W9-BXZ-007

BIOLOGICAL SCIENCE

SIXTH EDITION

JAMES L. GOULD
WILLIAM T. KEETON
WITH CAROL GRANT GOULD

BIOLOGICAL SCIENCE

SIXTH EDITION

W.W. NORTON & COMPANY NEW YORK LONDON

Copyright © 1996, 1993, 1986, 1980, 1979, 1978, 1972, 1967 by W. W. Norton & Company, Inc.

ALL RIGHTS RESERVED

PRINTED IN THE UNITED STATES OF AMERICA

This book is composed in New Aster, Composition by New England Typographic Service, Inc. Manufacturing by R. R. Donnelley & Sons, Company. Book design by Martin Lubin. Cover illustration: Wendell Minor.

SIXTH EDITION

Library of Congress Cataloging-in-Publication Data

Gould, James L., 1945–
Biological science / by James L. Gould, William T. Keeton with Carol Grant Gould.—6th ed.
p. cm.
Keeton's name appears first on the earlier edition.
Includes bibliographical references and index.
1. Biology. I. Keeton, William T. II. Gould, Carol Grant. III. Title.
QH308.2.G69 1996
574—dc20 95-44962

ISBN 0-393-96920-7

W. W. Norton & Company, Inc., 500 Fifth Avenue, New York, N.Y. 10110
W. W. Norton & Company, Ltd., 10 Coptic Street, London WC1A 1PU

1 2 3 4 5 6 7 8 9 0

■ ABOUT THE COVER

The painting on the cover illustrates a central theme of modern biology: the evolution of species by natural selection. The imaginary scene is based on the work of Peter and Rosemary Grant of Princeton University and their colleagues, whose long-term study of natural selection in Darwin's finches has been called "a modern classic in the study of evolution in action" (by Jonathan Weiner, author of *The Beak of the Finch,* winner of the 1995 Pulitzer Prize for nonfiction). The researchers, who have spent extended periods every year for more than two decades observing the finches in their native habitat in the Galapágos Islands, have noted that changes in the supply of seeds available as food can produce measurable changes in beak size (an inherited trait) from one generation to the next—a much shorter time than Darwin's original field observations had suggested. In one particularly striking case, represented here, the sequence of a typical wet year (left) followed by an unusually dry one (right) resulted in an evident change not only in the island's vegetation but also in the beaks of successive generations of the medium-sized ground finch *Geospiza fortis.* Normally, the finches prefer to eat small, soft seeds, altering the proportion of small and large seeds left on the ground. In the drought year, few or no new seeds were produced, leaving only the previous years' larger, harder seeds for food. Birds with more massive beaks, such as the one on the right, were better adapted to break these tougher seeds open, and hence were more likely to survive, while smaller-beaked birds such as the one on the left perished. The population evolved when the offspring of these survivors inherited the large-beak trait from their parents. The evolutionary significance of these findings is discussed further in the "How We Know" box on page 506.

Cover design and painting by Wendell Minor, based on photographs by Peter T. Boag, Queens University

NEW

In This Edition

Biological Science has long been recognized as the standard for accurate, up-to-date coverage of introductory biology. To provide easier access to the high-quality contents of this classic work, we are pleased to introduce a number of novel pedagogical elements in this edition, including a coordinated set of "navigational aids," designed to help students find their way around in the book more easily.

NEW *Topic outlines* now open every chapter, providing students with a handy overview of that chapter's contents.

CHAPTER 27

WHAT HETEROTROPHS NEED TO EAT

WHY ISN'T SUGAR ENOUGH?

WHAT VITAMINS DO

WHY MINERALS ARE NECESSARY

THE UNIQUE FEEDING STYLE OF FUNGI

INTERNALIZED DIGESTION: ANIMALS AND PROTOZOANS

HOW PROTOZOANS CAPTURE AND DIGEST FOOD

THE ANIMAL AS STOMACH: COELENTERATES

THE ALL-PERVADING STOMACH OF FLATWORMS

DISASSEMBLY LINES: THE COMPLETE DIGESTIVE TRACT

HOW THE VERTEBRATE DIGESTIVE SYSTEM WORKS

HOW FOOD IS DIGESTED

HOW CELLS HARVEST ENERGY ■ 173

now move through the cytochrome series (left) to cytochrome a_3. This last enzyme group uses the energy of the electrons to split molecular oxygen (O_2) and catalyze the following reaction:

$$1/2O_2 + 4H^+ + 2e^- \xrightarrow{\text{cytochrome } a_3} H_2O + 2H^+$$

This reaction reduces by four the number of hydrogen ions in the inner compartment. Two of these hydrogen ions are exported to the outer compartment, while the other two are incorporated into the water molecule. About 100 electrons can pass through each transport chain each second.

Chemiosmotic synthesis (Stage V of aerobic respiration) All the energy stored in the NAD_{re} has now been used up without generating any new ATP. However, an electrostatic and osmotic concentration gradient has built up across the inner membrane: the inner compartment, having lost H^+, has become negatively charged, while the outer compartment—the space between the inner and outer membrane—is filled with H^+ ions and therefore is positively charged. The result is a molecular battery that uses the difference in charge across the inner mitochondrial membrane to make ATP from ADP. Most of the H^+ ions return from the outer to the inner compartment by way of the F_1 complex (Fig. 6.12), passing down a ***chemiosmotic gradient***. As they do so, the energy of this electrochemical gradient—the combined osmotic and electrostatic gradients—is harvested (Figure 6.13). The abundant energy generated by this elegant system is stored in the phosphate groups of ATP. The same H^+ gradient probably also powers the pump that exports ATP to the cytosol. Peter Mitchell won the Nobel Prize in 1978 for working out this indirect route by which mitochondria convert the energy of NAD_{re} into ATP.

It is a continuing challenge to explain how such efficient, highly refined examples of biological engineering as the electron-transport chain might have evolved. How do the first elements in such a complicated array, which seem useless without the rest, ever come into existence? To answer the question, biologists look at a wide variety of species for preadaptations—fragments of the system as we know it now that are employed in other ways and which might have been present in ancestral organisms and available to natural selection for "remodeling."

Consider the most intricate element of the mitochondrial system, the F_1 complex, which uses the H^+ gradient to make ATP. Relict species of anaerobic bacteria living today still lack an electron-transport chain but nevertheless have an F_1 assembly that works in reverse. These anaerobes burn ATP to pump H^+ out of their cells, eliminating the acidic by-products of fermentation. There is similar evidence to suggest that the cytochrome complexes evolved later in ancient bacteria, where they served similar detoxification functions. Natural selection must then have favored combining these various systems into a rudimentary electron-transport chain using sulfur, carbon, or nitrogen as the ultimate electron acceptor.

Summary of respiration energetics Now that we have looked at all five stages in the aerobic breakdown of glucose, we can summarize their combined energy yield. The overall flow of energy in aer-

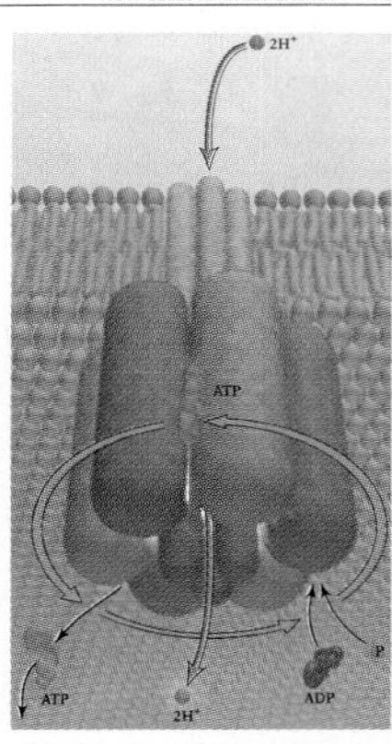

6.13 Operation of the F1 complex
The F_1 complex is composed of at least 18 subunits. The core group carries the H^+ to one of three protein pairs of the head region where ATP is formed; each of the three pairs is undergoing one of the three steps in the cycle: ADP and phosphate binding, ATP formation, and ATP release. Many researchers believe that the head rotates during ATP synthesis.

NEW *Figures* illustrate both basic and advanced concepts in biology better than ever. The entire art program has been carefully reviewed and, wherever possible, simplified for greater clarity.

NEW *"How We Know"* boxes show students how scientists approach interesting and important questions, reinforcing the emphasis on science as a process, not just a collection of facts.

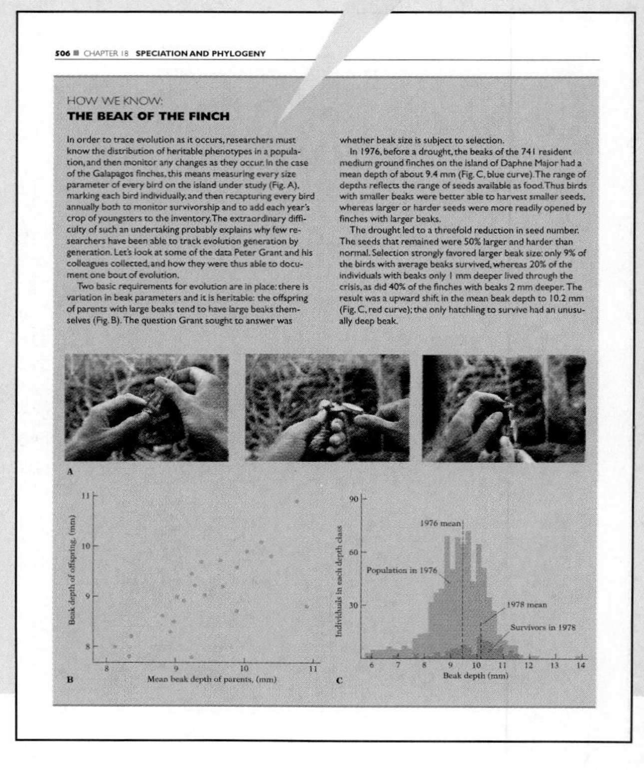

506 ■ CHAPTER 18 SPECIATION AND PHYLOGENY

HOW WE KNOW:
THE BEAK OF THE FINCH

In order to trace evolution as it occurs, researchers must know the distribution of heritable phenotypes in a population, and then monitor any changes as they occur. In the case of the Galapagos finches, this means measuring every size parameter of every bird on the island under study (Fig. A), marking each bird individually, and then recapturing every bird annually both to monitor survivorship and to add each year's crop of youngsters to the inventory. The extraordinary difficulty of such an undertaking probably explains why few researchers have been able to track evolution generation by generation. Let's look at some of the data Peter Grant and his colleagues collected, and how they were thus able to document one bout of evolution.

Two basic requirements for evolution are in place: there is variation in beak parameters and it is heritable: the offspring of parents with large beaks tend to have large beaks themselves (Fig. B). The question Grant sought to answer was whether beak size is subject to selection.

In 1976, before a drought, the beaks of the 741 resident medium ground finches on the island of Daphne Major had a mean depth of about 9.4 mm (Fig. C, blue curve). The range of depths reflects the range of seeds available as food. Thus birds with smaller beaks were better able to harvest smaller seeds, whereas larger or harder seeds were more readily opened by finches with larger beaks.

The drought led to a threefold reduction in seed number. The seeds that remained were 50% larger and harder than normal. Selection strongly favored larger beak size: only 9% of the birds with average beaks survived, whereas 20% of the individuals with beaks only 1 mm deeper lived through the crisis, as did 40% of the finches with beaks 2 mm deeper. The result was a upward shift in the mean beak depth to 10.2 mm (Fig. C, red curve); the only hatchling to survive had an unusually deep beak.

ANIMAL NUTRITION AND DIGESTION

Unlike plants and other autotrophs, heterotrophs cannot synthesize high-energy compounds from low-energy inorganic raw materials. To survive, they must feed on the high-energy molecules of other organisms. There are four main groups of heterotrophs: nonphotosynthetic bacteria, fungi, nonphotosynthetic protozoans, and animals. This chapter focuses mainly on how animals obtain the nutrients they need.

Bacteria and fungi lack internal digestive systems and hence depend mainly on absorption for feeding. They are usually either ***saprophytic*** (living and feeding on dead organic matter) or ***parasitic*** (living on or in other organisms and feeding on them). By contrast, the principal mode of feeding for animals and protozoans is ingestion—taking in and digesting particulate or bulk food. Animals and protozoans may be ***herbivores***, in which case they obtain high-energy compounds by eating plants or other photosynthesizers, or they may be ***carnivores***, eating animals or protozoans that have fed on plants. Some animals, the ***omnivores***, need both plant and animal material to survive. Much of the morphological diversity among living things reflects their different ways of employing the three major modes of nutrition—photosynthesis, absorption, and ingestion.

WHAT HETEROTROPHS NEED TO EAT

■ WHY ISN'T SUGAR ENOUGH?

Carbohydrates, fats, and proteins are 'he main energy sources for heterotrophs. Of these, carbohydrate: lone would suffice if organic nutrients functioned only as an e· gy source. In fact, adult bees and hummingbirds can survive s· on carbohydrate-rich nectar until they attempt to reproduce ·. 27.1). Because all amino acids and nucleotides incorporate · en, which carbohydrates lack, reproduction and growth requ· ource of fixed nitrogen. Many bacteria and fungi, as well as · protozoans, can flourish and reproduce on a diet consisting s· carbohydrates, fixed nitrogen, and minerals; like plants, the· ynthesize for themselves all the other classes of compounds · for life.

Animals are especially deficient in synthetic · ong their

27.1 A honey bee gathering pollen and nectar
Adult worker honey bees can survive on the sugars in the nectar they collect from plants; the pollen (which is packed into pollen baskets on the rear legs) is used as a protein source, and is fed to growing larvae and the egg-laying queen.

NEW *Chapter introductions* have been extensively rewritten to allow the chapters to stand alone better, making it easier for instructors to assign the chapters in any order. The introductions now serve primarily as a preview of the material to follow in that chapter.

NEW *Section headings* raise questions and issues rather than simply state conclusions, thereby emphasizing the active nature of scientific inquiry.

26.18 Leaf of sundew *(Drosera intermedia)*
A damselfly is caught in the sticky fluid on the ends of the glandular hairs.

long stiff teeth line the margin of each lobe. When an insect touches small sensitive hairs on the surface of the leaf, the lobes quickly change shape and come together with their teeth interlocked. When the trigger hairs are stimulated, they initiate a rapid pumping of H^+ ions through the cell membrane, consuming about 30% of the cells' available ATP in 2 sec. The pH change in turn leads to a quick osmotic movement of water from the intercellular spaces into the cells at the base of the trap; their consequent rapid enlargement causes the leaf to close on its victim. The trapped animal is then digested by enzymes secreted from glands on the leaf surface, and the resulting amino acids are absorbed.

The leaves of sundews (Fig. 26.18) bear numerous hairlike tentacles, each with a gland at its tip. Small insects, attracted by the plant's odor, become trapped in the sticky fluid secreted by the glands. The stimulus from a trapped insect causes nearby tentacles to bend over toward the animal, further entangling it. The proteins of the insect are digested by enzymes, and the amino acids are absorbed.

CHAPTER SUMMARY

AUTOTROPHS VERSUS HETEROTROPHS
Autotrophs can synthesize organic compounds from inorganic materials; heterotrophs require energy-rich organic molecules in their diet. (p. 743)

WHAT PLANTS ARE MADE OF
EARTH, AIR, OR WATER? The carbon and oxygen—together the elements that contribute most to a plant's mass of organic molecules—come from CO_2. The CO_2 comes from the air, most of the hydrogen comes from water, while the minerals used by plants mostly come from the earth. (p. 744)
WHY PLANTS NEED MINERALS Minerals are needed in amino acids and nucleic acids, at the active sites of many enzymes, and as components of coenzymes. (p. 745)

HOW PLANTS FEED
HOW ROOTS WORK Water containing minerals enters roots through root hairs or along cell walls; it crosses through endodermal cells, and then enters the xylem. The water usually enters because of a water-potential gradient generated by the evaporation of water from leaves. (p. 747)
WHERE PLANTS GET THEIR NITROGEN Nitrogen is obtained either as a dissolved soil compound (e.g., NH_3) or from symbiotic nitrogen-fixing bacteria, which convert N_2 into NH_3. (p. 755)
WHY DO SOME PLANTS EAT ANIMALS? Most often to obtain nitrogen that is lacking in the soil. (p. 756)

NEW *Chapter summaries* review the most important topics covered in each chapter. The summaries are keyed to both the topic outlines that open the chapters and the section headings that appear throughout the chapters.

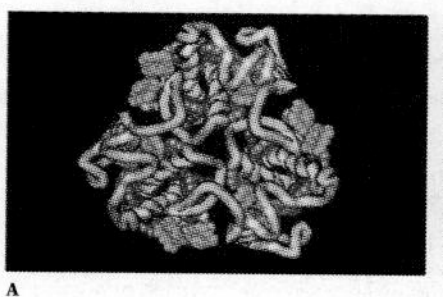

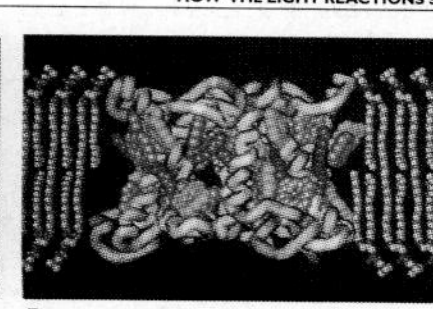

A B

7.7 How chlorophyll is mounted in the membrane
The antennas are organized as five to ten protein-bound units called light-harvesting complexes (LHC). Each LHC includes three subunits, viewed here from the stromal side of the membrane (A) and from the plane of the membrane (B). Each of these subunits, in turn, has approximately seven molecules of chlorophyll *a* (blue), five molecules of chlorophyll *b* (green), and several carotenoids (red); all are held in a protein matrix (tan). (The tails of the chlorophylls have been omitted for clarity.)

plants, *c* in brown algae, and *d* in red algae), and carotenoids. One group of four pigment molecules in each unit is distinct from all the rest; it is part of a complex that acts as a ***reaction center*** where energy conversion occurs. The other pigment molecules function somewhat like antennas for collecting light energy and transferring it to the reaction center.

When a photon strikes a chlorophyll (or a carotenoid) molecule and is absorbed, its energy is transferred to an electron of the pigment molecule; the excited electron moves up to a higher, relatively unstable energy level (Fig. 7.8). Not just any photon will do: because electrons occupy discrete energy levels, the photon must have a particular amount of energy to excite the pigment electron from its ground state to the higher level. This explains the well-

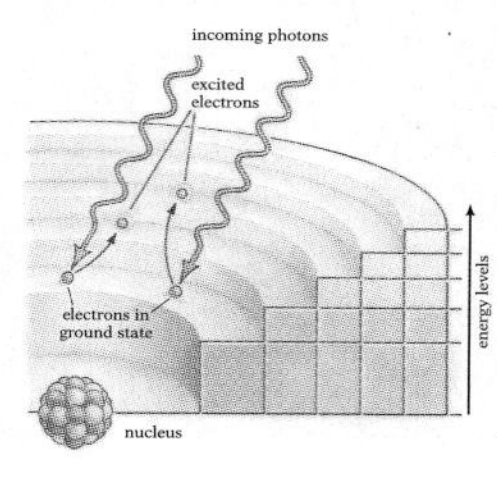

7.8 Effect of light on chlorophyll
When a photon is absorbed by a chlorophyll molecule, the photon's energy raises an electron to a higher energy level. This simplified representation shows a typical distribution of electrons at their lowest available energy levels, and the distribution as it is altered by absorption of a photon of red light, which raises one electron from its ground state in the first energy level to its first excited state in the second energy level. As we saw in Figure 7.6A, the absorption spectrum of a particular pigment can have several peaks. These exist because electrons can be excited from their ground state to their second excited state, two energy levels higher. In this illustration, another electron at the first level is shown being excited up to the third level—its second excited state—by a blue photon, which has enough more energy than a red photon that it can lift electrons two levels rather than one.

NEW *Scientific findings,* in this case the latest model of the structure of the light-harvesting chloroplast antenna complex, are woven seamlessly into the text and illustrations.

Biological Science is accompanied by a full range of electronic and print ancillaries:

FOR STUDENTS
- BioXplorer Plus *(interactive student software)*
- Study Guide

FOR TEACHERS
- Instructor's Manual with Test Bank
- Norton TestMaker *(electronic test-making program)*
- Overhead Transparencies
- Norton Presentation Maker and CD-ROM Resource Bank *(multi-media lecture aids)*

For more information on any of the ancillaries listed above, contact your local Norton representative or call (800) 233-4830.

CONTENTS IN BRIEF

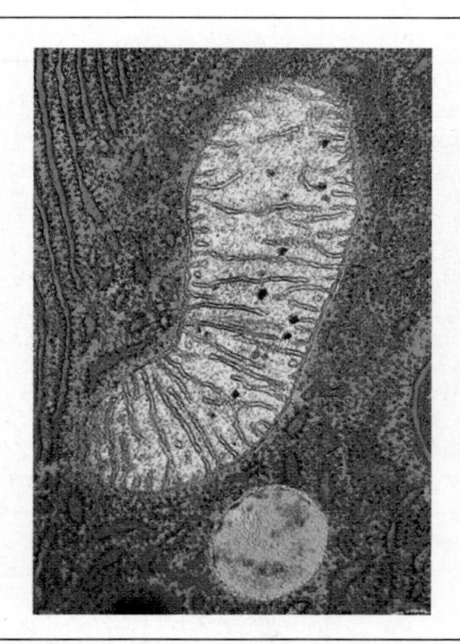

PART I ■ THE CHEMICAL AND CELLULAR BASIS OF LIFE

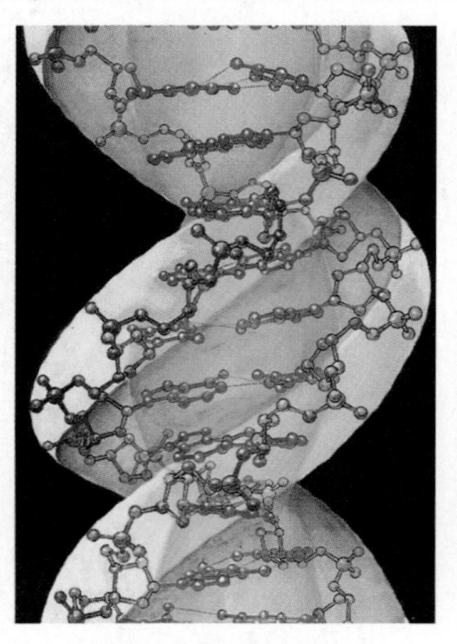

PART II ■ THE PERPETUATION OF LIFE

PART III ■ EVOLUTIONARY BIOLOGY

PART IV ■ THE GENESIS AND DIVERSITY OF ORGANISMS

PART V ■ THE BIOLOGY OF ORGANISMS

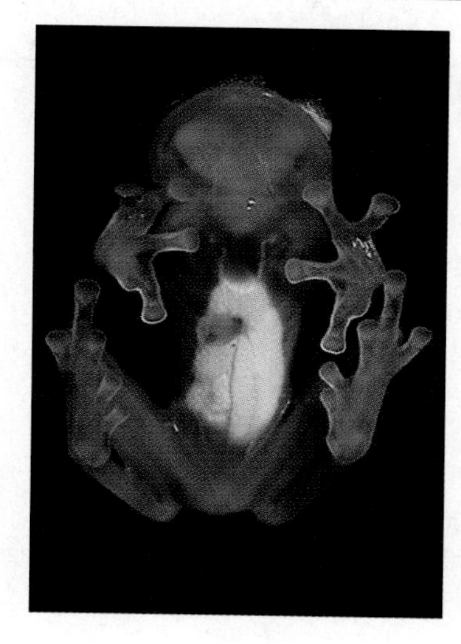

PART VI ■ ECOLOGY

CONTENTS

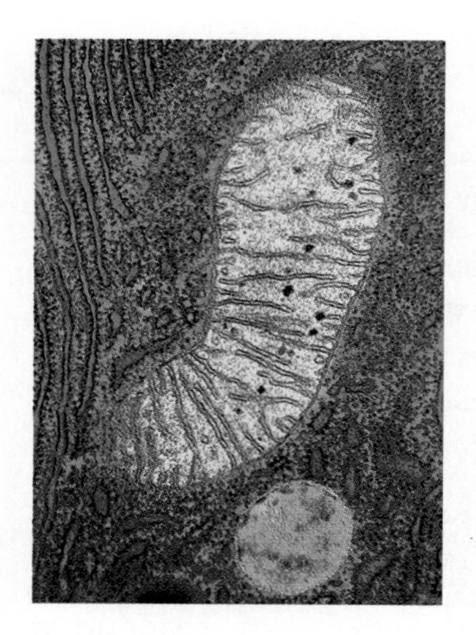

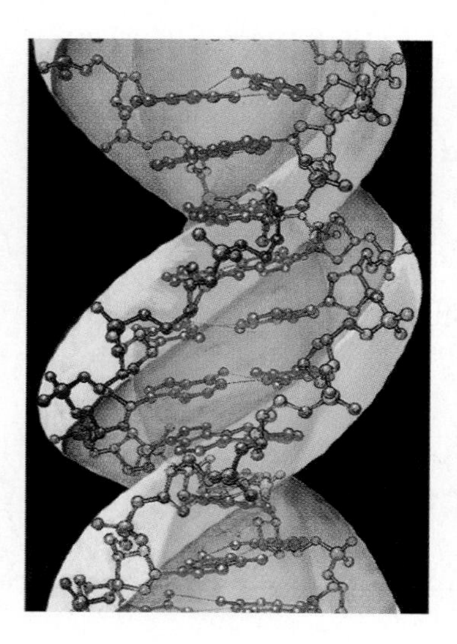

PART III

EVOLUTIONARY BIOLOGY

PART IV

THE GENESIS AND DIVERSITY OF ORGANISMS

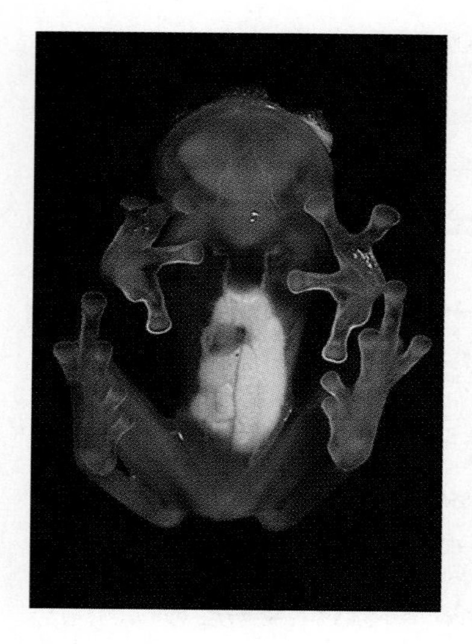

PART V

THE BIOLOGY OF ORGANISMS

CHAPTER 34 ■ HORMONES AND VERTEBRATE REPRODUCTION 969

CHAPTER 35 ■ NERVOUS CONTROL 991

CHAPTER 36 ■ SENSORY RECEPTION AND PROCESSING 1029

CHAPTER 37 ■ MUSCLES 1067

PART VI
ECOLOGY

ABOUT THE AUTHORS

James Gould, who took over as sole author of *Biological Science* following William Keeton's untimely death in 1980, is professor of biology at Princeton University. A graduate of the California Institute of Technology, where he majored in molecular biology, Jim became interested in animal behavior as a graduate student, obtaining his Ph.D. in ethology from Rockefeller University in 1975. He has since gone on to become an internationally recognized expert on various aspects of animal behavior—communication, navigation, learning and cognition, and sexual selection—all research topics that demand a thorough mastery of physiology, ecology, and evolution.

In short, Jim's own scientific training and special areas of expertise encompass the core contents of all six parts of *Biological Science*.

An enthusiastic user of the classic Keeton text for his introductory biology course at Princeton even before his involvement in the task of updating and revising this book, Jim also teaches a popular undergraduate course in animal behavior there; his first major textbook for W. W. Norton, *Ethology*, was a natural outgrowth of this course. In addition, he is a frequent contributor to both *Science* and *Nature*, and his work has been featured in three BBC Horizon films, which he wrote and narrated: "What Makes an Animal Smart," "Making Sex Pay," and "In the Company of Ants and Bees."

Carol Grant Gould has contributed substantially to the clarity, grace, and educational value of the three editions of *Biological Science* she and Jim have worked on together, as she has on their many other joint projects, including three popular books for the Scientific American Library series (*The Honey Bee*, *Sexual Selection*, and *The Animal Mind*) as well as numerous coauthored articles for both professional and general-interest publications. Her own background includes a Ph.D. in English literature from New York University and wide-ranging experience as a science writer, editor, and teacher.

Jim and Carol live in Princeton, New Jersey with their two children, both of whom have been intimately involved in the preparation of the software ancillaries that accompany this book. Grant Gould, now an undergraduate at the Massachusetts Institute of Technology, was the coauthor (with his father) of *Norton Presentation Maker*, the versatile lecture creation and presentation tool that is being introduced with this edition. Clare Gould worked on the scanned images, diagrams, and other graphic material found on the CD-ROM that serves as a resource bank for users of *Biological Science*.

A final note: Since Jim Gould has now had primary responsibility for as many editions of this work as his predecessor, the sequence of names in the byline has been reversed with this edition. Henceforth all of us—authors, editors, and longtime users of the book included—will have to get used to the idea of referring to it familiarly not as "Keeton/Gould," but as "Gould/Keeton."

PREFACE

This newest edition of *Biological Science* continues Bill Keeton's vision of biology as an integrated science. In his first edition, published in 1967, Keeton revolutionized the teaching of introductory college biology by combining biochemistry, cell biology, microbiology, botany, zoology, and ecology into a single integrated text, with evolution as its central unifying theme. In time, most other texts followed his lead, putting all of biology between two covers. Even so, few caught the essential spirit of his vision, and to this day most pay little more than lip service to Keeton's theme of "unity in diversity." (Indeed, many continue to segregate plants so completely from animals that students might well be led to believe they have nothing whatever in common!) Outside of the chapters on evolution per se, most other texts simply pass over the many wonderful examples of how natural selection works at all levels to solve the universal problems of living faced by a diverse range of organisms.

In preparing the Sixth Edition we had three main objectives in mind: (1) to update every chapter, both to reflect new discoveries and to respond to the changing needs of students who plan to go on to upper-level courses; (2) to clarify and simplify the presentation wherever possible; (3) to revise existing material and add new examples that help illustrate the *process* of science. We have also made it easier to present the chapters in different sequences, and to teach plant biology separately (while continuing to make illuminating comparisons and contrasts between plant and animal solutions to analogous problems of living).

In the course of the revision, we have actually managed to shorten the text significantly, reversing the seemingly inexorable trend toward greater and greater verbosity in books for the introductory course. The space created by this efficient presentation has allowed us to enlarge the type size for better readability, without substantially increasing the total number of pages. We have also added a topics outline at the beginning of each chapter and a detailed summary at the end (replacing the Concepts for Review section) to help students with their review of the material, and to enable them to find their way around in the book more easily. To emphasize that science is a process and not just a collection of facts, there are many new How We Know boxes, in addition to the popular Exploring Further boxes, introduced in earlier editions. In the same spirit, most of the section headings have been recast to suggest questions and issues rather than conclusions; for easy orientation, the revised headings are repeated verbatim in both the chapter-opening topics outline and the concluding summary.

SOME SPECIFIC CHANGES

We have added much new material on a variety of important topics. These additions are far too numerous to list separately, but here are a few of them:

In Part I (The Chemical and Cellular Basis of Life) we have added discussions of atomic-force microscopy, caveolae, the internal structure of the nucleus, the structure of the F_1 complex in mitochondria, and the structure of the light-harvesting complexes of chloroplast antennas and the associated reaction centers.

In Part II (The Perpetuation of Life) we have included new data on the structure of DNA polymerase, the process of replication in *E. coli*, the tertiary structure of rRNA, the operation of self-splicing introns, the nature of cancer, the operation of the cell cycle, early steps in development, the role of DNA repair in longevity, the biology of AIDS, and the molecular basis of blood groups. In addition, the discussion of recombinant DNA has been greatly expanded. There are new figures and discussions illustrating recombinant transduction, recombinant screening, how recombinant fragments are employed to inactivate viruses, the use of anti-sense DNA, engineered disease resistance in plants, the DNA particle-gun and microinjection techniques, the results of microinjection of growth-hormone genes, and DNA fingerprinting.

In Part III (Evolutionary Biology) we've added material on frequency-dependent selection, directly observed natural selection in populations (the subject of the cover), the role of parasites and parasitoids, intriguing theoretical work on the evolution of the camera eye, as well as the consequences of minigenes and horizontal transmission. There is more on speciation via developmental changes, and field studies of selection and speciation. In particular, we have expanded and revised the section on phylogenetic methods, adding simple examples and figures illustrating and comparing modern phenetic and cladistic techniques.

In Part IV (The Genesis and Diversity of Organisms) we have revised the phylogenetic trees with the most recent and reliable data available. There is now more on prions; the presentation of protists has been reordered to correspond to phylogenetic reality; and the treatment of fungi has been expanded. Human evolution has been brought up to the minute with the latest results on Y-chromosome dating.

In Part V (The Biology of Organisms) we have added material on the role of mycorrhizae, the mechanisms of uptake in intestinal villi, and the effects of plant toxins on digestion. The discussion of the structure and mechanisms of exchange in alveoli has been expanded, as have the presentations on the evolution of lungs, the heart cycle, molecular movements underlying ion-regulation of the fluids of invertebrates, and the life histories of parasitic plants. On the human level, we have added recent work on the feedback control of testosterone levels and menstruation, and the molecular biology of RU 486.

In Part VI (Ecology) we have made numerous revisions, including a new section on the Law of Constant Yield. There is more on biological control of pests, the causes and effects in population growth and population cycles, the interaction between economics and birth rates, and more on the issue of diversity and stability. We have reordered Chapter 40 (Ecosystems) for a more effective presentation, and added discussions of the relationship between latitude and climate (rainfall and temperature) as well as the role of ocean currents, followed by a discussion of Lieth's and other models of world-wide terrestrial and aquatic primary productivity. There is now a more quantitative treatment of energy flow in ecosystems, more on the role of decomposers, more on the effects of human activity on cycles of materials (including the effects of deforestation on local climate and succession), and more on the consequences of polar melting, the effects of erosion, the changing carbon cycle, and the enormous impact of agriculture on the nitrogen cycle.

ANCILLARIES

The Sixth Edition continues to be available in a one-volume hardcover or two-volume paperback format. It is accompanied by a CD-ROM that contains not only all of the figures found in the book, but also a variety of animations, user-controlled simulations, and QuickTime movies. The two dozen simulations are particularly effective in illustrating how dynamic processes depend on relevant parameters. Two examples will illustrate the general strategy: One simulation graphs the oxygen-binding curve of hemoglobin for any combination of temperature, pH, and body size; the instructor can change any parameter and plot the new curve in one of several contrasting line styles. A more complex simulation allows the teacher to select any combination of population-growth parameter values (setting the life expectancy, age of first reproduction, number of offspring per female, birth interval, and offspring survival), adjust the distribution of individuals in various age classes in the starting population, and then see the population profile and the total population change as the years pass.

The simulations and animations will be useful in lecture, lab, and student review—indeed, I use the three population-growth simulations as the basis for one of the weekly labs in my course. To encourage the creation of additional simulations and animations, the publisher is sponsoring a competition. Entries (simulations in either HyperCard or ToolBook format; animations in QuickTime for Macintosh or Windows) will be judged twice a year (in January and July). Submissions by teachers and students will be evaluated separately by a panel that includes one of the authors, an editor from W. W. Norton, the developer of the BioXplorer software ancillaries, and another teacher who uses *Biological Science.* Winners will receive their choice of a modestly priced piece of computer hardware (a CD-ROM drive, for instance) or the equivalent value in software, computer supplies, or a generous number of Norton books. Details can be found in the Contest file on the CD-ROM, as well as on the publisher's World Wide Web home page (http://web.wwnorton.com). We will incorporate the best animations and simulations into future editions of the CD ancillary. (We will also create a version of each selected simulation in the other programming format.)

The increasing number of users with LCD-projection facilities will find the remarkable *Norton Presentation Maker* program on the CD of special value. This program was designed by Grant F. Gould and the authors specifically for college teaching. It allows the lecturer to assemble slides, animations, simulations, and movie clips from the CD, plus similar material from other sources, as well as sound files, text, and videodisk stills or clips. These items can be added to a presentation quickly and in any order. It then creates a lecture file capable of displaying these elements either in the original sequence, or in any other desired order. It has many additional useful features too numerous to detail here. It can also be used by instructors to create custom tutorials, incorporating the teacher's own notes and commentary in combination with the material from the CD or the instructor's own digital material.

ACKNOWLEDGMENTS

We are grateful to the many instructors and specialists who carefully and critically reviewed the Fifth Edition and the revised draft of this edition. Our special thanks go to Marvin R. Alvarez, University of South Florida; Ron Ash, Washburn University of Topeka; J. Barber, Imperial College of Science, Technology, and Medicine; Robert Beckmann, North Carolina State University; Annalisa Berta, San Diego State University; Domenic Castignetti, Loyola University; Thomas Cavalier-Smith, University of British Columbia; Edward C. Cox, Princeton University; Barbara Demmig-Adams, University of Colorado; Harold B. Dowse, University of Maine; Paul R. Elliott, Florida State University; Gary L. Galbreath, Northwestern University; Emily Giffin, Wellesley College; Lindsay Goodloe, Cornell University; Peter R. Grant, Princeton University; H. Ernest Hemphill, Syracuse University; Linda Kohn, Erindale College, University of Toronto; William Z. Lidicker, Jr., University of California, Berkeley; Axel Meyer, State University of New York, Stony Brook; Kevin Moses, University of Southern California; Stephen W. Pacala, Princeton University; David J. Patterson, University of Sydney; William Rumbach, Central Florida Community College; Fred Wilt, University of California, Berkeley; Charles Wyttenbach, University of Kansas.

We are also grateful to the many individuals at W. W. Norton who have contributed to this revision. In particular we thank our tireless editors Joe Wisnovsky and Jim Jordan, as well as Stephen King, who did much to facilitate the development of Norton Presentation Maker. We also thank Susan Middleton, our gifted copy editor; Mary T. Kelly, our able and attentive project editor; Rachel Warren, our ever-resourceful editorial assistant and photo researcher; Susan Crooks, our excellent layout artist; and Michael Reingold, Michael Goodman and John McAusland, whose remarkable artwork helps bring these pages to life.

Finally, we appreciate the many thoughtful and helpful comments and criticisms that users of the book provide. Although letters to the publisher and authors continue to be very welcome, we invite you to communicate by e-mail to the senior author (gould@princeton.edu) or the editor (jwisnovsky@wwnorton.com).

JLG
CGG

Princeton
August, 1995

CHAPTER 1

SCIENCE AND EVOLUTION

Biology is the science of living things—what they are, how they work, how they interact, and how they change or evolve. Before we begin our exploration of biology, we must be clear about what we mean by life, science, and evolution.

WHAT IS LIFE?

Every day the sun radiates vast amounts of energy to the earth and our moon, and to all the other planets in our solar system. And every day the planets reradiate that energy into space. On earth there is a unique delay: for a brief moment, a minute portion of the sun's energy is trapped and stored (Fig. 1.1). This minor delay in the flow of cosmic energy powers life.

Plants capture sunlight and use its energy to build and maintain stems, leaves, and seeds, while animals secure the energy of sunlight by eating plants or by eating other animals that have eaten plants. Each stage in the many processes of living and dying produces waste heat, which joins the pool of energy being relayed back into space.

Biology is the study of this very special category of energy utilization—it is the science of living things. The world is teeming with life: millions of species of organisms of every description inhabit the earth, feeding directly or indirectly off lifeless sources of energy like the sun's light. What is there in this maze of diversity that unites all biology into one field? What do amoebae, redwoods, and people have in common that differentiates them from the nonliving? What, in short, is life?

Most dictionaries define *life* as the property that distinguishes the living from the dead, and define *dead* as being deprived of life. These singularly circular and unsatisfactory definitions give us no clue to what we have in common with protozoans and plants. The difficulty for the scientist as well as for the writer of dictionaries is that life is not a separable, definable entity or property; it cannot be isolated on a microscope slide or distilled into a test tube. To early mechanistic philosophers like Aristotle and Descartes, life was wholly explicable in terms of the natural laws of chemistry and physics. The vitalists, philosophers of the opposing school, were convinced there was a special property, a vital force, that was

1.1 Sunlight bursts through a forest canopy

A minute portion of the sun's energy powers all life's processes.

absent in inanimate objects and thus unique to life. Though some scientists continue to think in terms of an unnamed and intangible special property, vitalism has been discredited in biology for at least half a century. The more we learn about living things, the clearer it becomes that life's processes are based on the same chemical and physical laws we see at work in a stone or a glass of water.

If life is not a special property, what is it? One answer may be found by comparing living and nonliving things. Organisms from bacteria to humans have several attributes in common: all are chemically complex and highly organized; all use energy (metabolize), organize themselves (develop), and reproduce; all change (evolve) over generations. As far as we know, no nonliving thing possesses all these attributes. In addition, and perhaps most important, only living organisms have a set of instructions, or program, resident in the genes, that directs metabolism, organization, and reproduction, and is the raw material on which evolution acts.

HOW THE SCIENTIFIC METHOD WORKS

Science is based on disciplined curiosity. Children are notoriously curious and undisciplined. Professional scientists retain much of their youthful curiosity—springing in most cases from a profound respect for nature—but bring to their work a degree of discipline that can solve difficult problems, plus a basic belief that there is an underlying order to natural phenomena waiting to be uncovered.

The discipline that must accompany curosity in successful scientists is the rigorous and creative application of the ***scientific method***. Like so many truly great ideas, the scientific method is a basically simple concept and is used to some extent by almost everyone every day. As the English biologist T. H. Huxley (1825–1895) put it, the scientific method is "nothing but trained and organized common sense." Its power in the hands of a scientist stems from the consistency and ingenuity of its application.

■ THE NEED FOR HYPOTHESES

Science is concerned with the material universe, seeking to discover facts about it and to fit those facts into conceptual schemes, called theories or laws, that will clarify the relations between them. Science must therefore begin with observations of objects or events in the physical universe. The objects or events may occur naturally, or they may be the products of planned experiments; the important point is that they must be *observed,* either directly or indirectly. Science cannot deal with anything that cannot be observed.

Science rests on the philosophical assumption (well justified by its past successes) that virtually all events of the universe can be described by physical theories and laws, and that we get the data with which to formulate those theories and laws through our senses. Needless to say, natural laws are descriptive rather than prescriptive; they do not say how things *should* be, but instead how things are and probably will be.

Scientists readily acknowledge the imperfection of human sensory perception: the major alterations our neural processing imposes upon our picture of the world around us are themselves a subject of scientific study. In addition, experience has shown that there is often an interaction between phenomenon and observer: however careful we may be, our preconceived notions and even our physical presence may affect our observations and experiments. But to recognize the imperfection of sensory perception and observation is not to suggest that we can get scientific information from any other source (spiritual revelation, for example). No other route is open to scientists.

The first step in the scientific method is to formulate the question to be asked. This is not as simple as it sounds: scientists must decide which of the endless series of questions our escalating knowledge inspires are important and worth asking; the question must then be honed into a form that will generate a clear answer. The next step is to make careful observations in an attempt to answer the question. The difficulties here are in deciding what to observe and (since measuring everything is impossible) what to ignore. The scientist must also decide how to make the measurements and how to record the data, so that any patterns will emerge and nothing will be lost that might answer future questions. This is no trivial matter: an oversight or a mistake can render years of work useless.

The data that emerge from a set of observations must be analyzed and fitted into some sort of coherent pattern or generalization. A formal generalization, or ***hypothesis,*** is a tentative causal explanation for a group of observations. The step from isolated bits of data to generalization can be taken with confidence only if enough observations have been made to give a firm basis for the generalization, and then only if the individual observations have been reliably made. But even when data have been carefully collected, a hypothesis does not automatically follow. Often data can be interpreted in several ways or may appear senseless.

■ TESTING HYPOTHESES

Coming up with a general statement, or hypothesis, is not the end of the process. Scientists must devise ways of testing their hypotheses by formulating predictions based on them and checking to see if these predictions are accurate. Again, this is often not easy: a hypothesis may supply the basis for many predictions, but probably only one can be tested at a time. A scientist must decide which predictions can be most readily tested, and of those, which one provides the toughest challenge for the hypothesis (Fig. 1.2).

Perhaps most difficult of all, when researchers find a discrepancy between a hypothesis and the results of their tests, they must be ready to change their generalization. If, however, all evidence continues to support the new idea, it may become widely accepted as probably true and be called a theory. It is important to realize that

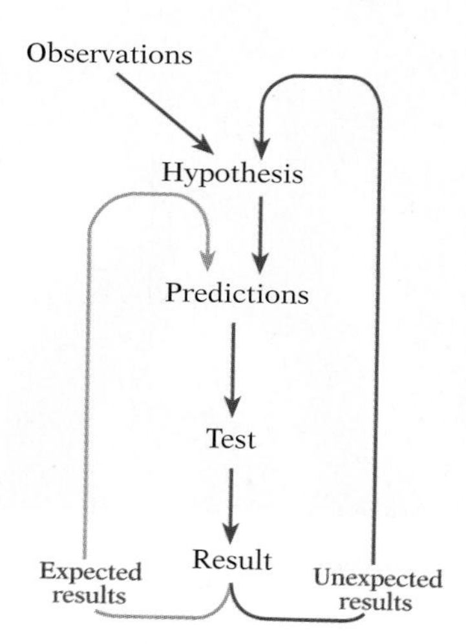

1.2 The scientific method

As this diagram suggests, experimental results are used, when necessary, to modify a hypothesis and to test predictions; the cycle of testing never really ends.

scientists do not use the term *theory* the way the general public does. To many people a theory is a highly tentative statement; when scientists dignify a statement by the name of theory, they imply that it has a very high degree of probability and that they have great confidence in it.

A ***theory*** is a hypothesis that has been repeatedly and rigorously tested. It is supported by all the data that have been gathered, and helps order and explain those data. Many scientific theories, like the cell theory, are so well supported by essentially all the known facts that they themselves are "facts" in the nonscientific application of that term. But the testing of a theory never stops. No theory in science is ever absolutely and finally proven. Good scientists must be ready to alter or even abandon their most cherished generalizations when new evidence contradicts them. They must remember that all their theories, including the physical laws, are dependent on observable phenomena, and not vice versa. Even incorrect theories, however, can be enormously valuable in science. We usually think of mistaken hypotheses as just so much intellectual rubbish to be cleared away before science can progress, but tightly drawn, explicitly testable hypotheses, whether right or wrong, catalyze progress by focusing thought and experimentation.

1.3 Louis Pasteur

■ CONTROLLING EXPERIMENTS

In its simplest form the scientific method begins with careful observations, which must be shaped into a hypothesis. The hypothesis suggests predictions, which must be tested. The tests, furthermore, must be *controlled,* a matter of vital importance since only a controlled test has any hope of illuminating anything. Controlling a test, or experiment, means making sure that the effects observed result from the phenomenon being tested, not from some other source.

The most familiar way of controlling an experiment is to perform the same process again and again, varying only one discrete part of it each time, so that if a difference appears in the result, we can easily trace the cause. When Louis Pasteur, for instance, took his memorable stand against the prevailing theory of spontaneous generation (the idea that life can arise spontaneously from nutrients), he designed his experiment so that there could be no doubt about the outcome, whatever it might be (Fig. 1.3). He took identical flasks, filled them with identical nutrient solutions, subjected them to identical processes, but left some of them open to the air, while the others were effectively sealed. In time bacteria and mold developed in the open flasks, but the liquid in the sealed flasks remained clear. Since exposure to the outside air was the only variable in the procedure, it was obvious that the bacteria and the mold-producing organisms, rather than being spontaneously generated from the broth itself, must have come from the air.

■ THE ROLE OF INTUITION

In actual practice science is neither easy nor mechanical. Most scientists will readily admit that every stage in the scientific process requires not just careful thought but a large measure of intuition and good fortune. At the end of the last century, for instance, physicists saw many small problems, or anomalies, in Newtonian physics. Experience justifies living with small anomalies when a theory works well, and most minor problems prove in time to be irrelevant or mistaken—the result of observations that were faulty or interpretations that missed a point. Sometimes, though, anomalies in a theory signal a conceptual error.

Albert Einstein sensed that two of the many irritating anomalies in Newtonian physics were crucial difficulties, and rearranged the pieces of the puzzle to create a revolutionary new theory that accounted for those two anomalies. Like most new hypotheses, Einstein's theory of relativity did not win over the world of science at once. For one thing, it seemed far more complex than the Newtonian mechanics with which everyone was familiar, and for another, it left at least as many (though different) loose ends dangling as had the theory it sought to replace. The potential appeal of Einstein's theory, however, was great: it made some very unlikely but testable predictions—that the light from stars, for instance, should bend as it passed near the sun. When these were subsequently proven to be correct, the theory of relativity was accepted rather quickly, loose ends and all, where a far more plausible but less dramatic hypothesis might well have been ignored.

Scientific investigation, then, depends on a combination of subjective judgments and objective tests, a delicate mixture of intuition and logic. Done well, scientific research is truly an art: the ability to make insightful guesses and imagine clever and critical ways in which to test them is usually the distinguishing characteristic of great scientists. But in the final analysis, the basic rules are the same for all: observations must be accurate and hypotheses testable. And as testing proceeds, hypotheses must be altered when necessary to conform to the evidence.

THE ORIGINS OF MODERN BIOLOGY

■ WHEN DID BIOLOGICAL SCIENCE BEGIN?

Although naturalists and collectors had been at work for centuries observing living things, the application of the scientific method to biology is a relatively recent innovation. The earliest scientific biologist was Andreas Vesalius (1514–1564), who made the first serious studies of human anatomy by dissecting corpses. He discovered that the body is composed of numerous complex subsystems, each with its own function, and he pioneered the comparative approach, using other animals to work out the purpose and organization of these anatomical units. A typical (if grisly) example is his demonstration that the nerve from the brain to the throat, common to so many animals, is responsible for controlling vocalization. When he took a squealing pig, dissected out the nerve, and cut it, the animal instantly became mute even though its vocal apparatus remained intact (Fig. 1.4).

1.4 Cutting the vocal nerve
The experimenter pictured in the initial letter in Vesalius' *Fabric of the Human Body* (1542) is cutting the vocal nerve of a pig.

This powerful style of comparative and experimental study was carried forward by the English physician William Harvey

1.5 William Harvey

(1578–1657), who showed conclusively that the heart pumps the blood and the blood circulates (Fig. 1.5). The heart, in short, is not in some metaphysical sense the seat of emotions but a mechanical device with a clear function. As a result of these studies and the anatomical work that followed, an increasingly mechanistic view of life began to develop.

The third of the pioneers was Antonie van Leeuwenhoek (1632–1723; Fig. 1.6). Leeuwenhoek, a cloth merchant, had the idea of using the powerful lens with which he inspected cloth to look at living things. The most important of his many discoveries were microorganisms (including bacteria), sperm and the eggs they fertilized, and the cells of which all living things seemed to him to be composed.

Centuries of painstaking observation were required to establish the science's fundamental generalizations. The cell theory, for instance, was not given its essentially modern form until 1858, and only about 130 years ago, in 1862, did Pasteur disprove the theory of spontaneous generation. With the realization that Leeuwenhoek's microorganisms might be responsible for disease, the English surgeon Joseph Lister proved the effectiveness of antiseptics (from *anti-*, "against," and *sepsis*, "decay") in 1865, and Pasteur greatly expanded the use of vaccination.

By far the most important figure in the history of biology, however, is Charles Darwin (1809–1882; Fig. 1.7). The publication in 1859 of his *Origin of Species*, presenting the theory of evolution by natural selection, suddenly provided a coherent, organizing framework for the whole of biology. His work sparked the explosive growth of biological knowledge that continues today. As the most important unifying principle in biology, the theory of evolution underlies the logic of every chapter of this book.

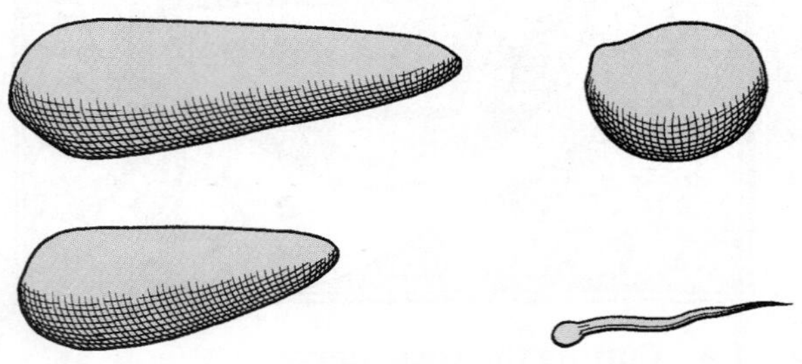

1.6 Antonie van Leeuwenhoek

The drawings are reproductions of his 1683 sketches of the intestinal protozoa of frogs.

■ DID DARWIN DISCOVER EVOLUTION?

The theory of evolution by natural selection, as modified since Darwin, will be treated in detail in Chapter 17. Since we will need to use it in earlier chapters, however, we will examine the essence of the theory here. It consists of two major parts: the concept of evolutionary change, for which Darwin presented a great deal of evidence, and the quite independent concept of natural selection as the agent of that change.

Do species change? Until only 200 years ago, it seemed self-evident that the world and the animals that fill it do not change: robins look like robins and mice like mice year after year, generation after generation, at least within the short period of written history. This commonsense view is very like our untutored impression that the earth stands still and is circled by the sun, moon, planets, and stars: it accords well with day-to-day experience, and until evidence to the contrary appeared, it provided a satisfying picture of the living world. The idea of an unchanging world also corresponded to a literal reading of the powerfully poetic opening of the Book of Genesis, in which God is said to have created each species independently, simultaneously, and relatively recently—a little over six thousand years ago by traditional scriptural reckonings.

But problems with the commonly held scriptural theory of creation arose from many sources. Scientists attempted first, quite naturally, to discount the evidence as ambiguous and then, when that proved impossible, to construct a new explanation. Let's look at the evidence for evolution that confronted Darwin and his contemporaries.

The most dramatic findings came from geology. By the eighteenth century, a picture of a changing earth had begun to emerge. Extinct volcanoes and their lava flows were discovered; most geological strata were found to represent sedimentary deposits, laid down layer upon layer a millimeter at a time in columns 3000 m (meters) or more deep; the gradual erosive action of wind and water were seen to have levelled entire mountains and carved out valleys, and unknown forces had caused mountains to rise where ocean floors had once been. This last fact in particular was impressed upon Darwin when he discovered fossilized seashells high in the Andes. Each of these phenomena implied continuous change during vast periods of time.

Another problem for the static view of life was presented by the New World fossils themselves. Many represented plants and animals wholly unknown in Europe, and though theologians had argued that the organisms these fossils represented were alive in the New World, increasingly intensive exploration of the Americas indicated that the hundreds of species of dinosaurs, for instance, were certainly extinct. In addition, many previously unknown and often bizarre animals inhabited the Americas (Fig. 1.8).

1.7 Charles Darwin

A

B

1.8 Challenges to traditional ideas of the origin of species

The discovery of fossils of now-extinct species brought into question the static view of life. Shown here are the remains of a baby mammoth (A) that had been preserved in the permafrost in Siberia. The discovery in the New World of organisms unfamiliar to Europeans, such as the anteater (B), also required a reinterpretation of traditional ideas of the origin of species.

1.9 Jean-Baptiste de Lamarck

The realizations grew that the number of animal species alive ran at least into the hundreds of thousands, and that the extinct species greatly outnumbered the living. New constellations of species had come and gone several times in the past. Moreover, the lowest, oldest rocks contained only the most primitive fossils—seashells, for instance—and these were followed in order by the more modern forms: fish appeared later, for example, reptiles still later, and then birds and mammals. The hypothesis of a young earth populated almost overnight by a single bout of creation began to seem very unlikely.

Jean-Baptiste de Lamarck (1744–1829; Fig. 1.9) was the first to offer the major alternative explanation of the fossil record: evolution. Lamarck had arranged fossils of various marine molluscs in order of increasing age; he saw clearly that certain species had changed slowly into others, and concluded that this process of slow change had continued right to the present day. As Lamarck put it in 1809, "it is no longer possible to doubt that nature has done everything little by little and successively," over a nearly infinite period of time. In Lamarck's view, the living world had begun with simple organisms in the sea, which eventually moved onto the land, and evolution had culminated in the appearance of our species, the inevitable result of the gradual trend toward change and "increasing perfection."

Lamarck was basically on the right track, but his mechanism for evolutionary change was incorrect. He was ignored for the very understandable reason that he could not offer sufficient evidence for the *fact* of evolution. Darwin, only 50 years later, was in a far better position: there was much more evidence of the sort Lamarck had pointed to, and Darwin, a respected geologist, was well acquainted with it. Furthermore, he had the ability to spot important data in the midst of apparent chaos.

One of the most important lines of evidence put forward by Darwin was the existence of morphological resemblances among living species (the findings of what we today call comparative anatomy). If we look at the forelimbs of a variety of different mammals, we see essentially the same bones arranged in the same order (Fig. 1.10). The basic bone structure of a human arm, a cat's front leg, and a seal's flipper is identical; the same bones are present even in a bird's wing. True, the size and shape of the individual bones vary from species to species, and some bones may be missing entirely in one species or another, but the basic construction is unmistakably the same. To Darwin the resemblance suggested that each of these species was descended from a common ancestor from which each had inherited the basic plan of its forelimb, modified to suit its present function.

The observation that structures with important functions in some species appear in vestigial, nonfunctional form in others further convinced Darwin of the reality of evolutionary change. Why otherwise would pigs, which walk on only two toes per foot, have two other toes that dangle uselessly well above the ground? Why would certain snakes, such as the boa constrictor, and many species of aquatic mammals, such as whales, have pelvic bones and small, internal hind-limb bones (Fig. 1.11)? Why would flightless birds such as penguins, ostriches, kiwis, and the cormorants of

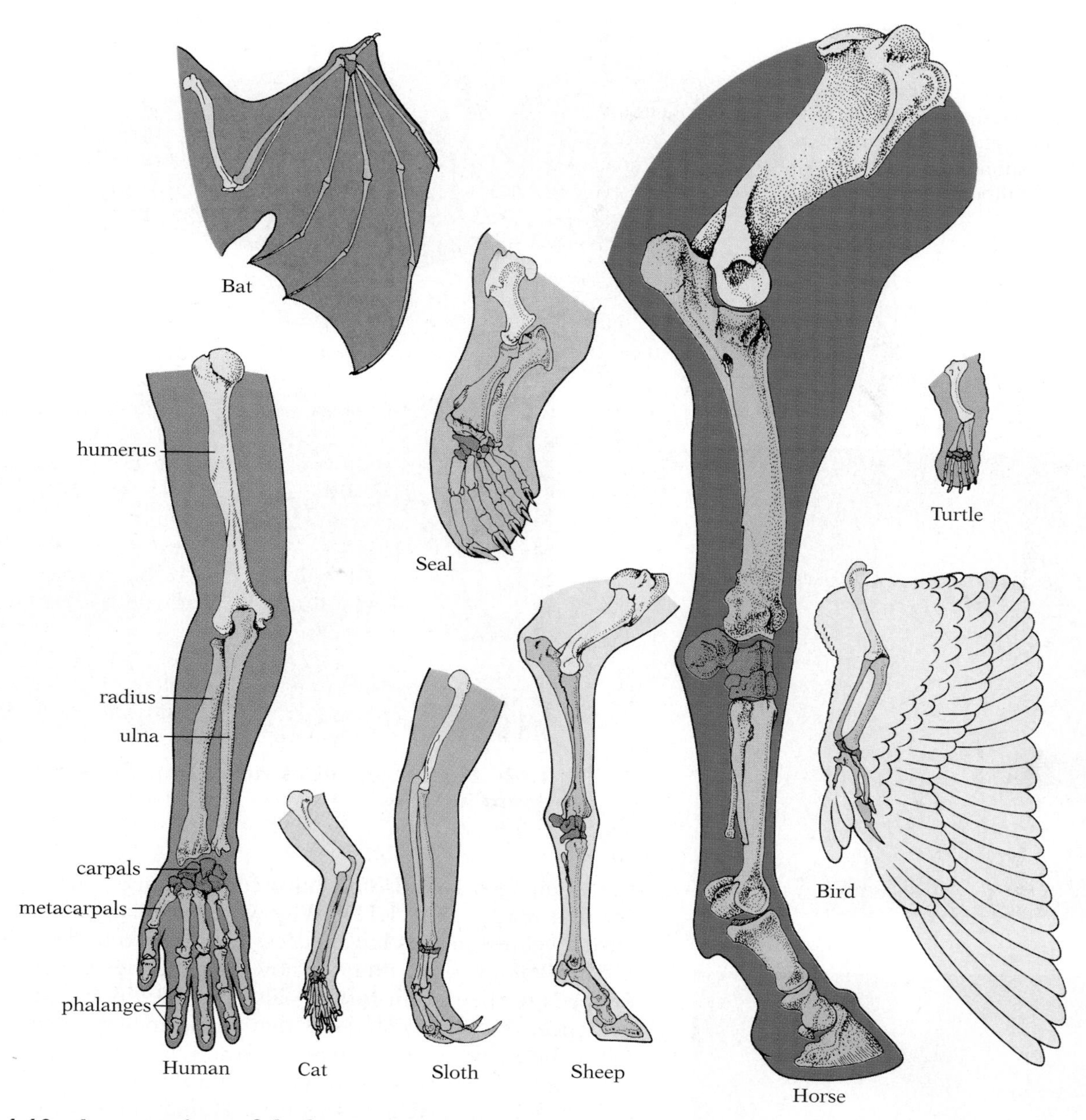

1.10 A comparison of the bones in some vertebrate forelimbs

The labeled and color-coded bones of the human arm at left permit identification of the same bones in the other forelimbs depicted. In the bat the metacarpals (hand bones) and phalanges (finger bones) are elongated as supports for the membranous wing. In the seal the bones are shortened and thickened in the flipper. The cat walks on its phalanges (toes), the metacarpals having come to form a part of the leg. The sloth normally hangs upside down from tree limbs—hence its recurved claws. The horse walks on the tip of one toe, which is covered by a hoof (a specialized claw), while the sheep walks on the hoofed tips of two toes (and therefore cloven-hoofed, though only one hoof can be seen in this side view). The carpals (wrist bones) of both the horse and the sheep are elevated far off the ground, because the much-elongated metacarpals (hand bones) have become a section of the leg. Small splintlike bones that are vestiges of other ancestral metacarpals can be seen on the back of the upper portion of the functional metacarpals of both horse and sheep. All the animals mentioned so far—human, bat, seal, cat, sloth, horse, and sheep—are mammals, but the same bones can also be seen in the leg of a turtle and the wing of a bird. (All limbs are drawn to the same scale except that of the turtle, which is enlarged.)

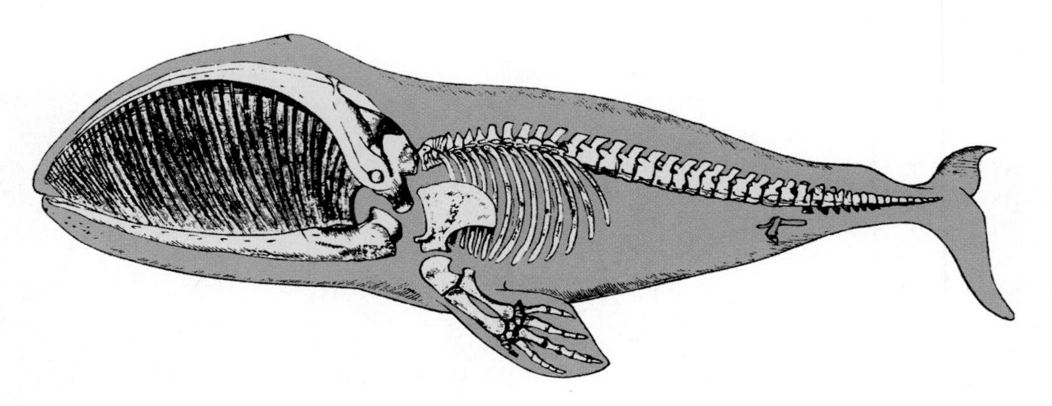

1.11 Rudimentary hind limb of a whale

Whales lost their hind limbs long ago, when they returned to the sea from the land, but they retain rudimentary bones that correspond to the pelvic girdle and the thighbone (red).

1.12 Flightless male ostriches from the Namib desert in Africa

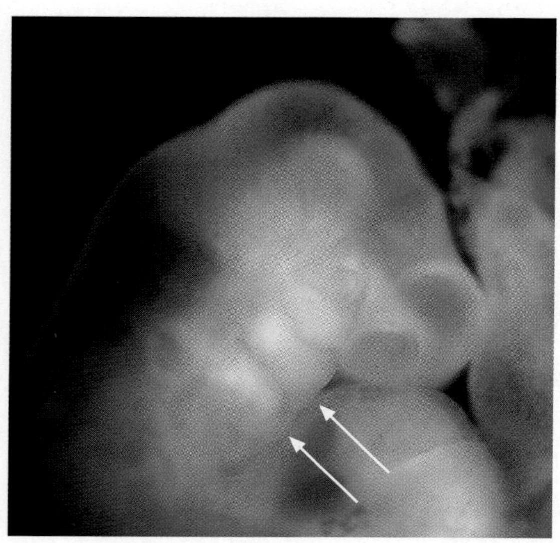

A

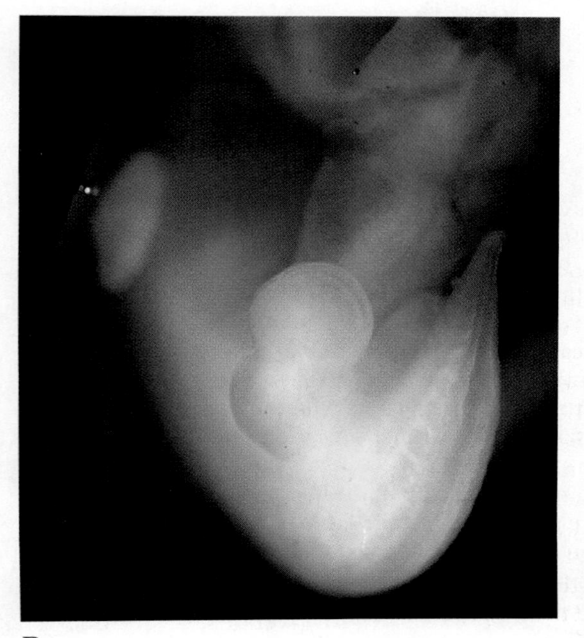

B

1.13 Embryological evidence of evolution

(A) Pharyngeal ("gill") pouches (arrows) in a 4-week human embryo.
(B) Tail in a 5-week human embryo.

the Galápagos Islands still have rudimentary wings—or feathers, for that matter (Fig. 1.12)? Why would so many subterranean and cave-dwelling species have useless eyes buried under their skin?

Embryology—the study of how living things develop from eggs or seeds to their adult forms—also provided powerfully suggestive evidence. Darwin pointed out that in marine crustaceans as different as barnacles and lobsters the young larvae are virtually identical, implying a common origin. Telltale traces of their genealogy are obvious in vertebrates as well. Human embryos, for instance, have gill pouches and well-developed tails that disappear before the time of birth (Fig. 1.13). It seemed clear to Darwin that such inappropriate structures are inherited vestiges of structures that functioned in ancestral forms, and that may still function in other species descended from the same ancestors.

Another particularly convincing line of evidence offered by Darwin was the ability of breeders to produce dramatic changes in both plants and animals. Artificial selection can bring about vast changes given sufficient time. Great Danes, sheepdogs, Irish setters, Yorkshire terriers, poodles, bulldogs, and dachshunds, for instance, are all members of the same species, bred from tamed wolves to look like almost anything breeders have fancied.

Similarly, cabbage, brussels sprouts, cauliflower, broccoli, kohlrabi, rutabaga, collard greens, and savoy have all been bred from the same species, the wild form of which looks nothing like its domesticated progeny (Fig. 1.14). The many varieties of chickens, cattle, horses, flowers, grains, and so on have been bred to fill different needs over the years. Who could compare the colors and shapes of the wild rose or jonquil with the many colors and shapes of the far larger domesticated roses or daffodils and doubt that a species has the capacity to change enormously even in a mere century? In everything Darwin looked at—fossils, anatomy, embryology, and breeding—he saw the same message: species can and do change (Fig. 1.15).

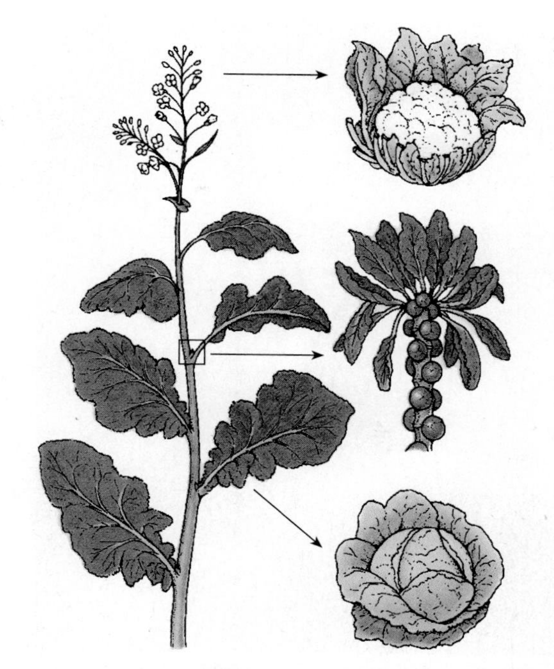

1.14 Selective breeding of wild cabbage
The wild species *Brassica oleracea* (left) has been bred to create cauliflower (upper right), brussels sprouts (middle right), and cabbage (lower right). Each represents a selective exaggeration of one part of the wild plant—the flower heads for cauliflower, the side buds for brussels sprouts, and the leaves for cabbage. Despite the extreme morphological differences, however, the three domestic vegetables can be interbred.

What causes species to change? Lamarck's now-discredited hypothesis was one of the first attempts to explain why species evolve. Lamarck was impressed by how well suited each animal was to its particular position in the web of life, even though the environment had changed enormously again and again over countless millions of years. To account for this ability to adapt, he imagined God had given each species a tendency toward perfection that allowed for small alterations in morphology, physiology, and behavior to accommodate changes in the environment, and that these alterations, once made, could be inherited by the offspring.

Belief in Lamarck's idea of a natural tendency toward perfection and the inheritance of acquired characteristics required no more faith in his day than did belief in other invisible everyday forces, such as gravity and magnetism. But where in the vestigial legs of whales or the dangling toes of pigs was there evidence of perfection? Instead the clear mark of compromise was everywhere. Plants and animals were well adapted to their places in the environment, but they were by no means perfect.

Darwin proposed a different mechanism—***natural selection***—requiring no internal tendency other than the one toward variation so obvious in nature. Darwin had conceived the idea of natural selection two years after his return from his voyage to the Americas

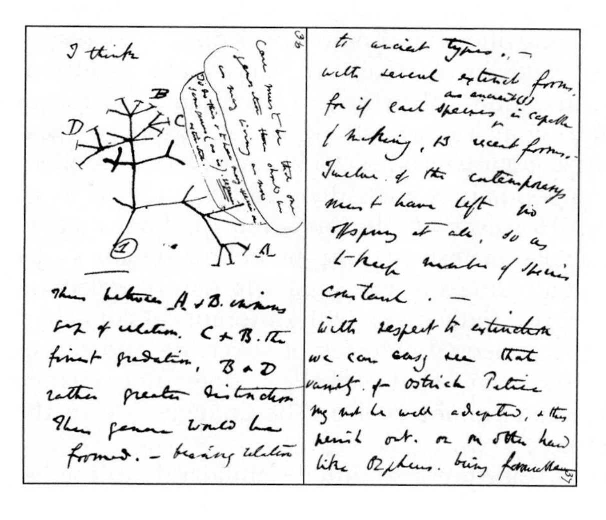

1.15 Pages from one of Darwin's notebooks
Scientists of the nineteenth century and earlier, among them Charles Darwin, often recorded in notebooks and journals their experiments, observations drawn from field trips, or simply their thoughts about hypotheses in the making. In addition to the many books he wrote, Darwin produced notebooks on a wide variety of subjects, including his trip to South America and the Galápagos (1831–1836), later described in the *Journal of Researches during the Voyage of the Beagle* (1840). The notebook from which this excerpt is taken was written in 1837, one year before Darwin saw that natural selection was the likely mechanism for evolution. These pages show Darwin's first recorded drawing of an evolutionary tree, a metaphor for the diversity and interrelatedness of species that has continued in the scientific literature to the present time. The trunk represents a common ancestor, the limbs major groups, and the twigs particular species, either extinct (indicated by a crossbar) or living.

1.16 Thomas Robert Malthus

on the *Beagle,* but was only goaded into publishing it 20 years later when in 1858 he received A. R. Wallace's manuscript proposing essentially the same theory. (We normally associate Darwin's name with the theory because of the impressive evidence he presented—he had been collecting it for two decades—and because of his thorough exploration of the theory's many ramifications.)

In essence, Darwin put together two ideas. The first was that numerous variations exist within species and that these variations are largely heritable. Immersed as he was in the Victorian preoccupation with plant and animal breeding, Darwin knew that, while cuttings produce plants identical to the parent, sexual reproduction produces individual offspring that differ both from their parents and from each other. Variation is a fact of life: breeders, as we know, are able to select for desirable traits and create new, morphologically distinct lines of plants and animals.

Darwin's second inspiration came when he reread the *Essay on the Principle of Population* by the economist Thomas Robert Malthus (Fig. 1.16). Malthus pointed out that humans produce far more offspring than can possibly survive; population growth always outruns any increase in the food supply and is held in check largely by war, disease, and famine. Vast numbers of people thus live perpetually on the edge of starvation.

Both Darwin and Wallace were struck by the consequence of applying the gloomy Malthusian logic to plants and animals: like humans, the creatures of each overpopulated generation must compete for the limited resources of their environment, and some—indeed most—must die. Each female frog, for example, produces thousands of eggs per year, and a fern produces tens of millions of spores, yet neither population is growing noticeably. Any organism with naturally occurring heritable variations that increase its chances in this life-or-death contest will be more likely than others to survive long enough to have offspring, some of which will inherit these variations. The offspring in turn will have an above-average chance to survive the struggle and so will form an increasingly large part of the population. As a result of this selection, the population as a whole will become better adapted—that is, it will evolve—and the never-ending struggle for existence will turn on the possession of still better adaptations. To distinguish this process from the sort of directed, artificial selection practiced by agriculturalists, Darwin called it *natural* selection.

The contrast between artificial and natural selection that served Darwin so well provides an instructive summary of the evolutionary process. In both, far more offspring are born than will reproduce; in both, differential reproduction, or selection, occurs, causing some inherited characteristics to become more frequent and prominent in the population and others to become less so as the generations pass. However, in the breeding of domesticated plants and animals, selection results from the deliberate choice by the breeder of which individuals to propagate. In nature, selection takes place simply because individuals with different sets of inherited characteristics have unequal chances of surviving and reproducing. Notice, by the way, that selection does not change individuals. An individual cannot evolve. The changes are in the makeup of populations.

Artificial and natural selection also differ significantly in the *de-*

1.17 Selective breeding of pigeons
By practicing rigorous selection, breeders can achieve major changes in relatively few generations. The ancestral rock dove is shown in the center. The domestic breeds (clockwise from upper left) are fantail, Silver Jacobin, Male Ice, Magpie Kormoner Tumbler, Norwich cropper, Voorburg Shield cropper, German Longface Tumbler.

gree of selection and its effect on the rate of change. Breeders can practice rigorous selection, eliminating all unwanted individuals in every generation and allowing only a few of the most desirable to reproduce. They can thus bring about very rapid change (Fig. 1.17). Natural selection, which involves a large measure of chance, is usually much less rigorous: some poorly adapted individuals in each generation will be lucky enough to survive and reproduce, while some well-adapted members of the population will not. Hence evolutionary change is usually slow; major changes may take thousands or even millions of years, depending on the degree of selection pressure imposed by the environment and by other species.

Darwin's evidence for evolutionary change and the common descent of at least the major groups of organisms was widely accepted in his time, but the idea of natural selection by small steps remained controversial until the 1930s. Some biologists had difficulty seeing how an elaborate and specialized structure like an eye, for instance, could evolve, since the first rudimentary but necessary steps might lack obvious survival value. An expanded understanding of the nature and organization of genes and their role in development has now made it clear that natural selection does explain most evolutionary change. Darwin extolled the beauty and simplicity of such a system in the final sentence of later editions of *The Origin of Species:* "There is grandeur in this view of life, with its several powers, having been originally breathed by the Creator into a few forms or into one; and that, whilst this planet has gone cycling on according to the fixed law of gravity, from so simple a beginning endless forms most beautiful and most wonderful have been, and are being evolved."

In summary, then, Darwin's explanation of evolution in terms of natural selection depends upon five basic assumptions:

1 Many more individuals are born in each generation than will survive and reproduce.

2 There is variation among individuals; they are not identical in all their characteristics.

3 Individuals with certain characteristics have a better chance of surviving and reproducing than individuals with other characteristics.

4 Some of the characteristics resulting in differential survival and reproduction are heritable.

5 Vast spans of time have been available for change.

All the known evidence supports the validity of these five assumptions.

■ THE TREE OF LIFE

Darwin's insights into evolution and its mechanisms, together with a variety of techniques (including, most recently, molecular analyses of genes themselves), have permitted a fairly reliable reconstruction of the course of evolution of life on earth and of the relationships between the species alive today. A diagram of the relationships among the eight major groupings (called kingdoms) of living species recognized at present is shown in Fig. 1.18. The organisms with the longest histories are the two kingdoms of bacteria, which are fundamentally different from the other groups because they lack nuclei; their chromosomes are mixed with the

1.18 Relationships among the eight kingdoms of life

In this representation, kingdoms are shown in color, and the width of the line indicates the relative number of species in each kingdom. The distance from the bottom of the figure at which a group's line departs from the central "trunk" of this evolutionary tree indicates the chronological order in which the kingdoms are thought to have arisen. Common names are used for most groups; technically, animals are in Kingdom Animalia, plants are in Plantae, brown algae are in Chromista, and "true" bacteria are in Eubacteria.

rest of the cellular contents instead of being segregated into a separate compartment. As a result, the bacteria are called ***procaryotes*** ("before nuclei"), while the other six kingdoms make up the ***eucaryotes*** (having "true nuclei").

The eucaryotes comprise the two kingdoms that now dominate the earth—the plants and the animals—as well as three smaller but significant kingdoms, the protozoans (most unicellular organisms), the fungi (mushrooms, for example), and the brown algae (like kelp); they also include one tiny, little-known group, the archezoans, which are by far the most primitive of the eucaryotes.

As we will see, the organisms in these kingdoms are marvelously diverse, having evolved to exploit a bewildering variety of habitats, and yet all share a common set of genetic and biochemical processes that unify the study of life.

CONTEMPORARY BIOLOGY

The work in the latter half of the 19th century by scientists like Darwin, Pasteur, Gregor Mendel (the monk who discovered the basic principles of inheritance), and a number of developmental biologists combined to set the stage for the emergence of the range of biological studies that go on today.

Although no one event formally marks the beginning of contemporary biology, the discovery of the structure of DNA in 1953 was an especially important foundation for later work. The double-helix model proposed by James D. Watson and Francis Crick (Fig. 1.19) provided a physical basis for Mendel's genes, which are the basic units of Darwinian selection. The structure immediately suggested how DNA could act as a template for its own reproduction and as a means for issuing instructions for building and operating cells.

Watson and Crick's model of DNA has helped fuel four decades of research on the messages encoded in this remarkable molecule; understanding the code is proving critical to every biological discipline. In relation to cell biology (our focus in Part I), research on DNA has led to a detailed appreciation of how cells are organized, both physically and chemically. One of the more impressive feats of investigators has been the deciphering of the chemical codes used as "mailing labels" on DNA products—codes that tell the elaborate transport system in the cell where to deliver each molecule.

In relation to genetics (Part II), we will see how research on DNA has been used to construct a nearly complete picture of how cells process information to regulate their metabolism, growth, and reproduction. This knowledge is critical in modern treatments of sickness and inborn defects, in the manufacture of chemicals of potentially enormous clinical and commercial importance like synthetic hormones and vaccines, and in the endowment of some plants with new genes that confer benefits such as disease resistance. An equally impressive by-product of DNA research is progress toward understanding the immune system—the body's main line of defense against pathogens—as well as the precisely orchestrated development of a complex, trillion-celled organism like a human being from a single, unspecialized cell. Another is the invention, since 1953, of many techniques for manipulating the genetic material and its protein products. Together the development of these techniques and the investigation of biological processes at the level of molecules are usually referred to as molecular biology.

The benefits of the new molecular biology for whole-organism studies are even more remarkable because they are less obvious. In Part III we will have a glimpse of the enormous impact molecular

1.19 Watson and Crick with their model of the DNA double helix

techniques are having on the study of evolution. Biologists now understand the chemical nature of the variation that is the basis of natural selection, and so are coming to grips with the question of how different species arise in nature and how they adapt to novel conditions. Small changes in genes controlling development, for example, can produce new species almost instantly.

We will see in Part IV that an understanding of how DNA and its chemical collaborators work is providing valuable insights into the origin and early evolution of life, as well as a reliable and often surprising reconstruction of the evolution of the major groups of organisms that populate the earth today. Molecular techniques now allow precise determination of species relationships; they reveal, for instance, that superficial similarities notwithstanding, pandas are not bears but overgrown raccoons.

In Part V, we will see how the molecular approach has exposed the workings of many physiological systems, including those responsible for circulation, gas exchange, movement, and internal coordination. For example, we can now analyze at the molecular level how neurons communicate and how hormones and other chemical signals (like insulin) regulate the activity of cells, tissues, and organs. So, too, the failures of organs to operate properly—breakdowns that lead to degenerative diseases—are beginning to be understood (and corrected) at the molecular level.

Finally, in Part VI, we will see how molecular techniques are aiding our understanding of ecological relationships. For example, the pesticide contaminant dioxin exerts its devastating ecological effects because the detoxification enzymes that animals use to dispose of unfamiliar (and thus potentially dangerous) chemicals alter it to an inappropriate form. And, of course, the potential of molecular engineering for solving or at least ameliorating environmental problems, like water pollution and even greenhouse-effect warming, is one avenue of hope for the future of our planet.

Now more than ever, as we stand able to read the genetic programs of organisms word for word, biology is the most exciting, intellectually stimulating, and promising discipline among the sciences.

CHAPTER SUMMARY

WHAT IS LIFE?

Living things have genetic instructions that direct their metabolism and reproduction, and they can evolve.

HOW THE SCIENTIFIC METHOD WORKS

THE NEED FOR HYPOTHESES The scientific method begins with formulating hypotheses that can account for observations. (p. 2)

TESTING HYPOTHESES Hypotheses are tested to see how well they predict events; failures lead us to revise our hypotheses. (p. 3)

CONTROLLING EXPERIMENTS A test usually compares two cases differing in only one variable; the hypothesis predicts the change that variable will cause compared with the "control" situation. (p. 4)

THE ROLE OF INTUITION Good judgment is required to select likely hypotheses for testing, and to design tests that are well controlled. (p. 4)

THE ORIGINS OF MODERN BIOLOGY

WHEN DID BIOLOGICAL SCIENCE BEGIN? The most important early biologists were Vesalius and Harvey, who studied functional anatomy, and Leeuwenhoek, who discovered microorganisms. Each helped make biology more experimental and mechanistic. (p. 5)

DID DARWIN DISCOVER EVOLUTION? Darwin collected much supporting data for evolution, but the idea was not his. Darwin discovered natural selection, which is the major mechanism of evolution: excess offspring are produced in each generation, and those with characteristics that are the most adaptive tend to live and reproduce, leading to an increase in the frequency of those characteristics in the next generation. (p. 6)

THE TREE OF LIFE The tree of life reconstructs the path evolution took to yield the diversity of organisms that inhabit the earth today. (p. 14)

CONTEMPORARY BIOLOGY

The discovery of DNA and the genetic code has led to a fuller understanding of the mechanistic bases of nearly every aspect of biology. (p. 16)

STUDY QUESTIONS

1 What roles do anomalies and intuition play in the scientific method? (pp. 3–5)

2 How might evolution occur in the absence of natural selection? (pp. 11–14)

3 What are three important requirements in a population for selection to take place? Can evolution occur in the absence of any one of these? (pp. 11–14)

4 How did Darwin's intimate knowledge of plant and animal breeding help him formulate his theory? (pp. 10–12)

SUGGESTED READING

COMROE, J. H., 1977. *Retrospectroscope.* Von Gehr Press, Menlo Park, Calif. *A fascinating study of how important scientific discoveries are made. The author concludes that great advances usually arise out of research directed at wholly unrelated problems.**

DARWIN, C., 1859. *The Origin of Species. Of the many reprints of this classic work, the edition by R. E. Leaky (Hill and Wang, New York, 1979) provides perhaps the best introduction and illustrations.**

* Available in paperback.

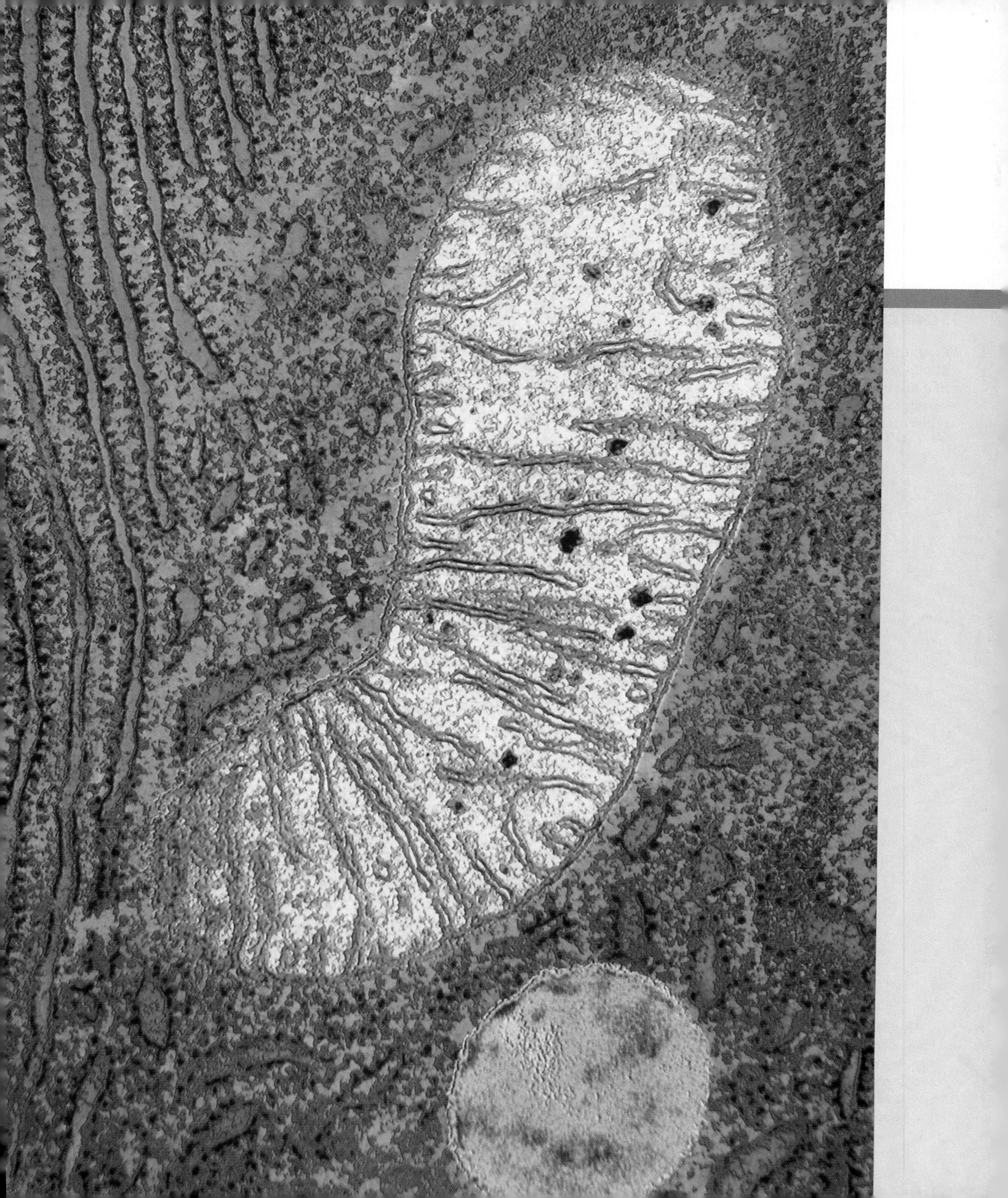

PART I

THE CHEMICAL AND CELLULAR BASIS OF LIFE

◄ **Powerhouse of the cell,** the elongated mitochondrion highlighted in yellow and red in this color-enhanced transmission electron micrograph (TEM) provides energy to the cell by completing the digestion of glucose, thereby releasing molecules of the cell's all-purpose energy currency: ATP (adenosine triphosphate). The black and blue membranous material surrounding the mitochondrion is mostly endoplasmic reticulum. The rounded, pinkish bodies are vesicles.

CHAPTER 2

SOME SIMPLE CHEMISTRY

All life processes obey the laws of chemistry and physics. In the behavior of molecules, atoms, and subatomic particles lies the key to such complex biological phenomena as the trapping and storing of solar energy by green plants, the extraction of usable energy from organic nutrients, the growth and development of organisms, the patterns of genetic inheritance, and the regulation of the activities of living cells. The study of biology, then, begins with the basic laws of chemistry and physics.

THE ELEMENTS ESSENTIAL TO LIFE

All the matter of the universe is composed of 92 basic substances called elements, some of which were formed in the original Big Bang that created the universe; others are released by the exploding stars called novas (Fig. 2.1). Each element is designated by one or two letters that stand for its English or Latin name. Thus H is the symbol for hydrogen, O for oxygen, C for carbon, Cl for chlorine, Mg for magnesium, K for potassium (Latin, *kalium*), Na for sodium (Latin, *natriu*), and so on. Only a few of the 92 elements are important to life (Table 2.1).

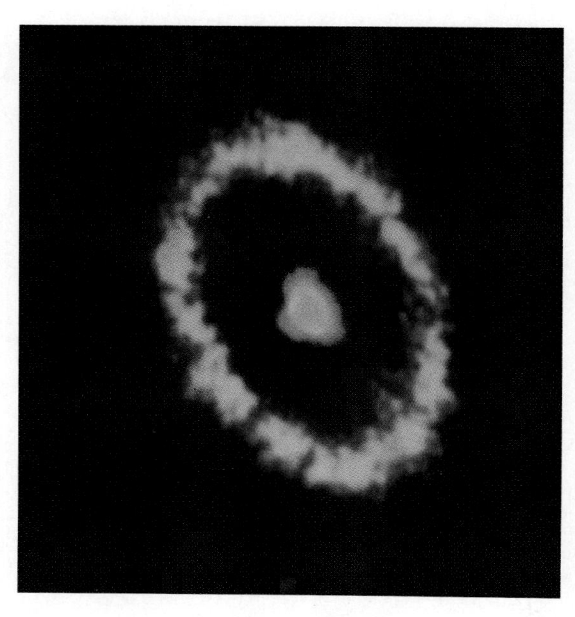

2.1 The Supernova of 1987

This nova consists of enormous amounts of gas produced when an aging star exploded. The gas includes many heavy elements created inside the star out of lighter elements. When a new star like the sun condenses out of this gas, the surrounding planets that may form are enriched in the heavy elements, many of which are essential to life.

■ THE STRUCTURE OF ATOMS

Matter is composed of ***atoms.*** The atoms of a particular element differ in many measurable ways from the atoms of all other elements. Atoms are themselves composed of still smaller particles, three of which—the proton, the neutron, and the electron—play a central role in determining the activity of elements. In their interactions lie the power and the cohesion that make life possible.

Inside the nucleus All the positive charge and almost all the mass of an atom are concentrated in its nucleus, which contains two kinds of primary particles, the ***proton*** and the ***neutron.*** Each proton carries an electric charge of +1. The neutron has no charge. The proton and the neutron have roughly the same mass (Table 2.2).

The number of protons in the nucleus is unique for each element. This ***atomic number*** can be found in Table 2.2 for many ele-

TABLE 2.1 Elements important to life

SYMBOL	ELEMENT	ATOMIC NUMBER/ TYPICAL MASS NUMBER	APPROXIMATE PERCENTAGE OF EARTH'S CRUST BY WEIGHT	APPROXIMATE PERCENTAGE OF HUMAN BODY BY WEIGHT
H	Hydrogen	**1**/1	0.14	9.5
B	Boron	**5**/11	trace	trace
C	Carbon	**6**/12	0.03	18.5
N	Nitrogen	**7**/14	trace	3.3
O	Oxygen	**8**/16	46.6	65.0
F	Fluorine	**9**/19	0.07	trace
Na	Sodium	**11**/23	2.8	0.2
Mg	Magnesium	**12**/24	2.1	0.1
Si	Silicon	**14**/28	27.7	trace
P	Phosphorus	**15**/31	0.07	1.0
S	Sulfur	**16**/32	0.03	0.3
Cl	Chlorine	**17**/35	0.01	0.2
K	Potassium	**19**/39	2.6	0.4
Ca	Calcium	**20**/40	3.6	1.5
V	Vanadium	**23**/51	0.01	trace
Cr	Chromium	**24**/52	0.01	trace
Mn	Manganese	**25**/55	0.1	trace
Fe	Iron	**26**/56	5.0	trace
Co	Cobalt	**27**/59	trace	trace
Ni	Nickel	**28**/59	trace	trace
Cu	Copper	**29**/64	0.01	trace
Zn	Zinc	**30**/65	trace	trace
Se	Selenium	**34**/79	trace	trace
Mo	Molybdenum	**42**/96	trace	trace
Sn	Tin	**50**/119	trace	trace
I	Iodine	**53**/127	trace	trace

TABLE 2.2 Fundamental particles

PARTICLE	MASS (DALTONS[a])	ELECTRIC CHARGE
Electron	0.001	−1
Proton	1.013	+1
Neutron	1.014	0

[a] One dalton equals 1.650×10^{-24} g, which is 1/16 of the mass of an oxygen atom. For definitions of units of measurement, see Glossary.

ments. The atomic number of hydrogen is 1; its nucleus contains only one proton. In contrast, each oxygen nucleus contains eight protons.

The total number of protons and neutrons in a nucleus is the ***mass number;*** it approximates the total mass (commonly called the ***atomic weight***) of the nucleus. The mass number is usually written as a superscript immediately preceding the chemical symbol. For example, most atoms of oxygen contain eight protons and eight neutrons; the mass number is therefore 16, and the nucleus can be symbolized as ^{16}O.

Though the number of protons is the same for all atoms of the same element, the number of neutrons is not always the same, and neither, consequently, is the mass number. For example, most oxygen atoms contain eight protons and eight neutrons and have a

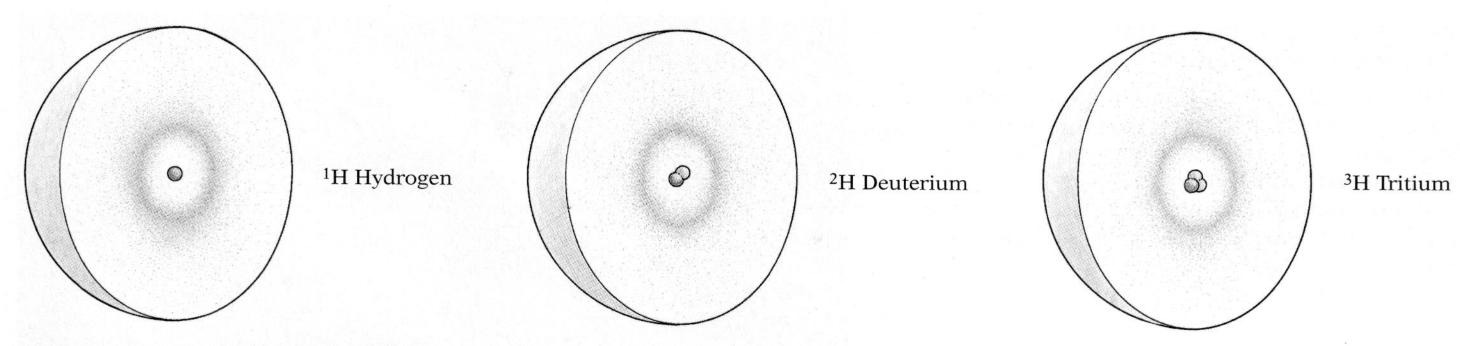

2.2 Three principal isotopes of hydrogen

Each of the three isotopes has one proton (blue) in its nucleus and one electron orbiting the nucleus. The isotopes differ in that ordinary hydrogen (^{1}H) has no neutrons in its nucleus, deuterium (^{2}H) has one, and tritium (^{3}H), which is unstable, has two. The volume within which the single electron can be found 90% of the time (in this case, a sphere) is indicated by stippling; the denser the stippling in this cross-sectional view, the greater the likelihood that at any given moment the electron will be found in that portion of the sphere.

mass number of 16; some, however, contain nine neutrons and therefore have a mass number of 17 (symbolized as ^{17}O), and still others have ten neutrons and a mass number of 18 (symbolized as ^{18}O). Atoms of the same element that differ in mass, because they contain different numbers of neutrons, are called ***isotopes;*** ^{16}O, ^{17}O, and ^{18}O are three isotopes of oxygen.

Figure 2.2 illustrates three different isotopes of hydrogen: ^{1}H, which is the usual form of the element; ^{2}H, a stable isotope called deuterium; and ^{3}H, an unstable isotope called tritium. Isotopes of hydrogen and other elements are invaluable research tools for biologists (Fig. 2.3).

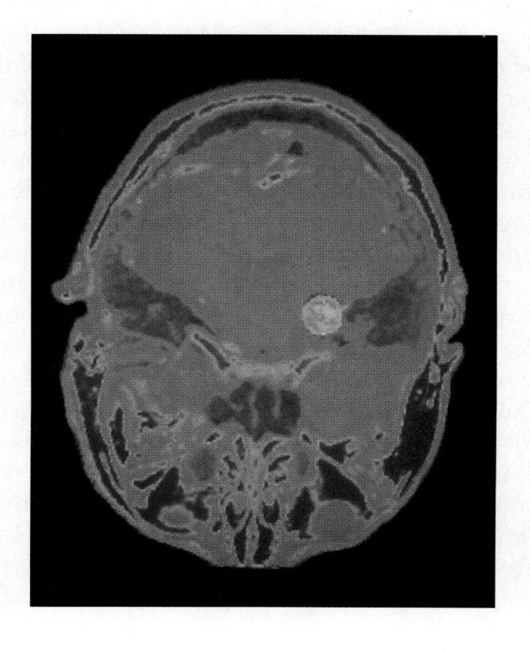

2.3 Magnetic Resonance Imaging technique

Isotopes whose atomic mass is an odd number (like ^{31}P) are weakly magnetic. When exposed to a strong magnetic field, a given isotope will absorb radio waves of a characteristic frequency and then reemit them when the signal is removed. In this image, the strong signal emitted by the relatively high concentration of organic compounds in a growing brain tumor stands out as a green area against the relatively watery cells of the healthy tissues. The tumor is growing in an auditory area, and is causing progressive deafness.

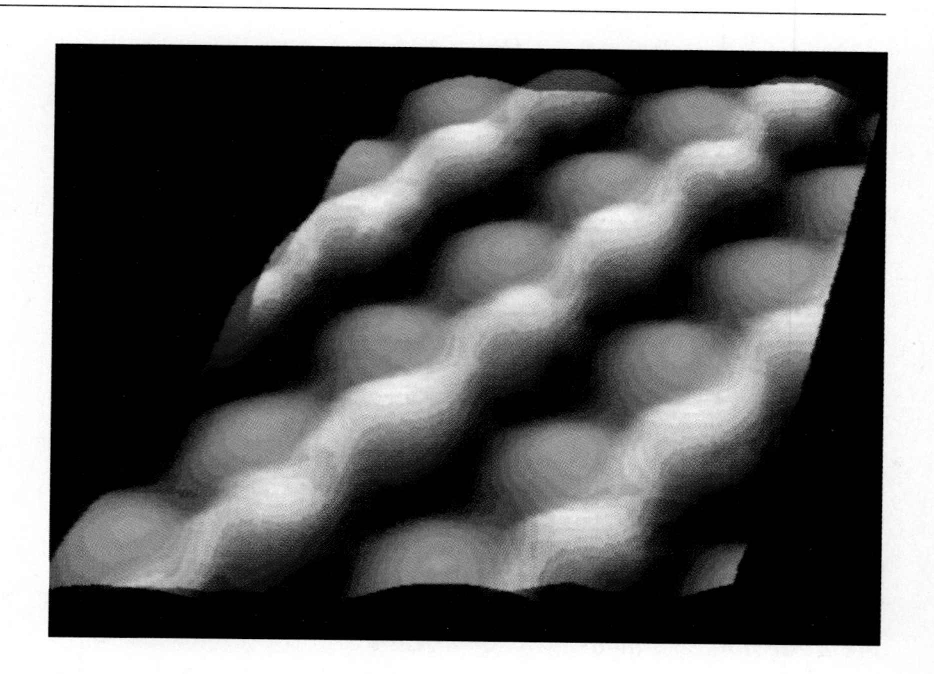

2.4 Atoms in a crystal of the semiconductor gallium arsenide

High-speed electrons create an image of hazy outer electron shells in this scanning tunneling micrograph. Color has been added to distinguish gallium atoms (blue) from arsenic atoms (red). (The operation of scanning tunneling microscopes, along with other sorts of microscopes, is described in Chapter 4.)

The electron cloud The portion of the atom outside the nucleus contains the third kind of primary particle—the ***electron.*** Though electrons have very little mass (see Table 2.2), their behavior is the single most crucial factor in the chemistry of life. Each electron carries a charge of -1, exactly the opposite of a proton's charge. In a neutral atom, the number of electrons around the nucleus is the same as the number of protons in the nucleus. The positive charges of the protons and the negative charges of the electrons cancel each other, making the total atom neutral.

The electrons are not in fixed positions outside the nucleus. Each is in constant motion, making 10^{15} to 10^{16} orbits of the nucleus each second. Hence, it is impossible to know exactly where any electron is; photographs made with the latest imaging technology record only a wispy shell boundary (Fig. 2.4). For this reason some illustrations of atoms, such as Fig. 2.2, do not show the electron itself but indicate the region where the electron is likely to be. All illustrations of atomic structure exaggerate the size of the nucleus. If a proton were the size indicated in Fig. 2.2, the outer edge of the electron cloud would extend 150 meters in all directions.

The average distance of an electron from the nucleus is a function of its energy; the higher its energy, the greater its probable distance from the nucleus. But in any particular atom, only certain discrete amounts, or levels, of energy—like steps on a staircase—are possible. To occupy a certain step, or energy level, an electron must possess a specific amount of energy. To achieve a higher energy level an electron must absorb additional energy from some outside source. Conversely, when an electron falls into the next lower level, it emits the same amount of energy it previously took to move up from that level (Fig. 2.5). We refer to an electron occupying the lowest step available to it in the atom as being in the ground state. Once it has absorbed enough energy to move up to the next energy level, it is said to be in an excited state.

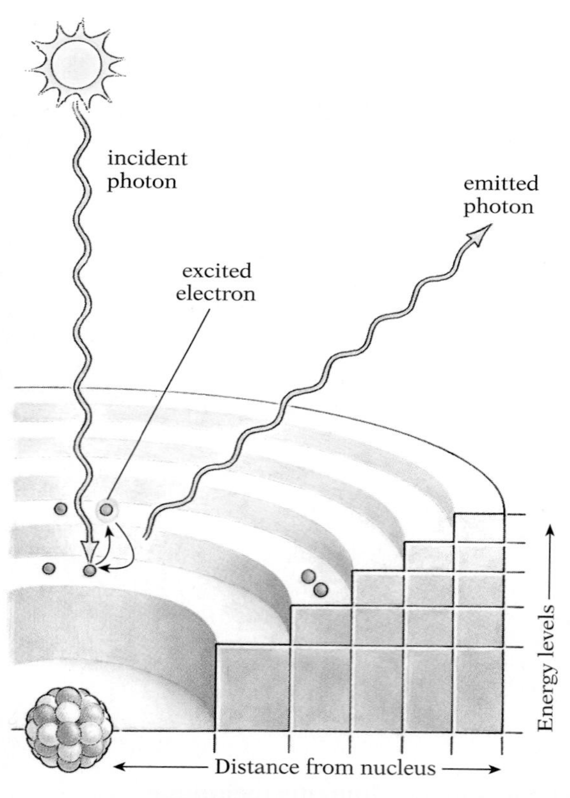

2.5 Energy levels of electrons

The electrons in an atom occupy discrete energy levels. If an electron absorbs the right amount of energy (shown as a photon, a discrete particle of light) and there is a vacancy in a higher energy level, the electron can move up the energy staircase to this level and so become excited. Usually an excited electron quickly reemits the absorbed energy (here again shown as a photon) and returns to its original energy level.

An electron in the excited state has a strong tendency to return to its ground state by emitting, in some form, the additional energy just acquired. Most often the energy is released as light. It is the decay of excited electrons in the lining of fluorescent tubes, for instance, that produces light. As we will see, this fleeting moment of excitation, lasting 10^{-8} sec, is critical to living things: life is based on the ability of specialized molecules in photosynthetic organisms to capture the energy of excited electrons before they drop back down the energy staircase.

The volume within which an electron can be found 90% of the time is known as its ***orbital.*** In illustrations of atoms the orbital is often represented by a circle, with the electrons sometimes shown as round spots on the circle (Fig. 2.6).

The energy level of the hydrogen electron is the level nearest the nucleus (often referred to as the *K* level). The *K* level can contain only two electrons. In an atom such as oxygen, with eight electrons, only two can be accommodated at the *K* level; the other six occupy a higher energy level, farther from the nucleus, called the *L* level. A maximum of eight electrons can fit into the *L* level. Since the most stable configuration for an atom is one in which its electrons have minimum energy, the six electrons outside the *K* level in an oxygen atom are all at the *L* energy level. None are found in the next higher energy level—the *M* level. Only when elements have more than 10 electrons, or when one of the lower-level electrons has been excited beyond the *L* level, are electrons found at the *M* level.

The most likely distance from the nucleus of each of the six *L*

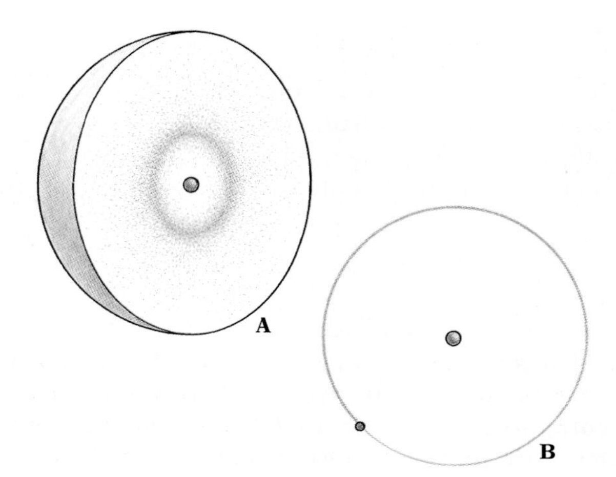

2.6 Two ways of representing the hydrogen atom

Since no one has ever seen the particles that make up an atom, all our knowledge of what atoms look like is indirect; we can only picture them as models that fit the data. (A) The nucleus is shown here as a central blue area, with the "cloud" around it in cross section representing the region where the electron is likely to be. The circle encloses the orbital of the electron—the spherical volume within which the electron will be found 90% of the time (see also Fig. 2.2). (B) Sometimes, for convenience, only the circle indicating the circumference of the orbital is shown; the electron may be represented by a small ball on the circle.

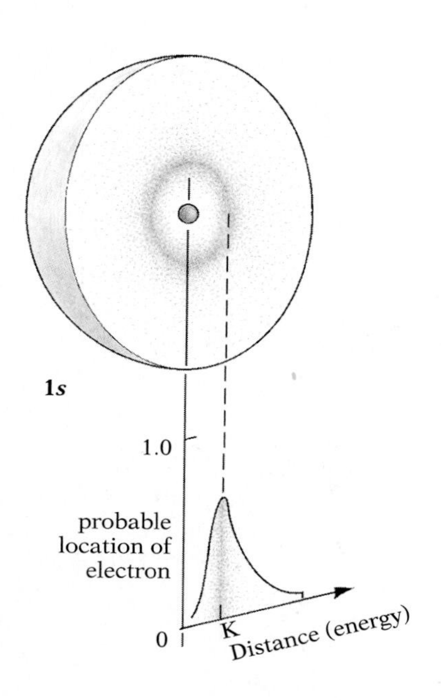

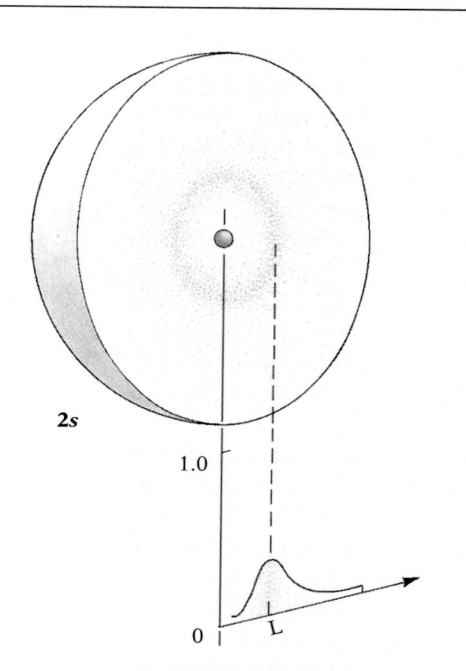

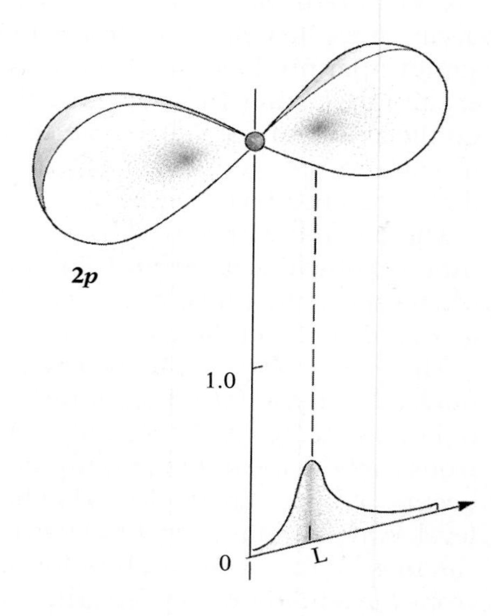

2.7 Electron orbitals

Top: The orbitals of *s* electrons are approximately spherical, while those of *p* electrons are roughly dumbbell-shaped. The numerals before *s* and *p* indicate the energy level. Thus the 1*s* electron is at the first energy level (*K*), nearest the nucleus; the 2*s* and 2*p* electrons are at the *L* level, a higher energy level, and hence are at a greater average distance from the nucleus than the 1*s* electron. Note that, despite the very different shapes of their orbitals, the 2*s* and 2*p* electrons are at the same energy level—their most probable distances from the nucleus are the same, as shown in the graphs (bottom).

2.8 Three 2*p* electron orbitals

Each of the dumbbell-shaped orbitals is oriented in a different dimension of space, at right angles to the other two.

electrons of an oxygen atom is somewhat greater than the most likely distance of the *K* electrons (Fig. 2.7). While the orbital of two electrons at the *K* level is circular, the other three pairs occupy dumbbell-shaped orbitals oriented at right angles to one another (Fig. 2.8).

■ HOW ELECTRONS DETERMINE CHEMISTRY

When the elements are arranged in sequence according to their atomic numbers—beginning with hydrogen (atomic number 1) and proceeding to uranium (number 92)—elements with very similar properties are found at regular intervals in the list (Fig. 2.9). For example, fluorine (number 9) is more like chlorine (number 17), bromine (number 35), and iodine (number 53) than it is like oxygen (number 8) or neon (number 10), the two elements immediately adjacent to it in the list. This tendency for chemical properties to recur periodically throughout the sequence of elements is called the Periodic Law.

The explanation for this periodicity is that the reactivity and other chemical properties of elements are largely determined by the number of electrons in their outer shell (the highest energy level). If that shell is complete, as in helium (atomic number 2),

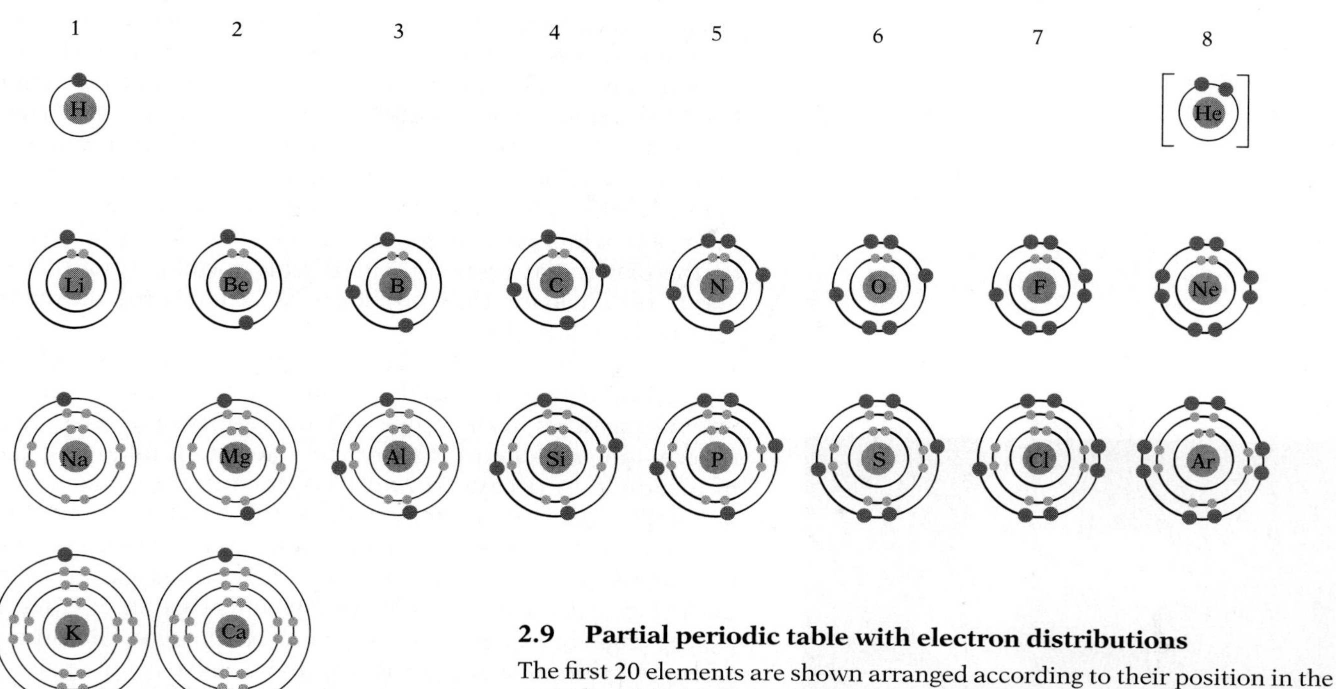

2.9 Partial periodic table with electron distributions
The first 20 elements are shown arranged according to their position in the periodic table. Elements in the same column share many chemical properties because they have the same number of electrons in their outer shell. (Helium is placed in column 8 even though it has only two outer electrons because, like neon, argon, and the other so-called noble gases, its outer shell is full, and its chemical properties are therefore those of a noble gas.)

neon (10), or argon (18), the element has very little tendency to react chemically with other atoms (Fig. 2.9). If the outermost shell has one electron fewer than the full complement, as in fluorine, chlorine, bromine, and iodine, the element has certain characteristic chemical properties; if, like oxygen, it lacks two electrons, the element has somewhat different properties.

A convenient way to represent the electron configuration of the outer shell is to symbolize each electron by a dot placed near the chemical symbol for the element under consideration. Thus fluorine and chlorine, which have seven electrons at their outer energy level, would have identical electron symbols:

$$:\underset{\cdot\cdot}{\ddot{F}}\cdot \qquad :\underset{\cdot\cdot}{\ddot{Cl}}\cdot$$

Similarly, hydrogen with one electron in its single shell, carbon with four in its outer shell, nitrogen with five, and oxygen with six are represented as follows:

$$H\cdot \qquad \cdot\underset{\cdot}{\dot{C}}\cdot \qquad \cdot\underset{\cdot}{\ddot{N}}\cdot \qquad \cdot\underset{\cdot\cdot}{\ddot{O}}\cdot$$

■ WHEN ATOMS DECAY

Though the various isotopes of an element carry different numbers of neutrons, their identical electron distributions give them the same chemical properties. Their physical properties, however, differ in two ways that are important to biological research and to human health.

Unusual isotopes are taken up by tissues just as well as the more common forms, but since isotopes differ significantly in atomic weight, they can be distinguished from each other by weight-sensitive techniques. For example, there was once controversy over whether the oxygen gas (O_2) released by plants comes from carbon dioxide (CO_2) or water (H_2O). The issue was settled by supplying plants with water containing a heavy isotope of oxygen (^{18}O rather than ^{16}O), while providing normal CO_2. The mass of the O_2 released by the plants was then compared with normal O_2 and found to be about 12% heavier, thus proving that water is the ultimate source for the oxygen in the air.

The other biologically significant physical property of some isotopes is their tendency to decay into a more stable form, giving off various particles in the process. The stability of an isotope is measured by the ***half-life*** of the isotope: the time it takes half the atoms in a sample to decay. Tritium (^3H), for instance, has a half-life of about 12 years; ^{32}P, roughly 14 days; ^{14}C, 5700 years; and so on.

Radioactive isotopes are extraordinarily useful in biology, since an isotope added to a sample emits radiation that acts as a label scientists can track (Fig. 2.10). With a labeled isotope of carbon in CO_2, for instance, we can trace how plants use carbon to build sugars. Because they are taken up by tissue as readily as their more stable counterparts, radioactive isotopes can also be used as tracers to help doctors locate circulatory blockages and pinpoint tumors. Naturally occurring isotopes make possible the dating of many rocks and fossils. The ratio of the radioactive isotope ^{14}C to the stable ^{12}C, for instance, is relatively constant in the CO_2 of the atmosphere, but once a plant has captured a CO_2 molecule and built it into a product like cellulose, the decay of ^{14}C atoms causes the ratio of ^{14}C to ^{12}C to decline steadily with time. Hence, the $^{14}C : {}^{12}C$ ratio in a sample provides a moderately accurate measure of age.

2.10 Autoradiography of a soybean plant

A radioactive isotope, ^{32}P, was taken up by the plant's roots and transported to sites of nucleic acid synthesis. To trace this movement, the plant was laid on a photographic emulsion, which recorded the sites of radioactive decay.

Isotopes can be dangerous as well. A radioactive atom in a living cell poses two potential threats. First, it can, like uranium, decay into another element by losing protons, thereby altering the chemistry of its molecule completely. More often, radioactive isotopes produce highly reactive molecules that have too many or too few electrons to balance the electrical charge of their protons, and thus have a net charge. Since the behavior of electrons, as we will see, determines the chemistry of life, such unpredictable and uncontrolled movement of electrons can disrupt the precisely ordered and carefully regulated workings of the cell. For instance, a change in a critical part of a cell's DNA can trigger the complicated chain of events that leads to cancer. Changes of this sort arise in all of us every day from exposure to the sun's radiation and from the natural decay of radioactive elements in the earth's atmosphere and crust, and each cell has a battery of defense mechanisms to reduce the damage.

THE GLUE OF LIFE: CHEMICAL BONDS

The arrangement of electrons in the outer shell of the atoms of most elements gives those atoms an ability to bind to others. When two or more atoms are bound together in this fashion, the force of attraction that holds them together is called a chemical bond. The atoms of each particular element can form only a limited number of such bonds; the arrangement of its electrons gives each element its own characteristic bonding capacity.

■ BONDS BETWEEN IONS

Atoms are particularly stable when the outer electron shell is complete—that is, in most cases, when it contains eight electrons. There is consequently a general tendency for atoms to form complete outer shells by reacting with other atoms.

Consider, for example, an atom of sodium (atomic number 11). This atom has two electrons in its first shell, eight in the second, and only one in the third. One way sodium might gain a complete outer shell would be to acquire seven more electrons from some other atom or atoms. But the sodium atom would then have an enormous excess of negative charge, and since like charges repel each other, the electrons would tend to push each other away from the sodium nucleus. Instead it becomes an ***electron donor,*** giving up the lone electron in its third shell to some other atom (called the ***electron acceptor),*** which leaves the complete second shell as the new outer shell (Fig. 2.11A).

Next, consider an atom of chlorine (atomic number 17). This atom has two electrons in its first shell, eight in its second shell, and seven in its third shell. Its outer shell lacks only a single electron. It cannot lose the seven electrons in its outer shell because it would become too highly charged. However, by gaining an extra electron from some electron donor, chlorine acquires a complete outer shell.

If a strong electron donor like sodium (an atom with a strong tendency to lose an electron) and a strong electron acceptor like chlorine (an atom with a strong tendency to acquire an extra electron) come into contact, an electron may be completely transferred from the donor to the acceptor. The result, in the present example, is a sodium atom with one electron fewer than normal and a chlorine atom with one more electron than normal. Once it has lost an electron, the sodium is left with one more proton than it has electrons, and it therefore has a net charge of +1. Similarly, the chlorine atom that gained an electron has one more electron than it has protons and has a net charge of −1. Such charged atoms are called ***ions,*** and are symbolized by their chemical symbol with a superscript indicating the charge. Sodium and chlorine ions are written Na^+ and Cl^-.

A sodium ion with its positive charge and a chlorine ion (usually called a chloride ion) with its negative charge tend to attract each other, since opposite charges attract. This important kind of electrical interaction is known as ***electrostatic attraction.*** In this instance electrostatic attraction holds the two ions together to form the compound we know as table salt, or sodium chloride, NaCl (Fig. 2.11B). Such a bond, involving the complete transfer of an electron and the mutual electrostatic attraction of the two ions thus formed, is an ***ionic bond.***

Ionic bonding may entail the transfer of more than one electron,

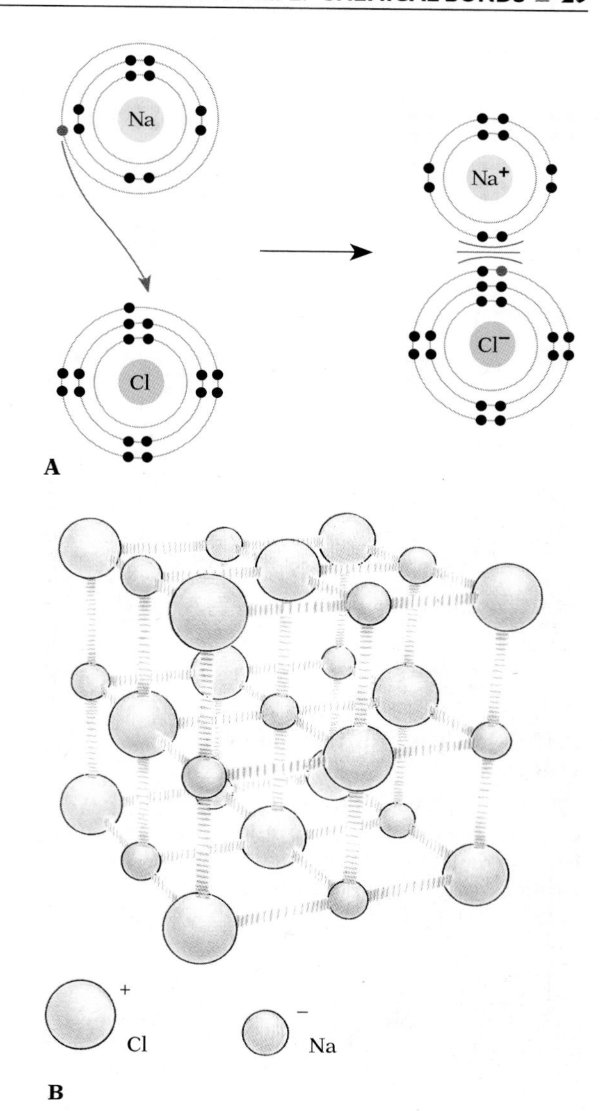

2.11 Ionic bonding of sodium and chlorine

Sodium has only one electron in its outer shell, while chlorine has seven (A). Sodium acts as an electron donor, giving up the one electron in its outer shell, whereupon the complete second shell functions as its new outer shell. Chlorine acts as an electron acceptor, picking up an additional electron to complete its outer shell. After donating an electron to chlorine, the sodium, left with one more proton than it has electrons, now has a positive charge. Conversely, the chlorine, with one more electron than it has protons, now has a negative charge. The two charged atoms, called ions, are attracted to each other electrostatically by their unlike charges. The result is sodium chloride (NaCl), or ordinary table salt, which in its crystalline state takes the form of a lattice-like structure (B).

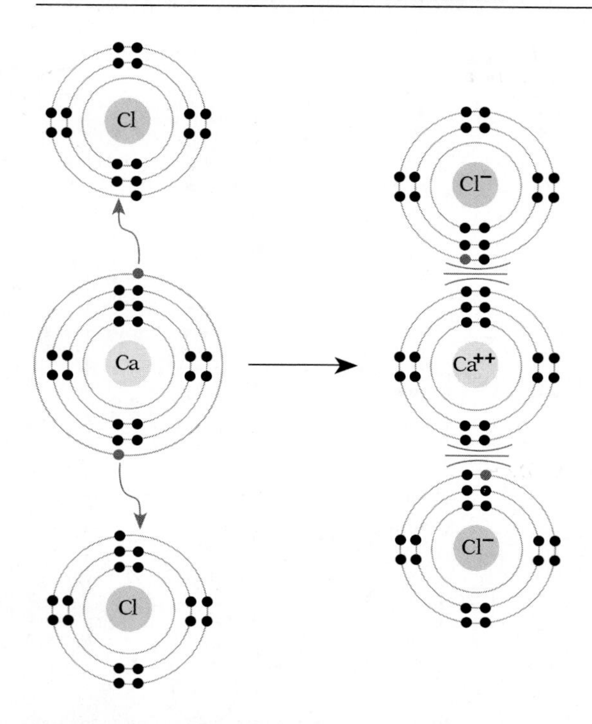

2.12 Ionic bonding of calcium and chlorine

Calcium has two electrons in its outer shell. It donates one to each of two chlorine atoms, and the two negatively charged chloride ions thus formed are attracted to the positively charged calcium ion to form calcium chloride ($CaCl_2$).

as in calcium chloride, another common salt. Calcium (atomic number, 20) has two electrons in its outermost shell, and it loses both to form the calcium ion, Ca^{++} (Fig. 2.12). Chlorine, however, requires only one electron to complete an octet in its outer shell. Hence it takes two chlorine atoms to act as acceptors for the two electrons from a single calcium atom; three ions bond together to form calcium chloride, symbolized as $CaCl_2$. Calcium has a bonding capacity, or ***valence,*** of +2, while sodium has a valence of +1 and chlorine a valence of −1.

Ionic bonding occurs between strong electron donors and strong electron acceptors. It is not common between configurations that have intermediate numbers of electrons in the outer shells, or between two strong electron donors or electron acceptors.

The binding of a calcium atom to two atoms of chlorine produces one molecule of calcium chloride. A ***molecule*** is generally defined as an electrically neutral aggregate of atoms bonded together strongly enough to be regarded as a single entity. In many instances, however, ***ionization*** (the transfer of one or more electrons from one atom to another to form ions) occurs without true molecular formation. Substances like sodium chloride (NaCl) and calcium chloride ($CaCl_2$), in which the bonds are ionic, tend to dissociate into separate ions when in solution. (We will look at why this happens shortly.) When they are ionized in solution, then, they exist as separate ions rather than molecules; NaCl, for instance, dissociates into a Na^+ ion and a Cl^- ion (Fig. 2.13).

Since ions are charged particles, they behave differently from neutral atoms or molecules, and substances wholly or partly ionized in water play many important roles in the functioning of biological systems. In later chapters we will see the effects of charge on the movements of materials through the membranes of cells.

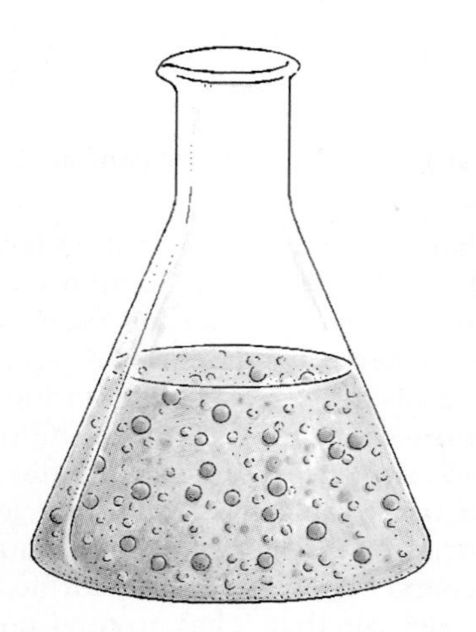

2.13 Ionization of sodium chloride

When in solution, the NaCl dissociates into separate Na^+ (yellow) and Cl^- (green) ions.

■ ACIDS AND BASES

Cells are extemely sensitive to the chemistry of the fluids that bathe them. The delicate balance of acids and bases, in particular, is crucial to most tissues. An ***acid*** is a substance that increases the concentration of hydrogen ions (H^+) in water; a ***base*** is a substance that decreases the concentration of hydrogen ions, which in water is equivalent to increasing the concentration of hydroxyl ions (OH^-).

The degree of acidity or basicity (usually called alkalinity) of a solution is commonly measured in terms of ***pH,*** which reflects the concentration of hydrogen ions. The pH scale generally ranges from 0 on the acidic end to 14 on the alkaline end. A solution is neutral (containing equal concentrations of H^+ ions and OH^- ions) if its pH is exactly 7. Solutions with a pH of less than 7 are acidic (with a higher concentration of H^+ ions than of OH^- ions). Conversely, solutions with a pH higher than 7 are alkaline (with a higher concentration of OH^- ions than of H^+ ions). A change of one pH unit means a 10-fold change in the concentration of hydrogen ions. Thus the concentration of H^+ ions in the solution of a very strong acid (pH 0) is 100 trillion (10^{14}) times greater than in the solution of a very strong base (pH 14). Figure 2.14 illustrates the range of pH values we normally encounter.

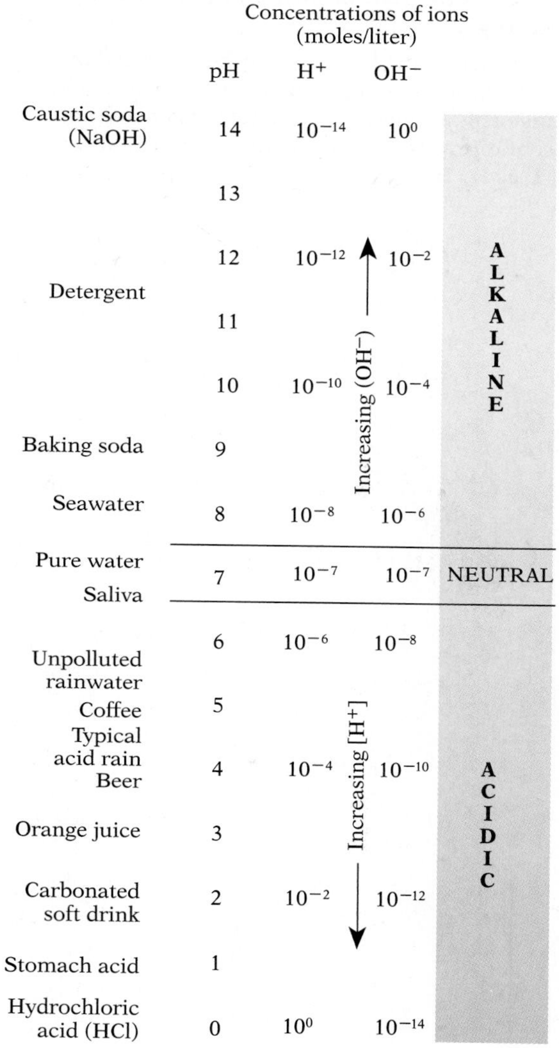

Except in parts of the animal digestive tract and a few other isolated areas, most cells function best when conditions are nearly neutral. Most of the interior material of living cells has a pH of about 6.8. The blood plasma and other fluids that bathe the cells in our own bodies have a pH of 7.2–7.3. Special mechanisms stabilize the pH of these fluids. In particular, chemicals known as buffers have the capacity to bond to H^+ ions, removing them from solution whenever their concentration begins to rise and, conversely, to release H^+ ions into solution whenever their concentration begins to fall. Buffers thus help minimize fluctuations in pH, which would otherwise be considerable since many of the biochemical reactions in cells either release or use up H^+ ions. The most important biological buffer in vertebrates is carbonic acid, which stabilizes the pH of the blood as it circulates through the tissues.

2.14 The pH scale

The concentration of hydrogen ions in a solution is measured by pH. At pH 7, the concentration of hydrogen ions (H^+) exactly balances the concentration of hydroxyl ions (OH^-), and so the solution is neutral. At lower pH values (corresponding to higher H^+ concentrations) solutions are acidic; at higher pH values (corresponding to lower H^+ concentrations) solutions are alkaline, or basic. Notice that the pH number matches the concentration of H^+ in moles per liter—for example, pH 8 corresponds to a H^+ ion concentration of 10^{-8}. A mole of any compound contains the same number of atoms—Avogadro's number, or 6.02×10^{23}—as a mole of any other substance, and its weight in grams corresponds to the molecular weight (the summed atomic masses) of the substance.

pH is the negative logarithm of the hydrogen ion concentration:

$$\text{pH} = \log\left(\frac{1}{[H^+]}\right) = -\log[H^+]$$

Concentrations of ions (moles/liter)

	pH	H^+	OH^-	
Caustic soda (NaOH)	14	10^{-14}	10^{0}	ALKALINE
	13			
Detergent	12	10^{-12}	10^{-2}	
	11			
	10	10^{-10}	10^{-4}	
Baking soda	9			
Seawater	8	10^{-8}	10^{-6}	
Pure water, Saliva	7	10^{-7}	10^{-7}	NEUTRAL
Unpolluted rainwater	6	10^{-6}	10^{-8}	ACIDIC
Coffee	5			
Typical acid rain, Beer	4	10^{-4}	10^{-10}	
Orange juice	3			
Carbonated soft drink	2	10^{-2}	10^{-12}	
Stomach acid	1			
Hydrochloric acid (HCl)	0	10^{0}	10^{-14}	

Increasing (OH^-) ↑

Increasing [H^+] ↓

■ **HOW STRONG BONDS ARE FORMED**

Ionic bonds, as we have seen, involve the complete transfer of electrons from one atom to another. However, in most cases bonding occurs not by complete transfer, but by a sharing of electrons between the atoms involved. Bonds of this sort, based on shared electrons, are called ***covalent bonds.*** These may be nonpolar or polar.

Nonpolar bonds To see how covalent bonds are formed, consider an atom of hydrogen. A complete first shell for hydrogen would contain two electrons, one more than each atom has normally. Two atoms of hydrogen can bond to each other and form molecular hydrogen (H_2):

$$H\cdot + H\cdot \longrightarrow H{:}H$$

In this molecule, each atom shares its electron equally with the

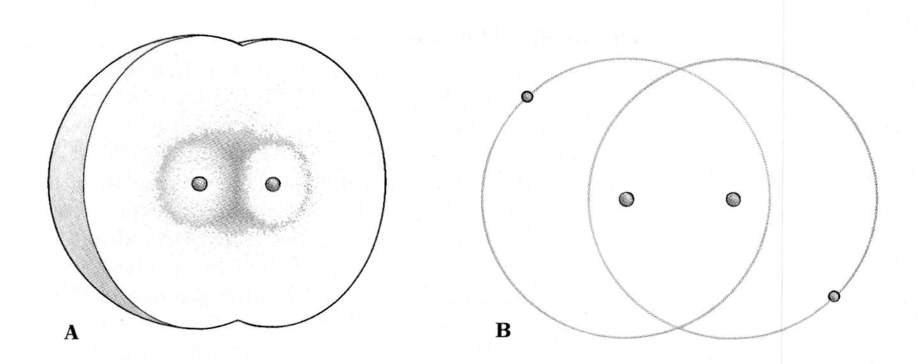

2.15 Covalent bonding of two hydrogen atoms

The sharing of electrons may be indicated by overlapping electron clouds (A) or by interlocking orbital rings (B). The latter system of representation parallels the textual representation of H_2 as H : H.

other atom, so that each hydrogen has, in a sense, two electrons (Fig. 2.15).

Covalent bonds are not limited to the sharing of one electron pair between two atoms. Sometimes two atoms share two or three electron pairs and form double or triple bonds. When two atoms of oxygen bond together, they form a double bond (since each oxygen atom needs two electrons to complete its outer shell), and when two atoms of nitrogen (atomic number 7) bond together, they form a triple bond, because each nitrogen atom needs three additional electrons to fill its outer shell:

$$:\ddot{\underset{\cdot}{O}}::\ddot{\underset{\cdot}{O}}: \qquad :N\vdots\vdots N:$$

A covalent bond may be represented simply by a line between two atoms, instead of a pair of dots; the other electrons in the outer shells are then ignored:

$$\mathrm{H-H} \qquad \mathrm{O=O} \qquad \mathrm{N\equiv N}$$

The number of covalent bonds an atom can form is the same as its valence (ignoring the sign): hydrogen atoms tend to form only one bond, oxygen two bonds, and nitrogen three bonds. This covalent bonding capacity corresponds to the number of vacancies in the outer shell of an atom with one to three vacancies, and to the number of sharable electrons in an atom with one to three outer electrons. The maximum covalent bonding capacity is 4: in an atom with four outer electrons, this is the number both of sharable electrons *and* of vacancies. Carbon's bonding capacity of 4 confers on it the ability to form the maximum number of bonds.

Polar bonds Suppose that instead of being bonded to each other, two hydrogen atoms are covalently bonded to an oxygen atom, forming water (H_2O):

$$\mathrm{H}\cdot + \mathrm{H}\cdot + \cdot\ddot{\underset{\cdot}{O}}: \longrightarrow \mathrm{H}:\ddot{\underset{\mathrm{H}}{\underset{\cdot\cdot}{O}}}:$$

Oxygen, with six electrons in its outer shell (Fig. 2.9), needs two more. By sharing electrons with two hydrogen atoms, the oxygen atom can obtain a full outer octet, while at the same time each hydrogen obtains a complete first shell of two electrons. A covalent bond between a hydrogen atom and an oxygen atom is somewhat different from one between two hydrogen atoms or between two oxygen atoms, however. No two elements have exactly the same

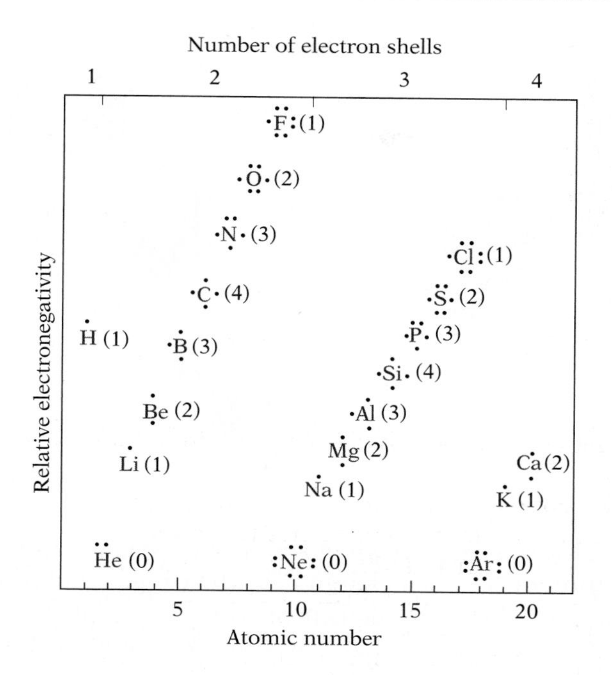

2.16 Electronegativity

The relative tendency of an atom to attract electrons depends on the number of spaces in the outer shell left to be filled and on the distance of the outer shell from the nucleus. Hence, lithium (Li), with seven vacancies, is less attractive to electrons (that is, less electronegative) than carbon (C), which has four. Oxygen (O), with only two missing electrons, is yet more electronegative. This graph also helps explain why in methane (CH_4) the shared electrons will be nearer the carbon atom, while in carbon dioxide (CO_2) they will be nearer the oxygens: carbon is more electronegative than hydrogen but less electronegative than oxygen. The electronegativities of the noble gases (which have filled outer shells) are estimated. This knowledge of relative electronegativity permits us to make important predictions about many biochemical reactions. (The covalent bonding capacity of each atom is shown in parentheses.)

affinity for electrons. Consequently, when a covalent bond forms between two different elements, the shared electrons tend to be pulled closer to the more attractive element. Such a bond is called a ***polar*** covalent bond because the charge is distributed asymmetrically.

The formal measure of an atom's attraction for free electrons is its ***electronegativity;*** this depends on the number of vacancies in the outer shell—an atom like oxygen with only two electron openings is generally more electronegative than one with three or more—and on the distance of the outer shell from the nucleus (Fig. 2.16).

The phenomenon of polarity helps explain many of the properties of various molecules in living systems. Whole molecules can be polar as a result of the polarity of bonds within them. One example is the water molecule. We can imagine how, even though the two hydrogen-oxygen bonds are polar, the atoms in the water molecule might be aligned in a straight line so that the charge would be distributed symmetrically within the molecule, which would then be nonpolar:

H : Ö : H

However, this is not the actual arrangement. The covalent bonds formed by atoms are maximally separated from one another, and thus are oriented to the four corners of a tetrahedron (Fig. 2.17). The three atoms of water form a bent-chain, or V-shaped, structure, with the oxygen at the apex of the V and the two hydrogen atoms as the arms:

(–)
(+) H : Ö :
H
(+)

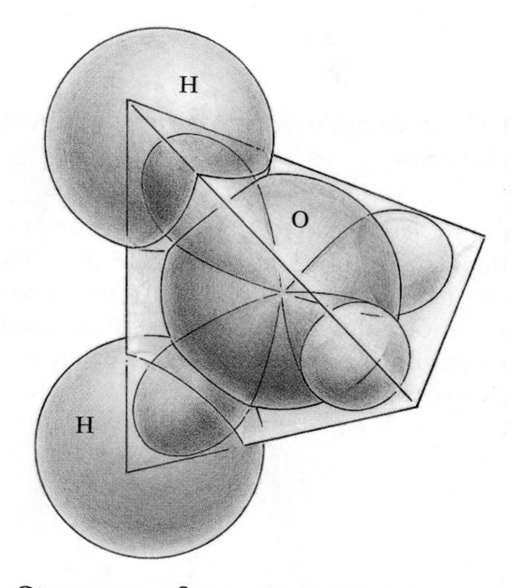

2.17 Structure of a water molecule

When oxygen bonds covalently with two hydrogen atoms, its second-level electrons become oriented to the four corners of a tetrahedron. As a result, the angle between the two hydrogens is neither 90° nor 180°, as might be expected from the perpendicular arrangement of the 2*p* orbitals, but rather is 104.5°.

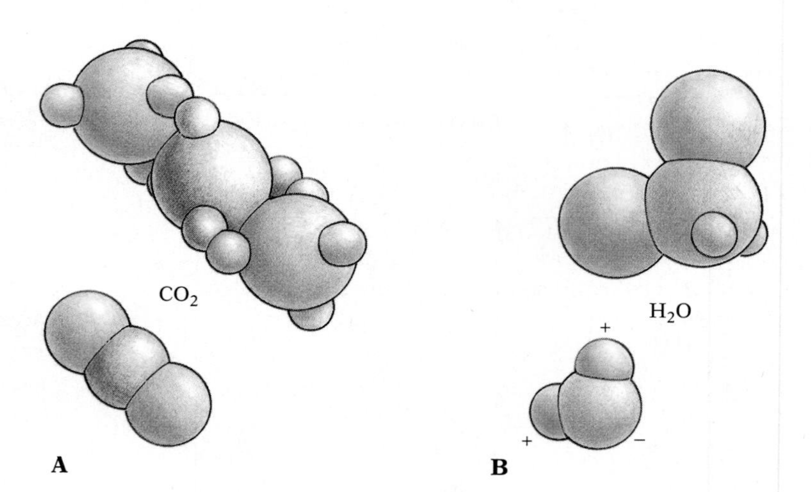

2.18 Polarities of two biologically important molecules

(A) In carbon dioxide, each of the four electrons of the carbon's outer orbital is shared between the carbon (gray) and an oxygen (red). The result is a linear molecule with no electrical polarity. (B) In water, two hydrogen atoms (blue) are bonded covalently to one oxygen atom, but the shared electrons are pulled closer to the oxygen because of its higher electronegativity. Because the atoms in water are arranged at an angle of 104.5°, the charge distribution is asymmetrical, with negative charge concentrated at the oxygen end; as a result the molecule as a whole is polar.

Since the electrons are drawn closer to the oxygen atom, there is a concentration of negative charge near the oxygen end of the molecule. Therefore, the molecule is polar (Fig. 2.18, right).

The carbon dioxide molecule, on the other hand, exhibits no polarity: its double bonds hold its atoms in rigid linear alignment (Fig. 2.19, left). Hence CO_2 is nonpolar.

As we will see shortly, the polarity of certain molecules has crucially important biological implications.

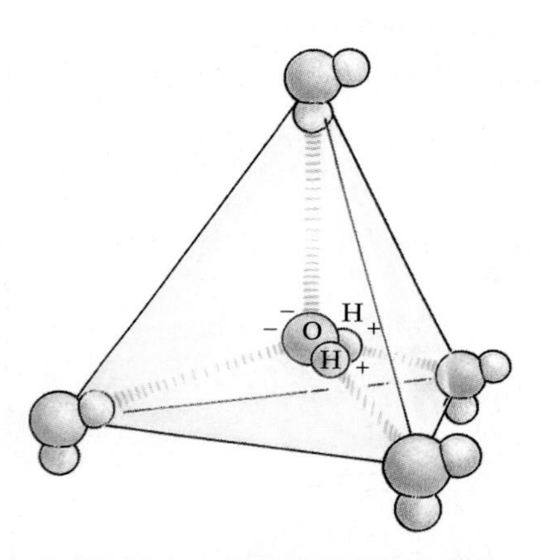

2.19 Hydrogen bonding between water molecules

Like the central H_2O molecule shown here, each water molecule can form hydrogen bonds (red bands) with four other water molecules. The array then assumes the shape of a tetrahedron. Water molecules near the edge of this imaginary tetrahedron can simultaneously form hydrogen bonds with two or three other water molecules, creating an interlocking array of tetrahedrons.

■ WHY ARE WEAK BONDS IMPORTANT?

Strong versus weak bonds To maintain internal stability, or ***homeostasis,*** living organisms must be able to adjust to the fluctuations of their environments. The changes all begin at the molecular level, and are powered by the liberated energy derived from strong, energy-rich covalent bonds. Covalent bonds are called strong because breaking them is hard, usually requiring between 50 and 110 kilocalories of energy per mole.[1] Bond breakage usually results from collisions of rapidly moving molecules. But since the energy from even the most rapidly moving molecules at physiological temperatures is almost never above 10 kcal/mole, covalent bonds are stable and show little tendency to rupture spontaneously.

However, life depends on a capacity for change, as well as on stability. The crucial sources of this ability to change are weak noncovalent bonds, which can readily be broken and re-formed. Ionic bonds in aqueous solutions, for instance, are relatively weak, averaging about 10 kcal/mole. The average duration of an ionic bond (the interval between formation of the bond and a collision with a

[1] A mole is the amount of a substance, in grams, that equals the combined atomic mass of all the atoms in a molecule of that substance; there are approximately 6 × 10^{23} molecules in a mole. A calorie (spelled with a small *c*) is defined as the quantity of energy, in the form of heat, required to raise the temperature of one gram of water from 14.5°C to 15.5°C. One ***kilocalorie*** (kcal) is 1000 calories. Nutritionists use a different scale to measure energy; their Calorie (spelled with a capital C) is equal to one kilocalorie on the standard scale.

BIOXPLORER PLUS

Student Software

DOWNLOAD a sample chapter from the World Wide Web
http://web.wwnorton.com/bioxplus.html

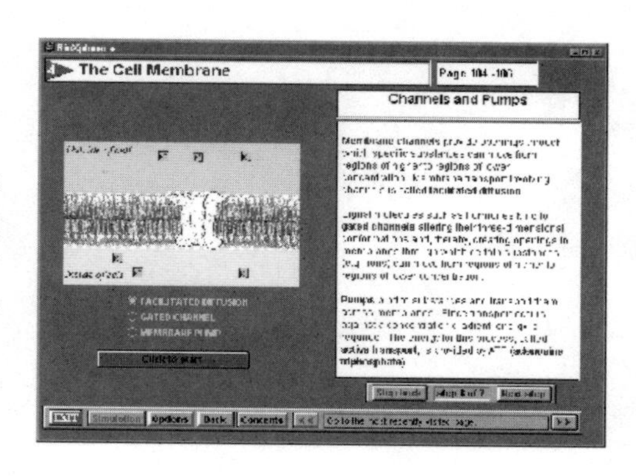

This enjoyable and easy-to-use program uses animation, simulated interactive experiments, quizzes, matching games, and an on-line glossary to help you understand and remember important concepts. **BioXplorer Plus** is carefully keyed to *Biological Science*, Sixth Edition, and includes page references to the text on every card.

REVIEW key text material, summarized on more than 250 brief notecards.

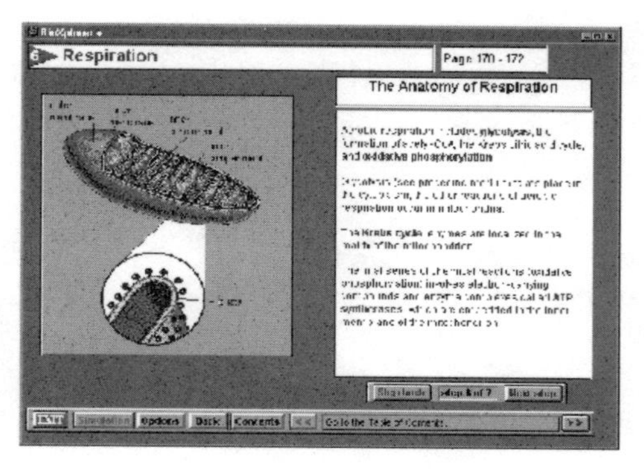

ANIMATIONS of complex biological processes, such as active transport across biological membranes, the mechanisms of photosynthesis and respiration, and recombinant DNA technology, make difficult concepts easier to grasp. Interactive **SIMULATIONS** allow you to manipulate variables within set experiments and see the results on-screen.

Test your understanding with the **CHAPTER QUIZ** feature. **BioXplorer Plus** includes more than 550 questions (with answers and feedback) covering every chapter of the text.

A **GLOSSARY**, complete with search function, provides handy definitions of key terms.

Detach and mail to:
W.W. Norton & Company
c/o National Book Company
800 Keystone Industrial Park
Scranton, PA 18512-4601
Attn: Mary C

System Requirements: The Macintosh version will run on any system that has sufficient memory to run Hypercard. The Windows version requires a 386 or higher processor operating under Windows 3.1 or higher and at least 8MB of memory.

YES, please send me BioXplorer Plus:

BioXplorer Plus	$17.95
Tax (in CA, NY, PA, IL)	______
Total	______

Choose one: ❑ Windows® (3.5″) ❑ Macintosh®

❑ My check for $17.95 (plus tax, where applicable) is enclosed. Make checks payble to W. W. Norton & Company.

Please charge to my: ❑ Visa ❑ MasterCard ❑ American Express

Acct. # ____________ Exp. date ________ Telephone ____________

Name ____________ Signature ____________

Ship to: Address ____________

City ____________ State/Prov. ________ Zip/PC ________

Billing Address (if different): Address: ____________

City: ____________ State/Prov. ________ Zip/PC ________

molecule moving rapidly enough to break it) is quite short. As a result, several weak bonds must act in concert to produce molecular stability sufficient for most of life's metabolic processes. The weak bonds of biological significance include ionic bonds (discussed earlier), hydrogen bonds, and van der Waals interactions.

Hydrogen bonds The electrostatic attraction between oppositely charged portions of neighboring polar molecules creates ***hydrogen bonds*** (also called ***polar bonds***). Water molecules provide an excellent example. The hydrogen atoms in each water molecule are covalently bonded to the oxygen atom, but because of the polarity of the bond—the electrons being closer to the oxygen end than to the hydrogen ends—each hydrogen has a net positive charge. The hydrogen atoms are therefore attracted to other oxygen atoms, with their net negative charge, in nearby water molecules. Specifically, each hydrogen can form a weak attachment with the oxygen of another water molecule, and the oxygen can form a weak attachment with two hydrogens in other water molecules. Thus each water molecule has the potential for being simultaneously linked by hydrogen bonds to four other water molecules (Fig. 2.19), which are themselves weakly bonded to yet other water molecules.

The distinction between hydrogen bonds and ionic bonds is clear: hydrogen bonds result from the electrostatic attraction between polar but electrically neutral molecules like water, while ionic bonds result from the electrostatic attraction between oppositely charged atoms (ions). There is also a hybrid of these two kinds of electrostatic attraction: ions can attract polar molecules, as we will see when we discuss, later in the chapter, the ***hydration sphere***—the shell of polar water molecules drawn around an ion in solution. In aqueous solutions, hydrogen bonds usually have a bonding energy of about 4–5 kcal/mole; ionic bonds, about 10 kcal/mole; and hybrid polar-ionic bonds such as those of the hydration sphere, about 7–8 kcal/mole.

Van der Waals interactions Much weaker than ionic or hydrogen bonds are the linkages known as ***van der Waals interactions***, which have bonding energies of only 1–2 kcal/mole. These linkages occur between electrically neutral molecules (or parts of molecules) when they are a precise distance from each other that allows the electrons in their outer orbitals to begin a synchronous, mutually avoiding motion. The normal repulsion between the two sets of outer electrons is lessened, and the atoms are able to bond weakly to each other. Van der Waals interactions play a crucial role in the enzymatic reactions that control virtually all the processes of life. They are also important in stabilizing aggregations of nonpolar molecules in cells.

The strength in weakness Weak bonds play a crucial role in stabilizing the shape of many of the large molecules—DNA and proteins in particular—and they often hold together groups of such molecules in orderly arrays. Like the minute hooks and eyes of a Velcro fastening, the bonds are individually quite weak (their average duration is only 10^{-11} sec), but they can be strong and stable

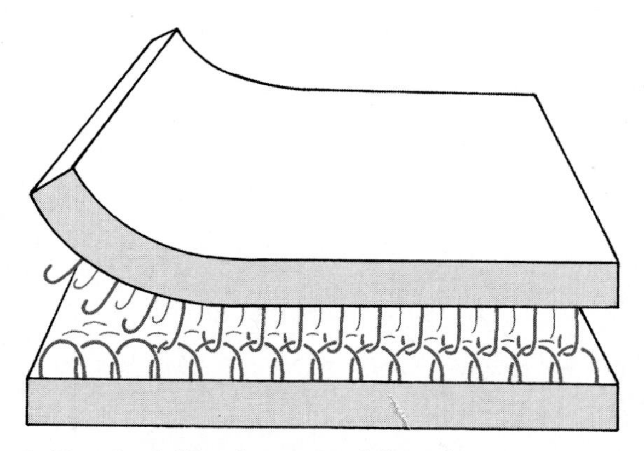

2.20 Stability from weak bonds

An array of many individually weak bonds—here represented as the many hooks and eyes of a Velcro fastener—can be surprisingly strong as a unit.

when many act together (Fig. 2.20). However, just as a Velcro patch is easy to unfasten from one end, a few "bonds" at a time, so too an array of molecules held together by weak bonds (as opposed to the more gripperlike covalent bonds) can be disassembled and rearranged with relative ease by forces that dislodge the fastening bonds one by one. An enormous number of life processes depend on just these sorts of changes.

WHY LIFE NEEDS INORGANIC MOLECULES

Chemists have traditionally referred to complex molecules containing carbon as ***organic*** compounds. All other compounds are called ***inorganic.*** Despite their name, many inorganic (non-carbon-based) substances are basic to the chemistry of life.

■ THE REMARKABLE CHEMISTRY OF WATER

Life on earth is totally dependent on water. Between 70% and 90% of all living tissue is water, and the chemical reactions of life all take place in an aqueous medium.

Water as a solvent One of the main reasons water is so well adapted as the medium for life is that it is a superb solvent for many important classes of chemicals. This advantage arises from the marked polarity of water molecules and the corresponding propensity of water to form hydrogen bonds. Thanks to this polarity, both ionic and polar molecules are soluble in water (Fig. 2.21). Let us consider the effect of water on each.

2.21 Dead Sea

Because water is polar and NaCl is ionic in solution, relatively large amounts of salt can be dissolved in water. In the Dead Sea, nearly 30% of the fluid is dissolved salt, and the density is so great that humans float high in the water.

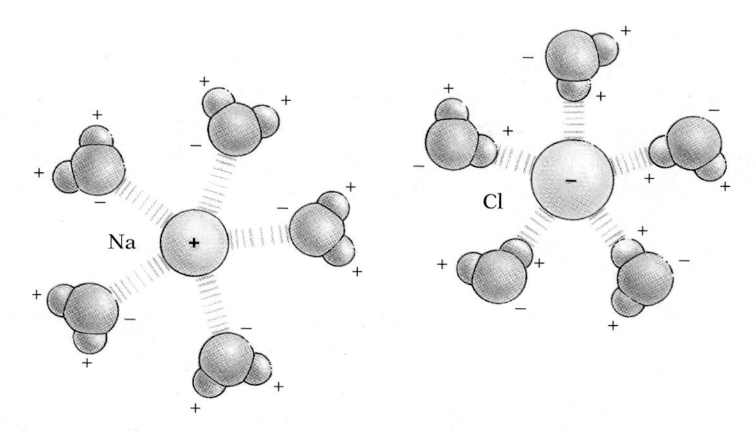

2.22 Hydration spheres of Na^+ and Cl^-

When dissolved in water, each Na^+ and Cl^- ion is hydrated—that is, surrounded by water molecules electrostatically attracted to it. Note that the oxygen (red) of the water molecules is attracted to the positively charged Na^+, while the hydrogen (blue) of the water molecules is attracted to the negatively charged Cl^-. Water molecules in a hydration sphere are called bound water. This bonding between ion and polar molecules (red bands) makes evident the common electrostatic basis of ionic bonds and hydrogen (polar) bonds.

The ionic bonds linking the atoms of NaCl are relatively weak when the salt is in an aqueous medium, but within a dry crystal of the same salt the bonds are comparatively strong. Why the difference? When the crystal is put into water, the negatively charged oxygen ends of the water molecules are attracted to the positively charged sodium ions: similarly the positively charged hydrogen ends of the water molecules are attracted to the negatively charged chloride ions. By force of numbers these attractions are able to overcome the strong mutual attraction between the Na^+ and Cl^- ions. In water, then, the ionic bonds are broken with extreme ease because of the competitive attraction of the water. The Na^+ and Cl^- ions dissociate, and each of the ions is surrounded by a hydration sphere of regularly arranged water molecules electrostatically attracted to it—a process called ***hydration*** (Fig. 2.22). The ionic bonds between Na^+ and Cl^- atoms are now weaker simply because the ions are kept far apart, and the strength of electrostatic attraction decreases exponentially with distance.

Water is also an excellent solvent for nonionic molecules if they are polar. Such molecules—ethyl alcohol, for instance—are called ***hydrophilic*** ("water-loving"). They dissolve in water because of the electrostatic attraction between the charged parts of the solute molecules and the oppositely charged parts of the water molecules. This occurs especially when the molecule has an oxygen with a hydrogen attached to it (−OH). As in water molecules, the hydrogen in such a group has a net positive charge and is therefore attracted by the negatively charged oxygen end of a nearby water molecule, with the result that a hydrogen bond is formed. The dissolved

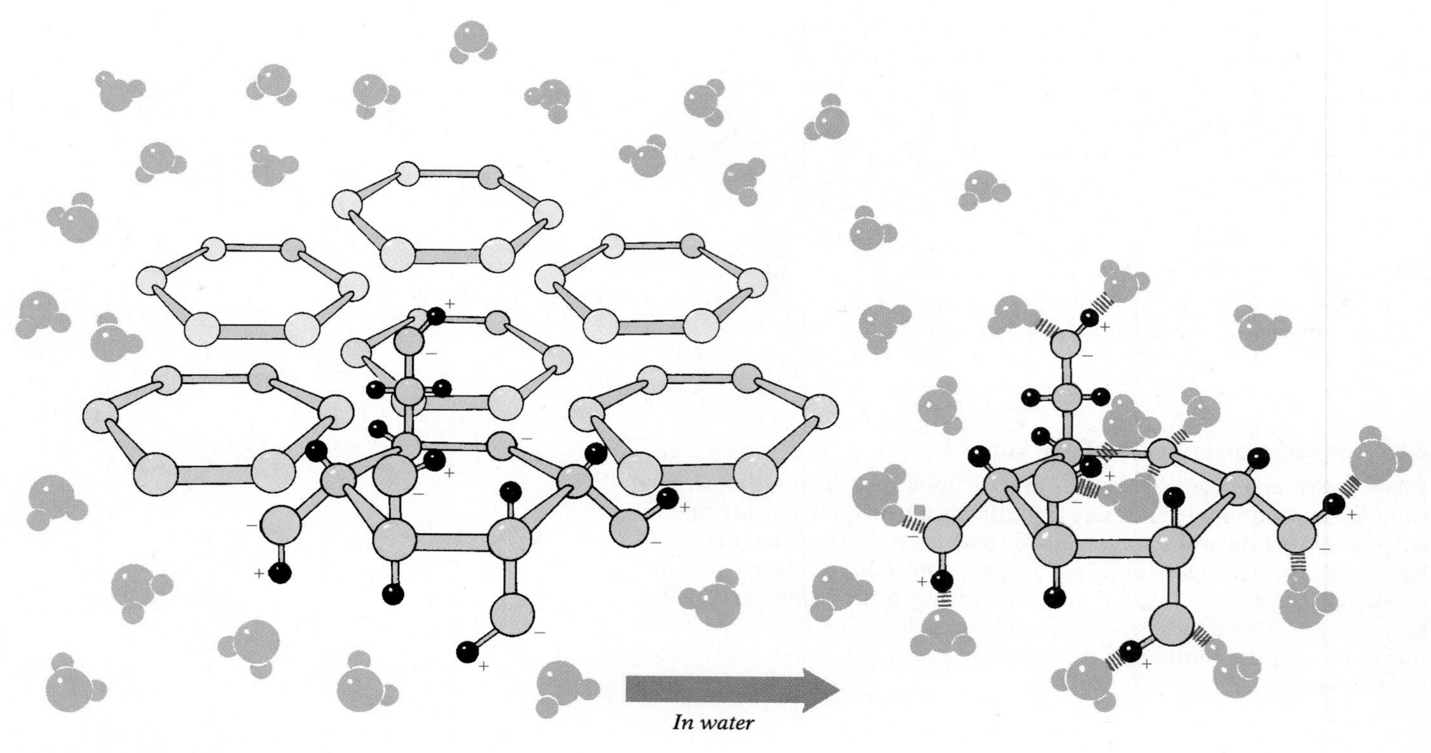

2.23 Polar basis of solubility
When a polar substance such as glucose, an energy-rich sugar (left), is placed in contact with water, the water molecules are attracted to the polar atoms of the sugar. (For clarity the polar –OH groups—green and black spheres—are shown for only one of the sugar molecules.) The water forms hydrogen bonds with the substance, surrounding it with water molecules, and so dissolves it (right).

(solute) molecules and the water molecules thus become weakly linked to each other (Fig. 2.23).

Substances that do not dissolve in water are electrically neutral and nonpolar. They therefore show no tendency to interact electrostatically with water. A ***hydrophobic*** ("water-fearing") substance like vegetable oil, when stirred into water, will soon begin to separate out, because the water molecules tend slowly to reestablish the hydrogen bonds broken by the physical intrusion of the insoluble material, and thus push it out. As a result, the nonpolar molecules tend to coalesce to form droplets, which usually fuse and form a separate layer outside the water (Fig. 2.24). As we will see, this tendency of hydrophobic molecules to be driven out of water is the basis for the spontaneous formation of the cell membranes that protect organisms as they grow and develop.

The polar nature of water also helps explain the mundane but important workings of soaps and other detergents. Stains left by water-soluble chemicals readily dissolve away when washed in water, but grease and other hydrophobic materials leave stains that cannot be lifted so easily. Detergents (Fig. 2.25) are able to remove grease from a natural fabric because their molecules have ionic heads and hydrophobic tails. The heads make them soluble in water, while the tails tend to be driven into tight-packed company with the grease molecules and dissolve them by incorporating them into hydrophobic detergent droplets. Nonpolar fabrics (like most synthetics) pose a special problem because water cannot easily wet them, which it must if the detergent is to be carried into the fibers; even if the water does penetrate well, the water will often "herd" the stain molecules into close association with the fabric rather than the detergent. Molecules of fluids used by dry cleaners stand a better chance because they readily wet nonpolar materials and dissolve grease, yet because they are small enough to evaporate, will not simply remain with the fabric after displacing the stain. Covalently bonded stains like dried blood are notoriously difficult to remove; they require chemically active agents like bleach or digestive enzymes.

The stickiness of water Water molecules in the hydration spheres of ions are arranged in orderly arrays; such water, referred to as ***bound water,*** is essentially immobilized. The same is true of water molecules around polar groups of nonionic compounds. The orderly arrays of bound water are very different from those of pure water (see Fig. 2.19), and the physical properties of bound water are consequently different from those of free water; the greater the proportion of bound water in a given volume, the lower the freezing point and the higher the boiling point of that liquid. Since much of the water inside cells is bound water, the physical properties of the cell contents are very different from those of pure water.

The strong ordering of water molecules by hydrogen bonding has important implications for life. For example, water has a high ***surface tension***: the surface of a volume of water is not easily broken. The effects of surface tension are evident when a water strider

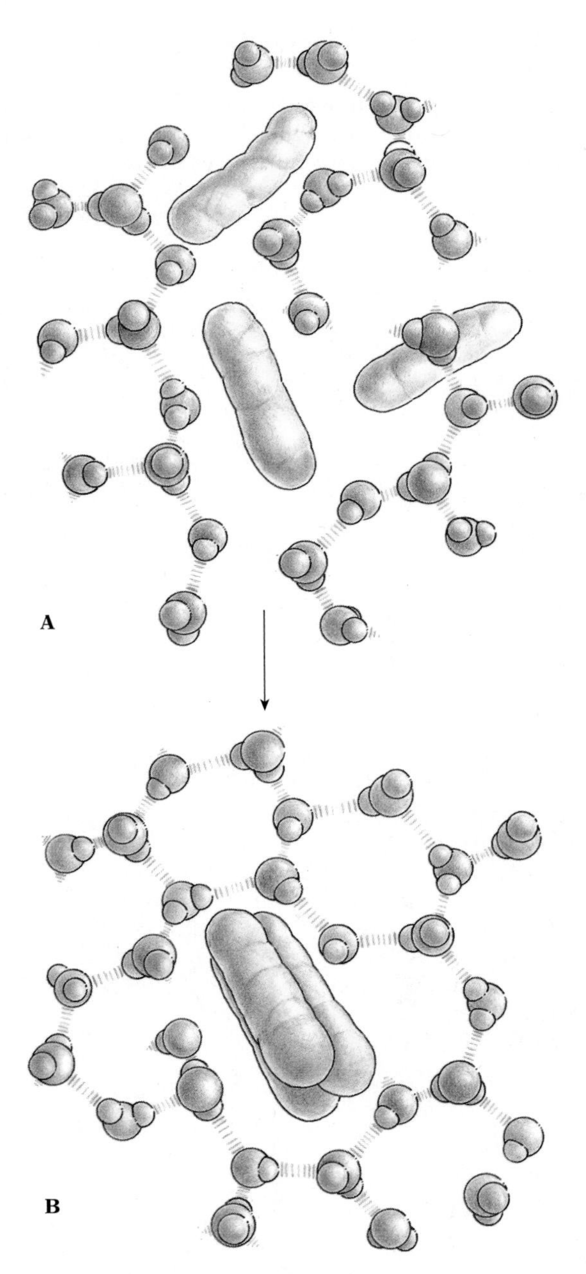

2.24 Water-induced clumping of hydrophobic molecules

Dispersed hydrophobic molecules disrupt the polar bonding pattern of pure water, so that few hydrogen bonds can form in the solution (A). As hydrophobic molecules (represented as brown ovals) encounter one another randomly in a solution of water, they tend to become trapped in clumps by polar bonding of water molecules to one another (B). Because there is more polar bonding when hydrophobic molecules are clumped, the solution becomes stabilized in this form.

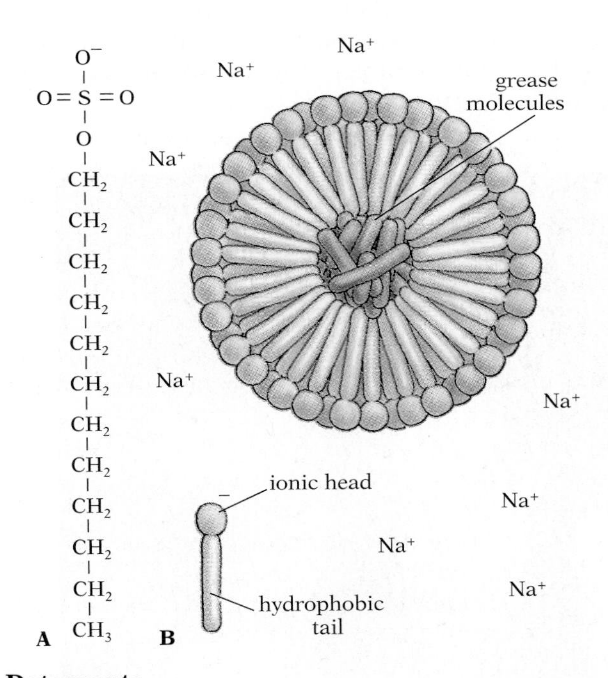

2.25 Detergents

Sodium dodecyl sulfate is a strong detergent often used to dissolve cell membranes and other hydrophobic molecules in experiments requiring the separation of these components for further analysis. It has a long, straight hydrophobic tail and, because it ionizes in water, a charged head (A). Water molecules dissolve the heads and drive the tails into tightly packed clumps that dissolve hydrophobic grease molecules (B). During washing, whether in a laboratory preparation or a home washing machine, the entire assembly is rinsed out.

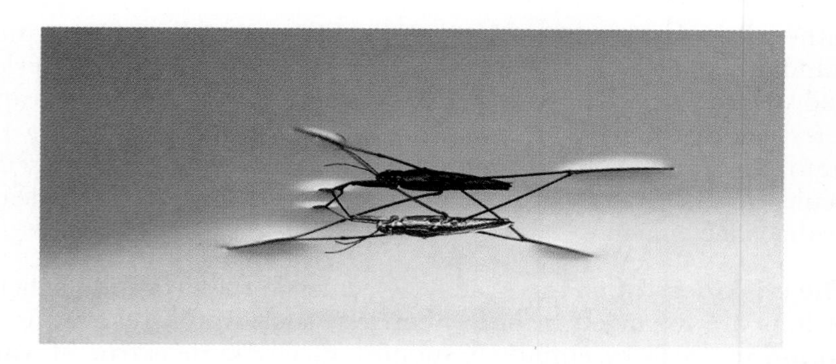

2.26 A water strider on the surface of the water

Water striders can move rapidly across the surface of still water, where they hunt for prey. Note the dimples in the water surface where each foot rests.

or other insect walks on the surface of a pond without breaking the surface (Fig. 2.26) or when the water in a glass that is filled slightly above the rim does not spill. Water has a high surface tension because hydrogen bonds link the molecules at the surface to each other and to the molecules below them. Before the legs of the water strider (or any other object, for that matter) can penetrate the water's surface, they must break some of these hydrogen bonds and deform the orderly array of water molecules.

Just as water molecules are attracted electrostatically to areas of charge on dissolved molecules, so also are they attracted to the charged groups that characterize hydrophilic surfaces. Consequently such surfaces are ***wettable***—that is, water spreads over them and binds loosely to them. By contrast, hydrophobic surfaces—those of most plastics and waxes, for example—lack surface charge and are not wettable; water on them will form isolated droplets, but will not spread out (Fig. 2.27).

2.27 Water beading

The polar bonding between water molecules causes droplets to form on hydrophobic surfaces like this feather.

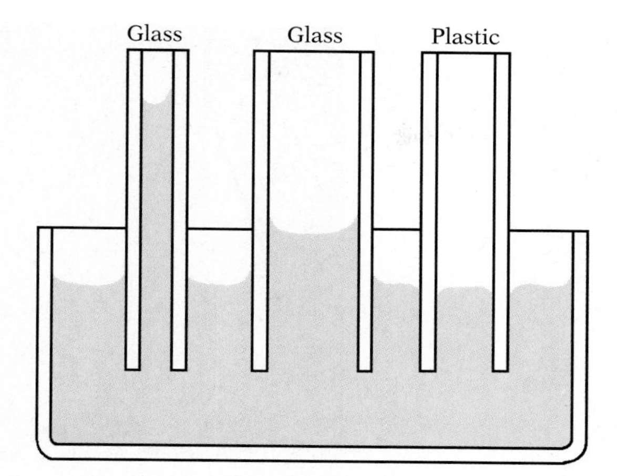

2.28 Capillarity

Water rises higher in a glass tube of small bore (left) than in one of large bore (center) because in the smaller tube a higher percentage of the water molecules are in direct contact with the glass and can form hydrogen bonds with charged groups on the glass. By contrast, water cannot "stick" to the surface of a plastic tube (right) because plastic is uncharged.

The propensity of water to bind to hydrophilic surfaces explains the phenomenon of ***capillarity***—the tendency of aqueous liquids to rise in narrow tubes. If the end of a narrow glass tube is inserted below the surface of a volume of water, water will rise in the tube to a level well above that of the water outside (Fig. 2.28). Because glass is hydrophilic, the water molecules are electrostatically attracted to the numerous charged groups on its surface and tend to creep up the inside of the tube. As the ring of water molecules in contact with the inner surface creeps upward, it pulls along other water molecules to which it is linked by hydrogen bonds. The water level stops rising when the pull of gravity just counteracts the electrostatic forces that contribute to capillarity. The larger the diameter of the tube, however, the smaller the percentage of water molecules in direct contact with the glass and, correspondingly, the smaller the rise in the water. Even though the relatively few molecules in contact with the glass have a tendency to creep upward, they are held back by their cohesion via the network of polar bonding with the rest of the water in the tube.

Capillarity is by no means restricted to glass tubes. Water will climb any charged surface. We are all familiar with the way it climbs up the fibers of paper towels and spreads through the fibers of many kinds of cloth.

Unlike most other substances, which become increasingly dense as the temperature falls, water first becomes denser as it is chilled to 4°C, and then begins to expand again below 4°C. This means that ice, being less dense than cold water, floats and—furthermore—that ponds and streams freeze from the top down rather than from the bottom up. The crust of ice that forms at the surface insulates the water below it from the cold air above and thereby often prevents the pond or stream from freezing solid, even in very cold weather. This special property of water makes life possible in the many ponds and streams that would otherwise freeze solid in the winter.

This curious behavior of cold water is a consequence of polar bonding. Though water molecules have the potential for forming hydrogen bonds with four other water molecules (see Fig. 2.19), this potential is not fully realized because molecular motion prevents stabilization. Hydrogen bonding reaches its full potential in

A

B

2.29 Molecular structure of ice

Because of the tetrahedral arrangement around each water molecule, the lattice is an open one, with considerable space between molecules (A). In liquid water the arrangement is not quite so rigid, and the packing of molecules is therefore slightly denser, but the general lattice arrangement is nonetheless largely preserved. (Planes have been added to help show the three-dimensional disposition of the molecules.) Note the hexagons created by the hydrogen bonding in this bit of ice. This conformation is the basis of the hexagonal shape of most snow crystals (B). Each snowflake contains about 10^{16} water molecules.

water that has frozen into ice. When all four possible bonds have formed, each is oriented in space at the greatest possible distance from the other three. Consequently the bonds are directed toward the four corners of the tetrahedron. The resulting three-dimensional lattice of water molecules in ice is an open one (Fig. 2.29); the rigidly tetrahedral structure maintains space between the molecules, so they can be only loosely packed compared to liquid water.

Water as a temperature regulator The hydrogen bonds in water give it a high internal cohesion, which enables it to absorb much heat without undergoing a large increase in temperature and to release much heat energy without undergoing a great drop in temperature. When most substances absorb heat, their molecules move more rapidly in relation to one another; temperature is an indication of the amount of such molecular motion. In water, by contrast, much of the absorbed heat energy is dissipated in increased vibration of the hydrogens, each of which is shared between the oxygen to which it is covalently bound and the oxygen of another water molecule, to which it is electrostatically bound. As a result, relatively little of the added heat energy is expressed as movement of whole water molecules, so the temperature increase is therefore modest. The high ***heat capacity*** of water (the amount of heat energy that must be added or subtracted to change the temperature by 1°C) and its high ***heat of vaporization*** (the amount of heat energy required for evaporation, or turning water from liquid to vapor) enables it to act as an effective buffer against extreme temperature fluctuations in the environment (Fig. 2.30). In this way water helps stabilize the earth's temperature within the range favorable to life.

2.30 Water vapor
The three states of water are visible in this volcanic pool in Yellowstone Park during the winter.

■ CARBON DIOXIDE AND THE CYCLE OF LIFE

Carbon has four electrons in its outer electron shell and thus a covalent bonding capacity of 4. Carbon dioxide (CO_2) is formed when two atoms of oxygen bond to one atom of carbon. Though this substance contains carbon, it is generally considered inorganic because it is simpler than the compounds classified as organic.

Only a very small fraction of the atmosphere (0.033%) is CO_2; yet atmospheric CO_2 is the principal inorganic source of carbon, and carbon is the principal structural element of living tissue. Before CO_2 can take part in chemical reactions, it must usually first dissolve in water, which it does very readily on the thin aqueous films that coat most cells. Carbon dioxide then reacts with the water to form carbonic acid (H_2CO_3):

$$CO_2 + H_2O \longrightarrow H_2CO_3$$

This reaction involves so little change in energy that it is easily reversible, and CO_2 can readily be released from water solution when conditions are appropriate:

$$H_2CO_3 \longrightarrow CO_2 + H_2O$$

Carbon dioxide and water are the raw materials from which plants manufacture many complex organic compounds essential to life. When these complex compounds are metabolized, they are broken down again to carbon dioxide and water, and the carbon dioxide is eventually released into the atmosphere. Carbon dioxide, then, is the beginning and the end of the immensely complex carbon cycle in nature.

■ OXYGEN AND LIFE

Molecular oxygen (O_2) constitutes approximately 21% of the atmosphere. It is necessary for life in most organisms. It is utilized by both plants and animals in the process of extracting energy from nutrients. Its role, as we will see, is to serve as the ultimate acceptor of electrons. This is a crucial task: without oxygen to accept electrons, most cells can run at only 5% of their normal efficiency. Oxygen is not very soluble in water, but it dissolves enough to supply the needs of aquatic organisms provided (1) that the water is not too hot and (2) that the water's surface is exposed to the air or, alternatively, plants are growing in it, releasing oxygen as they photosynthesize. Photosynthetic bacteria and plants are the source of virtually all atmospheric oxygen.

Though water, oxygen, and carbon dioxide are truly basic to life as we know it, still other compounds are used to capture, store, transport, and utilize the energy that fuels life. In the next chapter we will examine these complex, carbon-based compounds in an effort to understand the chemical reactions that make life possible.

CHAPTER SUMMARY

THE ELEMENTS ESSENTIAL TO LIFE

THE STRUCTURE OF ATOMS Negatively charged electrons occupy cloud-like regions around the positively charged protons of the nucleus. Electrons can have only certain energy levels, and they decay spontaneously to any vacancy in the lowest available level. (p. 21)

HOW ELECTRONS DETERMINE CHEMISTRY The chemistry of an element—particularly which other elements it will react with—is largely determined by the number of electrons in its outermost (highest) energy level. (p. 26)

WHEN ATOMS DECAY Atoms with unstable combinations of neutrons and protons tend to decay; the particles thrown off by such decay can be used to trace reactions and date fossils. (p. 27)

THE GLUE OF LIFE: CHEMICAL BONDS

BONDS BETWEEN IONS Atoms with very few outer electrons or very few outer vacancies tend to lose or gain electrons, respectively, becoming ions. Oppositely charged ions are attracted to one another electrostatically. (p. 29)

ACIDS AND BASES Acids increase the concentration of H^+ ions in a solution; bases increase the concentration of OH^- ions. Acidity is measured on a pH scale. (p. 30)

HOW STRONG BONDS ARE FORMED Atoms with complimentary numbers of outer electrons and outer-electron vacancies can share electrons, cooperating to complete each other's outer level and thus forming a strong covalent bond. When the shared electrons are attracted more strongly by one atom (the more electronegative atom), and the geometry of the resulting molecule does not cancel out this electron asymmetry, the distribution of charge in the molecule is unequal, generating a polar molecule. (p. 31)

WHY ARE WEAK BONDS IMPORTANT? Electrostatic bonds between polar molecules are weak, allowing flexibility in bonding but great strength when many of these weak hydrogen (polar) bonds cooperate. (p. 34)

WHY LIFE NEEDS INORGANIC MOLECULES

THE REMARKABLE CHEMISTRY OF WATER Water is polar and thus dissolves other polar molecules and ions while forcing nonpolar molecules into clumps. Hydrogen bonds between water molecules give water its surface tension, as well as its ability to climb polar surfaces and the capacity to retard temperature changes. (p. 36)

CARBON DIOXIDE AND THE CYCLE OF LIFE Carbon is the basis of all organic compounds; it comes initially from CO_2 captured by plants. (p. 43)

OXYGEN AND LIFE Oxygen is a by-product of photosynthesis and is essential to the efficient use of energy by both plants and animals. (p. 43)

STUDY QUESTIONS

1 Compare and contrast hydrogen (polar) bonds, ionic bonds, and covalent bonds. (pp. 29–34)

2 Why are electrons so important to the chemistry of life? (pp. 26–34)

3 What is electronegativity, and why is it important? (pp. 33–34)

4 What sorts of chemicals dissolve well in water? Explain their solubility. (pp. 36–39)

5 Do hydrogen bonds have to involve hydrogen? (p. 35)

SUGGESTED READING

ATKINS, P. W., 1987. *Molecules.* Scientific American Library, New York. *A beautifully produced "molecular glossary" illustrating the chemical formula, three-dimensional structure, and biological action of many common molecules.*

DICKERSON, R. E., AND I. GEIS, 1976. *Chemistry, Matter, and the Universe.* W. A. Benjamin, Menlo Park, Calif. *An excellent introduction to chemistry from a biological perspective; a real classic.*

CHAPTER 3

WHY CARBON?

THE ROLE OF CARBOHYDRATES

THE ROLE OF LIPIDS

THE ROLE OF PROTEINS

THE ROLE OF NUCLEIC ACIDS

HOW CHEMICAL REACTIONS WORK

WHY FREE ENERGY ISN'T FREE

PREDICTING REACTIONS: THE EQUILIBRIUM CONSTANT

HOW ACTIVATION ENERGY GUIDES REACTIONS

HOW CATALYSTS BREAK THE RULES

BIOLOGICAL CATALYSTS: ENZYMES

THE CHEMISTRY OF LIFE

Chemistry does what all good science does: it makes complex phenomena easier to predict. Chemistry explains why certain reactions among molecules will take place, and why particular molecular combinations will be stable. Biochemistry applies this understanding to organic compounds and the reactions critical to life. This chapter examines the molecular actors and how they interact in cells.

WHY CARBON?

The source of the vast chemical diversity in living things is the bonding capacity of just one element—carbon. Carbon's power lies in its versatile structure: four unpaired electrons in its outer shell, which allow it to form covalent bonds with up to four other atoms. These four bonding options make possible an almost endless variety of carbon-based, and thus organic, molecules. We will look first at the important kinds of organic compounds—carbohydrates, lipids, proteins, and nucleic acids—and then at how some of the crucial well-ordered chemical changes inside cells are orchestrated and controlled.

Though carbon can and does bond to a variety of elements, its four unpaired electrons are most commonly bonded to hydrogen, oxygen, nitrogen, or more carbon. Compounds containing only carbon and hydrogen, the ***hydrocarbons***, are of central importance in organic chemistry; the number of different compounds of this kind is immense. The readiness with which carbon-carbon bonds can form and produce chains of varying lengths and shapes generates a great variety of hydrocarbons. Hydrocarbon chains may be simple, branched, or may form circles of varying numbers of carbons (Fig. 3.1). The more atoms a molecule contains, the more different arrangements of those atoms will be possible. Compounds with the same atomic content and molecular formula but different atomic arrangements are called ***isomers*** (Fig. 3.2). Very large organic molecules may have hundreds of isomers, with many differing physical properties.

Another source of variety in hydrocarbons is the capacity of ad-

straight chains

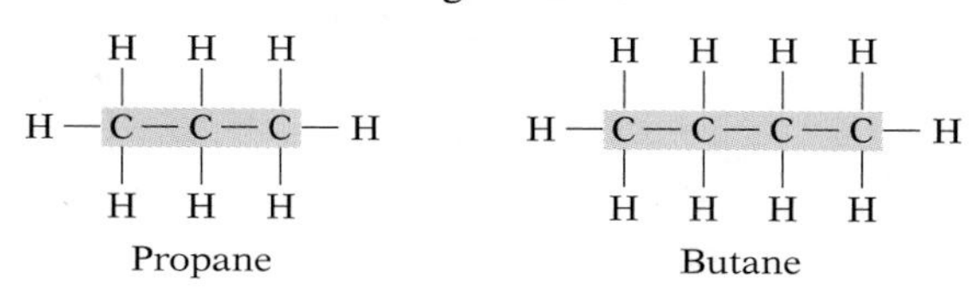

branched chains

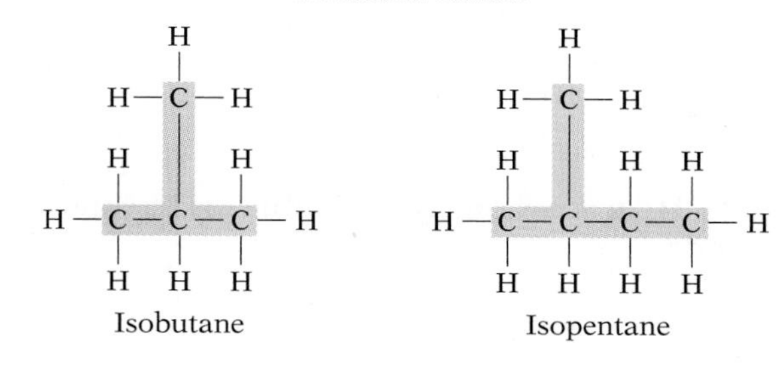

closed chains

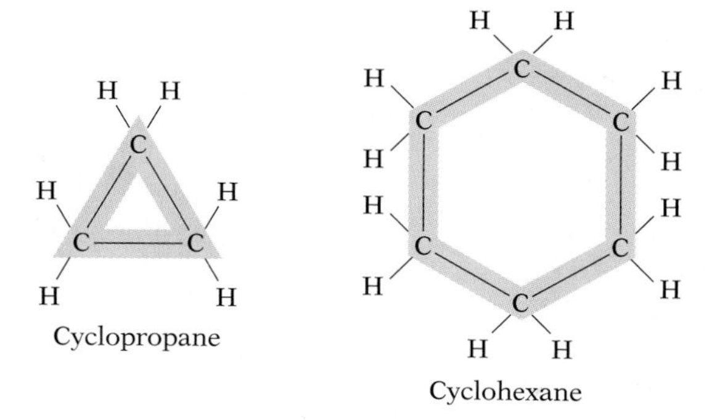

3.1 Examples of hydrocarbons

The molecules appear flat in these conventionally drawn structural diagrams, though they are, in fact, three-dimensional. The bonds around a carbon atom that forms only single bonds are oriented toward the four corners of a tetrahedron.

Glucose Galactose Fructose

3.2 Three isomeric hexoses

Each of these six-carbon sugars has the same molecular formula, $C_6H_{12}O_6$; hence each is an isomer of the others.

jacent carbon atoms to form single, double, or triple bonds (Fig. 3.3). Substitution of other elements or groups of elements for hydrogen atoms in hydrocarbons makes possible an almost infinite number of ***derivative hydrocarbons***. The total number of hydrocarbons and derivatives that form in nature exceeds half a million. This great capacity for diverse atomic organization makes hydrocarbons ideal for building chemicals with unique properties.

The four major classes of complex organic compounds are carbohydrates, lipids (both derivative hydrocarbons), proteins, and nucleic acids. Molecules in each of these classes are often identified on the basis of the subgroups they contain. Each subgroup, or ***functional group***, has its own characteristic properties, which help determine solubility, reactivity, and other traits of the chemical "personality" of the whole molecule. We will refer to many of the groups listed in Table 3.1 in this and later chapters.

double bond **triple bond**

$H_2C{=}CH_2$ $H{-}C{\equiv}C{-}H$

Ethylene Acetylene

3.3 Hydrocarbons with double and triple bonds

Table 3.1 Important functional groups

GROUP	NAME	PROPERTIES	IONIC FORM
$-OH$	**Hydroxyl**	Polar (soluble, because it is able to form hydrogen bonds)	
$-CH_2OH$	**Alcohol**	Polar (soluble)	
$-COOH$	**Carboxyl**	Polar (soluble); often loses its hydrogen, becoming negatively charged (an acid)	$-COO^-$
$-NH_2$	**Amino**	Polar (soluble); often gains a hydrogen, becoming positively charged (a base)	$-NH_3^+$
$-CHO$	**Aldehyde**	Polar (soluble)	
$>C{=}O$	**Ketone**	Polar (soluble)	
$-CH_3$	**Methyl**	Hydrophobic (insoluble); least reactive of the side groups	
$-PO(OH)_2$	**Phosphate**	Polar (soluble); usually loses its hydrogens, becoming negatively charged (an acid)	$-PO_3^{2-}$

■ THE ROLE OF CARBOHYDRATES

Carbohydrates are derivative hydrocarbons composed of carbon, hydrogen, and oxygen. In simple carbohydrates the hydrogen and oxygen are characteristically present in the same proportions as in water: there are two hydrogen atoms and one oxygen atom for each carbon atom. Here is its actual molecular arrangement:

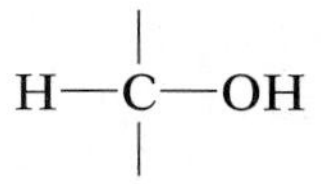

$$\mathrm{H{-}\overset{|}{\underset{|}{C}}{-}OH}$$

Some carbohydrates, such as starch and cellulose, are very large complex molecules. However, like most large organic molecules, they are composed of many simpler "building-block" compounds bonded together (Fig. 3.4). Understanding the constituent building blocks is the first step toward understanding more complex substances.

Ready energy: the simple sugars The basic carbohydrate molecules are simple sugars, or ***monosaccharides***. All sugars when in straight-chain form contain a —C=O group (Fig. 3.2). If the double-bonded O is attached to a terminal C of a chain, the combination is called an aldehyde group; if it is attached to a nonterminal C, the combination is called a ketone group (Table 3.1). The —OH (hydroxyl) groups are polar. Hence sugars readily form hydrogen bonds with water and are soluble, unlike simple nonpolar hydrocarbon molecules, which tend to clump together in water.

3.4 A moderately complex organic compound

One of the largest of the moderately complex organic compounds is starch, the main energy-storage molecule in plants. Despite its apparent complexity, the starch molecule is actually composed of a repetitive string of glucose units, each represented here as a hexagon. Only a small part of one starch chain is shown. A branched molecule composed of units of several different compounds would be far more complex.

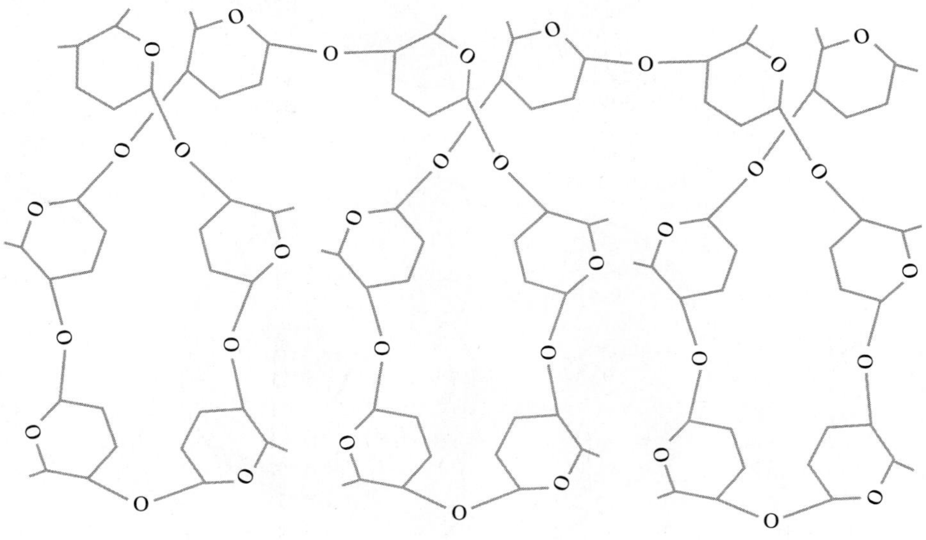

STRUCTURAL ISOMERS

H—C—C—OH (each C bearing two H)

Ethyl alcohol

H—C—O—C—H (each C bearing two H)

Dimethyl ether

GEOMETRIC STEREOISOMERS

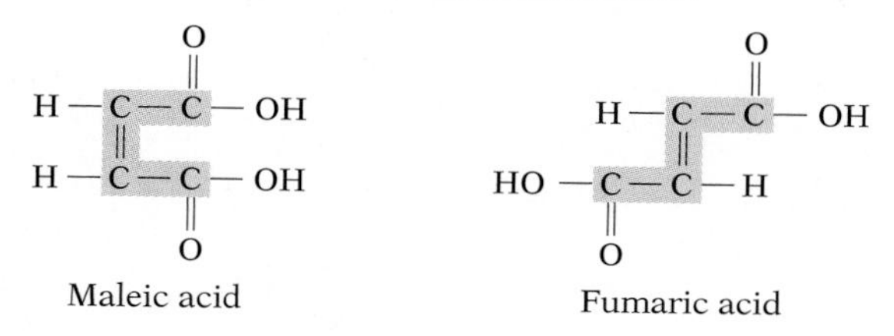

OPTICAL STEREOISOMERS

3.5 Three types of isomerism

The two structural isomers differ in the basic grouping of their constituent atoms, one being an alcohol (characterized by an —OH group) and the other an ether (characterized by an oxygen bonded between two carbons). The two geometric stereoisomers are fixed in different spatial arrangements by their inability to rotate around the double bond between the middle two carbons. As Figure 3.6 shows, the two optical stereoisomers are asymmetric molecules that cannot be superimposed on each other.

The six-carbon sugars (hexoses) are the most important building blocks for more complex carbohydrates. There are many six-carbon sugars, ***glucose*** and fructose being two of the most common. All have the same molecular formula, $C_6H_{12}O_6$, and are therefore isomers. Glucose and fructose are structural isomers (see Fig. 3.2); the basic groupings of their constituent atoms are different, making one an aldehyde and the other a ketone sugar.

In addition to structural isomerism, there is another, more subtle, kind of isomerism called stereoisomerism. In a given pair of stereoisomers, identical groups are attached to the carbon atoms, but the spatial arrangements of the attached groups are different. The two middle compounds shown in Figure 3.5 are stereoisomers; if the carbon-carbon bonds in these two molecules were single, the two would in fact be the same compound, because free rotation is possible around a single bond. A double bond, however, holds its atoms in a rigid configuration.

Despite the way they are usually drawn, stereoisomers (Fig. 3.5, bottom) are not flat. In a carbon atom the four unpaired electrons form the corners of a tetrahedron; if in two molecules the groups attached to corresponding electrons are different, the resulting molecules will be different (Fig. 3.6). Glucose and galactose are stereoisomers because the position of one —OH group is different in each (see Fig. 3.2). The stereoisomers of a compound usually

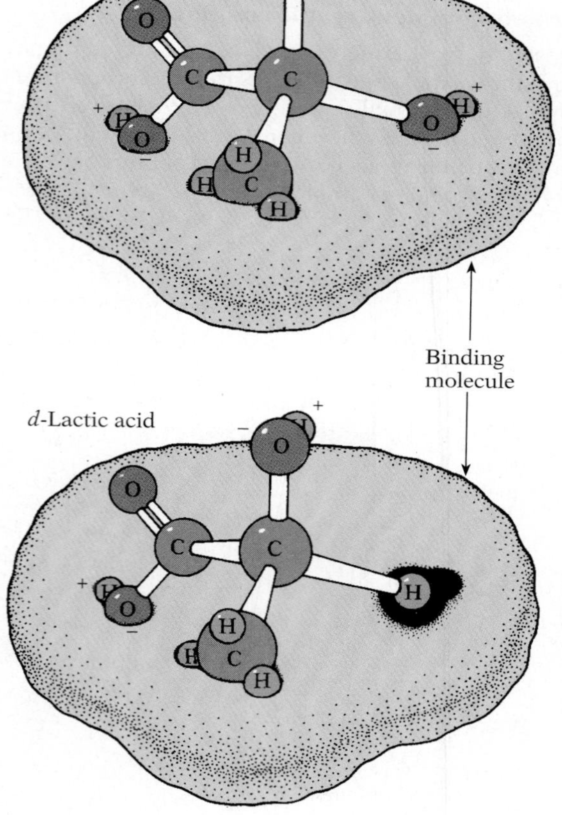

3.6 Optical stereoisomers and chemical specificity

Of these two stereoisomers of lactic acid, only one (top) can fit all the holes in the hypothetical schematic molecule to which it binds. The other will not fit the holes no matter which way it is turned. Subtle distinctions of this kind are crucially important: they enable molecules to recognize and bind certain other particular molecules as part of the network of critical chemical reactions necessary for life.

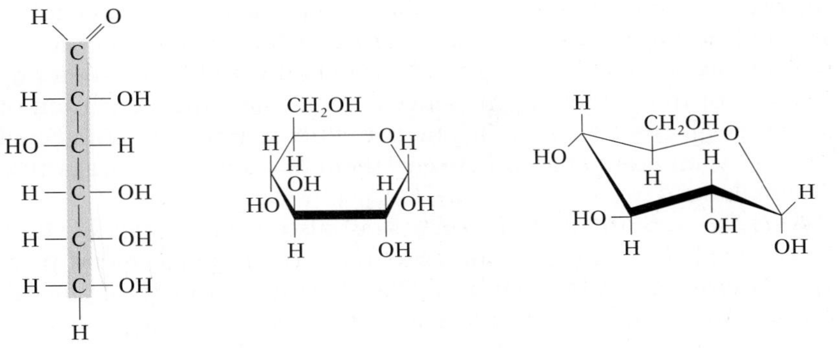

3.7 **Two forms of glucose**

Glucose may exist in the straight-chain aldehyde form (left) or as a ring structure (center). The ring structure is the most common. (By convention, the unmarked corners of the hexagon signify carbon atoms.) Both representations fail to convey the true shape of the molecules, since the four bonds of each carbon atom are directed to the four corners of a tetrahedron. The illustration at right is a more realistic representation of the ring form, but such realism is possible for only the simplest organic molecules.

have quite different biological properties and behavior; subtle differences in shape may determine which isomer can bind to or react with a particular molecule and which cannot (Fig. 3.6). Hence, different isomers can play very different roles in the chemistry of cells.

Glucose, which usually exists in a ring form composed of five carbons and one oxygen (Fig. 3.7), plays a unique role in the chemistry of life. As the primary product of photosynthesis in plants, glucose becomes the ultimate source of all the carbon atoms in both plant and animal tissue. Moreover, the energy stored in its covalent bonds is usually, directly or indirectly, the source of the energy that powers cells. Other six-carbon monosaccharides, among them fructose and galactose, are constantly being converted into glucose or synthesized from glucose. Even fats and proteins can be converted into glucose or synthesized from glucose in cells.

In addition to ordinary monosaccharides composed only of carbon, oxygen, and hydrogen, there is a variety of derivative monosaccharides containing other elements. For example, some have a phosphate group attached to one of the carbons, and others an ***amino*** group: a nitrogen with two hydrogens, $-NH_2$ (Fig. 3.8).

Double sugars The ***disaccharides*** are composed of two simple sugars bonded through reactions that remove a molecule of water. This kind of reaction series is consequently called a ***condensation*** or ***dehydration reaction***. The disaccharide ***maltose***, or ***malt sugar***, for instance, is synthesized by a condensation reaction between two molecules of glucose. The overall reaction (omiting intermediate steps) is:

$$2C_6H_{12}O_6 \longrightarrow C_{12}H_{22}O_{11} + H_2O$$

The hydrogen atom from a hydroxyl group ($-OH$) of one glucose molecule combines with a complete hydroxyl group from another

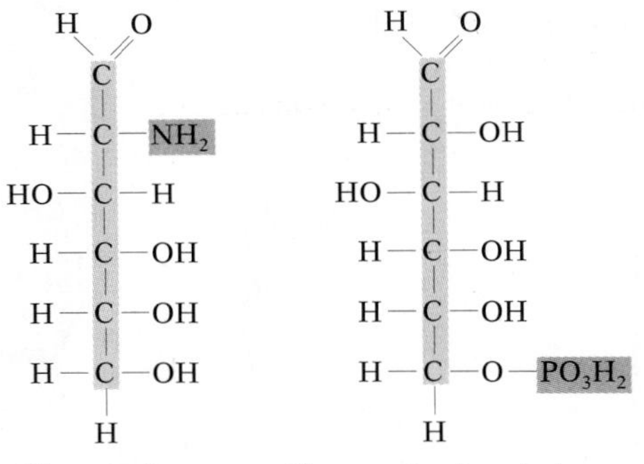

Glucosamine

Glucose-6-phosphate

3.8 **Two examples of derivative monosaccharides**

Glucosamine, a chemical used in the synthesis of some protein building blocks as well as in insect exoskeletons, is a glucose molecule with an amino group ($-NH_2$) substituted for an $-OH$ group. Similarly, glucose-6-phosphate, which is used in an important step in the harvesting of energy from glucose, is glucose with a phosphate group added.

glucose molecule to form water (Fig. 3.9A). The oxygen valence −1 (created by the removal of hydrogen) and the carbon valence +1 (created by the removal of —OH) are neutralized by the bonding together of the oxygen of one glucose molecule with the carbon of the other glucose. As a result, the two glucose units are connected by an oxygen atom shared between them, producing the disaccharide maltose.

Sucrose—common table sugar—is also a disaccharide. It is synthesized by a condensation reaction between glucose and fructose. ***Lactose***, or milk sugar, is a disaccharide composed of glucose and galactose joined by a condensation reaction (Fig. 3.9B). In fact, the synthesis of complex molecules from simpler units almost always produces water.

Disaccharides can be broken down to their constituent simple

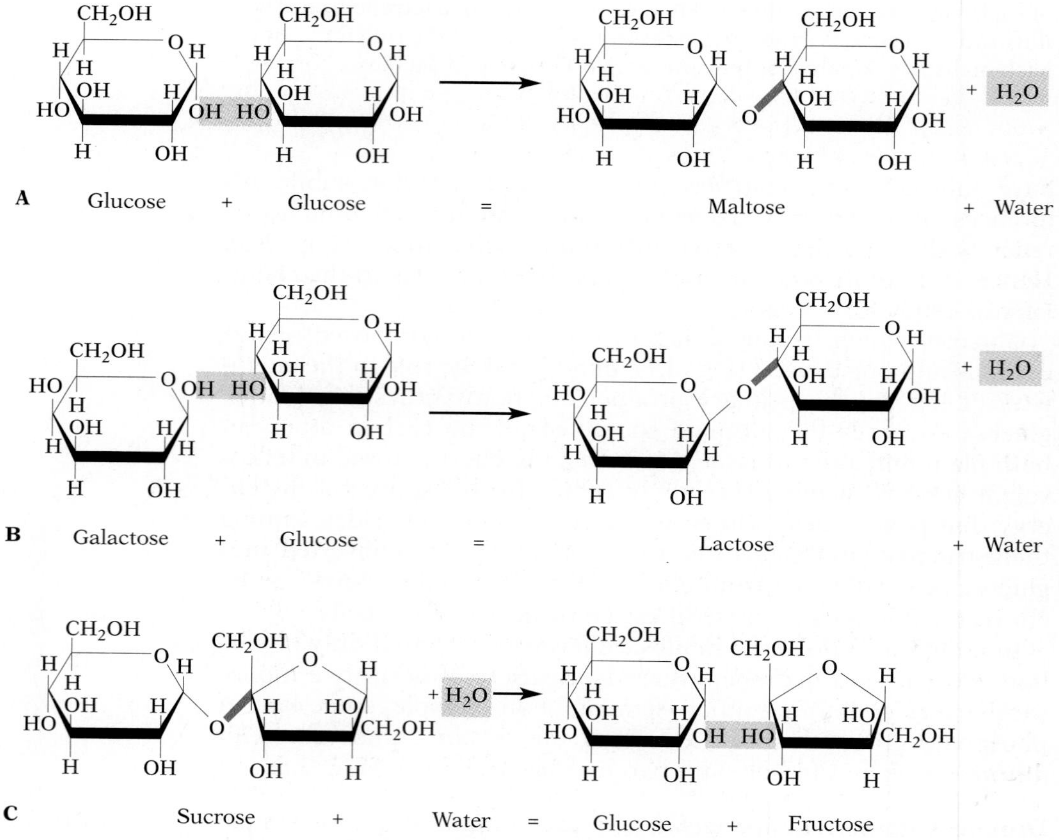

A Glucose + Glucose = Maltose + Water

B Galactose + Glucose = Lactose + Water

C Sucrose + Water = Glucose + Fructose

3.9 Synthesis and digestion of disaccharides

Removal of a molecule of water (blue) between two molecules of sugar (a condensation reaction) results in the formation of a bond (red) between the two. In the first two examples shown here, intermediate steps are omitted. (A) A bottom-to-bottom (α) linkage forms between two glucose molecules, yielding maltose. (B) A top-to-bottom (β) linkage forms between galactose and glucose, yielding the milk sugar lactose; some adults suffer from milk intolerance because they cannot digest this linkage. (C) The hydrolysis reaction leading to the breakdown of sucrose involves adding back a water molecule.

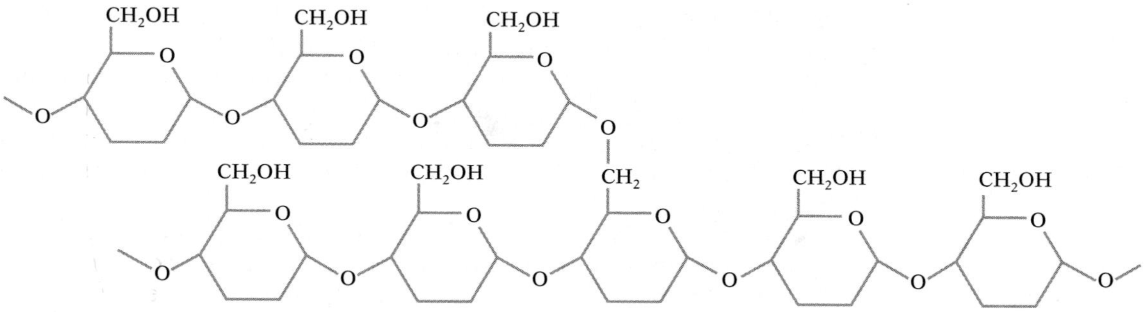

3.10 Branched starch

Shown here is a small segment of a starch molecule. This starch is branched, but some forms are unbranched. Like the cellulose of plants, starch is a polymer of glucose, but in cellulose the glucose is connected by β linkages (see Fig. 3.9), whereas in starch the glucose is connected by α linkages.

sugars by the reverse process: adding back a water molecule. This reaction, called ***hydrolysis***, splits a water molecule into a hydrogen atom and a hydroxyl group, and then adds them to the subunits through a series of steps. The summary reaction is:

$$C_{12}H_{22}O_{11} + H_2O \longrightarrow 2C_6H_{12}O_6$$

Hydrolysis reactions are particularly important in digestion, which breaks down complex molecules into simple building blocks, ready for subsequent use (Fig. 3.9C).

Storage and structure: the polysaccharides The prefix *poly-* means "many." Thus ***polysaccharides*** are complex carbohydrates composed of many simple-sugar building blocks bonded together in long chains (Fig. 3.10). They are synthesized by condensation reaction and can be broken down into their constituent sugars by hydrolysis.

A number of complex polysaccharides are of great importance in biology. ***Starches*** are the principal carbohydrate storage products of higher plants. They are composed of many hundreds of glucose units bonded together. ***Glycogen*** is the principal carbohydrate storage product in animals. Its molecules are much like those of starch; they have the same type of bond between adjacent glucose units. ***Cellulose*** is the most common carbohydrate on earth. It is a highly insoluble polysaccharide used by plants as their major supporting material. The bonds between its glucose units are β linkages rather than α linkages (Fig. 3.11); animals can digest the bonds of starch and glycogen, but most animals are unable to hydrolyze those of cellulose. ***Chitin*** (Fig. 3.11), the polysaccharide that serves as the major structural component of insect exoskeletons and fungal cell walls, is functionally equivalent to cellulose (Fig. 3.12).

All reactions in which small molecules (called ***monomers***) bond together to form long chains are called polymerization reactions; polymerization of monosaccharides, for example, creates a polysaccharide. The products of polymerization are called ***polymers***. Polymers play a critical role in biology, as we will see.

CH_2OH CH_2OH CH_2OH CH_2OH

Cellulose

CH_2OH CH_2OH CH_2OH

$NHCOCH_3$ $NHCOCH_3$

Chitin

3.11 Cellulose and chitin

Cellulose is composed of long chains of β-linked glucoses. Chitin is composed of β-linked acetylglucosamines (glucoses with a combined acetylamino side group, $—NHCOCH_3$).

A

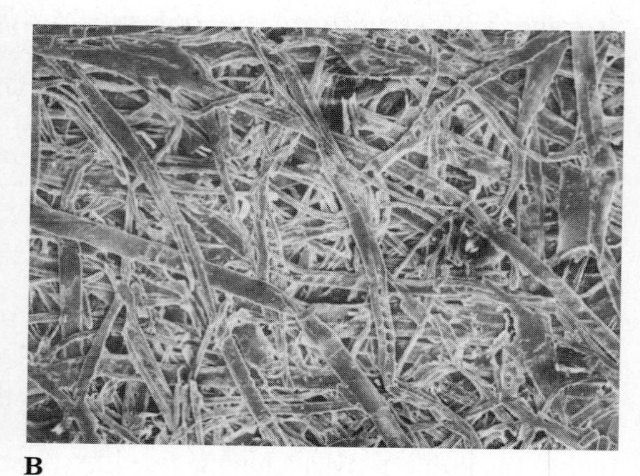

B

3.12 (A) Mantis exoskeleton and (B) scanning electron micrograph (SEM) of paper fibers

The chitin of insect exoskeletons (visible here in the shed skin of a mantis) is chemically similar and functionally equivalent to the cellulose of plants (seen here as fibers in uncoated paper). Books like this one have a coating of clay over the fibers; this coating, which gives the paper a shine, prevents absorption and spreading of the ink, and thus permits the printing of high-resolution photographs.

■ THE ROLE OF LIPIDS

Like carbohydrates, a second major group of derivative hydrocarbons, the lipids, are composed principally of carbon, hydrogen, and oxygen, but they may also contain other elements, particularly phosphorus and nitrogen. In their simplest form, ***lipids*** are hydrocarbons with a carboxyl group (—COOH) at one end (Fig. 3.13). Such lipids are primarily nonpolar, by virtue of their long hydrocarbon tails. They are therefore relatively insoluble in water but will dissolve in organic solvents such as ether. Most lipids, as we will see, are more complex, having an ionic group attached to the carboxyl end; the long hydrophobic tails are universal features.

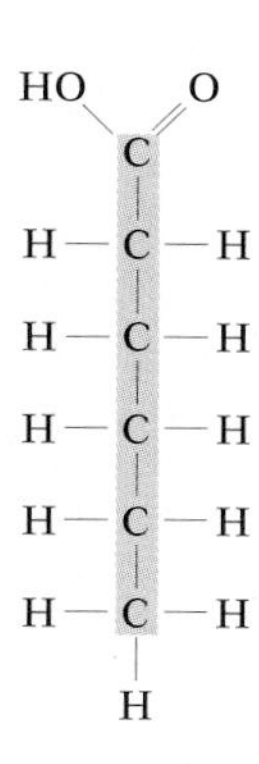

3.13 A simple lipid

Storage and insulation: the fats and oils Among the best-known lipids are the neutral fats, which are called oils when liquid. Important as energy-storage molecules, the fats also provide insulation, cushioning, and protection for various parts of the body (Fig. 3.14). Each molecule of fat is composed of fatty acids joined together by glycerol. Glycerol (also called glycerin) has a backbone

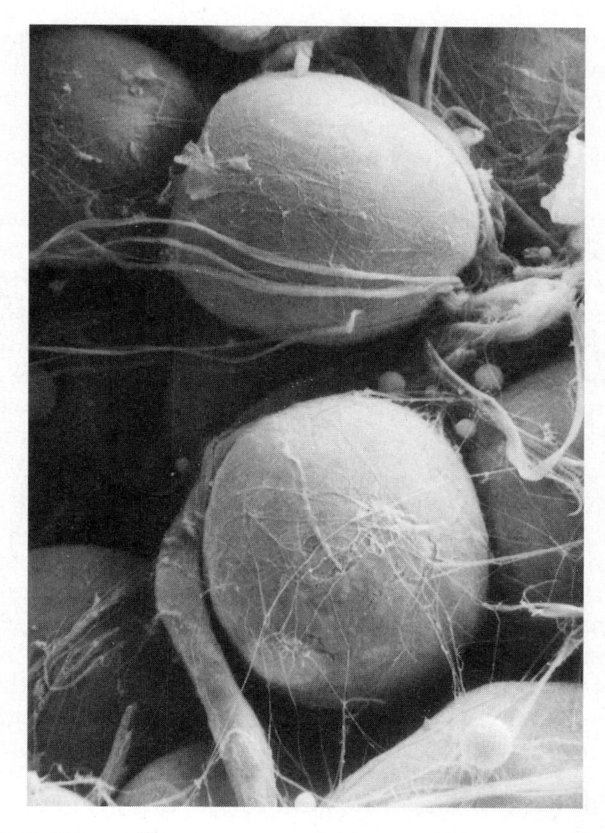

3.14 Fat-storage cells

Lipids are stored in spherical fat cells called adipocytes, seen here. Small blood vessels (capillaries) and support fibers (collagen) hold these cells in place. In the same manner as the styrofoam beads they resemble, these human adipocytes cushion and insulate underlying parts of the body. (From *Tissues and Organs: A Text-Atlas of Scanning Electron Microscopy* by Richard G. Kessel and Randy H. Kardon. Copyright 1979 W. H. Freeman and Company. Used with permission.)

of three carbon atoms, each carrying a hydroxyl —OH group (Fig. 3.15).

Fatty acids, like all organic acids, contain a carboxyl group. When both a double-bonded oxygen and an —OH group are attached to the same carbon atom, the double-bonded oxygen tends to cause the —OH part of this carboxyl to lose its hydrogen, making the group ionic and causing the compound to behave as an acid (see Table 3.1).

There are many different fatty acids, which vary in carbon-chain length, in the number of single or double carbon-carbon bonds, and in other characteristics. The fatty acids in edible fats and oils contain an even number of carbon atoms, and most of them have from 4 to 24 carbons in their backbones; three of the most common are stearic acid (18 carbons), palmitic acid (16 carbons), and linoleic acid (18 carbons; Fig. 3.16).

Organic acids and alcohols have a tendency to combine through condensation reactions. Since glycerol is an alcohol (by virtue of having a $—CH_2OH$ group) and has three hydroxyls, it can combine with three molecules of fatty acid to form a molecule of fat (Fig. 3.15). Hence fats are sometimes also called triglycerides.

The various fats differ in the specific fatty acids, or types of fatty acids, composing them. You have doubtless read of the concern over saturated fats. ***Saturated fats*** are simply those incorporating fatty acids with the maximum possible number of hydrogen atoms attached to each carbon, and hence no carbon-carbon double bonds (Fig. 3.16). The fatty acids in ***unsaturated fats*** (actually oils, since they are usually liquid at room temperature) have at least one carbon-carbon double bond—that is, they are not completely saturated with hydrogen. The double bond induces a "kink" in the otherwise linear structure, which prevents solidification. There is now good evidence that an elevated intake of saturated fats is one of many factors predisposing humans to atherosclerosis—a disease of the arteries in which fatty deposits in the arterial walls cause partial obstruction of blood flow, which can lead to strokes and heart failure.

Since fats are synthesized by condensation reactions, they, like complex carbohydrates, can be broken down into their building-block compounds by hydrolysis, as happens in digestion. In addition, because fat, though slow to be metabolized, contains 2.5 times as much usable energy per gram as monosaccharides, it is a good substance for long-term energy storage. In fact, if all the energy the average individual stores in fat (about a month's supply) were to be maintained in the form of sugar, we would each be 25–30 kg heavier.

The key to membranes: phospholipids Various lipids contain a phosphate group at the carboxyl end of the chain. Among the most common of these phospholipids are those composed of one unit of glycerol, two units of fatty acid, and a phosphate group often linked with a nitrogen-containing group (Fig. 3.17). The phosphate group is bonded to the glycerol at the point where the third fatty acid would be in a fat. Because the phosphate group tends to lose a hydrogen ion, one of the oxygens usually becomes negatively charged; similarly, the nitrogen, being electronegative, tends to attract a hydrogen ion and thus to become positively charged. In

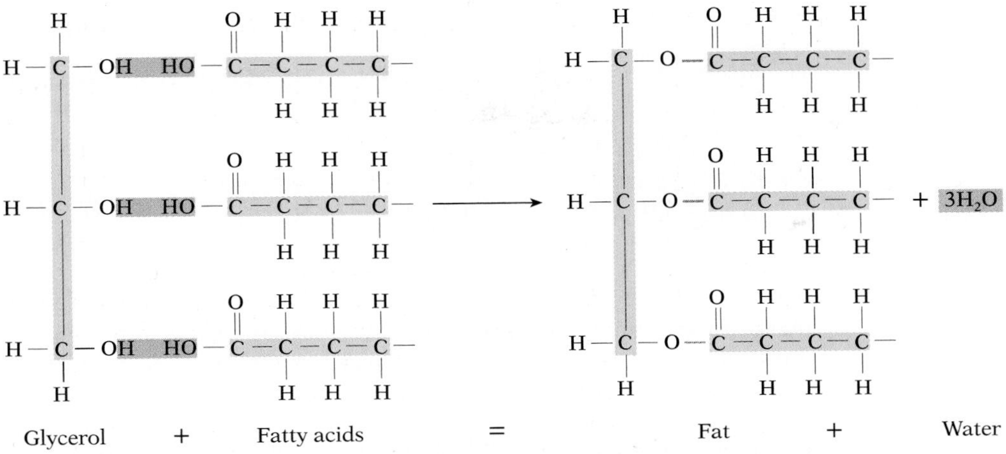

Glycerol + Fatty acids = Fat + Water

3.15 Synthesis of a fat

Removal of three molecules of water by condensation reactions results in the bonding of three molecules of fatty acid to a single molecule of glycerol. (Intermediate steps are omitted in this example.) Conversely, three molecules of water will be added by hydrolysis when this molecule of fat is digested. The carbon chains of the fatty acids are usually longer than shown here.

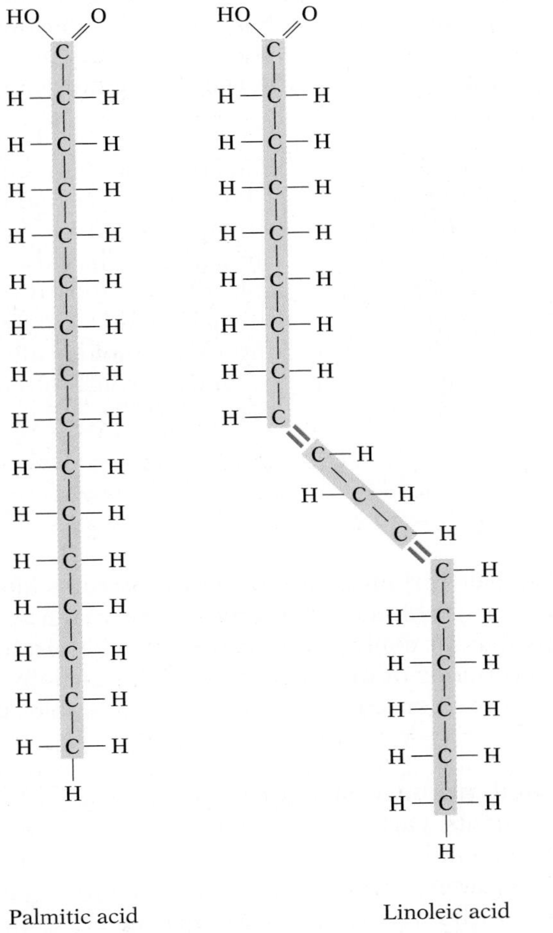

Palmitic acid

Linoleic acid

3.16 Examples of saturated and unsaturated fatty acids

Palmitic acid is saturated with hydrogen—that is, it contains the maximum number of hydrogens possible. By contrast, linoleic acid, with its two inflexible carbon-carbon double bonds, accommodates four fewer than the maximum number of hydrogens.

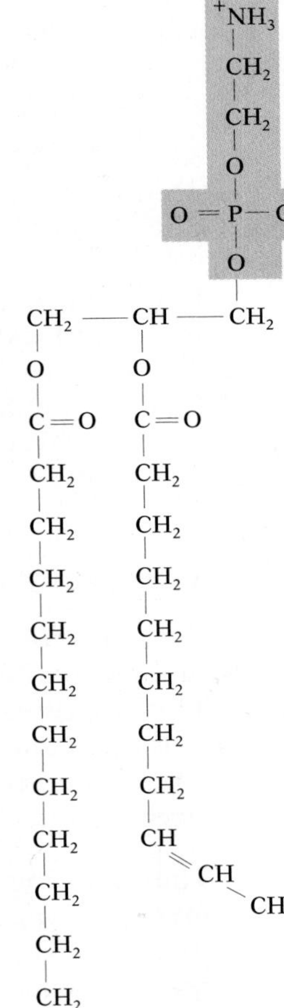

3.17 A phospholipid

The portion of the molecule with the phosphate and nitrogenous groups (brown) is soluble in water, whereas the two hydrocarbon chains are not. This particular phospholipid, ethanolamine phosphoglyceride, is one of the two most abundant in the cell membranes of higher plants and animals.

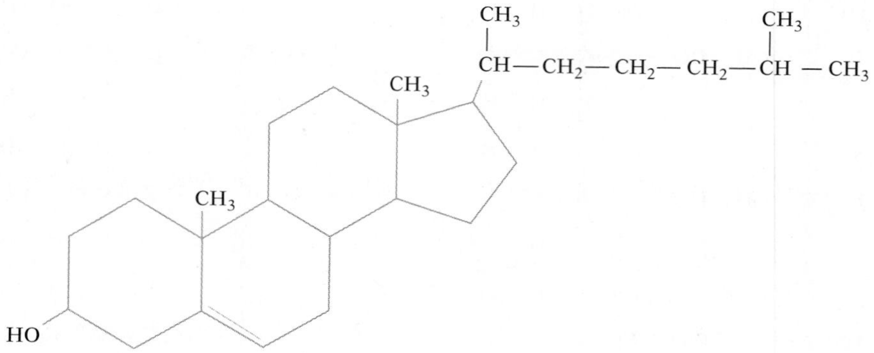

3.18 A steroid

All steroids have the same basic unit of four interlocking rings, but they differ in their side groups. This particular steroid is cholesterol. (By convention, a hexagon signifies a six-carbon ring with its valences completed by hydrogens; see cyclohexane in Figure 3.1. A pentagon signifies a five-carbon ring, also with hydrogens attached to the carbons.)

CH_3 CH_3
$CH—CH_2—CH_2—CH_2—CH—CH_3$
CH_3
CH_3
HO

short, the end of the phospholipid molecule with the phosphate and nitrogenous groups is strongly ionic and hence soluble in water, whereas the other end, composed of the two long hydrocarbon tails of the fatty acids, is nonpolar and insoluble. This curious property of solubility at one end but not at the other makes phospholipids especially well suited to function in cellular membranes, as we will see in Chapter 4.

Steroids Though commonly classified as lipids because their solubility characteristics are similar to those of fats, oils, waxes, and phospholipids, the steroids differ markedly in structure from the other lipids we have discussed (Fig. 3.18). They are not based on fatty acid–alcohol bonds, but are complex molecules composed of four interlocking rings of carbon atoms, with various side groups attached to the rings. Steroids are very important biologically. Some vitamins and hormones are steroids, and steroids often occur as structural elements in living cells, particularly in cellular membranes.

■ THE ROLE OF PROTEINS

Far more complex than either carbohydrates or lipids, proteins are fundamental to both the structure and function of living material. They are directly responsible for controlling the delicate chemistry of the cell and exist in thousands of different forms. Like carbohydrates and lipids, proteins are composed of simple building-block compounds.

The building blocks and primary structure of proteins All ***proteins*** contain four elements: carbon, hydrogen, oxygen, and nitrogen; most proteins also contain some sulfur. These elements are bonded together to form subunits called amino acids, which, being organic acids, contain the carboxyl (—COOH) group (Fig. 3.19). In addition, they each have an amino (—NH_2) group. Finally, each amino acid has a side chain, designated R:

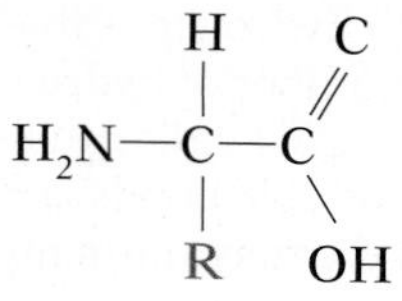

H C
$H_2N—C—C$
R OH

NONPOLAR R GROUPS

Glycine (Gly) | Alanine (Ala) | Valine (Val) | Leucine (Leu) | Isoleucine (Ile)

Methionine (Met) | Phenylalanine (Phe) | Tryptophan (Trp) | Proline (Pro)

POLAR R GROUPS

Serine (Ser) | Threonine (Thr) | Cysteine (Cys) | Tyrosine (Tyr) | Asparagine (Asn) | Glutamine (Gln)

IONIC R GROUPS

Aspartic acid (Asp) | Glutamic acid (Glu) | Lysine (Lys) | Arginine (Arg) | Histidine (His)

3.19 The 20 amino acids common in proteins classified by R group

The amino acids are shown in their ionized form. All have the same arrangement of a carboxyl group and an amino group attached to the same carbon; they differ in their R groups (red or blue). Top: These nine amino acids have nonpolar R groups and are relatively insoluble in water. Glycine is an exception. Its R group, a single hydrogen atom, is nonpolar but is too small to outweigh the charge of the amino and carboxyl groups. The molecule therefore behaves more like a polar amino acid and is water-soluble. Proline is also unusual. It is technically not an amino acid because the nitrogen is bonded to part of the R group. However, it is included because it is regularly incorporated into proteins along with the true amino acids. Middle: These six amino acids have polar R groups and are soluble. Bottom: These five amino acids, with R groups ionized at intracellular pH levels, are electrically charged and thus water-soluble; the first two, being negatively charged, are acidic, whereas the last three, with a positive charge, are alkaline.

The side chains of the various amino acids can be very simple, as in glycine (only a hydrogen atom), or they may be very complex, as in tryptophan, which includes two ring structures. Twenty different amino acids are commonly found in proteins (Fig. 3.19). The various R groups give each of the amino acids different characteristics, which in turn greatly influence the properties of the proteins incorporating them. For example, some amino acids are relatively insoluble in water, owing to R groups that are nonpolar at a pH between 6.5 and 7 (Fig. 3.19, top), whereas other amino acids are water-soluble because their R groups are polar (middle) or ionic (that is, electrically charged; bottom row).

Proteins are long and complex polymers of the 20 common amino acids. The amino acid building blocks bond together by condensation reactions between the —COOH groups and the —NH_2 groups (Fig. 3.20). The covalent bonds between amino acids are called ***peptide bonds***, and the chains they produce are called ***polypeptide chains***. The amino acid units incorporated into a chain are called peptides. The number of peptides in a single polypeptide chain within a protein molecule is usually between 40 and 500; for any given protein, however, the chain length is constant. Proteins have a three-dimensional shape that is largely determined by the distribution of their polar, charged, and nonpolar R groups: the chain will tend to fold so that the nonpolar (hydrophobic) groups are inside the protein, while the hydrophilic groups are exposed at the surface, where they interact with nearby polar molecules, particularly water.

Protein molecules often consist of more than one polypeptide chain. The chains may be held together by numerous weak bonds, especially hydrogen bonds; for example, a single molecule of hemoglobin, the red oxygen-carrying protein in blood, has four polypeptide chains linked by hydrogen bonds. Insulin, an important hormone secreted by the pancreas in vertebrates, exemplifies a protein with polypeptide chains held together by both hydrogen and covalent bonds. The covalent bonds, called ***disulfide bonds***,

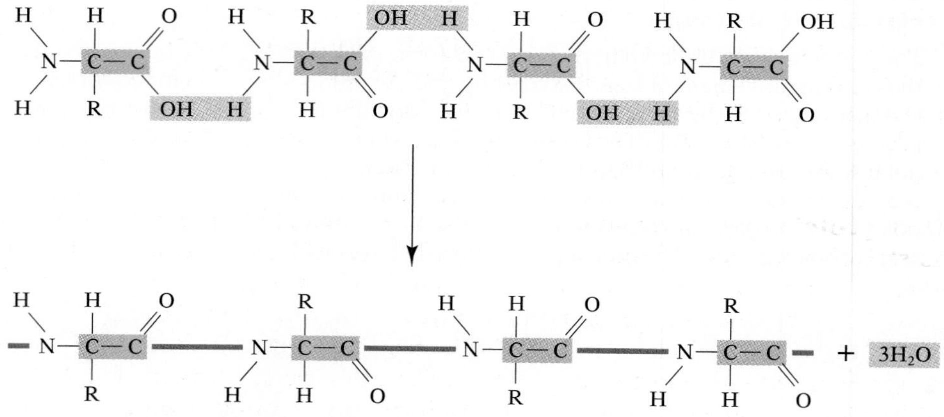

3.20 Synthesis of a polypeptide chain

Condensation reactions between the —COOH and —NH_2 groups of adjacent amino acids result in peptide bonds (red) between the acids. Notice again that the process of combining units releases water.

3.21 Structural formula of a cystine bridge

A cystine bridge is formed when two cysteine peptides (orange) from different parts of a protein are linked by a disulfide bond.

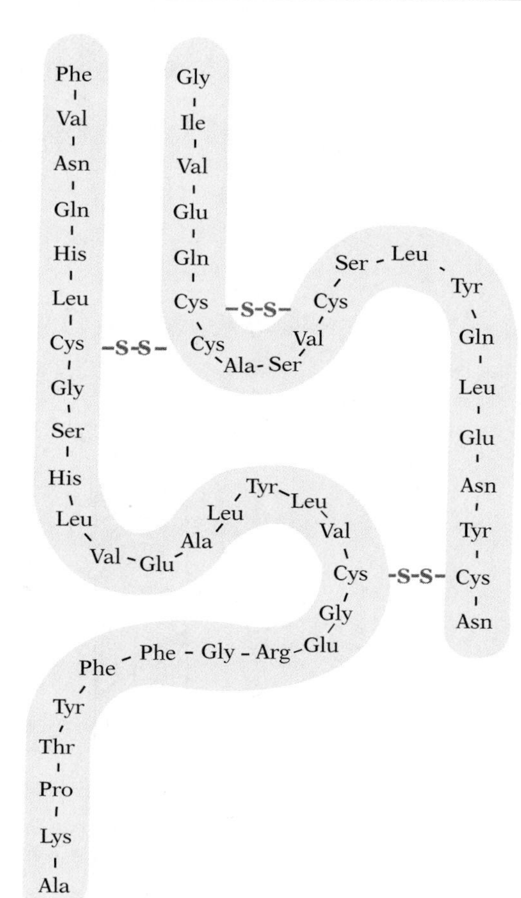

3.22 Structure of bovine insulin

The molecule consists of two polypeptide chains joined by two disulfide bonds. There is also one disulfide bond within the shorter chain (right). Hydrogen bonds (not shown) between the chains and between segments of the same chain are also present.

are between the sulfur atoms of two units of the amino acid cysteine (Fig. 3.19), which can react with each other to form a cystine bridge (Fig. 3.21). Disulfide bonds can also link two parts of a single polypeptide chain, maintaining it in a bent or folded shape (Fig. 3.22).

The number and sequence of amino acids is the ***primary structure*** of a protein molecule. The primary structure is different for every kind of protein; the potential number of different proteins is enormous. For a relatively short polypeptide chain of 100 amino acids, for instance, 20^{100} sequences are possible. Most organisms have between 1000 and 50,000 different proteins.

How proteins get their shape Proteins are not laid out simply as straight chains of amino acids. Instead, they coil and fold into complex spatial conformations, which play a crucial role in determining the distinctive biological properties of each protein. Much of this three-dimensional character is a consequence of weak interactions between peptides in the protein. Certain precise degrees of coiling allow internal hydrogen bonds to form and stabilize what is called an ***alpha (α) helix*** (Fig. 3.23). A helix can be visualized as a ribbon wound around a regular cylinder (Fig. 3.23A). In a protein, each complete turn of the helix takes up approximately 3.6 amino acid units of the polypeptide chain (Fig. 3.23B). The chain is held

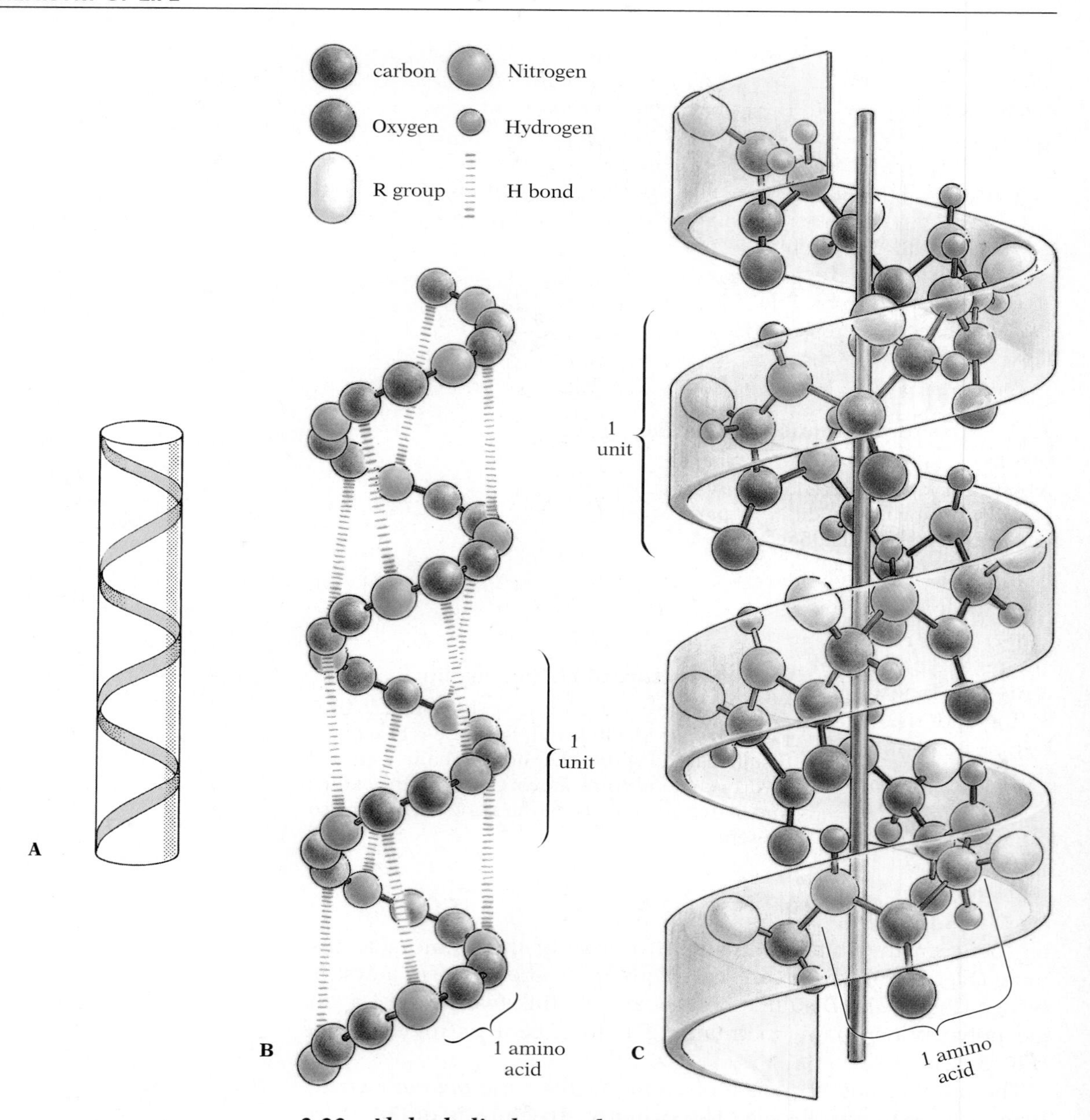

3.23 Alpha-helical secondary structure of some proteins

(A) The helix may be visualized as a ribbon wrapped around a regular cylinder. (B) The backbone of a polypeptide chain (the repeating sequence of N–C–C–N–C–C–N–C–C–) is shown coiled in a helix (with all other atoms and R groups omitted). It takes approximately 3.6 amino acid units (N–C–C) to form one complete turn of the helix. The hydrogen bonds shown extend between the amino group of one amino acid and the oxygen of the third amino acid beyond it along the polypeptide chain. (C) A ball-and-stick model of an α-helical section of a protein.

in this helical shape by hydrogen bonds formed between the amino group of one amino acid and the oxygen of the third amino acid beyond it along the polypeptide chain—which is the amino acid next to it in the axial direction of the helix (Fig. 3.23C). The ***conformation***, or arrangement in space, of amino acids in a peptide chain is called ***secondary structure***.

The helical pattern is seen at its simplest in some ***fibrous proteins***. One category of these insoluble proteins includes the ***keratins***, which provide the structural elements for many of the specialized derivatives of skin cells. Keratins with extensive α-helical secondary structure, such as nails, hooves, and horns, are hard and brittle. The hardness results from an extraordinarily large number of covalent cystine bridges: as many as 1 amino acid in 4 is cysteine. Others, such as hair and wool, are soft and flexible and can easily be stretched (especially when moistened and warmed). The stretching is possible because there are many fewer cystine bridges. The intrachain hydrogen bonds are easily broken, and the polypeptide chains can then be pulled out of their compact helical shapes into a more extended form. The chains tend to contract to their normal length, with re-formation of the hydrogen-bonded α helix, when the tension on them is released (or when they are dried and cooled).

Another stable arrangement of peptides within a polypeptide chain is the beta (β) structure, often called the ***pleated sheet***. In this conformation, also seen at its simplest among some of the keratins, many side-by-side polypeptide chains are cross-linked by interchain hydrogen bonds (Fig. 3.24). The resulting arrangement is flexible and strong but resists stretching because the polypeptide chains are already almost fully extended. The best-studied β-keratin is silk; other examples include spider webs, feathers (Fig. 3.25), and the scales, claws, and beaks of reptiles and birds. A number of important variants of the β structure give other distinctive shapes to protein segments.

Another kind of fibrous protein, with its own distinctive secondary structure, is ***collagen***, the most abundant protein in higher

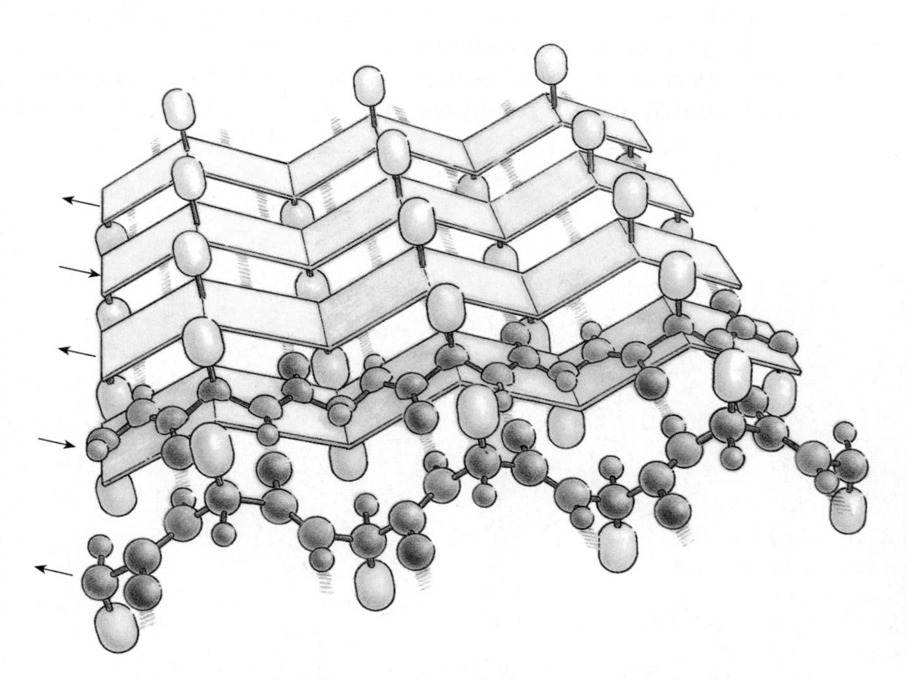

3.24 Pleated-sheet secondary structure of some proteins
Diagrammatic representation of five parallel polypeptide chains in β conformation, with the imaginary pleated sheet shown for four of the chains.

3.25 The feather of a bird, a β-keratin
This SEM shows the base of a parakeet's tail feather.

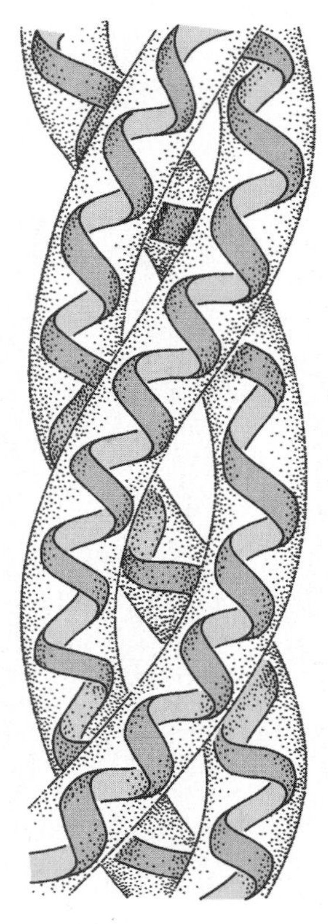

3.26 Model of a portion of a molecule of collagen

Three polypeptide chains, each helically coiled, are wound around one another to form a triple helix. The "sheaths" here and in Figures 3.27 and 3.30 are intended as a reminder that each molecule consists not merely of a backbone, but also of R groups, which give it volume.

vertebrates. Collagen can constitute one-third or more of all the body protein and is especially abundant in skin, tendons, ligaments, bones, and the cornea of the eye. A molecule of collagen comprises three polypeptide chains, each first helically coiled and then wound around the other two to form a triple helix (Fig. 3.26). Collagen fibers are exceedingly strong and very resistant to stretching.

The polypeptide chains of globular proteins are folded into complicated spherical or globular shapes (Fig. 3.27). Because of charged and polar R groups on their exposed surfaces, globular proteins (which include organic catalysts known as enzymes), proteinaceous hormones, the antibodies of the immune system, and most blood proteins are usually water-soluble. Typically, they are made up of sections of α helix and β sheet interspersed with nonhelical regions, strengthened by disulfide bonds (Fig. 3.28). The protein myoglobin, which is the oxygen-storage protein in muscles, provides a typical example of a helix-dominated protein. It consists of one polypeptide chain containing eight α helices connected by short regions of irregular (nonhelical) coiling. At each nonhelical region, the three-dimensional orientation of the polypeptide chain changes, giving rise to the protein's characteristic folding pattern. This three-dimensional folding pattern, which is superimposed on the secondary structure, is called ***tertiary structure*** (Fig. 3.27). In practice, tertiary structure is difficult to determine. The protein must first be crystallized. Then X rays are beamed through the crystals; deflected by the electrons of the thousands of atoms, they form a pattern that is then deciphered by a computer. This process is called X-ray crystallography (Fig. 3.29).

When a globular protein is composed of two or more independently folded polypeptide chains loosely held together (usually by weak bonds), the manner in which the already folded subunits fit together is called ***quaternary structure*** (Fig. 3.30).

Several aspects of a protein's primary structure (its amino acid sequence) contribute to producing its tertiary and quaternary

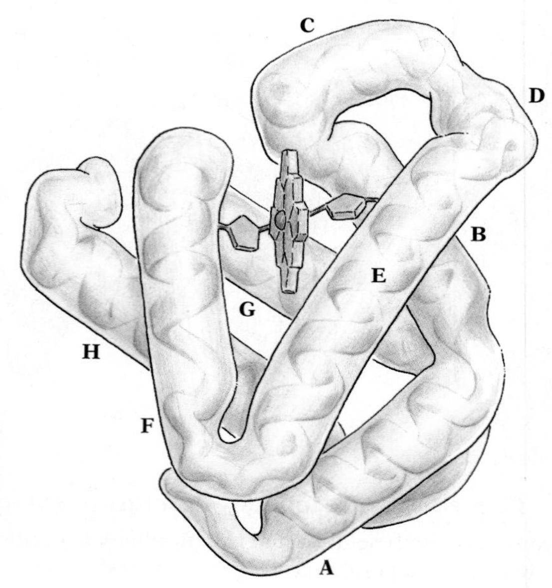

3.27 Tertiary structure of myoglobin

Myoglobin is a globular protein related to hemoglobin and, like hemoglobin, is characterized by a strong affinity for molecular oxygen. It has a single, complexly folded polypeptide chain of 151 amino acid units, and attached to the chain is a nonproteinaceous group called heme (represented by the disk). The polypeptide chain consists of eight sections of α helix (labeled A through H), with nonhelical regions between them. These nonhelical regions are a major factor in determining the tertiary structure of the molecule—that is, the way the helical sections are folded together. Section D is oriented perpendicular to the plane of the page.

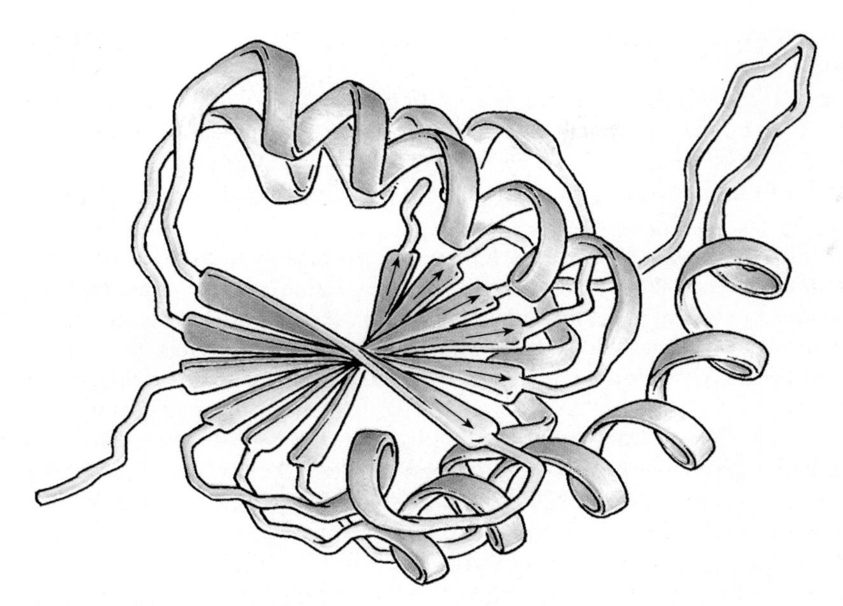

3.28 Mixture of helices and sheets in an enzyme

The enzyme lactate dehydrogenase incorporates five α helices and six β sheets, indicated schematically here as coils (red) and ribbons (blue).

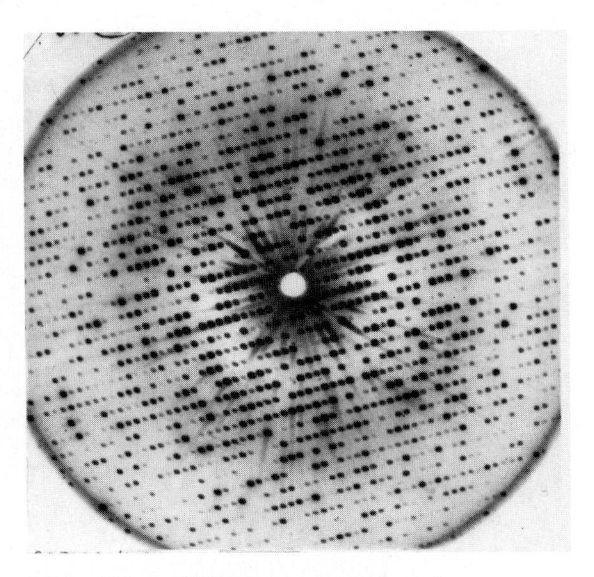

3.29 X-ray diffraction pattern of myoglobin

The intensity of each dot in this photograph of an X-ray diffraction pattern of sperm whale myoglobin provides information about the location of atoms in the molecule.

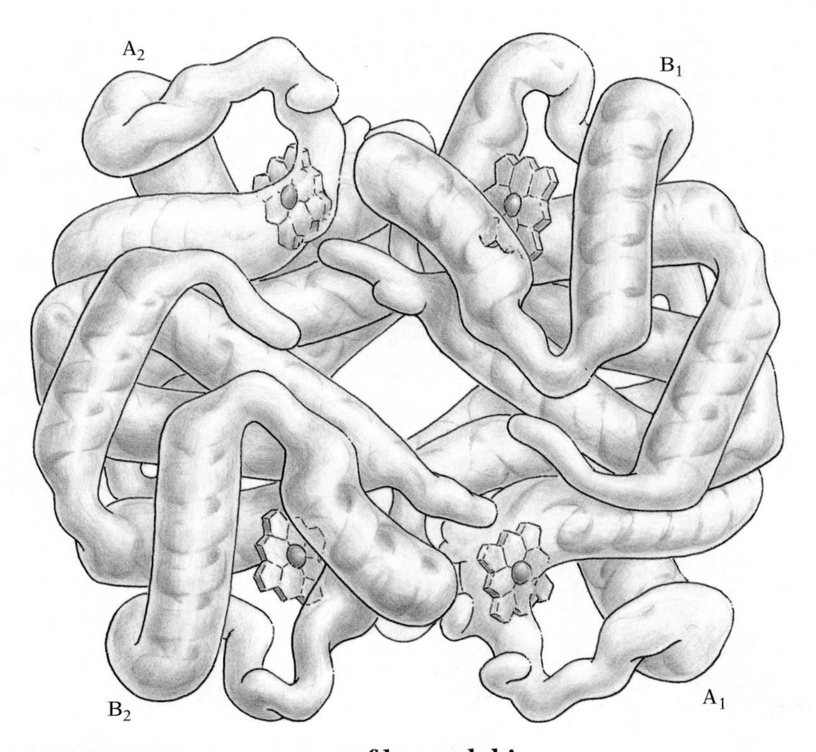

3.30 Quaternary structure of hemoglobin

A single molecule of hemoglobin is composed of four independent polypeptide chains, each of which has a globular conformation and its own prosthetic group. The spatial relationship between these four—the way they fit together—is called the quaternary structure of the protein. Chains A_1 and A_2 are identical, as are B_1 and B_2.

HOW WE KNOW:
CHROMATOGRAPHY

Chromatography is one of the most valuable techniques available for separating the substances found in blood or cells, or separating a single substance into its constituent parts. It is chromatography that first demonstrated that bacteria have more than 1000 different proteins, and allowed most of them to be individually isolated. After isolating a single protein using chromatography, we can hydrolyze it and then use chromatography again to separate the amino acids from one another. The many kinds of chromatography all share the same fundamental mechanism—the simultaneous exposure of the mixture being studied to two different substances, such as two solvents that will not mix, or a solvent and an adsorbent solid. (An adsorbent is so named because molecules of a gas, dissolved solute, or liquid adhere to its surface.) Each molecule of each solute in the mixture will diffuse back and forth between the two substances; the relative affinity of a solute molecule for those two solvents will determine how long, on average, it remains in each. For example, if solute A, a material highly soluble in water but only minimally soluble in phenol, is shaken in a jar with water and phenol (which do not mix), each molecule of A will divide its time between the two solvents in such a fashion that it will more often be in the water.

Paper chromatography is one of the simplest kinds of chromatography. Several drops of a mixture of unknown molecules are placed near one bottom corner of a piece of filter paper moistened with water (A, in the figure). The bottom edge of the paper is then dipped into a nonaqueous solvent such as phenol. As the solvent migrates up the paper by capillary action, those solutes in the mixture that have a much higher affinity for the solvent than for the water in the filter paper travel freely up the paper with the solvent. By contrast, those solutes that have a much higher affinity for the water do not travel far, but instead quickly transfer from the flowing solvent and bind to the stationary film of water on the filter paper. Materials with intermediate affinities for the solvent, compared with the water, travel intermediate distances. At the end of a measured time interval, the various solutes in the original mixture have come to rest at different places along the filter paper (B). Frequently, further separation is achieved by using a second solvent and allowing the solutes to travel across the paper in a new direction (C and D). The technique is then known as two-axis chromatography.

Another way to separate substances is electrophoresis. After chromatographic separation along one axis, an electric field is set up along the other; molecules with different ionic strengths and polarities migrate at different speeds toward one end or the other.

Column chromatography works like simple paper chromatography, except that instead of filter paper, a glass column packed with some hydrated adsorbent material, such as starch or silica gel, is used. The mixture to be tested, in a nonaqueous solvent, is poured into the top of the column and allowed to filter downward (or is pulled down by a pump). Each component in the mixture tends to move at its own rate, which depends on its relative affinity for the flowing solvent and the stationary beads of adsorbent material. When enough solvent is poured into the column to keep all the materials moving, each substance emerges from the bottom of the column at a different time, and they are collected in separate containers, ready for further analysis.

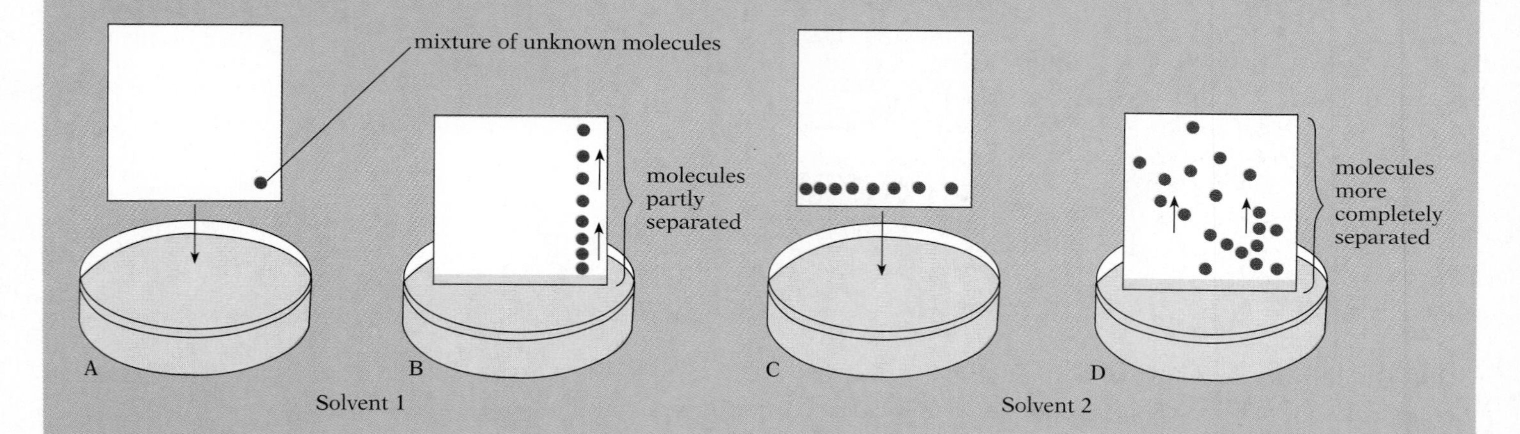

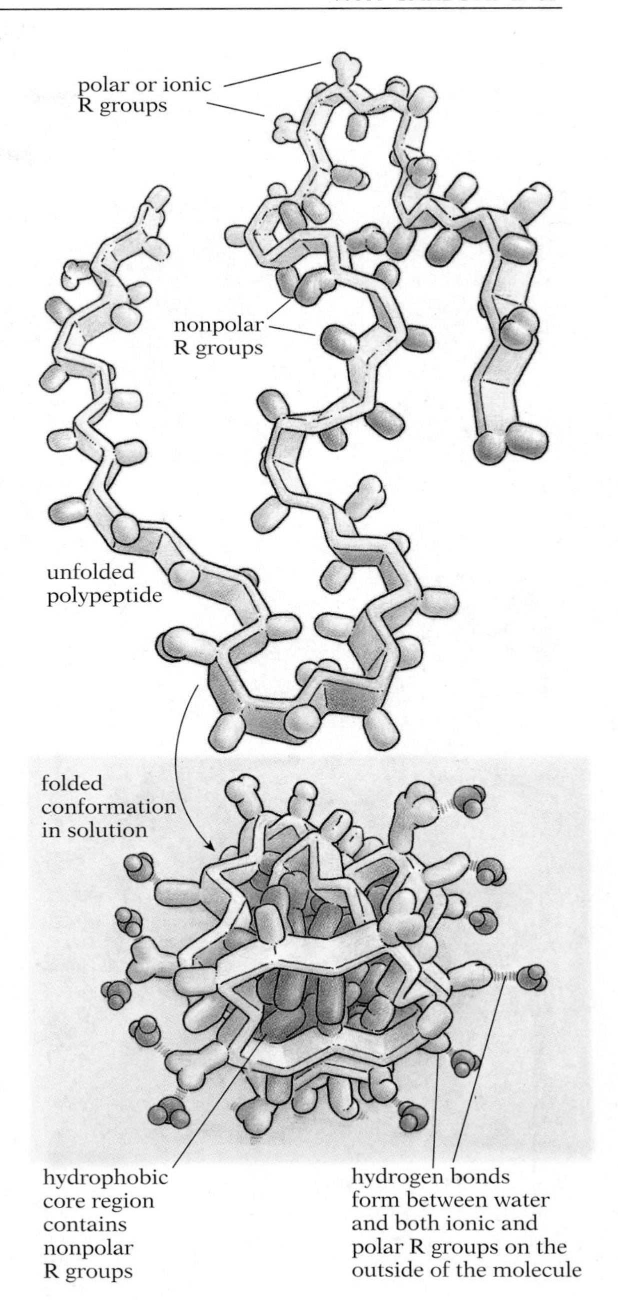

3.31 Folding of a globular protein in solution

As water molecules form hydrogen bonds with polar and ionic side groups, polypeptide chains can fold spontaneously so that the nonpolar groups are herded into the middle.

structure. If, for example, a polypeptide chain contains two cysteine units, the intrachain disulfide bond joining them may introduce a fold in the chain or stabilize one created in other ways (Fig. 3.22). The most common cause of folding is proline. Wherever there is a proline, a kink or bend occurs, because the structure of proline cannot conform to the geometry of an α helix. Four of the eight bends in globular myoglobin, in fact, result from the presence of prolines in the chain.

The distinctive properties of the various R groups of the amino acids also impose constraints on the shape of the protein. For example, hydrophobic groups tend to be close to each other in the interior of the folded chains—as far away as possible from the water that suffuses living tissue—whereas hydrophilic groups tend to be on the outside, in contact with the water (Fig. 3.31). In myoglobin, too, all the hydrophobic peptides are in the interior, and all but two of the hydrophilic peptides are on the outside. (The two exceptions, both ionic amino acids, hold the central, iron-containing heme group in place.) Thus the various kinds of weak bonds discussed earlier play crucially important roles in forming and stabilizing the tertiary structure of proteins.

Ultimately, the primary structure of a protein determines its spatial conformation. This is because there is one energetically most favorable, and therefore most stable, possible arrangement of the polypeptide chains. Further support for this comes from studies of ***denatured*** proteins—proteins that have lost most of their secondary, tertiary, and quaternary structure (and with it their normal biological activity) through exposure to high temperature or extreme pH. Under favorable test-tube conditions, some denatured proteins can spontaneously refold, regaining their native three-dimensional conformation, and thus recovering their normal biological activity. Since only the primary structure is available to dictate the folding pattern in such cases, it alone must be sufficient to determine all other aspects of protein structure (Fig. 3.32). In the crowded conditions within cells, however, many proteins need help from other molecules (called chaperones) to fold into a biologically active shape; most chaperones bind to the hydrophobic regions of unfolded protein chains and delay folding long enough to allow the protein to "discover" its most stable tertiary structure.

Conjugated proteins Attached to some proteins are nonproteinaceous groups called ***prosthetic groups***; an example is the heme group of myoglobin, a disklike structure with an iron atom in its center (see Fig. 3.27). Prosthetic groups may be as simple as a single metal ion bonded to the polypeptide chain; they may be sugars or other carbohydrate entities, or they may be composed of lipids. The prosthetic group alters the properties of the protein in important ways. Without their heme groups, for example, myoglobin and

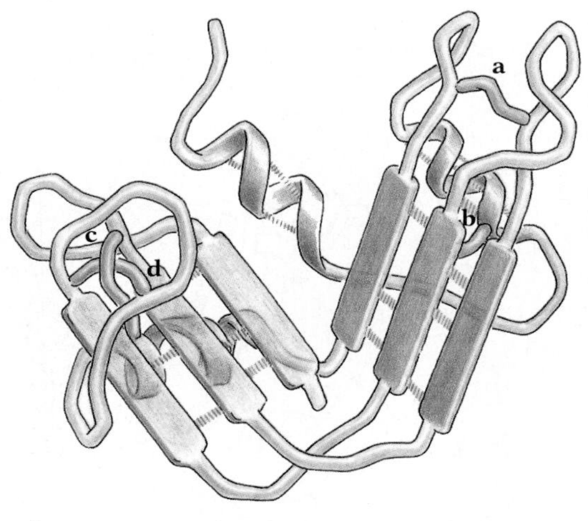

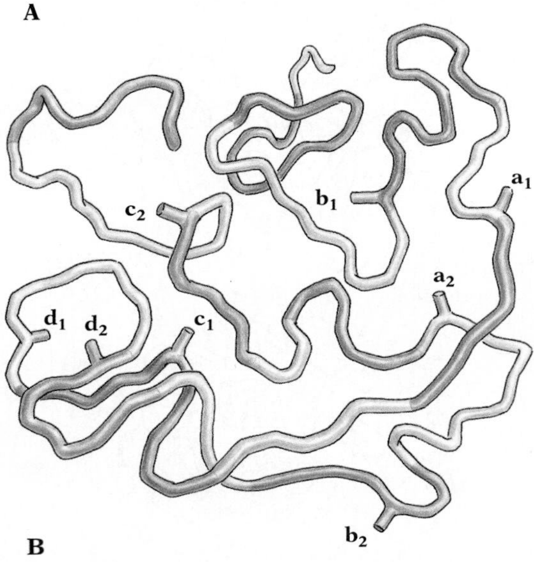

3.32 Denaturation and renaturation of ribonuclease

When ribonuclease, a normally globular protein (A), is denatured, with both its weak bonds and its four intrachain disulfide bonds (green) broken, it unfolds into an irregularly coiled state (B). In this denatured condition the ribonuclease lacks its usual ability to digest RNA. When the denaturing agents are removed and favorable conditions restored, the protein spontaneously refolds into its native conformation (A) and regains its capacity for biological activity as an enzyme. Even the four disulfide bonds re-form correctly. Since there are 105 possible ways to join the cysteines but the enzyme folds only to bring the correct pairs together, the conformation does not result just from the positions of the cysteines; most disulfide bonds serve to stabilize, rather than to determine, structure.

hemoglobin lose their high affinity for molecular oxygen. All proteins that contain nonproteinaceous substances are called ***conjugated proteins***.

■ THE ROLE OF NUCLEIC ACIDS

Nucleic acids constitute a fourth major class of organic compounds crucial to all life. They are the materials of which genes, the units of heredity, are composed. They are also the messenger substances that convey information from the genes in the nucleus to the rest of the cell.

Nucleic acid molecules are long polymers of smaller building blocks called ***nucleotides***, which are themselves composed of smaller units: a five-carbon sugar, a phosphate group, and an organic nitrogen-containing base. Both the phosphate group and the nitrogenous base are covalently bonded to the sugar (Fig. 3.33).

Information storage: deoxyribonucleic acid This nucleic acid, commonly called ***DNA***, is the one genes are made of. Four different kinds of nucleotide building blocks occur in DNA. All have deoxyribose as their sugar, but they differ in their nitrogenous bases. Two bases, ***adenine*** and ***guanine***, are double-ring structures of a class known as purines; the other two, ***cytosine*** and ***thymine***, are single-ring structures known as pyrimidines (Fig. 3.34).

The nucleotides within a DNA molecule are bonded so that the sugar of one nucleotide is attached to the phosphate group of the next nucleotide in the sequence (Fig. 3.35). Thus a long chain of alternating sugar and phosphate groups is established, with the nitrogenous bases oriented as side groups off this chain. The sequence in which the four different nucleotides occur is essentially constant in DNA molecules of the same species but differs between species. It is this sequence that determines the specificity of each type of DNA that encodes hereditary information, which is expressed largely through protein synthesis.

DNA molecules do not ordinarily exist in the single-chain form shown in Figure 3.35. Instead, two such chains, oriented in opposite directions, are arranged side by side like the uprights of a ladder, with their nitrogenous bases constituting the cross rungs of the ladder (Fig. 3.36). The two chains are held together by hydro-

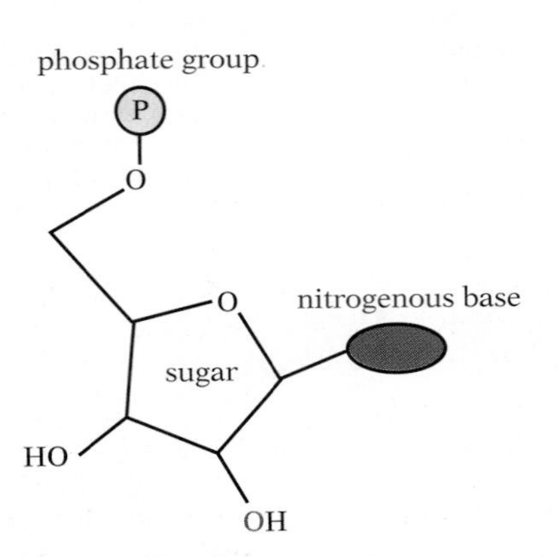

3.33 Diagram of a nucleotide

A phosphate group and a nitrogenous base are attached to a five-carbon sugar. The sugar shown is a ribose, part of RNA; the sugar in DNA (deoxyribose) lacks the oxygen (red) at the bottom.

3.34 Nitrogenous bases in DNA and RNA

The three single-ring bases—uracil, thymine, and cytosine—are pyrimidines; the two double-ring bases—adenine and guanine—are purines. Adenine, guanine, and cytosine are used in both DNA and RNA; thymine is restricted to DNA, while uracil takes its place in RNA.

PYRIMIDINES

PURINES

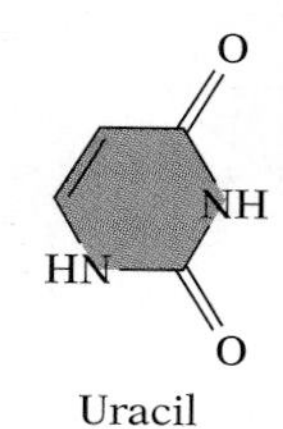

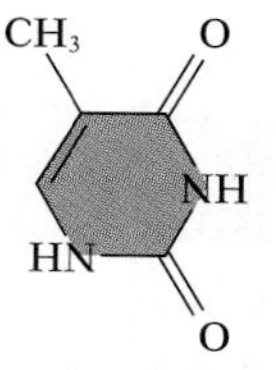

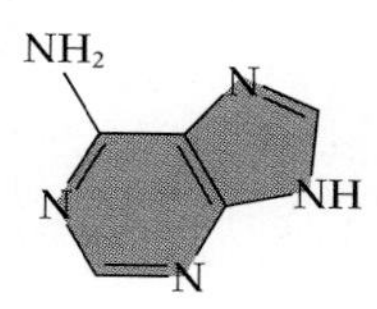

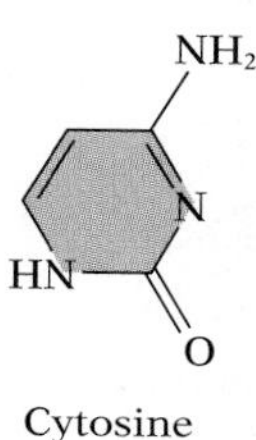

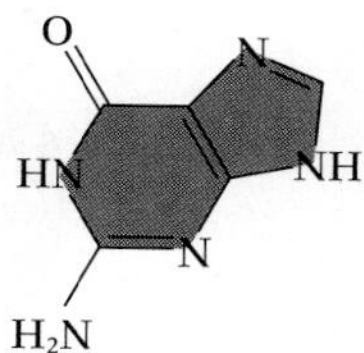

3.35 Portion of a single chain of DNA

Nucleotides are linked by bonds between the sugars and the phosphate groups. The nitrogenous bases (G, guanine; T, thymine; C, cytosine; A, adenine) are side groups.

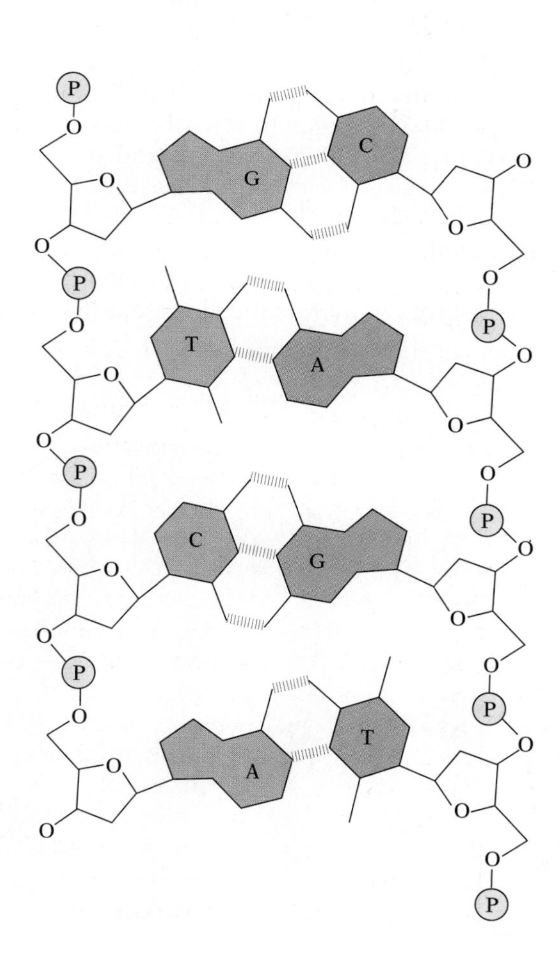

3.36 Portion of a DNA molecule uncoiled

The molecule has a ladderlike structure, with the two uprights composed of alternating sugar and phosphate groups and the cross rungs composed of paired nitrogenous bases. Each cross rung has one purine base and one pyrimidine base. When the purine is guanine (G), the pyrimidine with which it is paired is always cytosine (C); when the purine is adenine (A), the pyrimidine is thymine (T). Adenine and thymine are linked by two hydrogen bonds (red bands), guanine and cytosine by three. Note that the two chains run in opposite directions—that is, the free phosphate is at the upper end of the left chain and at the lower end of the right chain.

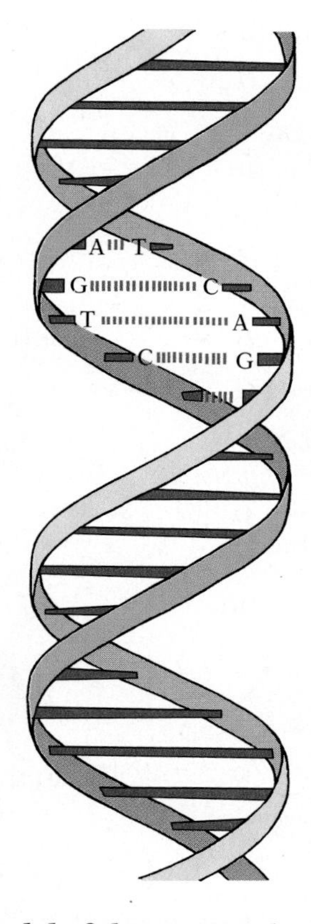

3.37 A model of the DNA molecule

The double-chained structure is coiled in a helix. As shown in detail in the second segment, it consists of two polynucleotide chains held together by hydrogen bonds (red bands) between their adjacent bases.

gen bonds between adjacent bases. Finally, the entire double-chain molecule is coiled into a double helix (Fig. 3.37).

The regular helical coiling and the hydrogen bonding between bases impose two important constraints on how the cross rungs of the ladderlike DNA molecule can be constructed. First, each rung must be composed of a purine (double ring) and a pyrimidine (single ring); only in this way will all cross rungs be of the same length, allowing the formation of a regular helix. Second, if the purine is adenine, the pyrimidine must be thymine, and, similarly, if the purine is guanine, the pyrimidine must be cytosine; only these two pairs are capable of forming the required hydrogen bonds (Fig. 3.36). Since it does not matter in which order the members of a pair appear (A–T or T–A; G–C or C–G), the double-chain molecule can have four different kinds of cross rungs (Fig. 3.36). The biological significance of this arrangement is that the base sequence of one chain uniquely specifies the base sequence of the other, so that the two strands can be separated and exact copies made every time a cell divides.

Jack of all trades: ribonucleic acid A second important category of nucleic acids comprises the ribonucleic acids, or ***RNA***. There are several types of RNA, each with a different role in protein synthesis. Some act as messengers carrying instructions from the DNA to the sites of protein synthesis in the cell. Others are structural components of subcellular structures, called ribosomes, on which protein synthesis takes place. Still others transport amino acids to the ribosomes for use in synthesis. A few even direct chemical reactions. RNA differs from DNA in three principal ways: (1) The sugar in RNA is ribose, whereas that in DNA is deoxyribose (Fig. 3.33); (2) instead of the thymine found in DNA, RNA contains a very similar base called ***uracil*** (Fig. 3.34); (3) RNA is ordinarily single-stranded, whereas DNA is usually double-stranded.

The relative contributions of DNA, RNA, proteins, and lipids to the volume of a typical cell are compared in Figure 3.38.

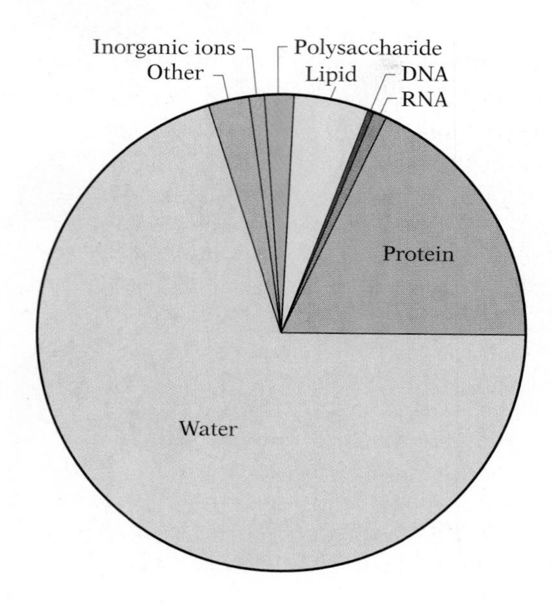

3.38 Chemical composition of a typical mammalian cell

Values are expressed as a percent of weight.

HOW CHEMICAL REACTIONS WORK

All the processes of life depend on the ordered flow of energy, and in particular on the behavior of electrons. Virtually all the energy for living things comes as light from the sun and is captured by electrons, which are thereby excited into higher-energy orbitals. The energy released by such electrons as they move to more highly electronegative atoms is harvested to fuel all the processes of life. To understand biology, then, it is essential to understand how the transfer of energy in chemical reactions takes place as one set of covalent bonds is replaced by another, an important subject in the field known as thermodynamics.

■ WHY FREE ENERGY ISN'T FREE

Suppose that two substances, A and B, can react with each other in solution to produce two new compounds, C and D:

$$\underset{\text{reactants}}{A + B} \longrightarrow \underset{\text{products}}{C + D}$$

What determines whether a reaction will take place spontaneously? The answer is that ***energy***, defined as *the capacity to do work*, must be available. ***Free energy*** (as the term is used in chemistry and biology) is the energy in a system available for doing work under conditions of constant temperature and pressure. Where there is energy to be tapped—whether in the weight of the water stored behind a dam, in the covalent bonds of sugars like glucose, in an electron that has been excited into a higher orbital by sunlight, or in the tightly bound nuclei of the atoms in a nuclear reactor—the potential for work is present (Fig. 3.39).

The ***First Law of Thermodynamics***—the Law of Conservation of Energy—tells us that *the total energy in the universe is constant*: if the energy needed to do work in a particular system—in a cell, for example—is not already available internally, it must be obtained from a source outside that system, which thereby loses a corresponding amount of energy. The ***Second Law of Thermodynamics*** states that *in the universe as a whole the total amount of free energy is declining*. This is because practically every energy transfer generates heat that is then no longer available for doing work. The magnitude of this waste is enormous, as the need for cooling towers in power plants and radiators in car engines to dissipate unused (and unusable) thermal energy makes evident. Even in our own bodies, waste heat can be a serious problem, and highly specialized mechanisms have evolved for releasing this waste energy to the environment.

The net change in free energy that accompanies a reaction determines whether or not it can proceed spontaneously. In other words, the course of the reaction depends on whether the total free energy of the covalent bonds in the reactants is greater or less than that of the covalent bonds in the products. We quantify any free-energy change as ΔG for a defined set of conditions: a temperature of 25°C, a pressure of 1 atm (atmosphere), pH 7.0, and with both

3.39 Bald eagle swooping down on prey
Many birds of prey use gravitational potential energy to power a rapid dive toward the animals they hunt. This eagle is braking from its dive, talons extended to grasp its prey.

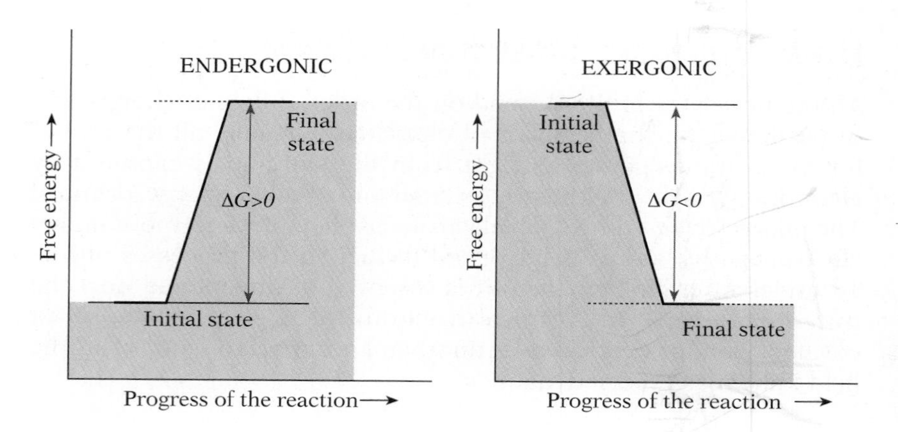

3.40 Exergonic versus endergonic reactions

Reactions either liberate some of the free energy of the bonds within the reactants or consume free energy from outside the system. Those that release energy (in the amount represented by ΔG) and thus go downhill from the initial state to the final state are said to be exergonic (right), while those that require energy and go uphill are said to be endergonic (left). (As we will see, these diagrams describe ideal conditions and are therefore oversimplified.)

the reactants and products at a concentration of 1 mole/liter.[1] Any reaction from a thermodynamic perspective involves the free energy of the initial reactants (G_i), the free energy of the final products (G_f), and the change in free energy engendered by the reaction (ΔG):

$$\begin{array}{cccc} & \text{initial state} & \text{final state} & \\ \text{(reactants)} & A + B & \longrightarrow\ C + D & \text{(products)} \\ & G_i & \longrightarrow\ G_f & (\Delta G = G_f - G_i) \end{array}$$

The term ΔG (pronounced "delta G"), the change in free energy as a result of the reaction, is the crucial variable. *If the reaction results in products with less free energy in the covalent bonds than the reactants possessed, the reaction is "downhill" and can proceed spontaneously.* Since the free energy liberated from the covalent bonds of the reactants is usually released as heat, such reactions are said to be exothermic (heat-producing) or, more generally, ***exergonic*** (energy-releasing). If, on the other hand, the reaction requires an input of external energy, it cannot proceed spontaneously. An "uphill" reaction of this kind, in which the covalent bonds of the products have more free energy than those of the reactants, is said to be ***endergonic***. These two alternatives are illustrated schematically in Figure 3.40.

All spontaneous reactions are downhill, or exergonic. This explains why all systems, living or inanimate, tend to lose free energy and ultimately reach a state in which their free energy is as low as possible, just as rocks on a hill tend to move down to the bottom rather than up to the top.[2] Throughout this book we will indicate the free-energy change, ΔG, resulting from biological reactions to

[1] In thermodynamics texts the formal symbol for changes in free energy under these conditions is $\Delta G'^{\circ}$. Since this is the only version of ΔG we will discuss, we can dispense with the superscripts.

[2] Physicists and chemists frequently relate free energy and order, which is defined as an arrangement of components in a system that is unlikely to occur by chance. Consider the unlikely circumstance in which all the molecules of air in a room are on one side, with a vacuum on the other. This orderly arrangement disintegrates rapidly as the molecules of air spread throughout the room, with a corresponding loss of free energy. The more orderly a system is, the more free energy it possesses; however, since increasing disorder is inevitable, so also is the spontaneous loss of free energy. The amount of disorder in a system—the energy unavailable for doing work—is called entropy.

the right of the reaction formula. Remember that when ΔG is negative (meaning that the covalent bonds of the reactants had more energy than those of the products), the reaction is exergonic; when ΔG is positive, the covalent bonds of the products have gained energy, so the reaction is endergonic. This relationship holds even when factors like temperature do not correspond to the standard conditions listed earlier: though the exact magnitude of ΔG will shift as conditions change, the *relative* magnitude of the ΔG values of different reactions in the same cell normally will not. Hence the ΔG values of different reactions under the same conditions can be directly compared, and many aspects of cell chemistry can then be predicted.

■ PREDICTING REACTIONS: THE EQUILIBRIUM CONSTANT

Our discussion so far has implied that two reactant molecules can combine spontaneously to create products only if free energy is liberated, and that if a reaction is exergonic, all the reactants in a mixture will ultimately be turned into products. However, we have ignored the possibility of a ***back reaction***, with the products C and D combining to regenerate the reactants A and B:

$$\mathrm{C + D \longrightarrow A + B}$$

Though this reaction is endergonic (uphill), it can take place—very slowly—because there is a source of energy within every system. This source, which we have ignored so far, is the energy of motion: the ***kinetic*** (or ***thermal***) ***energy*** of the molecules. Every molecule has both a certain characteristic amount of stored energy in its bonds and energy of motion, which depends on its speed. In any solution, some molecules move very fast, others more slowly. In our example, when two fast-moving product molecules, C and D, collide, their energy of motion will sometimes be converted into covalent bond energy to produce two (slow-moving) reactants, A and B. Though this back reaction is rare, it is very important when the forward reaction is near completion. At this point the reactants A and B are so scarce that the rare, kinetic-energy-dependent back reaction may be just as likely as the forward reaction.

When the forward reaction, slowed in consequence of the increasing scarcity of reactants, is just counterbalanced by the rare back reaction, the two processes are in equilibrium and no further net change in the concentration of substances takes place. This ultimate stable ratio of products to reactants is the ***equilibrium constant*** **(K_{eq})**:

$$K_{eq} = \frac{\text{product concentration}}{\text{reactant concentration}}$$

This relationship succinctly summarizes the results of any reaction, and so complements ΔG in describing chemical reactions. For instance, an equilibrium constant of 10 means that when equilibrium has been reached under conditions of stable temperature and pressure, the products of the reaction outnumber the reactants by a factor of 10:

$$\mathrm{A + B} \xrightarrow[\leftarrow]{K_{eq} = 10} \mathrm{C + D} \qquad (\Delta G = -1.4 \text{ kcal/mole})$$

Table 3.2 Relationship between K_{eq} and ΔG at 25°C

ΔG (KCAL/MOLE)	K_{EQ}	
4.1	0.001	Endergonic reactions
2.7	0.01	
1.4	0.1	
0	1.0	
−1.4	10.0	Exergonic reactions
−2.7	100.0	
−4.7	1000.0	

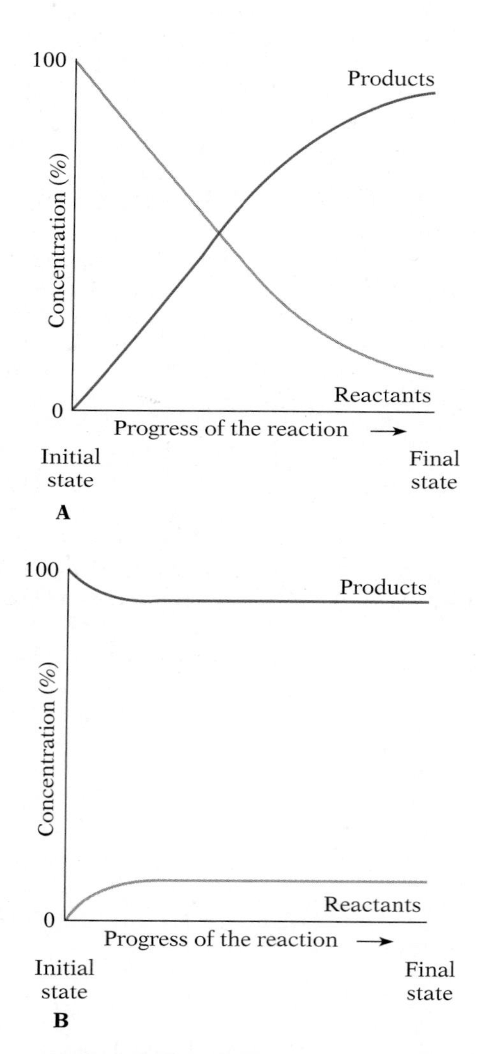

3.41 Relationship between the equilibrium constant and concentration

The equilibrium constant, specifying the final ratio of products to reactants, is independent of the starting concentration. In the example shown here, the equilibrium constant is 10, which means that products (red curves) will be 10 times more common than reactants (blue) when the reaction ends. This is the outcome whether we begin with pure reactants (A) or pure products (B).

Such a reaction is downhill: the products are more abundant than the reactants, and the reaction liberates energy. The predominance of the forward reaction is indicated by the longer arrow pointing from the reactants to the products. In a reaction with an equilibrium constant of 0.1, on the other hand, the reactants outnumber the products by 10 to 1:

$$E + F \underset{}{\overset{K_{eq} = 0.1}{\longleftarrow\!\rightleftharpoons}} G + H \qquad (\Delta G = +1.4 \text{ kcal/mole})$$

Such a reaction is uphill; it consumes energy, and so relatively few of the initial reactant molecules will be converted into products. And finally, an equilibrium constant of 1.0 indicates that the products and the reactants are equally abundant:

$$I + J \overset{K_{eq} = 1.0}{\rightleftharpoons} K + L \qquad (\Delta G = 0)$$

As we would expect, there is no free-energy change in this reaction. The numerical relationship between K_{eq} and ΔG at 25°C is summarized in Table 3.2.

Keep in mind that the equilibrium constant of a reaction is a ratio—a simple empirical description of the outcome of the reaction—and depends, ultimately, on the ΔG of the reaction. Hence, the ultimate ratio of products to reactants in no way depends on the starting conditions. Whether we start with a great deal of reactant or only product, we wind up with the ratio specified by the equilibrium constant (Fig. 3.41).

■ HOW ACTIVATION ENERGY GUIDES REACTIONS

How do particular compounds come to follow particular reaction pathways? Suppose the cell manufactures compounds D and Z, as follows:

$$A + B \rightleftharpoons C + D \qquad W + X \rightleftharpoons Y + Z$$

What is to prevent B from reacting with W, X, Y, or Z instead of with A? Some of these combinations may well have negative ΔG values and so be energetically favorable. What prevents all but the "correct" reactants from combining?

For two molecules to combine, they must be brought unusually close to each other in a particular orientation, and frequently one or more preexisting bonds must be broken. This requires energy—specifically known as ***activation energy*** **(E_a)**—so even an exergonic reaction has an endergonic first step (Fig. 3.42). The only source of energy for this "priming" is the kinetic energy of collision.

Because covalent bonds are strong, substantial energy may be necessary to break the preexisting bonds of reactants—often far more than the amounts listed for the ΔG in Table 3.2. Consider this enormously exergonic reaction:

$$2H_2 + O_2 \longrightarrow 2H_2O$$

Even though this combination of oxygen and hydrogen can be explosive—for example, it provides much of the energy that pushes the space shuttle into orbit—the two reactants can coexist as a stable mixture indefinitely. A single spark, however, will initiate an explosive reaction. The same stability is a property of most reactants in living systems: the energy necessary to bring the reactants together and break their covalent bonds is far greater than the energy of all but the very few most rapidly moving molecules in a solution. The activation-energy barrier therefore prevents most reactions from taking place at a significant rate (Fig. 3.42). Without such a barrier the complex high-energy molecules on which life depends would be unstable and would break down.

Once a reaction does get started, the combination of one pair of reactants may release enough energy (usually in the form of heat) to activate the next pair, and so on, in a chain reaction. This is precisely what happens in a rocket engine and in the combustion of a dry piece of firewood. The wood, as we all know, can lie in a woodpile for years without bursting into flames spontaneously; once set on fire, it literally consumes itself as the free energy liberated by the combining of carbon and oxygen into CO_2 supplies enough activation energy to continue the burning. Put quite simply, heating a mixture will increase the *rate* of reaction (though it cannot affect the ultimate ratio of products to reactants, since that is a constant—K_{eq}). Cells, however, literally cook at temperatures much above the internal 37°C of most mammals and birds, so they cannot use added heat as a way of overcoming the activation-energy barrier. Moreover, cells must use some method that lowers the barrier selectively, so that some exergonic reactions run, while others do not. How is this crucial task managed?

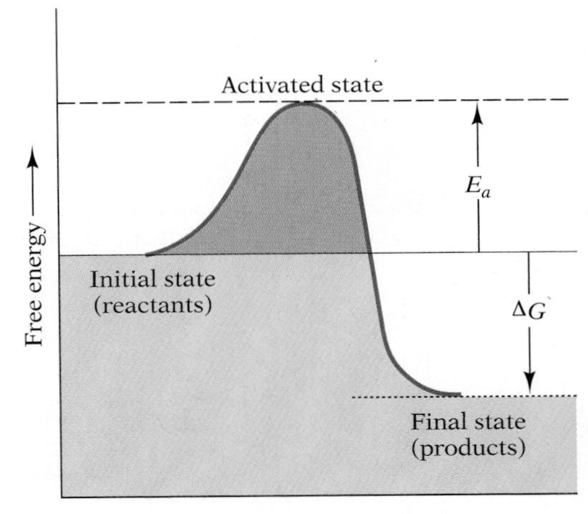

3.42 Energy changes in an exergonic reaction

Though the reactants are at a higher energy level than the products, the reaction cannot begin until the reactants have been raised from their initial energy state to an activated state by the addition of activation energy (E_a). It is the need for activation energy that ordinarily prevents high-energy substances like glucose from breaking down, and hence makes them stable; the higher the activation-energy barrier, the slower the reaction and hence the more stable the substance. When activation energy is available, the reactants form a temporary and unstable activated complex, which breaks down to yield the end products of the reaction; in the process both activation energy and free energy (ΔG) are released.

■ HOW CATALYSTS BREAK THE RULES

Certain chemicals speed up reactions between other chemicals. As we have seen, a simple mixture of hydrogen and oxygen does not react. If we add a small quantity of platinum, however, the mixture will explode. After the reaction is over, the platinum will still be present, unchanged.

A substance that speeds up a reaction but is itself unchanged when the reaction is over is a ***catalyst***. Catalysts affect only the *rate* of reaction; they simply speed up reactions that are thermodynamically possible to begin with. A catalyst cannot alter the direction of

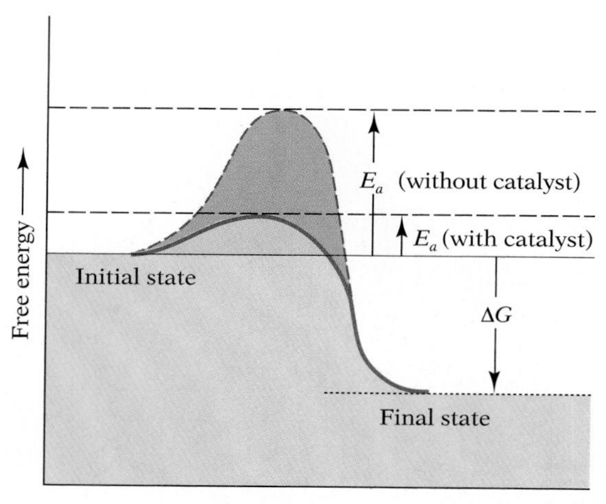

3.43 Reduction of necessary activation energy by catalysts

The activation energy (E_a) necessary to initiate the reaction is much less in the presence of a catalyst than in its absence. It is this lowering of the activation-energy barrier by enzyme catalysts that makes possible most of the chemical reactions of life. Note that the amount of free energy liberated by the reaction (ΔG) is unchanged by the catalyst—it is the same for both the catalyzed and the uncatalyzed reaction—and that only the activation energy is changed.

a reaction, its final equilibrium, or the reaction energy involved.

Catalysts work by decreasing the activation energy needed for the reaction to take place (Fig. 3.43), thereby increasing the proportion of reactants energetic enough to react (Fig. 3.44). A catalyst binds reactants in an intermediate state in which they are correctly oriented and important internal bonds are weakened (Fig. 3.45). As a consequence of binding, then, conditions are highly favorable for the reaction.

An inorganic catalyst like platinum is relatively unselective about the reactants it "helps"; the spacing of its atoms on a crystal face happens to match the distance between pairs of small covalently bonded atoms, and its loosely bound outer electron readily interacts with the outer shells of other atoms. However, cells need catalysts that only promote specific reactions; other combinations of potential reactants are undesirable. Lacking the necessary activation energy, and without specific catalysts to lower this thermody-

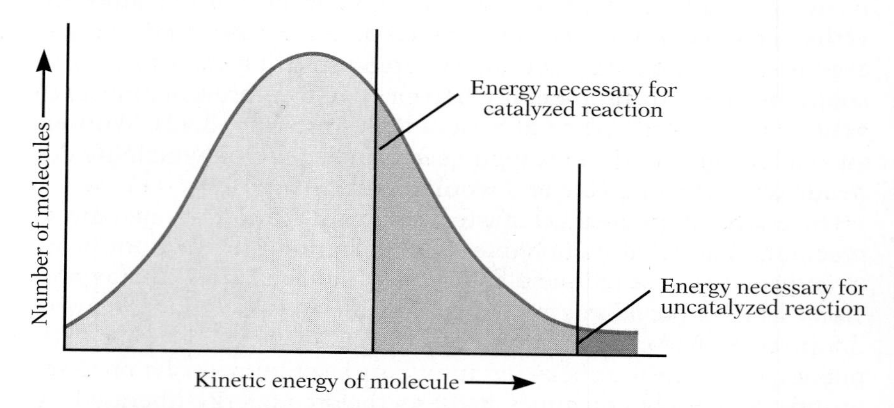

3.44 Effect of a catalyst on the ability of kinetic energy to activate a reaction

As indicated by this bell curve, reactant molecules exhibit a wide range of kinetic (thermal) energy. Only a minute fraction (right) have enough energy to overcome the activation-energy barrier for most reactions. In the presence of a catalyst, such as an enzyme, the barrier is lower, so a much larger proportion of the reactants (middle) can combine to form products.

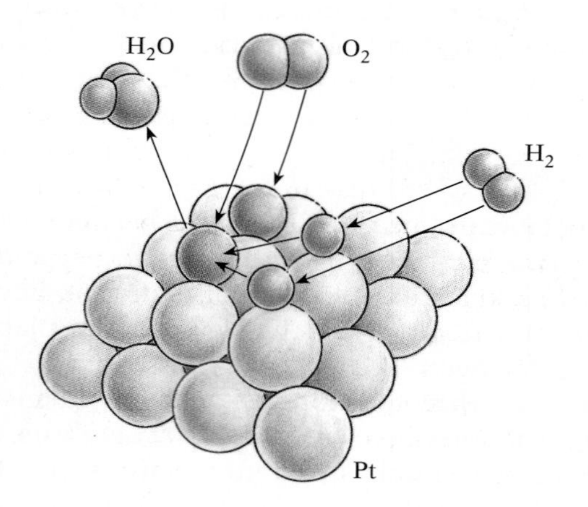

3.45 Action of platinum as a catalyst

Because of its loosely bound outer electron, platinum (Pt) is able to form weak temporary bonds with molecules of both hydrogen (blue) and oxygen (red). This binding draws the hydrogen and oxygen electrons away from their covalent positions, thus weakening the bonds within their respective molecules. In addition, the spacing of the platinum atoms tends to align the hydrogen and oxygen atoms in such a way that new bonds between hydrogen and oxygen can be more easily formed. Platinum is a catalyst in that it facilitates the reaction without being itself altered.

namic barrier, these reactants are unable to combine with one another. Cells have evolved an enormous variety of highly specialized organic catalysts, called ***enzymes***, which direct cellular chemistry along useful pathways. Virtually all enzymes are globular proteins.

■ BIOLOGICAL CATALYSTS: ENZYMES

Enzymes make use of the basic facts of biological thermodynamics: (1) A chemical reaction can proceed spontaneously if it releases free energy. (2) Because the activation energy is relatively high, biological reactions occur only very slowly without catalysts. (3) Catalysts, including enzymes, alter neither the equilibrium constant of a reaction nor the net change in free energy—they cannot, by themselves, cause a reaction to run uphill, but they can make specific downhill (exergonic) reactions occur quickly.

Why enzymes are so specific Enzymes are highly selective: a particular enzyme generally inteacts with only one type of reactant or pair of reactants, customarily called the ***substrates***. The enzyme thrombin, for instance, acts only on certain proteins and only at very specific sites. It "recognizes" the bond between the amino acids arginine and glycine, which it then hydrolyzes (an important step in the formation of blood clots). Like all catalysts, enzymes lower the activation energy required (see Fig. 3.43), thereby increasing the proportion of substrate molecules energetic enough to react (see Fig. 3.44). As a result, enzymes vastly speed up the reactions they catalyze; a single molecule of enzyme may cause thousands or even hundreds of thousands of molecules of reactant to combine into product each second.

Because of their efficiency and specificity, enzymes both steer specific substrates into particular reaction pathways and block them from others, thus guiding the chemistry of life with great precision. This specificity depends on the enzyme's three-dimensional molecular conformation; a given enzyme interacts only with substrates whose molecular configurations (their conformation and location of charged groups) "fit" its surface. This explains why when proteins are denatured—when their three-dimensional conformation is disrupted—their enzymatic properties vanish (Fig. 3.46). It also explains why enzymes are highly sensitive to changes in pH (Fig. 3.47). A change in pH results in the breakage of many of the weak bonds that help stabilize the conformation of proteins, thus changing the shape of the protein.

Enzyme and substrate fit together like a lock and key, or like pieces of a puzzle. The reactive portion of the substrate molecule and the portion of the enzyme known as the ***active site*** must fit together in space intimately enough to become temporarily bonded, like the platinum with the hydrogen and oxygen in Figure 3.44. In this way they form a transient enzyme-substrate complex:

$$E + S \longrightarrow ES \longrightarrow E + P$$

(E stands for enzyme, S for substrate, and P for product.) But enzymes and their substrates do not always have to fit together exactly *before* the enzyme-substrate complex (ES) can form; many

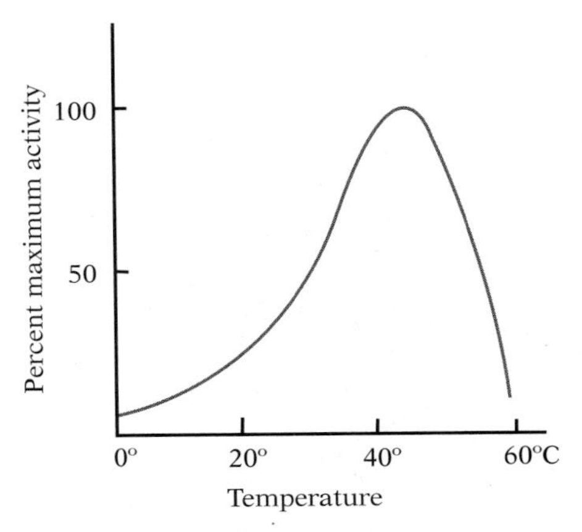

3.46 Enzyme activity as a function of temperature

Though temperature sensitivity varies somewhat from one enzyme to another, this curve applies to typical enzymes. Its activity rises steadily with temperature (approximately doubling for each 10°C increase) until thermal denaturation causes a sudden sharp decline, beginning between 40° and 45°. The enzyme becomes completely inactivated at temperatures above 60°, presumably because its three-dimensional conformation has been severely disrupted. Bacteria living in volcanic pools, by contrast, have well-stabilized enzymes with activity optima near 100°; below 60°, however, their enzymes have almost no activity.

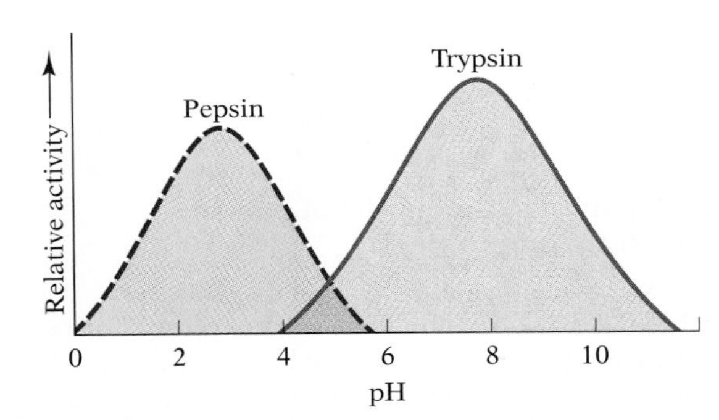

3.47 Enzyme activity as a function of pH

Most enzymes are very sensitive to pH, but they differ markedly in the pH value at which they are most active. Pepsin and trypsin are both enzymes that digest protein, but the pH ranges within which they are active overlap only slightly. Pepsin, a stomach enzyme, is most active under strongly acidic conditions, while trypsin, an enzyme secreted into the small intestine, is most active under neutral and slightly alkaline conditions.

(perhaps most) enzymes undergo conformational changes in the course of bonding, which create an ***induced fit*** (Fig. 3.48).

Spatial complementarity is only one of the prerequisites for enzyme-substrate interaction. Another is that enzyme and substrate be chemically compatible and capable of forming numerous and precise weak bonds with each other. The bonds between enzyme and substrate are usually the same types of weak bonds that stabilize protein conformation. These bonds can be made and broken rapidly in the collisions that result from random thermal motion at normal temperatures. The type of substrate to which a given enzyme molecule can become bonded depends on the exposed R groups of the amino acids at the active site and on their arrangement relative to one another. Suppose the active site of a particular enzyme is a curving groove and that most of the exposed R groups in this groove are electrically charged. The reactive portion of the substrate must have a complementary shape and charge or polarity; an electrically neutral nonpolar substrate molecule, or one with the same charge as the active site, could not bond to the active site of this enzyme. Conversely, only a hydrophobic substrate could interact with an active site made up largely of hydrophobic R groups.

Figure 3.49 shows what is currently known about the active site

SUBSTRATE

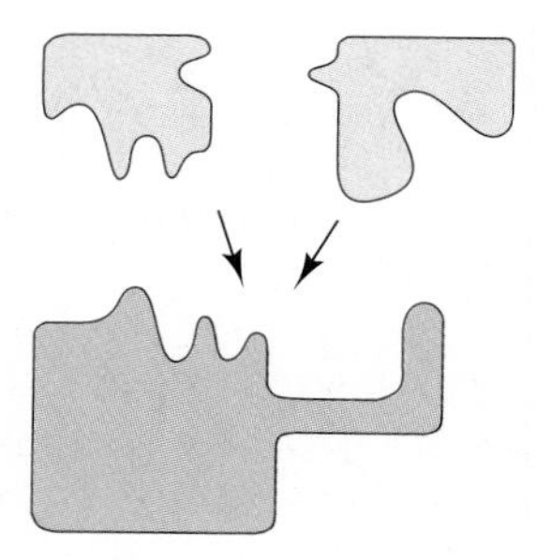

ENZYME

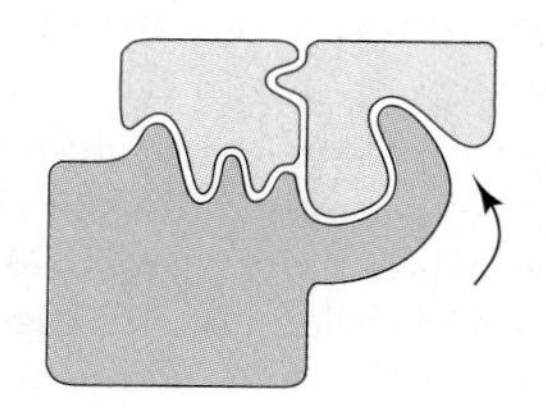

Enzyme-substrate complex

PRODUCT

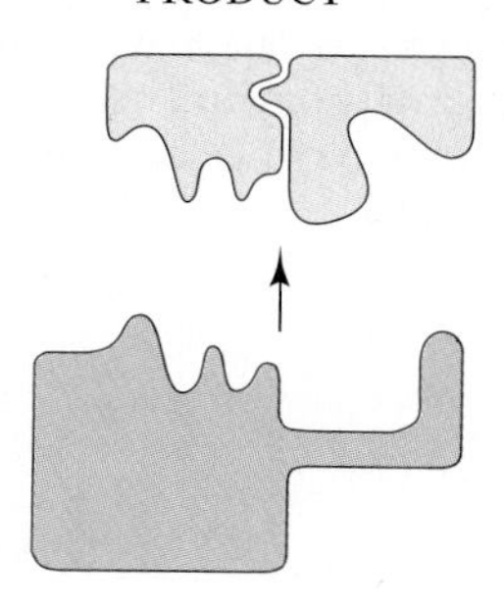

Enzyme resumes original configuration

3.48 Induced-fit model of enzyme-substrate interaction

The enzyme molecule has an active site onto which the substrate molecules can fit (top), forming an enzyme-substrate complex (middle). The binding of the substrate induces conformational changes in the enzyme that maximize the fit and force the complex into a more reactive state. The enzyme molecule reverts to its original conformation when the product is released (bottom).

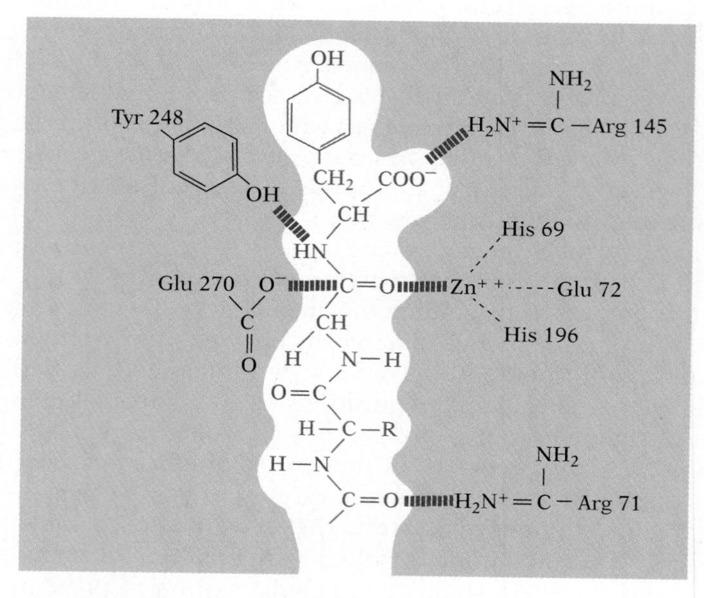

3.49 Active site of an enzyme

Shown here in schematic form is the base of the cleft (white) where the active site of carboxypeptidase (green) is located. (The hydrophobic entrance to the cleft is out of the drawing, to the bottom. The entire enzyme, with the active site highlighted, is depicted in Figure 3.50). Part of a substrate molecule (red) is shown in the cleft, linked to the enzyme by five weak bonds (red bands). Seven of the amino acids of the active site are indicated by their abbreviated names; the numbers beside the names refer to the positions of the amino acids in the enzyme polypeptide chain. The function of this enzyme is to separate the terminal amino acid (top) from the rest of the substrate amino acid chain (extending down out of the figure) at the covalent bond indicated by the arrow. The highly electronegative zinc and the charged oxygen draw away the electrons of this bond and so initiate its rupture.

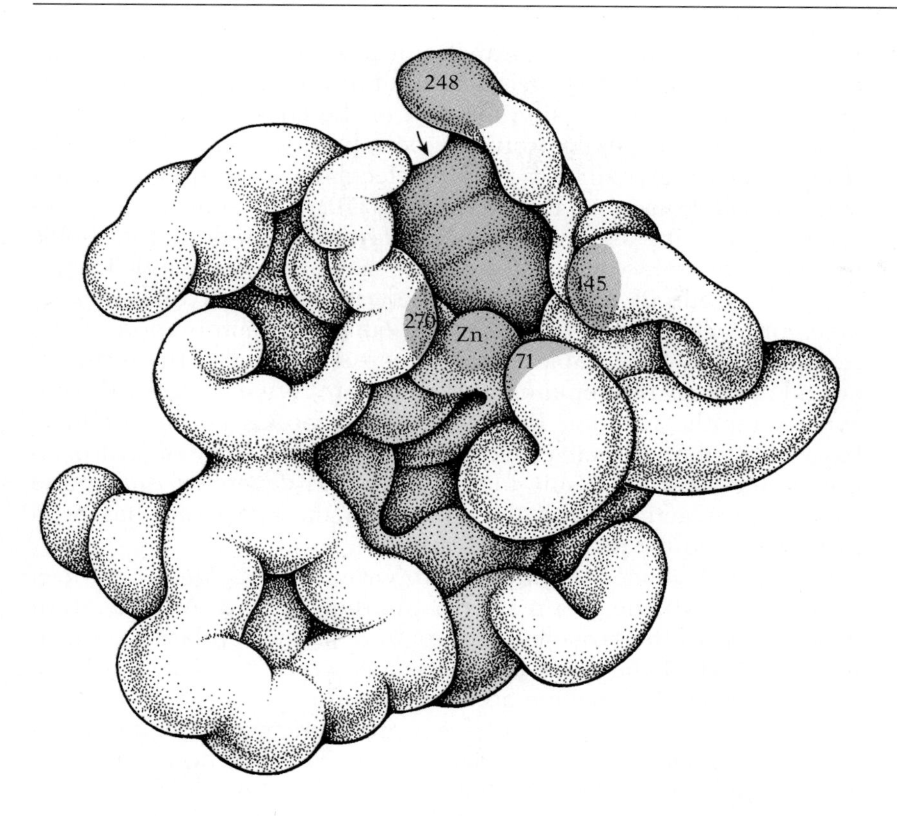

3.50 Location of the active site in an enzyme

The folding of the long chain of amino acids of carboxypeptidase brings together the zinc atom and three of the four amino acids that bind to the substrate at the active site, even though—as their numbers indicate—they are located at different places in the chain. The zinc atom itself is held in place by three different amino acids: nos. 69, 72, and 196 (not shown). The fifth part of the site folds into place (arrow) when the substrate binds to the enzyme. The region at the top center in dark color is the active site, while the area in light color is the hydrophobic entrance to the cleft. Most of the length of the polypeptide chain serves to (1) create an appropriately shaped cleft, (2) precisely position the R groups at the active site, (3) provide for the hingelike action of the induced-fit arm (containing amino acid 248), and (4) supply a second location for enzyme regulation (discussed later).

of the enzyme carboxypeptidase,[3] which catalyzes the removal of the terminal amino acid from one end of polypeptide chains during digestion. The enzyme's active site is a cleft into which the end of the substrate molecule can fit. The substrate forms several weak bonds with the R groups of amino acids at the active site. In addition, the substrate binds to a zinc held in place by the R groups of three other amino acids. Note that the critical amino acids in the active site (nos. 71, 196, 72, 69, 145, 248, and 270) are not adjacent to each other in the polypeptide chain, which means that the complex folding of the protein—its tertiary structure—has brought amino acids from several regions of the protein close together to form the active site (Fig. 3.50). Active sites nearly always include some nonadjacent amino acids. This helps explain why elevated temperature or a major pH change may greatly reduce an enzyme's activity: anything that changes the precise folding pattern of the polypeptide chain is likely to alter the critical arrangement of amino acids in the active site.

Carboxypeptidase is also typical in that the entrance to the cleft in which the active site is located is hydrophobic, so that the water molecules that surround the substrate are stripped away as it enters the groove. The binding energy of the weak bonds is just strong enough to stabilize the enzyme-substrate complex. The ability of carboxypeptidase to form these particular bonds accounts in

[3] The suffix *-ase* designates an enzyme, and is derived from the name *diastase*, coined in 1833 by two French chemists from the Greek *diastasis*, "to separate". The function of most enzymes can be guessed from their names. In the case of carboxypeptidase, *-peptid-* is from the Greek *pepsis*, "to digest," and *carboxy-* indicates that the digestion is from the carboxyl end of the protein.

part for its specificity. The functioning of this enzyme also illustrates the induced-fit strategy: as the binding begins at the other four parts of the active site, the part of the chain containing tyrosine (no. 248) moves in from the periphery to trap the substrate (the polypeptide chain). The highly electronegative zinc atom and the charged oxygen of glutamate (no. 270) help activate the substrate by drawing away the electrons of the N—C bond that holds the terminal amino acid to the rest of the chain (Fig. 3.49). The resulting cascade of electron shifts between atoms of differing degrees of electronegativity makes the outcome—elimination of the terminal bond—inevitable. Finally, the two products, the terminal amino acid and the remainder of the chain, separate from the enzyme after the reaction because the new bonds in the products have redistributed their electrons. The tenuous array of hydrogen bonds and van der Waals interactions that depended on precise electron interactions with the enzyme at the active site has been disturbed, and the products drift free.

Figure 3.49 illustrates another important point. Many enzymes contain a prosthetic group essential to their activity. A metal atom is often part of the prosthetic group; in carboxypeptidase the metal is zinc. Most of the trace elements our bodies require (see Table 2.1) are needed for enzyme prosthetic groups.

Some enzymes that do not have a prosthetic group require a cofactor to which they bond briefly during the reactions they catalyze. The cofactors may be metal ions, or they may be nonproteinaceous organic molecules called ***coenzymes***. Coenzyme molecules are much smaller and less complex than proteins. Like enzymes, they are not used up or permanently altered by the reactions in which they participate, and hence they can be used over and over again. Though only tiny amounts are needed, if the supply falls below normal, the health or even the life of the organism may be endangered. This is why vitamins, which act as parts of essential coenzymes, are so necessary in the diet.

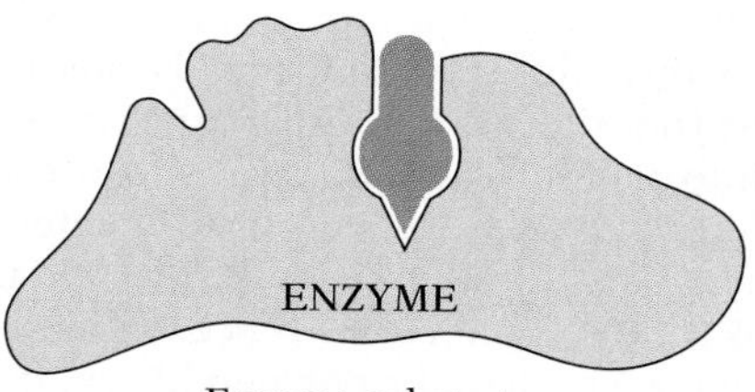

Enzyme-substrate complex

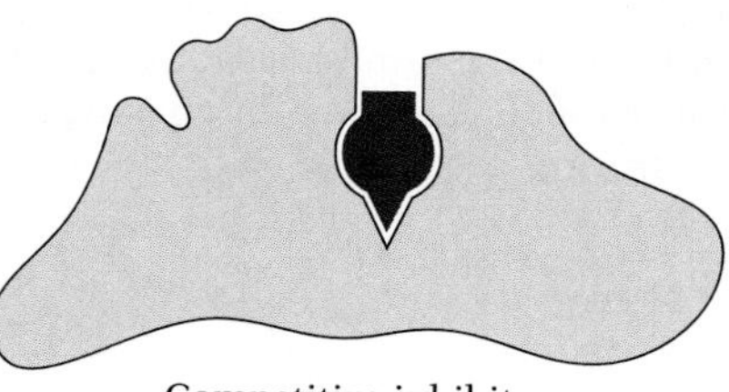

Competitive inhibitor bound to enzyme

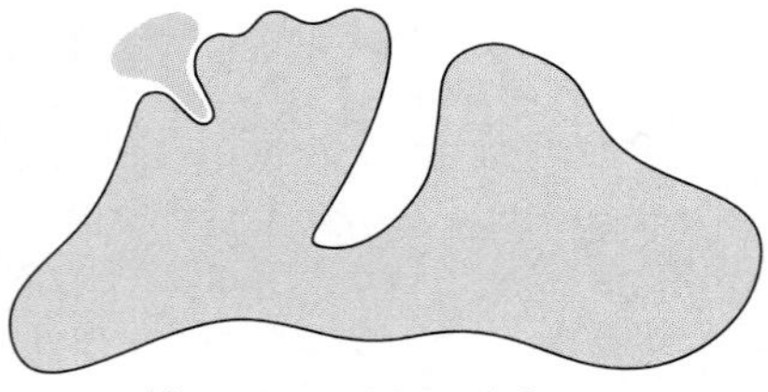

Noncompetitive inhibitor bound to enzyme

3.51 Competitive and noncompetitive inhibition of an enzyme

Top: The substrate is bound to the active site of the enzyme. Middle: The binding of a competitive-inhibitor molecule to the catalytic site prevents the substrate from binding. Bottom: A noncompetitive inhibitor bound to a different site on the enzyme induces an allosteric change that prevents the active site from catalyzing reactions.

How enzyme activity is controlled A variety of mechanisms have evolved for controlling the activity of enzymes. These mechanisms often depend on chemical agents that mask, block, or alter the active sites of the enzymes they help regulate.

One common form of enzyme control, ***competitive inhibition***, uses an inhibitor substance similar to the normal substrate of the enzyme. It binds reversibly to the enzyme's active site but is not chemically changed in the process; a bond normally broken in the substrate molecule, for instance, may in the inhibitor be too strong or too well "insulated" by nearby bonds to be severed. By binding to the active site, the inhibitor (I) masks the site and prevents the normal substrate molecules from gaining access to it (Fig. 3.51 middle). Thus the reaction

$$E + I \longrightarrow EI$$

competes with the reaction

$$E + S \longrightarrow ES \longrightarrow E + P$$

because both involve the same enzyme, which is present only in very small quantities.

Which of the two reactions will predominate depends on their relative energetics and, even more, on the relative concentrations of I and S. If there is much inhibitor and a low concentration of substrate molecules, a high percentage of the enzyme will be bound as EI and therefore unavailable. On the other hand, if there is much S and only a small concentration of I, then most of the enzyme molecules will be free to catalyze the reaction of substrate molecules to form the product.

Carbon monoxide poisoning is an extreme example of competitive inhibition. The carbon monoxide competes with oxygen for the active sites in hemoglobin (see Fig. 3.30), the molecule in the blood of vertebrates that carries oxygen to the body's cells. Carbon monoxide binds so strongly to the active sites that oxygen is excluded. Moreover, the inhibition is essentially irreversible: once bound, carbon monoxide remains in place. As a result of the oxygen deprivation that ensues, living tissue, particularly brain tissue, can be damaged or destroyed. Most competitive inhibitors, however, bind to their target only briefly; this reversible inhibition means that inhibitor and substrate molecules are always competing: a change in the concentration of either is quickly reflected by an alteration in enzyme activity.

A second category of reversible inhibition, ***noncompetitive inhibition***, depends on two kinds of binding sites in the same enzyme molecules: the usual active sites to which substrate can bind, and other sites to which inhibitors can bind. Most noncompetitive inhibition is ***allosteric inhibition***. An allosteric enzyme can exist in two distinct spatial conformations, which usually reflect alterations of tertiary structure. When the molecule is in one conformation, the enzyme is active; in the other conformation it is inactive (or less active) because the substrate-binding site is disrupted. The binding of inhibitor molecules—usually called negative ***modulators***—stabilizes the enzyme in its inactive conformation (Fig. 3.51 bottom). Quite often the product itself (or the product of a later reaction in the same biochemical pathway) is the modulator: when present in such high concentrations that no more is needed, it turns off the process responsible for its own synthesis. This self-limiting strategy is known as ***feedback inhibition***. Other allosteric enzymes have binding sites for positive modulators, which induce conformational changes enhancing enzyme reactivity.

Instead of having different kinds of binding sites, some allosteric enzymes have two or more sites of a single kind, and so can bind substrate at two or more locations simultaneously. The binding of substrate at one active site causes conformational changes that make the remaining sites more reactive. This phenomenon, called ***cooperativity***, is exemplified in hemoglobin. A single molecule of hemoglobin is capable of carrying four oxygen molecules. The binding of the first oxygen molecule induces changes in the quaternary structure of the hemoglobin molecule that give the other three binding sites a higher affinity for oxygen. Thus in cooperativity, the first substrate molecule functions as the modulator. It stabilizes the allosteric protein in one of its possible conformations—in the case of hemoglobin, its most reactive conformation.

The techniques of molecular biology now permit selective substitution or deletion of individual peptides in enzymes. As a result,

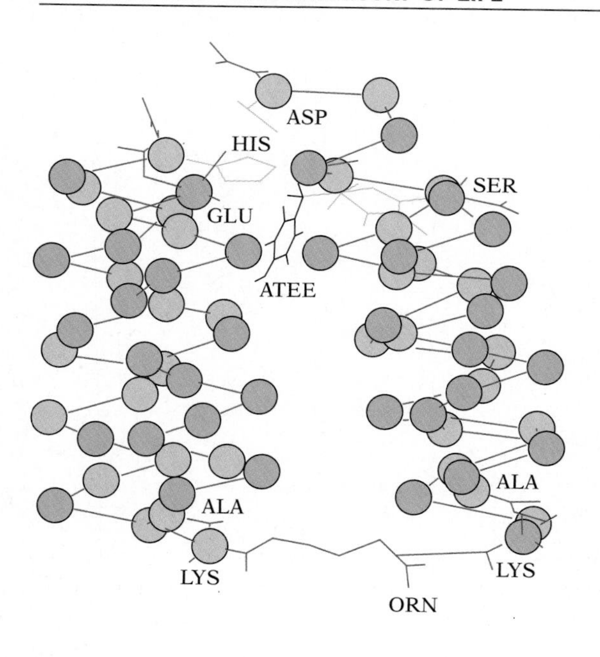

3.52 A synthetic enzyme

This 73-peptide designer enzyme mimics the action of chymotrypsin, a protein-digesting enzyme 245 amino acids long that is secreted by the pancreas. The active site with the R groups shown in blue is just as specific as the natural enzyme, is blocked by the same competitive inhibitors, and increases the reaction rate 100,000 times over that of an enzyme-free solution. Nevertheless, this abbreviated version of chymotrypsin is 1000 times slower than its natural counterpart, indicating that the complex globular structure imparted by the other 172 amino acids probably serves to optimize catalytic activity. (The part of the substrate actually digested is shown in black.)

the action of any amino acid in a protein can be explored experimentally. One consequence of the knowledge this technique has brought is that biochemists can now plan and synthesize artificial enzymes (Fig. 3.52). These "designer enzymes" can be made with specificities and activities unknown in nature. Such enzymes could have nearly limitless potential in treating disease or even solving problems of environmental pollution.

CHAPTER SUMMARY

WHY CARBON?

Carbon is a versatile backbone for organic molecules because it can form four covalent bonds. Many properties of organic compounds are determined by the functional groups attached to carbon skeletons. (p. 47)

THE ROLE OF CARBOHYDRATES Carbohydrates include simple sugars (used for energy) and chains of sugars (polysaccharides) used for energy storage and structural strength. (p. 49)

THE ROLE OF LIPIDS Lipids include fats (used for insulation and energy storage), phospholipids (the major component of cell membranes), and steroids. (p. 55)

THE ROLE OF PROTEINS Proteins are composed of chains of amino acids. Amino acids have specific characteristics depending on their side group. Proteins fold into active shapes that depend on polar, ionic, and covalent interactions between the side groups of their amino acids, and between the side groups and water. The most common structures for parts of a protein are α helices and β pleated sheets. Proteins can act as structural elements or as enzymes. (p. 58)

THE ROLE OF NUCLEIC ACIDS DNA stores information in the sequence of its nucleic acids. RNA is used in a variety of ways; most is involved in carrying the information from genes to the sites of protein synthesis. (p. 68)

HOW CHEMICAL REACTIONS WORK

WHY FREE ENERGY ISN'T FREE Cells depend on free (biological potential) energy to fuel the processes of life. Energy-consuming endergonic reactions must be powered by energy-releasing exergonic ones. (p. 71)

PREDICTING REACTIONS: THE EQUILIBRIUM CONSTANT The ratio of products to reactants when a reaction reaches equilibrium—the equilibrium constant—is directly related to the free energy the reaction produces (or consumes). (p. 73)

HOW ACTIVATION ENERGY GUIDES REACTIONS Most energy-releasing reactions do not occur spontaneously. Weakening the bonds involved in the reaction requires more activation energy than is available from the thermal motion of the molecules involved. Selectively lowering the activation energy for a reaction causes that reaction to occur instead of other reactions the reactants could participate in. (p. 74)

HOW CATALYSTS BREAK THE RULES Catalysts lower activation energy by binding reactants in an intermediate state. (p. 75)

BIOLOGICAL CATALYSTS: ENZYMES Enzymes act as highly specific biological catalysts. Reactants bind at an active site, bonds are loosened, and the reaction occurs. Enzyme activity, and thus the reactions they catalyze, is regulated to match the needs of the cell. (p. 77)

STUDY QUESTIONS

1 How might a change in pH affect the "personality" of polar and ionic amino acids, and the activity or solubility of an enzyme? (pp. 77–78)

2 Why is an α-helix more elastic than a β-pleated sheet? (pp. 61–64)

3 Calculate the approximate bonding energy holding the two 1,000,000-nucleotide-long strands of *E. coli* DNA together. How do you suppose the strands are separated to allow duplication? (pp. 34–36, 68–70)

4 What effect does raising the temperature or increasing the concentration of the relevant enzyme have on an equilibrium constant? (pp. 73–75)

5 Why is activation energy a blessing for organisms? (pp. 74–75)

6 Why is it essential for a cell to be able to control enzyme activity? What are some ways this task is accomplished, and why are some more appropriate for certain sorts of chemical pathways? (pp. 80–82)

SUGGESTED READING

ATKINS, P. W., 1987. *Molecules.* Scientific American Library, New York. *A beautifully produced "molecular glossary" illustrating the chemical formula, three-dimensional structure, and biological action of many common or unusually interesting organic molecules.*

CRAIG, E. A., 1993. Chaperones: helpers along the pathway to protein folding. *Science* 260, 1902–1903. *A brief overview of chaperone-assisted folding.*

DOOLITTLE, R. F., 1985. Proteins, *Scientific American* 253 (4). *Reviews the properties of amino acids and the structure of proteins, and discusses the evolution of different modern proteins from common ancestral enzymes.*

DRESSLER, D., AND H. POTTER, 1991. *Discovering Enzymes.* W. H. Freeman, New York. *A well-written and illustrated history of the study of enzymes, with particular emphasis on how the digestive enzyme chymotrypsin works.*

STROUD, R. M., 1974. A family of protein-cutting proteins, *Scientific American* 231 (1). *A good discussion of how enzymes like chymotrypsin work.*

STRYER, L., 1988. *Biochemistry,* 3rd ed. W. H. Freeman, San Francisco. *A beautifully produced, clearly written, but highly technical exposition of biochemistry.*

CHAPTER 4

THE CELL MEMBRANE

Like the departments in a giant corporate conglomerate, the organic molecules that make up life depend on precise compartmentalization and organization. If we put a population of amoebae through a blender, the resulting stew of organic molecules will not reorganize itself spontaneously into living entities. A complex ***plasma membrane*** protects the interior of the cell—the ***cytoplasm***—from assault by the external environment. The cytoplasm itself is intricately organized: it contains not just organic fluids (the ***cytosol***), but many internal membranes and a variety of specialized, self-contained entities known as organelles, which separate themselves from the rest of the cell by means of their *own* membranes. In this way the cell's chemical processes are partitioned off—food, for example, is digested in membrane-enclosed compartments that prevent digestion of the rest of the cell. This chapter will examine the organization and operation of membranes, which are critical to life.

THE DISCOVERY OF CELLS

■ THE CELL THEORY

The discovery of cells and of their structure came with the development of magnifying lenses. In the 17th century Antonie van Leeuwenhoek and other interested amateurs constructed simple microscopes, and in 1665 Robert Hooke reported to the Royal Society of London "the first microscopical pores I ever saw, and perhaps, that were ever seen," in a piece of cork (Fig. 4.1). Hooke's pores were, in fact, cells.

The idea that all living things are composed of cells—the ***cell theory***—was not proposed until about 1840. An important extension of the cell theory came 20 years later: all living cells arise from preexisting living cells. This theory of ***biogenesis***, life coming from life, contradicted the prevailing belief in spontaneous generation. In 1862 Louis Pasteur proved that living cells cannot arise from nonliving matter (see *How We Know: Spontaneous Generation Disproved*, p. 87).

These two components of the cell theory—that all living things are composed of cells and that all cells arise from other cells—pro-

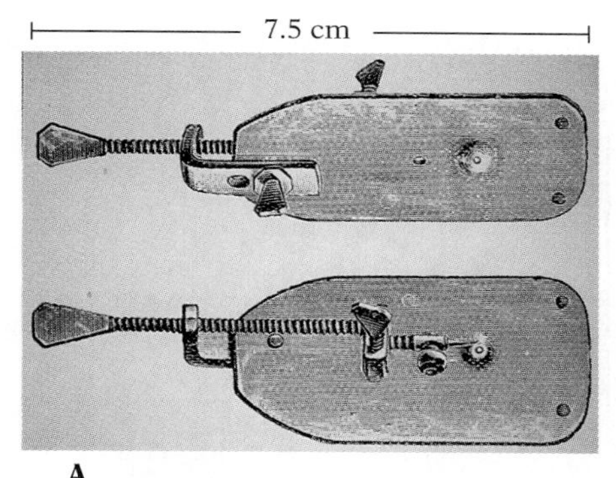

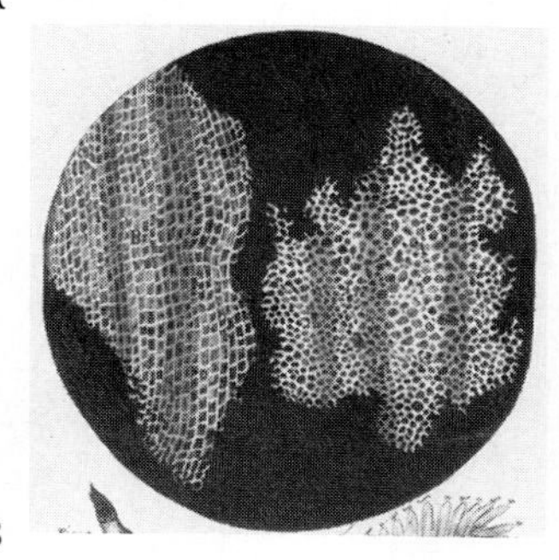

4.1 Van Leeuwenhoek's microscope

(A) This magnifier was first developed to inspect samples of cloth but was soon adapted for viewing living things. (B) Drawings of cork that appeared in Robert Hooke's ***Micrographia,*** published in 1665.

vide a working definition of life: living things are chemical organizations composed of cells and are capable of reproducing themselves.

■ HOW CELLS ARE STUDIED

Much of our knowledge of subcellular organization has been made possible by the development of better and more powerful microscopes. In the detailed analysis of subcellular structure, three attributes of microscopes are of particular importance: magnification, resolution, and contrast. Magnification is a means of increasing the apparent size of an object. Resolution is the capacity to make adjacent objects distinct. Contrast is important in distinguishing one part of a cell from another.

In the ordinary compound light microscope, light passes through a specimen and is captured, bent, and brought into focus by lenses (Fig. 4.2A). Depending on the magnification and the size of the specimen, a whole cell or only a tiny part may be in the field of view at any one time.

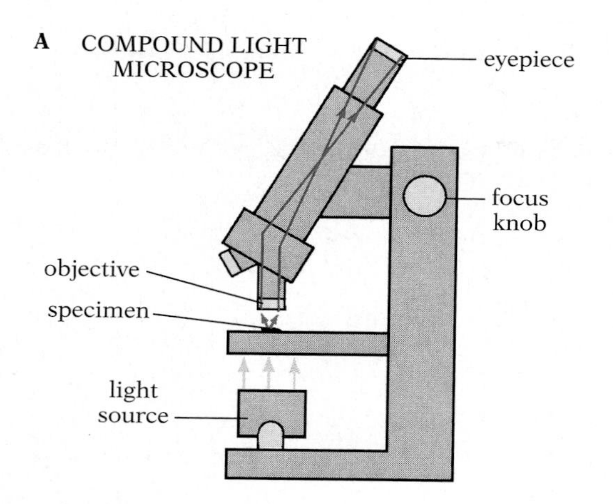

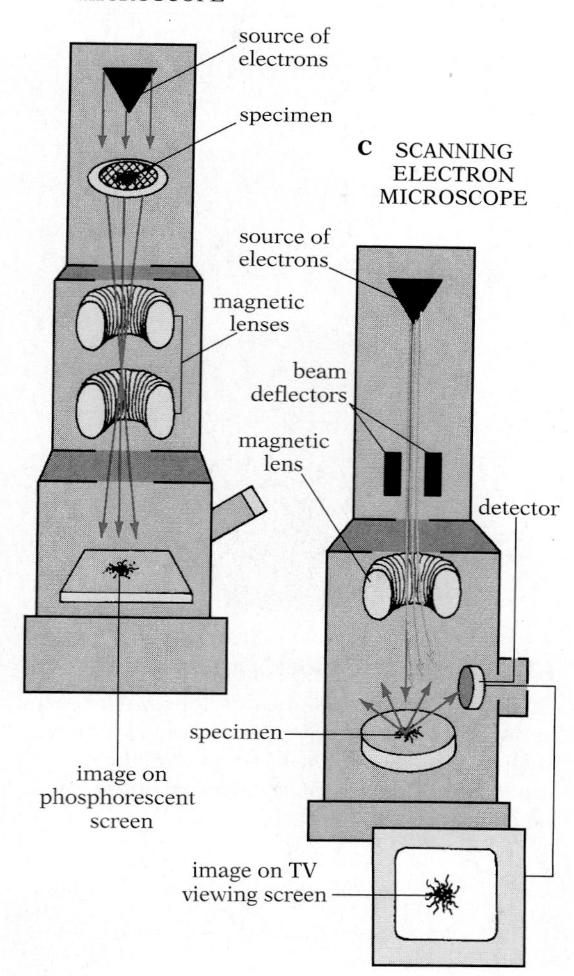

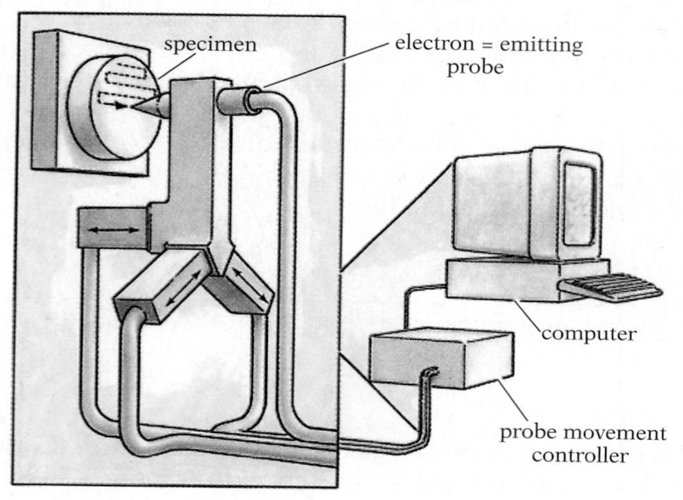

4.2 Microscopes

(A) In a compound light microscope, light passes through a specimen to the objective lens. Here the light is refracted and directed to the eyepiece, where it is focused for a camera or for the eye. (B) In a transmission EM, electrons provide the illumination, which passes through the specimen and is then focused by magnets to form an image for photographic film or on a phosphorescent screen. (C) In a scanning EM, a focused beam of electrons moves back and forth across the specimen, while a detector monitors the consequent emission of secondary electrons and reconstructs an image. (D) The scanning tunneling microscope (STM) moves an electron-emitting tip over the sample, constantly adjusting its height to maintain a fixed distance above the sample; these variations in tip elevation allow reconstruction of the specimen's contours. The minute movements necessary are accomplished by a mounting arm attached to piezoelectric crystals. Since the dimensions of such crystals change as different amounts of voltage are applied, it is possible to control the position of the tip exactly.

HOW WE KNOW:
SPONTANEOUS GENERATION DISPROVED

Pasteur's experiments on spontaneous generation are classically simple, well-controlled tests of a hypothesis. He placed various nutrient broths in long-necked flasks and then bent the necks into curves (see figure). Next he boiled the broths in the flasks to kill any microorganisms that might be in them. Then he left the flasks standing. Any microbe-laden dust particles in the air moving into the flasks were trapped in the films of moisture on the humid curves of the necks; the curved necks thus acted as filters. The broths in the swan-neck flasks remained unchanged indefinitely. On the other hand, identical broths boiled in flasks with straight necks—the control solutions—were soon teeming with life. Similarly, if the swan neck was broken off, the experimental broth rapidly developed colonies of molds and bacteria. Since the handling of the control solutions differed in only one respect from the experimental flasks—exposure to airborne microbes—the changed outcome had to be attributed to that difference. In the same way, Pasteur showed that the source of the microorganisms that ferment or putrefy milk, wine, and sugar-beet juice is the air; there was no spontaneous generation in the nutrient media.

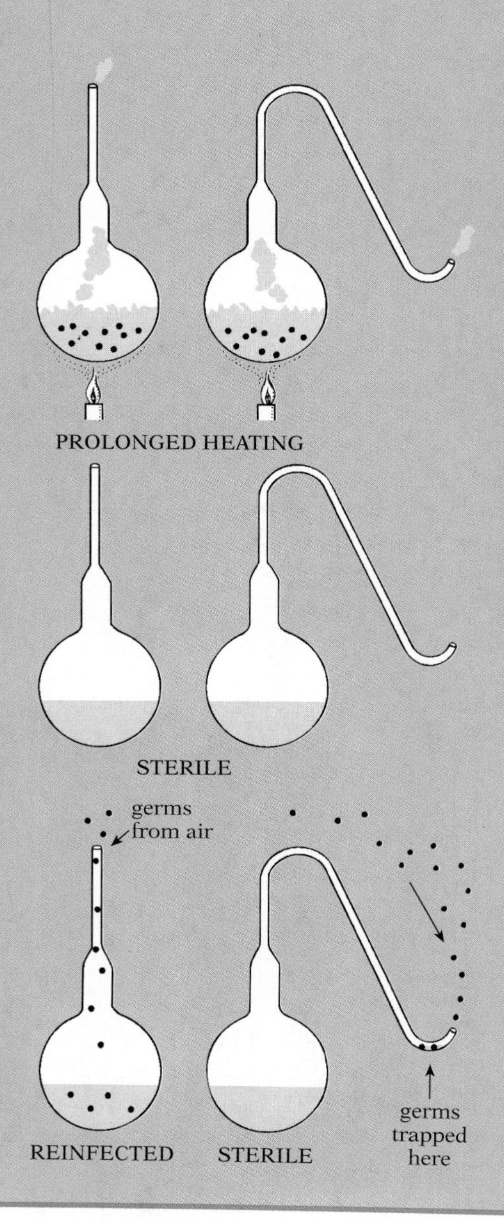

Pasteur's experiment

Nutrient broths in two kinds of flasks, one with a straight neck, the other with a bent neck, were boiled to kill any microbes they might contain (top). The sterile broths were then allowed to sit in their open-mouthed containers for several weeks (middle). Microorganisms entering the straight-necked flask contaminated the broth, but those entering the bent neck of the other flask were trapped in films of moisture in the curves of the neck (bottom).

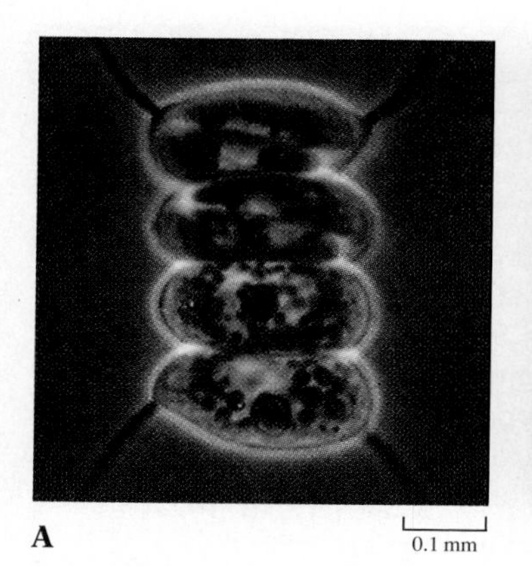

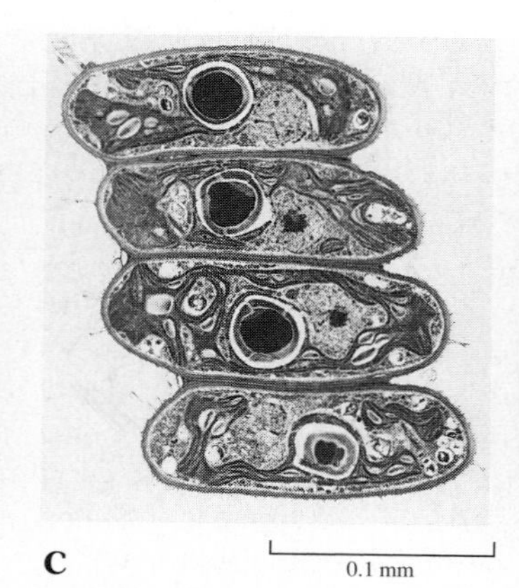

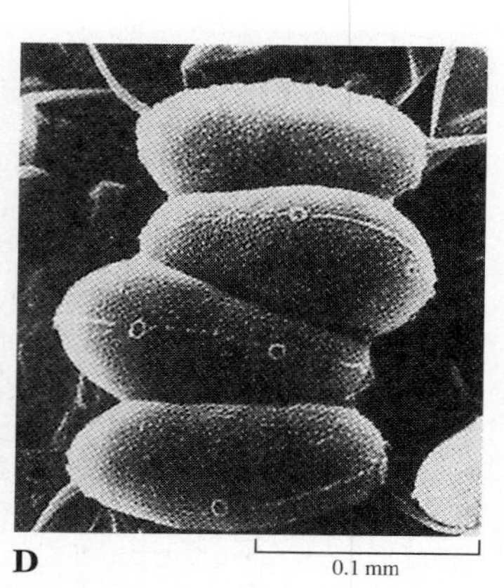

4.3 Views of the green alga *Scenedesmus* obtained with microscopes of various types

(A) Photograph of an unstained specimen as seen with a phase-contrast light microscope. (B) Photograph taken by the Nomarski process. (C) Transmission electron micrograph. Many of the membranous and particulate intracellular organelles show up much more clearly here than under the light microscope. They are made visible by a stain containing heavy-metal atoms, which combines differentially with various structures. (D) Scanning electron micrograph, providing a three-dimensional view of surface features. The scale bars below the images in this and later figures indicate relative dimensions.

We vary the magnification by using lenses with different shapes that accept light from larger or smaller sections of the specimen: the higher the magnification, the smaller the amount of light that reaches the eyepiece. The limit on useful magnification in light microscopes is not a matter of exhausting the illumination, however; but arises from the tendency of light to bend as it passes near an edge. This phenomenon, known as diffraction, spoils images by bending light out of the straight-line path as it moves from the source, through the specimen, and to the eyepiece, thus blurring the image. For ordinary light, useful magnification is limited to about 1000 times the actual size of the object in focus. Though magnification of 1000x is an enormous improvement over the unaided eye, it is still not enough to let us see many of the smaller subcellular structures.

Contrast is as important as magnification. Most cellular components are colorless and have essentially the same texture; they do not contrast with one another. However, different parts of the cell often differ in their affinities for various dyes, so when these areas are stained, they stand out from one another. Unfortunately, staining usually kills the cell and may thereby change its internal structure. Techniques such as phase-contrast and Nomarski optics, both of which depend on elaborate optical manipulation, greatly increase the value of the light microscope because they create contrast optically, without staining (Fig. 4.3A-B). Another technique uses small, nonlethal amounts of one or more fluorescent stains to create contrast inside cells. The cell being studied is illuminated with a highly focused laser beam of just the wavelength to cause fluorescence, and the focal point is systematically scanned back and forth through a specimen; a second highly focused optical system is used to pick up only the light emitted from the focal point of the scanning beam, and the resulting output is used to reconstruct a crisp cross-sectional image with a computer.

The electron microscope (EM) uses a beam of electrons instead of light as its source of illumination. Because resolution improves (that is, diffraction decreases) as the wavelength of the illumination becomes shorter, and because electron beams have much shorter wavelengths than visible light, electron microscopes can resolve objects about 10,000 times better than light microscopes (Fig. 4.3C-D).

In a transmission EM (Fig. 4.2B) the electrons, focused by magnets rather than by lenses, pass through the thinly sliced "sectioned" specimen and fall on a photographic plate or a phosphorescent screen, where they produce an image of the specimen. Because cells are essentially transparent to electrons, a specimen being prepared for the transmission EM must be stained with an electron-dense chemical that binds to specific cell structures. Many stains contain heavy-metal atoms, which bind to particular cellular structures, blocking the passage of electrons in these places (Fig. 4.3C). Another approach is to tilt the specimen to allow atoms of the electron-dense substance to fall onto it (Fig. 4.4). The shadows and highlights in the resulting EM picture create a three-dimensional effect—a kind of topographical map of the specimen that can reveal important surface details.

A scanning EM (SEM) can also produce a three-dimensional view. A specimen coated with atoms of metal is scanned from above by a moving beam of electrons (Fig. 4.2C). This focused probe does not penetrate the specimen but instead causes so-called secondary electrons to be emitted from the surface. The intensity of the emission of secondary electrons depends on the angle at which the probe beam strikes the surface, and therefore varies with the contours of the specimen. Hence a point-by-point recording of the emission produces a three-dimensional picture (Fig. 4.3D). Though the resolution of the scanning EM does not approach that of the transmission EM, its ability to create a better three-dimensional effect is an advantage for many applications. Since shadowing is not required, the same specimen can be turned repeatedly and observed from various perspectives.

A further elaboration of the electron microscope, the scanning tunneling microscope (STM), uses an ultrafine electron-emitting tip, which is moved back and forth over a metal-coated sample, scanning it systematically from top to botton (Fig. 4.2D). The

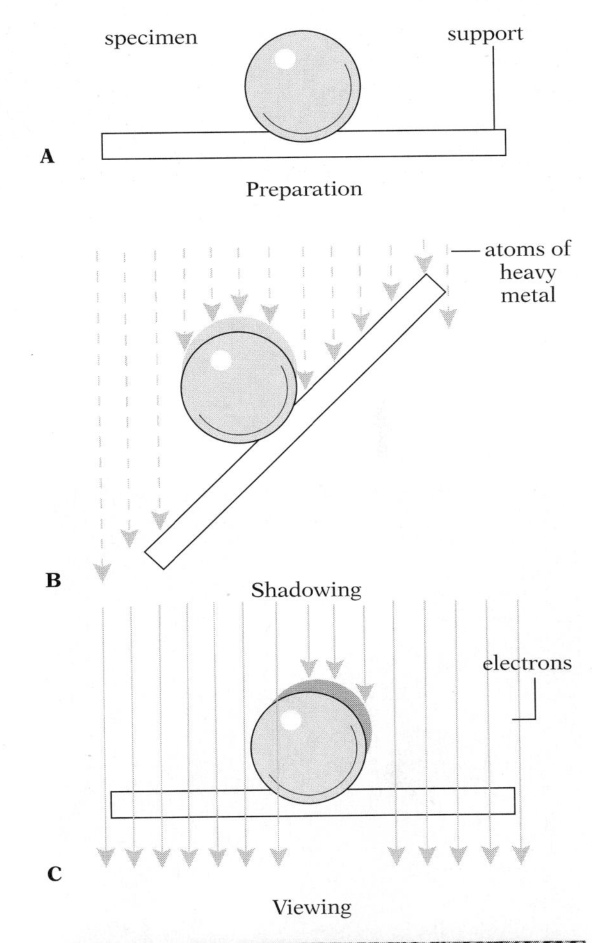

4.4 Shadow staining for the transmission EM

The specimen (A) is tilted and dusted, or "shadowed," with atoms of a heavy metal such as platinum (B). (Another technique is to "shoot" the metal atoms from an angled source onto a horizontal specimen.) When an electron beam is aimed at the coated specimen (C), the unevenly distributed metal plating prevents most of the electrons from penetrating the shadowed areas, and produces a three-dimensional view of the surface of the specimen, as in the electron micrograph of polio virus (D).

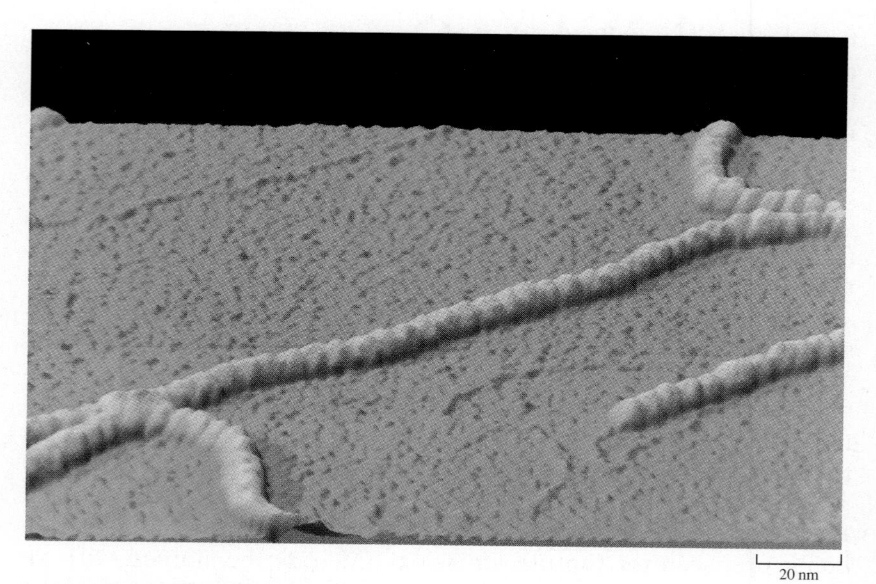

4.5 STM image of DNA

This computer-enhanced picture shows both ordinary DNA (thin coils) and DNA complexed with the protein SSB (discussed in detail in Chapter 8), which holds the double helix in a much more open configuration during the process of chromosomal duplication (thick coils).

4.6 Atomic-force image of a moving cell

This living cell is in the process of flattening itself, part of its normal behavior in the nervous system where it envelopes neurons. The flattened outer edges of the membrane are searching for the molecules that identify the class of cells that it envelopes. The thickened center (N) contains the nucleus; the linear structures (F) lie just below the membrane and are part of the machinery that moves the cell.

amount of current that flows from the tip to the specimen depends on the gap between them; by recording the constant adjustments of tip height necessary to maintain a given current flow, a computer can reconstruct the topography of the sample and create a picture on a video monitor (Fig. 4.5). An analogous technique uses atomic forces to map surfaces (Fig. 4.6).

HOW CELLS EXPLOIT DIFFUSION

At one time, the cell membrane was considered little more than a bag to hold in all the organic chemicals that somehow combine to produce life. In fact, the cell membrane is critical to cell movement, and to the cell's sampling of the extracellular environment; moreover, it bears the primary responsibility for regulating the chemical traffic between the precisely ordered interior of the cell and the essentially unfavorable and potentially disruptive outer environment. All substances moving into or out of a cell must pass through a membrane barrier, and the membrane of each cell can be quite specific about what is to pass, at what rate, and in which direction. The cell membrane exercises this control in two ways: by utilizing natural processes such as diffusion, and by transporting specific substances in and out.

■ DIFFUSION AS A CHEMICAL REACTION

As we have seen, temperature affects the rates of chemical reactions by increasing the kinetic energy (the average velocity) of the molecules involved. Imagine a box containing marbles in a cluster near one end (Fig. 4.7A). When we shake the box, the marbles disperse almost evenly over the bottom (Fig. 4.7B). This obvious result is worth a closer look, because the marbles can be thought of as molecules, and the shaking as adding kinetic, or thermal, energy to the system.

Of all the possible directions in which a given marble might move, more lead *away* from the center of the cluster than toward it. Hence *random movement will tend to disrupt the cluster rather than to maintain it.* In the absence of any counteracting external influence, the Second Law of Thermodynamics—in particular, the principle of ***entropy***—dictates that a dynamic system will tend to move toward the more probable, disorderly state rather than toward the less likely, orderly state. This is what happens when a lump of sugar dissolves in a cup of warm coffee: the sugar molecules move from the region of high concentration (the sugar crystal) to regions of lower concentration; eventually the sugar molecules disperse throughout the liquid. The warmer the liquid, the more kinetic energy the molecules in solution will have on average, and the faster diffusion will take place.

We can now make an important generalization: all other factors being equal, *the net movement of the particles of a substance is from regions of higher concentration to regions of lower concentration of that substance.* Note that we speak of *net* movement. There will always be some particles moving in the opposite direction, but overall the movement will be away from the centers of concentration. An obvious result is that the particles of a given substance tend to become equidistant from one another within the available space. When this uniform density has been reached, the system is in equilibrium; the particles continue to move, but there is little net change in the system.

The random, unrestrained movement of molecules from one place to another is called ***diffusion***. Diffusion is fastest in gases, where there is much space between the particles and hence relatively little chance of collisions that retard movement. Diffusion in liquids is much slower (Fig. 4.8); in the absence of convection currents a substance can take a very long time—years, in fact—to move in appreciable quantity only 1 m through cold water. Diffusion in solids is, of course, much slower still: there is very little space between the molecules of a solid, and collisions occur almost before the molecules get going. In all these instances, however, regardless of the rate of diffusion, the net effect is movement away from regions of higher concentration, as long as all regions are at the same temperature and pressure. In living organisms, where molecules are generally in a warm aqueous solution and the distances involved are measured in fractions of a millimeter, diffusion is a highly significant process; an amino acid in an aqueous medium will typically diffuse about one cell diameter (10-50 μm; also called microns or micrometers) in less than 0.5 sec.

So far, we have discussed diffusion in terms of movement from a higher to a lower concentration along a gradient. In the living world, however, diffusion is rarely a function of concentration alone; differences in temperature and pressure, for instance, affect diffusion. It is more useful to look at diffusion in terms of the free

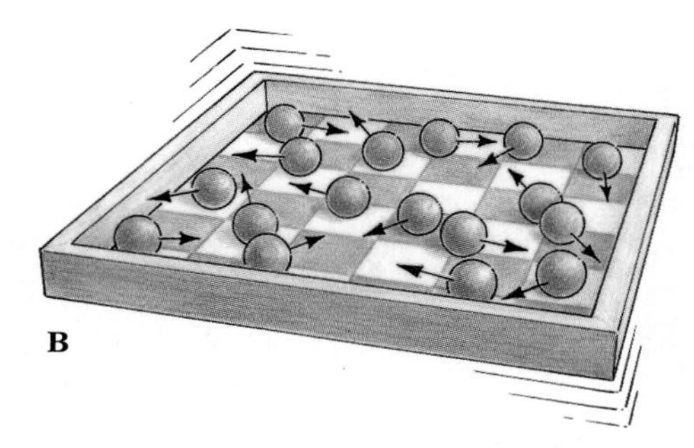

4.7 Mechanical model for diffusion

(A) All 17 marbles are placed in a cluster at one end of a rectangular box. (B) When the box is shaken to make the marbles move randomly, they become distributed throughout the box in nearly uniform density.

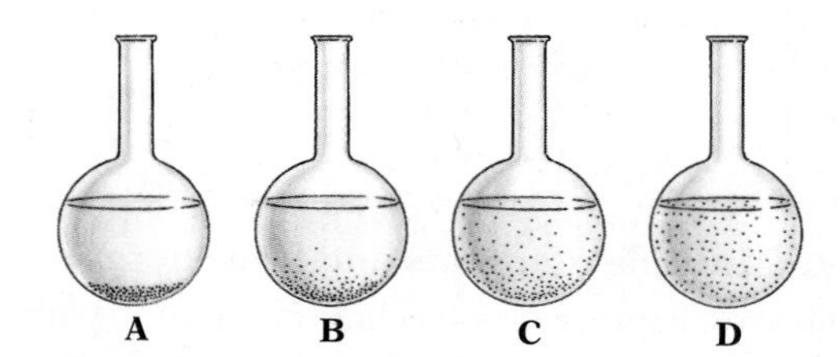

4.8 Diffusion in a liquid

Particles of solute are at the bottom of a flask of water (A). The particles slowly diffuse away from the cluster (B, C) until they are distributed with nearly uniform density through the water (D). If the water is cold and there are no convection currents to help move the particles, it may take a long time to reach uniform distribution.

energy of the particles involved. A local concentration of a substance is a relatively orderly and unlikely arrangement. We can see, for instance, that energy (among other things) would be needed to change a mixture of sugar molecules and coffee back into the original lump of sugar and unsweetened coffee. A random, disorderly arrangement of molecules has less potential for doing work than an orderly one, and accordingly it has less free energy. As we know from the Second Law of Thermodynamics, the amount of free (useful) energy in the universe is always decreasing; entropy is always increasing.

Diffusion, then, is spontaneous because orderly molecules, concentrated together, have greater free energy than dispersed molecules: it is a downhill reaction from order to disorder. The mixture (or product) has less free energy than the separate original substances (the reactants). If we begin with two pure substances, the rate of diffusion is fastest at the outset, and slows as the equilibrium point (complete mixture) is approached. If we could observe diffusion at the molecular level, we would see the sugar molecules speeding away from the lump during the early part of the reaction. Later, however, as the substances became more evenly mixed, the frequency of the back reaction returning sugar molecules to the location of the dissolving lump would rise until equilibrium was reached. In fact, diffusion is a chemical reaction with its own free energy, which depends on the characteristics of the substances involved:

$$\text{sugar} + \text{coffee} \rightleftarrows \text{sweetened coffee} \qquad (\Delta G = -x)$$

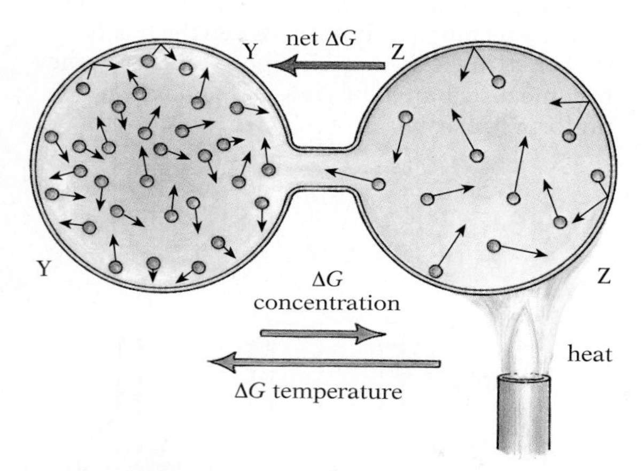

4.9 Multiple gradients and free energy
This two-chambered vessel has one side (Y) with a slightly higher pressure than the other (Z), but the temperature in Z is much higher than that in Y. If concentration alone were important, the net diffusion would be from Y, the region of higher concentration, to Z, the region of lower concentration. But the higher the temperature, the greater the thermal motion of the particles in that system; and the greater the thermal motion, the greater the free-energy content. Because the difference in free energy associated with the temperature gradient from Z to Y outweighs the difference in free energy associated with the concentration gradient from Y to Z, net diffusion will be from Z to Y.

Free energy tells us how a substance will diffuse; concentration gradients are only one component. Consider a situation in which there is a slight concentration gradient in one direction and a pronounced temperature gradient in the reverse direction (Fig. 4.9). The opposing effects of these two gradients produce a net movement of molecules that depends on the relative free energies of the two gradients; in this case it is from the region of higher temperature to the area of higher concentration.

The importance of diffusion and its basis in free energy with respect to cells is clear: the concentration of organic molecules and a select group of ions inside a cell is a very unlikely arrangement. Without the cell membrane, the free energy of the cellular chemistry would be lost as the contents diffused into the environment. There must be a barrier between the inside and the outside of the cell to maintain the integrity of the cellular chemistry. Moreover, the free-energy gradient across the cell membrane is available to do work.

■ MANIPULATING DIFFUSION: OSMOSIS

To envision the way cell membranes can function, imagine a U-shaped chamber divided by a membrane (Fig. 4.10). Suppose that molecules of water can pass through the membrane while molecules of sugar cannot. How will this ***differentially permeable*** (or ***selectively permeable***) membrane affect the diffusion of materials between the two halves of the chamber? Suppose that side A of the

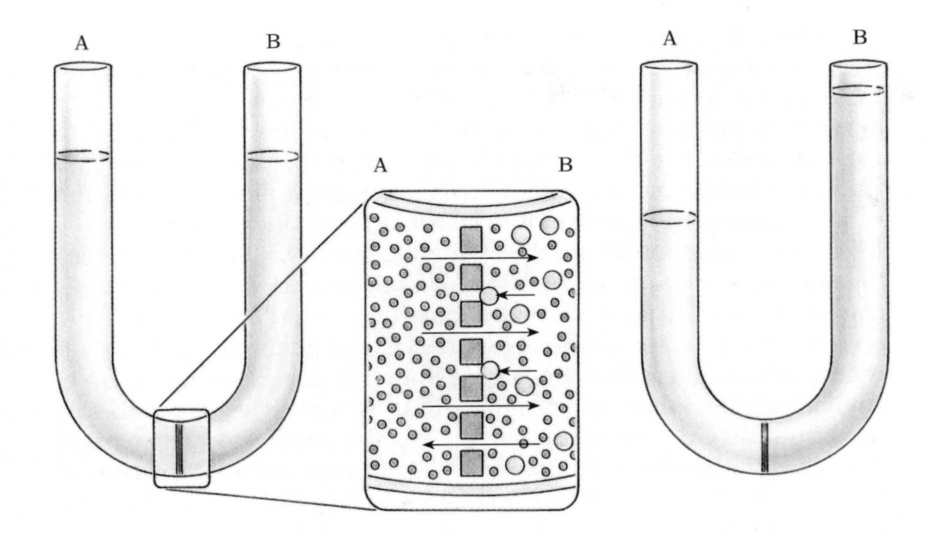

4.10 U-tube divided by a selectively permeable membrane

The membrane at the base of the U-tube is permeable to water but not to sugar molecules (yellow balls). Left: Side A contains only water; side B contains a sugar solution. Initially the quantity of fluid in the two sides is the same. Center: A larger number of water molecules (blue balls) bump into the membrane per unit time on side A than on side B. Right: Because more water molecules move from A to B than from B to A, the level of fluid on side A falls while that on side B rises.

U-tube contains pure water and side B an equal quantity of sugar solution (sugar dissolved in water), and that both sides are at the same temperature and pressure. If the membrane is permeable to water but not to sugar, water molecules will be able to pass in both directions, from A to B and from B to A.

This movement of a solvent (usually water) through a selectively permeable membrane is called ***osmosis***. Since water is already present on both sides of the membrane in the U-tube, it might seem that the movement of water molecules across the membrane should have no net effect. Consider the differences between the pure water and the sugar solution more carefully. We have seen that substances diffuse from regions of higher concentration to regions of lower concentration. Water, being more concentrated on side A, will tend to diffuse to side B; sugar molecules, trapped by the membrane on side B, will not cross.

We can also see why water molecules move from A to B by picturing to ourselves the events at the membrane itself. On side A, all the molecules that bump into the membrane during a given interval are water molecules, and because the membrane is permeable to water, many of these will pass through the membrane from A to B. By contrast, on side B, some of the molecules bumping into the membrane during the same interval will be water molecules, which may pass through, and some will be sugar molecules, which cannot pass through because the membrane is impermeable to them. Hence more water molecules will move across the membrane from side A to side B per unit time than in the opposite direction; the net osmosis will be from A to B.

We can also think of the matter more abstractly, in terms of entropy. The arrangement of water molecules in pure water is orderly, in that every molecular location is occupied by a water molecule, whereas the arrangement in the sugar solution is disorderly, in the sense that any given molecular location may be occupied by either a water molecule or a sugar molecule. Since an orderly system possesses more free energy than a disorderly one, it follows that the orderly water molecules in the pure water (side A) have more free energy than the disorderly water molecules in the sugar solution (side B). There is a free-energy gradient for water from side A to side B, so there will be a net movement of water down this gradient, from A to B. No matter how you choose to view it, remember that in the absence of temperature or pressure gradients, water moves from a region of high water concentration (that is, pure water) to one of lower water concentration.

We can make make another generalization about osmosis. *The free energy of water molecules is always decreased if osmotically active substances* (dissolved or colloidally suspended particles) *are present in the water.* (Colloidal particles are generally larger than the separated individual molecules of a dissolved substance, yet small enough so that—unlike the still larger particles of a true suspension—they do not settle out at an appreciable rate but remain dispersed within the fluid medium.) The ***osmotic concentration*** of a fluid—the number of osmotically active particles per unit volume—thus bears a direct relationship to that fluid's free energy. In the U-tube example, for instance, *the difference in the free energy of water molecules in a solution versus pure water is proportional to the osmotic concentration of the solution.* Since the osmotically active particles disrupt the orderly three-dimensional array of the water molecules, the fluid has less free energy.

Each solution, then, has a certain free energy, depending on its osmotic concentration. Under conditions of constant temperature and pressure, this free energy can be calculated; it is called ***osmotic potential***. (Pure water is arbitrarily assigned an osmotic-potential value of zero. Since osmotic potential decreases as osmotic concentration increases, all solutions have values of less than zero.) If two different solutions are separated by a membrane permeable only to water, and temperature and pressure are constant, *the net movement of water will be from the solution with the lower osmotic concentration to the solution with the higher osmotic concentration*. The steeper the osmotic-concentration gradient, the more rapid the movement: the water flows from regions of higher osmotic potential to regions of lower potential at a rate proportional to the degree of difference in osmotic potential. These generalizations are summarized in Table 4.1.

Table 4.1 Some rules of osmotic diffusion

I. The free energy of water molecules is always decreased if osmotically active substances (dissolved or colloidally suspended particles) are present in the water.

II. The decrease in the free energy of water molecules is proportional to the osmotic concentration.

III. The net movement of water will be from the solution with the lower osmotic concentration to the solution with the higher osmotic concentration.

If the net movement of water in the U-tube is from side A to side B, the volume of liquid will increase on side B and and decrease on side A. Does this process continue until side A is dry? Clearly, if the membrane is completely impermeable to sugar molecules, conditions on the two sides will never be equal, no matter how many water molecules move from A to B: the fluid in B will remain a sugar solution, though an increasingly weak one, and the fluid in A will remain pure water. Nevertheless, under normal conditions, the fluid level in B will rise to a certain point and then stop. Why? The column of fluid is, of course, being pulled downward by gravity. As the column rises, its weight exerts increasing downward hydrostatic pressure. As the pressure increases, the free energy of the water in the sugar solution rises, because pressure too is a form of free (useful) energy. Eventually the column of sugar solution becomes so high, and its pressure and free energy so great, that water molecules are pushed across the membrane from B to A as fast as they move into B from A.

When water is passing through the membrane in opposite direc-

EXPLORING FURTHER

OSMOTIC POTENTIAL, OSMOTIC PRESSURE, AND WATER POTENTIAL

As we have seen, osmotic potential is useful for thinking about how two solutions with differing osmotic concentrations interact. It is easily related to the underlying difference between them in free energy, which results in the movement of the solvent. Many researchers, however, prefer to think in terms of the pressure that must be exerted on a solution to keep it in equilibrium with pure water when the two are separated by a selectively permeable membrane. In our U-tube example, this pressure corresponds to the hydrostatic pressure exerted by the sugar solution at equilibrium; it is known as ***osmotic pressure***. Clearly, *the osmotic pressure of a solution is a measure of the tendency of water to move by osmosis into it.* The more dissolved particles in a solution, the greater the tendency of water to move into it and the higher the osmotic pressure of the solution. Thus, under constant temperature and pressure, water will move from the solution with the lower osmotic pressure to the solution with the higher osmotic pressure.

While the terms *osmotic pressure* and *osmotic potential* are regularly used by physiologists studying animals, plant physiologists more often refer to ***water potential***, which is essentially the same as the free energy of water. At a pressure of 1 atm (atmosphere), pure water is assigned a water potential of zero. Since the water potential decreases as the osmotic concentration increases, all solutions have values of less than zero. In this sense, water potential is like osmotic potential. Unlike osmotic potential, which is a function of solute concentration alone, water potential (like free energy) is also a function of temperature and pressure. When two solutions are separated by a selectively permeable membrane, water will move from the solution with the higher water potential to the solution with the lower water potential. Throughout this book we will discuss fluid movement in plants in terms of water potential, while the analogous descriptions of fluids in animals will be in terms of osmotic potential.

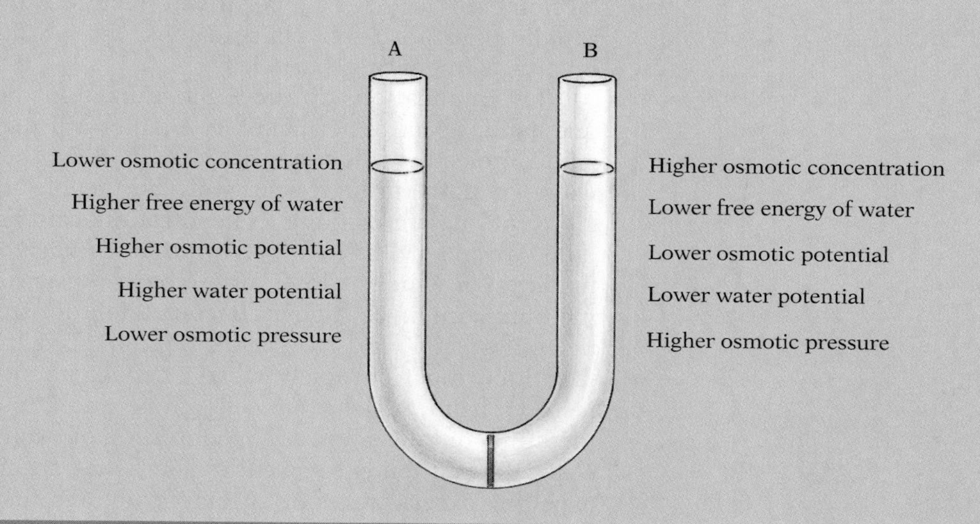

tions at the same rate, the system is in dynamic equilibrium, with the free energy—the osmotic potential—of the pure water on one side of the membrane just matching the free energy—the osmotic potential and hydrostatic pressure—of the column of solution on the other side. The greater the concentration difference across the membrane, the higher the column of solution will rise before this osmotic difference is counterbalanced by the difference in hydrostatic pressure.

Osmotic concentration is not concentration by weight, it is molecular or ionic concentration—the total *number* of solute particles per unit volume. If there are several kinds of solutes in the same solution, then the osmotic concentration of that solution is determined by the total (per unit volume) of *all* the particles of all kinds. If a dissolved substance ionizes, then each ion functions osmotically as a separate particle: 1 mole of sodium chloride (NaCl) dissolved in water produces 2 moles of particles: 1 mole of Na^+ ions and 1 mole of Cl^- ions. Colloidal particles may also contribute to the total osmotic concentration.

■ BETWEEN SWELLING AND COLLAPSE

We have discussed diffusion and osmosis at such length because the cell membrane is selectively permeable, and the processes of diffusion and osmosis are fundamental to cell life. Though the membranes of different types of cells vary widely in their permeability characteristics—the membrane of a human red blood corpuscle,[1] for instance, is over 100 times more permeable to water than the membrane of *Amoeba*—we can make a few generalizations: Most cell membranes are relatively permeable to water and to certain simple sugars, amino acids, and lipid (fat and oil)-soluble substances. They are relatively impermeable to polysaccharides, proteins, and other very large molecules. In short, cell membranes pass the building blocks of complex organic compounds, not the compounds themselves. The permeability of cell membranes to small ions varies greatly, depending on the particular ion, but in general, negatively charged ions can cross more rapidly than positively charged ions, though neither can do so as readily as uncharged particles.[2]

What implications do these generalizations hold for life? On the one hand, selective permeability enables cells to retain the large organic molecules they synthesize. On the other, the tendency of water to pass through selectively permeable membranes into regions of higher osmotic concentration can be harmful or even fatal. When a cell is in a medium that is ***hypertonic*** relative to it (a medium to which it loses water by osmosis, usually because the medium contains a higher concentration of osmotically active particles), the cell tends to shrink (Fig. 4.11); if the process goes too far, it may die. Conversely, when a cell is in a medium ***hypotonic*** to it (one from which the cell gains water, usually because the medium contains a lower concentration of osmotically active particles), the cell tends to swell; unless it has special mechanisms for expelling excess water or special structures that prevent excessive swelling (as most plant cells do), it may burst. A cell in an ***isotonic*** medium (one with which the cell is in osmotic balance, usually because it contains the same concentration of osmotically active particles) neither loses nor gains much water by osmosis.

The osmotic relationship between the cell and the surrounding medium can be critical. Some cells are normally bathed by an isotonic fluid and therefore have no serious osmotic problems. Human red blood corpuscles are an example; they are normally

[1] A red blood corpuscle begins as a cell but (in mammals) loses its nucleus as it matures and becomes specialized to transport oxygen. However, it remains sufficiently like true cells to be used as an example of many cellular properties.

[2] The process in which, in addition to solvent, some solutes selectively cross the membrane is often called dialysis.

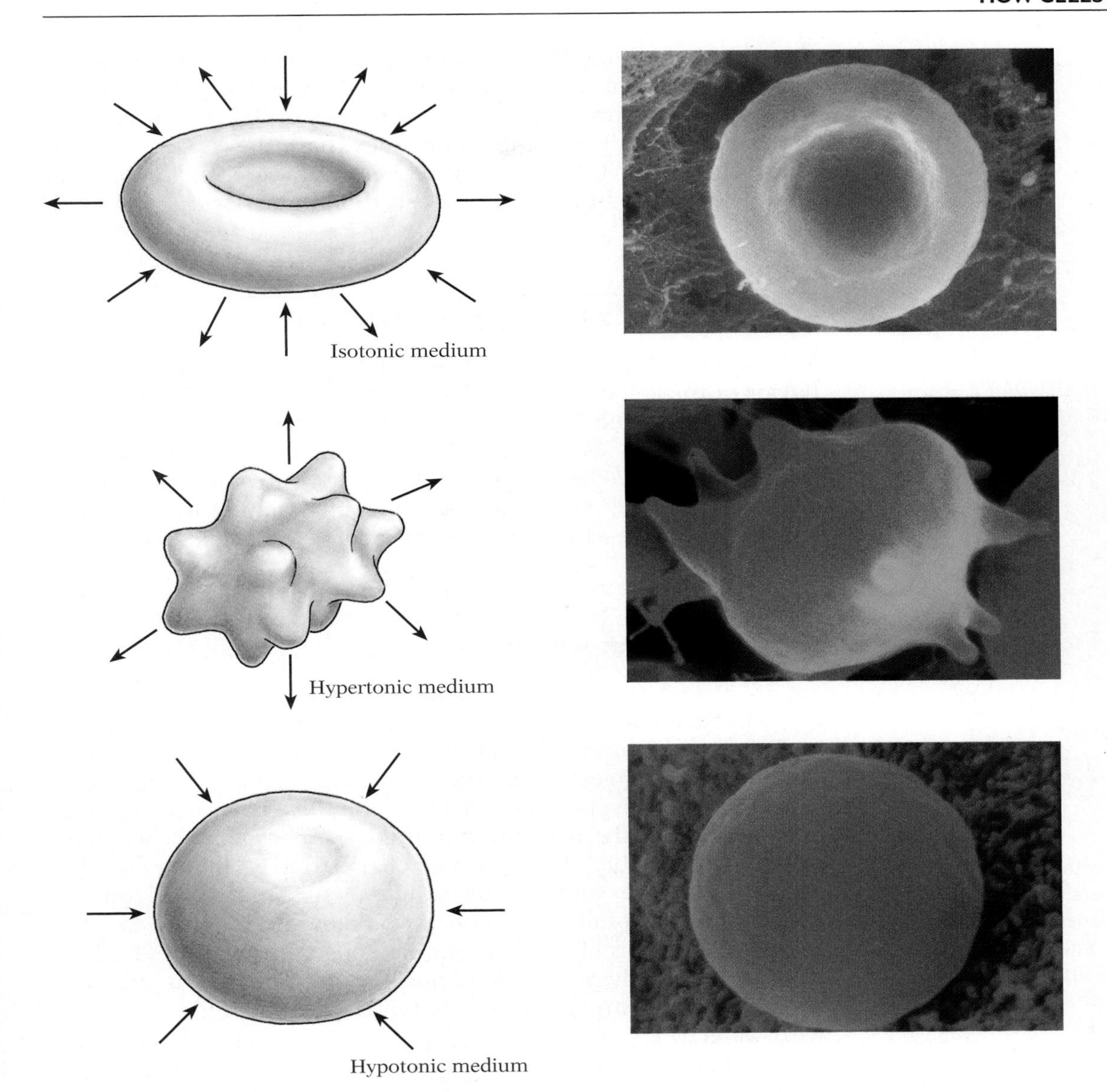

4.11 Osmotic relationships of a cell

In an isotonic medium, water gain and water loss are equal, and so the cell neither shrinks nor grows. In a hypertonic medium, there is a net loss of water from the cell, and the cell shrinks. In a hypotonic medium, the cell swells as water moves from the medium inside. The accompanying photographs are of human red blood cells.

bathed by blood plasma, with which they are in relatively close osmotic balance. The cells of most simple oceanic plants and animals have an osmotic concentration close to that of sea water. All cells, however, have a higher osmotic concentration than fresh water. Freshwater organisms therefore live in a hypotonic medium and have had to evolve ways either to avoid taking in excess water or to dispose of it when it too much does get in.

Controlling the flow of water is only one problem. Though the selective permeability of the membrane effectively traps large molecules inside, it does not provide any mechanism for concentrating the organic building blocks necessary for constructing substances like DNA, proteins, and polysaccharides in the first place. For nutrients to be captured and retained, wastes expelled, and cell volume controlled, the cell membrane must have the capacity to pass many chemicals in only one direction. The secret of this critical ability lies in the structure of the cell membrane itself.

HOW THE CELL MEMBRANE WORKS

Researchers struggled for decades to explain the remarkable behavior of the cell membrane. The key is the membrane's structure, which was inferred from its various properties. For instance, permeability studies showed that lipids and many substances soluble in lipids can move between the cell and the surrounding medium. Therefore the cell membrane must contain lipids to allow fat-soluble substances to cross. But how could a lipid membrane be stable?

4.12 A liposome
Mixing phospholipids with water produces spherical phospholipid bilayers, each enclosing a droplet of water. The spontaneous formation of these spheres, called liposomes, is a result of the energetically favorable interaction of the hydrophilic ends of the phospholipids with the water molecules. Cell membranes are structured in exactly this way. Hence, they are basically stable, forming almost automatically and requiring no energy to maintain.

■ LIPID BILAYERS: MEMBRANES HELD TOGETHER BY WATER

A workable model was proposed in the late 1930s by J. F. Danielli of Princeton University and H. Davson of University College, London. They suggested that the membrane might be composed of two layers of phospholipids oriented with their polar (hydrophilic) ends exposed at the two surfaces of the membrane and their nonpolar (hydrophobic) hydrocarbon chains buried in the interior, hidden from the surrounding water. A structure based on hydrophobic-hydrophilic interactions would be very stable because of the polar bonding of the hydrophilic ends with water. Indeed, as we now know, spheres—called ***liposomes***—composed of phospholipid bilayers form spontaneously when phospholipids are mixed with water (Figs. 4.12, 4.13). Electron micrographs of cell membranes (Fig. 4.14) reveal a structure nearly identical to that of liposomes.

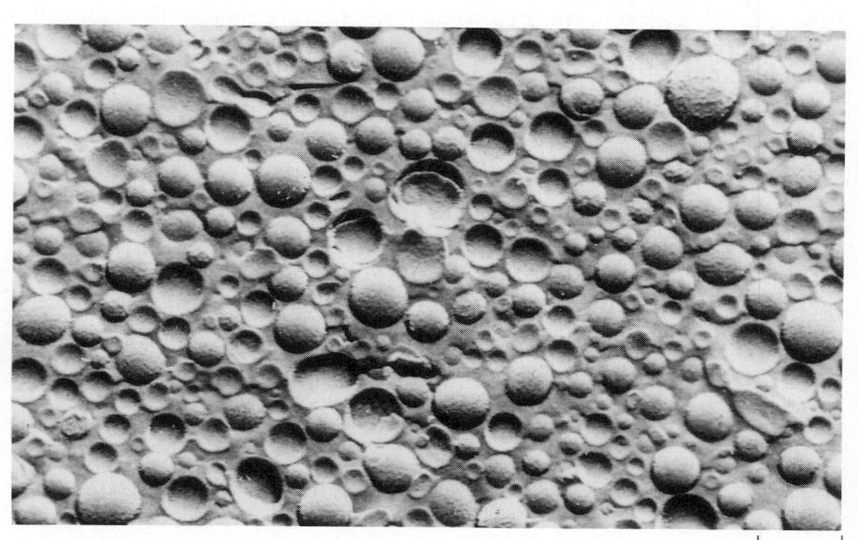

4.13 Single-layer liposomes
Liposomes can be manufactured to deliver concentrated dosages of drugs to sites of infection, inflammation, or cancerous growth. This electron micrograph shows a freeze-fracture of artificial liposomes in a cell membrane.

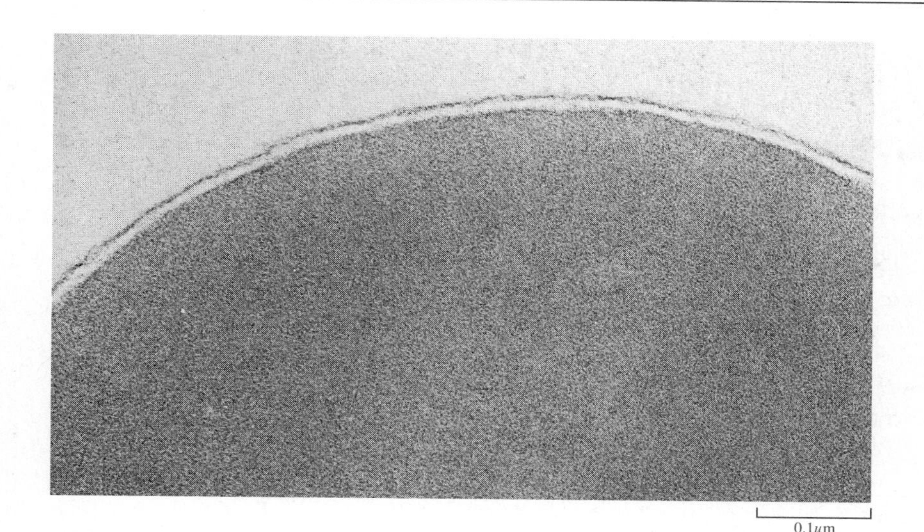

4.14 Membrane of a red blood corpuscle

The cytoplasm of the corpuscle is in the dark grey area of this electron micrograph. The membrane consists of two dark lines (the phosphate-group "heads") separated by a lighter area (the hydrocarbon tails).

The phospholipid-bilayer model accounts for the stability, flexibility, and lipid-passing characteristics of the membrane, but it does not explain the selective permeability of the membrane to certain ions and chemicals. In 1972 S. J. Singer of the University of California at San Diego and G. L. Nicolson of the Salk Institute proposed the ***fluid-mosaic model***, a hypothesis that is now almost universally accepted. The model incorporates the Davson-Danielli conception of a bilayer of phospholipids oriented with their hydrophobic tails directed toward the interior and their hydrophilic heads exposed to the aqueous environment on both surfaces. To explain the selective permeability, the fluid-mosaic model imagines various specialized proteins inserted in the membrane that act as transmembrane pores (Fig. 4.15).

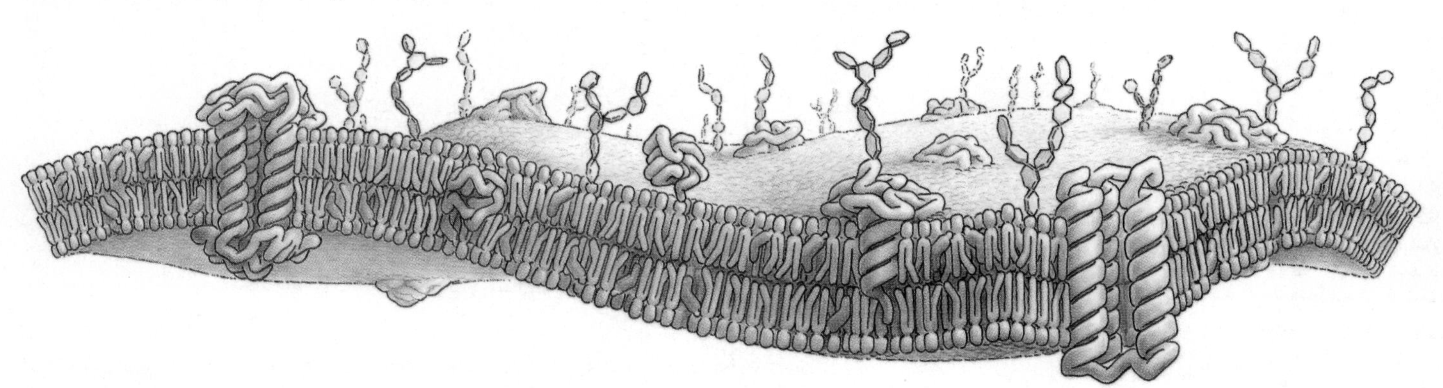

4.15 The fluid-mosaic model of the cell membrane

A double layer of lipids forms the main continuous part of the membrane; the lipids are mostly phospholipids, but in plasma membranes of higher organisms cholesterol (brown) is also present. Proteins occur in various arrangements. Some, called peripheral proteins, lie entirely on the surface of the membrane, to which they are anchored by a covalent bond with a membrane lipid. Others, called integral proteins, are wholly or partly embedded in the lipid layers; some of these may penetrate all the way through the membrane. In the front right corner, note the three protein units, joined by covalent bonds (not shown) to form part of a single protein molecule bounding a membrane-spanning pore. Proteins make up about half of the weight of membranes. The hexagons represent carbohydrate groups.

HOW WE KNOW:

FREEZE-FRACTURE AND FREEZE-ETCHING

Freeze-fracture etching, a technique for preparing specimens for electron microscopy, has become an indispensable tool for studying the details of membrane structure. The specimen is first rapidly frozen and then fractured along the plane of its bilipid membrane (A-B). Some of the ice is then removed from the specimen by sublimation (conversion directly to vapor), which exposes the inside surface of the membrane and gives the specimen an etched appearance (C). Carbon and a metal, usually platinum, are then applied at an angle to the specimen (D) so as to shadow any irregularities in the membrane. Next the original specimen is removed from the platinum cast or surface replica thus formed (E). The replica can now be examined by microscopy. EMs of freeze-etched cell membranes or other structures in cells usually have a three-dimensional appearance (Fig. 4.13).

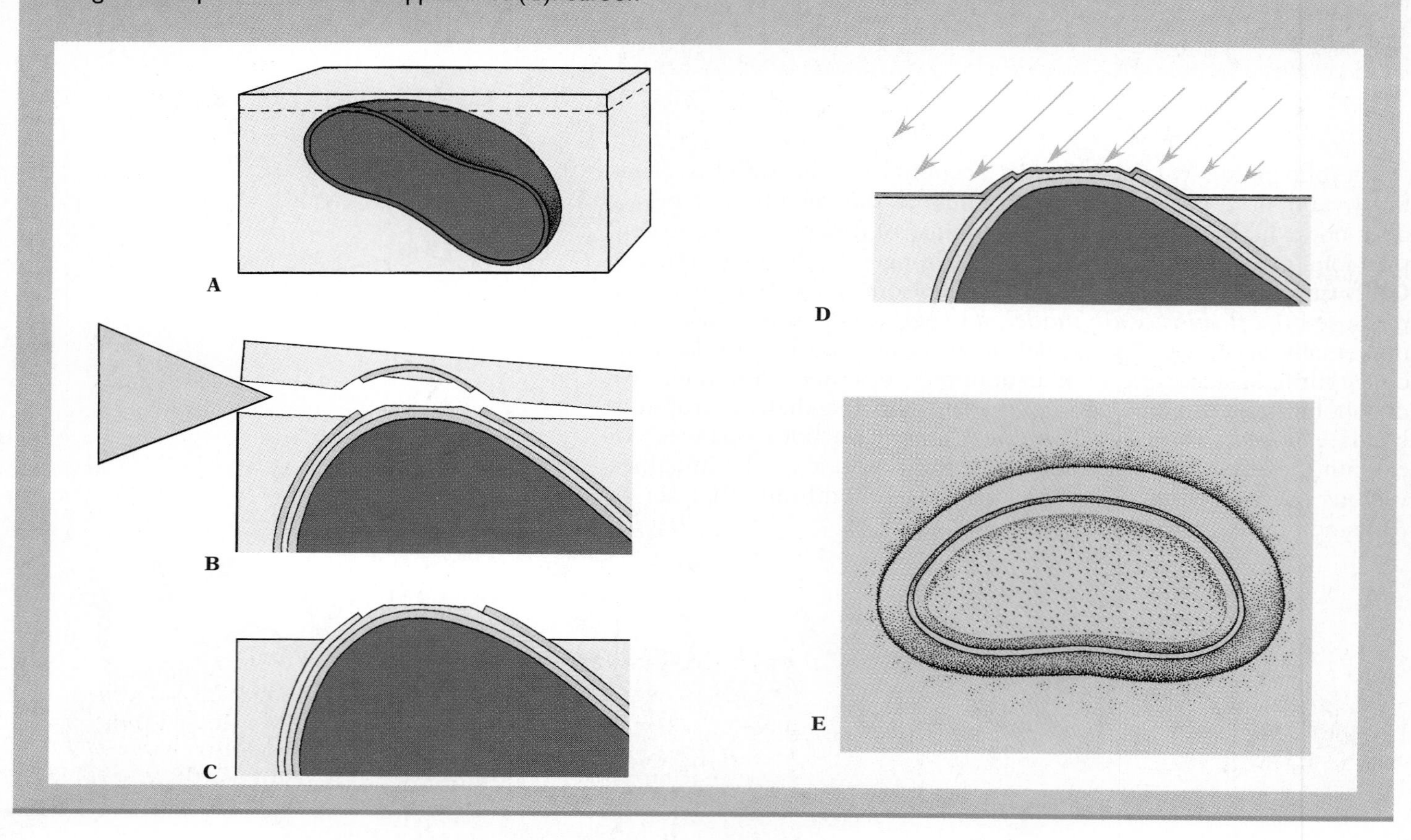

Some membrane proteins (***peripheral proteins***) are confined to the surfaces; others, the ***integral proteins***, are located wholly or largely within the lipid bilayer (Figs. 4.15, 4.16). Integral proteins may be buried within the bilayer, or have parts that project through the surface; some are confined to the outer half of the lipid core, and others to the inner half, while some extend entirely through the bilayer, projecting into the watery medium on both sides. Hydrophilic amino acids (those with polar or electrically charged R groups) predominate in the portions of the protein molecules that project out into the water, whereas hydrophobic (nonpolar) amino acids are abundant in the portions buried in the bilayer (Fig. 4.17). Indeed, the location of the hydrophilic and the hydrophobic amino acids in a membrane protein determines

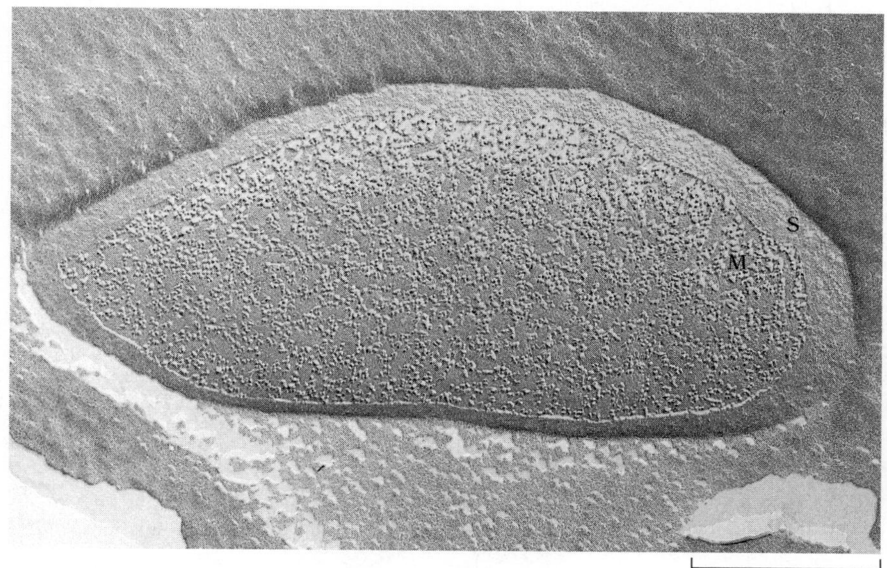

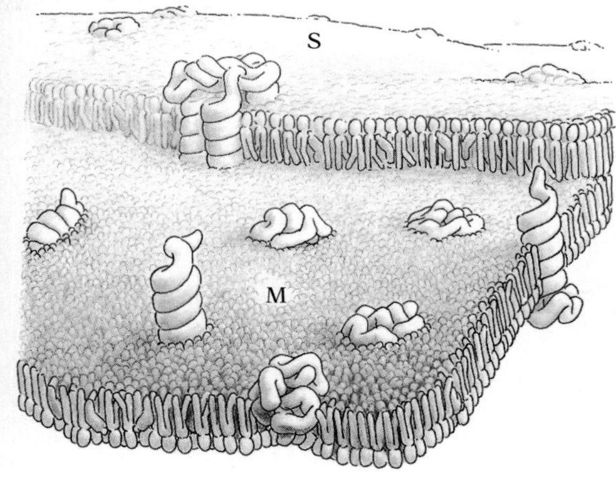

4.16 Plasma membrane of a red blood corpuscle

In this electron micrograph the plasma membrane has been frozen and then fractured along the plane between the two layers of lipids—that is, along the middle of the bimolecular lipid core (see sketch). The numerous spherical particles visible in the micrograph are proteins (see the gray coiled entities in sketch). S, outer surface of membrane; M, interior of fractured membrane.

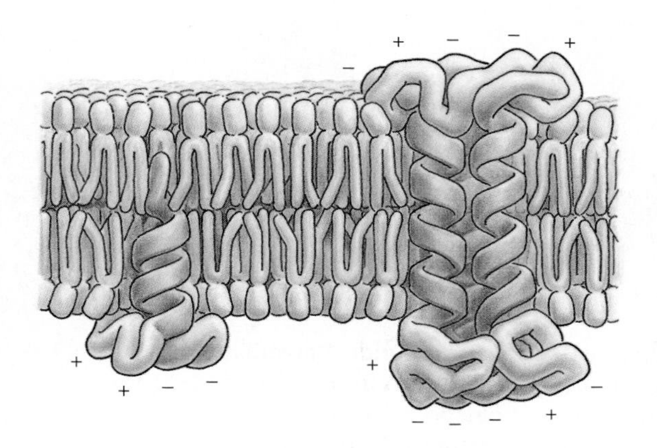

4.17 Orientation of proteins within membranes

The parts of the polypeptide chain containing most of the hydrophilic amino acids (polar or charged R groups; blue) tend to project into the watery medium outside the lipid layers, whereas the parts of the chain with hydrophobic amino acids (brown) tend to be folded into the inner, lipid portion of the membrane. The relative diameter of the protein strands has been reduced for clarity.

which part of the protein will be anchored in the membrane and whether the protein will be integral or peripheral.

The structure of the membrane is not static. Most lipid molecules can move in the plane of the membrane, so a particular molecule found in one position at a given moment may be in an entirely different position only seconds later. Mobility of the lipids is greatest in membranes that are high in unsaturated phospholipids (Fig. 4.18) and contain no cholesterol. Speeds of 2μm (micrometers)/sec are possible—an astonishing mobility considering that many organisms (the bacterium *Escherichia coli*, for example) are only about 2μm long.

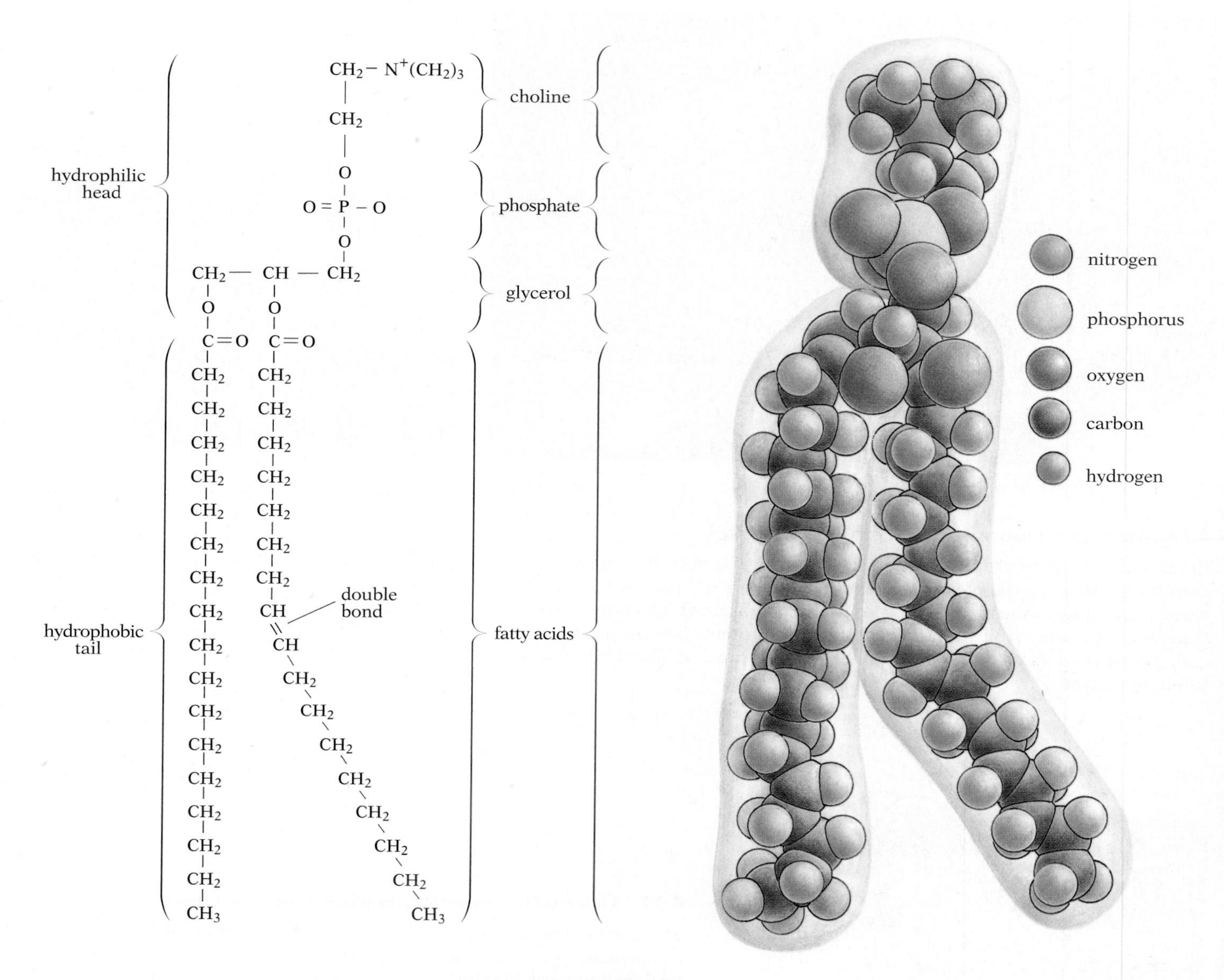

4.18 A phospholipid

The cell membrane is made up mostly of phospholipids. Phosphatidylcholine, a common membrane phospholipid found almost exclusively in the outer half of bilayer membranes, consists of a polar head (a positively charged choline, a negatively charged phosphate, and an uncharged glycerol) joined to two hydrophobic fatty acid chains. The kink in the right tail is created by a double-carbon bond. Because this tail is unsaturated—that is, not every carbon has its full complement of hydrogen atoms—the phospholipid will be less tightly packed in the membrane and hence will be more mobile.

When cholesterol is present, it binds weakly to adjacent phospholipids and reduces the permeability of the membrane to small molecules. It can also have one of two other effects. If the membrane phospholipids are mostly saturated, cholesterol prevents them from packing so closely and regularly that they crystallize into a stiff and solid layer. When, on the other hand, the phospholipids are mostly unsaturated (so the "kinks" in their hydrophobic tails keep them loosely packed), cholesterol can fill these gaps, bind to adjacent phospholipids, and thus join them together (Fig. 4.19). Plant membranes lack cholesterol and instead depend on the cell wall for stability.

The cholesterol concentration and the degree of saturation vary enormously between species and even among tissues in the same organism, depending on the amount of flexibility needed. Though medical opinion is far from unanimous, there is great concern that too much cholesterol and saturated fat in the human diet can lead to hardening of cell membranes, particularly those lining the arteries. Hardening of the arteries (atherosclerosis) is a major cause of stroke and heart disease.

Some proteins in the cell membrane can also move laterally to some extent, but others are anchored in place, thus limiting the fluidity of the membrane.

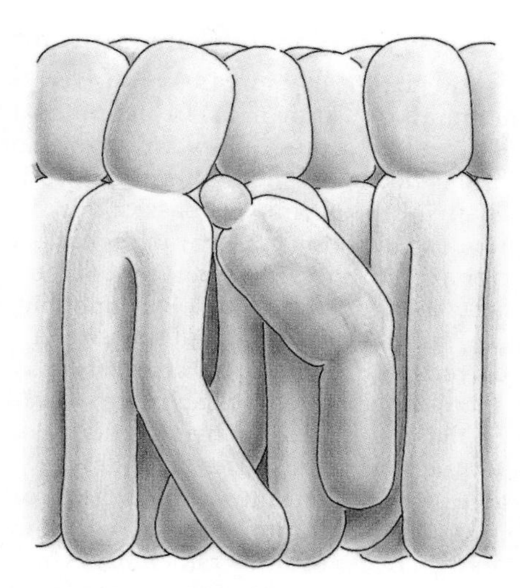

4.19 Cholesterol in the membrane
Cholesterol (brown) binds weakly but effectively to two adjacent phospholipids, thereby partially immobilizing them. The result is a less fluid and mechanically stronger membrane. The amount of cholesterol varies widely according to cell type; the membranes of some cells possess nearly as many cholesterol molecules as phospholipids, while others lack cholesterol entirely. For the structural formula of cholesterol, see Figure 3.18, p. 58.

■ HOLES IN THE MEMBRANE: CHANNELS AND PUMPS

The lipid bilayer forms spontaneously from phospholipids manufactured by the cell, creating a flexible but effective barrier between the inside of the cell and the world outside. However, the need to pass specific substances through the membrane—sometimes actively pumped against their osmotic-concentration gradients—requires specialized pores. The pores in the membrane are actually channels through one or (more often) a group of protein molecules (Fig. 4.15). The distinctive properties of the various R groups of the amino acids in the proteins give the pores some selectivity; not all ions or molecules small enough to fit in the pores can actually move through them. Since they enable specific chemicals to permeate the membrane, channels and pumps are known collectively as ***permeases.***

How channels control membrane traffic The simplest of the permeases, the ***membrane channels,*** provide openings through which specific substances can diffuse across the membrane. These channels, though selective, are passive: they simply permit particular chemicals to move down their concentration gradients. This ***facilitated diffusion*** is the basis of the membrane's highly specific permeability. The protein channel for potassium ions (Fig. 4.20A) provides a good example. Cellular processes result in the accumulation of K^+ inside most cells. As a charged particle, K^+ is insoluble in the hydrophobic interior of the lipid membrane, but the channel specific for K^+ allows it to leak out slowly at a controlled rate. Without such leakage the internal K^+ concentration would become too high for the cell to function. The specificity of the K^+ channel results from its internal shape and its charge, though the details are not understood.

More complex channels can move two specific substances in concert. For example, ion-exchange channels, called ***antiports***, work by passively trading two similarly charged ions. As one moves into the cell, the other exits, thereby maintaining an electrical balance. The many integral membrane proteins in Fig. 4.16 are channels that exchange Cl^- for HCO_3^- (dissolved carbon dioxide), as part of the red blood corpuscle's job of carrying waste CO_2 from cells to the lungs.

4.20 Strategies of membrane transport

Many strategies for moving substances across the cell membrane are known. (A) In facilitated diffusion of the simplest kind, a protein channel, or pore, embedded in the membrane provides a direct path for the chemical it passes "down" its osmotic-concentration gradient. The channel's diameter and the chemical environment the channel creates (hydrophilic or hydrophobic, for example) serve to prevent all but the correct substance from passing. (B) Other channels pass two substances cooperatively, or exchange two substances. Illustrated here is a hypothetical mechanism for the channel that uses the highly favorable osmotic-concentration gradient for bringing sodium ions into the cell to overcome the osmotic-concentration gradient working against glucose (labeled G). When Na^+ binds to the channel (middle), it induces an allosteric change that enables glucose also to bind to the channel. This binding of glucose causes the channel to close to the outside and open to the inside. This change releases the glucose as well as the Na^+ into the interior (right). Having lost the Na^+ and glucose, the channel can reopen to the outside. (C) An allosteric interaction between a signal molecule (red) and the gated channel causes the gate to open, so that diffusion can take place down a favorable concentration gradient. Other molecular systems, not shown, then inactivate the signal molecule so that the channel can close again. (D) A mobile carrier would not provide transmembrane channels, but would itself migrate back and forth from one surface to the other. The existence of mobile carriers is controversial.

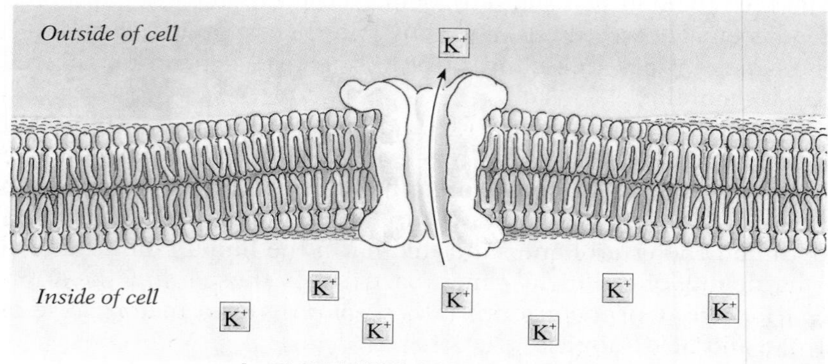

A FACILITATED-DIFFUSION CHANNEL

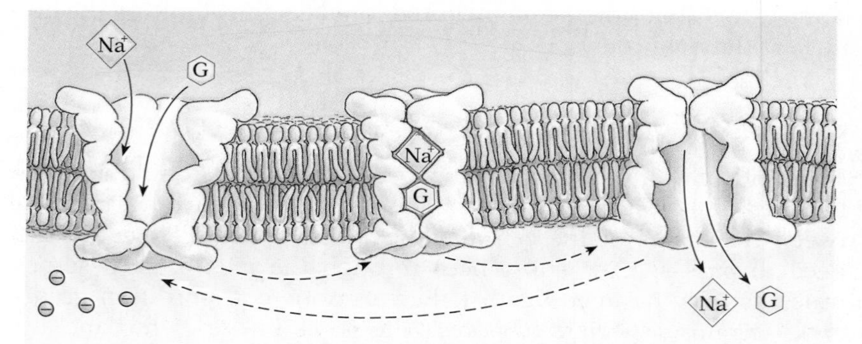

B COOPERATIVE ION CHANNEL (SYMPORT)

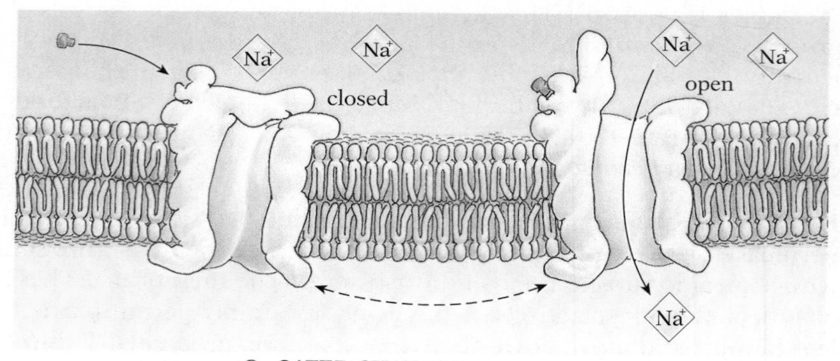

C GATED CHANNEL

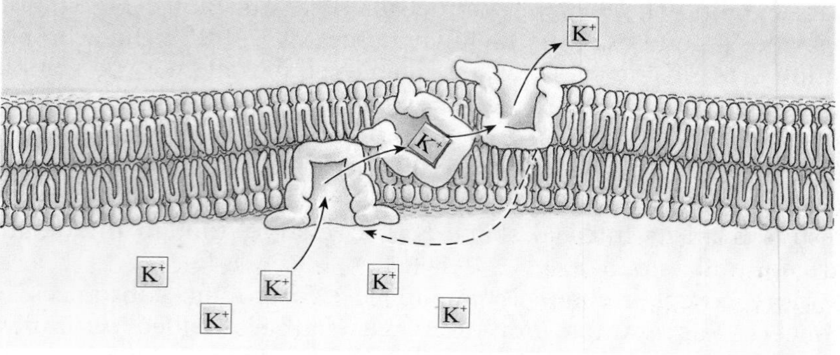

D MOBILE CARRIER

Other cooperative channels called ***symports*** move two substances in the same direction. Coordinated movement of this sort (called ***cotransport***) moves glucose (the molecule most cells use as their major energy source) through the membrane in company with Na^+. Sodium ions are 11 times more concentrated outside the cell, and are therefore subject to a highly favorable osmotic gradient to the inside. Yet they must be accompanied by glucose to pass through the appropriate channel (Fig. 4.20B). Both Na^+ and glucose must bind to the outside of the channel before this special membrane pore will open. Thus the free energy of the osmotic-concentration gradient of Na^+ is exploited to overcome the smaller, unfavorable concentration gradient of glucose. In thermodynamic terms, the two diffusion "reactions" are linked: moving Na^+ "downhill" releases more free energy than is consumed in moving glucose "uphill."

The osmotic-concentration gradient of Na^+ accounts for only part of the energy used in moving glucose across the membrane. As you know, oppositely charged ions attract one another electrostatically, while ions with the same charge repel one another. As a result, if a cell has more negative than positive ions, positive ions will be attracted to it from the surrounding fluid. (Most cells actually have a net negative charge of about 70 mv (millivolts) relative to the fluids surrounding them.) The difference in charge across the membrane generates an ***electrostatic gradient***, and when appropriate channels are open, positive ions tend to flow into the cell, while negative ions tend to flow out. For ions like Na^+, which are far more concentrated outside, the osmotic and electrostatic potentials combine to create a strong ***electrochemical gradient*** (Fig. 4.21). It is the free energy of these combined potentials that accounts for the particular effectiveness of Na^+ in moving glucose into cells.

Gates across membrane channels are another way to control movement across the membrane. One use is to convert molecular signals specialized for carrying information between cells into other signals more suitable for communicating inside the cell. When a molecular signal—a hormone or one of the transmitter substances that carry messages from one nerve to another—binds to an exposed part of a transmembrane ***receptor***, an allosteric change (a change in shape) takes place. This conformational change allows the gate to open, and the second signal, usually an ion like Na^+ or Ca^{++}, then moves across, carrying the message into the cell (Fig. 4.20C). This ***gated channel*** strategy underlies the transmission of many chemical messages in both plants and animals, and of the nerve impulses by which animals sense the outside world and move their muscles. This strategy is also used to respond to internal signals: for instance, internal gating regulates flow through Cl^- channels, which are important in regulating the secretion of fluids from cells. A defect in this channel can cause cystic fibrosis.

Another way molecules might get across the membrane would be for the permease to act as a ***mobile carrier***, taking them through one by one (Fig. 4.20D). No definite example of such a permease is yet known, but valinomycin acts in just the way we would expect a mobile carrier to act. Valinomycin is a ring-shaped

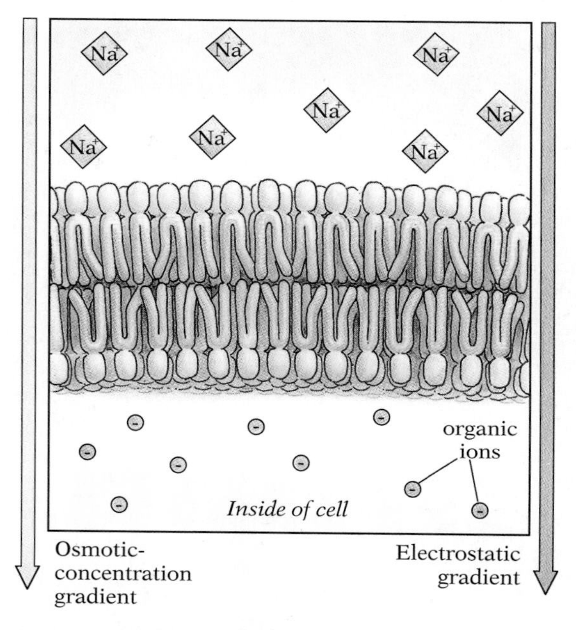

4.21 An electrochemical gradient

Cells have an electrical potential of about 70 mv negative with respect to the fluids that surround them. This gradient arises mainly because large numbers of negatively charged organic ions are trapped inside the cells, while a relatively high concentration of positively charged Na^+ exists outside. Sodium ions are therefore subject to both a strong osmotic-concentration gradient and a sizable electrostatic gradient. The effects of these two gradients combine to create an electrochemical gradient.

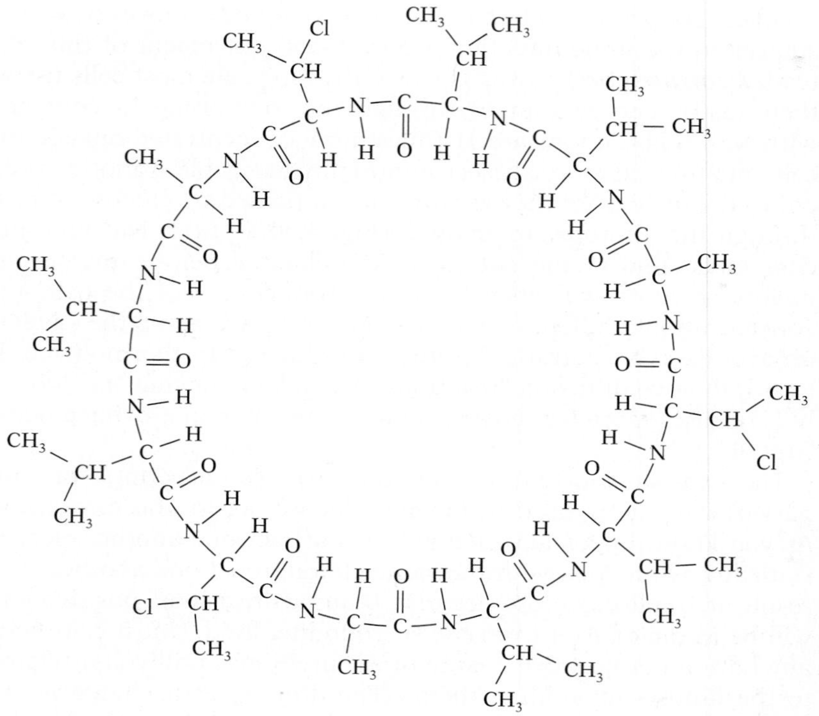

4.22 Valinomycin

This permease, which is synthesized by some bacteria, kills cells that take it up. It consists of three kinds of peptides: alanine, valine, and a chlorine-substituted valine. The potassium ion is held to the center by the oxygens while it is carried across the host membrane.

polymer with a hydrophobic exterior and a polar interior (Fig. 4.22). The polar pocket, lined with six oxygen atoms, can hold a single potassium ion. Apparently the complex bobs back and forth randomly, carrying K^+ ions in both directions. The net transfer of K^+ is a statistical consequence of the concentration gradient: valinomycin more often picks up K^+ inside the cell and releases it outside simply because there is more K^+ inside than out. Valinomycin is actually an antibiotic produced by certain bacteria to poison competing microorganisms by altering the selective permeability of their membranes. The evidence for mobile carriers in normal membranes is equivocal.

Moving molecules "uphill" Other permeases, known as ***pumps***, do not depend on free-energy gradients. Instead, pumps use the cell's store of energy to move substances against their gradients. This ***active transport*** is important in ridding the cell of undesirable molecules. Pumps also transport many essential building blocks into the cell. The best-understood example of a membrane pump, however, is the ***sodium-potassium pump***, which maintains the electrochemical gradient across the membrane: (Fig. 4.23). The free-energy source for this pump, and for many cellular processes, is the cellular energy carrier ATP (adenosine triphosphate), which we will discuss in detail in later chapters. The pump utilizes this energy to exchange potassium and sodium ions across the membrane, thereby maintaining the electrochemical gradient.

The effects of this pump are far-reaching: it is responsible for the electrical activity of nerves and muscles; it indirectly supplies the free energy for many osmotic transport systems, including the one for glucose; it supplies the gradient that permits a Na^+-H^+ antiport

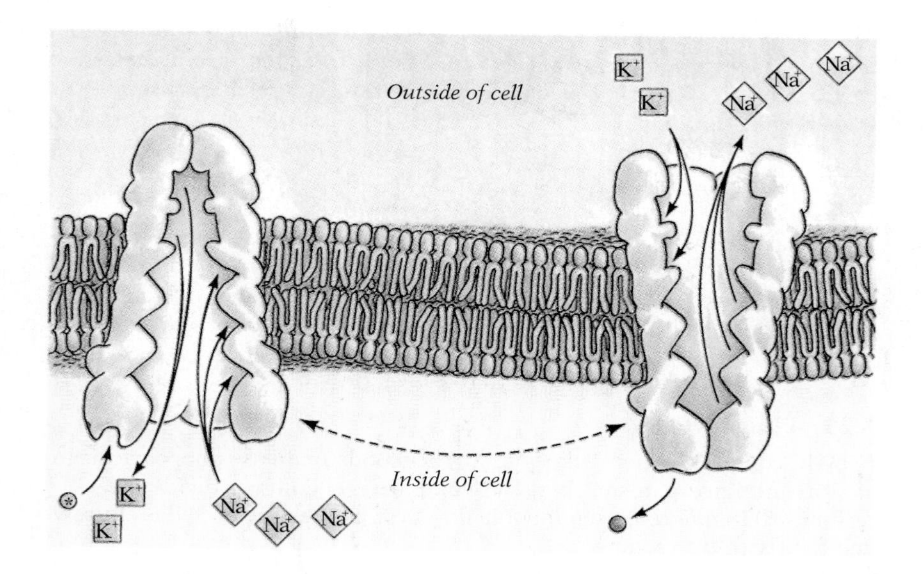

4.23 The sodium-potassium pump

In contrast to other methods of membrane transport, the pump strategy uses energy from the cell (rather than the free energy of a concentration gradient) and provides active transport of a substance against its gradient. In this case, three sodium ions are exchanged for two potassium ions; both kinds of ions are already more concentrated on the side to which they are being moved. In the model shown here, the release of K^+ ions brought in during the previous cycle is followed by the binding of three Na^+ ions and an energy source, ATP (circle), on the inside. The resulting conformational changes in the protein open it to the outside, reduce its affinity for Na^+, which is then released, and increase its affinity for K^+ ions. The binding of K^+ then causes the channel to open to the inside, increase its affinity for Na^+, and decrease its affinity for K^+, and the cycle begins again. The net ionic effect is to pump positive charges out of the cell, and the inside of the cell becomes negatively charged with respect to the outside. The electrical and osmotic potential created by the sodium-potassium pump ultimately makes possible the cooperative transport of glucose illustrated in Figure 4.21B.

channel to control cellular pH; it facilitates the uptake of water by plant roots; and it helps regulate the volume of many cells by controlling osmotic potential. Indeed, when the pump is destroyed by the poison ouabain, cells swell uncontrollably with water until they burst.

In addition to the engineering feats of channels and pumps, cells control osmotic potentials across the membrane by chemical means. We saw how the Na^+ gradient can be used to "pull" glucose inside because the gradient against glucose is very small. Yet the world outside most cells has very little glucose to begin with, while most cells use it in large amounts. How can the concentration of glucose inside the cell be kept so low? The cell manages this trick by binding the glucose into another compound as soon as it is inside (Fig. 4.24). As a result, the free-glucose concentration of glucose remains artificially low inside, and the osmotic potential against glucose does not get out of hand.

■ LARGE-SCALE MOVEMENT: ENDOCYTOSIS AND EXOCYTOSIS

As we have seen, permeases are the means by which substances enter and leave cells through the cell membrane. Cells also have ways of admitting substances, usually in larger quantities, without having them pass through channels. Through ***endocytosis***, a cell encloses the substance in a membrane-bounded vesicle that is pinched off from the cell membrane. There are three types of endocytosis, all of which depend on specialized membrane proteins:

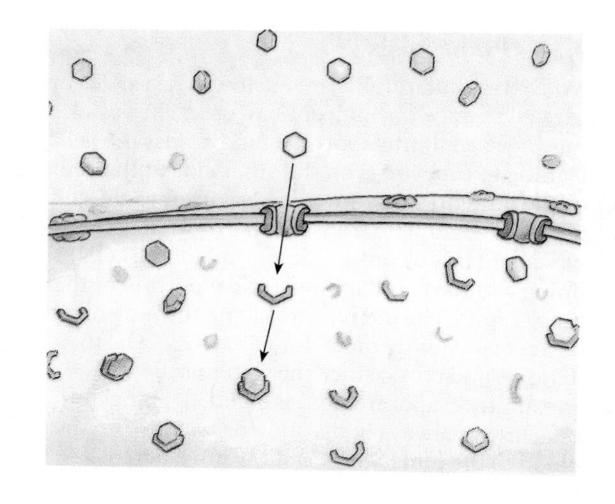

4.24 Formation of a complex inside the cell

As fast as glucose molecules (yellow hexagons) enter the cell, they combine with acceptor molecules (brown) already in the cell to form a new substance. The concentration of free glucose inside the cell therefore remains low, so glucose continues to diffuse inward.

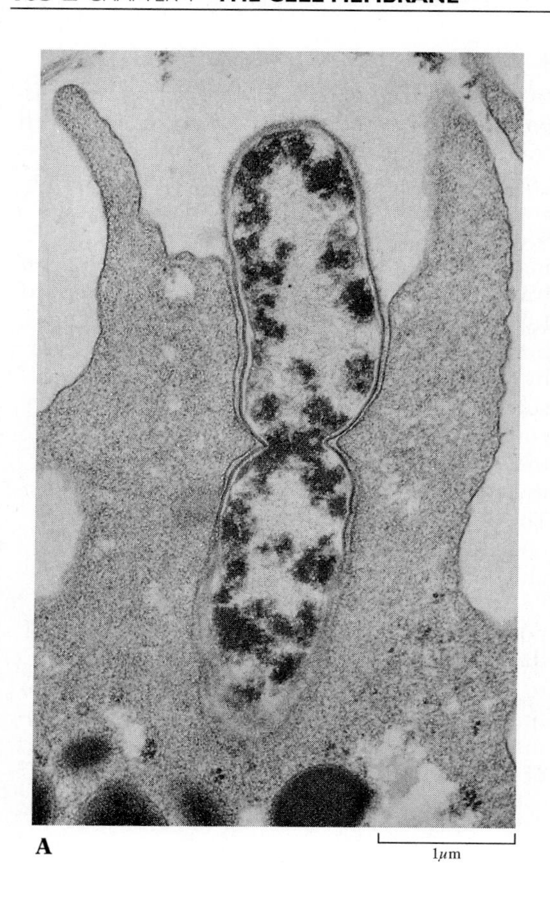

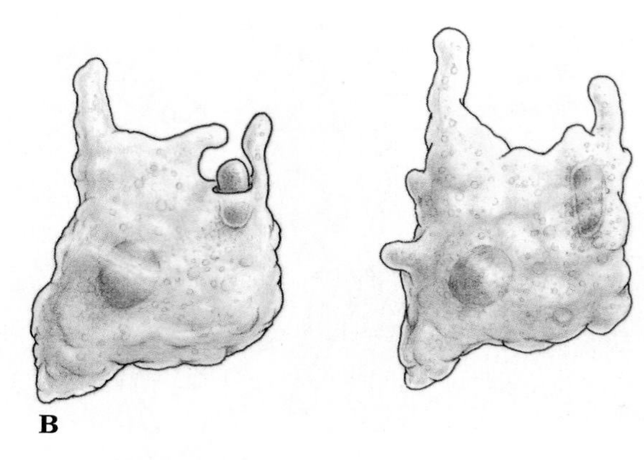

4.25 Phagocytosis

(A) White blood cells, or leukocytes, in the bloodstream use phagocytosis to capture foreign organisms; here the leukocyte is engulfing a dividing bacterium. (B) In *Amoeba,* pseudopodia flow around the prey until it is entirely enclosed within a vacuole.

1 When the material engulfed is in the form of large particles, the process is called ***phagocytosis***, or "cellular eating" (Fig. 4.25). Usually, armlike ***pseudopodia*** ("false feet") flow around the material, enclosing it within a vesicle, which then detaches from the plasma membrane and migrates to the interior of the cell. Phagocytosis is triggered only when special membrane-mounted receptor proteins bind to a suitable target; this binding is analogous to enzyme-substrate binding. In vertebrates, phagocytosis is generally restricted to scavenger cells in the blood, which ingest debris and invading microorganisms (Fig. 4.25B).

2 When the engulfed material is liquid the process is called pinocytosis, or "cellular drinking." Pinocytosis is used to ingest extracellular fluid or to transport it across a cellular barrier (Fig. 4.26).

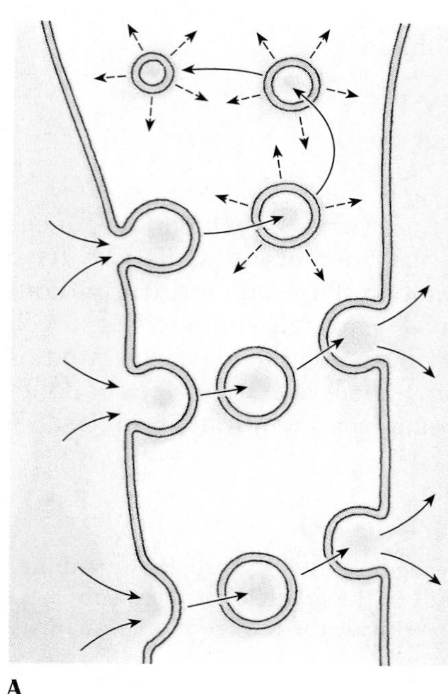

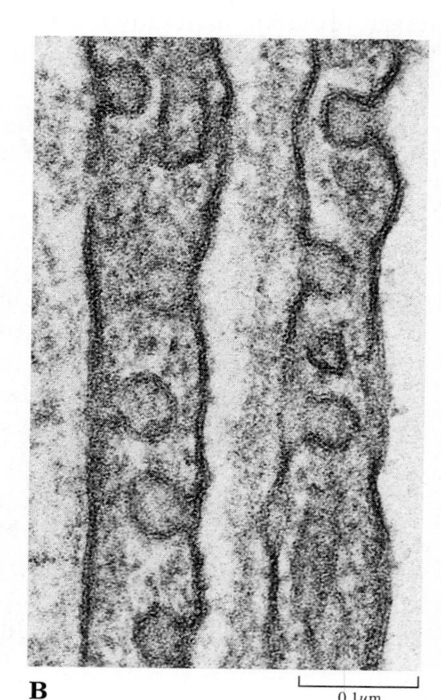

4.26 Pinocytosis

(A) Extracellular fluids are enclosed in vesicles on the cell surface through endocytosis. The vesicles may be used to transport the fluid across the cell, leading to exocytosis on the other side of the cell, or the fluid may be allowed to escape into the cell itself. (B) Complete transcellular movement can be seen in this electron micrograph of the cells lining a blood capillary (left) and a portion of the lung (right). The membrane facing the lung must be kept constantly moist for gases (O_2, CO_2) to dissolve and cross; since the water used constantly evaporates and is exhaled, pinocytotic vesicles are always in use moving water from the blood to the inner surface of the lung, across a basement membrane.

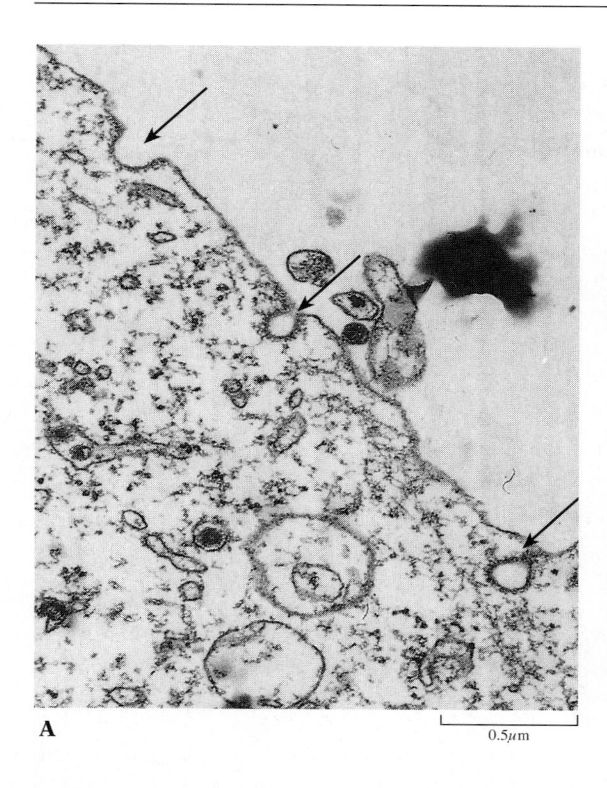

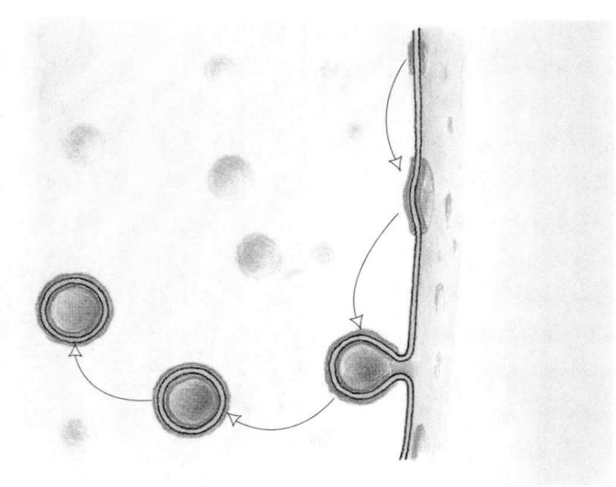

4.27 Receptor-mediated endocytosis

(A) Three stages of receptor-mediated endocytosis of a short polypeptide used in cell-to-cell communication are seen taking place in a cultured nerve cell. (B) Molecules bind to receptors on the cell surface, triggering vesicle formation and endocytosis.

3 When the material to be ingested is adsorbed on the cell membrane at selective binding sites, the process is termed ***receptor-mediated endocytosis.*** The loaded vesicles are formed and detach from the membrane at the cell surface (Fig. 4.27). In most cases, the site that collects a particular substance before trapping it in a vesicle appears as a coated pit, with receptor molecules clustered in one spot in the membrane (Fig. 4.28).

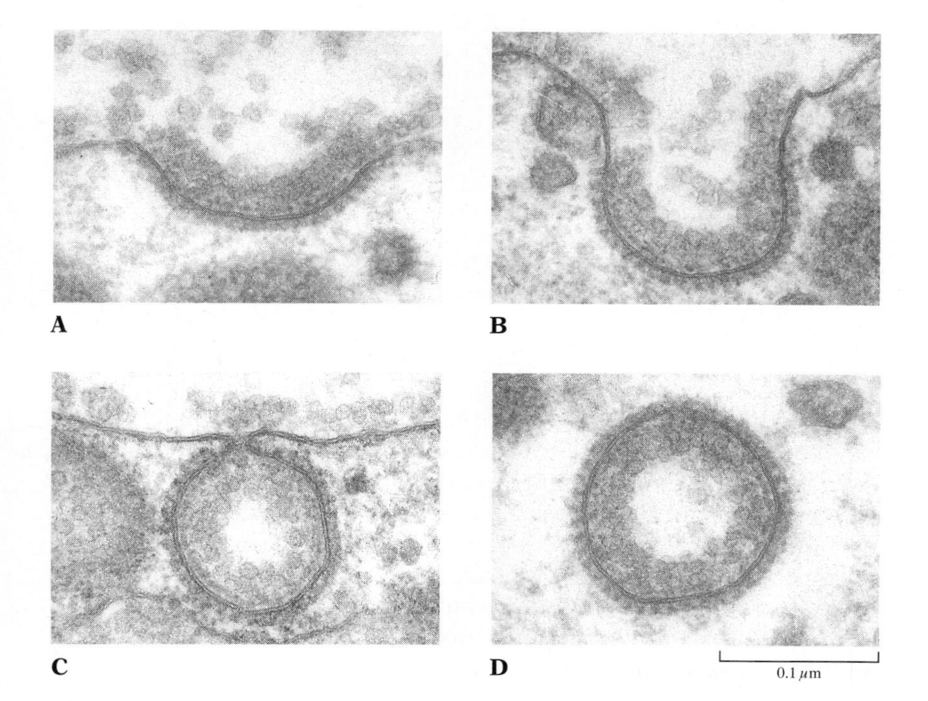

4.28 Endocytosis by means of a coated pit

(A) Specialized receptors for lipoproteins have aggregated to form a coated pit in the membrane of an egg cell. (B–D) The pit is subsequently pinched off to form a vesicle. Most endocytotic vesicles are transported to lysosomes, organelles within the cell where the contents are enzymatically altered. The lipoprotein in the vesicle shown here will become part of the yolk.

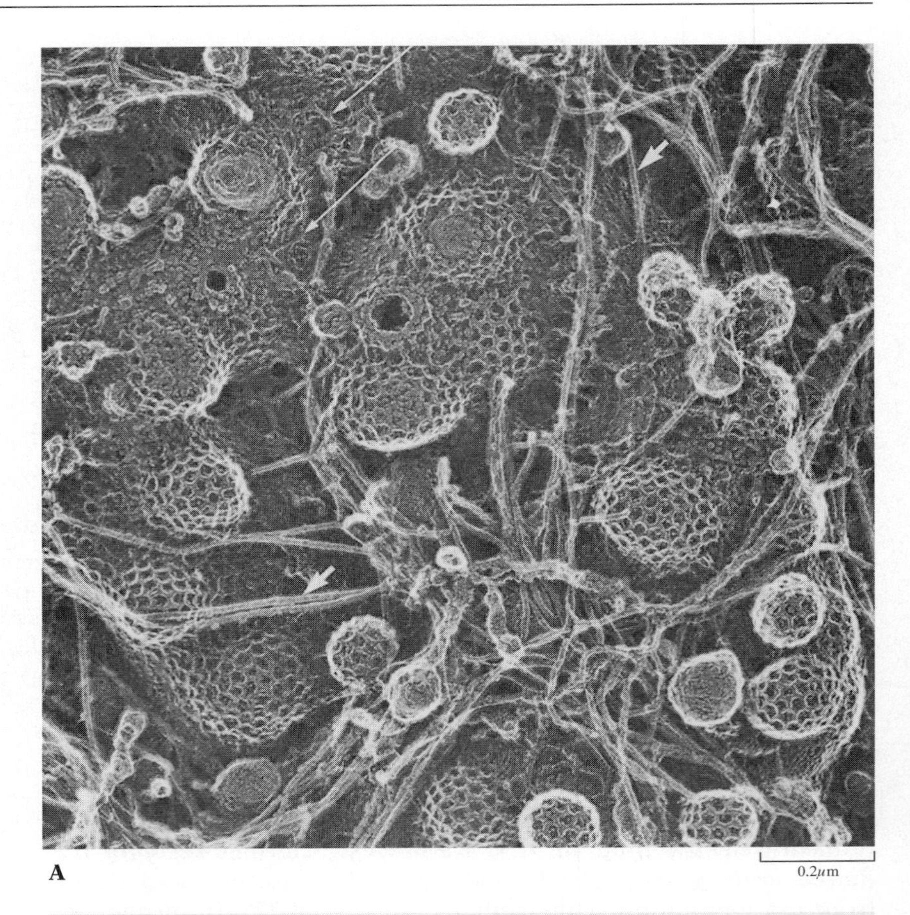

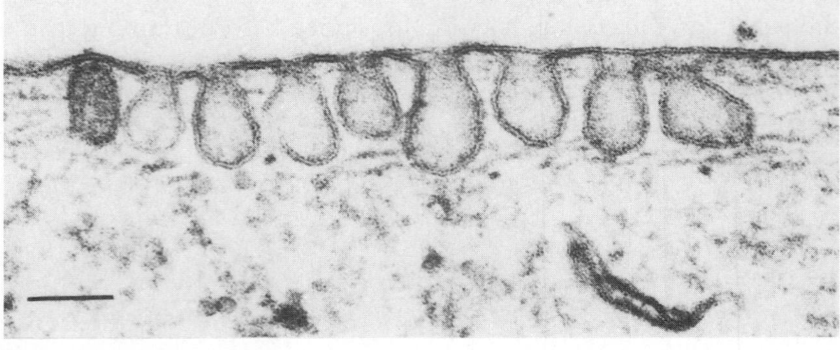

4.29 Clathrin-coated pits versus caveolae

(A) Clathrin-coated pits, seen from the inside of a liver cell, are in the process of budding off from the membrane to form intracellular vesicles. The cablelike structures are part of the cytoskeleton, discussed in the next chapter. (B) Caveolae, seen from the inside of a fibroblast, lack the characteristic clathrin coating; one clathrin-coated pit is visible on the upper left-hand side.

These outward-facing receptors (and the substances they bind) produce what looks like a haze on the extracellular surface of the membrane (Fig. 4.28). A similar darkening appears on the intracellular surface of the membrane at the spot destined to form a vesicle. The best-understood component of this internal patch is the structural protein ***clathrin***, which begins assembling itself into the foundation for a membrane indentation, or pit, and then into the vesicle's coating. The clathrin scaffolding is clearly visible surrounding newly formed vesicles (Fig. 4.29A). Smaller invaginations, called caveolae, have a separate scaffolding protein and a different range of membrane-mounted receptors (Fig. 4.29B); they may form vesicles, or merely concentrate their target molecules and then pass them to the cytoplasm.

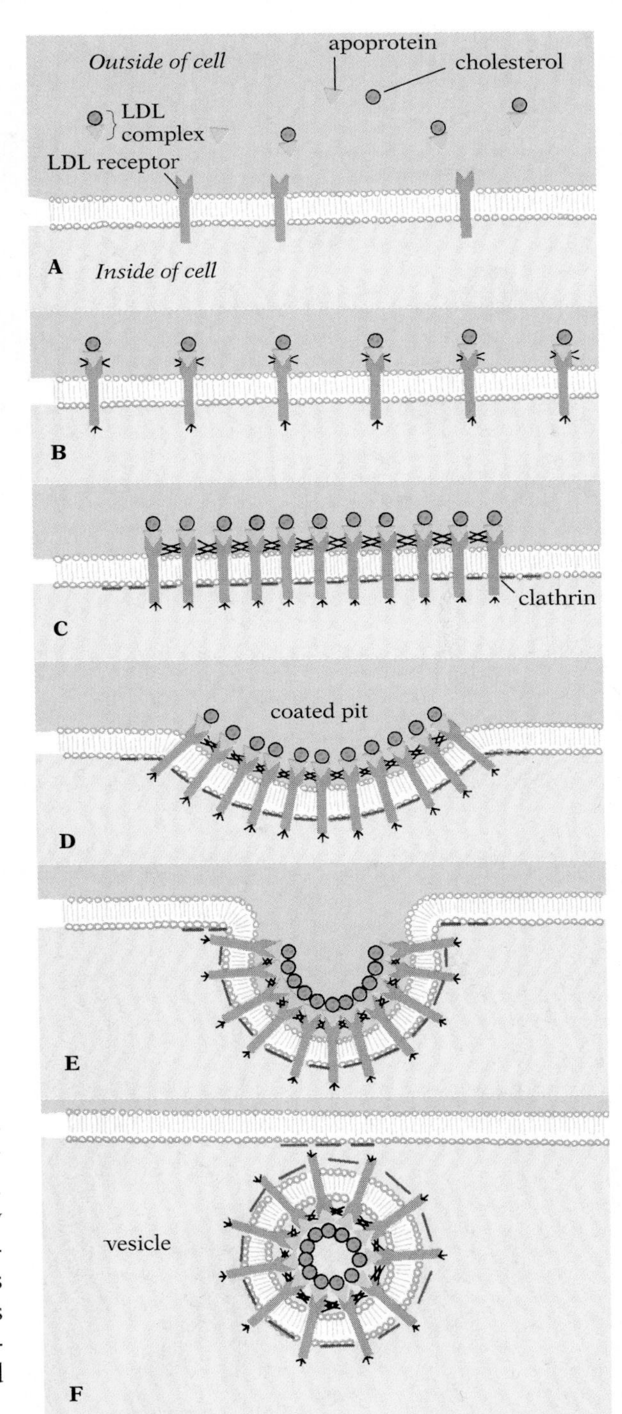

4.30 Endocytosis of cholesterol

(A) When a cell needs cholesterol, it synthesizes receptors for low-density lipoprotein (LDL) and incorporates them into the cell membrane, where they are free to migrate. (B) The receptors soon bind LDL, a carrier complex that transports cholesterol in the blood. The LDL complex includes some 2000 cholesterol molecules and an associated protein (called apoprotein) that binds to the LDL receptor. (C) Having bound the LDL, the receptors stop drifting in the membrane and stick to one another over a clathrin-rich patch of membrane. Even in the absence of bound LDL, many of the receptors eventually aggregate spontaneously. (D–E) Aggregations of LDL receptors trigger endocytosis, the first step of which is the formation of a coated pit. (F) The resulting vesicle is transported to the site of membrane synthesis. This method of cholesterol uptake may demonstrate the usual strategy by which cells obtain nutrients that cannot pass through the membrane directly.

One example that illustrates the importance and specificity of receptor-mediated endocytosis involves cholesterol uptake by cells. Cholesterol is transported in the blood to cells by a carrier complex of low-density lipoprotein (LDL). When the cell needs cholesterol—usually for use in the manufacture of new membrane—LDL receptors are synthesized and incorporated into the cell membrane (Fig. 4.30). The LDL receptors aggregate spontaneously over membrane patches rich in clathrin, particularly once they have bound LDL. The cholesterol is then transported in endocytotic vesicles, known as ***endosomes***, for use in the cell's vast complex of membranes. One cause of atherosclerosis involves failure of the receptors to bind LDL; another involves failure of the receptors, having bound LDL, to aggregate and to initiate endocytosis. Either defect can allow cholesterol-rich plaques to form on artery walls.

Material enclosed within endocytotic vesicles has not yet become part of the cytoplasm. It is still separated by a membrane; it must eventually cross that membrane (or the membrane must disintegrate) if it is to become incorporated into the cell. Normally the vesicle membrane forms from cell membrane. Meanwhile, vesicles full of digestive enzymes form within the cell. These ***lysosomes*** fuse with the endosomes to create miniature cellular stomachs, which break down the endosomal contents. After digestion, many of the products can cross the lysosome membrane into the cytoplasm, while others remain trapped inside. As a result the cell is able to ingest the substances it requires, while the unwanted parts of the meal, and the lysosome's destructive enzymes, remain segregated from the cell's delicate chemical interior. Endosomes and lysosomes will be described more fully in the next chapter.

Exocytosis is essentially the reverse of endocytosis. Materials contained in vesicles are conveyed to the periphery of the cell, where the vesicular membrane fuses with the cell membrane and then bursts, releasing the materials to the surrounding medium (Fig. 4.31). Many glandular secretions are released from cells in this way; the hormone insulin, for example, is released by exocytosis from the pancreatic cells that synthesize it. Exocytosis also exports waste products. The undigested remains of materials brought in by endocytotic vesicles, for example, are normally disposed of through exocytosis. In some cases a coupling of endocytosis and

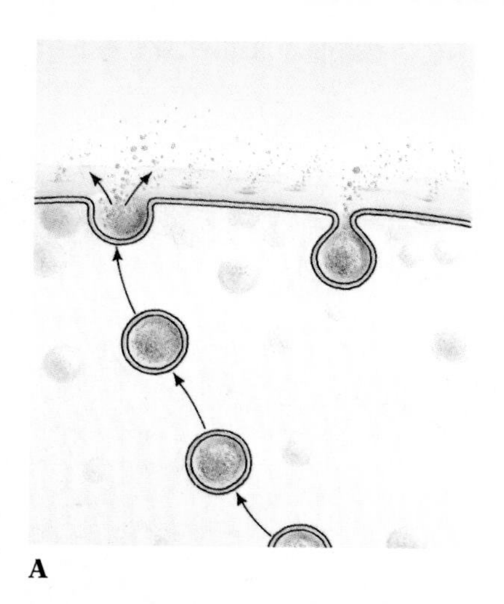

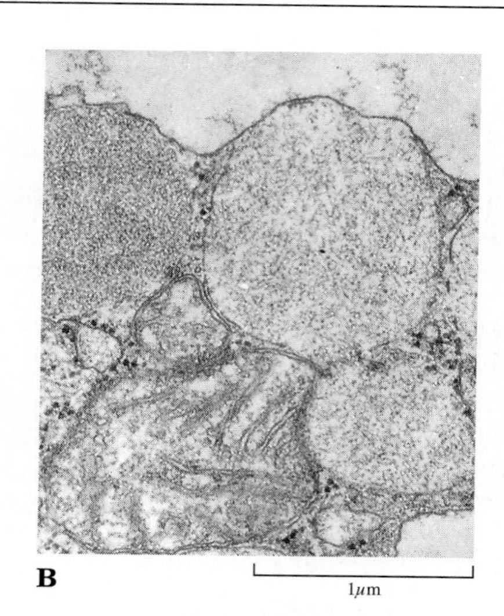

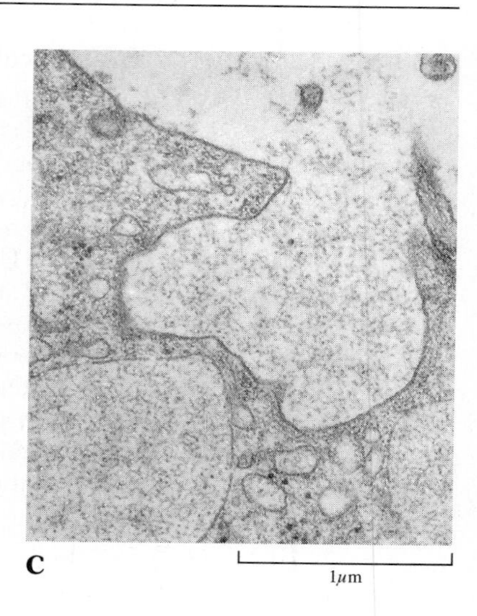

4.31 Exocytosis

(A) A membranous vesicle moves to the periphery of the cell, where it bursts, releasing its contents to the exterior. (B) The final steps of exocytosis are seen here, as a vesicle containing tear fluid fuses with the plasma membrane and bursts.

exocytosis moves a substance entirely across a cellular barrier, such as the wall of a blood vessel. The substance is picked up by endocytosis on one side of the cell, and the vesicle moves through the cell to the other side, where the substance is released by exocytosis (see Fig. 30.16, p. 855).

Exocytosis is not simply a matter of a vesicle colliding with a membrane, fusing, and thus dumping its contents; if this were the case, vesicles containing toxic waste might deliver their contents to any number of membrane-enclosed intracellular organelles, with fatal results. Instead, vesicles have membrane molecules that allow them to attach only to their proper targets; even then, fusion proteins must be recruited and cooperate to form an initial pore that allows the vesicle membrane to fuse with the cellular membrane. The potent nerve toxins responsible for botulism and tetanus apparently exert their fatal effects by blocking the fusion of the specialized exocytotic vesicles of nerve cells responsible for communicating with other nerve and muscle cells.

OUTSIDE THE MEMBRANE: CELL WALLS AND COATS

Unlike animal cells, plant cells are encased in conspicuous cell walls. These walls, which are located outside the plasma membrane, are composed primarily of carbohydrates. Fungi and most bacteria also have strong, thick walls rich in carbohydrates. Animal cells, too, have carbohydrates on the outer surface of their membranes, but they do not form a wall; instead, they are attached as side groups to some lipids and membrane proteins. This cell "coat" plays an important role in determining certain properties of the cells.

■ HOW CELL WALLS ARE BUILT

The principal structural component of the cell wall that encloses plant cell membranes is the complex polysaccharide ***cellulose***, which is generally present in the form of long threadlike structures called fibrils. The cellulose fibrils are cemented together by a matrix of other carbohydrate derivatives, including pectin and hemicellulose. The spaces between the fibrils are not entirely filled with matrix, however; they generally allow water, air, and dissolved materials to pass freely through the cell wall. The membrane located within the cell wall, rather than the wall itself, usually determines which materials can enter the cell.

The first portion of the cell wall laid down by a young growing cell is the ***primary wall***. As long as the cell continues to grow, this somewhat elastic wall is the only one formed. Where the walls of two cells abut, a layer between them, known as the ***middle lamella***, binds them together. ***Pectin***, a complex polysaccharide generally present in the form of calcium pectate, is one of the principal constituents of the middle lamella. If the pectin is dissolved away, the cells become less tightly bound to one another. That is what happens, for example, when fruits ripen: the calcium pectate is partly converted into other, more soluble forms, the cells become looser, and the fruit becomes softer. Many of the bacteria and fungi that produce soft rots in the tissues of higher plants do so by first dissolving the pectin, reducing the tissue to a soft pulp that they can absorb. In cooking, extra pectin is added to thicken fruit preserves.

Cells of the soft tissues of the plant have only primary walls and intercellular middle lamellae. After ceasing to grow, the cells that eventually form the harder, more woody portions of the plant add further layers to the cell wall, forming the ***secondary wall***. Since this wall, like the primary wall, is deposited by the cytoplasm of the cell, it is located inside the earlier-formed primary wall, lying between it and the membrane (Fig. 4.32). The secondary wall is

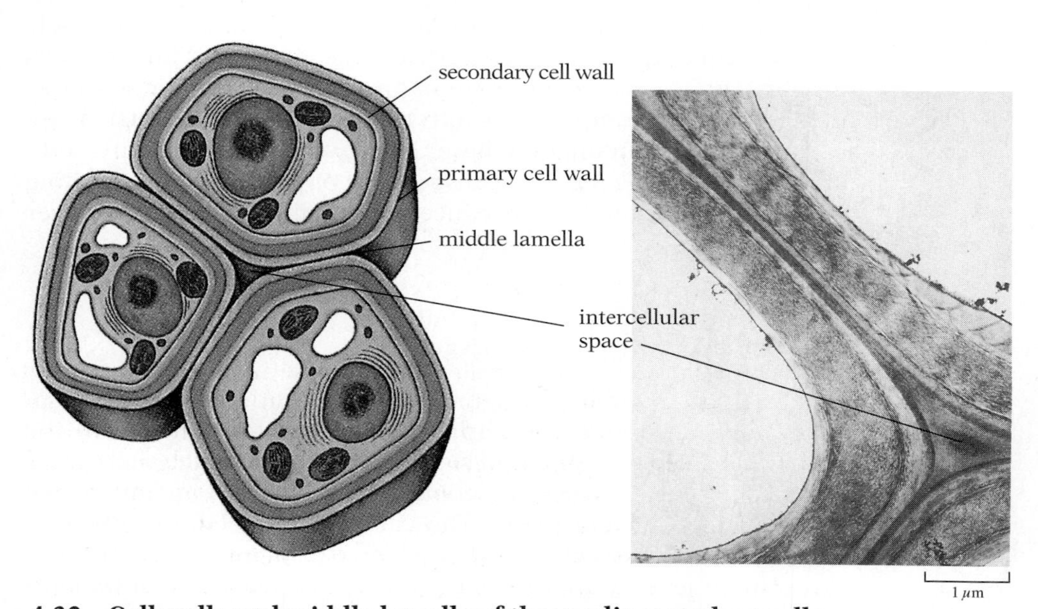

4.32 Cell walls and middle lamella of three adjacent plant cells

4.33 Cellulose microfibrils from the cell wall of a green alga

The microfibrils in this electron micrograph are laid out in parallel lines in two directions; each is about 20 nm (nanometers) wide. Water and ions move freely through this meshwork.

often much thicker than the primary wall and is composed of a succession of compact layers, or lamellae. The cellulose fibrils in each lamella lie parallel to each other and are generally oriented at angles of 60-90° to the fibrils of the adjacent lamellae (Fig. 4.33). This arrangement gives added strength to the cell wall. In addition to cellulose, secondary walls usually contain other materials, such as ***lignin***, which make them stiffer. Once deposition of the secondary wall is completed, many cells die, leaving the hard tube formed by their walls to function in mechanical support and internal transport for the body of the plant.

The cellulose of plant cell walls is commercially important as the main component of paper, cotton, flax, hemp, rayon, celluloid, and, obviously, wood itself. Lignin extracted from wood is sometimes used in the manufacture of synthetic rubber, adhesives, pigments, synthetic resins, and vanillin.

The cell walls of both fungi and bacteria differ from those of plant cells. In most fungi the main structural component of the wall is not cellulose but ***chitin***, a polymer that is derived from the amino sugar glucosamine (see Fig. 3.8, p. 51); chitin is also the major component of insect exoskeletons. In bacteria the cell walls contain several kinds of organic substances, which vary from subgroup to subgroup. (One small group of bacteria lack cell walls altogether). The distinctive responses of these organic substances to diagnostic stains help identify bacteria in the laboratory. Structurally, however, the cell walls of all bacterial groups are alike in one respect. Part of each bacterial wall has a rigid framework of polysaccharide chains cross-linked covalently by short chains of amino acids.

The presence of cell walls means that the cells of plants, fungi, and bacteria can withstand exposure to fluids with low osmotic concentrations without bursting. In such media the cells are in a condition of ***turgor*** (distention). Water tends to move into them by osmosis, as a result of the high osmotic concentration of the cell contents. The cell swells, building up ***turgor pressure*** against the cell walls. Equilibrium is reached when the resistance of the wall is so great that no further increase in the size of the cell is possible and, consequently, no more water can enter the cell. Thus the cells of plants, fungi, and bacteria are not as sensitive as animal cells to the difference in osmotic concentration between the cellular material and the surrounding medium. Moreover, turgor pressure actually strengthens the mechanical structure of plants, just as inflating an initially limp balloon produces a much stiffer and stronger structure.

■ WHAT THE CELL COAT DOES

In plants, fungi, and bacteria, the cell wall is entirely separate from the membrane; if the cell shrinks in a hypertonic medium, the membrane shrinks with it and separates from the much more rigid wall (see Fig. 31.2, p. 881). By contrast, the "coat" of an animal cell is not an independent entity. The carbohydrates (short chains of sugars called oligosaccharides) of which it is composed are part of the membrane: they are covalently bonded to nearly 5% of the protein and lipid molecules in the plasma membrane (Fig. 4.34). The resulting complex molecules are termed glycoproteins and glycolipids, and the cell coat itself is often called the ***glycocalyx***.

The glycocalyx provides the recognition sites on the surface of the cell that enable it to interact with other cells. For example, if individual liver and kidney cells are mixed in a culture medium, the liver cells will recognize one another and reassociate; similarly, the kidney cells will seek out their own kind and reassociate. The nature of the carbohydrate markers varies consistently from tissue to tissue and from species to species. Cell recognition in the process of embryonic development must also depend, at least in part, on the glycocalyx, and the same is probably true for the control of cell growth. When normal cells grown in tissue culture touch each other, they cease moving, and their growth slows down or stops altogether. This phenomenon of ***contact inhibition*** is absent in most cancer cells, which continue growing without restraint because defective glycocalyces prevent them from interacting normally.

As a source of identity, the glycocalyx is also important in a variety of infectious diseases: malaria parasites, for instance, recognize their host (the red blood corpuscle) by a distinctive carbohydrate marker that the corpuscle manufactures for a completely different purpose. The recognition of host cells by invading viruses also often depends on the carbohydrate markers of the glycocalyx. The markers in the glycocalyces of foreign cells probably provide the cues that the immune system's antibody molecules use to recognize invaders.

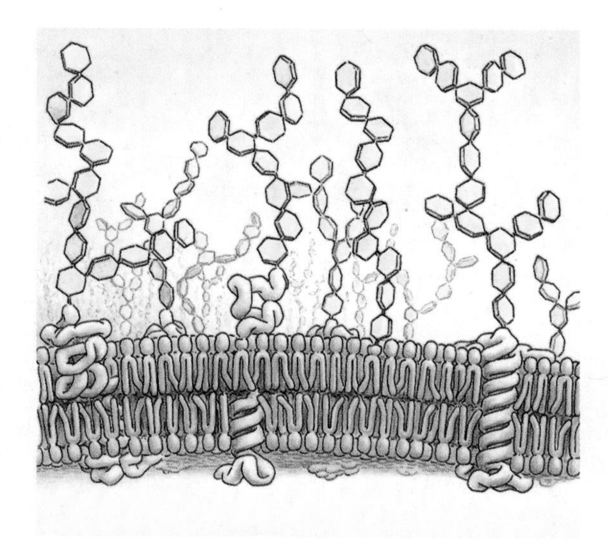

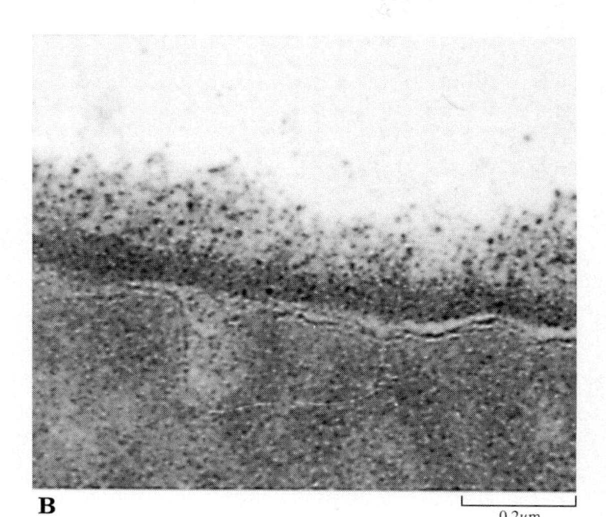

4.34 Plasma membrane with glycocalyx

(A) The glycocalyx of an animal cell is composed of oligosaccharides (branching carbohydrates) attached to some of the protein and lipid molecules of the outer surface of the membrane. (B) In this electron micrograph the glycocalyx of a red blood corpuscle gives the outer surface of the membrane a fuzzy appearance.

HOW CELLS STICK TOGETHER

Having completed our examination of the structure and operation of individual cell membranes, we need to look briefly at how cells can bind to one another to create multicellular organisms like ourselves.

■ LIMITS TO CELL SIZE

There are enormous benefits to being multicellular. Larger size enhances an organism's ability to capture or harvest smaller organisms efficiently, to move farther and faster, and so on. However, large size cannot be achieved by simply increasing the size of a single cell indefinitely. A cell must take in its nutrients and oxygen across the membrane. As a cell triples in volume, so does its need for nutrients and oxygen, and yet its membrane surface area does not even double. Since metabolic needs increase faster than the surface area through which they are satisfied, a point arrives at which the membrane can no longer support its contents. The need for efficient diffusion, therefore, puts a strict limit on the surface-to-volume ratio of a cell and consequently limits cell size.

Many single-cell organisms are extraordinarily complex, for the one cell must do everything needed for survival. Although simple aggregations of identical cells do exist in nature, such as the 32-cell disks of some green algae and the amoeboid clusters of slime molds, in most assemblies of cells more sophisticated specialization becomes possible. The course of evolution has demonstrated that arrangements in which certain cells concentrate on particular functions (propulsion, feeding, reproduction, and so on) can be far more effective than those in which each cell pursues a jack-of-all-trades strategy.

■ ALTERNATIVE WAYS TO CONNECT CELLS

There are two dramatically different strategies for giving form to multicellular aggregations, which would otherwise be amorphous lumps of cells. In certain animal tissues specialized cells called ***fibroblasts*** secrete fibrous proteins, among them collagen (see Fig. 3.27, p. 64) and elastin, which are components of an ***intercellular matrix***. Cells are held in place by this structural network, and they grow and function there.

The other strategy for providing shape and strength is for cells to adhere to one another. The cells must specifically recognize which other cells are appropriate partners and then secure their membranes to them. This recognition, which is especially critical during embryological development, remains largely a mystery. At least certain types of cells have characteristic carbohydrate markers on the lipids and proteins of their outer membranes that are recognized by specialized receptors on other cells. Other cells have special glycoproteins, called cell-adhesion molecules, which bind directly to one another. A curious class of plant proteins, called lectins, recognize cells of specific tissue types on the basis of their carbohydrate identification tags. The role of lectins is unclear; apparently they help many plant cells adhere to one another, but there is also evidence that they glue together, thereby immobilizing the bacterial and fungal cells that otherwise cause plant diseases.

However cells locate each other, they frequently form strong junctions of several sorts. Most of the cellular attachment junctions utilized in multicellular organisms other than plants can be seen in the lining of the small intestine (Fig. 4.35). Intestinal cells are arranged with microvilli projecting into the intestine, where they absorb the nutrients released by digestion. They must adhere to one another not only to form the tubular channel of the gut but also to prevent the digestive enzymes (against which they are specifically armored) from leaking out and digesting the rest of the organism.

The general mechanical joining of two cells is accomplished by structures known as ***spot desmosomes***. A spot desmosome consists of two cytoplasmic plaques, one just inside each of two adjacent cells (Fig. 4.35B). The outside faces of the plaques are joined by intercellular filaments, acting like rivets, while the inside faces are firmly attached to the cellular cytoskeleton, which will be described in the next chapter. There is a superficial resemblance between spot desmosomes and ***belt desmosomes***, which are also composed of plaques and filaments (Fig. 4.35). Belt desmosomes, however, have no role in cell-to-cell adhesion. Instead, the circumferential plaque containing contractile fibers provides internal support for the cell.

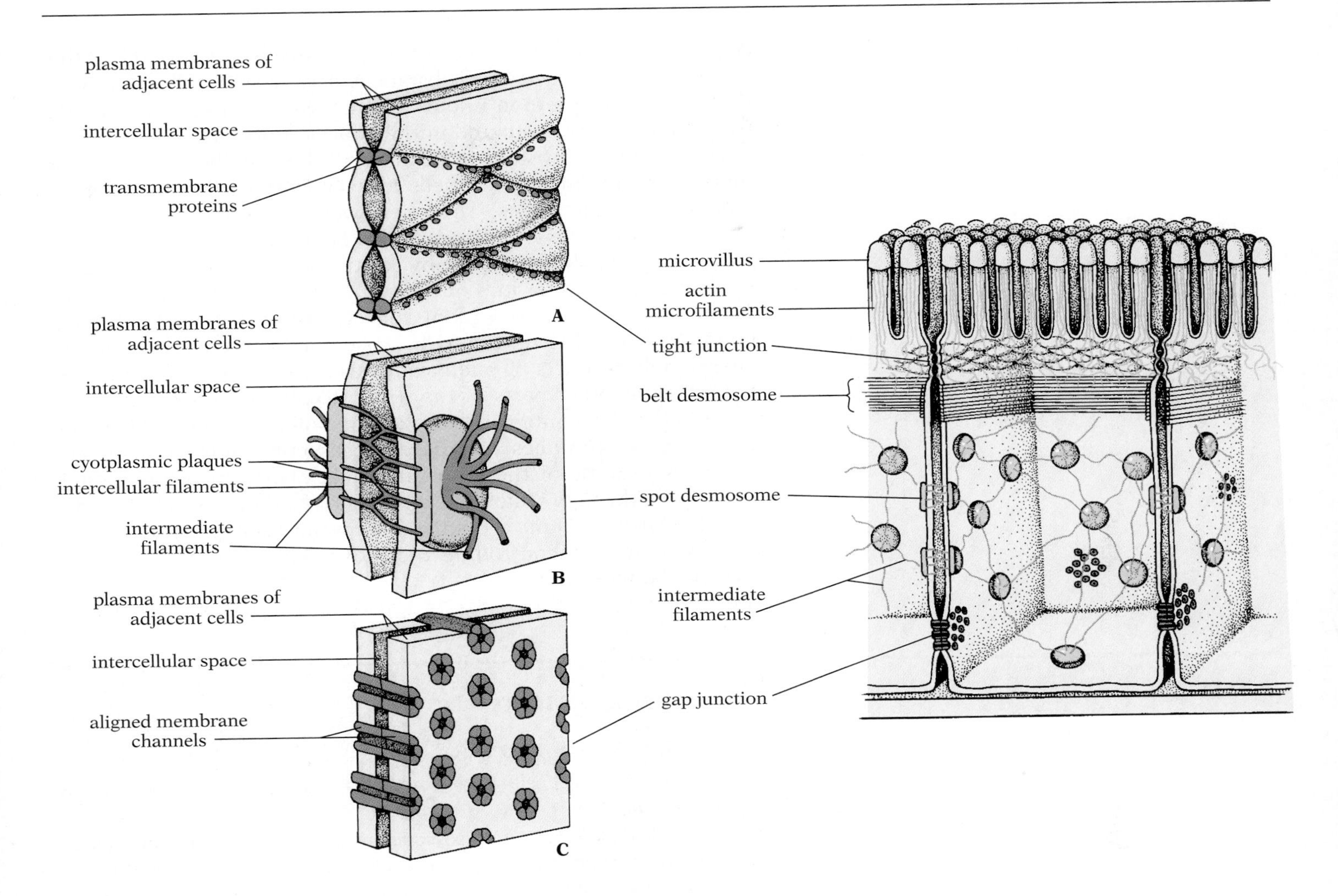

4.35 Varieties of cellular junctions

The cells lining the mammalian small intestine are attached by several types of specialized junctions. An example of each is shown in enlarged detail. (A) A tight junction is composed of rows of transmembrane proteins in adjacent cells that bind to each other. (B) A spot desmosome consists of a pair of cytoplasmic plaques, each just inside the cell membrane and connected to the other across the intermembrane space by specialized intercellular filaments. Each plaque is also attached to fibers of the cytoskeleton within its cell. (C) A gap junction is formed by a pair of membrane channels aligned and bound together to create a specialized pathway between cells. The chemical structure of intermediate filaments and actin microfilaments is described in Chapter 5.

Cells in the intestine are also joined by ***tight junctions***, in which specific transmembrane proteins attach directly to their counterparts in the adjacent cell (Fig. 4.35A). The cells are thus drawn together in such intimate association that there is no intercellular space and hence no possibility of leakage.

Finally, cells may be connected by ***gap junctions***. A junction of this type is apparently formed by a pair of identical membrane channels in the two cells that line up with and bind to each other

(Fig. 4.35C). The result is both mechanical strength and the ability to share certain particular substances between the cells. Gap junctions are most common in developing tissue, and they may play a role in the arrangement and initial adhesion of cells.

The problems involved in achieving cellular adhesion and communication for most plant cells are very different from those for animal cells. In plants, rigid cell walls intervene between the plasma membranes of adjacent cells. Hence adhesion must be accomplished for the most part by relatively simple cross-linking between the polysaccharides of the cell walls. Effective adhesion and communication between plant cells is crucial, however, if water and inorganic nutrients are to be passed up from the roots, and the energy-rich products of photosynthesis are to be transmitted from the leaves to other parts of the plant. To accommodate this need, plant cell walls contain specialized openings—***plasmodesmata***—where the membranes of adjacent cells come into contact with each other. Some of these openings take the form of membrane-lined holes through which the cytoplasm of adjoining cells can mix directly. Others retain a double-membrane barrier; these play an important role in controlling the movement of both solutes and solvents between cells.

CHAPTER SUMMARY

THE DISCOVERY OF CELLS

THE CELL THEORY All cells arise from other cells; there is no spontaneous generation. (p. 85)

HOW CELLS ARE STUDIED Microscopes magnify objects and help resolve adjacent structures; contrast, often supplied by stains, helps distinguish objects from each other. (p. 86)

HOW CELLS EXPLOIT DIFFUSION

DIFFUSION AS A CHEMICAL REACTION Diffusion is the tendency of a substance to move from a region of high concentration to one of lower concentration; like ordinary reactions, it has an equilibrium and an associated free energy. (p. 90)

MANIPULATING DIFFUSION: OSMOSIS Osmosis occurs when a semipermeable membrane allows some but not all substances in a solution to diffuse across. (p. 92)

BETWEEN SWELLING AND COLLAPSE Cells in a hypotonic medium—one with less osmotically active solutes—tend to gain water and swell, while those in a hypertonic medium tend to lose water and collapse. (p. 96)

HOW THE CELL MEMBRANE WORKS

LIPID BILAYERS: MEMBRANES HELD TOGETHER BY WATER The phospholipids of the membrane form a bilayer with their hydrophilic ends exposed to the water of the cytoplasm and surrounding fluids, and their hydrophobic ends hidden together in the interior of the membrane. (p. 98)

HOLES IN THE MEMBRANE: CHANNELS AND PUMPS Channels are openings in the membrane that pass specific substances. Facilitated-diffusion channels simply pass their particular substance at a controlled rate; gated channels open in response to chemical or electrical stimuli. Antiports are channels that trade substances going in opposite directions; symports are channels that pass two substances the same direction cooperatively. Pumps use chemical energy to move substances against their osmotic gradient. (p. 103)

LARGE-SCALE MOVEMENT: ENDOCYTOSIS AND EXOCYTOSIS Large quantities of material can be brought into cells by trapping it endocytotically in vesicles; material can be expelled from cells through vesicle-mediated exocytosis. (p. 107)

OUTSIDE THE MEMBRANE: CELL WALLS AND COATS

HOW CELL WALLS ARE BUILT Cell walls, found in bacteria, some protists, fungi, and plants, can provide enough rigidity to keep cells from bursting in hypotonic media. (p. 113)

WHAT THE CELL COAT DOES The cell coat consists of membrane-mounted carbohydrates that help in cell adhesion and recognition. (p. 114)

HOW CELLS STICK TOGETHER

LIMITS TO CELL SIZE Cells cannot grow indefinitely since larger cells have proportionally less surface area, and thus have access to smaller relative quantities of nutrients and can dispose of proportionally less waste material. (p. 115)

ALTERNATIVE WAYS TO CONNECT CELLS Cells can be attached by a variety of junctions that hold them in place; some junctions permit adjacent cells to share contents or information. (p. 116)

STUDY QUESTIONS

1 How do water, saturated and unsaturated phospholipid tails, and cholesterol contribute to membrane stability and flexibility? (pp. 98–103)

2 Why does the sodium-potassium pump require chemical energy from the cell? Why does the sodium-glucose symport not require any? (pp. 103–107)

3 List five different cross-membrane transport strategies, and explain how each one works. (pp. 103–112)

4 What are the costs and benefits of the plant strategy of using cell walls to stabilize membranes? (pp. 113–114)

5 What changes in conditions would serve to increase the rate of diffusion of an ionic substance? (The list is not short.) (pp. 90–92)

6 Explain osmosis in terms of probability at the molecular level, and then in terms of thermodynamics. (pp. 92–96)

SUGGESTED READING

BRETSCHER, M. S., 1985. The molecules of the cell membrane, *Scientific American* 253 (4). *Reviews the bilayer plasma membrane and membrane proteins and the process of endocytosis.*

BROWN, M. S., AND J. L. GOLDSTEIN, 1984. How LDL receptors influence cholesterol and atherosclerosis, *Scientific American* 251 (5).

DAUTRY-VARSAT, A., AND H. F. LODISH, 1984. How receptors bring proteins and particles into cells, *Scientific American* 250 (5). *The life cycle of coated pits.*

GOODSELL, D. 1992. A look inside the living cell. *American Scientist* 80 (5). *What the interior of a cell might look like if we could see the molecules themselves.*

SHARON, N., AND H. LIS, 1993. Carbohydrates in cell recognition. *Scientific American* 268 (1). *On the cell-surface markers that play a role in cell adhesion, development, and the immune response.*

STOSSEL, T. P. 1994. The machinery of cell crawling. *Scientific American* 271 (3). *On how cells move in amoeboid fashion.*

TAYLOR, D. L., M NEDERLOF, F. LANNI, AND A. S. WAGGONER, 1992. New vision of light microscopy. *American Scientist* 80 (4). *On modern methods of creating contrast.*

UNWIN, N., AND R. HENDERSON, 1984. The structure of proteins in biological membranes, *Scientific American* 250 (2).

CHAPTER 5

INSIDE THE CELL

The cell membrane creates a general-purpose chemical environment favorable to life—a comfortable organic home in a hostile world. A human home may have, in addition to a heating and cooling system to maintain a suitable general temperature, an oven for tasks that require unusually high temperatures, a refrigerator with freezer for low-temperature needs, and other specialized tools and appliances to help it operate more efficiently. Cells too need special structures to create the unusual internal conditions necessary for particular tasks, as well as a set of tailor-made tools. Many of these subcellular entities are enclosed in their own membranes, and are called ***organelles***. The term is also applied to the more complex of the other subcellular units. This chapter examines these specialized cellular components.

THE ROLES OF SUBCELLULAR COMPARTMENTS

■ THE NUCLEUS: THE CELL'S CONTROL CENTER

The largest internal compartment in most organisms (it is absent in bacteria) is the membrane-bounded ***nucleus*** (Fig. 5.1). The nucleus is the control center of the cell; its genetic material issues the "instructions" that guide the life processes of the cell.

Though bacteria too possess genetic material that controls the cell's activities, they lack a membrane-bounded nucleus. Similarly, this group has few of the subcellular structures found in other organisms. These differences are so fundamental that bacteria are classified into separate kingdoms of their own. They are called ***procaryotes*** ("before any nucleus"), whereas all other organisms are ***eucaryotes*** ("having a true nucleus"). The characteristics of procaryotic cells will be discussed later.

The eucaryotic nucleus contains two large structures, the chromosomes and the nucleolus. Both are embedded in a matrix of filaments that seem to provide structure and paths along which processing and transport can occur (Fig. 5.2). The entire nucleus is bounded by a closely associated pair of membranes called the ***nuclear envelope***.

The ***chromosomes*** (Fig. 5.3) are long, threadlike bodies clearly visible only when they condense in preparation for cell division; at

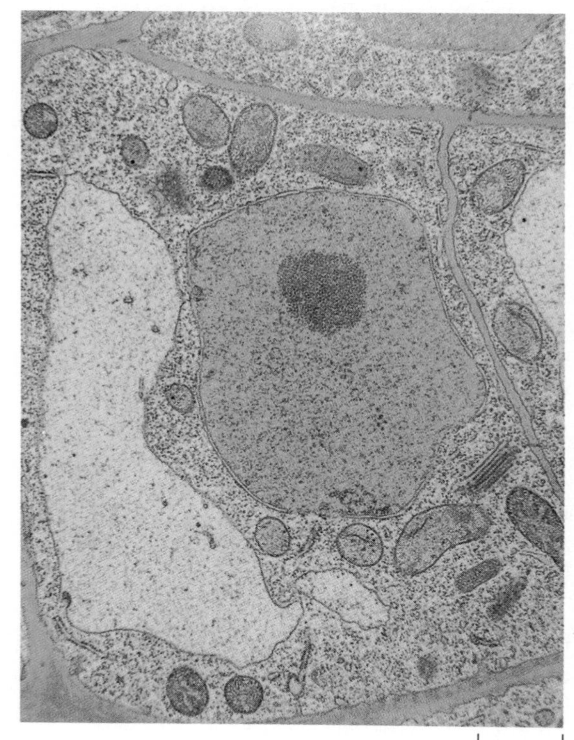

5.1 A plant cell nucleus

The nucleus of most cells is about one-third the diameter of the whole cell, thus occupying 3–4% of cellular volume. In this case, the nucleus is the large orange body at the center, with the darker nucleolus embedded in it.

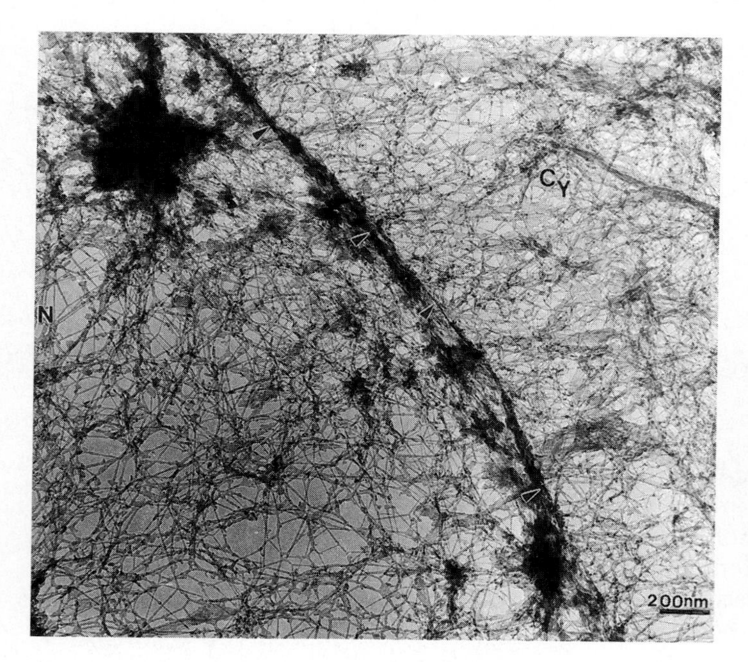

5.2 Internal matrix of the nucleus

The nucleus contains a meshwork of fibers along which molecules are transported.

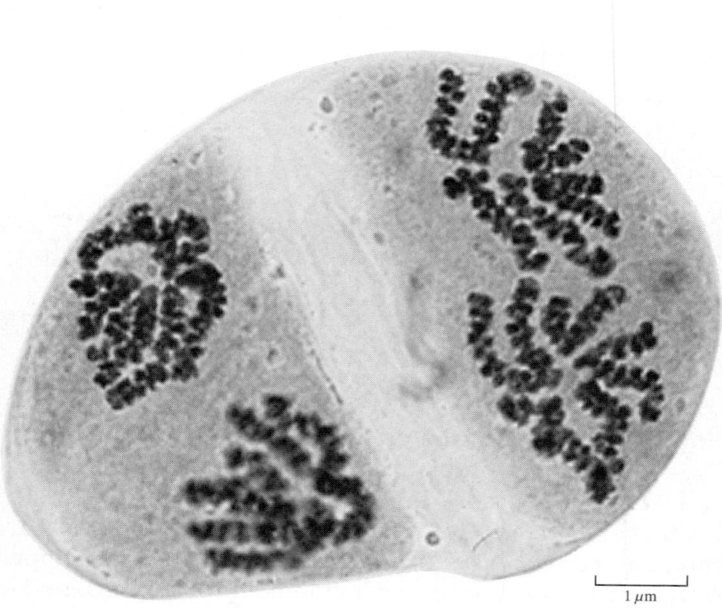

5.3 Chromosomes in a dividing cell of *Trillium*

The separate chromosomes in the two nuclei of this spring-flowering plant can easily be distinguished.

all other times the chromosomes are in an uncondensed conformation that, when stained, appears as a dark, hazy material called chromatin. The chromosomes are composed of DNA and protein; the DNA is the substance of the basic units of heredity, called ***genes***, while the protein provides spool-like supports, or cores, on which the DNA is wound to form ***nucleosomes*** (Fig. 5.4). The genes are duplicated whenever a cell divides, and a copy is passed to each new cell. The genes determine the characteristics of cells and control the day-to-day activities of living cells.

The hereditary information carried by the genes is written in the sequence of the nucleotide building blocks of the DNA molecules. The genes remain in the nucleus, while most of the processes they control take place in the cytoplasm. The mechanism for conveying the genetic information from the nucleus to the cytoplasm is transcription (detailed in Chapter 9): the nucleotide sequence in the DNA gives rise to a corresponding nucleotide sequence in RNA. This RNA sequence apparently moves along a nuclear filament on which it is modified into messenger RNA (mRNA), which leaves the nucleus and moves to the sites of protein synthesis in the cytoplasm. There amino acids are linked by peptide bonds in a sequence corresponding to that of the mRNA nucleotides to form proteins, including enzymes (Fig. 5.5). This process is known as translation (discussed in Chapter 9). The sequence of amino acids—the primary structure of a protein—determines the three-dimensional conformation of the protein and the biological activity that this conformation bestows. The genes encode all the information necessary for the synthesis of the enzymes regulating the myr-

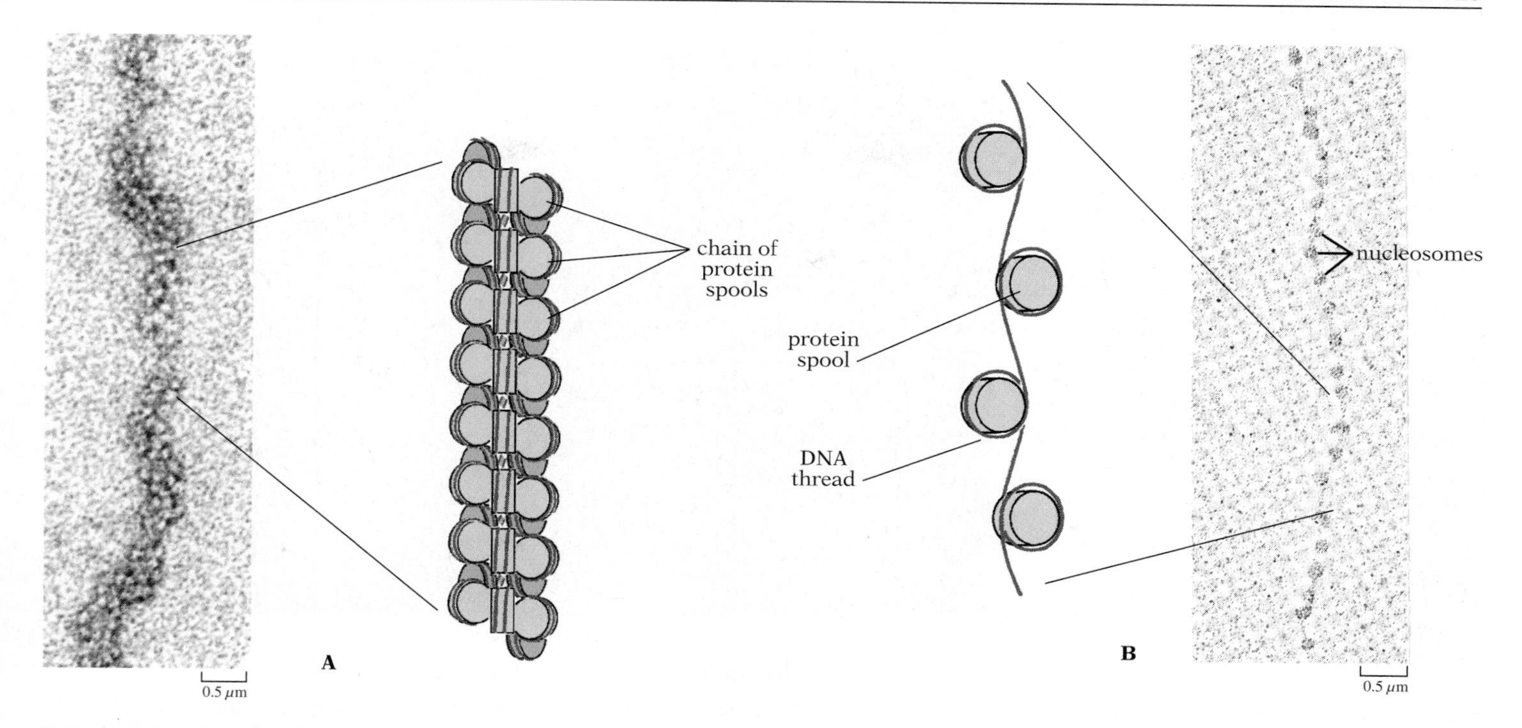

5.4 DNA on nucleosomes

The chromosomal DNA of eucaryotes is wound on protein spools, or cores, to form structures called nucleosomes. Normally the spools adhere to one another in a regular way, giving the DNA the appearance of a piece of yarn (A). When the DNA is treated to break the connections between nucleosomes, the individual protein spools and the thread of DNA can be seen (B).

5.5 Flow of information from nucleus to cytoplasm

Information encoded in a region of the chromosome (a gene) is transcribed by an enzyme complex to create an RNA copy. This messenger RNA is then exported to the cytoplasm, where it is decoded by a ribosome and the protein specified by the gene is synthesized.

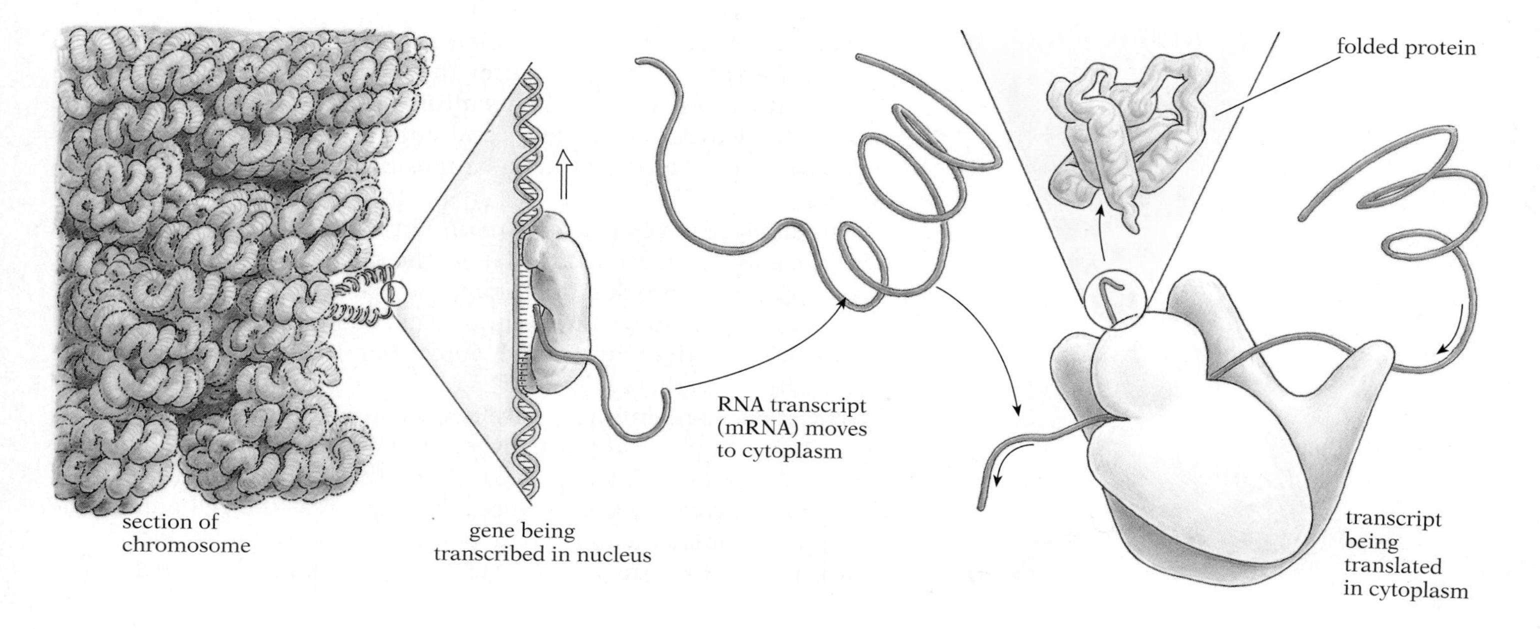

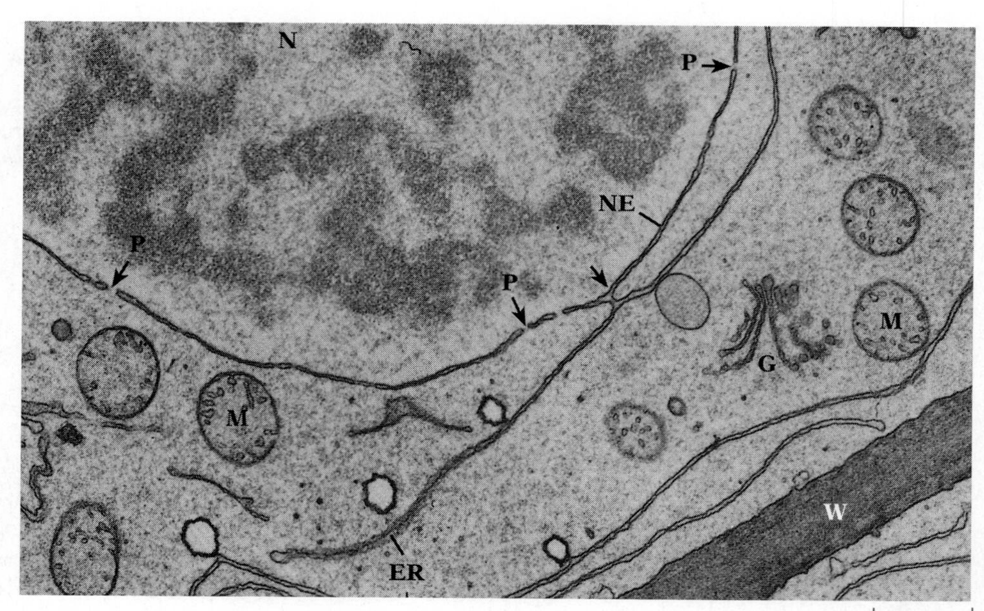

5.6 The nuclear envelope of a corn root cell

The large structure filling the upper left quarter of this electron micrograph is the nucleus. The unlabeled arrow indicates a point where the endoplasmic reticulum and the double nuclear membrane interconnect. ER, endoplasmic reticulum; G, Golgi apparatus; M, mitochondrion; N, nucleus; NE, nuclear envelope; P, pore in nuclear envelope; W, cell wall.

iad interdependent chemical reactions that determine the characteristics of cells and organisms.

The other structure in the nucleus is the ***nucleolus***, one or more dark-staining, generally oval areas usually visible within the nuclei of nondividing cells. Nucleoli are, in fact, simply specialized parts of the chromosome; like the rest of the chromosome, they are composed of DNA and protein. The DNA of the nucleoli includes multiple copies of the genes from which a type of RNA called ribosomal RNA (rRNA) is transcribed. After the rRNA is synthesized, it combines with proteins, and the resulting complex detaches from the nucleolus, leaves the nucleus, and enters the cytoplasm, where it becomes a part of the protein-synthesizing complexes called ***ribosomes***.

The nuclear envelope helps maintain a chemical environment in the nucleus different from that of the surrounding cytoplasm. It also appears to provide anchorage for the two ends of each chromosome. Unlike the cell membrane, the complete nuclear envelope consists of distinct inner and outer membranes, with space enclosed between them (Figs. 5.1, 5.6).

This double-membrane envelope is interrupted by fairly large and elaborate pores at points where the outer and inner membrane are continuous (Figs. 5.6–5.8). Nevertheless, the membrane is highly selective: many substances that can cross the cell membrane into the cytoplasm cannot pass through the nuclear envelope into the nucleus. Even molecules much smaller than the pores fail to

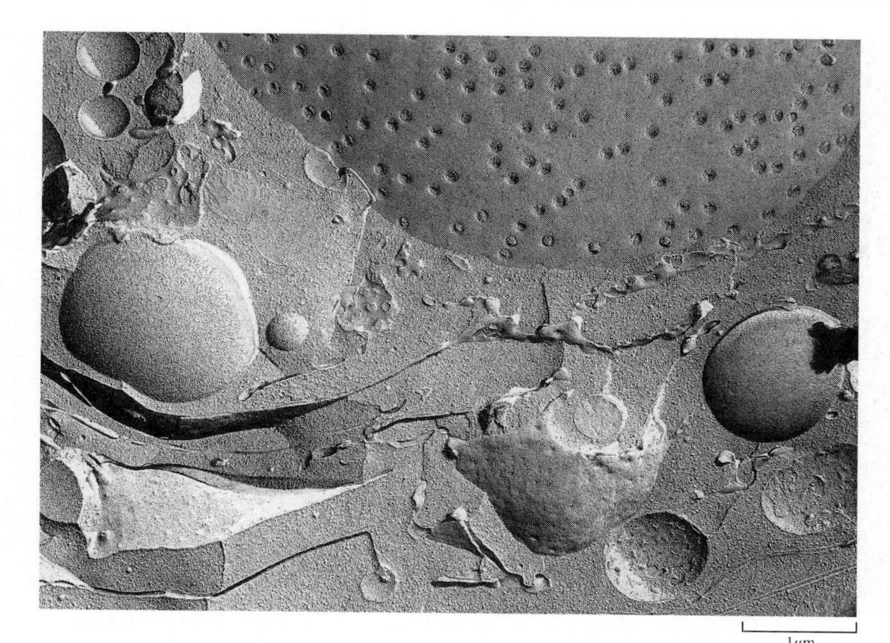

5.7 Freeze-fractured cell from the tip of an onion root

At upper right of this electron micrograph is the surface of the nuclear envelope, with numerous pores; a typical nucleus has a few thousand pores. A variety of vesicles can be seen in the cytoplasm.

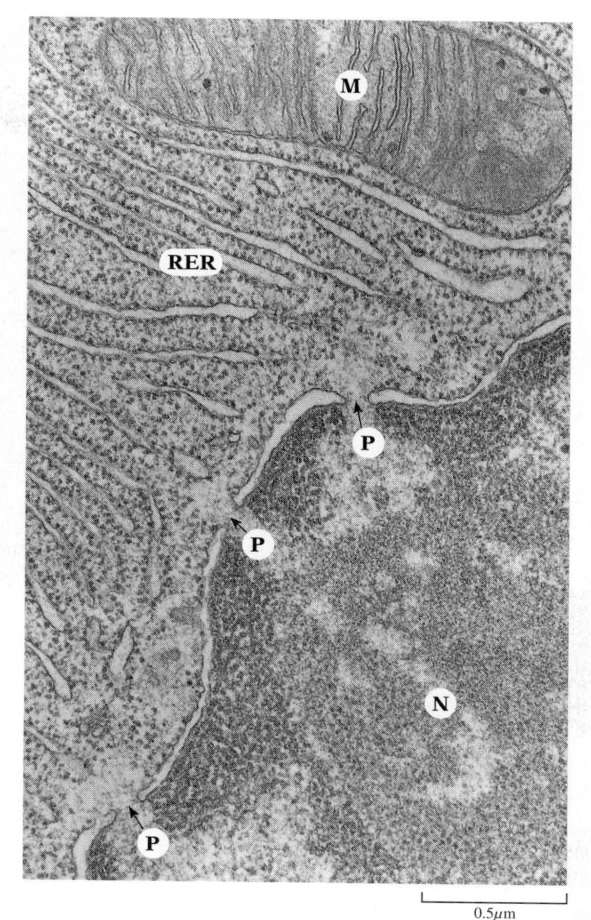

5.8 The rough endoplasmic reticulum

This electron micrograph of a thin section of a pancreatic cell from a bat shows many flattened cisternae of rough ER; the ribosomes lining the RER membranes can be clearly distinguished. In the lower right portion of the micrograph is part of the nucleus (N); note the very prominent pores (P) in the double membrane of the nuclear envelope. A mitochondrion (M) is at the top.

move through these openings, yet some large molecules pass readily. These macromolecules are primarily substances produced on the genes (such as mRNA) that are moving out of the nucleus, proteins moving into the nucleus to be incorporated into nuclear structures or to catalyze chemical reactions in the nucleus, and various substances from the cytoplasm that move into the nucleus and help regulate gene activity.

This highly selective and directional exchange through the nuclear pores between the nucleus and the cytoplasm is based on "passwords," chemical signal sequences attached to one end of the various molecules that tell the pores whether to allow the larger molecule through, and if so in which direction. Certain viruses (including HIV, the AIDS virus) have broken the molecular code used by nuclear pores; they are able to smuggle in a copy of their chromosome, which can then be incorporated into the cell's DNA library.

The nuclear envelope has another interesting elaboration: it is directly connected to an extensive cytoplasmic membrane system called the endoplasmic reticulum (Fig. 5.6).

■ THE ENDOPLASMIC RETICULUM: SYNTHESIS AND SORTING

Attempts to separate subcellular organelles from the cytoplasm often make use of centrifugation. Cells are first lysed (broken) by detergents that disrupt the lipids in the membrane, and then they are layered on top of a test tube containing a viscous solution of, say, glucose; finally, the samples are spun in a centrifuge to sort out the cell parts by weight. Each component travels through the solution at a characteristic rate, the densest moving most rapidly.

Using these techniques, Albert Claude isolated some unusual cytoplasmic components in 1938. These "microsomes" (as he called them) made up 15–20% of the total cell mass and had a very high nucleic acid content. They also contained a high percentage of the cytoplasmic phospholipids. In 1945 Keith R. Porter found that the microsomes were the fragments of a complex network of membranes in the cytoplasm plus associated ribosomes. This system, which Porter named the ***endoplasmic reticulum (ER)*** (*reticulum* being Latin for "network"), exists in large part to synthesize new membrane phospholipids, to package proteins into vesicles for transport to other parts of the cell, and to store calcium ions.

There are two anatomically distinct forms of endoplasmic reticulum, each with a different role in synthesis and packaging. One, the ***rough endoplasmic reticulum (RER)*** exists as flattened, fluid-filled, membrane-enclosed sacs called ***cisternae***. The RER probably forms a series of interconnected sheets, so that all the cisternae interconnect. The rough appearance of the RER membrane arises from its association with the numerous protein-synthesizing ribosomes (Fig. 5.8). Ribosomes also bind to the outer membrane of the nuclear envelope, to which the RER is connected.

The association of the ribosomes with the RER membrane appears to be necessary for the proteins they produce to penetrate the membrane to the cisternae or to become embedded in the membrane. Most or all of the proteins to be packaged and transported by the ER are synthesized by the ribosomes of the RER. Proteins synthesized on free ribosomes in the cytoplasm are apparently released to function as enzymes in the ***cytosol*** (the more fluid part of the cytoplasm). The mRNA from the nucleus that is to produce enzymes for the ER binds first to free ribosomes. The mRNA, however, signals the ribosome (with the help of signal-recognition particles) to bind to special channels in the RER. Once the ribosome binds, the translation begins and the enzyme being synthesized on the ribosome is threaded through the channels into the lumen (the interior) of the RER.

As soon as a ***signal sequence*** in this mRNA is synthesized as part of the growing polypeptide chain, and this sequence is threaded into the lumen of the ER, enzymes in the RER membrane attach a small sugar complex (an oligosaccharide) that will aid in subsequent sorting (Fig. 5.9): some proteins remain in the ER, whereas others are exported to a variety of sites. Modifications of this molecular "mailing label" specify where the protein is to go. Errors in the tagging system can have serious consequences.

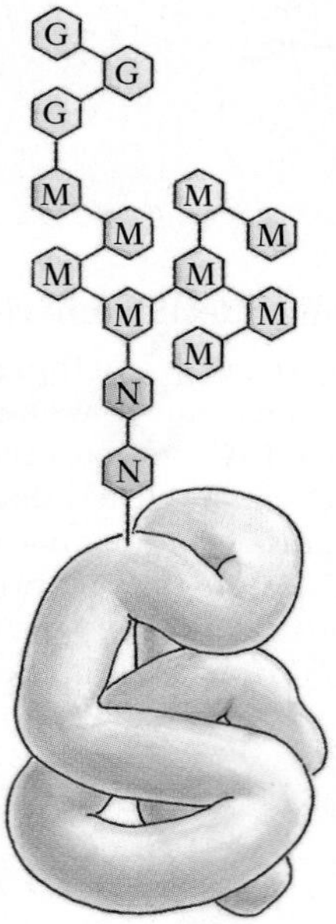

Beyond its importance in intracellular transport, the RER functions as a cytoplasmic framework, providing catalytic surfaces for some of the biochemical activity of the cell. Its complex folding provides an enormous surface for such activity.

The other form of endoplasmic reticulum, the ***smooth endoplasmic reticulum (SER)***, not only lacks ribosomes but also has a much more tubular appearance (Fig. 5.10) and a very different set of characteristic membrane proteins. The most obvious functions

5.9 Oligosaccharide mailing label attached to proteins in the ER

A 14-sugar side chain is attached to nearly all proteins synthesized on the RER, and serves as a "mailing label." Proteins lacking this label remain in the RER. When the four terminal sugars are removed, the protein is exported to the Golgi apparatus in vesicles. The size of the tag in this drawing is exaggerated compared with the protein. (G, glucose; M, mannose; N, *N*-acetylglucosamine)

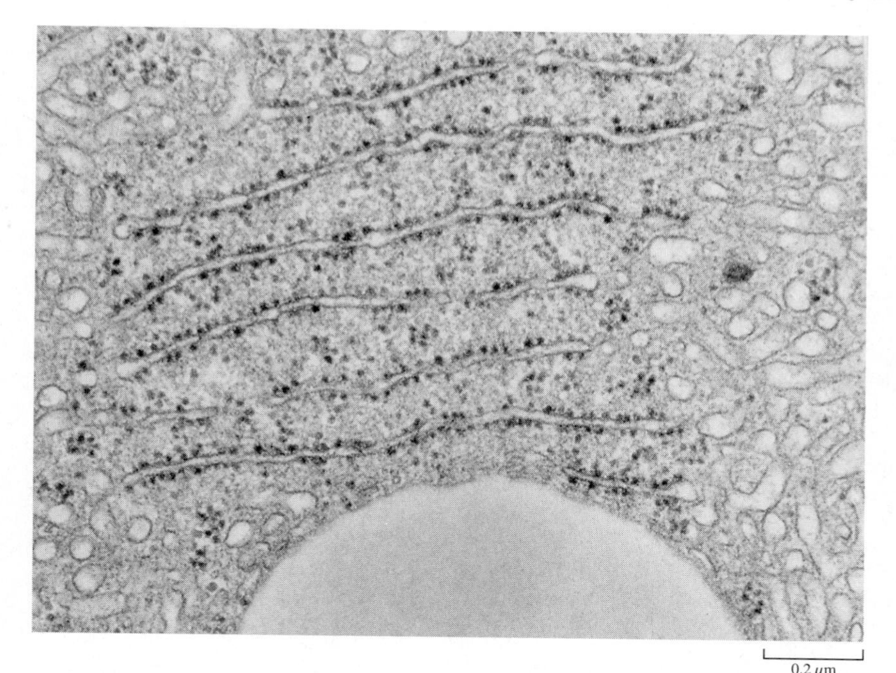

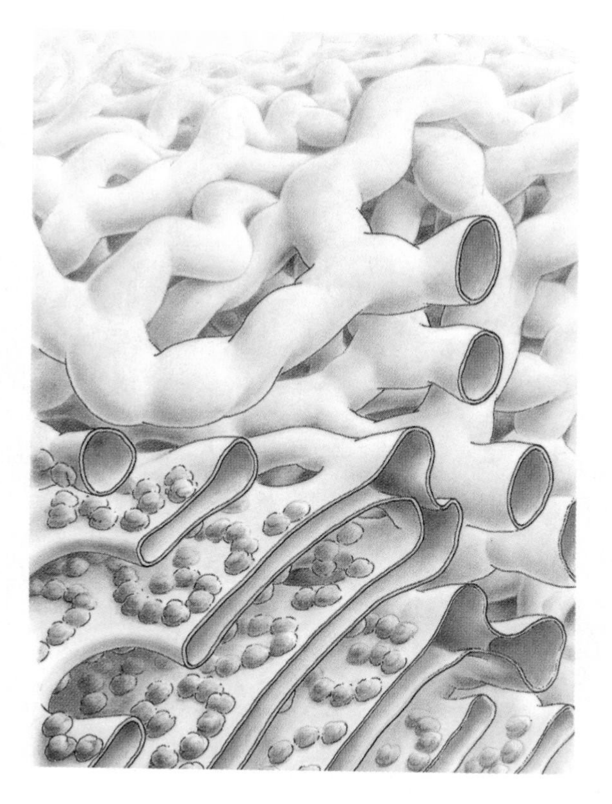

5.10 Rough and smooth endoplasmic reticulum

The endoplasmic reticulum from a steroid-producing cell of a guinea pig testis, shown in this electron micrograph, consists of a complex system of membranes including the RER, with associated ribosomes, and SER. The relationship between RER and SER varies from cell to cell; the sketch shows an association more typical of a cell in which macromolecules are synthesized in the RER and transported to the SER. This representation assumes that the RER and SER are physically continuous; they may actually be separate organelles that communicate via vesicles.

of the SER are to synthesize membrane phospholipids and to package certain proteins in the cisternae into membrane-bounded vesicles for transport to other locations in the cell. The vesicles also carry membrane-mounted proteins with them. Though most of the proteins in the cisternae and the SER membrane are synthesized first in the RER, it is not yet clear how they get to the SER. It may be that the RER and SER are continuous, as was once universally assumed; alternatively, the RER may send membrane-embedded proteins and cisternal proteins to the SER via vesicles.

Only liver, muscle, and hormone-secreting cells have large amounts of SER. The SER of liver cells holds enzymes that detoxify many poisons, including barbiturates, amphetamines, and morphine. Poisons that appear in the bloodstream are transported to the liver. There the rapid synthesis of more phospholipids can cause the surface area of the liver cells' SER to double. The positioning of detoxification enzymes in the ER takes maximum advantage of the cell's potential for compartmentalization to protect the cytosol from the poisons. The products of the detoxification process are probably packaged in vesicles that have been formed from the ER itself, and are then transported to other organelles, where they are further broken down.

The SER also stores and releases calcium ions, whose concentration, as we will see, is important in controlling the building or disassembly of elements of the cytoskeleton. The SER in muscles holds enormous quantities of calcium ions, which are used to coordinate contraction. The relative sizes of some of the subcellular components discussed so far are shown in Figure 5.11.

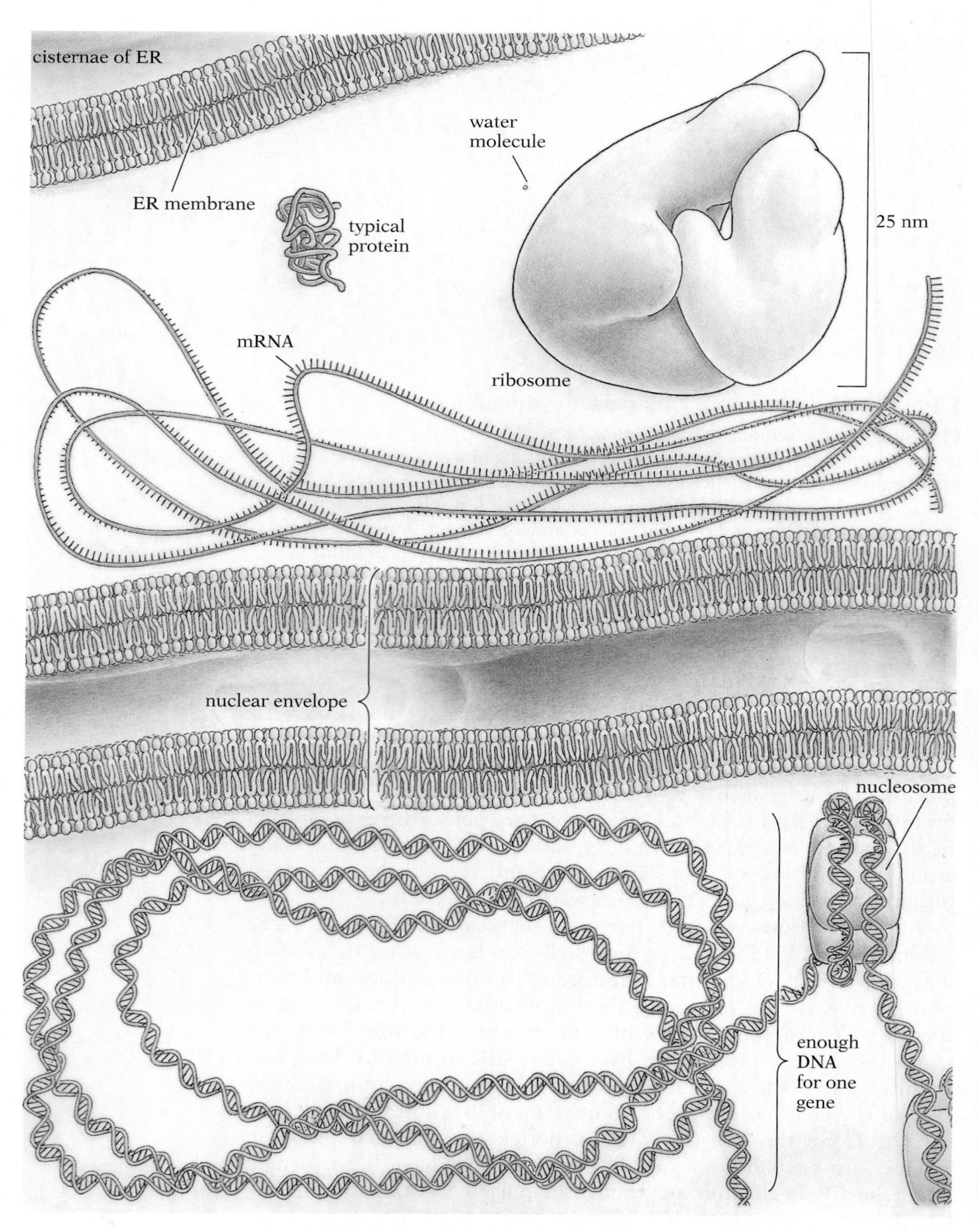

5.11 Relative sizes of subcellular components

■ THE GOLGI APPARATUS: SORTING AND PACKAGING

In 1898 Camillo Golgi discovered a different system of membrane-delimited cellular compartments, arranged approximately parallel to one another; this system is now called the ***Golgi apparatus*** (Fig. 5.12).

The Golgi apparatus is particularly prominent in cells involved in the secretion of chemical products; as the level of secretory activity of these cells changes, corresponding changes occur in the morphology of the organelle. For example, in certain cells of the pancreas of guinea pigs, a zymogen (the inactive precursor of an enzyme) synthesized on the ribosomes moves into the channels of the ER; it reaches the Golgi apparatus in vesicles that have budded off from the ER. The zymogen arrives at the compartment nearest the nucleus (the most highly curved cisterna in Figure 5.12, known as the *cis* compartment), and then moves from one layer to the next until it reaches the farthest layer (the *trans* compartment). Movement between layers is by means of vesicles that bud off from one cisterna and then fuse with another, farther from the nucleus. During this *cis-trans* movement the zymogen is modified, and in the final cisterna it is concentrated and stored; it is eventually released from the cell via secretory vesicles that are produced by this outer compartment of the Golgi and move to the cell surface. In this way the Golgi apparatus stores, modifies, and packages secretory products.

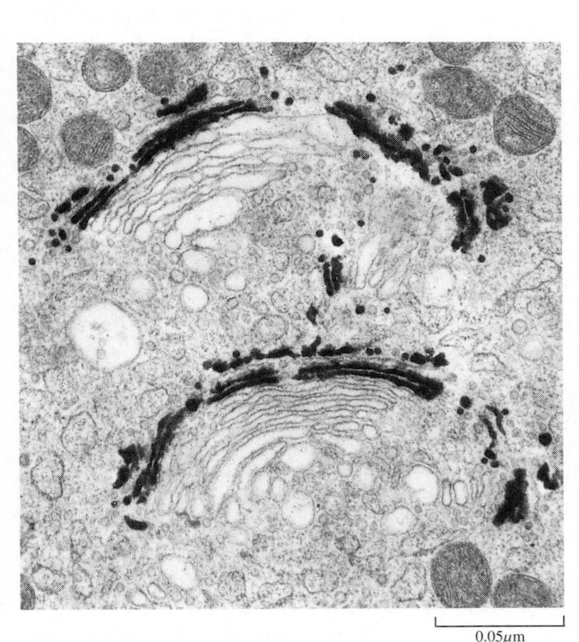

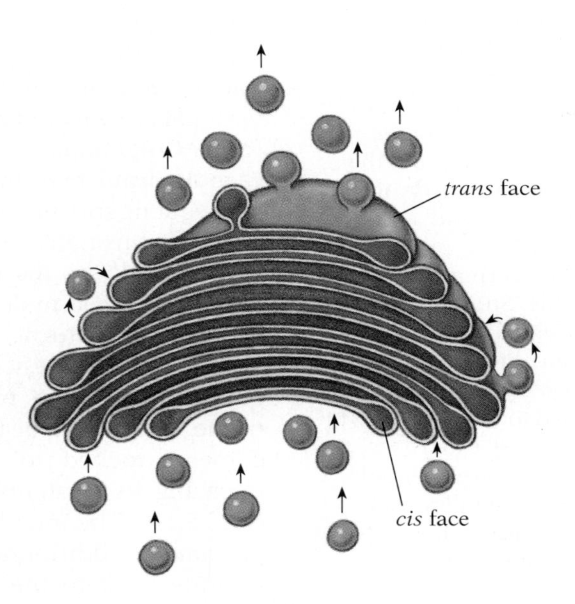

5.12 The Golgi apparatus

Electron micrograph of Golgi apparatus from an amoeba. Vesicles forming at the ends of some of the cisternae can be seen in the EM and in the interpretive drawing. The movement is from the *cis* layer, through the various intermediate compartments, to the *trans* cisterna. *Cis* (meaning "this side of") and *trans* ("other side of") are defined with respect to the nucleus. The Golgi sorts, modifies, relabels, and packages molecules into vesicles for further transport. In this EM, the *trans* compartment and the vesicles they have released are labeled with an electron-dense stain.

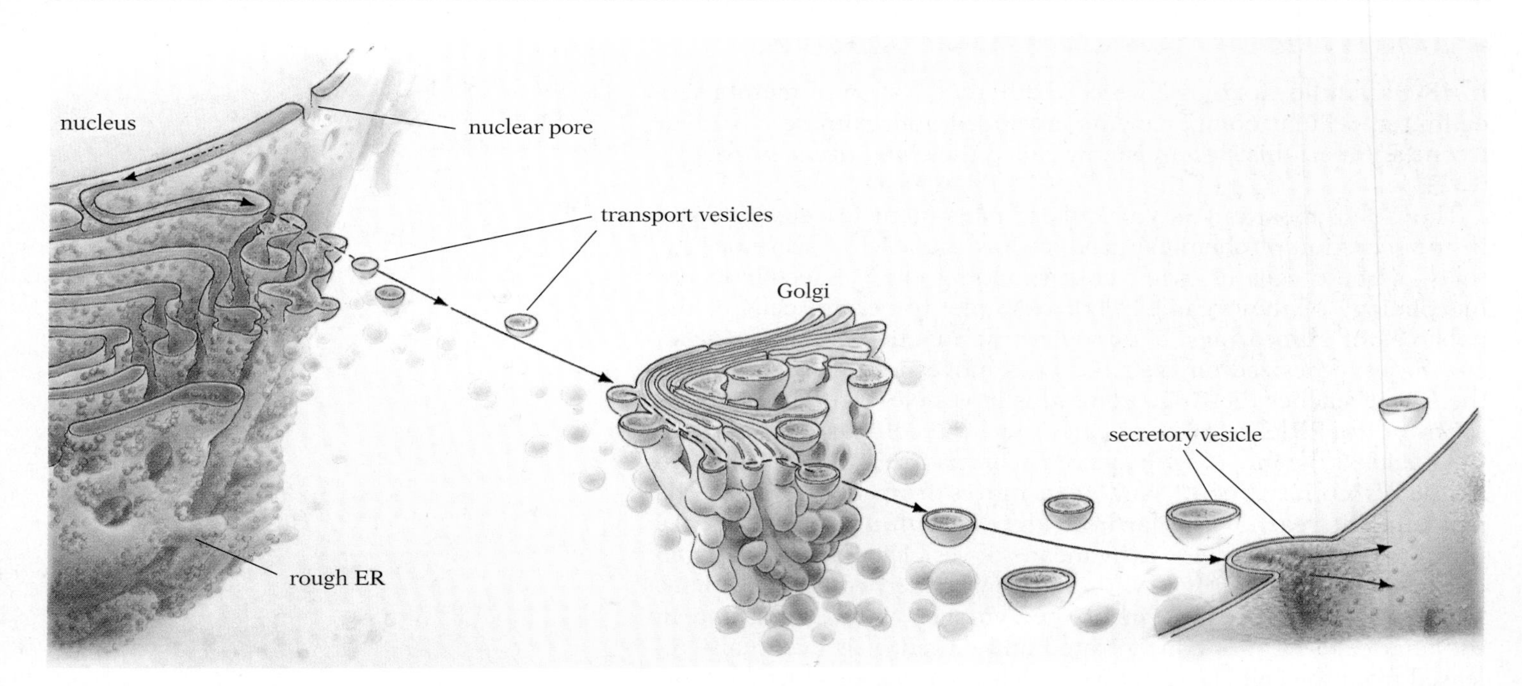

5.13 One path of phospholipid movement in the cell

Structural molecules of cellular membranes are constantly on the move. The path traced here shows the movement of a phospholipid from a point of synthesis in the nuclear envelope until it is incorporated into the cell membrane. (Phospholipids are also synthesized on the ER.) The lipid first moves through the lumen of the RER and (either directly or via vesicles) into the SER. There it receives a specific carbohydrate marker, which makes it a glycolipid, and becomes part of a vesicle transporting proteins to the Golgi apparatus; it is then incorporated into the membrane of that organelle. Subsequently the same molecule, perhaps with a new or modified carbohydrate marker, moves to the outer layer of the Golgi, where it becomes part of a secretory vesicle that is carried to the plasma membrane. There, in the process of exocytosis, the vesicle membrane (including the glycolipid molecule) fuses with the plasma membrane of the cell. Other membrane molecules follow different paths. (Note that the most highly curved Golgi compartment in this illustration is the *trans* cisterna, whereas in Fig. 15.12 it was the *cis* compartment. This serves to emphasize the point that *cis* and *trans* in this flexible organelle are defined by their relative distance from the nucleus.)

The Golgi apparatus is also the major director of macromolecular transport in cells. Though no protein synthesis takes place in the Golgi, polysaccharides are synthesized there from simple sugars and attached to proteins and lipids to create glycolipids and glycoproteins. Some of these are transported to the glycocalyx as part of vesicle membranes. In addition, proteins already marked with carbohydrate groups (true for nearly all proteins transported to the Golgi apparatus from the ER) usually have their sugar-based carbohydrate tags modified there. How the modified tags aid in subsequent sorting and packaging is still a mystery.

Traffic through the Golgi is not entirely one way: some proteins synthesized into the RER get into Golgi-directed vesicles by error; these misplaced molecules are captured in the Golgi, packaged into vesicles, and returned to the sender.

The secretory vesicles produced by the Golgi apparatus probably play an important role in adding surface area to the cell membrane. When one of these vesicles moves to the cell surface, it becomes attached to the plasma membrane and then ruptures, releasing its contents to the exterior in the process of exocytosis (Fig. 5.13). The membrane of the ruptured vesicle may remain as a permanent addition to the plasma membrane, or it may eventually migrate back to the Golgi apparatus or some other organelle as part of an empty vesicle. Recycling of membrane phospholipids back to the Golgi and ER is essential if the outer membrane is not to grow indefinitely.

■ LYSOSOMES: DIGESTIVE ORGANELLES

First described in the 1950s by Christian deDuve, ***lysosomes*** are membrane-enclosed bodies that function as storage vesicles for many powerful digestive (hydrolytic) enzymes (Fig. 5.14). The lysosome membrane contains an ionic pump that maintains a highly

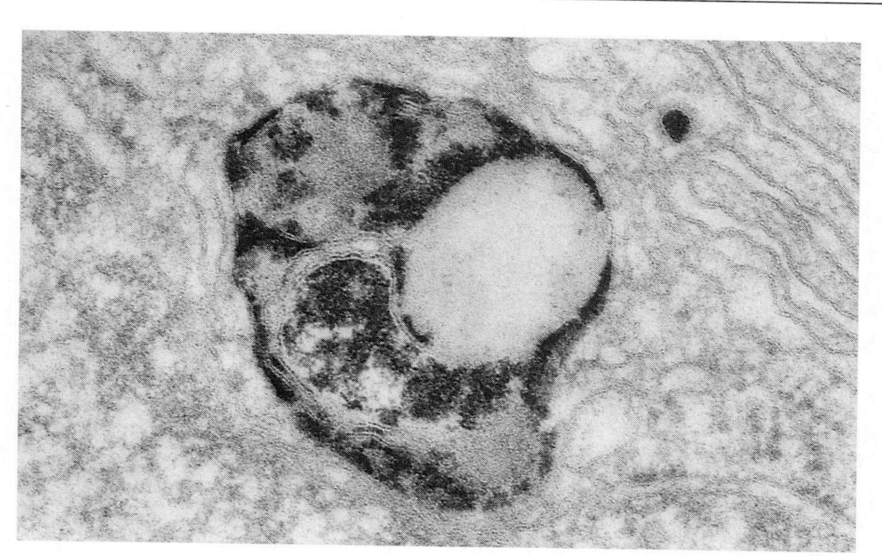

5.14 Lysosomes in a connective-tissue cell from the vas deferens of a rat

The small dark body at upper right is a primary lysosome. The much larger body at left is a secondary lysosome (digestive vacuole) formed by fusion of a primary lysosome with a phagocytic or pinocytic vesicle. (The dark appearance of the lysosomes results from staining for acid phosphatase, a digestive enzyme whose presence is used as the definitive test for these organelles.)

acidic internal environment. The membrane permits desirable reaction products to pass through to the cytosol, but it is impermeable to the hydrolytic enzymes and capable of withstanding their digestive action. If the lysosome membrane is ruptured, the hydrolytic enzymes—no longer safely confined—are released into the surrounding cytoplasm and break down the interior of the cell.

Lysosomes act as the digestive system of the cell, enabling it to process some of the bulk material taken in by endocytosis. Their hydrolytic enzymes are synthesized in the form of zymogens in the RER, packaged into transport vesicles in the SER, and carried to the Golgi apparatus. There, receptor proteins on the inside of the Golgi membrane form coated pits to attract and trap these zymogens, which are recognized by their characteristic carbohydrate markers. When a region has received an appropriate supply of the zymogens, it buds off from the Golgi membrane as a lysosome, and the enzymes are activated. Proteins mounted on the exterior of the lysosome's membrane serve as recognition sites to assure that the enzymes it carries are delivered to the appropriate target. The targets of these primary lysosomes include endocytotic vesicles (endosomes) newly arrived from the cell surface, and secondary lysosomes, also known as digestive vacuoles, which are the digestive vesicles already produced by the fusion of primary lysosomes and endocytotic vesicles. When digestion is complete, the useful products pass into the cytosol, while the residue is discharged by exocytosis, and the vesicle membrane and any receptors it contains are reincorporated into the cell membrane (Fig. 5.15).

Several diseases are caused by lysosome disorders. In human inclusion cell disease, for instance, the enzyme that adds a distinctive sugar to the mailing label is defective; as a result most of the 40 or so hydrolytic enzymes are mismarked for secretion. Undigested "food" builds up in cells, resulting in general debilitation, followed by death when one or more of the body's organs fail. In the devastating nervous-system disorder known as Tay-Sachs disease, lipid-digesting lysosomes lack a particular enzyme. When these deficient lysosomes fuse with lipid-containing vesicles, they do not fully digest their contents. The resulting defective secondary lysosomes

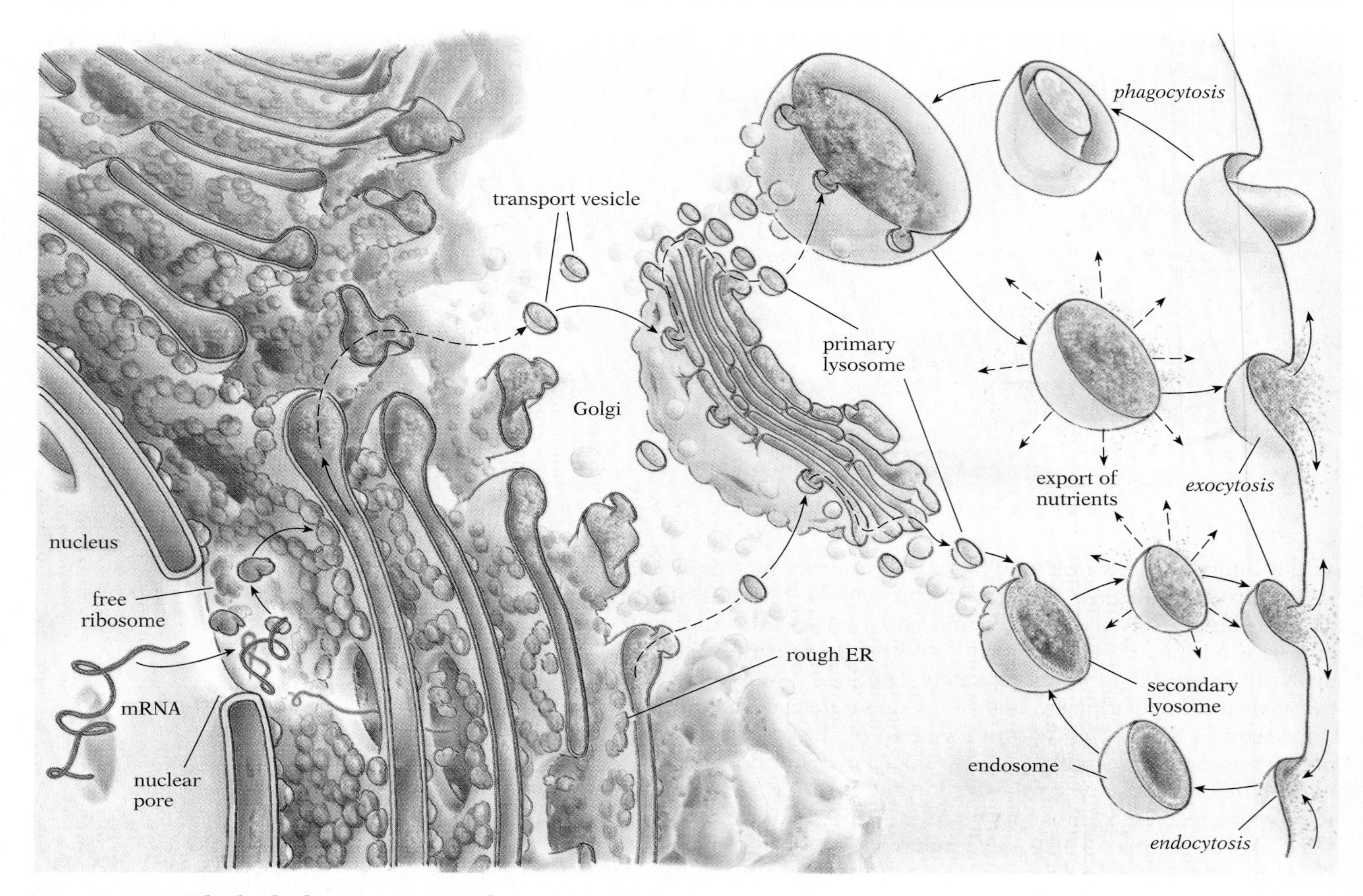

5.15 The hydrolytic enzyme cycle

The role of the ER and the Golgi apparatus in cellular digestion is representative of their functions in the cell. DNA in the cell nucleus produces the mRNA that encodes the hydrolytic enzymes. The mRNA is transported to ribosomes, which then bind to the RER. The enzymes are synthesized there, passed into the ER, and marked with signal sequences for later transport. The enzymes are next collected by receptors in the SER and packaged into transport vesicles. Markers on the outside of the transport vesicles cause them to fuse with the membrane of the Golgi apparatus, where new vesicles (primary lysosomes) are created. These packets of specific mixtures of hydrolytic enzymes fuse with appropriately marked endocytotic vesicles (endosomes) or with secondary lysosomes (produced by previous fusions of this kind) and help digest the contents. Useful products of this digestion pass through the membranes of the secondary lysosomes into the cytosol, while the residue is disposed of through exocytosis. The same pattern of synthesizing and marking enzymes in the ER, transporting them to the Golgi apparatus, sorting, chemically modifying, and repackaging them there, and then dispatching them to various intracellular targets seems to be the general strategy for managing macromolecules in cells.

can accumulate and block the long thin parts of nerve cells responsible for transmitting nerve impulses.

The elaborate organization of membrane movement in cells is vulnerable to many infectious diseases. Semliki Forest virus (SFV), for instance, infects a wide range of hosts from invertebrates to humans. The virus carries a marker on its surface that mimics a substance for which one kind of coated pit has specific receptors. SFV is carried into the cell by endocytosis, after which its chromosome escapes into the cytosol. The virus uses the host cells' ribosomes,

ER, Golgi apparatus, and the whole system of carbohydrate tags to reproduce itself and direct the subsequent exocytosis of its viral progeny. The virus thus ensures its own propagation by exploiting much of the specialized membrane machinery of the cell.

■ PEROXISOMES: DETOXIFICATION ORGANELLES

Cells also contain a variety of small organelles known collectively as microbodies. The most thoroughly understood are the ***peroxisomes*** (Fig. 5.16). Like lysosomes, peroxisomes contain an assortment of powerful enzymes. However, where the enzymes of the lysosomes are hydrolytic (water-splitting), these enzymes catalyze condensation reactions. Examples are the oxidative removal of amino groups from amino acids, the detoxification of alcohol, the oxidation of the dangerous compound hydrogen peroxide into water and oxygen, the breakdown of fatty acids into useful building blocks, and the reactions involved in the production of macromolecules used in respiration and other biochemical pathways.

Peroxisomes are independent organelles that multiply by fissioning: a cell that fails to inherit at least one peroxisome from the cytoplasm of its parent cell cannot make its own from scratch and will die. Peroxisomes are not self-sufficient, however: their membrane phospholipids are synthesized in the SER, recognized and picked up by transfer proteins, carried through the cytoplasm, and deposited in the peroxisomal membrane; the precursors of their enzymes are produced in the cytoplasm (rather than the ER) and transported to these tiny organelles. Just as proteins synthesized on the ER must be supplied with labels to assure accurate delivery, so too proteins built in the cytoplasm must have signals to assure proper routing, whether to peroxisomes or other destinations. In the cytoplasm, however, the signals are embedded in the amino acid sequence at one end of the protein rather than in a carbohydrate side chain.

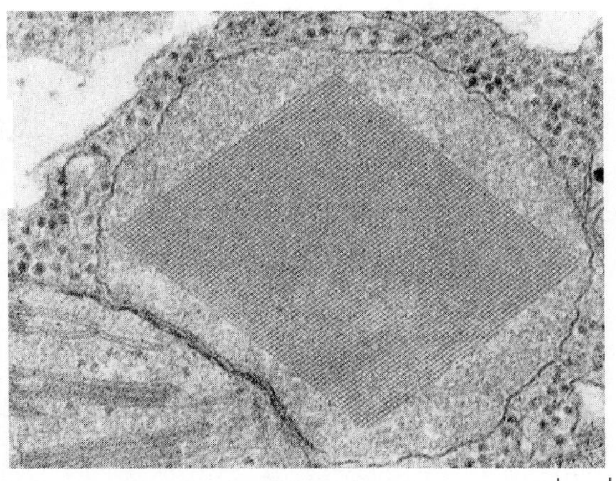

5.16 A peroxisome

This electron micrograph shows part of a tobacco-leaf cell, which has been treated with stain to reveal the crystalline core of a peroxisome and the presence of oxidizing enzyme within it. In mammals, peroxisome enzymes neutralize hydrogen peroxide, alcohol, and other potentially harmful substances.

■ MITOCHONDRIA: CELLULAR POWER STATIONS

Mitochondria carry out respiration, the chemical reactions that extract energy from food and make it available for innumerable energy-demanding activities. Each mitochondrion is bounded by a double membrane; the outer membrane is smooth, while the inner membrane has many inwardly directed folds called ***cristae*** (Fig. 5.17). Energy-rich reactants are concentrated in this organelle, and here, with the help of the appropriate enzymes, they are combined with oxygen to generate water, carbon dioxide, and the energy that runs the cell. The operation of mitochondria is fully described in Chapter 6.

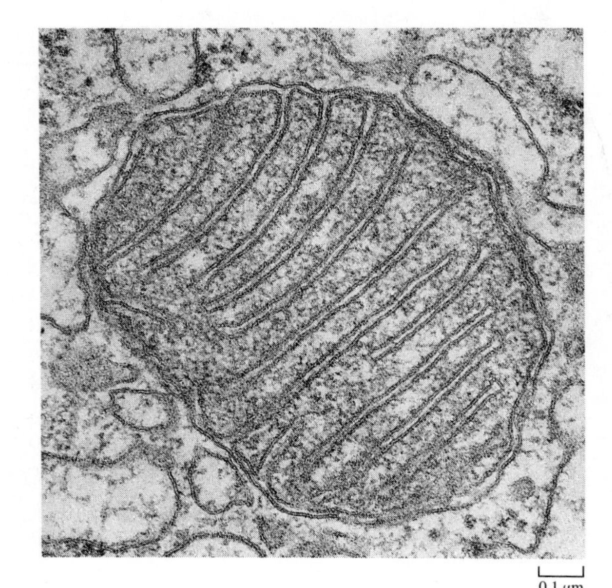

5.17 A mitochondrion from a rat epithelial cell

Note the double outer membrane and the numerous cristae, which can be seen to arise as folds of the inner membrane in this electron micrograph.

■ CHLOROPLASTS: SOLAR ENERGY RECEPTORS

Plastids are large cytoplasmic organelles found in the cells of plants but not in those of fungi or animals. Plastids are clearly visible through an ordinary light microscope. There are two principal

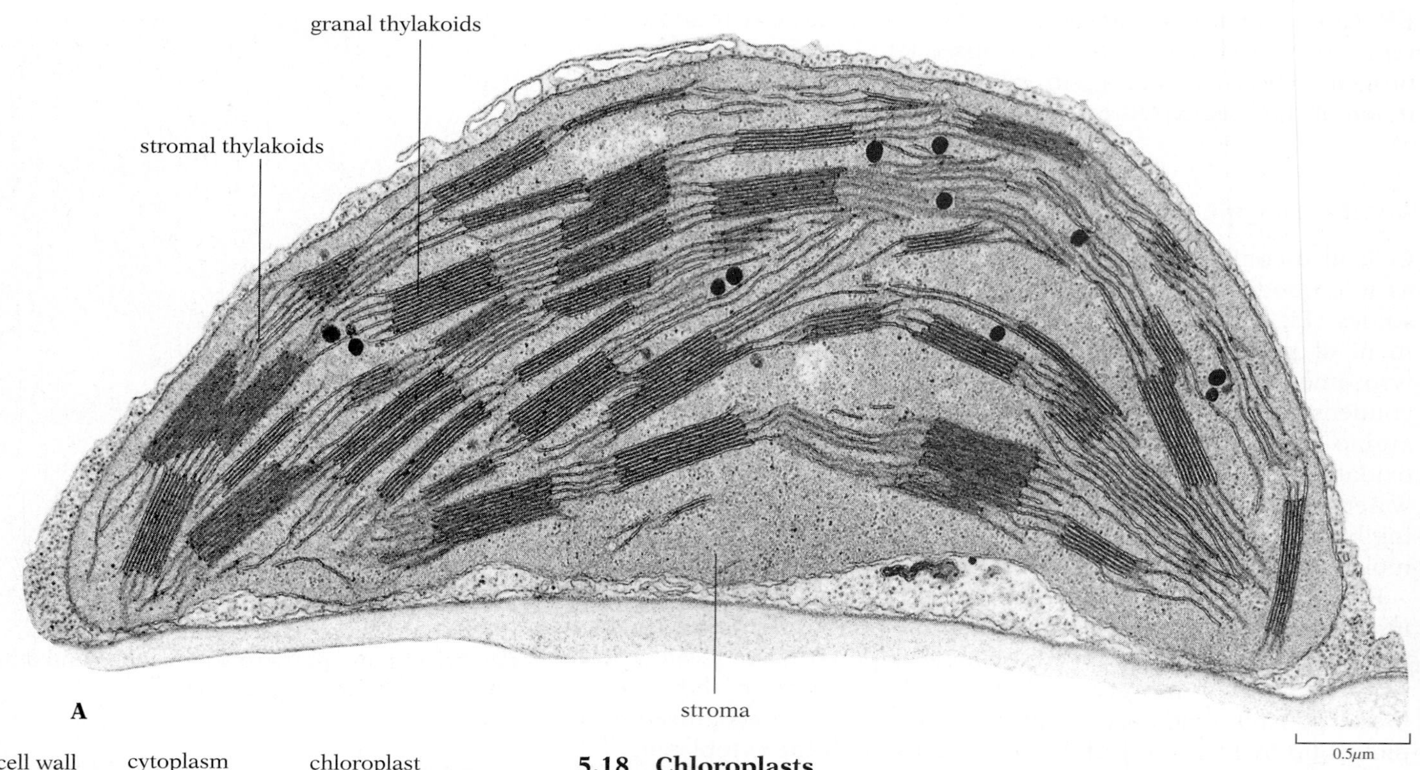

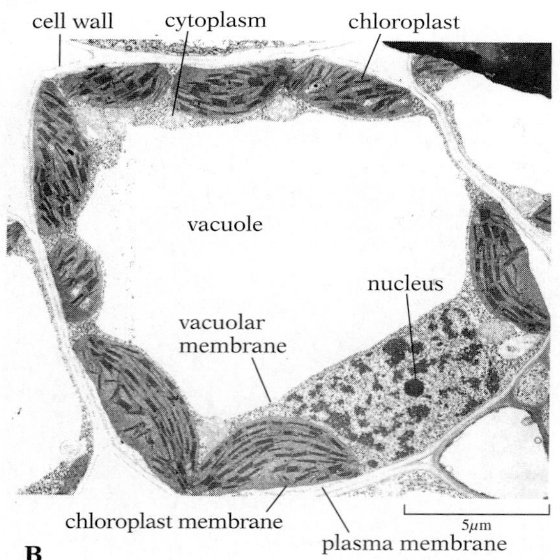

5.18 Chloroplasts

(A) Stacks of disklike thylakoids, forming grana, can be seen in this chloroplast of a corn leaf. (B) Numerous chloroplasts lie close to the perimeter of a mature leaf cell of timothy grass.

categories: ***chromoplasts*** (colored plastids) and ***leucoplasts*** (white or colorless plastids).

Chloroplasts are chromoplasts that contain the green pigment ***chlorophyll***, along with various yellow or orange pigments. Chapter 7 fully describes how the radiant energy of sunlight is trapped in the chloroplasts by molecules of chlorophyll and is then used in the manufacture of complex organic molecules (particularly glucose) from simple inorganic raw materials such as water and carbon dioxide.

Typical chloroplasts are bounded by two concentric membranes and have a complex internal membranous organization (Fig. 5.18). The fairly homogeneous internal proteinaceous matrix is called the ***stroma***. Numerous flat compartments, called ***thylakoids***, are embedded in it. In most higher plants these thylakoids come in two varieties: separate thylakoids that run through the stroma, and stacks of platelike thylakoids known as ***grana*** (Fig. 5.18). Most of the important reactions of photosynthesis occur on membranes of the thylakoids.

■ LEUCOPLASTS: STORAGE ORGANELLES IN PLANTS

The colorless plastids, or leucoplasts, are primarily used to store starch, oils, and protein granules. Plastids filled with starch, called

amyloplasts, are particularly common in seeds and in storage roots and stems, such as carrots and potatoes. Starch is an energy-storage compound in plants, and is deposited as a grain or group of grains in plastids (Fig. 5.19); no starch is found in other parts of the cell.

All types of plastids form from small colorless bodies called proplastids. Once formed, many kinds of plastids can be converted into other types under appropriate conditions. For example, under certain conditions leucoplasts exposed to light develop chlorophyll and are converted into chloroplasts, complete with the characteristic internal thylakoids, which develop from invaginations of the inner membrane of the plastid.

■ VACUOLES: MULTIPURPOSE ORGANELLES

Membrane-enclosed, fluid-filled spaces called ***vacuoles*** are found in both animal and plant cells, though they have their greatest development in plant cells. There are various kinds of vacuoles, with a corresponding variety of functions. In some protozoans, specialized vacuoles, called contractile vacuoles, play an important role in expelling excess water and some wastes from the cell. Many protozoans also have food vacuoles, chambers that contain food particles. They are similar to the vesicles formed by many cells when they take in material by endocytosis.

The distinction between vesicles and vacuoles, both of which are membrane-bounded, is hard to draw with any precision, particularly since vesicles may fuse with or bud off from vacuoles. The most obvious differences are in permanence, activity, and size: vesicles are relatively short-lived transport vehicles, while vacuoles tend to be long-lived; vesicles usually move quickly, while vacuoles are relatively static; finally, vesicles are usually small, while vacuoles are most often quite large. These hazy distinctions are underscored by the observation that many plant vacuoles develop from lysosomes.

In most mature plant cells, a large vacuole occupies much of the volume of the cell. The immature cell usually contains many small vacuoles. As the cell matures, the vacuoles take in more water and become larger, eventually fusing to form the very large definitive vacuole of the mature cell (Fig. 5.20). This process pushes the cyto-

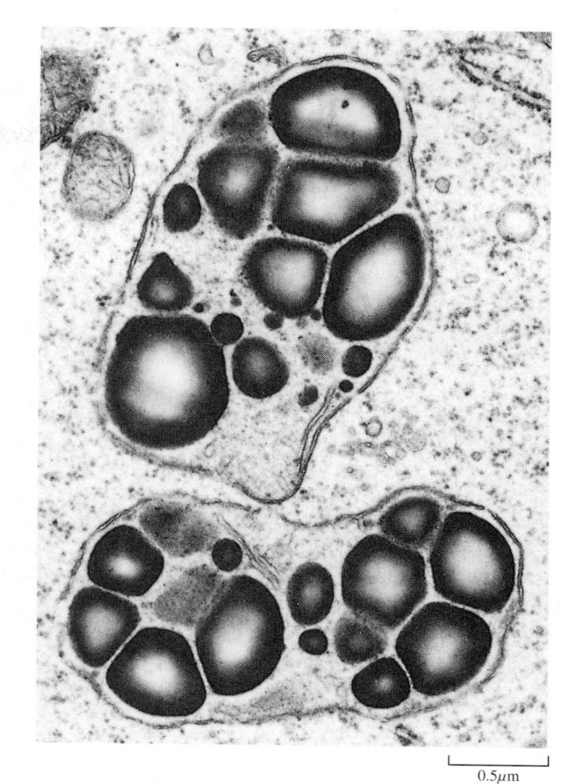

5.19 Leucoplasts from the root tip of *Arabidopsis*

Because they contain numerous prominent starch grains, these leucoplasts from the small desert plant *Arabidopsis* are called amyloplasts.

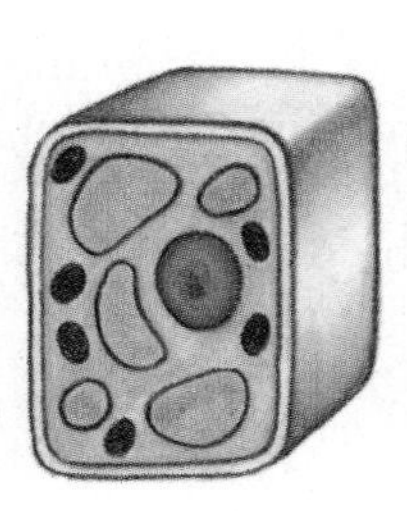

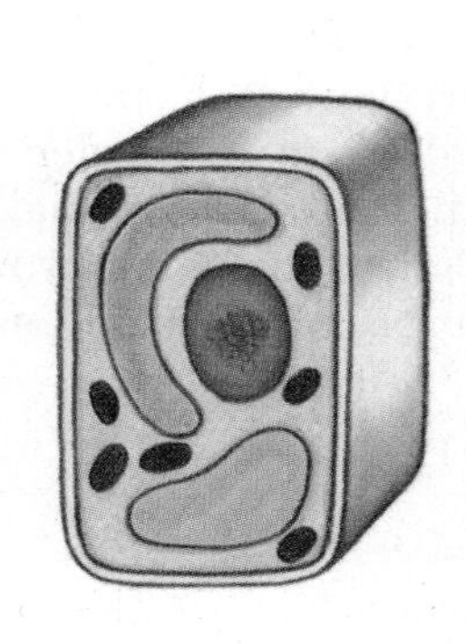

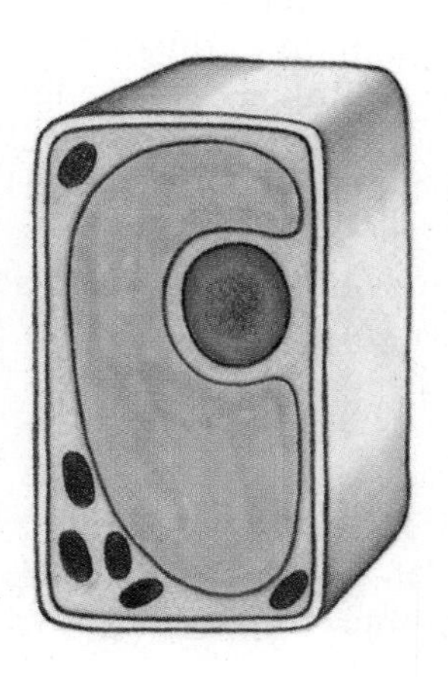

5.20 Development of the plant-cell vacuole

The immature cell (left) has many small vacuoles, which arise from lysosomes. As the cell grows (middle), these vacuoles fuse and eventually form a single large vacuole, which occupies most of the volume of the mature cell (right). Note the cytoplasm has been pushed to the periphery.

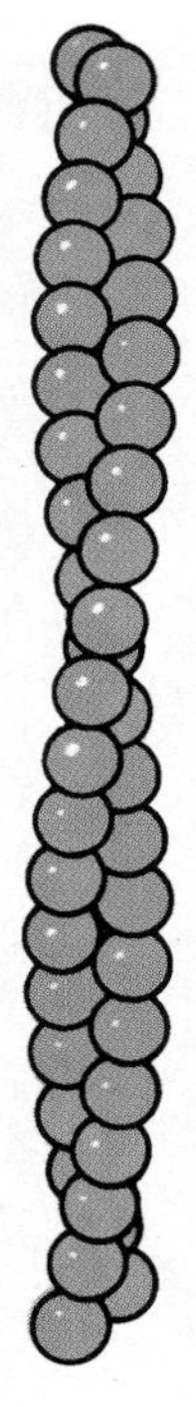

5.21 **An actin microfilament**
This portion of an actin microfilament shows the helically intertwined chains of protein subunits.

plasm to the periphery of the cell, where it forms a relatively thin layer.

The plant vacuole contains a liquid called cell sap, which is primarily water with a variety of substances dissolved in it. Since the cell sap is generally hypertonic relative to the external medium, the vacuole tends to take in water by osmosis. As the vacuole swells, the vacuolar membrane (or tonoplast, as it is often called) pushes outward against the cytoplasm, which transmits the pressure to the cell wall. The wall is strong enough to limit the swelling and prevent the cell from bursting; The outward push of the vacuolar membrane thus maintains cell turgidity and stiffness.

Many substances of importance in the life of the plant cell are stored in the vacuoles, among them high concentrations of soluble organic nitrogen compounds, including amino acids; vacuoles also store sugars, various organic acids, and some proteins. The vacuoles also function as dumping sites for noxious wastes. Enzymes secreted into them degrade some of these wastes into simpler substances that can be reabsorbed into the cytosol and reused.

THE CYTOSKELETON: STRUCTURE AND MOVEMENT

As we have seen, the cell is full of specialized, membrane-lined organelles that mediate much of cellular chemistry. Cells also need a variety of important protein-based components that help organize movement, not only within the cell but by the cell itself, and that aid in defining and controlling cell shape.

■ MICROFILAMENTS: CELLULAR ROPES AND CABLES

Molecules of a protein called ***actin*** polymerize spontaneously under conditions of elevated Ca^{++} and Mg^{++} concentration, or normal levels of K^{+} and Na^{+}. The resulting long, extremely thin polymers, helically intertwined, form actin microfilaments (Fig. 5.21), a general-purpose cabling that accounts for 2–20% of the protein in most cells. A filament spanning the entire length of a cell can form in a matter of minutes.

Actin microfilaments can play a purely structural role; when cross-linked for strength, they are a component of the cytoskeleton, the complex weblike array of molecules that helps maintain cell shape. This network is densest just under the cell membrane (Fig. 5.22), to which it is anchored by special proteins. Parallel, cross-linked arrays of actin microfilaments provide reinforcement for various stiff cellular protuberances, including the dense forests

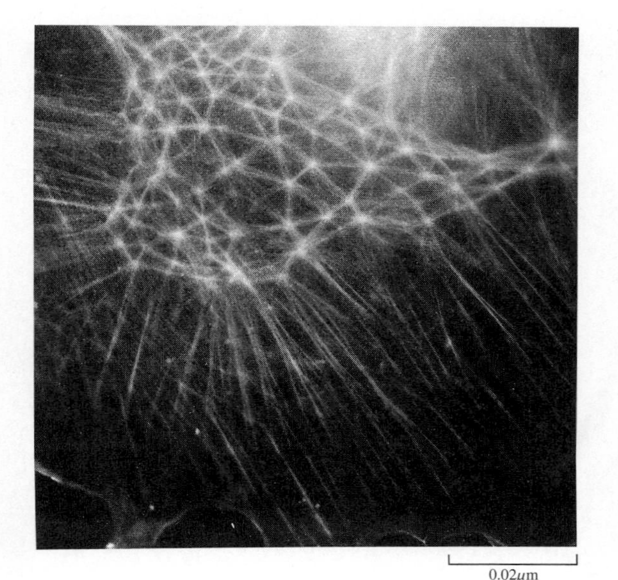

5.22 **Actin in a rat fibroblast cell**
Actin microfilaments appear to run between distinct anchor points on the membrane, providing structure and a scaffolding for movement.

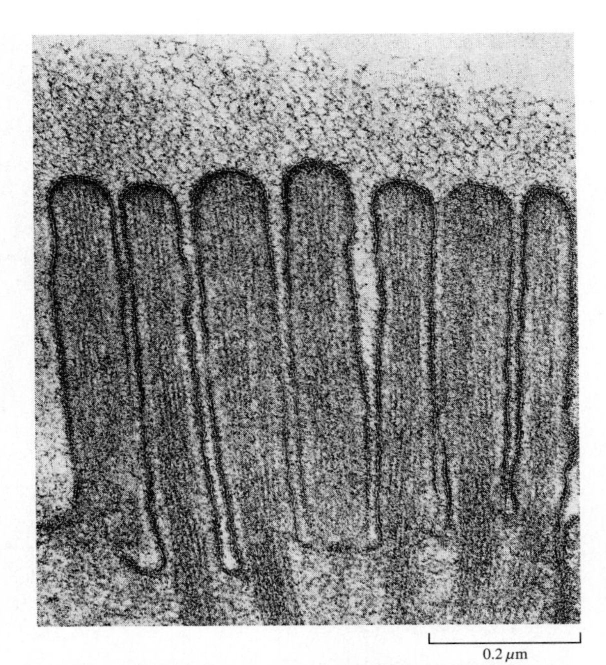

5.23 Microfilaments in the microvilli

Cross-linked into tight bundles, actin microfilaments in the microvilli provide rigidity.

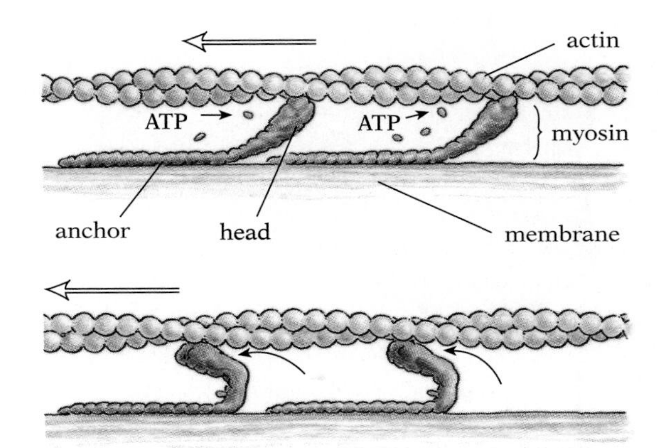

5.24 Interaction of actin and myosin

Some (perhaps all) nonmuscle myosin is anchored in membranes and, when fueled by ATP, can bind to actin microfilaments and undergo a conformational change. The resulting movement of the myosin head (described in detail in Chapter 37) moves the membrane (to which the anchor is permanently connected) and the microfilament (to which the head is temporarily attached) past one another in opposite directions. Myosin may also be able to anchor itself to vesicles or actin filaments, and so move them along the microfilament network. Myosin can create movement of 1–9 μm/sec.

of rodlike projections known as microvilli that line the intestine (Fig. 5.23).

In addition to its structural role, actin is involved in cell movement through interactions with a second cytoskeleton protein, myosin. ***Myosin*** consists of a two-part head plus an anchor. The tip of the head can bind to actin, while the base can accept energy from ATP, which causes the head to twist relative to the anchor. This twisting is used to pull actin cables (Fig. 5.24).

In muscle cells (described in detail in Chapter 37) the anchor is actually a long tail that binds to other myosin tails to create a highly specialized bundle that makes coordinated contractions possible. In nonmuscle cells, the myosin anchor is relatively short and is bound to cell or vesicle membranes or to actin filaments.

The best-understood case of myosin-mediated movement in nonmuscle cells occurs during cell division, when the daughter cells are created. Actin filaments are formed along the midline of the parent cell, and myosin molecules (perhaps mounted on the membrane) pull on the actin to "pinch off" the two cells from one another.

Similar actin-myosin interactions are responsible for cell movement (Fig. 5.25). At the trailing edge of the cell the membrane is being constantly tightened; the myosin pulls the actin cables, thus squeezing the cytoplasm toward the front. Vesicles produced from the rear membrane are sent forward to enable the leading edge to grow. Myosin-mediated movement of actin filaments is believed to cause extension of the pseudopodium at the front, though the details of this process are still a mystery. While most of the cytoplasm in the cell, a region often called ectoplasm, contains highly cross-

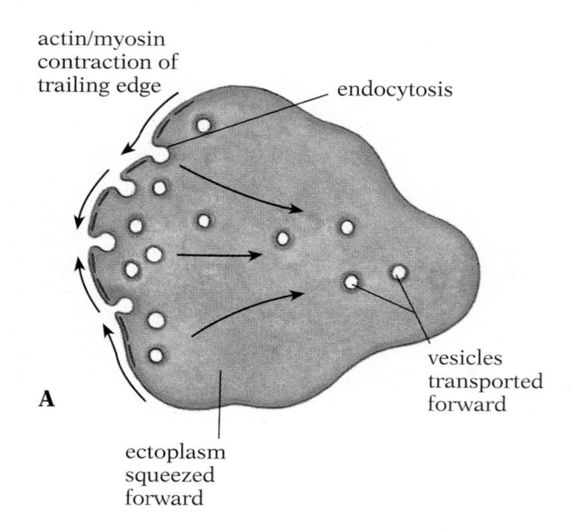

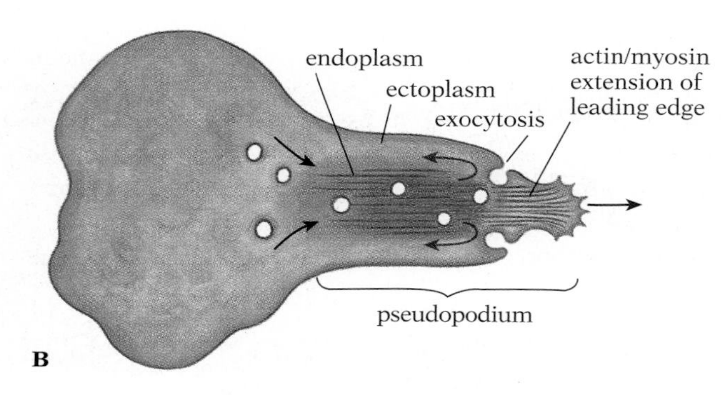

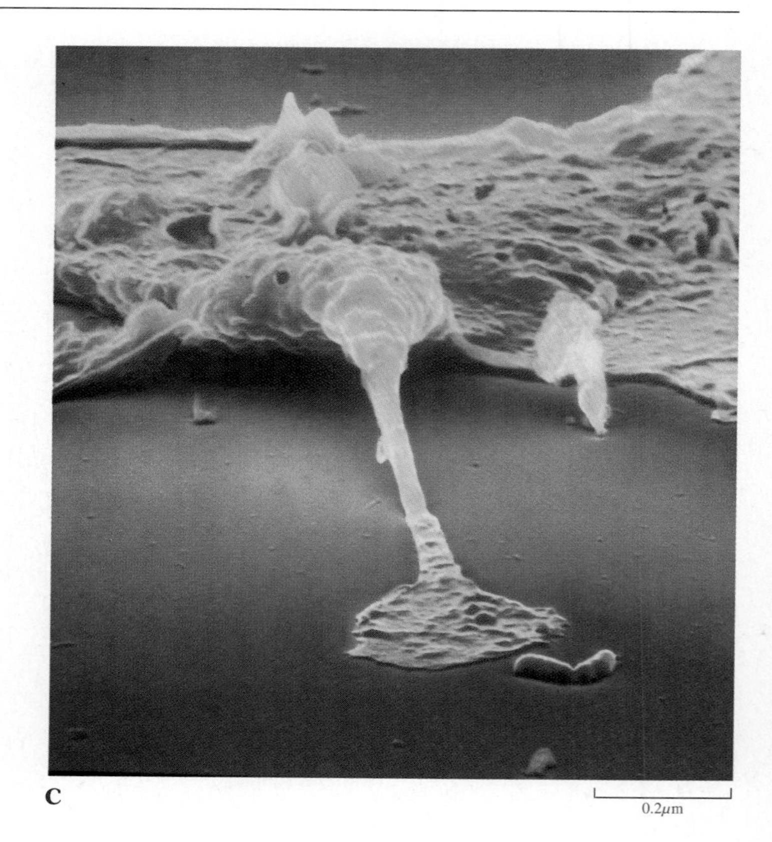

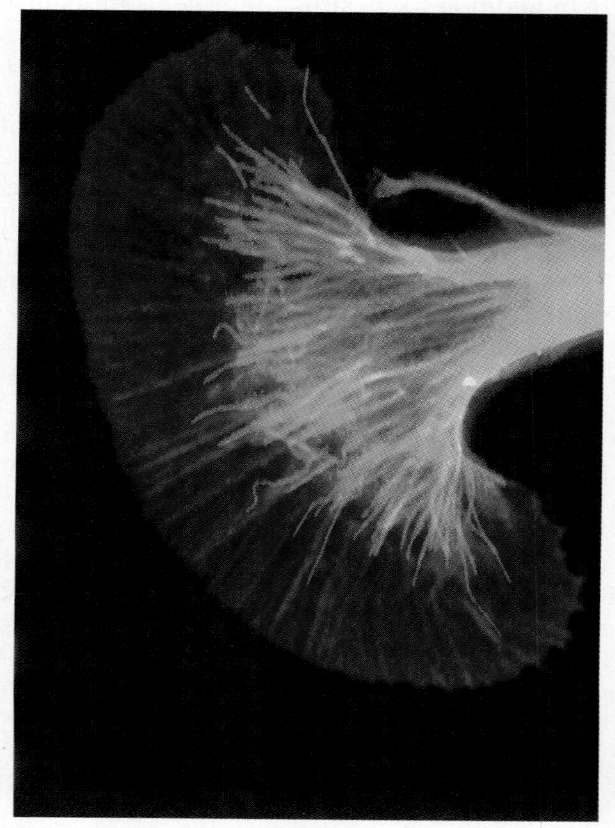

5.25 Amoeboid movement

(A) As the trailing edge of the cell contracts, squeezing the cytoplasm forward, the cytoplasm moving into the pseudopodium loses its actin cross-linking, and becomes more fluid (endoplasmic). This endoplasmic core slides into the pseudopodium between the peripheral layers of stationary ectoplasm. (B) The pseudopodium extends as peripheral actin filaments (straight lines) push it forward and stiffen its cortex; vesicles transported from the rear provide more membrane. Endoplasm is pushed down the middle of the pseudopodium. As the core endoplasm moves forward, ectoplasm at the rear of the pseudopod and from the rest of the cell is converted into endoplasm, while endoplasm at the front of the pseudopod is converted into ectoplasm. (C) In this dramatic example of actin-based cell movement, a phagocyte (an amoeboid cell in the blood that captures foreign material by endocytosis) is extending a pseudopodium toward a bacterium, which it has detected on the basis of a waste chemical released by the bacterium. (D) Actin bundles are visible in the tip of the pseudopodium of this migrating nerve cell.

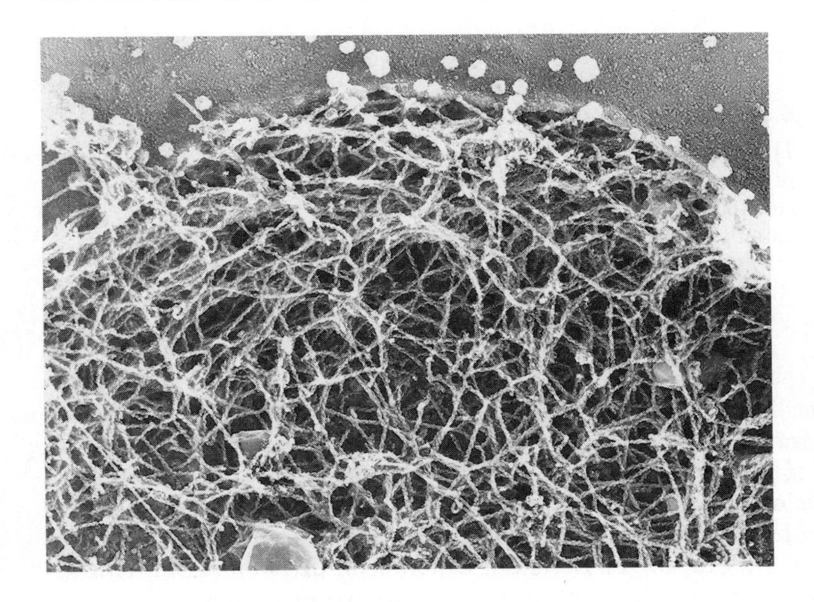

5.26 Cross-linked actin filaments

This electron micrograph shows the internal structure of the leading edge of a moving white blood cell.

linked actin filaments and is relatively rigid (Fig. 5.26), the cytoplasm flowing into the leading edge, the endoplasm, is unlinked, and so is much more fluid. Once this endoplasm has been squeezed into the extension provided by the myosin-moved actin filaments, the actin becomes cross-linked again; other regions of ectoplasm are then converted into endoplasm to feed further movement.

■ MICROTUBULES: CELLULAR GIRDERS AND RAILWAYS

Microtubules can be thought of as heavy-duty versions of microfilaments. They are long, hollow, cylindrical structures (Fig. 5.27) that, like microfilaments, form spontaneously in response to the

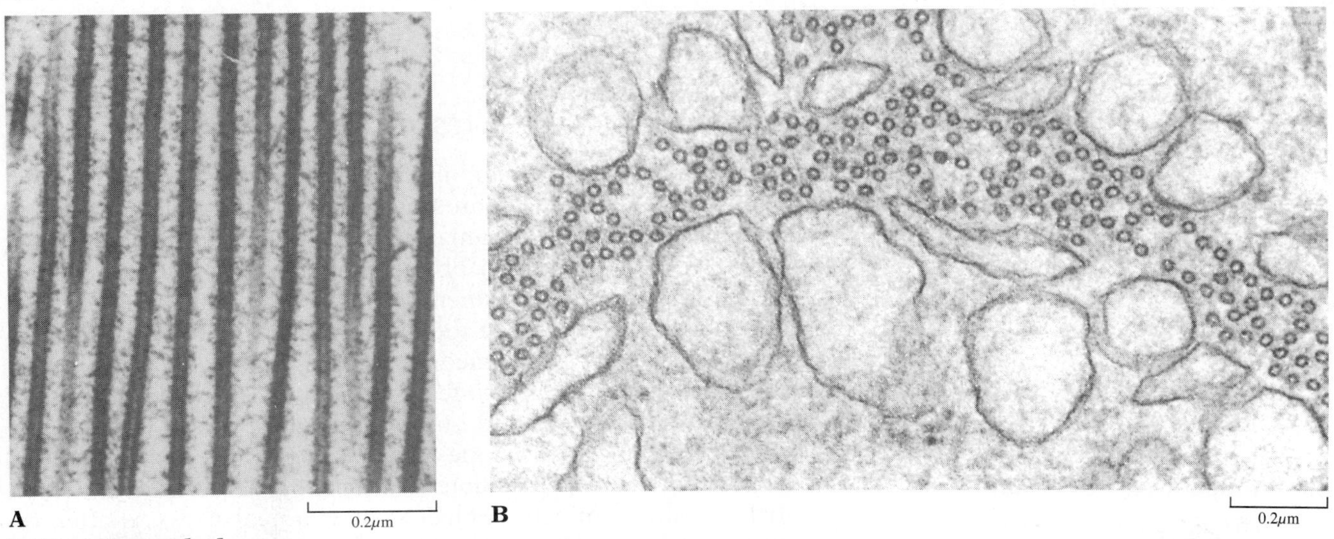

5.27 Microtubules

(A) Longitudinal section, from bovine brain. (B) Cross section, from hamster spermatid.

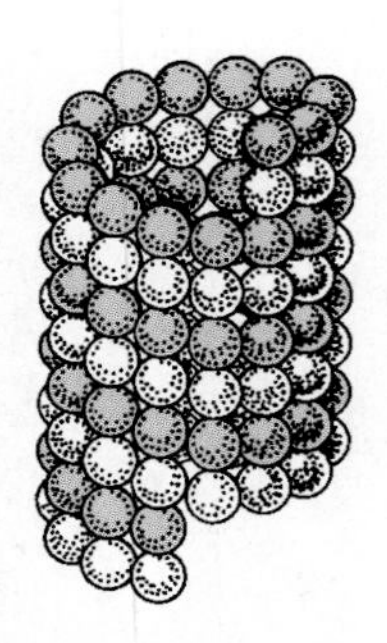

5.28 Structure of portion of a microtubule
The subunits of tubulin, each consisting of two proteins (shown colored and white), are helically stacked to form the wall of the tubule, which is usually 13 subunits in circumference.

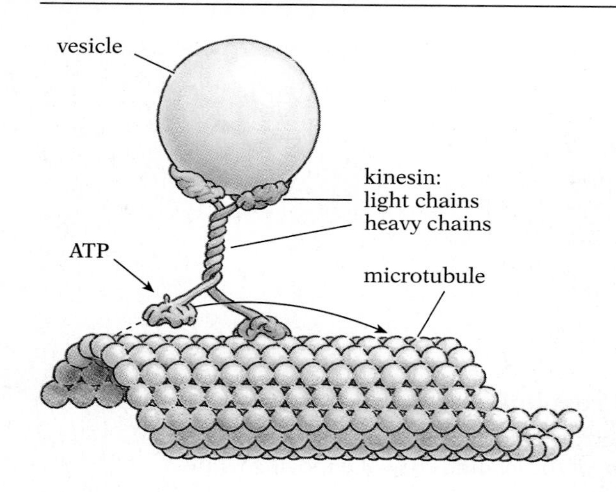

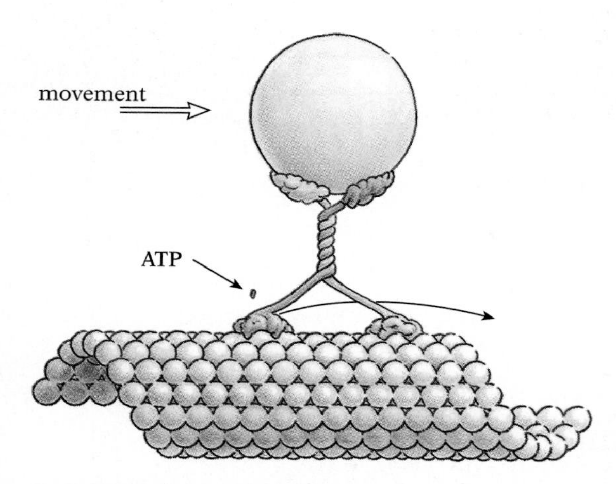

5.29 Kinesin-mediated movement
Kinesin has two identical heavy chains (which bind to the microtubule and "walk" along it in a particular direction) and two identical light chains (which bind the vesicle to the heavy chains).

proper ionic signals. Molecules of the globular protein ***tubulin***, each molecule consisting of two proteins (α and β), polymerize to form a helical stack (Fig. 5.28). Microtubules radiate from ***microtubule organizing centers*** (usually near the Golgi) and play a critical role in general cell structure, in vesicle and organelle movement, and in cell division. During cell division prominent arrays of microtubules radiate from ***centrosomes***, areas near structures called centrioles at each end of the cells in many species, to form a basketlike arrangement (the spindle), which is instrumental in moving the chromosomes to the locations of the new nuclei.

Special proteins (called kinesins) move vesicles away from the organizing center; one end binds to the vesicle while the other, which has two "feet," steps along the microtubule at speeds of 1–3 μm/sec (Fig. 5.29). Other kinesin-like proteins move mitochondria and other cargo. Yet other molecules (especially dynamin) move various loads *toward* the organizing center (Fig. 5.30). The positioning of the organizing center of this elaborate transportation network near the cellular mailroom—the Golgi—is no accident. Like microfilaments, microtubules also help provide shape and support for the cell and its organelles as part of the cytoskeleton. In fact, the microtubules appear to actively stretch the ER out into the farthest reaches of the cell: when the connections between microtubules and ER are broken, the ER collapses into a compact clump around the nucleus.

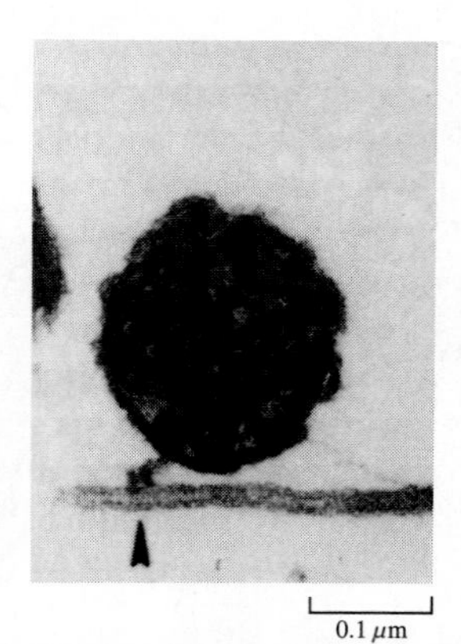

5.30 Dynamin-mediated movement
A mitochondrion from the giant amoeba *Reticulomyxa* is being moved along a microtubule. The dynamin molecule attaching the organelle to the microtubule is clearly visible.

■ INTERMEDIATE FILAMENTS: A CELLULAR LATTICE?

There are numerous tubular fibers larger in diameter than actin but smaller in diameter than microtubules, with (in general) a much lower degree of organization. Although there are several chemically distinct varieties of these fibers, they are lumped together as ***intermediate filaments*** (Fig. 5.31). (Had their hollow structure been evident when they were named, they would doubtless have been called "intermediate tubules" instead.) Other intermediate filaments are associated with the cell membrane (where they provide structure and aid in the formation of cell-to-cell junctions) and the nuclear envelope (Fig. 5.32).

The collection of intermediate filaments in the cytoplasm, which is most prominent in cells subject to mechanical stress, is often referred to as a lattice (Fig. 5.33). The information we have about the cytoplasmic lattice can be interpreted in two quite different ways. For example, the proteins in the cytosol are concentrated on these

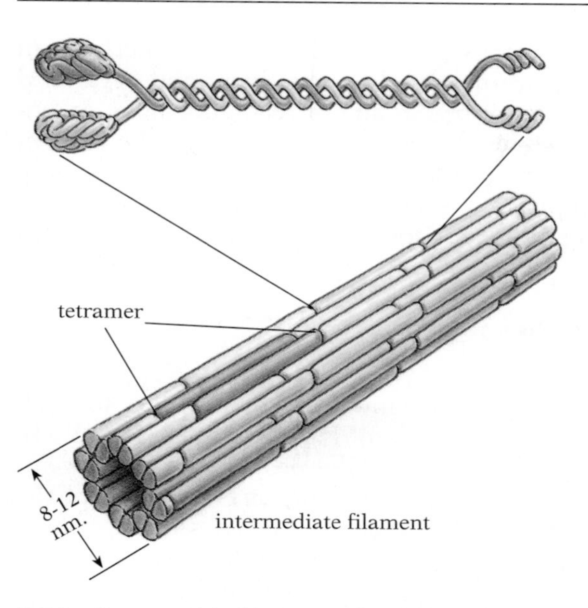

5.31 Structure of intermediate filaments

Top: The basic subunit is a protein dimer formed from two identical molecules coiled together. Bottom: each dimer is paired with another, slightly offset, dimer to form a tetramer. The tetramers bind head to tail to create strands. Eight strands bound together in the form of a hollow tube generate the structure typical of intermediate filaments.

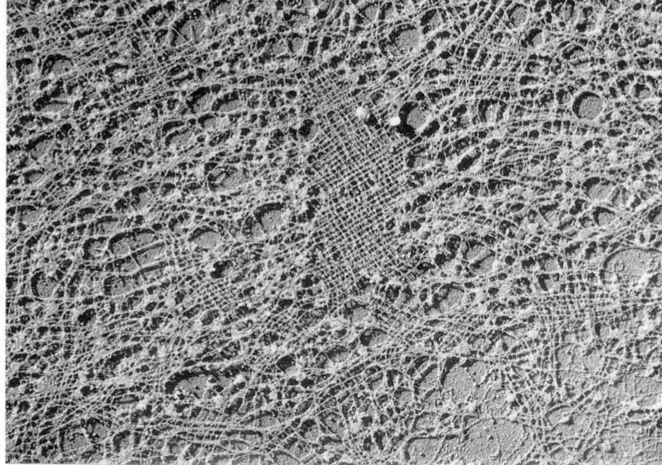

5.32 Nuclear lamina

This meshlike array of intermediate filaments lies just inside the inner membrane of the nuclear envelope in frog eggs.

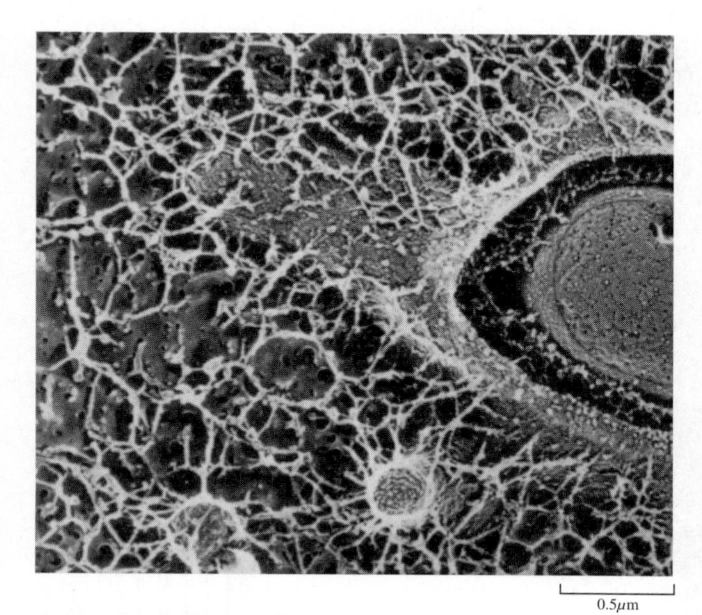

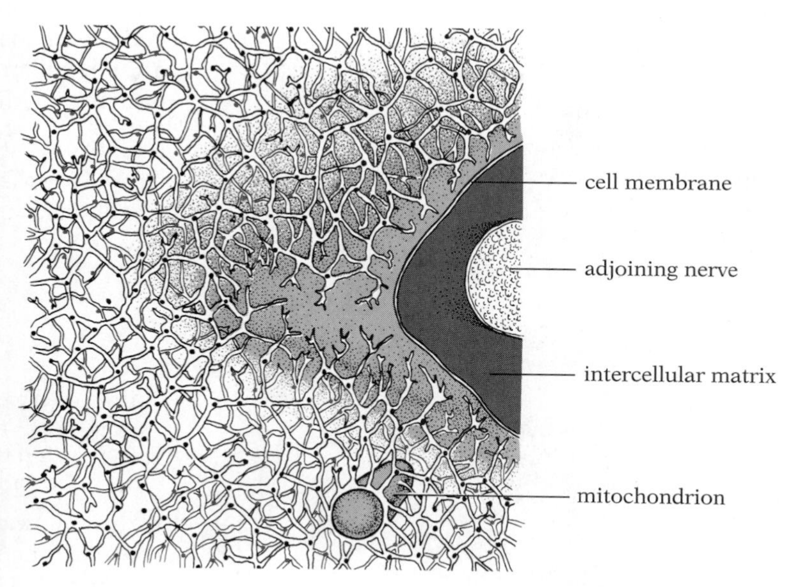

5.33 Cytoplasmic lattice

The cytoplasmic lattice is thought by many researchers to be composed of a network of fibers that help give shape to the cell and hold various cellular organelles in position. The hypothesis suggests that fibers are anchored to the cell membrane, as shown in the interpretive drawing, and are also linked to cellular microtubules and microfilaments (not shown). Other researchers argue that the lattice is merely an artifact of the fixation technique used to prepare the specimen for the EM.

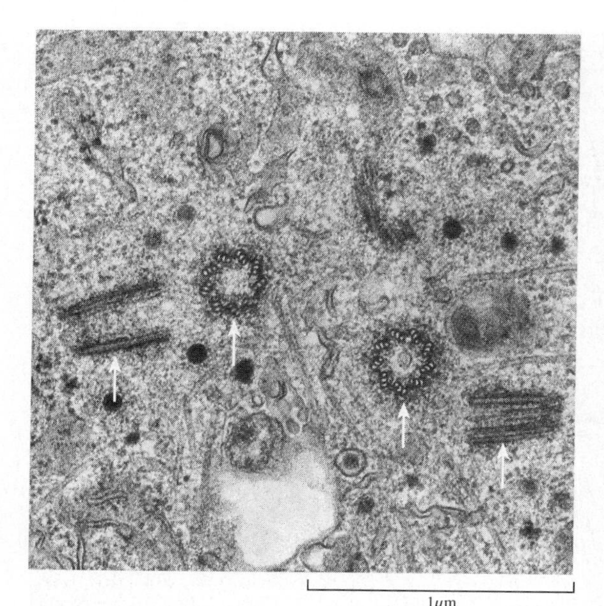

5.34 Centrioles

This electron micrograph shows newly replicated centrioles (arrows). Since the centrioles of each pair lie at right angles to each other, the sectioning of the specimen results in one from each pair being cut longitudinally and one being cut in cross section.

fibers, while the surroundings are basically aqueous. Perhaps in living cells the proteins are mounted on the fibers and organized into arrays for the sequential processing of reactants along an enzymatic pathway, or perhaps they merely stick to the fibers when the specimen is undergoing fixation for viewing. Similarly, the free ribosomes of the cytoplasm are concentrated at lattice intersections, perhaps in living cells, perhaps as a result of fixation.

■ CENTRIOLES AND BASAL BODIES: STRUCTURAL ORGANIZERS

Centrioles are found in pairs, oriented at right angles to each other, just outside the nuclei of many kinds of cells (Fig. 5.34). From the neighborhood of these organelles projects an array of microtubules. When centrioles are present, they are the focus of the microtubule spindle during cell division. In cross section, centrioles display a uniform structure, with nine triplets arranged in a circle and each triplet composed of three fused microtubules (Fig. 5.35B). Centrioles do not appear to be necessary for cell division; they may simply be the templates for making basal bodies.

Basal bodies anchor the many hairlike cilia and flagella (discussed next) to the cell membrane. As Figure 5.35A indicates, basal bodies have exactly the same structure as centrioles, and are probably the same organelle since basal bodies can become centrioles and vice versa. During cell division in the mobile alga *Chlamydomonas*, for example, the basal bodies of the two flagella abandon their posts and migrate to the poles of the cell, where they take up their positions in the spindle. Similarly, the basal body of the flagellum in many kinds of sperm becomes one centriole of the egg after fertilization. Also, the centrioles of cells that differentiate to line the oviduct multiply to become the basal bodies of the cilia

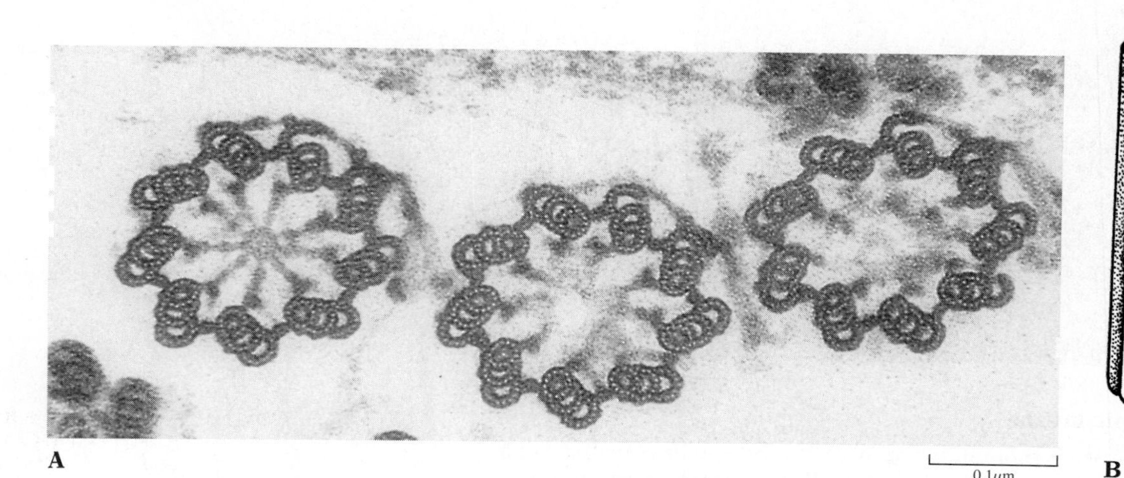

5.35 Basal bodies and centrioles

(A) Basal bodies, such as the three shown in cross section in this electron micrograph of a protozoan, are essentially identical to centrioles in structure. (B) Centrioles are composed of nine triplet microtubules.

whose rhythmic beating moves eggs from the ovary to the site of fertilization.

■ CILIA AND FLAGELLA: CELL MOVERS

Some cells of both plants and animals have one or more movable hairlike structures projecting from their free surfaces. If there are only a few of these appendages and they are relatively long in proportion to the size of the cell, they are called ***flagella***. If there are many and they are short, they are called ***cilia*** (Fig. 5.36). The basic structure of flagella and cilia in eucaryotes is the same, and the terms are often used interchangeably. Both usually function either in moving the cell or in moving liquids (or small particles) across the surface of the cell. They occur commonly on unicellular and small multicellular organisms and on the male reproductive cells of most animals and many plants, and can be the principal means of locomotion. They are also common on the cells lining many internal passageways and ducts in animals, where their beating aids in moving materials through the passageways. In the trachea they reach densities of a billion per square centimeter.

The flagella and cilia of eucaryotic cells are an extension of the cell membrane, containing a cytoplasmic matrix, with 11 groups of microtubules embedded in the matrix. Almost invariably, nine of these groups are fused pairs arranged around the periphery of the cylinder, while the other two are isolated microtubules lying in the center (Fig. 5.37). Each cilium and flagellum is anchored to the cell by a basal body; the similarities between these organelles and the

5.36 Ciliated surface of the trachea of a hamster

Coordinated movements of the many cilia shown in this SEM function to sweep dust particles and other foreign material from the respiratory surfaces to the mouth.

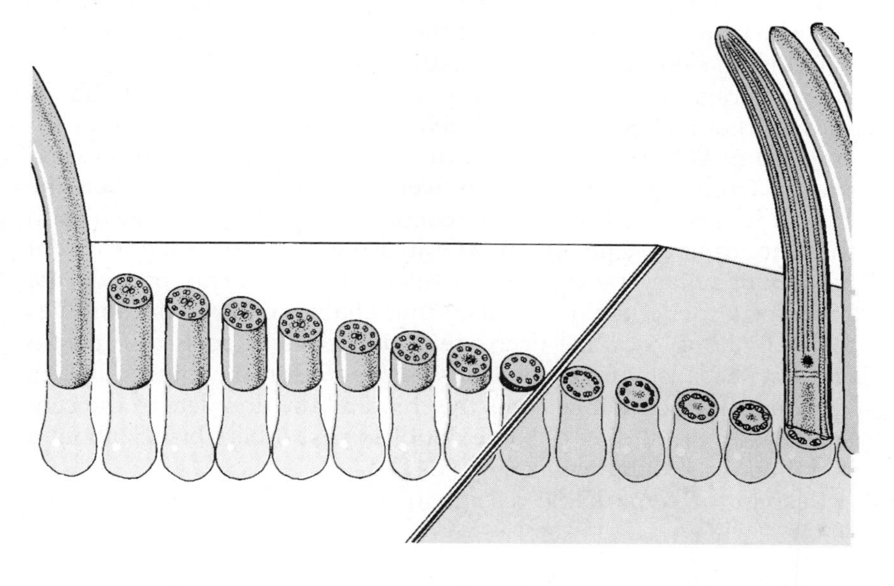

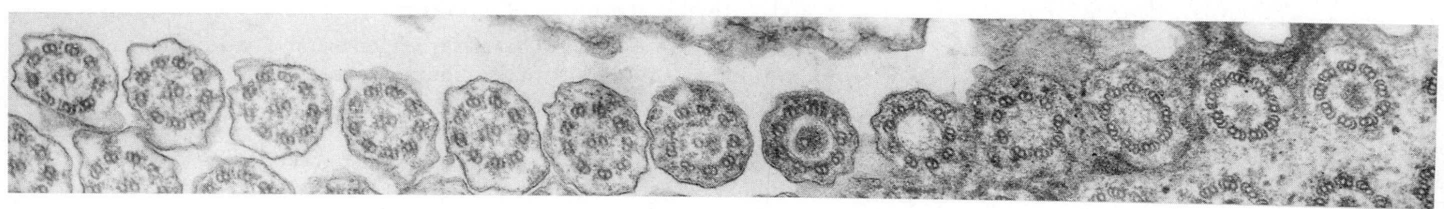

5.37 Cross section of cilia from the protozoan *Tetrahymena*

Most of this electron micrograph of an oblique section of surface tissue reveals the "9 + 2" arrangement of microtubules in cilia; the two cross sections at the far right show the nine triplet microtubules characteristic of basal bodies. The interpretive drawing shows the plane of the section and the position of the microtubules in a lateral section of an adjacent cilium.

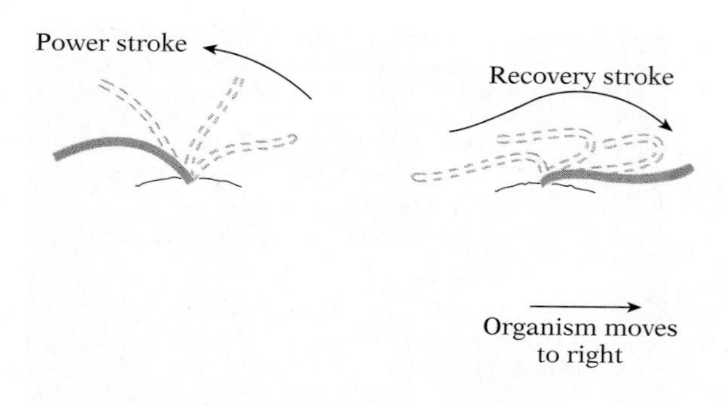

5.38 Stroke cycle of a cilium

In the power stroke the stalk is extended fairly rigidly and swept back by bending at its base. The recovery stroke brings the cilium forward again, as a wave of bending moves along the stalk from its base. Note that at no time during the recovery stroke is much surface opposed to the water in the direction of movement.

5.39 Flagellum movement in a sea urchin spermatozoon

Successive waves of bending that move along the flagellum, from its base toward its tip, push against the water and propel the sperm cell forward. This photograph was taken using four light flashes 10 msec (milliseconds) apart. In some protozoans, the flagella have lateral "hairs" that reverse the thrust; as a result, the waves move in the opposite direction, and so pull the cell along.

basal body in the arrangement of microtubules is obvious. The beating of cilia and flagella (Figs. 5.38 and 5.39) depends on two clawlike arms that allow each of the nine outer doublets to interact with an adjacent doublet (Fig. 5.40).

In a series of definitive experiments in the 1970s, Ian Gibbons showed that a protein called ***dynein***, found in cilia, can extract energy from ATP. He also found that when all the dynein was extracted from cilia, the doublets were left armless, and when the dynein was restored, the arms reappeared—a clear demonstration that the arms are the site of ATP hydrolysis for the cilia. Current models of ciliary movement postulate that the arms provide the basis for a rachetlike mechanism that enables ciliary microtubules to "walk along" or slide over one another. The models also postulate that because of the shear resistance within the cilium, due in large part to the radial spokes that bind all the doublets to the central sheath, the sliding of some doublets past others brings about a bending of the ciliary stalk (Fig. 5.41).

Despite the remarkable resemblance in mode of action between

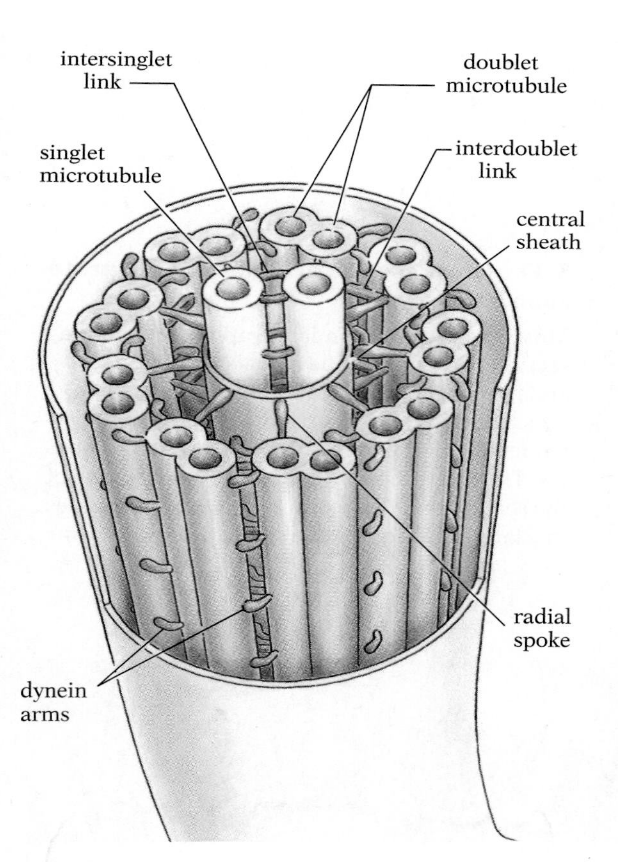

5.40 Internal structure of a cilium

The nine microtubule doublets are arranged around two singlet microtubules in the center of the cilium. Movement is generated when the dynein arms of one subset of doublets begin to "walk" along the length of other doublets. The doublets moving the most thus force the other doublets to bend. The intersinglet links may prevent lateral bending, and so explain why ciliary beats are constrained to a fixed plane.

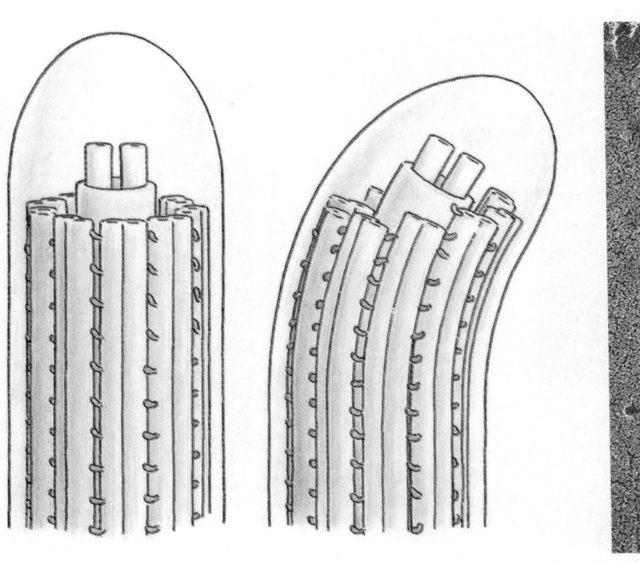

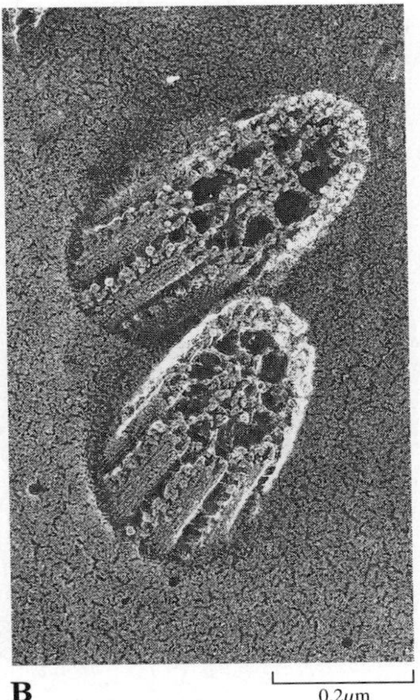

5.41 Bending in cilia produced by microtubules sliding past each other

(A) The microtubules on the concave side of the bend (green, right) have slid tipward. Because the tubules are all interconnected, their changes in relative position can be accommodated only if the stalk of the cilium bends. (B) Rows of outer dynein arms are visible in this electron micrograph of two *Tetrahymena* cilia that have been stripped of their outer cell membranes. The cilia were first frozen and then fractured obliquely to their axes. The surrounding matrix was subsequently etched back to expose the internal structures.

the tubulin-dynein system and the actin-myosin system, the two are unrelated: the amino acid sequences in tubulin and dynein show no similarities to those of actin and myosin. These two systems evolved their similar mechanisms for producing ratchet-driven sliding motion independently; this is a truly impressive example of what is called convergent evolution—the independent evolution of two very similar solutions to the same problem.

ARE EUCARYOTES COOPERATIVES OF PROCARYOTES?

There are two basic kinds of cells: eucaryotes (most organisms) and procaryotes (bacteria). Nearly all the organelles we have surveyed are unique to eucaryotes. Yet there is good reason to believe that some of these organelles may have evolved from procaryotes. To examine this argument we must first summarize eucaryotic organization.

■ HOW A "TYPICAL" EUCARYOTIC CELL IS ORGANIZED

The "typical" cell is a useful fiction; in fact, plant and animal cells differ from one another in dramatic ways, cells of particular plants or animals differ from those of other plants or animals almost as much, and within the body of any one plant or animal the various cells often differ strikingly in shape, size, and function. Now that the number of known cellular components has grown so large and their great variability has been so well demonstrated, it becomes even more obvious that no single diagram, or even series of diagrams, can really portray a typical cell. Nevertheless, to help summarize and visualize the arrangement of the organelles discussed in the preceding pages, two diagrams of relatively unspecialized cells (Figs. 5.42 and 5.43) are given here.

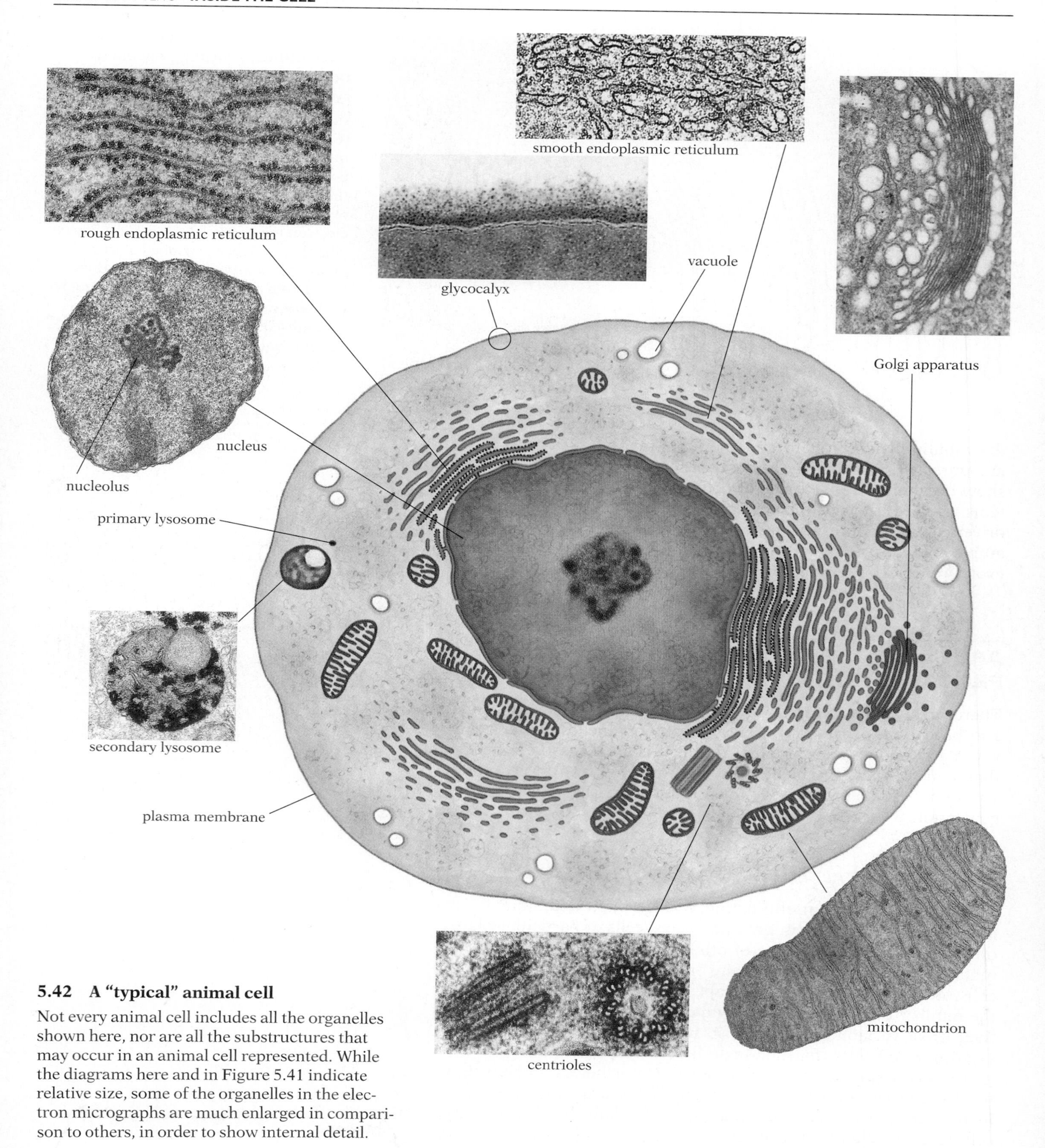

5.42 A "typical" animal cell

Not every animal cell includes all the organelles shown here, nor are all the substructures that may occur in an animal cell represented. While the diagrams here and in Figure 5.41 indicate relative size, some of the organelles in the electron micrographs are much enlarged in comparison to others, in order to show internal detail.

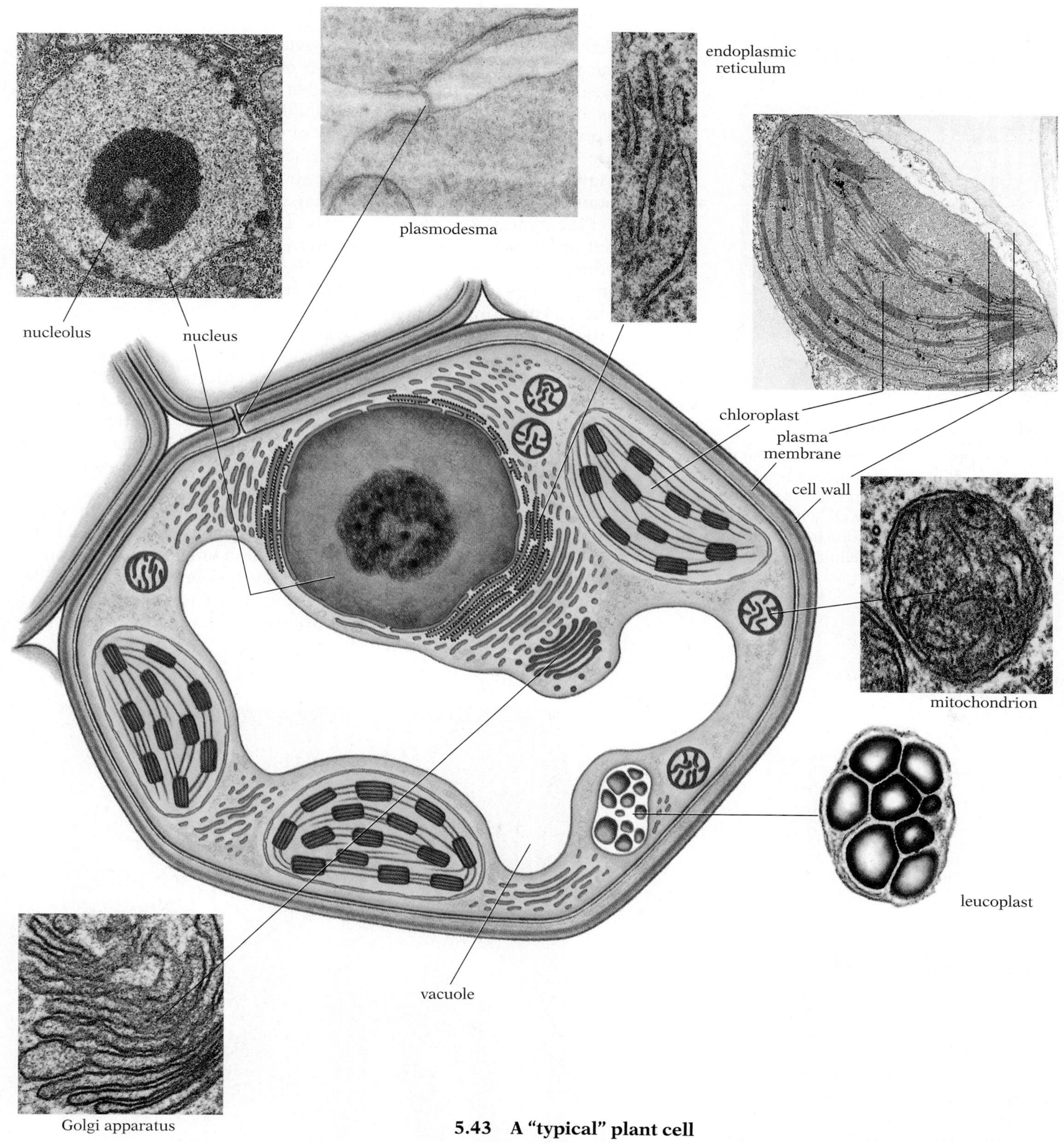

5.43 A "typical" plant cell

The organelles shown here do not occur in every plant cell, and some plant-cell substructures are not represented. The small amount of inner membrane compared with that of animal mitochondria is typical of plants.

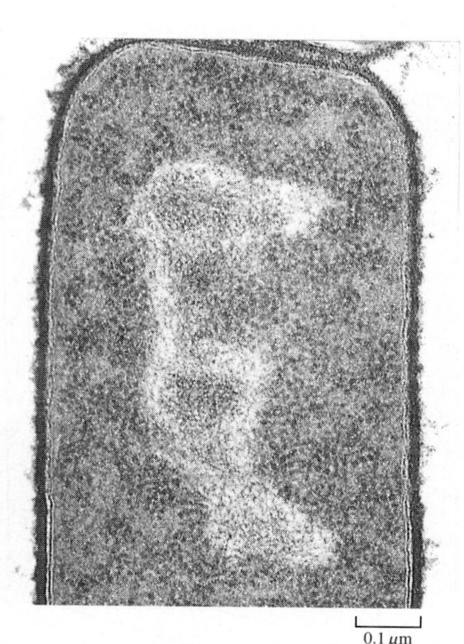

5.44 Part of a bacterial cell

The light area in the center of the cell in this electron micrograph is called the nucleoid; it contains DNA but is not bounded by a membrane. Note the prominent cell wall, with the plasma membrane visible just inside it, and the absence of membrane-bounded cellular organelles.

■ WHAT MAKES PROCARYOTES DIFFERENT?

Procaryotic cells lack most of the cytoplasmic organelles present in eucaryotic cells (Fig. 5.44). As their name implies, they have no nuclear membrane; they also lack other membranous structures such as an endoplasmic reticulum, a Golgi apparatus, lysosomes, peroxisomes, and mitochondria. (Many of the functions of mitochondria are carried out by the inner surface of the plasma membrane.) Many photosynthetic bacteria, however, do have separate membranous vesicles or lamellae containing chlorophyll.

Procaryotic cells contain a large DNA molecule, which, though not tightly associated with proteins as DNA is in eucaryotic cells, is nonetheless a true chromosome (Fig. 5.45). Often there are also small independent pieces of DNA, called plasmids. Unlike eucaryotic chromosomes, which are usually linear, the procaryotic chromosome and plasmids are ordinarily circular. Like eucaryotic chromosomes, the procaryotic chromosome bears, in linear array, the genes that control both the hereditary traits of the cell and its ordinary activities. The DNA functions by directing protein synthesis on ribosomes via mRNA, in the way already described for eucaryotic cells. Ribosomes are prominent in the cytoplasm of both eucaryotic and procaryotic cells. Those of procaryotic cells, however, are somewhat smaller than those of eucaryotic cells.

Some bacterial cells possess hairlike organelles, used in swimming, that have traditionally been called flagella, but they lack microtubules; instead they are composed of a single kind of protein, flagellin. Flagellin forms a stiff helix that is rotated like a propeller by a special structure in the membrane at its base. Table 5.1 gives a summary of some of the most important differences between procaryotic and eucaryotic cells.

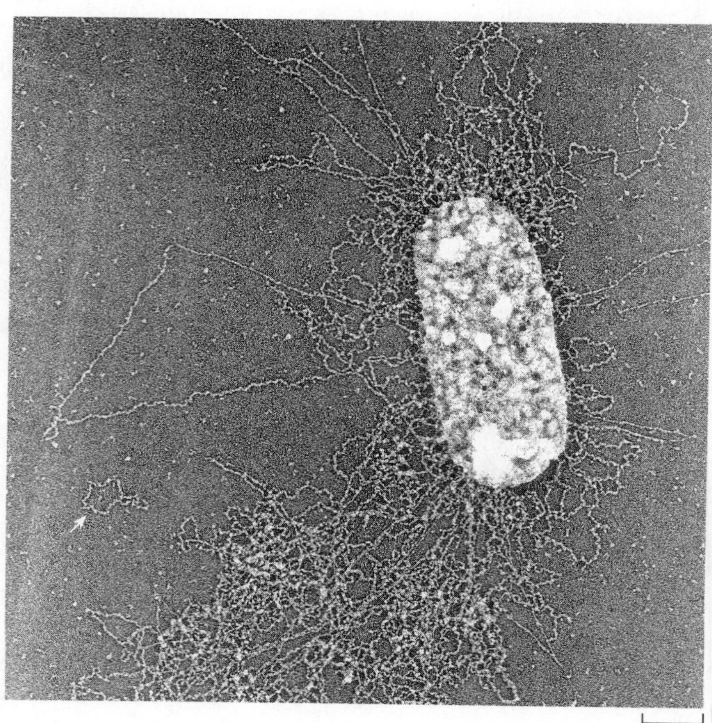

5.45 A disrupted cell of *Escherichia coli*

Exposure to detergent releases the DNA of the common intestinal bacterium *E. coli,* most of which can be seen outside the cell in this electron micrograph. Though not readily recognizable here, the main chromosome and the small accessory chromosomes (plasmids; arrow points to one) of a bacterium each form a circle rather than individual strands in the cell. The mottled appearance of the surface of the bacterium results from the effects of alcohol, drying, and shadowing with platinum.

Table 5.1 A comparison of typical procaryotic and eucaryotic cells, and certain eucaryotic organelles

CHARACTERISTIC	PROCARYOTIC CELLS	EUCARYOTIC CELLS	MITOCHONDRIA AND CHLOROPLASTS
Size	1–10 μm	10–100 μm	1–10 μm
Nuclear envelope	Absent	Present	Absent
Chromosomes	Single, circular, with no nucleosomes	Multiple, linear, wound on nucleosomes	Single, circular, with no nucleosomes
Golgi apparatus	Absent	Present[a]	Absent
Endoplasmic reticulum, lysosomes, peroxisomes	Absent	Present[a]	Absent
Mitochondria	Absent	Present[a]	
Chlorophyll	Not in chloroplasts	In chloroplasts	
Ribosomes	Relatively small	Relatively large	Relatively small
Microtubules, intermediate filaments, microfilaments	Absent	Present	Absent
Flagella	Lack microtubules	Contain microtubules	

[a] Certain parasites and anaerobic organisms lack mitochondria, having lost them either because high-energy compounds are available directly from the host cell, or because they cannot use the oxygen-dependent enzymatic pathways unique to mitochondria. One kingdom of primitive eucaryotes (discussed in Chapter 21) lacks mitochondria, ER, and Golgi.

■ THE ENDOSYMBIOTIC HYPOTHESIS: A PROCARYOTIC CORPORATION

Scientific opinion today is moving toward the view that at least two organelles found only in eucaryotes—mitochondria and chloroplasts—are the descendants of procaryotic organisms that took up residence in the "hospitable" precursors of eucaryotes. There are several lines of evidence for this ***endosymbiotic hypothesis*** (from *endo-*, "within," and *symbiosis*, "state of living together"); some of these are based on the features summarized in Table 5.1.

1 Many symbiotic associations of procaryotes and eucaryotes are known. For instance, many present-day photosynthetic bacteria live inside eucaryotic hosts, providing them with food in return for shelter. Similarly, certain nonphotosynthetic bacteria live symbiotically within eucaryotes, extracting and sharing energy from foods that their hosts cannot themselves metabolize. Such associations provide a clear starting point for the evolution of ***obligate*** symbioses, in which two organisms depend on one another for their mutual survival.

2 Both mitochondria and chloroplasts contain their own ribosomes and chromosomes; the chromosomes code for their ribosomal RNA and ribosomal proteins, and for some (though not all) of their enzymes. Mitochondria and chloroplasts also build their own membranes.

3 The organelle chromosomes resemble those of procaryotes in that they are circular, not wound on special protein spools, and not enclosed by a nuclear envelope.

4 The internal organization of the organelle genes is similar to that of procaryotes but very different from that of eucaryotes.

5 Cell division in eucaryotes involves a spindle apparatus, whereas bacteria, mitochondria, and chloroplasts divide by fissioning.

6 The ribosomes of mitochondria and chloroplasts are more similar to those of procaryotes than to the ribosomes in the cytosol of the cells in which they live.

In summary, the endosymbiotic hypothesis assumes that a bacterium with the unique metabolic capabilities exhibited by mitochondria was captured through endocytosis about 1.5 billion years ago, resisted digestion, lived symbiotically inside its host, and divided independently of its host. Later, some of the symbiont's genes were moved to the host nucleus, which took control from its guest. From this partnership all present-day plants, animals, fungi, and unicellular eucaryotic organisms must have evolved. Subsequently, according to this hypothesis, various photosynthetic bacteria met the same fate, the result being the chloroplasts found in algae and plants.

Some current speculation, less well supported but thought-provoking, holds that peroxisomes, basal bodies and centrioles, and nematocysts (the devices by which jellyfish and similar creatures harpoon potential prey and careless swimmers) may also have had an endosymbiotic origin.

CHAPTER SUMMARY

THE ROLES OF SUBCELLULAR COMPARTMENTS

THE NUCLEUS: THE CELL'S CONTROL CENTER The nucleus is bounded by a double membrane, the nuclear envelope, and contains the chromosomes. (p. 121)

THE ENDOPLASMIC RETICULUM: SYNTHESIS AND SORTING The RER binds the ribosomes synthesizing proteins destined for the ER. Some of these proteins are used in the ER; the rest are transferred to the SER, where they are packaged into vesicles for transport elsewhere. The ER also synthesizes new membrane. (p. 125)

THE GOLGI APPARATUS: SORTING AND PACKAGING The Golgi apparatus modifies and sorts molecules arriving in vesicles, and dispatches new vesicles including secretory vesicles. (p. 129)

LYSOSOMES: DIGESTIVE ORGANELLES Lysosomes contain digestive enzymes; they fuse with endocytosed vesicles containing food. (p. 130)

PEROXISOMES: DETOXIFICATION ORGANELLES Peroxisomes contain powerful oxidative enzymes. (p. 133)

MITOCHONDRIA: CELLULAR POWER STATIONS Mitochondria are complex organelles that carry out respiration. (p. 133)

CHLOROPLASTS: SOLAR ENERGY RECEPTORS Chloroplasts are complex organelles that carry out photosynthesis. (p. 133)

LEUCOPLASTS: STORAGE ORGANELLES IN PLANTS Leucoplasts store high-energy compounds for later use. (p. 134)

VACUOLES: MULTIPURPOSE ORGANELLES Vacuoles help give plant cells support and can store some molecules. (p. 135)

THE CYTOSKELETON: STRUCTURE AND MOVEMENT

MICROFILAMENTS: CELLULAR ROPES AND CABLES Actin microfilaments give structural support and, in conjunction with myosin, are involved in cell movement. (p. 136)

MICROTUBULES: CELLULAR GIRDERS AND RAILWAYS Microtubules give cells support and provide channels along which vesicles can move. Many radiate from an organizing center, the centrosome. (p. 139)

INTERMEDIATE FILAMENTS: A CELLULAR LATTICE? Intermediate filaments are tubules with at least some structural role. (p. 140)

CENTRIOLES AND BASAL BODIES: STRUCTURAL ORGANIZERS Centrioles are found near the centrosomes, and generate the basal bodies found at the base of cilia and flagella. (p. 142)

CILIA AND FLAGELLA: CELL MOVERS Cilia and flagella are mounted on the membrane and are used either to propel a cell or move material past it. (p. 143)

ARE EUCARYOTES COOPERATIVES OF PROCARYOTES?

HOW A "TYPICAL" EUCARYOTIC CELL IS ORGANIZED Eucaryotes have the organelles discussed in this chapter. (p. 145)

WHAT MAKES PROCARYOTES DIFFERENT? Procaryotes lack the organelles described in this chapter, though some are organized like mitochondria or chloroplasts. (p. 148)

THE ENDOSYMBIOTIC HYPOTHESIS: A PROCARYOTIC CORPORATION This hypothesis proposes that many organelles were either once free-living procaryotes or internal components of procaryotes. Some organelles even have their own DNA organized along procaryotic lines and use procaryote-like ribosomes. (p. 149)

STUDY QUESTIONS

1 Trace the pathway of information flow from the chromosomes to the outer membrane during the secretion of a digestive enzyme. (pp. 121–133)

2 Are the carbohydrate markers placed on proteins that are destined to be mounted in the cell membrane ever exposed to the cytosol? (pp. 129–131)

3 What would happen to a cell if it lost its Golgi apparatus? its nucleolus? its peroxisomes? (pp. 123–124, 129–130, 133)

4 List five important differences between procaryotes and eucaryotes. How do these contrasts bear on the endosymbiotic hypothesis? (pp. 148–150)

SUGGESTED READING

ALBERTS, B., D. BRAY, J. LEWIS, M. RAFF, K. ROBERTS, AND J. D. WATSON, 1994. *Molecular Biology of the Cell,* 3rd ed. Garland, New York. *This massive tome is the most complete and up-to-date summary of cell biology available.*

ALLEN, R. D., 1987. The microtubule as an intracellular engine, *Scientific American* 256 (2).

DEDUVE, C., 1986. *The Living Cell.* Scientific American Library, New York. *This two-volume set provides a well-illustrated, up-to-date tour of the cell.*

GOVER, D. M., C. GONZALEZ, AND J. W. RAFF, 1993. The centrosome. *Scientific American* 268 (6). *On how these organelles help determine cellular structure.*

ROTHMAN, J. E., 1985. The compartmental organization of the Golgi apparatus. *Scientific American,* 253 (3). *An incisive analysis of the fine structure of this important organelle.*

SIMMONS, K., H. GAROFF, AND A. HELENIUS, 1982. How an animal virus gets into and out of its host cell, *Scientific American* 246 (2). (Offprint 1511) *A fascinating account of how Semliki Forest virus subverts the membrane system of its many vertebrate hosts.*

STOSSEL, T. P., 1994. The machinery of cell crawling. *Scientific American* 271 (3).

WEBER, K., AND M. OSBORN, 1985. The molecules of the cell matrix, *Scientific American* 253 (4). *Reviews the structure and function of microfilaments and microtubules.*

CHAPTER 6

THE FLOW OF ENERGY

HOW METABOLISM EVOLVED

HOW ENERGY IS TRANSFERRED IN REACTIONS

ATP: THE CURRENCY OF ENERGY EXCHANGE

HOW CELLS HARVEST ENERGY

RESPIRATION WITHOUT OXYGEN

RESPIRATION WITH OXYGEN

HOW FATS AND PROTEINS ARE METABOLIZED

RESPIRATION

Energy is essential to life. It is used by cells to fuel growth and reproduction, and even just to maintain the status quo. On the earth at present, stored energy is usually released through aerobic respiration, in which, ultimately, oxygen and glucose are combined to yield CO_2 and water; the energy can then be used to run the cell's enzyme-mediated reactions (Fig. 6.1). These reactions, collectively known as ***metabolism***, belong to one of two phases—***anabolism***, the processes by which complex organic molecules like glucose are assembled, and ***catabolism***, the processes by which living things extract energy from food. This chapter will examine the cell's strategy of energy extraction and the machinery it uses.

THE FLOW OF ENERGY

■ HOW METABOLISM EVOLVED

Metabolism evolved under conditions very different from those that exist over most of the earth today. In particular, the ancient atmosphere had no oxygen. Relict anaerobic species found today in bogs, volcanic pools, and on the ocean bottom provide a glimpse of

6.1 Summary of biological energy flow
Today virtually all the energy for life originates in the sun, where hydrogen is converted by fusion into helium, and light is produced. In the process of photosynthesis, green plants convert the radiant energy of sunlight into chemical energy, which is most often stored initially in glucose. When cells in most organisms need energy, the glucose is broken down and some of its chemical energy is recovered by the process of aerobic respiration; the resulting product—ATP—supplies the energy in a more manageable form, making it available for muscular contraction, nerve conduction, active transport, and other work. Aerobic respiration utilizes oxygen, a by-product of photosynthesis, and photosynthesis utilizes carbon dioxide and water, by-products of respiration. With each of the transformations shown, much energy is lost as waste heat.

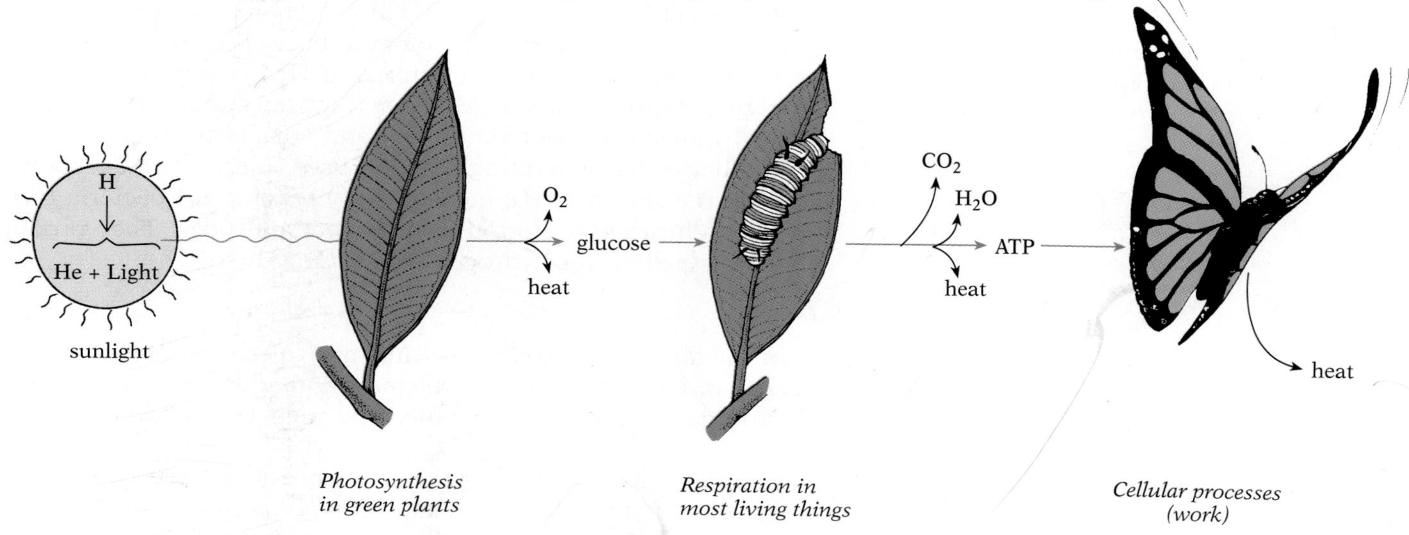

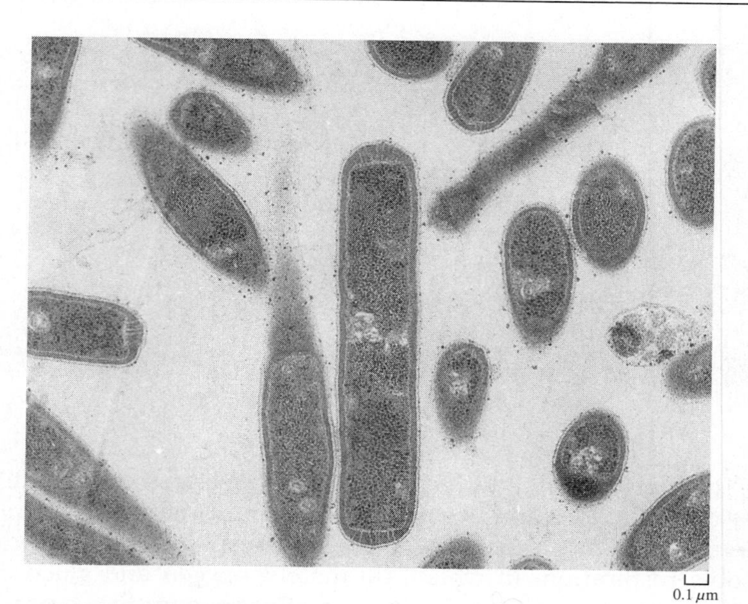

6.2 Chemosynthetic bacteria

These chemosynthetic bacteria, *Methanothermus fervidus,* live at temperatures up to 97°C in Icelandic hot springs. They combine H_2 and CO_2 to produce methane and energy.

the earliest metabolic pathways. The first organisms appeared about 3.5 billion years ago and subsisted on a limited supply of energy-rich organic molecules generated by inorganic processes. However, natural selection strongly favored organisms capable of synthesizing their own food from scratch. To create such organic compounds, this second group of early organisms used energy from inorganic sources, a process known as chemosynthesis.

Some early chemosynthetic organisms combined naturally occurring CO_2 with energy-rich H_2 to produce methane and water:

$$CO_2 + 4H_2 \longrightarrow CH_4 + 2H_2O + \text{energy}$$

As a result of this reaction, in which carbon and (especially) oxygen act as electronegative acceptors of energy-rich electrons, some of the energy stored in the electrons of H_2 is released and can be used to do work in the cell. Methane ("swamp gas") is still produced today by anaerobic bacteria in bogs and hot springs (Fig. 6.2).

Another energy-liberating reaction uses sulfur as an acceptor to liberate the energy in H_2; indeed, many present-day bacteria thrive in the sulfur-rich waste found in sewers and bogs. They produce the foul-smelling gas hydrogen sulfide (H_2S):

$$H_2 + S \longrightarrow H_2S + \text{energy}$$

In the absence of oxygen, organisms evolved the ability to use a variety of less electronegative elements: in addition to sulfur, some microorganisms still use arsenic, iron, manganese, and even uranium.

Early chemosynthetic organisms also evolved the capacity for long-term energy storage. The primary storage substance was probably glucose, or polysaccharides that could be converted into

EXPLORING FURTHER

THE HIGH ELECTRONEGATIVITY OF OXYGEN: ITS ROLE IN ENERGY TRANSFORMATIONS

Life depends on the efficient management of electrons. With the aid of highly specialized enzymes, each cell harvests the energy of electrons as they move from a position of relatively great potential energy to one of less potential energy. In biochemical pathways, an electron's potential energy can be unlocked in two ways. In photosynthesis, as we will see, a photon can energize an electron to a higher energy level within an atom—from level *L* to *M*, for example—and then this extra potential energy can be captured and used later. More often, however, biological reactions make use of the difference in potential energy between the energy of an outer electron in one atom and a vacancy with a lower potential energy in another atom. When an electron is shifted between atoms, energy is released. This movement need not be—indeed, usually is not—between different electron energy levels; instead, because the energy of, for instance, the *L* level of carbon is significantly higher than that of the *L* level of oxygen (Fig. A), movement of an electron from one to the other is highly exergonic. The same holds for electrons in covalent bonds within a molecule: the shared electrons in a C-H bond, for example, have about 11 kcal/mole more potential energy than those in an O-H bond. Transferring a hydrogen from a carbon to an oxygen, either within a single covalently bonded molecule or between two different molecules, thus releases a substantial amount of energy.

The difference in potential energies of outer-shell electrons accounts for the differences in electronegativities (see Fig. B). Of the elements commonly found in living things, oxygen is the most electronegative; hence, moving electrons to oxygen releases more free energy than shifting them to any other element. In the early stages of the evolution of life, free oxygen was exceedingly rare, so electrons usually had to be shifted to other electron acceptors. Even when oxygen that was part of a larger molecule could be used, the result was not optimal: though a substitution of electrons can result in a net liberation of energy—replacing a C-O bond with an O-H bond, for instance, is exergonic—it releases only about 25% of the energy available from the direct formation of an O-H bond from free oxygen (Fig. B). Using sulfur as an alternative electron acceptor, to produce an S-H bond rather than an O-H bond, is also exergonic but yields less than 30% of the electron's potential energy. Some bacteria still use sulfur, nitrogen, or even carbon as their electron acceptors.

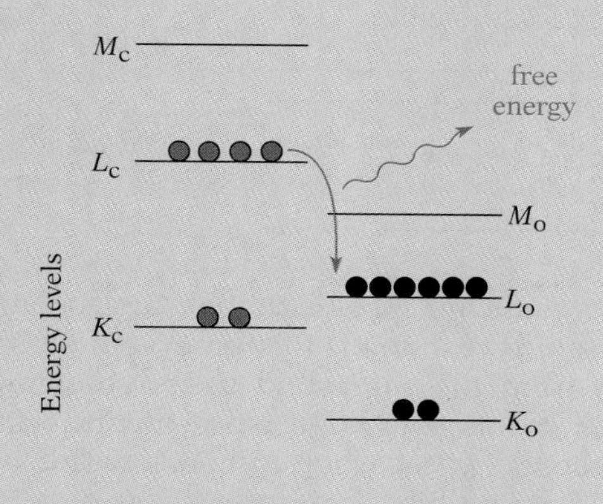

A Relative energy levels

The corresponding electron levels in two elements never have the same energy. As a result, an electron moving from the L level of carbon to the L level of oxygen will lose (liberate) energy in the process.

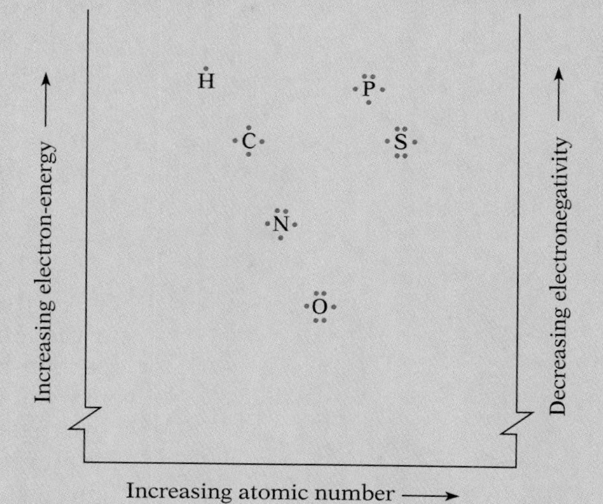

B Electronegativity and potential energy

The electronegativity chart has been inverted to show differences in electron potential energy—the essential factor controlling whether a reaction is endergonic (requiring that electrons move up on the graph) or exergonic (in which electrons move down to lower energy values). Only the six most important elements in living tissue are shown. Note that hydrogen is the least electronegative of these elements; electron transfers from hydrogen are therefore inevitably exergonic. Oxygen, on the other hand, is the most electronegative; transfers of electrons to it always release energy.

glucose. When needed, some of the energy of the glucose was extracted, as energy still is today, through glycolysis—a series of reactions that we will examine in a moment.

About 3 billion years ago certain organisms acquired a primitive ability to capture the sun's energy directly and use it to synthesize glucose and other important organic compounds. Then, about 2.5 billion years ago, the forerunner of aerobic respiration evolved. This pathway, which includes glycolysis as its first stage, extracts large amounts of energy from the end products of glycolysis, but it must use molecular oxygen (O_2) to be maximally efficient. Since molecular oxygen was in short supply, less electronegative acceptors (sulfur, nitrogen, and carbon) had to suffice, and so cells wasted much of the potential energy in the chemical bonds of storage compounds.

Oxygen became increasingly available when the descendants of the early photosynthetic organisms began producing it as a by-product of a more advanced form of photosynthesis; thus, about 2.3 billion years ago, oxygen began to accumulate in the atmosphere. Since oxygen (in the form of ozone, O_3) filters out destructive X rays and ultraviolet light from the sun, organisms could now emerge from beneath the radiation shielding provided by earth and water to colonize the surface of the land. In addition, the abundance of oxygen made aerobic respiration the dominant catabolic pathway.

■ HOW ENERGY IS TRANSFERRED IN REACTIONS

Sugars like glucose are energy-rich compounds, while carbon dioxide is energy-poor. Through aerobic respiration glucose is "burned" to produce CO_2 and energy for the cell:

$$C_6H_{12}O_6 + 6O_2 + 6H_2O \longrightarrow 6CO_2 + 12H_2O + \text{energy}$$

As the colored arrows indicate, the essential molecular rearrangement that occurs in this reaction is the removal of all of the carbon-bound hydrogens from the sugar and their rebonding to oxygen. Energy is released in this reaction because any transfer of hydrogen electrons from bonds with carbon to bonds with oxygen is exergonic. The enzymes of aerobic respiration use some of this energy to do work in the cell; the rest of the energy is lost as heat. This transfer of hydrogen atoms from carbon to oxygen is an example of ***oxidation***. Stated more generally, a substance is oxidized when it loses electrons (almost always hydrogen electrons)—and sometimes the rest of the atom to which the electrons are attached—to a more electronegative substance (usually oxygen); as a result, oxidation inevitably liberates energy.

The converse process—storing energy in a compound—is called, oddly enough, ***reduction***. Photosynthesis, for example, works to reduce CO_2, basically reversing aerobic respiration: solar energy is used to power the synthesis of high-energy sugar from energy-poor CO_2. The reduction of CO_2 occurs through the redeployment of hydrogens from low-energy bonding with electronegative oxygen to high-energy bonding with carbon:

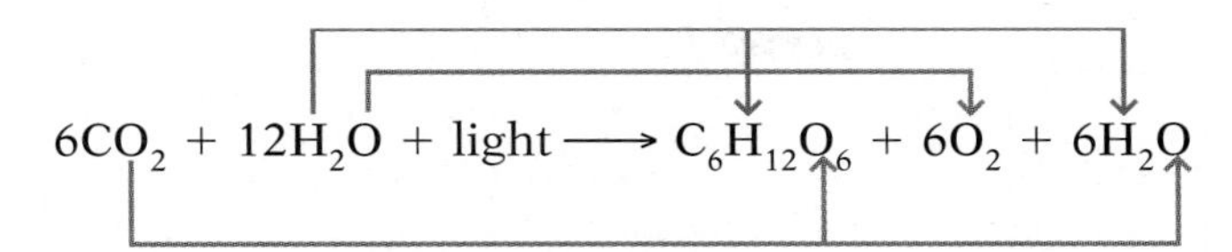

Though reduction in cellular reactions usually involves the transfer of hydrogen atoms from oxygen to carbon, reduction in general occurs whenever an electron (with or without the rest of its atom) is moved to a less electronegative atom. This broader definition is important because we will see cases in both photosynthesis and aerobic respiration in which oxidation and reduction entail the movement of isolated electrons, or involve elements other than oxygen and hydrogen.

We can see that oxidation and reduction are mirror-image processes, one lowering, the other raising an electron's potential energy. Since all reactions involve making or rearranging bonds, it follows that if one atom or molecule gains energy, another must have lost energy—that is, whenever one substance is reduced, another is oxidized. We can follow this so-called ***redox*** reaction from the perspective of an electron, as it moves from one atom to another:

$$A^{e-} + B \longrightarrow \underset{\substack{\text{has been}\\ \text{oxidized}\\ \text{(lost energy)}}}{A} + \underset{\substack{\text{has been}\\ \text{reduced}\\ \text{(gained energy)}}}{B^{e-}}$$

More often though, the electron undergoing a change in potential is part of an atom that is rebonded; no ions are created. Just as the most common reduction reactions involve the removal of oxygen or addition of hydrogen, oxidation usually involves the addition of oxygen or removal of hydrogen:

$$A + BO \longrightarrow \underset{\substack{\text{has been}\\ \text{oxidized}}}{AO} + \underset{\substack{\text{has been}\\ \text{reduced}}}{B}$$

or

$$AH + B \longrightarrow \underset{\substack{\text{has been}\\ \text{oxidized}}}{A} + \underset{\substack{\text{has been}\\ \text{reduced}}}{BH}$$

In cells, removal or addition of an electron derived from hydrogen is the most frequent mechanism of redox reactions. A summary of redox reactions appears in Table 6.1.

It now becomes clear why the synthesis of sugar from carbon dioxide constitutes reduction of the CO_2: hydrogen obtained by splitting water molecules is added to the CO_2, thereby storing energy that is later recovered through respiration. The main product of this subsequent respiration is ATP, the major energy currency of the cell.

Table 6.1 Redox reactions

OXIDATION	REDUCTION
Adds oxygen	Removes oxygen
Removes hydrogen	Adds hydrogen
Removes electron(s)	Adds electron(s)
Liberates energy	Stores energy

■ ATP: THE CURRENCY OF ENERGY EXCHANGE

One of the essential substances of life is the compound ***adenosine triphosphate***, or ***ATP***, which plays a key role in nearly every energy transformation in living things (Fig. 6.3).

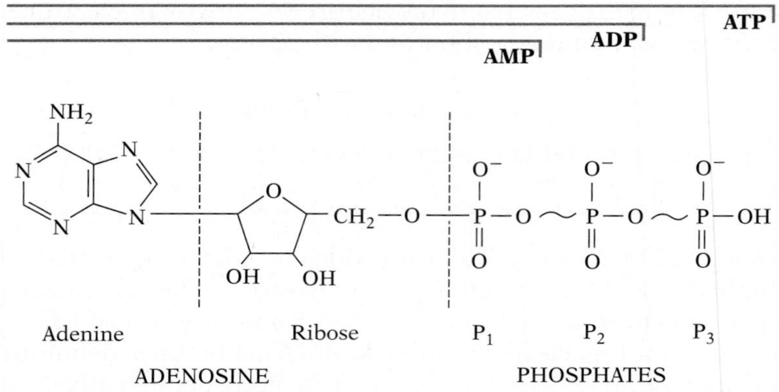

AMP ADP ATP

NH_2 N N N N O CH_2—O—P—O ~ P—O ~ P—OH

OH OH

Adenine Ribose P_1 P_2 P_3

ADENOSINE PHOSPHATES

6.3 The ATP molecule

Adenosine triphosphate (ATP) is composed of an adenosine unit (a complex of adenine and ribose sugar) and three phosphate groups arranged in sequence. The last two phosphates are attached by so-called high-energy bonds (wavy lines). The cell stores energy by adding a phosphate group to ADP (adenosine diphosphate) to make ATP, and later recovers some of this energy by hydrolyzing ATP into ADP and inorganic phosphate. Occasionally ADP is hydrolyzed to make AMP (adenosine monophosphate), one of the nucleotides of which RNA is composed. The widely used term *high-energy bonds* is misleading: the energy is actually stored within the phosphate group itself.

The ATP molecule is composed of a nitrogen-containing compound (adenosine) plus three phosphate groups bonded in sequence:

adenosine – Ⓟ~ Ⓟ~ Ⓟ

According to convention, Ⓟ stands for the entire phosphate group, and the wavy lines between the first and second and the second and third phosphate groups represent so-called high-energy bonds.[1]

Actually, it is often only the terminal phosphate bond of ATP that is involved in energy conversions. The exergonic reaction by which this bond is hydrolyzed and the terminal phosphate group removed leaves a compound called ***adenosine diphosphate,*** or ***ADP*** (adenosine plus two phosphate groups), and inorganic phosphate (symbolized by P_i):

$$ATP + H_2O \xrightarrow{\text{enzyme}} ADP + P_i + \text{energy}$$

If both the second and third phosphate groups are removed from ATP, the resulting compound is ***adenosine monophosphate,*** or ***AMP***.

New ATP can be synthesized from ADP and inorganic phosphate

[1] The widely used term *high-energy bond* does not mean that energy is stored in the phosphate bonds themselves. Instead, it means that the bonds between the negatively charged phosphate groups are very unstable (and so are thermodynamically inexpensive to break) and that substantially more energy is released when the phosphate groups form new bonds.

if adequate energy is available to force a third phosphate group onto the ADP. Addition of phosphate is termed ***phosphorylation***:

$$\text{ADP} + \text{P}_i + \text{energy} \xrightarrow{\text{enzyme}} \text{ATP} + \text{H}_2\text{O}$$

ATP is often called the universal energy currency of living things. Cells initially store energy in the form of carbohydrates, such as glucose, and of lipids, such as fat. But the amount of energy in even a single glucose molecule—670 kcal/mole—is inconveniently large for driving most reactions. The hydrolysis of ATP to ADP releases a more useful amount of energy: 7.3 kcal/mole. Glucose molecules are the hundred-dollar bills of the cellular economy, while ATP molecules are the everyday denominations, sufficient for making and breaking covalent bonds. Though some other compounds can supply energy, ATP is the one most often used by cells in the various kinds of work they perform: it powers synthesis of more complex compounds, muscular contraction, nerve conduction, active transport across cell membranes, light production, and so on. The price of the work is paid through the energy-releasing hydrolysis of ATP to ADP.

HOW CELLS HARVEST ENERGY

The energy stored in lipids and carbohydrates is liberated gradually through a series of reactions, each catalyzed by its own specific enzyme. Certain steps release small amounts of energy that are transferred to ATP by the phosphorylation just described. Let us look at the most important steps in the catabolism of glucose, the central energy-liberating pathway in all organisms.

■ RESPIRATION WITHOUT OXYGEN

Glycolysis The complete catabolism of glucose involves five stages, divided between anaerobic and aerobic series of reactions. Stage I, ***glycolysis***, is the anaerobic portion of the process, and breaks glucose down to pyruvic acid. Glycolysis is the most ancient part of the pathway; it evolved long before free oxygen became available. We will trace the steps of glycolysis in detail because they illustrate how reaction pathways are organized into small, enzyme-mediated steps and are made to work through the careful management of free energy changes. It is these principles and the overall sequence of steps (summarized in Fig. 6.4) that are important, not the exact thermodynamics and the structural details of the molecular intermediates.

Glucose is a stable compound, one with little tendency to break down spontaneously. If its energy is to be harvested, glucose must first be activated by the investment of a small amount of energy. The first steps of glycolysis, therefore, are preparatory, enabling the later steps to extract the stored energy.

The energy for initiating glycolysis (Fig. 6.4) comes from ATP. The initial reaction, like the succeeding ones, is made possible by a step-specific enzyme, which binds (via weak bonds) to the reactants, activates them (by causing the electrons of the bound molecules to redistribute themselves), and then joins or rearranges the

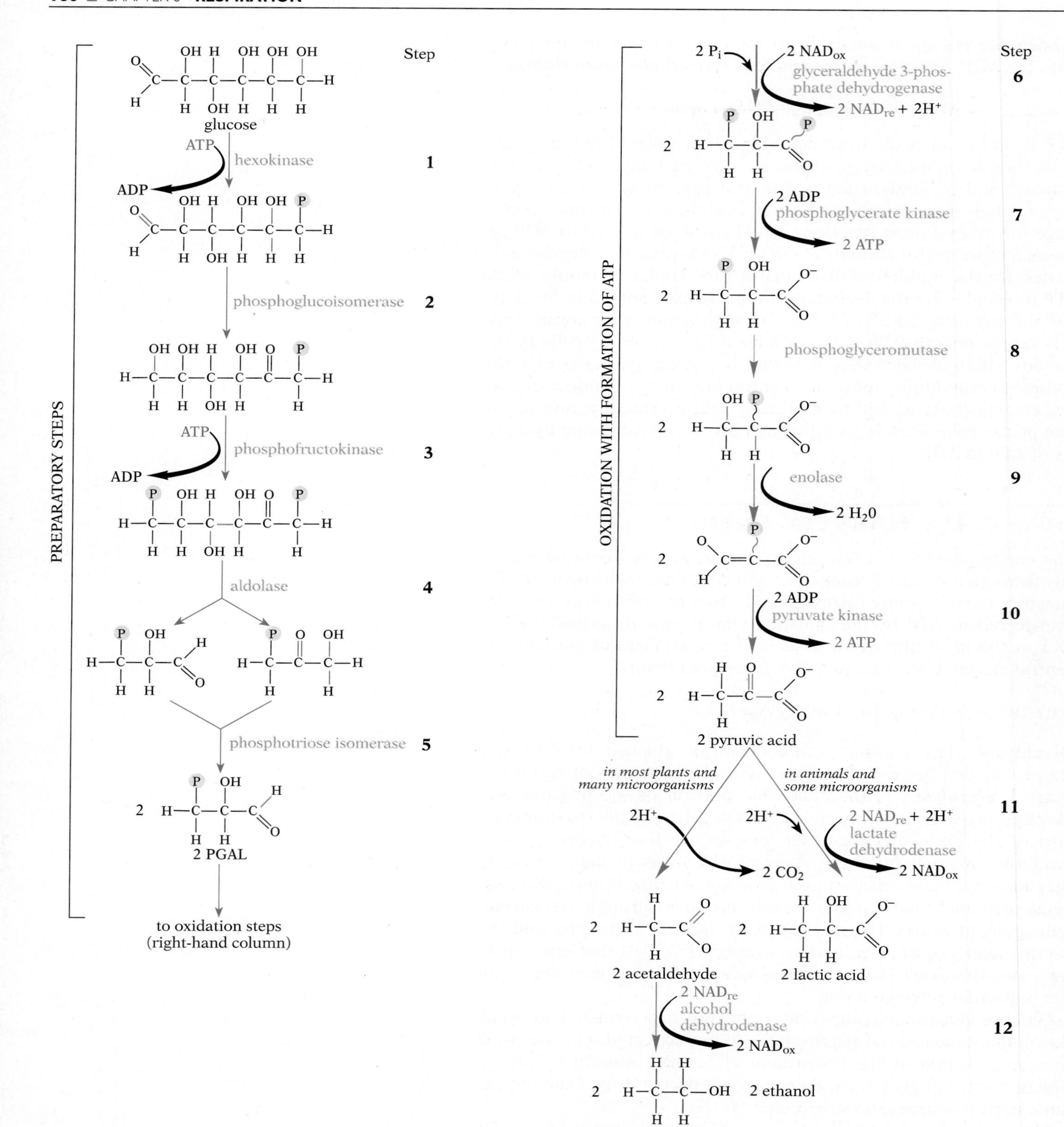

Step
glucose
ATP
hexokinase
ADP
1
phosphoglucoisomerase
2
ATP
phosphofructokinase
ADP
3
aldolase
4
phosphotriose isomerase
5
2 PGAL
to oxidation steps
(right-hand column)
PREPARATORY STEPS
OXIDATION WITH FORMATION OF ATP
2 Pi
2 NADox
glyceraldehyde 3-phosphate dehydrogenase
2 NADre + 2H+
6
2 ADP
phosphoglycerate kinase
2 ATP
7
phosphoglyceromutase
8
enolase
2 H20
9
2 ADP
pyruvate kinase
2 ATP
10
2 pyruvic acid
in most plants and many microorganisms
in animals and some microorganisms
2H+
2H+
2 NADre + 2H+
lactate dehydrodenase
2 NADox
2 CO2
11
2 acetaldehyde
2 lactic acid
2 NADre
alcohol dehydrodenase
2 NADox
12
2 ethanol

6.4 Glycolysis and fermentation

The entire reaction series for glycolysis is shown to illustrate how biochemical pathways work; there is no reason to try to memorize each step. Glucose in solution normally exists as a ring structure, but the straight-chain form is adopted here for clarity. Energy to initiate the breakdown of glucose is supplied by two molecules of ATP (Steps 1–3). The resulting compound is then split into two molecules of PGAL (5). This completes the preparatory reactions. Next, the PGAL is oxidized by the removal of hydrogen, and inorganic phosphate is added to each of the three-carbon molecules (6). A series of reactions then results in the synthesis of four new molecules of ATP, for a net gain of two (7–10). The pyruvic acid produced by this anaerobic breakdown can be further oxidized in most cells if O_2 is present (by reactions not shown here). In the absence of sufficient O_2, the pyruvic acid may be converted to lactic acid in some kinds of organisms, or to CO_2 and ethanol in others (11–12). At each step a particular enzyme catalyzes a specific redistribution of electrons and thereby brings about changes in bonding. This step-by-step strategy is the only one by which enzymes can guide reactions. (In each diagram, bonds to be altered by enzymatic actions in the next step are shown in green. The fate of particular oxygens and hydrogens cannot always be traced from step to step because they are sometimes incorporated into units, such as phosphate groups, for which full diagrams are not given.) Figure 6.4 provides a more schematic summary of glycolysis.

reactants before releasing the products. In the first reaction, a molecule of ATP donates its terminal phosphate group to the glucose.

$$(1)\ \underset{\text{glucose}}{\text{C—C—C—C—C—C}} + \text{ATP} \xrightarrow{\text{enzyme}} \underset{\text{glucose-6-phosphate}}{\text{C—C—C—C—C—C—}\text{Ⓟ}} + \text{ADP}$$

$$(\Delta G = \text{-}4.0 \text{ kcal/mole})$$

(The simplified equations given here show only the carbon skeleton; the more complete molecular structures are shown in Fig. 6.4.) The name of the product tells us that it is a glucose with a phosphate group attached to its sixth carbon atom.

Let's look carefully at what has happened in this reaction. An enzyme, hexokinase, has bound glucose and ATP, catalyzed the transfer of a phosphate group to the glucose, and released the products. The overall change in free energy of the electrons rearranged in this reaction is -4.0 kcal/mole; the free energy is liberated primarily as heat. As always, the minus sign means this is an exergonic (downhill) reaction.[2] In fact, the reaction has an equilibrium constant (K_{eq}) of about 1000—that is, the products outnumber the reactants by 1000 to 1. The free energy for this reaction comes from ATP: the energy available from the terminal phosphate of the ATP is 7.3 kcal/mole. Only 4.0 kcal/mole is liberated to drive this first step of glycolysis, and the other 3.3 kcal/mole from the ATP is stored in the electrons of the product—the activated glucose.

[2] As noted in Chapter 3, we use ΔG (rather than the formal symbol $\Delta G'^0$) for the change in the free energy of a reaction under standard conditions. In addition, we divide the ΔGs value simply between products and liberated free energy, even though, in fact, a more complicated but poorly understood division is probably occurring. The resulting approximation is reasonably accurate. The equilibrium constant (K_{eq}) is being treated for simplicity as though there is only a single product. This is possible because glycolysis has relatively little effect on the concentration of water, ATP, ADP, and so on; the concentration of these compounds is maintained through various homeostatic mechanisms in the cell.

The next step in glycolysis converts glucose-6-phosphate into the nearly identical compound fructose-6-phosphate:

(2) C—C—C—C—C—C—Ⓟ $\xrightarrow{\text{enzyme}}$ C—C—C—C—C—C—Ⓟ
glucose-6-phosphate → fructose-6-phosphate
(ΔG = +0.4 kcal/mole)

The positive ΔG indicates that this is an uphill, endergonic (energy-requiring) reaction, one that cannot proceed spontaneously. How is it, then, that glycolysis does not grind to a halt? The answer lies in the operation of ***coupled reactions:*** two reactions that share a common intermediate molecule—in this instance, glucose-6-phosphate, the product of Step 1 *and* the reactant of Step 2—can proceed as a single reaction. The 1000:1 ratio of products to reactants in Step 1 supplies an enormous number of reactant molecules to Step 2; this abundance insures that, even though the equilibrium constant of the second reaction is only about 0.2 (that is, a 1:5 ratio of products to reactants), many product molecules are nevertheless created to feed Step 3.

From the perspective of thermodynamics, these two steps can be treated as one reaction. The -4.0 kcal/mole liberated by Step 1 is combined with the +0.4 kcal/mole consumed by Step 2 to yield a net ΔG of -3.6 kcal/mole. Taken together, the two steps are strongly exergonic, and so the reaction proceeds. The glycolytic pathway is a series of such coupled reactions, in which exergonic steps push or pull endergonic steps, with the favorable *net* free-energy change of the steps taken together enabling the sequence of reactions to proceed.

The conversion of glucose-6-phosphate into fructose-6-phosphate also provides a good illustration of how enzymatic pathways work. Recall that when a substrate binds to an enzyme by means of weak bonds, a slight shift in the electron distribution of the substrate is induced, which lowers the activation energy required for a particular change in bonding and thereby catalyzes the reaction. The result of the redistribution of electrons can be seen in Step 2, in which two hydrogens, bonded to the fifth carbon and its oxygen, are transferred to the first carbon and its oxygen. This trivial change is essential to prepare for the next step in glycolysis. Once the change has taken place, the substrate no longer "fits" the enzyme, and so the substrate drifts away to be captured by the next enzyme in the series. Each step in the glycolytic pathway is therefore very small and is mediated by a highly specific enzyme.

After the formation of fructose-6-phosphate in Step 2, another molecule of ATP is consumed, to add a phosphate to the other end of the molecule. Of the ATP's energy, 3.9 kcal/mole is stored in the product, while the remaining 3.4 kcal/mole is liberated as heat; hence the reaction is exergonic:

(3) C—C—C—C—C—C—Ⓟ + ATP $\xrightarrow{\text{enzyme}}$
fructose-6-phosphate
Ⓟ—C—C—C—C—C—C—Ⓟ + ADP
fructose-1,6-bisphosphate
(ΔG = -3.4 kcal/mole)

Since this reaction in its turn is coupled to the previous one (with which it shares the intermediate compound fructose-6-phosphate), we can add the free energy of the separate steps along the way. The overall reaction chain so far has liberated 7.0 kcal/mole, and so it has a highly favorable K_{eq} of more than 10^5.

Next, the fructose-1,6-bisphosphate is split between the third and fourth carbons, forming two essentially similar three-carbon molecules (Step 4). One is ***PGAL*** (phosphoglyceraldehyde), and the other, an intermediate compound, is usually converted immediately to PGAL, in Step 5. (The cell can also use PGAL to synthesize fat if conditions warrant.) PGAL, a phosphorylated three-carbon sugar, is a key intermediate in both glycolysis and photosynthesis.

$$(4\text{–}5)\quad \underset{\text{fructose-1,6-bisphosphate}}{Ⓟ\text{—C—C—C—C—C—C—}Ⓟ} \xrightarrow{\text{enzyme}} 2\ \underset{\text{PGAL}}{\text{C—C—C—}Ⓟ}$$

$$(\Delta G = +7.5 \text{ kcal/mole})$$

To this point, instead of releasing energy from glucose to form new ATP molecules, glycolysis has actually cost the cells two ATPs. Indeed, Steps 4 and 5 are so unfavorable energetically that the net change in free energy is now +0.5 kcal/mole. For subsequent reactions to proceed, significant amounts of free energy must be liberated to pull the reactants past the five preparatory steps.

The next reaction actually involves two separate molecular changes, which we summarize for simplicity in one step. The first change is oxidation of the PGAL by reduction of ***NAD*** (nicotinamide adenine dinucleotide). (The characteristic function of NAD is temporary storage of high-energy electrons; it transports energy from one pathway to another, or from one step in a pathway to another step, elsewhere in the pathway.) Each NAD_{ox} accepts two hydrogens, keeping one and the electron of the other to produce NAD_{re} and an H^+ ion. The second change is phosphorylation of the PGAL:

$$(6)\quad 2\ \text{C—C—C—}Ⓟ + 2\ NAD_{ox} + 2P_i \xrightarrow{\text{enzyme}}$$

$$2\ Ⓟ\sim\text{C—C—C—}Ⓟ + 2\ NAD_{re} + 2\ H^+$$

$$(\Delta G = +3.0 \text{ kcal/mole})$$

The oxidation phase of this reaction, taken alone, is strongly exergonic, while the phosphorylation phase is strongly endergonic. Since the two processes occur together, the energy that would have been released by the oxidation (more than 100 kcal/mole) is conserved in the reduced NAD (NAD_{re}) and the phosphorylated PGAL. The consequence of Step 6 is to make the net change in free energy for the overall reaction chain even more unfavorable (+3.5 kcal/mole), but the next downhill reaction, to which Step 6 is coupled, begins once again to turn the balance, as the high-energy phosphate bond is broken; some of the free energy is harvested by transferring the phosphate group to the ADP, to form ATP:

$$(7)\quad 2\ Ⓟ\sim\text{C—C—C—}Ⓟ + 2\text{ADP} \xrightarrow{\text{enzyme}} 2\ \text{C—C—C—}Ⓟ + 2\text{ATP}$$

$$(\Delta G = -9.0 \text{ kcal/mole})$$

At this point, then, the cell regains the two ATP molecules invested to activate the glucose in Steps 1 and 3, and the overall net change

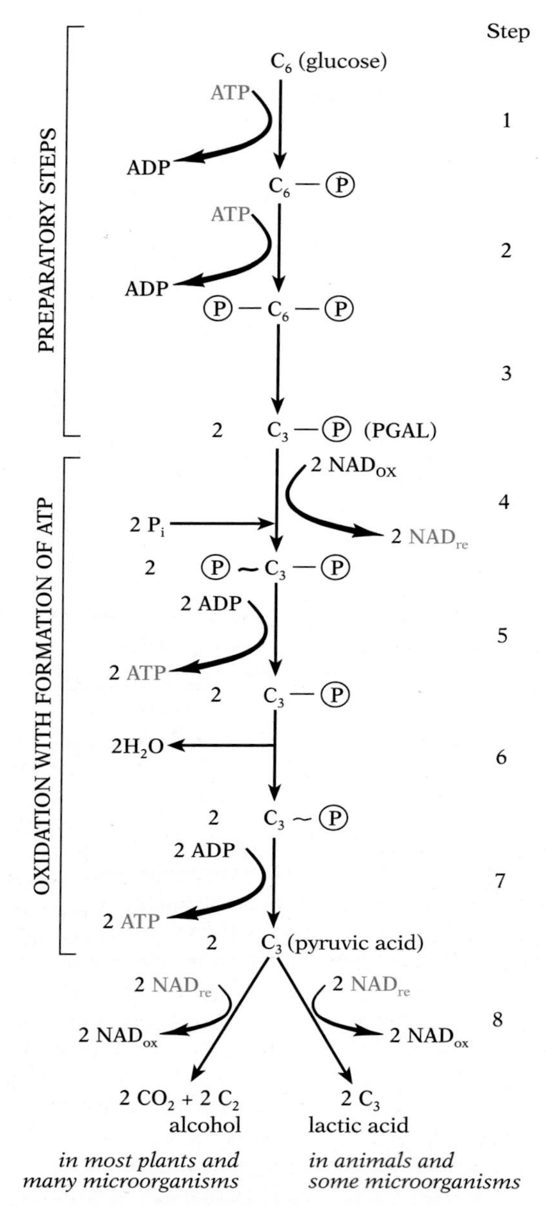

6.5 Summary of glycolysis and fermentation

in free energy is again favorable: -5.5 kcal/mole. Moreover, a great deal of energy has been stored in NAD_{re}.

Next comes a reaction that energizes the remaining phosphate groups:

$$(8)\quad 2\ C—C—C—Ⓟ \xrightarrow{\text{enzyme}} 2\ C—C—C{\sim}Ⓟ + H_2O \qquad (\Delta G = +1.5\ \text{kcal/mole})$$

After a reaction that rearranges the substrate for a ΔG of -0.4 kcal/mole (Step 9), these energized phosphate groups are transferred to ADP; the products are ***pyruvic acid*** (also referred to as pyruvate) and ATP:

$$(10)\quad 2\ C—C—C{\sim}Ⓟ + 2\ ADP \xrightarrow{\text{enzyme}} \underset{\text{pyruvic acid}}{2\ C—C—C} + 2\ ATP \qquad (\Delta G = -15.0\ \text{kcal/mole})$$

Obviously this last reaction is overwhelmingly favorable; indeed, the K_{eq} is greater than 10^9. Moreover, a profit of two ATPs is generated by this step of glycolysis. Because the two ATP molecules used in Steps 1 and 3 have already been regained (in Step 7), the two additional molecules formed here represent a net gain in ATP for the cell. The highly exergonic last step results in an overall ΔG of -19.4 kcal/mole. It is the liberation of this energy, primarily as heat, that causes this series of coupled reactions to proceed.

Figure 6.5 summarizes the essential steps of glycolysis; Figure 6.6 shows graphically both the thermodynamics of the reaction steps and the changes in free-energy content at each successive step in glycolysis, from glucose to pyruvic acid (as well as in the process of fermentation, which will be discussed shortly).

We can now summarize the most important features of glycolysis:

1 Each molecule of glucose (a six-carbon compound) is broken down into two molecules of pyruvic acid (a three-carbon compound).

2 Two molecules of ATP are used to initiate the process. Later, four new molecules of ATP are synthesized, for a *net* gain of two molecules of ATP from each glucose.

3 Two molecules of NAD_{re} are formed.

4 Glycolysis can occur whether or not O_2 is present. Glycolysis is found in the cytoplasm of all living cells.

Fermentation We have seen that in glycolysis two molecules of NAD_{ox} are reduced to NAD_{re}, and that NAD functions in the cell as a shuttle for high-energy electrons. Thus NAD is only a temporary acceptor of electrons, promptly passing its extra electrons to some other compound and then going back for another load. The cell has only a limited supply of NAD molecules, and these must be used over and over again. If the NAD_{re} molecules formed in glycolysis could not quickly unload electrons (that is, be oxidized back into NAD_{ox}), all the cell's NAD would soon be tied up. With Step 6 of glycolysis thus blocked, the process would come to an end.

As we noted in the discussion of Step 6, more than 50 kcal/mole of free energy is stored in each of the two molecules of NAD_{re} produced during the glycolysis of a glucose molecule. For the cell to harvest this energy, the NAD_{re} must transfer its electrons to a lower energy level in some more electronegative acceptor molecule. In most cells, molecular oxygen is the ultimate acceptor of the transferred electrons. Under anaerobic conditions, however, with no oxygen present, the pyruvic acid formed by glycolysis must accept the electrons from NAD_{re}. This reduction of pyruvic acid results in the formation of ***lactic acid*** in animal cells and some unicellular organisms, and of ***ethanol*** (ethyl alcohol) and carbon dioxide in most plants and many unicellular organisms:

$$(11)\ \text{pyruvic acid} + NAD_{re} + H^+ \xrightarrow{\text{enzyme}} \text{lactic acid} + NAD_{ox}$$

or

$$(11)\ \text{pyruvic acid} \xrightarrow{\text{enzyme}} \text{acetaldehyde} + CO_2$$

$$(12)\ \text{acetaldehyde} + NAD_{re} + H^+ \xrightarrow{\text{enzyme}} \text{ethanol} + NAD_{ox}$$

Thus, under anaerobic conditions, NAD shuttles back and forth, picking up electrons (becoming NAD_{re}) in Step 6 and giving up the electrons (becoming NAD_{ox}) in Step 11 or 12. The process that begins with glycolysis and ends with the transformation of pyruvic acid into ethanol or lactic acid is called **fermentation**.[3]

Fermentation enables a cell to continue synthesizing ATP by breaking down nutrients under anaerobic conditions. However, because the electrons transferred from NAD_{re} remain at a relatively high energy level in the reduction of pyruvic acid, fermentation extracts only a very small portion (about 2%) of the energy present in the original glucose.

Fermentation by yeast cells and other microorganisms is the basis for the industries that produce breads and alcohol. Bacterial fermentation is also essential to the production of most cheeses, yogurt, and a variety of other dairy products.

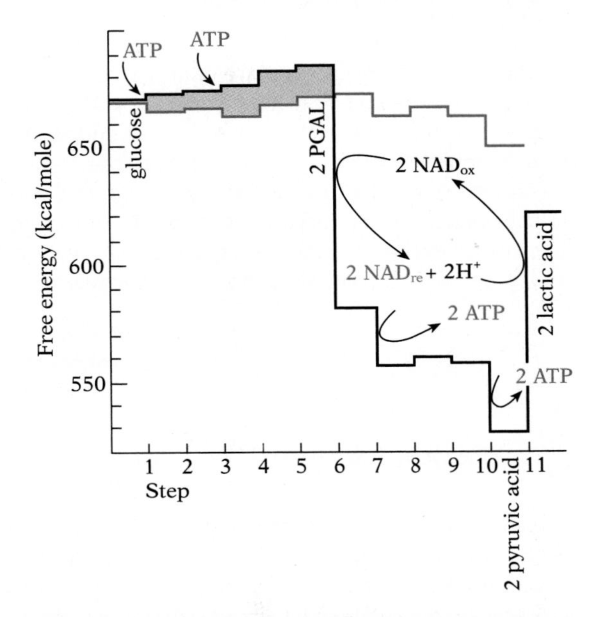

6.6 Changes in free energy at successive steps in glycolysis and fermentation

The graph shows two very different free-energy summaries. The first (red) traces only the free-energy changes resulting from each reaction; these values correspond to the ΔG values cited in the text, and show the net change of -19.4 kcal/mole that accompanies glycolysis. The black line represents the energy in the chemical intermediates; it includes the cost of the ATPs invested in Steps 1 and 3 to activate the glucose. The large drops in free energy correspond to steps in which energy is removed from those intermediates and stored in ATP or NAD_{re}. In the five preparatory steps, which convert glucose into PGAL, the free-energy content of the intermediate substances is slightly increased owing to the investment of two molecules of ATP. In Step 6 there is a sharp drop in free energy associated with the formation of two molecules of NAD_{re}. There are also major drops in Steps 7 and 10, each associated with the formation of two molecules of ATP. The space between the two free-energy summaries indicates the cellular energy profit; the process runs at a loss until Step 6.

■ RESPIRATION WITH OXYGEN

In eucaryotic cells the more efficient process of energy extraction that occurs in the presence of abundant molecular oxygen, aerobic respiration, takes place exclusively in the mitochondria. When O_2 is present, it acts as the electron acceptor; mitochondrial enzymes move the transient electrons to the oxygen, thereby releasing the free energy of NAD_{re}:

$$O_2 + 2\ NAD_{re} + 2H^+ \longrightarrow 2H_2O + 2\ NAD_{ox} \quad (\Delta G = -52.4\ \text{kcal/mole})$$

[3] The term *fermentation* has been used in countless ways in the scientific literature. It is often restricted to the breakdown of glucose to ethanol. It is also applied to the production of either ethanol or lactic acid by microorganisms—lactic acid production in animal cells being called glycolysis. Both these uses lead to confusion between the terms *fermentation* and *glycolysis,* and both tend to obscure the general occurrence of the same basic fermentation process in all living cells. Accordingly, *fermentation* is here applied to any process in which glucose is catabolized and organic molecules are used as the electron acceptors of glycolysis. The glycolytic pathway to pyruvic acid is taken both as a preparatory reaction sequence leading to the Krebs citric acid cycle when sufficient oxygen is present, and as the initial portion of fermentation in the absence of sufficient oxygen.

EXPLORING FURTHER

COUPLED REACTIONS

Many important reactions are endergonic; outside energy is required to build DNA, RNA, proteins, and other molecules, to form structures like cell membranes, and to activate energy sources such as glucose and fat so their stored energy can be utilized. This outside energy is supplied by a coupling of the unfavorable endergonic reactions to one or more strongly exergonic reactions. The first two reactions of glycolysis provide an example of how coupling works:

$$\text{(1)}\quad \text{glucose} + \text{ATP} \xrightleftharpoons{K_{eq} = 610} \text{glucose-6-phosphate} + \text{ADP} \quad (\Delta G = -4.0 \text{ kcal/mole})$$

$$\text{(2)}\quad \text{glucose-6-phosphate} \xrightleftharpoons{K_{eq} = 0.54} \text{fructose-6-phosphate} \quad (\Delta G = +0.4 \text{ kcal/mole})$$

Reaction 1 is downhill: the products eventually outnumber the reactants by 610:1, while reaction 2 is uphill, as the reactant eventually outnumbers the product by about 2:1. Because one of the products of reaction 1 is the reactant of reaction 2, the reactions are coupled—they share an ***intermediate compound***, glucose-6-phosphate. The table, a molecular scorecard, shows these coupled reactions set in motion by the combination of 10^5 molecules of glucose and an equal number of molecules of ATP (*a*). The equilibrium constant of reaction 1 (610) means that the reaction will proceed until a ratio of roughly 99,836 molecules of glucose-6-phosphate to 164 of glucose has been reached (*b*). Since glucose-6-phosphate is a reactant of reaction 2, some of it will be converted into fructose-6-phosphate. The equilibrium constant of this reaction is 0.54, so about 35,007 molecules of product will then be produced (*c*).

With the consequent reduction in the total amount of glucose-6-phosphate, reaction 1 is no longer at equilibrium, so additional glucose and ATP react to produce glucose-6-phosphate (*d*), only to have some of this intermediate promptly converted into fructose-6-phosphate (*e*). At this point equilibrium is reached, with the ratios of product to reactant for both reactions 1 and 2 approximating the corresponding equilibrium constants. The addition of more glucose and ATP (which happens almost continuously in cells) or the removal of fructose-6-phosphate (which also happens continuously, as it is converted into another compound in the third step of glycolysis) will cause more molecules to move along this reaction pathway (*f–j* and *k–n*).

Thus favorable and unfavorable reactions are coupled in the cell: reactants are pushed through the uphill steps as long as appropriate intermediate compounds are present and the *net* ΔG is negative. In fact, cells function by coupling literally thousands of reactions into long chains in this manner.

	REACTANTS	REACTION 1 (K_{eq} = 610)	INTERMEDIATE COMPOUND	REACTION 2 (K_{eq} = 0.54)	PRODUCT
	glucose + ATP	⇆	glucose-6-phosphate	⇆	fructose-6-phosphate
a	100,000		0		0
b	164		99,836		0
c	164		64,829		35,007
d	106		64,887		35,007
e	106		64,866		35,028
Addition of 10^4 molecules of glucose and 10^4 molecules of ATP					
f	10,106		64,866		35,028
g	123		74,849		35,028
h	123		71,349		38,528
i	117		71,355		38,528
j	117		71,353		38,530
Removal of 10^4 molecules of fructose-6-phosphate					
k	117		71,353		28,530
l	117		64,857		35,026
m	106		64,868		35,026
n	106		64,864		35,030

The free energy is then utilized in the formation of ATP. Moreover, the pyruvic acid (which still has 590 kcal/mole of free energy at Step 10) can be broken down to yield additional energy for the synthesis of still more ATP. (If lactic acid has already been formed, it can be reconverted into pyruvic acid—thus regenerating the lost NAD_{re}—when sufficient oxygen becomes available.)

The process of aerobic breakdown of nutrients with accompanying synthesis of ATP is called ***aerobic respiration***. Whereas anaerobic respiration comprises fermentation (glycolysis, followed by the transfer of electrons to pyruvic acid, producing lactic acid or ethanol), aerobic respiration consists of a longer sequence of events: glycolysis (Stage I of aerobic respiration), oxidation of pyruvic acid to acetyl CoA, the reactions of the Krebs citric acid cycle, a series of reactions involving an electron-transport chain, and finally the processes that culminate in the synthesis of ATP. It is the transfer of electrons from NAD_{re} to highly electronegative oxygen atoms—a transfer mediated by the electron-transport chain—that ultimately produces most of the ATP from glucose.

Oxidation of pyruvic acid (Stage II of aerobic respiration) The net effect of the aerobic oxidation of pyruvic acid is to break down the three-carbon pyruvic acid to CO_2 and the two-carbon compound acetic acid, which is connected by a high-energy bond to a coenzyme called CoA; the complete compound is acetyl-coenzyme A, or ***acetyl-CoA***. When a molecule of pyruvic acid is oxidized to acetyl-CoA and CO_2, hydrogen is removed and a molecule of NAD_{re} is formed. Since two molecules of pyruvic acid were formed from each glucose molecule, two molecules of NAD_{re} are formed here. This complicated series of reactions can be summarized by the following equation:

$$2 \text{ pyruvic acid} + 2 \text{ CoA} + 2 \text{ NAD}_{ox} \longrightarrow 2 \text{ acetyl} \sim \text{CoA} + 2CO_2 + 2 \text{ NAD}_{re} + 2H^+$$

The Krebs cycle (Stage III of aerobic respiration) The acetyl-CoA is next fed into a complex series of reactions called the Krebs citric acid cycle, after the British scientist Sir Hans Krebs, who was awarded a Nobel Prize for working out this system. Each of the two two-carbon acetyl-CoA molecules formed from one molecule of glucose is combined with a four-carbon compound (oxaloacetic acid) already present in the cell, to form a new six-carbon compound called ***citric acid***. Each of the citric acid molecules is then oxidized to a five-carbon compound plus CO_2. The five-carbon unit, in turn, is oxidized to a four-carbon compound plus CO_2. This four-carbon compound is then converted into the four-carbon compound: oxaloacetic acid. It can now pick up more acetyl-CoA, forming new citric acid and beginning the cycle again (Fig. 6.7). The cycle is simplified in Figure 6.8.

We see, then, that two carbons are fed into the Krebs cycle as the acetyl group, and two are released as CO_2. Since each glucose molecule being oxidized yields two molecules of acetyl-CoA, two turns of the cycle are required, and four carbons are released as CO_2 during this stage of glucose breakdown. With the two carbons already released as CO_2 during the oxidation of pyruvic acid to acetyl-CoA, all six carbons of the original glucose are accounted for.

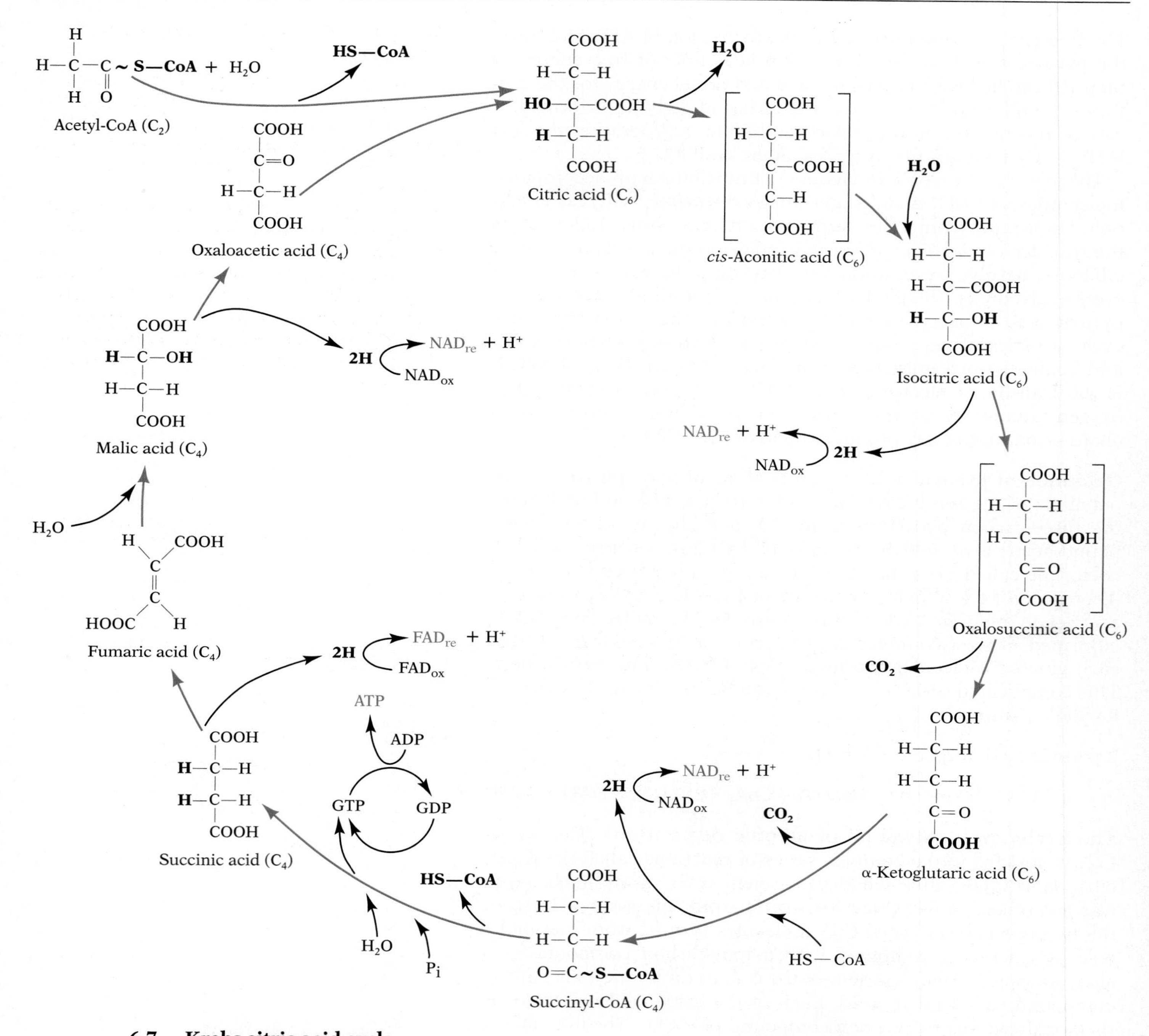

6.7 Krebs citric acid cycle

The complete cycle is shown here only to illustrate the characteristic complexity of metabolic pathways; there is no point in trying to learn all the reactions involved. As shown at upper left, the acetyl group (two carbons) from acetyl-CoA enters the cycle by combining with oxaloacetic acid (four carbons) to form citric acid (six carbons). During subsequent reactions two of the carbons are removed as CO_2 (between oxalosuccinic acid and α-ketoglutaric acid, and between α-ketoglutaric acid and succinyl-CoA), and a total of eight hydrogens are removed. These hydrogens are picked up by NAD (or by the related acceptor molecule FAD). One molecule of ATP is synthesized (bottom). Finally, oxaloacetic acid is regenerated and can combine with a new acetyl group to start the cycle over again. The cycle is completed twice for each molecule of glucose oxidized. (The atoms removed at each step are shown in boldface in the structural formulas. The two substances in brackets—*cis*-aconitic acid and oxalosuccinic acid—are enzyme-bound intermediates that seldom exist as free compounds.)

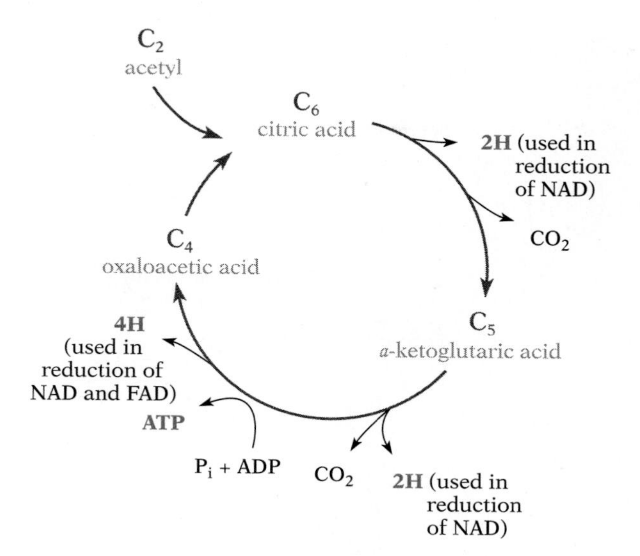

6.8 Simplified version of the Krebs citric acid cycle

The two carbons of the acetyl group combine with a four-carbon compound to form citric acid, a six-carbon compound. Removal of one carbon as CO_2 leaves a five-carbon compound. Removal of a second carbon as CO_2 leaves a four-carbon compound, which can combine with another acetyl group and start the cycle over again. In the course of the cycle, one molecule of ATP is synthesized and eight hydrogens are released, which are used in the reduction of NAD and FAD. Since one molecule of glucose gives rise to two acetyl units, two turns of the cycle occur for each molecule of glucose oxidized, with production of four molecules of CO_2, two molecules of ATP, and 16 hydrogens.

The oxidative breakdown of each molecule of acetyl-CoA via the Krebs cycle also involves the removal of eight hydrogens, which are picked up by NAD_{ox} (or by FAD_{ox}, the oxidized form of a related electron-carrier protein called flavin adenine dinucleotide); four units of reduced carrier are thus formed (Fig. 6.7). Since the breakdown of one molecule of glucose leads to two turns of the Krebs cycle, a total of eight molecules of reduced carrier (six NAD_{re} and two FAD_{re}) are formed during this stage of the breakdown of glucose. Two molecules of ATP are also synthesized in two turns of the Krebs cycle.

Figure 6.9 summarizes the yield of ATP, NAD_{re}, FAD_{re}, and CO_2 from the three stages of the breakdown of glucose.

The electron-transport chain (Stage IV of aerobic respiration) So far we have seen a net gain of only four new ATP molecules (two in glycolysis and two in the Krebs cycle). These represent just a small fraction of the energy originally available in the glucose. Of the remainder, some is liberated during the first three stages (mostly as heat, which is essential for the reactions to proceed); the rest is stored in the high-energy intermediates NAD_{re} and FAD_{re}. Twelve of these molecules are synthesized in the breakdown of each molecule of glucose (Fig. 6.9).

How is this energy used to synthesize ATP? We said earlier that under aerobic conditions the regeneration of NAD_{ox} from NAD_{re} is

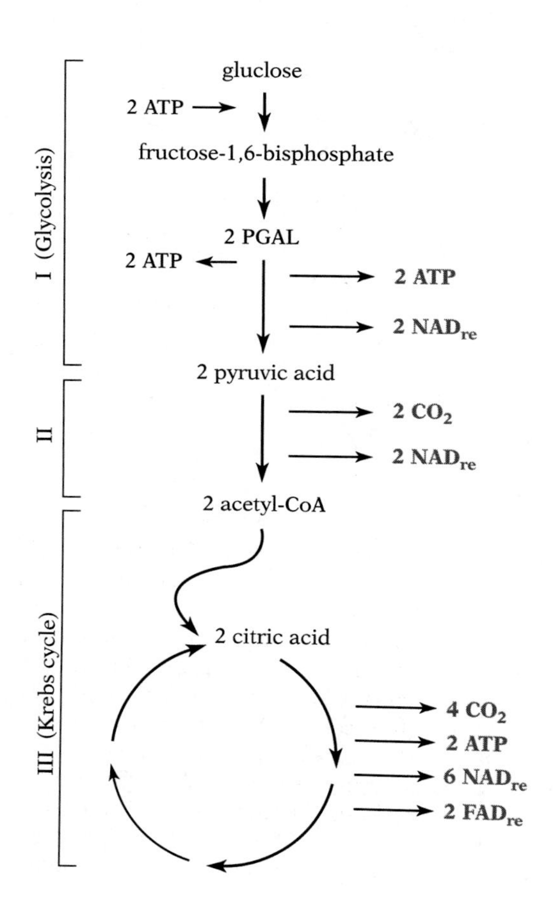

6.9 Summary of the most important products of Stages I, II, and III in the complete breakdown of one molecule of glucose

Stage I (glycolysis) begins with the expenditure of two molecules of ATP to activate glucose and produce two molecules of PGAL. The two PGAL molecules are then broken down to two molecules of pyruvic acid, in a process that first pays back the two ATP molecules originally invested and then yields two molecules each of ATP and NAD_{re} (red). Stage II yields two molecules each of CO_2 and NAD_{re}. Stage III, in which the two molecules of acetyl-CoA are fed into the Krebs cycle and further broken down, yields four CO_2, two ATP, six NAD_{re}, and two FAD_{re} molecules. (The H^+ ions liberated in the production of NAD_{re} and FAD_{re} are not shown.)

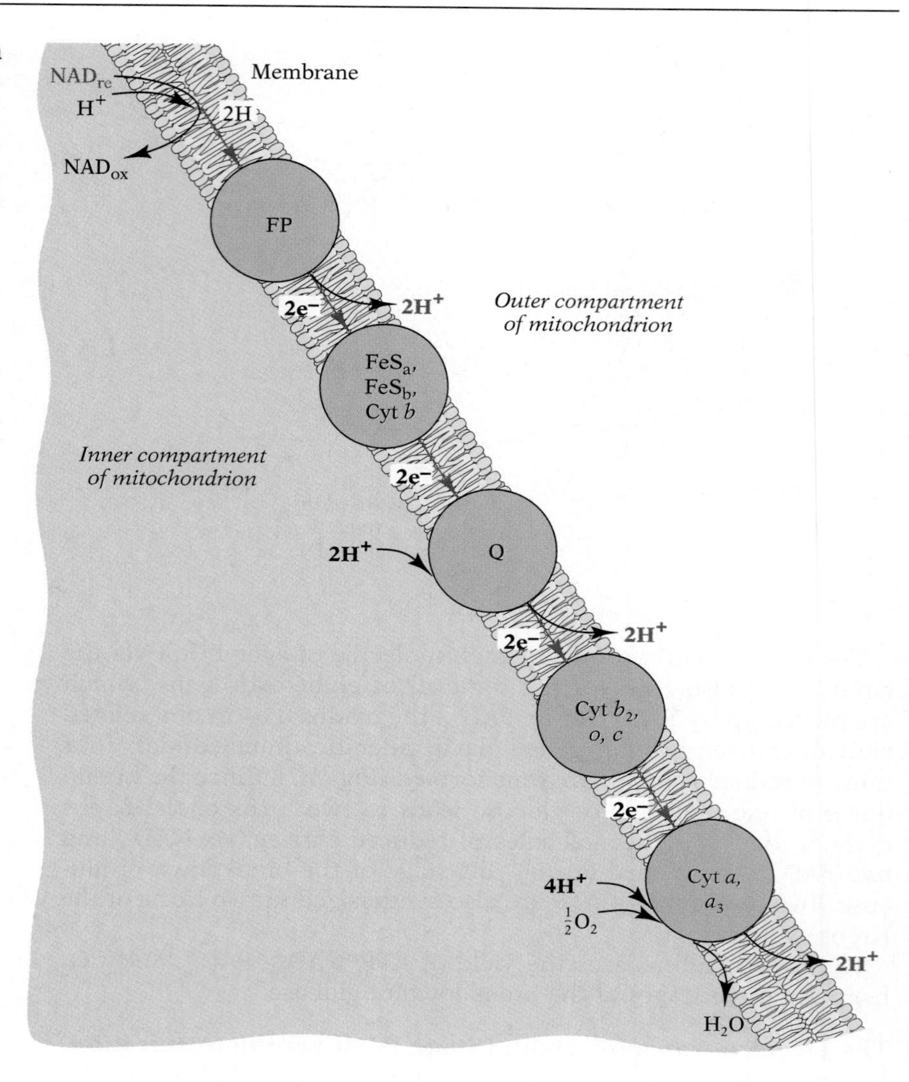

6.10 Respiratory electron-transport chain The reactions summarized in this diagram take place on the inner membrane of the mitochondrion. NAD_{re} donates two electrons and a proton to the electron-transport chain; a second hydrogen ion is drawn from the medium. The electrons are passed from one acceptor substance to the next, step by step down an energy gradient from their initial high energy level in NAD_{re} to their final low energy level in H_2O. (When available, molecules of FAD_{re} can also donate their electrons to the electron-transport chain; since these electrons have less energy than those of NAD_{re}, they enter lower down on the chain, at Q.) Each successive acceptor molecule is cyclically reduced when it receives the electrons, and then oxidized when it passes them on to the next acceptor molecule. At three sites along the chain, some of the free energy released is used to pump H^+ ions into the compartment outside the inner membrane. Later the H^+ gradient generated by the electron-transport chain is used in the synthesis of ATP. The electron acceptors are a flavoprotein (FP); coenzyme Q; cytochromes a, a_3, b, b_2, c, and o; and two proteins containing iron and sulfur, FeS_a and FeS_b. For simplicity, steps have been combined wherever possible. A more complete sequence is shown as part of Figure 6.12.

achieved by the passage of electrons from NAD_{re} to O_2, with oxygen thus acting as the ultimate acceptor of electrons:

$$O_2 + 2NAD_{re} + 2H^+ \longrightarrow 2H_2O + 2NAD_{ox}$$

The NAD_{re} does not, however, pass its electrons directly to the oxygen, as this summary equation suggests. The electrons and their associated protons reach their ultimate targets indirectly. In particular, the hydrogen electrons are passed down a "respiratory chain" of electron-transport compounds, many of which are iron-containing enzymes called ***cytochromes*** (Fig. 6.10). The electrons move to ever-lower energy levels with each transfer. As we will see, the energy extracted in the electron-transport chain is then used for the production of ATP.

The anatomy of respiration The elaborate internal structure of the mitochondrion plays a crucial role in respiration. The extensively folded inner membrane divides each mitochondrion into two compartments (Fig. 6.11). This membrane separates two very different chemical environments and serves as scaffolding for gates,

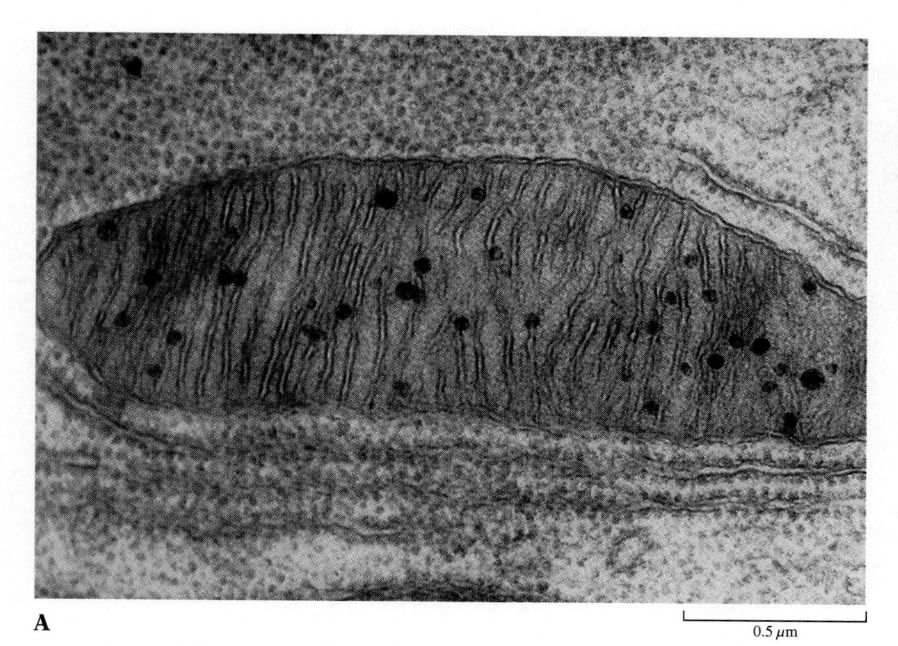

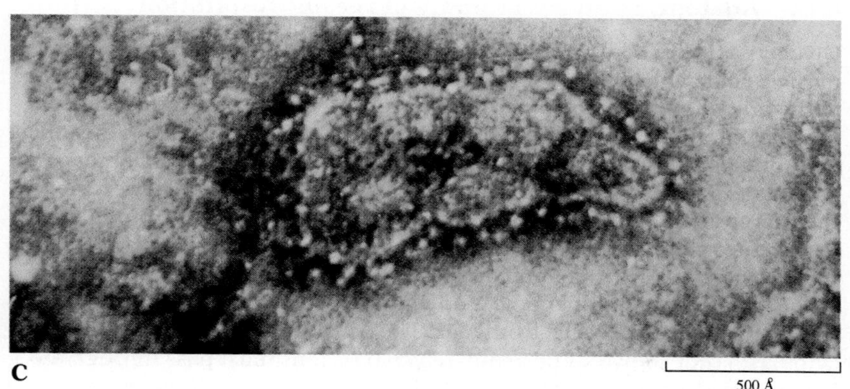

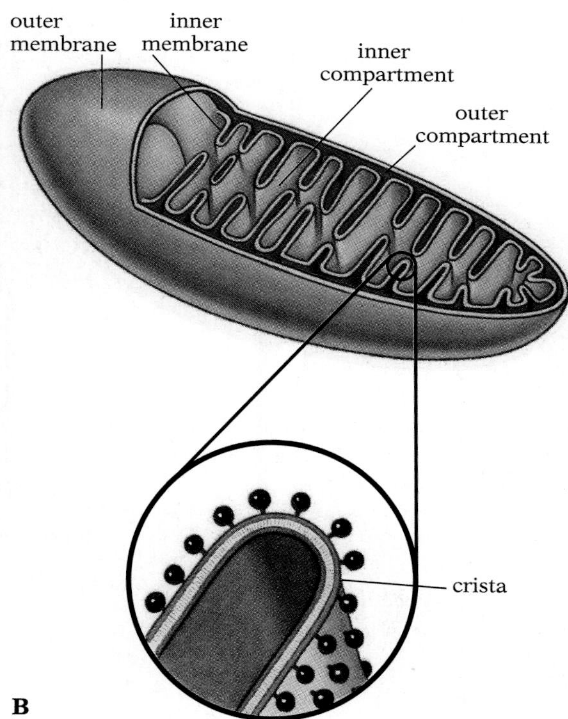

6.11 Structure of a mitochondrion
(A) A typical animal mitochondrion. (B) In this interpretive drawing, much of the outer membrane has been cut away, and the interior has been sectioned to show how the inner membrane folds into cristae. The mitochondria of metabolically very active cells have more cristae than those of less active cells. The inner compartment is within the inner membrane, while the outer compartment is the space between the two membranes. (C) A close-up of the inner membrane. As shown in Figure 6.12, much of the activity of cellular respiration, such as electron transport, occurs across the inner membrane between the inner and outer compartments. The knoblike structures are the F_1 enzyme complexes responsible for synthesizing ATP; as such, they are often called ATP synthases.

pumps, and organized enzyme arrays. While the enzymes of the Krebs cycle are contained in the fluid of the inner compartment, the enzymes of the electron-transport chain are found in the inner membrane.

The anatomy of the electron-transport chain is shown in Figure 6.12. The major raw material, NAD_{re}, was generated by the sequence of events already described: glycolysis (which takes place outside the mitochondrion in the cytosol) $\longrightarrow$ pyruvic acid (which is transported through both the outer and inner membranes to the inner compartment of the mitochondrion) $\longrightarrow$ acetyl–CoA $\longrightarrow$ Krebs cycle $\longrightarrow$ NAD_{re} (as well as CO_2, ATP, and FAD_{re}). The electron-transport chain must now harvest this energy.

The high-energy compound NAD_{re} carries its electrons to the inner membrane. Here NAD_{re} is oxidized by losing a hydrogen and an electron to a flavoprotein (FP), which is reduced; simultaneously, FP accepts a H^+ ion from the medium of the inner compartment. The hydrogen from NAD_{re} is split into a hydrogen ion (H^+) and an electron (e^-). The H^+ ion, along with the H^+ accepted from the medium, is deposited in the outer compartment; the two elec-

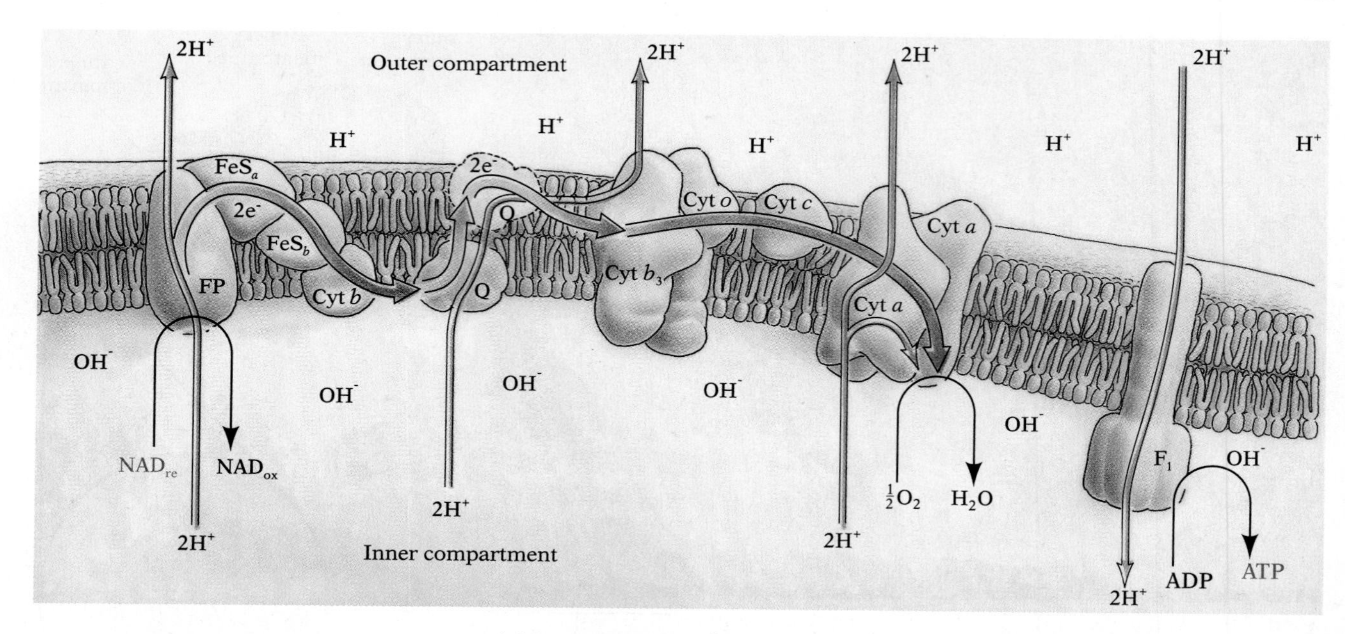

6.12 Anatomy of Stages IV and V of aerobic respiration

Glycolysis (not shown) takes place in the cytosol of the cell, and supplies pyruvic acid, which is converted into acetyl-CoA in the inner compartment of the mitochondrion. The Krebs cycle (not shown) operates in the inner compartment, producing CO_2, ATP, $NAD_{re,}$ and FAD_{re}. Mounted in the inner membrane of the mitochondrion, which separates the inner and outer compartments, are the enzymes of the electron-transport chain. They extract the energy from NAD_{re} and FAD_{re}, and use it to pump H^+ ions from the inner to the outer compartment. Some of the ions are carried across by Q, which shuttles between the inner and outer surface of the membrane. The energy of the resulting electrochemical gradient is then used by the enzyme complex F_1 (right) to make ATP from ADP. The ATP is then exported from the cytoplasm (not shown). Approximately two H^+ ions must pass through the F_1 complex to create one ATP from ADP. Since the energy of the two electrons donated by each NAD_{re} molecule is used to transport six H^+ into the outer compartment, the oxidation of an NADre produces about three ATPs. Two electrons from a molecule of FAD_{re}, which have less energy than those of NAD_{re}, can enter the electron-transport chain at Q; the oxidation of the lower-energy compound FAD_{re} produces about two ATPs.

trons are passed immediately to another enzyme in the sulfur-containing complex FeS_a. The result is that the H^+ concentration of the outer compartment is raised, the H^+ concentration of the inner compartment is lowered, high-energy-level electrons (two from each NAD_{re}) are inserted into the transport chain (center) where they will be used to do work, and NAD_{ox} is returned to the citric acid cycle to be "recharged."

The transfer of electrons from one acceptor to another proceeds because free energy is liberated (that is, lost by electrons) at each step. For each electron reaching enzyme Q (which shuttles back and forth across the membrane), another H^+ enters the chain from the inner compartment and is deposited in the outer compartment. (At this stage, the lower-energy electrons of FAD_{re} can also enter the electron-transport chain.) The electrons, whatever their origin, now move through the cytochrome series (left) to cytochrome a_3. This last enzyme group uses the energy of the electrons to split molecular oxygen (O_2) and catalyze the following reaction:

$$1/2O_2 + 4H^+ + 2e^- \xrightarrow{\text{cytochrome } a_3} H_2O + 2H^+$$

This reaction reduces by four the number of hydrogen ions in the inner compartment. Two of these hydrogen ions are exported to the outer compartment, while the other two are incorporated into the water molecule. About 100 electrons can pass through each transport chain each second.

Chemiosmotic synthesis (Stage V of aerobic respiration) All the energy stored in the NAD_{re} has now been used up without generating any new ATP. However, an electrostatic and osmotic concentration gradient has built up across the inner membrane: the inner compartment, having lost H^+, has become negatively charged, while the outer compartment—the space between the inner and outer membrane—is filled with H^+ ions and therefore is positively charged. The result is a molecular battery that uses the difference in charge across the inner mitochondrial membrane to make ATP from ADP. Most of the H^+ ions return from the outer to the inner compartment by way of the F_1 complex (Fig. 6.12), passing down a ***chemiosmotic gradient***. As they do so, the energy of this electrochemical gradient—the combined osmotic and electrostatic gradients—is harvested (Figure 6.13). The abundant energy generated by this elegant system is stored in the phosphate groups of ATP. The same H^+ gradient probably also powers the pump that exports ATP to the cytosol. Peter Mitchell won the Nobel Prize in 1978 for working out this indirect route by which mitochondria convert the energy of NAD_{re} into ATP.

It is a continuing challenge to explain how such efficient, highly refined examples of biological engineering as the electron-transport chain might have evolved. How do the first elements in such a complicated array, which seem useless without the rest, ever come into existence? To answer the question, biologists look at a wide variety of species for preadaptations—fragments of the system as we know it now that are employed in other ways and which might have been present in ancestral organisms and available to natural selection for "remodeling."

Consider the most intricate element of the mitochondrial system, the F_1 complex, which uses the H^+ gradient to make ATP. Relict species of anaerobic bacteria living today still lack an electron-transport chain but nevertheless have an F_1 assembly that works in reverse. These anaerobes burn ATP to pump H^+ out of their cells, eliminating the acidic by-products of fermentation. There is similar evidence to suggest that the cytochrome complexes evolved later in ancient bacteria, where they served similar detoxification functions. Natural selection must then have favored combining these various systems into a rudimentary electron-transport chain using sulfur, carbon, or nitrogen as the ultimate electron acceptor.

Summary of respiration energetics Now that we have looked at all five stages in the aerobic breakdown of glucose, we can summarize their combined energy yield. The overall flow of energy in aer-

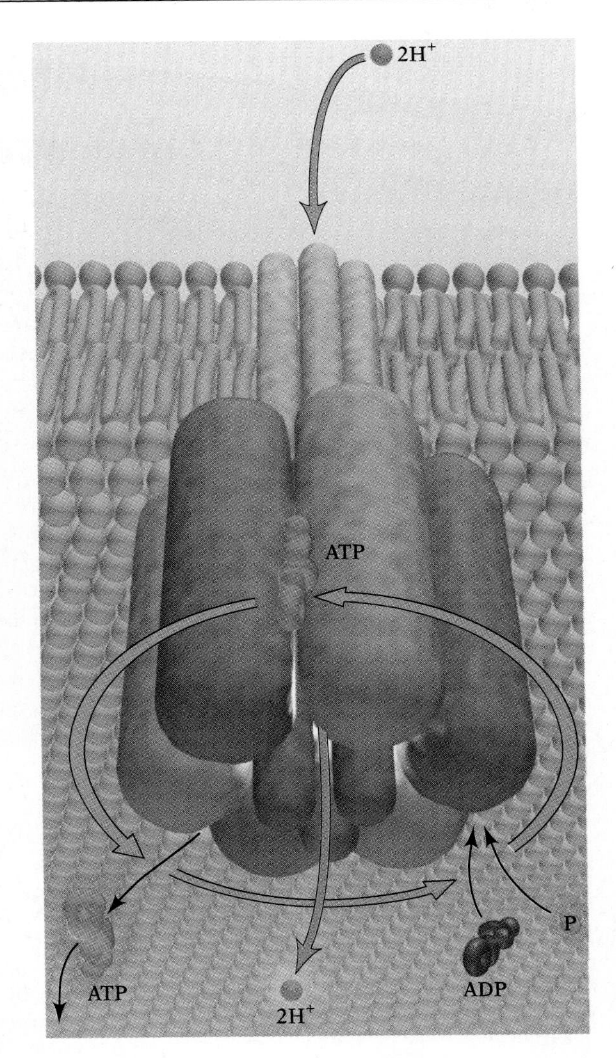

6.13 Operation of the F1 complex

The F_1 complex is composed of at least 18 subunits. The core group carries the H^+ to one of three protein pairs of the head region where ATP is formed; each of the three pairs is undergoing one of the three steps in the cycle: ADP and phosphate binding, ATP formation, and ATP release. Many researchers believe that the head rotates during ATP synthesis.

HOW WE KNOW: TESTING THE MITCHELL HYPOTHESIS

Convincing evidence for Peter Mitchell's brilliant chemiosmotic hypothesis comes from several researchers, including (besides Mitchell) Efraim Racker, Andre T. Jagendorf, and Peter C. Hinkle. Much of their research employs artificially reconstructed vesicles. The technique takes advantage of the tendency of phospholipids in an aqueous solution to form spherical liposomes. When mitochondria are exposed to ultrasound (sound of very high frequency), both the outer and inner membranes break into sheetlike fragments; the inner membranes subsequently "heal" by forming spheres about 100 nm (nanometers) in diameter, as shown in the accompanying figure. These submitochondrial vesicles display 9-nm spheres that are thought to be parts of the ATP-synthesizing F_1 complexes. Researchers disrupt the mitochondria, let the submitochondrial vesicles form in one solution—trapping that chemical milieu inside the vesicles—and then transfer the vesicles to a different chemical environment. In this way, the chemistry of the two sides of the membrane can be altered at will.

According to Mitchell's chemiosmotic battery model, in a normal mitochondrion the necessary gradient exists because the H^+ concentration is higher on the outside of the inner membrane than on the inside. If the model is correct, the inside-out patches of membrane should function when the H^+ concentration is higher on the inside. And in fact, when the vesicles are created with more H^+ inside than out, they begin producing ATP, but when the H^+ concentration is higher on the outside, they do nothing. The model also predicts that if these vesicles are supplied with NAD_{re} on the outside, they will increase the gradient by raising the H^+ concentration on the inside. This too happens. In addition, the model predicts that the insertion into these vesicles of channels permeable to H^+ should short-circuit the battery by allowing H^+ to cross the membrane freely, and should thereby prevent ATP synthesis. Again, this is exactly what is observed.

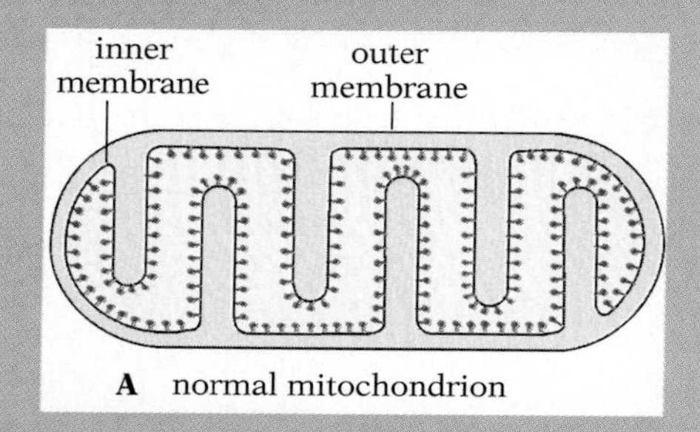

A normal mitochondrion

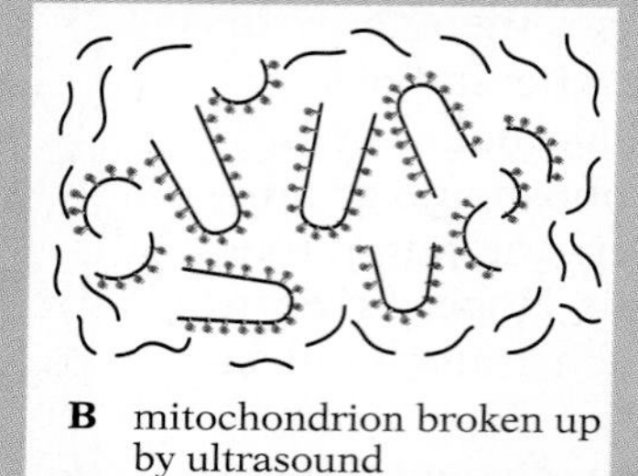
B mitochondrion broken up by ultrasound

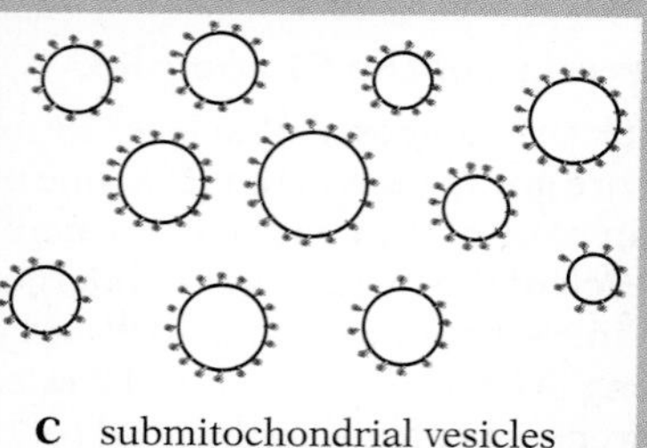
C submitochondrial vesicles formed when fragments "heal"

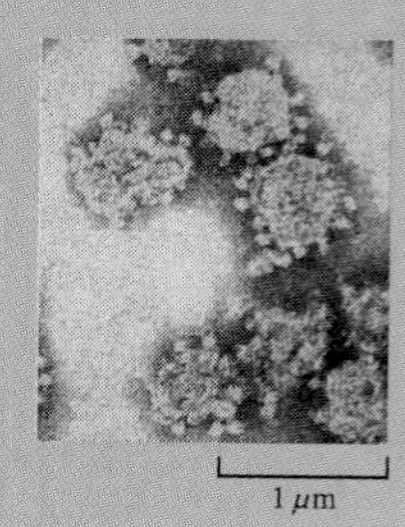

Formation of submitochondrial vesicles

Submitochondrial vesicles are made by breaking the mitochondrion (A) into pieces with ultrasound (B), and then allowing the bits to "heal" (C). By controlling the medium in which vesicle formation takes place, and the solution into which the resulting inside-out membrane spheres are then put, the researcher can create virtually any desired combination of internal and external chemical environments.

obic respiration is summarized in Figure 6.14. As we have seen, glycolysis generates two molecules of ATP per molecule of glucose, along with two NAD_{re}; the conversion of pyruvic acid into acetyl-CoA yields another two NAD_{re}; the Krebs citric acid cycle produces an additional two ATPs, six NAD_{re}, and two FAD_{re}; finally, and most critically, the charging of the mitochondrial battery—the expulsion of H^+ across the inner membrane by means of the electron-transport chain (driven by oxidation of the NAD_{re} and FAD_{re})—stores enough energy to synthesize roughly another 28 ATPs. Each NAD_{re}

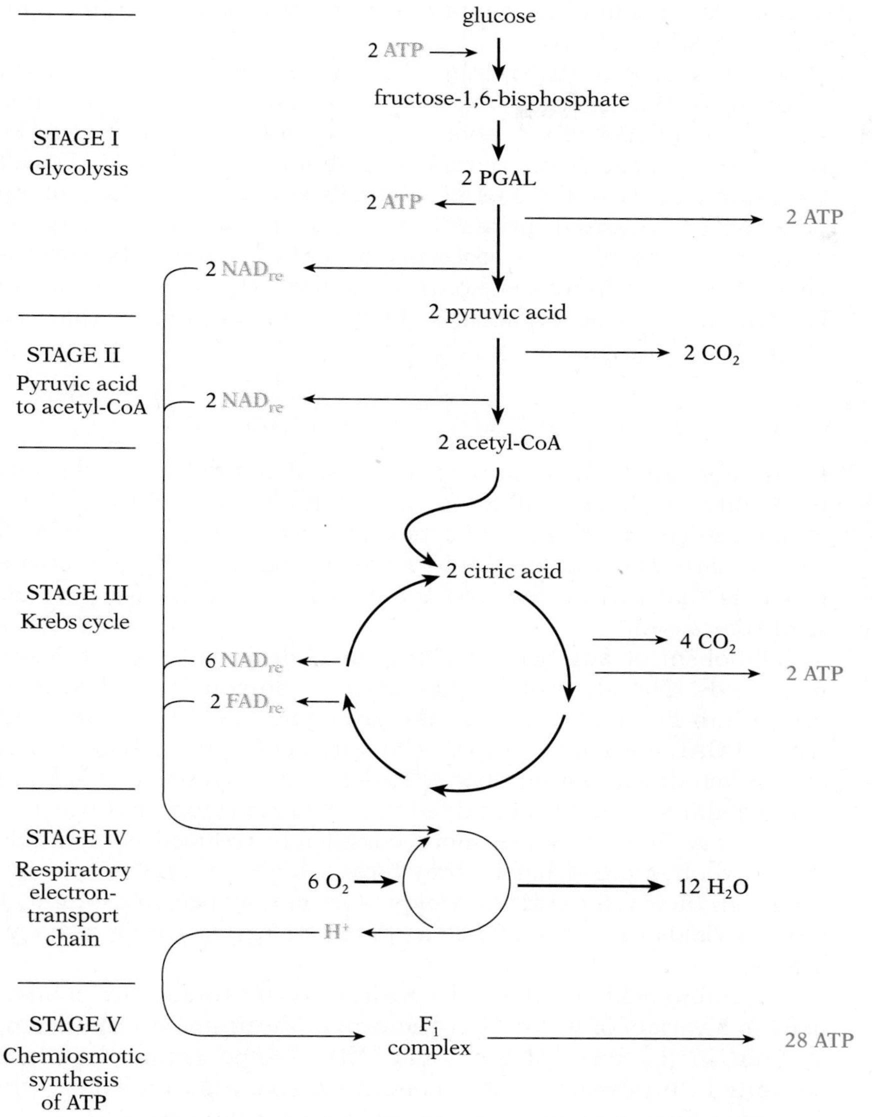

6.14 The ATP yield from complete breakdown of glucose to carbon dioxide and water

In the first three stages, four molecules of ATP are directly synthesized, along with ten of NAD_{re} and two of FAD_{re}. In Stage IV the NAD_{re} and FAD_{re} are fed to the electron-transport chain, and their stored energy is utilized to create a difference in charge across the inner membrane of the mitochondrion. Finally, in Stage V the F_1 enzyme complex uses the energy of this chemiosmotic gradient across the membrane to make approximately 28 more molecules of ATP.

molecule provides the energy to pump enough H^+ ions for the synthesis of almost three ATPs ($10NAD_{re} \longrightarrow 25ATP$), while each FAD_{re} pumps enough H^+ to power the synthesis of two ATPs ($2FAD_{re} \longrightarrow 3ATP$).

Now it is clear why atmospheric oxygen is so important to life as we know it: aerobic respiration extracts 16 times as much energy from glucose as does anaerobic metabolism (glycolysis followed by fermentation). Metabolic poisons like cyanide (which binds irreversibly with a cytochrome and thereby blocks the electron-transport chain) are fatal because the cell is suddenly denied 92% of its normal energy. However, parts of our bodies *can* operate anaerobically for short periods. During intensive exercise, for instance, muscles often need so much energy that the oxygen supplied from breathing is insufficient. In such cases glycolysis and fermentation provide the needed energy for a time, but they are inefficient and fatigue soon results. Later, the oxygen debt is paid back by deep breathing or panting, and the lactic acid that accumulated in the muscles as a result of fermentation is removed to the liver and converted back into glucose.

We can calculate the overall efficiency of aerobic respiration: a molecule of glucose has a free energy of about 670 kcal/mole, while a molecule of ATP stores about 7.3 kcal/mole. Since the 32 ATPs that are generated therefore represent about 234 kcal/mole, the cell has retained 35% of the energy originally stored in the glucose; the other 65% is released, primarily as heat. This liberated energy is essential to shuttling the reactants smoothly through the various chemical chains. Moreover, some of the inevitable "waste" heat can be used to raise an organism's internal temperature so that the chemistry of life runs more efficiently.

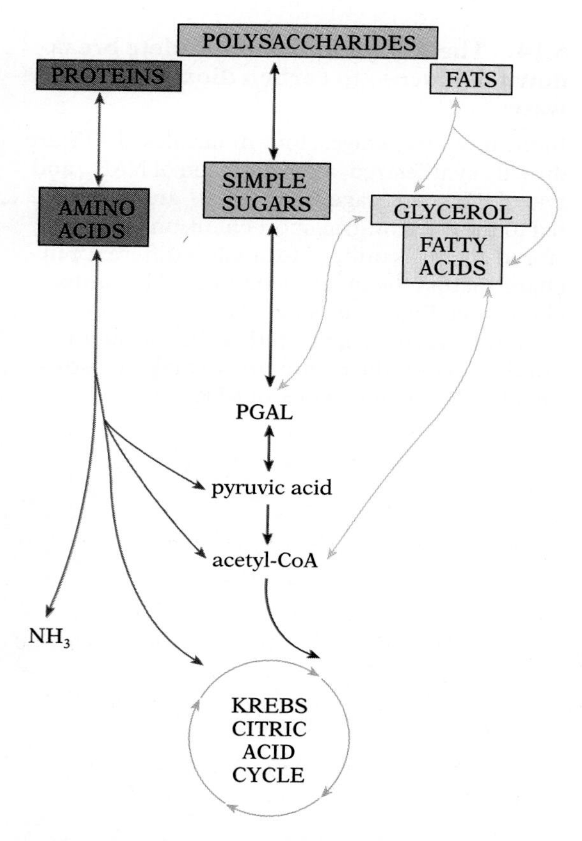

6.15 Relationships of the catabolism of proteins and fats to the catabolism of carbohydrates

■ HOW FATS AND PROTEINS ARE METABOLIZED

Cells can extract energy in the form of ATP, not only from the carbohydrates we have focused on so far but also from the two other major categories of nutrients: fats and proteins. As Figure 6.15 shows, early steps in the breakdown of fats and proteins create products that can be fed into the enzyme pathways we have already discussed.

Catabolism of fats begins with their hydrolysis to glycerol and fatty acids. The glycerol (a three-carbon compound) is then converted into PGAL and fed into the glycolytic pathway at the point where PGAL normally appears. The fatty acids, on the other hand, are broken down to a number of two-carbon fragments, which are converted into acetyl-CoA and fed into the respiratory pathway farther along. Since fats are more completely reduced compounds than carbohydrates (that is, they have a higher proportion of hydrogens), their full oxidation yields more energy per unit weight; 1 g of fat yields more than twice as much energy as 1 g of carbohydrate.

The amino acids produced by hydrolysis of proteins are catabolized in a variety of ways. After removal of the amino group (deamination) in the form of ammonia (NH_3), some amino acids are converted into pyruvic acid, some into acetyl-CoA, and some into one or another compound of the citric acid cycle. Complete oxidation of a gram of protein yields roughly the same amount of energy as that of a gram of carbohydrate.

Such compounds as pyruvic acid, acetyl-CoA, and the compounds of the citric acid cycle, which are common to the catabolism of several different types of substances, not only play a crucial role in the oxidation of energy-rich compounds to carbon dioxide and water but also function in the anabolism of amino acids, sugars, and fats. They serve as biochemical crossroads, at which several enzyme pathways intersect. By investing energy, the cell can reverse the direction in which substances move along some of these pathways; for example, the PGAL and acetyl-CoA produced at different points in the breakdown of carbohydrate can be moved up the pathways to glycerol and fatty acids, for use in building fats. Similarly, many amino acids can be converted into carbohydrate via the common intermediates in their metabolic pathways. Not all pathways are two-way, however; most higher animals lack a reversible pyruvic acid–to–acetyl-CoA enzyme, as well as any alternative enzyme, and so cannot convert fatty acids into carbohydrate.

CHAPTER SUMMARY

THE FLOW OF ENERGY

HOW METABOLISM EVOLVED Early organisms harvested energy from inorganic molecules. Later they evolved the ability to store energy as glucose for later use. The evolution of photosynthesis allowed massive conversion of low-energy inorganic molecules into high-energy organic compounds. (p. 153)

HOW ENERGY IS TRANSFERRED IN REACTIONS Electron energy is transferred in chemical reactions; molecules losing energy are oxidized, while those gaining energy are reduced. (p. 156)

ATP: THE CURRENCY OF ENERGY EXCHANGE Most energy transfers between biochemical pathways in the cell are made via ATP. (p. 157)

HOW CELLS HARVEST ENERGY

RESPIRATION WITHOUT OXYGEN During glycolysis the cell harvests some of the energy in glucose, storing it in ATP and NAD_{re}. In the absence of oxygen, NAD is oxidized during fermentation, and its energy is lost. (p. 159)

RESPIRATION WITH OXYGEN In the presence of oxygen, the pyruvic acid produced by glycolysis is oxidized and enters the Krebs cycle, which extracts additional energy and generates CO_2. Much of the extracted energy is in the form of high-energy electrons in NAD and FAD, which enter the electron-transport chain where they create a strong electrochemical gradient in the mitochondrion. The energy in this gradient is then used to synthesize ATP. Aerobic respiration extracts 16 times more energy from glucose than does fermentation. (p. 165)

HOW FATS AND PROTEINS ARE METABOLIZED Fats and proteins are metabolized by being converted into molecules found at some step in aerobic respiration; they enter the pathway at these points. (p. 176)

STUDY QUESTIONS

1. What is the source of energy in the glucose molecule that is extracted in fermentation? How exactly can you localize it? (pp. 159–165)
2. In the absence of oxygen, which element is the best to use as an electron acceptor in order to extract the most energy: carbon, nitrogen, or sulfur? Why? (p. 155)
3. Why are there two separate entry points into the electron-transport chain, one for electrons from NAD and one for electrons from FAD? (pp. 170–172)
4. What is the purpose of the space between the inner and outer membranes of the mitochondrion? Why not have just one membrane, as the procaryotic precursor of this organelle must have had? (pp. 169–173)
5. Why does reaction coupling work? How does it help explain the operation of fermentation? (pp. 159–166)

SUGGESTED READING

CLOUD, P., 1983. The biosphere, *Scientific American* 249 (3). *On the combined evolution of the earth, life, and the atmosphere, with particular emphasis on the role of oxygen concentration.*

HINKLE, P. C., AND R. E. MCCARTY, 1978. How cells make ATP, *Scientific American* 238 (3). *A difficult but rewarding explanation of how ATP is made.*

MORAN, L. A., K. G. SCRINGEOUR, H. R. HORTON, R. S. OCHS, AND J. D. RAVEN, 1994. *Biochemistry,* 2nd ed. Prentice Hall, Englewood Cliffs, N.J. *Traces in great detail the biochemical pathways of respiration, and their regulation.*

TUNNICLIFFE, V. 1992. Hydrothermal-vent communities of the deep sea. *American Scientist* 80 (4). *How chemosynthetic bacteria manage to extract energy from sulfurous 250°C water at deep sea vents, and support an exotic ecosystem of other creatures.*

CHAPTER 7

THE PHOTOSYNTHETIC REACTION

HOW THE LIGHT REACTIONS STORE LIGHT ENERGY

HOW CHLOROPHYLL CAPTURES THE ENERGY OF LIGHT

HARVESTING THE ENERGY OF EXCITED ELECTRONS

THE ROLE OF CHLOROPLAST STRUCTURE

HOW THE DARK REACTIONS FIX CARBON

CARBOHYDRATE SYNTHESIS BY THE CALVIN CYCLE

THE MYSTERY OF PHOTORESPIRATION

THE LOGIC OF LEAF DESIGN

THE ANATOMY OF ORDINARY LEAVES

LEAVES WITH KRANZ ANATOMY

AVOIDING PHOTORESPIRATION: C_4 PHOTOSYNTHESIS

AVOIDING PHOTORESPIRATION: CRASSULACEAN ACID METABOLISM

PHOTOSYNTHESIS

The earliest forms of life on earth survived by metabolizing high-energy inorganic molecules. Without oxygen, these anaerobic species could harvest only a tiny fraction of the available energy, and they were also in danger of extinction when their limited supply of inorganic food ran out. For eons, life on earth was a marginal phenomenon.

About 3 billion years ago, prospects suddenly changed: some primitive organisms evolved the ability to capture the sun's energy. Photons energized the electrons in special pigments, and these activated molecules passed their energy to other molecules, leading eventually to the synthesis of glucose and other organic molecules (Fig. 7.1).

This primitive form of photosynthesis—the transformation of light into chemical energy—freed organisms from their dependence on inorganic food. Further evolution resulted in a more efficient form of photosynthesis that also produced molecular oxygen as a by-product. As oxygen began to accumulate in the atmosphere 1.5–2 billion years ago, its high electronegativity allowed high-efficiency aerobic respiration to become the main mechanism for extracting energy from food. Thus the two major chemical path-

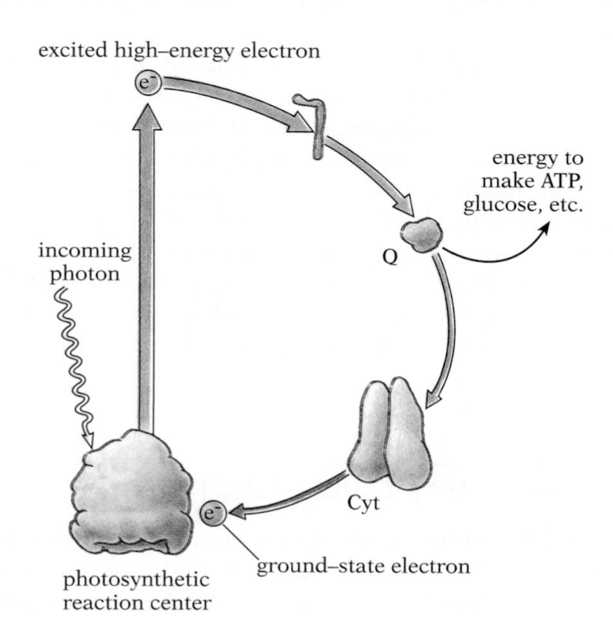

7.1 Energetic basis of life

The basic energy equation of life begins when a photon from the sun excites an electron in a pigment molecule by moving the electron up to a higher level. The energy thus acquired ultimately excites an electron in an acceptor molecule. Subsequently, as the excited electron is returned to a low-energy level in a pigment molecule by way of a series of acceptors, its free energy is used for generating an energy-storage compound (usually glucose or ATP). The pathway shown here is the oldest and most basic form of photosynthesis.

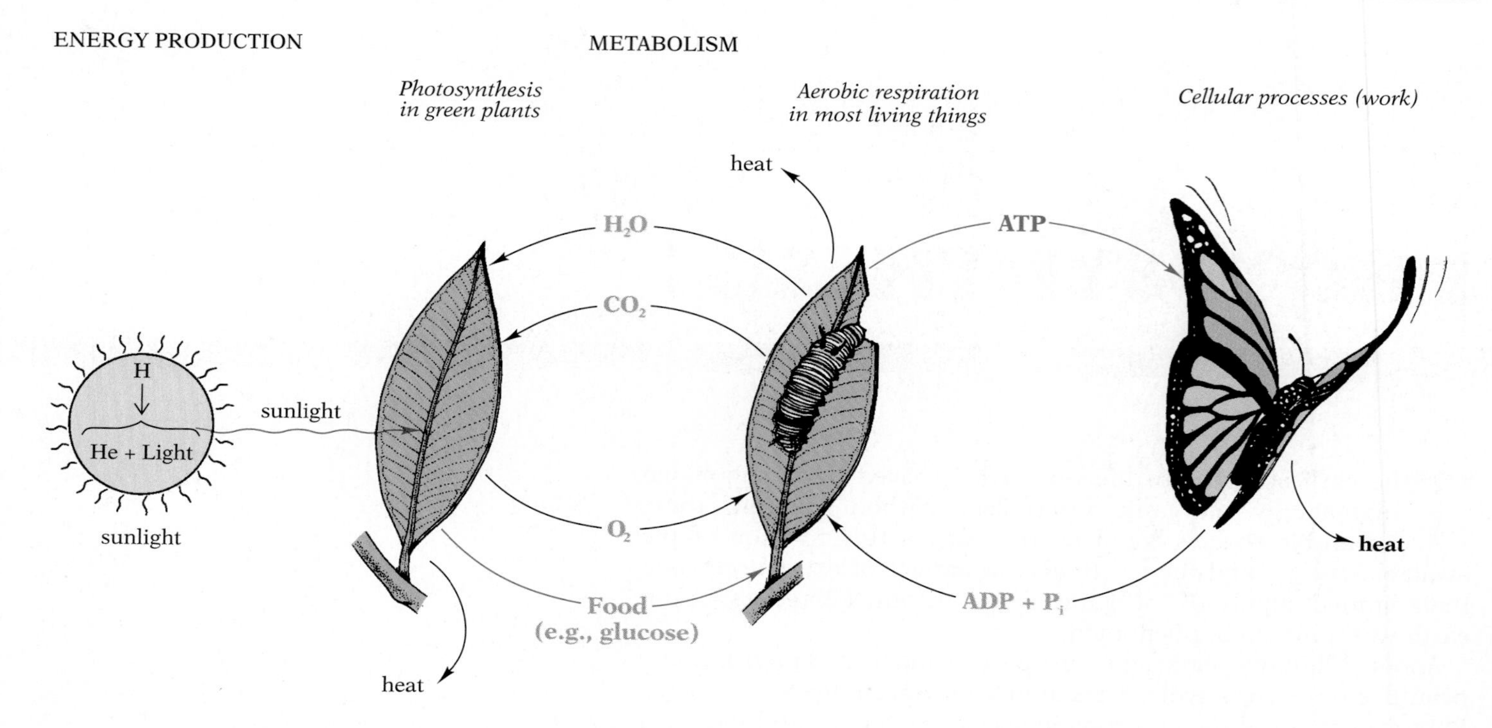

7.2 Energy flow in aerobic pathways

With the evolution of the modern form of photosynthesis approximately 2.3 billion years ago, oxygen, the electronegative element essential to the efficient operation of catabolic processes, began to accumulate in the atmosphere. This led (about 1.5 billion years ago) to the spread of organisms capable of high-efficiency aerobic respiration, and to the unification of the two great metabolic pathways—photosynthesis and respiration—that is characteristic of life processes today. The pathways are unified by shared products and by-products: organic food, H_2O, CO_2, and O_2. Red arrows trace the flow of energy. Note that both the caterpillar and the leaf respire.

ways—photosynthesis and respiration—were joined through shared by-products (Fig. 7.2). Life became the dominant feature of the earth.

Today, virtually all organisms depend directly or indirectly on photosynthesis. ***Autotrophs***, which are organisms (such as plants) that are capable of making organic nutrients from inorganic materials, depend almost exclusively on photosynthesis. ***Heterotrophs*** are organisms (such as animals) that must obtain organic nutrients from the environment; they depend indirectly on photosynthesis since they consume autotrophs, heterotrophs that have eaten autotrophs, or both. Photosynthesis, then, is life's single most important biochemical process. This chapter explores how autotrophs harvest the energy of the sun.

THE PHOTOSYNTHETIC REACTION

By the early 19th century, biologists knew that plants use light to combine water and carbon dioxide to generate organic compounds and oxygen. For cases in which glucose is the organic product, the

HOW WE KNOW: THE INGREDIENTS OF PHOTOSYNTHESIS

For most of recorded history scientists had no idea that the sun supplies the surface of the earth with virtually all its energy or that green plants trap that energy and produce the invisible gas we breathe. Only in 1772 did the English clergyman Joseph Priestley demonstrate that plants affect air in such a way as to reverse the effects of burning or of breathing. As he reported it:

> One might have imagined that, since common air is necessary to vegetable, as well as to animal life, both plants and animals had affected it in the same manner, and I own I had that expectation, when I first put a sprig of mint into a glass-jar, standing inverted in a vessel of water; but when it had continued growing there for some months, I found that the air would neither extinguish a candle, nor was it at all inconvenient to a mouse, which I put into it.
>
> Finding that candles burn very well in air in which plants had grown a long time, and having had some reason to think, that there was something attending vegetation, which restored air that had been injured by respiration, I thought it was possible that the same process might also restore the air that had been injured by the burning of candles.
>
> Accordingly, on the 17th of August, 1771, I put a sprig of mint into a quantity of air, in which a wax candle had burned out, and found that, on the 27th of the same month, another candle burned perfectly well in it. This experiment I repeated, without the least variation in the event, not less than eight or ten times in the remainder of the summer.

Priestley was the first to show that plants produce oxygen, though he did not realize that this was happening. Seven years later, the Dutch physician Jan Ingenhousz demonstrated the necessity of sunlight for oxygen production, and he also showed that only the green parts of the plant could photosynthesize. Just five years later a Swiss pastor, Jean Senebier, showed that the process depended on a gas that he called "fixed air" (carbon dioxide). Finally, in 1804, another Swiss researcher, Nicolas Théodore de Saussure, found that water is necessary for the photosynthetic production of organic materials. Thus, in a span of only 32 years, the qualitative formula of photosynthesis was worked out:

$$\text{light} + \text{water} + \text{carbon dioxide} \longrightarrow \text{organic matter} + \text{oxygen}$$

formula of the most important reaction for life on earth is:

$$6CO_2 + 12H_2O + \text{light} \longrightarrow 6O_2 + C_6H_{12}O_6 + 6H_2O \quad (\Delta G = -1300 \text{ kcal/mole})$$

What the formula does not reveal, however, is how this reaction actually occurs—that is, how a special combination of enzymes and cellular architecture manages to harvest energy from light. Nor does it explain why water appears on both sides of the equation. In fact, it does not even tell us whether the oxygen we breathe comes from H_2O or CO_2.

For many years, most scientists believed that light energy splits carbon dioxide (CO_2), and that the carbon is then combined with water (H_2O), to form the group $—CH_2O$, from which carbohydrates like glucose ($C_6H_{12}O_6$) are in turn constructed. According to this view, the oxygen released by the plant during photosynthesis comes from CO_2. In 1930, however, C. B. van Niel showed that photosynthetic bacteria that use hydrogen sulfide (H_2S) instead of water as a raw material for photosynthesis, give off sulfur instead of oxygen as a by-product. It seemed clear that these bacteria must be using sulfur from H_2S as their electron acceptor instead of the oxygen from CO_2; thus the oxygen produced by plants during photosynthesis must come from H_2O rather than CO_2. This supposition was later proven by using a heavy isotope of oxygen (^{18}O instead of the usual ^{16}O) to label the oxygen in water; $^{18}O_2$ rather than $^{16}O_2$ was liberated. Thus we can now trace the fates of all the atoms involved in generating the basic organic building block —CH_2O (shown below in parentheses because it is a component rather than an independent molecule):

$$CO_2 + 2H_2O + \text{light} \xrightarrow{\text{chlorophyll}} O_2 + (-CH_2O) + H_2O$$

The reason water appears on both sides of the equation is that the water produced by the photosynthetic process is new; it is not the water used as a raw material.

Now that we have sorted out the fates of the various atoms, we can look at how the energy for making —CH_2O is harvested. The process occurs in two basic steps: "light" reactions convert and store the energy from light in specialized energy-transfer molecules like ATP, while "dark" reactions use that stored energy to convert (or fix) carbon dioxide into carbohydrates such as glucose.

HOW THE LIGHT REACTIONS STORE LIGHT ENERGY

Light is used to drive two processes: the temporary storage of energized electrons in electron acceptors and the phosphorylation of energy-storage compounds (usually ADP). The term ***photophosphorylation*** is often used to describe the light-dependent conversion of ADP into ATP:

$$ADP + P_i + \text{energy} \xrightarrow{\text{enzyme}} ATP + H_2O$$

■ HOW CHLOROPHYLL CAPTURES THE ENERGY OF LIGHT

Visible light constitutes one small region of the spectrum of electromagnetic radiation (Fig. 7.3). Each form of radiation in this spectrum has a specific wavelength and energy content. These two characteristics are inversely related: the longer the wavelength, the smaller the energy content. Within the narrow band visible to humans, the shortest light waves produce the sensation of violet and

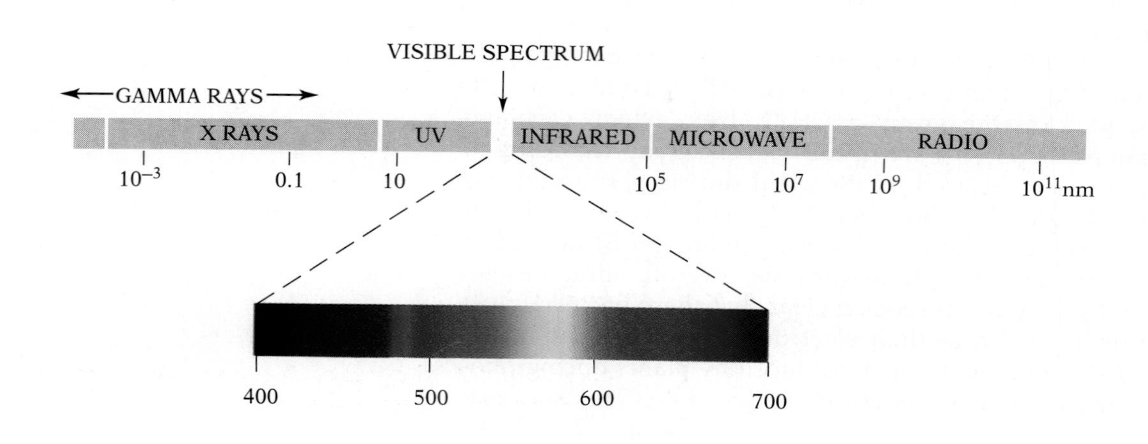

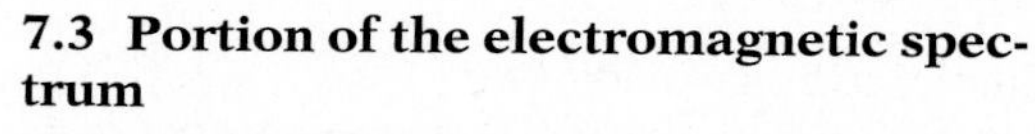

7.3 Portion of the electromagnetic spectrum

Light visible to humans constitutes only a very small portion of the total spectrum. Within the visible spectrum, light of different wavelengths stimulates different color sensations in us. Besides vision and photosynthesis, other radiation-dependent biological processes rely on this same small portion of the electromagnetic spectrum (sometimes extended into the ultraviolet or infrared). Light of wavelengths shorter than about 300 nm is absorbed by the atmosphere, while wavelengths longer than 1000 nm have too little energy to drive biological reactions.

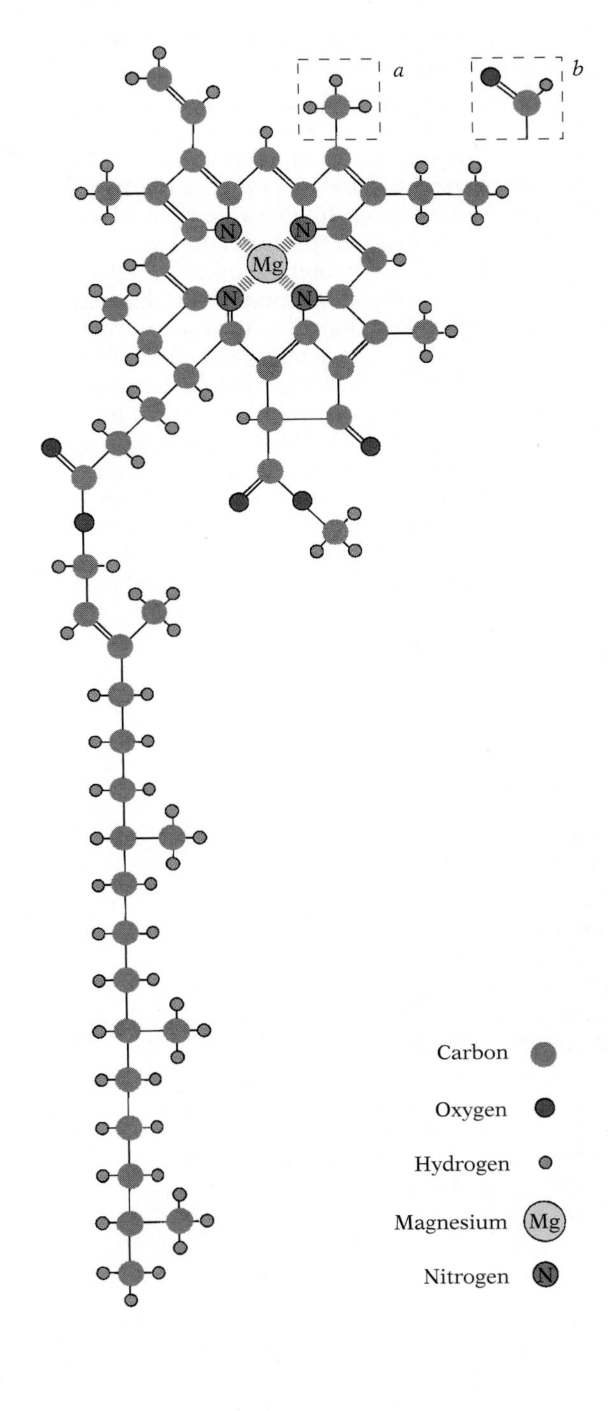

7.4 Molecular structure of chlorophyll

The structure of chlorophyll *a* is shown. The structure of the other major type of chlorophyll in green plants, chlorophyll *b*, differs only in the side group shown in the inset: a formyl group (—CHO) is substituted for the methyl group (—CH_3) of the *a* form. The electron that is excited by light energy is in the "head" region, near the magnesium atom. Note the similarity of chlorophyll's prosthetic group to the heme of hemoglobin.

the longest produce the sensation of red. The combination of all visible wavelengths creates the sensation of white. Electromagnetic waves shorter than those of violet light, such as ultraviolet (UV) radiation, X rays, and gamma rays, are invisible to us; waves longer than those of red light—infrared, microwave, and radio or TV rays, for instance—are also invisible.

Not all light is equally effective for photosynthesis. The all-important green pigment ***chlorophyll***, which traps light energy and helps convert it into chemical energy, captures only certain wavelengths. There are several slightly different kinds of chlorophyll; we will look first at the light-capturing ability of the most widespread form: chlorophyll *a* (Fig. 7.4).

Light falling on an object may pass through the object (be transmitted), be absorbed by it, or be reflected from it (Fig. 7.5). We can see transmitted and reflected light, but not absorbed light. Since chlorophyll is green, then it follows that chlorophyll cannot be absorbing much green light or there would be little radiation of that color reaching our eyes. On the other hand, chlorophyll must be absorbing radiation of other wavelengths within the visible part of the spectrum, or the light transmitted or reflected to us would appear white. Thus chlorophyll must be absorbing some light at the red and blue ends of the spectrum.

Chlorophyll can be extracted from a leaf and exposed to light to determine more precisely the amount of absorption at each wave-

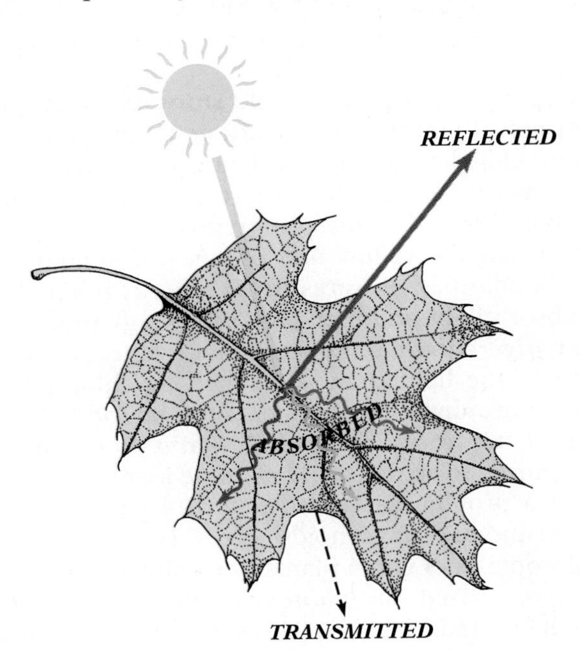

7.5 Light striking a leaf

Light striking an object, such as a leaf, may be reflected, absorbed, or transmitted.

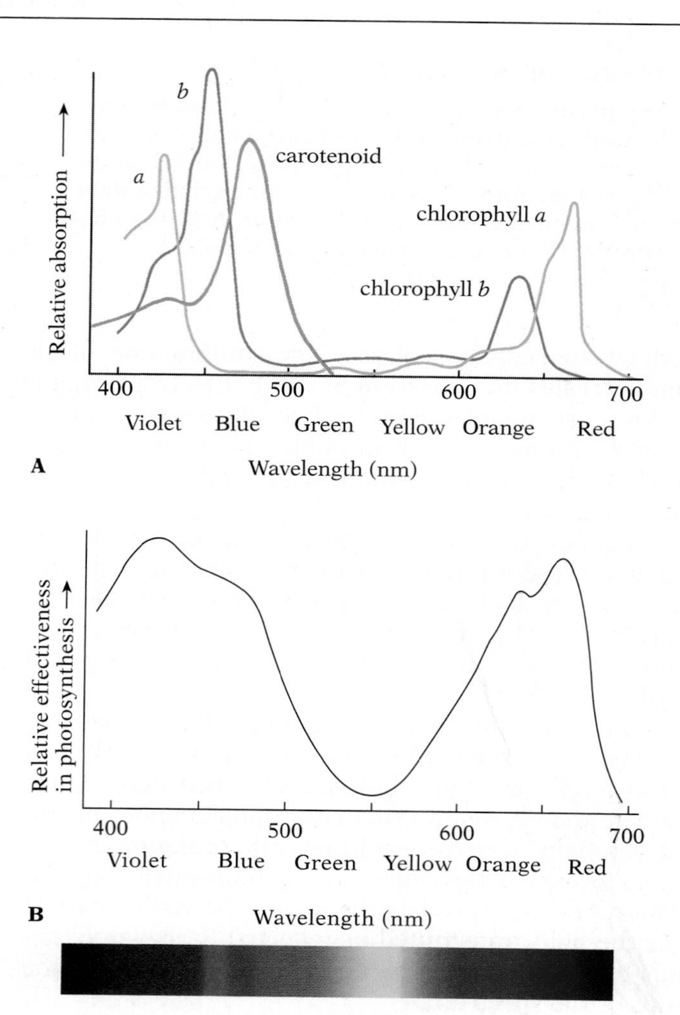

7.6 Absorption and action spectra of photosynthetic pigments

(A) Absorption spectra of chlorophyll *a*, chlorophyll *b*, and a carotenoid from a P680 complex—one of the two pigment complexes in plants. (The absorption spectra in the other type of complex, P700, are shifted about 20 nm to the right.) Taken together, the absorption spectra of the two chlorophylls and the carotenoid in each kind of complex cover more of the range of wavelengths available to the plant than does the spectrum of chlorophyll *a* alone. (B) Action spectrum of P680-dominated photosynthesis, indicating the relative effectiveness of different colors of light in driving photosynthesis. Light of intermediate wavelengths is more effective in driving photosynthesis than would be predicted on the basis of the absorption spectrum of chlorophyll *a* alone. Other pigments, such as carotenoids, absorb light of these intermediate wavelengths and pass the energy to the photosynthetic pathway.

length. The resulting absorption spectrum of chlorophyll *a* (Fig. 7.6A) shows that light in the violet, blue-violet, and red regions is absorbed, while light from the green, yellow, and orange regions is mostly transmitted.

But chlorophyll *a* is not the whole story: if we measure which wavelengths contribute to photosynthesis in an intact leaf, the resulting graph, called an action spectrum, shows that there is high activity in parts of the spectrum where chlorophyll *a* absorbs very little light (Fig. 7.6B). The reason for this difference is that other pigments—principally yellow and orange carotenoids, as well as other forms of chlorophyll—are also present in green plants. These molecules absorb light and then pass the energy to the chlorophyll *a*. ***Accessory pigments*** like the carotenoids thus enable the plants to use more of the light than is trapped by chlorophyll *a* alone. Most of the chlorophylls and pigments are bound by proteins in a highly ordered structure called a light-harvesting complex (LHC). The most common form, LHC II, has three identical subunits, each consisting of a protein, seven molecules of chlorophyll *a*, five of chlorophyll *b*, and two carotenoids (Fig. 7.7).

In the chloroplasts of green plants, the chlorophyll and accessory pigments are organized into ***photosynthetic units***. Each unit contains between five and 10 LHCs, with some 300 pigment molecules, including chlorophyll *a*, another variety of chlorophyll (*b* in green

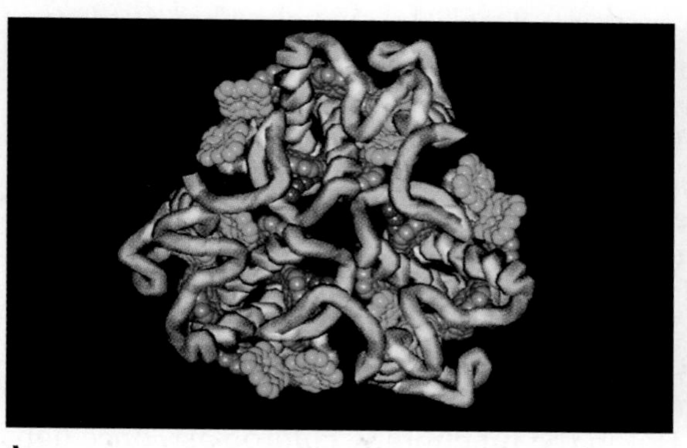

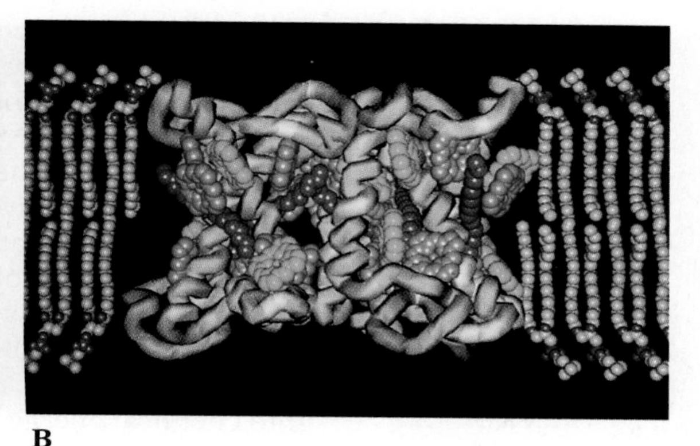

7.7 How chlorophyll is mounted in the membrane

The antennas are organized as five to ten protein-bound units called light-harvesting complexes (LHC). Each LHC includes three subunits, viewed here from the stromal side of the membrane (A) and from the plane of the membrane (B). Each of these subunits, in turn, has approximately seven molecules of chlorophyll *a* (blue), five molecules of chlorophyll *b* (green), and several carotenoids (red); all are held in a protein matrix (tan). (The tails of the chlorophylls have been omitted for clarity.)

plants, *c* in brown algae, and *d* in red algae), and carotenoids. One group of four pigment molecules in each unit is distinct from all the rest; it is part of a complex that acts as a ***reaction center*** where energy conversion occurs. The other pigment molecules function somewhat like antennas for collecting light energy and transferring it to the reaction center.

When a photon strikes a chlorophyll (or a carotenoid) molecule and is absorbed, its energy is transferred to an electron of the pigment molecule; the excited electron moves up to a higher, relatively unstable energy level (Fig. 7.8). Not just any photon will do: because electrons occupy discrete energy levels, the photon must have a particular amount of energy to excite the pigment electron from its ground state to the higher level. This explains the well-

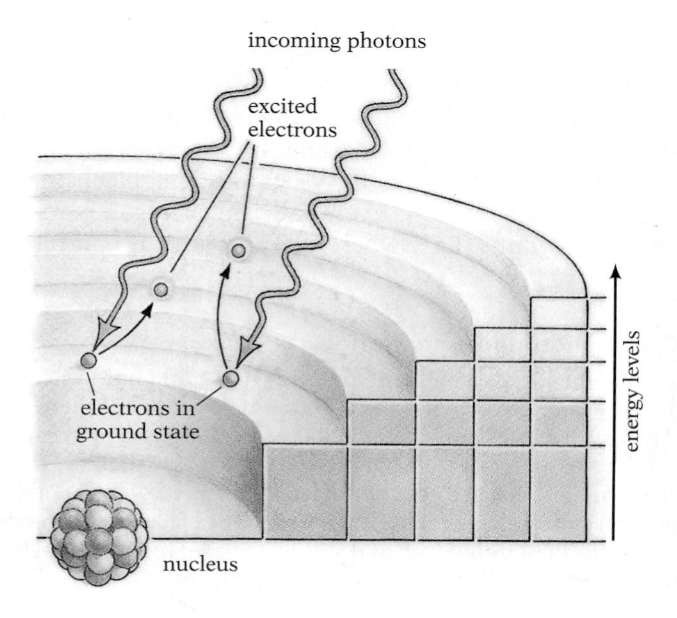

7.8 Effect of light on chlorophyll

When a photon is absorbed by a chlorophyll molecule, the photon's energy raises an electron to a higher energy level. This simplified representation shows a typical distribution of electrons at their lowest available energy levels, and the distribution as it is altered by absorption of a photon of red light, which raises one electron from its ground state in the first energy level to its first excited state in the second energy level. As we saw in Figure 7.6A, the absorption spectrum of a particular pigment can have several peaks. These exist because electrons can be excited from their ground state to their second excited state, two energy levels higher. In this illustration, another electron at the first level is shown being excited up to the third level—its second excited state—by a blue photon, which has enough more energy than a red photon that it can lift electrons two levels rather than one.

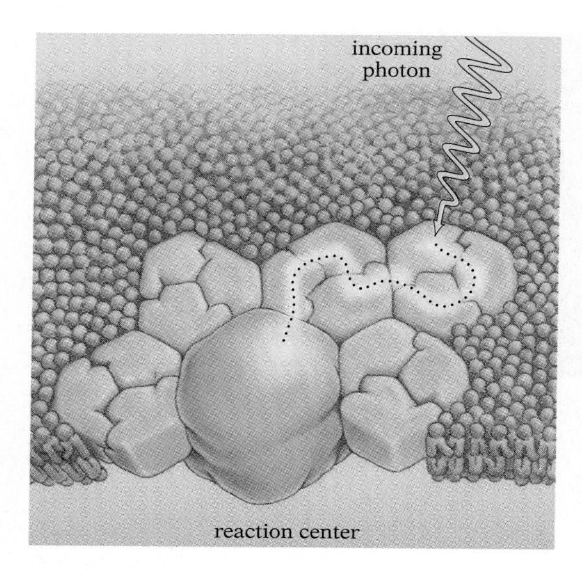

7.9 Flow of excited state within a photosynthetic unit

A photon strikes one of the antenna pigments in one of the three subunits of an LHC, raising an electron in the pigment to a higher energy level. This excited state is then passed from one pigment molecule to another in a random sequence (dotted pathway) until it eventually reaches the reaction center where it is trapped.

defined absorption peaks in Figure 7.6A: since the energy of a photon is related to its wavelength, photons with the proper amount of energy come only from light within a certain range of wavelengths (Fig. 7.6B).

An excited and therefore unstable electron will return spontaneously to its inactive state almost immediately, giving up its absorbed energy. Isolated chlorophyll in a test tube, for instance, promptly loses part of the energy it captures by reemitting it as visible light—a process known as fluorescence. (The rest of the energy is lost as heat.) Chlorophyll molecules alone, separated from their photosynthetic units, are incapable of converting light energy into chemical energy. But in the functioning chloroplast, once light energy has raised an electron in an antenna molecule to a high-energy state, the excited state is passed from one pigment molecule to another and eventually reaches the reaction-center complex, which traps it (Fig. 7.9).

The excited state is captured in the reaction center because the free energy of this complex is *lower* than that of the antenna molecules. Hence, once the excited state has reached the reaction center trap, it cannot easily escape. In this molecule, the energized electron that characterizes the excited state is passed quickly to an acceptor molecule and enters a series of redox reactions that convert the energy into a form more readily used by the cell.

■ HARVESTING THE ENERGY OF EXCITED ELECTRONS

There are two general pathways by which the energy from excited electrons is harvested: the cyclic pathway and the noncyclic pathway. The cyclic pathway involves only one of the two types of photosynthetic units found in most plants; the noncyclic pathway involves both. The cyclic pathway is the simpler and more ancient of the two.

Cyclic electron flow and photophosphorylation In cyclic photophosphorylation the reaction center is designated P700 because its chlorophyll absorbs maximally at 700 nm. Once the reaction center has received energy from the antenna, the electron thus activated is passed immediately to an acceptor molecule—a redox enzyme (FeS) containing an iron and sulfur prosthetic group. The acceptor is reduced and P700 is oxidized by this electron transfer. The electron is then passed along by a series of membrane-mounted molecules (Fig. 7.10) very like the electron-transport chain of the mitochondrion. Eventually the electron reaches a molecule of plastocyanin (PC), where it waits until it can fill an open-

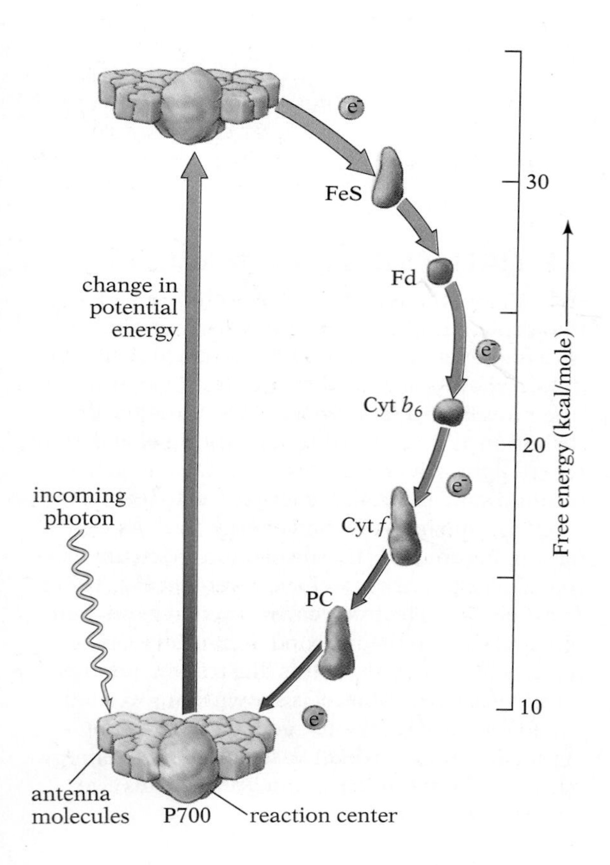

7.10 Cyclic photophosphorylation

A photon of light strikes a pigment molecule in the antenna system. The excited state eventually reaches the P700 complex, which is thereby energized. An energized electron from P700 then begins passage from one acceptor molecule to the next, each more electronegative than the previous one. Free energy is released at each step, and the electron ultimately returns to the ground state at which it began in P700. The electron acceptors are FeS, a molecule containing iron and sulfur; ferredoxin (Fd); cytochrome b_6; cytochrome f; and plastocyanin (PC).

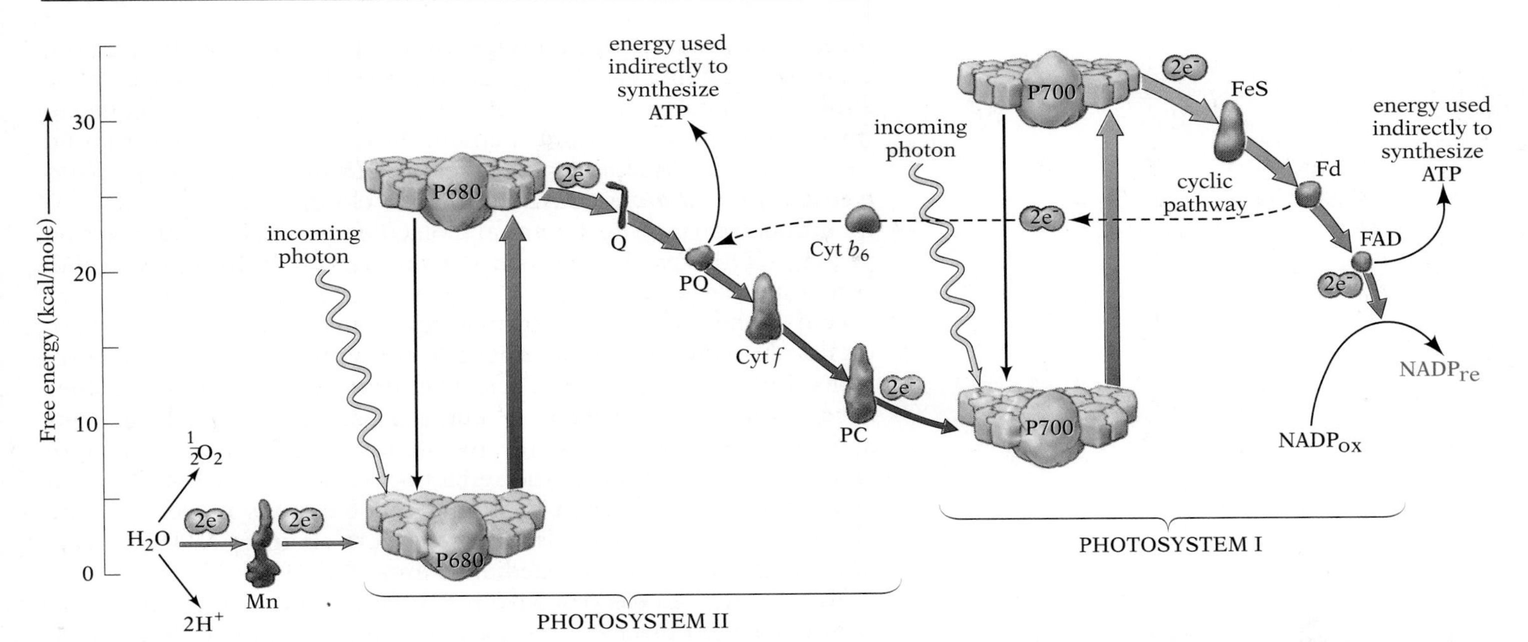

ing in the P700 complex. When another reaction-center electron is energized and transferred to FeS, the slot opens and the first electron fills it, thereby completing the cycle.

When a light-energized electron leaves a chlorophyll molecule, it is energy-rich; when it finally returns, it is energy-poor. The transition is gradual. As the electron is passed from transport molecule to transport molecule in the chain, it releases some of its extra energy with each transfer. Hence when the electron finally falls back into the chlorophyll, it has discharged all its extra energy, but it has not released it all at once, as happens with isolated chlorophyll in a test tube. Instead, the energy has been released in a series of small portions of manageable size. The process is called ***cyclic photophosphorylation*** because the electron is returned to the chlorophyll and some of its energy is used (indirectly, as we will see) to phosphorylate ADP into ATP.

Cyclic photophosphorylation was the first elaborate photosynthetic pathway to evolve, and it is still the dominant form found in bacteria. As Figure 7.10 indicates, however, the cyclic system is not very efficient: of the 25 kcal/mole of energy gained when P700 is excited by a photon, only the energy liberated in the passage from cytochrome b_6 to cytochrome f—3.4 kcal/mole—is actually available to the cell. The energy released in the other steps is wasted. Though 3.4 kcal/mole is better than nothing (particularly since photons are free), most photosynthesis now follows a highly modified, noncyclic pathway, which is more efficient under many conditions. In most organisms today cyclic photophosphorylation merely augments the noncyclic pathway.

Noncyclic electron flow and photophosphorylation Like the cyclic process, noncyclic electron flow and photophosphorylation (Fig. 7.11) begins when a photon of light strikes an antenna molecule of chlorophyll and raises an electron to an excited state, which is later trapped by the reaction center P700. As before, an ener-

7.11 Noncyclic photophosphorylation

As in cyclic photophosphorylation, electrons move along the pathway one at a time. However, for convenience in illustrating reactions at the two ends of the noncyclic pathway, the passage of a pair of electrons is shown throughout the diagram. One of the two essential light events occurs when a photon of light strikes a pigment molecule in Photosystem II; the resulting excited state eventually reaches P680, which donates an energized electron to substance Q. The electron moves down the Photosystem II electron-transport chain from Q to PQ to cytochrome f to PC; as in cyclic photophosphorylation, the energy released in the passage from PQ (plastoquinone) to cytochrome f is used by the cell (by the chemiosmotic process described later). At PC the electron waits to fill an opening in the P700 complex, which occurs when a photon strikes Photosystem I and P700 transfers an energized electron to the Photosystem I electron-transport chain. This electron passes from FeS to Fd and then, with the help of the enzyme FAD, to NADP. Energy released in the step down from FAD is used by the cell, and for each two electrons transported, one NADP$_{re}$ is generated to supply energy for carbon fixation. The vacancy in P680 that was created by the passage of an electron to Q is filled by an electron released in the splitting of water (lower left) by a manganese-containing enzyme (Mn), with one molecule of water yielding two electrons, along with two H^+ ions and one atom of oxygen. As indicated by the changes in relative free energy, the electron that has moved through Photosystem II still has much of its original energy when it enters Photosystem I. The primitive cyclic pathway (which connects Fd to PQ via cytochrome b_6) is also shown.

gized electron is then led away from the P700 complex by electron acceptors, the first two being FeS and ferredoxin (Fd). But here the similarity to cyclic photophosphorylation ends. Instead of continuing down the cyclic transport chain, the electron is passed from Fd to a different acceptor molecule: ***nicotinamide adenine dinucleotide phosphate,*** or ***NADP,*** which is closely related to the NAD in mitochondria. The antenna molecules and the P700 reaction center, together with the electron-transport chain from FeS to Fd, constitute ***Photosystem I***.[1]

Unlike the reduced electron-acceptor molecules in cyclic photophosphorylation, $NADP_{re}$ does not promptly pass along the electrons it receives from the electron-transport chain to another acceptor molecule. Instead, it retains a pair of energized electrons and their associated protons. Eventually $NADP_{re}$ acts as a high-energy electron donor in the reduction of CO_2 to carbohydrate, a process known as carbon fixation. Thus electrons move from the chlorophyll, through an electron-transport chain to NADP, to carbohydrate (via other intermediate compounds).

Since energized electrons from P700 are retained by $NADP_{re}$ and eventually incorporated into carbohydrate, it follows that Photosystem I is left short of electrons. These "electron holes" are filled, indirectly, by electrons derived from water through a second light-driven process. This second light event involves a different type of photosynthetic unit, which, like Photosystem I, contains about 300 molecules of chlorophyll including about 200 molecules of chlorophyll *a*. In this case the reaction-center complex is called P680 because it absorbs light maximally at 680 nm. The antenna molecules and the P680 reaction center, plus its special set of electron-transport molecules, constitute ***Photosystem II***.

When light of a proper wavelength strikes a pigment molecule of Photosystem II, the energy is passed around within the photosynthetic unit until it reaches the P680 trap. Excited P680, in turn, donates a high-energy electron to an acceptor, designated Q (Fig. 7.11). Substance Q then passes the electron to a chain of acceptor molecules, which transport the electron, by some of the same steps we saw in cyclic photophosphorylation, down an energy gradient to the electron hole in the P700 molecule of Photosystem I. As the electron moves down the transport chain, some of the energy released along the way is used by the cell indirectly to synthesize ATP (via the chemiosmotic pathway described later).

Thus the electron holes created in Photosystem I by the first light event are refilled by electrons moved from Photosystem II by the second light event. However, this process alone simply shifts the holes from Photosystem I to Photosystem II. This difficulty is resolved when P680 (with the aid of a manganese-containing enzyme which we shall designate simply as Mn) pulls replacement electrons away from water, leaving behind free protons and molecular oxygen:

$$2H_2O \longrightarrow 4e^- + 4H^+ + O_2$$
$$\downarrow$$
$$\text{P680}$$

[1] Some workers do not consider the electron-transport chain to be part of the photosystem.

HOW WE KNOW:
THE P680 REACTION CENTER

Painstaking research by H. Michel, J. Deienhofer, and R. Huber (for which they shared a Nobel prize) has revealed the structure of a photosynthetic reaction center in purple bacteria. In the details lies the explanation of how the center complex manages to capture and keep electron energy that is transferred freely between the conventional chlorophyll molecules in the antenna. The reconstruction of a P680 chloroplast reaction center shown here is based on the structure of the evolutionary precursor in photosynthetic bacteria; the two are thought to be very similar. A photon is shown activating an antenna chlorophyll in the chloroplast membrane immediately adjacent to the reaction center (1); this molecule transfers its energy to one of the two central reaction-center chlorophylls (2). The energized electron is first moved to a peripheral reaction-center chlorophyll (3) and then to a pheophytin molecule (4); from there it moves immediately to a bound quinone (5), which is the molecule Q of the electron-transport chain. The entire process is then repeated (white arrow), with the resulting high-energy electron being transferred to the same quinone used by the first electron (6). Now two electrons are waiting to leave.

The electrons leave the reaction-center complex by way of the membrane quinone, PQ (7, 8). The electrons missing from the P680 chlorophylls are replaced (10) by the enzyme Mn when it splits a water molecule (9). Because the movement of only one electron is pictured here, the water-splitting reaction is shown as involving only half a water molecule. The eight molecules of the reaction-center complex are bound into a unit by a large protein, shown here in outline.

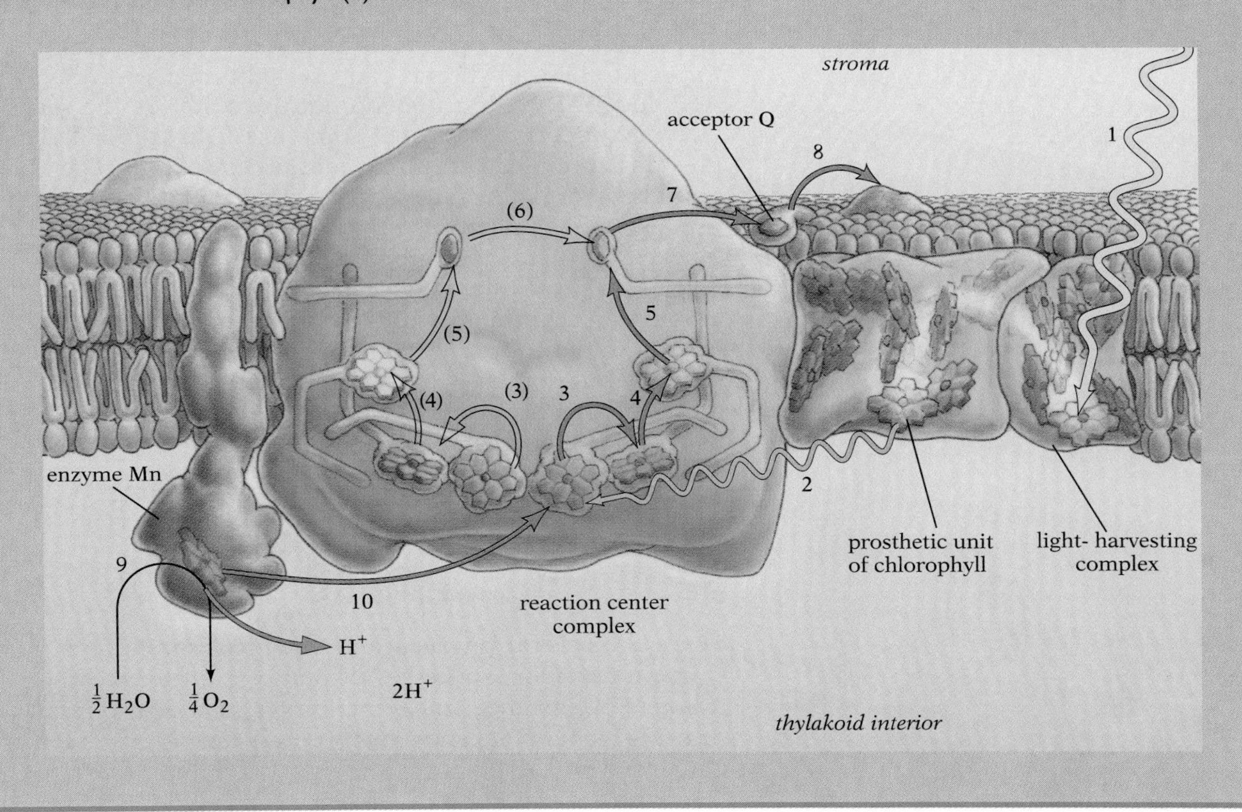

The oxygen from H_2O is released as a gaseous by-product. The protons, as we will see, also play an important role.

To summarize, the electrons involved in the second light event move from water to the P680 reaction center of Photosystem II, to Q, to the transport chain of Photosystem II, and to Photosystem I

(Fig. 7.11). Thus the sequence from water through the two photosystems to carbohydrate is as follows:

H_2O ⟶ P680 ⟶ Photosystem II transport chain ⟶ P700 ⟶
Photosystem I transport chain ⟶ $NADP_{re}$ ⟶ carbohydrate

This sequence shows that the electrons necessary to reduce carbon dioxide to carbohydrate come from water, and that the movement of electrons from water to carbohydrate involves the cooperation of a series of redox reactions similar to those occurring in mitochondria.

Since electrons are not passed in a closed loop, this series of reactions is termed noncyclic electron flow, which gives rise to ***noncyclic photophosphorylation***. The whole process results in the formation of both ATP and $NADP_{re}$ as well as the release of molecular oxygen. Though the events of NADP reduction and photophosphorylation have traditionally been known as the light reactions of photosynthesis, only two steps—the capture of photons by the antenna complexes, followed by the export of the electron from the reaction centers—are directly light-dependent.

The details given above apply only to photosynthesis by plants, algae, and the cyanobacteria. Other bacteria possess different forms of chlorophyll and can use light energy in the synthesis of ATP and $NADP_{re}$, but they do not use water as the source of electrons. As mentioned earlier, some use hydrogen sulfide (H_2S) and give off sulfur instead of oxygen; others use nitrogen-based compounds or organic acids. Even ordinary plants can be induced to use a source of electrons other than water. If oxidation of water is blocked by chemical inhibitors and a strong electron donor is provided as a substitute, noncyclic electron flow and photophosphorylation can continue without production of oxygen. However, oxygen is the most electronegative element among these alternatives, and its source—water—is abundant. It is not surprising that the most successful photosynthetic organisms, the plants, evolved using photosynthesis based on water, which is ubiquitous in the chemistry of life.

■ THE ROLE OF CHLOROPLAST STRUCTURE

There are several physical and chemical similarities between the electron-transport strategies of photosynthesis and those of respiration. For instance, the functioning of the respiratory electron-transport chain depends on ordered arrays of redox molecules embedded in the highly folded inner membrane of the mitochondrion; we will see something very similar in the chloroplast. We will also find that transport of H^+ across compartment membranes, which creates an electrochemical gradient, supplies the energy for the synthesis of ATP.

Like the mitochondrion, the chloroplast is a membrane-bounded organelle with its own genome, which probably derives from an ancient endosymbiotic bacterium. However, unlike the mitochondrion, which has a heavily folded inner membrane containing electron-transport chains, the chloroplast has an inner membrane that is relatively smooth and flat, and follows the contours of the outer membrane (Fig. 7.12); the inner membrane has many selective gates and pumps but no electron-transport complexes. It encloses the main volume, or ***stroma***, of the chloroplast. The antenna pigments, reaction centers, and electron-transport-chain molecules of the chloroplast are in fact embedded in a third membrane inside the stroma, which forms a series of flattened, interconnected compartments known as ***thylakoids***.

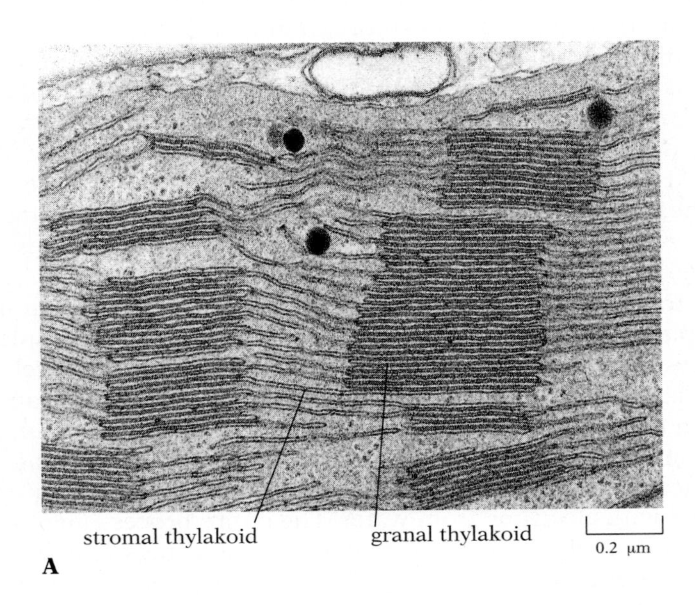

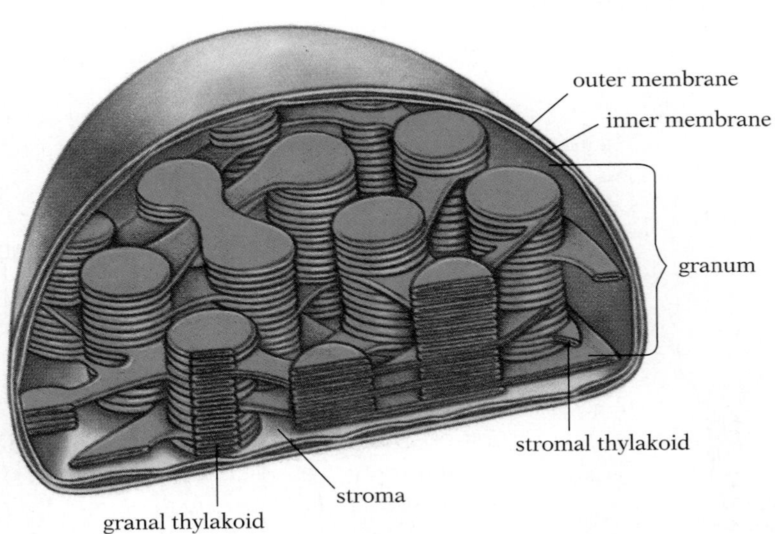

7.12 Structure of a chloroplast

(A) Electron micrograph of a section of a chloroplast of timothy grass showing several grana; note the continuity between granal and stromal thylakoids. (B) This cutaway view of a typical chloroplast shows the inner and outer membranes lying close together, enclosing the large compartment known as the stroma. Inside the stroma is a third distinct membrane, which forms the interconnected compartments called thylakoids. The stacks of flat, disklike thylakoids are grana. Chlorophyll molecules and most of the electron-transport-chain molecules are located in the thylakoid membrane.

Thylakoids occur in two distinct but interconnected arrangements, each with its own functional specializations (Fig. 7.12). When loose in the stroma, the proportion of the surface area of the ribbonlike stromal thylakoids that is in contact with stromal fluids is essentially 100%. When stacked neatly in piles called ***grana***, the densely packed, chlorophyll-laden membranes of granal thylakoids are highly efficient at capturing photons, but only a small proportion of the granal membrane is exposed to stromal fluids.

Stromal thylakoids are ideal for conditions of bright light, while granal thylakoids are well adapted to take advantage of moderate light. Like the inner mitochondrial membrane, the thylakoid membrane makes possible an electrochemical-potential gradient, which, functioning like a battery, supplies energy for the synthesis of ATP. However, in the chloroplast the H^+ ions accumulate in the innermost compartment of the organelle—the interior of the thylakoids—while the outer compartment, the stroma, becomes more alkaline.

Let's trace the sequence of events in photophosphorylation in the chloroplast. A photon is absorbed in the antenna complex of Photosystem II, exciting an electron to a higher energy level. This energy is transferred to the reaction-center complex. After another electron follows the same path, the pair is fed to an acceptor, plastoquinone (PQ), in the membrane (Fig. 7.13). The resulting electron vacancies in the normal chlorophylls in the reaction center are filled with electrons from water. The entire process involves abstracting four electrons from two water molecules (catalyzed by Mn) to produce one molecule of oxygen and four H^+ ions, and takes about a nanosecond (10^{-9} sec).

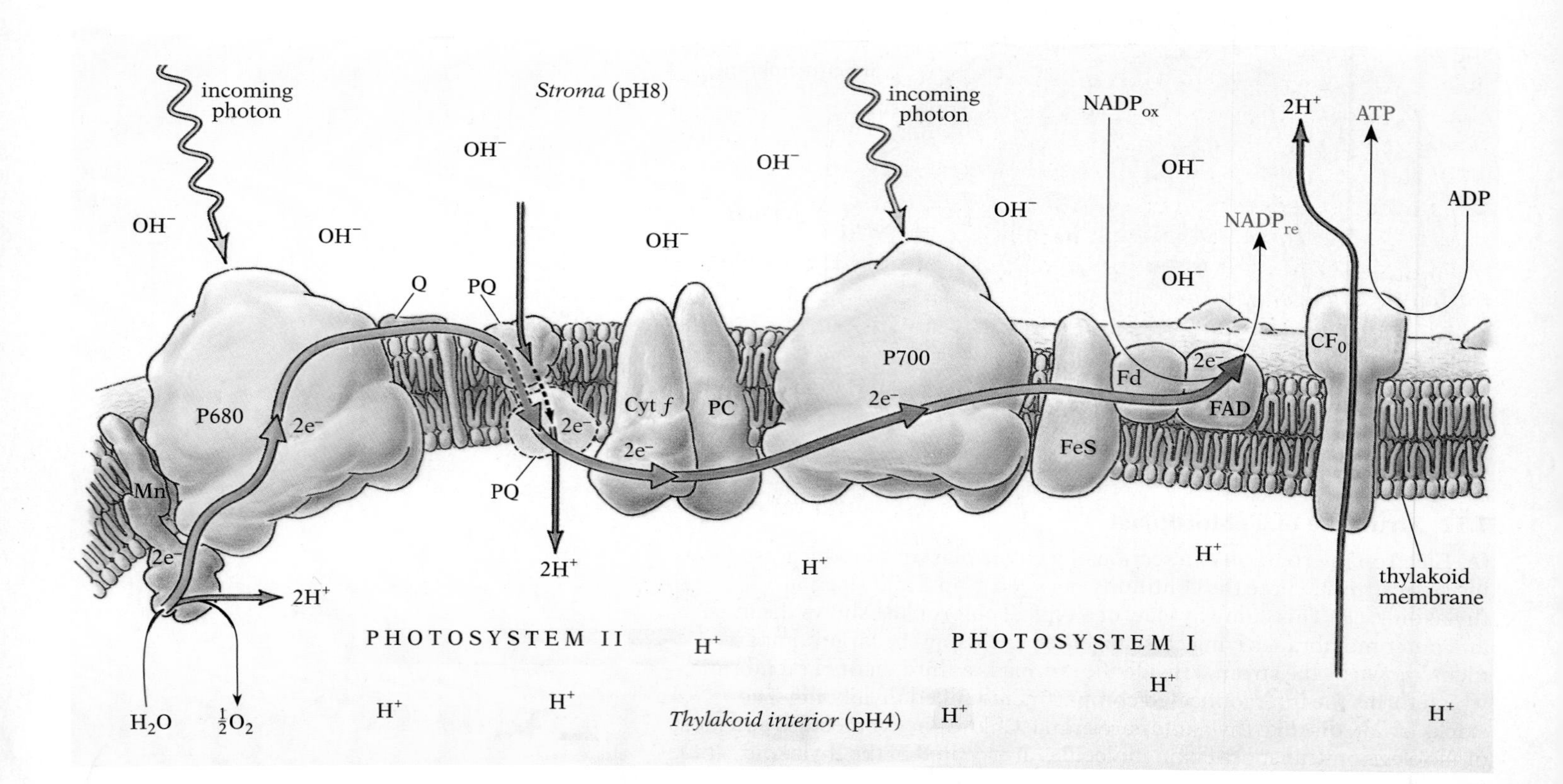

7.13 Anatomy of photophosphorylation

The enzyme molecules of photophosphorylation are shown here for simplicity as a linear chain; in the membrane the chain is thought to be folded back on itself, so that cytochrome b_6 connects Fd and PQ, thereby making possible cyclic photophosphorylation (not shown). The antenna molecules of the two photosystems are omitted. This representation makes it clear how electrons from water are moved along the electron-transport chains, and how the thylakoid battery is charged. Since the pH scale is logarithmic (see Fig. 2.14, p. 31), the values shown indicate that the concentration of H^+ ions is 10,000 times greater within the thylakoid than on the other side of the membrane. Note that PQ migrates from one side of the membrane to the other to transport H^+ ions. The CF_0-CF_1 complex, which uses the energy of the electrochemical-potential gradient to make ATP, is also shown.

The two pairs of energized electrons that left the reaction center via the acceptor (Q) are delivered, one electron at a time, to PQ, a molecule that can shuttle between the stromal side of the membrane and the thylakoid side. We will follow only one pair, and thus in our electrical accounting consider only two of the four H^+ ions produced when the two molecules of water were split. Some of the energy of each electron pair is used here to move two H^+ ions from the stroma to the thylakoid interior. Each electron now moves, one at a time, to cytochrome *f* and then to PC until there is a vacancy in P700.

When a photon is absorbed by an antenna molecule in Photosystem I and excites an electron, the excitation energy is passed to P700. The energized electron of P700 is immediately passed to an acceptor (FeS). The resulting vacancy in P700 can now be filled by an electron from Photosystem II waiting in PC. A second photon absorption is required to create a vacancy for the other electron in the pair we are following.

The two energized electrons of Photosystem I, meanwhile, move one at a time to another acceptor (Fd), from which each can follow one of two pathways. Under normal conditions, the most likely fate for the electrons is to help reduce $NADP_{ox}$. This process requires both electrons and consumes two H^+ ions from the stroma, so that in all, four H^+ ions are now gone from the stroma while four H^+ ions have been added to the thylakoid: two via PQ and two when the water molecule was split. The alternative (not shown in Fig. 7.13) is the cyclic pathway (dashed line in Fig. 7.11). The cyclic alternative represents a less efficient use of energy, but it is favored when $NADP_{ox}$ is in short supply or when the cell is more in need of ATP than of $NADP_{re}$.

The result of this highly ordered flow of electrons is twofold. On the one hand, it generates a supply of the high-energy carrier molecule $NADP_{re}$, whose role in carbohydrate synthesis we will examine shortly. On the other, it helps create, through the flow of H^+ ions, a powerful electrochemical gradient, which is then used to generate ATP: as in the mitochondrion, the membrane in the thylakoid contains numerous enzyme complexes (here called CF_0-CF_1) that can utilize the energy of the gradient to phosphorylate ADP.

The organization and anatomy of the electron-transport chains of photophosphorylation and of respiration are very similar. In fact, there is little doubt that the anabolic and catabolic mechanisms have a common evolutionary origin. Current evidence indicates that a low-efficiency anaerobic pathway similar to glycolysis followed by fermentation evolved first, enabling chemosynthetic cells to store energy in carbohydrates for later use. Then followed a higher-efficiency membrane-embedded electron-transport chain, which used electron acceptors like sulfur, carbon, and nitrogen to yield greater supplies of free energy for the cell.

Although primitive kinds of photosynthesis had evolved earlier, a major modification of the electron-transport system used in metabolism gave rise to cyclic phosphorylation. When the more versatile noncyclic system evolved, providing energy for carbon fixation, atmospheric oxygen was produced. For all its potential advantages, oxygen was toxic to much of the anaerobically evolved chemistry of the cell; the evolution of oxygen-tolerant enzymes must have

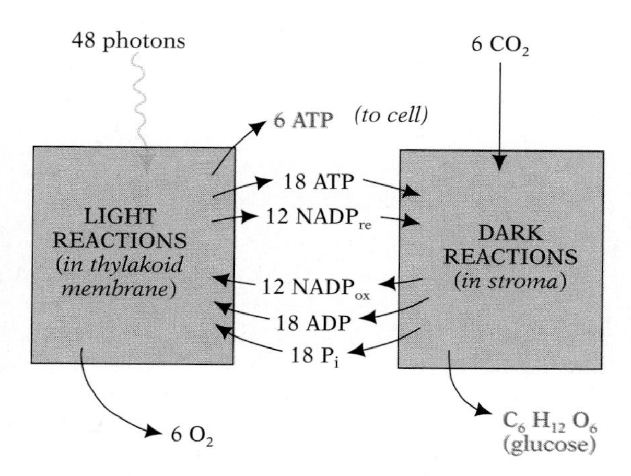

7.14 Light and dark reactions

Photosynthesis consists of two physically separate but interlocking sets of reactions. The light reactions use light to generate the energy intermediates ATP and $NADP_{re}$. These reactions, as we have seen, take place in the thylakoid membrane. The dark reactions—carbon fixation—use these energy intermediates to turn carbon dioxide into carbohydrates like glucose. Noncyclic photophosphorylation, summarized in the left half of this figure, produces roughly six ATPs in excess of those needed for the synthesis of a single molecule of glucose. Cyclic photophosphorylation goes on simultaneously to provide additional ATP for the cell, but since it does not produce $NADP_{re}$ for carbon fixation, we have ignored it here.

been the next step. With this, the evolution and rapid spread of the respiratory pathway using oxygen as its electron acceptor became possible.

HOW THE DARK REACTIONS FIX CARBON

So far, we have seen how the energy of photons is captured and used to make ATP and $NADP_{re}$, but not how this energy is used in turn to transform the low-energy substance CO_2 into high-energy compounds like glucose. The entire process of photosynthesis is frequently divided into the light reactions of electron-H^+ transfer and photophosphorylation on the one hand, and the dark reactions of carbon fixation on the other.

■ CARBOHYDRATE SYNTHESIS BY THE CALVIN CYCLE

Carbohydrates contain much chemical energy, while CO_2 contains very little. The reduction of CO_2 to form glucose proceeds by many small uphill steps, each catalyzed by a specific enzyme. The energy for this stepwise synthesis of carbohydrates from CO_2 comes from light via ATP and $NADP_{re}$ (Fig. 7.14). The series of dark reactions is called the ***Calvin cycle*** in honor of Melvin Calvin, who first worked out the many steps involved.

The Calvin cycle begins as CO_2 combines with the five-carbon sugar ***ribulose bisphosphate*** (***RuBP***) to form an unstable six-carbon compound, which is promptly broken into two three-carbon molecules of phosphoglyceric acid (PGA). Each molecule of PGA is

HOW WE KNOW:

THE CALVIN CYCLE

Since there are many sequential steps in the reduction of CO_2 to carbohydrates, and since many of the intermediate compounds occur also in other processes, leading to different end products, how was the exact sequence of steps ever discovered? The tool that made such discoveries possible was a radioactive isotope of carbon, ^{14}C. Samuel Ruben and Martin D. Kamen at the University of California, Berkeley, who discovered this isotope at about 1940, immediately recognized its potential as a research tool in photosynthesis. They showed that plants exposed to carbon dioxide containing the radioactive isotope (that is, $^{14}CO_2$ instead of the normal $^{12}CO_2$) incorporated the isotope into a variety of compounds.

In 1946, Melvin Calvin and his associates, also at Berkeley, began an intensive long-term investigation of carbon dioxide fixation in photosynthesis, using ^{14}C as their principal tool. They exposed algal cells to light in an atmosphere of $^{14}CO_2$ for a few seconds and then killed the cells by immersing them in alcohol. The alcohol also inactivated the enzymes that catalyze the reactions of photosynthesis. With the enzymes paralyzed, whatever amount of each intermediate compound existed in the cell at the moment of inactivation was, in effect, locked in. Calvin and his coworkers could then determine which of these locked-in intermediate compounds contained ^{14}C.

How long the algal cells were exposed to the $^{14}CO_2$ before being killed determined the number of compounds in which ^{14}C was detected: when the time was very short, the ^{14}C reached only the first few compounds in the synthetic sequence; when the time was longer, the isotope moved through more steps in the sequence and appeared in a great variety of compounds. Thus the order in which these compounds appeared in cells reflected their place in the reaction chain. After years of painstaking research, Calvin, who in 1961 was awarded the Nobel Prize for his critically important investigations, worked out the sequence of reactions now called the Calvin cycle.

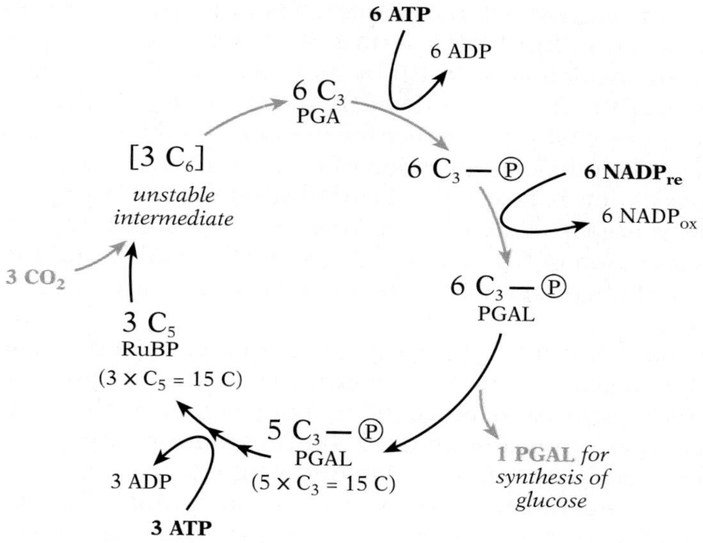

6 ATP
6 ADP
6 C_3 PGA
[3 C_6] *unstable intermediate*
6 C_3—Ⓟ
6 $NADP_{re}$
6 $NADP_{ox}$
3 CO_2
6 C_3—Ⓟ PGAL
3 C_5 RuBP ($3 \times C_5 = 15$ C)
5 C_3—Ⓟ PGAL ($5 \times C_3 = 15$ C)
1 PGAL *for synthesis of glucose*
3 ADP
3 ATP

7.15 Synthesis of carbohydrate by the Calvin cycle

Each CO_2 molecule combines with a molecule of ribulose bisphosphate (RuBP), a five-carbon sugar, to form a highly unstable six-carbon intermediate, which promptly splits into two molecules of a three-carbon compound called PGA. Each PGA molecule is phosphorylated by ATP and then reduced by $NADP_{re}$ to form PGAL, a three-carbon sugar. Thus each turn of the cycle produces two molecules of PGAL. Five of every six new PGAL molecules formed are used in the synthesis of more RuBP by a complicated series of reactions (not shown here) driven by ATP. The sixth new PGAL molecule can be used in the synthesis of glucose. The path of carbon from CO_2 to glucose is traced by blue arrows. Since it takes three turns of the cycle to yield one PGAL for glucose synthesis, the diagram begins with three molecules of CO_2; it would require a total of six turns to produce one molecule of glucose, a six-carbon sugar. Note that the cycle is driven by energy from ATP and $NADP_{re}$, both formed by the light reactions of photophosphorylation.

then phosphorylated by ATP and reduced by hydrogen from $NADP_{re}$. The resulting energy-rich three-carbon compound is PGAL (phosphoglyceraldehyde). This compound, which is also an intermediate in glycolysis, is a true sugar and, in a sense, is the stable end product of photosynthesis. Because PGAL is a three-carbon compound, as are the intermediate compounds leading to its formation, the Calvin cycle is often called ***C_3 photosynthesis*** (Fig. 7.15).

Five of every six molecules of PGAL are used in the formation of new RuBP (by a complicated series of reactions powered by ATP), with which more CO_2 can be processed. The sixth molecule can be combined by a series of steps with another molecule of PGAL (produced in another turn of the cycle) to form the six-carbon sugar glucose. Thus, it takes six carbon dioxide molecules and six turns of the Calvin cycle to produce one molecule of glucose.

Though glucose has traditionally been considered the end product of photosynthesis, free glucose is not present in significant amounts in most higher plants. Some of the PGAL produced by the Calvin cycle is at once utilized in the formation of lipids, amino acids, and nucleotides. Even when glucose is synthesized, it is normally used almost immediately as a building-block unit for double sugars (disaccharides like sucrose), starch, cellulose, or other polysaccharides. Carbohydrates are generally stored in higher plants in the form of starch, which, because it is insoluble in water, has much less osmotic activity than glucose. An excessive accumulation of carbohydrates in the form of sugars would raise the osmotic concentration of the cytoplasm relative to the environment and severely upset the osmotic balance between the cell and the surrounding fluid. The result would be the intake of too much water by the cell, leading to extreme swelling.

■ THE MYSTERY OF PHOTORESPIRATION

One property of the Calvin cycle is perplexing in that it has no obvious biological function: RuBP carboxylase, the enzyme that catalyzes the carboxylation of ribulose bisphosphate (that is, the addition of CO_2 to RuBP) at the start of the Calvin cycle, can also catalyze the oxidation of RuBP by molecular oxygen (the addition of O_2 to RuBP). In other words, CO_2 and O_2 are alternative substrates that compete with each other for the same binding sites on this enzyme. When the concentration of CO_2 is high and that of O_2 is low, carboxylation is favored and carbohydrate synthesis by the Calvin cycle proceeds. When the reverse conditions prevail—when the concentration of CO_2 is low and that of O_2 is high—oxidation is favored. Higher-than-normal temperatures also favor the oxidation pathway.

Oxidation of the high-energy photosynthetic intermediate RuBP leads to regeneration of low-energy CO_2, a wasteful process known as ***photorespiration***. Since it does not result in synthesis of ATP, as other types of respiration do, it appears to short-circuit the Calvin cycle to no purpose. The situation could be worse, however. By means of a complex series of reactions involving chloroplasts, mitochondria, and peroxisomes, plant cells salvage much of the energy they stand to lose from the breakdown of RuBP. Only one of every three carbons entering photorespiration is actually lost as CO_2. Still, photorespiration squanders useful energy.

Because photorespiration predominates over photosynthesis at low concentrations of CO_2, plants that depend exclusively on the Calvin cycle for CO_2 fixation cannot synthesize carbohydrates unless the CO_2 concentration in the air is above a critical level (commonly about 50 ppm, or parts per million); even at normal levels much of the production of photosynthesis is undercut by concurrent photorespiration. At atmospheric CO_2 concentrations, net photosynthesis by such plants could be increased by as much as 50% if photorespiration could be stopped. We will return to this question presently.

THE LOGIC OF LEAF DESIGN

As we have seen repeatedly, life depends both on the precisely catalyzed reactions of various biochemical pathways, with the accompanying interplay of their products, and on the particular anatomy of cells and their organelles. Nowhere is this intimate relationship between form and function more obvious than in the specialized tissues responsible for photosynthesis.

■ THE ANATOMY OF ORDINARY LEAVES

Photosynthesis can occur in all green parts of the plant, but in most plants the leaves expose the greatest area of green tissue to the light and are therefore the principal organs of photosynthesis.

Figure 7.16 shows leaves of a variety of familiar land plants. Most leaves consist of a stalk, or ***petiole***, and a flattened ***blade***. However, some leaves, particularly those of the grasslike plants, lack petioles, the base of the blade being attached directly to the stem. The blades of most leaves are broad and flat and contain a

7.16 Leaf types

A leaf usually consists of a blade and a petiole, which sometimes has stipules at its base. Veins run from the petiole into the blade. The main veins may branch in succession off the midvein (pinnate venation); or they may all branch from the base of the blade (palmate venation); or they may be parallel. The blade may be simple, or it may be compound—that is, divided into leaflets that may be pinnately or palmately arranged. Leaves with parallel venation (lower left) are characteristic of the monocot group, plants whose seedlings have only one "seed leaf" or cotyledon; they include grasses, grains, and spring bulbs, for example. The other leaves are from the dicot group (containing two cotyledons).

VENATION

LEAF TYPE

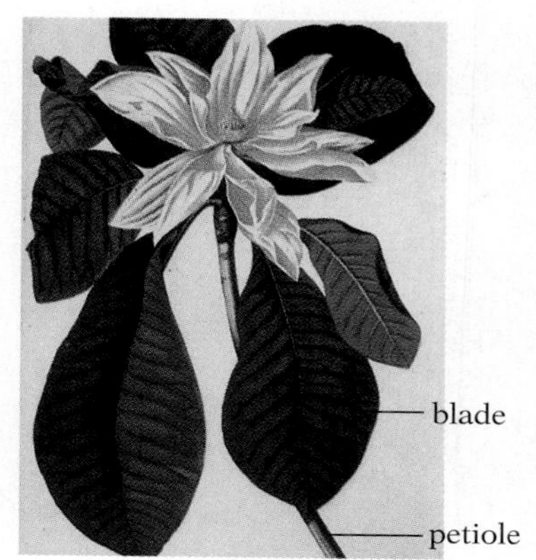

Pinnate venation

Simple

Palmate venation

Pinnately compound

Parallel venation

Palmately compound

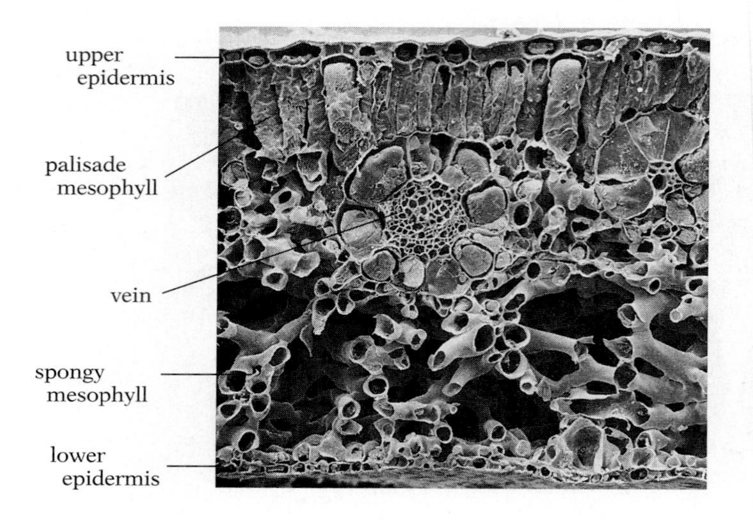

7.17 Leaf Anatomy

This scanning electron micrograph shows a cross section of a typical dicot leaf. Note that the mesophyll cells are loosely packed and have many spaces between them. The round structure in the center is a vein.

complex system of veins. Because of the flatness of the blade, the leaf exposes to the light an area that is very large in relation to its volume.

When we examine a transverse section of a leaf under the microscope (Fig. 7.17, 7.18), we see that the outer surfaces are formed by layers of epidermis, usually only one but sometimes two or more cells thick. A waxy layer, the ***cuticle***, usually covers the outer surfaces of both the upper and the lower epidermis, but is generally

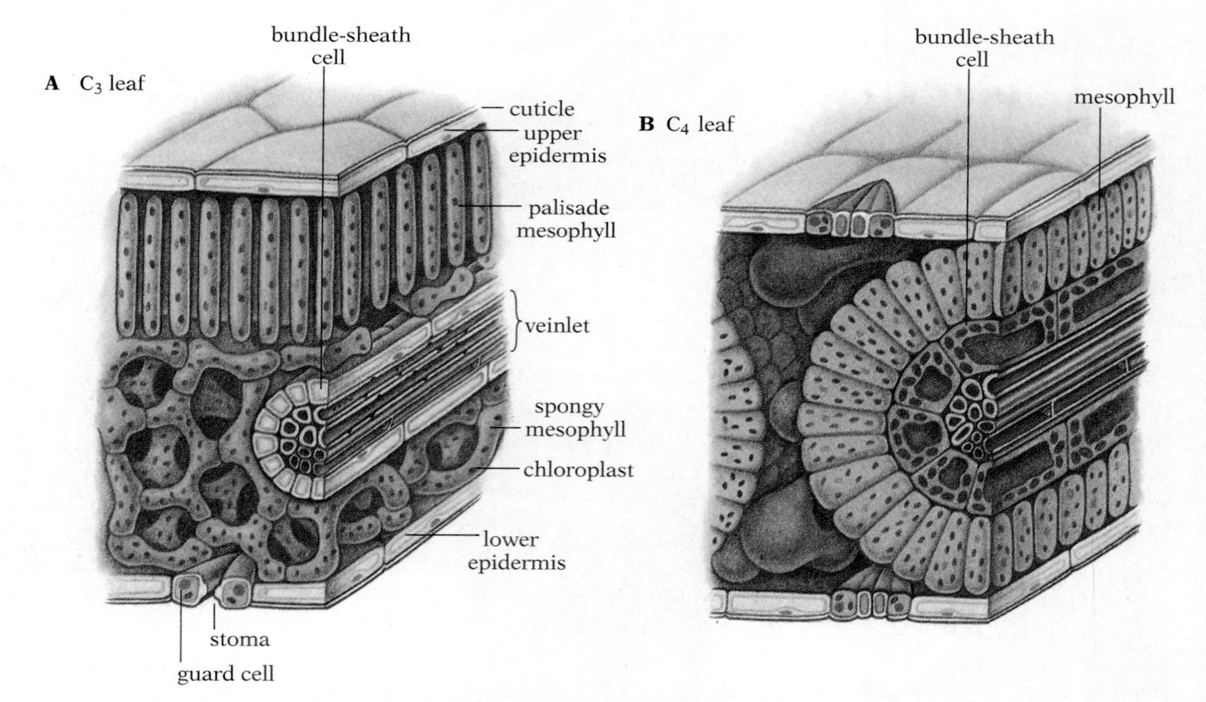

7.18 Anatomy of C_3 and C_4 (Kranz) leaves

In a C_3 leaf the palisade mesophyll cells typically form a layer in the upper part of the leaf; the corresponding mesophyll cells in a C_4 leaf are usually arranged in a ring around the bundle sheath. While the bundle-sheath cells of C_4 leaves have chloroplasts (dark green), those of C_3 leaves usually lack them.

thicker on the upper side. The chief function of the epidermis is to protect the internal tissues of the leaf from excessive water loss, from invasion by fungi, and from mechanical injury. Most epidermal cells do not contain chloroplasts.

The entire region between the upper and lower epidermis constitutes the ***mesophyll***. The mesophyll is usually divided into two fairly distinct parts: an upper palisade mesophyll, consisting of cylindrical cells arranged vertically, and a lower spongy mesophyll, composed of irregularly shaped cells (Fig. 7.18A). The cells of both parts of the mesophyll are very loosely packed and have many intercellular air spaces between them. These spaces are interconnected and communicate with the atmosphere outside the leaf by way of holes in the epidermis called ***stomata*** ("mouth" in Greek; *stoma* is the singular). The size of the stomatal openings is regulated by a pair of modified epidermal cells called ***guard cells***.

A conspicuous system of veins (also called vascular bundles) branches into the leaf blade from the petiole (Fig. 7.16). The veins form a structural framework for the blade and also act as transport pathways. Each vein contains cells of the two principal vascular tissues, xylem and phloem. Each vein is also usually surrounded by a ***bundle sheath***, composed of cells packed so tightly together that there are few spaces between them. In most cases the branching of the veins is such that no mesophyll cell is far removed from a vein; typically the veins have a combined length of 100 cm per square centimeter of leaf blade.

■ LEAVES WITH KRANZ ANATOMY

As early as 1904, German plant anatomists observed that the leaf anatomy of some plants of tropical origin—plants associated with bright, hot, but especially dry habitats—showed a combination of features not generally found in plants native to the temperate zones. This unusual complex of features came to be called ***Kranz anatomy***. *Kranz*, which means "wreath" in German, refers to the ringlike arrangement of photosynthetic cells around the leaf veins of these plants.

The bundle-sheath cells of plants with Kranz anatomy (also called C_4 plants, for reasons we will see shortly) contain numerous chloroplasts, whereas those of other plants (C_3 plants) often do not. In plants with Kranz anatomy the mesophyll cells that correspond to the palisade layer tend to be clustered in a ringlike arrangement around the veins, just outside the bundle sheaths (Fig. 7.18B). These mesophyll cells contain numerous chloroplasts, but the spongy mesophyll cells outside the rings often have few chloroplasts or none at all. In Kranz plants the chloroplasts of the bundle-sheath cells and mesophyll cells usually differ in a number of ways. In the bundle-sheath cells the chloroplasts are bigger, they accumulate large amounts of starch in the presence of light, and the grana are few and poorly developed; in the mesophyll cells the chloroplasts are smaller, they usually do not accumulate much starch in the presence of light, and they have numerous large grana (Fig. 7.19).

Though there is considerable variation in the structural details among species exhibiting Kranz anatomy, the common trends sug-

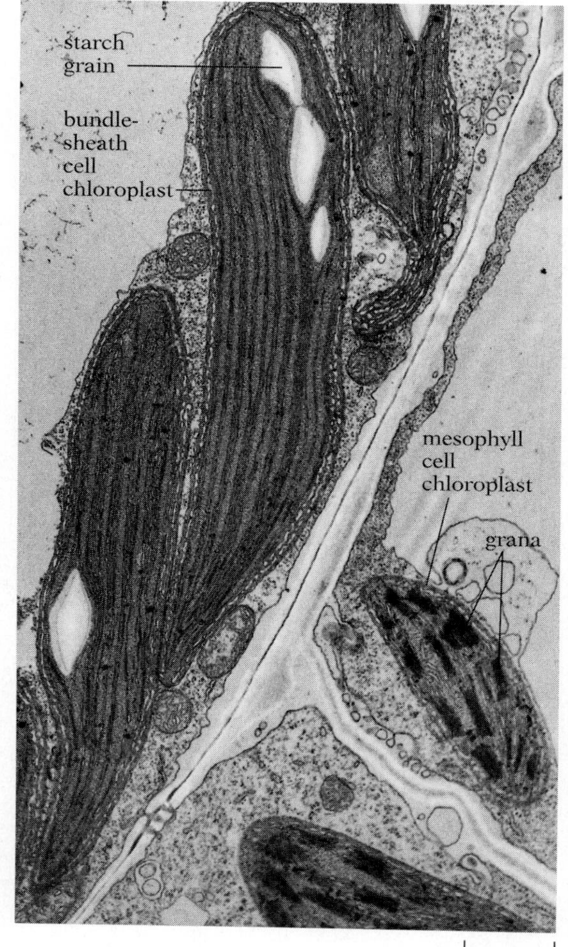

7.19 Two kinds of chloroplasts in a C_4 leaf

At left in this electron micrograph of a segment of a corn leaf is part of a bundle-sheath cell; its chloroplasts have small grana and contain starch grains (light areas). At right and bottom are parts of two mesophyll cells; their chloroplasts are smaller and contain numerous grana but no starch grains.

gest a special adaptation to conditions of high temperature, intense light, low moisture, and low CO_2 and high O_2 concentrations—all conditions far from optimal for plants that depend entirely on the Calvin cycle for CO_2 fixation. All these various structural specializations serve a common biochemical adaptation: the C_4 pathway of photosynthesis.

■ AVOIDING PHOTORESPIRATION: C_4 PHOTOSYNTHESIS

As Figure 7.20A shows, corn, which originated in the tropics and has Kranz anatomy, can carry out photosynthesis at very low concentrations of CO_2; bean plants, which are native to the temperate zone, fail to photosynthesize at CO_2 levels below about 50 ppm because of photorespiration. At CO_2 concentrations of 200 or 300 ppm, where corn approaches its maximum photosynthetic capacity, beans still perform well below their potential capacity. Again because of photorespiration, the concentration of O_2 that inhibits photosynthesis is far lower for beans than for corn (Fig. 7.20B). Kranz anatomy is correlated with the ability of a plant to function even when the CO_2 level in the leaf is low and O_2 concentration is high.

Low CO_2 and high O_2 result from exposure to great heat, dryness, and brilliant light, as in a desert or on a dry savanna. Under such conditions, when the moist walls of the mesophyll cells risk losing too much water by evaporation through the stomata, the guard cells close the stomata almost completely. Water loss is thus reduced, but now gases can no longer move freely between the atmosphere and the air spaces inside the leaf. As CO_2 is used up in photosynthesis, the nearly closed stomata prevent the supply inside the leaf from being fully replenished, and the CO_2 concentration in the air spaces around the mesophyll cells falls. Under such conditions, non-Kranz plants like the bean will carry out so much photorespiration that their ability to synthesize carbohydrate from CO_2 will be greatly reduced. By contrast, corn and other Kranz

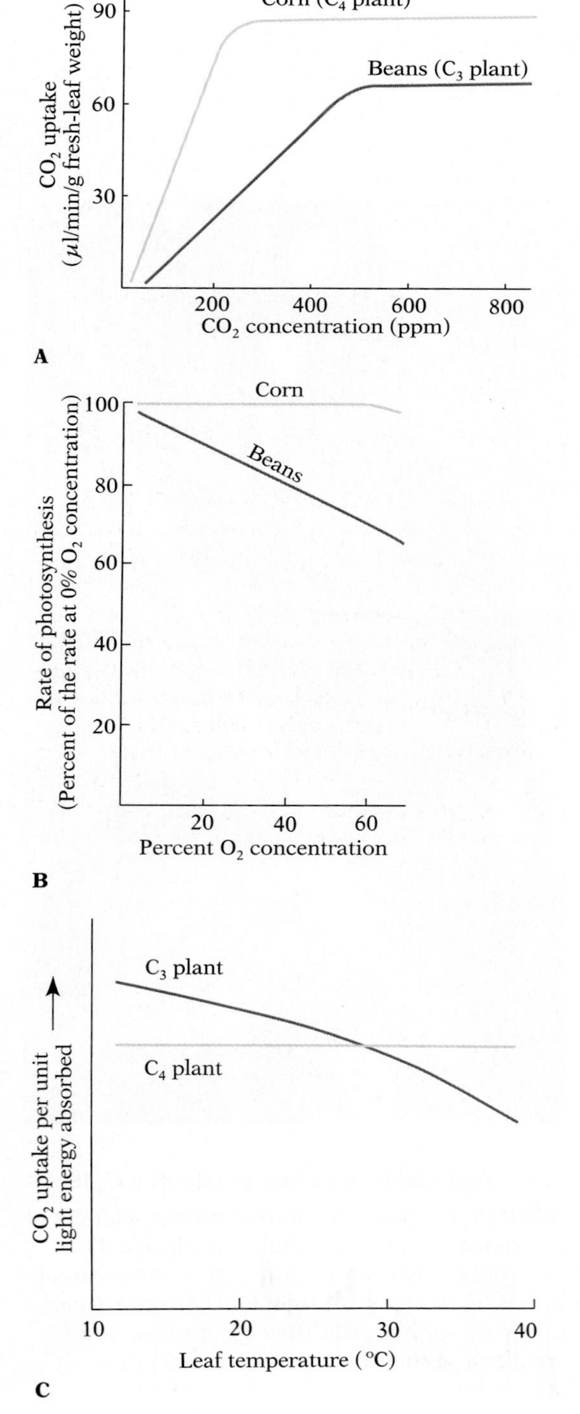

7.20 Comparison of photosynthetic efficiency in C_3 and C_4 plants

(A) Corn can fix carbon at CO_2 concentrations as low as 1 ppm, and it carries out photosynthesis at a very high rate at concentrations of 200–300 ppm. (A normal concentration of CO_2 in the atmosphere is about 330 ppm.) By contrast, beans perform no net carbon fixation at CO_2 concentrations below about 50 ppm, and their rate of photosynthesis at concentrations of 200–300 ppm is not very high. (B) Photosynthesis in corn shows no inhibition at all at O_2 concentrations below 65%, whereas the photosynthetic rate of beans falls steadily as the O_2 concentration rises. (A normal concentration of O_2 in the atmosphere is about 21%). Both A and B are for a temperature of 20°C and a light intensity of 2000 foot-candles. Obviously, C_4 photosynthesis is superior under these conditions. However, when the temperature or illumination varies (and while the O_2 and CO_2 concentrations remain at normal levels), a very different picture emerges. For example, as the temperature drops (C), C_3 plants clearly perform more efficiently, so they have the advantage in cooler (that is, more temperate) climates.

plants *can* synthesize carbohydrate under dry conditions; they are thus able to survive in climates that would be fatal to other plants. Even under less extreme conditions they can often carry out photosynthesis at a higher rate than other plants.

In the late 1960s M. D. Hatch and C. R. Slack worked out the biochemical pathway of photosynthesis unique to Kranz plants. They found that in the mesophyll cells—the cells richest in chloroplasts that are arranged in rings around the veins—CO_2 is combined, not with ribulose bisphosphate as in the Calvin cycle, but rather with a three-carbon compound called phosphoenolpyruvate or PEP, to form a four-carbon (C_4) compound. The enzyme that catalyzes this carboxylation of PEP, unlike the one that catalyzes the carboxylation of RuBP in the Calvin cycle, does not have O_2 as an alternative substrate and thus is not inhibited by high O_2 concentrations. Thus it enables Kranz plants (which are therefore also called C_4 plants) to fix CO_2 under conditions when photorespiration would predominate over photosynthesis in C_3 plants, which use only the Calvin cycle.

Curiously enough, the C_4 compound formed in the mesophyll cells is not used directly for growth or nutrition by the plant. Instead, it is passed in reduced form into the bundle-sheath cells (which in C_4 plants are very well developed and contain chloroplasts), where it is decarboxylated: the C_4 compound is broken down to CO_2 and a C_3 compound (Fig. 7.21). The C_3 residue moves back to the mesophyll cells, where it is reconverted into PEP and starts the C_4 cycle over again. The CO_2 remains in the bundle-sheath cells, where it is picked up by the RuBP carboxylase in these chloroplasts and incorporated into carbohydrate through the conventional Calvin cycle.

In both C_3 and C_4 plants the ultimate assimilation of CO_2 into carbohydrate is by the Calvin cycle. The difference is that in C_3 plants the Calvin cycle is the only pathway of CO_2 fixation, whereas in C_4 plants there is another, preliminary fixation pathway. It might seem strange that a plant would have evolved a special mechanism for fixing CO_2 as C_4 in the mesophyll, only to combine it with a mechanism for promptly breaking off the CO_2 again and refixing it as carbohydrate in the bundle-sheath cells. However, this "laundering" of CO_2 by C_4 plants gives them an advantage over C_3 plants under conditions of high temperature and intense light, when stomatal closure results in low CO_2 and high O_2 in the air spaces inside the leaf. Under such conditions, C_3 plants are unable to use CO_2 effectively because O_2 competes for RuBP. In contrast, C_4 plants *can* fix CO_2, because the mesophyll cells, acting as CO_2 pumps, can elevate the CO_2 concentration in the bundle-sheath cells to a level at which carboxylation of ribulose bisphosphate (leading into the Calvin cycle) exceeds its oxidation. The Kranz anatomy, with its concentric rings of mesophyll and bundle-sheath cells, makes possible the compartmentalization on which the process of CO_2 pumping depends.

The combination of Kranz anatomy and C_4 photosynthesis has evolved independently in a variety of unrelated plants, including corn, sugarcane, sorghum, crabgrass, saltbush, and portulaca. It is an impressive illustration of the intimate relationship between structure and function in living systems.

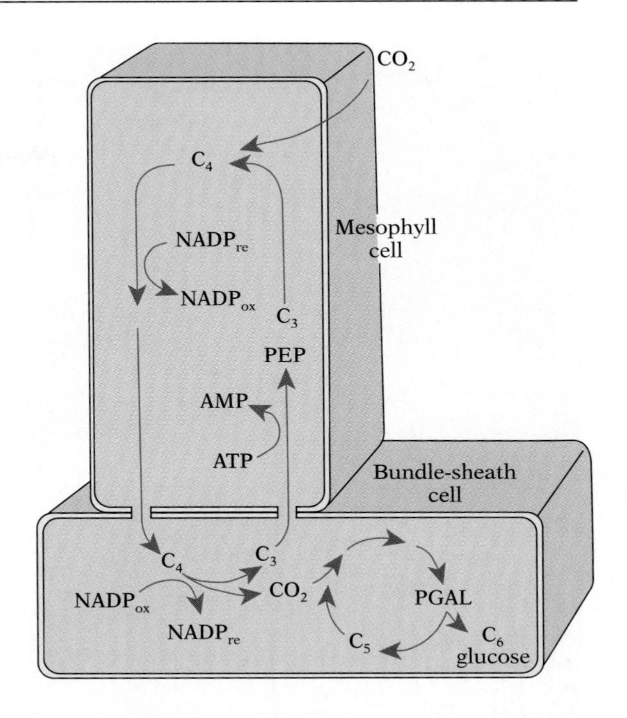

7.21 Hatch-Slack pathway of C_4 photosynthesis

The path of carbon is traced by red arrows. A mesophyll cell absorbs CO_2 from the intercellular air spaces (top). The CO_2 combines with PEP, a three-carbon compound, to form a four-carbon compound. After reduction by $NADP_{re}$, the C_4 substance moves into an adjacent bundle-sheath cell. There it is oxidized by $NADP_{ox}$ and split into a C_3 compound and CO_2. The C_3 compound moves back to the mesophyll cell and is converted (by a reaction probably driven by energy from ATP) into PEP. The CO_2 is fed into the Calvin cycle in the bundle-sheath cell and is incorporated into carbohydrate.

■ AVOIDING PHOTORESPIRATION: CRASSULACEAN ACID METABOLISM

Another remarkable dry-climate variation of photosynthesis is found in many succulents—plants that store water in fleshy leaves (Fig. 7.22)—and a few other plants, including pineapple, Spanish moss, and some cacti. Like C_4 plants, these organisms avoid water loss in their hot environment by closing their stomata during the day and opening them at night. The CO_2 necessary for photosynthesis is captured and stored at night in the form of malic acid and isocitric acid, and then slowly released in the cells during the day, to be fixed by C_3 photosynthesis. This process is called ***crassulacean acid metabolism (CAM)***, named for the genus of plants in which the process was discovered. Though CAM seems to have evolved independently of C_4 photosynthesis, it serves the same purpose. The essential difference is that Kranz plants solve the CO_2-buildup problem by segregating the two steps of carbon fixation anatomically, while CAM plants separate them temporally. The existence of these variations illustrates again how natural selection can lead to several solutions of the same problems.

7.22 Stonewort, a member of the Crassulaceae
The thick, fleshy leaves of this plant are typical of succulents.

CHAPTER SUMMARY

THE PHOTOSYNTHETIC REACTION

Photosynthesis combines water and carbon dioxide:
$CO_2 + 2H_2O + \text{light} \longrightarrow O_2 + (—CH_2O) + H_2O$ (p. 180)

HOW THE LIGHT REACTIONS STORE LIGHT ENERGY

HOW CHLOROPHYLL CAPTURES THE ENERGY OF LIGHT Chlorophyll absorbs red and blue light, exciting an electron to a higher energy level; accessory pigments extend the action spectrum. Groups of chlorophyll act as antennas, passing their excited electrons to a reaction center, from which the electrons enter an electron-transport chain. (p. 182)

HARVESTING THE ENERGY OF EXCITED ELECTRONS The cyclic pathway, used exclusively by anaerobic photosynthesizers but available as a backup in aerobic autotrophs, transfers excited electrons through an electron-transport chain back to the antenna, extracting energy used to create an electrochemical gradient in the chloroplast. In the noncyclic pathway the excited electron is passed through the electron-transport chain of Photosystem II to the antenna of Photosystem I, where it fills a vacancy created when an electron excited there passes into its electron-transport chain; the Photosystem I electron ends up in $NADP_{re}$. The vacancy in the antenna of Photosystem II is filled with an electron obtained from the splitting of water, which produces oxygen as a by-product. Energy extracted in the electron-transport chain is used to create an electrochemical gradient in the chloroplast. (p. 186)

THE ROLE OF CHLOROPLAST STRUCTURE The electrochemical energy is stored as a H^+ gradient across the thylakoid membrane, and is used to synthesize ATP. (p. 190)

HOW THE DARK REACTIONS FIX CARBON

CARBOHYDRATE SYNTHESIS BY THE CALVIN CYCLE The Calvin cycle uses ATP and the high-energy electron in $NADP_{re}$ to convert CO_2 into PGAL, which is then converted into other compounds, especially glucose. (p. 194)

THE MYSTERY OF PHOTORESPIRATION In the presence of too much oxygen, the enzyme that begins the fixation of CO_2 can instead bind O_2 and "unfix" CO_2. (p. 195)

THE LOGIC OF LEAF DESIGN

THE ANATOMY OF ORDINARY LEAVES Leaves have veins to bring in raw materials and export products, a cuticle and adjustable stomata to minimize water loss, and an internal mesophyll, where gases are exchanged and photosynthesis occurs. (p. 196)

LEAVES WITH KRANZ ANATOMY Leaves of Kranz plants have much of their photosynthetic mesophyll packed tightly around the veins; very little cell surface is exposed for gas exchange. Kranz anatomy is an adaptation for C_4 photosynthesis. (p. 199)

AVOIDING PHOTORESPIRATION: C4 PHOTOSYNTHESIS C_4 plants synthesize a four-carbon intermediate in mesophyll cells using an enzyme that is not sensitive to oxygen; this intermediate is then moved to bundle-sheath cells, where it is converted into a three-carbon molecule and CO_2. The three-carbon molecule is moved back to the photosynthetic cell, and the CO_2 is fixed via the Calvin cycle in the bundle-sheath cells. C_4 photosynthesis is most common in plants adapted for arid or hot conditions, where the need to conserve water by keeping the stomata closed leads to a buildup of waste oxygen in the leaves. (p. 200)

AVOIDING PHOTORESPIRATION: CRASSULACEAN ACID METABOLISM CAM metabolism solves the same problem by capturing CO_2 in intermediate compounds at night, when temperatures are lower; the CO_2 is released during the day for use in the Calvin cycle. (p. 202)

STUDY QUESTIONS

1. Make a list of the analogies between the electron-transport chain in respiration and that in photosynthesis. Be sure to include not only the strategies used by the two processes but also the enzymes and the chemicals involved in each. (pp. 169–173; 186–190)
2. Distinguish between C_3 photosynthesis, C_4 photosynthesis, and CAM photosynthesis. Why hasn't C_4 photosynthesis taken over? (pp. 195–196; 199–202)
3. If the predictions of some climatologists are borne out, human activity (especially the burning of trees and fossil fuels) will increase the CO_2 concentration in the atmosphere significantly and, through the greenhouse effect, raise the average temperature of the earth. What might this do for the various strategies of photosynthesis? (pp. 200–202)
4. Compare the dark reactions of photosynthesis with the Krebs citric acid cycle. Is there any hint of a common evolutionary origin? (pp. 167–169; 194–195)

SUGGESTED READING

GOVINDJEE, AND W. J. COLEMAN, 1990. How plants make oxygen, *Scientific American* 262 (2). *On the operation of the water-splitting enzyme of noncyclic photosynthesis.*

MILLER, K. R., 1979. The photosynthetic membrane, *Scientific American* 241 (4). *An excellent discussion relating the chemiosmotic theory of chloroplast function to the structure of thylakoid membranes as shown by freeze-etch microscopy.*

YOUVAN, D. C., AND B. L. MARRS, 1987. Molecular mechanisms of photosynthesis, *Scientific American* 256 (6). *An account of the molecular events occurring during the first 200m seconds following photon absorption.*

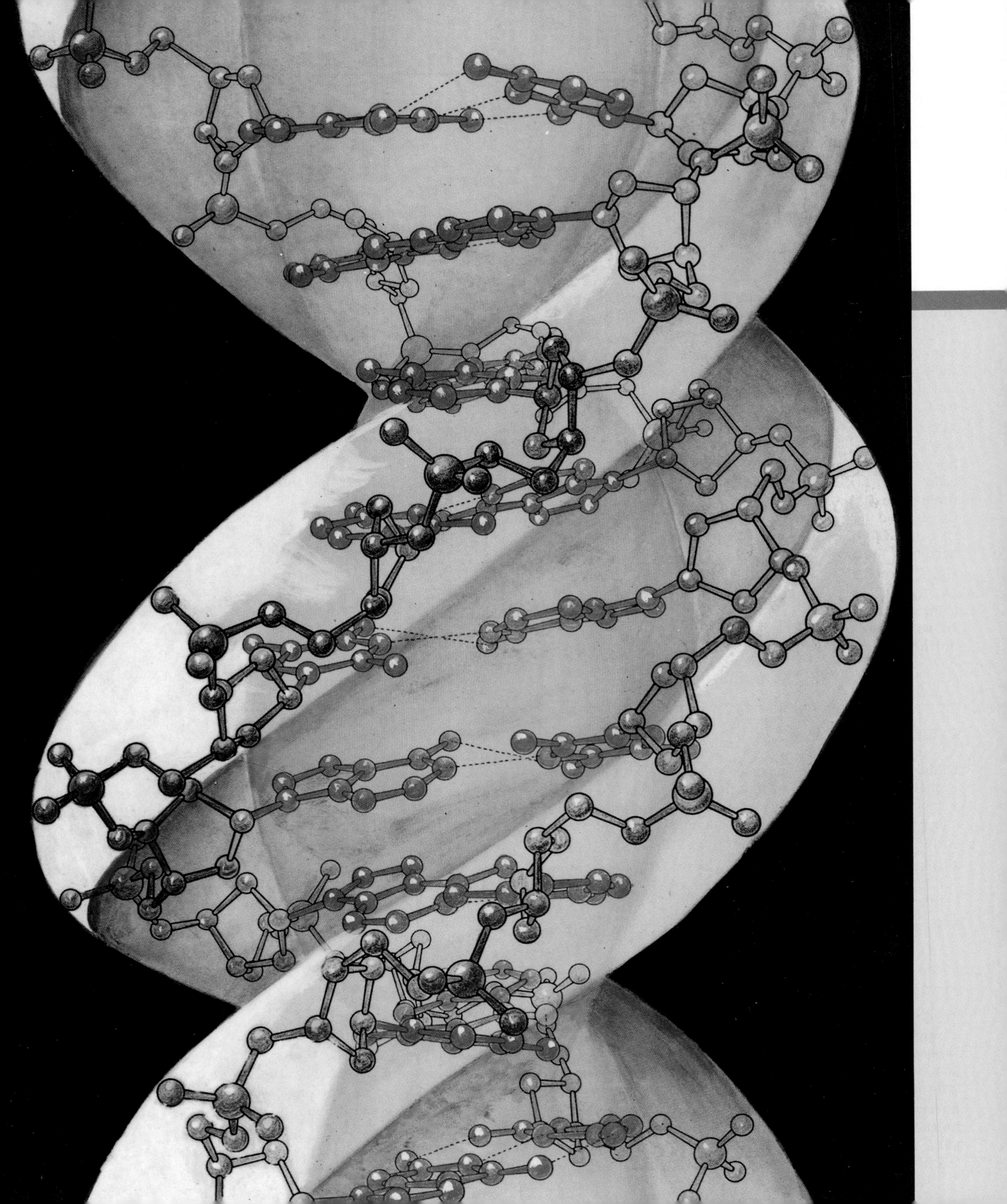

PART II

THE PERPETUATION OF LIFE

◀ **Molecule of inheritance,** the double helix of DNA is vividly portrayed in this painting by the noted scientific illustrator Irving Geis. The atomic-scale "ball and stick" model—based on single-crystal X-ray analysis—represents the dioxyribose backbones of the molecule as blue and the nucleotide bases as red; weak hydrogen bonds are indicated by dotted lines. The white ribbons accentuate the molecule's overall double-helical structure.

CHAPTER 8

WHAT ARE GENES MADE OF?

THE UNUSUAL CHEMISTRY OF CHROMOSOMES

GENES: DNA OR PROTEIN?

THE MOLECULAR STRUCTURE OF DNA

HOW DNA IS REPLICATED

THE TEMPLATE HYPOTHESIS

THE DIFFICULTIES OF REPLICATION

REPLICATION IN ORGANELLES

HOW ERRORS ARE REPAIRED

THE STRUCTURE AND REPLICATION OF DNA

The processes of life are guided by an elaborate and precise series of information transfers. The genes of an organism's DNA contain all the information necessary to build and operate the organism— information that orchestrates its development from a single unspecialized cell into a complex corporation of specialized tissues and organs, and also directs events ranging from glycolysis and the Krebs cycle to the organism's behavior. This chapter is concerned with how the genetic information as a whole is duplicated and transferred to each new cell, whether it be a cell added as the organism grows, or one of the ***gametes*** (sex cells, like sperm and egg) that unite to form an entirely new organism (Fig. 8.1C).

Genes are most conveniently defined as sequences of DNA that encode proteins (or, occasionally, encode special sequences of the nucleic acid RNA that are used directly in the construction of ribosomes and in other ways). The genes are organized in a linear sequence on one or more chromosomes. The chromosomes and the DNA in organelles, as well as the tiny chromosome-like elements called plasmids (discussed in Chapter 10), together comprise the ***genome***.

Modern molecular genetics begins with the study of DNA structure; we will look first at the structure of the genetic material and then go on to examine the process of chromosome duplication, the essential step of which is ***replication***. This process, which includes self-repair of any replication errors, enables each daughter cell produced during cell division to receive its own copy of each chromosome (and thus a complete set of genes) from the parental cell.

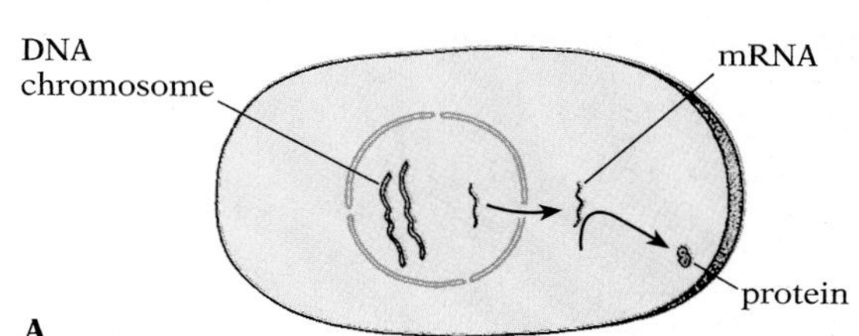

A

Instructions in the DNA are used to build cellular components and direct biochemical events. In most eucaryotes, there are two copies of each chromosome.

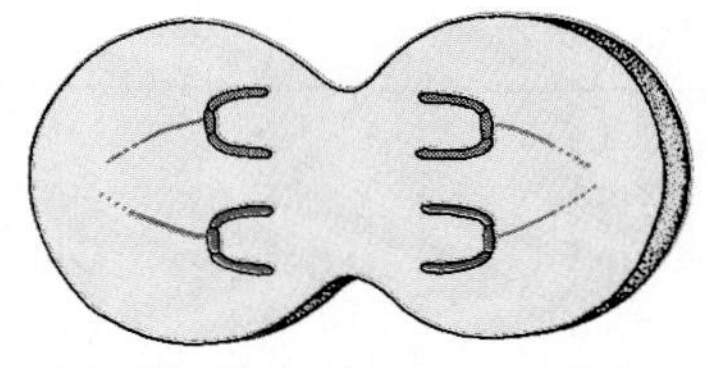

B

Chromosomes are duplicated prior to cell division and a complete copy of each pair is passed to each new cell.

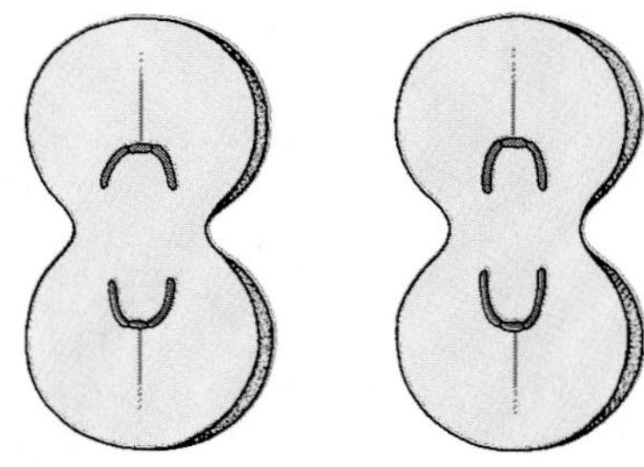

C

When gametes (sex cells) are formed, they receive only one chromosome from each pair. The normal number of chromosomes is restored by fusion of a male and female gamete.

8.1 Transfer of genetic information

The set of instructions in the DNA is transcribed and used to direct events within the cell (A), is duplicated and passed on to daughter cells during normal cell division (B), and is halved and passed to gametes in preparation for sexual reproduction (C).

WHAT ARE GENES MADE OF?

■ THE UNUSUAL CHEMISTRY OF CHROMOSOMES

The special chemistry of the nucleus came to light in 1868, when Friedrich Miescher showed that protein-digesting enzymes, though they destroy nearly everything else in cells, leave the contents of the nucleus largely intact. This meant that the nucleus contains large quantities of an unusual compound, now called nucleic acid.

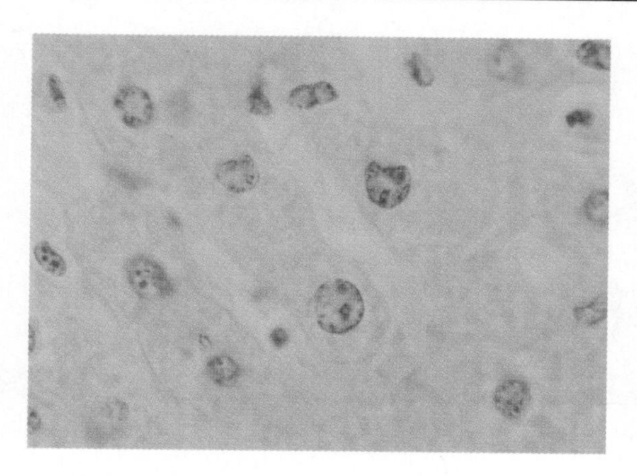

8.2 Feulgen staining

Feulgen staining was used to distinguish the DNA in this sample of liver cells.

In 1914 Robert Feulgen devised a method of selectively staining nucleic acids a brilliant crimson, and showed that nuclear DNA is restricted to the chromosomes (Fig. 8.2).

Staining and other techniques show that, in a given organism, all the ***somatic cells*** (all the cells except the gamete-producing ***germ cells***) ordinarily contain the same amount of DNA. This is true even though cells from such different tissues as liver, kidney, heart, nerve, and muscle differ drastically in the amounts of other substances they contain. Furthermore it became clear that egg and sperm cells contain only half as much DNA as the somatic cells. Since cell division distributes a complete set of genes to every somatic cell, regardless of its eventual role, while the process of gamete production distributes to every sperm and egg cell exactly half the amount of genetic material found in the somatic cells, these two discoveries indicated that either nucleic acid or the protein intimately associated with it is the essential material of genes.

■ GENES: DNA OR PROTEIN?

In 1928 Fred Griffith showed that mice injected with a virulent (disease-producing) strain of pneumococci, the bacteria that cause pneumonia, inevitably die. This S strain (so named because on artificial media it grew as smooth colonies) was compared with a nonvirulent R (for rough colonies) strain, which did not kill mice, and heat-killed S-strain bacteria, which were also harmless (Table 8.1). However, mice injected with a mixture of live strain-R and heat-killed strain-S bacteria died. How could a mixture of dead and nonvirulent live bacteria have killed the mice? Griffith found the dead mice full of live strain-S bacteria! Somehow the live strain-R bacteria had been transformed into live strain-S bacteria by material from the dead strain-S cells.

Table 8.1 Griffith's results

BACTERIA INJECTED	REACTION OF MICE
Live strain R	Survived
Live strain S	Died
Dead strain S	Survived
Live strain R + dead strain S	Died

In 1943 O. T. Avery, Colin MacLeod, and Maclyn McCarty showed that the essential agent in the process of bacterial ***transformation*** was DNA (Fig. 8.3). However, this did not prove that

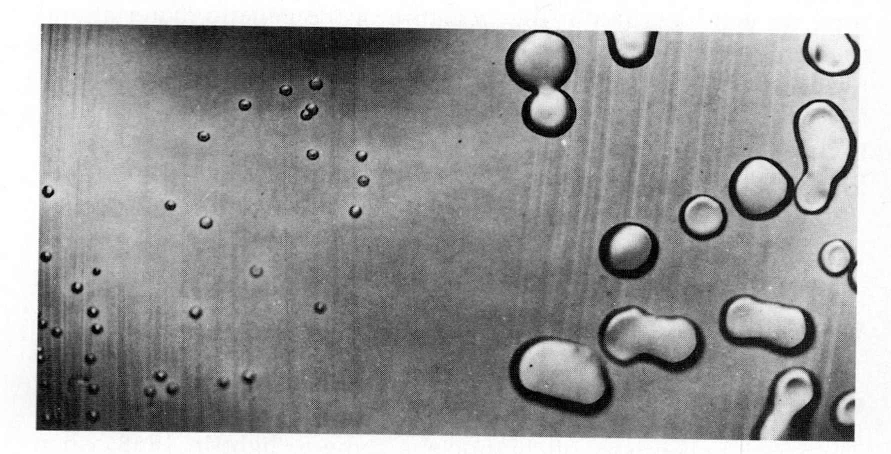

8.3 Rough versus smooth pneumococci

Rough (untransformed) colonies are on the left; smooth (transformed) colonies are on the right. (Photograph from the 1994 paper by Avery, MacLeod, and McCarty.)

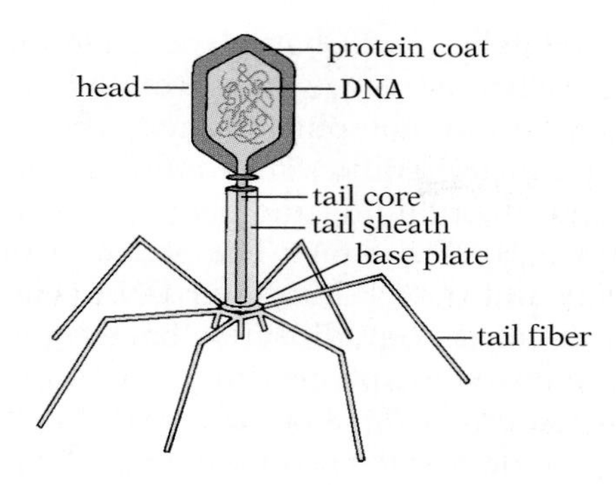

8.4 A complex bacteriophage

genes are made of DNA since the DNA from the virulent strain might merely have activated the protein-based genes in the nonvirulent strain.

In 1952 Alfred D. Hershey and Martha Chase showed that one of the viruses that attacks the bacterium *Escherichia coli*, which is abundant in the human digestive tract, does so by means of its DNA. Bacteria-destroying viruses are called ***bacteriophage,*** or *phage* for short (from the Greek *phagein*, "to eat"). These tiny parasites subvert the cellular machinery of host organisms to reproduce their own genes. They consist primarily of a protein coat and a nucleic acid core. Some phage viruses have a head region (containing the genetic material) and an elongate tail region made up of a hollow core, a surrounding sheath, and six distal fibers (Fig. 8.4). When such a phage attacks a bacterial cell, proteins on the tip of its tail fibers and its base plate bind to proteins on the wall of the bacterial cell (Fig. 8.5). The DNA of the phage enters the cell, but the protein coat is left behind. Later the host cell releases dozens of hundreds of new viruses.

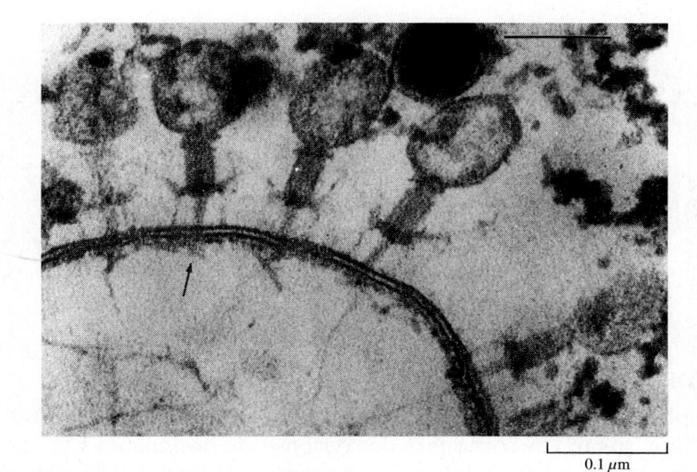

8.5 Bacteriophage replication

Each bacteriophage attaches to the bacterial cell wall by its tail fibers and base plate, and injects its genetic material into the cell. Once inside, the genetic material takes over the metabolic machinery of the cell and puts it to work making new phage.

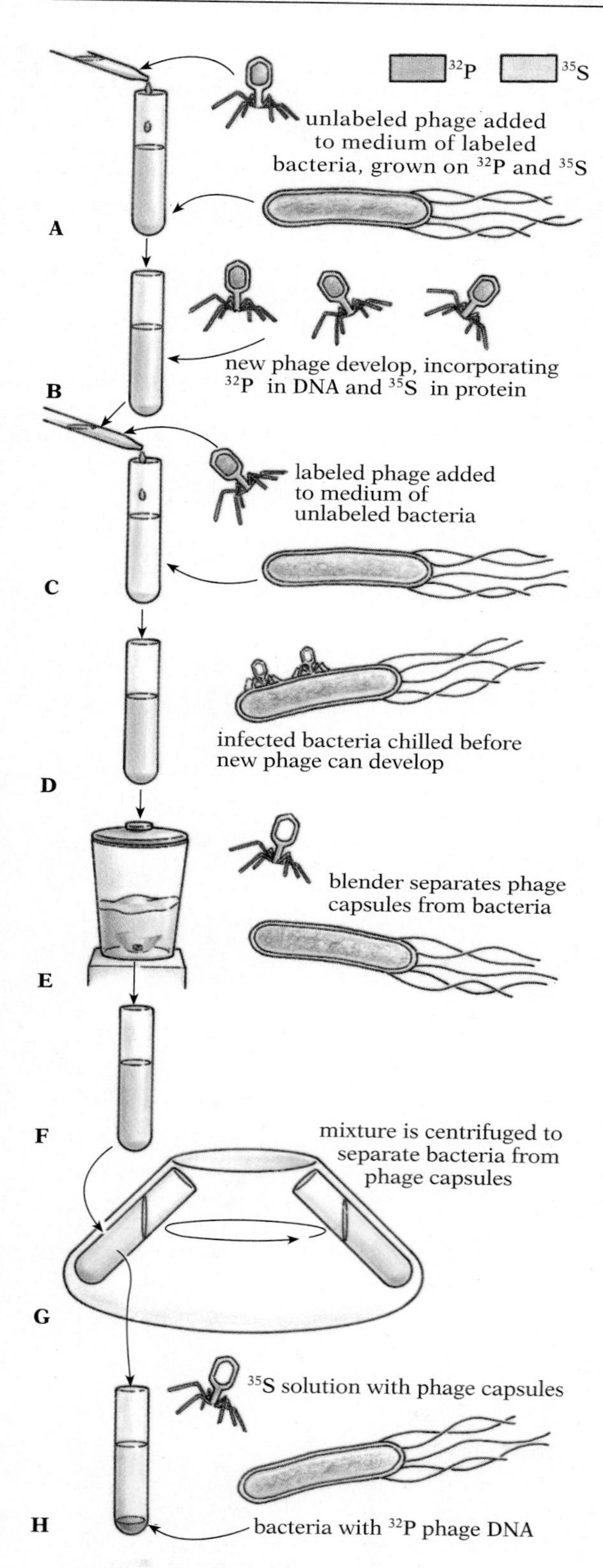

8.6 Hershey-Chase experiment

For details of this experiment, which demonstrated that DNA rather than protein carries genetic information, see the text.

The conclusive Hershey and Chase experiment took advantage of differences in the chemistry between proteins and nucleic acids: proteins have sulfur but not phosphorus, whereas nucleic acids have phosphorus but not sulfur. They cultured phage on bacteria grown on a medium containing radioactive phosphorus (^{32}P) and radioactive sulfur (^{35}S) (Fig. 8.6A). The phage incorporated the ^{35}S into their proteins and the ^{32}P into their DNA (Fig. 8.6B). Hershey and Chase then infected nonradioactive bacteria with the radioactive phage. They allowed the phage time to bind to the bacteria and inject hereditary material (Fig. 8.6C–D). Next they agitated the bacteria in a blender to detach what remained of the phage from their surfaces. They then centrifuged the mixture to separate the infected bacteria (reduced to a pellet at the bottom of the sample tube) from the phage coats (suspended in the fluid left behind; Fig. 8.6E–G). The fluid contained a substantial amount of ^{35}S but little ^{32}P, an indication that only the empty protein coat had been left outside the bacterial cell. The bacterial fraction contained much ^{32}P but no ^{35}S, meaning that only DNA had been injected into the bacteria by the phage (Fig. 8.6H). From these results they concluded that DNA alone was sufficient to transmit all the genetic information necessary to produce new viruses.

■ THE MOLECULAR STRUCTURE OF DNA

As detailed in Chapter 3, DNA is composed of building blocks called nucleotides, each of which is itself composed of a five-carbon sugar bonded to a phosphate group and a nitrogenous base (Fig. 8.7). By convention, the five carbons are designated by num-

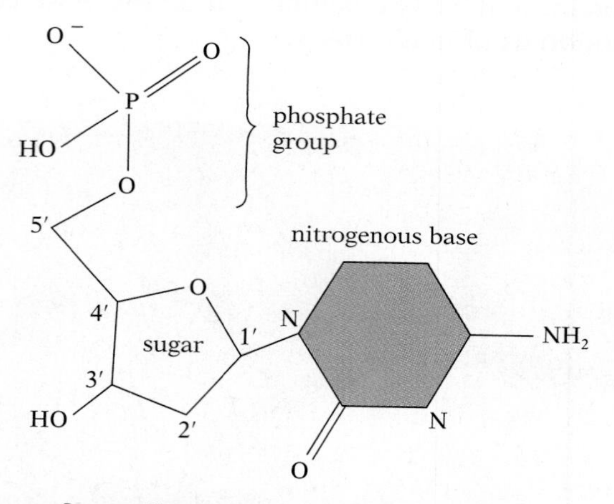

8.7 Diagram of a nucleotide from DNA

A phosphate group and a nitrogenous base are attached to deoxyribose, a five-carbon sugar. The carbons in deoxyribose are designated 1′–5′, as shown here, though the numbers are not normally included in molecular diagrams. The phosphate group is bound to the 5′ carbon of the sugar, while a hydroxyl group is bound to the 3′ carbon. Of the four different nitrogenous bases, cytosine is illustrated here. RNA nucleotides have a sugar identical to the one shown except that the 2′ carbon has a hydroxyl group bound to it.

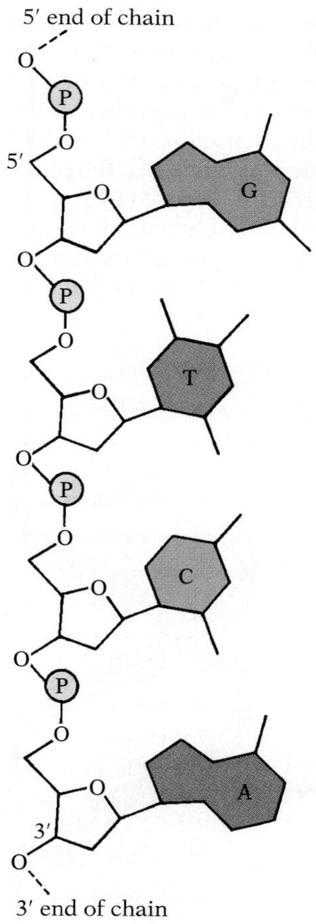

8.8 Portion of a single chain of DNA

Nucleotides are linked together by bonds between their sugar and phosphate groups. The nitrogenous bases (G, guanine; T, thymine; C, cytosine; A, adenine) are side groups. In this diagram P represents the main components of each phosphate group—the phosphorus atom with its hydroxyl group and the double-bonded oxygen; only the oxygen atoms in the connecting chain are shown separately.

bers and primes: 1′ to 5′. There are four kinds of nucleotides in DNA, which differ from one another in their nitrogenous bases. Two of the bases, ***adenine*** and ***guanine***, are purines, which are double-ring structures; the other two, ***cytosine*** and ***thymine***, are pyrimidines, which are single-ring structures (Fig. 8.8).

The nucleotides in a particular DNA molecule are arranged in a specific sequence; the sugars are held together by the phosphate groups that link the 3′ carbon of one sugar to the 5′ carbon of the next; the nitrogenous bases are arranged as side groups off the chains (Fig. 8.8). DNA molecules ordinarily exist as double-chain structures, with the two chains, or strands, held together by hydrogen bonds between their nitrogenous bases. Such bonding can occur only between cytosine and guanine or between thymine and adenine (Fig. 8.9). Thus the sequence of bases in one strand determines the complementary sequence in the other (Fig. 8.10). Notice that the polarities of the two strands are opposite: one runs from 5′ to 3′, while the other goes from 3′ to 5′.[1] Finally, the ladderlike double-chain molecule is coiled into a double helix (Fig. 8.11), and stabilized further by hydrogen bonds aligned with the chains; these bonds between separate rungs of the nucleotide ladder are

[1] This strand "polarity" has nothing to do with the unequal distribution of charge that gives rise to polar molecules.

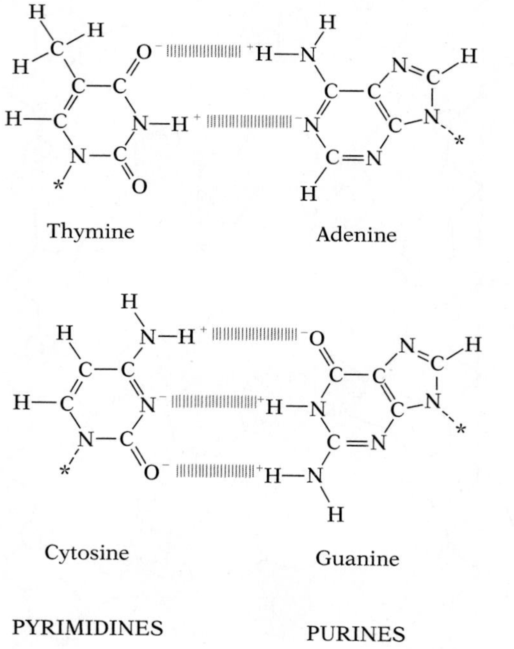

8.9 Bonding of nitrogenous bases in nucleotides

Because of the differing electronegativities of oxygen, hydrogen, nitrogen, and carbon, the nitrogenous bases of DNA have polar segments. The spacing and polarity of these segments allow thymine to form hydrogen bonds (red striped bands) with adenine, and cytosine to bond with guanine. (The asterisk marks the point where each base attaches to a sugar.)

8.10 Portion of a DNA molecule uncoiled

The molecule has a ladderlike structure, with the two uprights composed of alternating sugar and phosphate groups and the cross rungs composed of paired nitrogenous bases. Each cross rung has one purine base (a pentagon attached to a hexagon) and one pyrimidine base (a hexagon). When the purine is guanine (G), the pyrimidine with which it is paired is always cytosine (C); when the purine is adenine (A), the pyrimidine is thymine (T). Adenine and thymine are linked by two hydrogen bonds, guanine and cytosine by three. Note that the two chains run in opposite directions: the free phosphate is linked to the 5' carbon at the upper end of the left chain and to the corresponding carbon at the lower end of the right chain.

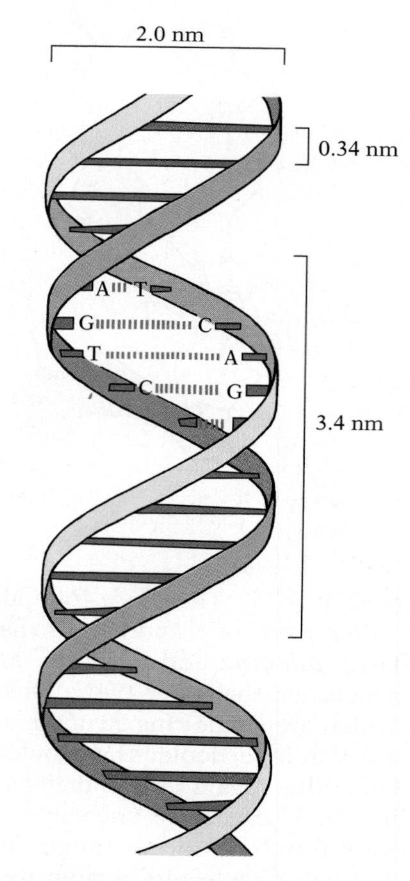

8.11 Watson-Crick model of DNA

The molecule is composed of two polynucleotide chains held together by hydrogen bonds between their adjacent bases. The double-stranded structure is coiled in a helix. The width of the molecule is 2.0 nm, the distance between adjacent nucleotides is 0.34 nm, and the length of one complete coil is 3.4 nm. Interactions between bases within each chain (not shown) help stabilize the molecule in the helical shape shown here.

analogous to those that stabilize the α helices of proteins (see Fig. 3.23, p. 62).

Determining the structure of so complicated—and important—a molecule as DNA had become an irresistible challenge to many scientists. X-ray diffraction analysis of DNA by Rosalind Franklin

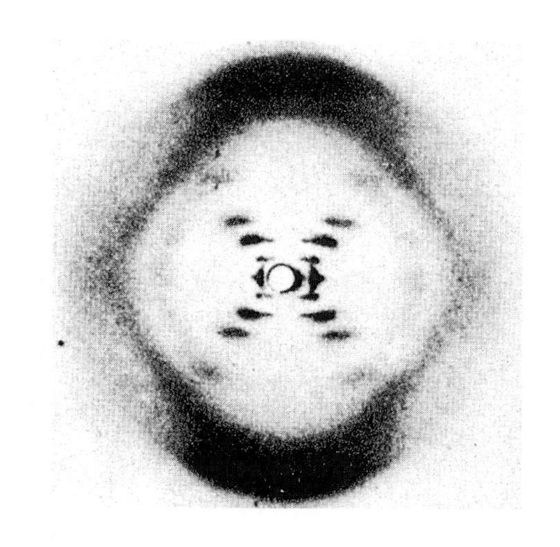

8.12 X-ray diffraction image of DNA

As Crick had discovered, the X-shaped cross radiating from the center of this pattern is diagnostic of a helix. The position of the strong bands at the top and bottom indicate a periodicity of 0.34 nm. More subtle patterns suggest a 3.4-nm period as well.

produced the first sharp X-ray diffraction patterns (Fig. 8.12). Francis H. C. Crick, who had developed mathematical methods for interpreting X-ray patterns of protein helices, used the Franklin photographs to deduce that DNA is a helix with three major periodicities: repeating patterns of 0.34, 2.0, and 3.4 nm.

James D. Watson (Fig. 8.13) and Crick decided to develop a model of the structure of the DNA molecule by combining (1) what was known about the chemical content of DNA, (2) the information gained from Crick's analysis of the X-ray diffraction studies, and (3) data on the exact distances between bonded atoms in molecules, the angles between bonds, and the sizes of atoms. Watson and Crick built scale models of the component parts of DNA and then attempted to fit them together in a way that would agree with the information from all these sources.

They were certain that the 0.34-nm periodicity corresponded to the distance between successive nucleotides in the DNA chain, the 2.0-nm periodicity to the width of the whole chain, and the 3.4-nm periodicity to the distance between successive turns of the helix. They found that a single chain of nucleotides coiled in a helix 2.0 nm wide with turns 3.4 nm long would have a density only half as great as that of DNA; from this they concluded that DNA molecules must have two nucleotide chains rather than one. They tried several arrangements of their scale model; the one that best fitted all the data had the two nucleotide chains twisted together, running in opposite directions, fitting within a hypothetical cylinder of appropriate diameter, with the purine and pyrimidine bases oriented toward the interior (Fig. 8.11).

With the bases oriented in this manner, hydrogen bonds between the bases of opposite chains could supply the force to hold the two chains together and to maintain the helical configuration. The DNA molecule, when unwound, would have a ladderlike structure, with the ladder uprights formed by the two long chains of alternating sugar and phosphate groups, and each of the cross rungs formed by two nitrogenous bases linked loosely by hydrogen bonds (Fig. 8.10).

Watson and Crick realized that each cross rung must be composed of one purine base and one pyrimidine base: two purines opposite each other occupied too much space (because each had two rings, for a total of four), and two pyrimidines opposite each other did not come close enough to bond properly (because each had only one ring). This left four possible pairings: A–T, A–C, G–T, and G–C. Further examination revealed that only A–T and G–C pairs had the right geometry to form hydrogen bonds. This pairing quite unexpectedly explained an earlier finding by the biochemist Erwin Chargaff. He had noted that while the proportion of cytosine in DNA varies from species to species, it is constant between individuals of a given species, and always matches the proportion of gua-

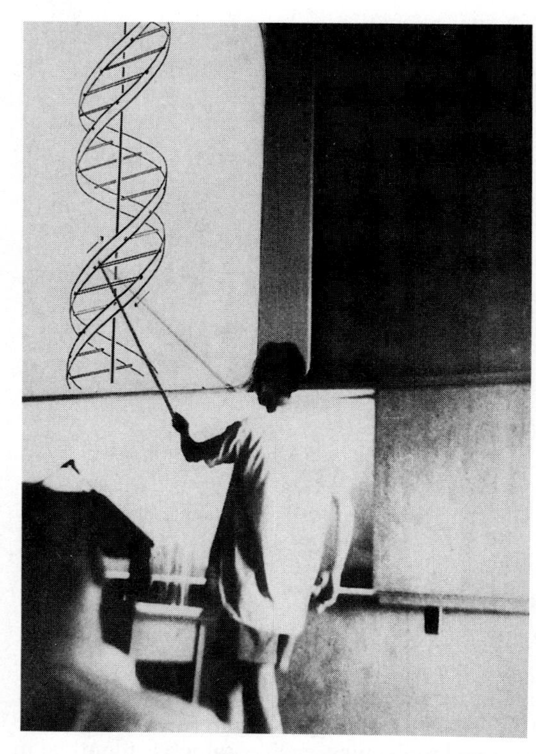

8.13 Watson describing the Watson-Crick model of DNA

This photograph was taken at a 1953 seminar at the Cold Spring Harbor Laboratory.

nine in that species. Similarly, the DNA of any given species always contains exactly equal amounts of adenine and thymine nucleotides. (Conversely, the amounts of adenine and thymine within a species usually differ from the amounts of guanine and cytosine.) Watson and Crick completed their deduction of DNA structure in 1953.

HOW DNA IS REPLICATED

■ THE TEMPLATE HYPOTHESIS

To be the genetic substance, DNA must have built into it the information necessary to replicate itself. One of the most satisfying things about the Watson-Crick model is that the basic strategy of replication is obvious from the structure of DNA.

The Watson-Crick model Since the DNA of all organisms is alike in being a polymer composed of only four different nucleotides, the essential distinction between the DNA of one gene and the DNA of another gene must be the sequence in which the four possible types of base-pair cross rungs (A–T, T–A, G–C, and C–G) occur. The basic question of genetic replication is, then, What tells the cell's biochemical machinery how to put these nucleotide building blocks together in exactly the sequence characteristic of the DNA already present in the cell?

Watson and Crick pointed out that if the two chains of a DNA molecule are separated by rupturing the hydrogen bonds between the base pairs, each chain provides all the information necessary for synthesizing a new partner identical to its previous partner. Since an adenine nucleotide must always pair with a thymine nucleotide, and a guanine always with a cytosine, the sequence of nucleotides in one chain (strand) precisely specifies what the sequence of nucleotides in its complementary strand must be. Thus, separating the two chains of a DNA molecule and using each chain as a template, or mold, against which to synthesize a new partner for it would result in two complete double-chained molecules identical to the original molecule (Fig. 8.14).

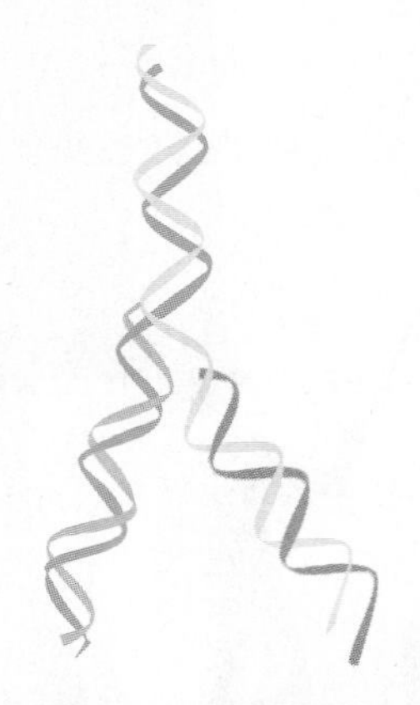

8.14 Replication of DNA

As the two polynucleotide chains of the old DNA (yellow and blue) uncoil, new polynucleotide chains (green) are synthesized on their surfaces. The process produces two complete double-stranded molecules, each of which is identical in base sequence to the original double-stranded molecule. For simplicity, replication is shown here beginning at the end of the molecule; in actuality it begins at specific internal points.

Experimental support Satisfying as it was, this template theory of DNA replication was pure speculation when first put forward by Watson and Crick. Since then, convincing evidence has come from the work of a number of investigators.

In 1957 Arthur Kornberg and his associates developed a method for synthesizing DNA in a test tube. They extracted an enzyme complex from the bacterium *E. coli* that catalyzes the synthesis of DNA—a ***DNA polymerase*** (an enzyme that creates a DNA polymer)—and combined it with a radioactive ATP-activated supply of the four nucleotides as raw material. (The activation consisted of a high-energy phosphate group that provides the energy for the later reaction steps; the isotope used was ^{14}C.) DNA was added to serve both as a primer—a starting point for the expected reactions—and as a potential template. Soon new DNA containing ^{14}C appeared.

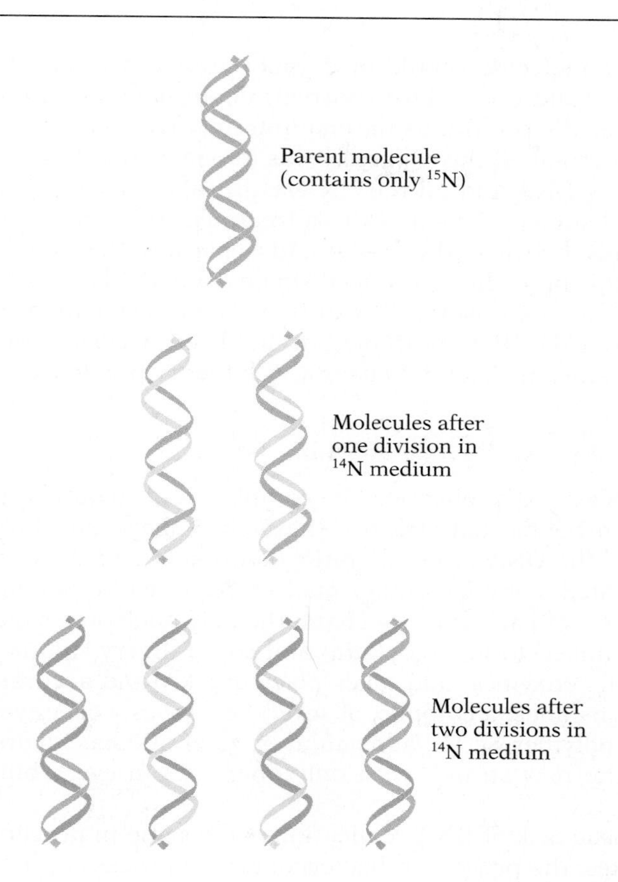

8.15 Results of the Meselson-Stahl experiment

The parent DNA molecule (blue chains) contained only heavy nitrogen. After one division, the DNA had an intermediate density, an indication that half the nitrogen in each molecule was heavy and half was light (normal); the two heavy parental chains had separated, and each had acted as the template for synthesis of a complementary normal chain (yellow and green). Even after several additional duplications, the two original heavy parental chains remained intact, though in two separate molecules of DNA.

The ratio of adenine and thymine to guanine and cytosine in the new DNA was precisely the same as that in the primer DNA.[2]

Direct support for the Watson-Crick model of replication came in 1958 from work by Matthew S. Meselson and Franklin W. Stahl. Meselson and Stahl grew *E. coli* for many generations on a medium whose only nitrogen source was the heavy isotope ^{15}N. Eventually all the DNA in these bacteria contained the heavy isotope instead of the normal isotope ^{14}N. Then the researchers changed the nitrogen source abruptly from ^{15}N to ^{14}N. Cell samples were removed at regular intervals thereafter, and the DNA was extracted from them and subjected to a complicated procedure designed to separate DNA of different densities.

The experiment showed that when cells containing only heavy (^{15}N) DNA were allowed to undergo one division in the normal (^{14}N) medium, the DNA of the new cells was intermediate in density between heavy DNA and normal DNA—the nitrogen in the DNA of the new cells was half ^{15}N and half ^{14}N. This is precisely what would be expected if the two chains of the heavy parental DNA separated and acted as templates for the synthesis of new partners from nucleotides containing only ^{14}N (Fig. 8.15). Each

[2] Ironically, we now know that there is more than one kind of DNA polymerase in cells, and that the one Kornberg isolated, now called DNA polymerase I to indicate its priority of discovery, is primarily used for repairing damage to chromosomes and cleaning up loose ends created during replication (see the *Exploring Further* box, pp. 216–217). The second DNA polymerase found is involved in creating the nucleolus (see pp. 124 and 286). The main enzyme complex responsible for replication, DNA polymerase III, was discovered last.

new DNA molecule should be composed of one heavy chain from the parent and one lighter (normal) chain newly synthesized; the new molecule would thus have an intermediate density.

To prove that all the ^{15}N really was in one chain of the intermediate-density DNA and all the ^{14}N was in the other chain, Meselson and Stahl subjected the DNA to a treatment that breaks the hydrogen bonds between the bases and separates the chains. As expected, this procedure produced single chains of heavy and normal density; the two isotopes had not been distributed randomly throughout the DNA molecule but had been localized, each in one of the chains, just as the Watson-Crick theory predicted.

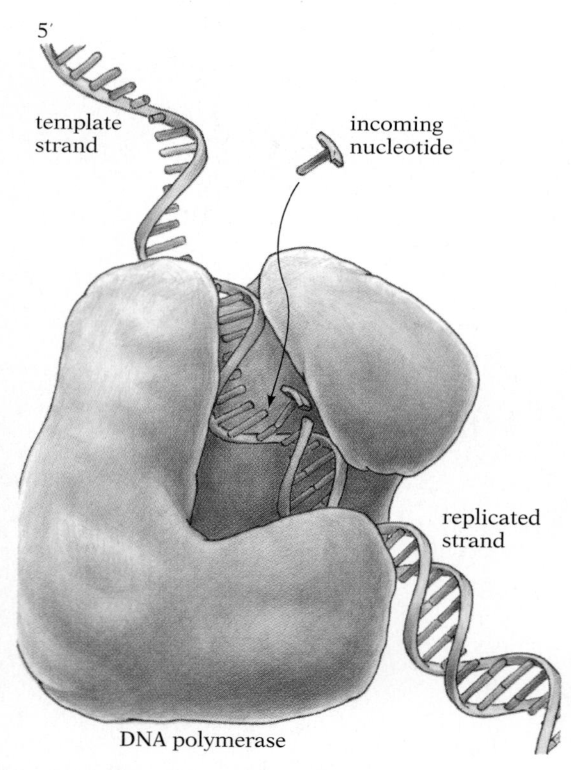

8.16 DNA polymerase

The approximate shape of DNA polymerase complex is known from X-ray crystallography. It is shown here replicating one strand of DNA, moving from upper left to lower right. Not shown is a ring-shaped protein that forms from two subunits and surrounds the DNA just behind the polymerase; by attaching itself to the replication enzyme, this protein keeps the polymerase firmly attached to the DNA it is copying.

■ THE DIFFICULTIES OF REPLICATION

The process of replication is complex and requires precision. Hydrogen bonds that stabilize the helical shape and link the two chains of the DNA molecule must be broken, and the chains must be separated; complementary nucleotides must be paired with the nucleotides of each existing chain; the new nucleotides must be covalently linked to form a chain; and so on. Every step is managed by specific enzymes, and takes place quickly and accurately. In *E. coli*, for instance, a complex of several enzymes, collectively known as DNA polymerase III, adds an average of 500 base pairs per second to the new strand, with only one error in every billion pairs copied.

The basic task of DNA replication is the same in procaryotes and eucaryotes: the process in bacteria and the process in humans display many similarities. For example, though the chromosomes differ in shape—bacterial chromosomes are circular while ours are linear—in both groups, replication begins at particular spots in the DNA and proceeds in both directions away from the initiation site; replication in eucaryotes never begins at an end. In both procaryotes and eucaryotes, replication enzymes can move along the parental strands only from the 3′ end to the 5′ end, generating complementary strands running from 5′ to 3′ (Fig. 8.16). This means that one chain is copied continuously while the other must be replicated "backward," in discontinuous segments (as described in the *Exploring Further* box). Both groups of organisms have mechanisms that locate and correct errors; both have special mechanisms that prevent the chains from tangling.

Along with these general similarities come differences in detail. For example, because eucaryotic chromosomes are packaged on protein spools to form nucleosomes (see Fig. 5.4, p. 123), replication proceeds at a rate of only 50 base pairs per second; it takes time for the DNA to unwind from the spools and, after a duplicate set of spools has been synthesized, to be rewound. In addition, because eucaryotic chromosomes are much larger than procaryotic chromosomes, replication is initiated at many independent sites on each chromosome simultaneously. Otherwise, it might take weeks or even months for a complex eucaryotic cell to divide.

Another major difference between procaryotes and eucaryotes comes in their strategies for ***segregation***, or sorting newly replicated chromosomes into two daughter cells. Procaryotes accomplish this task by attaching their single chromosome to the cell membrane; the replicated chromosome also attaches to the cell membrane near the same anchor point. After replication, a membrane grows in between these two anchor points, enclosing one copy in each daughter cell (Fig. 8.17A). For eucaryotes, which typically have 10–20 times as many genes, and two copies of each of their several chromosomes, the problem is more difficult. In these organisms, a special spindle structure aligns each replicated pair along the plane of the cell's coming division. Microtubules run from each pole to one member of each replicated pair, and serve to draw each new cell's complement of chromosomes away from the midline as cell division proceeds (Fig. 8.17B).

Another unique feature of nearly all eucaryotes is sexual reproduction. Most eucaryotes reproduce by means of gametes (eggs and sperm, for example), which contain only one copy of each chromosome; fusion of two gametes restores the normal number. Clearly, the pattern of replication or segregation necessary to produce gametes must be different from that used in normal cell division (Fig. 8.17C). Sexual reproduction serves, at least in part, to promote variation between parent and offspring. This variation occurs because the two copies of each chromosome in an individual are not absolutely identical, since one of the chromosomes came from the female gamete—the egg—while the other was contributed by the male gamete.

Although each member of a chromosome pair contains the same kinds of genes in the same order, the base sequence of a gene on one chromosome is often slightly different from that of the same gene on the other, and thus may encode a product with slightly different characteristics. Alternative forms of a gene are called ***alleles***. The precise set of alleles in one gamete of an individual usually differs from that in most of the other gametes, and usually differs substantially from the set of alleles in gametes produced by other individuals. The result is that when two gametes from different in

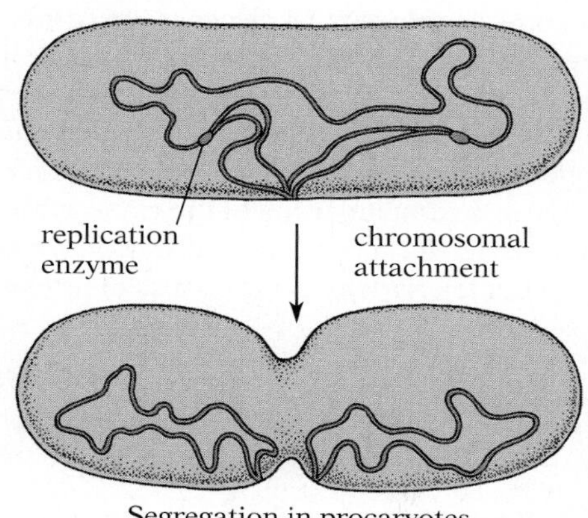

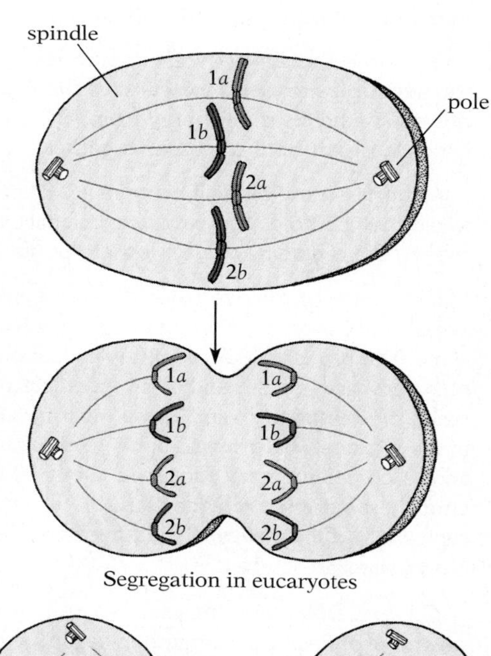

C
1a
2a
1b
2b
Gamete production

8.17 Alternative patterns of segregation after DNA replication

(A) In procaryotes, the chromosome remains attached to the cell membrane. After cell replication, the two copies are separated by the growth of the new membrane. (B) In most eucaryotes, there are two copies of each chromosome before replication, and each copy is used to produce a replica of itself; in this example, there was copy *a* and *b* of chromosome 1 and chromosome 2, each of which was then replicated to create the four pairs shown. As cell division proceeds, each pair of replicates becomes aligned along a plane, and one member of each pair is drawn to each end of the cell. (C) In preparation for the production of gametes, each copy of a chromosome is replicated, and then each two replicated pairs join to form a group of four—a tetrad. Two rounds of cell division follow, generating four gametes, each with only one of the two chromosomes in each pair.

EXPLORING FURTHER

REPLICATION OF THE *E. COLI* CHROMOSOME

The process of replication in *E. coli* illustrates the complex series of steps necessary to make an accurate copy of a chromosome. Highly specific enzymes are responsible for each event. Like all enzymes they recognize specific reactants by their complementary shape and pattern of polar charges. And as enzymes often do, they work together to rearrange the bonds in various reactants and produce a final product—in this case, a complete copy of the circular bacterial chromosome. Enzymes in complex pathways work nearly simultaneously, but for clarity we will discuss the events of this process in sequence.

1 In *E. coli*, the process begins when a protein, DNA B, recognizes an initiation site on the chromosome by its particular sequence of bases and binds to it (top figure).

2 Next, molecules of the enzyme group known as DNA gyrases (or topoisomerases, enzymes that change the shape of molecules) begin to relax the supercoiling of the chromosome on each side of the DNA B protein.

3 As the two DNA gyrase molecules move away from the initiation site, two molecules of the ***rep*** enzyme unzip the double helix, using energy from ATP to break the hydrogen bonds that hold the bases together.

4 Single-strand binding proteins (SSB) then form a scaffolding, which holds the two strands apart and prevents them from rebinding to each other spontaneously (Fig. A).

5 The last of the five steps that must occur before the actual replication of DNA can begin involves a primer enzyme (the primase). Because the replication enzyme can only add bases to the end of an incomplete strand as it reads the template strand, there must be at least a few new bases in place. The primase binds to the initiation site and adds a complementary sequence about 10 bases long. Oddly enough, the primer is made of RNA, which must later be replaced by DNA; we will see how this substitution is accomplished presently.

6 Next, ***DNA polymerase III***, a complex consisting of several proteins bound together, begins to replicate one of the two DNA strands by binding to it and adding complementary bases, creating, as its name implies, a polymer of DNA nucleotides. A complex is necessary because DNA polymerase must catalyze several different reaction steps and must be able to use four different nucleotides as reactants, depending on what it "reads" from the strand it is copying (the template strand). Presumably there are four active sites, one each for adenine, cytosine, guanine, and thymine. Once the complex has read the template strand and brought the complementary nucleotide into place, it catalyzes the binding of the nucleotide to the growing complementary strand. The nucleotides that are added have been activated by the addition of a high-energy phosphate group from ATP, which provides the energy for this step.

Because of its enzymatic specificity, DNA polymerase III can add nucleotides only to the 3′ end of a nucleotide strand—the end without the phosphate group. This creates a serious problem: since the two strands of the double helix have opposite polarities, with one running from 5′ to 3′ and the other from 3′ to 5′ (see Fig. 8.10), they must be copied in opposite directions. Only one strand can be copied by DNA polymerase III, following along behind the rep enzyme as it unzips the DNA (Fig. A); the other strand must somehow be copied backward. The DNA formed by the DNA polymerase that follows the rep enzyme is known as the ***leading strand***, while the DNA synthesized backward is known as the ***lagging strand***. The latter is formed bit by bit in a looping "backstitch" pattern.

7 Backstitching begins when the primer enzyme synthesizes short lengths of complementary RNA nucleotides at intervals along the single-stranded DNA (Fig. A).

8 The short segments created in Step 5 provide the 3′ free end for the DNA polymerase III; from here the polymerase works backward, copying the strand, until it reaches the preceding RNA primer segment (Fig. B). One strand, then, is copied continuously while the other is copied in sections, known as Okazaki fragments, which are 1000–2000 bases long and are flanked by the RNA primers.

9 Now a series of enzymes must patch the fragments together into a continuous strand. The repair complex ***DNA polymerase I*** removes the anomalous 10-base RNA primer segments and replaces then with DNA (Fig. C).

10 Finally, the fragments are welded together by ***DNA ligase*** (Fig. D), and the new strand is finished.

Current evidence suggests that the two DNA polymerase IIIs—the one on the leading strand and the one on the lagging strand—are physically attached. This means that the section (shown here as linear) that includes the SSB proteins and primer, running from the rep enzyme and the DNA polymerase III on the lagging strand, is actually a loop.

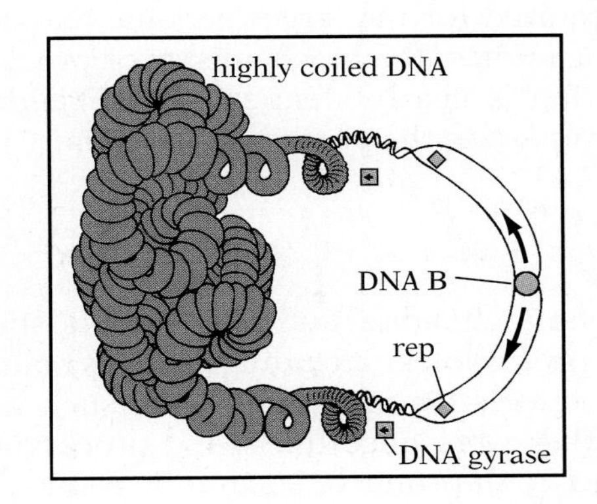

Step	*Symbol*	*Substance*	*Function*
1	●	DNAB	finds and marks initiation site
2	■	DNA gyrase	relaxes coiling
3	◆	rep	separates DNA strands
4	⬬	SSB	hold strands apart
5,7	○	RNA primase	primes lagging strand for replication
6,8	◁	DNA polymerase III	synthesizes complementary DNA
9	◊	DNA polymerase I	erases primer and replaces with DNA
10	▽	DNA ligase	welds gaps

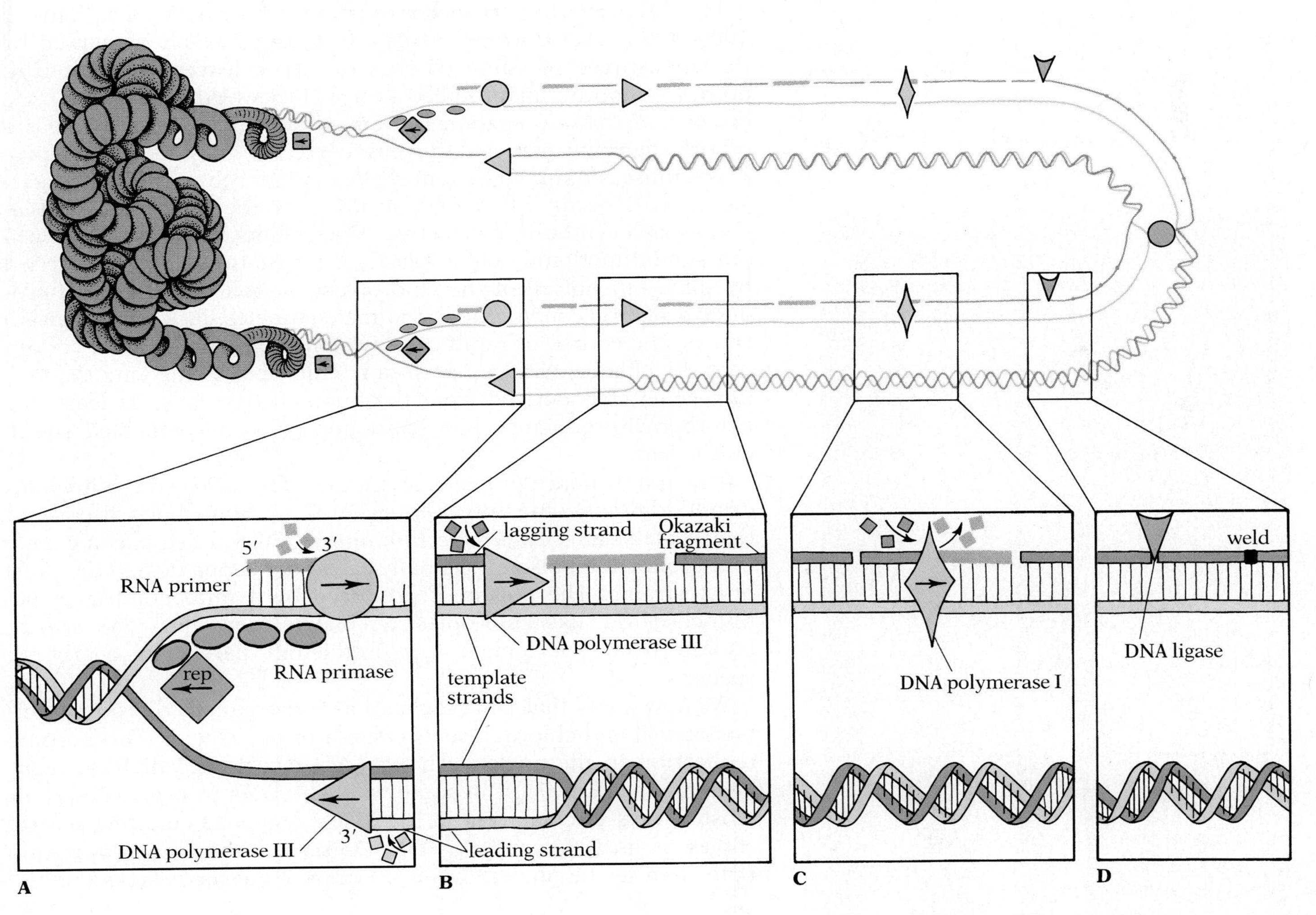

dividuals fuse and begin developing into an organism, the resulting individual is likely to be at least somewhat different from either of its parents, and even from its siblings.

Despite the differences required for the larger genome and sexual reproduction of most eucaryotes, the biochemistry of procaryotic and eucaryotes replication is nearly identical. The *Exploring Further* box on page 216 summarizes the process in bacteria.

■ REPLICATION IN ORGANELLES

As described in Chapter 5, mitochondria and chloroplasts share several characteristics with procaryotes, including (usually) circular chromosomes. The most interesting interpretation of such similarities is that these organelles were once free-living procaryotes that took up symbiotic residence in primitive eucaryotic cells.

The discovery that organelles must have their own genes came in 1909, when Carl Correns traced variegation, which is marked by the appearance of white patches on green leaves, in the bushy, evening-blooming plant called four o'clocks (*Mirabilis jalapa*). He discovered that variegation is transmitted through the cytoplasm of the maternal gamete. Correns correctly guessed that chloroplasts must contain genes and replicate themselves, and that variegation can occur when one or more of the genes involved in chlorophyll synthesis is defective. Since plants cannot survive without some functioning chloroplasts, the mutations must be carried by some but not all of the chloroplasts supplied by the seed-producing plant. (Pollen grains, the male gametes, lack chloroplasts.) During the course of rapid cell division in growing leaves, certain cells, by chance, wind up with only chloroplasts carrying the mutant gene; cells derived from the first cell that lacks at least one chlorophyll-producing chloroplast appear as an elongated streak on the leaf.

The first evidence of genes in another organelle—the mitochondrian—did not come until 1938, when T. M. Sonneborn discovered that some strains of *Paramecium aurelia* carry a cytoplasmic gene that produces a poison (originally called the kappa factor) that kills other strains. The poison is produced by a mitochondrial gene. Mitochondria, like chloroplasts, replicate themselves and are inherited from the female (egg-producing) parent in nearly all species.

We now know that DNA replication and organelle division in mitochondria and chloroplasts occur out of phase with chromosome replication in the nucleus, though the pace of cell division regulates organelle division in some way. The DNA in organelles is almost always circular (Fig. 8.18). Each organelle contains several copies of its DNA, and this DNA is more similar to procaryotic DNA than to the nuclear DNA of eucaryotic cells: it lacks the his-

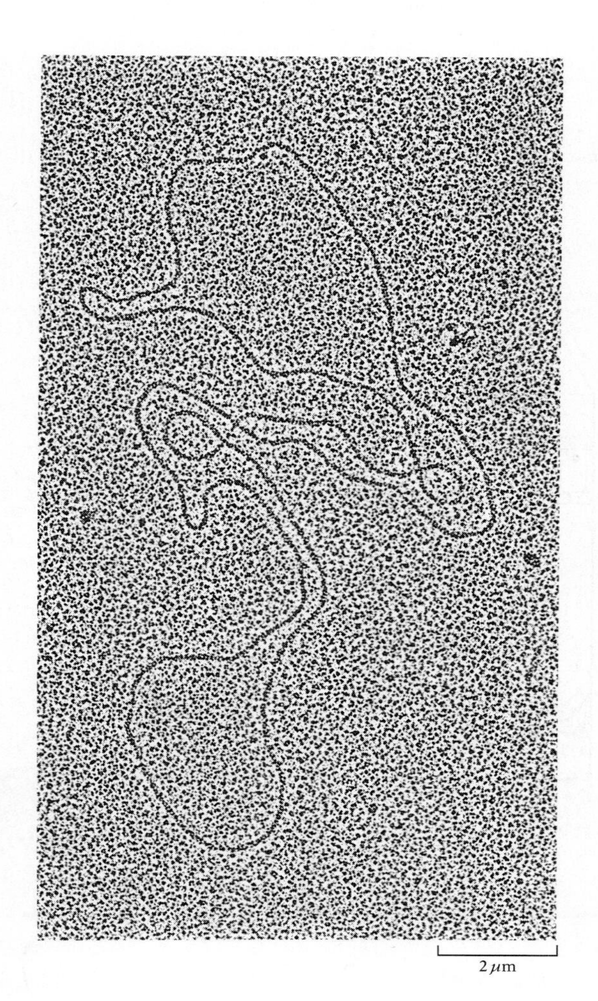

8.18 Organelle DNA

Electron micrograph of mitochondrial DNA from a liver cell of a chicken.

tone nucleosome cores characteristic of nuclear DNA, for example, and is attached to the inner membrane of the organelle much as the procaryotic chromosome is attached to the cell membrane. However, it differs from bacterial DNA in having considerably fewer base pairs: the *E. coli* chromosome encodes about 3000 products, while the genes of a typical animal mitochondrion encode about 40. Since these gene products are insufficient to carry out organelle synthesis and operation (even in bacteria at least 90 gene products are required just for replication, transcription, and translation), most of the genes necessary for organelle function must at some point in evolution have been transferred to the cell's nucleus.

■ HOW ERRORS ARE REPAIRED

The precise replication of DNA is essential for normal cell function. ***Mutations***—random changes in genes—are far more likely to disrupt a pathway than to improve it. If the delicate architecture of an essential enzyme is altered, it usually becomes less active, thus slowing the entire reaction chain of which it is a part. Since the genetic instructions for synthesis of the thousands of enzymes, structural proteins, regulatory proteins, and so on are each hundreds or thousands of bases long, an error rate as low as one in a thousand bases is far too great. Special enzymes have evolved in both procaryotes and eucaryotes to detect and repair mutations. Some enzymes keep uncorrected replication errors at a very low level, while other enzymes locate and repair most of the damage that occurs to the DNA *between* periods of replication. Repair enzymes attach themselves to faulty or damaged areas of the DNA that display the patterns of spacing and polar charges to which their active sites bind; the enzymes loosen specific bonds and catalyze the formation of new ones. Low rates of repair are correlated with susceptibility to cancer and premature aging.

The need for proofreading The initial error rate of the replication enzymes of both procaryotes and eucaryotes is about 3 in every 100,000 pairs.[3] If the errors produced at this rate were left uncorrected, the result would be a mistake in roughly 3% of each cell's proteins—perhaps 1000 changed proteins in every human cell after each replication.

Fortunately, the DNA polymerase complex includes one or more enzymes that successively proofread each base, clipping out mistakes. Other enzymes in the polymerase complex then substitute a rematched base for the excised unit; the complex then moves on without checking again. This second pass reduces the chance of an error to roughly 1 in 10^9. For humans, with our 50,000 functional genes, each of which has an average of about 1500 bases on each

[3] This value, like much of what is known about how DNA polymerases work, is obtained from studies of polymerase complexes with one or more inactive component enzymes. From the change in the operation of the complex, the function of the inactive enzyme can often be deduced. In this case replication proceeds when the enzymes under study are inactive, but most errors are not corrected.

of its two strands, this corresponds to an error in a gene somewhere in the chromosomes once in every 10 cell divisions.

Repairing mutations The integrity of the genetic message is also threatened by alterations in base sequences induced by heat, radiation, and various chemical agents. The rate at which these mutations occur is astoundingly high: thermal energy alone, for instance, breaks the bonds between roughly 5000 purines (adenine and guanine) and their deoxyribose backbones in each human cell every day. Cytosine is chemically converted to uracil (a nucleotide normally found only in RNA, and misread by DNA replication and transcription enzymes) at a rate of about 100 per cell per day. Ultraviolet (UV) radiation from sunlight fuses together adjacent thymines at a high rate in exposed epidermal cells (see Fig. 9.20, p. 242). And yet, because of the continuous operation of repair enzymes, the rate at which mutations actually accumulate in cells is even lower than the rate of uncorrected errors in replication.

The strategy for repairing mutations is basically the same as for replication errors: enzymes locate and bind to the faulty sequences and clip out the flaws, and the intact complementary strand guides repair. A remarkable array of specific enzymes is involved. For instance, chromosomes are scanned for chemically altered bases by fully 20 different enzymes, each specific for a particular class of problems. Another five or so enzymes are specific for finding faulty covalent bonding between a base and some other chemical, or between adjacent bases on one strand; the fusion of thymines induced by UV radiation is the most common error of this type. Other enzymes bind at the sites of missing bases, like the purines that are easily lost to thermal energy. In all, some 50 enzymes locate and correct errors.

Although elaborate, no enzymatic repair system is perfect, so some mutations manage to survive. The results can range from neutral effects like blue eyes (which result from a defective pigment gene) to genetic diseases. Mutations survive for several reasons: Some errors are missed in the repair process, and others are repaired incorrectly. When a mutation occurs just before or during replication, there may not be time for detection and repair. Still other mutations are not overlooked by the repair system, but are actually created by it.

One well-understood class of mutations generated by repair enzymes is ***misalignment deletion***. Small deletions (loss of a few base pairs) tend to occur at susceptible sequences. The basis of this differential susceptibility is the relative weakness of the base pairing between adenine and thymine: they are connected by only two hydrogen bonds, whereas cytosine and guanine are connected by three. The hydrogen bonds are continually breaking and re-forming, and it frequently happens, particularly in regions rich in adenine-thymine pairs, that by chance all the bonds in a small segment of DNA are broken simultaneously. The bonds re-form spontaneously, but occasionally the rebonding is incorrect. For example, when a region with an extended series of adenine-thymine pairs undergoes a transient separation, there is a small chance that the two strands will be misaligned when they pair again. DNA repair enzymes may then remove the unpaired bases and thereby in-

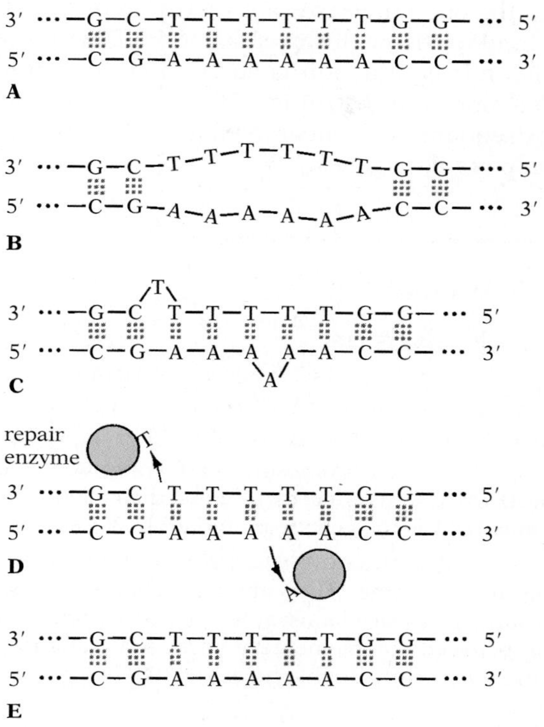

8.19 Misalignment deletion

(A) A sequence with an extended series of A–T pairs is especially susceptible to deletion. (B) When, by chance, a series of hydrogen bonds breaks simultaneously, the two strands of the helix separate transiently. (C) When they pair again, there is a small chance that they will be misaligned. (D) If the resulting distortion is detected by DNA repair enzymes, the two unpaired bases will be cut out by the enzymes . (E) The loose ends will then be reattached with DNA ligase, leaving each strand of DNA one nucleotide short.

troduce a misalignment deletion (Fig. 8.19). The repair is likely to alter the meaning of the gene by causing errors during translation of the gene's mRNA.

Another common mutation involves regions in genes that normally have the same three-base sequence repeated end to end several times (for example, . . . CAGCAGCAGCAGCAG . . .). Sometimes, either during replication or repair, extra copies of the triplet are added. When the number is large enough, the gene becomes defective: a massive CAG repeat is the cause of Huntington's disease, while one kind of mental retardation results from a hugely expanded CGG repeat.

Finally, some mutations simply cannot be detected by the repair enzymes. An example of this occurs in bacterial DNA when a cytosine that was modified by the addition of a methyl group ($-CH_3$) loses its amino group ($-NH_2$) after replication, producing a thymine (Fig. 8.20). The thymine is now mismatched with guanine, the partner of its predecessor. Since thymine is a normal base, there is no way for the repair enzymes to determine whether the incorrect base is the thymine in one chain or the guanine in the other.

The methylated-cytosine problem arises because the methylation of certain cytosines can be a useful adaptation. In bacteria, enzymes known as ***endonucleases***, present in the cytoplasm, are often able to break up the DNA of invading viruses. (As detailed in Chapter 10, the DNA-chopping ability of certain endonucleases is used in recombinant DNA research.) If the enzymes digest invading DNA, what keeps them from also cutting bacterial chromosome? In the bacterial DNA, cytosines in those sequences

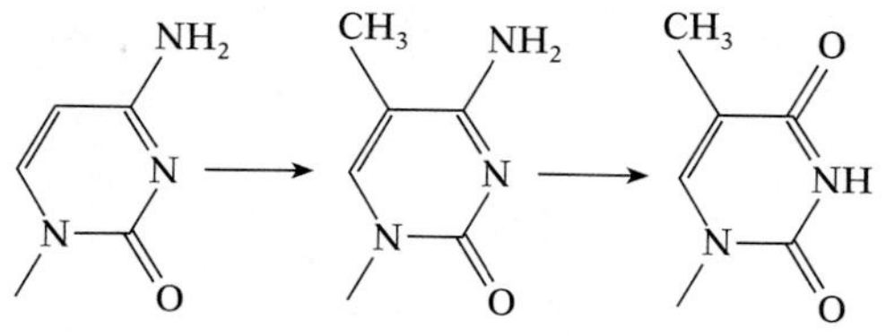

Cytosine → Methylated cytosine → Thymine

8.20 Deamination of methylated cytosine

Cytosine is sometimes methylated by enzymes. When a methylated cytosine undergoes deamination (loss of its amino group), usually as the result of oxidation by a mutagenic chemical, the result is a normal thymine. This change cannot be corrected reliably because the repair system cannot determine whether it is the thymine or the guanine (the base in the complementary strand to which the cytosine was originally paired) that is incorrect.

vulnerable to the endonucleases are methylated, so the endonucleases cannot bind to them. However, bacterial DNA protected in this way from endonuclease action is susceptible to mutation by conversion of cytosines into thymines. (There is active methylation of specific DNA sequences in eucaryotes as well, though its function there is to help regulate genes.)

CHAPTER SUMMARY

WHAT ARE GENES MADE OF?

THE UNUSUAL CHEMISTRY OF CHROMOSOMES DNA is the one substance whose quantity does not usually vary from one somatic cell to another, although it is reduced by half in gametes. (p. 205)

GENES: DNA OR PROTEIN? Chromosomes are composed of DNA and protein. Only the DNA of viruses needs to enter their host in order to begin multiplying; thus DNA is the genetic material. (p. 206)

THE MOLECULAR STRUCTURE OF DNA DNA has four bases: adenine, guanine, cytosine, and thymine. They are long chains of sugars linked by phosphate groups; pairs of chains are attached to one another through polar bonding between complementary bases—adenine to thymine, guanine to cytosine—to form a double helix. (p. 208)

HOW DNA IS REPLICATED

THE TEMPLATE HYPOTHESIS Replication occurs by separating the two strands of a double helix and using each strand as a template for synthesizing a complementary strand. (p. 212)

THE DIFFICULTIES OF REPLICATION The strands must be kept separate while the various enzymes do their work. The polymerase can only synthesize DNA in one direction, so one strand must be replicated in small backward steps, which must then be joined. In eucaryotes the nucleosomes must be removed and a new set synthesized. The daughter chromosomes must be segregated into separate progeny cells. (p. 214)

REPLICATION IN ORGANELLES Organelles have their own very small chromosomes, similar to bacterial DNA in structure and replication. (p. 218)

HOW ERRORS ARE REPAIRED Mutations occur in the DNA, either because of the action of chemicals or UV light on the chromosomes, or through errors in replication. Proofreading by the polymerase minimizes replication errors. Many kinds of special enzymes patrol the chromosomes looking for spontaneous mutations and repairing them. (p. 219)

STUDY QUESTIONS

1 Compare and contrast ATP and the nucleotide having adenine for its nitrogenous base (that is, adenosine monophosphate, or AMP). (pp. 158, 209)

2 Bacteria have only one copy of each gene, whereas most eucaryotes have two. If most serious mutations destroy the activity of the product encoded by a gene, what effect does having two copies of a gene have on the chance of suffering a complete loss of a gene?

3 The bacterial chromosome has about 4 million base pairs. Assuming all bases occur with equal frequency, calculate the approximate total energy of the hydrogen bonds holding the two strands together. (p. 35)

4 Assuming one ATP molecule is needed to prime each of the 8 million bases added during *E. coli* replication, how many glucose molecules must be metabolized to fuel a single such cell division? (Ignore the energy needed for the other steps.) (p. 175)

5 In the 1958 experiment of Meselson and Stahl, what pattern of DNA densities would be observed halfway through the first replication? halfway through the second replication? (pp. 212–214)

SUGGESTED READING

HOWARD-FLANDERS, P., 1981. Inducible repair of DNA, *Scientific American* 245 (5). *On how cells recognize when DNA has been damaged, how they "switch on" genes to generate repair enzymes, and how the enzymes work.*

LINDAHL, T., 1993. Instability and decay of the primary structure of DNA. *Nature*, 362, 709–714. *An excellent overview of spontaneous mutation, repair, and their connection with cancer and aging.*

RADMAN, M., AND R. WAGNER, 1988. The high fidelity of DNA replication, *Scientific American* 259 (2). *On how the proofreading component of DNA polymerase works.*

WANG, J. C., 1982. DNA topoisomerases, *Scientific American* 247 (1). *On the enzymes responsible for untangling DNA during replication.*

YUAN, R., AND D. L. HAMILTON, 1982. Restriction and modification of DNA by a complex protein, *American Scientist* 70, 61–69. *On how some endonucleases can cut DNA at a specific site if it is fully unmethylated, or finish methylating (and thereby protect) the same site if it is partially methylated.*

CHAPTER 9

TRANSCRIPTION AND TRANSLATION

In this chapter we will look at how information is actually encoded in DNA, and how the information is transcribed into ribonucleic acid (RNA). We will also see how a fraction of this RNA is used directly in structural and enzymatic roles, while the rest—messenger RNA (mRNA)—is translated into protein. Finally, we will look at some of the consequences of transcription and translation errors.

HOW TRANSCRIPTION WORKS

With a few exceptions, every cell has a full set of chromosomes, which are replicated, repaired, and passed on at every cell division. But at any given moment, only a small proportion of the thousands of genes in the cell are active. Which genes are needed depends not only on what the cell is doing—dividing, growing, resting, moving—but also on the cell's environment and, if it is part of a complex multicellular organism, on its specialty. Transcription enzymes must recognize the small number of active genes in a cell and ignore the rest, and must know where genes begin and end. (Replication, which is an all-or-none process, does not make any distinction between genes.) In Chapter 11 we will take up the important question of ***gene expression***: how genes are activated and inactivated according to the needs of the cell.

■ THE NEED FOR MESSENGER RNA

Genetic information is stored in the DNA of chromosomes, while the proteins encoded by genes are synthesized on ribosomes. How is information conveyed from the genetic library to the cellular building sites? By the early 1940s cells in tissues such as the vertebrate pancreas, where protein synthesis is particularly active, were known to contain large amounts of RNA. Radioactive-tracer experiments demonstrated that eucaryotic RNA is synthesized in the nucleus and then moves into the cytoplasm (Fig. 9.1). These two pieces of evidence suggested that RNA might be the chemical messenger between DNA and the ribosomes.

Though RNA and DNA are very similar, they differ in three important ways, described in Chapter 3: (1) The sugar in RNA is ribose, whereas in DNA it is deoxyribose (Fig. 9.2). (2) RNA has

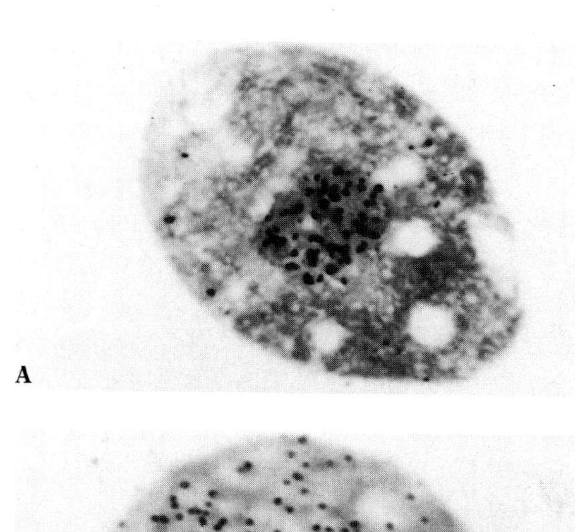

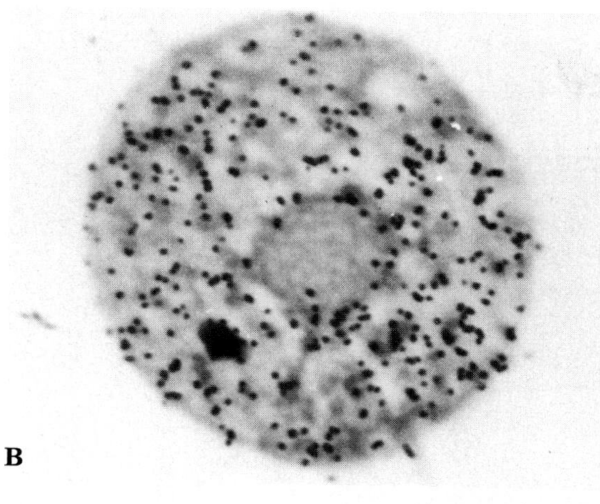

9.1 Radioactive-tracer experiment showing the movement of messenger RNA

(A) Cells were placed briefly in a medium containing radioactive RNA nucleotides. After 15 min the labeled nucleotides were incorporated into messenger RNA in the nucleus, where they appear as dark spots. (B) 75 min later, however, the labeled mRNA was found in the cytoplasm, demonstrating that mRNA moves from the nucleus to the cytoplasm.

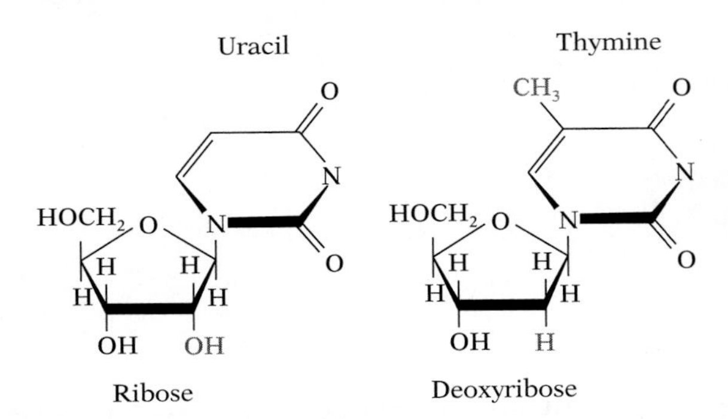

9.2 Comparison of uracil in RNA and thymine in DNA

The five-carbon sugars of RNA and DNA differ only at the site shown in color, where deoxyribose lacks an oxygen atom that is present in ribose. Uracil differs from thymine only in that it lacks a methyl group ($-CH_3$).

uracil where DNA has thymine;[1] (3) RNA is ordinarily single-stranded, whereas DNA is usually double-stranded.

These differences aside, DNA acts as a template for the production of RNA. The synthesis proceeds in essentially the same way as that of new DNA: the two strands of a DNA molecule uncouple, and RNA is synthesized along one of the DNA strands by RNA polymerase (Fig. 9.3). For every adenine in the DNA template, a uracil ribonucleotide, rather than a thymine, is added to the growing RNA strand; for every thymine in the DNA, an adenine ribonucleotide is added; for every guanine, a cytosine; and for every cytosine, a guanine (Fig. 9.4). In short, the synthesis of RNA—the process of ***transcription***—operates exactly like replication; the resulting strand of RNA, which is complementary to the transcribed DNA, will later act as an intermediary in protein synthesis.

There are three major types of RNA: ***messenger RNA (mRNA)***—the type just discussed—carries the information necessary to specify the sequence of amino acids in a protein to the ribosomes where proteins are synthesized; ***ribosomal RNA (rRNA)*** forms part of the ribosomes themselves; and ***transfer RNA (tRNA)*** brings amino acids to the ribosomes during protein synthesis.

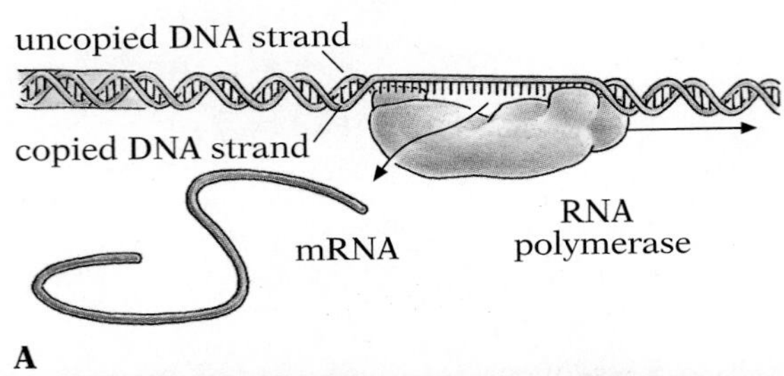

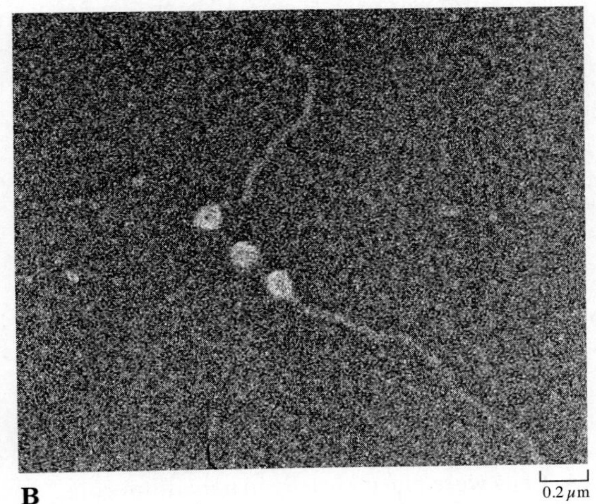

9.3 Transcription of a gene

(A) As the RNA polymerase complex moves along the DNA, it catalyzes transcription of only one of the two DNA strands. (B) An electron micrograph shows three polymerases transcribing a gene in *E. coli*.

■ CREATING MESSENGER RNA

Transcription of DNA into RNA is accomplished by a six-protein enzyme complex called ***RNA polymerase***, which both binds to the DNA and opens up the helix. Like DNA polymerase, the transcription complex moves from the 3′ end to the 5′ end of the DNA, and synthesizes a strand of the opposite polarity (5′ to 3′), in this case a strand composed of ribonucleotides. In procaryotes, a single kind of RNA polymerase is responsible for all RNA synthesis. In eucaryotes, RNA polymerase II is responsible for the synthesis of mRNA. Two other RNA polymerases synthesize rRNA, tRNA, and other RNAs that serve structural and, in some cases, enzymatic roles; all of these RNA polymerases work in much the same way.

Transcription in procaryotes RNA polymerases know where to start and stop synthesizing the messenger because specific control sequences in DNA mark the beginning and end of a gene. The first pair of these is called the ***promoter***; active sites on the RNA polymerase bind specifically to regions with the base sequences of the promoter. The promoter in *E. coli*, reading from 5′ to 3′ on the complementary strand, usually begins with TTGACA or a very similar sequence.[2] This is followed by roughly 17 bases with another

[1] The presence of uracil in place of thymine in RNA makes it possible for the active site of a wide variety of enzymes to distinguish DNA from RNA, and so to be selective in their activity. DNA polymerase, for instance, does not attempt to replicate or transcribe messenger RNA, nor can enzymes in the cytoplasm that digest the DNA of invading viruses chew up the cell's own RNA.

[2] Though the RNA polymerase works from the template strand of the DNA, the custom is to refer to the sequences in question as they exist on the complementary strand. This method of reference is convenient because the RNA sequences produced in transcription are identical to those of the corresponding regions of the complementary DNA strand—except, of course, that U is substituted for T. Thus, since the first three DNA bases transcribed are usually GTA (Fig. 9.5A), they are rendered as CAU in the mRNA, which corresponds to CAT in the complementary

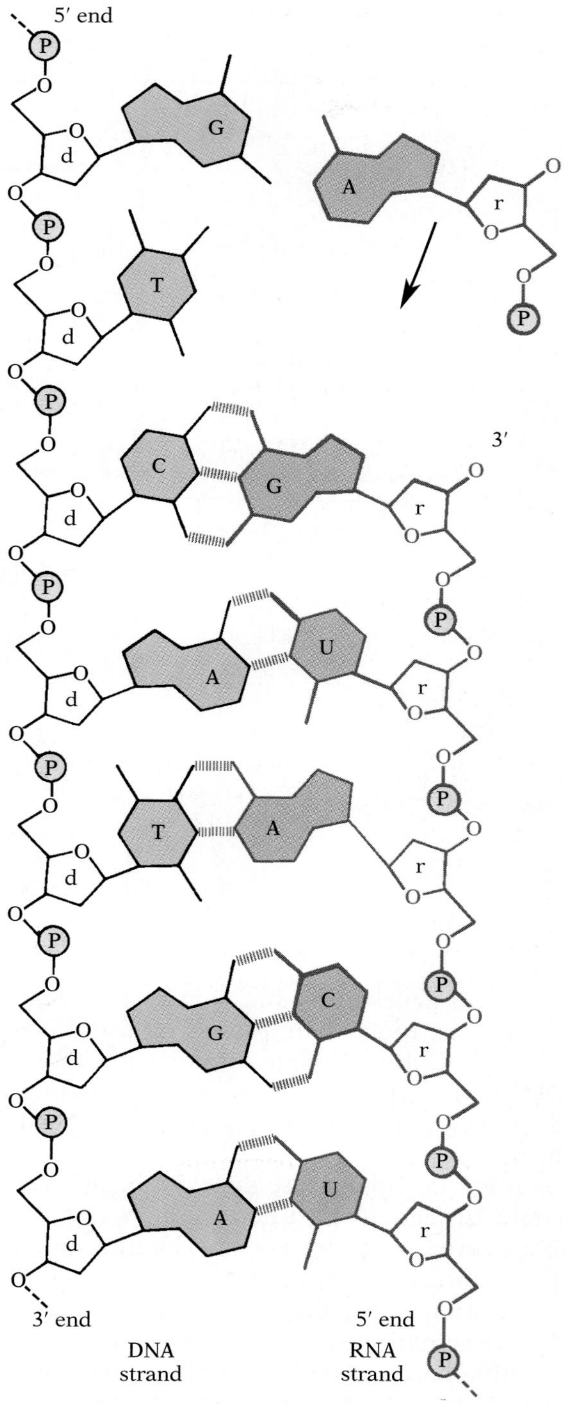

9.4 Synthesis of RNA by transcription of a DNA template

The sugar in RNA (ribose, r) is slightly different from that in DNA (deoxyribose, d); in addition, the uracil (U) in RNA takes the place of the thymine in DNA. Transcription is accomplished by the enzyme complex RNA polymerase (not shown).

function; then comes TATATT or a very similar sequence (Fig. 9.5A). The polymerase recognizes these two regions in the DNA and is large enough to bind simultaneously to both.

The exact promoter sequences shown in Figure 9.5 are called ***consensus sequences***, since each base shown is the one most commonly found at the promoter location in question in procaryotes. (Much less is known about eucaryotic promoters, but the sequence TATA appears to be important in them as well.) The promoter se-

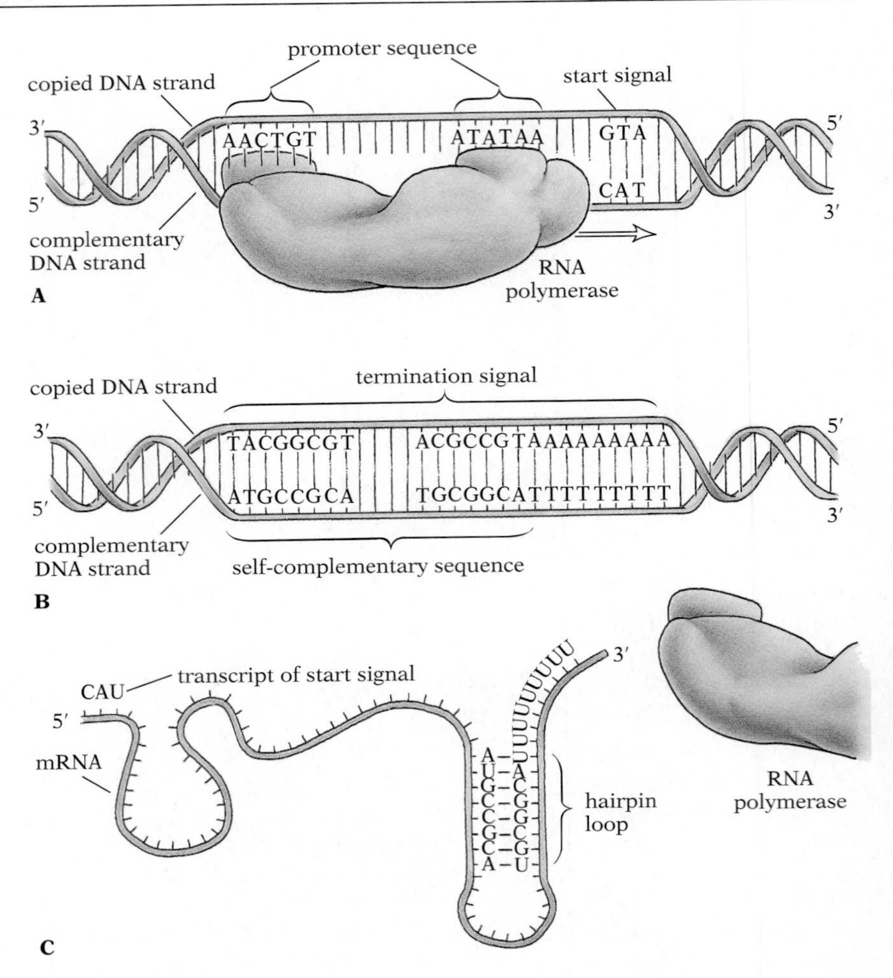

9.5 Transcription signals in *E. coli* DNA

(A) RNA polymerase binds to a region of DNA with the sequences shown. Once the polymerase is bound, transcription begins with the GTA signal (corresponding to the CAT sequence on the complementary strand) and proceeds until a termination signal is encountered. (B) The termination signal consists of a "self-complementary" sequence—one that can bind to itself—and four to eight adenines (corresponding to thymines on the complementary strand). (C) At this point, the tail of the mRNA forms a short double-helix hairpin loop, and the polymerase stops synthesizing. Note that RNA polymerase inserts a uracil wherever DNA polymerase would add a thymine.

quences of most genes differ in one way or another from the consensus sequences; more than a hundred variations have been discovered in *E. coli*, for example. Minor variations in the promoter sequences cause the polymerase to bind less strongly and less often to some of them, so some genes are transcribed less frequently than others.

Once bound, the polymerase does not begin synthesizing mRNA immediately. Instead, it must find the start signal—often CAT—located about seven bases beyond the binding point, toward the 3′ end of the complementary strand. In most procaryotes, synthesis of mRNA continues until the polymerase encounters a termination signal. The termination signal has two components. First, there is a region with a base sequence that allows the corresponding bases in the tail of the mRNA to pair off and bind together to form a small loop, known as a ***hairpin loop*** (Fig. 9.5C). This is followed on the copied strand by a run of four to eight adenines.

When the polymerase moves into the region of the adenine run, the hairpin loop forms in the RNA just produced and, apparently from the physical stress it puts on the enzyme complex, slows or temporarily halts transcription. Two things now happen that terminate the production of the RNA. First, because the loop sequence has pulled away from the DNA, that portion of the gene is able to re-form its double helix, adding yet more strain to the complex. Second, the weak bonding between the run of adenines in the DNA and the uracils of the RNA copy (only two hydrogen bonds between each base pair) is unable to anchor the RNA to the gene for long; the pause and the stress together allow the transcribed copy and its polymerase to drift away from the chromosome.

Transcription in eucaryotes: editing of mRNA The mechanism of mRNA synthesis just described is found in procaryotes, mitochondria, and chloroplasts. Transcription in eucaryotic nuclei is more complicated. One complication is that eucaryotic mRNA is "tagged" on both ends: a cap of 7-methylguanosine is added at the front (5′) end, while a tail of 100–200 adenines is affixed to the 3′ end. Among other things, these modifications are used later to identify the molecule as mRNA. The result—the ***primary transcript***—is still not a usable messenger.

The dramatic difference between a primary transcript and functional mRNA came in 1977 as a total surprise: Phillip A. Sharp discovered that though most primary transcripts are about 6000 bases long, the mRNA actually transcribed is only about one-third that length. Large specific regions *within* the primary transcript of most eucaryotic messengers must be removed in the nucleus to create a functional mRNA molecule. The regions of the primary RNA transcript that survive this processing and operate during protein synthesis (as well as the parts of the gene that gave rise to these sections) are called ***exons*** (because they are expressed), while the intervening sequences of the primary transcript, that are removed in the nucleus (and the corresponding regions of the gene) are referred to as ***introns*** (Fig. 9.6).

Despite their early removal, many introns are necessary to the functioning of RNA: most mRNA transcribed from artificially manufactured genes that lack introns fails to get into the cytoplasm, while mRNA from genes with some introns intact is often processed correctly and slips through the nuclear pores to the cytoplasm.

Intron removal is a formidable task (Fig. 9.7). Some genes have as many as 50 introns, and a mistake of even a single base in the excision process can render the mRNA useless. The precise beginnings and ends of introns on the primary transcripts are marked by signals so that they can be recognized and removed.[3] The intron-boundary signals are recognized by a short bit of RNA found in curious RNA-protein complexes known as the small nuclear ribonucleoprotein particles—snRNPs (sometimes pronounced "snurps"), for short. At least four different kinds of snRNPs cooperate in most splicing. The RNA in these particles is like ribosomal RNA in that it is used directly, and has both an enzymatic and a structural role.

The best known snRNP has a base sequence complementary to the boundary between the end of exons and the beginning of introns. SnRNP complexes catalyze the breakage of the bond between the primary-transcript nucleotides at each intron-exon

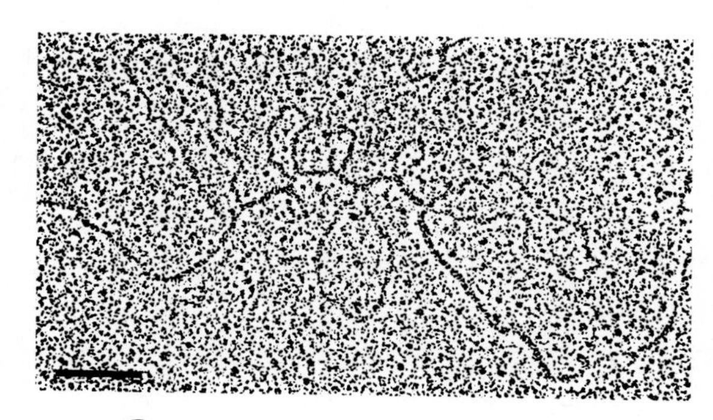

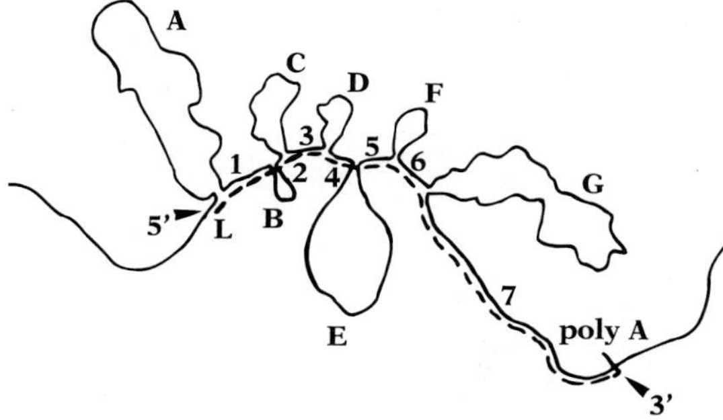

9.6 Visible evidence of introns

A gene for an egg protein, ovalbumin, was denatured to create single-stranded DNA, and then mixed with mature mRNA transcribed from this gene. Because the mRNA sequence is complementary to the DNA strand from which it is transcribed, the two molecules are able to form a double-stranded hybrid. Because the introns in the primary transcript have been removed by editing, the intron regions in the DNA form unmatched loops, revealing that this gene has six introns. The scale in the photograph is 1 μm.

[3] The sequence at the beginning of the intron is AG-GUAAGU; the hyphen marks the exon-intron boundary. At the end of the intron it is CAG-G.

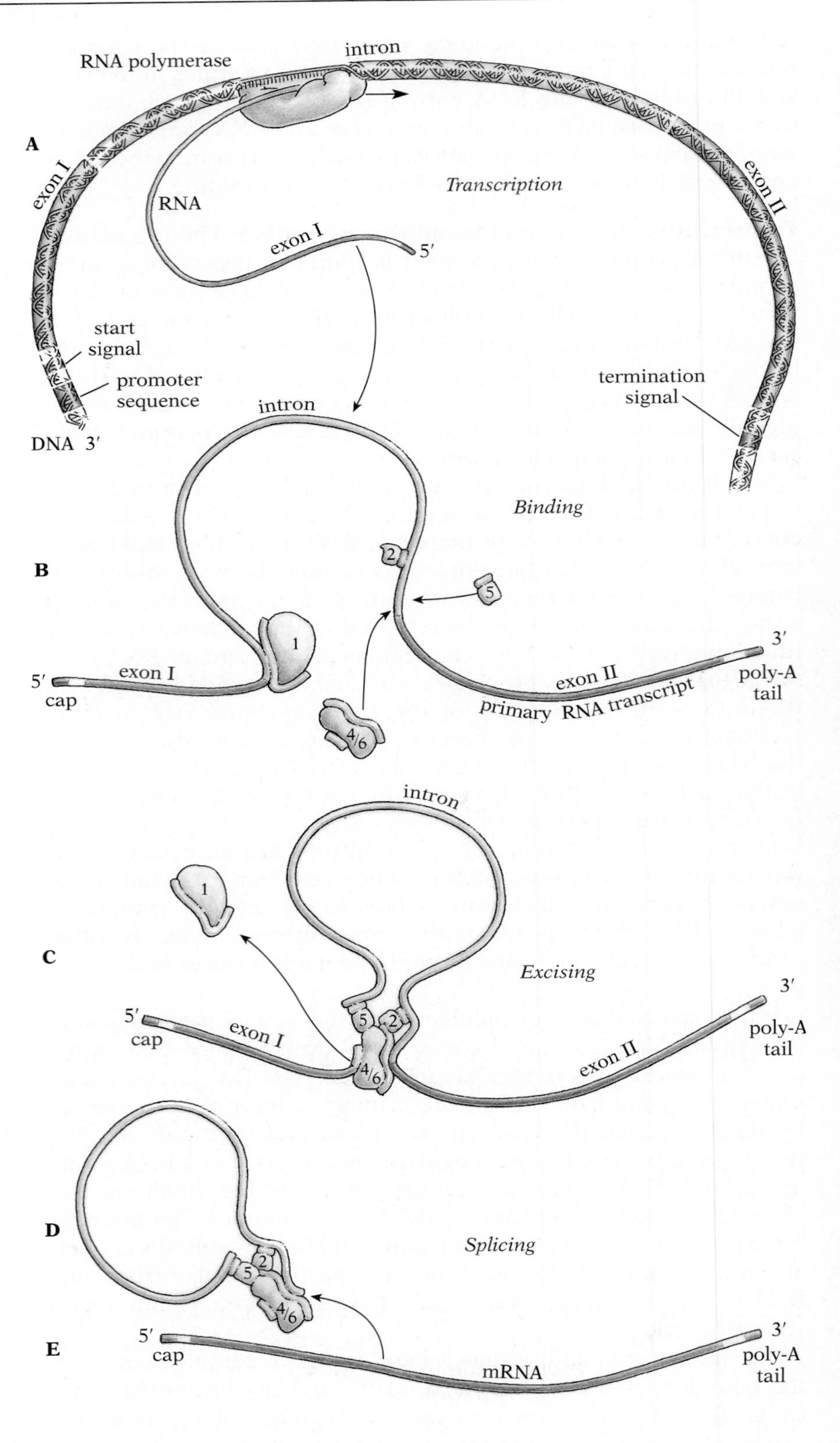
RNA polymerase
intron
A
exon I
Transcription
exon II
RNA
exon I
5′
start
signal
promoter
sequence
termination
signal
DNA 3′
intron
Binding
B
2
5
1
3′
5′
exon I
exon II
poly-A
tail
cap
primary RNA transcript
4/6
intron
1
C
Excising
3′
5′
cap
exon I
5
2
poly-A
tail
exon II
4/6
D
Splicing
2
5
4/6
5′
3′
E
cap
poly-A
tail
mRNA

9.7 Messenger RNA processing in eucaryotes

(A) RNA polymerase binds to a promoter sequence and moves along the DNA strand, synthesizing a complementary RNA strand. (B) A 7-methylguanosine cap is added at the beginning of the transcript by another enzyme. After the RNA polymerase transcribes the termination signal, the RNA between the last exon and the sequence of four to eight adenines is removed; a separate enzyme extends the poly-A tail by 100–200 bases, completing the primary RNA transcript. (B–E) With the aid of several small nuclear ribonucleoprotein particles (snRNPs), any introns are removed and the resulting transcript is spliced. For clarity, only one intron is shown. The process begins with recognition of the beginning of the intron by snRNP 1 (B). Next, snRNPs 2, 4/6, and 5 come together to form a complex at the end of the intron (C). The leading end of the intron is cut (D) and bound to the complex, which then cuts the trailing end of the intron and splices the two exons together (E). At this point the mature mRNA is ready for export to the cytoplasm.

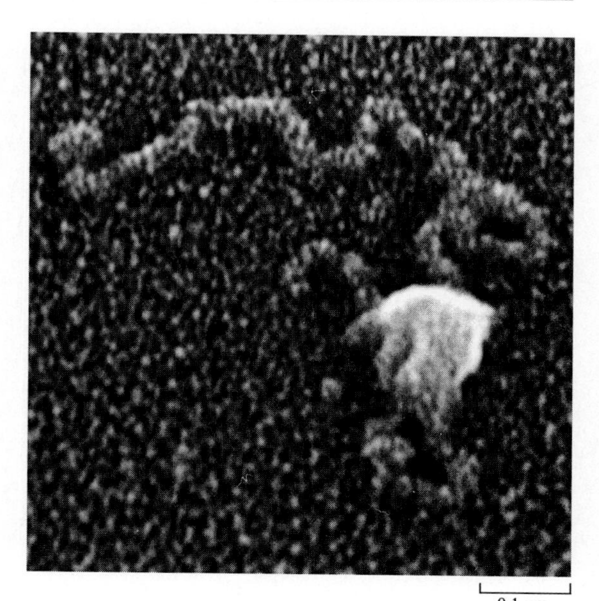

9.8 A snRNP-intron complex

boundary. The intron drifts away (Fig. 9.8) and is thought to be digested by other enzymes. Other components of the snRNP complex then splice the two exons. Once all the introns have been removed, the mature mRNA is exported to the cytoplasm. This splicing system probably evolved from a self-splicing mechanism, still seen in a few genes in certain species: the intron itself folds into a catalytic form, excises itself, and splices the exons together (Fig. 9.9).

HOW THE MESSAGE IS TRANSLATED

Messenger RNA contains the sequence of bases that specifies the identity and order of the amino acids that will be linked to form the protein encoded by the gene. This sequence, which is preceded by a short "leader" and followed by a "trailer" (analogous to the blank margins at the top and bottom of a page), is decoded on a ribosome, a process called ***translation***: information is translated from one molecular "language" to another. We will look first at the two languages, and then at the process of translation.

■ THE GENETIC CODE

The need for codons When Watson and Crick discovered that DNA is essentially a linear array of four nitrogenous bases, it became clear that the unique sequence of amino acids found in a particular protein must be encoded by *groups* of bases. After all, if each DNA base designated a particular amino acid, there would have to be 20 different bases instead of four. If the bases were taken two at a time—AA, AC, AG, AT, CA, CC, and so on—only 16 combinations would be possible. Since 20 amino acids are commonly present in proteins, the information has to be encoded in sets of three or more bases, called ***codons***.

9.9 A self-splicing intron

The intron portion of the transcript has formed internal pair bonds and folded into an enzymatic conformation that brings the end of one exon adjacent to the beginning of the next exon. In this drawing the 5′ end of the intron has just been cut free of the transcript.

Crick and his associates established the length of the codon in 1961. Bacteria were treated with compounds that insert or delete nucleotides from DNA. Crick reasoned that adding or deleting a single nucleotide would render a message meaningless because it would disrupt the translation process by shifting the apparent starting point in subsequent coding units. Suppose the message is

THE BIG RED ANT ATE ONE FAT BUG

Deletion of the first *E* will make it

TH~~E~~B IGR EDA NTA TEO NEF ATB UGX

A single deletion redefines all subsequent codons; if it occurs near the start of the message, the amino acid sequence of the protein is completely altered. Two deletions would have the same effect, but if the number of deletions corresponded to the codon length, and if they all occurred near the beginning, translation would produce an enzyme that would probably retain some activity:

T~~H~~E~~B~~I GR~~E~~D ANT ATE ONE FAT BUG

Crick and his associates made different numbers of nucleotide deletions, and determined for each concentration whether or not active enzymes were produced. Subjecting these results to elaborate statistical analyses, they concluded that codons were three bases long.

Deciphering the code The next problem to be solved in understanding translation was to determine the exact relationships between codons and amino acids. This was a long and difficult process, and was not completed until H. G. Khorana developed a technique for generating specific RNA sequences. Table 9.1 on page 242 summarizes the genetic code.

As the genetic dictionary of the table makes clear, all but two amino acids (methionine and tryptophan) are represented by more than one codon. The synonymous codons—those coding for the same amino acid—usually have the same first two bases, but differ in the third; thus CCU, CCC, CCA, and CCG all code for proline. In fact, U and C are always equivalent in the third position, and A and G are equivalent in 14 of 16 cases. Notice also that four codons—AUG, UAA, UAG, and UGA—serve as "punctuation," marking the beginning and end of the actual message within the mRNA.[4]

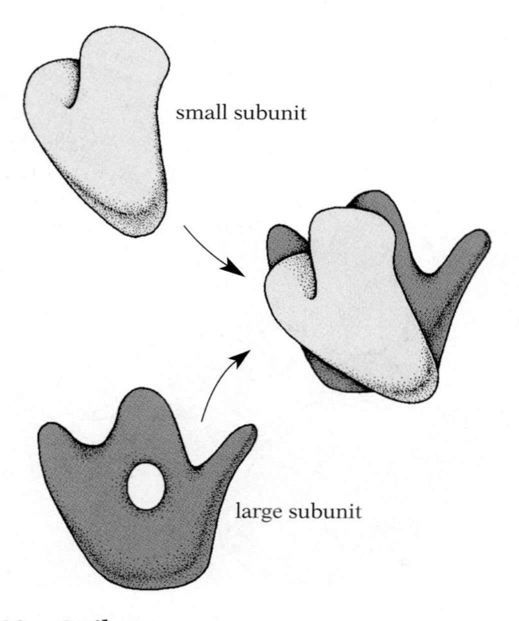

9.10 A ribosome

A functional ribosome is formed when the two kinds of independent subunits join. This joining can take place only after mRNA binds to the small subunit. The tunnel in the large subunit is about 2.5 nm in diameter.

■ HOW RIBOSOMES WORK

Translation occurs on ribosomes. Ribosomes have a large and a small subunit, each a complex of ribosomal RNA (rRNA), enzymes, and structural proteins (Fig. 9.10). When not carrying out protein synthesis, the ribosomal subunits are separate. In procaryotic ribo-

[4] Certain minor inconsistencies in the genetic code have been discovered. For example, in some ciliates the codons UAA and UAG, which are normally termination signals, instead code for glutamine; in some bacteria and most (perhaps all) eucaryotic mitochondria, UGA, which can also be a termination signal, codes for tryptophan.

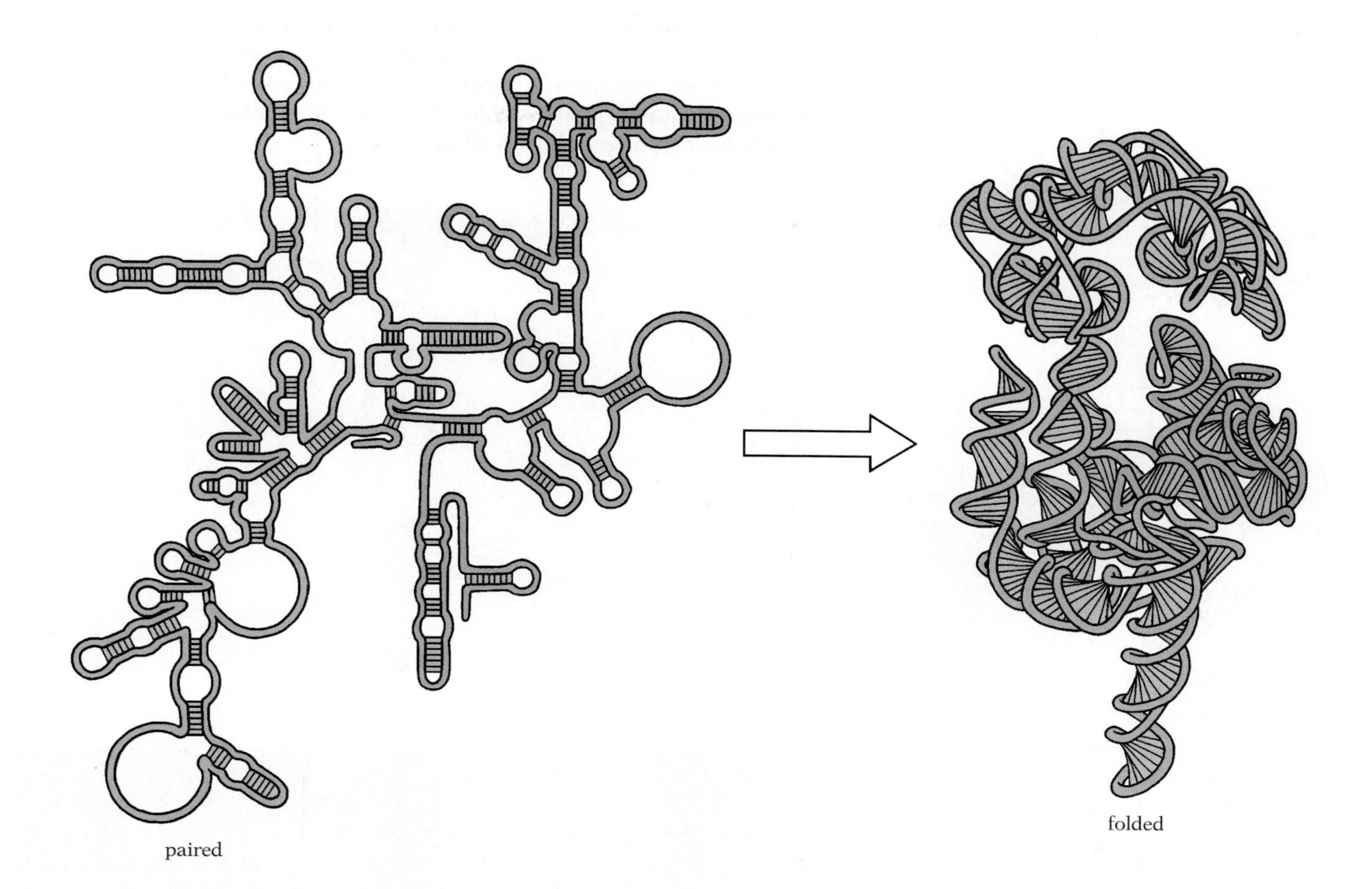

9.11 Ribosomal RNA

Ribosomal RNA has many stretches of complementary base pairs (color), which allow the chain to fold back on itself and form double-stranded helical arms. The open loops represent areas in which the bases are not complementary. This arrangement probably plays an important role in ribosomal function. The rRNA shown here is found in the small ribosomal subunit of a bacterium.

somes the large subunit is composed of two molecules of rRNA and roughly 35 proteins; the small subunit has one molecule of rRNA and approximately 20 proteins. Eucaryotic ribosomes are somewhat larger. In both eucaryotes and procaryotes, the rRNA itself is highly structured: long segments of complementary base pairs are held together in a helical chain by hydrogen bonds, creating a complex pattern of arms and loops (Fig. 9.11).

In procaryotes, the 5′ end of the mRNA binds to the small subunit, which is then able to bind the large subunit (Fig. 9.12). The binding of the mRNA involves base pairing between a section of the rRNA and a binding signal—usually AGGAGGU, near the end of the mRNA in the leader segment. This signal binds in the small subunit's cleft (Fig. 9.12B–C), along with the initiation codon (AUG) located just a few bases farther along on the mRNA. Special proteins called initiation factors aid in this process. When the binding of the large subunit is complete, translation can begin. AUG codes for methionine, so methionine is always the first amino acid incorporated during translation. Another special enzyme later removes the methionine from the end of most polypeptides.

Because procaryotes have no nuclear membrane, and the process of transcription is not segregated from the ribosomes and other translational machinery in the cytoplasm, ribosomes are able to bind one end of an mRNA molecule and begin translating it into protein while the RNA polymerase is still transcribing the rest of the message from the chromosome. (The same situation occurs in chloroplasts and mitochondria.) While the first ribosome to bind is translating later parts of the message, additional ribosomes can bind and begin translation (Fig. 9.13).

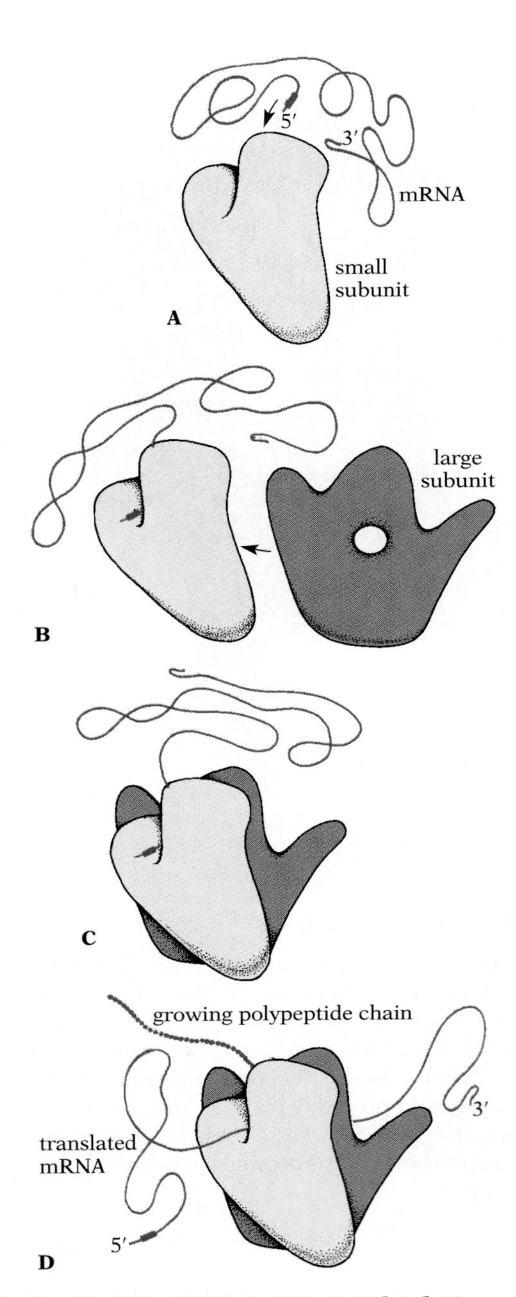

9.12 Synthesis of a polypeptide chain by a ribosome

Free ribosomes exist as two separate subunits (A–B). A signal sequence on the mRNA binds to the small subunit, causing a change that allows it to bind the large subunit (B–C). The ribosome then translates the mRNA into a polypeptide chain, producing a protein (D). The growing polypeptide chain is thought to be threaded out through the tunnel in the large subunit. The mRNA is much longer than shown here, and additional ribosomes can bind and begin translation while the first ribosome is still at work.

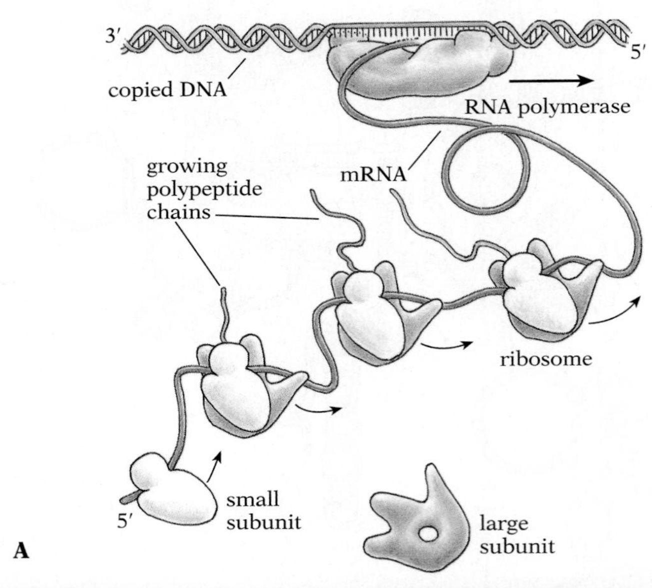

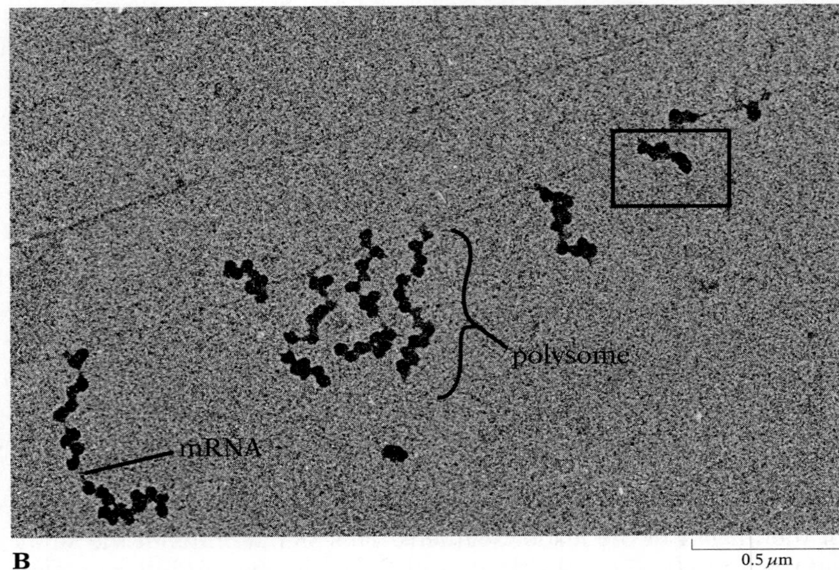

9.13 Simultaneous transcription and translation

In procaryotes and in eucaryotic organelles, translation can begin as soon as the first part of a message has been transcribed from the chromosome, and several ribosomes can be involved simultaneously. The complex of mRNA and two or more ribosomes is often called a polysome. In the micrograph, the two thin horizontal lines are DNA; the lower strand is being transcribed and, simultaneously, the RNA transcripts are being translated by ribosomes (dark spots). Note that the transcripts become longer from left to right as transcription of the gene proceeds. The growing polypeptide chains are not visible. The boxed area corresponds to the interpretive sketch.

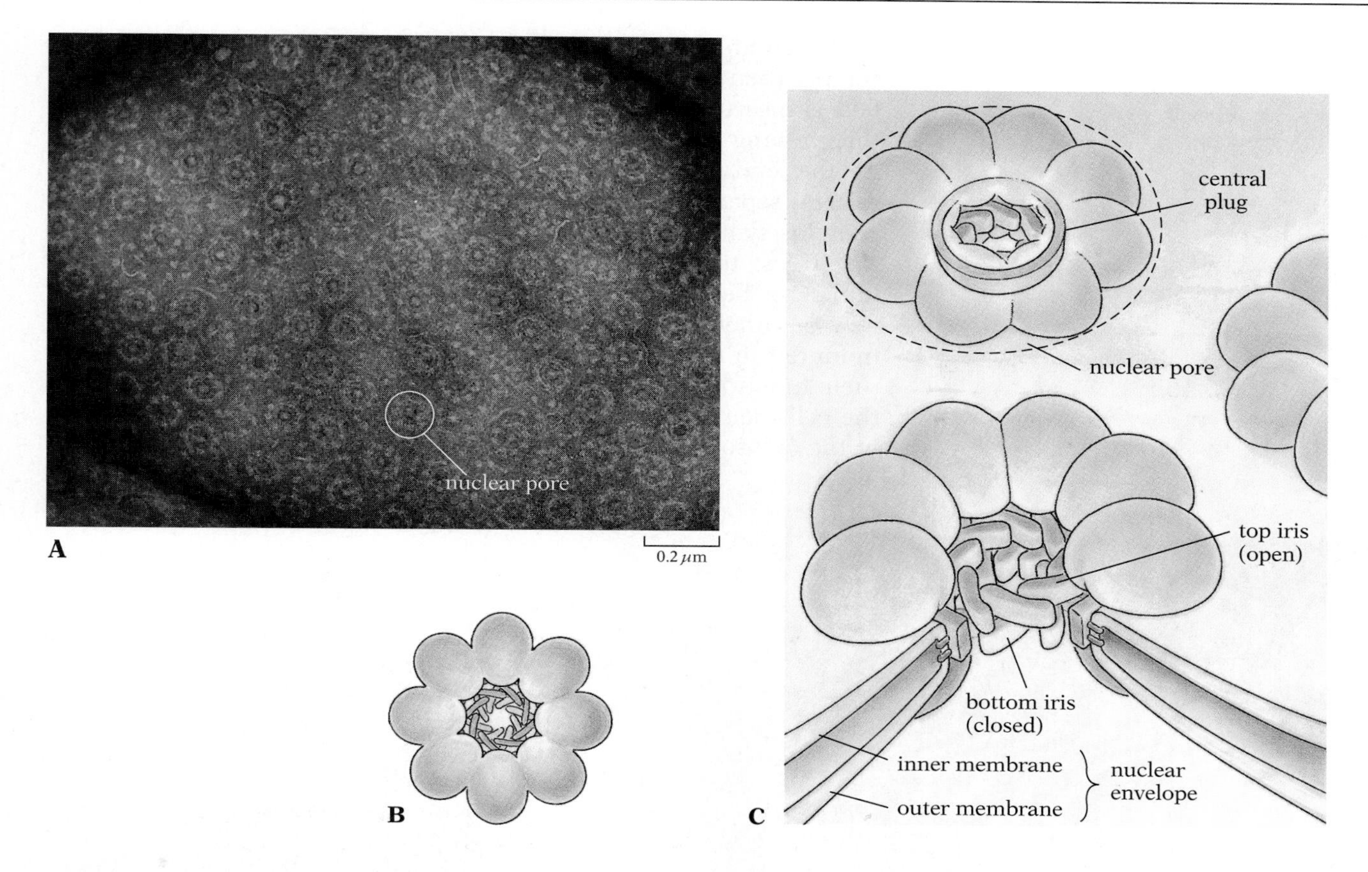

9.14 Nuclear pores

(A) The nuclear envelope contains many specialized pores, which provide a pathway from the nucleus to the cytoplasm for mRNA. The density of these portals is indicated by the micrograph of a small section of the nuclear membrane from a frog egg. (B) The probable structure of these octagonal pores includes a central plug, which appears to have two sets of eight protein arms, one on the inner face of the nuclear envelope, the other on the outer face. (C) The arms seem to act like an iris, opening to a diameter of 20 nm to admit correctly tagged macromolecules, and closing to 9 nm after they pass.

In eucaryotes, four steps intervene between transcription and translation. We have already mentioned three of these: modification of the transcript at the 5′ end, modification at the 3′ end, and splicing of the RNA after the removal of introns (see Fig. 9.7). The fourth step is movement through the pores of the nuclear envelope (Fig. 9.14) into the cytosol, where unbound ribosomes are available to do translation. Most RNA is translated by ribosomes in the cytosol, but the mRNA for certain proteins is always translated on the rough endoplasmic reticulum (ER). The mRNA of this type begins the process of translation in the normal way, by binding to a small ribosomal subunit in the cytosol and thereby making possible the binding of the large subunit and the start of translation. However, translation ceases almost immediately; the leading end of the partially synthesized protein contains a sequence that binds to a signal-recognition complex; this binding causes translation to stop.

The combined mRNA, recognition complex, and ribosome assembly then binds to the rough ER. The partially synthesized protein is inserted into a channel of the ER, translation resumes, and the growing chain of amino acids is fed into the lumen of the ER. At the end of translation, the ribosomal subunits release the mRNA, separate, and drift free of the ER. As detailed in Chapter 5, proteins synthesized in the rough ER are destined for noncytoplasmic roles: they may be soluble, and either drift free in the ER lumen, or be collected together in the ER and packaged into secretory or other vesicles (Fig. 9.15). Alternatively, they may become mounted in the ER membrane, either to remain there to perform their function, or to be packaged into vesicle membranes or into the cell membrane itself. Most proteins synthesized on ribosomes in the cytosol, by contrast, remain in the cytosol.

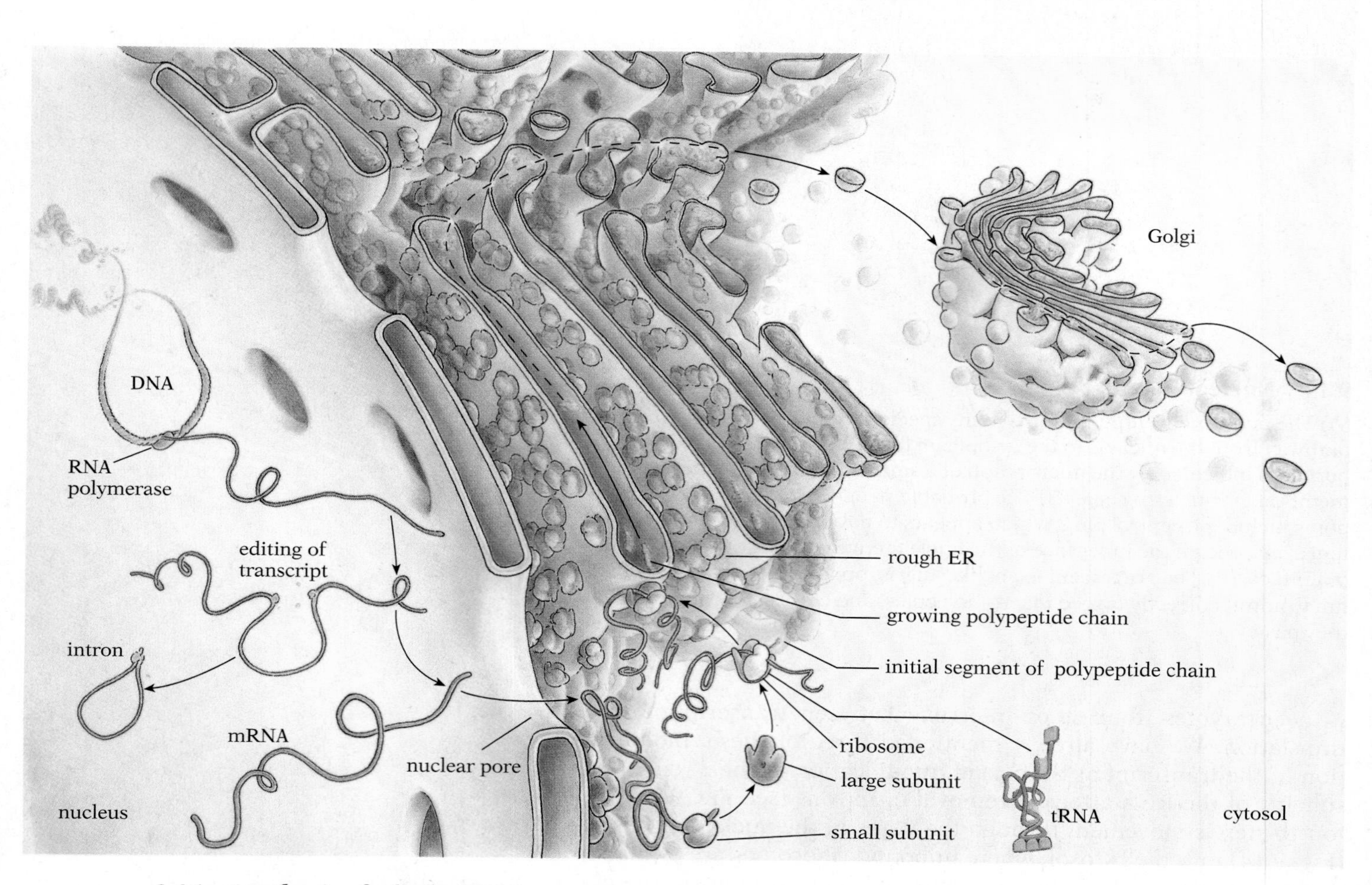

9.15 Synthesis of a hydrolytic enzyme

Most mRNA is translated freely in the cytosol, but when protein products are destined for the ER lumen, the ER membrane, or secretion, translation of mRNA occurs on the rough ER. A gene is transcribed, and the transcript is processed to form mRNA. Next, the mRNA passes through the nuclear pore (shown here without its plug) and binds to a small ribosomal subunit. A large ribosomal subunit then binds. Translation begins and continues until a signal sequence on the growing polypeptide chain binds to a signal-recognition complex that interrupts translation. The ribosome complex then binds to the rough ER and translation resumes, with the growing polypeptide chain passing into the lumen of the ER. Finally, the newly synthesized protein, in this case a hydrolytic enzyme, is packaged into a vesicle for transport to the Golgi apparatus.

■ THE NEED FOR TRANSFER RNA

In 1957 Mahlon Hoagland and his associates demonstrated that each amino acid becomes attached to some form of RNA *before* it arrives at the ribosome, where it is added to the growing polypeptide. This kind of RNA acts to transfer amino acids to the ribosomes; Hoagland and his colleagues called it transfer RNA. Each tRNA molecule binds a single molecule of amino acid, activates it with energy from ATP, and transports it to a ribosome.

At least one form of tRNA is specific for each of the 20 common amino acids; the amino acid arginine, for example, combines only with tRNA specialized to transport arginine. All forms of tRNA share certain structural characteristics: a length of 73–93 nucleotides, a structure consisting of a single chain that is folded into a cloverleaf shape, with internal base pairing (Fig. 9.16), and a CCA termination sequence. When an amino acid binds to its specific tRNA, it always binds at the CCA end.

Specific enzymes (aminoacyl tRNA synthetases) match and bind each amino acid with its appropriate tRNA (Fig. 9.17). The tRNA then carries its amino acid to a ribosome that has bound a strand of mRNA. There the tRNA becomes attached to the mRNA by complementary base pairing: each tRNA has an exposed portion containing an unpaired triplet of bases called an ***anticodon***, which is complementary to an mRNA codon for its particular amino acid. For example, the mRNA codon CCG codes for the amino acid proline. One type of tRNA for proline has the anticodon sequence GGC, which is complementary to CCG. When a molecule of the proline tRNA with an attached molecule of proline approaches a molecule of mRNA, its exposed GGC triplet can bind to the mRNA only where the mRNA has a CCG triplet.

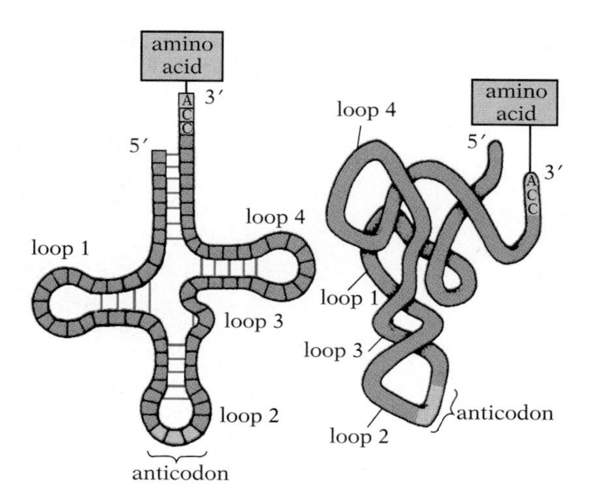

9.16 Structure of a molecule of tRNA

The single polynucleotide chain folds back on itself, forming five regions of complementary base pairing, four loops with unpaired bases, and an unpaired terminal portion to which the amino acid can attach. The unpaired triplet that acts as the anticodon is on loop 2. Loop 4 is thought to function in binding the tRNA to the ribosome (probably by binding to rRNA), and loop 1 probably binds an activating enzyme. The molecule is shown flattened at left; its tertiary structure is diagrammed at right.

■ THE TRANSLATION CYCLE

The actual binding of tRNA molecules and linkage of the amino acids they carry is managed in a stepwise fashion by the large subunit of the ribosome, but only after completion of the preliminaries: first, the tRNA with the anticodon sequence for methionine is bound to the small subunit; in eucaryotes as well as procaryotes this sequence is complementary to the initiation codon on the mRNA, which can now also bind to the small subunit; finally, the large subunit is bound to the small one. As translation then proceeds, the ribosome moves along the mRNA, with the part of the message being translated lying in the groove between the two subunits; two adjacent tRNA binding sites—the ***P-site*** and the ***A-site***—bring the appropriate nucleotide sequences together.

When the cycle begins, the anticodon of the first tRNA is in register with the initiation codon of the mRNA at the P-site (Fig. 9.18:4). The adjacent A-site then brings a molecule of tRNA with

9.17 Matching of an amino acid with its tRNA

The matchmaking enzyme responsible for loading a particular amino acid (glutamine in this case) onto its tRNA is shown in this computer-generated image in blue. It binds to the tRNA (red and yellow) at both the anticodon (bottom) and the acceptor arm (top center), the two regions of the tRNA responsible for its specificity. The green molecule is ATP.

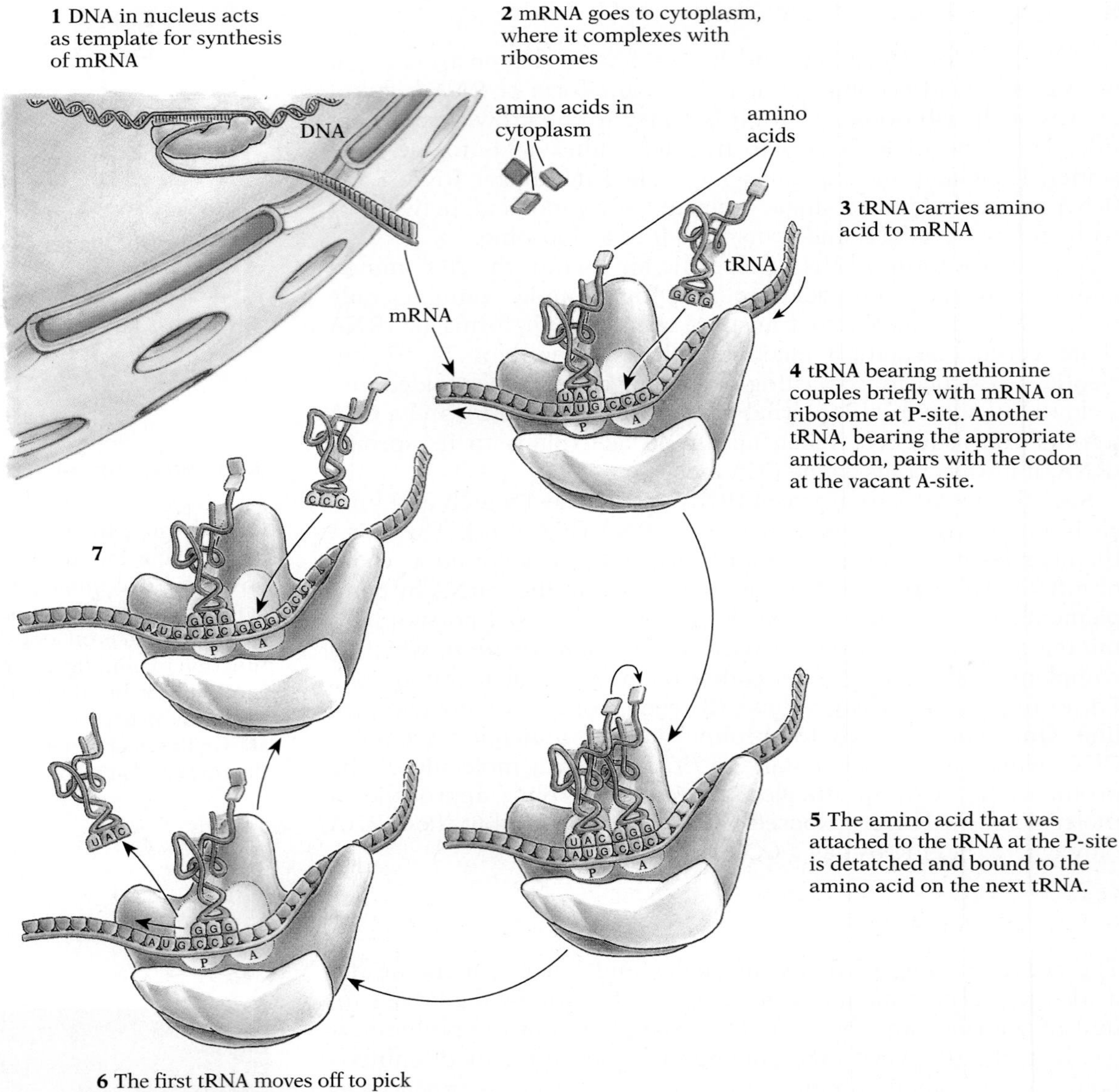

9.18 Translation cycle on the ribosome

After messenger RNA is created by transcription (1), it enters the cytoplasm (2), where it binds to a small ribosomal subunit. After a large subunit binds, translation can begin. At this point the tRNA with the anticodon for methionine (which was bound to the small subunit before the two subunits are combined) is already paired with the initiation codon on the mRNA at the P-site (4). Another tRNA (3), bearing the appropriate anticodon, pairs with the codon at the vacant A-site (5). Next the amino acid that was attached to the tRNA at the P-site is detached and bound to the amino acid on the tRNA at the A-site. The P-site tRNA then drifts free (6). Finally, the messenger is moved so that the remaining tRNA occupies the P-site, and the next codon is brought into position at the A-site, ready to accept another complementary anticodon (7). In these drawings part of the small subunit is shown cut away to reveal the mRNA, which lies in the groove between the subunits. The size of the tRNAs is greatly exaggerated.

the appropriate anticodon together with the next codon triplet on the mRNA (Fig. 9.18:5). Once the new tRNA is bound to the A-site, the enzyme peptidyl transferase moves the amino acid from the P-site tRNA and binds it to the amino acid at the A-site. The tRNA at the P-site, having relinquished its amino acid, is released (Fig. 9.18:6), and the cycle is completed as the ribosome moves along the mRNA by one codon, thereby bringing the codon that was at the A-site (and its tRNA, with the growing polypeptide chain) to the P-site (Fig. 9.18:7). As a result, the codon to be translated next now occupies the A-site.

The translation cycle adds about 15 amino acids per second to the polypeptide chain, with an error rate of about one mistake in every 30 chains. It comes to an end when the ribosome encounters one of the termination codons and, with the aid of a protein known as the release factor, causes the completed polypeptide to be released.

The ribosome complex's potential for producing protein is enormous. Since the average protein chain is 300–500 amino acids long, a ribosome can produce a chain in about 25–35 sec. Because many ribosomes can read a single mRNA at the same time, new peptide chains can be generated from each mRNA about every 3 sec. A typical active eucaryotic cell has on the order of 300,000 mRNA molecules in circulation; if sufficient raw materials are available, some 100,000 proteins can be synthesized every second. Cells specialized for secretion or rapid growth are far more active than this.

■ TRANSCRIPTION AND TRANSLATION IN ORGANELLES

Just as predicted by the endosymbiotic hypothesis (discussed in Chapters 5 and 19), transcription and translation in organelles are remarkably similar to the corresponding processes in procaryotes. For example, since there is no nuclear membrane to segregate a strand of mRNA from the ribosomes during its synthesis in an organelle, translation can begin at one end before transcription of the DNA has been completed at the other, just as it does in a procaryote (see Fig. 9.13). The resulting transcript receives neither the 7-methylguanosine cap nor the poly-adenine (poly-A) tail characteristic of mRNA produced in the nucleus.

Another similarity can be seen in the translation of the initiation codon, usually AUG. Eucaryotic tRNA pairs this codon with an unmodified methionine, whereas procaryotes begin translation with a tRNA carrying a special methionine—one to which a formyl group has been attached; organelles also begin translation with a formylated methionine. The rRNAs and ribosomal proteins of organelles are much closer in size and composition to those of procaryotes than they are to those elsewhere in the same cell (Fig. 9.19). Indeed, the ribosomes of *E. coli* and chloroplasts are interchangeable.

Despite their genetic similarity to procaryotes, organelles have many fewer genes. Mitochondrial DNA, for instance, usually encodes about 40 products, including two of the three rRNAs of the organelle ribosomes, all of their various tRNAs, one of the dozens of ribosomal proteins, one of the nine proteins of the F_1 complex

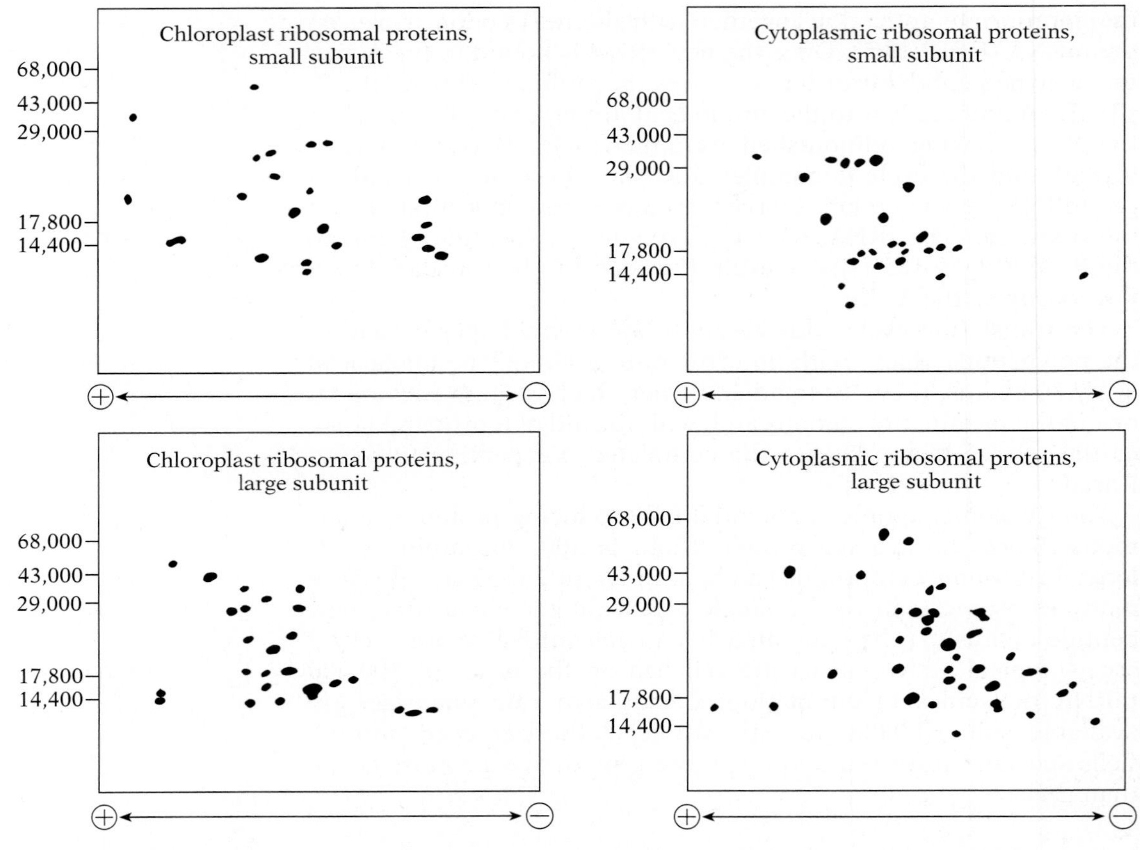

9.19 Proteins from chloroplast and cytoplasmic ribosomes of eucaryotes

The proteins from large and small subunits of organelle and cytoplasmic ribosomes have been dissociated and separated by molecular weight (vertical axis) and charge (horizontal axis). It is evident that organelle ribosomes (left) have fewer proteins than the cytoplasmic ribosomes of eucaryotes (right); furthermore, few if any of the proteins from the two sources are alike in weight and charge, an indication that they also differ in composition.

(which makes ATP, utilizing energy stored by the respiratory electron-transport chain), and several other organelle proteins. The other products used by mitochondria—the third rRNA; other ribosomal proteins; the replication, transcription, and translation enzymes; the electron-transport and other F_1 proteins, and so on—must be encoded by nuclear genes and transported to the organelles. The nucleus, therefore, contains not only all the genes for the cytoplasmic ribosomes, but nearly all the genes for mitochondrial ribosomes as well.

The abbreviated chromosome of mitochondria is probably an evolutionary consequence of intracellular competition. During early development, when a growing organism's cells are multiplying rapidly, any mitochondrion that can reproduce faster than others in the cell will contribute a greater proportion of mitochondria to the next generation of cells, including the ones destined to produce gametes in the adult. In time, natural selection favors such mitochondria, which come to dominate their slower-growing counterparts. The key to reproductive speed lies largely in reducing the number of genes that must be replicated. Selection, therefore, has favored mitochondria that have transferred as many genes as possible to the nucleus. The more puzzling question is why any genes at all are left in organelles.

HOW ERRORS ARISE

Mutations—alterations in the sequence of bases in the DNA—can change the information content of genes and thus produce new alleles. As discussed in Chapter 8, genes are subject to various mutational events. Here we will examine briefly some of the consequences of these events.

■ TYPES OF MUTATION

Two types of mutation that are particularly harmful are ***additions*** and ***deletions*** of single bases. These usually result in the production of inactive enzymes: at the point of the insertion or deletion the ribosomes begin to translate incorrect triplets, and the original meaning of subsequent codons is lost. Synthesis of the polypeptide chain may even be stopped too early if the shift in reading the codons has created a spurious termination signal, or stopped too late if the stop signal is misread. Alterations that result in the misreading of subsequent codons are called ***frameshift mutations***.

Another type of mutation is ***base substitution*** (also called point mutation), in which one nucleotide is replaced by another. For example, a codon that normally has the base composition CGG may be changed to CAG, which codes for a different amino acid. Base substitution is not as serious as addition or deletion, however, because in most codons a change in the third base does not alter the meaning (see Table 9.1). Roughly 30% of all base substitutions therefore have almost no effect.[5] Moreover, even when an amino acid is changed, the substitution usually has relatively little effect if, for example, one nonpolar amino acid replaces another. On the other hand, exchanges that occur between polar and nonpolar amino acids, or between positively charged and negatively charged amino acids, are likely to cause trouble. And alterations that create new prolines or cysteines, or remove old ones, are often serious since they may change the tertiary structure of the protein. All in all, just over half of all base substitutions are likely to affect protein function strongly.

Another very common type of mutation is ***transposition***. This phenomenon (discussed in detail in Chapter 10) results from the insertion of long stretches of DNA from one part of the genome into the middle of another. The majority of easily detectable spontaneous mutations in *Drosophila*, for example, result from transposition.

[5] There is some effect from a third-base mutation because the tRNAs for some alternative codons for an amino acid may be less numerous, less active, or less accurate in binding to the codon. In genes that are transcribed and translated at a high rate, selection favors the version of the codon that yields the best translation performance; for less frequently translated genes, there is little effect from a third-base substitution.

Table 9.1 The genetic code (messenger RNA)

FIRST BASE IN THE CODON	SECOND BASE IN THE CODON: U	C	A	G	THIRD BASE IN THE CODON
U	Phenylalanine	Serine	Tyrosine	Cysteine	U
	Phenylalanine	Serine	Tyrosine	Cysteine	C
	Leucine	Serine	*Termination*	*Termination*	A
	Leucine	Serine	*Termination*	Tryptophan	G
C	Leucine	Proline	Histidine	Arginine	U
	Leucine	Proline	Histidine	Arginine	C
	Leucine	Proline	Glutamine	Arginine	A
	Leucine	Proline	Glutamine	Arginine	G
A	Isoleucine	Threonine	Asparagine	Serine	U
	Isoleucine	Threonine	Asparagine	Serine	C
	Isoleucine	Threonine	Lysine	Arginine	A
	Methionine[a]	Threonine	Lysine	Arginine	G
G	Valine	Alanine	Aspartic acid	Glycine	U
	Valine	Alanine	Aspartic acid	Glycine	C
	Valine	Alanine	Glutamic acid	Glycine	A
	Valine	Alanine	Glutamic acid	Glycine	G

[a] Also *Initiation* when located at leading end of mRNA.

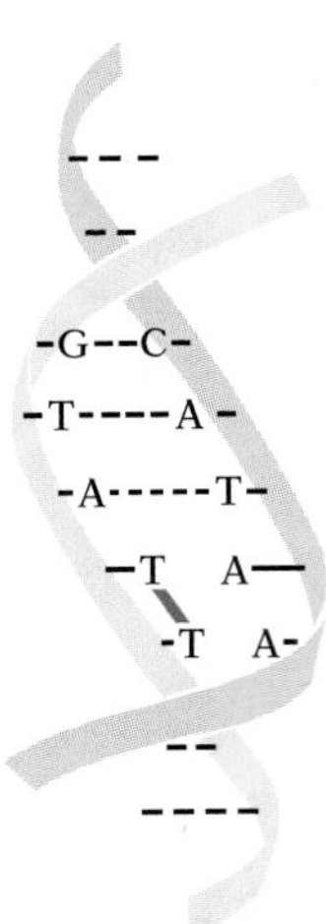

9.20 A thymine dimer within a DNA molecule

Two adjacent thymine nucleotides are bonded to each other covalently, so they cannot form hydrogen bonds with the adenine nucleotides of the complementary strand. Such a mutation, often induced by UV light, will inactivate the DNA if it is not repaired. Some species that normally live in the dark (for instance, the intestinal bacterium *Salmonella*) lack the enzyme that can repair thymine dimers.

■ ENVIRONMENTAL SOURCES OF MUTATION

High-energy radiation, including both ultraviolet light and ionizing radiation such as X rays, cosmic rays, and radioactive decay, can cause mutations. In addition, a variety of chemicals are mutagenic. A normal spontaneous mutation rate for a single gene is one mutation in every 10^6-10^8 replications, but the rate can be greatly increased by unusual exposure to mutagenic agents.

Ionizing radiation at various wavelengths sometimes induces simple base substitutions, but it also frequently produces large deletions of genetic material. It accomplishes the latter either directly by colliding with the DNA and causing breaks to occur, or indirectly by splitting water to produce highly reactive free radicals (single, unbonded, strongly electronegative oxygen atoms), which combine with and break the DNA.

Ultraviolet (UV) light most often exerts its mutagenic effect by causing abnormal bonding of thymine bases. When two adjacent thymine units absorb UV light and are thus energized, they can bond to each other, forming a ***thymine dimer*** (Fig. 9.20). An unrepaired dimer inactivates a strand of DNA; not only can no mRNA be transcribed from it, but DNA replication cannot take place.

Some mutagenic chemicals produce their effects by directly converting one base into another. For example, nitrous acid (HNO_2), a powerful mutagen found in cigarette smoke, deaminates (removes the NH_2 group from) cytosine, changing it into uracil; it also converts adenine into hypoxanthine and guanine into xanthine. Other chemical mutagens, called base analogues, are themselves some-

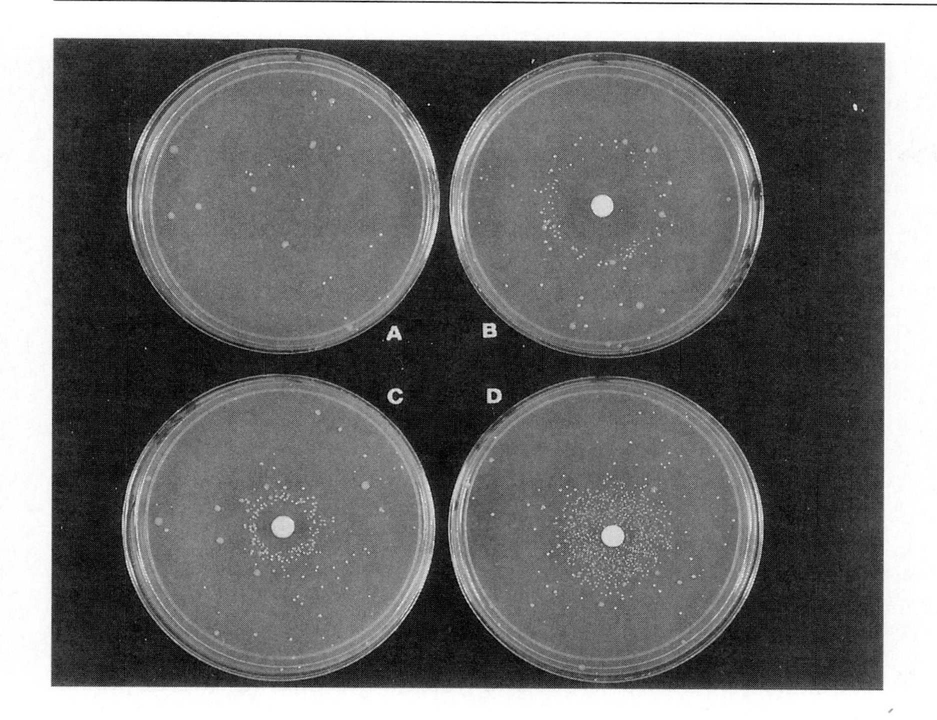

9.21 Ames test

Each of these Petri dishes has vast numbers of histadine-requiring bacteria in a layer of agar. The agar contains everything the bacteria need to grow except histadine. (A) Each white spot is a colony descended from a single bacterium in which a spontaneous mutation has restored the ability to synthesize histadine. (B–D) The large white disk in the center contains a mutagenic substance that is diffusing out from the disk: furylfuramide (B), aflatoxin (C), and 2-aminofluorene (D). The mutagenic potential of these chemicals is indicated by the large number of colonies able to grow as a result of a genetic change that restored their capacity to synthesize histadine.

times incorporated into nucleic acids in place of one of the normal bases. An example is 5-bromouracil, an analogue of thymine. When a strand of DNA contains a unit of 5-bromouracil, it is prone to errors of replication, because the 5-bromouracil will sometimes pair with guanine rather than with the requisite adenine. Virtually all of these mutations are detected and repaired before they can exert any effect.[6]

There is a close correlation between the mutagenicity of a chemical and its carcinogenicity—its cancer-inducing activity. This strongly suggests that many cancers are caused, at least in part, by mutations in somatic cells. Thus we see that alterations of the DNA are important not only in germ cells, where they may affect future offspring, but also in somatic cells, whose metabolism or growth they may disrupt, causing disease or degeneration.

The connection between mutagenicity and carcinogenicity is the basis for the test developed by Bruce N. Ames. Environmental pollutants, reagents used in industrial processes, proposed new drugs, and food additives are screened by the Ames test for potential carcinogens. The compound to be tested is added to a special mutant strain of *Salmonella* requiring the amino acid histidine as a nutrient. When the mixture of bacteria and chemical is incubated on a medium deficient in histidine, any cells that undergo a mutation that reverses their original mutation regain the ability to synthesize histidine and to grow on the deficient medium. A count of these so-called revertant colonies, each derived from one cell with a reverse mutation, provides a rough measure of the mutagenic potential of the chemical being tested (Fig. 9.21). An increase in the

[6] By flooding a cell with 5-bromouracil, it is possible to overwhelm the repair enzymes. This strategy is often used in cancer chemotherapy to kill rapidly dividing tumor cells. An unfortunate side effect is that some highly active normal cells—those producing hair, for instance—are also killed.

mutation rate in the bacteria over the normal spontaneous level predicts the likely cancer-inducing potency of the chemical in humans.

Many chemicals, though not themselves mutagens or carcinogens, are transformed in the body, especially by detoxification enzymes in the liver, into derivatives that are mutagenic and carcinogenic.[7] For this reason newer versions of the Ames test add liver homogenate to the bacterial culture, so that the bacteria will be exposed both to the original chemical and to its metabolic derivatives.

[7] As described in Chapter 31, the liver operates on the principle that unfamiliar chemicals are bad, and alters them with a variety of enzymes that prepare them for excretion; unfortunately, a few substances (like the pesticide contaminant dioxin) are thereby turned into carcinogens.

CHAPTER SUMMARY

HOW TRANSCRIPTION WORKS

THE NEED FOR MESSENGER RNA Messenger RNA (mRNA) carries genetic information from specific genes on the chromosome to sites of protein synthesis. (p. 225)

CREATING MESSENGER RNA RNA polymerase transcribes one strand of the DNA in a gene, creating an mRNA. The polymerase binds to a promoter region, begins transcription at a nearby start sequence, and stops at a termination sequence. In eucaryotes this creates a primary transcript whose introns must be removed before the mature mRNA can leave the nucleus. (p. 226)

HOW THE MESSAGE IS TRANSLATED

THE GENETIC CODE Nucleic acid base triplets (codons) encode the 20 common amino acids as well as signals for initiating and ending translation. (p. 231)

HOW RIBOSOMES WORK Ribosomes consist of two complex subunits, each a combination of rRNA and proteins. The smaller subunit binds to the mRNA, and then to the larger subunit. Translation begins with the first codon for methionine. The messengers for products destined for the ER have a sequence that causes their ribosomes to bind to the rough ER early in translation. (p. 232)

THE NEED FOR TRANSFER RNA Each amino acid is carried to the ribosomes while attached to a tRNA specific to that kind of amino acid; each tRNA has an anticodon sequence complementary to an mRNA codon for that amino acid. (p. 237)

THE TRANSLATION CYCLE The ribosome has adjacent P- and A-sites that bind sequential codons. The tRNA binds to the A-site; its amino acid binds to the amino acid attached to the tRNA already bound to the P-site; the P-site tRNA then detaches and the mRNA is shifted so that the codon formerly at the A-site is now attached to the P-site. (p. 237)

TRANSCRIPTION AND TRANSLATION IN ORGANELLES These two processes in organelles are very similar to those of procaryotes: there are no introns or nuclear membrane, and ribosomes can bind to mRNA while it is still being transcribed. (p. 239)

HOW ERRORS ARISE

TYPES OF MUTATION Additions and deletions of single bases create frameshift mutations that change the meaning of subsequent codons. Base substitutions can change individual codons. Transpositions rearrange pieces of genes. (p. 241)

ENVIRONMENTAL SOURCES OF MUTATION A variety of chemicals as well as ionizing radiation can induce mutations. Many cancers result from mutations. (p. 242)

STUDY QUESTIONS

1 Compare and contrast replication and transcription. (pp. 212–218, 226–229)

2 Compare the several types of RNA used in eucaryotic cells. (pp. 226, 229–233, 237–239)

3 Compare and contrast transcription and translation in procaryotes with the more complicated series of events in eucaryotes. How does the process in organelles fit into the picture? Is there any obvious reason for these differences? (pp. 225–236, 239–241)

4 Make a list of all the signal sequences (and the function of each) we have encountered so far, whether in DNA, RNA, or peptides. What others are likely to exist that have not been discussed yet? (pp. 226–239, 242)

5 In all, it takes four high-energy phosphate bonds to catalyze the several steps necessary to add each amino acid to a growing peptide chain. Given that the average protein contains 300–500 amino acids (the lower figure being more typical of procaryotes), how expensive (in terms of glucoses metabolized) is a protein, assuming that the amino acids are freely available? (p. 175)

SUGGESTED READING

CECH, T. R., 1986. RNA as an enzyme, *Scientific American* 255 (5). *On the enzymatic properties of ribosomal and especially snRNP RNA, and the possibility that RNA originally served the functions now taken over by DNA and protein.*

DARNELL, J. E., 1985. RNA, *Scientific American* 253 (4). *Reviews transcription, processing, translation, and transcriptional control.*

ALBERTS, B., D. BRAY, J. LEWIS, M. RAFF, K. ROBERTS, AND J. D. WATSON, 1994. *Molecular Biology of the Cell,* 3rd ed. Garland, New York. *Excellent detailed study of transcription and translation.*

LAKE, J. A., 1981. The ribosome, *Scientific American* 245 (2). *On the three-dimensional structure of the ribosome and the details of translation.*

STEITZ, J. A., 1988. "Snurps," *Scientific American* 258 (6). *On the enzymes that remove the introns from eucaryotic primary transcripts.*

CHAPTER 10

MOBILE GENES AND GENETIC ENGINEERING

Life depends on a dynamic balance between stability and change. Stability is essential for DNA replication, as well as for the transcription and translation of genes to produce RNA and protein. Change also is critical: natural selection can work only on the variation in a species, and a static genome would lead to the inevitable extinction of any species unable to evolve in the face of changing climate, resources, or competition.

Our emphasis in this chapter is on large-scale, dynamic alterations that can occur in the genome. By *large scale* we mean changes that involve an entire gene or major parts of it. Mutations involving single-base changes in functional genes play a smaller role in evolution, because such "tinkering" is unlikely to alter the gene product dramatically.[1] By contrast, the consequences of large-scale chromosomal changes range from cancerous growth in the individual to the potential for major evolutionary change in the species. Even for bacteria, whose sensitivity to mutational change is great and whose opportunities for major gene exchange and movement are limited compared with those of most eucaryotes, mutational alterations are thought to account for only a few percent of the evolutionary changes observed in the species.

Much of the work being done in genetic engineering today makes use of the natural agents of genetic mobility, especially the curious "accessory" chromosomes discussed presently. We will also look at viruses, whose unique reproductive strategies allow them to move new genes into target cells. Finally, we'll examine how gene mobility is orchestrated from within the genome, and how the enzymes responsible for mobile genes facilitate the techniques of genetic engineering, with their enormous potential for improving everything from food crops to human health.

THE DISCOVERY OF BACTERIAL PLASMIDS

Bacteria have one copy of their DNA instructions on a large circular chromosome; like all other organisms with only one set of instructions, they are classified as ***haploid.*** By contrast, sexually

[1] As described in Chapter 11, single-base changes in the sequences that *control* gene expression can have a significant effect on cell physiology by altering the *amount* of product synthesized.

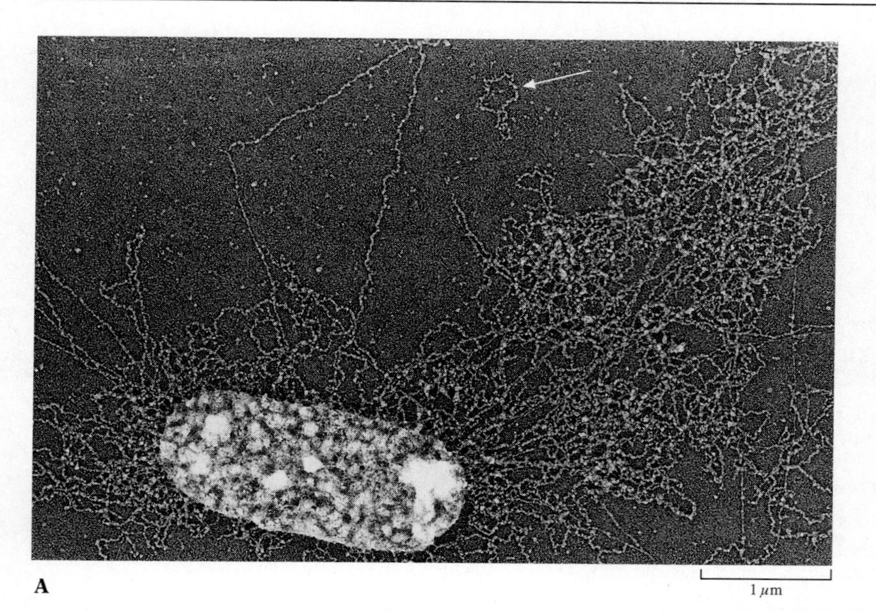

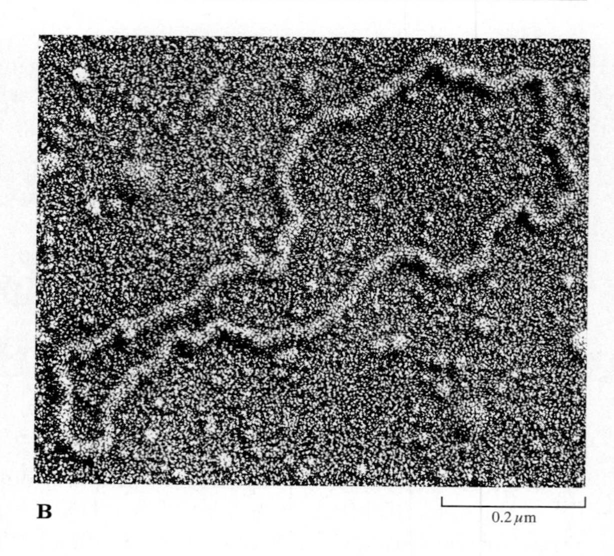

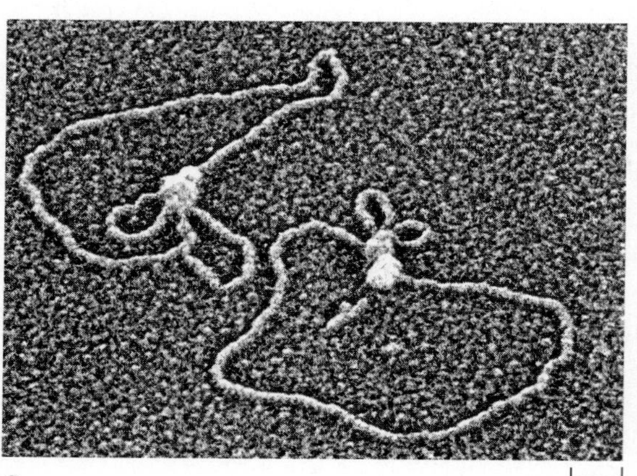

10.1 Electron micrographs of plasmids from *E. coli*

(A) A plasmid (arrow; top, center) is visible next to the main chromosome of this lysed *E. coli.* (B) The isolated plasmid seen here carries a gene encoding a product that confers resistance to the antibiotic tetracycline. (C) Plasmids are replicated independently of the main chromosome. The bacterial enzyme complex DNA polymerase can be seen copying each of these plasmids.

reproducing organisms, which have two copies of their genetic instructions in every cell except in their gametes, are ***diploid;*** gametes fuse to restore the duplicate set prior to fertilization. In 1946, however, Joshua Lederberg and Edward L. Tatum found that mating (of a sort) nevertheless does take place in bacteria. This led to the discovery of ***plasmids,*** small circular accessory molecules of DNA (Fig. 10.1), which are also found in eucaryotes ranging from yeast to mammals.

■ DO BACTERIA MATE?

Lederberg and Tatum isolated two mutant strains of *E. coli,* one of which was unable to synthesize the amino acid methionine or the vitamin biotin, while the second could not make the amino acids threonine or leucine. The mutants could be grown only on a medium that supplied the nutrients they could not synthesize. However, when the two mutant strains were mixed on a minimal medium (one lacking all four critical nutrients), a few healthy colonies grew on the deficient medium. The individuals in these colonies had somehow inherited both the first strain's ability to synthesize threonine and leucine and the second strain's ability to synthesize methionine and biotin.

The survival of these bacteria resulted from a ***recombination*** of traits from the two original mutant strains: genetic material from two different individuals had become combined in a single genome. Recombination occurs whenever two haploid gametes fuse to form a fertilized egg, which is thereby rendered diploid;

bacteria, however, are never diploid, and in most species a stiff cell wall precludes fusion. Bacterial recombination usually depends on ***conjugation.*** When two bacterial cells lie close to each other and other conditions are favorable, a narrow cytoplasmic bridge, or pilus, may be built from one to the other (Fig. 10.2). Genetic material passes through the bridge from one cell to the other.

Conjugation can occur only between cells of different mating types, designated F^+ ("male") and F^- ("female"). Unlike F^- cells, F^+ cells have a DNA plasmid containing a ***sex factor*** in their cytoplasm. Under most circumstances this factor replicates within nondividing bacteria, and a copy can be transferred from F^+ cells to F^- cells. The products of conjugation are always F^+. The population never becomes a pure culture of F^+ cells because conjugation is very time-consuming; thus an F^- cell can undergo one or more fissions while an F^+ cell engages in one conjugation. Hence, the proportion of F^+ cells can actually decline with conjugation.[2] Another problem limits the F^+ population: the pilus (which only F^+ cells can build) provides an attachment point for certain viruses, which thus kill only F^+ bacteria.

F^+ strains usually include a very few cells that, when they conjugate, may fail to transfer an intact sex factor to the F^- cells; instead they transfer chromosomal DNA. As a result, conjugation initiated by these high-frequency recombinant (*Hfr*) cells does not usually convert F^- cells into F^+ or *Hfr* cells. The difference between F^+ and *Hfr* strains is that in *Hfr* strains the plasmid is incorporated into the bacterial chromosome (Fig. 10.3). During conjugation, a copy of the main chromosome of the donor cell is moved through the pilus, but this delicate bridge rarely remains unbroken long enough for the sex factor (part of which is at the end of the chromosome) to be transferred. *Hfr* conjugation is important because it vastly increases chromosomal recombination.

The sex factor is unusual because at times it is free in the cytoplasm as a tiny circle of DNA, while at other times it is incorporated into the chromosome and behaves like other chromosomal genes. In a later section we will look at similar behavior in viruses and in movable elements in eucaryotic chromosomes called transposons.

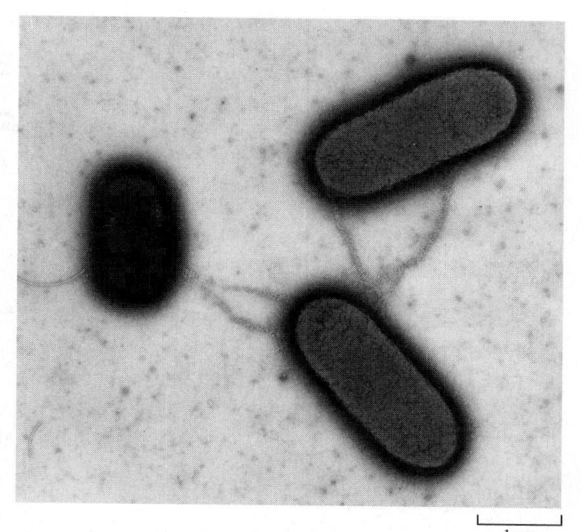

10.2 Conjugating bacteria

Long cytoplasmic bridges, or pili, connect the lower cell with two others with which it is conjugating simultaneously. The pili also provide attachment sites for certain viruses that can only infect F^+ (or "male") cells.

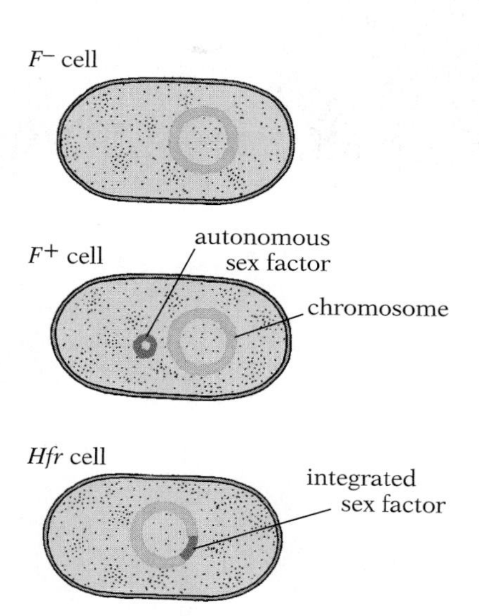

10.3 F^-, F^+, and *Hfr* cells of *E. coli* compared

F^- cells lack the sex factor. In F^+ cells the sex factor is a plasmid free in the cytoplasm. In *Hfr* cells the sex factor is integrated into the chromosome.

■ AUTONOMOUS PLASMIDS

Most bacterial cells contain some autonomous plasmids, which, unlike the sex-factor plasmid, are never integrated into the main chromosome. Some such plasmids carry only one or two genes, while others may be as much as one-fifth the size of the main chromosome and carry many genes.

Plasmids with genes for resistance to various antibiotics, including streptomycin, tetracycline (Fig. 10.1B) and ampicillin, are particularly important to modern medicine. When bacteria with resistance plasmids are exposed to one of these antibiotics, the plasmids begin replicating, and a cell that originally had only two

[2] This reproductive disadvantage is kept to a minimum because F^+ bacteria typically initiate conjugation only when the population is growing very slowly, perhaps because of a shortage of some critical nutrient.

or three plasmids may soon have a thousand or more. Plasmids may also carry genes that make the bacteria virulent, as in the case of the pneumococci that cause pneumonia. As we will see, recombinant DNA technology makes it possible to insert new genes into both autonomous plasmids and plasmids that are integrated into the host genome.

■ TRANSFORMATION

Some species of bacteria can take up pieces of DNA from their surroundings. (Many species, for instance, can be induced to take up foreign DNA when the cell walls of the bacteria are digested away and the bacteria are then maintained in an isotonic medium.) Imported DNA, often a fragment of the chromosome of a dead bacterium, may recombine with a similar region of the host genome. This necrophilic version of procaryotic sex, called ***transformation***, is a very important tool in genetic engineering.

HOW VIRUSES MOVE GENES

Viruses are tiny intracellular parasites with a miniature chromosome, a protective capsule, and (rarely) an enzyme. Their various reproductive strategies provide avenues for both natural and artificial transfer of genes.

■ HOW VIRUSES REPRODUCE

Lytic viruses In the 1940s Max Delbrück elucidated the reproductive cycle of a common bacteriophage. When he mixed phage with a culture of bacteria, the viruses seemed to vanish, but half an hour or so later a hundred times as many phage as he had started with suddenly appeared in the culture. The phage had entered the host cells, taken control of the cellular machinery, replicated themselves, and then caused the host cells to burst (lyse), releasing a new generation of infectious phage. The new viruses then attacked other bacterial cells and started the ***lytic cycle*** over again (Fig. 10.4A). Many techniques of genetic engineering take advantage of the willingness of viral-coat proteins to encapsulate nonviral DNA: the new genes to be delivered to other cells are introduced into infected cells; new viral caspules are built automatically around small pieces of DNA in the host, but in this case many of these genetic sequences are the nonviral fragments; and then the resulting viruses are used to infect new hosts.

Retroviruses A variety of different viral strategies are now understood. For example, many disease-causing viruses in animals have an RNA chromosome, which is both replicated and transcribed (Fig. 10.4B). The most extreme departure from the standard DNA- and RNA-based strategies is found in ***retroviruses.*** These parasites have an RNA chromosome, and carry an enzyme, ***reverse tran-***

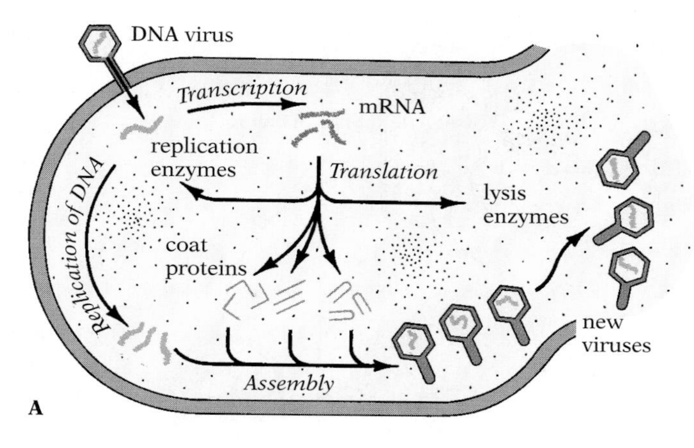

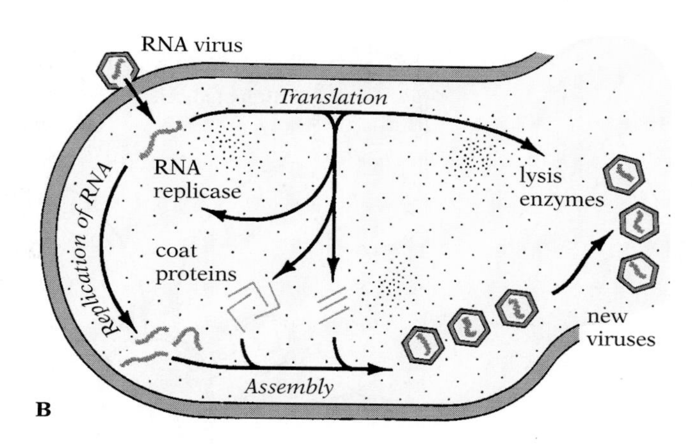

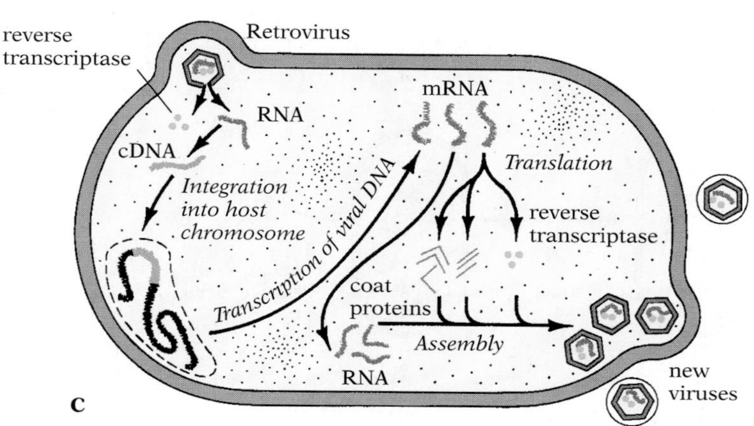

10.4 Three common strategies of viruses Viruses exploit their hosts in a variety of ways. The most frequent strategy is for a viral genome consisting of DNA to be both replicated and transcribed, and the transcripts translated, all by means of host enzymes. Transcription and translation generate coat proteins, enzymes that modify host-cell function, and enzymes that later lyse the host. (A) Other DNA viruses utilize replication enzymes produced through transcription and translation. (B) For RNA viruses the transcription step is unnecessary, but most encode a special enzyme, RNA replicase, to catalyze replication of the RNA. Some RNA viruses lyse their hosts, as shown, but many others escape from their hosts without lysing them. (C) Retroviruses are an unusual kind of RNA virus. Each retrovirus carries an enzyme, reverse transcriptase, that catalyzes the formation of a cDNA copy of the viral RNA; this copy is then incorporated into a chromosome of the host (often an animal cell). The host's enzymes then take over and accomplish replication (not shown), transcription, and translation of the viral genes. As the drawing indicates, retroviruses may enter the host cell by a kind of endocytosis, losing the protein coat only after entry; similarly, retroviruses generally do not lyse their hosts but instead leave by extrusion, a process whereby the virus is enveloped in a segment of the cell membrane as it emerges from the cell.

scriptase, that catalyzes the formation of a DNA copy of the RNA (known as ***cDNA*** because it is complementary to the RNA from which it is copied); thus they reverse the usual flow of information from DNA to RNA (Fig. 10.4C). Retroviral genes are usually inserted into the host chromosome, where they may become permanent residents.

Temperate viruses Some viral chromosomes act in a manner analogous to the F^+ plasmid: they can both enter and leave the host chromosome. This ability was first discovered when certain apparently normal strains of bacteria were exposed to ultraviolet (UV) light or X rays or various chemicals used to induce mutations. The bacteria would lyse within 1 hour, releasing large numbers of infectious phage: the bacteria had been carrying dormant viruses. Cells that harbor inactive viruses are said to be ***lysogenic.***

Viruses whose reproductive cycle includes an inactive stage inside their hosts are called ***temperate*** rather than virulent. Virulent phages invariably kill their hosts; temperate phages may or may not kill their hosts, depending on a variety of conditions. When they do not kill their hosts, their injected DNA usually becomes associated with the host chromosome at a particular location, and the ***lysogenic cycle*** begins (Fig. 10.5).

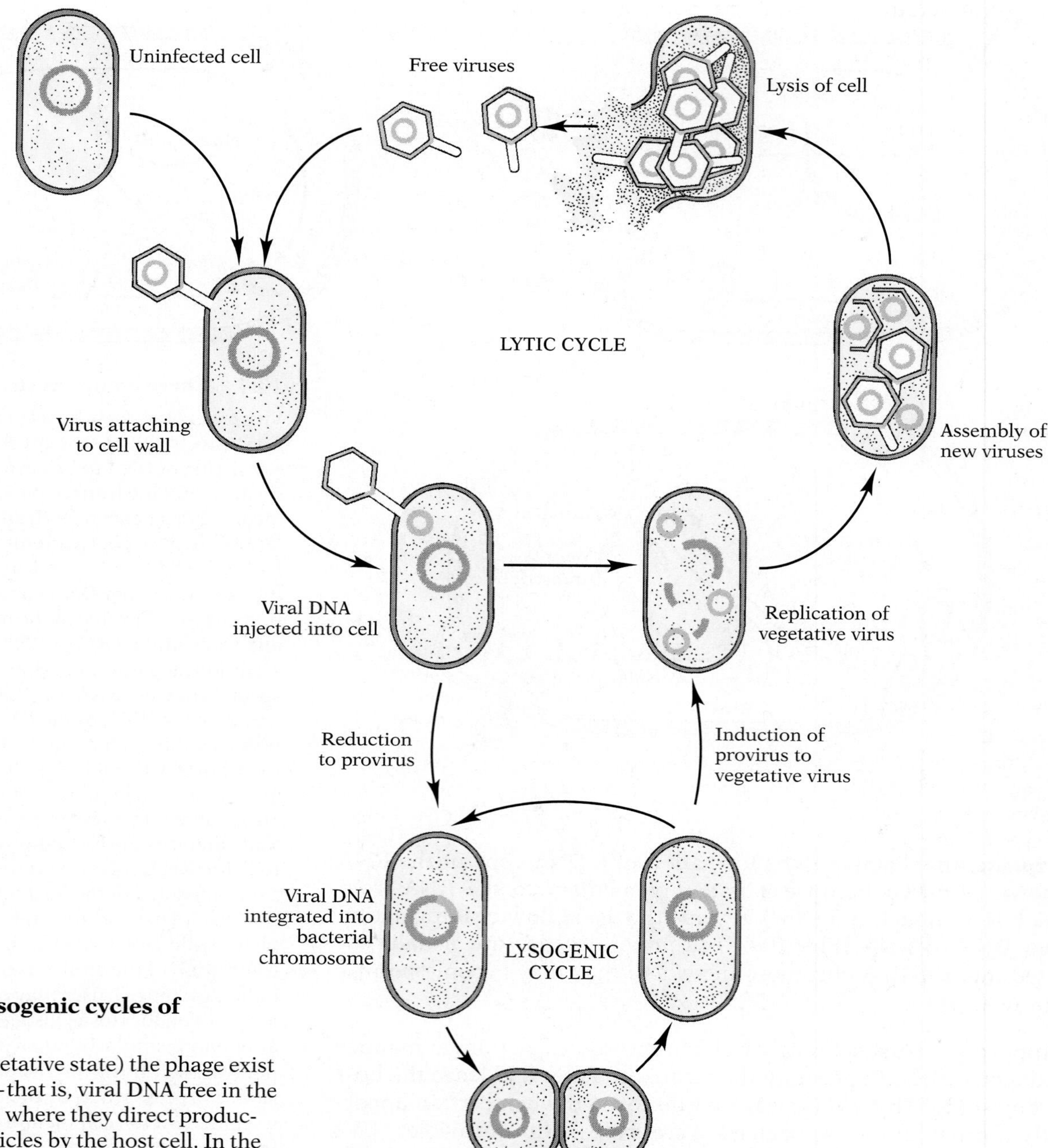

10.5 Lytic and lysogenic cycles of bacteriophage

In the lytic cycle (vegetative state) the phage exist only as free viruses—that is, viral DNA free in the host cell's cytoplasm, where they direct production of new viral particles by the host cell. In the lysogenic cycle the phage DNA is integrated into the host cell's chromosome, and only on occasion is it induced to break loose and initiate viral replication. (Bacteriophages are much smaller relative to their hosts than is indicated here.)

While integrated into the bacterial chromosome, the viral DNA, like the sex factor of *Hfr* strains, behaves as part of that chromosome. It is replicated with the rest of the chromosome, and it can be transferred from one cell to another during conjugation. Its genes can undergo recombination with bacterial genes, and they can even produce visible effects in the host bacterium, such as modifications of the morphology of the colony, changes in the properties of the cell wall, and changes in the production of enzymes. For example, diphtheria bacteria produce a toxin that causes the disease only if they are carrying a specific type of viral gene. Viral genes often make the bacterium in which they reside immune to further infection by the same type of virus, a phenomenon similar to the ability of the sex-factor genes to block the entry of additional F^+ DNA. Thus the DNA of temperate viruses can exist in an autonomous state, replicating independently and eventually destroying the cell, or it can exist in the integrated state, as a ***provirus,*** functioning and replicating as a part of the chromosome. Several human ailments, including cold sores and shingles, are caused by lysogenic viruses engaging in a bout of lytic reproduction.

■ VIRAL TRANSFER OF NONVIRAL GENES

Not only can some viral genes move into chromosomes, but the reverse can also occur. The phenomenon is best understood in bacteria. When temperate viruses are in the autonomous state and have put the bacterial cell to work making more viruses, small fragments of the bacterial chromosome may become enclosed in the new viral coats. If a temperate virus carrying bacterial DNA in this manner infects a new host, it injects bacterial DNA into the new host. Sometimes the injected bacterial genes undergo recombination with the new host's genes. The virus has thus acted as a vehicle for transferring genes from one bacterial cell to another, a process called ***transduction*** (Fig. 10.6).

Some of the genes that produce important observable effects in the human body were probably moved into human chromosomes by viruses. These genes may have been transferred from other humans or even from other species. This can happen because some viruses (notably those responsible for influenza) can infect other hosts, as in the "swine flu" epidemic of the 1980s. As described in the next chapter, some genes transported by viruses may be involved in cancer. Viruses may also have been involved in moving organelle genes into the nucleus of eucaryotes. In short, by moving entire genes, viruses can play a crucial role in creating the variation on which natural selection works in organisms.

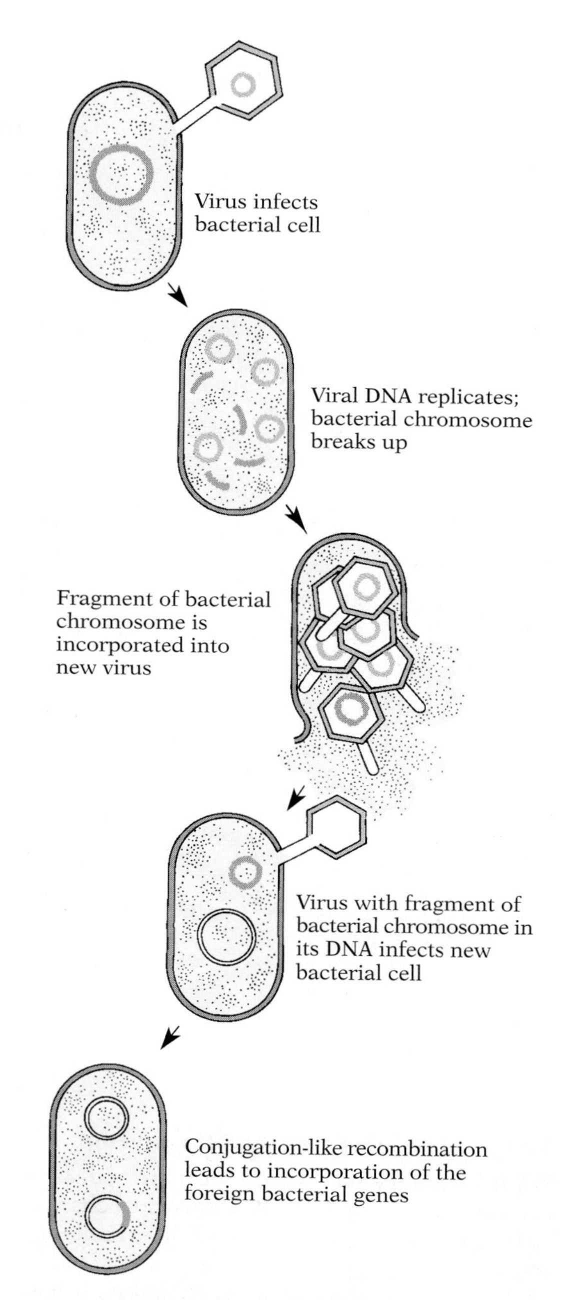

10.6 Model of transduction by bacteriophage

HOW GENES ARE MOVED WITHIN THE GENOME

While the movement of genes from one organism into the chromosomes of another is important, equally significant alterations can be caused by moving genes around *within* a genome. Since each gene's transcription is controlled by nearby sequences (including the promoter) that are not necessarily moved with the gene, the timing and degree of transcription can be significantly altered. Gene movement, therefore, can lead to dramatic changes.

■ THE BEHAVIOR OF TRANSPOSONS

Genetic units that move about in the genome, either by removing themselves to new locations or by duplicating themselves for insertion elsewhere, are called ***transposons***. They are found in all cells, procaryotic and eucaryotic, and can also insert themselves into plasmids. Transposition in eucaryotes was discovered more than four decades ago by Barbara McClintock. She found that in corn certain genetic elements will occasionally move, particularly after cells are subjected to trauma, such as exposure to intense UV radiation. These movements produced kernels with unusual colors that could not have resulted from normal recombination. Her results were so completely at odds with the prevailing concept of a static gene that the mobile genetic elements she had discovered were considered some sort of abnormality. By 1983, however, when she was awarded the Nobel Prize, many such transposons had been discovered, and their possible role in evolution was beginning to be recognized.

Transposons can consist of one or several genes, or just a control element, and can move in several ways, none of which is fully understood. Transposons within a chromosome are flanked by a pair of identical sequences, some of which are actually part of the transposon. Some transposons move from one site on the chromosome to another; others leave a copy behind; still others remain fixed but dispatch copies to other sites. A transposon that has physically moved leaves the flanking sequences as telltale scars (Fig. 10.7A); those that leave duplicates behind are copied in full by

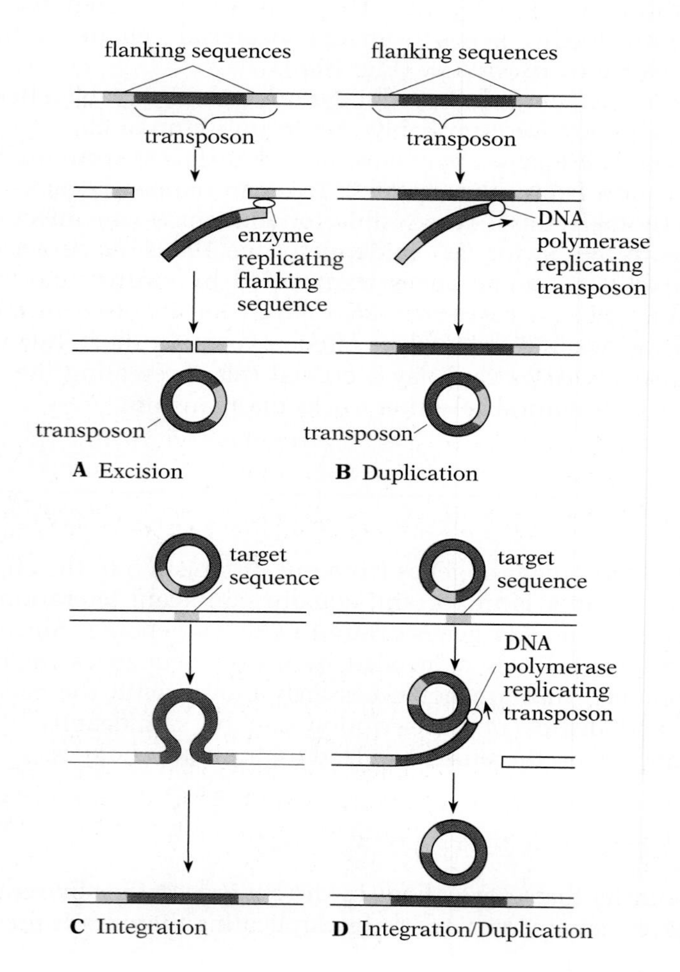

10.7 A model of transposition events
While integrated into a chromosome a transposon exists as a stretch of DNA flanked by characteristic sequences. At least one of these flanking sequences moved into the chromosome as part of the transposon, while the other may already have been there, and served as a target site. The movement of the transposon may involve either excision from the chromosome, leaving one or more target sequences behind (A), or duplication, with one copy of the transposon then left behind at the original location (B). In both cases, the mobile transposon probably exists as a circle, which can be integrated into any other part of the genome with (or, for certain transposons, without) the target sequence (C). The integration is frequently accompanied by a duplication, which produces another mobile transposon to continue the cycle (D).

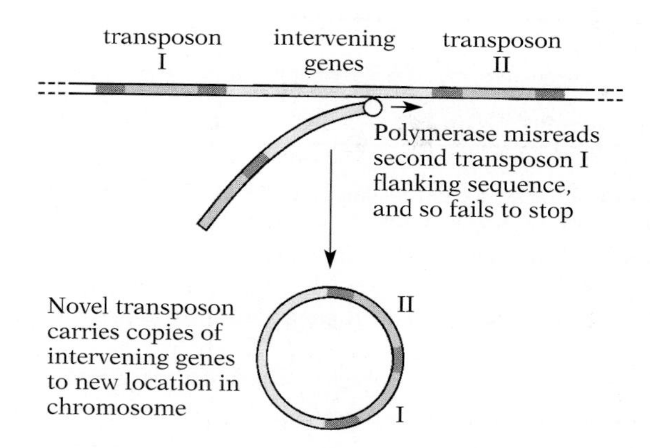

10.8 Model for capture of chromosomal genes by transposons

Chromosomal genes lying between two transposons are occasionally incorporated into a single hybrid transposon. A mutation in the second flanking sequence makes this incorporation inevitable.

DNA polymerase (Fig. 10.7B). Those that dispatch a copy can employ either DNA or RNA. In the case of an RNA copy the "retrotransposon" encodes reverse transcriptase. When such an RNA copy escapes to the cytoplasm, it is translated; the reverse transcriptase thus produced proceeds to make a cDNA copy for insertion back into the chromosome. Transposons are by no means rare: approximately 5% of the human genome consists of copies of a single kind—the L1 retrotransposon, present in at least 50,000 partial copies; this particular transposon, when it moves into a specific position on the X chromosome, appears to cause the most common form of hemophilia.

In many cases the receiving chromosome carries particular sequences that serve as targets for the mobile transposon. This sequence can pair with the one copy of the flanking sequence carried on the transposon. A special enzyme, usually encoded by the transposon itself, enables the transposon to recognize and act on the target sequence in the host chromosome. The transposon can then, with the aid of another enzyme usually encoded by the transposon, incorporate itself into the host chromosome. Other transposons have enzymes able to effect insertion anywhere in the genome. Like viruses and plasmids, transposons can sometimes pick up additional genes from the main chromosome (Fig. 10.8). There is an obvious similarity between transposons, temperate viruses, and those plasmids capable of incorporating themselves into a chromosome: each generally encodes the enzymes it needs to orchestrate its movements. The possible evolutionary relationships between these genetic entities are outlined in Figure 10.9.

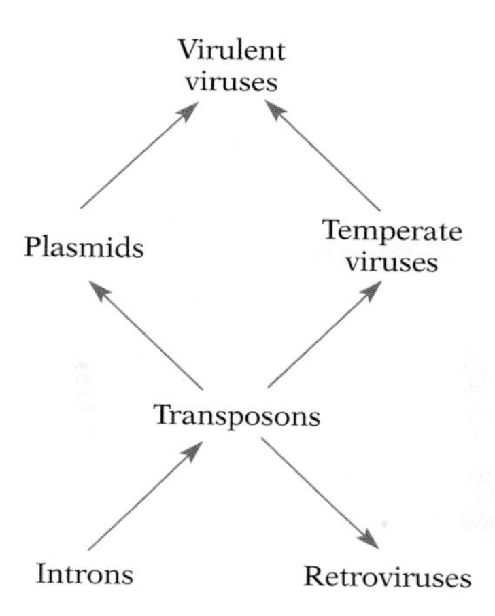

10.9 Possible evolutionary relationships between mobile genetic entities

Although it is not yet possible to trace the evolution of mobile genetic elements with certainty, it seems likely that the development of self-splicing introns may have been the first step. (The likely adaptive value of early introns is discussed in Chapter 17.) The next step would have been the evolution of a system to allow reinsertion of excised introns, producing transposons. Transposons could then have given rise to three new entities: (1) retroviruses, from the evolution of RNA transposons; (2) plasmids, from defective transposons, trapped in their independent circular state; and (3) the less benign entities we know as temperate viruses, from transposons that developed the ability to move between cells. Virulent viruses could have evolved from either plasmids or temperate viruses. Though not indicated here, at least some of these steps are reversible: temperate viruses, for example, could lose their ability to switch from one cell to another, and thus return to the status of transposons.

■ OTHER WAYS TO MOVE GENES

Although of less use in genetic engineering, there are other ways genes can move. These alternative mechanisms are important in understanding chromosomal organization, cancer, immunology, and evolution. Perhaps the most intriguing case is the occasional instance of retroinsertion, in which an mRNA is mistakenly bound by reverse transcriptase; it can then serve as a template for the creation of novel cDNA, which may then be incorporated back into the chromosome. The human genome has many such cDNAs, which are easily recognized from their lack of introns (missing in

Table 10.1 Sources of large-scale genetic change

PHENOMENON	MECHANISM	POSSIBLE CONSEQUENCES
INSERTION	Transformation	New genes from a dead cell imported from surrounding medium and incorporated into chromosome of bacterium
INSERTION	Transduction	New genes accidently picked up from previous host and imported into cell by a virus
INSERTION	Plasmid insertion	Existing gene or genes become integrated into genome, and subject to novel controls
INSERTION	Lysogenic insertion	Novel genes of temperate phage inserted into host genome
INSERTION	Retroviral insertion	cDNA copy of novel genes of retrovirus inserted into host genome
INSERTION	Intron insertion	Excised introns inserted into genome, mainly at exon-exon junctions in cDNA insertions
DUPLICATION, GENE MOVEMENT	Retroinsertion	cDNA copies of transcribed host DNA incorporated into genome, providing duplicate copies of genes
DUPLICATION, GENE MOVEMENT	Breakage and fusion	Part of one chromosome breaks off and fuses to the end of another during gamete formation; some gametes may obtain duplicate copies of genes on the broken fragment (discussed further in Chapter 16)
DUPLICATION, GENE MOVEMENT	Unequal crossing over	Chromosomes may be misaligned during a complex process called crossing over (described in Chapters 12 and 16); some gametes may obtain duplicate copies of some genes
GENE MOVEMENT	Transposition	Chromosomal DNA moved with genome, or both duplicated and moved

mRNAs after processing) and their telltale caps and poly-A tails (absent in normal genes, but added to mRNA during processing).

Genes can also be moved when chromosomes break and the pieces are reattached in the wrong places, or when errors occur during meiosis. Finally, introns removed during mRNA processing can sometimes act as templates for cDNA, and then be incorporated back into the genome. The range of possible mechanisms for gene movement is summarized in Table 10.1.

GENETIC ENGINEERING

■ ARTIFICIAL TRANSFORMATION

The agents of gene mobility, especially reverse transcriptase and plasmids, play a major role in ***recombinant DNA*** techniques for inserting selected genes from one kind of organism into cells of another kind. These techniques have great scientific, medical, and commercial potential. Specific genes can be isolated, introduced into bacteria, and used to produce large quantities of a desirable gene product such as insulin. Alternatively, new genes can be introduced into plants or animals that might benefit from them—genes for disease resistance in plants, for instance, or genes for growth hormones in domestic animals. However, recombinant DNA techniques have become a subject of public debate because of concern over the possibility of accidentally producing new pathogens, or developing "genetic monsters."

Plasmids, like DNA from bacterial chromosomes, can transform bacterial cells: if bacteria of certain species are placed in a medium containing free plasmids, some of the cells will pick up the plasmids. Thus nonresistant bacteria can be transformed into resistant ones by exposure to a medium in which bacteria with plasmids for antibiotic resistance have been killed. Recombinant DNA technology makes use of this transforming potential of plasmids: purified plasmids are modified by the addition of foreign genetic material, and when bacterial cells then pick up the modified plasmids, they acquire the foreign genes.

Let's follow the procedure in more detail: Bacterial cells containing plasmids are broken up, and their DNA is extracted. The DNA is then centrifuged to separate the plasmids from the main bacterial chromosomes. The purified plasmids are next exposed to a restriction endonuclease, one of a class of procaryotic enzymes that cleave the DNA circle at a particular nucleotide sequence (Fig. 10.10). The plasmid DNA is now linear, with "sticky" ends (unpaired bases) where it has been cleaved. The plasmid DNA is next mixed with fragments of foreign DNA prepared with the same restriction endonuclease and therefore equipped with sticky ends that are complementary to those of the plasmid DNA. In such a mixture, under the appropriate environmental conditions (temperature, pH, and so on), the plasmid DNA and the foreign DNA spontaneously anneal by complementary base pairing, re-forming a circle in the process (Fig. 10.11). The backbones (chains of alter-

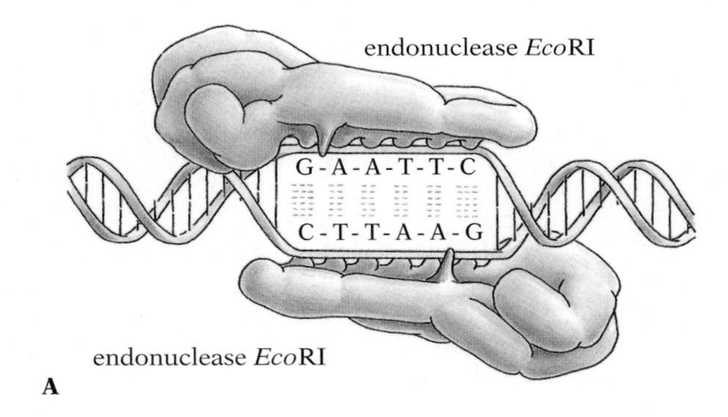

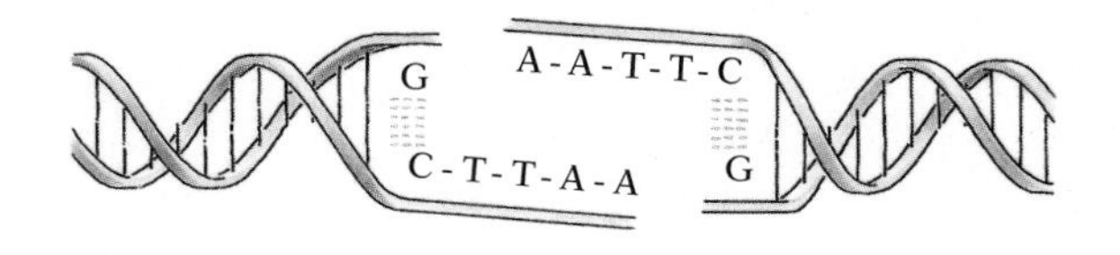

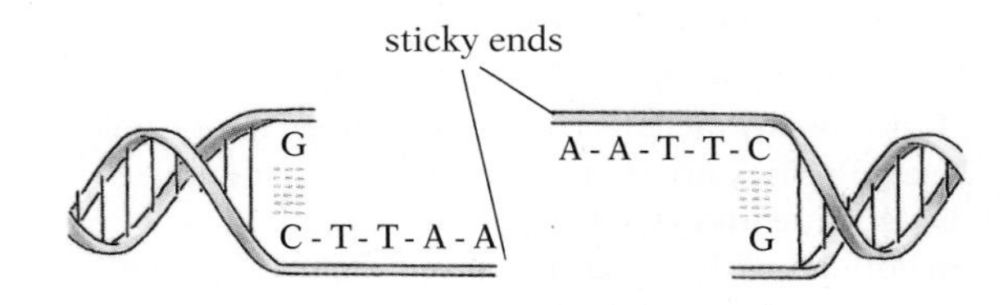

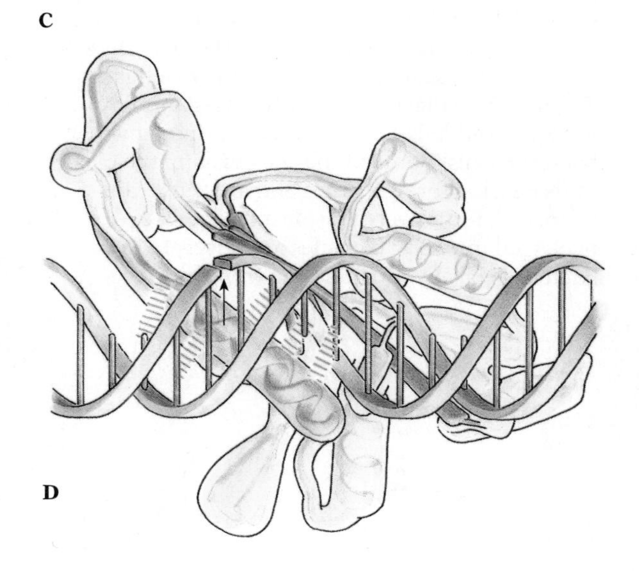

10.10 Endonuclease action

(A) Restriction endonucleases bind to a pair of target sequences—GAATTC in the case of the enzyme *Eco*RI. (B) They break a particular phosphate bond in the backbone of each strand of the DNA. (C) The resulting cut ends with unpaired bases are called sticky ends because under favorable conditions they will pair with cut ends having the complementary sequence of bases. (D) A more detailed look at a restriction endonuclease. One *Eco*RI attaches to a target sequence; the other endonuclease is omitted for clarity. The blue portion of the enzyme is important for recognizing the target sequence; the binding is indicated by dashed red lines. The solid red portion of the enzyme is involved in weakening the bonds in the DNA; the arrow indicates where the helix is cut. More than 100 endonucleases are known, each with its own specific target sequence. The normal function of endonucleases is to digest infecting viral DNA.

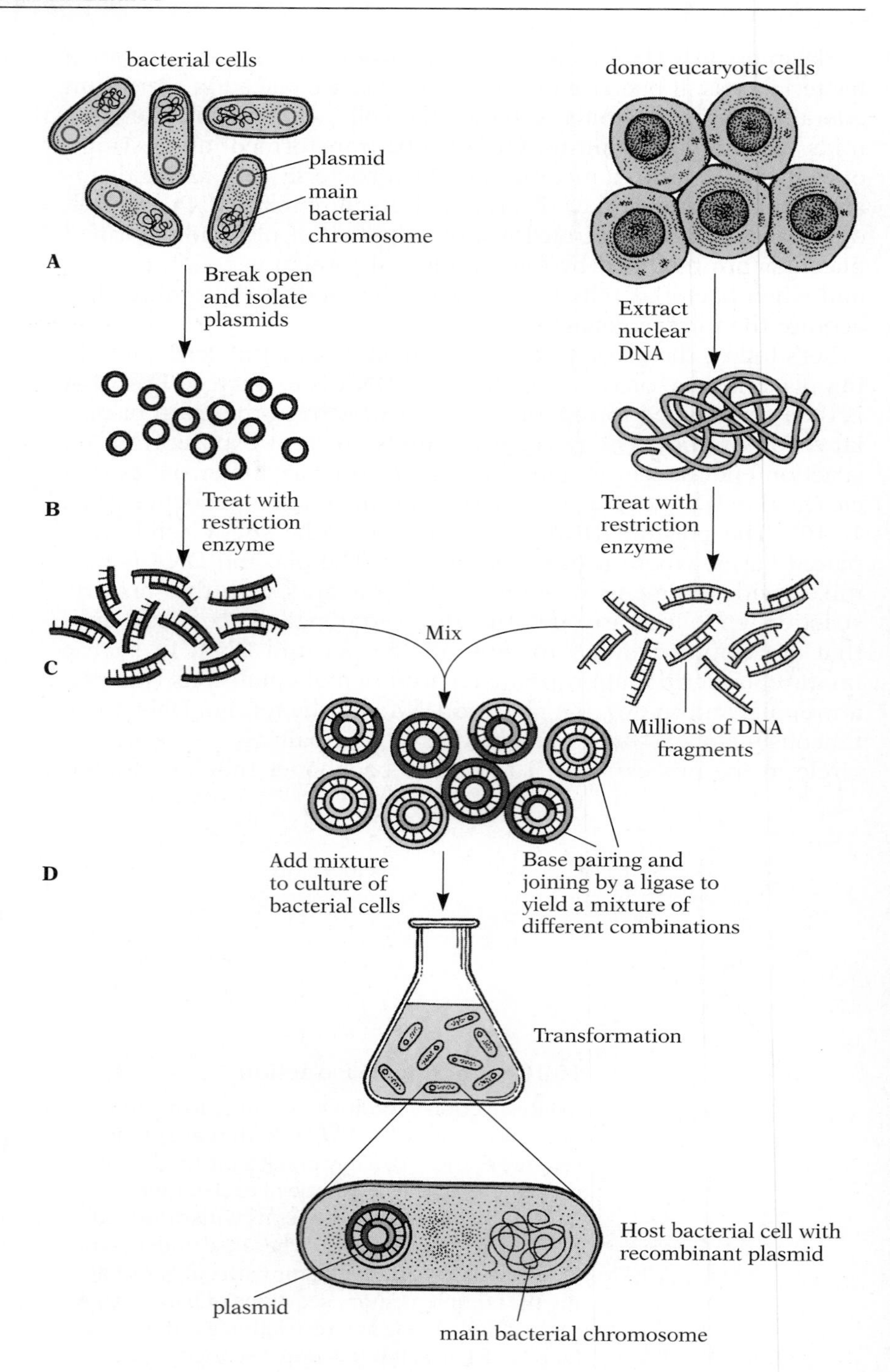

10.11 A recombinant DNA technique using transformation

Plasmids (shown here in red), removed from donor bacterial cells (A), are cut by an endonuclease (B). They are then mixed with fragments of DNA (blue and yellow) from other cells, produced with the same endonuclease (C). The plasmid DNA and the foreign DNA can join at their sticky ends by complementary base pairing, re-forming a circle in the process (D). The DNA ends are then sealed by treatment with a ligase. The modified plasmids, some bearing foreign genes, are added to a medium containing live bacterial cells. Some of the cells pick up the plasmids bearing foreign genes and are thus transformed. Alternatively, the plasmids can be packaged into viral capsules and inserted in the bacterial cells by transduction (not shown).

nating phosphate and sugar groups) of the DNA circle can then be sealed with the enzyme DNA ligase. The end products are plasmids that contain a graft of foreign genetic material.

These hybrid plasmids are then mixed with bacterial cells (usually treated with a calcium salt to make them more permeable). The bacterial cells pick up the modified plasmids. For more efficient transfer, transduction can be used: the hybrid plasmids are mixed with viral coat proteins, which will self-assemble around the plasmid if it is not too large; these plasmid-carrying viruses are

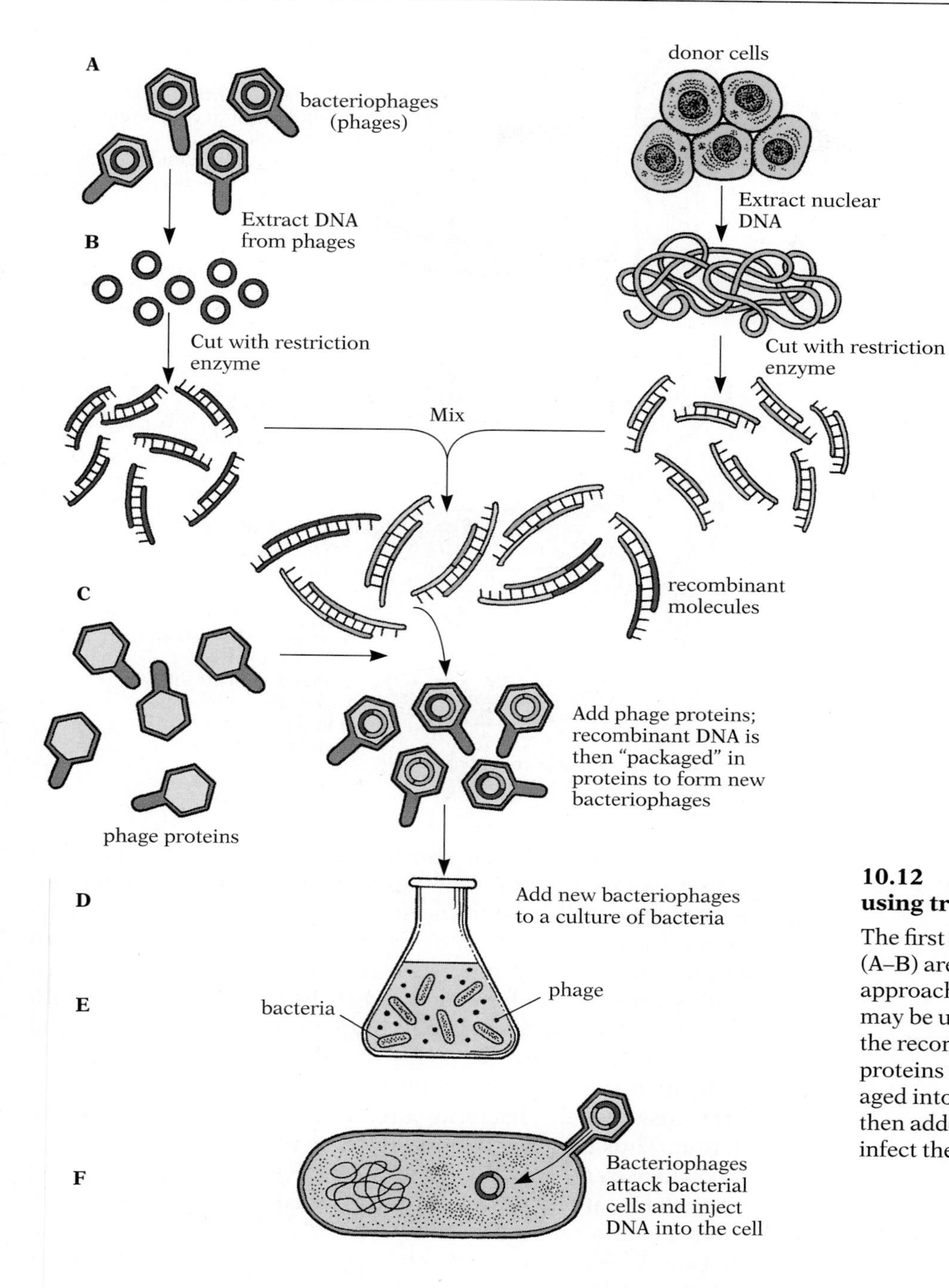

10.12 A recombinant DNA technique using transduction

The first steps of the transduction technique (A–B) are similar to those of the transformation approach (see Fig. 10.11), except that viral DNA may be used instead of a bacterial plasmid. After the recombinant molecules are obtained, viral proteins are added, and the hybrid DNA is packaged into new viral particles (C). The viruses are then added to the host culture (D) where they infect the target cells (F).

used to infect target cells (Fig. 10.12). In general, plasmids containing genes that confer antibiotic resistance are used; the experimenter then treats the bacterial cells exposed to the recombinant plasmids with antibiotic, and only the bacteria that have incorporated hybrid plasmids survive. As the plasmids reproduce, the foreign DNA is copied, or ***cloned***.

One disadvantage of this so-called genomic-cloning approach is that the desired combination of plasmid and foreign DNA is not the sole result. After endonuclease treatment, the foreign DNA is left as a mixture of various fragments, of which only one may be of interest. These genomic-clone fragments are frequently used as

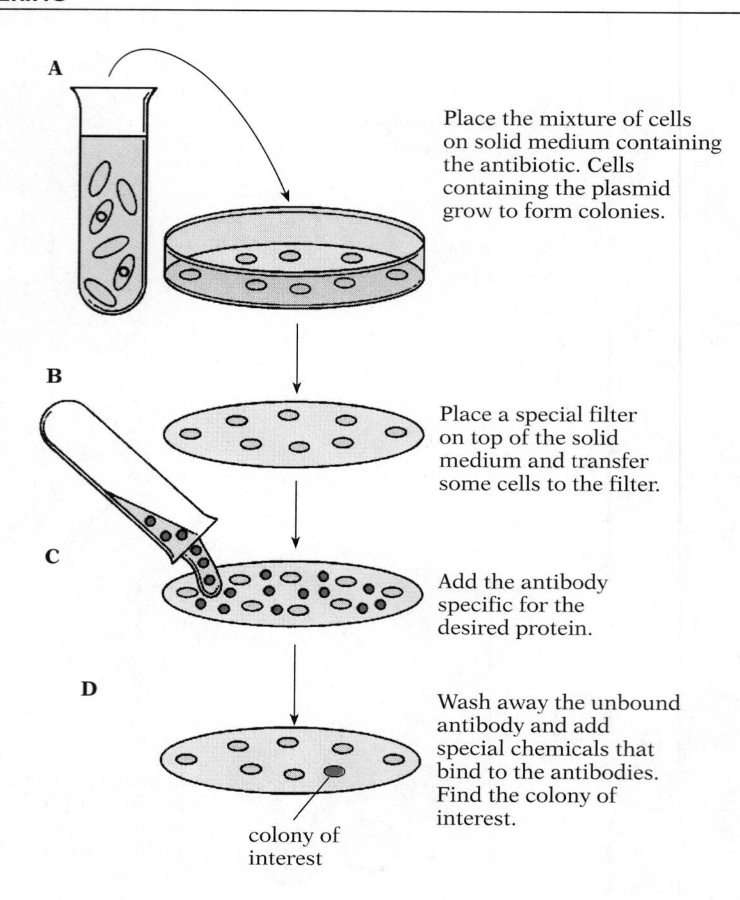

10.13 Screening for cells with recombinant DNA

There are several ways to locate colonies with recombinant plasmids carrying a desired gene. (A) In many of them the treated cells are plated into a dish and allowed to grow. (B) A piece of special filter paper is pressed against the dish, and thus picks up a few cells from each colony. (C) One method then requires the cells on this filter-paper "replica" to be exposed to antibodies specific to the gene product being sought. (D) After washing away the antibody molecules that fail to bind to the cells containing the desired gene DNA of the killed cells, chemicals are added that stain the antibodies. This stain indicates the location of the colonies in the original dish containing the desired gene; these colonies are removed from the dish and cultured. (The way in which antibodies bind to specific proteins is described in Chapter 15).

probes for locating genes on chromosomes, but are less useful in generating working genes. One potential problem is that the endonuclease may well have found its target sequence within a particular gene, and so cut the gene itself. Repeating the process with a different endonuclease may yield fragments with a gene intact. Rigorous screening is often required to isolate the bacteria with plasmids that have incorporated complete copies of the desired genes (Fig. 10.13). Worse yet, if the foreign DNA is from a eucaryote, as is often the case in recombinant DNA research, it will contain introns, which bacteria are unable to remove before translation. Thus the proteins produced by the bacteria include the introns, which renders the products inactive. To avoid these and other problems, many researchers use yeast, a simple eucaryote. Other researchers rely on more selective techniques to isolate and incorporate desirable genes into bacteria. One of these techniques is cDNA-based cloning.

■ HOW SPECIFIC GENES ARE CLONED

Recombinant techniques The crucial step in one method of mass-producing clones of a particular eucaryotic gene is to find cells that specialize in manufacturing that gene's product—pancreatic cells, for instance, if the desired product is insulin. The cytoplasm of such cells will have a high concentration of mRNA molecules coding for their special product, and these mRNAs will already have undergone intron removal in the nucleus. Many techniques exist to separate out the particular kinds of mRNA on the basis of physical characteristics like weight, so that the mRNA found in unusual abundance in the specialist cells can be isolated. A variety of other tricks are available to identify the mRNA of genes that are only rarely transcribed. (Another method of targeted gene cloning, the polymerase chain reaction, is described in the *Exploring Further* box on page 262).

Once the appropriate mRNA has been isolated, the next step is to produce from it the corresponding single-stranded DNA, using reverse transcriptase from a retrovirus. DNA polymerases are then used to replicate the DNA strand by complementary base pairing, a process that supplies the second strand for the double-helical structure of the transcript. The procedures already discussed for inserting foreign DNA into plasmids are then brought into play: a restriction endonuclease cuts open the plasmid, creating a pair of sticky ends; the cloned DNA with a complementary set of sticky ends is added, and the plasmid DNA and cDNA anneal; a ligase restores the bonds in the DNA backbone; and the plasmid is inserted into a bacterial host.

This procedure has two immediate advantages: the expensive process of sorting for bacteria bearing the desired gene is eliminated, since only the appropriate cDNA (in this case, the cloned DNA) is used; and introns, which cannot be removed by procaryotes, have already been eliminated in the production of the mRNA. The transformed bacterial cells grow and divide rapidly, creating limitless numbers of bacteria that may synthesize the desired product, particularly if the host plasmid has signals that enhance transcription of the cDNA. This technique is used to produce other hormones besides insulin that are difficult to synthesize, notably growth hormones. Recombinant bovine growth hormone is widely used to boost milk production by 10–40%, and synthetic human growth hormone is employed to prevent stunted growth in hormone-deficient children. A naturally produced but poorly understood agent called interferon, which sensitizes the immune system and so holds great promise in the treatment of various diseases, is also now widely available for research and medical applications as a result of recombinant DNA technology.

Recombinant techniques have also made possible several new approaches to vaccination. The one that seems most promising for finally defeating the common cold takes advantage of the need of viruses to locate and bind to the exposed portion of a particular membrane protein in their host's cells. Researchers can clone the portion of the gene that encodes the extracellular part of the target protein, and then mass-produce this fragment. When the body of a potential victim is flooded with these molecular decoys, the viruses attach themselves to the free-floating target segments, and so render themselves harmless (Fig. 10.14). Years of testing, however, will be needed before such treatments become widely available.

Gene cloning makes it possible not only to use the host cells as chemical factories that produce substances of medical or commercial importance, but also to study the sequencing and activity of genes from eucaryotic cells. Indeed, one of the many spinoffs from recombinant DNA technology is a method of mapping genes with enormous precision. A piece of single-stranded DNA is transcribed from an mRNA of known function in a medium containing ra-

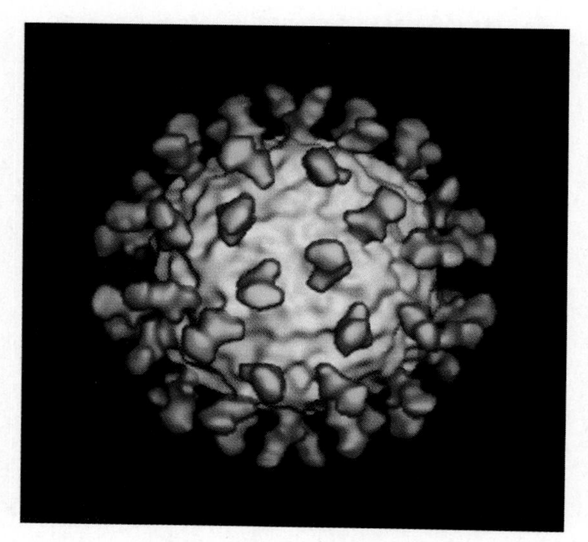

10.14 Inactivating a cold virus with recombinant proteins

Rhinoviruses, which are responsible for about half of the cases of colds, gain entry to the cells they infect by first attaching to cell-adhesion molecules (CAMs) mounted on the membranes of many types of cells in the body. (The role of CAMs in development is described in Chapter 14.) By flooding the bloodstream with the small part of the CAM to which the virus binds (red), the pathogen is rendered harmless.

EXPLORING FURTHER

THE POLYMERASE CHAIN REACTION

Another way of making a vast number of copies of a gene depends on knowing the base sequence flanking each end of the region of interest. Many copies of short DNA primers complementary to the two known flanking sequences are then synthesized, one for the 3′ end of the region on each strand. The gene's chromosome is cut into fragments with an endonuclease and heated to denature the strands, and the two primer sequences are added. Because the primers greatly outnumber the chromosomes, they are far more likely to bind to the 3′ flanking target sequence on a strand than is the complementary strand of the chromosome.

After the mixture has cooled slightly, a DNA polymerase from a species of bacterium adapted to life in the near-boiling waters of hot springs is introduced. The polymerases find the primers and begin synthesizing a complementary copy of each fragment. When enough time has passed for the polymerases to finish their work, the mixture is again heated to denature the two strands of each fragment. Once more, free primers bind to the flanking sequences, now present on four strands (the two original strands of the fragment with the gene the researchers want, plus two copies), and the mixture is again cooled to allow the polymerases to do their job. With each cycle of heating and cooling, the number of copies of the original fragment doubles, generating a million replicas in as

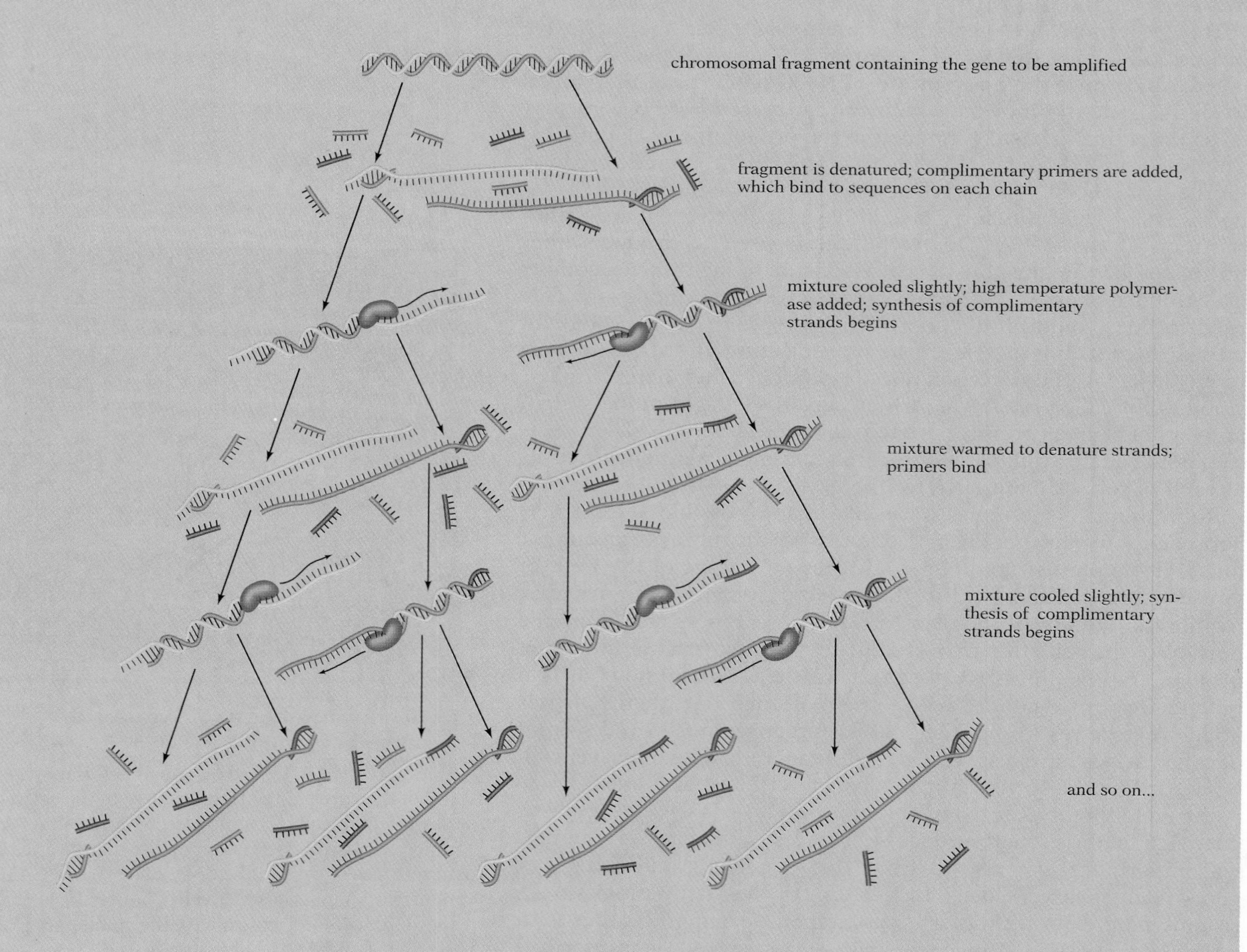

little as 8 hours.

This polymerase-chain-reaction (PCR) technique allows unlimited amplification and analysis of genes even if there is only a single copy present at the outset. In addition to the many obvious applications of the procedure to the study of genetic mechanisms, PCR has been used for a number of unusual tasks, ranging from the identification of criminals based on the DNA in a single hair, spot of blood, or drop of semen, to studying the evolutionary relationships of extinct organisms using nothing more than a minute quantity of surviving mitochondrial DNA in museum specimens or fossils. More recently, it was used to analyze whale meat sold in Japan; analysis revealed that a substantial proportion of the meat came from protected species, including fin, minke, and humpback whales—animals that had been killed and marketed illegally (and, until the invention of the PCR technique, with little risk of detection).

dioactive phosphorous (^{32}P). This cDNA is then mixed with a set of chromosomes treated to separate the double helix into single strands. When base pairing is made possible in the resulting mixture, the cDNA frequently pairs with the corresponding gene on the chromosome (Fig. 10.15). Because the cDNA carries a radioactive label, the exact location on the chromosome of the gene to which it binds is revealed (see Fig. 11.6, p. 284).

Prospects for the future Scientists have begun the complete mapping and sequencing of the human genome's 50,000 genes. (See *Exploring Further* box: DNA Sequencing.) This may make possible many kinds of therapeutic intervention, and cast considerable light on chromosomal organization and evolution. One of many uses of a complete map of the human genome may be to create DNA sequences complementary to a portion of the mRNA of a gene whose activity needs to be brought under control—a gene involved in cancer, for example (described in detail in Chapter 11). This cDNA—also known as antisense DNA—binds to the corresponding mRNA and blocks its translation (Fig. 10.16).

Other spinoffs of recombinant DNA research include techniques for (1) identifying nucleotide sequences in DNA quickly and precisely, (2) rearranging genes and their components to study gene interaction and gene control, and (3) creating specific mutations at specific sites in genes. This third procedure makes possible precise investigation of the roles of signal sequences in DNA and mRNA, and also enables the researcher to alter proteins at will to explore the bases of enzyme function.

Looking into the future, some researchers have suggested using cDNA technology to transfer genes for the fixation of atmospheric nitrogen to crop plants. This would be a formidable undertaking, but would eliminate the need to apply nitrogenous fertilizers. Researchers are also working to introduce the genes for proteins rich in essential amino acids into grains or beans. The result would

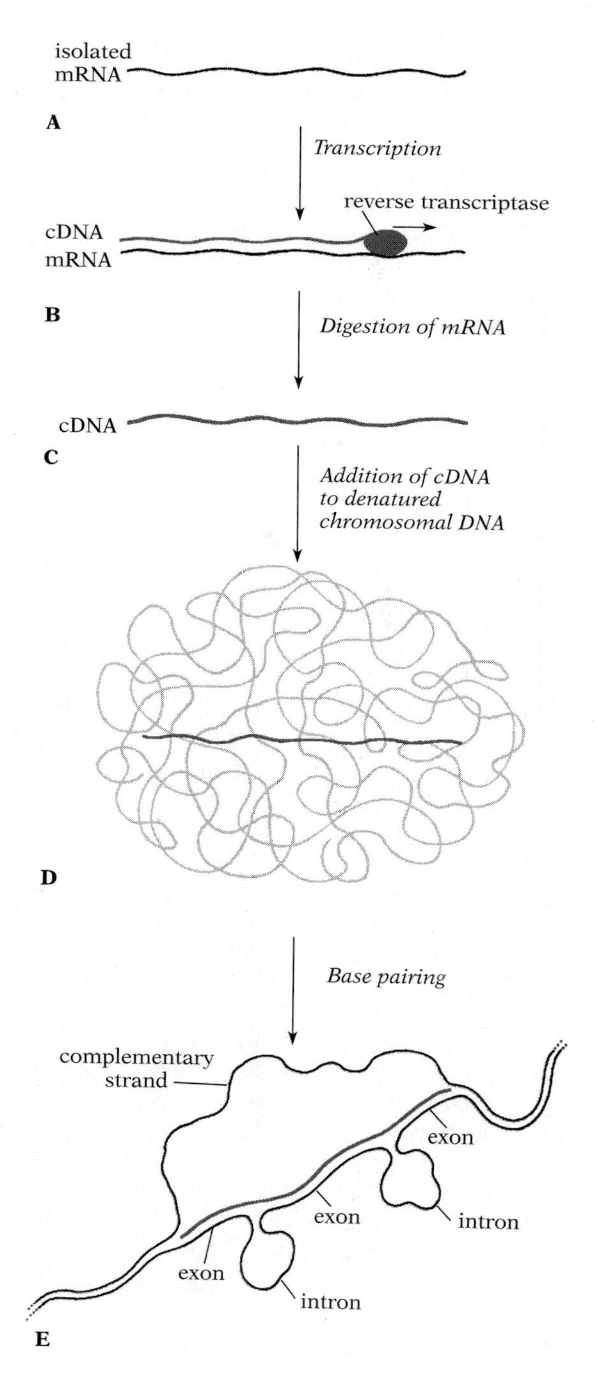

10.15 Locating a gene with a cDNA probe

(A) A particular kind of mRNA is isolated. (B) Reverse transcriptase is used to make a cDNA transcript incorporating a radioactive label, (C) then RNAse is added to digest the mRNA. (D) The cDNA is next mixed with denatured chromosomal DNA. (E) During base pairing, the cDNA binds to the gene coding for the mRNA, thus showing its location on the chromosome. Note that the intron segments loop out; they do not match up with the cDNA because they were absent from the mRNA.

EXPLORING FURTHER

DNA SEQUENCING

One of the cornerstones of recombinant DNA technology is sequencing—determining the order of bases in a gene. To obtain enough copies of the gene to be sequenced, a segment of the chromosome containing the gene is usually inserted into a plasmid and allowed to replicate many times. A widely used method for sequencing begins by breaking these multiple copies of the chromosome segment into pieces of manageable length by introducing an endonuclease. Because of the specificity of this enzyme, each copy of the chromosome segment is cut into an identical set of pieces. These pieces are then separated by any of a number of means (most often by molecular weight). All of the copies are tagged at one end with a radioactive marker. Next each collection of identical fragments is divided into four equal parts, each of which is treated with a low concentration of an enzyme that cuts out a specific nucleotide in the DNA; one part is exposed to a cytosine-specific enzyme, another to a guanine-specific chemical, and so on (Fig. A). The fragments are then denatured to separate the two strands; only one strand (the one labeled at the 5′ end) is used in the subsequent analysis.

Though the DNA in each part is cut next to a specific base, different copies of the DNA are severed at different places along the DNA. (Some, by chance, are even cut two or three times, but every break point is adjacent to a base of the kind for which the cutting enzyme is specific.) The result is a collection of labeled fragments of varying length; each labeled fragment extends from the labeled end to one of the bases in question. These fragments are next separated by weight (length); the weight of each labeled fragment appears as a radioactive band on a gel, and so each band specifies the position (that is, the distance from the labeled end) of one copy of the base. The fragments in the other three parts provide similar information about the location of the other three bases. When the resulting bands from each sample are aligned, the sequence of bases can be read out directly (photo).

The same procedure must be repeated for each sample in order to sequence the entire set of chromosome pieces. At this point, however, there is still no way to put the sequences into the proper order with respect to each other. This is done by treating another collection of fragments of the same chromosome with a different endonuclease, and then sequencing those fragments (Fig. B). Because the second endonuclease will cut the DNA at different spots, the break points of the segments from the first analysis will lie within the sequences worked out in the second. Thus, by looking for the beginnings and ends of the sequences from the first digestion *within* the sequences from the second digestion, the relative order of the fragments can be established.

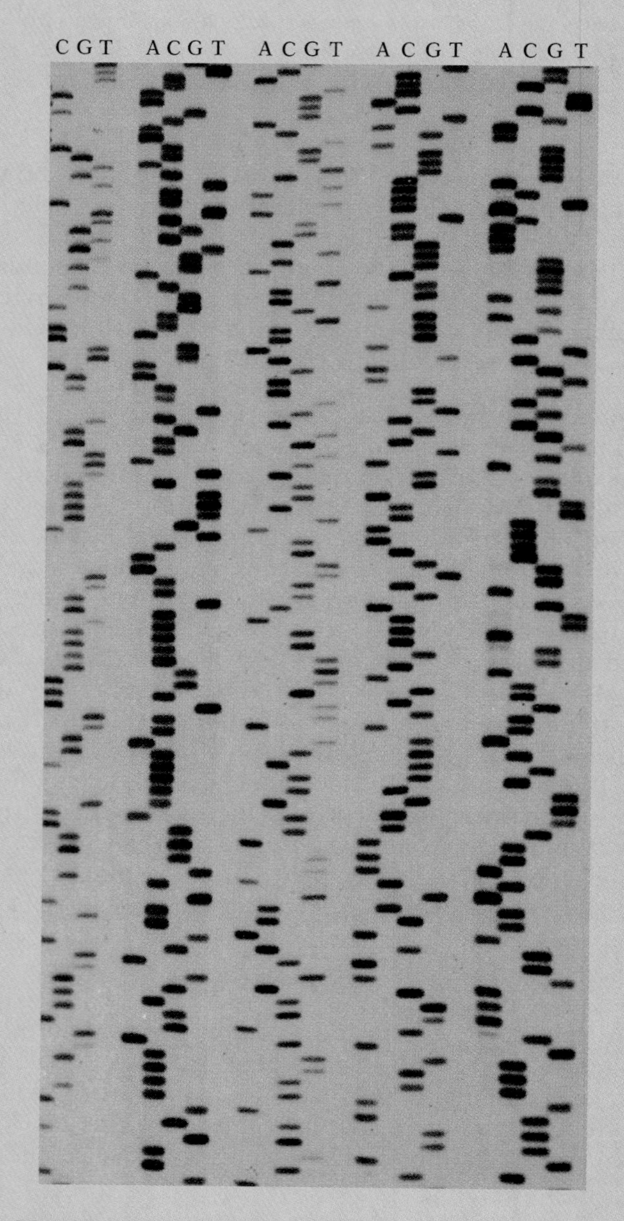

Nucleotide sequences produced by a sequencing technique: Each sample generates a group of four vertical bands on a gel.

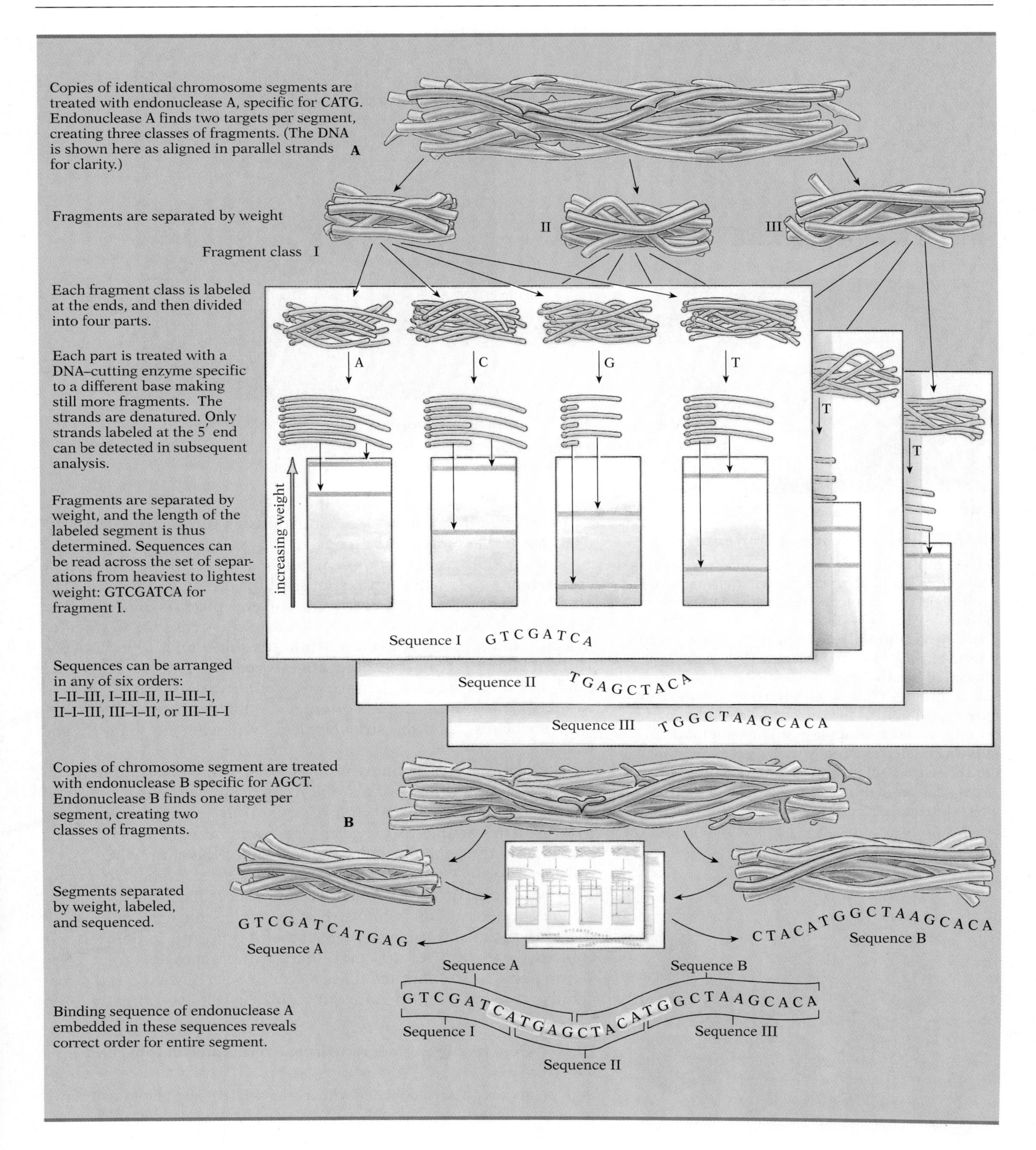

Copies of identical chromosome segments are treated with endonuclease A, specific for CATG. Endonuclease A finds two targets per segment, creating three classes of fragments. (The DNA is shown here as aligned in parallel strands for clarity.)
A
Fragments are separated by weight
Fragment class I
II
III
Each fragment class is labeled at the ends, and then divided into four parts.
Each part is treated with a DNA–cutting enzyme specific to a different base making still more fragments. The strands are denatured. Only strands labeled at the 5′ end can be detected in subsequent analysis.
A
C
G
T
increasing weight
Fragments are separated by weight, and the length of the labeled segment is thus determined. Sequences can be read across the set of separations from heaviest to lightest weight: GTCGATCA for fragment I.
Sequence I GTCGATCA
Sequences can be arranged in any of six orders: I–II–III, I–III–II, II–III–I, II–I–III, III–I–II, or III–II–I
Sequence II TGAGCTACA
Sequence III TGGCTAAGCACA
Copies of chromosome segment are treated with endonuclease B specific for AGCT. Endonuclease B finds one target per segment, creating two classes of fragments.
B
Segments separated by weight, labeled, and sequenced.
GTCGATCATGAG
Sequence A
CTACATGGCTAAGCACA
Sequence B
Sequence A
Sequence B
GTCGATCATGAGCTACATGGCTAAGCACA
Sequence I
Sequence II
Sequence III
Binding sequence of endonuclease A embedded in these sequences reveals correct order for entire segment.

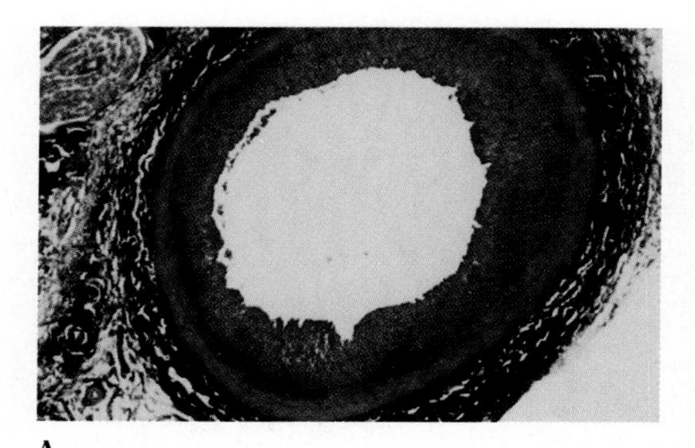

A

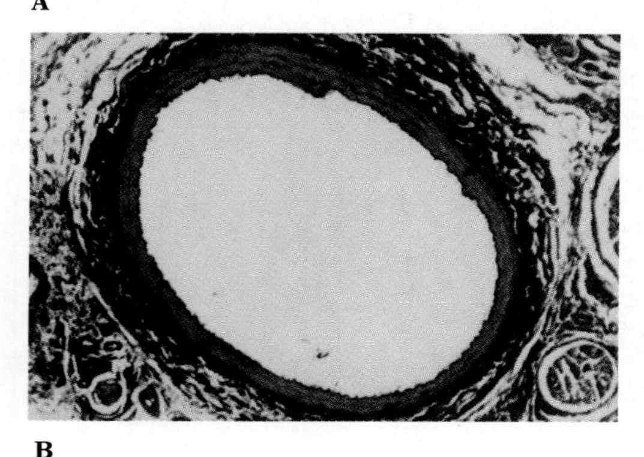

B

10.16 Therapeutic use of antisense DNA

DNA complementary to the mRNA from a gene involved in stimulating cell growth (a gene also involved in some cancers) can be used to block dangerous "healing" inside arteries after therapeutic procedures like balloon angioplasty, which is used to widen clogged arteries. (When the unwanted healing response is triggered, the resulting proliferation of cells in the artery wall can narrow and harden the vessel, thereby undoing the effects of the angioplasty). The vessel shown in B was treated externally with a gel containing the antisense DNA; a newer technique uses viruses delivered during the angioplasty itself, which introduce a gene that makes cells in the artery wall susceptible to a chemical that blocks cell division. Part A shows an untreated vessel.

10.17 Crown gall tumor

The bacterium *Agrobacterium tumefaciens* inserts a plasmid into its host's cells, leading the plant to produce the tumor.

be a complete human food, as nutritious as meat and dairy products, but far less fatty and expensive.

Genes for disease resistance might be transferred to susceptible crops. Using the bacterium responsible for crown gall disease (Fig. 10.17), a pathogen that inserts its 10-gene plasmid into the host genome, researchers have already transferred genes for herbicide-, insect-, and disease-resistance into petunias and tobacco (Fig. 10.18). Plants currently being engineered for disease resistance (and often herbicide tolerance as well) include alfalfa, apples, canola, canteloupes, corn, cotton, cucumbers, papaya, potatoes, rice, soybeans, squash, strawberries, sunflowers, tomatoes, and walnuts.

Vaccines against the many viruses for which conventional vaccination techniques fail can also be mass-produced. For example, if a virus cannot be cultured in large quantities, a gene coding for a coat protein capable of triggering an immune response might be cloned, and the gene product used instead of the entire virus.

Gene therapy—which adds normal copies of a gene to organisms with mutant alleles—has already succeeded in tests in *Drosophila,* rodents, and humans. For example, retroviruses or liposomes (which can fuse with the cell membrane and so deliver their contents) can be used to carry therapeutic genes to target cells. Cystic fibrosis, an inherited disease that blocks digestive and sweat glands, causing painful and dangerous cysts, has already been

10.18 Genetically engineered disease resistance in tobacco plants

Both plants have been inoculated with a microbial disease at two spots on each leaf. The plant on the right, with resistance genes introduced by *Agrobacterium,* shows no disease symptoms, whereas the unengineered plant on the left shows the characteristic halo of a spreading infection.

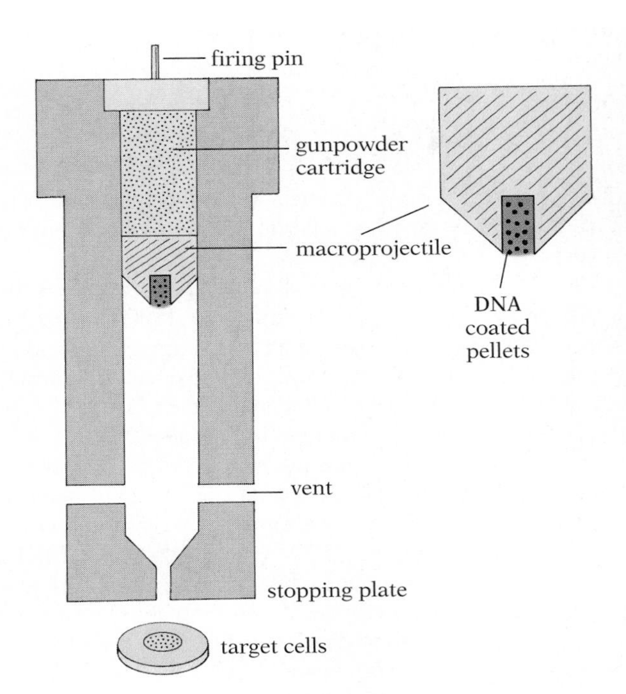

10.19 A DNA particle gun

DNA to be inserted into target cells is coated onto metal pellets, which are then loaded into a projectile in the particle gun. An explosive charge accelerates the projectile, which is stopped by a constriction at the end of the barrel. Inertia carries the coated pellets into the cells, where some of the DNA is incorporated into the chromosomes.

cured in cultured laboratory cells using this delivery system, opening the door to a dramatic new way of treating otherwise incurable diseases. More recently, the genes to be inserted have been coated onto microscopic gold pellets, which are then literally shot into cells (Fig. 10.19). Though developed for plants, this approach has been used successfully to insert genes that promote tissue repair into cells near wounds, leading to a dramatic reduction of healing times. Another delivery technique involves encapsulating engineered cells in a polymer shell, and then inserting the capsules into specific tissues: the foreign cells, safe from the host's immune system, secrete the desired product, which diffuses out of the capsule, while host nutrients diffuse into the capsule and maintain the caged cells.

Gene transfers to domesticated animals are clearly possible. The germ line is altered by effecting the change in the egg, either with a retrovirus, a pellet gun, or by microinjection (Fig. 10.20). Extra

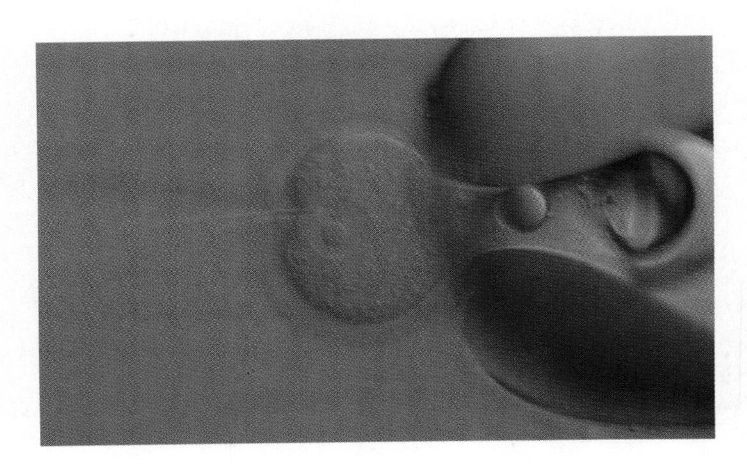

10.20 Microinjection of DNA

The very fine needle (barely visible at left) is being used to inject DNA into a male mouse's pronucleus (the nucleus of a sperm cell after it has entered an egg cell but before it has fused with the egg nucleus). Microinjection can also be performed after fertilization (that is, when the two gamete nuclei have fused), but the success rate is lower.

HOW WE KNOW:

DNA FINGERPRINTING

The techniques of genetic engineering have provided a new tool for identifying individuals based on samples of blood, skin, hair, or semen. These techniques are already being used for determining paternity, both in civil cases and in studying animal societies. The basic tool is the polymerase chain reaction, which generates enough copies of the DNA under study to permit quantitative analysis. But the key to individual identification lies in a particular class of genetic variation found in virtually all eucaryotes. As described in detail in Chapter 11, a large proportion of the human genome consists of untranscribed "junk" DNA, which includes hundreds of thousands of partial copies of transposons and still more copies of short nucleotide repeats—C-A-T, for instance. The repeats, which are concentrated near the centromeres and near the ends of the chromosomes, vary enormously in number and distribution from one individual to another (identical twins excluded). One person may have one copy of a particular eight-base repeat at a specific location on a given chromosome, while a sibling may have 20 copies of the same repeat at that spot.

To compare two samples to see if they are from the same individual, the DNA is cut with specific endonucleases, and the length and charge of the resulting fragments (as judged by the speed with which they move through a gel) are compared. This separation is very similar to the procedure used in gene sequencing (described in the box on p. 264). The position on the gel of fragments containing the so-called hypervariable regions of the genome differ greatly from one individual to another. In theory, the patterns are so specific that the chance of two persons having the same fragment lengths is less than 1 in 700 trillion—far more specific than a set of conventional fingerprints.

The DNA fingerprinting technique was developed by the British geneticist Alec Jeffreys in the early 1980s. The first use in the United States involved an accused rapist in Orlando, Florida, in 1987, but the technique is used widely in murder and assault cases as well. DNA from the defendant ("D" in the figure), the victim (V), and the blood found on the suspect's clothes (amplified with the PCR technique to yield enough material; labeled "jeans" and "shirt" in the figure) are treated with the endonuclease and run on a gel, along with some calibration samples ("TS," "1kb," and "γ"). A careful comparison of the columns on the gel show clearly that the blood on the suspect's clothes closely matches that of the victim, and is unrelated to the defendant's own blood. The defendant in this case was convicted on the DNA evidence.

DNA fingerprinting, like all forensic evidence, is only as reliable as the honesty of the investigators and the competence of the laboratory performing the analysis. In addition, DNA fingerprinting is not nearly as specific in practice as theory suggests. Compare the victim's blood in the figure with the blood found on the shirt in the "8μg" and "4μg" columns: the pattern is very similar, but the bands do not line up precisely. If the shirt sample were not available and you had only the very faint "jeans" column to compare with the victim's and defendant's blood, would you consider any of them a match? Clearly there is some degree of variation within a single gel, which amplifies a major problem with small sam-

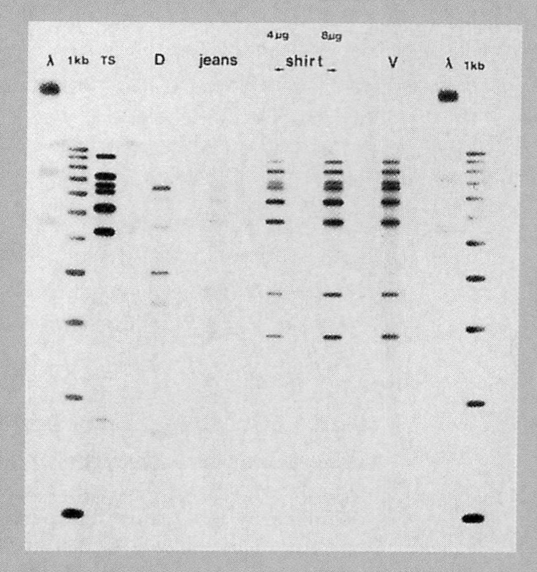

DNA fragments from a murder victim's blood (V), the defendant's blood (D), and the blood (amplified through the PCR technique) from the suspect's clothes were treated with endonuclease and separated on a gel. (The three columns at the left and the two at the right contain control fragments for calibration.) The fragments are made visible by applying a radioactive stain and then placing the gel (or a replica created by pressing a mesh onto the gel) on a photographic emulsion; radioactive decays create the dark areas visible here.

ples; each limits the certainty of the test. If the blood on the clothing had begun to degrade from exposure to light, air, chemicals, or simply through the passage of time, numerous random fragments of unpredictable lengths might have been created before the endonuclease was applied. This factor also reduces the potential resolution of any comparison.

A more difficult problem with DNA fingerprinting is that some of the original calculations of a chance match did not take into account the fact that the more related two individuals are, the more similar their basic pattern of fragment-length variability. Thus while the patterns of victim and suspect from two different ethnic groups will be very dissimilar, members of the same group, innocent or guilty, will be more alike—a similarity that ought to be greater if they are from the same interbreeding subgroup (the same town, for instance). Moreover, some ethnic groups simply have less variability, particularly when the founder population from which the group is descended was small. To take an extreme example, using DNA fingerprinting to solve a violent crime within an Amish colony in the United States—a remote possibility given the ethos of this sect—would be very difficult because the range of DNA variation is highly limited.

At present the exact probabilities in any given case, whether it entails animal research or human legal proceedings, are difficult to compute, and next to impossible for a jury to understand. In general, a worst-case scenario of maximum genetic similarity is used, which generates wildly conservative probabilities close to 1 in a million for obvious matches. How willing will juries be to convict on 1-in-a-million cases versus 1-in-a-trillion? Though this approach essentially eliminates the chance of false convictions, it raises the odds of mistaken acquittals. Increasingly, investigators are subdividing samples and analyzing them with additional endonucleases, generating extra patterns to compare. Each new set of digestion fragments reduces the probability of a chance match. Despite the lingering statistical controversy, and the resulting highly conservative interpretation of pattern matches, the DNA fingerprinting technique promises to revolutionize criminal investigation, and to increase the certainty with which juries can render judgments.

genes for growth hormone, paired with control regions that encourage high rates of transcription, have been added in this way, leading to strains of rodents, swine, cattle, and fish that grow larger and store less fat (Fig. 10.21). This shift in metabolic priorities overburdens some of the internal organs of these creatures, however, indicating that much remains to be learned and done to realize the full promise of the technique.

Recognition of the risk of undesirable side effects of recombinant DNA technology has led to strict regulation of how the technology is used. Laboratories must be equipped with facilities for sterile handling and for physical containment similar to those long used in the medical microbiology laboratories where dangerous pathogens are handled. In addition, there is a consensus that the microorganisms used in recombinant DNA research should be incapable of becoming pathogenic, or should be derived from strains with crippling mutations that make them incapable of surviving outside the laboratory.

10.21 Effects of microinjection of growth hormone genes

The eggs of coho salmon were microinjected shortly after fertilization with a gene for growth hormone and a control region that promoted its transcription. The five 14-month-old fish at the bottom are untreated; the five at the top were successfully injected with the gene and are on average 11 times heavier at the same age.

CHAPTER SUMMARY

THE DISCOVERY OF BACTERIAL PLASMIDS

DO BACTERIA MATE? Some kinds of bacteria can undergo conjugation, in which one bacterium donates genetic material to another. Usually the sex factor that allows a bacterium to initiate conjugation is found in a plasmid, which is all that is exchanged; when the plasmid has been incorporated into the chromosome, however, large sections of the donor's genome can be transferred. Donated chromosomal DNA can recombine with the recipient's DNA. (p. 248)

AUTONOMOUS PLASMIDS Most gene-carrying plasmids lack a sex factor; they reproduce autonomously and are passed to daughter cells. (p. 249)

TRANSFORMATION Certain bacteria can take up DNA from the surrounding medium; the DNA fragment may then recombine with the host chromosome. (p. 250)

HOW VIRUSES MOVE GENES

HOW VIRUSES REPRODUCE Some viruses simply inject DNA into hosts; the viral genes direct the replication of the viral chromosome and, via transcription and translation, the production of necessary enzymes and structural products. Lytic viruses reproduce and then lyse the host cell. In temperate viruses the DNA can be incorporated into the host chromosome and reproduced when the chromosome is replicated; later this lysogenic state can change into a lytic phase as the viral DNA becomes active and directs the production of new viruses. Retroviruses have RNA genes that are copied into DNA by reverse transcriptase; this copy is then incorporated into the host chromosome. (p. 250)

VIRAL TRANSFER OF NONVIRAL GENES Host genes sometimes become incorporated into new viruses, and can thus be carried to new hosts; this process is called transduction. (p. 253)

HOW GENES ARE MOVED WITHIN THE GENOME

THE BEHAVIOR OF TRANSPOSONS Transposons are genetic units that can move within the genome. Some move themselves, while others spread by making and dispatching RNA or DNA copies of themselves. (p. 253)

OTHER WAYS TO MOVE GENES In addition to breakage and fusion events, most other changes occur when cDNA copies of mRNA or other RNAs are made by reverse transcriptase and incorporated into the genome. (p. 255)

GENETIC ENGINEERING

ARTIFICIAL TRANSFORMATION Recombinant DNA techniques are used to incorporate new genes, or new versions of existing genes, into target genomes. Most methods utilize plasmids as gene carriers. (p. 256)

HOW SPECIFIC GENES ARE CLONED A variety of techniques are used to make copies of genes, including the polymerase chain reaction and the generation of cDNA copies of particular mRNAs. (p. 260)

STUDY QUESTIONS

1 Why can't cDNA genes—the coding sequence for insulin, for instance—be introduced directly into bacteria, thus dispensing with the plasmid and the several extra steps involved in using it to carry the gene? (pp. 256–269)

2 Construct a plausible scenario for the evolution of a functional RNA plasmid. How does it manage to maintain itself generation after generation? (pp. 249–256)

3 Now play the role of natural selection. How would you turn the plasmid into a working RNA virus, either single- or double-stranded?

4 What sort of evidence could convince you that plasmids evolved from viruses, not the reverse? (pp. 249–256)

5 Based on the evidence in Chapter 5 and in Chapters 8 and 9, summarize the case for the endosymbiotic hypothesis. If mitochondria and chloroplasts are really tame bacteria, why do you suppose that most but not all of their genes now reside in the nucleus (including many for products used only in organelles)? Surely it would have been simpler to have left them "on site," rather than to have to transport so many special products there. Given that most of their enzymes are being taken to the organelle anyway, why leave any genes behind? (pp. 149–150, 218–219, 239–241)

SUGGESTED READING

AHARONWITZ, Y., AND G. COHEN, 1981. Microbial production of pharmaceuticals, *Scientific American* 245 (3). *On how recombinant DNA techniques are used to make microbes produce antibiotics, hormones, and other drugs. Includes an explanation of how antibiotics work to destroy bacteria, which suggests how plasmid genes may confer resistance.*

AMÁBILE-CUEVAS, C.F., AND M.E. CHICUREL, 1993. Horizontal gene transfer, *American Scientist* 81 (4). *On gene movement between species.*

ANDERSON, W. FRENCH, 1995. Gene therapy, *Scientific American* 273 (3). *Concise review of gene therapy techniques and tests, with a look at future prospects.*

CAPECCHI, M.R., 1994. Targeted gene replacement, *Scientific American* 270 (3). *On the powerful technique that explores gene function by knocking out particular genes selectively.*

CHILTON, M.D., 1983. A vector for introducing new genes into plants, *Scientific American* 248 (6). *On bacteria that transduce host cells.*

COHEN, J. S., AND M. E. HOGAN, 1994. The new genetic medicines, *Scientific American* 271 (6). *On antisense DNA and related strategies for combating disease.*

FEDOROFF, N. V., 1984. Transposable genetic elements in maize, *Scientific American* 250 (6). *A modern interpretation of the transposition discovered by Barbara McClintock.*

GASSER, C.S., AND R.T. FRALEY, 1992. Transgenic crops, *Scientific American* 266 (6). *On the improvement of crops through genetic engineering.*

HOPWOOD, D. A., 1981. The genetic programming of industrial microorganisms, *Scientific American* 245 (3). *An excellent summary of how basic recombinant DNA techniques work.*

JOYCE, G. F., 1992. Directed molecular evolution, *Scientific American* 267 (6). *On how induced mutation and the polymerase chain reaction technique can be combined to select for novel forms of genes that encode products with more desirable properties.*

LASIC, D., 1992. Liposomes, *American Scientist* 80 (1). *On how liposomes can be used to deliver recombinant molecules to specific targets in the body.*

VARMUS, H., 1987. Reverse transcription, *Scientific American* 257 (3). *A detailed description of the process that reverses the usual direction of information flow.*

VERMA, I. M., 1990. Gene therapy, *Scientific American*, 263 (5). *On attempts to correct genetic defects.*

CHAPTER 11

ORCHESTRATING BACTERIAL GENES

HOW GENES ARE INDUCED

HOW GENES ARE REPRESSED

POSITIVE VERSUS NEGATIVE CONTROL

HOW EUCARYOTIC CONTROL IS DIFFERENT

THE ODD ORGANIZATION OF EUCARYOTIC CHROMOSOMES

HOW CHANGES IN GENE TRANSCRIPTION CAN BE SEEN

IMPRINTING: PRESERVING PATTERNS OF GENE ACTIVITY

EUCARYOTIC CONTROL: INDUCERS AND ENHANCERS

GENE AMPLIFICATION

CONTROL AFTER TRANSCRIPTION

THE IMPACT OF MUTATIONS

CANCER: WHEN GENE CONTROL FAILS

TISSUE CULTURE: A WAY OF STUDYING CANCER CELLS

HOW ARE CANCEROUS CELLS DIFFERENT?

THE STEPS THAT LEAD TO CANCER

THE ROLE OF ONCOGENES

ENVIRONMENTAL CAUSES OF CANCER

CONTROL OF GENE EXPRESSION

Every cell in the body of a multicellular organism receives the same set of genetic instructions—the organism's full evolutionary endowment. Yet individual cells may look entirely different from other cells, and may behave in entirely different ways. In fact, only a particular subset of all the genes a cell contains will ever generate proteins in that particular cell, and only a small percentage of this DNA is active at any one time. The selective activation and control of the degree of expression of each of a cell's genes is critical to life. In this chapter we will examine the logic and mechanisms of gene control, a process that involves chemicals that bind directly or indirectly to DNA or mRNA.

ORCHESTRATING BACTERIAL GENES

Because bacteria have only about 3000 genes (compared with the 50,000 of humans, for instance) and can be grown rapidly in huge numbers, they have been especially useful for study of the control of gene transcription. The first models of gene control emerged from research on *Escherichia coli*. They were based on the assumption (now known to be largely correct) that the availability of a reactant in a cell will turn on transcription of the genes encoding the enzymes that process the reactant, while excessive amounts of product will turn off transcription of the genes encoding the enzymes of the biochemical pathway that produce the product.

■ HOW GENES ARE INDUCED

In the course of an extended investigation of enzyme synthesis in *E. coli*, beginning in the late 1940s, the French biochemists François Jacob and Jacques Monod (Fig. 11.1) formulated a powerful model of gene regulation in bacterial cells. They concentrated on the enzyme β-galactosidase, which catalyzes the breakdown of lactose into glucose and galactose.

The operon Lactose is not continuously available to *E. coli*, and so the gene for β-galactosidase is normally transcribed at a very low rate. Jacob and Monod found that further production of this digestive enzyme is triggered by the presence of an ***inducer***, in this instance allolactose, a derivative of lactose automatically produced

A

B

11.1 François Jacob (A) and Jacques Monod (B)

in the cell when lactose is present. Normally, then, β-galactosidase is an ***inducible enzyme***.

Four genes are involved in the production of β-galactosidase and the two other enzymes involved in lactose breakdown: three ***structural genes***, each specifying the amino acid sequence of one of the three enzymes, and a ***regulator gene***, which controls the activity of the structural genes. The regulator gene, which is located at some distance from the structural genes, normally directs the synthesis of a ***repressor*** protein that inhibits transcription of the structural genes.

Jacob and Monod also discovered that a special region of DNA contiguous to the structural genes for β-galactosidase determines whether transcription of the structural genes will be initiated. This special region is the ***operator***, which itself is not a gene since it doesn't code for a specific product. The combination of an operator and its associated structural genes is an ***operon***. This particular operon, since it is responsible for the synthesis of enzymes involved in the breakdown of lactose, is called the *lac* operon.

The operator is located between the promoter—the region to which RNA polymerase binds—and the structural genes. Hence, when the repressor binds to the operator, RNA polymerase cannot bind to the promoter, and transcription is blocked (Fig. 11.2A). If an inducer is present, it will bind to the repressor, thus causing a conformational change in the repressor that forces it to dissociate from the operator; in short, the inducer deactivates the repressor (Fig. 11.2B). RNA polymerase is now free to bind to the promoter and can initiate transcription of the structural genes and the production of mRNA (Fig. 11.2C). The mRNA carries the instructions for all three structural genes. This messenger binds to ribosomes in the cytoplasm, where its information is translated and the three enzymes necessary for lactose metabolism are synthesized. The number of β-galactosidase molecules rises to about 5000 per cell when the operon is not repressed; during repression, there are only about 10 copies of that enzyme.

Summary of the Jacob-Monod model The condition of the operator region is one key to whether or not there will be activation of the *lac* operon. If repressor protein is bound to the operator, there will be no transcription. If no repressor is bound to the operator (because the repressor has been inactivated by inducer), transcription can proceed freely. Only a few molecules of inducer are required to bind all the repressor molecules in a cell.

The three jointly controlled structural genes of the *lac* operon specify enzymes with closely related functions. It is typical for the structural genes of an operon to determine the enzymes of a single biochemical pathway; thus the whole pathway can be regulated as a unit.

HOW GENES ARE REPRESSED

Not all operons are regulated in the same way as the *lac* operon, which is an inducible operon—that is, one that is inactive until it

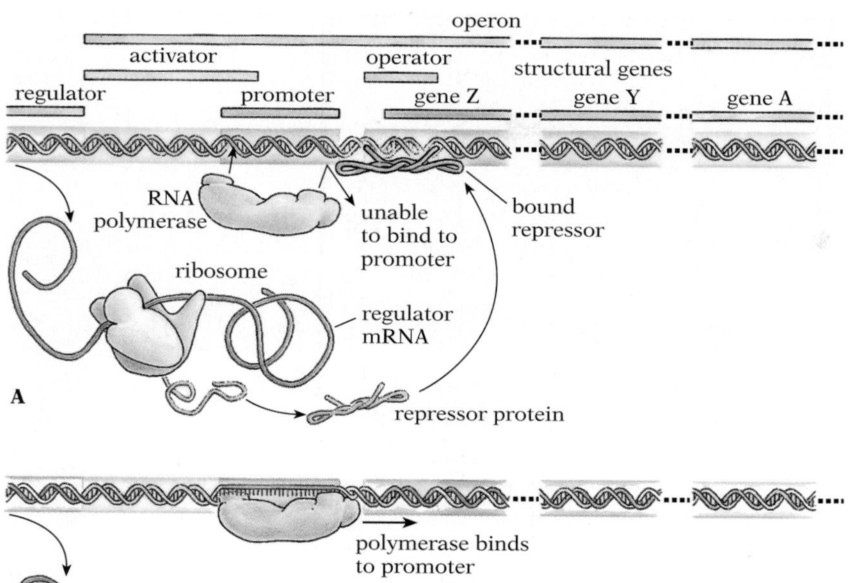

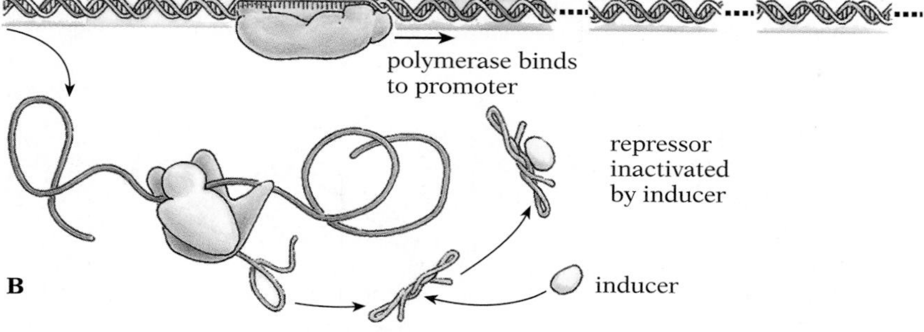

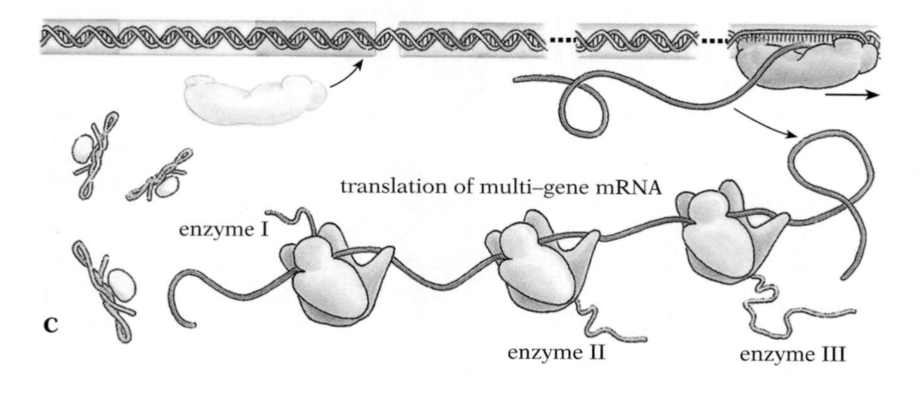

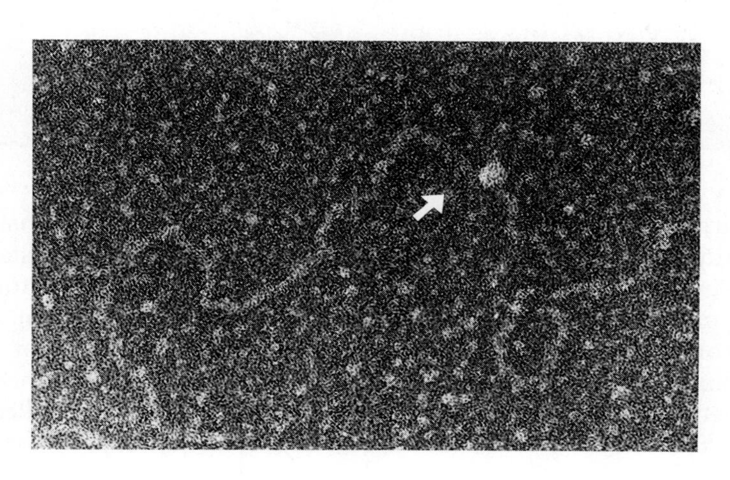

11.2 An inducible operon: the *lac* operon

(A) The operon consists of a promoter-operator region and three structural genes (Z, Y, and A). The operator sequence overlaps the beginning of the structural gene. The regulator gene codes for mRNA, which is translated on the ribosomes and determines synthesis of the repressor protein. When the repressor protein binds to the operator, it blocks one of the promoter's binding sites for RNA polymerase and thus prevents the initiation of transcription of the structural genes. (B) Binding of inducer to repressor deactivates the repressor, and the RNA polymerase can then bind to the promoter regions. (C) Polymerases initiate transcription of the structural genes, which are transcribed as a unit, producing an mRNA coding for three gene products. The mRNA then binds to ribosomes and is translated into three enzymes. Enzyme I is β-galactosidase; enzyme II is a permease, which helps transport lactose into the cell; and enzyme III is a transacetylase, whose role in lactose utilization is not understood. The transcribed strand of the structural genes has been emphasized. The function of the activator sequence will be discussed presently. (D) Electron micrograph of the *lac* repressor protein bound to the operator region.

EXPLORING FURTHER

HOW CONTROL SUBSTANCES BIND TO DNA

For the repressor to bind to a particular operator, it must bear active sites that match the DNA substrate exactly. Such matching involves polar amino acids on the repressor and complementary polar groups exposed on the *sides* of the base pairs of the DNA.

The deoxyribose backbones of DNA are attached to the bases slightly off center. As detailed in Chapter 8, the two chains have opposite polarities, with one running from 3′ to 5′ and the other from 5′ to 3′. The result is that one of the exposed sides, or grooves, of the helix is somewhat wider

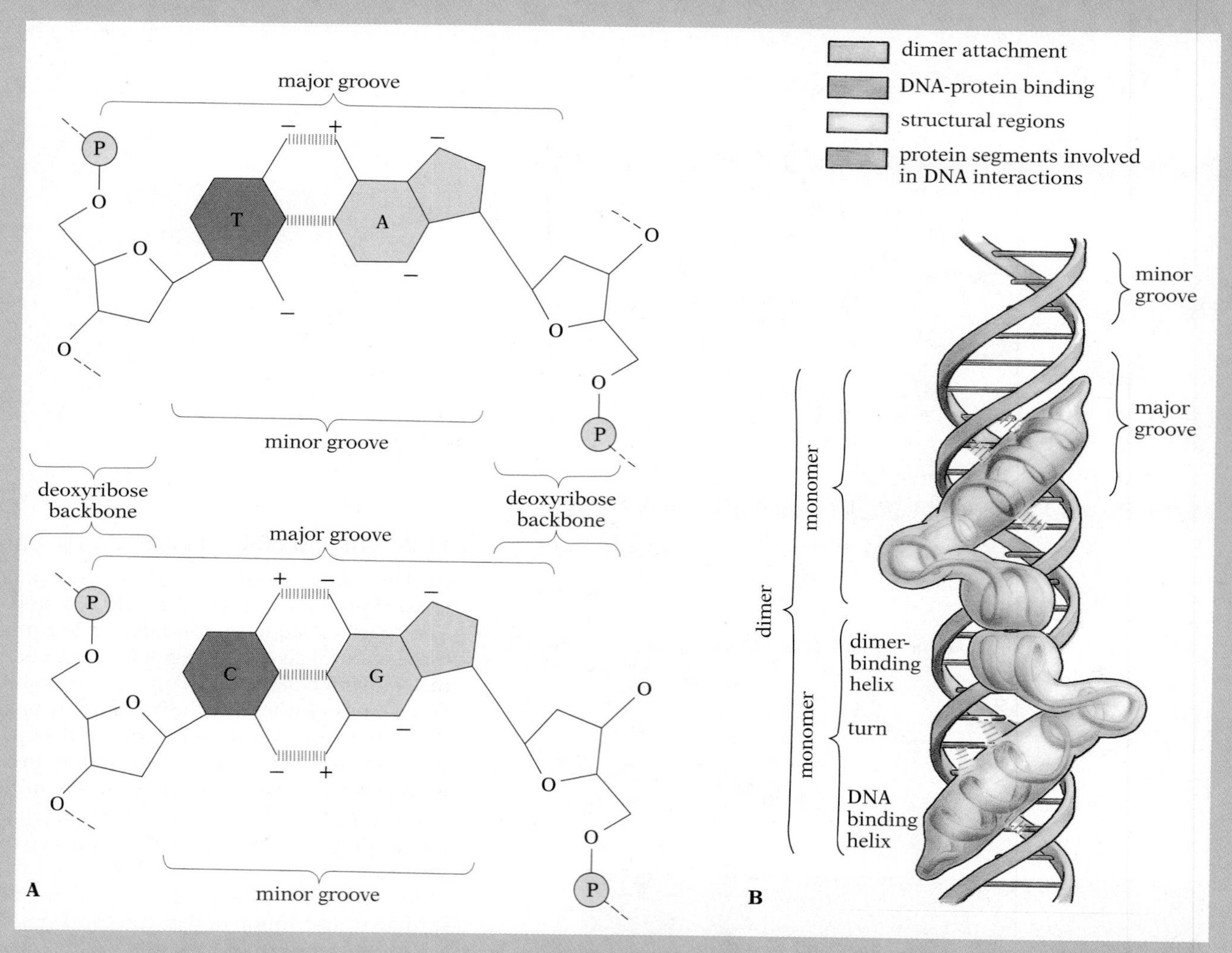

is turned on by an inducer substance. Many operons are, instead, continuously active unless turned off by a ***corepressor*** substance. One example is the operon whose five structural genes code for the enzymes necessary to synthesize the amino acid tryptophan. This operon is normally turned on, but when *E. coli* is grown in a medium containing tryptophan, it switches off. Enzymes encoded by genes that are usually active but can be repressed are called ***re-pressible enzymes***. Here the repressor protein encoded by the reg-

than the other (Fig. A). The wider opening—the ***major groove***—contains more sequence-specific information, and is therefore thought to be more important in binding repressors. For example, the major groove of a T–A pair has three polar groups—O, NH, and N, reading from T to A—their respective polarities being −, +, and −. The major groove of the C–G pair has the polar groups NH, O, and N, with polarities of +, −, and −. The ***minor groove*** is less useful: for T–A, the polar "code" is −, X, − (where "X" means no charge), while for C–G the sequence is −, +, −. Each particular repressor has a shape corresponding to the differing physical widths of the operator's major and minor grooves, and polarity patterns complementary to those of the operator's base pairs.

Several molecular morphologies are typical of DNA control substances. The "helix-turn-helix" motif (Fig. B) involves two α helices: one binds in the major groove, while the other attaches itself to a second copy of the binding chemical to create a dimer. The "leucine zipper" (Fig. B) is also a protein dimer with two helices. The shorter helix has a leucine at every seventh position, so that four of the leucine side groups project out on the same side of the short helix; these leucines interlock in a zipperlike fashion with the aligned leucine side groups of the other protein's short helix (Fig. C). The two long helices bind in the major groove of the DNA in what is called a "scissors grip." A third family of binding proteins also uses a dimer with a helix devoted to binding in the major groove (Fig. D). The "zinc-finger" proteins are stabilized by zinc ions, and (again) the dimers bind to each other as well as to the DNA. Other morphologies exist as well.

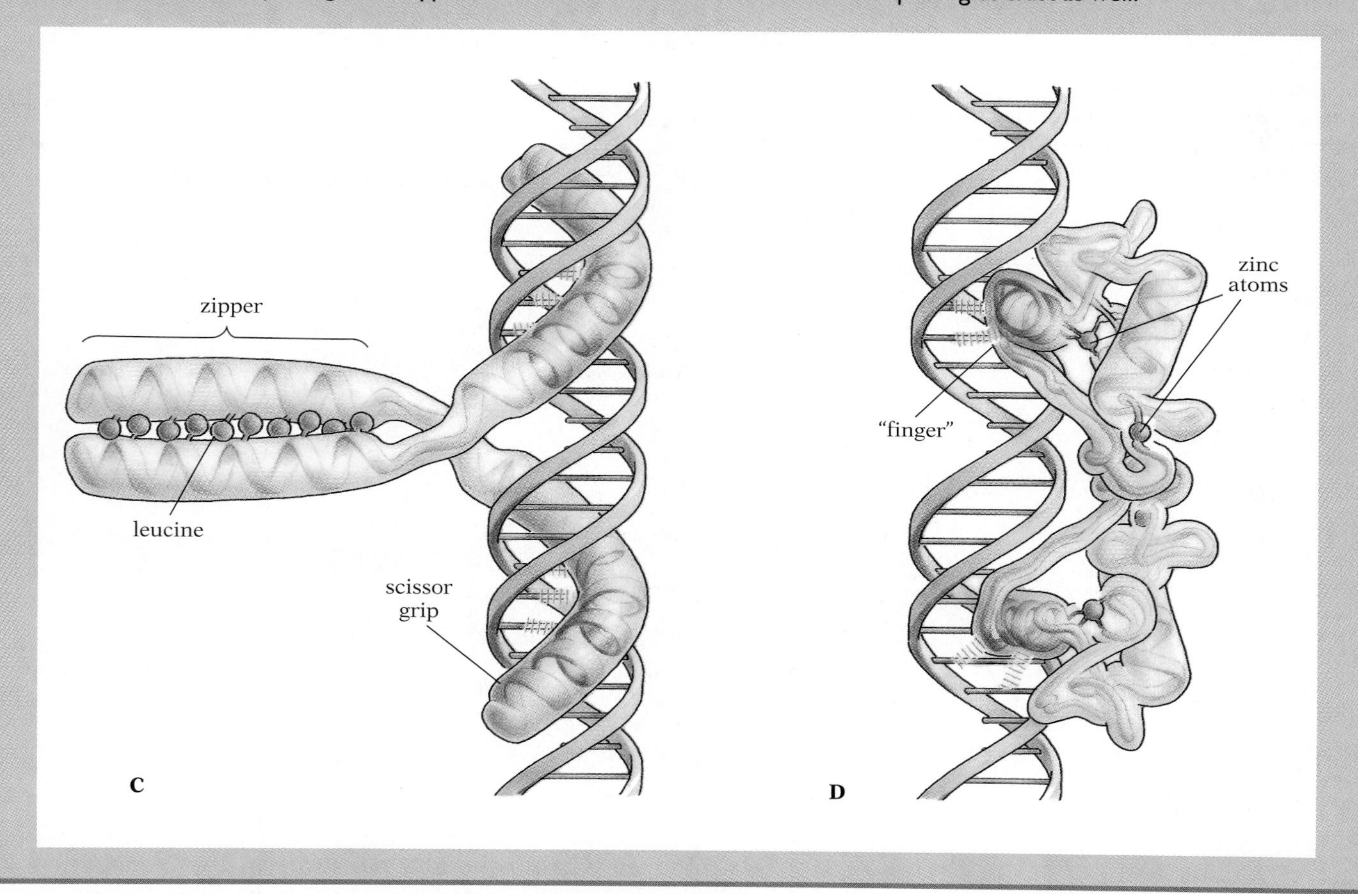

ulator gene is inactive when first produced. Only if a corepressor substance binds to and activates it can it bind to the operator and block RNA polymerase binding (Fig. 11.3). Unlike inducible enzymes, which are synthesized only if their operon is turned on by an inducer, repressible enzymes are automatically synthesized unless their operon is turned off by a corepressor. In tryptophan synthesis, the tryptophan itself activates the repressor protein, enabling it to bind to the operator.

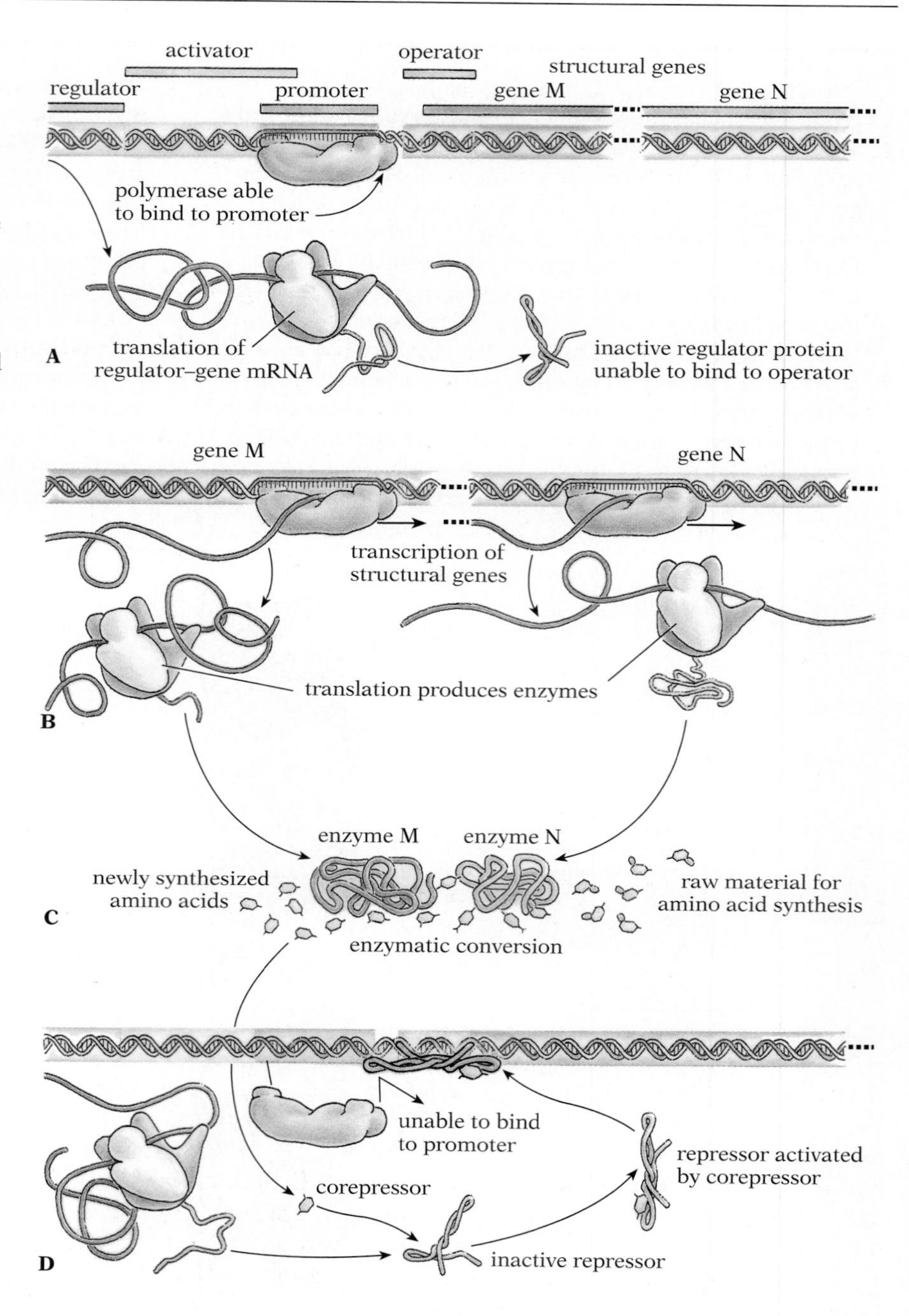

11.3 A repressible operon

The repressor protein encoded by the regulator gene is initially inactive (A). As a result, polymerases can bind to the promoter and transcribe the structural genes (B). These genes frequently encode enzymes that lead to the synthesis of an end product such as an amino acid (C). When the inactive regulator protein combines with a specific corepressor molecule (often the end product of the biochemical pathway served by the enzymes encoded by the operon), it can bind to the operator and block transcription of the structural genes (D). After the operon has been repressed, the concentration of the corepressor falls as it is used in cellular metabolism, but no more is produced. When the corepressor becomes scarce, the repressor tends to lose it to metabolic enzymes. As a result, the repressor can no longer bind to the operator, the RNA polymerase binds to the promoter, and transcription resumes. The operon shown here is responsible for the synthesis of an amino acid, which is incorporated into new proteins.

An inducer is often the first substrate in the biochemical pathway being regulated. Similarly, a corepressor is usually the end product of the biochemical pathway being regulated. In both substrate induction and end-product corepression, then, gene transcription is generally regulated by the cellular substances most affected by the transcription.

■ POSITIVE VERSUS NEGATIVE CONTROL

Both of the cases we have discussed are examples of ***negative control:*** a repressor bound to the operator turns off transcription.

Table 11.1 Summary of transcriptional control strategies in procaryotes

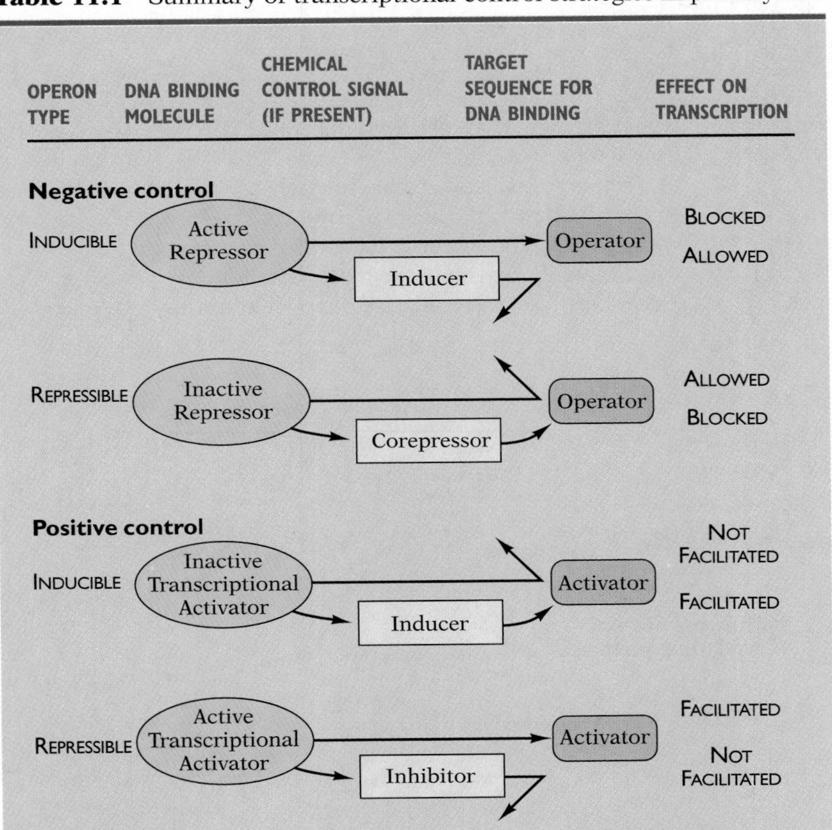

Control is effected in one case by deactivating the repressor with an inducer, and in the other case by activating the repressor with a corepressor (Table 11.1). Though negative control is the most common way of regulating gene expression in procaryotes, some systems are regulated by ***positive control***. In many of these cases a control protein binds directly to the DNA to activate the operon.

Positive control takes advantage of the variation in promoter sequences: RNA polymerase binds to the promoter sequence adjacent to the operator, but few genes have promoters with exactly the same nucleotide sequence. Genes with promoters that differ significantly from the optimal one are rarely transcribed, because the RNA polymerase will bind to them only weakly. In positive control of such genes, a control protein called a ***transcriptional activator (TA)*** binds to a region a little upstream from the promoter (Fig. 11.4), and then helps the polymerase bind. The presence of the bound activator protein compensates for the poor promoter sequence.

Two versions of positive control are known. In one, a chemical inducer binds to the TA, causing a conformational change that activates the TA, enabling it to bind to the DNA and facilitate transcription. A TA that works this way is the ***catabolic gene-activator protein (CAP)*** of *E. coli*, which helps control the transcription of

EXPLORING FURTHER

THE LAMBDA SWITCH AND CONTROL OF THE LYTIC CYCLE

The expression of genes is often regulated by systems that involve the interaction of three or more control substances, two or more separate operators, and sometimes other DNA-control regions in addition to the regulator genes, operators, and promoters already discussed. The most thoroughly investigated example of these more complex interactions involves the switching of the temperate virus lambda from its lysogenic to its lytic cycle; similar control systems are used by the host itself.

Temperate viruses are able to insert their DNA into a specific section of the host chromosome and remain there without lysing the bacterial cell. Under normal conditions (that is, steady growth of the bacterial colony), this passive kind of infection rarely occurs, but once the virus does enter

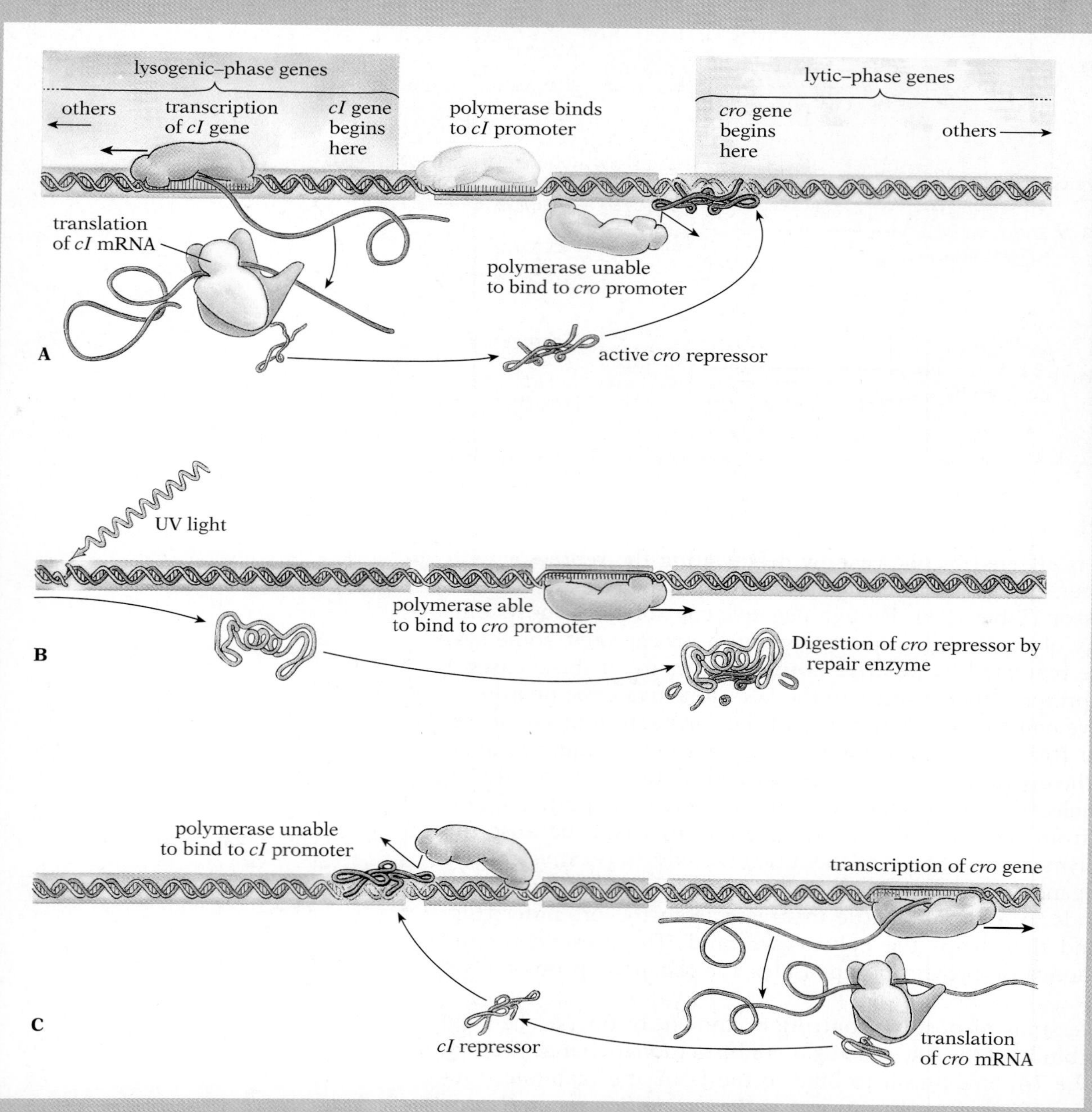

the host chromosome, it can remain there, in the lysogenic cycle, indefinitely. Exposure of an infected bacterium to ultraviolet radiation, however, may incite the viral DNA to leave the host chromosome and begin reproduction. In the lambda virus, which parasitizes *E. coli*, two interacting operons on complementary strands of the DNA control this switch from a lysogenic to a lytic cycle.

Excision of lambda DNA from the host chromosome and initiation of the lytic cycle are controlled by the *cro* gene and several other genes downstream from the operon. These genes are normally repressed: the *cI* gene on the complementary strand produces a repressor protein that binds to the operator region of the *cro* gene, and thereby blocks transcription (Fig. A). Repression of the *cro* and associated genes is stable until the host bacterium begins to synthesize unusually large quantities of DNA repair enzymes, often in response to UV damage (Fig. B). While fixing the DNA, one of the repair enzymes digests part of the repressor synthesized by the *cI* gene, freeing the *cro* operator to begin transcription of the normally repressed *cro* and associated viral genes. The *cro* gene codes for a second repressor, which binds to the *cI* operator and turns off transcription of *cI* (Fig. C). Transcription of lytic-phase genes can now proceed (Fig. D), leading to excision of the lambda genome from the host chromosome (Fig. E), the production of replicates of the viral DNA, synthesis of capsule proteins, viral assembly, and finally host lysis.

The lambda switch is irreversible: even the disappearance of the bacterial DNA repair enzyme does not stop production of the second repressor. The lambda virus reacts to the first sign of trouble (UV radiation, in this case) by initiating the synthesis of new phage, which will abandon the damaged bacterium and seek out other hosts.

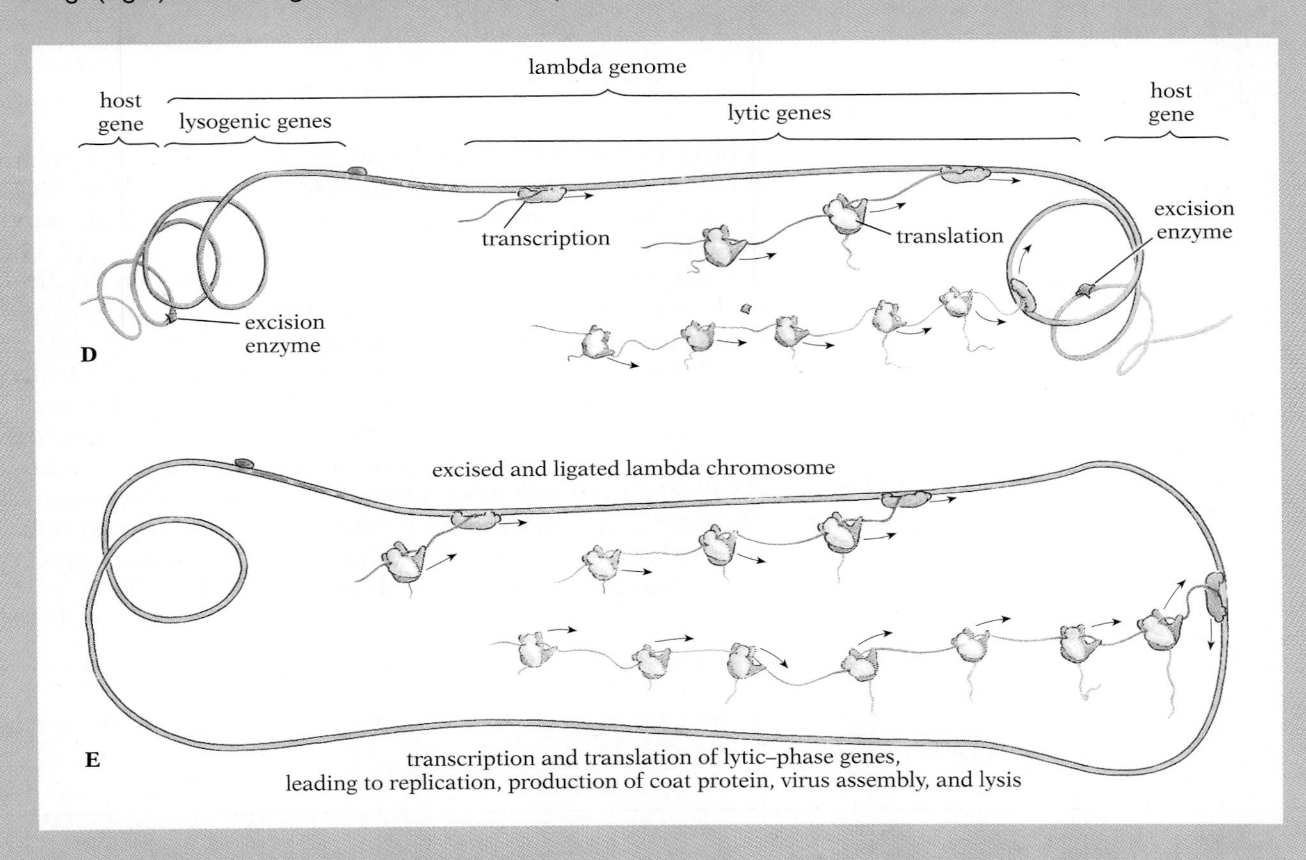

genes encoding enzymes that digest unusual nutrients. When glucose, the bacterium's usual food, becomes rare, a messenger substance—cyclic AMP—is produced. The cAMP binds to and activates CAP, greatly increasing transcription of the alternative nutrient-metabolism operons (Fig. 11.4; Table 11.1). CAP, then, can affect the activity of many operons at once.

The other positive-control strategy involves a normally active

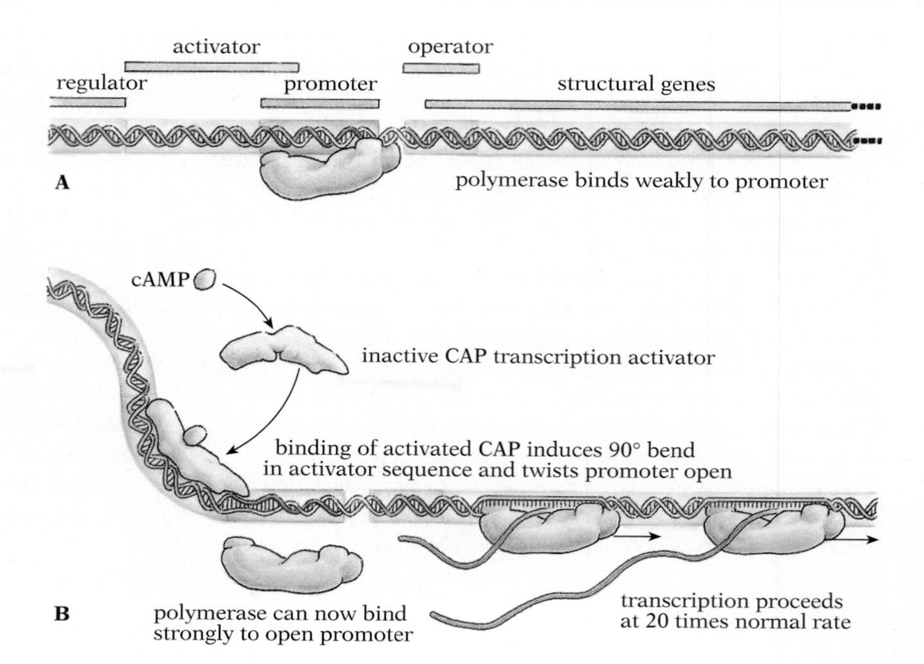

11.4 Induction of transcription by CAP (A) Many genes have promoters that bind polymerase at low rates. (B) When a CAP transcriptional activator protein, itself activated when bound by a molecule of cAMP, binds to a sequence adjacent to the promoter, it induces a 90° bend in the DNA; the resulting twist opens the promoter, and increases the polymerase-binding rate 20-fold.

control protein that is turned off by the binding of a small molecule. Once switched off, the protein cannot help the RNA polymerase bind to the poor promoter sequence, so transcription is minimized (Table 11.1). The cAMP signal that, via CAP, activates the operons for alternative nutrient-digestion genes, simultaneously *inactivates* the TA proteins that help the polymerases bind to and transcribe the genes for transporting glucose into the cell. The metabolic systems for alternative food sources are activated, and the now-useless glucose system is temporarily shut down.

The *lac* system is one of the most complex procaryotic operons; it involves both negative control (a repressor, which can be inactivated by an inducer when lactose is present) and positive control (the CAP transcription factor, which can be induced to bind when glucose is rare). As a result, the *lac* operon is most active when lactose is present and glucose is absent (Table 11.2).

Table 11.2 Summary of *lac* operon control

ENVIRONMENTAL CONDITIONS		OPERON STATUS		TRANSCRIPTIONAL ACTIVITY OF LACTOSE-METABOLIZING GENES
GLUCOSE PRESENT	LACTOSE PRESENT	cAMP-CAP TA COMPLEX BOUND TO ACTIVATOR	REPRESSOR BOUND TO OPERATOR	
yes	no	no	yes	none
yes	YES[a]	no	NO[a]	LOW
NO[a]	no	YES	yes	none
NO	YES	YES	NO	HIGH

[a] Uppercase entries indicate factors that help enhance lactose digestion.

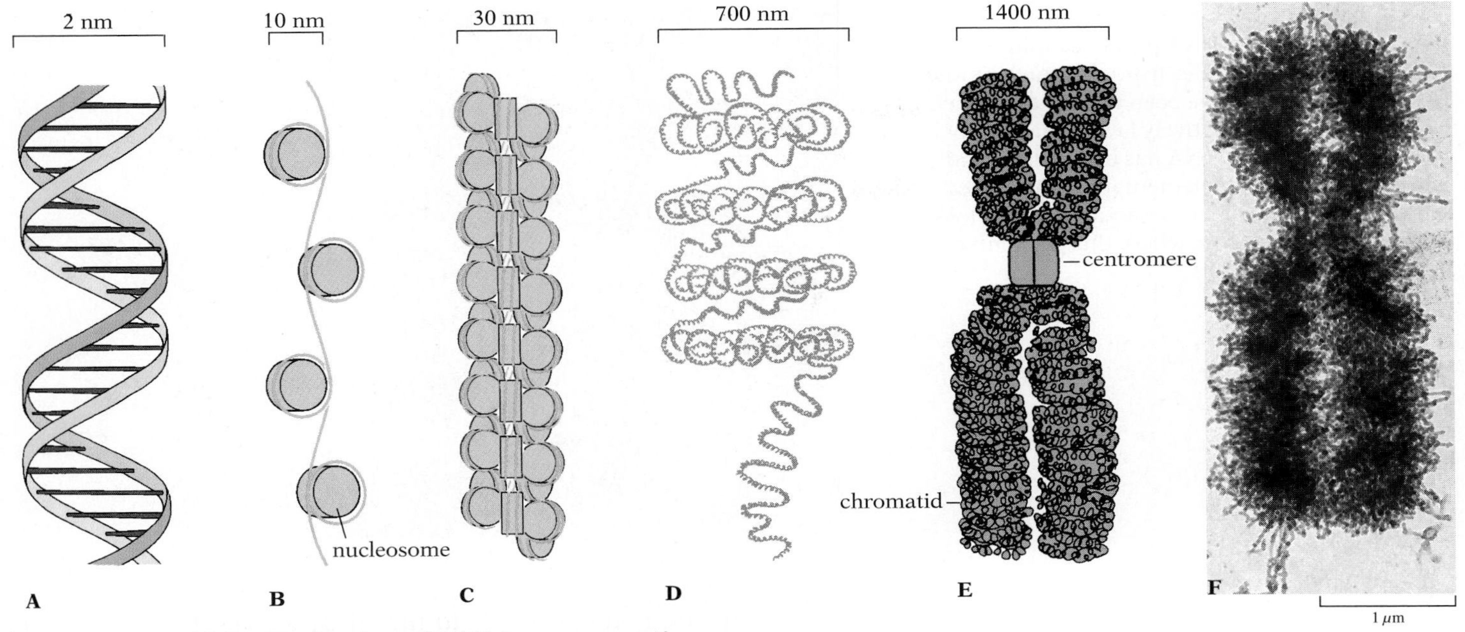

11.5 Packaging of DNA in a eucaryotic nuclear chromosome

The double-helical strand of DNA (A) is wound around nucleosomes about 10 nm in diameter (B). This chain of spools is itself coiled in some way to produce a thick strand about 30 nm in diameter (C). The arrangement shown here is hypothetical. Early in cell division the thick strands are collected into a long series of loops, which are wound into a helix (D). The result is the ragged appearance of the chromosome, as seen in drawing (E) and in the electron micrograph of a human chromosome (F). Some of the loops are tightly condensed while others are not, and can be seen extending out from the chromatid cores.

HOW EUCARYOTIC CONTROL IS DIFFERENT

The control of gene expression in eucaryotes is more complex, and it needs to be: even the simplest eucaryote has many more genes to regulate than any procaryote. In addition to functional complexity, there are also structural complications: as described in Chapter 5, eucaryotic DNA is wrapped on nucleosomes, wound into tightly packed coils, organized into loops, and condensed to form visible chromosomes (Fig. 11.5). Prior to transcription, a eucaryotic gene must be "unpacked."

■ THE ODD ORGANIZATION OF EUCARYOTIC CHROMOSOMES

Chromosomal proteins Eucaryotic chromosomes consist of DNA and protein, an association often called ***chromatin***. Chromosomal proteins include ***histones***, most of which are essential components of nucleosomes, and ***nonhistone proteins***. Histones may be involved in gene expression, but their role seems to be passive: when the nucleosome cores are in place, transcription is not possible, since RNA polymerase cannot gain access to the DNA.

Nonhistone proteins are much more important in gene regula-

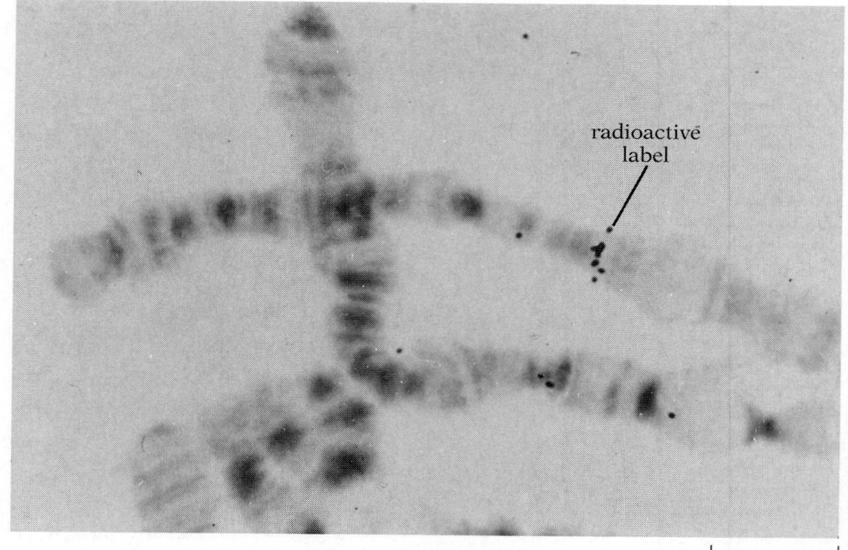

11.6 Locating a gene
In this version of the DNA probe technique, salivary-gland chromosomes from the much-studied fruit fly *Drosophila* have been treated to weaken base pairing; a radioactively labeled sample of previously identified DNA has been added and has bound to the complementary region. The dark dots reveal where radioactive decay is taking place, and therefore where this particular gene is located.

tion. Some are bound directly to the DNA, while others are linked to the nucleosome cores. They exhibit a rich diversity, not only from organism to organism, but also from tissue to tissue within a given organism, and even within a single cell at various times, depending on its developmental stage and its current functional condition. Hence, nonhistone proteins may have the specificity necessary for control elements. Their role, at least in some cases, is to bind to specific control regions in the DNA to cause a chromosomal loop to decondense (uncoil from its nucleosomes). They also play a role in the second step of eucaryotic gene control. Since most loops are about 100,000 base pairs long, each usually contains several genes, which are often selectively activated.[1] This level of control probably includes removing one or more nucleosomes from the promoter regions of the genes destined to be switched on. As we will see, activation rather than inhibition of transcription appears to be the rule in eucaryotes.

Highly repetitive DNA The organization of eucaryotic chromosomes has been revealed in large part through the technique of ***DNA hybridization***. The first step is to heat the chromosomal DNA under controlled conditions; this causes the two strands of each helix to separate. In one type of hybridization the next step is to add to the chromosomal DNA a labeled "probe" of a particular DNA sequence of interest, or an mRNA (or a cDNA made from it by reverse transcriptase). The mixture is then cooled very slowly. The RNA or DNA will frequently bind to the complementary region of the chromosomal DNA, and because of the labeling, the gene or sequence for it can then be located on the chromosome (Fig. 11.6). Clever variations on this basic method have produced a series of surprises for researchers, making it clear that the evolutionary histories of procaryote and eucaryote chromosomes have been very different. Perhaps the most consistent contrast is that eucaryotic

[1] The length of the average eucaryotic gene is 2000 nucleotides, of which only 1200 are in exons.

genomes are filled with sequences that are never transcribed, many of which exist in multiple copies. For instance, about 10% of most eucaryotic DNA contains base sequences that are found not once but thousands of times in the genome, constituting what is called ***highly repetitive DNA***.

There are at least four classes of highly repetitive DNA. The first consists of a vast number of copies of a few kinds of short sequences located at the ***centromere*** (the specialized region that holds the two copies of a replicated chromosome together until cell division) and in large blocks in chromosome arms, especially at the ends, where they are called ***telomeres*** (Fig. 11.7A). The function of this DNA, which is never transcribed, is to facilitate particular steps in the process of cell division (described in Chapter 12), and to maintain stability in the chromosome.

Another class of highly repetitive DNA consists of long units repeated in tandem. These areas contain genes that code for the smallest of the four ribosomal RNAs, called 5S RNA (Fig. 11.7B). (The name refers to the rate at which this RNA sediments through a sucrose solution; the other three rRNAs—6S, 18S, and 28S—are all larger and heavier by this measure.) In one well-studied frog species, each 5S gene is associated with a nonfunctional region known as a ***pseudogene*** (so called because its sequence is almost identical to that of a functional gene) and a very long "spacer" region. Neither the pseudogene nor the spacer is ever transcribed; most of this region of the chromosome appears to be functionless. Though the gene for 5S rRNA makes up only a small fraction of the region, there is a clear reason for the large number of copies of the 5S gene. Developing eggs need enormous numbers of ribosomes (perhaps 10^{12}) to handle all the protein synthesis necessary for rapid growth. Repeated transcription of a single copy of this gene would be inadequate to meet the needs of the cell for rRNA; consequently most eucaryotes have approximately 25,000 copies of this sequence.

A third class of highly repetitive DNA, known as "long interspersed elements," consists of sequences that are several thousand base pairs long and are found tens of thousands of times in the chromosomes of many eucaryotes (Fig. 11.7C). Their function, if any, is unknown.

The fourth class consists of relatively short (300 base-pair) segments scattered throughout the genome. Humans have about 500,000 copies of one such sequence, which is simply a transposon (derived from an RNA polymerase promoter) that reproduced and spread rampantly at some stage in our evolution; it appears to have no function.

Moderately repetitive DNA About 20% of the typical eucaryotic genome consists of ***moderately repetitive DNA*** whose sequences are repeated hundreds rather than thousands of times. Moderately repetitive DNA comes in one of two varieties. The first is a tandem repeat of certain genes. In particular, genes for the other three kinds of rRNA are repeated tandemly as a group: 18S, 6S, 28S, untranscribed long spacer; 18S, 6S, 28S, long spacer; and so on (Fig. 11.7D). Since every ribosome must have one of each of the four kinds of rRNA, how do cells manage to get by with only 100–1000

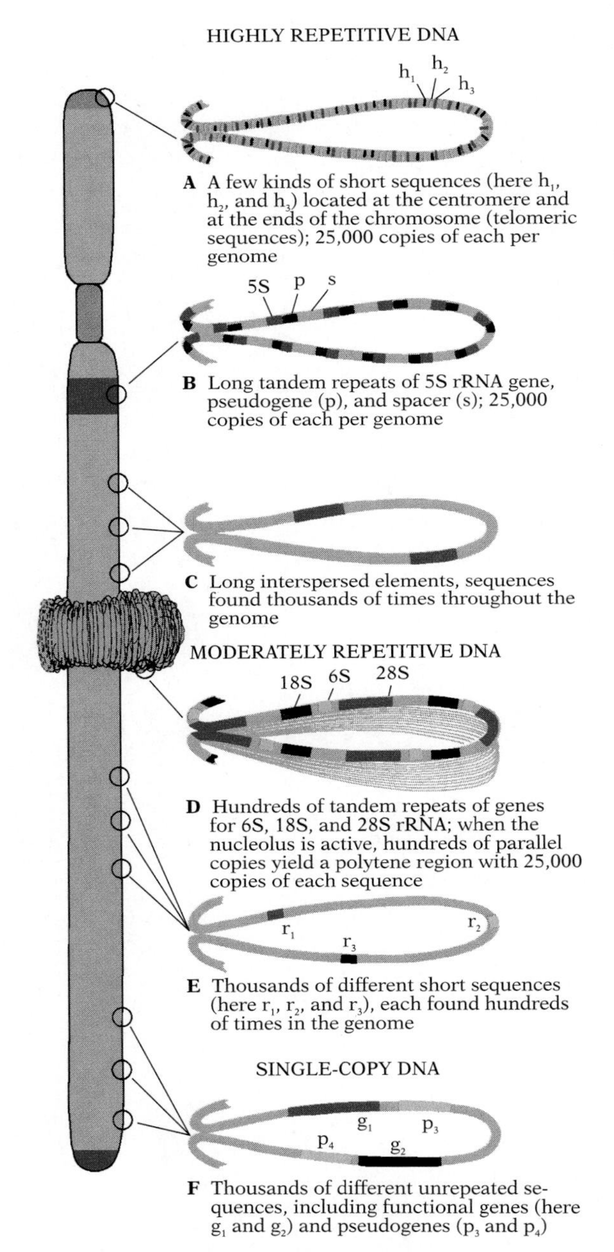

11.7 Organization of eucaryotic chromosomes

The short highly repetitive DNA (A), long tandem repeats (B), and polytene tandem repeats (D) are confined to different specialized regions of the chromosome. The long interspersed elements (C), moderately repetitive sequences (E), and single-copy DNA (F) are actually intermixed; for clarity, however, they are shown here in isolation on separate loops.

copies of these genes while there are many thousands of copies of the 5S gene? The answer is that when new ribosomes are needed, the 18S–6S–28S–spacer tandem repeat can be *replicated repeatedly, independent of the rest of the chromosome.* The result is a ***polytene*** region consisting of 25–250 parallel copies of the repeat region. Replicating the polytene region in turn produces up to 25,000 copies of each gene in active cells. The region of the chromosome with the many replicated repeating segments forms the nucleolus (described in Chapter 5).

The other class of moderately repetitive DNA is more mysterious, and varies widely in size and frequency between species. These sequences, of which there may be as many as 5000 different types, are only about 300–3000 bases long—as little as one-tenth the length of the average functional gene. Each type is usually scattered between functional genes throughout the chromosome in 30–500 different locations (Fig. 11.7E). Some of this DNA is clearly derived from mRNA containing the promoter sequence for reverse transcriptase; presumably some of the DNA copies were synthesized from mRNA by reverse transcriptase as an artifact of retroviral infection, and made their way into the host chromosome. One of the cell's tRNAs appears to be particularly susceptible to this process. Other cases, however, do not fit this model.

Single-copy DNA Though 70% of a typical eucaryotic genome consists of ***single-copy*** sequences (Fig. 11.7F), the majority of this DNA is *never* transcribed. In mammals, much of it exists as pseudogenes—nearly identical but untranscribed copies of functional genes. Only about 1% of eucaryotic DNA codes for mRNA that is subsequently translated. This is dramatically different from procaryotes, in which well over 90% of the genome is translated at one time or another. The high density of information in procaryote genomes—haploid and almost devoid of introns, spacers, and pseudogenes—is probably a result of intense selection for reproductive speed. In contrast, few eucaryotes find any reproductive advantage in high-frequency cellular doubling. This fact alone, however, seems insufficient to account for the eucaryotic cell's tolerance of countless introns, pseudogenes, and spacers, as well as transposons, retroinsertions, and other genetic debris. Some researchers believe that eucaryote genomes simply lack efficient ways of expunging genetic "junk," which therefore accumulates endlessly; others suspect that much of this vast proportion of untranscribed DNA is actively tolerated as a laboratory for evolutionary experiments in exon shuffling and mutation, or that it plays an undetected role in gene regulation. The logic of eucaryotic gene organization remains one of the most interesting and elusive questions in modern biology.

■ HOW CHANGES IN GENE TRANSCRIPTION CAN BE SEEN

Profound as the differences are in the organization of the procaryotic and eucaryotic genomes, it is not surprising that their systems of gene control differ as well. The eucaryotic genome, with its need to decondense tightly packed loops of DNA before transcribing them, must have an added level of control that procaryotes do not require. In addition, eucaryotes must orchestrate multiple constellations of genes, each specific for different tissues and different stages in their complex development from a single cell.

Heterochromatin versus euchromatin The internal structure of a chromosome is not uniform; much of its variation reflects gene activity. For example, some regions of chromosomes stain only faintly when treated with basic dyes, whereas other regions stain intensely. The nonstaining regions are called ***euchromatin*** and the staining ones ***heterochromatin***. Chromosomal mapping over the past few years has shown that euchromatic regions contain active genes whereas heterochromatic regions are inactive.

Some heterochromatic regions are simply devoid of genes, for instance, the large region of heterochromatin with highly repetitive base sequences located around the centromere. However, most heterochromatic regions do not lack genes; instead, the genes (often long series of genes) are simply inactive, and may be euchromatic at other stages of development.

Even within a region of euchromatin only a few widely spaced genes may be active; the clumping of a small set of related genes into an operon that is so characteristic of bacteria is rare in eucaryotes. Indeed, in many instances in which there appears to be simultaneous control of the synthesis of functionally related enzymes, the enzymes are encoded by widely separated genes. The genes that specify two polypeptide chains of a single protein (the α and β chains of hemoglobin, for example) are often on different chromosomes.

Lampbrush chromosomes and chromosomal puffs The developing egg cells of many vertebrates synthesize large amounts of mRNA for later use. The mRNA is then ready during the early stages of development, after fertilization, when the chromosomes are so busy with DNA replication in support of rapid cell division that they are largely unavailable for transcription. The parts of the chromosome bearing genes that are being repeatedly transcribed—the euchromatic regions—loop out laterally from the main chromosomal axis, while the parts bearing repressed genes remain tightly compacted; chromosomes with many looped-out regions are called ***lampbrush chromosomes*** (Fig. 11.8).

The giant chromosomes of the salivary glands of larval flies are lateral arrays of replicated chromosomes (polytene chromosomes) that have remained stuck together. Polytene chromosomes result from the same processes that, on a much smaller scale, create the polytene region called the nucleolus. About 10 cycles of replication are involved in the creation of a polytene chromosome, yielding 1024 (2^{10}) parallel copies of each gene. The result of this unselective gene amplification is a capacity for rapid synthesis of large

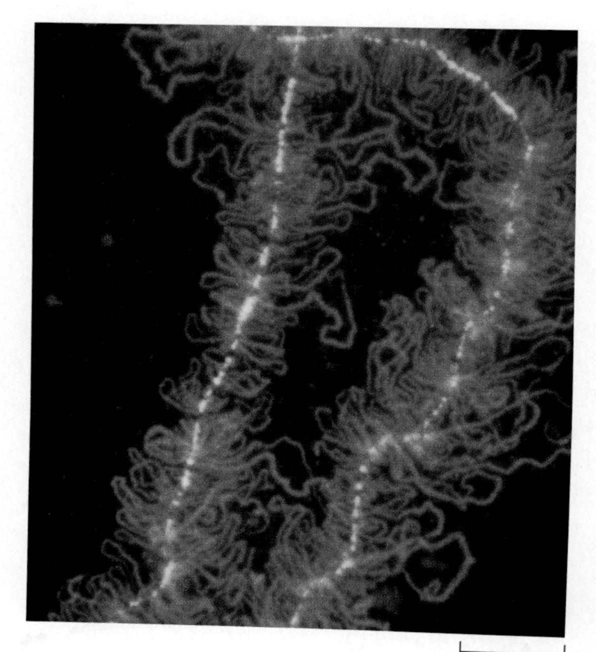

11.8 Lampbrush chromosome from a developing egg cell of the spotted newt (*Triturus viridescens*)

The many feathery projections from the chromosome are regions bearing genes that are being repeatedly transcribed.

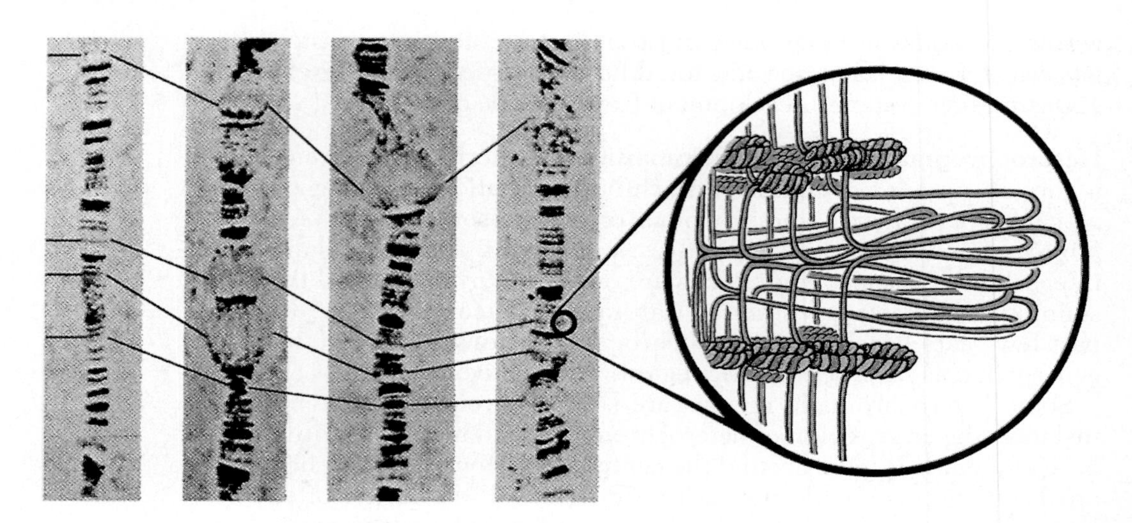

11.9 Chromosomal puffs

Puffs in polytene chromosomes consist of hundreds of parallel copies of the chromosomal DNA looped out to expose the maximum surface for synthesis of RNA. The interpretive drawing shows only nine of the loops, flanked by tightly coiled regions. The series of photographs indicates how the location of puffs in a *Drosophila* chromosome changes dramatically but predictably over time; we can clearly see the varying activation of four different genes or sets of genes on chromosome 3 (each identified by a series of connecting lines) over a period of 22 hours (left to right) in larval development.

amounts of RNA. When certain regions of a giant chromosome are especially active, all of its parallel DNA strands form brushlike loops (***chromosomal puffs***) in those regions (Fig. 11.9). The locations of puffs are different on chromosomes in different tissues. They are also different in the same tissue at different stages of development (Fig. 11.9), though at any given time all the cells of any one type in any given tissue show the same pattern of chromosomal puffing.

The puffs indicate the location of active genes, and thus the sites of rapid RNA synthesis. As such, the puffs provide a way of determining visually whether or not changes in the extranuclear environment—the cytoplasm or extracellular chemicals—can alter the pattern of gene activity. For example, if ecdysone, the hormone that causes molting in insects, is injected into a fly larva, the chromosomes rapidly undergo a shift in their puffing pattern, taking on the pattern characteristically found at the time of molting. If chromosomes from one type of cell are exposed to the cytoplasm of another type of cell or of the same type of cell at a different developmental stage, they quickly lose the puffs characteristic of their original cells and develop puffs characteristic of the type and stage of cells providing the new cytoplasm.

■ IMPRINTING: PRESERVING PATTERNS OF GENE ACTIVITY

Rapidly dividing cells in a developing eucaryote must have some way of preserving the patterns of loop decondensation and gene activity that give them the appropriate chemical "personality" for the tissue the cell is a part of. In fact, the patterns of euchromatin and heterochromatin can be passed directly from a growing cell to the two daughter cells it produces through a process called ***imprinting.*** One component of this imprinting system in vertebrates involves methylation of the cytosines in certain C–G sequences in inactive genes. Methylation can transmit activity patterns to daughter cells: Because only C–G sequences can be methylated by the enzyme involved, and since C–G is always paired with G–C on the other strand, the same pattern of methylation exists on both strands of the DNA:

```
                 m
                 |
:   :   : A : T : C : G : T : C : A :   :   :
:   :   : T : A : G : C : A : G : T :   :   :
                     |
                     m
```

(The colored *m* indicates the cytosine has been methylated by the addition of a CH_3 group.) Replication creates a hybrid structure wherever the parental DNA had been methylated:

```
                 m
                 |
:   :   : A : T : C : G : T : C : A :   :   :      parental strand
:   :   : T : A : G : C : A : G : T :   :   :      new strand
```

A special enzyme, maintenance methylase, scans the DNA for methylated cytosines in C–G sequences and methylates the corresponding cytosine on the new strand. Thus the pattern of methylation—gene inactivation—is passed on intact. If, however, both methyl groups at a site are lost by chance, cells are produced with altered gene activity. At least some of the problems associated with aging probably arise from the progressive loss of methylation over time, and hence the activation of inappropriate genes in cells. Demethylation in gametes can even pass mistaken gene-activation patterns on to offspring, and several diseases are now ascribed to such imprinting errors.

■ EUCARYOTIC CONTROL: INDUCERS AND ENHANCERS

Little is known about the selective decondensation of chromosomal loops—the necessary precursor to transcription—that creates polytene puffs and lampbrush projections. Certain master control chemicals like ecdysone can trigger major changes in looping, but how they act is a mystery. Despite this, knowledge about the basis of individual gene regulation on decondensed loops is growing rapidly.

Since eucaryotes rely mainly on positive control, transcription does not occur without active aid from a ***transcription factor* (TF)**, which helps the polymerase bind to the promoter. In addition, eucaryotes control TF binding in gene-specific ways from two other locations on the chromosome.

The first control locus is an inducer region just upstream of the TF sequence (Fig. 11.10A). It consists of one or (usually) more individual sites that specific control proteins can bind to. Some of these control molecules in turn help the TF protein bind (Fig.

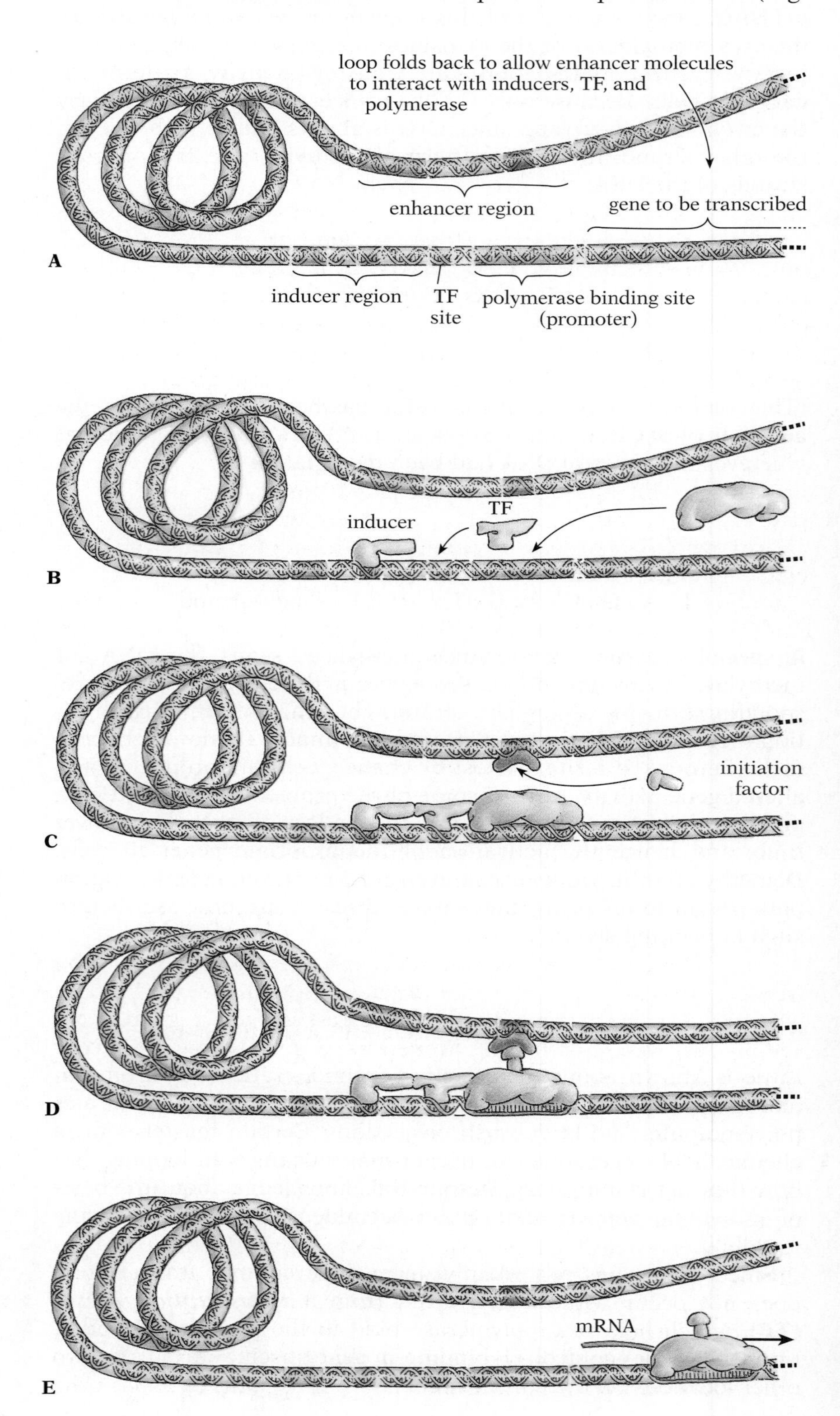

11.10 Eucaryotic gene control

(A) For this hypothetical gene there are four inducer sites adjacent to the transcription-factor (TF) binding sequence, and four enhancer sites upstream. (B) In this simplified positive-control model an activating inducer binds to one site and loads the TF. (C) This in turn helps the polymerase to bind. An activating enhancer binds upstream. (D) Once the loop folds into position, the enhancer loads transcription-enhancing factor onto the waiting polymerase. (E) This action initiates transcription. As described in the text, there are other ways inducers and enhancers can participate in starting or blocking transcription.

11.10B–C), while others inhibit it. For instance, the β-globin gene (which encodes one of the two kinds of protein chains that combine to form hemoglobin) has seven of these control sites in its inducer region.

The other locus of action is a few thousand nucleotides away on the same decondensed loop. It consists of a cluster of sites collectively called the ***enhancer region.*** These sites exert their effects when the loop folds back on itself to bring the entire region into contact with the promoter and inducer sites (Fig. 11.10). Some enhancers help load the TF protein, which then aids in polymerase binding; others directly help the polymerase to bind. Still others add a protein to the polymerase complex that can greatly enhance its efficiency in transcription. Other regulatory substances that can bind to the enhancer region inhibit one or another of these steps.

This action-at-a-distance system is nearly ubiquitous in eucaryotes, and a few bacterial genes have been found to be controlled in this way as well. Why do most eucaryotic genes have so many alternative ways to regulate transcription? Eucaryotes have much more complex life histories than procaryotes. Eucaryotes are usually multicellular, and so must grow in a controlled and directed way to achieve a species-specific morphology. Moreover, eucaryotes usually have distinct tissues and organs whose cells must develop particular attributes to fulfill their specialized roles. In short, gene activity depends not only on immediate metabolic needs, but on a cell's location and developmental age as well.

Studies of specific clusters of enhancer and inducer sites support this idea. Some sequentially active enhancer and inducer sites are found in many genes scattered throughout the chromosomes, indicating that those genes are expressed in synchrony. There are four versions of the β-globin gene, for instance: two encode hemoglobin subunits whose affinity for oxygen is appropriate early in development, before the fetal circulatory system matures; another version of the gene takes over for the interval during which oxygen must be obtained from the maternal placenta; and the fourth version leads to the production of a subunit adapted to the high-oxygen conditions in the lungs of children and adults. Each has a regulatory sequence shared with the many other genes active at similar ages.

Other sequences are less common, and seem to promote gene transcription only in particular tissues; the activity of β-globin genes, for instance, in addition to being restricted to particular developmental stages, is further limited to cells in the bone marrow, along with all the other specialized genes needed to produce red blood corpuscles and to maintain the marrow itself. It is only in the marrow cells that the necessary enhancers for these genes are found; the enhancers, in turn, are produced in response to earlier chemical signals from neighboring cells (signals that identify tissue type), and internal signals encoding developmental age.

Finally, there are control sequences in these same marrow cells that are unique, or are shared by only a few genes at most; at least some of these appear to adjust the concentration of hemoglobin to ensure an adequate supply. An active gene in a eucaryote, then, would probably need the cooperation of at least three control substances bound to the inducer and enhancer regions, as well as the

Table 11.3 Comparison of procaryotic and eucaryotic gene control

	PROCARYOTES	EUCARYOTES
Control of DNA	Usually negative, blocked by a repressor	Usually positive, possible only with aid of transcription factor (TF)
Direct control of TA/TF binding: inducer sites	Allosteric change after binding of transcriptional activator to activating (inducer) site	Inducer must bind to DNA before helping TF bind; inhibitors can bind to DNA and block TF or inducers
Indirect control: enhancer sites	Very rare; exerts negative effect	Very common: can exert positive or negative effect
Levels of control	Usually one; rarely two	Usually at least three; very often four or more

absence of any inhibitors, in order to be transcribed. The comparison of the main features of procaryotic and eucaryotic gene regulation are summarized in Table 11.3.

■ GENE AMPLIFICATION

Ordinarily a single copy of each gene per chromosome is sufficient to meet the needs of a cell for the substance coded by the gene. The rate of transcription of a single gene is such that in 4 days it may produce 100,000 mRNA molecules, which can lead to the synthesis of as many as 10 billion (10^{10}) protein molecules. However, in some cases the demand for a product is so great that a single gene cannot meet it. We've already seen that so much rRNA is required by a eucaryotic cell for construction of ribosomes that the 5S rRNA gene is repeated in tandem 25,000 times (see Fig. 11.7B). The other three rRNA genes are repeated only 100–1,000 times in tandem, but when large numbers of ribosomes are needed, many parallel copies of this region of the chromosome are synthesized (Fig. 11.7D). This polytene region usually loops out from the main axis of one of the chromosomes to form the nucleolus, and rRNA is produced there at a high rate (Fig. 11.11). The production of these extra copies of the rRNA genes from moderately repetitive DNA is one type of ***gene amplification.*** Inappropriate gene amplification is common in many kinds of cancer.

■ CONTROL AFTER TRANSCRIPTION

Cellular control mechanisms are not limited to regulating mRNA synthesis. For example, some primary transcripts can be spliced (processed) in more than one way, yielding slightly different products, depending on the cell's needs. In *Drosophila,* for example, the way one gene's mRNA is spliced controls the expression of a constellation of other genes that help orchestrate the development of either a male or female fly (Fig. 11.12). The critical event occurs when the splicing complex prepares to join the end of exon 3 to the beginning of the next exon. If a particular gene (*transformer,* or *tra*) is fully expressed, it causes exon 4 to be added, but blocks further

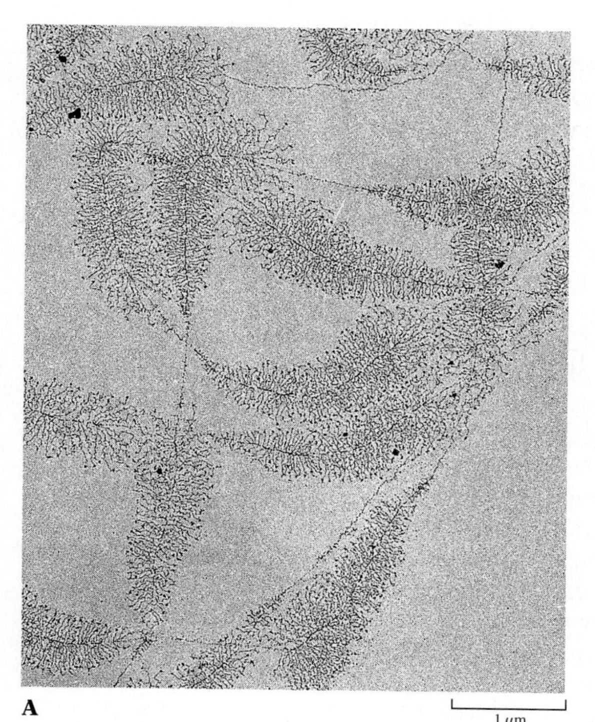

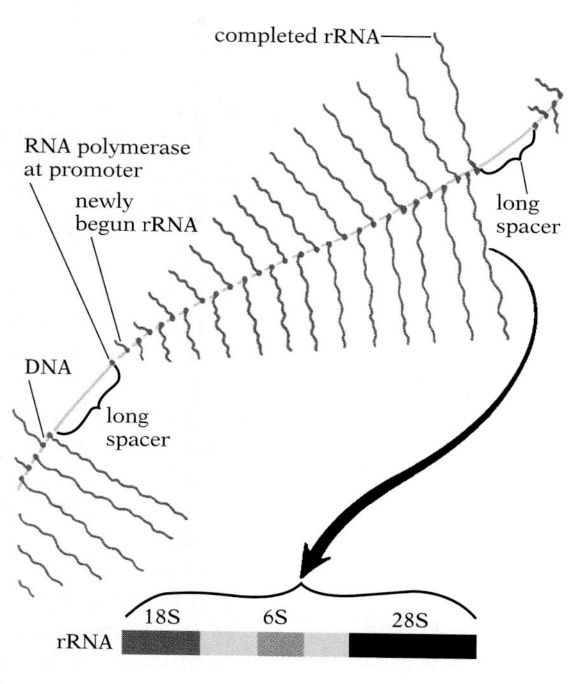

11.11 Transcription of rRNA on DNA in the nucleolus

(A) Electron micrograph of a portion of the nucleolus in an egg cell of the spotted newt. This continuous strand of DNA bears multiple copies of the genes for three RNAs. Strands of rRNA at progressive stages of synthesis feather out from each set of genes. Because the successive gene sets are separated by regions of intergenic DNA spacers where no rRNA is being synthesized, we can see more or less exactly where on the DNA molecule each set of genes begins and ends. (B) Diagram of rRNA synthesis on one gene set. Many molecules consisting of the three kinds of rRNA joined together in a single strand (red) are being synthesized, as RNA polymerase molecules (red circles) specialized for transcribing rRNA genes move along the DNA. Strands of rRNA attached to the DNA near the beginning of the gene set are still short, their synthesis having just begun.

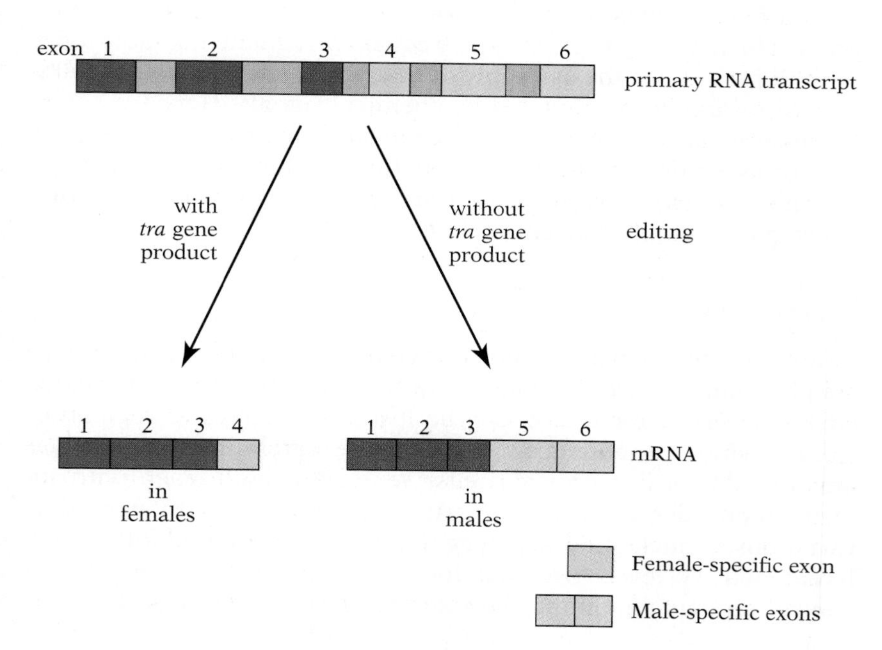

11.12 Alternative splicing in *Drosophila*

The primary transcript of the "double-sex" gene in *Drosophila* can be spliced in two ways, depending on whether the *tra* gene product is present to intervene in the process. If the exon sequence is 1-2-3-4, the protein translated from the mRNA leads to the expression of female-specific genes, presumably by binding to the appropriate places in the DNA. If the sequence is 1-2-3-5-6, the mRNA encodes a product that leads to the development of a male fly.

splicing; if active *tra* gene product is not available, exon 4 is skipped and all the remaining exons are spliced normally. (The *tra* protein mimics the portion of the splicing complex that binds to the leading end of exons, but how it alters splicing is not known.) The *tra* product itself is generated in response to signals in the nucleus that reflect whether the developing organism is a male or a female fly.

Even after splicing, almost half of the mature mRNA produced in the nuclei of most cells never reaches the cytoplasm for translation. There must be a system that can selectively block the export of specific mRNA molecules, probably in response to chemical signals that reflect the cell's needs. Even when mRNA reaches the cytoplasm, it may not be translated. Inhibitor substances can bind to specific messengers in the cytoplasm and block either ribosomal binding or complete translation. This strategy can reduce the cell's response lag to changing conditions by maintaining a pool of mRNAs ready to begin or complete translation when an appropriate chemical signal binds to the inhibitor and causes it to disassociate from the messenger. In most cases, the translation inhibitor is a protein, but in a few instances a segment of RNA with a sequence complementary to the mRNA binds though base pairing. This binding of ***antisense RNA*** is the mechanism by which herpes viruses are able to remain dormant in cells for extended periods. How antisense inhibitors come to be removed is not yet understood. Other control chemicals facilitate rather than inhibit translation; these bind to specific mRNAs and help them to bind ribosomes.

Finally, the expression of mRNA can be regulated by the rate at which the messenger is broken down: some types of mRNA have long life spans, while others are destroyed within minutes. An mRNA's degree of resistance to digestion by RNAse is encoded near the trailing end of the messenger. Control can also be exerted after translation: for example, protein longevity (that is, resistance to proteases) is written into the sequence of subunits, and ranges from a few minutes to days or, for some structural proteins, even years. The activity of enzymes is frequently regulated by activators or inhibitors. Even the assembly of many structural proteins (collagen, for example) is regulated by chemical signals. Hence, a gene's expression can in some cases be controlled at every step from before transcription until after translation. Failure of these control systems can have disastrous results, producing a variety of diseases, most notably cancer.

■ THE IMPACT OF MUTATIONS

In addition to the usually harmful changes in protein activity created by mutations, alterations in control sequences can have major effects: levels of regulation can be lost, and degrees of regulation can be reduced or enhanced. Consider the starch-digesting enzyme amylase. Many different variants were discovered, each with its own unique degree of activity. At first, researchers thought that variations in nucleotide sequence in the gene accounted for the different activity levels. In fact, the critical alterations are in the control regions regulating the rate of gene transcription. This observation has had a major impact on the way scientists now think about gene evolution.

Alterations of regulatory regions can also be important to human health. Consider, for example, the two steps involved in detoxifying alcohol in humans:

$$\text{ethyl alcohol} \xrightarrow{\text{alcohol dehydrogenase}} \text{acetyl aldehyde} \xrightarrow{\text{aldehyde dehydrogenase}} \text{acetic acid}$$

In many individuals, the transcriptional activity of the aldehyde dehydrogenase gene is unusually low, leading to a buildup of the toxic intermediate acetyl aldehyde; the result is a low tolerance for alcohol. Most individuals of Asian and Native American descent have characteristically low aldehyde dehydrogenase levels. Alcoholics, on the other hand, almost invariably have elevated levels of aldehyde dehydrogenase transcription; presumably, one of their control sequences regulating transcription of this gene is unusually active. Such patterns suggest a genetic basis for susceptibility to at least some diseases or behavior patterns. The most dramatic cases of altered gene regulation, however, come from studies of the many forms of vertebrate cancer.

CANCER: WHEN GENE CONTROL FAILS

Biologists hope to learn more about how normal cellular controls operate by studying how they fail—by investigating how the normal controls can become so deranged as to permit cancerous growth. The most distinctive feature of cancer cells is their unrestrained proliferation, which results in the formation of malignant tumors. These tumors grow, consuming the body's resources while also crushing nearby structures (the lungs, for instance), rendering them useless. Worse yet, the tumor cells often spread from the original site of growth to many other parts of the body, a process known as ***metastasis.***

■ TISSUE CULTURE: A WAY OF STUDYING CANCER CELLS

One of the most profitable ways of studying the properties of specific kinds of cells is to grow them in tissue culture in the laboratory. Embryonic cells and tumor cells grow best. Among nontumor cells, embryonic fibroblasts outgrow all other cells in culture. Fibroblasts (literally, "fibrous germinating cells") serve many regenerative functions such as producing scar tissue. The usual lab procedure is to seed a large number of cells onto (or into) a sterile culture medium to which a variety of nutrients and growth-stimulating factors have been added. The nutrients and other factors are frequently supplied by serum—the liquid portion of blood. Serum contains, for example, the platelet-derived growth factor that signals fibroblasts to initiate the growth of scar tissue. In the body, the growth factor escapes from the circulatory system and binds to receptors on fibroblasts whenever a wound causes a break in the blood vessels.

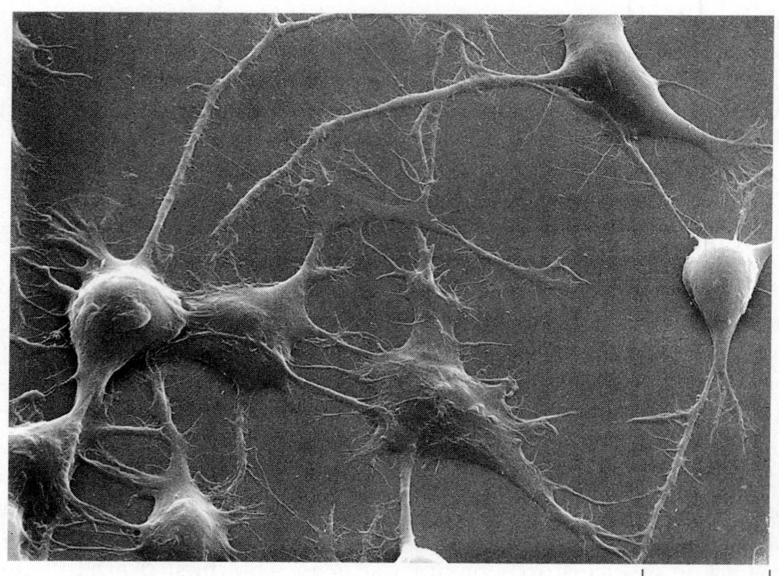

11.13 Cultured neuroblastoma cells from a mouse

These cells, from a malignant tumor of the nervous tissue (neuroblastoma), are growing on a solid surface.

A cell culture, no matter how elaborately set up and controlled, is far from the normal environment of the cells. Nevertheless, some of the cells may survive and engage in many of their usual developmental and functional activities, including cell division (Fig. 11.13). Most cultured lineages of animal cells stop dividing when they become crowded or, if crowding is prevented, after a certain number of cellular generations. The number of generations tends to be specific for the species and tissue of origin.

These two mechanisms limiting the growth of cells in culture are found in organisms as well. The first one—the cell's reaction to the effect of crowding—is called ***contact inhibition.*** When receptors in the cell membrane recognize markers in the cell coats (glycocalyces) of adjacent cells of the same type, the receptors signal the nucleus, and further cell division is inhibited. This inhibition is not absolute, however: adding high concentrations of growth factors, for example, can often stimulate a further round of cell division even in a crowded culture. The second mechanism—the automatic cessation of cell division after a set number of divisions—is an independent system for limiting cell proliferation. In most tissues it comes into play only if contact inhibition fails to work normally.

Cultured cells that continue to divide indefinitely fall into one of two categories: those that have lost the fixed-number-of-divisions control, and will go on growing as long as they are not crowded, and those that have lost both kinds of control. The cells of each category are potentially "immortal" in culture, but the former resemble ***benign*** (that is, self-limiting) tumors, while the latter resemble cancers: they will grow indefinitely regardless of crowding to create an ever-larger mass of tissue.

■ HOW ARE CANCEROUS CELLS DIFFERENT?

The loss of normal control systems can produce a variety of changes in a cell's anatomy, chemistry, and general behavior.

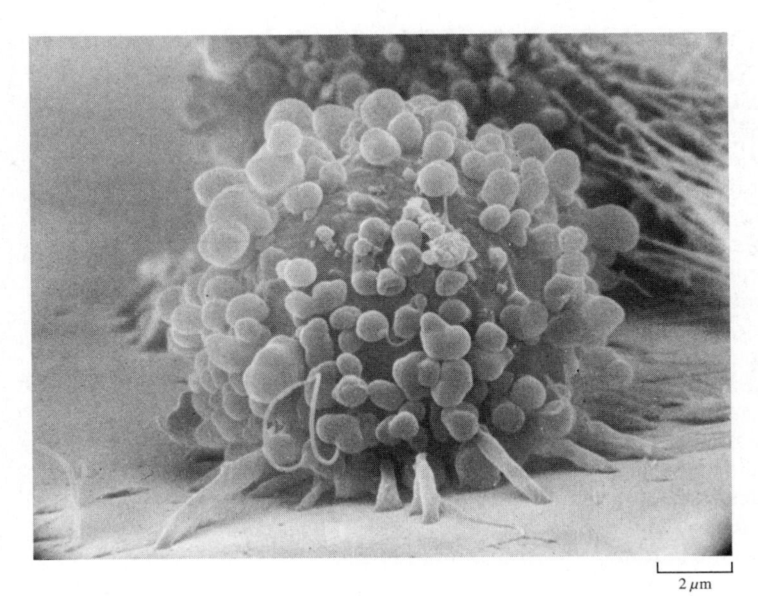

11.14 A typical cancer cell

This cell, from the HeLa line, is growing in tissue culture. The cell is rather spherical, and is covered with many small "blisters," called blebs, whose significance is not understood.

Studies of these changes tell us quite a bit about how cancer cells differ from normal ones, and sometimes suggest potential avenues of treatment.

One property of cultured cancer cells is that they almost always have an abnormal set of chromosomes. For example, the cells of the human cancer line called HeLa (Fig. 11.14), by far the most widely studied line of cultured human cells, typically have 70–80 chromosomes instead of the normal 46. Interestingly enough, noncancerous cell lines that become potentially immortal in culture also have extra chromosomes; hence it seems likely that possession of extra chromosomes somehow frees cultured cells from some of the normal constraints on proliferation.

In cancerous tissue in organisms, however, such an increase in chromosome number is not typical. Instead, there are frequently numerous self-perpetuating chromosomal segments called minute chromosomes, or specific rearrangements of the normal chromosomes. For example, in 90% of patients with Burkitt's lymphoma (a cancer of cells in the immune system), the tip of chromosome 8 has been moved to the end of chromosome 14 (see Fig. 12.4, p. 311). The insertion point in chromosome 14 is next to a gene coding for an immune-system polypeptide while the translocated region contains a gene that is frequently involved in cancers. The other 10% of patients with Burkitt's lymphoma show other translocations, from different sites on chromosome 8 to the same target site on chromosome 14. The implication is that the critical site in chromosome 8 has a control region that can contribute to cancer by stimulating the transcription of specific genes. Predictable translocations are also found in the cultured cells of several other sorts of cancer.

Besides differences within the nucleus, cancer cells and normal

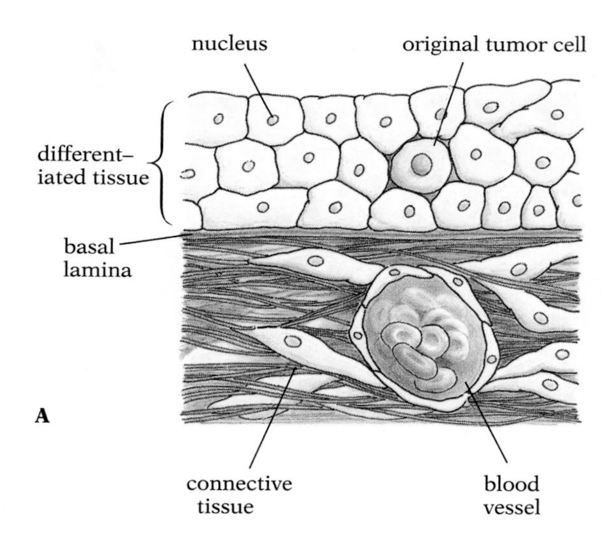

11.15 Development of a malignant tumor
(A) Normal tissue has differentiated cells with small, quiescent nuclei, separated from the circulatory system and loosely packed connective tissue by a tough basal lamina (the basement membrane). (B) When a cell incurs enough changes to begin proliferating, its nucleus enlarges, its shape changes, and it begins to replicate its DNA. As the cell reproduces, the clone it forms begins to push other cells back. If it lacks contact inhibition, it will grow regardless of crowding, and if it is not limited to a fixed number of divisions, the clone will grow as long as nutrients (blood supply) permit, but may still be contained by the lamina. (C) If a cell in the clone develops the ability to penetrate the basal lamina, its descendants will move across and proliferate. The tumor will metastasize widely when it penetrates the lamina protecting blood vessels.

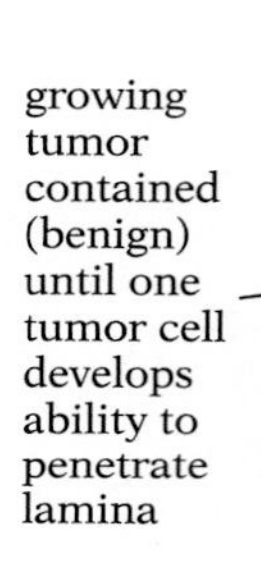

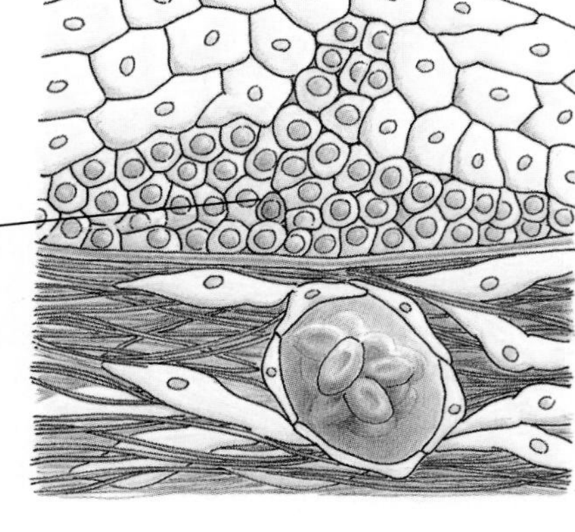

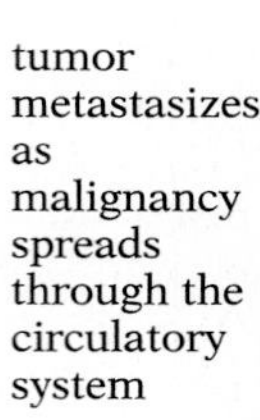

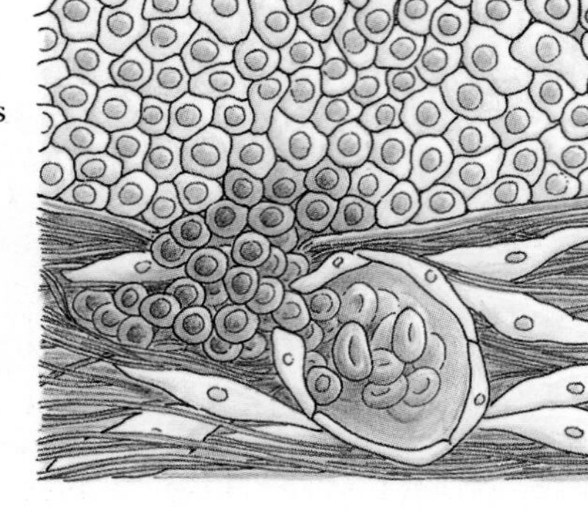

cells display a host of significant differences related to cell shape and to the nature of the cell surface. Cultured cancer cells, for instance, tend to have a spherical shape, one seen in normal cells only during a short period immediately following cell division. This peculiarity of shape probably results from their having an abnormally small number of functional structure-stabilizing microfilaments, a factor that also makes these cells more mobile than normal cells. This characteristic is also related to a striking feature of cultured cancer cells known as anchorage-independence: most cells must "cling" to a solid surface in order to grow—a useful constraint, since they will require support later in order to function properly. Cancer cells are not constrained in this way, and so they can grow in liquids or on soft surfaces.

In order to metastasize in an individual, and so escape from its tissue or organ of origin, a cancerous cell must usually be able to produce and transport to its own membrane a receptor called laminin, which enables cancer cells to bind to the basal membrane (a tough layer, or lamina, that underlies or surrounds many tissues and organs, including the vessels of the circulatory system). It must then be able to secrete collagenase, which digests the lamina and allows the cell to cross the barriers that otherwise contain the growth of tumors (Fig. 11.15). In particular, the capacity to break through the lamina surrounding capillaries allows cancer cells to spread throughout the body. Since most normal cells can neither produce laminin nor secrete collagenase, the majority of tumors are contained at their site of origin, and so are benign.

Cancer cells, with their abnormal surfaces, appear to lack the ability to react appropriately to environmental changes. Cancer cells typically have fewer and different glycolipids and glycoproteins in their cell coats; these differences are probably correlated both with the absence of normal contact inhibition during cell division, and with the apparent inability of cancer cells to recognize other cells of their own tissue type. When mixed in culture, normal cells from two different tissues (such as liver and kidney) tend to sort themselves out and reaggregate according to tissue type; cancer cells, however, don't do so. This absence of normal cellular affinities is probably one of the reasons why malignant cells of many cancers can spread.

■ THE STEPS THAT LEAD TO CANCER

The conversion of a normal cell into a cancer cell involves several changes: loss of fixed-number-of-division control, loss or reduction of contact inhibition, loss of anchorage-dependence and, sometimes, tissue-specific cell-surface changes. But in many tissues even these changes result only in benign tumors that grow slowly and then stop, apparently because the cells fail to produce the chemical signal or signals that cause vascularization (the establishment of a capillary system that supplies oxygen and nutrients to growing tissue). Even when it can recruit capillaries, the tumor cannot metastasize without being able to bind to and destroy the basement membrane. Do all these changes take place as a result of a single genetic event, or is each change the consequence of a separate event?

Health statistics suggested the answer to this question before the actual mechanisms began to be elucidated. If just a single genetic event were required, the probability of an individual's having contracted a particular type of cancer should increase proportionally with age. For example, if the chance of contracting skin cancer were 1% per year, then the chance of having contracted it would be 2% for a two-year-old, 3% for a three-year-old, and so on. In fact, however, cancer is normally a disease of old age (Fig. 11.16). To account for the exponential rise of the incidence of cancer, several independent genetic events are required. The curve would be the product of the individual probabilities of the separate events.

Some cancers require only two or three genetic events; examples of this group are leukemia and immune-system cancers because they involve tissues in which tissue-type affinity, vascularization, crossing of a lamina, and a solid surface for growth are not necessary for the normal cells. Other kinds of cancer require four to seven genetic events.

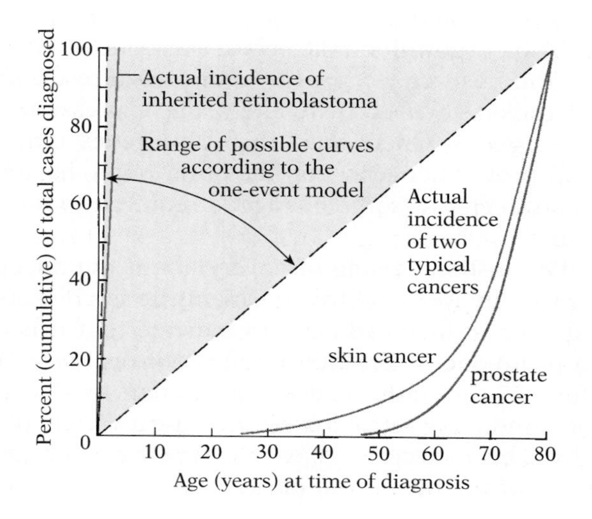

11.16 Incidence of representative cancers

If each type of cancer were caused by a single transforming event, the probability of developing a particular cancer would increase linearly with age; the shaded area indicates the range of possible curves. Actually, the incidence of most cancers increases exponentially, indicating that several contributing events are usually involved. The incidence curve for skin cancer closely approximates a three-event curve, while the curve for prostate cancer approximates a five-event curve. By contrast, the curve for inherited retinoblastoma is linear, approximating a single-event curve; this is because the other transforming event required for this form of cancer is already incorporated in a mutant gene carried by these individuals.

Additional evidence for a multistep model comes from studies of individuals who inherit a tendency to get certain cancers. Individuals with a genetic susceptibility to retinoblastoma (a cancer of the eye) usually develop it when young. The cumulative-incidence curve for this cancer shows a linear rise with age, rather than the more common exponential one. This suggests that in such individuals only a single transforming event is necessary to trigger the disease. (All other necessary changes must already be present in the form of mutant genes.) Because inheritance of genetic susceptibility in this kind of cancer behaves as a single-gene characteristic, developing retinoblastoma must require a total of *two* transforming events among individuals who are not genetically predisposed to it; genetically susceptible individuals require only one, because the other event is already built into the genome.

Statistical studies cannot tell us what constitutes a genetic event leading to cancer—among the possibilities are single-base changes, deletions, insertions, or chromosomal rearrangements. Nor do the studies tell us the precise effects of the events. Most steps are probably linked directly to the loss of a particular level of control. The best data on this come from studies of oncogenes.

■ THE ROLE OF ONCOGENES

The discovery of ***oncogenes***—genes that cause cancer—has a complex history. In 1910 Peyton Rous found that what is now known as Rous sarcoma virus could cause cancer in chickens; how it did so was then a mystery. We know now that the causative agent is a retrovirus, which, like other retroviruses, produces a cDNA copy of its RNA that is incorporated into the host chromosome (see p. 250). Some strains of Rous sarcoma virus cause cancer quickly, and at a high rate, because the cDNA brings with it an oncogene (known as *v-src*—the *v* identifies it as a viral gene while *src* specifies that it causes a sarcoma). The expression of this gene in the host chromosome, which does not depend on the site of incorporation, makes the cell cancerous.

Ordinary retroviruses—those not containing oncogenes—can also cause cancer, but only at a very low rate, and even then only after a long latency. Incorporation of cDNA into the host genome is apparently random, and cancer from ordinary retroviruses occurs only when the incorporation takes place near one of a few particular genes in the host, called ***proto-oncogenes.*** Presumably an active control region in the cDNA alters the expression of one or more of these genes. Oncogenes, then, are genes that inevitably cause cancer; proto-oncogenes, on the other hand, have the potential to cause cancer, but some change is required to convert them into oncogenes.

Perhaps the most surprising discovery about the oncogene *v-src* is that it is virtually identical to a normal gene in chickens; indeed, this normal gene (designated *c-src* to indicate that it is cellular in origin) is a proto-oncogene: incorporation of ordinary cDNA near *c-src* can (eventually) cause cancer. More than two dozen rapidly acting, cancer-inducing retroviruses infecting a variety of birds and mammals have been discovered, each carrying an oncogene almost identical to a normal host proto-oncogene.

Table 11.4 Events that can create oncogenes

I. Altered gene, resulting in altered product
- 1. Incorporation of retroviral oncogene
- 2. Mutation of normal proto-oncogene to create an oncogene

II. Altered gene expression, in which a gene is transcribed at the wrong time or at too high a rate
- 1. Insertion next to a proto-oncogene of retroviral cDNA containing an active control region
- 2. Translocation of a proto-oncogene to a position next to an active control region
- 3. Translocation of an active control region to a position next to a proto-oncogene
- 4. Mutation of a control region next to a proto-oncogene

III. Loss of anti-oncogene activity

The viral cancers are convenient to study, but as yet few forms of cancer in humans are definitely viral in origin. However, studies of cultured human and nonviral animal tumors reveal strong similarities to the observed behavior of retroviruses: some of the genetic changes that are responsible for making the cells in these cultures cancerous involve the same oncogenes also found in cancer-inducing retroviruses. Some of the nonviral oncogenes appear to have arisen through a mutation in a proto-oncogene that altered the base sequence through a base change, insertion, or deletion. Others appear to be proto-oncogenes that have been moved (translocated) from their usual locations, and perhaps released from their normal controls, to become oncogenes.

Conversely, either mutation or translocation involving a control region next to a proto-oncogene can convert a proto-oncogene into an oncogene. Indeed, when grown in culture, cells of several leukemias, one ovarian cancer, and several other kinds of cancer display predictable chromosomal translocations or losses. The translocations seem to involve moving a structural gene and its control region, which is responsible for its transcription in the tissue at a high rate, to a position next to a proto-oncogene, or moving the proto-oncogene next to a structural gene and its control region. Finally, there are ***antioncogenes***—genes that encode proteins that deactivate oncogene products. Failure of such an inhibitor gene is equivalent to activating a proto-oncogene. The gene involved in retinoblastoma is an antioncogene.

The various mechanisms that may result in the presence of an oncogene in the genome are summarized in Table 11.4.

Most oncogenes encode products that fall into one of four general categories (Fig. 11.17):

1 Growth factors (extracellular signal molecules that stimulate cell proliferation)

2 Receptors (for growth factors, contact inhibition, or surface adhesion)

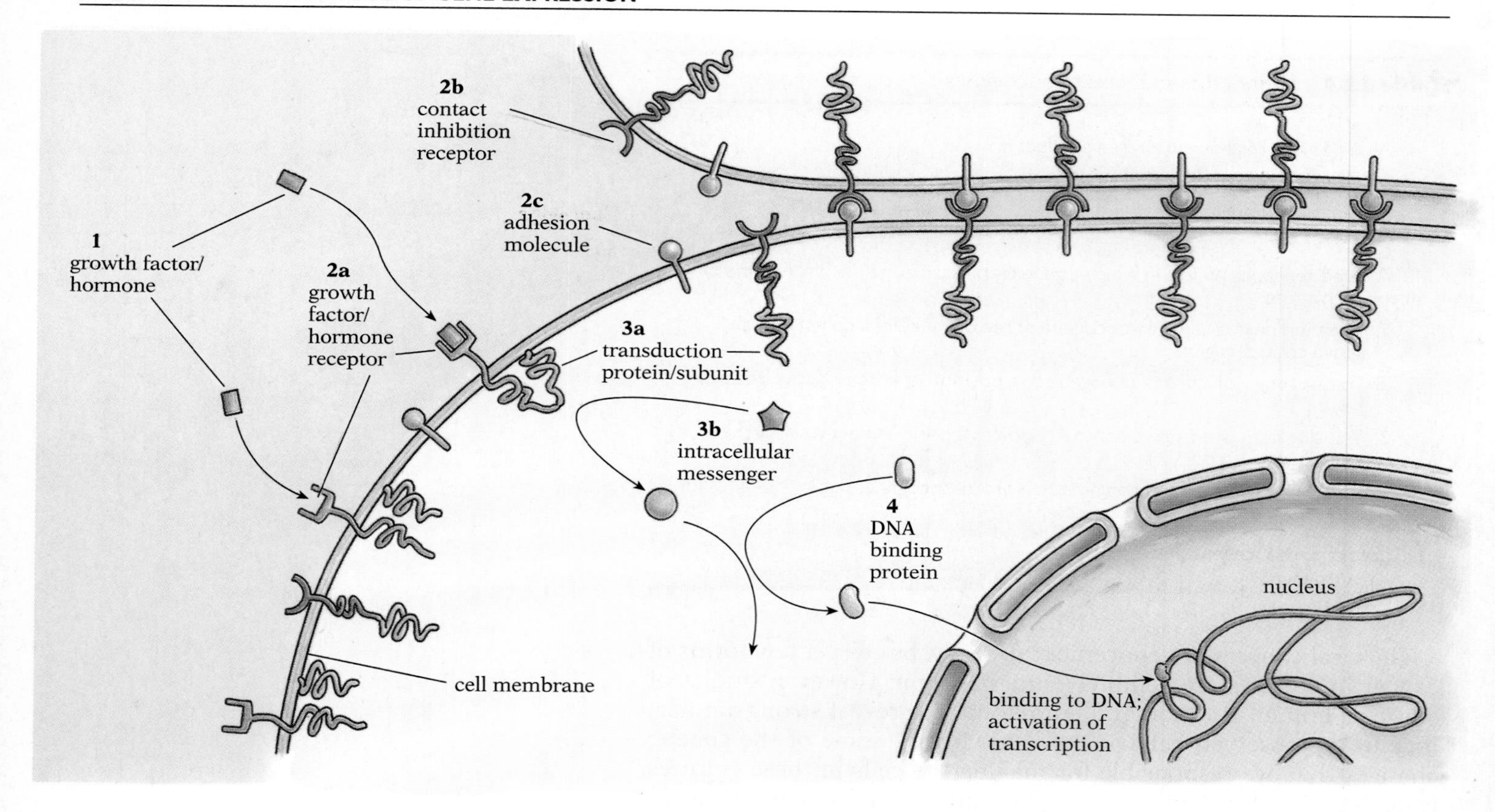

11.17 General sites of oncogene action

Oncogenes can exert their effect at any point in the flow of information from the extracellular environment to the chromosomes. Some act to create high levels of growth factor (1) or growth factors with unusual properties. Others alter the sensitivity of membrane receptors for growth factor or other relevant signals, like those involved in adhesion or contact inhibition (2). Still others alter the communication between the receptors and the intracellular messenger chemicals (3); some receptors incorporate domains that alter intracellular messengers, while others interact with an intermediate enzyme that makes the appropriate change. Finally, some oncogenes alter the DNA-binding proteins that are activated or repressed by the intracellular messengers (4).

3 Intracellular signalling systems (which communicate information from receptors to intracellular enzymes or binding proteins)

4 DNA binding molecules (which regulate transcription of specific genes or replication of the entire genome)

These categories mirror most of the array of normal cellular control strategies: cell division, contact inhibition, anchorage-dependence, tissue-type affinities, vascularization, and the expression of the genes needed to breach the basement membrane.

The first evidence concerning the actual operation of oncogenes came from work on the *src* oncogene by Raymond Erikson and Marc Collett. They found that the enzyme encoded by *src* is a type of intracellular signal molecule called a protein kinase—one of a class of enzymes that phosphorylate a particular protein or a particular component of proteins. Phosphorylation is frequently involved in controlling chemical pathways, usually serving to

Table 11.5 A partial list of classes of cancer-causing and cancer-suppressing genes

NAME	TYPE
Extracellular signals	
sis	Mutant growth factor (GF)
Receptors	
erbB, neu	Mutant GF receptor
ros, erbA	Mutant hormone receptor
C-myc	Mutant adhesion molecule
Transducers	
mos, raf	Mutant serine protein kinase (PK)
src, met	Mutant tyrosine PK
ras	Mutant guanine binding protein
Second messengers	
(none known)	
DNA binding	
fos, jun	Mutant transcription factor
myc, myb	Mutant inducer or enhancer
RB	Antioncogene; blocks binding at some stage in the control pathway

activate an enzyme; the *src* enzyme phosphorylates the amino acid tyrosine. In cells with *src* oncogenes, tyrosine phosphorylation increases to 10 or more times the normal level. Several other oncogenes are now known to code for tyrosine kinases, while a few others encode serine kinases.

Some oncogenes seem to encode products that bridge two steps in the pathway from extracellular signals to DNA binding. The epidermal growth factor (EGF) receptor, for instance, has an extracellular component that binds to EGF, and an intracellular section that acts as a kinase. At least one oncogene appears to code for an enzyme that is very similar to the kinase part of the activated EGF receptor, but lacks the extracellular portion that binds to EGF. This oncogene product may act by signalling the cell to divide continuously, whether or not EGF is present.

Understanding cancer in the near future may be a realistic hope: there appear to be only a limited number of ways cancer is triggered, only a limited number of proto-oncogenes in any cell (fewer than a hundred in humans), and an even smaller number of functions for oncogene products (Table 11.5). Understanding oncogenes will probably prove the most important step in finding effective treatments for cancer.

■ ENVIRONMENTAL CAUSES OF CANCER

Mutations, translocations, and retroviruses can each play a role in triggering cancer. Though many, if not all, mutations and translo-

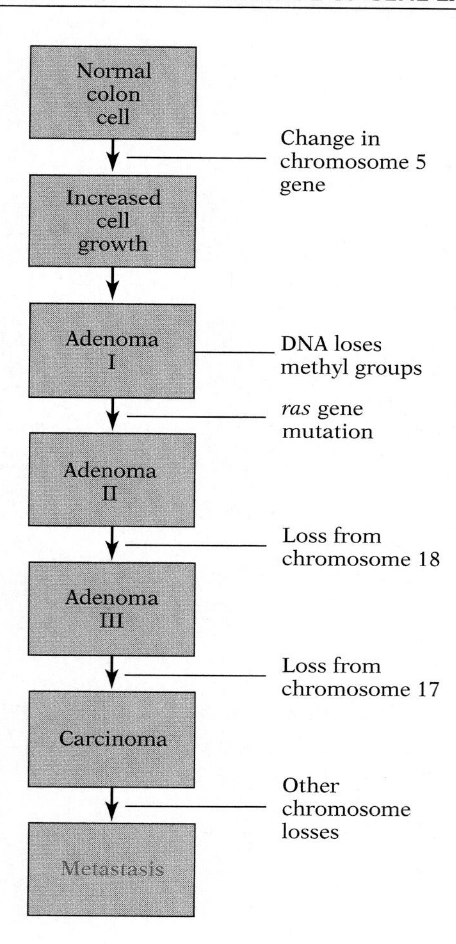

11.18 Chromosomal changes leading to colon cancer
Colon cancer usually begins with small growths called adenomas, which are categorized according to morphology. The cancer is unusual because some of the chromosomal changes involved are cytologically visible and occur more or less in a particular order. These peculiarities have made it possible to outline the steps involved. The *ras* oncogene encodes an intracellular signalling enzyme, which couples the response of growth-factor receptors to DNA-binding protein, which in turn regulates part of the cell-division control system.

cations are caused by external agents—radiation and mutagenic chemicals—linking particular chemicals or radiation sources to cancer in humans and evaluating the relative danger of each source is difficult. Not every mutation occurs in a cancer-causing location, and even a mutation that is in such a location will trigger only one of the several steps necessary for development of a cancer. Thus the final step may occur years after the first significant exposure to a mutagenic chemical or radiation. Only in the case of colon cancer are researchers even close to understanding the steps involved (Fig. 11.18).

The absence of strict cause-and-effect relationships between particular mutagenic agents and particular cancers has made the identification and evaluation of causative agents difficult, even in the most obvious cases. For example, though health statistics have shown that cigarette smoking is probably responsible for about 150,000 fatal cancers annually (and for 25% of fatal heart disease), the tobacco industry denies the link. While granting that smokers develop lung cancer and heart disease far more often than nonsmokers, they point out that many smokers never develop lung cancer, and that some nonsmokers do. An informed understanding of how oncogenes work and of the probabilistic nature of their development, however, reveals the weaknesses in these arguments: The genetic events necessary for inducing lung cancer must be greatly facilitated by the highly mutagenic tar in cigarette smoke. (Tar is a deadly combination of chemicals that on average kills two-pack-per-day smokers 8 years early.) However, the changes must occur in the correct locations (including a specific deletion on chromosome 3) in at least one cell. Thus, even with regular exposure to mutagenic smoke, some smokers will escape unscathed. At the same time, other sources of mutation, such as radiation, are also present, and though less potent, they will sometimes cause the changes necessary to trigger lung cancer even in nonsmokers.

Another way to evaluate the carcinogenic potential of chemicals and radiation is to expose animals—usually mice—to measured doses of these agents. This procedure is expensive and time-consuming, and in addition it assumes that what causes cancer in mice is equally carcinogenic in humans. Another assay is the Ames test, described on page 243, which involves exposing bacteria to potential carcinogens to see if mutations result. The assumption is that a mutagen for *E. coli* is a carcinogen for humans. The Ames test has identified many substances as potential carcinogens. Others having no initial effect are converted into mutagens when liver homogenate is added to the bacterial culture, and are presumably also thus converted in the liver of an organism. A few of the substances identified by the Ames test as possible carcinogens are hexachlorophene soaps, the coal-tar components of many hair dyes and cosmetics, the seared protein of grilled meat, the smoke from wood fires, and several chemicals in certain vegetables and spices. When tested on animals each of these substances has proven to be carcinogenic. It seems only sensible to assume that all mutagens are carcinogenic, but given the ubiquity of environmental mutagens, avoiding them all is impossible. The only sensible course, then, is to eliminate our exposure to the most potent ones—especially cigarette smoke and coal tars—and, weighing the costs and benefits (just as we do each time we decide to take the risk of driving a car), to minimize contact with the rest.

CHAPTER SUMMARY

ORCHESTRATING BACTERIAL GENES

HOW GENES ARE INDUCED The genes for inducible enzymes are normally not transcribed. An inducer binds to and deactivates a gene's repressor protein, which then can no longer bind to the gene's operator region; a deactivated repressor can no longer prevent RNA polymerase from binding to the promotor region and transcribing the gene. (p. 273)

HOW GENES ARE REPRESSED Genes that are normally transcribed are shut down when a corepressor substance binds to and activates the gene's repressor. (p. 274)

POSITIVE VERSUS NEGATIVE CONTROL Genes that use repressors are under negative control. Positive control usually depends on transcription factors binding to activator regions; this binding changes the ability of RNA polymerase to bind to the gene's promoter. (p. 278)

HOW EUCARYOTIC CONTROL IS DIFFERENT

THE ODD ORGANIZATION OF EUCARYOTIC CHROMOSOMES The nucleosomes onto which eucaryotic DNA is wound must be temporarily removed to allow transcription. Eucaryotic chromosomes have regions of highly repetitive DNA: in the centromeres and telomeres it has a structural role or aids in cell division; other repetitive regions encode rRNA; yet others have no known function. Some moderately repetitive DNA encodes rRNA, while the function of other sorts is unknown. Most single-copy DNA is never transcribed, and its function is not known. (p. 283)

HOW CHANGES IN GENE TRANSCRIPTION CAN BE SEEN DNA transcription requires decondensation of a chromosomal loop; this decondensation is sometimes visible as chromosomal puffs in polytene chromosomes and as a "lampbrush" pattern in certain chromosomes. (p. 286)

IMPRINTING: PRESERVING PATTERNS OF GENE ACTIVITY The activity of some genes is controlled by methylation of the bases. Special enzymes scan the parental strand after replication and methylate the daughter strand to match. Gametes containing methylated genes carry with them these methylation patterns. (p. 288)

EUCARYOTIC CONTROL: INDUCERS AND ENHANCERS Most eucaryotic genes utilize positive control, but the binding of transcription factors to a gene's inducer region is controlled by a multitude of other molecules—especially those that bind at its distant enhancer region. The large number of control elements seems to reflect the need for complex developmental control. (p. 289)

GENE AMPLIFICATION Genes are said to be amplified when multiple copies of a gene are created, thus generally increasing the transcription rate of the gene. (p. 292)

CONTROL AFTER TRANSCRIPTION Gene activity can be controlled after transcription through alternative ways of splicing its exons, which include sequences that can block or facilitate export of an mRNA to the cytoplasm, alter its translation rate, and determine its degree of resistance to proteases in the cytoplasm. (p. 292)

THE IMPACT OF MUTATIONS Mutations can alter the sequence of amino acids in the protein a gene encodes and the activity of control elements, including promoters, operators, inducers, and enhancers. (p. 294)

CANCER: WHEN GENE CONTROL FAILS

TISSUE CULTURE: A WAY OF STUDYING CANCER CELLS Most cells fail to become cancerous because they undergo only a fixed number of divisions, are inhibited from division by crowding (contact inhibition), cannot recruit additional capillaries to support extra tissue, and cannot bind to and penetrate the basement membrane that confines tissues. (p. 295)

HOW ARE CANCEROUS CELLS DIFFERENT? Cancerous cells have lost levels of control that keep normal cells from multiplying in their tissue type. Many cancer cells have extra chromosomes or translocations, which alter patterns of gene expression. (p. 296)

THE STEPS THAT LEAD TO CANCER Most cancers develop after several independent events—events that presumably remove each level of control that normally prevents unrestrained growth. (p. 299)

THE ROLE OF ONCOGENES Oncogenes are genes that cause cancer; proto-oncogenes are genes that can turn into oncogenes when their sequence or control regions are altered. Most oncogenes encode growth factors, cell-surface receptors, intracellular signalling components, and gene-control molecules. (p. 300)

ENVIRONMENTAL CAUSES OF CANCER Cancers are caused by mutations (either of genes or control regions), translocations, gene amplification, or the insertion of genes by retroviruses. Each change removes one of the defenses against uncontrolled growth of the cell. Most mutations and translocations are caused by ionizing radiation or chemicals in the cell. The chemicals may be normal, highly reactive by-products of cellular chemistry, or unusual molecules obtained through the diet. (p. 303)

STUDY QUESTIONS

1 Are there good reasons why eucaryotes should rely much more than procaryotes on positive gene control? (pp. 283–295)

2 Why do you suppose that skin cancer is so often controllable? (pp. 295–299)

3 The eucaryotic genome has 50,000 genes, each managed by one or more control genes, which in turn are orchestrated by thousands of middle-level executive genes, which themselves respond to hundreds of upper-management sequences, and so on. Why is the genome not strangled by this bureaucracy? (pp. 289–292)

4 Why bother keeping inactive regions of chromosomes condensed?

5 What sorts of mutations could affect the *lac* operon, and what effects would they have? What would happen if a cell had two of these mutations at once? (pp. 273–275, 278–282)

SUGGESTED READING

BISHOP, J. M., 1982. Oncogenes, *Scientific American* 246 (3). *An illuminating look at the relationship between cancer genes carried by viruses and the similar noncancerous genes in normal cells.*

CAVENEE, W. K., AND R. L. WHITE, 1995. The genetic basis of cancer. *Scientific American* 272 (3). *A clear presentation of the multiple steps that lead to cancer.*

CROCE, C. M., AND G. KLEIN, 1985. Chromosome translocations and human cancer, *Scientific American* 252 (3). *A clear discussion of the translocations involved in Burkitt's lymphoma.*

FELDMAN, M., AND L. EISENBACH, 1988. What makes a tumor cell metastatic? *Scientific American* 259 (5). *About the oncogenes that permit tumor cells to stop adhering to other cells or structures, and so spread to other parts of the body.*

FELSENFELD, G., 1985. DNA, *Scientific American* 253 (4). *The role of DNA structure in the regulation of gene expression.*

MCKNIGHT, S. L., 1991. Molecular zippers in gene regulation, *Scientific American* 264 (4). *On the operation of the leucine zipper.*

PTASHNE, M., 1989. How gene activators work, *Scientific American* 260 (1). *An excellent summary of how promoters work in bacteria and yeast.*

SAPIENZA, C., 1990. Parental imprinting of genes, *Scientific American* 263 (4). *On the inheritance of gene switches bound to the DNA of gametes.*

TIOLLAS, P., AND M. A. BUENDIA, 1991. Hepatitis B virus, *Scientific American* 264 (4). *On the life history of a cancer-promoting virus.*

TIJAN, R., 1995. Molecular machines that control genes, *Scientific American* 272 (2). *On the elaborate organization of transcription complexes.*

WEINBERG, R. A., 1988. Finding the anti-oncogene, *Scientific American* 259 (3). *On the genes that restrain cell growth, focusing on retinoblastoma.*

WEINTRAUB, H., 1990. Antisense RNA and DNA, *Scientific American* 262 (1). *On the use of antisense RNA to regulate mRNA activity.*

CHAPTER 12

CELLULAR REPRODUCTION

The previous four chapters trace the flow of genetic information within individual cells. This chapter describes how cells themselves reproduce, passing their genetic endowment from cell to cell and from parent to offspring. Chapters 13 and 14 describe how the genome organizes the development and cellular specializations typical of complex, multicellular organisms.

FISSION VERSUS DIVISION

Before a cell can divide to produce two new cells, it must first replicate all the genetic information in its nucleus and then make sure that a full set of information is given to each daughter cell. The process of DNA replication is detailed in Chapter 8; the emphasis here is on the larger-scale organizational task of duplicating the entire genome and accurately segregating the copies into two progeny cells.

■ PROCARYOTIC FISSION

Cell division is less complex in procaryotes than in eucaryotes. As the procaryotic cell elongates sufficiently to form two independent daughter cells, the single circular chromosome replicates and the resulting second chromosome attaches to a different point from the first one on the expanding plasma membrane (Fig. 12.1). Next, new plasma membrane and wall material form near the midpoint of the parental cell and grow slowly inward, cutting through both

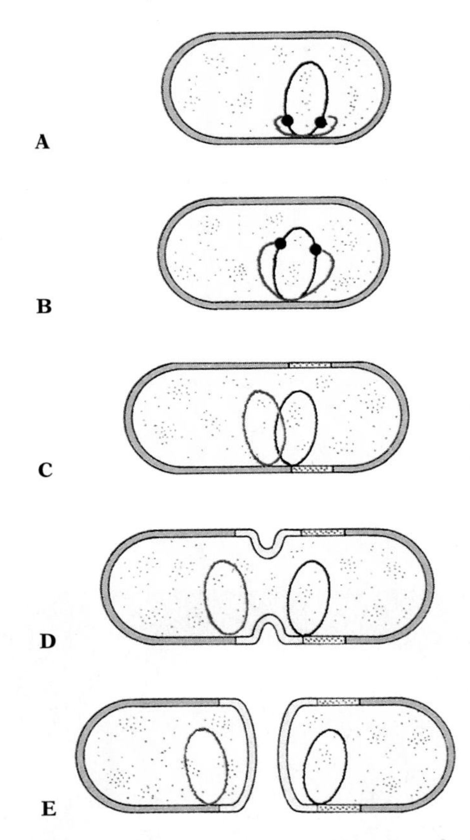

12.1 Binary fission of a procaryotic cell

(A) The circular chromosome of a cell is attached to the plasma membrane near one end; it has already begun replication (with the partially formed second chromosome shown in red). (B) Replication is about 80% complete. (C) Chromosomal replication is finished, and the second chromosome now has an independent point of attachment to the membrane. During replication additional membrane and wall (stippled) has formed. (D) More new membrane and wall (tan) has formed between the points of attachment of the two chromosomes. Part of this growth forms invaginations that will give rise to a septum cutting the cell in two. (E) Fission is complete, and two daughter cells have formed. (For clarity, the chromosomes are depicted here as small open circles; actually they are so long that they must be looped and tangled to fit into a cell; at the scale of this drawing, the true circumference of each chromosome would be about 30 m.)

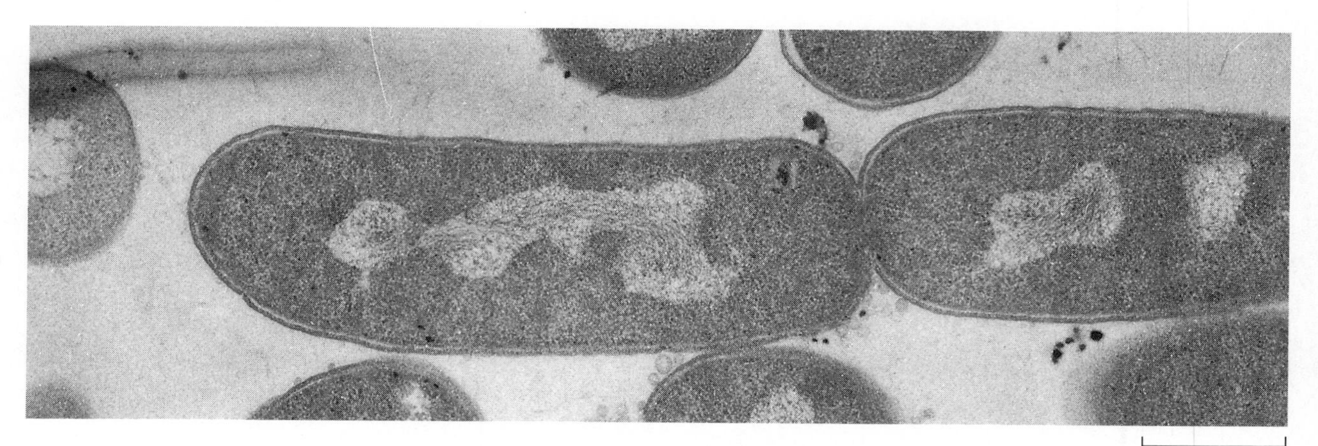

12.2 Electron micrograph of a dividing bacterial cell
New wall is growing inward, cutting this *E. coli* in two. The nucleoid (white areas inside cells) has already divided.

the cytoplasm and the nucleoid (Fig. 12.2). Each new daughter cell receives a complete chromosome. This process is known as transverse fission, or ***binary fission***.

■ WHY EUCARYOTIC CHROMOSOMES POSE PROBLEMS

As described in Chapter 11, the chromosomes of eucaryotes differ from those of procaryotes in several ways. First, except in mitochondria and other organelles, eucaryotic chromosomes are linear rather than circular. Second, eucaryotic chromosomal DNA in the nucleus is wound on nucleosome cores. Third, each chromosome is organized into loops, all of which remain in a highly condensed state during cell division, probably to prevent tangling (Fig. 12.3).

The final major difference between eucaryotic and procaryotic chromosomes arises because most eucaryotes receive chromosomes from two parental cells—from a sperm cell and an egg cell, for instance—rather than from just one. Most eucaryotes are diploid, and their chromosomes thus occur in ***homologous pairs***, each consisting of one chromosome from each parent bearing basically the same genes in the same order (Fig. 12.4). Procaryotes are haploid: their chromosome is unpaired. Though chromosomal number may vary enormously between species—cats, for instance, have 19 pairs, fruit flies 4 pairs, onions 8 pairs, and humans 23 pairs—the chromosomal number within a species is normally uniform.

12.3 Condensed chromosome showing highly condensed loops

When eucaryotic chromosomes condense and become visible early in cell division, each chromosome has already been replicated to produce compact twins, connected by a centromere (Fig. 12.5). Each member of this bound pair is referred to as a ***chromatid***.

There are two kinds of cell division in eucaryotes: division as part of the growth of an organism, also known as mitosis or somatic cell division, and division to produce the gametes that give rise to new individuals, also known as meiosis. They differ in important ways.

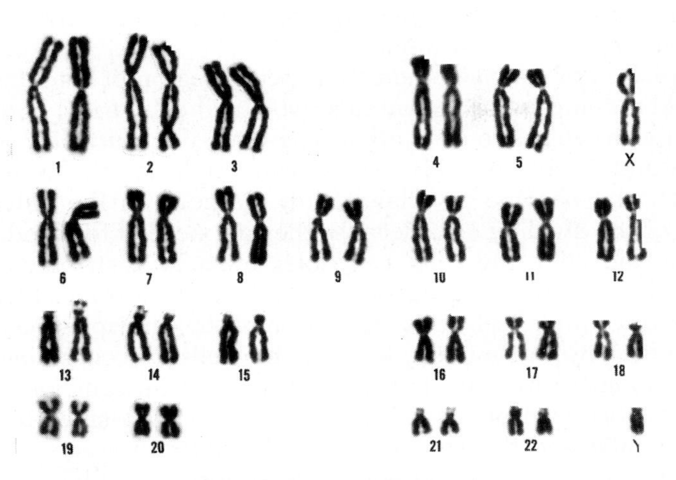

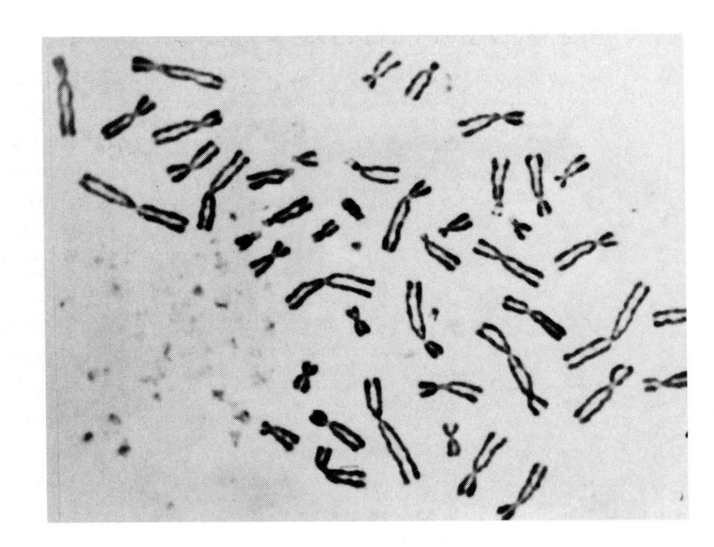

12.4 Chromosomes of a human male

At left, the chromosomes have been arranged as homologous pairs and numbered according to accepted convention. A human somatic cell (any cell except the egg or sperm cells) contains 23 pairs of chromosomes, including a pair of sex chromosomes (for a male, X and Y).

HOW SOMATIC CELLS DIVIDE

Cell division in eucaryotic cells involves division of the nucleus and division of the cytoplasm. The process by which the nucleus divides to produce two new nuclei, each with the same number of chromosomes as the parental nucleus, is called ***mitosis*** (from the Greek *mitos*, "thread"). The process of division of the cytoplasm is called ***cytokinesis***.

■ NUCLEAR DIVISION: MITOSIS

For convenience, each mitotic cycle, from one cell division to the next, is divided into a series of stages, each designated by a special name. Though each stage will be discussed separately here, the entire process is continuous rather than a series of discrete occurrences.

Interphase and the cell cycle The nondividing cell is in the interphase state. The nucleus is clearly visible as a distinct membrane-bounded organelle, and one or more nucleoli are usually prominent. Chromosomes are not visible in the nucleus; there are none of the distinct rodlike bodies that microscopes and sophisticated staining techniques reveal in a dividing cell (Fig. 12.6). Interphase chromosomes are so thin and tangled that they cannot

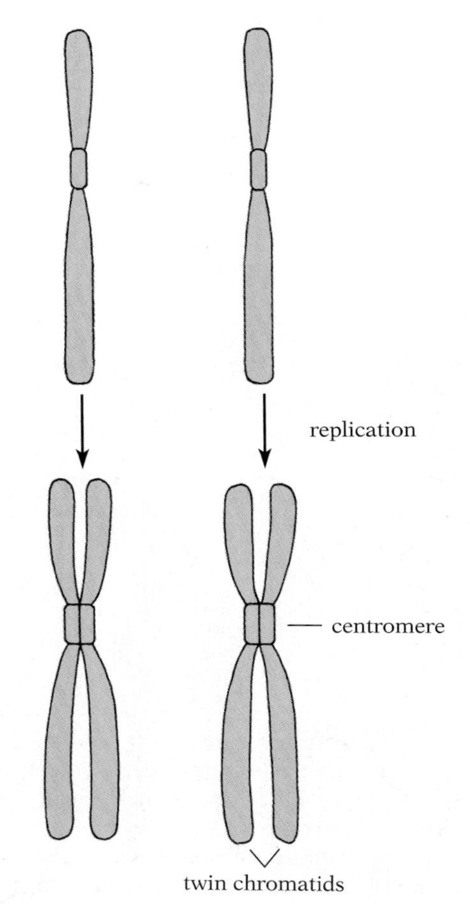

12.5 Production of chromatids through replication

At some time prior to cell division the genetic material of each chromosome duplicates itself. As a result, twin chromatids take form upon condensation, joined to each other by a centromere. For clarity, the unduplicated chromosomes are shown above in condensed form, even though condensation normally takes place only after duplication.

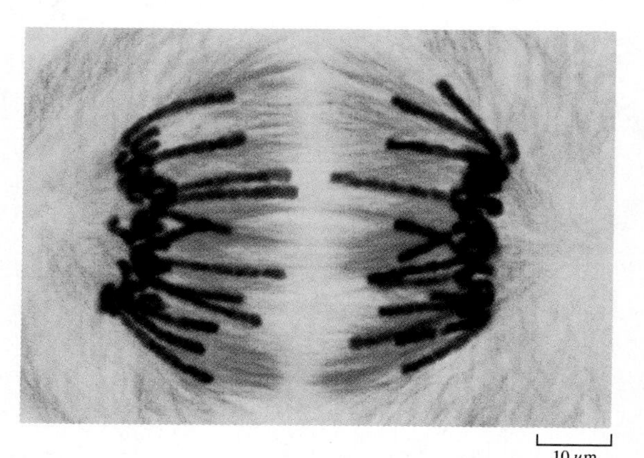

12.6 Chromosomes in a dividing cell of the African blood lily

The separating chromosomes for each of the new nuclei are easily distinguishable.

be recognized as separate entities. They appear only as an irregular granular-looking mass of chromatin.

In interphase animal cells, there is a special region of cytoplasm just outside the nucleus that contains two small cylindrical bodies oriented at right angles to each other. These are the ***centrioles*** (see Figs. 5.34 and 5.35B, p. 142), which will duplicate themselves, move apart, and become associated with the poles of the mitotic apparatus of the dividing cell. Because they are needed for producing flagella and cilia, the cells' centrioles must be distributed to daughter cells as reliably and accurately as the chromosomes themselves. In many animal cells this separation of the centrioles occurs just before the onset of mitosis, but in some cells it occurs during interphase, long before mitosis begins. There are no centrioles in the cells of most seed plants, which is not surprising since these species lack cilia and flagella; centrioles do occur in some algae, fungi, bryophytes (mosses, liverworts, and so forth), and ferns that produce motile sperm. Centrioles are not essential for cell division even in animal cells; they remain close to the spindle-organizing centers during mitosis, but do not participate in spindle formation.

Interphase cells carry out all the innumerable activities of a living, functioning cell—respiration, protein synthesis, growth, differentiation, and so forth. In addition, during interphase the genetic material is replicated and all the cellular machinery is duplicated in preparation for the next division sequence.

Replication of the genetic material does not begin immediately after completion of the last division sequence. There is a gap, designated the G_1 stage, before genetic replication; this is when ribosomes and organelles begin to be duplicated. Next comes the S stage, during which the synthesis of new DNA takes place, along with the further duplication of organelles. Another period, G_2, occurs after replication, during which time the cell prepares for mitosis. Then cytokinesis and mitosis proper (together referred to as the M stage) occurs. The three subdivisions of interphase, along with mitosis and cytokinesis, constitute the ***cell cycle*** (Fig. 12.7).

The duration of the complete cell cycle can vary greatly. Though usually lasting 10–30 hours in plants and 18–24 hours in animals, it may be as short as 20 min in some organisms or as long as several days or even weeks. All the stages can vary in duration to some degree, but by far the greatest variation occurs in the G_1 stage. At one extreme very rapidly dividing embryonic cells may pass so quickly through the G_1 stage that it can hardly be said to exist at all; at the other extreme some cell types become arrested in the G_1 stage. Differentiated skeletal-muscle cells and nerve cells, for example, are suspended in the G_1 stage and normally never divide again. There are a few cases in which cells are arrested in the G_2 stage: they have replicated their DNA but do not divide.

Arrest in either the G_1 or G_2 stages results from a failure to produce an essential control chemical. If the nucleus from a cell in G_1 arrest is transplanted into a cell that is just entering the S stage, the transplanted nucleus will promptly enter the S stage, apparently because it has been stimulated by a control substance present in the cytoplasm of the host cell. Similarly, when a cell in G_2 arrest is fused with a mitotic cell, its chromosomes soon begin to condense and the cell enters mitosis.

12.7 The cell cycle

This particular cycle assumes a period of 24 hours, but some cells complete the cycle in less than an hour and others take many days. Similarly, the ratios of the four stages of the cycle vary, with G_1 exhibiting the most variation.

The chemicals that control the cell cycle are called ***cyclins***. S-cyclin is involved in stimulating replication (DNA synthesis), while M-cyclin helps trigger mitosis. Each can bind to the same **cell-division-cycle protein (*cdc*)**, which in turn can activate a cellular messenger specific for either replication or mitosis.

To see how the system works, let's follow a complete cycle beginning with the G_1 phase (Fig. 12.8). As G_1 begins, the cdc protein is

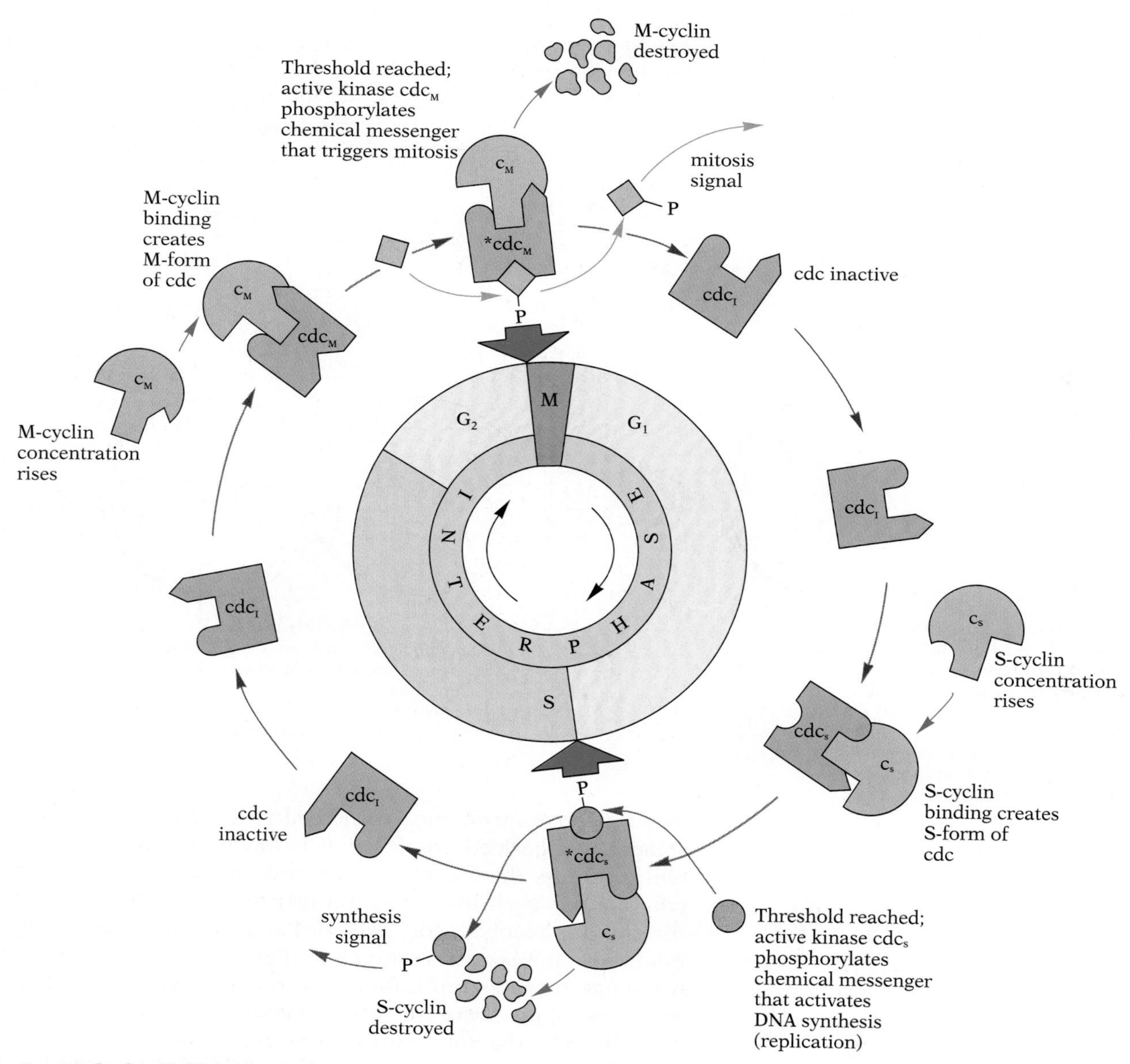

12.8 Control of cell division

A single protein kinase, cdc, is responsible for signalling the cell to replicate or undergo mitosis. It adopts the replication-promoting form as the concentration of S-cyclin rises in G_1 (lower right), becoming activated at a certain threshold (activation indicated here by an asterisk). It then phosphorylates a messenger that triggers replication and the destruction of S-cyclin. Similarly, as the M-cyclin concentration rises in G_2 (upper left), it binds to cdc, which adopts a mitosis-promoting form. Once the threshold has been reached, the cdc becomes activated and catalyzes the phosphorylation of a messenger that triggers mitosis and the destruction of M-cyclin.

2. EARLY PROPHASE
Centrioles begin to move apart.
Chromosomes appear as long thin threads.
Nucleolus becomes less distinct.

3. MIDDLE PROPHASE
Centrioles move farther apart.
Asters begin to form.
Twin chromatids become visible.

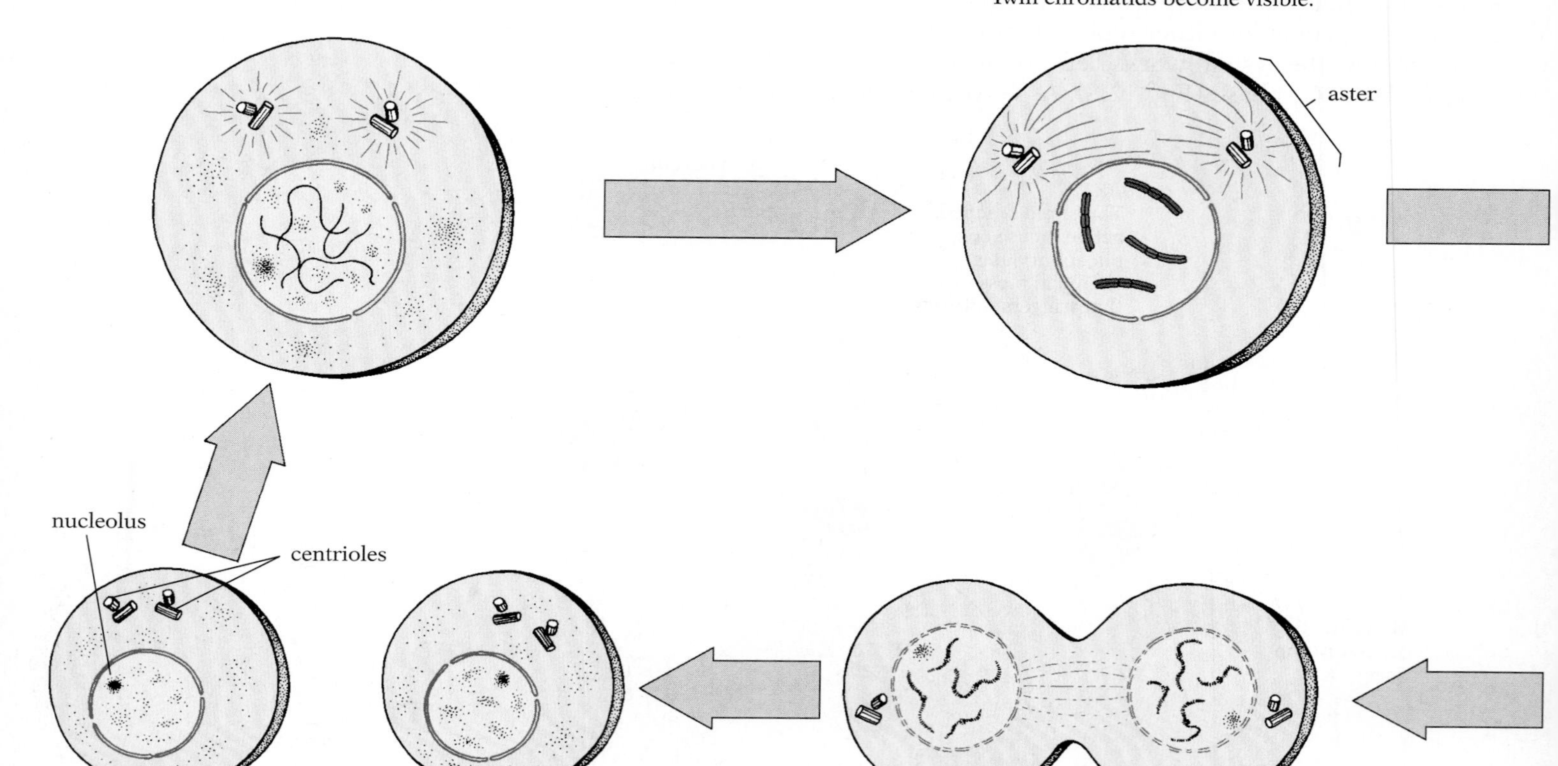

1/9. INTERPHASE
Cytokinesis is complete.
Nuclear membranes are complete.
Nucleolus is visible in each cell.
Chromosomes are not seen as distinct structures.
Replication of genetic material occurs before the end of this phase.
Centrioles are replicated.

8. TELOPHASE
New nuclear membranes begin to form.
Chromosomes become longer, thinner, and less distinct.
Nucleolus reappears.
Cytokinesis is nearly complete.

in its inactive form, cdc_i. If the cell is to divide again, S-cyclin is steadily synthesized; the faster it is made, the sooner the S phase will begin. As the S-cyclin concentration rises, it starts to bind to cdc_i, converting it into its synthesis-promoting form, cdc_s. When a threshold is reached, cdc_s becomes activated and begins phosphorylating a messenger compound (Fig. 12.8); thus activated, the messenger triggers replication, and other enzymes quickly destroy the S-cyclin, restoring cdc_s to its inactive cdc_i form.

During G_2, the mitosis-inducing cyclin—M-cyclin—is steadily synthesized. As its concentration rises, it binds to cdc_i and converts it into a mitosis-promoting form, cdc_m. Again, when a threshold is passed, cdc_m is activated and begins phosphorylating a different messenger compound; this new messenger triggers mitosis and the digestion of M-cyclin, thus turning cdc_m into cdc_i (Fig. 12.8).

The potential links between control of the cell cycle and the control failures that underlie cancer are becoming clear. Excess cyclin (leading to uncontrolled cell division) could be produced, either

4. LATE PROPHASE
Centrioles nearly reach opposite sides of nucleus.
Spindle begins to form and kinetochore microtubules project from centromeres toward spindle poles.
Nuclear membrane is disappearing.
Nucleolus is no longer visible.

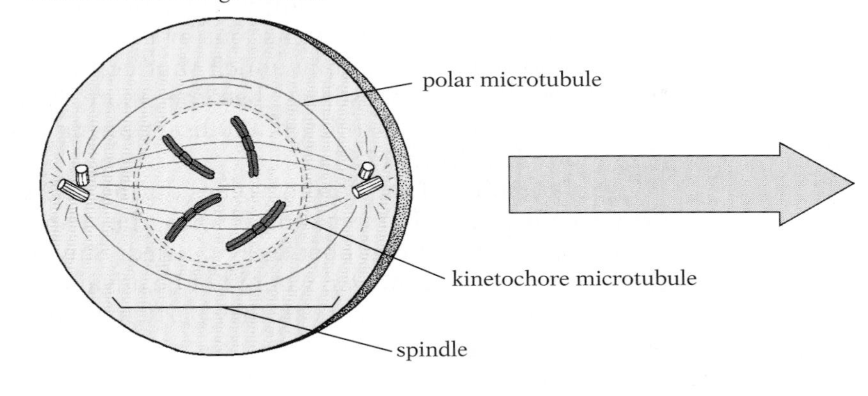

5. METAPHASE
Nuclear membrane has disappeared.
Kinetochore microtubules move each twin-chromatid chromosome to the midline; other spindle microtubules interact with spindle tubules from opposite pole.

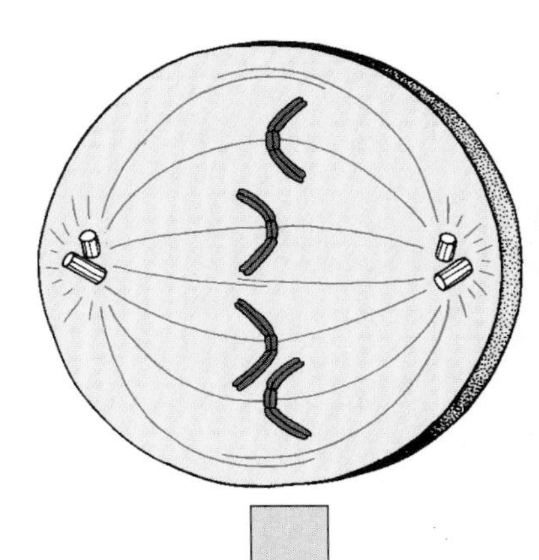

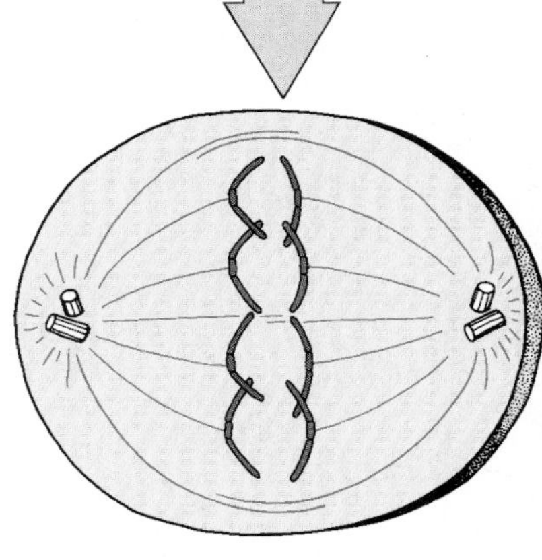

6. EARLY ANAPHASE
Centromeres have split and begun moving toward opposite poles of spindle.
Spindle microtubules from opposite poles force the poles apart.

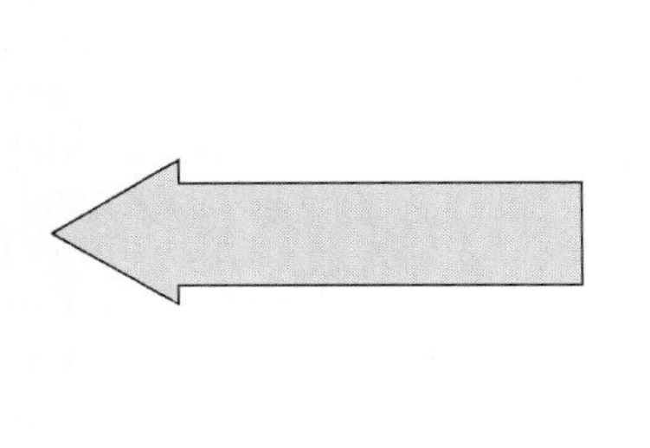

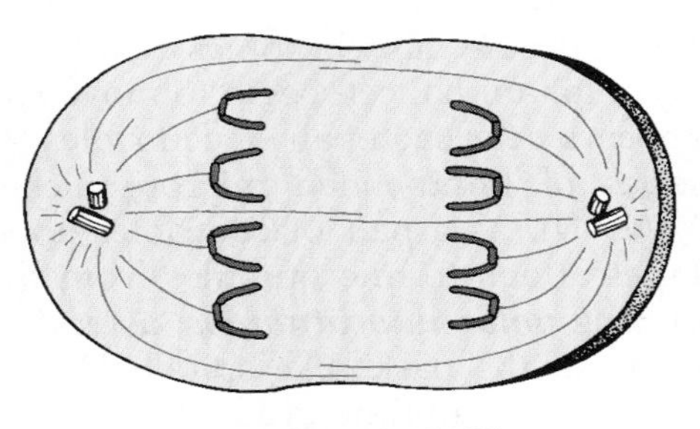

7. LATE ANAPHASE
The two sets of new single-chromatid chromosomes are nearing their respective poles.
The poles are being pushed apart.
Cytokinesis begins.

12.9 Mitosis and cytokinesis in an animal cell

In these drawings, purple and green are used to distinguish the two chromosomes inherited from each parent. For clarity, only a few microtubules are shown. For simplicity, only single kinetochore microtubules are shown pulling the chromatids toward the poles; in reality, each centromere is attached to a bundle of microtubules.

through alteration of cyclin transcriptional control, or (as in the case of several oncogenes mentioned in Chapter 11) through overstimulation of growth-factor receptors. Alternatively, a mutant form of cdc could remain constantly active, or be impossible to inactivate; the *RB* proto-oncogene of retinoblastoma, for instance, is thought to be involved in keeping cdc turned off. With the same effect, the messenger compound itself may be made permanently active. Finally, the DNA target sequence of a messenger could be altered, throwing the gene it regulates into a transcriptionally active form. Studies of the control of replication and mitosis may lead to techniques for restoring quiescence to tumor cells.

Let's assume now that the cell we are examining has passed through the G_1, S, and G_2 stages of interphase and is entering mitosis proper. Because mitosis is such a complex process, it is customarily divided into four stages: prophase, metaphase, anaphase, and telophase (Fig. 12.9).

Prophase During prophase, the cell readies the nucleus for the crucial separation of two complete sets of chromosomes into two daughter nuclei. As the two centrioles of an animal cell move toward opposite sides of the nucleus, the chromosomes begin to condense into visible threads. (During early prophase, they appear as long, thin, intertwined filaments, but by late prophase the individual chromosomes will become much shorter, thicker rodlike structures.) As the chromosomes become more distinct, the two nucleoli become less distinct, often disappearing altogether by the end of prophase (Fig. 12.9).

The shortening of the chromosomes during cell division has one obvious advantage. In their shorter form, they can be moved about freely without becoming hopelessly tangled. Staying in the shorter form permanently, however, would not be advantageous, since only in their long uncoiled form can they participate in the critical task of replication during interphase.

Each chromosome from a late-prophase nucleus consists of separate twin chromatids (Fig. 12.5). The replication that occurred during interphase creates two identical copies of the DNA molecules of the original chromosome. Hence, the two chromatids of a prophase chromosome are genetically identical.

The means by which chromosomes are separated during cell division, the mitotic ***spindle***, also becomes visible in late prophase. As the centrioles of an animal cell begin to move apart, a system of microtubules appears near each pair of centrioles (Fig. 12.9, stages 2–3). These series of blind-ended microtubules are called the ***asters*** (Fig. 12.10C–D). As a pair of centrioles approaches a pole, some microtubules attach to microtubules from the opposite pair of centrioles forming ***polar microtubules***; asters and polar microtubules together form the basketlike spindle (Fig. 12.9:4). In late prophase, the nuclear membrane gradually disappears and some aster microtubules bind to protein plates called ***kinetochores***, which form on the centromere of each chromatid; the kinetochore microtubules thus connect the centromeres to the poles.

Metaphase The stage of metaphase proper is preceded by a short period known as the prometaphase. The chromosomes begin to move toward the equator, or middle, of the spindle. This movement occurs because the two bundles of kinetochore microtubules that anchor each chromosome are able to alter their lengths relative to each other: the shorter microtubule bundle adds tubulin subunits at its aster-pole end, while the longer one digests (removes) them. In this way the position of each chromosome continues to shift until, by the end of prometaphase, it is positioned on the midline.

During the brief stage of metaphase proper, the chromosomes are arranged on the equatorial plane of the spindle, and in side view appear to form a line across the middle of the spindle (Figs. 12.9:5, 12.10D). Metaphase ends when the centromeres of each pair of twin chromatids split apart. Each chromatid then becomes an independent chromosome with its own centromere. By the end of metaphase the total amount of genetic material is unchanged though the number of independent chromosomes in the nucleus has doubled.

Anaphase Next comes the separation of the two complete sets of chromosomes. At the beginning of anaphase the centromeres that

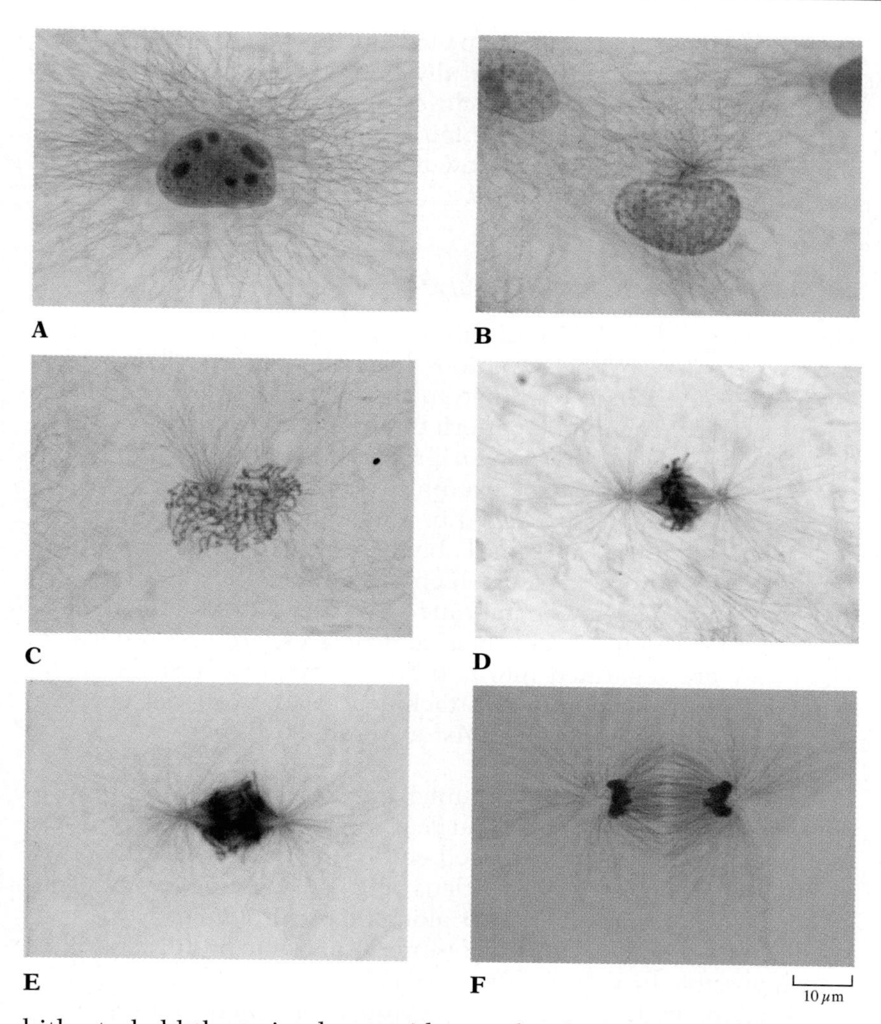

12.10 Mitosis in a cell from the endothelium of a frog's heart

(A) Interphase: a diffuse network of microtubules (stained red) is visible. (B) Early prophase: the chromosomes begin to condense; the centrioles have not yet separated. (C) Late prophase: prophase ends as the nuclear envelope disappears; the centrioles have moved apart, and the two asters are prominent. (D) Metaphase: the chromosomes are lined up along the equator of the spindle, midway between the two asters. (E) Middle anaphase: the two groups of chromosomes have begun to move apart from each other, toward their respective poles of the spindle. (F) Early telophase: the two groups of chromosomes are being organized into new nuclei; cytokinesis has begun, and the line dividing the two daughter cells is clearly visible.

hitherto held the twin chromatids together have just broken apart. The two new sets of single-chromatid chromosomes now begin to move away from each other, one going toward one pole of the spindle and the other going toward the opposite pole. This movement toward the respective poles is accomplished in two ways. A process similar to the one that lined up the chromosomes at the midline during prometaphase now pulls the centromeres, with their attached chromosomes, to the poles. (The only difference is that now the tubulin subunits are digested at *both* poles, and no tubulin is added.) Meanwhile, the polar microtubules from opposite ends of the dividing cell form cross bridges between themselves; by adding tubulin subunits to their ends, they increase their lengths, thus pushing the poles apart (Figs. 12.9:6, 12.10E). By late anaphase the cell contains two groups of chromosomes that are widely separated, the two clusters having almost reached their respective poles of the spindle (Fig. 12.9:7). Cytokinesis often begins during late anaphase.

Telophase Telophase (Figs. 12.9:8, 12.10F) is essentially a reversal of prophase. The two sets of chromosomes, having reached their respective poles, become enclosed in new nuclear membranes as the spindle disappears. Then the chromosomes begin to uncoil and to resume their interphase form, while the nucleoli slowly reappear. Cytokinesis is often completed during telophase. Telophase ends when the new nuclei have fully assumed the characteristics of interphase, thus bringing to a close the complete mitotic process. What was a single nucleus containing one set of twin-chromatid chromosomes in prophase is now two nuclei, each with one set of single-chromatid chromosomes.

■ THE EVOLUTION OF NUCLEAR DIVISION

The preceding description of mitotic division fits most but not all eucaryotes. Most of the deviations from the pattern illustrated in Figure 12.9 are seen in primitive eucaryotes, and they may indicate how the typical eucaryotic pattern of nuclear division evolved from the sequence of cell division in procaryotes. Some primitive eucaryotic dinoflagellates, for example, even though they have numerous chromosomes packaged in a membrane-bounded nucleus, still divide by a process remarkably like the binary fission of procaryotes. The chromosomes, which are attached to the inner surface of the nuclear membrane by very short kinetochore microtubules running from their centromeres, are first replicated. Then they are separated into two groups by growth of the membrane between the points of attachment of the original chromosomes and their replicates. As a result, two new nuclei are organized.

These dinoflagellates rely on microtubules in their cytoplasm to determine the direction of nuclear division. Bundles of these tubules run in parallel from each end of the cell toward the other, meeting in "tunnels" in the nucleus (Fig. 12.11A); the two new nuclei move away from each other along this scaffolding.

Advanced dinoflagellates and some fungi and protozoans illustrate a possible further evolutionary progression in their method of cell division, while still relying on cytoplasmic microtubules to orient the process. As in the primitive dinoflagellates, their chromosomes are attached to the nuclear membrane by very short kinetochore microtubules from the centromeres, but there is just a single tunnel. Additional spindle microtubules, originating at the same poles as the transnuclear bundles, run to the nuclear membrane and physically pull the two newly formed nuclei apart (Figs. 12.11B, 12.12).

As we have seen, in higher eucaryotes the nuclear membrane disappears before division, and there is no apparent bundling of microtubules. Instead of securing the chromosomes to the nuclear envelope, as in some dinoflagellates, kinetochore microtubules

A

B

C

12.11 Evolution of nuclear division

(A) In primitive dinoflagellates, parallel bundles of microtubules run from each end toward the other, meeting and interacting in tunnels in the nucleus. The chromosomes are attached to the nuclear membrane and separate as the nucleus divides, and the daughter nuclei slide along the microtubules. (B) In advanced dinoflagellates there is only a single microtubule tunnel through the nucleus. Additional microtubules run from the poles to the nuclear membrane and pull the daughter nuclei apart. (C) In higher eucaryotes the nuclear membrane disappears during division, and certain of the polar microtubules interact with the kinetochore microtubules to pull the individual chromosomes toward the poles.

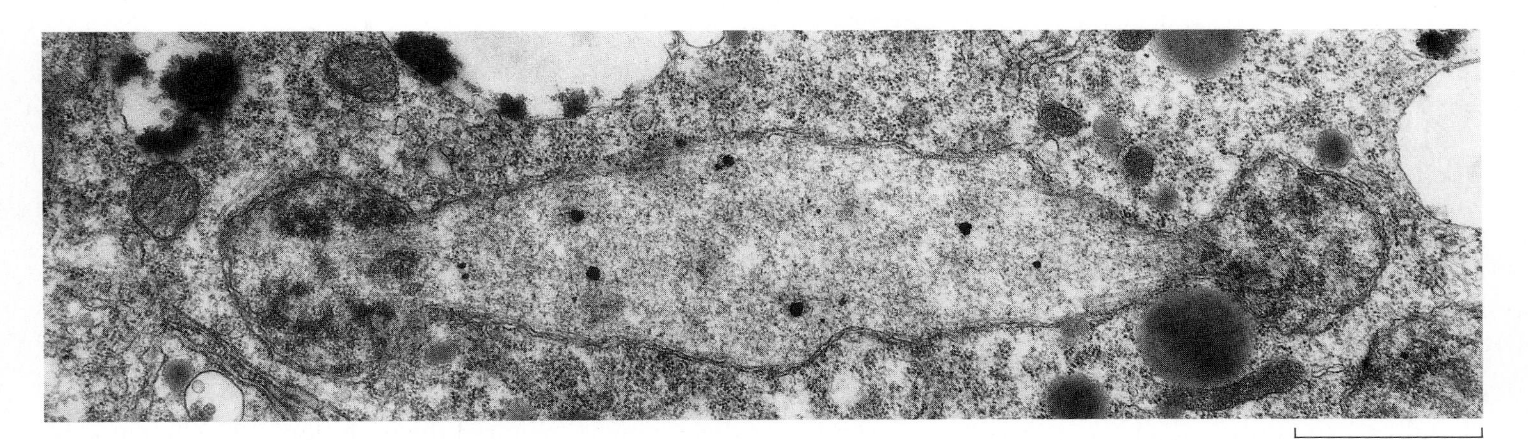

12.12 Persistent nuclear membrane during mitosis in the fungus *Catenaria*

In this fungus, part of the spindle runs through the nucleus. Here two new telophase nuclei, at opposite ends of the photograph, are being pinched off by constrictions of the nuclear envelope. The middle part of the envelope, between the new nuclei, will eventually disintegrate.

connect to polar microtubules or to the poles themselves. These tubule complexes exert a direct pull on the chromosomes, rather than on the daughter nuclei (Fig. 12.11C).

■ CELL DIVISION: CYTOKINESIS

Division of the cytoplasm usually accompanies division of the nucleus, but mitosis without cytokinesis is common in some algae and fungi, producing ***coenocytic*** plant bodies (bodies with many nuclei but few, if any, cellular partitions). It regularly occurs during certain phases of reproduction in seed plants and certain other vascular plants. It is also common in a few lower invertebrate animals with coenocytic bodies. During the early development of insect eggs, mitosis without cytokinesis produces hundreds of nuclei in a limited amount of cytoplasm; later, cytokinesis cuts up this cytoplasm to produce many new cells in a very short time.

How animal cells divide Division of an animal cell normally begins with the formation of a ***cleavage furrow*** running around the cell (Fig. 12.13). When cytokinesis occurs during mitosis, the location of the furrow is ordinarily determined by the orientation of

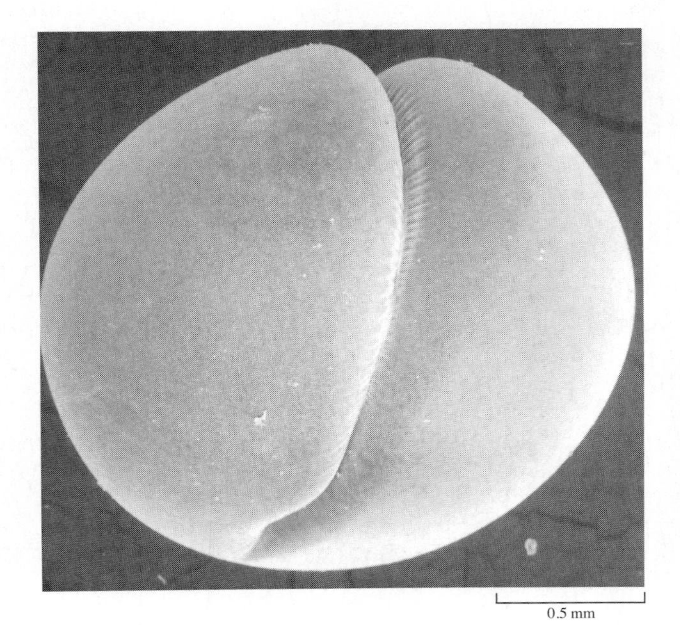

12.13 A dividing frog egg

The cleavage furrow is not yet complete (see bottom of cell). Note the puckered stress lines in the furrow visible in this scanning electron micrograph.

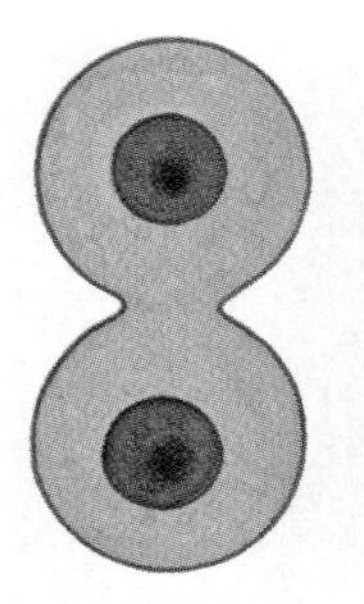
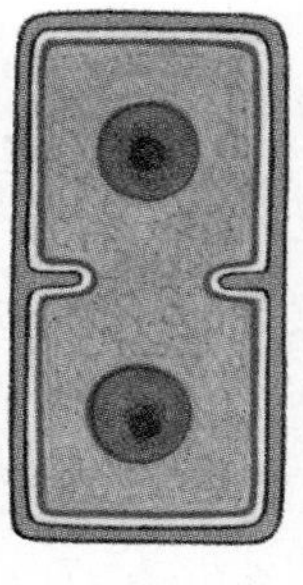
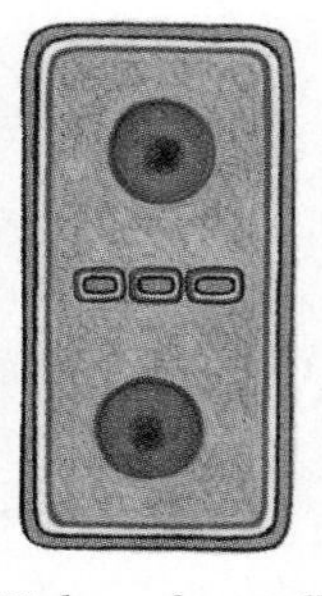

12.14 Three mechanisms of cytokinesis

Cytokinesis in animal cells typically occurs by a pinching-in of the plasma membrane. In many algal cells cytokinesis occurs by an inward growth of new wall and membrane. In higher plants cytokinesis typically begins in the middle and proceeds toward the periphery, as membranous vesicles fuse to form the cell plate.

the spindle, in whose equatorial region the furrow forms (see Fig. 12.9:7). The furrow becomes progressively deeper until it cuts completely through the cell (and its spindle), producing two new cells.

Since the location of the furrow is usually related to the position of the centrioles and spindle, a dense belt of actin and myosin microfilaments at the site of the cleavage furrow is probably responsible. Drugs like cytochalasin, which block the activity of microfilaments, stop cytokinesis. Since agents that bind actin and myosin also stop the cleavage process, the sort of actin-myosin interaction known to be involved in cell movement, as discussed in Chapter 5, is probably responsible for cytokinesis as well.

How plant cells divide The relatively rigid cell walls of plants, essential for their support, required the evolution of a different kind of cytokinesis. In many fungi and algae, new plasma membrane and cell wall grow inward around the wall midline until the growing edges meet and completely separate the daughter cells (Fig. 12.14B). In higher plants a special membrane, called the ***cell plate***, forms halfway between the two nuclei at the equator of the spindle if cytokinesis accompanies mitosis (Figs. 12.14C, 12.15D). The cell plate begins to form in the center of the cytoplasm and slowly becomes larger until its edges reach the outer surface of the cell and the cell's contents are cut in two. Thus cytokinesis of higher-plant cells progresses from the middle to the periphery, whereas cytokinesis in animal cells progresses from the periphery to the middle.

The cell plate forms from membranous vesicles that are carried

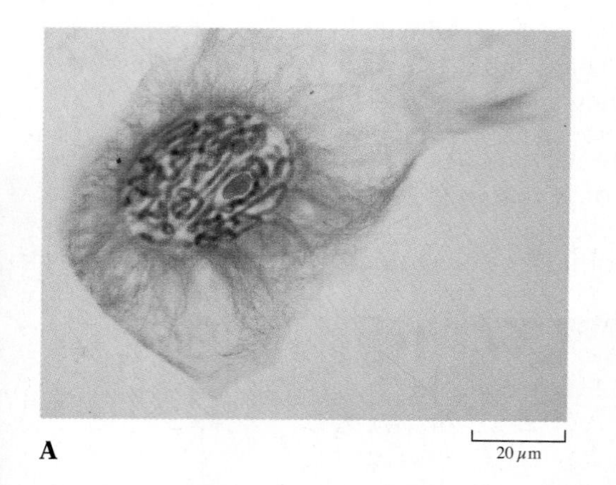

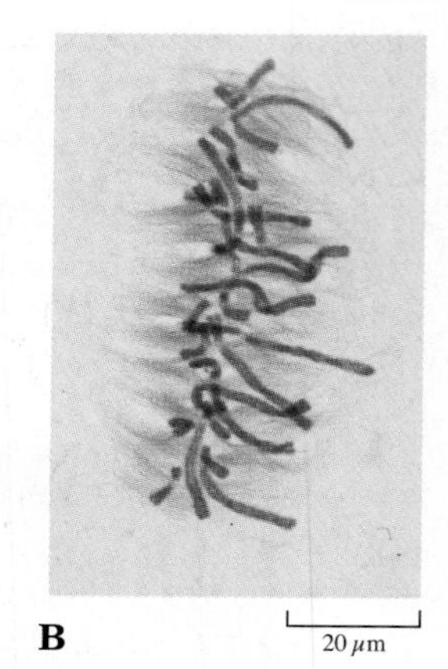

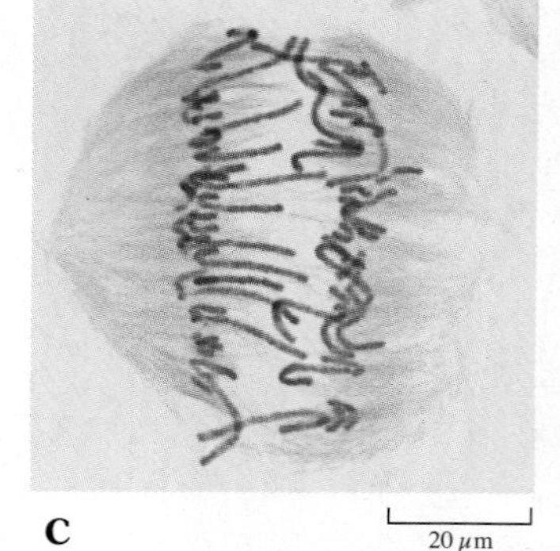

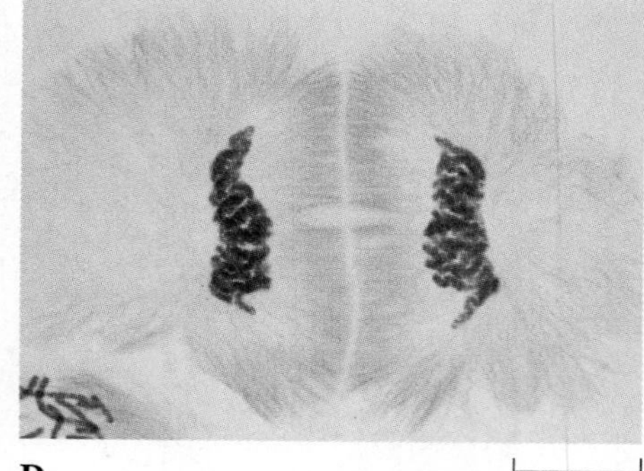

12.15 Cell division in a plant, the African blood lily

(A) Prophase: the chromosomes have condensed; microtubules (red) are visible. (B) Metaphase. (C) Anaphase: the two groups of chromosomes are moving to opposite poles of the cell. (D) Telophase: a cell plate has begun to form.

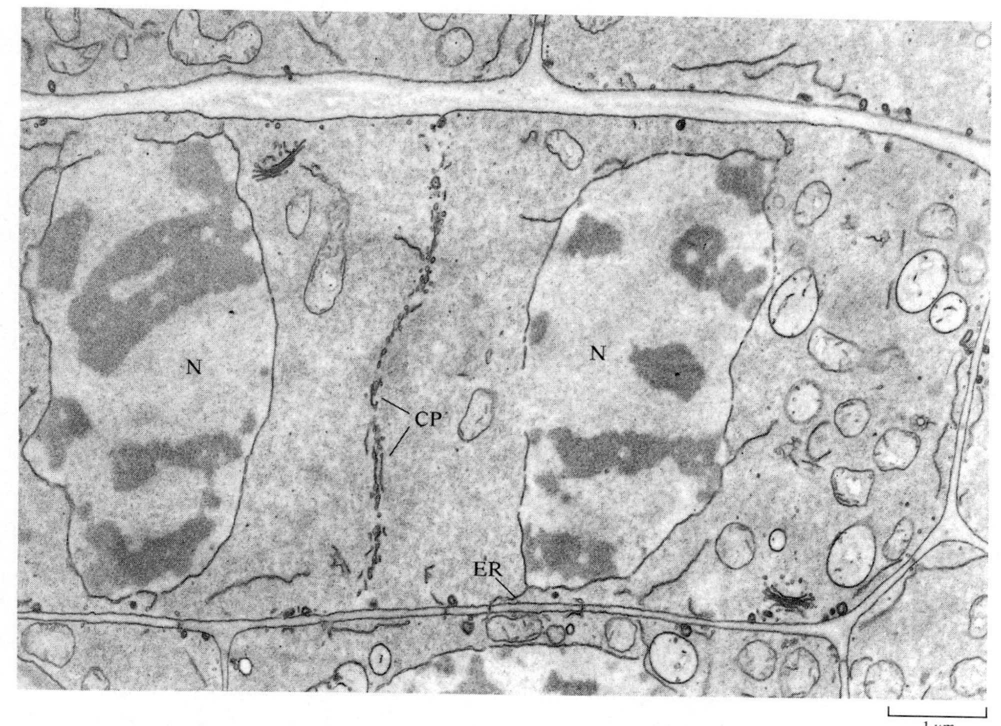

12.16 A late-telophase cell in corn root, showing formation of the cell plate

Mitosis has been completed, and the two new nuclei (N) are being formed; the chromosomes (dark areas in the nuclei) are no longer visible as distinct structures, but the nuclear membranes are not yet complete. A cell plate (CP) is being assembled from numerous small vesicular structures. At the lower end of the nucleus on the right, a length of endoplasmic reticulum (ER) can be seen in this electron micrograph that appears to run from the nuclear envelope to the cell wall, through the wall, and into the next cell.

to the site of plate formation by the microtubules that remain after mitosis. There they first line up and then unite (Fig. 12.16). The vesicles are derived mainly from the Golgi and, to a lesser extent, from the ER. As the vesicle membranes fuse with one another and, peripherally, with the old plasma membrane, they constitute the partitioning membranes of the two newly formed daughter cells. The contents of the vesicles, trapped between the daughter cells, give rise to the middle lamella and to the beginnings of primary cell walls.

MAKING GAMETES: MEIOSIS

Mitosis maintains a constant number of chromosomes in somatic cells. What would be its effect in reproductive cells? If two gametes (an egg and a sperm, for example) each carried the normal number of mitotic chromosomes, the ***zygote*** (the first cell produced by their union) would have double that number. At each successive generation the number would again double. Gamete production requires

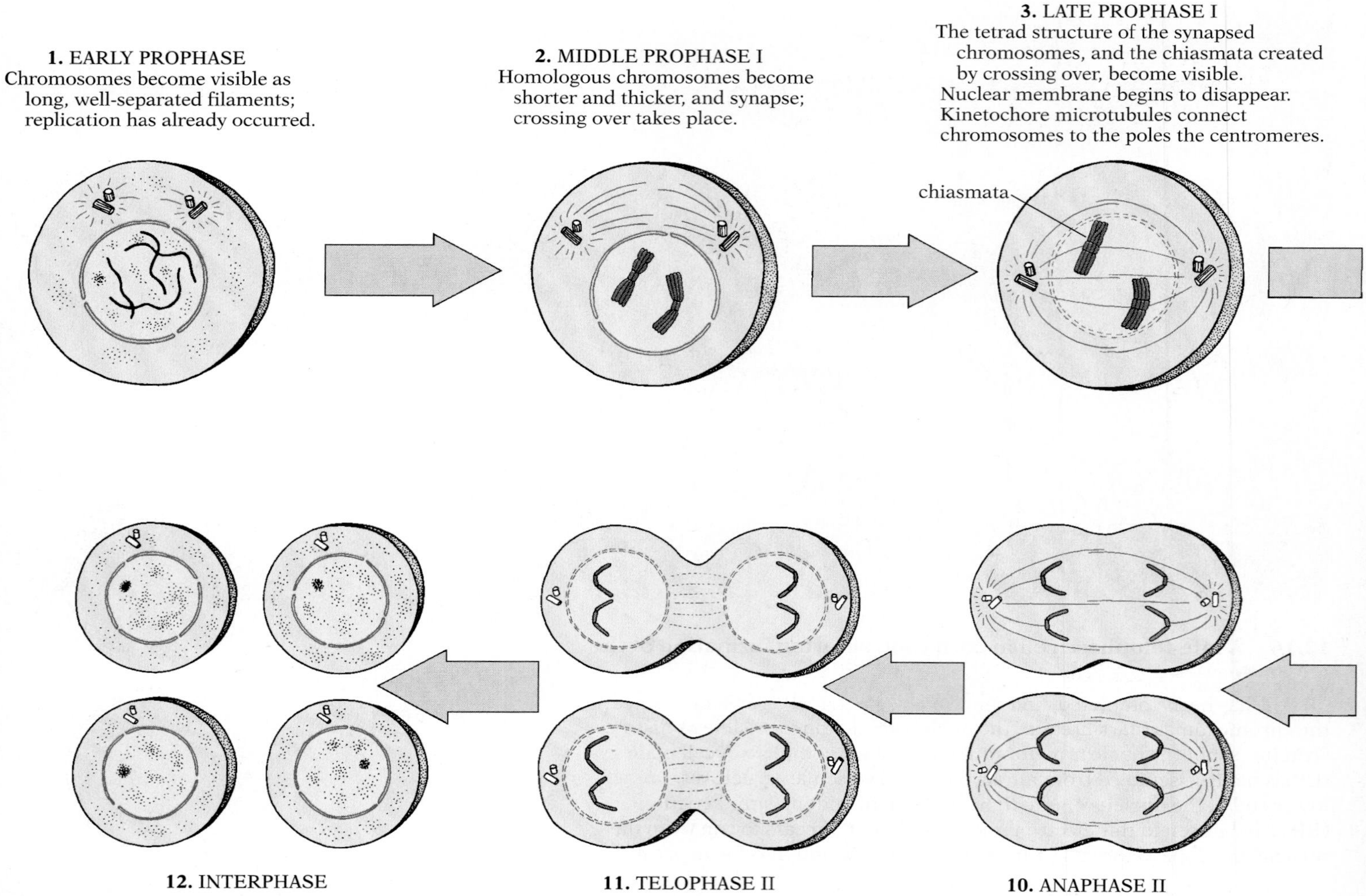

1. EARLY PROPHASE
Chromosomes become visible as long, well-separated filaments; replication has already occurred.

2. MIDDLE PROPHASE I
Homologous chromosomes become shorter and thicker, and synapse; crossing over takes place.

3. LATE PROPHASE I
The tetrad structure of the synapsed chromosomes, and the chiasmata created by crossing over, become visible.
Nuclear membrane begins to disappear.
Kinetochore microtubules connect chromosomes to the poles the centromeres.

chiasmata

12. INTERPHASE

11. TELOPHASE II

10. ANAPHASE II

a division strategy that reduces the number of chromosomes to half, so that when the egg and sperm unite in fertilization the normal diploid number is restored. This special process of reduction division is ***meiosis*** (from the Greek word for "diminution"). In all multicellular animals meiosis occurs at the time of gamete production. Consequently each gamete possesses only half the species-typical number of chromosomes.

Note that in the reduction division of meiosis, the chromosomes of the parental cell are not simply separated randomly into two halves. The diploid nucleus contains two of each type of chromosome, and meiosis partitions these chromosome pairs so that each gamete contains one of the two homologues. Such a cell, with only one of each type of chromosome, is therefore haploid. When two haploid gametes unite in fertilization, the resulting zygote is diploid, having received one of each chromosome type from the sperm of the male parent and one of each type from the egg of the female parent.

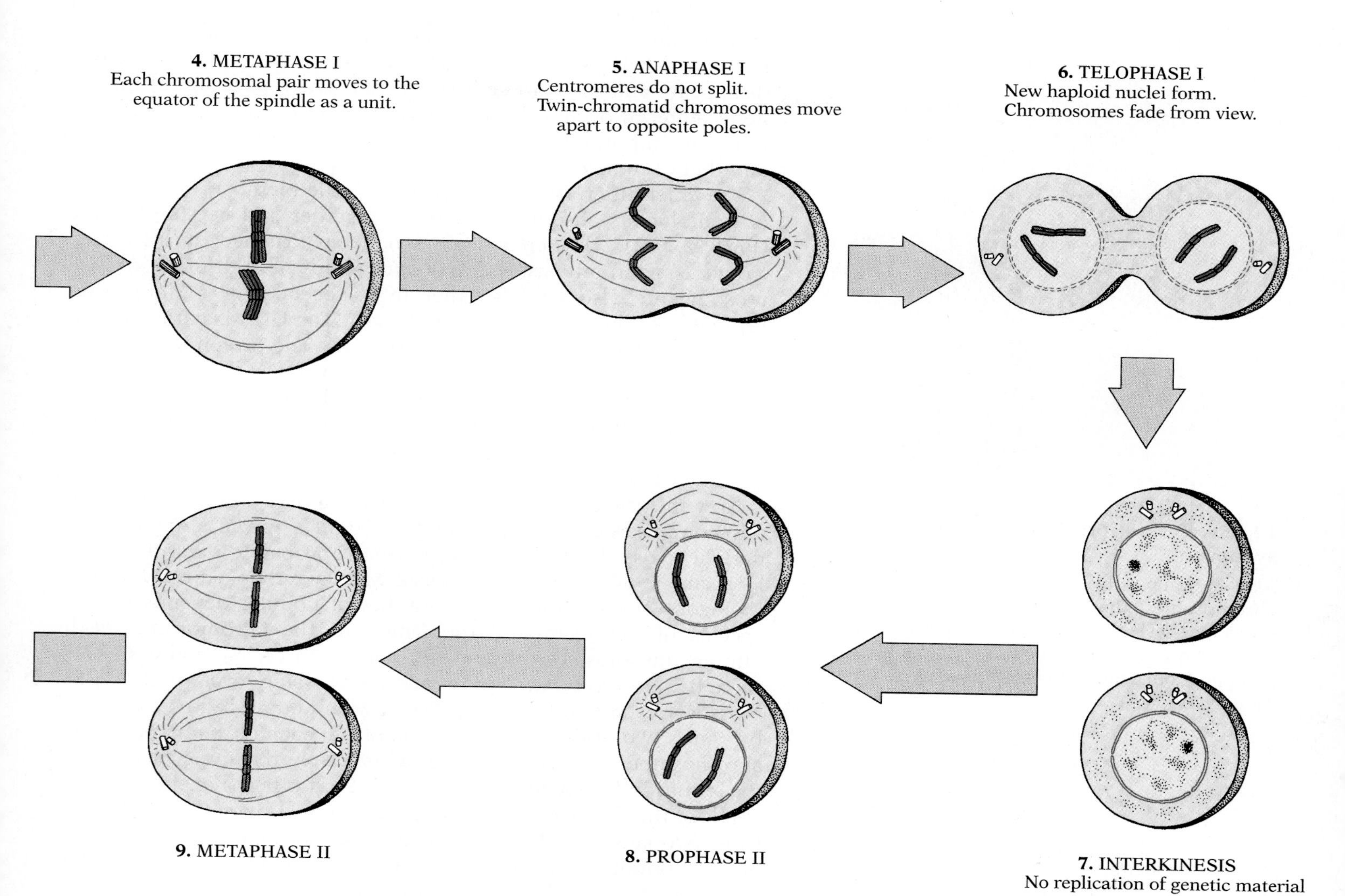

12.17 Meiosis in an animal cell

The chromosomes inherited from each parent are shown in different colors. As a result, each member of the homologous pair is a different color, which aids in visualizing the results of crossing over.

■ THE MEIOTIC CYCLE

Production of two haploid gametes from a diploid cell could be accomplished by a single division; instead, complete meiosis (Fig. 12.17) involves two division sequences, which result in four new haploid cells. As in mitosis, meiosis is preceded by replication, which creates chromatid pairs. This step seems unnecessary, and may exist solely to make possible the process called crossing over (to be discussed presently). The result is that the gamete-producing cell now has 4 times as much DNA as the gametes require, and meiosis proceeds to correct this excess. The first division sequence reduces the number of chromosomes; the second separates chromatids. The same four stages found in mitosis—prophase, metaphase, anaphase, and telophase—are recognized in each division sequence in meiosis.

Prophase I: crossing over Many of the events in prophase I of meiosis superficially resemble those in the prophase of mitosis.

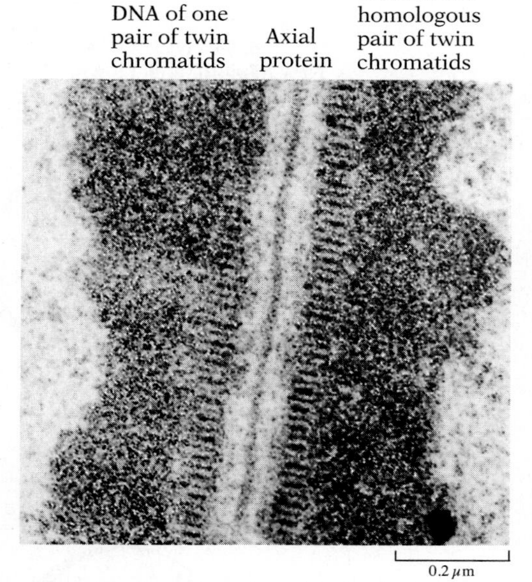

12.18 Formation of the chromosomal synapse

Axial proteins gather the replicated DNA of each chromosome—that is, the twin chromatids—into a long series of paired loops, as seen in this longitudinal section through a synaptonemal complex from a meiotic cell of the ascomycete *Neottiella*.

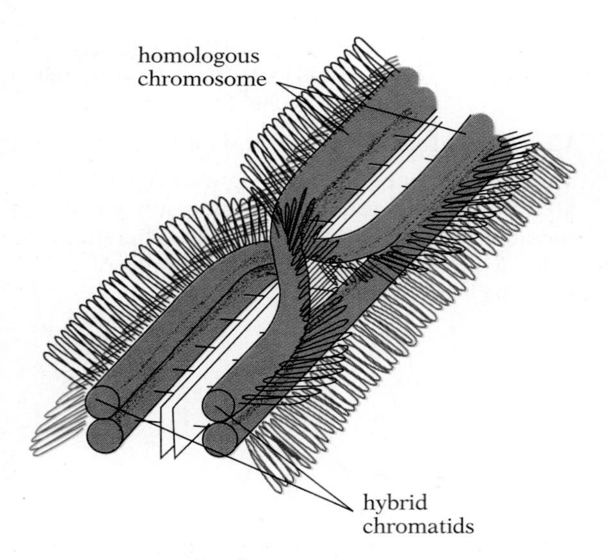

12.19 Crossing over

When two homologous chromosomes have synapsed, chromatids from one chromosome exchange fragments with chromatids from the other chromosome to create hybrid chromatids. One such exchange is shown here.

The individual chromosomes come slowly into view as they coil and become shorter, thicker, and more easily stainable. The two nucleoli slowly fade from view and, finally, the nuclear envelope disappears and the spindle is organized. The replication of the genetic material occurs during interphase, as in mitosis. There are, however, important differences between the prophase of meiosis and that of mitosis.

The chief difference is that in meiosis the members of each pair of homologous chromosomes move together and lie side by side (Fig. 12.17:2). Also, in addition to being fastened at their centromeres, as in mitosis, the twin chromatids of each meiotic chromosome are also held together by a special pair of long, thin protein axes that run their entire length. The DNA is gathered into loops as part of the condensation process. The protein axes of the two homologous chromosomes now join by means of protein cross bridges to form an intricate compound structure known as a ***synaptonemal complex***, which lines up the four chromatids (often referred to as a tetrad) in perfect register (Fig. 12.18). This process is known as ***synapsis***.

Now the remarkable process of ***crossing over*** begins. Large protein complexes called recombination nodules appear along the ladderlike cross bridges. These nodules probably determine the locations at which genetic material will be exchanged between homologous chromosomes. The number of nodules depends on the species of organism and the length of the chromosome; each human chromosome has an average of three. In many species there is also a sexual difference in crossing over, with fewer crossover events in male cells. Next, each nodule begins a process by which two of the chromatids, one from each of the homologous chromosomes, are clipped open at precisely the same place, and the resulting fragments are spliced to each other (Fig. 12.19). The result of this splicing is that the chromatids of the tetrad no longer form two sets of twins; the recombined chromatids are now hybrids, containing genetic material descended from both the mother's and the father's homologous chromosomes.

When the synaptonemal complex begins to break up in late prophase, the points at which crossing over has taken place become visible. The hybrid chromatids produced by crossing over link the two homologous chromosomes at these points, called ***chiasmata*** (Fig. 12.20). Each crossover event can involve a different pair of chromatids (Fig. 12.21). Crossing over is not rare or accidental: it is a frequent and highly organized mechanism.

As prophase ends, most chromosomes now consist of two hybrid chromatids rather than twin chromatids, since genetic material has been exchanged between the chromatids of homologous chromosomes. Spindle microtubules appear, radiating from the two poles of the cell, and kinetochore microtubules attach to the centromeres. There is a distinct difference between the activity of the centromeric microtubules in mitosis and in meiosis. In meiosis, microtubules from one centromere of each homologous pair attach to a pole so the two pairs are pulled toward opposite poles (Fig. 12.22).

Metaphase I In mitosis each chromosome consists of a pair of twin chromatids, and moves independently to the midline of the

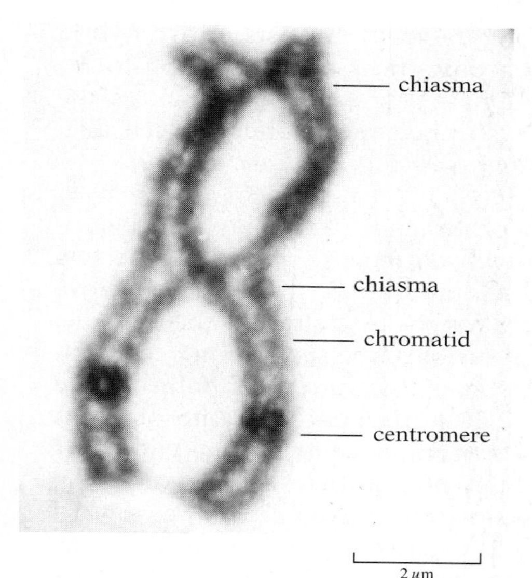

12.20 Two chiasmata between homologous chromosomes

Each chromosome is clearly recognizable as a pair of chromatids. The centromeres, too, are clearly visible. Crossing over has taken place at two points—the chiasmata. Note that the crossing over involves only one chromatid from each chromosome. These chromosomes of a Costa Rican salamander are from a spermatocyte in prometaphase I.

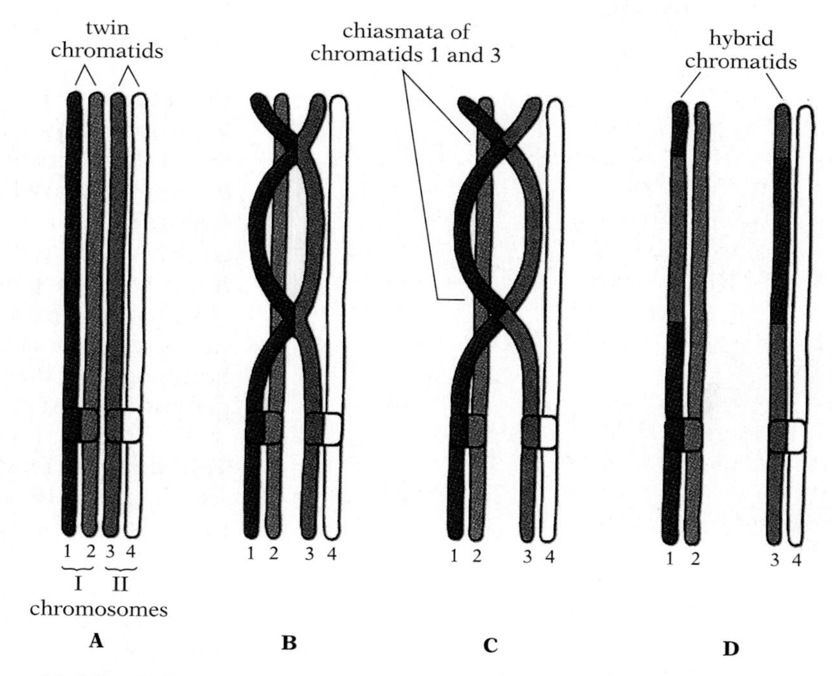

12.21 Schematic summary of synapsis and crossing over

(A) Crossing over begins as homologous chromosomes (I and II), each consisting of twin chromatids (1,2 and 3,4), are brought into register and synapse. (The synaptonemal complex has been omitted for clarity.) (B) Parts of separate chromatids are spliced together, while the protein axis (not shown) keeps each segment firmly attached to its twin. (C) When the synaptonemal complex breaks up, the homologous chromosomes begin to drift apart, but the chiasmata of spliced hybrid chromatids prevent them from separating. (D) The homologous chromosomes separate fully during anaphase.

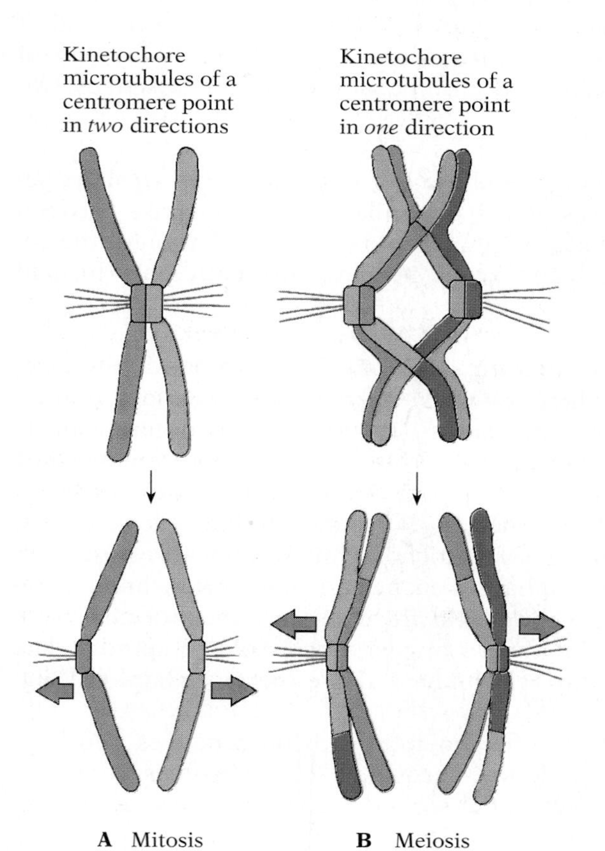

12.22 Kinetochore microtubules in mitosis and meiosis

(A) In mitosis, each centromere is attached by kinetochore microtubules to both poles. As a consequence, the two chromatids are pulled apart and wind up at opposite ends of the cell during anaphase. (B) In meiosis I, however, one centromere in each pair is attached to one or the other of the two poles. The result is that in anaphase I the chromatids remain joined, and homologous two-chromatid chromosomes wind up in separate cells.

cell. In meiosis each chromosome typically consists of two hybrid chromatids, and homologous chromosomes move as a unit to the midline in preparation for anaphase. As a result, in meiosis the number of independent units waiting at the midline for cell division is only half the number of mitosis (Fig. 12.17:4).

Anaphase I In mitosis, metaphase ends and anaphase begins when the centromere of each twin chromatid splits, and the two independent single-chromatid chromosomes thus formed move away from each other toward opposite poles of the spindle. However, in the first division of meiosis, the separation occurs instead between the two chromosomes that have been joined since the middle of prophase (Fig. 12.22). Since each chromosome has its own centromere, there is no splitting of centromeres. With their respective centromeres attached by microtubules to only one pole, the homologous chromosomes move away from each other toward opposite poles during anaphase. Thus in a four-chromosome organism, only two chromosomes, each with a hybrid chromatid, move to each pole (Fig. 12.17:5), in contrast to the four single-chromatid chromosomes that move to each pole in mitosis. Because the synaptic pairing was not random, but involved the two homologous chromosomes of each type, the two daughter nuclei get not just any two chromosomes, but rather one of each type.

Telophase I Telophase of mitosis and meiosis are essentially the same, except that in mitosis each of the two new nuclei has the same number of chromosomes as the parental nucleus, whereas each of the new nuclei formed in meiosis has half the chromosomes present in the parental nucleus (Fig. 12.17:6). At the end of telophase in mitosis, the chromosomes are single; at the end of telophase I of meiosis, the chromosomes are composed of two chromatids each.

Interkinesis Following telophase I of meiosis, there is a short period called interkinesis, which is similar to an interphase between two mitotic division sequences except that no replication of the genetic material occurs and hence no new chromatids are formed (Fig. 12.17:7).

The second set of meiotic divisions The second division sequence of meiosis, which follows interkinesis, is essentially mitotic from the standpoint of mechanics, though the functional result is different (Fig. 12.17:8–12). The chromosomes do not synapse; they cannot, since the cell contains no homologous chromosomes. Each two-chromatid chromosome moves to the midline independently, and its centromere sends kinetochore microtubules toward each pole. At the end of metaphase II each centromere splits, and during anaphase II the single-chromatid chromosomes thus formed move away from each other toward opposite poles of the spindle. The new nuclei formed during telophase II are therefore haploid (Fig. 12.23).

In summary, then, the first meiotic division produces two haploid cells containing double-chromatid chromosomes. Each of these cells divides again in the second meiotic division, producing

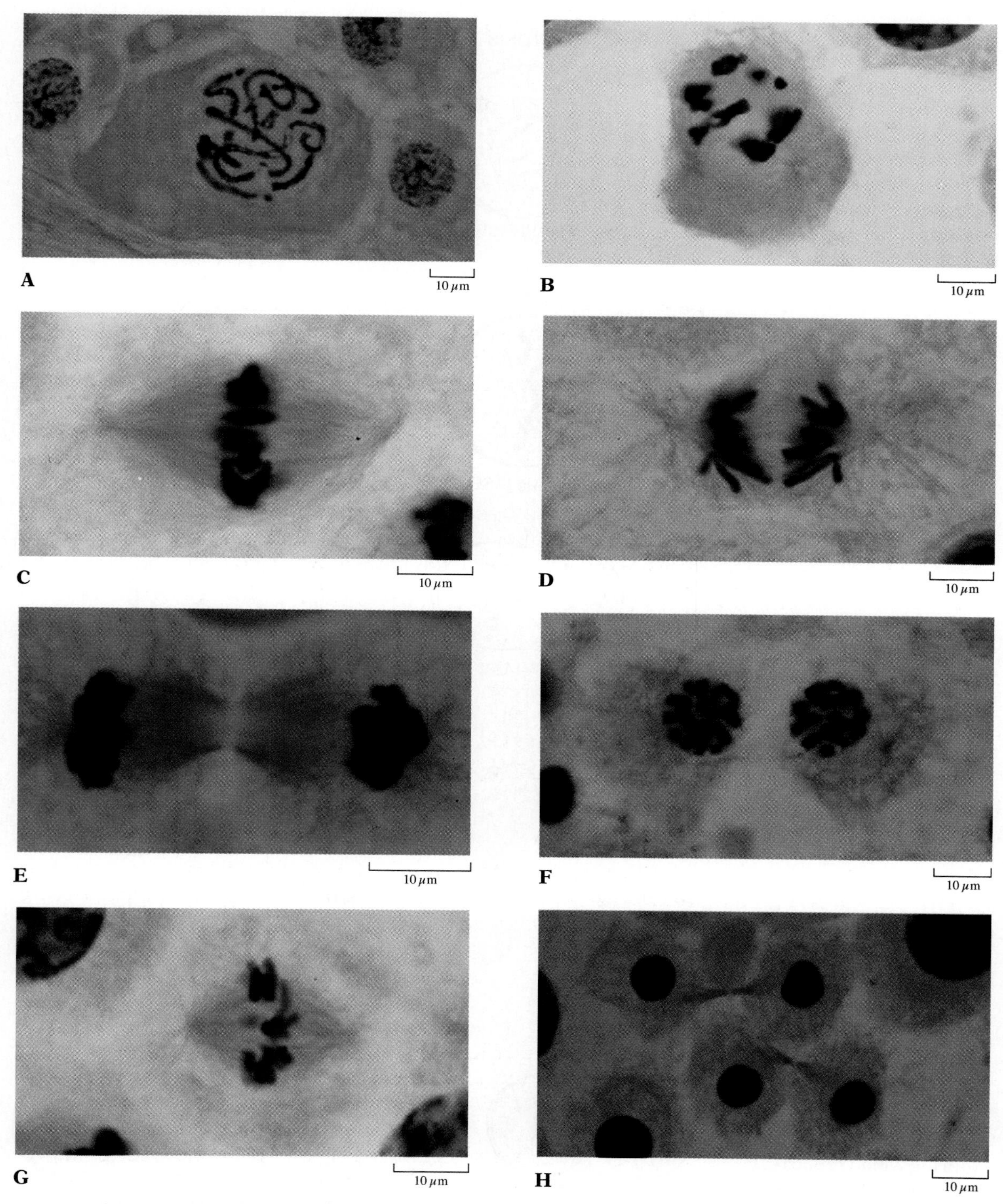

12.23 Meiosis in a cell from the grasshopper *Mongolotetix japonicus*

(A) Early prophase I: the chromosomes are seen as long filaments. (B) Middle prophase I: synapsing of homologous chromosomes takes place. (C) Metaphase I: chromosomal pairs line up at the equator. (D) Anaphase I: homologous chromosomes have separated and are moving to opposite poles. (E) Telophase I: division into two haploid nuclei has begun. (F) Prophase II: early in this stage, separate chromosomes are barely distinguishable. (G) Metaphase II: chromosomes in each cell are at the equator, and the spindles are evident. (H) Telophase II: four new haploid nuclei can be seen.

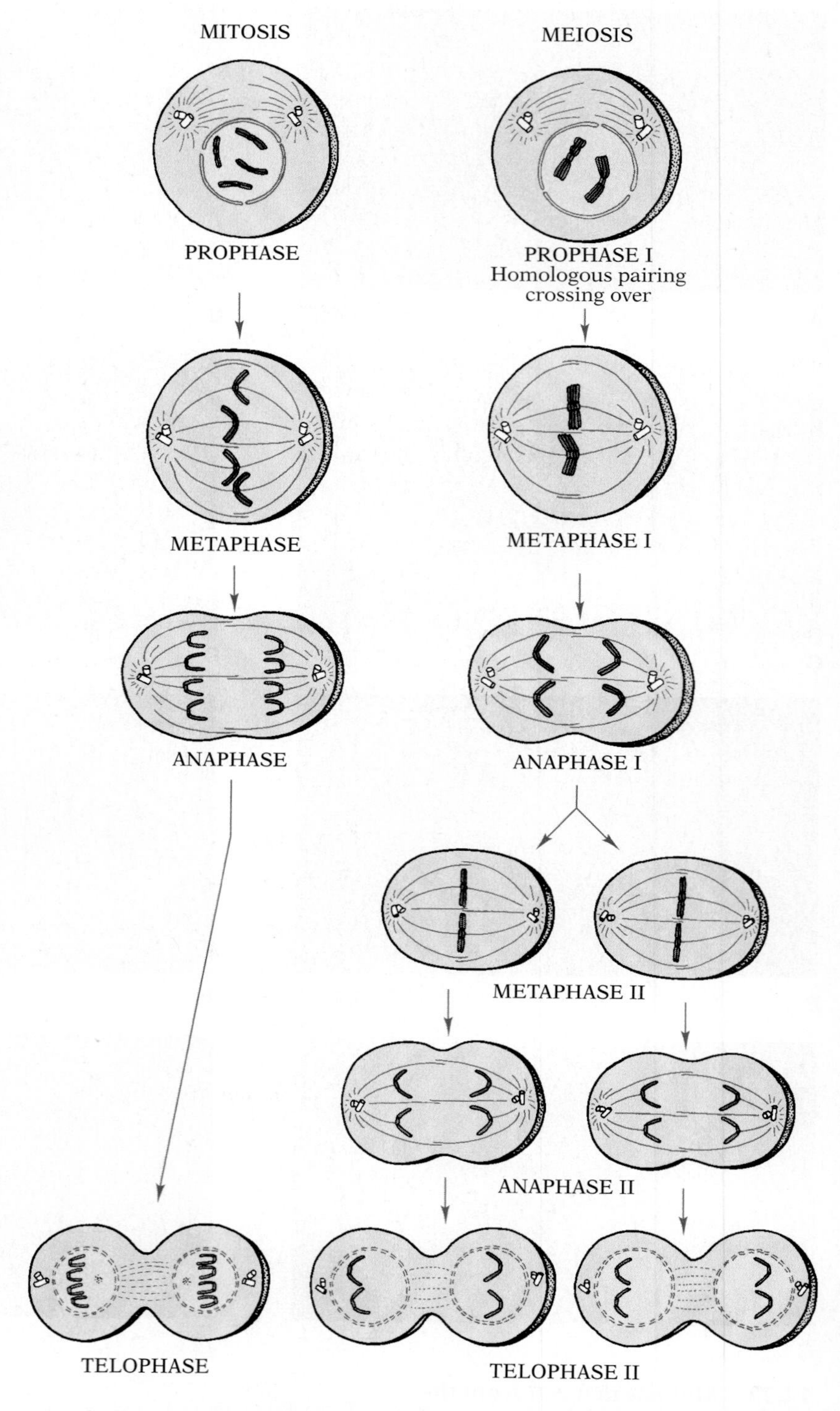

12.24 Comparison of mitosis and meiosis
The first step, interphase (during which replication occurs), is omitted, as are the stages between anaphase I and metaphase II of meiosis.

a total of four new haploid cells containing single-chromatid chromosomes. Mitosis and meiosis are compared in Figure 12.24.

■ WHAT IS THE PURPOSE OF SEX AND CROSSING OVER?

Life depends on an optimal balance between stability and change—maintaining the fidelity of the genetic message while at the same time providing enough variation to permit selection for improved genes, particularly in the face of unpredictable changes in competition, predation, habitat, and climate. Sexual organisms go to great lengths to produce haploid gametes for mating (***sexual recombination***): asexual cloning of diploid offspring is metabolically less expensive, and gamete-producing cells are put to considerable trouble to make crossing over possible. Most researchers believe these two processes serve the goal of generating change.

The most important consequence of sexual reproduction is that it creates offspring with new combinations of characteristics. Each member of a homologous pair of chromosomes comes from a different parent, and contains genes coding for the same kinds of RNA, structural proteins, and enzymes, but the exact base sequences in two homologous genes are not necessarily identical. Instead, the copy of a gene inherited from an organism's male parent is often at least slightly different from the copy inherited from the female parent. Different versions of the same gene are called ***alleles***, and the enzymes or structural proteins they produce are likely to have slightly (or even very) different activities. For example, blue eyes develop in people when both copies of their eye-pigment gene code for a defective (colorless) screening pigment; people with brown eyes have at least one copy of the allele that codes for a functional pigment. Since an individual organism is a mix of alleles from two different parents, its chromosomes are almost certain to contain an ensemble of alleles different from that of either parent, and many aspects of its morphology, physiology, and behavior will be correspondingly different.

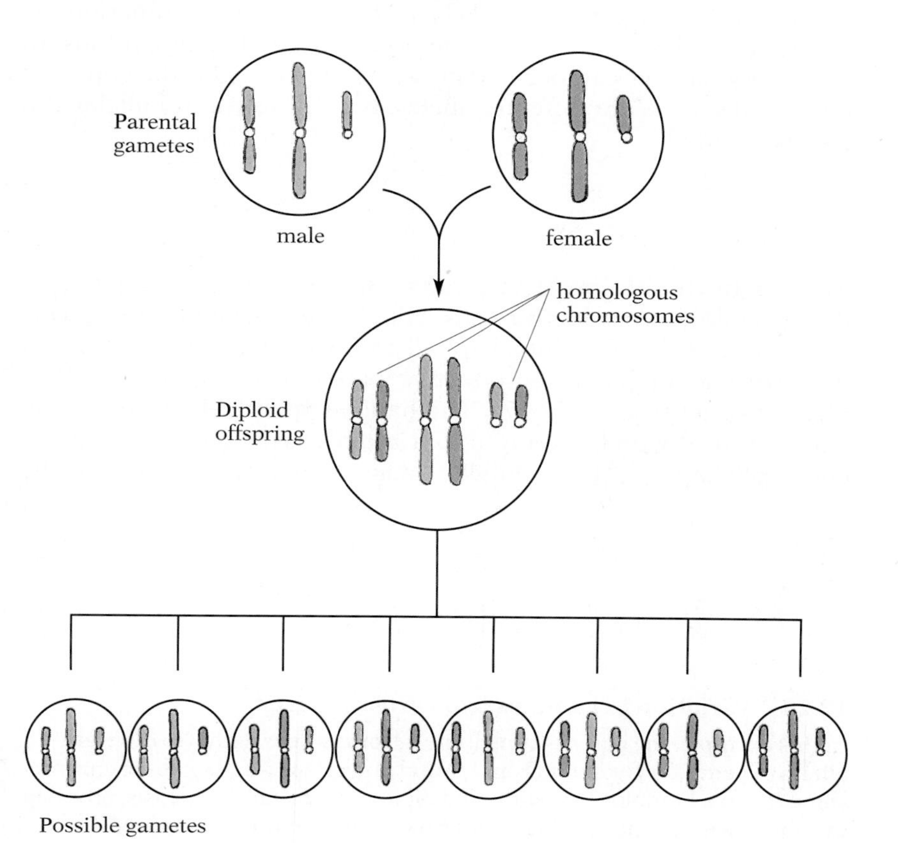

12.25 Variation in gametes

Even without crossing over, a diploid organism produces 2^n different kinds of gametes, where n is the number of pairs of homologous chromosomes. In this hypothetical organism, $n = 3$, so there are eight (2^3) possible kinds of gametes.

To understand why variation arises in sexual recombination, let's consider a hypothetical six-chromosome organism (Fig. 12.25). Three chromosomes came originally from the gamete of the male parent (the product of meiosis), and three from the gamete of the female parent, to form the three pairs of homologous chromosomes in this diploid organism. Each time meiosis occurs in an individual producing gametes for reproduction, all of that organism's maternal chromosomes *might* go into one gamete and all the paternal ones into another, but many other combinations are equally likely. If an organism has three pairs of chromosomes, eight different combinations may be found in the gametes. In the zygote formed by two gametes in such a species, 64 (8×8) different combinations of chromosomes are then possible. For humans, with our unusually large number of chromosomes, the number of different chromosome combinations possible in the offspring of the same two parents is about 7×10^{13}.

The highly ordered process of crossing over increases the variety of possible gametes astronomically, since crossover points are essentially random, and virtually every hybrid chromatid is spliced together at a unique set of points. To take a specific example, assume that the father contributes a gene coding for black fur (which we can call *F*) and a gene coding for a large body (*S*), while the alleles from the mother code for gray fur (*f*) and a small body (*s*). Though the *F* and *S* of the chromosome contributed by the father remain together (or ***linked***) throughout the *mitotic* life of a cell, as do the *f* and *s* on the maternal chromosome, crossing over in *meiosis* can rearrange things (Fig. 12.26). As a result, the chromosome in the offspring's gamete is likely to have a new combination—*F* and *s*, or *f* and *S*—rather than one of the parental combinations. In fact, crossing over can occur even within a gene, so if the copies of a particular gene are different alleles, two entirely novel alleles can also be created.

■ WHEN DOES MEIOSIS OCCUR?

Meiosis in the life cycle of plants Meiosis in plants usually produces haploid reproductive cells called *spores*, which often divide mitotically to develop into haploid multicellular plant bodies. At one extreme in plant reproductive patterns are some primitive plants like algae (Fig. 12.27). The haploid spore cells (stage 1) divide mitotically and develop into a rapidly growing, poorly differentiated, haploid, multicellular stage (stage 2) in which the

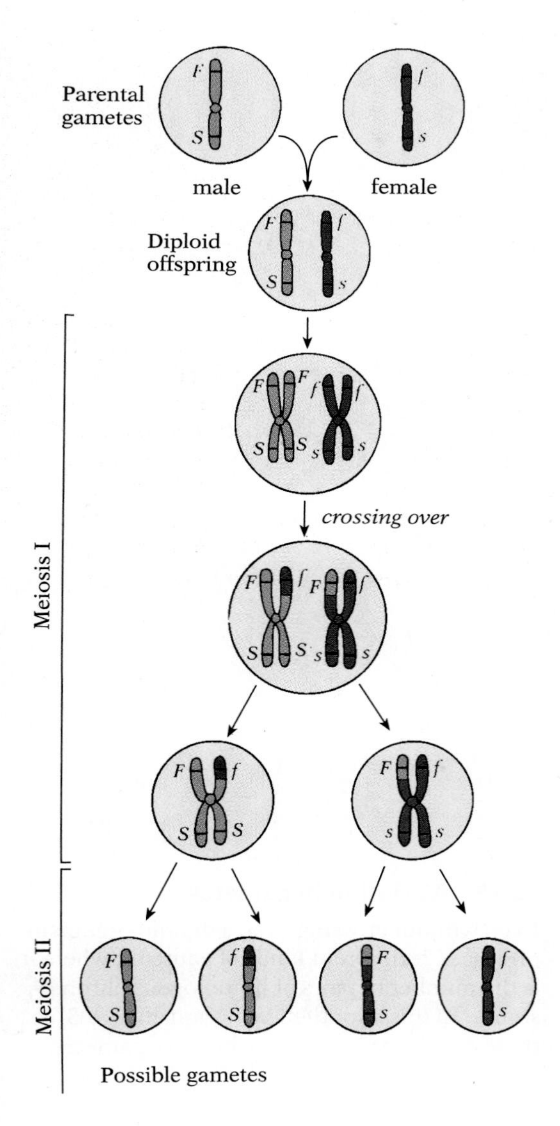

12.26 Variation in gametes from crossing over
In this example, the parents contribute chromosomes with two different alleles of genes for fur color (*F* and *f*) and body size (*S* and *s*); these chromosomes form a homologous pair in the diploid offspring. At meiosis, crossing over breaks up the paternal and maternal combinations, so some of the resulting gametes have hybrid chromosomes, unlike those of either parent.

A PRIMITIVE PLANT

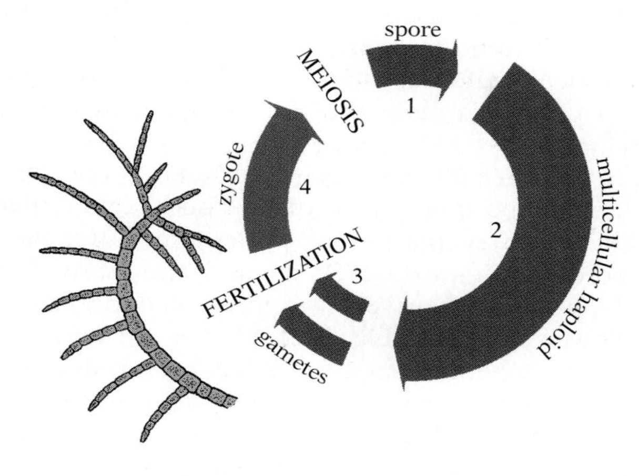

B INTERMEDIATE PLANT

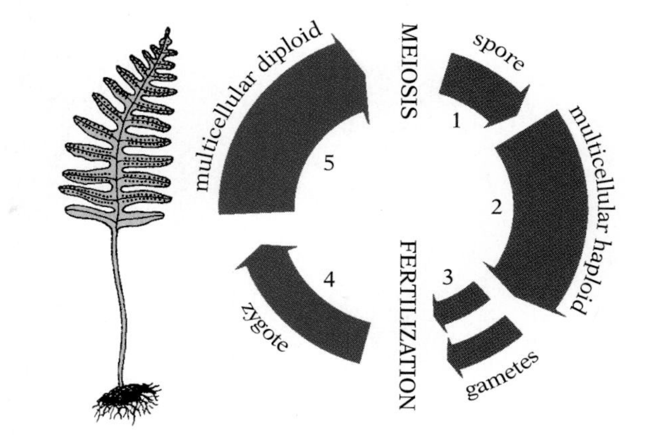

C ADVANCED PLANT

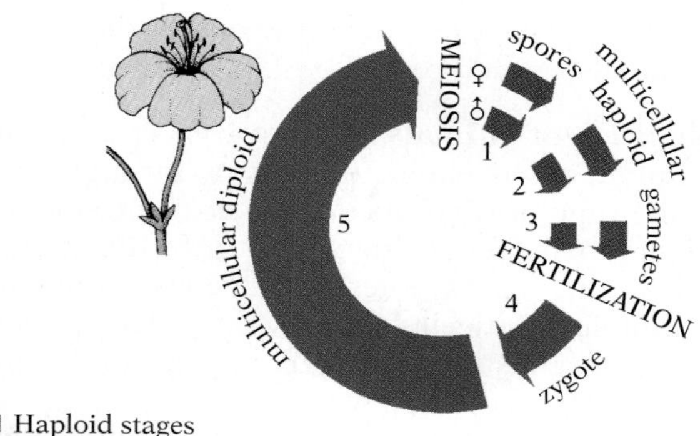

D ANIMAL

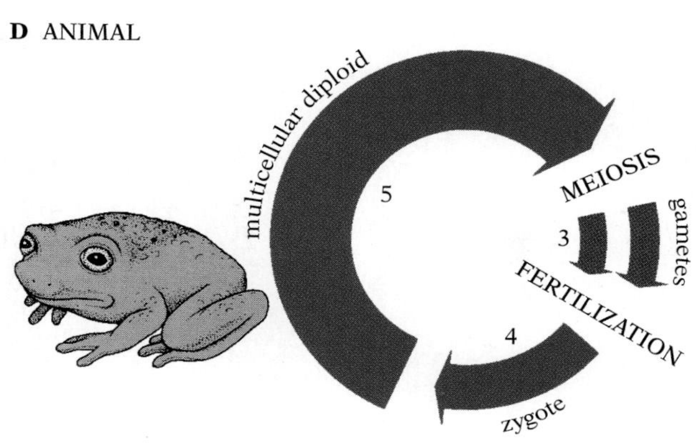

Haploid stages
Diploid stages

12.27 Four types of life cycles

(A) In some very primitive plants the diploid phase is represented only by the zygote, which quickly divides by meiosis to produce haploid spores, which may in turn divide mitotically to produce a multicellular haploid plant. (B) In most multicellular plants there are two multicellular stages, one haploid and one diploid (stages 2 and 5). The relative importance of these two stages varies greatly from one plant group to another; the cycle shown here is an intermediate one, in which stages 2 and 5 are nearly equal. (C) In flowering plants the multicellular diploid stage (stage 5) is the major one, and the multicellular haploid stage (stage 2) is much reduced, being represented by a tiny organism with very few cells. (D) Animals and a very few plants have a life cycle in which meiosis produces gametes directly—the spore stage and the multicellular haploid stage (stages 1 and 2) being absent.

organism passes most of its life. This multicellular plant eventually produces, by mitosis, cells specialized as gametes (stage 3), which unite to form the diploid zygote (stage 4); the zygote promptly undergoes meiosis to produce four haploid spores, thus beginning the cycle again. In such an organism, then, the haploid phase of the cycle (particularly stage 2) is dominant, and the only diploid stage—the zygote—is brief.

Most plants devote much of their life cycle to the diploid state. Many ferns, for instance, divide their time about evenly between haploid and diploid phases (Fig. 12.27). While the fern is in the diploid portion of its life cycle, certain cells in its reproductive organs divide by meiosis to produce haploid spores (stage 1). These spores divide mitotically and develop into haploid multicellular plants (stage 2). The haploid multicellular plant eventually produces cells specialized as gametes (stage 3). As with algae, the gametes are produced by mitosis, not meiosis, because the cells that divide to produce the gametes are already haploid. Two of these gametes unite in fertilization to form the diploid zygote (stage 4), which divides mitotically and develops into a diploid multicellular plant (stage 5). In time, this plant produces spores and the cycle starts over again.

As described in Chapter 22, the several groups of plants vary greatly in the relative importance of the diploid and haploid phases in their life cycles. A very few plants have cycles almost like the animal life cycle shown in Figure 12.27D; stages 1 and 2 are absent, and the haploid phase of the cycle is represented only by the gametes. In the flowering plants (Fig. 12.27C), stages 1 and 2 have not been abandoned altogether, but stage 2 has been reduced to a tiny three-to-eight-cell entity that is not free-living, and the plant spends most of its life as a multicellular diploid organism (stage 5). In general, haploid-dominated plants tend to be short-lived, to reproduce rapidly and prolifically, and have little differentiated tissue; diploid-dominated plants are at the other end of the scale, and more like higher animals in each regard.

Meiosis in the life cycle of animals With rare exceptions, higher animals exist as diploid multicellular organisms through most of their life cycle. At the time of reproduction, meiosis produces haploid gametes, which, when their nuclei unite in fertilization, give rise to the diploid zygote. The zygote then divides mitotically to produce the new diploid multicellular individual. The gametes—sperm and egg cells—are thus the only haploid stage in the animal life cycle (see Fig. 12.27D).

In male animals, sperm cells (spermatozoa) are produced by the germinal epithelium lining the seminiferous tubules of the testes (Fig. 12.28). When one of the epithelial cells undergoes meiosis, the four haploid cells that result are all quite small, but approximately equal in size (Fig. 12.29). All four differentiate into sperm cells with long flagella, but with very little cytoplasm in the head,

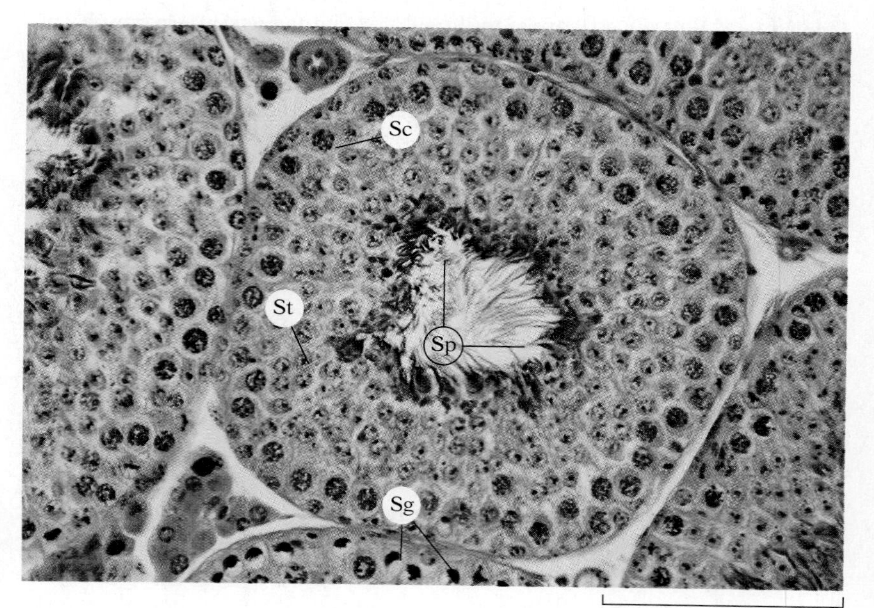

12.28 Cross section of rat seminiferous tubule, showing spermatogenesis

The dark-stained outermost cells in the wall of the tubule are spermatogonia (Sg), which divide mitotically, producing new cells that move inward. These cells enlarge and differentiate into primary spermatocytes (Sc), which divide meiotically to produce secondary spermatocytes and then spermatids (St). The spermatids differentiate into mature sperm cells, or spermatozoa (Sp), whose long flagella can be seen in the lumen of the tubule in this photograph.

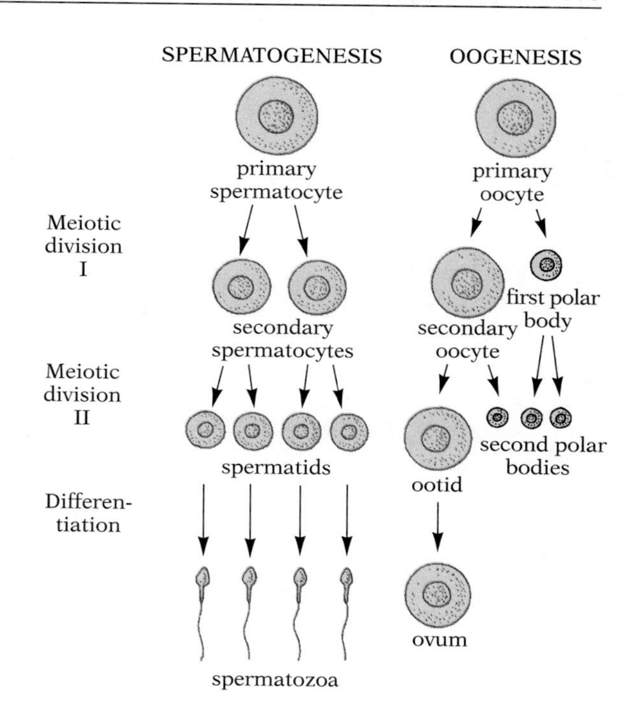

12.29 Schematic illustration of spermatogenesis and oogenesis in an animal

In some animals the first polar body does not divide.

which consists primarily of the nucleus. This process of sperm production is called ***spermatogenesis***. Meiosis in human males begins at puberty.

In female animals the egg cells are produced within the follicles of the ovaries through ***oogenesis***. When a cell in the ovary undergoes meiosis, the haploid cells that result are very unequal in size. The first meiotic division produces one relatively large cell and a tiny one called a first ***polar body***. The second meiotic division of the larger of these two cells (secondary oocyte) produces a tiny second polar body and a large cell that soon differentiates into the egg cell (or ovum). The first polar body may or may not go through the second meiotic division. If it does redivide, there are three polar bodies altogether (Fig. 12.30). Thus, when a diploid cell in the ovary undergoes complete meiosis, only one mature ovum is produced (Fig. 12.29). In human females, the oocytes complete the first meiotic prophase in the fetal ovaries, and then complete meiosis upon ovulation (when a mature egg is released from an ovary). The interval between these two events can exceed 40 years.

The unequal cytokinesis of oogenesis provides an unusually large supply of cytoplasm and stored food to the ovum for use by the embryo that will develop from it. In fact, the ovum provides almost all the cytoplasm and initial food supply for the embryo. The tiny, highly motile sperm cell essentially contributes only its genetic material.

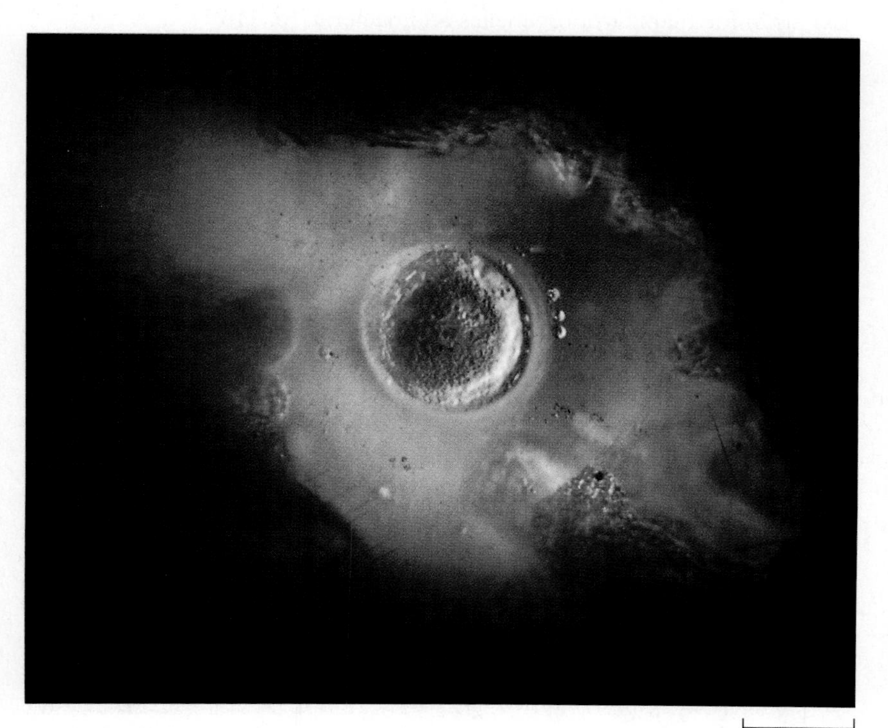

12.30 Human egg cell with polar bodies

The polar bodies are the three small circular structures at the right.

CHAPTER SUMMARY

FISSION VERSUS DIVISION

PROCARYOTIC FISSION Bacteria divide by replicating their single circular chromosome, and then dividing into two daughter cells, each containing a copy of the chromosome. (p. 309)

WHY EUCARYOTIC CHROMOSOMES POSE PROBLEMS Eucaryotic chromosomes are generally linear, organized as homologous pairs, wound on nucleosomes, and organized into condensed loops. All these factors make cell division more complicated. (p. 310)

HOW SOMATIC CELLS DIVIDE

NUCLEAR DIVISION: MITOSIS Eucaryotic cells must distribute a copy of each chromosome and one pair of centrioles to each daughter cell. During interphase the centrioles and chromosomes are duplicated; the duplicated chromatids of each chromosome are joined at the centromere. Interphase is divided into an initial gap, G_1 (when ribosomes and organelles are first duplicated), the synthesis phase, S (when the chromosomes are duplicated), and a second gap, G_2 (when the cell prepares for division). The interactions between cyclins and the cell-division-cycle protein (cdc) regulate interphase. During prophase the centioles move to the poles, the chromosomes condense, the nuclear membrane is disassembled, and the spindle forms. During metaphase the spindle attaches to microtubules from the centromeres, the chromosomes are moved to a plane equidistant between the poles, and the centromeres break. In anaphase the single-chromatid chromosomes move to the poles. In telophase the nuclear membrane reforms around the two daughter nuclei, the chromosomes decondense, and physical division of the daughter cells—cytokinesis—is completed. (p. 311)

THE EVOLUTION OF NUCLEAR DIVISION In some primitive eucaryotes the nuclear membrane remains intact throughout division, and the nucleus fissions in the manner of organelles and bacteria. (p. 318)

CELL DIVISION: CYTOKINESIS The cleavage furrow that marks where animal cells will divide forms equidistant between the two poles. Plants form a membraneous cell plate midway between the poles, which grows until it divides the two daughter cells. (p. 319)

MAKING GAMETES: MEIOSIS

THE MEIOTIC CYCLE Meiosis reduces the number of chromosomes by half, so that gametes have only one chromosome from each homologous pair. Meiosis proceeds in two steps: all chromosomes are duplicated and two diploid daughter cells are produced; then another round of division produces four haploid gametes. The first step differs from mitosis in that after duplication the homologous chromosomes pair up, and crossing over occurs; then one member of each homologous pair is moved into each daughter cell. Thus, the centromeres that connect the chromatids do not break until the second step of meiosis. (p. 323)

WHAT IS THE PURPOSE OF SEX AND CROSSING OVER? Sexual recombination occurs when haploid gametes combine to create a diploid zygote. Sexual recombination creates variation in three ways: crossing over recombines the homologous chromosomes inherited from an individual's two parents, creating novel hybrid chromosomes; a randomly selected member of each homologous pair is incorporated into each gamete, creating many different gametes; and gametes from different individuals combine to generate a unique diploid. (p. 328)

WHEN DOES MEIOSIS OCCUR? Plants alternate between a diploid spore-producing phase and a haploid gamete-producing phase. In most plants the spore-producing form is the dominant phase; the spores are produced through meiosis, and grow into the form that generates the gametes that recombine. Animals have no such alternation. Most plants and all animals produce small male gametes (pollen and sperm) and large female gametes (eggs). Animal eggs are produced by unequal meiotic division that concentrates all the cytoplasm in just one of the four haploid daughter cells; sperm are produced by equal division of the precursor cell entering meiosis. Egg production in many animals begins long before reproduction and is then arrested until eggs are needed; males generally produce sperm as needed. (p. 330)

STUDY QUESTIONS

1 It has been said that the only possible function of the odd-process of meiosis is to make crossing over possible. Compare mitosis and meiosis with this suggestion in mind. (pp. 314–315, 322–323)

2 Crossing over is less frequent in males than in females in many species—male *Drosophila*, for instance, never undergo crossing over in meiosis. Even in hermaphroditic species, in which every organism can have both male and female reproductive organs, the same pattern holds. What might be the evolutionary logic behind this curious asymmetry? (pp. 328–330)

3 Interpret the following observations and formulate a master hypothesis about what sorts of chemical signals must appear in the cell at various points, and what effect they have:

a When a cell just entering S phase is fused with a G_1-phase cell arrested in that state, a process that mixes their cytoplasms but leaves the nuclei separate, both nuclei enter S phase.

b When a mid-S-phase cell is fused with a G_1-phase cell, the S nucleus remains frozen until the G_1 nucleus catches up.

c When a G_1-phase cell and a G_2-phase cell are fused, the G_2 nucleus remains frozen until the G_1 nucleus catches up.

d When an M-phase cell is fused with an S-phase or either sort of G-phase cell, the M nucleus is unaffected, but the other nucleus loses its nuclear envelope and its chromosomes condense. (pp. 312–315)

4 Do animals with larger numbers of chromosomes but similar crossover rates have more, less, or the same general amount of variation in their progeny? (pp. 328–330)

SUGGESTED READING

GOULD, J. L., AND C. G. GOULD, 1989. *Sexual Selection.* Scientific American Library, New York. *A wide-ranging, nontechnical account of the many theories that seek to account for the evolution of sex and gender.*

MCINTOSH, J. R., AND K. L. MCDONALD, 1989. The mitotic spindle, *Scientific American* 261 (4).

MURRAY, A. W., AND M. W. KIRCHNER, 1991. What controls the cell cycle, *Scientific American* 264 (3). *On the biochemical control of cell division, with emphasis on cyclin and cdc.*

CHAPTER 13

THE COURSE OF ANIMAL DEVELOPMENT

Probably no aspect of biology is more amazing than the development of a complete new organism with perhaps more than 10^{12} cooperating cells arising from one fertilized egg. The process is so precisely controlled that the entire intricate organization of cells, tissues, organs, and organ systems that will characterize the functioning adult comes into being, each element in just the right place with respect to the others.

Development requires a precisely programmed and coordinated series of changes in gene expression, cell-surface proteins, cell shape, and cell motility, as cells interact to create a characteristic species morphology and to give rise to the lines of specialist cells that constitute the various tissues and organs. This chapter will describe the major morphological and physiological events of development in animals, from fertilization through embryonic development, birth, and postnatal development. The next chapter will examine the molecular and biochemical mechanisms that control and coordinate development.

We will concentrate in these chapters on the patterns and mechanisms of animal development; for a number of reasons plant development is treated separately (in Chapter 32). For one thing, the vast majority of plants are autotrophic, and are adapted to remain rooted during a life cycle to a specific locale, and to collect sunlight. Since plants are not adapted for locomotion, they can enjoy the structural advantages of more rigid cell walls—a specialization that fixes each cell in place, making impossible the dynamic aspect of animal development that is examined in the next chapter: the migration of cells during the growth of the embryo.

Furthermore, because few plants feed on other organisms, they do not need nervous systems and muscles to guide and power the capture or harvesting of food; nor do they need mouths, stomachs, or digestive systems, or the more than 300 specialized cell types that make up kidneys, bladders, rectums, and other specialized tissues and organs of animals. Because plants compete with each other by growing taller or broader to gather more light, and by extending their roots to obtain more water and minerals, they produce new organs like leaves and flowers whenever and wherever they are needed throughout their life cycle. Animals usually generate their entire array of tissues and organs early in development, and devote later energies to their maintenance and growth.

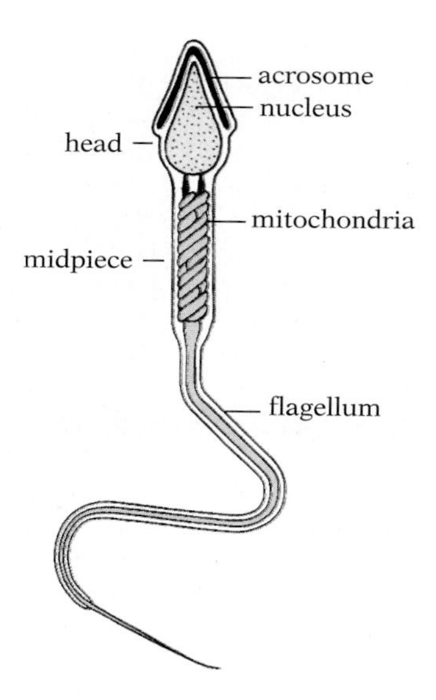

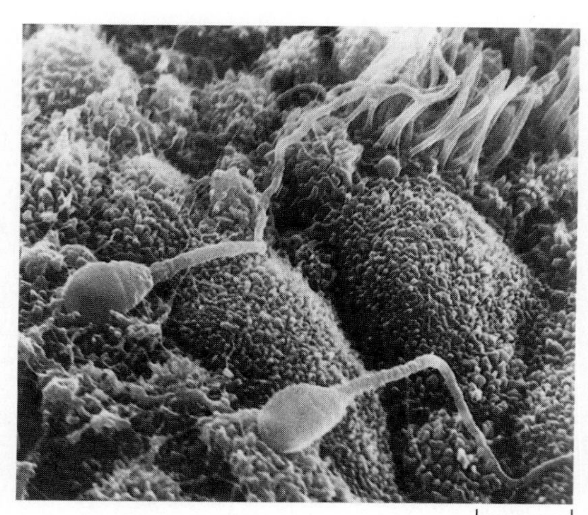

13.1 Human sperm

(A) Three sections are easily distinguishable: the head, containing the nucleus; the midpiece, tightly packed with mitochondria, which produce the ATP that fuels the sperm's swimming; and the long flagellum. The chromosomes in the nucleus are supercondensed into an almost crystalline, genetically inactive form. In the apex of the head, immediately in front of the nucleus, is the acrosome, a membrane-bounded vesicle derived from the Golgi apparatus. There is only a tiny amount of cytosol in the cell. (B) The photograph shows human sperm in a tube leading to the egg.

THE EVENTS OF FERTILIZATION

In the egg-producing organs, or ***ovaries***, certain cells are set aside early as egg primordia. These cells grow to an unusually large size, and when they then undergo meiosis, the divisions are unequal; almost all the cytoplasm is retained in the ripe ovum. The sperm, on the other hand, are unusually small cells with very little cytoplasm. The ovum thus furnishes most of the initial cytoplasm for the ***embryo***—the entity that develops from the fertilized egg and orchestrates its own conversion into a new individual. Moreover, even in eggs with relatively little stored food, or ***yolk***, the yolk granules are more abundant in some parts of the eggs than others. Often one side of the egg is richer in yolk; this side is then called the vegetal pole. As described in the next chapter, this uneven distribution of yolk reflects an underlying polarization that helps to provide order and direction in early development.

The fusion of sperm and egg cell creates the ***zygote***, which then immediately starts to develop into an embryo. However, it is the contact between the two cell membranes—not the fusion of the sperm nucleus with egg nucleus to form a diploid nucleus—that causes development to begin. Apparently ***fertilization***—the joining of two haploid sets of chromosomes—is not necessary to induce embryonic development in many animals. For example, it is easy to induce unfertilized frog eggs to begin development in the laboratory by pricking them with a fine needle dipped in blood. A few such eggs will develop into viable, apparently normal tadpoles.[1] Adult rabbits have been produced from unfertilized eggs by similar procedures. Unfertilized eggs can even be stimulated to begin development through a mild electric shock, a change in the salt concentration in the surrounding fluid, or a physical jolt. In most species, however, the development aborts after only a few divisions.

In many mammalian species the egg is initially enclosed in a thin protective layer of cells called the ***follicle***. This barrier may be loosened by enzymes released by the ***acrosome***, a vesicle located at the apex of the sperm in many species (Fig. 13.1). After the follicle-cell barrier has been loosened, the sperm encounters a thick jelly-like coat, called in mammals the ***zona pellucida*** (Fig. 13.2A). Receptors in the head of the sperm then bind to a long, filamentous glycoprotein embedded in the coat. Species-specific features of the glycoprotein prevent the sperm of other species from attaching.

When the acrosomal vesicle releases its enzymes, it also everts to form a long filament that penetrates the jelly coat (Fig. 13.2B–C). The membrane of the filament fuses with the membrane of the microvilli of the egg, thus fully fusing the two gametes (Fig. 13.2D–E). Once this fusion has occurred, the electrical potential of the egg-cell membrane of many species changes dramatically (thus making fusion with the other sperm difficult for the next minute or so), and the sperm nucleus moves into the cytoplasm of the egg. At the

[1] Most frog embryos developed from unfertilized eggs are haploids and die. The embryos that reach adulthood are the few that undergo spontaneous chromosomal doubling to become diploid (though the two sets of chromosomes are identical).

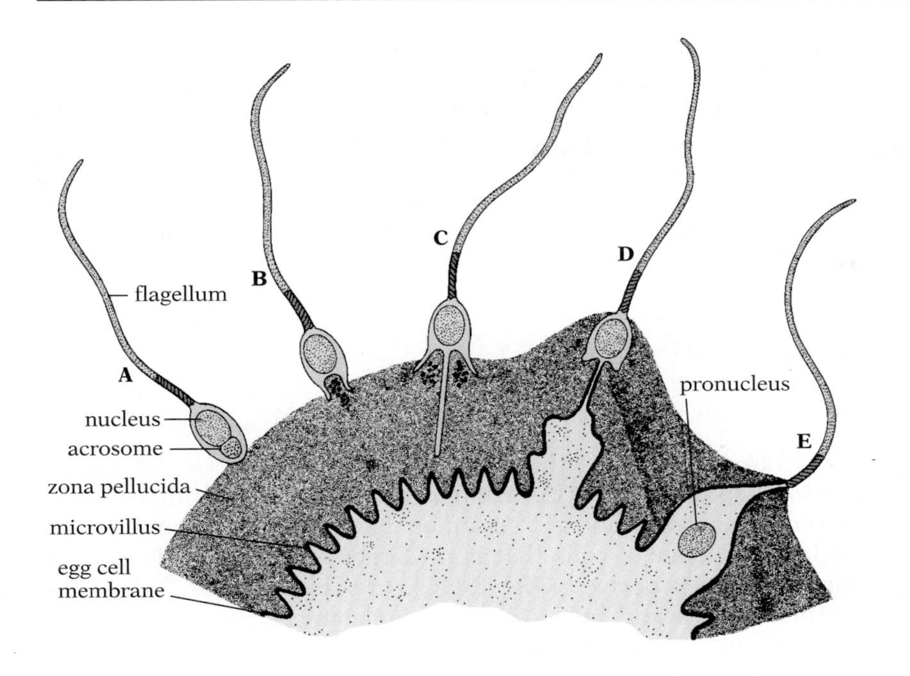

13.2 Fertilization process

(A) After the follicle-cell coat (not shown) has been broken, a sperm cell comes into contact with the zona pellucida surrounding the egg cell membrane. (B) If the recognition of species-specific markers on the two gametes is successful, the membrane of the acrosome fuses with the plasma membrane of the sperm and ruptures, releasing the acrosomal contents, which include enzymes that act on the jelly coat and on the membrane of the egg cell. (C) A tubular filament derived from the acrosome membrane pushes through the zona pellucida. (D) The tube fuses with an enlarged microvillus of the egg cell; there is now no membranous barrier between the contents of the sperm cell and the egg cell. (E) The sperm pronucleus moves into the egg cell.

same time, the structure of the egg-coat glycoprotein responsible for sperm binding is altered to prevent the attachment of additional sperm even after the membrane potential returns to normal.

As the sperm nucleus, now known as a pronucleus, moves into the egg, many changes begin. First, calcium-containing vesicles in the egg release their contents, triggering a wave of calcium release from vesicles throughout the egg, initiating the earliest steps in postfertilization development (Fig. 13.3). (The techniques for inducing development in unfertilized eggs, mentioned earlier, probably succeed by triggering calcium release.) Among the many other

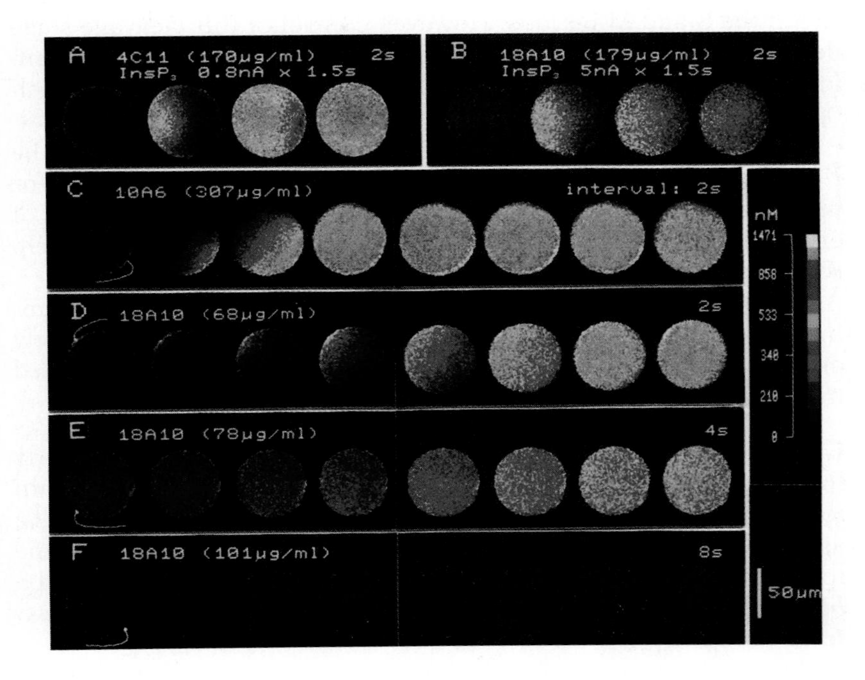

13.3 Calcium release spreading across an egg cell

The release of calcium (orange) across this hamster egg from the site of sperm fusion occurs over a period of just 2 sec.

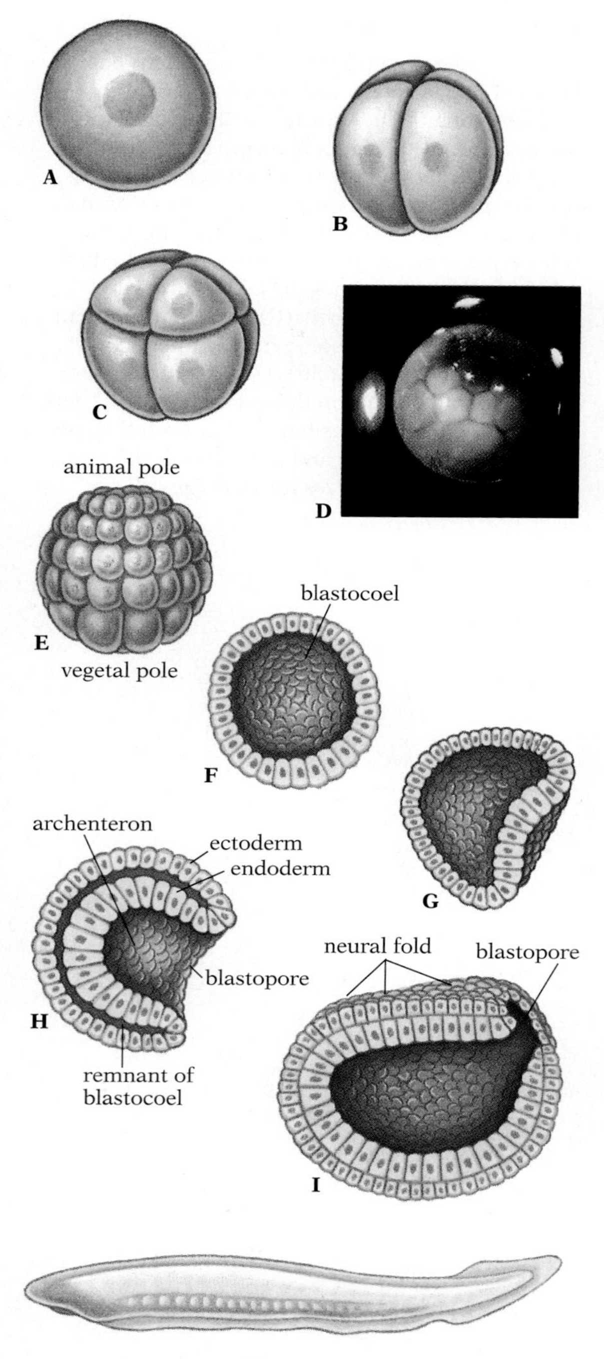

13.4 Early embryology of amphioxus

(A) Zygote. (B–D) Early cleavage stages, culminating in formation of a blastula (E). (F) Longitudinal section through a blastula, showing the blastocoel. (G–H) Longitudinal sections through an early gastrula and a late one. Notice that the invagination is at the vegetal pole of the embryo, where the cells are largest. (I) The blastopore becomes the anus of the gastrula, and a neural fold begins to form. (Subsequent steps are traced in Fig. 13.5.) An adult amphioxus is also shown at the bottom.

changes regularly seen in the egg after fusion is a striking alteration in the permeability of the plasma membrane.

True fertilization—the union of the two gamete nuclei—depends on some attraction of the sperm pronucleus by the mature egg pronucleus. The activated egg also induces the decondensation of the sperm chromosomes, and the preparations for DNA replication prior to the first mitotic division of embryonic development.

HOW EMBRYOS DEVELOP

■ EARLY CLEAVAGE AND MORPHOGENESIS

In normal development the zygote begins a rapid series of mitotic divisions soon after fertilization occurs. In most animals these early cleavages are not accompanied by cytoplasmic growth; they produce a cluster of cells that is no larger than the single egg cell from which it is derived (Fig. 13.4C). In some animals, however—notably reptiles and birds—some cytoplasmic growth does occur, as nutrients from the yolk (stored food) are consumed and transformed into cellular material.

Subdividing the ovum During this early cleavage stage of development in most eggs, the nuclei cycle very rapidly between chromosomal replication (the S period of the cell cycle, when DNA is synthesized) and mitosis (the M period); the G_1 and G_2 (gap) periods are practically absent. Such rapid cycling, in which G_1 and G_2 are skipped, is possible because the ovum already contains huge quantities of the DNA polymerase necessary for catalyzing repeated chromosomal replication, as well as most of the mRNA required for synthesis of proteins during early cleavage.

Perhaps the brevity of the interval taken up by transcription (because so little new mRNA is needed) allows the rapid cycling between the S and M periods. However, control of this cleavage stage depends largely on mRNA synthesized in the oocyte prior to fertilization, so the paternal genes have little input until later in development.

Ova are remarkably large cells. They are so large, in fact, that the ratio of nuclear to cytoplasmic material would be too low for proper control of ordinary cellular activities. The early cleavages of embryonic development, with their minimal cell growth, thus help restore a more normal ratio of nuclear to cytoplasmic material.

The one exception to the pattern outlined here is seen in mammals: the eggs are small, have little yolk, and divide slowly: they rely on RNA transcription from the outset. They probably do not need much yolk because the placenta will provide nutrients to the embryo.

Giving the embryo shape As cleavage continues, the newly formed cells (blastomeres) of many species begin to pump sodium ions into the intercellular spaces in the center of the mass of cells; as a result, water diffuses into this space and the blastomeres come to be arranged in a sphere surrounding a fluid-filled cavity called a ***blastocoel*** (Fig. 13.4F). An embryo at this stage is termed a ***blastula***.

Next begins a series of complex movements important in estab-

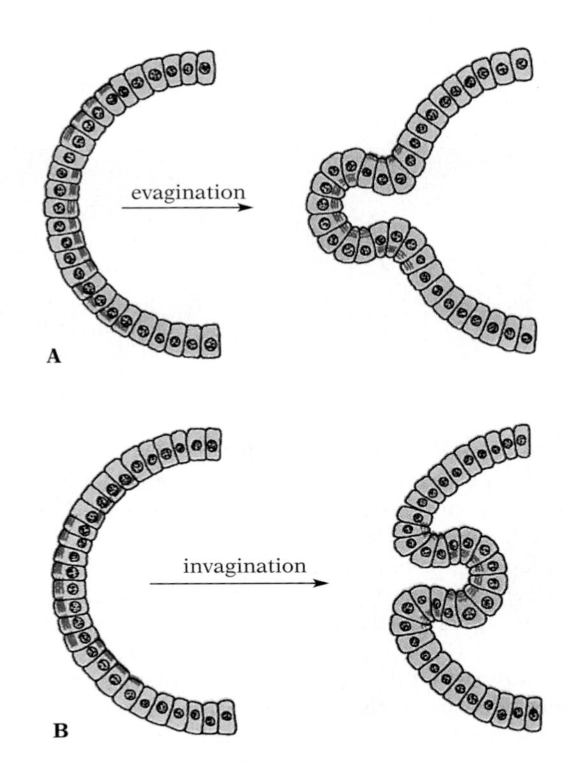

13.5 Mechanism of some morphogenetic movements in cells
Contraction of microfilament-microtubular complexes (red), asymmetrically positioned in the cells, may change the shapes of the cells and produce evaginations (A), invaginations (B), or other alterations of the arrangement of cells in a developing organ. The contractile interactions are of the same type as those involved in amoeboid movement and cytokinesis.

lishing the definitive shape and pattern of the developing embryo. The establishment of shape and pattern in all organisms is called ***morphogenesis*** (meaning "the genesis of form"). Morphogenesis in animals always involves movements of cells in large masses during the early stages. Often it also involves changes in cell shapes, probably effected by interactions between actin microfilaments and myosin. These changes may be relatively small (Fig. 13.5), or they may be extensive. Some cells actively move. Other movements involve changes in the adhesive affinities of the cells for neighboring cells or, in the case of epithelial (surface-covering) cells, for the basement membrane on which these cells sit.

Since the pattern of cleavages and cell movement is greatly influenced by the amount of yolk in the egg, we will examine separately the patterns in animals whose eggs have small, medium, and large amounts of yolk.

Development of zygotes with little yolk In amphioxus (Fig. 13.4), a tiny marine animal whose egg has very little yolk, the movements that occur after formation of the blastula (when it is composed of about 500 cells) convert it into a two-layered structure called a ***gastrula***. The process of ***gastrulation*** begins when a broad depression, or invagination, starts to form at a point on the surface of the blastula where the cells are somewhat larger than those on the opposite side (Fig. 13.4F). The smaller cells make up the ***animal hemisphere*** of the embryo, and the larger cells make up the ***vegetal hemisphere*** (Fig. 13.4E). It is at the pole of the vegetal hemisphere that the invagination of gastrulation occurs in amphioxus.

As gastrulation proceeds, the invaginated layer bends farther and farther inward, until eventually it comes to lie against the inside of the outer layer, nearly obliterating the old blastocoel (Fig. 13.4H). In some animals, pseudopods extend from the inner membranes of the cells where invagination began into the blastocoel. The pseudopods may adhere to the membranes of other blastula cells lining the blastocoel. In at least some cases contractions of these pseudopods assist invagination.

The resulting gastrula is a two-layered cup, with a new cavity that opens to the outside via the ***blastopore***. The new cavity, called the ***archenteron***, will become the cavity of the digestive tract, and the blastopore will become the anus. (Amphioxus is a chordate. In all animals except echinoderms, chordates, and two minor phyla, however, the blastopore becomes the mouth rather than the anus).

Gastrulation, as it occurs in amphioxus, first produces an embryo with two primary cell layers: an outer ***ectoderm*** and an inner layer. The inner layer subsequently separates in turn into two layers: the ***endoderm*** and the ***mesoderm;*** the mesoderm lies dorsally (toward the top side) between the ectoderm and the endoderm. In

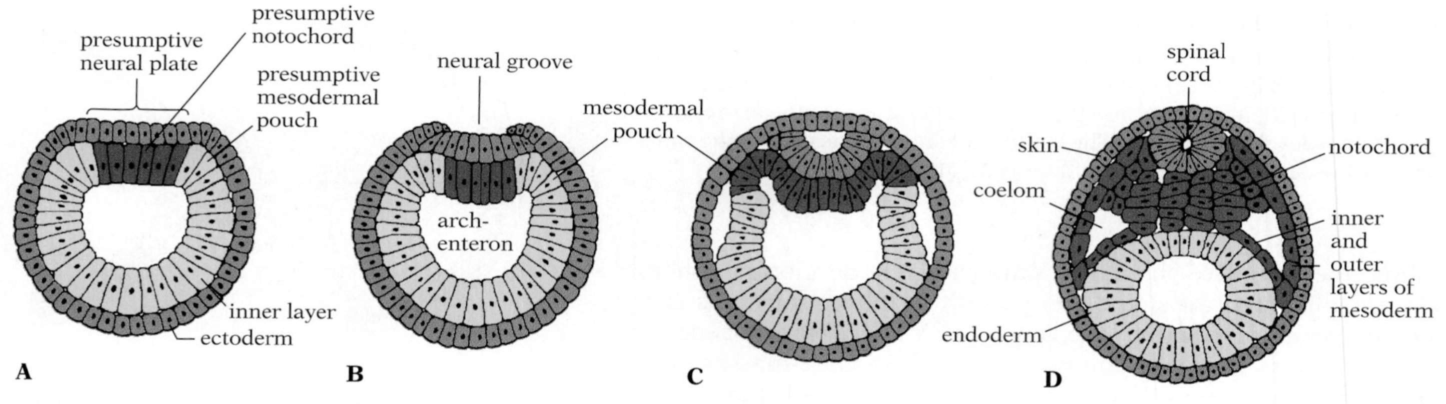

13.6 Neurulation in amphioxus

Cross sections through amphioxus embryos show the progressive formation of mesoderm and the neural tube. (A) When gastrulation is complete, the dorsal part of the inner layer has already been segregated as presumptive mesoderm—presumptive notochord and mesodermal pouches. (Presumptive tissue is tissue which, though not yet differentiated, is destined for a given developmental fate.) Similarly, part of the ectoderm has differentiated as presumptive neural plate. (B–C) The notochord and mesodermal pouches form as evaginations from the inner layer, and the neural plate invaginates from the ectoderm as it begins to form the spinal cord. (D) In this later embryo, called the neurula, both the spinal cord and the mesoderm are taking their definitive form. Notice that there is a cavity (the coelom) in the mesoderm.

amphioxus the mesoderm originates as pouches flanking the ***notochord***, a supportive central rod found in all chordates (at least in the embryo); these pouches all pinch off from the inner layer (Fig. 13.6). The remaining part, the endoderm, pinches together to form a tube that becomes the digestive tract. Figure 13.6 shows the further development of the embryo to the ***neurula*** stage, so called because it incorporates the beginnings of the nervous system (discussed in a later section). All these changes are orchestrated by well-timed changes in gene activity, described in the next chapter.

In the amphioxus egg, where the distinction between animal and vegetal hemispheres is only slight (owing to the small amount of yolk in the vegetal hemisphere), the early cleavages are nearly equal. The new cells are thus of nearly the same size, and gastrulation occurs in a direct and uncomplicated manner. The eggs of many organisms, however, have far more yolk in their vegetal hemisphere, and this deposit of stored food imposes complications and limitations on cleavage and gastrulation. Generally, the more yolk an egg contains, the more cleavage tends to be restricted to the animal hemisphere and the more gastrulation departs from the pattern in amphioxus.

Development when there is more yolk The frog egg contains far more yolk than that of amphioxus but much less than that of a bird, say. The first two cleavages, which are perpendicular to each other, cut through both the animal and vegetal poles, producing cells of roughly the same size (Fig. 13.7B). The next cleavage is equatorial (parallel to the egg's equator) and located decidedly nearer the animal pole (Fig. 13.7C); hence the four cells produced at the animal end of the egg are considerably smaller than the four at the vegetal end. From this stage onward, more cleavages occur in the animal hemisphere of the embryo than in the vegetal hemisphere as the blastula develops. As in amphioxus, there is no increase in total mass during these early cleavage stages (Fig. 13.8).

After the blastula has been formed, the frog embryo begins gastrulation. Simple invagination of the vegetal hemisphere is not mechanically feasible, because of the large mass of inert yolk. Instead, portions of the cell layer of the animal hemisphere apparently expand and move down around the yolk mass and then turn in at the edge of the yolk. This involution begins at what will be the dorsal side of the yolk mass, forming initially a crescent-shaped blasto-

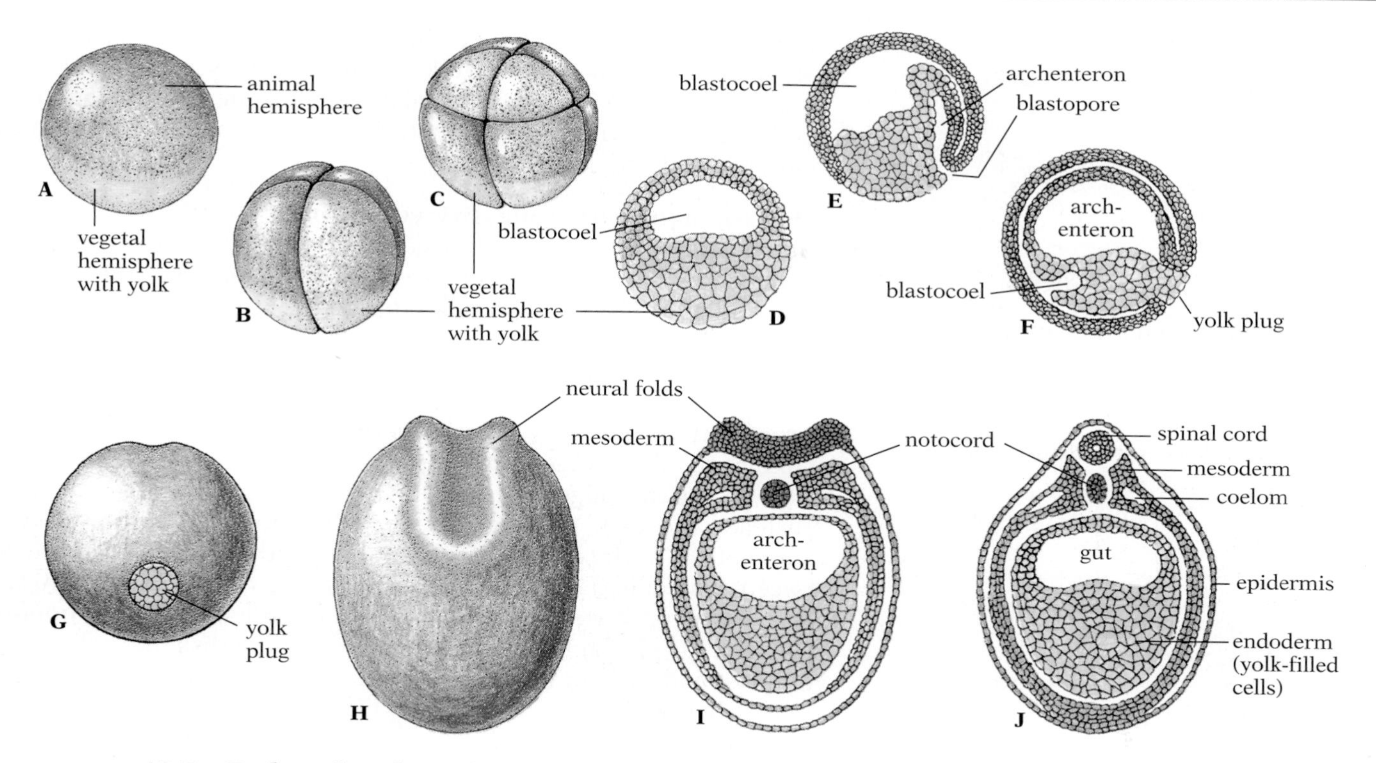

13.7 Early embryology of a frog

The large amount of yolk in the frog egg causes its pattern of gastrulation to differ from that in amphioxus. (A) Zygote. (B–C) Early cleavage stages. Note that because the first horizontal cleavage is nearer the animal pole, the cells at the vegetal pole are much larger. (D) Longitudinal section of a blastula. (E–F) Longitudinal sections of two late gastrula stages. (G) End view of late gastrula with first hint of developing neural fold at top. (H) End view of an early neurula, showing the neural folds and neural groove. (I) Cross section of a neurula after formation of mesoderm. (J) Cross section of a later embryo, showing definitive spinal cord; this stage is reached about 24 hours after fertilization.

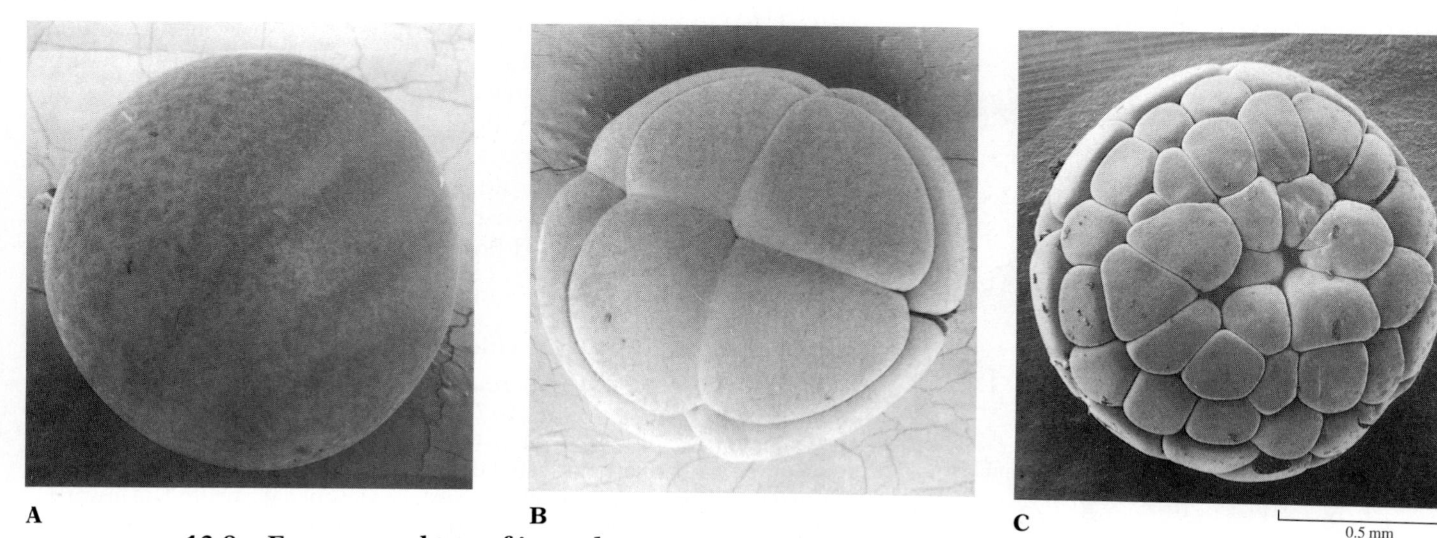

13.8 Frog egg and two of its early cleavage stages

(A) Unfertilized egg. (B) eight-cell stage. (C) 32–64-cell stage. All three scanning electron micrographs are at the same magnification: 46x. Note that there has been no overall growth in size during these cleavage stages—the 32–64-cell embryo is no larger than the egg cell.

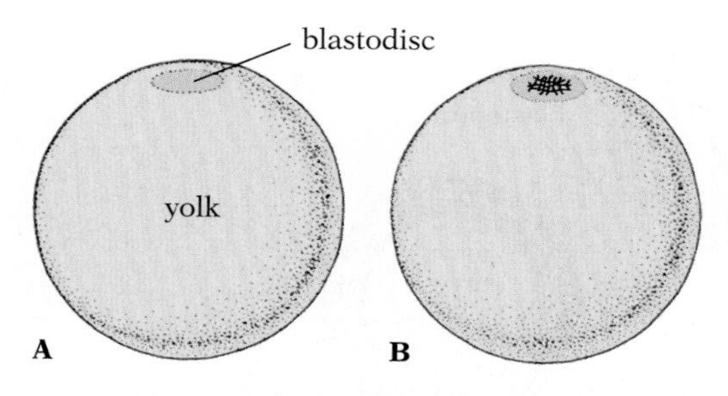

13.9 Egg and early-cleavage embryo of a chick

(A) The zygote. A small cytoplasmic disc—the blastodisc—lies on the surface of a massive yolk. (B) Early cleavage. There is no cleavage of the yolk.

pore at the edge of the yolk. The infolding of surface cells slowly spreads to all sides of the yolk, so that the crescent blastopore is converted into a circle. Movement of the other cells around the yolk eventually encloses this material almost completely within the cavity of the archenteron.

Development with a massive yolk supply Birds' eggs contain so much yolk that the small disk of cytoplasm on the surface is dwarfed by comparison. No cleavage of the massive yolk is possible, and all cell division is restricted to the small cytoplasmic disk, or ***blastodisc*** (Fig. 13.9). (Note that the yolk and the small lighter-colored disk on its surface constitute the true egg cell. The albumen, or "white," of the egg lies outside the cell.) The gastrulation process is of necessity greatly modified in such eggs. Neither invagination of the vegetal hemisphere (as in amphioxus) nor involution along the edge of the yolk mass (as in a frog) can occur. Instead, the disk floating on the yolk becomes multilayered as some of the surface cells move inward into the fluid-filled space between the cellular layer and the underlying cavity. Eventually two layers are produced, an outer ***epiblast*** and an inner ***hypoblast*** (Fig. 13.10A–C). In the posterior portion of the disk, the cells of the epiblast converge toward the midline, giving rise to a clearly visible line or depression on the epiblast (Fig. 13.10D); this line, called the primitive streak, is in effect a very elongate closed blastopore. Individual cells move downward from this region. Some of these cells stay between the epiblast and the hypoblast and give rise to the mesoderm, while others intrude into the hypoblast and help form the endoderm.

Developmental fates The fates of cells in different parts of the three primary layers of vertebrates have been traced by staining them with dyes of different colors, or by marking them with radioactive carbon or other isotopes, and then following their movements. The ectoderm eventually gives rise to the outermost layer of the body—the top layer of the skin, called the ***epidermis***—and to structures derived from the epidermis, such as hair, nails, eye lens, pituitary gland, and epithelium of the nasal cavity, mouth, and anal canal. The endoderm gives rise to the innermost layer of the body: the epithelial lining of the digestive tract and of other structures derived from the digestive tract, such as respiratory passages and lungs, liver, pancreas, thyroid, and bladder. The mesoderm gives rise to most of the tissues in between, such as muscle, connective tissue (including blood and bone), and the notochord. The origins

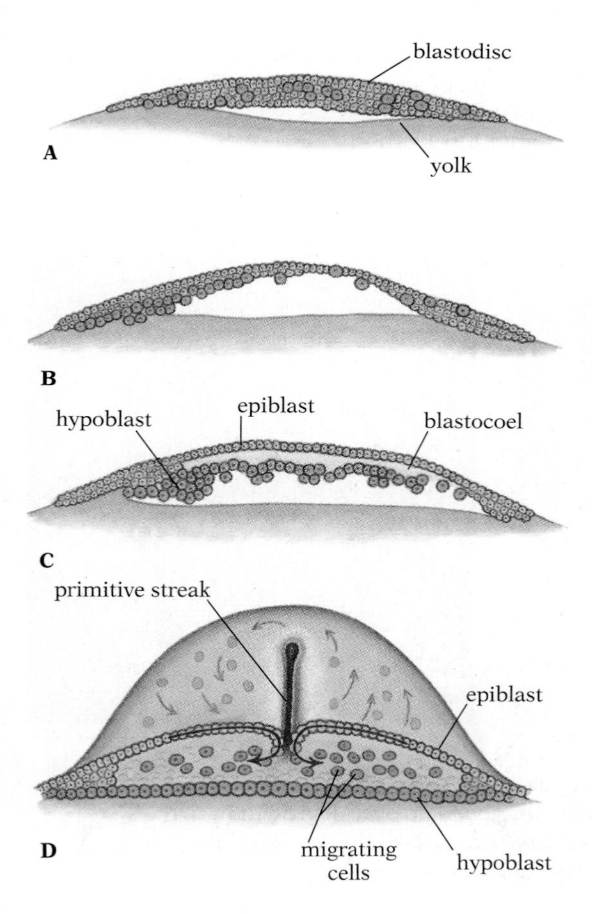

13.10 Gastrulation in the chick embryo

(A) Longitudinal section through a blastula. Larger yolk-laden cells are intermixed with smaller cells. (B) The larger cells begin to accumulate on the lower surface of the cell mass. (C) The layer of larger cells separates from the layer of smaller cells to become the hypoblast; the cavity between the two layers is the blastocoel. (D) Surface view of a gastrula. The inward movement of cells along the midline of the embryo during gastrulation (arrows) produces a clearly visible primitive streak, which is essentially a very elongate blastopore. Some epiblast cells of the primitive streak move downward to form the mesoderm; others invade the hypoblast to form the endoderm.

of various body organs and tissues are summarized in Table 13.1.

Oddly enough, nervous tissue is derived from the ectoderm. Soon after gastrulation, the ectoderm becomes divided into two components, the epidermis and the neural plate. A sheet of ectodermal cells lying along the midline of the embryo dorsal to the newly formed digestive tract and developing notochord bends inward in a process called ***neurulation***, and forms a long groove extending most of the length of the embryo (Figs. 13.6A–C and 13.7H–I). The dorsal folds that border this groove then move toward each other and fuse, converting the groove into a long tube lying beneath the surface of the back. This neural tube becomes detached from the epidermis dorsal to it, and in time differentiates into the spinal cord and brain (Figs. 13.6D, 13.7J).

Table 13.1 Origins of certain organs and tissues

PRIMARY LAYER	TISSUE OR ORGAN
Ectoderm	Epidermis
	Hair
	Nails
	Eye Lens
	Nervous System
	Lining of nose and mouth
Mesoderm	Muscle
	Bone
	Blood
	Notochord
	Kidneys
	Gonads
Endoderm	Lungs
	Liver
	Pancreas
	Lining of digestive tract
	Bladder

■ LATER EMBRYONIC DEVELOPMENT

Gastrulation and neurulation provide the organization that shapes the development of the early embryo. The late embryo must be converted from this promising beginning into a fully developed young animal ready for birth: the individual tissues and organs must be formed, and an efficient circulatory system must quickly come into being (Fig. 13.11). In a vertebrate the four limbs must develop, the elaborate system of nervous control must be established, and so on. The complexity and precision characterizing these developmental changes are staggering to contemplate: in each human arm and hand, for example, approximately 43 muscles, 29 bones, and many hundreds of nervous pathways must form. To function properly, all these components must be precisely correlated. Each muscle must have exactly the right attachments; each bone must be jointed to the next bone beyond it in a certain way; each nerve fiber must have all the proper connections with the central nervous system (CNS) and muscle cells.

Surprisingly enough, the developmental processes that produce these remarkably precise later embryonic changes are the same ones at work in the early embryo—cell division, cell growth, cell ***differentiation*** (as cells take on increasingly specialized roles), and morphogenetic movements. Bursts of mitotic activity in some areas and cessation of cell division in other areas alter the balance

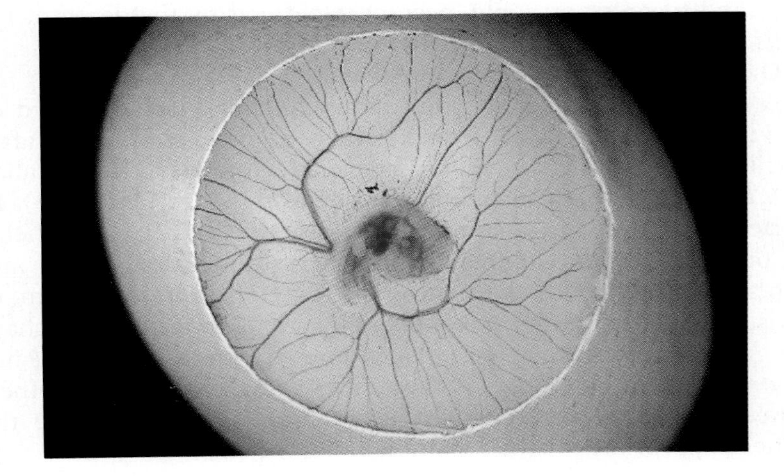

13.11 Chick embryo after 4 days of incubation

The tiny embryo lies on the surface of the yolk. It has a functional circulatory system, including a beating heart, even at this early stage of its development. Note the long branching blood vessels that run out of the embryo into the yolk; they transport nutrients to the embryo.

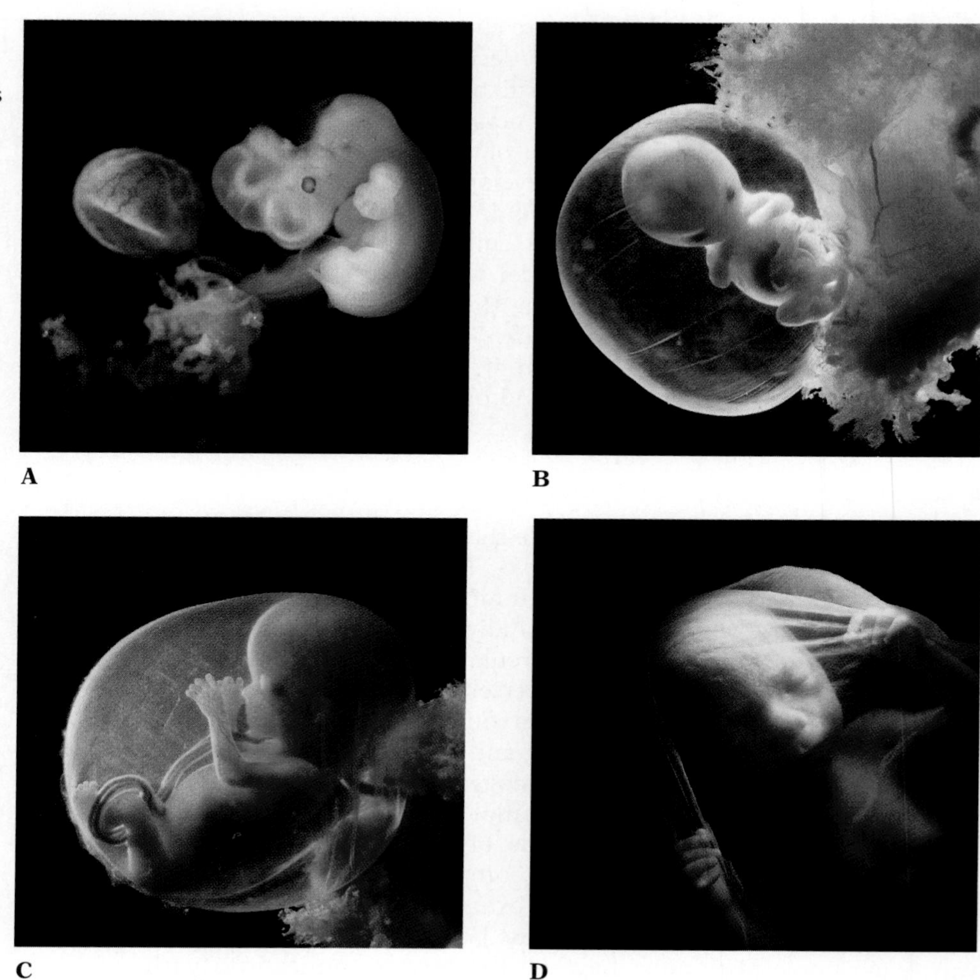

13.12 Human embryos at successive stages of development

(A) The 5-week embryo, 1 cm long, shows the beginnings of eyes but no distinctive face; note its mittenlike hands and feet, with no separation between the digits. (B) By contrast, the 7-week embryo, 2 cm long, has a distinct face, and there is separation between its fingers. (C) At 13 weeks the fetus is over 7 cm long and weighs about 30 g, 15 times more than it did at 7 weeks. (D) By 17 weeks the fetus is over 15 cm long and 7 times heavier than it was a month earlier; all the internal organs have formed.

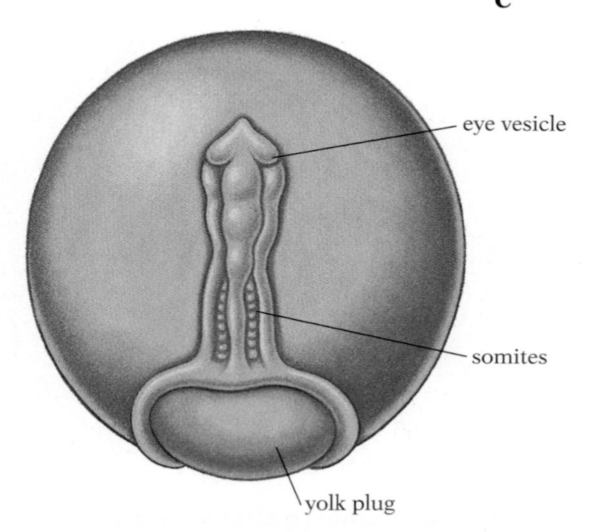

13.13 Formation of somites in the chick embryo

After the primitive streak forms, it develops into a neural groove, beginning at the anterior end and proceeding posteriorly. Somites begin to appear, each of which organizes the development of a separate segment of the body.

between the parts. Special patterns of cell growth produce important changes in size and shape. Through differentiation cells may lose particular capacities, but become more efficient at performing other functions. Folds and pouches establish the primordia of lungs and glands, of eyes and bladder. Even cell death plays an important role in the normal development of the living animal: fingers and toes, for example, become separated by the death of the cells between them (Fig. 13.12B).

One simplification in the organization of development is the repetitive organization of older embryos. Once the notochord or primitive streak is fully formed (which occurs about 1 day after fertilization in birds), regularly spaced clumps of cells called ***somites*** begin to appear along the dorsal midline (Fig. 13.13). In vertebrates each pair of somites gives rise to a vertebra, and helps to organize the development of the nerves, muscles, bones, and other nearby structures. Hence, the developmental program of most animals in compartmentalized. As detailed in the next chapter, specialized characteristics of segments (like those of somites that produce limbs) arise as these organizing centers "read" their anterior-posterior position in the embryo and then activate the genes appropriate for that position along the axis.

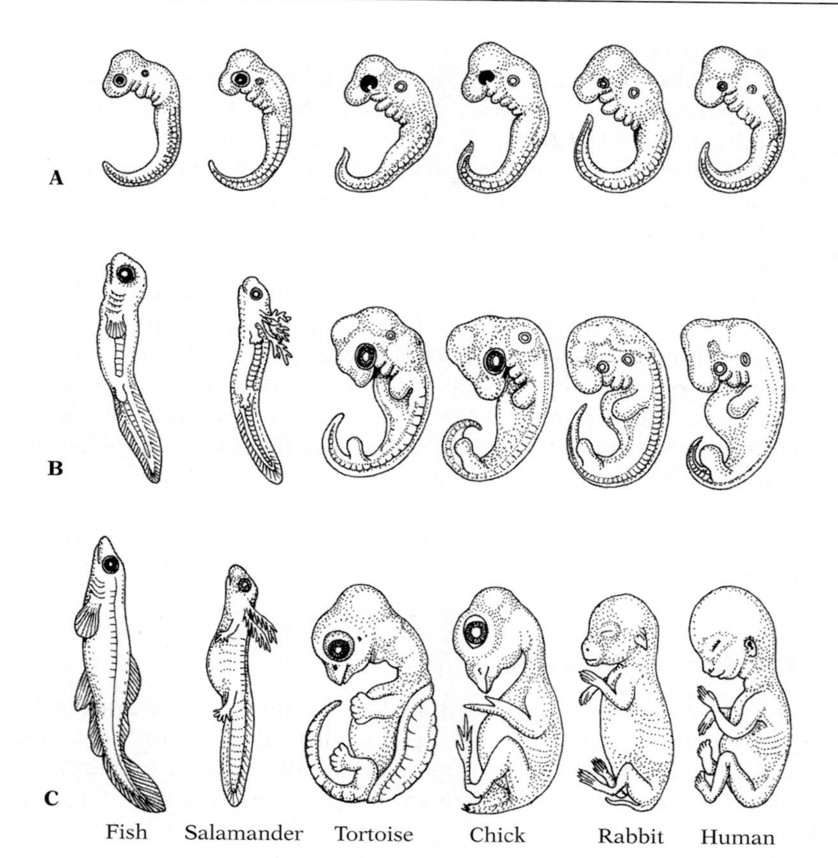

13.14 Vertebrate embryos compared at three stages of development

(A) All the embryos—whether fish, amphibian, reptile, bird, or mammal—strongly resemble one another. (B) At a later stage the fish and salamander are noticeably different, but the other embryos are still very similar; note the pharyngeal pouches in the neck region and the prominent tail. (C) By this stage, each embryo has taken on many of the features distinctive of its own species.

One fact of embryology that pushed Darwin toward the idea of evolution is that the early embryos of most vertebrates closely resemble one another. The early human embryo, with its well-developed tail and a series of pouches in the pharyngeal region (where the neck will appear), looks very much like an early fish embryo; it looks even more like an early rabbit embryo (Fig. 13.14). Only as development proceeds to later stages do the distinctive traits of each kind of vertebrate become apparent (Fig. 13.14B–C). About 100 years ago, Ernst Haeckel proposed that the development of each organism retraces in detail its evolutionary history—that "ontogeny recapitulates phylogeny." According to this hypothesis, early human embryos resemble fish because mammals are the evolutionary descendants of fish. It is certainly true that the suites of genes working together to control developmental patterns are conserved by natural selection, changing little over time as compared with superficial morphology. Nevertheless, we now know that the general developmental pattern of a species can skip the steps of ancestors, or create new structures *de novo*. There is even evidence to suggest that novel groups of organisms can arise through major and relatively rapid changes in developmental programs.

DEVELOPMENT AFTER BIRTH

The extent to which an animal has developed by the time of birth varies greatly among different species. Some young animals are entirely self-sufficient from the time they are born, and neither need nor receive parental care. Others are born while still at an early stage of development, and are nearly helpless and totally dependent on parental care. Newly hatched robins, for instance, are blind, almost devoid of feathers, and unable to stand.

The extent of development at birth is often (though not always) a reflection of the length of the embryonic period, which is usually correlated in animals that lay eggs with the amount of yolk the eggs contain. Among birds in particular, species that have a short incubation period for the eggs characteristically have altricial (incompletely developed) young (see Fig. 25.30, p. 724), while species that have a longer incubation period characteristically have precocial (well-developed) young. For example, robins (which are altricial) incubate their eggs for only 13 days, while precocial chickens have a 21-day incubation period. Regardless of their state of development at birth, all animals have a complete circulatory system, gastrointestinal tract, and respiratory system; nevertheless, even the most precocial young continue to undergo major developmental changes during their postembryonic life.

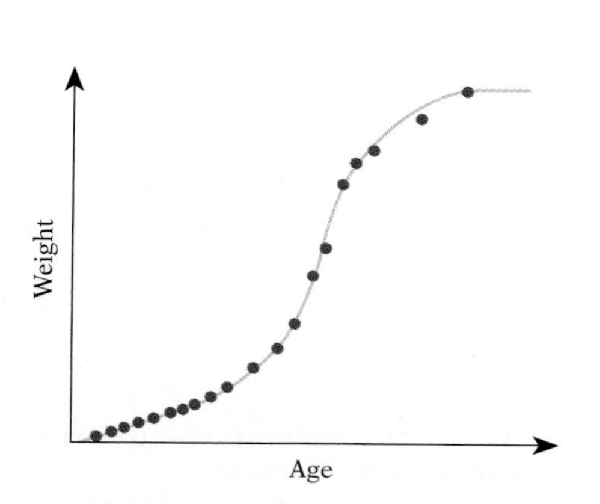

13.15 Typical S-shaped growth curve

■ PATTERNS OF GROWTH

Though development after birth seldom involves any major morphogenetic movements, there is some cell multiplication and cell differentiation. But the most obvious characteristic of postembryonic development by far in many animals is growth in size. Usually growth begins slowly, becomes more rapid for a time, and then slows down again or stops. This pattern yields a characteristic S-shaped growth curve (Fig. 13.15). The slope of the curve is different for different species, depending on whether they grow very rapidly for a shorter time or more slowly for a longer time. The shape of the curve is seldom as smooth as it appears in a generalized growth curve because many factors can affect the rate of growth. For example, in most mammals growth slows down for a while immediately after weaning, and it often varies greatly during puberty (Fig. 13.16). An especially marked departure from the smooth generalized curve is seen in the growth of arthropods, a group that includes the insects (Fig. 13.17). These animals can undergo only limited growth between molts, because the hard ex-

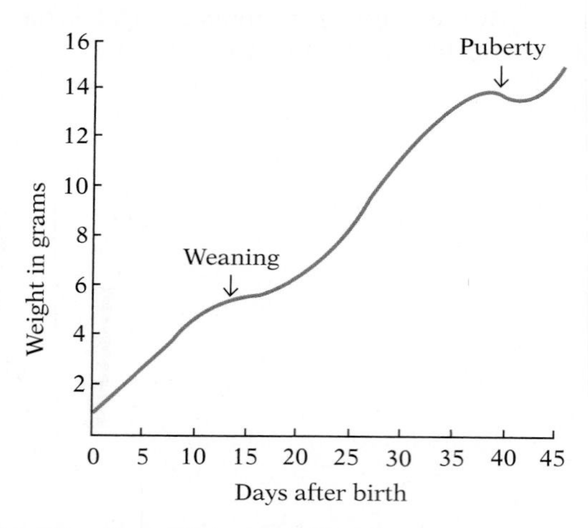

13.16 Growth in weight of a mouse

The rate of growth is slower at weaning and at puberty.

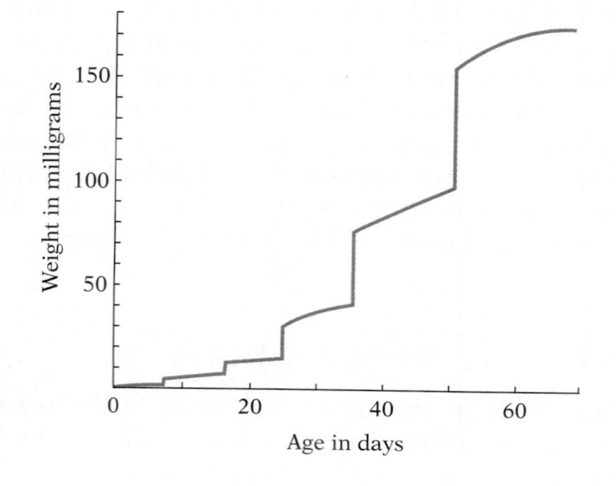

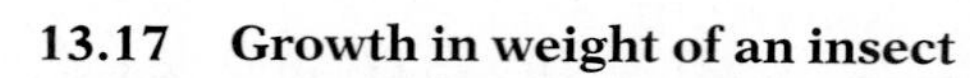

13.17 Growth in weight of an insect

The growth spurts of the water boatman, an aquatic insect, occur at the time of molt, when the old exoskeleton has been shed and the new one has not yet fully hardened.

oskeleton that encases their bodies can be stretched only slightly. At each molt, however, there is a sharp burst of growth during the short period after the old exoskeleton has been shed and before the new one has hardened.

Growth does not occur at the same rate or at the same time in all parts of the body. It is obvious to anyone that the differences between a newborn baby and an adult human are differences not only in overall size but also in body proportions. The head is far larger in relation to the rest of its body in a young child than in an adult, while a child's legs are disproportionately short. Normal adult proportions arise because the various parts of the body grow at quite different rates or stop growing at different times (Fig. 13.18).

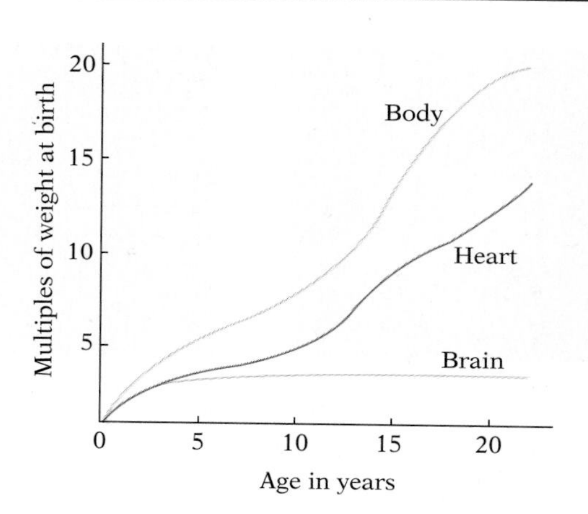

13.18 Differences in relative growth of body, heart, and brain of a human

■ HOW LARVAE DEVELOP

Growth in size is not always the principal mechanism of postembryonic development. Many aquatic animals, particularly those leading sessile (nonmobile) lives as adults, go through ***larval*** stages that bear little resemblance to the adult (Fig. 13.19). The series of sometimes drastic developmental changes that convert an immature animal into the adult form is called ***metamorphosis***. It often involves extensive cell division and differentiation, and sometimes even morphogenetic movement.

In many aquatic animals dispersal of the species depends on the larval stage; the tiny larvae either swim or are passively carried by currents to new locations, where they settle down and undergo metamorphosis into sedentary adults. In other species where the adult is not sedentary, the adaptive significance of the larval stage seems less a matter of dispersal than of exploiting alternative food sources. For example, frog larvae (tadpoles) feed primarily on microscopic plant material, while adults are carnivorous and take fairly large prey.

Though a larval stage occurs in the life history of many aquatic animals, probably the most familiar larvae are those of terrestrial insects, including butterflies and moths. The butterfly or moth larva is called a caterpillar. In the course of their larval lives, these insects molt several times and grow much larger, but this growth does not bring them any closer to adult appearance; they simply become larger larvae (Fig. 13.20B–D). Finally, after they have completed their larval development, they enter an inactive stage called the ***pupa***, during which they are usually enclosed in a case or co-

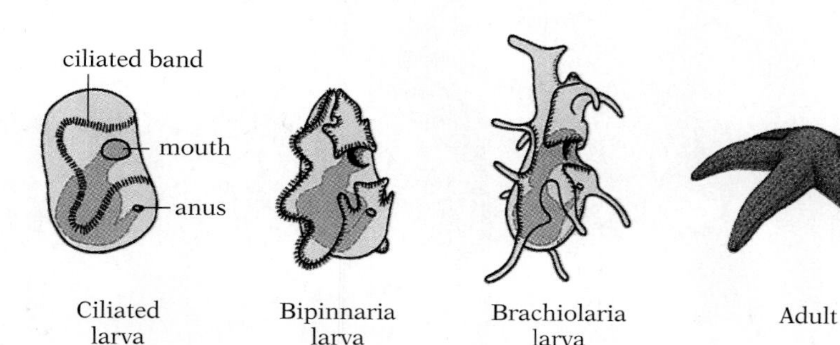

13.19 Three larval stages and adult of the sea star *Asterias vulgaris*

The gastrula develops into the ciliated larva, which changes into the bipinnaria larva, which changes into the brachiolaria larva, which metamorphoses into the characteristically shaped sea star with five arms.

A Monarch butterfly egg

B First-stage larva eating its own egg shell

C Later-stage larva eating its own molted skin

D Full-grown larva going into pupation

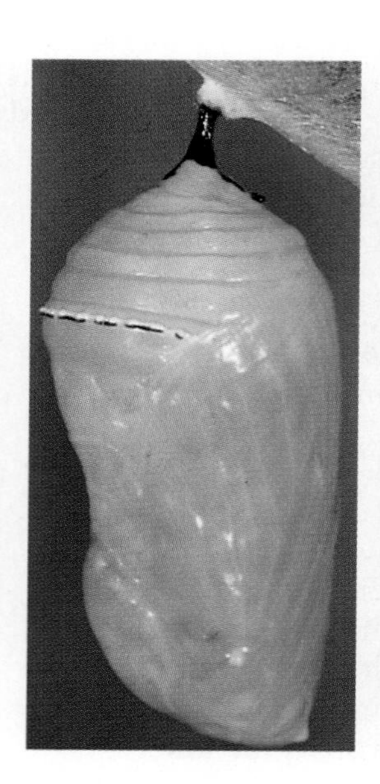

E Early chrysalis

F Late-stage chrysalis

G Inflation of wings by newly emerged butterfly

13.20 Developmental stages of a monarch butterfly, an insect with complete metamorphosis

(A–C) From the egg hatches a small larva which goes through a series of molts, growing larger at each. (D–E) Eventually the full-grown larva attaches to a plant and goes into pupation, forming a case around itself; in butterflies the encased pupa is called a chrysalis. (F) During pupation most of the larval tissues are broken down and new adult tissues are formed; in the monarch some of the adult structures can be seen through the case of the late chrysalis. (G) The newly formed adult emerges from the pupal case and rests while it inflates its wings. (H) It then flies off to feed at flowers before mating and laying eggs (almost always on milkweed) to start the cycle over again.

H Adult butterfly at flower

coon (Fig. 13.20E). During the pupal stage most of the old larval tissues are destroyed, and new tissues and organs develop from small groups of cells called ***imaginal discs,*** which were present in the larva but never underwent terminal development. The adult that emerges from the pupa (Fig. 13.20G–H) is therefore radically different from the larva; it is the product of an entirely different developmental program—one could almost consider it a new organism built from the raw materials of the larval body.

Insects with a pupal stage and the type of development just described are said to undergo ***complete metamorphosis.*** The sharp distinction in form between the larval and adult stages in such insects has meant evolution in two markedly different directions; in general, the larva is more specialized for feeding and growth, while the adult is more specialized for active dispersal and reproduction.

Many insects, such as grasshoppers, instead undergo ***gradual metamorphosis*** (Fig. 13.21). Young larvae of such insects resemble the adults, except that their body proportions are different (the wings and reproductive organs, especially, are poorly developed). They go through a series of molts during which their form gradually changes, becoming more and more like that of the adult, largely as a result of a differential growth of the various body parts. They have no pupal stage and experience no wholesale destruction of the immature tissues.

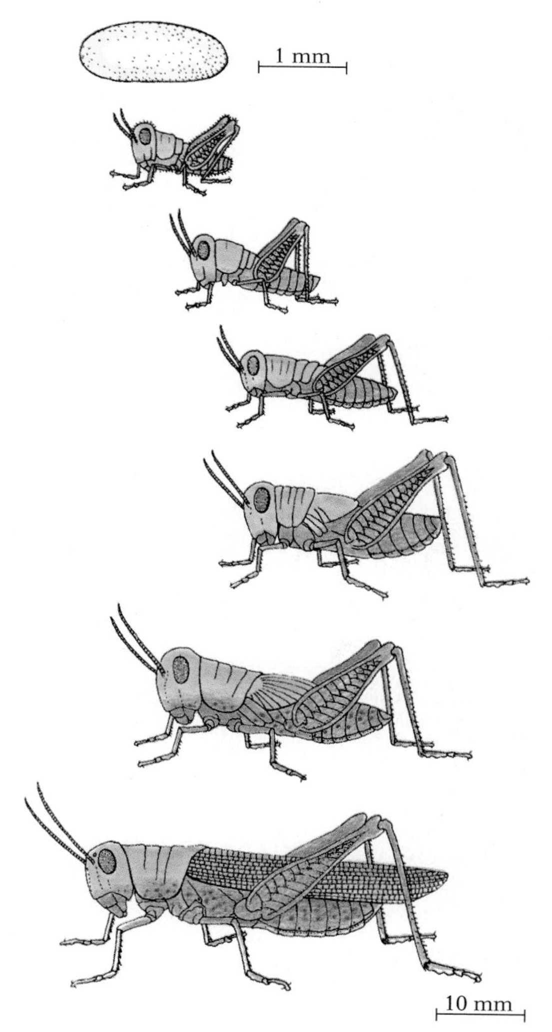

13.21 Gradual metamorphosis of a grasshopper

The insect that emerges from the egg (top) goes through several nymphal stages that bring it gradually closer to the adult form (bottom). The increase in size is more dramatic than it appears: note that the scale in this drawing changes gradually by 10-fold from egg to adult.

■ ARE AGING AND DEATH PROGRAMMED?

The adult organism is not a static entity; it continues to change, and hence to develop, until death brings the developmental process to an end. The term *aging* is applied to the complex of developmental changes that lead, with the passage of time, to the deterioration of the mature organism (Table 13.2) and ultimately to its death. Modern scientific progress and improved medical techniques have

Table 13.2 Average decline in a human male from ages 30 to 75

CHARACTERISTIC	PERCENT DECLINE
Weight of brain	44
Number of nerve cells in spinal cord	37
Velocity of nerve impulse	10
Number of taste buds	64
Blood supply to brain	20
Output of heart at rest	30
Speed of return to normal pH of blood after displacement	83
Number of filtering subunits in kidney	44
Filtration rate of kidney	31
Capacity of lungs	44
Maximum O_2 uptake during exercise	60

greatly increased our ability to protect ourselves against disease, starvation, and the destructive forces of the physical environment. More people are living to an advanced age. As the life expectancy increases and the proportion of the population in the upper age brackets rises, the changes associated with aging become more obvious and more important to all of us.

The process of aging seems to be correlated with specialization of cells for one or a few highly specific functions. Cells that remain relatively unspecialized and continue to divide do not age as rapidly (if at all) as cells that have lost the capacity to divide. Cancer cells divide continually, and so can live indefinitely. Bacteria and some other unicellular organisms do not age, for any cell that is not destroyed eventually divides to produce two young cells; division is thus a process of rejuvenation. Animals that continue to grow throughout their lives (many fish and reptiles, for example) show fewer obvious symptoms of aging than, say, mammals and birds, which cease growing soon after they reach maturity. Within a single species, individuals whose period of growth and development is slowed and extended by a very limited diet are usually older than normal before they begin to show signs of aging.

Clearly, then, the aging and death of individual cells and the aging and death of the multicellular organism as a whole are two different things. Paradoxical as it may seem at first, cell death is an essential part of life: the death of individual cells, as noted previously, plays an essential role in the development of the animal embryo and in the complete metamorphosis of some insects. Early death of individual red blood corpuscles and epidermal cells is entirely normal even in a young healthy mammal. Aging of the whole organism, therefore, is a matter not simply of the death of its cells, but of the deterioration and death of those cells and tissues that cannot be replaced.

Only a few of the factors that correlate with aging are understood:

1 When cells in most types of tissue die as a result of disease or injury, no new cells are formed to replace the ones lost. Wound healing in such cases involves the growth of connective (scar) tissue, which serves as a patching material but cannot function like the original cells. As more and more irreplaceable cells die, the increased burden placed on the remaining cells of that type may contribute to their aging.

2 The changing hormonal balance, such as that caused by a drop in the level of sex hormones, may disturb the function of a variety of tissues.

3 As cells become older, they tend to accumulate some metabolic wastes, and these wastes—particularly highly reactive products like hydrogen peroxide—may contribute to the eventual deterioration of the cells. At the same time, aging cells produce ever smaller amounts of the antioxidant molecules needed to detoxify these dangerous chemicals.

These factors are not really explanations but symptoms. The real question is why these changes occur. Perhaps somatic cells slowly cease to function and eventually die as a result of damage by radiation (particularly X rays and cosmic rays). In the laboratory, however, radiation damage is greatest in actively dividing cells—the cells that age most slowly. Furthermore, the amount of radiation damage should be proportional to the chronological age of the cells, but aging is a function of physiological age—a 5-year-old rat is physiologically very old indeed, and its tissues show pronounced symptoms of aging, whereas 5-year-old tissues in a human being are not yet even mature.

The changes characteristic of aging may be programmed in the genes just like the earlier developmental changes. Though extrinsic environmental factors doubtless influence aging, they do so only by speeding up or slowing down processes that would occur anyway. These processes may involve a decline in the production of important enzymes or an altered chemical balance or physical structure, with an ensuing loss of ability to perform certain functions. They could also result in development of autoimmune reactions (allergies against parts of the organism's own body) that result in destruction of essential tissues; or they may involve increased rupture of lysosomes and release of destructive hydrolytic enzymes within cells.

A particularly interesting proposal is a variation of two hypotheses already mentioned: that aging is a developmentally programmed termination of life that results from an organism's failure to repair somatic-cell mutations or other damage. According to this hypothesis, the different rates of aging in various species reflect different inherited capacities for DNA repair and the production of antioxidant molecules. Long-lived species are the ones with high levels of DNA repair enzymes, and this pattern holds true for the concentration of antioxidants as well (Fig 13.22). Conversely, species with short life spans and early aging invest little in either DNA repair or cellular detoxification. This does not appear to be a chance relationship: genetically engineered *Drosophila* having 50% higher concentrations of the enzymes that detoxify hydrogen peroxide and oxygen radicals live far longer than their unengineered

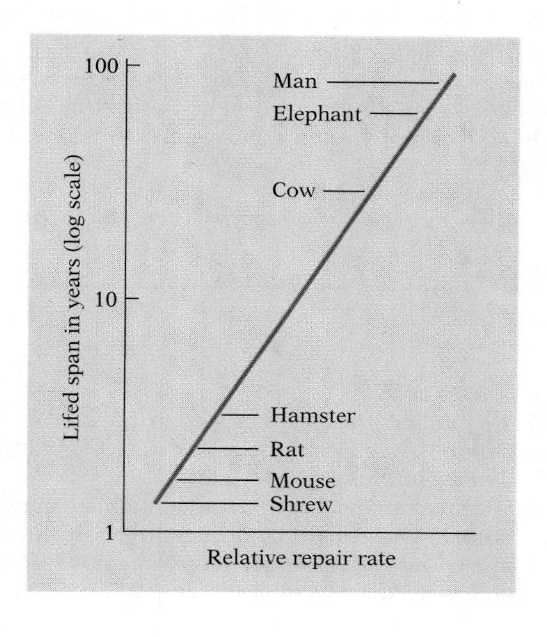

13.22 Activity of DNA repair systems in various animals

Longer-lived animals invest more energy and material in repair enzymes.

counterparts. If, as seems indicated by such data, aging is a genetically programmed trait (and a critical gene from carp has been isolated that significantly extends the life span of mice), it must have originated and been maintained because of a selective advantage over a longer average life span. As yet, however, there is no general explanation of why allowing itself to age and die should be to an individual's advantage. The opportunity to synthesize concepts and data from molecular biology and evolution is one of the most exciting challenges of modern biology.

CHAPTER SUMMARY

THE EVENTS OF FERTILIZATION

Fusion of egg and sperm creates a zygote. Fusion triggers calcium release in the egg, initiating the first steps of embryonic development. (p. 338)

HOW EMBRYOS DEVELOP

EARLY CLEAVAGE AND MORPHOGENESIS The zygote is divided into many smaller cells by cell division: it then forms a hollow blastula. Morphogenic movements cause part of the blastula to fold into itself, forming a two-layer gastrula. The inner layer becomes endoderm, which lines the gastrointestinal and respiratory tracts; the outer layer becomes ectoderm, which forms the skin, nervous system, and other structures; a new third layer, the mesoderm, forms from ingrowths of the endoderm, and develops into muscle and bone. (p. 340)

LATER EMBRYONIC DEVELOPMENT Organs and other structures are formed by progressive differentiation and morphogenesis. (p. 345)

DEVELOPMENT AFTER BIRTH

PATTERNS OF GROWTH Development after birth involves mostly differential growth of organs and structures. (p. 348)

HOW LARVAE DEVELOP Many animals have a larval stage that may be very different from the adult. Many insects go through a series of larval molts followed a pupal stage in which imaginal discs develop to produce the adult (complete metamorphosis). (p. 349)

ARE AGING AND DEATH PROGRAMMED? Aging is species-specific, and correlates with the organism's investment in repair and its capacity for continued growth and differentiation; thus aging seems to be an inborn property. (p. 351)

STUDY QUESTIONS

1 What evolutionary pressures and aspects of life history are likely to favor eggs with small versus large quantities of yolk? (p. 340)

2 What would be the consequence of having several sperm fertilize an egg?

3 Why do you suppose sperm are small and streamlined? Why should selection not have favored male gametes that bring large quantities of cytoplasmic resources for the zygote to use?

4 Why might the nervous system have evolved from ectodermal tissue? What does this suggest about the course of that evolution? (pp. 344–345)

5 Summarize the case for programmed aging and death. Explain what might be the adaptive value of such a system in a species whose life history you know well enough to be able to speculate about. (pp. 351–354)

SUGGESTED READING

PARTRIDGE, L., AND N. H. BARTON 1993. Optimality, mutation, and the evolution of aging, *Nature* 362, 305–311. *An evolutionary analysis of aging.*

SHARON, N., AND H. LIS, 1993. Carbohydrates in cell reognition, *Scientific American* 268 (1). *On the cell-surface markers that allow cell-to-cell recognition during development.*

WASSARMAN, P. M., 1988. Fertilization in mammals, *Scientific American* 259 (6). *On how eggs manage to be fertilized by only one sperm.*

CHAPTER 14

MECHANISMS OF ANIMAL DEVELOPMENT

A single fertilized egg cell contains all the genetic information necessary to orchestrate the precise series of changes from a blastula to a gastrula, then a neurula, and finally to a mature multicellular organism. The intricacies of eucaryotic gene control, discussed in Chapter 11, can account for the many changes in gene expression that successful development requires. Throughout development, cells need to know their location in the organism, what their immediate neighbors are doing, and the organism's developmental age. The major sources of all this information are chemical signals that modify each cell's pattern of gene expression.

This chapter will examine how genes control development: the differentiation of cells into ever more specific classes, the movement of cells at particular times, the formation of morphological patterns to create organs and limbs, and the repeating cycles of growth. Much remains to be learned about development, but the essential molecular mechanisms are becoming clear. In particular, the process of ***induction***, by which chemical signals alter the expression of a cell's DNA to specify or determine its developmental fate, is yielding to analysis. We will see in broad terms how, in the process of cell differentiation, each cell type follows a different avenue of determination leading to a distinctive cellular biochemistry and morphology.

HOW POLARITY ORGANIZES DEVELOPMENT

The early steps in the development of an embryo establish the basic body plan, a kind of coordinate system by which the blastula will orient during gastrulation and further development will proceed. The egg, as seen in Chapter 13, is often polarized into animal and vegetal hemispheres by the concentration of yolk (stored food) at the vegetal end. There are also crucial differences in the concentration of various proteins and messenger RNAs (mRNAs) between different parts of eggs; this molecular polarization may underlie, at least in part, the organization of the developing embryo. In frogs, for instance, the animal pole becomes the anterior part of the organism (the head), while the vegetal pole becomes the posterior.

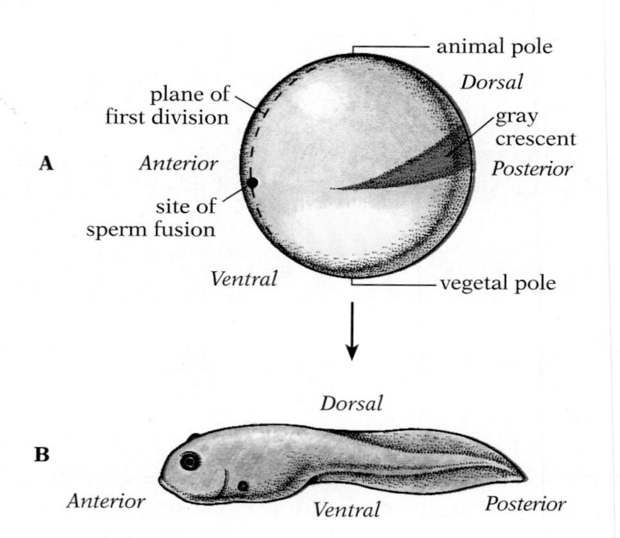

14.1 Polarity of the fertilized frog egg

After a sperm fuses with the egg, a gray crescent develops opposite the point of fusion (A). The crescent, in turn, helps define the anterior-posterior and dorsal-ventral axes of the tadpole (B). The first cleavage divides the embryo's left side from the right. Not all parts of the tadpole arise from corresponding sites on the egg: the anterior half of the egg contributes all of the ectoderm; cells developing from the posterior half fold back into the anterior and give rise to mesoderm.

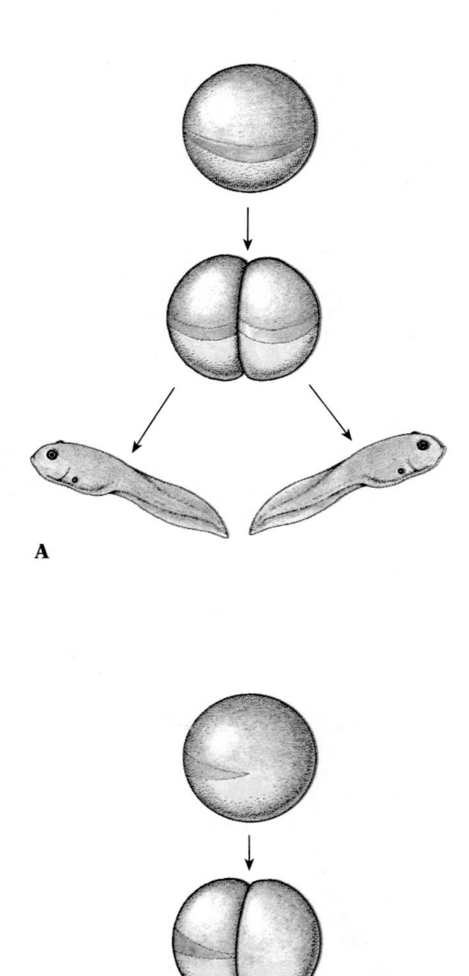

The site of sperm fusion can also provide a polarizing cue: as soon as the egg cell of a leopard frog fuses with a sperm cell, some of the contents of the egg shift position, and a crescent-shaped grayish area appears on the egg roughly opposite the point where the sperm entered (Fig. 14.1); these two points define the embryo's dorsal-ventral axis. This ***gray crescent*** is the site where gastrulation will later begin. The first cleavage of the frog zygote normally passes through the gray crescent, so that each daughter cell receives half (Fig. 14.2A). If these two cells are separated, each will develop into a normal tadpole, since each cell retains the information specifying both axes. If the plane of the first cleavage is experimentally made to shift to the side of the gray crescent, however, the result of separating the daughter cells will be very different; the cell that contains the gray crescent will develop into a normal tadpole, but the other cell will form only an unorganized mass of cells (Fig. 14.2B). In other words, the way in which the material of the egg, especially the material near the gray crescent, is distributed is critical to the development of the embryo.

14.2 Importance of the gray crescent in the early development of a frog embryo

(A) If the two cells produced by a normal first cleavage, which passes through the gray crescent, are separated, each develops into a normal tadpole. (B) If the first cleavage is experimentally oriented so that it does not pass through the gray crescent, and the two daughter cells are separated, the cell with the crescent develops normally, but the other cell develops into an unorganized cellular mass.

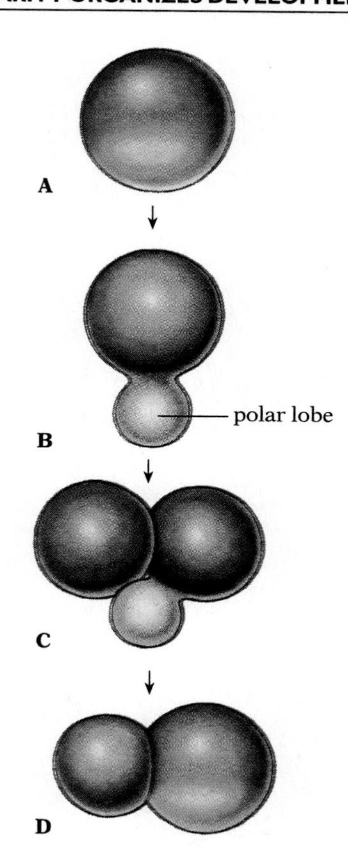

14.3 Cleavage of the fertilized egg of the sea snail *Ilyanassa*
(A) At one pole of the zygote is a region of clear cytoplasm (yellow). (B) Just before the first cleavage, this polar cytoplasm moves into a large protuberance, the polar lobe. (C) The first cleavage partitions the zygote in such a way that the entire polar lobe goes to one of the daughter cells. (D) The lobe recedes during interphase, but it will form again prior to the next division sequence. Only the cell that receives the polar-lobe material can give rise to the specific set of external structures seen in normal larvae.

After the first cleavage of the frog zygote, both of the new cells are ***totipotent:*** they have the full developmental potential of the original zygote (since the polarized cytoplasmic substances are equally distributed in each). In this respect their development is characteristic of sea stars and many of their relatives, and of most vertebrates, including humans. Mammalian blastomeres generally remain totipotent until at least the eight-cell stage. In fact, eight-cell embryos of two different strains of mice can be mixed to create a 16-cell ***chimera*** (after the Chimera, a monster in Greek mythology composed of parts from several different animals) that develops normally. This technique enables researchers to study how cells from one strain interact with those of another during development. Only animals with this kind of development can give birth to identical twins.[1]

In some other groups of animals, such as molluscs and segmented worms, the normal first cleavage partitions critical cytoplasmic constituents asymmetrically; as a result, when the daughter cells are separated, they do not have equivalent developmental potential. For example, in some molluscs, a protuberance called the polar lobe develops on the fertilized egg cell just before the first cleavage occurs. The plane of cleavage is oriented in such a way that one of the two daughter cells receives the entire polar lobe (Fig. 14.3). If the two daughter cells are separated, the one with the lobe material (which was drawn back into the main body of the cell soon after division was accomplished) will form a normal embryo possessing two prominent structures—a so-called apical organ and the posttrochal bristles—that are seen on normal larvae; the one with no lobe material forms an aberrant embryo lacking these structures. Something in the polar-lobe material must be essential for formation of the apical organ and the bristles.

In sea urchins (relatives of sea stars), cytoplasmic determinants promoting various kinds of differentiation are distributed along the animal-vegetal axis of the embryo. If the animal and vegetal halves of the embryo are separated (along the plane of the third cleavage), the animal half will give rise to an abnormal blastulalike larva with overdeveloped cilia, and the vegetal half will often give rise to a different type of abnormal larva with overdeveloped digestive organs (Fig. 14.4B). By contrast, if the eight-cell embryo is di-

[1] Identical twins develop from the same zygote, usually as a result of a double gastrulation event. Nonidentical (fraternal) twins develop from separate zygotes when two egg cells are released from the ovaries at the same time and are fertilized by different sperm cells.

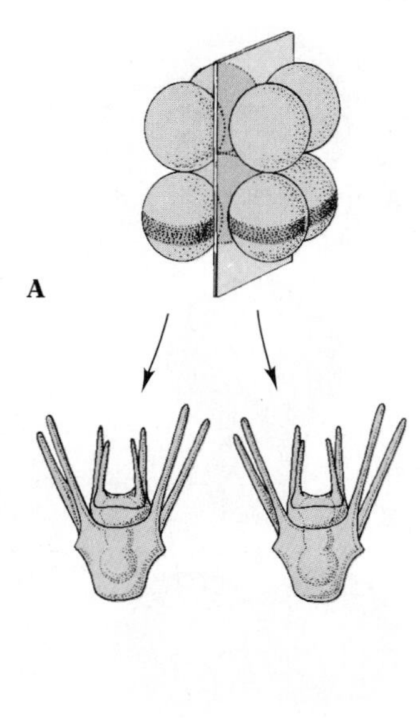

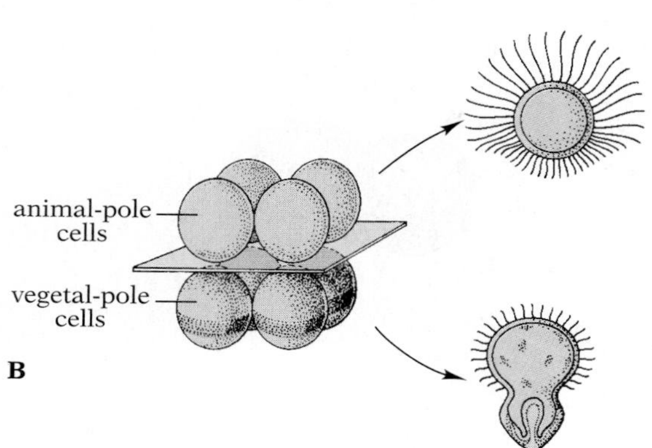

14.4 Experimental separation of cells after the third cleavage in the embryo of a sea urchin

(A) If the embryo is partitioned meridionally, so that each half receives both animal-pole and vegetal-pole cells, each half develops into a normal larva. (B) If the embryo is partitioned equatorially, so that one half receives only animal-pole cells and the other half only vegetal-pole cells, the animal half develops into an abnormal blastulalike larva with overdeveloped cilia, and the vegetal half develops into an abnormal larva with an overdeveloped digestive cavity.

vided along its animal-vegetal axis, each half will develop into a small but normal larva (Fig. 14.4A). In an intact embryo each of these halves would have developed into only half of the normal larva; evidently, as in the frog, interactions between the different regions of the embryo normally act to modify the course of differentiation of neighboring cells, integrating them into a single properly structured and proportioned larva.

Clearly, then, some unevenly distributed cytoplasmic substances must play a prominent role during early embryonic development. Some substances function by activating some genes and repressing others. Others are maternal mRNAs produced by genes expressed during oogenesis, as well as the products resulting from translation of such RNAs. The unequal distribution of these products of previous gene activity along the animal-vegetal or another axis leads to the expression of different traits in different parts of the embryo. Some polarization is present at oogenesis; additional polarization appears early in embryonic development (as early as the first cleavage in some organisms).

INDUCING DEVELOPMENTAL EVENTS

■ HOW GASTRULATION AND NEURULATION ARE INDUCED

The polarity of the embryo is first revealed when gastrulation takes place. The site of gastrulation (the blastopore of vertebrates) becomes the anus in vertebrates and echinoderms; in all other animals the gastrulation site becomes the mouth. In addition to knowing where to initiate gastrulation, the embryo also needs to know *when* the appropriate moment for gastrulation has arrived. This is not simply a matter of counting cell divisions until the right number is reached. In the frog embryo these developmental events are cued by the concentration of specific molecules, a kind of molecular developmental clock. Researchers have inferred these cues by studying a variety of so-called heterochronic mutations, which cause developing cells to misread the molecular clock, so that structures are produced at the wrong times.

Once the time has arrived for gastrulation in a frog embryo, cells in the region of the gray crescent begin the process of involution. In the early gastrula the cells derived from the gray-crescent portion of the egg become the ***dorsal lip*** of the blastopore. These cells move inward and form parts of the roof of the mouth and the

14.5 Spemann's experimental transplantation of the gray crescent

(A) When Spemann and Mangold transplanted dorsal-lip material from a light-colored salamander embryo to a dark-colored one, the blastula with two dorsal lips proceeded to form an embryo with two gastrulation zones—one at the original dorsal lip of the blastopore, and one at the transplanted dorsal lip. A double larva of mostly dark-colored tissue was the result. (B) A similar transplantation experiment in the frog *Xenopus laevis* produced an embryo with two body axes, including two heads and a second spinal cord.

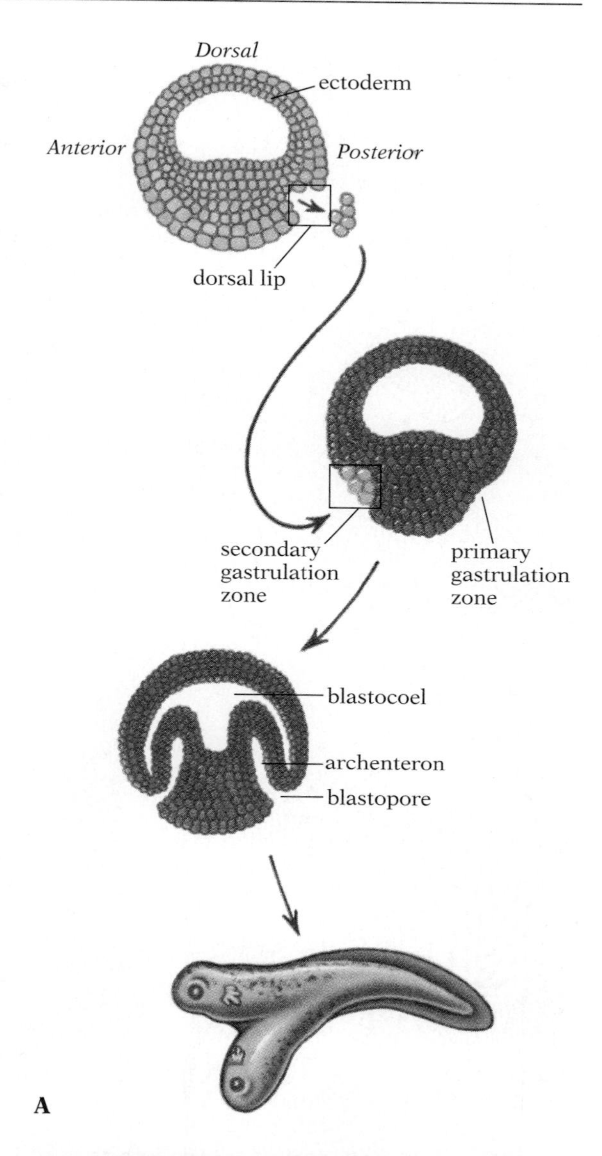

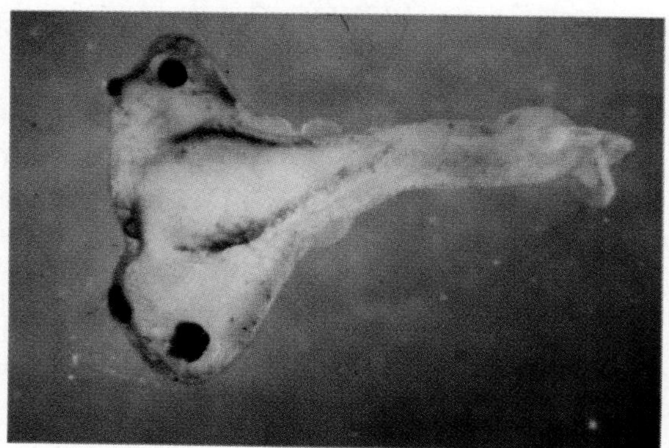

chordamesoderm, which is at first located in the roof of the newly forming archenteron (see Fig. 13.7E, p. 343). The chordamesoderm soon detaches from the roof of the archenteron to form the notochord and other mesodermal structures (see Fig. 13.7I–J). The chordamesoderm also induces the ectodermal tissue lying over it to fold inward during neurulation, forming the neural tube, which differentiates into the brain and spinal cord.

In 1924 Hans Spemann, who had previously demonstrated the importance of the gray crescent in early cleavage, and Hilde Mangold discovered another critical developmental structure. They transplanted the dorsal lip from its normal position on a light-colored salamander embryo to the belly region of a darker-colored embryo. After the operation, gastrulation occurred in two places on the recipient embryo—at the site of its own blastoporal lip and at the site of the implanted lip. Eventually two nervous systems were formed, and sometimes even two nearly complete embryos developed, joined together ventrally (Fig. 14.5). Most of the tissue in both embryos was dark-colored, an indication that the transplanted blastoporal lip had altered the course of development of cells derived from the host. Transplanting other tissues had little effect.

The dorsal lip of the blastopore plays a crucial role even after it takes the lead in gastrulation. Signals from those prospective chordamesodermal cells, as they make their way anteriorly under the dorsal ectoderm, induce the formation of the neural plate, which in turn is important in establishing the longitudinal axis of the embryo and in inducing formation of other structures.

■ HOW ORGANS ARE INDUCED

Some of the most definitive studies on embryonic induction were begun in 1905 by Warren H. Lewis. Lewis worked on the development of the eye lens in frogs. In normal development the eyes form as lateral outpockets (optic vesicles) from developing brain tissue. When one of these outpockets comes into contact with the epidermis on the side of the head, the contacted epidermal cells form a thick plate of cells that sinks inward, becomes detached from the epidermis, and eventually differentiates into the eye lens (Figs. 14.6, 14.7). Lewis cut the connection between one of the optic vesicles and the brain before the vesicle came into contact with the epidermis. He then moved the vesicle posteriorly into the trunk of

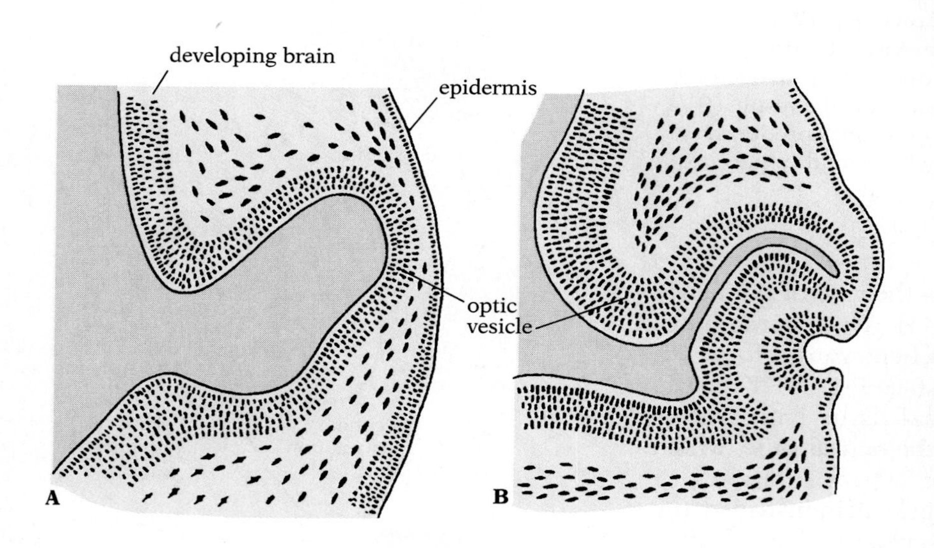

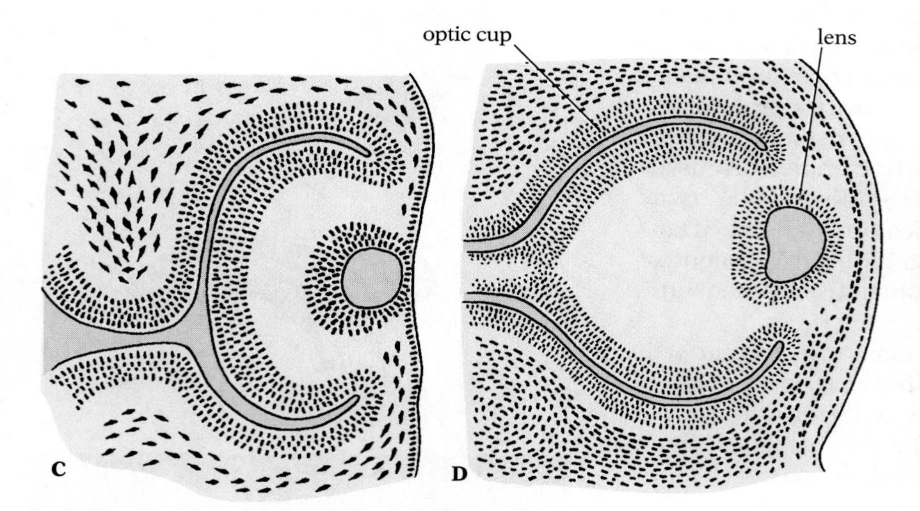

14.6 Development of optic vesicles and their induction of lenses in the vertebrate eye

(A–B) The proximity of the optic vesicle helps induce the nearby cells of the epidermis to fold inward and form a lens. (C–D) The infolding presumptive lens tissue, in turn, helps mold the optic vesicle into a two-layered structure called the optic cup. The cup cells then differentiate, the layer adjacent to the lens forming visual receptor cells and nerve cells.

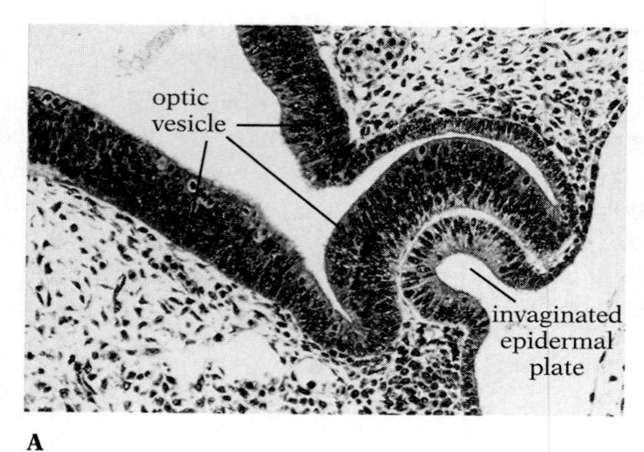

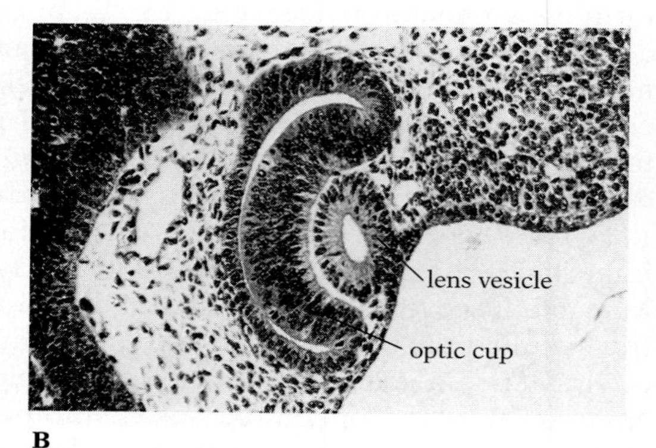

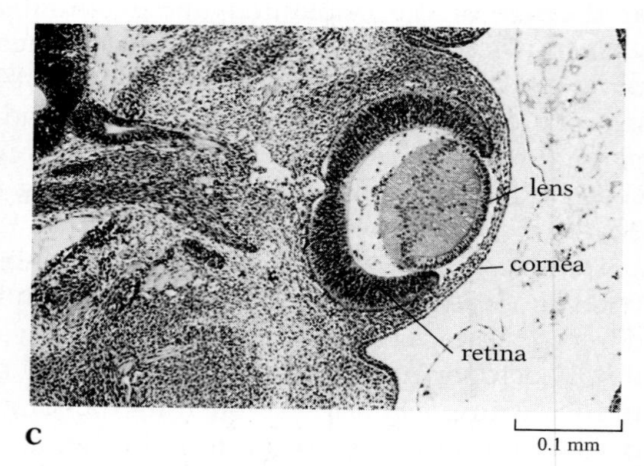

14.7 Development of a mammalian eye

Photographs of mammalian eye development over a 4-day period. (A) The contact with the tip of the optic vesicle causes the epidermis to invaginate. (B) The epidermal region differentiates to form the lens vesicle, while the optic vesicle develops into the optic cup. (C) The optic cup ultimately becomes the retina—the tissue at the rear of the eye that contains light receptors.

the embryo. Despite its lack of connection to the brain, the optic vesicle continued to develop, and when it came into contact with the epidermis of the trunk, that epidermis differentiated into a crude lens. The epidermis on the head that would normally have formed a lens usually failed to do so. Clearly, the differentiation of epidermal tissue into lens tissue depends on some inductive stimulus from the underlying optic vesicle. Later work showed that the ability to form a lens is triggered by a substance released from cells in the ectoderm dorsal to the usual site of lens formation. This chemical spreads over a limited area of the ectoderm, priming these cells for a brief period to form a lens at the site specified by the arrival of the developing optic vesicle.

The regulation is not all one-way; the lens, once it begins to form, also influences the further development of the optic vesicle. If epidermis from a species with normally small eyes is transplanted to the sides of the head of a species with large eyes, the eyes that are formed do not have a large optic cup (formed from the optic vesicle) and a small lens. Instead, both the optic cup and the lens are intermediate in size and correctly proportioned to each other. Obviously, each influences the other as they develop together.

Of the many chemicals that can act as intercellular inducers in embryonic development, some play an instructive role, others play a more permissive role. The ***instructive inducers*** restrict the developmental potential of the target cell and thus help determine the course of differentiation. The ***permissive inducers*** amplify potentialities already expressed. For example, a rudimentary organ like the pancreas may form fully committed cells during embryonic development, but it will not complete its development and become functional until acted on by a permissive inducer.

Instructive inducers do not give their target cells specific instructions about the design of the tissues or organs they are to form; they instruct only in the sense that, through repression of some gene expression and induction of expression in others, they tell the cells what part of their genetic endowment they are to use. For example, when Spemann and Oscar E. Schotté transplanted ectoderm from the flank of a frog embryo to the mouth region of a salamander embryo, they found that, though induced by salamander endoderm, the transplanted frog tissue formed the typical horny jaws of a frog. The constellation of genes for producing a jaw was activated by inducers from the adjacent tissue in the salamander, but the induced genes, being from frog ectoderm, encoded the morphology of a frog jaw.

■ HORMONES AS INDUCERS

The body uses many internal chemical signals, called ***hormones***, to pass messages over relatively long distances, often through the circulatory system; testosterone (a male sex hormone) and adrenalin (an alerting hormone) are examples. Hormone action is discussed in Chapter 33.

In vertebrates, hormones play a predominantly permissive role in development, helping differentiated tissue assume its full character. The essential interplay between instructive inducers and permissive hormones is clear in the gonads of amphibians. The ***gonads*** are the primary sex organs—the sperm-producing ***testes*** of the male, and the egg-producing ***ovaries*** of the female. Two distinct kinds of cells are set aside early to become gonads in amphibian embryos: peripherally located cortical cells and centrally located medullary cells (Fig. 14.8). The cortical cells have the potential for forming ovarian tissue, and the medullary cells for form-

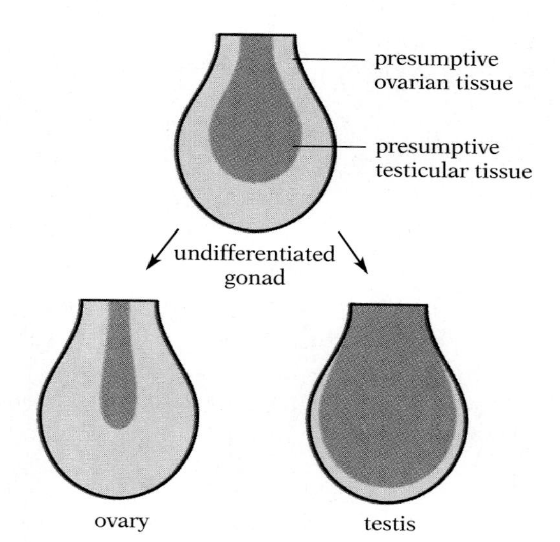

14.8 Development of ovary or testis from the undifferentiated gonad of an amphibian

Depending on the hormonal condition of the embryo, one or the other of the two types of tissue in the undifferentiated gonad gains developmental ascendancy and the other type is repressed.

ing testicular tissue. Only when sex hormone acts on the sexually uncommitted embryonic gonad does one or the other set of cells begin to express its potential.

Similarly, the same embryonic primordia give rise to the accessory sexual organs of both sexes in humans (Fig. 14.9). Whether these primordia form male or female structures depends on whether or not male sex hormone is present in the embryo at the critical stages of embryonic development. In birds, the situation is reversed; the sexual organs will follow a masculine developmental course unless female hormone is present at the critical embryonic stages.

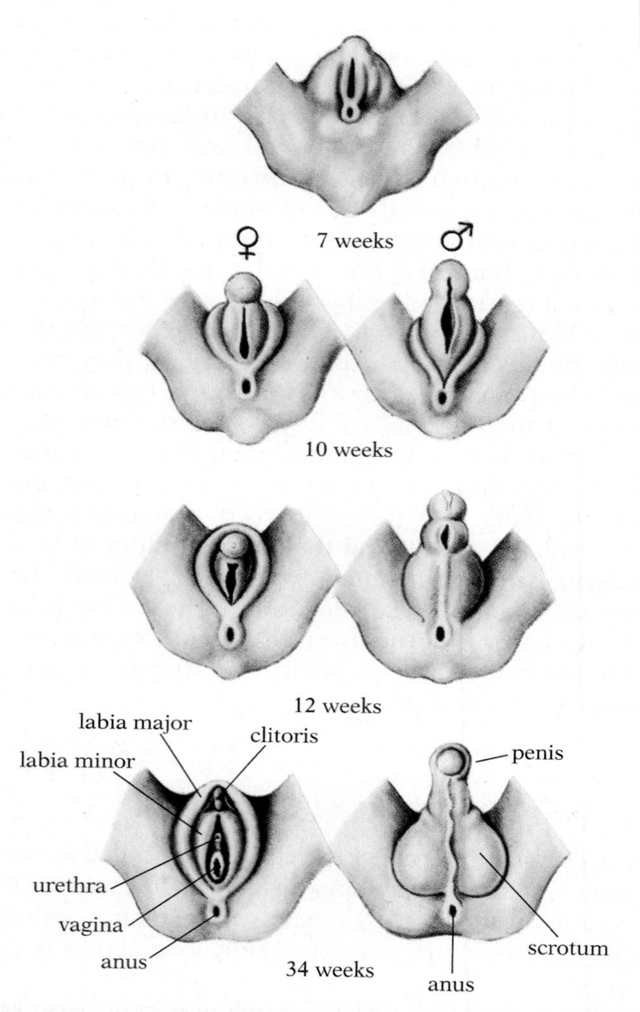

14.9 Development of the external genitalia of human beings
At 7 weeks the genitalia of male and female fetuses are virtually identical. At 10 weeks the penis of the male is slightly larger than the clitoris and labia minor, which form from the same primordium in the female. At 12 weeks these differences are more pronounced, and the male scrotum has formed from the tissue that becomes the labia major in the female. At 34 weeks the distinctive features of the genitalia of the two sexes are fully apparent. It is largely the concentration of male sex hormones like testosterone that determines which of these developmental pathways will be followed.

■ HOW MIGRATING CELLS KNOW WHERE TO GO

One of the most remarkable phenomena of development is the organized movement of cells and the folding of tissues into new shapes (morphogenesis). Dorsal-lip cells of amphibian gastrulas crawl in the right direction and stop at the correct spot; cells of the optic vesicle "know" when to begin their evagination outward toward the ectoderm, when they have reached their target, and then, how to create a suitable cup. What molecules provide directional information to migrating cells, and how are the coordinates read?

At least some orientation is to gradients of diffusible chemicals—molecular beacons—produced by landmark cells. Tissue-dissociation experiments also suggest another strategy: if cells from two different developing organs—the liver and spleen, say—are separated from one another and the two groups of cells mixed together, they slowly but accurately sort themselves out into two clumps. Each cell extends several pseudopods, touching and adhering to neighboring cells, and then pulls itself toward the most similar cells.

Each cell membrane is richly supplied with one or (usually) several kinds of ***cell-adhesion molecules*** (CAMs). Unlike the typical binder-receptor systems we have dealt with up to now, in which two different molecules—glucose and the glucose receptor, for instance—bind to one another, some kinds of CAMs attach themselves exclusively to other molecules of the same sort of CAM (Fig. 14.10), a phenomenon called ***homophilic binding***. The proportion of the several different classes of CAMs on the membranes of various liver cells is fairly consistent and is different from the corresponding ratio on spleen cells. As a result, liver cells stick only slightly to spleen cells, but more firmly to their own kind. It is possible that when a migrating liver cell attempts to withdraw its pseudopods from a spleen cell, the pseudopod loses its hold, but when it tries to withdraw from another liver cell, it remains stuck and instead drags itself toward the other liver cell. This differential affinity would permit the two classes of organ cells to sort themselves out.

The same thing may occur when the dorsal-lip cells begin to move. Their developmental clock tells them it is time for gastrulation, which may cause a new set of CAMs to be mounted in the membrane. General chemical gradients may provide some initial orientation, and the presence of other cells or structures to attach to must constrain the choice of routes available to migrating cells. However, many researchers believe CAMs to be the decisive factor in determining which among a cell's alternative routes to follow once the cell begins to grow. Pseudopods begin to form and seek attachments, and would find the strongest interactions in the anterior-dorsal direction. Pulled along by the discovery of even better

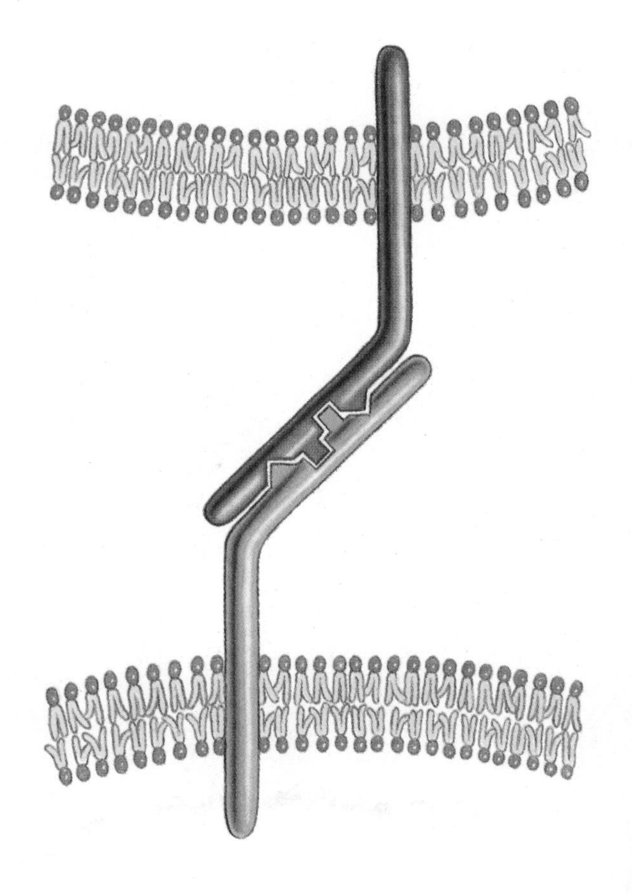

14.10 Homophilic binding of cell-adhesion molecules

The complementary structure of the arms of CAMs allows each type to bind to others of the same class. Several dozen kinds of CAM are thought to exist, though fewer than 10 have yet been characterized. Heterophilic CAMs (which bind to CAMs not of the same class) also exist.

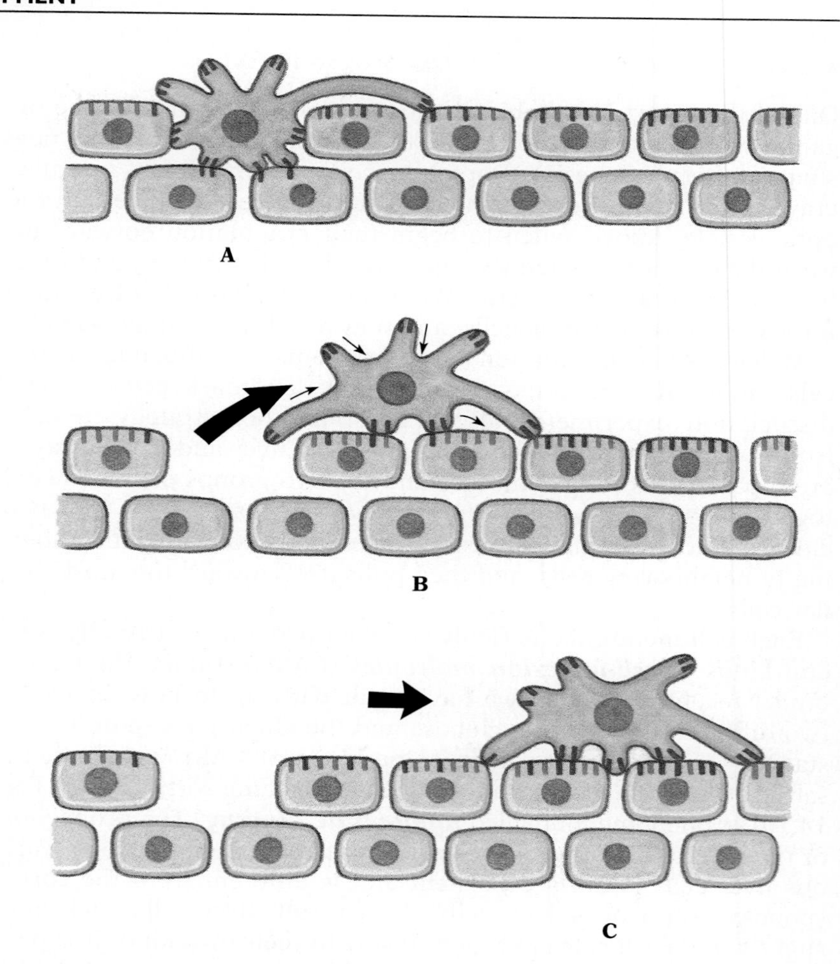

14.11 A model of cell migration

The CAM-based migration hypothesis: (A) A cell induced to begin searching for a better CAM match extends pseudopods and attaches more or less strongly to other cells. (B) When the pseudopods are periodically retracted, the cell is pulled in the direction of the best match. (C) After many such steps, the cell reaches a point from which there is no available improvement in CAM matching.

matches farther forward, the dorsal lip cells would drag themselves (and the posterior surface of the blastula) forward until their CAMs found the best matches available; when no direction offers the prospect of any stronger attachments, migration would stop (Fig. 14.11).

Analyses of the structure of CAMs have yielded a major surprise: they are closely related to specialized proteins of the immune system. Since CAMs are also found in organisms that lack immune systems such as the fruit fly *Drosophila* and lower vertebrates, this developmental recognition system probably provided the basis for the evolution of the molecules of the immune system of mammals (described in Chapter 15).

PATTERN FORMATION

Once the embryo has undergone gastrulation and neurulation, different organs begin to appear up and down its length. The strategy of further steps in development, from insects to humans, is one of subdivision of the embryo into a series of domains, followed by largely independent development of each domain.

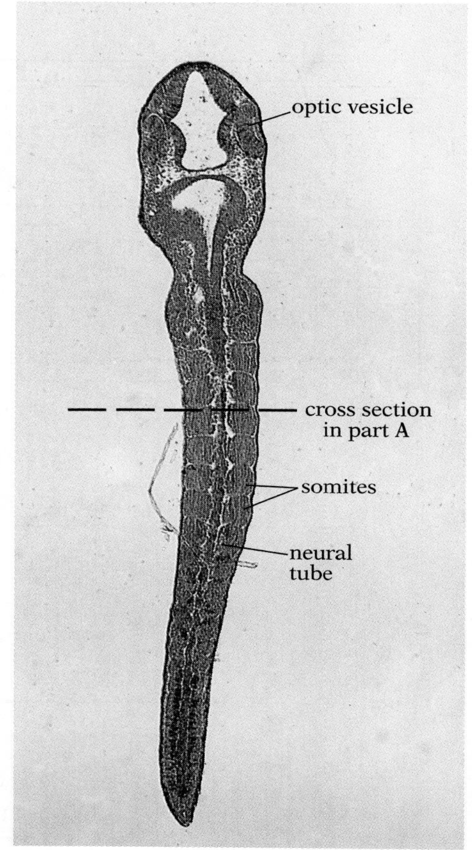

14.12 Somite formation

(A) The dorsal portion of embryonic mesoderm differentiates into a series of cell blocks, called somites, which give rise to vertebrae and dermis, as well as (depending on the somite) ribs and muscles of the limbs. (B) A photograph of an amphibian larva in frontal section with somites clearly visible.

■ WHY EMBRYOS ARE SEGMENTED

Vertebrate somites Once the spinal cord has fully formed in a vertebrate embryo, the most dorsal region of mesoderm begins to differentiate into blocks of tissue called somites (Fig. 14.12). This reorganization of the dorsal mesoderm may be based both on an anterior-posterior chemical gradient, which triggers formation of the somites, and on local interactions between somites, which assure proper grouping and segregation.

Each somite goes on to produce a vertebra, the pair of ribs (if any) associated with it, the muscles unique to that vertebra, and the dermis (the layer of cells just below the epidermis). Each somite "knows" which set of bones, nerves, and muscles to construct on the basis of its anterior-posterior location, and becomes ***determined***—that is, committed to that fate. Some embryologists believe that positions along axes in vertebrates are interpreted by the genes of somites and other units by measuring the concentration of one or more chemical ***morphogens***, which are secreted from a specific point in the embryo and diffuse away.

Morphogens and segmentation in *Drosophila* Morphogens are no mere hypothesis in invertebrates. The genetic and molecular bases of subdivision, and the use of morphogens for position finding, are well understood in *Drosophila,* where the segmentation in both the larva and adult is clearly visible externally. (In the adult, there are three segments in the head, three in the thorax—the middle section, to which the legs and wings are attached—and eight in the abdomen.) Two chemical axes are established in the blastoderm—one anterior-posterior, the other dorsal-ventral—which determine the overall organization of the larva and adult. The dorsal-ventral axis is established, and locations on it are interpreted, by a group of about 20 genes; another 30 or so genes are involved in the construction and reading of the anterior-posterior axis that generates segmentation.

Segmentation depends on gradients of two protein morphogens already being produced in the egg; one diffuses from the front toward the rear, while the other (or perhaps a pair) emanates from the posterior end forward. The result is two overlapping concentration gradients (Fig. 14.13A). The genes of blastula cells apparently respond to the ratio of these two chemicals. Even before segmentation begins, if cytoplasm is removed from the anterior end of the egg and replaced with posterior cytoplasm, the larva develops

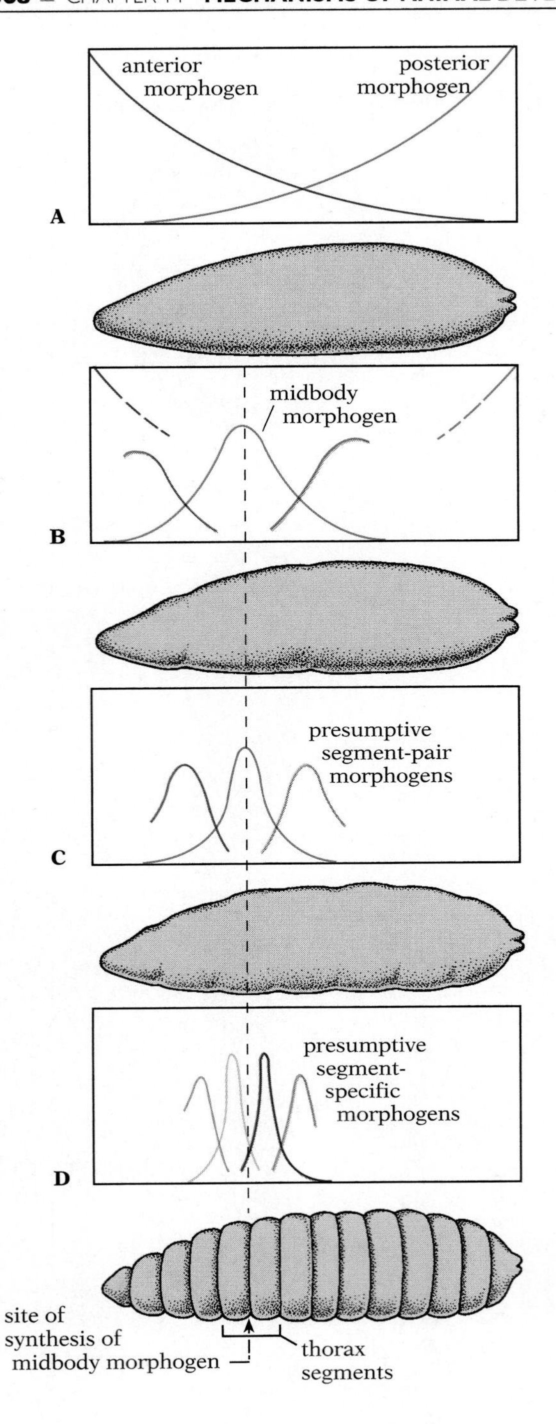

14.13 Model of morphogen concentration gradients responsible for segmentation in *Drosophila*

(A) Two morphogens important in establishing segmentation are produced in *Drosophila* eggs. One (blue curve), encoded by the gene *bicoid,* is synthesized near the anterior end and diffuses toward the rear; the other (red curve), the product of the gene *nanos,* is generated at the posterior end and diffuses toward the front. (A second posterior morphogen may also exist.) The result is a pair of overlapping concentration gradients. (The gradient pictured for the posterior morphogen may not be this regular; there is some evidence that it is actively transported anteriorly before beginning to diffuse.) (B) Additional gradients are created later in development. One, based on a morphogen encoded by the gene *Krüppel,* helps organize the middle of the larva. (C) These more local morphogens allow the larva to divide itself into seven segments. (D) These give rise to the final complement of 14, each of which then organizes itself internally.

without a head but with an abdomen at each end. Mutants lacking one of these polarizing morphogens can be "rescued" through an injection of the absent chemical into the appropriate end of the egg.

Though the overlapping concentration gradients inform cells of their approximate location in the embryo, this information is not precise enough to direct accurate development. At least one intermediate landmark provides a closer point of reference for cells in the long midregion of the larva. A tiny band of cells about a third of the way back along the larva, stimulated by the appropriate ratio of anterior-to-posterior morphogens, begins synthesizing and releasing a different morphogen that helps organize the middle of the embryo (Fig. 14.13B); other morphogens are produced by bands elsewhere along the axis.

The next step in organizing the embryo requires the genes in the various cells to interpret the anterior and posterior morphogen gradients and the local morphogen to determine which of seven segment pairs the cell is located in (Fig. 14.13C). Presumably this triggers the release of further chemicals, which enable cells to localize themselves to the anterior or posterior member of their segment pair (Fig. 14.13D). Finally, a set of 10 or so genes sets to work polarizing each segment individually. The sequential and hierarchical organization of whole-body morphogens, followed by ever-more-local morphogens, produces a gradual, step-by-step determination of cells toward increasingly specific fates.

The roles of homeotic genes At this point, as segment identity is fixed and the polarizing coordinates established, particular members of a family of about 20 ***homeotic genes*** (so called because the mutant alleles of these genes alter a segment or structure so that it resembles another) are activated according to which segment they are in. The products of homeotic genes, like the morphogens that induced them, are often control substances that bind to DNA and orchestrate the operation of many other genes; in this way, they cause each segment to express its unique character. Like the other

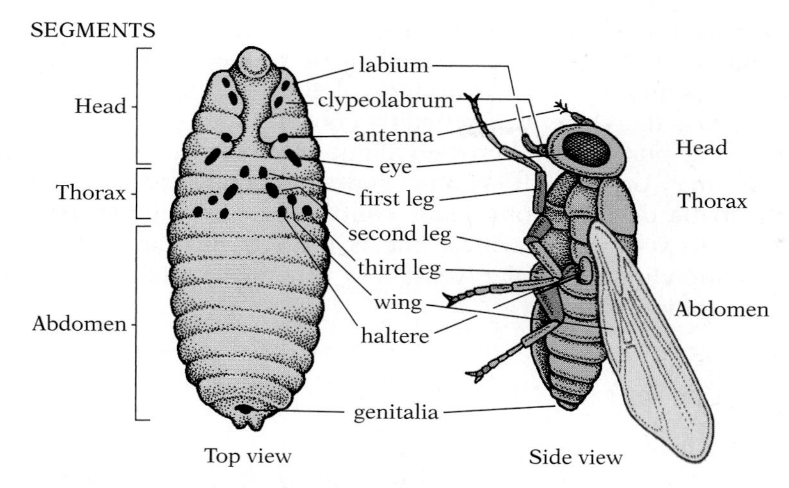

14.14 Segmentation of a *Drosophila* larva

Induced by morphogens, homeotic genes in each segment give rise to imaginal discs (shown as black spots) in the larva that control the development of particular structures in the adult.

control genes, homeotic genes make permanent changes in a cell's genome that are passed on to its daughters when it divides, thus inducing and stabilizing cellular determination.

The most dramatic consequence of the action of homeotic genes in insects is the creation of imaginal discs—the platelike clumps of cells mentioned in Chapter 13 that differentiate during the pupal stage to form adult structures (Fig. 14.14). Although differentiation occurs in the pupal form, determination took place much earlier, in the larval stage, soon after segmentation, when homeotic genes first became active: if a presumptive leg disc is exchanged with an antenna disc, the resulting adult will have a leg on its head and an antenna on its thorax.

Most of the homeotic genes are organized on the chromosome in two tight groups, the bithorax and antennapedia complexes. A gene's position within a complex correlates with its site of action on the embryo: beginning with those in the antennapedia complex that operate at the extreme anterior end of the embryo, genes farther along the chromosome operate on ever-more-posterior structures, and the pattern holds for the bithorax group as well. The logic of this organization is not yet understood.

Another intriguing discovery is the 180-nucleotide sequence called the ***homeobox***, which is found in each homeotic gene as well as in many other developmental-control genes. Although the exact sequence of the homeobox varies a little from gene to gene, a strong similarity is clear. Homeobox gene products are transcription factors, which suggests that the variation in basic homeobox sequence accounts for the specificity of homeotic genes in activating the particular constellation of genes needed in each segment. One particular mutation, for example, causes a gene normally transcribed only in the thorax to become active in the head, leading to the production of legs in place of antennae (Fig. 14.15); this mimicking of the effects of disc transplants described earlier is almost certainly the consequence of a mutation in a control element that regulates the activity of a homeotic gene.

A

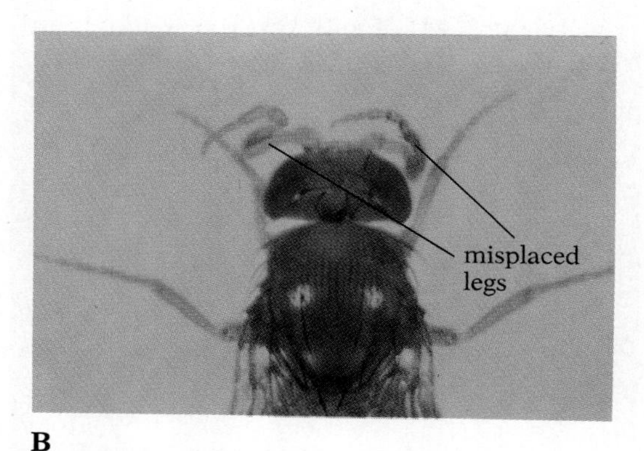

B

14.15 Two homeobox mutations in *Drosophila*

(A) Bithorax. (B) Antennapedia.

Studies of other species have turned up homeoboxes in all kinds of multicellular organisms, from plants to roundworms to humans. An understanding of the workings of the segmentation genes in *Drosophila,* therefore, will probably contribute to our understanding of how our own embryos go about organizing themselves. The evolutionary origin of homeobox genes is equally intriguing. They seem to be derived from genes controlling very simple developmental events and polarities: one is involved in mating-type determination in unicellular yeast and others in anterior-posterior differentiation in hydra.

■ HOW LIMBS FORM

Once a particular somite or imaginal disc is told to help produce a limb or an organ, there comes the major developmental task of orchestrating its construction. One popular model to account for vertebrate limb growth and patternings is based on investigations into wing development in chicks. The model assumes a bicoordinate system: a proximo-distal axis (from the body to the end of the extremity) and an anterio-posterior (front-to-rear) axis.

The hypothetical proximal-distal coordinate is linked with the ***progress zone*** of the wing, the area in which new cells are produced as the bud grows. The progress zone is associated with an ectodermal ridge running across the tip of the limb bud (Fig. 14.16A). The new cells produced in the progress zone are left behind as the area is pushed farther and farther away from the body. A cell's proximal-distal positional values might be determined by the time it spent in this progress zone (Fig. 14.16B). Cells left behind by the progress zone very early in development of the wing would have a low positional value, which might cause them to develop into the basal part of the wing (the humerus portion; see Fig. 1.10). Cells left behind a bit later would have an intermediate posi-

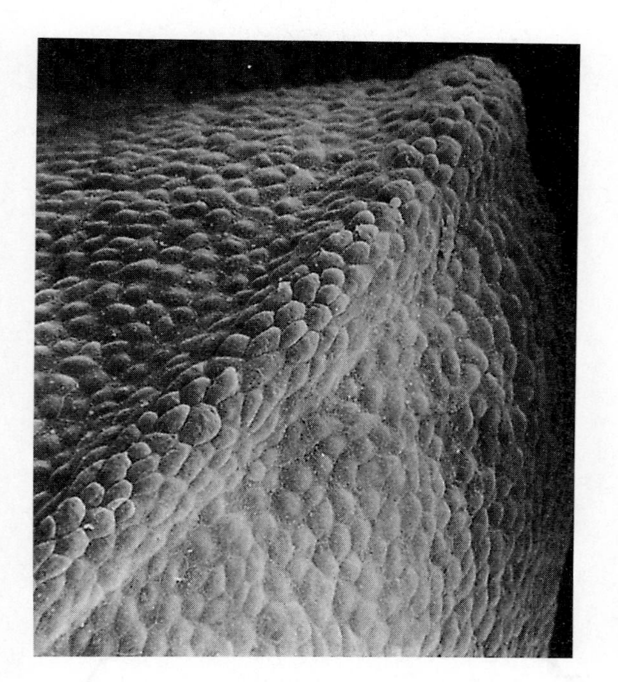

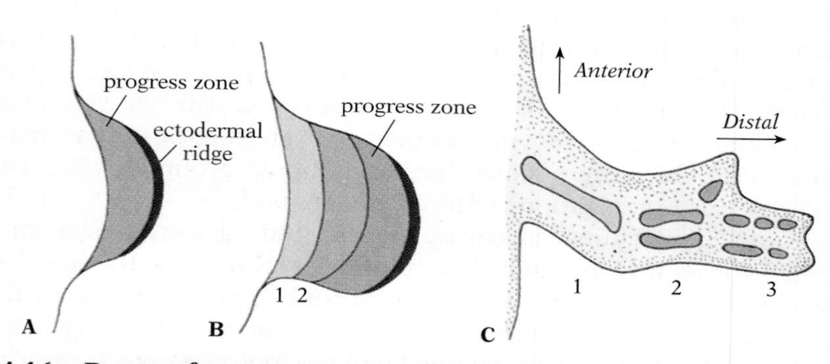

14.16 Pattern formation in the development of the chick wing
Left: Scanning EM of the ectodermal ridge of a chick wing bud. Right: Three stages of development of the wing bud, according to the progress-zone model. (A) Just behind the ectodermal ridge is a progress zone, where new cells are produced. (B–C) The first band of cells derived from the progress zone will become the humerus section of the wing; the second band will become the section containing the radius and ulna; and a third band, derived from the progress zone late in the development of the wing bud, will become the distal part of the wing.

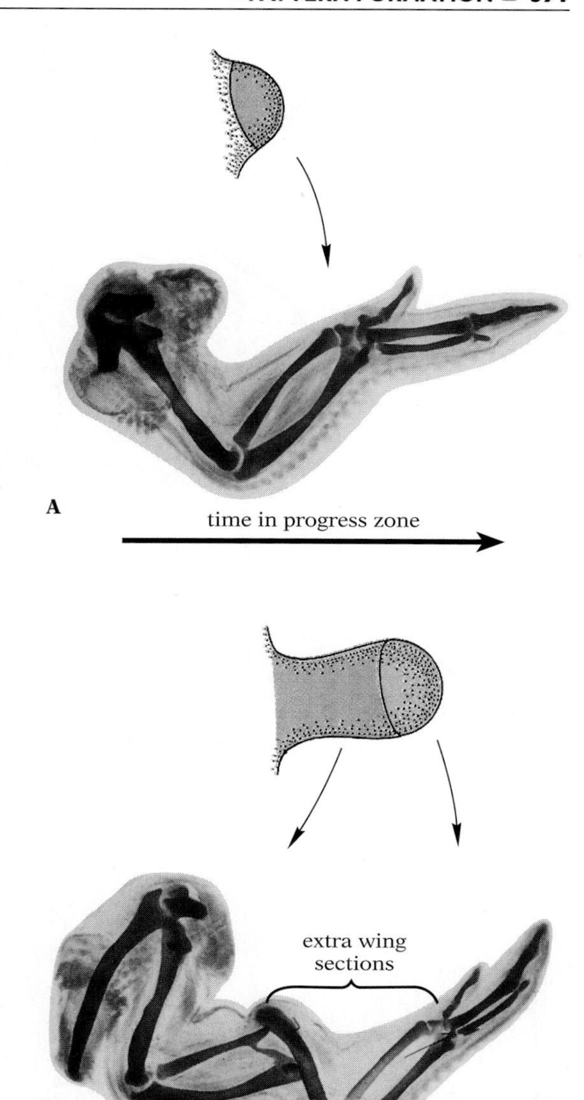

14.17 Experiment with a grafted progress zone of a chick wing bud

(A) A normal wing developed from an intact wing bud. The tan area in the bud is the progress zone. (B) The wing that developed when an early wing bud (lighter tan) was grafted onto the original bud (darker tan) after the basal part of the wing had already begun to develop. The wing has extra humerus and radius-ulna sections.

tional value, appropriate to formation of the middle part of the wing (the radius and ulna portion). Cells left behind late in the development, having spent a long time in the progress zone, would have a high positional value, appropriate to formation of the distal part of the wing. This information must somehow be stored in the cells as they are left behind, enabling them to regenerate new limbs from any point of amputation.

For the anterior-posterior coordinate, the model imagines that a diffusion gradient of a morphogen (which may be retinoic acid or triggered by it) is secreted by a small group of cells at the rear margin of the wing bud and polarizes the bud. With these two coordinates, the cells could "read" their position with sufficient accuracy to ensure their differentiation into an appropriate structure.

One test of the model involved grafting the progress zone (the ectodermal ridge and the associated area of actively dividing cells) from an early wing bud onto the end of an older bud whose own progress zone had been removed. The result was development of a wing with two humerus sections and two radius-ulna sections (Fig. 14.17). The cells of the graft had no way of telling they were so far out on the wing that they should form only its distal parts. Because they had spent so little time in the progress zone, some of these cells were left behind, programmed to read their position as being near the wing base, and developed into structures appropriate to such a position. The converse experiment of grafting the progress zone from an older bud onto an early bud produced a wing with only the distal parts, the phalanges: the transplanted tissue had no indication that the humerus, radius, and ulna had not yet developed. Having spent a long time in the progress zone, it developed structures appropriate to the wing tip.

To test the anterior-posterior gradient part of the hypothesis, part of the supposed polarizing region of the bud from the rear (posterior) edge was transplanted to the front (Fig. 14.18A–C). The results were dramatic: a partial set of mirror-image phalanges developed on the front edge. Their orientation to the transplanted part of the polarizing region was identical to the orientation of the normal set of bones to the polarizing tissue at the rear of the wing. Evidently this is because the morphogen from the transplanted region polarized the front half of the bud, diffusing from front to rear, just as the morphogen from the intact polarizing region on the rear edge polarized the rear half of the bud, diffusing from rear to front. Implants of retinoic acid mimic these transplants; moreover, retinoic acid receptors in wing-bud cells, once bound, are known to move to the nucleus and bind to the DNA, just as we

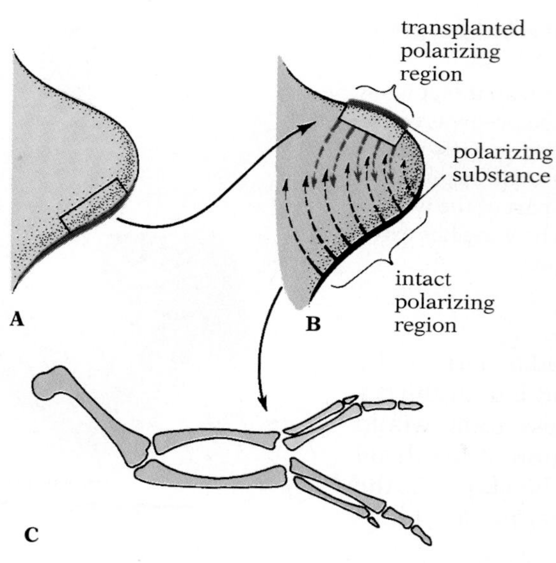

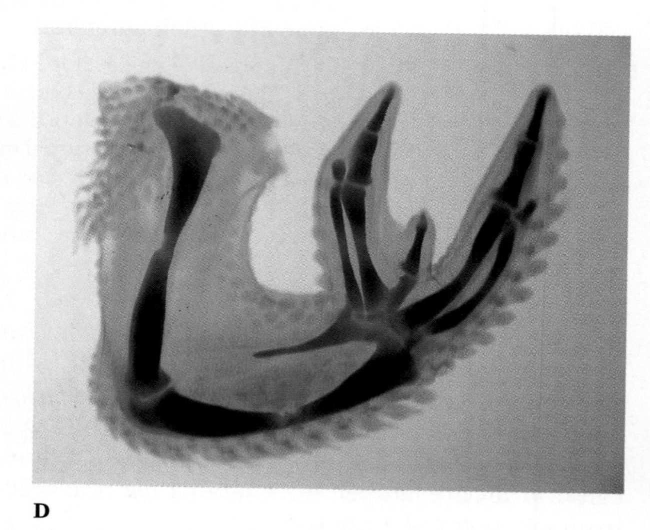

14.18 Experiment with a transplanted polarizing region
Tissue from the rear margin of one wing bud (A) is transplanted to the leading edge of another (B). There the morphogen diffusing from the transplant causes cells to differentiate into a second set of phalanges (C). The same effect can be achieved by treating the leading edge of a limb bud with retinoic acid (D).

would expect from a gene-control substance (Fig. 14.18D). In addition, which homeotic gene is expressed correlates with the concentration of retinoic acid. Whether retinoic acid is actually the primary morphogen, however, is controversial.

REVERSING DEVELOPMENT

The usual sequence of development, as we have seen, is for a morphogen or other chemical (often from an adjacent cell) to act as an inducer, altering the pattern of gene expression of a cell. This responding cell has now become to some degree determined. Subsequent experience with additional morphogens or other control substances further focuses a cell's fate, until it may become fully determined. At the same time, a cell's morphology and chemistry are responding to the activity of the genes in its nucleus, so that the cell differentiates (though often after considerable delay). Determination and differentiation, therefore, are usually gradual and proceed in that order.

Though differentiation usually follows determination, certain cells become differentiated *before* determination fully fixes their fate; they are capable of some degree of ***dedifferentiation***—that is, they can regress to less differentiated states, and then begin differentiating into a new type of tissue. In mammals and birds, dedifferentiation is usually restricted to embryos. In amphibians, however, this ability to maintain incomplete determination in at least some cells into adulthood leads to remarkable powers of re-

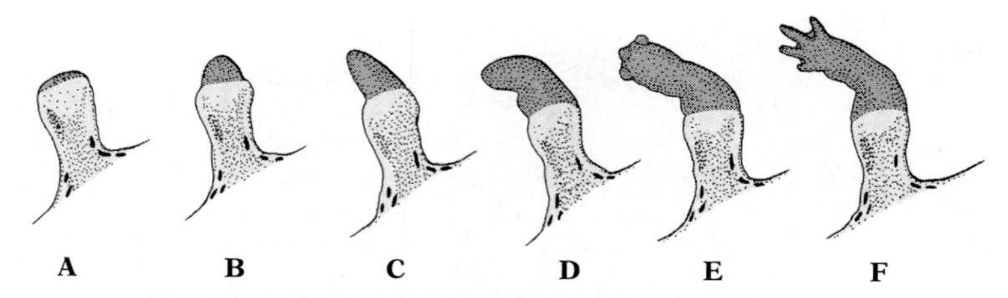

14.19 Regeneration of a salamander arm

generation. For example, if the leg of a salamander is amputated, cells near the wound begin to dedifferentiate under the epidermis at the tip of the stump. Gradually, as mitotic activity and cellular redifferentiation take place within this area, the mound comes to look more and more like the normal limb bud of an embryonic salamander (Fig. 14.19). It slowly elongates, and after several weeks a distinct elbow and digits appear, complete with muscle, tendon, bone, connective tissue, etc.

Dedifferentiation is most common in animals that grow throughout their life cycles; many fish, reptiles, and amphibians have no typical adult size, but instead simply grow larger until they die. With the need to be able to grow indefinitely, the developmental options of at least some cells may need to be kept open.

Although dedifferentiation is possible for some cells in some species, full determination is seldom reversed. Cancerous cells are one exception: they may suffer mutations that remove some of the determining constraints, begin dividing again, and produce cells with a different identity (most often that of an earlier, less differentiated cell type). Obviously this requires profound changes in patterns of gene expression.

HOW NERVOUS SYSTEMS DEVELOP

Most aspects of development are illustrated with particular clarity in the nervous system. Nerve cells, or ***neurons***, are "born," migrate to their proper places, send out fibers called axons and dendrites to specific target locations, and so come to form a highly integrated functional network that is more complex than any other organ system in the body.

■ NERVE-CELL MIGRATION

The first step in the life of a newly formed presumptive neuron, like that of many other cells in a developing organism, is usually movement from where it was formed to where it is destined to be. The cells that give rise to the retina, for instance, must grow out from the developing brain to where the eyes are to be located, and then form the optic vesicle, while the cells of the outer layer of the brain, the cerebral cortex, must move through layers of older cells to get from the core of the brain, where they were engendered, to the outside layer of the cortex, where they end up.

The movement of neuronal cells during development is quite

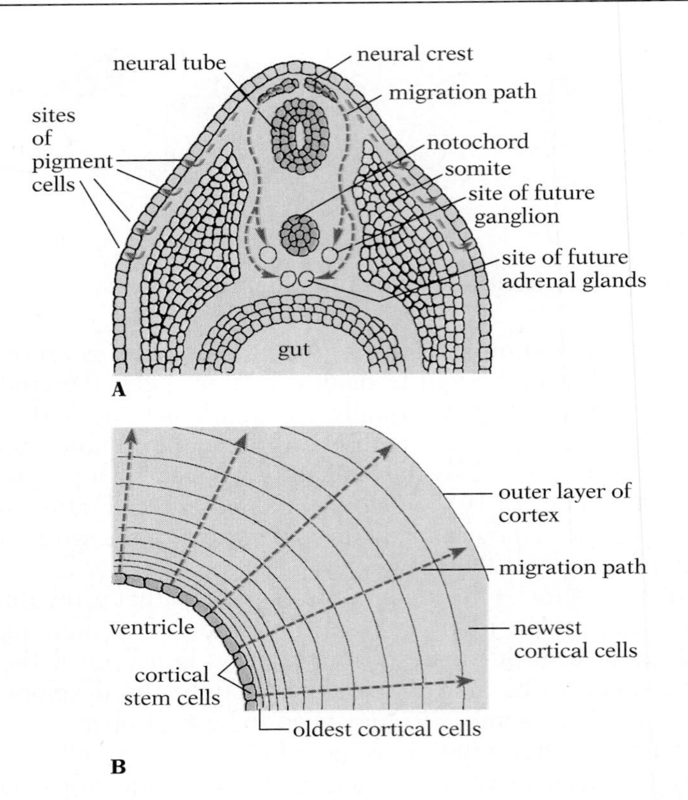

14.20 Migration of nerve cells

Before nerve cells join together in a network, many must move from their place of origin to their final location. (A) Cells from the neural crest move down and along glycoprotein filaments to positions at the future sites of spinal ganglia and the adrenal glands. Other neural-crest cells encounter another set of filaments, leading to positions just under the ectoderm; cells following this pathway become pigment cells in the skin. (B) New cells of the cerebral cortex are generated by a basal layer of stem cells lining each fluid-filled cavity, or ventricle, and migrate out along filaments to form a layer on the outside of the cortex. Cells produced later will move through this layer to positions still farther out.

well oriented. Consider the cells that give rise to the adrenal glands of the kidney and to nearby clumps of nerve cells (ganglia). They originate in the neural crest—the region just above the developing spinal cord—and migrate down to specific spots near the notochord (Fig. 14.20A). They move in the proper general direction from the outset, follow highly predictable pathways, and stop at precise spots. Similarly, cells of the cerebral cortex are generated in a basal layer of tissue and then migrate outward through other, older cortical cells until they reach the outer layer of the cortex (Fig. 14.20B). The three stages the neural-crest and cortical cells pass through—determining the initial direction, following a path, and determining the stopping point—are characteristic of the migration of developing cells. At least three mechanisms are utilized: diffusing chemicals, CAMs, and tactile cues.

The gradients of the diffusing chemicals help guide neuronal cells by causing them to move in an amoeboid fashion toward (or, in some cases, away from) the source of the chemical, leading neural-crest cells "down," cortical cells "outward," and other cells in an anterior or posterior, or dorsal or ventral, direction. One current model proposes that most path finding also involves the other two mechanisms: CAMs and tactile cues. Neural-crest cells (guided by filaments of glycoproteins that lead around the spinal cord and past the notochord) and cortical cells (following a radial array of filaments up through the cortex) apparently recognize their respective pathways by means of the particular ratios of CAMs they encounter on the other cells along the pathway. They partially envelop the guide cells, and then move along them, maintaining intimate tactile contact, finding ever-better matches between self- and substrate-borne CAMs.

According to this model, when a migrating cell encounters the optimum CAM correspondence on the cells it has touched, it stops moving and proceeds to form the cell-to-cell attachments that will anchor it in place. The molecular specificity of nerve-cell surface markers is so precise that each class of neurons probably has distinctive molecules in its membrane.

■ AXON GROWTH AND TARGETING

Once a neuronal cell has reached its permanent place in the nervous system, it must send axons (the long, thin processes specialized for transmitting information) and, in the case of some sensory cells, dendrites to specific target cells. Here again, both chemical and tactile information seem to play a role. The leading edge of the developing axon displays an unusual type of structure known as a ***growth cone*** (Fig. 14.21). The growth cone continually extends and retracts spikelike pseudopods called filopodia, which probably sample the environment for specific chemicals and for the actual presence of certain guide cells. If the cone encounters the chemical or tactile stimulus of such a guide cell (usually the axon of another nerve), it partially envelops it and grows along it. Other chemicals can repel growing axons, and thus help prevent mistargeting.

In vertebrates, as many as 1 million axons may project information from the body to one of several large arrays of target cells like those found in the visual, auditory, and tactile areas of the cortex. These connections of axons to target cells are spatially organized and genetically predetermined. It seems unlikely that 10^6 specific molecular labels exist in the visual system to assure that all axons find their correct target cells. Nor does it seem plausible to imagine that a simple bicoordinate morphogen gradient could be sufficiently precise. In fact, the initial wiring is not very exact; the final pattern is determined dynamically through neural competition.

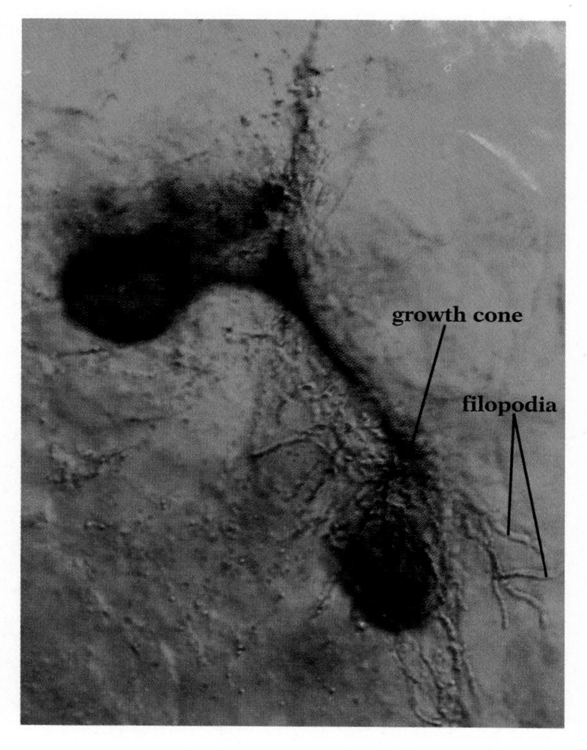

14.21 Growth cone of a developing axon in a grasshopper

The filopodia have made contact with another neuron (lower right).

■ DEATH AND COMPETITION IN DEVELOPMENT

In most animal nervous systems, many more cells are born than are actually needed. In vertebrates, for example, identical ganglia containing vast numbers of cells develop next to each of the vertebrae. And yet only the ganglia serving the many muscles and sensory receptors of the arms and legs require so many cells; in the other ganglia, the extra cells die. Apparently it is easier or more efficient for the developmental program to build all segments alike initially, and then to allow functionless cells to die.

In nematodes (roundworms) a similar pattern is evident. Their simple nervous system consists of a long ventral nerve cord containing 12 ganglia. Each ganglion arises during development from one of 12 precursor cells, each a part of a clump of embryonic cells. The ganglia mature to control rhythmic swimming movements and mating. The pattern of cell division is the same in all 12 ganglia even though the eventual organization of each ganglion may be very different. Cells that become neurons in only some (or even just one) of the ganglia are nevertheless formed in all 12, and

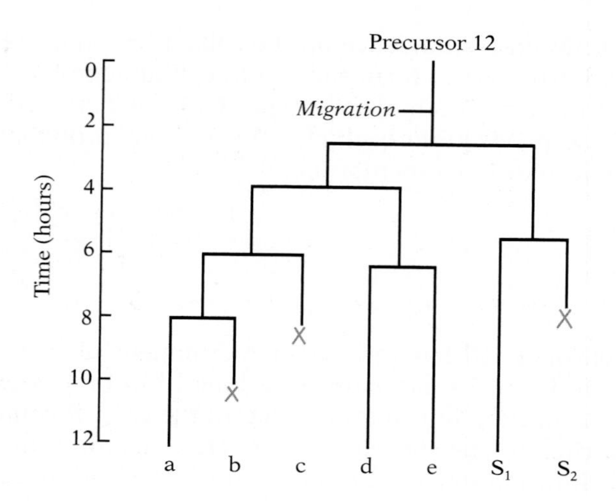

14.22 Cell death during the formation of nematode ganglion 12

This family tree of the cells in the last ganglion of a nematode suggests that the developmental program for the neurons involves a fixed series of divisions regardless of which neurons are needed by the ganglion. For example, the ganglion at the tip of the tail (illustrated here) does not need cells b and S_2, which in other ganglia connect to the next ganglion to the rear. Cell c is also unnecessary except in the middle ganglia of males, where it controls some of the animal's reproductive behavior. Nevertheless, a cell can apparently be produced only if cells b and c are produced as well.

those not needed are then allowed to atrophy and die (Fig. 14.22).

An analogous phenomenon occurs in many animals at the level of neuron-to-neuron connections: many more of the axonal endings specialized to communicate neural activity from one cell to another—structures called synapses—are formed than survive. When cells wired to a particular target are prevented from responding to external sensory stimuli (either by experimental manipulation or because the cells are misplaced or defective), their synapses seem to get crowded out on the target neuron by synapses from other cells that do fire normally. In animals as diverse as crickets and mammals, cells that lose in this competition seem to atrophy and disappear. This process serves to fine-tune the connections in large spatial arrays.

CHAPTER SUMMARY

HOW POLARITY ORGANIZES DEVELOPMENT

Egg cells are polarized by virtue of having different concentrations of critical molecules in different parts of the egg. This polarity, the site of sperm fusion, and the plane of an early division can orient later development. (p. 357)

INDUCING DEVELOPMENTAL EVENTS

HOW GASTRULATION AND NEURULATION ARE INDUCED Developmental timers tell the embryo when to gastrulate. In frogs the cells derived from the gray crescent organize gastrulation and much of early development. (p. 360)

HOW ORGANS ARE INDUCED Many organs are formed in response to inducer chemicals produced by cells near where the organ is destined to lie. (p. 361)

HORMONES AS INDUCERS Some hormones—particularly the sex hormones—act as organ inducers. (p. 363)

HOW MIGRATING CELLS KNOW WHERE TO GO The migration of cells from their point of origin to their ultimate location may be guided by cell-adhesion molecules (CAMs). Moving cells appear to seek the best CAM match with the other cells they encounter until they reach a location from which no better matches are within reach. (p. 365)

PATTERN FORMATION

WHY EMBRYOS ARE SEGMENTED Most embryos develop as a series of segments that are initially very similar, but specialize later based on their anterior-posterior position—information that they read from concentrations of morphogens. Misreading the correct location creates homeotic changes that transform one region into another. (p. 367)

HOW LIMBS FORM Limb formation requires organization along two axes: a distal-proximal one (which seems to depend on developmental age) and an anterior-posterior axis (which relies on a morphogen released from the rear edge of the developing limb). (p. 370)

REVERSING DEVELOPMENT

Some cells that appear to be fully differentiated are nevertheless able to dedifferentiate, divide, and take up new roles under special circumstances. Some cancerous cells mimic this pattern. (p. 372)

HOW NERVOUS SYSTEMS DEVELOP

NERVE-CELL MIGRATION Many nerve cells migrate from their site of birth to a ganglion along specific routes guided by chemical gradients; CAMs and tactile cues may also be involved. (p. 373)

AXON GROWTH AND TARGETING Axon growth shows the same pattern as nerve-cell migration, except that the distances are often far longer and the precision much greater. (p. 375)

DEATH AND COMPETITION IN DEVELOPMENT Many more nervous-system connections are made than ultimately survive. Competition between connections based on the degree of neural activity seems to fine-tune the connections. Each ganglion has many more neurons than it eventually needs; these excess cells, which play a role in other ganglia with different specializations, atrophy and die. (p. 375)

STUDY QUESTIONS

1 When the anterior half of a frog egg was separated from the posterior half (Fig. 14.2B), one segment became a normal tadpole while the other turned into a disorganized mass of cells. What might happen if you were to perform the same experiment on a *Drosophila* egg or blastula? (pp. 367–369)

2 How does the overlapping two-gradient strategy of *Drosophila* allow for greater accuracy in reading anterior-posterior location than the use of a single morphogen? Is there any evidence that *Drosophila* needs this degree of precision? (pp. 367–370)

3 Can you think of any reason why it might make sense to arrange the homeotic genes in anatomical order?

4 Think of five different ways unnatural chemicals in the diet of a pregnant mammal might disrupt the development of an embryo. (pp. 361–372)

5 Why might incomplete determination be essential to organisms with indefinite growth? Why might incomplete determination be a disadvantage to species with fixed growth—species that usually live longer and grow larger? (pp. 372–373)

SUGGESTED READING

COOKE, J., 1988. The early embryo and the formation of body pattern, *American Scientist* 76 (1). *A wide-ranging review looking for common mechanisms.*

EDELMAN, G. M., 1984. Cell-adhesion molecules: A molecular basis for animal form, *Scientific American* 250 (4). *On the likely molecular basis of cell-to-cell adhesion and changes in adhesion during embryonic development.*

GILBERT, S. F., 1994. *Developmental Biology, 4th ed.* Sinauer, Sunderland, Mass. *Excellent, highly detailed text.*

GOODMAN, C. S., AND M. J. BASTIANI, 1984. How embryonic nerve cells recognize one another, *Scientific American* 251 (6). *An excellent description of how axons of invertebrates employ the stepping-stone strategy, following first gradients and then one preexisting axon after another to reach their targets.*

MCGINNIS, W., AND M. KUZIORA, 1994. The molecular architects of body design, *Scientific American* 270 (2). *On how the homeobox genes orchestrate development in both insects and vertebrates.*

CHAPTER 15

IMMUNOLOGY

Once we have had measles or mumps, we cannot contract them again. This immunity develops because the cells that fight the disease while it is present are able to recognize and destroy the disease-causing ***pathogen*** far more rapidly in the future. The number of foreign cells and unfamiliar chemicals the immune system can learn to recognize is essentially infinite, yet only a few genes are involved. How does this remarkable defense system work? The answer, as we will see in this chapter, involves the cell-adhesion and cell-surface molecules discussed in Chapter 14, as well as powerful mechanisms of genetic variation.

HOW THE IMMUNE RESPONSE WORKS

Nearly all animals have phagocytic cells that ingest bacteria and dead cells. These ***macrophages*** are attracted by chemicals released by damaged tissue and many foreign cells; they create a localized inflammation where they are active, as well as a liquid mass of dead cells and other debris (commonly called pus). This slow and unselective line of defense has been elaborated in vertebrates into a highly specific immune system adaptated to larger body size and longer life.

The best-known receptor molecules of the immune system are the ***antibodies***, which bind to molecules that are foreign to the organism. Foreign substances, collectively called ***antigens*** (short for "antibody generating"), are almost always large molecules (usually proteins or polysaccharides). Antigens may be free in solution, as are the toxins secreted by some pathogenic microorganisms. Antigens may also be built into the outer surfaces of the pathogens themselves (as with some viruses and bacteria), or they may be part of the coating of otherwise innocuous entities like grains of pollen. The antigens stimulate certain cells in the immune system to produce highly specific antibodies—proteins that bind to these antigens exactly as enzymes bind to reactants. The binding of antibodies to toxins inactivates them, often by simply covering the toxin's active site, and therefore altering their toxicity. Antibodies inactivate viruses by binding to the receptors by which they recognize their hosts. When antibodies bind to microorganisms, however, they target these pathogens for subsequent destruction by one of several mechanisms described in the following sections.

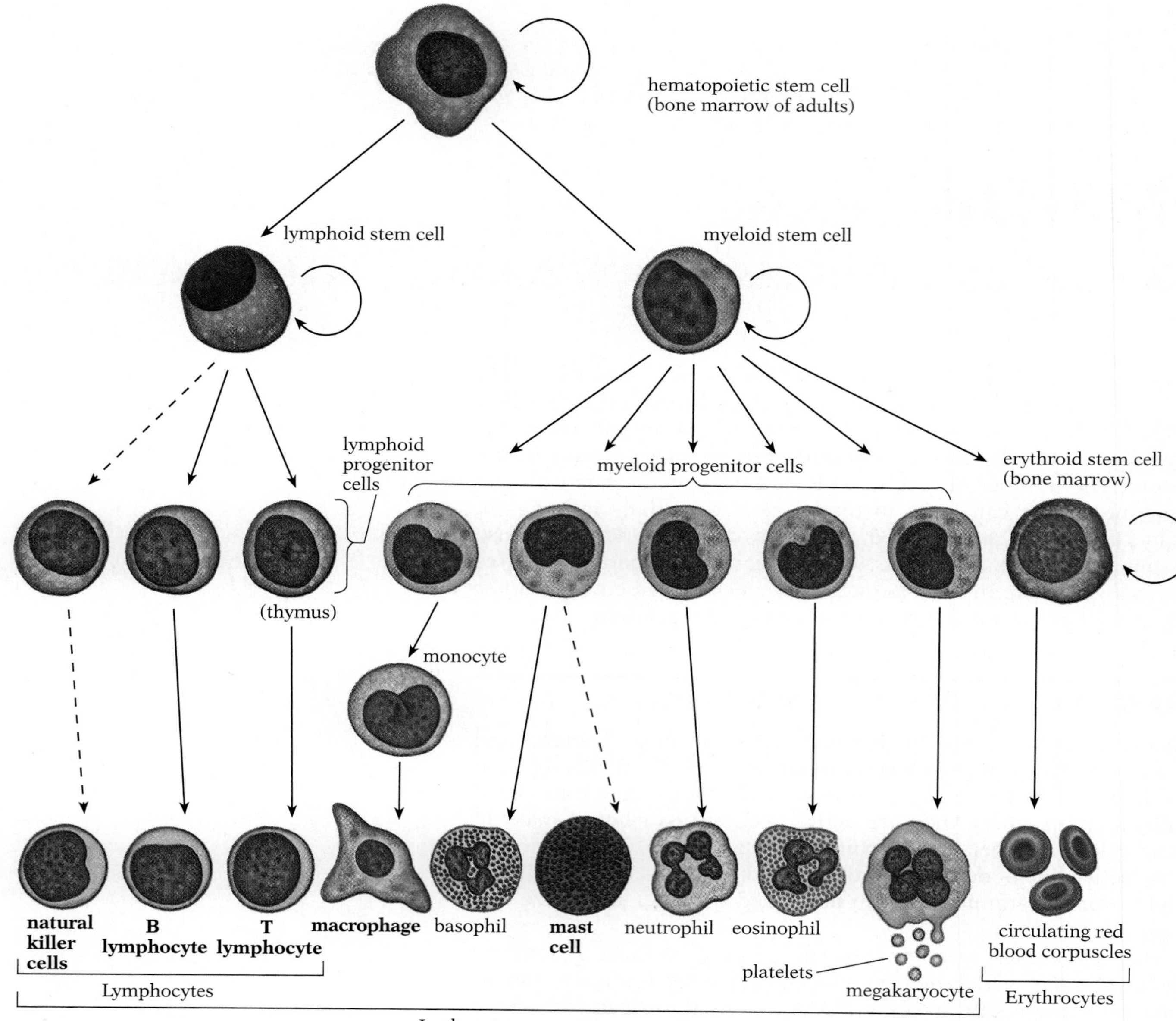

15.1 Hematopoiesis: the origin of blood cells

Hematopoietic stem cells give rise to at least two kinds of more highly determined stem cells: lymphoid and myeloid stem cells. Both kinds of stem cells have the ability to regenerate themselves. The more highly determined stem cells in turn give rise to more specialized cells. Lymphoid stem cells give rise to three still more highly determined, progenitor cell types, which in turn produce the natural killer cells, B lymphocytes, and T lymphocytes that we will discuss in this chapter. Myeloid stem cells produce at least six classes of more highly determined cells, which themselves go on to generate the more specialized monocytes (which mature into the cell-scavenging macrophage discussed later), as well as red blood corpuscles, basophils and (perhaps) mast cells. (Myeloid stem cells also give rise to neutrophils and eosinophils, which attack bacteria and larger parasites, as well as megakaryocytes, which produce platelets—substances crucial to clotting, discussed in Chapter 30.) The presentation cells (which we will encounter later in this chapter) are derived from hematopoietic stem cells along an unknown pathway. The text discussion focuses on the interactions of the best-understood immunocytes, designated here with boldface type.

■ THE CAST: CELLS AND ORGANS OF THE IMMUNE SYSTEM

Where immune-system cells come from In most vertebrates, all ***immunocytes***—cells with immunological function—derive from precursors that form in the yolk sac of the early embryo. These hematopoietic stem cells (so named because they have become developmentally determined as progenitors of blood cells) migrate to specific tissues and organs, where they give rise to the red and white blood cells (Fig. 15.1). Red cells carry oxygen in the blood and have no immunological role. A class of white cells that comes from the thymus gland and from bone marrow gives rise to ***lymphocytes;*** it is these cells that respond to the presence of foreign antigens or kill microorganisms tagged by antibodies. Lymphocytes derive their name from the high proportion of time they spend in the lymphatic system, where dead cells and debris tend to collect, and into which toxins and infectious organisms usually find their way.

The role of the lymphatic system Blood consists of a solid portion (cells) and a liquid portion (plasma). As blood flows through the many fine capillary beds in the body, a portion of the plasma leaks out between capillary cell walls into surrounding tissue. This fluid, supplemented by liquid transported endocytotically or lost osmotically, brings nutrients to tissue cells outside the bloodstream, and picks up waste products. As blood moves from the capillaries into larger vessels, most but not all of the lost plasma and metabolic wastes of tissues is reabsorbed. The rest is collected as ***lymph*** into a parallel system of lymphatic veins and funneled into a large thoracic duct in the chest, from which it rejoins the blood (Fig. 15.2). Fluid and unanchored cells (generally dead or foreign) from body tissues can also enter the lymphatic system.

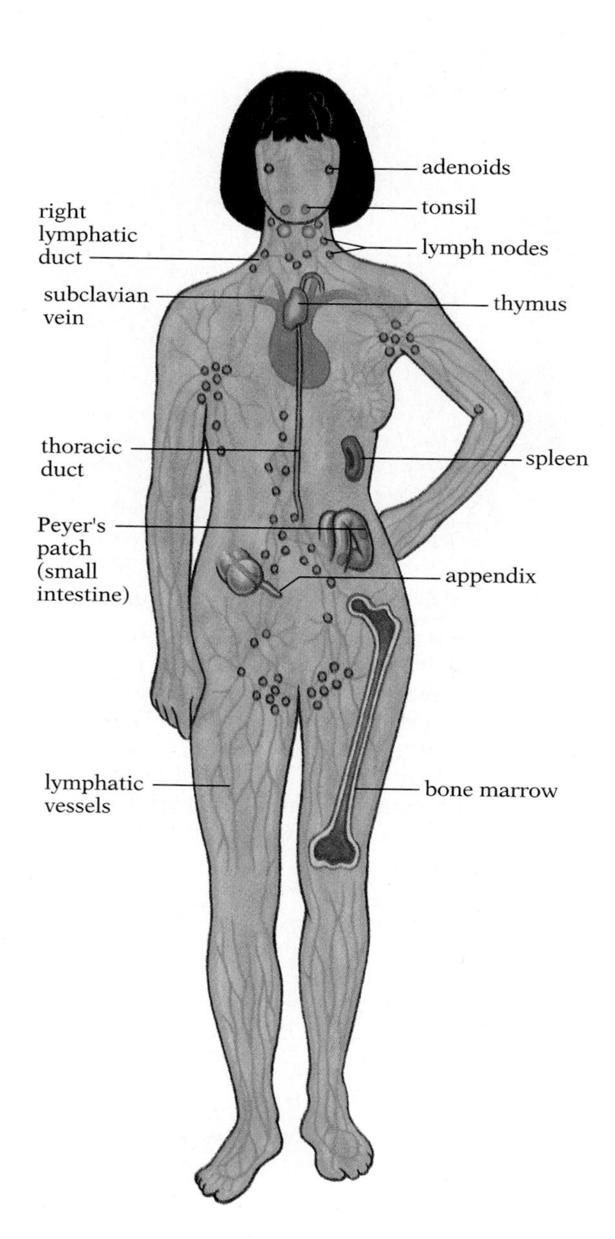

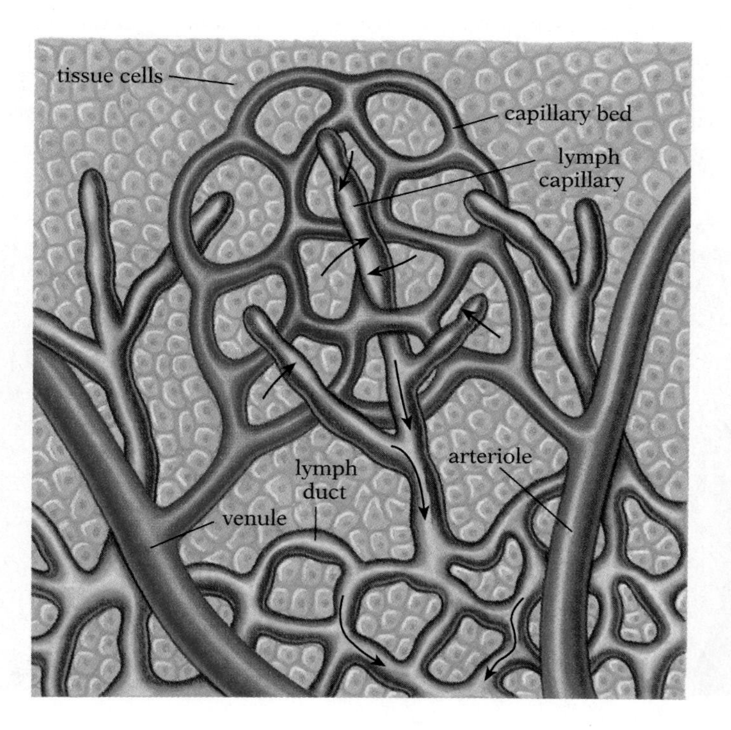

15.2 Human lymphatic system

Above: The primary organs of the lymphatic system—the thymus and bone marrow (only one bone is shown)—produce lymphocytes that circulate from extremely fine blood vessels, called capillaries, into the lymphatic system. Other tissues and organs with connections to the lymphatic system are shown. Left: The source of lymph is fluid forced from the blood capillaries, which is picked up by lymph capillaries; these merge to form lymph ducts, which in turn empty into the lymph vessels shown above.

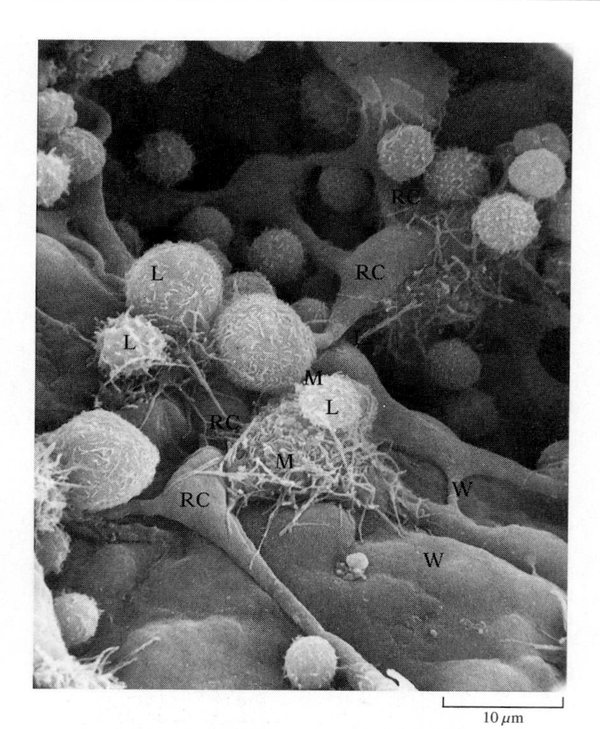

15.3 Lymph node seen from the inside

Reticular cells (RC) and their extensions create a meshwork that acts as a crude filter and provides a support surface for lymphocytes (L) and macrophages (M). The wall of the node (W) is also visible. (From *Tissues and Organs: A Text-Atlas of Scanning Electron Microscopy* by Richard G. Kessel and Randy H. Kardon. Copyright ©1979 W.H. Freeman and Company. Reprinted with permission.)

The transport of lymph is passive: body movements squeeze lymph past one-way check valves on its slow journey to the chest.

Lymphatic veins from capillary beds in the chest contain one or more ***lymph nodes***; these are regions with a fine tissue mesh that filters dead cells and other large fragments from the lymph (Fig. 15.3). The nodes are home to many phagocytic cells (which consume the material trapped in the mesh) and lymphocytes. Like the spleen, through which all of the body's blood is frequently filtered, the nodes provide a convenient place for immunocytes to monitor the plasma for foreign antigens, and immune responses are frequently localized in the nodes, particularly those in the tonsils.

The diversity of immunocytes The two main types of lymphocytes are B cells (which mature in the bone marrow) and T cells (which mature in the thymus). B cells manufacture and secrete antibodies, and are responsible for the ***humoral immune response***—the response triggered by antibodies circulating in the body fluids, blood, and lymph. Because their antibodies can recognize and bind to cell-surface markers, B cells are particularly effective against bacteria, fungi, parasitic protozoans, and viruses, as well as toxins free in the blood and plasma. The T cells, on the other hand, are responsible for the ***cell-mediated response***, which kills those of the organism's own cells that have become infected. T cells can also bind to and modulate the activity of B cells.

The major phagocytic cells of vertebrates are the macrophages. The macrophages of vertebrates can be far more selective than those of invertebrates because they look specifically for cells to which antibodies have been bound, thus identifying them as foreign. Another kind of white cell, the natural killer (NK) cell, is also found in invertebrates, but like the macrophages, it has evolved in vertebrates to home in on and attack antibody-tagged cells. The discussion that follows will emphasize the best-understood actors in immune reactions: macrophages, B cells, T cells, NK cells, and mast cells. Other white blood cells with immunological functions include neutrophils (Fig. 15.4), eosinophils (which attack bacteria and larger parasites), megakaryocytes (which produce platelets crucial to clot formation), basophils (which, like mast

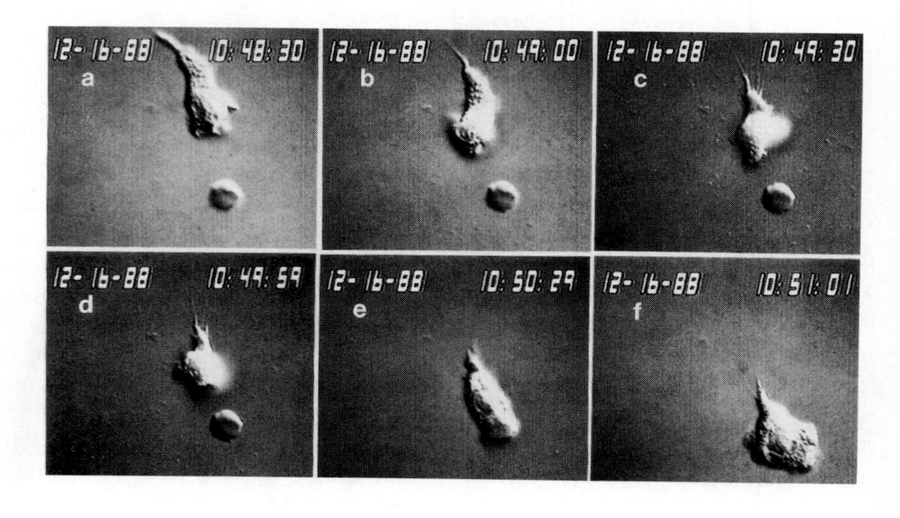

15.4 Neutrophil

Neutrophils home in on invading organisms (usually bacteria) by detecting some of their waste products. The neutrophil seen here crawls toward its target, releases toxic chemicals and enzymes onto it, and then envelopes it by phagocytosis.

cells, release a chemical—histamine—that signals injury or infection, and thus create inflammation), and presentation cells. Basophils are specialized for circulating in the blood, while mast cells reside in tissues. Presentation cells are derived from hematopoietic stem cells along an unknown pathway.

■ WHAT THE HUMORAL RESPONSE ACCOMPLISHES

Most B-cell lymphocytes migrate back and forth from the blood to the lymph, taking up temporary residence in the spleen and lymph nodes where, using their membrane-mounted antibodies, they are well placed to monitor the body's fluids or dispense free antibodies. When needed, they can disperse to the circulation and tissues.

The B-cell antibody Each antibody molecule consists of four polypeptide chains—two identical "heavy" chains and two identical shorter "light" chains, linked by disulfide bonds (Fig. 15.5). There are five classes of heavy chains—A, D, E, G, and M—that differ in the amino acid sequence at the COOH (tail) ends. The tail regions play no role in antigen specificity; they determine which reaction of the humoral antibody response will take place. For example, after an antigen has been bound, heavy chains with a G tail undergo an allosteric change in conformation that allows them to be recognized by macrophages, which can then ingest the antibody along with whatever the bound antigen is attached to—a virus, for instance.

Antibodies with other heavy chains are used to activate other parts of the immunological reaction: antibodies with E tails become mounted on the membranes of mast cells, signalling these early-warning cells to begin releasing histamine when an antigen is encountered. Others are specialized for tasks to be described presently, like conferring maternal immunity on newborns, or activating a series of enzyme reactions known as the complement system. Together, all the antibodies of all classes are called ***immunoglobulins (Ig)***, and the individual classes are often referred to with added letters, such as IgG, IgE, and so on.

Regardless of which class of heavy or light chain is incorporated into an antibody, most of each chain has a constant amino acid sequence and structure; the variability crucial to antigen specificity lies mostly at the free amino ends. The binding sites for antigens (two identical sites on each antibody molecule) are at the ends of the variable portions. Each binding site is a pocket bounded partly by the heavy chain and partly by the light chain (Fig. 15.5). This region can bind to roughly six amino acids or carbohydrate units of an antigen.

How the humoral response develops Before an organism has encountered a particular antigen, its ***B lymphocytes*** which can bind that antigen are small, metabolically quiescent ***virgin cells*** that move freely between the blood and the lymphatic tissues. Each of the millions of virgin cells generated during embryonic development has thousands of identical antibodies mounted in the membrane, but no two virgin cells are likely to display antibodies with the same antigenic specificity. When a cell's antibodies begin

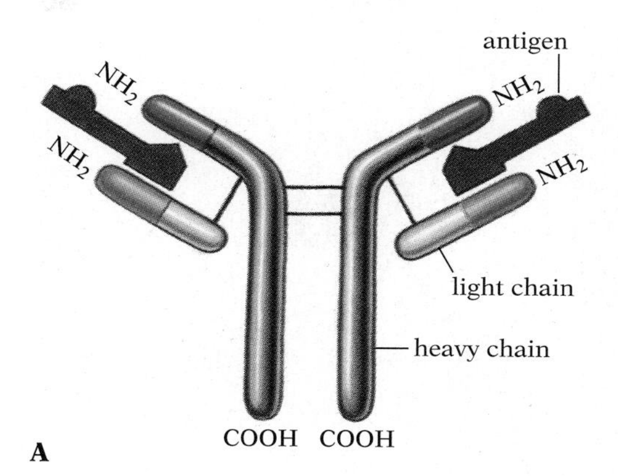

15.5 B-cell antibody molecule

(A) B-cell antibodies consist of two identical pairs of polypeptides; each pair has a heavy chain and a light chain, which is readily seen in this schematic representation. The sections shown in gray have relatively constant sequences, while the colored portions vary greatly from one B cell to another. Antigens bind in the cleft between the heavy and light chain of each pair. B-cell antibodies can exist as freely circulating molecules (as shown here) or mounted in the membrane of B lymphocytes or mast cells. (B) A space-filling model showing the three-dimensional shape of antibodies. Each sphere represents one amino acid. The antibody shown here is IgG.

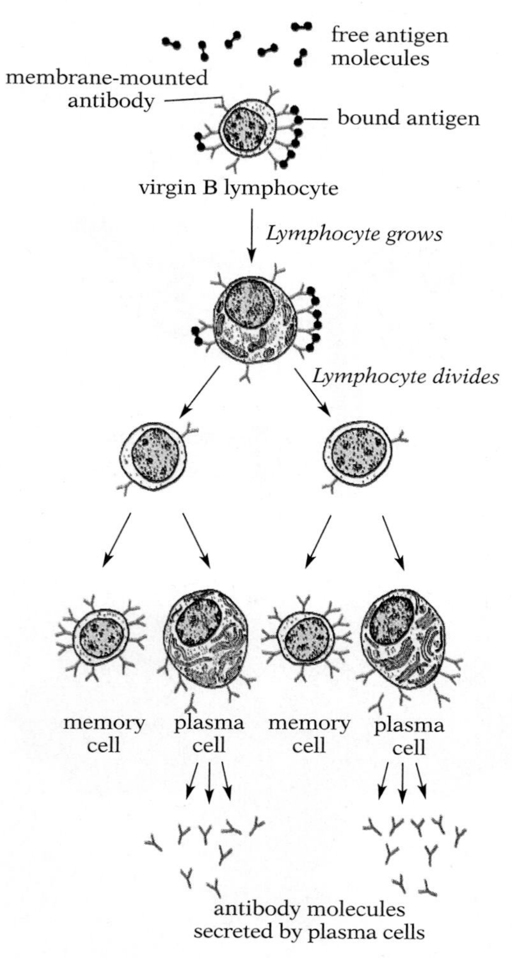

15.6 Stimulation of B lymphocyte by antigen

When the membrane-mounted antibodies of a virgin B cell are able to bind a particular antigen, the lymphocyte first grows larger and then begins a series of cell divisions (only two being shown here). Some of the cells produced by this proliferation are memory cells, which resemble the original lymphocyte; others become specialized as plasma cells, which secrete antibodies. The antigen in this example is a toxin.

binding antigen, these small lymphocytes start to grow larger and divide (Fig. 15.6).

Cell division of a stimulated B lymphocyte gives rise over a period of several days to numerous ***plasma cells***, and it is primarily these cells that secrete antibody molecules. A stimulated B lymphocyte also gives rise to other lymphocytes like itself; these serve as ***memory cells***, which will make possible a more rapid response in the future should that particular antigen be encountered again. (How this is accomplished will be described later in the chapter.) It is the rapidity of this subsequent response that confers immunity.

Each of the millions of kinds of antibodies has a different set of active sites, so each binds to one or more different antigens or antigen regions. Some antibodies are so well matched to an antigen that this bonding is rapid and strong; other antibodies bind with a lower affinity. Because each antibody molecule can bind to two antigen molecules, the antibodies can agglutinate (lump together) any microorganisms or viruses bearing the antigen (Fig. 15.7).

This agglutination in turn can trigger three reactions. First, large phagocytic macrophages (see Fig. 4.25, p. 108) in the lymph recognize antigen-bound antibodies, and engulf them and their targets (Fig. 15.8A). This reaction is effective against toxins, viruses, and most bacteria. Second, ***natural killer (NK) lymphocytes*** recognize the bound antibodies, bind to them, and destroy any foreign eucaryotic cells they have marked (Fig. 15.8B). The mechanism by which NK lymphocytes kill is not yet fully understood, but it involves making holes in the membrane of the target cell; osmotic entry of extracellular water then kills the cell.

Finally, the bound antibodies can trigger the ***complement system***, a cascade reaction involving more than 20 plasma proteins (mostly zymogens—enzymes that exist in a nonfunctional conformation until activated). In the cascade reaction, each protein in turn catalyzes the activation of another. The complement-system proteins assemble to create membrane channels in the invading cell; the channels allow water to enter by osmosis into the microor-

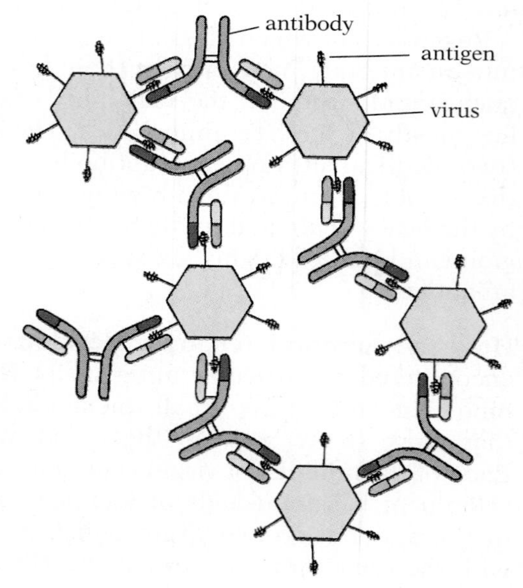

15.7 Agglutination by antibodies bound to antigens

Each antibody molecule can bind to two antigen molecules; hence the microorganisms or viruses bearing the antigens can be held together in large clumps. Here the antigens are surface proteins or carbohydrates on invading viruses. (For clarity, the antibodies and antigens have been enlarged; they are in fact much smaller than viruses.) This agglutination aids in the destruction of antibody-bearing pathogens by macrophages, killer cells, or complement-system proteins.

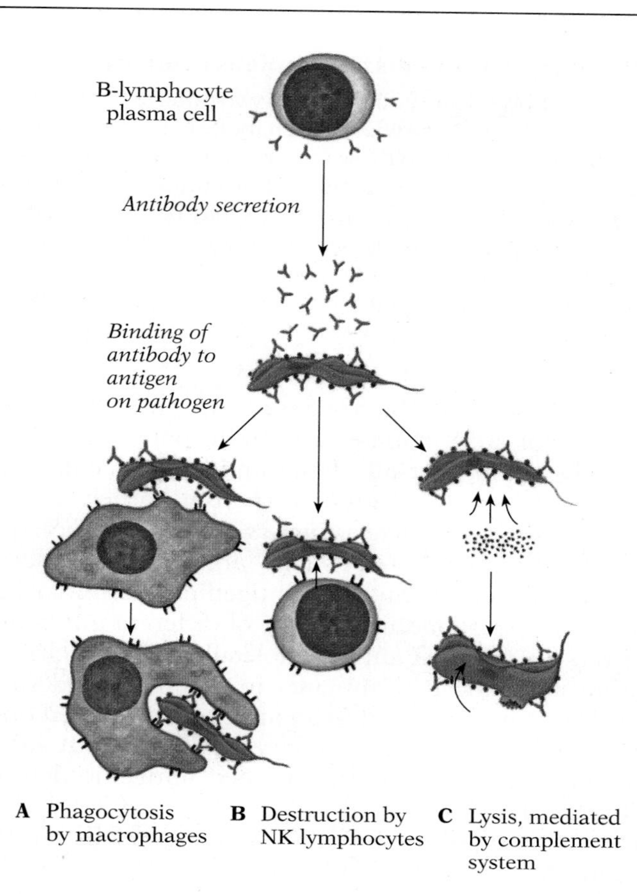

15.8 How humoral antibodies facilitate the destruction of pathogens

Once bound to antigens, the antibodies secreted by B lymphocytes can trigger three sorts of reactions. (A) Bound antibodies, and the pathogen to which they are attached, are ingested by phagocytic macrophages; this is also the primary mechanism by which agglutinated toxins and viruses are eliminated. (B) Bound antibodies are recognized by NK lymphocytes, which destroy the marked pathogen by creating holes in the target's membrane. (C) Bound antibodies trigger a cascade reaction by which zymogens of the complement system are activated and catalyze the construction of a membrane channel in the invading microorganism; water moves into the invader through this channel by osmosis, causing the cell to swell and burst. (The action of circulating antibodies on mast cells is not shown in this drawing.)

ganism, which swells and bursts (Fig. 15.8C). Some membrane-enclosed viruses are neutralized in this way.[1] This system operates almost exactly like the hole-punching strategy of NK cells. Having detected an invader, the immune system clearly takes no chances; several redundant systems work simultaneously to attack anything foreign.

In addition to binding directly to antigens and thus targeting them for destruction, some circulating antibodies can become attached by their bases to ***mast cells***. The antibodies thus mounted on the cell membrane of the mast cell have their antigen-binding region free and exposed to the surrounding medium. When an antigen is bound by a mast-cell-mounted antibody, the mast cell is induced to release histamine and other chemicals. Histamine in turn causes nearby blood vessels to dilate and become "leaky," allowing the plasma, rich in antibodies and complement-system proteins, to reach the site of the histamine release. Lymphocytes and macrophages are also attracted. The mast cell–antibody system, then, acts as a sort of cellular burglar alarm to recruit the other elements of the immune system to concentrations of antigens. It is particularly important in the response to parasites embedded in tissue rather than moving through the blood or lymph.

[1] Some viruses seem to have evolved defenses against the complement system. Herpes and Epstein-Barr viruses, for instance, bear receptors that bind and inactivate the third protein in the cascade, bringing the reaction sequence to a halt; vaccinia virus (one of the pox viruses) protects itself by binding the fourth protein.

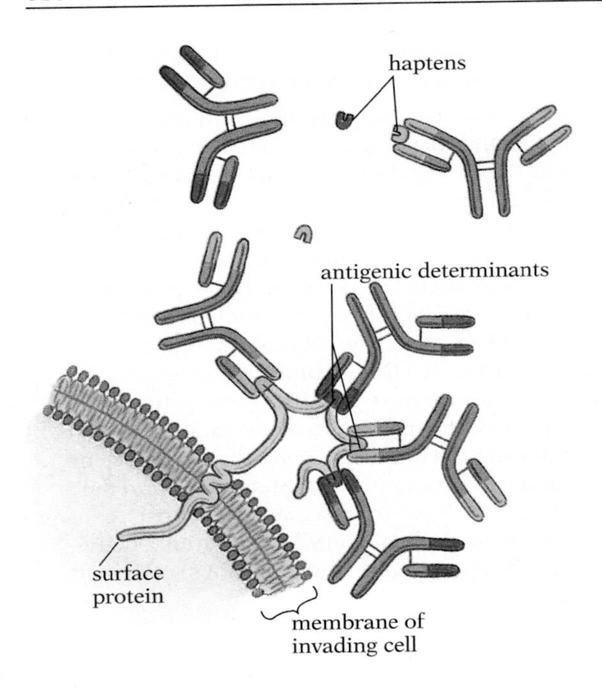

15.9 Antigenic determinants and haptens compared

The surface protein on this invading cell has two different antigenic determinants, each of which occurs twice in each molecule. Specific membrane-mounted antibodies can bind to these regions and trigger an immune response, which includes the production of free antibodies (shown). The surface protein and its antigenic determinants may be unique to this kind of invading cell, or they may be shared by cells of other kinds. Isolated antigenic determinants, or haptens, can also be bound by antibodies and stimulate an immune response if the immune system has been previously exposed to a large molecule bearing that particular determinant.

How antigens provoke the response An antigen is almost always a large molecule—usually a protein, polysaccharide, glycoprotein, or glycolipid. Not all of the antigen molecule stimulates lymphocytes to begin the immune response. Rather, restricted regions serve as ***antigenic determinants***—sites of interaction with the receptors on lymphocytes. A single large antigen molecule may include several different kinds of antigenic determinants, which are bound by a corresponding number of different antibodies (Fig. 15.9). Conversely, different antigen molecules may, by chance, have one or more antigenic determinants in common, and so "share" antibodies. The initial reaction to an antigen is triggered only if the antigenic determinant is part of a large molecule, but subsequent reactions can be initiated by the isolated antigenic determinant, called a ***hapten*** (Fig. 15.9).

Proper functioning of the immune system depends on the availability of an enormous number of slightly different lymphocytes, each specific for a particular antigenic determinant. An individual may have 10 billion (10^{10}) or more antibody types. Each antigen reacts only with those very few lymphocytes that have antibodies capable of binding to part of it, and this binding is necessary to induce proliferation of the appropriate lymphocyte types (Fig. 15.10).

Each stimulated lymphocyte gives rise to a clone of cells (a group of genetically identical cells descended from a common virgin cell). Hence the proliferation of the particular lymphocytes that react with a specific antigen is called ***clonal selection***. Each of the plasma cells to which a given B lymphocyte gives rise when stimulated may transcribe as many as 20,000 mRNA molecules from its genes for antibody, enabling each of the plasma cells to secrete 5 million identical antibody molecules per hour.[2] But the immune system is careful not to trigger this massive response by accident.

[2] A partial but rapid immune response can also be triggered in individuals directly inoculated with the appropriate antibodies. Though there are no memory cells to begin supplying new antibodies, the antibody molecules introduced from outside can deactivate pathogens, or mark them for phagocytosis or complement-mediated destruction. Breast feeding is one means of creating this "passive immunity": maternal antibodies specialized to cross the intestinal wall (by virtue of a special heavy-chain tail) and enter the fetal bloodstream provide a significant degree of temporary protection until the infant's own immune system gains experience. Another source of antibodies is an injection of gamma globulin, a protein extract from blood containing antibodies from individuals immune to a particular disease, though the extract contains many other kinds of antibodies as well. Gamma globulin treatment may reduce the severity of symptoms in someone who has been exposed to certain diseases, and can be useful in the absence of an effective vaccine, or when there is no time for an immunization to "take."

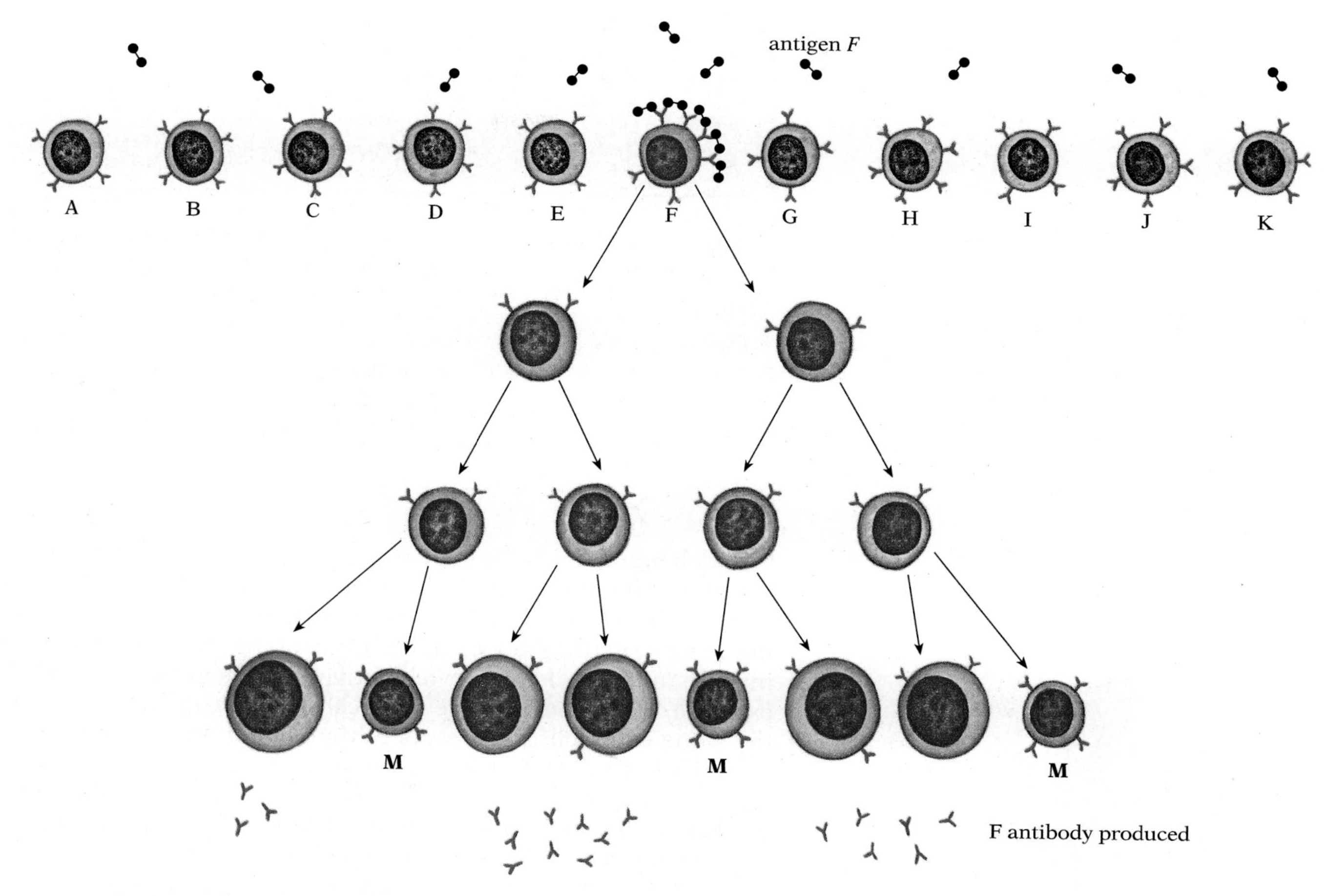

15.10 Clonal selection

Antigens are able to react with the membrane-mounted antibodies of only one or a few very specific lymphocytes from among the billions of kinds of lymphocytes in the organism's body. In this example antigen *F* can be bound only by the B lymphocyte of type F; it does not affect the other lymphocytes (top row). Lymphocyte F, stimulated by the binding of the antigen, proliferates to form a clone of genetically identical cells. Some cells of the clone are memory cells (M); others are plasma cells that actively secrete F antibody. The cross-linking of the antibodies on lymphocyte F greatly facilitates the proliferation of the F clone.

The cross-linking of receptors (which, by requiring at least two simultaneous antigen-binding events, serves to reduce false alarms, and probably explains why haptens cannot stimulate virgin cells) appears to be the event that induces the lymphocytes to begin dividing (Fig. 15.10).

■ WHAT THE CELL-MEDIATED RESPONSE ACCOMPLISHES

Another set of immune reactions is mediated by ***T lymphocytes***. One kind—the ***cytotoxic T cell***—kills infected cells of an individual's own body before the pathogen (usually a virus) can spread further. Cytotoxic cells, sometimes called killer T cells, are often confused with NK cells; the distinction between the two is made clear in Table 15.1. However, the B-cell and T-cell systems are not

Table 15.1 Simplified outline of immune-system cells

FUNCTION	B-CELL SYSTEM (HUMORAL RESPONSE)	T-CELL SYSTEM (CELL-MEDIATED RESPONSE)
Early-warning	Mast cells	Presentation cells
Diversity generation	Virgin B cell	Virgin T cells
Modulation		Helper T cells
Effector systems	Complement cascade	Cytotoxic T cells
	Natural killer cells	
	Macrophages	

as independent as Table 15.1 suggests. For example, other sorts of T cells modulate the activity of both humoral and cell-mediated systems, accelerating the response at the outset and later preventing it from getting out of hand. Controlling the immune reaction involves tuning a lymphocyte's activity and is antigen-specific. For example, a particular clone of T cells will manage the activity of the clone of B cells that reacts to the antigen in question. Each of these jobs requires T cells to recognize specific antigens, but to interact only with other cells of the body; free antigens or antigenic markers on pathogens are ignored. The mechanisms by which this dual recognition is accomplished will become clear when we look at the structure of the T-cell receptor.

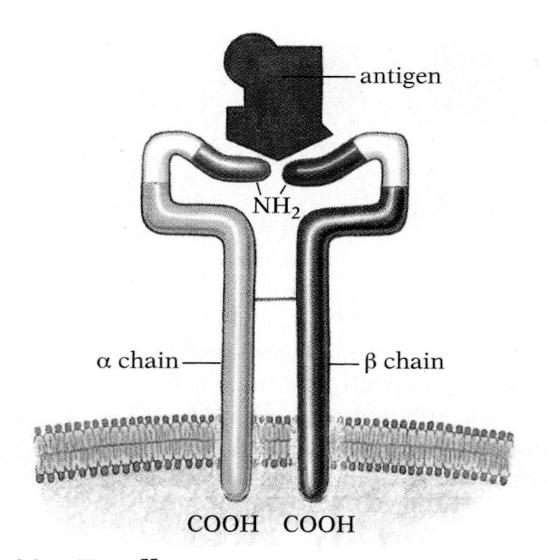

15.11 T- cell receptor

T-cell receptors consist of two peptide chains, each with its own sequence. Each chain has a relatively constant section (gray) and a highly variable region (purple) to which an antigen can bind. Another section in each chain (white) binds to a portion of complementary MHC molecules (described later in the text).

The T-cell receptor The receptor molecules of T lymphocytes are not secreted, but remain instead firmly attached by their tails to the lymphocyte membrane (Fig. 15.11). Like the B-cell antibody, each polypeptide arm of the T-cell receptor has a constant region at its base and a variable section farther out. The cleft lies between the two variable-region arms, which cooperate to bind the antigen. Like B cells, each T cell produces receptors specific to only one antigenic determinant, and nearly every T cell has a unique specificity.

T-cell receptors differ from B-cell antibodies in that they can bind only one antigen at a time. In addition, each arm has a region that is used to bind to cell-surface markers on other cells within the organism's body. It is this binding that keeps T cells from duplicating the work of B cells. The "self" markers involved in this recognition are membrane-mounted proteins produced by the genes of the ***major histocompatibility complex (MHC)***. As we will see, the MHC molecules are active participants in the T-cell response.

Self-protection: the MHC system There are two general types of MHC molecules: MHC-II proteins are found on the membranes of B cells and certain immune-system cells which are specialized for antigen presentation and are located in the tissues; MHC-I proteins are found on all other cells in the body. This dichotomy reflects the dual role of T cells as modulators of B-cell activity and assassins of disease-infected cells: cells bearing MHC-II proteins participate in regulation of the immune system, while those with MHC-I molecules can be killed by cytotoxic T cells.

MHC molecules, like the immune-system antibodies and receptors we have already discussed, are composed of two chains, and involve both constant and variable regions (Fig. 15.12A–B); T-cell receptors probably evolved from MHCs. MHC molecules bind antigens and then "present" them on the surface of the cell for binding

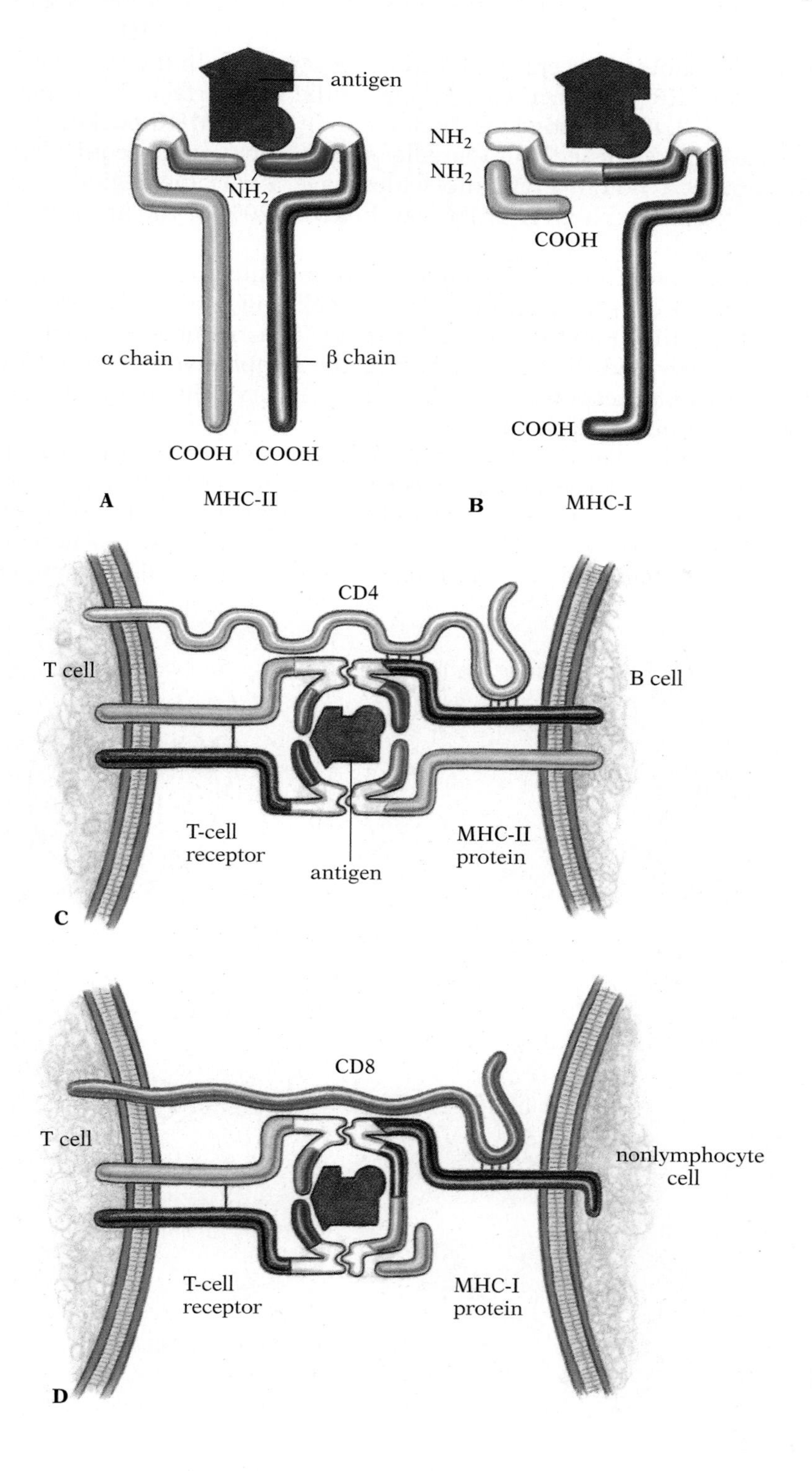

15.12 Structure and function of MHC molecules

(A) MHC-II molecules are similar to T-cell receptors: they have a pair of distinct chains, each with a constant region (gray), a variable section that binds an antigen (color), and a portion (white) that is complementary to a corresponding T-cell receptor (C). (B) MHC-I molecules are functionally almost identical to MHC-II, though structurally they are dissimilar. They too bind antigen (though the two variable domains are on the same chain) and a matching T-cell receptor (D). (C–D)When either MHC molecule binds to a T-cell receptor, the antigen-binding portions of the T-cell receptor and MHC protein each interact with different parts of the antigen they hold in common. The binding is stabilized by a cluster determinant protein (CD4 or CD8).

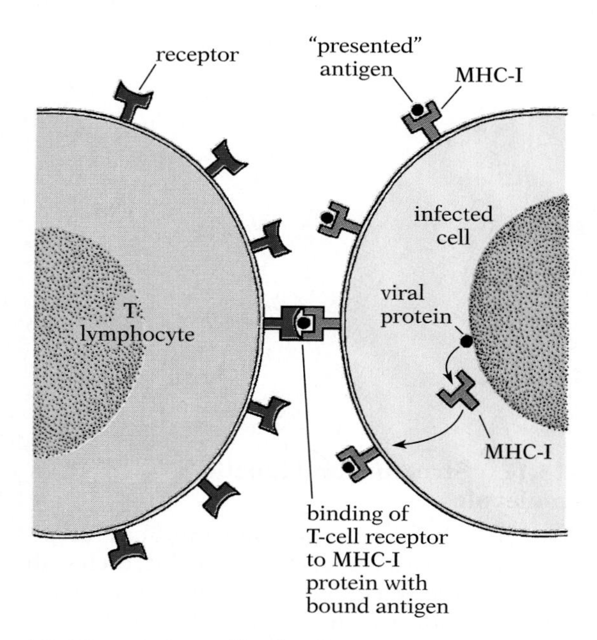

15.13 Antigen display by MHC-I molecules

In this model, MHC-I molecules in the cytoplasm of a cell bind antigens and carry them to the membrane for display. A T cell with a specific affinity for that antigen will bind to it and the MHC-I simultaneously, and become activated.

by appropriate T-cell receptors (Fig. 15.12C–D). This binding is then stabilized by a specific class of glycoproteins called cluster determinants: CD8 for MHC-I, and CD4 for MHC-II.

How the cell-mediated response develops Like their humoral counterparts, T lymphocytes begin as antigen-specific virgin cells. The first step in their activation occurs when MHC-I molecules in an infected cell begin presenting pathogenic antigens. The antigens are usually viral-coat proteins synthesized by the infecting organism in preparation for its further spread in the host. MHC-I molecules probably intercept these foreign compounds in the cytoplasm or on the ER, and then carry them to the cell surface for display (Fig. 15.13). If the antigen is too large to fit in the MHC pocket, it is "processed" (digested) into smaller pieces functionally equivalent to haptens. Though MHC molecules have a variable region, it is not very specific: a given MHC can bind 10–20% of the antigens it encounters.

Once the MHC-antigen complexes are mounted on the surface of an infected cell, a matching virgin T cell will eventually bind to both the antigen and the MHC-I protein. Thus stimulated, a virgin T cell grows and divides to produce the lymphocytes involved in the immediate response, as well as the memory cells that make future reactions more rapid.

The simplest variety of activated T lymphocyte is the cytotoxic T cell mentioned earlier. It destroys cells bearing the MHC-I–antigen complex to which its receptor can bind by perforating the infected cell (Fig. 15.14), which allows water to enter and burst it. Another class of specialized T lymphocytes—the helper cells—react to

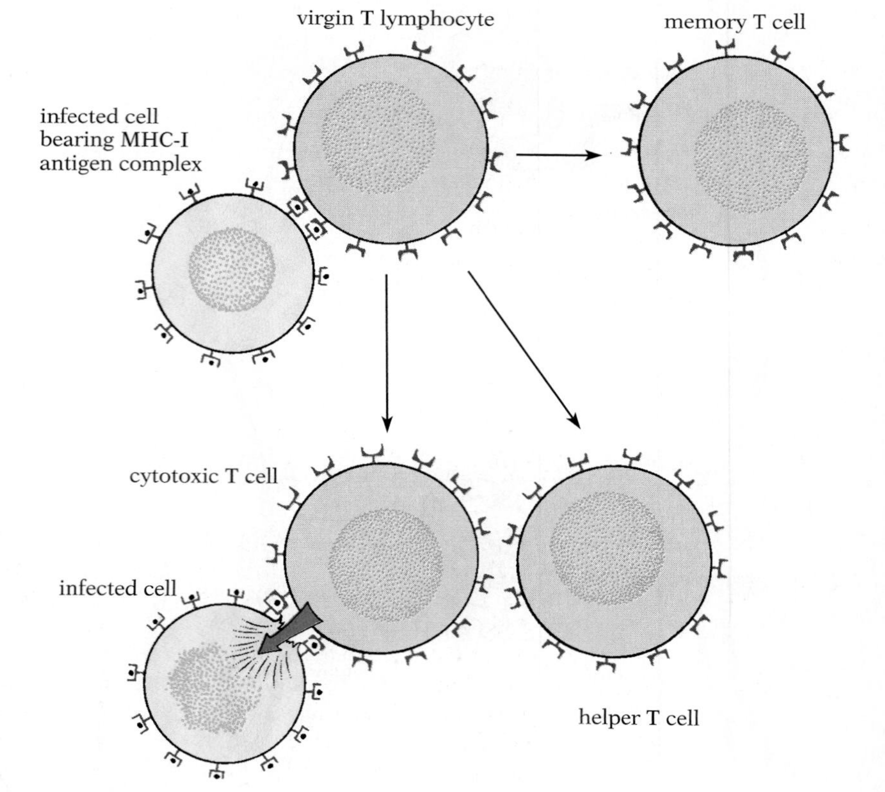

15.14 Cell-mediated immune response

When a virgin T lymphocyte binds to an infected antigen-bearing body cell, it becomes activated, grows, and begins dividing to generate memory cells (which permit faster responses when the antigen is encountered again), cytotoxic T cells (which kill infected cells), and helper T cells (which regulate the responses of B lymphocytes, cytotoxic T cells, macrophages, and other elements of the immune system to the same antigen).

MHC-II–antigen complexes, and are involved in modulating the immune response.

■ HOW T CELLS FINE-TUNE THE RESPONSE

So far we have seen several strategies for capturing antigens: cells with MHC-I complexes collect antigens found in the cytoplasm; the immune-system cells with MHC-II proteins actively import antigens, process them if necessary, and then display them on the MHC-II complex; B cells use their membrane-mounted antibodies to collect specific antigens (Fig. 15.15A); and T cells use the T-cell receptor. ***Presentation cells***, whose structure we have not yet explored, have neither receptors nor antibodies to capture antigens. These curious cells disperse from the bone marrow to the tissues, differentiate in tissue-specific ways, and take up their role as antigen presenters. They continually and unselectively endocytose material from the surrounding tissue fluid, process it, and then display the fragments on their MHC-II molecules.

Helper T cells begin their modulation of the immune response by binding to MHC-II proteins displaying the appropriate processed antigens on the cell surface of other immune-system cells (Fig. 15.15B). However, helper T cells do not attack the antigen or its source directly; instead, once bound to B cells or presentation

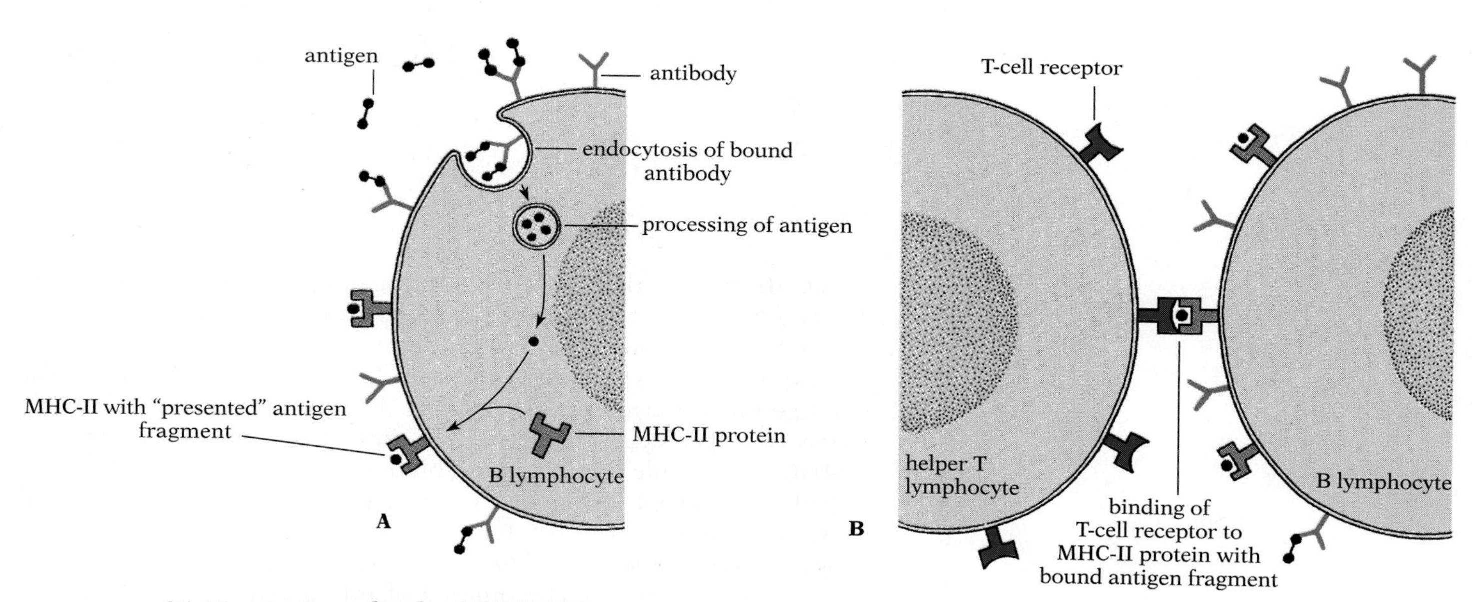

15.15 Antigen display by MHC-II molecules

MHC-II molecules are found in B lymphocytes and presentation cells. (A) When the membrane-mounted antibodies of a B cell (shown here) or the receptors of a cytotoxic T cell bind their particular antigen, the antibodies are taken in through endocytosis. Antigens are removed in special proteolytic (protein-digesting) vesicles and, if necessary, broken into convenient lengths; these fragments are then bound by MHC-II molecules in the cytoplasm and displayed on the cell surface. (B) When a helper T cell specific for that antigen is encountered, the two lymphocytes bind and the T cell becomes active. (Activation by a presentation cell, not shown, involves nonselective endocytosis of intercellular material, processing, and display on its MHC-II molecule.) This system of indirect activation is essential if the immune reaction is to be accurately modulated.

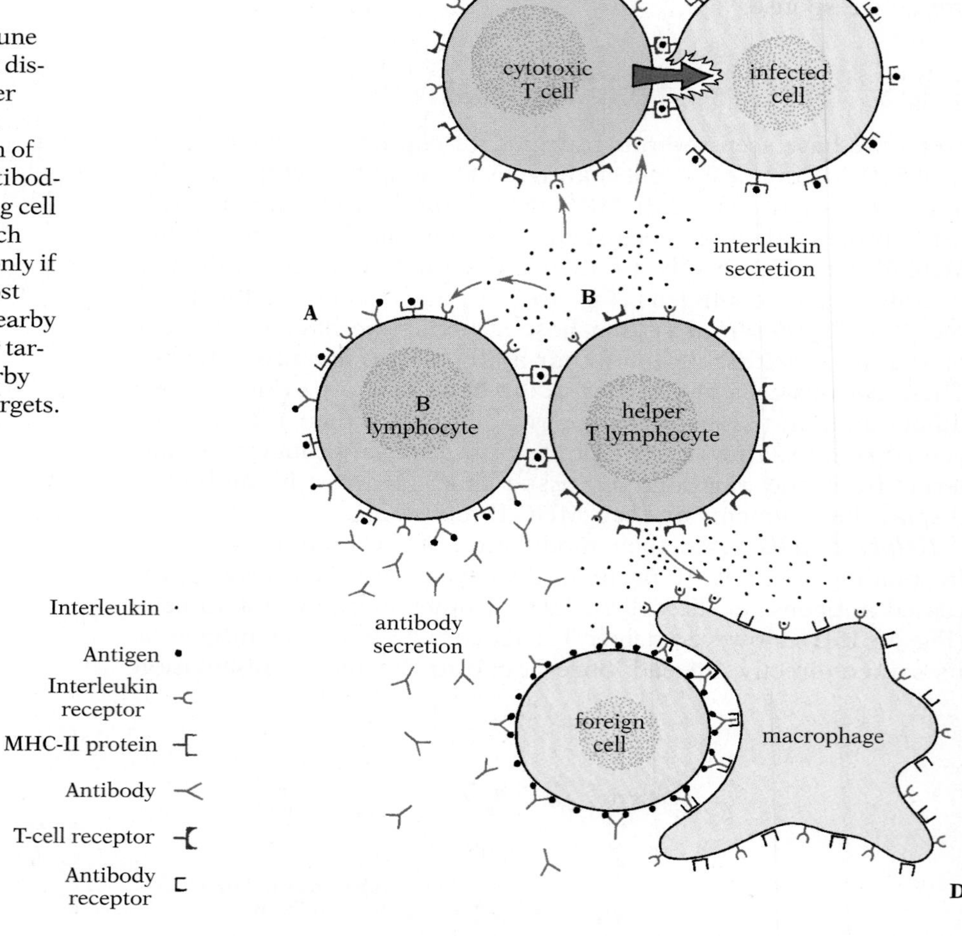

15.16 Modulatory action of T lymphocytes

(A) Helper cells regulate the humoral immune response by binding to B lymphocytes that display a unique antigen. (B) The bound helper secretes various interleukins, one of which causes the helper to multiply; another form of interleukin induces the B cell to secrete antibodies, which bind the antigens on the invading cell (A). (C) In the cell-mediated response (which normally would be active simultaneously only if the invading microorganisms can infect host cells), a third kind of interleukin induces nearby bound cytotoxic T lymphocytes to kill their targets. (D) A fourth interleukin activates nearby macrophages to ingest antibody-marked targets.

cells displaying the appropriate antigenic determinant, they help orchestrate the activation of these elements of the system.

Helper cells become aware of the presence of an antigen when it is exposed in the binding cleft of an MHC-II protein. Whether the antigen is displayed by a B cell (Fig. 15.16A) or a presentation cell, only a helper cell specific for the antigenic determinant *and* the MHC-II molecule holding it can bind and become activated. An activated helper can advance the campaign against the antigen and its source in several ways. First it installs receptors for a chemical-signal molecule, ***interleukin***, in its own membrane; then it begins secreting interleukin. The binding of interleukin to its own receptors causes the helper cell to proliferate (Fig. 15.16B). The secreted interleukin also induces multiplication of any activated cytotoxic T cells nearby that have recently encountered their specific antigens; more often than not, the helper cell and its cytotoxic neighbor will be responding to the same pathogen (Fig. 15.16C).

When bound to stimulated B lymphocytes, helper T cells produce an interleukin that encourages the B cells to secrete antibodies (Fig. 15.16A). Yet another kind of interleukin energizes nearby macrophages (Fig. 15.16D).

This two-step activation mechanism—in which B lymphocytes and cytotoxic T cells must both bind an antigen *and* be induced by a helper T cell specific for the same antigen—is probably a safety feature designed to prevent the potent cell-destruction capacity of the immune system from making mistakes. In particular, this double-checking strategy prevents the immune system from erroneously attacking the organism's own proteins, which can lead to a slow and sometimes fatal process of self-digestion. We will look at the problem of recognizing and ignoring friendly molecules in the next section.

One consequence of over stimulation by helper T cells is an excess of antigen-specific mast cells, and therefore a hypersensitivity to the antigen in question (Fig. 15.17). Even a slight exposure to the antigen under these conditions can lead to a massive release of histamine, triggering excessive loss of fluid from the blood. The result is an allergic reaction that, if extreme, can cause anaphylaxis: loss of consciousness as blood pressure falls, and even asphyxiation as fluid-induced swelling in the throat flattens the trachea (windpipe).

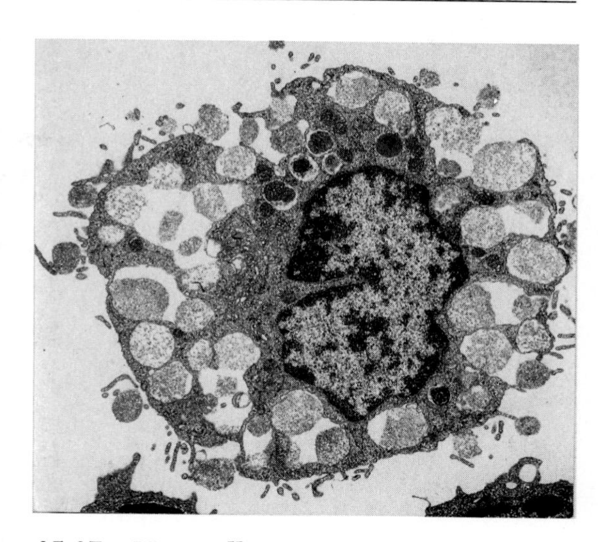

15.17 Mast cell

Massive exocytosis of histamine by a mast cell from a rat.

■ INITIAL CALIBRATION: RECOGNITION OF SELF

While the immune system of an organism is able to recognize and destroy almost any foreign antigen or antigen-bearing cell, it must not confuse its self antigens—the cell-surface proteins on its own cells—with foreign antigens. This recognition of self, or self-tolerance, is learned early in the embryonic development of the immune system: if a piece of tissue from one organism is transplanted into or grafted onto another adult animal, the recipient's immune system will almost always reject the tissues by mounting an immune reaction against the donor's cell-surface proteins. However, foreign tissue transplanted prenatally, while the recipient's immune system lacks experience, will be accepted; transplants from that same donor will be accepted later, even when the recipient is an adult.

Self antigens are generally membrane-anchored glycoproteins of the glycocalyx (see pp. 114–115). Some are specific to particular tissues, and seem to play a role in cell-type recognition and cell-to-cell adhesion. Others are encoded by the MHC, which devotes at least seven genes to the task, each of which can exist in as many as 100 versions. The diversity of MHC proteins is important if they are to bind and display the full range of processed antigens; it also has important consequences for cell recognition.

Inactivation of virgin cells Self-tolerance originates during fetal development when a process of selective lymphocyte inactivation takes place. Millions of virgin B cells are released into the prenatal blood at a time when no foreign antigens are present. Because any cells whose antibodies bind during this stage of development must be attaching themselves to normal proteins, the immune system is designed to inactivate B cells that bind prenatally. This still leaves vast numbers of virgin cells specific for nonself antigens capable of proliferating and mounting an immune response should the need arise.

The inactivation of T cells is more complicated. A T cell must be repressed not only if the receptor binds to a normal protein present during fetal development, but also if it *fails* to bind to one of the organism's MHC complexes. Some such T cells are actually killed, in a process known as ***clonal deletion***. Inactivation or deletion is essential if T cells are to perform their normal role of killing infected cells and modulating the activity of other immune-system cells.

Inactivation is an ongoing process that depends on the continued presence of each self antigen. Even with continued exposure, however, inactivation may not be permanent: should control over inactivation be loosened, as by infection with certain bacteria or viruses, an ***autoimmune disease*** can result. One example is myasthenia gravis, a neurological disorder involving the loss of tolerance of the body's billions of receptors for acetylcholine, a chemical that is used for communicating between nerve cells and muscles. The result is a debilitating deterioration of muscle control. Multiple sclerosis, rheumatoid arthritis, and Type I diabetes are among the more familiar autoimmune diseases.

Why transplants fail When a foreign cell is introduced into an organism as part of an organ transplant, it bears several different MHC-I proteins on its surface. The diversity of MHC proteins plays an important role in cell recognition, as we have said. Most of the MHC variants on the cells of the transplant are novel to the host's immune system, and so B-cell antibodies will bind to them and trigger a humoral response. At the same time, the overwhelming odds are that the foreign cell and the host share at least one MHC-I variable region; as a result, at least one class of T-cell receptor is able to bind to one region of this one kind of MHC protein. Unfortunately, the rest of that foreign MHC molecule almost certainly contains novel regions, and thus is seen as an antigen by the T cell; as a result, a cell-mediated response is also triggered.

The resulting two-pronged immune reaction leads to transplant rejection. This attack on transplanted tissue can be suppressed with certain drugs (like cyclosporin, which blocks an enhancer of interleukin genes), but not without cost: the recipient, whose entire immune system is now depressed, is at risk even from common colds. A recipient's lymphocytes may, with the passage of time, come to learn to ignore the foreign antigens, and then immunosuppression therapy can be relaxed or discontinued.

■ SUBVERTING THE IMMUNE SYSTEM: AIDS

The adaptive value of the vertebrate immune system is graphically illustrated by ***acquired immune deficiency syndrome (AIDS)***, a disease that destroys most of the immune response by eliminating the all-important helper T cells.

The AIDS virus The virus that causes AIDS (human immunodeficiency virus type 1, or HIV-1) is a complex retrovirus (Fig. 15.18). It consists of two copies of its RNA genome, each coated by a pair of proteins (called P7 and P9), and all enclosed by two very differ-

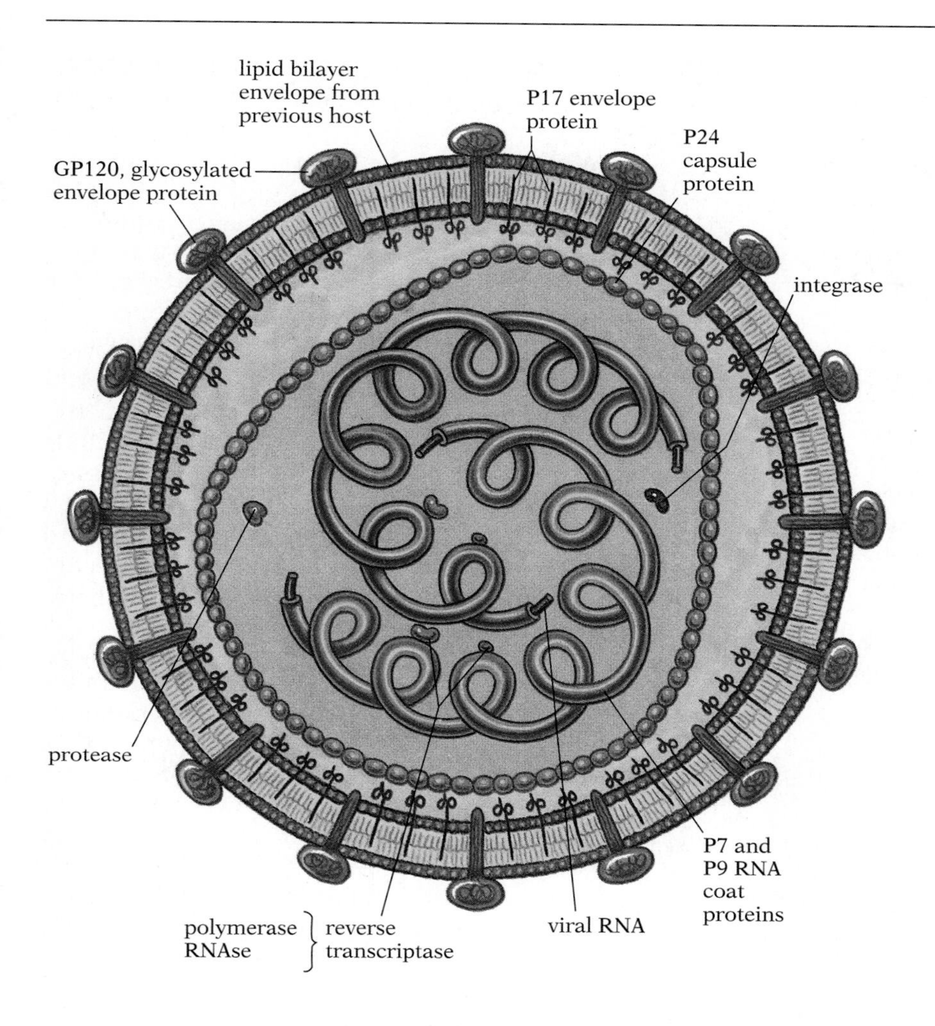

15.18 Structure of HIV

The two copies of the viral RNA are protected by a protein coat and enclosed in a capsule. Also contained in the capsule are reverse transcriptase enzymes (which convert the single-stranded RNAs into double-stranded DNA copies), integrase (which inserts the DNA version into the host genome), and protease (which catalyzes the budding off of the virus). The capsule is itself enclosed in a bilayer membrane obtained from the previous host cell; in the membrane are mounted P17 (the remains of a protein involved in budding off) and GP120 (the glycoprotein that binds the helper T-cell receptor and so enables the virus to gain entry into its host).

ent layers. The inner layer is a self-assembling structure of one kind of protein (P24). The outer envelope is more complex: its primary constituent is a lipid bilayer obtained from the host cell as the virus buds off (Fig. 15.19), but embedded in this vesicle are one sort of protein facing inward (P17) and a glycoprotein (GP120) directed outward. HIV locates and enters its host by means of GP120, which binds specifically to the CD4 protein of helper T cells.

The HIV-1 retrovirus carries four enzymes. The first and second work together as the reverse transcriptase: a polymerase synthesizes a complementary DNA copy of the single-stranded RNA genome, and an RNAse separates the viral template from the complementary DNA allowing the polymerase to complete the job of making the HIV DNA double-stranded for insertion into the host genome. The third enzyme, an integrase, cuts the host DNA and inserts the parasite's own instructions into the genetic library. The fourth enzymatic passenger, a protease, activates certain viral proteins when needed.

The viral genome contains several regulatory sequences. The virus can exist in a low-impact, wholly lysogenic or semilytic phase, during which it produces a slow stream of progeny. As with

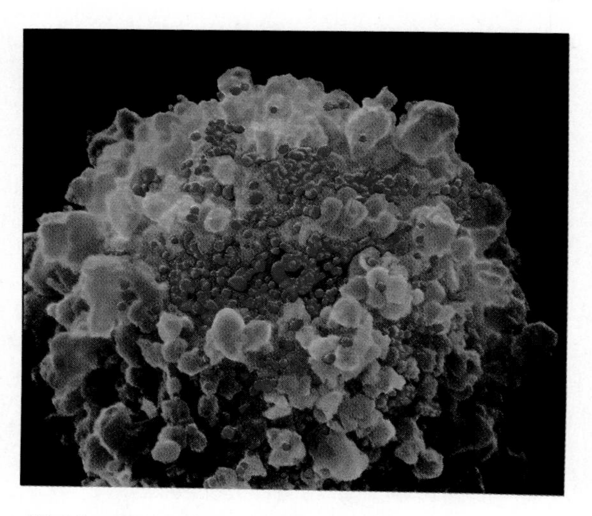

15.19 HIV viruses budding from a cultured lymphocyte

EXPLORING FURTHER

MONOCLONAL ANTIBODIES

A growing understanding of the workings of the immune system has enabled researchers to develop techniques for learning about cellular architecture—approaches that can provide new medical applications while they help reveal the structure of the cell. In one such technique, researchers select and clone a single type of lymphocyte—typically a B lymphocyte—whose antibodies bind to an antigenic region on a particular kind of cell, a specific structure in a cell, or some other particular substance. This ***monoclonal antibody technique***, as it is called, can be used to locate all the tubulin in a cell, or the sodium-potassium pumps, the F_1 complexes of mitochondria, the RNA polymerases, or any of the thousands of enzymes whose distribution and function in a cell are not yet well understood.

To clone antibodies specific to a particular substance, researchers first inject that substance into mice (see figure). Soon there is a substantial increase in the B lymphocytes producing various antibodies specific to the many different antigens on the foreign cell or chemical. Most of these lymphocytes are producing antibodies to antigens common to many different foreign cells or chemicals; however, a few may, by chance, be producing antibodies specific to an antigen unique to the foreign material. If just these lymphocytes could be selected and cultured, they would produce a supply of the single antibody desired.

There are two major problems: the specific antibody-producing cells must be separated from the others, and they must be propagated. In the actual cloning procedure, the second of these problems is solved first. Normal cells have a set number of cell divisions, after which they age and die, while cancer cells are immortal. If the lymphocytes were cancerous, they would multiply indefinitely. Hence the next step is to force cells of the mixed collection of lymphocytes to fuse with cells from a cancerous line of B cells. Those that fuse become immortal, and are called hybridomas.

This still leaves the problem of selection. It is necessary to eliminate all but the hybridoma cells, and then to select those producing the desired antibody. The elimination is relatively easy. The nonhybrid lymphocytes from the mouse simply die off after completing their set number of cell divisions. The more difficult problem of sorting out the nonhybrid cancerous lymphocytes is accomplished by using a line of cancer cells lacking a crucial synthetic pathway. When these cancer cells are cultured in a medium without the substance they cannot make, they die. Only the hybridomas will survive, since they are able to synthesize the missing nutrient because of genes from the mouse lymphocytes, and are immortal by virtue of genes from the cancer cells.

The next step is to separate the individual hybridoma cells, to culture each one separately, and then to test each culture by adding the antigen. If the cells of a particular group are producing the desired antibody, the test will result in antigen-antibody clumping. These cells can then be cultured further, to form a clone of immortal cells producing a continuous supply of a specific antibody. Researchers have even been able to

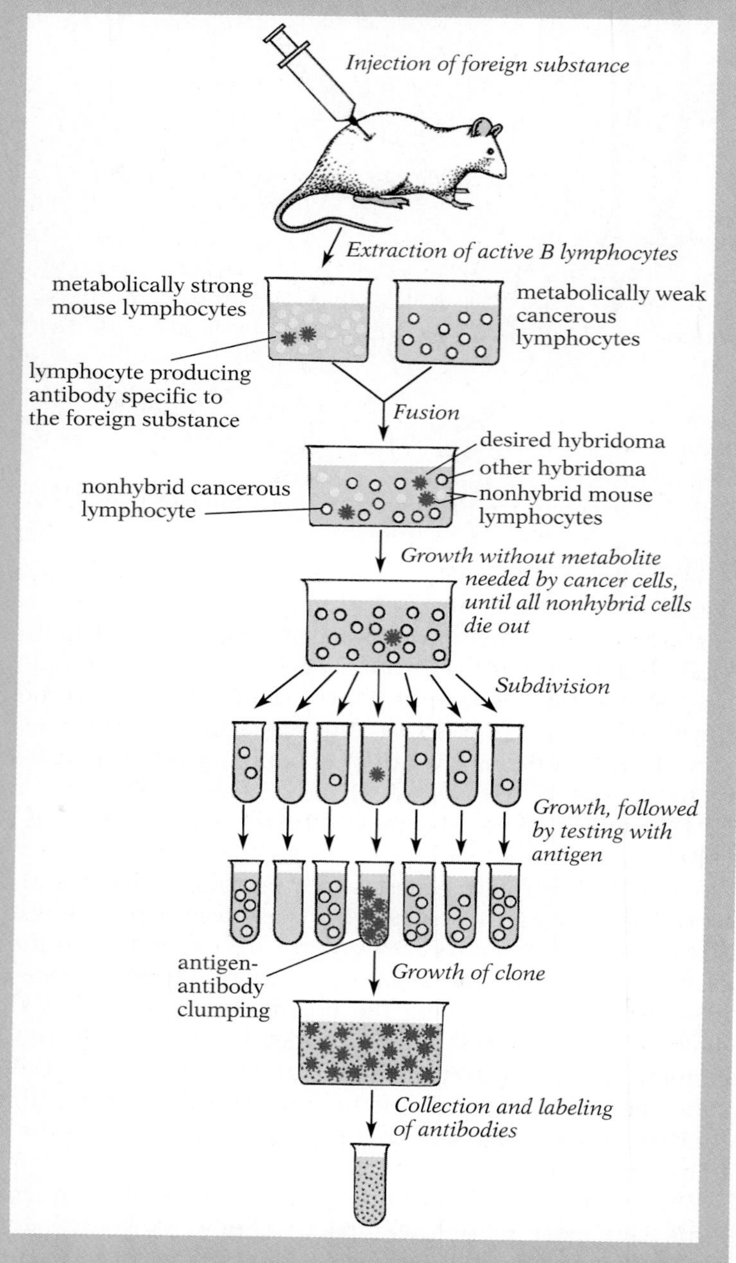

transplant the antibody genes from certain hybridoma cells into plants, and (via the phage lambda) into *E. coli*, where these immunoglobins are produced far more rapidly and cheaply.

Researchers use monoclonal antibodies to tag specific cells, cellular structures, or chemicals by labelling the antibodies with fluorescent dye, the electron-dense substance ferritin, or a radioactive marker. Once a labelled antibody has reached its target, the location of the antigen can be found with a light microscope, an electron microscope, or any of a variety of radioisotope analyses. A labelled antibody can even be used to mark certain kinds of cancer cells selectively for destruction; ovarian cancer cells, for example, can be bound by ferritin-linked antibodies and then destroyed by ferritin-specific T cells. The monoclonal antibody technique, then, has enormous potential both for basic research and for medical technology.

more conventional viruses, induction into the fully lytic state seems to require a shock or threat to the host cell, including, for example, its participation in an immune response. The virus reproduces both by replicating its entire genome and by producing mRNAs that subsequently direct synthesis of the envelope protein GP120, the "P" proteins, and the integrase and protease. The proteins and RNA aggregate in the host membrane, with only the GP120s facing out (Fig. 15.20). When a sufficient collection of membrane proteins accumulates, the membrane begins to bud, and the protease activates the enzymes destined to be packaged with the virus, as well as the proteins of the coat, or ***capsid***. Once the capsid is formed, the virus buds off and is free to bind another host cell.

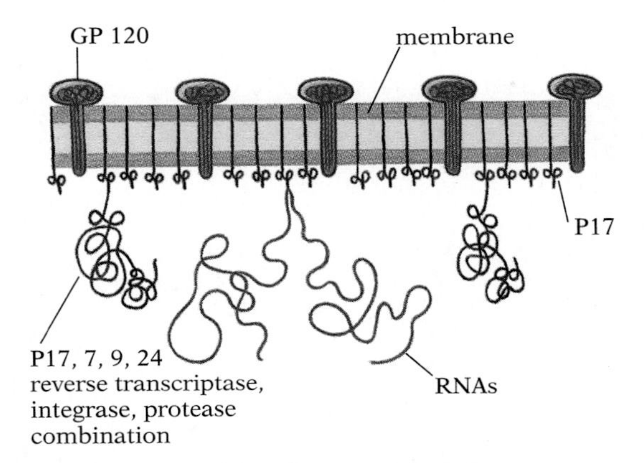

15.20 Assembly of a new HIV virus
Assembly occurs on the host-cell membrane where viral RNA, GP120, the P-protein precursor, and the long P-protein–enzyme precursors aggregate. When the collection of parts is sufficient, the piece of membrane begins to bud outward. Then a protease is freed and quickly cuts out the many enzymes and P-proteins, allowing coating of the RNA and capsid formation, steps that lead to complete budding off.

How AIDS attacks the immune system Initially, the immune system responds normally to HIV, with B cells releasing suitable antibodies, helper T cells amplifying the response, and macrophages consuming free viruses. But the virus lives on, hidden in helper T cells and, undigested, in macrophages. Though the total population of T cells remains normal for about a year (on average), infected cells slowly begin to be recruited into the lytic or semilytic phase. As GP120 antigens begin to appear on the surface of T cells, these hosts are destroyed by cytotoxic T cells and NK cells, but this is not of any real benefit. Not only is the body's arsenal of helper T cells being culled (dropping, on average, to 25% of normal 3 years after infection); GP120s are released into the blood and lymph, where they bind to other helper cells, which, though not actually infected, are then attacked and killed. Worse, an infected cell with GP120 on its membrane can bind to other T cells; just as the viral envelope fuses with the host, the membrane of the infected cell fuses with the bound cell. Dozens or hundreds of T cells are drawn into this fusion event, with the result that all die together, either at the hands of killer or cytotoxic cells, or because the resulting multinucleate cell becomes too large to function.

AIDS also affects the nervous system. Macrophages can cross from blood vessels into the nervous system where, as the viruses they harbor escape, HIV infects the glial cells that insulate neurons. As the insulating cells die, conduction becomes slower and less efficient, and the accuracy with which new neurons are routed to their targets declines.

Things go from bad to worse. Soon the number of helper cells is down to 5% of normal or below, and other infections that come along face only a minor immune response. AIDS patients generally die 5–10 years after infection, often of diseases that pose no particular threat to humans with healthy immune systems.

The epidemiology of AIDS Studies of how AIDS spreads indicate that HIV is almost always passed from the blood or seminal fluid of one individual to the circulatory system of another. Hence the disease spreads readily through transfusion of infected blood, reuse of hypodermic needles, and anal intercourse. The virus is also transmitted, though at a lower rate, through vaginal intercourse.

The spread of AIDS (like any epidemic) depends on the average incubation period of the infective agent, the length of the infectious period, the rate of contact (most often, in this case, the number of sexual partners or shared needles), and the efficiency of transferring the infection with each encounter. Most of this information can now be estimated for the United States. It appears that the average time from infection to death is about 8 years (the mortality rate as of 1995 was essentially 100%), and carriers are infectious throughout this period, though more so early and late in the cycle.

Among American homosexuals and intravenous drug users, the infection rates are as high as 70% in urban areas like San Francisco and New York City, where contact rates are very high; on average, each infected individual communicates the disease to several others. In less urban regions, where the contact rate is lower and the susceptible pool of potential victims smaller, the spread is slower. Overall rates of transmission in the United States are declining primarily because a large portion of the high-risk group is already infected; several million Americans are thought to carry the virus.

Among heterosexuals, both transmission rates and contact rates are lower. It is not clear whether there will be an epidemic among heterosexuals in Europe or the Americas: the statistic that defines an epidemic is whether the typical carrier will infect, on average, more than one other individual. In Africa, HIV transmission is primarily heterosexual, and risk correlates with the number of sexual partners, which is larger on average than in most other parts of the world. The World Health Organization estimates that nearly 60% of all AIDS deaths, which totalled 3.6 million through 1995 (and will rise beyond 20 million by the end of the decade), have occurred in Africa, where millions more are infected. Clearly a heterosexual epidemic *is* possible.

Can AIDS be cured? Early attempts to combat AIDS suffered from a lack of knowledge about the mode of transmission of this disease, its long latency, and ignorance of its elaborate life history. Now that more is known, one thing is immediately obvious: though nearly all conventional vaccines use a surface antigen as the means for teaching the immune system to recognize a particular disease, this approach will not work with AIDS since a surface-antigen vaccine merely prepares the body to destroy its own immune-system cells. Instead, some step in the life cycle must be blocked. Preventing binding by masking the CD4 molecules is useless, since this would inactivate the immune response. Blocking the GP120 active site is a possibility—unless, of course, it turns out to be an analogue of the MHC molecule to which the T-cell receptors normally bind, which seems all too likely.

Preventing a step in viral duplication inside the infected cell is more promising, though that step must be unique to the virus or the treatment will be toxic to the host. A good example of this trade-off is seen with 3'-azido-2',3'-dideoxythymidine (commonly called azidothymidine, or AZT; Fig. 15.21). This thymidine analogue is actually preferred by HIV's DNA polymerase over thymidine, whereas host-cell polymerases prefer conventional thymidine. When AZT is incorporated into a growing DNA chain, replication stops because the host-cell polymerase is unable to add the next nucleotide. At suitably low doses of AZT, host cells replicate fairly normally but HIV reproduces less successfully. There are side effects, however; these include substantial attrition in the bone marrow (where rapid production of blood cells is always under way) and often anemia. Nevertheless, AZT can delay, for about a year, the death of patients with clear symptoms.

One big risk with AZT is that the HIV polymerase gene, which has an unusually high error rate (one mispairing per 2000 bases in a genome only about 10,000 nucleotides long) will mutate to discriminate against AZT. The search for a more practical and effective treatment is being intensively pursued. An antisense RNA complementary to part of the HIV genome, for instance, would be a better treatment because it would be completely specific and nontoxic. To make it work, however, means must be found to deliver it efficiently to target cells and make it resistant to RNAse. Injecting the portion of CD4 to which the virus binds holds promise as well. At the moment, though, prevention through behavioral changes among those at risk seems the most practical alternative.

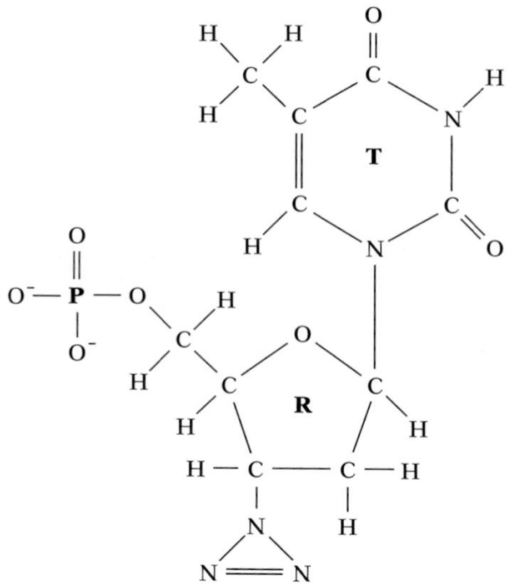

15.21 AZT

AZT is a DNA thymidine mimic. The base (T) is normal, but the deoxyribose unit (R) differs from DNA's sugar in that the hydroxyl group to which the next nucleotide is attached is replaced by N_3, which cannot be used by DNA polymerase.

THE SOURCE OF ANTIBODY DIVERSITY

■ SHUFFLING EXONS

How is the vertebrate genome able to produce an almost infinite variety of antibodies and T-cell receptors without employing millions or billions of different genes, one for each potential antigen? The answer to this question involves a remarkable strategy of exon selection, which operates initially during the fetal development of the immune system, and leads to the vast diversity of lymphocytes available to the organism during its lifetime. In B cells, the process involves selection from a pool of exons in only three genes—one for the heavy chains and one for each of the two classes of light chains. The strategy used for T cells is identical in most details.

Let's look at the chromosomes of a developing human B lymphocyte in the bone marrow before the cell specializes to produce a single type of antibody. In the human genome, the part of the DNA coding for the constant regions of the heavy chain is composed of 22 exons arranged in five sets, one corresponding to each class of heavy chain—A, D, E, G, and M. Each set contains exons for re-

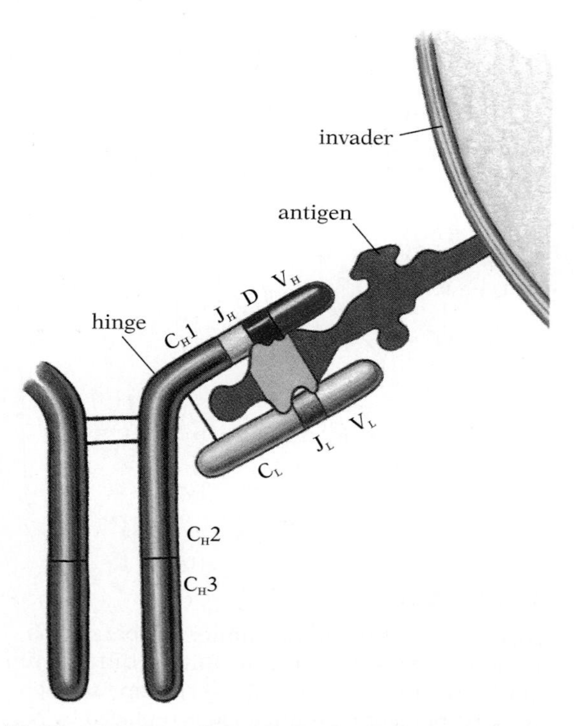

15.22 Antigen-binding site of an antibody molecule

The binding site is a pocket formed by the interaction of a heavy chain and a light chain. The sectors labeled C_H3, C_H2, hinge, C_H1, J_H, D, V_H, C_L, J_L, and V_L on this schematic representation are encoded by separate exons in the genes. The five different classes of heavy chains have different C_H2 and C_H3 regions; class E and class G chains have a fourth C_H region. Only one of the antigenic determinants of the antigen (beige) is bound by this antibody. Completely different antibodies may bind to separate sites on the same kind of antigen. (The left half of the antibody beyond the hinge is omitted.)

gions C_H1, hinge, C_H2, C_H3, and, in the sets that code for heavy chains of classes E and G, C_H4 (Fig. 15.22).

This arrangement of exons in the heavy-chain gene seems at first glance to be counterproductive, since so much has to be transcribed that is ultimately not used: thus, even if the virgin lymphocyte later produces only antibodies with class A chains, exons for the other classes would be transcribed as well. In the portion of the gene encoding the variable region, the organization seems even more counterproductive: there are four different exons for the J_H (joiner) region, even though each heavy chain has only one J_H segment. Still farther toward the 5' end, are approximately 12 different exons for the D region. Again, each antibody produced by a virgin cell has only one D segment. Still farther toward the 5' end lies a string of roughly 200 different V_H exons, only one of which is eventually translated.

How is it, then, that the roughly 240 exons in the original heavy-chain antibody gene give rise to the protein product encoded by only seven or eight exons? Exon removal seems to be accomplished in two stages. In the first step, which takes place during fetal development, large regions of the gene are cut out of the chromosome, and the remaining ends are spliced together. One such excision, for example, extends a random distance from the D region into the V_H exons (Fig. 15.23B). Any number of V_H exons from 0 to 199 may be removed. The first V_H exon not excised is the one ultimately trans-

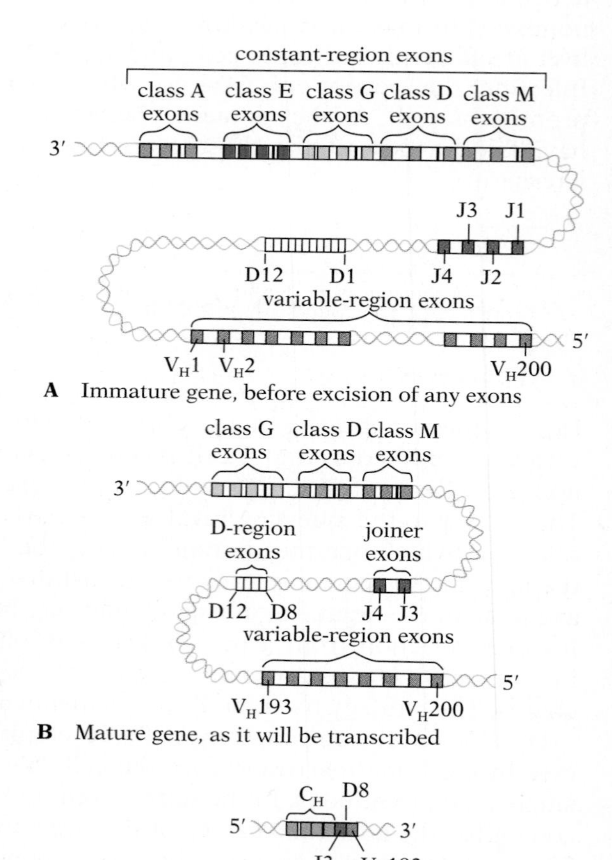

15.23 Organization of immature and fully mature heavy-chain antibody gene, and corresponding mRNA

(A) The immature heavy-chain gene originally contains five sets of constant-region exons (one set for each class of heavy chain), four different joiner exons, approximately 12 different D-region exons, and about 200 different variable-region exons. (B) When the gene is fully mature, many of the exons have been removed, and the remaining exons and introns are transcribed. (C) However, only the exon closest to the 5' end of the primary RNA transcript of each exon group survives processing and appears in the mature mRNA. The two steps of exon removal enable a B lymphocyte to produce a unique kind of antibody from millions of possible alternatives.

lated and expressed. During transcription, RNA polymerase copies all the remaining V_H exons on the 5' side, but these superfluous transcribed exons are removed during RNA processing to yield an mRNA with only the first V_H exon that survived excision. As a result, only the V_H exon closest to the D-region exon will be translated (Fig. 15.23C). All but one transcribed joiner exon and D-region exon are removed by means of two similar steps. A corresponding process is seen in the gene for the light chain, which has one C_L exon, four alternative joiner exons, and about 300 alternative V_L exons.

The diversity made possible by an active but random assembly of exons like the one just described is vast. For instance, since the gene for the heavy chain has four alternative joiner exons and 12 alternative D-region exons, there are 48 (4 x 12) possible joiner–D-region combinations. Moreover, since there are roughly 200 V_H alternative exons, a lymphocyte can produce antibodies with one of 9600 (48 x 200) forms of the variable region of the heavy chain. Similarly, the light chain, with its four alternative joiner exons and 300 V_L exons, can be produced in 1200 different forms. Since each antibody consists of one kind of randomly selected heavy chain and one kind of light chain, these different forms give rise to about 12 million (1200 x 9600) different antibodies during the development of the immune system. Additional variation is generated by a process of active mutation in the joiner and D-region exons and in parts of the V_L and V_H exons. Since the mutations take place in developing B cells, which are somatic cells, they are unique to each cell and are not perpetuated in the germ line.

■ HYPERMUTATION

As an immune response gets under way, the affinity of the B-cell antibodies produced by some cells in the activated clone actually increases. This fine-tuning involves yet another level of immunological learning. After a virgin cell is stimulated by an antigen, the exons for the variable regions of the antibody genes are replicated, during which time mistakes are permitted at a million times the normal rate. This error rate means that roughly half of the daughter cells will have slightly modified versions of the original antibody. Because the further cloning of cells from this first generation of daughter cells depends on how well each lymphocyte's version of the antibody binds the antigen on the cell membrane, a mutant with a better match will outreproduce its clone mates. As a result, the cells producing the better-binding antibody come to dominate the response. Here is a kind of natural selection in miniature.

HOW THE IMMUNE SYSTEM EVOLVED

The molecules of the immune system are related to cell-adhesion molecules (CAMs) described in Chapter 14. Of course, differential cell adhesion is critical for immune-system cells: not only must they attach themselves to their targets; lymphocytes must also be able to move to and from the blood, lymph, and any tissue that is

EXPLORING FURTHER

SLEEPING SICKNESS: HOW THE TRYPANOSOME CHANGES ITS COAT

Sleeping sickness is one of the most debilitating diseases and remains virtually impossible to treat. Its cause is a protozoan known as a trypanosome (see figure), which bloodsucking tsetse flies carry from an infected host to a potential host. The trypanosomes multiply in the blood of the host and later spread into the nervous system. Symptoms begin with fever and fatigue, and progress to drowsiness during the day and insomnia at night. The victim ultimately becomes too sleepy to eat, then comatose, and finally dies.

Since the trypanosomes are in the bloodstream of the host for weeks before causing death, how is it that the immune system, with its millions of antibodies generated by exon shuffling, fails to react effectively? The answer is that during these weeks the trypanosomes are shuffling their genes as well.

Trypanosomes are unusual in having only one kind of protein exposed on their cell membranes. Within a few days of infection the immune system of the host mounts a full-scale immune response. By the fifth day, however, all the surviving trypanosomes have changed their surface proteins. The trypanosome genome has hundreds—perhaps thousands—of surface-protein genes, and a new one is transposed into an active site in the genome roughly every 5 days. The number of trypanosome antigens increases by the fifth day from the few present at the time of infection to a great many, a different one on nearly every surviving trypanosome. As a result, the existing immune reaction becomes irrelevant, and a new one, responsive to a larger set of antigens, must begin from scratch: a new, different set of lymphocytes must now bind to the parasites, to begin again the process in which virgin B and T lymphocytes are activated and divide, antibodies are produced, and so on.

Again, even if the battle is being won, the surface-protein genes are shuffled. New surface proteins appear, and the number of new antigens multiplies. Eventually, the body's immune system is overwhelmed. Transposition in trypanosomes, then, seems to be a precisely controlled event. Transposons may be important elements in many diseases.

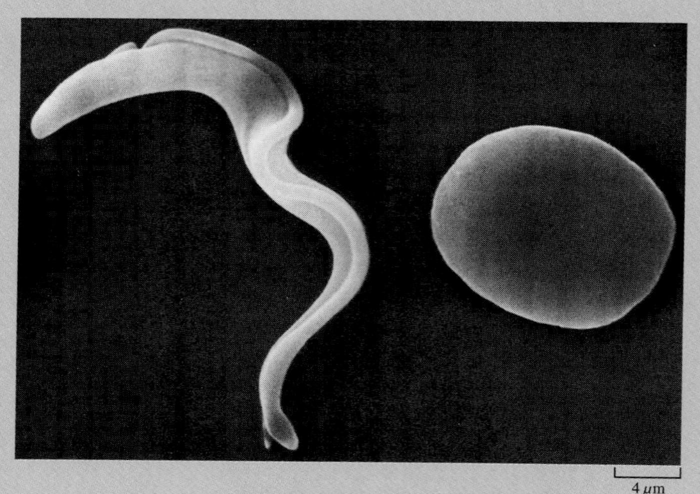

A trypanosome and a red blood corpuscle

releasing the chemical signals that draw these cells to the site of trouble. In fact, the molecules on which the lymphocytes rely evolved from proteins that animals have been using for hundreds of millions of years—long before the first immune systems appeared. For example, some vertebrate immunoglobulins are virtually identical to CAMs discovered in *Drosophila*, where they are involved in axonal guidance. As outlined in Chapter 17, the process involves repeated gene duplication, followed by the independent evolution of the different copies.

Let's look closer at a partial cast of these molecular actors (Fig. 15.24). One of the most telling similarities is between the light

15.24 Some molecules used by the immune system

Most of the proteins and protein complexes used by immunocytes for cell binding fall into three molecular "families," each of which is part of a larger family of developmental-control substances. The similarities are obvious at a glance: compare the structure of antibodies to MHC complexes, T-cell receptors, and CAMs. Other immune-system-specific members of developmental control families are shown in the lower half of the figure.

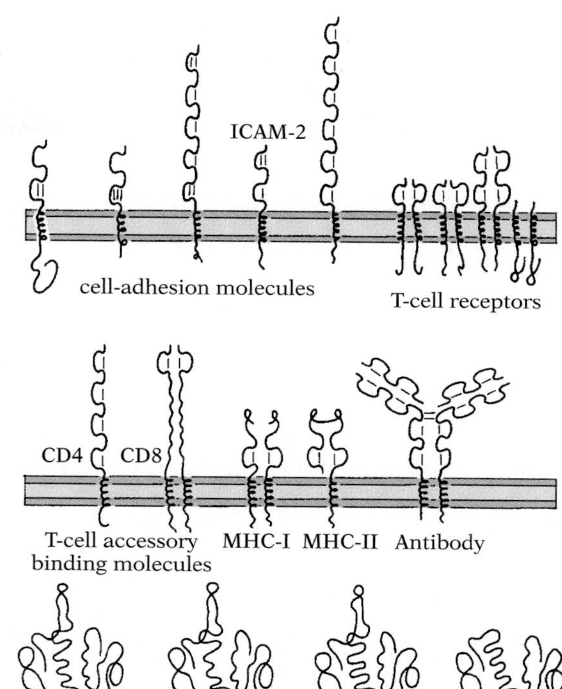

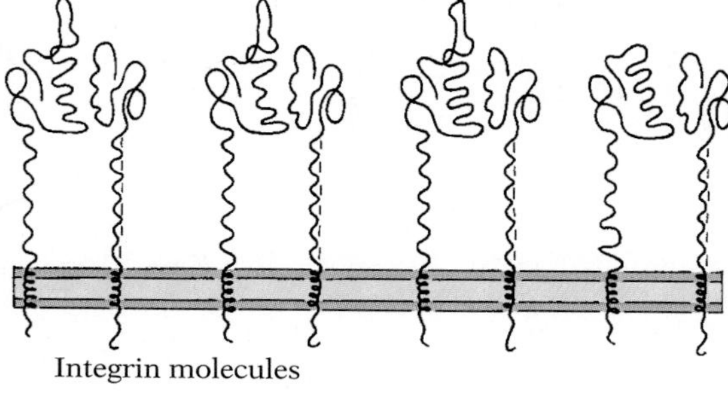

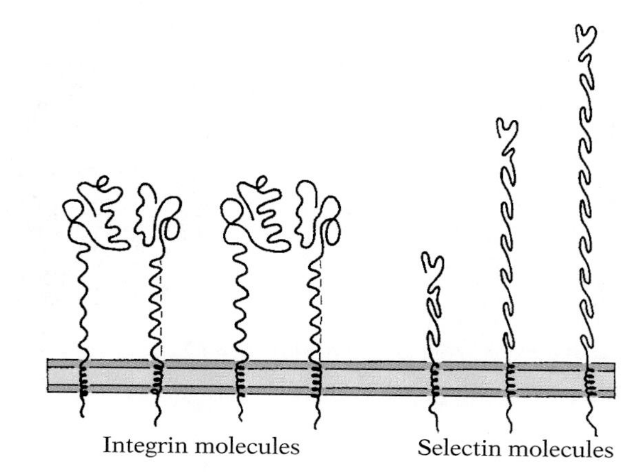

chains of antibody molecules, the α chain of the T-cell receptor, and the cell-adhesion molecule ICAM-2. This particular CAM is expressed on the membranes of cells in inflamed regions (in response to locally secreted chemicals like the interleukins) and thus encourages T-cell adhesion. ICAM-2 is also very similar to the MHC molecules, to CD2 (which helps T cells find antigen concentrations in lymph nodes), to LFA-3 (the molecule to which CD2 binds), and the β chain of the T-cell receptor. Except that they are missing a transmembrane anchor, the light chains of antibody molecules also strongly resemble the members of this ancient cell-adhesion family.

Three other components of the T-cell receptor (ε, γ, and δ, which communicate the fact of T-cell receptor binding to the interior of the T cell) are just truncated versions of ICAM-2, while CD8 is very like ICAM-2 with a long insertion. CD4, the heavy chains of antibody molecules, and two other immune-system CAMs are essentially ICAM-2s with a doubled outer segment.

Two other families of developmental CAMs are also involved in the immune response. The immune system's ***integrins*** help lymphocytes home in on infection sites, and modulate T-cell proliferation. ***Selectins***, for their part, regulate lymphocyte binding to tissue at sites of inflammation. In short, lymphocytes have evolved to use the selective CAMs of morphogenesis to track down infections, where they bind both normal cell receptors and unfamiliar antigens prior to exerting their specific effects. In a way, many lymphocytes are like undifferentiated stem cells in early development, migrating to their proper location by means of CAMs and chemical gradients. The cells of the immune system, however, come equipped with dozens of CAMs and other morphogenic binding molecules, and respond to any of a large number of gradients; hence, they have the potential to go anywhere in the body in search of cells broadcasting chemical cries for help. Selection has acted to transform the existing genetic systems of organismal development into a self-guided internal security system.

CHAPTER SUMMARY

HOW THE IMMUNE RESPONSE WORKS

THE CAST: CELLS AND ORGANS OF THE IMMUNE SYSTEM Antibodies bind to foreign antigens, marking them for engulfment by macrophages or destruction by killer cells; binding to antibodies on mast cells causes the release of histamine and other signalling molecules. These cells are found in both the bloodstream (particularly the spleen) and lymph (particularly the nodes). (p. 381)

WHAT THE HUMORAL RESPONSE ACCOMPLISHES The humoral response involves the B cells, each of which produces a unique antibody. When part of an antigen (an antigenic determinant, called hapten when it is isolated from the rest of the antigen) binds to a surface-mounted antibody on a B cell, the cell begins dividing and releasing antibodies into the blood. Some cells produced by this clonal selection specialize in antibody secretion while others become memory cells, which allow a faster reaction when the antigen is next encountered. Some of the secreted antibodies become mounted on mast cells. Others bind to the antigen directly, and trigger various reactions, including engulfment by macrophages, perforation by complement enzymes, and attack by natural killer cells. (p. 383)

WHAT THE CELL-MEDIATED RESPONSE ACCOMPLISHES The cell-mediated response involves the T cells. Each T cell mounts in its membrane a receptor molecule with unique specificity. The version of the receptor found on cytotoxic T cells can bind to haptens presented by cell-surface markers (MHC-I molecules) on the body's cells; the MHC-I molecules in each body cell constantly scavenge molecular fragments from within the cell and carry them to the surface for display. Cytotoxic T cells can kill infected cells displaying MHC molecules with antigen haptens. Clonal selection leads to the proliferation of T cells that bind antigen haptens. (p. 387)

HOW T CELLS FINE-TUNE THE RESPONSE Immune-system cells use their surface-mounted antibodies (B cells) or receptors (T cells) to bind circulating antigens; these are imported into the cell, where they are processed, and then displayed on the cell surface by the MHC-II molecules unique to the cells of the immune system. Helper T cells stimulate B cells with the same antigen specificity—that is, with MHC molecules displaying haptens to which the T-cell receptor can bind. (p. 391)

INITIAL CALIBRATION: RECOGNITION OF SELF Early in development, B and T cells that bind molecules in the body are deactivated or repressed (clonal deletion) to prevent an autoimmune response; at the same time, T cells that fail to bind the body's MHC molecules are deactivated. (p. 393)

SUBVERTING THE IMMUNE SYSTEM: AIDS The AIDS virus infects helper T cells and also survives engulfment by macrophages. As the virus slowly reproduces, it spreads to more T cells, eventually disabling the ability of T cells to stimulate immune responses. (p. 394)

THE SOURCE OF ANTIBODY DIVERSITY

SHUFFLING EXONS The number of possible antibodies is nearly infinite. Much of the diversity is accomplished by reassembling the antibody-coding region in each B and T cell as it matures, using a randomly selected exon from among many alternatives for each of several segments of the binding region of the antibody. (p. 399)

HYPERMUTATION Additional diversity is generated when a stimulated B or T cell begins to multiply. Mutations are induced in the antibody-coding regions of the progeny cells. As a result, any of these that have a higher affinity for the antigen are favored in further cloning. (p. 401)

HOW THE IMMUNE SYSTEM EVOLVED

The immune system molecules are closely related to the molecules involved in cell-to-cell recognition during development. Apparently the high specificity of these molecules was used to create a system for recognizing foreign chemicals. (p. 401)

STUDY QUESTIONS

1 Compare and contrast clonal selection with clonal deletion. (pp. 386–387, 393–394)

2 What would be the physiological effect on an individual born with a defect in either histamine release or reception? (pp. 385, 393)

3 What happens when, by chance, the virgin B cell for an antigen is never generated (that is, the appropriate antibody gene is not created by recombination and hypermutation) but the helper T cells are present? (pp. 388–393)

4 What would be the consequences of having a B-cell system that produced hybrid antibodies—that is, every cell produced antibodies, one arm of which bound to one antigen while the other bound to another? (pp. 383–393)

5 What happens to a cell if it suffers a mutation in its MHC gene? (Consider both immunocytes and other cell types.) (pp. 388–394)

6 How could the MHC system be exploited by the body to avoid inbreeding (mating with close kin)? (pp. 388–394)

SUGGESTED READING

ANDERSON, R. M., AND R. M. MAY, 1992. Understanding the AIDS pandemic, *Scientific American* 266 (6). *On the epidemiology of the AIDS epidemic from an ecological and evolutionary point of view.*

GREY, H. M., A. SETTE, AND S. BUUS, 1989. How *T* cells see antigen, *Scientific American* 261 (5). *A clear discussion of how processed antigens are "presented" by MHC proteins, and recognized by T-cell antibodies.*

HASSELTINE, W. A., AND F. WONG-STAAL, 1988. The molecular biology of the AIDS virus, *Scientific American* 259 (4). *An excellent description of how HIV-1 works, emphasizing gene regulation. This same issue has nine related articles on various aspects of the AIDS epidemic.*

MARRACK, P., AND J. KAPPLER, 1986. The T cell and its receptor, *Scientific American* 254 (2). *An excellent summary of how T cells interact with the other elements of the immune system.*

MILLS, J., AND H. MASUR, 1990. AIDS-related infections, *Scientific American* 263 (2).

OLD, L. J., 1988. Tumor necrosis factor, *Scientific American* 258 (5). *On the many messenger chemicals that immune-system cells use to regulate each other's activity.*

YOUNG, J. D.-E., AND Z. A. COHN, 1988. How killer cells kill, *Scientific American* 258 (1). *Compares the action of the complement system with the analogous strategy of killer cells in perforating the membranes of target cells.*

CHAPTER 16

INHERITANCE

Many observable traits are inherited. In dogs, for example, in addition to such obvious characteristics as height and build, strong breed propensities, such as the drive toward herding, hunting, or retrieving, are the result of inheritance. Thousands of years before the discovery of DNA humans used this knowledge to develop better grains, vegetables, fruits, and domesticated animals. Without understanding inheritance, or why selective breeding sometimes worked, breeders knew that when plants produce runners (reproducing asexually by cloning), the progeny were exactly like the parents. When plants reproduce sexually, however, breeders have long known that the progeny can be very different, not only from the parents but from each other as well.

The most common hypothesis until the beginning of this century, that of ***blending inheritance***, predicted that progeny would always embody a recognizable blend of parental traits. Nearly everyone knew this hypothesis to be wrong, or at least highly incomplete: in populations of European descent, two brown-eyed parents sometimes have a child with blue eyes. In any group, traits absent for generations can mysteriously recur, while altogether new ones occasionally appear out of the blue.

In this chapter we will see how the discovery of the laws of inheritance made sense of these seeming contradictions, clearing the way for the discovery of their genetic basis. This understanding of how inheritance works makes possible a deeper appreciation of how selection operates to create evolution, the topic of Part III.

HOW SINGLE-GENE TRAITS ARE INHERITED

The Austrian monk Gregor Mendel was the first to make some sense of the conflicting phenomena of inheritance. In 1856, he began breeding garden peas. The experiments he performed over the next 12 years laid the groundwork for what is now called the ***Chromosomal Theory of Inheritance***.

■ WHAT MENDEL FOUND

Mendel began with several dozen strains of peas, mostly purchased from commercial sources. He raised each variety for several years to discover which strains had recognizable morphological varia-

Table 16.1 Mendel's results from crosses involving single character differences

P CHARACTERS	F_1	F_2	F_1 RATIO
1 Round × wrinkled seeds	All round	5474 round; 1850 wrinkled	2.96:1
2 Yellow × green seeds	All yellow	6022 yellow; 2001 green	3.01:1
3 Red × white flowers	All red	705 red: 224 white	3.15:1
4 Inflated × constricted pods	All inflated	882 inflated: 299 constricted	2.95:1
5 Green × yellow pods	All green	428 green: 152 yellow	2.82:1
6 Axial × terminal flowers	All axial	651 axial: 207 terminal	3.14:1
7 Long × short stems	All long	787 long: 277 short	2.84:1

tions that bred true (in other words, that reappeared consistently in each generation).

Dominant and recessive traits In the end, Mendel reported his work on seven of the characteristics he studied. Each of these seven traits occurred in two contrasting forms: the seeds were either round or wrinkled, the flowers were red or white, the pods were green or yellow, and so on (Table 16.1). When Mendel bred plants with contrasting forms of just one of these characteristics, he found that all the offspring (usually referred to as the ***F_1***, or first filial, generation) were alike and resembled only one of the two parents (the ***P***, or parental, generation). When these offspring were crossed among themselves, however, some of their offspring (the ***F_2***, or second filial, generation) showed one of the original contrasting traits and some showed the other (Table 16.1). Thus, a trait that had been present in one of the parents, but not in any of their children, reappeared in the third generation, just as blue eyes can reappear after a generation or more of brown eyes.

Mendel's cross of red-flowered with white-flowered plants, for example, produced red flowers in all plants of the F_1 generation (Fig. 16.1). Apparently one form of each characteristic takes precedence over the other: red takes precedence over white in the flowers. Mendel referred to the traits that appear in the F_1 offspring of such crosses as ***dominant characters***, and the traits that are latent in the F_1 generation (white flowers, for example) as ***recessive characters***.

When Mendel allowed the F_1 peas from the cross involving flower color, all of which were red, to breed freely among themselves, their offspring (the F_2 generation) consisted of 705 plants with red flowers and 224 with white flowers. The recessive character had reappeared in approximately one-fourth of the F_2 plants. The same pattern held true for each of the other six characters he described crossing (Table 16.1); in each case the recessive character disappeared in the F_1 generation, but reappeared in approximately one-fourth of the plants in the F_2 generation. We can

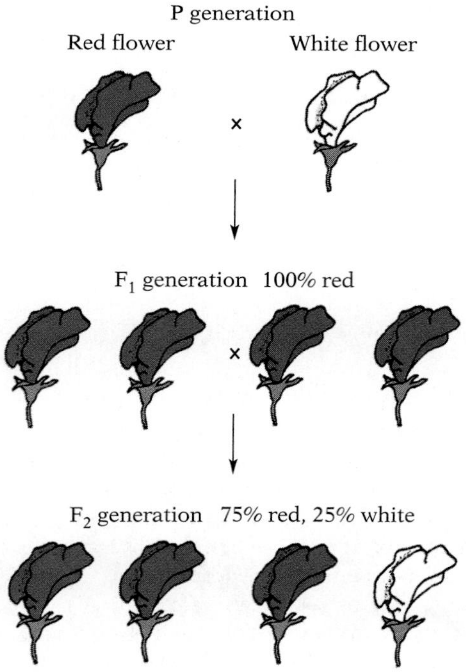

P generation
Red flower White flower
F_1 generation 100% red
F_2 generation 75% red, 25% white

16.1 Results of Mendel's cross of red-flowered and white-flowered garden peas

summarize the flower-color experiment as follows:

P	red × white
	↓
F_1	all red
	↓
F_2	3/4 red 1/4 white

Segregation These experiments led Mendel to formulate and test a series of hypotheses. Finally, he drew the revolutionary conclusion that each pea plant possesses two hereditary "factors" for each character, and that when gametes are formed the two factors segregate and pass into separate gametes. Each gamete, then, possesses only one factor for each character. Each new plant thus receives one factor for each character from its male parent and one for each character from its female parent. That two contrasting parental traits, such as red and white flowers, can both appear in normal form in the F_2 offspring indicates that the hereditary factors must exist as separate entities in the cell; they do not blend or fuse with each other. Thus the cells of an F_1 pea plant from the cross involving flower color contains one factor for red color and one factor for white color, the factor for red being dominant. These two factors remain distinct, and segregate unchanged when new gametes are formed. Mendel referred to this pattern of inheritance as the ***Principle of Segregation.***

Mendel's conclusions are consistent with what we now know about the chromosomes and their behavior in meiosis: the diploid nucleus contains two of each type of chromosome, one homologous chromosome from each parent. Each of the two chromosomes in any given pair bears genes for the same characters; hence the diploid cell contains two copies of each type of gene; these are the two hereditary factors for each character that Mendel described. Since homologous chromosomes segregate during meiosis, gametes contain only one chromosome of each type and hence only one copy of each gene, just as Mendel deduced. At the time, however, no one had even observed chromosomes.

Mendel may have doubted his own conclusions: some of his unpublished results could not be reconciled with his model, for, as we will see, not all genetic characters can be categorized as simply dominant or recessive, while others do not segregate independently. Mendel abandoned his efforts in 1868. In 1900, after cell division and chromosome segregation had been observed and described in detail, three biologists—Hugo De Vries in Holland, Carl Corren in Germany, and Erich von Tschermak-Seysenegg in Austria—independently rediscovered the phenomenon of segregation and, almost immediately, Mendel's original paper.

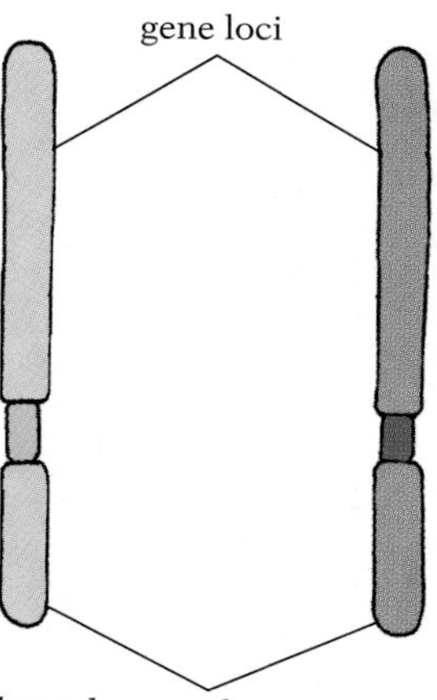

16.2 Anatomy of segregation

Somatic cells have pairs of homologous chromosomes, a pair consisting of one chromosome from each parent. Each gene contains two copies, one on each chromosome of the homologous pair at corresponding loci. When the copies at the corresponding loci are different, they are called alleles. In meiosis, each gamete receives a copy of only one chromosome from each homologous pair, and hence only one of the two alleles.

A modern interpretation of Mendel's results Let's consider Mendel's results on flower color from a modern perspective. In the cells of a pea plant, a gene for flower color exists at the same position, or ***locus***, on two homologous chromosomes (Fig. 16.2). The gene can exist in many different forms, or ***alleles***, but no individual can have more than two—one on each homologous chromosome containing that gene. Genes are designated by letters, using capital letters for dominant alleles and small letters for recessives. Flower color comes from the presence of pigment molecules, whose structure is determined by the genetic information in the chromosomes. Mendel's red allele (*C*) encodes a functional pigment that absorbs all wavelengths of light except the red that we see reflected, while the white allele (*c*) encodes a nonfunctional pigment molecule.

Since a diploid cell contains two copies of each gene, it may have two copies of the same allele or one copy of one allele and one copy of another allele (Fig. 16.2). Thus the cells of a pea plant may contain two copies of the allele for red flowers (*C/C*), or two copies of the allele for white flowers (*c/c*), or one red allele and one white allele (*C/c*). Cells with two copies of the same allele (*C/C* or *c/c*) are said to be ***homozygous*** for that trait. Those with two different alleles (*C/c*) are said to be ***heterozygous***. (The slash between letters indicates that the alleles are on separate chromosomes.)

We cannot tell by visual inspection whether a given pea plant is homozygous dominant (*C/C*) or heterozygous (*C/c*), because each type of plant has red flowers. Where one allele is dominant over another, the dominant allele takes full precedence over the recessive allele, and heterozygous organisms exhibit the dominant trait; one copy of the dominant allele is as effective as two. For this reason, there is often no one-to-one correspondence between the different possible genetic combinations, or ***genotypes***, and the possible appearances, or ***phenotypes***, of the organisms. In the example of flower color in peas, there are three possible genotypes—*C/C, C/c,* and *c/c*—but only two possible phenotypes, red and white. The heterozygote (*C/c*) produces a red phenotype because the red allele is dominant.

Just as we often cannot tell whether an individual is heterozygous or homozygous for a dominant allele, natural selection cannot act on genotypes directly. Hidden recessive traits are immune to selection; it is an organism's phenotype—the sum of its expressed traits—that matters in the struggle to survive and reproduce. Only when recombination produces a homozygous recessive, and the trait it encodes is therefore expressed, can selection operate on that allele.

The functional-nonfunctional dichotomy is the most common basis of dominant and recessive phenotypes. Wrinkled peas, for instance, result from a gene encoding an enzyme that polymerizes sugar to make starch. The recessive allele encodes a defective version of the enzyme. In the homozygous condition, this leads to a buildup of sugar, which draws extra water in osmotically; the developing seed thus becomes abnormally large. Water is removed from seeds when they mature; because these swollen seeds have more water and wall area than normal, wrinkles develop when the water is removed.

Using the modern conception of gene function, we can rewrite Mendel's pea cross to show both the genotypes and phenotypes:

P		*C/C* red	×	*c/c* white	
			↓		
F_1		*C/c* red	×	*C/c* red	
			↓		
F_2	*C/C* red	*C/c* red		*c/C* red	*c/c* white

Mendel began with a cross in the parental generation between a plant with a homozygous dominant genotype (red phenotype) and a plant with a homozygous recessive genotype (white phenotype). All of the F_1 progeny had red phenotypes, because all of them were heterozygous, having received a dominant allele for red (*C*) from the homozygous dominant parent and a recessive allele for white (*c*) from the homozygous recessive parent. However, when the F_1 individuals were allowed to cross freely among themselves, the F_2 progeny were of three genotypes and two phenotypes; 1/4 were homozygous dominant with the red phenotype, 1/2 (2/4) were heterozygous and showed red phenotypes, and 1/4 were homozygous recessive and produced white phenotypes. The ratio of genotypes in F_2 is 1 : 2 : 1; the ratio of the phenotypes is 3 : 1.

Figuring out the possible genotypic combinations in F_2 is simple in such a ***monohybrid cross*** (a cross involving only one character). All individuals in the F_1 generation are heterozygous (*C/c*); that is, they have one of each of the two types of alleles. Each allele is located on a separate homologous chromosomes and thus the two alleles segregate in meiosis. This means that half the gametes produced by a heterozygous individual will contain the C allele and half the c allele. When two such individuals are crossed (Fig. 16.3), there are four possible combinations of their gametes:

C from female parent, *C* from male parent
C from female parent, *c* from male parent
c from female parent, *C* from male parent
c from female parent, *c* from male parent

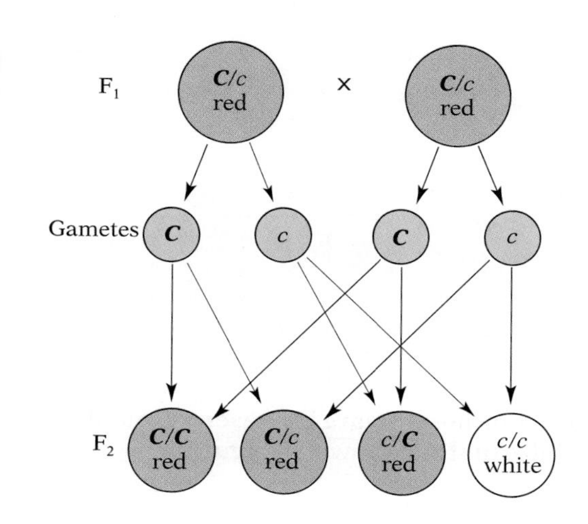

16.3 Gametes formed by F_1 individuals in Mendel's cross for flower color, and their possible combinations in F_2

Notice that the first combination produces homozygous dominant offspring (red), the second and third produce heterozygous offspring (also red), while the fourth produces homozygous recessive offspring (white). Since each of the four combinations is equally probable, we would expect, if a large number of F2 progeny are produced, a genotypic ratio close to 1 : 2 : 1 and a phenotypic ratio close to 3 : 1, just as Mendel found.

An easy way to figure out the possible genotypes produced in F_2 is to construct a ***Punnett square*** (named after the English geneticist R. C. Punnett). Along a horizontal line, write all the possible kinds of gametes the male parent can produce; in a vertical column to the left, write all the possible kinds of gametes the female parent can produce; then draw squares for each possible combination of these:

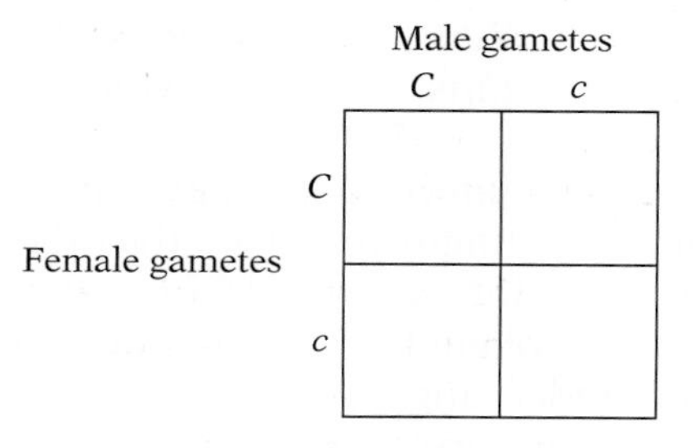

Next, write in each box the allele letters for the female and male gamete (in that order) separated by a slash. Each box then contains the symbols for the genotype of one possible zygote combina-

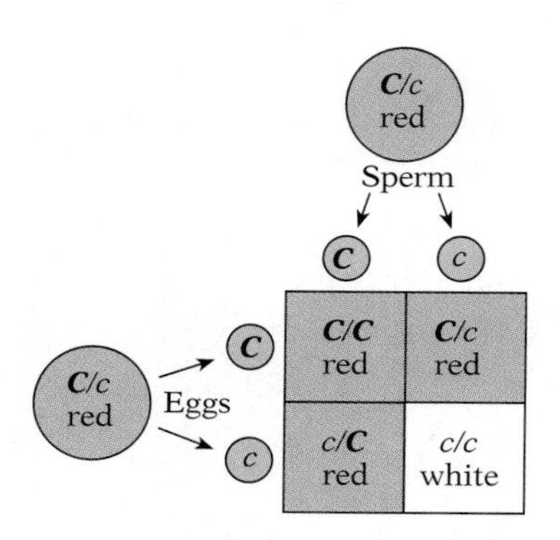

16.4 Punnett-square representation of the information shown in Figure 16.3

tion from the cross. The completed Punnett square in Figure 16.4 shows that the cross yields the expected 1 : 2 : 1 genotypic ratio and, since dominance is present, the expected 3 : 1 phenotypic ratio.

Whenever there is dominance, the expected phenotypic ratio is 3 : 1. If the samples are large and these ratios are not obtained, some complicating condition must be present. The failure of monohybrid crosses to yield the expected results has often led to the discovery of genetic phenomena that are critical to the understanding of evolution. Two of these phenomena—linkage and sex-specific effects—will be discussed presently. Others include strict maternal inheritance (which led to the discovery of organelle genes), genetic "imprinting" (which can also lead to early expression of only maternal or paternal alleles; see pp. 288–289), and alleles that eliminate any heterozygous allele on the homologous chromosome, thereby assuring their own transmission in all gametes.

■ PARTIALLY DOMINANT TRAITS

The seven characters of peas that Mendel reported on involved an allele, known as a full dominant, that showed complete dominance over the other. Some characteristics do not show this pattern, including some that Mendel himself studied but did not discuss. When a heterozygous individual clearly displays both alleles, the alleles are referred to as partial (or incomplete) dominants.

In many cases of partial dominance, heterozygous individuals have a phenotype that is intermediate between the phenotype of individuals homozygous for one allele and the phenotype of individuals homozygous for the other—a kind of blending inheritance. For example, crosses between homozygous red snapdragons or sweet peas and homozygous white varieties yield plants producing pink blossoms (Fig. 16.5). When these pink-flowering plants are crossed among themselves, they yield red, pink, and white offspring in a ratio of 1 : 2 : 1, as follows:

P		R/R ×	R'/R'	
		red	white	
		↓		
F_1		R/R' ×	R/R'	
		pink	pink	
		↓		
F_2	R/R	R/R'	R'/R	R'/R'
	red	pink	pink	white

When there is partial dominance, both alleles are designated by a capital letter, and one is distinguished from the other by a prime, as here, or by a superscript (for example, C^r for red, C^w for white). Since both alleles affect the phenotype, each is subject to selection.

The chemical events underlying the production of pink flowers are not fully understood. Heterozygous pea flowers, as we have seen, appear red. In pea flowers the red pigment is so intense that it probably masks the white pigment totally, while the less saturated red pigment of snapdragons does not. You can see how this is

16.5 Partial dominance in sweet peas

When plants homozygous for red flowers are crossed with plants homozygous for white blossoms, the heterozygous F_1 sweet-pea plants have pink flowers.

possible by noticing how little pure red pigment it takes to turn a gallon of white base paint bright red.

In other cases of partial dominance, the heterozygous phenotype is not necessarily intermediate between the two homozygous phenotypes. For example, in certain strains, a mating between a black chicken and a so-called splashed white chicken produces offspring all of which have a distinctive appearance called blue Andalusian. A cross between two Andalusians produces black, blue Andalusian, and splashed white offspring in a ratio of 1 : 2 : 1:

P	black	×	white	
		↓		
F_1	blue	×	blue	
		↓		
F_2	black	blue	blue	white

Here the gene products of two alleles, one for black and one for white, interact to produce a phenotype that is not intermediate between black and splashed white.

HOW MULTIPLE AND MULTIGENIC TRAITS ARE INHERITED

We have limited our discussion so far to monohybrid crosses, those in which a single phenotypic character is studied. Moreover, our examples have involved a single character controlled by a single gene. Such a character is called a ***monogenic*** trait. However, most characters are controlled by several different genes at once, and so involve ***multigenic*** inheritence. In addition, organisms have many different contrasting characters, and so all crosses are actually ***multihybrid crosses***.

■ INHERITING TWO TRAITS AT ONCE

Mendel's experiments sometimes involved two or more of the characters listed in Table 16.1. For example, he crossed plants bearing round yellow seeds with plants producing wrinkled green seeds. The resulting F_1 plants all had round yellow seeds—all showed dominant phenotypes. When these plants were crossed among themselves, the F_2 progeny showed four different phenotypes:

315 had round yellow seeds
101 had wrinkled yellow seeds
108 had round green seeds
32 had wrinkled green seeds

These numbers represent a ratio of about 9 : 3 : 3 : 1 for the phenotypes.

This experiment produced a new and interesting result: even if full dominance is present, a dihybrid cross can produce new plants phenotypically different from either of the original parental plants; here the new phenotypes were wrinkled yellow and round green

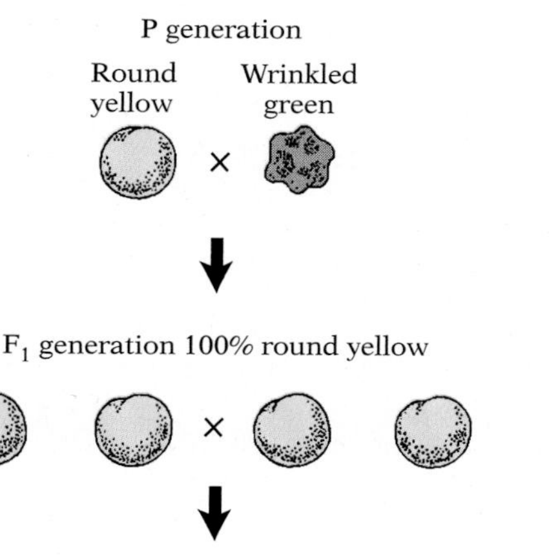

F₂ generation four phenotypes in ratio 9 : 3 : 3 : 1

Round yellow

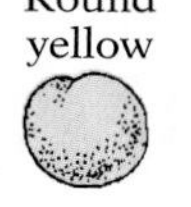

Wrinkled yellow

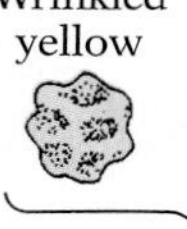

Round green

Wrinkled green

Novel phenotypes

16.6 Expression of novel phenotypes
Crosses between plants homozygous for round yellow seeds and plants homozygous for wrinkled green seeds produce two novel phenotypes in the F_2 generation: wrinkled yellow seeds and round green seeds.

(Fig. 16.6). This confirms that during meiosis the genes for seed color and the genes for seed form do not necessarily remain paired as they were in the parents, but can separate and reassemble in different allelic combinations in the gametes. This implies that the genes for seed color are on one homologous pair of chromosomes while the genes for seed form are on another pair. Consequently, the genes for the two characters segregate independently during meiosis, sometimes producing new phenotypes on which selection can operate.

The 9 : 3 : 3 : 1 phenotypic ratio is characteristic of the F_2 generation of a dihybrid cross (with dominance) in which the genes for the two characters are ***independent*** (located on nonhomologous chromosomes). Each independent gene behaves in a dihybrid cross exactly as in a monohybrid cross. For example, if we view Mendel's F_2 results as the product of a monohybrid cross for seed color (ignoring seed form), we find that there were 416 yellow seeds (315 + 101) and 140 green seeds (108 + 32), which closely approximates the 3 : 1 F_2 ratio expected in a monohybrid cross. Similarly, if we treat the experiment as a monohybrid cross for seed form and ignore seed color, the F_2 results also show a phenotypic ratio of approximately 3 : 1. The dihybrid F_2 ratio of 9 : 3 : 3 : 1 is simply the product of two separate and independent 3 : 1 ratios.

We can summarize this cross using the following symbols: (with *R* symbolizing the round-seed allele and *r* for the wrinkled-seed allele, *G* for the yellow-seed allele and *g* for the green-seed allele[1]; the dash means that it does not matter phenotypically whether the dominant or the recessive allele occurs in the spot indicated):

P	*R/R G/G* round yellow	×	*r/r g/g* wrinkled green	
		↓		
F_1	*R/r G/g* round yellow	×	*R/r G/g* round yellow	
		↓		
F_2	9 *R/– G/–* round yellow	3 *r/r G/–* wrinkled yellow	3 *R/– g/g* round green	1 *r/r g/g* wrinkled green

The round yellow parent could produce gametes of only one genotype, *RG*. The wrinkled green parent could produce only *rg* gametes. When *RG* gametes from the one parent united with *rg* gametes from the other parent in the process of fertilization, all the resulting F_1 offspring were heterozygous for both characters (*R/r G/g*) and showed the phenotype of the dominant parent (round yellow). Each of these F_1 individuals could produce four different types of gametes, *RG, Rg, rG,* and *rg*. When two such individuals were crossed, there were 16 possible combinations of gametes (4 × 4). These 16 combinations included nine genotypes, which determined four phenotypes in the ratio of 9 : 3 : 3 : 1 (Fig. 16.7).

One way to determine the genotypic and phenotypic ratios in a dihybrid cross is to construct a Punnett square and then count the

[1]Mendel used *G* and *g* to denote color factors because the typical color of peas is green; he was surprised to discover later that green is recessive.

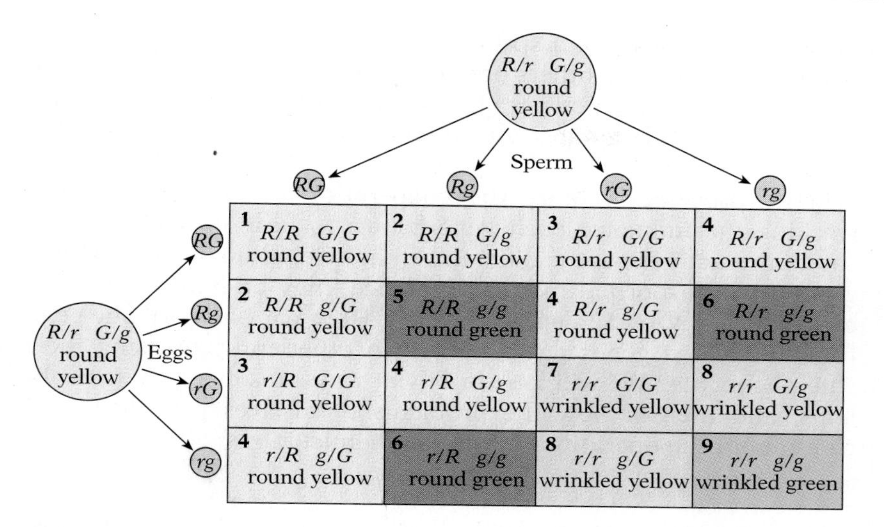

16.7 Punnett-square representation of an F_1 dihybrid cross

When two individuals heterozygous for both seed color and shape were crossed, for every plant that produced wrinkled green seeds, Mendel obtained roughly nine plants that produced round yellow seeds, three that produced round green seeds, and three that produced wrinkled yellow seeds. (Combinations of alleles that give rise to identical phenotypes have the same shading.) Underlying these results were nine different genotypes, as indicated by the numbers in the boxes. (Combinations of alleles that represent identical genotypes are shown with the same number.)

number of boxes representing each genotype and each phenotype. This method becomes prohibitively tedious in any cross involving more than two characters. An alternative procedure is much easier. It is based on the principle that *the chance that a number of independent events will occur together is equal to the product of the chances that each event will occur separately,* a principle known as the ***Product Law***. It is best explained by an example.

Suppose we want to know how many of the 16 combinations in Mendel's cross will produce the wrinkled yellow phenotype. We know that wrinkled is recessive; hence it is expected in one-fourth of the F_2 individuals in a monohybrid cross. We know that yellow is dominant; hence it is expected in three-quarters of the F_2 individuals. Multiplying these two values (1/4 × 3/4) gives us 3/16; three of the 16 possible combinations will produce a wrinkled yellow phenotype. Similarly, if we want to know how many of the combinations will produce a round yellow phenotype, we multiply the expectancies for two characters (both of which are dominant) (so 3/4 × 3/4) to get 9/16.

The Product Law applies equally well to more complex examples. Suppose we want to find out, for a trihybrid cross involving the two seed characters and flower color, what fraction of the F_2 individuals (produced by allowing *C/c R/r G/g* individuals to cross among themselves) will exhibit a phenotype combining red flowers, wrinkled seeds, and yellow seeds. The separate probability for red flowers in a monohybrid cross is 3/4, that for wrinkled seeds is 1/4, and that for yellow seeds is 3/4. Multiplying these three values (3/4 × 1/4 × 3/4) gives us 9/64.

■ HOW GENES INTERACT

We have seen that different alleles of the same gene can interact when both are partially dominant. Separate genes can also interact. Indeed, most traits or characters (phenotypic characteristics), whether they are morphological, like flower color, or more subtle chemical characteristics, such as enzyme pathways, are controlled by several genes. Glycolysis, for instance, is a dozen-step process. Each step requires a different enzyme, and each enzyme is pro-

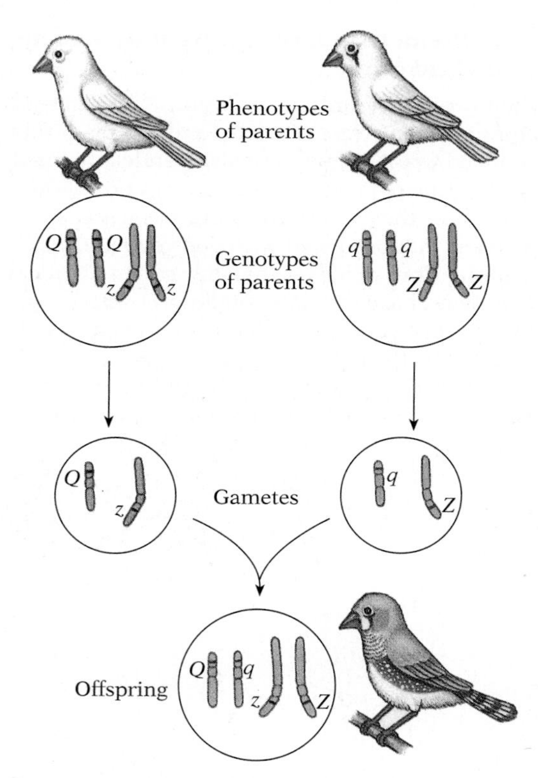

16.8 Complementation test with zebra finches

There are two varieties of albino zebra finch. When these are crossed, the offspring have the slate-gray bodies and colorful markings of normal wild zebra finches. The genetic basis of this phenomenon is shown here. Finches of one strain (B, left) are homozygous for the recessive allele at the *Q* locus (genotype *q/q*), and therefore fail to produce functional pigment. Those of the other strain (A, right) are homozygous recessive at the *Z* locus (genotype *z/z*), and lack normal color because of flaws in the mechanisms controlling gene activity. The offspring of the cross are heterozygous at both loci, and so can produce the normal amount of pigment. In the example shown here, both parents are homozygous for both genes. If one or both of the parents were heterozygous for one gene (with the genotype *Q/q z/z* or *q/q Z/z*), we would expect some of the offspring to be normal and some to be albino.

duced by the action of a specific gene. Most of these enzymes depend on one or more others operating earlier in the chemical chain reaction to provide a substrate with which they can interact, and so in turn provide a substrate suitable for the next enzyme in the series.

Many processes require the cooperation of many genes, but tracking the influence each gene brings to bear in such complicated interactions can be difficult. An individual unable to perform glycolysis at a normal rate, for instance, may be homozygous recessive for any of 12 separate genes. How can we sort out gene functions when genes depend on each other for expressing their true phenotype? We will look briefly at ways of solving these problems, and then at how the problems themselves provide useful clues about the operation of cellular chemistry.

Complementary genes Genes that are mutually dependent are normally detected during crosses between individuals with similar recessive phenotypes. This ***complementation test*** can show whether the trait in question is produced by the same gene in both individuals, or by two separate genes. For example, a normal zebra finch is brightly colored, but there are also two forms of albino finch: one form (strain A), though almost totally white, has a slight brown tinge near its eyes; the other (strain B) is completely white. When these two forms are crossed, their offspring are normal ***wild types*** (the most common naturally occurring form). This result indicates that each form of albinism is produced by a separate gene, at a different locus on the chromosome; the offspring of the cross, being heterozygous in both genes, display the normal phenotype (Fig. 16.8).

Failure in the normal pigmentation system in zebra finches can occur at two known points. The pigment protein can be defective, which is the problem in strain B; or the control mechanisms that turn on pigment production and control the amount made can be defective, which is the problem in strain A. A finch that is homozygous recessive for either element of the pigment system will be an albino. If two white individuals with the *same* defect are crossed, all the progeny will be white, but if two white individuals with different defects are crossed, some or all of the offspring will be normal (Fig. 16.9). This reversion to the normal phenotype in offspring heterozygous at both loci exemplifies the phenomenon known as complementarity, and the two mutually dependent genes are said to ***complement*** each other.

Complementarity can lead to some unusual phenotypic ratios in crosses. Consider the following cross between zebra finches:

P		*Q/Q Z/Z* normal	×	*q/q z/z* albino
			↓	
F_1		*Q/q Z/z* normal	×	*Q/q Z/z* normal
			↓	
F_2	9 *Q/– Z/–* normal	3 *Q/– z/z* albino	3 *q/q Z/–* albino	1 *q/q z/z* albino

Were this a dihybrid cross, involving two traits, each at a separate locus, we would have expected four phenotypes in a ratio of 9 : 3 : 3 : 1. Instead, the cross involves two loci but only *one* trait, and hence is digenic but monohybrid. In this example, the 9 : 7 ratio arises because the last three genotype combinations yield the albino phenotype.

Epistasis Complementary genes find expression in the dominant phenotype if and only if the organism is heterozygous or homozygous dominant at both loci. Each gene in some sense has veto power over the other, because their products must cooperate. In ***epistasis***, by contrast, these two genes do not have equal "votes": because of the biochemistry of the interaction of their products, only one of the genes can be vetoed by the other. When one gene has the effect of suppressing the phenotypic expression of another gene but not vice versa, the first gene is said to be epistatic to the second. (The Greek root of *epistasis,* logically enough, means "standing upon.")

In guinea pigs, for example, a gene for the production of the skin pigment melanin is epistatic to one for the *deposition* of melanin. The first gene has two alleles: *C,* which causes pigment to be produced, and *c,* which does not; hence a homozygous recessive individual, *c/c,* is albino. The second gene has an allele *B* that causes deposition of much melanin, which gives the guinea pig a black coat, and an allele *b* that causes deposition of only a moderate amount of melanin, producing a brown coat. However, neither *B* nor *b* can cause deposition of melanin if *C* is not present to make the melanin:

P	*C/C B/B* black	×	*c/c b/b* albino	
		↓		
F_1	*C/c B/b* black	×	*C/c B/b* black	
		↓		
F_2	9 *C/– B/–* black	3 *C/– b/b* brown	3 *c/c B/–* albino	1 *c/c b/b* albino

Instead of an F_2 phenotypic ratio of 9 : 3 : 3 : 1, this monohybrid cross yields a ratio of 9 : 3 : 4.

Notice that epistasis and dominance are quite different. Dominance is the phenotypic expression of one member of a pair of alleles at the expense of the other. Epistasis is the suppression of one gene for a phenotypic effect by another entirely different gene; the epistatic gene prevents natural selection from acting on *either* allele of the second gene. In other words, dominance involves the interaction between alleles, while epistasis refers to the interaction between nonallelic genes.

Collaboration Sometimes two genes interact to produce a novel phenotype. Such collaborative interaction is seen in the control of comb form in chickens (Fig. 16.9). One gene, *R,* produces a rose comb, while its recessive allele, *r,* produces a single comb. Another

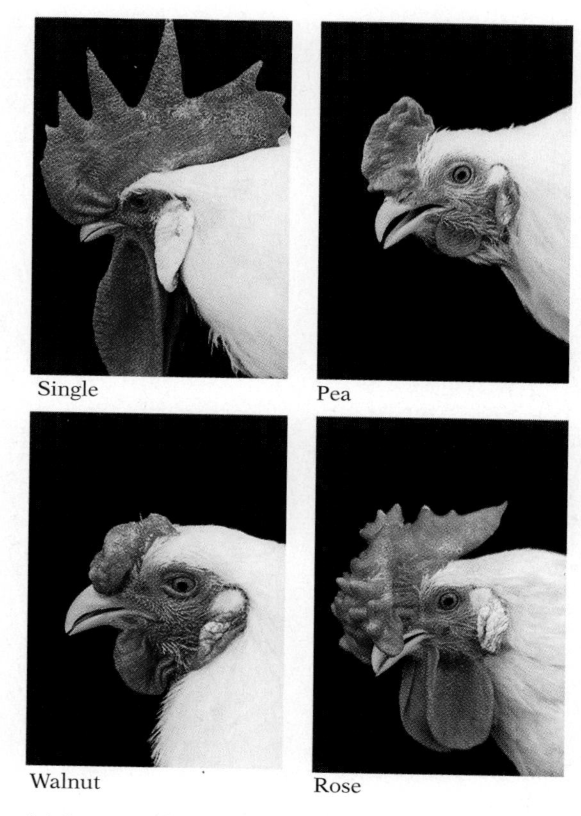

16.9 Comb types in chickens

gene, *P,* produces the "pea" comb, while its recessive, *p,* also produces a single comb. When *R* and *P* occur together, they collaborate to produce the "walnut" comb. When rose-comb Wyandotte chickens are crossed with pea-comb Brahma chickens, four phenotypes result:

P	*R/R p/p* rose	×	*r/r P/P* pea	
		↓		
F_1	*R/r P/p* walnut	×	*R/r P/p* walnut	
		↓		
F_2	9 *R/– P/–* walnut	3 *R/– p/p* rose	3 *r/r P/–* pea	1 *r/r p/p* single

Notice that when collaboration occurs, it is not possible to identify the action of each gene on the basis of its phenotypic effects in the clearcut way Mendel could. The action of selection on any one allele, then, may depend on the alleles present at another locus.

Modifiers Probably no inherited character is controlled exclusively by one gene pair, though for convenience we often analyze crosses as though only a single gene were important for a particular trait. However, even when only one principal gene is involved, its expression is usually influenced to some extent by other genes. An example is eye color in humans.

Human eye color is largely controlled by one gene with two alleles—a dominant allele, *B,* for brown eyes, and a recessive, *b,* for blue eyes. Brown-eyed people (*B/B* or *B/b*) have pigment cells containing melanin in the *front* layer of the iris. Blue-eyed people (*b/b*) lack melanin; the blue is an effect of the black pigment on the *back* of the iris seen faintly through the semiopaque front layer.

Though most eyes can be categorized as brown or blue, eyes can be green or gray (both genetically forms of blue), hazel or black (both forms of brown), as well as shades in between. Obviously, modifier genes are also involved, some affecting the amount of pigment in the iris, some the tone of the pigment (which may be light yellow, dark brown, and so on), some its distribution (uniform over the whole iris, or in scattered spots, or in a ring around the outer edge, and so forth). In fact, in rare cases two blue-eyed people can have a brown-eyed child, because one of them, in whom the lack of pigmentation is a consequence of the action of modifier genes, actually carries the genotype *B/b* instead of *b/b.*

Multiple-gene inheritance Not all characters have a limited number of relatively distinct phenotypes. While pea flowers are either red or white, and eyes blue or brown (with intermediates generated by modifier genes) other traits have no distinct boundaries. Human height, skin pigmentation, and IQ are three examples. What accounts for the absence of distinct phenotypes?

The most common reason is that two or more separate genes can affect the same character in the same way, in an additive fashion. This kind of inheritance is called ***polygenic.*** The first clear demon-

16.10 Nilsson-Ehle's trigenic cross

A line of wheat with pure white kernels and a line with dark red kernels were crossed to produce hybrid medium red F_1 offspring; these were then crossed to produce offspring with the distribution of phenotypes shown here. Nilsson-Ehle concluded from his observations that (1) there are three genes for kernel color in wheat, each with a partially dominant allele for red (*A, B, C*) and a partially dominant allele for white (*A′, B′, C′*); (2) that the genotype of the dark red line is *A/A B/B C/C*, while that of the white line is *A′/A′ B′/B′ C′/C′*; and (3) that the F_1 hybrids therefore have the genotype *A/A′ B/B′ C/C′*. A gradation of phenotypes is seen in the offspring because all the alleles are additive. Eight different kinds of gametes occur because each of the genes for color is on a different chromosome. The existence of the seven distinct phenotypes of F_2 offspring in the ratio 1 : 6 : 15 : 20 : 15 : 6 : 1 is predicted by this Punnett square of a trigenic cross postulating partial dominance in kernel color.

stration came in 1909, when Herman Nilsson-Ehle showed that the color of wheat kernels, which vary from white through various shades of pink and red to a very dark red, results from the interactions of three genes. Each gene has two alleles, a partially dominant allele for red and a partially dominant allele for white. Dark red kernels are homozygous for the red allele in all three genes, while pure white kernels are homozygous for the white allele in all three genes. All the phenotypes in between result from different heterozygous mixtures of the alleles (Fig. 16.10).

If we graph the probable frequency of colors in F_2, we obtain a jagged approximation of the bell-shaped curve that represents the so-called normal distribution of most continuously varying traits in a population, traits like weight, height, IQ, and the like (Fig. 16.11). The greater the number of phenotypes, the less jagged the distribution should appear. In most cases, however, the genes involved in multiple-gene effects do not all contribute equally to the phenotype, and the effects of modifier genes and of the environment blur the boundaries. As a result, it is usually impossible to determine the number and nature of the genes involved simply by scrutinizing the distribution of phenotypes in F_2. On the other hand, asymmetries in plots of actual phenotypic variation in a population can suggest that selection is operating against certain ranges of the phenotypic spectrum.

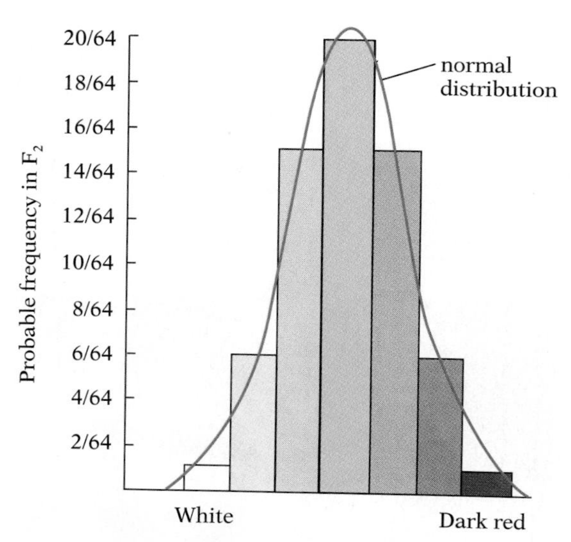

16.11 Frequency distribution of kernel color in wheat

As Nilsson-Ehle showed, a cross of heterozygous wheat kernels results in offspring with seven classes of kernel color, their relative frequencies producing a so-called normal distribution centered around light red.

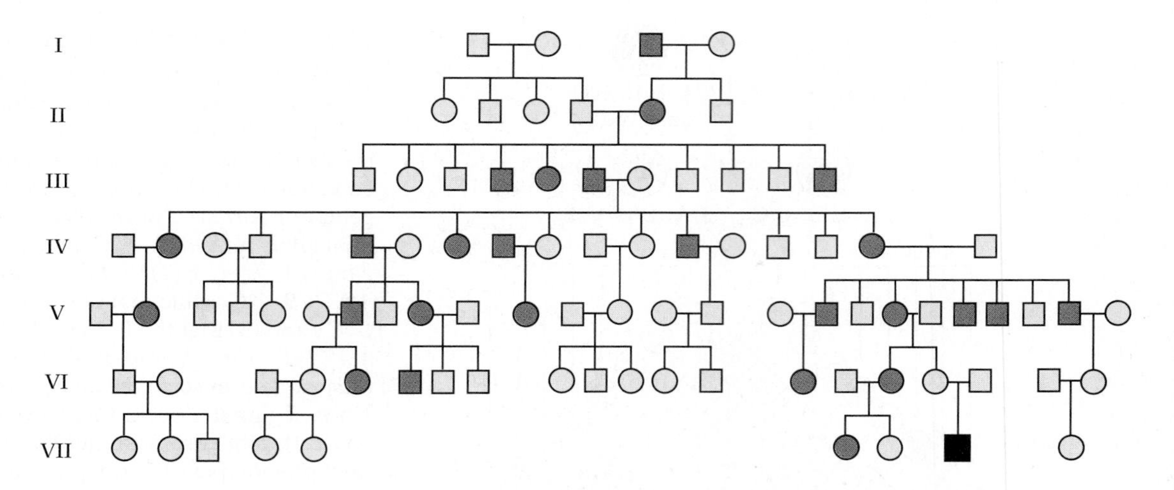

16.12 Pedigree of syndactyly in seven generations of one family
Squares represent males, and circles represent females. Red indicates syndactyly and gray indicates the normal phenotype. The character appears to be inherited as a simple dominant: there is no correlation with sex, and individuals with syndactyly have a parent who also shows this trait. There is one exception: an individual (black square) who has syndactyly even though neither of his parents shows the trait. Presumably the gene was present in his mother without expression.

■ WHEN GENES ARE ONLY PARTIALLY EXPRESSED

Because genes interact, complementary or epistatic genes, or a combination of modifier genes, may prevent the expression of a dominant allele. And even when an allele is expressed, the effects of other genes may modify the intensity of phenotypic expression. Certain alleles thus have incomplete penetrance and variable expressivity: ***penetrance*** refers to the percentage of individuals in a population carrying a given allele that actually express the phenotype; ***expressivity*** denotes the manner in which the phenotype is expressed.

Incomplete penetrance and variable expressivity are illustrated by an allele in humans that causes blue sclera: the whites of the eyes appear bluish. This allele is dominant, but only about 9 out of 10 people who have the allele actually show the phenotype. Thus the penetrance of this gene is about 90% (or 0.9). Among those that do show the phenotype, the expressivity is variable, with the intensity of the bluish coloration ranging from very pale whitish blue to very dark blackish blue.

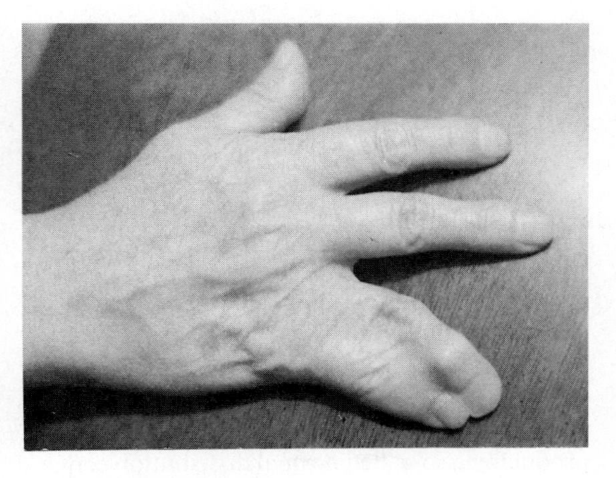

16.13 Syndactyly

Figure 16.12 shows a portion of the pedigree of a family in Virginia that for generations has had many members with syndactyly of the ring and little fingers—the two fingers are joined by a web of muscle and skin (Fig. 16.13). This character exhibits variable expressivity: a few individuals have three fingers webbed; most have two fingers fully webbed; some have two fingers partially webbed; and a few have a crooked finger with no webbing.

Syndactyly also exhibits incomplete penetrance. Of the more than 50 individuals in this family known to have had syndactyly, all but one had a parent who also had syndactyly—almost exactly the pattern of a pure dominant. (By contrast, a recessive character often appears in a person neither of whose parents showed the character, because they were heterozygous.) The abnormality in the one man having two parents with normal fingers demonstrates that the allele for syndactyly has incomplete penetrance: the man's mother must have been carrying the allele, but did not exhibit its phenotype because of the allele's incomplete penetrance.

Both examples demonstrate that penetrance and expressivity are aspects of the same phenomenon. Lack of penetrance of the gene for blue sclera produces white color, which is simply one extreme of the expressivity gradient from white through very pale blue to dark blue. Lack of penetrance of the gene for syndactyly is simply one extreme of the expressivity gradient whose other extreme is the full webbing of three fingers.

Penetrance and expressivity are often also affected by the environment. For example, *Drosophila* that are homozygous for the allele for vestigial wings have wings that are only tiny stumps when reared at normal room temperatures (about 20°C), but their wings grow almost as long as normal wings when reared at temperatures as high as 31°C. Himalayan rabbits are normally white with black ears, nose, feet, and tail (Fig. 16.14), but if a patch of fur on a patch on the rabbit's back is plucked and an ice pack is kept on the patch, the new fur that grows there will be black: the allele for black color can express itself only if the temperature is low, which it normally is only at the body extremities.

The same phenomenon underlies the coloring of Siamese cats. An enzyme required for pigment synthesis in this breed retains its normal active conformation only at low temperatures, and denatures—changes to an inactive conformation—when heated to body temperature.

Thus the expression of a gene can depend both on the other genes present (the genetic environment) and on the physical environment (temperature, sunlight, humidity, diet, and so forth). We don't inherit characters. We inherit only genetic potentialities; other factors govern whether the potentialities are realized. All organisms are products of both their inheritance and their environment.

16.14 Effect of temperature on expression of a gene for coat color in the Himalayan rabbit

(A) Normally, only the feet, tail, ears, and nose are black. (B) Fur is plucked from a patch on the back, and an ice pack is applied to the area. (C) The new fur grown under the artificially low temperature is black. Himalayan rabbits are normally homozygous for the gene that controls synthesis of the black pigment, but the enzyme encoded by the gene is active only at low temperatures (below about 33°C).

■ WHEN GENES HAVE SEVERAL ALLELES

Mendel simplified his analysis by concentrating on just two alleles at each locus. And of course, the maximum number of alleles for each gene that any diploid individual can possess is two, because the organism has only two copies of each gene. However, within a population to which that individual belongs, many other alleles may be present for a given gene, so multiple alleles can play a major role in evolution.

Eye color in *Drosophila* One of the first examples of multiple alleles was discovered in the fruit fly *Drosophila melanogaster*—a species whose genetics have been extensively studied. Though normally red-eyed, fruit flies may have white-, eosin- (a brightly fluorescing red), wine-, apricot-, ivory-, or cherry-colored eyes. Each of these colors is controlled by a different allele of the same gene. The allele for the wild-type eye (red) is dominant over all the rest—that is, the normal red pigment masks the pigment produced by any other allele. When two of the other alleles occur together in a heterozygous fly, however, they produce an intermediate eye-color phenotype.

Blood types A well-known example of multiple alleles in humans is that of the A-B-O blood series, in which four blood types are generally recognized: A, B, AB, and O. The red blood corpuscles (erythrocytes) in type A blood bear antigen A on their surface; in type B, antigen B; in type O, neither A nor B; and in type AB, both A and B.

An antigen is a chemical capable of triggering an immune reaction by which antibodies are produced that bind to and help destroy that particular antigen and the cell that bears it (see Chapter 15). Because the immune system soon becomes insensitive to antigens present from birth, an individual whose red blood corpuscles bear antigen A has no antibodies—called anti-A—to that antigen. Similarly, an individual with antigen B has no anti-B antibodies. The person with type AB blood, therefore, having corpuscles that bear both A and B antigens, will have neither anti-A nor anti-B in the blood plasma; the person with type O blood, on the other hand, having corpuscles that bear neither antigen, will have both anti-A and anti-B in the blood plasma (see Table 16.2).

Table 16.2 Antigen and antibody content of the blood types of the A-B-O series

BLOOD TYPE	BLOOD CONTAINS	
	CELLULAR ANTIGENS	PLASMA ANTIBODIES
O	None	anti-A and anti-B
A	A	anti-B
B	B	anti-A
AB	A and B	None

The presence of these antigens and antibodies in the blood has important implications for transfusions. Because the antibodies present in the plasma of blood of one type tend to react with the antigens on the red blood cells of other blood types and cause clumping, it is always best to obtain a donor who has the same blood type as the patient. Blood of another type may be used, provided that the *plasma of the patient* and the *erythrocytes of the donor* are compatible. The reason is that, unless the transfusion is to be a massive one or is to be made very rapidly, the donor's plasma is sufficiently diluted during transfusion so that little or no agglutination occurs.

This means that type O blood can be given to anyone, because its red blood corpuscles have no antigens and hence are obviously compatible with the plasma of any patient; type O blood is sometimes called the universal donor. But type O patients can receive transfusions only from type O donors, because their plasma contains both anti-A and anti-B and hence is obviously not compatible with the erythrocytes of any other class of donor. Luckily for those of us with type O blood, it is the most common variety in most parts of the world. Conversely, people with the rare type AB, whose plasma contains no anti-A or anti-B antibodies, are universal recipients, but cannot act as donors for any except type AB patients (Table 16.3).

At first glance, you might suppose that two independent genes are involved in the A-B-O system, one determining whether the A antigen is present and another whether the B antigen is present. Actually the inheritance of the A-B-O groups is for the most part controlled by three alleles of the same gene, I^A, I^B, and i. Both I^A and I^B are dominant over i, but neither I^A nor I^B is dominant over the other. Accordingly the four blood-type phenotypes correspond to the genotypes indicated in Table 16.4. I^A and I^B are alleles that code for different functional proteins, while i codes for a nonfunctional protein. In fact, the protein is an enzyme that modifies the carbohydrate groups on a membrane component of the glycocalyx (Fig. 16.15).

Blood typing is often used as a source of evidence in paternity cases. For example, a man with type O blood could not be the father of a child with type A blood whose mother is type B. The child's true father must be either type A or type AB, because the child must have received its I^A allele from its father; an O man has no such allele. Similarly, a man with type AB blood could not be the father of a type O child, because the child must have received an i allele from each parent, but an AB man has no such allele. Blood-type analysis can only determine who could *not* be the father. Genetic "fingerprinting" techniques, which usually measure the number of copies of each type of repetitive DNA (a pattern that is similar between related individuals, but unique in detail for each

Table 16.3 Transfusion relationships of the A-B-O blood groups

BLOOD GROUP	CAN DONATE BLOOD TO	CAN RECEIVE BLOOD FROM
O	O, A, B, AB	O
A	A, AB	O, A
B	B, AB	O, B
AB	AB	O, A, B, AB

Table 16.4 Genotypes of the A-B-O blood types

BLOOD TYPE	GENOTYPE
O	i/i
A	I^A/I^A or I^A/i
B	I^B/I^B or I^B/i
AB	I^A/I^B

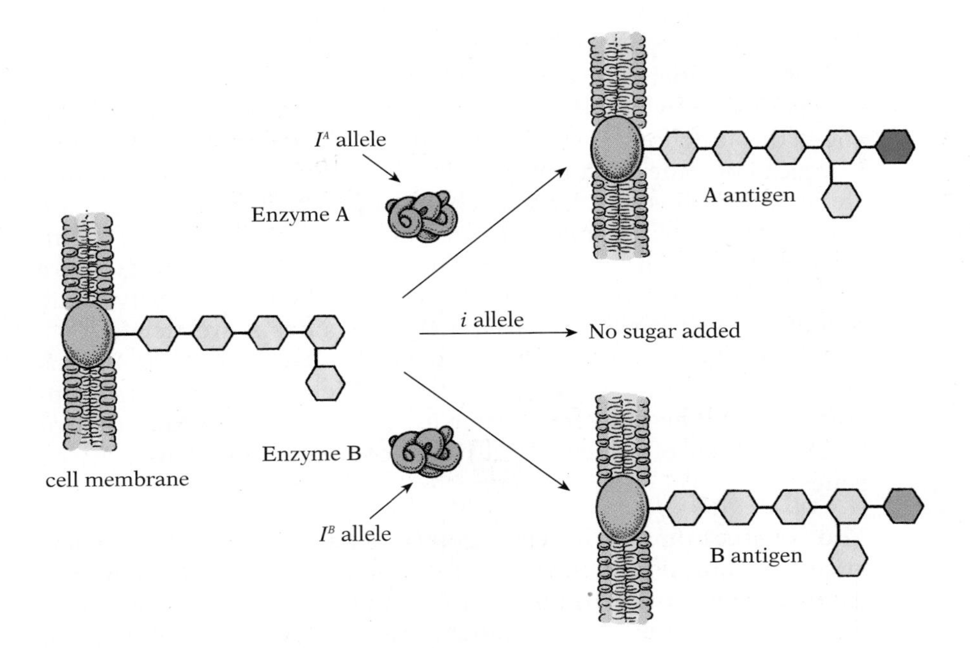

16.15 Molecular basis of the A-B-O blood group antigens

Blood cells produce a carbohydrate chain with five sugar groups that is mounted on a cell-surface lipid or protein. Cells with an I^A allele produce an enzyme that adds a particular sixth sugar (red) to the chain, producing the type A antigen; cells with an I^B allele produce an enzyme that adds a different sixth sugar (blue) to the chain, producing the type B antigen. Cells with both alleles produce both kinds of six-sugar chains, leading to the AB phenotype; cells homozygous for the i allele fail to modify any of the surface-mounted carbohydrates.

Table 16.5 Frequencies of A-B-O blood groups in selected populations

POPULATION	O	A	B	AB
U.S. whites	45%	41%	10%	4%
U.S. blacks	47	28	20	5
African Pygmies	31	30	29	10
African Bushmen	56	34	8	2
Australian aborigines	34	66	0	0
Pure Peruvian Indians	100	0	0	0
Tuamotuans of Polynesia	48	52	0	0

person) can be more reliable, though it is very time consuming to perform.

As indicated in Table 16.5, the frequencies of the various A-B-O blood types vary in populations of different ancestral extraction. Since the most frequent phenotype in most human populations is type O, which corresponds to the homozygous recessive phenotype, the *i* allele is more common than the I^A or I^B alleles. This illustrates an important point: whether an allele is dominant or recessive does not determine whether it will be common or rare in the population. "Dominant" and "recessive" describe the way the alleles interact when they occur together in a heterozygous individual; they do not indicate which allele determines the more advantageous phenotype.

Rh factors The A and B antigens are not the only surface proteins on red blood corpuscles. Rh factors, named for antigens first identified in rhesus monkeys, are produced by alleles of the Rh gene. Individuals whose two copies of the Rh allele coding for nonfunctional products are said to be Rh-negative (Rh^-). Individuals with at least one functional Rh allele (and there are many, including four common ones) are said to be Rh^+. Another gene for surface proteins on red corpuscles has two alleles, *M* and *N*, giving rise to the genotypes *M/M*, *M/N*, and *N/N*. Each of these groups of blood antigens can create immunological problems during transfusions.

■ HOW DELETERIOUS ALLELES ARISE AND SURVIVE

As discussed in Chapters 8, 9, and 11, a variety of influences can cause mutations in genes. Because cells have elaborate DNA-repair mechanisms, the rate at which any particular gene undergoes mutation is ordinarily extremely low. However, every individual organism has a large number of different genes, and the total number of genes in all the individuals of a species is vast indeed. Hence mutations are constantly occurring within a species.

Every living organism is the product of billions of years of evolution and is a finely tuned, smoothly running, astoundingly intricate mechanism, in which the function of every part in some way influences the function of every other part. By comparison, the most complex computer is simple indeed. If you were to take such a machine and make some random change in its parts, the chances are great that you would make it run worse rather than better: a random change in any delicate and intricate mechanism is far more likely to damage it than to improve it. Thus the vast majority of the mutations that have obvious phenotypic effects are deleterious. Only very rarely is a mutation beneficial. Nevertheless, mutations are a major source of the variation on which natural selection operates. (The other sources include crossing over and sexual recombination, each of which creates new combinations of preexisting genes.)

The protection of heterozygosity When a deleterious allele arises by mutation, natural selection can act against it only if it causes some change in the organism's phenotype. Because dominant deleterious mutations are expressed, they can be eliminated from the population rapidly by natural selection; thus most mutations that persist in a population are recessive to the normal alleles. Deleterious alleles that are not dominant may be retained in the population in a heterozygous condition for a long time. Clearly, organisms that are diploid throughout their lives are much less sensitive to mutation than those that have extended haploid stages. This is probably the main reason diploidy predominates among long-lived organisms.

When a mating occurs between two diploid individuals carrying the same deleterious recessive allele in the heterozygous condition, about one-fourth of the progeny will be homozygous for the deleterious allele, and these homozygous offspring will have the harmful phenotype. Selection (death before reproduction) culls out such organisms, thereby reducing the frequency of the deleterious allele in the population. A recessive allele whose phenotype is fatal is called a ***lethal***. The occurrence of lethals can greatly modify the phenotypic ratios obtained in the progeny of some crosses.

In numerous instances alleles harmful or even lethal when homozygous are actually beneficial when heterozygous. The Dexter cattle breed is a good beef producer; however, it is impossible to establish a pure-breeding herd because some of its most desirable characteristics are caused by the heterozygous expression of an allele that is lethal when homozygous.

An example in humans is the allele for ***sickle-cell anemia***. In an individual homozygous for the sickle-cell allele, the mutant hemoglobin tends to crystallize under acidic conditions (as when blood levels of CO_2 rise during exercise). When the hemoglobin in a corpuscle crystallizes, the corpuscles become curved like a sickle and bear long filamentous processes (Fig. 16.16). These abnormal corpuscles tend to form clumps and to clog the smaller blood vessels. The resulting impairment of circulation leads to severe pains in the abdomen, back, head, and extremities, and to enlargement of the heart and atrophy of brain cells. In addition, the tendency of the deformed corpuscles to rupture brings about severe anemia. Homozygous victims of sickle-cell anemia usually suffer an early death. Individuals heterozygous for the sickle-cell allele sometimes

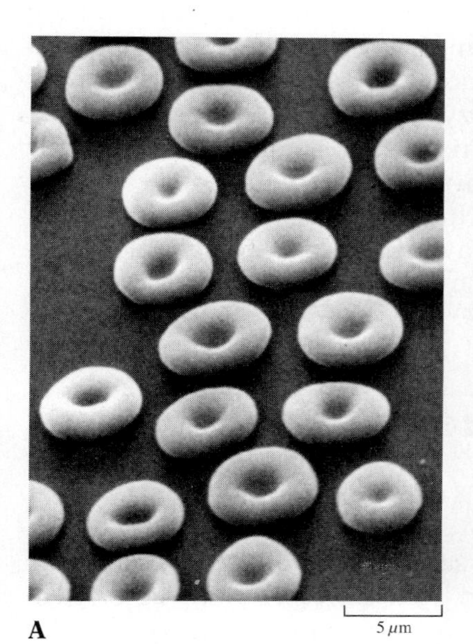

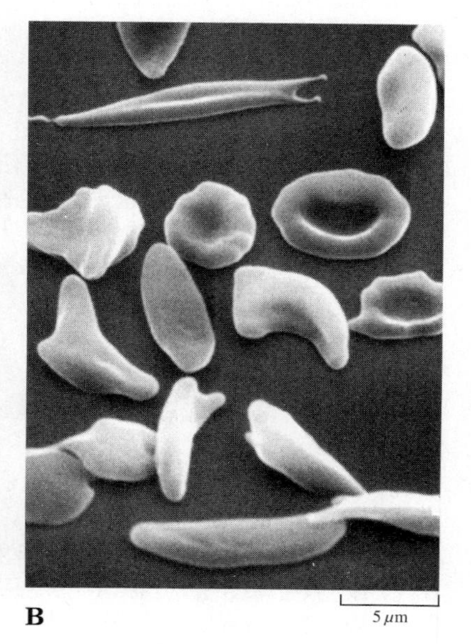

16.16 Normal and sickled red blood corpuscles

Normal erythrocytes (A), which are biconcave disks, look dramatically different from sickled cells (B). Some of the sickled cells seen here in this scanning electron micrograph bear the filamentous processes that may cause clogging in the body's smaller blood vessels.

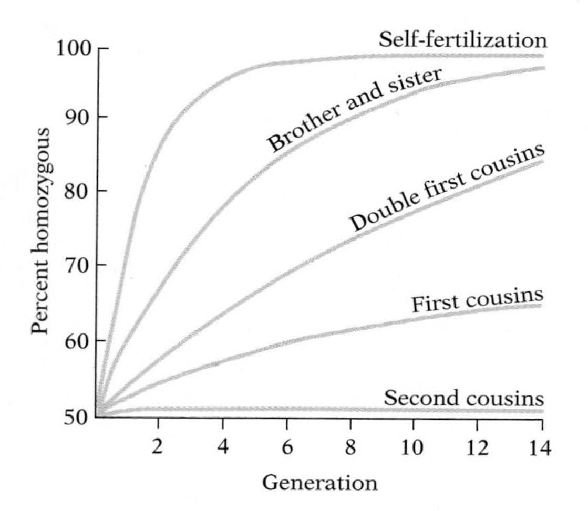

16.17 Percentage of homozygotes in successive generations under different degrees of inbreeding

It is assumed in this graph that the initial condition is two alleles of equal frequency (hence the 50% homozygosity at the start). If a similar graph were drawn for very rare recessive alleles, the initial rises in homozygosity would be much steeper, particularly in the three curves for cousin-to-cousin matings. Thus matings between first or second cousins, though of little effect in the situation graphed here, may greatly increase the percentage of homozygosity of rare, perhaps deleterious recessive alleles. (Double first cousins result when siblings of one family marry siblings of another family—when two brothers marry two sisters, for example. The offspring of such marriages are first cousins through *both* of their parents rather than through just one of them.)

show mild symptoms of the disease, but the condition is usually not serious.

You might suppose that natural selection would operate against the propagation of any allele so obviously harmful and that such an allele would be held at very low frequency in the population. But the allele is surprisingly common in many parts of Africa, being carried by up to 20% of the black population. In fact, individuals heterozygous for this allele have an unusually high resistance to malaria. Since malaria is very common in many parts of Africa, the sickle-cell allele is beneficial when heterozygous. Hence, in Africa there is selection for the allele when it is heterozygous and selection against it when it is homozygous. In other parts of the world, however, malaria is infrequent, and the benefits of the sickle-cell allele are outweighed by its costs. The frequency of the sickle-cell allele in the descendants of African blacks living in the United States is steadily declining, a clear instance of evolution.

Sickle-cell anemia is a dramatic example of an allele that has more than one effect. Such alleles are ***pleiotropic***. Pleiotropy is very common (though its effects may be subtle). A striking example involves the temperature-sensitive coat-color allele of Siamese cats mentioned earlier. This same allele also causes a mysterious misrouting during development of the axons carrying visual information from the eye to the brain. Many Siamese cats are cross-eyed to compensate for their visual miswiring.

The dangers of inbreeding The conditions under which genes cause deleterious phenotypes explain the danger of matings between closely related humans. Everyone carries in heterozygous combination many alleles that would cause harmful effects if present in homozygous combination, including some lethals. Because most of these deleterious alleles originated as rare mutations, however, and are limited to a tiny percentage of the population, the chances are slight that two unrelated individuals will be carrying the same deleterious recessive alleles and produce homozygous offspring that show the harmful phenotype.

The chances that two closely related persons are carrying the same harmful recessives, having received them from common ancestors, are much greater. If they mate, they may have children homozygous for the deleterious traits. In short, close inbreeding increases the percentage of homozygosity (Fig. 16.17). Matings between first cousins cause slight increases in homozygosity; while brother-sister matings and matings between double first cousins (those sharing two sets of grandparents) cause rapid increases. Many species have evolved behavioral or physiological mechanisms that prevent close relatives from mating.

HOW SEX ALTERS INHERITANCE

■ HOW SEX IS DETERMINED

The sex chromosomes We have said that diploid individuals have two of each type of chromosome, and hence two copies of each gene in every cell. In fact, this is not quite correct: in most species with distinct sexes, the chromosomal endowments of males and females are slightly different. In general, one of the two sexes has one pair of nonmatching chromosomes. These are the ***sex chromosomes***, which play a fundamental role in determining the sex of the individual; they exhibit their sex-specific effects primarily in the brain, in certain hormone-producing organs, in cells in sexually dimorphic portions of the external and internal anatomy, and in the gonads. All other chromosomes are called ***autosomes***.

Let's look first at the chromosomes of *Drosophila* and humans. In each species, one chromosome—the ***X chromosome***—bears many genes; the other, which is much smaller—the ***Y chromosome***—has only a few genes. Normal females have two X chromosomes, while males have one X and one Y. The diploid number in *Drosophila* is 8 (four pairs); females have three pairs of autosomes and one pair of X chromosomes, while males have three pairs of autosomes and a pair of sex chromosomes consisting of one X and one Y (Fig. 16.18). The diploid number in humans is 46 (23 pairs); females have 22 pairs of autosomes and one pair of X chromosomes, while males have 22 pairs of autosomes plus one X and one Y (see Fig. 12.4, p. 311).

When a female produces egg cells by meiosis, all the eggs receive one of each type of autosome plus one X chromosome. When a male produces sperm cells by meiosis, half the sperm cells receive one of each type of autosome plus one X chromosome, while the other half receive one of each autosome plus one Y chromosome (Fig. 16.18). When fertilization takes place, the chances are approximately equal that the egg will be fertilized by a sperm carrying an X chromosome or a sperm carrying a Y chromosome. If fertilization is by an X-bearing sperm, the resulting zygote will be XX and will develop into a female. If fertilization is by a Y-bearing sperm, the resulting zygote will be XY and will develop into a male. The sex of an individual is normally determined at the moment of fertilization and depends on which of the two types of sperm fertilizes the egg.

This XY-male system, though characteristic of many plants and animals (including all mammals), is not universal. Birds, butterflies and moths, and a few other animals have just the opposite system, where XX is male and XY is female. (To distinguish this from the usual XY system, the symbols Z and W are often substituted: ZZ for male and ZW for female.) In still other species—many reptiles and marine fish, for example—sex is environmentally determined: the sex of alligators depends on the temperature of the eggs during development, while some coral-reef fish reverse sex as adults to take advantage of variations in food and competition. Many species of animals (ranging from insects to mammals) have the ability to manipulate the sex ratio of their offspring to exploit changes in habitat conditions, competition, or individual social status that maximize the reproductive potential of their progeny.

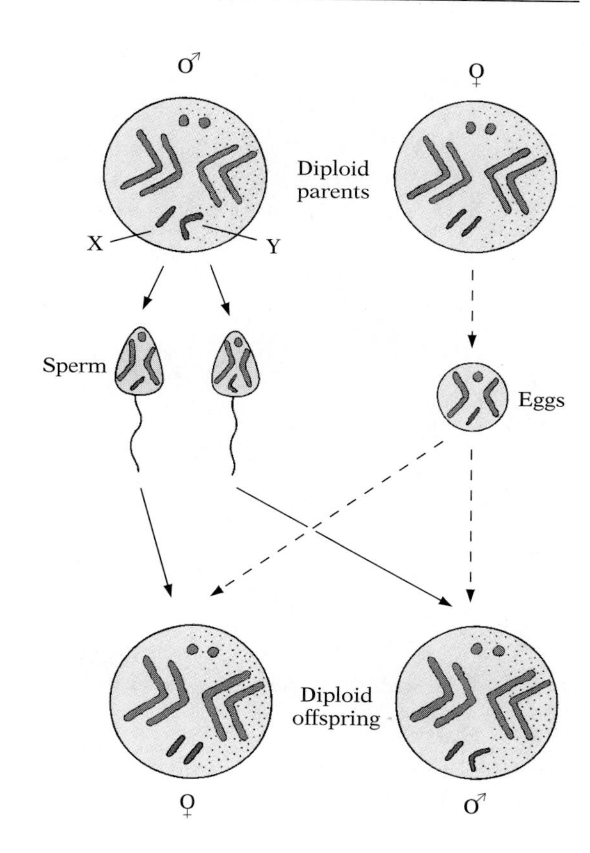

16.18 Chromosomes of male and female *Drosophila melanogaster*

There are three pairs of autosomes and one pair of sex chromosomes. Males (♂) have one X chromosome and one Y chromosome; since these separate at meiosis, half the sperm carry an X and half carry a Y. Since females (♀) have two X chromosomes, all eggs have an X. The sex of the offspring depends on which type of sperm fertilizes the egg.

What does the Y chromosome do? Occasionally the members of a homologous pair of chromosomes fail to separate properly in meiosis, so both move to the same pole. The effect of such ***nondisjunction*** is that one of the daughter cells receives one too many chromosomes while the other cell receives one too few. If a human gamete carrying an extra chromosome is involved in fertilization,

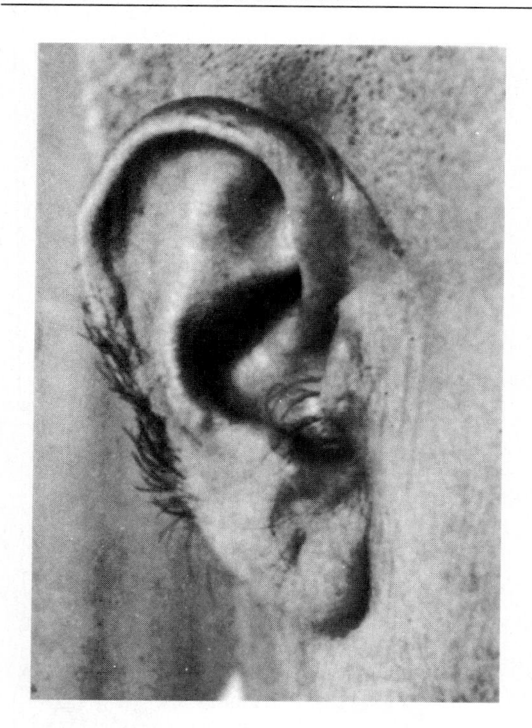

16.19 Hairy pinna
This trait is thought to be determined by a gene on the Y chromosome.

the zygote produced has 47 chromosomes (22 pairs plus three of one type) instead of 46 (23 pairs).

When the sex chromosomes fail to separate in meiosis, an XXY individual may result. In *Drosophila,* XXY individuals are essentially normal females, which suggests that only the X chromosomes function in sex determination: one produces a male and two a female, regardless of any Y chromosome. And indeed, in *Drosophila,* sex is determined by the ratio of certain gene products produced by the X chromosomes to certain products produced by the autosomes. A low ratio results in a male, while a high ratio produces a female.

This X-based model accounts for sex determination in grasshoppers and other animals that lack the Y chromosome entirely—females have two X chromosomes and males a single unpaired X. Such a system of sex determination is known as the XO system, with "O" denoting the absence of a chromosome.

The X-based system is not universal, however. The human Y chromosome, for example, bears a gene with strong male-determining properties. Its product is a DNA-binding protein that, early in fetal life, binds to numerous sites on the autosomes, initiating the developmental program that will produce a male. In the absence of this gene—even in XY individuals in whom this gene is deleted—a female develops. Genes unique to the Y chromosome are termed ***holandric***. The phenotypic traits they control appear, of course, only in males (Fig. 16.19). There are very few genes on the human Y chromosome.

■ THE GENETICS OF SEX-LINKED CHARACTERS

Most genes on the X chromosome are absent from the Y chromosome. Such genes are said to be sex-linked. The inheritance of characteristics controlled by sex-linked genes are completely different from those controlled by autosomal genes: females have two copies of each sex-linked gene, one from each parent, but males have only one copy of each sex-linked gene, and that one copy always comes from the mother. Furthermore, since the male has only one copy of each sex-linked gene, recessive alleles cannot be masked; selection acts directly on deleterious recessives on the X chromosome. Recessive sex-linked phenotypes, such as red-green color blindness, occur much more often in males than in females.

Sex linkage was discovered in 1910 by Thomas Hunt Morgan. It was Morgan who began the systematic use of *Drosophila* in genetic studies. With fruit flies it is possible to perform in a few months experiments that would have taken Mendel years with peas. *Drosophila* can be cultured in large numbers in the laboratory. It produces a new generation in 10–12 days and is subject to many easily detectable genetic variations. Most of the modern knowledge of eucaryotic genetics derives from work on this insect.

The first sex-linked trait discovered by Morgan was white eye color in *Drosophila*. This mutation, like dozens discovered since (including many others affecting eye color) arose spontaneously; it is controlled by a recessive allele *r.* The normal red eye color is controlled by the dominant allele *R.* If homozygous red-eyed females are crossed with white-eyed males, all the F_1 offspring, regardless of sex, have red eyes, since they receive from their mother an X chromosome bearing an allele for red. In addition, the F_1 females receive from their father an X chromosome bearing an allele for white eyes, but the red allele masks its presence. The F_1 males, like the females, receive from their mother an X chromosome bearing an allele for red eyes. Unlike the females, however, they receive no gene for eye color from their father, who contributes a Y chromosome instead of an X. (In writing the genotype of a male for a sex-linked character, the Y is shown to indicate that no second X chromosome is present and hence there is no second copy of the sex-linked gene.) We can summarize this cross as follows (♀ denotes females, ♂ males):

P		*R/R* red-eyed ♀	×	*r/Y* white-eyed ♂	
			↓		
F_1		*R/r* red-eyed ♀	×	*R/Y* red-eyed ♂	
			↓		
F_2	*R/R* red-eyed ♀	*r/R* red-eyed ♀		*R/Y* red-eyed ♂	*r/Y* white-eyed ♂

Notice that when the F_1 flies of this cross are allowed to mate among themselves, the F_2 flies show the customary 3 : 1 phenotypic ratio of a monohybrid cross where dominance is present. This 3 : 1 ratio is different, however, from the 3 : 1 ratio obtained in a cross involving autosomal genes. In an autosomal cross there is no correlation of phenotype with sex, but in this cross all F_2 individuals showing the recessive phenotype are males. This asymmetry illustrates the potential for selection to operate differentially on the two sexes.

Two well-known examples of recessive sex-linked traits in humans are red-green color blindness and hemophilia (a failure to form blood clots, which can lead to uncontrolled bleeding). Color blindness occurs in about 8% of white males in the United States and in about 4% of black males. It occurs in only about 1% of white females and about 0.8% of black females. More men than women will show such a trait since men need only one copy of the allele to show the phenotype. For a woman to be color-blind, she must have two copies of the allele, and so be homozygous; not only must her father be color-blind, but her mother must be either color-blind or a heterozygous carrier of the allele. Since the allele is not very common in the population, it is not likely that two such people will marry.

Though human females have two copies of each sex chromosome, one of the X chromosomes in each somatic cell condenses into a tiny dark object called a ***Barr body*** (Fig. 16.20). Most of the genes on this condensed X chromosome are inactive. Hence, a normally functioning female cell contains only one active copy of most sex-linked genes. Why, then, are sex-linked recessive traits expressed in females only when they are homozygous?

The explanation is that it is not always the same X chromosome

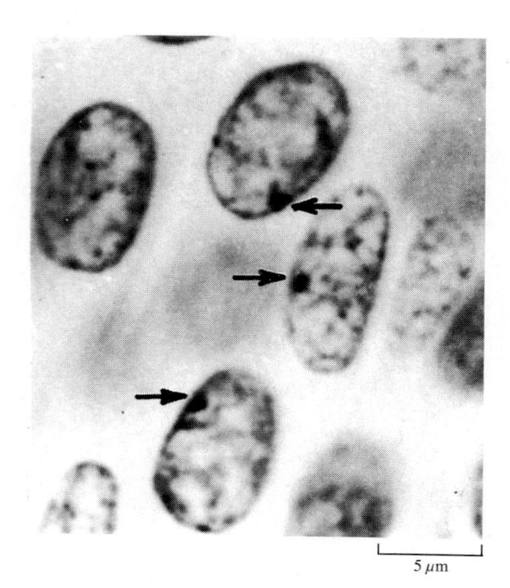

16.20 Nuclei from epidermal cells of a human female

The arrows indicate the Barr bodies. Since Barr bodies are present in the cells of female fetuses, the sex of an unborn child can be ascertained by examination of the nuclei of cells sloughed off into the fluid of the mother's womb.

16.21 X-chromosome inactivation in cats

One of the most common natural demonstrations of X-chromosome inactivation is the coat color of tortoiseshell and calico cats. A gene for color found on the cat's X chromosome has two common alleles—black and yellow (or orange). Males can have only one allele, so (in the absence of modifier genes) they are either yellow or black, usually with various white markings. Females, however, can have both alleles, and after inactivation some cells will express the black allele and others the yellow allele. Large patches of cells with one allele or the other develop, and the result is a mosaic of yellow, black, and (usually) white patches. Except for rare XXY individuals, all cats with both yellow and black fur are females.

that condenses into a Barr body in the different somatic cells, with no discernible pattern as to which chromosome is active in which cell. For example, consider the distribution of alleles of the sex-linked gene for the enzyme glucose-6-phosphate dehydrogenase; suppose one of the alleles codes for an active version of the enzyme, the other for a defective version. When the red blood corpuscles of women heterozygous for these two alleles are examined, roughly half the cells have normal enzyme activity while the other half have none. Women—and the females of most other mammals—are, in a sense, genetic mosaics for sex-linked traits. For some of these traits, the mosaic pattern finds phenotypic expression; an example is the coat color of tricolor (tortoiseshell and calico) cats (Fig. 16.21). For others, it does not. It seems, for instance, that as long as half the cells are normal in a woman heterozygous for red-green blindness or for hemophilia, she will be phenotypically normal.

The phenomenon of X-chromosome inactivation explains why the sex of mammals depends on the presence or absence of the Y chromosome rather than on the ratio of X chromosomes to autosomes, as in *Drosophila:* because of inactivation each mammalian somatic cell has only one functional X chromosome, and so is either XO (female) or XY (male); both sexes have the same number of X-chromosome genes available for transcription. In *Drosophila,* on the other hand, females have twice the number males have. The resulting difference in the amount of certain gene products determines the sex of the individual. For many other products, various mechanisms have evolved to achieve ***dosage compensation***—to equalize the amount resulting from transcription of the two X chromosomes in females with the amount resulting from transcription of the single chromosome in males. On average, genes on the single X chromosome of *Drosophila* males do double duty: they are transcribed twice as often as the corresponding genes on any one female X chromosome.

■ THE EXPRESSION OF SEX-DEPENDENT CHARACTERS

As we have seen, sex-linked genes may control characters not customarily regarded as sexual. Conversely, not all genes commonly associated with sex are sex-linked; many are located on the autosomes. For example, a number of genes that control growth and development of the sexual organs, as well as those that control distribution of body hair, size of breasts, pitch of voice, or other secondary sexual characteristics, are autosomal and are present in individuals of both sexes. That their phenotypic expression is different in the two sexes indicates that they are sex-*limited,* not sex-linked. The sex chromosomes determine what hormones are to be synthesized in each sex, and these hormones influence the activity of the sex-limited autosomal genes secondarily, either inhibiting or stimulating them.

There are other instances of sex-correlated inheritance. One, discussed in Chapter 10, involves the inheritance of mitochondria: in most higher organisms, sperm and pollen only rarely contribute mitochondria to the zygote, so mitochondrial genes are essentially always inherited from the female parent. The other case is imprinting: as described in Chapter 11, progeny can occasionally inherit genes with different patterns of methylation; if the gene in question is permanently inactivated in the gamete of one parent, the phenotype of the zygote is determined by the allele contributed by the gamete from the other parent.

HOW GENES ON THE SAME CHROMOSOME ARE INHERITED

Mendel formulated two generalizations. The first was that each individual carries two copies of every gene, and these copies segregate during gamete formation; this Principle of Segregation is often called Mendel's first law. The second generalization was that when several genes are involved in a cross (as in a dihybrid cross), they sort out into gametes independently of one another; this ***Principle of Independent Assortment*** is frequently referred to as Mendel's second law. All seven of the traits Mendel reported on from garden peas did indeed assort independently, and were free of the complications of modifiers, partial dominance, sex linkage, multiple alleles, and so on.

In fact, independent assortment describes only the simplest genetic interactions—the behavior of genes on separate chromosomes. Mendel almost certainly knew that some pairs of factors do not assort independently to produce offspring in neat 9 : 3 : 3 : 1 ratios, and that some do not show a simple pattern of recessiveness and dominance. He probably chose to ignore such anomalies. However, these nagging problems—particularly the failure of many crosses to exhibit independent assortment—were well known to geneticists and students of evolution. They were not understood until the discovery and description of chromosomes.

■ HOW GENES COME TO BE LINKED

In 1900, shortly after the rediscovery of Mendel's paper, W. S. Sutton pointed out the striking accord between Mendel's conclusion that hereditary factors (genes) occur in pairs in somatic cells and separate in gametogenesis, and the recent cytological evidence that somatic cells contain two of each kind of chromosome and that these chromosomes segregate in meiosis. Sutton interpreted this agreement as powerful evidence that the chromosomes are the bearers of the genes.

Now that Mendel's ideas could be directly related to observations made with microscopes, scientists could focus on the anomalies. In time it became apparent that traits assort independently only when their respective genes occur on two different chromosomes. Genes that occur on the same chromosome cannot assort independently during meiosis unless separated by crossing over. Genes on the same chromosome are said to be ***linked.*** One of the first examples of linkage was reported in 1906 by R. C. Punnett and William Bateson. They crossed sweet peas that had purple flowers and long pollen with ones that had red flowers and round pollen. All the F_1 plants had purple flowers and long pollen, as expected. (They knew that purple is dominant over red and that long is dominant over round.) The F_2 plants from this cross did not show the expected 9 : 3 : 3 : 1 ratio, however, but a highly anomalous one. Next, Bateson and Punnett crossed the F_1 plants back to homozygous recessive plants (with red flowers and round pollen). These results were equally anomalous. Using the symbols *B* for purple, *b* for red, *L* for long, and *l* for round, we can summarize Punnett and Bateson's results as follows:

BbLl × *bbll*
purple long   red round

↓

7 *BbLl*	1 *Bbll*	1 *bbLl*	7 *bbll*
purple long	purple round	red long	red round

According to the Principle of Independent Assortment, the heterozygous purple long parent should have produced four kinds of gametes (*BL*, *Bl*, *bL*, and *bl*) in equal numbers. When united with the *bl* gametes from the homozygous recessive parents, *BL* gametes should have given rise to purple long offspring, *Bl* gametes to purple round, *bL* to red long, and *bl* to red round, and these four phenotypes should have occurred in equal numbers, in a 1 : 1 : 1 : 1 ratio. But the result Bateson and Punnett actually obtained—a ratio of 7 : 1 : 1 : 7—makes it appear that the heterozygous parent produced far more *BL* and *bl* gametes than *Bl* and *bL* gametes.

Only in 1910 did Thomas Hunt Morgan, who had obtained similar results from *Drosophila* crosses, conceive of the answer. He postulated that the anomalous ratios were caused by linkage. Hence we should write the genotypes of the parents in Bateson and Punnett's second cross with slashes as *BL/bl* and *bl/bl,* to show that *B* and *L* are on one chromosome and *b* and *l* on the other. (We would write these genotypes as *B/b L/l* and *b/b l/l* if the genes were not linked).

If in Bateson and Punnett's cross the genes for purple and long and the genes for red and round were linked, the *BL/bl* parent in the second cross should have produced only two kinds of gametes, *BL* and *bl,* and the second cross should have yielded offspring of only two phenotypes, purple long and red round, in equal numbers. Yet the cross also yielded some purple round and red long offspring. Morgan suggested that some mechanism occasionally breaks the original linkages between purple and long and between red and round and establishes in a few individuals new linkages between purple and round and between red and long, making possible the production of *Bl* and *bL* gametes. The mechanism of this recombination is, of course, crossing over.

From an evolutionary perspective, crossing over is important because it increases the number of genetic combinations a cross can produce, and therefore the number of phenotypes on which selection can operate. Linkage can also have a major impact. Imagine that there were no crossing over. In such a case, an allele *a* on a chromosome containing a deleterious allele *Z* would be culled by selection operating against *Z*, even though *a* may be innocuous or even beneficial. Crossing over allows *a* to escape from its association with *Z*, but the more tightly linked two genes are—that is, the nearer they are to one another on the chromosome—the less often escape can occur (Fig. 16.22). Two unrelated genes located next to each other may be so tightly linked that selection for or against one is, in practice, selection for or against the other. Thus, it is possible for the frequency of one allele to be strongly affected by selection on an unrelated locus. At the other extreme, genes at opposite ends of a long chromosome in species with high crossover rates will be inherited virtually independently.

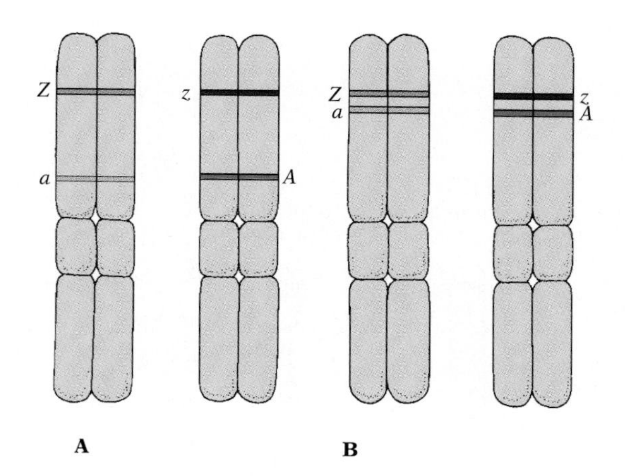

16.22 Effects of linkage on crossing over
When alleles *Z* and *a* are located far apart on a chromosome (A), the odds of a crossover event occurring between the two loci (and thus separating the alleles, leading to a *ZA* or *za* combination on a chromosome) are greater than when the two loci lie close together (B). In the latter case, selection operating on the allele at one locus can strongly affect the allele at neighboring loci.

■ MAPPING GENES ON CHROMOSOMES

Alfred H. Sturtevant—then an undergraduate working in Morgan's lab—pointed out that if the probability of crossing over is approximately equal at any point along the length of a chromosome, then the greater the distance between two linked genes, the greater the frequency with which they will be separated: the frequency of recombination between any two linked genes will be proportional to the distance between them. Sturtevant realized that the percentage of recombination can serve as a tool for mapping the location of genes on chromosomes. We speak in terms of the percentage of re-

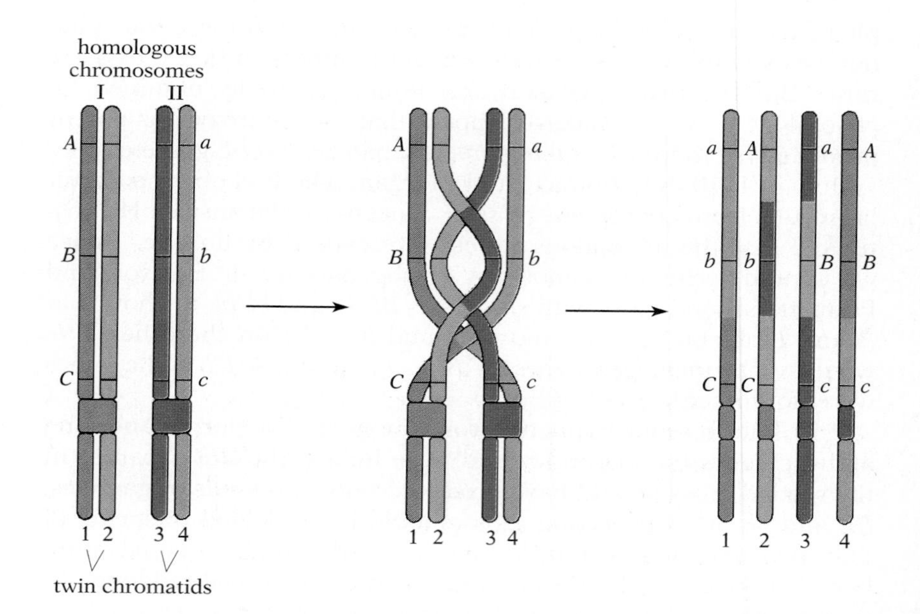

16.23 Linkage and recombination

After this typical round of crossing over, the three genes originally linked as *ABC* and *abc* are combined in four ways: *abC*, *AbC*, *aBc*, and *ABc*. Notice that the farther apart two genes are, the more likely it is that a crossover event will take place between them. In this example, crossing over between *A* and *C* takes place in every instance, whereas crossing over between *A* and *B* occurs only half as often. Notice also that crossing over between two genes does not always recombine them: because chromatids 2 and 3 have two compensating crossover events, *A* and *C* remain together, as do *a* and *c*, even though segments between the two genes have been interchanged. As a result, recombination frequency is always lower than crossover frequency.

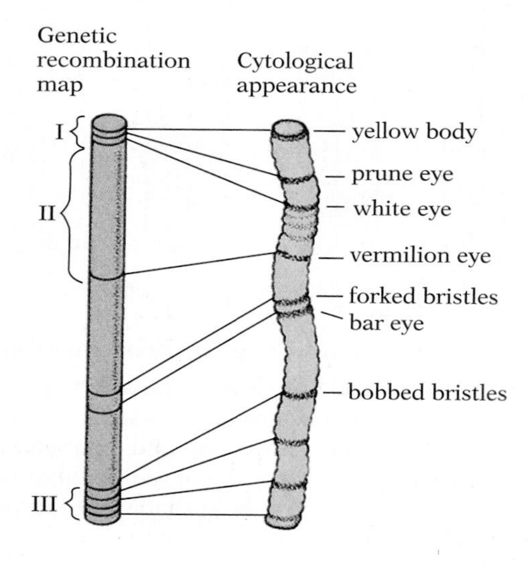

16.24 Comparison of a genetic-recombination map and cytological appearance of a portion of the X chromosome in *Drosophila melanogaster*

Staining and photographic techniques now exist to localize the genes on chromosomes directly. When these cytological results are compared with crossover frequencies (linkage maps), the effects of variations in crossover rates in different parts of the chromosome become apparent. In this case, crossing over is suppressed within two groups of genes (I and III), and greatly enhanced between the loci for white eyes and vermilion eyes (II). Suppression results from a relative lack of possible crossover sites while enhancement is a consequence of an overabundance of them.

combination rather than the number of crossover events because most chromosomes cross over at more than one place; therefore, two crossover events can cancel each other and so go undetected (Fig. 16.23).

The percentage of recombination gives no information about the absolute *distances* between genes (Fig. 16.24). Crossover events occur primarily at special DNA sequences, and selection can act to increase or decrease the number of such sites in any part of the chromosome if unusual crossover rates between particular genes are adaptive. On the other hand, crossover rates do reveal gene *order.* By convention, one unit of map distance on a chromosome is the distance within which recombination occurs 1% of the time. In Bateson and Punnett's test cross, two out of 16 of the offspring were recombinant products of crossing over. Two is 12.5% of 16; hence the genes controlling flower color and pollen shape in the sweet peas of this cross are located 12.5 map units apart.

Suppose we know that linked genes *B* and *L* are 12.5 map units apart. Further, suppose we find another gene, *A*, linked with these, that crosses over with gene *L* 5% of the time. How do we determine the order of the genes? The order could be *B–A–L:*

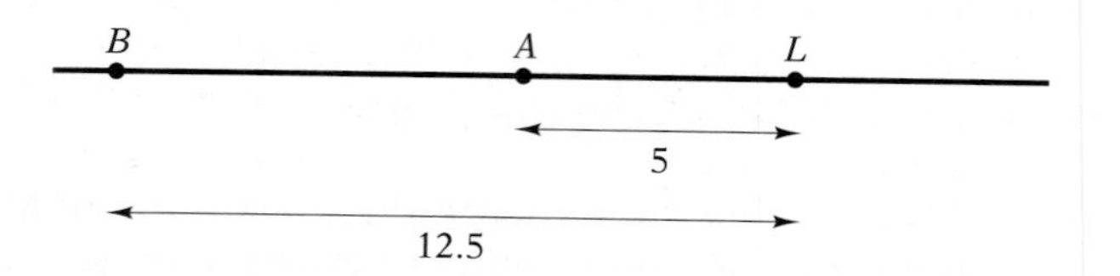

or it could be *B–L–A:*

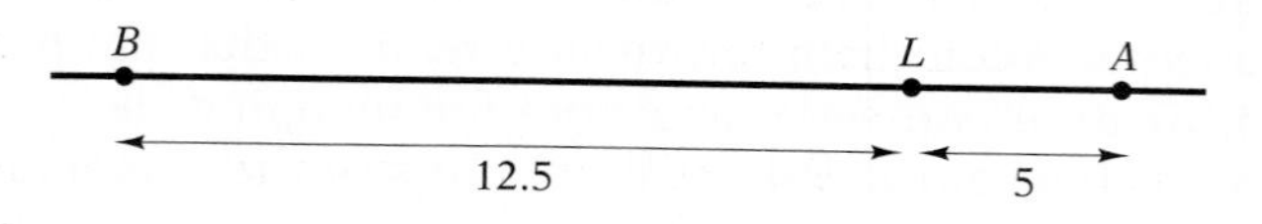

The way to decide between these two alternatives is to determine the frequency of recombination between *A* and *B*. If this frequency is 7.5% (12.5 − 5), then the first alternative is correct; if it is 17.5% (12.5 + 5), then the second alternative is correct. In this way, by determining the frequency of recombination between each gene and at least two other known genes, the arrangement of genes on a chromosome can be mapped. Such maps reveal which traits are so closely linked that selection acts on them as a unit (Fig. 16.24).

■ LINKAGE AND VARIATION

Genetics is critical to understanding evolution because it is the major basis of heritability and variation, two crucial requirements for natural selection. As detailed in Chapter 12, the recombination of chromosomes during gamete formation and fertilization generates enormous diversity (Fig. 16.25). In our species, for instance, even without crossing over, each individual can produce more than 8 million different kinds of gametes, and thus, when two unrelated individuals pair, 70 trillion different zygotes. Crossing over increases these values enormously. The average human gamete, for instance, carries a set of chromosomes that has undergone an average of 30 crossover events during meiosis, so that most chromosomes have recombined internally at least once. The result is that any individual can generate an essentially infinite number of distinct gametes. Recombination and crossing over provide far more variation for natural selection to operate on than does mutation.

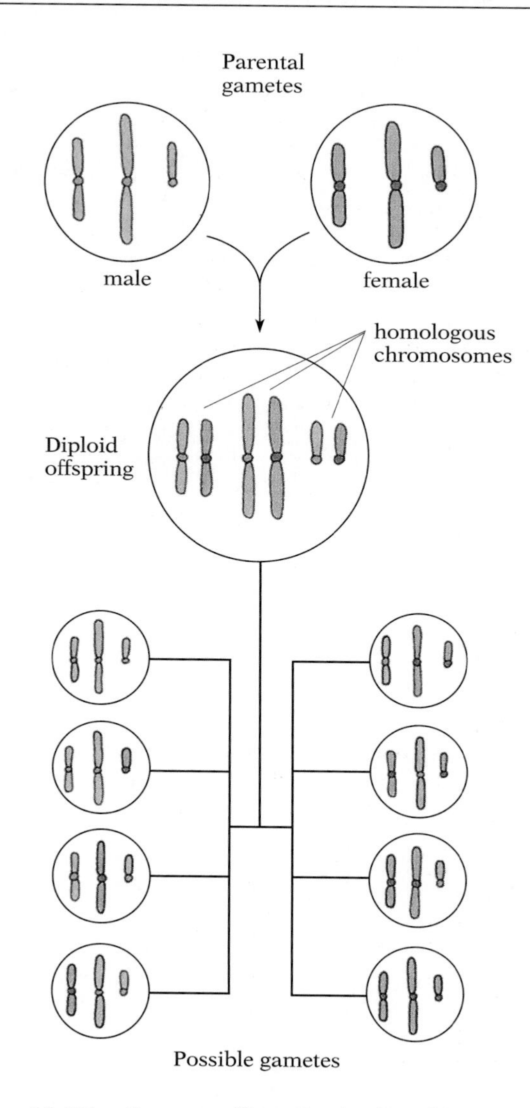

16.25 Gamete diversity in the absence of crossing over

Even in the absence of crossing over, each individual is capable of generating 2^n different kinds of gametes (where n is the number of chromosomes). Hence this hypothetical three-chromosome species produces eight classes of gametes, while our own species, with 23 chromosomes, can generate 2^{23}, or more than 8 million.

THE EFFECTS OF CHROMOSOMAL ALTERATIONS

A number of genetic alterations can have important consequences for selection and the evolution of new species—mutation and transposon movement, for instance, discussed in Chapters 8 through 11. In this section we will look at a few of the other chromosomal events that affect variation, linkage, and the emergence of new species.

■ STRUCTURAL CHANGES

In one form of alteration, called ***translocation***, portions of two nonhomologous chromosomes are interchanged, with a consequent modification in linkage groups. Suppose, for example, that a pair of bar-shaped chromosomes in a certain species bear the genes *ABCDEFG* and that a pair of J-shaped chromosomes bear the genes *lmnopqrst*. If the *EFG* end of one bar-shaped chromosome and the *st* end of one J-shaped chromosome were interchanged, the result would be a shorter bar chromosome bearing only the genes *ABCDst* and a longer J-shaped chromosome bearing the genes *lmnopqrEFG*. By changing the linkage relationships of genes, translocations can have important effects on phenotypes and selection.

Sometimes a piece breaks off one chromosome and fuses onto

the end of the homologous chromosome. Such an alteration is a kind of ***duplication***. An example would be loss of the *ABC* portion from one bar chromosome and fusion of this portion onto the homologous chromosome. The chromosome in which the loss occurred would thus bear only the genes *DEFGH*, while the chromosome undergoing the duplication would bear the genes *ABCABCDEFGH*. Translocation and duplication are most often caused by X-ray-induced damage to chromosomes.

Since translocations and duplications do not result in a net loss of genes, they usually have little effect on somatic cells. But these chromosomal rearrangements create problems in meiosis, when homologous chromosomes first synapse and then separate in the formation of gametes: If one chromosome of a pair has lost genes in a duplication event, half the gametes produced will lack these genes. If one chromosome of a pair has undergone a translocation exchange with a nonhomologous chromosome, synapsis with its homologue may be difficult, and meiosis may be blocked altogether. When meiosis does occur, the two nonhomologous chromosomes that have exchanged genes in translocation (losing some and gaining others) will segregate together only half the time. When they do not, half the resulting gametes will have nonhomologous duplications, while the other half will have two chromosomes with genes missing. A gamete that has lost essential genes may be inviable. Any zygote that does result from fusion involving such a gamete will lack the second copy of the genes in question; as a result, deleterious recessive alleles carried by the other gamete may be "exposed" (expressed in the organism's phenotype).

■ CHANGES IN CHROMOSOME NUMBER

As we saw earlier, the separation of chromosomes in cell division does not always proceed normally, and chromosomes that should move to opposite poles of the spindle may move instead to the same pole and become incorporated into the same daughter nucleus. The result of this nondisjunction may be an organism with one or more extra chromosomes. The presence of three chromosomes of one type in an otherwise diploid individual is called ***trisomy***.

Occasionally cell division may be so aberrant that all the chromosomes move to the same pole, giving rise to a daughter cell with twice the normal number of chromosomes. If this happens during meiosis, the gamete produced is diploid instead of haploid. If such a gamete unites at fertilization with a normal haploid gamete, a triploid zygote results; if it unites with another diploid gamete, also produced by aberrant meiosis, a tetraploid zygote results. Cells or organisms that have more than two complete sets of chromosomes—that are triploid, tetraploid, hexaploid, and so on—are said to be ***polyploid***.

Polyploidy is fairly common in plants. It has sometimes given rise to new species that are adaptively superior to the original diploid species under certain environmental conditions. Polyploidy can be stimulated in the laboratory by treating plants with special chemicals that cause nondisjunction during cell division. This procedure has been used in the production of many of the new strains

EXPLORING FURTHER

TRISOMY IN HUMANS

In humans most trisomies are lethal. Trisomy-18 (Edwards' syndrome) and trisomy-13 (Patau's syndrome), for example, produce physical malformations and mental and developmental retardation so severe that most afflicted infants die within a few weeks after birth. Because trisomies of most other autosomes result in spontaneous abortion, they are not found in live births. Two kinds of trisomy—trisomy-21 (Down's syndrome) and trisomies of the sex chromosomes—are exceptional in that the infants may survive.

Down's syndrome, in which three chromosomes of type 21 occur in the individual's cells, was the first clinical condition ever linked to a chromosomal abnormality. It is associated with a variety of characteristic physical features (broad head, rounded face, perceptible folds in the eyelids, a flattened bridge of the nose, protruding tongue, small irregular teeth, and short stature) and also mental retardation (IQ of about 40). The incidence of Down's syndrome is often related to the age of the mother. It occurs in fewer than one out of 1000 births in women under 20; it is over 7 times more common in women 35–39 years old, more than 20 times more common when the mother is 40–44, and more than 50 times more common when she is 45 or older. A similar association with the age of the mother is seen in Edwards' and Patau's syndromes.

Trisomy of the sex chromosomes can take several forms. In one, called Klinefelter's syndrome, the chromosomal makeup is XXY, and the individuals are males. Most victims show a variety of physical abnormalities and mental retardation; furthermore, they often suffer from thyroid dysfunction, chronic pulmonary distress, and diabetes.

Males with a second type of sex-chromosome trisomy, the XYY syndrome, generally show fewer and less severe abnormalities, though they often have subnormal intelligence. Because the incidence of XYY individuals is often significantly higher in penal institutions than in the general population, some investigators have advanced the controversial suggestion that men with the XYY condition are predisposed to aggressive behavior.

Women with triple-X syndrome (XXX) usually have underdeveloped sexual characteristics and sometimes subnormal intelligence, but their abnormalities are not debilitating.

These trisomic conditions, as well as many other kinds of genetic or chromosomal diseases, can be detected during embryonic development by the process of ***amniocentesis***, in which fluid containing sloughed-off epidermal cells from the fetus is withdrawn from the uterus with a long needle inserted through the mother's abdominal wall. The fetal cells are then cultured and examined for abnormalities. A new technique, in which embryonic tissue cells are obtained directly through the maternal vagina, promises earlier and safer detection of these and other chromosomal conditions.

of cultivated plants (Fig. 16.26). Polyploidy is rare in animals and has probably not been an important factor in the origin of new animal species. There are, however, at least two exceptions: there exist several species of triploid lizards, which are of special interest because they are asexual. In addition, polyploidy may have played

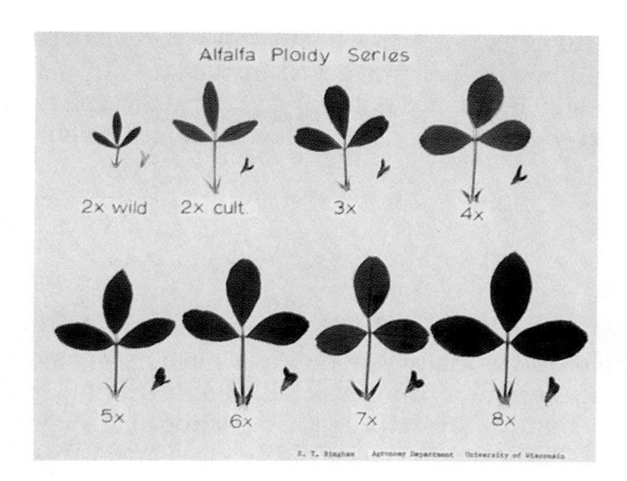

16.26 Induced polyploidy in alfalfa

Tetraploidy and octoploidy in alfalfa and other commercially valuable crops can be induced experimentally. Tetraploid alfalfa (4×) is the most stress-tolerant, and is the type cultivated by farmers; cultivated alfalfa is a naturally occurring tetraploid. Octoploid alfalfa (8×) grows well in the greenhouse but is sensitive to the stress induced by lack of water in the field. (Diploid alfalfa, 2×, is shown at the upper left.) The polyploids shown here were obtained by a process known as sexual polyploidization, in which gametes with unreduced numbers of chromosomes are produced during meiosis. The union of such gametes results in tetraploid plants; repetition of the process in the next generation yields octoploids. This process is also the principal source of polyploidy of plants in nature.

a role in the evolution of bees and related insects, among which the various species usually have 4, 8, 12, or 16 chromosomes; a similar pattern exists in certain groups of fish and some frogs.

HOW EXPERIMENTAL RESULTS ARE TESTED

■ WHY STATISTICAL ANALYSIS IS NECESSARY

So far we have mentioned Mendelian ratios such as 1 : 2 : 1, 3 : 1, 1 : 1, and 9 : 3 : 3 : 1—all commonly expected results of various types of crosses. Geneticists frequently use the phenotypic ratios obtained in breeding experiments to deduce the underlying genetic phenomena. Evolutionary biologists, on the other hand, look for evidence of evolution—that is, changes in gene frequencies—in the form of departures from the expected Mendelian ratios. For example, when selection is operating against a homozygous dihybrid recessive, the size of the fourth class of the 9 : 3 : 3 : 1 distribution is reduced. However, chance deviations from a predicted distribution are common: although four coin tosses have an expected distribution of two heads and two tails, a 3 : 1 ratio is hardly surprising. However, if the four-toss sequence is repeated 100 times, yielding 300 heads and 100 tails, we would suspect that the coin was biased in some way. To judge whether a phenotypic distribution indicates the operation of selection (or of some other unexpected genetic process), we therefore must consider both the degree of discrepancy from the expected ratio *and* the sample size.

Scientists constantly encounter the same fundamental question—whether the deviations they observe in their experimental results (from a prediction or from another set of results) are significant or not. They cannot rely simply on a subjective impression. To help them decide, they can compute the mathematical probability that any observed deviation in their sample could have occurred by chance alone. The calculated probability is the chance of obtaining the observed results (or results even more extreme) by drawing data at random from the predicted set, or from the other set of results being compared. Statisticians have devised many tests for evaluating experimental or observational data. Though these tests differ in their form and in the sorts of data to which they can validly be applied, all are ways of calculating the probability that any deviations of the observed values might be due to chance alone.

■ THE CHI-SQUARE TEST

One test of statistical significance, the chi-square (χ^2) test, is particularly applicable to many genetic experiments. This test measures whether any deviation from the predicted distribution that occurs in experimental results exceeds the deviation that might occur by chance. It is designed for categorical data—green versus red, for example—rather than continuous data, such as 2.1 cm versus 2.4 cm. The formula for chi-square is

$$\chi^2 = \Sigma\, d^2/e$$

where d is the deviation from the expected value, e is the expected value, and Σ means "the sum of."

Consider two hypothetical crosses in which we expect that the phenotypic ratio should be 1 : 1 in the absence of selection for or against one of the phenotypes. In one cross we get values of 45 and 55 instead of 50 and 50, and in the other we get values of 5 and 15 instead of 10 and 10. We want to know in each case whether the deviation of the observed values from the expected values can reasonably be attributed to chance, or implies that selection is at work.

First we must determine the chi-square value for the two crosses (Table 16.6). In each of these experiments the absolute deviations of the observed values from the expected values are the same: a deviation of 5 in each phenotype. Note, however, that the chi-square values obtained in the two crosses are very different: the one based on a sample of 20 is 5 times as large as the one based on a sample of 100. This illustrates how sensitive the chi-square test is to sample size: the difference in sample size alone has made the great difference in these two chi-square values.

Each of these crosses involved only two classes, in this case two different phenotypes. Hence their chi-square values were calculated on the basis of only two squared deviations. Now suppose we had been analyzing a cross involving three different phenotypes. Then the chi-square would have been calculated on the basis of three squared deviations, and it is only reasonable to expect that the chi-square value obtained would have been higher than one based on only two. It is clear, then, that in evaluating chi-square values we must also take into account the number of classes on which they are based. The number of independent classes in a chi-square test is termed the ***degree of freedom***. The number of independent classes is one fewer than the total number of classes in the cross.

In our crosses involving two phenotypes, there is only one independent class (and so one degree of freedom), while in a cross involving three phenotypes there are two independent classes and two degrees of freedom. A moment's thought will tell you why this is so. In our cross based on a sample of 100, once we know that 45 offspring show the first phenotype, we automatically know that 55 must show the other phenotype. Since we know the total, the number in one class automatically tells us the number in the other class. In other words, the number in the second class is dependent on the number in the first class. Therefore, only the first class is an independent class. The same reasoning applies if we perform a cross involving three different phenotypes, and the total number of observations in our sample is 100; once we know the number showing the first and second phenotypes, we automatically know the number showing the third phenotype, because the number in the third class is dependent on the number in the first two classes.

We now know the chi-square values (1.0 and 5.0) and the degrees

Table 16.6 Chi-square analysis of two crosses

	FIRST PHENOTYPE	SECOND PHENOTYPE
45:55 experiment:		
Observed values	45	55
Expected values (e)	50	50
Deviation (d)	−5	+5
Deviation squares (d^2)	225	25
d^2/e	25/50 = 0.5	25/50 = 0.5
$\chi^2 = \Sigma d^2 e = 0.5 + 0.5 = \mathbf{1.0}$		
5:15 experiment		
Observed values	5	15
Expected values (e)	10	10
Deviation (d)	−5	+5
Deviation squared (d^2)	25	25
d^2/e	25/10 = 2.5	25/10 = 2.5
$\chi^2 = \Sigma d^2 e = 2.5 + 2.5 = \mathbf{5.0}$		

Table 16.7 Probabilities for certain values of chi-square

DEGREES OF FREEDOM	P = 0.20 (1 IN 5)	P = 0.10 (1 IN 10)	P = 0.05 (1 IN 20)	P = 0.01 (1 IN 100)
1	1.64	2.71	3.84	6.64
2	3.22	4.60	5.99	9.21
3	4.64	6.25	7.82	11.34
4	5.99	7.78	9.49	13.28
5	7.29	9.24	11.07	15.09
6	8.56	10.64	12.59	16.81
7	9.80	12.02	14.07	18.48
8	11.03	13.36	15.51	20.09
9	12.24	14.68	16.92	21.67
10	13.44	15.99	18.31	23.21
15	19.31	22.31	25.00	30.58
20	25.04	28.41	31.41	37.57
30	36.25	40.26	43.77	50.89

Source: Based on a larger table in R.A. Fisher, 1946. *Statistical Methods for Research Workers*, 10th ed., Oliver & Boyd, London.

of freedom (one for each experiment) for our two hypothetical crosses. The next step is to consult a table of chi-square values. Table 16.7 gives four different chi-square values for each of a series of different degrees of freedom, and gives the probability (*P*) that a deviation as great as or greater than that represented by each chi-square value would occur simply by chance.

We can now evaluate the results obtained in our hypothetical crosses. In the first cross the deviation from the expected result yielded a chi-square value of 1.0. The cross had one degree of freedom. According to the table, a value as high as or higher than 1.64 has a chance probability of 0.20 (20%); that is, deviation from the expected as great as or greater than that represented by 1.64 will occur about once in five trials by chance alone. Our chi-square value is less than 1.64; hence the deviation in the experiment can be expected to occur by chance even more often than once in five trials. Most biologists agree that deviations having a chance probability as great as or greater than 0.05 (5%, or 1 in 20) are not statistically significant. Thus the deviation in our experiment is not regarded as statistically significant; it is presumed to be a chance deviation, which can be disregarded.

In our second experiment, the chi-square value representing the deviation from the expected results turned out to be 5.0. Again there was one degree of freedom. Looking at the listings in the table for one degree of freedom, we find that the value of 5.0 is greater than 3.84, which has a probability of 0.05 (5%), but less than 6.64, which has a probability of 0.01 (1%). Hence the probability that the deviation in this cross resulted purely from chance is between 1% and 5%, and thus is significant. Some factor other than chance was involved in producing the disagreement between result and prediction.

At this point, a geneticist would begin the search for a reasonable explanation: the original observations are always open to scrutiny. Selection may have acted against one of the phenotypes, so that some of those individuals died, thus leading to fewer representatives of this class than were expected; or the assumptions concerning the genetics involved in this cross may have been wrong. In any event, something interesting is probably occurring.

CHAPTER SUMMARY

HOW SINGLE-GENE TRAITS ARE INHERITED

WHAT MENDEL FOUND Mendel discovered that some traits are dominant, and thus mask other (recessive) traits. Thus the phenotype reflects only the dominant traits; any recessives in the genotype are hidden. Each organism carries two factors for a trait, which can be the same (so the organism is homozygous) or different (heterozygous). In reproduction the factors segregate independently, with one going into each gamete. Factors are now called alleles, and each allele is one form of a gene that occupies a particular locus on a chromosome. (p. 407)

PARTIALLY DOMINANT TRAITS When neither of two alleles is masked by the other, they are said to be partially dominant, and the phenotype is often a blend of the two traits. (p. 412)

HOW MULTIPLE AND MULTIGENIC TRAITS ARE INHERITED

INHERITING TWO TRAITS AT ONCE Traits are inherited completely independently when their loci are on different pairs of chromosomes. The distribution of genotypes can be computed with the Product Law (p. 413)

HOW GENES INTERACT Many genes depend on other genes for their full expression. A complementation test can determine whether a given recessive phenotype in two individuals arises from alleles at a single locus. Some genes can modify the expression of other genes. An epistatic gene determines whether or not one or more other genes will be expressed. Traits controlled by many separate genes often interact to produce a graded range of phenotypes; in such cases the effect of an individual gene is often impossible to discern. (p. 415)

WHEN GENES ARE ONLY PARTIALLY EXPRESSED Not all individuals with a particular genotype will express the associated phenotype; such genes have only partial penetrance. The genetic or physical environment may control the expressivity—the manner in which a genotype is expressed. (p. 420)

WHEN GENES HAVE SEVERAL ALLELES While an individual can only have two alleles of a gene, the population may harbor a large number, yielding a variety of genotypes and phenotypes. Examples are eye color in fruit flies and blood type in humans. (p. 422)

HOW DELETERIOUS ALLELES ARISE AND SURVIVE Recessive deleterious alleles at low frequency are generally hidden in heterozygous individuals. Some deleterious alleles actually confer an advantage in the heterozygous state. Inbreeding vastly increases the probability of generating offspring that are homozygous for a deleterious allele. (p. 424)

HOW SEX ALTERS INHERITANCE

HOW SEX IS DETERMINED Many species have sex-specific chromosomal differences, which result from an absent or much-reduced copy of one member of one chromosomal pair. (p. 426)

THE GENETICS OF SEX-LINKED CHARACTERS When one sex has only one full-length copy of the sex chromosome, it generally expresses any recessive alleles on that chromosome. In many species, the second sex chromosome in the sex with two full-sized sex chromosomes condenses to form a Barr body. (p. 428)

THE EXPRESSION OF SEX-DEPENDENT CHARACTERS Most sex-dependent characters are determined by genes on the autosomes. Some signal that is dependent directly or indirectly on the sex chromosomes determines which sex-dependent genes will be active. (p. 431)

HOW GENES ON THE SAME CHROMOSOME ARE INHERITED

HOW GENES COME TO BE LINKED When genes lie on the same chromosome, they usually do not segregate independently. Their degree of independence depends on the probability of a crossover event occurring between them during meiosis; in some species the crossover rate is so high that genes at opposite ends of long chromosomes are inherited essentially independently. (p. 432)

MAPPING GENES ON CHROMOSOMES The probability of genes on a chromosome becoming unlinked during meiosis can be used to map the approximate distance between them and to establish their order. (p. 433)

LINKAGE AND VARIATION Crossing over enormously increases the number of unique gametes meiosis can produce, and thus increases the range of variation in offspring. (p. 435)

THE EFFECTS OF CHROMOSOMAL ALTERATIONS

STRUCTURAL CHANGES Chromosomes sometimes break and are rejoined incorrectly. Occasionally pieces of chromosomes are copied mistakenly and incorporated into the genome, leading to duplications. Both translocations and duplications make pairing during meiosis difficult. (p. 435)

CHANGES IN CHROMOSOME NUMBER Occasionally chromosomes fail to segregate properly, and a cell winds up with too many or too few copies of a chromosome. Down's syndrome is caused by three copies of chromosome 21. (p. 436)

HOW EXPERIMENTAL RESULTS ARE TESTED

WHY STATISTICAL ANALYSIS IS NECESSARY Experiments either compare two sets of results, or a set of results with a prediction. Statistical techniques determine the probability that the degree of difference observed could have occurred by chance—that is, by drawing the data for one set at random from the other set. (p. 438)

THE CHI-SQUARE TEST The chi-square test is used when the data being collected are categorical. (p. 438)

STUDY QUESTIONS

1 In squash an allele for white color (*W*) is dominant over the allele for yellow color (*w*). Give the genotypic and phenotypic ratios for the results of each of the following crosses:

$W/W \times w/w$

$W/w \times w/w$

$W/w \times W/w$

2 A heterozygous white-fruited squash plant is crossed with a yellow-fruited plant, yielding 200 seeds. Of these, 110 produce white-fruited plants, while only 90 produce yellow-fruited plants. Using the chi-square test, would you conclude that this deviation is the result of chance, or that it probably represents some complicating factor? What if there were 2000 seeds, and 1100 produced white-fruited plants while 900 produced yellow-fruited individuals?

3 In humans, brown eyes are dominant over blue eyes. Suppose a blue-eyed man marries a brown-eyed woman whose father was blue-eyed. What proportion of their children would you predict will have blue eyes?

4 If a brown-eyed man marries a blue-eyed woman and they have 10 children, all brown-eyed, can you be certain that the man is homozygous? If the 11th child has blue eyes, what will that show about the father's genotype?

5 The litter resulting from the mating of two short-tailed cats contains three kittens without tails, two with long tails, and six with short tails. What would be the simplest way of explaining the inheritance of tail length in these cats? Show the genotypes.

6 When Mexican hairless dogs are crossed with normal-haired dogs, about half the pups are hairless and half have hair. When, however, two Mexican hairless dogs are mated, about a third of the pups have hair, about two-thirds are hairless, and some deformed puppies are born dead. Explain these results.

7 In peas an allele for tall plants (*T*) is dominant over the allele for short plants (*t*). An allele of another independent gene produces smooth peas (*S*) and is dominant over the allele for wrinkled peas (*s*). Calculate both phenotypic ratios and genotypic ratios for the results of each of the following crosses:

$T/t \quad S/s \times T/t \quad S/s$

$T/t \quad s/s \times t/t \quad s/s$

$t/t \quad S/s \times T/t \quad s/s$

$T/T \quad s/s \times t/t \quad S/S$

8 In some breeds of dogs a dominant allele controls the characteristic of barking while trailing. In these dogs an allele of another independent gene produces erect ears; it is dominant over the allele for drooping ears. Suppose a dog breeder wants to produce a pure-breeding strain of droop-eared barkers, but he knows that the genes for silent trailing and erect ears are present in his kennels. How should he proceed?

9 In Leghorn chickens, colored feathers are produced by a dominant allele, (*C*); white feathers are produced by the recessive allele (*c*). The dominant allele of another independent gene (*I*) inhibits expression of color in birds with genotypes *C/C* or *C/c*. Consequently both *C/– I/–* and *c/c –/–* are white. A colored cock is mated with a white hen and produces many offspring, all colored. Give the genotypes of both parents and offspring.

10 If the dominant allele *K* is necessary for hearing, and the dominant allele *M* of another independent gene results in deafness no matter what other genes are present, what percentage of the offspring produced by the cross *k/k M/m* × *K/k m/m* will be deaf?

11 What fraction of the offspring of parents each with the genotype *K/k L/l M/m* will be *k/k l/l m/m?*

12 Suppose that an allele, *b*, of a sex-linked gene is recessive and lethal. A man marries a woman who is heterozygous for this gene. If this couple had a large number of normal children, what would be the predicted sex ratio of these children?

13 Red-green color blindness is inherited as a sex-linked recessive. If a color-blind woman marries a man who has normal vision, what would be the expected phenotypes of their children with reference to this character?

14 In cats, short hair is dominant over long hair; the gene involved is autosomal. An allele B^1 of another gene, which is sex-linked, produces yellow coat color; the allele B^2 produces black coat color; and the heterozygous combination B^1/B^2 produces tortoiseshell and calico coat color. If a long-haired black male is mated with a tortoiseshell female homozygous for short hair, what kind of kittens will be produced in F_1? If the F_1 cats are allowed to interbreed freely, what are the chances of obtaining a long-haired yellow male?

15 In *Drosophila melanogaster* there is a dominant allele for gray body color and a dominant allele of another gene for normal wings. The recessive alleles of these two genes result in black body color and vestigial wings, respectively. Flies homozygous for gray body and normal wings are crossed with flies that have black bodies and vestigial wings. The F_1 progeny are then crossed, with the following results:

Gray body, normal wings	236
Black body, vestigial wings	253
Gray body, vestigial wings	50
Black body, normal wings	61

Would you say that these two genes are linked? If so, how many units apart are they on the linkage map?

16 The recombination frequency between linked genes *A* and *B* is 40%; between *B* and *C*, 20%; between *C* and *D*, 10%; between *C* and *A*, 20%; between *D* and *B*, 10%. What is the sequence of the genes on the chromosome?

SUGGESTED READING

HOLLIDAY, R., 1989. A different kind of genetic inheritance, *Scientific American* 260 (6). *On genetic "imprinting."*

WHITE, R., AND J. M. LALOUEL, 1988. Chromosome mapping with DNA markers, *Scientific American* 258 (2). *On modern methods for mapping the human genome.*

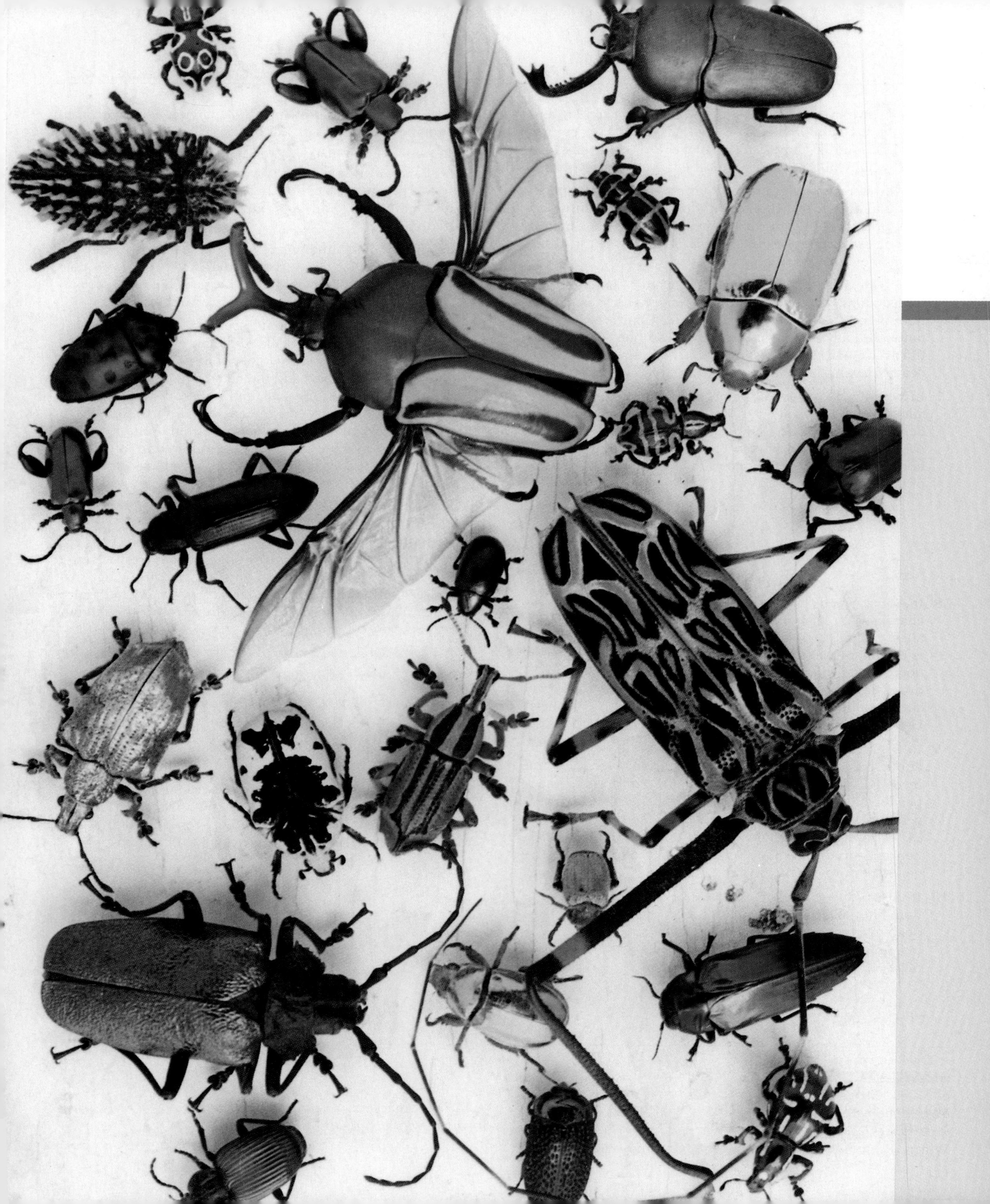

PART III

EVOLUTIONARY BIOLOGY

◄ **Diversity of species** produced by natural selection is suggested by this colorful array of beetles. The species represented here constitute a minute fraction of the estimated 400,000 extant species in the order Coleoptera, one of the 28 orders that make up the vast class of animals known as Insecta. (Actually, one of the specimens shown is a true bug—a member of the order of insects called Hemiptera.)

CHAPTER 17

VARIATION, SELECTION, AND ADAPTATION

Evolution, by definition, is the change in allelic frequencies in populations across generations. Because an individual's genotype is fixed at conception, an individual cannot evolve; only populations can. At its simplest, evolution can occur through the emigration or immigration of members of a population. In this chapter we will concentrate instead on the roles played by chance and natural selection in evolution. After we look at how these two processes generate changes in allele frequencies, we will examine some especially compelling cases of adaptation.

HOW SELECTION OPERATES

Evolution by natural selection depends on five factors:

1 More offspring are produced than can survive to reproduce (excess progeny).

2 The characteristics of living things differ among individuals of the same species (variability).

3 Many differences are the result of heritable genetic differences (heritability).

4 Some differences affect how well adapted an organism is (differential adaptedness).

5 Some differences in adaptedness are reflected in the number of offspring successfully reared (differential reproduction).

The excess reproductive capacity of organisms is self-evident. Darwin made the point incisively:

> The elephant is reckoned the slowest breeder of all known animals, and I have taken some pains to estimate its probable minimum rate of natural increase; it will be safest to assume that it begins breeding when 30 years old and goes on breeding until 90 years old; if this be so, after a period from 740 to 750 years there would be nearly 19 million elephants descended from this first pair.

After about 1200 years, this hypothetical elephant population would cover the entire land area of the earth, Antarctica included, shoulder to shoulder and head to tail (Fig. 17.1). Clearly, not all offspring survive long enough to reproduce.

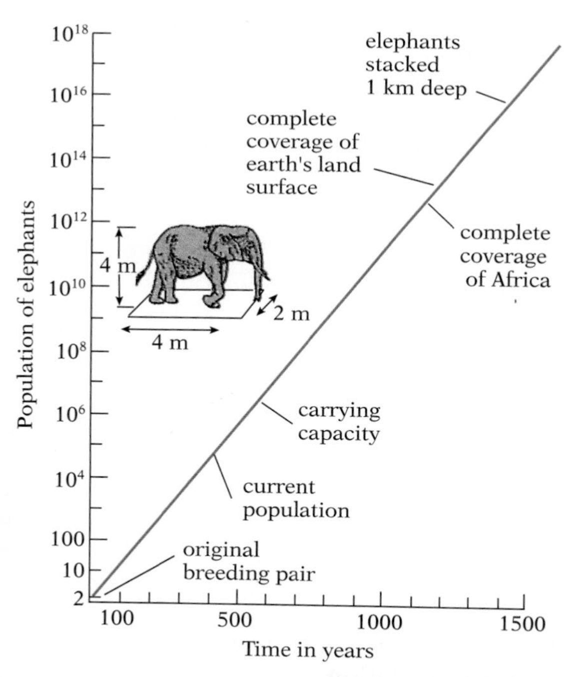

17.1 Unrestrained growth of a hypothetical elephant population

When Darwin's calculations for the slowest reproducing animal known are graphed and extrapolated, his point that not all offspring can survive becomes clear. (Darwin overestimated both the reproductive life span and the age of sexual maturity in elephants, but these minor errors cancel one another out.)

The nature of heritability of variant alleles was described in Chapter 16. As pointed out there, most variants are recessive, and therefore are unexpressed and hidden from selection when present in only one copy. Here we will look at the sources of variation and their consequences, and then consider mechanisms of selection. The second major section of the chapter will explore several types of adaptation.

■ WHAT CAUSES VARIATION?

A population is composed of many individuals. With rare exceptions, no two of these are exactly alike. We are well aware of the uniqueness of individual humans; we know from experience that each has distinctive anatomical and physiological characteristics, as well as distinctive abilities and behavioral traits. We are also fairly well aware of individual variation in dogs, cats, and horses. However, we tend to overlook the similar individual variation in less familiar species, like robins, squirrels, earthworms, sea stars, dandelions, and corn. Yet even though this variation may be less obvious to our unpracticed eye, it exists in all species.

The members of a population, then, share some important features but differ from one another in numerous ways. It follows that, if there is selection for or against certain heritable variants within a population, the makeup of that population may change with time.

17.2 Sexual recombination as a source of variation

Offspring of the same parents frequently do not resemble each other or either of their parents. Shown here with their mother are the various kittens of a single litter.

Genetic variation Variation in sexual organisms arises from three main sources: crossing over in meiosis, the union of unrelated haploid gametes (sexual recombination; Fig. 17.2), and mutation (Fig. 17.3). The first two processes, discussed in Chapters 12 and 16, do not lead to new alleles but instead to a recombination of existing ones. Although beneficial in the long run, crossing over and sexual recombination are part of the metabolically expensive and geneti-

A

B

17.3 Spontaneous mutation as a source of variation

(A) Most wisterias, like the one shown here, have lavender flowers. (B) Several decades ago, however, the famous white-flowering Eno wisteria appeared behind a biology building on the Princeton campus. It is almost certainly the result of a spontaneous mutation in a pigment gene of an offspring of one of the many lavender wisterias nearby.

cally chancy process of sexual reproduction. Since selection operates on individuals, the existence of recombination suggests that sex persists because it creates variation. Mutation, which includes point mutations as well as large-scale exon recombinations, is a potential source of entirely new alleles; we will look at how new alleles arise in more detail in a later section. Taken together, these processes generate new alleles and new combinations of alleles, and they provide the genetic variability on which natural selection can act to produce evolutionary change.

Purely phenotypic variation Some kinds of variation are immune or irrelevant to selection. Natural selection can act on genetic variation only when it is expressed in the phenotype. A completely recessive allele never occurring in the homozygous condition would be totally shielded from the action of natural selection.

Another potential complication is that any phenotypic variation within a population may give rise to reproductive differentials between individuals, whether or not the variation reflects corresponding genetic differences. Thus variations produced by exposure to different environmental conditions during development, or produced by disease or accidents, are subject to natural selection. In general, such selection has no effect on the overall genetic makeup of the population: there is no correlation between an individual's alleles and any chance events that lead to its lowered viability and reproductive potential. However, there are cases in which luck *can* affect evolution. For example, if the population is small and the one carrier of a rare allele is killed by an accident, gene frequencies will change. We will return to the role of chance in small populations in both this chapter and the next.

Variation produced by somatic mutations is not raw material for evolutionary change. It would be possible, for example, for an important mutation to occur in an ectodermal cell of an early animal embryo. All the cells descended from the mutant cell would be of the mutant type. The result might be a major change in the animal's nervous system, but the change could not be passed on to the animal's offspring. The ectodermal cells are not the ones that give rise to gametes. Hence selection that acts on variations produced by somatic mutations cannot result in evolutionary change in sexually reproducing organisms.

The theory of ***evolution by natural selection*** had a rival during the 19th century in the concept of ***evolution by the inheritance of acquired characteristics***—an idea often identified with Jean-Baptiste Lamarck. Lamarck thought that somatic characteristics acquired by an individual during its lifetime could be transmitted to its offspring. Thus the characteristics of each generation would be determined, in part at least, by all the modifications—including those caused by experience, use and disuse of body parts, and accidents—to members of the preceding generations. Evolutionary changes would be the gradual accumulation of such acquired modifications over many generations. The classic example is the evolution of the long neck of the giraffe (Fig. 17.4).

According to the Lamarckian view, ancestral giraffes with short necks tended to stretch their necks as much as they could to reach the tree leaves they fed on. This frequent neck stretching caused

A

B

17.4 An okapi and a giraffe, two related African herbivores

The okapi (A) and giraffe (B) are thought to have had a common ancestor with a relatively short neck. The long-necked giraffe of today can reach food unavailable to shorter individuals.

EXPLORING FURTHER

THE LOGIC OF SEX

At first glance, sex in higher organisms makes no evolutionary sense. Half of a female's reproductive effort goes into sons, who will bear no offspring and so contribute little to her reproductive output. If males could be eliminated, each female could produce twice as many daughters and far more grandchildren. This is readily observed in groups of higher plants and animals that have both sexual and asexual species. In nearly every case, however, the asexuals arose relatively recently from sexual precursors.

In addition to reducing reproductive output, sex tampers with success. A constellation of alleles that has worked well enough to allow an organism to reproduce is broken up, and half are assigned arbitrarily to each gamete, rather like taking cards randomly from a royal flush and a full house and recombining them to form a new hand. The novel collection of cards is unlikely to be as good as either "parent" hand. Some asexual species are quite successful: dandelions are, despite their flowers, incapable of sexual reproduction.

The evolution of sex presumably began with haploid asexual species such as bacteria that normally reproduce asexually. *E. coli* can reproduce every 20 min if enough food is available; since it takes 18 min to replicate the chromosome, the contest to generate offspring while conditions are good is a race against the clock.

Diploidy is one step toward sex. A bacterium with two copies of its its chromosome to replicate would be at an enormous reproductive disadvantage, having to collect enough nutrients to synthesize twice as much DNA as its haploid counterpart. The cost of haploidy, however, is the immediate expression of any mutation: since most genetic changes are deleterious, this penalty can be large. This is the most likely explanation of the evolution of diploid dominance in plants and animals.

Diploidy by itself does not entail sexual recombination (dandelions, for example, are diploid), but even without it diploidy confers several advantages. Most obviously, deleterious mutations are usually hidden. In addition, the intact chromosome can be used as a template for the repair of double-stranded damage to the other. Diploids can also accumulate and harbor variants on the "spare" copy that may prove useful when conditions change.

Sexual reproduction adds the features of crossing over and genetic recombination. Since crossing over involves the same enzymes that asexual diploids use for repair and that bacteria use for their rare bouts of conjugation, it seems likely that the first sexual species took advantage of crossing over as a way of correcting errors just prior to gamete formation. But changes bring new opportunities, and crossing over now creates more variation than it repairs.

There are two general hypotheses concerning the role of variation. The first, the ***tangled-bank model***, is based on a reference to a diverse, multi-species environment discussed near the end of *The Origin of the Species*. Darwin realized that if conditions were everywhere the same, a few species would come to dominate any habitat. Yet when he surveyed a square meter of his own lawn, he counted more than 20 species of plants. Darwin's description of a tangled bank emphasized the many different microhabitats found in even a small area. According to the tangled-bank model, asexual species are at a disadvantage because they are superior in only one microhabitat, whereas sexual species produce a range of offspring-microhabitat matches. Because of their adaptive narrowness, asexual siblings may even wind up competing primarily with each other, whereas the many unique sexual progeny of one pairing might be different enough to seek out their own most appropriate sites.

A Alice and the Red Queen

They were unable to outrun their environment, no matter how hard they tried.

The second hypothesis, the ***red queen model***, takes its inspiration from the scene in Lewis Carroll's *Through the Looking Glass* in which Alice and the Red Queen are running as quickly as they can and yet make no progress: however fast they move, they cannot outrun their surroundings (Fig. A). The red queen theory suggests that variation is needed to keep up with a rapidly evolving background of predators, prey, and parasites as well as with changing conditions. The emphasis is on variation in time rather than variation in space.

The potential problem from parasites is especially easy to understand: most parasites are strain-specific and can devastate an asexual or inbred population. New strains of barley in Britain, for example, have a useful life of only 3–5 years before fungal parasites adapt enough to ruin crops. Clones of thrips (juice-sucking insects) on long-lived hosts regularly develop specializations for that one plant. This degree of specialization is possible because parasites have a short generation time compared with their hosts; they can go through many rounds of selection before the host reproduces.

One of the predictions of the red queen hypothesis is that crossing over should be most common among the longest-lived species; this would enable the host to compensate for rapid parasite evolution by making its offspring as different as possible. Crossover rates increasing with life span seem to be the pattern at least among mammals. What is hard to understand is why parasites have not wiped out dandelions and other asexual organisms.

With regard to habitat variation, both hypotheses are consistent with pattern seen in many species that can reproduce either sexually or asexually. For instance, the first aphid to find a suitable new plant shoot reproduces asexually, creating a clone of daughters equally well adapted to this local habitat. When conditions deteriorate, as when the shoot becomes overcrowded or suffers from drought, the founder begins producing winged reproductive offspring that leave the shoot, mate with other aphids, and search out a suitable host (Fig. B). Clearly the aphid life cycle can be interpreted as an adaptation to unpredictable spatial *or* temporal variation. There is every reason to believe that the degree and nature of the advantage of sexual reproduction can differ between species.

B Sexual (winged) and asexual aphids on a yarrow stem

The common thread, however, is that sex has been selected for because it creates variation. Without the capacity to change faster than mutations alone allow, long-lived species would be in trouble.

their offspring to have slightly longer necks. Since these also stretched their necks, the next generation had still longer necks. As a result of neck stretching to reach higher and higher foliage, each generation had slightly longer necks than the preceding generation.

Natural selection, on the other hand, proposes that ancestral giraffes probably had short necks, but that neck length varied from individual to individual because of their different genotypes. If the supply of food was limited, individuals with longer necks had a better chance of surviving and leaving progeny than those with shorter necks. This does not mean that all individuals with shorter necks perished or that all with longer necks survived to reproduce, but simply that a higher proportion of those with longer necks survived and left offspring. As a result, the proportion of individuals with genes for longer necks increased in the succeeding generation. As the proportion of individuals with somewhat longer necks rose, the increased competition for food higher up on the trees resulted in a selective advantage for those with yet longer necks, and so evolution continued.

■ HOW DO NEW GENES AND ALLELES ARISE?

One of the troubling mysteries of evolution for Darwin and his early supporters was how a complex organ or other structure could arise through selection on small variants of existing traits. How, for example, could the elaborate camera eye of vertebrates evolve? Wouldn't all the intermediate steps, as well as the first, be useless or even maladaptive, altering some important preexisting structure? In fact, new genes, new alleles, new morphological structures generally arise out of previous ones by a combination of well-documented genetic mechanisms.

Point mutations—deletions, substitutions, and additions of one or a few bases—can have two useful effects. First, they can "fine-tune" one allele of an existing gene, safe from negative selection during intermediate steps by virtue of the protection afforded by diploidy. (The "orthodox" version of the gene is on the other copy of the chromosome, performing its job.) Second, when small mutations occur in control regions—the binding sites for transcription factors, for example—they can dramatically alter the specificity, timing, and degree of responsiveness of gene activity in one or a group of alleles on that copy of a chromosome.

Though these are potentially useful sources of variation, the origin of much major evolutionary change is based on two other mechanisms: (a) gene duplication followed by the independent evolution of the spare copies of the gene, and (b) recombinations of preexisting exons.

How new genes arise Three observations suggest a novel molecular basis for evolution:

1 Eucaryotic chromosomes contain introns, exons, seemingly functionless pseudogenes, and repetitive DNA (Chapter 11). More than 90% of eucaryotic DNA is apparently functionless and dormant.

2 The realization that the fetal immune system actively disposes of redundant gene segments (Chapter 15) indicates that the genome is not necessarily as stable as had been thought. The discovery of transposons (Chapter 10) underscores this point.

3 The unique structure and organization of antibody genes, MHC proteins, and T-cell receptors (discussed in Chapter 15) implies that the exons of the variable and constant regions of both the light and heavy chains of antibodies arose from the duplication and rearrangement of some prehistoric gene sequence—probably one of the cell-adhesion molecules. (The cell-adhesion molecules are themselves part of a closely related family believed to be derived by duplication from a still more ancient gene.)

Duplication of genes or exons followed by small-scale evolutionary change in the superfluous copies is more likely to lead to functional genes with novel properties than random changes alone would be. Imagine how rarely we could generate a meaningful sentence by randomly arranging letters and spaces, whereas if we began with a meaningful sentence (a gene for a functional protein) consisting of words and spaces (exons and introns) and changed a few existing letters or words, the odds of ending up with an intelligible sentence with a new meaning would be fairly high. We will look at some documented examples of duplication, and then at the evidence that new genes are created by recombining the exons of old genes.

Evidence for gene duplication The evidence for the role of duplication is particularly clear in the genes for myoglobin and hemoglobin. Myoglobin, the oxygen-storage protein in muscles, consists of a single polypeptide chain, while hemoglobin, which carries oxygen in the blood, has two pairs of chains, for a total of four. The three-dimensional conformations of the α and β chains in hemoglobin are nearly identical, and they also closely resemble the conformation of the single chain in myoglobin; the genes for all are thought to have evolved by duplication from a single ancestral gene (Fig. 17.5). This conclusion is reinforced by the discovery that introns are located in the same places in all these genes. (Myoglobin itself evolved from a bacterial gene whose product is used to capture cellular oxygen that would otherwise destroy a nitrogen-fixing enzyme complex.)

Hemoglobin is synthesized in slightly different forms at different times in an organism's life—during embryonic development and during adult life, for instance—and is thus specialized for differing conditions of pH and oxygen concentration. Again, the genes for these alternative forms of hemoglobin, which in humans lie near one another on chromosome 7, are thought to have originated by duplication and then to have followed independent evolutionary pathways. The same region contains many pseudogenes with sequences very similar to those of the hemoglobin genes. Perhaps the pseudogenes are duplications that never evolved into improved functional gene products.

The same pattern is seen in many groups of enzymes. Elastase, the digestive enzymes trypsin, and chymotrypsin, and the blood-clotting enzyme thrombin all have different functions, but the genes for them have base sequences and intron locations that are nearly identical. It is unlikely that each evolved independently into a near duplicate of the others. The genes for these enzymes probably began separately as duplications of the gene for some primordial enzyme, and then went their separate evolutionary ways.

Gene evolution thus depends in part on gene duplication, followed by changes in base sequence that give rise to functionally different products. Such changes occur slowly, and most mutations result in genes with products of reduced function—or no function at all—rather than of altered specificity. For every new gene that produces a functional protein, there may be tens or hundreds of incomplete or failed "experiments" involving duplications. Unless some process is at work to edit out useless duplications, the chromosomes should be full of nonfunctional base sequences with clear similarities to those of functional genes. And indeed, well over 90% of the mammalian genome does not code for functional products. It may be that the enormous number of pseudogenes in eucaryotic chromosomes is evidence of past duplications that never evolved into functional genes.

Another source of genes is ***horizontal transmission***—importation of a gene from another species. This interspecific transfer can occur when a parasite—especially a virus—picks up a host gene, incorporates it into its capsule, and then releases it into its next victim. Lysogenic viruses and retroviruses (discussed in Chapter 10) are especially efficient mediators of horizontal exchange, since they incorporate their genetic material into the host's genome.

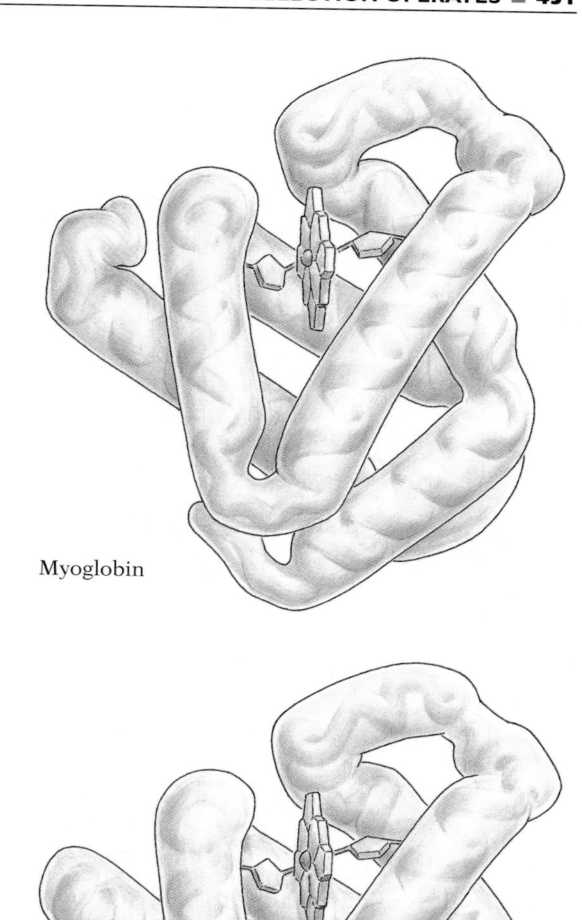

17.5 Myoglobin and the β chain of hemoglobin compared

The similarity in conformation between these peptide chains is evident from this representation. The genes for these two chains are thought to have arisen from a duplication event.

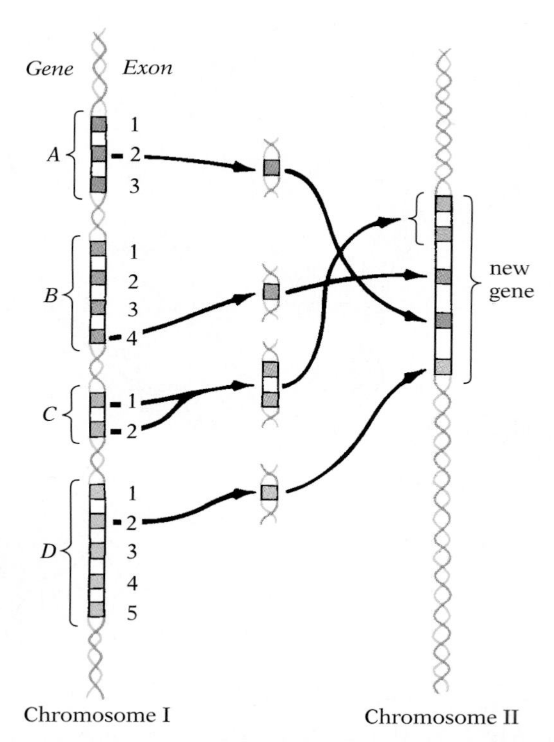

17.6 Hypothetical recombination of duplicate exons to create a new gene

Duplicates of five exons (colored blocks), from four different genes on a single chromosome (designated chromosome I), are imagined to be inserted near one another in a different chromosome (chromosome II). The intron regions (white blocks) flanking each exon that is being inserted provide the signals necessary for the new gene's transcript to be processed correctly. Since many exons code for protein subunits, the newly combined exons are more likely to encode a functional product than would have been the case if they began and ended at random points. If their product is at least partially functional, the new gene will survive in the germ line (the gametes transmitted to offspring) and be subject to improvement through natural selection.

Exon recombination Walter Gilbert and Colin Blake have suggested that individual exons from different genes can be brought together to produce new combinations. This widely entertained proposal makes excellent sense if exons code for parts of the resulting protein called "domains"—distinct subunits that, like building blocks of various shapes, can form new structures when put together in new ways. Careful examination of the genes for dozens of proteins confirms that this is sometimes the case. For instance, introns in some myoglobin genes occur between regions that code for sections joined at a major turn of this highly folded globular protein. Thus each intron defines a boundary of sorts between compact domains, and each myoglobin exon can be thought of as coding for one of these domains, or subunits.

In genes for some other proteins, introns fall at the boundaries between regions coding for sections of α helix and regions coding for sections of the β pleated sheet. Alternatively, the introns sometimes flank regions coding for sections of the protein containing the active site. Because the regions encoded by exons form distinct subunits, a recombination of exons would have a real chance of generating a working enzyme with novel properties. A new gene could evolve by combining, say, exon 2 of gene *A*, exon 4 of gene *B*, exons 1 and 2 of gene *C*, and exon 2 of gene *D* at a new site (Fig. 17.6). Such a recombination could be effected by a simple movement of the exons from their original genes or from duplicates. It might involve movement of the exons themselves, or of copies of them. In either case, insertion of exons (with their flanking introns) within an intron region of the chromosome would improve the chances of generating a functional new gene.

The best evidence for exon shuffling comes from the genes involved in blood clotting. Intron positions in many other genes do not seem to fit neatly with the domain hypothesis. Perhaps, as some suggest, modern exons are the descendants of ancient minigenes that have been strung together to create larger and more efficient proteins; some have taken on domainlike identities, while others have not. Still other researchers believe that many introns are actually dormant transposons (described in Chapter 10) that have inserted themselves at random into modern genes; many exons, therefore, are random slices of genes rather than functional domains.

Despite the controversy over the multiple sources of exons and introns, comparative study of DNA sequences argues strongly for a major role of exon recombination in the evolution of many genes. Gilbert has estimated that all the genes that exist today (50,000 in humans alone) evolved from perhaps as few as 1000 unique exons or minigenes. The processes of duplication and intragenic recombination have almost certainly been efficient mechanisms of major evolutionary innovation. Molecular biologists have recombined existing exons to create hybrid "designer" genes; these hybrid genes can generate working enzymes with novel properties.

■ HOW DO MUTATIONS SPREAD IN POPULATIONS?

Because mutation rates are low, when a novel mutation occurs, it is necessarily rare in the population. The mutation may go extinct (particularly if it is harmful), achieve a stable intermediate frequency (as part of a balanced polymorphism, discussed later in this chapter), or become the most common form of the gene. A rare mutation could spread and become a common allele in two ways: (1) in the absence of any selective advantage, it could nevertheless increase in frequency by chance, or (2) by conferring an advantage on its bearers, it could be favored by natural selection. Let us look at how the frequency of an allele is traced, and then at how chance and selection can drive that frequency up or down.

The gene pool Chapter 16 discusses genetics in terms of the genotype, which is the genetic constitution of an individual. The genetics of populations—the level at which evolution operates—depends on the ***gene pool***, which is the sum total of all the genes possessed by all the individuals in the population.

As detailed in Chapter 16, the genome of a diploid individual contains no more than two alleles of any given gene.[1] But the gene pool of a population can contain any number of different allelic forms of a gene. The gene pool is characterized with regard to any given gene by the frequencies of the alleles of that gene in the population. Suppose that gene *A* occurs in only two allelic forms, *A* and *a*, in a particular sexually reproducing population. And suppose that allele *A* constitutes 90% of the total of both alleles, while allele *a* constitutes 10%. The frequencies of *A* and *a* in the gene pool of this population are therefore 0.9 and 0.1. If those frequencies were to change with time, the change would be evolutionary. Hence it is possible to determine what factors cause evolution by determining what produces shifts in allelic frequencies.

For our alleles *A* and *a*, with frequencies of 0.9 and 0.1 in a population, how can we calculate the genotypic frequencies in the next generation? If we assume the population is large and that all genotypes have an equal chance of surviving, this calculation is not hard. If the frequency of *A* in the entire population is 0.9 and the frequency of *a* is 0.1, the alleles carried by sperm and eggs will also appear at these frequencies. Using this information, we can set up a Punnett square analogous to those introduced in Chapter 16:

		Sperm	
		0.9*A*	0.1*a*
Eggs	0.9 *A*	0.81*A*/*A*	0.09*A*/*a*
	0.1 *a*	.09*a*/*A*	.01 *a*/*a*

Notice that the only difference between this and a Punnett square for a cross between individuals is that here the sperm and eggs are not those produced by a single male and a single female, but those produced by all the males and females in the population, with the frequency of each type of sperm and egg shown on the horizontal and vertical axes respectively. Filling in the square (by combining

[1] There is one obvious exception to this generalization: because the genome contains multiple copies of rRNA genes, it is almost inevitable that each individual will carry more than two alleles of these genes.

the indicated alleles and multiplying their frequencies) tells us that the frequencies of the homozygous dominant genotype (*A/A*) in the next generation of this population will be 0.81, of the heterozygous genotypes (*A/a* and *a/A*) will be 0.18, and of the homozygous recessive genotype (*a/a*) will be 0.01. If the frequencies we have computed change in successive generations, the population has evolved.

Evolution versus equilibrium It is easy to believe that the more frequent allele (*A*) would automatically increase in frequency while the less frequent allele (*a*) would decrease and eventually be lost; thus variation—the raw material for evolution—would inevitably vanish. If there were little variation, evolution would be a minor phenomenon in nature. In fact, however, the rarity of a particular allele in a large population does not doom it to automatic disappearance.

As noted above, the genotypic frequencies in the gene pool of the second generation of our hypothetical population will be 0.81, 0.18, and 0.01. We can use these figures to compute the allelic frequencies of *A* and *a* in this generation. Since the frequency of the *A/A* individuals is 0.81, the frequency of their gametes in the gene pool will be 0.81. All these gametes will contain the *A* allele. Likewise, the frequency of the gametes of the *a/a* individuals will be 0.01, and each gamete will contain an *a* allele. The frequency of the heterozygous (*A/a* and *a/A*) individuals is 0.18, and the frequency of their gametes in the gene pool will be 0.18, but their gametes will be of two types, *A* and *a*, in equal numbers. The frequency of the *A* and *a* alleles in the gametes of the population can be calculated:

Frequency of genotypes	Frequency of *A* gametes	Frequency of *a* gametes
0.81 *A/A*	0.81	0
0.01 *a/a*	0	0.01
0.18 *A/a* + *a/A*	0.09	0.09
	0.9	0.1

The allelic frequencies are 0.9 and 0.1, the same frequencies we started with in the preceding generation.

Since the allelic frequencies are unchanged, the genotypic frequencies in the succeeding generation will again be 0.81, 0.18, and 0.01; the allelic frequencies in turn will be 0.9 and 0.1. In populations large enough to swamp chance effects, variation is retained. Moreover, evolution can occur only when something disturbs this genetic equilibrium. This was first recognized in 1908 by G. H. Hardy and W. Weinberg. According to the ***Hardy-Weinberg Law***, *under certain conditions of stability both phenotypic and allelic frequencies remain constant from generation to generation in sexually reproducing populations.*[2]

[2] For this statement to hold true for phenotypes, the initial genotypic frequencies must be in equilibrium. If these frequencies are not in equilibrium, they will change in successive generations until the equilibrium is achieved. For example, if genotypes *A/A* and *a/a* were present in a population but *A/a* was missing (say, because of human intervention), all three genotypes would appear in the next generation, and thereafter the genotypic and phenotypic frequencies would remain constant.

For a population to be in genetic equilibrium, five conditions must be met:

1 The population must be large enough to make it unlikely that chance could significantly alter allelic frequencies.

2 Mutations must not occur, or there must be mutational equilibrium.

3 There must be no immigration or emigration that alters allelic frequencies in the population.

4 Mating must be random with respect to genotype.

5 Reproductive success—the number of offspring and the number of their eventual offspring—must be random with respect to genotype.

The Hardy-Weinberg Law demonstrates that variability and heritability, two bases of natural selection, cannot alone cause evolution. Because the five conditions are *never* completely met, however, evolution does occur. The value of the Hardy-Weinberg Law is that it provides a baseline against which to judge data from actual populations. By defining the criteria for genetic equilibrium, it indicates when a population is not in equilibrium and helps to isolate the causative agents of evolution. The role of the investigator is then to discover the relative contribution of each factor in the evolution of a population.

With regard to the first condition, chance is always a factor: it is unlikely that an allele will appear in the next generation in exactly the same proportion of the individuals that carry it in the current generation; variation arises because gametes carrying any allele tend to contribute to zygotes more or less often than the exact number of times their proportion in the population would predict. This is analogous to the familiar observation that a coin flipped 10 times has only a 25% chance of yielding exactly five heads and five tails. But when the number of flippings is large, the average chance deviation from 50:50 is very small. Similarly, many natural populations are large enough that chance alone is not likely to cause any appreciable alteration in the allelic frequencies in their gene pools. However, in small isolated populations of fewer than 100 breeding-age members, allelic frequencies are highly susceptible to random fluctuations. Moreover, these fluctuations can easily lead to loss of an allele from the gene pool even when that allele is adaptively superior. In the absence of immigration or mutation, such an allele is lost forever. In such populations, in fact, there are relatively few alleles with intermediate frequencies; most alleles are either lost or become fixed as the only allele present. Small populations tend to have a high degree of homozygosity, while large populations tend to be more variable. But since chance changes, called ***genetic drift***, are not much influenced by the adaptiveness of alleles, it is an indeterminate evolution, as likely to take one direction as another (Fig. 17.7).

The second condition for genetic equilibrium—either no mutation or mutational equilibrium—is rarely met in populations. Mutations are always occurring. Most genes undergo mutation probably once every 1 million to 100 million replications; the rate of mutation for different genes varies greatly. As for mutational equilibrium, very rarely, if ever, are the mutations of alleles for the

17.7 Possible genetic drift in a cichlid fish

Pseudotropheus zebra, one of hundreds of species of cichlid fish living in the rift lakes of Africa, is divided into numerous isolated populations, many of which have evolved their own distinctive morphology. As there is no known selective force that accounts for this diversity, the varied colors are thought by many researchers to be the result of genetic drift in each small population.

EXPLORING FURTHER

THE HARDY-WEINBERG EQUILIBRIUM

On p. 452 we used a Punnett square to calcualte the frequencies of the genotypes produced by alleles *A* and *a*, whose respective frequencies in a hypothetical population were given as 0.9 and 0.1. The same results can be obtained more rapidly algebraically.

Expansion of the binomial expression $(p + q)^2$ where p is the frequency of one allele (in our case, *A*) and q is the frequency of the other allele (*a*), yields the formula for the Hardy-Weinberg equilibrium:

$$p^2 + 2pq + q^2 = 1$$

Substituting the allelic frequencies 0.9 and 0.1 for p and q respectively, we obtain

$$\begin{array}{ccccccc} p^2 & + & 2pq & + & q^2 & = & 1 \\ (0.9)(0.9) & + & 2(0.9)(0.1) & + & (0.1)(0.1) & = & 1 \\ 0.81 & + & 0.18 & + & 0.01 & = & 1 \end{array}$$

The three terms of the Hardy-Weinberg formula indicate the frequencies of the three genotypes:

$$\begin{array}{lllll} p^2 & = & \text{frequency of } A/A & = & 0.81 \\ 2pq & = & \text{frequency of } A/A + a/A & = & 0.18 \\ q^2 & = & \text{frequency of } a/a & = & 0.01 \end{array}$$

These are the same results we obtained using the Punnett square.

In this example we have assumed that we know the allelic frequencies and want to compute the corresponding genotypic frequencies. The Hardy-Weinberg formula allows many other sorts of calculations as well. Suppose, for example, we know a certain disease caused by a recessive allele *d* occurs in 4% of a certain population and we want to find out what percentage are heterozygous carriers of the disease. Since the disease occurs only in homozygous recessive individuals, the frequency of the *d/d* genotype is 0.04. Letting q^2 stand for the frequency of *d/d* in the formula, we can write

$$q^2 = 0.04$$

The frequency of allele *d*, then, is the square root of 0.04:

$$q = \sqrt{0.04} = 0.2$$

If the frequency of allele *d* is 0.2, the frequency of allele *D* must be 0.8, because the two frequencies must always add up to 1 (that is, $p + q = 1$). Substituting the frequencies of both alleles in the Hardy-Weinberg formula, we can compute the frequencies of the genotypes:

$$\begin{array}{ccccccc} p^2 & + & 2pq & + & q^2 & = & 1 \\ (0.8)(0.8) & + & 2(0.8)(0.2) & + & (0.2)(0.2) & = & 1 \\ 0.64 & + & 0.32 & + & 0.04 & = & 1 \end{array}$$

Since the term $2pq$ stands for the frequency of the heterozygous genotype, which is what we want to know, our answer is that 0.32, or 32%, of the population are heterozygous carriers of the allele *d* that causes the disease we are studying. Powerful as this method is, however, it is important to remember that these equations apply only to populations in Hardy-Weinberg equilibrium.

This type of reasoning can be used to calculate changes in allelic frequencies when only phenotypes can be measured directly. Suppose in a large population of freely interbreeding plants we find 59% with yellow blossoms (the dominant phenotype) and 41% with white (a recessive phenotype). In the spring of the following year, after a very severe winter, we find 64% with yellow blossoms and 36% with white blossoms. Clearly the plants with the dominant allele survived better, but exactly how much have the allelic frequencies changed?

Since white blossoms indicate a recessive phenotype, the frequency of the genotype *y/y* was 0.41 initially. Setting $q^2 = 0.41$, we calculate that $q = 0.64$ (approximately), which means that the frequency of allele *y* was 0.64. The frequency of dominant allele *Y* was therefore 0.36 ($1 - 0.64 = 0.36$). The next spring the frequency of white blossoms has fallen to 36%, which gives $q^2 = 0.36$ and $q = 0.60$. The frequency of allele *y* is therefore 0.60, and the frequency of *Y* must be 0.40. Thus the frequency of allele *y* has changed from 0.64 to 0.60 and the frequency of *Y* from 0.36 to 0.40.

Similar procedures, even if considerably more complicated mathematically, can be used for situations involving multiple alleles. Thus the Hardy-Weinberg formula for a triallelic (three-allele) situation requires expansion of the trinomial $(p + q + r)^2$, where r is the frequency of the third allele. A quadriallelic (four-allele) situation uses $(p + q + r + s)^2$.

same character in exact equilibrium: the number of forward mutations per unit time is rarely exactly the same as the number of back mutations.[3] The result of this difference is a ***mutation pressure*** tending to cause a slow shift in the allelic frequencies in the population. More stable alleles increase in frequency while more muta-

[3] By convention, the mutation from the more common allele to the less common one is called the forward mutation; the reverse is the back mutation.

ble alleles decrease, unless some other factor offsets the mutation pressure. Eventually, of course, the frequency of the more stable allele will become so high that it will undergo the same number of mutations per unit time as the more mutable allele, despite its lower mutation rate, and equilibrium will be achieved. This requires so much time, however, that other events almost always change allelic frequencies before mutational equilibrium is reached. But even though mutation pressure is almost always present, it is seldom a major factor in producing changes in allelic frequencies in a population in the short run. On the other hand, as we saw earlier, gene duplication, exon recombination, and mutations in gene-control regions do provide much of the basis for long-term genetic change.

According to the third condition for genetic equilibrium, a gene pool cannot have immigrants from other populations that introduce new alleles or different allelic frequencies, and it cannot undergo changes in allelic frequencies by emigration. Most natural populations, however, experience at least a small amount of gene migration, generally called ***gene flow***, which tends to upset the Hardy-Weinberg equilibrium and lead to evolution. Such evolution—that is, a change in allelic frequencies—need not, at least in the short run, involve natural selection. On the other hand, there are doubtless populations—those on distant islands or in isolated habitat patches, for instance—that experience no gene flow, and in many instances where flow does occur it is probably so slight as to be negligible. Thus the third condition for genetic equilibrium is sometimes met in nature.

The final two conditions for genetic equilibrium are that mating and reproductive success be random. Among the vast number of factors involved in mating and reproduction are choice of a mate, physical efficiency and frequency of mating, fertility, total number of zygotes produced at each mating, percentage of zygotes that develop successfully, survival of the young to reproductive age, fertility of the young, and even survival of postreproductive adults when their survival affects either the chances of survival or the reproductive efficiency of the young. For mating and reproductive fitness to be random, all these factors must be independent of genotype, so that natural selection cannot operate.

These conditions are probably never met in any real population. An organism's genotype almost always influences each of these factors. To take an obvious case, female guppies do not mate at random but instead choose showy males with large tails (Fig. 17.8). In short, few aspects of reproduction are totally uncorrelated with genotype. Nonrandom reproduction is normal. Aside from mate choice, which is considered part of sexual selection (discussed later in this chapter), nonrandom reproduction is a component of natural selection. Natural selection, then, is almost always operating in populations; there is always ***selection pressure*** acting to disturb the Hardy-Weinberg equilibrium and cause evolution, even if selection serves merely to limit the frequency of deleterious mutations.

While we will concentrate on natural selection as the most influential mechanism in causing evolution, it is useful to think about Lamarck's long-necked giraffes again in the context of the Hardy-Weinberg factors. Though the Lamarckian concept of inheritance of acquired characteristics is wrong, the Darwinian explanation of

A

B

17.8 Female-choice sexual selection in guppies

In most species of vertebrates, including guppies, the female chooses the mate. (A, B) Male guppies exhibit great variety in the size, color, and location of their spots, and in the size and patterning of their tails. These features are largely heritable. Females choose males on the basis of their conspicuousness. (C) Given a choice between males with tails of two different sizes—large and small (bars at left), large and medium (the latter produced by surgical shortening; bars at middle), or medium and small (bars at right)—females prefer to be near males with larger tails. Subsequent findings indicate that they also preferentially mate with them. Females also prefer high display rates and greater coloration (particularly orange spots).

longer necks as a result of natural selection, which would favor them if they offer an adaptive advantage, is by no means the only possibility. There are several other evolutionary scenarios that might conceivably have given rise to the present-day giraffe.

The most plausible alternative to natural selection is that giraffes have long necks as a result of chance. Suppose that the ancestral population had a wide variety of neck lengths, or that a mutation caused a small subset of the population to have unusually long necks. If we assume that long necks are adaptively neutral—neither advantageous nor disadvantageous, or, more likely, that the benefits they confer are balanced by physiological costs—neck length in the population will not change. Now suppose that the population suddenly declined because of some environmental factor like disease or bad weather, so only a few individuals survived the crisis. If, by chance, a disproportionate number of the survivors were long-necked, the trait could become established without the intervention of natural selection (Fig. 17.9).

As this hypothetical example indicates, evolution does not depend exclusively on any one mechanism, whether natural selection, genetic drift, mutation, or migration. It also underscores the potential error of assuming that all the traits of the living things around us are the adaptive result of natural selection. Throughout this chapter and the next we will consider the alternatives to natural selection. Nevertheless, natural selection is apparently the most important factor in the evolution of most populations.

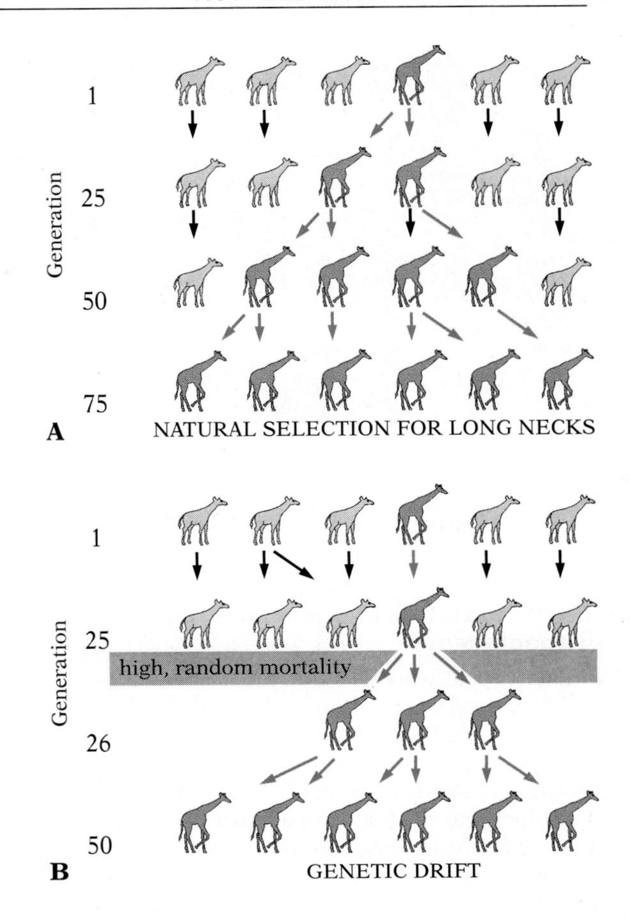

17.9 Natural selection versus chance
In this hypothetical example, a small group of long-necked individuals arises in an otherwise stable population. (For simplicity, each animal in the drawing represents a fraction of the total population.) In evolution by natural selection (A), long-necked individuals become prevalent because they are able to reach more vegetation, and so survive to have proportionately more offspring in succeeding generations. In evolution by chance (B), the frequency of long-necked individuals does not change until a chance catastrophe—fire, flood, heavy predation, or disease, for example—kills most members of the population, or the population size becomes small in some other way. Because only long-necked individuals happen to survive, their offspring multiply to fill the habitat even though their distinctive trait is of no net selective advantage.

■ NATURAL SELECTION

How selection changes allelic frequencies Our hypothetical population had alleles *A* and *a*, with frequencies 0.9 and 0.1; the genotypic frequencies were 0.81, 0.18, and 0.01. Let's look now at the effects of selection pressure on these alleles.

Suppose that selection acts against the dominant phenotype, and that this negative selection pressure is strong enough to reduce the frequency of *A* in the present generation from 0.9 to 0.8 before reproduction occurs. (Note that there will be a corresponding increase in the frequency of *a* from 0.1 to 0.2 among the survivors, since the two frequencies must total 1.) If we set up a Punnett square, we can calculate the genotypic frequencies in the zygotes of the second generation:

		Sperm	
		0.8*A*	0.2*a*
Eggs	0.8 *A*	0.64*A/A*	0.16*A/a*
	0.2 *a*	0.16*a/A*	0.04*a/a*

The genotypic frequencies of these zygotes are different from those in the parental generation; instead of 0.81, 0.18, and 0.01, the frequencies are 0.64, 0.32, and 0.04. If selection now were to act against the dominant phenotype in this generation, and thereby again reduce the frequency of *A*, the frequency of *A/A* will be still lower and that of *a/a* higher. Eventually the frequency of *A/A* would fall to a very low level and the frequency of *a/a* would rise. Natural selection would cause a change from a population in which 99% of the individuals showed the dominant phenotype, to one in which most showed the recessive phenotype. This evolutionary change from the prevalence of one phenotype to the prevalence of another would have occurred without any new mutation, simply as a result of natural selection operating on preexisting mutant alleles.

Selection has been observed to produce radical shifts in allelic frequencies. For example, soon after the discovery of penicillin, *Staphylococcus aureus* (a bacterial species that can cause numer-

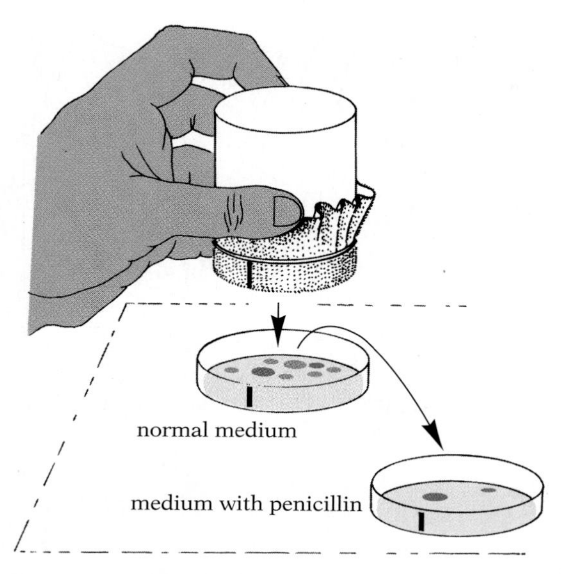

17.10 Mutation for resistance to penicillin is spontaneous

Bacterial cells were cultured on a normal agar medium, and many colonies developed (upper culture dish). Then a block wrapped with velveteen was pressed against the surface of the culture to pick up cells from each of the colonies. The block was next pressed against the surface of a second culture dish, containing sterile medium to which penicillin had been added; care was taken to align the transfer block and the culture dishes according to markers on the block and dishes (black lines). The cells from most of the colonies on the original dish failed to grow on the penicillin medium, but those of a few colonies (two are shown here) did grow. Had the cells of those two colonies spontaneously become penicillin-resistant before being transferred to the penicillin medium, or did exposure to the penicillin induce a mutation for resistance?

Because the transfer block had been aligned with each dish in the same way, according to the markers, it was possible to tell precisely which original colonies had given rise to the two colonies on the penicillin medium. Cells could therefore be taken from those original colonies, which had never been exposed to penicillin, and tested for resistance. They were found to be resistant. Hence the mutation for resistance must already have arisen; it was not induced by exposure to the drug.

ous infections, including boils and abscesses) developed resistance to the drug. Higher and higher doses of penicillin were necessary to kill the bacteria, and the resistant bacteria became a serious problem in hospitals. The bacterial population evolved in the face of the strong selection exerted by the penicillin. The drug itself does not induce mutations for resistance; it simply selects against susceptible bacteria by killing them (Fig. 17.10). Apparently some alleles determining metabolic pathways that confer resistance to penicillin are already present in low frequency in most populations, having arisen earlier as a result of random mutations. Individuals possessing these alleles are thus ***preadapted*** to survive the antibiotic treatment, and, since they are the ones that reproduce and perpetuate the population, the next generation shows a marked resistance to penicillin. If such alleles were not already present in a population exposed to penicillin, no cells would survive and the population would be wiped out.

The importance of selection pressure does not mean that new mutations are irrelevant; in fact, continued selection with penicillin usually leads to gradually increased resistance, some of which results from new mutations that enhance resistance. However, mutations beneficial in an environment containing penicillin arise by chance whether or not the drug is administered; they are simply not selected for in the absence of penicillin. They can nevertheless persist if they have no adverse effect.[4]

Evolution in haploid organisms can proceed faster than in conventional biparental diploid organisms. Nevertheless, even very small selection pressures can produce major shifts in gene frequencies in biparental populations over an evolutionarily brief period. J. B. S. Haldane showed that if the individuals carrying a given dominant allele consistently benefit by as little as 0.001 in their capacity to survive (that is, if 1000 *A/A* or *A/a* individuals survive to reproduce for every 999 *a/a* individuals), then the frequency of the dominant allele would increase from 0.00001 to 1.0 in fewer than 24,000 generations. This may sound like a large number of generations, but many plants and animals have at least one generation a year—*Drosophila*, for example, has more than 30. In very few species is the generation time more than 10 years. Hence 24,000 generations often means fewer than 2400 years and rarely more than 240,000 years. These are short time spans when measured on the geological time scale. Moreover, selection pressures in nature are usually much larger than 0.001; hence major changes in allelic frequencies sometimes take less than a century, perhaps less than a decade (and less than a year for microorganisms).

Directional selection of polygenic characters So far, we have discussed idealized situations involving only two clearly distinct phenotypes determined by two alleles of a single gene. In reality, as

[4] There is evidence that in the face of adverse conditions, some species of bacteria can increase the rate at which their genomes mutate, either by neglecting to repair some errors, or by actively creating mistakes (as in the hypermutation process at work on the antibody genes, described in Chapter 15). In the face of declining reproductive potential, such a system would increase the variability in the population. With variability comes the chance of obtaining a favorable mutation that could "rescue" at least one individual, thus allowing it to found a new clone able to thrive despite altered conditions.

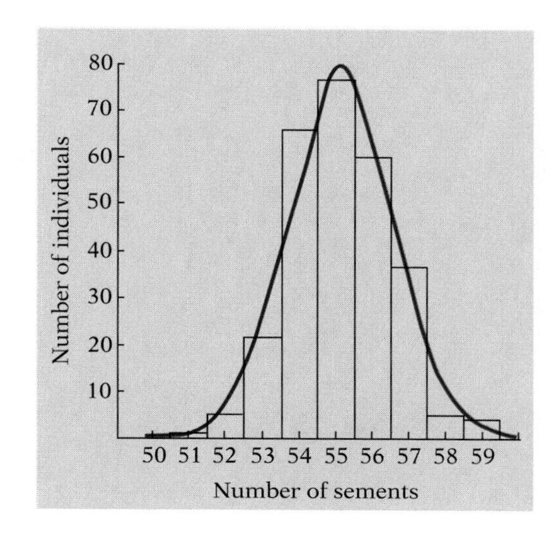

17.11 Frequency distribution of number of body segments in a population of the millipede *Narceus annularis*

The pattern of variation in number of segments (shown by the vertical bars) approximates, but does not exactly fit, the bell-shaped normal curve of probability.

discussed in Chapter 16, the vast majority of characters on which natural selection acts are influenced by many different genes, most of which have multiple alleles in the population; the expression of many characters, moreover, is influenced considerably by environmental conditions. Consequently such characters as height usually vary continuously over a wide range. When graphed, the frequency distribution often approximates the normal, or bell-shaped, curve (Fig. 17.11).

If the environmental conditions change, creating a shift in the selection pressure, we would expect the curve of phenotypic variation to shift as a result of changing allelic frequencies. For example, assume that the conditions under which a certain plant grows best are genetically determined (Fig. 17.12). The first curve (Fig. 17.12A) shows the annual rainfall at which the various plants in a particular population would grow best. The actual rainfall in the area where this population occurs averages 40 cm (arrow 1), though it will vary around this mean from year to year. The population contains a few plants (S) that would grow best if the annual rainfall were about 32 cm and a few (W) that would grow with about 48 cm of rain. Plants that would grow best if the annual rainfall were about 36 cm (T) or 44 cm (V) are fairly common in the population. This phenotypic diversity is maintained because of the variability of the rainfall about the mean, so that in some unusually dry years, for example, the group S plants would have the advantage. In any given year, then, there is selection for an optimum phenotype, but the optimum varies. Since the average rainfall is 40 cm, the U plants are best adapted in the long run and thus are the most common.

Suppose that the average annual rainfall in the area slowly increases until it is 44 cm (arrow 2). Under these new environmental conditions, the V plants (which grow best when the rainfall is about 44 cm) will do better than before; a higher percentage of them can be expected to survive and reproduce, and their frequency should increase. Similarly, the W plants (which grow best with about 48 cm of rain) will now grow better than formerly, and they too should increase in frequency. Conversely, the T plants and

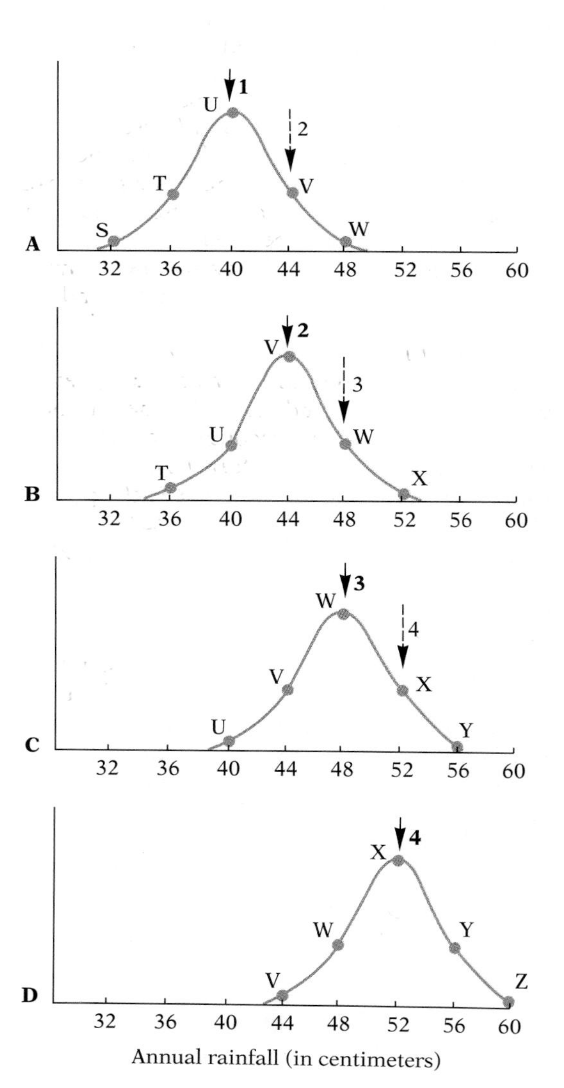

17.12 Evolutionary change of a hypothetical plant population in response to directional selection by changing rainfall

The various phenotypes (S, T, U, etc.) in a hypothetical plant population reflect different genetically determined growth responses to annual rainfall. The four curves (A–D) show the frequency distributions of the phenotypes at different times, under conditions of average annual rainfall (solid black arrows). As described in the text, a systematic change in average rainfall (toward the values shown by dashed arrows) exerts directional selection on the plant population, shifting the curve of relative phenotypic frequencies more and more to the right.

the U plants will not grow as well as formerly, so they will decrease in frequency. Finally, the S plants, only a few of which manage to survive with less than 40 cm of rain, will now be so poorly adapted to the prevalent conditions that none can survive. These changing frequencies, produced by the shift in the selection operating on the population, give rise a new curve (Fig. 17.12B).

If the average annual rainfall continues to increase over the years until it reaches 48 cm (arrow 3), the W and X plants will increase in frequency, the U and V plants will decrease in frequency, and the T plants will disappear (Fig. 17.12C). If the average annual rainfall increases to 52 cm (arrow 4), further shifts in frequencies will occur (Fig. 17.12D).

Changing environmental conditions, then, give rise to ***directional selection***, which causes the population to evolve in a particular way. If the population has not been sufficiently variable to respond to the environmental changes, it will be much reduced and might even become extinct.

How selection can create novel phenotypes Notice in Figure 17.12 that directional selection does not simply shift the *peak* of the population curve to the right, as might happen when the variation in the gene pool is small. Instead, it causes the *entire curve*—the extremes as well as the peak—to shift to the right. The shift is eventually so great, in fact, that a class of plants (X) not even present in the original population becomes the largest class. If X, Y, and Z plants are not present initially, how do they arise in the descendant populations? One possibility is that new genes or alleles arise that make their possessors grow better in wetter habitats; such novel genes or alleles would be strongly selected for and would spread rapidly through the population. However, if moisture preference is influenced by many different genes, as is highly likely, new phenotypes such as X, Y, and Z could arise without the necessity of any new genetic variation, simply through the separate increase in frequency of particular alleles already present, which would then be more likely to occur together and produce a new phenotype.

Haldane calculated how long it would take for a new phenotype to be created in this way. He showed that if one particular allele of each of 15 independent genes is present in 1% of the individuals of a population, then all 15 alleles will occur together in only one of 10^{30} individuals. But there has never been a population of higher organisms containing anywhere near 10^{30} individuals. Hence the chance that all 15 alleles would occur together in even one individual in a real population is essentially zero. However, if there is moderate natural selection for each of the 15 alleles, it would take only about 10,000 generations for the frequency of each allele to increase from 1 to 99%. Once each allele is present in 99% of the population, 86% of the individuals in it will have all 15 alleles and hence will show the phenotype that was previously nonexistent in the population. Thus recombination and selection alone can produce new phenotypes by combining old genes in new ways and systematically favoring certain combinations.

This process is illustrated by a long-term experiment performed on corn. Researchers at the University of Illinois systematically se-

HOW WE KNOW:
HOW COMPLEX STRUCTURES EVOLVE

While it is easy enough to imagine directional selection for longer necks leading to the evolution of the giraffe, it is much more difficult to see how a complex structure like the camera eye of vertebrates could be selected for. After all, what good is any part of it—retina, lens, cornea, iris, ciliary body, sclera, choroid, or the suspensory ligament—without the other components already in place? And given the enormous spans of time required for the evolution of even simple structures, what could ever be learned by experimentation?

An increasingly fruitful approach in evolutionary biology involves modeling selection mathematically and then allowing the computer to compress millions of generations into a few minutes of digital activity. Such electronic experiments can determine whether or not a model is plausible and provide a rough idea of the time required for particular changes to occur. Dan Nilsson and Susanne Pelger took this approach in an attempt to see whether camera eyes can evolve from a patch of light-sensitive cells.

They began with a light-detecting patch sandwiched between a transparent covering and a dark supporting layer (Fig. A). In each "generation" they allowed tissue deformations or changes of density (refractive index) of 1% in any direction, and permitted the "progeny" to survive only if the visual resolution of the structure was an improvement over that of the previous generation. Thus every intermediate step had to represent progress.

The evolutionary route that led to a camera eye in this simulation involved a steadily increasing invagination of the bottom layer to create an eyeball over the course of about a thousand generations (Figs. A–E). At this point further improvement came only with a localized increase in the optical density of what had been the protective layer; after another 800 generations of 1% steps, selection led to a lens capable of forming a sharply focused image (Figs. E–H).

Thus, by a series of exceedingly tiny selective events, each required to increase the fitness of the individual involved and constrained to operate on a very small level of morphological variation, the essentials of the most advanced optical system in nature appear spontaneously in fewer than 2000 generations—2000–20,000 years for a typical vertebrate species. So rapid is this process when modeled that the perennial question of how eyes could evolve must now be replaced with an equally puzzling problem: how could eyes have taken so long to evolve?

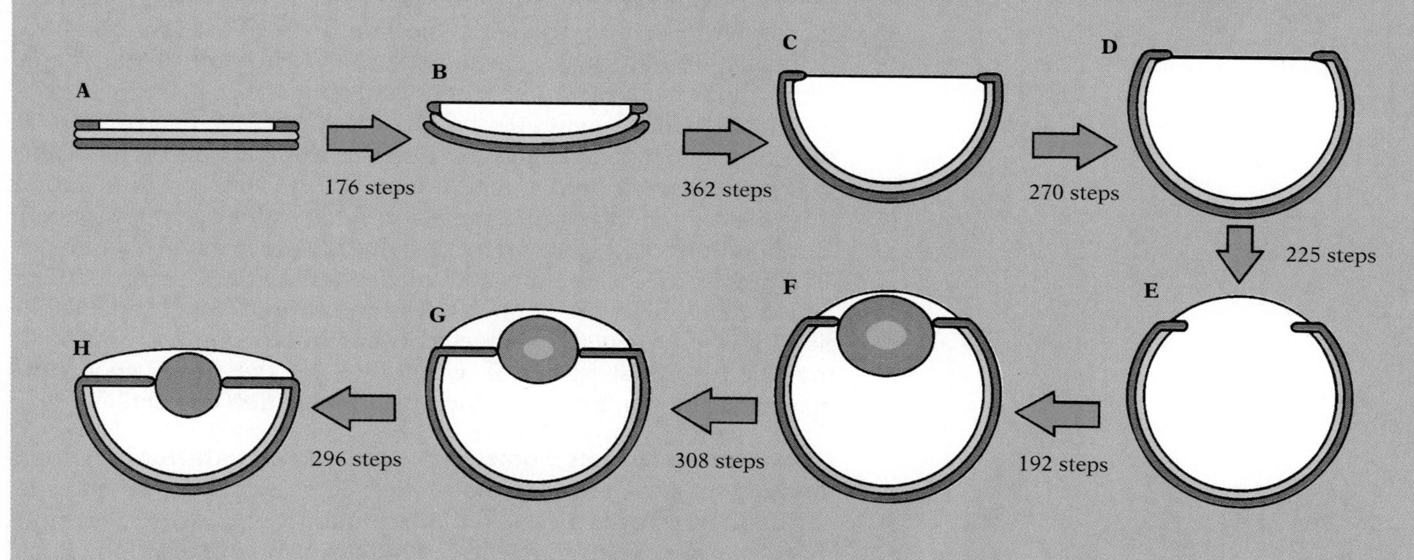

lected for high oil content in the kernels across 50 generations. There was a steady increase in oil content throughout most of this period (Fig. 17.13). The kernels of the original stock of corn plants averaged about 5% oil; those of the plants in the 50th generation after selection averaged about 15% (higher than any individuals in the first generation), and there was no indication that a maximum had been reached.

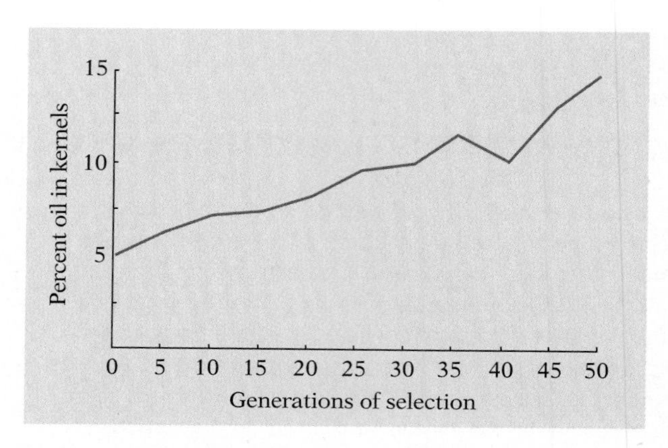

17.13 Results of 50 generations of selection for high oil content in corn kernels

This steady change through 50 generations probably resulted mainly from the formation of new genetic combinations through selection rather than from new mutations. In 50 generations the researchers raised between 10,000 and 15,000 plants. The usual rate of mutation per gene in corn is lower than one in 50,000 plants. It is unlikely that a mutation contributing to an increase in oil content occurred in any *particular* gene affecting this phenotype during the experiment. Since the trait is polygenic, the chance of a mutation in one of the relevant genes is greater; if, say, 20 genes were involved, 2 to 10 mutations are likely to have occurred over the brief interval studied, one or more of which could have been favorable. Therefore, though some small part of the increase in oil content may have arisen from mutation, nearly all of the change came about from recombination of existing alleles.

Thus selection—whether natural or artificial—determines the direction of change largely by altering the frequencies of alleles that arose usually many generations before, through duplication, transposition, and random mutation. This establishes new genetic combinations and gene activities that produce new phenotypes. The processes that create new alleles and genes are not usually major *directing* forces in short-term evolution; the principal evolutionary role of genetic changes consists in creating and replenishing the essential store of variability in the gene pool, thus providing the potential on which future selection can act.

Disruptive selection Sometimes a polygenic character in a population is subject to two (or more) directional selection pressures favoring the two extremes of the distribution. Suppose, for example, that a certain population of birds shows much variation in bill length. Suppose, further, that as conditions change, there are increasingly good feeding opportunities for the birds with the shortest bills and also for those with the longest, but decreasing opportunities for birds with bills of intermediate length. This might happen if the population of plants producing fruits suitable for intermediate-sized bills began to decline, or if a competing species more efficient at harvesting such fruits were to immigrate and become established. The effect of such selection, at least in the short run, would be to divide the population into two distinct

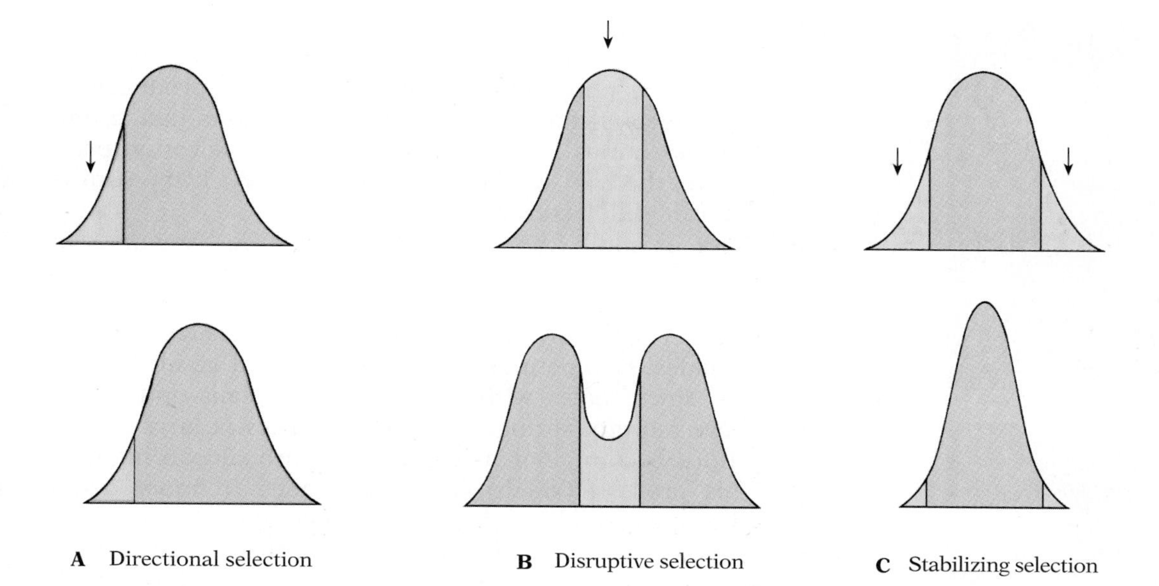

types, one with short bills and one with long bills. The combined action of the opposing directional pressures would thus disrupt the smooth curve of phenotypes in a population (Fig. 17.14B); hence this sort of selection is called ***disruptive selection.***

The most important presumed example of disruptive selection is gamete dimorphism: large eggs and tiny sperm or pollen. If we assume that gametes were originally all the same size, as is still the case in many primitive species, any that were slightly larger would have enhanced the ultimate survival of the zygotes they contributed to by providing more food to fuel initial development. At the same time, organisms that produced small gametes would have gained a reproductive advantage by virtue of making more gametes with the same amount of food, thereby outnumbering the competition from other organisms. In addition, smaller gametes can swim faster because they encounter less resistance, so where speed is important, they would have been at an advantage. If this widely accepted scenario is correct, then all the vast diversity of sex-specific morphology, physiology, and behavior is a consequence of disruptive selection.

Stabilizing selection Most often, when a polygenic character in a population is subject to two or more opposing directional selection pressures operating simultaneously, the pressures select against the two extremes. Plants that are too tall to resist high winds, for example, and those of the same species that are so short that they are shaded by other plants will both be selected against. Similarly, unusually severe winter storms often kill a disproportionate number of the largest and smallest birds in a population. When selection operates against individuals at the two ends of the distribution for a polygenic trait, the process is called ***stabilizing selection*** (Fig. 17.14C).

Costs and benefits: net selection Many characteristics benefit the organisms that possess them in some ways and harm them in

17.14 Directional, stabilizing, and disruptive selection

Each graph indicates the relative abundance of individuals of various heights in a population—in the original condition (above) and later after the specified selection (below). (A) Directional selection acts (arrow) against individuals exhibiting one extreme of a character (here the shortest individuals, represented by the blue area under the upper curve). The eventual result (bottom curve) is that the distribution of heights in the population has shifted to the right, indicating that the population has evolved in the direction of greater height. (B) By contrast, disruptive selection acts against individuals in the middle part of a distribution, thereby favoring both extremes; in our example both the shortest and the tallest individuals would be favored, but individuals of medium height would be selected against. The result is a tendency for the population to split into two contrasting subpopulations. (C) Stabilizing selection acts against both extremes, culling individuals that deviate too far from the mean condition and thus decreasing diversity and preventing evolution away from the standard condition.

17.15 Male and female guppies

The showy coloration of the male (top) contrasts with the drab gray of the female.

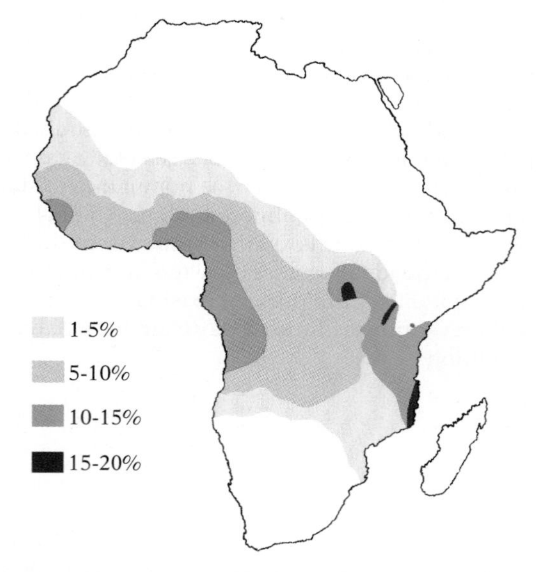

17.16 Distribution of sickle-cell anemia in Africa

The various colors indicate the percentage of the population in each area that has the disease.

others. The evolutionary fate of such characteristics depends on whether the various positive selection pressures produced by their advantageous effects outweigh the negative pressures produced by their harmful effects. If the algebraic sum (an addition taking into account plus and minus signs) of all the separate selection pressures is positive, the trait will increase in frequency; if the sum is negative, the trait will decrease in frequency.

Consider, for example, the selection pressures on showiness in male guppies, a fish native to the freshwater streams of South America (Fig. 17.15). As we have already seen, males with larger tails and brighter spots are more likely to mate because females prefer such males. Competition for females should result in very showy males, since males with alleles for large tails and bright spots will leave more offspring; and, indeed, males in large aquarium populations become increasingly showy with succeeding generations. This process, which will be described in more detail shortly, is known as female-choice sexual selection.

No such selection pressure has operated on the females. These gray, nondescript fish closely resemble one another and are much less easily seen by predators than the males, whose bright markings and showy tails make them more subject to predation. This liability causes strong selection against such features; indeed, in laboratory situations, predators capture the showiest males first. As we might expect, males found in the wild are much less conspicuous than their counterparts in predator-free aquaria. The showiness of males, then, exemplifies stabilizing selection—a balance between selection for greater showiness (by the females, who confer reproductive success on the most "attractive" males) and selection for inconspicuousness (by the predators, who can put an end to further mating if a male is too visible).

Just as polygenic traits often have both advantageous and disadvantageous effects, the alleles of a single gene also usually have multiple effects (a condition called pleiotropy), and it is unlikely that all of them will be advantageous. For example, *Drosophila* carrying alleles that disrupt the activity of specific systems involved in communication between nerve cells also have altered patterns of wing veins; those individuals with a muscle defect causing them to hold their wings upright also will not walk toward light, though this is a powerful response in normal fruit flies. *Whether an allele increases or decreases in frequency is determined by whether the sum of the various selection pressures favoring it is greater or smaller than the sum of the selection pressures acting against it.*

Many instances are known in which the effects of a given allele are more advantageous in the heterozygous than in the homozygous condition. As discussed in Chapter 16, for example, in some parts of Africa the allele for sickle-cell anemia occurs in humans much more often than we might expect in view of its lethal effect when homozygous (Fig. 17.16). This is because the allele, when heterozygous, gives the possessor a partial resistance to malaria. The equilibrium frequency of the sickle-cell allele is thus determined by at least four separate selection pressures: (1) the strong selection against the recessive homozygotes, who suffer the full debilitating effects of sickle-cell anemia; (2) the weaker selection against the heterozygotes as a result of their mild anemia; (3) the selection against the dominant homozygotes, who are more susceptible to malaria; and (4) the fairly strong selection favoring the heterozygotes as a result of their resistance to malaria. A similar pattern seems to underlie the cystic fibrosis gene: heterozygotes are less susceptible to cholera.

Balanced polymorphism Polymorphism is the occurrence in a population of two or more distinct forms, or ***morphs***, of a genetically determined character. These phenotypes do not grade into one another like those represented by a bell-shaped curve for height, where "short" and "tall" designate the extremes of a continuum. Instead, individuals fall into separate categories known as ***discontinuous phenotypes,*** and intermediates are rare or absent. For example, Mendel's peas had polymorphic flower color: plants had either red or white flowers but not pink ones. Polymorphism is common in wild populations: several species of snails, for instance, occur in banded and unbanded forms, and the red fox has both red and silver-colored morphs.

When the relative frequencies of the different morphs in a population are stable over time, we speak of ***balanced polymorphism***. Sometimes the balance is maintained because the polymorphism itself is advantageous. Thus if a polymorphic species lives in an environment subdivided into many local areas where different conditions prevail, one of the morphs may do better in one area, while others are more likely to succeed in other areas. Sometimes one morph is adaptively superior at one time of year or in one habitat, and another is superior at another time or in another place; hence an individual's descendants may have a better chance of survival if they are polymorphic than if all belong to a single form. Individuals that produce a pure culture of one form—putting all their eggs in one basket—run the risk of having none survive.

There are many instances of balanced polymorphism in which some of the contrasting morphs, instead of being themselves selected for, are the unavoidable by-products of selection for a heterozygous phenotype. As in the case of sickle-cell anemia in Africa, the heterozygotes are sometimes better adapted than the homozygotes—a condition known as ***heterozygote superiority*** (or sometimes heterosis or overdominance).

Heterozygote superiority favors balanced polymorphism, because it leads to the retention in the population of both alleles of a given gene at frequencies higher than would be predicted from the selection acting on the homozygous phenotypes. Thus, if *A/a* individuals are adaptively superior to both *A/A* and *a/a* individuals, both allele *A* and allele *a* will be retained in fairly high frequency in the population; neither will be eliminated, as might tend to happen if one of the homozygotes were superior. Therefore all three possible genotypes—*A/A*, *A/a*, and *a/a*—will occur frequently in each generation, and if each of these produces a noticeably different phenotype, the population will be polymorphic. The relative frequencies of *A* and *a* and of the three morphs they produce will be determined, as in sickle-cell anemia, by the balance between the several selection pressures acting on the system.

Playing the odds: frequency-dependent selection Balanced polymorphism can also arise when the selection pressure on an individual displaying one morph (whether physiological, anatomical,

17.17 Handedness in scale-eating cichlids

The ratio of the two morphs fluctuates near 1:1; this balanced polymorphism is maintained by frequency-dependent selection: the rarer phenotype enjoys higher success in attacking prey.

or behavioral) depends on the frequency of that morph compared with the alternatives. For example, there are seven species of cichlids in Lake Tanganyika that feed exclusively on the scales they bite off other cichlids. Each species consists of right-mouthed and left-mouthed forms (Fig. 17.17). Right-mouthed cichlids always approach from the rear and attack the left flanks of their victims; left-mouthed cichlids always attack from the right. This behavior, which becomes apparent in the fry while they are still feeding on plankton, is controlled by a simple single-gene, two-allele system in which right-mouthedness is dominant. Field observations show that the proportion of left-mouthed individuals varies from 0.4 to 0.6. The prey species are extremely watchful (only one attack in 500 is successful), and they focus their attention on the flank most often attacked. In any given year, therefore, the scale eaters attacking from the less common side enjoy greater success, driving the phenotypic ratio in the next generation far in the other direction; then the situation is reversed, and the previously less-successful morph does better. Thus the ratio of the two morphs oscillates around 1:1.

■ SEXUAL SELECTION

Darwin distinguished between selection that affects physical survival and selection that operates on traits used exclusively to attract and keep mates. In evolutionary terms, the failure to reproduce is just as fatal as early death, but since selection for characters that enhance an individual's chance of mating are often very different for the two sexes, he considered *sexual selection* to be nearly independent of natural selection.

Contests Darwin identified two basic varieties of sexual selection. The more common form involves contests between members of one sex (usually the males) for access to the other sex. These duels can take the form of dominance fights, which establish a male hierarchy and thus access to reproductively ready females (Fig. 17.18). More often the males fight for the best territories in the habitats fe-

17.18 A dominance ritual in mountain sheep

Male mountain sheep work out a dominance hierarchy through a ritualized duel in which a run on hind legs culminates in a loud collision, followed by a head-turning display. Dominant males obtain unchallenged access to females.

males favor; the strongest males obtain the highest-quality territories and therefore the largest number of matings (or, in monogamous species, the female best able to gain access to the territory). Male-specific features like large size, offensive weapons (including the horns of mountain sheep), and defensive structures (like the lion's mane, which protects his neck from bites inflicted by other males) are each the result of sexual selection.

Choices Darwin proposed that other uniquely male features exist only to attract females. The elaborate tail of the peacock, for instance, is not used to attack males but rather is displayed whenever a peahen is nearby. Experiments in which these dimorphisms have been altered demonstrate that females of some species do indeed select males on the basis of, for example, the length of their tails (Fig. 17.19; see also Fig. 17.8). Since many male dimorphisms exaggerate a species-specific recognition sign, the female may benefit from the increased certainty of mating with a male of the right species, who also carries genes that will enhance the attractiveness (and therefore the reproductive potential) of her sons.

She may also benefit from genes that confer the sort of physiological superiority that have enabled the male to survive despite the burden of his morphological "handicap"—genes that might contribute to the physiological health of her daughters. In other cases the female's attraction to a male's dimorphism seems to have been favored by selection because the presence of bright colors and the like indicate that the male is free from the sorts of parasites that threaten the population, and the progeny would benefit from this immunity. The mating systems of many species include elements of both male-contest and female-choice sexual selection.

17.19 A normal male widow bird in flight
Experimental shortening or lengthening of tails demonstrates that females select males with the longest tails.

ADAPTATION

Every organism is, in a sense, a complex bundle of thousands of ***adaptations***. In biology, an adaptation is any genetically controlled characteristic that increases an organism's fitness. ***Fitness***, as the term is used in evolutionary biology, is an individual's (or allele's or genotype's) probable genetic contribution to succeeding generations. An adaptation, then, is a characteristic that increases an organism's chances of perpetuating its genes, usually by leaving descendants. Adaptations do not necessarily increase the organism's chances of survival: although it ordinarily enhances prereproductive survival, it need not affect postreproductive survival. In many species it is, in fact, more adaptive for the adults to die soon after they have reproduced.

Adaptations may be structural, physiological, or behavioral. They may be genetically simple or complex. They may involve individual cells or subcellular components, or whole organs or organ systems. They may be highly specific, beneficial only under very limited circumstances, or they may be general and of value under many and varied circumstances.

A population may become adapted to changed environmental conditions very rapidly. W. B. Kemp reported a classic example in

1937. The owner of a pasture in southern Maryland had seeded the pasture with a mixture of grasses and legumes. Then he divided the pasture into two parts, allowing one to be heavily grazed by cattle, while protecting the other from the livestock and letting it produce hay. Three years after this division, Kemp obtained specimens of blue grass, orchard grass, and white clover from each part of the pasture and planted them in an experimental garden where all the plants were exposed to the same environmental conditions. He found that the specimens of all three species from the heavily grazed half of the pasture exhibited dwarf, rambling growth, while specimens of the same three species from the ungrazed half grew vigorously upright.

In only three years the two populations of each species had become markedly different in their genetically determined growth pattern. The grazing cattle in the one half of the pasture had cropped most of the upright plants, and only plants low enough to be missed had survived and set seed. There had been intense selection against upright growth in this half of the pasture and correspondingly intense selection for the adaptively superior dwarf, rambling growth pattern. In the half of the pasture where there was no grazing, by contrast, upright growth was adaptively superior, and dwarf plants were unable to compete effectively. Analogous results are observed when animals are subjected to various degrees of predation (Fig. 17.20).

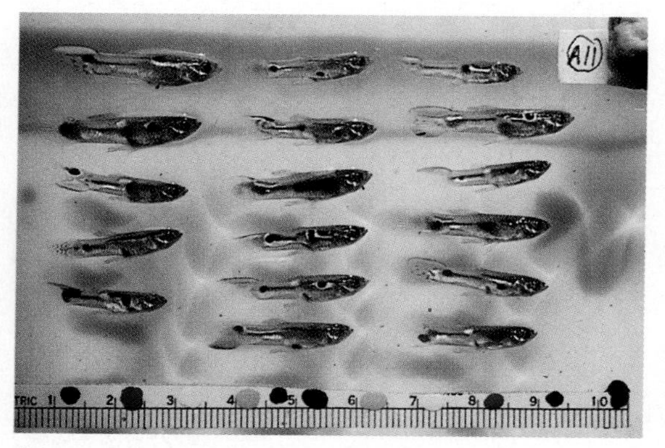

A

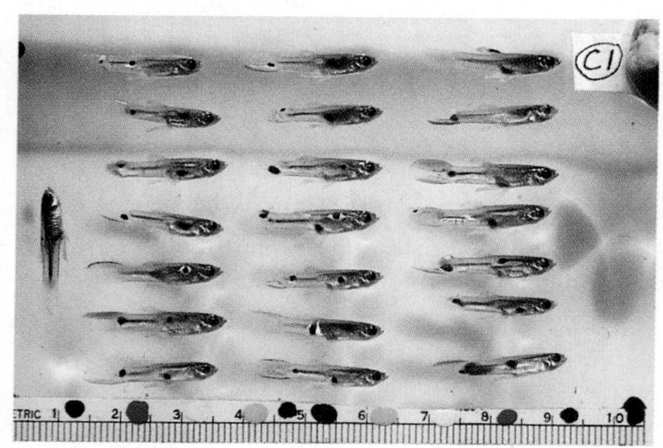

B

17.20 Effects of predation on male guppies

When John Endler kept guppies in a predator-free environment, they evolved brighter coloration and larger tails (A), presumably as a result of female-choice sexual selection; when predators were introduced into the same stock for a few generations, males evolved smaller tails and less dramatic colors (B), probably as a result of predator pressure.

Experimental tests of adaptiveness are usually not as easy to design as straightforward laboratory evaluations of cause and effect. There may be several alternative explanations for why a trait may be adaptive. It is important to remember, furthermore, that not all characters present in an organism need be adaptive in the first place. Some will be incidental pleiotropic effects of genes whose main effects—those that make them adaptive—are quite different. Other traits are essentially historical: we have five fingers because the vertebrate developmental program produces five (or, rarely, six) digits; five may or may not be the optimal number for every kind of amphibian, reptile, bird, and mammal, but selection can only operate on existing variation. As illustrated in Chapter 1, selection has acted to customize digit size rather than number, enlarging or reducing the size of individual digits; the basic five-digit plan remains remarkably consistent. Apparently, some changes are simply not genetically feasible. This sort of evolutionary constraint is often called ***phylogenetic inertia***.

Devising tests to discover the selective pressures that have led to particular traits requires ingenuity and persistence. Niko Tinbergen was one of the first scientists to insist on putting his evolutionary theories to the test. When Tinbergen wondered, for instance, why ground-nesting gulls meticulously remove broken eggs from their nests (Fig. 17.21A), he formulated a variety of possible explanations: the damaged eggs might be a source of disease, which would infect the newly hatched young; the jagged edges of the broken shells might endanger the chicks; or the unrelieved white of the exposed interiors of the broken shells might nullify the camouflage provided by the olive-drab exteriors.

Tinbergen solved this problem (and many similar ones) in two steps: first, by making interspecific comparisons and then by con-

A

B

17.21 Eggshell removal by gulls

Ground-nesting gulls remove broken eggshells and other debris from their nests and carry them at least a meter away (A), while cliff-nesting species like the kittiwake do not remove eggshells (B). The nesting habit of the kittiwakes protects them from predation, while ground-nesting gulls must work to keep their nests inconspicuous.

ducting experimental tests. The species-comparison step allowed him to isolate the most likely hypothesis for testing. He observed that kittiwake gulls, which live on cliffs and are therefore virtually immune to predation, do not remove broken eggshells (Fig. 17.21B). Since disease and cuts, if they posed significant threats, ought to be as dangerous for the kittiwakes as for ground-nesting gulls, Tinbergen decided to test the predation hypothesis first. He did this by setting out an array of nests containing normal eggs, with broken eggs placed at varying distances from them. The results were clear-cut: broken eggshells nearby called the attention of predators to an otherwise inconspicuous nest (Table 17.1). In the absence of direct experimental tests or compelling species comparisons, even the most plausible explanations of adaptiveness remain speculative and should serve as working hypotheses to stimulate research.

Let's look at some examples of adaptations, which help clarify the processes of evolution.

Table 17.1 Survival value of eggshell removal[a]

DISTANCE FROM EGG TO EGGSHELL (CM)	PERCENT EGGS TAKEN BY PREDATORS
5	65
15	42
100	32
200	21
No eggshell	22

[a]SOURCE: N. Tinbergen et al., 1963. Egg Shell Removal by the Blackheaded Gull, *Behaviour* 19.

■ ADAPTATIONS FOR POLLINATION

Flowering plants depend on external agents to carry pollen from the male parts in the flowers of one plant to the female parts in the flowers of another plant (Fig. 17.22). The flowers of each species are adapted in shape, structure, color, and odor to the particular pollinating agents on which they depend, and they provide an especially clear illustration of the evolution of adaptedness. Evolving

A

B

C

D

17.22 A variety of plant pollinators

(A) A bat at a flower, its face liberally dusted with pollen. (B) A honey possum feeding on nectar from flowers. (C) A hummingbird hovering under nectar-bearing flowers. (D) A butterfly inserting its long proboscis into a flower to obtain nectar. The proboscis is likely to be dusted with pollen, which the butterfly may carry to the next flower it visits.

together, the plants and their pollinators became more finely tuned to each other's peculiarities—a process often termed ***coevolution***.

There are indeed striking correspondences between the pollinators and the species they pollinate. Bees are attracted innately to bright colors, UV bull's-eye patterns, and sweet, aromatic, or minty odors. They are active only during the day, and they usually alight on a petal before moving into the part of the flower containing the nectar and pollen. Bee-pollinated flowers have showy, brightly colored petals that are usually blue or yellow (colors bees can see) but seldom red (which bees cannot see at all). In addition, most bee-pollinated flowers have a UV bull's-eye, a sweet, aromatic, or minty fragrance, daytime opening or nectar production, and a special protruding lip or other suitable landing platform. However, these observations do not indicate how these correspondences may have come about—whether the preferences of the pollinators provided the exclusive selective force on flower morphology and odor, or whether early flowers provided the selection pressures leading to the innate preferences of modern bees, or both. A look at other species of pollinators provides some clues.

Hummingbirds, for example, can see red well but blue only poorly; they have a weak sense of smell, and they ordinarily do not land on flowers but hover in front of them while sucking the nectar. Flowers pollinated primarily by hummingbirds are usually red or yellow, are nearly odorless, and lack a landing platform. Since flowers of the same genus can have very different morphologies to suit different pollinators (Fig. 17.23), it is probably the flowers that have done most of the adapting. However, pollinators have probably been adapting to flowers too, though to a lesser extent. Different species of bees, for example, can have very different tongue lengths, suitable for different flower morphologies.

This pattern of coadaptation between pollinators and flowers extends to other nectar-feeding species as well. In contrast to both bees and hummingbirds, moths and bats are generally most active at dusk and during the night. The flowers they pollinate are mostly white, are open only during the late afternoon and night, and often have a heavy fragrance that helps guide the moths and bats to them.

Moths play a role in an interesting adaptation of plants to their pollinators. The flowers produced by scarlet gilia plants near Flagstaff, Arizona, range from red through pink to white. The dark-red flowers are most effective in attracting hummingbirds, but these pollinators emigrate a month after the season begins; the white flowers are most effective in attracting hawk moths, the pollinators available throughout the blooming season. The plants compensate for this shift in relative pollinator abundance by doubling the production of white flowers late in the season and stopping production of red ones (Fig. 17.24).

Unlike bees and moths, the short-tongued flies (which feed primarily on carrion, dung, humus, sap, and blood) are attracted by rank rather than sweet odors, and they rely very little on vision in locating food. The flowers of plants that depend on these flies for pollination are usually dull colored and ill smelling.

A particularly dramatic example of adaptation for pollination is seen in some species of orchids, whose flowers resemble in shape, odor, and color the females of certain species of wasps, bees, or

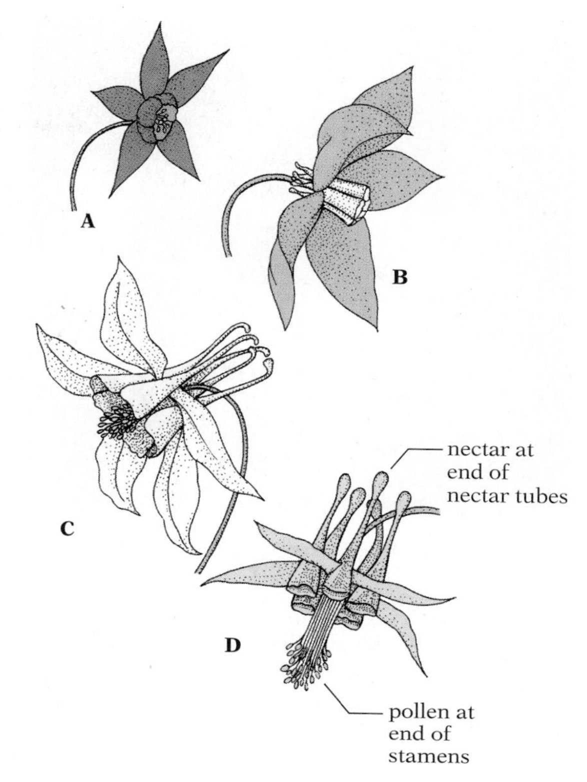

17.23 Characters of columbine flowers correlated with their pollinators

(A) *Aquilegia ecalcarata*, pollinated by bees. (B) *A. nivalis*, pollinated by long-tongued bees. (C) *A. vulgaris*, pollinated by long-tongued bumble bees. (D) *A. formosa*, pollinated by hummingbirds. The length and curvature of the nectar tubes of the flowers are correlated with the length and curvature of the bees' tongues and the hummingbirds' bills. The length of the pollen-bearing stamens is suited to the size of the pollinator, and the reduced petal width of *A. formosa* reflects the hummingbird's lack of need for a perch.

17.24 Pollinator tracking by scarlet gilia (*Ipomopsis aggregata*)

During the early part of the summer, when these plants are pollinated by both hummingbirds and hawk moths, they produce about twice as many red flowers (preferred by hummingbirds) as white flowers (A). Later in the summer, as the hummingbird populations leave the area, scarlet gilia plants shift over to producing pink and especially white flowers, which are more attractive to the sole remaining pollinator, a local species of hawk moth (B).

17.25 An orchid flower that resembles a fly

The flowers of this species (*Ophrys insectifera*) look enough like female flies to attract some male flies to land on them. The males thus become dusted with pollen, which they may carry to other flowers

flies (Fig. 17.25). The male insect is stimulated to attempt to copulate with the flower and becomes covered with pollen in the process. When he attempts to copulate with another flower, some of the pollen from the first flower is deposited on the second.

Flowers pollinated by wind or water rather than animals characteristically lack bright colors, special odors, and nectar. In fact, most of them have no petals, and their sexual parts are freely exposed to the air currents.

The characteristics of flowers are, therefore, not simply pleasing curiosities of nature. They are important adaptations that have evolved in response to fundamental selection pressures.

■ DEFENSIVE ADAPTATIONS

Cryptic appearance Many animals blend into their surroundings so well as to be nearly undetectable. Frequently their color matches the background almost perfectly (Fig. 17.26). In some cases animals even have the ability to alter the condition of their own pigment cells and change their appearance to harmonize with their background (Figs. 17.27 and 17.28). Often, rather than match the color of the general background, the animals may resemble

17.26 Cryptic coloration of a crab

The sargassum crab, which lives in dense growths of the brown alga *Sargassum* off Bermuda, is the same color as the alga; its rounded body resembles the algal floats.

A

B

17.27 Flounders on two different backgrounds

These fish can change color to match the background, whether light-colored (A) or dark (B).

A B

17.28 Color change by the frog *Hyla versicolor*

Individuals of this species are able to change color to match either a tree trunk (A) or vegetation (B).

inanimate objects commonly found in their habitat, such as leaves (Fig. 17.29) or twigs (Fig. 17.30). When the shape or color of an animal offers concealment against its background, it is said to have a ***cryptic*** appearance.

Cryptic appearance is an adaptive characteristic that helps animals escape predation. In one study, F. B. Sumner invested the interaction of Galápagos penguins and mosquito fish (*Gambusia partuelis*). Mosquito fish contract or expand their pigment cells to become lighter or darker, depending on their background. Sumner showed that the penguins catch 70% of the fish that contrast with their background but only 34% that resemble their background. In a similar experiment, F. B. Isely studied predation of grasshoppers of various colors, on differently colored backgrounds, by chickens,

17.29 Leaflike mantis

The mantis (at top in picture) looks strikingly like the green leaves below it. Even the relationship between its thorax and abdomen reflects the way the leaves often occur in pairs.

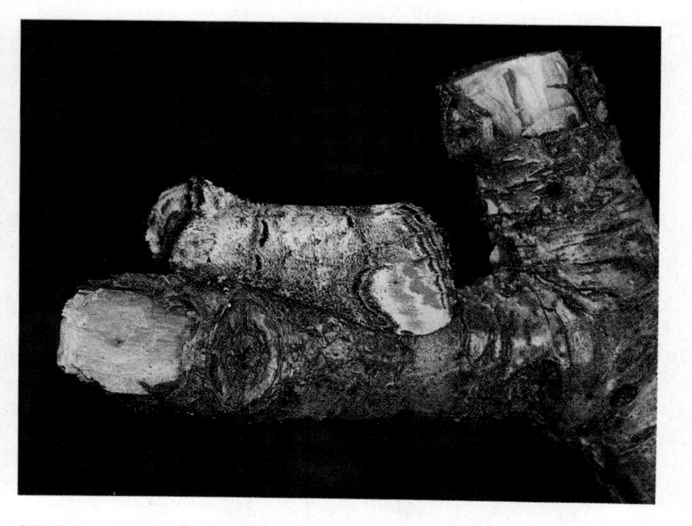

17.30 A moth that looks like a broken twig

The perfection of the cryptic appearance of this moth (*Phalera bucephala*) is a triumph of evolutionary adaptation.

turkeys, and native birds. He found that 88% of the nonprotected grasshoppers were eaten, but only 40% of the cryptically colored ones.

One of the most extensively studied cases of cryptic coloration is the so-called industrial melanism of moths. Since the mid-1880s many species of moths have become decidedly darker in industrial regions. This is actually a case of polymorphism in which the less frequent of two forms has become the more frequent; the originally predominant light form in these species of moths has given way in industrial areas to the dark (melanic) form. In the Manchester area of England, the first black specimens of the species *Biston betularia* were caught in 1848; by 1895 melanics constituted about 98% of the total population in the area. For such a remarkable shift in frequency to have occurred in so short a time, the melanic form must have had at least a 30% advantage over the light form.

The various species of moths exhibiting the rapid shift to melanism, though unrelated to one another, all habitually rest during the day in an exposed position on tree trunks or rocks, being protected from predation only by their close resemblance to their background. In former years the tree trunks and rocks were light colored and often covered with light-colored lichens. Against this background the light forms of the moths were difficult to see, whereas the melanic forms were quite conspicuous (Fig. 17.31). Under these conditions it seems likely that predators captured melanics far more easily than the cryptically colored light moths. The light forms would thus have been strongly favored, and they would have occurred with much higher frequency than melanics. With the advent of extensive industrialization, tree trunks and rocks were blackened by soot, and the lichens, which are particularly sensitive to such pollution, disappeared. In this altered envi-

A

B

17.31 Cryptic coloration of peppered moths

(A) Light and dark morphs of *Biston betularia* at rest on a tree trunk in unpolluted countryside. (B) Light and dark morphs on a soot-covered tree trunk. Here the light form is easier to see.

ronment the melanic moths resembled the background more closely than did the light moths. Thus selection should now have favored the melanics, which would explain why they increased in frequency.

This scenario was put to a test in the mid-1950s, when H. B. D. Kettlewell released equal numbers of the light and melanic forms of *Biston betularia* onto trees in a rural area in Dorset, England, where the tree trunks were light colored, lichens were abundant, and the wild population of the moth was about 94.6% light colored. Of 190 moths observed to be captured by birds, 164 were melanics and only 26 were light forms. Of hundreds of marked individuals of each form released in another experiment, roughly twice as many light moths as melanic moths were recaptured in traps set up in the woods, an indication that more of the light moths had survived.

Taken alone, however, these experiments did not prove that the factor favoring the light moths over the melanics was their resemblance to their background. Perhaps the birds preferred the melanics because of some difference in flavor. Kettlewell duplicated the experiments under the reverse environmental conditions—in woods near Birmingham, England, where the tree trunks were blackened with soot, lichens were absent, and the wild population of the moth was about 85% melanic. The results of these experiments were the reverse of those in the Dorset experiments. Now birds captured nearly three times as many light moths as melanics, and roughly twice as many melanics were recaptured in traps. Recently, the imposition of pollution controls has led to a shift back toward the light morph.

Warning coloration Whereas some animals have evolved cryptic coloration, others have evolved colors and patterns that contrast boldly with their background and thus render them clearly visible to potential predators. Such warning appearance is called ***aposematic.*** (Fig. 17.32). Nearly all of these animals are in some way disagreeable to predators; they may taste bad, smell bad, sting, or secrete poisonous substances. They are animals that a predator will usually reject after one or two unpleasant encounters. Such animals benefit by being gaudy and conspicuous because predators that have experienced their unpleasant features learn to recognize and avoid them more easily in the future.

The warning is sometimes so effective that, after unpleasant experiences with one or two warning-colored insects, some vertebrate predators simply avoid all flashily colored insects, whether or not they resemble the ones they encountered earlier. G. D. H. Carpenter demonstrated this by offering over 200 different species of insects to an insectivorous monkey. The monkey accepted 83% of the cryptically colored insects but only 16% of those with the warning coloration, even though many of the insects belonged to species the monkey had probably not previously encountered. It is possible, therefore, that avoidance of aposematic insects does not depend entirely on learning, and that some predators have a genetic predisposition to avoid brightly colored prey. They would probably have an adaptive advantage over predators that waste time and energy pursuing inedible prey; hence innate conservatism could evolve.

17.32 Aposematic coloration

The bright color of the poison-arrow frog (from which South American Indians obtain poison for their arrow tips) makes it easily recognizable by predators, which carefully avoid it.

Mimicry Species not naturally protected by some unpleasant character of their own may closely resemble (mimic) in appearance and behavior some dangerous or unpalatable aposematic species (called models). Such a resemblance can be adaptive: the mimics may suffer little predation because predators cannot distinguish them from their models, which the predators have learned are unpleasant. This phenomenon is called ***Batesian mimicry*** (Fig. 17.33A,B).

Convincing evidence for the potential effectiveness of this type of mimicry in protecting the mimic species comes from the elegant experiments of Jane van Z. Brower. Brower produced an artificial model-mimic system with starlings as the predators and mealworms, which starlings ordinarily eat voraciously, as prey. She painted a tasteless color band on the mealworms and dipped some in a distasteful solution. She then presented various ratios of noxious (dipped) models and mimics to different groups of birds. After a few unpleasant encounters with the models, the birds learned to recognize and avoid the painted worms, with the result that the mimics also escaped predation, particularly when the percentage of mimics presented to the starlings was 60% or less.

A few species pursue the opposite approach: these "aggressive mimics" provide lures of apparently palatable prey to attract potential victims (Fig. 17.33C).

In addition to Batesian mimicry, which is based on deception—mimicry of a distasteful or dangerous species by individuals of a species that is neither—there is a second kind of mimicry, called

A

B

17.33 Examples of Batesian and aggressive mimicry

(A) The prominent imitation eyes and mouth of the spicebush swallowtail larva give it the appearance of a predator to be avoided. (B) The markings of the harmless syrphid fly (top) resemble those of a stinging bumble bee. (Upon close inspection, flies are readily distinguished from bees because their antennae are very short and they have only a single pair of wings.) (C) The warty frogfish is an aggressive mimic: its body, which resembles an algae-covered rock, is merely cryptic, but it displays a lure that resembles a small fish.

C

Müllerian mimicry, which involves the evolution of a similar appearance by two or more distasteful or dangerous species. In this type of mimicry, individuals of each species act as both model and mimic. The members of each species have some defensive mechanism, but if each species had its own characteristic appearance, the predators would have to learn to avoid each of them separately; the learning process would thus be more demanding, and would involve the death of some individuals of each prey species. Selection favors evolution toward one appearance; the various protected species thus come to constitute a single prey group from the standpoint of the predators, which learn avoidance more easily.

One striking case of Müllerian mimicry involves the monarch butterfly *(Danaus plexippus)* and the unrelated viceroy *(Limenitis archippus)*. These two species look very much alike (Fig. 17.34) and are each distasteful to birds, though in different ways: monarchs sequester poisons from the milkweed plant upon which their caterpillars feed, while viceroys synthesize their own bad-tasting chemical. Until recently, the viceroy was thought to be a Batesian mimic of the monarch because blue jays, the predators used in early tests, are one of the few species of birds that find viceroys only slightly distasteful. Other kinds of birds that taste one member of either species learn to avoid them both.

The selective advantage of Müllerian mimicry may explain the similar markings of many unrelated species of wasps and bees, or of the group of poisonous reptiles known as coral snakes. If avoidance involves any genetic predisposition, then resemblances among the prey animals would facilitate more rapid selection for improved prey-recognition mechanisms in the predators. In fact, there is evidence that some predators recognize coral snakes innately; perhaps only Müllerian mimicry can provide a strong enough selection pressure to cause the evolution of such specialized recognition—a recognition that benefits individuals of both predator and prey species.

A

B

17.34 Müllerian mimicry in butterflies
The monarch butterfly (A) and the viceroy (B) which are not closely related, have evolved strikingly similar color patterns.

■ SYMBIOTIC ADAPTATIONS

The term *symbiosis* is used in a variety of ways in the biological literature. Some authors apply it only to cases in which two species live together to their mutual benefit. Etymologically, however, symbiosis simply means "living together," without any implied value judgments. This is the meaning it was given when it was first introduced into biology, and this is the meaning we use here. There are four categories of symbiosis: ***commensalism***, a relationship in which one species benefits while the other receives little or no benefit or harm; ***mutualism***, in which both species benefit; ***parasitism***, in which one species benefits and individuals of the other species are harmed; and ***enslavement*** in which one species is enslaved by another (Table 17.2).

Table 17.2 Varieties of symbiotic relationships

RELATIONSHIP	SPECIES A	SPECIES B
Commensalism	+	0
Mutualism	+	+
Parasitism	+	–
Enslavement	+	–

Commensalism The advantage derived by the commensal species from its association with the host often involves shelter, support, transport, food, or several of these. For example, in tropical forests numerous small plants, called epiphytes, grow on the branches of the larger trees or in forks of their trunks (see Fig. 40.35B, p. 1186). These epiphytes, including species of orchids and bromeliads, are

17.35 An example of commensalism

An anemone fish living among the tentacles of a sea anemone from the Palau Islands

often not parasites. They use the host trees as a base of attachment and do not obtain nourishment from them. They apparently do no harm to the host except very rarely, when so many of them are on one tree that they stunt its growth or cause limbs to break. A similar type of commensalism is the use of trees as nesting places by birds.

Sometimes it is difficult to tell what benefit is involved in a commensal relationship. For example, certain species of barnacles occur nowhere except attached to the heads of whales, and other species of barnacles occur nowhere except attached to the barnacles that are attached to whales. Just what advantages either of these groups of barnacles enjoys is not clear.

In some cases of commensalism, however, the benefit is dramatically obvious. For example, certain species of fish live in association with sea anemones, deriving protection and shelter from them and sometimes stealing some of their food (Fig. 17.35). These fish swim freely among the tentacles of the anemones, even though the tentacles quickly paralyze other fish that touch them. The anemones regularly feed on fish, yet the particular species that live as commensals with them sometimes enter the gastrovascular cavity of their host and emerge later with no apparent ill effects. The physiological and behavioral adaptations that make such a commensal relationship possible are extensive.

Another striking example is a small tropical fish that lives in the respiratory tree of a particular species of sea cucumber. The fish emerges at night to feed and then returns to its curious abode by first poking its host's rectal opening with its snout and then quickly turning so that it is drawn tail first through the rectal chamber into the respiratory tree. Still another example is a tiny crab that lives in the mantle cavity of oysters. The crab enters the cavity as a larva and eventually grows too big to escape through the narrow opening between the two valves of the oyster's shell. It is thus a prisoner of its host, but a well-sheltered prisoner. It steals a few particles of food from the oyster but apparently does it no significant harm.

Mutualism Symbiotic relationships beneficial to both species are common. Figure 17.36 illustrates two instances of the widespread phenomenon called cleaning symbiosis, which is patently mutual-

A

B

17.36 Cleaning symbiosis

(A) A giant seabass being cleaned by a cleanerfish. (B) Yellow-billed oxpeckers search for parasitic insects on a black rhinoceros. In both cases the symbiosis is mutualistic: the cleaner obtains food, and the host gets rid of parasites that could endanger its health.

istic. Other examples of mutualism include the relationship between a termite and the cellulose-digesting microorganisms in its digestive tract, or between a human being and the bacteria in the intestine that synthesize vitamin K. The plants we call lichens are actually formed of an alga or a cyanobacterium and a fungus united in such close symbiosis that they give the appearance of being one plant (see Fig. 23.9, p. 650). The fungus benefits from the photosynthetic activity of its "guest," and in at least some cases the alga or bacterium benefits from the water-retaining properties of the fungal walls.

The division of symbiosis into four subcategories is in many ways arbitrary. In particular, commensalism, mutualism, and parasitism are all parts of a continuous spectrum of possible interactions. The categories are only devices to help us organize what we know about nature and form testable hypotheses. What is important is to keep in mind how commensalism, mutualism, and parasitism grade into each other; each case of symbiosis is different from all others and must be studied and analyzed on its own.

Parasitism There is no strict line between parasitism and predation. Mosquitoes and lice both suck mammalian blood, yet we usually call only the latter parasites. Foxes and tapeworms may both attack rabbits, but foxes are called predators and tapeworms are called parasites. The usual distinction is that a predator eats its prey quickly and then goes on its way, while a parasite passes much of its life on or in the body of a living host, from which it derives food in a manner harmful to the host. Parasites generally do not kill their hosts; those that eventually do so are called ***parasitoids*** (Fig. 17.37).

17.37 A caterpillar (tomato hornworm) with numerous pupae of a parasitoid wasp attached to its body

The wasp laid her eggs inside the body of the host caterpillar, and the larvae fed on it until they emerged and pupated.

Parasites are customarily divided into two types: external parasites and internal parasites. The former live on the outer surface of their host, usually either feeding on the hair, feathers, scales, or skin of the host or sucking its blood. Internal parasites may live in the various tubes and ducts of the host's body, particularly the digestive tract, respiratory passages, or urinary ducts; or they may bore into and live embedded in tissues such as muscle or liver; or, in the case of viruses and some bacteria and protozoans, they may actually live inside the individual cells of their host.

Internal parasitism is usually marked by more extreme specializations than external parasitism. The habitats available inside the bodies of living organisms are completely unlike those outside, and the unusual problems the host environments pose have resulted in evolutionary adaptations entirely different from those seen in free-living forms. Internal parasites, for example, have often lost organs that would be essential in a free-living species: tapeworms have no digestive system, but they live in their host's intestine and are bathed by the products of its digestion, which they absorb directly across their body wall.

Because of their frequent evolutionary loss of structures, certain parasites are often said to be degenerate. "Degenerate" implies no value judgment but simply refers to the lack of many structures present in their free-living ancestors. This condition is common in internal parasites, like the tapeworms mentioned above, but also can occur in external parasites. For example, the ant species

17.38 A parasitic ant

Queens of the "ultimate ant," *Teleutomyrmex*, ride on the host queen and are fed by a host worker (lower right). They dispense their eggs along with those of the host queen, and their offspring, all reproductives, mate in the colony.

Teleutomyrmex (Fig. 17.38) lives in the nests of another species of ant. It is degenerate to the extent that it cannot survive outside its hosts' nest: it has lost many of its glands, its sting, its pigmentation, and its ability to digest anything but liquid fed to it by its hosts. Even its brain is degenerate. From an evolutionary point of view, however, loss of structures that are useless in a new environment is an instance of specialized adaptation.

Specialization, then, does not necessarily mean increased structural complexity; it only means the evolution of characteristics particularly suited to some special situation or way of life. In internal parasites—and cave animals, which frequently lack eyes—the development and maintenance of structures that no longer serve a useful function would require energy that the organism might use to more advantage in some other way. Moreover, some useless structures, such as eyes in internal parasites, deep-sea fish, and cave animals, might well be a handicap in these special environments, because they would be a likely point of infection. Natural selection often favors those individuals in which such useless organs are either relatively small or lacking entirely. Alternatively, the unused structures could have been adaptively neutral and lost through genetic drift.

As parasites evolve, the host species is also evolving, and there is strong selection pressure for the evolution of more effective defenses against the ravages of parasites (Fig. 17.39). This constant interplay between host and parasite is at the heart of the red queen hypothesis. (see "Exploring Further: The Logic of Sex," p. 448). Those individuals of the host species with superior defenses will be better able to survive and reproduce. Correspondingly, those individuals of the parasite species with the best ways of counteracting the defenses will be most likely to prosper. In turn, their counteractions will lead to pressure on the host to evolve still better defenses, against which the parasite may then evolve new means of surviving, and so forth.

17.39 A bumble bee defense against parasitoids

Up to 70% of foraging bumble bees are parasitized by fly larvae. Infected bees respond by visiting the hive only to deliver their loads of pollen and nectar; they spend the nights in the field on flowers rather than in the colony. This counterploy not only slows parasitoid development by exposing it to cold nighttime temperatures (thus lengthening the useful life of the forager), it also greatly lowers the chance that the parasitoid will chance to emerge and spread within the colony.

Probably most long-established host-parasite relationships are balanced ones. Relationships that result in serious disease in the host tend to be relatively new, or they are relationships in which a new and more virulent form of the parasite has recently arisen or in which the host showing disease symptoms is not the main host of the parasite. Native Americans, for example, suffered severely when exposed to diseases first brought to the Americas by European colonists, even though some of these same diseases caused only mild symptoms in the Europeans, who had been exposed to the disease-causing organisms for many centuries and in whom the host-parasite relationship had nearly reached a balance. Many examples are known in which humans are only occasional hosts for a particular parasite and suffer severe disease symptoms, though the wild animal that is the major host shows few ill effects.

It is worth keeping in mind, however, that benign interactions between the host and a parasite need never evolve regardless of how much time passes. The reason is that the optimal strategy for a parasite depends critically on the life histories of itself and its host, and for some combinations there is no advantage to the parasite in achieving a balance and no way for the host to impose one. In some diseases—for example, rabies (which is fatal to humans and many other mammals) and smallpox (which, untreated, kills about 30% of its victims)—the parasite apparently benefits from a massive attack on the host, which enables it to spread its offspring rapidly. A slower, longer-lasting release of progeny produces lower reproductive success. An evolutionary perspective on host-parasite dynamics—particularly life histories and transmission efficiencies—can lead to optimal designs for disease treatment and prevention, whereas intuitively attractive alternatives that fail to consider these interactions often turn out to be futile.

Enslavement Enslavement occurs when an individual of one species maintains members of another species, controlling their reproduction and using their effort to its own ends. Some cases are clear-cut. Several species of ants enslave workers of other species, either by raiding the host colonies for pupae (which hatch into workers in the slave-making colony), or by killing the host queen and establishing one of their own queens in her place. The algae in some lichens are enslaved in that they cannot escape. Mitochondria and chloroplasts are in the same position, having been enslaved by eucaryotes more than a billion years ago. Other instances shade into parasitism or predation.

CHAPTER SUMMARY

HOW SELECTION OPERATES

Evolution is a change in allelic frequencies. Natural selection operates on heritable variation, favoring better adapted individuals. Chance (genetic drift) can also cause evolution. (p. 444)

WHAT CAUSES VARIATION? Variation arises from crossing over, sexual recombination, and mutation. (p. 445)

HOW DO NEW GENES AND ALLELES ARISE? Alleles can be modified through point mutations. Larger-scale changes arise from gene or exon duplication, exon recombination, and horizontal transfer of genes from other species. (p. 448)

HOW DO MUTATIONS SPREAD IN POPULATIONS? In the absence of immigration and emigration (gene flow), nonrandom mating (sexual selection), natural selection, mutation, and genetic drift, the allelic frequencies in a population (gene pool) will not change from one generation to another. (p. 451)

NATURAL SELECTION Selection operates by increasing the reproductive fitness—the number of viable offspring produced—of individuals carrying well-adapted alleles relative to individuals lacking these alleles. Selection can be directional (favoring one end of a range of variation), stabilizing (favoring a certain value in the continuum of variation), or disruptive (favoring two—or, rarely, more—alternative values). Crossing over and recombination can create novel allelic combinations. Net selection pressure on a trait usually depends on the difference between the trait's costs and benefits. Frequency-dependent selection depends on the relative frequency of a trait in the population: in many cases, rarer traits are favored. (p. 458)

SEXUAL SELECTION Selection favors an ability to increase the quality and quantity of offspring, in part through optimizing the quality and quantity of mates. One sex (usually males) often competes for mates, either through dominance contests, or through courtship displays and sex-specific morphology. (p. 467)

ADAPTATION

Adaptations help an organism survive and reproduce. The most spectacular cases are those that are unique to a particular species and suit it for making its living in a highly specific way. (p. 468)

ADAPTATIONS FOR POLLINATION Plants and their pollinators have coevolved so that each is well suited to the other. (p. 470)

DEFENSIVE ADAPTATIONS Many species are cryptic in their particular habitat, thus avoiding predation. Toxic or dangerous species often have aposematic (warning) colors that may be innately recognized or readily learned by predators. Batesian mimics benefit by having the same warning signals without paying the costs of defense. Müllerian mimics are each toxic but, by resembling each other, "share" the cost of teaching predators to avoid them. (p. 473)

SYMBIOTIC ADAPTATIONS One species in a commensal pair benefits from its association with the other, while the second is not helped or harmed. Mutualistic species each benefit. In a parasitic relationship, one species benefits at the expense of the other. Finally, one species can enslave members of another species. (p. 478)

STUDY QUESTIONS

1 Imagine a relatively large, long-lived animal (say at least 5 kg, living a decade on average) which, because of its habitat and lifestyle, does not benefit from sexual recombination. Derive a list of the rare conditions that would favor cloning in such an animal. (pp. 447–448)

2 Preadaptations can enable a species to expand into a novel niche. Many species—rats, mice, roaches, silverfish, sea gulls, pigeons, and so on appear to have occupied the niches created by humans. What sorts of adaptations, already useful or neutral in their normal habitat, enable these species to exploit new opportunities? Is there a common theme? Is it just a coincidence that several of these species are standard laboratory animals? (pp. 458–459)

3 What sorts of ecological factors might favor male contests over female choice? How might such factors tip the balance toward harems rather than monogamy or bigamy? (pp. 467–468)

4 Some researchers believe that human females are, on average, inherently better at some tasks (verbal expression, making subtle distinctions, and so on) than males, while males are, on average, better at mathematical and geometric problems. (Others, of course, believe the differences that have been observed are the result of conditioning.) Is it possible for natural selection to work in different directions on the two sexes of one species? Cite examples to support your conclusion. If the answer is yes, what is the likely logic of differential specializations? (pp. 463–464, 467–468)

5 What, if anything, keeps parasites from driving their hosts to extinction? (There are several possibilities.) (pp. 480–482)

SUGGESTED READING

AMÁBILE-CUEVAS, C. F., AND M. M. CHICUREL, 1993. Horizontal gene transfer. *American Scientist* 81 (4). *On how gene transfers between different species can fuel evolution.*

DOOLITTLE, R. F., AND P. BORK, 1993. Evolutionary mobile modules in proteins. *Scientific American* 269 (4). *Thoughtful alternatives to Gilbert's model for exon duplication and recombination emphasizing horizontal transmission.*

GOULD, J. L., AND C. G. GOULD, 1989. *Sexual Selection.* Scientific American Books, New York. *Well-illustrated treatment of the evolution of sex and mate choice.*

LAWN, R. M., 1992. Lipoprotein (a) in heart disease. *Scientific American* 266 (6). *Latter half of article details how this protein evolved from exon recombination.*

LEWONTIN, R. C., 1978. Adaptation. *Scientific American* 239 (3). *Points out that most features are compromises between different selection pressures, and that chance plays a role in evolution when more than one solution to a problem is possible.*

LI, W. H., AND D. GRAUAR, 1991. *Fundamentals of Molecular Evolution.* Sinauer Associates, Sunderland, Mass. *An excellent account of how gene evolution can occur through duplication and exon recombination.*

RIDLEY, M., 1993. *Evolution.* Blackwell, Oxford. *Excellent text on the subject.*

CHAPTER 18

SPECIATION AND PHYLOGENY

HOW SPECIATION OCCURS

The previous chapter discussed how populations change through time. Eventually such a population may split, giving rise to two or more different descendant populations, perhaps generating new species in the process. This chapter will examine the processes underlying speciation, as well as the ways in which evolutionary history is reconstructed.

■ WHAT ARE SPECIES?

Demes To understand speciation, we need to look at population structure in more detail. With reference to sexually reproducing organisms, we can define a ***population*** as a group of individuals that interbreed and so share a common gene pool. A ***deme***, in contrast, is a small local population, such as all the deer mice or all the red oaks in a certain woodland, or all the perch or all the water striders in a given pond (Fig. 18.1). Though no two individuals in a deme are exactly alike, the members of a deme usually resemble one another more closely than they resemble the members of other demes. There are at least two reasons for this: (1) the individuals in a deme are more closely related, because pairings occur more frequently between members of the same deme than between members of different demes; and (2) the individuals in a deme are exposed to more similar environmental influences and hence to more nearly the same selection pressures.

Demes are not clear-cut permanent units of population. Though the deer mice in one stand of trees are more likely to mate among themselves than with deer mice in nearby woods, there will almost certainly be occasional matings between mice from different woods. Similarly, though the female parts of a particular red oak tree are more likely to receive pollen from another red oak tree in the same stand, they will sometimes receive pollen from a tree in another nearby woods. And woods themselves are not permanent ecological features. They have only a transient existence as separate and distinct ecological units; neighboring stands may fuse after a few years, or a single woods may become divided into two or more separate smaller ones. Such changes in ecological features will produce corresponding changes in the demes of deer mice and oaks. Demes, then, are usually temporary units of population that intergrade with other similar units.

18.1 Aquatic demes
Each of these ponds contains a separate deme of each species of fish and aquatic insect.

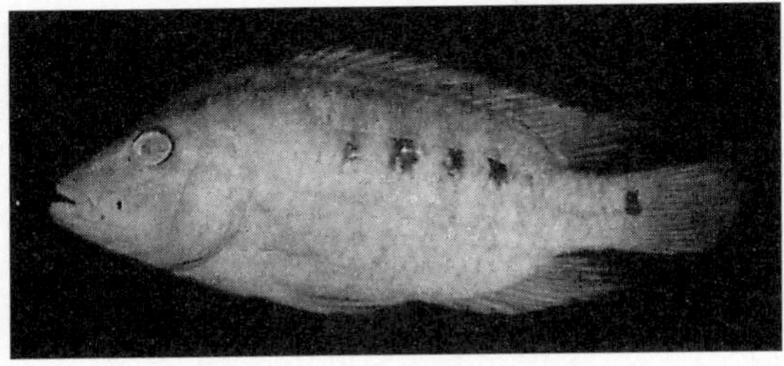

A

B

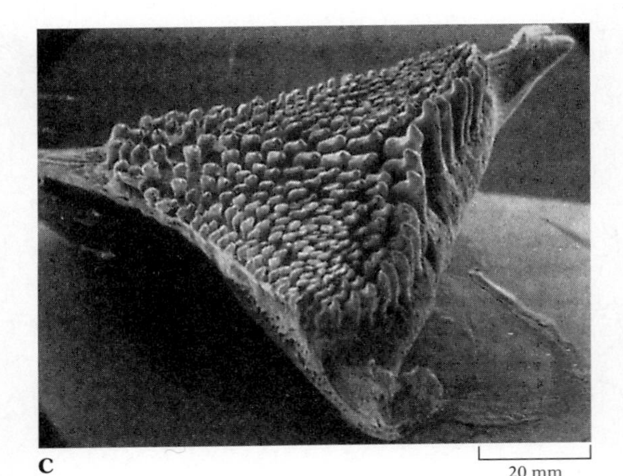

C

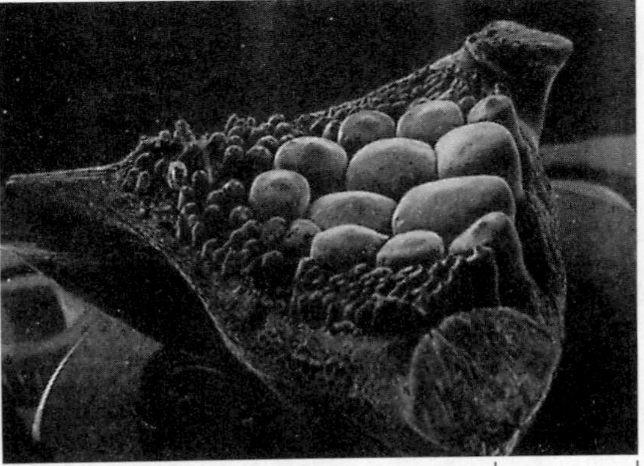

D

18.2 Variations in morphology between populations within a species

In some species, populations within a single area show easily observable morphological variations, which may be continuous (such as size) or discrete (distinct alternative morphs). A male of the cichlid species *Cichlasoma minckleyi,* found in the Cuatro Ciénegas basin in Mexico, exhibits the deep-bodied morph (A), while another male of the same species from the same basin exhibits the slender-bodied morph (B); there are no intermediates in this character. An independent variation of the food-grinding structures in the lower pharyngeal jaw of males of this species also exists. Some males, regardless of body form, display the papilliform morph (C), while others display the molariform morph (D). Despite these major differences, individuals of the various morphs readily interbreed and produce fertile offspring; hence they are all considered members of the same species.

Species For centuries it has been recognized that plants and animals seem to fall naturally into many separate and distinct "kinds," or species. This does not mean that all the individuals of any one species are precisely alike—far from it. It does mean that all the members of a single species share certain biologically important attributes and that, as a group, they are genetically separated from other such groups.

A ***species***, in the modern view, is a genetically distinctive population, a group of natural demes that share a common gene pool and that are reproductively isolated from all other such groups. A species is the largest unit of population within which effective gene flow occurs or can occur. The key word here is *effective;* we will see later why two species whose members mate but produce infertile hybrids are not classified as a single species.

The modern concept of species says nothing about how different from each other two populations must be to qualify as separate species. Most species can be separated on the basis of fairly obvious anatomical, physiological, or behavioral characters. However, the final criterion for living species is always reproduction—whether or not there is actual or potential gene flow.[1] If there is complete reproductive isolation between two outwardly almost identical populations—that is, if there can be no gene flow between them—then those populations belong to different species despite their similarity. On the other hand, if two populations show striking differences, but there is effective gene flow between them, those populations belong to the same species (Fig. 18.2). Anatomical, physiological, or behavioral characters simply serve as clues toward the identification of reproductively isolated populations; they do not in themselves determine whether a population constitutes a species.

Variation within species Though there is significant variation between individuals in a deme, there is often a superimposed variation correlated with geographic distribution. This leads to

[1] Because paleontologists deal almost exclusively with fossils, they must rely to a great extent on morphological criteria in distinguishing between species.

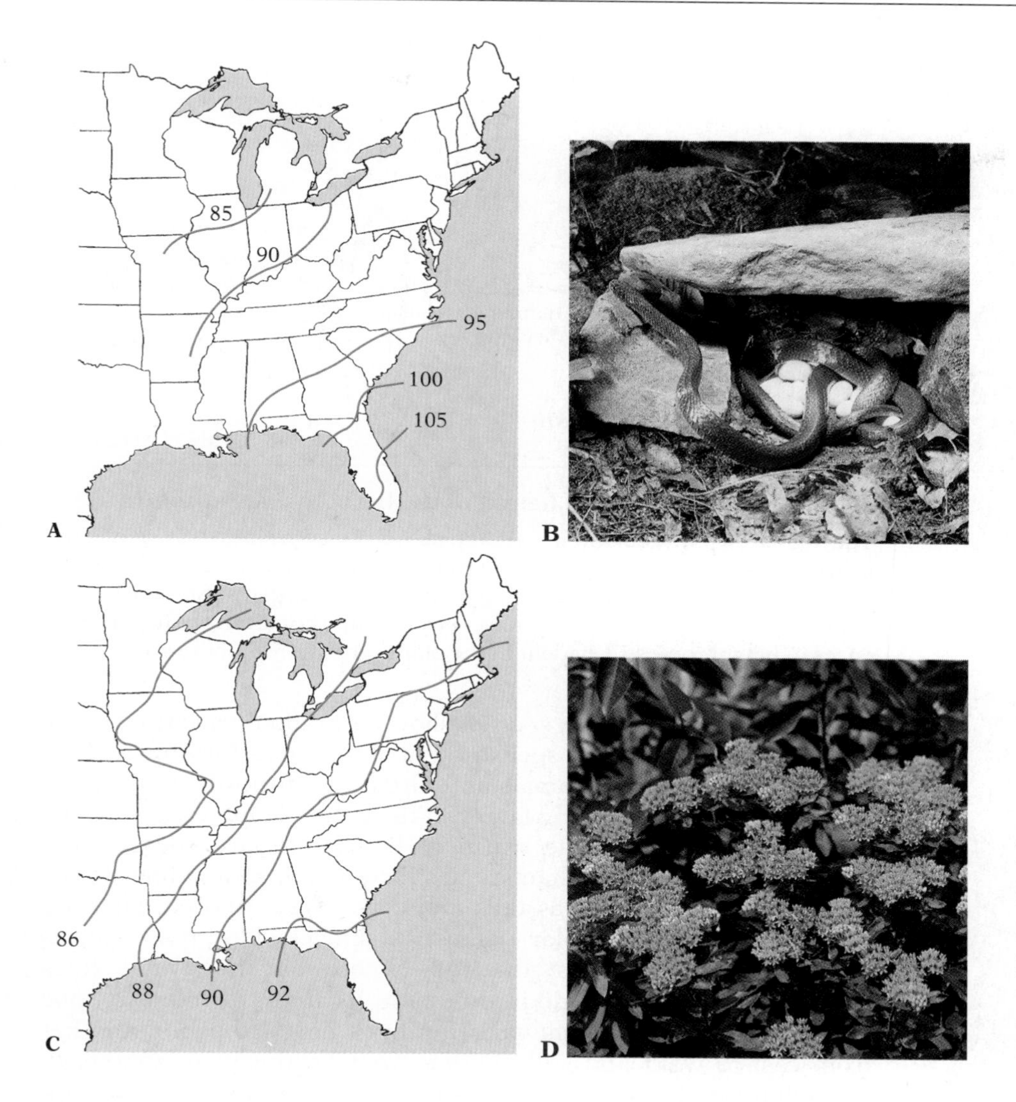

18.3 Clinal variation

(A) Map showing, by means of isophene lines (lines connecting equal values), the geographic variation in the mean number of subcaudal scales of the snake *Coluber constrictor* (the racer; B). Isophene map (C) showing geographic variation in the apical taper of leaves of the milkweed *Asclepias tuberosa* (D). The numbers represent degree of apical taper.

larger-scale differences between demes. Some of this pattern may result from genetic drift, but much of the geographic variation reflects differences in selection pressures operating on the populations, resulting from differences in environmental conditions. Each local population or deme tends to evolve adaptations to the specific environmental conditions in its own small portion of the species range. Such geographic variation is found in the vast majority of animal and plant species.

Environmental conditions often vary geographically in a more or less regular manner. There are changes in temperature with latitude or with altitude on mountain slopes, or changes in rainfall with longitude, as in many parts of the western United States, or changes in topography with latitude or longitude. Such environmental gradients are usually accompanied by gradients of allelic frequency in the species of animals and plants that inhabit the areas involved. Most species show north-south gradients in many characters; east-west gradients (Fig. 18.3A,C) and altitudinal gradients (Fig. 18.4) are not uncommon.

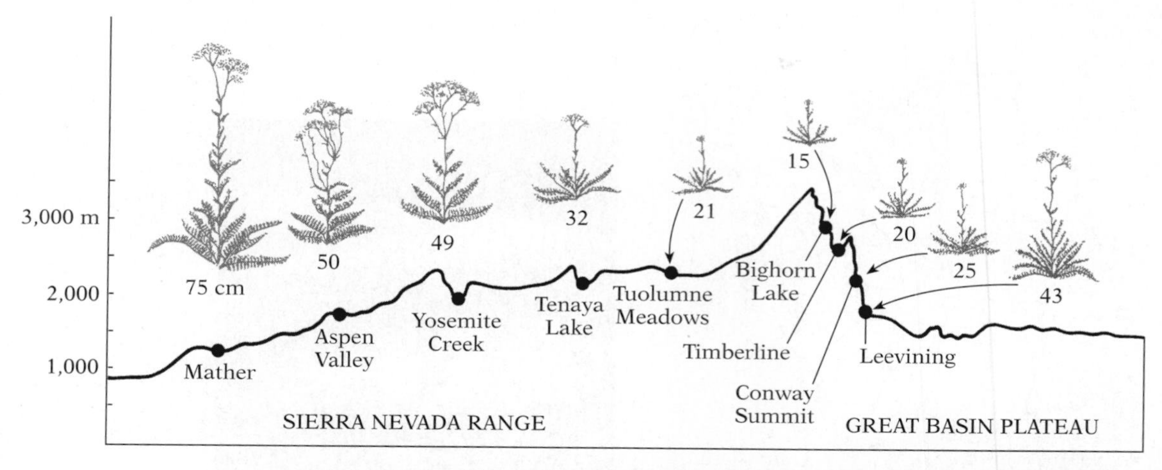

18.4 Altitudinal cline in height of the herb *Achillea lanulosa*

The higher the altitude, the shorter the plants. This variation was shown to be genetic (as opposed to merely environmental) by collecting seeds from the locations indicated and planting them in a test garden at Stanford where all were exposed to the same environmental conditions. The differences in height were still evident in the plants grown from these seeds.

When a character of a species shows a gradual variation correlated with geography, we speak of that variation as forming a ***cline***. For example, many mammals and birds exhibit north-south clines in average body size, being larger in the colder climates toward the poles and smaller in the warmer climates toward the equator. Similarly, many mammalian species show north-south clines in the size of such extremities as tails and ears: these exposed parts are smaller in the more polar demes.[2] A single widespread species often has many characters that vary clinally, but the several clines frequently do not coincide in direction, location, or intensity; one character may show clinal variation from north to south, another from east to west, and still another from northwest to southeast. Though most clines reflect adaptation to local conditions, some appear to be chance patterns.

Sometimes geographically correlated genetic variation is not as gradual as in clines. When an abrupt shift in a genetically determined character occurs in a geographically variable species, the populations involved may be called ***subspecies***. This term is also sometimes applied to more isolated populations—such as those on different islands or in separate mountain ranges or, as in fish, in separate rivers—when the populations are recognizably different genetically but are potentially capable of interbreeding freely. Subspecies, then, are groups of natural populations within a species that differ genetically and that are partly isolated from each other because they have different ranges (Fig. 18.5).

Two subspecies of the same species cannot long occur together geographically, because it is only the limitation on interbreeding imposed by distance that keeps them genetically distinct. If they occur together, they will interbreed and any distinction between

[2] Increase in average body size with increasing cold, called Bergmann's rule, is very common in warm-blooded animals; the tendency toward decrease in the size of the extremities with increasing cold is known as Allen's rule. The adaptive significance of these clines reflects the role of surface-to-volume ratios in heat loss.

A

B

18.5 Two subspecies of Canada goose

These two subspecies of Canada goose have different breeding grounds and different ranges. *Branta canadensis maxima* (A) breeds in the central and south-central United States. *Branta canadensis moffitti* (B) breeds largely in central and western Canada.

them will disappear. Some biologists have argued against recognizing subspecies, mainly because the distinctions are often made arbitrarily on the basis of one morphologically obvious character, while other less obvious characters may form different patterns of variation that are ignored (Fig. 18.6). Nevertheless, assigning names to distinguish separate populations can be a great convenience.

As a result of intraspecific geographic variation, two populations belonging to the same species but occurring in two widely separated localities often show no more resemblance to each other than to populations belonging to other species. Such intraspecific dissimilarity underscores the point made earlier: it is not the degree of morphological resemblance that determines whether or not two populations belong to the same species, but rather whether they are reproductively isolated. There are even instances where two widely separated populations are regarded as belonging to the same species even though the respective individuals, when brought together, are incapable of producing viable offspring. They are members of the same species because the populations are connected by an unbroken chain of intermediate populations that can interbreed.

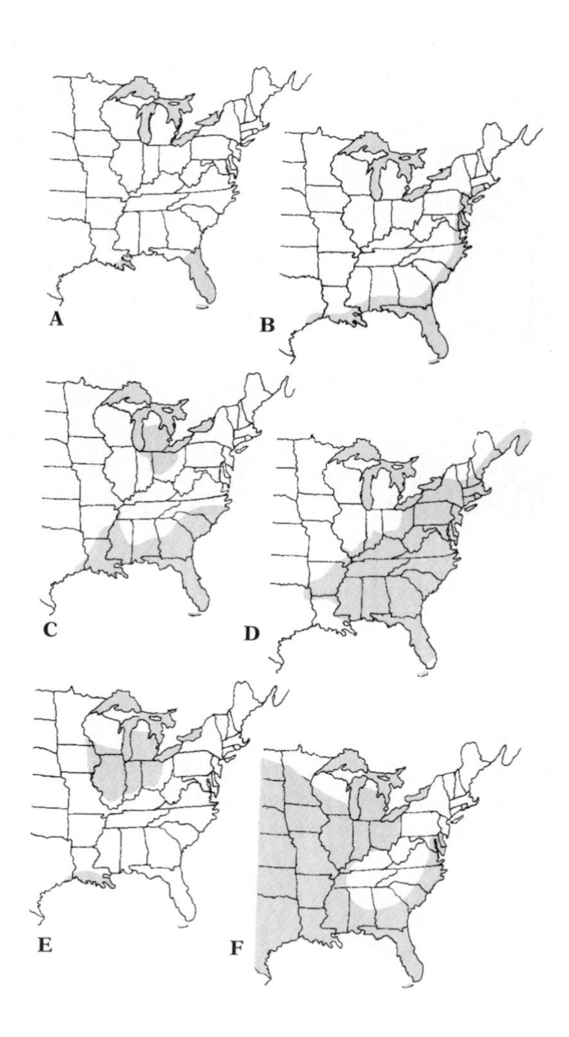

18.6 Discordant geographic variation in six characters of the snake *Coluber constrictor* in the eastern United States

Since no two of the characters vary together, selection of any one of them as a criterion for recognition of subspecies is largely arbitrary. (A) Red eyes are found in juveniles. (B) Red ventral spots are found on juveniles. (C) The loreal scale is in contact with the first supralabial scale in at least 10% of these specimens. (D) Black adults are found. (E) Dark postocular stripes are found. (F) Full-grown adults have white chins.

A

B

18.7 Squirrels of the Grand Canyon

Two populations of squirrels that live in different ranges of the Grand Canyon in Arizona are morphologically distinct. The Kaibab squirrel (A), which lives on the Kaibab Plateau on the northern rim of the canyon, is darker than the Abert squirrel (B), of the related population that inhabits a range on the southern rim. Whether these two groups are separate species is not known.

■ THE PROCESS OF SPECIATION

New species can arise as an existing species changes, or as it splits into two or more species. We will focus on the latter phenomenon, known as divergent speciation.

The role of geographic isolation Since species are defined in terms of reproductive isolation, the fundamental question of divergent speciation is: How do two populations that initially share a common gene pool come to have completely separate gene pools?

In the majority of cases the initiating factor in speciation is geographic separation. As long as all the populations of a species are in direct or indirect contact, gene flow will normally continue throughout the system, and splitting will not occur. Various populations within the system, however, may diverge in numerous characters and thus give rise to much intraspecific variation. If the continuous system of populations is divided by some geographic barrier to individual dispersal, then the separated populations will no longer be able to exchange genes, and further evolution will be independent (Fig. 18.7).

Isolated populations are said to be ***allopatric*** (literally "of different groups"); populations that share a habitat, on the other hand, are called ***sympatric***. Given sufficient time, the two separate populations may become quite unlike each other as each evolves in its own way. At first, the only reproductive isolation between them will be geographic—isolation by physical separation—and they will still be potentially capable of interbreeding. Eventually, however, they may become genetically so different that there would be no effective gene flow between them even if they should again come into contact. When this point in their divergence has been reached, the two populations constitute separate species. This is not to say that separation inevitably leads to speciation, though that is likely: the sycamores in Europe have been isolated from the population in China for more than 30 million years, and yet they can still interbreed.

There are at least three factors that can make geographically separated populations diverge:

1 The chances are good that the two populations will have different initial gene frequencies. Because most species exhibit geographic variation, it is unlikely that a geographic barrier would divide a variable species into portions that are genetically alike; more likely it would separate populations that are already genetically different, with different alleles adapting them to the local conditions in their geographic regions. In fact, geographic isolation often cuts off the terminal portions of a cline. Alternatively, small numbers of individuals frequently manage to cross existing barriers and found new geographically isolated colonies.

Regardless of how the population becomes divided, if one group is relatively small, its members will carry with them in their own genotypes a relatively small percentage of the total genetic variation present in the gene pool of the parental population. The new colony will have allelic frequencies different from those of the parental population; this is a special form of genetic drift called the ***founder effect***. If from the moment of their separation two populations have different genetic potentials, their future evolution may follow different paths. Most cases of geographic isolation do involve small populations that exhibit the founder effect and genetic drift.

2 Separated populations will experience different mutations. Mutations are random; some mutations will occur in one of the populations and not in the other, and vice versa. Since there is no gene flow between the populations, a new mutant allele arising in one of them cannot spread to the other.

3 Isolated populations will almost certainly be exposed to different environmental selection pressures, since they occupy different ranges.

The barriers that cause initial spatial separation leading to speciation can differ from one species to another. A mountain range is a barrier to species that can live only in lowlands, a desert is a barrier to species that require a moist environment, and a valley is a barrier to montane species. On a grander scale, oceans and glaciers have played a role in the speciation of many plants and animals.

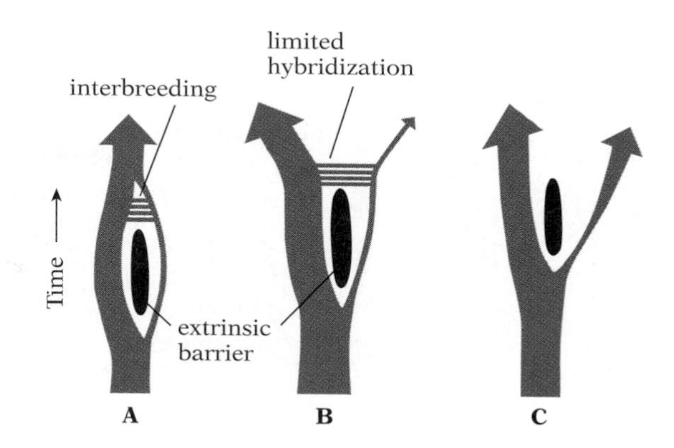

18.8 Model of geographic speciation

An ancestral population is split by an extrinsic (geographic) barrier for a time, and diverges with regard to various traits (only one of which is considered in these hypothetical examples). (A) The barrier breaks down before the two subpopulations have been isolated long enough to have evolved intrinsic reproductive isolating mechanisms, so the populations will interbreed and fuse back together. (B) Two populations are isolated by an extrinsic barrier long enough to have evolved incomplete intrinsic reproductive isolating mechanisms. When the extrinsic barrier breaks down, some hybridization occurs, but the hybrids are not as well adapted as the parental forms. Hence there is a strong selection pressure favoring forms of intrinsic isolation that prevent mating, and the two populations diverge more rapidly until mating between them is no longer possible. This rapid divergence is called character displacement. (C) Two populations are isolated by a geographic barrier so long that by the time the barrier breaks down they are too different to interbreed. In most cases, one population has many fewer members than the other, as indicated by the width of the "branches" where they separate. The smaller population—often quite small—usually diverges more from the common ancestor than does the larger population. This greater divergence results from the founder effect, a greater tendency for genetic drift, and the need to adapt to a smaller and perhaps more specialized habitat.

Reproductive isolation The initial factor preventing gene flow between two closely related populations is ordinarily an extrinsic one—geography. As two populations diverge, they accumulate differences that will lead, given enough time, to the development of intrinsic isolating mechanisms—biological characteristics involving morphology, physiology, chromosomal compatibility, or behavior that prevent the two populations from occurring together or from interbreeding effectively when (or if) they meet again (Fig. 18.8). There are several kinds of intrinsic isolating mechanisms that can evolve:

1 ***Ecogeographic isolation*** Two populations, initially separated by some extrinsic barrier, may in time become so specialized for different environmental conditions that even if the original barrier is removed they may never become sympatric, because neither can survive the conditions where the other lives. Two well-known species of the genus *Platanus*—*P. occidentalis* (the buttonwood tree), which occurs in the eastern United States, and *P. orientalis* (the Oriental plane tree), which occurs in the eastern part of the Mediterranean region—can be artificially crossed, and the hybrids are vigorous and fertile. But each species is adapted to the climate in its own native range, and the climates in the two ranges are so different that neither species will long survive in the range of the other. Thus there are genetic differences that under natural conditions prevent gene flow between the two species.

2 ***Habitat isolation*** When two sympatric populations occupy different habitats within their common range, the individuals of each population will be more likely to encounter and mate with members of their own population than with members of the other population. Their genetically determined preference for different habitats thus helps keep the two gene pools separate.

There are numerous examples of such habitat isolation. *Bufo woodhousei* and *B. americanus* are two closely related toads that can cross and produce viable offspring. However, in those areas where the ranges of the two toads overlap, *B. woodhousei* elects to breed in the quieter water of streams, while *B. americanus* breeds in shallow rain pools. The dragonfly *Progomphus obscurus* lives in northern Florida, while its close relative *P. alachuensis* lives in southern Florida. When the ranges of the two species overlap in north-central Florida, the two species occupy different habitats, *P. obscurus* being restricted to rivers and streams and *P. alachuensis* to lakes. In California the ranges of *Ceanothus thyrsiflorus* and *C. dentatus*, two species of buckthorn shrubs, overlap broadly, but *C. thyrsiflorus* grows on moist hillsides with good soil, while *C. dentatus* grows on drier, more exposed sites with poor or shallow soil.

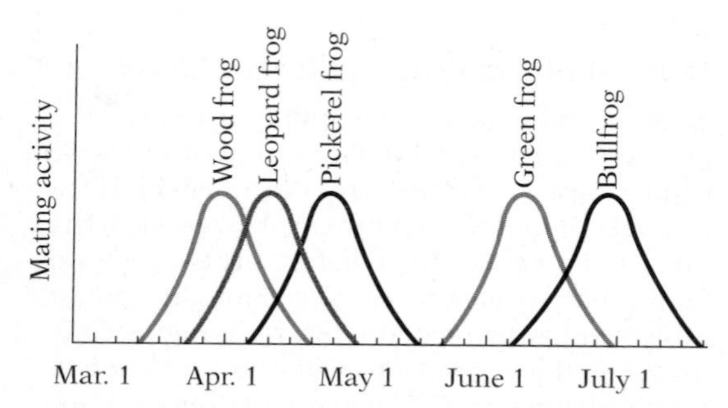

18.9 Mating seasons at Ithaca, New York, for five species of frogs of the genus *Rana*

The period of most active mating is different for each species. Where the mating seasons for two or more species overlap, different breeding sites are used.

3 *Seasonal isolation* If two closely related species are sympatric, but breed during different seasons of the year, interbreeding will be impossible. *Pinus radiata* and *P. muricata*, two species of pine, are sympatric in some parts of California. They are capable of crossing, but rarely do so under natural conditions, because *P. radiata* sheds its pollen early in February while *P. muricata* waits until April. *Reticulitermes hageni* and *R. virginicus*, two closely related species of termites, are sympatric in southern Florida, but the mating flights of the former occur from March through May while those of the latter occur in the fall and winter months. Five species of frogs belonging to the genus *Rana* are sympatric in much of eastern North America, but the period of active mating is different for each species (Fig. 18.9).

4 *Behavioral isolation* Chapter 38 discusses the immense importance of behavior in courtship and mating, particularly with respect to species recognition. In many cases, species have unique courtship patterns, and where two species are sympatric, crosses rarely occur because a courtship between members of different species involves so many wrong responses that it is unlikely to proceed all the way to copulation.

A particularly interesting example of the role of visual displays in species recognition was discovered by Jocelyn Crane. She found 12 species of fiddler crabs of the genus *Uca* actively courting on the same small beach (only about 56 m^2) in Panama. Each species had its own characteristic display, which included waving the large claw (cheliped), elevating the body, and moving around the burrow (Fig. 18.10). Crane found that the displays were so distinctive that she could recognize each species from a considerable distance merely by the form of its display; presumably the female crabs do likewise.

18.10 A male fiddler crab (*Uca*)

The animal waves its large cheliped in the air as a courtship display. Critical details of the display differ among the various species of fiddler crabs.

Auditory stimuli are important in species recognition among many animals, particularly birds and insects, and help prevent mating between related species. In several instances specialists have noticed that two or three very different calls are produced by what had been considered members of a single species of cricket. On investigation, they found that each call was, in fact, produced by a different species so similar morphologically that no one had previously distinguished among them. They do not hybridize in nature even when they are sympatric, because females do not respond to the sounds made by a male of a different species. ***Sibling***

species—species so closely related that humans can hardly distinguish them—may be a fairly common phenomenon.

5 *Mechanical isolation* If structural differences between two closely related species make it physically impossible for matings between males of one species and females of the other to occur, the two populations will obviously not exchange genes. If, for example, one species of animal is much larger than the other, matings between them may be very difficult, if not impossible. Likewise, if the genital organs do not fit, mating will be prevented.

Mechanical isolation is probably more important in plants than in animals, particularly in plants that depend on insect pollinators. Consider *Salvia apiana* and *S. mellifera*, two closely related species of sages with overlapping ranges in California. They are reproductively isolated by differences in habitat and flowering season and by the behavior of their pollinators. In addition, mechanical features play a role. Whereas *S. mellifera* is pollinated by relatively small bees, the flowers of *S. apiana* can be entered only by very large bees whose weight is sufficient to cause the landing platform (lower lip of the corolla) to unfold and permit free entrance into the flower; a similar isolating mechanism is observed in other plants as well, including Scotch broom (Fig. 18.11).

6 *Gametic isolation* If individuals of two animal species mate or if the pollen from one plant species gets onto the stigma of another, fertilization still may not take place. Some 68 interspecific combinations are known in tobacco in which no cross-fertilization will occur; the sperm nucleus from the pollen is unable to grow to the egg cell in the ovary. In *Drosophila*, if cross-insemination occurs between *D. virilis* and *D. americana*, the sperm are immobilized by the unsuitable environment in the reproductive tract of the female. In other species of *Drosophila*, interspecific matings cause an antigenic reaction in the female genital tract, which kills the sperm before they reach the eggs.

The mechanisms discussed so far exact little cost from the individual organisms: the machinery of isolation is either imposed from without, or generated internally from existing behavior, physiology, and morphology. As a result, no great loss of fitness is involved in maintaining isolation. The remaining four mechanisms, though effective, exact a significant cost from the individual, and this cost acts as a selection pressure favoring evolution of the more efficient mechanisms of reproductive isolation already mentioned.

7 *Developmental isolation* Even when cross-fertilization occurs, embryonic development is often irregular and may cease before birth. The eggs of fish can often be fertilized by sperm from a variety of other species, but development usually stops early on. Crosses between sheep and goats produce embryos that die before birth.

8 *Hybrid inviability* Hybrids are often weak and malformed and frequently die before they reproduce; hence there is no gene flow through them from the gene pool of one parental species to the gene pool of the other. Certain tobacco hybrids, for example, develop tumors in their vegetative parts and die before they flower.

A

B

C

18.11 Pollination of Scotch broom (*Cytisum scoparius*) by a bumble bee

The nectar and pollen are inaccessible to lighter pollinators like honey bees, whose weight is insufficient to trip release of the reproductive structures. (A) An untripped bloom, (B) A bumble bee gathering nectar, (C) A flower after tripping.

9 *Hybrid sterility* Some interspecific crosses produce vigorous but sterile hybrids. The best-known example is the cross between a female horse and a male donkey, which produces a mule. Mules have many characteristics superior to those of both parental species, but they are sterile.

10 *Selective hybrid elimination* The members of two closely related populations may be able to cross and produce fertile offspring. If those offspring and their progeny are as vigorous and well adapted as the parental forms, then the two original populations will not remain distinct if they are sympatric. But if the fertile offspring and their progeny are less well adapted than the parental forms, then they will be eliminated. There will be only limited gene flow between the two parental gene pools through the hybrids. The parental populations are consequently regarded as separate species.

These last four mechanisms isolate species from one another effectively, but their cost is significant: valuable metabolic resources are invested in doomed embryos or in frail or possibly sterile young, and the seasonal nature of many reproductive cycles may offer no second chance to mate and rear young. Mistaken matings waste gametes, whether fertilization takes place or hybrids are viable. Selection, therefore, will strongly favor individuals whose behavior, morphology, or physiology reduces the chance of a mismatch in the first place. If the parental populations are sympatric, there will be strong selection for the evolution of more effective intrinsic isolating mechanisms. Gene combinations that lead to correct mate selection will increase in frequency until eventually all hybridization ceases. The tendency of closely related sympatric species to diverge rapidly in characteristics that reduce the chances of hybridization and/or minimize competition between them is called ***character displacement*** (see Fig. 18.8). Table 18.1 summarizes the different intrinsic isolating mechanisms.

Situations in which only one of the 10 isolating mechanisms is operative are extremely rare. Ordinarily several contribute to keeping two species apart. For example, closely related sympatric plant species often exhibit habitat and seasonal isolation in addition to some form of hybrid incapacity.

Table 18.1 Intrinsic isolating mechanisms

EFFECT	MECHANISM	INDIVIDUALS AFFECTED
Mating prevented	1. Ecogeographic isolation	**Parents:** fertilization prevented
	2. Habitat isolation	
	3. Seasonal isolation	
	4. Behavioral isolation	
	5. Mechanical isolation	
Production of hybrid young prevented	6. Gametic isolation	
	7. Developmental isolation	**Hybrids:** success prevented
Perpetuation of hybrids prevented	8. Hybrid inviability	
	9. Hybrid sterility	
	10. Selective hybrid elimination	

Speciation by polyploidy and chromosomal change Speciation through the divergence of geographically separated populations is not the only way new species can arise. Speciation not involving geographic isolation is called ***sympatric speciation***. One important example is speciation by polyploidy, the condition of having more than two sets of chromosomes. This condition can arise so quickly that it is possible for a parent to belong to one species and its offspring to another. (Other forms of sympatric speciation will be discussed later.) Speciation by polyploidy has been common in plants: by some estimates, 70% of the flowering plant species arose this way.

One type of polyploid speciation, called ***autopolyploidy***, involves a sudden increase in the number of chromosomes, usually as a result of the nondisjunction of chromosomes during meiosis. An example of this type of polyploidy was discovered by the Dutch geneticist Hugo De Vries, while he was making extensive studies of the evening primrose, *Oenothera lamarckiana*. This diploid species has 14 chromosomes. During De Vries's studies, a new form suddenly arose (Fig. 18.12). The new form, *Oenothera gigas*, has 28 chromosomes. This tetraploid is reproductively isolated from the parental species because hybrids between *O. lamarckiana* and *O. gigas* are triploid (they received one of each type of chromosome from their *O. lamarckiana* parent and two of each type from their *O. gigas* parent); triploid individuals, because of the highly irregular distribution of their chromosomes at meiosis, are sterile. Hence polyploid populations fulfill all the requirements of the modern definition of species—they are genetically distinct and reproductively isolated—though botanists do not always give each polyploid population a formal species name.

The reproductive isolation of the polyploid daughter species from the ancestral stock sometimes permits adaptive divergence that would not otherwise be possible. For example, new polyploid plant species have become adapted to mineral soils like mine tailings and serpentine outcrops. If the plants were not reproductively isolated, gene flow from the large surrounding population of normal diploids would prevent local adaptation to these special soils. Polyploidy, then, substitutes for geographic isolation.

A second type of polyploid speciation, called ***allopolyploidy***, involves a multiplication (usually a doubling) of the number of chromosomes in a hybrid between two species. The hybrid has one set of chromosomes from each of the two species. Unless these two species are so closely related that they have homologous chromosomes capable of pairing in meiosis, the hybrid will almost certainly fail to produce gametes. However, if the hybrid undergoes chromosome doubling before meiosis, it will have a complete diploid set of chromosomes from each species, and viable gametes are more likely. This type of polyploidy has probably been more important in speciation than autopolyploidy. Allopolyploid individuals are able to breed freely among themselves, but they cannot cross with either of the parental species. Consequently the allopolyploid population is a distinct species.

Allopolyploid plants are only rarely more vigorous than the parental diploid plants, probably because each parental species has a well-balanced mixture of genes, while the allopolyploid combines two sets of gene products, metabolic pathways, control systems, and, especially, developmental instructions. But sometimes allopolyploids, because of this mix of genes, can grow in habitats that neither parental species can colonize. Allopolyploid speciation has played an important role in the perpetuation of some plant groups during periods of widespread environmental change.

Both allopolyploidy and autopolyploidy have been proved to be of great importance in the production of valuable new crop plants. As soon as plant breeders realized that many of our most useful plants—such as oats, wheat, cotton, tobacco, potatoes, bananas, coffee, and sugarcane—are polyploids, they began trying to stimulate polyploidy, and obtained many new varieties. The chemical colchicine readily induces polyploidy. One of the first artificially produced allopolyploids came from a cross made in 1924 between the radish and the cabbage. Unfortunately it had the root of the

18.12 Evening primrose

This species arose in a single generation from an autopolyploidy event in 1904.

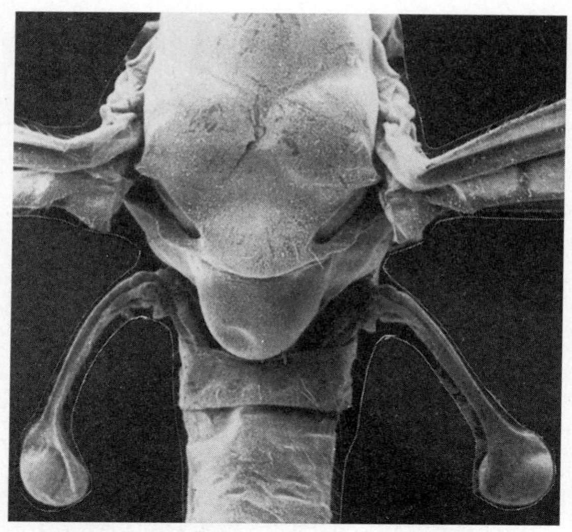

A

B

18.13 Speciation through a developmental mutation?

Strepsiptera is an insect order closely related to Diptera, the true flies. The order probably arose through a developmental mutation (a homeotic mutation, discussed in Chapter 14) that reversed two thoracic segments. In flies (A) the club-shaped structures called halteres are located on the segment behind the wings and spin during flight to provide a gyroscopic stability; in strepsipterans (B) the haltere- and wing-bearing segments are reversed.

cabbage and the shoot of the radish. Other crosses have yielded more desirable results (see Fig. 16.26, p. 437).

Chromosome doubling creates an obvious problem: if tetraploids cannot breed with diploids, they are restricted to breeding with each other. Hence, the rare diploid pollen grain must find an equally rare diploid egg cell to produce a fertile tetraploid. Most likely, the resulting plant must then fertilize itself, since other tetraploids will be exceedingly rare. The same is true of allopolyploids: the sole hybrid is extremely unlikely to find a genetically compatible plant, and so will probably have to fertilize itself. The result in both cases will be severe inbreeding of the new species, a strong founder effect, and rapid genetic drift. The new population will persist only if it has a strong competitive advantage in the habitat.

Some researchers believe that chromosomal rearrangements—whether the result of a few major breakage-and-fusion events or of smaller but more numerous transpositions, duplications, and deletions—might give rise to individuals or small populations genetically incompatible with the rest of their species. This would be especially likely if a developmental control gene were to be altered (Fig. 18.13). If the variant organisms are at a competitive advantage, selection would favor further intrinsic isolation, and sympatric speciation might occur as a result. There is some evidence for this mechanism of speciation, at least among the *Drosophila*.

Nonchromosomal sympatric speciation Though most speciation not involving gross chromosomal changes is certainly allopatric (requiring a period of geographic isolation), there is increasing evidence that sympatric speciation can occur without polyploidy or other major chromosomal rearrangements. Reproductive isolation is just as essential to these types of sympatric speciation, but is effected by other means. For example, relatively small changes in habitat preference may produce habitat isolation. Thus a species of clover adapted to soil containing mine tailings is now reproductively isolated from the widespread and closely related species from which it evolved, not by polyploidy but by virtue of having a different flowering season.

Another alternative to geographic isolation is related to host specificity. Treehoppers often mate on the plants on which they feed, so subspecies that are adapted for a particular species of host plant will tend to inbreed. Selection should favor those individuals that breed strictly on the host species; consequently, intrinsic isolating mechanisms might develop that would serve to isolate tree hoppers with adaptations to different host plants (Fig. 18.14).

Yet another mechanism that may contribute to sympatric speciation, particularly in birds, is sexual imprinting (detailed in Chapter 38). While young, the members of one sex of many species memorize features displayed by their parents or siblings. The features may be visual, auditory, or olfactory. Memorization enables these individuals later to identify suitable mates. If the offspring imprint on a parent displaying a mutation in the feature being committed to memory (a brightly colored eye ring, for example, or novel elements in the courtship song), and if the young are later able to locate mates displaying the same mutation, the mutant individuals are likely to pair with each other. The result could be the founding of a population that does not interbreed with the rest of the species. In birds, then, instant reproductive isolation based on imprinting may play an important role in shaping populations.

Imprinting can also create a kind of habitat preference that leads to reproductive isolation. The picture-wing fruit fly, *Rhagoletis pomenella*, courts and deposits its eggs on hawthorn berries; the larvae feed on the fruit, pupate, and emerge to continue the cycle. At some stage in its life, the fly imprints on the odor of hawthorn, and uses that food-based memory to locate a suitable host plant. Jeff Feder has shown that the larvae are well adapted to hawthorn berries, growing and pupating about twice as well as on any other kind of tree fruit. However, there are two specialized species of parasitic wasps that seek out hawthorn berries and insert their eggs into the growing fruit fly larvae, killing about 90% of them.

About 150 years ago, land was cleared and large commercial apple orchards were established in the Hudson Valley of the northeastern United States; hawthorns became less common, and subsequently (probably by mistake) a few *Rhagoletis* began to lay their eggs on apples. Imprinting committed the next generation to apples, so reproductive isolation must have been nearly complete from the outset. Although *Rhagoletis* larvae grow only about half as well on apples, they suffer far less parasitism: not only are the wasps not looking for them on apples, but also the fruits appear earlier (so the larvae probably enter the pupal stage before the height of the season for parasites) and apples are larger, allowing the larvae to burrow so far in that the wasps cannot reach them. The result is that six times as many of the eggs laid on apples produce adult flies as compared with hawthorn. Whether full speciation has yet taken place is unknown, but the potential for this pathway to sympatric speciation is clear.

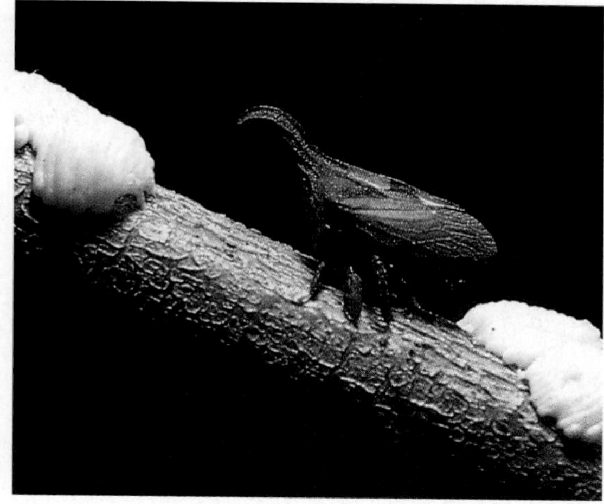

A

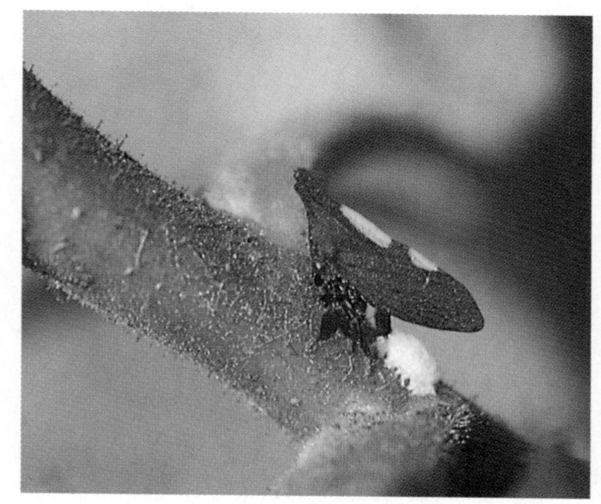

B

18.14 Sympatric speciation in treehoppers

Two sympatric populations of the treehopper *Enchenopa binotata* have evolved adaptations to different host plants. The treehopper in A lives on bittersweet, while the one in B lives on butternut. Host specificity may take the place of physical separation (allopatry) in preventing these two populations from interbreeding.

Adaptive radiation One of the most striking aspects of life is its extreme diversity. A bewildering array of species now occupies this globe. And the fossil record shows that, of the species that have existed at one time or another, those now living represent probably less than 0.1%, all the other species being now extinct. Clearly, then, divergent evolution—the evolutionary splitting of species into many separate descendant species—has been as frequent as

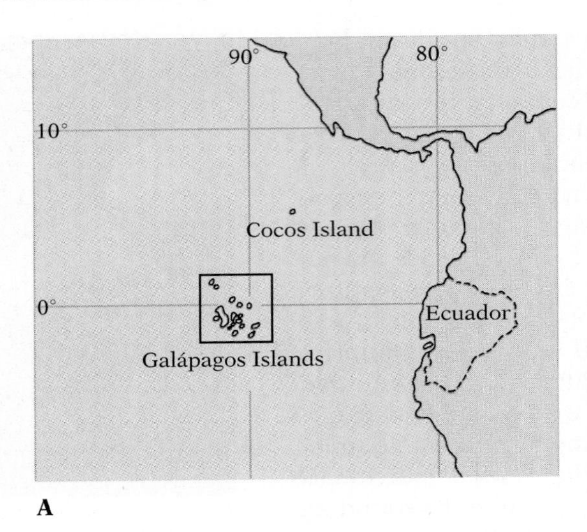

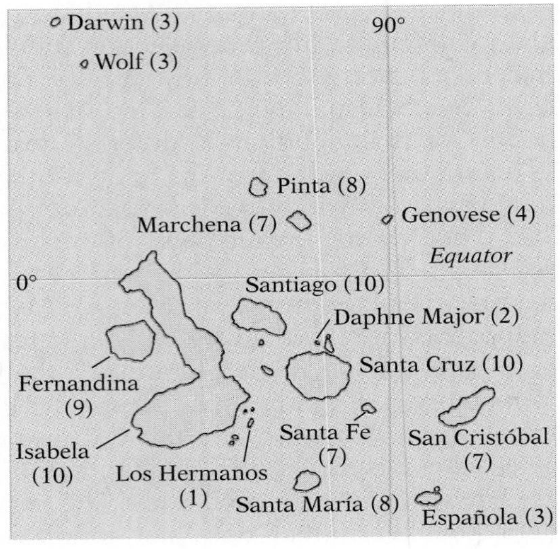

18.15 Galápagos Islands

Left: The islands are located about 950 km off the coast of Ecuador. Cocos Island is about 700 km northeast of the Galápagos. Right: The islands shown in greater detail. The number in parentheses after each name indicates the number of species of Darwin's finches that occur on the island.

extinction. How could opportunities for geographic isolation have been sufficient to lead to most of this speciation? Perhaps the best documented example of rampant divergent evolution—often called ***adaptive radiation***—involves the finches on the Galápagos Islands, which (along with tortoises and mockingbirds on those islands) played a major role in leading Darwin to formulate his theory of evolution by natural selection.

The Galápagos Islands lie astride the equator in the Pacific Ocean roughly 950 km west of the coast of Ecuador (Fig. 18.15). The islands have never been connected to one another. They apparently arose from the ocean floor as volcanoes approximately 7 million years ago. At first they were completely devoid of life, and were therefore open to exploitation by the few species that chanced to reach them from the mainland. The only land vertebrates present on the islands before human beings arrived were at least seven species of reptiles (one or more snakes, a species of huge tortoise, and at least five species of lizards, including two very large iguanas), seven species of mammals (five rats and two bats), and a limited number of birds (including two species of owls, a hawk, a dove, a cuckoo, a warbler, two flycatchers, a martin, four mockingbirds, and Darwin's finches).

The 14 species of Darwin's finches constitute a separate subfamily found nowhere else in the world. Thirteen of them are believed to have evolved on the Galápagos Islands, and one on Cocos Island, from some unknown finch ancestor that colonized the islands from the South or Central American mainland. The descendants of geographically isolated colonizers often undergo so much evolutionary change as to become, in time, very unlike their mainland ances-

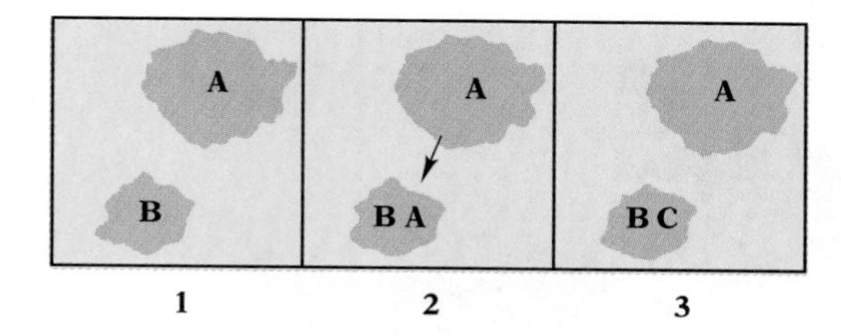

18.16 Model of speciation on the Galápagos Islands

An ancestral form colonized the larger of these two hypothetical islands. Later, part of the population dispersed to the smaller island. (1) Eventually the two populations, being isolated from each other, evolved into separate species A and B. (2) Some individuals of A dispersed to B's island. The two species coexisted, but intense competition between them led to rapid divergent evolution. (3) This rapid evolution of the population of A on B's island caused it to become increasingly different from the original species A, until eventually it was sufficiently distinct to be considered a full species, C, in its own right. At the same time, the selection pressure imposed by the small invading population caused the large population of species B to evolve to a small degree as well.

tors. How did the descendants of the original immigrants split into today's 14 species?

The Galápagos group is a cluster of more than 25 separate islands. Finches will not readily fly across wide stretches of water, and they show a strong tendency to remain near their home area. Hence a population on any one of the islands is effectively isolated from the populations on the other islands. The initial colony was probably established on one of the islands where the colonizers, perhaps blown by high winds, chanced to land. Later, stragglers from this colony wandered or were blown to other islands and founded new colonies. Because of the founder effect the allelic frequencies in the new colonies differed from those in the original colonies from the moment they started. In time the colonies on the different islands diverged even more, because of different mutations, different selection pressures, and, in such small populations as some of these must have been, genetic drift. We might expect, therefore, a different species or subspecies on each of the islands. In fact, most of the islands have more than one species of finch, and the larger islands have 10 (Fig. 18.15, right). What has happened?

Let us suppose that form A evolved originally on the island of Santa Cruz and that the closely related form B evolved on Santa María. If form A later spread to Santa María before the two forms were isolated long enough to evolve any but minor differences, they might have interbred freely and merged. However, if A and B were separated long enough to evolve major differences before A invaded Santa María, then A and B might have been intrinsically isolated from each other, having developed into separate species, and they might have been able to coexist on the same island without interbreeding (Fig. 18.16). Selection would have favored individuals that mated only with their own kind, and this selection pressure would have led rapidly to more effective, intrinsic isolating mechanisms. In fact, Darwin's finches readily recognize members of their own species and show little interest in other species.

It would be highly unlikely that A and B could coexist on Santa María indefinitely if they used the same food supply. The ensuing competition would be severe, and the less well adapted species would tend to be eliminated by the other unless it evolved differences that minimized the competition. In short, wherever two or more very closely related species occur together, competition will lead either to the extinction of one or to character displacement—in this instance to the evolution of different feeding specializations.

Character displacement is indeed what we find in Darwin's finches (Fig. 18.17). The 14 species form four groups (genera). One

18.17 Darwin's finches

Darwin's finches fall into four genera: birds 1, 3, 4, 5, 6, and 10 are the tree finches (*Camarhynchus*); birds 7, 8, 11, 12, 13, and 14 are the ground finches (*Geospiza*); bird 2 is the unfinchlike warbler finch; bird 9 is the one species inhabiting Cocos Island.

1. Vegetarian tree finch (*Camarhynchus crassirostris*)
2. Warbler finch (*Certhidea olivacea*)
3. Large insectivorous tree finch (*Camarhynchus psittacula*)
4. Medium insectivorous tree finch (*C. pauper*)
5. Mangrove finch (*C. heliobates*)
6. Small insectivorous tree finch (*C. parvulus*)
7. Large cactus ground finch (*Geospiza conirostris*)
8. Cactus ground finch (*G. scandens*)
9. Cocos finch (*Pinaroloxias inornata*)
10. Woodpecker finch (*C. pallidus*)
11. Large ground finch (*G. magnirostris*)
12. Sharp-beaked ground finch (*G. difficilis*)
13. Medium ground finch (*G. fortis*)
14. Small ground finch (*G. fuliginosa*)

group contains six species that live primarily on the ground; of these, some feed primarily on seeds and others mostly on cactus flowers. Of the species that feed on seeds, some feed on large, some on medium-sized, and some on small seeds. These feeding preferences result from the morphological specialization of the beaks: small beaks are more efficient at handling small seeds, while larger beaks can crack large seeds. From a series of careful beak measurements, David Lack was able to find clear evidence of character displacement. For example, where the small and medium ground finches coexist on the larger islands, their beak sizes are widely separated, with depths averaging about 8.4 and 13.2 mm, respectively. On small islands where only one of the two species exists, the birds of that species tend to have beaks of intermediate size, on the order of 9.7 mm (Fig. 18.18).

The second group of finches contains six species that live primarily in trees. Of these, one is vegetarian and the others eat insects, but the insect eaters differ from one another in the size of their prey and in where they catch them (Fig. 18.19). A third group contains only one species, which has become very unfinchlike and strongly resembles the warblers of the mainland. The fourth group also contains one species, restricted to Cocos Island, which is about 700 km northeast of the Galápagos Islands and about 500 km from Panama. Correlated with the differences in diet among the species are major differences in the size and shape of their beaks. (Noteworthy variations in beak size and shape among birds thought to share a common ancestry can be found in other island

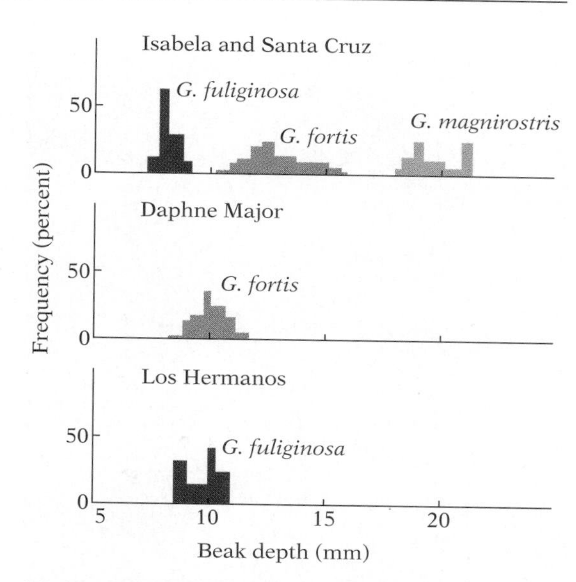

18.18 Beak sizes of ground finches

On large islands like Isabela and Santa Cruz, three species of ground finch coexist. Though the beak sizes of individuals within each species show the sort of variation essential for evolution by natural selection, they fall into three separate ranges, specialized respectively for efficient feeding on small, medium-sized, and large seeds. However, on small islands where only one species exists (either the small or the medium ground finch), the beak sizes fall into an intermediate range regardless of the size specialization found in the same species on the larger islands. This shows that, in the face of competition for food on the larger islands, character displacement has taken place.

18.19 Tool-using finch

One of the insectivorous tree finches of the Galápagos has evolved a most unusual feeding habit. Sometimes called the woodpecker finch, it chisels into wood after insects, but it lacks the long tongue that a woodpecker uses to probe insects out of a crack. Instead, it pokes into the crack with a cactus spine or twig that it holds in its beak. The mangrove finch does the same thing in a different habitat. These are two of the few known cases of tool use by birds.

18.20 Beak differences in Hawaiian honeycreepers

Differences in beak size and shape are apparent in related species of Hawaiian honeycreepers, which, like Darwin's finches, are thought to have evolved from a common finchlike ancestor.

habitats as well, as shown, for example, in Figure 18.20. These characteristics of the beak are one important means by which the birds recognize other members of their own species; song is another.)

If selection on Santa María favored character displacement between species A and B, the population of species A on Santa María would become less and less like the population of species A on Santa Cruz (Fig. 18.16). Eventually these differences might become so great that the two populations would be intrinsically isolated from each other and would thus be separate species. We might now designate as species C the Santa María population derived from species A. The geographic separation of the two islands would thus have led to the evolution of three species (A, B, and C) from a single original species. The process of island hopping followed by divergence could continue indefinitely, producing many additional species. It was doubtless such a process that led to the 14 species of Darwin's finches.

The sorts of processes seen on islands can account also for radiation in a continental area, where intervening habitats can limit the frequency of movement as much as an expanse of ocean. Adaptive radiation on islands, as in the case of Darwin's finches, is dramatic and lends itself particularly well to analysis, but it does not differ in principle from adaptive radiation under other conditions.

It should be clear, as Darwin himself pointed out, that the rate of evolutionary divergence is not constant. When the first colonizing finches reached the Galápagos Islands, they would have encountered environmental conditions quite unlike those they had left behind in Central or South America. The selection pressures to which they were subjected would probably have been quite different from those in their former home; differences in the resources available, for example, may have led to selection for different morphology, physiology, and behavior. On the other hand, if there was initially little or no competition from other species, the result might have been a temporary relaxation of selection pressures. Only when the new habitat became saturated with finches would intraspecific competition for the available resources have become important. And so, sooner or later, selection pressures must have led to rapid divergence from the ancestral population. Later, as the finches became increasingly well adapted to conditions on the Galápagos, the rate of evolutionary change probably slowed down.

In general, when conditions change radically and organisms have new evolutionary opportunities for which they are at least modestly preadapted, they may undergo an evolutionary burst—a period of rapid adaptational change—which may then be followed by a more stable period, during which any further evolutionary changes are merely a fine-tuning of their characteristics. Such bursts of rapid evolutionary divergence probably characterized the tremendous radiation of amphibians when they moved onto land for the first time, and the explosive radiation of mammals when the demise of the dinosaurs left many niches—that is, ways of surviving—unoccupied. (The concept of niche is detailed in Chapter 39.)

How constant and important is competition? Chapter 17 contrasted the roles played by genetic drift (chance) and natural selection in the evolution of populations. The discussion there emphasized that evolution is a vital, ongoing phenomenon, while drift (potentially very important in small populations) and selection are two contributing—and clearly demonstrated—mechanisms by which it occurs. In our discussion of speciation here we have touched on a similar theme: barriers that contribute to reproductive isolation can arise by chance (primarily genetic drift), by natural selection, or by a combination of the two. We have emphasized the role of competition in the formation of the species: even reproductively isolated populations may not be able to coexist indefinitely if they compete for precisely the same food, since even a slight but systematic superiority of one will tend to lead to the extinction of the other.

G. F. Gause, who first observed this phenomenon in the laboratory in the 1930s, formulated the competitive exclusion principle, often called the ***Niche Rule***: no two species occupying the same niche can long coexist (Fig. 18.21). Only the character displacement that natural selection produces, resulting in changes in food preference, habitat choice, and the like, will allow closely related species to coexist. This Darwinian interpretation of how separate species form, with its emphasis on reproductive isolation and character displacement in the face of competition, probably accounts for most speciation.

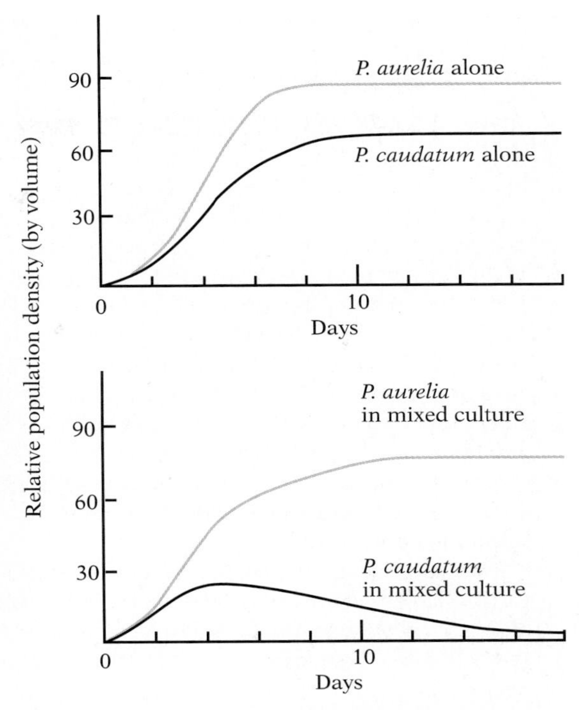

18.21 Competition and extinction

When a few individuals of either species of *Paramecium* are introduced into a tank alone (top), they multiply until they reach a limiting density. When individuals of both species are introduced together (bottom), they multiply independently for the first 3 days, and then begin to compete for resources. *P. aurelia* is in some way more efficient under these conditions, and drives *P. caudatum* to extinction in only 3 weeks.

In the last 20 years, however, the nature and role of competition in speciation have come under critical scrutiny. Competition is no longer thought to be as overwhelmingly important or constant; chance is now seen as a greater force than Darwin thought. There are at least two ways in which chance can take precedence over competition in causing species to diverge. The first we have already discussed: in small populations, genetic drift can be very powerful, driving alleles to extinction before the generally more gradual process of natural selection can produce stability. The other process, however, is more fundamental to our understanding of evolution. Most Darwinian analyses tacitly assume that selection pressures change relatively slowly, and that large continuous populations have ample time to evolve specialized adaptations to their environments. Do conditions really remain sufficiently constant for a species to achieve a stable set of allelic frequencies in the face of all the selection pressures affecting it?

Apparently, many populations are not always at equilibrium. Two things in particular can disrupt the balance: ***crises***, which greatly reduce population size, and ecological changes, which alter the competitive "landscape." Both factors are evident in one particularly thorough study by Peter Grant and his colleagues. They tagged hundreds of medium-sized ground finches on one of the smaller Galápagos islands, Daphne Major. In 1977 only 20% of the normal amount of rain fell, and the plants on the island produced many fewer seeds than usual. Relatively dry or wet seasons, a surprisingly early or late hard frost, or especially warm or cold years are familiar conditions that organisms can generally take in stride, but the effects of extreme events on plants and animals can be drastic.

On Daphne Major the consequences of the reduction in food were dramatic for the medium ground finch (*Geospiza fortis*). With only 35% of the usual amount of food available, 387 of the 388

HOW WE KNOW:
THE BEAK OF THE FINCH

In order to trace evolution as it occurs, researchers must know the distribution of heritable phenotypes in a population, and then monitor any changes as they occur. In the case of the Galapagos finches, this means measuring every size parameter of every bird on the island under study (Fig. A), marking each bird individually, and then recapturing every bird annually both to monitor survivorship and to add each year's crop of youngsters to the inventory. The extraordinary difficulty of such an undertaking probably explains why few researchers have been able to track evolution generation by generation. Let's look at some of the data Peter Grant and his colleagues collected, and how they were thus able to document one bout of evolution.

Two basic requirements for evolution are in place: there is variation in beak parameters and it is heritable: the offspring of parents with large beaks tend to have large beaks themselves (Fig. B). The question Grant sought to answer was whether beak size is subject to selection.

In 1976, before a drought, the beaks of the 741 resident medium ground finches on the island of Daphne Major had a mean depth of about 9.4 mm (Fig. C, blue curve). The range of depths reflects the range of seeds available as food. Thus birds with smaller beaks were better able to harvest smaller seeds, whereas larger or harder seeds were more readily opened by finches with larger beaks.

The drought led to a threefold reduction in seed number. The seeds that remained were 50% larger and harder than normal. Selection strongly favored larger beak size: only 9% of the birds with average beaks survived, whereas 20% of the individuals with beaks only 1 mm deeper lived through the crisis, as did 40% of the finches with beaks 2 mm deeper. The result was a upward shift in the mean beak depth to 10.2 mm (Fig. C, red curve); the only hatchling to survive had an unusually deep beak.

Despite the strong selection, the overall shift in mean beak depth was just 0.8 mm. The reason is that there were very few individuals with longer beaks for selection to act on; if the variation in the 1976 population had been greater, the shift in beak depth would doubtless have been more dramatic. (Note, however, the evident role of chance: as Fig. C shows, despite the strong selection, the bird with the longest beak died, while the one with the shortest beak survived.)

The effect of this bout of intensive selection on the next generation of finches was clear: in 1978 the mean depth of hatchling beaks was about 0.8 mm longer. Evolution—a change in allele frequencies—had occurred. By 1978, however, the average seed size and hardness had fallen back to normal levels. In 1978, therefore, many finches, young and old, had beaks that were awkwardly large compared to many of the seeds that were being produced once the rainfall returned to normal levels. This change in the food supply created a less intense directional selection for smaller beaks, and the mean beak depth began to fall; another precisely documented bout of evolution was underway.

nestlings of 1976 died, as did the majority of adults. Furthermore, the adult survivors were not a representative cross section of the former population: roughly 180 of the 500 adult males survived, but only about 30 of the 500 females did so, and the birds that did survive the year of starvation were significantly larger than those that succumbed. Exceptionally intense natural selection in favor of large size—especially larger beaks—had occurred. At the same time, the role of chance was also clear: the individual with the shortest beak in the population nevertheless managed to survive.

Thus one of several rare and unpredictable crises sharply altered the selection pressures for a year, imposed a large element of chance into the struggle to survive, and reduced the population to a level at which genetic drift (and thus chance) became a real possibility. What had seemed to be an adaptive equilibrium was upset by a chance disturbance, and a population with altered characteristics emerged. The finch population had been forced through a period of environmental crisis. When a crisis is so severe as to cause major changes in allelic frequencies in the surviving population, it is called an ***evolutionary bottleneck***.

Is it possible that the role of interspecific and intraspecific competition, leading gradually to character displacement, is dwarfed by chance crises that upset whatever stability exists in the allelic frequencies within a species? Crises can be caused by any environmental factor—pestilence or extreme weather, for example—that severely affects an isolated population, or that serves to isolate one part of a population from the remaining body. Because of their small population size, endangered species are in this position. During such a crisis, as an isolated population passes through the evolutionary bottleneck, one character or another may gain ascendancy. Having been selected by a founder effect, such a character is still at risk if the surviving population is small, because it may decline in frequency or even become extinct through genetic drift. Alternatively, if a character that was not adaptive prior to the crisis proves adaptive thereafter, it may increase in frequency in the population. In the case of the finches, the effects of the drought in 1977 (and another in 1982) were canceled by another freak climatic event: record rainfall in 1983. Unprecedented numbers of seeds were produced, and selection strongly favored smaller beaks.

Though most researchers still believe that competition is the predominant force leading to speciation over time, many studies now

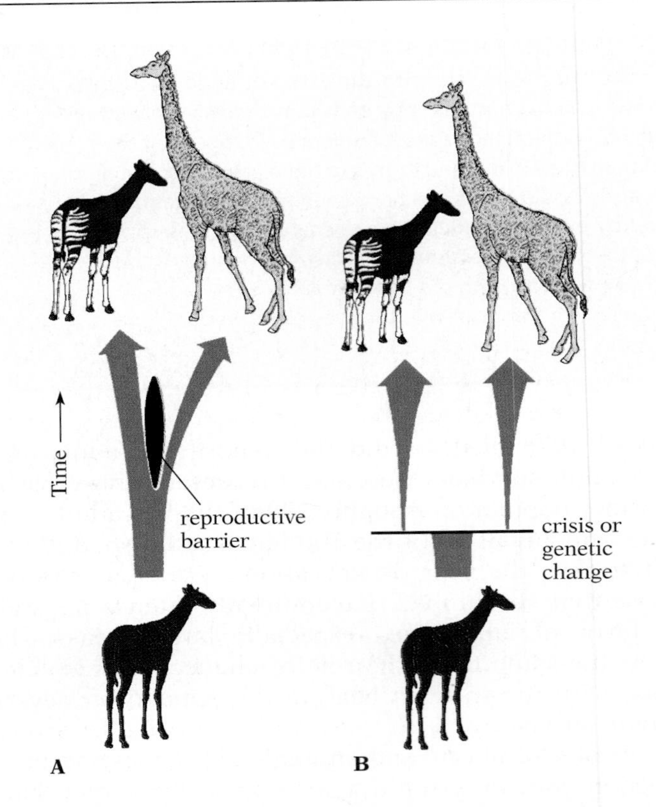

18.22 Gradualism versus punctuated equilibrium

These drawings represent two hypotheses of how different species—here, the okapi and the giraffe—might have evolved from a common ancestor, the "pre-okapi." Forms intermediate between the pre-okapi and the okapi and giraffe are not shown. (A) In gradualism, the conventional interpretation of natural selection, a population is imagined to evolve slowly (as suggested by the gentle leftward movement of the lower part of the arrow, indicating a slow change in one trait) until a reproductive barrier (usually geographic isolation) separates two parts of the population. Generally one of the subgroups is very small, as indicated by the narrowness of the right-hand branch; it may diverge slowly from the larger subgroup through selection, genetic drift, or both. (B) According to the punctuated equilibrium model, there is little or no gradual change in a trait over time. Instead, a crisis or genetic event selects or gives rise by chance to one or more populations with traits very different from those of the ancestral population.

suggest that rare boom-or-bust events have been important in the evolution of at least certain populations.

Punctuated equilibrium One consequence of the increasing evidence that natural catastrophes can affect the allelic frequencies of populations has been a growing interest in ***punctuated equilibrium***. This much-debated hypothesis concerning the mode and tempo of speciation was originally formulated by Niles Eldredge and Stephen Jay Gould. Based on careful study of certain fossil records, their hypothesis suggests that most allopatric speciation events are the result of crises or major genetic alterations that punctuate long periods of equilibrium, or ***stasis***, in which the morphology of the species remains relatively constant. Gould and Eldredge maintain that the fossil record does not support ***gradualism***—the post-Darwinian view that speciation occurs as a gradual accumulation of morphological and physiological changes (Fig. 18.22). The changes they emphasize are *geologically* instantaneous.

"Instantaneous" speciation, however, is by no means as rapid as it sounds. The time scale of geological layering can be measured in millennia. Gould and Eldredge believe speciation events may have taken thousands of generations, up to 100,000 years, to complete. Perhaps most important, because the population that survives and evolves after a crisis would seem to arise suddenly in the fossil record, punctuated equilibrium provides an explanation of the gaps that exist in the record. The more gradual processes of change we are used to seeing in evolution are, from this point of view, of minor importance in speciation, and serve mostly to fine-tune a species to its environment.

The debate between punctuated equilibrium and the conventional perspective is fueled to a great extent by the many differences in the way the fossil record can be interpreted. In some species, for instance, certain morphological features may change rapidly in response to selection pressures while others may remain relatively constant. A researcher who examines or is only able to discern the characters that remain constant may overlook this ***mosaic evolution*** and assume that the species exhibits stasis—no change at all. On the other hand, supporters of gradualism often assume that slow morphological or physiological change is occurring even when it is not obvious from the record. Of course, neither physiological nor morphological changes in most soft body parts are preserved in fossils.

Another difference in interpretation concerns the great discontinuities in the record. The conventional view of these gaps is that they are, in most cases, just anomalies, and that the missing pieces in the picture of gradual evolutionary change might, if found, provide valuable information about transitions between the species preceding and following the gaps. The supporter of punctuated equilibrium, by contrast, regards the gaps as the norm, and as proof of the rapid speciation events. However, gaps in the record may frequently result from the kinds of environmental crises already discussed. Consider what happens when a lake, whose sediments have provided an excellent record of its organic inhabitants for hundreds of thousands of years, suddenly dries up as a result of a major climatic change. The formation of the fossil record in this location is terminated just at the time when the inhabitants of the lake are faced with extraordinary selection pressures and potential extinction.

The punctuated equilibrium hypothesis holds that besides crises, punctuational change could be generated by major genetic alterations—polyploidy, hybridization, translocations, and deletions—particularly if they affect developmental-control genes, which can change the expression of entire constellations of other genes. One widely cited example is the idea that vertebrates evolved from the larval form of an aquatic invertebrate (p. 704). A change in developmental control could have blocked transformation into the adult morphology; this would have been followed by an elaboration of the larval phase and the evolution of the first fish. (Another possible example, involving the evolution of arthropods, is discussed in the next section.) Of course, genetic anomalies face the same problems as polyploids: with only siblings to mate with (at best), inbreeding and genetic drift become serious problems and founding a healthy population becomes difficult.

The Cambrian explosion One of the most spectacular cases of rapid evolutionary radiation occurred in the geologic time division called the Cambrian period, about 550 million years ago. The "sudden" appearance of complex animals (older fossils are nearly all the remains of single-celled creatures) is a potentially compelling instance of punctuation. Two sites with well-preserved remains have been discovered, each about the size of a city block. The more extensively studied one is in the Burgess shale deposit in the mountains of western Canada; the other one is near Chengjiang in southern China. Because many of the fossils are similar, the animals probably had a worldwide distribution, and must have evolved over the course of a few million years. Until these discover-

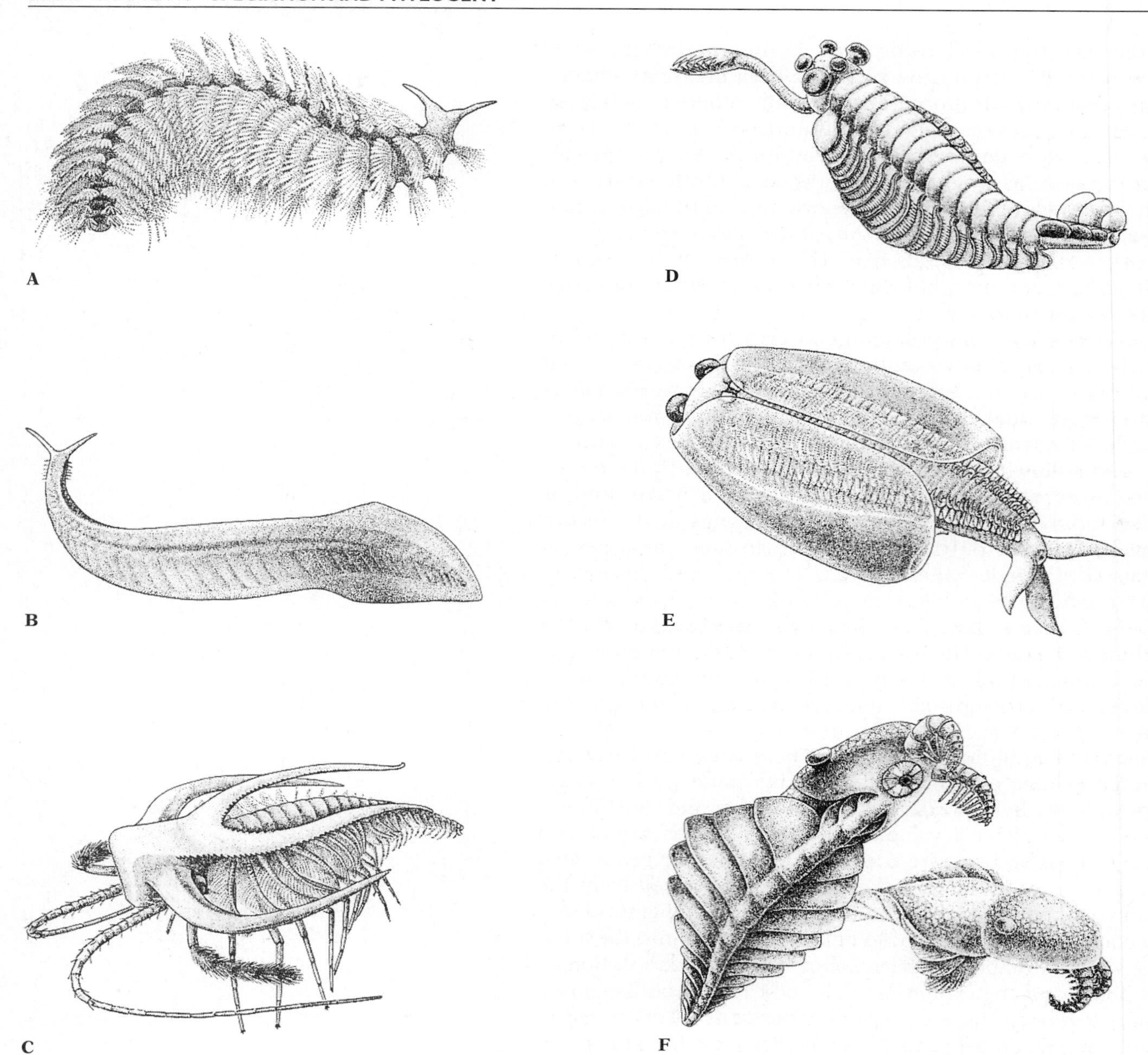

18.23 Some species from the Burgess shale

The Burgess fauna includes annelids (segmented worms) like the polychaete worm shown here (A) and the oldest preserved chordate (B), whose notochord and somite-produced muscle bands are clearly visible. Most of the remaining species, of which only a small fraction are illustrated here (C–F), represent extinct groups of arthropods.

ies, we had no clear picture of the kind of diversity that existed before selection "pruned" the tree of life.

The animals that typify a given place or a particular period are known collectively as a ***fauna***. The Burgess fauna contains a group of polychaetes (a kind of segmented worm, whose physiology will be discussed in Part V) and the oldest known chordate (Fig. 18.23A,B), but mostly it consists of arthropods. These include some trilobites (an extinct group), a few uniramians (the group of invertebrates with mandibles—that is, chewing mouthparts—to which crustaceans and insects belong), and cheliceratans (the group that includes spiders, scorpions, and horseshoe crabs). However, none of these Burgess arthropods look like any modern species. In the same deposits there are at least 20 (and perhaps 30) other classes of arthropods, all now extinct (Fig. 18.23C–F).

The diversity of design is staggering, and poses two evolutionary problems: how did they come into being over such a relatively brief period, and why is it that 80–90% of these creatures have left no modern descendants, while other groups, with no obvious advantages, continue to thrive? We will deal with this second question in the next section. A potential answer to the first, favored by many biologists, is that the explosion in diversity resulted from a biological breakthrough: the invention of a novel developmental strategy. As described in Chapter 14, the embryonic development of animals involves the formation of a blastula, followed by gastrulation. The breakthrough came with the appearance of compartmentalization: the embryo divides along the anterior-posterior axis into a number of nearly identical units, each of which can differentiate to produce specialized organs and appendages. It is likely that the evolution of the first multicellular creatures was followed quickly by the evolution of this more efficient multisegmental plan.

The remarkable diversity that followed would have been a consequence of the absence at that point of the highly redundant and hierarchical system of morphogens that now exist to assure precision and stability during development; as a result, variation would have been generated at an extraordinary rate, while competition in an environment with nothing occupying the many niches available to multicellular organisms must have been relatively lax initially. As the number of different species began to saturate the environment, however, selection would necessarily become more intense, and less efficient designs and less stable developmental programs probably went extinct. This scenario, therefore, postulates a kind of punctuation, followed by more gradual evolution and selective pruning. As we will see, there is another school of thought—an alternative to selection—to explain how many species may go extinct.

Whatever the fate of punctuation theory, the evolution of organisms can be less uniform and gradual than many biologists have supposed. Indeed, Darwin himself rejected simple gradualism, pointing out that both local and global environmental phenomena must redefine and magnify selection pressures enormously and hence alter the direction and rate of evolution. The usual tempo of speciation probably lies somewhere between the gradual-change and the punctuated equilibrium models.

How important is chance? As we saw, in the Galápagos one species of finch underwent enormous evolutionary radiation into a variety of available lifestyles. The first group of birds to reach the islands might just as well have been some other species of finch or some other sort of bird altogether. Chance events led to the creation of the new habitat and the immigration of the first birds. Later avian arrivals failed to thrive because the habitat was already being efficiently exploited by the (now) well-adapted first arrivals.

On a larger scale, there have undeniably been massive bouts of extinction, probably resulting from chance worldwide calamities such as asteroid impacts or other catastrophic events. These random prunings of the tree of life have cleared away thousands of species at a time, leaving the survivors the opportunity to radiate and exploit novel lifestyles in the absence of competition. Perhaps this is what happened to most of the Burgess species. As described in Chapter 24, the enormous radiation of mammals after the Cretaceous-Tertiary boundary probably occurred because a global disaster eliminated the dinosaurs, which had dominated the terrestrial large-animal opportunities. In some sense, therefore, the evolution of humans depended on this accident, and subsequent crises—warmings, ice ages, the continental movement that opened the Rift Valley in Africa (where the earliest human fossils are found), and so on—may also have been critical to the history of our species. We may be where we are now because distant ancestors happened to be in the right place at the right time. If many large-scale evolutionary developments depend on chance, simple luck may often be more important than selection. Had the cards fallen differently, the dominant species of animals today might be something quite different.

18.24 Flower and seeds of the dandelion, an asexual organism

Despite appearances, the dandelion reproduces asexually at all times. The flower (right) of the dandelion, which was inherited from a sexually reproducing precursor, now has infertile pollen and diploid eggs that go on to form seeds without fertilization.

■ PROBLEMS IN IDENTIFYING SPECIES

The modern definition of species cannot always be applied without difficulty. Though most biologists accept the major ideas on which the definition rests, and though most of them, if they were to study the same set of natural populations, would probably agree in the great majority of instances about which populations represent full species and which do not, there would be a small percentage of populations on which they could not agree. Let us examine a few such cases.

Asexuals Since the modern definition of species assumes interbreeding, it cannot apply to asexual organisms. Though most so-called asexual organisms actually do have provision for occasional sexual recombination, a few lack sexual mechanisms altogether. Dandelions, for instance, despite their flowers, are totally asexual (Fig. 18.24). Can such truly asexual organisms be said to form species in any sense?

Asexual organisms do seem to form recognizable groups or kinds, even though the members of a group cannot exchange genes. Gaps, or discontinuities in the variation, occur between the various kinds just as they do between sexual species. One possible explanation for the groupings is that each group of asexual organisms evolved from a sexual species. The flowers of dandelions, which now have infertile pollen and diploid eggs that go on to form seeds without fertilization, originated in sexually reproducing ancestors. Since not all the variations that occur over the course of time are likely to be equally well adapted, asexual organisms like dandelions would continue to form recognizable groups: only those individuals whose genotypes produced well-adapted phenotypes would survive in significant numbers. Hence there would be a limited number of types, and all individuals falling within the bounds of one such type would constitute a natural group that could be called a species. In the long run, the inability of asexual organisms to exploit the potential of genetic recombination makes them less able to survive the vicissitudes of environmental change. And, indeed, nearly all macroscopic multicellular asexual species are of recent evolutionary origin. Among short-lived asexuals, however, the combination of mutations and high reproductive rate seem to create sufficient variation for natural selection to operate on, and some of these species are extremely ancient.

Fossil species The modern definition of species can be applied formally only to organisms that coexist, since the criterion of interbreeding cannot be used when comparing an organism with its likely ancestors of a million years earlier. Therefore, paleontologists can compare organisms from different periods in the earth's history only by using morphological criteria and geographic distribution. Two forms can be classified as separate species when they differ to about the same degree as related organisms from the present day that are known to constitute reproductively isolated species. For practical purposes, paleontologists usually regard gaps in the fossil record as breaks between species, even though they are fully aware that no gaps actually occurred in the lineages of the organisms.

Populations versus species Geographically isolated populations can slowly diverge until they have reached the level of full species. The intrinsic reproductive isolation that makes them full species itself usually evolves gradually. Hence there is no precise point at which the diverging populations suddenly reach the level of full species. However, the definition of species makes no provision for such intermediate stages. Consequently, intermediate stages, when they are encountered, pose a problem to any biologist intent on rigid categorization of what is in nature a fluid system.

Allopatric species One of the most obvious and frequently encountered problems in applying the modern definition of species arises when two populations are closely related and completely allopatric. Since they are allopatric, they are obviously not exchanging genes, but lack of actual effective gene flow is not sufficient: there must be no *potential* gene flow for the two populations to be regarded as separate species. How can potential gene flow be determined? One way that immediately comes to mind is to release a large sample of individuals from one population in the range of the other, and then see whether free interbreeding takes place and, if it does, whether the hybrids are as viable as the parents.

The wholesale introduction of foreign plants and animals is seldom desirable or even legal. An alternative would be to bring individuals from the two allopatric populations together in the laboratory to see whether they will interbreed. If the individuals will breed freely with other members of their own population but not with members of the other population, then the two populations are probably intrinsically isolated and are therefore separate species.

Interbreeding between members of different populations in the laboratory, however, does not prove two groups are part of the same species. Instead it demonstrates that certain types of intrinsic isolation do not exist between the populations. It says nothing about other types of intrinsic isolation. For example, behavioral isolation may operate in nature but not in the laboratory. Many species of animals that will have nothing to do with each other in the wild, because of important differences in their behavior, will mate in the laboratory. Lions and tigers, for example, are distinct allopatric species in the wild, and never mate in their small region of range overlap; in the unnatural environment of a zoo, however, they mate and produce living offspring. Clearly, when members of two different allopatric populations mate in the laboratory and produce viable offspring, the question of whether they belong to the same or to different species remains unanswered. The same ambiguity exists for the many organisms that will not breed at all under laboratory conditions: after all, males and females of the *same* species often refuse to mate outside of their natural habitats.

In many cases, then, there is no good test for determining whether two allopatric populations belong to the same or to different species. Although character displacement among sympatric species adds a level of complication, the usual practice in such cases is to determine the extent to which the two populations differ, and then to compare this degree of difference with that seen in related sympatric species. If the differences between the allopatric populations are of the same order of magnitude as (or greater than) those that distinguish sympatric species, the allopatric populations are considered separate species. If the difference is less, the two allopatric populations are regarded as belonging to the same species.

THE CONCEPT OF PHYLOGENY

Evolution implies that many different species have a common ancestor and that all forms of life probably stem from the same remote beginnings. Hence one of the tasks evolution sets for biologists is to discover the relationships among the species alive today and to trace the ancestors from which they descended (Fig. 18.25). Once these relationships are understood, they are summa-

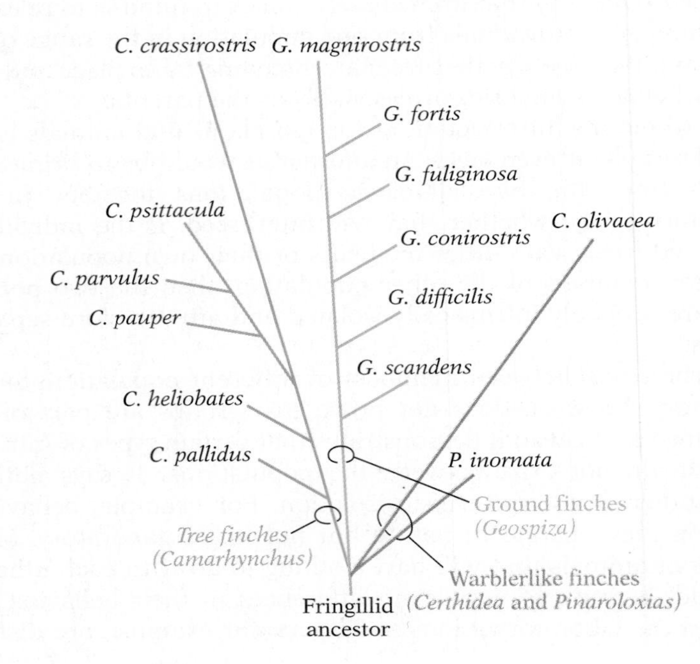

18.25 A phylogenetic tree for Darwin's finches

This tree is based on both the degree and the nature of the morphological differences between species. The distances along branches from any one species to another reflect inferred degrees of relatedness.

rized by grouping species into ***taxa*** (collections of related organisms, ranging from genera to kingdoms). There is lively disagreement about the best ways to determine ancestry and define groups.

■ DETERMINING PHYLOGENETIC RELATIONSHIPS

When systematists,[3] also known as taxonomists, set out to reconstruct the ***phylogeny*** (evolutionary history) of a group of species that they think are related, they have before them the species living today and the fossil record. To reconstruct a phylogenetic history as closely as possible, they must make inferences based on observational and experimental data. The difficulty is that what can be measured is *similarity*, whereas the goal is to determine *relatedness*. There are four major approaches to systematics—classical evolutionary taxonomy, phenetics, cladistics, and molecular taxonomy. Each uses different techniques to infer relatedness from similarity.

Classical evolutionary taxonomy Classical systematics depends more than any other approach on experience and judgment. The usual procedure in reconstructing phylogenies by the classical method is to examine as many independent characters of the species in question as possible, and to determine in which characters the species differ and in which they are alike. The assumption is that the differences and resemblances will reflect, at least in part, their true phylogenetic relationships. Ordinarily, as many different types of characters as possible are used in the hope that misleading data from any single character will be detected by a lack of agreement with the data from other characters.

The most easily studied and widely used characters are morphological—external morphology, internal anatomy and histology (tissue types), and the morphology of the chromosomes in cell nuclei. It is particularly helpful when morphological characters of living species can be compared with those of fossil forms. The characters chosen for comparison must be ones that vary within the group being analyzed; especially useful are characters that are unique to the group, and so have a common and relatively recent evolutionary origin. Among Darwin's finches (Fig. 18.17), for example, morphological characters that could also be checked in fossils include beak depth, ratio of beak depth to beak width, angle of the beak relative to the head, linear dimensions of the various bones that fuse to create the skull, relative areas of the skull bones, pattern of muscle insertions (visible as small indentations in bone) in the vocal apparatus, and so on. Characters that are usually lost in fossils but can be measured in live finches include tail length, markings of the young (degree of streaking, for instance), changes in markings with age, and the like. Of these, the wide range of beak sizes, the depressed angle of the beaks, and the relatively short tails set Darwin's finches apart from their cousins on the South American mainland.

[3] Systematics, or taxonomy, in the words of G. G. Simpson, is "the scientific study of the kinds and diversity of organisms and of any and all relationships among them."

Characters preserved in fossils are of special importance because the fossil record is the most direct source of evidence about the stages through which ancient forms of life passed. Unfortunately, that record is usually incomplete, and for many groups of organisms there is no suitable record at all. At best, fossils suggest the broad outlines of the evolution of major groups. In some groups, notably the horses, fossils have provided phylogenetic information that could have been obtained from no other source (Fig. 18.26).

Another frequently used source of information is embryology. Morphological characters are often easier to interpret if the manner in which they develop is known. For example, if a particular structure in organism A and a structure of quite different appearance in organism B both develop from the same embryonic primordium, the resemblances and differences between those structures in A and B provide information about the phylogeny of

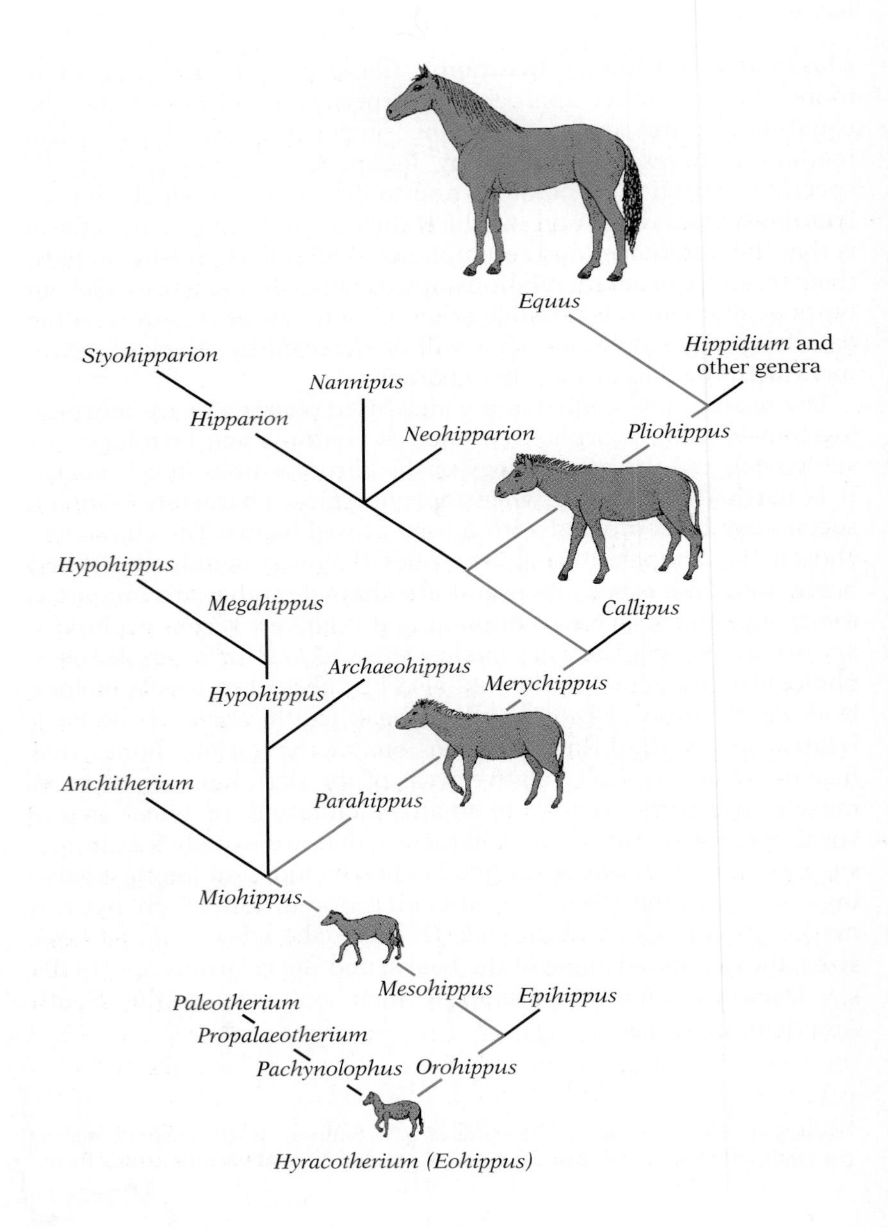

18.26 Presumed evolution of horses

The fairly complete fossil record of horses has enabled paleontologists to work out a reasonable picture of the evolutionary history of the group. The emphasis here is on the direct ancestors of the modern horse; many major branches left no modern descendants. *Hyracotherium* lived in the Eocene epoch about 55 million years ago. It was a small animal, weighing only a few kilograms. It had four toes on each front foot and three on each rear foot. It was a browser, feeding on trees and bushes. *Mesohippus,* which lived during the Oligocene epoch about 35 million years ago, was a bit larger, and its front feet, like the rear feet, had only three toes. *Merychippus,* a grazer, lived during the Miocene about 25 million years ago. It had three toes on each foot, the middle one much larger than the others, which were short and thin and did not reach the ground. *Pliohippus,* of the Pliocene, often had only one toe on each foot, though in some individuals tiny remnants of other toes persisted. *Equus,* the modern horse, is much larger than the ancestors shown here. It has only one toe on each foot.

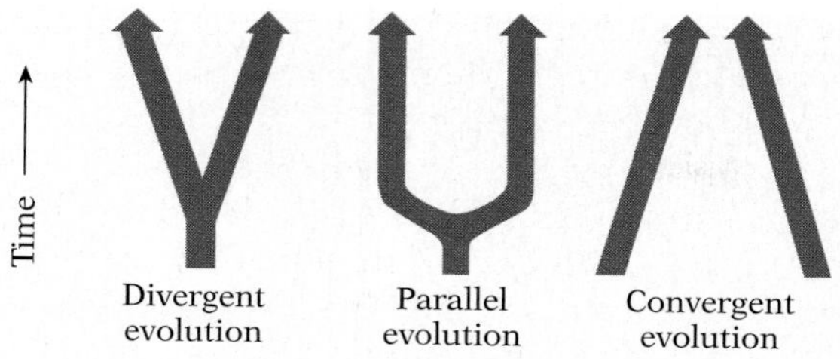

Time ⟶

Divergent evolution

Parallel evolution

Convergent evolution

18.27 Patterns of evolution

In divergent evolution one stock splits into two, which become less and less like each other as time passes. In parallel evolution two related species evolve in much the same way for a long period of time, probably in response to similar environmental selection pressures. Convergent evolution occurs when two species that are not closely related come to resemble each other more and more as time passes; this is usually the result of living in similar habitats and adopting similar lifestyles; as a result, they experience similar selection pressures.

the organisms. The message would be different if they had developed from entirely separate embryonic structures. Embryological evidence often allows biologists to trace the probable evolutionary changes that have occurred in important structures, and helps them reconstruct the probable chain of evolutionary events that led to the modern forms of life. For example, the brief appearance of pharyngeal gill pouches during the early embryology of mammals, including humans, is thought to indicate that the distant ancestors of land vertebrates were aquatic.

Life histories have also played an important role in classical phylogenetic studies. The stages through which plants pass during their life cycles are particularly important sources of information.

The problem of convergence The classical approach, then, evaluates similarities in a range of characters. However, similarity by itself does not necessarily indicate common descent. A particular similarity might reflect similar adaptations to the same environmental situation. This is common in nature and is a serious source of confusion in phylogenetic studies.

When organisms that are not closely related become more similar in one or more characters because of independent adaptation to similar environmental situations, they are said to have undergone convergent evolution, and the phenomenon is called ***convergence*** (Fig. 18.27). Whales, which are mammals, have evolved flippers from the legs of their terrestrial ancestors; those flippers superficially resemble the fins of fish, but the resemblances result from convergence; the flippers conceal well-developed finger bones. The "moles" of Australia are not truly moles but marsupials (mammals whose young are born at an early stage of embryonic development and complete their development in a pouch in their mother's abdomen, rather than in a placenta inside the womb). They occupy the same habitat in Australia as do the true moles in other parts of the world and have, as a result, convergently evolved many startling similarities to the true moles. Many Australian marsupials are strikingly convergent with placental mammals of other continents (Fig. 18.28).

As a result, when classical systematists find similarities between two species, they must try to determine whether the similarities are probably ***homologous*** (inherited from a common ancestor) or merely ***analogous*** (similar in function and often in superficial structure, but of different evolutionary origins). Thus the wings of robins and those of bluebirds are considered homologous, since both were derived from the wings of a common avian ancestor. But the wings of robins and those of butterflies are only analogous;

A

B

C

D

18.28 Marsupials that are convergent with placental mammals of other continents

(A) A marsupial mouse. (B) A marsupial glider, convergent with placental flying squirrels. (C) The tiger cat, a marsupial carnivore. (D) A cuscus, a marsupial monkey with a prehensile tail.

they evolved independently and from different ancestral structures.

Two structures can be homologous and analogous at the same time. For example, the wings of birds and of bats are not homologous as wings, for they evolved independently, but they contain homologous bones: both types of wings evolved from the forelimbs of ancient terrestrial vertebrates that were ancestors to both birds and mammals. In short, the wings of birds and bats are analogous as wings and homologous as forelimbs.

Phenetics and molecular techniques The degree of subjectivity evident in classical taxonomy has motivated many systematists to attempt to develop more objective methods. One, which later fell into disfavor, is phenetics. This approach to taxonomy used as many morphological characters as possible, weighted all characters equally, and ignored the issue of analogy versus homology. Since evidence from fossils, embryology, and behavior is difficult to quantify, it was generally not included. The expectation was that if enough characters were compared, the subjective judgements necessary for making relative weightings and identifying cases of analogy would be unnecessary; any errors would be canceled out, or would be swamped by the mass of other data. Cladistic analysis, described in the next section, has replaced simple phenetic analysis by applying an objective rule for excluding many potentially misleading traits.

However, phenetics is not dead: phenetic techniques have been revived now that molecular sequences of genes are available, and "molecular taxonomy" is enjoying enormous popularity. In essence, each nucleotide in a gene (or each amino acid in a protein) is treated as a trait without regard to analogy or homology. The potential number of these "traits" in a genome runs into the millions. Each species' characters (base sequences) are compared with those of each other species, generating a value for the degree of difference for each pairing. A branch diagram is then constructed that places each species at a distance from each other species that corresponds to the calculated difference values (Fig. 18.29, top). Most of the phylogenetic diagrams in Part IV are derived in this way using the sequence of nucleotides in ribosomal RNA.

The analysis is complicated by two problems: First, such a tree is "unrooted"—that is, there is no indication of which species are most like the ancestral form, or the order of branching points in the evolutionary tree. To determine each species' distance from the common ancestor, a suitable outgroup—a species not closely related to those being analyzed—must be added to the calculations to orient the tree correctly (Fig. 18.29, bottom). Second, since errors, uncertainties, and chance effects enter into the measurements and the evolutionary changes they attempt to quantify, the various branch lengths are never entirely consistent from pair to pair. Sometimes these inevitable anomalies are so small that they create no problem: only one branching diagram is consistent with the data, and discordant distances inferred from various pairings can simply be averaged. However, when many species are included and branch points are close together, ambiguities are possible.

In theory, the branch lengths reflect differences that have accu-

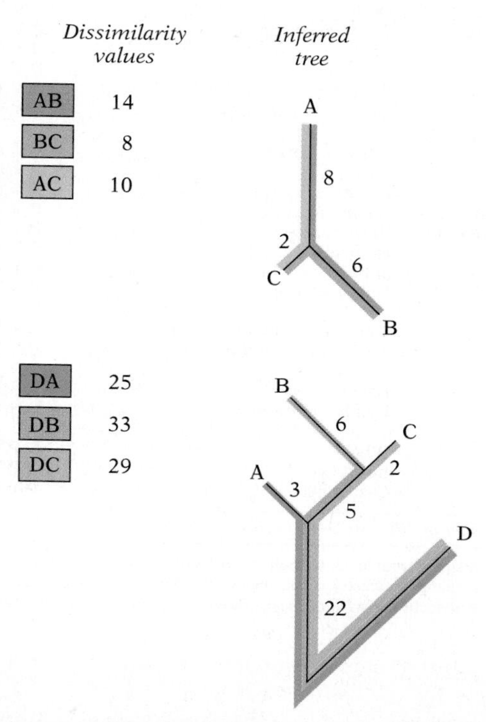

18.29 Construction of evolutionary trees from phenetic data on similarity

Species are compared in pairs, and dissimilarity values are obtained (top left). In the simplest case these values are used to construct a branching diagram (top right). In this example, since AB is 4 units greater than AC, the AB side of the BC branch must be 4 units longer than the AC side, which yields this unrooted tree. To locate the point of common ancestry of this group, an outgroup D must be included (bottom left). From the differences in path lengths to each species from the outgroup, the deepest branch among the species being analyzed can be determined (bottom right); in this case, it falls between species A and the branch point between B and C. (The branch point between the ABC group and the outgroup cannot be calculated on the basis of these data; it is placed here arbitrarily as a reminder that the outgroup and the species being analyzed also have a common ancestor.) As the text points out, the measurements usually are slightly inconsistent; you can see how this complicates matters by attempting to recompute this tree after substituting, for instance, a value of 13 for AB.

Sequence of amino acids in cytochrome c for **28** *organisms*[a]

Group	Organism	Sequence, positions 1–50
Mammals	Human, chimpanzee	GDVEKGKKIFIMKCSQCHTVEKGGKHKTGPNLHGLFGRKTGQAPGYSYTA
	Rhesus monkey	GDVEKGKKIFIMKCSQCHTVEKGGKHKTGPNLHGLFGRKTGQAPGYSYTA
	Horse	GDVEKGKKIFVQKCAQCHTVEKGGKHKTGPNLHGLFGRKTGQAPGFTYTD
	Donkey	GDVEKGKKIFVQKCAQCHTVEKGGKHKTGPNLHGLFGRKTGQAPGFSYTD
	Cow, pig, sheep	GDVEKGKKIFVQKCAQCHTVEKGGKHKTGPNLHGLFGRKTGQAPGFSYTD
	Dog	GDVEKGKKIFVQKCAQCHTVEKGGKHKTGPNLHGLFGRKTGQAPGFSYTD
	Rabbit	GDVEKGKKIFVQKCAQCHTVEKGGKHKTGPNLHGLFGRKTGQAVGFSYTD
	California gray whale	GDVEKGKKIFVQKCAQCHTVEKGGKHKTGPNLHGLFGRKTGQAVGFSYTD
	Great gray kangaroo	GDVEKGKKIFVQKCAQCHTVEKGGKHKTGPNLNGIFGRKTGQAPGFTYTD
Other vertebrates	Chicken, turkey	GDIEKGKKIFVQKCSQCHTVEKGGKHKTGPNLHGLFGRKTGQAEGFSYTD
	Pigeon	GDIEKGKKIFVQKCSQCHTVEKGGKHKTGPNLHGLFGRKTGQAEGFSYTD
	Pekin duck	GDVEKGKKIFVQKCSQCHTVEKGGKHKTGPNLHGLFGRKTGQAEGFSYTD
	Snapping turtle	GDVEKGKKIFVQKCAQCHTVEKGGKHKTGPNLNGLIGRKTGQAEGFSYTE
	Rattlesnake	GDVEKGKKIFTMKCSQCHTVEKGGKHKTGPNLHGLFGRKTGQAVGYSYTA
	Bullfrog	GDVEKGKKIFVQKCAQCHTCEKGGKHKVGPNLYGLIGRKTGQAAGFSYTD
	Tuna	GDVAKGKKTFVQKCAQCHTVENGGKHKVGPNLWGLFGRKTGQAEGYSYTD
	Dogfish shark	GDVEKGKKVFVQKCAQCHTVENGGKHKTGPNLSGLFGRKTGQAQGFSYTD
Insects[b]	Tobacco hornworm moth	GNADNGKKIFVQRCAQCHTVEAGGKHKVGPNLHGFFGRKTGQAPGFSYSN
	Fruit fly (*Drosophila*)	GDVEKGKKLFVQRCAQCHTVEAGGKHKVGPNLHGLIGRKTGQAAGFAYTN
Fungi[b]	Baker's yeast	GSAKKGATLFKTRCELCHTVEKGGPHKVGPNLHGIFGRHSGQAQGYSYTD
	Red bread mold	GDSKKGANLFKTRCAECHGEGGNLTQKIGPALHGLFGRKTGSVDGYAYTD
Plants[b]	Wheat	GNPDAGAKIFKTKCAQCHTVDAGAGHKQGPNLHGLFGRQSGTTAGYSYSA
	Sunflower	GDPTTGAKIFKTKCAQCHTVEKGAGHKQGPNLNGLFGRQSGTTAGYSYSA
	Castor bean	GDVKAGEKIFKTKCAQCHTVEKGAGHKQGPNLNGLFGRQSGTTAGYSYSA
Number of different amino acids		1 3 5 5 4 1 3 3 4 1 4 3 2 1 3 3 1 1 2 3 3 4 2 3 4 2 1 4 1 1 2 1 5 1 3 2 1 1 3 2 1 3 3 6 1 2 3 1 2 4

[a] Adapted from M. O. Dayhoff, ed., *Atlas of Protein Sequence and Structure* (Washington, D.C.: National Biomedical Research Foundation, 1972), vol. 5; and R. E. Dickerson, The structure and history of an ancient protein, *Sci. Am.*, April 1972, copyright © 1972 by Scientific American, Inc.; all rights reserved.

[b] In cytochrome *c* from insects, fungi, and plants, a few (4–8) amino acids are usually ahead of what is here labeled Position 1; these are omitted from the table.

HOW WE KNOW:
NUCLEIC ACIDS AND PROTEINS AS TAXONOMIC CHARACTERS

A mutation that changes a single base in a gene affects the gene product to varying degrees. At one extreme, it may not affect the gene product at all; if the change is in the third base of a codon, the new codon will often specify the same amino acid as the old one (see Table 9.1, p. 242), and thus be "silent." In most cases, however, the change will result in a codon for either a similar amino acid (for example, one hydrophobic amino acid instead of another, a neutral mutation) or a very different one. Less often, a change creates a termination codon, which will cause translation to end prematurely, or convert a termination codon into a coding sequence so that the transcription continues into the next gene. Base changes that survive are of two major types: silent or neutral mutations that do not significantly alter the activity of the gene product, and mutations that are fixed by selection because they improve the gene product. Distinguishing between silent/neutral and selectively advantageous changes is not easy.

Mutations that do not alter the meaning of the codon are probably nearly neutral, and most of them probably accumulate randomly with time after two groups of organisms have diverged from a common ancestor. By comparing the extent of single-base changes of this type in the sequences of genes common to a wide range of species—genes for the enzymes of glycolysis, the citric-acid cycle, or the electron-transport chain, for example—we might get some measure of the time elapsed since two groups diverged. As a tool for dating evolutionary events, therefore, neutral single-base changes have great potential.

Another approach is to compare the amino acid sequences of the gene products. Consider, for example, cytochrome *c*, an essential component of the respiratory chain in mitochondria. The complete amino acid sequence of this enzyme has been worked out for a variety of organisms. The table above shows the sequence for some of the species so far examined, with the various functional groupings of amino acids indicated by a color code.

Perhaps the most obvious feature of this table is that cytochrome *c* is remarkably similar in all the species, even

```
        55        60        65        70        75        80        85        90        95       100     104
A N K N K G I I W G E D T L M E Y L E N P K K Y I P G T K M I F V G I K K K E E R A D L I A Y L K K A T N E
A N K N K G I T W G E D T L M E Y L E N P K K Y I P G T K M I F V G I K K K E E R A D L I A Y L K K A T N E
A N K N K G I T W K E E T L M E Y L E N P K K Y I P G T K M I F A G I K K K T E R E D L I A Y L K K A T N E
A N K N K G I T W K E E T L M E Y L E N P K K Y I P G T K M I F A G I K K K T E R E D L I A Y L K K A T N E
A N K N K G I T W G E E T L M E Y L E N P K K Y I P G T K M I F A G I K K K G E R E D L I A Y L K K A T N E
A N K N K G I T W G E E T L M E Y L E N P K K Y I P G T K M I F A G I K K T G E R A D L I A Y L K K A T K E
A N K N K G I T W G E D T L M E Y L E N P K K Y I P G T K M I F A G I K K K D E R A D L I A Y L K K A T N E
A N K N K G I T W G E E T L M E Y L E N P K K Y I P G T K M I F A G I K K K G E R A D L I A Y L K K A T N E
A N K N K G I I W G E D T L M E Y L E N P K K Y I P G T K M I F A G I K K K G E R A D L I A Y L K K A T N E

A N K N K G I T W G E D T L M E Y L E N P K K Y I P G T K M I F A G I K K K S E R V D L I A Y L K D A T S K
A N K N K G I T W G E D T L M E Y L E N P K K Y I P G T K M I F A G I K K K A E R A D L I A Y L K Q A T A K
A N K N K G I T W G E D T L M E Y L E N P K K Y I P G T K M I F A G I K K K S E R A D L I A Y L K D A T A K
A N K N K G I T W G E E T L M E Y L E N P K K Y I P G T K M I F A G I K K K A E R A D L I A Y L K D A T S K
A N K N K G I I W G D D T L M E Y L E N P K K Y I P G T K M I F T G L S K K K E R T N L I A Y L K E K T A A
A N K N K G I T W G E D T L M E Y L E N P K K Y I P G T K M I F A G I K K K G E R Q D L I A Y L K S A C S K
A N K S K G I V W N N D T L M E Y L E N P K K Y I P G T K M I F A G I K K K G E R Q D L V A Y L K S A T S -
A N K S K G I T W Q Q E T L R I Y L E N P K K Y I P G T K M I F A G L K K K S E R Q D L I A Y L K K T A A S

A N K A K G I T W Q D D T L F E Y L E N P K K Y I P G T K M V F A G L K K A N E R A D L I A Y L K Q A T K -
A N K A K G I T W Q D D T L F E Y L E N P K K Y I P G T K M I F A G L K K P N E R G D L I A Y L K S A T K -

A N I K K N V L W D E N N M S E Y L T N P K K Y I P G T K M A F G G L K K E K D R N D L I T Y L K K A C E -
A N K Q K G I T W D E N T L F E Y L E N P K K Y I P G T K M A F G G L K K D K D R N D I I T F M K E A T A -

A N K N K A V E W E E N T L Y D Y L L N P K K Y I P G T K M V F P G L K K P Q D R A D L I A Y L K K A T S S
A N K N M A V I W E E N T L Y D Y L L N P K K Y I P G T K M V F P G L K K P Q E R A D L I A Y L K T S T A -
A N K N M A V Q W G E N T L Y D Y L L N P K K Y I P G T K M V F P G L K K P Q D R A D L I A Y L K E A T A -
1 1 2 5 2 3 2 6 1 6 4 3 2 2 5 3 1 1 3 1 1 1 1 1 1 1 1 1 1 1 3 1 5 1 2 2 1 6 9 2 1 7 2 2 2 2 2 2 1 6 4 3 5 4
```

Symbol	*Amino acid*
▭	NONPOLAR
G	glycine
A	Alanine
V	Valine
L	Leucine
I	Isoleucine
M	Methionine
F	Phenylalanine
W	Tryptophan
P	Proline
▭	POLAR
S	Serine
T	Threonine
C	Cysteine
Y	Tyrosine
N	Asparagine
Q	Glutamine
▭	ACIDIC
D	Aspartic acid
E	Glutamic acid
▭	BASIC
K	Lysine
R	Arginine
H	Histidine

G is printed in color, because glycine, despite its technically nonpolar R group, behaves like a polar amino acid.

though some of them have probably not had a common ancestor for more than 1 billion years. For example, all the cytochromes have the same amino acid sequence from positions 70 through 80. In fact, cytochrome *c* is an evolutionarily conservative protein; its amino acid sequence has changed at a considerably slower average rate (about 20 million years for a 1% change) than, for example, that of hemoglobin (5.8 million years) or fibrin (only 1 million years). The minimal change in cytochrome c suggests that only minor alterations can be tolerated if the enzyme is to continue functioning properly. Even at points along the chain where there are differences, the amino acids are often functionally similar (with one polar amino acid substituted for another, one nonpolar amino acid for another, and so on). Some of the alterations may be neutral, having essentially no effect on the activity of the gene product, while others may represent minor but adaptive species-specific modifications of the protein. Some, however, almost certainly indicate major changes in the gene product. Among these are substitution of a polar for a nonpolar amino acid, and alterations involving proline (which induces turns in polypeptide chains) or cysteine (which forms strong covalent bonds with other cysteines).

If we compare various species in the table with one another, we find that the number of differences in amino acids usually agrees reasonably well with the presumed evolutionary distances among the species. Thus the mammals differ less among themselves than any of them differ from the fish. Humans and chimpanzees do not differ at all; both differ by one amino acid from the rhesus monkey, by an average of 10.4 amino acids from the other mammals, 14.5 from the reptiles, 18 from the amphibian, and 22.5 from the fish. This is an accurate reflection of the generally accepted evolutionary age of vertebrate classes, with fish predating amphibia, which are more ancient than reptiles, which in turn gave rise to mammals.

Because cytochrome *c* is evolutionarily so conservative, its value as a taxonomic character is limited to studies of the relationships among evolutionarily distant organisms. More rapidly changing proteins, such as fibrin, may prove useful in cases of closely related species. In any comparison, reliable conclusions depend on using several different proteins; because of unusual selection pressures or an unusually large or small number of mutations in the gene, the rate of change in one particular protein may be too great or too small to be representative.

Species and their character traits:

1. A B C D
2. A B C D
3. A B C D
4. A B C D
5. A B C D
6. A C D

Random tree

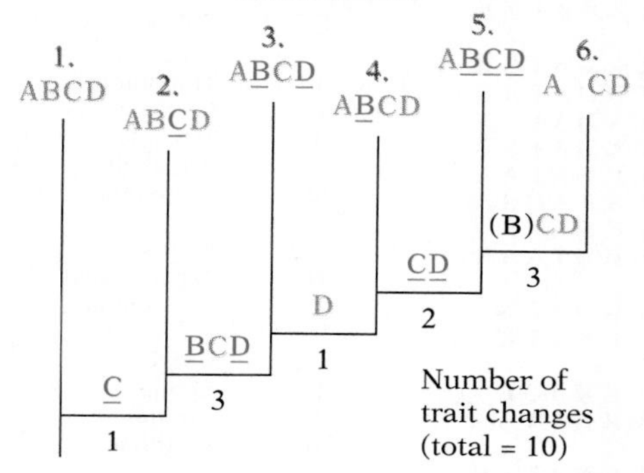

Parsimonious tree

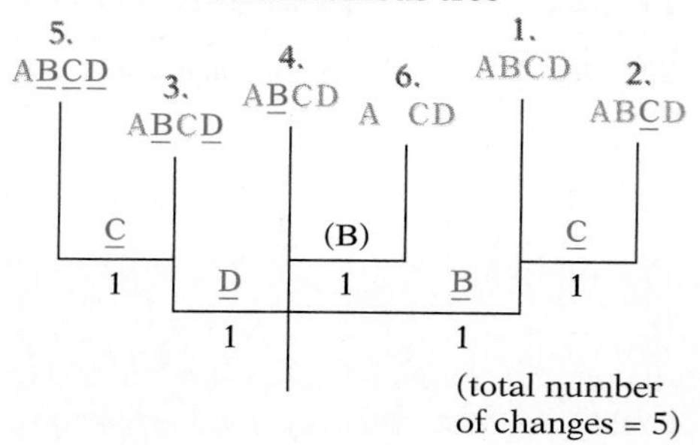

18.30 An example of cladistic analysis
Six species in a group of organisms with a set of traits A–D are compared with one another to construct a phylogenetic tree. Each trait can exist in two forms, indicated by blue versus red letters. Cladistic analysis treats the absence of B in one species as a possible case of secondary loss; character A is ignored since it is common to all members. Cladistic analysis tries many different arrangements of the species to find the tree with a minimum number of changes. Two of the possible 954 trees are shown; one, selected at random, has 10 changes; the other, much simpler one, has five. The tree with the minimum number is used to infer the primitive state (ABCD); thus the loss of B is taken to be secondary.

mulated as a result of drift and selection. If only gradual drift were at work, the total distance from the deepest branch point to each present-day species would be the same, reflecting the time that has passed since the last common ancestor. Deviations—that is, longer paths—imply unusual degrees of selective pressure, and suggest something about the tempo and mode of evolution in the group.

Cladistics Cladists seek to avoid confusing analogy with homology by focusing on ***shared derived characters***—traits that are common to most but not all of the several species in question and thus probably represent relatively recent rather than ancient adaptation.[4] Hence, in comparing two species of bats, the traits shared by mammals in general would not be considered. The cladistic approach is helpful in providing a rule for ignoring the absence (secondary loss) of traits—the hind limbs of aquatic mammals, for example—that causes many species to stand out as obvious exceptions to the general taxonomic patterns of their group. Cladists weight each of the traits they consider equally, but by ignoring features presumed to be shared at the point at which the speciation event in question occurred, they can be more selective than pheneticists: they have a criterion, or rule, for choosing the traits to be included.

While phenetic approaches focus on degrees of difference—the evolutionary "distance" that translates into branch lengths—cladists concentrate on discrete differences, or character states. They use the number of trait differences between each pair of species under consideration to create the simplest, or most parsimonious, branching diagram (Fig. 18.30). There are some drawbacks to cladistics. For one thing, evolution need not occur by the simplest route. For another, cladistic analysis usually ignores fossils, embryology, and behavior. The result is sometimes controversial. For example, cladistic analysis places the crocodiles with the birds, and places mammals close to snakes and turtles (Fig. 18.31). In addition, like phenetics and classical taxonomy, cladistic analysis can be fooled by convergences, such as those exhibited by cer-

[4] Cladistics takes its name from the term *clade*, which refers to any monophyletic group—that is, a group with a single common ancestor. Examples are a genus, a class, a kingdom (all described at the end of this chapter), or all living things.

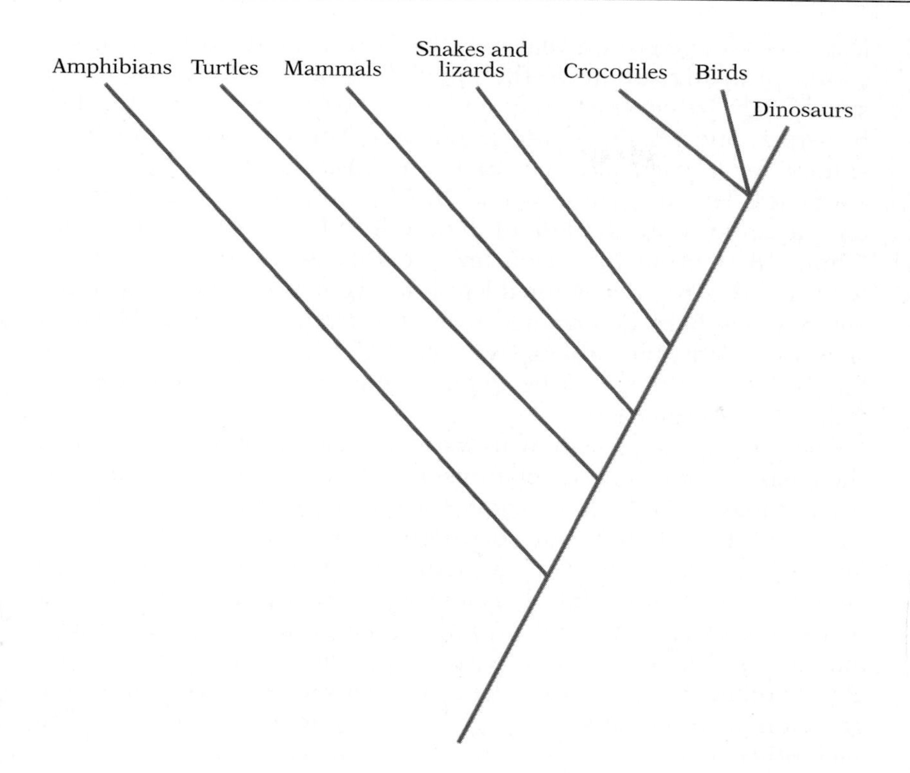

18.31 Cladistic analysis of the phylogeny of higher vertebrates

The major feature of this reconstruction is that reptiles do not belong to a single class. Depending on how a cladist chooses to group species, mammals, snakes, and turtles could form one class, with birds and crocodiles in the other; alternatively, there could be five different classes of reptiles, with birds and mammals in their own classes.

tain Australian marsupials. For analyzing the phylogeny of closely related genera or species, however, cladistics is now widely accepted. (Fig. 18.25, for instance, is based on cladistic techniques.) And cladistic analysis of molecular data is generally considered at least as reliable as phenetic calculations.

■ HOW ORGANISMS ARE CLASSIFIED

Over 1 million species of animals and over 325,000 species of plants are known (and many more are yet to be discovered). To deal with this vast array of organic diversity, we need a system by which species can be classified in a logical and meaningful manner. Many different kinds of classification are possible. We could, for example, classify flowering plants according to their color: all white-flowered species in one group, all red-flowered species in a second group, all yellow-flowered species in a third group, and so on. This sort of system has its usefulness in field guides, but the information such categories convey fails to set apart fundamentally different organisms. The classification system used in biology today attempts to encode the evolutionary history of the organisms; it is thus often a means of conveying information about many of their characteristics to those familiar with the various taxonomic groups (taxa) to which the organisms are assigned.

The classification hierarchy Suppose you had to classify all the people on earth on the basis of where they live. You would probably begin by dividing the world population into groups based on country. This subdivision separates inhabitants of the United States from the inhabitants of France or Argentina, but it still leaves very large groups that must be further subdivided. Next, you would probably subdivide the population of the United States by states, then by counties, then by city or village or township, then by street, and finally by house number. You could do something similar for Canada, Mexico, England, Australia, and all the other countries. This procedure would enable you to place every individual in an orderly system of hierarchically arranged categories (Table 18.2). Note that each level in this hierarchy is contained within and partly determined by all levels above it. Thus, once the country has been determined to be the United States, a Mexican state or a Canadian province is excluded. Similarly, once the state has been determined to be Pennsylvania, a city in New York or California is excluded.

Table 18.2 Classification hierarchies

BIOLOGICAL	POSTAL
Kingdom	Country
Phylum/division	State/province
Class	City
Order	Street
Family	Number
Genus	Last name
Species	First name

The same principles apply to the classification of living things on the basis of phylogenetic relationships. A hierarchy of categories is used (Table 18.2). Each category (taxon) in this hierarchy is a collective unit containing one or more groups from the next-lower level in the hierarchy. Thus a genus is a group of closely related species (Fig. 18.32), a family is a group of related genera, an order is a group of related families, a class is a group of related orders, and so on. The species in any one genus are believed to be more closely related to each other than to species in any other genus, the genera in any one family are believed to be more closely related to each other than to genera in any other family, and so on.

Ideally, all the organisms in a taxon share a common ancestor—a progenitor not in the ancestry of any other taxa at the same level of the classification hierarchy. Pure taxa of this sort, which include the common ancestor and all its descendants, are ***homophyletic*** (Fig. 18.32A–C). Sometimes species are grouped into two taxa on the basis of similarities and differences in characters, thus separating species with a common ancestor and mixing species with different lineages; the classification of birds and mammals as taxa distinct from reptiles (see Fig. 18.31) is a clear example of this. When some but not all descendants of a common ancestor are grouped into a taxon, separating other descendants on the basis of dissimilarity, the classification is ***paraphyletic*** (Fig. 18.32D). When species are grouped into a taxon on the basis of similarities in such a way that their common ancestor is excluded, the taxon is ***polyphyletic*** (Fig. 18.32E). Two of the kingdoms discussed in Part IV (Archaebacteria and Archezoa) are probably paraphyletic, while another (Protista) is polyphyletic.

Table 18.3 gives the classification of six species. The table shows us immediately that the six species are not closely related, but that a human and a wolf are more closely related to each other than either is to the herring gull. It also shows us that the mammals and the bird are more closely related to each other than to the house fly, which is in a different phylum, or to moss or red oaks, which are in a different kingdom.

The current system dates from the work of Carolus Linnaeus (1707–1778). His system used kingdoms, classes, orders, genera, and species; the phylum and family categories were added later. The rationale on which Linnaeus based his system was very different from the phylogenetic one employed today. He worked a century before Darwin; he had no conception of evolution, thinking of each species as an immutable entity. He simply grouped organisms

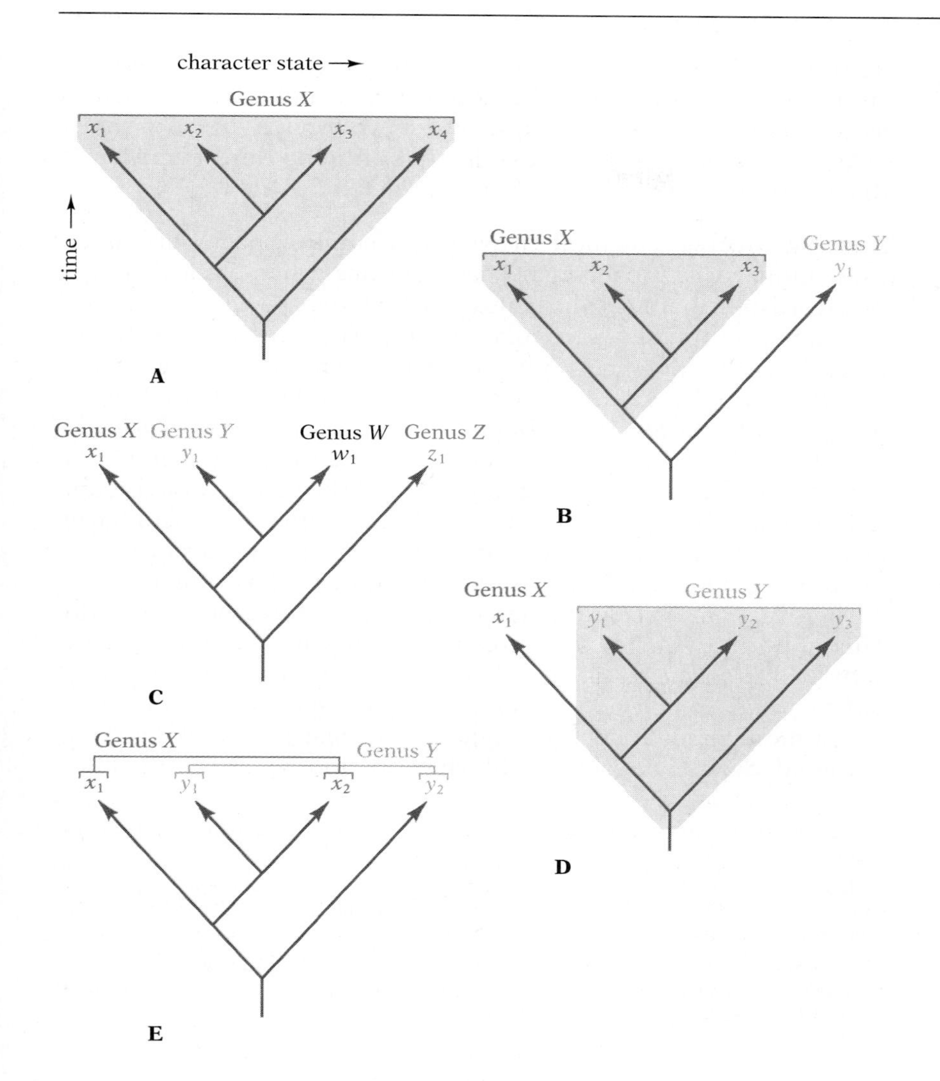

18.32 Alternative generic groupings for four related species

Biologists try to group species in a way that will indicate their phylogenetic relationships. Thus a genus is a group of related species. But how closely related? There is no absolute answer to this question. Some biologists (the "lumpers") like large genera containing many species (A); others (the "splitters") prefer small, compact genera containing only species that are very closely related (C). In the drawing, five alternative ways of grouping four related species are shown. The first recognizes only one genus, the second recognizes two, and the third recognizes four. All are homophyletic: genera include all species sharing common descent. The fourth classification is paraphyletic: three very similar species, y_1, y_2, and y_3, are placed in the same genera, but species x_1, though closely related to y_1 and y_2, is so different that it is placed in a separate genera. The fifth classification is polyphyletic because it excludes the common ancestor of y_1 and y_2 from genus Y.

Table 18.3 Classification of six species

CATEGORY	HAIRCAP MOSS	RED OAK	HOUSE FLY	HERRING GULL	WOLF	HUMAN
Kingdom	Plantae	Plantae	Animalia	Animalia	Animalia	Animalia
Phylum or Division	Bryophyta	Tracheophyta	Arthropoda	Chordata	Chordata	Chordata
Class	Musci	Angiospermae	Insecta	Aves	Mammalia	Mammalia
Order	Bryales	Fagales	Diptera	Charadriiformes	Carnivora	Primata
Family	Polytrichaceae	Fagaceae	Muscidae	Laridae	Canidae	Hominidae
Genus[a]	*Polytrichum*	*Quercus*	*Musca*	*Larus*	*Canis*	*Homo*
Species[a]	*commune*	*rubra*	*domestica*	*argentatus*	*lupus*	*sapiens*

[a]The name of a particular species consists of the genus name and the species designation, both customarily in italics, as shown in this table.

according to similarities, primarily morphological. His results were similar to those obtained today because morphological characters, as products of evolution, usually reflect evolutionary relationships.

An outline of a modern classification of living things is given in the Appendix and is discussed in some detail in Part IV.

Naming species The modern system of naming species also dates from Linnaeus. Linnaeus gave each species a name consisting of two words: the name of the genus to which the species belongs and a designation for that particular species. Thus one species of carnation is *Dianthus caryophyllus*. Other species in the genus *Dianthus* have the same first word in their names, but each has its own specific designation (*Dianthus prolifer, Dianthus barbatus, Dianthus deltoides*). No two species can have the same name.[5] The names are always Latin (or Latinized), and the genus name is capitalized while the species name is not.[6] Both names are customarily printed in italics. The correct name for any species, according to the present rules, is usually the oldest validly proposed name.[7]

The same Latin scientific names are used throughout the world. This uniformity of usage ensures that each scientist will know exactly which species another scientist is discussing. Thus the plant *Bidens frondosa* can be unambiguously designated even though it is commonly known, just in English, as beggar-tick, sticktight, bur marigold, devil's bootjack, pitchfork weed, and rayless marigold.

[5] More precisely, no two species of plants can have the same name, and no two species of animals can have the same name.

[6] This rule always holds for zoological names, but specific botanical names are sometimes capitalized when they are derived from the name of a person or from other proper nouns.

[7] For purposes of priority, botanical naming dates from the publication of *Species Plantarum* by Linnaeus in 1753, and zoological naming dates from the publication of the 10th edition of his *Systema Naturae* in 1758.

CHAPTER SUMMARY

HOW SPECIATION OCCURS

WHAT ARE SPECIES? A species is a group of organisms that are capable of breeding among themselves and are reproductively isolated from other populations. Species generally exist as groups of local demes that may be somewhat different genetically from one another; systematic differences that vary with habitat are called clines. (p. 487)

THE PROCESS OF SPECIATION The reproductive isolation of two populations required for speciation can occur quickly through chromosomal changes, or slowly as habitat or geographic isolation keeps the populations separate while reproductive differences evolve. Many speciation events involve the isolation of small founder populations in which genetic drift is likely to play a major role. Reproductive isolation can involve genetic or habitat separation, or behavioral isolation based on mating habitat, times, or displays, or genetic or developmental incompatibilities. Speciation often involves adaptive radiation, as populations of a single species become specialized for particular ways of making a living. Evolution can be gradual when selection pressure is mild, or relatively rapid when selection is intense, drift is important, or major genetic or developmetal changes occur. (p. 492)

PROBLEMS IN IDENTIFYING SPECIES Asexuals form distinct groups even though they do not interbreed. Whether allopatric populations are distinct species often cannot be established for certain since they have no opportunity to interbreed. (p. 512)

THE CONCEPT OF PHYLOGENY

DETERMINING PHYLOGENETIC RELATIONSHIPS Phylogeny uses similarity to infer relatedness. A major problem lies in distinguishing evolutionary convergence from true kinship—analogous versus homologous characters. The classical approach depends on subjective weighting of different kinds of similarity. Molecular taxonomy relies on the large numbers of bases in genetic sequences to swamp out chance convergence. Cladistics focuses on identifying shared derived characters to exclude convergence. (p. 515)

HOW ORGANISMS ARE CLASSIFIED Taxonomists use a hierarchical system of kingdoms, phyla or divisions, classes, orders, families, genera, and species; the system reflects evolutionary relatedness rather than mere similarity. (p. 523)

STUDY QUESTIONS

1 One problem with molecular techniques is the issue of whether a given difference in base sequence between two species is neutral or adaptive. After all, there can be subtle differences created by the substitution of one nonpolar amino acid for another, and even third-codon changes can have an effect since they use different tRNAs, which may have different abundances, specificities, and activities. How might one go about attempting to resolve this issue for one specific gene product—an rRNA or a particular enzyme, for example? Feel free to imagine plausible advances in molecular techniques. (pp. 519–522)

2 The argument is sometimes made (almost always in the context of humans) that the concept of subspecies is useless because the variation within a subspecies for most characters exceeds the difference between the two subspecies. Is this line of reasoning correct? Does it work for species within a genus? Could it be used to prove that there are no "important" differences between males and females of a species? (pp. 490–491, 512–514)

3 How could you use one of the taxonomic techniques described in this chapter to determine which if any species in a genus were likely to have been produced through an evolutionary bottleneck? (pp. 507, 515–523)

4 For what sorts of species (including factors like size, generation time, reproductive strategy, habitat, and niche) is chance more likely to have been important than competition? What kinds of patterns would you expect to find through molecular-taxonomic analysis in such cases? (pp. 504–509, 519–522)

SUGGESTED READING

GRANT, P. R., 1991. Natural selection and Darwin's finches. *Scientific American* 265(4). *On the effects of a season of drought on the finch population in the Galápagos. A well-written and modern analysis of speciation.*

LEVINTON, J. S., 1992. The big bang of animal evolution. *Scientific American* 267 (5). *On the apparent explosion of diversity at the beginning of the Cambrian.*

LI, W. H., AND D. GRAUR, 1991. *Fundamentals of Molecular Evolution.* Sinauer, Sunderland, Mass. *Excellent treatment of how molecular techniques can be applied to phylogeny.*

RIDLEY, M., 1993. *Evolution.* Blackwell, Oxford. *Excellent text on the subject.*

SCHOENER, T. W., 1982. The controversy over interspecific competition, *American Scientist* 70, 586–595.

SIBLEY, C. G., AND J. E. AHLQUIST, 1986. Reconstructing bird phylogeny by comparing DNAs, *Scientific American* 254 (2). *On the DNA hybridization technique.*

SIMPSON, G. G., 1983. *Fossils and the History of Life.* Scientific American Books, New York. *Clearly written and well-illustrated exposition of evolution.*

STEBBINS, G. L., AND F. J. AYALA, 1985. The evolution of Darwinism, *Scientific American* 253 (1). *A modern summary of evolutionary theory, with particular attention to the claims for punctuated equilibrium.*

WILSON, A. C., 1985. The molecular basis of evolution, *Scientific American* 253 (4). *On methods for tracing evolution through similarities in nucleotide or amino acid sequences.*

PART IV

THE GENESIS AND DIVERSITY OF ORGANISMS

◀ **Major milestones** in the early evolution of life on earth are exposed at low tide at Shark Bay in western Australia. The moundlike structures are stromatolites, the solidified remains of successive layers of bacteria. Fossilized stromatolite formations, some dating from as far back as the Precambrian era, are among the oldest known remains of procaryotes—single-celled organisms lacking a distinct membrane-bounded nucleus.

CHAPTER 19

HOW DID LIFE BEGIN?

FORMATION OF THE EARTH

THE FIRST ORGANIC MOLECULES

THE EARLIEST CELLS

PRECAMBRIAN EVOLUTION

WHAT FOSSILS TELL US

THE FIRST EUCARYOTES

THE KINGDOMS OF LIFE

ORIGIN AND EARLY EVOLUTION OF LIFE

Few problems have exercised the human imagination like the question of the origin of life. Not until the latter part of the 19th century was a truly scientific (that is, testable) explanation—the theory of evolution by natural selection—able to account for the origin of species. This chapter will examine how life might have begun, and consider the problem of how the kingdoms of life are best defined.

HOW DID LIFE BEGIN?

How is it that life arose spontaneously from nonliving matter under the conditions prevailing on the early earth? There is no direct evidence concerning the origin of life, but most researchers agree on a general outline of what probably happened.

■ FORMATION OF THE EARTH

The universe is about 10-20 billion years old, but the sun and its planets formed more recently, between 4.5 and 5 billion years ago, from a cloud of cosmic dust and gas. Most of this material condensed into a single compact mass, producing enormous heat and pressure; this initiated thermonuclear reactions, and converted the condensed mass into the sun. Within the remainder of the dust and gas cloud, which now formed a disk held in the gravitational field of the newborn sun, lesser clouds of condensation began to form (Fig. 19.1). These became the planets, of which the earth is one. Finally, beyond the orbit of Pluto, billions or even trillions of "dirty snowballs" up to a few kilometers in diameter are thought to surround the solar system. This distant swarm of objects, known as the Oort cloud, is the source of the comets that may have drastically altered the course of evolution.

As the earth condensed, its components stratified; the heavier materials, such as iron and nickel, moved toward the center, and the lighter substances rose to the surface. Among these lighter materials were hydrogen, helium, and the noble gases, which formed the first atmosphere. Unlike larger planets such as Jupiter and Saturn, the earth was too small, and its gravitational field too weak, to retain this first atmosphere; eventually all the gases es-

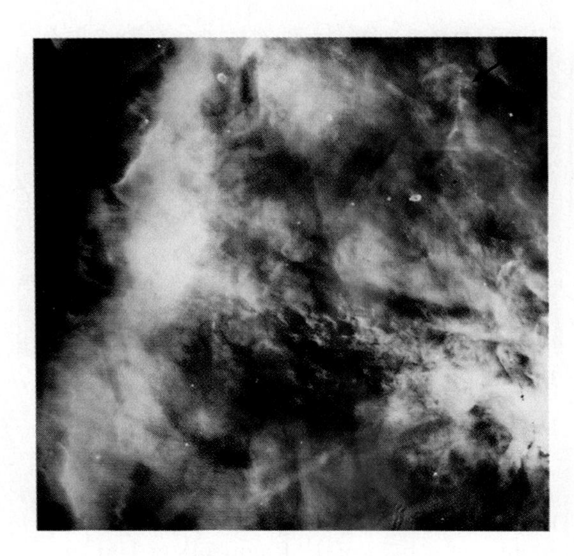

19.1 Star and planet formation

Stars and accompanying planets condense from the gas of nebulae. The arrow indicates the largest gas and dust disk visible in this photograph of the Great Nebula in Orion taken by the Hubble Space Telescope.

caped into space, leaving a bare rocky globe with neither oceans nor atmosphere.

As time passed, gravitational compression of the earth, together with radioactive decay, generated an enormous amount of heat, and the interior of the earth became molten. The result was further stratification into a core of iron and nickel, a mantle some 4700 km thick composed of dense silicates of iron and magnesium, and an outer crust 8–65 km thick composed mainly of lighter silicates. The intense heat in the interior of the earth also tended to drive out various gases, which escaped primarily by volcanic action. These gases formed a second atmosphere for the earth.

The present atmosphere contains about 78% molecular nitrogen (N_2), 21% molecular oxygen (O_2), and 0.033% carbon dioxide (CO_2), as well as traces of rarer gases such as helium and neon. When the atmosphere first formed, it contained virtually no free oxygen. Instead, the earth's early atmosphere was made up primarily of the gases that occur in the present-day outgassing of volcanoes: H_2O, CO, CO_2, N_2, H_2S, and H_2. In such a mixture, hydrogen cyanide (HCN) and formose (H_2CO) are easily formed and would also have been present in the atmosphere.

Initially, most of the earth's water was probably present as vapor in the atmosphere, a condition leading to torrential rains. These would have filled the low places on the crust with water and given rise to the first oceans. As rivers rushed down the slopes, they must have dissolved away and carried with them salts and minerals (iron and uranium being of particular importance), which slowly accumulated in the seas. Atmospheric gases probably also dissolved in the waters of the newly formed oceans. Whatever free oxygen might have existed in the atmosphere would have been rapidly removed by dissolved ions in the oceans; the extensive deposits of uraninite (UO_2) dating from that period must have formed in this way.

A

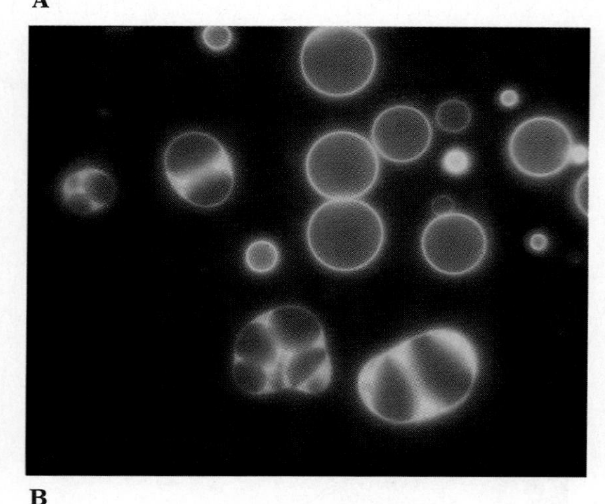

B

19.2 A carbonaceous chondrite meteorite

(A) This golf-ball-sized fragment is part of a meteorite that fell near Murchison, Australia, in 1969. Tiny particles of organic compounds, accounting for 1-2% of the fragment's weight, are scattered throughout the stone. (B) When the organic material is extracted, some of the molecules self-assemble into vesicles. The yellow-green color is produced by the fluorescence of polycyclic aromatic hydrocarbons, a class of extremely complex organic molecules.

The primitive seas contained a mixture of salts, CO_2, H_2S, HCN, H_2CO, and N_2, which are thermodynamically stable: there is no tendency for these substances to react with each other to form other compounds. Yet for life to have arisen, the critical building-block materials—particularly the amino acids and the purine and pyrimidine bases—would have been necessary at the very least. How might these compounds have been formed on the abiotic primitive earth?

Extraterrestrial molecules? There are two basic hypotheses to account for the accumulation of complex organic compounds on the early earth. One, developed in its modern form in 1990, notes that comets as well as many asteroids and meteors are rich in complex organics created during the formation of the solar system billions of years ago (Fig. 19.2). Given the denser atmosphere of the early earth, which would have been better able to slow incoming objects before they struck the surface, and the vast number of comets and meteors whose orbits intersected the earth's path, some astronomers estimate that from 10^6–10^7 kg of complex organic molecules could have survived impact annually. Allowing for the thinning of both the atmosphere and the population of comets and meteors, the accumulation of extraterrestrial organics would have declined to negligible levels within 1 billion years. The total mass that could have been accumulated by this scenario is $2–20 \times 10^{14}$ kg; the total organic mass of all living things at present is estimated to be no more than 6×10^{14} kg. It is possible, therefore, that the early earth had a vast supply of the complex molecules necessary for the evolution of life.

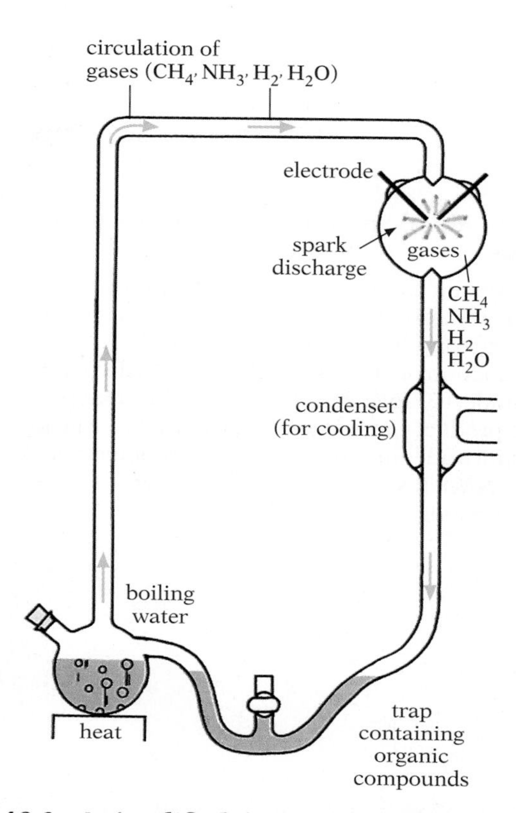

19.3 A simplified drawing of Miller's apparatus for synthesizing organic molecules under abiotic conditions

The closed system contained CH_4, NH_3, H_2, and H_2O, gases that would have been abundant if the early atmosphere was reducing. Water in the lower flask was boiled to circulate gases past the electrical sparks in the upper flask. As the products of the reaction passed through the condenser, they were cooled and condensed into liquid that accumulated in the trap, forming compounds like urea, hydrogen cyanide, acetic acid, and lactic acid. Similar results have been obtained with volcanic gases.

Terrestrial synthesis? The more conventional view of the development of life assumes that complex organics were generated from small compounds already present on the early earth. For the changes that could have initiated life to occur, however, some external source of energy must have been acting on the mixture. During the early history of the earth there was no shortage of such energy, which came from several sources: solar radiation (especially ultraviolet—UV—light), lightning, and heat from the earth's core and the sun.

In 1953 Stanley L. Miller showed that this sort of energy could cause reactions between simple inorganic molecules, which produce complex organic compounds. Miller set up an airtight apparatus in which a mixture of ammonia, methane, water, and hydrogen was circulated past electrical discharges from tungsten electrodes (Fig. 19.3). At the end of just one week he found that a variety of organic compounds had been synthesized. Among these were some of the biologically most important amino acids as well as urea, hydrogen cyanide, acetic acid, and lactic acid.

In the years since 1953, many investigators have used mixtures of gases characteristic of volcanic emissions, plus hydrogen cyanide—on the more widely accepted assumption that the early atmosphere contained these or similar gases—and have also achieved positive results (Fig. 19.4). Those who have turned to

19.4 Volcanoes and lightning

One effective combination for producing complex organic molecules abiotically is volcanic gas and electrical discharges. Here nature performs a similar experiment during the eruption in 1963 of Surtsey, off the coast of Iceland.

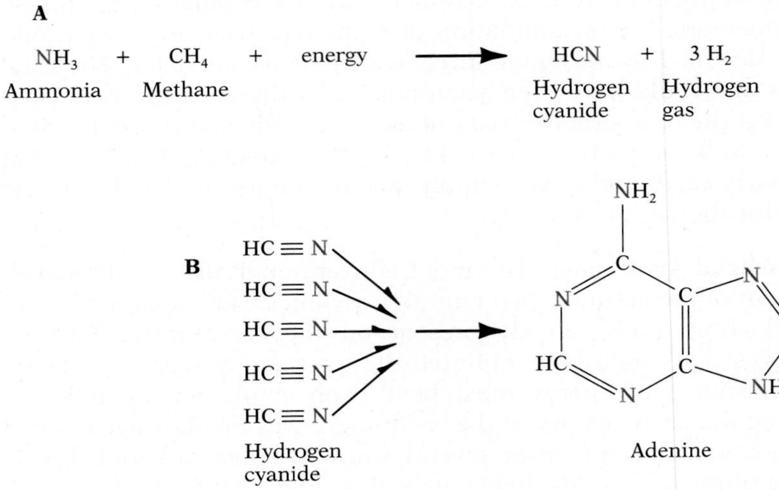

A

NH_3 (Ammonia) + CH_4 (Methane) + energy ⟶ HCN (Hydrogen cyanide) + $3\ H_2$ (Hydrogen gas)

B

$HC{\equiv}N$ ×5 (Hydrogen cyanide) ⟶ Adenine

19.5 Abiotic production of a nitrogenous base

(A) Mixing methane and ammonia in the presence of external energy of the sort available during the early history of the earth results in the production of, among other things, hydrogen cyanide. (B) Even in the absence of a catalyst, this product can react with itself at a low rate to produce adenine, an important component of ATP, NAD, RNA, and a variety of other compounds.

other energy sources, such as ultraviolet light, heat, or both, have also obtained large yields. Significantly, the amino acids most easily synthesized in all these experiments performed under abiotic conditions are the very ones that are most abundant in proteins today, and the most important nitrogenous base—adenine—is also the one most readily produced abiotically (Fig. 19.5).

The wide variety of conditions under which abiotic synthesis of the organic compounds has now been demonstrated makes it safe to conclude that, even if no extraterrestrial molecules survived impact, these compounds would inevitably have appeared and become dissolved in the waters of the seas.

Though organic compounds appeared on the primordial earth, most of these molecules are highly perishable under present conditions: they react with oxygen and become oxidized, or are consumed by microorganisms. Since the prebiotic atmosphere contained virtually no free oxygen, and there were no organisms, neither oxidation nor decay would have destroyed the organic molecules; they could have accumulated in the seas over hundreds of millions of years. No such accumulation would be possible today.

Formation of polymers Suppose, then, that a variety of hydrocarbons, fatty acids, amino acids, purine and pyrimidine bases, simple sugars, and other relatively small organic compounds slowly accumulated in the ancient seas. This combination would still not be a sufficient basis for life, which also needs macromolecules, particularly polypeptides and nucleic acids. How could these polymers have formed from the mixture of building-block substances present in the "soup" of the ancient oceans? This question is not easy to answer, and several hypotheses are currently being investigated.

Some think that the concentration of organic material in the seas was high enough for chance bondings between simpler molecules to give rise in time to considerable quantities of macromolecules. They point out that, even though each such polymerization reaction is rather unlikely in the absence of protein enzymes, on a vast time scale enough rare and unlikely events would probably occur to produce, collectively, a major change.

Other investigators suggest that concentration mechanisms must have speeded up chemical reactions. One such mechanism might have been adsorption of the building-block compounds on the surfaces of clay minerals. Another might have involved the accumulation of small amounts of dilute solutions of building-block compounds in puddles on the beaches of lagoons and ponds. The heat of the sun would have evaporated most of the water, thus concentrating the organic chemicals, and providing energy for polymerization reactions. The resulting polymers might then have been washed back into the pond. Such a process could slowly have built up a supply of macromolecules in the pond. Sidney W. Fox has shown that if a nearly dry mixture of amino acids is heated, polypeptide molecules are rapidly synthesized (particularly if phosphates are present). Alternatively, after condensation by evaporation, the energy for polymerization reactions in the puddles might have come from UV radiation rather than heat. In fact, in experiments of both the Miller and Fox types, polymers often form before monomers.

■ THE EARLIEST CELLS

Cells from droplets? How could the first cells have arisen from this ancient soup of salts and organic molecules? Under appropriate conditions of temperature, ionic composition, and pH, colloids (stable suspensions) of macromolecules tend to give rise to complex units called ***coacervate droplets***. Each such droplet is a cluster of largely hydrophobic macromolecules stabilized by a shell of water (see Fig. 2.24, p. 39). There is thus a definite interface between the droplet and the liquid in which it floats. In a sense, the shell of oriented water molecules forms a membrane around the droplet.

Coacervate droplets have a marked tendency to adsorb and incorporate various wholly or partially hydrophilic substances from the surrounding solution. In this way the droplets grow at the expense of the surrounding liquid. Also as a result of hydrophilic and ionic interactions, coacervate droplets have a strong tendency to form an internal structure—that is, the molecules within the droplet tend to become arranged in an orderly manner. As more and more different materials are incorporated into the droplet, a membrane consisting of surface-active substances (analogous to phospholipids) may form just inside the shell of oriented water molecules. Thus, though coacervate droplets are not alive in the usual sense of the word, in protecting their internal chemistry from the surrounding environment, they exhibit many properties ordinarily associated with living organisms. In fact, the droplets look so much like organisms that experienced biologists have on occasion mistaken them for bacteria and attempted to assign them to species.

Fox envisions the prebiological systems (prebionts) that led to development of the first cells as ***proteinoid microspheres*** rather than coacervate droplets. Fox's microspheres are droplets that form spontaneously when hot aqueous solutions of polypeptides are cooled. The microspheres exhibit many properties characteris-

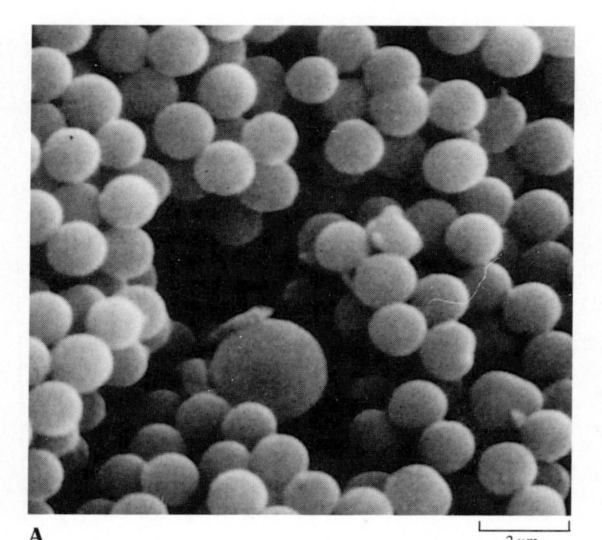

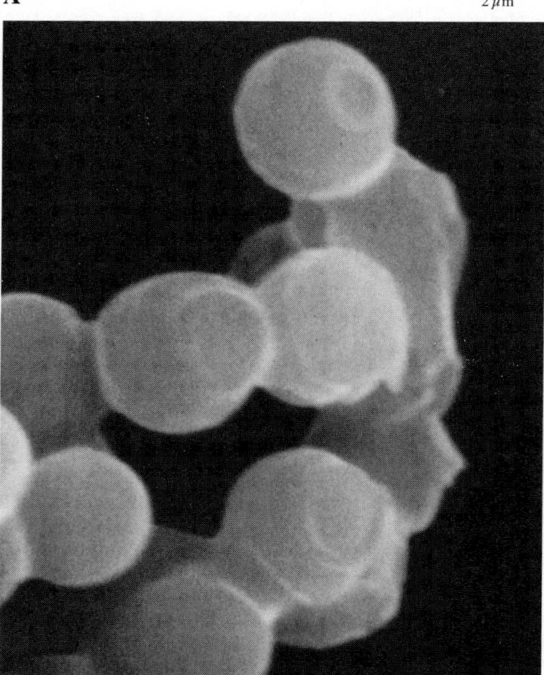

19.6 Proteinoid microspheres

(A) The spheres seen in this scanning electron micrograph are remarkably uniform in size, though a few are as much as 4 times larger than the others. (B) At higher magnification, connecting bridges between the spheres can be seen. The scars on the surface of some of the spheres indicate locations where bridges have been broken.

tic of cells: swelling in a hypotonic medium and shrinking in a hypertonic one; formation of a double-layered outer boundary; internal movement; growth in size and increase in complexity; budding in a manner superficially similar to the method of reproduction seen in yeasts; and a tendency to aggregate in clusters resembling those seen in many bacteria (Fig. 19.6). When these microspheres are illuminated with sunlight, they develop an electrical potential across their membranes not unlike the potential seen in essentially all living cells.

Since either type of droplet—complex coacervate or proteinoid microsphere—is structurally organized and sharply separated from the external medium, the chemical reactions that take place within the droplet depend not only on the conditions of the medium, but also on the organization of the droplet itself. Since various substances may be more concentrated in the droplet, the probability of chemical reactions is increased; because of the organization within the droplet, each reaction will influence other reactions in ways that are most unlikely when the substances are free in the external medium. Furthermore, catalytic activity of both inorganic substances like metallic compounds and organic ones like proteins is enhanced by the regular spatial arrangement of molecules within the droplet. In short, the special conditions within the droplet will exert selective and regulative influence over the chemical reactions taking place there.

Vast numbers of different prebiological systems of this kind may have arisen on the early earth. Most would probably have been too unstable to last long, but a few might have been more stable: some proteinoid microspheres have survived in the laboratory for more than 6 years, long enough for countless chemical reactions to occur, for many "generations" of growth and budding to take place, and for natural selection to operate. Some of the early droplets must have contained particularly favorable combinations of materials, especially complexes with catalytic activity, and may thus have developed harmonious interactions between the reactions occurring within them. As such droplets increased in size, they would have been more susceptible to fragmentation, which would have produced new smaller droplets with composition and properties essentially similar to those of the original droplet. These, in turn, would have grown and fragmented again. This primitive reproduction would not initially have been under the control of nucleic acids, even though these compounds could have been synthesized under abiotic conditions and may well have been incorporated into some of the prebionts.

The essential next step in this scenario is the evolution of organic catalysts and the ability to make copies of them for "offspring." The discovery of the RNA-based enzymes—ribozymes—that catalyze intron removal (Fig. 19.7) has led to the suggestion that RNA might originally have served this dual role as both enzyme and genetic library. In fact, a synthetic ribozyme has been created that can slowly replicate itself; moreover, one subunit of a naturally occurring three-RNA intron-splicing complex is also able to produce copies of itself. Chemical considerations make it likely that the first catalyst/gene combinations were only RNA analogues; a pure RNA-based system would probably have come sec-

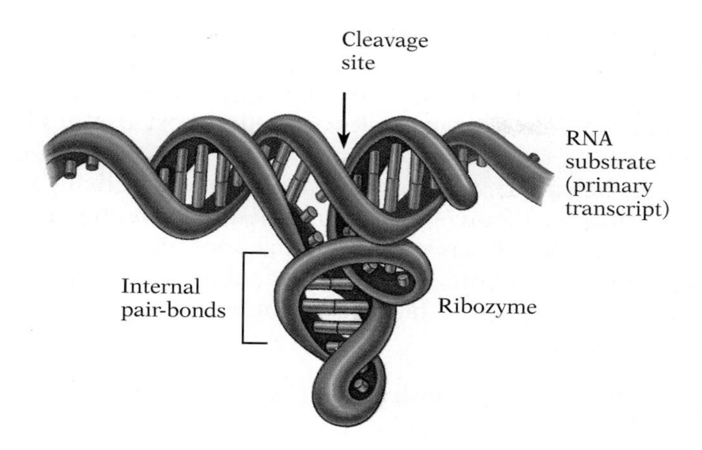

19.7 A ribozyme

The 39-nucleotide ribozymes that aid in intron editing bind to the target site on the primary transcript and then cleave a particular bond.

ond, supplanting its less efficient predecessor. Intense natural selection would have operated to cull the less-fit replication and metabolic-control ribozymes, and to select among the many variants that the high mutation rates of these primitive, repairless systems would inevitably have generated.

The next step—and the most difficult one for theorists—is the evolution of translation. Though the selective advantage to living systems of using proteins rather than RNA as enzymes and structural elements is clear, the gap between the processes of transcription and translation appears enormous. Just how a correlation between nucleotide sequences in nucleic acids and amino acid sequences in proteins could have arisen remains a mystery. There are, however, some suggestive leads. The first is that the genetic code is almost universal; with minor exceptions, the codons have the same amino acid translations in all organisms, from microorganisms to humans. This could mean that the code evolved only once and is arbitrary.

On the other hand, the fact that so little variation has arisen during the whole course of cellular existence leads one to believe that mere happenstance is unlikely to have dictated which codon would go with which amino acid. For example, a careful look at the dictionary of the genetic code (Table 9.1, p. 242) shows that all codons in which the second letter is U (uracil) code for hydrophobic amino acids, and that all five electrically charged amino acids (see Fig. 3.19, p. 59) have codons in which the middle letter is a purine (A or G). In short, some chemical system may underlie the pairing of codons and amino acids. Though the chemical rationale, if any, is unknown, the presumption is that, as it gradually came into being, it made possible both more accurate duplication during the reproductive process and more precise control over the chemical reactions taking place within the droplets.

The next step must have been that a small percentage of prebionts with particularly favorable characteristics slowly developed into the first primitive cells.

Not all biologists accept the sequence outlined here. Some think it more likely that the first "living" things were self-replicating macromolecules such as nucleic acids: in other words, "naked" genes. The first cells would then have arisen as these macromolecules slowly accumulated a shell of other substances (a primitive cytoplasm) around themselves. Another scenario imagines that life developed on the surfaces of clay particles. Clay has the ability to absorb certain molecules and weakly catalyze reactions. Selection would then have favored particles with the most useful chemical properties.

However they arose, these primitive precursors of living cells absorbed chemicals ("nutrients") from their environments, and at some point began converting these substances into the chemicals necessary for replication. As entities capable of "eating" and "reproducing," they began to fit at least some of the criteria that define living things.

Complex biochemical pathways The earliest organisms arose well over 3.5 billion years ago. They were heterotrophs using carbohydrates, amino acids, and other organic compounds that were free in the environment as nutrients. In other words, they depended on previous abiotic synthesis of organic compounds. However, as the organisms became more abundant and more efficient at removing preformed nutrients from the medium, they must have begun to deplete the supply of nutrients. The rate of spontaneous formation of organic matter must have been low. This supply was probably being used up at an ever-increasing rate, which must have led to increasing competition among organisms. Forms inefficient at obtaining nutrients doubtless perished, while the more efficient survived in greater numbers. Natural selection would have favored any new mutation that enhanced the ability of its possessor to obtain, process, or synthesize food.

At first, the primitive organisms probably carried out relatively few complex biochemical transformations. They could obtain most of the materials they needed ready-made. Yet these very materials—which could have been used directly with little alteration—would have dwindled most rapidly. Hence there would have been strong selection for any organisms that could use alternative nutrients. Suppose, for example, that a compound A that was necessary for the life of cells had been available in the medium initially, but its supply was being rapidly exhausted. Suppose some cells possessed a gene that coded for an enzyme a that catalyzed synthesis of A from another compound, B, which was in greater supply in the medium. Those cells would then have had a marked adaptive advantage over cells that lacked the gene. They could survive even when A was no longer available by carrying out the reaction

$$B \xrightarrow{a} A$$

However, now there would have been increasing demand for free B, and the rate of its utilization would soon have exceeded the rate of its abiotic synthesis. Thus the supply of B would have dwindled, and there would have been strong selection for any cells possessing a second gene that coded for enzyme b, catalyzing synthesis of B from C. These cells would not be dependent on a free supply of either A or B, because they could make both A and B for themselves as long as they could obtain enough C:

$$C \xrightarrow{b} B \xrightarrow{a} A$$

This general process of evolution of synthetic ability might have continued until most cells made all the complex compounds they required by carrying out a long chain of chemical reactions:

$$G \xrightarrow{f} F \xrightarrow{e} E \xrightarrow{d} D \xrightarrow{c} C \xrightarrow{b} B \xrightarrow{a} A$$

In this way, the primitive cells would slowly have evolved more elaborate biochemical capabilities.

It seems most unlikely that much synthetic ability could have arisen unless the cells possessed some mechanism for handling the chemical energy released from such catabolic reactions as splitting molecular hydrogen or hydrolyzing organic compounds. Since all living things, from bacteria to humans, employ ATP as their principal energy currency, the use of ATP was probably an early evolutionary development. In fact, ATP must have been available to early organisms. The two organic precursors of ATP are adenine and the five-carbon sugar ribose (Fig. 19.8). Both of these com-

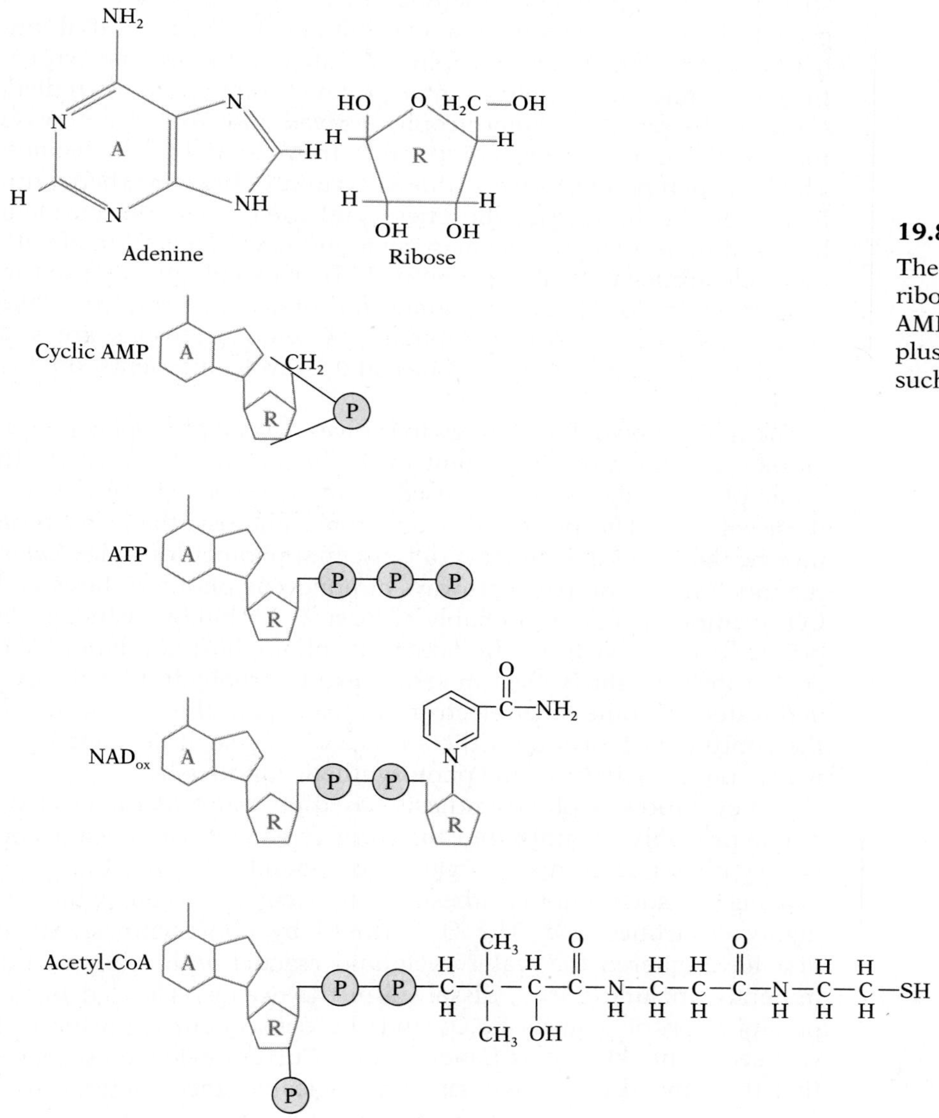

19.8 Adenine and its derivatives

The nitrogenous base adenine and the sugar ribose combine to form the structural basis for AMP, ATP, NAD, and acetyl-CoA (all shown here), plus other important biological intermediates such as FAD and NADP.

pounds also occur in nucleic acids, and both are synthesized abiotically under presumed prebiological conditions. Indeed, of the five nitrogenous bases that occur in DNA and RNA, adenine is the one that forms most easily. It is probably no accident, therefore, that adenine is the base found in a host of biologically critical compounds—not only ATP (and ADP, AMP, and cAMP), but also the electron carriers NAD, NADP, and FAD (Fig. 19.8). ATP would probably have been available for coupling exergonic and endergonic reactions as soon as the primitive cells could perform the reactions.

The earliest form of metabolism using ATP, which probably arose about 3.5 billion years ago, is likely to have been fermentation, the anaerobic process universal in living organisms today.

Evolution of autotrophy Primitive heterotrophs probably evolved increasingly elaborate biochemical pathways that enabled them to use a greater variety of the organic compounds that were free in their environment. Some of these organisms probably also evolved such purely heterotrophic methods of feeding as parasitism and predation. Nevertheless life would eventually have ceased if all nutrition had remained heterotrophic. Yet life did not become extinct. Instead, while the supply of free organic compounds dwindled, some of the primitive heterotrophs evolved autotrophic pathways that synthesized organic compounds from inorganic molecules. The first such pathways were almost certainly chemosynthetic, utilizing for the most part the energy released when the covalent bonds of molecular hydrogen are split and new bonds formed with more electronegative atoms (see p. 155). This energy was used to synthesize the organic compounds no longer available from the "soup," as well as novel compounds. Chemoautotrophs are still found today, particularly in bogs and at volcanic vents on the ocean floor.

The next autotrophic strategy to evolve—cyclic photophosphorylation—was far more important to the history of life on earth. In cyclic photosynthesis, light is used as an energy source in the synthesis of ATP. The present-day anaerobic photosynthetic bacteria may be the direct descendants of those first photoautotrophs. Later, the much more complex pathways of noncyclic photosynthesis and CO_2 fixation appeared, probably at least 2–2.5 billion years ago in primitive cyanobacteria. In noncyclic photophosphorylation, energy from sunlight is used in synthesis of carbohydrate from CO_2 and water (or some other electron source). From this time onward, the continuation of life on the surface of the earth depended primarily on the activity of the photosynthetic autotrophs.

The evolution of photosynthesis based on water as the electron donor probably administered the *coup de grace* to significant abiotic synthesis of complex organic compounds. An important by-product of such photosynthesis is molecular oxygen, which is highly electronegative. The O_2 released by photosynthesis must first have entered the water cycle and reacted with many of the minerals, including iron, dissolved in the oceans. This led to the precipitation of iron as Fe_3O_4 and the consequent deposition of vast sediments known as banded iron. (Current evidence suggests that the banded iron was formed biologically, since bacteria used iron as an an electron receptor in their metabolism.) With dissolved iron removed from the water, free oxygen would have been able to accumulate in the atmosphere. Once the layer of ozone (O_3) now present high in the atmosphere had been formed by some of the O_2, this layer effectively screened out most of the UV radiation from the sun and allowed very little high-energy radiation to reach the earth's surface. Thus living organisms changed their environment in a way that destroyed the conditions that had made the origin of life possible; they caused what has sometimes been called an oxygen revolution.

Once molecular oxygen became a major component of the atmosphere, both heterotrophic and autotrophic organisms could use the biochemical pathways of aerobic respiration, by which far more energy can be extracted from nutrient molecules than by fermentation alone.[1] The progressive increase in atmospheric O_2 and the formation of the ozone shield were probably the major factors in permitting organisms to leave the UV-absorbing oceans and to move onto the land. Unicellular organisms were on the land by at least 1.2 billion years ago.

PRECAMBRIAN EVOLUTION

■ WHAT FOSSILS TELL US

The oldest known fossils (as of this writing), which are from deposits in Western Australia, date back at least 3.6 billion years (Fig. 19.9). They appear to be bacteria. It was probably the evolution of cyanobacteria from these organisms between 2.5 and 3.5 billion years ago that initiated the oxygen revolution. The cyanobacteria, also called blue-green algae, are like true plants but unlike other photosynthetic bacteria in that they use water as the electron source in noncyclic photophosphorylation, and hence release molecular oxygen as a by-product; they can also fix nitrogen, which may be what allowed them to begin growing on land at least 1.2 billion years ago.

More and more fossils of procaryotic organisms from the first 2.5 billion years of life are being found, and some fossils of eucaryotic algae are thought to date back at least 2.1 billion years. The oldest geological period from which fossils of higher forms of life are fairly abundant is the ***Cambrian***, which began about 550 million years ago (Table 19.1). Many of the Cambrian fossils are of relatively complex organisms—most of the animal phyla extant today are represented. The few Precambrian fossils of eucaryotic organisms are mostly of simple algae, protists, and possibly a few invertebrates, whose relationships are poorly understood.

[1] As detailed in Chapters 6 and 7, the electron-transport chains of aerobic respiration, cyclic photophosphorylation, and noncyclic photophosphorylation are closely related. Which process evolved first is not known, but some elements of the respiratory electron-transport chain probably arose prior to the photosynthetic pathways: the ability to use the small quantities of electron acceptors—nitrates, sulfates, carbonates, and even the minute quantities of free oxygen—would have given heterotrophs with an electron-transport chain enormous advantages over obligate fermenters.

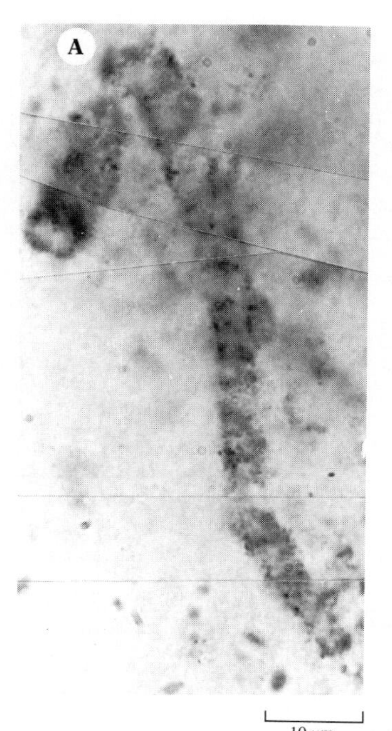

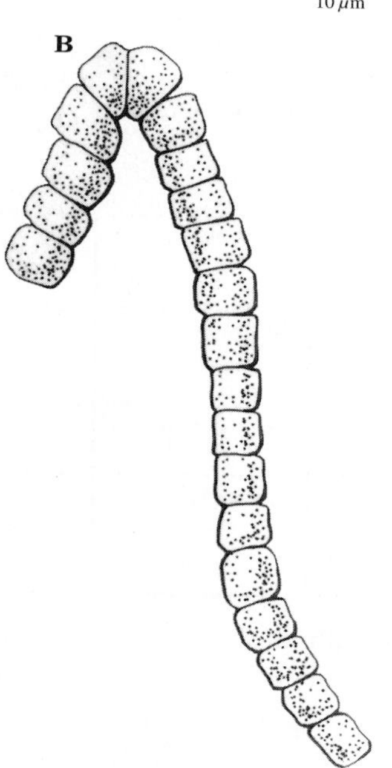

19.9 Filamentous procaryotic microfossil from Western Australia

The bacteriumlike organism shown in the photograph (A) and corresponding drawing (B) lived about 3.6 billion years ago.

Table 19.1 The geologic time scale

ERA	PERIOD	EPOCH	MILLIONS OF YEARS (APPROX.) FROM START OF PERIOD TO PRESENT	PLANT LIFE	ANIMAL LIFE
CENOZOIC	Quaternary	Recent Pleistocene	0.01 2	Increase in numer of herbs[a]	Rise of civilizations First humans
	Tertiary	Pliocene Miocene Oligocene Eocene Paleocene	5 25 40 55 65	Dominance of land by angiosperms[b]	First hominines Dominance of land by mammals, birds, and insects
			BUILDING OF ANCESTRAL ROCKY MOUNTAINS		
MESOZOIC	Cretaceous		140	Angiosperms[b] expand as gymnosperms[c] decline	Last of the dinosaurs; second great radiation of insects
	Jurassic		210	Angiosperms arise; gymnosperms[c] still dominant	Dinosaurs abundant; first birds
	Triassic		250	Dominance of land by gymnosperms[c]	First mammals First dinosaurs
			BUILDING OF ANCESTRAL APPALACHIAN MOUNTAINS		
PALEOZOIC	Permian		280	Precipitous decline of primitive vascular plants	Great expansion of reptiles; decline of amphibians; last of the trilobites
	Carboniferous[d]		360	Great coal forests, dominated at first by large but primitive vascular plants, and later also by ferns and gymnosperms[c]	Age of Amphibians; first reptiles; first great radiation of insects
	Devonian		410	Expansion of primitive vascular plants; origin of first seed plants toward end of period	Age of Fishes; first amphibians and insects
	Silurian		440	Invasion of land by the first vascular plants toward end of period	Fish radiate
	Ordovician		500	Marine algae[e] abundant	First vertebrates; invasion of land by a few arthropods
	Cambrian		550	Marine algae[e] and photosynthetic bacteria (especially cyanobacteria)	Marine invertebrates abundant (including representatives of most phyla)
			INTERVAL OF GREAT EROSION		
PRECAMBRIAN				Primitive marine life	

[a]Perennial plants whose aboveground growth dies back in winter or drought. [b]Flowering plants. [c]Conifers and their relatives. [d]In North America, the Lower Carboniferous is often called the Mississippi period, and the Upper Carboniferous is called the Pennsylvanian period. [e]Aquatic plants lacking any vascular system.

It is not clear why there are so few Precambrian fossils. Most Precambrian organisms were soft-bodied and hence did not readily form fossils; it is the hard parts of plants and animals that are most often fossilized, because these are the most likely to resist decay and become buried. It's possible that the many hard-bodied animals present at the start of the Cambrian arose suddenly; alternatively, they may have descended from Precambrian ancestors that gradually developed shells, exoskeletons, or other hard parts, possibly in response to predation. A few beds of Precambrian fossils, notably in Australia, suggest the existence of hard-bodied in-

19.10 Fossil of a Precambrian animal (*Dickinsonia costata*) from South Australia

This animal lived some 600 million years ago. Reproduced actual size.

vertebrates before the Cambrian (Fig. 19.10), but many experts classify these organisms as multicellular protists unrelated to later animals. In any event, it is possible that the ranges of these hard-bodied organisms were restricted to areas with environmental conditions that made fossilization unlikely. Furthermore, a high percentage of any fossils formed as long ago as the Precambrian is likely to have been destroyed before there were humans to study them.

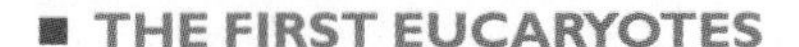

■ THE FIRST EUCARYOTES

The absence of cytological detail in fossils from the Precambrian era leaves us with practically no direct evidence concerning the evolution of the first eucaryotic cells.

As discussed in Chapter 5, there is growing anatomical and molecular evidence that certain subcellular organelles of eucaryotes evolved from procaryotes (Fig. 19.11). For example, chloroplasts and mitochondria are self-replicating bodies (Figs. 19.12, 19.13) that contain genetic material and carry out protein synthesis on ribosomes of their own making (see p. 149). They are about the size of procaryotes (1–10 μm), lack nuclear membranes, and have a single, usually circular chromosome that is not packaged on nucleosomes (Fig. 19.14). They are also like procaryotes in that their

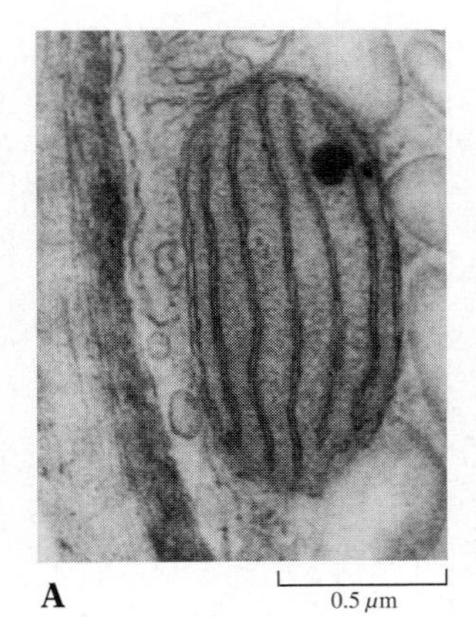

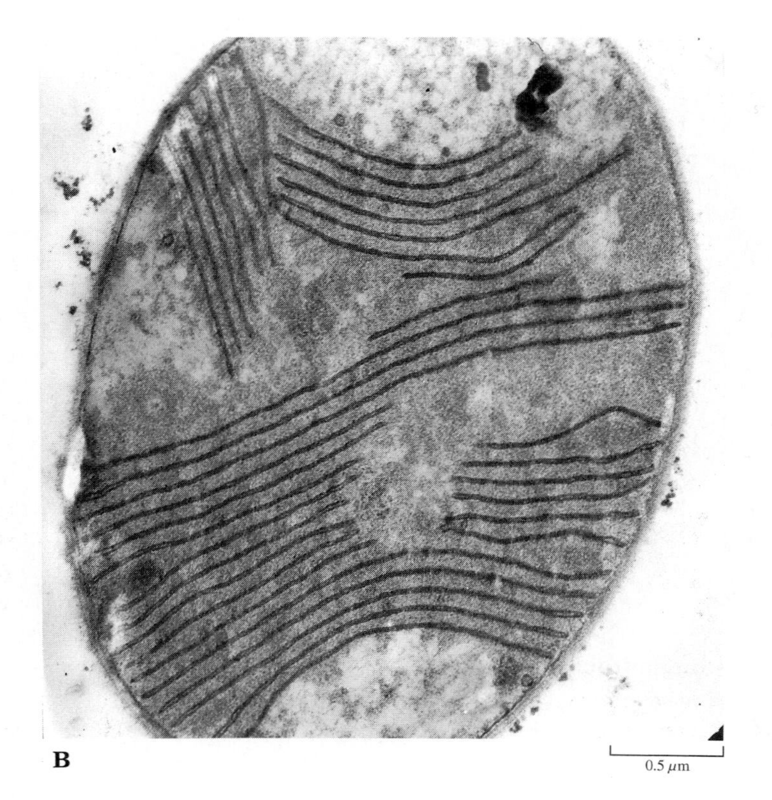

19.11 Chloroplast of a red alga (A) compared with a whole cyanobacterium (B)

(A) The photosynthetic lamellae in the chloroplasts of red algae are not arranged in the stacks (grana) that characterize the chloroplasts of most plants. (B) In this respect a red algal chloroplast resembles an entire cyanobacterium, whose lamellae also show no granum arrangement but are free in the cytoplasm, for there is no plastid membrane.

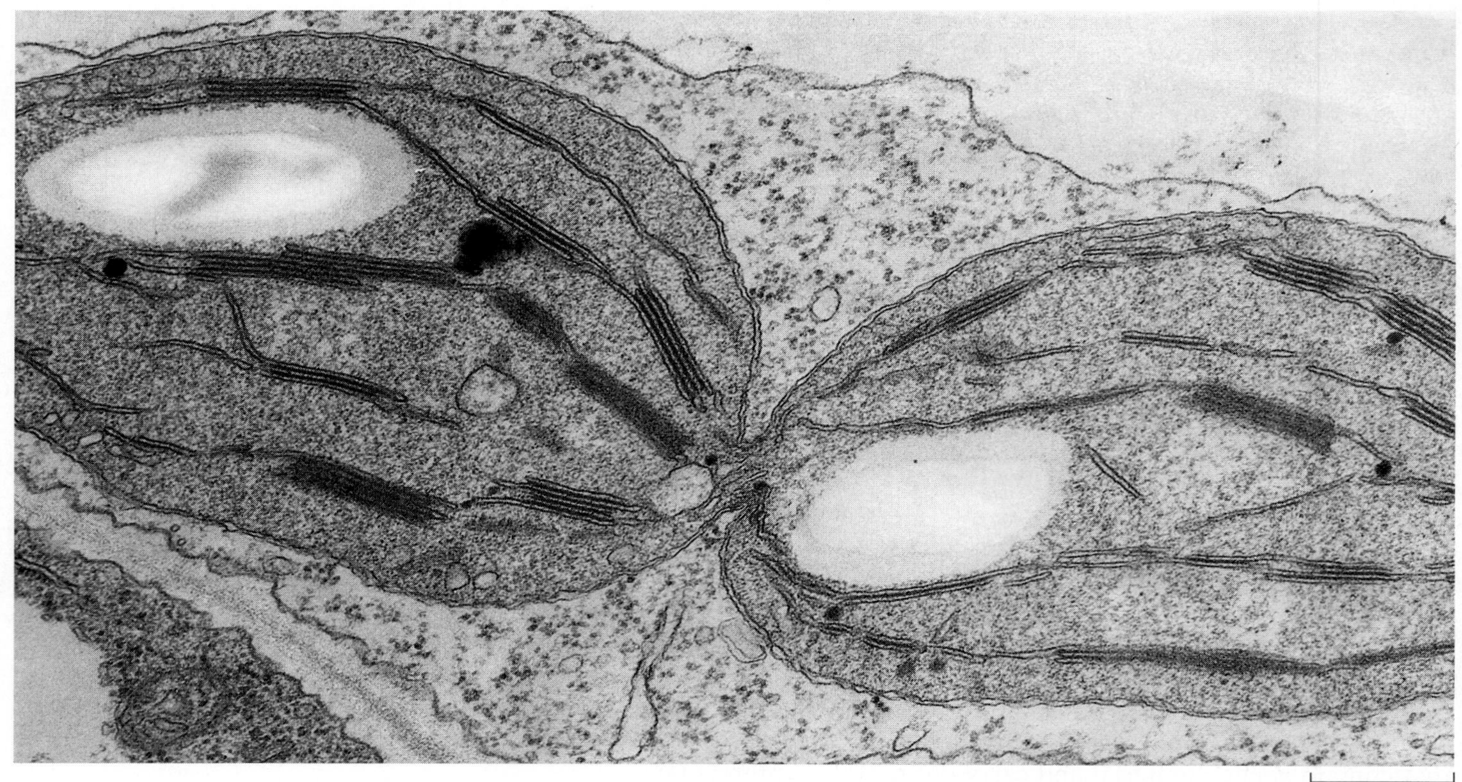

19.12 A dividing chloroplast in a tobacco leaf

The large white areas in each of the daughter chloroplasts shown in this electron micrograph are starch granules.

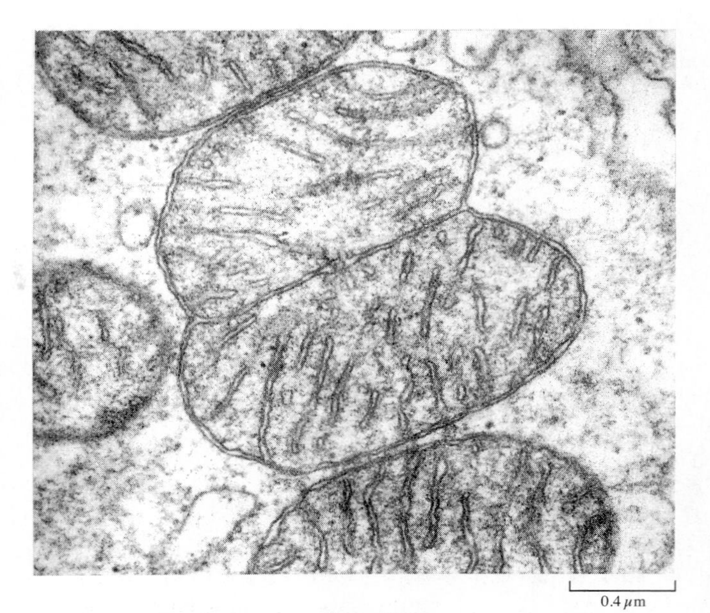

19.13 A dividing mitochondrion

The partition between the two daughter mitochondria seen in this electron micrograph is nearly complete.

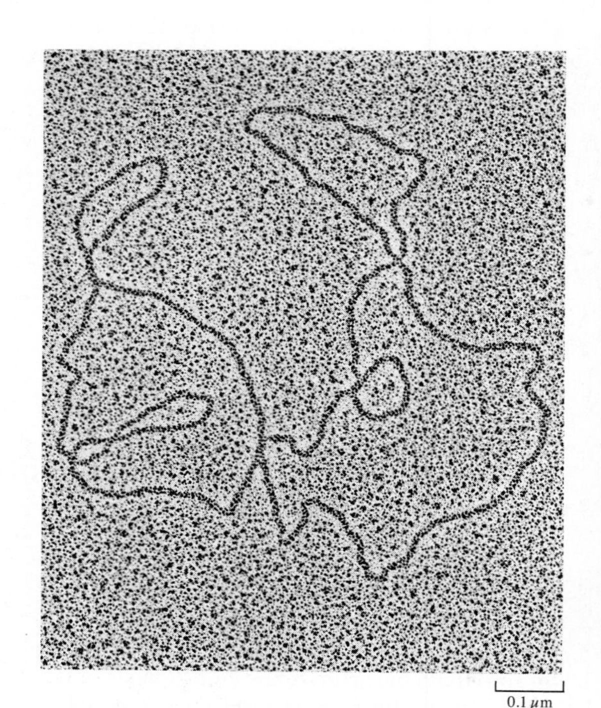

19.14 Mitochondrial chromosome

This electron micrograph shows the circular chromosome of a mitochondrion, from an oocyte of a frog, genus *Xenopus*.

ribosomes are small (even interchangeable with procaryotic ribosomes in some cases), and they use the same control-sequence codes and initial amino acid as most bacteria. Moreover, gene expression in both procaryotes and organelles is controlled primarily by negative control strategies; eucaryotes usually employ positive control. The RNA and DNA polymerases of procaryotes and organelles are similar, but qualitatively different from those of eucaryotes. Procaryotes and organelles both accomplish protein synthesis, metabolism, and internal transport without the ER (endoplasmic reticulum), Golgi bodies, lysosomes, microtubules, and subcellular organelles typical of eucaryotes.

Present-day eucaryotes, therefore, may have arisen from precursors, now extinct, that had developed a symbiotic relationship with certain procaryotes. The capacity of unicellular organisms to live as endosymbionts inside the cells of other organisms is illustrated by present-day examples. A variety of protozoans contain endosymbiotic single-celled algae, and the gastrodermal (stomach-lining) cells of Chlorohydra and of several species of corals and sea anemones also contain algal cells (see Fig. 40.42, p. 1191). Some heterotrophic protists inhabit the guts of termites, and harbor bacteria that digest the cellulose the termites eat; neither the termites nor the protists can live without them (Fig. 19.15). Several species of molluscs have intracellular algal symbionts, and intracellular symbiotic bacteria occur widely in both the plant and animal kingdoms. Some parasitic bacteria live in animal hosts and are passed from generation to generation in the cytoplasm of egg cells.

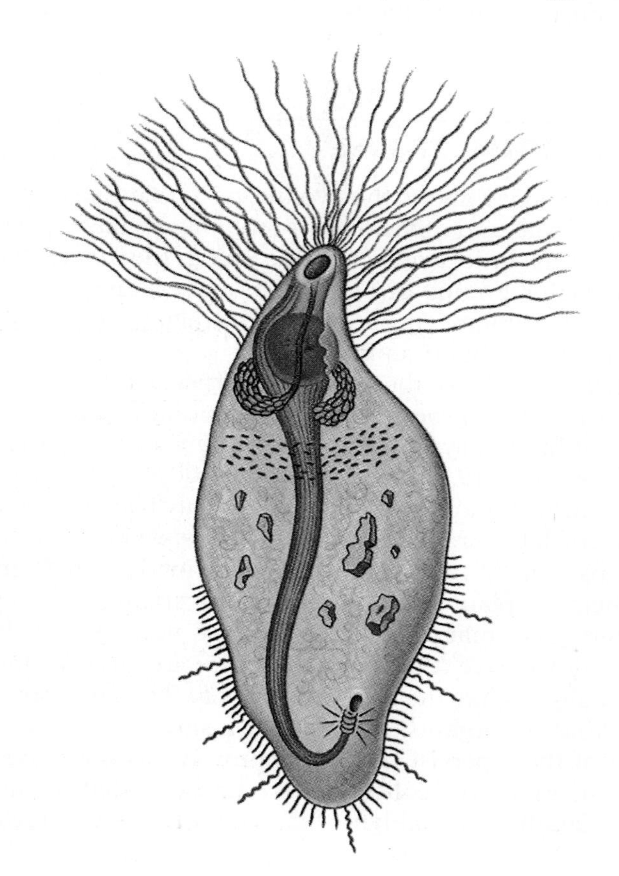

19.15 Endosymbiosis in the termite gut

The single-celled intestinal protist shown here lives in the termite gut, and harbors within it a bacterium capable of digesting the cellulose in the termite's food. Neither the protist nor the termite could survive without the presence of the bacterium. Spirochete bacteria attached to the cell membrane generate currents that bring food to the protist and enable it to move (see also Fig. 20.19, p. 565).

Though both anaerobic photosynthetic bacteria and cyanobacteria probably arose long before the first eucaryotic cells, the cyanobacteria seem a more likely progenitor of most chloroplasts, since they have the same chlorophyll *a* and use the same noncyclic photophosphorylation pathways. Whether chloroplasts have a single origin or multiple origins is hotly debated. Mitochondria are derived from aerobic bacteria. The inner mitochondrial membrane is biochemically similar to the bacterial plasma membrane, particularly in its capacity for electron transport.

Though these organelles contain DNA and divide, grow, and differentiate partly on their own, they are not autonomous entities. Most of their proteins are specified by nuclear genes. Presumably the symbionts gave up much of their genetic control to the host hundreds of millions of years ago.

In contrast to the wealth of data and speculation on chloroplasts and mitochondria, there is little conjecture about the origin of the nuclear membrane and the associated endoplasmic reticulum. Since there is membrane flow in eucaryotic cells from the nuclear membrane through the ER to the Golgi apparatus (see Fig. 5.13, p. 130), the nuclear membrane may have been the first part of the intracellular membrane system to evolve, with the other parts arising from it later. How the nuclear membrane itself arose is still a mystery. As discussed in Chapter 21, archezoans—a kind of missing link between bacteria and protozoans—do have a nuclear membrane and basal bodies; the Golgi apparatus and mitochondria, which they lack, are likely to have come later, with chloroplasts being the most recent additions.

THE KINGDOMS OF LIFE

Until recently, ideas about the evolutionary relationships between the major phyla and divisions of organisms have been rather vague. There is little evidence concerning the relationships between the major groups of algae. It is uncertain how some protists—the unicellular eucaryotes—are related to multicellular plants or to multicellular animals.

Nevertheless, at least the broad patterns of evolution are now clear, and the logical groupings and evolutionary relationships that any artificial human system of classification seeks are more tangible than ever (Fig. 19.16). The genealogy of life resembles a tree: few of the earliest branches survive, primarily because most die "in the shade" of later branches—that is, competition from more recent groups can drive earlier ones to extinction. This imperfect analogy helps explain some otherwise surprising features of the evolutionary tree: many major groups are now extinct; the older groups tend to have the fewest remaining species, and newer groups usually display the greatest diversity. The once-popular idea that new kinds of organisms appear at a slow and steady rate and that most of them persist (in ever-improved forms) to the present, is simply incorrect; the collection of organisms that present themselves for classification is highly biased toward newer phyla.

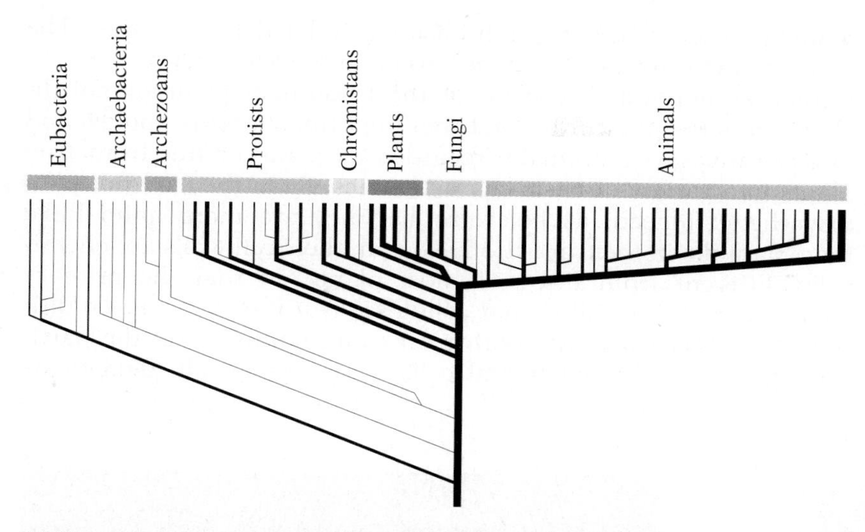

19.16 Schematic view of the evolution of major phyla

The basic outlines of the evolutionary history of modern phyla are indicated in this diagram. Extinctions create a bias toward more recently evolved diversity. The kingdom groupings used in this text are indicated at the top. Line thickness suggests the approximate number of species in each phylum. The details of the branching patterns are simplified and highly tentative; in particular, the branches leading to Archaebacteria and Archezoa may each have two origins rather than one.

One of the oldest attempts to fit nature into a consistent and convenient classification system recognizes only two kingdoms:—one for plants (formal name Plantae) and the other for animals (Animalia). This dichotomy works well as long as the organisms to be classified are the generally familiar ones. Things become a bit more difficult when the organisms in question are bread molds or sponges, which don't fit the intuitive concept of *plant* or *animal*.

With the advent of electron microscopy it became clear that bacteria differ from all other forms of life in a very fundamental way: they are procaryotic, whereas all other organisms are eucaryotic. Though bacteria had traditionally been assigned to the plant kingdom, researchers began to set them apart as a separate kingdom. The techniques of molecular biology, however, have provided data that force a yet–more–radical revision of our ideas about bacteria.

Evolutionary relationships have traditionally been inferred from structural, chemical, and developmental similarities. It is now also possible to compare directly the nucleotide sequences of genes. Since all organisms (except viruses) have ribosomes, the sequences encoding ribosomal RNA are used most often in interkingdom comparisons. In theory at least, the greater the difference between the rRNA of two species, the longer it has been since they shared a common ancestor. As with other sorts of phylogenetic comparisons, experience has revealed many problems with a simplistic application of sequence comparisons, but the technique has nevertheless become a powerful aid in reconstructing the course of evolution. On this basis, it is now clear that there are actually two major groups of bacteria only distantly related to the other. The two kingdoms are the Archaebacteria and the Eubacteria.

Another important revision of the traditional plant-animal dichotomy is the realization that the fungi (mushrooms, molds, and their relatives), traditionally regarded as plants, differ from true plants in a variety of basic characteristics. First, they are not photosynthetic; indeed, as heterotrophs, they are more animal-like than plantlike in their nutrition. Second, their cell walls are chemically different from those of most plants, in that the primary component of the walls is not cellulose. Third, they are not multicellular in the sense that both plants and animals are; the partitions between adjacent fungal cells, if present at all, tend to be

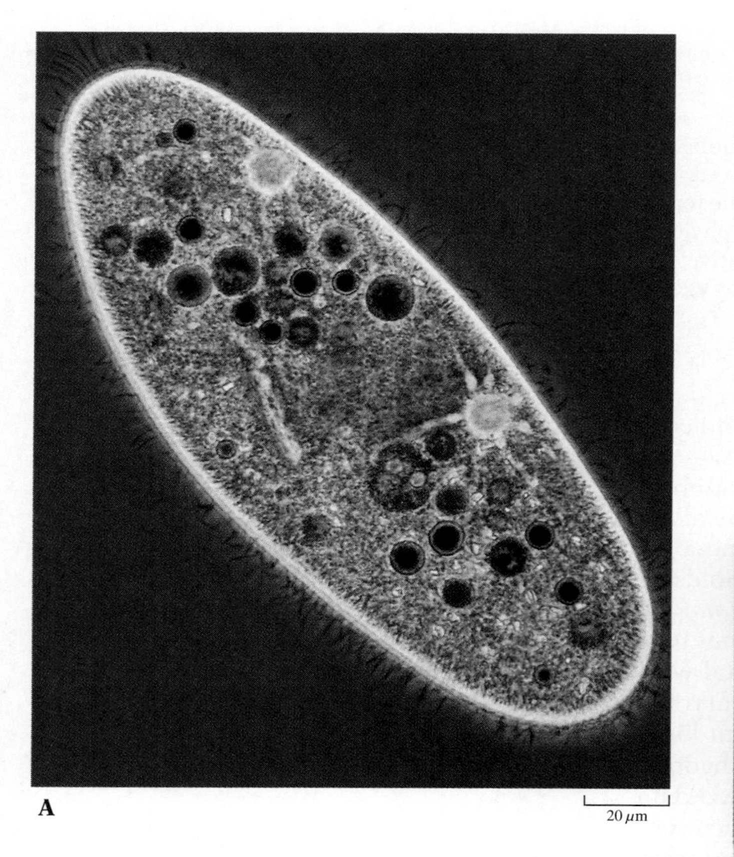

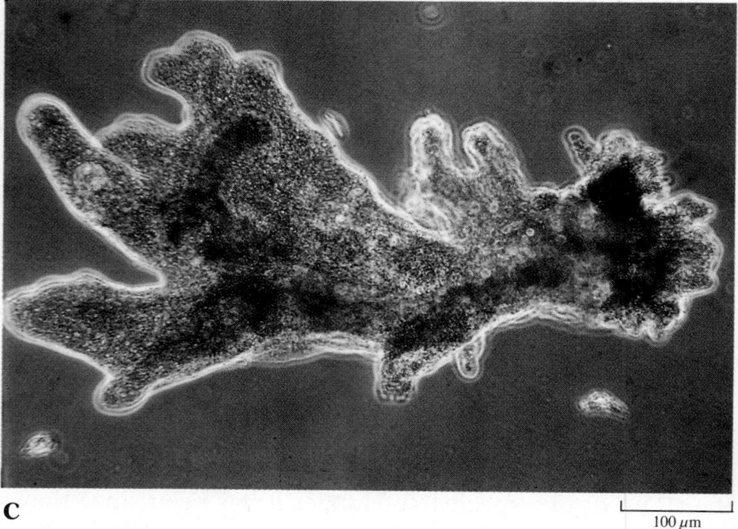

19.17 Protist diversity

Paramecium (A) has cilia and is heterotrophic, whereas *Euglena* (B) has flagella and chloroplasts. Other protists (C) move by means of amoeboid changes in shape.

incomplete and the cytoplasm is continuous. In short, the fungi represent a separate evolutionary development, and comparisons of DNA and RNA nucleotide sequences fully confirm this conclusion. They are thus placed in a separate kingdom: Fungi.

Recognizing the kingdom Fungi and two kingdoms of bacteria resolves two serious anomalies in the older classifications. But another major problem remains: the classification of eucaryotic unicellular organisms. The ones zoologists have traditionally called protozoans have been particularly troublesome, especially the group of protozoans known as flagellates. These creatures have long flagella that enable them to swim actively in a manner intuitively felt to be "animal-like." Yet some of them, such as *Euglena* (Fig. 19.17B), have chlorophyll and carry out photosynthesis. How can organisms like these be classified? One possibility would be to rule arbitrarily that all those with chlorophyll are to be considered plants, and all those lacking chlorophyll to be animals. However, some photosynthetic euglenoids are clearly very closely related to colorless species; yet the proposed rule would put the green ones in a different kingdom from their colorless close relatives. Furthermore, some species of euglenoids include both green and colorless subspecies; the suggested rule is unable to deal effectively with such species, and violates the whole purpose of modern taxonomy, which is to codify evolutionary relationships.

Whatever the criteria chosen, it is impossible to make a clean separation between plants and animals at the unicellular level. The reason is obvious. Unicellular organisms (and some multicellular ones relatives) are not part of the evolutionary lineage that led to plants and animals. How these protists should be classified to best capture their phylogenetic affinities is a contentious issue. Should they be grouped together as one kingdom, or divided into several? If they are to be separated, what groupings most accurately reflect their evolutionary history?

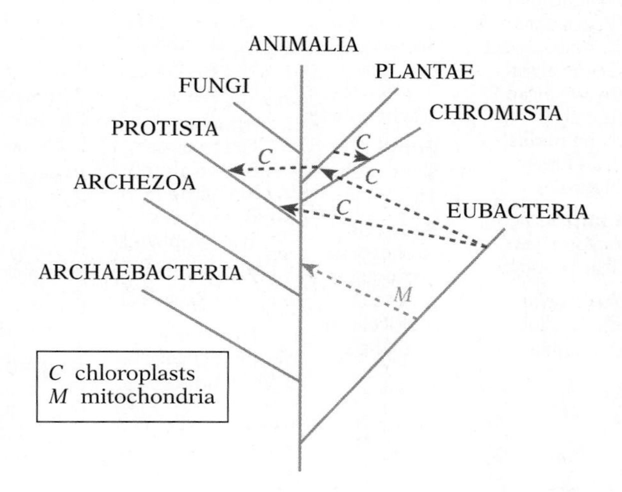

19.18 A simplified classification showing possible endosymbiotic events

The protists diverged from the rest of the eucaryotes before the divergence of plants, animals, and fungi. The tree also assigns origins (indicated by dashed lines) for eucaryotic mitochondria and chloroplasts. There is no consensus about how the various groups of photosynthetic protists obtained their chloroplasts; the most likely scenario is shown here.

For one group of protists, the appropriate grouping is obvious. Cytological evidence, now backed up by sequence analysis, makes it clear that three kinds of single-celled eucaryotes—the archaeamoebae, the metamonads, and the microsporidians—are radically unlike any other kind of organism. Though they have true nuclei, they lack mitochondria, peroxisomes, and a Golgi apparatus. They probably diverged from the rest of the eucaryotes very early, and are increasingly placed in a kingdom called Archezoa, a practice we will follow.

For the other unicellular (and primitively multicellular) eucaryotes, an alternative that has won much support is to assign them to a separate kingdom, designated Protista (or Protoctista); (Fig. 19.18). As Table 19.2 shows, the classification of protists varies greatly among the many different kingdom classifications encountered in textbooks today. Any particular set of boundaries for Protista gets around some of the difficulties inherent in the two-kingdom system, but creates other problems. As detailed in Chapter 21, the protists probably comprise six distinct kingdoms.

For the brown algae (and the other eucaryotes with chloroplasts located within the rough ER and containing chlorophyll *c*) there is an increasing tendency to put them into a separate kingdom called Chromista, and so resolve a major difficulty. We will adopt this practice here. (Chromista also include some nonphotosynthetic organisms, notably the pseudofungi.) There is also considerable support for a separate kingdom for red algae, or for reclassifying them as protists rather than plants; for now, pending more data, we will

Table 19.2 Various kingdom classifications

SYSTEM 1	SYSTEM 2	SYSTEM 3	SYSTEM 4	SYSTEM 5	SYSTEM 6	SYSTEM 7	SYSTEM 8
Plantae Bacteria Archezoans Euglenoids Chrysophytes Green algae Brown algae Red algae Slime molds True fungi Bryophytes Tracheophytes **Animalia** Protozoans Myxozoa Multicellular animals	**Monera** Bacteria **Plantae** Euglenoids Chrysophytes Green algae Brown algae Red algae Slime molds True fungi Bryophytes Tracheophytes **Animalia** Archezoans Protozoans Myxozoa Multicellular animals	**Protista** Bacteria Archezoans Protozoans Slime molds Myxozoa **Plantae** Euglenoids Chrysophytes Green algae Brown algae Red algae True fungi Bryophytes Tracheophytes **Animalia** Multicellular animals	**Protista**[a] Bacteria Protozoans Archezoans Euglenoids Chrysophytes Green algae Brown algae Red algae Slime molds True fungi Myxozoa **Plantae** Bryophytes Tracheophytes **Animalia** Multicellular animals	**Monera** Bacteria **Protista**[a] Archezoans Euglenoids Protozoans Chrystophytes Green algae Brown algae Red algae Slime molds True fungi Myxozoa **Plantae** Bryophytes Tracheophytes **Animalia** Multicellular animals	**Monera** Bacteria **Plantae** Euglenoids Chrysophytes Green algae Brown algae Red algae Bryophytes Tracheophytes **Fungi** Slime molds True fungi **Animalia** Archezoans Protozoans Myxozoa Multicellular animals	**Monera** Bacteria **Protista** Archezoans Protozoans Chrysophytes Slime molds Euglenoids Myxozoa **Plantae** Green algae Brown algae Red algae Bryophytes Tracheophytes **Fungi** True fungi **Animalia** Multicellular animals	**Archaebacteria** **Eubacteria** **Archezoa** **Protista** Protozoans Slime molds Euglenoids **Chromista** Chrysophytes Brown algae **Plantae** Red algae Green algae Bryophytes Tracheophytes **Fungi** True fungi **Animalia** Myxozoa Multicellular animals

[a]When multicellular groups are included, the Protista are sometimes rechristened Protoctista. Note that all kingdoms include at least a few one-celled species.

leave them with the green plants. Similarly, there is some support for a separate kingdom for slime molds, but our approach will be to place them in Protista until more data become available. (Some systems assign them to Fungi, but sequence analysis makes this affinity unlikely.) The eight-kingdom arrangement used in this book is summarized as System 8 in Table 19.2, and is shown diagrammatically in Figure 19.16.

As we said at the outset, classification systems are artificial human constructs created to serve the dual purposes of providing convenient groupings and indicating evolutionary relationships. If we look again at the pattern of evolution of the modern phyla (Fig. 19.19), we can see the inevitable bias toward the dominance of recently evolved diversity (inevitable since more-modern forms have in general survived at the expense of more-ancient ones, which are in consequence now extinct). This preponderance of recently evolved phyla leads most researchers to favor kingdom groupings that reflect current diversity rather than absolute evolutionary age. A few workers, however, favor a strict age-based approach to defining kingdoms; the consequences of this strategy are illustrated in Figure 19.19. Even the diversity-based approach is fraught with controversy and ambiguity (for example, how to classify sponges, red algae, and the separate groups of protists). These issues underscore the point that any classification system is tentative, and the ultimate test of its value is how well it serves the goal of providing convenient categories that are consistent with evolutionary relationships.

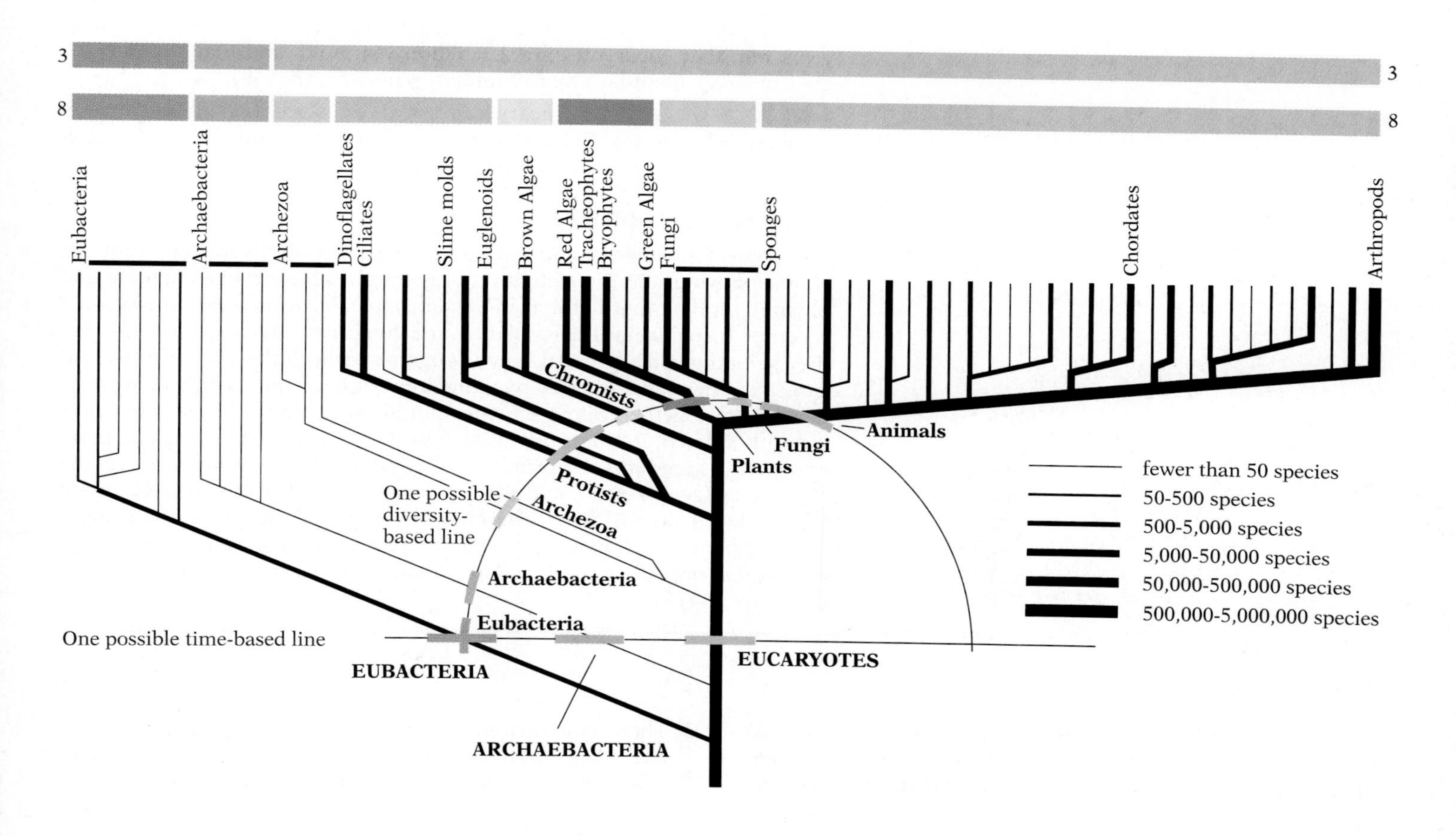

19.19 Diversity-based versus age-based classification

This simplified overview of the evolution of modern phyla illustrates two very different approaches to assigning phyla to kingdoms. If kingdoms are defined strictly by age, then a horizontal line can be drawn at any convenient point; in the case illustrated here, the line is drawn about a billion years ago, and divides the phyla into three kingdoms: eubacteria, archaebacteria, and eucaryotes. The colored bar next to the level "3" shows how the phyla are categorized in such a system. A more recent line might put the archezoans in a separate kingdom. If, as is more usual, kingdoms are defined by diversity, the imaginary line is more like an arc. The one shown here is just large enough to separate animals and fungi into separate kingdoms, which is the modern practice; a slightly larger arc would put sponges into a separate kingdom (which some workers advocate). Note the ambiguity about the status of red algae, and the clear division of protists into multiple kingdoms. The categorization of phyla that result from the eight-kingdom approach is shown by the colored bar next to the label "8."

CHAPTER SUMMARY

HOW DID LIFE BEGIN?

FORMATION OF THE EARTH The earth formed between 4.5 and 5 billion years ago with an atmosphere devoid of oxygen. (p. 529)

THE FIRST ORGANIC MOLECULES Organic molecules arrived in comets and meteorites, and were also synthesized on earth from inorganic precursors with heat, lightning, and UV radiation providing the necessary energy. (p. 530)

THE EARLIEST CELLS The earliest cells may have developed from spontaneously formed coacervate droplets or proteinoid microspheres, held together by the polar bonding of surrounding water in much the same way as modern cell membranes. The internal chemistry of these entities would have been buffered from the external medium, and could evolve. The original information-storage molecule was probably RNA or an analogue. DNA, protein-based enzymes, and the genetic code would have evolved later. Early organisms were chemoautotrophic and heterotrophic; the evolution of anaerobic respiration made it possible to store energy. Photoautotrophy, in the form of cyclic photosynthesis, provided a major new source of energy: light. Noncyclic photosynthesis evolved later; this more efficient process produces oxygen as a byproduct. Aerobic respiration, utilizing an electron-transport chain similar to that of photosynthesis, allowed respiration to be much more efficient when oxygen was available. Accumulating oxygen in the atmosphere led to the formation of the ozone layer that shields the earth from UV radiation. (p. 533)

PRECAMBRIAN EVOLUTION

WHAT FOSSILS TELL US Fossils indicate that life is at least 3.6 billion years old, that eucaryotes are 2.1 billion years old, and that complex organisms were abundant by 550 million years ago. (p. 539)

THE FIRST EUCARYOTES Eucaryotes probably evolved as symbiotic associations of procaryotes. Chloroplasts, mitochondria, and possibly other cellular organelles are descended from free-living bacteria. (p. 541)

THE KINGDOMS OF LIFE

The goal of phylogeny and classification is to objectively represent the course of evolution. The definition of hierarchical levels like kingdom and phylum, on the other hand, is more subjective. Most modern classification schemes recognize animals, plants, and fungi as separate kingdoms, but they sometimes disagree on how to organize unicellular and primitive organisms: There are certainly two kingdoms of bacteria, though many texts lump them together. Some schemes put all other species, no matter how distantly related, into a protist kingdom. More modern systems recognize as separate kingdoms the primitive missing link between procaryotes and eucaryotes (archezoans), and the separately evolved photosynthetic group that includes brown algae (the chromists). They also place green algae with the plants, and divide the remaining microorganisms into three to six groups that are either combined as protists or separated into distinct kingdoms. (p. 544)

STUDY QUESTIONS

1 Where, in the RNA-based scenario (in which RNA serves as the original genes and enzymes), would you draw the line between living and nonliving? How would you justify your dichotomy? (pp. 534– 536)

2 There is controversy over which group of photosynthetic bacteria gave rise to chloroplasts. How would you go about using sequence analysis to resolve the issue? (p. 545)

3 The archezoans were originally considered part of the kingdom Protista; since they are mostly parasitic, the Golgi apparatus was thought to have been lost because it was unnecessary. Based on your knowledge of how these organelles work, what sorts of chemical "scars" might you look for in the proteins of archezoans that would reveal a previous history of Golgi-based organization?

SUGGESTED READING

FINCHEL T., AND B.J. FINLAY, 1994. The evolution of life without oxygen. *American Scientist* 82 (1). *On the evolution of early eucaryotes.*

KNOLL, A. H. 1991. End of the Proterozoic eon, *Scientific American* 265 (4). *On the change in fauna and flora 550 million years ago when multicellular life began to flourish.*

ORGEL, L. E., 1994. The origin of life on earth. *Scientific American* 271 (4). *On the RNA-based origin of life.*

REBEK, J., 1994. Synthetic self-replicating molecules. *Scientific American* 271 (1). *How organic molecules can serve as templates for their own synthesis.*

SCHOPF, J. W., 1978. The evolution of the earliest cells, *Scientific American* 239 (3). *The fossil evidence for the appearance of cellular life on earth; the special metabolic characteristics of bacteria that enabled them to prosper under conditions that shut out most higher forms of life.*

CHAPTER 20

VIRUSES AND BACTERIA

This chapter will explore the lifestyles of the smallest genetic entities—the viroids, viruses, and bacteria. Bacteria are obviously living organisms, with a clear position in the evolution of life. Biologists are divided, however, over whether to consider viruses a form of life. On the one hand, viruses lack all metabolic machinery and cannot reproduce in the absence of a host. They are not in a kingdom because they are only rarely related to one another, and have no common ancestor; viruses did not evolve as a group, but rather piecemeal from many different procaryote and eucaryote cells. On the other hand, viruses do possess nucleic acid genes that encode sufficient information for the production of new viruses, and reproduction with gene-controlled heredity is a basic attribute of life. Regardless of whether viruses are classed as living, as nonliving, or as something in between, it is instructive to compare their minimalist lifestyle with that of the simplest free-living organisms: the bacteria.

VIRUSES AND VIROIDS

■ HOW VIRUSES WERE DISCOVERED

By the latter part of the 19th century Louis Pasteur and others had shown that many diseases are caused by microorganisms. For some diseases, however, biologists could find no causal microorganism. For instance, by 1796 Edward Jenner had discovered that smallpox could be induced by something in the pus from a smallpox victim, but no infective agent could be found. Similarly, in 1892 Dmitri Ivanovsky showed that the tobacco mosaic disease, which makes leaves mottled and wrinkled, is independent of conventional microorganisms: when juice extracted from an infected plant was rubbed on the leaves of a healthy one, the latter developed the disease. Even juice from an infected tobacco plant that had been passed through a very fine porcelain filter (which removes bacteria) remained infective.

Many other diseases of both plants and animals were found to be caused by infectious agents so small they could pass through porcelain filters and were invisible even under the best light microscopes. They were called ***viruses***, from the Latin word for "poi-

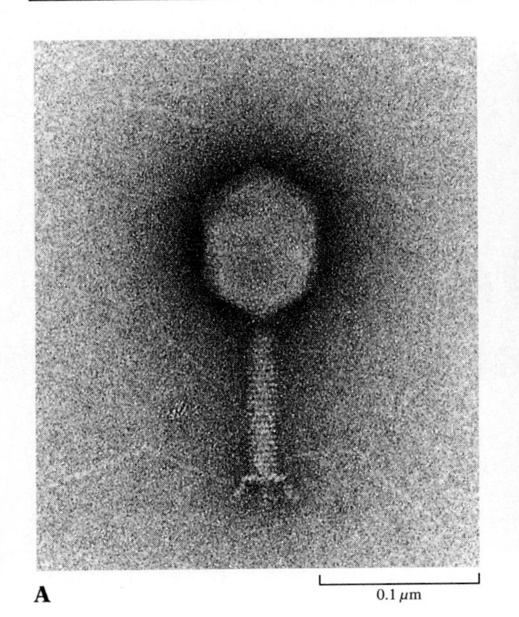

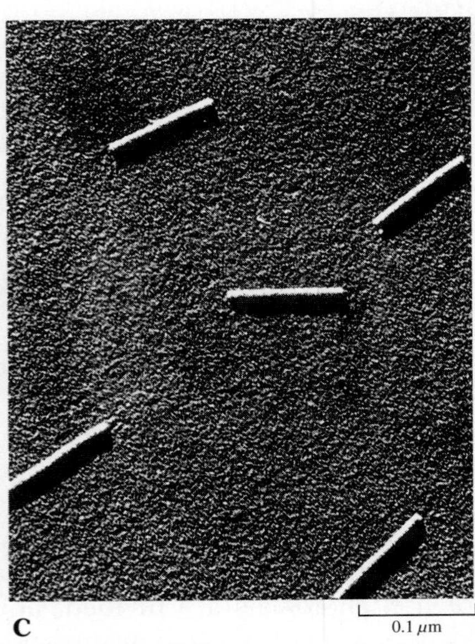

20.1 Electron micrographs of three types of virus

(A) T4, a complex virus with a "head" containing its DNA, a "tail," and six leglike fibers that aid in attaching it to a host cell (normally *E. coli*). (B) Adenovirus, a polyhedral virus that multiplies in the upper respiratory tract to produce coldlike symptoms. The form of adenovirus is an icosahedron, having 20 equal sides. (C) Tobacco mosaic virus, a rod-shaped virus.

son." Finally, in 1935 W. M. Stanley isolated and crystallized tobacco mosaic virus. The fact that viruses could be crystallized showed that they were not cells but must be much simpler chemical entities.

■ THE STRUCTURE OF VIRUSES

As described in Chapter 10, a free virus particle usually consists of a protein coat and a nucleic acid core. The protein coat, or ***capsid***, may be complicated, as in the bacteriophage (bacteria-infecting virus) designated T4 (Fig. 20.1B), with its tail and long leglike fibers; or it may be a simple polyhedron (Fig. 20.1C) or rod (actually a helix of protein molecules; Fig. 20.1A). Many animal viruses, some plant viruses, and a few bacteriophages have a membranous ***envelope*** surrounding the capsid; sometimes this envelope is derived from the plasma membrane of the host cell in which the virus was produced, and sometimes it is synthesized in the host cell's cytoplasm: in either case it usually contains some virus-specific proteins. In addition to the proteins of the capsid and envelope, some viruses—the retroviruses in particular—carry with them a limited number of enzymes.

The nucleic acid is usually a single molecule[1] consisting of as few as 3500 nucleotides[2] or as many as 600,000, depending on the kind of virus. If 1000 nucleotides is taken to be a reasonable esti-

[1] A few viruses have their nucleic acid in several pieces (six to eight for influenza virus); HIV (the AIDS virus) carries two copies of its chromosome.

[2] Some so-called defective viruses have even fewer nucleotides; they can reproduce only when the host cell is infected with another, "helper" virus.

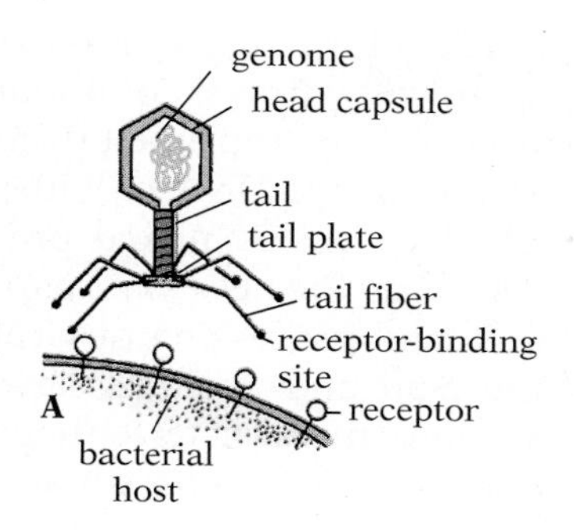

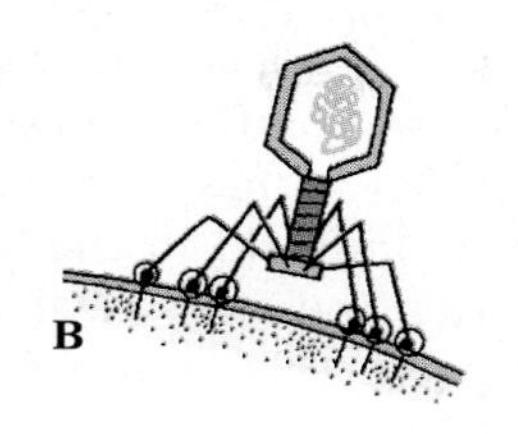

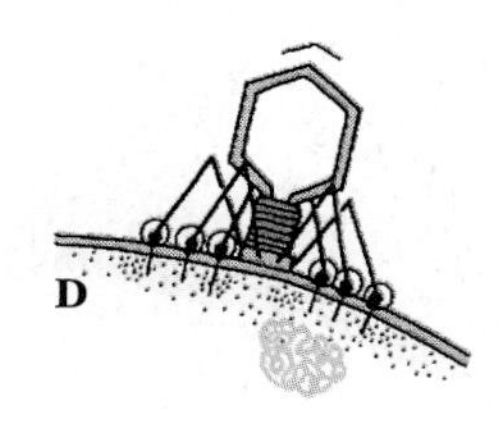

20.2 Lytic cycle of a typical bacteriophage

(A) The bacteriophage T4 has attachment sites at the tips of its tail fibers. (B) When the fibers encounter the host cell wall, they are bound by receptors on the bacterium and (C) bring the tail plate into contact with the host; the receptors normally bind useful substances like sugars or cell-surface markers on other bacteria. (D) This contact triggers contraction of the tail, and the viral genome is injected into the bacterium. (E) The phage DNA is then both replicated and transcribed by host enzymes. The viral mRNA is translated to make coat proteins, regulatory and assembly enzymes, and, at the end of the cycle, lysis enzymes such as lysozyme. (F) The head capsule is assembled around the replicated DNA, and the tail is added, followed by the tail plate and fibers. (G) Finally, the host is lysed, and mature bacteriophage are released.

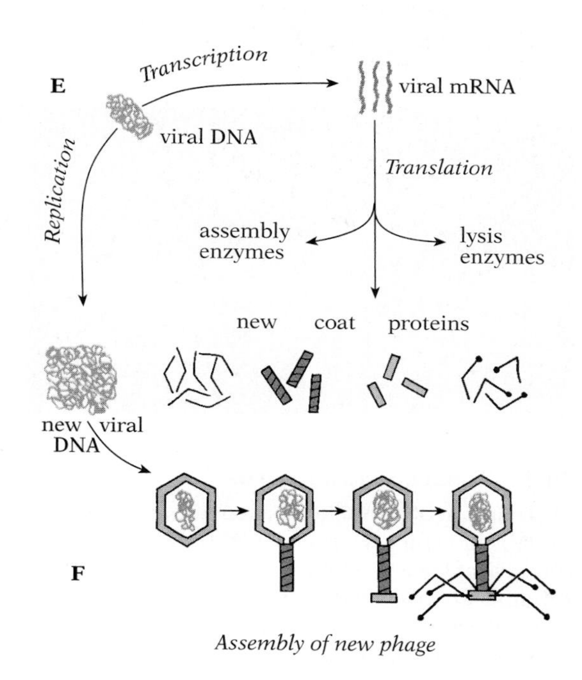

mate of the average length of a gene, then the total number of genes in a virus particle ranges from fewer than five to several hundred.

Some viruses have double-stranded DNA, while others have single-stranded DNA; in either case the DNA may be linear or circular. Many other viruses have RNA genes; in most instances the RNA is single stranded.

■ HOW VIRUSES REPRODUCE

Viruses lack multienzyme systems and cannot generate ATP; they lack raw materials for synthesis. Thus it is the host cell, not the virus, that replicates the viral genome, transcribes the viral genes, and translates the mRNA to produce the special enzymes and structural proteins the virus needs for its reproduction. Hence viruses cannot be cultured on artificial media. Pharmaceutical companies and research laboratories often grow them in bacteria, in fertilized chicken eggs, or in cells in tissue cultures.

Three of the diverse reproductive strategies of viruses are described in Chapter 10 (see Fig. 10.4, p. 251). Some viruses simply begin using the host cell's chemical machinery to make new viral particles which then exit the cell either one at a time by budding off, or all at once by lysing the cell. Other viruses insert their genetic material into the host chromosome, and then either produce a slow stream of progeny or lie dormant until something triggers active reproduction. In the case of retroviruses, the RNA genome must be copied into DNA by reverse transcriptase; the DNA version is then incorporated into the host genome.

The sequence of events in a relatively complex bacteriophage is summarized in Figure 20.2. The phage becomes attached by its tail

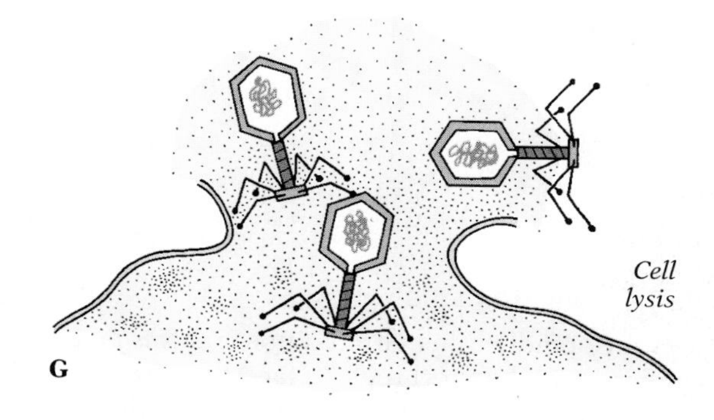

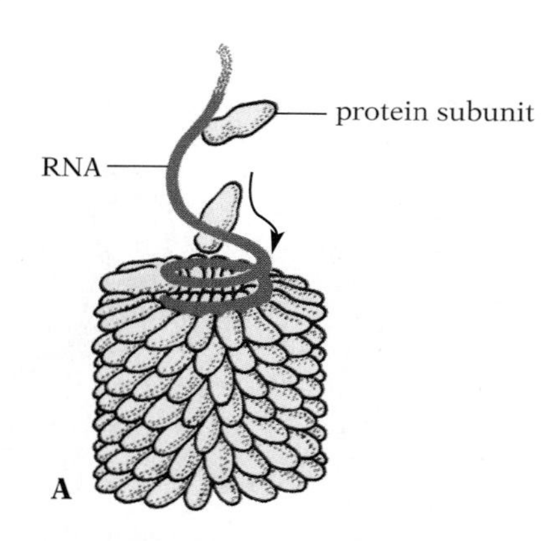

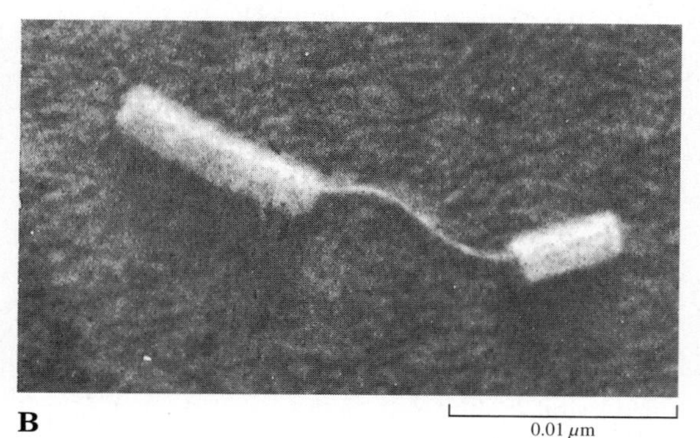

20.3 Tobacco mosaic virus

(A) Diagram of part of a tobacco mosaic virus (TMV), showing its structure. At the center of the cylindrical virus is a long RNA molecule. Protein subunits are packed in a helical array around the coiled RNA; they are added in sequence as a virus is assembled. (B) Scanning electron micrograph of the rod-shaped TMV. Part of the protein coat has been stripped away from the center portion of the virus, exposing the nucleic acid core.

fibers to the wall of a bacterium; its nucleic acid is injected into the host. The energy for this injection comes from hydrolysis of about 140 ATP molecules in the tail of the phage (ATP expropriated from the previous host). Once inside the host, the phage DNA provides genetic information for the synthesis of new viral DNA and protein. These proteins include structural proteins for new viral capsids and enzymes that aid in the synthesis and processing of viral components. After the new viral nucleic acid and proteins have been manufactured, they are assembled into new bacteriophage and released when a phage enzyme attacks the host's cell wall and lyses the bacterium.

The simplest viruses do not bother with assembly enzymes. Instead they use one or two types of coat proteins, which essentially crystallize around the viral genome. This process of self-assembly is exemplified by tobacco mosaic virus (TMV), whose 2130 identical protein subunits seem to form a helix spontaneously, with a long RNA molecule inside (Fig. 20.3).

■ VIROIDS: NAKED RNA

In 1971, T. O. Diener discovered that the agent responsible for potato spindle-tuber disease is a small circular strand of RNA that lacks a protein coat (Fig. 20.4). Since then more than a dozen diseases of higher plants have been shown to be caused by these remarkable naked bits of RNA, called ***viroids***. Because of extensive stretches of complementary bases, viroid RNA can pair internally; this self-pairing results in a stable, double-stranded conformation (Fig. 20.5), just as self-pairing helps stabilize the structure of tRNA, rRNA, and the tail of mRNA (see Figs. 9.5, 9.11, and 9.16, pp. 228, 233, 237).

The viroid genome is 10 times smaller than the shortest viral chromosomes, and there is no evidence that any enzyme or other product is generated by viroids. Apparently viroids contain only enough information to allow an RNA polymerase to copy them.

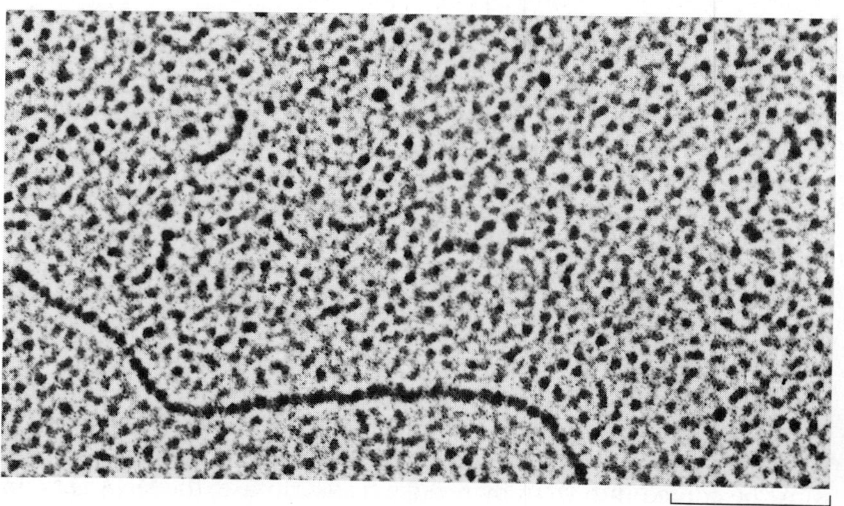

20.4 Viroids

Potato spindle-tuber viroids appear at this magnification to be short rods. In fact, they are circular, single-stranded bits of RNA; because they have several lengths of complementary base sequences, they are able to assume the extended double-stranded hairpin-turn form seen in the drawing. Their double-stranded form may protect them from digestion by certain host-cell enzymes. A segment of DNA from the bacterial virus T7 runs through the bottom left of the photograph and provides a sense of scale.

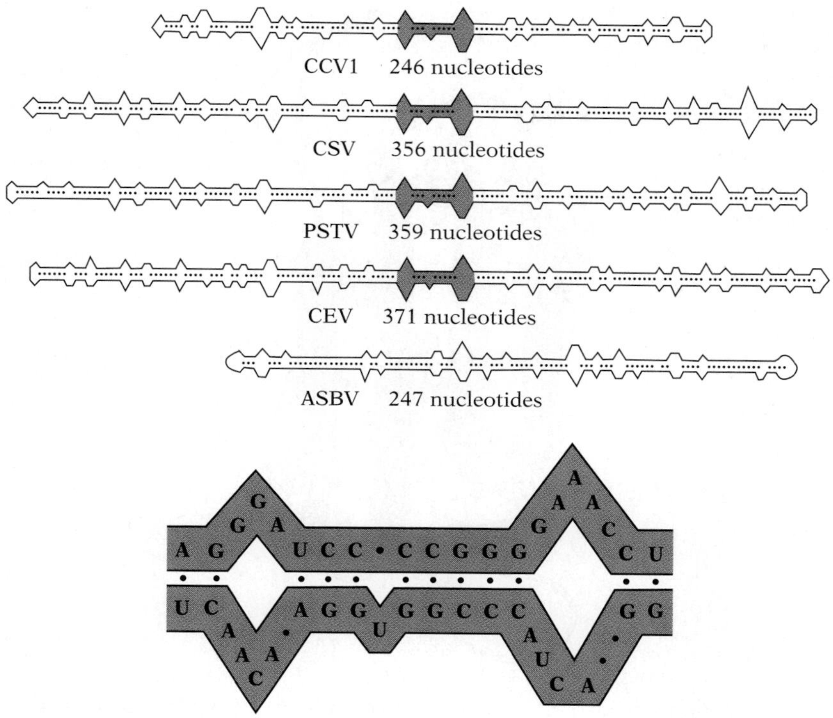

CCV1 246 nucleotides

CSV 356 nucleotides

PSTV 359 nucleotides

CEV 371 nucleotides

ASBV 247 nucleotides

20.5 Structure of viroids

All five of these viroids have a similar self-complementary structure that gives rise to their double-stranded conformation. In addition, a specific sequence in the central regions (shown here in red and enlarged) is common to four of the five, suggesting that it plays an important signaling role. The dots indicate bonds between complementary bases. CCV1 infects coconuts; CSV infects chrysanthemums; PSTV, potatoes; CEV, citrus trees; and ASBV, avocados. (Reproduced courtesy of T. O. Diener, 1983, *Amer. Sci.*, vol. 71.)

Since RNA polymerase is found only in the nuclei of eucaryotes, some mechanism, perhaps involving a signal sequence on the viroid RNA, must effect transport into and out of the nucleus. And then, since RNA polymerase normally recognizes a promoter on the DNA of the nucleus, a viroid sequence must somehow mimic the host's DNA promoter. The extensive double-stranded structure of viroids may protect them from enzymatic degradation and provide the physical structure that is necessary for polymerases to bind to.

In short, viroids are like an envelope correctly addressed and stamped, but empty. Of the viroids discovered so far, all are known only because, as a side effect of their presence, they cause visible disease symptoms in commercially valuable plants—tumors on potatoes, for example. Their pathogenic action may arise from a 40-base region that is largely complementary to a ribosomal RNA. Since viroids tend to become concentrated in nucleoli, where rRNA is synthesized, it seems likely that they bind to rRNA and interfere with the assembly of ribosomes.

Viroids may be widespread in nature, but since they do not lyse their host cells, their presence may go undetected. Very little is known about how they spread.

■ PRIONS

The infectious agent responsible for scrapie, a degenerative disease of the central nervous system of sheep and goats, appears to be even simpler than a viroid; in fact, it is almost certainly a protein. Two rare neurological diseases of humans—Creutzfeldt-Jakob dis-

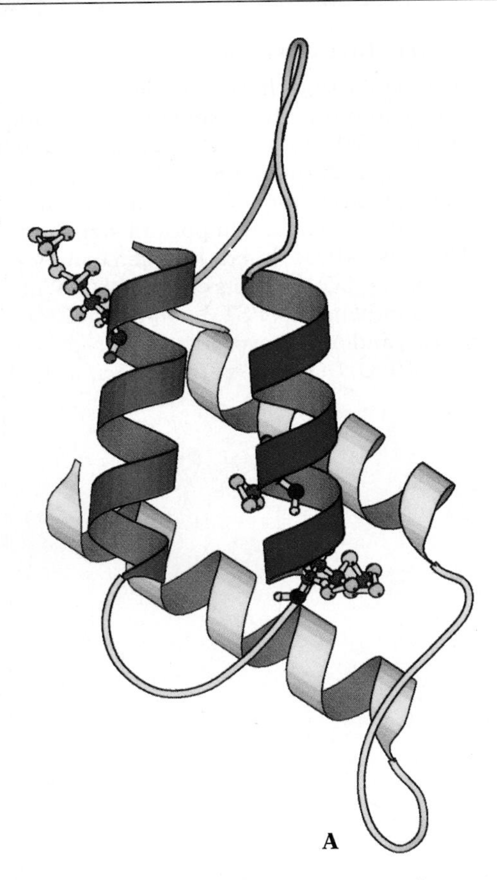

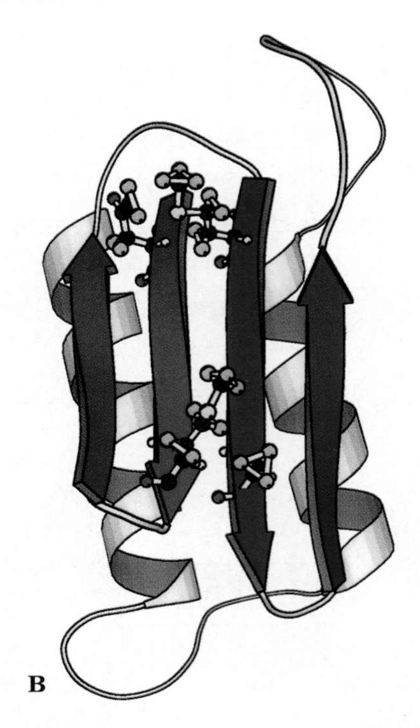

20.6 Structure of the prion protein

The prion-related protein (PrP^c) produced by cells is normally found on the surface of nerve cells as a dimer consisting of two identical subunits, each composed of four alpha helices (A). Mutations at one or more of the four sites whose amino acids are shown here in ball-and-stick form can cause two of the helices to lapse into four β sheets (B). The new conformation (PrP^{sc}) seems to create an enzymatically active protein that can catalyze the conversion of normal PrP^c into more PrP^{sc}.

ease and Gerstmann-Strassler syndrome—are also casued by proteins. How such a protein, now called a ***prion***, can cause the production of more infectious proteins is not fully understood. Apparently the prion protein modifies a preexisting product in the host cell (Fig. 20.6) to create new versions of itself. Scrapie-infected animals have just such a protein, identical to a product encoded by an exon in a host gene except for two to four amino acids. The prion polypeptide appears to catalyze the conversion of the normal protein (a receptor involved in neural regulation) into the prion version; this process must accelerate until the infected cell has no chance of getting a normal receptor safely into its membrane.

■ HOW DID VIRUSES EVOLVE?

An early hypothesis to account for the origin of viruses suggested that they are organisms that have reached the extreme of evolutionary specialization for parasitism. Internal parasites commonly lose unnecessary structures; perhaps these viruses have dispensed with everything but their nuclei. A second hypothesis suggested that the ancestors of modern viruses were free-living noncellular predecessors of cellular organisms. When the organic nutrients of the primordial seas disappeared, they survived by becoming parasites on the cellular organisms that had arisen by that time. According to this view, modern viruses represent an early stage in the origin of life.

The modern view is that viruses are fragments of genetic material derived from cellular organisms. They might have begun as bare nucleic acid similar to viroids or plasmids, and later evolved the capacity to incite their host cells to synthesize a protein shell. Some viruses may have arisen as fragments of bacterial DNA, others as fragments of plant nucleic acid, and still others as fragments of the genetic material of higher animals. The host specificity of viruses would then normally be a reflection of their origin; a given type of virus might be able to parasitize only species fairly closely related to the one from which it originally derived. In fact, sequence analysis indicates a close relationship between virus and host. Thus any attempt at a taxonomic classification of viruses based on evolutionary relationships is pointless, since most viruses have evolved independently of each other as "offspring" of their hosts. Thus viruses are not placed in a separate kingdom.

■ VIRAL DISEASES

Among the many human diseases caused by viruses are chicken pox, mumps, measles, smallpox, yellow fever, rabies, influenza, viral pneumonia, the common cold, poliomyelitis (infantile paralysis), fever blisters (herpes), several types of encephalitis, infectious hepatitis, and AIDS (acquired immune deficiency syndrome). The association of viruses with cancer is discussed in Chapter 11.

Some of the most pernicious virus diseases, notably smallpox and polio, can be prevented with vaccines; indeed, smallpox has been eradicated as a result of intensive worldwide vaccination programs.

Immune responses, though important in preventing reinfection, ordinarily appear too late to account for recovery from viral diseases. At least one important factor in recovery is a protein called ***interferon***, which is produced by host cells in response to invading viruses. The interferon cannot save those cells, but when it is released into the medium and encounters uninfected cells, it interacts with a receptor site on their membranes. This action helps to generate a resistance to viral infection, probably by inducing production of antiviral proteins that block translation of the viral mRNA.

Viruses infect a wide variety of organisms, often at remarkably high rates. For example, viruses are found at densities of 15×10^6 per millimeter 10 m below the surface in the North Atlantic, and at concentrations up to a dozen times higher in lakes. These viruses infect about 70% of the photosynthetic bacteria and algae that flourish in these waters, and are often more important than nutrients in limiting these populations.

THE TRUE BACTERIA

Bacteria are the most ancient of organisms. They have been evolving and diversifying for billions of years, and so span a range of habitats and lifestyles greater than that of any eucaryotic kingdom. Some are plantlike photosynthesizers (photoautotrophs)—***producers*** of organic material from light and inorganic chemicals; some are chemosynthesizers (chemoautotrophs), and have no counterpart in the rest of the catalogue of life. Some are heterotrophs—***consumers***, including funguslike saprophytes (which live on dead organic material), dangerous parasites (both inter- and intracellu-

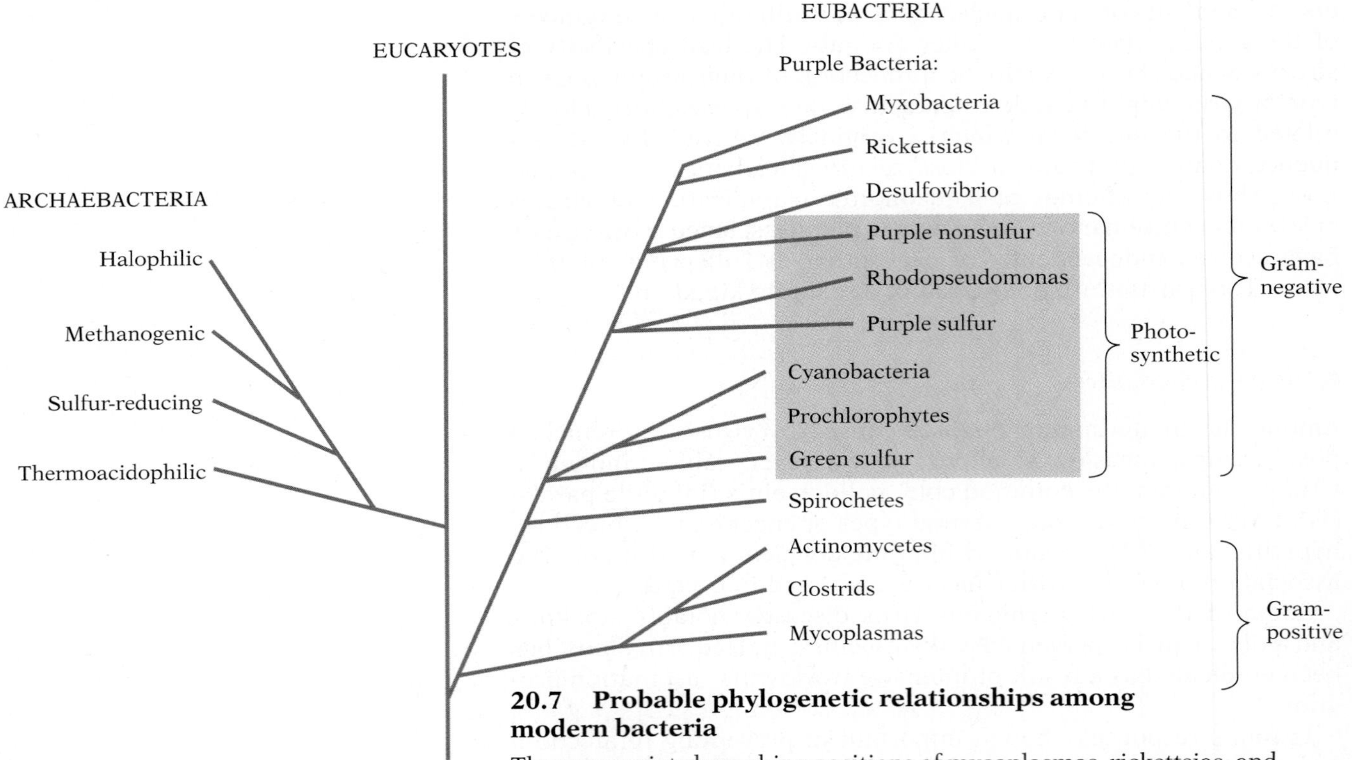

20.7 Probable phylogenetic relationships among modern bacteria

The appropriate branching positions of mycoplasmas, rickettsias, and gram-positive bacteria among the Eubacteria are especially uncertain. Archaebacteria are shown here as arising from a single branching event; many researchers believe this group contains organisms that arose from two independent branchings.

lar), harmless commensals (living within a host but doing no damage), and beneficial symbionts.

There are two bacterial kingdoms: Archaebacteria, or "ancient" bacteria, and Eubacteria, the much more common "true" bacteria (Fig. 20.7). Both are procaryotic—they lack a nuclear membrane, mitochondria, an endoplasmic reticulum, a Golgi apparatus, and lysosomes. Bacteria divide by fission rather than by mitosis or meiosis.

■ SYSTEMATICS

There is no consensus about the course of eubacterial evolution, and hence no single accepted scheme for classifying this diverse group of organisms. Sequence analysis, where it has been performed, permits a partial reconstruction. It is combined with major differences in cell-wall composition, chlorophyll type, and metabolic pathways to create the system of six divisions used here (Fig. 20.7).

One group stands out: the mycoplasmas, which are parasitic and have no cell wall. Nevertheless, they are placed with the gram-positive bacteria—bacteria that have exposed cell walls, and react with Gram's stain. The two other important groups of gram-positive bacteria are the clostrids and the actinomycetes. The five remaining divisions of eubacteria are gram-negative.

One division of gram-negative bacteria—the spirochetes—is unique because its members are the only bacteria with internal structures resembling microtubules. The spirochetes can be grouped with the next three divisions—prochlorophytes, cyanobacteria, and green sulfur bacteria—because they use NADP rather than NAD as their electron "shuttle."

Most of the organisms in the three divisions just mentioned, and the majority of the purple bacteria as well, are photosynthetic. Green sulfur bacteria have a chlorophyll similar to chlorophyll *a*, but are anaerobic, using H_2S rather than H_2O as their electron donor. Cyanobacteria, which have genuine chlorophyll *a*, are aerobic; water is their electron donor. The prochlorophytes are also aerobic and, like true plants, have both chlorophylls *a* and *b*.

The members of the last and largest division, the purple bacteria, use NAD. The photosynthetic forms have a pigment known as bacteriochlorophyll. Some purple bacteria are aerobic, using oxygen as an electron acceptor for respiration—this is true for the purple nonsulfur group and, when oxygen is available, the rhodopseudomonads. Others—the purple sulfur group—are anaerobic and use H_2S or other compounds as electron donors. Desulfovibrio bacteria are chemotrophic, obtaining energy directly from the reduction of sulfur. The purple bacteria also include two odd classes: the rickettsias, which are intracellular parasites, and myxobacteria, which propel themselves by secreting a slime over which they "glide" along.

■ ANATOMY

For the most part, bacteria are smaller than eucaryotic cells. Exceptions are the cyanobacteria, which were long mistaken for algae and are still called "blue-green algae," and one newly discovered symbiont from the gut of surgeonfish. In fact, the mycoplasmas, rickettsias, and some actinomycetes are almost as small as some of the largest viruses (Fig. 20.8). However, even the smallest bacteria always have ribosomes and integrated multienzyme sys-

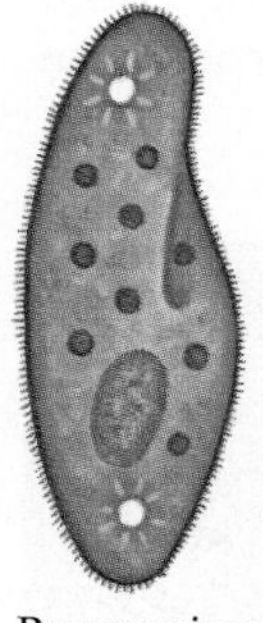

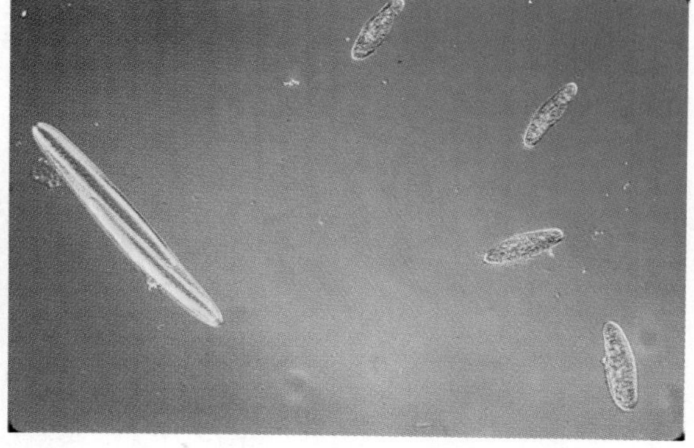

20.8 Sizes of viruses, bacteria, and eucaryotes compared

(A) Though bacteria are almost always smaller than eucaryotes (like *Paramecium* here), the enormous, 400 μm-long surgeonfish symbiont (B) is a notable exception. The photo includes four paramecia on the right to show scale.

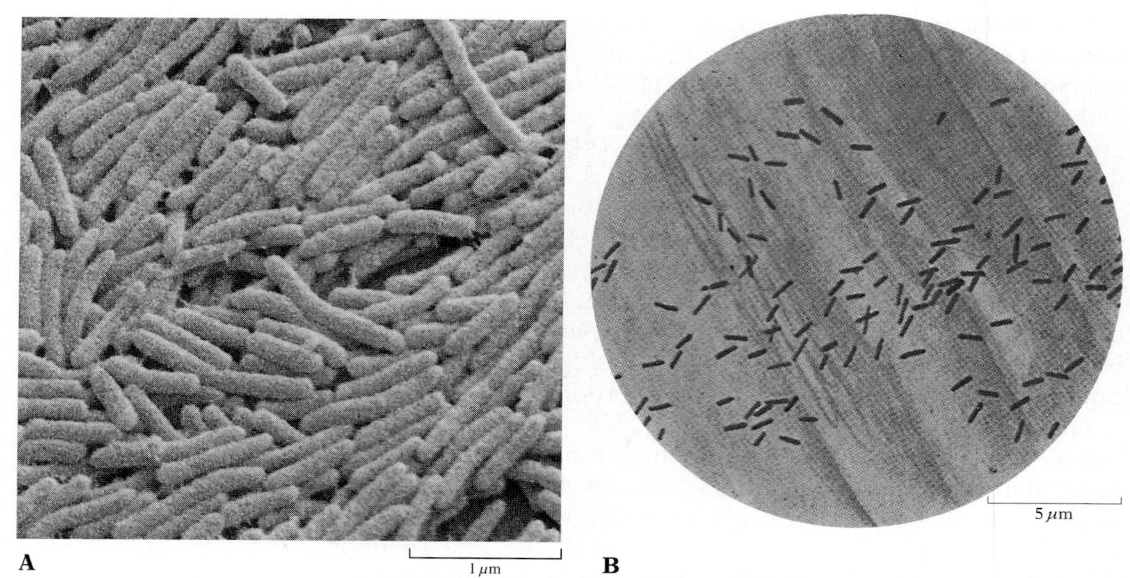

20.9 *Escherichia coli,* a rod-shaped purple bacterium found in the intestines of vertebrates

(A) Electron micrograph of a colony. (B) Stained cells as they appear under a high-power light microscope.

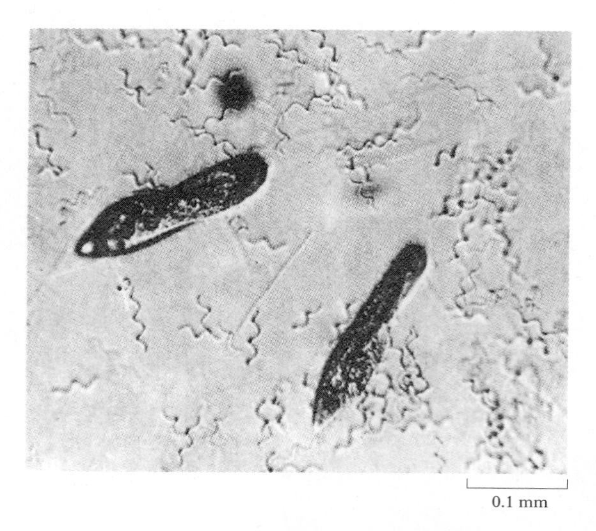

20.10 *Spirillum volutans*, a helically coiled bacterium

The two paramecia (which are protozoans) are included for size comparison.

tems, and provide both the raw material and all the metabolic machinery for their own reproduction.

Cell shape The cells of most bacteria have one of three fundamental shapes: spherical or ovoid, cylindrical or rod-shaped (Fig. 20.9), or helically coiled (Fig. 20.10). Spherical eubacteria are called ***cocci*** (singular, *coccus*); rod-shaped ones are called ***bacilli*** (singular, *bacillus*); and helically coiled ones are called ***spirilla*** (singular, *spirillum*).

When cell division takes place, the daughter cells of some species remain attached and form characteristic aggregates. Thus cells of the eubacterium that causes pneumonia are often found in pairs (diplococci). The cells of some spherical species form long chainlike aggregates (streptococci; Fig. 20.11), while others form grape-

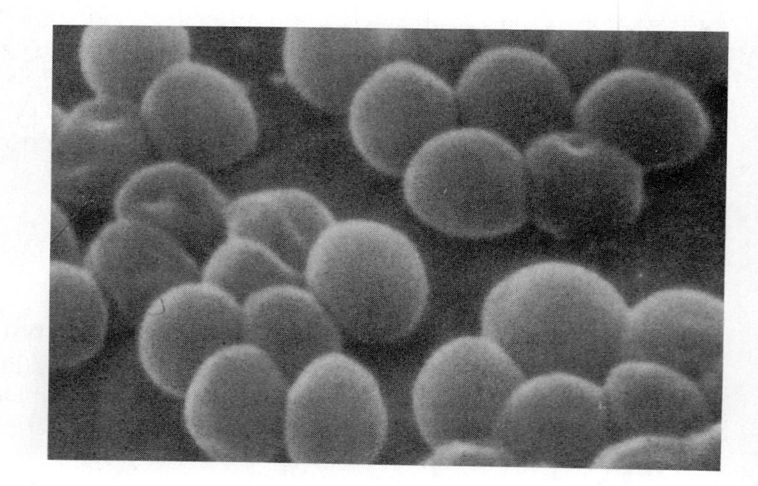

20.11 *Streptococcus salivarius*, a clostrid bacterium common in the human mouth

All streptococci are spherical (coccal) bacteria; they are normally grouped in chainlike clusters.

20.12 *Staphylococcus aureus*, a clostrid bacterium common in human skin

The round cells occur in grapelike clusters. They are the cause of pimples.

like clusters (staphylococci; Fig. 20.12). Each cell in a diplococcal, streptococcal, or staphylococcal aggregate is an independent organism, but some bacteria, notably some cyanobacteria, form either multinuclear or multicellular filaments. Adjacent cells of a filament have a common cell wall. Many actinomycetes resemble molds (which are fungi; Fig. 20.13).

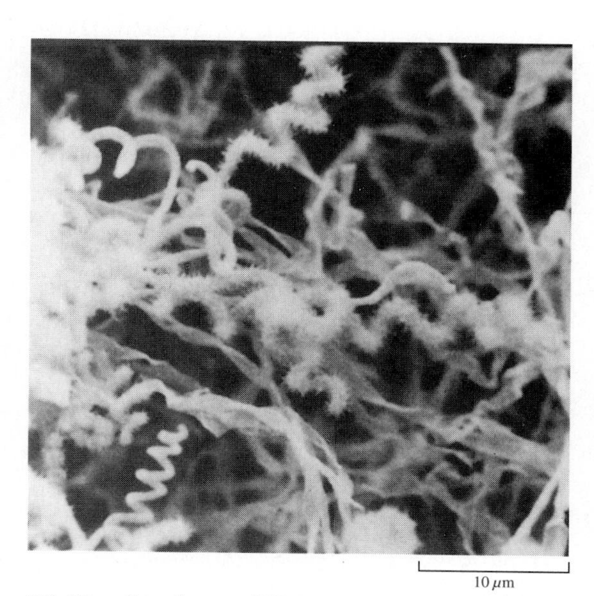

20.13 A colony of *Streptomyces*, one of the actinomycetes

The actinomycetes seen in this scanning electron micrograph have a much more complicated morphology than most other bacteria.

Cell walls Like plant cells, most eubacterial cells are enclosed in cell walls, which protect the cells from physical damage and osmotic disruption. The walls of eubacteria, however, differ from those of plants. The walls of eucaryotic cells derive their tensile strength largely from cellulose (chitin in fungi), whereas eubacterial walls derive theirs from peptidoglycan (also called murein), a huge polymer of polysaccharide chains covalently cross-linked by short chains of amino acids. Muramic acid, one of the principal polysaccharide constituents of peptidoglycan, never occurs in the wall of eucaryotic cells or archaebacteria. The walls can be thick and exposed, as in gram-positive bacteria, or thin and enclosed by an outer membrane of lipopolysaccharides, as in gram-negative bacteria (Fig. 20.14). Penicillin, which is harmless to plants and animals, is toxic to many growing eubacteria because it inhibits the formation of exposed peptidoglycan, and thus interferes with the multiplication of gram-positive bacteria.

Mycoplasmas lack walls and can live only as parasites of plants and animals—that is, in environments where their osmotic relations are such that the absence of a mechanically strong cell wall is not lethal. Mycoplasmas are the smallest known cellular organisms. They have less than half as much DNA as most other bacteria. Some of the mycoplasmas cause severe human diseases.

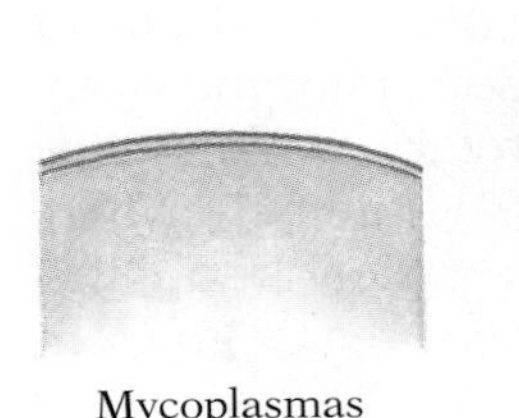

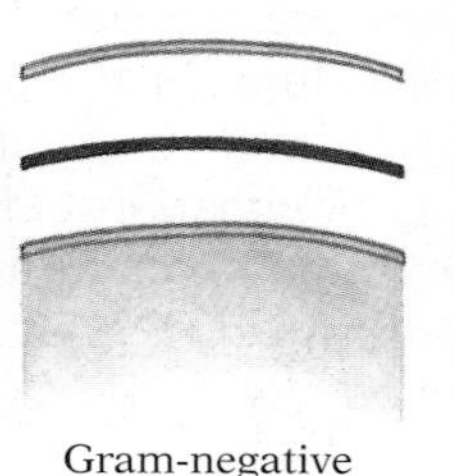

20.14 Cell walls of eubacteria

Mycoplasmas have only a plasma membrane. Other gram-positive bacteria have a thick layer of murein (also called peptidoglycan), which readily binds Gram's stain. Gram-negative bacteria have a thinner murein wall and an outer membrane of lipopolysaccharides. Many bacteria secrete a hard or sticky capsule outside of these structures.

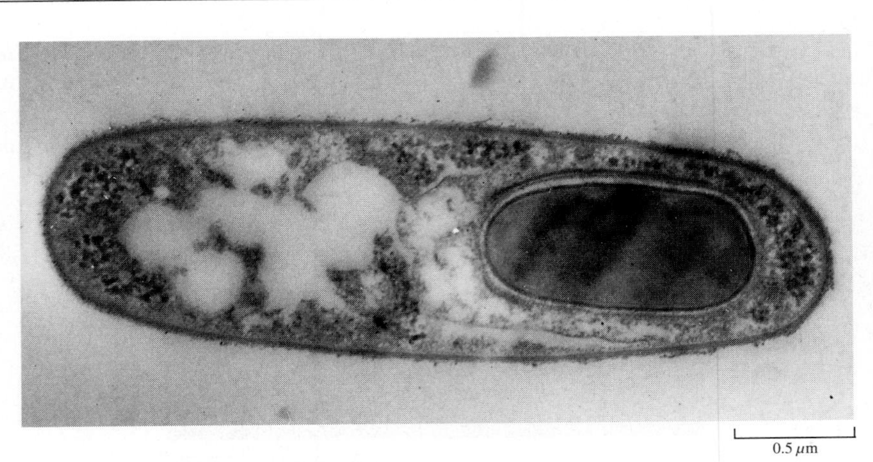

20.15 A sporulating bacillus

The spore is the dark oval at the right side of the cell in this electron micrograph. The developing spore coat is clearly visible. The white areas in the left portion of the cell are not vacuoles, but areas filled with fatty material.

Many eubacterial cells secrete mucoid materials (usually polysaccharides) that accumulate on the outer surface of the cell wall and form a ***capsule***. The capsule makes the cell more resistant to the defenses of host organisms.

Some eubacteria (mostly rod-shaped ones) can form special resting cells called ***endospores***, which enable them to withstand conditions that would kill the normal active cell. Each small endospore develops inside a vegetative cell and contains DNA plus a limited amount of other essential material (Fig. 20.15). It is enclosed in an almost indestructible spore coat. Once the endospore has fully developed, the remainder of the vegetative cell may disintegrate. Because of their very low water content and tough coats, spores of many species can survive an hour or more of boiling. They can be frozen for decades (perhaps for centuries) without harm. They can survive long periods of drying. When conditions again become favorable, the spores may germinate, giving rise to normal vegetative cells that resume growing and dividing.

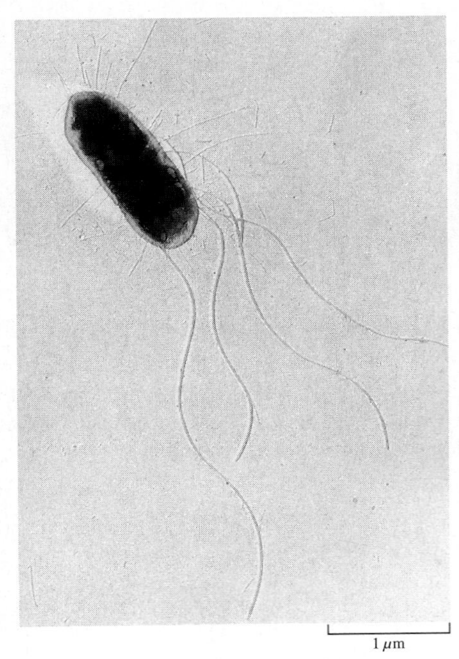

20.16 Long helical flagella of *E. coli*

Though the bacterial flagella seen in this electron micrograph are organelles of locomotion, they differ from the flagella of eucaryotic cells in lacking internal microtubules, and therefore use a different mechanism of movement.

■ HOW BACTERIA MOVE

Many bacteria are motile and move about actively. In most cases the motion is produced by the rotation of stiff helical flagella (Fig. 20.16). Virtually all bacterial flagella are structurally very different from the flagella of eucaryotic cells. They are not enclosed within the cytoplasmic membrane, do not contain the array of microtubules found in all flagella of eucaryotic cells, and lack tubulin. Instead they are formed from protein subunits called flagellin. A bacterial flagellum has approximately the same diameter as one of the tubules from a eucaryotic flagellum. Its motion is not the whiplike beating of eucaryotic flagella, but rather a rotary motion made possible by a complex attachment unlike anything seen elsewhere in the living world (Fig. 20.17). The energy for rotation is supplied by hydrogen ions moving down an electrochemical gradient

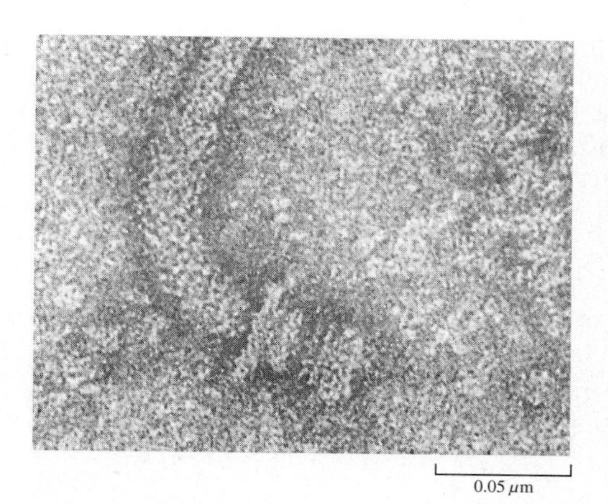

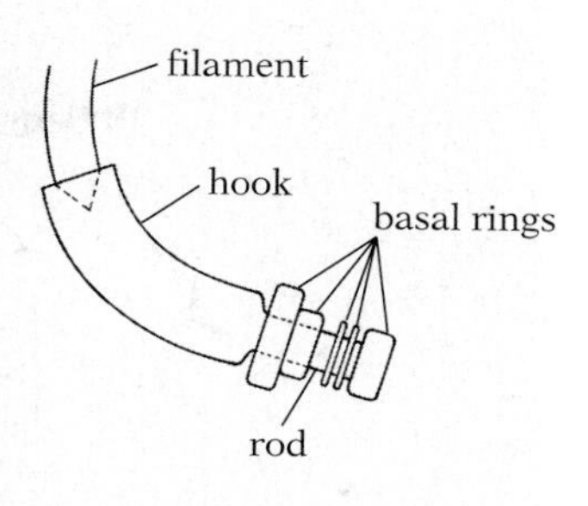

20.17 Basal portion of a typical eubacterial flagellum

The electron micrograph shows the base of a flagellum from *Caulobacter crescentus*. The interpretive drawing elucidates the complex system of rings, rod, hook, and filament. Unlike the flagella of eucaryotes, which beat back and forth, nearly all eubacterial flagella rotate. Their complicated structure provides the necessary anchor, joints, and couplings for such rotational motion. The flagella of *E. coli* rotate about 270 times per second (16,000 rpm); rotation rates of other bacterial flagella range from 170 to 1700 per second.

through channels at the base of each flagellum; the same gradient provides energy for the synthesis of ATP.

An exception to the characteristic bacterial flagella is found in certain species of spirochetes; these bacteria appear to contain internal microtubules that propel the organisms in some yet-to-be-understood way (Fig. 20.18). A spirochete or spirochete ancestor may have given rise to the eucaryotic flagellum (Fig. 20.19).

Some eubacteria that lack bacterial flagella (the myxobacteria) exhibit a peculiar gliding movement that does not involve any visible locomotor organelles; they secrete a slime that somehow aids in propulsion.

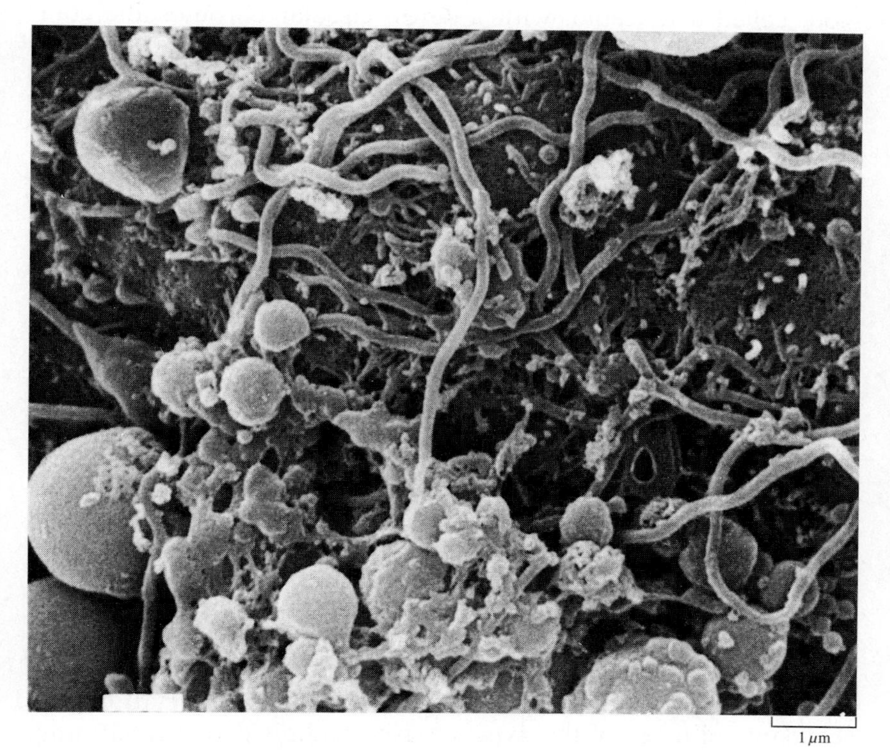

20.18 *Borrelia burgdorferi*, the spirochete responsible for Lyme disease

These long, slightly helical bacteria were taken from the midgut of a deer tick.

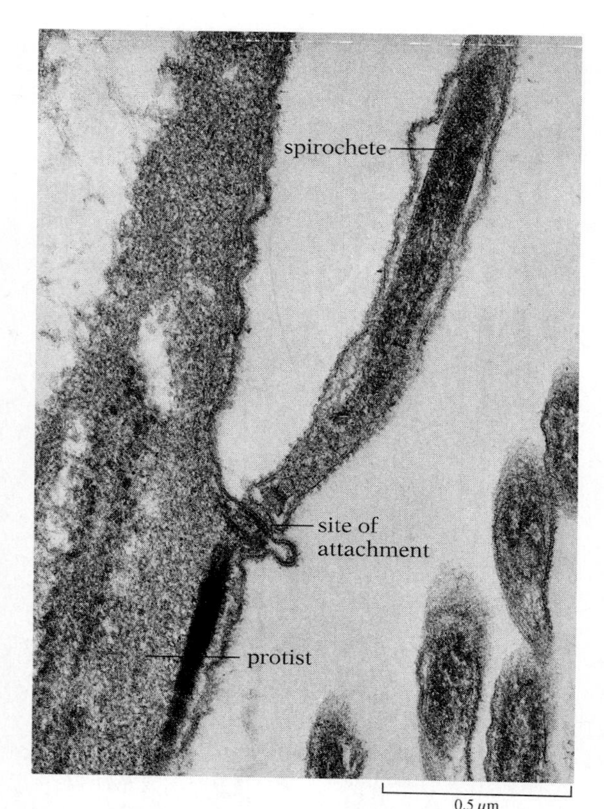

20.19 Symbiosis between a spirochete bacterium and a protistan

Spirochetes have curious tubules and move in a manner that is not understood. Here, a spirochete specialized for attachment to eucaryotic cells is bound to a nonmotile protistan that inhabits the gut of a termite; as a result of this attachment, the protist is able to move.

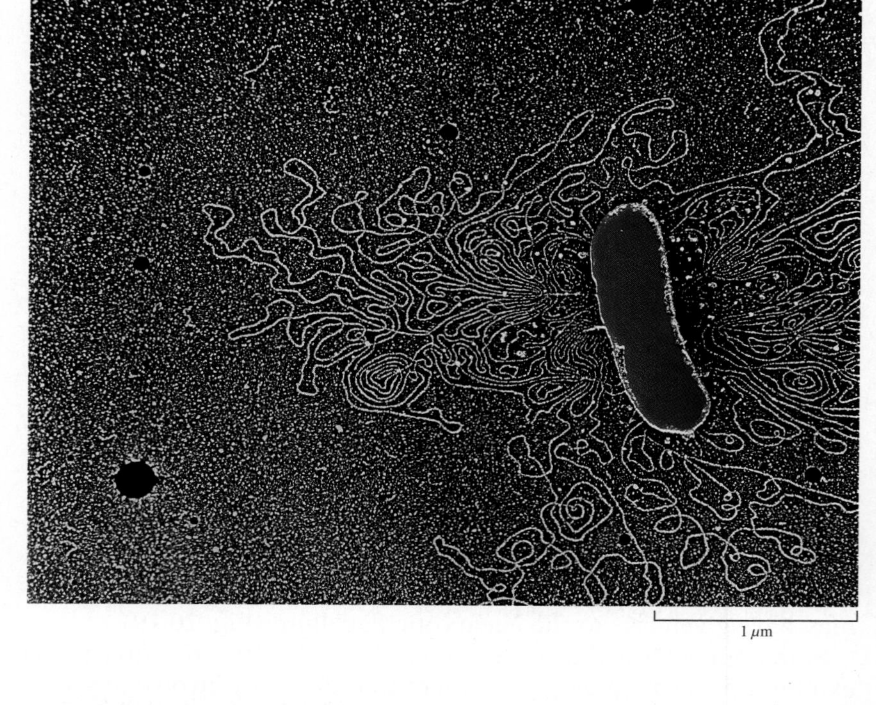

20.20 The DNA of a bacterial cell (*E. coli*) spread onto a surface

All the many coils of circular DNA seen in this electron micrograph are parts of a single chromosome. The supercoiling, which normally keeps the chromosome compact, has been relaxed.

■ HOW BACTERIA REPRODUCE

Bacteria lack membrane-bounded nuclei, but there is a nuclear region, called a nucleoid, with a single circular chromosome composed of double-stranded DNA (see p. 310). The single chromosome is surprisingly long—1000 times longer than the cell itself (Fig. 20.20).

Most bacteria reproduce by ***binary fission:*** two equal daughter cells with characteristics essentially like those of the parent cell are produced without the complex steps of mitosis (see Fig. 12.1, p. 309). When a bacterial cell undergoes fission, each daughter cell receives a complete chromosome with a full set of genes. The DNA, which is attached to the cell membrane, is replicated, and the two chromosomes move apart into separate nuclear zones before division of the cytoplasm occurs. When the plasma membrane grows inward, it partitions the binucleate parent cell into two daughter cells, each of which already has its own nucleoid. In many bacteria, a much-convoluted inward extension of the plasma membrane, called a mesosome, is seen at or near the site of division. Its role is not understood.

A major elaboration of the usual pattern of binary fission is seen in the myxobacteria, which can form elevated fruiting bodies and release spores (Fig. 20.21). A similar pattern of reproduction, seen in cellular slime molds, is described in Chapter 21.

20.21 Myxobacteria

This group of myxobacteria has formed a spore-releasing fruiting body.

Bacteria have enormous reproductive potential. Many species divide once every 20 min under favorable conditions. If all the descendants of a cell of this type survived and divided every 20 min, the single initial cell would have about 500,000 "offspring" at the end of 6 hours; by the end of 24 hours the total weight of its descendants would be nearly 2,000,000 kg. Though increases of this magnitude cannot actually occur, the real increases are frequently huge, which helps explain the rapidity with which food sometimes spoils or a disease develops.

Though the reproductive process itself is asexual, genetic recombination does occur occasionally, at least in some bacteria. As described in Chapter 10, three mechanisms of recombination are known: conjugation, in which part of a chromosome is transferred from a donor cell to a recipient; transformation, in which a living cell picks up fragments of DNA released into the medium from dead cells; and transduction, in which fragments of DNA are carried from one cell to another by viruses. When a normally haploid bacterial cell receives extra DNA, it becomes partly diploid; haploidy is soon reestablished by elimination of all genes that do not become incorporated into the new recombinant chromosome.

■ OBTAINING ENERGY AND RAW MATERIALS

Most well-known bacteria are heterotrophic, being either parasites or saprophytes. Like animals, many bacteria are aerobic—that is, they cannot live without molecular oxygen, which they use in the respiratory breakdown of carbohydrates and other food into carbon dioxide and water. Aerobic respiration is carried out with the help of an electron-transport system built into the cell membrane or its invaginations.

For some bacteria that obtain all their energy by fermentation, however, molecular oxygen is lethal. Such bacteria are called ***obligate anaerobes***. *Clostridium botulinum*, the causal agent of botulism, the most dangerous kind of food poisoning, is an obligate anaerobe; it grows well in tightly closed food containers that have not been properly sterilized before being filled. Other anaerobic bacteria live on inorganic compounds or organic fluids in the soil. Only in the early and mid-1990s, with the advent of sterile drilling techniques, has the immense number of such organisms living deep in the earth become apparent. Ocean drilling regularly recovers more than 10^7 bacteria per gram of sediment down to 500m, with no indication of any lessening in the numbers deeper still; terrestrial drilling often turns up densities of $10^5 - 10^8$ at 300m, and not much less even further down. If the values from these test drillings can be extrapolated to the earth as a whole, more than half of the mass of living things on the planet consists of subterranean bacteria. Many—perhaps the vast majority—may turn out to be archaebacteria.

Other bacteria, called ***facultative anaerobes***, can live in either the presence or absence of oxygen. Some of them are simply indifferent to O_2, obtaining all their energy from fermentation whether O_2 is present or not. Others obtain their energy by fermentation when O_2 is not available; when it is available, they carry out respiration (via the Krebs cycle and the electron-transport chain), and may grow faster under these conditions. Respiration is much more efficient than fermentation. Certain types of facultative anaerobes do not rely on fermentation when O_2 is absent; they continue to carry out complete respiration by using inorganic substitutes (such as nitrates, sulfates, or carbonates) in place of O_2.

Besides lactic and alcoholic fermentations—the most common fermentative processes (see Chapter 6)—at least 10 other types of

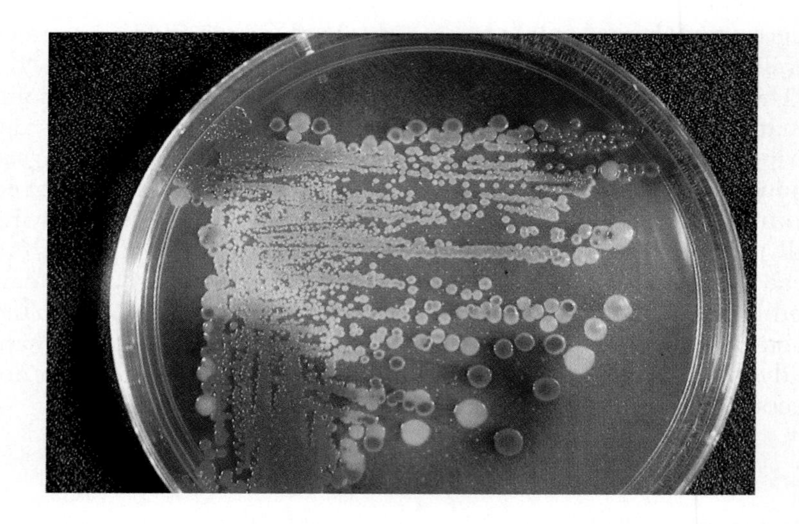

20.22 Colonies of two kinds of eubacteria growing on the same culture medium, with different effects on the medium

Colonies of *E. coli* on Endo agar carry out fermentation, making the colonies red, whereas *Salmonella* bacteria do not carry out fermentation and the colonies are colorless.

fermentation occur in different groups of bacteria. The products include acetic acid, butylene glycol, butyric acid, and propionic acid.

Bacteria differ considerably in the sorts of molecules they can use as energy sources and in the specific amino acids and vitamins they require. These differences provide valuable diagnostic characters for workers attempting to identify unknown bacteria (Figs. 20.22 and 20.23). Samples of the organisms to be identified are placed on a variety of nutrient media and cultured at standard temperatures. It is often possible to identify the unknown organisms by determining on which of the media the organisms grow and on which they do not, and when they do grow, by comparing the color, texture, and other characteristics of the colony they produce with data established for known species.

■ BACTERIAL PHOTOSYNTHESIS

Most eubacteria are chemoheterotrophs: they are parasites or decomposers that must obtain energy *and* carbon from organic compounds. Other bacteria are chemoautotrophs: they extract energy from inorganic compounds but fix CO_2 to make organic molecules. These chemosynthetic bacteria oxidize inorganic compounds, such as ammonia, nitrite, sulfur, hydrogen gas, or ferrous iron, and trap the released energy. The mechanisms of some of these reactions are discussed on pp. 153–156. The bacteria that oxidize ammonia or nitrite are the nitrifying bacteria, important in the nitrogen cycle (discussed in Chapter 40).

There are several kinds of photosynthetic bacteria. A few are

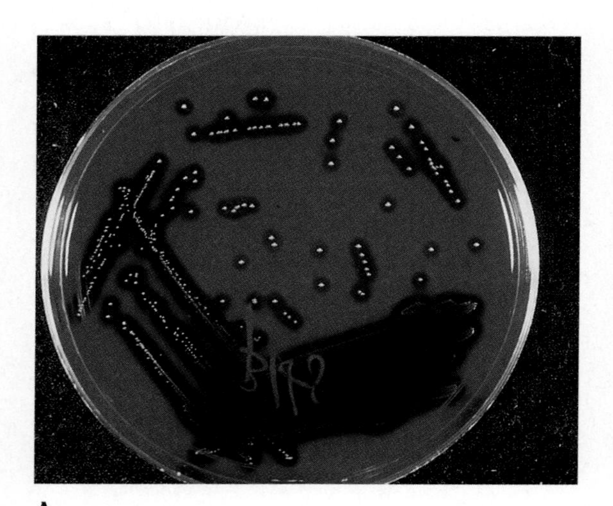

A

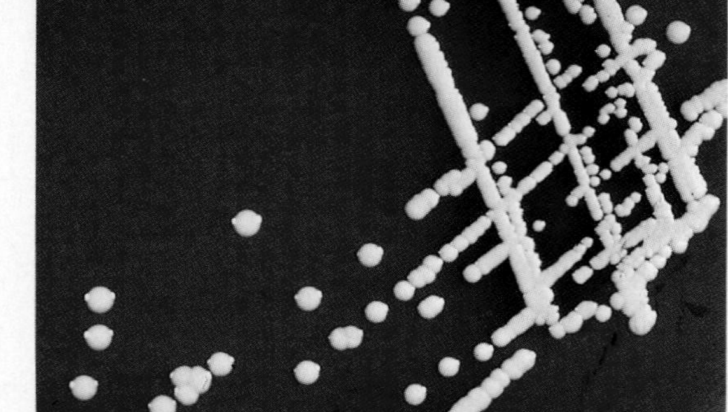

B

20.23 Different appearances of colonies of two kinds of eubacteria growing on blood agar

(A) Colonies of *Streptococcus* are surrounded by dark clear zones of hemolysis (breakdown of red blood corpuscles). (B) Colonies of *Sarcina lutea* have no surrounding zones of hemolysis.

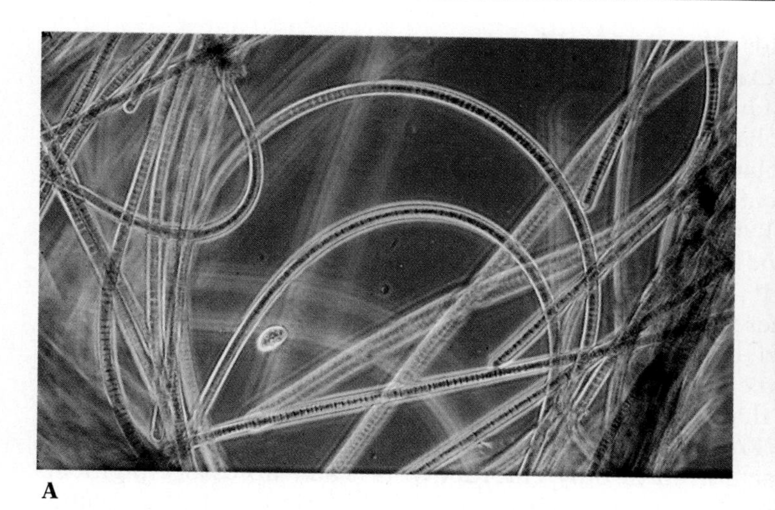

A

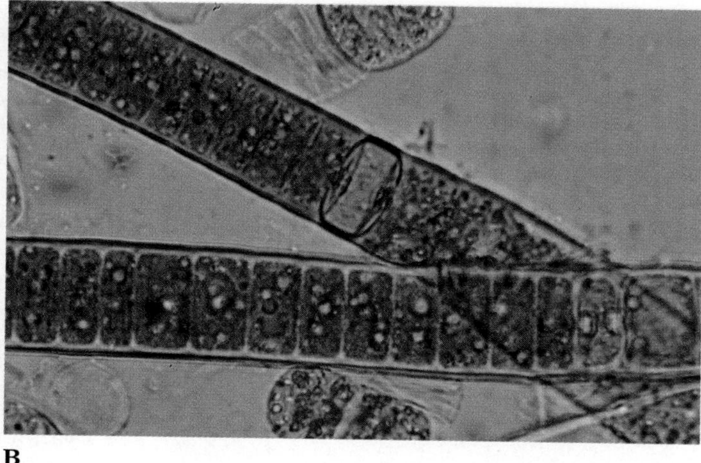

B

20.24 Some representative cyanobacteria

(A) *Oscillatoria*, a filamentous bacterium without a sheath. Members of this genus exhibit an odd oscillatory movement. (B) *Scytonema*, another filamentous bacterium, has oval, nearly rectangular cells. The lighter colored cell in the diagonal filament is a heterocyst, a cell specialized for carrying out nitrogen fixation (converting molecular nitrogen into a biologically useful form). (C) *Chroococcus*, from the Pine Barrens of New Jersey. Small groups of cells are enclosed within common gelatinous sheaths.

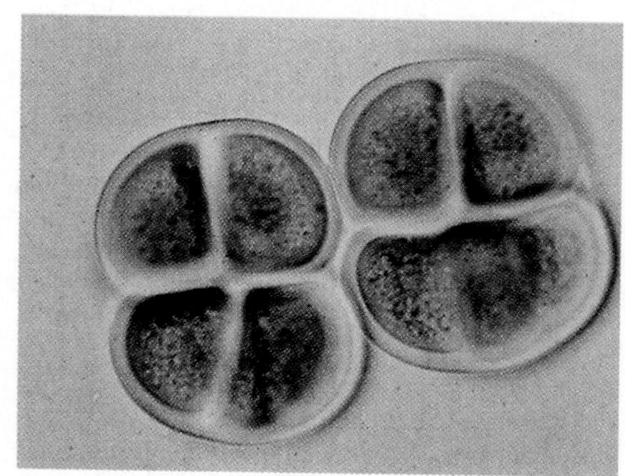

C

photoheterotrophs: they get energy from light but cannot fix carbon; as a result, they must "eat" organic compounds. Far more common are the groups of photoautotrophs, which differ from one another in the type of chlorophyll they possess. The green sulfur bacteria and the purple bacteria—both eubacteria—are similar in that neither group has chlorophyll *a*, the chief light-trapping pigment in higher plants. Unlike higher plants, neither group uses water as the ultimate electron donor in photosynthesis; hence neither produces molecular oxygen. Depending on the species of bacteria, the electron (and hydrogen) donor for reduction of NADP is molecular hydrogen, reduced sulfur compounds (such as H_2S), or organic compounds, but the oxidation of these substances is not light-driven. These photosynthetic bacteria possess only Photosystem I (and carry out only cyclic photophosphorylation).

The pigments and enzymes of the light-trapping process in anaerobic photosynthetic bacteria are located in chromatophores; these organelle-like structures are composed of vesicles or paired membranous lamellae, and differ between the purple and the green sulfur bacteria. The chromatophores are not contained within any membrane-bounded structure that could be interpreted as a chloroplast. Hence the enzymes involved in the dark reactions of photosynthesis (carbon fixation via the Calvin cycle), which are in the stroma portion of plant chloroplasts, are in the general cytoplasm of the bacterial cell.

The major group of aerobic photosynthetic bacteria are the cyanobacteria (Fig. 20.24). All cyanobacteria possess photosynthetic pigments located in flattened membranous vesicles called

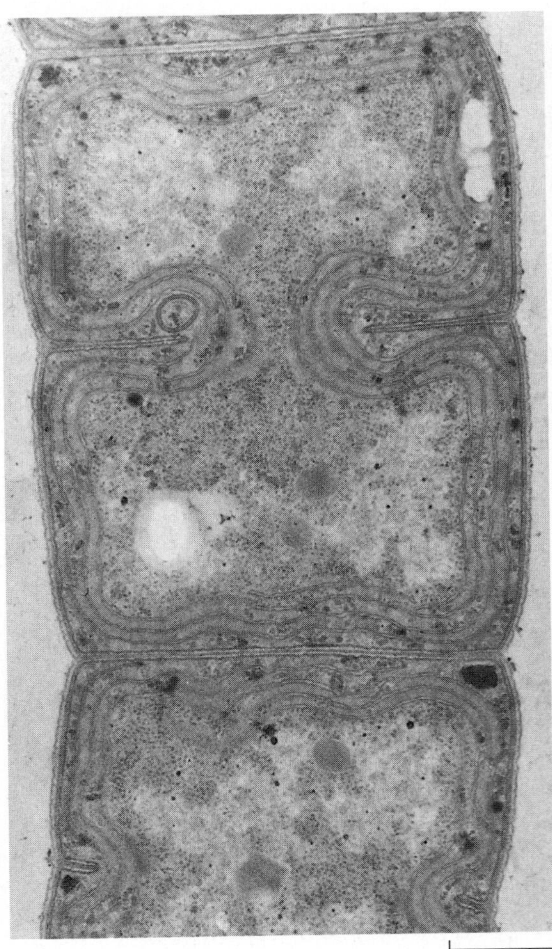

20.25 A cyanobacterium, *Plectonema boryanum,* undergoing cell division

The photosynthetic membranes (thylakoids) lie near the periphery of the cell. Walls have begun to grow inward at a point midway along the cell's length.

thylakoids (Fig. 20.25). These structures are similar to the chromatophores of the photosynthetic green and purple bacteria and, like the chromatophores, are not contained within chloroplasts.

Cyanobacteria have chlorophyll *a,* a pigment also found in higher plants. Like the higher plants, cyanobacteria generate molecular oxygen as a by-product of their photosynthesis. In addition to chlorophyll and various carotenoids, these organisms contain ***phycocyanin*** (a blue pigment) or sometimes ***phycoerythrin*** (a red pigment). It is the phycocyanin together with the chlorophyll that gives these bacteria their characteristic blue-green color. However, so-called blue-green algae can be black, brown, yellow, red, or grass green. The periodic redness of the Red Sea is caused by a species that contains large amounts of phycoerythrin.

The Prochlorophyta have both chlorophylls *a* and *b* whereas cyanobacteria have only chlorophyll *a*. Their accessory pigments correspond to the standard carotenoids found in higher plants, with no trace of the red or blue pigments of cyanobacteria. These organisms are thought by some to be the likely ancestor of most eucaryotic chloroplasts. Some sequence analyses support this view, and the morphological similarities between the prochlorophytes and eucaryotic chloroplasts are striking.

■ FIXING NITROGEN

One particularly important property of many cyanobacteria and some other anaerobic bacteria is their ability to fix atmospheric nitrogen (N_2). This process depends on the enzyme nitrogenase, which is inactivated by oxygen. Nitrogen-fixing species often grow symbiotically with plant roots; the plant host provides high-energy compounds that the bacteria use to power the conversion of N_2 into NH_3 (ammonia), and supply hemoglobin-like chemicals that trap any stray oxygen. Other species ferment cellulose for energy. Cyanobacteria are unique: they are oxygen-generating photosynthesizers; how can they fix nitrogen as well?

The species of cyanobacteria most active in N_2 fixation are, with few exceptions, filamentous forms that produce a few highly specialized cells called ***heterocysts***, where the N_2 fixation takes place (Fig. 20.24B). Heterocysts have exceptionally thick walls. Because O_2 is largely excluded from the interior of the heterocyst, nitrogenase can function there. Heterocysts lack Photosystem II, the oxygen-generating component of the photosynthetic apparatus. Equipped only with Photosystem I, the heterocyst can generate ATP in the light, but it cannot manufacture carbohydrate and must therefore depend on neighboring cells to supply it with this essential material. The heterocyst exports fixed nitrogen to the other cells of the filament. Heterocysts are rarely present when the bacteria are growing in an environment where a supply of already fixed nitrogen is ample, but they form quickly if the fixed nitrogen becomes limiting (Fig. 20.26).

Nitrogen fixation by cyanobacteria proceeds best when the heterocysts are in an atmosphere with less than 10% free O_2. At higher concentrations, some O_2 can leak into the heterocysts and begin to

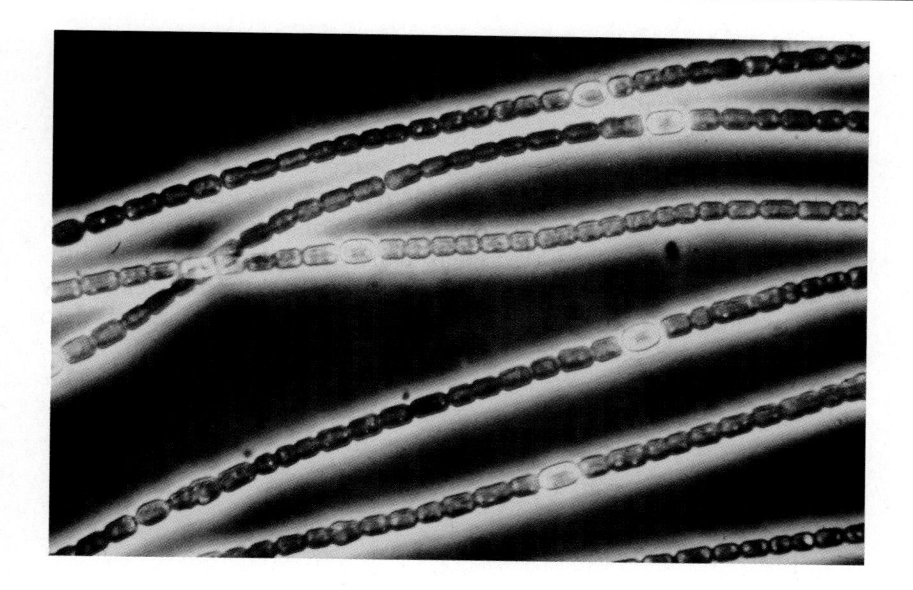

20.26 Effect of environmental nitrogen on heterocyst formation by a cyanobacterium, *Anabaena*

When grown in a culture containing sodium nitrate, the filaments form no heterocysts. On a medium in which N_2 is the only nitrogen source, the filaments form numerous heterocysts.

inhibit the nitrogenase. It is interesting that photosynthesis, too, is inhibited in cyanobacteria when the O_2 concentration is above 10%. Perhaps the metabolism of cyanobacteria functions best with low O_2 concentrations because these organisms are relicts of nearly 2 billion years of evolution under conditions of an oxygen-poor earthly atmosphere.

Because cyanobacteria are self-sufficient—able to trap solar energy and fix both CO_2 and N_2—they are important members of many food chains. They are particularly common in the upper few meters of oceans and lakes, and on barren surfaces like rocks, where they pave the way for less metabolically independent plants.

■ DISEASE-CAUSING BACTERIA

Perhaps the best-known bacteria are the ones that cause diseases in humans, domesticated animals, and cultivated plants. Eubacteria cause a long list of human diseases, including bubonic plague, cholera, diphtheria, syphilis, gonorrhea (Fig. 20.27), leprosy, scarlet fever, tetanus, tuberculosis, typhoid fever, whooping cough, bacterial pneumonia, bacterial dysentery, meningitis, strep throat, toxic shock, boils, abscesses, and peptic ulcers. In addition, rickettsias cause typhus fever and Rocky Mountain spotted fever, which are spread to humans by the bites of ticks, mites, and fleas. Chlamydias (which are actinomycetes) cause such diseases as ornithosis, lymphogranuloma, and trachoma. Both the rickettsias and the chlamydias are obligate intracellular parasites. Mycoplasmas, which are also obligate parasites, cause kidney stones and some types of pneumonia. One species, *Mycoplasma incognitus*, is a major killer of people with AIDS. This microorganism, which produces flulike symptoms initially, causes widespread tissue damage in the skin, spleen, liver, and brain; unchecked, it is invariably fatal. New variants of disease-causing bacteria can evolve quickly: beginning in 1986 a new form of the bacterium re-

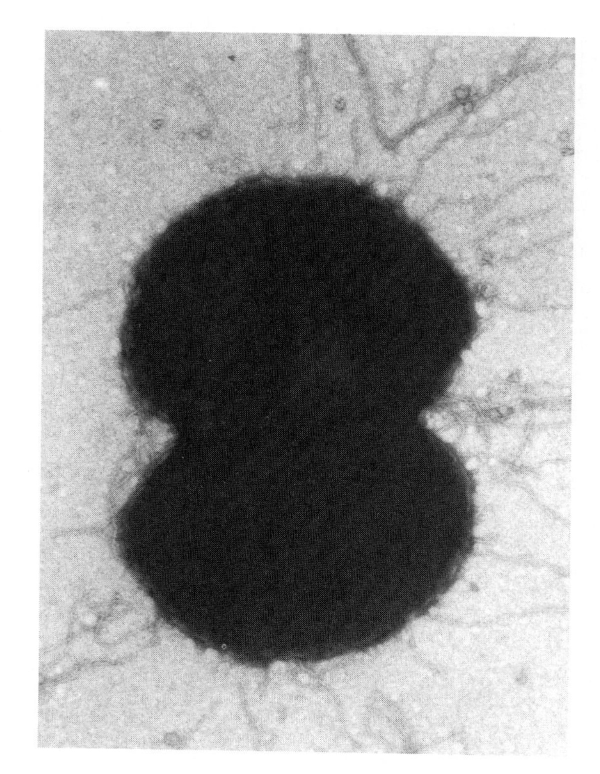

20.27 Gonorrhea bacterium

A diplococcal pair of the gonorrhea bacterium, *Neisseria gonorrhoeae*, showing many pili, which appear to play a role in the attachment of the bacterium to human mucosal cells. (As described in Chapter 10, pili can also function in bacterial recombination.) This bacterium infects at least 1 million people in the United States annually. Unchecked infections can cause blindness in newborns and sterility in adults.

sponsible for strep throat has been spreading; the novel strain can rapidly digest muscle.

Bacteria cause disease symptoms in a variety of ways. In some cases their immense numbers place such a tremendous material burden on the host's tissues that they interfere with normal function. In other cases the microorganisms destroy cells and tissues. In still other cases bacteria produce poisons, called ***toxins***. Exotoxins are toxins that are released by living bacterial cells into the host's tissues, as in diphtheria or tetanus; exotoxins released by diarrhea-causing bacteria, to cite another example, either bind to channels in the cells lining the intestinal wall, causing them to release water, or actually poke holes in the cells, creating massive leaks. In contrast, endotoxins are retained in the cells of the bacteria that produce them and are released into the host when the bacteria die and disintegrate.

■ BENEFICIAL BACTERIA

Beneficial bacteria far outnumber harmful ones. In particular, nitrogen-fixing bacteria are crucial to agriculture, and the organisms of decay not only prevent the accumulation of dead bodies and metabolic wastes but also recycle materials such as the nitrogen of proteins into a form usable by other living things. Eubacteria in the intestine synthesize essential vitamins absorbed by the body and aid in the digestion of certain materials. The breakdown of cellulose, whether in cattle or termites, is wholly dependent on bacteria in the digestive system.

Bacteria are also of great importance in many industrial processes. Manufacturers often find it easier and cheaper to use cultured microorganisms in certain difficult syntheses than to try to perform the syntheses themselves. Among the many substances manufactured commercially by means of bacteria are acetic acid (vinegar), acetone, butanol, lactic acid, and several vitamins. Bacteria are also used in the retting of flax and hemp, a process that decomposes the pectin material holding the cellulose fibers together; the fibers, once freed, may be used in making linen, other textiles, and rope. Commercial preparation of skins for making leather goods often involves the use of bacteria.

Many branches of the food industry depend on bacteria. Bacteria are essential in manufacturing yogurt and the various kinds of cheeses; the characteristic flavor of Swiss cheese, for example, is due in large part to propionic acid produced by bacteria, and its holes are produced by the carbon dioxide released during fermentation.

Particularly interesting is the use of bacteria in the production of antibiotics that can help control other bacteria. Most of the antibiotic drugs in use today are produced by various species of bacteria of the actinomycetes group or, if synthesized artificially, were discovered in these organisms. Among these drugs are streptomycin, Aureomycin (chlorotetracycline), Terramycin (oxytetracycline), and neomycin.

Most recently, bacteria—particularly *E. coli*—have become the workhorses of genetic engineering, turning out large quantities of enzymes or other products like insulin that are encoded by genes removed from other organisms and inserted into a bacterial host.

ARCHAEBACTERIA

Archaebacteria differ from eubacteria in a number of ways: their cell walls, if present, have a markedly different chemical composition; the lipids in their cell membranes are branched rather than straight like these in eubacteria and eucaryotes; their translation system differs in several small but consistent ways and is, if anything, more like that of eucaryotes. Though their name means "ancient bacteria," their kingdom is younger than the eubacterial kingdom.

The archaebacteria typically occupy very challenging habitats; some of them are thought to be similar to those in which life evolved. On the basis of sequence analysis, the archaebacteria are now usually classified into four groups: methanogens, halophiles, sulfur reducers, and thermoacidophiles.

■ METHANOGENS

The methanogens are anaerobic chemosynthesizers. They are found most often in oxygen-poor bogs and marshes, where they feed on decaying vegetation, producing methane ("marsh gas") as a by-product:

$$CO_2 + 4H_2 \longrightarrow CH_4 + 2H_2O$$

Some are also found as gastrointestinal symbionts in cellulose-fermenting herbivores. Others live deep in the oceans, at depths up to 11 km (where pressures are a thousand times that at sea level), feeding on organic matter that has sunk to the bottom after organisms have died. Still others live in the hot waters near undersea volcanic vents, enduring temperatures up to 110°C. How the DNA, proteins, and cell membrane are stabilized against such heat is not known; most proteins, for instance, denature above 45°C. These methanogens are so well adapted, however, that some species grow best at 98°C, and die below 84°C (183°F).

■ HALOPHILES

Halophiles (Fig. 20.28) live in extremely salty environments like drying brine pools near the ocean, the (otherwise) Dead Sea, and the Great Salt Lake (which is 25% salt, as compared with the oceans, which are 3%). Some halophiles do best in saturated solutions of salt (36%); such solutions are called saturated because no more salt will dissolve at room temperature. Added to the osmotic challenge of living in such salty water is the extremely alkaline pH of many pools, which ranges up to 11.5. Halophiles are photosynthetic. Their version of chlorophyll is called bacteriorhodopsin, a

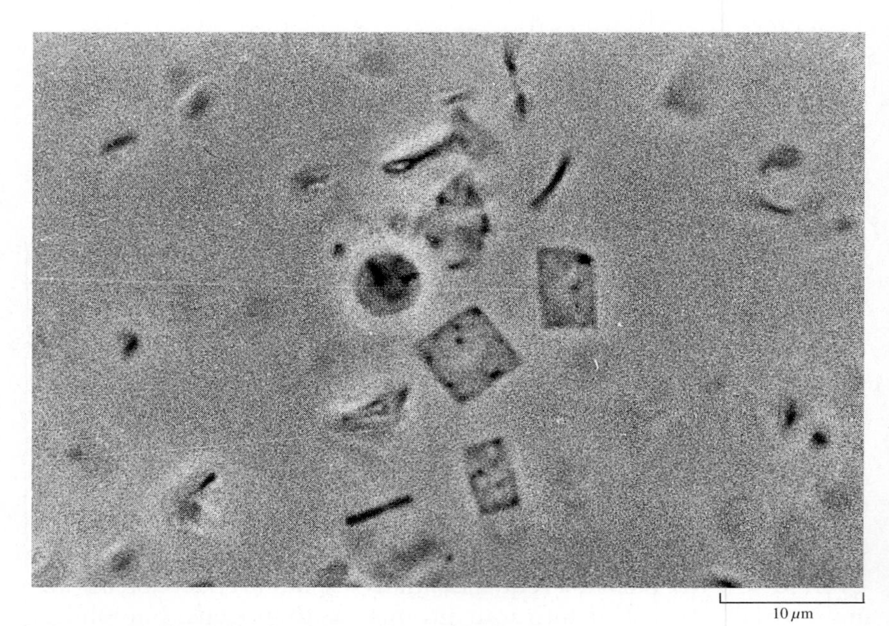

20.28 Halophilic bacteria from a brine pond

This species grows as thin square units that resemble miniature postage stamps.

purple substance nearly identical to the vertebrate visual-system pigment rhodopsin. Photons cause the cell to pump protons across the membrane; the electrical and chemical potential of these H^+ ions provides the chemiosmotic energy used to synthesize ATP. The enormous salt gradient is another important source of potential energy for halophiles.

■ SULFUR REDUCERS

Sulfur reducers use hydrogen and inorganic sulfur (usually from volcanic vents or pools) as their energy source:

$$H_2 + S \longrightarrow H_2S \qquad 6H_2S + 3O_2 \longrightarrow 6S + 6H_2O$$

Little is known about the members of this recently discovered group beyond their tolerance for temperatures up to 85°C.

■ THERMOACIDOPHILIC BACTERIA

Thermoacidophiles (Fig. 20.29) survive by oxidizing sulfur, not only near volcanic vents but in volcanic pools and hot sulfur springs:

$$S + O_2 + 2H_2O \longrightarrow H_2SO_4 + H_2$$

They prefer temperatures of 65-80°C and highly acidic environments with a pH as low as 1.0.

Because archaebacteria are mainly confined to extreme habitats,

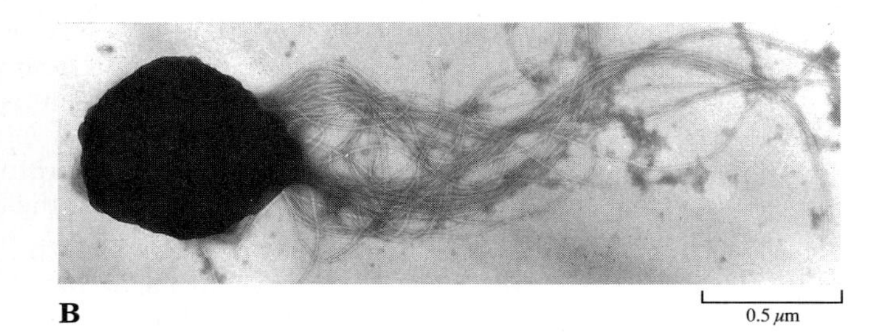

20.29 Thermoacidophilic archaebacteria

(A) Scanning electron micrograph of a dense mat of chemosynthetic bacteria growing on a mussel shell near an underwater volcanic vent in the Pacific Ocean. (B) Transmission electron micrograph of a single archaebacterium collected from the hot water of a submarine hydrothermal vent in the eastern Pacific.

they probably have a set of adaptations that protect them from their surroundings but put them at a disadvantage in competition with conventional organisms. As indicated earlier, the massive numbers of subterranean bacteria, living under conditions of intense pressure and (often) heat, may also be archaebacteria. The nature of their hardy chemistry remains to be discovered.

CHAPTER SUMMARY

VIRUSES AND VIROIDS

HOW VIRUSES WERE DISCOVERED Viruses were found when the search for a bacterial cause of certain diseases demonstrated that the agent responsible was far too small to be a bacterium. (p. 553)

THE STRUCTURE OF VIRUSES Viruses have a genome enclosed in a protein capsid; some are further enclosed in a membranous envelope. A few types carry one or more enzymes in the capsid. (p. 554)

HOW VIRUSES REPRODUCE Most DNA viruses use their genome as a template for replication and transcription; most RNA viruses use their genome for replication and translation. Some viruses incorporate their genome into the host chromosome; in the case of RNA viruses (retroviruses), this requires transcribing a DNA copy of the viral RNA. (p. 555)

VIROIDS: NAKED RNA These short pieces of RNA are too small to encode a conventional enzyme; their mode of replication and spread is not understood. (p. 556)

PRIONS Prions are proteins that cause more prions to be produced, apparently by causing normal versions of the protein to adopt the prion configuration. (p. 557)

HOW DID VIRUSES EVOLVE? Viruses do not form a kingdom; they are largely unrelated to each other. Most evolved from their hosts as genetic "escapees." (p. 558)

VIRAL DISEASES Viruses are important agents of disease in humans, other animals, plants, and protists. (p. 559)

THE TRUE BACTERIA

SYSTEMATICS Mycoplasms lack a cell wall. Gram-positive bacteria have a thick cell wall. Gram-negative bacteria have a second membrane outside their wall, which is thinner than that of the Gram-positive bacteria. Several groups of Gram-negative bacteria have distinctive features: spirochetes have microtubule-like internal structures; several are photosynthetic, including the cyanobacteria and purple bacteria. (p. 560)

ANATOMY Bacteria lack a nucleus and all of the subcellular organelles seen in eucaryotes with the exception of ribosomes. Nearly all bacteria have cell walls. (p. 561)

HOW BACTERIA MOVE Most motile bacteria have stiff rotating flagella with which they propel themselves. A few glide by an unknown mechanism. Spirochetes use their tubules to move about. (p. 564)

HOW BACTERIA REPRODUCE Bacteria propagate by fissioning. A few produce spores or can undergo conjugation. (p. 566)

OBTAINING ENERGY AND RAW MATERIALS Some bacteria are chemoautotrophs, using inorganic compounds as food. Others are photoautotrophs, using some form of photosynthesis to make organic material. Many others are parasitic or saprophytic, consuming organic food. Independent of the source of energy, many are anaerobic, not needing (or perhaps even tolerating) oxygen, while others are fully aerobic. (p. 567)

BACTERIAL PHOTOSYNTHESIS The photosynthesis of prochlorophytes appears very similar to that of green plants; cyanobacteria are not too different. Other groups use unusual oxygen acceptors like H_2S or unconventional chlorophylls. Many have unusual accessory pigments. (p. 568)

FIXING NITROGEN Only bacteria can fix nitrogen, and the ability is limited even in this group, in part because the enzymes responsible are inactivated by oxygen. The most important nitrogen fixers are cyanobacteria and certain plant-root mutualists. (p. 570)

DISEASE-CAUSING BACTERIA Bacteria cause a wide range of diseases in plants and animals. Some are a result of the metabolic burden imposed by these quickly growing organisms. Others are a consequence of direct damage to host cells or toxins released by the pathogenic bacteria. (p. 571)

BENEFICIAL BACTERIA Bacteria, along with fungi, consume dead organisms, form important mutualistic relationships with plants and animals, and perform a large number of important commercial tasks. (p. 572)

ARCHAEBACTERIA

METHANOGENS Methanogens use hydrogen gas as their energy source, producing methane as a by-product. They live in anaerobic environments ranging from the intestines of animals to vocanic vents on the ocean floor. (p. 573)

HALOPHILES Halophiles use photosynthesis to create a salt gradient across their membranes; they harvest the electrochemical energy of the gradient to power their chemistry. They live in intensely salty and alkaline water. (p. 573)

SULFUR REDUCERS Sulfur reducers combine sulfur and hydrogen gas to harvest energy. They live in intensely hot environments. (p. 574)

THERMOACIDOPHILIC BACTERIA Thermoacidophiles combine sulfur and oxygen to harvest energy. They live in environments that are both intensely hot and acidic. (p. 574)

STUDY QUESTIONS

1 Why might you expect more diversity of life histories in a bacterial kingdom than among, say, fungi, plants, or animals? (pp. 559–560)

2 What is the evidence that viruses are not the direct descendants of the first life? (pp. 558–559)

3 If viruses fatally infect 70 percent of the cyanobacteria in the ocean, and produce numerous progeny, why don't they drive these photosynthetic bacteria extinct? (p. 559)

4 Bacteria are often considered primitive relicts of the early stages in the evolution of life. Which divisions have to be more recent, and how can you be sure? (pp. 559–561)

5 If all bacteria were to go extinct today, how would the world change over the next month? The next year? Decade? Century? (p. 572)

SUGGESTED READING

BERG, H. C., 1975. How bacteria swim, *Scientific American* 223 (2). *The structure and mode of action of bacterial flagella.*

DIENER, T. O., 1981. Viroids, *Scientific American* 244 (1).

KOCH, A. L., 1990. Growth and form of the bacterial cell wall, *American Scientist* 78: 327-341.

PRUSINER, S. B., 1995. The prion diseases, *Scientific American* 272 (1).

STANIER, R. Y., E. A. ADELBERG, AND J. L. INGRAHAM, 1986. *The Microbial World*, 5th ed. Prentice-Hall, Englewood Cliffs, N.J. *Excellent general microbiology text.*

WOESE, C. R., 1981. Archaebacteria, *Scientific American* 244 (6).

CHAPTER 21

ARCHEZOANS AND PROTISTS

The members of the two kingdoms considered in this chapter are the descendants of the earliest eucaryotes. They are essentially single-celled (though some are multicellular or colonial), and many are parasites of higher organisms—where "higher" means species with at least some differentiated tissues, including nearly all the members of the kingdoms that evolved later. Archezoans and protists have suffered an evolutionary fate similar to that of many bacteria: just as procaryotes must once have dominated the living world but have now been forced into marginal lifestyles, so too these primitive eucaryotes have been crowded out by their multicellular competitors. Archezoans in particular seem to have been painted into this kind of evolutionary corner; protists, on the other hand, have retained a hold on certain biological "occupations" that reward small size and high reproductive rates. Both groups have exploited niches created by the evolution of multicellular organisms.

ARCHEZOANS: THE OLDEST EUCARYOTES

Archezoans were once considered to be protozoans that had lost their mitochondria and Golgi apparatus after having adopted a parasitic way of life. Sequence analysis, however, reveals that some of these tiny organisms are actually the oldest nucleated cells; they diverged from the mainstream of eucaryotic evolution over 2 billion years ago, before the endosymbiotic incorporation of mitochondria and the development of the endoplasmic reticulum (ER) and Golgi. The members of this kingdom also lack peroxisomes and have small ribosomes like those found in bacteria.

Like the kingdom Archaebacteria, Archezoa is a relict kingdom that may have much to tell us about the early evolution of life. There are 400 known species, but perhaps thousands or hundreds of thousands of undescribed species, many living in beetles. There are three phyla of archezoans: Archaeamoebae (or Pelobiontida), Metamonada, and Microsporidia. Archaeamoebae are primitive amoeboid organisms. The best-known species of archaeamoeba is *Pelomyxa*, a multinucleate, unicellular organism about 200 μm in diameter (Fig. 21.1). It inhabits the mud of freshwater ponds and

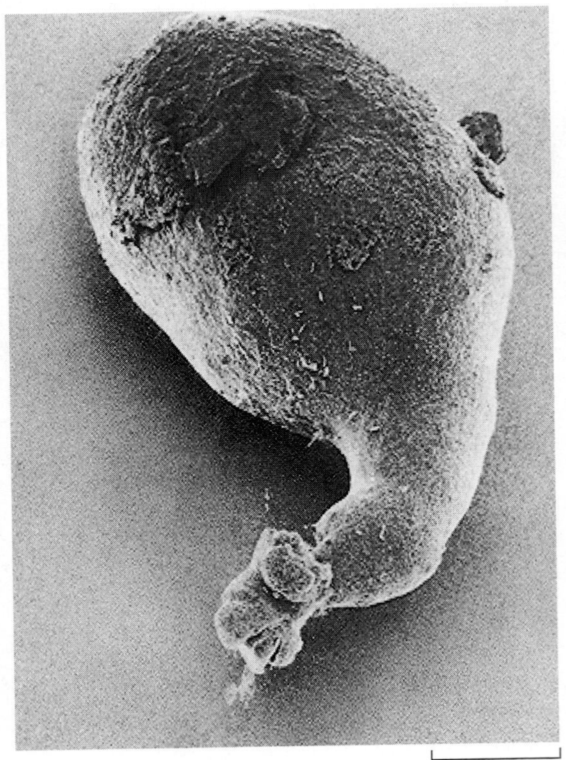

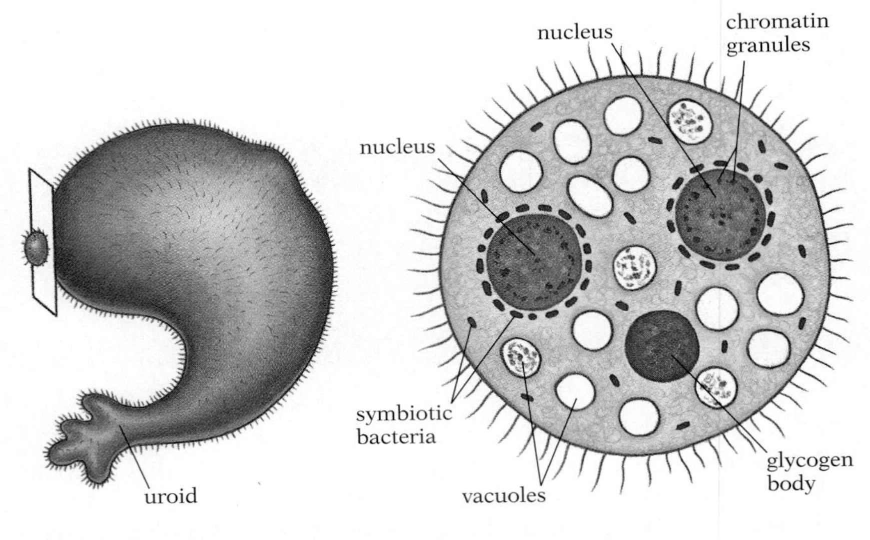

21.1 An archaeamoeba *(Pelomyxa palustris)*

Two of the three species of symbiotic bacteria found within the multinucleate cell are methanogens. One species always encircles the nuclei, while the other two are distributed throughout the cytoplasm. The cilia are nonmotile in *P. palustris*.

utilizes at least three species of endosymbiotic bacteria to make up for its lack of mitochondria.

Though a few metamonads are free-living, most are parasites that inhabit the intestines of animals, both vertebrate and invertebrate. The majority are harmless, but a few are pathogenic; one species, for instance, causes severe gastric distress in turkeys, and another (*Giardia*) is a common parasite of humans.

The microsporidians are entirely parasitic and infect new hosts by means of spores. Most of the hosts are invertebrates, in which the microsporidians infect fat bodies, vacuoles, and the like. Some microsporidians, however, attack the muscles of vertebrates—primarily fish, but also amphibians, reptiles, and even AIDS patients in a few cases (Fig. 21.2).

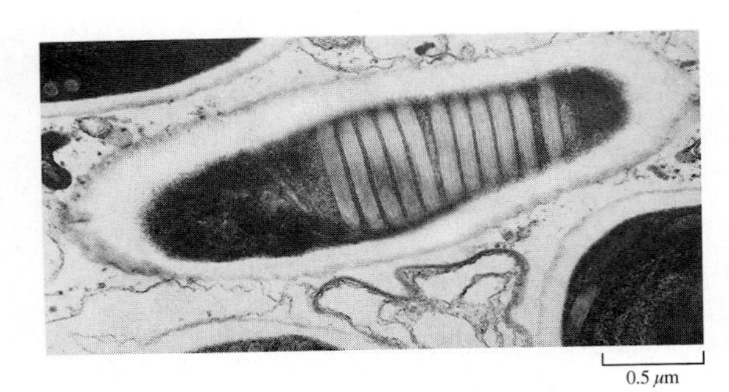

21.2 A microsporidian spore

This species is a parasite of flounders. The structure wound around and around just inside the membrane is the polar filament, which is used to attach the spore to its target.

PROTISTS: HOW SHOULD THEY BE CLASSIFIED?

Protozoans are usually unicellular. Instead of the organs characteristic of multicellular species, they have functionally equivalent subcellular structures. Protozoans occur in a variety of habitats, including the sea, fresh water, soil, and the bodies of other organisms—in fact, wherever there is moisture. Most are solitary. Many are free-living, but others are commensal, mutualistic, or parasitic. They are abundant, widespread, and dominate a number of ecosystems.

The heterotrophic protists, traditionally referred to as "protozoa," usually digest food particles in food vacuoles (discussed in Chapter 27). There are no special organelles for gas exchange; instead, the cell membrane serves as the exchange surface. Many species, particularly those living in hypotonic media such as fresh water, have contractile vacuoles, which function primarily in osmoregulation (described in Chapter 31). Most nitrogenous wastes are released as ammonia by diffusion across the cell surface. Locomotion occurs by formation of pseudopods, or by means of beating cilia or flagella. Where a single individual has many cilia, their action is coordinated by a system of fibrils connecting their basal bodies; these fibrils on the cell membrane have conductile properties like those of nerves in multicellular animals. Most freshwater and parasitic protists can encyst when conditions are unfavorable; they secrete a thick resistant case around themselves and become dormant.

Protists are separate evolutionary lines, most of which arose prior to the appearance of multicellular organisms. Comparisons of nucleic acid sequences yield a phylogeny that mixes species with chloroplasts (and thus are plantlike) and species that are exclusively heterotrophic (and thus animal-like) with no regard for human intuition (Fig. 21.3). For instance, dinoflagellates, which are often photosynthetic, are closely related to the nonflagellated animal-like ciliates (like *Paramecium*), but they are only distantly related to the euglenoids, the group that contains the well-known flagellated photosynthetic genus *Euglena*. The kinship is so remote, in fact, that the protists are often divided into six kingdoms, and up to 60 phyla. Molecular evidence shows that the group is far more diverse than plants, fungi, and animals taken together. Their cellular and subcellular organization is equally diverse. At the least, the group must be considered polyphyletic. For our purposes, the situation is less confusing than it could be because we place the single-celled algae most closely related to plants in the plant kingdom, and the parasitic unicellular myxosporians in the animal kingdom. The protists are divided into about 14 phyla and subphyla, which are grouped into heterogeneous assemblages of structurally similar but often unrelated taxa: Kinetoplastida, Alveolata, aerobic zooflagellates, and slime molds. The probable relationships of the major phyla are indicated in Figure 21.3.

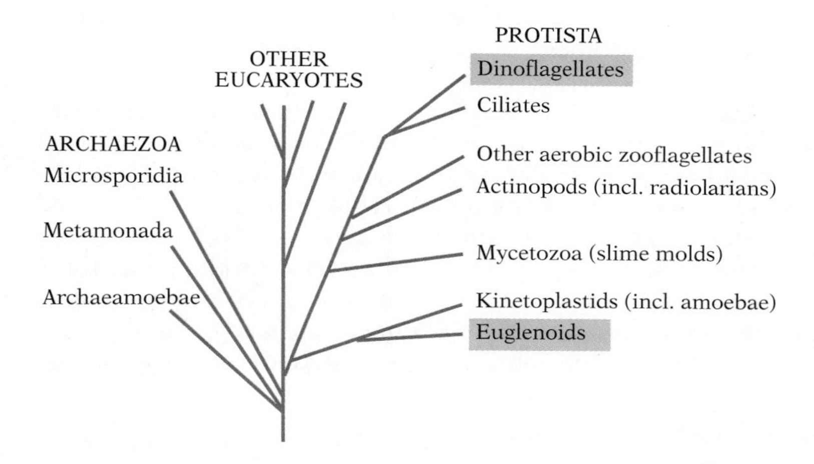

21.3 Probable phylogeny of archezoans and protists

This tree indicates multiple origins for the organisms now classified as Archezoa; the kingdom is therefore paraphyletic. The same pattern is shown for Protista, but the situation is more complex because the group is polyphyletic. Notice that the animal-like protists fall into two distantly related groups, each of which also has photosynthetic phyla (green). (See Fig. 18.32, page 525 for a comparison of homophyletic, paraphyletic, and polyphyletic groupings.)

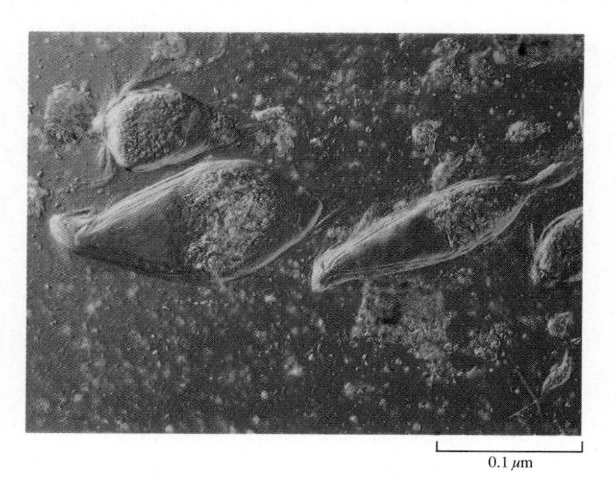

21.4 *Trichonympha*, a parabasalan flagellate that inhabits the gut of termites

This flagellate helps digest the cellulose in the termite's diet.

THE KINETOPLASTIDS

■ PARABASALA

The members of the phylum Parabasala have flagella as their principal locomotor organelles. They appear to be among the oldest of all the protozoans, and they may be derived from the ancestors of all higher organisms. A few parabasalans are free-living aquatic organisms, but most live as symbionts in the bodies of higher animals. Several species, for instance, live in the termite gut, where they work mutualistically to digest the cellulose the termite consumes (Fig. 21.4).

■ EUGLENOZOA: PROTOZOA WITH ODD MITOCHONDRIA

The Euglenozoa include two subphyla, Kinetoplasta and Euglenoida, which together are distinguished by their discoidal mitochondrial cristae—that is, their cristae take the form of flat disks joined to the inner membrane of the mitochondrion by a short stalk.

Kinetoplasta *Trypanosoma* is a genus of parasitic heterotrophic flagellates that cause several severe diseases in humans and domestic animals. *Trypanosoma gambiense* (Fig. 21.5) is the causative agent in African sleeping sickness.[1] The trypanosomes live in the blood of their host, where they multiply and release a poisonous by-product. In humans or domestic animals (but not in the native wild mammals of Africa), they eventually invade the nervous system, causing lethargy and finally death. The trypanosomes are spread from host to host by blood-sucking tsetse flies. When a tsetse fly draws blood from an infected animal, some of the trypanosomes are sucked into its intestine, where they multiply and undergo several developmental changes. They then migrate to the fly's salivary glands, where they undergo additional development and continue to multiply. If the fly now bites an uninfected vertebrate, some of the trypanosomes are injected from the salivary glands into the vertebrate host. Sleeping sickness, which makes large parts of Africa nearly uninhabitable for humans, is difficult to control because of the many wild animals that serve as a constant reservoir of trypanosomes.

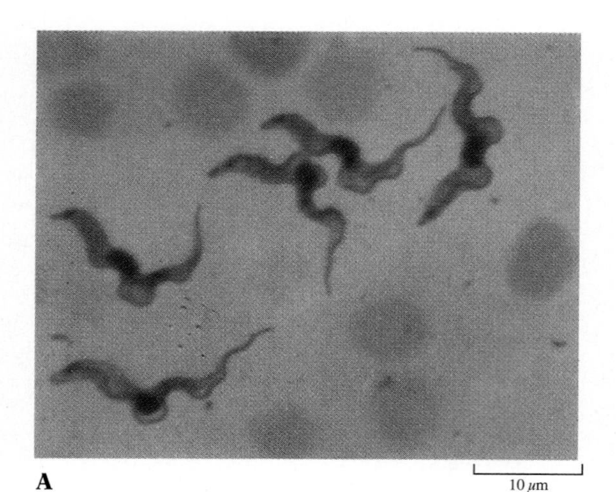

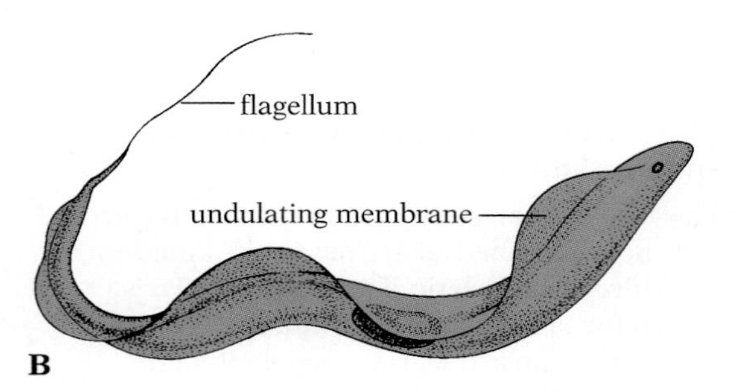

21.5 *Trypanosoma gambiense*, the cause of African sleeping sickness

(A) Photograph of the protozoans among red blood corpuscles. (B) Drawing showing the general structure of the trypanosome.

Euglenoida: the euglenoids Some euglenoids are photosynthetic, and thus plantlike. All are animal-like in lacking a cell wall and being highly motile; the species that lack chlorophyll are heterotrophic. Though their pigments (chlorophylls *a* and *b* and carotenoids) are like those of the green algae and land plants, they are not related to the algae. Their chloroplast envelope has three membranes compared with the two of plants. The original euglenoid chloroplast may have been obtained when a green algal cell was eaten by a heterotrophic euglenoid and the chloroplast be-

[1] African sleeping sickness is different from ordinary sleeping sickness, or encephalitis, which is caused by a virus.

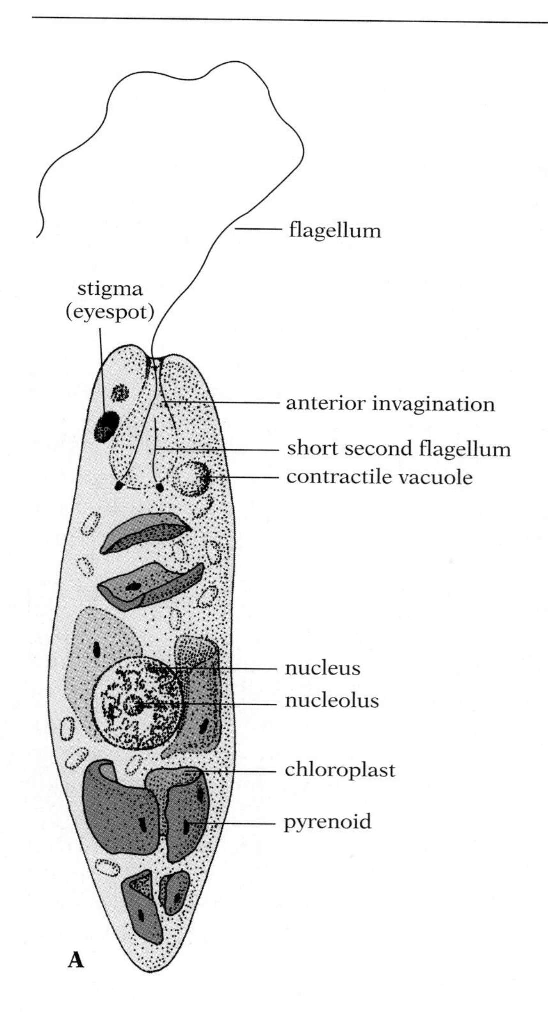

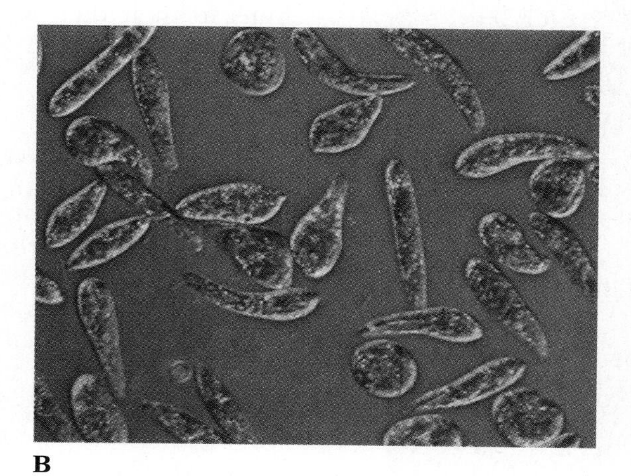

21.6 ***Euglena***

(A) For a full description of the structures shown in this drawing, see text. (B) Photograph of a group of Euglena.

came permanently incorporated into the cell rather than being digested. If correct, this scenario indicates that endosymbiosis can involve transfer of organelles *between* eucaryotes as well as the conventional pathway from procaryotes to eucaryotes. The carbohydrate storage molecule of euglenoids—a glucose polymer called paramylum—is unique to them. There are about 450 species of euglenoids. Most live in fresh water, but a few are found in soil, on damp surfaces, or even in the digestive tracts of certain animals.

A representative genus is *Euglena*. A typical cell is an elongate ovoid, with a long flagellum emerging from an anterior invagination (Fig. 21.6).[2] An orange granule, called the ***stigma*** (or eyespot), is located near the anterior end of the cell and functions in light detection and phototaxic responses—that is, movements that orient the organism toward or away from light. (There is no stigma in nonphotosynthetic euglenoid species.) Movement up during the day allows it to harvest more sunlight; movement down at night takes it into regions that usually have higher nutrient levels). Most green euglenoids have a special organelle, the ***pyrenoid***, which produces paramylum. A large contractile vacuole also lies near the anterior end of the cell.

[2] *Euglena* possesses a second shorter flagellum that does not emerge from the anterior invagination. In some other euglenoids both flagella are emergent.

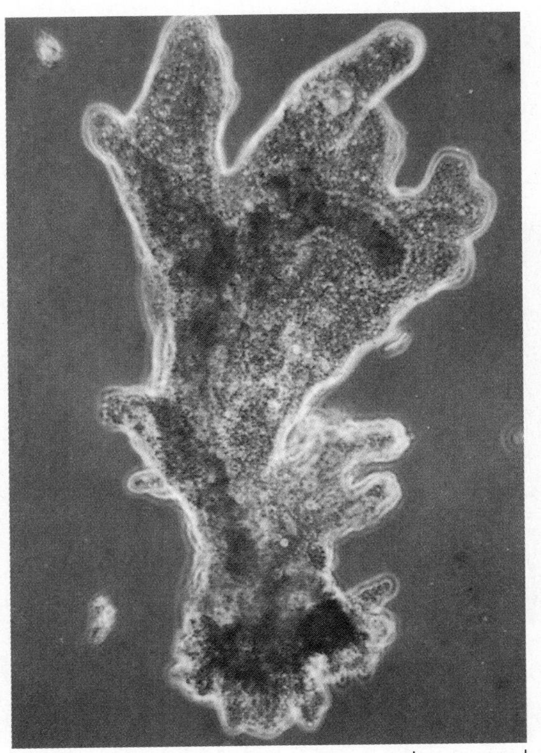

21.7 An amoeboid protozoan *(Chaos carolinensis)*

"Amoeboid"—from the Greek word for "change"—describes a cell that can alter its shape as it thrusts out or withdraws many armlike extensions.

21.8 A group of calcareous foraminiferan shells

The euglenoid species that lack chlorophyll are obligate heterotrophs—that is, they cannot produce their own nutrients. The species that have chlorophyll are facultative heterotrophs; although they are autotrophic, they can survive in the dark if they have a source of organic nutrients. Indeed, even in the light they are not entirely autotrophic, since they require one or more vitamins in their diet. Reproduction in euglenoids is by longitudinal mitotic cell division. Sexual reproduction has never been demonstrated.

■ THE AMOEBOID (SARCODINE) PROTOZOA

Sarcodines are the amoeboid protozoans. Their close relationship with the other kinetoplastids is underscored by the observation that many kinetoplastids undergo amoeboid phases while some primitive sarcodines have flagellated stages. Most sarcodines have tubular mitochondrial cristae.

The most familiar sarcodines are freshwater species of the genus *Amoeba* (Fig. 21.7), which have asymmetrical bodies that constantly change shape as new pseudopods are formed and old ones retracted. These pseudopods, which are large and have rounded or blunt ends, function both in locomotion and in feeding by phagocytosis. The food includes small algae, other protozoans, and even some small multicellular animals such as rotifers and nematode worms.

Some sarcodines secrete hard calcareous or siliceous shells (that is, shells containing calcium carbonate or silica) around themselves. These shells, often quite elaborate and complex, can be used in species identification. The pseudopods of the shelled sarcodines can be broadly lobed or very thin; in some forms they have no locomotor function, being exclusively feeding devices. This type of pseudopod does not flow around and engulf the prey, but instead functions as a trap. When prey organisms touch it, they become stuck in the mucoid adhesive secretion that coats the pseudopod surface. The secretion contains proteolytic enzymes that digest the prey, which is eventually enclosed in a food vacuole and drawn into the cell.

One group of shelled sarcodines, the Foraminifera (Fig. 21.8), has played a major role in the geological history of the earth. When the individuals die, their shells, which are calcareous, sink to the mud at depths of 2500–4500 m. Much of the earth's limestone and chalk were formed from deposits of foraminiferan shells.

ACTINOPODA

Instead of the labile pseudopods found in the sarcodines, these protozoa have stiff projectiles called axipods (Fig. 21.9) supported by a skeleton of regularly arranged and cross-linked microtubules. They include the shelled radiolarians (Fig. 21.10) and the Heliozoa, which are often naked. The bottom ooze in deeper parts of the ocean is composed chiefly of the siliceous shells of radiolarians. Radiolarian shells contribute to the formation of siliceous rocks such as chert.

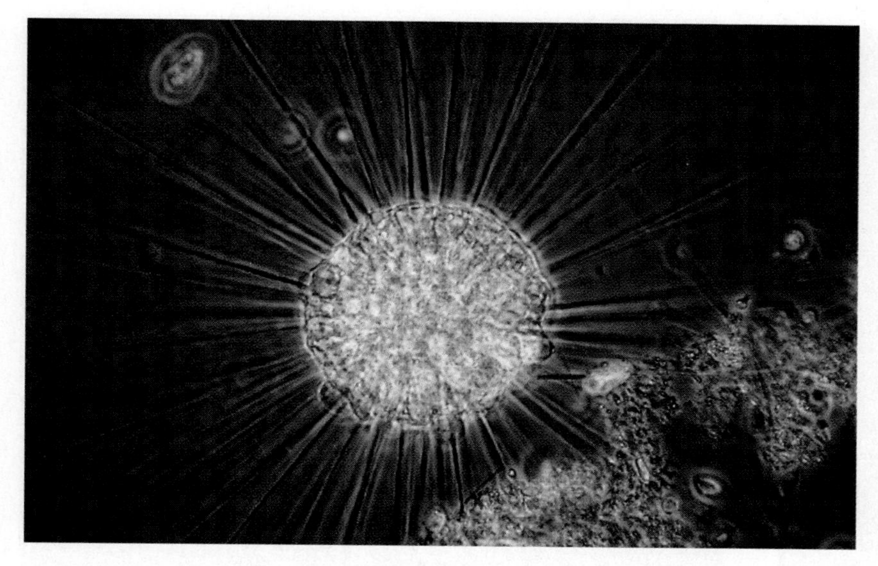

21.9 A sarcodine *(Actinosphaerium)* with long pointed axipods

21.10 A group of siliceous radiolarian shells

THE ALVEOLATE PROTISTS

This group contains three superficially different phyla: Ciliophora, Dinozoa (or Dinoflagellata), and Apicomplexa. They are characterized by membranous sacs, called alveoli, in their cell cortex. Sequence analysis confirms this phylogenetic grouping.

■ CILIOPHORA: PROTOZOA WITH CILIA

The ciliophorans include about 9000 species. As their name implies, ciliophorans use numerous cilia as locomotor organelles (Figs. 21.11, 21.12). In most species the cilia are present through-

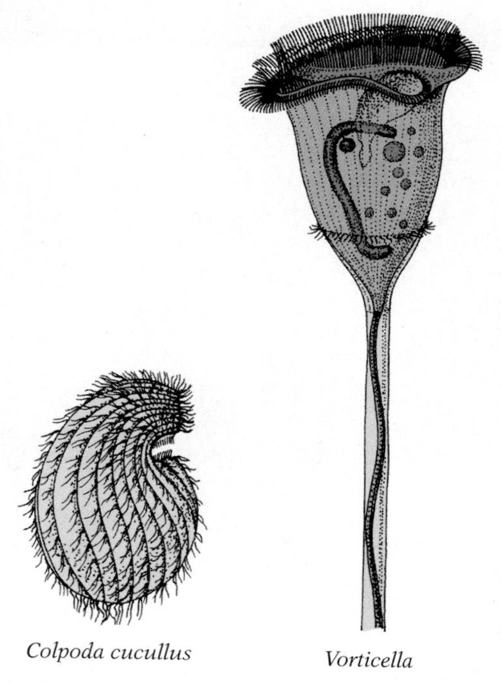

21.11 Two ciliate protozoans
The drawings illustrate the remarkable complexity of these unicellular organisms.

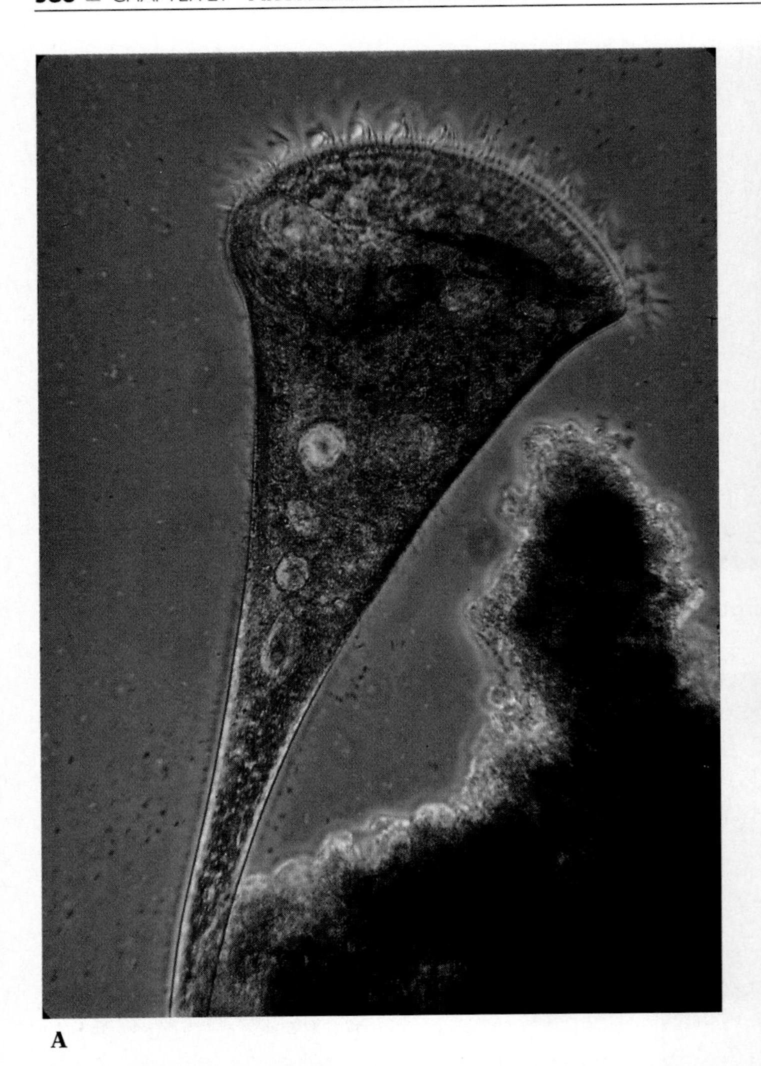

A

B

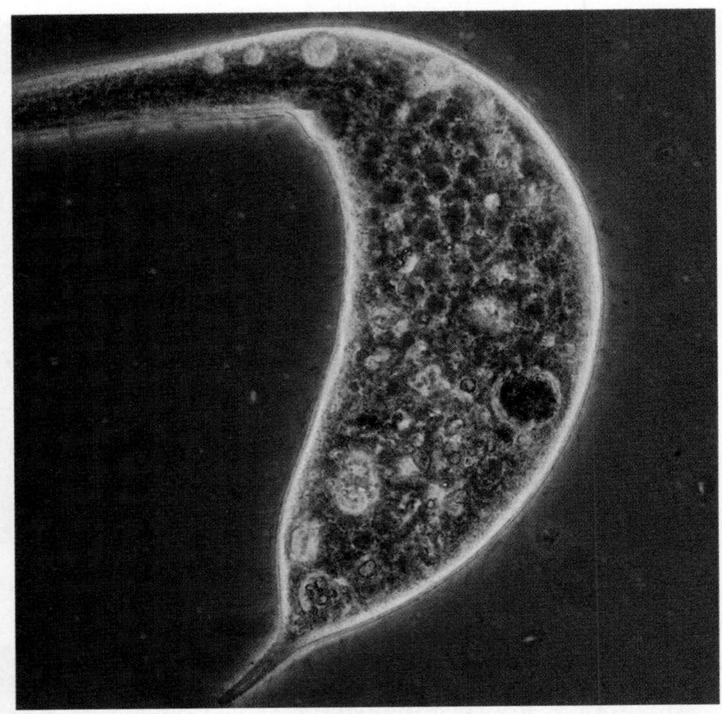

C

D

21.12 Four representative ciliates
(A) *Stentor*. (B) *Blepharisma*. (C) *Dileptus*. (D) *Paramecium* (with cilia specifically stained).

out life, but in Suctoria they are absent in the adult stages. The ciliates exhibit the greatest elaboration of subcellular organelles of any protozoan group (Fig. 21.13). Species such as *Paramecium* have a special oral groove and cytopharynx, into which food particles are drawn in currents produced by beating cilia; often there is also an anal "pore" through which indigestible wastes are expelled from food vacuoles. Skeletal fibrils connect the bases of the cilia, and there can be a system of contractile fibers (sometimes striated) analogous to the muscular system of multicellular animals.

21.13 A living ciliate viewed with Nomarski optics

This technique permits study of internal detail without damage to the organism. The large nucleus is visible near the center of the cell. Below the nucleus, and to the left, are contractile vacuoles of the sort filled by small fusion vesicles, some of which can be seen around the periphery of the largest vacuole. The cell contains numerous food vacuoles of various shapes containing several kinds of algae, including diatoms and filamentous forms. An array of undischarged trichocysts can be seen just inside the plasma membrane at right.

Stiffened plates occasionally found in the cell's pellicle (a proteinaceous layer just beneath the plasma membrane) together constitute a "skeleton." In some species there is a long stalk by which the individual may attach itself to the substrate (Figs. 21.11, 21.12A). A few species have tentacles for capture of prey. Some can discharge toxic threadlike darts called ***extrusomes*** (Fig. 21.13); these may function in defense against predators, in capturing prey, or in anchoring the organism to the substrate during feeding. The trichocysts of *Paramecium* are the best-known kind of extrusome.

Ciliates differ from nearly all other protozoans in having two types of nuclei: a large macronucleus and one or more small micronuclei. The macronucleus, which is polyploid (about 860-ploid in *Paramecium aurelia*), controls the normal metabolism of the cell. This is a special sort of gene amplification. The diploid micronuclei are involved only with reproduction and with giving rise to the macronucleus. During asexual reproduction the micronuclei divide mitotically.[3] The macronucleus, however, appears to divide by simple constriction.

Many ciliates occasionally engage in a sexual process called ***conjugation.*** Two individuals of appropriate mating types come together and adhere in the oral region; there is some fusion in the area of contact (Fig. 21.14). Next, the micronuclei divide by meio-

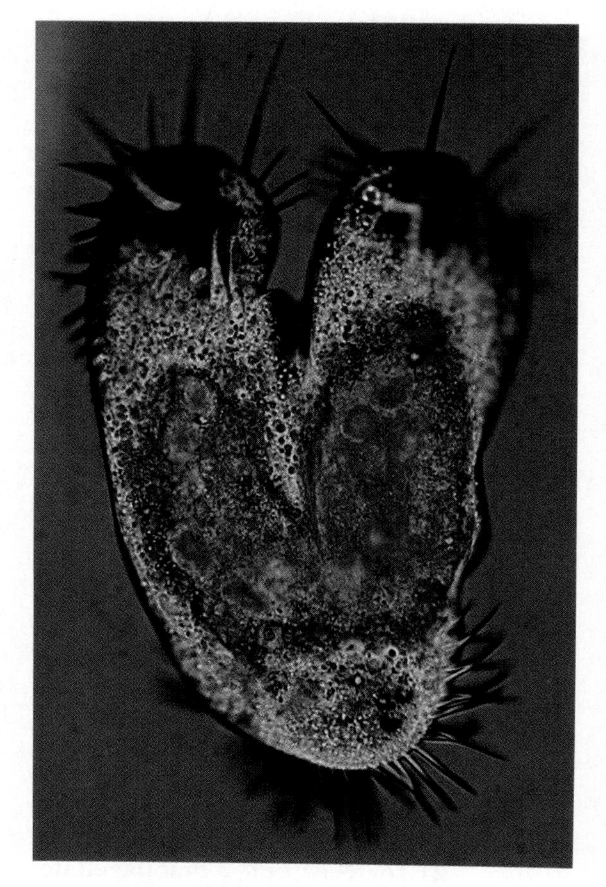

21.14 Two *Stylonychia* conjugating

[3] The nuclear envelope of the micronuclei of ciliates does not disappear during mitosis, and the spindle forms inside the membrane-bounded nucleus.

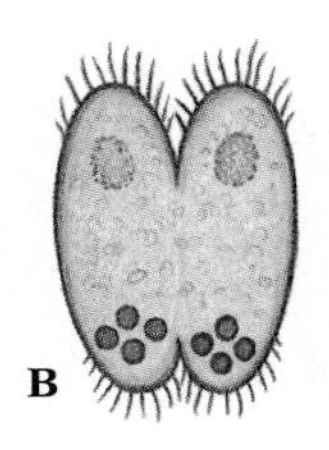

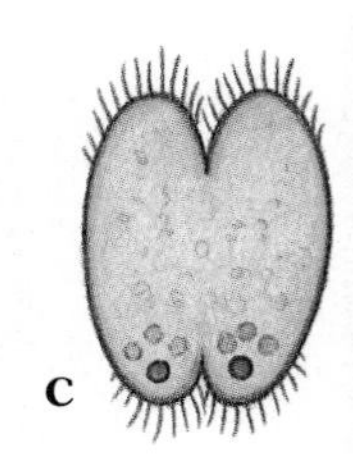

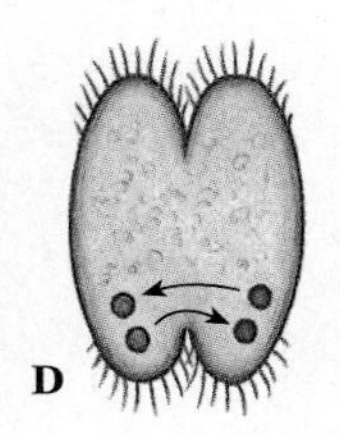

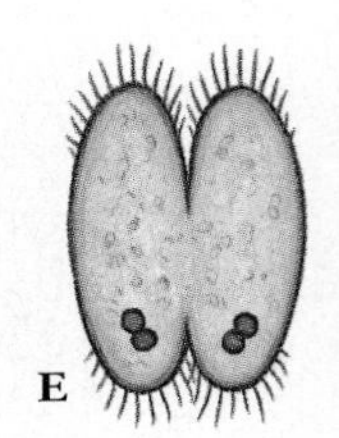

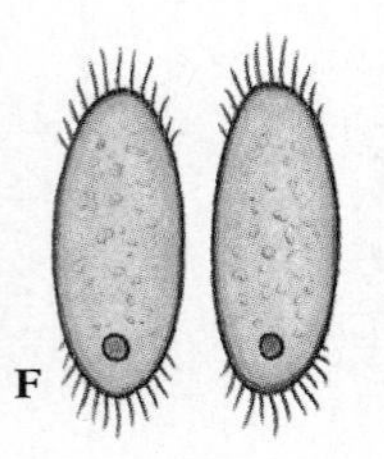

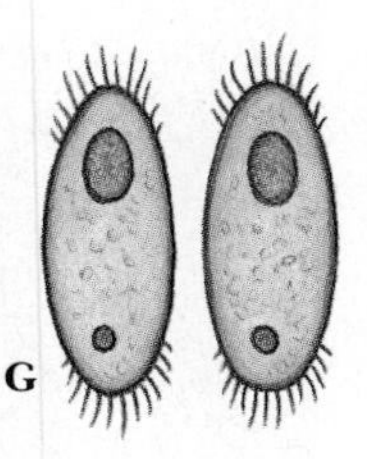

21.15 Conjugation in ciliates

(A–B) Each conjugating ciliate eliminates its macronucleus while its micronucleus undergoes meiosis. (C) All but one of these haploid nuclei in each organism is eliminated. (D-F) The survivors are duplicated, and one copy is transferred to the partner; the two now fuse to form a new diploid micronucleus. (G) The chromosomes are then repeatedly replicated to create a macronucleus.

sis (Fig. 21.15B). Then all but one of the resulting haploid micronuclei in each cell disintegrate; the macronucleus also usually disintegrates (Fig. 21.15C). The surviving "gamete" nucleus is duplicated through mitosis. One of the two nuclei in each cell remains stationary and functions as the female nucleus; the second moves into the other cell and fuses with that cell's female nucleus in the process of fertilization (Fig. 21.15D–E). Thus each cell acts as both male and female, donating a nucleus and receiving one in return. When the two cells part, each has a new recombinant diploid nucleus (Fig. 21.15F). This nucleus then undergoes one or more divisions, and some of the new nuclei thus produced develop into macronuclei (Fig. 21.15G).

■ DINOFLAGELLATES

The dinoflagellates are small, usually unicellular alveolate organisms (Fig. 21.16). When present (only in cysts), the cell wall is composed largely of cellulose; some species also build an "armor" of cellulose plates, each constructed by one of the cortical alveoli. Photosynthetic species possess chlorophylls *a* and *c*; they usually

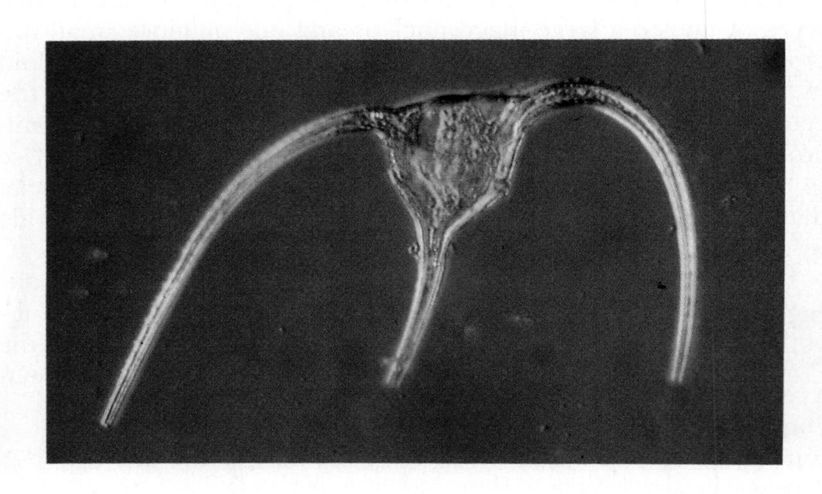

21.16 ***Ceratium*, a dinoflagellate**

have a yellowish-green to brown color as a result of an abundance of carotenoids and other accessory photosynthetic pigments called xanthophylls, several of which are unique to these organisms. There are many colorless species of dinoflagellates as well; some feed on particulate organic matter, and others live as parasites in marine invertebrates. The energy-storage material is either starch or oil.

Most species have two very unequal flagella, which are attached laterally. One of these runs along a groove to the posterior end of the cell and extends beyond the cell like a tail; the other lies in a groove that encircles the midportion of the cell like a belt (Fig. 21.17). Many dinoflagellates produce trichocysts similar to those of ciliate protozoans such as *Paramecium*. A few even produce organelles that resemble the nematocysts of hydra (p.773). Some species can produce light, and are responsible for much of the luminescence often seen in ocean water at night.

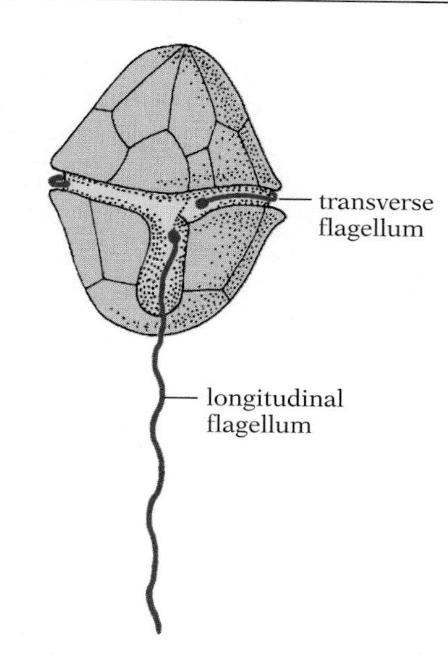

21.17 A freshwater dinoflagellate *(Glenodinium cinctum)*

The two flagella lie in distinctive grooves on the surface of the cell.

As plankton, dinoflagellates (along with the cyanobacteria and diatoms—see Chapter 22) are the primary producers of organic matter in the marine environment; they play a lesser role in fresh water. Not only are they important as food, but they also function as endosymbionts for an amazing variety of marine invertebrates. Some types of corals take in as much as 60% of the carbon fixed by their dinoflagellate symbionts (zooxanthellae); such corals would deposit calcium in their skeletons only one-tenth as fast if they lacked the symbionts (see Fig. 40.42, p. 1191).

A number of species of dinoflagellates are poisonous. Some of these contain red pigments. When the water is warm and upwelling currents bring extra nutrients up from the bottom, or heavy rains carry phosphates into the sea, these species can multiply rapidly until they turn the water red. They produce a potent nerve toxin—saxitomin—which blocks sodium channels. During one of these periodic "blooms," known commonly as a red tide, the toxin can kill many millions of fish. Red tides are fairly common in the Gulf of Mexico, off the coast of Florida, and along the shores of California. In at least some cases, the otherwise autotrophic dinoflagellates then consume the flesh of the fish they have killed.

■ APICOMPLEXA: SPORE-FORMING PROTISTS

Since many apicomplexans have a sporelike infective cyst stage in their life cycles, they were earlier called sporozoa. They were once considered primitive fungi, but their cell walls are composed of cellulose rather than chitin, the hallmark of true fungi. The apicomplexans lack special locomotor organelles (except in the male gametes). Nearly all are internal parasites, and they usually have complex life cycles. They cause a variety of serious diseases, including malaria in humans.

Malaria, which is caused by species of the genus *Plasmodium*, is transmitted from host to host by female *Anopheles* mosquitoes. When an infected mosquito bites a person and starts to suck blood, it releases saliva containing both a chemical that prevents coagulation of the blood and *Plasmodium* cells in a stage called sporozoites (Fig. 21.18; stage 5). The sporozoites enter the victim's

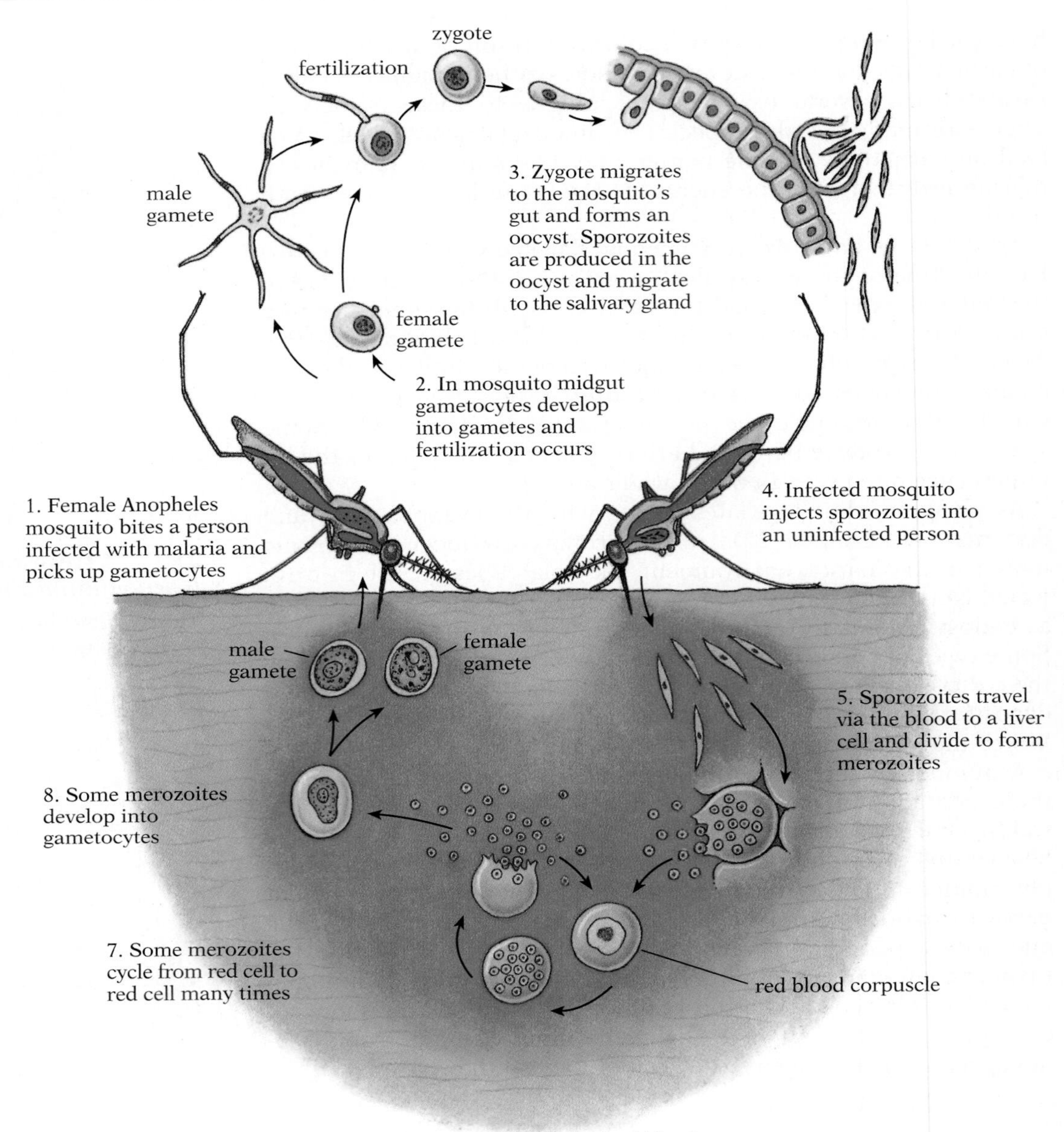

21.18 Life cycle of *Plasmodium*, the malarial parasite

bloodstream and are carried to the liver, where they enter liver cells and grow for 5-15 days. Then each sporozoite divides, producing a large number of new cells called merozoites. These are released from the liver cells and locate red blood corpuscles, attaching to a surface receptor involved in immune-system response. The merozoites enter the corpuscles, where they reproduce asexually, producing additional merozoites. At regular intervals (48 hours in some types of malaria), all infected red corpuscles burst, releasing the merozoites, which enter new blood corpuscles and repeat the asexual reproductive process. The host experiences attacks of chills and fever each time a release occurs.

Eventually some of the merozoites develop into special sexual cells capable of becoming either male or female gametes; however, they do not mature as gametes in the blood of the human host. If an anopheline mosquito sucks blood containing these cells, they complete their development in the midgut of the mosquito. The male gametes then fertilize the female gametes, and the zygote thus produced, which is amoeboid, works its way into the wall of the gut and encysts. Within the cyst, a series of divisions ultimately produces new sporozoites, which are released when the cyst ruptures. These sporozoites migrate to the salivary glands of the mosquito, from which they are discharged into a new vertebrate host when the mosquito feeds.

THE FLAGELLATED PROTOZOA

Apart from the numerous nonphotosynthetic dinoflagellates and euglenoids, flagellated protozoa are widespread in the oceans, fresh water, and soil. One of the several types of flagellated protozoa is the group called the choanoflagellates (phylum Choanozoa) which have a single flagellum surrounded by a collar (like the choanocyte of sponges; see Fig. 24.3, p. 656) and flat mitochondrial christae like those of animals and fungi. Very probably, the sponges and other animals evolved from choanoflagellates or a common ancestor; fungi too may owe their origin to this group.

Another important type of flagellated protist, phylum Opalozoa, has tubular mitochondrial cristae, and is much more diverse. Opalozoans usually have two flagella (except for the parasitic species, which have four or more). Some are commensals in the guts of amphibians, but most are free-living. They include the most common soil flagellates, and sometimes dominate aquatic ecosytems where their populations can exceed 1000 per milliliter.

MYCETOZOA: THE SLIME MOLDS

The slime molds are curious organisms decidedly protozoan at some stages in their life cycle and superficially funguslike at others. The three classes of free-living slime molds are generally found growing on damp soil, rotting logs, leaf mold, or other decaying organic matter in moist woods, where they look like glistening masses of slime. They are sometimes white, but are often red or yellow (Fig. 21.19).

In the so-called ***true slime molds*** (class Myxogastrea), the vegetative phase of the life cycle is a large diploid multinucleate amoeboid mass called a ***plasmodium,*** which moves slowly and feeds on organic material by phagocytosis. Under certain conditions, however, the plasmodium becomes stationary and develops ***fruiting bodies,*** which may be either simple rounded masses or elaborate stalked organs. Meiosis occurs within the ***sporangia*** (spore-producing structures) of the fruiting bodies, and the haploid cells thus formed are released as spores, whose walls contain cellulose. When the spores germinate, they produce flagellated gametes. These fuse

21.19 Plasmodium of a yellow slime mold *(Hemitrichia stipitata)*

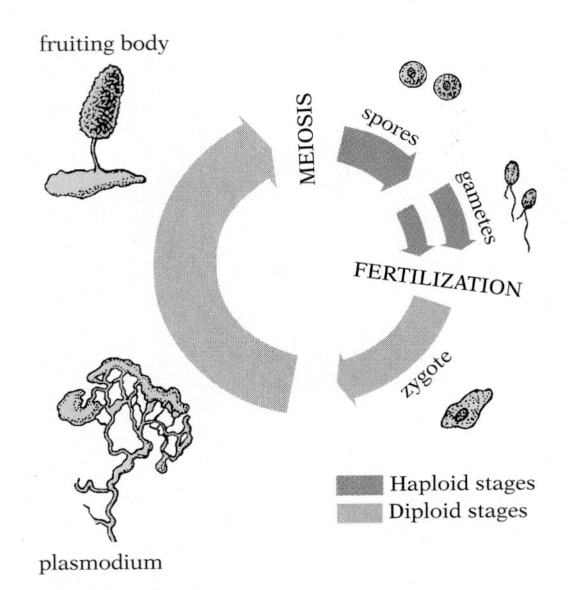

21.20 Life cycle of a true slime mold
See text for description.

in pairs to form zygotes, which become amoeboid. As this amoeboid form flows along the substrate, engulfing bacteria and other organic particles and digesting them in vacuoles, its diploid nucleus undergoes repeated mitotic divisions without accompanying cytokinesis. In this way the zygote develops into a multinucleate plasmodium, which may grow to a length of 25 cm. Some growth may also occur by fusion of the cytoplasms of two or more young plasmodia. In summary, the life cycle proceeds from diploid amoeboid plasmodium, to stationary spore-producing plasmodium, to haploid spores, to flagellated gametes, to zygote, and back to amoeboid plasmodium (Fig. 21.20).

The life cycle of the ***cellular slime molds*** (class Dictyostelea) is quite different from that of the true slime molds. The two groups are probably not closely related; they may have evolved from a common amoeboid ancestor that formed cysts on stalks like the members of the third class of the Mycetozoa, the protosteleans. In the cellular slime molds, the spores do not develop into flagellated gametes, but instead give rise to free-living soil-inhabiting amoeboid cells, each with a single haploid nucleus (Figs. 21.21, 21.22A). The amoebae feed on bacteria and other organic matter. During this feeding stage the amoebae divide repeatedly, producing independent uninucleate daughter cells. As the local food supply diminishes, the behavior of the amoebae changes. Two quite different cycles—one sexual and the other asexual—are possible, depending on environmental conditions. Most often the amoebae enter the

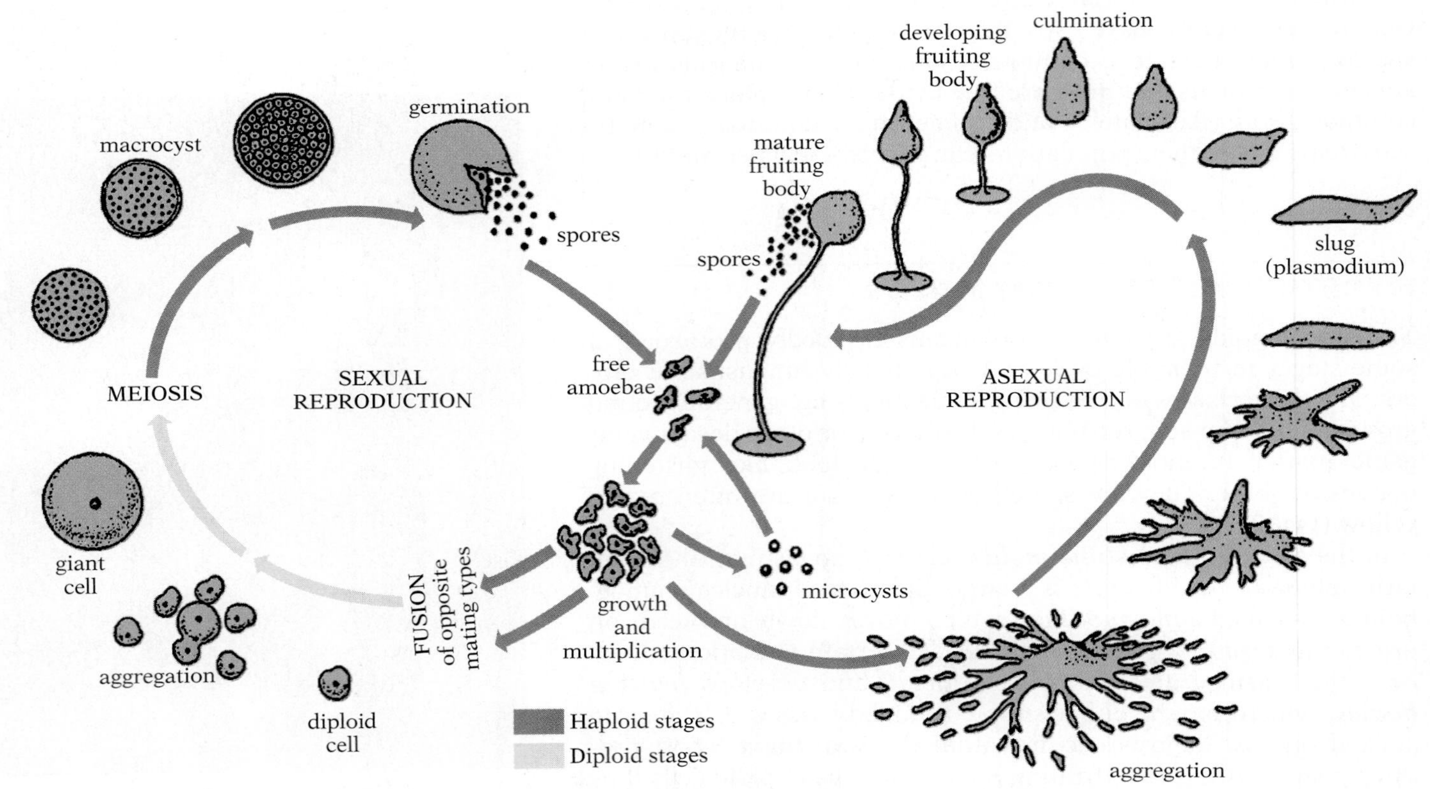

21.21 Life cycle of a cellular slime mold See text for discussion.

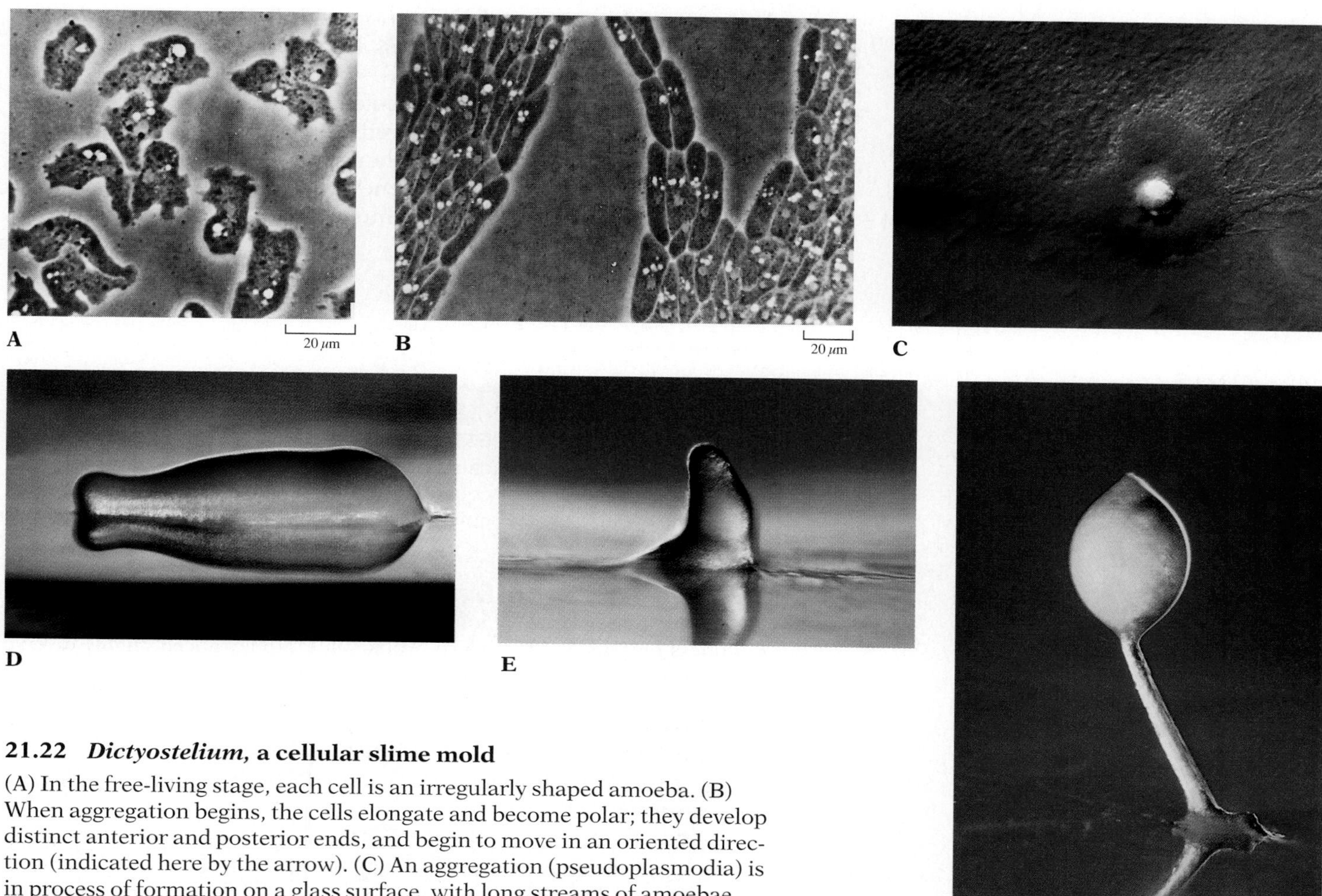

21.22 *Dictyostelium,* a cellular slime mold

(A) In the free-living stage, each cell is an irregularly shaped amoeba. (B) When aggregation begins, the cells elongate and become polar; they develop distinct anterior and posterior ends, and begin to move in an oriented direction (indicated here by the arrow). (C) An aggregation (pseudoplasmodia) is in process of formation on a glass surface, with long streams of amoebae funneling into the center of the aggregation. (D) A pseudoplasmodium moving on a glass surface. E) A pseudoplasmodium beginning to produce a stalk. (F) A fruiting body, the culmination of the asexual phase.

asexual cycle: They cease feeding and begin to aggregate at central collecting points where they clump together to form a sluglike ***pseudoplasmodium*** (Fig. 21.22C-D). The individual haploid cells do not fuse. The pseudoplasmodium of the cellular slime molds is thus very different from the diploid multinucleate plasmodium of the true slime molds. The slug may move around as a unit for a while, but eventually it becomes sedentary and forms a stalked fruiting body in which new spores are produced (Fig. 21.22E-F). This cycle does not include any sexual events.

Under very moist conditions, when the soil is so wet that the fruiting bodies would probably be under water, making spore dispersal difficult, the amoebae enter a different cycle, in which a sessile aggregation of free-living amoebae can form. If by chance the first two amoebae to begin this aggregation are of different mating types, sexual recombination occurs. The two nuclei combine, and as new cells aggregate, they are ingested by phagocytosis. The resulting giant cell grows and grows. Meiosis then restores the haploid state, and continued replication and division lead to a spore-filled macrocyst. When conditions become favorable, the cyst releases its spores.

The signal that cellular slime molds use to control aggregation is cAMP, the same substance involved in many types of intracellular control in animals. On the other hand, cAMP is a waste product of many bacteria, and is used by some nematodes as a homing signal to a potential meal; perhaps the hungry amoebae are simply zeroing in on a likely food source.

CHAPTER SUMMARY

ARCHEZOANS: THE OLDEST EUCARYOTES

Archezoans are the descendants of very early eucaryotes, having diverged before the endosymbiotic acquisition of mitochondria and peroxisomes, as well as the evolution (or acquisition) of ER and Golgi. Their ribosomes are also primitive. (p. 579)

PROTISTS: HOW SHOULD THEY BE CLASSIFIED?

While the ancestors of multicellular plants, animals, and fungi were doubtless protists, most modern protists fall into advanced, highly diverse groups not closely related to multicellular organisms. (p. 580)

THE KINETOPLASTIDS

Most kinetoplasts have mitochondria with disklike cristae. (p. 582)

PARABASALA These flagellated species are the most primitive members of the protists. (p. 582)

EUGLENOZOA: PROTOZOA WITH ODD MITOCHONDRIA This group is flagellated, and includes euglenoids, many of which are photosynthetic, lack cell walls, and have a unique energy-storage molecule. Another phylum, the kinetoplastans, include the agent of sleeping sickness. (p. 582)

THE AMOEBOID (SARCODINE) PROTOZOA This group contains amoeba, some of which are normally enclosed in shells. (p. 584)

ACTINOPODA

Members of this group have stiff axipods. (p. 584)

THE ALVEOLATE PROTISTS

Members of this group derive their name from the membranous sacs, called alveoli, in their cell cortex. (p. 585)

CILIOPHORA: PROTOZOA WITH CILIA Ciliates use cilia for movement and food capture. (p. 585)

DINOFLAGELLATES Most dinoflagellates are photosynthetic and have chlorophylls *a* and *c*. Though flagellated, they are closely related to the ciliates. Many have cellulose-based walls. They are a major component of plankton. (p. 588)

APICOMPLEXA: SPORE-FORMING PROTISTS These protists lack the chitinous walls that are diagnostic of true fungi, but they mimic the fungal lifestyle in various ways. (p. 589)

THE FLAGELLATED PROTOZOA

This widespread group is unrelated to any of the other groups; one may have been the precursor of fungi and animals. (p. 591)

MYCETOZOA: THE SLIME MOLDS

True slime molds have a multinucleate plasmodium lacking internal cellular divisions that engulfs food; cellular slime molds live as isolated cells, forming a plasmodium only during asexual reproduction; even then the individual cells remain distinct and do not feed. (p. 591)

STUDY QUESTIONS

1 If introns were once the rule, there must have been some selection pressure to dispense with both introns and nuclear envelopes in the lines that led to modern bacteria. Can you think what these might be, and what signs we might look for in current organisms? How about reproductive speed?

2 If your model of the loss of introns/envelopes is correct, what might this mean for our thinking about Archezoa? Does the fact that archezoan phyla lack mitochondria make a difference? (pp. 579–580)

3 To what extent can you match the niches and lifestyles of bacteria with those of protists? (pp. 559–575, 579–594)

4 Some protists, like bacteria, are haploid throughout their lives; others have brief diploid stages; in other species the diploid phase predominates. Draw up a list of costs and benefits for haploidy, diploidy, and a mixed strategy. Is there any obvious correlation with the life history of the various species of protists?

SUGGESTED READING

BONNER, J. T., 1983. Chemical signals of social amoebae, *Scientific American* 248 (4).

GRELL, K. G., 1973. *Protozoology*. Springer, Heidelberg. *Excellent comprehensive treatment of the Protozoa.*

KABNICK, K. S., AND D. A. PEATTIE, 1991. *Giardia:* A missing link between procaryotes and eucaryotes. *American Scientist* 79 (5). *On an archezoan that is an intestinal parasite.*

CHAPTER 22

CHROMISTS AND PLANTS

Plants and their relatives have traditionally been divided into algae and terrestrial plants. Algae are aquatic, obtaining water, oxygen, nutrients, and even mechanical support directly from the surrounding medium. As a result, algae need little specialization beyond leaves, structures for attachment to the substrate, and simple reproductive organs. The entire plant consists of a single tissue, known as the ***thallus***. Terrestrial plants, by contrast, face enormous problems: supporting their own weight, protecting the somatic tissue and reproductive organs from desiccation, obtaining water and nutrients from the soil and transporting them to the leaves, and so on. We begin this chapter by looking at groups of algae, taking them in the order in which they appear to have evolved (Fig. 22.1). Then, after seeing how terrestrial adaptations evolved, we look at the diversity of modern land plants.

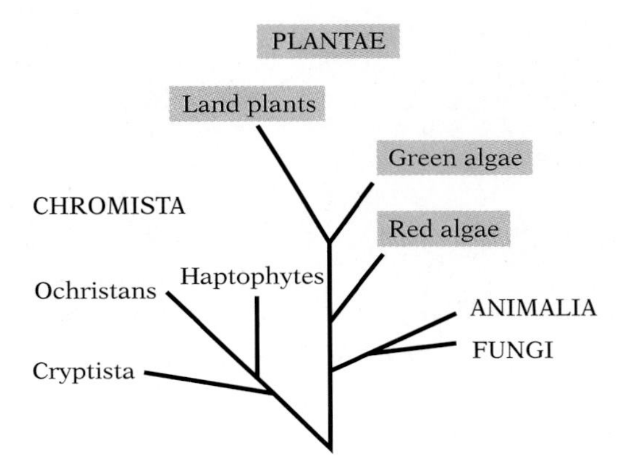

22.1 Probable phylogeny of plants and chromists

The brown algae (ochristans), cryptomonads (cryptista), and haptophytes (as well as a few nonphotosynthetic groups, which are not shown), constitute the kingdom Chromista; the remaining true algae are part of the kingdom Plantae. The Chromista are unrelated to plants, except that they obtained their chloroplasts by capturing green or red algae.

CHROMISTA: THE CHLOROPHYLL *C* ALGAE AND PSEUDOFUNGI

Sequence analysis indicates that, except for dinoflagellates, there are three major groups of algae with chlorophyll *c* (in addition to the usual chlorophyll *a*) that probably share a common descent: the cryptista (cryptomonads), haptophytes, and ochristans (which include the kelps). All three groups also have their chloroplasts *inside* the rough endoplasmic reticulum. Two of the three (cryptomonads and ochristans) have odd tubular hairs that project from one or both of their cilia and/or flagella (Fig. 22.2). This last characteristic is the basis for also incorporating several small groups of nonphotosynthetic species into this kingdom, known tentatively as Chromista.

Whether the chromistans are best considered a division of plants or protists, or deserve their own kingdom, is a point of contention. Sequence analysis places them almost halfway between the ciliate-dinoflagellate branch of the protists and the red algae line of the plants. On the one hand, it seems unreasonable to call them plants: unlike red and green algae (placed here in the plant kingdom), which obtained their chloroplasts by capturing a photosynthetic eubacterium, the chromists appear to have appropriated their chloroplasts much later, from the eucaryotic red and green

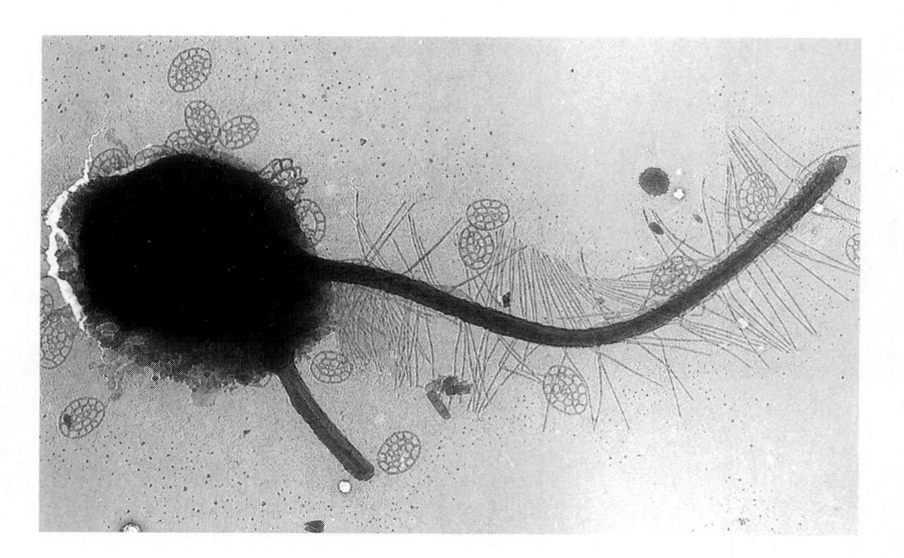

22.2 Hairs on a chromistan flagellum

These curious structures serve to reverse the thrust produced by such flagella; as a result, these flagella pull the cell along, while naked flagella push cells. The lacy oval structures are silica scales that have become detached from the cell.

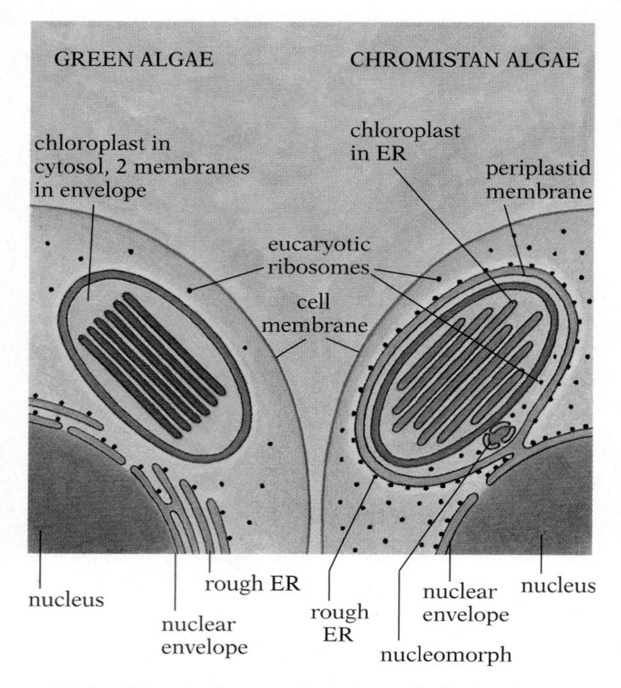

22.3 Unusual organization of chromistan algae compared with green algae

Green algae and plants have a conventional nuclear envelope with ER, and a chloroplast with a two-membrane envelope containing loose chloroplast DNA. Chromistan algae differ from green algae and plants in one or more of the following ways: (1) The chloroplast is enclosed in the ER. (2) The chloroplast has three membranes, the outermost of which (the periplastic membrane) is the former cell membrane of the endocytosed alga that housed the chloroplast. (3) The chloroplast has a nucleomorph, the remains of the nucleus of the endocytosed alga. This nucleomorph contains, for instance, the genes for an rRNA unique to the region that corresponds to the cytoplasm of the endocytosed alga. The equivalent cytoplasmic rRNA is encoded by the true nucleus, while the rRNA used inside the double-membrane-bounded region corresponding to the endocytosed algal chloroplast is found loose in that compartment.

algae that had already "tamed" these organelles (Fig. 22.3). On the other hand, it seems equally inappropriate to place them within the already polyphyletic Protista; given that brown algae are multicellular and can grow into plants dozens of meters long, calling them protists seems unreasonable. Other specialists divide them into two or three new, separate kingdoms on the basis of separate chloroplast-capture "events"; the independent-capture scenario is especially plausible for the cryptomonads, whose source of chloroplasts is more likely to have been a red alga. The many unique biochemical and ultrastructural features common to chromists certainly argue for separate kingdom status of some sort, but multiplying the number of kingdoms to accommodate minor branches is unwieldy and inconvenient. We have chosen the middle course of separating them into a single paraphyletic kingdom.

■ OCHRISTA: THE YELLOW-GREEN, BROWN, AND GOLDEN-BROWN ALGAE AND THE DIATOMS

The earliest classifications of algae were based on color, which depends on the sorts of pigments the cells contain. Fortunately, later study of other important characters—particularly flagellar structure, energy-storage materials, and the cell wall—showed that algae of like pigmentation usually share such characters as well. Thus most species in the six classes of algae in the phylum Ochrista are some shade of yellow or brown, in part from a predominance of carotenoids; they also resemble each other in possessing chlorophylls *a* and *c* but not *b*, and in using a polysaccharide other than starch as their storage material. The walls of many species are impregnated with silica or calcium. Two anteriorly attached flagella of unequal length are common (Fig. 22.2).

The majority of the ochristan classes are unicellular or colonial, though a few have large but structurally simple multicellular bodies. Reproduction is usually asexual. Most of the yellow-green and golden-brown algae live in fresh water. Diatoms are abundant in

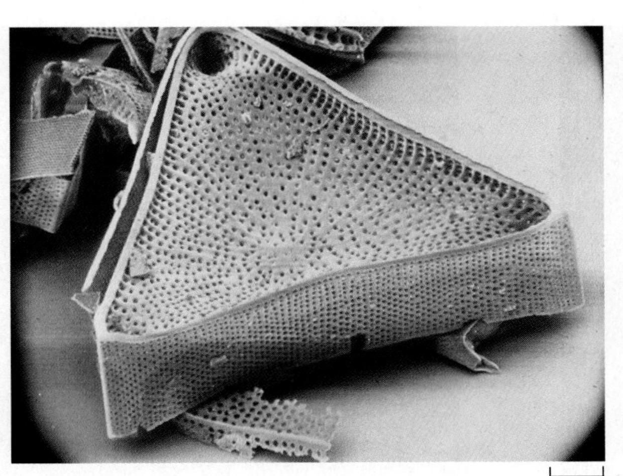

22.4 Scanning electron micrograph of a diatom (*Trinacria regina*)
The organism has a distinctly boxlike structure.

both freshwater and saltwater habitats. Some ochristans can both photosynthesize and capture prey.

Diatoms The diatoms are unusual in that the vegetative cells (cells not specialized for reproduction) are ordinarily diploid—not haploid, as is usual in primitive plantlike organisms. Unlike most other ochristans, diatoms lack flagella; some species, however, produce flagellated sperm cells. Silica-impregnated glasslike walls, composed of two pieces that fit together like a box with its lid (Fig. 22.4), often give the cells a jewellike appearance (Fig. 22.5). Classification of the diatoms is based almost entirely on the characters of the walls, or shells, as they are commonly called.

When diatoms die, their shells sink to the bottom, where they may accumulate in large numbers, forming deposits of diatomaceous earth. This material is used as an ingredient in many commercial preparations, including detergents, polishes, paint removers, decolorizing and deodorizing oils, and fertilizers. It is also used as a filtering agent (in swimming-pool filters, for example) and as a component in insulating and soundproofing products.

The diatoms play an extremely important role in aquatic food webs,[1] in both freshwater and marine habitats. They are the second most abundant component of marine plankton (after cyanobacteria); it is not unusual for a gallon of sea water to contain as many as 1–2 million diatoms. ***Plankton*** consists, by definition, of small organisms floating or drifting near the surface. Planktonic organisms are generally divided into two groups—phytoplankton (plant plankton) and zooplankton (animal plankton). The organisms constituting phytoplankton are the principal photosynthetic producers in marine communities.

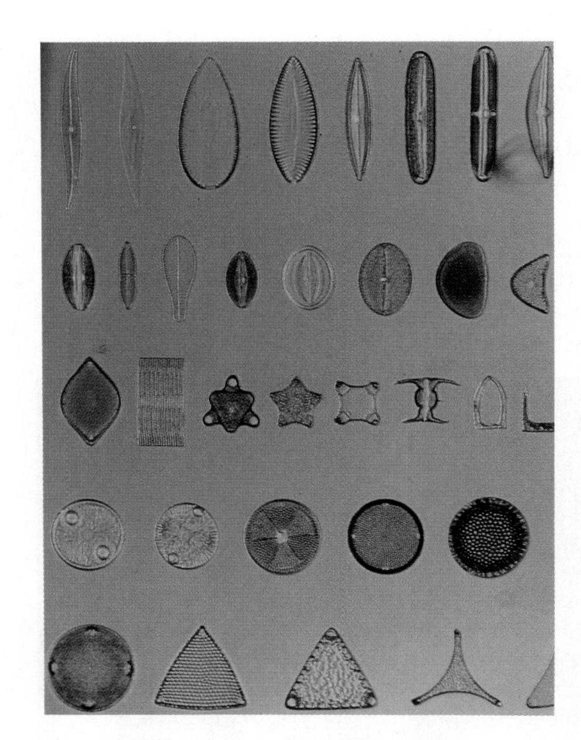

22.5 Some representative diatoms

Brown algae The brown algae (Phaeophyceae) are almost exclusively marine. Many of the seaweeds are members of this group. They are most common along rocky coasts of the cooler parts of the oceans, where they normally grow attached to the bottom by ***holdfasts*** in the littoral (intertidal) and upper sublittoral zones

[1] As discussed in Chapter 40, a food web is the network of energy transfers that depends on which kinds of organisms consume which other kinds.

22.6 A brown alga (*Saccorhiza polyschides*)

Brownish pigments mask the green pigment chlorophyll also present in this seaweed. The flattened leaflike blades are supported by the water in which the plant is growing; when uncovered at low tide, the blades lie flat on the substrate. Notice the holdfasts, which anchor this alga to the substrate.

22.7 Brown algae (mostly *Fucus*) growing on rocks exposed at low tide

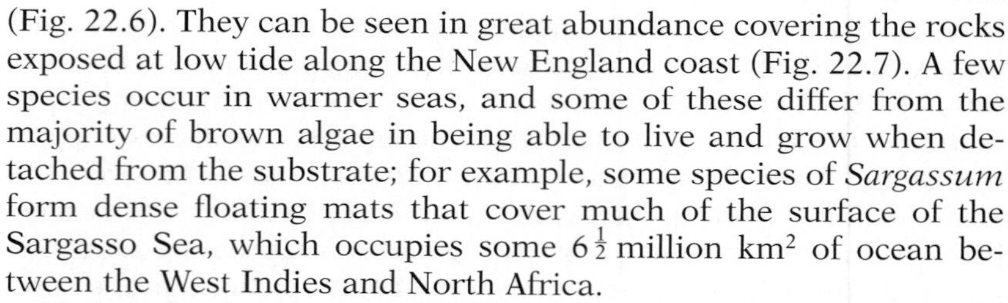

(Fig. 22.6). They can be seen in great abundance covering the rocks exposed at low tide along the New England coast (Fig. 22.7). A few species occur in warmer seas, and some of these differ from the majority of brown algae in being able to live and grow when detached from the substrate; for example, some species of *Sargassum* form dense floating mats that cover much of the surface of the Sargasso Sea, which occupies some $6\frac{1}{2}$ million km^2 of ocean between the West Indies and North Africa.

All brown algae are multicellular, and most are macroscopic, some growing as long as 45 m or more. The thallus (plant body) may be a filament, or it may be a large and rather complex three-dimensional structure (Fig. 22.8). The individual cells have cell walls composed of cellulose and a gummy carbohydrate called alginic acid. The cells usually contain a large vacuole, one or several plastids, and sometimes a pyrenoid—an organelle in the plastid that manufactures the storage products, a polysaccharide called laminaran. Unlike the cells of most higher land plants, brown algae usually have centrioles.

Like all photosynthetic eucaryotes, the brown algae possess chlorophyll *a*. However, they have chlorophyll *c* instead of the chlorophyll *b* found in green algae and in land plants. Large

22.8 A giant kelp (*Macrocystis pyrifera*)

At the base of each of the flattened blades is a gas-filled bladder that functions as a float.

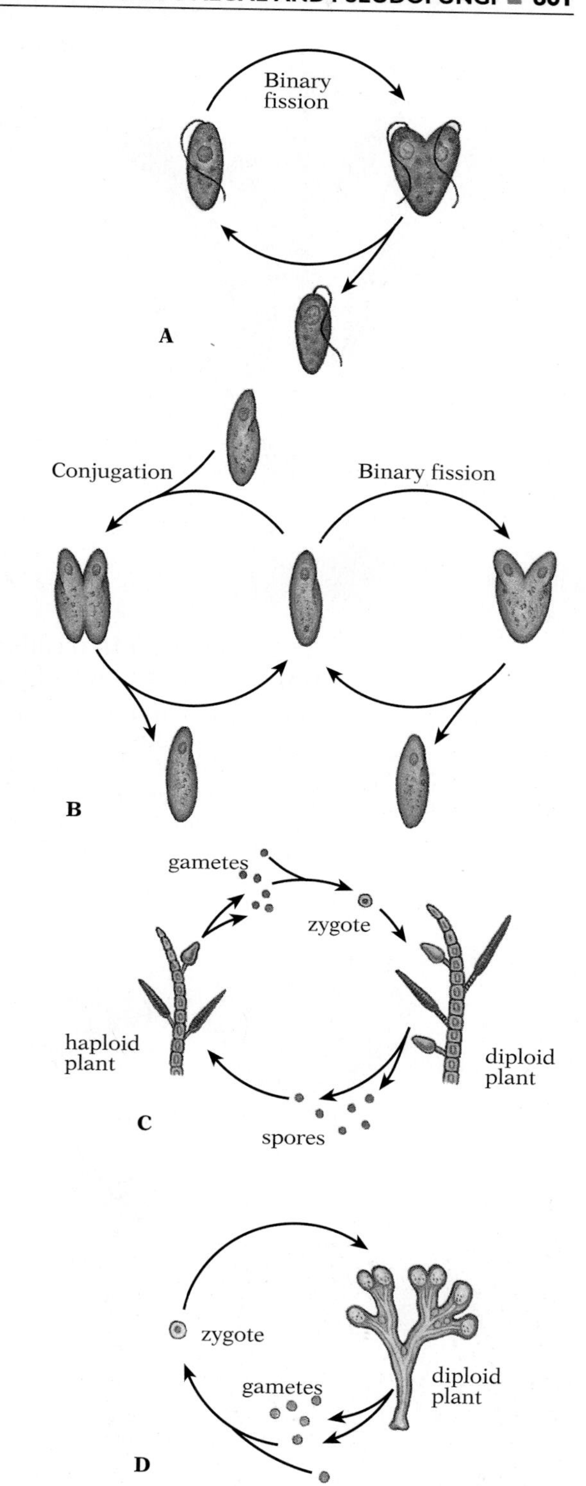

22.9 Comparison of common life histories

(A) Bacteria and many protists such as *Euglena* reproduce through an endless cycle of binary fission. (B) Protists such as *Paramecium* have the option of breaking the asexual fission cycle and conjugating. (C) Algae such as *Ectocarpus* have cycles that require alternation of asexual and sexual reproduction, with the growth of a full-size plant in each phase. (D) In most higher plants the haploid generation is shorter and the organism that develops is smaller than that of the diploid phase. *Fucus,* like animals, has no haploid growth at all; the diploid plant produces gametes that fuse to create a zygote.

amounts of a xanthophyll carotenoid called ***fucoxanthin*** give the characteristic brownish color to these algae.

Asexual reproduction most often occurs by flagellated ***zoospores***, which are motile asexual cells able to develop into a new plant. Sexual reproduction often involves specialized multicellular sex organs called ***gametangia***. In ***isogamous*** ("same gamete") species all the gametangia are alike, whereas in ***heterogamous*** species they are of the two kinds. The most familiar form of heterogamy, called ***oogamy***, involves large, sedentary eggs and small male gametes. The latter are flagellated sperm in animals and certain other organisms including brown algae. In oogamous species the gametangia that produce sperm are called ***antheridia,*** and those that produce eggs are ***oogonia***.

The life cycle of most brown algae illustrates the strategy of ***alternation of generations***. To appreciate this strategy, let's review others that we have described. The simplest life cycle is that of bacteria and many protists: asexual cloning. The organism grows, and then reproduces by mitosis and binary fission (Fig. 22.9A). There is no change in chromosome number: haploids stay haploid, diploids remain diploid. Organisms like *Paramecium* have a bimodal system, able to undergo meiosis and to conjugate for sexual reproduction (Fig. 22.9B). Cellular slime molds have a trimodal cycle: they can reproduce mitotically by binary fission, by producing spores mitotically, or by undergoing meiosis and conjugating (see Fig. 21.21, p. 592). These various reproductive alternatives to simple fission provide options that help organisms survive adverse conditions, disperse their offspring, and reap the benefits of occasional genetic recombination.

As organisms get larger, their reproductive strategy tends more and more to a single cycle in which haploid and diploid states alternate. In plants such as the brown alga *Ectocarpus,* the two states are equally prominent (Fig. 22.9C); they are also ***isomorphic:*** the haploid organism is virtually indistinguishable from the diploid form. From this obligatory alternation of generations the dominant evolutionary trend as body size increases has been to shorten or essentially eliminate the haploid stage (Fig. 22.9D). In the extreme, gametes fuse without further growth to form a diploid zygote, which then develops into the adult organism; this is the case with *Fucus* (Fig. 22.10), a brown alga whose reproductive strategy is similar to that of most animals (Fig. 22.11).

More typical is *Ectocarpus*, which has a branching filamentous thallus. The diploid sporophyte plants (stage 5 in Fig. 22.12) sometimes bear small unicellular sporangia, in which haploid zoospores

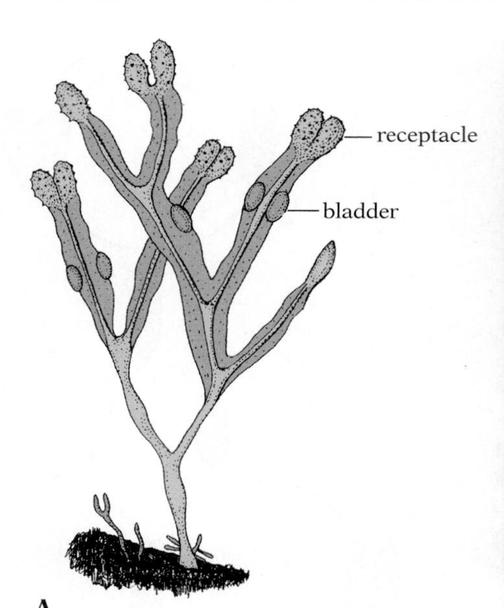

22.10 *Fucus* (often called rockweed), a brown alga common along northern coasts

(A) Each thallus is flattened and characterized by repeated dichotomous branching. Each younger axis consists of a midrib and thin paired wings. In some species (including the one shown here), there are bladders (floats) at intervals along the wings. The tips of fertile thalluses develop swollen reproductive structures called receptacles, whose surfaces are pocked by numerous tiny openings that lead into cavities (conceptacles) where the sex organs are located. In some species each individual has both male and female organs; in others the sexes are separate. (B) The receptacles can be seen especially well here.

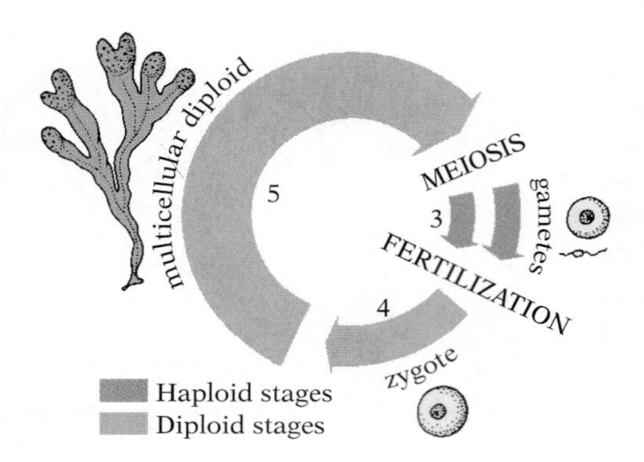

22.11 Life cycle of *Fucus*

Fucus and its close relatives have a very unusual life cycle. They are the only multicellular plants in which the multicellular haploid stage (the gametophyte) is completely absent and meiosis produces gametes directly (that is, both stage 1 and stage 2 are absent). In this respect their life cycle is like that of animals.

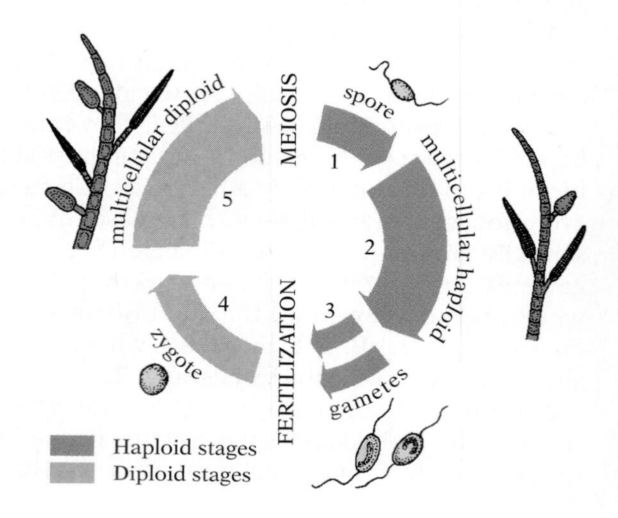

22.12 Life cycle of *Ectocarpus*

The gametophyte (multicellular haploid) and sporophyte (multicellular diploid) stages are equally prominent.

(stage 1) are produced by meiosis. After swimming about for a while, the zoospores settle down and develop into haploid multicellular gametophyte plants (stage 2). These plants may bear multicellular gametangia, in which morphologically isogamous gametes (stage 3) are produced. Two gametes (from different plants) may fuse in fertilization to form a zygote (stage 4). At first, the zygote is motile, but then it settles and germinates, giving rise to a new diploid multicellular sporophyte plant (stage 5) and thus completing the cycle.

Pelagophycus is one of the kelps. The sporophyte thallus is large (about 2 m long in this species and 45 m long or more in some kelps) and consists of a rootlike holdfast, a stemlike ***stipe***, and an expanded leaflike ***blade*** (see Fig. 28.2, p. 796). Although thallophyte plants usually lack tissue differentiation, the stipe of some kelps has an outer surface tissue (epidermis), a middle tissue (cortex) containing many plastids, and a central core tissue (medulla). It may even have a meristematic layer similar to the cambium of higher plants and, in a few genera, a conductive tissue in the medulla. In short, brown algae are complex plants that have convergently evolved a few tissues similar to those of terrestrial plants.

Pseudofungi Most pseudofungi are water molds, organisms long thought to be aquatic fungi. Some are distinctly nonaquatic, including *Phytophthora infestans*, the agent responsible for the Irish potato famine of 1845–1847. They are also distinctly different from true fungi: they have flagellated spores, while true fungi have identical nonmotile spores; they are oogamous whereas true fungi have identical gametes; they are usually diploid, unlike true fungi, which are usually haploid; and they lack chitin; the chief diagnostic criterion of fungi.

■ HAPTOPHYTA AND CRYPTISTA

The haptophytes are unicellular, motile, and (mostly) marine algae. Emerging from the same point as their paired flagella is a slender haptoneme—usually a tightly coiled thread—which the cell uses as a holdfast to attach itself to the substrate (Fig. 22.13A). The cell is covered with hundreds of tiny scales. This flagellated stage alternates with a resting phase, in which the cell covers itself with much larger calcium carbonate disks, often of elaborate design (Fig. 22.13B). Thus encased, the resting morph can survive extended periods of extreme conditions.

The cryptistans include freshwater photosynthetic species—most of the cryptomonads—and heterotrophs, some of which are parasitic. All are unicellular, have paired flagella, and possess a deep indentation called the gullet, which is surrounded, in the carnivorous species, by poison-secreting trichocysts. Some of the photosynthetic species also eat microbes.

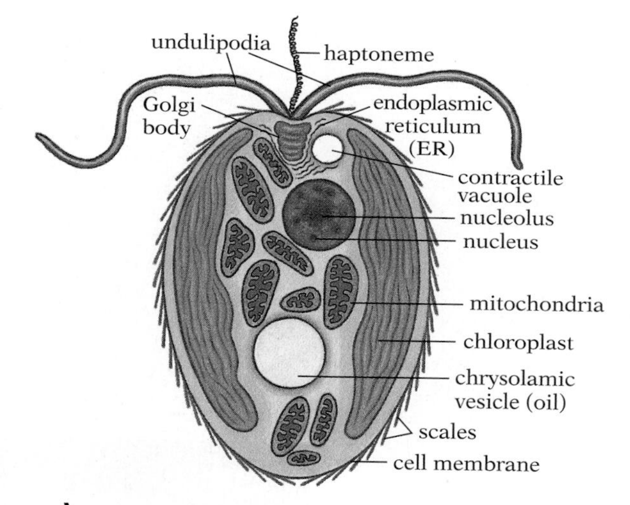

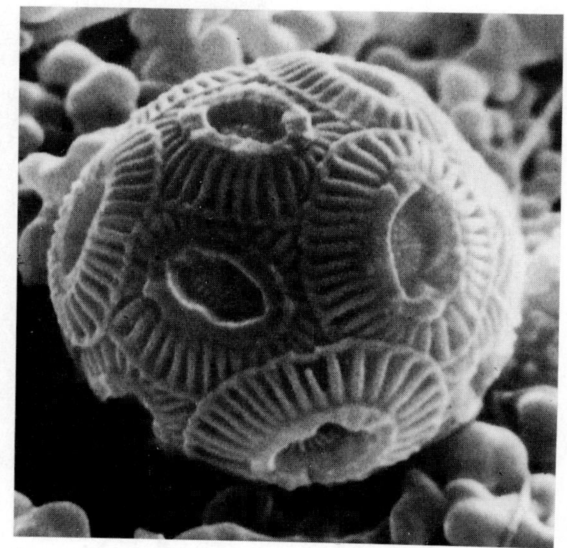

22.13 A haptophyte

(A) A haptophyte during its flagellated stage; note the haptoneme, which is used as a holdfast. (B) A haptophyte during its resting phase; the disks are called coccoliths.

LOWER PLANTS: RED AND GREEN ALGAE

The kingdom Plantae includes all the organisms we readily recognize as plants, plus the green and red algae. Like other plants, their chloroplasts have a two-membrane envelope (not three as in protists) and are free in the cytosol (rather than in the ER, as in the chromists). In this section we look at the fully or partially aquatic plants—organisms that form an evolutionary sequence that culminates in full independence from an aquatic medium, even for reproduction.

■ RHODOPHYTA: THE RED ALGAE

The red algae are mostly marine seaweeds (Figs. 22.14). They are smaller and often occur at greater depths than the brown algae. Most are multicellular and are attached to the substrate.

The cell walls contain cellulose and large quantities of mucilaginous (glue-like) material. The storage product is floridean starch; it is located in the cytosol, whereas conventional starch in other plants is stored in plastids. Red algae are an important source of commercial colloids, including agar used in culturing bacteria; suspending agents used in chocolate milk and puddings; stabilizers used in ice creams, some cheeses, and salad dressings; and moisture retainers used in icings, cosmetics, and marshmallows.

Rhodophytes use chlorophyll *a* in conjunction with several unique accessory pigments, including phycocyanins, phycoerythrins, and allophycocyanins. ("Phyco" is from the Greek *phykos*, meaning "seaweed.") It is the phycocyanins and allophycocyanins that give many of these algae their characteristic reddish color. Despite their common name, however, rhodophytes can also be black, violet, brownish, yellow, or even green.

A

B

22.14 Two examples of red algae

(A) *Kallymenia reniformis*, a species with a flattened bladelike thallus. (B) The seaweed *Scinaia furcellata*.

The life cycles of red algae usually involve an alternation of generations. Flagellated cells never occur; even the sperm cells lack flagella and must be carried to the egg cells by water currents.

■ CHLOROPHYTA AND CHAROPHYTA: THE GREEN ALGAE

The green algae are of particular interest because they include the group (the charophytes) from which the land plants arose. Like land plants, the green algae have chlorophylls *a* and *b* and carotenoids. It makes no sense to assign the unicellular green algae to one kingdom and their multicellular relatives to another. Sequence analysis indicates that the green algae are true plants. The majority of green algae live in fresh water, but some live in moist places on land, and there are many marine species.

Many divergent evolutionary tendencies can be traced in the green algae: (1) the evolution of motile colonies; (2) a change to nonmotile single cells and colonies; (3) the evolution of extensive tubelike bodies with numerous nuclei but without cellular partitions (coenocytic organization); (4) the evolution of multicellular filaments and even three-dimensional leaflike thalluses. The many unicellular and primitively multicellular members of the green algae illustrate this progression.

Chlamydomonas: a unicellular chlorophyte *Chlamydomonas* is the most thoroughly studied genus of unicellular green algae. Its many species are common in ditches, pools, and other bodies of fresh water and in soils. The individual organism is an oval haploid cell with a glycoprotein wall; unlike most green algae, it lacks cellulose. There are two anterior flagella of equal length and a single large cup-shaped chloroplast that fills the basal portion of the cell (Fig. 22.15). Inside the chloroplast are numerous chlorophyll-bearing lamellae, often arranged in stacks like the grana of higher

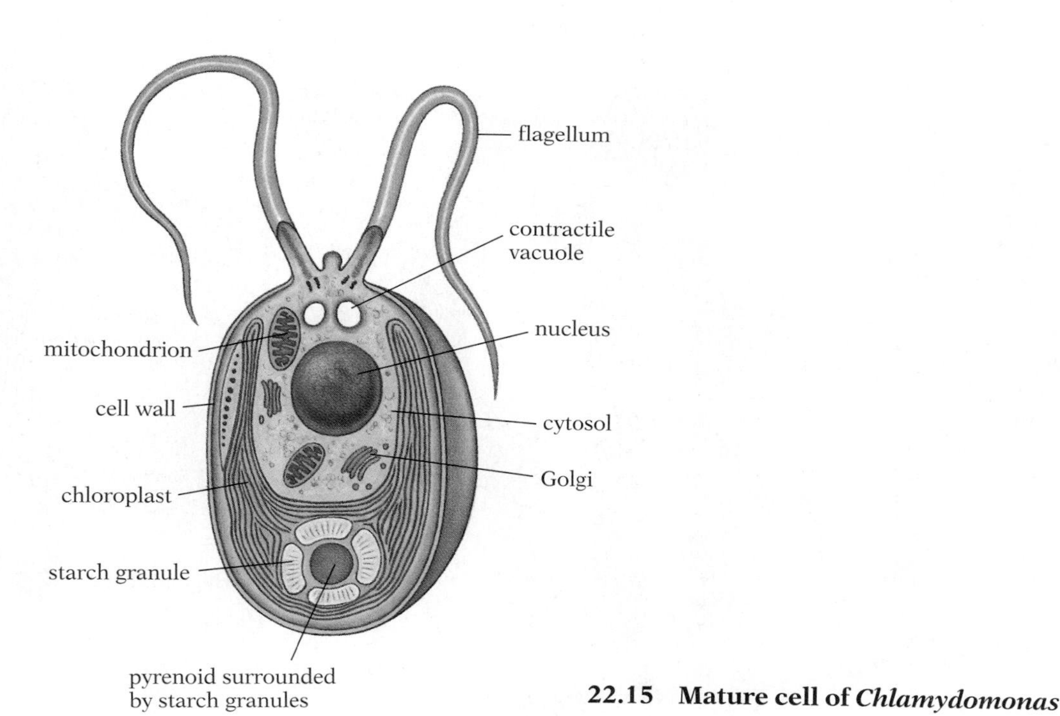

22.15 Mature cell of *Chlamydomonas*

plants. A conspicuous pyrenoid in the basal portion of the chloroplast functions as the site of starch synthesis. Two small contractile vacuoles lying near the base of the flagella discharge alternately and rhythmically.

Asexual reproduction is common in *Chlamydomonas* (Fig. 22.16). A vegetative cell resorbs its flagella; then mitotic division of the nucleus and cytokinesis occur. This gives rise to two daughter cells, which lie within the wall of the original cell. In some species the two daughter cells are promptly released by breakdown of the wall. In other species the daughter cells themselves divide while still inside the wall, and a total of 4, 8, 16, or more daughter cells is produced, depending on the species and conditions of growth. The daughter cells each develop a wall and flagella just before they are released as free zoospores. In *Chlamydomonas* the zoospores are smaller than mature vegetative cells, but otherwise they are indistinguishable from them; in many species of algae, however, there are noticeable morphological differences between the zoospores and the mature cells. The free zoospores grow to full size, completing the asexual reproductive cycle.

Under certain conditions, especially when the concentration of nitrogen in the medium is low, *Chlamydomonas* may reproduce sexually. A mature haploid vegetative cell divides mitotically to produce several gametes, which develop walls and flagella and are released from the parent cell. Gametes (usually of two different mating types) are attracted to each other and form large clusters. Eventually the clustered cells move apart in pairs. The members of a pair are positioned end to end, with their flagella, which bear species-specific and mating-type-specific binding sites at their tips, in close contact. The cells then shed their walls, and their cytoplasms fuse. Their nuclei unite, producing a single diploid cell, the zygote. The zygote sheds its flagella, sinks to the bottom, and develops a thick protective wall. The zygote can then withstand unfavor-

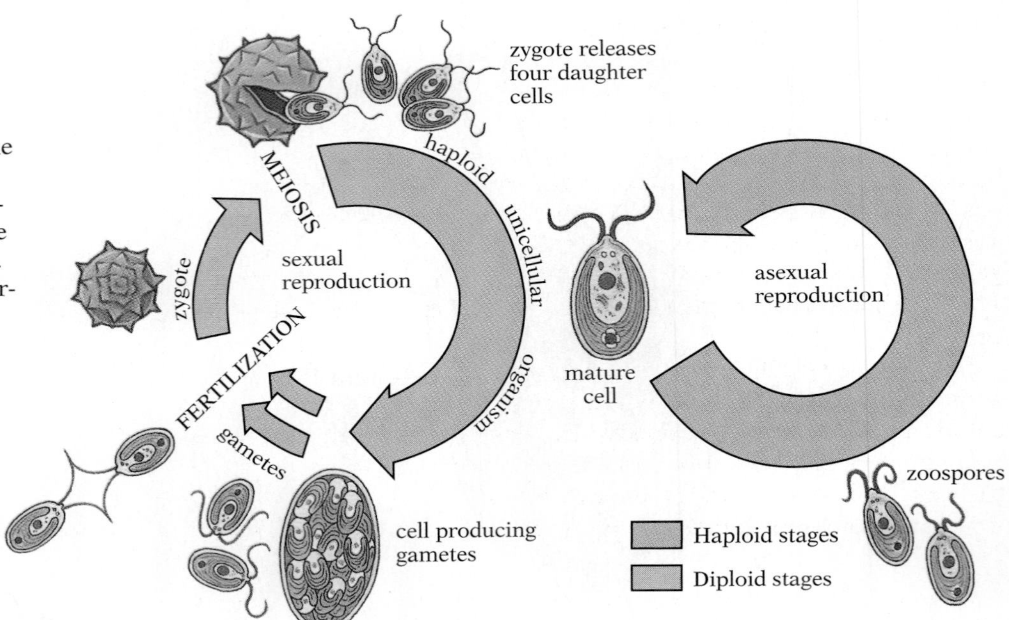

22.16 Life history of *Chlamydomonas*

The diagram shows all stages of both the sexual and asexual cycles. Schematic representation of life cycle for comparison with those of other organisms. Note that the zygote is the only diploid stage. This type of life cycle was probably characteristic of the first sexually reproducing unicellular organisms, and it may thus be the type from which all other types arose.

able environmental conditions, such as a pond drying up or a cold winter. When conditions are again favorable, it germinates, dividing by meiosis to produce four (or eight) new flagellated haploid cells, which are released into the surrounding water. The new cells mature, thus completing the sexual reproductive cycle.

Because sexual reproduction in most species of *Chlamydomonas* is relatively simple, it yields insights into the way sexuality probably arose. There are no separate male and female individuals. Furthermore, though the gametes usually differ in their mating-type-specific binding sites, they are usually morphologically indistinguishable. This isogamy is probably the primitive (ancestral) condition in plants. The isogametes of *Chlamydomonas* are indistinguishable from vegetative cells; they are simply small vegetative cells that tend to fuse and act as gametes under certain conditions.[2] This, too, is probably the primitive condition; the specialization of gametes as morphologically distinctive cells—a characteristic of most higher plants and animals—is probably a later evolutionary development.

The haploid stages of the life cycle of *Chlamydomonas* are the dominant ones; the only diploid stage is the zygote. Dominance of the haploid stages is characteristic of most very primitive plants, and it seems clear that this was the ancestral condition.

The volvocine series: evolution of multicellularity The evolutionary trend toward complex multicellular organization is illustrated by the volvocine (or motile-colony) series. This series of genera shows a gradual progression from the unicellular condition of *Chlamydomonas* to elaborate colonial organization.

Gonium is an example of the simplest colonial stage. Each colony of *Gonium* is made up of 4, 8, 16, or 32 cells (depending on the species), each of which is morphologically similar to *Chlamydomonas*. The cells are embedded in a mucilaginous matrix and are arranged in a flat or slightly curved plate. In some species delicate cytoplasmic strands run between the cells; these may provide a route for direct interaction and coordination between the cells of the colony. The flagella of all the cells beat together, thus enabling the colony to swim as a unit.

When asexual reproduction occurs in *Gonium*, all the cells in a colony divide simultaneously. When each cell of the parent colony has divided enough times to contain within its wall the same number of daughter cells as in the parent colony, its wall disintegrates and the daughter cells are released. The daughter cells remain together in a common matrix and mature into a new *Gonium* colony. Thus each cell of the parent colony gives rise to a complete new colony.

Sexual reproduction in *Gonium* is similar to that in *Chlamydomonas*. Individual free-swimming cells are released from a colony and function as gametes, fusing in pairs to form zygotes. The gametes are isogamous. As in *Chlamydomonas*, the zygote is the only diploid stage in the life cycle.

Colonies of genus *Pandorina* are more complex. Each is a hollow

[2] A few species of *Chlamydomonas* are heterogamous, and a few are even oogamous.

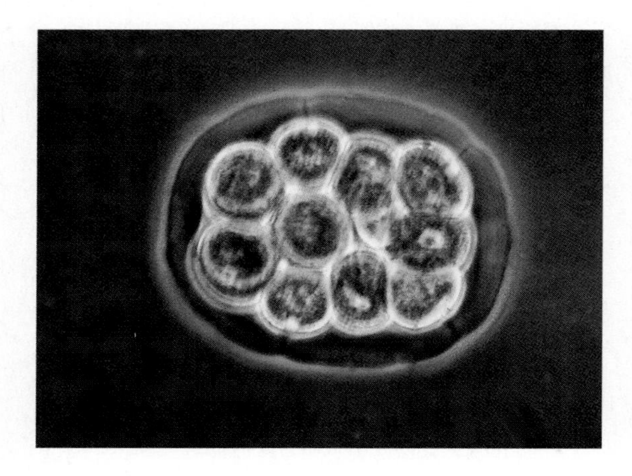

22.17 A *Pandorina* colony
The cells are embedded in a gelatinous matrix.

sphere of 8, 16, or 32 cells whose flagella are oriented to the outside (Fig. 22.17). Three main advances over *Gonium* are noticeable: (1) The colony shows some regional differentiation; it has definite anterior and posterior halves (detectable by the orientation of the colony when it is swimming). (2) The vegetative cells of the colony are so dependent on one another that they cannot live apart from the colony, and the colony cannot survive if disrupted. (3) Sexual reproduction involves two kinds of gametes; the male gametes are smaller than the female gametes, but both have flagella and are free-swimming. This type of heterogamy, where the only morphological difference is in size, is called ***anisogamy***.

Eudorina is a more advanced genus than *Pandorina*. The spherical colonies contain 16 or 32 cells. The differences between the anterior and posterior portions of the colony are greater, and the heterogamy is more pronounced: the large female gametes are not released, but remain embedded in the matrix of the colony and are fertilized there by the much smaller free-swimming male gametes.

A still more advanced genus is *Pleodorina* (Fig. 22.18), whose spherical colonies are composed of 32–128 cells. These large colonies exhibit considerable division of labor. The anterior cells are vegetative; the posterior cells, which function in both asexual and sexual reproduction, are much larger. Sexual reproduction is heterogamous. In some species reproduction is oogamous: the large female gametes lose their flagella and become nonmotile egg cells.

The culmination of this evolutionary series is represented by the genus *Volvox* (Fig. 22.19). Its spherical colonies are very large, consisting of about 500–50,000 cells. Delicate cytoplasmic strands between cells make intercellular communication possible (Fig.

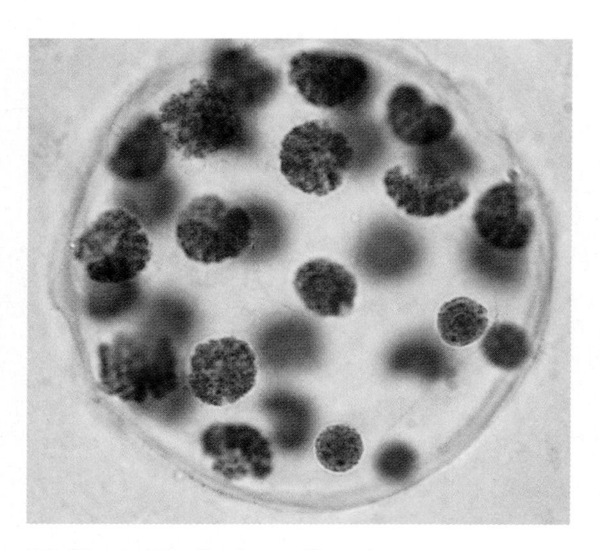

22.18 A *Pleodorina* colony
The individual clumps of cells are connected by strands of cytoplasm; without those connections the colony cannot reproduce.

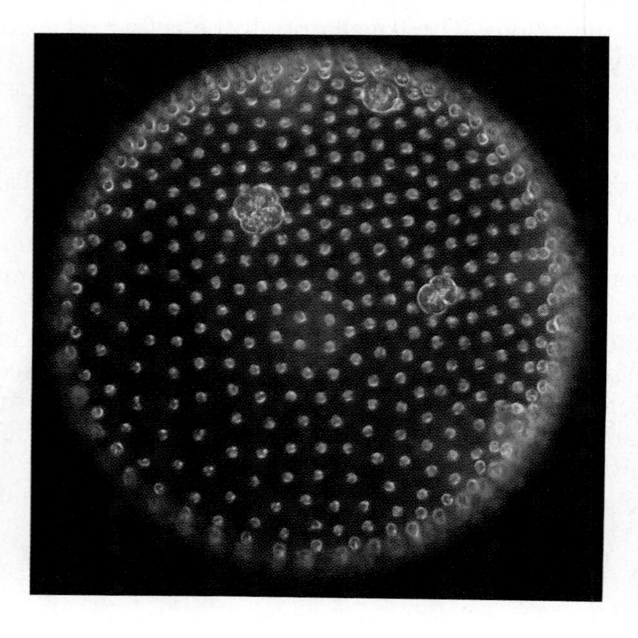

22.19 A *Volvox* colony
The colony is a sphere with a single layer of cells embedded in gelatinous material; the interior of the sphere is filled with a watery mucilage. Some small daughter colonies can be seen developing in this colony.

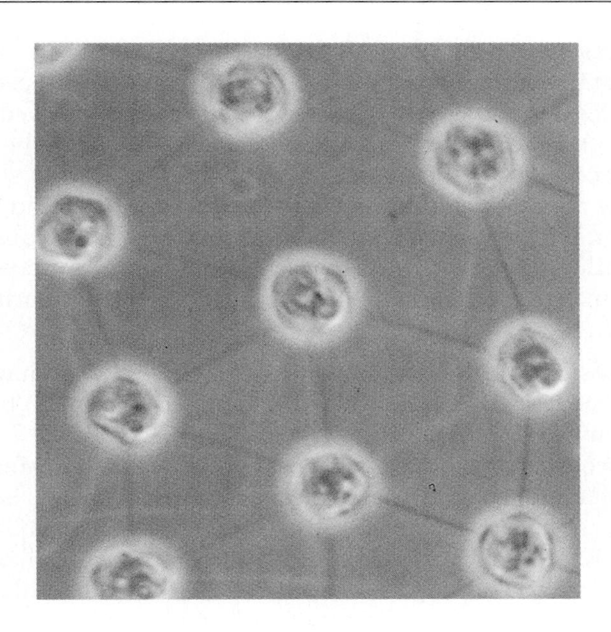

22.20 Cells of *Volvox* interconnected by cytoplasmic strands

22.20). Most of the cells are exclusively vegetative. A few cells (between two and 50), scattered in the posterior half of the colony, are much larger than the others and are specialized for reproduction. Each of the female reproductive cells can give rise to an entire new daughter colony (Figs. 22.19 and 22.21). Sexual reproduction is always oogamous.

The major lines of evolutionary changes manifest in this series include: (1) a change from unicellular to colonial life, and a tendency for the number of cells in the colonies to increase; (2) increasing coordination of activity among the cells; (3) increasing interdependence among the vegetative cells; (4) increasing division of labor, particularly between vegetative and reproductive cells; (5) a gradual change from isogamy to anisogamy to oogamy. In both the oogamous algae and the oogamous higher plants the female gametes are characteristically retained within the parental organism, and the meeting of gametes becomes less random. Hence fewer female gametes need to be produced, and more energy can be devoted to providing a large store of nutrients in those few.

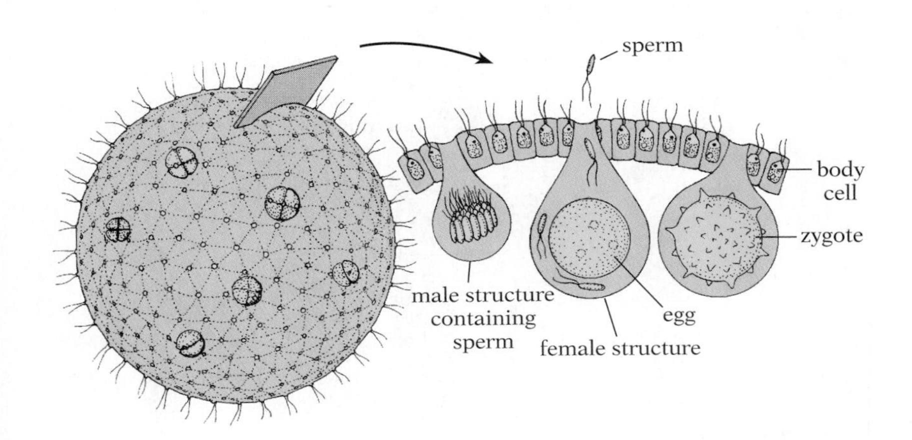

22.21 Reproduction in *Volvox*

Left: The colony is very large, containing 500–50,000 vegetative cells. Six daughter colonies at various stages of development can be seen still embedded in the matrix of the parent colony. Right: Section through the surface of a colony showing male and female reproductive structures. Sperm released by the male structure enter the female structure and fertilize the egg. After a period of inactivity, the zygote divides meiotically, and the haploid cells thus formed then divide mitotically, producing a new daughter colony, which is eventually released. The colony shown here is producing both male and female gametes; in some species or strains the sexes are separate, and a given colony produces only one kind of gamete.

22.22 *Stigeoclonium*, a branching filamentous green alga

The presentation of organisms in this series is not meant to imply that each genus necessarily evolved from the preceding one. Nevertheless, it is likely that each of these genera evolved from an ancestor that resembled in many important ways the modern genus placed just before it in this series. Study of this series suggests how complex colonial forms probably evolved, and indicates one way in which multicellularity may have arisen in plants. A similar evolutionary series, beginning with a nonwalled unicellular organism, may have been the beginning of multicellularity in the animal kingdom.

Life cycles of multicellular algae Many green algae have a multicellular stage in their life cycle. This stage is usually a branching or nonbranching filamentous thallus (Fig. 22.22).

The species of *Ulothrix* are unbranched filaments. Most live in fresh water (Fig. 22.23). The filament of each plant is a small threadlike structure attached to the substrate by a specialized holdfast cell. Except for the holdfast, all the cells of the filament are identical and are arranged end to end in a line. The filament increases in length as its cells grow and divide. Adjacent cells have common end walls—a basic step in the evolution of multicellularity in algae.

Ulothrix may reproduce by fragmentation (with each fragment growing into a complete plant), by asexually produced zoospores, or by sexual processes. In asexual reproduction any cell of the filament except the holdfast may act as a ***sporangium***, producing

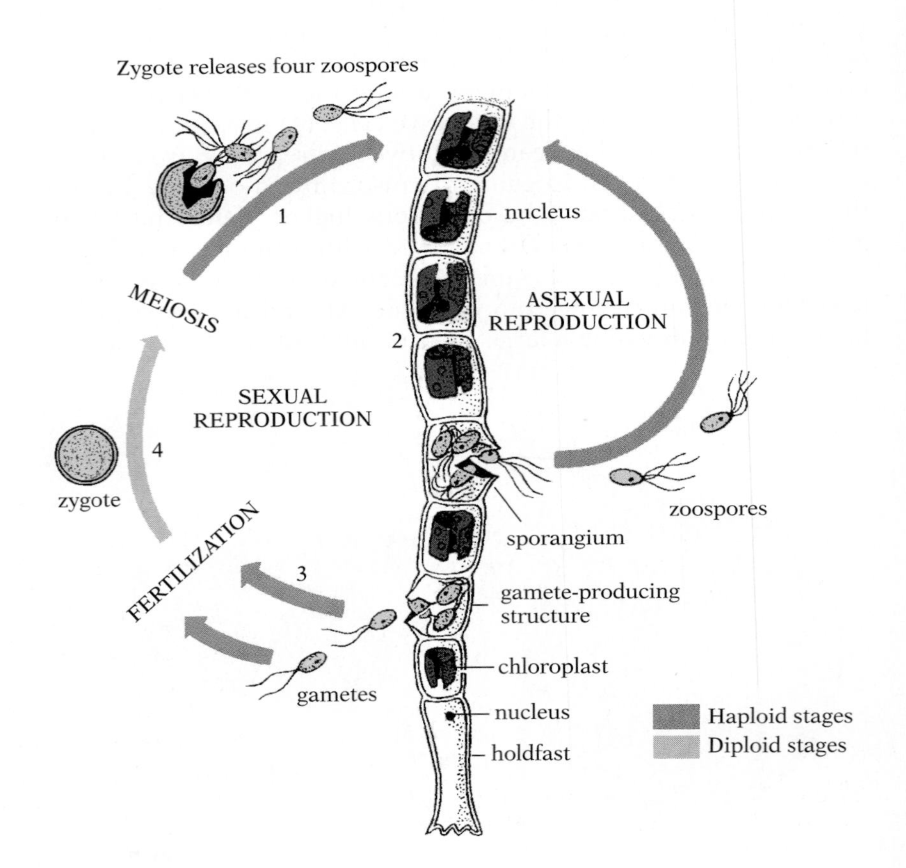

22.23 Life history of *Ulothrix*

The haploid plant may reproduce either asexually or sexually (though a single filament would never reproduce in both ways at once as shown here). Asexual reproduction is the more common; certain cells of the filament develop into sporangia (spore-producing structures) and produce zoospores, which settle down and develop into new filaments. Under certain environmental conditions the filament may cease reproducing asexually and begin reproducing sexually; a cell becomes specialized as a gametangium (gamete-producing structure) and produces isogametes. Two such gametes may fuse in fertilization, producing a zygote, which divides meiotically and releases zoospores.

zoospores, each of which has four flagella. After the zoospores are released, they swim about and then settle down and give rise to a new filament.

Sexual reproduction is isogamous. The zygote (stage 4 in Fig. 22.23), formed by the union of two of the biflagellate gametes, develops a thick wall and functions as a resting stage capable of withstanding unfavorable conditions. At germination the zygote divides by meiosis, producing haploid zoospores (stage 1), each of which grows into a new filament (stage 2). The main difference between this life cycle (Fig. 22.24) and that of *Chlamydomonas* is the addition of the haploid multicellular stage (stage 2).

Ulva, or sea lettuce, has an expanded leaflike thallus two cells thick (Fig. 22.25). Its sexual life cycle is more complex in that it includes both multicellular haploid and multicellular diploid stages (stages 2 and 5 in Fig. 22.26). Haploid zoospores (stage 1) divide mitotically to produce the haploid multicellular thalluses of stage 2. These may reproduce either asexually by means of zoospores or sexually by means of gametes (stage 3). Fusion of pairs of gametes (fertilization) produces diploid zygotes (stage 4). Upon germination, the zygotes divide mitotically (not meiotically as in the green algae previously discussed), producing diploid multicellular thalluses (stage 5). Eventually certain reproductive cells (sporangia) of these diploid plants divide by meiosis, producing haploid zoospores, which begin a new cycle.

Multicellularity in plants arose first in the gametophyte, and many green algae have no sporophyte stage. *Ulva* shows a more advanced life cycle in that both gametophyte and sporophyte stages are present. Furthermore, the two stages are equally prominent in *Ulva*; the haploid portion of the life cycle is no longer dominant.

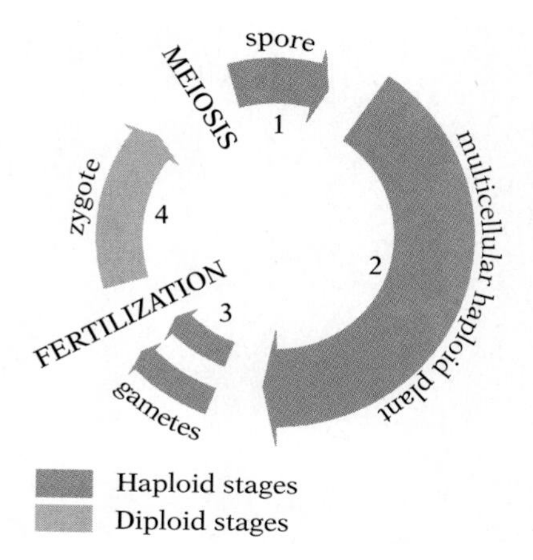

22.24 Life cycle characteristic of most multicellular green algae

Note that multicellularity is present only in the haploid phase.

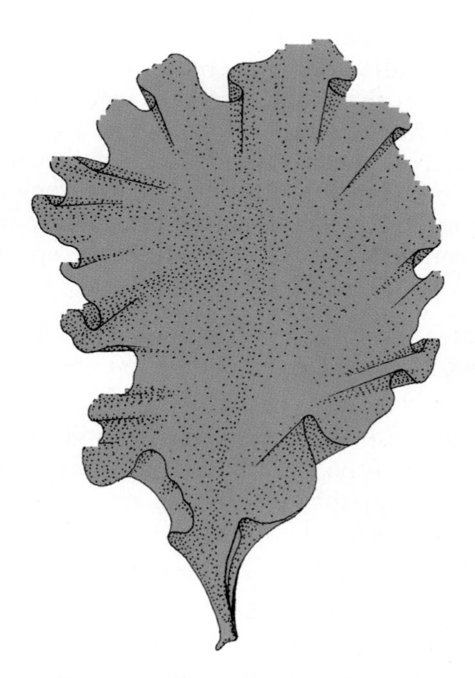

22.25 *Ulva*, a marine green alga with a leaflike thallus that is two cells thick

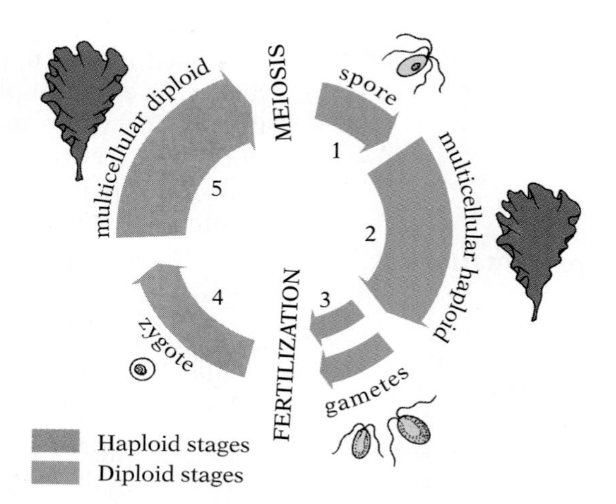

22.26 Life cycle of *Ulva*

The gametophyte (multicellular haploid) and sporophyte (multicellular diploid) stages are equally prominent. *Ulva* and its close relatives are unusual among the green algae in having a life cycle of this sort; most of the Chlorophyta have no alternation of generations, the sporophyte stage being absent.

22.27 Photograph of living *Spirogyra* in various stages of conjugation

Spirogyra: a common charophyte The most widely distributed member of the Charophyta is *Spirogyra* (Fig. 22.27), a genus of rather odd filamentous freshwater green algae. It has a sexual life cycle similar to that of *Ulothrix* except that the gamete cells are not flagellated and are not released from the plant that produces them. Instead, two filaments come to lie side by side and protuberances develop on the sides of the cells where they are in contact. The walls between the protuberances of each pair of cells disintegrate. Then one cell becomes amoeboid, moves through the conjugation tube, and fuses with the other cell, forming a zygote (Fig. 22.28). Relatively few green algae reproduce by this conjugation process. *Spirogyra* is a member of a group of filamentous green algae—the Charophyceae—of which most species utilize oogamy in reproduction. Sequence analysis indicates that it is from a complex member of this group of algae that land plants evolved.

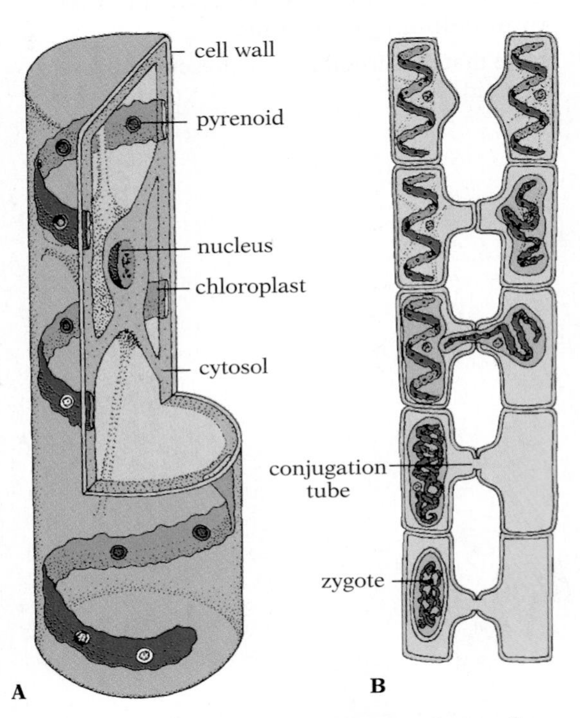

22.28 Diagrammatic representations of *Spirogyra*

(A) A single vegetative cell removed from the filament and partially sectioned. Note the unusual spiral chloroplast that runs the length of the cell. (Some species have more than one chloroplast.) Numerous pyrenoids are associated with the chloroplast. The cell has a large central vacuole, in which the nucleus is suspended by cytoplasmic threads; these threads connect to the peripheral cytosol, which forms a layer just inside the cell wall. (B) Conjugating filaments. The two filaments lie side by side, and a conjugation tube develops between each pair of cells. One cell acts as the sperm, moving through the tube to fuse with the other cell (see middle pair of cells). The zygote thus formed is the only diploid stage in the life cycle.

HOW PLANTS CONQUERED THE LAND

The terrestrial plants have evolved numerous adaptations for life on land. The few algae that live on land are not truly terrestrial, occurring as they do only in a film of water in very moist places. The evolutionary move from an aquatic existence to a terrestrial one was not simple. Most of the problems of living on land relate to the need for copious amounts of water. Water is much more important for plants than for any other group of organisms. For example, plants depend on obtaining raw materials—light, carbon dioxide gas, fixed nitrogen, minerals, and so on—which are normally very dilute. As a result, plants have evolved an enormous surface-to-volume ratio, which maximizes the area available to gather light and nutrients. To do this, they devote most of their energy to creating walls and membranes, and fill what little cellular volume they absolutely must have with about 99% water. A plant trying to live outside of an aqueous environment faces at least seven specific challenges:

1 Obtaining enough water (both for its dilute nutrients *and* as filler for new cells) when fluid no longer bathes the entire surface of the plant body.

2 Transporting water and dissolved substances from restricted areas of intake to other parts of the plant body, and transporting the products of photosynthesis to parts of the plant that no longer carry out photosynthesis for themselves.

3 Preventing excessive loss of water by evaporation.

4 Maintaining an extensive moist surface for gas exchange when the surrounding medium is air instead of liquid.

5 Supporting a large plant body against the pull of gravity when the buoyancy of an aqueous medium is no longer available.

6 Carrying out reproduction when there is little water through

Table 22.1 A comparison of the major chromist and plant phyla

CHARACTERISTICS	OCHRISTA	RHODOPHYTA	CHLOROPHYTA; CHAROPHYTA	BRYOPHYTA	TRACHEOPHYTA
Sperm usually flagellated	+	–	+	+	+ or –
Chlorophylls	a and c	a	a and b	a and b	a and b
Principal reserve material usually starch	–	–	+	+	+
Sporophyte equal or dominant to gametophyte in most species	+	*	*	–	+
Sex organs usually multicellular with jacket cells	–	–	–	+	+
Embryo development within archegonium	–	–	–	+	+
Cuticle usually present	–	–	–	+	+
Complete internal transport system (both xylem and phloem) present in most species	–	–	–	–	+

* Among Chlorophyta and Rhodophyta species for which the full life cycle is known, some have a dominant gametophyte and some have a dominant sporophyte. Since many species have not yet been studied, it is not possible to say which condition is the more usual.

which flagellated sperm may swim, and when the zygote and early embryo are in severe danger of desiccation.

7 Withstanding the extreme fluctuations in temperature, humidity, wind, and light to which terrestrial organisms are often subjected, and which are greatly moderated in lakes and seas because of the high heat capacity of water.

Much of the evolution of the terrestrial plants can be understood as adaptations to solve these problems.

Terrestrial plants characteristically have multicellular sex organs with an outer layer of ***jacket cells*** that helps protect the enclosed gametes from desiccation; male and female sex organs of this type are known as antheridia and archegonia.[3] The sporangia are also multicellular, and they too have a layer of jacket cells. They are oogamous, and the egg cells are fertilized while they are still contained within the archegonia. Each zygote develops into a multicellular diploid embryo while still inside the archegonium. The embryo obtains some of its water and nutrients from the parent plant and is thus a parasite. This type of embryonic development, which permits the stages of development most susceptible to desiccation to occur in a moist microenvironment, is strongly reminiscent of the internal gestation of mammals.

The surfaces of the aerial parts of the plant bodies of terrestrial species are usually covered by a waxy cuticle, which waterproofs the epidermis and helps prevent excessive water loss.

The principal pigments in land plants are chlorophylls *a* and *b* plus carotenoids, and the storage material is starch (Table 22.1). Thus land plants are biochemically similar to the charophytes, from which they almost certainly arose. We will consider the

[3] Like *oogonium*, *archegonium* denotes an organ producing female gametes, but most botanists restrict *archegonium* to the jacketed female reproductive organs of higher plants, and *oogonium* to the unjacketed ones of algae.

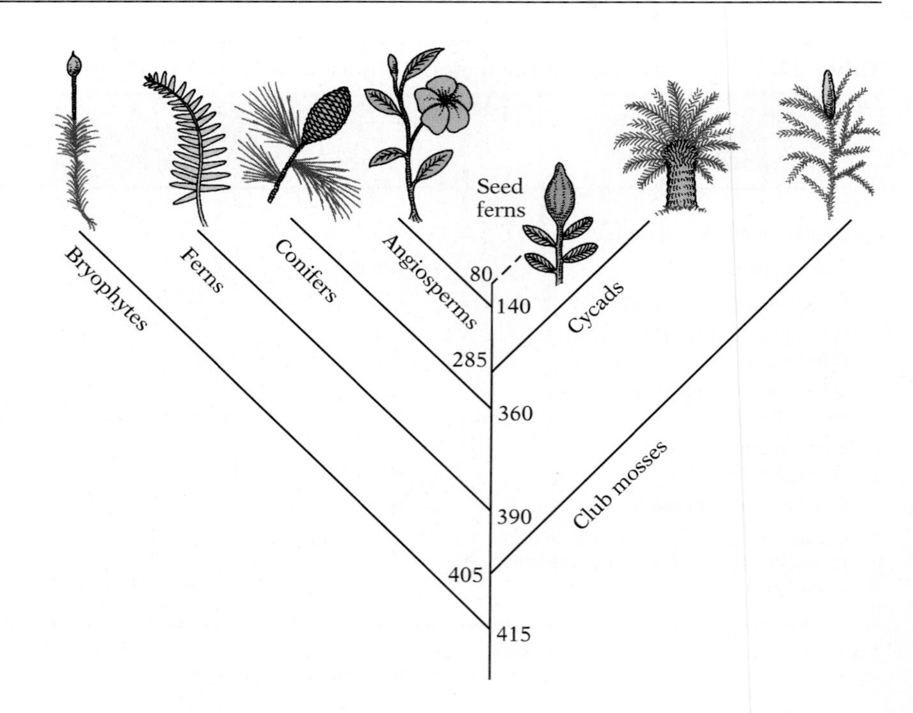

22.29 Phylogeny of modern land plants
The existence of fossils makes this tree certain and permits rough dating of the branch points (indicated in millions of years before the present). The dead end near the top represents the extinction of seed ferns about 80 million years ago.

22.30 A young moss plant
The spore (yellow) gives rise to a filamentous plant (called a protonema) that strikingly resembles a green alga. The protonema develops into the mature moss plant.

groups of terrestrial plants in the order in which they evolved (Fig. 22.29).

■ BRYOPHYTA: LIVERWORTS, HORNWORTS, AND MOSSES

The bryophytes are relatively small plants that grow in moist places on land—on damp rocks and logs, on the forest floor, in swamps or marshes, or beside streams and pools. Some species can survive periods of drought by becoming dormant. Thus though the bryophytes live on land, they have never freed themselves from their ancestral aquatic environment. Their strict dependence on a moist environment is linked to two characteristics: (1) They retain flagellated sperm cells, which must swim to the egg cells in the archegonia. (2) Most lack well-developed vascular tissues (tissues specialized as ducts or tubes) and hence the means for efficient long-distance internal transport of fluids. The absence of cells with secondarily thickened walls, which function as major supportive elements in vascular plants, has probably also limited their size.

The bryophytes may have arisen from filamentous green algae. Indeed, a very young moss plant, called a protonema (Fig. 22.30), often resembles a green algal filament. As the plant grows, it forms some branches (rhizoids) that enter the ground and function like roots, anchoring the plant and absorbing water and nutrients; other branches form upright shoots with stemlike and leaflike parts.

In the bryophytes, the haploid gametophyte (Fig. 22.31, stage 2)

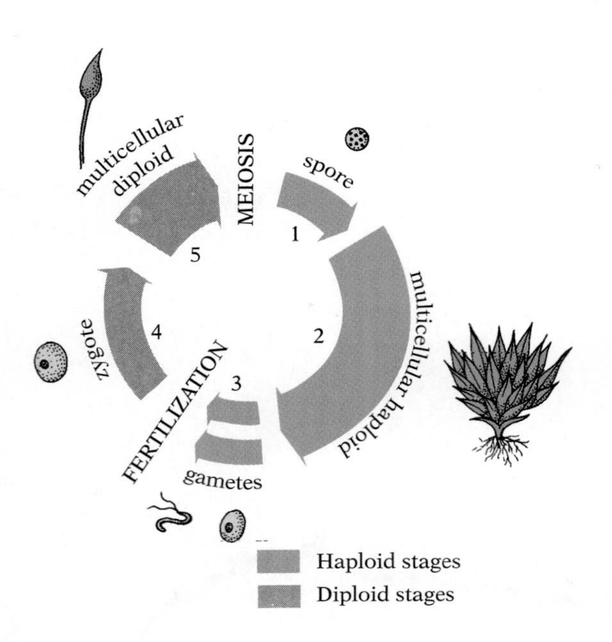

22.31 Life cycle of a bryophyte
Both gametophyte (stage 2) and sporophyte (stage 5) are present. The former is dominant.

22.32 Gametophyte and sporophyte stages of a moss
The haploid gametophyte (green) is the lower "leafy" plant; the "leaves," except at their midrib, are only one cell thick. The diploid sporophyte plant (yellow and brown), is attached to the gametophyte and is to some degree parasitic on it. The sporphyte consists of a foot (not visible here), a stalk, and a capsule. (Here the capsule is shown covered by a cap—the calyptra—derived from the archegonium of the gametophyte; in time it will fall away, leaving the capsule fully exposed.)

is clearly the dominant stage in the life cycle. The "leafy" green moss plant or liverwort is the gametophyte. These plants bear antheridia and archegonia in which gametes (stage 3) are produced by mitosis. The flagellated male gametes (sperm) are released from the antheridia and swim through a film of moisture, such as rain or heavy dew. Responding to chemical attractants, they swim to archegonia, where they fertilize the egg cells, producing zygotes (stage 4). Each zygote then divides mitotically, producing a diploid sporophyte (stage 5).

In a moss this sporophyte is a relatively simple structure consisting of three parts: a foot embedded in the "leafy" green gametophyte, a stalk, and a distal capsule, or sporangium (Fig. 22.32). The sporophyte has chloroplasts and carries out some photosynthesis, but it also obtains nutrients parasitically from the gametophyte to which it is attached. Meiosis occurs within the mature capsule of the sporophyte, producing haploid spores (stage 1), which are released (Fig. 22.33). These spores, encased in walls that are extremely tough, may remain inactive for a long time (sometimes many years) if conditions are unfavorable. When they do germi-

A

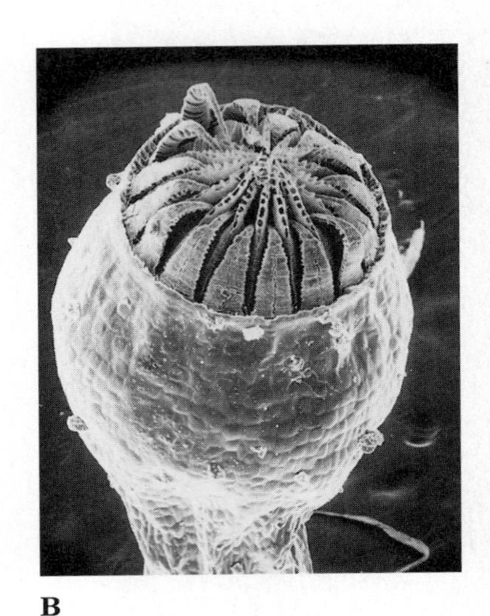

B

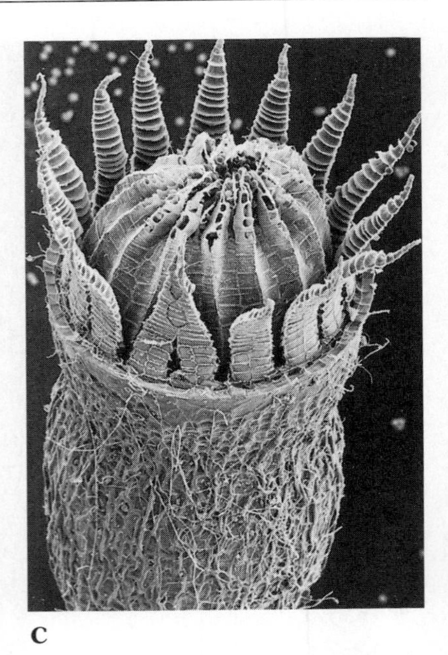

C

D

22.33 Spore release by a moss

The sporophytes of some species release their spores explosively. (A) Before this can happen, the cap must be shed. (B) A set of "teeth" in the capsule retains the spores until conditions are right. (C) Then they spring back to release the spores, which are carried by breezes to new sites (D). (A, B, C, from *Living Images* by Gene Shih and Richard G. Kessel. Copyright © 1982 by Science Books International.)

nate, they develop into protonemata (plural for *protonema*) and eventually into mature gametophyte plants (stage 2), thus completing the life cycle.

The gametophyte plants of some liverworts resemble mosses except that the "leaves" are scaly, and the "stem" is prostrate. Other liverworts grow as flat green structures lying on the substratum. In some species the antheridia and archegonia are borne in receptacles located at the top of stalks (Fig. 22.34). In other species there is no receptacle or stalk (sometimes only the stalk is missing), and the reproductive structures are embedded in the upper portion of the prostrate "leaf." The life cycle is much like that of mosses, except that the sporophyte is even simpler (Fig. 22.35). Asexual reproduction sometimes occurs by production of special clusters of cells called gemmae; these usually occur as cuplike structures located on the surface of the flat gametophyte (Fig. 22.36). When detached from the parent plant, the gemmae can grow into new gametophytes.

■ LOWER TRACHEOPHYTES: THE EARLY VASCULAR PLANTS

The ***tracheophytes*** have evolved a host of adaptations to the terrestrial environment that have enabled them to invade all but the most inhospitable land habitats. In the process, they have diverged into five subphyla:

22.34 Liverworts (*Marchantia*) with stalked receptacles bearing archegonia (yellow)

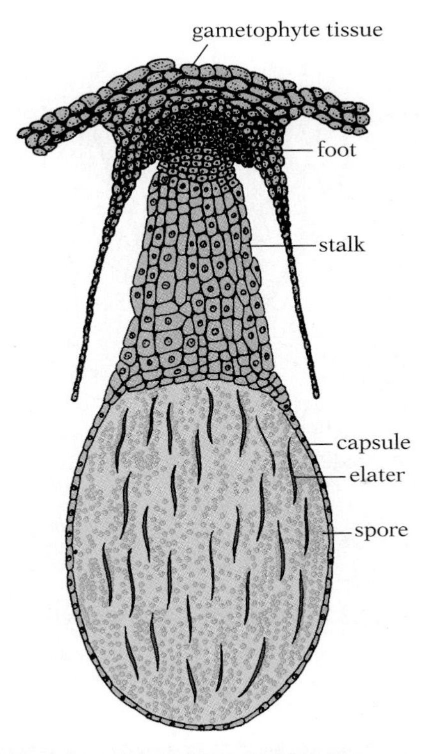

22.35 Sporophyte of *Marchantia*

The sporophyte of this liverwort is a small structure consisting of a foot, a short stalk, and a capsule. The foot remains embedded in the gametophyte plant, in the tissue of the undersurface of the umbrella-shaped receptacle (see Fig. 22.33). The mature capsule contains spores and elaters, which are elongate cells with spirally thickened walls. Eventually the wall of the capsule dries and bursts, releasing the spores. Ejection of the spores is aided by the elaters, which twist and jerk as they dry, thus throwing the spores from the capsule.

22.36 Gemmae

Several gemmae (cups) can be seen here on the gametophyte of *Marchantia*. Gemmae function in asexual reproduction in some liverworts.

Phylum Tracheophyta
- Subphylum Psilopsida (psilopsids)
- Subphylum Lycopsida (club mosses)
- Subphylum Sphenopsida (horsetails)
- Subphylum Pteropsida (ferns)
- Subphylum Spermopsida (seed plants)

Virtually all members of this phylum possess four important attributes absent in even the most advanced algae: a protective layer of sterile jacket cells around the reproductive organs; multicellular

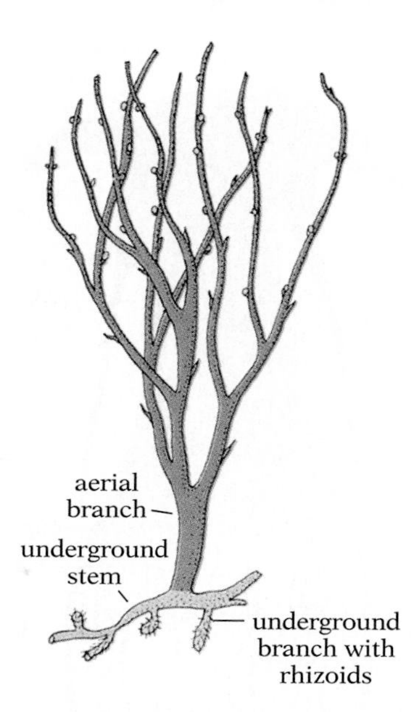

22.37 *Psilotum* sporophyte

embryos retained within the archegonia; cuticles on the aerial parts; and, most important, a complete internal transport system, which allows movement of water and nutrients up from the soil, as well as organized transfer of organic chemicals within the plant. This network of "plumbing" is found only in the sporophyte, and has naturally led to an increasing domination of the sporophyte phase of the life cycle at the expense of the gametophyte (see Table 22.1).

All four are clear adaptations for terrestrial existence. Many other such adaptations, absent in the earliest tracheophytes, appear in more advanced members of the division. These trends, which we trace below, reflect the increasingly extensive exploitation of the terrestrial environment by vascular plants.

Psilopsida: the first tracheophytes The oldest undisputed fossil representatives of the vascular plants appeared more than 395 million years ago late in the Silurian period (see Table 19.1, p. 540). They are classified in the subdivision Psilopsida, most of whose members have become extinct.

The psilopsid sporophytes are simple dichotomously branching plants that lack leaves[4] and have no true roots, although they do have underground stems with unicellular rhizoids similar to root hairs (Fig. 22.37). The aerial stems are green and carry out photosynthesis. Sporangia develop at the tips of some of the aerial branches. Within the sporangia, meiosis produces haploid spores. The psilopsids evolved from certain branching filamentous green algae. Some researchers believe that a few of the modern plants classified as psilopsids are actually degenerate forms of higher plants.

Lycopsida: the club mosses The first representatives of the subdivision Lycopsida appeared early in the Devonian period, almost 10 million years after the psilopsids. During the late Devonian and the Carboniferous periods these were among the dominant plants on land. Some of them were very large trees that formed the earth's first forests. Toward the end of the Paleozoic era, however, the group was displaced by more advanced types of vascular plants, and only five genera are alive today. One of these, *Lycopodium* (often called running pine or ground pine), is common in many parts of the United States and is frequently used in Christmas decorations (Fig. 22.38).

22.38 *Lycopodium*
The pale, conelike structures at the tops of stems are strobili (see Fig. 22.38).

Unlike the psilopsids, lycopsids have true roots. These probably arose from branches of the ancestral algae that penetrated the soil and branched underground. Lycopsids also have true leaves, which may have arisen as simple scalelike outgrowths (emergences) from the outer tissues of the stem. Certain leaves, called ***sporophylls***, are specialized for reproduction; they bear sporangia on their surfaces. In many lycopsids the sporophylls are congregated on a short length of stem and form a conelike structure (strobilus; Fig. 22.39). The cone is club-shaped, which gives rise to the common name *club mosses* for this group. (Note, however, that lycopsids are not related to the true mosses, which are bryophytes).

[4] Some psilopsids have scalelike structures that superficially resemble leaves.

The spores produced by *Lycopodium* are all alike, and each can give rise to a gametophyte that will bear both archegonia and antheridia. However, some lycopsids (such as *Selaginella*) have two types of sporangia, which produce different kinds of spores. One type of sporangium produces very large ***megaspores***, which develop into female gametophytes bearing archegonia; the other type produces small ***microspores***, which develop into male gametophytes bearing antheridia. The megaspores of *Selaginella* begin embryonic development within the protection of the sporangium, a strategy greatly elaborated in seed plants. Plants like *Lycopodium*, which produce only one kind of spore, and hence have only one kind of gametophyte bearing both male and female organs, are said to be ***homosporous***. Plants such as *Selaginella*, which produce both megaspores (female) and microspores (male), with separate sexes in the gametophyte generation, are said to be ***heterosporous***.

Sphenopsida: the horsetails The sphenopsids first appear in the fossil record late in the Devonian period. They became a major component of the land flora during the Carboniferous period and then declined. Members of the one living genus, *Equisetum*, are commonly called horsetails or scouring rushes. Though most of these are small (less than 1 m), some of the ancient sphenopsids were large trees (Fig. 22.40). Much of the coal we use today was formed from these plants.

Like the lycopsids, sphenopsids possess true roots, stems, and leaves. The stems are hollow and jointed like bamboo, an arrangement that makes them strong and light. Whorls of leaves occur at

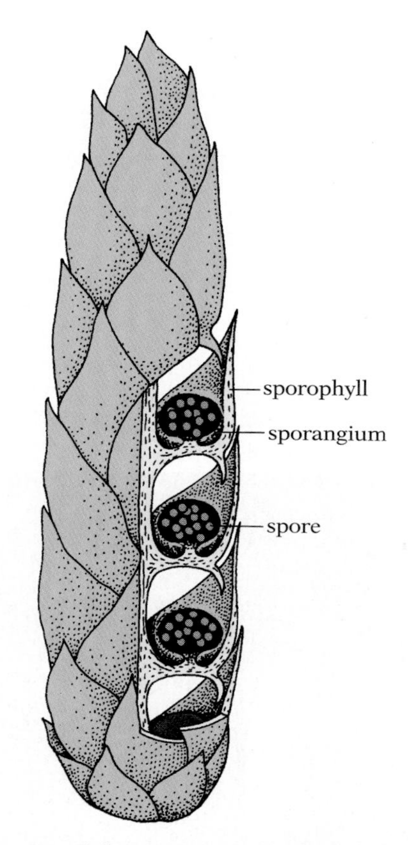

22.39 Strobilus of a lycopsid (club moss)

The strobilus is partially sectioned to show the arrangement of sporangia on the sporophylls.

22.40 Carboniferous swamp forest

This museum diarama reconstructs the probable environment of a swamp forest about 300 million years ago. The trees with jointed stems and whorls of leaves are sphenopsids, while the trunk at the right is a lycopsid.

22.41 A fossil of a sphenopsid
A whorl of leaves is located at each joint of the stem.

22.42 *Equisetum*
Three stalks bearing mature reproductive cones can be seen on these field horsetails.

each joint (Fig. 22.41). Spores are borne in terminal cones (strobili; Fig. 22.42). In *Equisetum* all spores are alike (since the plants are homosporous), and, since the sexes are not separate, they give rise to small gametophytes that bear both archegonia and antheridia.

Pteropsida: the ferns The ferns probably evolved from the psilopsids. They first appeared in the Devonian period and greatly increased in importance during the Carboniferous. Their decline late in the Paleozoic era was much less severe than that of the psilopsids, lycopsids, and sphenopsids, and there are many modern species.

The ferns are fairly advanced plants with very well-developed vascular systems and with true roots, stems, and leaves. The leaves probably arose in a different way from those of the lycopsids. Instead of emergences, they seem to be flattened and webbed branch stems; a group of small terminal branches probably became arranged in the same plane (planated), and the interstices filled with tissue.[5] Such leaves are larger, and provide a much greater surface area for photosynthesis, than emergence leaves.

The leaves of ferns are sometimes simple, but more often they are compound, being divided into numerous leaflets that may give the plant a lacy appearance.[6] In a few ferns, like the large tree ferns of the tropics, the stem is upright, forming a trunk. In most ferns, however, especially those of temperate regions, the stems are prostrate on or in the soil, and the large leaves are the only parts normally seen.

The large leafy fern plant is the diploid sporophyte phase (Fig. 22.43). Spores are produced in sporangia located in clusters on the underside of some leaves (sporophylls; Fig. 22.44). In some species the sporophylls are modified relatively little and look like the nonreproductive leaves. In other species the sporophylls look quite different from vegetative leaves; sometimes they do not look like leaves at all, forming spikelike structures instead (Fig. 22.45).

Most modern ferns are homosporous. After germination, the spores develop into gametophytes that bear both archegonia and antheridia (Fig. 22.46). These gametophytes are tiny (less than 1

[5] Leaves arising as emergences are called microphylls. Those arising as planated and webbed branch systems are called megaphylls.

[6] When a fern leaf, or frond, is divided into leaflets, the leaflets are called pinnae. The pinnae may themselves be subdivided into pinnules.

22.43 Marsh ferns in autumn
The large leafy fern plant is the diploid sporophyte phase.

22.44 Undersurface of fertile leaf (sporophyll) of polypody fern
Each round dot is a sorus, which is a cluster of many tiny sporangia.

22.45 Leaves of sensitive fern
The sterile leaves (left) have expanded blades, but the fertile leaves (sporophylls, to the right) are spikes bearing grapelike clusters of reproductive organs.

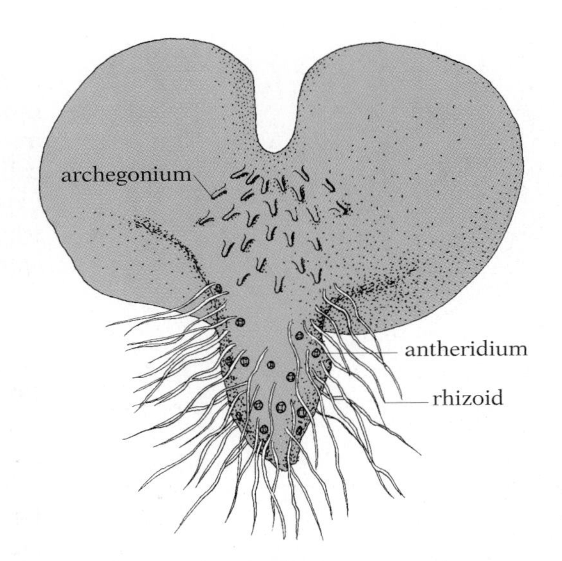

22.46 Fern gametophyte
This much-magnified view shows the undersurface of the tiny heart-shaped organism.

cm wide), thin, and often heart-shaped. Although most people are familiar with the sporophytes of ferns, few would recognize their gametophytes. Small and obscure as it is, however, the fern gametophyte is an independent photosynthetic organism. All five principal stages are present in this life cycle, but the multicellular

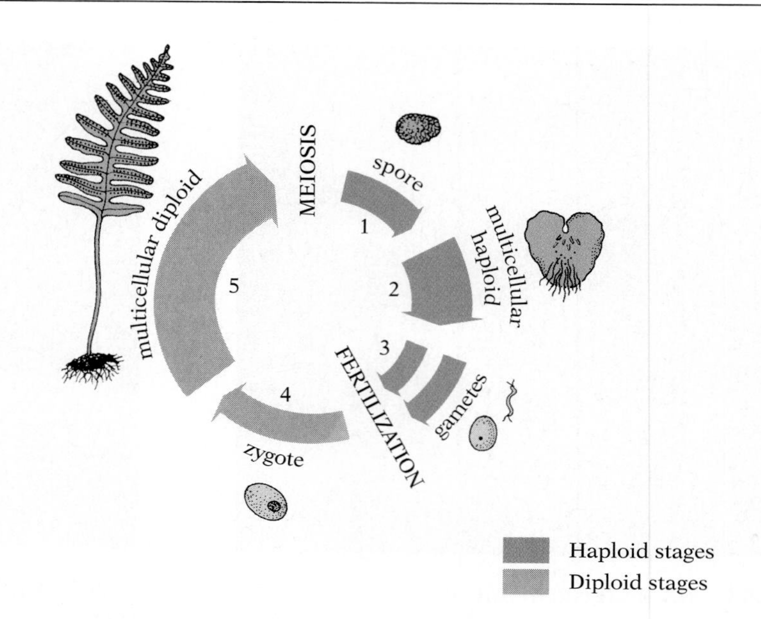

22.47 Life cycle of ferns

Both gametophyte (stage 2) and sporophyte (stage 5) are present; the latter is much the more prominent. Compare this life cycle with that of bryophytes (Fig. 22.31).

haploid stage has been much reduced and the multicellular diploid stage emphasized (Fig. 22.47).

In some respects, the primitive tracheophytes discussed here are no better adapted for life on land than the bryophytes. Their vascularized sporophytes can live in drier places and grow bigger, but their nonvascularized free-living gametophytes can survive only in moist places. In addition, since their sperm are flagellated and must have a film of moisture through which to swim to the egg cells in the archegonia, and the young sporophyte develops directly from the zygote without passing through any protected seedlike stage, these plants are most successful in habitats where there is at least a moderate amount of moisture.

HIGHER (FULLY TERRESTRIAL) PLANTS

■ HOW PLANTS ARE ORGANIZED INTERNALLY

The algae, as we saw, have no tissue specialization; because of their aqueous environment, they have no need for support, respiratory apparatus, or internal transport. The movement to land, however, made such tissues essential. The full range of tissue types is found in the seed plants; this summary is based on that group.

The tissues of vascular plants have been classified in a variety of ways by botanists. The lack of agreement on any one classification arises because different cell types are not perfectly distinct, but rather intergrade, and a given cell may even change from one type to another during the course of its life. Consequently, the tissues formed from such cells may share structural and functional characteristics. Furthermore, plant tissues may contain cells of only one type, or they may be complex, containing a variety of cell types. Thus plant tissues cannot be fully characterized or distinguished on the basis of a single criterion such as structure, function, location, or mode of origin (Table 22.2).

Table 22.2 Plant Tissues

TYPE OF TISSUE	DISTINGUISHING CHARACTERISTICS	FUNCTION
MERISTEMATIC TISSUE	Embryonic cells capable of cell division; small, thin walled cells	Produce new cells
Apical 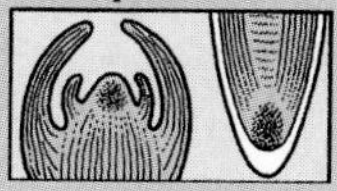	Located at tips of shoots and roots	Increase in length of plant body; produce primary tissues
Lateral 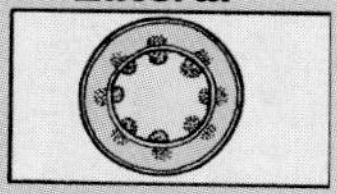	Located near periphery of roots and stems. Examples: vascular and cork cambia	Increase in girth of plant body; produce secondary tissues
PERMANENT TISSUE	Composed of more mature, differentiated cells	Protection of underlying tissues
Surface tissue	Cover the outer surfaces of plant body	Protection in young plants
Epidermis 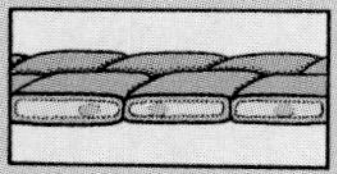	Flat cells, often with thicker outer walls; aerial parts often covered with waxy cuticle	
Periderm 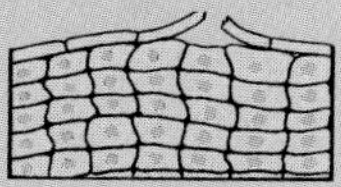	Waterproofed cells with thick cell walls that are dead at maturity	Forms outer bark in trees
Fundamental tissue	Simple tissues composed of a single type of cell	
Parenchyma 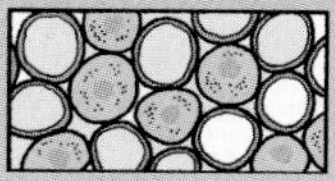	Unspecialized cells with thin primary walls	Photosynthesis, secreation, storage
Collenchyma 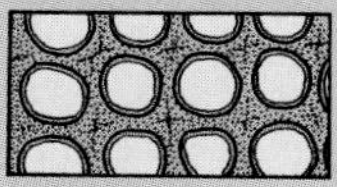	Elongated cells with unevenly thickened primary walls. Living at maturity.	Support in young leaves and stems
Sclerenchyma 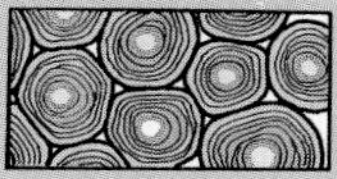	Elongated cells with very thick secondary walls. Dead at maturity	Support
Vascular tissue	Elongated cells, specialized for conduction	Transport of material throughout the plant body
Xylem 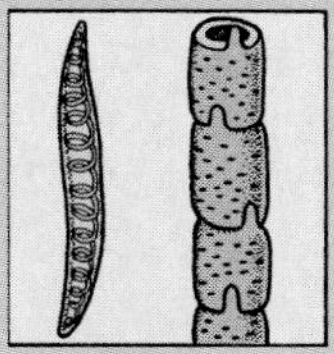	Conductive cells dead at maturity; thick cell walls arranged end to end to form empty passages	Transport of water and minerals upward from roots to leaves, support
Phloem 	Conductive cells living at maturity; sieve cells elongated, arranged end to end for conduction	Transport of organic materials up and down the plant body

All plant tissues can, however, be divided into two major categories: meristematic tissue and permanent tissue. ***Meristematic tissues*** are composed of immature cells and are regions of active cell division; ***permanent tissues*** are composed of more mature, differentiated cells without active cell division. This distinction is not absolute: some permanent tissues may revert to meristematic activity under certain conditions.

The permanent tissues fall into three subcategories: surface tissues, fundamental tissues, and vascular tissues. Each of these, in turn, contains several different tissue types. The classification used here is summarized in Figure 22.48.

Growth: meristematic tissue Meristematic tissues are composed of embryonic, undifferentiated cells capable of active cell division. Cell division occurs throughout the very early embryo, but as the

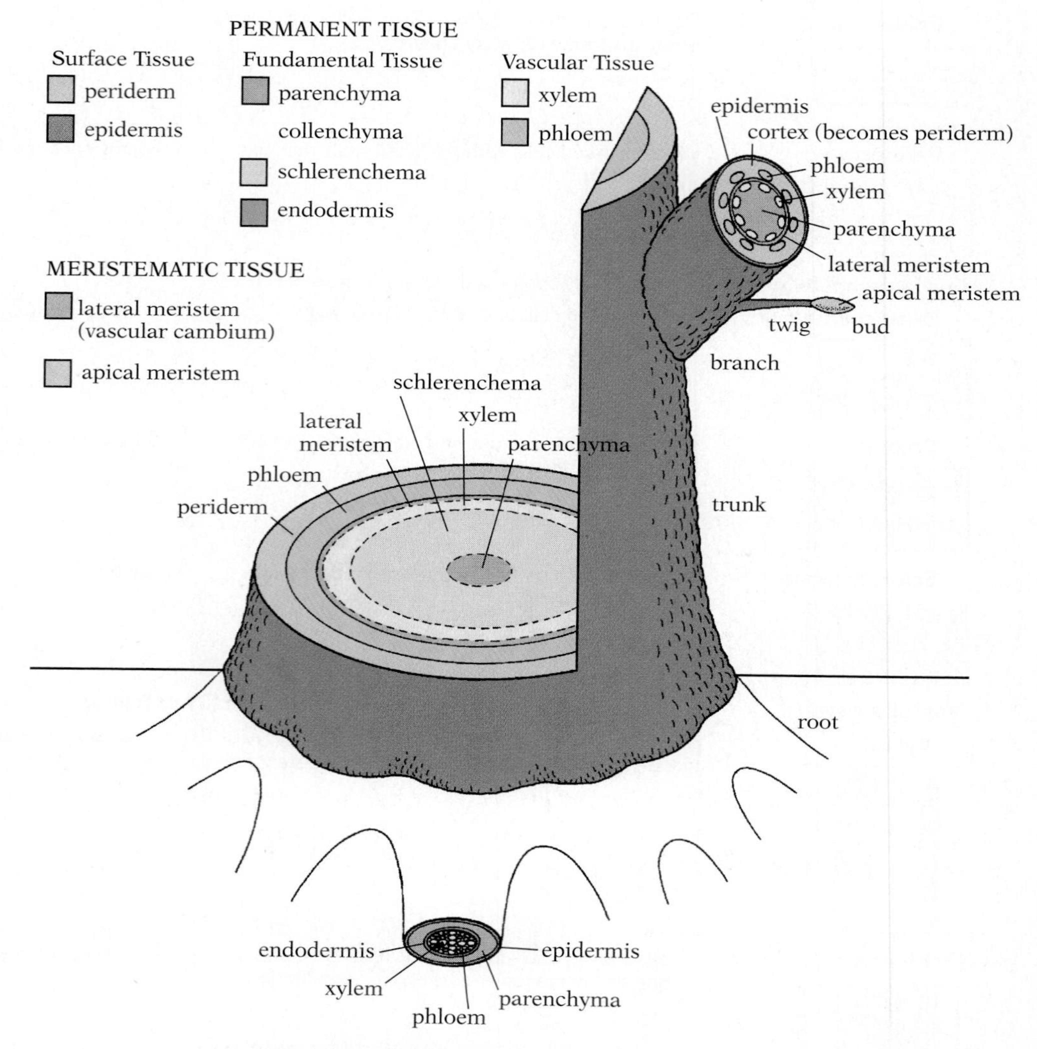

22.48 Classification of plant tissues

The example here is a young tree. Collenchyma, found in leaves and young shoots, is not shown.

EXPLORING FURTHER

EVOLUTION OF SPOROPHYTE DOMINANCE

An obvious question, in view of the tendency toward increasing dominance of the sporophyte generation in so many major groups of plants, is why this shift of emphasis in the life cycle should have occurred. Why did diploidy become adaptively superior to haploidy?

This issue is part of the larger question of why sexual recombination evolved at all, which is discussed in Chapters 12 and 17. The likely answer involves the dynamic balance between stability and change that organisms must maintain. Stability is essential in the short run so the genetic program that has built and operated a successful organism can accurately reproduce its design. At the same time, external conditions inevitably change over time, and modified enzymes or regulatory systems will sooner or later be needed to enable the organism to adjust to these changes.

Haploidy was the ancestral condition. Bacteria and many protists are nearly permanent haploids. Haploidy's main disadvantage is that unrepaired mutations slowly accumulate and any changes in the genome are directly expressed. In most cases these mutations sap the fitness of the individuals affected. However, with a doubling time of only 20 min, bacteria are able to "outrun" mutation, producing scores of letter-perfect offspring for every mutant. Moreover, the high reproductive rate of bacteria assures a constant stream of variants potentially able to accommodate change.

Many protists and algae share to a lesser degree the basic elements of bacterial lifestyle that make haploidy satisfactory: small size, simple body organization, short generation time, and high reproductive output. As organisms grow larger and more complex, however, their prereproductive exposure to mutation begins to increase, along with the number of genes and control sequences at risk. Also increasing are the odds that environmental conditions will have changed before gametes or offspring are produced.

Diploidy helps on all fronts. A second copy of each gene not only provides a backup in the event one is damaged; it even allows an individual to have two slightly different versions of a gene, and so perhaps to make two versions of an enzyme, each optimized for different conditions. This phenomenon is very common and, as described in Chapter 16, it can provide an advantage to the heterozygote (called hybrid vigor). Crossing over and sexual recombination enhance these diploid advantages, and confer higher fitness in the face not only of mutation, but also of competition, predation, and ecological change. In fact, sexual reproduction in simpler organisms is often triggered by environmental hardship, a likely sign that an organism's constellation of genes may no longer be the best combination of alleles. Sporophytes, therefore, are increasingly essential as plants become larger and more complex, and must find other ways to outrun the risks they face.

young plant develops, many regions become specialized for other functions and cease producing new cells. Cell division becomes restricted largely to certain undifferentiated tissues in localized regions called meristems. The cells of higher animals generally differentiate, committing themselves to a specific role, and so there is no equivalent to meristem in animals.

There is no such thing as a typical meristematic cell. However, they tend to be small, have thin walls, and be rich in cytoplasm (that is, to have only small vacuoles), and meristematic tissues tend to lack intercellular spaces. New cells produced by a meristem are initially like those of the meristem itself, but as they grow and mature, they become differentiated as components of other plant tissues.

There are regions of meristematic tissue at the growing tips of roots and stems. These ***apical meristems*** are responsible for increase in length of the plant body. In many plants, there are also meristematic areas toward the periphery of the roots and stems, and these ***lateral meristems*** are responsible for increases in girth.

Protection: surface tissue Surface tissues form the protective outer covering of the plant body. In young plants, the principal sur-

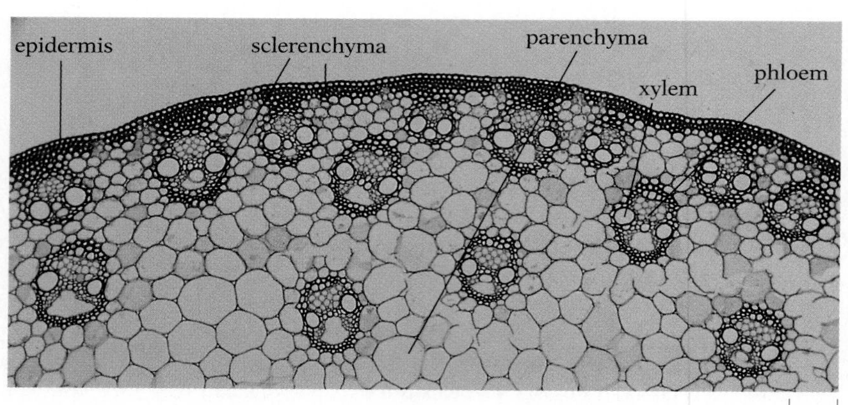

22.49 Partial cross section of corn stem

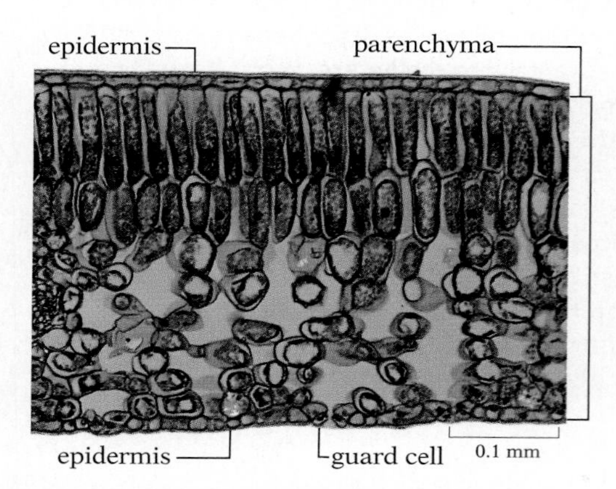

22.50 Photograph of cross section of an ivy leaf

Guard cells regulate the flow of gases into and out of the leaf.

face tissue of roots and stems is the ***epidermis*** (Fig. 22.49); epidermis is also the surface tissue of all leaves (Fig. 22.50). Often the epidermis is only one cell thick, though it may be thicker, as it is in some plants living in very dry habitats, where protection against water loss is critical.

Most epidermal cells have a very large vacuole and only a thin layer of cytoplasm. Often the outer and side walls of epidermal cells are thicker than the wall that faces the inside of the plant. Epidermal cells on the aerial parts of the plant often secrete a waxy, water-resistant cuticle on their outer surface; this, combined with the thick outer cell wall, aids in protecting against water loss, mechanical injury, and invasion by parasitic fungi.

Epidermal tissues of the aerial parts often give rise to unicellular or multicellular hairs, spines, or glands. Some epidermal cells, particularly of the leaves, are specialized as guard cells and regulate the size of small holes (stomata) in the epidermis through which gases can move into or out of the leaf (Fig. 22.50). As detailed in Chapter 26, epidermal root cells have no cuticle, and they function in water absorption; their outer walls often develop into long filaments (hairs) that greatly increase the total absorptive surface area.

As the stems and roots of plants with active lateral meristems increase in diameter, the epidermis is slowly replaced by another surface tissue, the ***periderm***. This tissue constitutes the corky outer bark so characteristic of old trees. Functional cork cells are dead; their waterproof cell walls act as the protective outer covering of the plant.

Structure and metabolism: fundamental tissue Most fundamental plant tissues are simple, since each is usually composed of only one type of cell. These same types of cells often occur as components of complex tissues such as those of the vascular system. Fundamental tissues form from the same embryonic regions. They are often defined as those tissues that are neither surface tissues nor vascular tissues. There are several kinds of fundamental tissues.

1 *Parenchyma* Parenchyma tissue occurs in roots, stems, and leaves. The parenchyma cells are relatively unspecialized, like those that make up almost the whole body of lower plants, and are mitotically dormant. However, they retain the ability to break out of dormancy by beginning to divide actively, and to differentiate into specialized tissue—that is, they can take on meristematic activity or undergo further specialization, forming other cell types. Parenchyma cells usually have thin primary walls and no secondary walls. They generally have a large vacuole surrounded by a peripheral layer of cytoplasm. The cells are ordinarily loosely packed; consequently intercellular spaces are abundant in parenchyma tissue (Figs. 22.49 and 22.50). Most of the chloroplasts of leaves are in the cells of parenchyma tissue. Parenchyma of stems and roots stores nutrients and water. When swollen with water, parenchyma gives support and shape to the plant.

2 ***Collenchyma*** Like parenchyma, collenchyma is a simple tissue whose cells remain alive during most of their functional existence. Collenchyma cells are characteristically more elongate, and their walls are irregularly thickened. The thickening is usually most prominent at the edges (the "corners" when viewed in cross section; Fig. 22.51). Collenchyma is an important supporting tissue in young plants, in the stems of nonwoody older plants, and in leaves.

3 ***Sclerenchyma*** Sclerenchyma, like collenchyma, functions in support. However, sclerenchyma cells are far more specialized than collenchyma cells. At functional maturity most are dead, and their thick, hardened secondary walls give strength to the plant body. Often these walls are so thick that the cell has scarcely any lumen (internal space; Fig. 22.52).

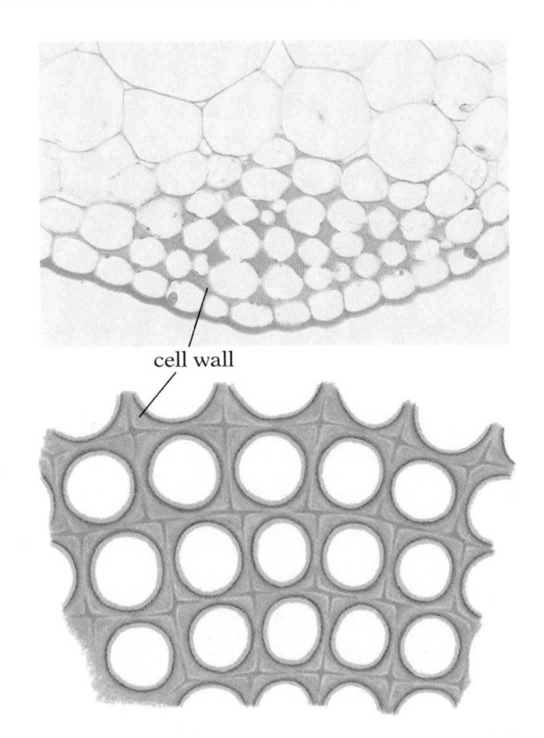

22.51 Collenchyma cells from a petiole of a beet leaf

Notice the particularly thick walls at the corners of the cells.

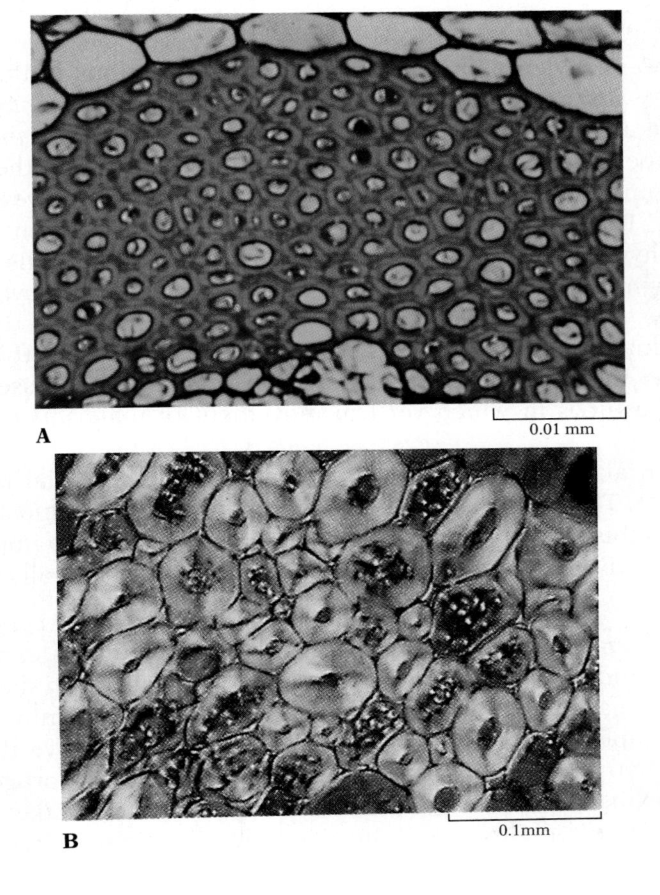

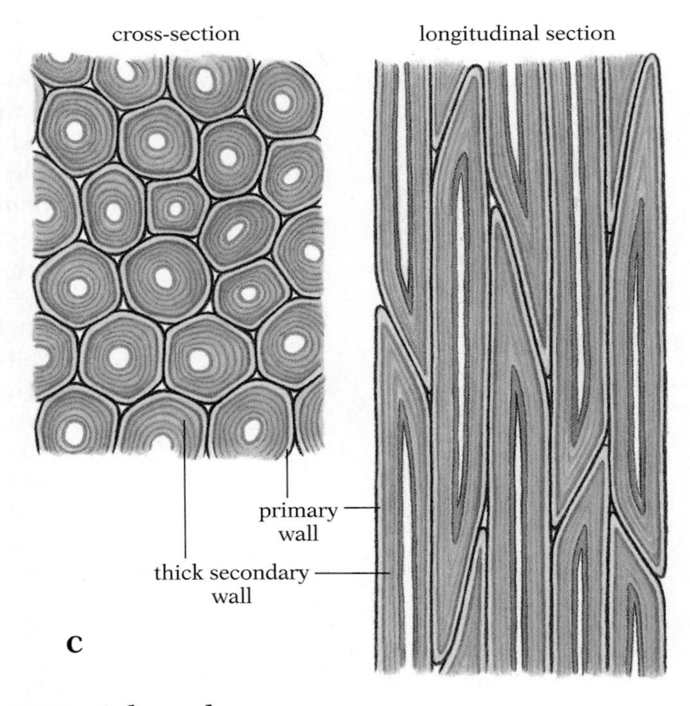

22.52 Sclerenchyma

(A) Cross section of fibers (red) from a corn stem. (B) Stone cells (sclereids) of pear fruit. (C) Cross section and longitudinal section of schlerenchyma tissue.

Sclerenchyma cells are customarily divided into two categories: fibers and sclereids. ***Fibers*** are very elongate cells with tapered ends. They are tough and strong, but flexible; commercial flax and hemp are derived from strands of sclerenchyma fibers. ***Sclereids*** are of variable, often irregular, shape. The simpler, unbranched sclereids are frequently called stone cells; they are common in nutshells and the hard parts of seeds, and are scattered in the flesh of hard fruits. The gritty texture of pears, for instance, comes from small clusters of stone cells (Fig. 22.52B).

4 *Endodermis* Endodermis is difficult to classify. It occurs as a layer surrounding the vascular-tissue core of roots and, less frequently, of stems. Young endodermal cells are much like elongate parenchyma cells, except that a band of chemically distinct thickening runs around each cell on its radial (side) and end walls. This reinforced, waterproof band is called the ***Casparian strip***. It prevents the movement of fluids into or out of the root except through metabolically active cells, where the passage of water and solutes can be controlled. The cells of endodermal tissue occur in a single layer and are compactly arranged without intercellular spaces.

Fluid movement: vascular tissue Vascular, or conductive, tissue is a distinctive feature of the higher plants, one that has made possible their extensive exploitation of the terrestrial environment. It incorporates cells that function as tubes through which water and nutrients move from one part of the plant body to another. There are two principal types of vascular tissue: xylem and phloem. Both are complex tissues: they consist of more than one kind of cell. The operation of the vascular system of terrestrial plants is detailed in Chapter 29.

1 *Xylem* Xylem is a vascular tissue that functions in the transport of water and dissolved substances upward in the plant body. It forms a continuous pathway running through the roots, the stem, and appendages such as leaves. In the flowering plants the xylem cells come in two specialized varieties: ***tracheids*** and ***vessel elements***. The xylem of flowering plants also includes numerous parenchyma and sclerenchyma cells. The parenchyma cells are the only living cells in mature functioning xylem: both the cytoplasm and the nuclei of tracheids, the vessel elements, and the sclerenchyma cells disintegrate at maturity, leaving their thick cell walls as the functional structures. The tracheid and vessel walls form the tubes in which vertical movement of materials can take place.

Xylem also provides support, particularly of the aerial parts of the plant. The numerous fibers in the xylem function almost exclusively in this way, and the thick-walled tracheids are also important as supportive elements. In trees, nearly all of what we call wood is xylem.

2 *Phloem* Phloem is unlike xylem in that materials can move both up and down in it. Phloem functions particularly in the transport of organic materials such as carbohydrates and amino acids. For example, newly synthesized organic molecules move through the phloem from the leaves to the stem and roots for storage or to the growing points of the plant for immediate use. Like xylem, phloem is a complex tissue, and contains both parenchyma and

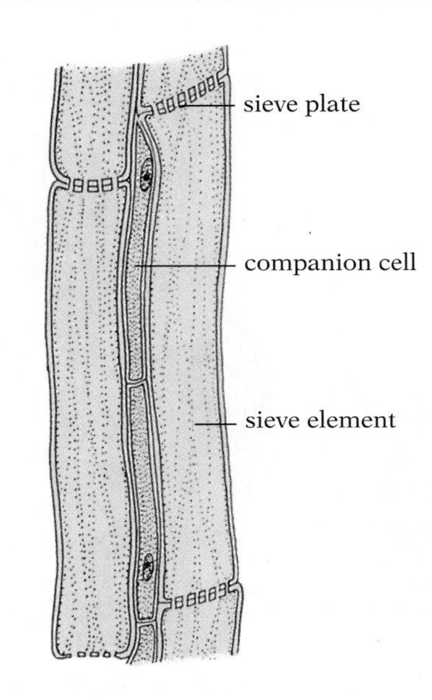

22.53 Longitudinal section of sieve elements and companion cells

The sieve elements lose their nuclei at maturity, but retain their cytoplasm.

sclerenchyma cells in addition to the sieve elements and companion cells that are unique to it. The ***sieve elements*** (Fig. 22.53) are the vertical transport units of phloem; at maturity their nuclei disintegrate, but their cytoplasm remains. ***Companion cells***, which retain both their nuclei and their cytoplasm at maturity, are closely associated with the sieve elements in the most advanced plants.

■ PLANT ORGANS

The body of the higher land plants consists of two major parts: the ***root*** and the ***shoot*** (Fig. 22.54). Many of the tissues are essentially continuous throughout both root and shoot. For example, the vascular tissue of root and shoot, despite a somewhat different arrangement in each, forms an uninterrupted transport system.

The roots of a plant function mainly in procurement of inorganic nutrients such as minerals and water, in transport, in nutrient storage, and in anchoring the plant to the substrate.

The shoot, which is structurally somewhat more complex than the root, consists of the stem and its appendages, particularly the leaves and the reproductive organs (cones and flowers). The stem functions in support and in internal transport, while the leaves are the main organs in which the critical process of photosynthesis takes place.

The number of distinct organs—root, stem, leaf, and reproductive organs—is much smaller than in higher animals. The plant body is not so clearly subdivided into readily distinguishable functional components, or organs, as the animal body; the parts of the plant grade more imperceptibly into each other and, in some ways, form a more continuous whole. Since most plants are autotrophs that remain rooted to one spot, they do not need most of the tissues and organs essential for animals, which must move about, ingest food, digest it, dispose of wastes, and so on.

■ SPERMOPSIDA: THE SEED PLANTS

The seed plants have been by far the most successful in fully exploiting terrestrial environments. They first appeared in the late Devonian; in the Carboniferous they replaced the lycopsids and sphenopsids as the dominant land plants, a position they still hold today. In these plants the gametophytes are even more reduced than in the ferns—they are not photosynthetic or free-living—and the sperm of most modern species are not independent free-swimming flagellated cells. In addition, the young embryo, together with a rich supply of nutrients, is enclosed within a desiccation-resistant seed coat and can remain dormant for extended periods if en-

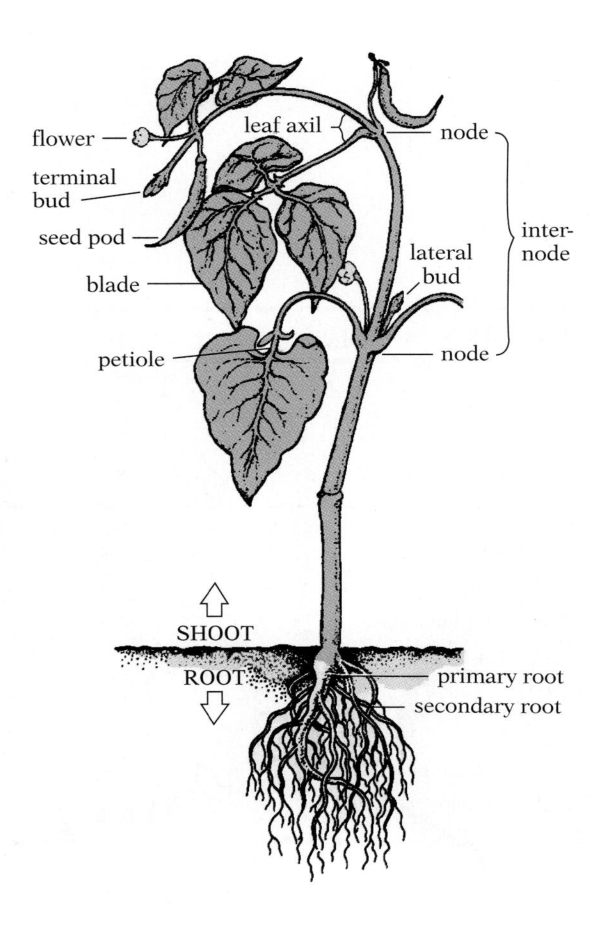

22.54 Diagram of flowering plant body

The vascular tissue (not visible in this external view) is continuous through all parts of the plant.

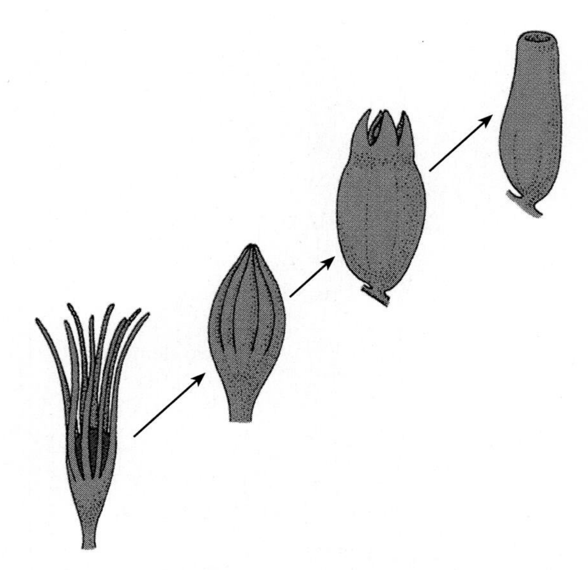

22.55 A model for the possible evolution of the seed

Shown here are the seeds of four species of extinct seed ferns (pteridosperms); note the progressive development of the integument. This sequence is thought to be similar to the one by which seeds of other plants evolved.

vironmental conditions are unfavorable (Fig. 22.55). In short, the aspects of the reproductive process that are most vulnerable in more primitive vascular plants have been eliminated in the seed plants (Table 22.3). In addition, seed plants produce secondary wood, an adaptation that allows them to grow to a greater size.

The seed plants have traditionally been divided into two classes, the Gymnospermae and the Angiospermae. In recent years, however, it has become increasingly clear that the relationships among the five groups bracketed together as the gymnosperms are not particularly close and that these groups differ from one another at least as much as they differ from the angiosperms. Consequently many modern classifications recognize each of the gymnosperm groups as a separate class. We have adopted this procedure in the technical classification given in the Appendix and outlined below, but will discuss the gymnosperm groups together.

Subphylum Spermopsida
- GYMNOSPERMS
  - Class Pteridospermae
  - Class Cycadae
  - Class Ginkgoae
  - Class Coniferae
  - Class Gneteae
- ANGIOSPERMS
  - Class Angiospermae

Conifers and other gymnosperms The first gymnosperms appeared in the late Devonian, some 350 million years ago. Many of those first seed plants had bodies that closely resembled the ferns. Today these fossil plants—usually called the seed ferns—are grouped together as the class Pteridospermae of the subdivision Spermopsida. No members of this class survive today.

Another ancient group, the cycads and their relatives (class Cycadae), may have arisen from the seed ferns. These plants first appeared in the Permian period and became abundant during the

Table 22.3 A comparison of the subphyla of *Tracheophyta*

CHARACTERISTICS	*PSILOPSIDA*	*LYCOPSIDA* (CLUB MOSSES)	*SPHENOPSIDA* (HORSETAILS)	*PIEROPSIDA* (FERNS)	*SPERMOPSIDA* GYMNOSPERMS (CONIFERS)	*SPERMOPSIDA* ANGIOSPERMS (FLOWERING PLANTS)
Vascular tissue	+	+	+	+	+	+
True roots and leaves	–	+	+	+	+	+
Megaphyllous leaves	–	–	–	+	+	+
Gametophyte retained in sporophyte tissue	–	–	–	–	+	+
Sperm cells without flagella	–	–	–	–	+ (– in primitive groups)	+
Production of seeds	–	–	–	–	+	+
Flowers and fruit	–	–	–	–	–	+

22.56 A living cycad

Though often called "sago palms," these plants are not really palms at all, but members of an ancient gymnosperm group.

22.57 Branch of a larch tree in spring

Note the needlelike leaves. At this time of year the new female cones have a pinkish coloration.

Mesozoic era. They had large palmlike leaves. The cycads declined after the rise of the angiosperms in the Cretaceous period, but nine genera containing over 100 species are in existence today (Fig. 22.56). They are generally called sago palms and are fairly common in some tropical regions.

The class Ginkgoae comprises still another group that was once widespread but is now nearly extinct. There is only one living species, the ginkgo or maidenhair tree, often planted as an ornamental tree but almost unknown in the wild.

By far the best-known group of gymnosperms is the conifers (class Coniferae), which include pines, spruces, firs, cedars, hemlocks, yews, and larches. The leaves of most of these plants are small evergreen needles or scales (Fig. 22.57), with an internal arrangement of tissues (Fig. 22.58) that differs somewhat from that in angiosperms. This group first arose in the Carboniferous period and was very common during the Mesozoic era, but only about 500 species survive today.

The life cycle of a pine tree illustrates the seed method of reproduction. The large pine tree is the diploid sporophyte stage (stage 5 in Fig. 22.61). This tree produces reproductive structures called ***cones***, of which there are two kinds: large female cones (Fig. 22.59A), in whose sporangia meiosis gives rise to haploid megas-

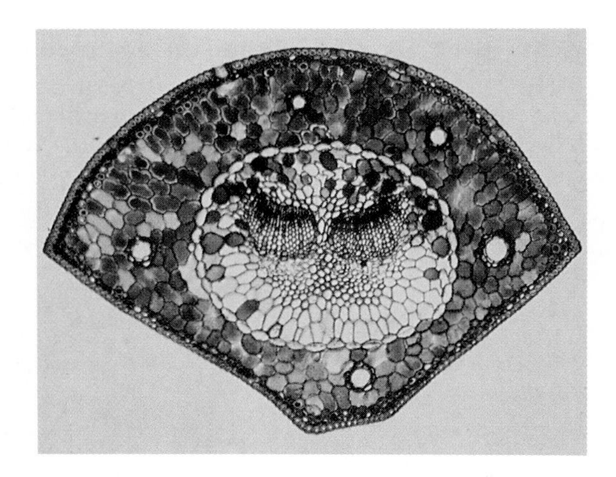

22.58 Cross section of a pine needle

Large resin ducts (white surrounded by blue) can be seen outside the prominent endodermis (circular array of oval cells) that bounds the large central stele.

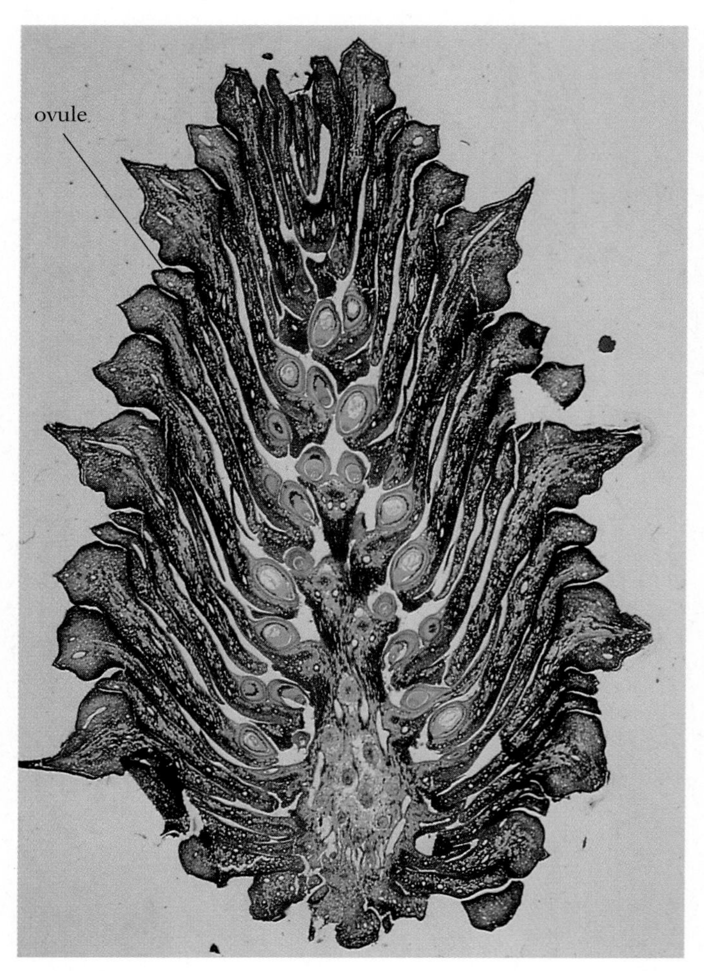

A

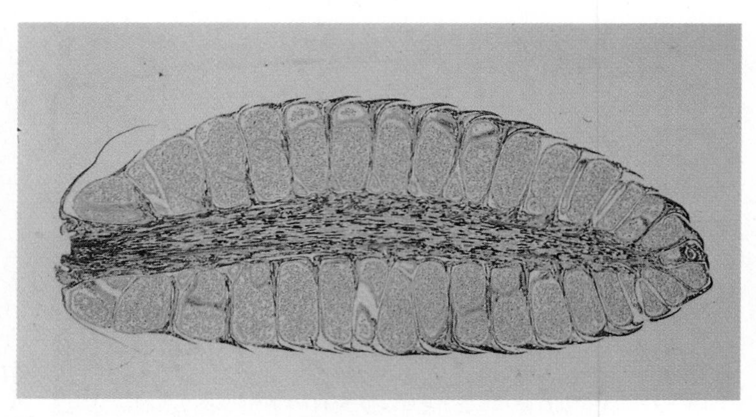

B

C

22.59 Sections of female and male pine cones

(A) Female cone. Ovules can be seen on the surface of the sporophylls near their base. (B) Male cone. Each sporophyll (cone scale) bears a large sporangium that becomes a pollen sac. (C) Male and female cones can occur together. A female cone, or strobilus, is seen at the bottom in this photo, with male cones just above it.

pores (stage 2), and small male cones (Fig. 22.59B), in whose sporangia meiosis gives rise to haploid microspores. (Production of distinctive male and female spores—heterospory—is characteristic of all seed plants, both gymnosperms and angiosperms.) In both kinds of cones the sporangia are produced by highly modified leaves (sporophylls).

Each scale of a female cone bears two sporangia on its upper (adaxial) surface (Fig. 22.60A). Each sporangium is encased in an integument, or shell, with a small opening, the ***micropyle***, at one end (Fig. 22.60B). Meiosis takes place inside the sporangium, producing four haploid megaspores, three of which disintegrate. The single remaining megaspore gives rise, by repeated mitotic divisions, to a multicellular mass, which is the female gametophyte (megagametophyte). The female gametophyte produces two to five tiny archegonia at its micropylar end. Egg cells develop in the

EXPLORING FURTHER

EVOLUTION OF PINE CONES

A male pine cone is a spiral cluster of cone scales, which are highly modified reproductive leaves (sporophylls) on a short section of stem (Fig. 22.59B). The same description was generally thought to apply to the larger female cone, too, until the fossil evidence forced a revision of that assumption.

Apparently the earliest female cones were compound structures composed of a section of stem bearing a series of modified nonreproductive leaves called bracts (Fig. A). In the axil where each bract joined the stem was a very short budlike lateral branch. A few of its spirally arranged scalelike leaves bore sporangia—that is, they were sporophylls. In the course of evolution, all the dwarf lateral branches moved close together, and each was reduced to one to three sporophylls, which fused with several tiny sterile leaves to form a single compound structure called an ovuliferous scale. Each ovuliferous scale, in turn, partly fused with the bract in whose axil it developed (Fig. B). Thus each "scale" of a modern female pine cone (Fig. 22.59A) consists of a bract (which is a modified leaf) and an ovuliferous scale (derived from a dwarf branch consisting of fused sporophylls and sterile leaves).

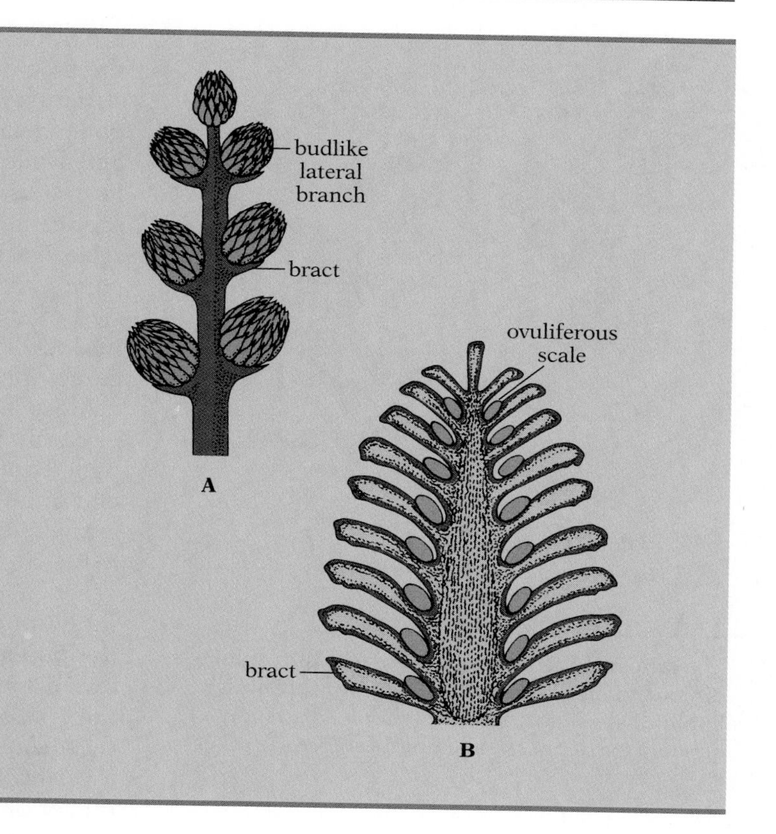

archegonia. Note that the megaspore is never released from the sporangium, and that the female gametophyte derived from it remains embedded in the sporangium, which is still attached to the cone scale. The composite structure consisting of integument, sporangium, and female gametophyte is called an ***ovule***.

Each of the many microspores produced by meiosis in a spo-

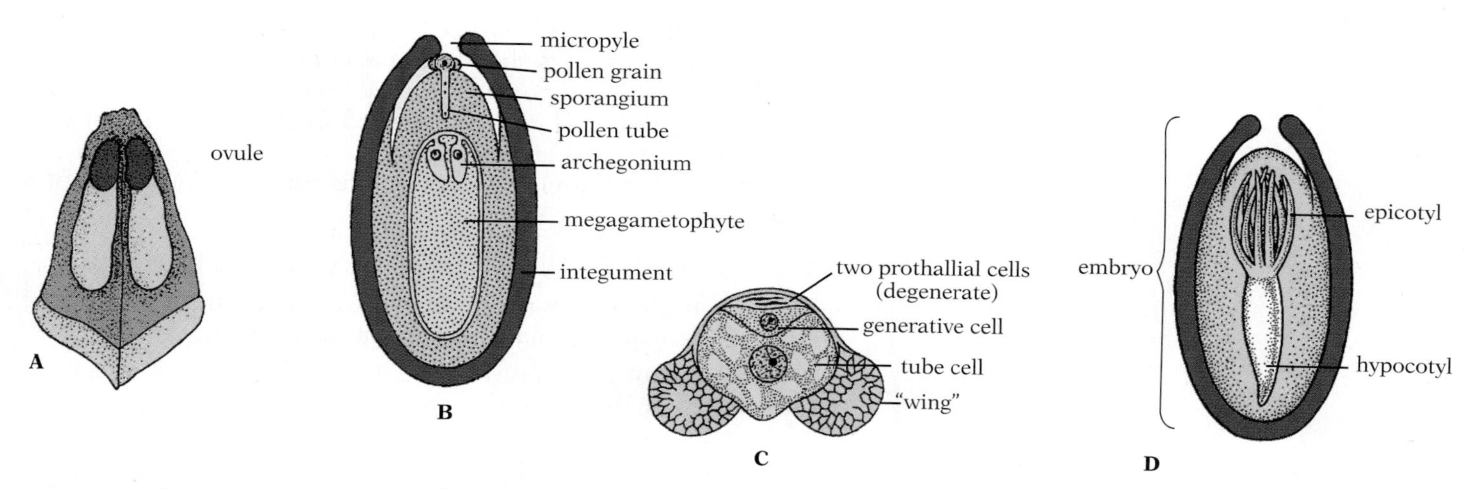

22.60 Ovules and pollen grains of pine

(A) Scale from female cone. The two ovules, each containing a sporangium, lie on the surface of the scale near its point of attachment to the cone axis. (B) Section of an ovule with germinating pollen. C) Pollen grain, composed of four cells, two of which are degenerate. (D) Developing embryo.

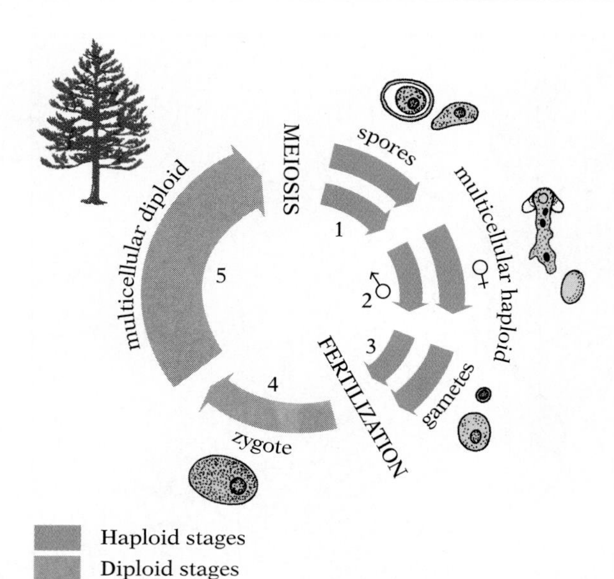

22.61 Life cycle of conifers

The familiar tree is the sporophyte (multicellular diploid) stage. The gametophytes (multicellular haploid) are very tiny and cannot lead an independent existence. In all haploid stages the sexes are separate.

rangium of a male cone becomes a ***pollen grain***. It develops a thick coat, which is highly resistant to loss of water, and winglike structures on each side, which aid its dispersal by wind. Within the pollen grain the haploid nucleus divides mitotically several times, and walls develop around each nucleus (Fig. 22.60C). Two of the four cells (called prothallial cells) degenerate; the two that remain are the generative cell and the tube cell. The mature pollen grain is released from the cone when the sporangium bursts. A single male cone may release millions of tiny pollen grains, which may be carried many kilometers by the wind. Note that the pollen grains are multicellular haploid structures and that they constitute the male gametophyte (microgametophyte; stage 2 in Fig. 22.61).

Most of the millions of pollen grains released by a pine tree fail to reach a female cone. A few sift down between the scales of a female cone and land in a sticky secretion near the open micropylar end of an ovule. As this secretion dries, it is drawn through the micropyle, carrying the pollen grains with it. The arms of the integument around the micropyle then swell and close the opening. When a pollen grain comes in contact with the end of the sporangium just inside the micropyle, it develops a tubular outgrowth, the ***pollen tube***. The nucleus of the tube cell enters the tube, followed by the generative cell. The generative cell then divides, giving rise to a sterile cell and a spermatogenous cell. The latter of these daughter cells divides again, producing two sperm cells. Thus a germinated pollen grain contains four active nuclei plus the two nuclei of the degenerate cells; this six-nucleate condition is as far as the male gametophyte of pine ever develops toward multicellularity.

The pollen tube grows down through the tissue of the sporangium and penetrates into one of the archegonia of the female gametophyte.[7] There it discharges its sperm cells, one of which fertilizes the egg cell. Fertilization usually occurs about 1 year after pollination. The resulting zygote (stage 4) then divides mitotically to produce a tiny embryo sporophyte consisting of a hypocotyl and an epicotyl (Fig. 22.60D). The embryo is still contained in the female gametophyte, which is itself contained in the sporangium. Finally, the entire ovule is shed from the cone as a ***seed***, which consists of three main components: a seed coat derived from the old integument, stored food material derived from the tissue of the female gametophyte, and an embryo.[8]

The life cycle of the pine shows several advances over that of a typical fern:

1 The sporophylls are more highly modified and less leaflike.

2 There are two types of sporangia, which produce two types of spores: microspores (male) and megaspores (female).

3 Two kinds of gametophytes are derived from the two types of spores; that is, the sexes are separate in the gametophyte stage.

[7] Since an ovule contains several archegonia, several embryos may begin development, but usually only one completes it.

[8] As a rule, the sporangium (or nucellus, as it is called in the ovule) eventually disintegrates and is not present in the seed, but in a few species it may be preserved, usually as the inner layer of the seed coat.

4 The gametophytes are much further reduced than those of ferns; they do not possess chlorophyll and are not free-living. A male gametophyte consists only of the six-nucleate pollen grain and tube. A female gametophyte is only a mass of haploid tissue in the sporangium.

5 There are usually no flagellated sperm cells.[9]

6 The young embryo is contained within a seed.

Flowering plants: the angiosperms Angiosperms appeared in the early Cretaceous. The group underwent great expansion, and became the dominant land flora of the Cenozoic era, as they are today. They include nearly 300,000 species. One reason for their success may be their larger vascular elements, which allows more rapid movement of fluids in both xylem and phloem. Another may be their much greater ability to adopt a deciduous life history, shedding leaves and entering dormancy during periods unfavorable for growing.

Finally, their novel reproductive structures, and in particular their ability to employ insect pollinators, are important. The reproductive structures of gymnosperms are cones; the ovules, which become the seeds, are borne naked on the surface of the cone scales. The reproductive structures of angiosperms, by contrast, are flowers, and the ovules are enclosed within modified leaves called ***carpels***.

A flower is a short length of stem with modified leaves attached. The modified leaves of a typical flower (Fig. 22.62) occur in four sets attached to the enlarged end (receptacle) of the flower stalk: (1) The ***sepals*** enclose and protect all the other floral parts during the bud stage. They are usually small, green, and leaflike; in some species, however, they are large and brightly colored. All the sepals together form the ***calyx.*** (2) Internal to the sepals are the ***petals***, which together form the ***corolla***. The calyx and corolla together constitute the ***perianth***. In flowers pollinated by insects, birds, or other animals, the petals are usually quite showy, but in those pollinated by wind they are often small or even absent.

(3) Just inside the circle of the corolla are the ***stamens***, which are the male reproductive organs—that is, they are the sporophylls that produce the microspores.[10] Each stamen consists of a stalk, called a ***filament***, and a terminal ovoid pollen-producing structure called an ***anther***. (4) In the center of the flower is one or more female reproductive organs, the ***pistils***. Each pistil consists of an ***ovary*** at its base, one or more slender stalks called ***styles***, which rise from the ovary, and an enlarged apex called a ***stigma***. The pistil is derived from one or more sporophylls, which in flowers are carpels.[11] All four kinds of floral organs—sepals, petals, stamens,

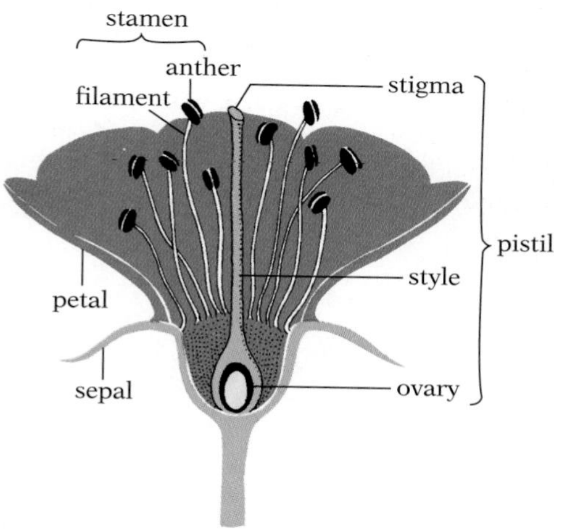

22.62 Major parts of a flower

[9] The pollen tubes of primitive gymnosperms, such as ginkgoes and cycads, produce flagellated sperm cells, but these have only a very short distance to swim within the cytoplasm of the pollen tube to reach the egg cells, since the pollen grains have already been carried to the ovules by the wind.

[10] It is not entirely certain that the stamens and carpels evolved directly from individual leaves.

[11] A simple pistil is composed of only one sporophyll, or carpel. A compound pistil is composed of several fused carpels.

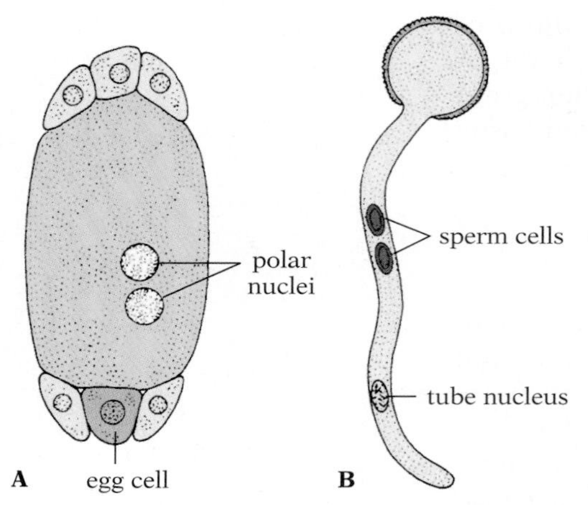

22.63 Gametophytes of an angiosperm

(A) Female gametophyte (embryo sac), which is composed of seven cells. One cell is much larger than the others and contains the two polar nuclei. (B) Male gametophyte (pollen grain and tube).

22.64 Pollen chambers of a lily anther

When the pollen chambers (derived from a sporangium) open, the numerous pollen grains (dark paired cells on circular white background) within it are released.

and pistils—are present in so-called complete flowers, whereas "incomplete" flowers lack one or more of them.[12]

Within the ovary are one or more (at least one for each carpel) sporangia, called ***ovules***, which are attached by short stalks to the wall of the ovary. Meiosis occurs once in each ovule, with formation of four haploid megaspores, three of which usually disintegrate. The remaining megaspore then divides mitotically several times, producing in most species a structure composed of seven cells; one is much larger than the others and contains two nuclei, called polar nuclei (Fig. 22.63A). This haploid seven-celled, eight-nucleate structure is the much-reduced female gametophyte (often called an embryo sac).[13] One of the cells located near the micropylar end will act as the egg cell.

Each anther has four sporangia in each of which many cells undergo meiosis, producing numerous haploid microspores. The wall of each microspore thickens, and the nucleus divides mitotically, producing a generative nucleus and a tube nucleus. The resulting thick-walled two-nucleate structure is a pollen grain—a male gametophyte—which is released from the anther when the mature sporangium (Fig. 22.64) splits open.

A pollen grain germinates when it falls (or is deposited) on the stigma of a pistil, which is usually rough and sticky. A pollen tube begins to grow, and the two nuclei of the pollen grain move into it.

[12] In some species, such as corn, willow, oak, and walnut, the stamens and pistils are in separate flowers. Incomplete flowers of this type, in which only one of the two kinds of reproductive structures is present, are called imperfect flowers. Flowers with both stamens and pistils (whether complete or incomplete) are called perfect flowers.

[13] The embryo sac of some species has more than eight nuclei, and that of a limited number of other species has fewer than eight. Furthermore, in some species no cytokinesis occurs, and all the nuclei lie in the same mass of cytoplasm. However, the most common sort of embryo sac is the seven-celled, eight-nucleate type described here.

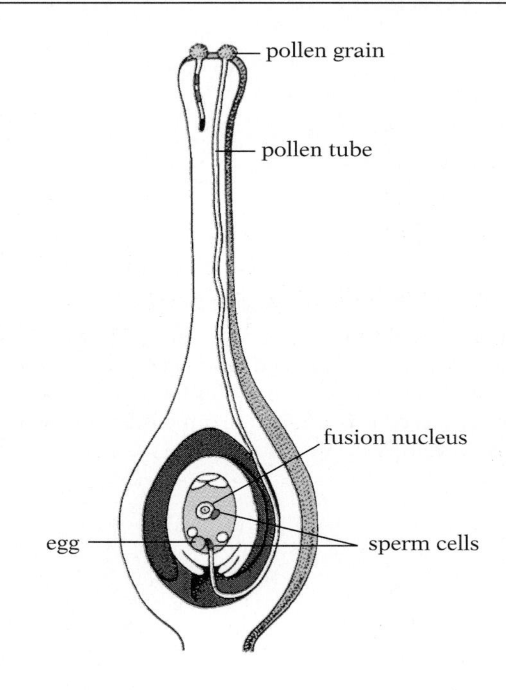

22.65 Fertilization of an angiosperm

Pollen grains land on the stigma and give rise to pollen tubes that grow downward through the style. One of the pollen tubes shown here has reached the ovule in the ovary and discharged its sperm cells into it. One sperm will fertilize the egg cell; the other sperm will unite with the diploid fusion nucleus (derived from the two polar nuclei) to form a triploid nucleus, which will give rise to endosperm.

The generative nucleus[14] then divides, giving rise to two sperm cells (Fig. 22.63B).[15] The pollen tube grows down through the tissues of the stigma and style and enters the ovary (Fig. 22.65). When the tip of the pollen tube reaches an ovule, it enters the micropyle and then discharges the two sperm cells into the female gametophyte (embryo sac). One of the sperm fertilizes the egg cell, and the zygote thus formed develops into an embryo sporophyte. By the time fertilization occurs, the two polar nuclei of the female gametophyte have combined to form a diploid ***fusion nucleus***, with which the second sperm unites to form a triploid nucleus. This nucleus undergoes a series of divisions, and a triploid tissue called ***endosperm*** is formed. The endosperm functions in the seed as a source of stored food for the embryo.

After fertilization, the ovule matures into a seed with a seed coat, stored food, and embryo. The angiosperm seed differs from that of pine in being enveloped by the ovary. It is the ovary that develops into the ***fruit***, usually enlarging greatly in the process. Sometimes other structures associated with the ovary, such as the receptacle, are incorporated into the fruit. The ripe fruit may burst, expelling the seeds, as in peas (where the pod is the fruit), or the ripe fruit with the seeds inside may fall from the plant to the ground, as in tomatoes, squash, apples, and acorns. The fruit not only helps protect the seeds from desiccation during their early development, but often also facilitates their dispersal by various means—the wind, say (Fig. 22.66), or an animal that, attracted by the fruit, carries it to other locations or eats both fruit and seeds and later releases the seeds unharmed in its feces.

The main features in which the angiosperm life cycle (Fig. 22.67) differs from that of gymnosperms include the following:

1 The reproductive structures are flowers instead of cones. The sporophylls (stamens and pistils) of flowers are less leaflike than those of cones.

2 The ovules are embedded in the tissues of the female sporophylls instead of lying bare on their surface.

22.66 Seed pod of the milkweed

The milkweed seed pod is a fruit and bears many seeds, each of which is attached to a set of fibers that act as a wind-borne sail.

[14] The generative nucleus is surrounded by a plasma membrane and is thus technically a cell, though with virtually no cytoplasm.

[15] The pollen grains in most species are released in the two-nucleate condition, and the division of the generative nucleus does not take place until germination. In some species, however, this division occurs earlier, and the pollen grains are released from the anthers in the three-nucleate condition.

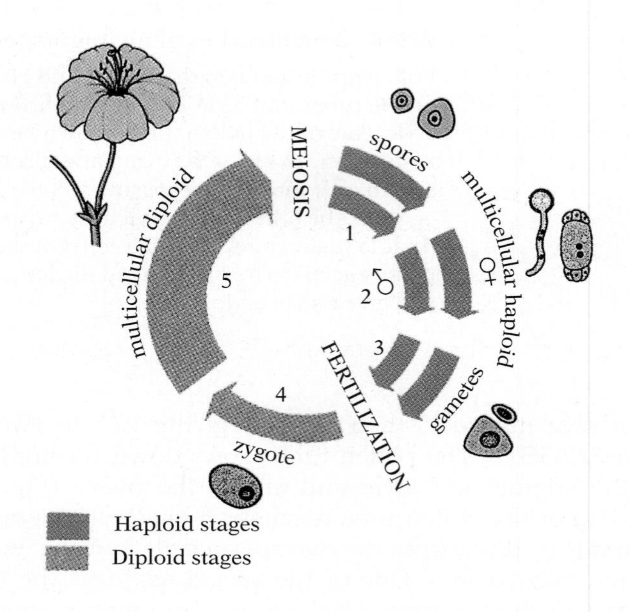

22.67 Life cycle of angiosperms

The mature diploid plant produces gametes—pollen, which is dispersed, and eggs, which are retained in the ovary. Once fertilization has occurred, a seed is produced, from which the new plant develops.

3 The gametophytes are even more reduced than those of gymnosperms. The male gametophyte (pollen grain and tube) has only three nuclei. The female gametophyte usually has only eight nuclei.

4 In pollination the pollen grains are deposited on the stigma; the pollen tube thus has much farther to grow in angiosperms.

5 Angiosperms have "double fertilization," one sperm fertilizing the egg cell and the other uniting with the fusion nucleus to produce a triploid endosperm.[16] Gymnosperms have single fertilization. The stored food in the seed of gymnosperms is the haploid tissue of the female gametophyte and is thus quite different from the triploid endosperm of angiosperms.

6 The seeds of angiosperms are enclosed in fruits that develop from the ovaries and associated structures; gymnosperms have no fruit.

The enormous length of the style means that the pollen grain must metabolize food reserves in the style itself to support its growth. This puts the pollen genome to a severe test: about half of the grain's 40,000 genes are active during this growth, and, because the pollen is a haploid gamete, any defective member of this basic constellation of metabolism and synthesis genes will be directly expressed. The result is a weeding out of many mutant alleles. In insect-pollinated plants, the effects are even more extreme. Instead of a few pollen grains arriving on the wind over the course of days, many hundreds may be delivered simultaneously whenever a pollinator arrives. The result is both a test of genome viability and a race between rival pollen grains down the long style; only the first arrivals, superior in their ability to use the resources provided by the style to grow rapidly and efficiently, succeed in fertilizing the limited number of eggs.

[16] In some angiosperms—lilies, for example—the endosperm is pentaploid ($5n$).

The class Angiospermae is customarily divided into two subclasses, the Dicotyledoneae (***dicots***) and the Monocotyledoneae (***monocots***). The dicots include oaks, maples, elms, willows, roses, beans, clover, tomatoes, asters, and dandelions. The monocots include grasses, corn, wheat, rye, onions, daffodils, irises, lilies, and palms. Most grains and spring-flowering bulbs are monocots. There are certain basic differences between the two groups:

1 The embryos of dicots have two embryonic leaves—cotyledons—whereas those of monocots have only one.

2 Dicots often have vascular cambium and secondary growth; monocots usually do not.

3 The vascular tissue in the stems of young dicots is arranged in a circular bundle, or fused to form a tubular vascular cylinder; monocots have scattered vascular bundles.

4 The leaves of dicots usually have net venation; those of monocots usually have parallel venation.

5 Dicot leaves generally have petioles; monocots generally do not.

6 The flower parts of dicots usually occur in fours or fives or multiples of these (for example, four sepals, four petals, four stamens); those of monocots usually occur in threes or multiples of three (Fig. 22.68).

A

B

22.68 Monocot and dicot flowers compared

(A) In a trillium flower (a monocot), the main parts occur in threes. (B) In a nasturtium flower (a dicot), the parts occur in fives.

Plants provide humans worldwide with more than 70% of our protein, and 70% of that fraction comes from monocots; the remaining 30% is mostly derived from legumes.

EVOLUTIONARY TRENDS IN PLANTS

One major trend among plants is the progressive decrease in the size and importance of the gametophyte (Fig. 22.69). Along with the reduction in size has come a gradual loss of independence. At the same time, the sporophyte has become increasingly larger and more independent. This is an adaptation to a longer life cycle and increased structural complexity.

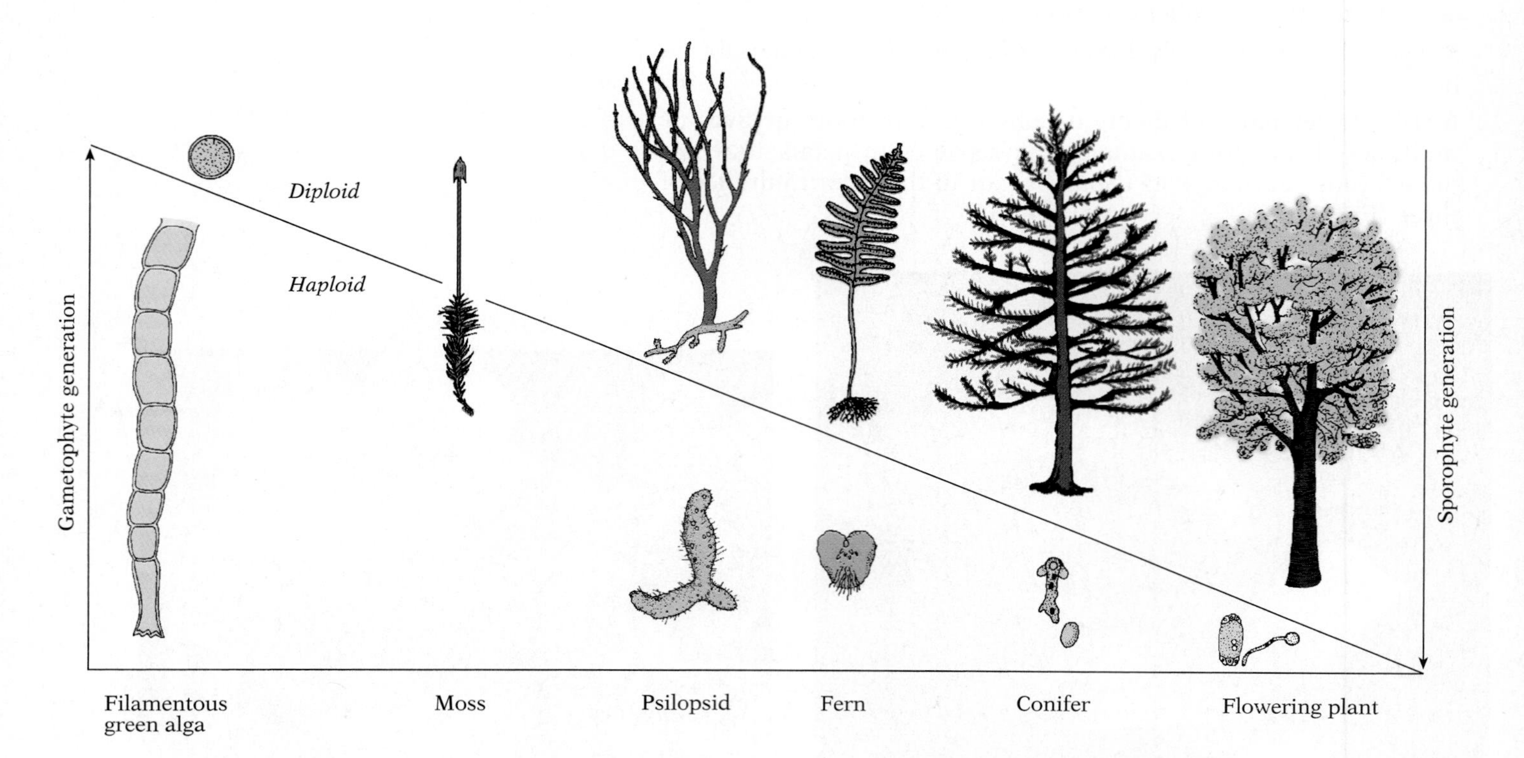

22.69 Transition from gametophyte to sporophyte dominance as shown by representative plant groups

The gradual decrease in the size and importance of the gametophyte and the corresponding increase in the size and importance of the sporophyte are shown. In filamentous green algae and in mosses, the gametophyte state is dominant and the sporophyte state is much reduced. In the other plant groups shown, the sporophyte is dominant and the gametophyte becomes progressively smaller. Psilopsids have a large underground gametophyte, ferns have a small but independent (photosynthetic) gametophyte, and conifers and flowering plants have tiny female gametophytes and male pollen grains, both dependent on the sporophytes. (The organisms are not drawn to scale.)

Another evolutionary trend in higher plants has been the change from the homosporous to the heterosporous condition. The early vascular plants were largely homosporous, producing only one type of spore, which developed into gametophytes containing both archegonia and antheridia. The more advanced vascular plants produce two types of spores: microspores and megaspores. The microspores give rise to male gametophytes, and the megaspores to female gametophytes.

Most of the evolutionary advances are adaptations to life on land. The evolution of vascular tissue in the primitive tracheophytes enabled plants to grow larger and survive in drier environments. However, water was still required for reproduction. The angiosperms and gymnosperms solved this problem by evolving pollen and seeds. Pollen grains don't need external water for fertilization, and the seed, with its coat and stored food, protects the embryo and enables it to withstand unfavorable conditions.

CHAPTER SUMMARY

CHROMISTA: THE CHLOROPHYLL *C* ALGAE AND PSEUDOFUNGI

They are distinguished by the presence of chlorophyll *c*, chloroplasts inside the ER, and lateral projections from their flagella. (p. 597)

OCHRISTA: YELLOW-GREEN, BROWN, AND GOLDEN-BROWN ALGAE AND THE DIATOMS This group includes simple algae and the planktonic diatoms. Especially important are the brown algae, which are seaweeds. Many brown algae are extremely long and plantlike in appearance. Most have specialized sex organs that produce morphologically identical (isogametic) gametes, or two distinct (heterogametic) gametes. Most have alternating generations of haploid gametophyte (gamete-producing) and diploid sporophyte (spore-producing) cells. (p. 598)

HAPTOPHYTA AND CRYPTISTA Most members of these groups are unicellular algae, though a few are parasites. (p. 603)

LOWER PLANTS: RED AND GREEN ALGAE

RHODOPHYTA: THE RED ALGAE Red algae are marine seaweeds; their color results from their unique accessory pigments. (p. 604)

CHLOROPHYTA AND CHAROPHYTA: THE GREEN ALGAE Green algae have the same chlorophylls (*a* and *b*) as land plants. Some live as motile unicellular individuals with both sexual and asexual reproductive alternatives. Others exist as colonies, some of which have shared cytoplasm, coordinated swimming, and even specialized reproductive cells. The most advanced forms have alternating generations. (p. 605)

HOW PLANTS CONQUERED THE LAND

Movement onto land requires adaptations to obtain and distribute water, mineral nutrients, and gases; to avoid desiccation; to reproduce; to withstand extreme environmental fluctuations; and to provide mechanical support. (p. 612)

BRYOPHYTA: LIVERWORTS, HORNWORTS, AND MOSSES Bryophytes live low to the ground in damp places. They lack vascular tissue. The sporophyte grows on the gametophyte. (p. 614)

LOWER TRACHEOPHYTES: THE EARLY VASCULAR PLANTS The lower tracheophytes have at least limited vascular tissue, and thus some transport of water and nutrients within the plant. Most also have roots. All require moist conditions for reproduction and early growth. The ferns are the most fully terrestrial members of this group, which also includes club mosses and horsetails. All have alternation of generations. (p. 616)

HIGHER (FULLY TERRESTRIAL) PLANTS

HOW PLANTS ARE ORGANIZED INTERNALLY The tissues of higher plants are either meristematic (capable of cell division) or permanent, though some permanent tissues can revert to meristematic growth. Apical meristem is found at growing tips; lateral meristem is usually located between the xylem (the vascular tissue that transports water and minerals from the roots) and phloem (vascular tissue that transports organic nutrients). Permanent tissues include the protective epidermis and periderm, loosely packed parenchyma, the more densely packed collenchyma, the supporting sclerenchyma, the endodermis that surrounds the vascular tissue in roots, as well as xylem and phloem. (p. 622)

PLANT ORGANS Plants consist of roots and shoots. Shoots in turn have distinct stems, leaves, and reproductive organs. (p. 629)

SPERMOPSIDA: THE SEED PLANTS They include 700 species of gymnosperms (mostly conifers) and nearly 300,000 species of angiosperms (mostly dicots). These plants have advanced vascular systems and are fully adapted to terrestrial reproduction: the female gamete is retained in the parent plant and later enclosed in a desiccation-resistant seed; the male gametes (pollen) resist desiccation while being transferred (by wind or insects) to the female reproductive organ; once there they grow a pollen tube to the ovule or ovary. The haploid stage of the alternation-of-generations cycle is tiny and brief. Monocots have a single leaf at germination, parallel venation, and flowers with three or six petals; dicots have a pair of first leaves, net venation, and flowers with petals in multiples of 4 or 5. (p. 629)

EVOLUTIONARY TRENDS IN PLANTS

As the sporophyte gets larger and takes longer to mature, the gametophyte grows smaller, and the time devoted to its part of the cycle becomes briefer. (p. 640)

STUDY QUESTIONS

1 Oogamy is almost universal among animals and higher plants. What might be the reproductive benefits to a multicellular organism in producing unusually large gametes, at least in certain habitats or with certain life cycles? What might be the advantage of producing unusually small gametes? If both kinds of benefits apply in a population, is there any advantage to medium-sized gametes? Would there be any selection against small-small or large-large fusion that could explain the observed pattern? (pp. 601–603, 608–609)

2 Besides mutation and climatic change, another risk that higher plants face over the course of their relatively long lives is fungal infection. Given that the fungi that attack cereal grains are strain-specific, can the long-generation-time argument for diploidy outlined in the Exploring Further box on sporophyte dominance (p. 625) be expanded to include disease and parasitism? Given the usual advantage of being big, and given the inevitability that selection will often favor long lifetimes, what are some of the defensive adaptations (at the genetic or life-history level) you might expect to evolve in order to counter such a threat? (p. 628)

3 How and to what degree is the task of growing a pollen tube through the style of an angiosperm different from the challenge that faces the male gametophyte in ferns and lower plants? Is the angiosperm female gametophyte facing less selective pressure than its evolutionarily more ancient counterparts? (pp. 622, 635–638)

4 In the case of insect-pollinated angiosperms, are plants in a position to use the style to do a little selective breeding of their own? How might they exploit such a potential? (pp. 636–638)

SUGGESTED READING

JACOBS, W. P., 1994. Caulerpa. *Scientific American* 271 (6). *On the largest single-celled organism in the world—a tropical alga.*

NIKLAS, K. J., 1989. The cellular mechanics of plants, *American Scientist* 77, 344–349. *On how land plants support themselves against the force of gravity.*

RAVEN, P. H., R. F. EVERT, AND S. E. EICHHORN, 1992. *Biology of Plants*, 5th ed. Worth, New York. *A well-written broad coverage of botanical science.*

CHAPTER 23

FUNGI

Fungi are primarily multicellular. Their multicellularity is unusual in that partitions between nucleated compartments, or "cells," are generally either absent or only partial; the cytoplasm is continuous. Compartments may have more than one nucleus. For many fungi, therefore, "multinucleate" is perhaps a more accurate description than "multicellular." The "cells" are usually organized into branched filaments called ***hyphae***, which form a mass called a ***mycelium***. The basic component of the cell wall is chitin, a modified polysaccharide containing nitrogen, which is also found in the exoskeletons of arthropods. Chitin is highly resistant to microbial attack. (Chitin and cellulose, which lacks nitrogen, are compared in Fig. 3.11, p. 54.)

Most fungi are parasitic or saprophytic. Saprophytic fungi usually secrete digestive enzymes onto their food and absorb the products through rootlike filaments called ***rhizoids*** or haustoria. Parasitic fungi may carry out extracellular digestion, or they may directly absorb materials produced by their host.

Some fungi are parasitic on or in animals, including humans; many skin diseases, including ringworm and athlete's foot, are caused by fungi. The pneumonia caused by *Pneumocystis carinii*, a relative of yeast, is a destructive disease agent infecting AIDS victims. Other fungi are parasitic on plants, and cause enormous losses when they attack agricultural crops. The most damaging are corn smuts, wheat rusts and ergots, and rice blasts. (Fungi are the only organisms with enzymes that can break down lignin, a major component of plant cell walls.) Still others cause spoilage of bread, fruit, vegetables, and other foodstuffs, and deterioration of leather goods, fabrics, paper, lumber, and other valuable products.

Many fungi are beneficial. For instance, the vast majority of plant roots have symbiotic fungi living in them which send hyphae into the soil. The hyphae, which typically constitute 15–20% of the plant's root mass, significantly enhance the plant's ability to obtain water and nutrients by greatly increasing the surface area available for absorption (Fig. 23.1). Many researchers believe that the movement of plants onto the land was only possible because of such mutualistic relationships, a theory supported by fossils of early plants. These associations of fungi and plant roots are called mycorrhizae.

23.1 Mycorrhizae

Yeasts are used extensively in the manufacture of alcoholic products and to make bread dough rise. The antibiotic penicillin is ob–

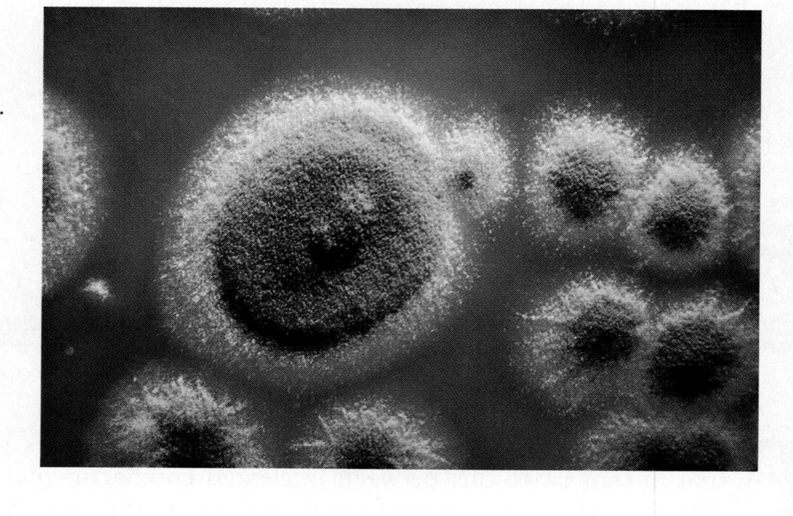

23.2 Colonies of *Penicillium chrysogenum* growing on culture medium in a Petri dish

Almost the entire world's supply of commercial penicillin is produced by this fungus in pharmaceutical laboratories.

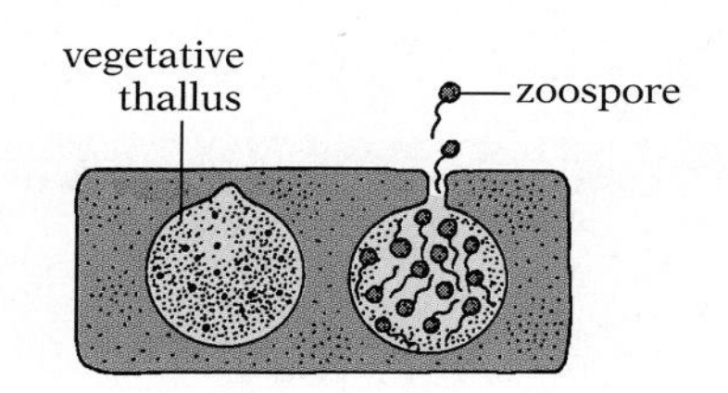

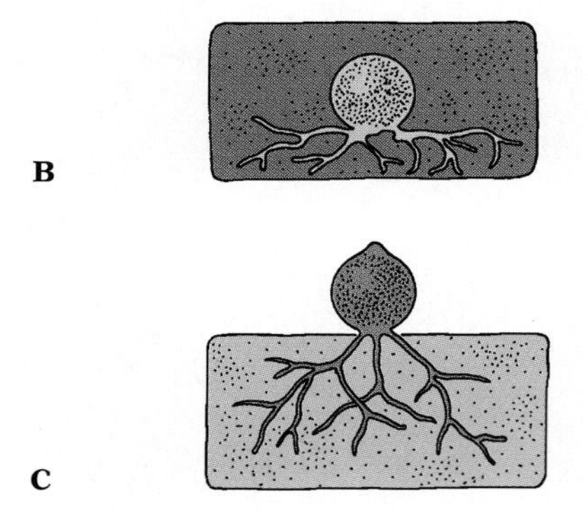

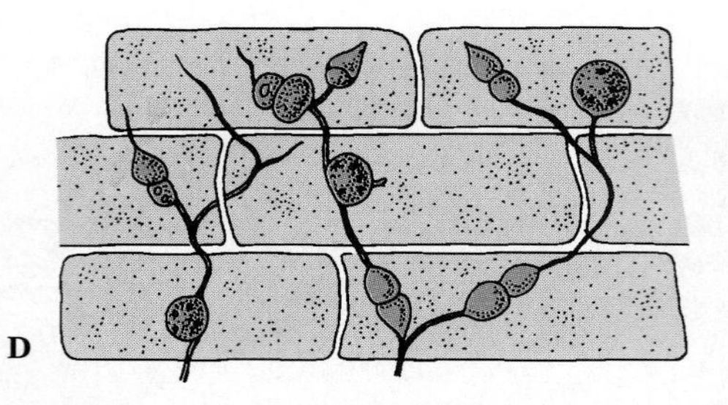

23.3 Thallus forms among chytrids

(A) Some species have a simple saclike thallus (body) located entirely within the host cell. When flagellated zoospores are produced, a tube grows out to the membrane of the host cell and then ruptures, and the zoospores are released to the exterior. (B) Other species that live entirely inside their host cells have rootlike rhizoids for nutrient absorption. (C) Still other species, in which the main part of the thallus is outside the host cell, send rhizoids into the host. (D) Some species exhibit a simple form of multicellularity, and their threadlike bodies exploit several host cells simultaneously.

tained from a fungus (Fig. 23.2). Even the destructive ergots are useful, producing chemicals used to treat migraines. Fungi are important in the manufacture of many cheeses, and certain mushrooms are highly prized as food. Fungi, together with bacteria, decompose vast quantities of dead organic matter that would otherwise accumulate rapidly and make the earth uninhabitable.

Reproduction in the fungi may be either asexual or sexual, but in both cases the haploid stages are usually dominant. Sexual reproduction can occur when haploid hyphae of opposite mating types encounter one another. The characteristics of sexual reproduction are especially important in distinguishing the four divisions of true fungi currently recognized.[1]

CHYTRIDIOMYCOTA: THE PROTIST FUNGI

The chytrids are small organisms often classified as protists. Many are aquatic, living as parasites or saprophytes in or on algae or other plants; some are found in soil, and a few are internal parasites of animals.

The haploid bodies of these organisms may be simple sacs living entirely inside a cell of the host (Fig. 23.3A), sacs with nutrient-absorbing rhizoids extending from their surface (Fig. 23.3B–C), or filamentous forms with sac-like reproductive structures (Fig. 23.3D). Reproduction begins when the nucleus divides mitotically but without cytokinesis, so the sacs become multinucleate. Eventually the cytoplasm is partitioned so that each nucleus receives some. The newly formed cells, each of which develops a single posterior flagellum, are then released into the surrounding medium. Under some circumstances these flagellated cells function as gametes in sexual reproduction. More often they act as asexual reproductive cells called zoospores, which settle down in a suitable location and develop into new sacs or filaments.

[1] It is often difficult to classify species for which sexual stages have not been found. Such species are customarily assigned to the category Fungi Imperfecti; the "imperfection" being in our understanding.

ZYGOMYCOTA: THE CONJUGATION FUNGI

The hyphae of members of the division Zygomycota characteristically lack internal cell walls; internal walls appear only during the formation of reproductive structures. Neither the spores nor the gametes are motile. Sexual reproduction involves conjugative fusion of cells from hyphae of two different mycelia of opposite mating types (Fig. 23.4C).

A typical example of a member of this division is the common black bread mold, *Rhizopus*. The hyphae of this mold form a whitish or grayish mycelium on the bread. The mycelium includes three types of hyphae: stolons, which form a network on the surface of the bread; rhizoids, which penetrate into the bread, both anchoring the fungus and absorbing nutrients; and sporangiophores, which grow upright from the surface and bear globular sporangia (Fig. 23.4B). Thousands of asexual spores are produced in each sporangium. The spores are very tiny and light, and when liberated (by disintegration of the wall of the sporangium) they can be carried long distances by wind or animals. If a spore lands in a warm and moist location, it germinates and gives rise to a new mass of haploid hyphae, thus completing the asexual cycle.

Sexual reproduction in *Rhizopus* resembles that of the green alga Spirogyra (see Fig. 22.28, p. 612). Short branches from two hyphae of different mating types contact each other at their tips (Fig. 23.3C). Cross walls form just behind the tips of these hyphal

A

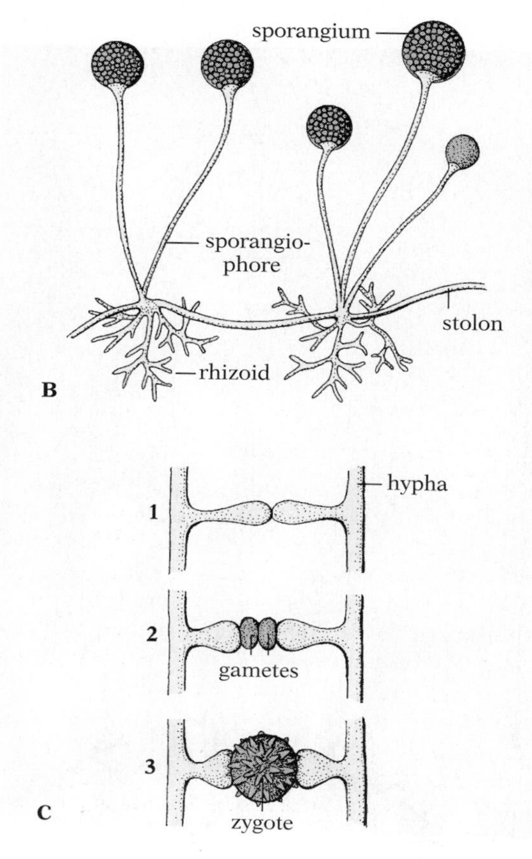

23.4 *Rhizopus* (black bread mold)

(A) *Rhizopus* is made up of numerous threadlike filaments hung with what look like black and white balls. These are sporangia; the black ones are ripe and ready to release their spores. (B) Interpretive drawing showing hyphae with sporangia. (C) Sexual reproduction by conjugation: (1) short branches from two different hyphae meet; (2) the tips of the branch hyphae are cut off as gametes; (3) the gametes fuse in fertilization to form a zygote with a thick spiny wall.

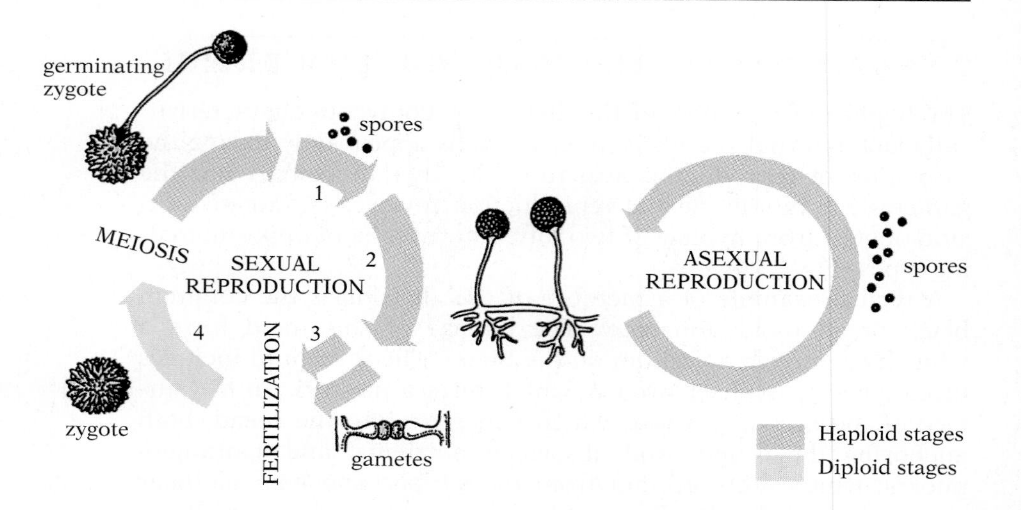

23.5 Life cycle of *Rhizopus*
See text for description.

branches. The gamete cells thus created then fuse to form a zygote, which develops a thick protective wall and enters a period of dormancy usually lasting from 1–3 months. At germination the nucleus of the zygote undergoes meiosis, and a short hypha grows from the zygote. This haploid hypha produces a sporangium, which releases asexual spores that grow into new mycelia. The only diploid stage in the entire sexual cycle is the zygote (Fig. 23.5).

Members of the division Zygomycota are widespread as saprophytes in soil and dung. Some, however, are parasites on plants, animals, or other fungi. More than 100 species are plant-root symbionts.

ASCOMYCOTA: THE SAC FUNGI

The members of this large division are very diverse, varying in complexity from unicellular yeasts through powdery mildews and cottony molds to complex cup fungi. This last group forms a cup-shaped structure composed of many hyphae tightly packed together (Fig. 23.6). The vegetative hyphae of Ascomycota, unlike those of Zygomycota, are septate—that is, they possess internal walls—but these walls, or septa, are perforated and the cytoplasm is continuous.

Though their vegetative structures differ, all Ascomycota resemble each other in forming a reproductive structure called an ***ascus*** during their sexual cycle. An ascus is a sac within which eight haploid spores are usually produced (Fig. 23.7). Sexual reproduction begins when hyphae of opposite mating types ("+" and "−") encounter one another. The "−" hyphae form a small structure, arbitrarily called the male organ, while the "+" hyphae produce a larger organ. Two haploid nuclei, one from each of the two parental hyphae, combine to form a diploid zygote, which then generates the haploid spores of the ascus. Most Ascomycota also

23.6 The cup fungus (*Peziza*) growing on a log in a rain forest in Central America

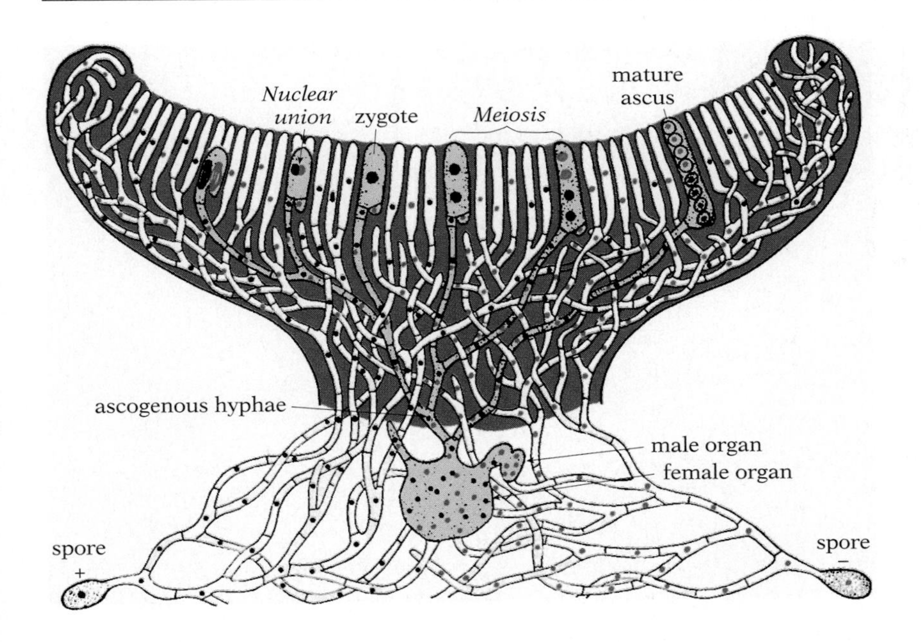

23.7 Section through a cup fungus

Hyphae of two haploid mycelia, one derived from a female (+) spore and the other from a male (–) spore, participate in forming the cup structure and in producing the spores. The female mycelium bears a female organ (ascogonium), and the male mycelium bears a male organ (antheridium). A tube grows from the female organ to the male organ, and then male nuclei (red dots) move into the female organ and become associated with female nuclei (black dots). Next, hyphae grow from the female organ; each cell in these hyphae contains two associated nuclei, one female and one male. A cell containing two nuclei is said to be dicaryotic; when, as here, the nuclei are genetically different (heterocaryotic), the cells are effectively diploid even though the individual nuclei are haploid. The terminal cells of the dicaryotic hyphae eventually become elongate, and their nuclei unite, forming a zygote nucleus (dark color). The zygote nucleus promptly divides meiotically, and each of the four haploid nuclei thus formed then divides mitotically. The result is eight small spore cells that are still contained within the wall of the old zygote cell, now called an ascus. When the mature ascus ruptures, the spores are released.

reproduce asexually by means of special spores called ***conidia*** (Fig. 23.8). Conidia are produced in chains at the end of conidiophore hyphae. Each conidium can grow into a new fungal plant.

It may seem strange that yeasts are considered members of the Ascomycota. They are unicellular, and they reproduce asexually by ***budding*** (Fig. 23.9). However, under certain conditions a single yeast cell may function as an ascus, producing four spores. The spores are more resistant to unfavorable environmental conditions than are vegetative cells, and they enable yeasts to survive temperature extremes or periods of prolonged drying. Sequence analysis confirms that yeast are sac fungi.

Lichens are plants composed of a fungus and a photosynthetic microorganism growing together in a symbiotic relationship (Fig. 23.10). The fungal components of most lichens are members of

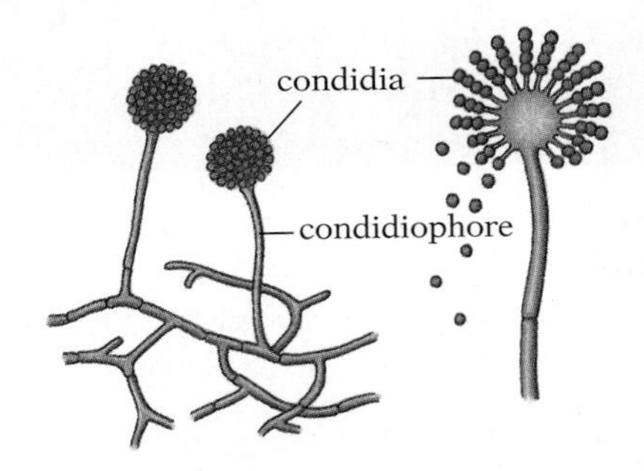

23.8 Asexual reproductive structures of *Aspergillus*, a member of the Ascomycota

Two conidiophores arise from a mycelium (left). An enlarged section through a conidiophore shows that the structure bears numerous spores, called conidia, arranged in chains (right).

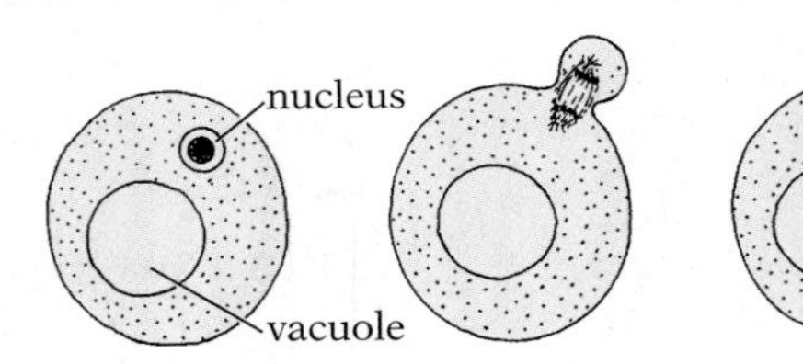

23.9 Budding in brewer's yeast

A small new cell is pinched off the larger parent cell.

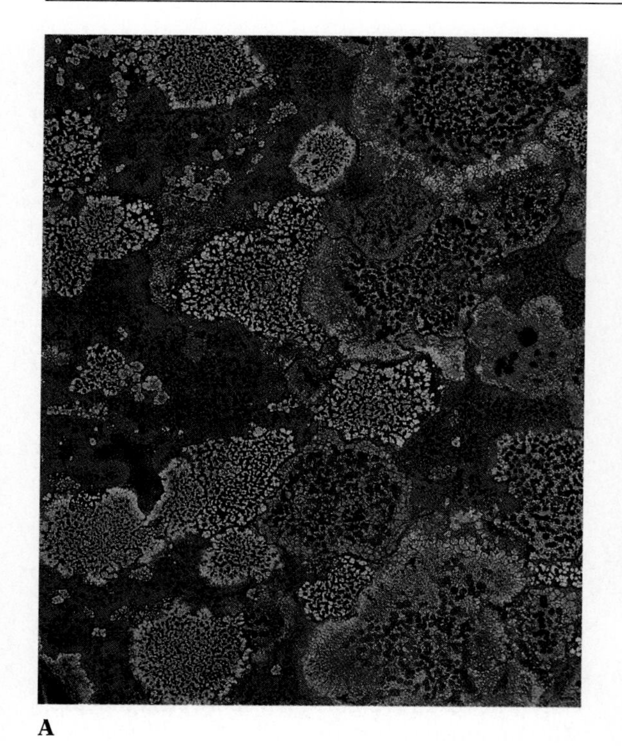

A

B

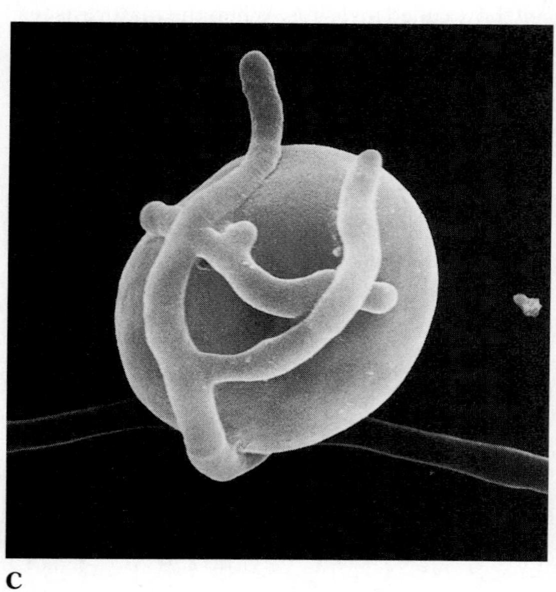

C

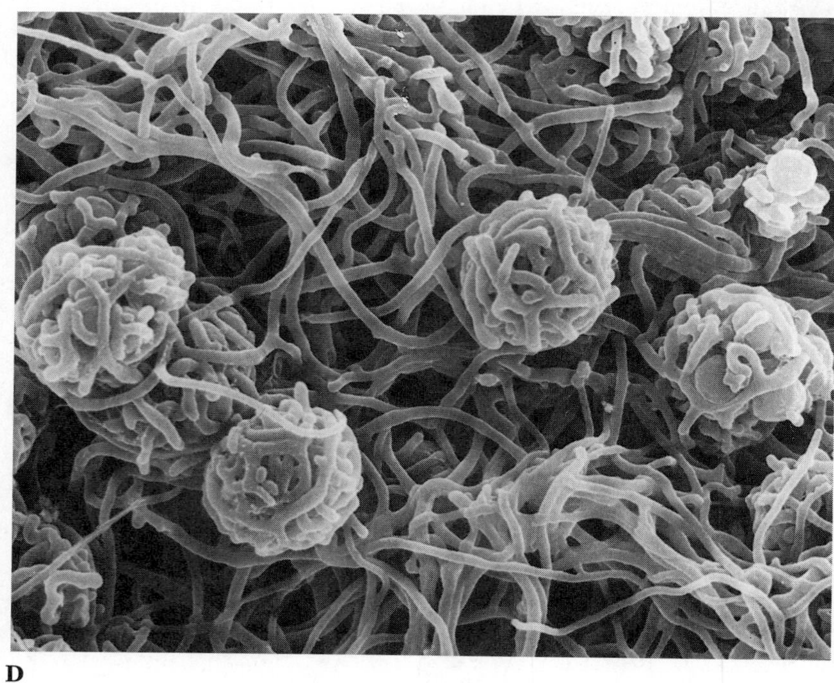

D

23.10 Two types of lichens

(A) Both *Lecidea atrata* (rust-red) and *Lecidea lithophila* (gray-black) grow as a crust, shown here on a boulder in the Cairngorms, Scotland. (B) British soldiers, an association of the fungus *Cladonia* and the alga *Trebouxia*, with their bright red tops, have a more upright growth. (C) A resynthesis of the algal-fungal association can be generated through the culturing of British soldiers. Here a fungal hypha encircles an algal cell. (D) In a later stage of development, many algal cells are held together by a network of hyphae.

A

B

23.11 Two representative Basidiomycota

Left: A tree fungus, *Cariolus versicolor*. Right: A puffball, *Lycopoerdon perlatum*, expelling spores after being struck by a drop of rain.

Ascomycota, though in a few tropical forms the fungus is a member of the Basidiomycota (discussed in the next section). The photosynthetic components are usually green algae (chlorophytes), though they may be cyanobacteria (in which case, since cyanobacteria can fix nitrogen, the lichen can grow on stone).

BASIDIOMYCOTA: THE CLUB FUNGI

Many of the largest and most conspicuous fungi—puffballs, mushrooms, toadstools, and bracket fungi—are Basidiomycota (Fig. 23.11). Though the above ground portion of these plants—the fruiting body—looks like a solid mass of tissue, it is nevertheless composed of hyphae. The fruiting body of many mushrooms is only a small part of the total plant; there is an extensive mass of hyphae in the soil. The hyphae are septate (divided by walls), but the cross walls are perforated. Some club fungi, particularly the rusts and smuts, are parasitic on plants; the more specialized rusts produce structures that mimic flowers and attract pollinators, which then spread the spores to other plants (Fig. 23.12).

23.12 A fungus that mimics a flower

The host is a herbaceous mustard. The rust inhibits the formation of the plant's own flowers, and induces instead the growth of a very different structure that produces nectar and spores.

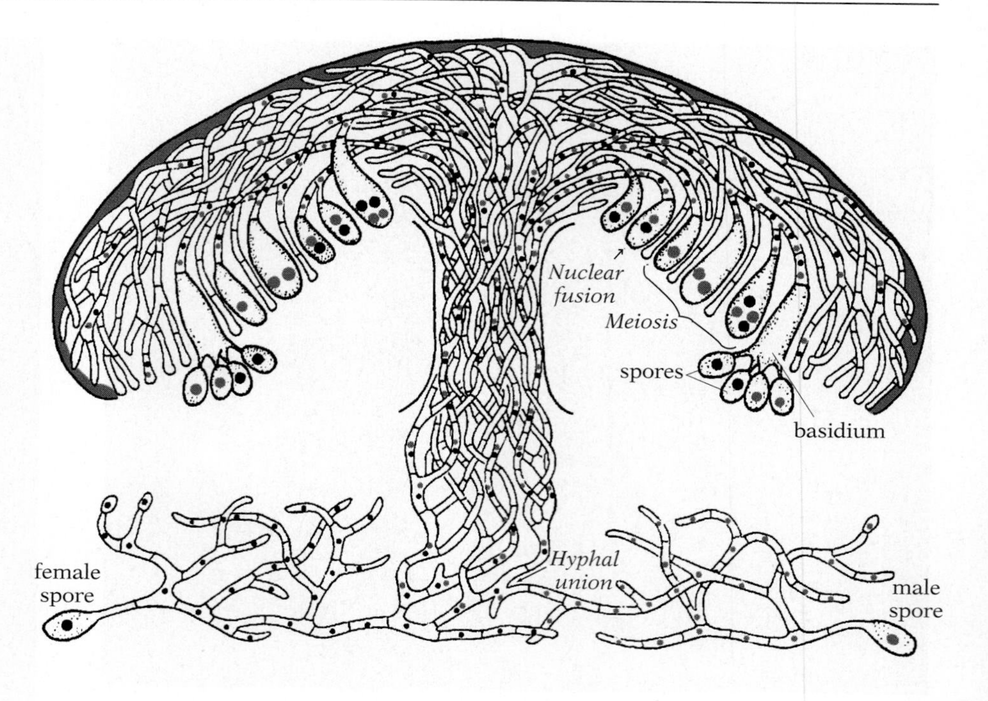

23.13 Section through a mushroom
Hyphae from two uninucleate mycelia—one female, the other male—unite and give rise to dicaryotic hyphae, which then develop into the above ground part of the mushroom. The entire stalk and cap are composed of these hyphae tightly packed together. Spores are produced by basidia on the undersurface of the cap.

The members of this class are distinguished by their club shaped reproductive structures called ***basidia*** (Fig. 23.13). The cells of the hyphae that produce basidia are binucleate (dikaryotic), containing one haploid "+" nucleus and one haploid "−" nucleus. Certain terminal cells of these hyphae, located in rows, or "gills," on the undersurface of the cap of gill fungi (Fig. 23.13), become zygotes when their two nuclei fuse in fertilization. The zygote then becomes a basidium. Its diploid nucleus divides by meiosis, producing four new haploid nuclei. Four small protuberances develop on the end of the basidium, and the haploid nuclei migrate into these. The tip of each protuberance then becomes walled off as a spore, which is usually ejected from the basidium. Each spore may give rise to a new mycelium. Some can reach astonishing sizes: one mycelium of a tree root colonizer is known to cover 15 hectares and weigh more than 10,000 kg; it is about 1500 years old.

CHAPTER SUMMARY

CHYTRIDIOMYCOTA: THE PROTIST FUNGI

Though very simple in organization, this group is like all fungi: they have chitinous walls, usually with incomplete separation of cells; most are parasitic or saphrophytic; reproduction can be sexual or asexual, depending on circumstances; and the diploid phase is brief. (p. 646)

ZYGOMYCOTA: THE CONJUGATION FUNGI

Conjugation fungi have more structural complexity: like nearly all higher fungi, they grow as filaments (hyphae) variously specialized into rootlike supports (rhizoids), vegetative extensions (stolons), spore stalks (sporangiophores), or a mass (mycelium). (p. 647)

ASCOMYCOTA: THE SAC FUNGI

The filaments in the more complex sac fungi grow vegetatively until they encounter hyphae of the opposite mating type. They produce a mating organ, and then, still as filaments, they form an elaborate reproductive cup that looks like an inverted mushroom. This group also includes the yeasts. (p. 648)

BASIDIOMYCOTA: THE CLUB FUNGI

All the club fungi are structurally complex; their hyphae "mate" in the manner of sac fungi, and most form either conventional mushrooms or puffballs. (p. 651)

STUDY QUESTIONS

1 Given the mating system of most fungi, and the observation that many species also have cellulose in their cell walls, develop a scenario for the evolution of this kingdom. What alteration in Figure 19.19 would be necessary to reflect your hypothesis? (pp. 591–594, 612, 647–648)

2 Compare the macronucleate strategy of ciliates with the coenocytic nature of fungi. (pp. 587–588, 646–649, 652)

3 What are some likely factors that keep fungi small and, moreover, favor the low-density tissues characteristic of mushrooms and the other largest fungi?

4 What are the fungi in lichens likely to be providing to the symbiotic algae? (pp. 649–651)

5 If cyanobacteria are so self-sufficient, what help could fungi provide that would make cooperation to form a lichen pay off? (pp. 569–571, 649–651)

SUGGESTED READING

BALL, D. M., J. F. PEDERSEN, AND G.D. LACEFIELD, 1993. The tall-fescue endophyte. *American Scientist* 81 (4). *On the symbiotic fungus that helps the most popular lawn grass thrive, in part by poisoning grazers.*

COOKE, R. C., 1978. *Fungi, Man and his Environment.* Longman, London. *Short but fascinating treatment of the biology of fungi, with emphasis on the many ways these organisms affect humans.*

LITTEN, W., 1975. The most poisonous mushrooms, *Scientific American* 232 (3). *The members of the genus Amanita and the highly toxic compound they produce.*

RAVEN, P. H., R. F. EVERT, AND S. E. EICHHORN, 1992. *Biology of Plants,* 5th ed. Worth, New York. *Includes an excellent treatment of the fungi.*

CHAPTER 24

INVERTEBRATE ANIMALS

In the kingdom Animalia there are at least five times as many species as there are in the other seven kingdoms combined. Moreover, the degree of specialization in tissues and organs in most animals dwarfs that of even the vascular plants. The diversity of the animal kingdom is so great that the focus here will be on the largest and most important groups, with only a brief mention of some of the smaller phyla. This chapter will deal with all the animal phyla except Chordata; the next chapter will examine the chordates. Part V discusses the comparative physiology of these major groups.

A formal classification of the animal kingdom is given in the Appendix. This classification, though widely used, is not accepted in detail by all biologists.

PORIFERA: THE SPONGES

Sponges are aquatic, mostly marine, animals (Fig. 24.1). Though the larvae are ciliated and free-swimming, the adults are always sessile (living in fixed spots) and are usually attached to rocks or shells or other submerged objects. They are multicellular, but show few of the features ordinarily associated with multicellular animals. They lack organs, having no digestive, nervous,[1] or circulatory systems.

[1]In some sponges a few cells of the body wall have elongate processes that may have special conductile properties.

24.1 Colonial sponges

Water, carrying microscopic food particles and oxygen, flows into the central body cavity of each of the sponges through the numerous smaller openings (pores) in their body walls, and is discharged through the larger openings (oscula).

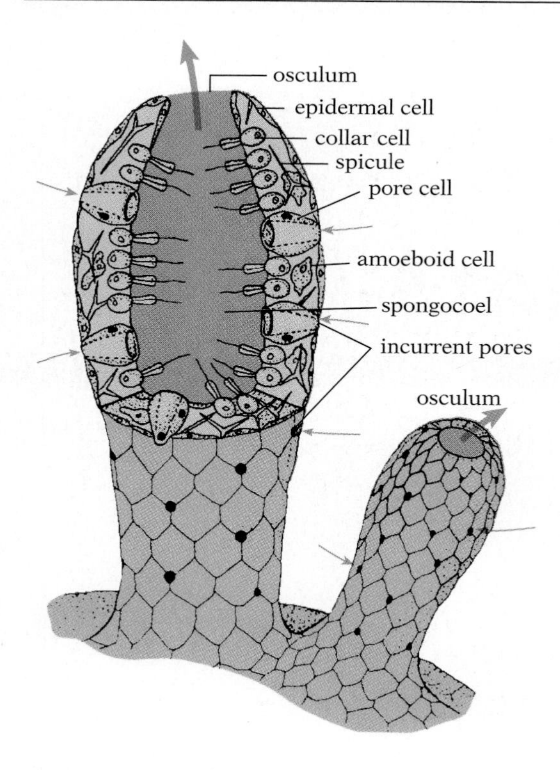

24.2 Detailed view of two colonial sponges

One sponge is shown partially sectioned, the other in exterior view. The blue arrows indicate the path of water through the sponges. These sponges have a simple tubular body. The majority of sponges have complexly folded walls in which water passes through a network of channels, but their basic body plan may be considered an elaboration of the tubular plan.

The body of a typical sponge is a perforated sac. Its wall has three layers: an outer layer of flattened epidermal cells, a gelatinous middle layer with wandering amoeboid cells, and an inner layer of flagellated ***collar cells*** (Fig. 24.2).

The wall of a sponge is perforated by numerous pores, each surrounded by a single pore cell. Water currents flow through the pores into the central cavity (spongocoel) and out through a larger opening (osculum) at the end of the body; the flow is enhanced by the beating of the flagella of the collar cells, which makes possible a higher metabolic rate, faster growth, and increased reproduction. Microscopic food particles brought in by water currents adhere to the collar cells and are engulfed; the food may be digested in food vacuoles of the collar cells themselves or passed to the amoeboid cells for digestion. The water currents also bring oxygen to the cells and carry away carbon dioxide and nitrogenous wastes (largely ammonia). As in the case of algae, the surrounding water and the organism's thin tissues make organs unnecessary in sponges. (One rare, highly modified group has secondarily lost both sac and collar cells: members of this group feed by capturing prey on hooks mounted on projecting filaments.)

Sponges characteristically possess an internal skeleton secreted by the amoeboid cells. This skeleton is composed of crystalline ***spicules*** or proteinaceous fibers or both. The spicules are made of calcium carbonate or siliceous material (chiefly silicic acid); their chemical composition and their shape are the basis for sponge classification. The fibrous skeletons of the bath sponges (*Spongia*) are cleaned and sold for many uses. A living bath sponge, which looks rather like a piece of raw liver, bears little resemblance to the familiar commercial object.

Some free-living flagellated protists, the choanoflagellates (phylum Choanozoa; Fig. 24.3), resemble the collar cells of sponges—cells found in no other organisms. Sequence analysis places sponges midway between the choanoflagellates and the most primitive radiate animals. Modern protists are only distantly related to any of these groups—in fact, fungi and even plants are far closer to animals than to any of the protozoans other than the choanoflagellates. Sponges are so unusual that they are often assigned to a separate subkingdom, the Parazoa; some specialists even propose an independent kingdom for sponges.

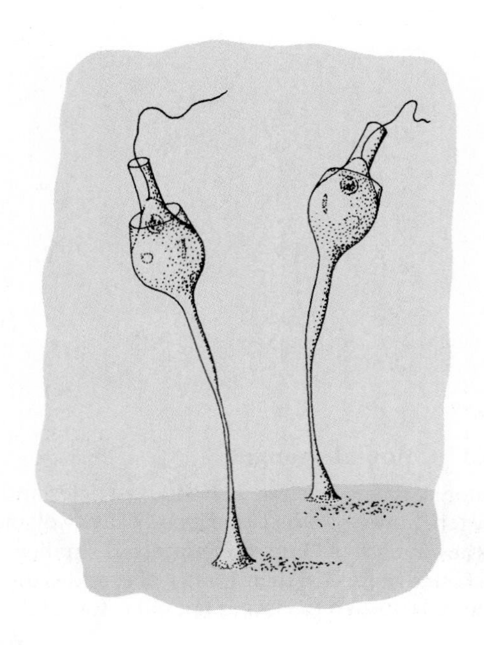

24.3 Two choanoflagellates (*Codosiga*)

These single-celled organisms strikingly resemble the collar cells of sponges (See Fig. 24.2). They have traditionally been classified as protozoan animals by zoologists and as colorless members of the algal group Chrysophyta by botanists.

THE RADIATE PHYLA: A CIRCULAR BODY PLAN

The two ancient radiate phyla—Cnidaria and Ctenophora—comprise radially symmetrical animals such as jellyfish with bodies at a relatively simple level of construction.[2] They have definite tissue layers, but no distinct internal organs (see Fig. 35.9, p. 998). There is a digestive cavity, but it has only one opening, which must serve as both mouth and anus—that is, it is a gastrovascular cavity. Tentacles are usually present. There is no coelom or other internal space between the wall of the digestive cavity and the outer body wall.

The bodies consist of three layers of cells—an outer epidermis (ectoderm), an inner gastrodermis (endoderm), and (usually) a mesoglea (mesoderm) in between (see Fig. 27.11, p. 773). The mesodermal layer is not well developed in the radiate phyla: it is usually gelatinous and has a few scattered cells, which may be amoeboid or fibrous. Sequence analysis reveals that the radiate phyla are not closely related to any other group of animals; some authorities suggest that they too may represent a separate kingdom.

■ CNIDARIA: THE COELENTERATES

The phylum Cnidaria (also called Coelenterata) include hydras, jellyfish, sea anemones, and corals. The hydras live in fresh water, but most other coelenterates are marine.

The coelenterate body shows some cell specialization and division of labor: the outer epidermis contains sensory-nerve cells, gland cells, special cells that produce nematocysts, small interstitial cells, and epitheliomuscular cells. These last are the main structural elements of the epidermis and consist of a columnar cell body with several contractile basal extensions.

The contractile elements of coelenterates are parts of cells that also have other important functions; there are no separate muscle cells. Moreover, these contractile elements are ectodermal, not mesodermal as in most higher animals. The gastrodermis also contains contractile elements; these are basal extensions of cells whose cell bodies constitute the bulk of the lining of the gastrovascular cavity and function in digestion. Thus there is some division of labor among cells in coelenterates, but it is not as complete as in most higher multicellular animals. Moreover, most functions performed by tissues derived from mesoderm in other animals are performed by ectodermal or endodermal cells in coelenterates.

The phylum Cnidaria is divided into four classes: Hydrozoa, Scyphozoa, Anthozoa, and Cubozoa. The class Cubozoa, comprising the sea wasps, has only 20 species, and will not be discussed separately.

Hydrozoa and cnidarian reproduction The best-known members of this class are the freshwater hydras (Fig. 24.4). In many ways, hydras are not typical members of their class. Many hydro-

A

B

24.4 Hydra, feeding

The green hydra *Chlorhydra* owes its color to the cells of a green alga that it incorporates into its own cells, thus benefiting from the alga's photosynthetic activity. Note the buds with tentacles on the upper portion of the stalks; hydras reproduce asexually by budding. (A) The hydra is manipulating a small crustacean with its tentacles. (B) The food has been taken into the digestive cavity, where it makes a prominent bulge.

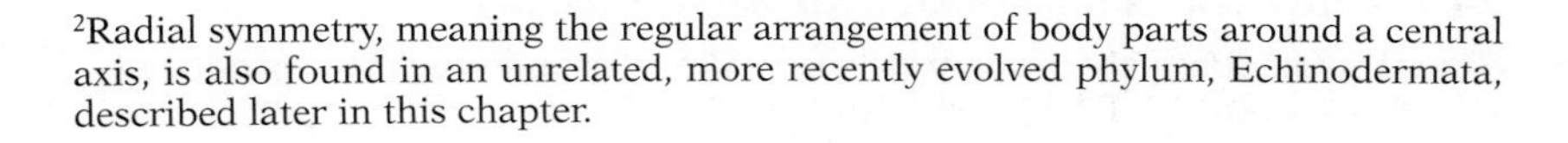

[2]Radial symmetry, meaning the regular arrangement of body parts around a central axis, is also found in an unrelated, more recently evolved phylum, Echinodermata, described later in this chapter.

24.5 A branching colonial hydrozoan, *Plumularia setacea*

The colony is composed of a stalk with many lateral branches, each bearing six to eight polyps.

zoans are colonial (Fig. 24.5), and have a complex life cycle in which a sedentary hydralike ***polyp*** stage alternates with a free swimming jellyfishlike ***medusa*** stage (Fig. 24.6). By contrast, hydras are solitary and have only a polyp stage (which is not completely sedentary).

Obelia is a typical example of a colonial hydrozoan (Fig. 24.7). Much of the life of *Obelia* is passed as a sedentary branching colony of polyps. The colony arises from an individual hydralike polyp by asexual budding: the buds do not separate, and the new polyps remain attached by hollow stemlike connections. The gastrovascular cavities of all the polyps are interconnected via the cavity in the stems. The cells lining the stem cavity have long flagella that circulate the fluid in the cavity. Both the stems and the polyps (mouth and tentacles excepted) are enclosed in a hard chitinous case secreted by the ectoderm.

A mature *Obelia* colony consists of two kinds of polyps: feeding polyps with tentacles and nematocysts, and reproductive polyps without tentacles. The reproductive polyps, which regularly bud off tiny transparent free-swimming medusas, are nourished with food captured by the feeding polyps and passed to them through the common gastrovascular cavity in the connecting stems.

Each medusa is umbrella-shaped or bell-shaped, with numerous tentacles hanging from the margin of the bell (Fig. 24.8). A tube with a mouth at its end hangs from the middle of the underside of the bell. Medusas represent the dispersal and sexual stage in the life cycle. They swim feebly by alternately contracting and relaxing the contractile cells in the bell, but mostly they drift with water currents.

Certain cells in mature medusas undergo meiosis and give rise to either sperm or eggs, which are released into the surrounding water. Fertilization takes place, and the resulting zygote develops into a hollow blastula. Gastrulation does not take place by the

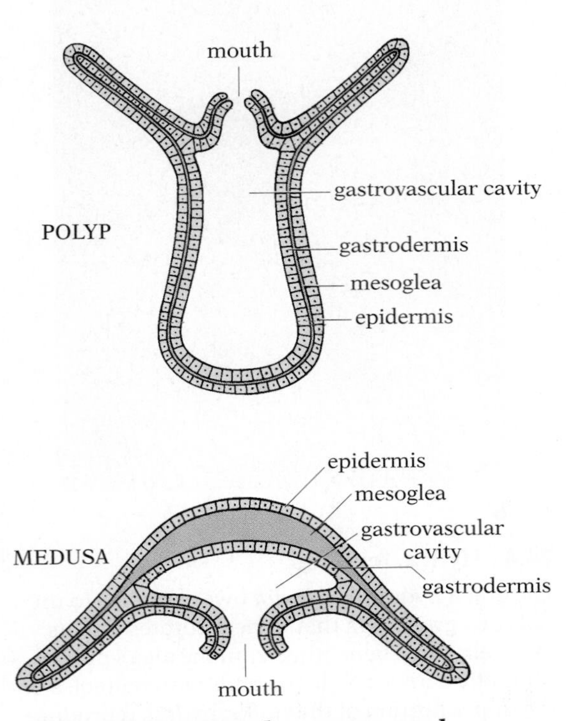

24.6 Polyp and medusa compared

The basic structure of these two forms is the same. A medusa is like a flattened polyp turned upside down.

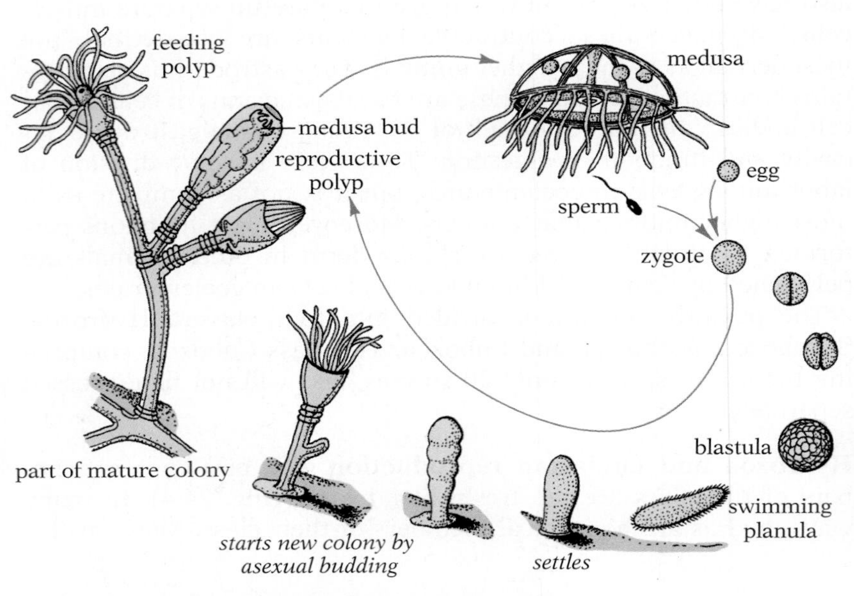

24.7 Life cycle of a colonial hydrozoan (*Obelia*)

Since the medusas are of separate sexes, the eggs and sperm are produced by different individuals. See text for description.

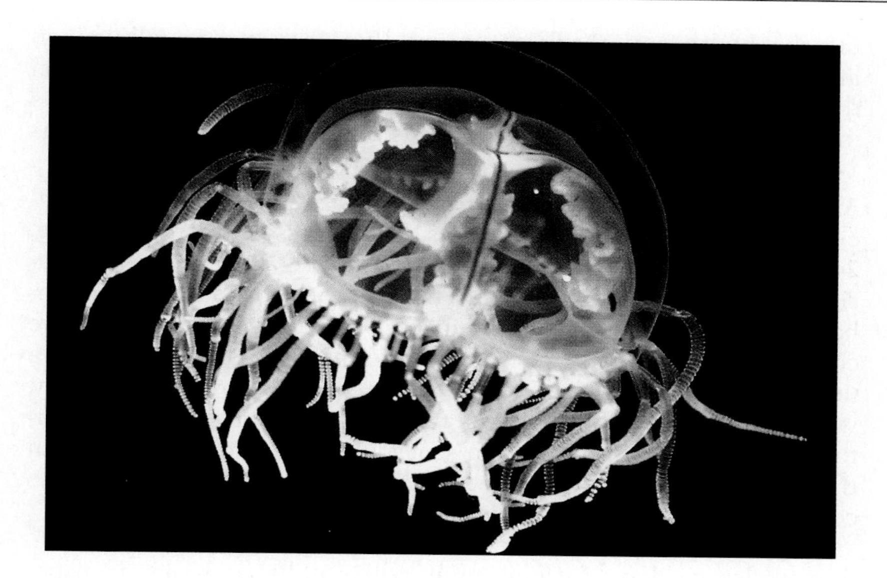

24.8 A hydrozoan medusa (*Gonionemus*)

This genus spends most of its life as a medusa—a weakly swimming bell-like creature with numerous tentacles.

process of invagination described in Chapter 13. Instead, endodermal cells generated by the wall (ectoderm) of the blastula wander into the blastocoel until they completely fill it. This solid gastrula then develops into an elongate ciliated larva called a ***planula***. The planula eventually settles to the bottom, attaches by one end to some object, and develops a mouth and tentacles at the other end, becoming a polyp that gives rise to a new colony. The alternation of polyp and medusa stages in a coelenterate differs from the alternation of generations in plants in that both polyp and medusa are diploid; the only haploid stage in the life cycle is the gamete.

Scyphozoa: the jellyfish The scyphozoans are the true jellyfish. The medusa is the dominant stage in the life cycle; the polyps are small larvae, which develop from the planula and promptly produce medusas by budding (Fig. 24.9). Scyphozoan medusas resem-

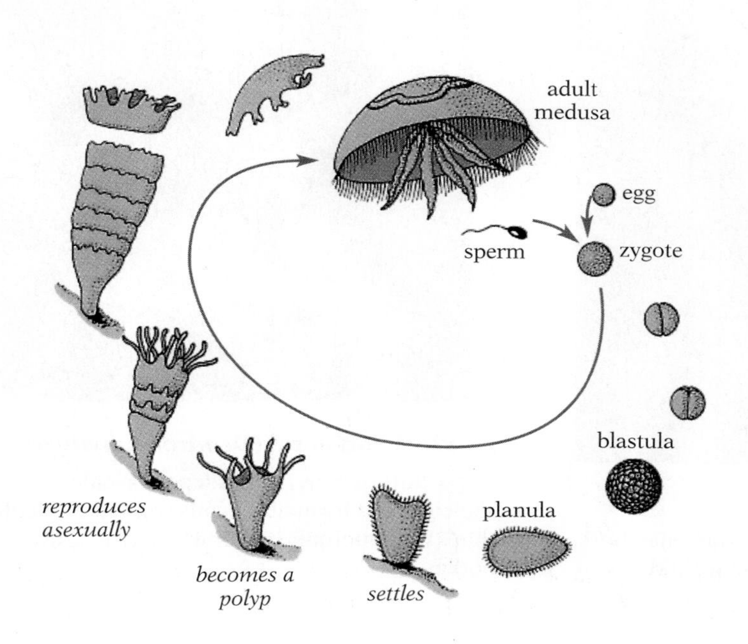

24.9 Life cycle of a jellyfish (*Aurelia*)

The polyps are shown much enlarged. Since the medusas are of separate sexes, the eggs and sperm are produced by different individuals. See text for description.

24.10 A jellyfish with oral arms and long marginal tentacles (Pelagia)

ble the hydrozoan medusas except that they are usually much larger and have long oral arms (endodermal tentacles) arising from the margin of the mouth; the marginal tentacles may be small, as in *Aurelia* (Fig. 24.9), or quite long, as in *Pelagia* (Fig. 24.10).

Anthozoa: sea anemones and corals The class Anthozoa includes sedentary polypoid forms such as sea anemones (Fig. 24.11), sea fans, and hard and soft corals (Fig. 24.12). All are marine. There is no medusa stage. They are the most advanced members of the Cnidaria, and their body structure is much more complex than that of simple polyps like the hydras. They possess a tubular ***pharynx*** (throat) leading into the gastrovascular cavity, which is divided into numerous radiating compartments by longitudinal septa; their mesoderm (mesoglea) is much thicker than that of other coelenterates and is often elaborated into a fibrous connective tissue; their muscles are also much better developed.

The skeletons of hard corals—anthozoans that secrete a hard, limy skeleton—have formed many reefs, atolls, and islands, especially in the South Pacific (see Figs. 4.32, 4.33). The Great Barrier Reef, a coral ridge many kilometers wide that extends more than 1600 km along the eastern coast of Australia, shows how these lowly animals can change the face of the earth and create a favorable environment for an exceptionally diverse collection of organisms. Most of the large oil deposits of the world are thought to be derived from the decay and burial of vast quantities of organic matter in enormous coral reefs.

24.11 A sea anemone (*Epiactis*)

A dense thicket of tentacles surrounds the animal's mouth. Many newly budded young anemones are attached to the stalk portion of its body.

24.12 A soft coral (*Dendronephythya*)

These anthozoans do not deposit a calcareous skeleton, but form numerous spinelike spicules that give structure to their otherwise fleshy bodies.

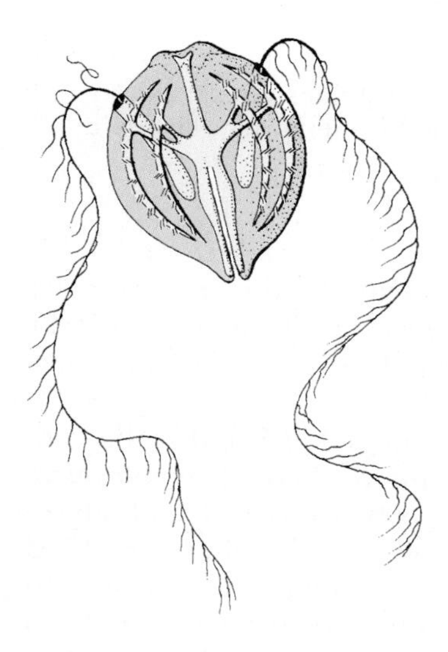

24.13 A common ctenophore (*Pleurobrachia*)

Note the rows of ciliary plates, as well as the very long antennae, that emerge from sheaths extending deep into the body.

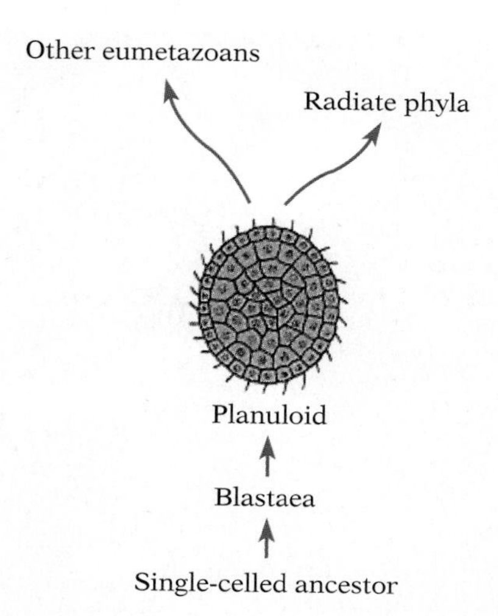

24.14 Origin of the eumetazoans according to the planuloid hypothesis

■ CTENOPHORA: THE COMB JELLIES

Like the coelenterates, members of the phylum Ctenophora, known as comb jellies or sea walnuts, are radial animals with a saclike body composed of epidermis, mesoglea, and gastrodermis; they also have a gastrovascular cavity and lack a coelom and organ systems. Unlike the coelenterates, however, they have independent mesodermal muscle cells, lack nematocysts, do not have a polymorphic life cycle; characteristically they have eight rows of ciliary plates (combs) that run across the surface (Fig. 24.13). The cilia in the eight rows beat in unison, enabling the animal to swim feebly. Most ctenophores float near the surface of the sea, chiefly near shore, and are carried about by currents and tides.

HOW DID EUMETAZOANS EVOLVE?

The subkingdom Eumetazoa includes essentially all multicellular animals except the Porifera, or sponges. Speculation concerning the origin of the eumetazoans has long centered on the radiate phyla just discussed. The early evolution of animals is inferred in large part from a precise study of their developmental patterns. The most widely accepted hypothesis for the evolution of the Eumetazoa focuses on the ***planuloid*** stage of cnidarians (Fig. 24.14). It supposes that the early ancestors of higher animals were planulas that had had a creeping mode of life, and slowly evolved gastrulation (creating a specialized gastrovascular cavity) and later a bilaterally symmetrical form.[3]

But why did gastrulation evolve in the first place, and why did the endodermis become specialized for nutritional activities? One early hypothesis suggests that the first eumetazoan was a ***plakula***, a two-layered, flattened creature creeping about on the sea bottom (Fig. 24.15). The ventral cell layer of such an organism would be

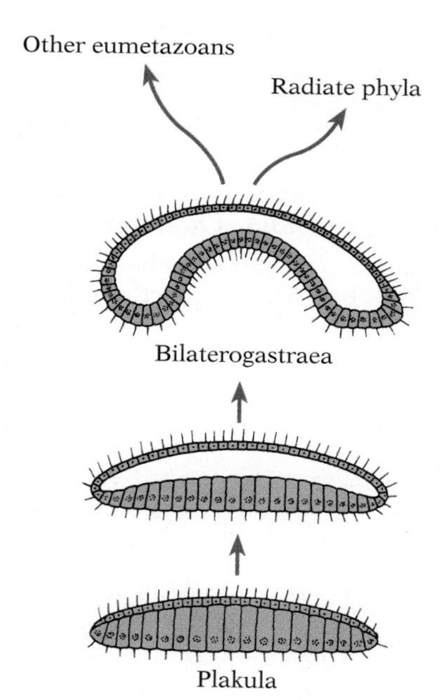

24.15 Origin of the eumetazoans according to the plakula-bilaterogastraea hypothesis

In this interpretation the earliest form of plakula was a two-layered organism with no internal space. The two layers gradually separated, and the organism became bilateral as it differentiated front and rear ends appropriate to its creeping mode of locomotion.

[3]Bilateral symmetry is the property of having two similar sides. A bilaterally symmetrical animal has definite dorsal (upper) and ventral (lower) surfaces and definite anterior (head) and posterior (tail) ends.

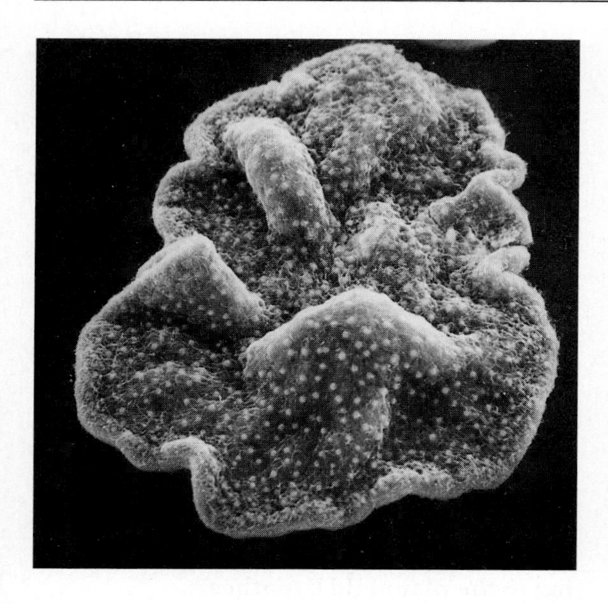

24.16 *Trichoplax adhaerens*

The animal has a flat platelike appearance. Of irregular shape, it has no constant symmetry. The many spherical particles in its surface layer are globules of a fatty material.

the layer that would come in contact with food particles most often. Therefore, it would be reasonable to expect the ventral layer to evolve nutritive specializations and the dorsal cell layer to evolve protective and perhaps locomotory ones.

If food items were captured beneath the plakula, and if the ventral epithelium was charged with digesting the food, it would have been advantageous for the organism to elevate part of its body to form a temporary digestive chamber from which neither the food nor the digestive enzymes secreted onto it could easily escape. In this manner, then, the plakula might sometimes have become a temporary ***gastraea***—a two-layered cavity. If the plakula crept rather than swam, bilateral symmetry would have been more adaptive than radial symmetry, leading to a ***bilaterogastraea*** (Fig. 24.15).

In 1969 K. G. Grell discovered a flattened, two-layered creature that crept about on the bottom (first described 85 years earlier and called *Trichoplax adhaerens*) on some algae sent to him from the Red Sea (Fig. 24.16). He found that it does, in fact, often rear up to form a temporary digestive cavity (Fig. 24.17). Moreover, he showed that this simple organism, with only two principal tissue layers and no organs (Fig. 24.18), is not just a larval form: it reproduces itself—asexually when conditions are not crowded, and sexually when they are.

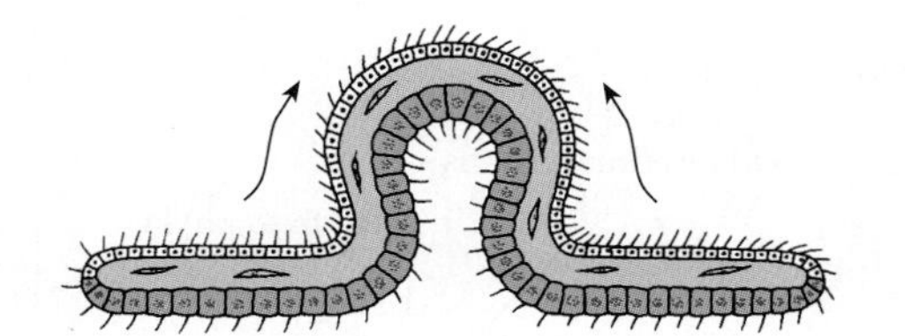

24.17 Formation of a temporary digestive chamber by *Trichoplax adhaerens*

The animal elevates part of its body to form a cavity. It thus becomes, in effect, a temporary gastraea.

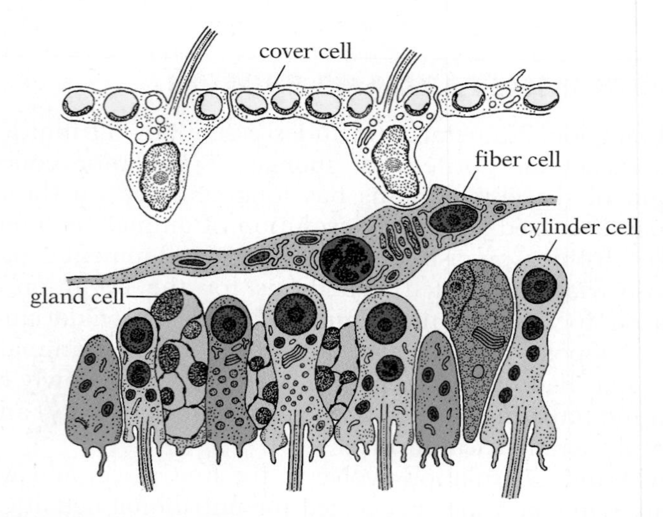

24.18 Histology of *T. adhaerens*

The thin dorsal epithelium, which is transparent, consists of cells of a single type, called cover cells by Grell; each bears a flagellum, and some incorporate large spherical vesicles containing fatty material. The much thicker ventral epithelium contains flagellated cylinder cells and nonflagellated gland cells. In the space between the two epithelia are some fiber cells, which probably function in locomotion.

Molecular evidence supports the view that *Trichoplax* is the most primitive multicellular animal known. Because *Trichoplax* has no bilateral symmetry, however, it is uncertain whether the evolutionary stage following the plakula was a gastraea or a bilaterogastraea (Fig. 24.19).

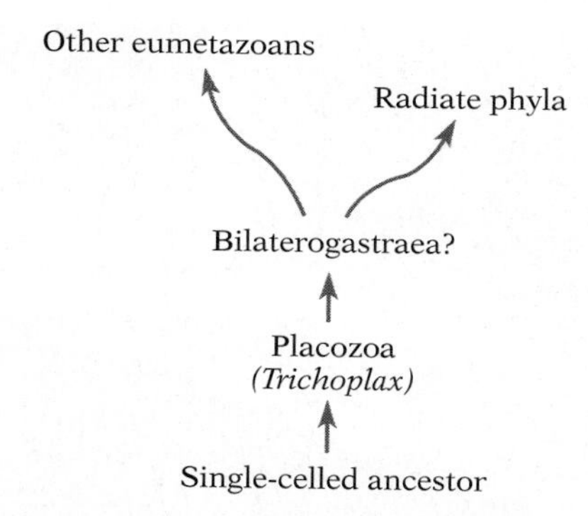

24.19 Origin of the eumetazoans according to the Placozoa version of the plakula hypothesis

Grell suggests that the placozoans may be ancestral to the sponges in addition to the eumetazoans; however, since the evidence for such a relationship is not convincing, it is not included in this diagram.

ACOELOMATE BILATERIA: PRIMITIVE BILATERAL ANIMALS

There are three phyla—the Platyhelminthes, the Gnathostomulida, and the Nemertea—that contain the most primitive bilaterally symmetrical animals. In each phylum the body is composed of three well-developed tissue layers—ectoderm, mesoderm, and endoderm—and the mesoderm is a solid mass that fills what once was the embryonic blastocoel. Thus there is no coelom—no cavity between the digestive tract and the body wall (see Fig. 24.31A). For this reason, these phyla are known as the acoelomate bilateria.

■ PLATYHELMINTHES: THE FLATWORMS

The flatworms are dorsoventrally flattened, elongate animals.[4] When present, their digestive cavity resembles that of coelenterates: it is a gastrovascular cavity with a single opening. However, there is a muscular pharynx leading into the cavity, and the cavity itself is often profusely branched. As in coelenterates, the amount of extracellular digestion is limited, most of the food particles being phagocytized and digested intracellularly by the cells of the wall of the gastrovascular cavity. Respiratory and circulatory systems are absent; these organisms are flat and thin enough to survive by exchanging O_2 and CO_2 directly with the water. However, there is an excretory system (described in Chapter 31), indicating that these species are too large and have too low a surface-to-volume ratio to dispose of excess salts adequately through direct exchange. There are also well-developed reproductive organs (usually both male and female in each individual).

The presence of both an excretory system and the reproductive organs signifies that the flatworms have advanced beyond the tissue level of construction of the radiate phyla to an organ level of construction. The more extensive development of mesoderm, leading to greater division of labor, was probably a major factor in making this advance possible. Mesodermal muscles are well developed. Several longitudinal nerve cords running the length of the body and a tiny "brain" ganglion located in the head constitute a central nervous system.

The phylum is divided into three classes: Turbellaria, Trematoda, and Cestoda. The last two are entirely parasitic.

Turbellaria The members of this class, which include freshwater planarians, are free-living flatworms ranging from microscopic size

[4]The term *worm* is applied to a great variety of unrelated animals. It is a descriptive, not a taxonomic, term that denotes possession of a slender elongate body, usually without legs.

24.20 An aquatic turbellarian (*Prosthecereus vittatus*)

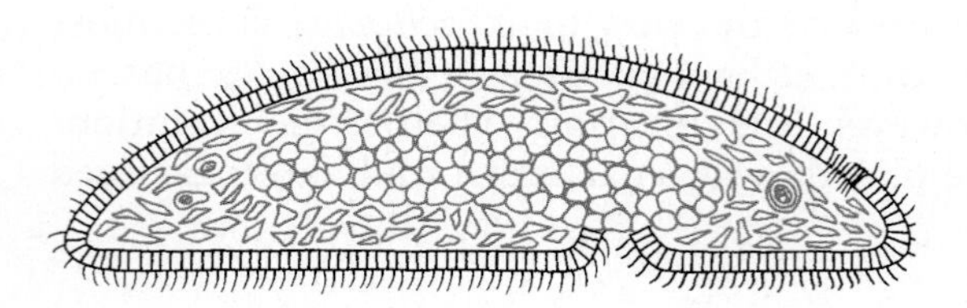

24.21 A flatworm, a typical acoelomate

There is a ventral mouth but no digestive cavity. The entire interior of the animal is filled with an almost solid mass of tissue (color).

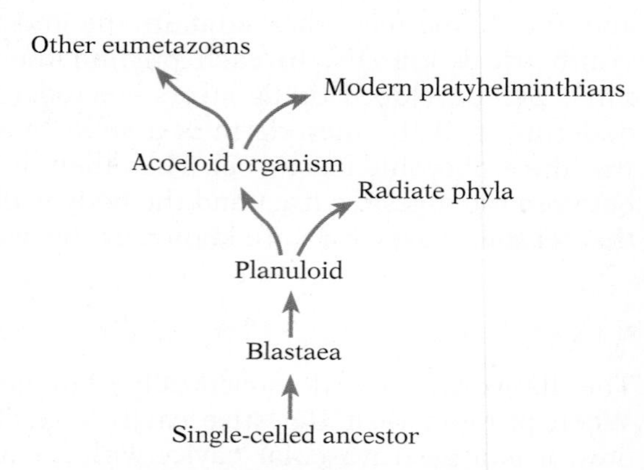

24.22 Origin of the bilateral eumetazoans according to the blastaea-planuloid-acoeloid hypothesis

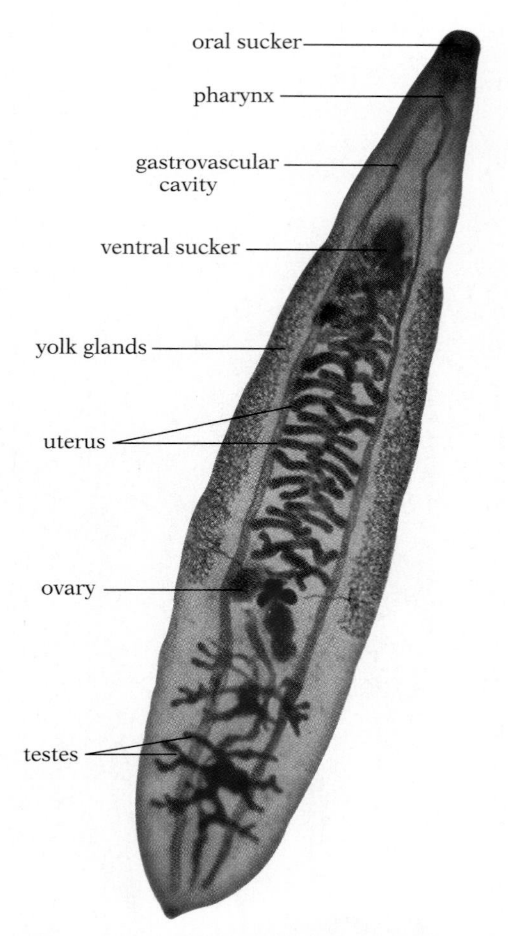

24.23 A Chinese liver fluke *Opisthorchis sinensis*

to a length of several centimeters. The body is clothed by an epidermal layer, which is usually ciliated (at least in part). Most turbellarians are aquatic (Fig. 24.20).

Turbellarians usually have a gastrovascular cavity, but most members of one small order, the Acoela, do not (Fig. 24.21). The acoels may be the most primitive bilateral animals. A primitive ***acoeloid*** organism might have arisen from a planuloid ancestor. The more complex flatworms and the other eumetazoan phyla could both have evolved from such an acoeloid organism (Fig. 24.22).

Trematoda: the flukes The flukes are parasitic flatworms. They lack cilia, and in place of the cellular epidermis of their turbellarian ancestors, they have a thick cuticle secreted by the cells below. This cuticle is highly resistant to enzyme action and is thus an important adaptation to a parasitic way of life. Flukes characteristically possess suckers, usually two or more, by which they attach themselves to their host (Fig. 24.23). They have a two-branched gastrovascular cavity that does not ramify throughout the body like that of turbellarians. A large proportion of their bodies is occupied by reproductive organs, including two or more large testes, an ovary, a long, much-coiled uterus in which eggs are stored prior to laying, and yolk glands.

The members of one order of flukes are ectoparasites (external parasites) on the gills or skin of freshwater and marine fishes. A

24.24 *Schistosoma mansoni*, a blood fluke responsible for schistosomiasis

few of these flukes sometimes wander into the body openings of their hosts, and it may have been from such a beginning that the endoparasitism (internal parasitism) typical of the other two orders arose.

Endoparasitic flukes often have very complicated life cycles involving two to four different kinds of hosts. The three species of blood fluke (genus *Schistosoma*), common in the tropics, has two sequential hosts. The adult blood fluke (Fig. 24.24) inhabits blood vessels near the intestine of a human host. When ready to lay its eggs, it pushes its way into one of the very small blood vessels in the wall of the intestine. There it deposits so many eggs that the vessel ruptures, discharging the eggs into the intestinal cavity, from which they are carried to the exterior in the feces. If there is a modern sewage system, that is the end of the story.

In many tropical countries, however, human feces are regularly used as fertilizer. Thus the eggs get into water in rice fields, irrigation canals, or rivers, where they hatch into tiny ciliated larvae. A larva swims about until it finds a particular kind of snail; it dies if it cannot soon locate the correct species. When it finds such a snail, it bores into the body of this host and feeds on its tissues. It then reproduces asexually, and the new individuals thus produced leave the snail and swim about until they come in contact with the skin of a human, such as a farmer wading in a rice paddy or a child swimming in a pond. They attach themselves to the skin and digest their way through it and into a blood vessel. Carried by the blood to the heart and lungs, they eventually reach the vessels of the intestine, where they settle down, mature, and lay eggs, thus initiating a new cycle.

Schistosomas cause the serious disease ***schistosomiasis***, which leads to severe dysentery and anemia. Victims become weak and emaciated, and often die of other diseases to which their weakened condition makes them susceptible. Schistosomiasis is one of the most widespread and debilitating human diseases in the world today; it is confined to warmer, less-developed regions of the earth.

The Chinese liver fluke, *Opisthorchis sinensis*, has three hosts. The adult lives in the human liver, where it lays its eggs. The eggs

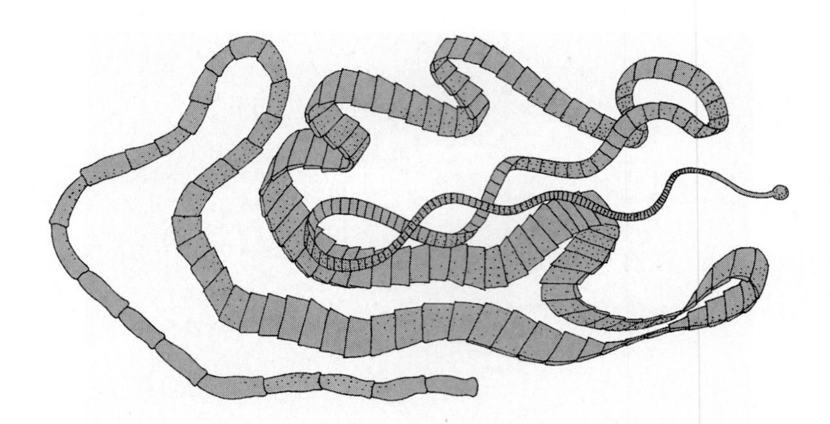

24.25 A tapeworm

The body is composed of a small head (scolex) and neck, followed by a large number of segments called proglottids. As the proglottids ripen, they break off and pass with the host's feces to the outside. New proglottids are produced just back of the neck.

pass into the intestine with the bile and are carried out in feces. If they get into water and are eaten by a certain species of snail, they hatch, and the larvae bore into the lymph spaces of the snail, where they reproduce asexually. The progeny leave the snail, burrow through the skin of a fish, and encyst in the fish's muscles. If a person eats the raw or insufficiently cooked fish, his digestive enzymes weaken the walls of the cysts, and the flukes emerge in his intestine, migrate up the bile duct to the liver, and settle down to start a new cycle (causing incurable enlargement of the liver, stomach distress, and the buildup of fluid in the abdomen). All three hosts—snail, fish, and human—are necessary for completion of the reproductive cycle of this fluke.

Cestoda: the tapeworms Adult tapeworms (Fig. 24.25) live as internal parasites of vertebrates, almost always in the intestine. The life cycle usually involves one or two intermediate hosts. The most familiar of these parasites is the beef tapeworm, whose intermediate host is a cow and final host is a human.

Tapeworms exhibit many special adaptations for their parasitic way of life. Like the flukes, they have a resistant cuticle instead of the epidermis of their free-living ancestors. They have secondarily lost the mouth and digestive tract. Bathed by the food in their host's intestine, they absorb digested nutrients across their body surface.

The head of a tapeworm is a small knoblike structure called a ***scolex***, which usually bears suckers and often hooks by which the worm attaches to the wall of the host's intestine (Fig. 24.26B). Immediately behind the scolex is a neck region, followed by a long ribbonlike body (beef tapeworms occasionally grow to 23 m). This long body is usually divided by transverse constrictions into a series of segments called ***proglottids*** (Fig. 24.26A).

Each proglottid is essentially a reproductive sac; each is ***hermaphroditic***, containing both male and female organs. Sperm cells, usually from a more anterior proglottid of the same animal, enter the genital pore and fertilize the egg cells, which are then combined with yolk from a yolk gland and enclosed in a shell. Unless there is another tapeworm infecting the same intestine, there is no chance of true sexual recombination. The fertilized eggs, already undergoing development, are stored in a uterus, which may become so engorged with eggs that it occupies most of

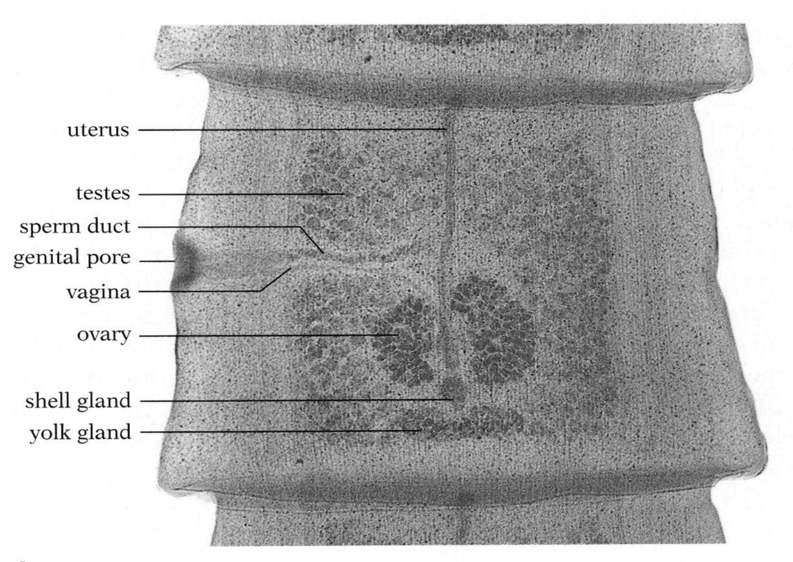

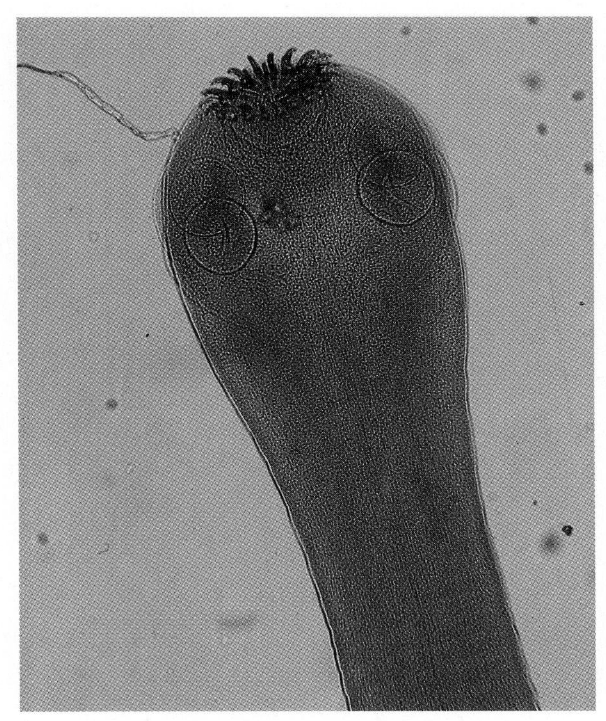

24.26 Scolex and proglottids of the dog tapeworm (*Taenia pisiformis*)

(A) A mature proglottid. (B) Scolex, neck, and some proglottids. Note the hooks and suckers on the scolex.

the volume of a mature proglottid. Eventually all the sexual organs except the uterus degenerate, and the proglottid, now "ripe," detaches from the worm and passes out of the host's body with the feces. As ripe proglottids are released from the end of the worm, new ones are produced just back of the neck. A single ripe proglottid may contain more than 100,000 eggs; the annual output of one worm can exceed 600 million.

If an appropriate intermediate host eats food contaminated with feces containing tapeworm eggs, its enzymes digest the shells of the eggs. The embryos thus released bore through the wall of the host's intestine, enter a blood vessel, and are carried by the blood to the muscles, where they encyst. If a person eats the raw or undercooked meat of this intermediate host, the walls of the cysts are digested away; the young tapeworms attach to the intestinal wall, begin to grow and produce eggs, thus starting a new cycle.

Tapeworms illustrate the compartmentalized developmental strategy of most higher animals: a series of initially similar segments or somites is produced, each of which gives rise to a section of the body along the anterior-posterior axis, which then specializes as necessary to produce the appropriate appendages and organs.

■ NEMERTEA: THE PROBOSCIS WORMS

The members of the phylum Nemertea are slender worms (Fig. 24.27) with a long muscular proboscis enclosed in a tubular cavity at the anterior end of the body. The proboscis is eversible—that is, it turns itself inside out as it is extruded. It is used in capturing

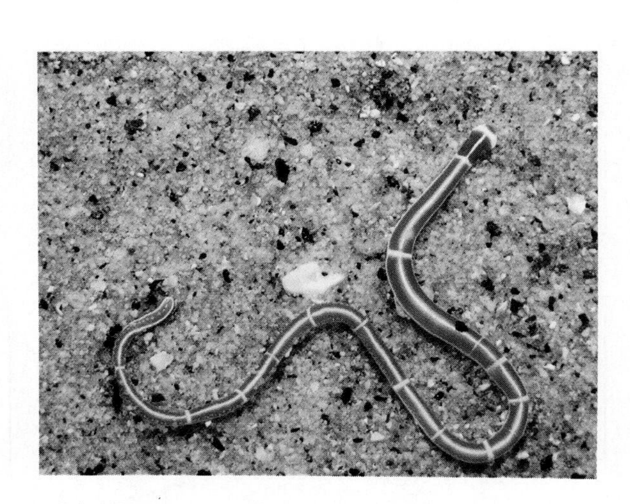

24.27 A nemertine worm (*Tubulanus annulatus*)

prey and in defense. It is often two times the length of the worm's body and is somewhat coiled when enclosed in its sheath. The worms are common along both the Atlantic and Pacific shores of the United States. They are usually found sheltered under stones, shells, or seaweeds, or burrowing in the sand or mud in shallow water.

Nemertines resemble turbellarian flatworms (their probable ancestors) in many ways, but they have two important features not encountered in the groups discussed thus far. First, they have a ***complete digestive system***—one that has two openings, a mouth and an anus. This makes possible specialization of sequentially arranged chambers for different digestive functions, thus permitting efficient assembly-line processing of food. Second, they have a simple blood circulatory system, which facilitates transport of materials from one part of the body to another.

■ GNATHOSTOMULIDA: THE JAW WORMS

The members of the phylum Gnathostomulida are small, free-living marine worms. These hermaphrodites live in detritus-rich sand and have specialized paired jaws.

HOW PROTOSTOMES AND DEUTEROSTOMES DIFFER

As described in Chapter 13, the embryonic archenteron cavity, formed during gastrulation, becomes the digestive tract of an adult. The archenteron has only one opening to the outside, the blastopore. In animals like coelenterates and flatworms, where the digestive tract is a gastrovascular cavity, the blastopore becomes the combined mouth and anus. In nemertines, however, the site of the embryonic blastopore becomes the mouth; the anus is an entirely new opening. This is also the case in many other animals, including nematode worms, molluscs, and annelids. In a few phyla, including Echinodermata and Chordata, the situation is reversed: the embryonic blastopore becomes the anus, and a new opening becomes the mouth.

This fundamental difference in embryonic development suggests that a major split occurred in the animal kingdom soon after the origin of a bilateral ancestor. One evolutionary line led to all the phyla in which the blastopore becomes the mouth; these phyla are called the ***Protostomia*** (from the Greek *protos*, "first," and *stoma*, "mouth"). The other evolutionary line led to the phyla in which the blastopore becomes the anus and the mouth is formed later: the ***Deuterostomia*** (from *deuteros*, "second," or "later," and *stoma*).

The protostomes and deuterostomes differ in a number of fundamental characteristics. For example, the early cleavage stages are usually determinate in protostomes and indeterminate in deuterostomes. That is, the developmental fates of the first few cells of a protostome embryo are usually already at least partly determined; if these cells are separated, not one of them can form a complete

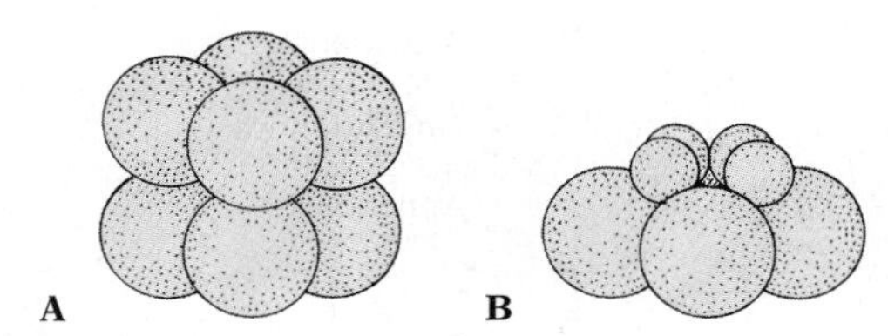

24.28 Radial and spiral cleavage patterns

(A) Radial cleavage, characteristic of deuterostomes. The cells of the two layers are arranged directly above each other. (B) Spiral cleavage, characteristic of protostomes. The cells in the upper layer are located in the angles between the cells of the lower layer.

individual. In contrast, the fates of the first few cells of a deuterostome embryo are not determined, and each cell, if separated, can develop into a normal individual. Furthermore, the two groups exhibit strikingly different patterns of cleavage: the early cleavages in protostomes are usually oblique to the polar axis[5] of the embryo and thus give rise to a spiral arrangement of cells (Fig. 24.28B); the early cleavages in deuterostomes are either parallel or at right angles to the polar axis and thus give rise to a so-called radial arrangement of cells (Fig. 24.28A). The basic larval types are also different in the two groups, as we shall see later in this chapter.

Another fundamental difference between the protostome and deuterostome phyla is seen in the origin of the mesoderm. The mesoderm of the radiate phyla arises from in-wandering cells derived from the ectoderm (Fig. 24.29A). A small amount of mesoderm forms this way in the protostomes also, though not usually in the deuterostomes. Most of the mesoderm in protostomes and all of it in deuterostomes is derived from endoderm instead of ectoderm. However, in protostomes this mesoderm arises as a solid ingrowth of cells from a single initial cell located near the blastopore (Fig. 24.29B), whereas in deuterostomes (except vertebrates) it arises from two saclike outfoldings of the gut wall (Fig. 24.29C; also described in Chapter 13).[6]

Another difference is in the formation of the coelom, if one is present. A true coelom is a cavity enclosed entirely by mesoderm and located between the digestive tract and the body wall. In the coelomate protostomes this cavity usually arises as a split in the initially solid mass of mesoderm. In the deuterostomes, by contrast, the coelom arises as the cavity in each of the mesodermal sacs as they evaginate from the wall of the archenteron.[7]

The protostomes and the deuterostomes also differ in many nondevelopmental traits. To give but one example, the visual receptor cells of the protostomes generally have a receptor organelle of specialized microvilli, whereas the analogous organelle in the deuterostomes is composed of specialized cilia.

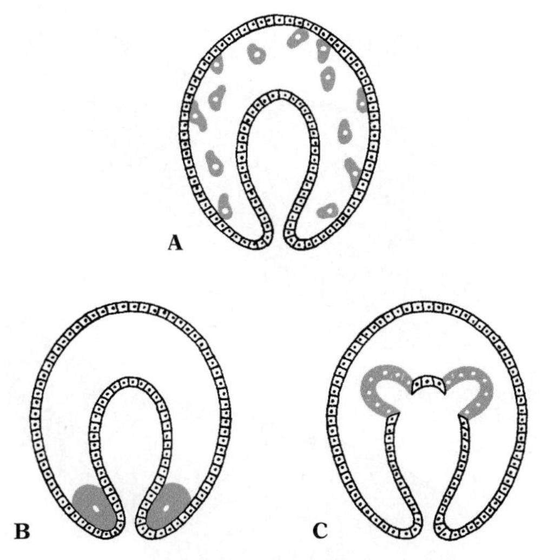

24.29 Different modes of origin of mesoderm

(A) In the radiate phyla, mesoderm (pink cells) arises from in-wandering cells derived from the ectoderm. A small amount of the mesoderm of the protostome phyla arises in this way also. (B) In most protostome phyla, a single initial cell (not shown) divides to form two primordial mesodermal cells. The bulk of the mesoderm arises from divisions of these cells, which are located near the blastopore, at the junction between the ectoderm and the endoderm. (C) In the deuterostome phyla, the mesoderm arises as a pair of pouches from the endodermal wall of the archenteron.

[5]The polar axis runs between the animal and vegetal poles.

[6]There are a variety of other ways in which mesoderm may arise from endoderm, but embryologists usually interpret them as variants of one or the other of the two processes described here.

[7]A coelom that arises as a split in an initially solid mass of mesoderm is called a schizocoelom. One that forms as the cavity in a pouch of mesoderm is called an enterocoelom.

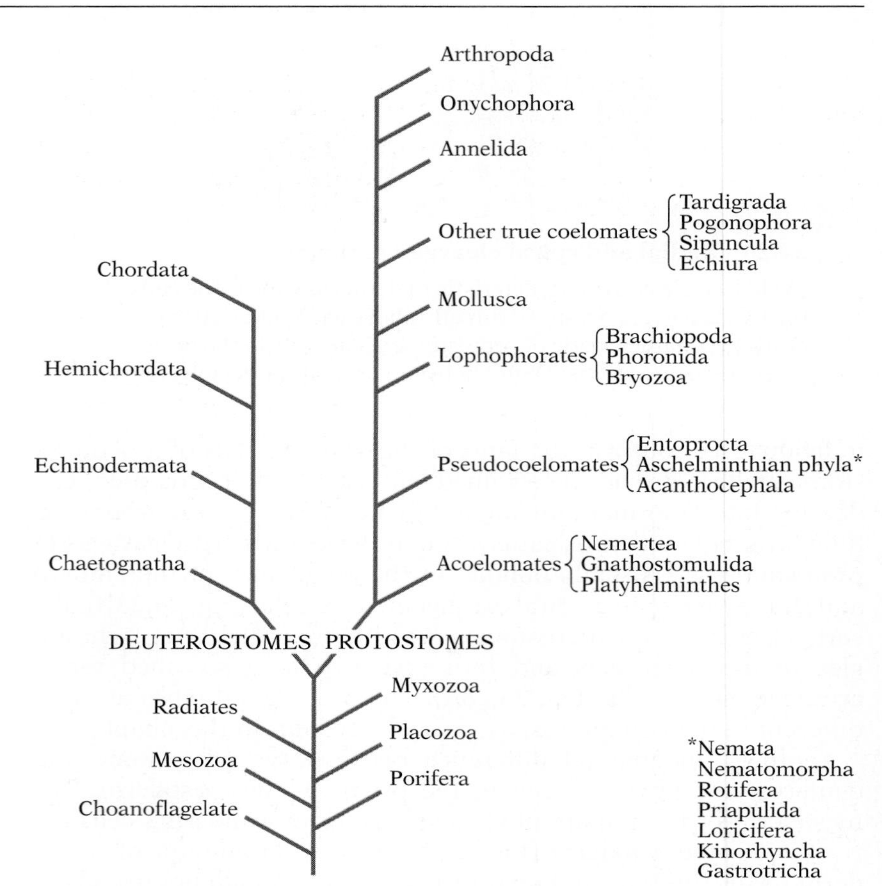

24.30 One interpretation of evolutionary relationships among the animal phyla

The Myxozoa are unicellular parasites formerly classed as fungal protists, but now known to be highly specialized animals. There is some evidence that the proboscis-holding cavity of nemertines is a highly modified true coelom; if so, the nemertines are misplaced on this tree.

Figure 24.30, drawn in the traditional form of a phylogenetic tree, shows a simplified interpretation of evolutionary relationships among the animal phyla.

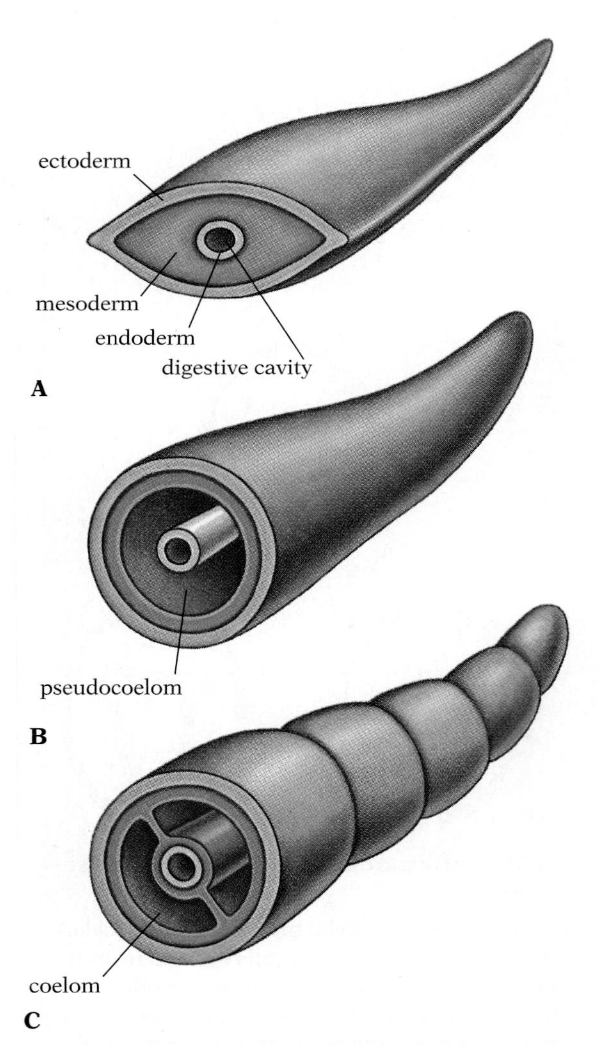

24.31 Acoelomate, pseudocoelomate, and coelomate body types

(A) Acoelomate body. There is no body cavity; the entire space between the ectoderm and endoderm is filled by a solid mass of mesoderm. (B) Pseudocoelomate body. There is a functional body cavity, but it is not entirely bounded by mesoderm. (C) Coelomate body. The body cavity is completely bounded by mesoderm.

THE PSEUDOCOELOMATE PROTOSTOMES

In several protostome phyla the body cavity is functionally analogous to a coelom but differs from a true coelom, which is entirely enclosed by mesoderm, in being partly bounded by ectoderm and endoderm. Such a cavity is called a ***pseudocoelom*** (Fig. 24.31B). It is actually the remnant of the embryonic blastocoel.

Three phyla of pseudocoelomate protostomes are recognized by many biologists: Acanthocephala, Entoprocta, and Aschelminthes. However, the current trend is to divide the traditional phylum Aschelminthes into six separate phyla. For simplicity, we will discuss these phyla as a group.

■ ACANTHOCEPHALA AND ENTOPROCTA

Adult members of the phylum Acanthocephala are endoparasites in the digestive tracts of vertebrates; the larvae live in invertebrates. These animals are often called spiny-headed worms because they have a proboscis armed with rows of recurved hooks. They have no digestive tract.

Members of the phylum Entoprocta are tiny, sessile, mostly marine animals that live attached by a stalk to rocks, shells, pilings, or animals such as crabs or sponges. Most are colonial. Their digestive tract is U-shaped, and both mouth and anus open inside a circle of ciliated tentacles. They feed on small plankton such as diatoms and protozoans.

■ THE ASCHELMINTHIAN PHYLA

The largest and most important pseudocoelomate phyla are those of the aschelminthian group. These phyla include an array of generally small, wormlike animals that lack a definitely delimited head, and that have a straight or slightly curved complete digestive tract and a cuticle. There is no respiratory or circulatory system. We will discuss only two of the phyla: Rotifera and Nemata. (The other four phyla are Priapulida, Loricifera, Kinorhyncha, and Gastrotricha.)

Rotifera: the rotifers The rotifers are microscopic, usually free-living aquatic animals with a crown of cilia at the anterior end (Fig. 24.32). The cilia are generally arranged in a circle, and when beating they often give the appearance of a rotating wheel. When feeding, rotifers attach themselves by a tapering posterior "foot," and the beating cilia draw a current of water into the mouth. Very small protozoans and algae are swept into a complicated muscular pharynx, where they are ground up by seven hard jawlike structures. This complex pharynx, a distinctive feature of the rotifers, is called a mastax.

The freshwater rotifers are abundant. Anyone examining a drop of water for protozoans under a microscope is likely to see one or more of these interesting animals. Most are no larger than protozoans, and it is often difficult to believe that they are multicellular.

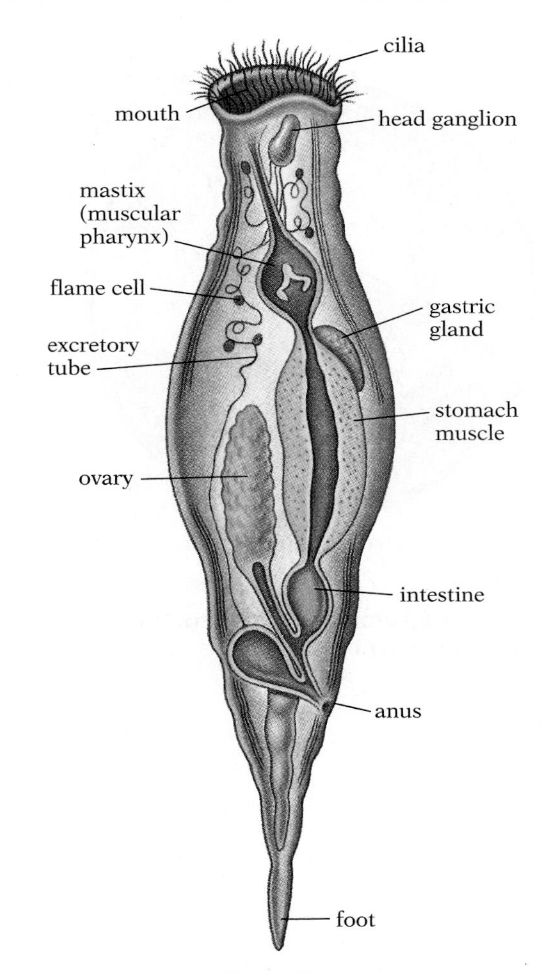

24.32 Section of a rotifer

Nemata: the roundworms Nematode worms have round elongate bodies that usually taper nearly to a point at both ends (Fig. 24.33). Unlike flatworms, they have no cilia. The body is enclosed

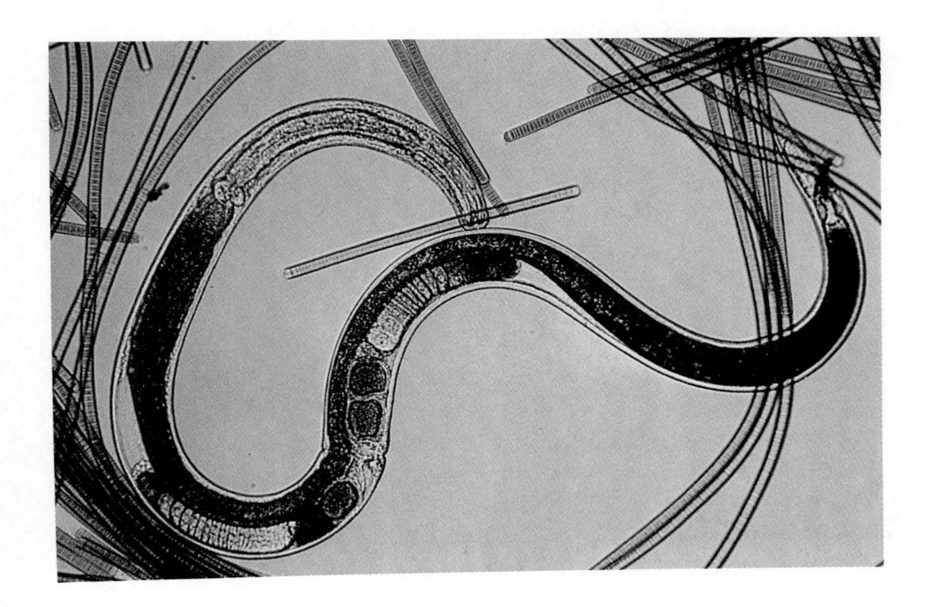

24.33 A living nematode worm viewed among the cyanobacteria it eats

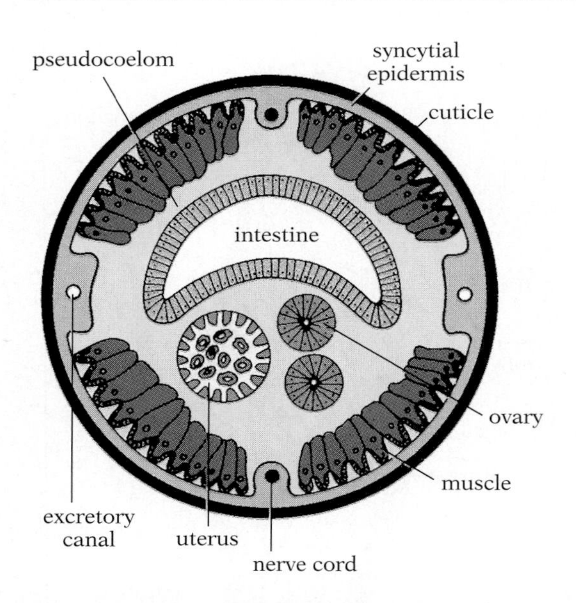

24.34 Diagrammatic cross section of a nematode worm

in a tough cuticle (Fig. 24.34). Just under the epidermal layer of the body wall are bundles of longitudinal muscles; there are no circular muscles. The lack of circular muscles and the stiff cuticle severely limit the types of movement possible for the worms.

The wall of the digestive tract consists of a single layer of endodermal cells; there is usually no muscle layer around the intestine, except sometimes at its posterior end. Between the intestine and the body wall is a fluid-filled cavity (Fig. 24.34). This cavity is bounded internally by the endodermal wall of the intestine, and externally by the bands of mesodermal muscle and, between the muscle bands, by the ectodermal layer of the body wall. Since the cavity is not entirely enclosed by mesoderm, it is a pseudocoelom.

Nematodes are extremely common, and are found in almost every type of habitat. Of the many free-living nematodes in soil or water, most are very tiny, often microscopic. A single spadeful of garden soil may contain a million or more individuals, and a bucket of water from a pond usually contains comparable numbers. Many other nematodes are internal parasites of both plants and animals; these also are often small, but some may attain a length of 1 m.

Nematodes parasitic on humans cause some serious diseases. *Trichinella spiralis*, for example, causes the disease ***trichinosis***, often contracted by eating insufficiently cooked pork. Adult *Trichinella* worms inhabit the small intestine of numerous species of mammals, among them hogs. Impregnated females bore through the wall of the intestine and deposit young larvae in the lymphatic vessels of their host. The larvae are carried by the lymph and blood to all parts of the body. They then bore out of the vessels, eventually entering every organ and tissue. However, only those that bore into skeletal muscles (especially the muscles of the diaphragm, ribs, tongue, and eyes) survive. In the muscles they grow in size (to about 1 mm) and then curl up and encyst (Fig. 24.35). If a human eats undercooked pork containing such cysts, the walls of the cysts are digested away and the worms complete

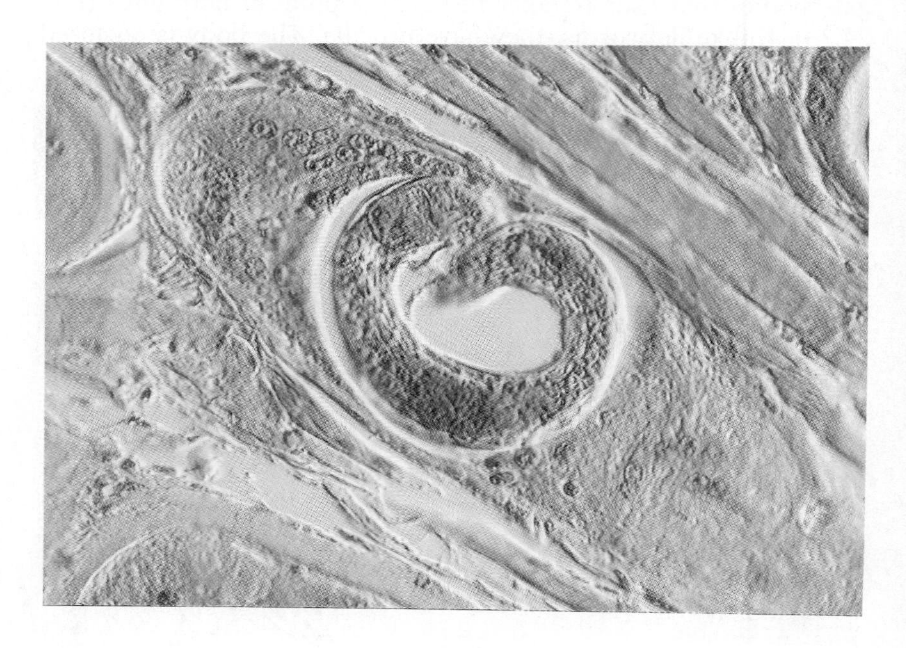

24.35 *Trichinella spiralis* encysted in muscle

their development in the host's intestine. The adult worms then deposit larvae in the lymph vessels in the wall of the intestine, from which they move through the host's body as they do through a hog's, eventually encysting in muscles.

Most of the damage of trichinosis occurs during the migration of the larvae, when half a billion or more may simultaneously bore through the body. Symptoms include excruciating muscular pains, fever, anemia, weakness, and sometimes localized swellings. Some victims die; survivors may sustain permanent muscular damage. Prevention of the disease is simple: pork must be thoroughly cooked to kill the encysted larvae. Fortunately, *Trichinella* is exceedingly rare in developed countries.

Among other nematodes that parasitize humans are (1) *Ascaris*, a large worm (up to 30 cm long), which lives in the digestive tract, lays eggs that pass to the outside with the host's feces, and infects new hosts when vegetables grown in soil contaminated with feces are eaten without adequate washing; (2) hookworms, tiny worms widespread in warm climates, that cause a severely debilitating disease usually contracted by going barefoot on soil contaminated with feces containing eggs; (3) pinworms, tiny worms common in schoolchildren, who usually get infected by putting unclean fingers with eggs on them in their mouths; (4) filaria worms, which are spread by the bite of certain mosquitoes in tropical and subtropical areas. Filaria worms live in the lymphatic system, where they may accumulate in such numbers that they block the flow of lymph, causing accumulation of fluid and often enormous swelling (elephantiasis; see Fig. 30.21, p. 858).

THE COELOMATE PROTOSTOMES

All the protostome phyla except the three discussed earlier possess true coeloms that in most groups arise as a split in an initially solid mass of mesoderm. All have a complete digestive tract, and most have well-developed circulatory, excretory, and nervous systems.

■ THE LOPHOPHORATE PHYLA

There are three small phyla, Phoronida, Bryozoa, and Brachiopoda, that resemble one another in having a ***lophophore***—a fold, usually horseshoe-shaped, that encircles the mouth and bears numerous ciliated tentacles. The lophophore is a feeding device; its tentacular cilia create water currents that sweep plankton and tiny particles of detritus into a groove leading to the mouth.

All members of the lophophorate phyla are aquatic, and most are marine. Adults are usually sessile and secrete a protective case, tube, or shell around themselves, but the larvae are ciliated and free-swimming. The digestive tract is U-shaped in phoronids and bryozoans and in some brachiopods; the anus lies outside the crown of tentacles.

The phylum Phoronida contains only about 10 species of wormlike animals that inhabit a tube of their own secretion (Fig. 24.36).

24.36 Phoronid worms (*Phoronis hippocrepia*)

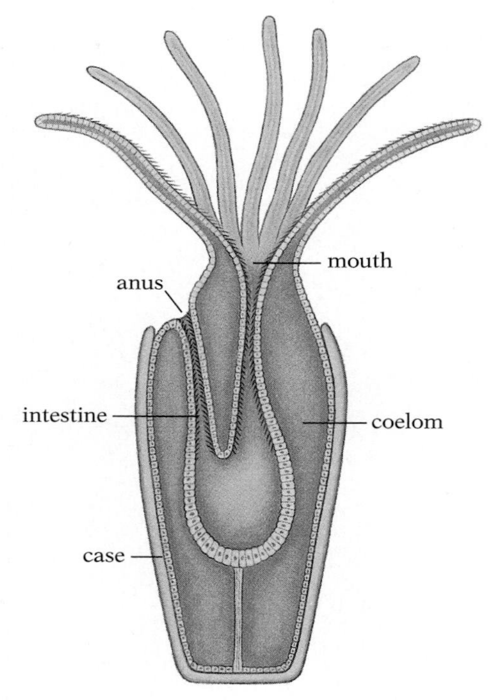

24.37 Section through the body of a bryozoan, showing the U-shaped digestive tract

Note the anus is located outside the ring of tentacles.

24.38 A brachiopod (*Terebratulina*)

The shape of this northern lamp shell from Maine is reminiscent of old Roman lamps. Brachiopods usually attach themselves to the substratum by means of a stalk.

They are found either buried in the sand or attached to rocks, shells, or other objects in shallow seas.

The bryozoans (or ectoprocts) are often called moss animals. They are very tiny (usually less than 0.5 mm long), colonial, sessile animals enclosed in a case open only at the lophophore end (Fig. 24.37). Unlike most other coelomate protostomes, they lack both excretory and circulatory systems. They superficially resemble entoprocts, which were formerly included in the same phylum with them, but entoprocts have a pseudocoelom instead of a coelom and their anus lies inside the ring of tentacles,[8] which is not a true lophophore.

The brachiopods, shelled animals that superficially resemble molluscs, are often called lamp shells because they are shaped like a Roman oil lamp (Fig. 24.38). They are usually permanently attached to the ocean bottom by a fleshy stalk. There are only about 300 living species of brachiopods, but more than 30,000 fossil species are known; they were among the most common animals in Paleozoic seas.

■ MOLLUSCA: THE MOLLUSCS

The phylum Mollusca is the second largest in the animal kingdom; it contains nearly 35,000 living and 35,000 fossil species. Among the best-known molluscs are snails (including slugs), clams, squids, and octopuses. Compared with the phyla examined so far, molluscs have relatively large and compact bodies. At this size there is insufficient surface area to exchange enough oxygen and carbon dioxide to keep the internal tissues alive; moreover, diffusion through the many layers of cells between the core and the surface would be too slow. Because of these factors, molluscs have evolved efficient internal plumbing and other specialized systems.

[8]*Entoprocta* means "internal anus"—an anus inside the crown of tentacles—and *Ectoprocta* means "external anus"—outside the crown of tentacles.

The various groups of molluscs differ considerably in outward appearance, but most have fundamentally similar body plans.[9] The soft body consists of three principal parts: (1) a large ventral muscular ***foot***, which can be extruded from the shell (if present) and functions in locomotion; (2) a ***visceral mass*** above the foot, which contains the digestive system, the excretory organs, the heart, and other internal organs; and (3) a heavy fold of tissue called the ***mantle***, which covers the visceral mass and which in most species contains glands that secrete a shell. The mantle often overhangs the sides of the visceral mass, thus enclosing a ***mantle cavity***, which frequently accommodates gills (see Fig. 24.46).

Molluscs have an open circulatory system: during part of each circuit the blood pools in large open sinuses, where it bathes the tissues directly rather than through a network of capillaries. Blood drains from the sinuses into vessels that run out into the gills, where the blood is oxygenated. From the gills, the blood goes to the heart, which pumps it into vessels that lead it back to the sinuses; a typical circuit, then, is heart–sinuses–gills–heart. This design is analogous to the oil-circulation system of an automobile: an oil pump forces the lubricant to the top of the engine; from there the oil drips down over various mechanical parts and is collected in the oil pan, from which it is pumped back up again.

Most marine molluscs pass through one or more ciliated free-swimming larval stages, but freshwater and land snails usually complete the corresponding developmental stages while still in the egg and hatch as miniature editions of the adult.

The phylum Mollusca is divided into eight classes: Caudofoveata, Solenogastres, Polyplacophora, Monoplacophora, Gastropoda, Scaphopoda, Bivalva, and Cephalopoda. The first three classes are exclusively marine; they are probably the most primitive living members of the Mollusca. The first two classes are so rare we will not discuss them separately.

Polyplacophora: the chitons The polyplacophorans ("bearing many tablets") have an ovoid bilaterally symmetrical body with a shell consisting of eight serially arranged dorsal plates (Fig. 24.39).

[9]The generalized body plan described here characterizes all living molluscs except members of class Solenogastres, which are very primitive wormlike molluscs having no head, mantle, foot, or shell.

24.39 A chiton

Note the series of plates composing the chiton's shell.

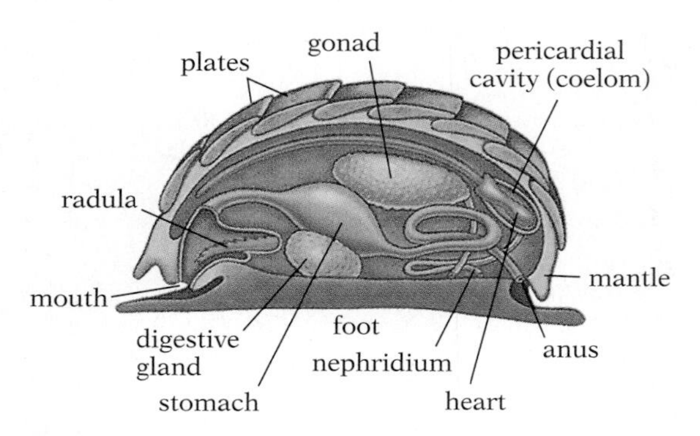

24.40 Lateral view of a chiton

They have an anterior mouth and a posterior anus (Fig. 24.40). The coelom is reduced to a small cavity surrounding the heart.[10]

Chitons lead a sluggish, nearly sessile life. They creep about on the surface of rocks in shallow water, rasping off fragments of algae with a horny toothed organ called a ***radula***. Their broad flat foot can develop tremendous suction, and when disturbed they clamp down so tenaciously to the rock that they can hardly be pried loose.

Monoplacophora: living fossils Members of this class have long been known as fossils, and for many decades it was thought that all had been extinct for about 350 million years. In 1952, however, 10 living specimens (genus *Neopilina*) were dredged from a deep trench in the Pacific Ocean off the coast of Costa Rica. They show a kind of internal segmentation, a characteristic seen in no other members of the phylum. The early cleavage pattern and larval type of molluscs show striking similarities to the corresponding developmental stages in the segmented worms (phylum Annelida); the "segmentation" of *Neopilina* suggests that ancestral molluscs may have been primitive annelids.

Gastropoda: the snails and their relatives Most gastropods have a coiled shell (Fig. 24.41). In some cases, however, the coiling is minimal. Some species, such as *Aplysia* and the other nudibranchs, have lost the shell as adults (Fig. 24.42).

[10]The lumina of the gonads and nephridia are also thought to be remnants of the coelom.

24.41 A marine gastropod mollusc (*Trivia monaca*)

The animal moves along the substratum by means of the large muscular foot, here plainly seen. Notice also the head region, bearing prominent antennae, eyes, and a long siphon through which water is brought to the gills.

24.42 Two nudibranch gastropods

As the derivation of *nudibranch* suggests, the animals have no shells. (*Nudus* is Latin for "naked"; *branchia* is Greek for "gills".)

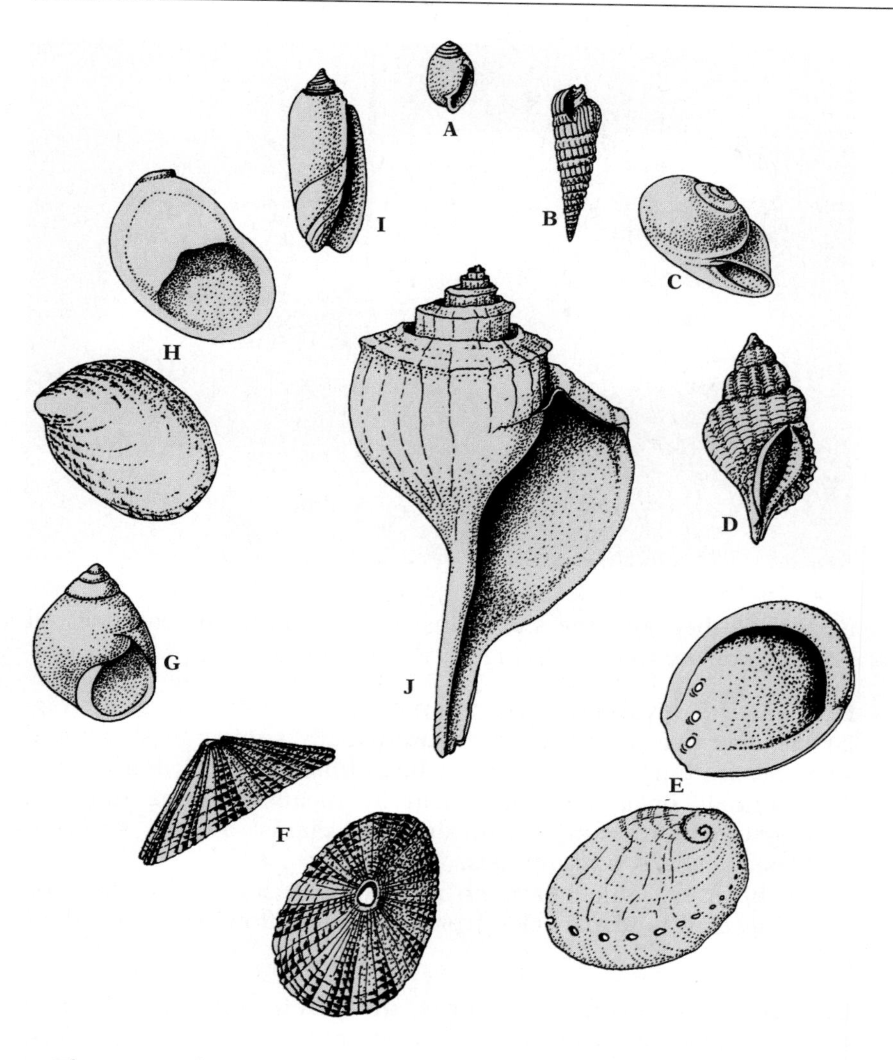

24.43 Shells of some representative gastropods

(A) Salt-marsh snail (*Melampus*). (B) Auger shell (*Terebra*). (C) Moon shell (*Neverita*). (D) Oyster drill (*Urosalpinx*). (E) Abalone (*Haliotis*), ventral and dorsal views. (F) Californian keyhole limpet (*Diodora*), dorsal and lateral views. (G) Periwinkle (*Littorina*). (H) Boat shell (*Crepidula*), ventral and dorsal views. (I) Olive shell (*Oliva*). (J) Channeled conch (*Busycon*).

The young larva in gastropods is bilateral. As it develops, however, the digestive tract bends downward and forward until the anus comes to lie close to the mouth. Then the entire visceral mass rotates through an angle of 180°, coming to lie dorsal to the head in the anterior part of the body. Most of the visceral organs on one side (usually the left) atrophy, and growth proceeds asymmetrically, producing the characteristic spiral.

Gastropods have a distinct head with well-developed sense organs, and most have a strong radula and feed on bits of plant or animal tissue that they grate, rasp, or brush loose with this organ.

Gastropods occur in a great variety of habitats. The majority are marine, often with large, decorative shells (Fig. 24.43), but there are also many freshwater species and some that live on land. The land snails are among the few groups of fully terrestrial invertebrates. In most the gills have disappeared, but the mantle cavity has become highly vascularized and functions as a lung. Such snails are said to be pulmonate. Some pulmonate snails have secondarily returned to the water and must periodically come to the surface to obtain air.

24.44 A scaphopod mollusc (*Dentalium vulgare*)

The shell is a long tapering tube open at both ends.

Scaphopoda: the tusk shells Scaphopods have a long tubular shell, open at both ends (Fig. 24.44). One end is usually smaller

24.45 A scallop (*Chlamys opercularis*), a representative bivalve
The shell is composed of two hinged valves. Note the numerous small eyes around the edges of the mantle.

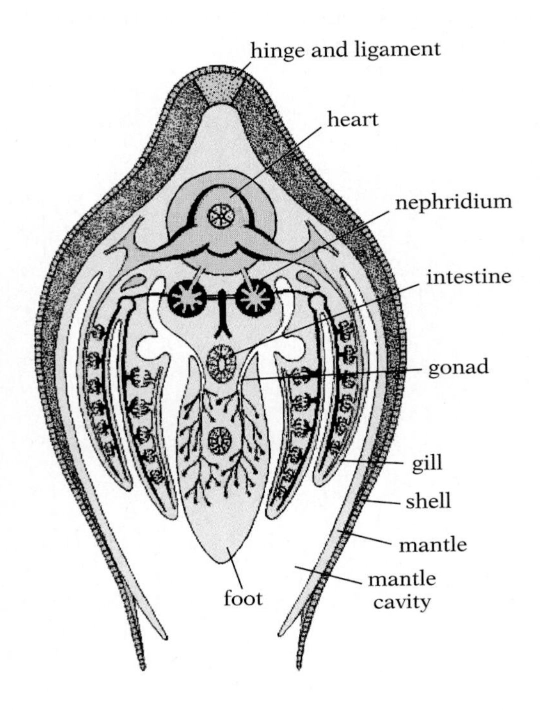

24.46 Cross section of a clam

24.47 A swimming octopus (*Octopus vulgaris*) Note the large suckers on the tentacles.

than the other, and the shell thus has a tusklike appearance. All scaphopods are marine, living buried in mud or sand.

Bivalva: the bivalves These animals have a two-part shell (Fig. 24.45). The two parts, or valves, are usually similar in shape and size and are hinged on one side (the animal's dorsum; Fig. 24.46). The animals open and shut them by means of large muscles. Among the more common bivalves are clams, oysters, scallops, cockles, file shells, and mussels.

The bivalves, which have no radula, are usually filter feeders, straining tiny food particles from the water flowing across their gills.

Cephalopoda: squids, octopuses, and their relatives Many of the cephalopods bear little outward resemblance to other molluscs. Unlike their sedentary relatives, they are often specialized for rapid locomotion and a predatory way of life—for killing and eating large prey such as fish and crabs. Though fossil cephalopods often have large shells, these are much reduced or absent in most modern forms. (The chambered nautilus, with its well developed shell, is a familiar exception; see Fig. 36.16, p. 1038.) The body is elongate, with a large and well-developed head encircled by long tentacles.

Some species grow up to several meters long. The giant squids (*Architeuthis*) of the North Atlantic are the largest living invertebrates; so far, the biggest recorded individual was 17 m long (including the tentacles) and weighed approximately 2 tons. Octopuses (Fig. 24.47) never grow anywhere near this size (except in Hollywood).

Cephalopods, particularly squids, have convergently evolved many similarities to vertebrates. For example, squids have internal cartilaginous supports analogous to the vertebrate skeleton, including a cartilaginous braincase rather like a skull. They have a well-developed nervous system with a large and complex brain. The most striking similarity is their large camera-type eye, which works almost exactly the way ours does.

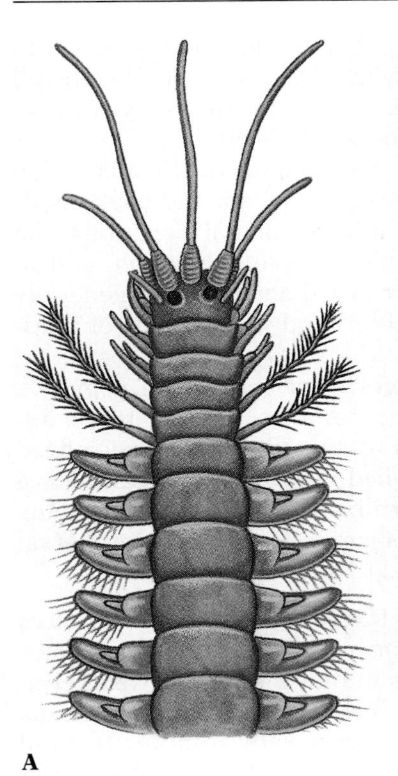

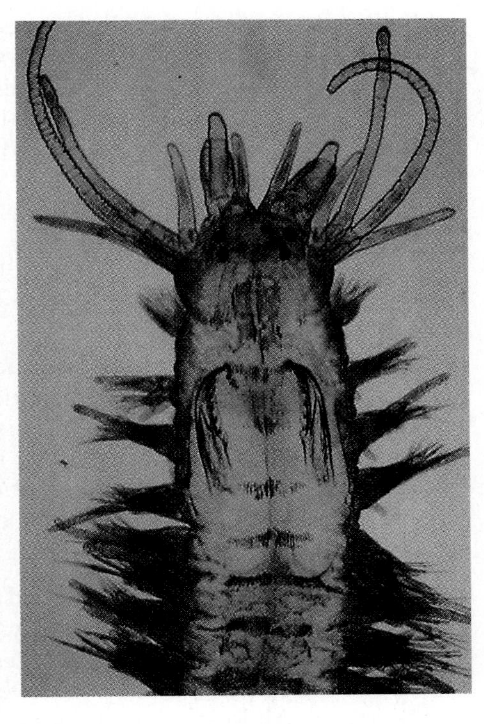

A B

24.48 Head and anterior gill-bearing segments of two polychaete worms

(A) *Diopatra*. (B) *Nereis*.

■ ANNELIDA: THE SEGMENTED WORMS

The physiology of the annelids, or segmented worms, is discussed in Part V. Included there are descriptions of their digestive system, gas exchange, circulatory system (which, like ours, consists of closed, pressurized, fluid-filled tubes), fluid-balance organs (nephridia), nervous system, as well as their hydrostatic skeleton, muscle arrangement, and locomotion. There are three classes: Polychaeta, Oligochaeta, and Hirudinoidea.

Polychaeta: the marine worms Polychaetes are marine annelids with a well-defined head bearing eyes and antennae (Fig. 24.48). Each of the numerous serially arranged body segments usually bears a pair of ***parapodia*** that function in both locomotion and gas exchange. There are numerous stiff setae (bristles) on the parapodia. (*Polychaeta* means "many setae.")

Some polychaetes actively swim or crawl; others are more sedentary, usually living in tubes they construct in the mud or sand of the ocean bottom. These tubes may be simple mucus-lined burrows, membranous structures, or elaborately constructed dwellings composed of sand grains cemented together. The tubes of some species are straight, while those of others are U-shaped and have two openings (Fig. 24.49). The beating parapodia keep water currents flowing through the tubes; these currents bring oxygen, and in some cases food particles, to the worm. Many of the

24.49 The polychaete worm *Phragmatopoma lapidosa*

This gregarious species builds sand tubes; large aggregations of these tubes can create reefs. Parts of three individuals shown here.

24.50 A fanworm (*Sabella crassicornis*)
The processes on this "feather-duster worm" help bring food and gases to the organism.

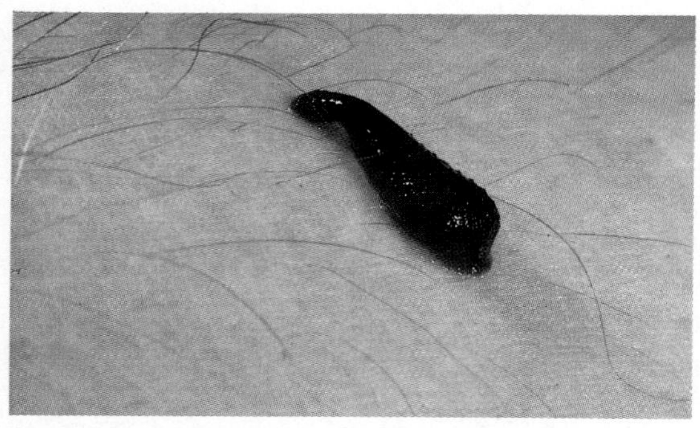

A

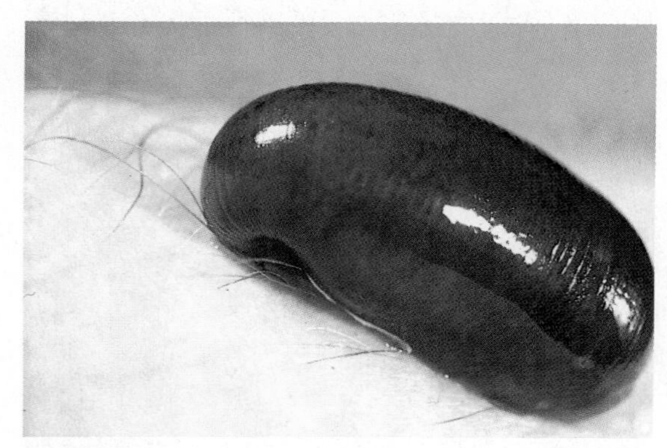

B

24.51 A terrestrial leech of Australia, sucking blood from a human arm
(A) The leech has just begun feeding. (B) Somewhat later its body has become engorged with blood.

tube dwellers are beautifully colored. Among the most striking are the fanworms and peacock worms, which have a crown of colorful, much-branched fanlike or featherlike processes that they wave in the water at the entrance to their tubes (Fig. 24.50).

All of the segments of the body are usually much alike. The coelom of each segment is partly separated from the coeloms of adjacent segments by membranous intersegmental partitions; the partitions of many polychaetes are not complete, however, and in some species they have been entirely lost. Each segment generally has its own local nerve center (ganglion) and its own pair of fluid-regulation organs (nephridia).[11]

The sexes are separate in the majority of species. In primitive polychaetes most segments produce gametes, but in more advanced species gamete production is restricted to a few specialized segments. The gametes are usually shed into the coelom and leave the body through the nephridia. Fertilization is external. In many species development includes a ciliated, free-swimming larval stage called a ***trochophore*** (see Fig. 24.77).

Oligochaeta: the freshwater and terrestrial worms The class Oligochaeta contains the earthworms and many freshwater species. They differ from polychaetes in that they lack a well-developed head and parapodia, and have few setae (*Oligochaeta* means "few setae"). Male and female organs are usually combined within the same individual, and they usually have nearly complete intersegmental partitions. Earthworm physiology is detailed in Part V.

Hirudinoidea: the leeches The leeches, which probably evolved from oligochaetes, are the most specialized annelids. Their body is dorsoventrally flattened and often tapered at both ends. The first and last segments are modified to form suckers, of which the posterior one is much the larger; the suckers are used in locomotion. Leeches show almost no internal segmentation; the intersegmental partitions have been completely lost except in very primitive species.

Some leeches are predaceous, capturing invertebrates such as worms, snails, and insect larvae and swallowing them whole. More familiar are the bloodsuckers, which attack a variety of vertebrate and invertebrate hosts (Fig. 24.51). When a leech of this type attacks a host, it selects a thin area of the host's integument or skin, attaches itself by its posterior sucker, applies the anterior sucker very tightly to the skin, and either painlessly slits the skin with small bladelike jaws or dissolves an opening by means of enzymes. It then secretes into the wound a substance (hirudin) that prevents coagulation of the blood, and begins to suck the blood, usually consuming an enormous quantity at one feeding and then not feeding again for a fairly long time. (Some leeches have been known to go unfed for more than a year without apparent harm.)

■ ONYCHOPHORA: AN EVOLUTIONARY MILESTONE?

There are only about 65 living species of this small phylum, all restricted to tropical regions or to the temperate parts of the south-

[11]In a few species of polychaetes, there is only one pair of nephridia for the whole animal.

24.52 An onychophoran

Note the numerous short unjointed legs and the prominent antennae.

ern hemisphere (Australia, New Zealand, South Africa, and the Andes). They are mostly confined to very moist habitats on land, living beneath leaves, logs, or stones in forests, and are nocturnal. Members of this group were very numerous in the past; many soft-bodied marine fossils from the Cambrian are onychophorans.

Looking rather like caterpillars, these velvet worms have a segmented wormlike body with 14–43 pairs of short unjointed legs (Fig. 24.52). Velvet worms are of special interest because they have a combination of annelid and arthropod characters and are regarded by some researchers as an early evolutionary offshoot of the line leading from an ancient annelid-like ancestor to the arthropods (see Fig. 24.54). They have a thin, flexible, permeable cuticle that is more like the cuticle of annelids than the exoskeleton of arthropods, and, like the annelids, they have a pair of nephridia in each segment. However, they resemble arthropods in having claws and an open circulatory system. The respiratory system consists of a highly ramified system of tubes called ***tracheae***, opening to the outside and penetrating into all parts of the body. Despite their intermediate morphology, preliminary sequence analysis places them among the arthropods. (Until sequencing and other evidence is more complete, however, we will treat them as a separate phylum.)

■ ARTHROPODA: THE LARGEST PHYLUM

The phylum Arthropoda is by far the largest of the phyla. Nearly a million species have been described, and there are doubtless hundreds of thousands more yet to be discovered. Probably more than 80% of all animal species on earth belong to this phylum.

Arthropods have jointed chitinous exoskeletons and legs. The exoskeleton, which is secreted by the epidermis, functions both as a point of attachment for muscles and as protective armor (particularly against desiccation). It has the disadvantage, however, of imposing limitations on growth, and must be molted periodically if the animal is to increase much in size (Fig. 24.53; see also Figs. 13.17 and 13.20, pp. 348 and 350). The arthropod cuticle is not restricted to the exterior surface of the body; long rod-shaped processes that often project from the surface deep into the interior of the animal function as bases for muscle attachment, and both

24.53 A molting centipede

Its new yellow-orange exoskeleton glistening, the animal is backing out of its old exoskeleton.

the anterior and posterior portions of the digestive tract (and the tracheae of land arthropods) are lined with cuticle. These internal extensions of the exoskeleton are also shed at each molt.

Together with their elaborate exoskeleton, arthropods have evolved a complex musculature unlike that of most other invertebrates. It comprises not only longitudinal and circular bands, as in many invertebrates, but also separate muscles that, running in myriad directions, make possible an extensive repertoire of movements, including flight. This phylum, to a far greater extent than any other invertebrate group, has solved the constellation of problems that come with a fully terrestrial life cycle.

The nervous system is very well developed. It consists (like that of the annelids) of a dorsal brain and a ventral double nerve cord (see Fig. 35.13, p. 1001). In primitive arthropods there are ganglia in each segment, but in many groups the ganglia have moved forward and fused into larger ganglionic masses. Sensory organs are many and varied.

Arthropods have an open circulatory system. There is usually an elongate dorsal vessel called the heart, which pumps the blood forward into arteries. From the arteries, the blood goes into open sinuses, where it bathes the tissues directly. Eventually the blood returns to the posterior portion of the heart.

The body spaces through which the blood moves constitute the ***hemocoel***—not a true coelom but a cavity derived from the embryonic blastocoel. Though arthropods almost certainly descended from an annelid-like ancestor with well-developed coelomic cavities, and though such cavities develop in arthropod embryos, they are not retained as the functional body cavity in the adult.[12]

In most aquatic arthropods, excretion of nitrogenous wastes (primarily ammonia) occurs principally by way of the gills. Aquatic species usually also have special saclike glands, located near or in the head, that play a minor role in excretion; these glands (usually called coxal glands or green glands) have their own ducts leading to the outside. The excretory organs in most groups of terrestrial arthropods are called Malpighian tubules; these organs are bathed by the blood of the open circulatory system, from which they collect dissolved nitrogenous wastes. As detailed in Chapter 31, the tubule system then recovers essentially all of the water, and excretes nearly dry wastes. The tubules, therefore, are an excellent adaptation for conserving water. The sexes are usually separate. Fertilization is internal in all terrestrial and in most aquatic forms—an adaptation that both conserves gametes and prevents their desiccation. Along with the cuticle and Malpighian tubules, internal fertilization helped make possible the radiation of early arthropods onto the land, a process well underway by 415 million years ago, when spiders and centipedes were abundant.

Arthropods probably evolved either from a polychaete annelid or from the ancestor of the polychaetes. The arthropod body plan is an elaboration of the segmented body of annelids. The first arthropods had long wormlike bodies composed of many nearly identical segments, each bearing a pair of legs. Among the host of modifications of this ancestral body plan that have arisen in the various

[12]The cavity of the gonads (and that of the excretory ducts in some arthropods) may be a remnant of the true coelom.

groups of arthropods during the millions of years of their evolution, four tendencies stand out:

1 Reduction in the total number of segments

2 Grouping of segments into distinct body regions, such as a head and trunk, or a head, thorax, and abdomen

3 Increasing cephalization—that is, incorporation of more segments into the head and concentration of nervous control and sensory perception in or just behind the head

4 Specialization of the legs of some segments for a variety of functions other than locomotion, and complete loss of legs from many other segments.

The arthropods are often divided into five subphyla: Trilobita (now extinct), Crustacea (which includes crabs, barnacles, and shrimp), Uniramia (a group that includes the insects), Chelicerata (a smaller group that includes spiders and horseshoe crabs), and Pentastomida (a tiny group of parasites not discussed here). Figure 24.54 shows one interpretation of the relationships among the arthropod subphyla and classes. The distinction between Uniramia and Chelicerata is considered dubious by some authorities.

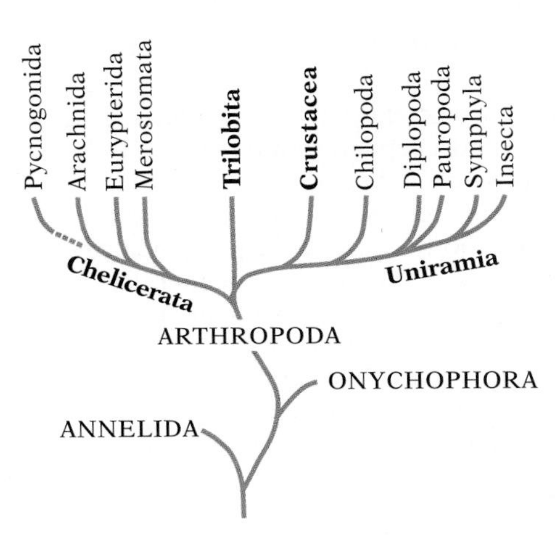

24.54 Possible relationships between the phyla Annelida and Onychophora, and the major subphyla and classes of the phylum Arthropoda

Members of Trilobita and Eurypterida are extinct, as are many unnamed groups (see Fig. 18.23, p. 510). Preliminary sequence analysis places the Onychophora among the Chelicerata. The distinction between Uniramia and Chelicerata is controversial.

Trilobita: the trilobites Arthropods were abundant in the Paleozoic seas. Particularly common in rocks of the first half of the Paleozoic are the fossils of an extinct group, the Trilobita (Fig. 24.55). The fossils show a usually oval and flattened shape and three body regions: (1) a head, apparently composed of four fused segments, that bore a pair of slender antennae and often compound eyes (composed of many independent facets, each with its own lens); (2) a thorax consisting of a variable number of separate segments; and (3) an abdomen (pygidium), composed of several fused segments. However, the name Trilobita refers to a different tripartite division: two prominent longitudinal furrows running along the dorsum separate the body into a median lobe and two lateral lobes.

24.55 A fossil of a trilobite

Note the two longitudinal furrows that partition the animal's body into a median lobe and two lateral lobes. It is this tripartite arrangement that suggested the name Trilobita.

24.56 A whipscorpion (*Mastigoproctus giganteus*)

Whipscorpions have two body regions: an abdomen and a cephalothorax. The cephalothorax has six pairs of appendages: a pair of fanglike chelicerae (not visible in the photograph); a pair of stout, toothed pincerlike pedipalps; and four pairs of legs. (The long and slender first pair have a sensory-tactile function and are not used in walking.) The posterior knob with its "whip" (not visible here) has slits through which the animal can spray a poisonous secretion. This female is carrying her eggs under her abdomen.

Chelicerata: spiders and their relatives The chelicerate body is usually divided into two regions: a cephalothorax (prosoma) and an abdomen (Fig. 24.56). There are no antennae. The appendages corresponding to the first pair of legs in ancestral arthropods and trilobites are modified as mouthparts called ***chelicerae***, which may be either pincerlike or fanglike. The cephalothorax usually bears five other pairs of appendages besides the chelicerae; in some groups these are all walking legs, while in others only the last four pairs are legs, the first pair being modified as feeding devices called ***pedipalps***, which are often much longer than the chelicerae. The legs arising from the abdominal segments in ancestral arthropods have either been lost or modified into respiratory or sexual structures.

The subphylum Chelicerata includes four classes. One class (Eurypterida) consists entirely of animals extinct since the Paleozoic era, and the members of another (Pycnogonida, the sea spiders) are very rare marine animals.

A third class (Merostomata) is composed of the horseshoe crabs (Fig. 24.57). These are not true crabs but living relicts of an ancient

24.57 Horseshoe crabs spawning

24.58 A spider (*Araneus marmoreus*)

The four pairs of legs characteristic of the Arachnida are easy to make out. Note how the animal places its first pair of legs on strands of its web; it can detect vibrations when prey touch the web, and can often even distinguish, by the type of vibration, what sort of prey it is.

chelicerate class, most members of which have been extinct for millions of years.

The fourth class of chelicerates (Arachnida) includes the spiders, ticks, mites, daddy longlegs, scorpions, whipscorpions (see Fig. 24.56), and their relatives. Though the various groups of arachnids differ structurally in many ways, most have two body regions: a cephalothorax and an abdomen (not distinguishable in ticks, mites, or daddy longlegs). There are often simple eyes on the cephalothorax but never any compound eyes or antennae. The cephalothorax bears six pairs of appendages: a pair of chelicerae, a pair of pedipalps, and four pairs of walking legs (Fig. 24.58). In most groups, prey is seized and torn apart by the pedipalps. The chelicerae may also function in manipulating prey, or they may be modified as poison fangs, as in spiders. The abdomen of arachnids may be long, as in scorpions, or short, as in spiders. In some (including spiders), the bases of one or two pairs of abdominal appendages are retained as much-modified lungs; in others, no trace of abdominal appendages remains. Many arachnids have tracheae, and some respire by means of tracheae only. Several arachnid groups have glands that secrete silk.

Crustacea: crabs and their relatives The members of this subphylum and the next (the enormous subphylum Uniramia, which includes the insects) differ from chelicerates in having antennae (two pairs in Crustacea versus one pair in Uniramia) and ***mandibles*** instead of chelicerae as their first pair of mouthparts. (Formerly Crustacea and Uniramia were combined in a single subphylum called Mandibulata.) The mandibles are modified from the basal segment (coxa) of the ancestral legs, and they function in biting and chewing, though in some species they are modified for

A

B

24.59 Representatives of the larger, better-known crustaceans

(A) The crab *Neolithodes grimaldii*. (B) A freshwater prawn, or crayfish.

piercing and sucking. They are never clawlike or pincerlike, as chelicerae often are. In most mandibulates there are two additional pairs of mouthparts called ***maxillae***.

Some representatives of this subphylum, such as crayfish, lobsters, shrimps, and crabs (Fig. 24.59), are familiar. Many other species bear little superficial resemblance to these; among them are fairy shrimps, water fleas, brine shrimps, sand hoppers, barnacles, and sow bugs (Fig. 24.60).

Crustaceans characteristically have two pairs of antennae, a pair of mandibles, and two pairs of maxillae. But the rest of the appendages vary greatly from group to group. The crustaceans are an enormously diverse group. Some have a cephalothorax and an abdomen; others have a head and a trunk, or a head, thorax, and abdomen, or even a unified body. Most are free-living, but some are parasitic. Most are active swimmers, but some, like barnacles, se-

A

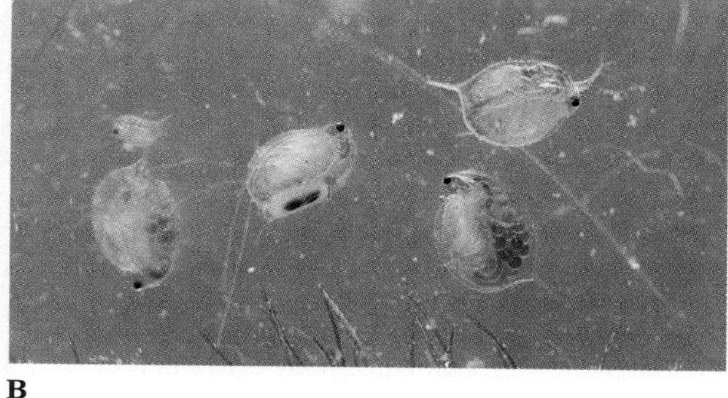

B

C

D

24.60 Representatives of the smaller crustaceans

(A) A marine amphipod (*Gammarus*). (B) The water flea (*Daphnia*). (C) A group of sow bugs (also called wood lice or pill bugs), one of the very few terrestrial crustaceans. (D) A freshwater crustacean (*Asellus aquaticus*).

crete a shell and are sessile (Fig. 24.61). The majority are marine, but there are many freshwater species, and a few, such as sow bugs, are terrestrial and have a simple tracheal system. This is a subphylum in which the basic arrangement of a segmented body with numerous jointed appendages has been modified and exploited in countless ways as the members have diverged into different habitats and adopted different modes of life.

Uniramia: centipedes, millipedes, and insects This vast subphylum comprises five classes. We briefly describe three of them here: Chilopoda, Diplopoda, and Insecta.

24.61 Goose barnacles (*Lepas*)

These animals are sedentary and secrete a protective shell. Most species of barnacles lack the long stalk so prominent in the goose barnacles.

24.62 Centipede

Centipedes have a segmented body with a pair of jointed legs on each segment. This Australian species is found under tree bark.

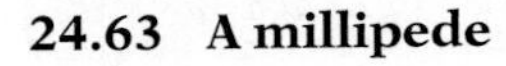

24.63 A millipede

Each segment (except a few at the front and rear) bears two pairs of legs.

1 *Chilopoda* (centipedes) The centipede body is divided into two regions—a head and a trunk (Fig. 24.62). The trunk is elongate and often somewhat flattened. The head bears a single pair of antennae and three pairs of mouthparts (mandibles and two pairs of maxillae). The animals are carnivorous, and the legs of the first trunk segment are modified as large poison claws. Each of the other trunk segments bears a single pair of walking legs. All centipedes are terrestrial.

2 *Diplopoda* (millipedes) These animals superficially resemble centipedes, but the two groups are not closely related.

The millipede body is divided into a head and a trunk (Fig. 24.63). The head bears a pair of antennae, but only two pairs of mouthparts (mandibles, and a pair of maxillae fused to form a platelike underlip). The animals are not carnivorous, feeding largely on decaying organic matter of various types. Each of the first four or five trunk segments bears a single pair of legs, but each of the other segments bears two pairs of legs (and also two pairs of spiracles). Clearly, each double-legged segment is formed by the fusion of two segments. In most millipede orders the legs (one or both pairs) of the seventh segment in the males are highly modified and function as organs for inserting sperm into the female.

3 *Insecta* (insects) This enormous group of diverse animals occupies almost every conceivable habitat on land and in fresh water. If numbers are the criterion by which to judge biological success,

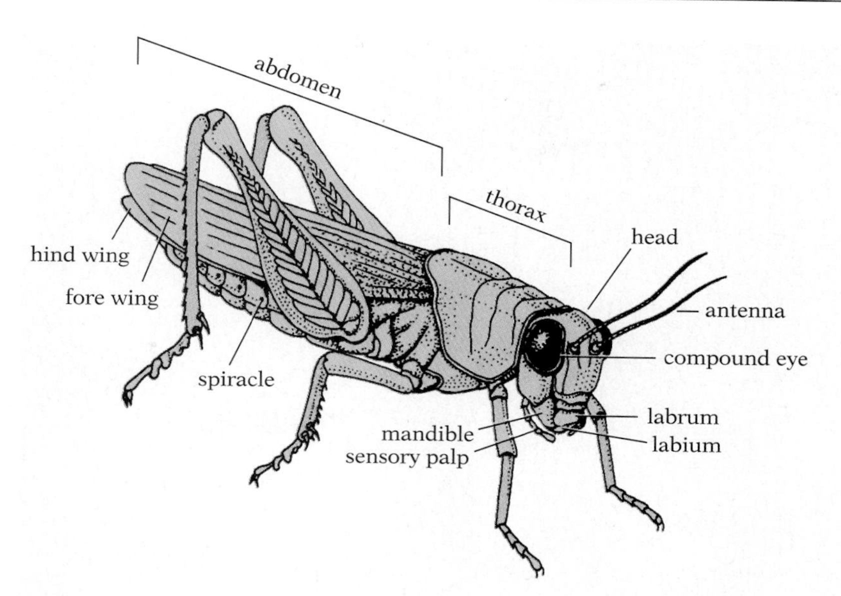

24.64 A grasshopper

Spiracles are openings to the tracheal system.

then the insects are the most successful group of animals that has ever lived; there are more species of insects than of all other animal groups combined. However, they do not occur in the sea (although a few species move on the ocean and pond surfaces or live in brackish water); the role played by insects on land is played in the sea by crustaceans.

There are a few insect fossils from the Devonian, but it was in the Carboniferous and Permian periods that insects took their place as one of the dominant groups of animals (see Table 19.1, p. 540). By the end of the Paleozoic era, many of the modern orders had appeared, and the number of species was enormous. A second great period of evolutionary radiation began in the Cretaceous and continues to the present time; this second radiation is correlated with the rise of flowering plants.

The insect body is divided into three regions: a head, a thorax, and an abdomen (Fig. 24.64). The head segments are completely fused. The head bears numerous sensory receptors, usually including compound eyes, one pair of antennae, and three pairs of mouthparts derived from ancestral legs. The mouthparts include a pair of mandibles, a pair of maxillae, and a lower lip, or ***labium***, formed by fusion of the two second maxillae (Fig. 24.65). The upper lip, or ***labrum***, may also be derived from ancestral legs.

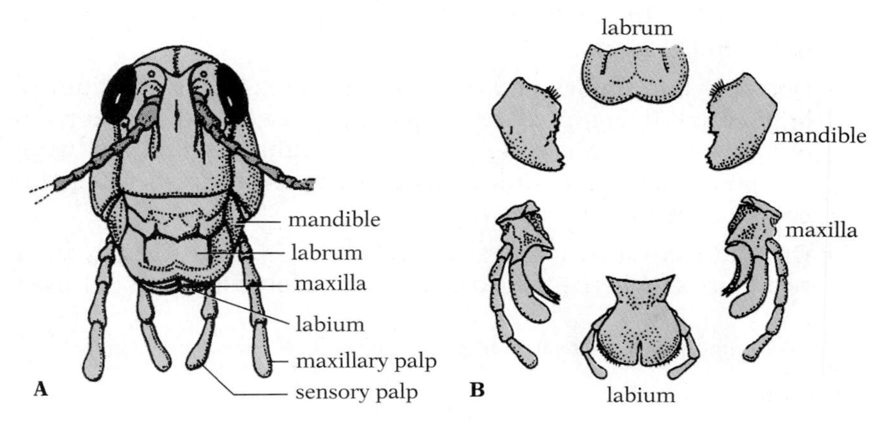

24.65 Mouthparts of a grasshopper

(A) Front view of head, with mouthparts in place. (B) The mouthparts removed from the head, but kept in their proper relative positions. The mandibles and probably the labrum (upper lip) are derived from the basal segments of ancestral legs; all the other segments of those legs have been lost. The maxillae and labium (lower lip) retain more of the segments of the legs from which they are derived; the basal segments are enlarged, but the distal segments form slender leglike structures called palps, which bear many sensory receptors.

24.66 A desert locust in flight

Like most flying insects except the flies, the locust has two pairs of wings. The front wings are leathery and, when the animal is at rest, provide a protective covering for the fragile pleated rear wings, which are important for flight.

The thorax is composed of three segments, each of which bears a pair of walking legs. In many insects (but not all), the second and third thoracic segments each bear a pair of wings (Fig. 24.66).

The abdomen is composed of a variable number of segments (12 or fewer). Abdominal segments are devoid of legs, but highly modified remnants of the ancestral appendages may be present at the posterior end, where they function in mating and egg laying.[13]

Insect physiology and behavior is discussed in detail in Part V. The insects are classified in approximately 28 orders. The following 11 orders are among the more familiar:

THYSANURA Bristletails and silverfish. Small, primitive, wingless; have chewing mouthparts, and long taillike appendages on rear of abdomen. Incomplete metamorphosis. Common in houses, particularly in kitchens and bathrooms; sometimes damage books in libraries.

ODONATA Dragonflies and damselflies (Fig. 24.67A). Medium to large, rapid-flying, predaceous on other insects. Have two pairs of long membranous wings, chewing mouthparts, and very large compound eyes. Immature stages (nymphs) in fresh water; incomplete metamorphosis.

ORTHOPTERA Grasshoppers, crickets (Fig. 24.67B). Usually have two pairs of wings—the coarse-textured forewings are not used

[13]A few primitive insects retain vestiges of appendages on many abdominal segments.

in flight, but function (when animal is at rest) as covers for the folded, fanlike hind wings. Have chewing mouthparts. Gradual metamorphosis.

ISOPTERA Termites. Highly social insects. Wings briefly present only on members of the reproductive castes; usually have chewing mouthparts. Gradual metamorphosis.

HEMIPTERA True bugs (Fig. 24.67C). Usually have two pairs of wings—basal half of forewings thick and leathery, distal half membranous, hind wings membranous. Have piercing-sucking mouthparts. Gradual metamorphosis.

ANOPLURA Sucking lice (Fig. 24.67D). External parasites. Wingless; have piercing-sucking mouthparts and legs and claws adapted for clinging to host. Gradual metamorphosis.

COLEOPTERA Beetles (Fig. 24.67E). Have two pairs of wings—forewings hard, meeting along middorsal line, forming a protective case for the folded membranous hind wings when animal is at rest. Have chewing mouthparts. Complete metamorphosis.

LEPIDOPTERA Moths and butterflies. Have two pairs of large scale-covered wings; have chewing mouthparts in larvae, and sucking (not piercing) mouthparts in adults. Complete metamorphosis.

DIPTERA True flies: mosquitoes, gnats, midges, house flies, horse flies, etc. (Fig. 24.67F). Have one pair of membranous wings; highly modified hind wings acting as tiny balancing organs. Have piercing-sucking or sponging mouthparts. Complete metamorphosis.

SIPHONAPTERA Fleas (Fig. 24.67G). Intermittent ectoparasites. Small, with body laterally compressed; have no wings; have piercing-sucking mouthparts; have long legs adapted for jumping. Complete metamorphosis.

HYMENOPTERA Sawflies, ants, bees, wasps (Fig. 24.67H). Usually have two pairs of membranous wings, interlocked in flight. Have chewing or chewing-lapping mouthparts. Thorax and abdomen connected by a very narrow waist. Complete metamorphosis.

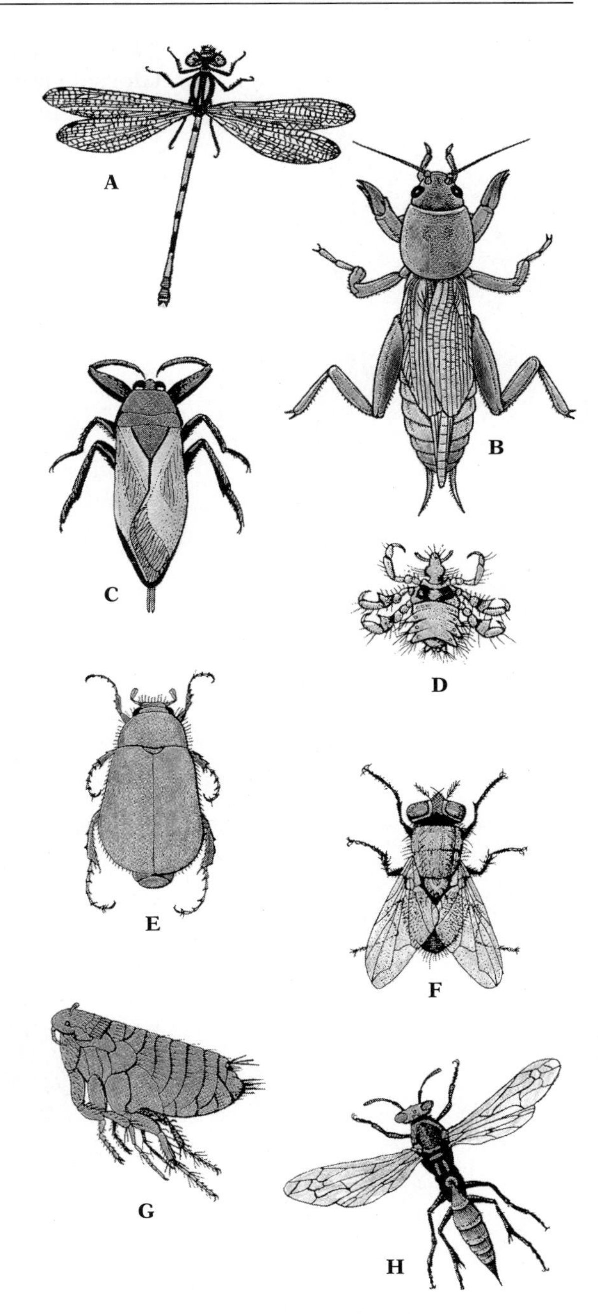

24.67 Representatives of some of the major insect orders

(A) Damselfly (order Odonata). (B) Mole-cricket (Orthoptera). (C) Bug (Hemiptera). (D) Louse (Anoplura). (E) Beetle (Coleoptera). (F) Fly (Diptera). (G) Flea (Siphonaptera). (H) Wasp (Hymenoptera). (Not to scale.)

THE DEUTEROSTOMES

The three phyla of Deuterostomia mark the transition from the invertebrate animals to the vertebrates. One phylum is clearly invertebrate: Echinodermata, whose most familiar representatives are the sea stars (starfish). A second phylum, Chordata, contains all of the vertebrates; the next chapter is devoted to this group. The third phylum, Hemichordata, includes the primitive marine animals known as acorn worms. We discuss this small but evolutionarily important phylum in this chapter.

■ ECHINODERMATA: SEA STARS AND THEIR RELATIVES

The echinoderms are exclusively marine, mostly bottom-dwelling animals. They are common in all seas and at all depths from the intertidal zone to the ocean deeps. Included in this distinctive phy-

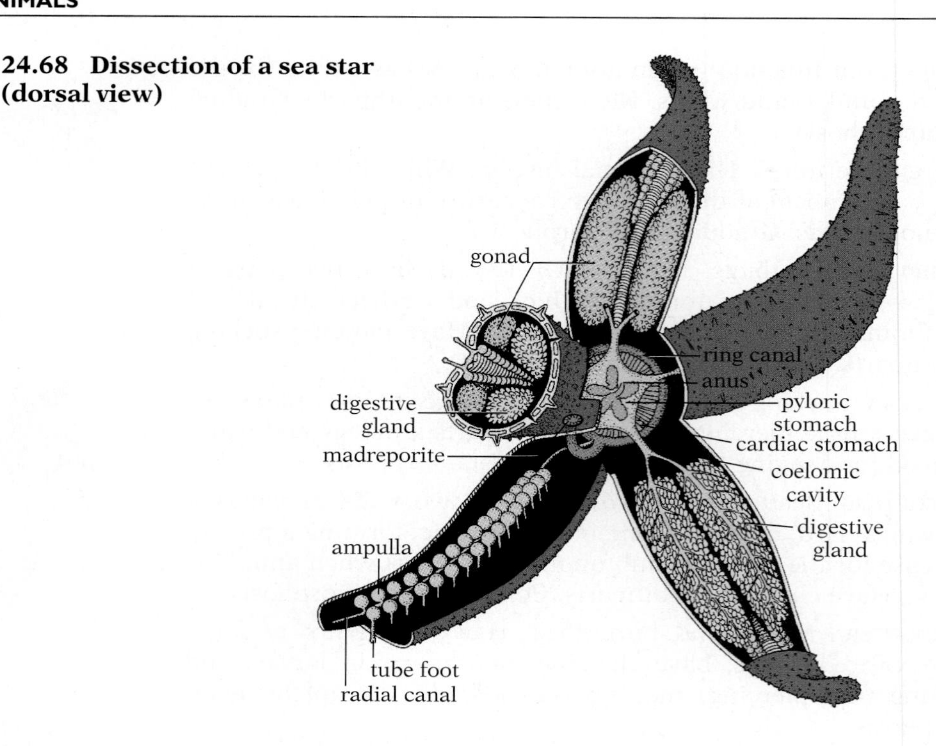

24.68 Dissection of a sea star (dorsal view)

lum are the sea stars (Fig. 24.68), brittle stars, sea urchins, sand dollars, sea cucumbers, and sea lilies.

The adults are radially symmetric, but the larvae are bilateral. Echinoderms probably evolved from bilateral ancestors; the radial symmetry probably arose as an adaptation to a sessile way of life. Most of the modern echinoderms (with the exception of sea lilies) move about slowly, but ancient echinoderms were completely sessile.

Almost all members of this phylum possess an internal skeleton composed of numerous calcareous plates embedded in the body wall. These plates may be separate, or they may be fused to form a rigid boxlike structure. The skeleton often bears many bumps or spines, particularly noticeable in sea urchins, that project from the surface of the animal (see Fig. 24.72). It is this characteristic that gives the animals the name Echinodermata (from the Greek *echino*, "spiny," and *derma*, "skin").

Echinoderms have a well-developed coelom in which the various internal organs are suspended. The complete digestive system is the most prominent of the organ systems. There is no special excretory system, and the blood circulatory system, though present, is poorly developed. The nervous system is radially organized, consisting of nerve networks that connect to ringlike ganglionated nerve cords running around the body of the animal; there is no brain.

A characteristic unique in echinoderms is their ***water-vascular system***. This is a system of tubes (usually called canals) filled with watery fluid. Water can enter the system through a sievelike plate, called a madreporite, on the surface of the animal. A tube from this plate leads to a ring canal that encircles the esophagus. Five radial canals branch off the ring canal and run along symmetrically spaced grooves or bands on the surface of the animal. Many short side branches from the radial canals lead to hollow ***tube feet*** that project to the exterior. Each tube foot is a thin-walled hollow cylinder with a sucker on its end.

At the base of each tube foot is a muscular ampulla containing fluid. When the ampulla contracts, the fluid, prevented by a valve from flowing into the radial canal, is forced into the tube foot, which is thereby extended. After the foot attaches to the substratum by its sucker, longitudinal muscles in the foot wall contract, shortening it and pulling the animal forward (while forcing the water back into the ampulla). This cycle of events, repeated rapidly by the many tube feet of an animal like a sea star, enables it to move slowly. The tube feet also enable it to hold tightly to a rock or other object by applying suction, and to pull apart shells of clams and oysters, on which the sea star feeds.

Stelleroidea: the sea stars The body of a sea star (starfish) consists of a central ***disk*** and (usually) five rays, or ***arms***, each with a groove bearing rows of tube feet running along the middle of its lower surface. The outer surface of the animal is studded with many short spines and numerous tiny skin gills, which are thin fingerlike evaginations of the body wall that protrude to the outside between the plates of the endoskeleton (Fig. 24.69). The cavity of each skin gill is continuous with the general coelom. Scattered between the spines and skin gills are often numerous small jawlike structures called pedicellariae, which are used for protection and for capturing very small animals.

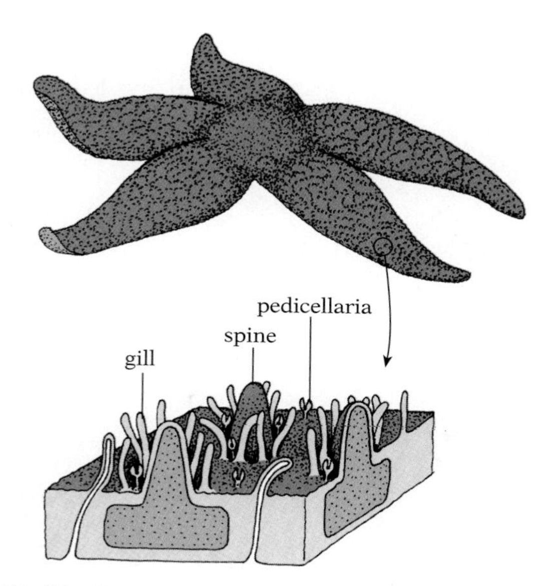

24.69 Sea star

The tiny skin gills are protected by the spines and by the pincerlike pedicellariae, which repel (or capture) small animals that might otherwise settle on the surface of the sea star.

The mouth is located in the center of the lower surface of the disk and the anus in the center of the upper surface. The digestive tract of a sea star is straight and very short, consisting of a short esophagus, a broad stomach that fills most of the interior of the disk, and a very short intestine. The stomach is divided by a constriction into two parts: a large eversible part (cardiac stomach) at the esophageal (lower) end, and a smaller, noneversible part (pyloric stomach) at the intestinal (upper) end. Attached to the noneversible part are five pairs of large digestive glands; each pair of glands lies in the coelomic cavity of one of the arms (Fig. 24.68).

When the sea star feeds, it pushes the lower part of the stomach out through the mouth, turning it inside out and placing it over food material such as the soft body of a clam or oyster. The stomach secretes digestive enzymes onto the food, and digestion begins. The partly digested food is then taken into the upper part of the stomach and into the digestive glands, where digestion is completed and the products are absorbed.

Sea stars have amazing regenerative abilities. Even a single detached arm can regenerate an entire new animal (Fig. 24.70).

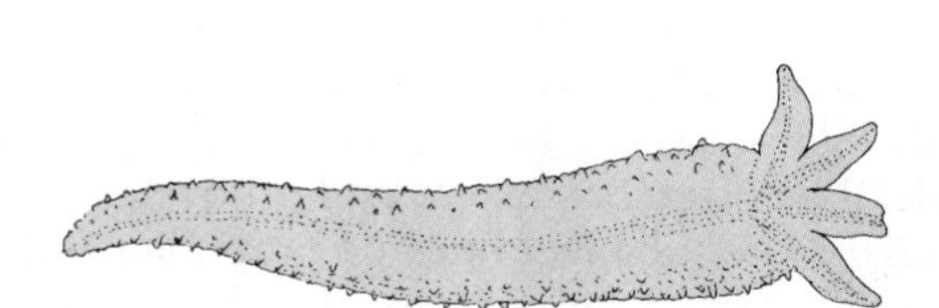

24.70 The detached arm of a sea star regenerating a new body

A

B

24.71 A brittle star (*Ophiopholis aculeata*) and a basket star (*Gorgonocephalus eucnemis*)

(A) The body disk of the brittle star is relatively small, and the arms are long and slender. (B) The arms of the basket star branch repeatedly to produce a mass of coils resembling tentacles.

In addition to sea stars, the class Stelleroidea includes brittle stars, serpent stars, and basket stars (Fig. 24.71). The organisms have five arms like sea stars, but the arms are longer, much more slender and flexible, often branched, and grooveless. The body disk is relatively small. The tube feet have no ampullae and are not used in locomotion. There is a large stomach, but no intestine and no anus. Gas exchange occurs in invaginated pouches in the periphery of the disk.

24.72 Giant red sea urchin (*Strongylocentrotis franciscanus*)
Sea urchins and their relatives have numerous large spines on their hard boxlike shells, or "tests."

Other echinoderms Although members of the three other echinoderm classes (described below) often show little superficial resemblance to sea stars and brittle stars, their structure is fundamentally similar.

1 *Echinoidea* These are the sea urchins, sand dollars, and heart urchins. They have no arms, but they do have five bands of tube feet. The body is spherical, or flattened and oval, and is covered with long spines (Fig. 24.72). The plates of the endoskeleton are fused to form a rigid case. There is a complex chewing apparatus just inside the mouth, and the intestine is long and coiled. Gas exchange occurs in small but highly branched gills or in modified tube feet.

2 *Holothuroidea* The sea cucumbers (Fig. 24.73) differ from the other echinoderms in having much-reduced endoskeletons and

24.73 A sea cucumber (*Pseudocolochirus*) in the Coral Sea
Notice the row of yellow tube feet. There are five of these rows, which correspond to the five arms of a sea star.

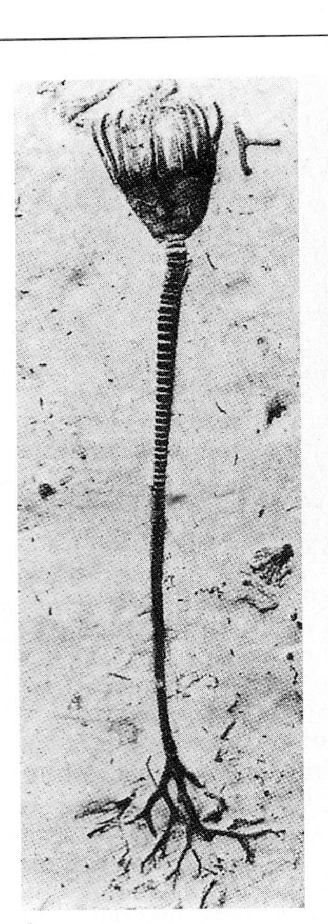

A

B

24.74 Sea lilies

(A) A fossil sea lily. (B) Diorama showing what the ocean floor might have looked like when sea lilies thrived.

leathery bodies. They lie on their sides rather than on the oral surface. The mouth is surrounded by tentacles attached to the water-vascular system. There is a very long coiled intestine, and gas exchange usually occurs in branched respiratory trees attached to the cloaca (a passage serving as both the anus and the site of reproduction; see Fig. 28.15, p. 805).

3 *Crinoidea* This is the most primitive of the living classes of echinoderms. The sea lilies, as most members of Crinoidea are called, are attached to the substratum by a long stalk and are thus sessile (Fig. 24.74). They have long feathery arms (often branched) around the mouth, which is on the upper side. Some modern crinoids—the feather stars—lack a stalk and are not sessile.

■ HEMICHORDATA AND THE PHARYNGEAL SLIT

The hemichordates, many of which belong to the class Enteropneusta (acorn worms; Fig. 24.75), are entirely marine. Often found living in U-shaped burrows in sand or mud along the coast, acorn worms are fairly large, ranging from 6 to 43 cm in length. Their bodies consist of an anterior conical proboscis

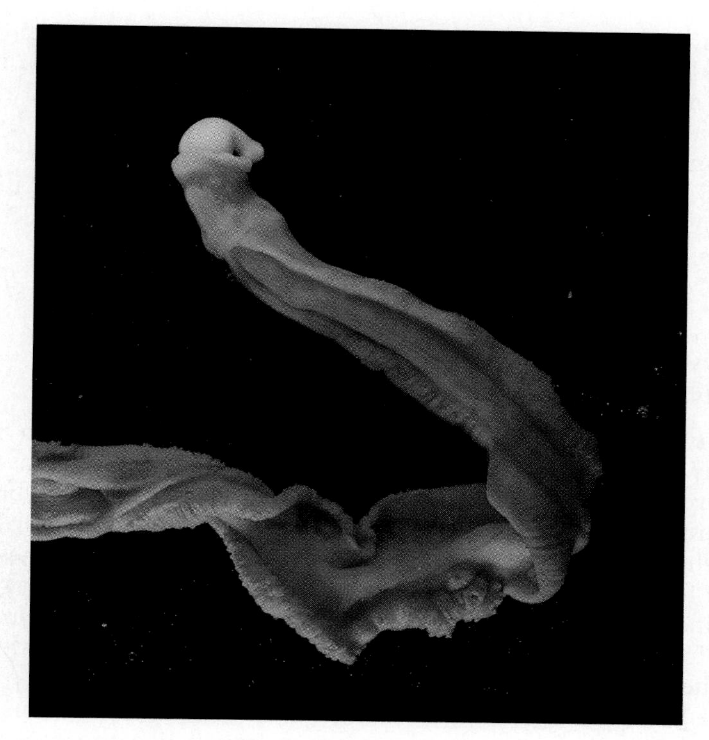

24.75 An acorn worm (*Balanoglossus*)
The short rounded proboscis is partially enveloped by the collar behind it. Only a part of the long trunk is shown.

(thought by some to resemble an acorn), a collar, and a long trunk (Fig. 24.76). The mouth is situated ventrally, at the junction between the proboscis and the collar.

A notable feature in the hemichordates is a series of ***pharyngeal slits*** in the wall of the pharynx. Water drawn into the mouth is forced back into the pharynx and out through these slits into sacs,

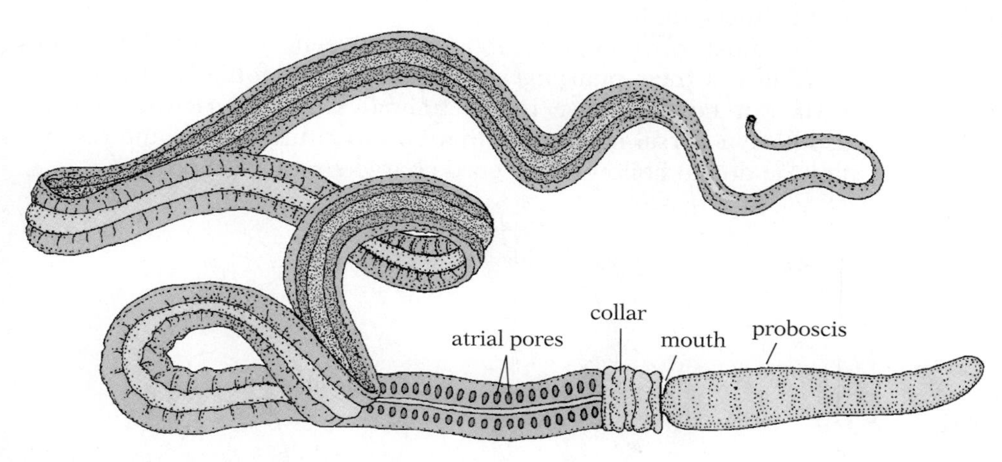

24.76 An adult acorn worm (*Saccoglossus*)
This particular genus has a more elongated proboscis than *Balanoglossus*, shown in Figure 24.75.

which open to the exterior via atrial pores. Oxygen is removed from the in-drawn water, and carbon dioxide is released into it, by blood in capillary beds in the septa between the slits. Food particles carried by the water into the pharynx are strained out by the slits and passed back into the esophagus.

Another important characteristic of hemichordates is a ciliated larval stage that strikingly resembles the larvae of some echinoderms.

■ HOW ECHINODERMS, HEMICHORDATES, AND CHORDATES ARE RELATED

It may seem strange that Echinodermata is the phylum most closely related to our own phylum, Chordata. After all, sea stars, sea urchins, and sea cucumbers don't look at all like vertebrates. Certain characteristics, however, link echinoderms, hemichordates, and chordates and set them apart from the protostomes. These include formation of the anus from the embryonic blastopore, radial and indeterminate cleavage, origin of the mesoderm as pouches, and formation of the coelom as cavities in the mesodermal pouches.

The hemichordates show clear affinities to the echinoderms. The ciliated larvae of hemichordates are so much like those of some echinoderms that they were mistaken for echinoderms when first discovered. This larval type, sometimes called a ***dipleurula*** (Fig. 24.77), is found only in the echinoderms and hemichordates. It has a band of cilia that forms a ring encircling the mouth. It thus differs from the trochophore larva found in many protostomes, which has a band of cilia encircling the body anterior to the mouth. The similar larvae of hemichordates and echinoderms, as well as the similarities in early embryology, indicate that these two groups must stem from a common ancestor. In view of the complicated transformation that in echinoderms produces a radial adult from a bilateral larva, it seems likely that echinoderms have deviated greatly from the ancestral type; hemichordates are probably closer to that ancestral form.

The most obvious resemblance between hemichordates and chordates is their pharyngeal slits, which are found in these two phyla but nowhere else in the animal kingdom. Hemichordates also have a dorsal nerve cord that is sometimes hollow and resembles the dorsal hollow nerve cord characteristic of chordates.

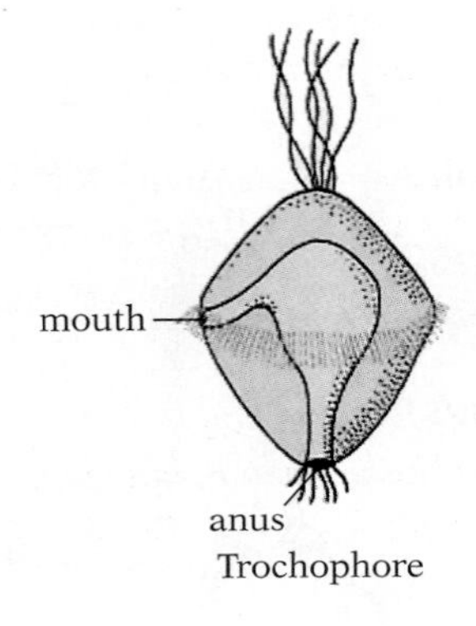

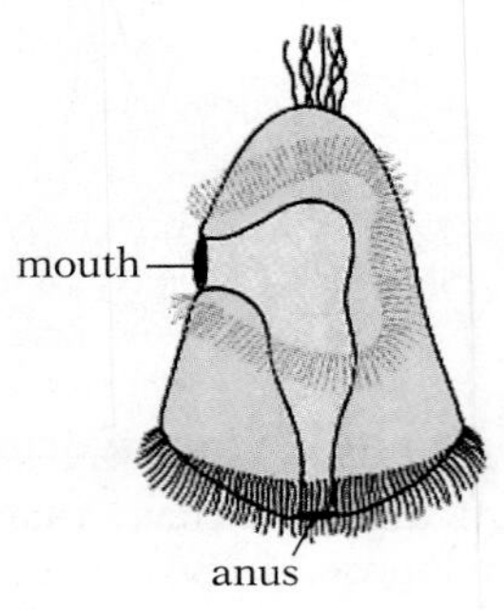

24.77 Trochophore and dipleurula larval types

The band of cilia (red) of the trochophore is located anterior to the mouth, whereas the corresponding band of the dipleurula encircles the mouth.

Table 24.1 A comparison of some of the major animal phyla

CHARACTERISTIC	PHYLUM								
	CNIDARIA	PLATYHELMINTHES	ASCHELMINTHIAN PHYLA	MOLLUSCA	ANNELIDA	ARTHROPODA	ECHINODERMATA	HEMICHORDATA	CHORDATA
Symmetry	Radial	Bilateral					Secondarily radial	Bilateral	
Cleavage	Determinate						Indeterminate		
Body cavity	None		Pseudocoelom	Coelom much reduced	Coelom	Hemocoel (coelom, degenerate)	Coelom		
Digestive tract	Gastrovascular cavity		Complete, with mouth from blastopore				Complete, with anus from blastopore		
Circulatory system	Absent			Open	Closed	Open	A special type; often poorly developed	Open	Closed (except in tunicates)
Ciliated larva	Planula	Trochophore-like in some	None or a unique type	Trochophore		None	Dipleurula		None
Segmentation	Absent	Absent or correlated with reproduction	Absent		Present		Absent		Present

Although echinoderms are chordates' closest relatives at the phylum level, chordates did not evolve from echinoderms; the two groups diverged from a common ancestor at some remote time.

Some of the important characteristics of the major animal phyla are compared in Table 24.1.

CHAPTER SUMMARY

PORIFERA: THE SPONGES

Sponges are the most primitive animals. They circulate water through the three-layer body sac; gases are exchanged directly, while collar cells capture and ingest food. (p. 655)

THE RADIATE PHYLA: A CIRCULAR BODY PLAN

CNIDARIA: THE COELENTERATES Radiates have no anterior-posterior axis. They have three-layer bodies surrounding a body sac, and show more division of labor than is seen in sponges. Cnidarians have some muscles and capture prey with nematocysts. Many species have a free-swimming larval medusa stage and a sedentary adult form. (p. 657)

CTENOPHORA: THE COMB JELLIES Comb jellies lack the nematocysts of cnidarians, have a different kind of muscle tissue, and live as free-swimming individuals. (p. 661)

HOW DID EUMETAZOANS EVOLVE?

Eumetazoans probably evolved from a plakula that formed a cavity over food. The ventral surface became specialized for digestion, while the dorsal surface provided protection. Eumetazoans used cavity formation to accomplish gastrulation. (p. 661)

ACOELOMATE BILATERIA: PRIMITIVE BILATERAL ANIMALS

Acoelomates have no coelom between the endoderm and ectoderm. (p. 663)

PLATYHELMINTHES: THE FLATWORMS Flatworms lack a complete digestive tract, but do have discrete excretory and reproductive organs. The group includes planarians, flukes, and tapeworms. (p. 663)

NEMERTEA: THE PROBOSCIS WORMS Proboscis worms have complete digestive and circulatory systems. (p. 667)

GNATHOSTOMULIDA: THE JAW WORMS These small marine worms have specialized, paired jaws. (p. 668)

HOW PROTOSTOMES AND DEUTEROSTOMES DIFFER

The blastopore created by gastrulation becomes the mouth in protostomes and the anus in deuterostomes; a complete digestive tract requires that a second opening be created in the embryo. The two groups differ in the geometry of early cleavages and the way in which the mesoderm arises and the coelom develops. (p. 668)

THE PSEUDOCOELOMATE PROTOSTOMES

Pseudocoeloms are internal cavities only partly bounded by mesoderm; the endoderm and/or ectoderm form part of the enclosure. (p. 670)

ACANTHOCEPHALA AND ENTOPROCTA Acanthocephalans are parasitic worms. The adults live in the digestive tracts of vertebrates; the larvae prey on invertebrates. Entoprocts are generally sessile marine creatures with the mouth and anus located side by side within a ring of cilia. (p. 670)

THE ASCHELMINTHIAN PHYLA The most important phyla in this group are Rotifera (rotifers) and Nemata (nematodes, or roundworms). Rotifers are tiny aquatic animals with a crown of cilia at the head that sweep food into the mouth. Roundworms are extremely abundant. They are characterized by a stiff cuticle and the absence of circular muscles—properties that severely limit movement. Several species are serious parasites of humans. (p. 671)

THE COELOMATE PROTOSTOMES

True coeloms are fully enclosed in mesoderm. (p. 673)

THE LOPHOPHORATE PHYLA Lophophores are named for the U-shaped fold that encircles the mouth and bears ciliated tentacles. (p. 673)

MOLLUSCA: THE MOLLUSCS The molluscs include snails, clams, and squids. They have a foot, a viceral mass with several specialized organs, and a protective mantle that usually produces a shell. (p. 674)

ANNELIDA: THE SEGMENTED WORMS The annelid worms have a segmented organization and elaborate organ systems. Each segment is a separate unit in a hydrostatic skeleton, and has its own set of muscles, neural ganglia, and fluid-balance organs. This group includes marine (polychaete), freshwater and terrestrial (oligochaete) worms, and the leeches. (p. 679)

ONYCHOPHORA: AN EVOLUTIONARY MILESTONE? The velvet worms are intermediate between annelids and arthropods; they may be descendants of the precursor of arthropods, or simply highly modified arthropods. (p. 680)

ARTHROPODA: THE LARGEST PHYLUM The majority of known species of all kingdoms taken together are arthropods. The success of this phylum, particuarly on land, is due in large part to its water-retaining jointed exoskeleton, which provides support and firm attachments for muscles; the phylum's strong hinged jaws are also critical. The group includes spiders, crabs, centipedes, millipedes, and the largest of all living classes, the insects. (p. 681)

THE DEUTEROSTOMES

ECHINODERMATA: SEA STARS AND THEIR RELATIVES Echinoderms are bilaterally symmetric as larvae but radially symmetric as adults. They retain a complete digestive tract. Included are sea stars and sea urchins. (p. 691)

HEMICHORDATA AND THE PHARYNGEAL SLIT Hemichordates, like the chordates, have a series of pharyngeal slits that serve both in gas exchange and filter feeding. (p. 696)

HOW ECHINODERMS, HEMICHORDATES, AND CHORDATES ARE RELATED In addition to the many features shared by all deuterostomes, the larvae of hemichordates greatly resemble the larvae of echinoderms, while the pharyngeal slits and (in some species) hollow dorsal nerve cord of hemichordates resemble those of chordates. (p. 698)

STUDY QUESTIONS

1 Why are so many aquatic invertebrates larger than the largest terrestrial invertebrates?

2 Are there any obvious (or possible) reasons insects have not been successful in marine environments? (pp. 688–691; you may also want to look at the discussions of insect and invertebrate physiology on pp. 812–814, 880, 884–885, and 891–892.)

3 The life histories of two-, three-, and four-host parasites seem unnecessarily complex. What purpose for the intermediate hosts seems most likely—that they are needed for nutritional reasons, as launching sites from which the parasite can reencounter the main host, or for some other reason that you can suggest? (pp. 664–667)

4 One of the major advances in the evolution of animals was the development of a complete digestive system—having a separate entrance and exit. Why do you suppose that lungs—the gas-exchange organs of terrestrial animals—remain essentially blastulalike, with new air and waste air sharing the same orifice? (pp. 668–669)

SUGGESTED READING

BUCHSBAUM, R., 1987. *Animals without Backbones,* 3rd ed. University of Chicago Press, Chicago. *One of the most readable and fascinating discussions of the invertebrates ever written.**

EVANS, H. E., 1984. *Insect Biology.* Addison-Wesley, Reading, Mass. *Probably the best entomology textbook available.*

MCMENAMIN, M. A. S., 1987. The emergence of animals, *Scientific American* 256 (4). *On the first invertebrates.*

*Available in paperback.

CHAPTER 25

CHORDATE ANIMALS

Much of our species' morphology and physiology is a heritage from our earlier vertebrate ancestors. This chapter will trace the evolution of vertebrates in general, and humans in particular. We are part of the phylum Chordata, which includes the vertebrates—animals with backbones. However, not all chordates are vertebrates: two of the three chordate subphyla—Tunicata and Cephalochordata—include only invertebrates. They provide important hints about the origin of the third subphylum, Vertebrata (the vertebrates).

All members of Chordata share three important characteristics:

1 All have, at least during embryonic development, a ***notochord***. This is a flexible supportive rod running longitudinally through the dorsum of the animal just ventral to the nerve cord.

2 All have pharyngeal slits (or pouches) at some stage in their development. (These slits are often called "gill slits," which may be misleading since they probably originated as feeding devices. They have become modified for hearing, reproduction, or gas exchange—in which case they are properly called gills—over the course of evolution.)

3 All have a hollow nerve cord just dorsal to the notochord.

The tunicates and cephalochordates have no backbones, and so are invertebrates.

THE INVERTEBRATE CHORDATES

■ TUNICATA: THE TUNICATES

In the best-known class of tunicates (sometimes called sea squirts), the adults are sessile marine animals that little resemble other chordates except in having pharyngeal slits.[1] Water taken in through the mouth (also called the incurrent siphon) goes into a large pharynx, and then filters through the pharyngeal slits into a chamber called the atrium, from which it passes to the exterior

[1] Members of two smaller classes of tunicates are free-swimming planktonic organisms.

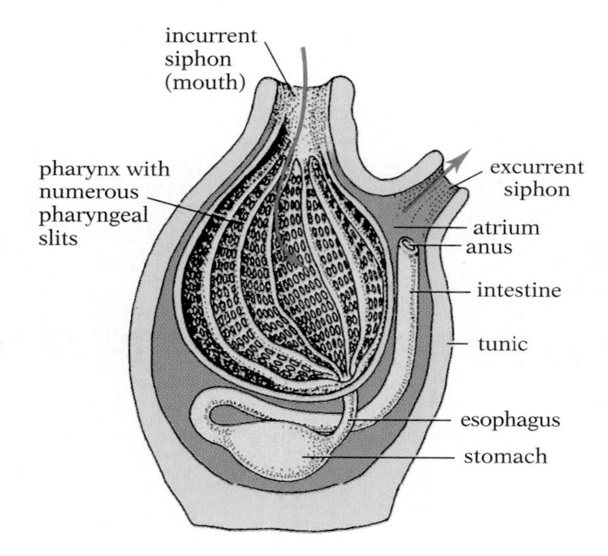

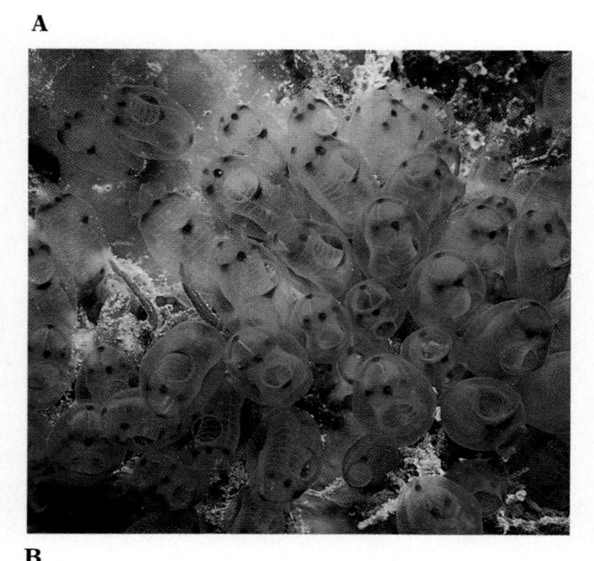

25.1 Adult tunicate

(A) The arrows show the path of respiratory water, which is drawn into the pharynx through the incurrent siphon, passes through the pharyngeal slits into the atrium, and then exits through the excurrent siphon. Oxygen is absorbed from the water across the walls of the pharyngeal slits, which thus serve as gills. Food particles drawn into the pharynx with the water do not pass through the slits, but instead are carried through the pharynx into the esophagus. (B) A colony of sea squirts (*Podoclavella moluccensis*) from the Great Barrier Reef in Australia.

through the excurrent siphon (Fig. 25.1). The pharyngeal slits function in both gas exchange and feeding, acting as a strainer for removing small food particles from the water flowing through them. The food particles become caught in a layer of mucus in a ciliated groove of the pharynx, called the endostyle, and are carried by the mucus into the esophagus, which leads to the stomach.

Larval tunicates, which are motile, show much more resemblance to other chordates. With their elongate bilaterally symmetrical bodies and long tails, they look like tadpoles. They possess a well-developed dorsal hollow nerve cord and a notochord beneath it in the tail region (Fig. 25.2). When the larvae settle down and undergo metamorphosis to the adult form, the notochord and most of the nerve cord are lost.

Some biologists hold that the tunicates and higher chordates descended from a common ancestor that was free-swimming and resembled a modern tunicate larva. If so, the sessile structure of modern adult tunicates is a later specialization. An alternative hypothesis is that the common ancestor was sessile, and that vertebrates evolved from its motile larva.

■ CEPHALOCHORDATA: THE LANCELETS

There are about 30 species of these small marine animals. Though capable of swimming, they spend most of their time partially buried tail down in sand in shallow water. They are filter feeders, taking in water through the mouth and straining it in the pharynx in the same manner as the tunicates. The water passes through pharyngeal slits into a large chamber, the atrium, and thence to the exterior through an atrial pore. Oxygen is removed from the water as it passes through the slits; additional oxygen is obtained through gas exchange across the rest of the body surface.

The genus of lancelets most commonly studied is *Branchi-*

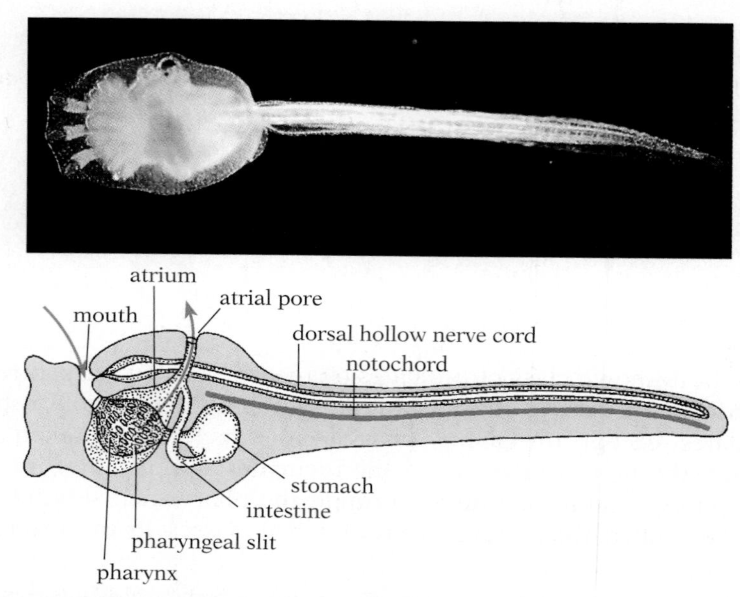

25.2 A larval tunicate

The arrows indicate the path of inflowing and outflowing water.

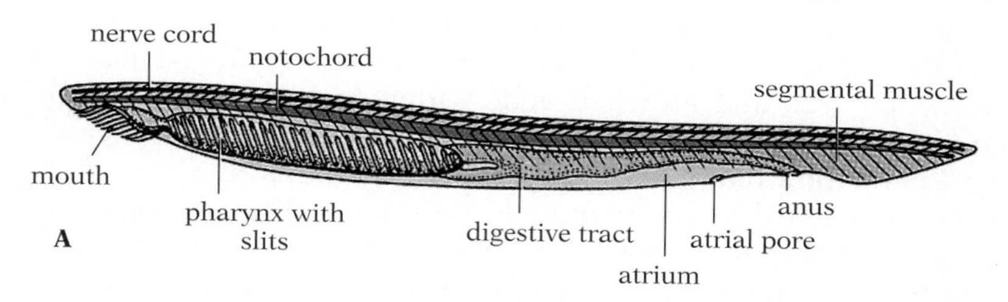

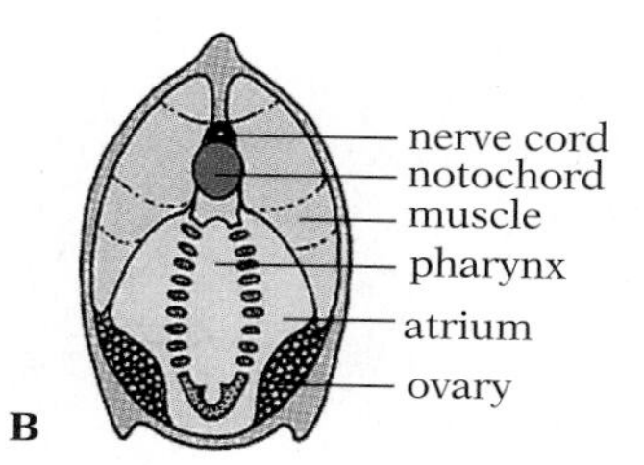

25.3 An adult lancelet (amphioxus)

(A) Longitudinal view. (B) Cross section, showing the relationship between the pharynx, the pharyngeal slits, and the atrium.

ostoma, usually called amphioxus. The body is typically about 5 cm long, translucent, and shaped like a fish (Fig. 25.3). Both the dorsal hollow nerve cord and the notochord are well developed and are retained through life. A feature not seen in tunicates but characteristic of both cephalochordates and vertebrates is segmentation. In lancelets this segmentation is most noticeable in the muscles, which are arranged in a line of V-shaped bundles.

THE TISSUES OF VERTEBRATES

The animals in this subphylum have an internal skeleton that includes a backbone composed of a series of vertebrae. The vertebrae develop around the notochord, which in most vertebrates is present in the embryo only. As in amphioxus, some parts of the vertebrate body are segmented longitudinally—in particular, the vertebrae and the main body musculature.

Vertebrates, like the arthropods, include many species that are fully terrestrial. In both cases, elaborate ***tissues*** (groups of any of several specialized types of cells) and ***organs*** (organized assemblies of two or more kinds of tissues) have developed to provide the necessary support and mobility, and to transport nutrients, gases, and wastes. Animal tissues are divided into four categories: epithelium, connective tissue, muscle, and nerve. Each of these contains numerous subtypes. The subtype classification used here is based primarily on mammals; attempts to apply it to other animals, particularly the invertebrates, are often not very useful. This classification of animal tissues is summarized in Table 25.1.

Table 25.1 Classification of animal tissues

MAJOR TYPE	SUBTYPE
Epithelial	Simple epithelium
	Squamous
	Cuboidal
	Columnar
	Stratified epithelium
	Stratified squamous
	Stratified cuboidal
	Stratified columnar
Connective	Vascular tissue
	Blood
	Lymph
	Connective tissue proper
	Loose connective tissue
	Dense connective tissue
	Cartilage
	Bone
Muscle	Skeletal muscle
	Smooth muscle
	Cardiac muscle
Nerve	

■ COVERING BOTH INSIDE AND OUT: EPITHELIUM

Epithelial tissue forms the covering or lining of all free body surfaces, both external and internal. The outer portion of the skin, for example, is epithelium, as are the linings of the digestive tract, the lungs, the blood vessels, the various ducts, the body cavity, and so on. Epithelial cells are packed tightly together, with only a small amount of cementing material between them and almost no intercellular spaces. Thus they provide a continuous barrier protecting the underlying cells from the external medium. Because anything entering or leaving the body must cross at least one layer of epithelium, the permeability of the cells of the various epithelia are exceedingly important in regulating the exchange of materials between the body and the external environment.

Since one surface of an epithelium is generally exposed to air or fluid and the opposite surface rests on other cell layers, it is no surprise that epithelial cells should show significant differences between their free and attached ends. Often highly specialized, the free ends commonly bear cilia, hairs, or short fingerlike processes (Fig. 25.4); they may also have deep depressions and are sometimes covered with waxy or mucous secretions. The portion of the plasma membrane on the outer surface of the cell, which is exposed to the extracellular environment, is quite different from the portions of the membrane adjacent to other epithelial cells (Fig. 25.4).

Epithelial cells are usually divided into three categories: squamous, cuboidal, and columnar (Figs. 25.5, 25.6). ***Squamous cells*** are much broader than they are thick and have the appearance of thin flat plates. ***Cuboidal cells*** are roughly as thick as they are wide and have a rather square shape when viewed in a section perpendicular to the tissue surface; in surface view, however, they look like polygons, often with six sides. ***Columnar cells*** are much longer than they are wide and, in vertical section, look like rectangles set on end.

Epithelial tissue, if only one cell thick, is called ***simple epithelium***, and if two or more cells thick is known as ***stratified epithelium***. In a third category, called ***pseudostratified epithelium***, the tissue looks stratified but actually is not. In true stratified epithelium only the cells in the lowest layer are in contact with the underlying membrane, whereas in pseudostratified epithelium all the cells are in contact with it. The various types of epithelia are named on the basis of cell type and degree of stratification. Epithelial tissue may be simple squamous, simple cuboidal, simple columnar, stratified squamous, stratified cuboidal, stratified columnar, and so on. (In stratified epithelia the cells of the outermost layer determine the name.) Epithelium is usually separated from the underlying tissue by an extracellular ***basement membrane*** containing collagen fibers (Fig. 25.5).

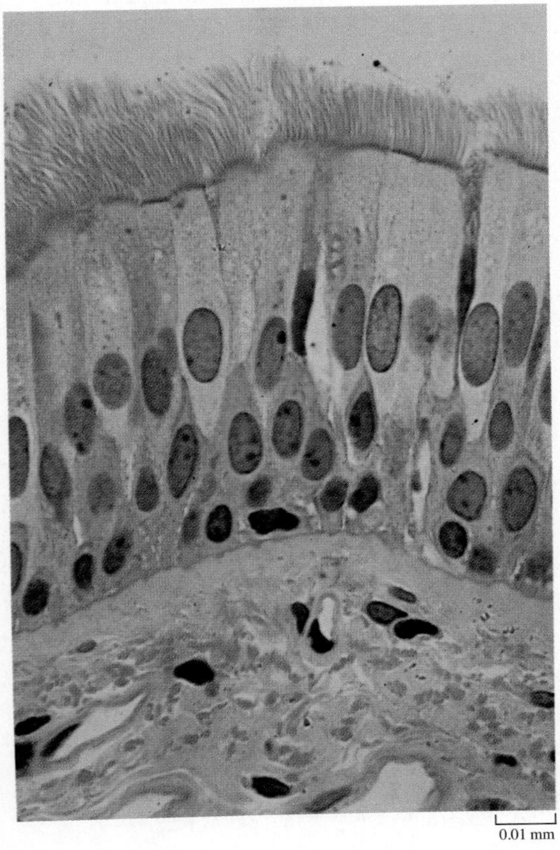

0.01 mm

25.4 Pseudostratified columnar epithelium from human trachea

The basal bodies of the cilia (dark band where the cilia protrude from the cells) are plainly visible. Note the basement membrane on which the epithelial cells rest.

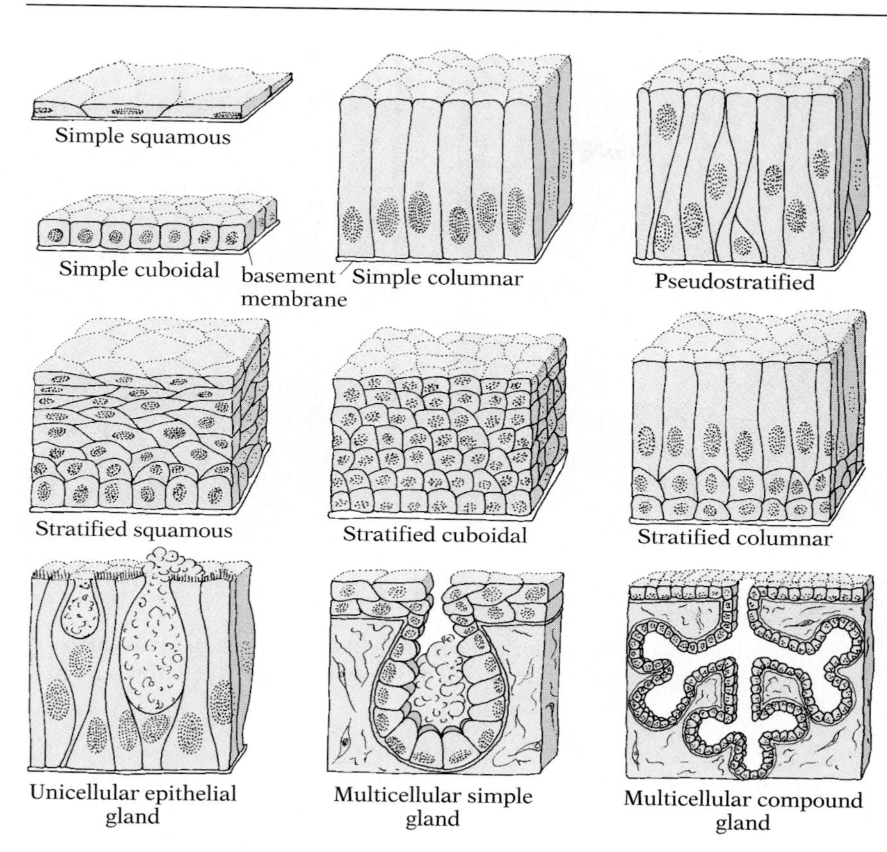

25.5 Varieties of epithelial tissue

Epithelial cells often become specialized as ***gland*** cells (cells specialized for producing secretions). The substances secreted by epithelial gland cells include sweat, body oil, and mucus (Figs. 25.6, 25.7). Sometimes a portion of the epithelial tissue becomes invaginated, and a multicellular gland is formed.

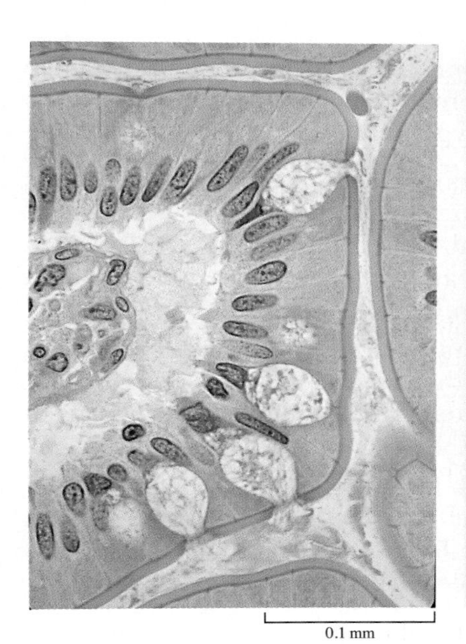

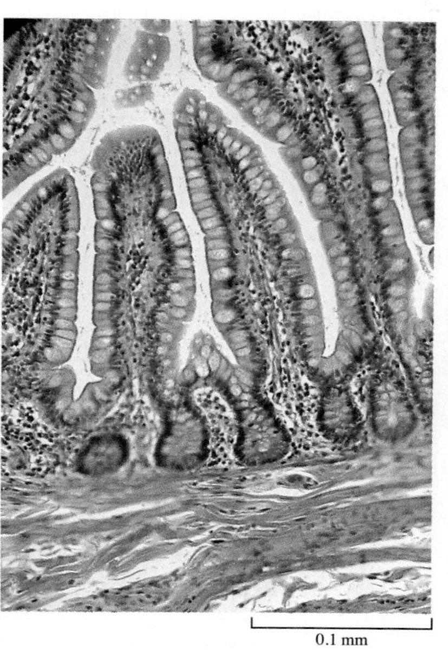

25.7 Gland tissues

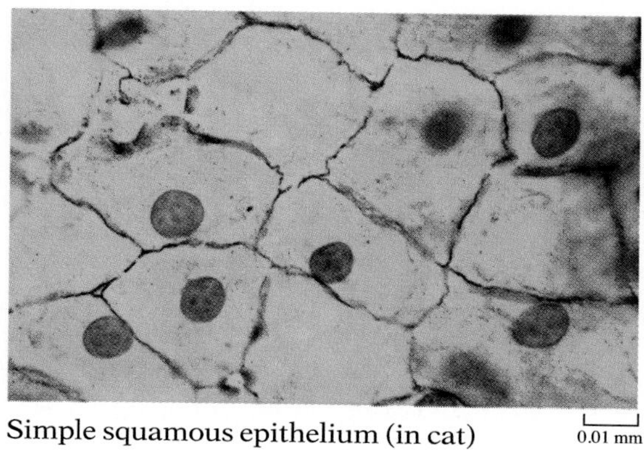

Simple squamous epithelium (in cat)

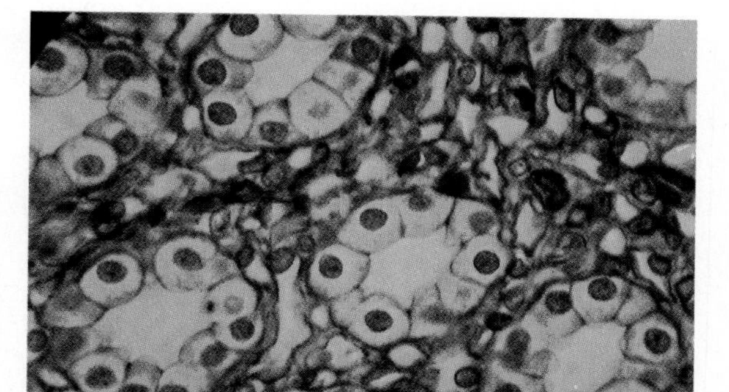

Simple cuboidal (in tubules of cat kidney)

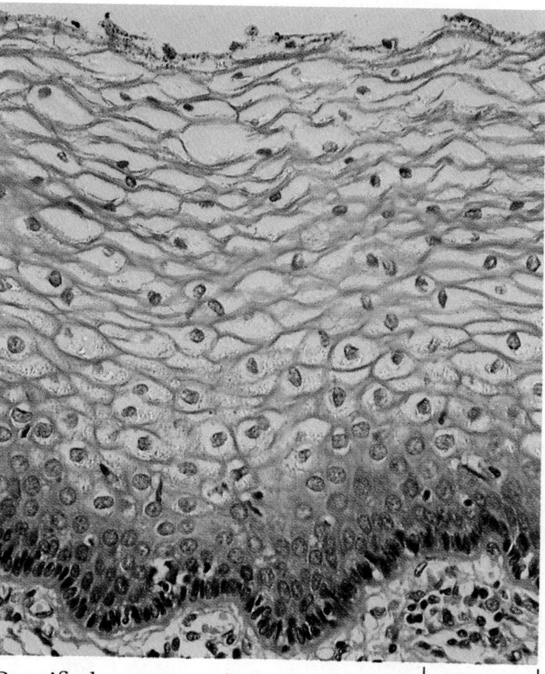

Stratified squamous (in human vagina)

25.6 Epithelial tissues

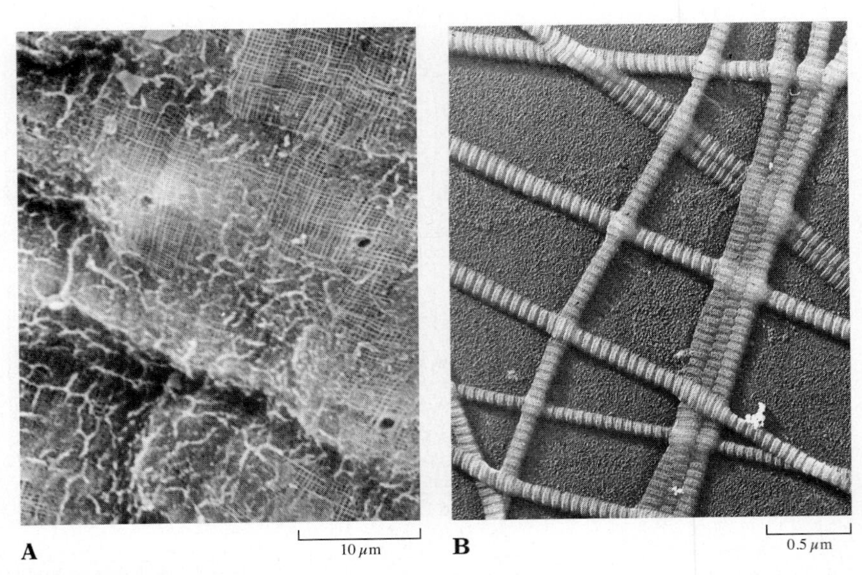

25.8 Collagen

(A) Scanning EM of the network of collagen fibers in the skin of an earthworm. (B) Higher-magnification transmission EM of collagen fibrils from calf skin.

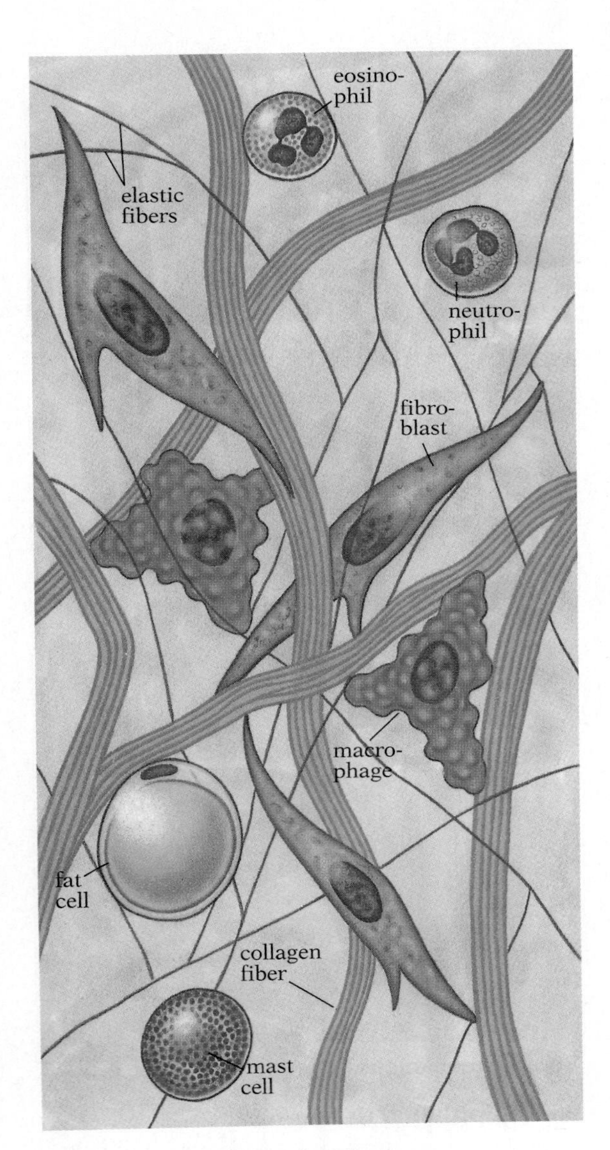

25.9 Loose connective tissue

The several varieties of cells are embedded in an extensive intercellular matrix of fibers and ground substance.

■ CONNECTIVE TISSUE: CELLS IN A MATRIX

In connective tissue the cells are always embedded in an extensive intercellular matrix. Much of the total volume of connective tissue is matrix, the cells themselves being often widely separated. The matrix may be liquid, semisolid, or solid. Connective tissue is often divided into four main types: (1) blood and lymph, or vascular tissue; (2) connective tissue proper; (3) cartilage; and (4) bone. The last three are called supporting tissues.

Blood and lymph Blood and lymph are rather atypical connective tissues with liquid matrices. They are discussed in Chapter 30.

Connective tissue proper Connective tissue proper is variable in structure, but its intercellular matrix always contains numerous fibers. These fibers are of three types:

Collagen fibers (or white fibers) are composed of numerous fine fibrils of collagen, a protein that constitutes a high percentage of the total protein in the animal body (Fig. 25.8). Such fibers are flexible, but resist stretching and confer considerable strength on the tissues containing them.

Elastic fibers (or yellow fibers) can easily be stretched. When the stretching force ceases, the fibers return to their former length. Elastic fibers are often much thinner than collagen fibers. They are composed of the protein elastin.

Reticular fibers branch and interlace to form complex networks. They are important at points where connective tissues and other tissues join, particularly in the basement membrane between epithelium and connective tissue.

Several kinds of cells are generally found in connective tissue proper (Fig. 25.9). They perform a variety of functions:

1 ***Fibroblasts*** secrete the proteins from which fibers form.

2 ***Macrophages,*** irregularly shaped cells particularly common near blood vessels, become mobilized when there is an inflammation. As described in Chapter 15, they move by amoeboid motion and actively engulf particles such as dead red blood corpuscles and bacteria.

3 ***Mast cells*** produce a substance (heparin) that tends to prevent blood clotting and another substance (histamine) that increases the permeability of blood capillaries.

4 ***Fat cells*** are cells highly specialized for fat storage. When they are very numerous in a region of connective tissue, the tissue is often called adipose tissue.

5 The various kinds of ***white blood cells*** (such as eosinophils and neutrophils) help fight infection. Some can move between the blood or lymph and connective tissue proper.

Both cells and fibers are embedded in an amorphous ***ground substance***, which is a mixture of water, proteins, carbohydrates, and lipids. Associated with the ground substance is the ***tissue fluid***, a liquid derived from the blood.

Connective tissue proper is customarily subdivided into two basic types—loose connective tissue and dense connective tissue.

1 ***Loose connective tissue*** is characterized by an open, irregular arrangement of its fibers, a large amount of ground substance, and the presence of numerous cells of a variety of types (Fig. 25.9). It is widely distributed in the animal body. Much of the framework of the lymph glands, bone marrow, and liver is loose connective tissue. Loose connective tissue also supports, surrounds, and connects the elements of all other tissues. For example, it binds muscle fibers together; attaches the skin to underlying tissues; forms the membranes that line the heart and abdominal cavities; forms the membranes, called mesenteries, that suspend the internal organs in their proper position and the membranes that bind together the parts of internal organs or that bind various organs together; functions as a packing material in the spaces between organs; and forms a thin sheath around blood vessels, consequently penetrating with them into the interior of most organs. The flexibility of loose connective tissue allows movement between the units it binds.

2 ***Dense connective tissue*** is characterized by the compact arrangement of its many fibers, the limited amount of ground substance, and the relatively small number of cells. The fibers may be irregularly arranged in an interlacing network, as in the dermis of the skin or the sheaths of bone; or they may be arranged in a definite pattern—usually parallel bundles oriented to withstand tension from one direction, as in tendons connecting muscle to bone (Fig. 25.10, A) or ligaments connecting bone to bone.

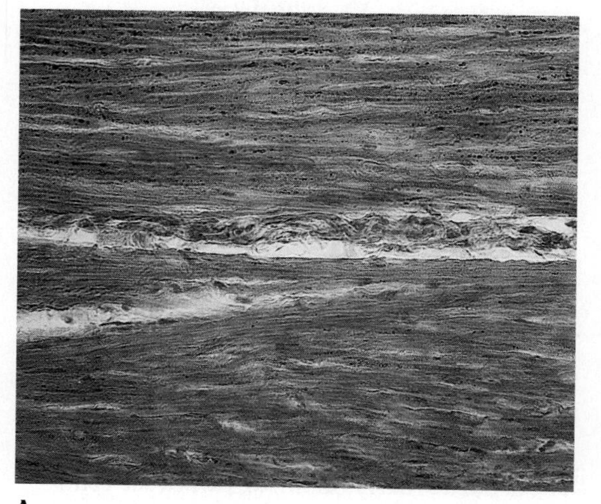

A

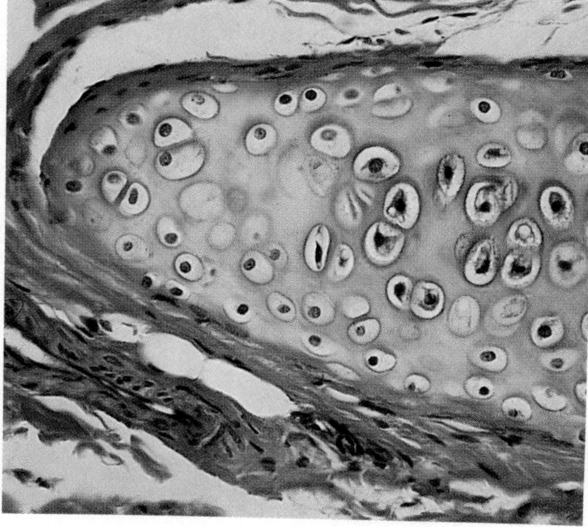

B

25.10 Tendon and cartilage

(A) Longitudinal section of a human tendon, surrounded by supporting tissue. (B) Section of human hyaline cartilage, embedded in the wall of the trachea.

Cartilage Cartilage (gristle) is a specialized form of dense fibrous connective tissue in which the intercellular matrix has a rubbery consistency (Fig. 25.10, B). The relatively few cells are located in cavities in the matrix. Cartilage can support great weight, yet it is often flexible. It can vary in texture, color, and elasticity.

Cartilage is found in the human body in such places as the nose and ears (where it forms pliable supports), the larynx ("voicebox") and trachea (you can feel the rings of cartilage in the front of your throat), intervertebral disks, surfaces of skeletal joints, and ends of ribs. Most of the skeleton of the early vertebrate embryo is composed of cartilage; the developing bones follow this model and slowly replace it. Some vertebrate groups—the sharks, for example—retain a cartilaginous skeleton even as adults.

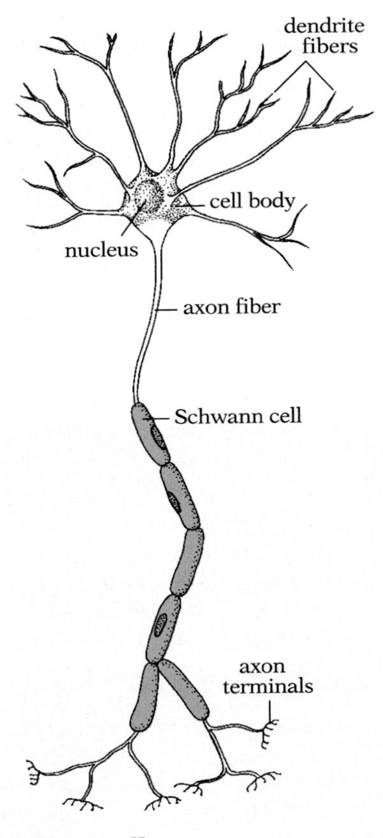

25.11 Nerve cell
The dendrites carry impulses toward the cell body; the axon carries impulses away from the cell body. The axons of vertebrates are often sheathed along much of their length by Schwann cells, which speed signal conduction.

Bone Bone has a hard, relatively rigid matrix that contains numerous collagen fibers and a surprising amount of water, and is impregnated with inorganic salts such as calcium carbonate and calcium phosphate. This inorganic material may constitute as much as 65% of the dry weight of an adult bone. The few bone cells are widely separated and are located in spaces in the matrix (see Fig. 37.7, p. 1072). The histology (tissue structure and anatomy) of bone is detailed in Chapter 37.

■ MUSCLE

Muscles are responsible for most movement in higher animals. The individual muscle cells are usually elongate and are bound together into sheets or bundles by connective tissue. Vertebrates have three principal types of muscle tissue (see Fig. 37.9, p. 1074): (1) skeletal or striated muscle, which is responsible for most voluntary movement; (2) smooth muscle, which is involved in most involuntary movements of internal organs; and (3) cardiac muscle, the tissue of which the heart is composed.

■ NERVE

Nerve cells are easily stimulated and can transmit impulses very rapidly. Each cell is composed of a cell body, containing the nucleus, and one or more long thin extensions up to 1m in length (Fig. 25.11). Nerve cells can conduct messages over long distances. Nerve fibers bound together by connective tissue constitute a nerve.

The functional combination of nerve and muscle tissue is important for all multicellular animals except sponges. The functioning of nerves and muscles is detailed in Chapters 35–37.

VERTEBRATE ORGANS

The complex integration of different types of cells and tissues to form an animal organ is illustrated by human skin (Fig. 25.12). Skin is remarkably complex. It contains elements of all four primary animal tissue types: epithelium, connective tissue, muscle,

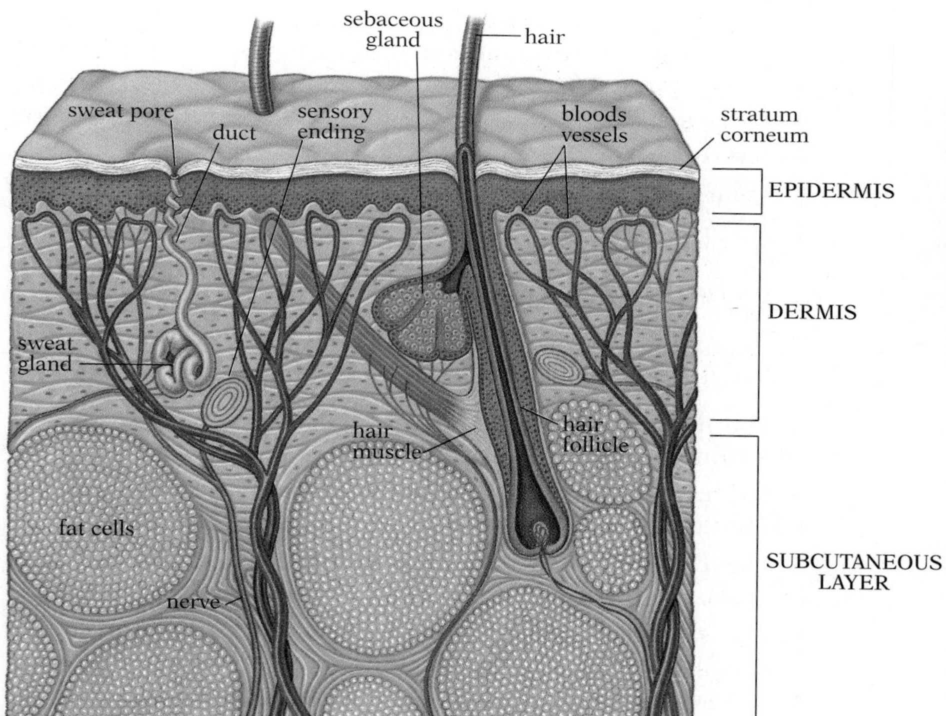

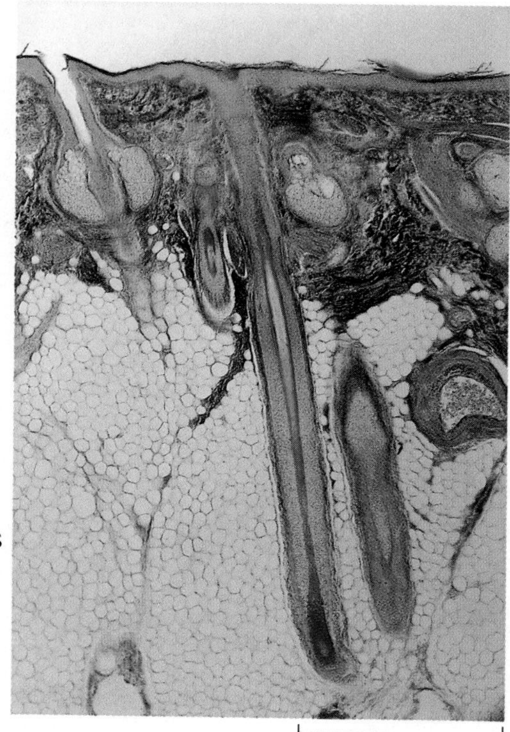

25.12 Human skin in cross section

The skin shown in the photograph is from the human scalp. The outer portion of the skin, the epidermis, is composed of stratified squamous epithelial tissue. (Individual epithelial cells are not visible.) The outermost layer (stratum corneum) consists of hardened dead cells that are constantly being sloughed off. Active cell division in the deeper layers of the epidermis produces new cells that are pushed outward and take the place of those that are shed. Beneath the epidermis is the dermis, which is composed chiefly of connective tissue (stained blue in the photograph). Blood vessels penetrate into the dermis but not into the epidermis. Sweat glands are embedded in the deeper layers of the dermis, and their ducts push outward through both dermis and epidermis to open onto the surface of the skin through sweat pores. Both the glands and their ducts are derived from the epidermis; they form initially as invaginations of the epidermis that push downward into the connective tissue of the dermis.

Hairs, and the inner layers of the hair follicles in which they are encased, are also derived from the epidermis and also develop asinvaginations downward. When the hair follicle is fully developed, the bulbous base of the follicle and the hair root lie in the subcutaneous layer; the shaft of the hair extends at a slant from the root to the surface of the skin and beyond. A small muscle runs diagonally from the upper portion of the dermis to the hair follicle near its lower end; when this muscle contracts, it pulls the hair erect. One or more sebaceous (oil) glands empty into the hair follicle.

Numerous nerves penetrate into the dermis, and a few even penetrate into the epidermis. Among them are nerves to the hair muscles, sweat glands, and blood vessels, and also nerves terminating in the sensory structures for detecting touch, temperature, and pain.

Beneath the dermis, and not sharply delimited from it, is the subcutaneous layer, which is not considered a part of the skin itself. This is a layer of very loose connective tissue, usually with abundant fat cells. It is this layer that binds the skin to the body. The extent and form of its development determine the amount of possible skin movement.

and nerve. Portions of these tissues, in turn, are organized into relatively complex structures such as glands, ducts, hairs, blood vessels, and sensory devices. All these structural elements are integrated to form the functional organ. This organ functions in many ways:

1 As a protective covering for the body that resists penetration by many harmful substances and disease-producing organisms, and helps prevent excessive water loss

2 As a mechanism for the excretion of several different waste materials

3 As an aid in regulating the temperature of the body

4 As an instrument whose sensory nerve endings receive impulses from the outside environment

5 As a depot in which reserve nutrients are stored

The numerous organs are commonly grouped into organ systems, of which the following are discussed in Part V:

1 The ***digestive system***, which procures and processes nutrients from food.

2 The ***respiratory system***, which exchanges gases; oxygen is taken into the body and waste carbon dioxide released.

3 The ***circulatory system***, the internal transport system of animals.

4 The ***excretory system***, which not only disposes of certain metabolic wastes, but regulates the chemistry of the body fluids.

5 The ***endocrine system***, whose glands and the hormones they produce play an important role in internal control.

6 The ***nervous system***, which helps coordinate the myriad functions of a complex multicellular animal.

7 The ***skeletal system***, which provides support and determines shape in most animals.

8 The ***muscular system***, which mediates movement.

9 The ***reproductive system***, which helps produce new individuals.

HOW VERTEBRATES EVOLVED

The earliest vertebrates were fish; mammals and birds are the most recently evolved groups.

■ AGNATHA: THE JAWLESS FISH

The oldest complete vertebrate fossils are from the Ordovician period, which began some 500 million years ago. These fossils are of bizarre fishlike animals covered by thick bony plates. None of them had true paired fins. They lacked an important character found in essentially all later vertebrates—jaws. These ancient fish constitute the class Agnatha (which means "jawless"). Most were filter feeders, straining food material from mud and water flowing through their gill systems in more or less the same manner as tunicates and amphioxus.

The class Agnatha continued as an important group through the Silurian period. By the end of that period the class had begun to decline, and its members disappeared from the fossil record by the end of the Devonian (Fig. 25.13).

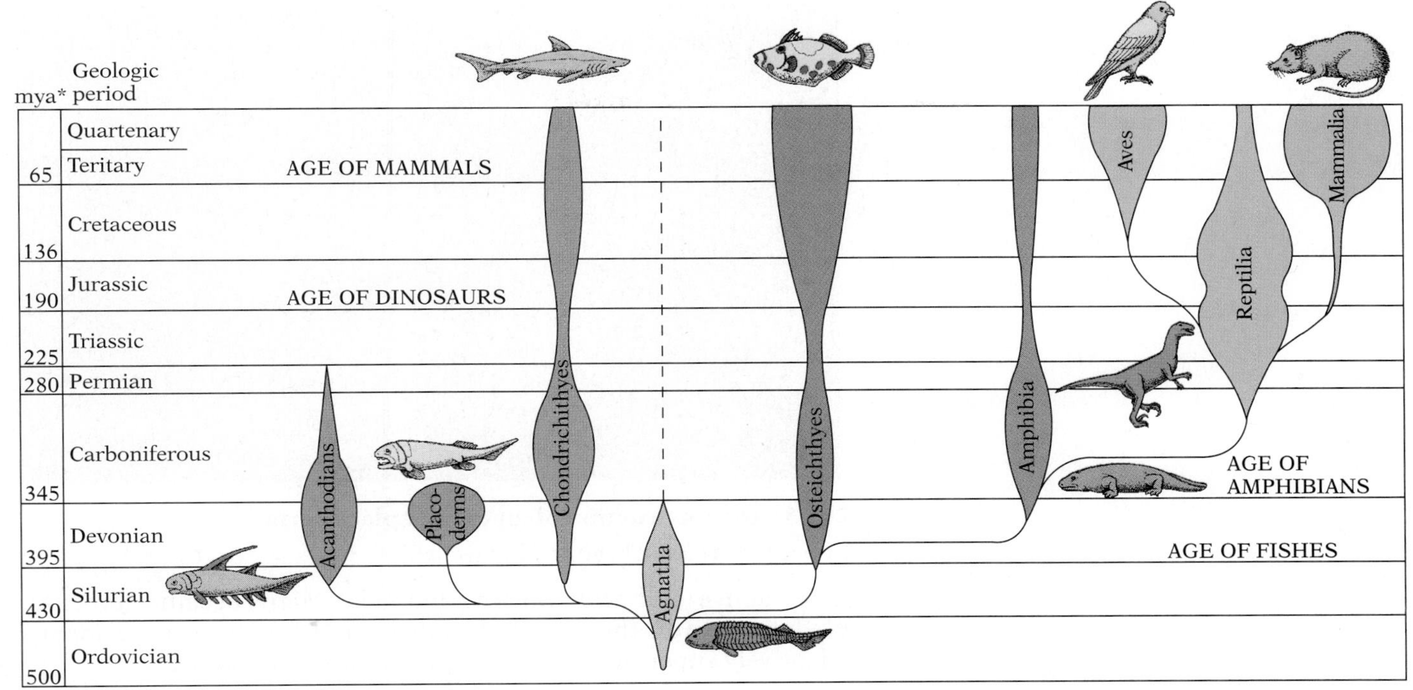

25.13 Evolution of the vertebrate classes

Approximate dates of period boundaries (in millions of years) are shown in parentheses.

The two kinds of jawless vertebrates living today, the lampreys (Fig. 25.14) and the hagfish, are generally classified as Agnatha, although they diverged from a common, cartilaginous ancestor;[2] they are also different from their armored Paleozoic relatives in that they have a soft body without either armor or scales, and their jawless mouth is modified as a round sucker that is lined with many horny teeth and accommodates a rasping tongue.[3] They feed by attaching themselves by their sucker to other fish, rasping a hole in the skin of the prey, and sucking blood and other body fluids. Lampreys have a larval filter-feeding stage that resembles amphioxus.

■ FISH WITH JAWS

The decline of the ancient Agnatha coincided with the rise of three classes of fish: the Placodermi (an armored fish), the Chondrichthyes (cartilaginous fish), and the Osteichthyes (bony fish). All evolved from a common ancestor that had hinged jaws. The acquisition of jaws was one of the most important events in the history of vertebrates. It made possible a revolution in the method of feeding and hence in the entire mode of life of early fish. They became

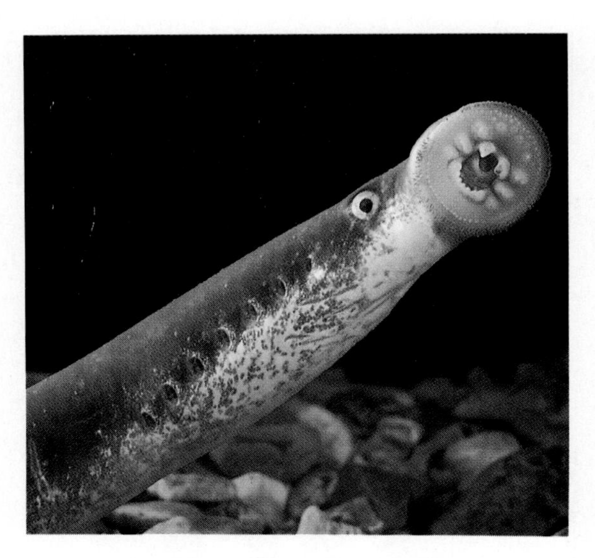

25.14 Head and pharyngeal region of a lamprey

The animal has a large oral sucker instead of jaws, and seven prominent external gill openings.

[2] Lampreys and hagfish are so different from the Paleozoic members of the Agnatha that some biologists erect a separate class for them.

[3] Horny materials are reinforced with protein, while bony materials are reinforced with minerals—usually calcium phosphate ($CaPO_4$).

25.15 Reconstruction of an extinct placoderm

Notice the armored head and hinged jaws.

more active and wide-ranging animals. Many became ferocious predators. Even those that remained mud feeders were evidently adaptively superior to the ecologically similar agnaths.

Placodermi: Armored Fish With Jaws The placoderms were an important group of armored fish during the Devonian; most became extinct by the end of that period (Fig. 25.15).

The hinged jaws of the placoderms probably developed from a set of gill support bars (Fig. 25.16). Although they are functionally analogous, hinged jaws arose independently in the arthropods and the vertebrates—in the one case from ancestral legs and in the other from skeletal elements in the wall of the pharyngeal region.

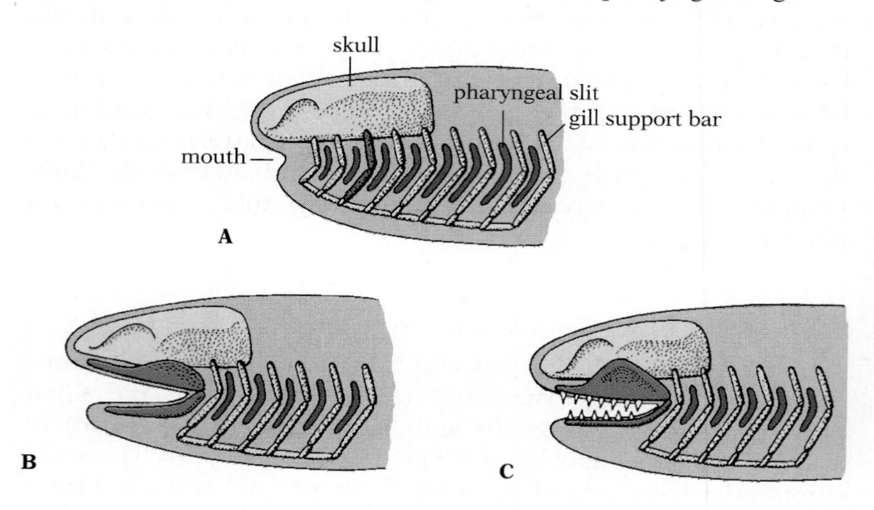

25.16 Evolution of the hinged jaws of vertebrates

(A) The earliest vertebrates had no jaws. The structures (dark brown) that in their descendants would become jaws were gill support bars. (B) A pair of gill support bars were modified into weak jaws, and the two most anterior support bars (shown in A) were lost. (C) The jaws became larger and stronger.

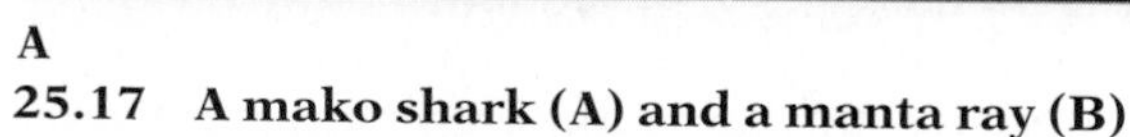

A

B

25.17 A mako shark (A) and a manta ray (B)

Rays have flattened bodies adapted for life on the bottom. They swim by "flying" through the water, using the thin lateral parts of their bodies as wings. The manta ray has a further adaptation for its more mobile lifestyle: fleshy appendages on both sides of its mouth direct the microscopic food it eats into its mouth as it moves through the water.

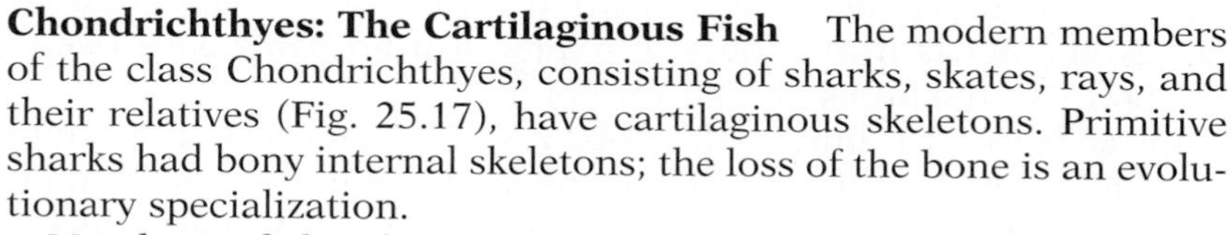

Chondrichthyes: The Cartilaginous Fish The modern members of the class Chondrichthyes, consisting of sharks, skates, rays, and their relatives (Fig. 25.17), have cartilaginous skeletons. Primitive sharks had bony internal skeletons; the loss of the bone is an evolutionary specialization.

Members of the class Chondrichthyes have neither swim bladders nor lungs. Osmoregulation in the subclass Elasmobranchii is unusual, involving retention of high concentrations of urea in the body fluids (see p. 886–887). Fertilization is internal, and the eggs have tough leathery shells. Most species are predatory.

Osteichthyes: The Bony Fish The last class of jawed fish—the Osteichthyes—includes most modern fish. This is a large class, whose members are the dominant vertebrates in both fresh water and the oceans, as they have been since the Devonian—the so-called Age of Fishes (see Table 19.1, p. 540). More than 18,000 species are known, and many remain to be discovered, particularly in the tropics and deeper parts of the oceans. The total number of living species may be as high as 30,000. They range from organisms 1 cm long when mature to giants more than 6 m long. They assume a host of different shapes, many of them bizarre. Some are sluggish and sedentary, while others can swim at speeds as great as 80 km/hr. Almost every type of food is used by some species of fish.

The earliest members of this class probably lived in fresh water. In addition to gills, they had lungs, which they probably used as

25.18 Coelacanth (*Latimeria*), a modern lobe-finned fish

supplementary gas-exchange devices when the water was stagnant and acidic, making it difficult to obtain O_2 and dispose of CO_2. As described in Chapter 28, the ventral lungs in most modern bony fish are modified into a swim bladder, which is used to control buoyancy. Such fish rely for gas exchange almost exclusively on their gills.

Soon after the class Osteichthyes arose, it split into two groups. One, the ray-finned fish (whose fins are supported by rays) diversified enormously, giving rise to nearly all the bony fish alive today. The other, common in the Paleozoic, is represented today by only six relict species—five species of lungfish and one species of lobe-finned fish. The latter, called a coelacanth, was once thought to have been extinct for 75 million years. In 1939 it was found in the deep waters off the southeastern coast of Africa (Fig. 25.18). The fins are mounted on lobes, which themselves are supported by bones; the bones in the lobes of related fish probably gave rise to the bones in the limbs of land vertebrates.

In addition to lungs, the lobe-fins had another important preadaptation for life on land—the large fleshy bases of their paired pectoral and pelvic fins. At times, especially during droughts, lobe-fins living in fresh water probably used these leglike fins to pull themselves onto land (Fig. 25.19), or to crawl to a new pond or stream when the one they were in dried up.

By the Devonian the land had already been colonized by plants, but it was still nearly devoid of animal life (the first insects and millipedes appeared in the Devonian, but they did not become common until the Carboniferous). Hence any animal that could survive on land would have had a whole new range of habitats open to it without competition. Any fish that had appendages better suited for land locomotion than did those of their fellows would have been able to exploit these habitats more fully; through selection pressure exerted over millions of years, these lobed fins

A

B

25.19 Movement of vertebrates onto land

(A) A model of a Devonian lobe-finned fish (*Eusthenopteron*), which probably pulled itself out of the water onto mud flats and sandbars. (B) A painting of an early amphibian (*Ichthyostega*). Its legs were better suited for locomotion on land than the lobe fins of *Eusthenopteron*, but it too probably spent most of its time in the water.

evolved into legs. By the end of the Devonian, along with a host of other adaptations for life on land that evolved at the same time, one group of ancient lobe-finned fish evolved into the first amphibians.

■ LEAVING THE WATER: AMPHIBIA AND REPTILIA

The Age of Amphibians The first amphibians were still quite fishlike (Fig. 25.19, bottom). In fact, they probably spent most of their time in the water. As they progressively exploited the ecological opportunities open to them on land, they slowly became a large and diverse group. So numerous were they during the Carboniferous that that period is often called the Age of Amphibians. The amphibians were still abundant in the Permian, but they slowly declined as a new class, the Reptilia, partially replaced them.

The end of the Permian, which also marked the end of the Paleozoic era, was a time of great change, both geological and biological. The last trilobites and the last placoderms disappeared; the once common brachiopods declined; and older types of corals, molluscs, echinoderms, crustaceans, and fish were replaced by more modern representatives of those groups. This Permo-Triassic crisis also witnessed the extinction of most groups of amphibians. By the end of the Triassic, the only members of this class that survived were the immediate ancestors of the modern amphibians—the salamanders (order Caudata), the rare, wormlike apodes (order Gymnophiona), and the frogs and toads (order Anura; Fig. 25.20).

The Age of Reptiles The first reptiles evolved from primitive amphibians by the late Carboniferous. The class expanded during the Permian, replacing its amphibian predecessors in most terrestrial niches. They became a huge and dominant group during the Mesozoic era, which is often called the Age of Reptiles.

A

25.20 Two modern amphibians

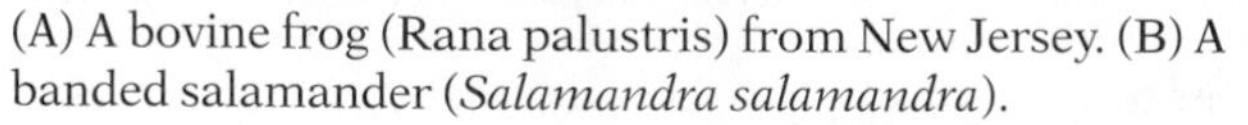

(A) A bovine frog (Rana palustris) from New Jersey. (B) A banded salamander (*Salamandra salamandra*).

B

25.21 Eggs of a toad (*Bufo boreus*) in Napa Creek, California

The reptiles were able to displace the once-dominant amphibians because, unlike amphibians, reptiles were fully terrestrial. Amphibians continued to use external fertilization and to lay fish-like eggs (Fig. 25.21)—eggs that had neither a tough, protective membrane (***amnion***) nor a shell, and hence had to be deposited either in water or in very moist places on land, lest they dry out. Larval development remained aquatic. Amphibians were thus largely bound to the ancestral freshwater environment by the necessities of their mode of reproduction. In addition, the Mesozoic, during which the reptiles gained ascendancy, was generally much dryer and warmer; this greatly reduced the reliable freshwater habitat suitable for reproduction. Furthermore, even adult amphibians probably had thin moist skin and were in danger of desiccation if conditions became very dry.[4]

Reptiles, on the other hand, use internal fertilization, lay amniotic shelled eggs (Fig. 25.22), have no larval stage, and have dry, scaly, relatively impermeable skin. Reptiles evolved the amniotic egg—often called the "land egg"—which provides a fluid-filled chamber in which the embryo may develop even when the egg itself is in a dry place. The amniotic egg was an advance as important in the conquest of land as the evolution of legs by the amphibians.

The reptiles had many other characteristics that made them better suited for some modes of terrestrial life. The legs of the ancient amphibians were small, weak, attached far up on the sides of the body, and oriented laterally. Hence they were unable to support much weight, and the belly of the animal often dragged on the ground; while swimming would have been quite rapid, walking would have been slow and labored, as it is in salamanders today. The legs of reptiles were usually larger and stronger and could thus support more weight and move more rapidly; in many (though not all) species the legs were also attached lower on the sides and oriented more vertically, so that the animal's body cleared the ground. Whereas the lungs of amphibians were well adapted to an aquatic or semiaquatic life, those of reptiles were larger, and stronger rib muscles made their breathing well adapted to a more terrestrial life. As described in Chapter 30, the amphibian heart retains a compromise design that, though it limits efficiency on land, allows skin breathing in the water; the reptile heart, by contrast, is fully terrestrial.

The class Reptilia is represented in our modern fauna by members of four groups: turtles (order Testudines), crocodiles and alligators (order Crocodylia), lizards and snakes (order Lepidosauria), and the tuatara (order Rhynchocephalia; Fig. 25.23). The tuatara, which is found on a few islands off the coast of New Zealand, is the sole surviving member of its ancient order. Members of the other three orders are fairly abundant, totaling about 6500 living species.

25.22 Baby lizards hatching from their eggs

Unlike the eggs of amphibians, those of reptiles are surrounded by a tough case that prevents the embryo from drying out. Note, however, that the shells of reptilian eggs are usually leathery, not brittle like birds' eggs, as can be seen here from the way the shells have buckled and bent.

[4] All modern amphibians have thin moist skin that functions as a respiratory organ (in addition to the gills and/or lungs), but this may not have been true of all ancient amphibians.

A

B

C

D

25.23 Representatives of the main groups of modern reptiles
(A) A tuatara (*Sphenodon*).
(B) A lizard (*Chlamydosaurus*).
(C) A snake (*Vipera*).
(D) A turtle (*Pseudemys*).
(E) A crocodile (*Crocodylus*).

E

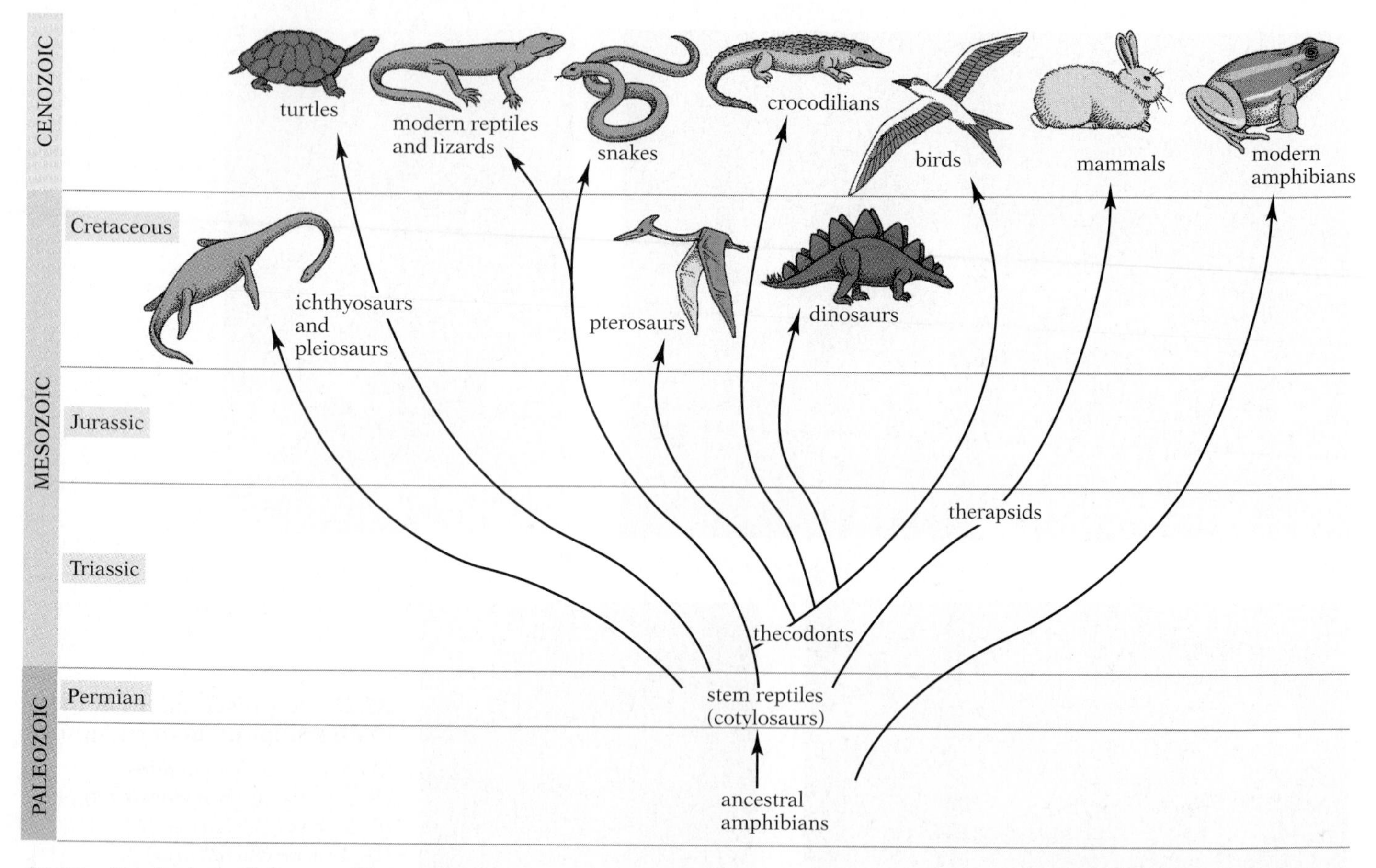

25.24 Evolution of the reptilian groups

The location of the branch point leading to crocodiles is controversial.

All the living reptiles except the crocodilians are directly descended from an important Permian group called the stem (or root) reptiles (Fig. 25.24). This group—the cotylosaurs— also gave rise to several other lineages: one of these (therapsids) ultimately led to the mammals (Fig. 25.25). Two groups (ichthyosaurs and

25.25 Restoration of a late Permian therapsid reptile from South Africa

Fossil evidence indicates that the mammals evolved from the therapsids, which may have had hair.

A

B

25.26 Two extremes of saurian adaptation

The plesiosaur (A) was representative of an ancient reptilian lineage that returned to an aquatic environment. The pterosaur (B) was a giant flying reptile. Both were descended from the Permian stem (or root) reptiles known as cotylosaurs. The coloration of the pterosaur, *Quetzalcoatlus northropi*, is conjectural.

A

B

25.27 Terrestrial dinosaurs

Great variation in body type among the fully terrestrial dinosaurs is evident in these two reconstructions. The scene at the left shows an encounter between two gracile (delicate-boned) species, a solitary *Oviraptor mongoliensis* and a trio of *Troodon mongoliensis*. The scene at the right shows a lumbering herd of single-horned, rhinoceros-like dinosaurs (*Monoclonius apertus*), crossing a stream.

plesiosaurs) returned to the water (Fig. 25.26A). A fourth group (thecodonts) in turn gave rise to crocodilians, the flying reptiles called pterosaurs (Fig. 25.26B), the dinosaurs, and the birds. The dinosaurs were abundant and varied during the Jurassic and Cretaceous periods (Figs. 25.27, 25.28).

25.28 A gathering of dinosaurs
The various types of dinosaurs depicted here are all representative of fossils found in the rocks of the Morrison formation in the western United States, dating from the late Jurassic period (approximately 145 million years ago). Included among the larger species are a herbivorous camptosaur, standing in the trees at the far left, a carnivorous allosaur, perched on the body of a fallen apatosaur at the center, and a herd of long-necked camarasaurs, coming on the scene from the right rear. In the foreground are a primitive crocodilian (left) and a bizarrely outfitted stegosaur (right).

Recent research has led some investigators to propose a radical revision in the long-established conception of dinosaurs. They think that, unlike modern reptiles, which are cold-blooded, these animals may have been warm-blooded—that is, they may have had the ability to maintain a high and constant metabolic rate, and hence great activity, despite fluctuations in environmental temperature. This energetically expensive strategy is often reflected in morphological specializations. The bones of at least some dinosaurs show indications of the extensive vascularization characteristic of warm-blooded animals, and it appears that at least some of them may have had heat-conserving body coverings. For example, the particular dinosaurs from which the birds arose probably had feathers as insulation before the first true birds appeared. Other investigators, however, reject the idea that the dinosaurs were characteristically warm-blooded.

By the end of the Cretaceous (which was also the end of the Mesozoic era), all the plesiosaurs and pterosaurs had disappeared. The dinosaurs, too, had disappeared, except for one group of specialized modern descendants, the birds. Besides birds and mammals, only members of the four groups of modern reptiles remained as representatives of this once enormous class.

The extinction of the dinosaurs After millions of years of slow decline, the rather abrupt extinction of the dinosaurs was a dramatic event in the history of life on earth; it was particularly crucial for mammals, which radiated extensively to fill the ecological vacancies left when these large reptiles vanished. (Indeed, our species might never have evolved had the dinosaurs survived.) Extinction was not limited to reptiles; many invertebrates also disappeared. Yet many other groups did not undergo significant change. In all, about 70% of the earth's species died out at the end of the Cretaceous.

In recent years, interest in mass extinctions has been greatly stimulated by the work of Walter Alvarez and his associates. They discovered that the concentration of specific elements, particularly iridium, is 30 times higher than normal in a thin layer of sediments at the Cretaceous-Cenozoic (K-T) boundary; they argue that the only plausible source for this element is extraterrestrial. They calculate that a 10-km asteroid or comet must have struck the earth and exploded to disperse enough iridium into the atmosphere to account for its sudden worldwide distribution in sediments.

One or more asteroids or comets might well have contributed to mass extinctions; if either fell into the ocean, an enormous tidal wave perhaps 8 km high would have swept around the world and devastated coastal life; there is some evidence for such an event. An impact on land might well have darkened the skies with enough dust to reduce photosynthesis below the level necessary to support most plants and animals. A greenhouse effect could have elevated temperatures by as much as 10°C. The acidity of the earth's rain might have exceeded pH 1 for months or years, which would help explain why species with calcium-based shells, which dissolve in acid, suffered far more than those with silica-based shells.

There is even some evidence of a massive fire at the time of one of the extinctions; the amount of what is apparently soot would have blackened the skies for a very long time. Moreover, dust or soot would have had a major effect on temperature, perhaps screening out the sun's warming rays and plunging the earth into a prolonged and extremely severe winter. The most likely impact site is a (now buried) crater 180 km in diameter straddling the shoreline of Mexico (Fig. 25.29). An impact in such a location could have produced both a huge tidal wave and an enormous dust cloud. A 35-km crater of about the same age is buried in Iowa, a coincidence that implicates a comet: comets frequently break up, as in the case of the one that separated into eight fragments that struck Jupiter in 1994; the chance of two large asteroids striking the earth at about the same time is remote.

All the large-scale extinctions in the fossil record need not have had an extraterrestrial cause: widespread volcanic activity could also produce enough dust and smoke to alter world climate dramatically and reduce photosynthesis; massive volcanic activity in Siberia is thought by many to have triggered the largest mass extinction in the earth's history, which occurred about 250 million years ago. An asteroid collision, comet impact, or cycle of intense volcanic activity would almost certainly have created an evolutionary crisis of some sort, and would have initiated a series of widespread changes in the tempo and direction of evolution.

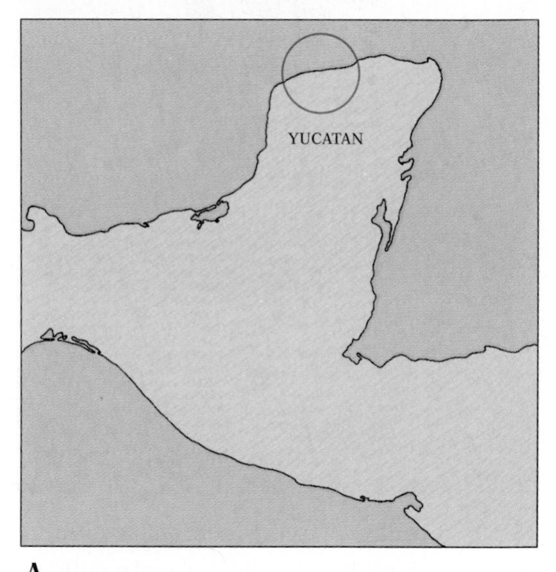

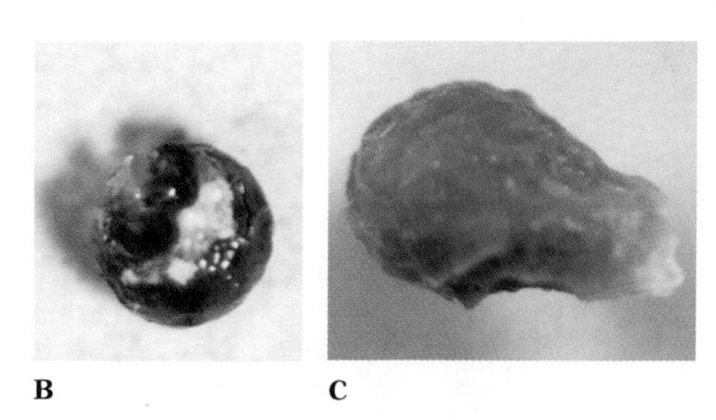

25.29 Evidence of impact at end of the Cretaceous

(A) Site of the Chicxulub crater in Mexico. (B–C) Glass fragments produced by an impact at this time.

■ AVES: THE BIRDS

By the late Triassic or early Jurassic, at least two lineages of reptiles, descended from the thecodonts, had developed the power of flight. One of these, the pterosaurs, included animals with wings consisting of a large membrane of skin stretched between the body and the enormously elongated arm and fourth finger; some species had wingspreads as great as 8 m. They eventually became extinct. The other lineage developed wings of an entirely different sort, in which many long feathers, derived from scales, were attached to the modified forelimbs. This line eventually became sufficiently different from the other reptiles to be designated as a separate class—Aves—the birds. Some authorities think the birds still belong in the class Reptilia as surviving dinosaurs; yet others prefer a new class that would include both the dinosaurs and the birds.

The oldest known fossil bird (*Archaeopteryx*), from the middle Jurassic, still had many reptilian characters including teeth and a long jointed tail. Modern birds have a beak instead of teeth and only a tiny remnant of the ancestral tail bones (the tail of a modern bird consists only of feathers).

Birds evolved from a group of reptiles with a host of adaptations for their very active way of life. One of the most important was warm-bloodedness (endothermy). An anatomical feature that helped make possible the metabolic efficiency necessary for endothermy was the complete separation of the four chambers of the heart (see Fig. 30.8, p. 848), a characteristic also seen in mammals and crocodiles. Unique to birds is their insulation against heat loss; in modern birds all the scales except those of the feet are modified as feathers. Some feathers are modified for flying, and complement other flight adaptations including light hollow bones and an extensive system of air sacs attached to the lungs (see Figs. 28.23, 28.24, p. 810).

The newly hatched young of birds are usually not yet capable of complete temperature regulation, and they cannot fly. In many species, in fact, they are featherless, blind, and virtually helpless. Accordingly, most birds exhibit elaborate nest-building and parental-care behavior (Fig. 25.30).

25.30 A cedar waxwing feeding its young
Many baby birds have brightly colored mouth linings, which act as visual triggers for the parental feeding response.

■ MAMMALIA: THE MAMMALS

Both birds and mammals evolved from reptiles that were probably at least partly endothermic, and both became highly successful groups. But they did not arise from the same ancestral reptilian stock. The line leading to the mammals split off from the stem reptiles early in the Permian (Fig. 25.24), while that leading to birds probably diverged from the dinosaur branch in the Triassic.

The mammals themselves did not appear in the Permian. The Permian saw the rise of the therapsid reptiles, some of which became very mammal-like (Fig. 25.25); they may even have had hair. It is impossible to say at what point therapsids became mammals.

Mammals have a four-chambered heart and are endothermic. They have a muscular structure—the diaphragm—that divides the upper chest (containing the lungs) from the rest of the body; downward movement of the diaphragm helps suck in air, thereby increasing breathing efficiency. The body is usually covered with an insulating layer of hair. The limbs are oriented ventrally and lift the body high off the ground. The lower jaw is composed of only one bone (compared with six or more in most reptiles), and the teeth are differentiated for various functions. There are three bones in the middle ear (versus one in reptiles and birds). The brain is much larger than in reptiles. No eggs are laid (except by monotremes); embryonic development occurs in the uterus of the mother, and the young are born alive. (These reproductive characteristics appear sporadically in other vertebrate classes—live-bearing fish, for example.) After birth, the young are nourished on milk secreted by the mammary glands of the mother.

One small group of mammals—the monotremes—are fundamentally different from all other members of the class. They lay eggs; yet they secrete milk. In many other ways they are a curious blend of reptilian traits, mammalian traits, and traits peculiar to themselves. They are an offshoot of the mammalian lineage from the days when most mammals still laid eggs. The only living monotremes are the echidna (spiny anteater) and the duck-billed platypus (Fig. 25.31); both are found in Australia, and echidnas also occur in New Guinea.

25.31 Duck-billed platypus, an egg-laying mammal

Though the platypus is well adapted for aquatic life, it lays its eggs on land.

The main stem of mammalian evolution split into two parts very early, one leading to the marsupials and the other to the placentals. The characteristic difference between them is that marsupial embryos remain in the uterus for a relatively short time and then complete their development attached to a nipple in an abdominal pouch of the mother (Fig. 25.32), whereas placental embryos complete their development in the uterus.

25.32 A kangaroo with young in pouch

The young kangaroo (called a joey) seen here is hundreds of times larger than when it first entered its mother's pouch; it is no longer attached to a nipple, and it often comes out of the pouch for extended periods.

The living placental mammals are classified in approximately 16 orders. A few of the most important orders are listed here in roughly the sequence they appeared:

EDENTATA Sloths, anteaters, armadillos
LAGOMORPHA Rabbits, hares, pikas
RODENTIA Rats, mice, squirrels, gophers, beavers, porcupines
INSECTIVORA Moles, shrews
PRIMATA Lemurs, monkeys, apes, humans
CHIROPTERA Bats
CARNIVORA Cats, dogs, bears, raccoons, weasels, skunks, minks, badgers, otters, hyenas, seals, walruses
CETACEA Toothed whales, dolphins, porpoises, baleen whales
ARTIODACTYLA Even-toed ungulates (hoofed animals): pigs, hippopotamuses, camels, deer, giraffes, antelopes, cattle, sheep, goats, bison
PERISSODACTYLA Odd-toed ungulates: horses, zebras, tapirs, rhinoceroses
PROBOSCIDEA Elephants

The oldest fossils identified as placental mammals are from the Jurassic. They are of small creatures that probably fed primarily on insects. They remained a relatively unimportant part of the fauna until the end of the Mesozoic. Of the modern orders, Insectivora is closest to this ancient group. The great radiation from the insectivore ancestors dates from the beginning of the Cenozoic era, as the mammals rapidly filled the many niches left open by the disappearance of the dinosaurs. The Cenozoic, which includes the present, is termed the Age of Mammals.

HOW PRIMATES EVOLVED

Primates arose from an arboreal stock of small shrewlike insectivores very early in the Cenozoic (Fig. 25.33). The group soon split into several evolutionary lines that have had independent histories ever since, though the modern representatives of these evolutionary lines generally share the following characteristics:

1 Retention of the clavicle (collarbone), which is greatly reduced or lost in many other mammals

2 Development of a shoulder joint, permitting relatively free movement in all directions, and an elbow joint permitting some rotational movement

3 Retention of five functional digits on each foot

4 Enhanced individual mobility of the digits, especially the thumb and big toe, which are usually opposable

5 Modification of the claws into flattened nails

6 Development of sensitive tactile pads on the digits

7 Abbreviation of the snout or muzzle

8 Elaboration of the visual apparatus and development of binocular vision

9 Expansion of the brain

10 Usually only two nipples

11 Usually only one young per pregnancy

Most of these traits are associated with an arboreal way of life.

In quadrupedal terrestrial mammals the limbs function as props and as instruments of propulsion for running and galloping; they have tended to evolve toward greater stability at the expense of freedom of movement. In the forelimbs of a dog or a horse, for example, the clavicles are greatly reduced or lost, the limbs are positioned close together under the animal, and their movement is restricted largely to one plane. By contrast, in an animal leaping

25.33 A Malay tree shrew

The living tree shrews, which are intermediate in many of their traits between the Insectivora and the Primata, are thought to resemble the early ancestors of the modern Primata. They are not closely related to squirrels (members of the Rodentia), which they resemble superficially.

25.34 A ring-tailed lemur

The animal has a long snout and a bushy tail—both uncharacteristic of the higher primates—but its hands and feet have opposable first digits.

about in the branches of a tree, the limbs function in grasping, clasping, and swinging. Mobility at the shoulder, elbow, and digits facilitates such activities, as does attachment of the limbs (braced by the clavicles) at the sides of the body instead of underneath.

The eyes of many quadrupedal terrestrial mammals, like horses and cows, are located on the sides of the head. As a result, they can survey a very wide total visual field, but the fields of the two eyes overlap only slightly; consequently they have little binocular (three-dimensional) vision. Binocular vision aids in localizing nearby objects, and early primates are thought to have been insectivores, searching for and grabbing insects off vegetation in trees. In addition, an animal that jumps from limb to limb must be able to gauge the position of the next limb accurately. The insectivorous and arboreal way of life of the early primates probably led to selection for binocular vision and, consequently, for eyes directed forward rather than laterally. This change, in turn, would have led to the distinctive flattened, forward-directed face of most higher primates.

Many of the other traits most important to humans first evolved because our distant ancestors lived in trees and hunted moving insects to supplement the standard simian diet of leaves and fruit.

■ PROSIMIANS: THE FIRST PRIMATES

The living primates are usually classified in two suborders: the prosimians (Stepsirhini or Prosimii) and the anthropoids (Haplorhini or Anthropoidea). The prosimians ("premonkeys") are a miscellaneous group of more or less primitive primates, including the lemurs, aye-ayes, lorises, pottos, and galagos.

The living lemurs and aye-ayes are found only on the island of Madagascar off the east coast of Africa. Their relatives—the lorises, pottos, and galagos—inhabit southern Asia and tropical Africa. Most lemurs are small arboreal animals with bushy coats. They have long foxlike snouts and bushy tails, and hardly resemble the higher primates (Fig. 25.34). However, they do have opposable first digits, and the digits are usually provided with flattened nails.

■ ANTHROPOIDS: THE ADVANCED PRIMATES

The second suborder, the anthropoids (Haplorhini), comprises tarsiers, New World monkeys, Old World monkeys, gibbons, and the hominoids: apes and humans. By the Oligocene epoch, after the first members of the anthropoids had evolved from the prosimian stock, the tarsiers arose from an ancestral line that led eventually to two or more lines of anthropoids: the New World monkeys (including the marmosets), and the Old World monkeys, apes, and hominoids (Fig. 25.35).

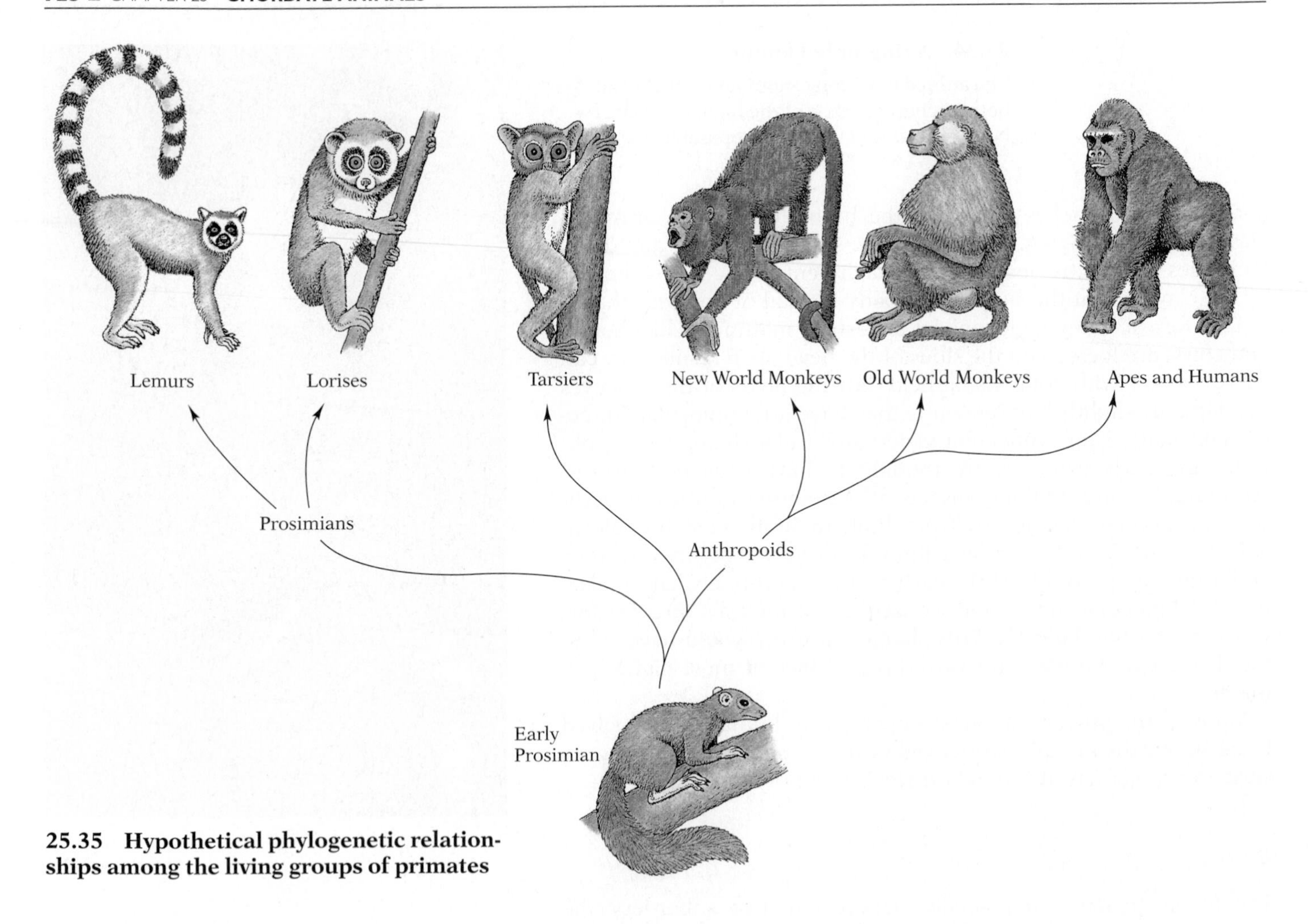

25.35 Hypothetical phylogenetic relationships among the living groups of primates

The anthropoid suborder is classified as follows:

Suborder Anthropoidea (or Haplorhini)
  Superfamily Tarsii
    Family Tarsiidae
  Superfamily Platyrrhini (or Ceboidea)
    Family Cebidae: New World monkeys
    Family Callithricidae: marmosets
    Family Callimiconidae: Goeldi's monkey
  Superfamily Catarrhini
    Family Cercopithecidae: Old World monkeys and baboons
    Family Hylobatidae: gibbons
    Family Pongidae: great apes
    Family Hominidae: humans

Tarsiers The tarsier (Tarsiidae), which is a small crepuscular (dusk-foraging) animal found in the Philippines and the East Indies, is a more advanced and specialized primate than the lemurs (Fig. 25.36). It shows more superficial resemblance to monkeys, though it differs in many ways from all other primates. It has a much shorter muzzle than a lemur. The eyes are enormous and are directed more completely forward than in lemurs. The hind

25.36 A tarsier

Note the distinct face and the large forward-directed eyes.

limbs are long and specialized for leaping. The long tail is naked except at the end.

Monkeys The New World (Cebidae) and Old World (Cercopithecidae) monkeys differ in many ways. Three differences are especially obvious: (1) Most New World monkeys have prehensile tails (used almost like another hand for grasping branches); the tails of Old World monkeys are not prehensile. (2) The nostrils of New World monkeys are separated by a wide partition and are thus oriented in a lateral direction; the nostrils of Old World monkeys are not widely separated and are directed forward and down. (3) New World monkeys lack the naked, brightly colored areas on the buttocks (ischial callosities) common in Old World monkeys.

Among the best-known New World monkeys are capuchins (the traditional organ-grinders' monkeys), howlers (Fig. 25.37A), spider monkeys, and squirrel monkeys. Examples of Old World monkeys are macaques, mandrills, baboons, proboscis monkeys, mona monkeys, and the sacred hanuman monkeys of India.

A

B

25.37 New World and Old World Monkeys

(A) A howler (*Alouatta seniculus*), a New World monkey, has a long prehensile tail. (B) The red patas (*Erythrocebus patas*), an Old World monkey, lacks a prehensile tail and has nostrils that are close together and directed forward and down.

Gibbons Several species of gibbons (Hylobatidae) are found in Southeast Asia. They are smaller than the great apes—about 1 m tall when standing. Their arms are exceedingly long, reaching the ground even when the animal is standing erect. The gibbons are amazing arboreal acrobats and spend almost all their time in trees (Fig. 25.38). Like all apes, gibbons are tailless.

The great apes The living great apes (Pongidae) fall into three groups: orangutans, gorillas, and chimpanzees. All are fairly large animals that have a relatively large skull and brain and very long arms. All have a tendency, when on the ground, to walk semierect.

The one living species of orangutan is native to Sumatra and Borneo (Fig. 25.39). Though orangs are fairly large (males average about 75 kg), and their movements are slow and deliberate, they spend most of their time in trees.

Gorillas, of which there are two forms in Africa, are the largest of the apes (Fig. 25.40); wild adult males may weigh as much as 200 kg and stand nearly 2 m tall. Their arms, while proportionately much longer than those of humans, are not as long as those of gibbons and orangs. Unlike gibbons and orangs, gorillas spend most of their time on the ground. They are not usually aggressive.

25.38 A gibbon (*Hylobates moloch*) from Sunda Island, Borneo

Notice the extremely long arms and lack of tail.

25.39 An orangutan (*Pongo pygmaeus*) with young

25.40 A young male lowland gorilla (*Gorilla gorilla*) thumping his chest

25.41 A chimpanzee using a twig as a tool to get termites out of their nest

The two species of chimpanzees, both of which are native to tropical Africa, are the most human-looking of the living apes (Fig. 25.41). They are about the same size as orangs, but their arms are shorter. Although they spend most of their time in trees, they descend to the ground more frequently than orangs, and sometimes even adopt a bipedal position. (Their usual locomotion, however, is quadrupedal, with the knuckles of the hand used for support.) They are quite intelligent and can learn to perform a variety of tasks. There has been modest success in teaching them to communicate with sign languages and symbols.

THE EVOLUTION OF HUMANS

The earliest humans, members of the family Hominidae (not to be confused with the hominoids, an informal grouping of apes and humans), almost certainly arose from the same pongid stock that produced the gorillas and chimpanzees. Both paleontological evidence and biochemical and serological data indicate that chimpanzees and humans are more closely related, in terms of recentness of common ancestry, than either is to gorillas, orangutans, or gibbons. Indeed, comparisons of the amino acid sequences of proteins and base sequences in DNA indicate that humans and chimps share about 99% of their genes (see table, pp. 520–521). Along with paleontological evidence, the sequencing data suggest that the separation of the line that led to chimpanzees from the hominid line took place no more than 4–6 million years ago (Fig. 25.42).

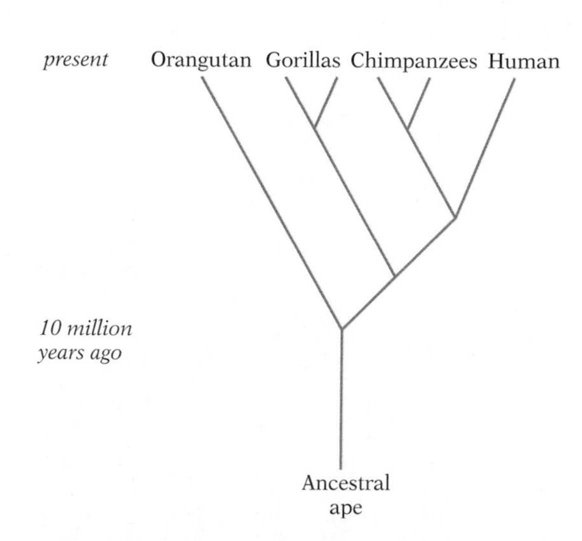

25.42 Hypothetical evolutionary relationships among the living apes and humans

As the drawing indicates, there are two species each of gorillas and chimpanzees.

There is an anomaly between the standard classification of humans into a family separate from the apes and the evolutionary tree in Figure 25.42. Technically, lineages leading to separate families ought, of logical necessity, to diverge before any divergences occur leading to separate genera within a family. It follows, therefore, that if orangutans are to be classed in the same family as chimpanzees, and if, as the evidence strongly indicates, the lineages leading to these genera diverged 12 million years ago, then the divergence 4–6 million years ago between the lineages leading to humans and chimpanzees unquestionably places our species in the family Pongidae. However, taxonomists consider humans to be so different from the apes as to require a separate family. (It is doubtful, however, that an objective extraterrestrial taxonomist would consider these two families. Indeed, humans and chimpanzees would probably be placed in the same genus.)

■ EVOLUTIONARY TRENDS

The current conception of the common ancestor of modern apes (at least of gorillas and chimpanzees) and humans is founded largely on fossils of several species assigned to the genera *Proconsul* and *Dryopithecus*, which first appeared some 25 million years ago (during the early Miocene) and ranged widely over Europe, Africa, and Asia. These animals had a skull with a low rounded cranium, moderate supraorbital ridges, and moderate forward projection of face and jaws. The arms were only modestly specialized for brachiation (swinging from branch to branch), and the feet indicate some tendency toward bipedal posture.

A few of the many anatomical changes that occurred in the course of evolution from ape ancestor to modern humans are as follows:

1 The jaw became shorter (making the muzzle shorter), and the teeth became smaller.

2 The point of attachment of the skull to the vertebral column shifted from the rear of the braincase to a position under the braincase, the skull thus becoming balanced more on top of the vertebral column (Fig. 25.43).

3 The braincase became much larger, and, as it did, a prominent vertical forehead developed.

4 The eyebrow ridges and other keels on the skull were reduced as the muscles that once attached to them became smaller.

5 The nose became more prominent, with a distinct bridge and tip.

6 The arms (though probably never as long as in the modern apes) became shorter.

7 The feet became flattened, and an arch developed.

8 The big toe moved back into line with the other toes and ceased being opposable.

Fossil humans are intermediate in these characters.

One of the most distinctive traits of humans is their bipedal locomotion and upright posture. How might this trait have evolved? When our ancestors moved from the forest to the savanna, their forelimbs, adapted to an arboreal existence, may by that very fact have been preadapted for uses other than locomotion. Their locomotion may well have been of the kind used by gorillas and chimpanzees today, in which the knuckles rather than the palms are on the ground. Knuckle walking enables a quadrupedal animal like the chimpanzee to carry an object such as a tool or weapon in one hand as it walks, using the other arm like a cane. This represents an important advance in transport capacity.

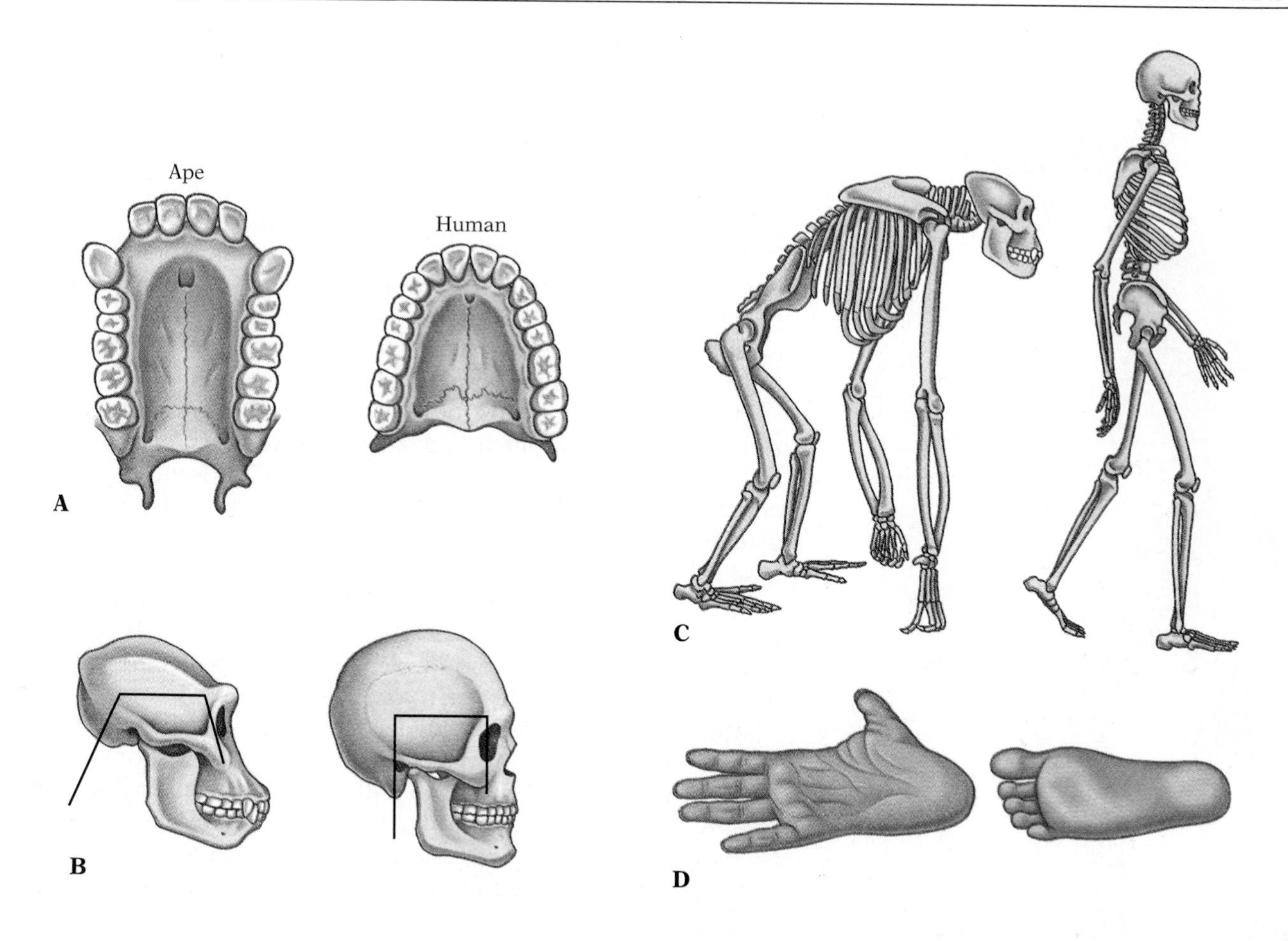

There is, however, considerable controversy over the selection pressures that led to the evolution of bipedalism at least 3 million years ago. The traditional assumption that early hominids were primarily hunters led many theorists to believe that having both hands free for the manipulation of weapons might have been the most important selection pressure for bipedalism. Current opinion holds that hunting arose later and that early hominids were mostly herbivorous. Bipedalism may have have arisen incrementally as a series of adaptations to ground feeding on plants. However, adaptation to a terrestrial way of life need not require the evolution of bipedalism: the ancestral baboons also moved into the savanna, and their descendants remain essentially quadrupedal.

The hands of early hominids, having opposable thumbs, must have been preadapted not only for carrying but also for the preparation and manipulation of tools. Jane Goodall has observed both capabilities in chimpanzees, which use sticks for poking and prying, blades of grass for collecting and eating ants and termites (Fig. 25.41), leaves for personal grooming, and stones for cracking nuts. If such behavior had become increasingly important to the early representatives of the hominid line, then there might have been selection for keeping the hands free, and hence for evolution of upright posture.

Another advantage of upright posture would have been the increased ease of maintaining surveillance—an important requirement for an animal living on the ground in the open country, if not for a chimpanzee living chiefly on the forest floor.

25.43 Anatomical changes between apes and humans

(A) Ape jaws are longer and more U-shaped compared to the shorter and bow-shaped human jaw. (B) A shift in the point of attachment from the rear of the braincase (apes) to a position under the braincase (humans) enables the skull to become balanced more on top of the vertebral column. Also note that the human face is flatter, with a prominent vertical forehead, and that the braincase is larger. (C) The stance of apes is quadrupedal whereas the stance of humans is upright, and bipedal. Note the differences in the shape and orientation of the pelvis, curvature of the vertebral column, and relative arm lengths. (D) The human foot is flat for bipedal walking and the big toe has moved back in line with other toes, ceasing to be opposable.

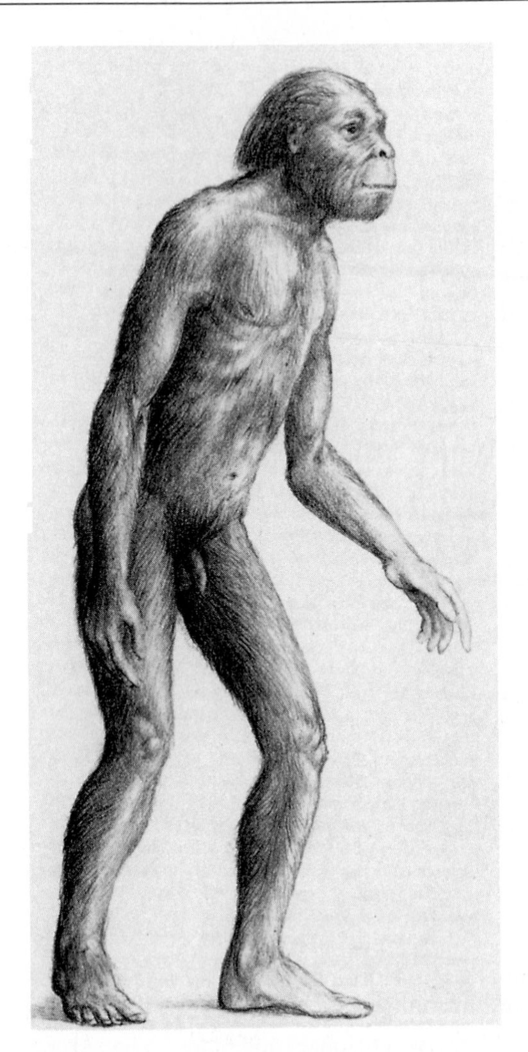

25.44 A reconstruction of *Australopithecus africanus* standing

Though this primitive human was fully bipedal, his stance was not as erect as that of our own species.

■ FOSSIL HUMANS

The first human fossil bones were found in 1856 in Germany by Johann Karl Fuhlrott. They excited lively debate, with Fuhlrott and his supporters maintaining that they were remnants of an ancient organism quite different from modern humans, and his opponents objecting that they were simply the remains of a person who had suffered several deformities. It was many years before enough similar fossils were found to establish the validity of Fuhlrott's claim.

There is considerable controversy over how fossil data on human evolution should be interpreted; the brief sketch given here is the most likely of a number of possible interpretations.

The first true hominids found in the fossil record are usually assigned to the genus *Australopithecus* (from *australis,* "southern"—because the first specimens were found in South Africa—and *pithecus,* "ape"). The oldest known fossils of *Australopithecus*, now usually assigned to the species *Australopithecus afarensis*, are about 4 million years old; this species must have originated 4–6 million years ago. *A. afarensis* was soon replaced by *A. africanus* (Fig. 25.44).

Apparently, at least three species of australopithecines lived contemporaneously for at least a million years. *A. africanus* had smaller teeth and a more slender form than its major contemporary, *A. robustus*, which was distinguished by large bony crests on its skull, and its greater height—40 cm taller on the average (Figs. 25.45A–B and 25.46). The third species of contemporaneous australopithecine, *A. boisei,* was also smaller than *A. robustus*; all three species are believed to have had overlapping ranges.

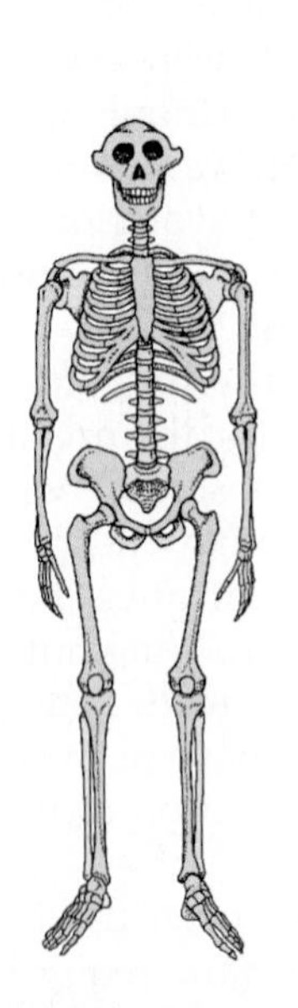

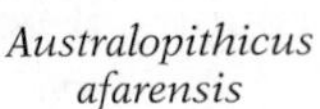

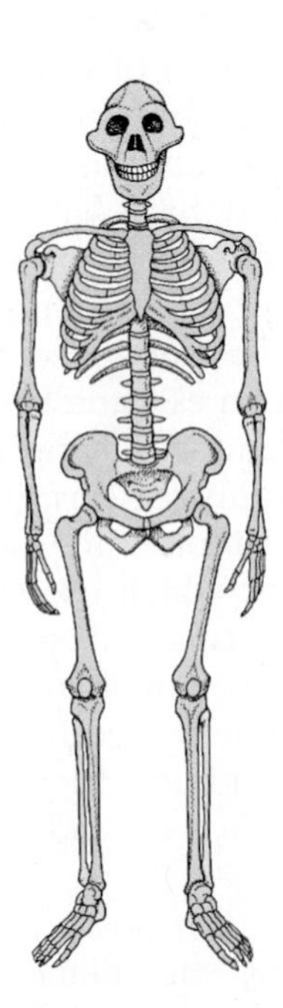

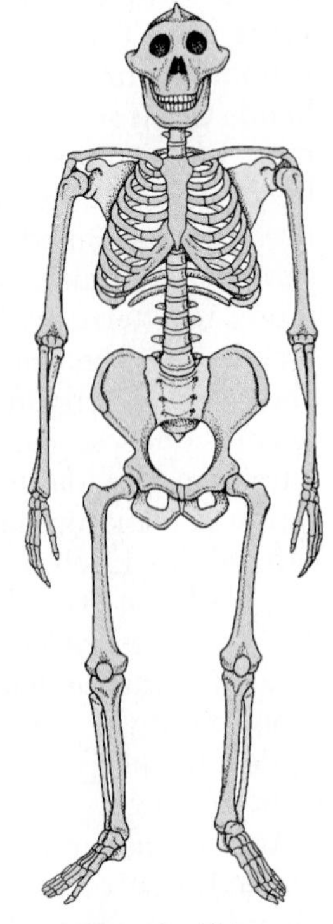

A

B

C

D

25.45 Skulls of prehistoric humans

(A) The skull of a five-year-old *Australopithecus africanus*. (B) *A. robustus*. (C) *Homo habilis*. (D) *H. erectus*, often called Java man.

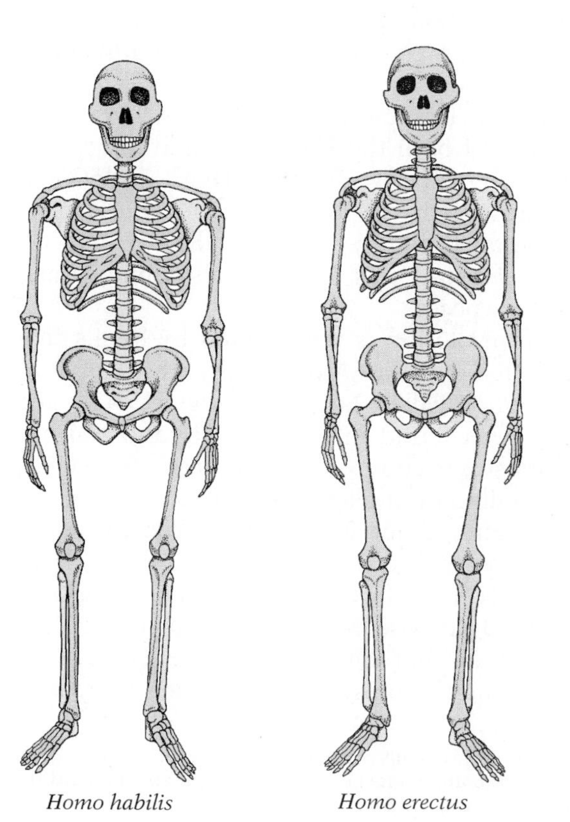

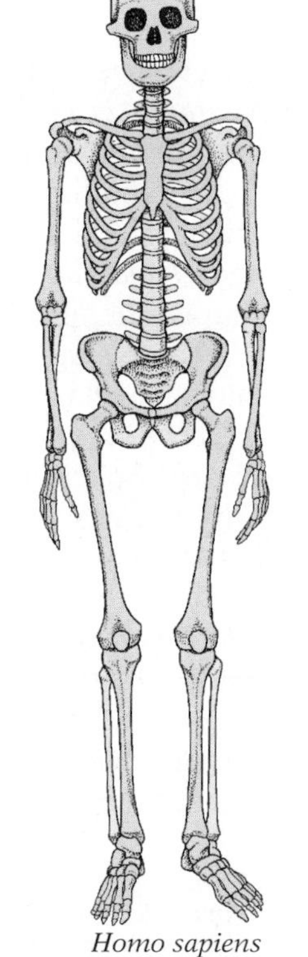

Homo sapiens

25.46 A comparison of reconstructed skeletons of six hominid species

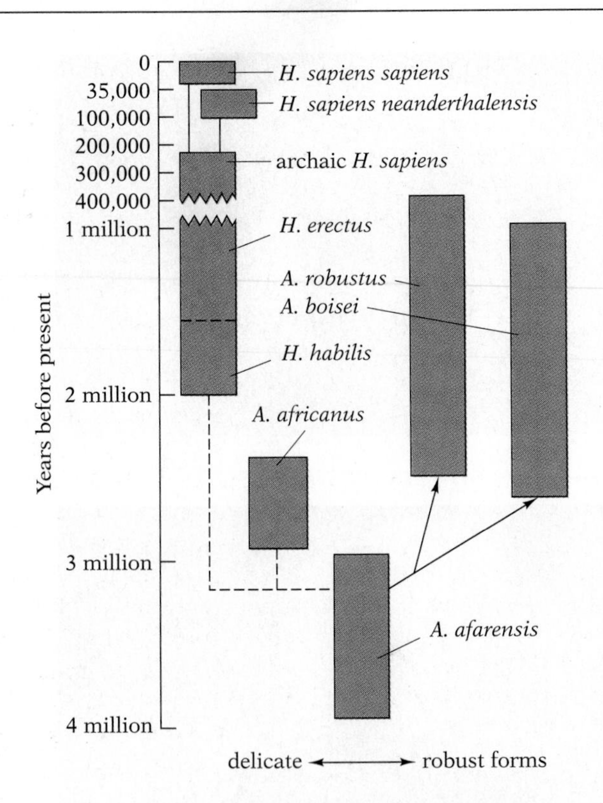

25.47 Hypothetical evolutionary relationships among the known hominids

There is still no universal consensus upon which to base a phylogenetic tree of hominids. This figure shows one of several possible arrangements. The relationships among the early hominids is unclear, which is indicated by the dashed lines. The boxes represent the approximate length of time each species was thought to exist, based on fossil evidence. Notice that there are several periods in which there were two or more different species of hominids living at the same time.

All australopithecine species, whose fossil remains have been found in South and East Africa, were apparently fully bipedal, though their stance may not have been as upright as that of modern humans. They had large jaws, but almost no forehead or chin, and their cranial capacity was only about 450–550 cc, compared with 1200–1600 cc (average about 1360 cc) for modern humans, and 350–450 cc for normal chimps. They probably used unworked stones and bits of wood as tools.

Evidence collected over the last few years strongly suggests that *A. afarensis* is the ancestor of both the *Homo* line and the rest of the *Australopithecus* line (Fig. 25.47). *H. habilis* (Figs. 25.45C and 25.46), the first clear representative of the *Homo* line, had a larger cranial capacity than the australopithecines, usually 650–775 cc, and *H. habilis* not only used stones as tools but also chipped and shaped them for various purposes.

A later stage in human evolution is represented by fossils that may be classified as *Homo erectus*[5] (originally described as *Pithecanthropus erectus*, often called Java man; Figs. 25.45D and 25.46). Specimens have been found in Asia, Africa, and Europe. This species, which almost certainly descended from *H. habilis*, first appeared about two million years ago. Its cranial capacity was considerably larger, averaging about 900 cc. However, the facial features remained primitive, with a projecting massive jaw, large teeth, almost no chin, a receding forehead, heavy bony eyebrow ridges, and a broad low-bridged nose. Not only did the members of this species make and use tools, but they also used fire. Casts of the interior of the skulls indicate the presence of the speech areas of

[5] *Homo erectus* includes the forms originally assigned to *Pithecanthropus, Sinanthropus, Telanthropus,* and *Atlanthropus*. Some authorities also prefer to regard the fossils we have called *Homo habilis* as early *Homo erectus*.

the brain; we have no way of knowing whether or how language was used.

Modern humans are given the Latin name *Homo sapiens* ("wise man"). Early representatives of this species first appeared about 250,000 years ago, apparently in Africa or southwestern Asia. Analysis of sequence variation in both mitochondrial DNA (which is inherited maternally) and Y-chromosome DNA (which is inherited paternally) indicates that the small group of individuals from whom all modern humans are descended dates from almost exactly this time. They spread to Europe, eastern Asia, and the Pacific islands; each group made many kinds of tools and buried their dead. The European form, the neanderthals, is the one Fuhlrott discovered. What happened next is controversial. One proposal is that each group evolved in parallel into the present-day humans in the various areas. Another hypothesis, more consistent with sequence analysis, supposes that parts of the African group migrated to Asia and Europe (and later to the Pacific islands and the Americas) about 150,000 years ago, coexisting with and then replacing the earlier humans. Indeed, more modern forms of *H. sapiens* lived in Europe at the same time as the neanderthals, with whom they did not interbreed; in this view, the neanderthals are a separate species, *H. neanderthalensis*. The African-origin hypothesis, however, fails to explain the DNA-sequence patterns of certain Asian and Pacific island populations: archaeological evidence suggests that a few specific groups are relatively recent, while their sequences indicate they are relatively old.

■ THE HUMAN RACES

Widespread species often tend to become subdivided into geographic subspecies, or races. Humans are no exception. *Homo sapiens* is an extremely variable species, and regional populations are often recognizably different (or were, prior to the great mobility of the last few centuries). Thus Scandinavians tend to have blue eyes and fair complexions, while southern Europeans tend to have brown eyes and darker complexions. Eskimos look different from Mohawk Indians, who in turn look different from Apaches, and yet all three groups are derived from a common founder stock that arrived in the Americas about 12,000 years ago. Many of the differences probably reflect adaptations to different environmental conditions. Thus, for example, the prevalence of darker skin in tropical and subtropical regions may be a protective adaptation against damaging solar radiation.

Races, by definition, are regional populations that differ genetically but can readily interbreed. There are seldom sharp boundaries between them, and populations can shift gradually from one set of characteristics to another over a wide area. Designation of races in most species is therefore an arbitrary matter, and there is no such thing as a "pure" race. Some authorities have chosen to recognize as many as 30 races of humans, while others recognize only three: Caucasoid, Mongoloid, and Negroid. Another widely used system recognizes five: the traditional three, plus American Indians and Australian aborigines. Sequence analysis suggests this latter grouping is more accurate than any other.

None of these systems has much biological significance, since human races are far more similar than the recognized subspecies of most other organisms. The geographic variation within our species will surely break down as people move about more. The main barriers to interbreeding in many parts of the world are now cultural or social rather than geographic; it seems very unlikely that such barriers will approach the effectiveness of the original geographic barriers. Hence, whatever races are recognized now, it seems probable that they will become less distinct as time goes on.

■ DO CULTURAL AND BIOLOGICAL EVOLUTION INTERACT?

Early hominids used tools long before their brains became much larger than those of apes. Thus the old idea that a large brain and high intelligence were necessary prerequisites for the use of tools has been discredited. The early hominids' use of tools may, in fact, have been an important factor in the evolution of higher intelligence. Once the use and making of tools began (Fig. 25.48), individuals that excelled in these endeavors would surely have had an advantage. There would thus have been strong selection for neural mechanisms that facilitated tool use. Thus perhaps, instead of considering culture the crown of the fully evolved human intelligence, we should think of early cultural development and increasing intelligence having evolved together synergistically; in a sense, the highly developed brain of the modern human may be as much a consequence as a cause of culture.

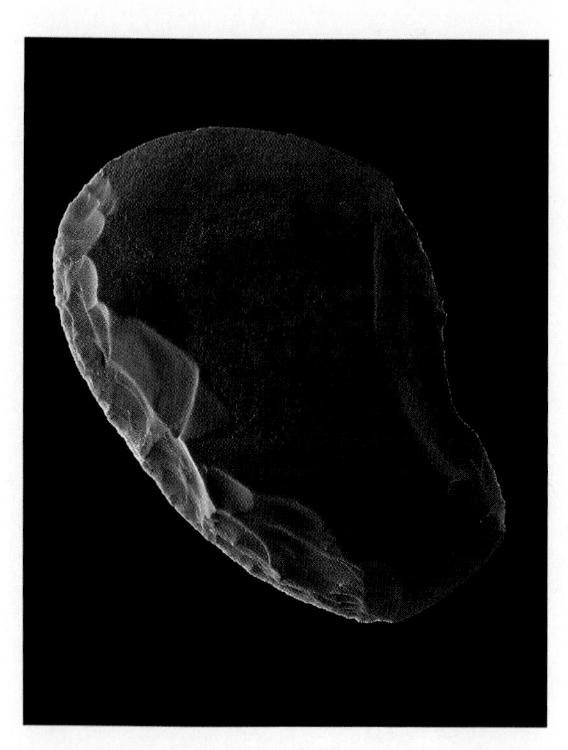

25.48 A primitive hide scraper

This artifact represents a type of stone tool used by early humans, to dress hides, for perhaps half a million years.

Cultural evolution can proceed at a far more rapid pace than biological evolution. Words as units of inheritance are much more effective than genes in spreading new developments and in giving dominance to new approaches originating with a few individuals. But the two types of evolution continue to be interwoven just as they were in the use of tools; they may be even more so in the future.

We humans, by our unrivalled ability to alter the environment, are influencing in profound ways the evolution of all species with which we come in contact. Thus, new strains of bacteria have evolved in response to the use of antibiotics; many species of insects have evolved new physiological and behavioral traits as a consequence of the intense selection resulting from the use of insecticides; the clearing of forests for agricultural purposes has led to drastic declines in the population densities of some species and to increases in others; long-established balances between prey species and their predators and parasites have been destroyed, often with far-reaching consequences for the entire ecosystem.

While some human actions on the environment have been deliberate, many have been unintentional. Regardless of intent, our behavior has precipitated a period of wholesale and rapid change rarely matched since life began. As detailed in Chapter 40, human activity already threatens to destroy the protective layer of ozone in the atmosphere, and to increase the CO_2 concentration enough to create major climatic shifts. Since disruption of ecosystems will unavoidably grow as civilization expands, the great challenge to our species is to use the resources of our knowledge and technology to guide the change in ways that will benefit both our own species and the other organisms around us.

Our ability to control our own evolution may eventually no longer depend primarily on regulating reproduction. Now that the genetic code has been deciphered and recombinant DNA techniques developed, some gene combinations that lead to deformity or disease can already be detected; the day will come when the DNA of genes can be deliberately manipulated to design, at least in part, new human beings. When that day comes, how do we decide what to design? What do we look for in human beings? We might all agree to rid our species of the genes for muscular dystrophy or sickle-cell anemia (at least in areas where there is no malaria), but once the techniques for achieving these apparently worthy ends are mastered, suggestions will surely be put forward for other alterations to which we cannot all agree.

More immediately pressing, perhaps, is the problem of regulating the *size* of human populations, now that we have interfered with the action of the many former natural regulating factors. Already some people are asking whether we should abandon the campaigns to eradicate malaria, to cure cancer and heart disease, or to slow the aging process. They point out that the current population problem is the result of major advances in the technology of death control, which societies have generally been as eager to accept as they have been reluctant to accept compensatory birth control. No species can continue indefinitely with its birth and death rates unbalanced, and there is little evidence that most populations of our species will consent soon enough to critical population-control measures.

These complex and unnerving questions—at once biological, economic, political, and ethical—must be faced, and soon. History provides only unpalatable precedents for population control. The answers the next few generations give to these questions may well have as profound an influence on the future of life as anything that has happened since the first cells materialized in the primordial seas.

CHAPTER SUMMARY

THE INVERTEBRATE CHORDATES

Like all chordates, invertebrate chordates at some stage in their life cycle possess a notochord, pharyngeal slits, and a hollow nerve cord. (p. 703)

TUNICATA: THE TUNICATES Larval tunicates use their pharyngeal slits for gas exchange and filter feeding. Adults lose their notochord and nerve chord. (p. 703)

CEPHALOCHORDATA: THE LANCELETS Adult lancelets use their pharyngeal slits for gas exchange and filter feeding, and retain the notochord and nerve cord throughout life. As with all invertebrates, the muscles are segmented. (p. 704)

THE TISSUES OF VERTEBRATES

COVERING BOTH INSIDE AND OUT: EPITHELIUM Epithelium provides a barrier that covers the body and lines the digestive tract, lungs, blood vessels, glands, and body cavity. It is usually separated from other tissues by a basement membrane. (p. 706)

CONNECTIVE TISSUE: CELLS IN A MATRIX Connective-tissue cells are found in a matrix, and include blood and lymph, bone, cartilage, and connective tissue proper (which has a matrix of collagen fibers, elastic fibers, or reticular fibers). (p. 708)

MUSCLE Muscle cells are specialized for contraction. (p. 710)

NERVE Nerve cells are specialized for signalling and information processing. (p. 710)

VERTEBRATE ORGANS

Most vertebrate organs are parts of organ systems, including the digestive, respiratory, circulatory, excretory, endocrine, nervous, skeletal, muscular, and reproductive systems. (p. 710)

HOW VERTEBRATES EVOLVED

AGNATHA: THE JAWLESS FISH Like the invertebrate chordates, the jawless fish (Agnatha) were filter feeders. (p. 712)

FISH WITH JAWS Jaws in fish probably evolved from gill supports in filter feeders. There have been three major groups of jawed fish: the placoderms (Placodermi) were armored fish (now extinct). In cartilaginous fish (Chondrichthyes, including sharks and rays), the muscles attach to cartilage rather than bone; though bone develops from cartilage, cartilaginous fish evolved from bony ancestors. The bony fish (Osteichthyes) form the largest class of vertebrates; early groups had lungs and heavy fins, and one of these gave rise to amphibians. (p. 713)

LEAVING THE WATER: AMPHIBIA AND REPTILIA Amphibians require water for their eggs and larvae; most also breathe through moist skin. Reptiles have dry skin and lay eggs with shells that can survive out of water. (p. 717)

AVES: THE BIRDS Birds are endotherms, with modified scales (feathers) for insulation; the feathers later evolved into flight aids. (p. 724)

MAMMALIA: THE MAMMALS Nearly all mammals rear their embryos internally and feed the young from mammary glands. Mammals are endotherms with hair for insulation. (p. 724)

HOW PRIMATES EVOLVED

The original primates were insectivores that evolved the now distinctive collarbone, arm joint, long digits with an opposable thumb and toe, and binocular vision—all adaptations for arboreal life. (p. 726)

PROSIMIANS: THE FIRST PRIMATES This group includes lemurs, lorises, pottos, and galagos. (p. 727)

ANTHROPOIDS: THE ADVANCED PRIMATES The advanced primates include tarsiers, monkeys, gibbons, and apes; this last group includes the orangutan, lowland and mountain gorillas, pigmy and common chimpanzees, and humans. Nearly all are arboreal, and the larger species are generally herbivores. (p. 727)

THE EVOLUTION OF HUMANS

EVOLUTIONARY TRENDS Human characteristics—shorter jaw, smaller teeth, larger skull positioned farther back over the spine, shorter arms, and feet adapted for bipedal walking—evolved gradually. (p. 732)

FOSSIL HUMANS Several species of hominids coexisted at one time. The first true humans—*Homo habilis*—made and used tools. *H. erectus* appeared about 2 million years ago and used fire. *H. sapiens* dates from about 250,000 years ago. (p. 734)

THE HUMAN RACES The concept of race has little biological meaning in the context of human evolution; it is primarily a nonscientific system for categorizing peoples on the basis of a limited range of morphological variables. (p. 737)

DO CULTURAL AND BIOLOGICAL EVOLUTION INTERACT? Cultural evolution in humans is far more rapid than biological evolution. Some elements of our precultural behavioral repertoire may be important forces predisposing our species toward overpopulation and environmental destruction. (p. 738)

STUDY QUESTIONS

1 If photosynthesis were to cease for 1 year—one of the possible consequences of an asteroid impact or extreme volcanic activity—what sorts of plants and animals would probably survive, and which would die out? What if the darkness lasted 3 years? (p. 723)

2 Some researchers suggest that without the crisis that led to the extinction of dinosaurs, most mammals (including primates) would never have evolved. How plausible is this argument? Do reptiles (or did dinosaurs) have the capacity to exploit efficiently many of the niches that are now held by mammals, and so exclude other phyla? (pp. 717–725)

3 The term *Age of Fish* implies that fish design is so greatly superior to older invertebrate body plans that fish were able to radiate extensively, exploiting previously empty niches and displacing preexisting invertebrate species from their niches. What was so great about being a fish? (pp. 712–717)

4 The sequence of first the Age of Amphibians and then the Age of Reptiles corresponds to an ever-greater ability to exploit terrestrial niches. This sequence presumably reaches its climax with birds and mammals. Yet no one has written of an Age of Birds. What are the costs and benefits of the avian body plan and lifestyle that make them superior to reptiles and yet unable to dominate the mammals? (p. 724)

5 If the earth were to suffer another crisis that wiped out most of the large species, it might result in a new "evolutionary age." Construct a plausible design for the new vertebrate class that could supplant the mammals.

SUGGESTED READING

ALVAREZ, W., AND F. ASARO, 1990. What caused the mass extinctions? An extraterrestrial impact, *Scientific American* 263 (4).

CHINSAMY, A., AND P. DODSON, 1995. Inside a dinosaur bone. *American Scientist* 83 (2), 174–180. *A very objective analysis of the argument for warm-blooded dinosaurs.*

CONROY, G., 1990. *Primate Evolution.* W. W. Norton, New York.

COPPENS, Y., 1994. East side story: the origin of humankind, *Scientific American* 270 (5). *On the evolution of humans in the Rift Valley of East Aftica.*

DUELLMAN, W. E., 1992. Reproductive strategies of frogs. *Scientific American* 267 (1). *How some species have circumvented the need for water for reproduction.*

FOREY, P., AND P. JANVIER, 1994. Evolution of the early vertebrates. *American Scientist* 82 (6), 554–565. *On the jawless fish.*

GLEN, W., 1990. What killed the dinosaurs? *American Scientist* 78, 354–370.

LOVEJOY, C. O., 1988. Evolution of human walking, *Scientific American* 259 (5). *Fascinating analysis of the "Lucy" fossil showing how anthropologists are able to conclude that our ancestors were bipedal 3 million years ago.*

STRINGER, C. B., 1990. The emergence of modern humans, *Scientific American* 263 (6). *A comparison of the DNA of various human groups is used to reconstruct and date the origin of our species.*

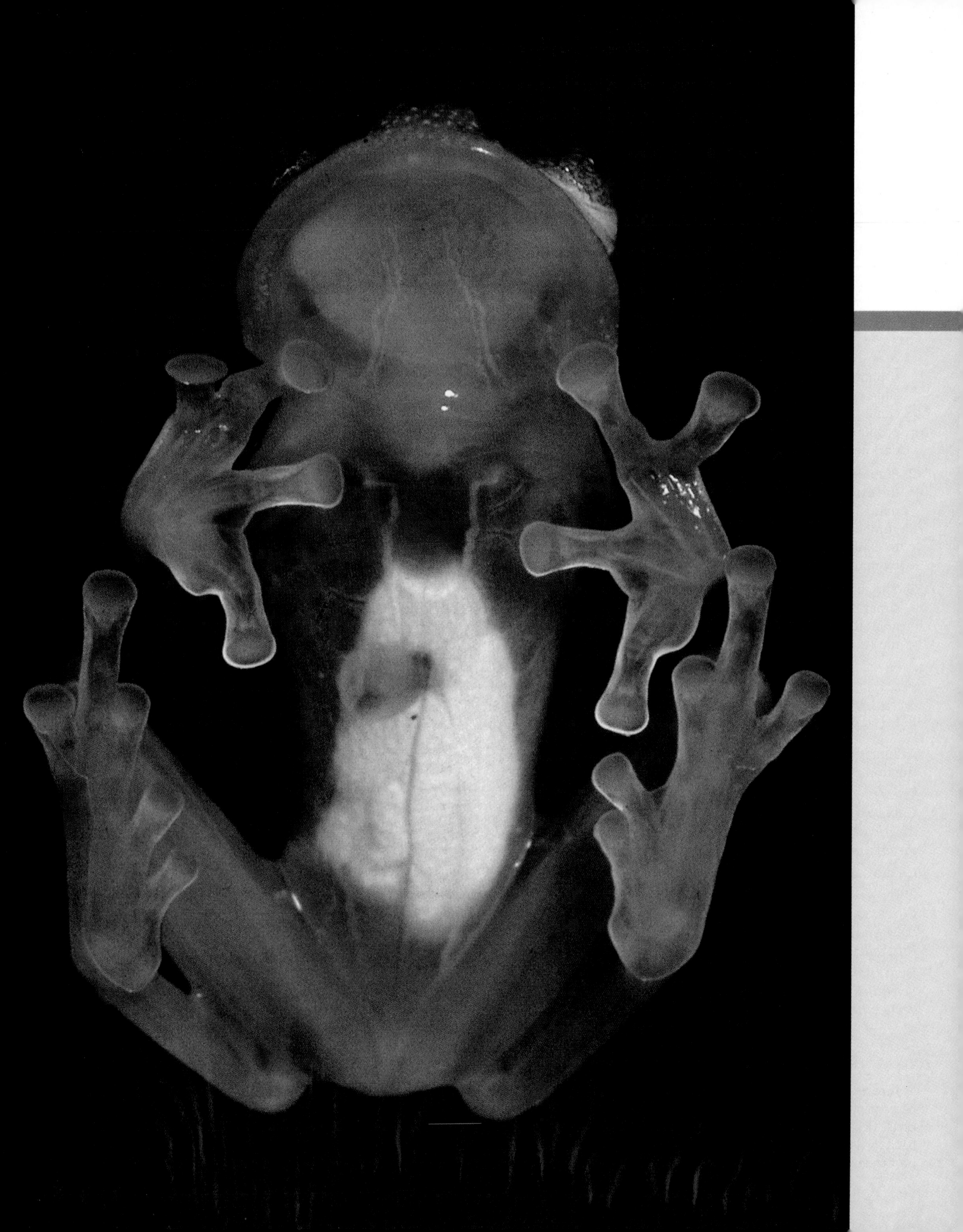

PART V

THE BIOLOGY OF ORGANISMS

◄ **Internal organs,** including the heart and a major blood vessel, are visible in this photograph of a transparent frog. The amphibian, a native of Costa Rica, utilizes a closed circulatory system. Like all organisms, it has evolved its own distinctive solution to the basic physiological challenges of life—nutrient procurement, gas exchange, internal transport, chemical control, information processing, etc.

CHAPTER 26

PLANT NUTRITION

To get the nutrients they need, most organisms need to synthesize specific organic molecules and break down energy-rich compounds. In addition to the oxygen that aerobic organisms use in cellular respiration, the environment provides either already synthesized high-energy compounds, or the raw materials from which those compounds and new organic molecules can be synthesized. This chapter and the next will deal with the procurement and processing of such nutrients.

AUTOTROPHS VERSUS HETEROTROPHS

Organisms can be divided into two classes on the basis of their method of nutrition. ***Autotrophs*** can manufacture their own organic compounds from inorganic raw materials taken from the surrounding medium. Since the molecules of these raw materials are small enough and soluble enough to pass through cell membranes when dissolved in water, autotrophic organisms do not need to digest their nutrients before taking them into their cells. Most autotrophs are photosynthetic; the few that aren't are chemosynthetic. The autotrophs include photosynthetic and chemosynthetic bacteria, photosynthetic protozoans, brown algae, and plants. The plants are by far the most important of the earth's autotrophs.

Heterotrophs, by contrast, are unable to manufacture energy-rich organic compounds from simple inorganic nutrients, and so must obtain prefabricated organic molecules from the environment; most bacteria, fungi, protozoans, and animals are heterotrophs. Many of the organic molecules found in nature are too large to be absorbed through cell membranes, and so must first be digested into smaller, more easily absorbable molecular subunits. Some of the products of digestion will supply the energy to alter and reassemble the others into organic macromolecules such as lipids, proteins, and nucleic acids.

Clearly, then, autotrophic and heterotrophic organisms will differ both in their nutrient requirements and in the problems associated with nutrient procurement. They have evolved radically different adaptations in response to the different selection pressures acting upon them. At the same time, as we will see, there are parallels in processing and procurement strategies that are intriguing and enlightening.

WHAT PLANTS ARE MADE OF

The higher photosynthetic organisms need light, carbon dioxide, and water. These raw materials supply the carbon, oxygen, and hydrogen that are the predominant elements in the organic molecules they manufacture, as well as the energy needed to drive the synthetic reactions. Where do plants get these elements?

■ EARTH, AIR, OR WATER?

For centuries it was assumed that the structural material of the plant body comes from the soil. By 1700 most researchers believed that water rather than the earth was the actual source (see the *How We Know* box). Later, when it became clear that about 40% of a plant's dry weight is in the form of carbon, which is virtually absent from soil and water, attention turned to the air as a source of some of the mass. Isotopic tracer techniques, first developed in the present century, demonstrated conclusively that CO_2 gas contributes not only the carbon but also the oxygen that is used in the synthesis of glucose, the substance from which most of a plant's organic compounds are directly or indirectly synthesized. The im-

HOW WE KNOW:

PLANT MASS DOES NOT COME FROM THE SOIL

About 1450, the German Cardinal Nicholas of Cusa advanced the revolutionary thought that the weight gained by a growing plant comes from water, not earth. He was sure that if seeds were planted in a pot with a carefully measured quantity of earth and allowed to germinate and grow large and heavy, the soil measured afterwards would weigh very much the same.

The experiment suggested by Nicholas of Cusa was finally performed by Jan Baptista van Helmont, a Flemish physician, and the results were published in 1648:

> **. . . I took an earthenware vessel, placed in it 200 pounds of soil dried in an oven, soaked this with rainwater, and planted in it a willow branch weighing 5 pounds. At the end of five years, the tree grown from it weighed 169 pounds and about 3 ounces. Now, the earthenware vessel was always moistened (when necessary) only with rainwater or distilled water, and it was large enough and embedded in the ground, and, lest dust flying be mixed with the soil, an iron plate coated with tin and pierced by many holes covered the rim of the vessel. I did not compute the weight of the fallen leaves of the four autumns. Finally, I dried the soil in the vessel again, and the same 200 pounds were found, less about 2 ounces. Therefore 164 pounds of wood, bark, and root had arisen from water only.**

Though van Helmont clearly demonstrated that most of the material of a plant's body does not come from the soil, he had not proved that its source was the water, though he could think of no other plausible alternative. Technically, the scientific method can only *disprove* hypotheses. Proof is possible only by exclusion: all possible alternatives must be eliminated. The trouble is, of course, that it is impossible to be absolutely certain that every alternative has been imagined. It was 1727, for example, before Stephen Hales, an English clergyman, suggested that plants might get at least some of their nourishment from the air. His improbable idea opened up the novel line of thinking that led to our present understanding: the vast majority of the dry mass of plants comes from carbon dioxide gas in the atmosphere.

plausible idea that a tree's massive substance could come mainly from something as insubstantial as air was for centuries a serious impediment to research.

The formula for glucose, the central carbohydrate in living things, is $C_6H_{12}O_6$. The atomic weight of carbon is 12, that of hydrogen is 1, and that of oxygen is 16; consequently the molecular weight of glucose is 180 (really 180.162 if you average the weight of all the isotopes). All the carbon and oxygen incorporated into glucose by photosynthesis comes from carbon dioxide, which in turn comes from air. Since the combined weight of the six atoms of carbon and six atoms of oxygen in glucose is 168, which is about 93% of the total weight of glucose, it follows that about 93% of the dry weight of a large, heavy tree comes initially from the air (Fig. 26.1). The other 7%—the hydrogen in glucose—comes from water.

■ WHY PLANTS NEED MINERALS

Despite their importance, carbon dioxide and water provide only three elements: carbon, oxygen, and hydrogen; yet we know that other elements enter into the composition of plants. Nitrogen, for example, is always present in the amino acids of enzymes and structural proteins; two amino acids—cysteine and methionine—also contain sulphur. Phosphorus is present in ATP and nucleic acids. Chlorophyll, the essential molecule of photosynthesis, contains magnesium (see Fig. 7.4, p. 183), and the cytochromes, critical in electron transport, contain iron and other metal ions. Where does the green plant obtain the nitrogen, sulphur, phosphorous, magnesium, iron, and other elements it needs? Obviously not from carbon dioxide or water. Here, finally, we see the role of the soil itself as a source of the minerals essential to plants.

By 1900, research on crops had discovered that plants need seven minerals: nitrogen, sulphur, phosphorous, potassium, calcium, magnesium, and iron. Three of them—nitrogen, phosphorous, and potassium—are particularly important, and modern commercial fertilizers are often designated by their N-P-K percentages. The widely used garden fertilizer called 5-10-5 contains 5% nitrogen, 10% phosphoric acid, and 5% soluble potash (a potassium compound) by weight. These three are the elements most rapidly removed from the soil, and they must be replenished if crops are to flourish.

Much of the important research on the mineral requirements of plants makes use of plants grown hydroponically in distilled water

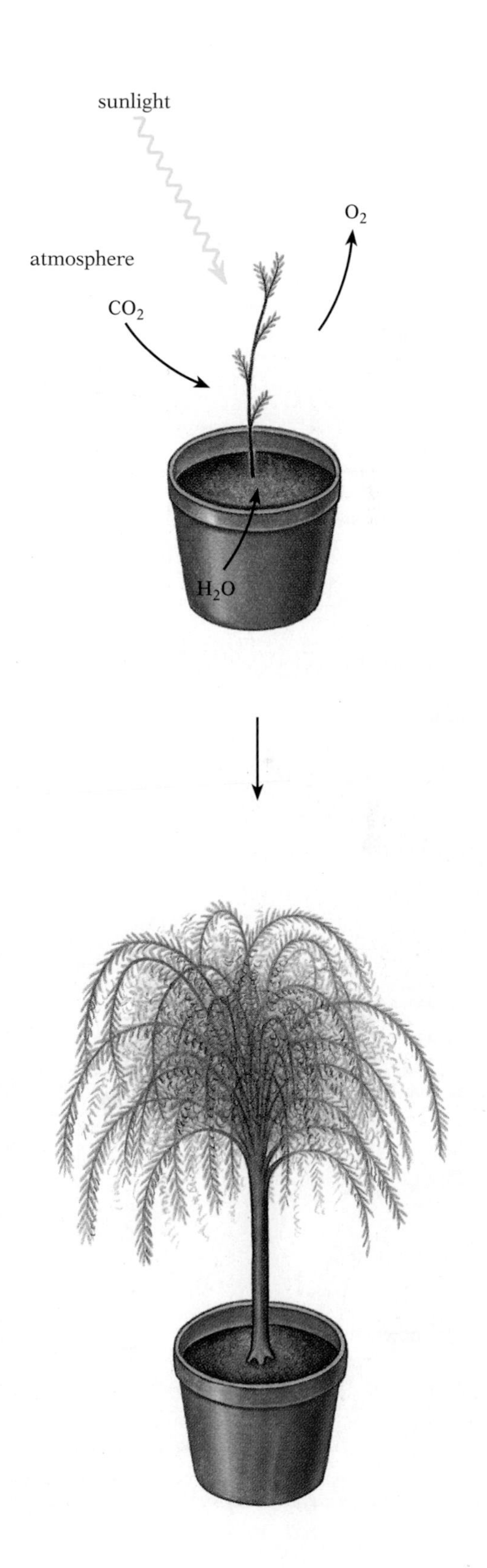

26.1 Major sources of plant mass

Most of the dry mass of plants is in the form of carbohydrates (or their derivatives) generated originally by photosynthesis. A growing seedling uses the energy of light to combine the carbon and oxygen of carbon dioxide with hydrogen split off from water to synthesize carbohydrates; the oxygen of water in an actively growing plant is mostly discharged into the air. Thus, more than 90% of the dry mass of a plant is obtained from the atmosphere; the remainder (including, in addition to hydrogen, essential minerals like nitrogen and phosphorus) comes from the soil.

to which precise quantities of minerals can be added. Hydroponic research has demonstrated that plants require manganese, boron, chlorine, zinc, copper, molybdenum, and nickel in very small amounts. Such elements, sometimes toxic in excess, are called micronutrients.

Table 26.1 lists all the essential nutrients known at present. The functions listed there make it clear why the trace elements are needed in only minute amounts. Most of them are present in the active sites of enzymes or coenzymes; since enzymes and coenzymes can be used over and over again, only a small number of mineral atoms is required for their synthesis.

Table 26.1 Essential minerals for higher plants[a]

ELEMENT	APPROXIMATE NUMBER OF GRAMS NEEDED TO GROW ONE CUBIC METER OF CORN	FUNCTION
MACRONUTRIENTS		
Nitrogen (N)	21	Structural component of amino acids, many hormones and coenzymes, etc.
Phosphorus (P)	5	Structural component of nucleic acids, phospholipids, ATP, coenzymes, etc.
Potassium (K)	16	Plays a role in the ionic balance of cells; cofactor for enzymes involved in protein synthesis and carbohydrate metabolism
Sulfur (S)	10	Structural component of two amino acids (cysteine and methionine) and of several vitamins
Magnesium (Mg)	7	Structural component of chlorophyll; cofactor for many enzymes involved in carbohydrate metabolism, nucleic acid synthesis, and the coupling of ATP with reactants
Calcium (Ca)	7	Influences permeability of membranes; component of pectic salts in middle lamellae and necessary for wall formation; activator for several enzymes
Iron (Fe)	0.3	Structural component of iron-porphyrins (hemes), which are incorporated into cytochromes, peroxidases, catalases, and some other enzymes; plays a role in the synthesis of chlorophyll
MICRONUTRIENTS		
Manganese (Mn)	0.04	Cofactor of many enzymes involved in cellular respiration, photosynthesis, and nitrogen metabolism
Boron (B)	0.008	Function unknown; may play a role in translocation of sugar; perhaps necessary for utilization of calcium in wall formation
Chlorine (Cl)	0.008	Plays an essential role in photosynthesis
Zinc (Zn)	trace	Necessary for synthesis of tryptophan (a precursor of auxin); activator of many dehydrogenase enzymes; may play a role in protein synthesis; found in some DNA-binding proteins
Copper (Cu)	trace	Structural component of many enzymes that catalyze oxidation reactions, and of plastocyanin, which is important in electron transport in chloroplasts
Molybdenum (Mo)	trace	Structural component of the enzyme that reduces nitrate to nitrite; essential for fixation of N_2 by nitrogen-fixing bacteria.
Nickel (Ni)	trace	Structural component of the enzyme that allows plants to use urea as a nitrogen source

[a]An element must meet three criteria to be regarded as essential: (1) it is needed for normal growth and reproduction in several different plants; (2) it is not replaceable by other elements; (3) its function is a direct one—in that it is not needed simply to correct a toxic condition induced by other substances.

HOW PLANTS FEED

We have seen that plants need three classes of nutrients: carbon dioxide, water, and minerals. In both aquatic and terrestrial plants, carbon dioxide is absorbed by the leaves. Since aquatic plants such as algae can also absorb the water and minerals they need from the water directly through their leaves, these plants usually lack specialized procurement organs like roots. Some nonaquatic plants that grow on other plants, often on the branches of large trees, also absorb all their nutrients through their leaves. Among these ***epiphytes*** are Spanish moss and many orchids. Some plants that normally obtain minerals through their roots can also absorb minerals through their leaves, but most higher land plants take in water and minerals primarily through roots.

■ HOW ROOTS WORK

Root structure The first root formed by the young seedling is called the ***primary root.*** Later, ***secondary roots*** (also called lateral roots) branch from the primary root, and a root system is formed. If the branching results in a system of numerous slender roots with no single root predominating, as in clover and many grasses, the plant is said to have a ***fibrous root system*** (Fig. 26.2A). If, however, the primary root remains dominant, with smaller secondary roots branching from it, the arrangement is called a ***taproot system*** (Fig. 26.2B). Dandelions, beets, and carrots have taproots. Taproots are frequently specialized as storage organs for the products of photosynthesis, though all roots store nutrients to some extent.

Roots also serve to anchor the plant. Usually the substrate is soil, but in addition to their normal root system in the ground, climbing vines commonly have short specialized roots arising from the stem that fasten the plant to a vertical surface such as a tree trunk or the side of a building. These aerial roots of vines are examples of ***adventitious roots***; the term *adventitious is* applied to any root that grows after the embryo stage from a structure that is not part of

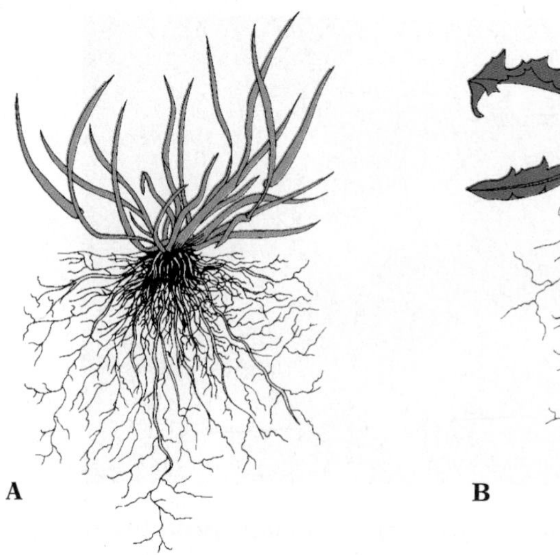

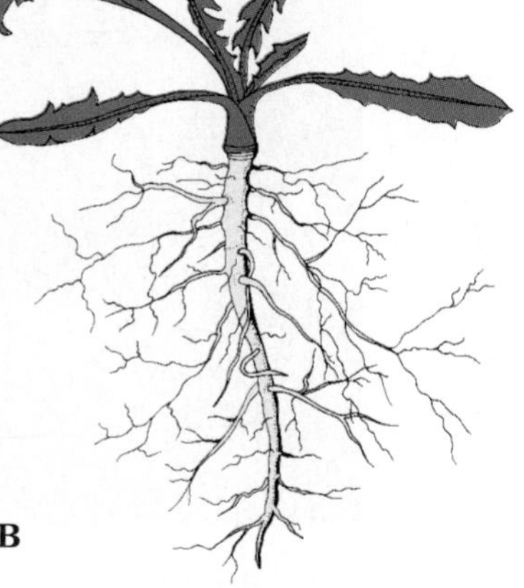

26.2 Two types of root system

(A) Fibrous root system of grass. (B) Taproot system of dandelion.

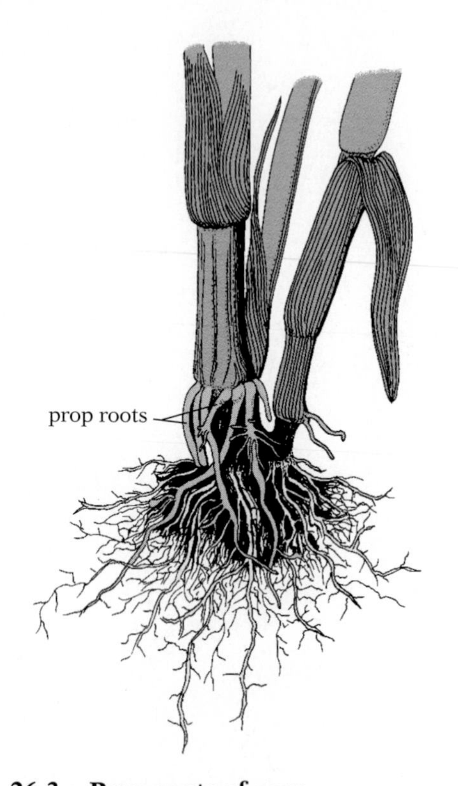

26.3 Prop roots of corn

These roots arise from a portion of the stem and are therefore adventitious roots.

the root system. The prop roots of corn are also adventitious roots; they arise from the lower portion of the stem (Fig. 26.3), penetrate the soil, and become important components of the root system.

The root system of a plant is normally far more extensive than you might realize. When you pull up a plant, you seldom get anything even approaching the entire root system, since most of the smaller roots are so firmly embedded in the soil that they break off and are lost. An extensive root system is important both in anchoring the plant and in providing sufficient absorptive surface. As a cell or organism gets bigger, its volume increases much faster than its surface area. A large multicellular organism therefore faces a serious problem: it needs an absorptive surface large enough to admit all the nutrients required to support its large volume. As an adaptation toward solving this problem, many organisms have a flat or hollow body plan, so that each cell is in direct contact with its surroundings, or at most only one or two cells away. This option allows kelp, for example, to absorb enough nutrients without true roots or any internal transport system. Nearly all terrestrial plants, on the other hand, have evolved extensively subdivided absorptive surfaces: a rye plant less than 1 m tall has some 14 million branch roots with a combined length of over 600 km.

Roots have another adaptation that increases their absorptive capacity. Just behind the growing tip of each rootlet, there is usually an area bearing a dense cluster of tiny hairlike extensions of the epidermal cells (Fig. 26.4). The zone of these ***root hairs*** on each rootlet may be anywhere from 1 cm long in some species to over 1 m in length. It is in this region that most absorption of water and minerals takes place. Even if the root-hair zone on any one rootlet is not very long, the number of root hairs on all the many rootlets is so vast that the total absorptive surface they provide is enormous. For example, the rye plant just mentioned may have as many as 14 billion root hairs with a total surface area of more than 400 m^2.

If the root of a typical young plant is viewed in cross section, the different tissue layers that form it become visible (Fig. 26.5). On the outer surface of the root is a layer of ***epidermis*** one cell thick.

26.4 Root of radish seedling with many prominent root hairs

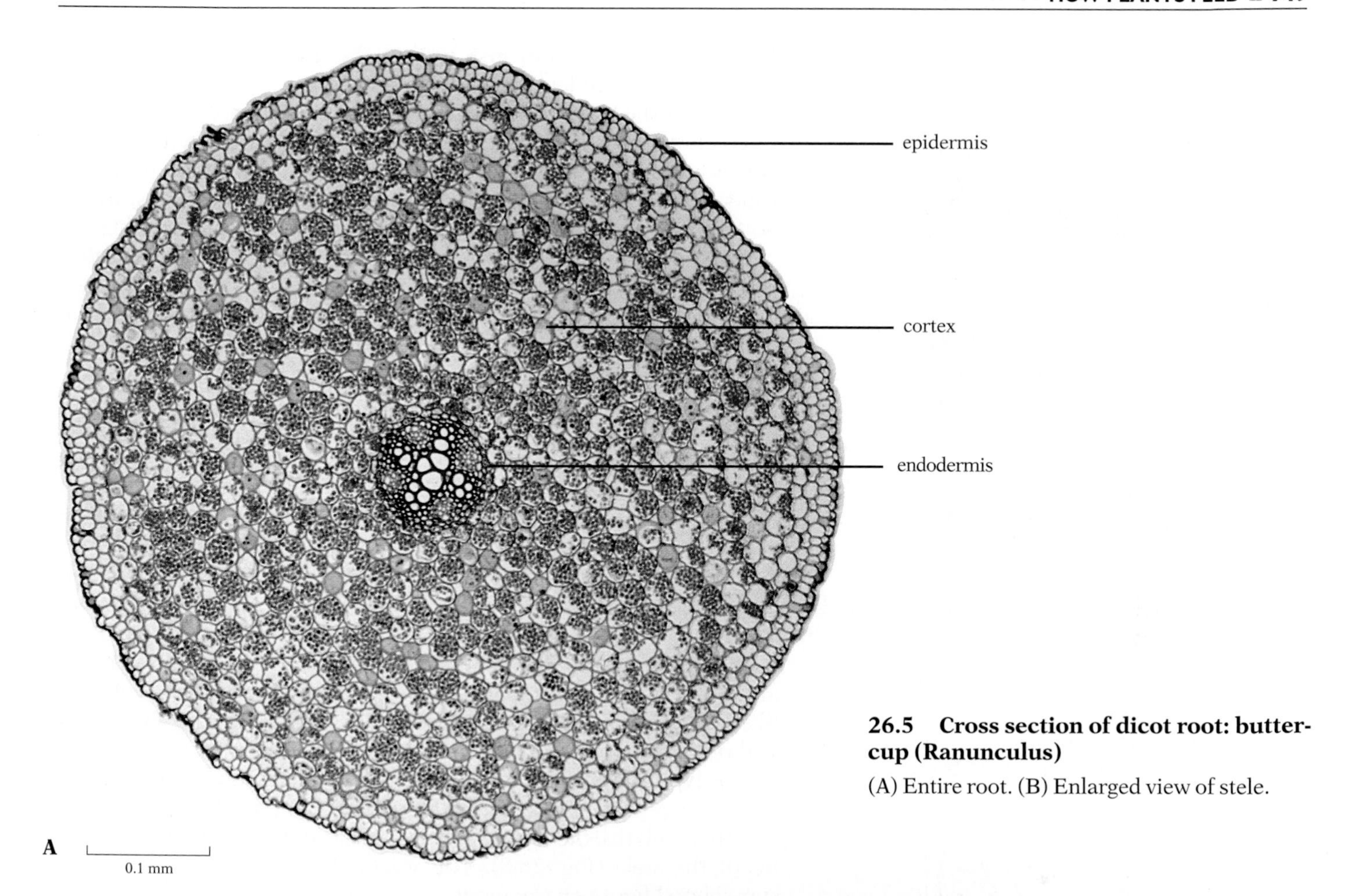

26.5 Cross section of dicot root: buttercup (Ranunculus)

(A) Entire root. (B) Enlarged view of stele.

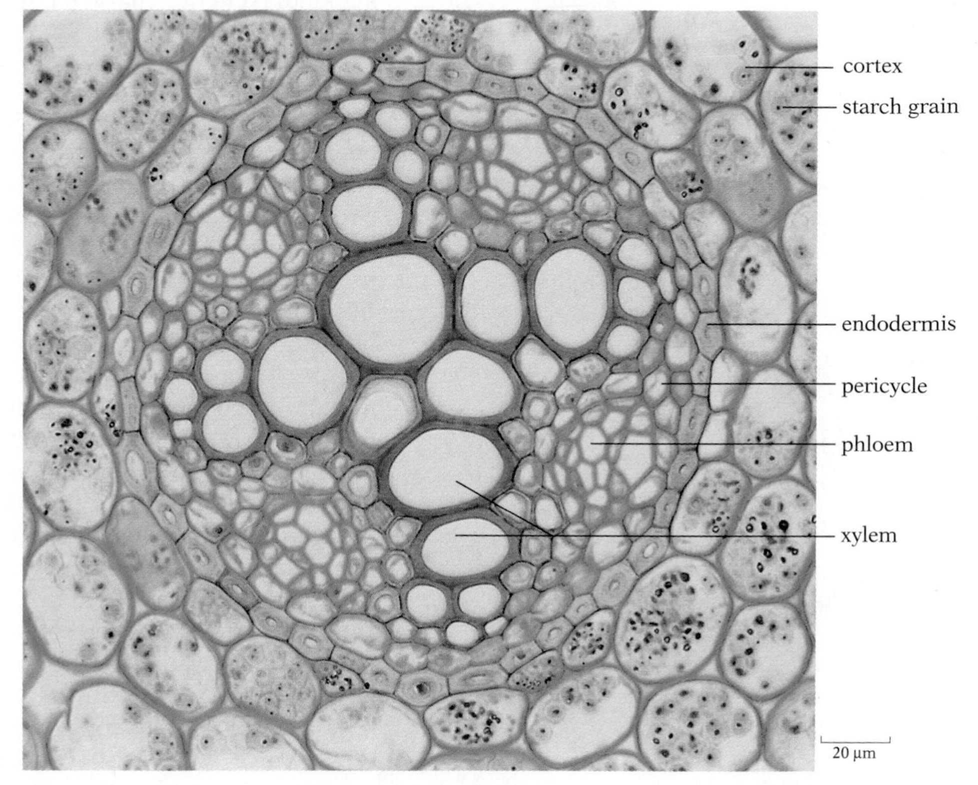

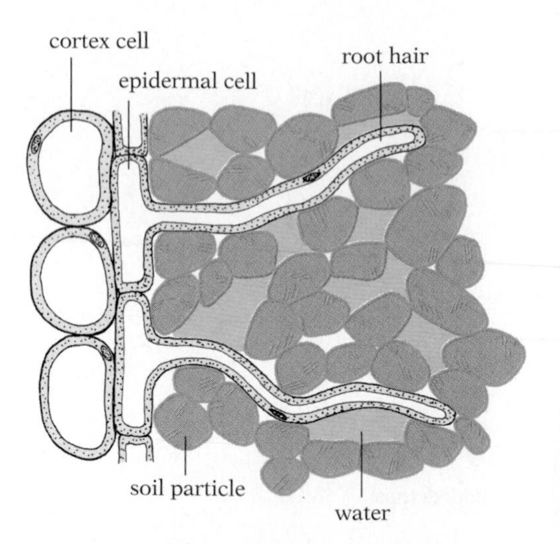

26.6 Root hairs penetrating soil

Each root hair, which is an extension of a single epidermal cell, is in contact with many soil particles (brown) and with soil spaces, some of which contain air, some water (blue).

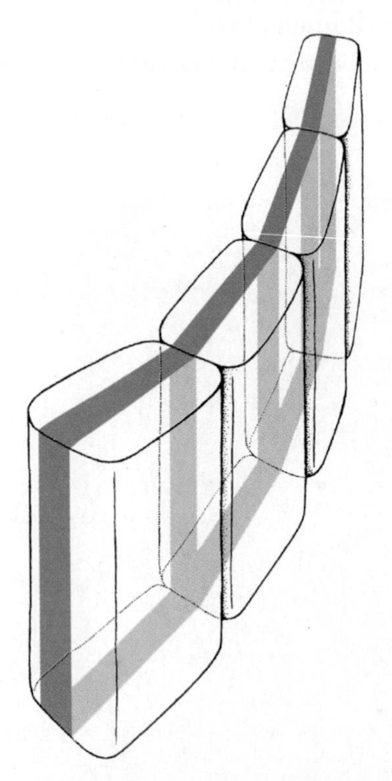

26.7 Endodermal cells with Casparian strip

The Casparian strip (color), located in the radial and end walls of each cell, forms a watertight barrier.

Each of the root hairs just behind the growing tip of a rootlet arises from an epidermal cell (Fig. 26.6).

Beneath the epidermis is the ***cortex***, a wide area with numerous intercellular spaces. Large quantities of starch are often stored in cortex cells. The cortex, so prominent and important in young roots, is frequently much reduced or even lost in older roots, where both cortex and epidermis may be replaced by a corky periderm.

The innermost layer of the cortex, one cell thick, is the ***endodermis*** (Fig. 26.5). Endodermal cells are characterized by a waterproof band, the ***Casparian strip***, which runs through their radial (side) and end walls (Fig. 26.7). All roots and some stems have a well-differentiated endodermis.

The endodermis forms the outer boundary of the central core of the root that contains the vascular cylinder. This core is called the ***stele***. Just inside the endodermis is a layer (often only one cell thick) of thin-walled cells, called the ***pericycle***. Cells of the pericycle can initiate new growth and give rise to lateral, or secondary, roots.

The central portion of the stele, surrounded by endodermis and pericycle, is filled with the two vascular tissues, ***xylem*** and ***phloem***. The thick-walled xylem cells often form a cross- or star-shaped figure (Fig. 26.5). Bundles of phloem cells are located between the arms of the xylem. Thus, instead of forming a continuous cylinder like the epidermis, cortex, endodermis, and pericycle, the xylem and phloem alternate in this part of the stele.

The typical root just described is characteristic of dicots, the largest group of plants. The large roots of monocots (the grasses and their relatives) commonly have an area of ***pith*** at the very center of the stele (Fig. 26.8). The xylem therefore cannot form the star-shaped figure characteristic of dicots, but the bundles of xylem and phloem still alternate.

Absorption of nutrients In the soil, rainwater generally becomes available to plants as a loose film around soil particles, known as capillary water. The roots, and particularly the root hairs, are in contact with this water. Capillary water can move into the roots in two ways. The water can cross a root-hair membrane, and then move from cell to cell through the cortex and across the endodermis to the core of the root. Once across the root hair membrane, this intercellular movement is usually unimpeded by further membranes because the cytoplasm of adjacent plant cells often form a ***symplast***, an association in which open channels (the plasmodesmata) interconnect the contents of neighboring cells. The symplast pathway is represented in Figure 26.9 as a light-blue arrow.

The other (and probably more important) route for water is the ***apoplast***, a network of cell walls and intercellular spaces that leads through the cortex (shown as a dark-blue line in Fig. 26.9). The Casparian strip prevents further apoplastic flow, so this water must now cross an endodermal cell membrane.

Whether it reaches the root core via the symplastic or apoplastic pathway, the water must enter a cell at some point. We can see how this could occur through simple osmosis: water flowing from regions of low solute concentration to areas with higher concentrations. Since cytoplasm typically has very high concentrations of

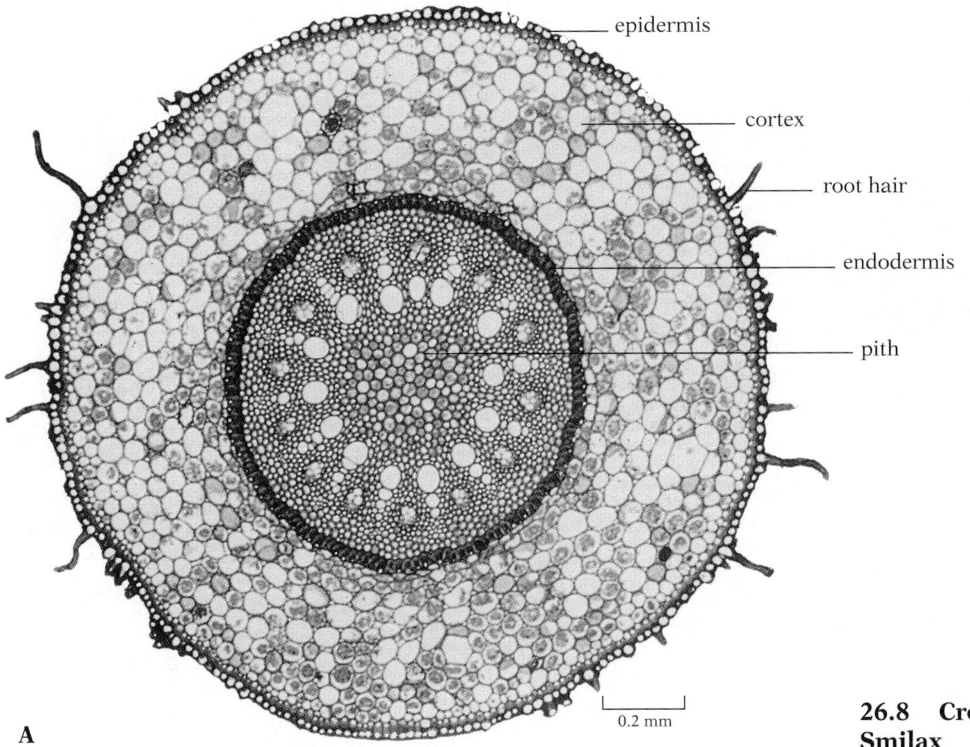

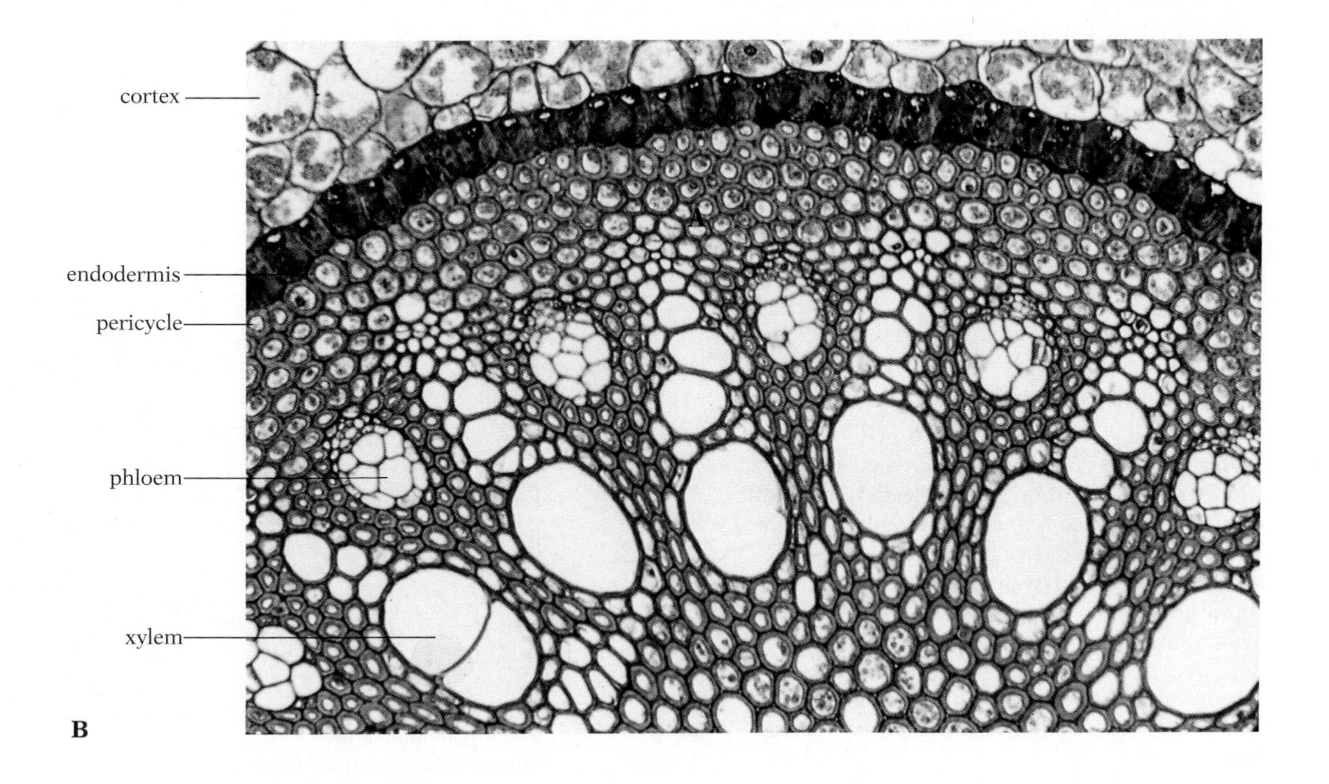

26.8 Cross section of monocot root: Smilax

(A) Entire root. (B) Enlarged portion of stele. The pericycle of this root, unlike that of the buttercup, is many layers thick. Note the very thick walls of the cells in the endodermal layer.

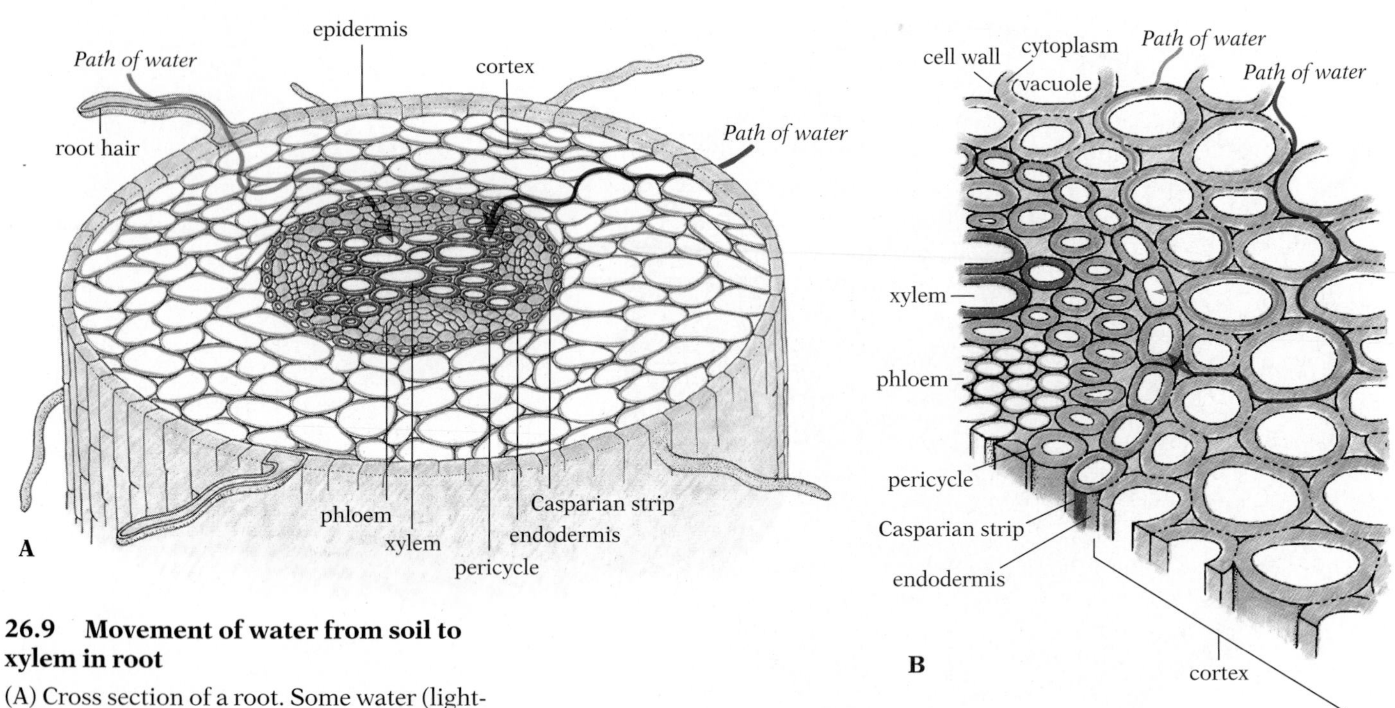

26.9 Movement of water from soil to xylem in root

(A) Cross section of a root. Some water (light-blue arrow) is absorbed by the epidermal cells, particularly the root hairs, and moves from cell to cell either by osmosis or by diffusion through plasmodesmata. Most of the water (dark-blue arrow) flows along cell walls and does not cross membranes of living cells until it reaches the endodermis. The Casparian strip (B, dark brown) of the endodermal cells prevents the flow of water along their radial and end walls; hence all water entering the stele must move through the living cells of the endodermis. (B) A region near the endodermis of a root. The pathway along cell walls (dark blue arrow) weaves through intercellular spaces.

dissolved substances like ions, sugars, and other hydrophilic organic molecules, water will tend to enter roots.

Plant scientists view the movement of water in terms of water potential—that is, the potential energy of the water on one side of a membrane compared with that on the other. From this perspective, all other factors being equal, water with fewer solutes has a higher potential than water with more dissolved substances, and thus the water flows from regions of higher potential to areas of lower potential (Fig. 26.10). This way of looking at water movement is especially appropriate in plants because factors other than relative osmotic concentration are often important. In particular,

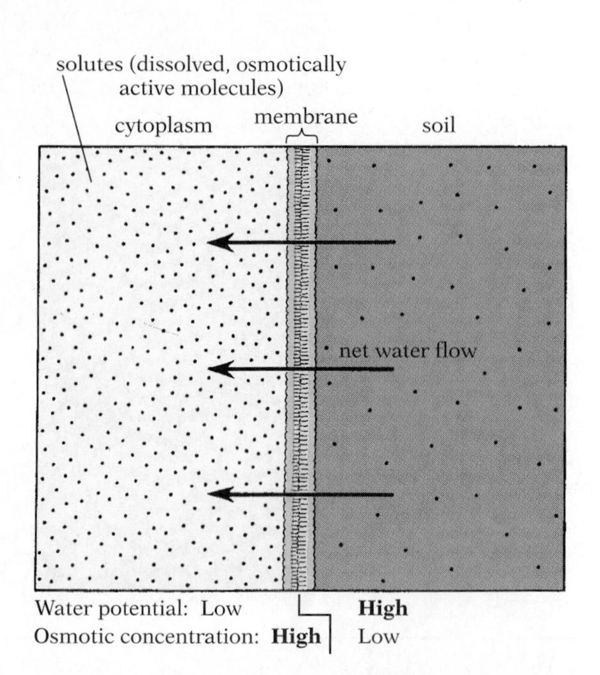

26.10 Water potential across plant membranes

Water potential is the pressure, created across a semipermeable membrane, that leads to the flow of water. It is the combined result of the osmotic concentration difference between the two sides *and* the difference in water pressure; only osmotic effects are considered in this illustration. Water flows from areas of higher potential (corresponding to lower osmotic concentration) to regions of lower potential. This terminology is described in more detail in the *Exploring Further* box on p. 95.

because plants have rigid cell walls, they can accept only a limited influx of water before the ***turgor pressure*** in the cell creates a potential energy difference that balances the osmotic pressure, producing a water potential of zero. Since it is essential for plants to maintain a net inward flow of water through the roots, mechanisms have evolved to help prevent zero or negative water potential in the roots.

Once water has entered an epidermal cell, it must be moved to where it is needed—usually the leaves and growing parts of the plant. This happens slowly but automatically, because plants maintain a gradient of ever-lower water potential from the outer roots to the xylem. The potential is set up in two ways. The dominant factor is the evaporation of water from leaf cells, which creates a very low water potential in leaves; in other words, there is little turgor pressure and a very high osmotic concentration. As a result, water is drawn in from the xylem, creating a lower water potential in the vascular system, which thus draws water in from the roots. The other factor occurs during cool, humid weather when the sky is cloudy or dark, and the evaporation rate is low; then molecular pumps transport ions into the xylem, creating a lower water potential (though at a significant metabolic cost). This strategy can also help when the soil is dry, and thus has a low water potential.

Now that we have seen how plants maintain a suitable gradient of water potential from leaves to roots, let's look at how nutrients are brought in with the water. Plants usually absorb minerals in ionic form: nitrogen is absorbed as nitrate (NO_3^-) or ammonium (NH_4^+) ions, phosphorous as dihydrogen phosphate ($H_2PO_4^-$) or monohydrogen phosphate (HPO_4^-), sulfur as sulfate ions (SO_4^-), and potassium, calcium, magnesium, and iron as their simple ions (K^+, Ca^{++}, Mg^{++}, and Fe^{++} or Fe^{+++}).

The ions available to plants for absorption are in solution in the soil water; their concentration varies according to the fertility and acidity of the soil and other factors. When the soil minerals are not in solution but bound by ionic bonds to soil particles, they are useless. Agricultural soil management often involves changing soil acidity to free bound minerals for absorption by roots. For example, adding lime to very acid soil in order to raise the pH may increase the availability of phosphorous, potassium, and molybdenum, but an excess of lime may decrease the available iron, copper, manganese, and zinc (Fig. 26.11).

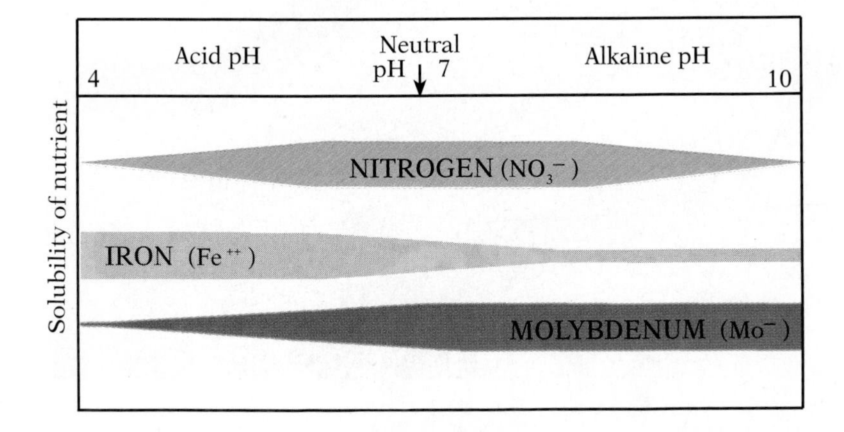

26.11 Solubility of three mineral nutrients as a function of pH

The changing width of each band indicates the relative solubility of the mineral between pH 4 and pH 10. Nitrogen (as NO_3^-) is most soluble, and hence most available to plants, in the neutral pH range; iron is most soluble under acid conditions, and molybdenum under alkaline conditions. Other minerals have their own distinctive solubility functions. Since it is obviously impossible for soil to have a pH at which all minerals will be maximally available to plants, farmers or gardeners must adjust the soil pH according to the nutrient requirements of the particular plants they wish to grow.

The rate of absorption of each mineral is independent of that of water and other minerals. Each nutrient moves into the root at a rate determined by such factors as its concentration both inside and outside the root, and the ease with which it can penetrate cell membranes through ion channels. When the concentration gradient favors inward movement, the rate of absorption is often greater than would be possible by passive diffusion alone—which means that facilitated diffusion is taking place. More often, however, plants take in minerals that are actually more concentrated inside the root cells than in the soil solution, and that would therefore move the other way if simple diffusion alone were involved. The plant clearly expends energy to transport its mineral nutrients actively.

To help minerals enter roots, plants must prevent their concentration in the root cells from becoming too high. As a result, a mineral may be transported to some other part of the plant as soon as it enters the root, or it may be utilized rapidly in the synthesis of a different compound. For example, nitrogen is quickly built into nitrogen-rich organic compounds such as amino acids, which are then transported and stored. Much of the storage is in cell vacuoles, where the concentration of the nitrogen compounds is often much higher than in the cytoplasm itself. This differential is an indication that the vacuolar membrane acts selectively, admitting the compounds into the vacuole but preventing their escape from it.

Efficient as the root systems of plants have become, there are still many soils in which plants cannot thrive unaided because proper nutrients either are not present, or exist in forms that are inaccessible to the plants. Fungi and plant roots sometimes form intimate associations known as mycorrhizae, which can greatly facilitate mineral uptake by the plants (Fig. 26.12). The fungi, some of which are visible on the soil surface as mushrooms, actually penetrate the root cells of their hosts, and transport nutrients in directly. Apparently the fungi also obtain nutrients from their hosts, so that the relationship is mutually advantageous, or mutualistic.

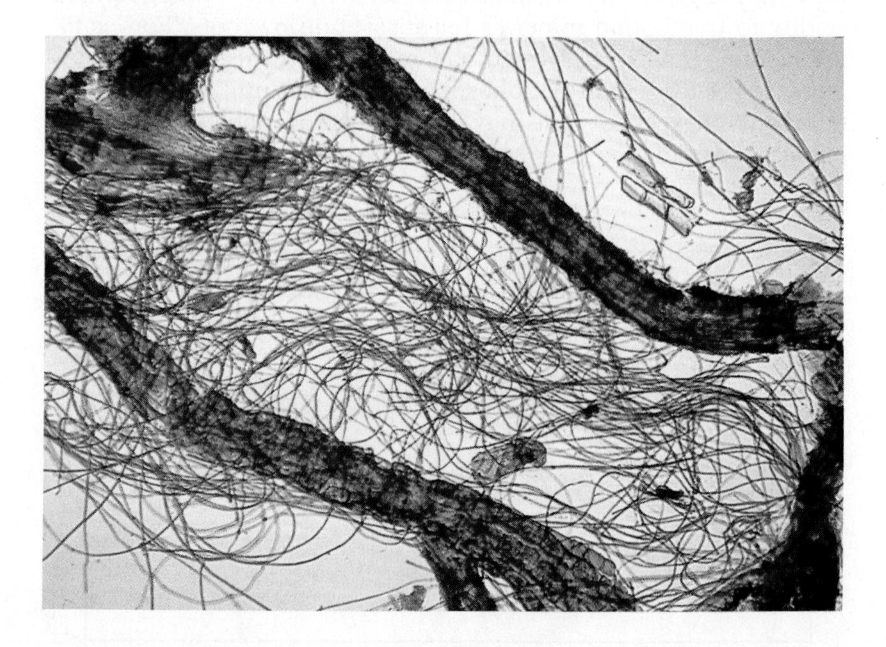

26.12 Mycorrhizae

■ WHERE PLANTS GET THEIR NITROGEN

Another mutualistic relationship between plant roots and other organisms is better understood. Nitrogen, which is essential to plants, is abundant in the atmosphere: air consists of about 78% nitrogen, 21% oxygen, and 0.03% carbon dioxide. Plants are amazingly successful at harvesting the minute amount of carbon dioxide in the air; however, they have never evolved the ability to capture the much more abundant atmospheric nitrogen. Instead, most plants depend on nitrogen-containing compounds absorbed from the soil. When the soil is poor, farmers must add nitrogenous fertilizer for their crops to grow well.

The plants of one group, the legumes (which include the lupines, clover, peas, peanuts, and alfalfa), have overcome dependence on soil nitrogen through mutualistic association with bacteria of the genus *Rhizobium*. These bacteria thrive in nodules on the plants' roots (Fig. 26.13). Molecular signals from the bacteria alter the development of the host plant's root hairs, leading to the formation of the nodules. The bacteria have powerful enzymes capable of breaking the triple bonds of molecular nitrogen and trapping (fixing) the nitrogen as ammonia (NH_3):

26.13 Roots of a legume (pea) with nodules

$$N_2 + 3H_2 \longrightarrow 2N + 3H_2 \longrightarrow 2NH_3 \quad (\Delta G = 147 \text{ kcal/mole})$$

The main enzyme involved, nitrogenase, is remarkably complex; it consists of two subunits incorporating 30 metal ions—28 irons and two molybdenums. Exposure to oxygen destroys this complex.

Since this highly endergonic reaction can take place only under anaerobic conditions, most oxygen-producing photosynthetic organisms like plants are unable to fix nitrogen directly. Even *Rhizobium* is not entirely anaerobic: it requires oxygen for one step in its metabolism. The way in which the oxygen problem is solved illustrates how the bacterium and its host legume have coevolved. The legume manufactures a protein called leghemoglobin, which binds oxygen in the vicinity of the bacterium's nitrogen-fixing enzymes, thereby maintaining nearly anaerobic conditions; the leghemoglobin then delivers the oxygen to the bacterial enzyme requiring it. Much of the fixed nitrogen is donated to the legume, which grows far better than plants without the benefit of bacteria (Fig. 26.14); the legume, in turn, supplies the bacteria with photosynthetically produced sugars. Modern agriculture makes use of this mutualism. Legumes such as clover are planted in rotation with other crops and plowed under to restore depleted nitrogen supplies, and legume crops like soybeans and peanuts are raised in less fertile soil.

26.14 Use of bacteria to improve nitrogen-fixing by legumes

In Africa, legume seeds are being inoculated with *Rhizobium* bacteria to increase the nitrogen-fixing capacity of the plants. In this test patch, inoculated plants are green, uninoculated (nitrogen-starved) plants, yellow.

Several groups of bacteria fix both nitrogen and carbon themselves, but the two processes are not connected. Many fix carbon using energy from cyclic photophosphorylation, which does not generate O_2 as a by-product. The most self-sufficient nitrogen-fixing bacteria are certain cyanobacteria, which can survive in almost any habitat, from volcanic springs to barren rocks. Some cyanobacteria can fix nitrogen *and* perform high-efficiency, oxygen-generating noncyclic photophosphorylation. These two chemically incompatible processes can be reconciled if the nitrogen-fixation system is sequestered in separate (anaerobic) compartments

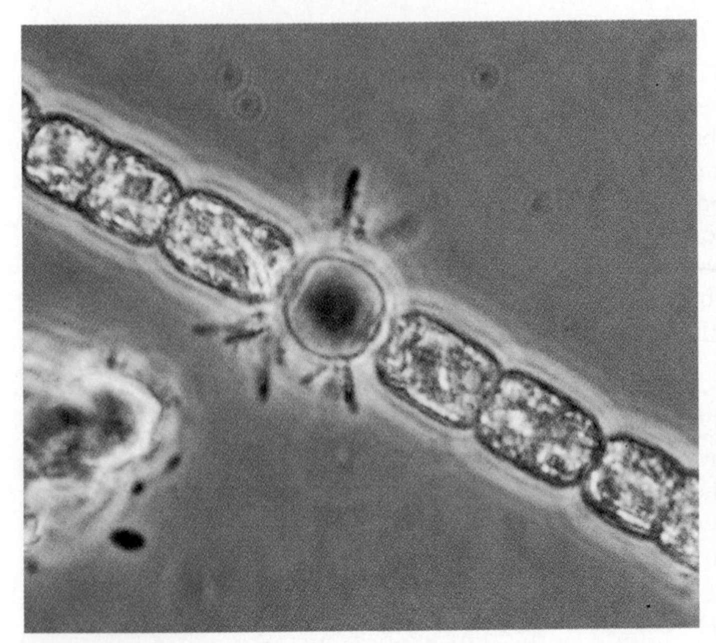
A

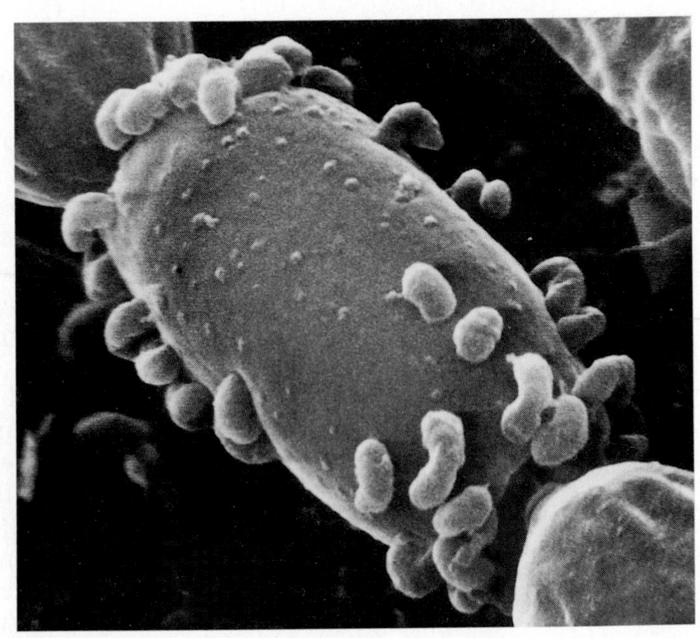
B

26.15 Nitrogen-fixing cyanobacteria

Some species of cyanobacteria form chains (A) consisting of photosynthetic cells and an occasional anaerobic nitrogen-fixing compartment known as a heterocyst. Such chains begin as a single cell. Planktonic cyanobacteria benefit from the sharing of products between the photosynthetic cells and the heterocysts. The heterocysts metabolize the bonds in molecular nitrogen anaerobically to produce ammonia, which is then utilized to produce amino acids and proteins, while the photosynthetic cells produce sugars that the heterocysts utilize as a source of energy. In some species of cyanobacteria, nitrogen fixation can be augmented indirectly by other bacteria, which congregate around the active heterocyst and ingest the excess amino acids that leak from it (B). This relationship appears to be of mutual benefit—in "return" for the amino acids, the metabolic activity of the bacteria consumes the oxygen in the immediate environment of the heterocyst, thus increasing its efficiency in nitrogen fixation.

known as ***heterocysts*** (Fig. 26.15). Other systems accomplish the necessary chemical separation by performing noncyclic photosynthesis during the day and fixing nitrogen at night. Up to 80% of the photosynthetic plankton of the oceans (which provide much of the oxygen in the atmosphere and virtually all the food for marine organisms) are aerobic, nitrogen-fixing cyanobacteria. One of the most challenging goals for modern applied biology is to use the techniques of genetic engineering to develop crop plants that can fix atmospheric nitrogen themselves.

■ WHY DO SOME PLANTS EAT ANIMALS?

A few photosynthetic plants supplement their inorganic diets with organic compounds that they obtain by trapping and digesting bacteria, insects, and other small organisms. Such plants are true autotrophs, since they can survive without capturing any prey;

26.16 Pitcher plant *(Sarracenia purpurea)*
Several fruit flies are trapped within the leaf.

when they do capture prey, however, the nutrients stimulate more rapid growth. In nutrient-poor lakes, for instance, deep-living algae make up for the low intensity of light by capturing and digesting bacteria. A typical algal cell consumes 30-40 bacteria per hour, and may obtain more than half its carbon in this way.

Insectivorous plants, which often grow in nitrogen-poor soils—particularly acid bogs and heavy volcanic clays—use the nitrogen compounds of their prey as nutrients. Pitcher plants (Fig. 26.16) trap their prey in leaves that are modified into tubes or sacs, partly filled with water. Insects that fall into the sacs are kept from climbing out by stiff downward-pointing hairs. Enzymes secreted into the water digest the proteins of the trapped insects, and the products of this digestion are absorbed by the inner surface of the leaf.

The leaf of the Venus flytrap (Fig. 26.17) takes a more active role in prey capture. It is composed of two lobes hinged by a midrib;

26.17 Leaf of Venus flytrap *(Dionaea muscipula)*
(A) One of the fly's legs is about to touch a trigger hair on the leaf surface. (B) The two halves of the leaf have quickly moved together, interlocking their marginal teeth and thus trapping the fly.

long stiff teeth line the margin of each lobe. When an insect touches small sensitive hairs on the surface of the leaf, the lobes quickly change shape and come together with their teeth interlocked. When the trigger hairs are stimulated, they initiate a rapid pumping of H^+ ions through the cell membrane, consuming about 30% of the cells' available ATP in 2 sec. The pH change in turn leads to a quick osmotic movement of water from the intercellular spaces into the cells at the base of the trap; their consequent rapid enlargement causes the leaf to close on its victim. The trapped animal is then digested by enzymes secreted from glands on the leaf surface, and the resulting amino acids are absorbed.

The leaves of sundews (Fig. 26.18) bear numerous hairlike tentacles, each with a gland at its tip. Small insects, attracted by the plant's odor, become trapped in the sticky fluid secreted by the glands. The stimulus from a trapped insect causes nearby tentacles to bend over toward the animal, further entangling it. The proteins of the insect are digested by enzymes, and the amino acids are absorbed.

26.18 Leaf of sundew *(Drosera intermedia)*

A damselfly is caught in the sticky fluid on the ends of the glandular hairs.

CHAPTER SUMMARY

AUTOTROPHS VERSUS HETEROTROPHS

Autotrophs can synthesize organic compounds from inorganic materials; heterotrophs require energy-rich organic molecules in their diet. (p. 743)

WHAT PLANTS ARE MADE OF

EARTH, AIR, OR WATER? The carbon and oxygen—together the elements that contribute most to a plant's mass of organic molecules—come from CO_2. The CO_2 comes from the air, most of the hydrogen comes from water, while the minerals used by plants mostly come from the earth. (p. 744)

WHY PLANTS NEED MINERALS Minerals are needed in amino acids and nucleic acids, at the active sites of many enzymes, and as components of coenzymes. (p. 745)

HOW PLANTS FEED

HOW ROOTS WORK Water containing minerals enters roots through root hairs or along cell walls; it crosses through endodermal cells, and then enters the xylem. The water usually enters because of a water-potential gradient generated by the evaporation of water from leaves. (p. 747)

WHERE PLANTS GET THEIR NITROGEN Nitrogen is obtained either as a dissolved soil compound (e.g., NH_3) or from symbiotic nitrogen-fixing bacteria, which convert N_2 into NH_3. (p. 755)

WHY DO SOME PLANTS EAT ANIMALS? Most often to obtain nitrogen that is lacking in the soil. (p. 756)

STUDY QUESTIONS

1 Kelps have thin leaves to maximize exposure to nutrients in the water. Land plants get most of their nutrients from roots, and yet still have thin leaves. What reasons might favor retention of the thin-leaf design? (p. 747)

2 Are there any waste products that plants need to void themselves of, and if so, how do they manage it?

3 One obvious trend in plant evolution is increasing size. What are the costs and benefits of being bigger?

4 Why do you suppose that most of the mass of a tree is dead?

5 One of the challenges imposed by moving out of the water is the lack of temperature buffering: water temperatures vary little and only slowly compared with air temperatures. Is there any obvious way terrestrial plants could still use water (taken up by the roots) to moderate internal temperature?

SUGGESTED READING

BEARDSLEY, T., 1991. A nitrogen fix for wheat, *Scientific American* 244 (3). *On inducing nodule formation on wheat.*

EPSTEIN, E., 1973. Roots, *Scientific American* 228 (5). *The mechanisms by which roots take up nutrients from the soil.*

HESLOP-HARRISON, Y., 1978. Carnivorous plants, *Scientific American* 238 (2).

CHAPTER 27

ANIMAL NUTRITION AND DIGESTION

Unlike plants and other autotrophs, heterotrophs cannot synthesize high-energy compounds from low-energy inorganic raw materials. To survive, they must feed on the high-energy molecules of other organisms. There are four main groups of heterotrophs: nonphotosynthetic bacteria, fungi, nonphotosynthetic protozoans, and animals. This chapter focuses mainly on how animals obtain the nutrients they need.

Bacteria and fungi lack internal digestive systems and hence depend mainly on absorption for feeding. They are usually either ***saprophytic*** (living and feeding on dead organic matter) or ***parasitic*** (living on or in other organisms and feeding on them). By contrast, the principal mode of feeding for animals and protozoans is ingestion—taking in and digesting particulate or bulk food. Animals and protozoans may be ***herbivores***, in which case they obtain high-energy compounds by eating plants or other photosynthesizers, or they may be ***carnivores***, eating animals or protozoans that have fed on plants. Some animals, the ***omnivores***, need both plant and animal material to survive. Much of the morphological diversity among living things reflects their different ways of employing the three major modes of nutrition—photosynthesis, absorption, and ingestion.

WHAT HETEROTROPHS NEED TO EAT

■ WHY ISN'T SUGAR ENOUGH?

Carbohydrates, fats, and proteins are the main energy sources for heterotrophs. Of these, carbohydrates alone would suffice if organic nutrients functioned only as an energy source. In fact, adult bees and hummingbirds can survive solely on carbohydrate-rich nectar until they attempt to reproduce (Fig. 27.1). Because all amino acids and nucleotides incorporate nitrogen, which carbohydrates lack, reproduction and growth require a source of fixed nitrogen. Many bacteria and fungi, as well as a few protozoans, can flourish and reproduce on a diet consisting solely of carbohydrates, fixed nitrogen, and minerals; like plants, they can synthesize for themselves all the other classes of compounds necessary for life.

Animals are especially deficient in synthetic ability. Among their

27.1 A honey bee gathering pollen and nectar

Adult worker honey bees can survive on the sugars in the nectar they collect from plants; the pollen (which is packed into pollen baskets on the rear legs) is used as a protein source, and is fed to growing larvae and the egg-laying queen.

many dietary requirements are at least some of the amino acids of which proteins are composed. Other amino acids can only be synthesized if nitrogen-containing organic molecules are available.

The loss of a synthetic pathway may seem maladaptive. But organisms can be at a competitive disadvantage if they use up their valuable raw materials and energy fabricating organic compounds that are readily available in their diet. Thus, the loss of unnecessary chemical pathways can be adaptive, and mutations in the genes coding for one or more of the enzymes necessary to synthesize common dietary compounds are not deleterious. Such mutations do, however, commit a species to an increasingly specific range of food. Hence, an organism's dependence on particular organic compounds implies that at some point in its evolution the compounds in question were readily available. Since nutritional needs tend to reflect a species' past diet, they provide valuable clues about the course of evolution.

Most animals have lost the ability to synthesize certain amino acids and must get them from their diet. These are called ***essential amino acids***. (All amino acids are necessary, but only a few are essential *in the diet;* the others can be synthesized by the organism itself from organic nitrogen compounds in the diet.) The essential amino acids vary for different species of animals, and even for different stages in the life history of the same species. Nine amino acids (histidine, isoleucine, leucine, lysine, methionine, phenylalanine, threonine, tryptophan, and valine) are essential for almost all animals; others may be essential for some, but not for all.

27.2 A child suffering from kwashiorkor

Kwashiorkor is a severe protein-deficiency disease caused by eating protein from a single source, such as rice or corn. The most obvious external symptom is inflammation of the skin, evident here on the child's arms and legs.

To accumulate enough of all the essential amino acids, animals must normally ingest several different proteins, since a single protein may not contain them all. Zein, the main protein in most varieties of corn, for example, is deficient in tryptophan and lysine. An animal that depended exclusively on zein would suffer from a deficiency not only of these two amino acids but of other essential amino acids as well, since protein synthesis requires that all the essential amino acids be available for use simultaneously. If there is not enough of one, utilization of the others is reduced proportionately, and since amino acids cannot be stored, they are lost through excretion. Specialist species like the corn moth, whose larvae dine exclusively on one plant, retain the ability to synthesize the amino acids missing in their host.

Nutritionists recommend that an average adult human male include in his daily diet at least 70 g (about 2.5 ounces) of protein, of which at least half should be of animal origin. The proportions of the various amino acids in plant proteins are often quite different from those in animal proteins; hence plant proteins are less reliable than animal proteins as a source of essential amino acids for humans. ***Kwashiorkor*** (Fig. 27.2), a protein-deficiency disease characterized by degeneration of the liver, severe anemia, and inflammation of the skin, is particularly common among children in countries where the diet consists primarily of a single plant material—as in Indonesia, where rice is the main food, and in parts of Africa, where corn is the principal staple.

A human following a vegetarian diet must take care to select a combination of plant proteins that will complement one another, making up for one another's deficiencies. For example, the proteins in beans are deficient in methionine, whereas those in wheat are deficient in lysine; if both beans and wheat are eaten daily (preferably at the same meal), they will complement each other and there will be no deficiency of either peptide. Since amino acids cannot be stored in the body, eating beans one day and wheat the next would be futile.

Some animals can survive and grow with little or no fat in their diets because they can interconvert carbohydrate and fat. Other animals (including humans) cannot synthesize enough fatty acids, no matter how many other organic compounds are available in the diet. Severe disease symptoms or even death may result from a lack of what are, for some species, ***essential fatty acids***.

The nutrient requirements for our species reinforce most current theories of human evolution, which picture early humans as omnivorous hunter-gatherers. Regularly ingesting animal protein and fat along with plant material, these humans experienced no selection pressure to maintain the synthetic pathways for substances provided ready-made by the diet. Because these pathways have been lost, a strict vegetarian diet is now inherently artificial in our species, and care must be taken to balance the amino acids and to include animal fats (usually from dairy products).

■ WHAT VITAMINS DO

Vitamins are organic compounds required in small quantities in the diet by organisms that cannot synthesize them. A compound may be a vitamin for one species and not for another, because the second species can synthesize it. Vitamins are necessary only in very small quantities. They usually function as coenzymes or as parts of coenzymes, catalysts that can be reused many times and hence are not needed in large amounts.

That certain diseases are connected with dietary deficiencies, now identified as vitamin deprivation, was recognized long ago. By 1750 fresh fruit was known to prevent ***scurvy***, a painful disease common among sailors long at sea; the symptoms are bleeding gums and loosening teeth, anemia, delayed healing of wounds, and painful and swollen joints. Lime and lemon juice were made standard parts of rations for British sailors—hence their nickname "limeys." The need for the vitamin C (Fig. 27.3A) provided by these citrus fruits reflects our evolution as omnivores: our species has always been able to ingest fresh vegetables, fruits, and berries. Carnivores do not suffer from scurvy despite the lack of fruit in their diets. Since they have never obtained much vitamin C from external sources, natural selection has worked to prevent the loss of the metabolic pathway for synthesizing this vitamin.

Beriberi is another dietary-deficiency disease; it came to light among Dutch troops in the East Indies, and is characterized by muscle atrophy, paralysis, mental confusion, and sometimes congestive heart failure. A team of investigators discovered that chickens that were fed primarily on polished rice accidentally dropped in the kitchen and dining area of the military quarters developed symptoms similar to those of human beriberi. Because it was so cheap, polished rice was the main food supplied to the troops. When unpolished rice was added to the diet, neither chickens nor

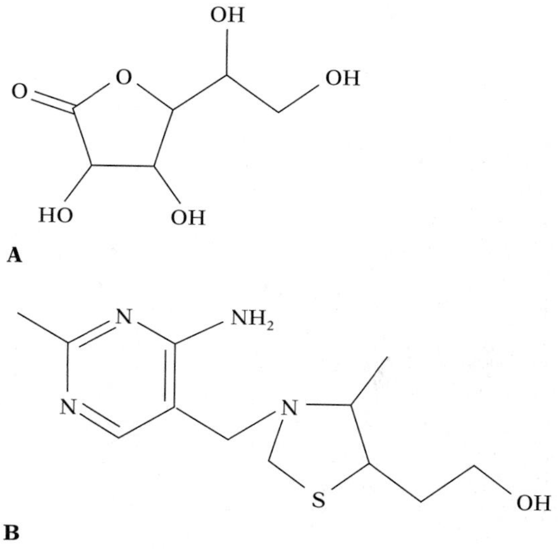

OH O OH O HO OH
A

N NH_2 N N S OH
B

27.3 Two important vitamins

(A) Vitamin C is a powerful reducing agent. The symptoms of scurvy occur when there is too little of the vitamin to keep the iron atom of a collagen-synthesizing enzyme in its reduced (ferrous) form; the result is insufficiently hydroxylated collagen, which is structurally weak. (B) Vitamin B_1, in its active form, has a chain of two phosphate groups in place of the hydroxyl group at the far right. One of the vitamin's functions is to act as a coenzyme in the conversion of pyruvate (the end product of glycolysis) to acetyl-coenzyme A (the starting point of the citric acid cycle). Lack of vitamin B_1 therefore reduces the rate at which the citric acid cycle can operate. Since this cycle harvests far more of the energy of foodstuffs than glycolysis alone, metabolic energy falls to the low levels that lead to the symptoms of beriberi.

humans developed the disease. The antiberiberi factor is water-soluble and is removed by polishing; it is now called vitamin B_1, or thiamine (Fig. 27.3B)

Determining reliable minimum daily requirements for vitamins is very difficult, and the established requirements are only rough approximations. Very little is known about how these needs change with age or health. One thing can be asserted with reasonable confidence: healthy people who eat a varied diet including meats, fruits, and vegetables will probably get all the vitamins they need.

The relation between the pathological symptoms of a vitamin deficiency and the actual biochemical function of the vitamin is often obscure. For example, the symptoms of beriberi give no indication that vitamin B_1 functions in the conversion of pyruvic acid into acetic acid and carbon dioxide. In fact, the exact biochemical roles of many vitamins are still unknown, despite extensive clinical information on the symptoms that a lack of them will create.

Water-soluble vitamins Vitamins can be classified as water-soluble or fat-soluble. For humans the water-soluble vitamins include vitamin C and the vitamins of the B complex, all of which tend to occur together. These substances often function as coenzymes in metabolic reactions that take place in almost all animal cells. Some animals can synthesize one or more of these coenzymes, and for them, of course, such coenzymes are not vitamins.

One of the most important functions of vitamin C, or ascorbic acid, is its role in the formation of collagen fibers, which are the chief components of connective tissue. When the diet is severely deficient in ascorbic acid, collagen formation ceases and scurvy results. Severe scurvy is rare among adults in developed countries, but occurs occasionally in infants; very mild cases, which are difficult to recognize, are more frequent. Since fresh fruits or vegetables in the diet provide an ample supply of ascorbic acid, supplements are usually advisable only for infants, pregnant women, and the seriously ill. Citrus fruits are a good source of vitamin C but not the best. They got their reputation because they travel well—they last much longer than richer sources of vitamin C like fresh cabbage, peas, and beans—and during a long sea voyage or a long winter, endurance counts.

The B complex includes a large number of compounds, unrelated chemically, that are somewhat similar in function. Several of them are components of coenzymes functioning in cellular respiration. Thiamine (vitamin B_1) is a part of the coenzyme that catalyzes the oxidation of pyruvic acid. The B vitamin pantothenic acid is a component of coenzyme A, which plays an essential role in carrying the acetyl group into the Krebs cycle. Riboflavin (vitamin B_2) is one of the carrier compounds in the respiratory electron-transport system. Pyridoxine (vitamin B_6) is a component of a coenzyme involved in transaminations—reactions transferring amino groups from one compound to another. Nicotinamide, another B vitamin, is a major component of both NAD and NADP. (Commercial vitamin preparations often contain niacin, which is converted into nicotinamide in the body.) The nicotinamide-deficiency disease ***pellagra*** is a severe problem in many poor areas, but it can also be caused by chronic alcoholism.

Some of the B vitamins—particularly B_{12} (cobalamin), which contains cobalt—seem to be involved in the formation of red blood corpuscles. Vitamin B_{12} deficiency results in ***pernicious anemia,*** a chronic disease more common in older people. This vitamin, like several others (vitamins E and K, niacin, pantothenic acid, and folic acid), is usually synthesized in mammals by microorganisms in the digestive tract, where it is absorbed. When humans develop pernicious anemia, the problem is usually an inability to absorb or convert vitamin B_{12} into an active form.

Folic acid, another B vitamin, is apparently also involved in the formation of red blood corpuscles. Its primary role, however, is in the synthesis of some of the nucleotides, which are building blocks for nucleic acids, making it essential for cell division (Fig. 27.4). Unlike the fat-soluble vitamins discussed next, water-soluble vitamins are not stored: we must replenish our supply of these substances frequently.

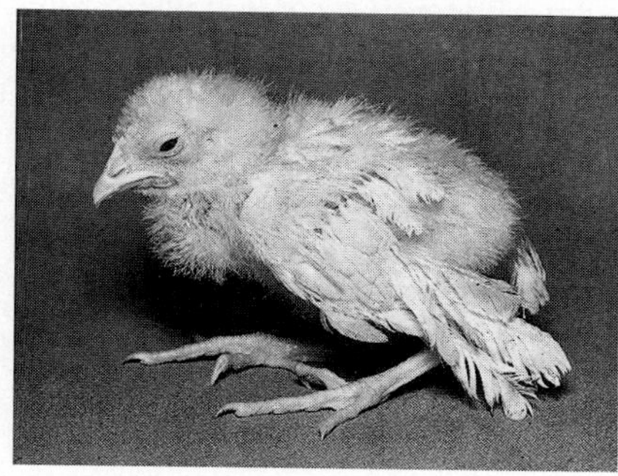

27.4 Folic-acid–deficient chick

Both birds are 4 weeks old. The one below was fed a diet deficient in folic acid, while the one above received a plentiful supply of the vitamin.

Fat-soluble vitamins The vitamins A, D, E, and K are the principal fat-soluble coenzymes. Symptoms of vitamin A deficiency include retarded growth, excessive keratinization of epithelia (hardening by deposition of keratins, the chief components of claws, nails, and horns), degeneration of epithelia, and ***xerophthalmia***—a keratinization of tissues of the eye that can lead to permanent blindness. A less extreme manifestation of vitamin A deficiency is night blindness: since vitamin A is a component of the light-sensitive pigment in the rod cells of the eye, lack of adequate vitamin A can cause a marked impairment of vision in dim light (Fig. 27.5). Vitamin A can be synthesized from the carotenoids in green and yellow vegetables and fruits (such as β-carotene, which gives carrots their characteristic color). These precursors are also abundant in butter, cheese, milk, and egg yolk.

Vitamin D is involved in calcium absorption and metabolism. In children a deficiency results in ***rickets***. The growing skeleton be-

A

B

27.5 Night blindness

(A) Road as seen by a normal individual. (B) Road approximately as seen under the same lighting conditions by an individual with a vitamin A deficiency. That person cannot see the road sign at all.

HOW WE KNOW:
VITAMINS FIGHT BACK

Vitamins were originally defined as complex molecules found in the diet that are needed in small quantities for general health. When the early vitamins were found to be coenzymes, or precursors of other important molecules (like the visual pigments), the assumption crept into biological orthodoxy that vitamins are passive building blocks in the body's chemistry. When an otherwise well-fed animal is sickly, we presume that a critical mineral or vitamin is missing, and therefore some building block is unavailable. Only recently has a more active and aggressive role for certain vitamins become apparent.

The most prominent examples of this unconventional role are provided by the antioxidant vitamins A, E, and C, which can detoxify reactive compounds produced in the course of normal cellular metabolism. They also help protect animals against certain compounds produced by plants that have evolved specifically to kill herbivores. Perhaps the first well understood example of this interaction involved St. John's wort *(Hypericum)*. This weed thrives in Europe and the Americas because most grazing animals and herbivorous insects avoid it. And no wonder: mammals that feed on it become ill and develop serious skin rashes; caterpillars eating the leaves generally die.

In the 18th century, scientists first noticed that black sheep are relatively immune to the toxic effects of St. John's wort, whereas white sheep are susceptible. Later, it became clear that most of the herbivorous insects that specialize on this weed hide themselves during the day and forage only at night, or burrow inside the plant. This observation suggested that the toxin depends on light, and led to the discovery that the chemical responsible, now called hypercin, is a receptor for blue light.

Hypercin uses captured solar energy to create free radicals from O_2 in the epidermal cells of its victims. Hyperacin molecules rapidly dispose of the light energy they capture by using it to split an oxygen molecule; they are then ready to repeat the process. The free radicals thus produced react with the first molecule they encounter, destroying it or altering its chemical properties. Slowly the cell is poisoned. Only dark-pigmented animals or insects that shun the light are safe from this phototoxin. On spotted animals, only the light patches are susceptible; dark spots are safe.

But there are a few caterpillars that feed on St. John's wort and other phototoxic plants while exposed to sunlight. This observation suggested that the caterpillars have some chemical defense in place. Scientists at the University of Ottawa isolated the major detoxicant in one species: β-carotene, the precursor of vitamin A in vertebrates. In another species, the molecular protection comes in the form of vitamin E, which had been thought until then to contribute to some

comes deformed because the bones, lacking sufficient calcium, are very soft. Exposure to sunlight is the best preventive for rickets, since the ultraviolet radiation in sunlight acts to produce vitamin D in the skin. Rickets is therefore largely confined to the temperate zones, where people spend much time indoors and, when outdoors, wear clothing that shields most of the body from sunlight. It is almost unknown in the tropics, where our species evolved and where, therefore, it is technically not a vitamin. Vitamin D is particularly abundant in egg yolk, milk, and fish oils.

Vitamin E is important in rats for maintaining good muscle and nerve condition, normal liver function, and male fertility, and for preventing rupture of red blood corpuscles. Since rats are omnivores, their dietary requirements are almost identical to ours, and it can be argued that vitamin E is likely to be equally important to human beings. However, a deficiency of this vitamin occurs only rarely in humans, and no clear deficiency symptoms are known.

Essential or not, vitamin E has one clearly beneficial effect. Many mutagenic and potentially cancer-causing substances have an exposed oxygen atom with a vacancy in its outer orbital. The most common example is the unstable compound hydrogen perox-

compound associated with male fertility in laboratory rodents, and whose importance to humans was considered a figment of the health-food industry's imagination. In fact, excessive quantities of vitamin E are probably essential to the health of actively reproducing albino rats simply to keep them safe from light-induced mutations.

Where do these defensive chemicals come from? They originate mainly in plants. (Although they are found in meat and other animal products, the animals obtain them originally from plants in their diets.) It seems likely that plants, in turn, synthesize these antioxidants to detoxify the oxidants produced by the high-energy photons that relentlessly irradiate their leaves. Even St. John's wort must defend itself against the sun by generating a steady supply of antioxidants. It survives its own phototoxins only by keeping them carefully sequestered in safe (but digestible) packaging, ready to poison foraging herbivores.

ide (H_2O_2). Because oxygen is highly electronegative, such compounds are highly reactive and therefore potentially disruptive. Worst yet, the oxygen can separate from the larger molecule ($H_2O_2 \rightarrow H_2O + O^*$), producing an unbonded singlet oxygen (O^*), which is the most reactive entity found in cells. Dangerous molecular subunits like singlet oxygen are called ***free radicals***. Vitamin E, which is fat-soluble, reacts with and detoxifies free radicals in lipid membranes; water-soluble vitamin C performs this function in the cytoplasm and extracellular fluids. The carotenoids found in vegetables are also able to detoxify free radicals, particularly singlet oxygen.

Vitamin K is essential for the formation of one of the chemicals necessary for blood clotting. In humans, enough vitamin K is normally synthesized by bacteria in the digestive tract, but a deficiency may develop if anything interferes with the absorption of fats and fat-soluble material in the intestine. Large doses of antibiotics can cause a temporary deficiency of vitamins synthesized by intestinal bacteria: such nonspecific drugs kill not only pathogenic bacteria but also the intestinal flora that keep us functioning.

Table 27.1 lists the main vitamins for humans and indicates deficiency symptoms and important sources.

Table 27.1 Some vitamins needed by humans

VITAMIN	SOME DEFICIENCY SYMPTOMS	IMPORTANT SOURCES
Fat-soluble		
Vitamin A (retinol)	Dry, brittle epithelia of skin, respiratory system, and urogenital tract; xerophthalmia and night blindness	Green and yellow vegetables and fruits, dairy products, egg yolk, fish-liver oil, liver, kidney, animal fat
Vitamin D (calciferol)	Rickets or osteomalacia (very low blood-calcium level, soft bones, distorted skeleton, poor muscular development)	Sunlight; egg yolk, milk, fish oils
Vitamin E (tocopherol) Need in humans not defintely established	In rats, malfunction of muscular and nervous systems; anemia (from rupture of red blood corpuscles); male sterility	Widely distributed in both plant and animal food such as meat, egg yolk, green vegetables, seed oils, grains; intestinal bacteria
Vitamin K (phylloquinone)	Slow blood clotting and hemorrhage	Green vegetables; intestinal bacteria
Water-soluble		
Thiamine (B_1)	Beriberi (muscle atrophy, paralysis, mental confusion, congestive heart failure)	Whole grains, yeast, nuts, liver, meat
Riboflavin (B_2)	Vascularization of the cornea, conjunctivitis, and disturbances of vision; sores on the lips and tongue; disorders of liver and nerves in experimental animals	Milk, cheese, eggs, yeast, liver, wheat germ, leafy vegetables, grains
Pyridoxine (B_6)	Convulsions, dermatitis, impairment of antibody synthesis	Whole grains, fresh meat, eggs, liver, fresh vegetables
Pantothenic acid	Impairment of adrenal cortex function, numbness and pain in toes and feet, impairment of antibody synthesis	Present in almost all foods, especially fresh vegetables and meat, whole grains, eggs; intestinal bacteria
Biotin	Clinical symptoms (dermatitis, conjunctivitis) extremely rare in humans, but can be produced by great excess of vitamin-consuming raw egg white in diet	Present in many foods, including liver, yeast, fresh vegetables
Nicotinamide	Pellagra (dermatitis, diarrhea, irritability, abdominal pain, numbness, mental disturbance)	Meat, yeast, grains; intestinal bacteria
Folic acid	Anemia, impairment of antibody synthesis, stunted growth in young animals	Leafy vegetables, liver, meat; intestinal bacteria
Cobalamin (B_{12})	Pernicious anemia	Liver, meat; intestinal bacteria
Ascorbic acid (C)	Scurvy (bleeding gums, loose teeth, anemia, painful and swollen joints, delayed healing of wounds, emaciation)	Fresh vegetables and fruits

■ WHY MINERALS ARE NECESSARY

Like the autotrophs, heterotrophs require certain minerals, which are usually absorbed as ions. Some, like sodium, chlorine, potassium, phosphorus, magnesium, and calcium, are needed in relatively large amounts. In humans the minimum daily requirement for these minerals varies from about 0.35 g for magnesium to nearly 3 g for sodium chloride. We need other minerals, like iron, manganese, and iodine, in much smaller quantities. Copper, zinc, molybdenum, selenium, and cobalt, though essential to life, are needed only in trace amounts—no more than a few milligrams per day. Some elements, like vanadium, barium, tin, silicon, and nickel, are necessary in some species of animals, but have not been proven essential for humans.

The function of some of the minerals is obvious. Calcium is a major constituent of bones and teeth in vertebrates and plays a variety of other roles in most organisms. Phosphorus is a component of nucleic acids and many high-energy organic compounds of critical importance. Iron is a constituent of the cytochromes and of hemoglobin. Sodium, chlorine, and potassium are important components of the body fluids. Iodine is found in the hormones produced by the thyroid gland.

Minerals needed only in trace amounts are usually components of enzymes or coenzymes; selenium, for example, is found at the active site of an enyzme that helps destroy mutagenic substances like hydrogen peroxide. Other minerals may be cofactors that help catalyze reactions without actually being physically incorporated into enzymes or coenzymes. The minerals an animal needs usually come from the normal diet. A few species of birds and mammals apparently compensate for mineral deficiencies by eating soil; other instances of consuming dirt, however, are driven by a need for clay, which helps detoxify certain poisonous foods.

THE UNIQUE FEEDING STYLE OF FUNGI

The fungi are sedentary heterotrophs that live on or in their food supply. Bread mold is a familiar example. The bread on which it grows is composed mostly of starch, a rich source of energy. Recall that starch is a polysaccharide, whose very large and insoluble molecules cannot move across the cell membranes of the mold. Before absorption can take place, the starch must be broken down to its constituent sugars. This digestion is nothing more than enzymatic hydrolysis of the bonds linking chemical subunits. For bread mold the hydrolysis takes place outside the cells. The enzymes involved in this ***extracellular digestion*** are synthesized inside the cells of the mold and released onto the bread, where they hydrolyze the starch. The simple sugars that are the products of this digestion are then absorbed, often by rootlike structures called ***rhizoids*** (Fig. 27.6A).

Mold living on bread is saprophytic, since it obtains its nutrients from dead organic matter. However, many fungi are parasitic. Indeed, bread mold itself can also grow on living fruit and vegetables. The various parasitic fungi differ in their relationships to their hosts. Some small fungi grow between the cells of their host, but send out rootlike structures, called ***haustoria***, that make deep invaginations in the host's cell membranes, through which they absorb nutrients from the host's cytoplasm (Fig. 27.6B). Still other filamentous types occupy many host cells simultaneously, penetrating through the cell membranes that divide one cell from the next. The intense itching caused by athlete's foot fungus is produced when the haustoria grow past sensory endings in the skin.

All parasitic fungi employ basically the same mode of nutrition as saprophytic fungi: enzymes are secreted into the food supply on (or in) which the fungus lives, digestion takes place extracellularly, and the products of digestion are absorbed by the fungus. Fungi have no internal cavity where bulk food can be digested; they sim-

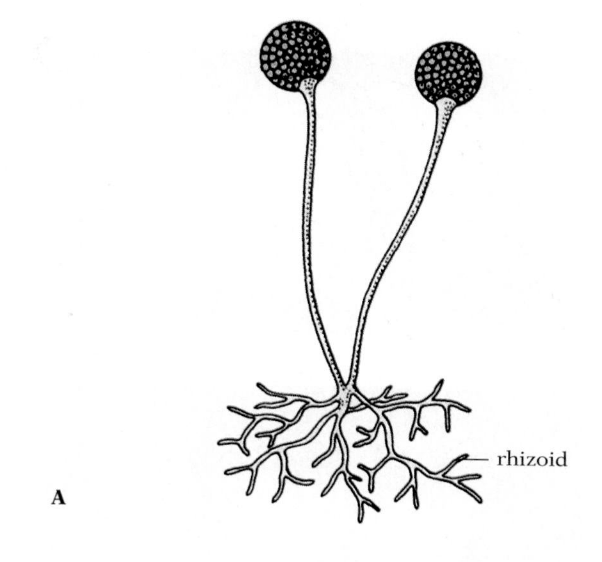

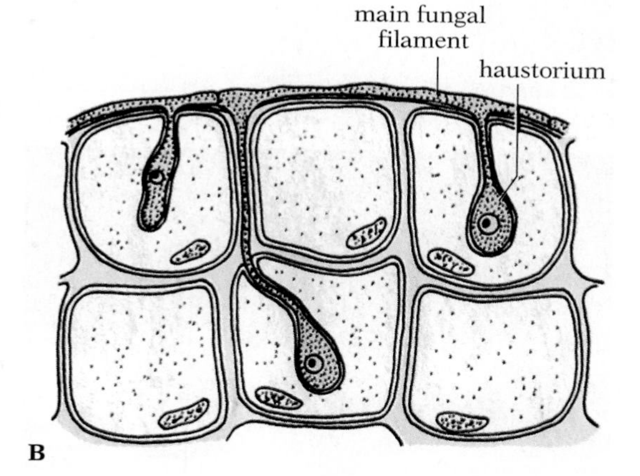

27.6 Nutrient-procurement structures of fungi

(A) Rootlike rhizoids of a saprophytic fungus. (B) Haustoria of a fungus parasitic on a multicellular plant. The body of the fungus (gray) is filamentous and can grow between the cells of the plant host. The haustoria penetrate the cell walls and make deep invaginations in the membranes of host cells, through which they absorb nutrients. Note that the haustoria are not in direct contact with the cytoplasm of the host cells, because the haustorial invaginations are lined with host-cell membrane.

ply absorb digested organic nutrients across the body surface, much as plant roots absorb inorganic nutrients.

A few fungi supplement their diets by trapping small animals such as nematode worms (Fig. 27.7). When the prey has been trapped, branches of the fungus penetrate the victim's body and release digestive enzymes; extracellular digestion takes place, and the products are absorbed.

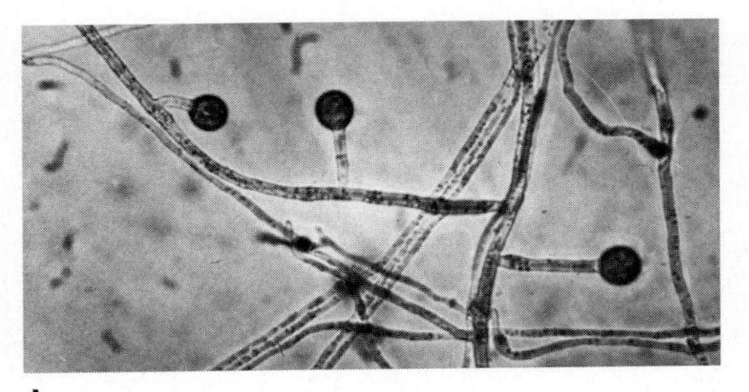

A

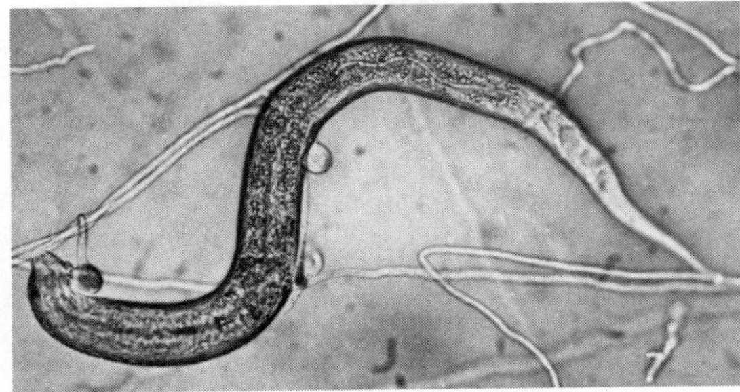

B

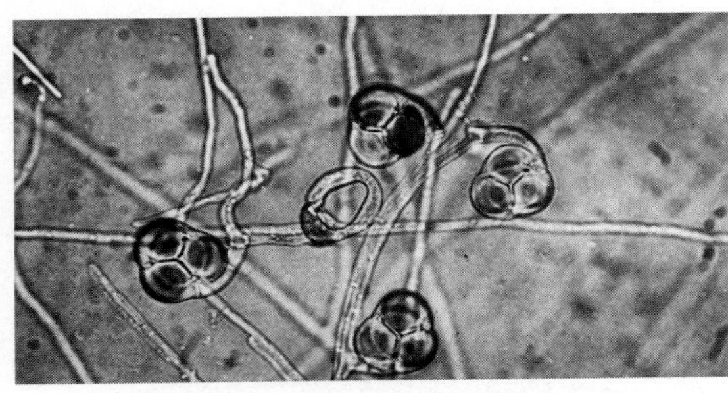

C

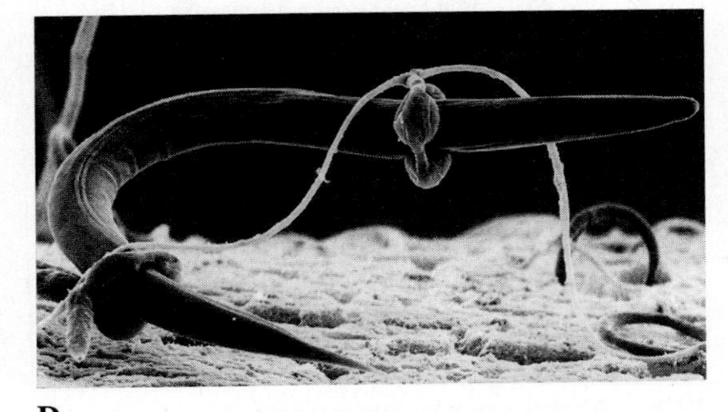

D

27.7 Two fungi that trap nematode worms

Dactylella drechsleri has sticky knobs (A), which hold a worm that contacts them (B). *Arthrobotrys dactyloides* has rings formed of three cells (C), which can be seen plainly in the closed rings. When a worm enters a ring, the cells swell and constrict the opening, trapping the worm (D).

INTERNALIZED DIGESTION: ANIMALS AND PROTOZOANS

Nutrient procurement usually involves much more activity in animals and protozoans than it does in plants and fungi. Animals must often resort to elaborate methods of locating and trapping their food. Their extremely varied feeding habits may be classified in any number of ways. One useful classification distinguishes carnivores, herbivores, and omnivores. Another classification is based on the size of the food: microphagous feeders strain microscopic organic materials from the surrounding water by an array of cilia, bristles, legs, nets, or the like, whereas macrophagous feeders break up larger masses of food by means of teeth, jaws, pincers, or gizzards, or solely by the action of enzymes. Other feeding variations are exhibited by sucking animals, adapted to extract fluid from plants or from animal prey, and parasitic animals and protozoans that are bathed in the nutrients of the host and absorb them directly through the body surface.

Like fungi, most animals must digest their food before it can cross the membranes of their cells. Food is likely to be in the form of large molecules such as polysaccharides, fats, and proteins, which must be hydrolyzed. Unlike the fungi, however, few animals secrete digestive enzymes directly onto their food. (Spiders and some insects are among these few. They inject their prey with digestive enzymes and drink the resulting "soup.") The vast majority ingest particles of food into some sort of digestive structure in which enzymatic action takes place.

The digestive structure itself is usually extracellular. In mammals, for instance, digestion takes place in the intestinal tube, and the nutrients released by the enzymatic breakdown are then absorbed by the surrounding cells. In some other organisms—the protozoans in particular—food is taken up directly by a cell by phagocytosis or a similar process, and then digested in a food vacuole. Though this process is classified as ***intracellular digestion***, the food material is actually separated from the rest of the cellular material by a membrane that it cannot cross until after digestion has occurred. Thus extracellular ingestion and intracellular ingestion are alike in that digestion always precedes the actual absorption of nutrients across a membrane.

Though the nutritional requirements and the basic processes of digestion are similar in protozoans and animals (from worms to humans), the body plans of these organisms vary so greatly that the structures involved in food processing and the details of that processing are often very different.

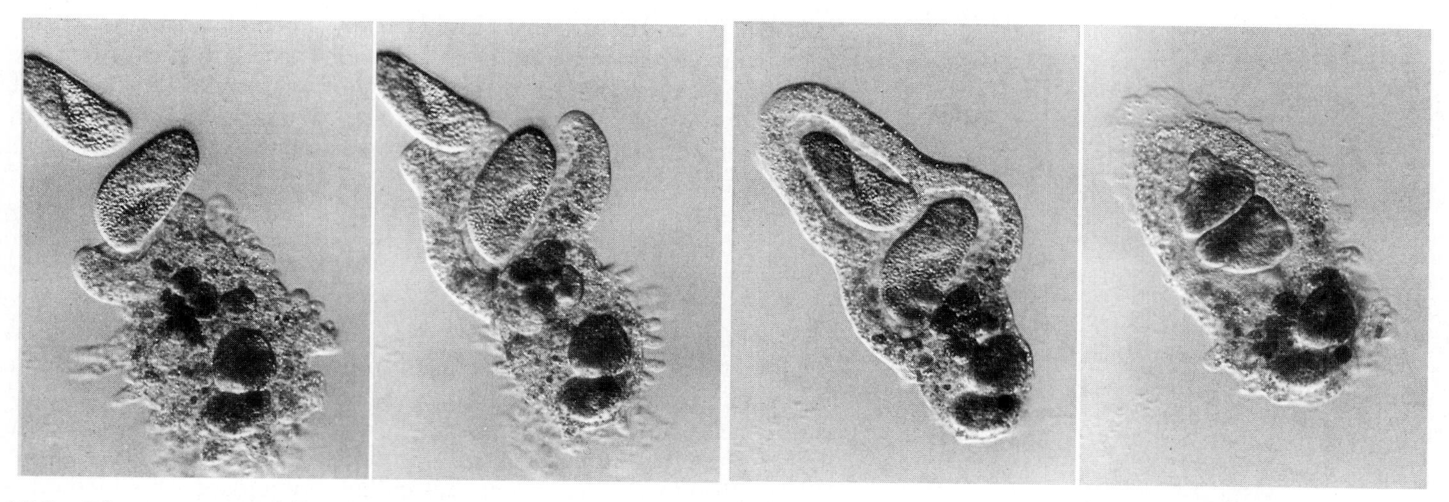

27.8 Phagocytosis of food by *Amoeba*

Pseudopods flow around the prey (two cells of *Paramecium*) until it is entirely enclosed within a vacuole.

■ HOW PROTOZOANS CAPTURE AND DIGEST FOOD

Amoeboid protozoans constantly change shape as the cytoplasm flows along; new armlike pseudopods are pushed out while others are withdrawn. When an amoeba is stimulated by nearby food, some of the pseudopods may flow around and completely surround the food—the process of phagocytosis. The food is completely engulfed by the cytoplasm and enclosed in a ***food vacuole***, where it will be digested (Fig. 27.8).

The ciliates are characterized by hairlike cilia that cover the surface of their bodies. Though they lack actual subdivision into recognizable cellular units, the more complex ciliates show much of the internal specialization usually associated with multicellular organisms. Unlike the amoeba, the complex ciliate *Paramecium* has a permanent organelle that functions in feeding (Fig. 27.9). Food particles are swept into the outer portion of the organelle; this ***oral groove*** is a ciliated channel located on one side of the cell. Water currents produced by the beating of the cilia move food down the groove into the inner portion of the organelle, the ***cytopharynx***.

As food accumulates at the lower end of the cytopharynx, a food vacuole forms around it. Eventually the vacuole breaks off and begins to move toward the anterior end of the cell. Digestive enzymes are secreted into the vacuole and digestion begins. As digestion proceeds, the products (simple sugars, amino acids, and the like) diffuse across the membrane of the vacuole into the cytoplasm, and the vacuole begins to move back toward the posterior end of

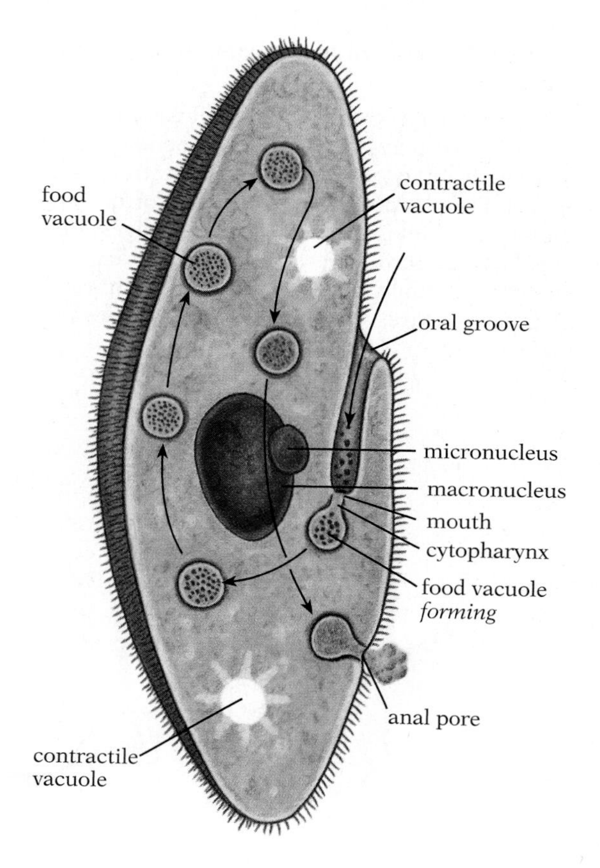

27.9 Digestive cycle in *Paramecium*

A food vacuole forms at lower end of the cytopharynx, then breaks off and moves toward the anterior end of the cell while enzymes are secreted into it; digestion takes place, and the products of digestion are absorbed into the general cytoplasm. The vacuole then moves toward the posterior end, attaches to the anal pore, and expels digestive wastes. The vacuole undergoes several changes in size and appearance as it moves.

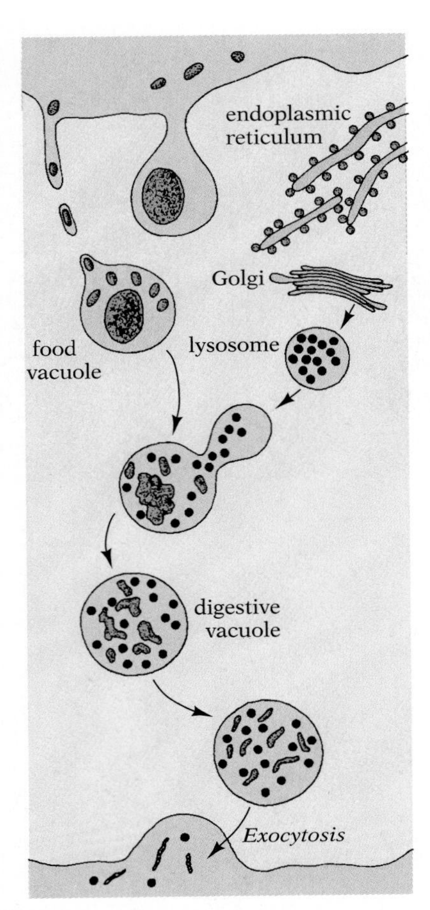

27.10 The role of lysosomes in intracellular digestion

Food material (brown) that the cell takes in by phagocytosis is enclosed in a food vacuole, which fuses with a lysosome containing digestive enzymes. The composite structure thus formed is a digestive vacuole, and the products of digestion are absorbed across the vacuolar membrane. The vacuole eventually fuses with the cell membrane and then ruptures, expelling digestive wastes to the outside.

the cell. When the vacuole reaches a tiny specialized region of the cell surface called the anal pore, it attaches there and ruptures, expelling by exocytosis any remaining bits of indigestible material. Not only does the vacuole function as a digestive chamber, but by its movement it helps distribute the products of digestion to all parts of the cell.

How are powerful digestive enzymes, capable of hydrolyzing such compounds as polysaccharides, fats, proteins, and nucleic acids, prevented from digesting the very organisms that employ them? A partial answer is given in Chapter 5: digestive enzymes are packaged in lysosomes, vesicles whose membranes are both impermeable to the enzymes and immune to their hydrolytic action. Whether in protozoans or animal cells, the digestive enzymes are synthesized on ribosomes, move through the endoplasmic reticulum to the Golgi apparatus, and there are packaged in a lysosome. When a food vacuole (sometimes called a phagosome) is formed, a lysosome fuses with it (Fig. 27.10), and food materials and the digestive enzymes are mixed in the resulting digestive vacuole. As this vacuole circulates in the cytoplasm, the products of digestion are absorbed, and indigestible materials are eventually expelled from the cell by exocytosis.

■ THE ANIMAL AS STOMACH: COELENTERATES

With the evolution of multicellular organisms came a corresponding evolution of cellular specialization—a division of labor among cells. Coelenterates have a small degree of specialization. Their saclike body is composed of two principal layers of cells—an outer epidermis and an inner gastrodermis—with a jellylike mesoglea between them (Fig. 27.11). The cells of the epidermis function as protective and sensory epithelium, while the gastrodermis is the nutritive epithelium. Some cells of both layers are specialized as muscle fibrils, and others as nerves. The central body cavity functions as a digestive cavity. It has only one opening to the outside, which is surrounded by mobile tentacles. A digestive cavity of this sort—with a single opening that functions in both ingestion and excretion—is called a gastrovascular cavity.

Coelenterates are strictly carnivorous. Embedded in their tentacles are numerous stinging ***nematocysts*** (Fig. 27.12). Each nematocyst consists of a slender hollow thread coiled within a capsule, with a tiny hairlike trigger penetrating to the outside. When prey comes into contact with the trigger, the nematocyst fires, the thread turns inside out, barbs on its surface unfold, and it either penetrates the body of the prey or entangles it in sticky loops. The nematocyst ejects a paralyzing poison, and the coelenterate's tentacles draw its prey toward the mouth, which opens wide to receive it. Once the food is inside the gastrovascular cavity, digestive enzymes are secreted into the cavity by gastrodermal cells, and extracellular digestion begins. As soon as the food has been reduced to small fragments, gastrodermal cells engulf them by phagocytosis, and digestion is completed intracellularly in digestive vacuoles. Indigestible remains of the food are expelled from the gastrovascular cavity via the mouth.

If phagocytosis and intracellular digestion are going to take place anyway, what adaptive advantage does extracellular digestion have? Intracellular digestion limits the size of the food to items no larger than a cell; extracellular digestion is limited only by the size of the mouth.

Once a coelenterate has eaten, the gastrodermal cells must distribute the products of digestion to the other cells. Since no cells are far removed from the gastrodermal layer, this distribution does not require any specialized transport system.

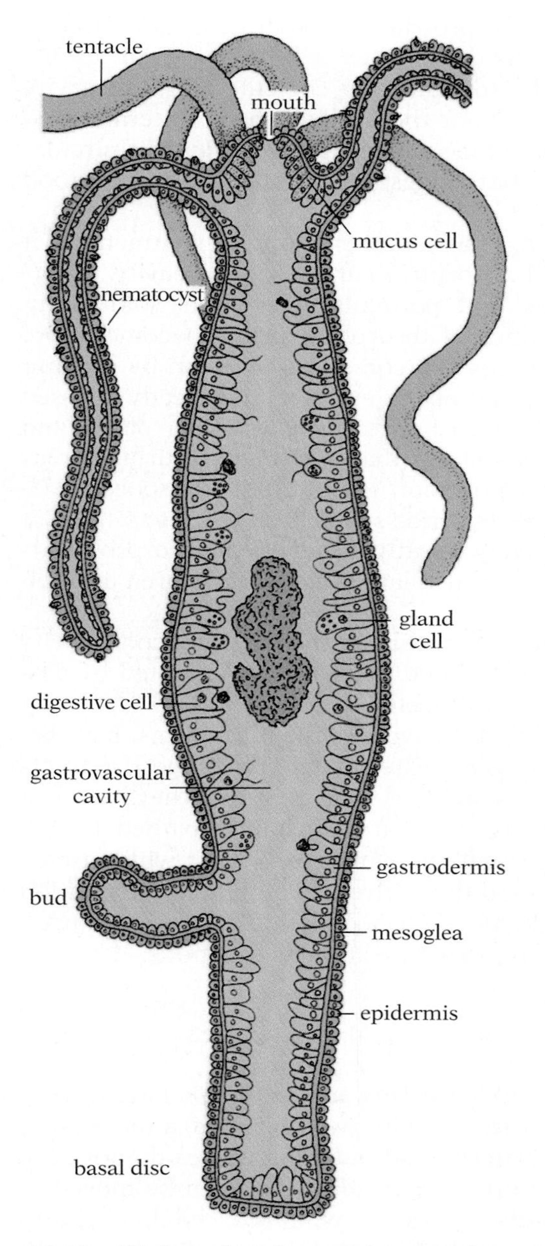

27.11 Hydra, showing gastrovascular cavity

The cavity contains food material (red). The mesoglea layer in the body wall is much more extensively developed in some other coelenterates.

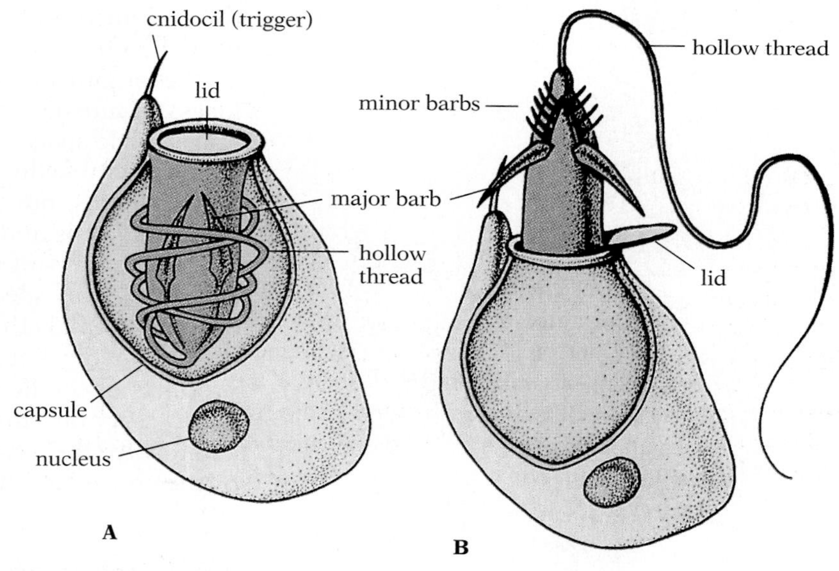

27.12 Nematocyst

Nematocysts are organelles contained in specialized cells known as cnidoblasts (light blue). This example of a nematocyst (yellow and dark blue), shown before (A) and after discharge (B), represents only one of more than a dozen general morphological types. The minor barbs are part of a hollow thread that is folded back inside itself before discharge. Most nematocysts fire only when the trigger has been stimulated simultaneously in two ways—for example, by touch and by exposure to chemicals characteristic of the coelenterate's prey. The exploding nematocyst propels the major barb at the target, and after it has struck, the hollow thread begins to be everted, and the minor barbs are exposed and dig themselves into the target. The rest of the hollow thread is similarly ejected, and the poison flows through it into the victim. This is the source of the jellyfish's painful sting.

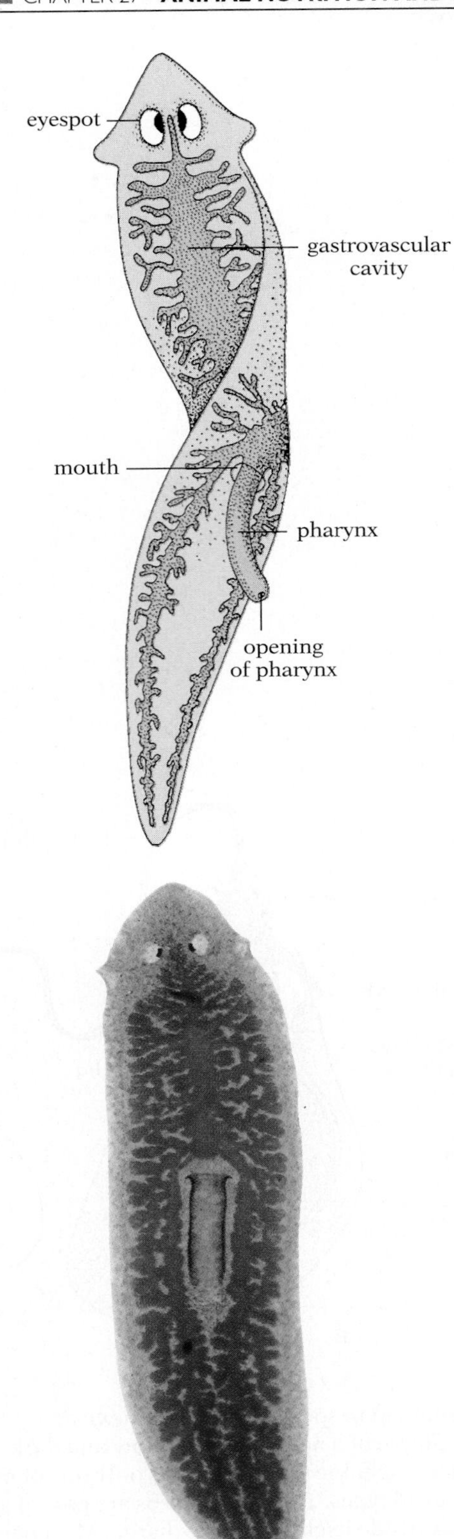

27.13 Planarian

Note the much-branched gastrovascular cavity and extruded pharynx.

■ THE ALL-PERVADING STOMACH OF FLATWORMS

Unlike the radially symmetrical coelenterates, the platyhelminths, or flatworms, are bilaterally symmetrical: they have distinct anterior and posterior ends, and also distinct dorsal and ventral surfaces. Their bodies are composed of three well-formed tissue layers. Many flatworms are parasitic on other animals, but some, like the planarians, are free-living (Fig. 27.13).

The mouth of a planarian is located on the ventral surface near the middle of the animal. It opens into a muscular, tubular ***pharynx***, which the planarian can protrude through its mouth directly onto its prey. The pharynx leads into a gastrovascular cavity. Though functionally similar to that of the coelenterates, this cavity branches elaborately throughout the animal's body. Literally gastrovascular (*gastro-* refers to the stomach and *vascular* to a circulatory vessel), it functions in both digestion and the transport of food to all parts of the body.

The extensive branching has another important function: it greatly increases the total absorptive surface of the cavity. As organisms increase in size, and particularly as their volume increases, the problem of sufficient absorptive surface becomes more acute. Kelps and coelenterates overcome this problem by remaining thin, flat, or hollow, so that nearly every cell is directly exposed to nutrients. The more compact body design of higher plants and animals, however, requires systems that create large surface areas for exchange and, usually, provision for internal transport to deliver the nutrients absorbed by these specialized tissues to the rest of the organism. By evolving greatly subdivided absorptive surfaces, organisms can concentrate a large total surface area into relatively little space.

Some extracellular digestion occurs in the gastrovascular cavity of planarians, but most of the food particles are engulfed by gastrodermal cells and digested intracellularly.

The members of one class of flatworms, the tapeworms, have become so highly specialized as parasites living in the digestive tracts of other animals that in the course of their evolution they have lost their own digestive systems. They are constantly bathed by the products of the host's digestion and can absorb them without having to carry out any digestion themselves. Evolutionary adaptation can involve the loss of structures, the acquisition of new structures, or the conversion of existing structures to new functions.

■ DISASSEMBLY LINES: THE COMPLETE DIGESTIVE TRACT

Unlike the coelenterates and flatworms, most animals have a ***complete digestive tract***—that is, one with two openings, a mouth and an anus. Incoming food material and outgoing wastes do not have to pass through the same opening. Instead, food can be moved in one direction through a tubular system, which can be divided into a series of distinct sections or chambers, each specialized for a different function. As the food passes along this assembly line, it is acted upon in a different way in each section. The sections may be variously specialized for mechanical breakup of bulk food, temporary storage, enzymatic digestion, absorption of the products of di-

27.14 Digestive system of an earthworm

gestion, reabsorption of water, storage of wastes, and so on. The result is a much more efficient digestive system.

The digestive system of earthworms is a good example of division into specialized compartments (Fig. 27.14). Food, in the form of decaying organic matter mixed with soil, is drawn into the mouth by the sucking actions of its muscular pharynx. After passing from the mouth through a short passageway into the pharynx and then through a connecting passage called the ***esophagus***, the food enters a relatively thin-walled ***crop***, which functions as a storage chamber. Next it enters a compartment with thick, muscular walls (the ***gizzard***), where it is ground up by a churning action; the grinding is often facilitated by small stones in the gizzard. The pulverized food, suspended in water, passes into the long ***intestine***, where enzymatic digestion and absorption take place. Finally, in the rear of the intestine, some of the water involved in the digestive process is reabsorbed, and the indigestible residue is eliminated from the body through the anus.

Earthworms use extracellular digestion. Glandular cells in the epithelial lining of the intestine secrete hydrolytic enzymes into the intestinal cavity, and the products are absorbed. To maximize the absorptive surface area of the intestine, a large dorsal fold, the ty-

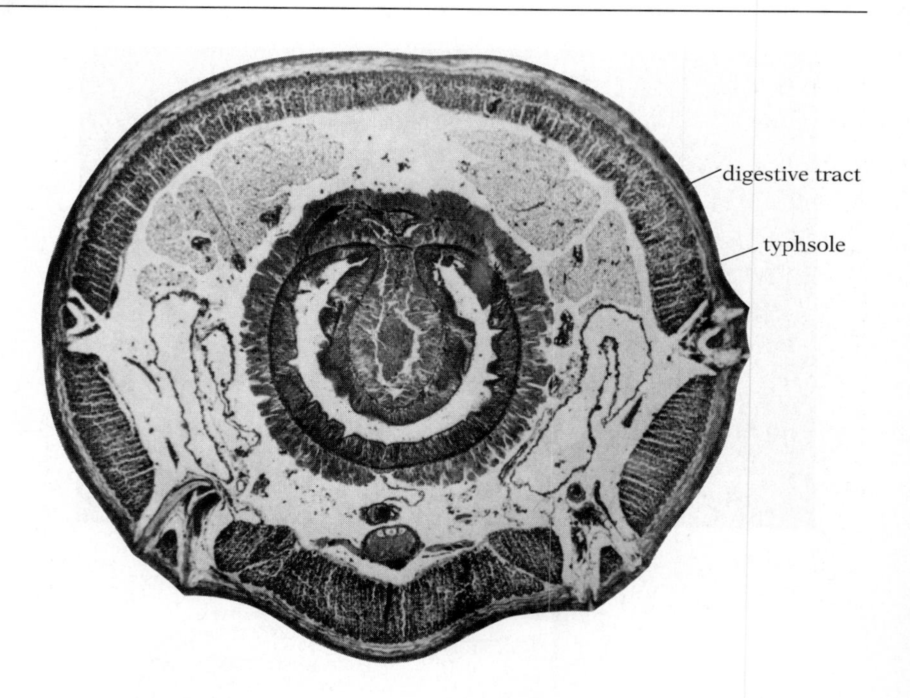

27.15 Cross section in the intestinal region of an earthworm

The typhlosole, which projects into the cavity of the intestine, greatly increases the surface area available for absorption of food.

phlosole, projects downward into the digestive cavity (Fig. 27.15). The typhlosole is functionally analogous to the root hairs of higher plants and to the branching of the gastrovascular cavity in planaria.

Capturing and grinding food The gizzard of the earthworm is critical to breaking up the bulk food it consumes. A variety of structures have evolved in different animals. In our own case, food is torn and ground by the teeth. Many snails have a hard toothed pharyngeal plate, the radula, with which they rasp off small particles from larger pieces of food. Cockroaches and many other insects that feed on solid food have a chamber (the proventriculus) similar to the earthworm's gizzard except that its inner wall often bears several very hard ridges and teeth.

The grinding or chewing device need not be in the first section of the digestive tract, as it is in humans; in both earthworms and cockroaches the grinding chamber comes after the crop, which is in some ways like our stomach, and mechanical breakup thus follows temporary storage instead of preceding it. A similar arrangement exists even in some vertebrates; the muscular gizzard of birds, in which hard food is ground with rocks and pebbles (often called grit), is posterior to the stomach (Fig. 27.16). The stomach of birds, though relatively unspecialized, is a true stomach in that it is muscular and secretes digestive enzymes, but does not function in absorption.

Not all complex multicellular animals break up and eat large pieces of food. Some, such as the bloodsucking and sap-sucking insects, have liquid diets. Others are ***filter feeders***, straining small particles of organic matter from water. Clams and many other molluscs filter water through tiny pores in their gills; microscopic food particles are trapped in streams of mucus that flow along the gills and enter the mouth, kept moving by beating cilia.

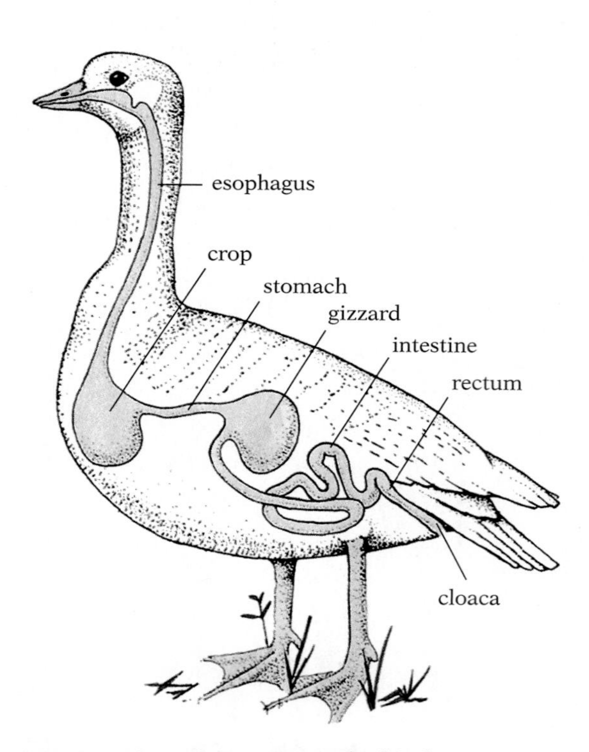

27.16 Digestive system of a bird

The chamber for mechanical breakup (gizzard) is located posterior to the stomach.

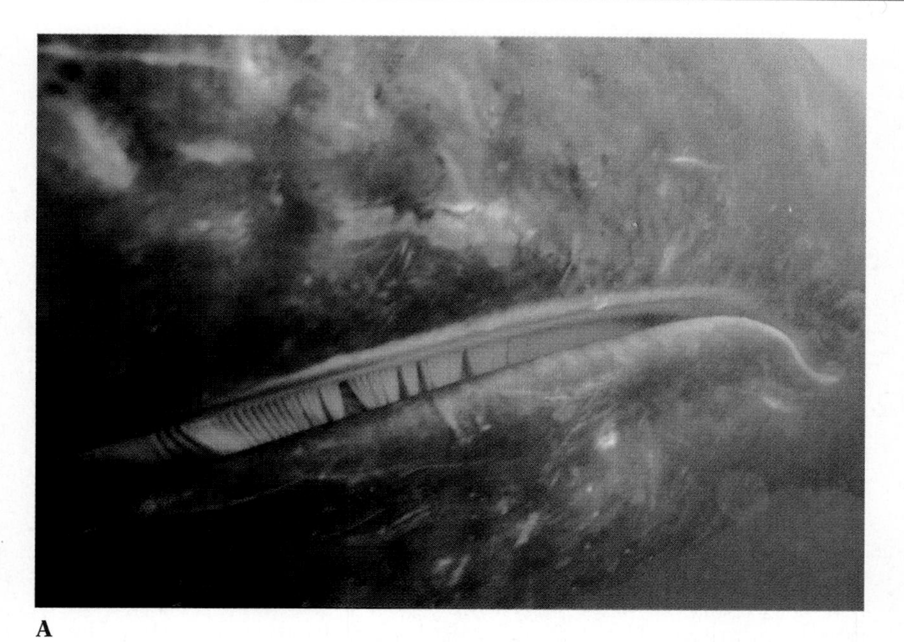

A

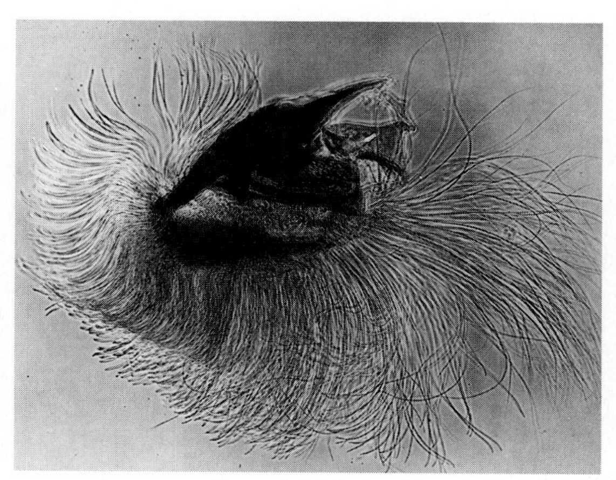

B

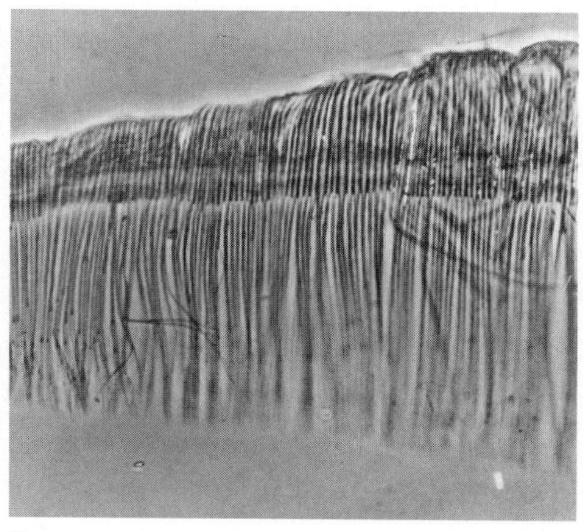

C

27.17 Structures used for filter feeding

(A) Strips of baleen hanging down from the top jaw of an Atlantic right whale, one of the filter-feeding whales. The baleen allows the cetacean to take a mouthful of water and strain out the organisms—primarily shrimp and small fish—that form its diet. (B) On a much smaller scale, the isolated food brush of a mosquito larva. Its motion causes water to flow into the mouth. (C) Much-enlarged portion of comb from the pharyngeal filter of a mosquito larva. It strains food particles from water.

The larvae of mosquitoes are also filter feeders (Fig. 27.17B–C). They eat bacteria and other small particles of organic matter in the water. Two small hair-covered brushes near the mouth beat in a circular scooping motion, setting up water currents toward the mouth. The particles and water pass through the mouth and into the pharynx. Muscles in the wall of the pharynx contract, expelling the water through two small canals. Tiny combs in the canals strain out the food while the water passes through; the larva swallows the food that remains.

Some of the largest present-day vertebrates—certain species of whales—are filter feeders, straining small plants and animals from the vast quantities of water they take into their mouths (Fig. 27.17A).

Food-storage chambers The food-storage crop of animals enables them to take in large amounts of food in a short time, when it is available and can be eaten in safety, and then to metabolize this food later at their leisure. Such discontinuous feeding allows an animal to devote much of its time to other activities, such as searching for a mate, mating, and, in some cases, caring for young. Our own stomachs function as storage organs analogous to the earthworm's crop; they enable us to live comfortably on only three meals a day and to devote the rest of our time to other pursuits.

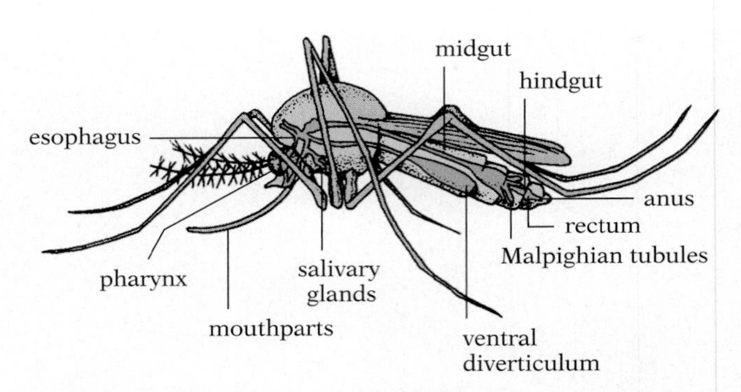

27.18 Digestive system of adult female mosquito
Notice the large diverticulum.

Though discontinuous feeding is nearly universal, tapeworms, nematode worms, and some other animals lack storage ability; they must be almost constantly in contact with food. It would be a mistake, however, to assume that such animals are unsuccessful or poorly adapted. Their long evolutionary history and their large numbers today testify to the contrary. They are simply adapted—successfully—for a different way of life. Biological success is not measured by structural complexity or by the possession of any particular organ. The earthworm with its crop and the nematode worm without one are both successful from the biological point of view, which equates success with survival.

Different kinds of food-storage organs occur in different species of animals. In many birds an expanded region of the esophagus anterior to the stomach forms a thin-walled chamber, the crop, which is functionally analogous to the earthworm's crop (Fig. 27.16). Some birds also use the crop in carrying food to their young; they fill it with seeds, berries, fish, or whatever their food may be, and then fly to the nest, where they disgorge the food for their young. In many animals storage organs take the form of blind sacs, or diverticula, branching off the digestive tract. A good example is seen in adult female mosquitoes, which have a very large diverticulum (Fig. 27.18) that occupies much of the abdominal cavity. The female mosquito locates a suitable animal, pierces its skin with long needlelike mouthparts, and sucks blood until this diverticulum is filled. A single large blood meal may suffice to carry the female through the entire process of locating an egg-laying site and laying her eggs—a matter of 4 or 5 days.

■ HOW THE VERTEBRATE DIGESTIVE SYSTEM WORKS

The mouth The first chamber of the human digestive tract is the oral cavity (mouth), where the teeth break up food by both biting and chewing. The internal structure of a human tooth is shown in Figure 27.19. Human teeth are of several different types, each adapted to a different function (Fig. 27.20A–B). In front are the chisel-shaped ***incisors***, which are used for biting. Then come the more pointed ***canine*** teeth, which are specialized for tearing food. Behind each canine are two ***premolars*** and three ***molars*** in adults; these have flattened, ridged surfaces, and function in grinding and crushing food. A child's first set of teeth lacks some of these; the

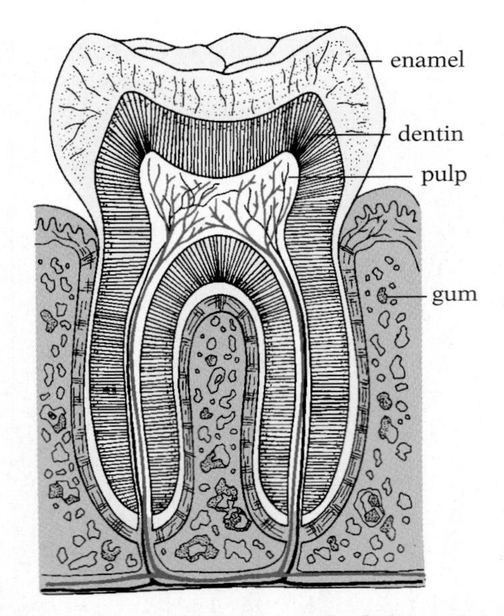

27.19 Internal structure of a human tooth
Blood vessels (red) and nerves penetrate into the pulp, but not into the outer harder layers.

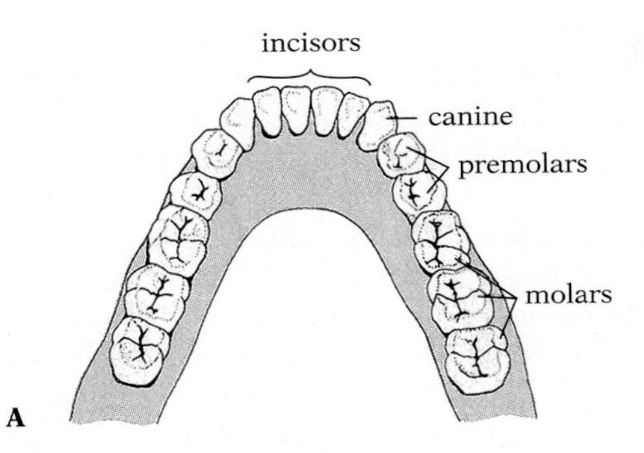

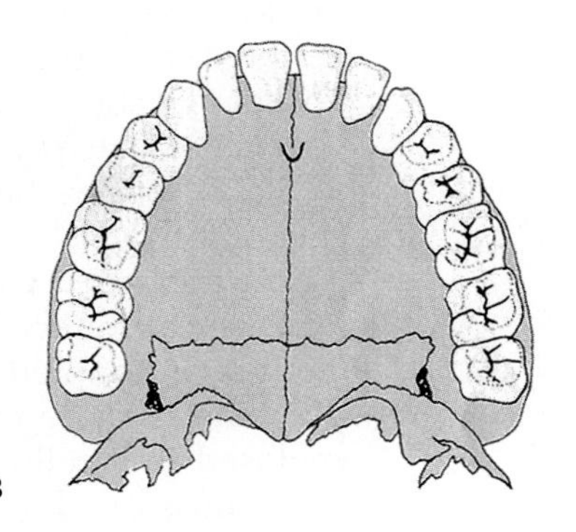

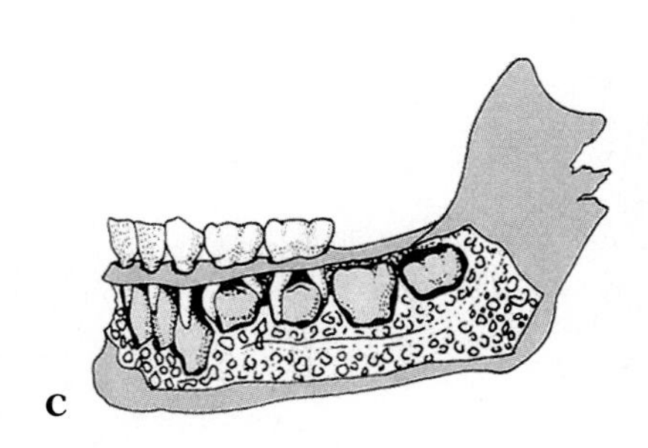

27.20 Human teeth

(A) Lower jaw of adult. (B) Upper jaw of adult. (C) Lower jaw of child, showing permanent teeth in gums below milk teeth.

first (or milk) teeth are lost as the child gets older and are replaced by permanent teeth that have been growing in the gums (Fig. 27.20C).

The teeth of different species of vertebrates are specialized in a variety of ways and may be quite unlike human teeth in number, structure, arrangement, and function. For example, the teeth of snakes are very thin and sharp (Fig. 27.21A) and usually curve backward; they function in capturing prey, but not in mechanical breakup; snakes swallow their food whole. Carnivorous mammals such as cats and dogs have pointed teeth (Fig. 27.21C); the canines

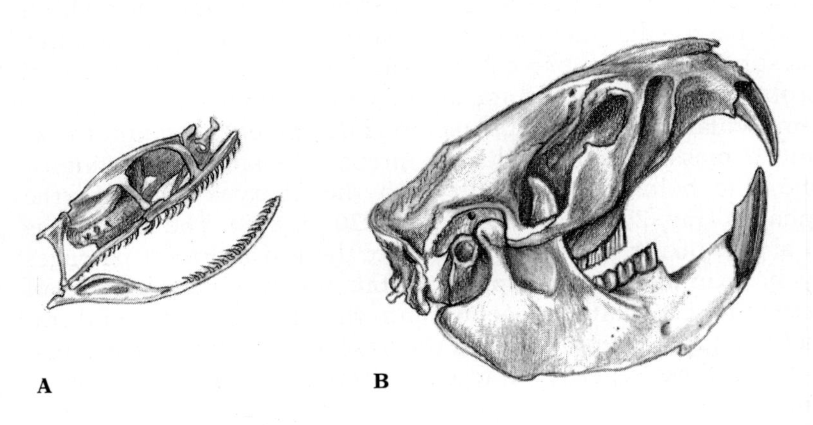

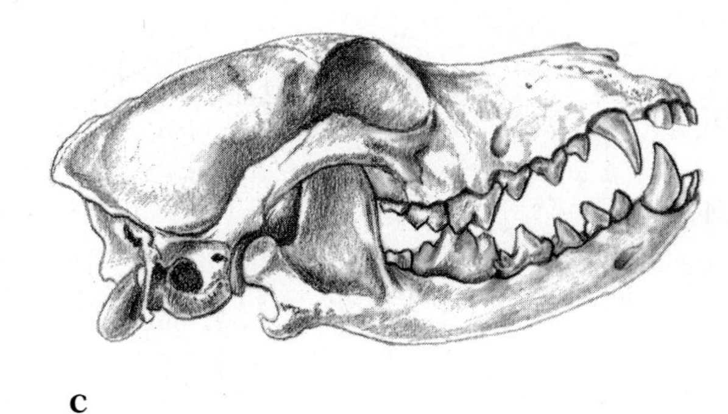

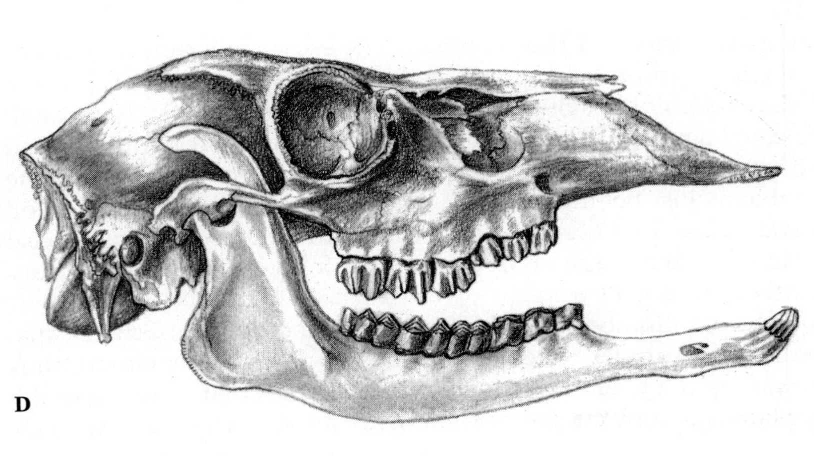

27.21 Structure and arrangement of teeth in different animals

(A) The snake has thin, sharp backward-curved teeth with no chewing function. (The snake skull is here shown disproportionately large in relation to the other three.) (B) The beaver (a gnawing herbivore) has few but very large incisors, no canines, and premolars and molars with flat grinding surfaces. (C) The dog (carnivore) has large canines, and its premolars and molars are adapted for cutting and shearing. (D) The deer (a grazing and browsing herbivore) has six lower incisors (three on each side) but no upper incisors (which are functionally replaced by a horny gum); the premolars and molars have very large grinding surfaces. Notice the large gap between the incisors and premolars.

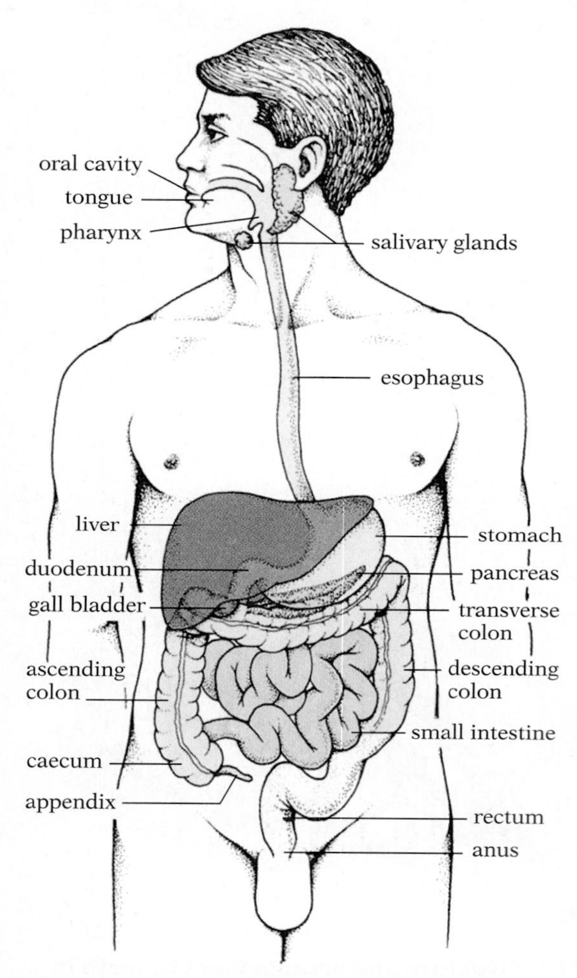

27.22 Human digestive system

The small intestine has been shortened for clarity.

are long, and the premolars lack flat grinding surfaces, being more adapted to cutting and shearing; the more posterior molars may be absent. Herbivores like cows and deer, on the other hand, have very large, flat premolars and molars, and may lack canines (Fig. 27.21D) Beavers have massive incisors that are used to gnaw trees (Fig. 27.21B).

Sharp pointed teeth poorly adapted for chewing seem to characterize meat eaters like snakes, dogs, and cats, whereas broad flat teeth well adapted for chewing seem to characterize vegetarians. Chewing is more important for herbivores because plant cells are enclosed in a cellulose cell wall. Few animals can digest cellulose; most must break up the cell walls of the plants before the cell contents can be exposed to digestive enzymes. Animal cells, including those in meat, lack walls, and thus can be digested directly. Because chewing is not as necessary for carnivores, dogs can gulp down their food, while cows must spend much of the time chewing. But carnivores must capture and kill their prey, and for this, sharp teeth capable of piercing, cutting, and tearing are well adapted. Humans, being omnivores, have teeth that belong, functionally and structurally, somewhere between the teeth of carnivores and herbivores.

The oral cavity is where food is tasted and smelled prior to acceptance, and it is here the food is mixed with saliva from the salivary glands. The saliva dissolves some of the food and acts as a lubricant, facilitating passage through the next portions of the digestive tract. Human saliva contains a starch-digesting enzyme, which initiates the process of enzymatic hydrolysis. It also contains an antimicrobial agent, the thiocyanate ion, together with a special enzyme that facilitates entry of the ion into microbial cells; these substances help prevent infection by the potentially harmful microbes that constantly come along with the food.

A muscular tongue manipulates food during chewing and forms it into a mass, called a bolus. In preparation for swallowing, it pushes the bolus backward through the pharynx and into the esophagus (Fig. 27.22; see also Fig. 28.20, p. 808). The pharynx is also a part of the respiratory passage; the air and food passages cross here, in fact. Swallowing, therefore, involves a complex set of reflexes that close off the opening into the nasal passages and trachea (windpipe), thereby forcing the food to move into the esophagus. When these reflexes fail to occur in proper sequence, choking results.

The esophagus and the stomach The esophagus runs downward through the throat and thorax to the stomach in the upper portion of the abdominal cavity (Fig. 27.22). Food moves quickly through the esophagus, pushed along by waves of muscular contraction, a process called ***peristalsis***. Circular muscles in the wall of the esophagus just behind the food bolus contract, squeezing the food forward (Fig. 27.23). As the food moves, the muscles it passes also contract, so that a region of contraction follows the bolus and constantly pushes it forward.

At the junction between the esophagus and the stomach is a special ring of muscles called a ***sphincter***. It is normally closed, thus preventing the contents of the stomach from moving back into the esophagus when the stomach moves during digestion. It opens

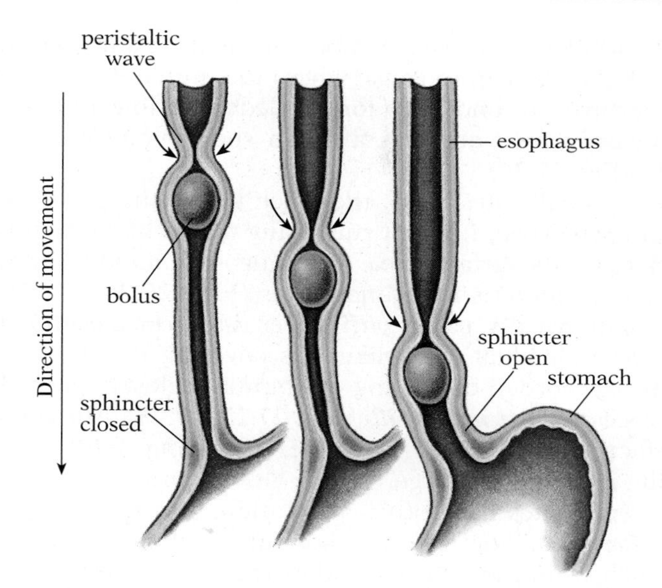

27.23 Peristalsis

The wave of muscular contraction pushes the bolus of food ahead of it. The sphincter muscle at the junction between the stomach and esophagus normally remains closed, and opens when a wave of peristaltic contraction coming down the esophagus reaches it.

when a wave of peristaltic contraction reaches the end of the esophagus.

The stomach is a large muscular sac that functions in part as a storage organ, making possible discontinuous feeding. Its thick walls are composed of three layers: an inner mucous membrane composed of connective tissue and columnar epithelium with many glands, a thick middle layer of smooth muscle, and an outer layer of connective tissue. The muscle layer contains fibers running around the stomach, others running longitudinally, and still others oriented diagonally. Hence the stomach is capable of a great variety of movements. When it contains food, it is swept by powerful waves of contraction which churn the food, mixing it and breaking up larger pieces. In this manner the stomach supplements the action of the teeth in the mechanical breakup of food. The glands of the stomach lining secrete mucus, which covers the stomach lining—hence the name ***mucosa*** or mucous membrane for the inner layer of the stomach wall—as well as ***gastric juice***, a mixture of hydrochloric acid and digestive enzymes. Enzymatic digestion, then, is a third important function of the human stomach.

The small intestine Food leaves the stomach as a soupy mixture. It passes through the pyloric sphincter into the small intestine, which is the part of the digestive tract where most of digestion and absorption takes place. The first section of the small intestine, attached to the stomach, is the ***duodenum*** (see Fig. 27.22). It leads into a long coiled section lying lower in the abdominal cavity. The entire small intestine of an adult male is about 7 m long and 2–3 cm in diameter.

The length of the small intestine varies among species. The intestine is usually long and coiled in herbivores, much shorter in carnivores, and of medium length in omnivores such as humans. These differences, like those of the teeth, result from the difficulty of digesting plant material. Even if the cellulose has been well broken up, it remains mixed with the digestible portions of the cells and tends to mask them from the digestive enzymes. This interference makes digestion and absorption of plant material much less efficient than the processing of animal material. A longer intestine

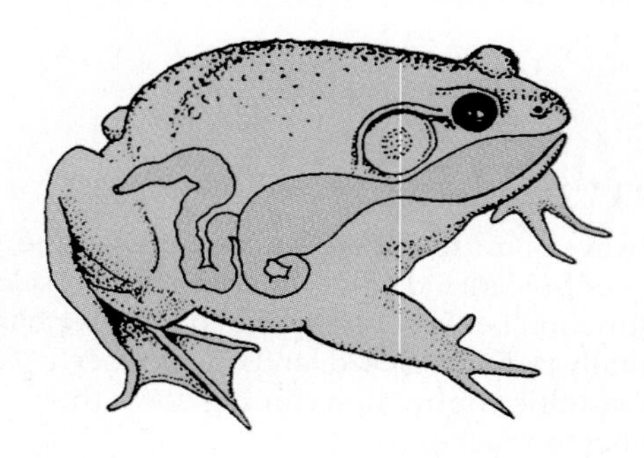

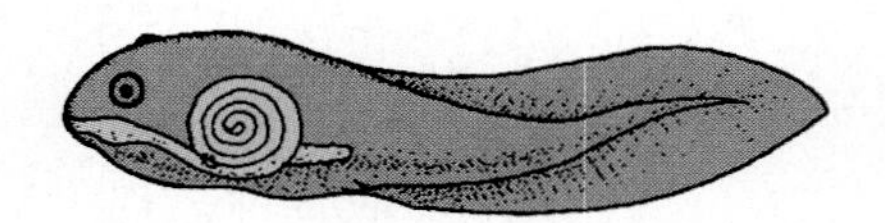

27.24 Intestines of adult frog and of tadpole

The much-coiled intestine of the tadpole is far longer relative to the size of the animal than is the intestine of the adult frog.

enables herbivores to extract more nutrients from their food. A striking change is seen in frogs, where the immature stage, or tadpole, is herbivorous and has a long coiled small intestine, while the adult is carnivorous and has a much shorter one relative to its body size (Fig. 27.24).

Since the small intestine is the place where the products of digestion are absorbed, it is not surprising that it has structures that vastly increase its surface area. The mucosa lining the intestine is arranged in numerous folds and ridges (Figs. 27.25, 27.26). Small fingerlike outgrowths, called ***villi***, cover the surface of the mucosa. Finally, the individual epithelial cells covering the folds and villi have a brush border, consisting of countless closely packed cylindrical processes, the ***microvilli*** (Fig. 27.27, 27.28). The total internal surface of the small intestine, including folds, villi, and microvilli, is incredibly large.

Some vertebrates have other adaptations for increasing absorptive surface area. For instance, special blind sacs, called ***caeca***, may branch from the anterior end of the small intestine; in many fish such caeca are present in the pyloric region. Another example is the spiral valve of many primitive fish and of sharks. The spiral valve is an epithelial fold extending the length of the intestine and forming a spiral within it (Fig. 27.29). Food must follow the spiral of the valve and thus contact much more epithelial surface than it would by moving straight through a tubular intestine of the same length.

The large intestine In humans the caecum is a blind sac at the junction between the small intestine and the large intestine (colon) (see Fig. 27.22). In humans the caecum is small, plays no role in nutrient absorption—indeed, is functionally unimportant. It may well be a relic of our long line of herbivorous primate ancestors (see next paragraph). A small fingerlike process, the ***appendix***, extends from the top of the caecum. The appendix sometimes becomes infected and must be surgically removed.

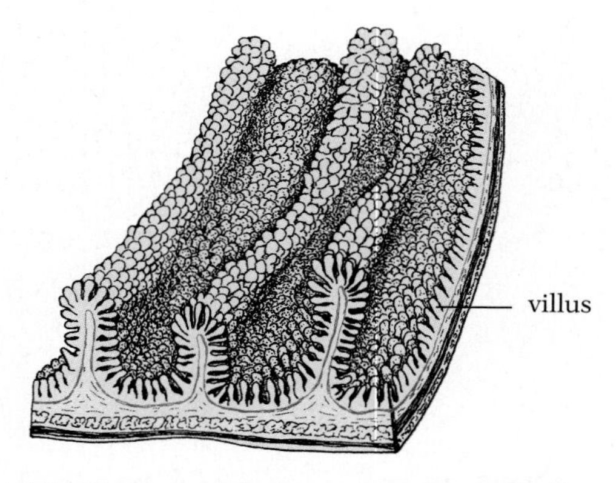

27.25 Longitudinal section through the wall of the human intestine

Three folds of the lining of the intestine are shown, each bearing numerous villi, which greatly increase the absorptive surface area. The villi are not stationary; they have smooth muscle fibers that enable them to move back and forth. Such movement increases after a meal.

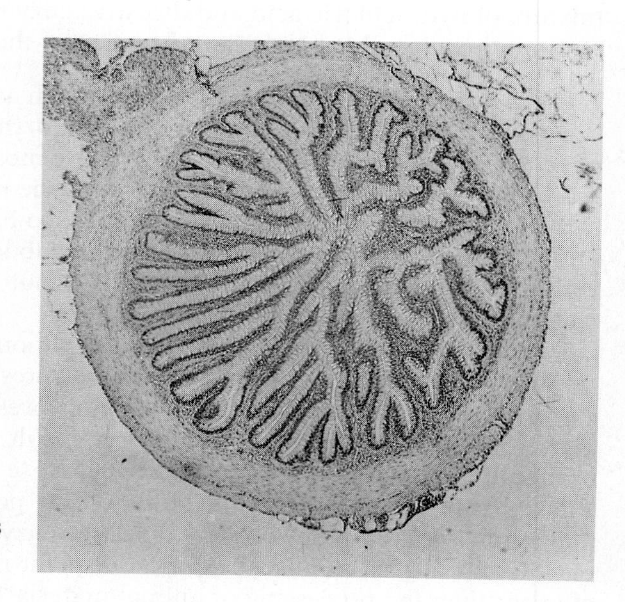

27.26 Cross section of intestine of calico bass

Notice the extensive folding.

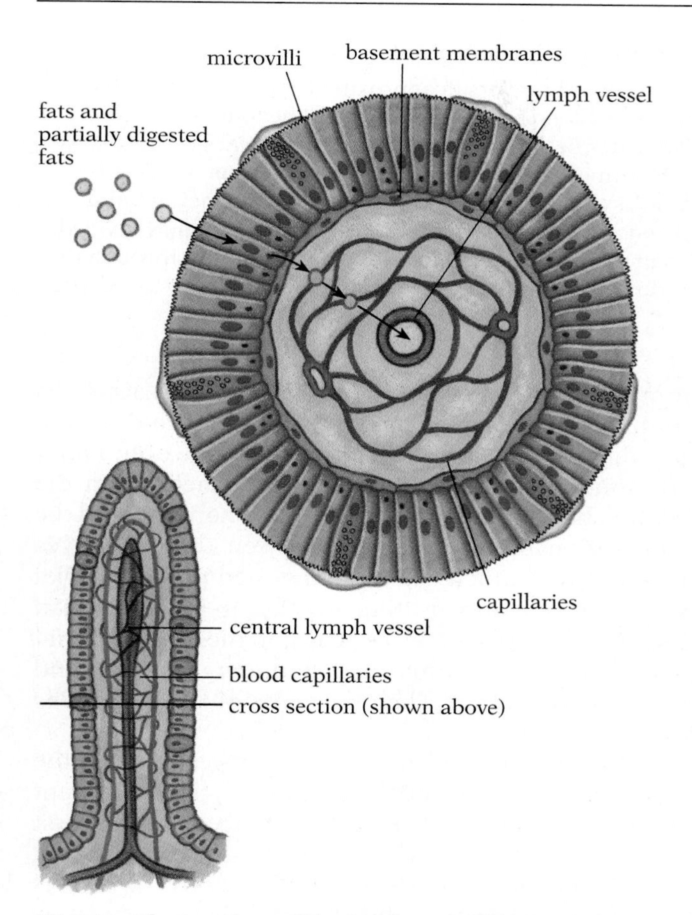

27.27 Absorption of fat in the small intestine

Shown here are longitudinal (bottom) and cross sections (top) of a villus in the small intestine. Lipase digests fat into monoglycerides and fatty acids. Some fat remains undigested. Because all these are lipid-soluble, they can diffuse through the membrane of the microvilli of the intestinal cell and enter the cytoplasm. There the partially digested fats are taken up by the smooth endoplasmic reticulum and new fats are synthesized. These combine with cholesterol and phospholipids, are coated with proteins, and are released from the cell by exocytosis. From there the lipoproteins move into the lymphatic vessels (green).

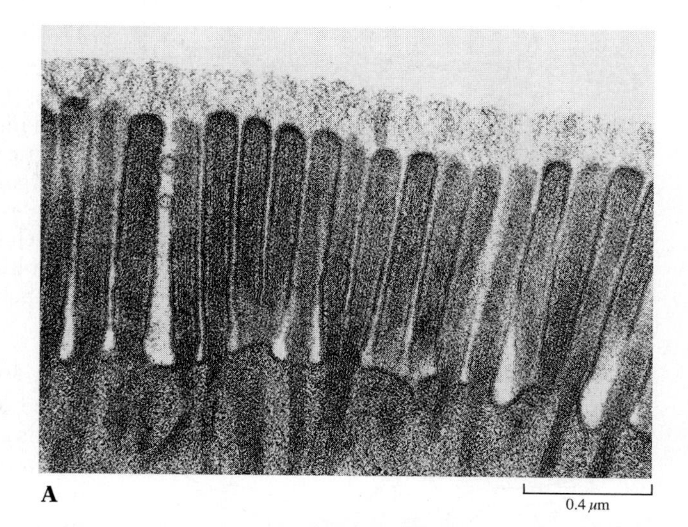

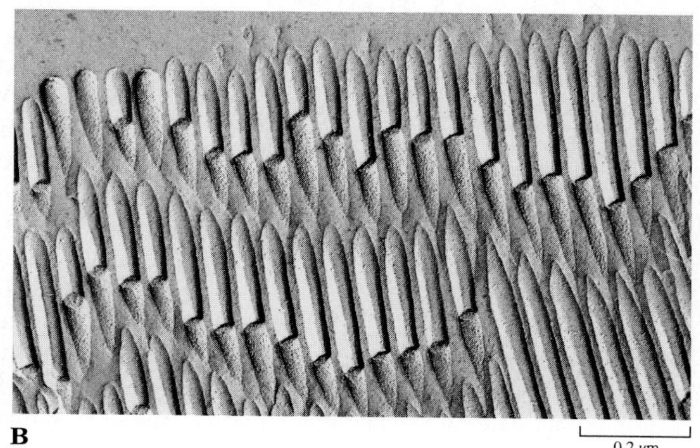

27.28 Microvilli on an epithelial cell of the intestinal lining of a cat

Notice the prominent glycocalyx covering the ends of the microvilli. (A) A bundle of microfilaments (composed of actin) runs into each microvillus. (B) Several rows of obliquely fractured intestinal microvilli are illustrated in this freeze-fracture replica. (From *Freeze Fracture Images of Cells and Tissues* by Richard L. Roberts, Richard G. Kessel, and Hai-Nan Tung, Oxford University Press, New York, ©1991.)

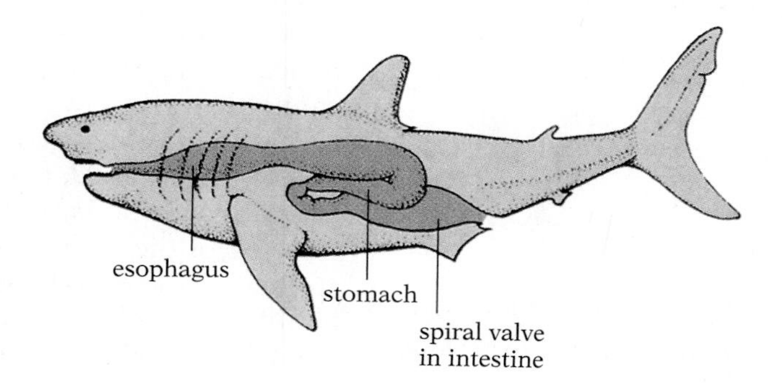

27.29 Digestive system of a shark

Because of the spiral valve of the intestine, food material must follow a winding course and is thus exposed to more surface area.

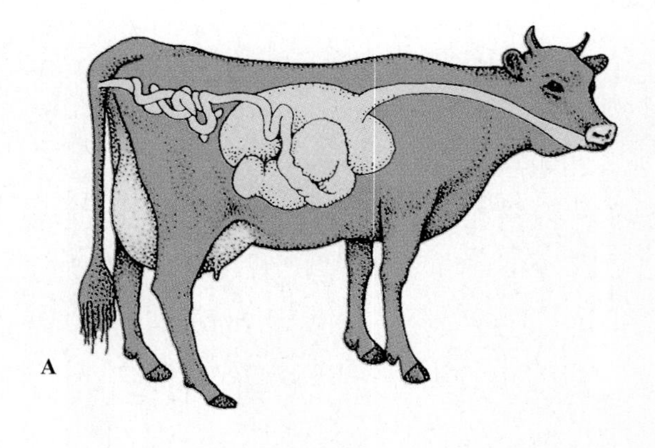

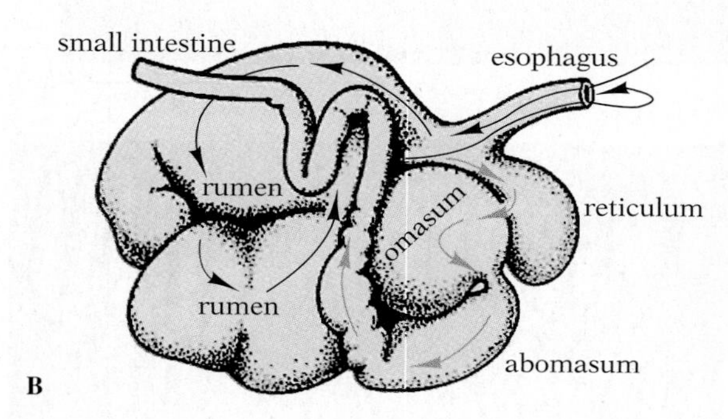

27.30 Digestive system of a ruminant

(A) A cow with the various chambers in approximately their correct locations. (B) Detailed path of food as it passes through the four "stomachs." Food moves initially into the rumen and the reticulum (black arrows show it going only into the rumen), where it is digested and fermented by microorganisms. Then, as the cud, it is regurgitated for more chewing and later reswallowed and returned to the rumen. The fluids and finely divided particles produced in this second phase (colored arrows) move through the reticulum into the omasum and then into the abomasum, which is the true stomach.

In some mammals, particularly herbivorous ones, the caecum is large and contains many microorganisms (bacteria and protozoans) capable of digesting cellulose. The caecum, however, is not located where the mammal can derive maximal benefit from the microbial action. It is behind the small intestine, where most of the digestion and absorption takes place. Hence even though horses have an enormous caecum, much coarse undigested plant material is expelled in their feces. A compensating adaptation has evolved in rabbits, which form two types of feces. One of these is material from the caecum; they reingest this, and it undergoes a second cycle of digestion and absorption. The other type of feces contains waste to be eliminated.

Ruminants such as cows also use microbes for digestion, but the microorganisms are held in front of the intestines. Four different stomachlike chambers are involved (Fig. 27.30), of which the first three are thought to be expanded sections of the esophagus. Vast numbers of bacteria and protozoans live in the ***rumen***, which is the largest of the four chambers, and in an adjacent chamber called the reticulum. Swallowed food enters the rumen and the reticulum, where the microbes begin digesting and fermenting it, breaking down not only protein, polysaccharides, and fats, but cellulose as well.

The larger, coarser material is periodically regurgitated for further chewing, as the animal chews its cud. Slowly the products of the microbial action and some of the microbes themselves move on into the true stomach (the abomasum) and intestine, where the more usual type of digestion and absorption takes place.

The rumen strategy is efficient but not perfect. Plant material must remain in the "fermentation vat" for a considerable period—hours, or even days—to give the microorganisms sufficient time to do their work. If the food is low quality, a cow or an antelope can literally starve to death waiting. The less efficient "afterburner" strategy of horses and zebras seems better adapted to handling large volumes of poor-quality food quickly. It enables zebras in Africa to thrive in desertlike habitats where ruminant antelope cannot live.

Digestion of cellulose by symbiotic microorganisms is not limited to mammals. A tropical leaf-eating bird, the hoatzin, ferments vegetation in a large, highly muscular crop that accounts for 15–20% of the animal's total weight (Fig. 27.31). Part of the inside surface of the crop is covered with cornified horny ridges that act as internal molars for chewing cud. Though the hoatzin benefits from the extra nutrients its symbiotic microbes extract for it, the

27.31 A tropical leaf-eating bird, the hoatzin
The hoatzin has an enlarged crop that, like the rumen of cattle and antelope, is used to ferment vegetation.

added weight of this avian version of the rumen makes the bird a poor flier. A more familiar example of a nonmammalian fermentor is the termite, one of several groups of insects that feed on wood.

In humans, the large intestine ascends from the caecum on the right side to the midregion of the abdominal cavity, then crosses to the left side and descends again (see Fig. 27.22). The three sections thus formed are frequently termed the ascending, transverse, and descending colons. One of the chief functions of the colon is reabsorption of much of the water used in digestion. If all the water secreted into the digestive tract with the enzymes were lost with the feces, most terrestrial animals would desiccate. Water recovery is accomplished mainly by active pumping of sodium; water follows osmotically.

Occasionally the intestine becomes irritated, and peristalsis moves material through it too fast for enough water to be reabsorbed; this condition is known as diarrhea. Conversely, if material moves too slowly, too much water is reabsorbed and constipation results. A proper amount of roughage (indigestible material, primarily cellulose) in the diet provides the bulk needed to stimulate enough peristalsis in the large intestine to prevent constipation. Our need for roughage is probably an artifact of our herbivorous ancestry.

A second function of the colon is the excretion of certain salts when their concentration in the blood is too high. The large intestine also contains great numbers of bacteria, which live on the undigested food that reaches the colon. The significance of most of these bacteria (those that do not synthesize vitamins) is unknown. Approximately half the dry weight of the feces is made up of masses of these bacteria.

The last portion of the large intestine, the ***rectum***, functions as a storage chamber for feces until defecation. Feces are eliminated from the rectum through the anus.

■ HOW FOOD IS DIGESTED

The role of saliva As food passes through the human digestive system, it changes chemically from the plant or animal (or collection of synthetic products) it once was into nutrients. Enzymatic digestion starts in the mouth. The saliva contains the enzyme ***amylase***[1] (also known as ptyalin), which begins the hydrolysis of starch to glucose. Though amylase produces some glucose, it yields primarily the disaccharide maltose (Fig. 27.32) and, in lesser amounts, fragments three or four glucose units long, which must be digested further in the intestine.

[1] The names of most enzymes end with the suffix *-ase*, which designates enzymes by international agreement. The first part of the name usually indicates the substrate on which the enzyme acts; thus *amyl-* (from *amylum*, Latin for "starch") indicates that amylase acts on starch.

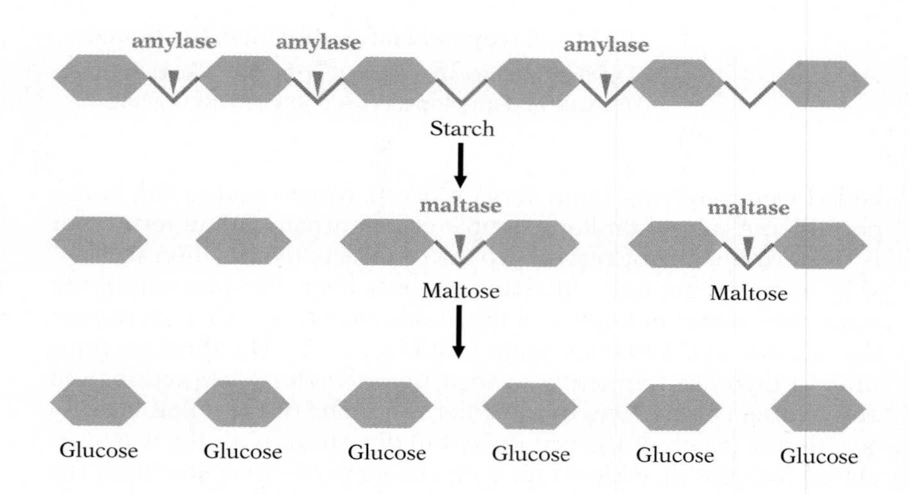

27.32 Digestion of starch

Amylase in the saliva and in the pancreatic juice hydrolyzes some of the bonds between glucose units, producing small amounts of free glucose, but much larger quantities of the disaccharide maltose. The maltose is then digested to glucose by maltase secreted by intestinal glands.

Since the food remains in the mouth only a short time, the amylase has little opportunity to work there. It acts mainly inside each bolus after the food is swallowed and moved into the stomach. The acid of the stomach soon inactivates the enzyme, however, and salivary amylase actually digests only a small percentage of the starch in the food. The saliva of many mammals contains no amylase at all.

Digestion in the stomach Once in the stomach, food is exposed to the action of gastric juice secreted by the gastric glands of the stomach wall. This juice contains hydrochloric acid (HCl) and several enzymes. The acid makes the contents of the stomach very acidic (with a pH of about 1.5–2.5).

The principal enzyme of the gastric juice is ***pepsin***, which digests protein. Unlike most proteolytic (protein-digesting) enzymes, it will function only in a strongly acid medium. Pepsin does not completely hydrolyze protein; it splits the peptide bonds adjacent to only a few amino acids, particularly tyrosine and phenylalanine (Fig. 27.33). The specificity of proteolytic enzymes is readily understandable. Since proteins are composed of a variety of amino acids, the structural configuration around the various peptide bonds depends on which two amino acids the bond joins; some of the bonds may fit in the active site of a particular enzyme, and others will not. Pepsin, for example, has an active site complementary to peptide bonds at the amino end of amino acids whose R groups include a six-carbon ring.

The wall of the digestive tract is covered with mucus, which shields it from digestive enzymes. When this defense breaks down, the enzymes begin to eat away a small portion of the lining; the resulting sore is known as an ulcer. Occasionally, an ulcer is so severe that a hole develops in the wall of the digestive tract, and the contents of the tract spill into the abdominal cavity. Most ulcers appear to develop from infections of a particular species of bacteria.

The wall of the stomach is protected in a second way: the gastric glands secrete not the active enzyme pepsin, but an inactive precursor called pepsinogen. Inactive enzyme precursors of this sort are known as ***zymogens***. Pepsinogen has no proteolytic activity, and as long as it is stored in the glands of the stomach wall, it poses no threat to the wall. It can only be changed into active

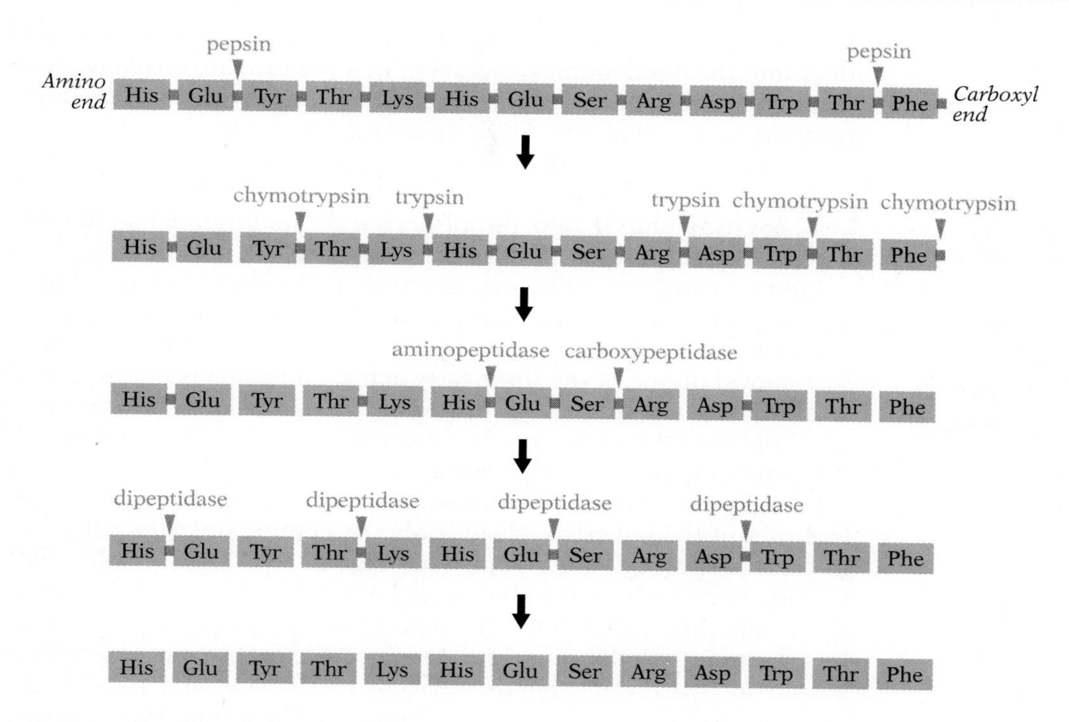

27.33 Digestion of protein

Pepsin in the stomach hydrolyzes peptide bonds at the amino ends of tyrosine (Tyr) and phenylalanine (Phe). Then the food moves into the intestine, where trypsin from the pancreas hydrolyzes bonds at the carboxyl end of lysine (Lys) and arginine (Arg), and chymotrypsin from the pancreas hydrolyzes bonds at the carboxyl end of tyrosine, tryptophan (Trp), and phenylalanine (and also bonds adjacent to methionine and leucine, when they are present). The remaining short chains of amino acids are digested in three ways: Aminopeptidase splits the bond connecting the peptide at the amino end of a chain to the rest of the fragment, while carboxypeptidase from the pancreas splits the bond connecting the peptide at the carboxyl end. In this way these two enzymes "nibble off" peptides from each end until only a dipeptide—a fragment consisting of two amino acids—remains. Bonds between these pairs of amino acids are then split by dipeptidases. With digestion now complete, the amino acids may be absorbed through the cells of the intestinal wall.

pepsin after exposure to HCl or the pepsin already active in the stomach. When 42 amino acids are removed from one end of the pepsinogen molecule, the shortened polypeptide chain that remains is pepsin.

Digestion in the small intestine Most digestion takes place in the small intestine. When partially digested food passes from the stomach into the duodenum, its acidity stimulates the release of a large number of different digestive enzymes into the interior (the lumen) of the intestine. These enzymes are secreted from two principal sources, the ***pancreas*** and the ***intestinal glands***. The pancreas, a large glandular organ lying just below the stomach (see Fig. 27.22), forms during fetal development as an outgrowth of the digestive tract; it retains a connection to the duodenum called the ***pancreatic duct***. When food enters the duodenum, the pancreas

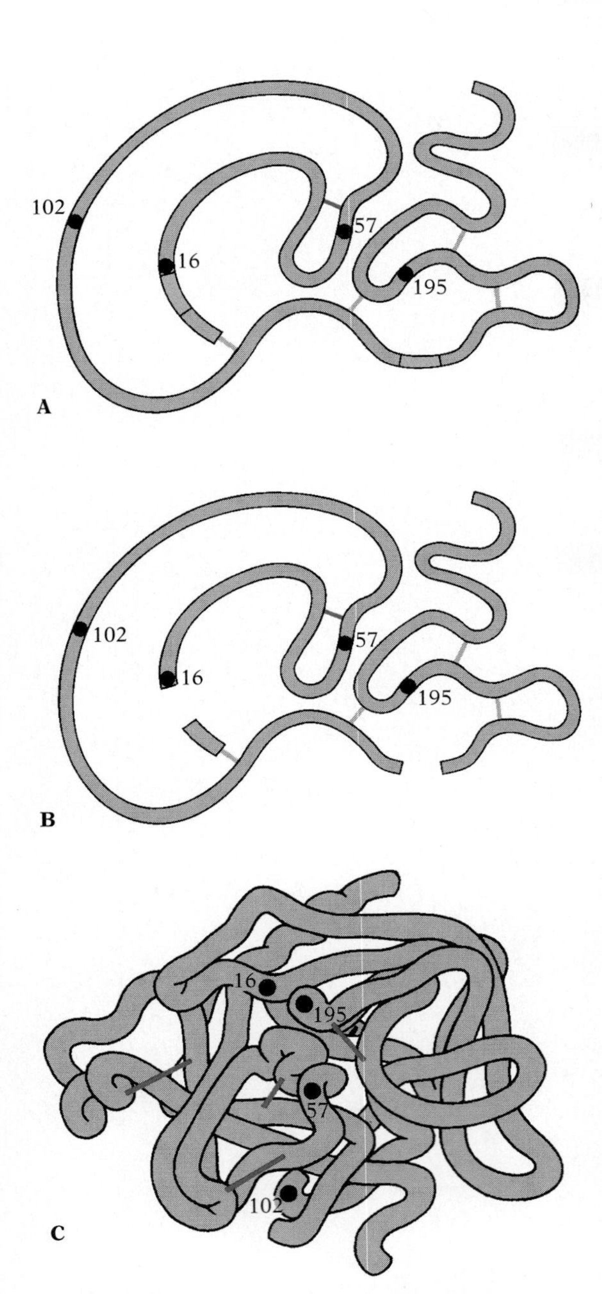

27.34 Conversion of chymotrypsinogen into chymotrypsin

(A) The single polypeptide chain constituting a molecule of the enzyme precursor chymotrypsinogen (shown unfolded here), is stabilized by disulfide bonds (red bars). The solid black circles indicate important parts of the catalytic site of the enzyme. Two portions of the polypeptide chain (blue) are removed (B), after which the enzyme refolds slightly, assuming its active configuration, chymotrypsin (C).

secretes a mixture of enzymes that flows through the pancreatic duct into the duodenum. Included in this mixture are enzymes that digest all three major classes of foods—carbohydrates, fats, and proteins—as well as some that digest nucleic acids.

One of the ***pancreatic enzymes*** is pancreatic amylase (sometimes called diastase or amylopsin), which acts like salivary amylase, splitting starch into the disaccharide maltose. This enzyme digests far more starch than salivary amylase.

Lipase, also secreted by the pancreas, is the body's principal fat-digesting enzyme, but it hydrolyzes a relatively small percentage of the fat to glycerol and fatty acids. Some of the fat is partly digested by removal of one of the three fatty acids, and some is not digested at all. Since fats, and the products of the partial digestion of fats, are lipid-soluble, they can be absorbed across cell membranes without being fully broken down.

Like pepsin, ***trypsin*** and ***chymotrypsin***, two of the proteolytic enzymes of the pancreas, cleave only the peptide linkages adjacent to certain specific amino acids (see Fig. 27.33). Trypsin splits the peptide bonds adjacent to lysine and arginine; chymotrypsin splits those adjacent to tyrosine, phenylalanine, tryptophan, and, to a lesser extent, methionine and leucine. Chymotrypsin resembles pepsin in hydrolyzing bonds adjacent to tyrosine and phenylalanine, but while pepsin cleaves the bonds on the amino side of tyrosine and phenylalanine, chymotrypsin breaks those on the carboxyl side.

Again like pepsin, both trypsin and chymotrypsin are secreted in inactive (zymogen) forms, called trypsinogen and chymotrypsinogen respectively. In the intestine, trypsinogen is converted into active trypsin. Then the trypsin thus formed removes two small internal pieces to produce active chymotrypsin (Fig. 27.34). Note that all three zymogens we have discussed—pepsinogen, trypsinogen, and chymotrypsinogen—have polypeptide chains longer than those of the active enzymes; in each case activation simply involves cutting off one or more pieces of the chain so that the polypeptide can refold to create an active site.

Studies of the chemical structure of trypsin and chymotrypsin have helped clarify both their probable evolution and the basis for their catalytic specificity. These two enzymes are sufficiently alike—in amino acid sequence, in conformational folding pattern, and even in their catalytic sites—to make it reasonable to believe that they evolved from the same ancestral proteolytic enzyme. There is, however, a functionally crucial difference between them in the pocketlike portion of the binding site that holds the R group of the substrate amino acid. In trypsin one of the amino acids that form this pocket is aspartic acid (which has a negatively charged R group); in chymotrypsin the same position is occupied by serine (which has an uncharged R group). This seemingly small variation accounts for the affinity of the two enzymes for different amino acids. That it should do so is a dramatic illustration of the critical importance of amino acid sequence (primary structure) in determining the functional properties of enzymes.

In summary, then, the action of the pepsin in the stomach and of trypsin and chymotrypsin from the pancreas results in splitting proteins into fragments of varying lengths, but does not produce many free amino acids. These three enzymes are known as ***en-***

EXPLORING FURTHER

THE DIGESTION OF LACTOSE BY HUMAN ADULTS

Digestive capabilities vary not only among different species, but also within particular species. The digestion of lactose, a sugar found only in milk, provides a striking example.

Secretion of milk by the mammary glands of female mammals evolved as a way of feeding the young. The only source of nourishment for very young mammals, milk is a nearly complete food; in most species it comprises carbohydrate, fat, and protein, as well as important minerals. Except for humans (and some of our domesticated pets), adult mammals do not use milk as a food. It is not surprising, then, that secretion of lactase, the lactose-digesting enzyme, usually greatly diminishes or even ceases altogether once an animal is weaned.

This pattern applies also to most humans; over much of the world, humans more than 4 years old secrete little or no lactase. Only people of European ancestry and those belonging to a few pastoral tribes in Africa have been found to secrete enough of the enzyme to be able to digest the lactose in large quantities of milk (see graph). When people of other ancestries, or even many adults of European heritage, drink much milk, they often develop cramps and diarrhea. They may have little or no trouble with milk from which the lactose has been removed, or with milk products like yogurt and cheese in which much of the lactose has been broken down by microbial action. One reason for the illness is that the undigested lactose sugar in the intestine upsets the normal osmotic balance, and an excessive amount of water moves into the intestinal lumen from the cells; another is that fermentation of the lactose by bacteria in the large intestine produces large quantities of acids and carbon dioxide. In Europeans and pastoral Africans, continued production of lactase in adults must have evolved during the roughly 10,000 years since domestic animals began to be kept for milk.

How widely peoples living near one another may differ is shown by the major tribes of Nigeria. The Ibo and Yoruba live in the southern part of the country, where conditions are unfavorable for cattle; milk has not traditionally been a part of their diet after weaning, and they cannot tolerate lactose. By contrast, the nomadic Fulani in northern Nigeria have been raising milk cattle for thousands of years, and they are lactose-tolerant (see graph). Most African-Americans are descended from nonpastoral tribes of western Africa, and are relatively intolerant of lactose.

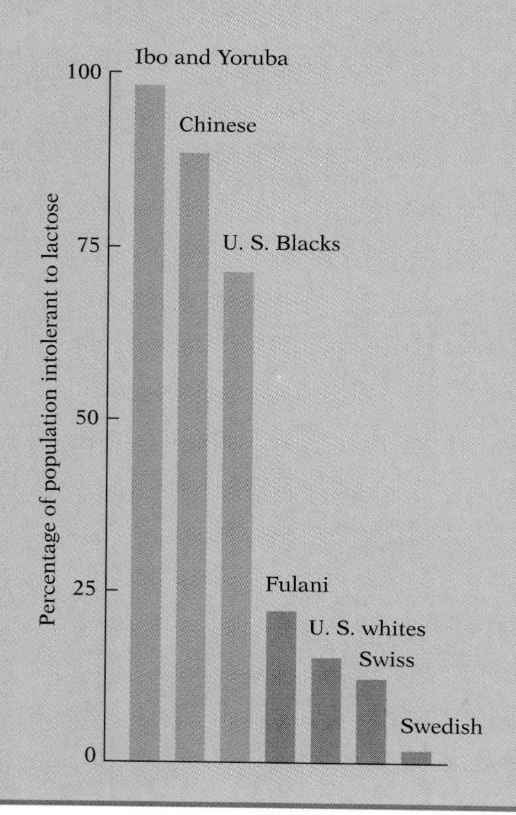

dopeptidases—that is, enzymes that hydrolyze peptide bonds between amino acids located within the protein, not bonds linking terminal amino acids to the original chain. Another class of enzymes, called ***exopeptidases***, hydrolyze off the terminal amino acids of the fragments produced by the endopeptidases, completing the digestive process.

There are many exopeptidases, each highly specific in its action. Carboxypeptidase, for example, hydrolyzes the linkage binding the terminal amino acid at the free carboxyl end of fragments with three or more amino acids. Another (aminopeptidase) hydrolyzes the linkage of the terminal amino acid at the free amino end of similar fragments. Dipeptidases break apart the resulting pairs of amino acids; one breaks only the bond of a fragment consisting of glycine linked to leucine, another only the bond of a fragment consisting of two molecules of glycine linked together, and so on. Most exopeptidases are secreted by intestinal glands; carboxypeptidase is produced in the pancreas.

The chemical action of the various proteolytic enzymes has been described at some length, not because it is important to remember in detail what bonds each enzyme hydrolyzes, but because the proteolytic mechanism provides a good example of enzyme specificity and of the way enzymes must often work in teams.

Just as certain enzymes from the intestinal glands complete the digestion of protein, other intestinal enzymes complete the digestion of starch begun by salivary and pancreatic amylase. The amylases split starch molecules primarily into disaccharides. The intestinal enzymes split disaccharides into simple sugars. For example, ***maltase*** splits maltose (see Fig. 27.32), ***sucrase*** splits sucrose, and lactase splits lactose. Digestion of nucleic acids is accomplished in an analogous manner: ***Nucleases*** secreted by the pancreas split nucleic acids into nucleotides. Nucleotides are themselves digested by ***phosphatases*** from the intestinal glands, which break their phosphate bonds, and by ***N-glycosylases***, which remove the ribose sugars. For an overview of the action of digestive enzymes, see Fig. 27.35.

Absorption of nutrients The products of digestion in the lumen of the small intestine must be absorbed into the blood if they are to be utilized. There is a capillary bed within each villus that picks up nutrients that have crossed the intervening layer of absorptive cells (Fig. 27.34). Some molecules simply diffuse from the lumen into the absorptive cells and then into the capillaries. For others, however, active transport supplements diffusion, leading to more rapid and complete removal of certain rare or especially valuable nutrients, including glucose and most amino acids. These nutrients enter the cell via cotransport channels (described in Chapter 4), which couple the movement of sodium ions (Na^+) down a steep electrochemical gradient into the cell with transport of the nutrient molecule. Transport out of the absorptive cell into the capillary space depends on ATP-powered pumps. From there the nutrients enter the capillaries by diffusion.

Bile The ***liver***, a critically important organ about which much will be said in later chapters, produces bile, which aids in fat digestion. The liver is very large, occupying much of the space in the upper part of the abdomen. On its surface is a small storage organ, the ***gall bladder*** (Fig. 27.22), which collects bile from the liver. When food enters the duodenum, the muscular wall of the gall bladder contracts, and bile is forced into the duodenum.

Bile is not a digestive enzyme. Instead, it is a complex solution of salts, pigments, and cholesterol. The bile salts act as emulsifying agents, causing large fat droplets to be broken up into many tiny droplets suspended in water. This action is much like that of a detergent. The many small fat droplets expose much more surface area to the digestive action of lipase than a few large droplets would. Bile salts apparently aid also in the absorption of fats. When insufficient bile salts are present in the intestine, both fat di-

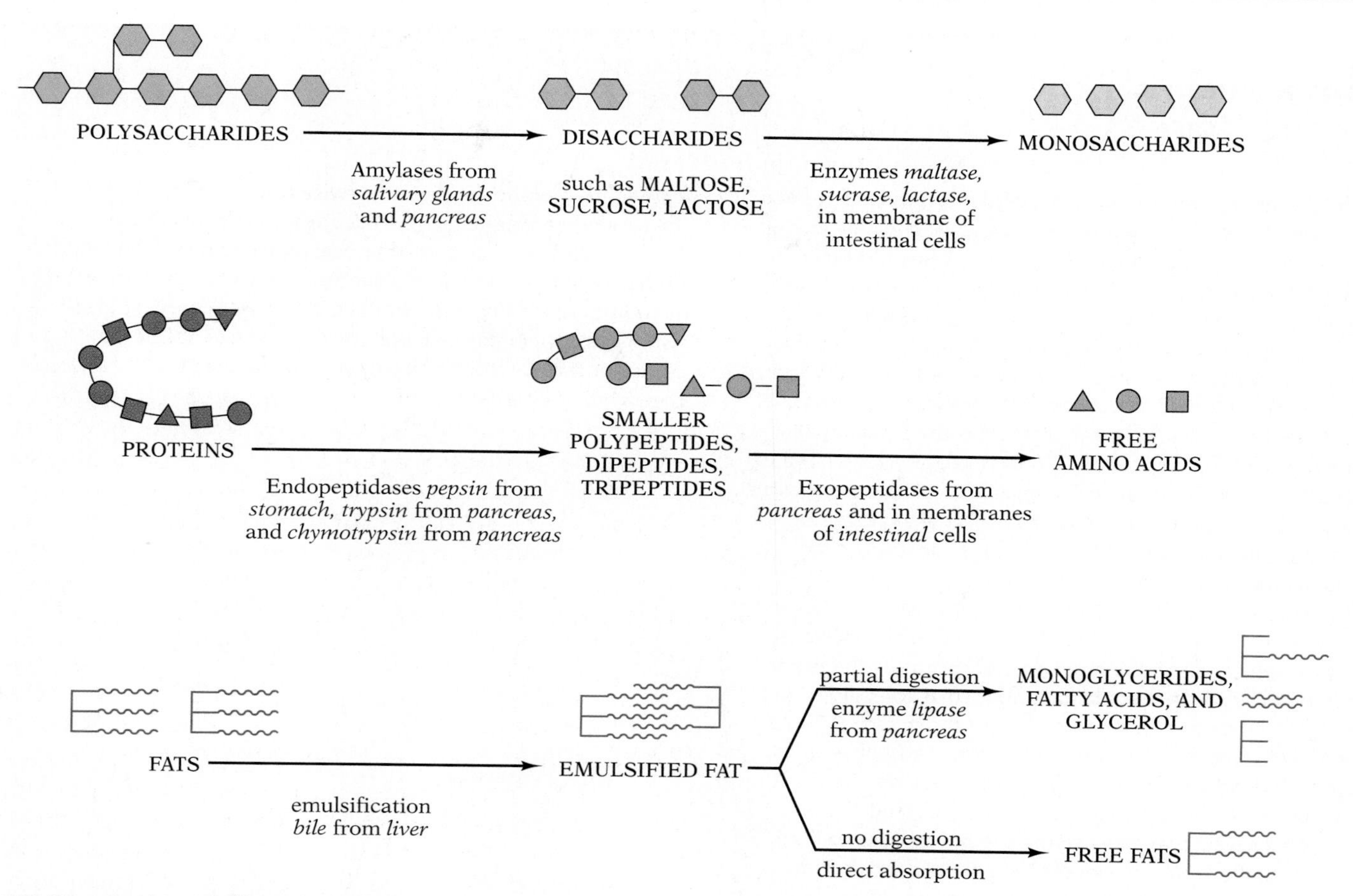

27.35 Action of digestive enzymes

gestion and absorption are seriously impaired. The bile salts are reabsorbed by the large intestine, transported back to the liver, and used again.

Bile pigments and cholesterol play no role in digestion. The pigments result from the destruction of red blood cells in the liver and give the characteristic brown color to feces. Cholesterol, a relatively insoluble compound, sometimes causes trouble by becoming concentrated into hard gallstones, which may block the bile duct and interfere with the flow of bile.

Wastes Plants, of course, do not produce feces. The reason is that plants can be highly selective about what nutrients they take in through their roots by either active transport or osmosis. Since plant nutrients need not be digested, the only waste products are gases. Even plants that feed through their leaves (tropical epiphytes and carnivorous plants, for example) are selective about which substances to admit.

For animals that eat bulk food, it is much more difficult to balance the diet at the mouth. Selectivity is largely imposed in the intestines, where microvilli act like plant roots, transporting in the appropriate products of digestion and letting the rest pass by. Even after the nutrients are absorbed, however, metabolism frequently produces new imbalances—an excess of nitrogen is especially common—which must be addressed with separate waste-manage-

EXPLORING FURTHER

TOXIC NUTRIENTS

One of the most common defenses of plants against potential herbivores is to synthesize one or more toxic substances that interfere with the digestion or metabolism of would-be grazers. When we see specialist herbivores like koalas (which eat only eucalyptus) or silkworms (which dine exclusively on mulberry), we can be pretty sure that the plants have a strong toxin that prevents most herbivores from feeding, and that the specialists have evolved a counterstrategy for evading the plants' line of defense. A typical example is the vinelike legume *Dioclea megacarpa*, whose only predator is the larva of the beetle *Caryedes brasiliensis*. The plant's trick is to synthesize an unusual amino acid, canavanine, a structural analogue of arginine that, in nearly all species, is readily substituted for arginine by the appropriate transfer RNA.

Canavanine is nonpolar, while arginine is positively charged. As a result, the enzymes produced by any normal herbivore feeding on the seeds of this vine and incorporating the apparent nutrient canavanine would have abnormal solubilities and a strong tendency to fold incorrectly—in short, some would be inactivated, and the development of the predator would be disrupted. Indeed, other insects fed on the legume never mature properly, but the *Caryedes* beetle is unaffected. Careful experiments reveal that the enzyme the beetle uses to load arginine onto its tRNA has become specialized to discriminate against canavanine, and a separate enzyme system has evolved to metabolize this otherwise toxic amino acid. This is the same defense the legume uses against its own poison. Another 250 nonstandard amino acids are known from plants, each probably the result of selection for herbivore resistance. In addition, certain plants produce other well-known herbivore toxins, including caffeine, cocaine, cannabinol, opium compounds, plus the molecules that make peppers hot and cocoa beans bitter.

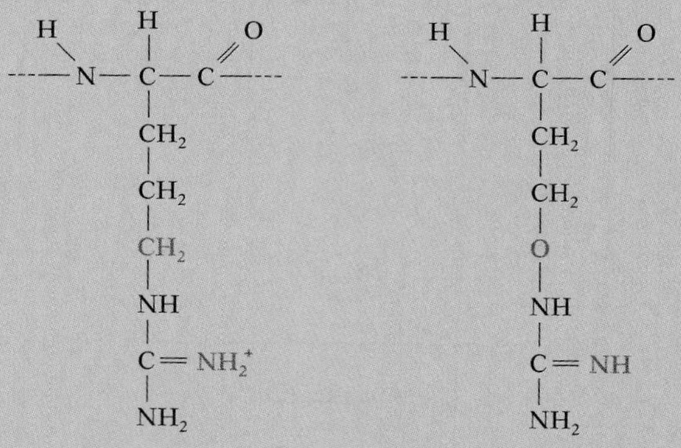

H, H, O; N—C—C; CH_2; CH_2; CH_2; NH; $C=NH_2^+$; NH_2

Arginine

H, H, O; N—C—C; CH_2; CH_2; O; NH; C=NH; NH_2

Canavanine

ment systems. (We look at how nitrogen is disposed of in Chapter 31.) Not only is waste inevitable for animals; it is, as we have seen, essential for the proper workings of the gastrointestinal tract: if a certain minimum amount of indigestible matter is not part of the diet, the flow of material through the digestive system ceases.

CHAPTER SUMMARY

WHAT HETEROTROPHS NEED TO EAT

WHY ISN'T SUGAR ENOUGH? In addition to an energy source (usually carbohydrates), most animals need certain essential amino acids and fatty acids that they cannot synthesize but that are regularly found in their diet. (p. 761)

WHAT VITAMINS DO Most vitamins act as coenzymes or parts of coenzymes. (p. 763)

WHY MINERALS ARE NECESSARY Minerals are needed to synthesize nucleic acids, amino acids, and the active sites of many enzymes and to maintain the ionic balance of fluids. (p. 768)

THE UNIQUE FEEDING STYLE OF FUNGI

Most fungi have rootlike structures (rhizoids) that secrete digestive enzymes and absorb the products of this extracellular digestion. (p. 769)

INTERNALIZED DIGESTION: ANIMALS AND PROTOZOANS

HOW PROTOZOANS CAPTURE AND DIGEST FOOD Some protozoans simply engulf their food; others have special localized structures for ingesting food. (p. 771)

THE ANIMAL AS STOMACH: COELENTERATES Coelenterates draw prey into a gastrovascular cavity—a stomach with a single opening—where digestion and absorption occur. (p. 772)

THE ALL-PERVADING STOMACH OF FLATWORMS Flatworms have a highly branched gastrovascular cavity that delivers the products of digestion to all parts of the body and greatly increases the surface area available for absorption. (p. 774)

DISASSEMBLY LINES: THE COMPLETE DIGESTIVE TRACT Complete digestive tracts have a separate entrance and exit. Food is usually ground into finer pieces at the entrance and digested in a sequential fashion; the products of digestion are absorbed, excess water is recovered, and wastes are expelled. (p. 774)

HOW THE VERTEBRATE DIGESTIVE SYSTEM WORKS Mammals have specialized teeth for capturing, cutting, and grinding food. The stomach mixes the food with digestive enzymes; the anterior portion of the intestine adds additional enzymes. The intestine has a high surface-to-volume ratio to aid in absorbtion of nutrients. Water is usually recovered in the posterior intestine. Ruminants have a pregastric chamber in which cellulose is digested and fermented; many other herbivores have a postgastric caecum that serves the same purpose. (p. 778)

HOW FOOD IS DIGESTED Enzymes that break large molecules like starch and protein into smaller segments are usually employed first, followed by enzymes that specialize in digesting the segments from one end or breaking specific bonds between subunits. (p. 785)

STUDY QUESTIONS

1. Though all nectars are mostly sugar and water, there are usually low concentrations of some minerals and certain amino acids as well. How would you go about discovering whether these "contaminants" are artifacts of a sloppy nectar-secretion system or the result of selective pressure in the competition for pollinators?
2. Draw up a list of the costs and benefits of the digestive strategy of fungi, coelenterates, and animals with complete digestive tracts. (pp. 769–770, 772–778)
3. The trend in antelopes, which are pregastric ruminants, is for the largest species to be able to live on the poorest vegetation—dry grass as opposed to growing leaves, for example. What might be the physiological and behavioral logic of this pattern? (pp. 784–785)
4. What effect might loss of the gall bladder have on a human's metabolism and physiology? Might loss of this organ offer potential benefits for modern humans? (pp. 790–791)
5. From their diet, which species of vertebrates would you predict would have the shortest large intestines? The longest? (pp. 781–782)
6. Why do carnivores have a greater variety of essential amino acids than herbivores? (pp. 761–762)

SUGGESTED READING

DEGABRIELE, R., 1980. The physiology of the koala, *Scientific American* 243 (1). *On the nutritional problems of this postgastric fermentor, whose need to maximize nitrogen extraction while minimizing water loss can cause it to starve with a full stomach.*

ROSENTHAL, G. A., 1983. A seed-eating beetle's adaptations to a poisonous seed, *Scientific American* 249 (5). *Ploy and counterploy as plants attempt to protect themselves by sabotaging the digestion and metabolism of herbivores.*

SANDERSON, S. L., AND R. WASSERSUG, 1990. Suspension-feeding vertebrates, *Scientific American* 262 (3). *On filter feeders.*

CHAPTER 28

GAS EXCHANGE

The most critical nutrients for plants during photosynthesis are water and carbon dioxide, a gas; the most important waste product, oxygen, is also a gas (Fig. 28.1). For animals (as well as for plants when they are metabolizing the high-energy molecules they produce during photosynthesis), the reverse is true: oxygen is crucial to efficient respiration, and metabolism generates waste carbon dioxide, as well as water (Fig. 28.1). Thus a basic problem for nearly all organisms is the procurement of one gas and the elimination of another. We will see in this chapter how these problems are solved.

THE PROBLEMS

Gas exchange between a cell and its environment always takes place by diffusion across a moist membrane. The gases must be in solution if they are to cross the membrane. In unicellular organisms and many small multicellular ones this poses no serious problem, because each cell is either in direct contact with, or only a few cells removed from, the surrounding medium. Hence these organisms have usually not evolved special systems for gas exchange.

Even a larger organism can still use direct diffusion for gas exchange if its body is essentially two-dimensional. Some brown algae, the kelps, may grow to a length of 50 m, but the blades remain very thin (Fig. 28.2). As a result, no cell is far from the surface of a blade, and the total gas-exchange area is fairly large in relation to the volume of the plant. Even the thicker stipe that connects the blades has intercellular spaces filled with water that is continuous with the external medium. In these large two-dimensional plants, no special gas-exchange mechanism is needed.

When increasing body size involves three dimensions, as it generally does, maintaining a respiratory surface of adequate size relative to the organism's volume becomes a problem, because surface area (a square function) increases much more slowly than volume (a cube function). The problem is most acute for the more active animals, whose rapid utilization of energy demands a large amount of oxygen per unit of body volume per unit time.

A further complication is that many organisms have evolved relatively impermeable body coverings. The waxy epidermis of the

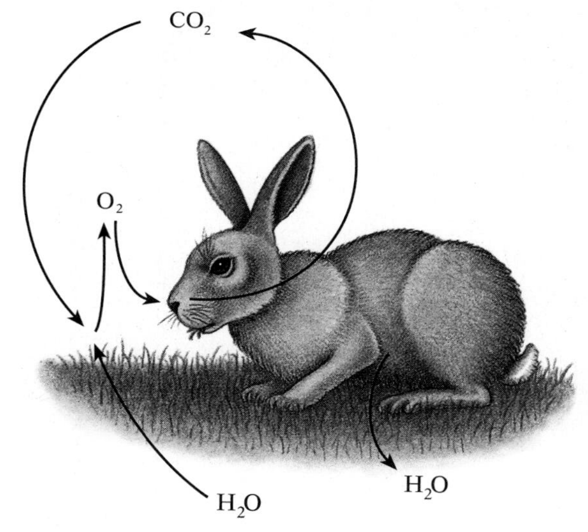

28.1 Basic pattern of gas exchange in plants and animals

Most heterotrophs require oxygen for respiration and produce waste carbon dioxide (and water), whereas most autotrophs require carbon dioxide (and water) for photosynthesis, generating waste oxygen. Of course, plants also respire (not shown), but in general the carbon dioxide consumed (and oxygen liberated) in photosynthesis exceeds the amount involved in respiration. Also not indicated here is the extra water that plants need for vascular transport, and the additional water that animals need to aid in temperature regulation and waste elimination (discussed in subsequent chapters).

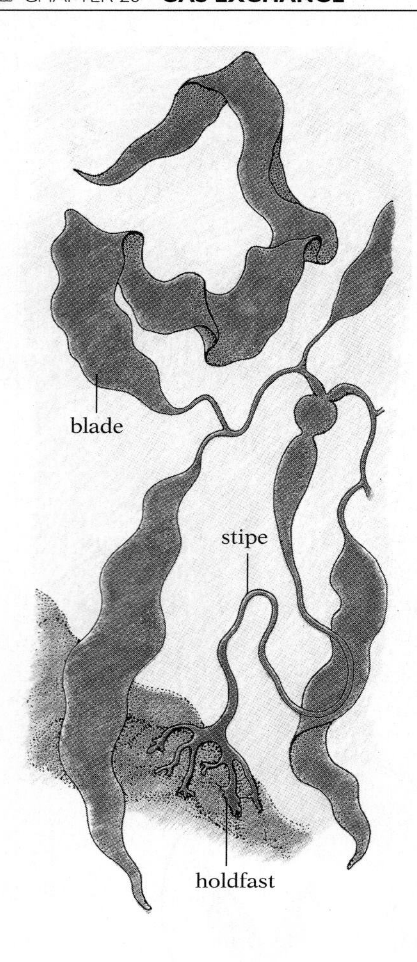

28.2 A kelp (*Pelagophycus*), one of the large brown algae
In a simple solution to the problem of gas exchange, the broad, flat kelp blades provide a relatively large surface for the diffusion of gases, with every cell in the organism close to the surface.

leaves of terrestrial plants, for example, and animal skin with its scales, feathers, or hair function as protective barriers between the fragile internal tissues and the often hostile outer environment. The presence of these coverings, though it confers many advantages, limits the gas-exchange surface to restricted regions of the body, making the problem of providing adequate exchange area even more critical.

Another problem for large three-dimensional organisms is that many cells are deep within the body, far from the gas-exchange surface. Diffusion alone is incapable of moving gases in adequate concentrations across the immense number of cells that may intervene between these more distant cells and the exchange surface. In general, simple diffusion suffices for movement of substances through aqueous media only when the distances are less than 1 mm. Some other mechanism for conveying gases to the individual cells of the organism therefore becomes essential.

The need for direct contact between the moist membranes across which gas exchange occurs and the environmental medium also poses serious difficulties for terrestrial organisms. The moist membranes must be exposed to the environment, but at the same time the chances of desiccation must be kept to a minimum. Since a large, thin, moist surface is often fragile and easily damaged, protective devices have evolved in many species, particularly when the respiratory surfaces protrude outward.

In general, specialized respiratory surfaces may be grouped into two categories: inward-oriented and outward-oriented extensions of the body surface (see Fig. 28.9). Each category embraces a diversity of form and detail, but each represents a way of meeting the basic needs discussed above: (1) a respiratory surface of adequate dimensions; (2) for many organisms, methods of transporting gases between the area of exchange with the environment and the inner cells; (3) means for protecting the fragile respiratory surface from injury; and (4) ways of keeping the surface moist without incurring excessive water loss.

HOW PLANTS EXCHANGE GASES

■ LEAVES AS PASSIVE LUNGS

Gas exchange associated with both photosynthesis and cellular respiration takes place at a particularly high rate in green leaves, organs strikingly adapted for this process.

Most of the visible outer surface of a leaf is covered by a waxy cuticle; it is relatively dry and impermeable, and hence ill suited for diffusion of gases. Exchange must therefore take place elsewhere. The mesophyll in leaves has large intercellular spaces (Figs.

28.3, 28.4). A high percentage of the total surface of each mesophyll cell is exposed to the air in these spaces, which are interconnected and continuous with the external atmosphere by way of openings in the epidermis: the ***stomata***. Gases can thus move easily between the surrounding atmosphere and the internal spaces of the leaf. The actual gas exchange—the diffusion of gases into and out of cells—takes place across the thin moist membranes of the cells inside the leaf.

The structure of the leaf helps it meet four critical requirements for respiratory systems:

1 Because of the large number of free-standing mesophyll cells, the surface area available for gas exchange in the leaf is enormous compared with the outer area of the leaf. The microvilli in the small intestine and the root hairs of plants represent analogous solutions to the problem of increasing surface area within a small volume.

2 Internal transport of gases occurs in leaves without any special adaptations. Gases can reach each cell directly via the intercellular spaces.

3 The danger of mechanical injury is relatively minor for an internal exchange surface. The epidermis functions as a protective covering for the entire leaf.

4 The exchange surfaces remain moist because they are exposed to air only in intercellular spaces. With the humidity within those spaces at nearly 100%, the membranes of the mesophyll cells always retain a thin film of water on their surfaces. Gases dissolve in this water before moving into the cells. The epidermal tissues and their waxy cuticle act as barriers between the dry outside air and the moist inside air.

These barriers cannot be complete or gas exchange between the mesophyll and the air would be impossible. The stomata provide thousands of essential openings in every square centimeter of leaf.

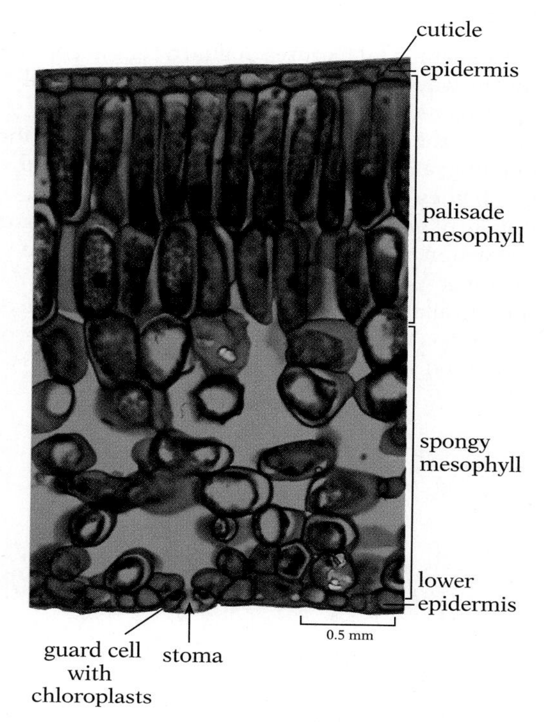

28.3 Cross section of part of a privet leaf

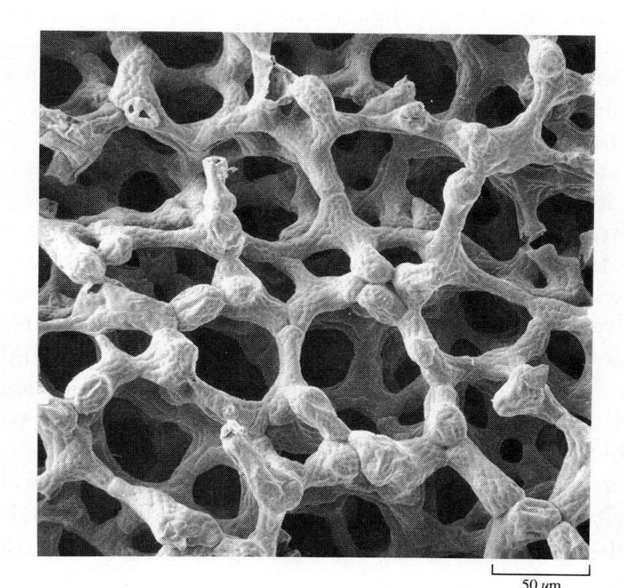

28.4 Spongy mesophyll in a bean leaf
This scanning electron micrograph gives an especially clear view of the extensive system of interconnecting air spaces.

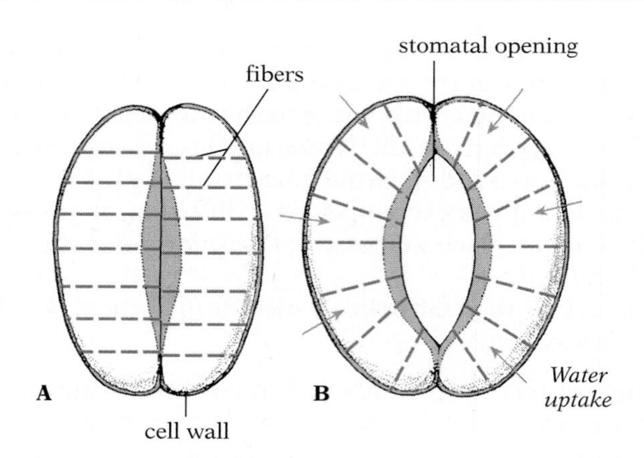

28.5 Guard cells

(A) Two unusual features of guard cells account for the way in which they regulate the stomatal opening: the cell walls are thicker on the side close to the stoma than on the side away from the opening; and bands of inelastic fibers—dashed green lines—run around each cells. (B) When uptake of water causes the cells to become turgid, expansion is limited to the sides away from the opening. As a result the guard cells buckle, pulling apart, and the stoma opens. The degree of swelling is largely controlled by the availability of water and sunlight.

The problem, then, is to balance the need for gas flow—supplying CO_2 and disposing of excess O_2—with the need to retain moisture.

When all three raw materials for photosynthesis—light, water, and CO_2—are available, the stomata must be open to allow gas exchange, but when one or more of them are missing, these openings must be closed to prevent water loss. Each opening, or stoma, in the epidermis is bounded by two specialized ***guard cells***, which are joined at their ends. Unlike most other epidermal cells, these bean-shaped cells contain chloroplasts (Fig. 28.3). The walls of each guard cell are of unequal thickness, being considerably thicker on the side next to the stoma than on the side away from it (Fig. 28.5A). Moreover, the guard cells are wrapped by bundles of inelastic fibers that prevent them from expanding radially.

Opening of the stomata is caused by the flow of water into guard cells, generally in response to the active pumping of potassium ions into the cell, which lowers their water potential. When the guard cells swell with water (which normally occurs only if sufficient water is available from the roots), the expansion of the cell surface to accommodate this increase in volume is restricted to the lengthening of the sides away from the stoma, where the wall is relatively thin. As a result the guard cells bend (Fig. 28.5B), and the stoma opens (Fig. 28.6). Gas exchange can now take place, and the leaf can obtain carbon dioxide for photosynthesis. When less water is available, or is being lost from the plant too quickly, the guard cells become flaccid, which causes the stoma to close, slowing the loss of water. Of course, this closing also limits diffusion of carbon dioxide into the leaf, and thus limits photosynthesis.

At night, when the problems of water loss are rarely as severe, the stomata nevertheless usually narrow, since sunlight is not available. How the rapid changes in guard-cell turgidity are tied to the availability of sunlight and water is still a matter of dispute, but there is good evidence that a yet-to-be-discovered blue-light receptor controls the activity of the potassium pump in at least many guard cells, causing potassium ions to be imported only when light is available.

Another hypothesis takes its cue from the unusual starch metabolism of guard cells. By contrast with most cells, which convert starch into sugar mainly in the dark, guard cells do so in the light.

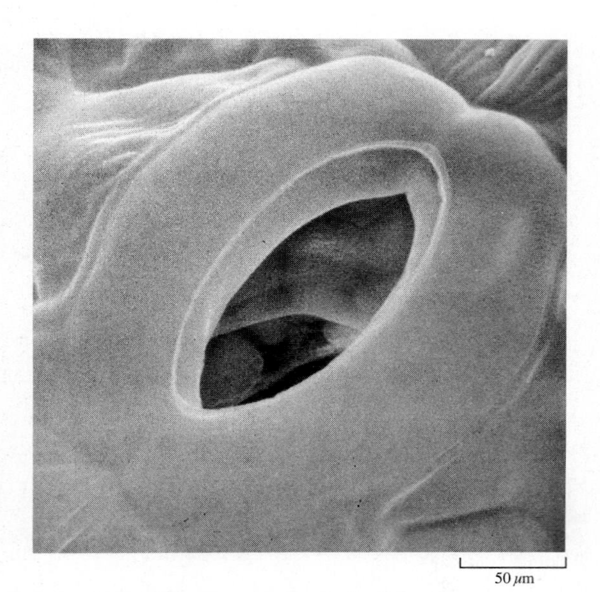

28.6 A stoma in a cucumber leaf

The stoma is open because the two large crescent-shaped guard cells have pulled away from each other. Mesophyll cells in the interior of the leaf can be glimpsed through the opening in this scanning electron micrograph.

The decreasing acidity (rise in pH) that results from the consumption of carbon dioxide in photosynthesis seems to make the enzyme responsible for converting starch into sugar in the guard cells progressively more active. The conversion of stored starch into sugar presumably brings about a rapid decrease in the water potential of the guard cells' cytoplasm, so that water then moves into the guard cells by osmosis, causing them to swell and open the stomata. There is also evidence that abscisic acid, a plant hormone produced in response to stress, can induce stomatal closing. The various factors known to affect stomatal opening are summarized in Table 28.1.

Table 28.1 Environmental factors affecting stomatal opening and closing

CONDITIONS FAVORING OPENING	CONDITIONS FAVORING CLOSING
Abundant water	Lack of water
Abundant light	Darkness
Low internal CO_2	High internal CO_2
	Presence of abscisic acid

Most plants lose large quantities of water by evaporation through the stomata every day in the process called ***transpiration***. Indeed, despite their obviously sedentary lifestyle, each day plants lose 10–20 times more water per unit of weight than animals. The reason is that plants have a far larger area of gas-exchange tissue per unit of mass, and water is constantly evaporating from these damp membranes. The extra surface area is needed because the atmospheric concentration of the CO_2 essential in photosynthesis is much lower than the concentration of the O_2 needed by animals (0.03% versus 21%), and so a proportionally larger "capture surface" is required.

Water lost by transpiration must be replaced; if the mesophyll tissues dry out, gas exchange ceases. Replacement water is drawn up through the xylem from the roots. Terrestrial plants have made a virtue of this necessity by using the flow of replacement water as a means of transporting the inorganic nutrients gathered by the roots, carrying them to the stems, leaves, and buds. Transpiration also serves as a valuable cooling mechanism, preventing excessive buildup of heat inside the leaves even when they are exposed to the direct rays of the sun.

Many plants have unusual stomata particularly suited to their lifestyle. In most plants the stomata are located primarily in the lower epidermis—the side of the leaf usually turned away from the sun's rays, where the drying tendency is less severe. Furthermore, the lower epidermis of many plants is covered with short hairs, which trap insulating air. This boundary layer of air functions in reducing the flow of direct air currents across the stomatal openings and thereby slows down desiccation. In the oleander (*Nerium*), which lives in a very dry habitat, the stomata are located in deep hair-lined depressions in the lower epidermis, an adaptation that eliminates convection currents across the stomata. In the pondweed (*Potamogeton*), by contrast, the stomata of the floating leaves are located only in the very thin upper epidermis (Fig. 28.7).

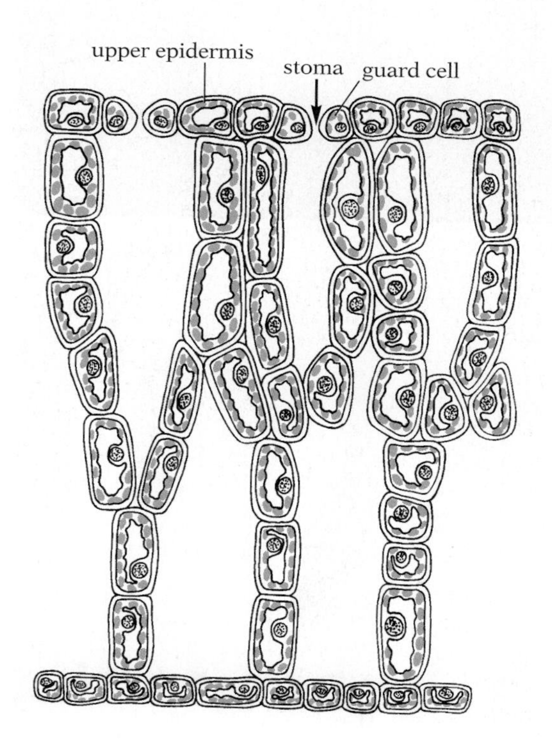

28.7 Cross section of portion of a floating leaf of pondweed (*Potamogeton*)

The stomata are in the upper surface, and the mesophyll has very large intercellular air spaces. Unlike terrestrial plants, where the epidermal cells have no chloroplasts and function as barriers to water loss, the epidermal cells of pondweed leaves are richly supplied with chloroplasts.

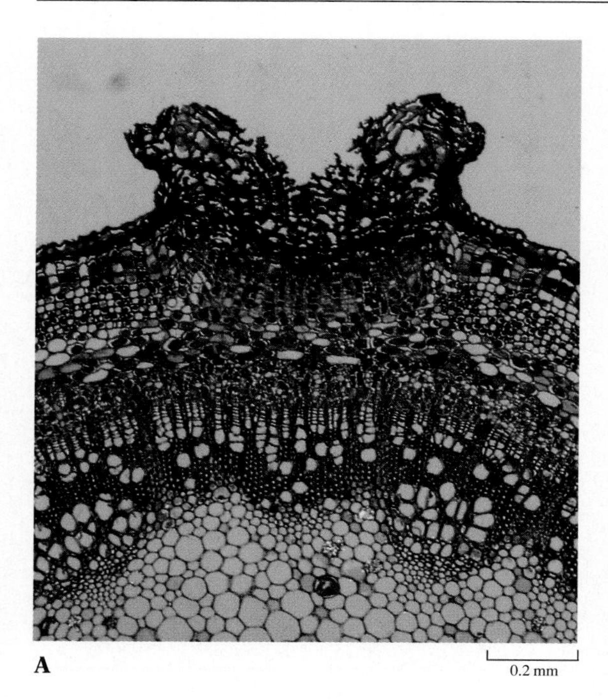

A

B

28.8 Lenticels

(A) The waterproof outer bark (layer of dark cells on the surface) on this section of elderberry (*Sambucus*) stem is interrupted at the center of the lenticel. Thus the more loosely arranged cell layers beneath, with their numerous intercellular air spaces, are exposed to the atmosphere.
(B) The individual lenticels can be seen as white streaks on the surface of a young sycamore maple (*Acer*) stem.

■ CAN ROOTS BE ASPHYXIATED?

Leaves are the major gas-exchange organs of plants, but stems and roots are also involved. The relatively impervious layer of bark on many old stems would effectively cut off most of their oxygen supply were it not for the development of numerous small areas of loosely arranged cells, called ***lenticels*** (Fig. 28.8), through which gases can move freely. Since most of the cells in the inner layers of large stems are dead, oxygen does not need to penetrate very deep.

Roots also carry out gas exchange. Gases can diffuse readily across the moist membranes of root hairs and other epidermal cells. For roots to obtain enough oxygen, however, the soil in which they grow must be well aerated. Different types of soils vary in their aeration characteristics, which depend on the amount of pore space, the affinity of the soil particles for water, and many other factors. Soils with very high percentages of clay particles, for example, have many pore spaces, but the tiny clay particles are so hydrophilic and absorb so much water during wet periods that the air spaces become filled with water. Frequent waterlogging may reduce the air content of the soil to the point where a plant cannot grow at all because its roots cannot "breathe."

Plants adapted to life in waterlogged soil or in standing water generally have specific structures for getting oxygen to their roots. Deep-water rice, for example, grows small tubes that rise out of the water to allow air to diffuse down; many other species of plants produce aerial roots that, in a less specialized way, accomplish the same end.

Plants, unlike animals, do not usually need any special gas-transporting organs. Most of the intercellular spaces in the tissues of land plants are filled with air; those in animal tissues are filled with fluid. The air-filled spaces of plants are interconnected to form a continuous system that opens to the outside through the stomata and lenticels. Incoming gases can therefore move directly from the air to the internal parts of the plant without having to cross membranes. Moreover, they don't have to diffuse long distances through water or cell liquids because they do not go into solution until they reach the film of water on the surfaces of the individual cells. Since oxygen can diffuse some 10,000 times faster through air than through liquids, the intercellular air-space system ensures that even the innermost cells are adequately supplied.

HOW AQUATIC ANIMALS EXCHANGE GASES

Unicellular animals have no need for special gas-exchange devices; simple diffusion across their cell membranes is sufficient (Fig. 28.9A). Some of the smaller and simpler multicellular animals like jellyfish, hydra, and planarians show little further development. Their multipurpose gastrovascular cavities facilitate the exposure of the inner cells to the environmental water (containing dissolved oxygen) that they draw in through the mouth; no cell in these animals is far from the water. A few larger aquatic animals, particularly some of the marine segmented worms, lack special respiratory systems; they use their skin, which is usually richly supplied with blood vessels (Fig. 28.9B). Most larger multicellular animals, however, have evolved true respiratory systems.

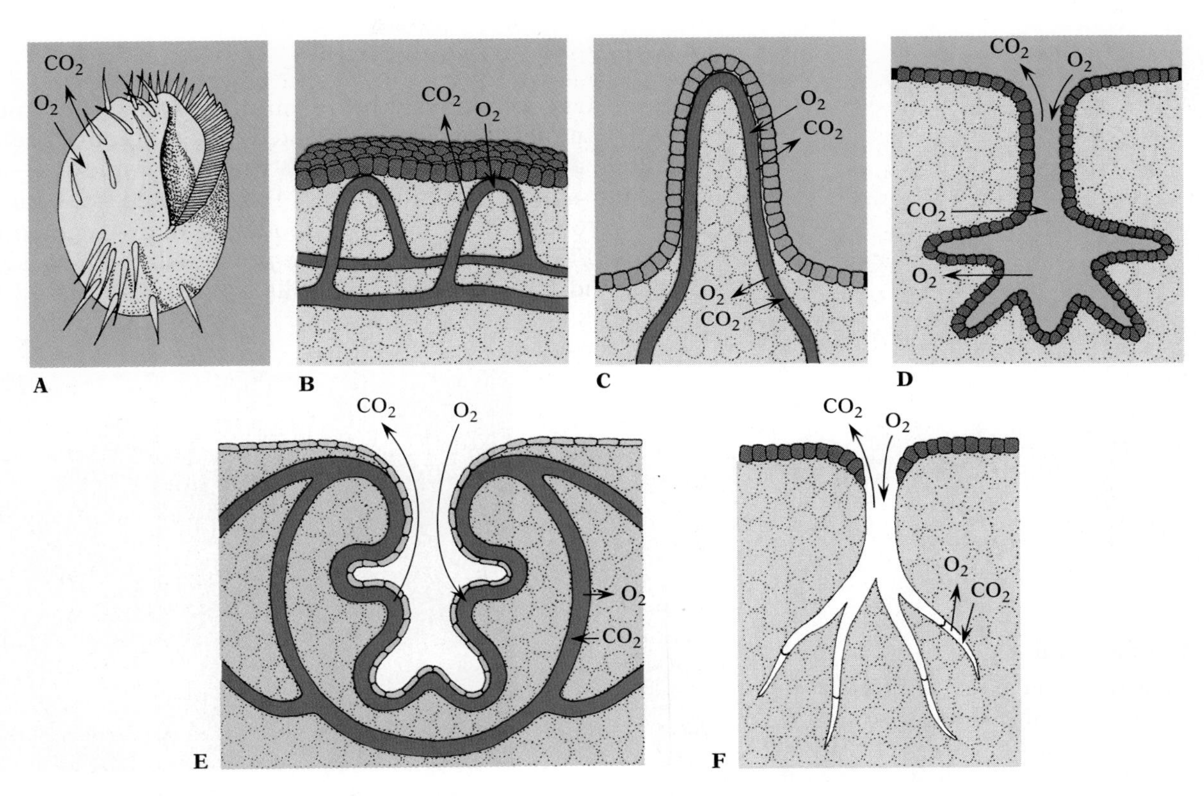

28.9 Types of gas-exchange systems in animals

(A) Unicellular organisms exchange gases with the surrounding water directly across the cell membrane. (B) Some multicellular aquatic animals use the body surface as an exchange surface; the blood (red) transports gases to and from the surface. (C) Many multicellular aquatic animals have specialized evaginated gas-exchange structures (gills). (D) A few aquatic animals, such as the sea cucumber, use invaginated exchange areas. (E) Most true air breathers have lungs, specialized invaginated areas that depend on a blood transport system. (F) Most land arthropods have tracheal systems, invaginated tubes that carry air directly to the tissues without the intervention of a blood transport system.

28.10 Gills of marine segmented worm (*Nereis*)

As the cross section shows, the flaplike parapodia (two on each segment) are richly supplied with blood vessels.

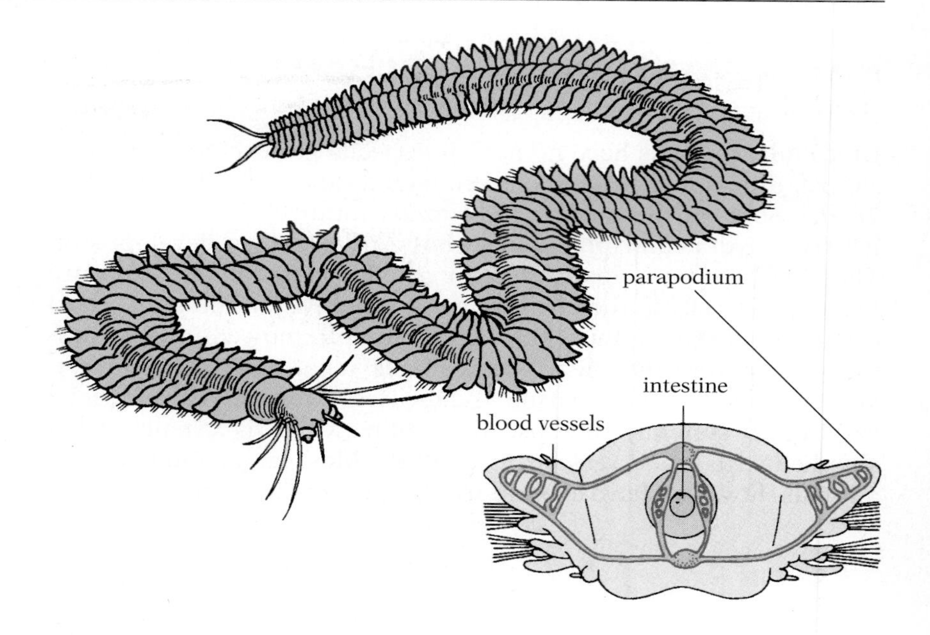

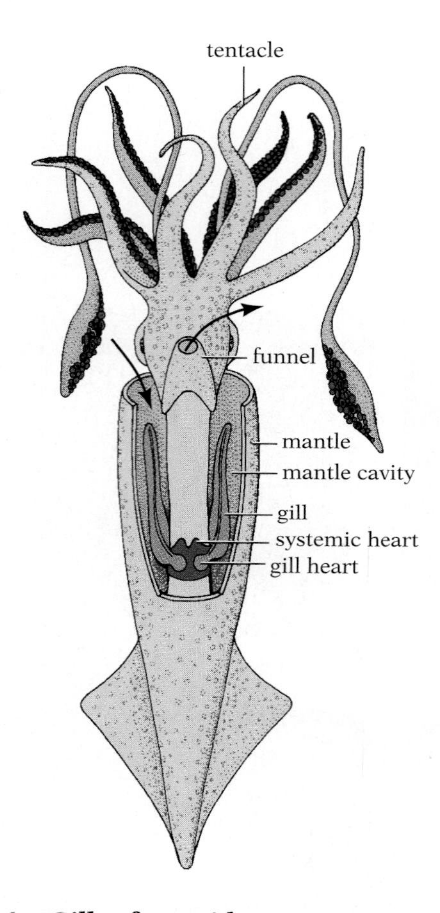

28.11 Gills of a squid

Part of the protective mantle has been cut away to expose the gills within the mantle cavity. Water (black arrows) containing dissolved oxygen flows into the mantle cavity when the mantle is relaxed. When the mantle contracts, its collar seals the opening and the oxygen-depleted water is forced out of the funnel. The jetlike expulsion of water when the mantle is contracted also propels the animal backward with great force.

■ WHY GILLS WORK SO WELL

With a few exceptions, the respiratory systems of multicellular aquatic animals involve evaginated (protuberant) exchange surfaces, usually known as ***gills***. Gills vary in complexity—all the way from the simple bumplike skin gills of some sea stars (see Fig. 24.69, p. 693), the flaplike parapodia of many segmented marine worms (Fig. 28.10), the mantle-protected gills of squids (Fig. 28.11), and the shell-protected gills of bivalves (Fig. 28.12), to the

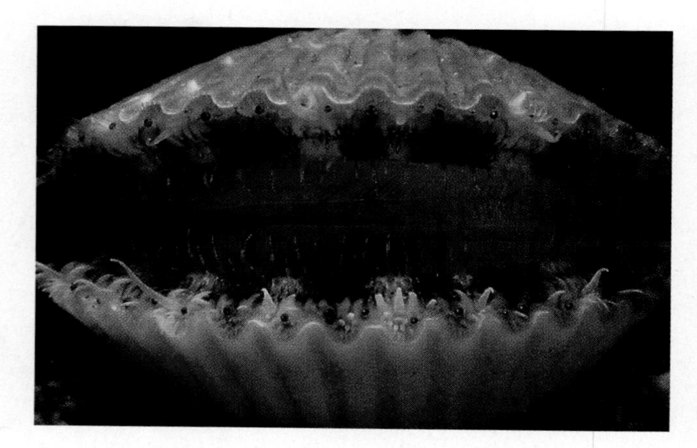

28.12 Calico scallop (*Argopecten gibbus*), a filter-feeding bivalve

The cilia on the gill move water through this bivalve. As the water crosses the gill, gas exchange—uptake of O_2 and release of CO_2—takes place. At the same time, a second set of cilia traps food particles and moves them to the labial palps near the mouth. Here yet another set of cilia separates out the coarse material, which is ejected; the remaining fine particles are ingested. (The small blue spheres along the shell rims are simple eyes.)

minutely subdivided gills of fish (Fig. 28.13). The gill system has evolved independently countless times; animals as diverse as clams, lobsters, sharks, and salamanders use gills for respiration.

Most gills, particularly those of very active animals, have such finely subdivided surfaces that a few small gills may expose an immense total exchange surface to the water. Thus, though the gas-exchange surface takes up a very limited part of the animal body, most of which can be protected by relatively impermeable coverings, the surface-to-volume ratio of the exchange surface is high.

Another characteristic of most gills is that they contain a rich

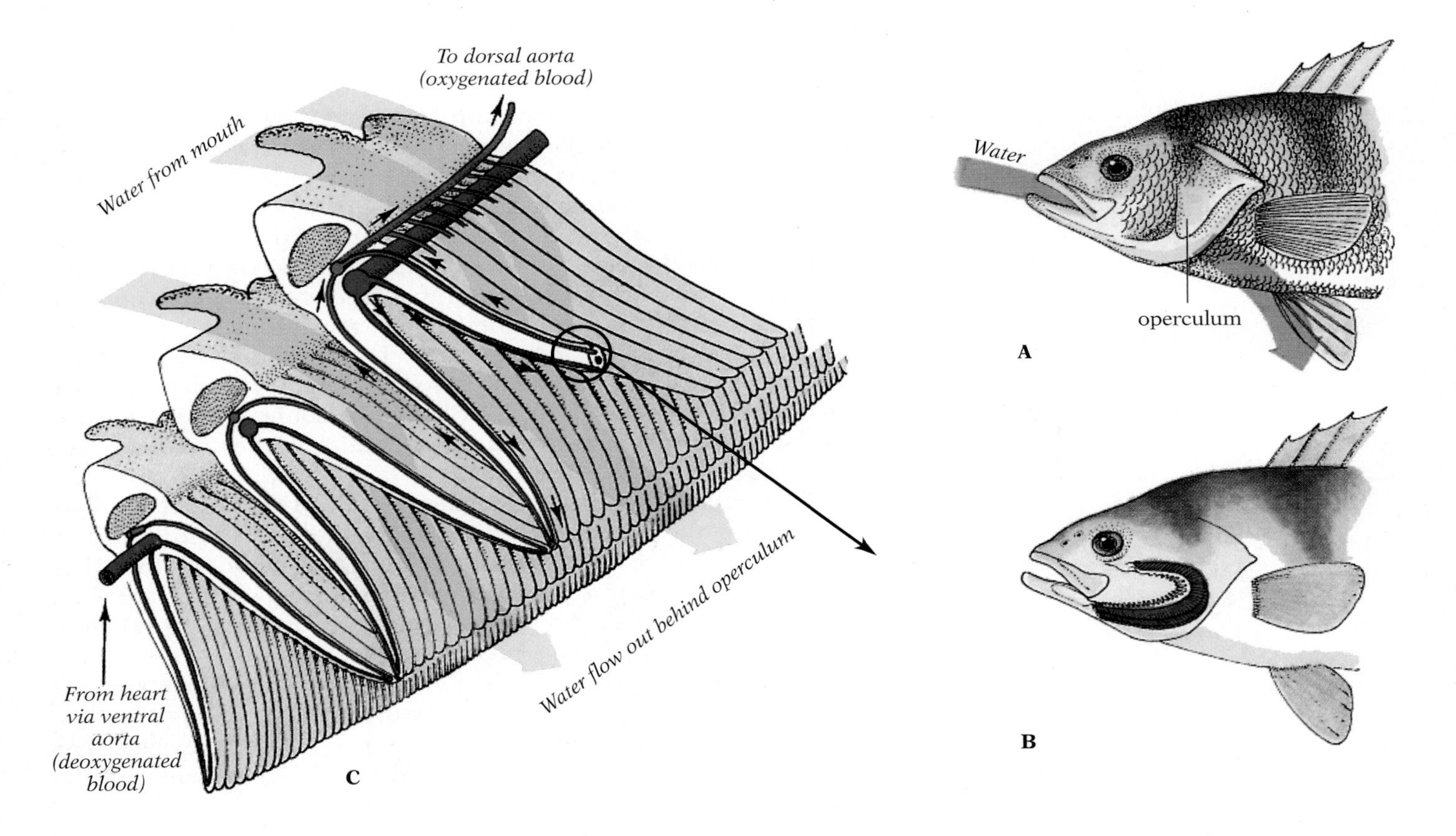

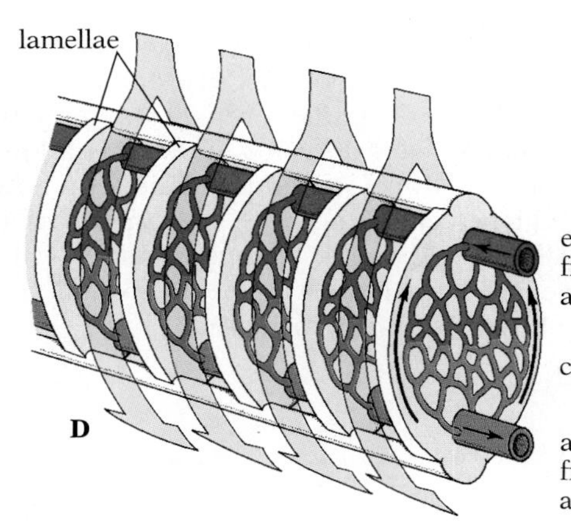

28.13 Gills of fish

(A) Head with operculum covering gills. Water carrying O_2 is drawn in through the mouth, flows across the gills, and exits behind the operculum. (B) Head with the operculum cut away and the gills exposed. (C) Portions of three adjacent gill arches. Each arch bears two rows of primary filaments. The main paths of blood flow from the heart to the filaments are shown in dark blue (afferent arteries); those from the filaments to the heart are in red (efferent arteries). The broad, light-blue arrows trace the path of water across the gills. Each primary filament bears many disklike lamellae, which contain capillaries that run from the afferent artery to the efferent artery. The end of one filament has been cut off and enlarged (D) to show the lamellae more clearly. The blood and the water flow between the lamellae as countercurrents, moving in opposite directions. The lamellae are the actual sites of gas exchange.

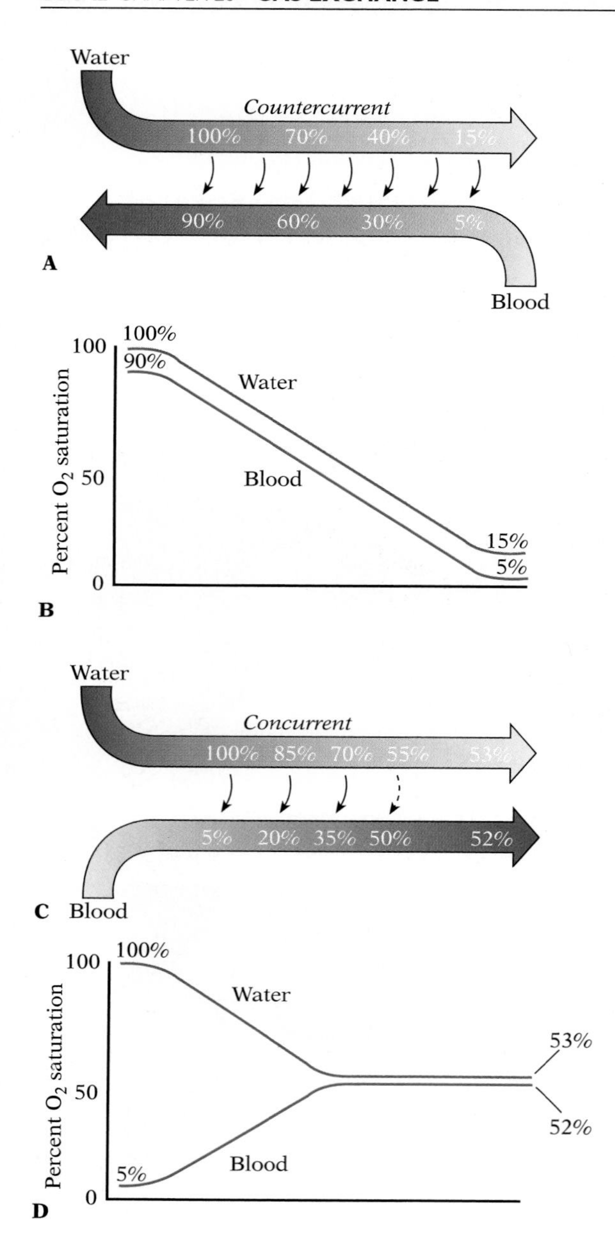

28.14 Countercurrent exchange system in lamella of fish gill

Percentages indicate the degree of oxygen saturation for both water and blood. Because of the counterflow (A), the O_2 gradient between water and blood always favors diffusion of O_2 (vertical arrows) from water to blood, and the blood can extract a high percentage of the O_2 from the water (B). If the flow were parallel, or concurrent (C), the blood could extract much less O_2 and would leave the gills far from fully loaded (D).

supply of blood vessels. Often the blood in these vessels is separated from the external water by only two cells: the single cell of the vessel wall and a cell of the gill surface. Sometimes even the vessel wall is eliminated, and only one cell remains between the blood and the water. Oxygen moves by diffusion from the water, across the intervening cell or cells, and into the blood, where it is ordinarily picked up by a carrier. The blood then distributes the oxygen throughout the body to the individual cells. Carbon dioxide produced by cellular metabolism is transported to the gills and discharged into the surrounding water.

The water flows over the surface of a gill lamella in a direction opposite to the flow of blood in the vessels of the lamella (Fig. 28.14A). Consequently, blood just about to leave the gill, and already almost fully loaded with O_2, encounters water that has not yet given up any of its O_2 because it is just reaching the gill. The O_2 gradient therefore favors pickup of more O_2 by the blood. At the other end of the lamella, oxygen-poor blood entering the gill encounters water that has already lost much of its O_2, but because even this water contains more O_2 than blood that has not yet picked up any, the gradient here too favors diffusion of O_2 from the water to blood (Fig. 28.14B). In short, this ***countercurrent exchange system***, thanks to a favorable gradient between blood and water at every point along the lamella, maximizes the amount of O_2 the blood can pick up from water. This would not be the case if the two fluids had the same direction of flow (concurrent exchange; Fig. 28.14C–D). As we will see, the countercurrent strategy is also employed in temperature regulation, water recovery, and fluid regulation.

The fragile gills are easily damaged, and various structures have evolved for their protection. Frequently these are hard coverings like the carapace of lobsters and the operculum of fish (Fig. 28.13A), but sometimes they take other forms, such as the fearsome array of spines and pedicellariae that surround the skin gills of sea stars (see Fig. 24.69, p. 693).

Obtaining enough oxygen is a greater problem for aquatic animals than for air breathers, for two reasons: First, O_2 has a low solubility in water, constituting only about 0.45% of seawater (and usually slightly more in fresh water, although the percentage is far more variable) as compared with approximately 21% in air. Second, the diffusion of O_2 is many thousands of times slower in water than in air. Most aquatic animals must therefore move water across the exchange surfaces. If the water remained still, the O_2 in the vicinity of the exchange surfaces would soon be depleted, and it would not be renewed by diffusion fast enough to sustain the animal. Most fish actively pump water into their mouths, across the gill filaments, and out behind the operculum. Many fast-swimming fish keep their mouths open as they swim, so that their forward motion forces water across the gills; some species are so dependent on this method of ventilation that they will die for lack of O_2 if prevented from moving. Because water is far more viscous than air and therefore much harder to move, some aquatic organisms use up as much as 20% of their metabolic energy just to move water across their gills.

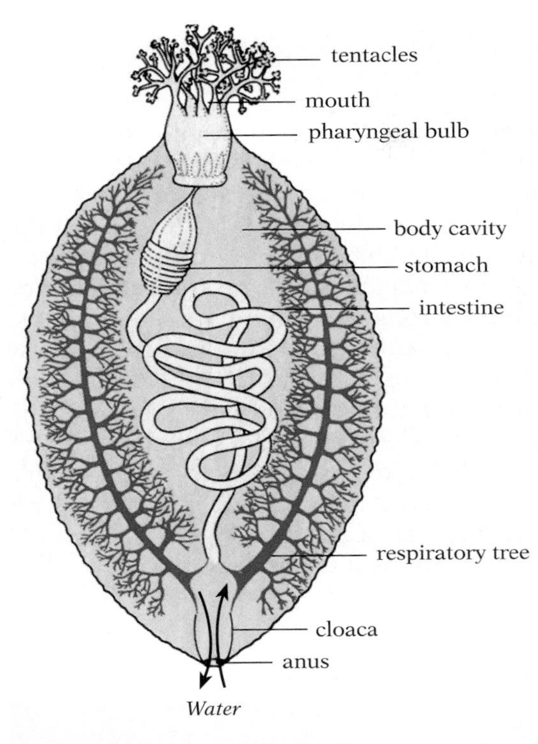

28.15 Respiratory tree of sea cucumber

28.16 Water spider *Argyroneta aquatica*

The male spider is beneath his underwater web, which holds a large bubble of air and to which he has returned to breathe. He transports the air there by trapping it among hairs on his abdomen.

■ ALTERNATIVES TO GILLS

Not all aquatic animals with special respiratory systems use evaginated gills. For example, sea cucumbers—relatives of sea stars—have specialized invaginated (infolded) systems called ***respiratory trees*** (Fig. 28.15). This system is an elaboration of a simpler one used by a few animals, in which gas exchange takes place as water is alternately drawn in and expelled through the enlarged thin-walled posterior portion of the digestive tract.

Many insects that live in water are not fully aquatic; they must periodically come to the surface to breathe air. Diving beetles (*Dytiscus*) store air under their hard shell-like forewings when they surface, then dive with the bubble and breathe from it. Some spiders construct an underwater web in which they store a large bubble of air (Fig. 28.16). You might think the O_2 in the bubble would soon be depleted, but this is not the case. Remember that air contains high concentrations of other gases besides O_2. Since these gases are not used up, a gas bubble remains, and as the partial pressure[1] of O_2 in the bubble falls, O_2 from the surrounding water tends to diffuse into the bubble, renewing the supply. In short, the bubble acts as a gill. In a few arthropods this mechanism is so refined that their store of gases is permanent and they do not have to surface to renew it.

[1] The partial pressure of a gas is the total pressure of the mixture of gases in which it occurs multiplied by the percentage of the total volume that it occupies. Thus, if the total pressure of all the atmospheric gases is about 760 mm Hg (millimeters of mercury, a measure of barometric pressure), and if O_2 is about 20% of this mixture, then the partial pressure of O_2 in the atmosphere is equal to 760 × 20%, or 152 mm Hg.

HOW TERRESTRIAL ANIMALS EXCHANGE GASES

A few land animals have evolved highly modified gill-like respiratory structures that function in air. An example is the book lungs of arachnids, in which the gills are evaginated into an open body cavity and therefore resemble the pages of a book (Fig. 28.17). But the hazards of desiccation are considerable for most exposed surfaces, and major structural problems plague any array of filaments or branched structure sufficiently strong to maintain its shape against surface tension and gravity, yet sufficiently thin-walled to allow easy passage of gases. It is not surprising, therefore, that most terrestrial animals have evolved invaginated respiratory systems. (Fig. 28.18) These systems are of two principal types, lungs and tracheae. In both, the air inside the system is kept moist, and the cells of the exchange surface are covered by a film of water in which gases can dissolve. Thus the process of gas exchange has remained essentially aquatic in land animals, as it has in leaves.

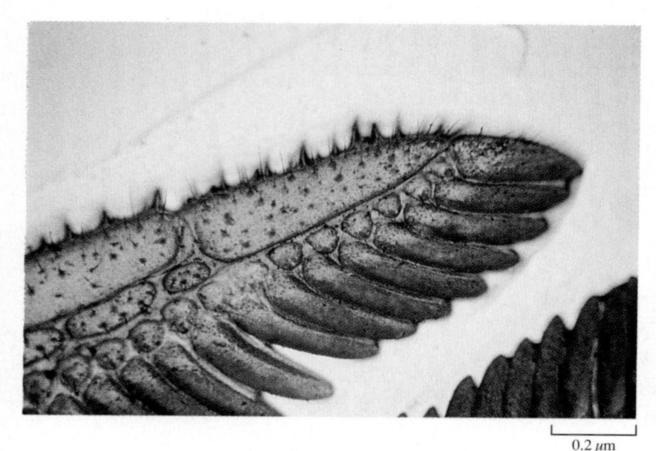

28.17 Book lungs, or pectines, of *Centruroides*

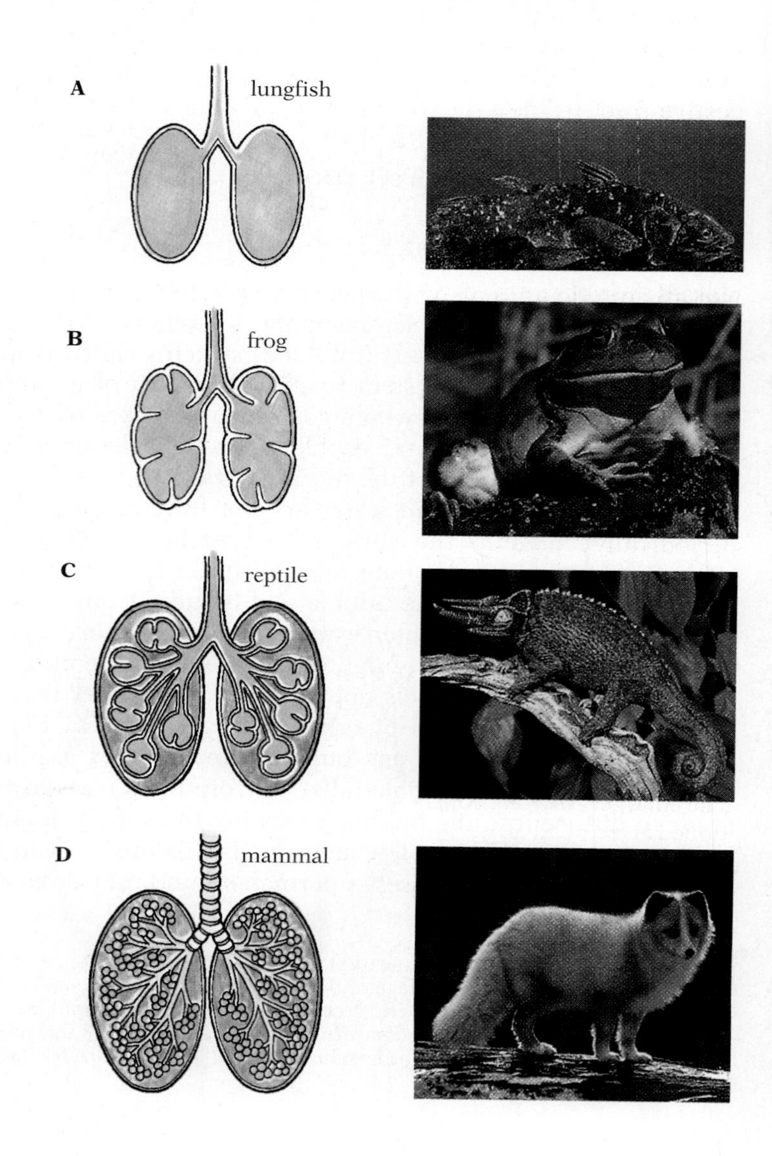

28.18 Evolution of vertebrate lungs

The lungs of vertebrates show a progressive increase in surface area from lungfish to amphibians, reptiles, and mammals. (A) The ancestral vertebrates had lungs that were little more than sacs supplied with blood. Some fish and amphibians have lungs of this type. (B) Other amphibians such as the frog, have increased partitioning within the lung, which gives them a greater gas-exchange surface. (C) Reptiles, which evolved from amphibians, have lungs with still more partitions. (D) Mammals, which evolved from reptiles, have lungs that are very finely divided into alveoli, which provide an enormous surface area for gas exchange.

■ HOW LUNGS WORK

Lungs are invaginated gas-exchange organs dependent on a blood-transport system for distributing oxygen to the tissues. They are typical of two unrelated groups, land snails and higher vertebrates, including some fish, most amphibians, and all reptiles, birds, and mammals. In their simplest form, lungs are little more than chambers with slightly increased vascularization in their walls and with some sort of passageway leading to the outside. This simple type of lung is found in some snails that inhabit the lower levels of the ocean beach, where the necessity for air breathing is seldom pressing because oxygen is available from the water. Evolution of lungs has tended toward a greatly increased surface area, by subdivision of its inner surface into many small pockets or folds, and toward increased vascularization of its exchange surfaces.

It is not surprising that terrestrial vertebrates have lungs, but it is odd that some relict species of fish (that is, species surviving essentially unchanged in structure from ancient times) have them also. The ancestral fish from which both modern fish and the land vertebrates evolved probably had lungs that enabled them to live in stagnant, poorly aerated water when necessary. These primitive lungs were simple sacs that arose as ventral evaginations of the digestive tract in the pharyngeal region behind the gills. A few salamanders still have such simple lungs, but in most terrestrial vertebrates the inner surface has become increasingly folded and subdivided, providing a far higher surface-to-volume ratio for the exchange process (Fig 28.18). This evolutionary tendency reaches its culmination in mammals and birds, the two warm-blooded classes, which must expend vast amounts of metabolic energy to maintain their stable high body temperatures, and hence have exceedingly high oxygen demands.

The human respiratory system (Fig. 28.19) provides a good example of the mammalian type. Air is drawn in through the ***external nares***, or nostrils, and enters the ***nasal cavities***, which function in warming and moistening the air, filtering out dust particles, and sensing odors. Bony ridges in the cavities cause eddies in the air stream, facilitating these processes; the mucous layer on

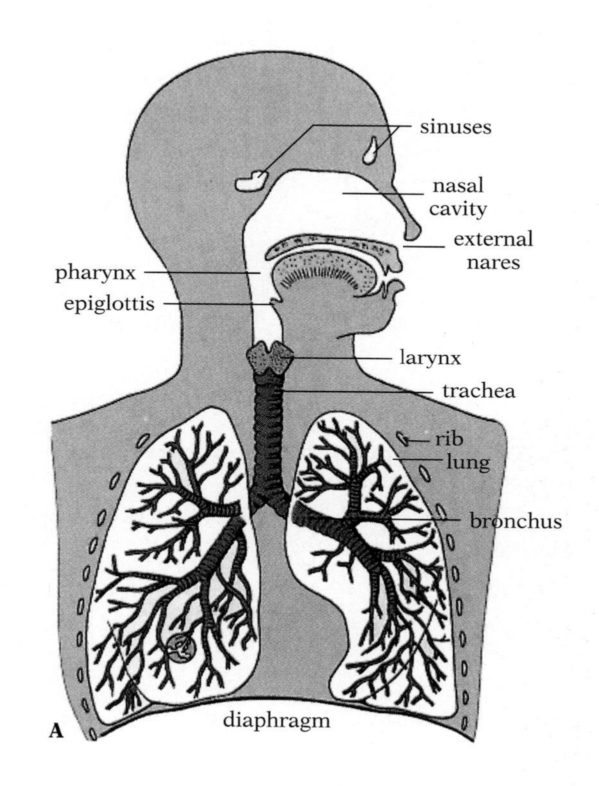

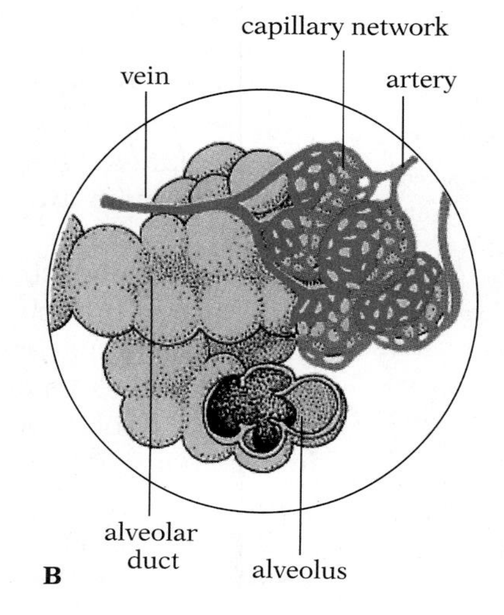

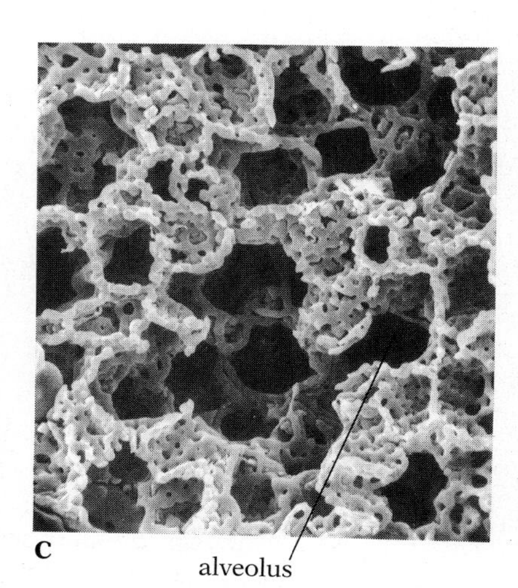

28.19 Human respiratory system

The enlarged drawing (B) shows a few sectioned alveoli and other alveoli surrounded by blood vessels. Actually, all alveoli, including those lying along alveolar ducts, are surrounded by networks of capillaries. The corresponding photograph of alveoli (C) is from *Tissues and Organs: A Text-Atlas of Scanning Electron Microscopy* by Richard G. Kessel and Randy H. Kardon. Copyright 1979 W.H. Freeman and Company. Used with permission.

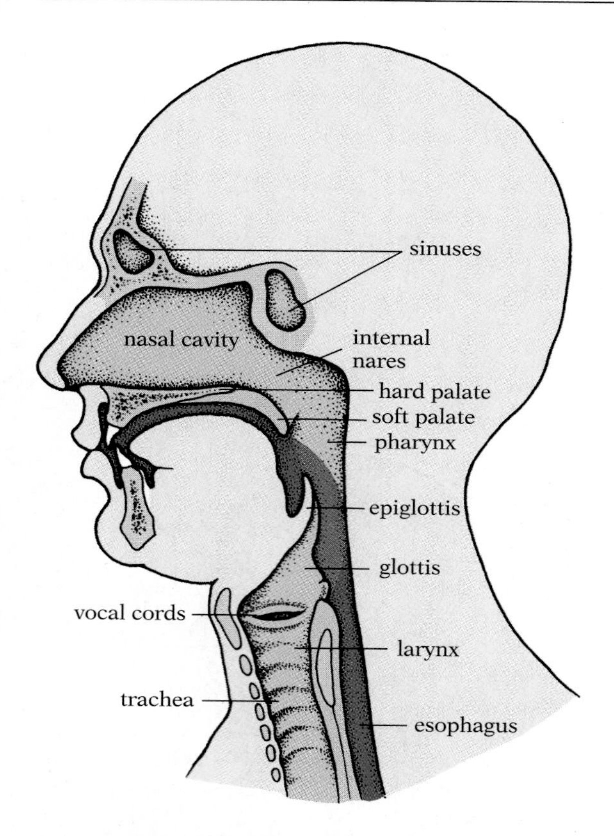

28.20 Upper portion of human respiratory and digestive system

The path of air is pink; the path of food is brown. The two paths cross in the pharynx.

the epithelium of the nasal passages also helps, as do the cilia of many of the epithelial cells. Curiously enough, of all the functions of this system, olfaction (odor detection) is the most primitive one; the external nares and the nasal passages originally had nothing to do with respiration, but evolved as odor-sensing devices. Fish use their nostrils for smelling but not breathing; they take in water for respiratory purposes through the mouth. Since smelling and feeding are so intimately related, it is not surprising that in fish, amphibians, and many reptiles, little or no separation exists between the nasal cavity and the mouth cavity. The separation has been developed furthest in mammals. In them a "roof of the mouth" has evolved that consists of an anterior bony palate and a posterior soft palate (Fig. 28.20).

Even in mammals, however, the air and food passages ultimately join in the region known as the ***pharynx***. During inhalation, air leaves the pharynx via a ventral opening, the ***glottis***, which leads into the larynx. Since air enters the pharynx dorsally and exits ventrally, and since food enters ventrally and exits dorsally into the esophagus, the air and food passages not only join but actually cross in the pharynx. (This rather inefficient arrangement results from natural selection's modification of the already-existing olfactory apparatus into respiratory passages. It is typical of much evolution: the new is built from the old.) Elaborate mechanisms help ensure that when food is forced back into the pharynx, it will not enter the nasal cavity or the larynx, but must be swallowed into the esophagus. The internal nares, which connect the nasal cavities with the pharynx, are closed by the soft palate, and the glottis is closed by a flap of tissue called the ***epiglottis*** when the larynx is raised against it during swallowing.

After leaving the pharynx through the glottis, air enters the ***larynx***, a chamber surrounded by a complex of cartilage (commonly called the Adam's apple). In many animals, including humans, the larynx functions as a voice box. It contains a pair of vocal cords—elastic ridges stretched across the laryngeal cavity that vibrate when air currents pass between them; changes in the tension of the cords result in changes in the pitch of the sounds emitted.

The ***trachea*** is an air duct leading from the larynx into the thoracic cavity. Its epithelial lining is ciliated; the cilia beat in waves that carry foreign particles and mucus up the trachea away from the lungs. A series of C-shaped rings of cartilage embedded in the walls of the trachea prevent it from collapsing upon inhalation. At its lower end, it divides into two ***bronchi***, tubes that lead toward the two lungs. (It is at the lower end of the trachea, where the bronchi branch away, that the voice box of birds, the syrinx, is located.)

Each bronchus branches and rebranches, and the bronchioles thus formed branch repeatedly in their turn, forming smaller and smaller ducts that ultimately terminate in tiny air pockets, each of which has a series of small chamberlike bulges in its walls termed ***alveoli***. The total alveolar surface in a pair of human lungs is enormous: about 100 m^2—many times greater than the total area of the skin.

The walls of the alveoli are usually only one cell thick, and each alveolus is surrounded by a dense bed of capillaries. The alveoli are the site of the actual gas exchange. Oxygen entering an alveolus dissolves in the film of water on its wall and then moves by diffusion across the intervening cells to the blood. The movement of O_2 and CO_2 requires nothing more than diffusion; no active transport across membranes is involved. O_2 is more concentrated in the air of the alveolus than in the blood, and CO_2 is more concentrated in the blood than in the alveolus; each gas simply moves from the region of its higher concentration to the region of its lower concentration. Because conventional lungs lack the countercurrent design of gills, this transfer is incomplete (Fig. 28.21). The air entering the lungs contains about 21% O_2 and almost no (0.04%) CO_2; the blood of a person at rest entering the capillaries of the aveoli contains the equivalent of 5% O_2 and 6% CO_2. Complete transfer would fully discharge the blood of CO_2 and raise the O_2 concentration to the equivalent of 21%; instead the blood leaving the lungs has the equivalent of 13% O_2 and 5% CO_2.

Air is drawn into and expelled from the lungs by the mechanical process called ***breathing***. In mammals, breathing generally involves muscular contractions of two regions, the ***rib cage*** and the ***diaphragm***. The latter is a muscular partition separating the thoracic and abdominal cavities (Fig. 28.22). Inhalation, or inspiration, occurs whenever the volume of the thoracic cavity, in which the lungs lie, is increased; such an increase reduces the air pressure within the chest to less than the atmospheric pressure and draws air into the lungs. The increase in thoracic volume is accomplished by contractions of the intercostal muscles, which draw the rib cage up and out, and by contraction, or downward pull, of the normally upward-arched diaphragm; the first mechanism is called chest breathing, while the second is called abdominal breathing.

Normal exhalation, or expiration, is a passive process; the muscles relax, allowing the rib cage to fall back to its resting position and the diaphragm to arch upward. This reduction of thoracic vol-

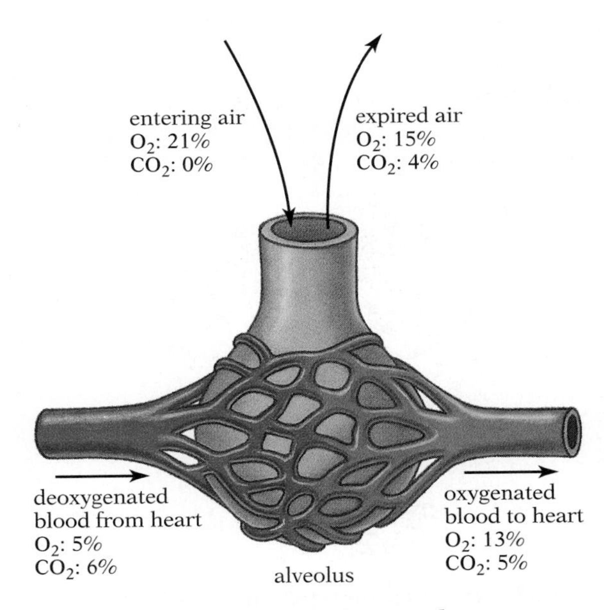

28.21 Gas exchange in an aveolus

Gas exchange in conventional lungs is less complete than in gills. Gills typically exchange about 90% of the differential between the medium and blood; blood leaving the aveoli of an individual at rest, by contrast, has picked up only half of the available oxygen differential and discharged only about a fifth of the carbon dioxide.

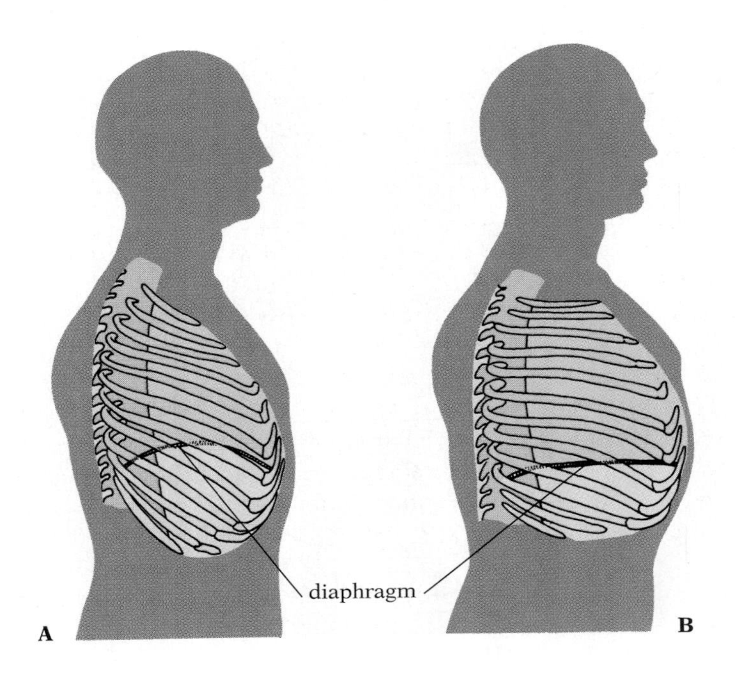

28.22 Mechanisms of human breathing

(A) Resting position. (B) Inhalation. The rib cage is moved up and out by the intercostal muscles, and the diaphragm is pulled downward. Both of these motions increase the volume of the thoracic cavity. Consequently the reduced air pressure in the cavity causes more air to be drawn into the lungs.

28.23 Respiratory system of a bird

Attached to the lungs are many air sacs (light brown), some of which even penetrate into the marrow cavities of the wing bones. As in mammals, the respiratory system has bilateral symmetry (not obvious in this side view).

ume, combined with the elastic recoil of the lungs themselves, causes a rise in the pressure inside the lungs to a level above that of the outside atmosphere and drives out the air.

The air moved by a single normal breath—the tidal air—represents only a small fraction of the total capacity of the lungs. Additional air (complementary air) can be forcibly inhaled, and, similarly, forcible exhalation can expel not only tidal and complementary air, but also additional air known as reserve air. The total breathing capacity, or ***vital capacity***, is the sum of tidal plus complementary plus reserve air; though it varies greatly from person to person, 4 liters is typical. Trained athletes usually develop larger lungs, and their vital capacity is usually substantially greater than normal.

The pattern of air flow in the respiratory system of birds differs fundamentally from that of mammals. In addition to paired lungs, birds possess several thin-walled air sacs that occupy much of the body cavity and even penetrate into the interior of some of the bones (Fig. 28.23). The air sacs are poorly supplied with blood vessels and do not themselves absorb O_2 or release CO_2. Their arrangement and bellowslike action, however, make possible continuous unidirectional flow of air through the lungs.

Like mammals, birds suck in air by increasing the volume of the body cavity. As Figure 28.24A shows, most of the air drawn in during inhalation does not go directly to the lungs, but flows through the bronchus to the posterior air sacs; simultaneously, air already in the lungs moves forward into the anterior air sacs via connecting passages called recurrent bronchi. During exhalation, air from the posterior sacs moves into the lungs, primarily via recurrent bronchi, while air from the anterior sacs empties to the outside (Fig. 28.24B). Thus air moves forward through the lungs during both inhalation and exhalation. Instead of alveoli, whose dead-end chambers would be incompatible with unidirectional air flow, bird lungs have tiny air ducts (parabronchi) running through the lung

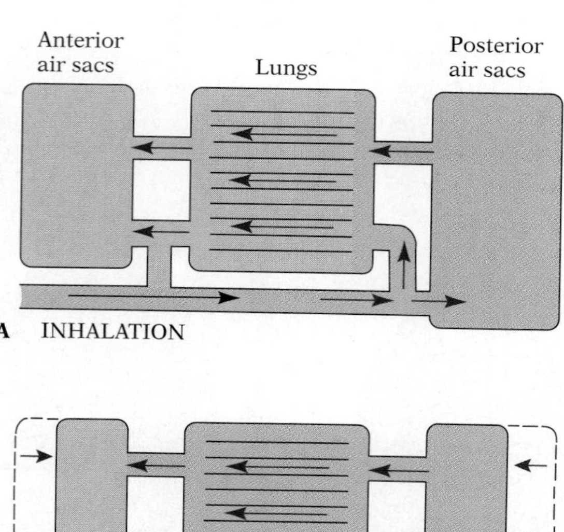

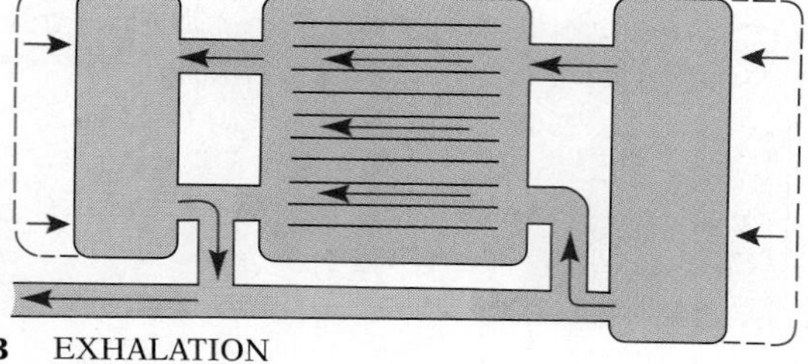

28.24 Respiratory cycle of a bird

(A) During inhalation, new air (pink) is drawn into the posterior air sacs; a small amount also enters the posterior portion of the lungs. Air already in the system (gray) is simultaneously moved forward through the lungs and into the anterior air sacs. (B) As air is exhaled from the anterior air sacs, air from the posterior air sacs moves forward into the lungs and across the gas-exchange surfaces; this movement is effected by a contraction of the air sacs. The portion of air (pink) inhaled in (A) will be exhaled during the following respiratory cycle. Note that during both inhalation and exhalation oxygen-rich air is moving unidirectionally through the lungs. In the parabronchi (Fig. 28.25) the blood moves from left to right, opposite the flow of air, which allows the gas exchange to benefit from a countercurrent design.

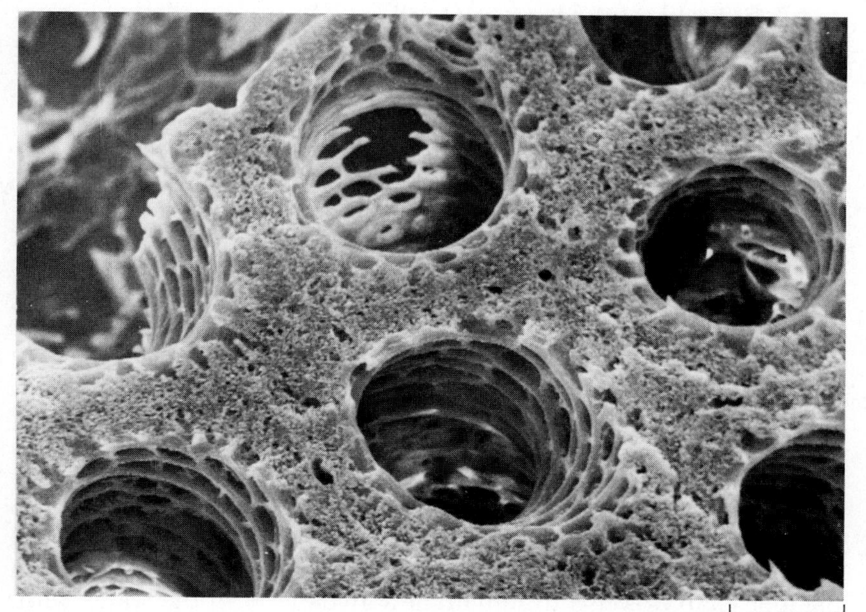

28.25 Cross section of parabronchi in a 2-week-old chick

It is through the parabronchial tubes that the unidirectional flow of air in the bird lung occurs. Exchange of gases takes place across the walls of these tubes, each less than 0.5 mm in diameter. Note the resemblance to the spongy mesophyll of a leaf (see Fig. 28.4).

tissue, and it is across their walls that gas exchange takes place (Fig. 28.25).

Birds are far more efficient than mammals in extracting O_2 from air, both because of the continuous unidirectional flow of air through their lungs, and because blood in the capillaries associated with the parabronchi moves against the flow of air and so provides some of the same benefits as the countercurrent exchange system of fish gills. This superior efficiency enables birds to fly actively at high altitudes, where the partial pressure of O_2 is low. Vance Tucker experimentally exposed sparrows and mice to an atmosphere simulating that at 6000 m altitude (350 mm Hg), and found that the sparrows could fly vigorously while the mice were unable to stand up and could barely crawl. Geese and swans can fly at altitudes up to at least 9,000 m.

The mammalian and avian method of breathing, in which air is drawn into the lungs, is known as ***negative-pressure breathing***. By contrast, in ***positive-pressure breathing*** air is forced into the lungs. Both methods are used by adult frogs. Mouth closed and nostrils open, the frog lowers the floor of its mouth, sucking air into the mouth cavity (negative-pressure method). Then it closes its nostrils and raises the mouth floor; this reduction in the volume of the mouth cavity exerts pressure on the imprisoned air and forces it into the lungs (positive-pressure method). In fact, frogs rarely need their lungs during cool weather; most of their gas exchange under these conditions occurs across the creature's soft moist skin.

Negative-pressure breathing works because the thorax, in which the lungs are suspended, is a sealed compartment filled with fluid that does not expand. As a result, when the thorax expands during inhalation, the walls of the lungs are pulled along, the volume of the lungs increases, and the resulting negative pressure draws in outside air. If the seal separating the thoracic cavity from the outside or from the rest of the body is broken, as in some human chest injuries, the expansion of the thorax may draw in air or abdominal fluid and the lung may collapse. Positive pressure is then needed to reinflate the lung, and, of course, the thoracic cavity must be sealed to restore breathing.

■ DO FISH HAVE LUNGS?

Many researchers believe that the primitively ventral lung of ancient fish evolved into the swim bladder of modern fish by gradually shifting to a dorsal position (Fig. 28.26). Its attachment to the esophagus also gradually shifted to a dorsal position; it is still there in some species, but in other modern species the connecting duct has been lost and direct entrance or exit to the swim bladder no longer exists.

Cytoplasm is significantly denser than fresh water and somewhat denser than salt water; as a result, aquatic organisms will tend to sink unless they expend energy swimming upward or have low-density organs to give them buoyancy. Since water density also changes with temperature and depth, any buoyancy system based on gases is strongly affected by water pressure (depth). The swim bladder enables a fish to remain at a given level in the water without sinking by adjusting the fish's overall density to that of the surrounding water. (Bottom-dwelling fish seldom have swim bladders.) If the fish swims upward, the swim bladder will swell in response to the reduced pressure, and gases must be removed from it to restore it to its normal size; otherwise the fish will rise ever faster to the surface. Conversely, if the fish swims downward, gases must be added. The addition of gases is accomplished by a region of specialized glandular cells (the gas gland), associated with many blood vessels, in the walls of the swim bladder. The removal of gases generally occurs elsewhere in the bladder, or even in a separate specialized area, the oval lobe (Fig. 28.27).

■ HOW INSECTS "BREATHE"

The second principal type of invaginated respiratory strategy is the tracheal system of most terrestrial arthropods. There is no localized respiratory organ and little or no transport of gases by the blood. Instead, the system is composed of many small tubes, called ***tracheae***, that branch throughout the body (Fig. 28.28). The tracheae and the smaller tracheoles into which they branch carry air directly to the individual cells, where diffusion across the cell membranes takes place (Fig. 28.29).

Air enters the tracheae by way of ***spiracles***, apertures in the body wall that usually open and close by valves (Fig. 28.30). Some

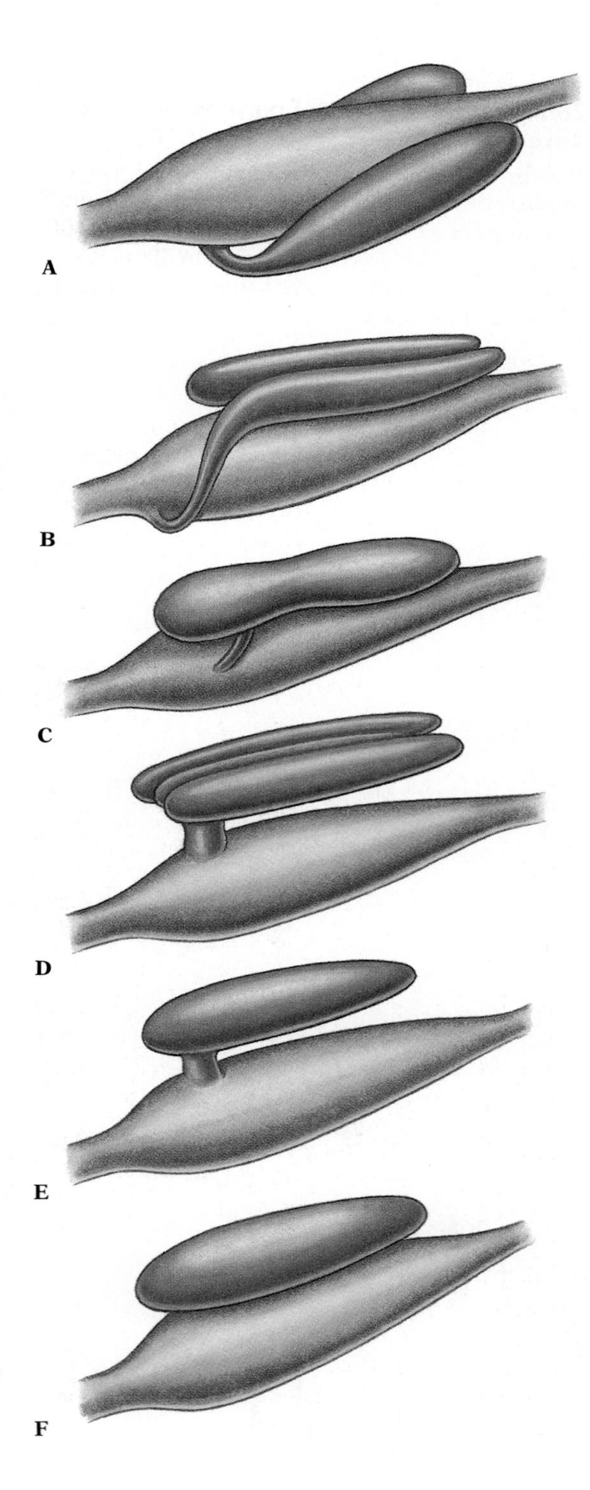

28.26 Evolution of swim bladders

In each drawing, the swim bladder or lung is blue, and the esophagus is red. According to one hypothesis, the swim bladder of modern fish evolved from a primitive ventral lung (A). The lung may first have moved to a dorsal position while retaining a ventral attachment to the esophagus, as in some living lungfish (B). The attachment to the esophagus may then have moved to the side (C) and finally become dorsal, as in typical modern swim bladders (D–E). In some modern fish the connection to the esophagus has been lost entirely (F).

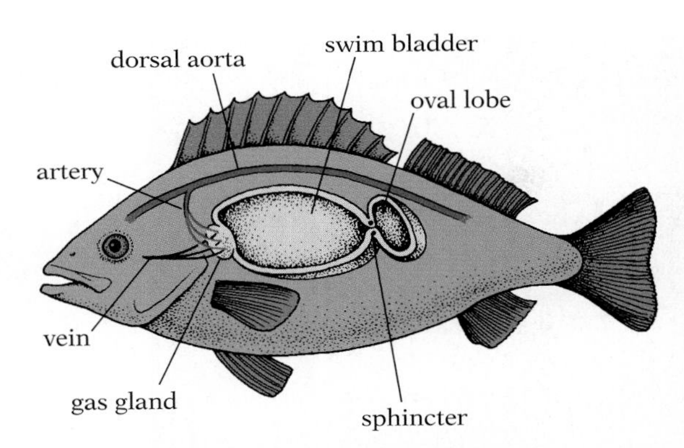

28.27 Functioning of a typical swim bladder

Gas is added to the bladder from the blood by means of the gas gland. Gas is removed from the bladder by absorption through the walls of the oval lobe. These two processes control the volume of the bladder, and hence the buoyancy of the fish. The sphincter is generally open only when gas is being removed.

28.28 Tracheae and spiracle of an insect

At top center is a spiracle (brown), from which numerous branching tubes—the tracheae—can be seen running to many parts of the insect's body. The spiracles are usually located on the sides of the body segments, the number varying in different kinds of insects. See Figure 28.29 for a drawing of terminal branches, or tracheoles.

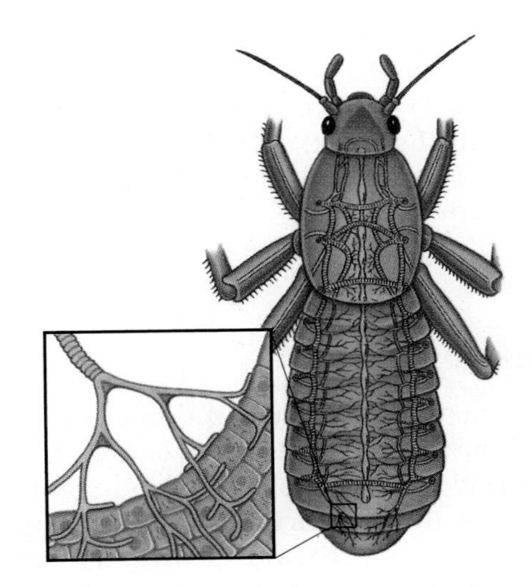

28.29 The tracheal system of insects

The tracheal system consists of many small internal tubes called tracheae that branch throughout the insect body. The tracheae open to the outside through holes called spiracles. The tracheae, and the smaller tracheoles (insert) into which they branch, carry air directly to the individual cells, where diffusion across the cell membranes takes place.

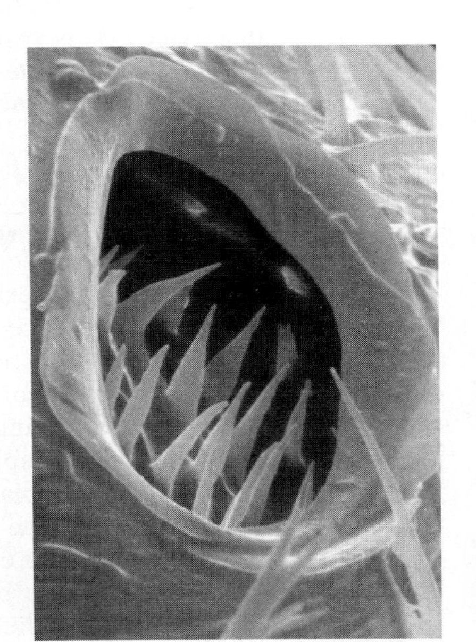

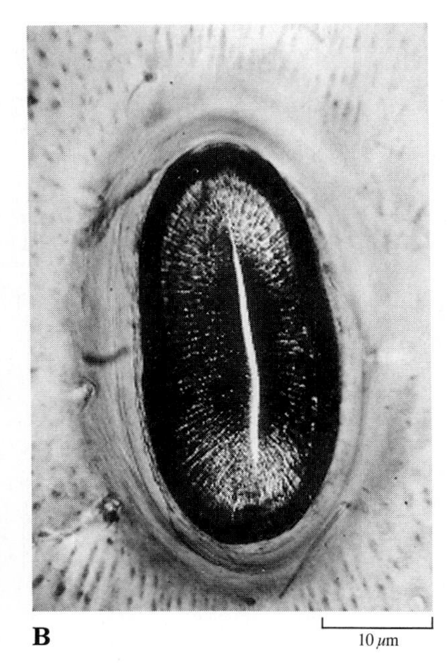

28.30 Spiracles of two insects

(A) Scanning electron micrograph of a fully open ant spiracle. The pointed projections are sensory hairs that monitor external conditions and can trigger spiracle closing when necessary. (B) A nearly closed grasshopper spiracle; the black areas are the valves. Note the resemblance to the stoma of a leaf (See Fig. 28.6).

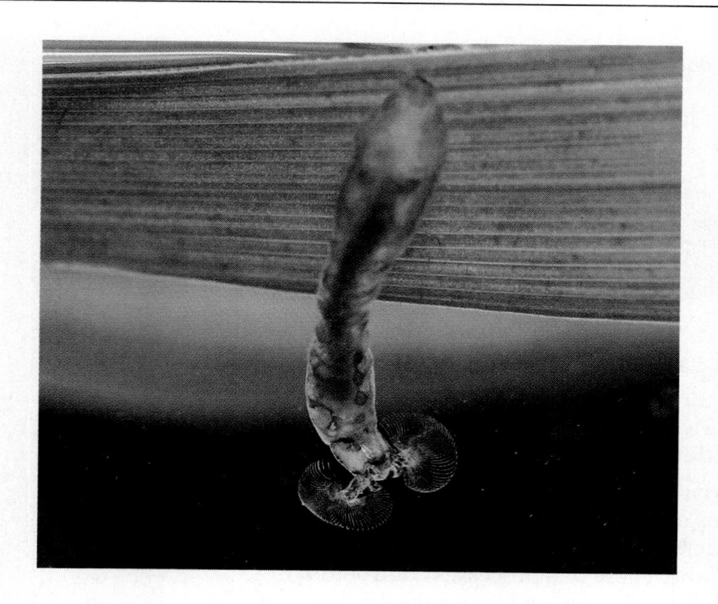

28.31 Larva of a buffalo fly (*Siphona exigua*) with tracheal gills extended

Like this juvenile buffalo fly, many aquatic insect larvae breathe by means of tracheal gills. These structures serve a dual purpose: they provide a fine surface for the absorption of dissolved oxygen, and also function as filter-feeding combs, sifting food from the environment.

of the larger insects actively ventilate their tracheal systems by muscular contraction, but most small insects do not. The diffusion of oxygen in air is rapid enough to maintain at the tracheal endings an O_2 concentration only slightly below that of the external atmosphere. This type of respiratory system, however, probably helps limit the size of insects.

The aquatic larvae of some insects, such as damselflies and mayflies, have tracheal gills—platelike or feathery structures richly supplied with tracheae (Fig. 28.31). In the other gills that we have discussed, O_2 is transported in the blood, but the O_2 absorbed by tracheal gills comes out of solution and moves in gaseous form into the general tracheal system.

CONSERVING WATER

One of the most serious problems organisms faced in evolving to a terrestrial existence was the loss of the continuous supply of water necessary to keep tissue surfaces moist. For most land plants and animals, specializations for water conservation are a necessity. Because gas exchange depends on a large, moist surface, measures for minimizing respiratory water loss are critical if an organism is to survive. When plants run short of moisture, they can block most water loss from the mesophyll by closing the stomata. As a result net photosynthesis ceases, but the plants are protected.

For animals the problem is more severe: as heterotrophs, they cannot produce their own O_2 for use in metabolism; they must breathe. Water loss from exchange surfaces is a particular problem for homeotherms (animals that maintain a constant elevated body temperature), whose oxygen needs remain high, especially at low temperatures.[2]

[2] Temperature control and waste disposal, which are discussed in later chapters, can also result in considerable fluid loss.

The difficulty of water conservation becomes clear when we trace the movement of air in the respiratory system. When air is inhaled, it passes along the respiratory tract into the lungs, where it is exposed to the enormous surface area of the aveoli. The air is in close contact with the respiratory tissues, which themselves are kept moist and warm by virtue of intense vascularization. As a result, inspired air becomes warmed to body temperature and fully saturated with moisture, as it must be to exchange gases effectively. If this moist air were to be exhaled, both heat and water would be lost.

The trick for recovering respiratory moisture that is most common among terrestrial creatures involves countercurrent exchange. The strategy is most obvious in animals from arid habitats, where selection for water recovery has been most intense. Inhaled air enters the nose and passes over a convoluted series of moist tissues as it moves to the lungs (Fig. 28.32). This transit accomplishes much of the necessary warming and moistening of the air that

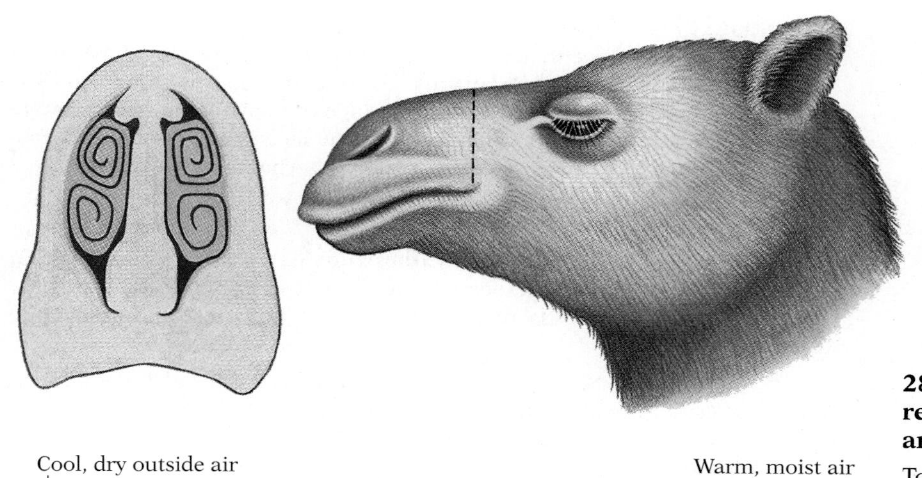

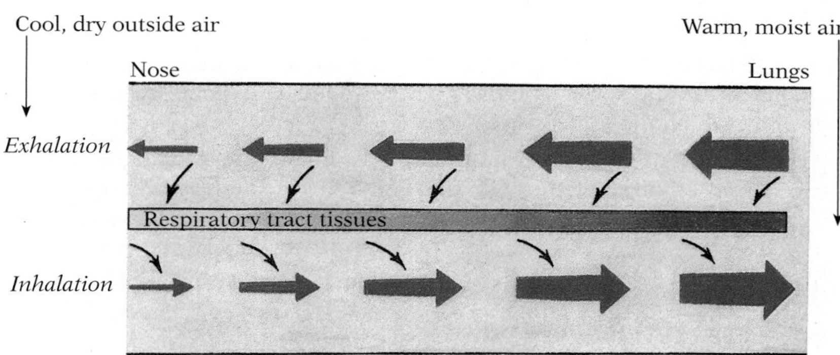

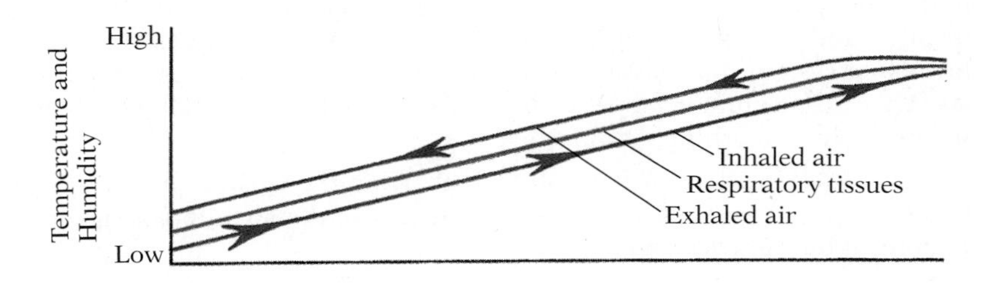

28.32 Countercurrent exchange in the respiratory tract of an animal adapted to arid climates

Top: This cross section through the nose of a camel shows how foldings of the nasal membranes serve to increase the surface area within the nose. Middle: Air inhaled through the nose (lower arrows) is dry and often cooler than the animal's body temperature. Exchange with the tissues lining the respiratory tract transfers moisture and heat to the incoming air, fully saturating the breath by the time it reaches the lungs, and establishing a nose-to-lung temperature and humidity gradient. When the air is exhaled (upper arrows), this gradient cools the air and causes the moisture to condense on the tissues, much as humid summer air condenses on a chilled glass. The air leaving the nose carries only a little more moisture than the incoming air. (Inhalation and exhalation are shown following adjacent paths for clarity.) Bottom: A corresponding graphic representation of the countercurrent exchange of heat and moisture in the respiratory tract.

would otherwise occur only in the lungs. The water and temperature exchange involved sets up a gradient from nose to lung. Because the tissues nearest the nose are exposed to the direct and (usually) coolest air, they lose much moisture and heat. As the air moves along into the system, exchange continues and the differential becomes ever smaller. Air reaching the lungs is almost fully warmed and humidified.

The air exhaled from the lungs encounters a countercurrent-like gradient on the way out (Fig. 28.32). This interaction between the departing air and the respiratory surfaces results in a highly efficient return of moisture to the nasal tissues. In camels, for instance, totally dry desert air is fully humidified en route to the lungs, but the countercurrent arrangement removes 95% of the moisture from the air being exhaled.

CHAPTER SUMMARY

THE PROBLEMS

Gas exchange must occur across moist membranes, creating a potential for water loss in terrestrial organisms. Every active cell must have access to the gases it needs, and the exposed surface area available for exchange must be sufficient to support the needs of the cells; thus specialized gas-delivery systems may be necessary to supply internal cells. (p. 795)

HOW PLANTS EXCHANGE GASES

LEAVES AS PASSIVE LUNGS Photosynthesizing leaves need to take in CO_2 and dispose of O_2. Exchange occurs mainly across the membranes of the spongy mesophyll. Stomata regulate the rate of exchange and associated water loss. (p. 796)

CAN ROOTS BE ASPHYXIATED? Active stem and root cells also need to exchange gases; poorly aerated soils can suffocate roots. (p. 800)

HOW AQUATIC ANIMALS EXCHANGE GASES

WHY GILLS WORK SO WELL Gills are highly subdivided to maximize exchange area, and water is moved across them. The blood flow is countercurrent, which maximizes exchange. (p. 802)

ALTERNATIVES TO GILLS Some aquatic animals are thin enough to exchange gases directly through the body surface; others use respiratory trees, analogous in structure to the highly ramified gastrovascular cavity of flatworms. Some spiders and beetles create underwater air supplies. (p. 805)

HOW TERRESTRIAL ANIMALS EXCHANGE GASES

HOW LUNGS WORK Gas exchange in most lungs occurs in tiny, highly folded alveoli. Air is usually drawn in through negative-pressure breathing in birds and mammals. Birds have a high-efficiency countercurrent parabronchial system that passes air in one direction across the gas-exchange surfaces. Some animals force air into their lungs, a strategy known as positive-pressure breathing. (p. 807)

DO FISH HAVE LUNGS? A few fish take air from the surface and exchange gases in a lung; in most, however, this structure, the swim bladder, serves to control buoyancy. (p. 812)

HOW INSECTS "BREATHE" Insects have a highly elaborated respiratory structure called the tracheal system, which brings air to all parts of the body; body movement helps bring fresh air in. (p. 812)

CONSERVING WATER
Incoming air is warmed to body temperature and fully humidified; countercurrent systems help recover this heat and condense the moisture when gas is exhaled. (p. 814)

STUDY QUESTIONS

1. Is the alveolar capillary system arranged to take at least partial advantage of countercurrent exchange? If not, could it be? (pp. 804, 806–809)
2. Of terrestrial vertebrates, why might birds alone have evolved unidirectional lung flow? What are the likely costs and benefits? (pp. 804, 810–811)
3. Even the lungs of birds are not fully unidirectional: inhalation and exhalation share a common channel, which reduces the "duty cycle" of breathing. Fish, on the other hand, can maintain a constant flow of oxygenated water across their gills. Is a fully unidirectional respiratory system possible for a terrestrial animal? What are the advantages and disadvantages of such a design? (pp. 801–804, 810–811, 814–816)
4. Assuming some constant value for respiratory exchange per unit surface area in the lung, but an oxygen need that parallels metabolic rate so that the respiratory surface required per unit of volume of animal follows the curve in Figure 30.24, what is the consequence for the evolution of large and small homeotherms? How does your conclusion relate to the argument in this chapter about the effects of gas-exchange strategies on insect size?

SUGGESTED READING

FEDER, M. E., AND W. W. BURGGREN, 1985. Skin breathing in vertebrates, *Scientific American* 253 (5). *On how some vertebrates supplement or even replace lungs or gills in obtaining oxygen and eliminating carbon dioxide.*

RAHN, H., A. AR, AND C. V. PAGANELLI, 1979. How bird eggs breathe, *Scientific American* 240 (2). *On the unique gas-exchange problems faced by developing embryos in eggs.*

SCHMIDT-NIELSEN, K., 1971. How birds breathe, *Scientific American* 225 (6). *On the remarkable phenomenon of unidirectional flow in the respiratory system of birds.*

SCHMIDT-NIELSEN, K., 1981. Countercurrent systems in animals, *Scientific American* 244 (5). *Excellent analysis of how noses conserve water during breathing.*

CHAPTER 29

LIVING WITHOUT PLUMBING

HOW VASCULAR PLANTS MOVE THEIR FLUIDS

THE MAIN THOROUGHFARE: STEMS

FROM ROOTS TO LEAVES: THE ASCENT OF SAP

MOVING SOLUTES IN THE PHLOEM

DO PLANTS HAVE A CIRCULATORY SYSTEM?

INTERNAL TRANSPORT IN PLANTS

Every cell, whether living on its own or as part of a multicellular organism, must have access to a medium from which it can extract raw materials and into which it can dump wastes. In unicellular organisms and some of the structurally simple multicellular ones, each cell is either in direct contact with the environmental medium or only a short distance from it. But in structurally complex plants and animals, many internal cells are far from any environmental medium. Nutrient procurement, gas exchange, and waste expulsion take place in specialized regions of the body. The next problem is to transport nutrients to internal cells from these specialized structures, and move wastes to equally specialized disposal areas.

This chapter will describe how transport is accomplished in unicellular organisms, simple multicellular organisms, and vascular plants. Chapter 30 examines transport in multicellular animals.

LIVING WITHOUT PLUMBING

Diffusion plays an important role in moving materials within bacteria, protozoans, and unicellular algae, as well as within individual cells in multicellular organisms. Individual particles of all substances within the cell tend to become evenly distributed when they are not blocked by intracellular membranes. Diffusion is also important in moving materials from cell to cell within the body of a multicellular organism. In plants, intercellular diffusion is facilitated by the plasmodesmata, which connect adjacent cells.

Diffusion is a slow process in large cells. It is not surprising, therefore, that even in unicellular and small multicellular organisms diffusion is supplemented by other transport mechanisms. Food vacuoles, for instance, commonly move along regular circuits within cells, distributing the products of digestion through the cytoplasm. Even the cytoplasm itself is seldom motionless; it frequently exhibits rapid massive flow within the cell, as in an active amoeba. The cytoplasm of many plant cells flows in definite currents along the surface of the cell vacuole, a process called ***cytoplasmic streaming*** (Fig. 29.1). Sometimes the streaming is local, but at other times most of the cytoplasm becomes involved. This mass flow is accomplished by the active movement of the endo-

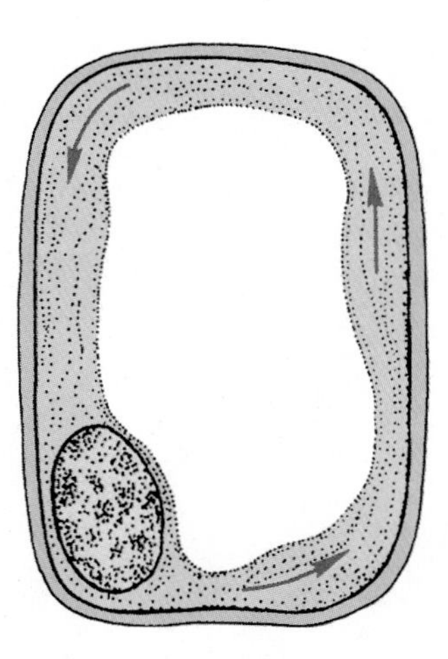

29.1 Cytoplasmic streaming in a plant cell
The cytoplasm flows around the large central vacuole.

plasmic reticulum along actin microfilaments; myosin bound to the ER interacts with the actin and pulls these membranes, and with them the cytoplasm. Streaming can transport substances many times faster than diffusion.

Virtually all multicellular plants have a specialized internal transport system to distribute water, essential gases, and nutrients to their cells. However, many algae lack conducting tissue. The cells of such plants are seldom far from the surrounding water or from water in intercellular spaces that connect to the outside. Consequently, each cell gets ample supplies locally, and long-distance transport is rarely necessary.

HOW VASCULAR PLANTS MOVE THEIR FLUIDS

Vascular plants are so named because their transport systems depend, like ours, on vessels. They incorporate two principal conducting tissues: the xylem (from the Greek *xylon*, "wood") and the phloem (from the Greek *phloios*, "bark," a reference to the location of this tissue in the outer growing perimeter of the plant). Their vessel systems have enabled vascular plants to evolve large bodies with many specializations and highly integrated control. Water and minerals are taken up by roots and distributed to the rest of the organism by the xylem; photosynthesis is restricted largely to the leaves, and its products are transported throughout the plant by the phloem. In some trees the distance between the roots and the leaves may be enormous, yet the xylem and phloem form continuous pathways between them, and they can exchange materials with relative ease (see Fig. 29.8).

■ THE MAIN THOROUGHFARE: STEMS

Stems of plants serve many functions. Some contain chlorophyll and carry out photosynthesis. Most also store nutrients in their pith; potato tubers, which are underground stems, are an extreme example. But stems (usually called trunks in larger plants) are mainly organs of transport and support.

From the standpoint of transport strategies, there are three basic kinds of higher plants: monocots, herbaceous dicots, and woody dicots.[1] ***Monocots***, of which there are about 50,000 species, are mostly annuals (that is, they live for only a single season). All the grasses (which include wheat, oats, rice, and corn) are monocots, as are tulips, lilies, daffodils, and palms. Monocots can be recognized by their morphology: the major veins in their leaves are roughly parallel, whereas those of most dicot leaves have a branching or netlike arrangement; monocot flowers have petals in threes or multiples of three, rather

[1] The vascular plants, or tracheophytes, also include about 650 species of conifers, such as pine trees, which have vascular systems similar to those of dicots, and about 12,000 species of ferns and their relatives.

than the fours or fives of dicot flowers; and germinating monocot ("one cotyledon") seeds have a single leaf rather than the two of dicots.

Herbaceous dicots are plants whose stems remain soft rather than becoming woody. They too are mostly annuals: the entire plant dies after one season of growth. In perennial herbaceous dicots, the part of the plant that remains above ground dies each year, but the roots survive; the plant regrows a new stem each year.

Woody dicots are all perennials, and both their top growth and their root systems survive for many seasons. Most trees and many flowering plants, such as the raspberry and the rose, are woody dicots. Altogether there are about 200,000 species of herbaceous and woody dicots.

How monocots are organized The simplest vascular organization is that of monocots. The nutrient- and water-conducting (vascular) tissue is organized in discrete vertical bundles scattered throughout the structural and supportive cells of the stems (Fig. 29.2, 29.3). Each bundle contains both xylem and phloem, and is

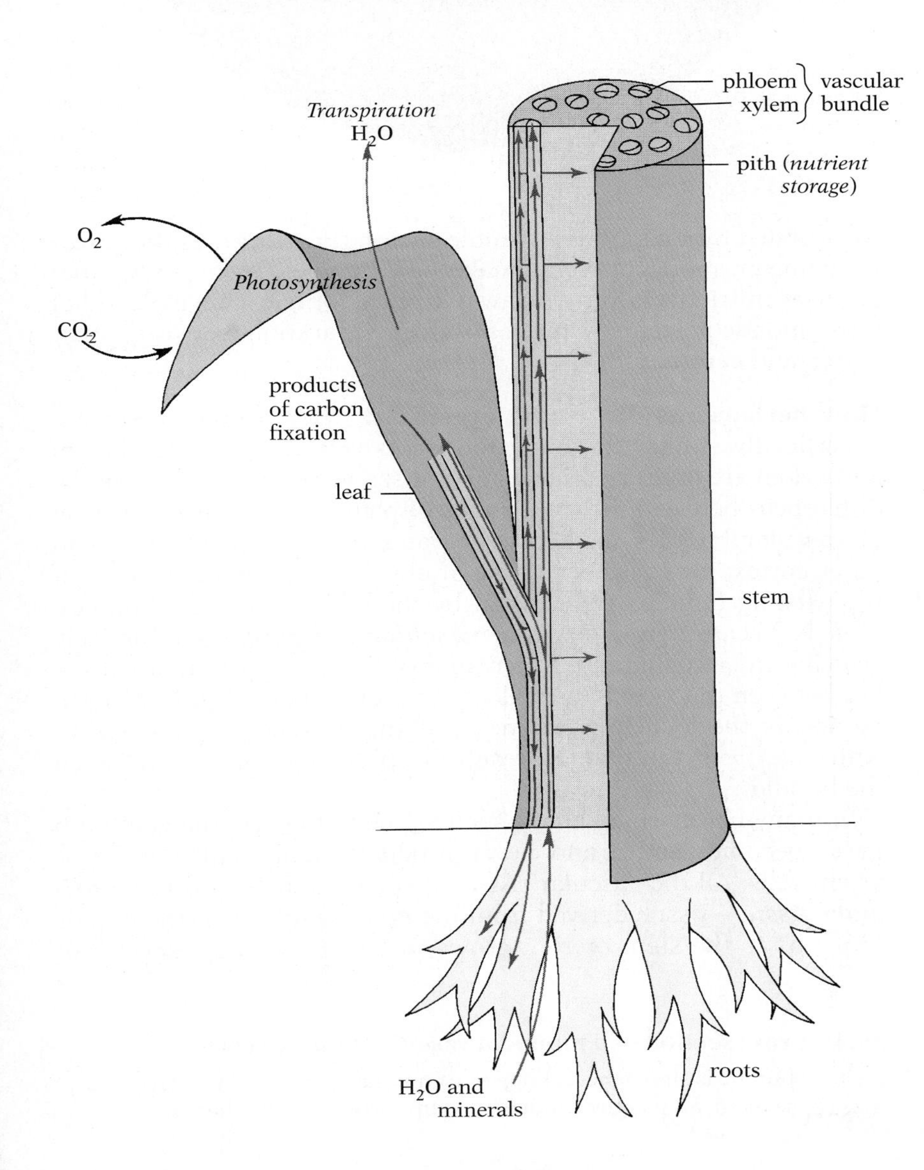

29.2 Summary of fluid flow in a monocot

In this stylized model of a grass plant, water and minerals enter the roots and are pulled up the xylem to the leaves by a process to be described later. Some of the water and the minerals are then used in photosynthesis and for making organic compounds. The nutrients and other materials thus generated are transported through the phloem to build or nourish cells, or to be stored in the pith. The phloem and xylem of monocots are organized into discrete bundles (one of which is shown here in vertical section) surrounded by pith. Nutrients to be stored simply diffuse out of the phloem into the pith, as indicated by the red arrows. There is no distinct outer cortex in monocots.

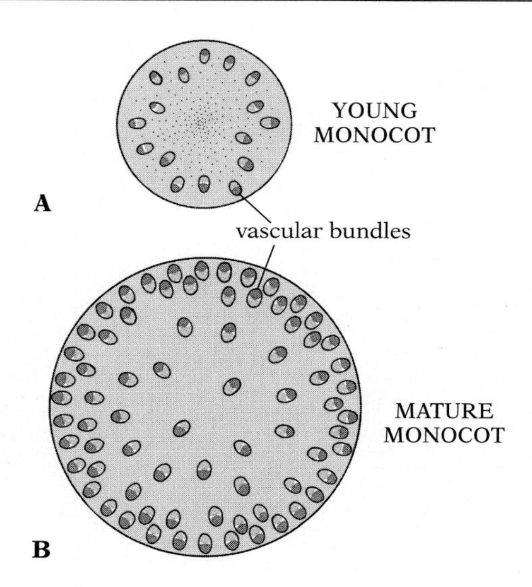

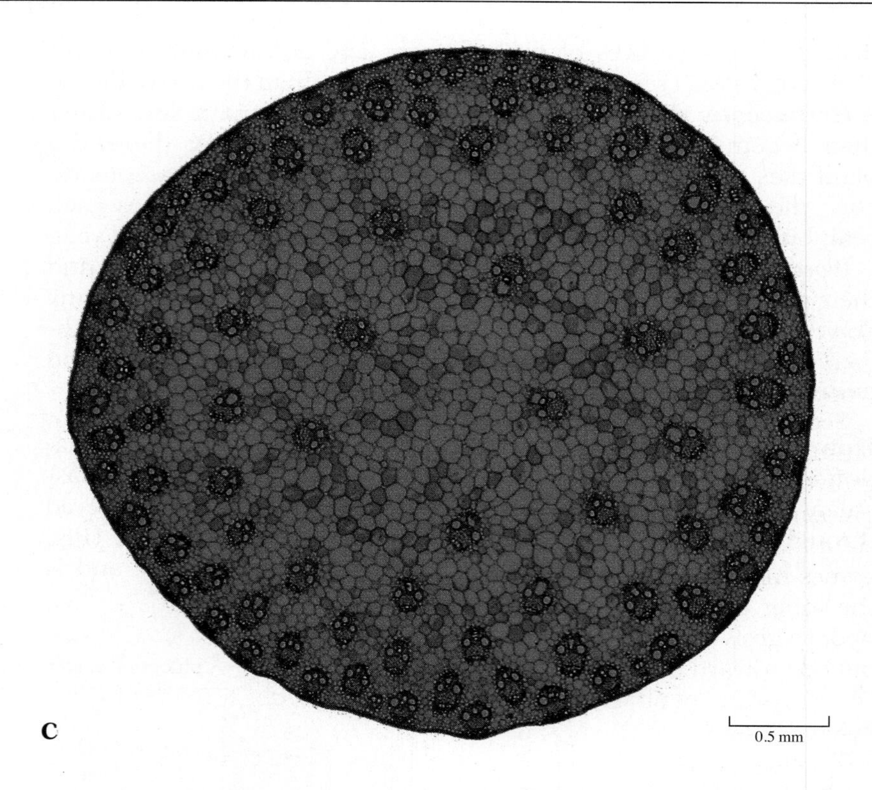

29.3 Cross sections of monocot stems

A young monocot has a ring of discrete vascular bundles, each with phloem and xylem vessels (A). As it grows, new vascular bundles develop, until they are distributed throughout the stem (B). The rest of the stem is pith, the tissue used for nutrient storage. (C) is a photograph showing a cross section of a monocot stem (corn).

surrounded by a supportive bundle sheath (Fig. 29.4). As they grow in diameter, monocot stems make new bundles, so no tissue that requires nutrients is very far away from a source. This means that most monocot stems exhibit no clear separation between outer cortex and central pith.

How herbaceous dicots are organized Herbaceous dicots are superficially similar to young monocots (Fig. 29.5A); the phloem and xylem are organized in a ring of discrete bundles. However, the differences between the two groups of plants are striking. The ring of vascular bundles in dicots separates the central pith from an outer cortex; new vascular tissue, if any, is added by the growth of the existing bundles rather than by the addition of new bundles. This is because in dicots the ***vascular cambium***—the actively growing (meristematic) tissue responsible for new vascular cells—lies between the xylem and phloem in each bundle (Fig. 29.5B). (In monocots the analogous tissue is at the periphery of the stem.) Nutrients from the phloem reach the pith through gaps between the bundles.

In many herbaceous dicots, such as the buttercup, the cambium never becomes active and never produces additional phloem or xylem cells. All the vascular tissue in such plants is said to be ***primary tissue***—tissue derived from the ***apical meristem*** (the bud or root tip) as the stem or root grows in length. In some species of

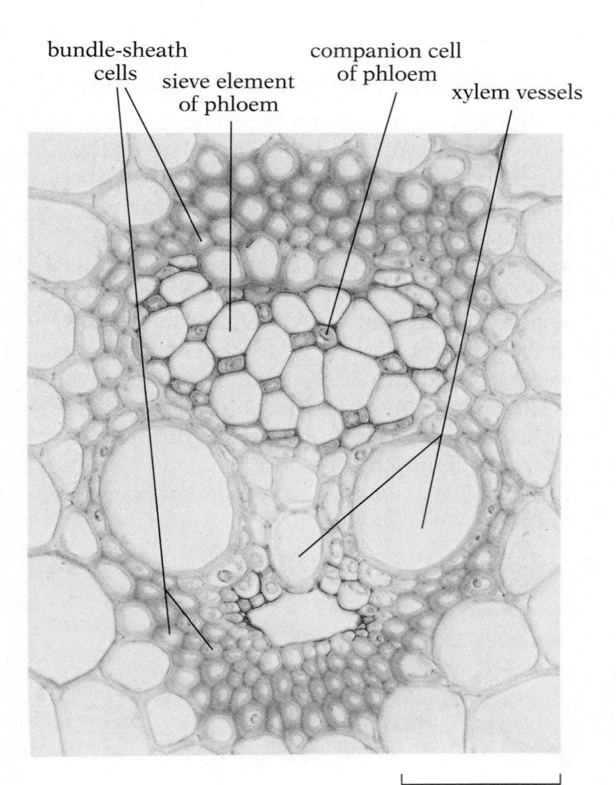

29.4 Cross section of a monocot vascular bundle (corn)

A single bundle is shown here. Each one contains a few xylem vessels in a mass of phloem, and is surrounded by a supportive bundle sheath.

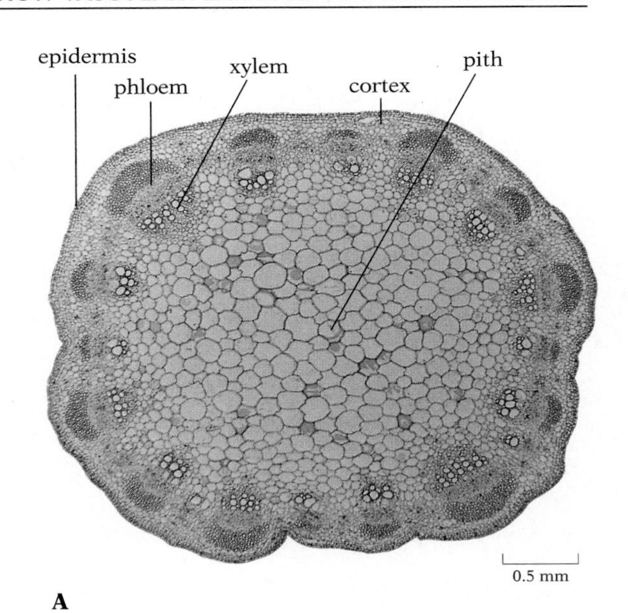

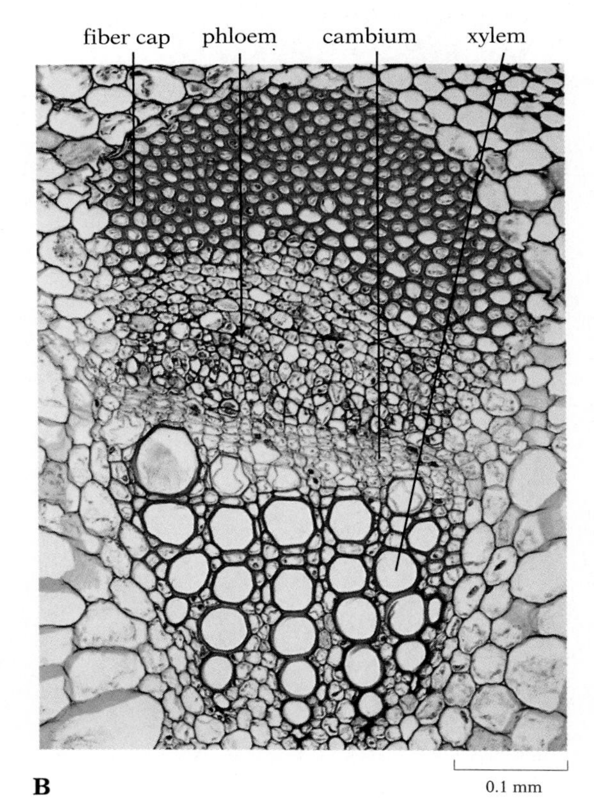

29.5 Cross sections of herbaceous dicot stems

(A) Whole stem of sunflower (*Helianthus*). Note that the vascular bundles are arranged in a circle. (B) Closeup of a single vascular bundle of sunflower.

herbaceous dicots, such as alfalfa, the cambium does become active. As the cambial cells divide, they give rise to new cells both to the inside and to the outside. The new cells formed on the outer side of the cambium differentiate as ***secondary phloem***; those formed on the inner side of the cambium differentiate as ***secondary xylem***.

Secondary vascular tissue, then, is tissue derived from the cambium, and its production results in growth in diameter rather than growth in length. The secondary phloem pushes the older, primary phloem farther away from the cambium toward the outside of the stem. Similarly, as secondary xylem is produced, the cambium becomes increasingly distant from the primary xylem, which is left in the inner portion of the stem. In a stem that has undergone secondary growth, therefore, the sequence of tissues (moving from the outside toward the center through a vascular bundle) is: epidermis, cortex, primary phloem, secondary phloem, cambium, secondary xylem, primary xylem, pith (Fig. 29.6).

The organization of the tissues in the central vascular core, or stele, of a dicot stem differs in two obvious ways from that in a typical young dicot root (see Fig. 26.5, p. 749): (1) The bundles of the

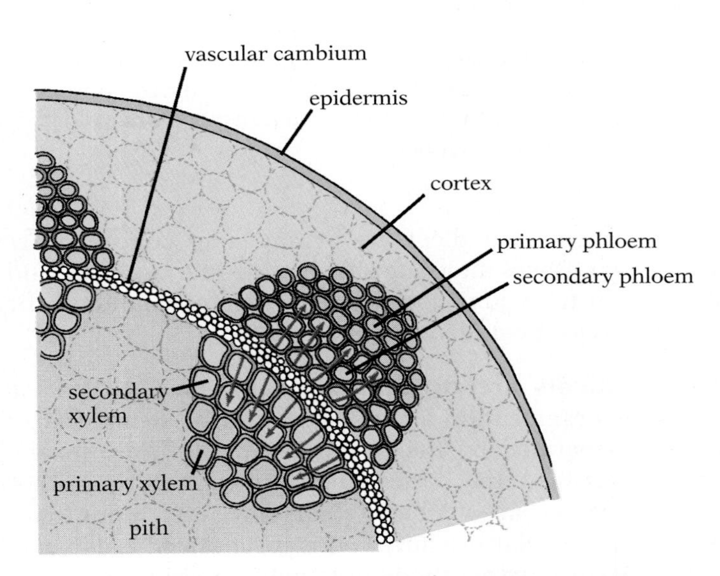

29.6 Cross section of part of a herbaceous dicot stem after secondary growth

The vascular tissue of herbaceous dicots is arranged in discrete bundles, with the xylem separated from the phloem by vascular cambium. As the stem grows, the cambium produces new (secondary) xylem and phloem cells, which push away the original (primary) xylem and phloem. Brown arrows indicate the directions of growth.

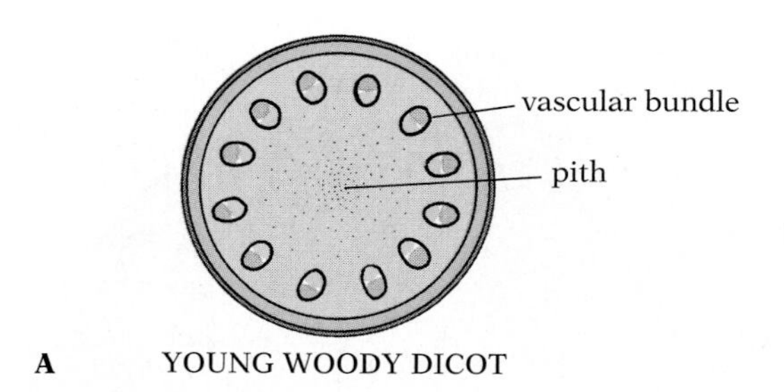

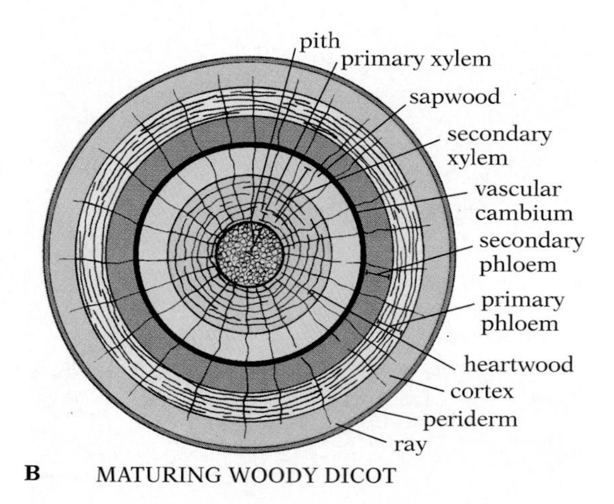

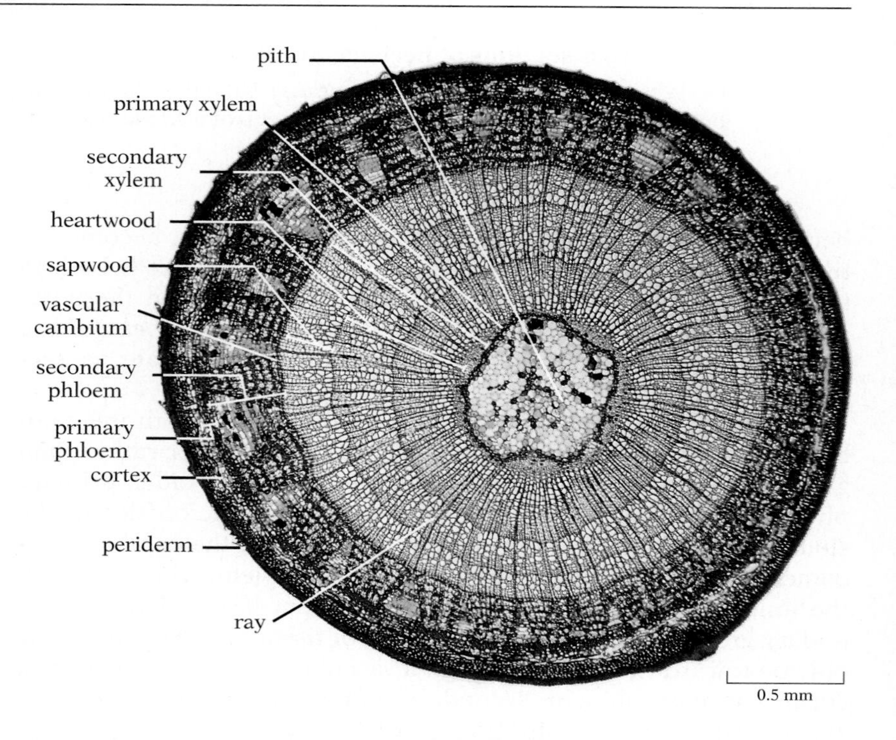

29.7 Cross sections of woody dicot stems

(A) In a young woody dicot the phloem and xylem are located in discrete vascular bundles. (B) After several seasons the primary xylem and phloem of the first year have been supplanted by new cells—secondary xylem and phloem—produced by the cambium, and the vascular bundles have fused to form a cylinder. Rays carry nutrients from the phloem to other parts of the stem. (C) Photograph showing a cross section of a woody dicot stem (basswood) at the end of 3 years of growth. Only the most recently produced xylem (the sapwood) is still active in transport, while older xylem (the heartwood) functions in support.

phloem and xylem in a dicot are arranged in concentric rings, whereas the two tissues alternate side by side in a circle around the central xylem in the young root. (2) Dicot stems characteristically have pith, whereas most dicot roots do not.

How woody dicots are organized Young woody dicot stems have the same organization as herbaceous dicot stems: the xylem and phloem, separated by the vascular cambium, are located in discrete bundles near the cortex and surrounding the pith (Fig. 29.7, top). Since herbaceous dicot stems are never more than one season old, this vascular organization is regenerated each year. But woody dicot stems survive the winter, and with continued growth their vascular anatomy changes. As additional tissue becomes necessary to transport the increasing quantities of water and nutrients required by the growing plant, some of the cells lying between the vascular bundles become vascular cambium. Like the cambium within the bundles, this new cambium produces xylem and phloem; eventually the vascular bundles fuse, forming concentric rings of xylem, cambium, and phloem (Fig. 29.7, bottom right).

As in herbaceous dicots, the youngest tissues lie closest to the

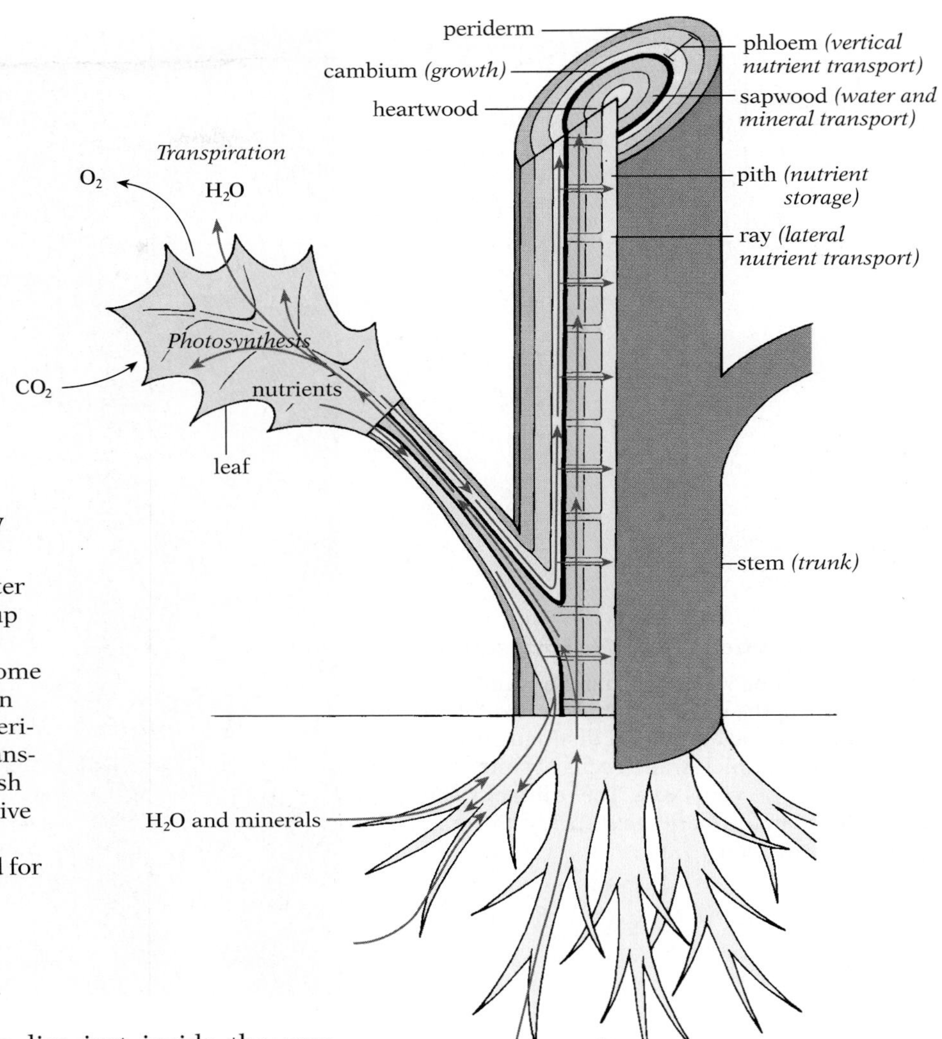

29.8 Summary of fluid flow in a woody dicot

In this stylized model of a tall forest tree, water and minerals enter the roots and are pulled up the xylem to the leaves by a process to be described later. As in monocots (Fig. 29.2), some of the water and the minerals are then used in photosynthesis and for making new cell materials. The nutrients and other materials are transported through the phloem to build or nourish cells. Rays carry the nutrients laterally to active cells and to storage areas in the plant. (The widths of some layers have been exaggerated for clarity.)

cambium: the newest secondary xylem lies just inside the cambium, while the newest secondary phloem is added just outside it. The secondary xylem becomes thicker and thicker, until almost the entire stem of an older plant is xylem tissue—commonly called wood. All but the newest phloem cells are crushed as the tree expands, so there is no buildup of phloem tissue. A horizontal system of vessels, called ***rays***, develops to carry out lateral transport (Fig. 29.7, right), moving nutrients from the phloem to the pith for storage and back out when needed (Fig. 29.8).

Xylem cells produced early in the growing season, when conditions are best, grow larger than cells produced later in the season. This growth pattern results in a series of concentric annual rings, clearly visible in cross sections of the stem. Each ring is made up of an area of spring wood with large cells and an area of summer wood with smaller cells (Fig. 29.9). A fairly accurate estimate of the age of a tree can be made by counting the annual rings. The width of the rings may be affected by such factors as the vigor of the tree and the climatic conditions during the growing season. Since variations in ring size from year to year tend to reflect climatic changes during the life of a tree, study of the rings of a large sample of very old trees can give a clue to the climate of an area in

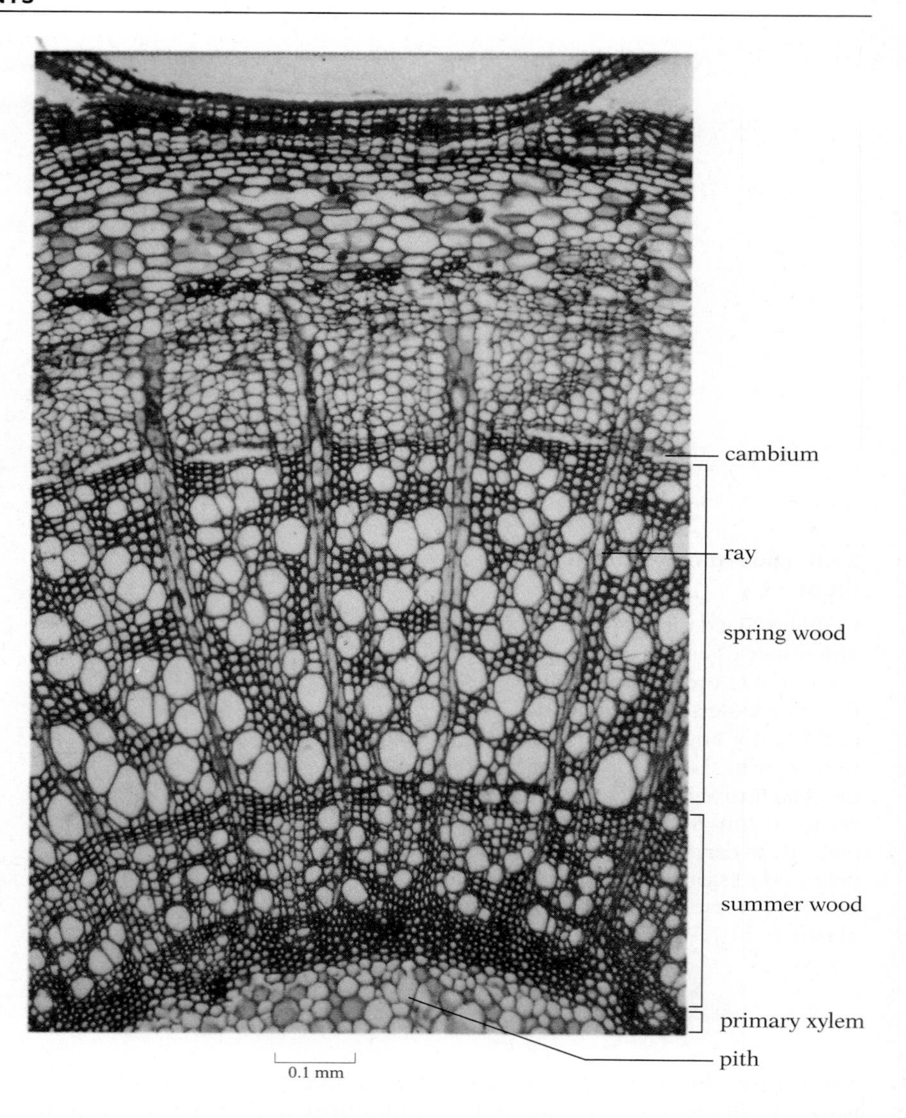

29.9 Cross section of ivy (*Hedera*) stem
In this photograph (taken during the plant's spring growth), the larger, thinner-walled xylem cells of spring wood are readily distinguished from the smaller, thick-walled cells of summer wood from the previous year. The primary xylem was produced in the plant's first spring. Several rays can also be seen.

past ages. Trees commonly live for many centuries, and some, like the bristlecone pine, may be as much as 4500 years old.

In older trees of most species, chemical and physical changes occur in the older rings of xylem toward the center of the stem. The conducting cells become plugged, and pigments, resins, tannins, and gums (all of which provide protection against insect pests) are deposited. The older xylem ceases to function in transport; the current year's xylem performs most or all of the transport from the roots. The older inactive xylem cells, known as ***heartwood***, support the tree physically, while the newest outer ring or rings—which still function in transport—constitute the ***sapwood***, or active xylem (Fig. 29.7). A tree can continue to live after its heartwood has burned or rotted away, but it is much weakened and cannot withstand strong winds.

The stem outside the cambium also undergoes changes as the tree ages. As woody stems (or roots) grow in diameter, a layer of cells outside the phloem takes on meristematic activity of its own and becomes the ***cork cambium***. In the second year this cambium forms ***cork cells*** just under the epidermis, and this growth causes the original epidermis and cortex to flake off. In subsequent years,

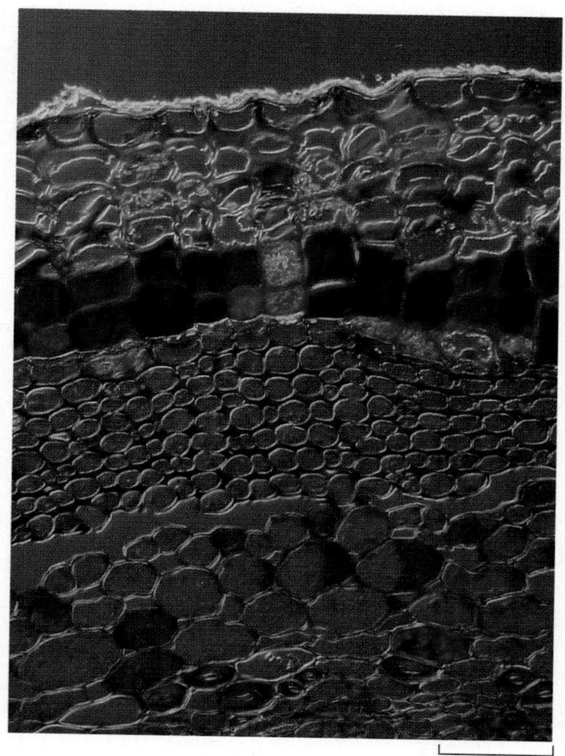

29.10 Periderm of young stem of elderberry (*Sambucus*)

The periderm layer, here consisting of several rows of flattened cork cells (the outermost reddish cells) and the cork cambium (the dark cells just below the cork), provides a protective layer on roots and stems that undergo secondary growth.

the cork cambium can either continue to form outside the primary phloem, or form outside the newest ring of secondary phloem cells. The cork cambium and the cell layers derived from it are collectively termed the ***periderm*** (Fig. 29.10). The walls of most cork cells usually develop layers of the same fatty substance—suberin—that coats the root, and often layers of waxes as well; together they provide a waterproof coating for the plant. As more and more layers of cork cells are produced, the cells in the older, outermost layers usually die and may begin to flake off. Because the breaking of the cork tends to be patchy, the outer bark of some species of trees is very rough and uneven.

In summary, then, the old woody stem of a tree has no epidermis or cortex. Its surface is covered by an outer bark of cork tissue. Beneath the cork cambium is the thin layer of crushed phloem cells, then active phloem, and then the vascular cambium, which is usually only one cell thick. The rest of the stem is mostly xylem, of which only the outer annual rings, or sapwood, still function in transport. It should be clear why girdling a tree (cutting a ring around it) will kill it: the thin phloem system, which encircles the tree, is severed, so nutrients produced by the photosynthetic activity of the leaves cannot reach and sustain the roots.

The design of xylem cells Xylem is a complex of conducting and supporting tissue formed by several types of cells. Two of these, tracheids and vessel elements, are important in the transport of water and minerals. When tracheids and vessel elements mature, their cellular contents, both cytoplasm and nucleus, disintegrate, leaving only their support structure. The main transport in the xylem occurs in these tubular remnants. The distinction between active and inactive xylem—between sapwood and heartwood—depends only on whether they transport water from the roots; each consists of dead cells.

Tracheids are elongate and tapering with thick, hard, secondary cell walls; the walls are particularly heavy in summer wood and are important as supportive elements. Tracheids of the first-matured primary xylem are stretched during their development, and their secondary walls usually take the form of rings or spirals (Fig. 29.11A–B). Tracheids of secondary xylem arise after all lengthwise growth has ceased, and they are not stretched; their secondary walls are more continuous, but are interrupted by numerous ***pits***

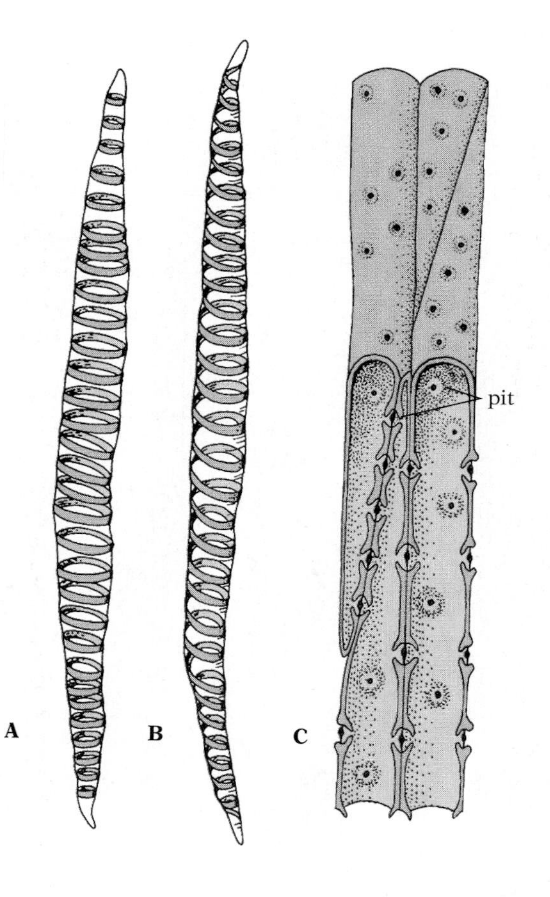

29.11 Tracheids

(A) Primary tracheid with secondary walls in the form of rings. (B) Primary tracheid with spiral secondary wall. (C) Secondary tracheids. Parts of four cells are shown, with one wall cut away from portions of three of the cells to expose their lumina and give a clearer view of the junction between cells. Notice the pits, which are particularly abundant along the tapering ends of the cells.

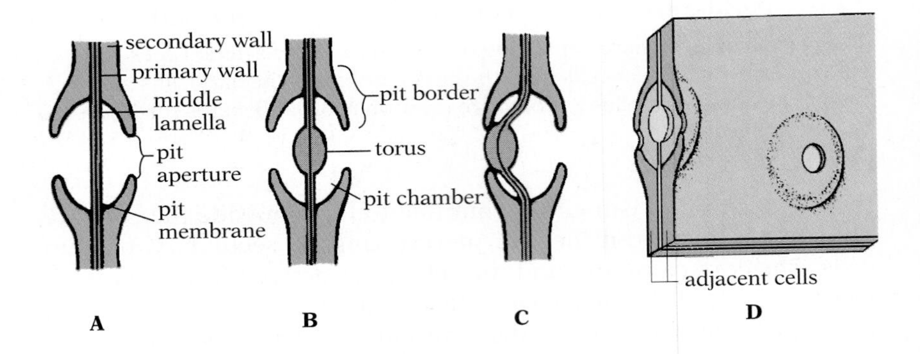

29.12 Pit structure

(A) Bordered pit pair without torus. The secondary walls overhang the pit chamber. The primary walls of any two adjacent cells, and the middle lamella between them, are continuous through the pit and constitute the pit membrane. (B–C) Bordered pit pair with torus, in pine. (C) Here, the torus has been pushed against the pit borders on one side, so movement of materials through the pit is impeded. (D) Three-dimensional representation, showing one bordered pit in section and another in surface view.

(Fig. 29.11C), which are a type of plasmodesmata. They may occur anywhere on the cell wall, but they are often particularly numerous on the tapered ends of the cell, where it abuts on the next cell beyond it. Water and dissolved substances move from tracheid to tracheid through the pits.

Most tracheids have intricate ***bordered pits*** (Fig. 29.12). At these pits, the secondary walls of two adjacent cells are interrupted and their edges overhang the pit chamber, forming the pit borders. The primary walls and middle lamella are continuous through the pit and form the pit membrane, which is generally very thin and permeable to water and dissolved substances[2]. The bordered pit provides extra tissue surface for permeability without reducing structural rigidity. The bordered pits of conifers and a few other plants are particularly interesting in that the pit membrane, though very thin toward its edges, is thickened centrally to form a buttonlike ***torus***. If the pressure in one of the cells becomes much greater than in the adjacent cell—perhaps because air entering the first cell has formed a bubble—the torus, forced against the pit borders of the adjacent cell, blocks the pit aperture and obstructs the flow of materials. In such plants the pit membrane with its torus functions as a check valve between the cells, stopping the flow of sap from a wound, for instance, or preventing moisture from ebbing in a drought.

Vessel elements are more highly specialized for transport than tracheids; fluid can flow directly through them, as through a pipe. They are characteristic of the flowering plants and do not occur in most conifers. In general, vessel elements are shorter and wider than tracheids. They have bordered pits in their sides, through which some lateral movement of substances may take place, but movement occurs chiefly through their ends, which are extensively perforated or may even be entirely open (Figs. 29.13, 29.14). Since the perforations lack the membrane found in pits, material moving vertically from one vessel element to the next forms a continuous column, and transport is more efficient. A vertical series of vessel elements is called a ***vessel***.

Some cells of the xylem are grouped together to form the rays that radiate through the xylem horizontally, perpendicular to the vessels (see Fig. 29.7), and they function as pathways for the lateral

[2] The pit membrane is not the kind of bilayer membrane common to cells; instead, it is part of the cell wall; it is composed of cellulose and pectin, and contains no selectively permeable membrane channels.

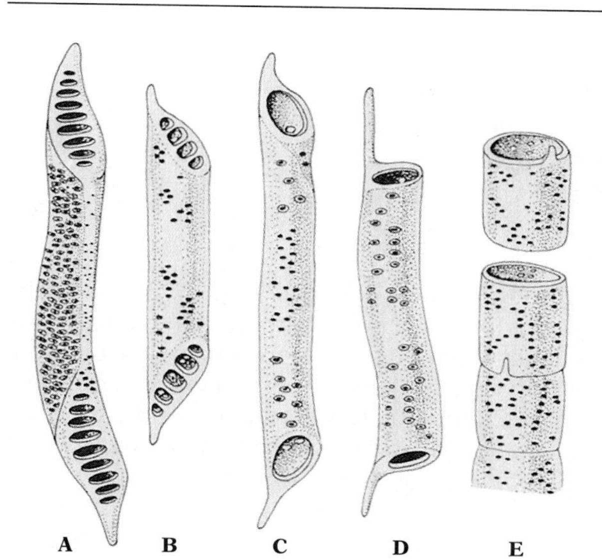

29.13 Vessel elements

Five different types of xylem vessel elements are shown—those thought to be the more primitive on the left, those thought to be the more advanced on the right. The last example (E) shows a single vessel element on top, three elements linked in sequence to form a vessel below. The evolutionary trend seems to have been toward shorter and wider elements, larger perforations in the end walls until no end walls remained, and less oblique, more nearly horizontal ends.

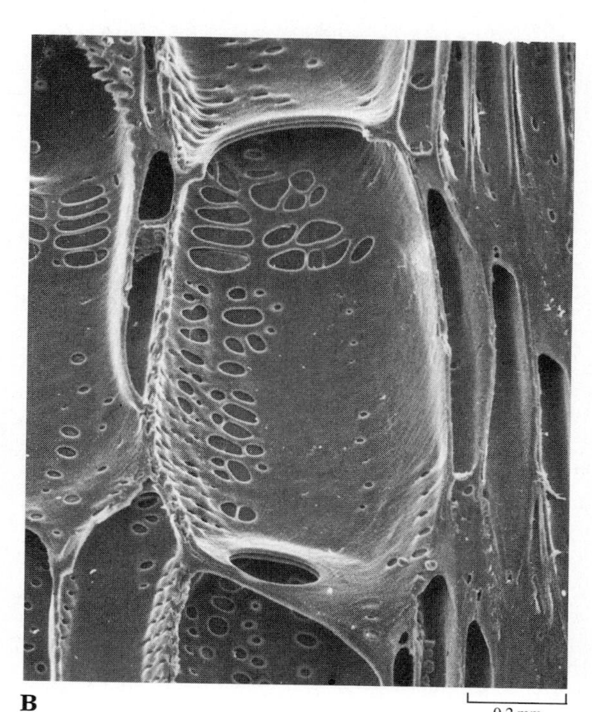

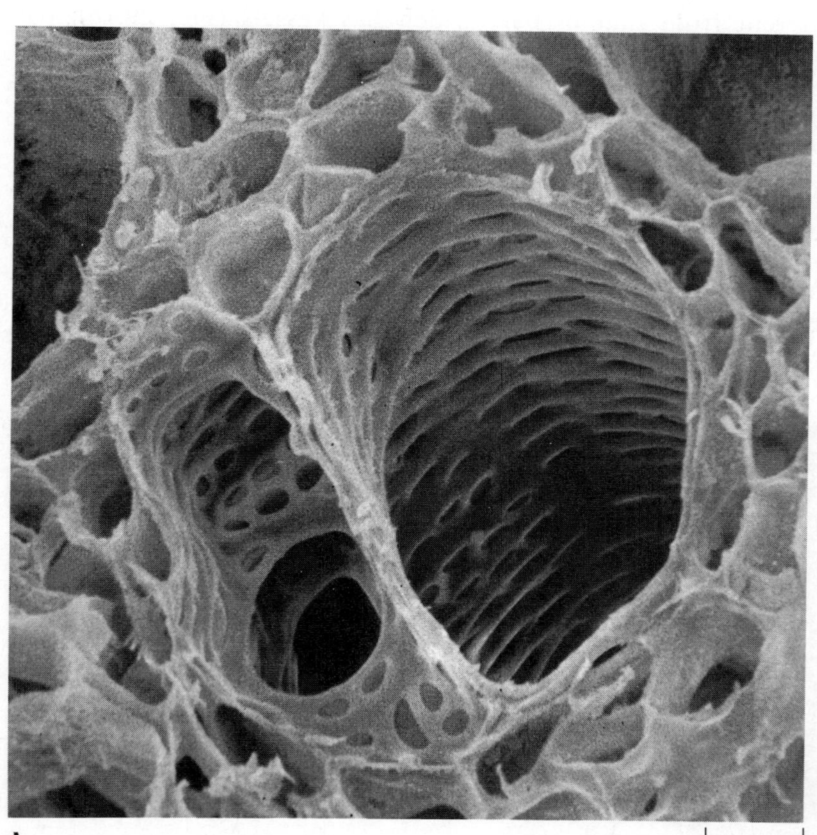

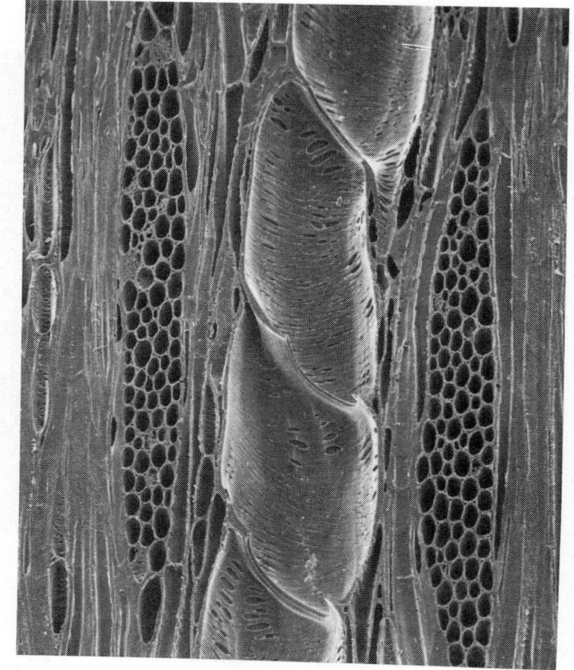

29.14 Xylem vessels

(A) Portion of cross section of corn root showing two large xylem vessels. (B) Inner surface of a single vessel element. Numerous pits can be seen in the side walls. The large opening at the lower end is the perforation through which water moves vertically from one element to the next. (C) Lateral view of four vessel elements stacked one above the other. (All images are scanning electron micrographs.)

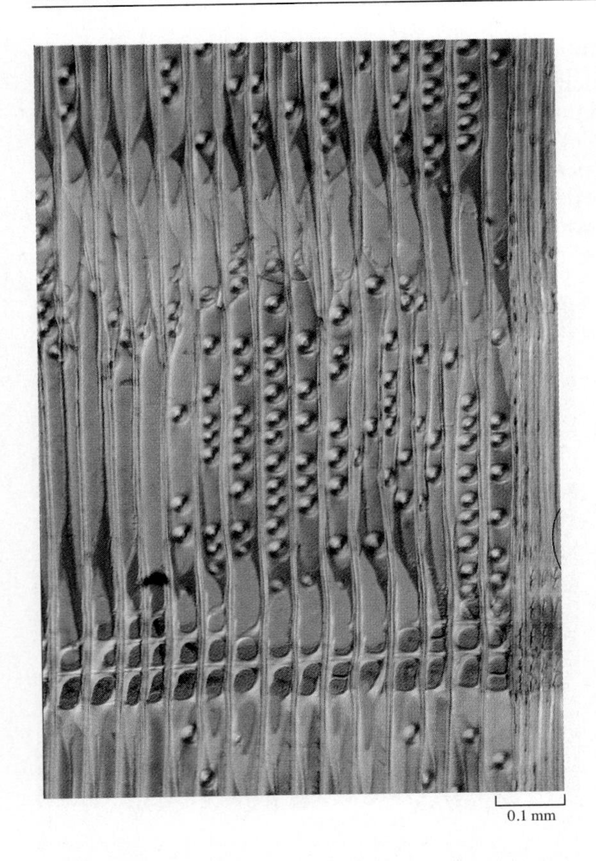

29.15 Vertical section of pine wood

Note the oblique junctions between successive tracheids. Bordered pits are clearly visible. Part of a ray about three cells high is shown in the lower quarter of the photograph.

movement of materials. Rays may be small (Fig. 29.15) or large, depending on the species of plant. Without the rays to transport nutrients to them, cells far from the active phloem cells would die. Rays are particularly important in older trees, where a continuous layer of cambium separates the active xylem (sapwood), from the phloem cells, and layer after layer of inactive xylem (heartwood) intervenes between the phloem and the pith.

The number, form, and distribution of tracheids, vessels, fibers, and other cells vary from species to species, and this variation causes the woods of different species to differ in appearance and properties. The rays in oak, for instance, are typically macroscopic, so their presence greatly affects the appearance of the wood. The wood of pine, a conifer, has rays but lacks vessels, and is thus very different from that of oak, which has vessels (Fig. 29.16); oak wood, with its relatively few vessels, is, in turn, different from elm wood or tulip-tree wood, both of which are very porous and have numerous vessels. These differences have both aesthetic and mechanical consequences when the wood is used in construction.

The design of phloem cells The principal function of the phloem is to transport the nutrient products of photosynthesis throughout the plant. The principal vertical conductive elements in phloem are the ***sieve elements***, which in vertical series form ***sieve tubes***. In their most advanced form, sieve elements are elongate cells with specialized areas on their end walls called ***sieve plates***

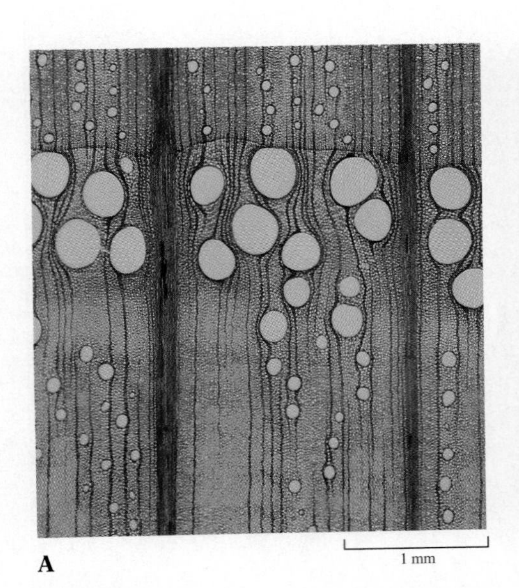

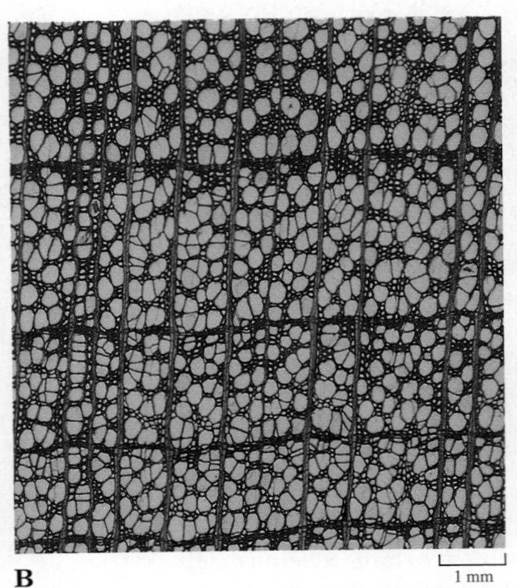

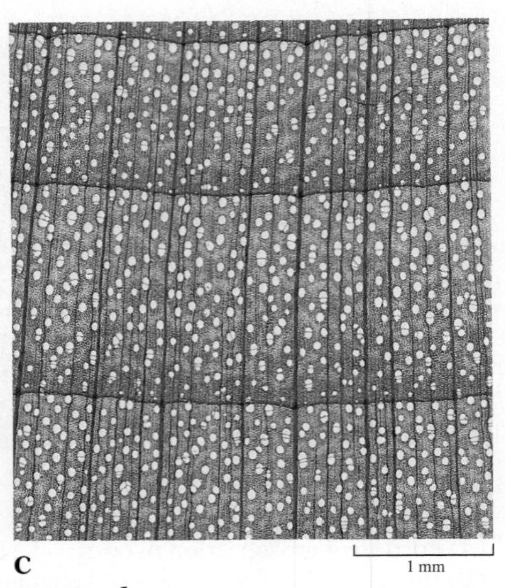

29.16 Cross sections of different woods

(A) Red oak. (B) Tulip tree. (C) Sugar maple. Note the differences in the number and size of vessels.

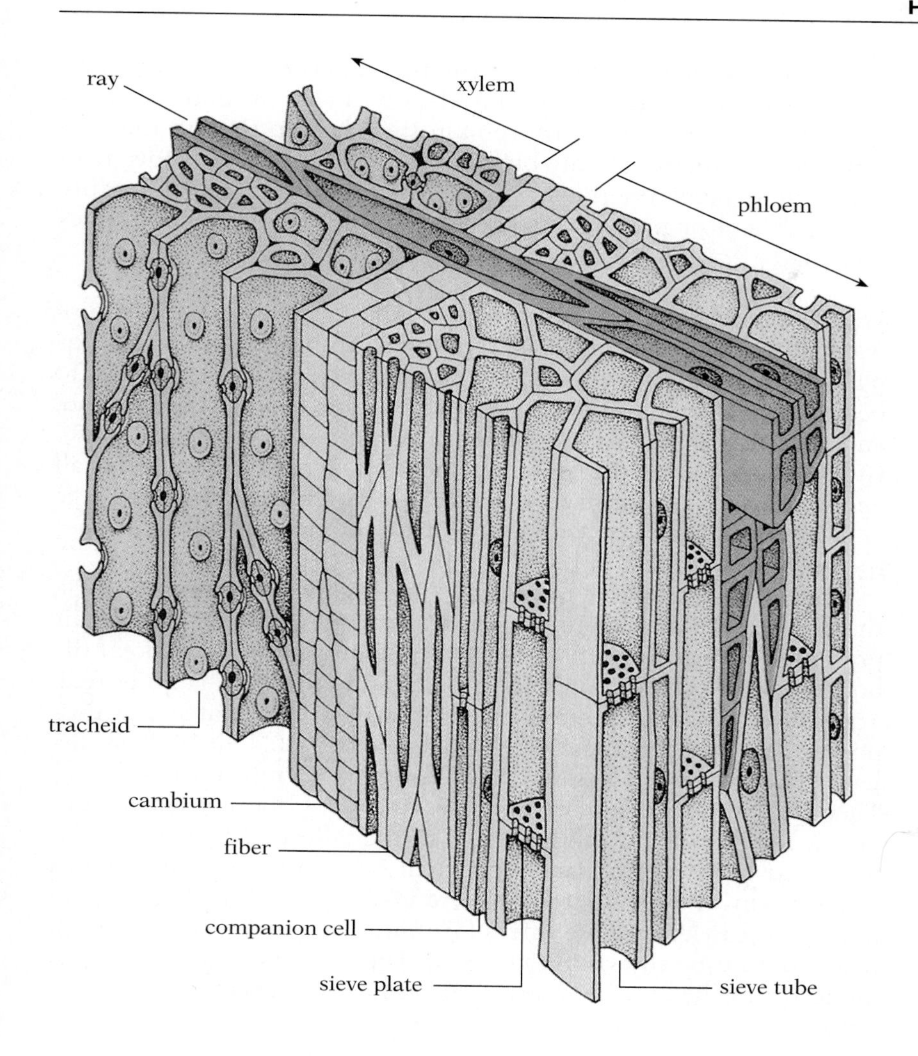

29.17 Phloem
Some nearby xylem tissue is also shown.

with numerous perforations or pores (Fig. 29.17). These openings are created in the cell's end as it matures; strands of cytoplasm connect the contents of one cell with those of the next. Unlike the tracheids and vessels of xylem, the sieve element retains its cytoplasm at maturity, though the nucleus disintegrates.

Closely associated with the sieve elements of most flowering plants is usually one or more mysterious, elongate ***companion cells*** (Fig. 29.17), which are derived from the same cell that gave rise to the associated sieve element and which retain both cytoplasm and nucleus. The nucleus of the companion cell may control both its own cytoplasm and that of the adjoining sieve element; this could help explain why the mature sieve element can continue to carry out many of the normal activities of a living cell even though it has no nucleus of its own.

■ FROM ROOTS TO LEAVES: THE ASCENT OF SAP

Sap—the water and dissolved elements that the roots absorb—moves upward through the plant body in the mature tracheids and vessels of the xylem. That the upward movement is primarily in the xylem can be easily demonstrated by ringing experiments; if the cork, phloem, and cambium are removed from a ring around the trunk of a tree, the leaves will still remain turgid, even though they are connected to the roots only by xylem. Furthermore, the movement does not depend on the few cells that remain alive in mature xylem; if heat or poisons kill these cells, the rise of sap continues unabated.

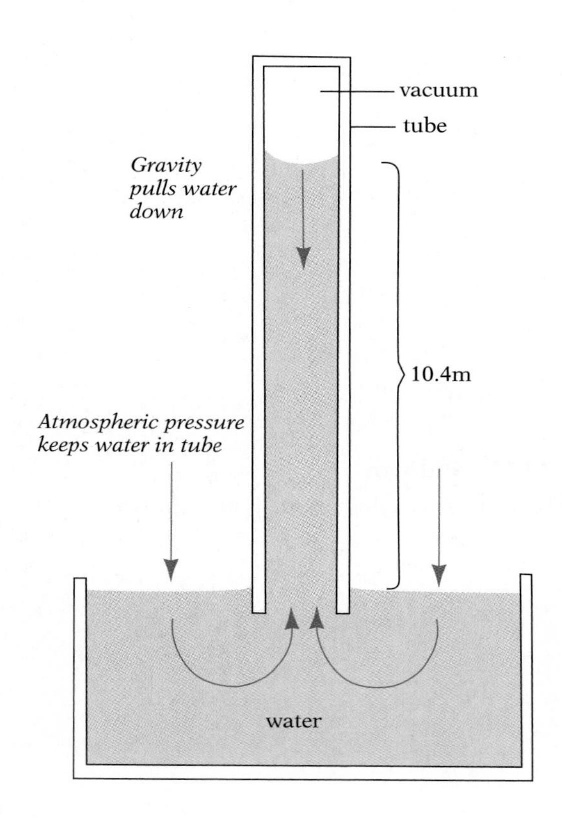

29.18 How atmospheric pressure supports a column of water

Atmospheric pressure—the weight of the air above us pressing down—is sufficient to support a column of water 10.4 m high. In the same way, the weight of the atmosphere supports the column of mercury in the familiar mercury barometer, but because mercury is so much denser than water, the column reaches only about 76 cm.

Where could the energy come from? What force could raise water to the tops of trees 90–120 m high? This question, which has puzzled researchers for decades, does not have a simple answer. To begin with, imagine filling a tube with water, sealing the top end, and leaving the bottom in an uncovered pool of water (Fig. 29.18). Gravity pulls down the water in the tube, and if the tube is tall enough, a vacuum forms at the top. The sap in the xylem of a tree is subject to this same gravitational force and yet still nourishes the highest branches.

What keeps the water in the tube at all is the pressure of air on the surface of the surrounding pool of water. The existence of air pressure is difficult for most of us to imagine, living as we do at the bottom of an ocean of air. Nevertheless, the pressure is quite real. The weight of miles of atmosphere above us—designated 1 atmosphere, or atm—supports a column of water 10.4 m high at sea level. (At higher elevations the column is significantly shorter.) Therefore, a pressure of at least 12 atm would be needed to support a column of 120 m, the height of the tallest trees. But the column must be more than supported: the fluid must be moved upward 1m or more per minute, and this movement must take place in the xylem, which offers far more frictional resistance than the smooth tube shown in Fig. 29.18. The movement of fluid to the highest branches of the tallest trees requires a force of at least 30 atm.

Capillarity? For a long time capillarity (the tendency of water to rise in a thin hydrophilic tube) was a popular candidate as the source of sap movement. Only an accurate assessment of the magnitude of the forces involved proved that capillarity, by itself, is far too weak to account for the massive phenomenon of the ascent of sap.

Root pressure? Another proposed explanation was root pressure. When the stems of certain species of plants are cut, sap flows from the surface of the stump for some time, and if a tube is attached to the stump, a column of fluid a meter high or more may rise in it. Similarly, when conditions are optimal for water absorption by the roots, but the humidity is so high that little water is lost by transpiration, water under pressure may be forced out at the ends of the leaf veins, forming droplets along the edges of the leaves. This process of water secretion is called ***guttation*** (Fig. 29.19). When the fluid in the xylem is under pressure, as in these instances of bleeding and guttation, the pushing force involved is apparently in the roots, and is called ***root pressure***. How is root pressure generated, and how strong is it?

Root pressure depends on energy from ATP, which drives the active transport of ions that establishes a decreasing water-potential gradient; ions in the root cells external to the endodermis are pumped inward, while ions in the pericycle are pumped out into the xylem. As a result, water following the symplast path (see Chapter 26) is drawn into the root, through the endodermis, and into xylem vessels; water moving via the apoplast route is drawn into the xylem by the low water potential of the endodermis followed by the still lower potential in the xylem. However, like capillarity, root pressure cannot be the whole answer or even a major part since it rarely exceeds 1 or 2 atm. At best it may be involved in the ascent of sap in some short plants, particularly very young plants in early spring.

Hydrogen bonds and evaporation? As water in the walls of mesophyll cells in the leaves (or other parts of the shoot) is lost by transpiration, it is replaced by water from the cell contents. With the removal of water the water potential of leaf cells falls, and they take up water from adjoining cells—which, in turn, withdraw water from cells adjacent to them. In this way a water-potential gradient extends to the xylem in the veins of the leaf, which enables the cells next to the xylem to withdraw water from the column in the xylem; in the process, this helps pull the column upward. This is not a matter of pulling the column up by suction: air pressure could raise water only 10.4 m. What is required is a continuity between the water on the evaporating surfaces of the cell wall and the water in the xylem, and a continuity between the water at the top of the xylem and that in the roots. If this continuity of water all the way from leaf cell to root were broken by the entrance of air into the system, that particular xylem pathway would cease to function.

The idea that sap could be pulled up by transpiration depends on the great cohesion between individual polar water molecules, and their tendency to adhere to the hydrophilic walls of the xylem tubes and the mesophyll cells in the leaves. Water molecules moving out of the top of the xylem must pull other water molecules behind them; there can be no break, no separation between the molecules of water, or between the sap and the walls of first the vessels and then the walls of mesophyll cells.

In theory, the cohesive strength of the entire column of water—a continuous system of hydrogen-bonded molecules, extending from roots to transpiration surfaces in the leaves, and enclosed in very narrow, highly hydrophobic tubes—could reach 15,000 atm. Actual experimental values approach 300 atm. If the ascent of sap requires a pressure of 30 atm or more, these values are compatible with the so-called ***transpiration-adhesion-tension-cohesion*** (TATC) theory, also known as the transpiration theory. (The *adhesion* here refers to the binding of water to vessel walls and, more importantly given the enormous surface area involved, the walls of leaf mesophyll cells; *tension* is the pull of water out of the xylem into the leaves, down the gradient of water potential; *cohesion* is the hydrogen bonding between water molecules in the xylem column and in the mesophyll.) The source of energy for the transpiration system is the sun, which causes the evaporation of water from the leaves that pulls replacement water up the xylem.

29.19 Guttation by strawberry leaves

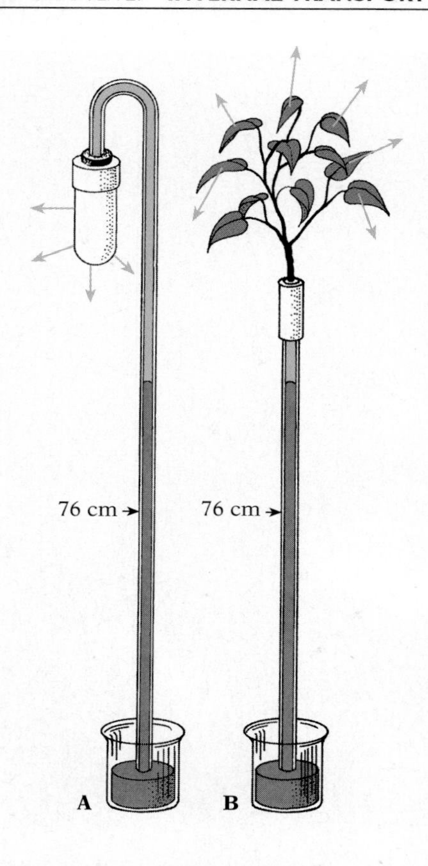

29.20 Demonstrations of water rising by pull from above

(A) Water is evaporated from a clay pot attached to the top of a thin tube whose lower end is in a beaker of mercury. The water (blue) in the tube rises and pulls a column of mercury (gray) to a point well above the 76 cm that atmospheric pressure can support. (B) The same results are obtained when transpiration from the leaves of a shoot is substituted for evaporation from a clay pot. The highly porous hydrophilic clay in A substitutes for the mesophyll cells of leaves in B.

The TATC theory has not been tested under conditions duplicating those in very tall trees, but researchers more than a century ago showed that if water is evaporated from the top of a thin glass tube whose lower end is immersed in mercury, a column of mercury will be pulled up the tube by the evaporating water. If the base of a cut branch is inserted tightly in the upper end of the tube and the leaves of the branch become the site of evaporation, the same thing happens. Clearly, the water molecules adhere to each other (and to the walls of the tube) tightly enough to pull up a heavy column of mercury (Fig. 29.20).

Support for the TATC theory comes from other types of experiments too. When sap is warmed in the early morning at one point, and then the time is measured until the sap reaches a thermocouple higher on the tree, it turns out that the water in the xylem begins to move in the upper parts of the tree earlier in the morning than it does in the lower parts of the trunk—an indication that the upward movement of the sap is initiated at the top of the tree, not at the bottom.

More evidence comes from precise measurements of the girth of tree trunks at different times of day. During the daylight hours, when the xylem should be under maximum tension because of high rates of transpiration, the trunk diminishes in girth—that is, it shrinks, just as any closed, flexible, liquid-filled container will do when its contents are being pulled upward. If the xylem were under pressure from the bottom, it should expand.

Though the TATC theory has gained wide acceptance among plant physiologists, it leaves a number of problems unresolved. The theory requires the maintenance of continuity within the water column in the xylem; yet breaks in the column occur frequently. For example, during times of drought some of the gases dissolved in the sap may form gas bubbles; when sap freezes in winter the dissolved gases are forced out, forming bubbles that break the water column. As these breaks in the water columns progressively affect a larger proportion of the xylem elements, a stem's total conductance declines. Plants have several strategies for repairing or isolating these air bubbles. In herbaceous plants, for example, root pressure generated at night (when the tension in the vessels is relaxed) can rejoin broken water columns. Gases forced out of solution during freezing can redissolve when temperatures rise. In addition, the pits of tracheids probably allow fluids to bypass gas-filled or damaged parts of the xylem. Finally, newly synthesized rings of xylem replace any unrepairable columns in the stem.

■ MOVING SOLUTES IN THE PHLOEM

Are all organic solutes moved in the phloem? Two principal classes of solutes are transported within the plant body: organic solutes and inorganic solutes. We can divide the organic solutes conveniently into two principal types: carbohydrates (usually transported as sucrose) and organic nitrogen compounds. (Plant hormones are also transported.)

As we have seen, xylem serves to move sap—water and inorganic ions—up from the roots, while the phloem transports carbohydrates produced through photosynthesis to regions of growth and storage, whether up or down from the site of production.

The situation is less clear for organic nitrogen compounds. It was once thought that nitrogen, absorbed by the roots primarily as nitrate, was carried upward in inorganic form through the xylem to the leaves, to be used in the synthesis of organic compounds that were then transported through the phloem. This sequence probably holds true for some plants. However, there is now good evidence that many species promptly incorporate incoming nitrogen into organic compounds such as amides and amino acids in the roots. In some species, especially herbaceous ones, these organic nitrogen compounds move upward in the phloem, but in other species, especially trees, they move in the xylem.

Are all inorganic solutes moved in the xylem? Inorganic ions are moved upward from the roots to the leaves primarily through the xylem. However, experiments using radioactive isotopes of these minerals indicate that some ions are able to travel rapidly back down in the phloem; they can move out of older leaves through the phloem to newer, actively growing leaves. This movement can be visualized by growing a plant briefly in a solution of radioactive phosphorus ions: placing the plant against a photographic plate shows the younger leaves to be highly radioactive (Fig. 29.21); a day later, after growth in nonradioactive water, pictures show that the radioactive phosphorus from the previous day has moved to newly developed leaves. Calcium, on the other hand, cannot move from old leaves to newer ones. Consequently, plants must obtain a steady supply of new calcium from the soil and move it up the xylem to where it is needed. Well-designed fertilization programs take into account such differences in the mobility of different minerals.

What powers the movement of phloem? Most organic solutes move through the phloem, but the force that powers this movement is still the subject of debate.

Any hypothesis about phloem transport must account for several facts:

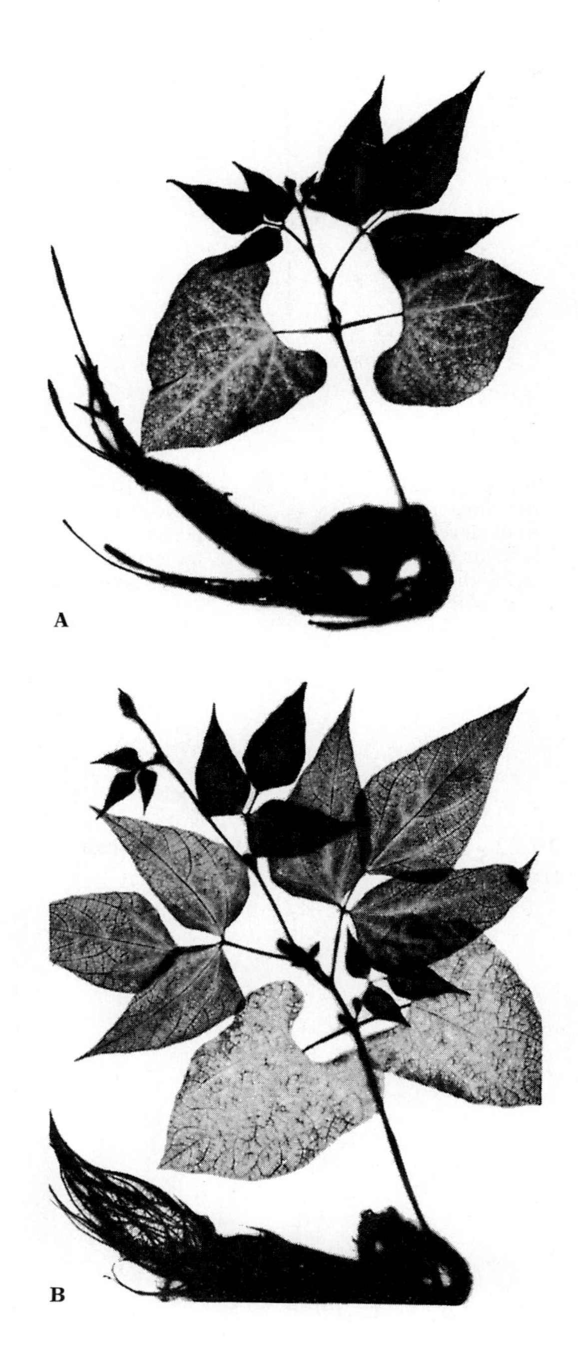

29.21 Movement of radioactive phosphorus in a growing plant

The plant was grown for 1 hr in a nutrient solution containing ^{32}P. It was then removed to a nonradioactive solution. At the end of 6 hr (A), the ^{32}P was particularly concentrated in the youngest leaves. At the end of 96 hr (B), much of the ^{32}P had moved from the leaves in which it was formerly most concentrated to new leaves that had developed above them. (The darker the area, the more ^{32}P it contains.)

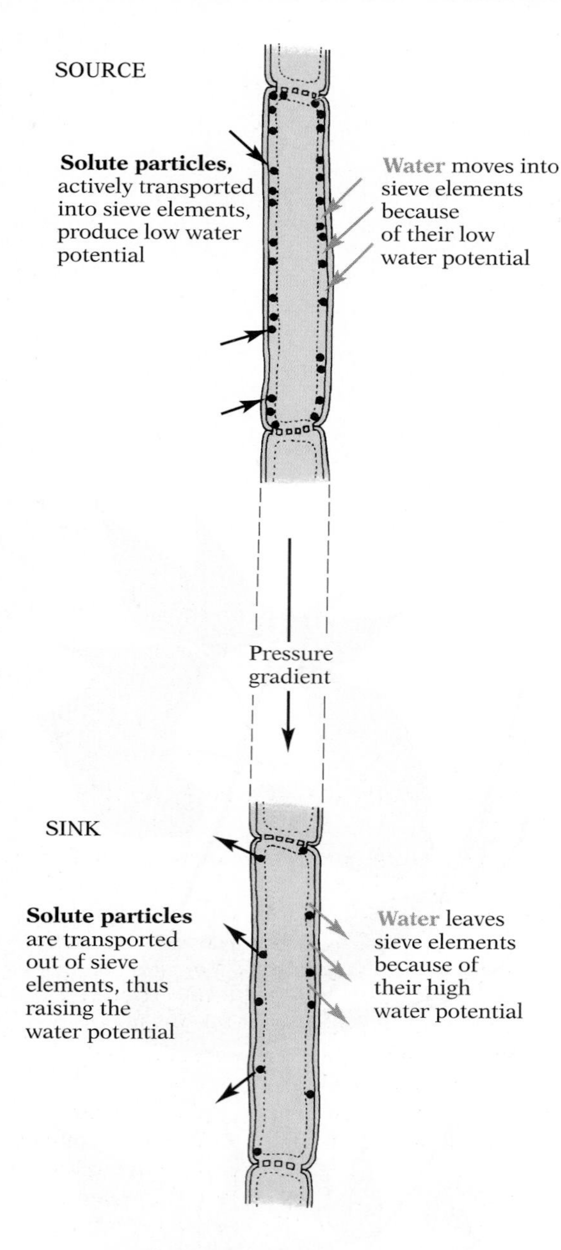

29.22 Pressure-flow model of phloem transport

1 The movement is often rapid (up to 1 m/hr); sugar moves through the phloem of a cotton plant more than 40,000 times faster than it diffuses in a liquid.

2 The speed of movement through the sieve tubes differs for different substances.

3 The direction of movement may be reversed periodically within a given sieve tube.

4 The movement in neighboring sieve tubes may be in opposite directions.

5 The movement takes place through sieve elements that, unlike xylem, retain their cytoplasm.

6 Unlike the ends of xylem vessels, the ends of the individual sieve elements are penetrated only by the tiny pores of the sieve plates.

Clearly, we are dealing with transport through active cells, not merely with movement through dead tubes by purely mechanical processes, as in mature xylem.

The most widely accepted hypothesis for phloem movement involves ***pressure flow***, or bulk flow—the mass flow of water and solutes through the sieve tubes along a turgor-pressure gradient. The process begins with active transport of photosynthetically produced sugars into phloem cells in the leaves. Because these cells then contain high concentrations of sugar, they have a low water potential; water tends to diffuse into them. This passive osmotic movement of water following the sugar causes the turgor pressure of the cells to rise. The pressure then tends to force substances from these cells into the cells next to them. Thus, under hydrostatic pressure, substances are forced en masse from cell to cell along the sieve tubes. At the same time, in storage organs or actively growing tissues, where sugars are being used up and actively removed from the sieve tubes, the water potential rises. The tubes therefore tend to lose water, and their turgor pressure drops.

The contents of the sieve tubes, then, are under considerable turgor pressure in one portion of the plant (the "source" or "producer") and under lower turgor pressure in another portion of the plant (the "sink"). The source or producer region—in which hydrostatic pressure is high—is usually the leaves, but sometimes includes storage organs when reserves are being mobilized for use, as is the case in early spring. Regions of low pressure (the sinks) are usually actively growing "consumer" regions or storage depots. The result is a mass flow of the contents of the sieve tubes from the source to the sink. The whole process depends on massive uptake of water by cells at the source or producer end, because of their low water potential, and massive loss of water by cells at the sink or consumer end, because their water potentials are raised by their loss of sugar (Fig. 29.22). Phloem transport is therefore push-powered by hydrostatic pressure, while xylem flow is pull-powered by transpiration.

■ DO PLANTS HAVE A CIRCULATORY SYSTEM?

Gram for gram, plants require 15–20 times as much water as animals. Some of this water is needed for growth: water is the main constituent of new cytoplasm, and a raw material for photosynthesis. More than 90% of the H_2O that ascends in the xylem, however, simply evaporates from the gas-exchange surfaces of the leaves and is lost. A small amount, as we have just seen, is actually circulated, passing from the xylem to the phloem (and later back) to assist in the movement of sugars. Without a constant supply of new water, plants cannot translocate inorganic nutrients or photosynthetically produced compounds: they cannot grow or photosynthesize beyond the minimal level of their own metabolism; they cannot even cool their delicate leaves (Fig. 29.23).

Water acquisition and conservation are major challenges for many species. Stomatal opening is keyed to the availability of

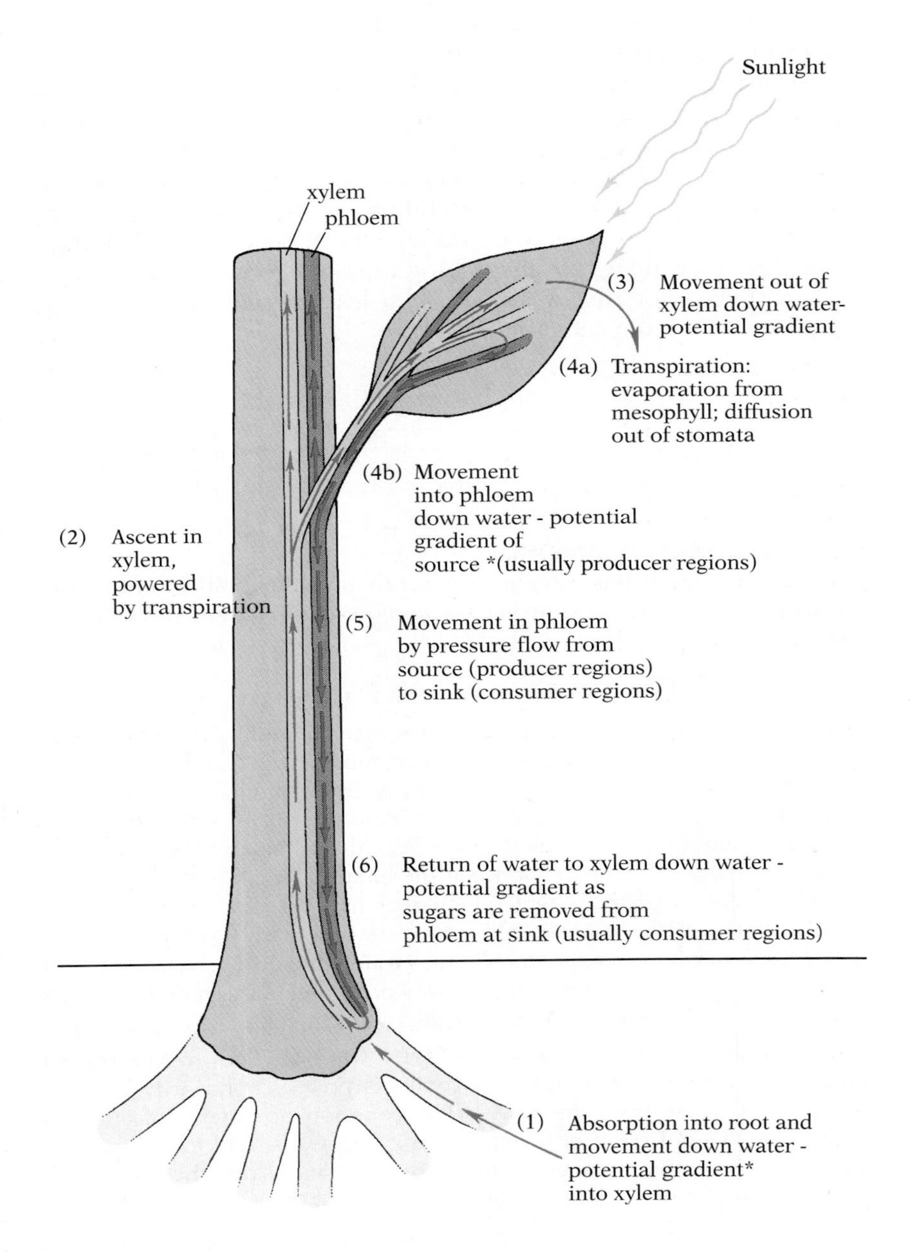

29.23 Summary of water movement in plants

Water enters roots by moving down a gradient of water potential (1) and is pulled up the xylem (2) by transpiration. Water moves from the xylem in leaves by flowing down another water-potential gradient (3), where it either evaporates from the gas-exchange surfaces (4a) or is drawn down the water-potential gradient into the phloem (4b). Water in the phloem is pushed away from the source (producer regions) of the solutes (mainly sucrose) by the continued entry of more water down the potential gradient (5). When the solutes are transported out at a sink (consumer region), water moves down a gradient of water potential back into the xylem (6). The energy for moving water is generated by active transport of ions in the roots (1), solar-powered evaporation of water in the leaves (4a), active transport of sugars into the phloem at sources (4b), and active transport of sugars into sinks (6).

water: when the soil cannot supply enough water, the stomata close and transpiration virtually ceases. As described in Chapter 7, some arid-adapted plants even have special carbon-fixation pathways (C_4 and CAM) that permit photosynthesis with the stomata closed; these gas-exchange orifices are opened at night to allow gas diffusion when lower temperatures retard transpiration.

Water loss is also increased by dry air and wind. Leaf shape is often a good clue to a plant's usual habitat. The leaves of forest trees, protected by other trees and leaves from wind and drying heat, are usually broad and flat; for these species, capturing the little light filtering through higher leaves is more important than conserving water. For plants adapted to grow in the open windswept plains, on the other hand, low, thicker, narrow leaves are the rule; light is abundant while water is usually limiting.

The most important long-term investment a plant can make, however, is in roots. The larger and (usually) deeper the root net, the greater a plant's potential for finding the water needed to support its growth. A corn plant, though adapted to arid conditions, still needs about 200 liters of water over the course of a season; it solves this supply problem by extending its roots up to 50 mm/day until the combined root and rootlet length reaches 1000 km or more. It's no wonder that disturbing a plant's roots can kill it in a matter of days, or that a drought can lead to wilting as water is withdrawn from the cells it holds rigid.

CHAPTER SUMMARY

LIVING WITHOUT PLUMBING

Small or thin organisms can sometimes rely on simple diffusion to move nutrients and wastes; individual cells can move their contents via cytoplasmic streaming. (p. 819)

HOW VASCULAR PLANTS MOVE THEIR FLUIDS

THE MAIN THOROUGHFARE: STEMS Monocots and young dicot stems have vascular bundles containing xylem, which moves water and minerals up from the roots, and phloem, which moves water with nutrients from areas of production or storage to areas for storage or use. In older dicots the vascular tissue forms a complete circle near the outside of the stem, with the xylem toward the inside; between the band of xylem and phloem is a band of vascular cambium, which generates new vascular tissue. Ray vessels run horizontally from the phloem to the xylem and to the center, or pith, where nutrient storage can occur. Monocots and herbaceous dicots have their vascular tissue in bundles; in woody dicots the bundles fuse to form a ring, and new rings are added each year. (p. 820)

FROM ROOTS TO LEAVES: THE ASCENT OF SAP The movement of sap—water with minerals from the roots—is powered primarily by transpiration of water from the leaves. The cohesion of water molecules to each other, the adhesion of water to the xylem walls, and tension generated by the water-potential gradient between the leaves and the xylem—all contribute to the process that pulls water up from the roots. (p. 831)

MOVING SOLUTES IN THE PHLOEM The nutrient-rich fluid of the phloem moves from areas of high solute concentration and water pressure (sources, including sites of photosynthesis and nutrient storage) to areas of low solute concentration and water pressure (sinks). (p. 835)

DO PLANTS HAVE A CIRCULATORY SYSTEM? Most water entering the xylem is lost in the leaves through transpiration; much of the rest enters the phloem and travels to sinks, where some is picked up by the xylem. (p 836)

STUDY QUESTIONS

1. Given that root pressure and capillarity can move water through xylem a few centimeters, is it possible to imagine a water conserving plant with no transpiration, and thus complete fluid circulation? What would be the costs and benefits of such a design in competition with conventional plants in various environments? (pp. 832, 836–838)
2. How might plants be different if the earth's atmosphere were 21% CO_2 and only 0.03% O_2, instead of the other way around? How about animals?
3. What happens to a typical plant's physiology on a muggy day? (Consider both sunny and cloudy alternatives.) (pp. 833–834, 836–838)
4. Aphids, which tap the juices of plants for a living, are far more common on growing tips of plants than on mature stems leading to photosynthetically active sources. Does this make sense? If the active transport systems of plants could be redesigned, how might you better defend a plant against aphids without producing any new chemicals or altering the geometry of the plumbing? (pp. 835–836)

SUGGESTED READING

RAVEN, P. H., R. F. EVERT, AND S. E. EICHHORN, 1992. *Biology of Plants*, 5th ed. Worth, New York. *An authoritative textbook.*

ZIMMERMAN, M. H., 1963. How sap moves in trees, *Scientific American* 208 (3).

CHAPTER 30

INTERNAL TRANSPORT IN ANIMALS

WHY CIRCULATION IS NECESSARY

Animals face the problem of getting nutrients to every cell in their bodies and carrying off metabolic wastes. Single-celled organisms can rely on simple diffusion, supplemented with active transport across membranes, because their surface area is large relative to their cellular volume. Even multicellular organisms may find diffusion and active transport sufficient if they are so thin that every cell is near the surrounding medium. In contrast, most larger animals must have specialized internal-transport systems.

The more rapid metabolism required by the active movement of most animals makes them correspondingly less able to rely on passive diffusion. Moreover, animals are much less likely than plants to have flat, thin bodies. Sea fans, a curious group of large, thin coelenterates, are an exception. Tapeworms are another example; they may be 20 m long or more, but they are always flat and thin, and no cell is far from the food supply that surrounds them in the host's digestive tract.

In general, however, even small, thin animals have some sort of specialized system to supplement diffusion. The hydra's body wall (see Fig. 27.11, p. 773) is basically only two cells thick, but even so the creature is so active that the cells of the inner layer must be exposed directly to water containing dissolved oxygen, which is drawn into the gastrovascular cavity. Branches of the gastrovascular cavity penetrate deep into each tentacle, and food particles can be absorbed directly from this cavity by the tentacle cells. The gastrovascular cavity of planarians serves the same role (see Fig. 27.13, p. 774).

Animal circulatory systems usually include some sort of pumping device, called a ***heart***. There may be only one heart, as in the case of humans, or a number of separate hearts, as in earthworms, where five blood vessels on each side of the animal pulsate, pumping blood from the main dorsal longitudinal vessel into the main ventral longitudinal vessel (Fig. 30.1). Many insects have both a large general heart and a series of smaller accessory hearts at the bases of their legs and wings.

The one-way pumping action of the heart, usually combined with a system of one-way valves, moves the blood in a regular fashion through the circuit. This circuit may flow exclusively through well-defined channels or vessels, in which case it is called a ***closed***

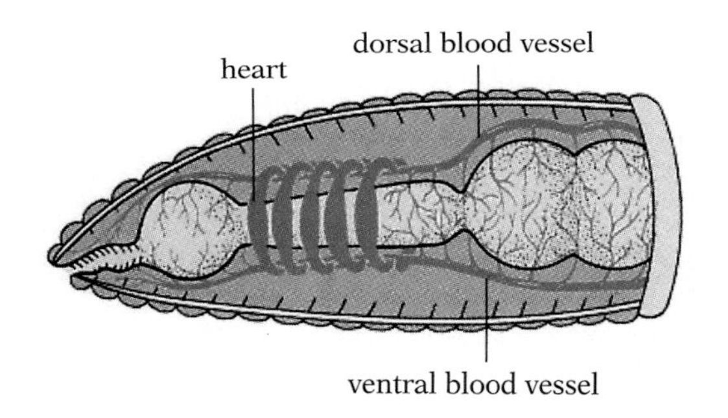

30.1 Circulatory system of an earthworm
Ten hearts, five on each side, pump blood through the longitudinal vessels, which themselves pulsate and help move the blood.

circulatory system. Or the circuit may have some sections where vessels are absent and the blood flows through large open spaces known as sinuses; such a system is called an ***open circulatory system***. Closed circulatory systems are found in many animals, including the earthworms and all vertebrates. Open circulatory systems are characteristic of most molluscs (snails, oysters, clams, etc.) and all arthropods (insects, spiders, crabs, crayfish, etc.).

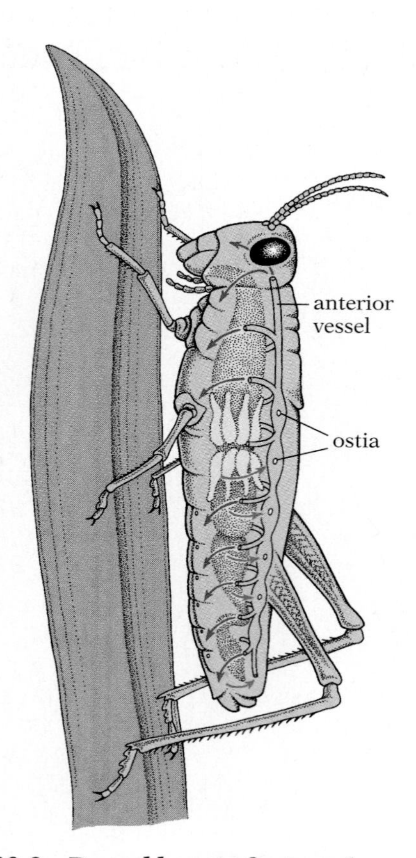

30.2 Dorsal heart of a grasshopper
Blood enters the vessel-like heart through the ostia and is pumped forward and out at its open anterior end.

THE INSECT STRATEGY: OPEN CIRCULATION

The movement of the blood through an open system is not as fast, orderly, or efficient as through a closed system. It may seem surprising that insects, which are very active, should have open circulatory systems. However, insects do not rely on the blood to carry oxygen to their tissues; for this they use a tracheal system. The insect circulatory system acts to deliver nutrients and carry off waste, and thus the blood need not flow very quickly.

The only definite blood vessel in most insects is a longitudinal vessel, often designated the heart, which runs through the dorsal portion of the thorax and abdomen (Fig 30.2). The posterior portion of the heart is pierced by a series of openings, called ostia, each regulated by a valve that allows movement of blood only into the vessel. When the heart contracts, it forces blood out of its open, anterior end into the head region. When the heart relaxes again, blood is drawn in through the ostia. Once outside the heart, the blood is no longer in vessels. The blood simply fills the spaces between the internal organs of the insect, bathing them directly.

The action of the insect heart causes the blood to move sluggishly through the body spaces from the anterior end, where it was released, to the posterior end, where it will again enter the heart. The movement of the blood is accelerated by the stirring and mixing action of the muscles of the body wall and gut during activity. Thus when the animal is most active, as in running or flying, and its organs are in most need of rapid delivery of nutrients and removal of wastes, the blood moves relatively fast because of the activity itself.

THE VERTEBRATE STRATEGY: CLOSED CIRCULATION

All vertebrates have a closed circulatory system, which consists basically of a heart and numerous arteries, capillaries, and veins. ***Arteries*** are blood vessels that carry blood away from the heart, while ***veins*** carry blood back toward the heart. These definitions are not based on the condition of the blood carried. Though most arteries carry oxygenated blood and most veins carry deoxygenated blood, oxygen content is not always a reliable way to distinguish them; for example, the vein that leads from the lungs to the heart carries the most highly oxygenated blood in the body. ***Capillaries*** are tiny blood vessels that interconnect the arteries and the veins;

they usually run from very small arteries, called arterioles, to very small veins, called venules. It is across the thin walls of the capillaries that most of the exchange of materials between the blood and the other tissues takes place.

■ THE CIRCUIT IN HUMANS

Let's trace the movement of blood through the human circulatory system. Blood returning from the legs and arms enters the upper right chamber of the heart, called the ***right atrium*** (Figs. 30.3, 30.4) and the relaxed ***right ventricle,*** the lower right chamber of the heart. The atrium then contracts, forcing extra blood through the tricuspid valve (see Fig. 30.9) into the right ventricle. This blood, having just returned to the heart from its circulation through the tissues, contains little oxygen and much carbon dioxide. The next step must be to obtain O_2 and dispose of CO_2.

Contraction of the right ventricle sends the blood through a valve (the pulmonary semilunar valve) into the ***pulmonary artery,*** which soon divides into two branches, one going to each lung. In the lungs the pulmonary arteries branch repeatedly, and each terminal branch connects with dense beds of capillaries in the walls of the alveoli. Gas exchange takes place as carbon dioxide is discharged from the blood into the air of the alveoli and oxygen is

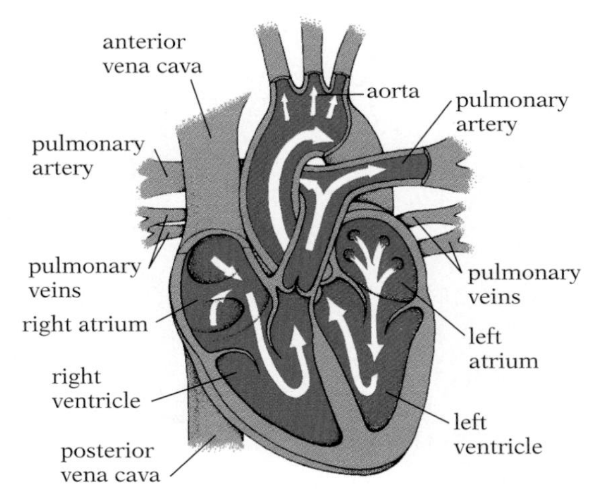

30.3 Human heart

The arrows show the direction of blood flow (oxygenated blood, red; deoxygenated blood, blue). Note that "left" and "right" refer to the sides of the organism's body; in this frontal view, the person's left as seen from the front is on the right, and vice versa.

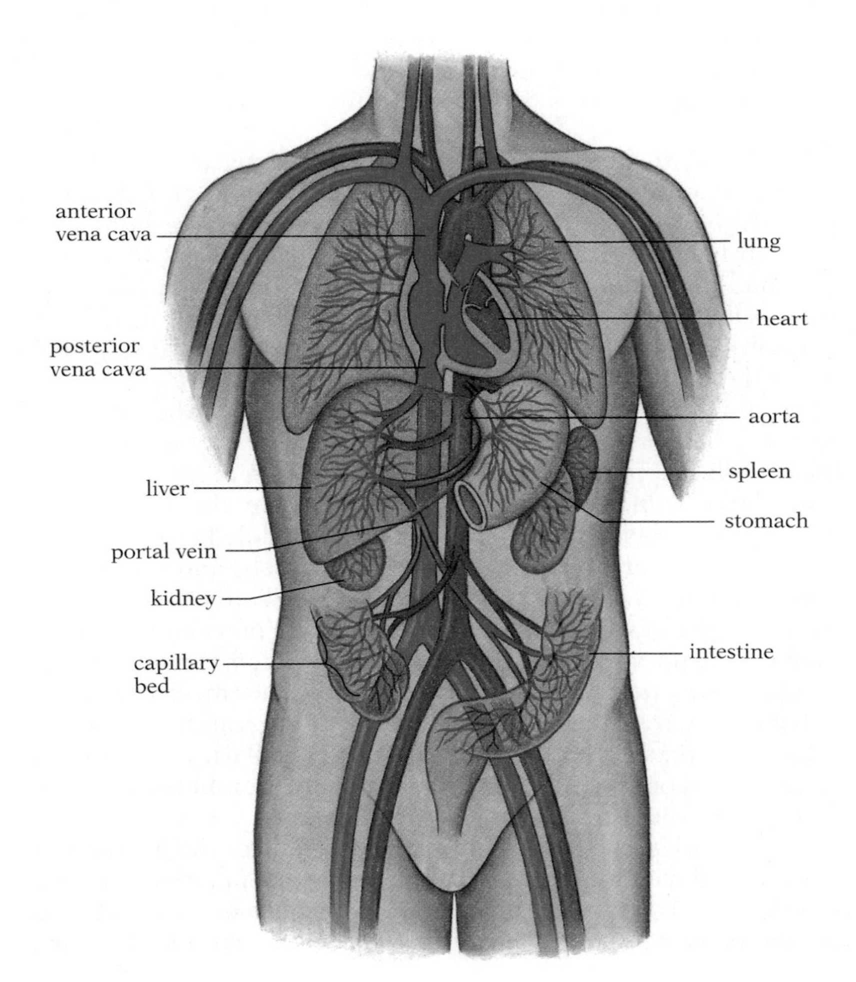

30.4 Human circulatory system

Red vessels contain oxygenated blood; blue vessels, deoxygenated blood. The transfer of oxygen and nutrients from the blood to tissues, as well as the transfer of carbon dioxide and other wastes from cells to the blood, takes place in each of the capillary beds. The capillaries, which are represented here as being relatively thick, are actually microscopic; also, only a very few of the vast number of arteries and the capillary beds they supply are shown.

picked up by the hemoglobin in the ***red blood corpuscles*** of the blood. (Though these bodies are often referred to as red blood cells or erythrocytes, the term *corpuscles* is the most accurate, since mammalian corpuscles lack nuclei and organelles. We will use all three terms, however.)

From the capillaries, the blood passes into venules, which join to form the large ***pulmonary veins*** that run back toward the heart from the lungs. The pulmonary veins empty into the ***left atrium***, the upper left chamber of the heart. When the left atrium contracts, it forces the blood through a valve (the bicuspid, or mitral, valve; see Fig. 30.9) into the ***left ventricle***, which is the lower left chamber of the heart. When the left ventricle contracts, it pushes the blood through a valve (the aortic semilunar valve) into a very large artery called the ***aorta***.

After the aorta emerges from the anterior portion of the heart (the upper portion, in humans standing erect), it arches over and runs posteriorly along the middorsal wall of the thorax and abdomen (Fig. 30.4). Numerous branch arteries arising from the aorta along its length carry blood to all parts of the body. For example, the first branch of the aorta is the coronary artery, which carries blood to the muscular wall of the heart itself. Each artery branches into smaller vessels, until eventually the smallest arterioles connect with the numerous tiny capillaries embedded in the tissues. Here oxygen, nutrients, hormones, and other substances move out of the blood into the tissues; waste products such as carbon dioxide and nitrogenous wastes are picked up by the blood, and substances to be transported, such as hormones secreted by the tissues, or nutrients from the intestine and liver, are also taken up.

The blood then runs from the capillary beds into tiny veins, which fuse to form larger and larger veins, until eventually one or more large veins exit from the organ or region in question. These veins, in turn, empty into one of two very large veins that empty into the right atrium of the heart: the ***anterior vena cava*** (sometimes called the superior vena cava), which drains the head, neck, and arms, and the ***posterior vena cava*** (inferior vena cava), which drains the rest of the body (Fig. 30.5).

Very little, if any, exchange of materials between the blood and the other tissues occurs across the walls of the arteries and veins. These walls are composed of three layers: (1) an outer, connective-tissue layer with numerous fibers, which give the vessels their characteristic elasticity and strength; (2) a middle layer of smooth muscle, which can change the size of the vessels; and (3) an inner layer of connective tissue lined with endothelial cells (Fig. 30.6). The two outer layers and the connective-tissue portion of the inner layer terminate at the ends of the arterioles and venules, leaving the capillaries with walls composed of only the one-cell-thick endothelium. Across these very thin walls of the capillaries the exchange of materials takes place. The capillaries provide an enormous amount of absorptive surface: the combined length of the capillary routes in humans is over 100 km.

To summarize, the blood enters the right side of the heart and is pumped to the lungs, where it picks up oxygen and gives up carbon dioxide; then it returns to the left side of the heart. This portion of the circulatory system is called ***pulmonary circulation*** (Fig.

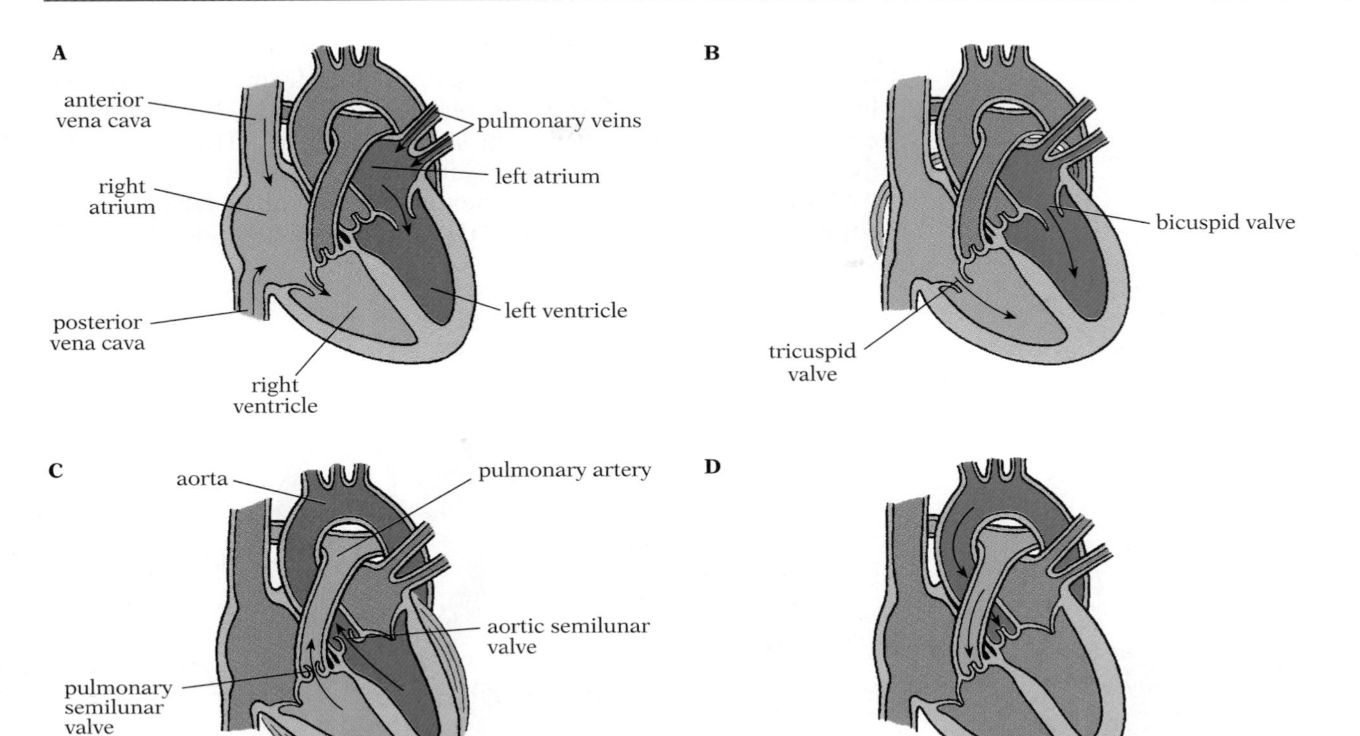

30.5 A summary of the events of the cardiac cycle

(A) Following diastole, the atria and ventricles are relaxed. Blood enters the right side of the heart from the two venae cavae and the left side from the pulmonary veins. Both the tricuspid and bicuspid valves are open, so blood flows from the atria into the ventricles. (B) Additional blood is forced into the ventricles when the atria contract. (C) The bicuspid and tricuspid valves close and the valves in the base of the pulmonary arteries and aorta open as the two ventricles contract simultaneously, forcing blood into the pulmonary arteries and the aorta. (D) The ventricles relax and the valves in the base of the pulmonary arteries and aorta close to prevent backflow.

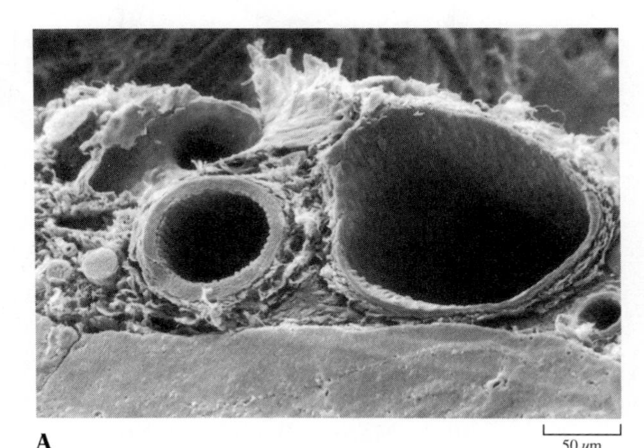

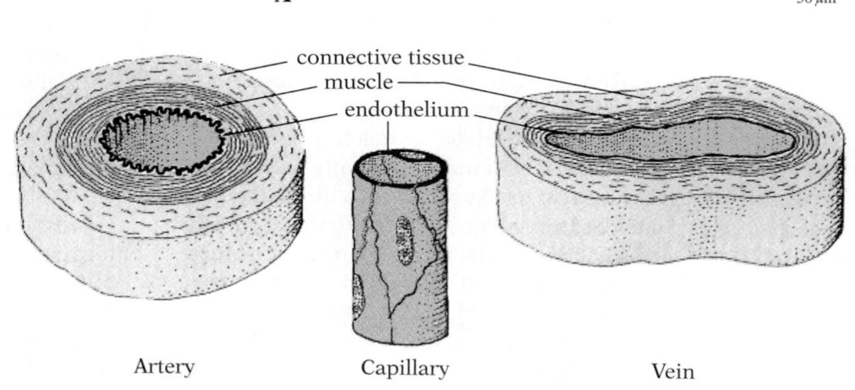

30.6 Walls of artery, vein, and capillary compared

Arteries and veins have the same three layers in their walls, but the walls of veins are much less rigid and they readily change shape when muscles press against them. Capillaries have walls composed only of a thin endothelium. In the scanning electron micrograph (A), a medium-size artery (left) and a medium-size vein (right) can be seen embedded in connective tissue surrounding a human vas deferens. Note that the vein has a thinner wall and a larger lumen than the artery. (From R. G. Kessel and R. H. Kardon, *Tissues and Organs*, W. H. Freeman: San Francisco, 1979.)

30.7A). From the left side of the heart, the blood is pumped into the aorta and its numerous arterial branches, from which it moves into capillaries, then into veins, and finally back in the anterior or posterior vena cava to the right side of the heart. This portion of the circulatory system is called ***systemic circulation***.

How can the developing tissues of a mammalian fetus be supplied with oxygenated blood in the absence of functioning lungs?

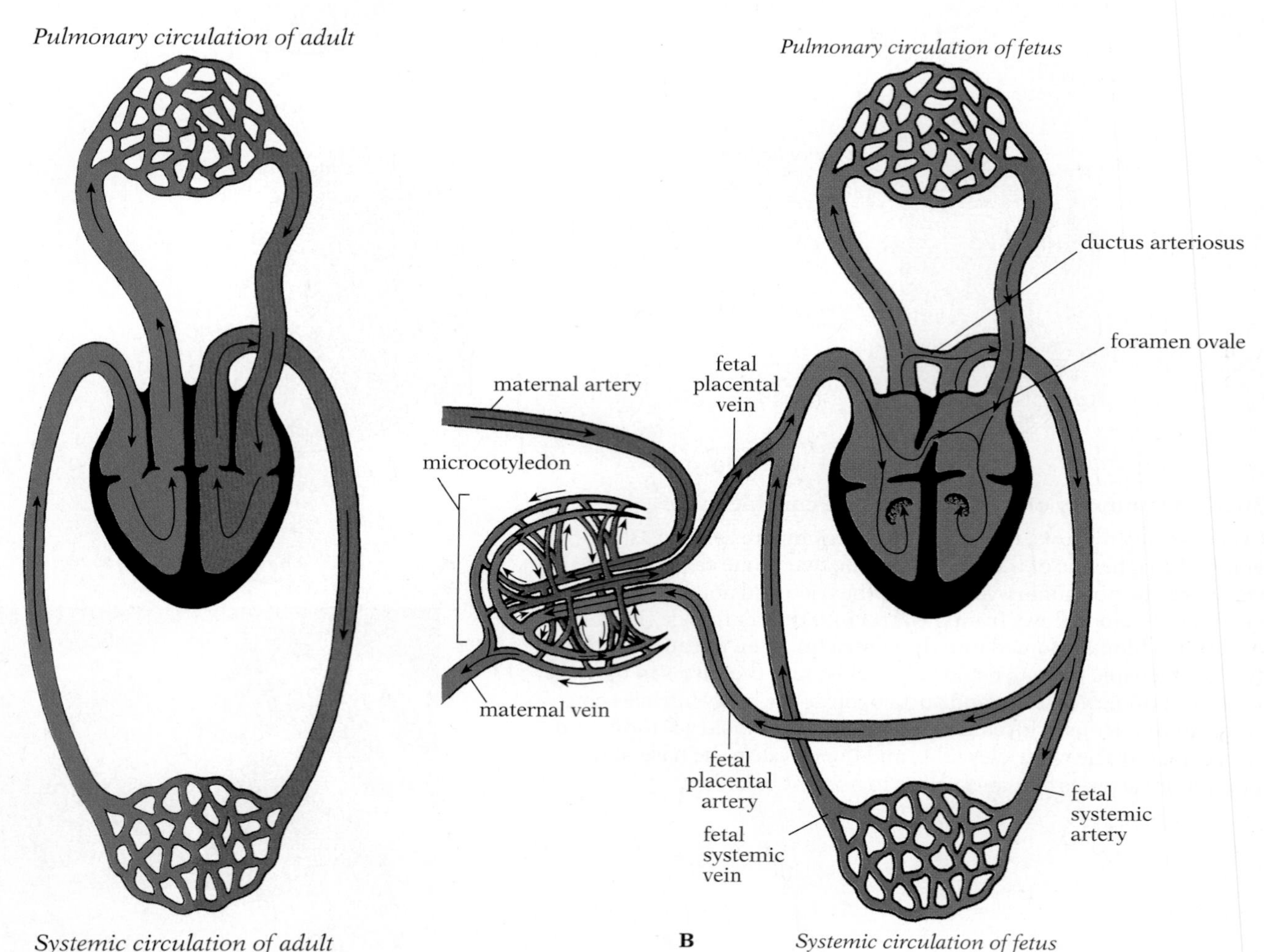

30.7 Mammalian circulation

Red vessels contain oxygenated blood; blue vessels, deoxygenated blood. (A) In an adult, blood passes from the right ventricle to the lungs, where it picks up oxygen. It then returns to the left side of the heart, from which it is pumped into the systemic circulation. Finally, having largely given up its oxygen to cells lining the capillary beds, the blood returns to the right side of the heart. (B) In a fetus, by contrast, the blood is oxygenated by a countercurrent system in the placenta, in which fetal blood passes through capillaries alongside those containing maternal blood. The transfer takes place in thousands of cuplike structures known as microcotyledons; this figure shows a diagrammatic version of a single cup. As discussed later in the chapter, a special sort of hemoglobin is needed to accomplish efficient gas exchange. The oxygenated fetal blood then joins the systemic veins leading to the heart, and forms a mixture (purple) with deoxygenated blood from the systemic circulation. Once the blood has reached the right atrium, the foramen ovale and the ductus arteriosus shunt most of it away from the nonfunctional lungs and into the systemic arteries; the other major pathway is to the placenta via the fetal placental artery.

Gas, nutrient, and waste transfers between the mother and the fetus take place in the placenta, a large adjoining structure that contains a specialized pair of capillary beds, one carrying fetal blood and the other carrying maternal blood (Fig. 30.7B). The umbilical cord brings part of the fetal blood supply to the fetal capillaries in the placenta, where CO_2 and other wastes diffuse into the maternal capillaries, while O_2 and nutrients move into the fetal system. Since very little of this freshly oxygenated blood is needed by the lungs, two alternate pathways, unique to fetuses, transfer most of the oxygenated blood to the systemic arteries: the first is the foramen ovale, a valve that joins the two atria; the second is the ductus arteriosus, which connects the pulmonary artery to the aorta.

At birth the placental exchange with the maternal capillaries comes to an end, and these two connections must be closed. This is accomplished within seconds of birth as a result of the expansion of the lungs with the first breath. As the capillary beds in the lungs of the newborn open out, blood suddenly begins to flow in large quantities for the first time through the pulmonary arteries and veins. As a result, the pressure in the right atrium abruptly drops, while the pressure in the left atrium rises sharply, and the flap between the two is forced shut. At the same time, the increased flow of blood into the pulmonary artery causes it to expand, and the now-useless ductus arteriosus contracts. Subsequently, both the duct and the flap are permanently sealed by new cell growth.

■ THE CIRCUIT IN OTHER VERTEBRATES

In effect, the human heart is two hearts in one, since blood in the left side of a normal heart is completely separated from blood in the right side. This type of heart—four-chambered, with complete separation of sides—is characteristic of mammals and birds, the two groups of endothermic (warm-blooded) vertebrates. These animals, which maintain a relatively constant high body temperature, have a high metabolic rate. Constant perfusion of the tissues with blood rich in oxygen is essential to them. Keeping the systemic and pulmonary circuits separate maximizes the oxygen level in blood entering the systemic system.

The first vertebrate hearts did not have separate systemic and pulmonary circuits. They had only one atrium and one ventricle, arranged linearly. Modern fish display an elaboration of this design (Fig. 30.8A). Blood aerated in the capillaries of the gills goes straight from the gills to the systemic circulation without first returning to the heart. This one-pump strategy has a drawback: because of the high resistance of the gill capillaries, blood leaving the gills to enter the systemic circulation is under relatively low pressure, and so moves through the systemic capillaries sluggishly. Since gills are 80–90% efficient at picking up oxygen from water, however, the sluggish flow is offset by the high concentration of oxygen in the blood.

This problem is overcome in air-breathing vertebrates by the addition of a second pump, between the gill/pulmonary and systemic circuits, to boost the pressure. In amphibians and reptiles, in contrast with birds and mammals, the two pumps are not fully sepa-

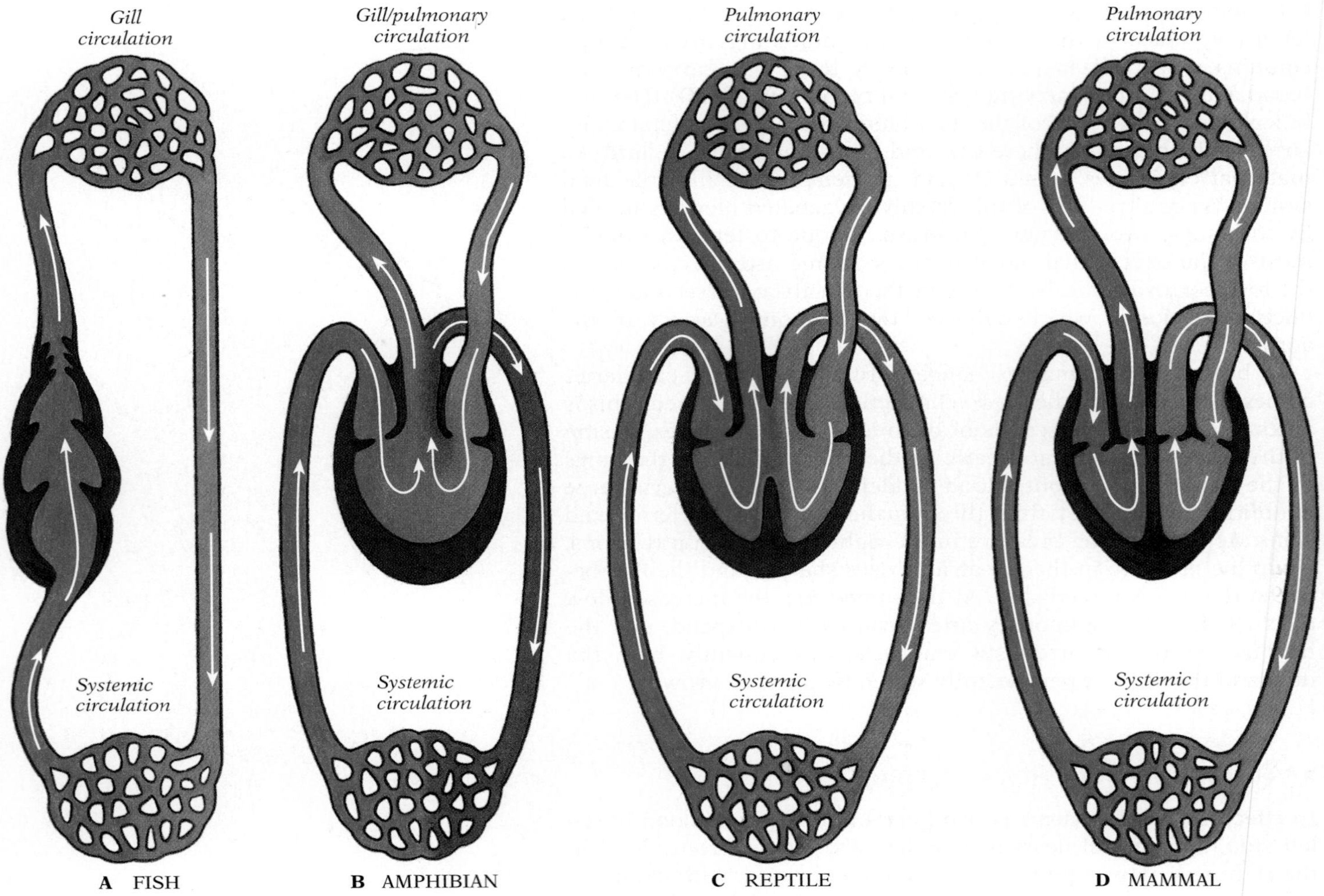

Gill circulation
Gill/pulmonary circulation
Pulmonary circulation
Pulmonary circulation

Systemic circulation
Systemic circulation
Systemic circulation
Systemic circulation

A FISH **B** AMPHIBIAN **C** REPTILE **D** MAMMAL

30.8 A comparison of vertebrate hearts

(A) In modern fish the blood is pumped through a linear, multichambered heart to the gills, where it picks up oxygen. The oxygenated blood (red) then passes without further pumping to the systemic circulation, where it gives up its oxygen before returning to the heart. (B) In the amphibians the blood that has picked up oxygen in the gills and/or lungs returns to the heart, from which it is pumped into the systemic circulation. Since the ventricle is not divided, some mixing of the pulmonary and systemic flows occurs in the heart. The colors used here represent the situation when the animal is out of the water. In most reptiles (C) the ventricles are partially divided, so less mixing takes place. (D) In mammals and birds the two halves of the heart are effectively separated.

rated: there are distinct right and left atria, but the ventricle is undivided in amphibians (Fig. 30.8B), and in reptiles the ventricles, though connected, are more distinct (Fig. 30.8C). Though mixing of oxygenated and deoxygenated blood is inevitable in amphibians, the amphibian heart is well adapted. Because most of these creatures spend much of their time in the water, the capillaries in the skin are able to perform gas exchange independently of the lungs. Muscular ridges in the ventricles of reptiles allow relatively little mixing to take place between pulmonary and systemic blood.

■ HOW THE HEART WORKS

Controlling the beat The two halves of the human heart beat essentially in unison. The beating is inherent in the heart itself; no stimulation from the central nervous system is necessary. In fact, vertebrate hearts continue to beat even after removal from the animal's body if placed in a solution with the proper ionic composition. Although the initiation of the beat and the beat itself are intrinsic properties of the heart, the rate of beat is partly regulated by stimulation from two sets of nerves; one set tends to accelerate the heartbeat and the other to decelerate it (see pp. 1022–1023).

The initiation of the heartbeat normally comes from the sinoatrial node, or ***S-A node***, the pacemaker of the heart; it is a small mass of ***nodal tissue*** on the wall of the right atrium. Nodal tissue is unique to the heart; it has the contractile properties of muscle and can transmit impulses like nerve fibers. A second mass of nodal tissue, the atrioventricular node, or ***A-V node***, is located in the septum (partition) between the right atrium and ventricle. A bundle of nodal-tissue fibers (the bundle of His) extends from the A-V node into the walls of the two ventricles, with branches (***Purkinje fibers***) in all parts of the ventricular musculature (Fig. 30.9).

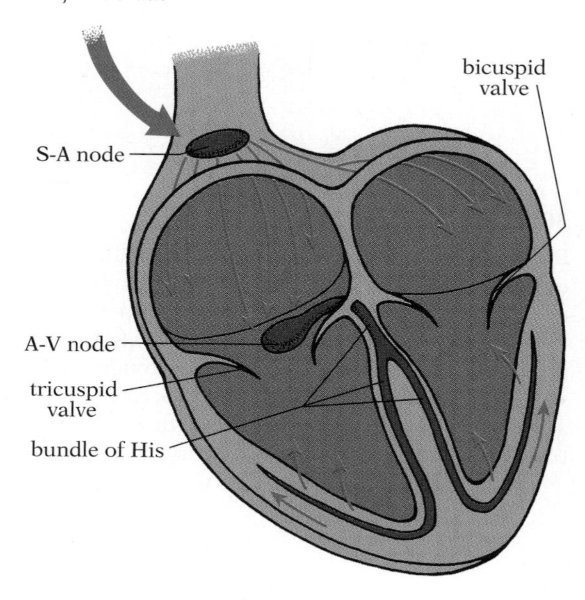

30.9 Electrical control of the heart

The pacemaker of the heart is the S-A node. Its spontaneous rate can be accelerated or decelerated by nerve impulses from the brain. Each time the S-A node fires, a signal spreads across the two atria, generating a beat and stimulating the A-V node. From there, after a brief delay, the signal is relayed rapidly down the fibers of the bundle of His and out through the network of Purkinje fibers (not shown) to generate a beat in the ventricles.

At regular intervals, a wave of excitation spreads from the S-A node across the walls of the atria. When this wave of excitation reaches the A-V node, the node is stimulated to send excitatory impulses rapidly through fibers of the bundle of His to all parts of the ventricles. These impulses stimulate the ventricles to contract.

The heart rate—the number of systoles (contractions) per minute—is inversely related to body size and thus varies from species to species. In the Asiatic elephant, a normal rate is 30 beats/min; in the tiny masked shrew the average is 780 beats/min; the rate for a normal human at rest is roughly 70 beats/minute.

In the course of the beat, the heart emits several characteristic sounds that can be heard through a stethoscope placed against the chest. First, there is a long, low-pitched sound produced by the closing of the valves between the atria and the ventricles and by the contraction of the ventricles. Then there is a shorter, louder, higher-pitched sound produced by the closing of the valves between the ventricles and the arteries leading from them. Changes in these sounds can indicate that the heart is defective. A normal heart valve opens when the pressure behind it is greater than the pressure in front of it. For example, when the atria start contracting, they put pressure on the blood they contain, and as soon as this pressure is greater than the pressure in the ventricles, the tricuspid and bicuspid valves are forced open and the blood can flow into the ventricles (Fig. 30.9). If a valve has been damaged and cannot shut completely, a hissing or murmuring sound can be heard as blood leaks backward through the damaged valve; this condition is called a diastolic heart murmur. Sometimes a damaged valve partly obstructs blood flow during systole, and the resulting sound (due to the turbulence in the blood flow) is called a systolic murmur. The more extensive the damage to the valve, the less efficient the heart action and the greater the strain placed on the heart. Sometimes a valve is so damaged that it must be replaced with an artificial valve (Fig. 30.10).

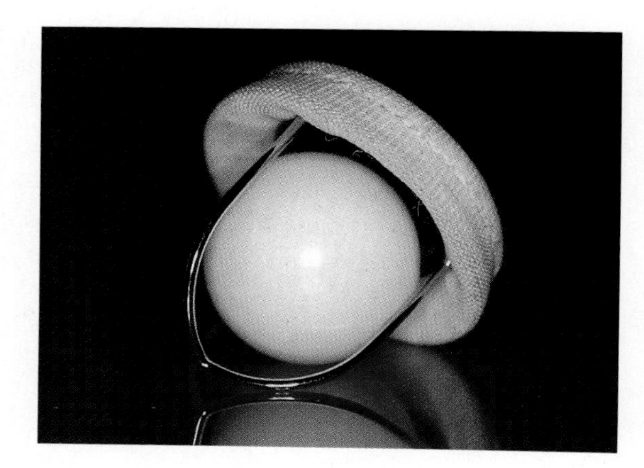

30.10 An artificial heart valve

Severely damaged heart valves can be replaced with artificial ones such as this ball-and-cage valve. When the heart contracts, the blood can pass by forcing the ball against the cage. When the heart relaxes, the blood starts to flow back and forces the ball against the ring, closing the valve.

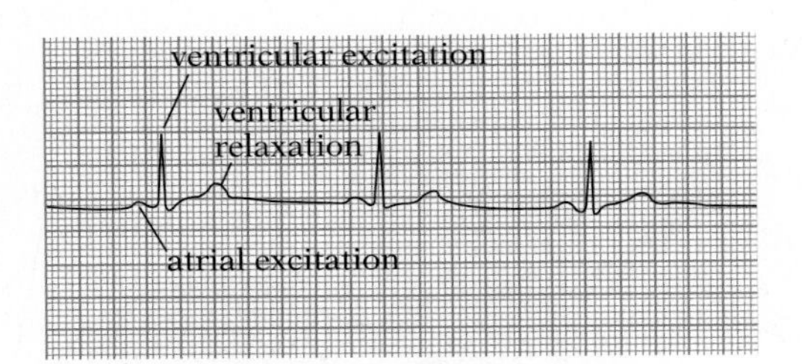

30.11 An electrocardiogram

By recording the electrical activity of the heart under various conditions of rest and exercise, electrocardiograms (EKGs) can sometimes provide important information about potential heart problems. Notice the regular order, strength, and spacing of events in this normal EKG. Since atrial relaxation usually takes place during ventricular excitation, it is not visible in this record.

In the course of contraction, the heart muscle undergoes a series of electrical changes. These changes can be detected by electrodes attached to the skin and can be graphed by an electrocardiograph (Fig. 30.11). Abnormalities in heart action alter the pattern of the resulting electrocardiogram.

The heart of a resting human adult pumps about 5 liters of blood every minute, which is approximately equal to the total amount of blood in the body. This does not mean, of course, that each individual drop of blood passes through the heart every minute; blood that happens to flow into one of the shorter circuits, such as those supplying the neck or chest, may return to the heart quickly and make several rounds in a minute, while blood going to more distant parts of the body, such as the legs, may take several minutes to return to the heart of a resting person. During exercise both the rate of contraction and the amount of blood pumped per beat (the stroke volume) increase greatly. The combination of elevated heart rate and increased stroke volume may raise cardiac output (total amount of blood pumped per minute) to a level 4 to 7 times the resting level.

Controlling the flow When the left ventricle contracts, it forces blood under high pressure into the aorta, and blood surges forward in each of the arteries. The walls of the arteries are elastic, and the pulse wave stretches them. During diastole, the relaxation phase of the heart cycle, the heart is not exerting pressure on the blood in the arteries, but elastic recoil of the previously stretched artery walls maintains some pressure on the blood. There is thus a regular cycle of pressure in the larger arteries that reaches its high point during systole and its low point during diastole.

In humans arterial blood pressure in the systemic circuit is usually measured in the upper arm, where systolic values of about 120 mm Hg and diastolic values of about 80 mm are normal in young adult males at rest. Blood pressure decreases continuously as the blood moves farther away from the heart, and falls even more rapidly in the arterioles and capillaries. It continues to decline, though more slowly, in the veins, reaching its lowest point in those nearest the heart (Fig. 30.12). The decline of the blood pressure in successive parts of the circuit is the result of friction between the flowing blood and the walls of the vessels. Such a gradient of pressure is essential, of course, if the blood is to continue to flow; the fluid can move only from a region of higher pressure toward a region of lower pressure. The problem with the original, linear vertebrate heart was the extreme drop in pressure across the capillaries in the gills with another set of high-resistance capillaries in the body tissues still to come. For such a system to work, the heart had to be quite powerful, and the capillaries had to be relatively wide (to provide less resistance) and hence less efficient for the exchange of gases, nutrients, and wastes. These constraints continue to impose an upper limit on metabolic activity in animals with linear hearts.

Several other changes occur along the route of human blood flow. First, with increased distance from the heart, the difference between systolic and diastolic pressures diminishes, because the elasticity of the artery walls tends to damp out the fluctuations in

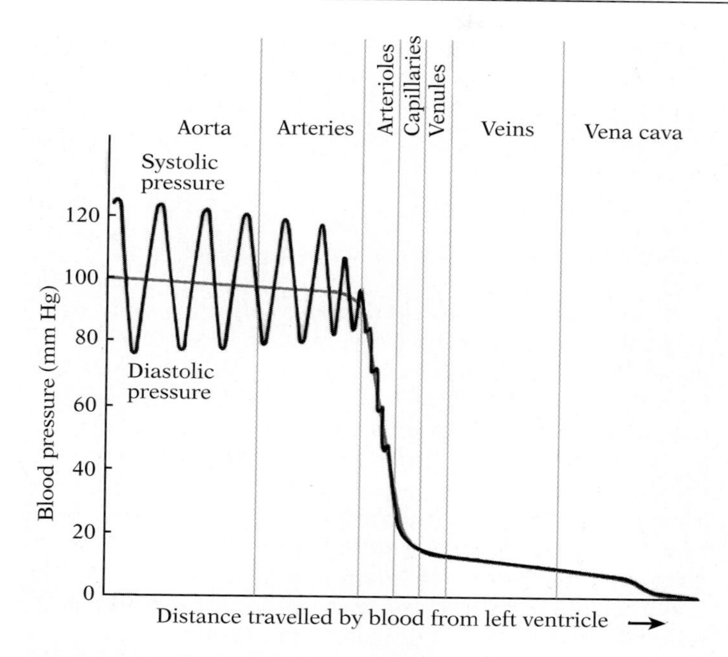

30.12 Blood pressure in different parts of the human circulatory system

The red curve traces the mean pressure values. As shown by the black curve, there is considerable fluctuation between the systolic and the diastolic pressure in the arteries. This fluctuation diminishes in the arterioles and no longer occurs in the capillaries and veins, because the elasticity of the vessel walls has effectively damped out the oscillations. The most rapid fall in pressure is in the arterioles and capillaries.

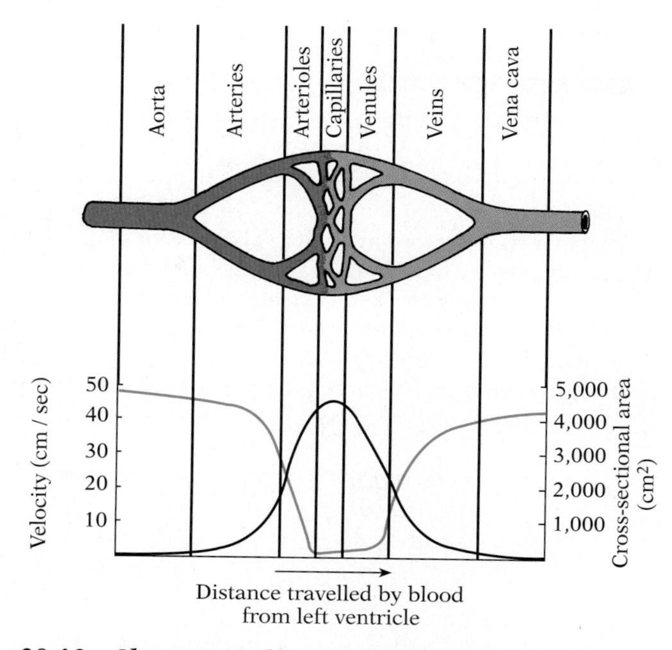

30.13 Changes in the velocity of blood flow in a systemic circulatory pathway

As the color curve shows, the velocity falls precipitously as the blood flows through the arterioles, where the total cross-sectional area (black curve) rises. The velocity remains low in the capillaries and venules, but rises again in the veins as the total cross-sectional area falls.

blood pressure. This means that the cyclic, surging type of flow characteristic in the arteries is replaced by a constant rate of flow by the time the blood reaches the capillaries and veins. Second, the linear velocity tends to fall as the blood moves through the branching arteries and arterioles; the velocity is lowest in the capillaries and increases again in the venules and veins. The reason is that the volume of blood pumped by the heart with each contraction is pushed quickly through the relatively narrow aorta, but the same volume is moved at a much more leisurely pace through the greater *total* cross-sectional area of the capillaries. As the capillaries unite to form venules and these join to form veins, the total cross-sectional area diminishes again, and the velocity necessarily increases (Fig. 30.13).

By the time the blood reaches the veins, the hydrostatic pressure is too low to move the blood efficiently. The walls of veins are easily collapsible; nearby muscles, which contract as the body moves, put pressure on the veins, compressing their walls and forcing the fluid in them forward. The fluid can move only toward the heart, because numerous one-way valves in the veins prevent it from flowing back (Fig. 30.14). When you stand still for a long period, your feet begin to swell because the muscle action in your legs is not sufficient to push the blood upward against gravity.

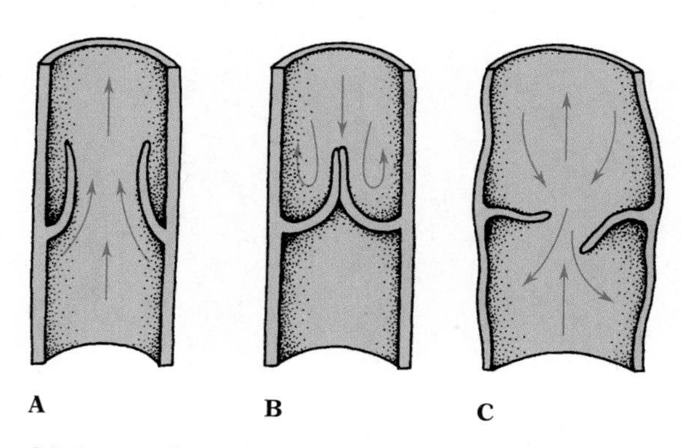

30.14 Valves in the veins

(A) When blood is flowing toward the heart, it forces the flaps open and moves through. (B) When the pressure declines and the blood is pulled down by gravity, the flaps automatically act as a check valve, closing to prevent a backward flow. (C) Damaged valves fail to prevent this backflow; the result is varicose veins, which are especially common in the legs.

EXPLORING FURTHER

THROMBOEMBOLIC AND HYPERTENSIVE DISEASE IN HUMANS

Thrombosis is the formation of a solid mass or plug of blood constituents in a blood vessel. The mass, or ***thrombus***, may wholly or partially block the vessel in which it forms (see Fig. A), or it may become dislodged and be carried to some other location in the circulatory system, in which case it is called a ***thromboembolus***. (An embolus is any large object moving through the circulatory system.) Thromboembolisms are a leading cause of serious illness and death in Western societies.

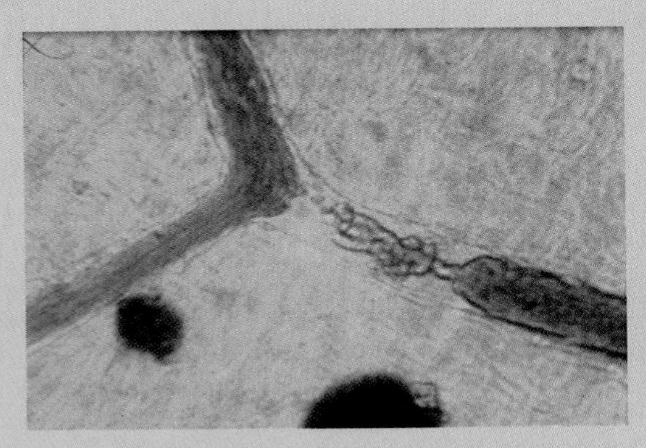

A A thrombus in a small blood vessel

The thrombus (tangled red mass) has blocked blood flow near a point where the vessel branches. The blood has pulled away from the left end of the thrombus and is beginning to pull away from the right end also.

Many factors can encourage thrombus formation, including irritation or infection of the lining of a blood vessel, or a reduced rate of blood flow through a vessel, which may result from disease or merely from long periods of inactivity. For example, formation of a thrombus in a vein, especially of the leg (a condition known as thrombophlebitis) is particularly common in postoperative patients who remain immobilized in bed; it is also common in the elderly, whose leg muscles have lost much of their pumping action. The thrombus often leads to an inflammatory reaction and pain.

A thrombus that becomes detached from its site of formation and moves in the bloodstream as an embolus is extremely dangerous, because it may become lodged in a vessel of some essential organ such as a lung, the heart, or the brain, and cut off its blood supply. Such emboli often lodge in the lung (pulmonary embolism) and cause death (necrosis) of a portion of the lung tissue.

When either an embolus or a locally formed thrombus blocks a blood vessel in the brain and causes necrosis of the surrounding neural tissue from lack of oxygen, the condition is known as a ***stroke***, or ***cerebral infarction***. The symptoms of stroke vary depending on the part of the brain that has been damaged. There can be some loss of muscular control in part of the body, sensory impairment, or some loss of language ability.

Contracting and relaxing the leg muscles minimizes this problem. To prevent pooling of blood in the lower extremities, hospitalized patients are encouraged to begin walking as soon as they can, often only a day or two after surgery; pooling of blood may lead to formation of a potentially fatal clot in a leg vein (a condition called thrombophlebitis).

The motions of the chest during breathing also aid in moving blood in the veins. When the chest expands during inhalation, the pressure in the thorax falls. Thus there is a pressure gradient from other parts of the body to the thorax, and blood tends to be drawn into the large vessels of the thorax and into the heart. In addition, the ventricles exert some degree of suction during diastole.

■ HOW THE CAPILLARIES WORK

No cell is far removed from at least one capillary: muscle tissue contains as many as 60,000 capillaries per square centimeter of

The structure of the circulatory system helps guard against damage from any small embolus becoming trapped as a thrombus in a capillary bed. Most parts of the body are reached by capillaries from two or more arterioles—a strategy known as collateral circulation. Hence, to cause damage an embolus must be of sufficient size to block the circulation upstream, in the relatively large blood vessel that has not yet branched into alternative pathways, or separate small emboli must block each of the branches independently. However, some parts of the body lack effective collateral circulation; these include the retina and, unfortunately, the heart.

Blockage of a blood vessel in the heart by an embolus (or by a locally formed thrombus) causes necrosis of a portion of the heart muscle, a condition known as ***myocardial infarction***, or heart attack. A high percentage of the deaths that occur in the first few hours after a heart attack result from disruptions of the control system of the heart, with accompanying arrhythmias, especially ventricular fibrillation.

In the vast majority of cases of cerebral or myocardial infarction, the patient already suffers from ***atherosclerosis***, a condition in which fatty deposits in the arteries and thickening of their walls diminish the size of the lumen (Fig. B); both the reduced size and the consequent reduced blood flow make it easier for an embolus to become lodged in the vessel. Directly or indirectly, atherosclerosis is responsible for half of the deaths in Western countries. Medical authorities recommend minimizing cholesterol and saturated fat in the diet to avoid atherosclerosis. The most reliable indicator of risk is the circulating concentration of LDL (low-density lipoprotein, a cholesterol-transport complex discussed in Chapter 4). A less common cholesterol complex, high-density lipoprotein (HDL), seems to aid in removing cholesterol deposits. Another common precursor to heart attacks is damage to the lining of the arteries from prolonged high blood pressure (hypertension). Hypertension can also lead eventually to weakening of the heart muscle (which thickens from the continuing strain imposed on it) and to declining efficiency of its pumping action. Blood may then back up in the heart and lungs, an often fatal condition called congestive heart failure.

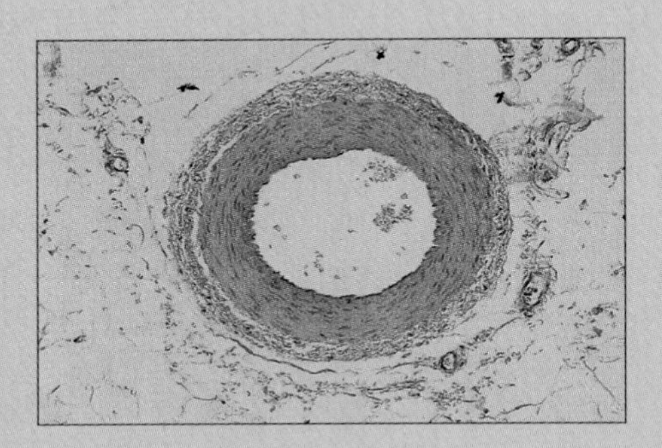

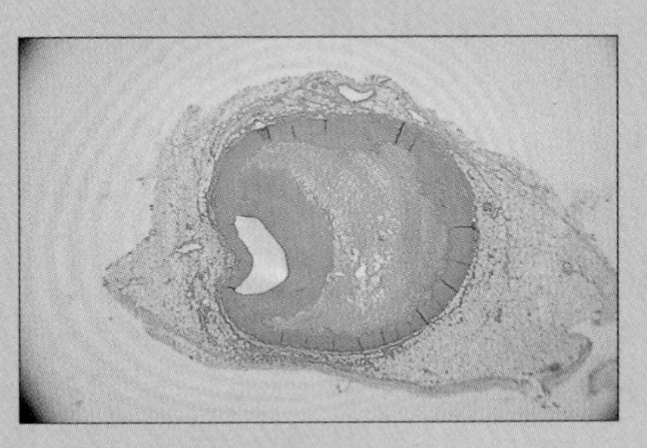

B Atherosclerosis

Top: Normal artery. Bottom: An artery clogged with fatty deposits in the wall.

cross section. The diameters of the capillaries are seldom much larger than those of the blood corpuscles that must pass through them (Fig. 30.15). The extensive branching and the small diameters of individual capillaries ensure that all portions of the tissues will be supplied with a large capillary surface area through which nutrient, gas, and waste exchange can take place. Because the branching causes the blood to flow more slowly in the capillaries (Fig. 30.13), there is more time for exchange. The considerable drop in blood pressure in the capillary bed also plays an important role, as we will see.

Exchange of materials between the blood in the capillaries and the tissue fluids outside the capillaries can occur in at least three ways: (1) The materials may move by diffusion through the membrane of an endothelial cell in the wall of the capillary, across the cytoplasm of that cell, and out through the cell membrane on the other side. (2) Vesicles in the endothelial cells of capillaries can pick up materials by endocytosis on one side of the cell, move

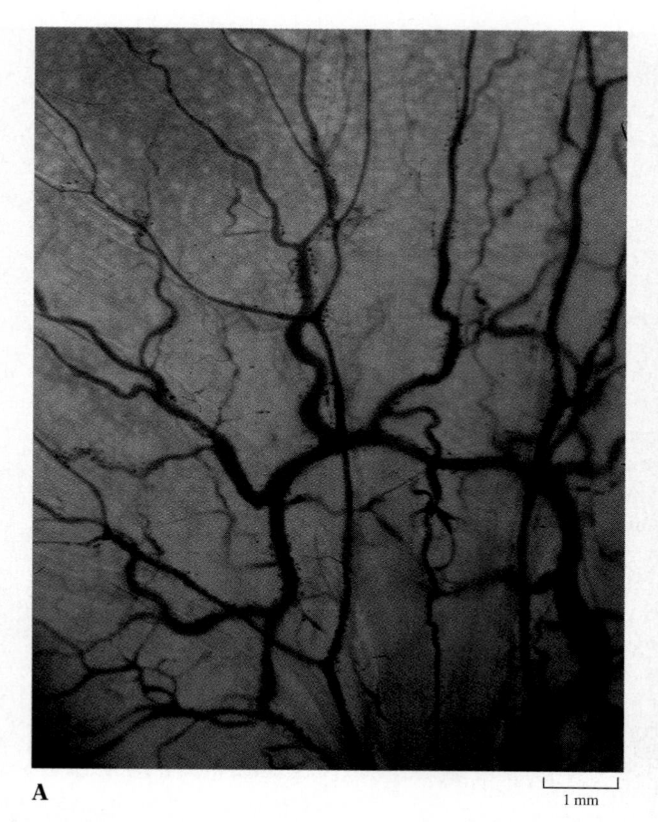

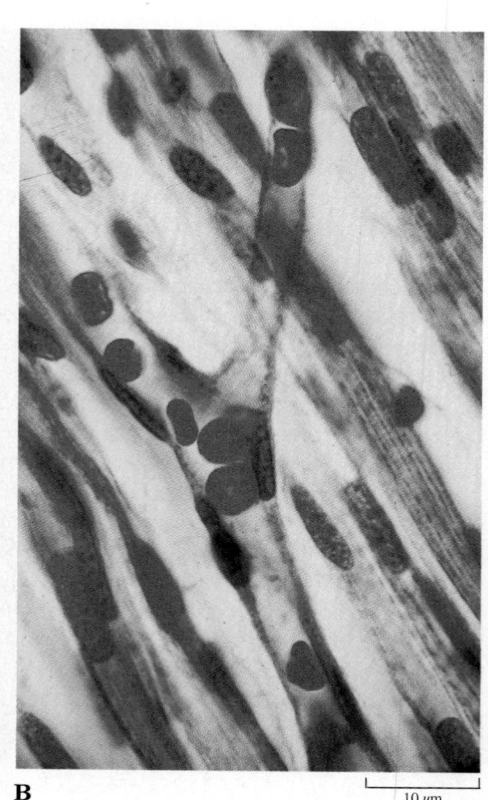

30.15 Blood vessels and capillaries

(A) The vessels in the web of a living frog's foot branch repeatedly. The smallest capillaries (seen here as very thin, faintly brown lines) reach all cells. (B) In this longitudinal section of a human capillary, individual erythrocytes are seen moving single file.

across the cell, and expel the materials by exocytosis on the other side (Fig. 30.16). (3) Clefts between adjacent endothelial cells in the capillaries in most parts of the body (the central nervous system being an exception) are wide enough to permit filtration of water and most dissolved molecules, but not proteins (Fig. 30.16).

The third mechanism is especially important for water, though most other small molecules leave the capillaries by diffusion. At the arteriole end of most capillary beds, the hydrostatic blood pressure is about 36 mm Hg higher than the hydrostatic pressure of the tissue fluid outside the capillaries (Fig. 30.17). The pressure differential falls to about 15 by the time the blood reaches the venule end of the capillary bed. The hydrostatic blood pressure tends to force materials out of the capillaries into the surrounding tissue fluid. If this were the only force involved, there would be a steady loss from the blood by filtration of both water and small dissolved substances. Normally, however, there is relatively little net loss of water from the blood in the capillaries. Clearly, a second force must be involved.

This other force derives from the difference in osmotic concentration between the blood and tissue fluid. The blood of mammals contains a relatively high concentration of proteins, and these large molecules cannot easily pass through the capillary walls. There is a much lower concentration of these proteins in tissue fluids. Thus the blood and tissue fluids have different osmotic pressures. Normally, the osmotic pressure of the blood is about 25 mm Hg higher than that of the tissue fluid, with the result that water

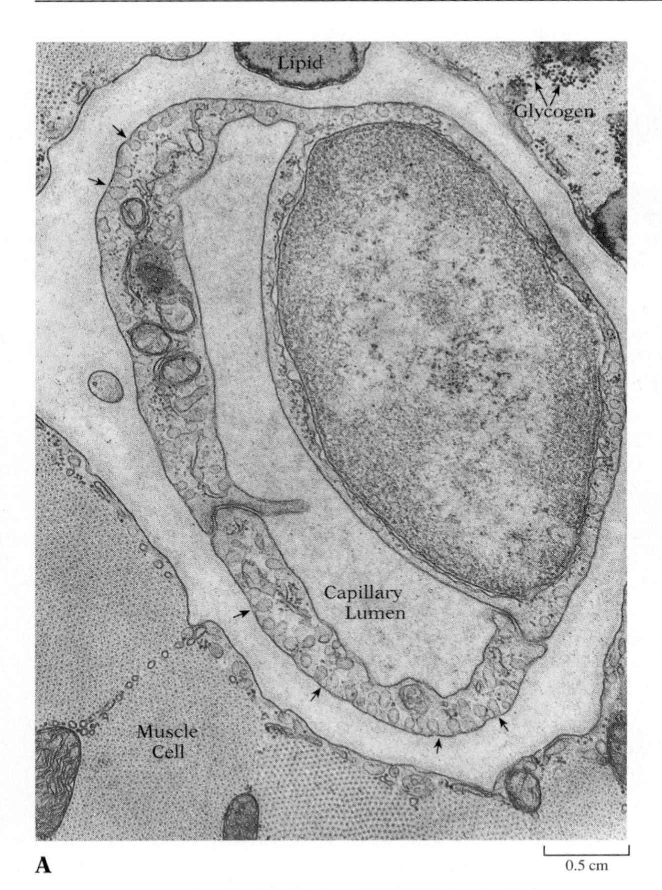

B

0.1 μm

30.16 Electron micrographs of cross section of capillary

(A) Two endothelial cells (the section shows the large nucleus of one of them) make up the capillary wall. Note the clefts at each of the two junctions between the cells and the numerous pinocytotic vesicles (arrows) in the cytoplasm. These vesicles may transport materials from outside the capillary, across the endothelial cell, into the lumen of the capillary, or they may take the reverse route. (B) Enlarged view of wall of a capillary, showing the cleft where two endothelial cells join (arrow), as well as numerous pinocytotic vesicles opening on the outer face of the lower cell.

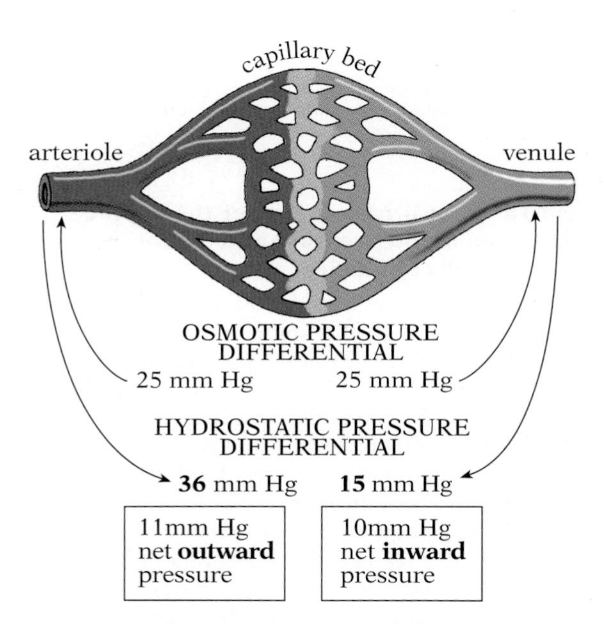

30.17 Forces involved in the filtration of materials across capillary walls

The blood in the capillaries has both a greater hydrostatic and a greater osmotic pressure than the surrounding tissue fluid. The hydrostatic pressure differential tends to force water and dissolved materials out of the capillaries into the tissue fluid; the osmotic pressure differential has the opposite effect, causing the capillaries to take up water and dissolved materials from the tissue fluid. At the arteriole end of a characteristic capillary bed, the hydrostatic pressure differential of 36 mm Hg is greater than the osmotic pressure differential of 25 mm; the difference of 11 mm favors the outflow of materials. At the venule end, the osmotic pressure differential, which remains 25 mm, is greater than the hydrostatic pressure differential, which has dropped to 15 mm; here the difference of 10 mm favors the pickup of materials from the tissue fluid.

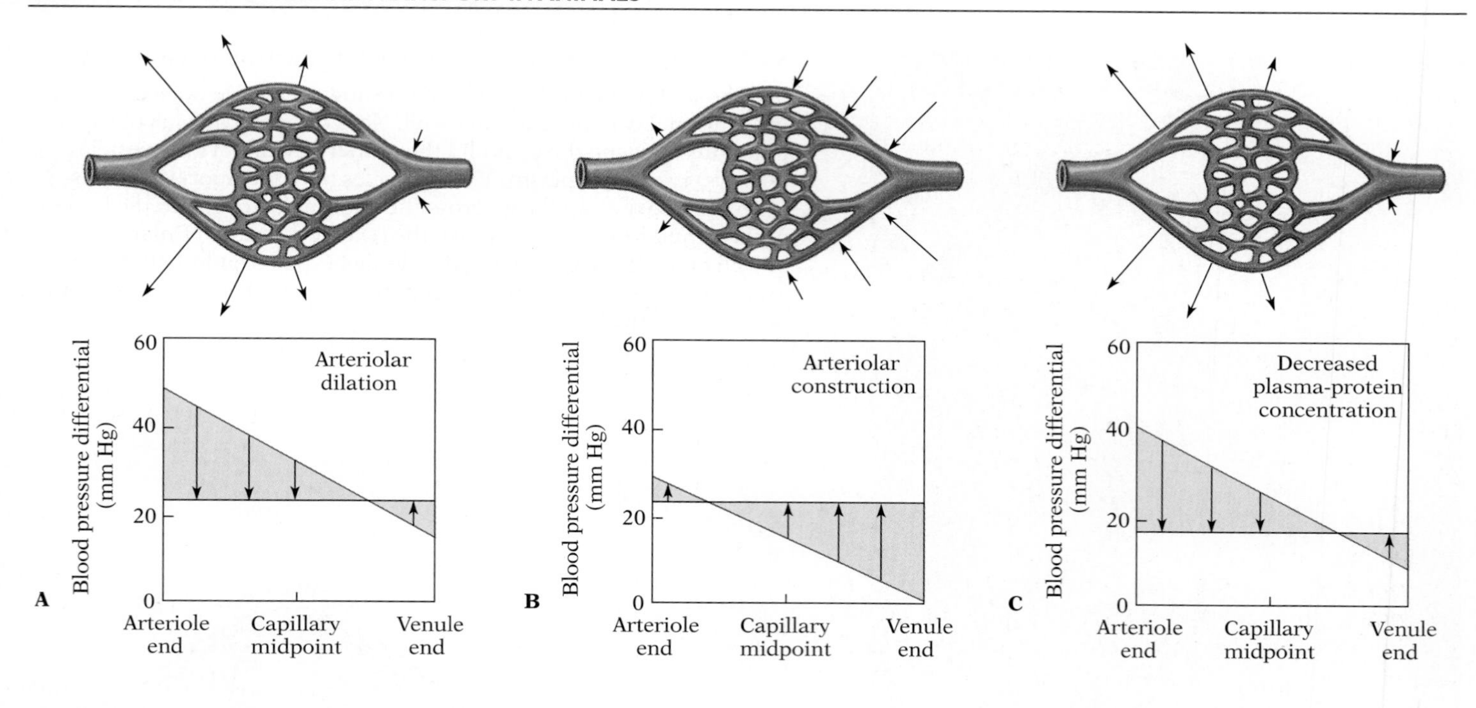

30.18 Conditions that alter the balance between blood pressure and osmotic pressure in the capillaries

The hydrostatic pressure differential (the difference between the hydrostatic pressure of the blood and that of the fluid in the surrounding tissue) is represented by the diagonal green line; the osmotic pressure differential is represented by the horizontal black line. (A) When the arterioles are dilated and much blood is flowing into the capillary bed, the pressure differential is much greater at the arteriole end of the capillary. Hence flow of water out of the capillary (left color area) greatly exceeds reabsorption near the venule end (right color area). (B) When the arterioles are constricted and little blood is entering the capillary bed, the hydrostatic blood pressure in the capillary is low; hence reabsorption of water into the blood greatly exceeds outward filtration. (C) When the plasma-protein concentration is low, the osmotic pressure of the blood falls (in this case lowering the differential—horizontal line—to below 20 mm Hg); hence the balance is shifted toward outward filtration of water.

tends to move into the capillaries from the tissue fluid by osmosis.

Thus, while hydrostatic pressure forces water out of the capillaries, osmotic pressure forces water into the capillaries. The net movement of water depends on the relative magnitudes of these two opposing forces. At the arteriole end of many capillary beds the hydrostatic differential is 36 and the osmotic differential is 25 (Fig. 30.17). Thus there is a net pressure of 11 tending to force water out. At the venule end of the bed, the hydrostatic differential has fallen to 15, while the osmotic differential has not changed greatly.[1] Therefore, there is now a net pressure of at least 10 tending to force water into the capillaries. Thus water is forced out of the capillaries at the arteriole end and into the capillaries at the venule end. The net effect is that about 99% of the water filtered out at the arteriole end is reabsorbed at the venule end. The fluid leaving the capillaries at the arteriole end carries nutrients and oxygen, which it exchanges in the tissues for wastes that it then carries back into the capillaries at the venule end. But because the total volume of liquid taking this extracellular route through the capillary bed is much smaller than the volume passing through the capillaries themselves, most gas and small-molecule exchange occurs across capillary walls.

The balance of hydrostatic and osmotic pressures in the capillaries is very delicate. Since it plays such an important role in the exchange of materials between the blood and the tissue fluid, any disturbance to it may have profound effects on the condition of the organism. For example, an increase in blood pressure in a capillary would tend to increase loss of fluid from the blood, while a decrease in blood pressure would have the opposite effect (Fig. 30.18A–B). Such changes in blood pressure can be produced by a

[1] Loss of water from the blood, of course, slightly increases the concentration of protein in the blood and raises the osmotic pressure accordingly, but this change is slight.

change in rate or strength of heart action, an increase or decrease in total blood volume, a change in the elasticity of the walls of the arteries, or any increased dilation or constriction of arterioles and capillary sphincters. (A capillary sphincter is a ring of muscle found at the base of most capillaries.)

The degree of dilation or constriction of capillary sphincters is also important in determining how much blood flows to any given tissue at a given time. Blood normally never circulates through all the capillaries of the body at the same time; there is simply not enough blood volume to go around, and when, as a result of trauma, the body attempts this redistribution, blood pressure falls drastically and shock results. Circulation is therefore selective, as blood is shunted preferentially to the regions of the body with the greatest current need. Constriction of a sphincter muscle can close off local circulation. In a resting muscle, for example, only certain capillaries—called thoroughfare channels, or metarterioles—are generally open; but once the muscle becomes active, and its need for oxygen and nutrients increases, the numerous precapillary sphincters in the muscle open, allowing blood to flow through the previously unused capillaries, and the local blood supply is greatly increased (Fig. 30.19). Similarly, flow through the capillaries in the wall of the intestine is greatly increased following a large meal. Increased flow through skin capillaries often gives the skin a reddish hue, seen in blushing, while constriction of the sphincters controlling these same capillaries gives the skin a bleached, whitish look. Large amounts of heat are lost from the body by radiation from the superficial capillaries of the skin (Fig. 30.20); as we will see, changes in the amount of blood flow in these capillaries help regulate heat loss.

When blood pressure falls because of vasodilation (extensive dilation of the vessels, producing a condition known as vascular shock) or because of loss of blood by hemorrhage, the consequent increased absorption of tissue fluid into the capillaries increases

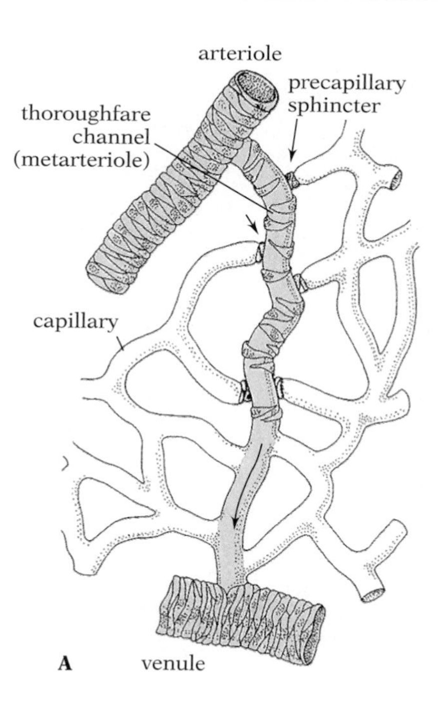

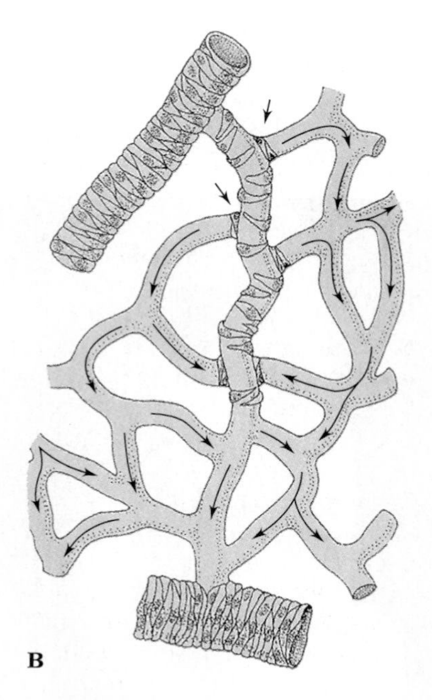

30.19 Vessels of a capillary bed

When the precapillary sphincters are closed (A), blood flows only through the capillaries known as thoroughfare channels. When the sphincters are open (B), blood flows through all the capillaries.

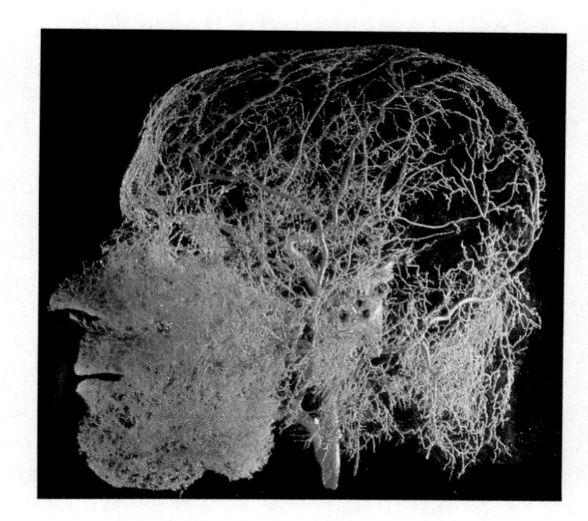

30.20 Blood vascular network in the human head

The human face, particularly the area around the lips, contains a dense array of capillaries.

the total blood volume and compensates partially for the deficiency. At such times, the supply of circulating blood is also augmented by reserve blood stored in the ***spleen***. Contractions of the smooth muscles in the walls of the spleen can expel this blood into the general circulation.

Changes in the relative concentration of proteins in the blood and in the tissue fluid can also alter the balance of forces operating in the capillaries (Fig. 30.18C). Increasing the protein concentration of the blood decreases loss of fluid from the blood and increases absorption of tissue fluid. Conversely, decreasing the protein concentration increases loss of fluid from the blood and decreases reabsorption of fluid from the tissues. The result is an abnormal accumulation of fluid in the tissues that causes swelling, a condition known as edema.

■ WHAT DOES THE LYMPH DO?

Fully 99% of the water leaving the capillaries at the arteriole end is normally reabsorbed at the venule end; what about the remaining 1%? Vertebrates have a special system of lymph vessels that return materials from the tissues to the blood. The lymph capillaries, which like the blood capillaries are distributed throughout most of the body, are closed at one end. Their walls are composed of a single layer of endothelium. Tissue fluid is absorbed into the lymph capillaries (at which point it is called ***lymph***) and slowly flows into small lymph veins, which unite to form larger and larger veins until finally two very large lymph ducts empty into the blood in the upper portion of the thorax near the heart. There are no lymph arteries.

30.21 Elephantiasis

Elephantiasis is a condition of extreme edema that occurs when lymph vessels become blocked by filarial worms. Here the left leg is swollen with the fluids accumulated in the tissues as a result of the blockage.

Though the walls of the lymph capillaries are structurally similar to those of blood capillaries, their permeability is different. The small amounts of protein lost from the blood capillaries move easily across the walls of the lymph capillaries. Returning these proteins to the blood is critical if normal osmotic balance is to be maintained. If a major lymph vessel becomes blocked, the protein concentration in the tissue fluid rises steadily, and the difference in osmotic concentration between it and the blood steadily diminishes. This means that less and less fluid is reabsorbed by the blood capillaries until the result is severe edema (Fig. 30.21).

Nodes are present in the lymphatic systems of mammals and some birds. The nodes are composed of a meshwork of connective tissue harboring many phagocytic cells; they act as filters and are sites where immune-system cells monitor the passing fluid for signs of infection. As the lymph trickles through the nodes, dead cells, cell fragments, and invading bacteria are engulfed. Nondigestible particles such as dust and soot, which phagocytic cells cannot destroy, are stored in the nodes. Since the nodes are particularly active during an infection, they often become swollen and sore, as the lymph nodes at the base of the jaw are apt to do during a throat infection.

Lymph is not moved by pressure generated by the heart. In mammals its movement, like that of blood in the veins, results from the contractions of skeletal muscles that press on the lymph

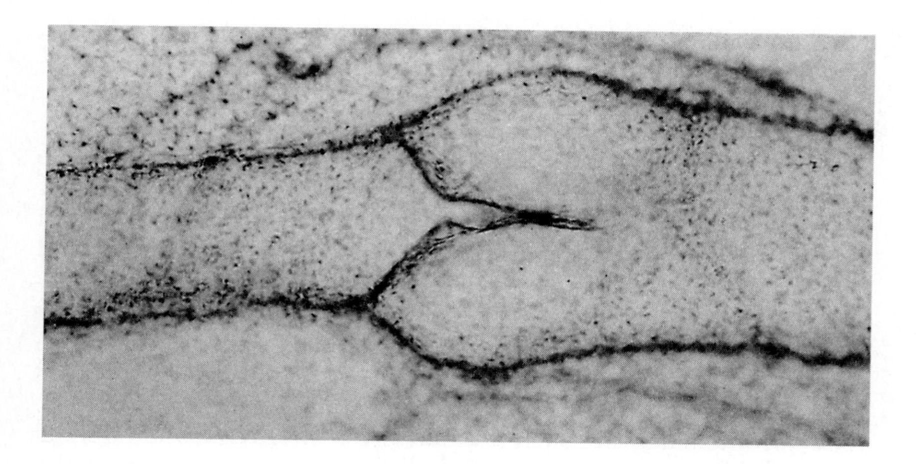

30.22 **Valve in a lymph vessel**

vessels and push the lymph forward past one-way valves (Fig. 30.22). Many other vertebrates have lymph hearts, which are pumping devices located along major lymph vessels.

HOW TEMPERATURE IS REGULATED

One of the major roles of the circulatory system in animals is temperature regulation. The role of the blood as a heating or cooling fluid depends greatly on whether the organism in question is warm-blooded—that is, attempts to maintain an elevated temperature near the optimum for most enzyme activity—or cold-blooded. As we will see, much of the chemistry of the blood and the dynamics of gas exchange in tissues depends on this feature of an animal's physiology. We will look first at the costs and benefits of attempting to maintain a constant temperature, and then at the circulatory mechanisms involved.

■ WHY BOTHER?

As described in Chapter 6, cellular metabolism captures some of the energy released by the oxidation of carbohydrates, fats, and proteins and converts it into the energy of high-energy phosphate groups in ATP. However, this process fails to recover roughly 60% of the energy—energy that serves instead to make metabolism highly favorable thermodynamically. Most of this energy is released as heat. The vast majority of animals (and plants) promptly lose most of this thermal energy to their environment. Such animals are known as cold-blooded, or more accurately, ***poikilothermic*** ("of variable temperature") or ***ectothermic*** ("externally heated"). Because the animals' heat comes largely from external sources, their body temperature fluctuates with the environmental temperature; when they are at rest, it is nearly the same as that of the surrounding medium, particularly if the medium is water.

The metabolism of an organism depends on the activity of its enzymes, which in turn is closely tied to temperature. Metabolic rate—the organism's rate of oxygen consumption and/or carbon

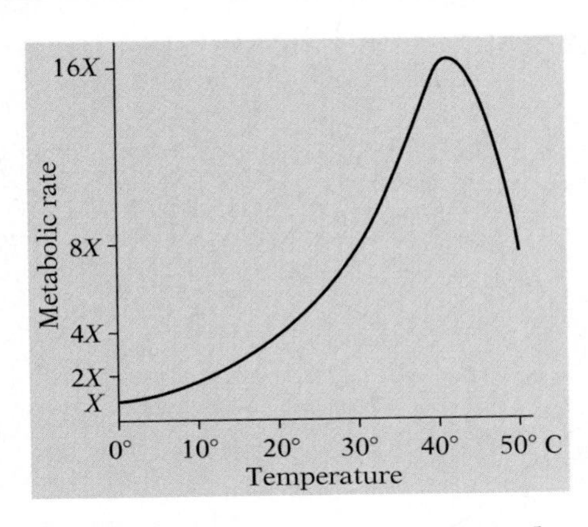

30.23 Changes in metabolic rate with changes in temperature

The hypothetical organism has a Q_{10} of 2—that is, its metabolic rate doubles with every 10°C rise in temperature. This rise is the result of the greater thermal energy of the reactants in the cell and the increasing effectiveness of the cellular enzymes. The abrupt decline above 40° represents the point at which the weak bonds that hold enzymes in their specific active conformations begin to break. As a result the enzymes become denatured and metabolic activity is severely disrupted.

Table 30.1 Insulating value of various materials

MATERIAL	RELATIVE INSULATION VALUE
Copper	1
Water	660
Muscle	835
Fat	1,800
Dry fur	10,000
Trapped air	16,000

dioxide production under standard conditions—increases with increasing temperature in a very regular fashion. The relationship between metabolic rate and temperature is often expressed as Q_{10}, which measures the rate increase for each 10°C rise in temperature. For example, if the rate triples for each 10° rise, then the value of Q_{10} is 3. Metabolic rates frequently have Q_{10} values of about 2. If the metabolic rate of a given animal at 0°C is X, and Q_{10}=2, then at 10°C the rate will be $2X$; at 20°, $4X$; at 30°, $8X$; and at 40°, $16X$. Notice that the rate increases more and more rapidly as the temperature increases (Fig. 30.23).

The activity of ectothermic animals is radically affected by temperature changes in their environment. As the temperature rises (within limits), they become more active; as the temperature falls, they become sluggish. Such animals are restricted in the habitats they can effectively occupy because they are at the mercy of the temperatures in those habitats. Of course, many have evolved behavioral adaptations such as basking on warm rocks or in the sun to increase their internal temperature relative to the environment at least briefly, and so extend their range somewhat.

A few animals, notably mammals and birds (plus a few fish, including tuna and swordfish), can make use of the heat produced during the exergonic reactions of their metabolism, because they have evolved mechanisms—often including insulation by fat, hair, feathers, etc. (Table 30.1)—whereby heat loss to the environment is retarded. Such animals are commonly called warm-blooded; biologists prefer the term ***endothermic*** ("internally heated"). The body temperature of animals that have internally produced heat is fairly high—usually higher than the environmental temperature—and relatively constant (and thus ***homeothermic***) even when the temperature fluctuates widely. The metabolic rate of endotherms can accordingly be maintained at a uniformly high level, and they remain active. They are thus less dependent on environmental temperatures than ectotherms, and are freed for successful exploitation of more varied habitats. They typically grow and reproduce faster as well.

If so many advantages accrue to endotherms, then why hasn't selection favored the evolution of endothermy in all animals? The answer lies in the costs of maintaining an elevated temperature.

In both endothermic and ectothermic animals, the normal metabolic rate is inversely related to body size; the smaller the organism, the higher the *relative* metabolic rate—that is, the higher the metabolic rate per gram of body tissue (Fig. 30.24). The reason for this is easily understood in the case of endotherms: smaller animals have a greater surface-to-volume ratio, and consequently a larger relative heat loss to the environment per unit time. To maintain a constant high body temperature despite rapid heat loss across its body surface, a small animal must oxidize food at a very high rate. Because the relative amount of food consumed and the pace of digestion, respiration, and so on must rise with decreasing size, there is a lower limit on the size of endotherms. The smallest mammals are shrews, which weigh only about 4 g. They must eat nearly their own body weight of food every day, and can starve to death in a few hours if deprived of food. In addition, animals of any size in habitats with relatively little food may simply not be able to afford the costs of endothermy.

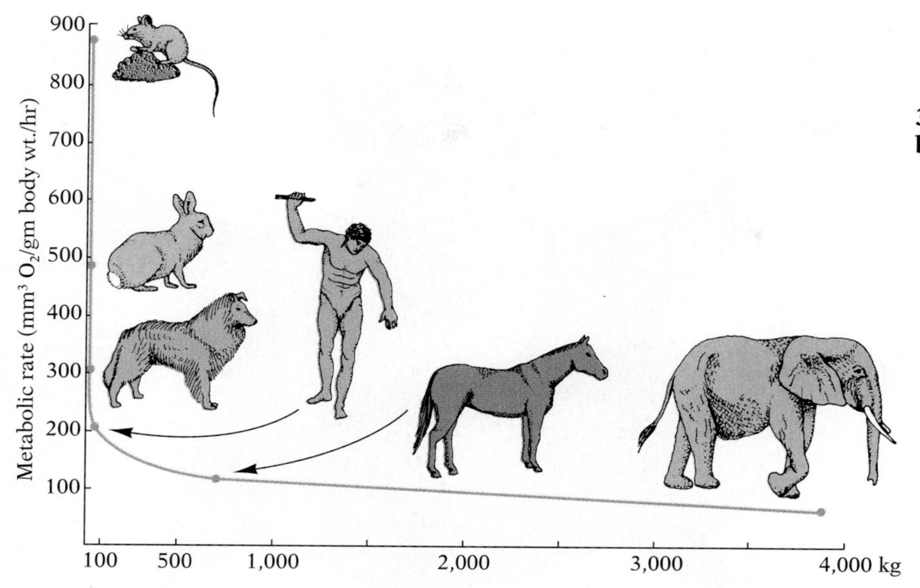

30.24 Inverse relationship between metabolic rate and body size in mammals

Oddly enough, there is an inverse relationship between size and relative metabolic rate in ectothermic animals. Since cold-blooded organisms lose their metabolic heat to the environment and do not normally respond to heat loss by increased metabolism, larger size and its concomitant smaller surface-to-volume ratio should retard heat loss somewhat, and the conserved heat ought then to speed up metabolism. In fact, why larger size in ectotherms tends to be correlated with *lower* relative metabolic rates has never been fully explained. One likely factor is that increasing size generally involves a disproportionate increase in the mass of skeletal and other connective tissues in animals—an alligator requires a great deal more inactive support structure than a salamander. Since these tissues are relatively inactive metabolically, the average metabolic rate per unit weight for the organism as a whole may fall as the proportion of these less active but necessary structural tissues rises.

■ GETTING WARM

In most places on earth, the ambient temperature is lower than 39°C. This means that most endotherms must actively heat themselves. If the environmental temperature is not too low and the creature is well insulated, the waste heat generated by its own metabolism may be sufficient, provided the circulatory system can distribute that warmth efficiently from sites of production to the rest of the body. If basal heat production is inadequate, most animals respond by shivering. The result of this involuntary version of isotonic exercise is that blood is heated as it passes through the capillaries serving the muscles; the warmth is carried by the circulatory system to the rest of the body.

Certain small animals, including shrews, bats, and hummingbirds, cannot fully afford the luxury of homeothermic endothermy: their large surface-to-volume ratio results in a disproportionally high "heating bill" (Fig. 30.24). Instead they adopt a kind of temporal ***heterothermy*** (nonuniform temperature), allowing their body

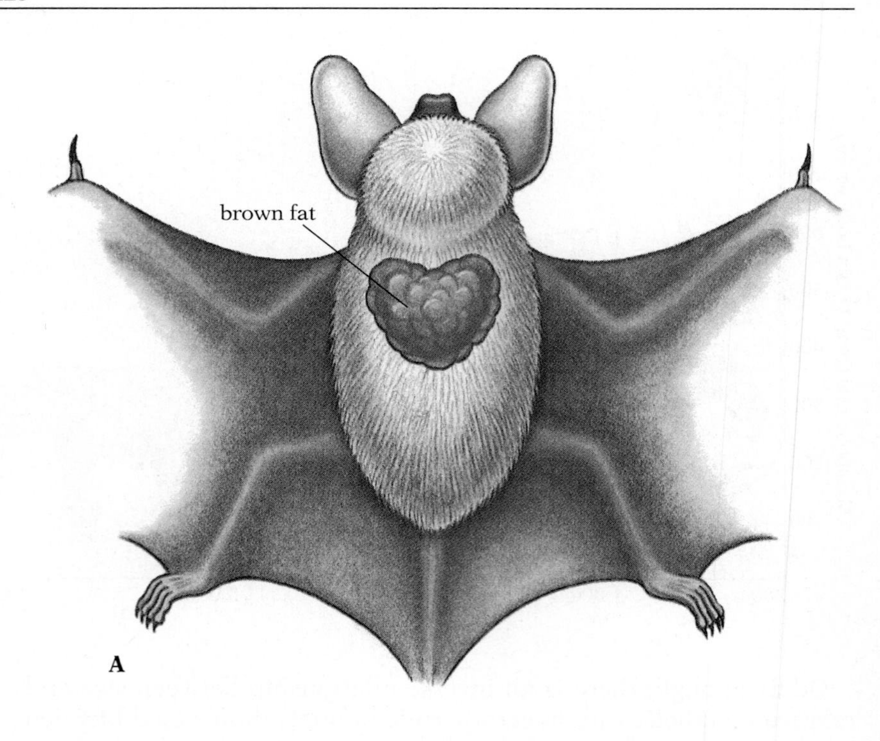

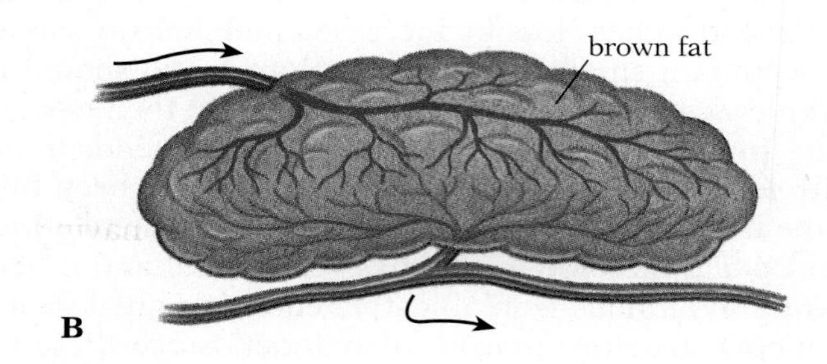

30.25 Brown fat in a bat

(A) The deposit of brown fat is located next to the heart in the upper thorax. (B) The heavy vascularization of this deposit carries the heat from oxidation rapidly to all parts of the body.

temperature to drop between bouts of feeding. Hibernators do the same, on a scale of weeks rather than hours. The problem in either case is for the animal to rouse itself out of its low-temperature stupor. One common strategy is to have a central heater richly supplied with capillaries. Bats, for instance, use deposits of "brown fat"—fat cells with many mitochondria (which impart the distinctive color) and the necessary oxidative enzymes (Fig. 30.25). (Normal fat cells do not metabolize their own lipids, but secrete them into the circulation when needed; the target cells then take the lipids up for their own use.) When the torpid animal needs to become active, it begins oxidizing the fat, a process that leads to massive heating of the blood.

■ STAYING WARM

When the ambient temperature is much below the body temperature of an animal, some heat loss is inevitable. Insulation—fur, feathers, and fat—can minimize but not altogether stop heat loss. Indeed, the presence of such an insulating layer can create a differ-

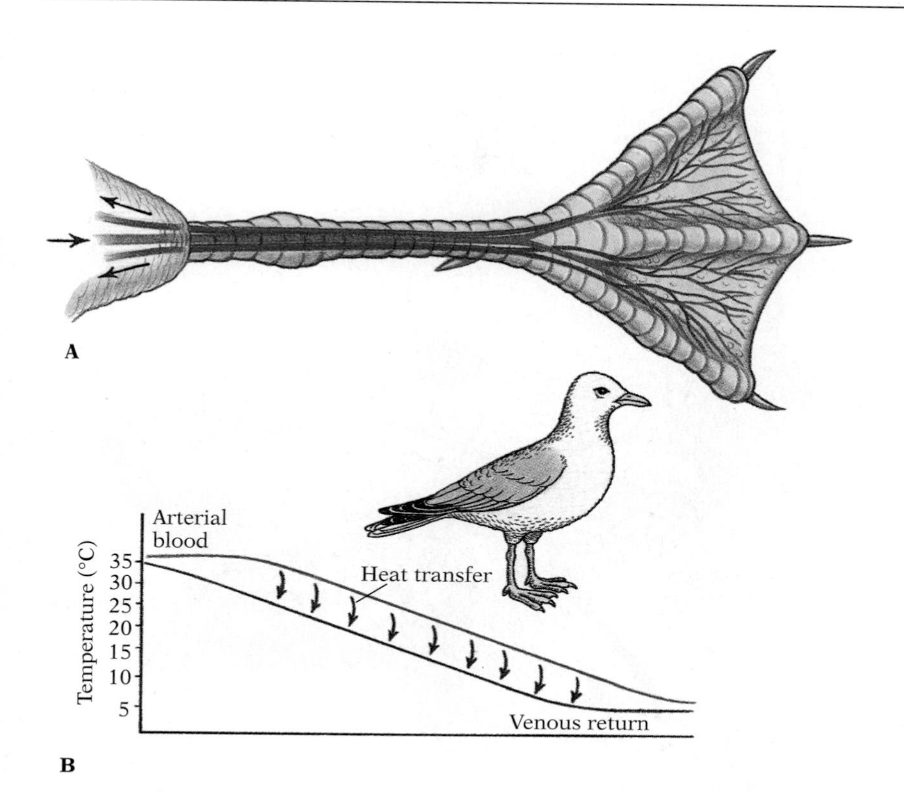

30.26 Use of countercurrent exchange to minimize heat loss

In cold weather, the temperature of the extremities of an animal like the Arctic gull is allowed to fall far below the animal's core temperature, a strategy known as spatial heterothermy. The gull further conserves heat through countercurrent exchange in its extremities between warm blood in the arteries and cool blood returning in the veins. As the graph shows, this arrangement sends much of the heat back into the body before it can be lost in the leg and foot.

ent set of problems on warm days or during heavy exercise, when excess heat must be disposed of.

The most severe challenges for minimizing heat loss are met in the limbs and other extremities of Arctic animals having a high surface-to-volume ratio. Most of these creatures resort to spatial heterothermy, maintaining a high core temperature while allowing legs and ears to grow cold (Fig. 30.26). The problem then is to conserve the heat in the blood that must circulate to the extremities to keep them alive. Their strategy is countercurrent exchange: arteries carrying warm blood out from the body core lie against the veins returning with cold blood, and most of the heat of the arterial blood is transferred to the veins.

■ COOLING OFF

Most animals cool themselves by evaporating water. For humans this means sweating. Blood is shunted to the skin where it can radiate heat directly into the air and benefit simultaneously from evaporative cooling. In species covered with insulating fur or feathers, however, evaporation must be more localized. The most frequently used surface in these cases is the tongue. The panting of dogs provides a familiar example.

Surprising as it may seem, a panting dog takes in air through its nose, not its mouth. As shown in Chapter 28, the respiratory tissues evaporate water into the inspired air, warming and humidifying it. But instead of being exhaled through the nose, and thereby recovering much of the warmth and moisture, the outgoing air during panting is blown across the wet tongue, carrying with it air

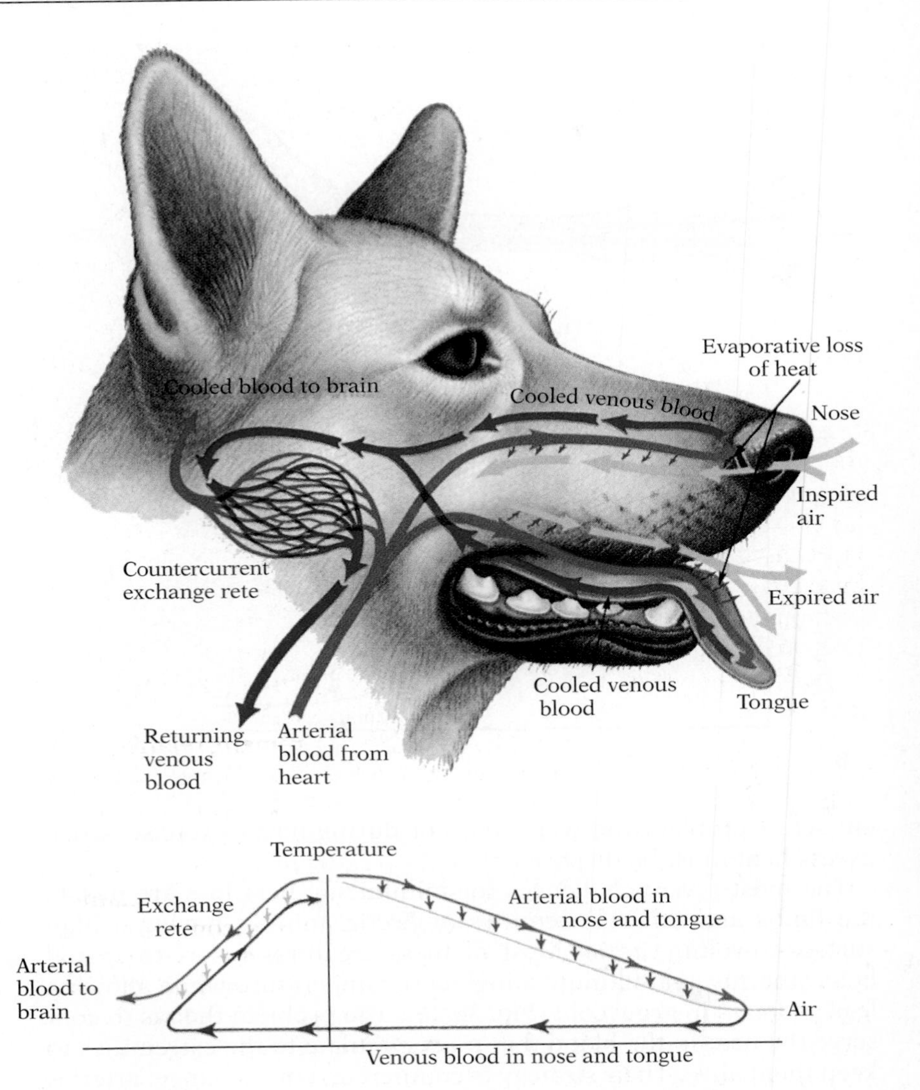

30.27 Panting and rete-exchange cooling

Animals can cool the blood in the capillaries of the respiratory system evaporatively by inhaling through the nose and exhaling over the tongue. The cooled blood can then exchange heat with blood en route to the brain, protecting that delicate organ from overheating. The anatomy of this three-step exchange is shown at top; the effects on blood temperature are illustrated in the graph (bottom).

from the sides of the mouth and evaporating still more water (Fig. 30.27). As a result, the blood is cooled in the capillaries of both the nose and the tongue. In many species—particularly those adapted to warm climates—this chilled blood is not returned directly to the body. Instead, it flows to another set of capillaries, the ***carotid rete***. (A rete is a network.) This bed of capillaries is intermingled with arterial capillaries[2] carrying blood to the brain, and cools that blood via countercurrent exchange. In this way the brain, which is the organ most easily damaged by heat, is preferentially cooled.

THE NATURE OF BLOOD

The intercellular liquid matrix of blood is called ***plasma***. Suspended in the plasma of vertebrates are three major types of cells and solid components: (1) red blood corpuscles, or erythrocytes; (2) white blood cells, or ***leukocytes***; (3) ***platelets***, which are small

[2] These are not true capillaries in that they do not exchange gases, nutrients, or wastes with the adjacent tissues or each other; they exist only to transfer heat.

disk-shaped bodies that arise as cell fragments.[3] Normally, the cells and platelets constitute about 40–50% of the volume of whole blood, while the plasma constitutes the other 50–60%.

■ WHY IS PLASMA IMPORTANT?

Plasma is roughly 90% water. Six main classes of solutes are dissolved in this water: (1) inorganic ions and salts, (2) plasma proteins, (3) organic nutrients, (4) nitrogenous waste products, (5) hormones, and (6) dissolved gases. We will describe each of these classes in this section.

1 The principal inorganic cations (positively charged ions) in the plasma are sodium (Na^+), calcium (Ca^{++}), potassium (K^+), and magnesium (Mg^{++}). The chief inorganic anions (negatively charged ions) are chloride (Cl^-), bicarbonate (HCO_3^-), phosphate (HPO_4^- and $H_2PO_4^-$), and sulfate (SO_4^-); of these, chloride and bicarbonate are by far the most abundant. Together, the inorganic ions and salts (ionic compounds composed of positively and negatively charged ions) make up about 0.9% of the plasma of mammals by weight; more than two-thirds of this amount is sodium chloride (NaCl, or table salt).

The concentrations of the individual ions remain relatively stable, regulated as they are by a variety of agencies, particularly the kidneys and other excretory organs, as well as a number of hormones. This maintenance of equilibrium—***homeostasis***—is essential. Any appreciable shift in the concentrations of sodium chloride (NaCl) and sodium bicarbonate ($NaHCO_3$), for example, would have severe effects on the cells, since these compounds (together with plasma proteins) help determine the osmotic balance between the plasma and the fluids bathing the cells. Even if the total concentration of dissolved substances remains the same, shifts in the concentrations of particular ions in the plasma, leading to corresponding shifts in the tissue fluids, can create serious disturbances. Nerves and muscles, for example, are highly sensitive to changes in the concentrations of K^+ and Ca^{++}. Similarly, the integrity of cell membranes depends on proper balance of Ca^{++}, Mg^{++}, K^+, and Na^+ in the intercellular medium. The concentrations of certain ions are also very important in determining the pH of the body fluids, and even slight changes in pH (normally slightly alkaline in plasma) can be fatal.

2 The plasma proteins, which constitute 7–9% by weight of the plasma, are of three types: fibrinogen, albumins, and globulins. Most of these proteins are synthesized in the liver, though some of the globulins are synthesized in lymphoid tissue or by circulating B cells.

We've seen that proteins help determine the osmotic pressure of the plasma and thus the movement of water between blood and tissue. These proteins are also instrumental in stabilizing the pH of the plasma and controlling its viscosity (a measure of the internal

[3] True platelets are found only in mammals. The blood of most other vertebrates contains cells called thrombocytes, which function in blood clotting in a manner similar to platelets.

friction, or "thickness," of a fluid). The heart can maintain normal blood pressure only if the viscosity of the blood is nearly normal.

Some plasma proteins can bind to certain hormones, fatty acids and other lipids, some vitamins, and various minerals, thereby greatly facilitating the transport of such substances by the blood. In addition, fibrinogen and certain globulins function in blood clotting (see box, p. 868); other globulins participate in the immune response.

3 Organic nutrients in the blood include glucose, fats, phospholipids, amino acids, and lactic acid. Some of these are picked up from the intestine; others enter the blood from storage areas such as the liver and the fat deposits. Lactic acid is a product of glycolysis, especially in muscles; it is transported by the blood to the liver, where some of it may be used in resynthesis of carbohydrates, and some may be further oxidized to carbon dioxide and water.

Plasma also contains cholesterol. It is metabolized to some extent as a source of energy and is a stiffening component of the cell membrane; it plays a major role as the precursor of most other important steroids, such as the bile acids and the steroid hormones.

4 The plasma carries nitrogenous wastes from their sites of formation to such organs of excretion as the kidneys.

5 Hormones are also carried by the plasma. These substances, synthesized by the endocrine tissues, are important regulatory chemicals.

EXPLORING FURTHER

SPECIAL ADAPTATIONS IN DIVING MAMMALS

When humans descend even to modest depths of a few dozen meters in water, they must take special precautions to avoid a variety of problems, the best known of which is "the bends." With each additional 10 m a diver ventures below the surface, the pressure on the body increases by 1 atm—the pressure felt in the air at sea level. This increased pressure compresses the body and forces the air present in compressible and collapsible structures (given enough depth, even sinuses will collapse) into the blood and tissues. The compressed air that divers carry provides pressure to the lungs to help inflate the chest and permit continued breathing, but that extra pressure also drives more gas molecules into tissues. As the diver returns to the surface, the lower pressure on the body allows nitrogen gas to form bubbles in the tissues, which causes the painful muscular cramps that characterize the bends.

Diving mammals have a number of adaptations that minimize the dangers of high pressure. One is behavioral: they exhale before diving, minimizing the volume of air to be forced into tissues. Many can even completely collapse their lungs. Disposing of air while diving, however, means the animal must carry sufficient oxygen to continue to metabolize while diving. The Weddell seal, for example, can dive to 500 m for more than an hour. It manages to supply itself with oxygen over this period by a combination of specializations. One adaptation involves maintaining about twice the concentration of myoglobin—the oxygen-storage protein of tissues—in its muscles. Another specialization is increased blood volume: seals have about twice as much blood in circulation per unit of mass as humans, and another supply of about the same volume stored in the spleen. Yet a third adaptation involves maximizing the number of corpuscles; seals have about twice as many per unit of blood volume as terrestrial mammals have.

Along with these tricks for increasing the oxygen supply, diving mammals have a number of strategies for minimizing oxygen consumption during deep dives in search of prey. Inessential organs and tissues are shut down, and many others (including, during long dives, the muscles) are allowed to run anaerobically to conserve oxygen for the brain (and, in pregnant females, the placenta). That such a highly efficient diving machine could evolve from a fully terrestrial ancestor is a testament to the power and potential of natural selection.

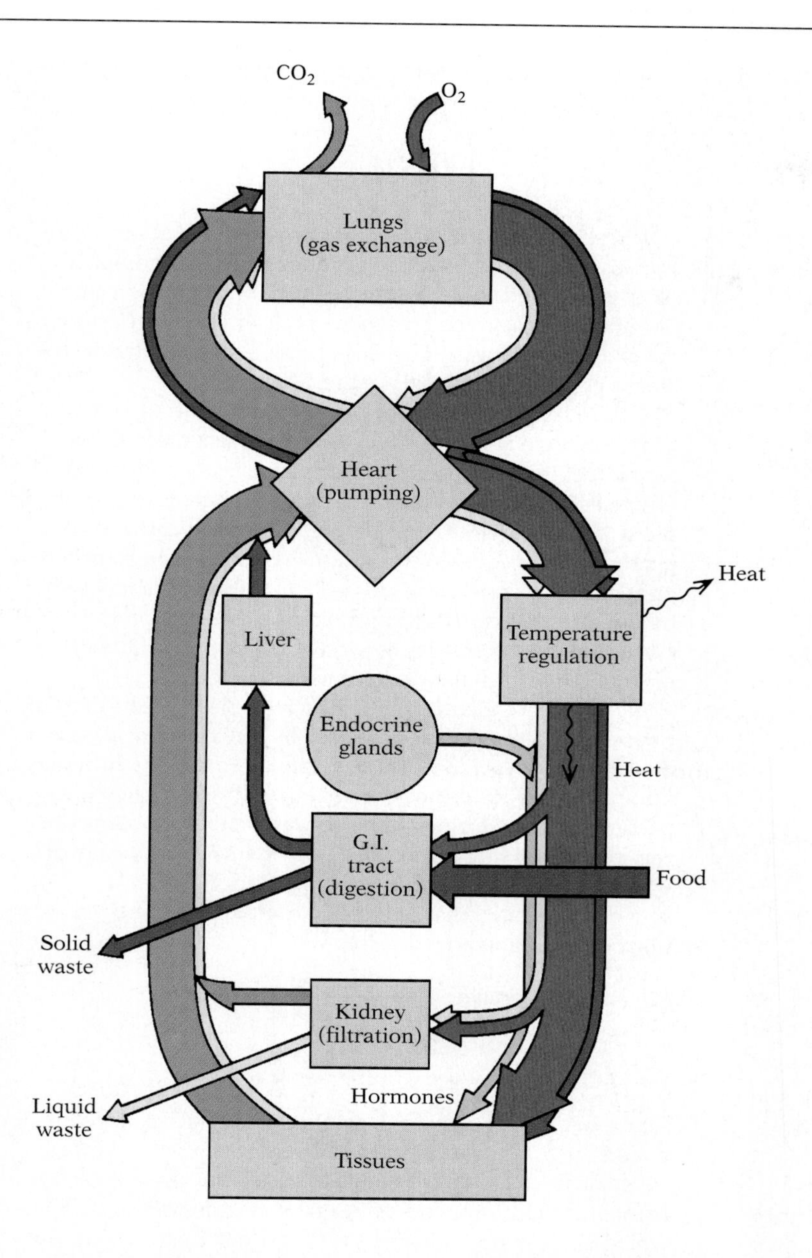

30.28 Major functions of the circulatory system

This highly schematic diagram traces the movement of gases, hormones, nutrients, wastes, and heat through the circulatory system. Oxygen (red) enters through the lungs, is carried by the blood, and is exchanged for CO_2 in the tissues; the CO_2 (blue) is carried to the lungs, where it is exchanged for more O_2. Nutrients (brown) enter from the digestive system and are carried to the liver, where most are removed, and then to the tissues, where the remainder are delivered. The tissues also load nitrogenous wastes (yellow) into the blood for transport to the kidneys. (The nutrients removed by the liver are added back to the blood between meals to keep a relatively constant level of sugars and other substances in circulation.) The circulatory system also transports hormones (green) and is used for heat exchange.

6 Three principal gases are found dissolved in the plasma. One of these, nitrogen, which diffuses into the blood in the lungs, is physiologically inert, and can be disregarded except in deep-diving mammals (see box). The other two, oxygen and carbon dioxide, are of critical importance, and we will discuss the details of their transport in a later section. In vertebrates most of the oxygen and much of the carbon dioxide are transported in the red blood corpuscles.

The multipurpose functioning of the circulatory system is indicated schematically in Figure 30.28.

■ HOW BLOOD CLOTS

Normally, the plasma of the circulating blood remains a liquid. When a blood vessel has been ruptured or otherwise damaged, or

EXPLORING FURTHER

HOW DOES BLOOD CLOT?

Understanding how blood clots form eluded researchers for decades. You can see why by considering the following series of experiments.

1 Blood removed from a blood vessel, but carefully prevented from touching the portion of the vessel that has been damaged, is put into an open dish lined with plastic, which is hydrophobic. The blood is thus simultaneously subjected to two conditions traditionally thought to result in clotting—exposure to air and cessation of flowing—yet it remains liquid for hours. Nor does it clot when it is held stationary in a portion of a blood vessel that has been tied off.

2 When the same procedure is repeated, but the blood is allowed to come into contact with the damaged vessel wall through which it is removed, clotting begins at once. Since clotting occurs when a vessel is damaged, and since it is initiated at the site of the damage, might the damaged tissue release some chemical that initiates the clotting process?

3 If we prepare an extract from tissue cells and add this to the liquid blood in the plastic-lined dish of the first experiment, a clot promptly starts to form. But when we remove blood from a blood vessel as we did in the first experiment and put it into a dish made of glass which is *hydrophilic*, the blood promptly clots even without the addition of tissue extract. Might there be some other clot-initiating factor?

4 If we allow fresh blood to touch glass and watch it under a microscope, we see blood platelets disintegrating on contact with the glass. Perhaps the disintegrating platelets release a clot-initiating chemical similar to that released by damaged tissues. Platelets seem to disintegrate readily on contact with hydrophilic surfaces like glass; it turns out that platelets also disintegrate when they touch damaged tissue.

The chemical basis of these observations is that both damaged tissues and disintegrating platelets release a complex substance, called ***thromboplastin***, that initiates blood clotting, though only in the presence of calcium ions. A calcium-dependent anticoagulant binding protein—thrombomodulin—blocks accidental triggering of the system.

A third factor essential for normal blood clotting is the plasma protein fibrinogen. However, if we mix these three substances in a dish, no clotting occurs. Clearly, something else must be involved. That something else seems to be one of the globulin proteins of the plasma, known as ***prothrombin***. If prothrombin is added to the mixture of fibrinogen, thromboplastin, and calcium ions, a clot will form. Prothrombin itself has no effect on clotting; it must first be converted into ***thrombin***, the substance that converts fibrinogen into its crystallized form, fibrin, which makes the clot.

Now we have identified the main ingredients in the clotting process. Thromboplastin, produced by disintegrating platelets or damaged tissue, converts the plasma protein prothrombin into thrombin; this is the reaction in which the calcium ions participate. The thrombin then converts another plasma protein, fibrinogen, into fibrin. The fibrin fibers form a meshwork, which begins to shrink; finally, the fluid blood serum is squeezed out, and a hardened clot is left in place. The reactions can be summarized as follows:

$$(1) \quad \text{prothrombin} \xrightarrow[\text{Ca}^{++}]{\text{thromboplastin}} \text{thrombin}$$

$$(2) \quad \text{fibrinogen} \xrightarrow{\text{thrombin}} \text{fibrin}$$

These simplified summary equations show the relations between the essential substances. Numerous other substances—accelerators, inhibitors, and the like—also play roles in the clotting process. A reaction series of this sort, in which the first step releases another, which then triggers yet another, and so on, is referred to as a cascade reaction.

when certain kinds of foreign substances have gained entrance into the circulating blood, or when the blood has been removed from the body, one of the plasma proteins, ***fibrinogen***, comes out of solution and converts into ***fibrin***, which forms a hard lump, or clot (Fig. 30.29). In this way a small hole in a vessel may be plugged, or a weakened place in a vessel wall may be strengthened. Blood clotting is a powerful evolutionary adaptation for emergency repair of the circulatory system and for preventing excessive blood loss (see box). Clotting occurs in all vertebrates and in certain invertebrates. Some invertebrates have an alternative adaptation serving the

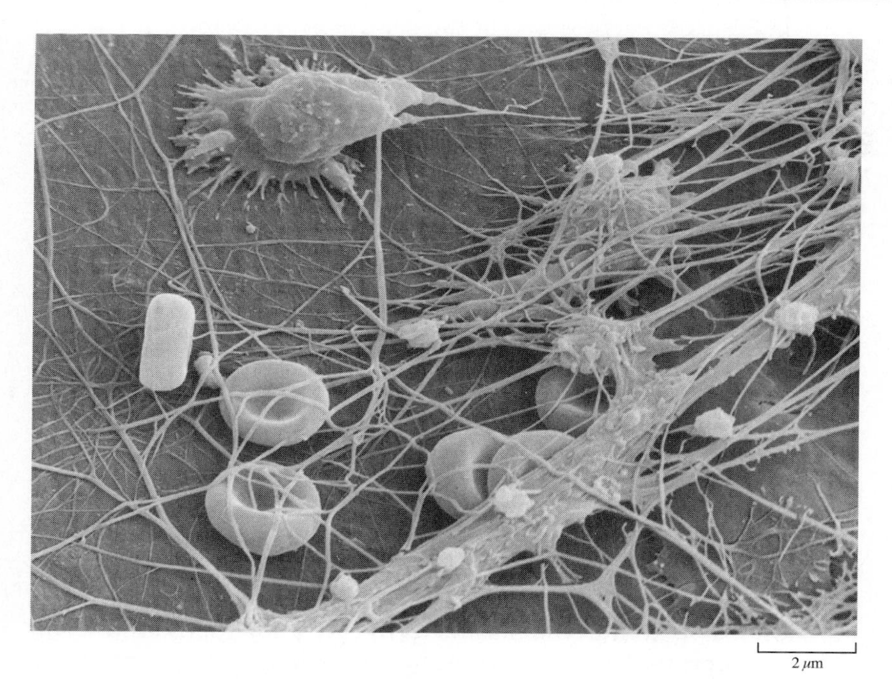

30.29 Blood clotting

Blood clots as the various components of the blood become trapped in a network of fibrin. Several erythrocytes and a mobile macrophage (top left), which plays a central role in eliminating dead erythrocytes and bacteria, are visible in this electron micrograph.

same basic function: powerful muscles contract and close off any hole or damaged area.

■ THE ROLE OF RED BLOOD CORPUSCLES

The two general classes of cells in human blood, leukocytes and erythrocytes, serve very different purposes. The white blood cells are the body's primary defense against toxins, infections, and cancers. They cooperate in elaborate ways to carry out their mission. Erythrocytes, on the other hand, are far simpler, and function only as mediators of efficient gas exchange.

Human erythrocytes, or red blood corpuscles, are biconcave, disk-shaped bodies without nuclei that were once cells (Fig. 30.30). There are normally about 5 million of them per cubic millimeter of blood. Though the number of corpuscles remains constant from

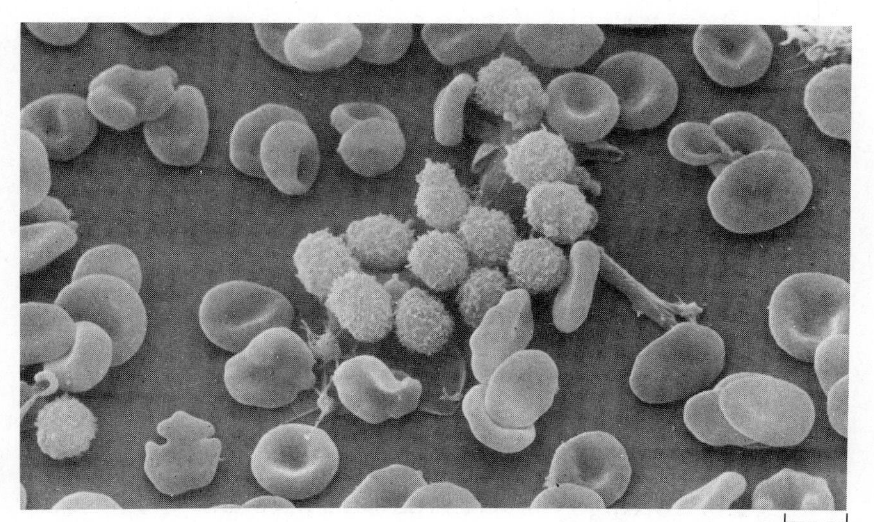

30.30 Blood cells and corpuscles

The erythrocytes are biconcave disks. In the center of this scanning electron micrograph is a cluster of lymphocytes. A large phagocytic macrophage can be seen under the lymphocytes.

day to day, destruction of some corpuscles and formation of new ones goes on continually; the normal survival time of an erythrocyte is 120 days. More than 2 million erythrocytes are destroyed every second, chiefly in the liver and the spleen, where they are engulfed by large phagocytic cells. Phagocytic cells in the lymph nodes destroy any erythrocytes that escape from the blood and get into the lymph.

The erythrocytes of adults are formed in the red bone marrow, which fills the interior of the upper ends of the long bones and the shafts of flat bones like those of the skull, ribs, and pelvis. Toward the end of their development, developing erythrocytes lose their nuclei and acquire the red oxygen-carrying pigment ***hemoglobin***, a protein with iron-containing prosthetic (nonproteinaceous) groups. The erythrocytes then enter the circulating blood. In vertebrates other than mammals the mature erythrocytes retain their nuclei.

Not all invertebrates have hemoglobin. In some of those that do, it is located within cells, as in vertebrates, but in many it is simply dissolved in the plasma. Though hemoglobin molecules might function just as well in the plasma as in erythrocytes, their location in cells or corpuscles is important for animals with high metabolic rates; more pigment molecules can then be carried per unit volume of blood, and the oxygen-transporting capacity is correspondingly increased. A single human erythrocyte usually contains about 280 million molecules of hemoglobin. If all this hemoglobin were loose in the plasma, the concentration of plasma protein would be about 3 times higher than at present, with profound effects on the osmotic balance between the blood and the tissue fluid. Erythrocytes, then, are a convenient method of packaging large amounts of hemoglobin with relatively little disturbance of the osmotic concentration of the blood.

Many of the invertebrates that lack hemoglobin have different oxygen-transporting pigments that, like hemoglobin, combine a metal with protein. For example, many molluscs and arthropods have a pigment called hemocyanin, which contains copper instead of iron; when oxygenated, it is blue instead of red. Hemocyanin never occurs in cells, but is dissolved in the plasma.

■ CARRYING THE OXYGEN: HEMOGLOBIN

How hemoglobin works The hemoglobin molecule is a globulin protein composed of four independent polypeptide chains (see Fig. 3.30, p. 65). Each of the four chains enfolds a complex prosthetic group called heme, which has an iron atom at its center (Fig. 30.31). The tetrameric hemoglobin of animals appears to have evolved from monomeric and dimeric forms found in certain bacteria and in the roots of some (perhaps all) plants. Some plant hemoglobin is used to remove oxygen from the vicinity of nitrogen-fixing enzymes in symbiotic bacteria in root nodules. How bacterial hemoglobin functions in oxygen uptake or regulation is yet to be discovered.

All vertebrate hemoglobins have essentially the same three-dimensional structure. In some hereditary blood diseases, however,

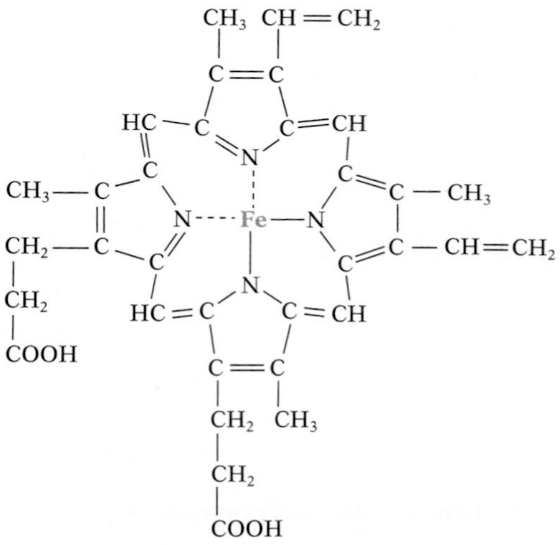

30.31 Structure of the heme group

A single molecule of hemoglobin has four of these iron-containing prosthetic groups.

changes in only one or two amino acids are accompanied by alterations that severely impair the oxygen-transporting capability of the hemoglobin molecule. The dysfunction that causes sickle-cell anemia results from a single amino acid substitution: valine, which is hydrophobic, replaces glutamate, which is polar. This reduces the solubility of the hemoglobin, so it tends to crystallize at certain levels of pH. Once crystallized, the hemoglobin can no longer load and unload oxygen efficiently; worse yet, the red corpuscles tend to collapse, taking on a sickle shape and forming clumps that can obstruct the capillaries.

As would be expected, the more closely related two animals are, the more similar their hemoglobins tend to be; those of humans and apes, for instance, are much more alike than those of humans and fish. For reasons that will become clear presently, the hemoglobin of an embryo is slightly different from that of an adult of the same species.

Each of the four iron atoms in a hemoglobin molecule can, by virtue of its structural relationships within the molecule, combine loosely with one molecule of oxygen. The compound formed by the union of one molecule of hemoglobin (Hb for short) with four molecules of oxygen is called oxyhemoglobin ($Hb + 4O_2$).

Hemoglobin exemplifies cooperativity in an allosteric molecule (see p. 81). When it binds with the first O_2 molecule, conformational changes occur that enhance its affinity for oxygen and greatly facilitate the binding of the other three O_2 molecules. An incidental effect of this conformational change is that the blood becomes redder; thus arterial blood in the systemic circulation is more crimson than venous blood.

Why dissociation has to be nonlinear The combination of hemoglobin with O_2 is a loose one. Under certain conditions the combination will form, and under other conditions it will break down. Clearly, conditions in the lungs must favor formation of oxyhemoglobin, and conditions in the capillary beds of the systemic circulation must favor release of O_2 and re-formation of hemoglobin. The critical condition in determining whether hemoglobin will load or unload O_2 is the partial pressure[4] of O_2 in the medium to which the hemoglobin is exposed. When the partial pressure of O_2 is high—that is, when oxygen is abundant—the hemoglobin picks up O_2; when the partial pressure of O_2 is low (oxygen is rare), the hemoglobin releases O_2. There is, of course, a relatively high partial pressure of O_2 in the air in the alveoli of the lungs and a relatively low partial pressure of O_2 in the tissues serviced by the systemic circulation, where O_2 is being consumed in cellular respiration. Consequently, hemoglobin in the capillaries of the lungs tends to pick up O_2 and to release O_2 in the capillaries of the systemic circulation.

Figure 30.32 shows the percentage of O_2 saturation of human hemoglobin at different partial pressures of O_2 in the blood; the lower the partial pressure, the greater the tendency for oxyhemo-

[4] The partial pressure of a gas dissolved in a liquid is the amount of pressure that must be exerted to force it to go into solution in the concentration observed. For a general definition of the partial pressure of a gas, see p. 805.

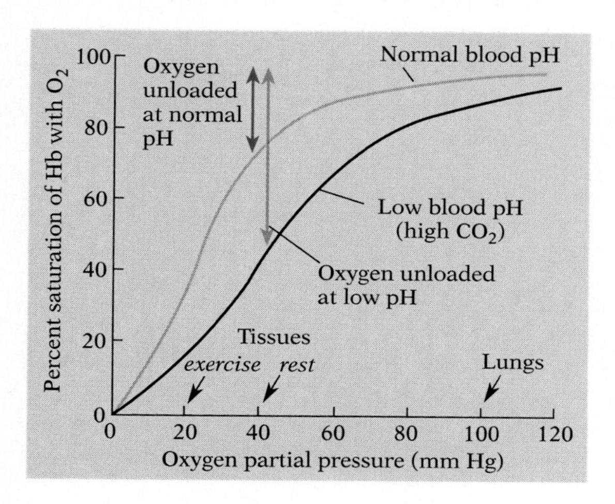

30.32 Dissociation curves of adult human hemoglobin at normal and low blood pH

The nonlinear shape of the curves is a consequence of cooperativity, which results in the enhanced affinity of the hemoglobin molecule for oxygen once it has bound its first molecule of O_2. In the lungs the hemoglobin becomes fully loaded with oxygen, whereas oxygen is unloaded in the tissue capillaries. As both curves make clear, when the oxygen concentration in the tissues is especially low, as during exercise, more of the oxygen is unloaded to meet the needs of the tissues. Notice also that when CO_2 rises, and so acidifies the blood (black curve), the change in pH alters the binding properties of hemoglobin, causing it to deliver more oxygen than usual.

globin to dissociate into hemoglobin and O_2—hence the name *dissociation curve* for a graph of this type. As you can see, when the blood pH is normal (brown curve), the hemoglobin is about 98% saturated with O_2 at the partial pressure of O_2 typical for the tissue fluid of the lungs (100 mm Hg), but only about 68% saturated at the partial pressure of O_2 typical of the fluid of tissues at rest (40 mm). The difference (30%) represents the approximate percentage of the O_2 carried by hemoglobin that is actually released to the tissues at rest. During exercise the more rapid utilization of O_2 by the muscle tissues causes a drop in its partial pressure in the tissues to a level of 20 mm or even lower. As a result, oxyhemoglobin releases more of its O_2 to the tissues, and the saturation of venous blood may fall so low that fully 75% of the oxygen is delivered.

The S shape of the dissociation curve of hemoglobin is a result of the cooperativity factor in the binding of O_2. At very low O_2 pressures the curve rises slowly, because the binding of the first O_2 molecule to a molecule of hemoglobin is difficult. After those first molecules are bound, the binding of additional O_2 is easy, and the curve rises steeply until it finally levels off when the hemoglobin is nearly saturated with O_2. This S-shaped curve has important implications for the release of O_2 to the tissues. Notice that the steepest portion of the curve lies in the range of O_2 partial pressures prevalent in tissue fluid. Only a slight drop in the pressure here, as occurs in the transition from rest to exercise, results in a very sizable increase in the amount of O_2 released from the oxyhemoglobin. A drop in pressure from 40 to 20 mm—from rest to exercise—means an increase in the total percentage of O_2 released from about 30% to about 73%.

The affinity of hemoglobin for O_2 is influenced markedly by pH, because H^+ ions act as negative allosteric modulators for hemoglobin. CO_2, produced as metabolic waste by the cells, is abundant in the tissue fluid of most parts of the body; it combines with water to form carbonic acid (H_2CO_3), which in turn dissociates into H^+ ions and bicarbonate ions (HCO_3^-). Corpuscles contain the enzyme carbonic anhydrase, which vastly speeds the reaction between water and CO_2; initially most of the bicarbonate is concentrated in the erythrocytes (Fig. 30.33). The result is increased acidity—that is, an increased concentration of H+ in the corpuscles:

$$CO_2 + H_2O \rightleftarrows H^+ + HCO_3^-$$

Hence hemoglobin in capillaries of the systemic circulation is exposed to an acid environment, where its affinity for O_2 is reduced, and it therefore readily unloads O_2 (Fig. 30.32, black curve). Conversely, in the capillaries of the lung, where CO_2 is released to the lungs and acidity is consequently lower, the hemoglobin is in an environment in which its affinity for O_2 is high, and it therefore readily loads O_2. Thus the waste product CO_2 plays an important regulatory role in shifting the condition of the hemoglobin back and forth between a propensity for loading O_2 in the lungs and for unloading it in the other tissues.

Temperature, as well as pH, affects the O_2 affinity of blood pigments. Hemoglobin releases O_2 more easily at higher temperatures, such as those that occur in muscles during strenuous exercise, when extra O_2 is needed (Fig. 30.34).

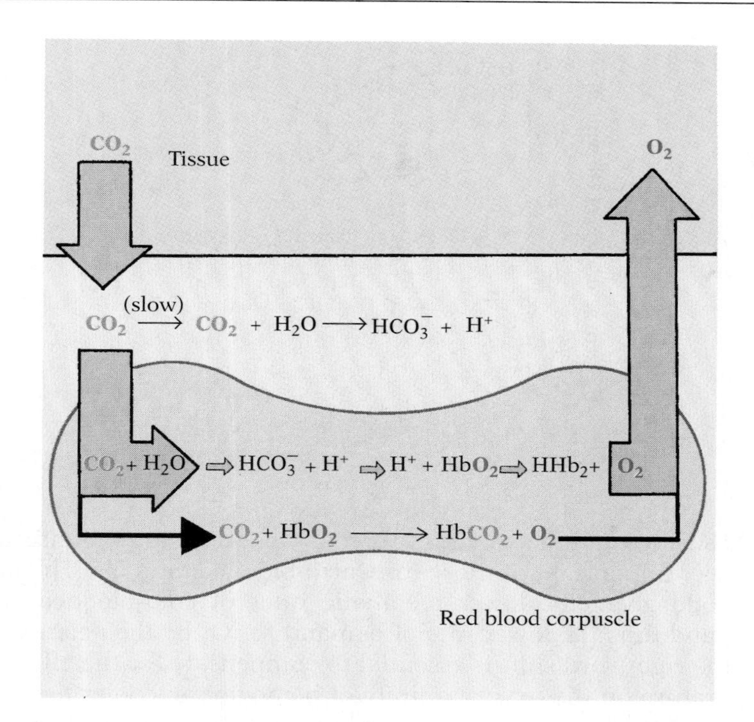

30.33 Summary of gas-exchange chemistry in tissue capillaries

In tissues, where the partial pressure of CO_2 is high (and hence the pH is low) and the partial pressure of O_2 is low, CO_2 enters the plasma (yellow). Most of this CO_2 then diffuses into corpuscles (pink), where it is primarily converted into bicarbonate, most of which diffuses back into the plasma (not shown). The hydrogen ions thus liberated help alter the quaternary structure of hemoglobin from that of ordinary hemoglobin (Hb) to that of acid hemoglobin (hemoglobin bound to a hydrogen ion, or HHb). As a result, the molecule loses much of its affinity for oxygen; O_2 is therefore released and diffuses into the plasma and from there into the tissues. In the pulmonary capillaries of the lung, where the partial pressures and pH are reversed, all of these reactions run the other way. A small quantity of CO_2 binds directly to hemoglobin (large black arrow), and an even smaller amount of CO_2 dissolves directly in the blood (yellow).

Why are there different hemoglobins? Species-specific differences appear in the hemoglobin dissociation curves of mammals, the curves for smaller animals generally being to the right of those for larger animals (Fig. 30.35). It is advantageous to the smaller animals, which have a higher metabolic rate and whose tissues therefore require more O_2 per unit time, that their oxyhemoglobin should dissociate more readily.

The curve for birds, which have very high metabolic rates (as would be expected in such active animals), tend to be to the right

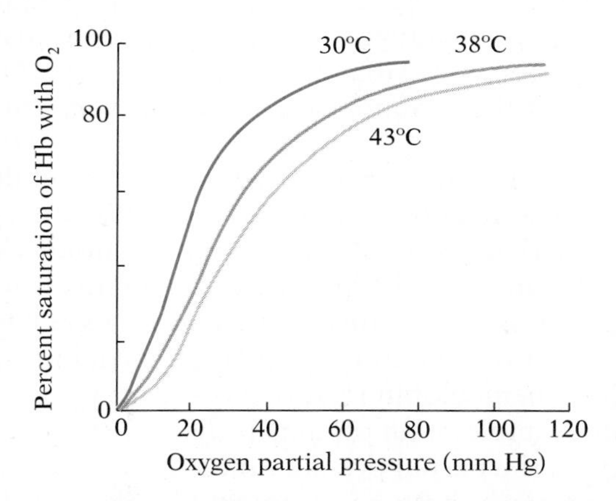

30.34 Effect of temperature on the affinity of human hemoglobin for oxygen

At higher temperatures the dissociation curve of hemoglobin is displaced to the right. In physiological terms, the hemoglobin has a lower affinity for O_2 and will therefore unload more of it at any given partial pressure.

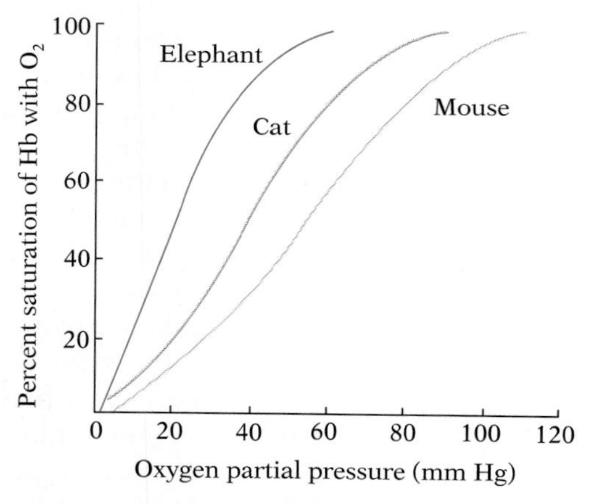

30.35 Dissociation curves for the hemoglobin of different animals

In general, the smaller the animal, the farther to the right its curve will be located. This means that small animals, with high metabolic rates and correspondingly high O_2 requirements, have hemoglobin that tends to unload more readily.

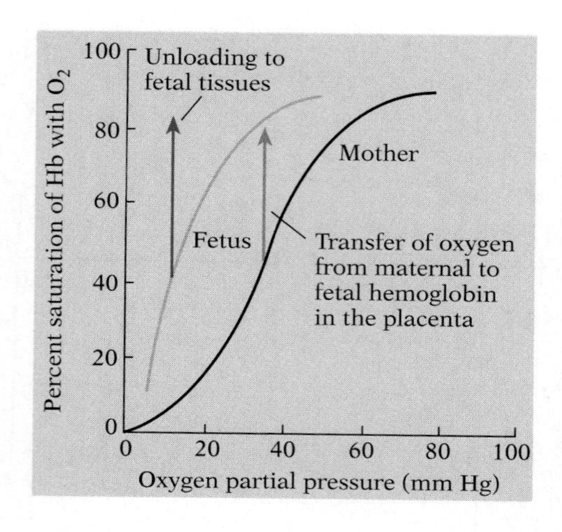

30.36 Dissociation curves for fetal and maternal hemoglobin in the cow

Because fetal hemoglobin has a higher affinity for O_2 than maternal hemoglobin, it can take O_2 from the maternal hemoglobin.

of those for mammals. Those for cold-blooded animals are also generally to the right of those for warm-blooded animals. This may seem odd given the lower metabolic rates of cold-blooded creatures, and thus the lower rate of demand for O_2 by their tissues. In fact, the rightward shift is essential to compensate for the effects of temperature on the oxygen affinity of hemoglobin; cooler temperatures shift the binding curve to the left (Fig. 30.34). Indeed, at 20°C (68°F, a comfortably warm temperature for most organisms) human hemoglobin has about the same oxygen affinity as myoglobin; thus, were our bodies ever to allow our internal temperature to fall this low, our hemoglobin would be useless for transferring oxygen to muscles, and we would be unable to move. But most cold-blooded animals must deal with temperatures of 20°C and far lower much of the time, especially those that live in the water; hence, the necessity of a rightward-shifted binding curve. However, this specialized binding curve creates another problem: at warmer temperatures, there can be only partial loading of the hemoglobin even at high partial pressures of oxygen. Thus cold-blooded creatures are almost always at a disadvantage relative to homeotherms; at low temperatures their chemistry runs slowly (see Fig. 30.23), while at higher temperatures they cannot obtain oxygen very efficiently.

Human fetal hemoglobin has a higher affinity for O_2 than adult hemoglobin (Fig. 30.36).[5] The adaptive significance of this difference is readily apparent. The fetus gets its O_2 from the mother's blood, not directly from the air. If the hemoglobin of the fetus is to take O_2 from the hemoglobin of the mother, it must have a greater affinity for O_2; that is, it must be able to take up O_2 at partial pressures at which the mother's hemoglobin readily gives up oxygen.

The human fetus is an example of an organism that must get its O_2 from a medium in which the partial pressure of O_2 is lower than in the atmosphere we ordinarily breathe. Animals like the South American llama and vicuña live at very high altitudes in the mountains, where the partial pressure of O_2 is appreciably lower than it is nearer sea level (as you know if you have ever experienced short-

[5] In isolation, adult and fetal hemoglobins have the same affinity for O_2. In the living animal, however, adult hemoglobin has a much greater tendency to bind a substance called 2,3-diphosphoglycerate and, as a result, a lesser tendency to bind O_2.

ness of breath at high altitudes). It is not surprising, therefore, to find that these animals, like the human fetus, have hemoglobin with a relatively high affinity for O_2; their dissociation curves are to the left of those of the average mammal, meaning that their hemoglobin loads O_2 more easily.

The llama and the vicuña have evolved a genetically determined type of hemoglobin that adapts them for life at high altitudes. It is also possible to become acclimatized to high altitudes, and thus no longer to experience the severe shortness of breath that occurred at first. As Table 30.2 shows, this acclimatization occurs because the blood at high altitudes contains more erythrocytes per unit volume than at sea level.

Carbon monoxide (CO) binds even more readily to hemoglobin than O_2. This gas is common in coal gas used for heating and cooking, in the exhaust from automobiles, and in tobacco smoke. It is a dangerous poison because, even when its partial pressure in the air is relatively low, such a high percentage of the hemoglobin may bind to it that the remainder cannot carry sufficient O_2 to the tissues. Severe symptoms of asphyxiation (impairment of vision, hearing, and thought) or even death may therefore result from exposure to carbon monoxide. Permanent brain damage is also a common consequence.

Table 30.2 Percentage of blood by volume occupied by corpuscles

ANIMAL	ALTITUDE	CORPUSCLES (%)
Human	Sea level	46
	5360 m	60
Sheep	Sea level	35
	4700 m	50
Dog	Sea level	35
	4540 m	50
Rabbit	Sea level	35
	5340 m	57
Vicuña	Sea level	30
	4700 m	32

SOURCE: Data from C. L. Prosser and F. A. Brown, 1961. *Comparative Animal Physiology*, Saunders, Philadelphia.

CHAPTER SUMMARY

WHY CIRCULATION IS NECESSARY

Circulation is needed to bring nutrients to cells and carry wastes away in organisms too large for simple diffusion to suffice. (p. 841)

THE INSECT STRATEGY: OPEN CIRCULATION

The blood is collected from pools and pumped to the top and front of the organism, and then allowed to percolate down and back through the tissues into collecting pools. (p. 842)

THE VERTEBRATE STRATEGY: CLOSED CIRCULATION

THE CIRCUIT IN HUMANS Blood is pumped by the right ventricle of the heart through the pulmonary arteries to the capillaries of the lungs (the sites of gas exchange); the pulmonary veins carry it back to the left atrium of the heart. The blood is then pumped to the left ventricle and into the systemic arteries. After reaching the systemic capillaries (the sites of gas, nutrient, and waste exchange), blood is collected by the systemic veins and returned to the right atrium of the heart. (p. 843)

THE CIRCUIT IN OTHER VERTEBRATES In fish, blood moves in a single circuit from the heart to the gill capillaries, then to the systemic capillaries, and then back to the heart. In amphibians and reptiles there is a separate pulmonary circuit, but the heart is not completely divided; as a result some degree of mixing of the pulmonary and systemic circulation can occur—no disadvantage when the organism is capable of a significant degree of skin breathing. (p. 847)

HOW THE HEART WORKS The heartbeat is produced spontaneously by the S-A node, where it spreads slowly across the atria (causing them to contract) to the A-V node, from which the fibers of the bundle of His carry the activity rapidly to the ventricles, initiating contraction there. One-way valves in the heart prevent blood from moving in the wrong direction. (p. 849)

HOW THE CAPILLARIES WORK The narrow capillary channels bring serum and blood cells within a few cell diameters of every cell in the body. Through diffusion, nutrients are exchanged for wastes, and oxygen is traded for carbon dioxide. Sphincters control blood flow so that only certain necessary capillaries receive maximal blood flow. (p. 852)

WHAT DOES THE LYMPH DO Much of the fluid in the blood is forced out of the capillaries, but nearly all is reabsorbed; most of the rest is collected by the lymph and returned to the venous circulation near the heart. The lymph is an important part of the immune system. (p. 858)

HOW TEMPERATURE IS REGULATED

WHY BOTHER? The activity of enzymes increases dramatically with temperature until they denature. Body chemistry is thus most efficient when the temperature is kept high but safely below the denaturing temperature. Endotherms can warm themselves; homeothermic endotherms maintain a constant, elevated internal temperature; ectotherms depend on external heat to warm them. (p. 859)

GETTING WARM Waste heat from metabolism is the major source of body warmth; shivering of muscles or the purposely inefficient metabolism of fat reserves can be used to warm the blood when necessary. (p. 861)

STAYING WARM Insulation, cutting off peripherial circulation, and countercurrent heat recovery are used to minimize heat loss. (p. 862)

COOLING OFF Shunting blood to peripheral capillaries, evaporating water, and countercurrent exchange between capillaries and airflow during panting are used to dispose of dangerous excess heat. (p. 863)

THE NATURE OF BLOOD

WHY IS PLASMA IMPORTANT? Plasma is the liquid part of the blood. It maintains the osmotic balance between cells and tissue fluid; it also carries antibodies and blood-clotting agents, transport proteins, nutrients, hormones, vitamins, gases, and metabolic wastes from one part of the body to another. (p. 865)

HOW BLOOD CLOTS When triggered by chemical signals, fibrinogen in the serum is converted into fibrin, which forms a clot. (p. 867)

THE ROLE OF RED BLOOD CORPUSCLES Corpuscles serve primarily as a means to increase the density of oxygen-carrying hemoglobin in the blood. (p. 869)

CARRYING THE OXYGEN: HEMOGLOBIN Hemoglobin is a tetrameric protein that can bind up to four oxygen molecules. Because its affinity for oxygen is lower at low pH values, it tends to release oxygen in regions of the body with high CO_2 concentrations. Since hemoglobin has a low affinity for the first oxygen, but binds the remaining three more rapidly, it has an S-shaped binding curve; as a result, it takes up oxygen faster in the oxygen-rich lungs and releases it more rapidly in the oxygen-poor tissues. (p. 870)

STUDY QUESTIONS

1 Redesign the distribution of oxygenated blood in the amphibian to allow for skin breathing. Considering the blood chemistry involved, would a fully separated heart work for an amphibian, or is the mixing tolerated simply because these creatures tend to be very inactive? (pp. 847–848)

2 Hemoglobin unloads oxygen to myoglobin, the oxygen-storage monomer in muscles. What must the myoglobin dissociation curve look like? (pp. 871–875)

3 Under what conditions should panting and/or sweating be ineffective or even dangerous? What behavioral countermeasures might animals need to take? (pp. 863–864)

4 Some athletes train at high altitudes for meets to be held at sea level. For what sorts of events might this be a smart strategy? (pp. 874–875)

5 Is there any plausible physiological use for concurrent exchange? (pp. 861–864)

6 How would acid rain affect the gas-exchange chemistry of fish? What about thermal pollution or greenhouse warming of ponds? (pp. 871–874)

SUGGESTED READING

HAJJAR, D. P. AND A. C. NICHOLSON, 1995. Atherosclerosis, *American Scientist* 83(5). *On the cellular and molecular bases of this increasingly common disease.*

LAWN, R. M., AND G. A. VEHAR, 1986. The molecular genetics of hemophilia, *Scientific American* 254 (3). *Insights into how clotting works.*

LILLYWHITE, H. B., 1988. Snakes, blood circulation, and gravity, *Scientific American* 259 (6). *Interesting comparison of circulatory adaptations of aquatic, terrestrial, and arboreal snakes.*

ROSS, R., 1993. The pathogenesis of atherosclerosis: a perspective for the 1990s. *Nature* 362, 801–808. *Focuses on the molecular biology of the disease.*

VOGEL, S., 1994. Nature's pumps, *American Scientist* 82 (5). *Biological pumps have much in common with mechanical pumps.*

ZAPOL, W. M., 1987. Diving adaptations of the Weddell seal, *Scientific American* 256 (6). *On the many specializations in this mammal's circulatory system that permit lengthy dives to great depths.*

CHAPTER 31

REGULATION OF BODY FLUIDS

Life originated in the ancient seas. The cytoplasm of the early cells and the internal fluids of primitive marine organisms were compatible with the stable composition of the sea water that bathed them; organisms depended on the stable conditions in sea water. But as evolution continued, the body fluids of different organisms evolved in different ways. Present-day marine animals, for example, differ noticeably in the chemical makeup of their body fluids, though these fluids are more similar to one another and to sea water than they are to the body fluids of freshwater or terrestrial organisms. Nonetheless, all these fluids have much in common (Table 31.1).

A constant internal fluid environment is essential to survival under varying external conditions. Hence as organisms have become more diverse, selection has necessarily led to the evolution of mechanisms for maintaining homeostasis in their body fluids. This chapter will explore these mechanisms.

HOW PLANTS CAN AVOID REGULATING THEIR FLUIDS

Multicellular marine algae differ markedly from multicellular marine animals in the sort of fluid environment to which their cells are exposed. Roughly 50% of the water in the body of a complex animal is extracellular tissue fluid, lymph, or blood plasma. Most of the cells are bathed in this extracellular fluid, which is separated from environmental water by cellular barriers; its composition differs both from that of the intracellular fluid and the surrounding water.

By contrast, most of the fluid content of a multicellular alga is intracellular, and continuous with the environmental water. The alga thus has no true tissue fluid analogous to the blood of an animal. Whereas animals must regulate the composition of both intracellular and extracellular fluids, the alga must regulate the composition of its intracellular fluids alone.

A similar contrast appears between an animal and a large vascular land plant. Such a plant obviously contains much extracellular fluid in the form of xylem sap and the water in the apoplast. But water can penetrate far into the cortex of a root by flowing along

Table 31.1 Concentrations of ions in sea water and in body fluids (millimoles/liter)

	Na^+	K^+	Ca^{++}	Mg^{++}	Cl^-
Sea water	470	9.9	10.2	53.6	548
Marine invertebrates					
Jellyfish (*Aurelia*)	454	10.2	9.7	51.0	554
Sea urchin (*Echinus*)	444	9.6	9.9	50.2	522
Lobster (*Homarus*)	472	10.0	15.6	6.8	470
Crab (*Carcinus*)	468	12.1	17.5	23.6	524
Freshwater invertebrates					
Mussel (*Anodonta*)	14	0.3	11.0	0.3	12
Crayfish (*Cambarus*)	146	3.9	8.1	4.3	139
Terrestrial animals					
Cockroach (*Periplaneta*)	161	7.9	4.0	5.6	144
Honey bee (*Apis*)	11	31.0	18.0	21.0	?
Japanese beetle (*Popillia*)	20	10.0	16.0	39.0	19
Chicken	154	6.0	5.6	2.3	122
Human	140	4.5	2.4	0.9	100

SOURCE: Based on a larger table in C. L. Prosser and F. A. Brown, 1961. *Comparative Animal Physiology*, Saunders, Philadelphia.

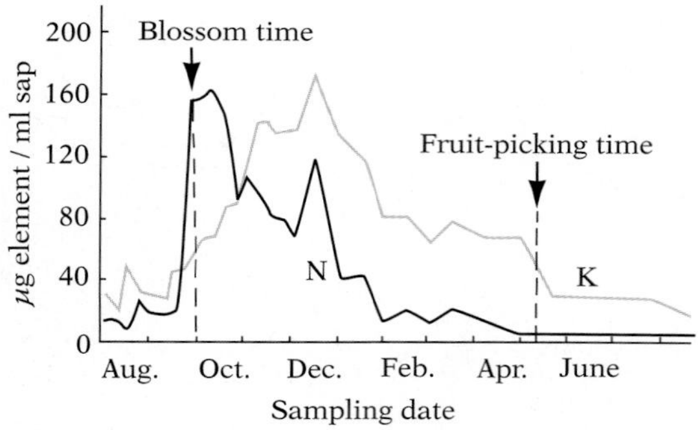

31.1 Yearly fluctuations in nitrogen and potassium concentrations in the xylem sap of apple trees in New Zealand

The fluctuations of both substances are far greater than could be tolerated by animal cells.

cell walls without having to cross any membranous barrier. Thus, much of the fluid that directly bathes plant cells, even those far inside the plant body, is essentially continuous with the environmental water. This means, of course, that the composition of much of the extracellular fluid of the plant cannot be as well regulated as the tissue fluid and blood of animals. Even the composition of sap fluctuates widely. Though separated from external water by the membranous barrier of the endodermis, sap composition depends on such factors as environmental conditions, health of the plant, and season of the year (Fig. 31.1).

Marine algae don't need to regulate the composition of the fluid that bathes their cells because it is essentially the same as sea water. What about the typical plant living in fresh water or on land, whose extracellular fluids fluctuate much more than the tissue fluids of animals and the seawater medium of marine algae?

How the cells of terrestrial and freshwater plants are able to withstand great fluctuations in the composition of the fluids bathing them is only partly understood. Unlike the typical animal cell, which thrives in fluids with an osmotic concentration like its own, the cell of a land or freshwater plant almost always exists in a medium that is much more dilute than the cell's contents. In such a situation an animal cell would take in so much water by osmosis that it would burst, unless it had some special mechanism for expelling the excess water. In contrast, the plant cell is surrounded by its cell wall; as the cell takes in more water and becomes turgid, the wall resists further expansion. Eventually the resistance of the wall is as great as the tendency of water to enter the cell by osmosis, so no further water enters.

While plant cells can function in fluids that are dilute (hypo-

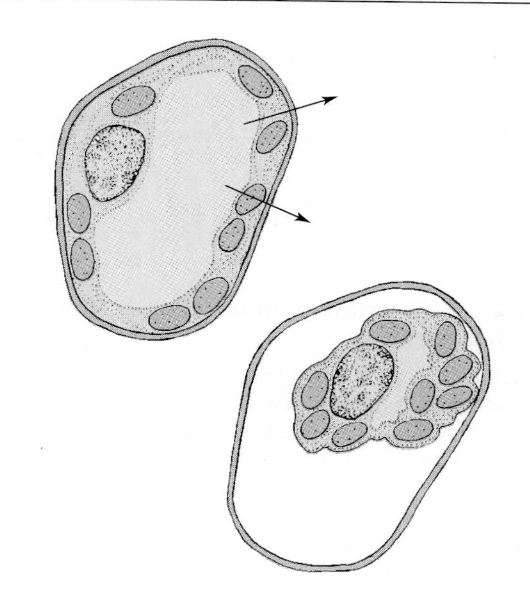

31.2 Plasmolysis

A plant cell in a hypertonic medium (left) loses so much water that, as it shrinks, it pulls away from its more rigid wall (right).

tonic) relative to the cell contents, when external fluids become hypertonic relative to the cell, it may lose much of its water and pull away from its wall; this phenomenon is called ***plasmolysis*** (Fig. 31.2). The presence of the cell wall in plants and its absence in animals thus make the problem of salt and water balance quite different in each.

Land plants, like land animals, are often at risk of excessive water loss by evaporation. Plants resist desiccation through a number of strategies: having cuticle on their exposed surfaces, regulating stomatal openings via guard cells, placing stomata in deep hair-lined pits (in some plants living in very dry regions), and so on. Some plants, including many that grow only in shade, show little tolerance of drought; when the soil dries out, the cells lose their turgidity, and the plant wilts and eventually dies.

Some plants, including many mosses, lichens, and ferns, are drought-tolerant because their cells can be dehydrated without permanent injury. Other plants can survive long periods of drought because they store large quantities of water and lose little by evaporation, thanks to having very thick cuticles, few stomata, and low surface-to-volume ratios. Cacti and other succulent desert plants are good examples. Most plants are only moderately well equipped to withstand drought; they depend on widespread or deep-penetrating root systems that increase absorptive capacity.

HOW THE LIVER REGULATES BLOOD CHEMISTRY

Consider the blood leaving the intestinal capillaries shortly after a meal. Digestion is taking place in the small intestine, and the products of digestion move in large quantities into the capillaries of the intestinal villi. The blood leaving these capillaries contains high concentrations of simple sugars and amino acids—concentrations considerably greater than those normally found in the blood in most parts of the circulatory system. The wholesale addition of these materials to the blood, if not controlled, would drastically alter the composition of the blood and other body fluids, and make the maintenance of a stable fluid environment for the cells impossible.

Vertebrates overcome this difficulty with the help of a large homeostatic organ, the liver. Blood from the capillaries of the intestine and stomach is collected in the ***portal vein***, which goes to the liver, where it breaks up into a network of capillaries (or, more precisely, sinuses; Fig. 31.3). The liver is one of only four places in the mammalian body where blood passes through a second set of capillaries before returning to the heart[1]; all other blood circuits involve only a single capillary bed.

[1] One is the heat-exchange rete discussed in Chapter 30. The others, to be discussed later, are in the kidney and the pituitary gland.

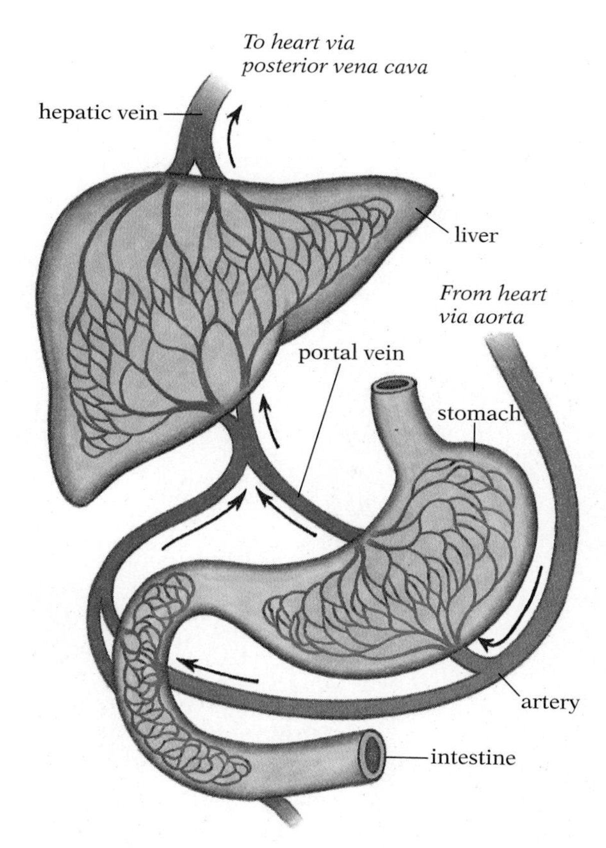

31.3 Hepatic portal circulation

Blood from the capillaries of the digestive tract and stomach is carried by the portal vein to a second bed of capillaries in the liver. It then flows via the hepatic vein into the posterior vena cava, which takes it to the heart.

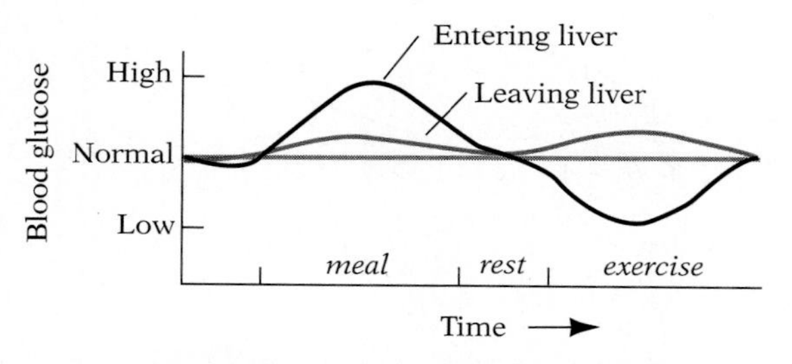

31.4 Regulation of blood sugar by the vertebrate liver

The liver acts to maintain constant levels of many substances in the blood. In this example, the liver removes and stores excess glucose carried by the blood from the intestine after a meal (left), but later liberates glucose into the blood during exercise, when muscles begin using up this source of energy at an unusually high rate (right). The slight excess of glucose in the blood leaving the liver is reduced to the normal level when it mixes in the heart with other returning blood.

■ REGULATING BLOOD-SUGAR LEVEL

After a meal, the blood coming to the liver has a higher-than-normal concentration of glucose (Fig. 31.4, left). The liver removes most of the excess, converting it into the insoluble polysaccharide glycogen, which is the principal storage form of carbohydrate in animal cells. Therefore, the blood leaving the liver via the ***hepatic vein*** (a vein leading into the posterior vena cava) contains a concentration of glucose only slightly higher than that normally found in the arteries. This blood then mixes in the vena cava with blood from other parts of the body, which contains a lower concentration of sugar because it has given up glucose to the tissues through which it has passed. As a result, the blood entering the heart now has a glucose concentration within the normal range.

If the incoming supply of glucose from the intestine exceeds all of the body's immediate needs, and the liver has stored its full capacity of glycogen, the liver begins converting glucose into fat, which can then be stored in the various regions of adipose tissue throughout the body. Thus, in spite of the great quantities of glucose absorbed by the intestine, the blood-sugar level in most of the circulatory system is not greatly raised and homeostasis is maintained.

The whole process is reversed when exercise depletes the supply of glucose in the blood. At such times, blood in the muscle capillaries gives up glucose to the muscle cells. This means that the blood reaching the liver is very low in glucose. The liver converts some of its stored glycogen into glucose and adds it to the blood, so that blood leaving the liver in the hepatic vein has at least its normal glucose concentration (Fig. 31.4, right). The liver functions in helping to maintain a homeostatic blood-sugar concentration by converting a highly variable input into a virtually constant output, which it accomplishes by actively adding and subtracting substances from the blood.

The human liver is capable of storing enough glycogen to supply glucose to the blood for a period of about 4 hours. If no new glucose has come to the liver from the intestine for 4 hours, tissues are at risk. In particular, brain cells cannot store much glucose themselves or use fats or amino acids as energy sources, and these cells are thus wholly dependent on a regular supply of glucose from the blood. Under such conditions the liver converts other substances, such as amino acids, into glucose, and in this way maintains the normal blood-sugar level. Some other tissues of the body, particularly muscle, can store glucose as glycogen, but muscle glycogen serves as a local fuel deposit only.

■ METABOLIZING LIPIDS AND AMINO ACIDS

The liver is responsible also for the processing and modification of fatty acids and other lipids. For example, much of the fatty acid mobilized from adipose tissues during periods of starvation is incorporated into lipoprotein by the liver before it is used as a source of metabolic energy by other tissues of the body.

The liver's role in amino acid metabolism is more complicated. Like glucose, amino acids absorbed by the villi of the intestine pass to the liver, where many are removed from the blood for temporary storage until they may be needed between meals; then they are returned to the blood, which carries them to the tissues that need them. But the usual diet contains far more amino acid than is needed in protein syntheses. Animals are unable to store amino acids, proteins, or other nitrogenous compounds for very long; those compounds used as an energy source when supplies of carbohydrates and fats are exhausted are not stored products but the actual structural material of living cells. Excess amino acids from the diet must be converted into other substances, such as glucose, glycogen, or fat. These conversions take place in the liver.

Amino acids differ from most carbohydrates and fats in that they contain nitrogen—an amino group (NH_2). The first step in converting amino acids into these other substances is ***deamination***, or the removal of the amino group. The amino group is converted into ***ammonia*** (NH_3).[2] In some animals the liver simply releases this waste ammonia into the blood, and it is soon removed from the blood and excreted. In many other animals, including humans, the liver first combines the ammonia with carbon dioxide to form a more complex but less toxic nitrogenous compound, called ***urea*** (Fig. 31.5), which is released into the blood. In still other animals, the liver converts the waste ammonia into an even more complex compound, ***uric acid***, which it releases into the blood. No matter what the nitrogenous waste product is, the liver dumps it into the blood, and another regulatory system of the body acts to prevent the wastes from reaching too high concentrations in the body fluids.

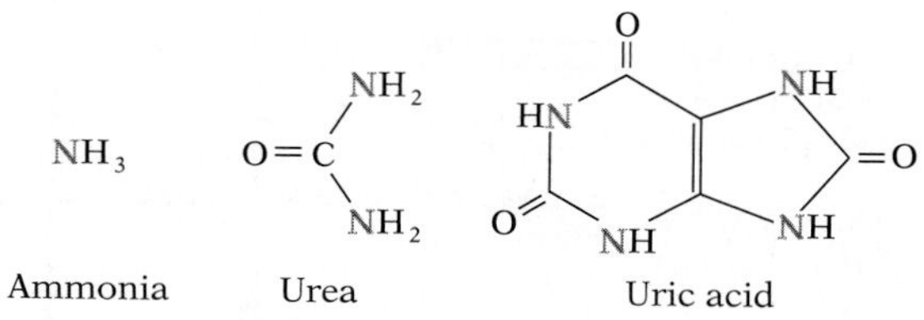

NH_3 — Ammonia; $O=C(NH_2)_2$ — Urea; Uric acid

31.5 Three important nitrogenous waste compounds

Because the ammonia produced as a result of deamination of amino acids can be toxic in relatively low concentrations and requires considerable water for its elimination, some animals combine it with CO_2 to produce urea, which is less toxic. Other animals go a step farther by converting ammonia into uric acid, which is insoluble and nontoxic. Though the conversion of ammonia to uric acid requires a considerably greater expenditure of ATP than conversion to urea, the greater energy investment may be worthwhile for certain land animals, since less water is required for uric acid elimination.

■ A SUMMARY OF THE LIVER'S FUNCTIONS

The range of liver functions is remarkable, as this partial list suggests:

1 The liver removes excess glucose from the blood and stores it as glycogen, and reconverts glycogen into glucose to maintain the blood-sugar level when the incoming supply is insufficient (Fig. 31.4).

2 It resynthesizes glycogen from some of the lactic acid produced by muscles during glycolysis.

3 It plays a major role in the interconversion of various nutrients, such as carbohydrates into fats, incoming fats into fats more typical of the organism's own body, and amino acids into carbohydrates or fats.

4 It deaminates amino acids; converts the ammonia thus obtained into urea, uric acid, or some other compound; and releases the nitrogenous wastes into the blood.

5 It detoxifies a great variety of injurious chemical compounds, and is thus one of the body's most important defenses against poisons such as alcohol and barbiturates.

6 It manufactures many of the plasma proteins, including fibrinogen, prothrombin, albumin, and some globulins.

[2] At physiological pH, most of the ammonia is rapidly converted into the ammonium ion (NH_4^+).

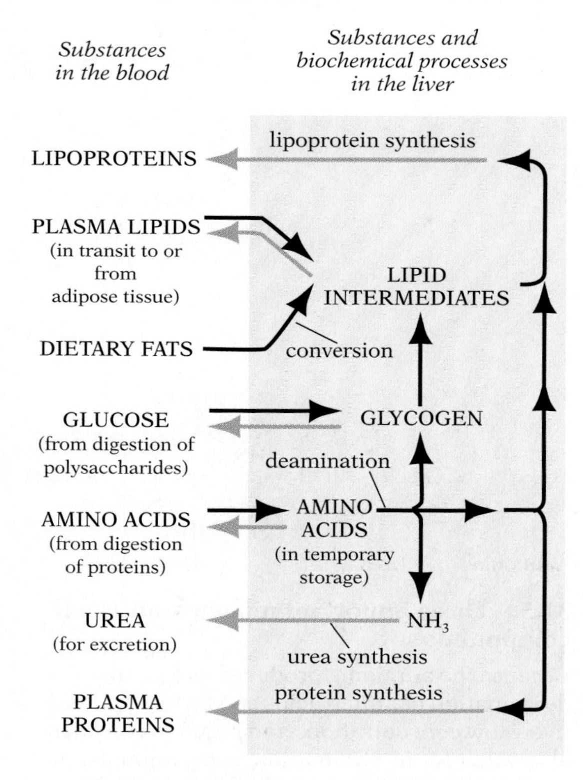

31.6 Chemical pathways in the human liver

The liver removes excess sugars, fats, and amino acids (among other things) from the blood when their concentration is too high, and releases them into the blood when their circulating levels fall too low. The liver also synthesizes lipoproteins, plasma proteins, and plasma lipids. Finally, the liver detoxifies many substances, and produces urea. Some of the chemical pathways involved are diagrammed here.

7 It manufactures some plasma lipids, including cholesterol, and plays an important role in processing and modifying lipids mobilized from adipose tissues.

8 It stores various important substances such as vitamins and iron.

9 It forms red blood corpuscles in the embryo.

10 It destroys aging red blood corpuscles.

11 It excretes bile pigments.

12 It synthesizes bile salts.

A few of these functions are summarized in Figure 31.6.

HOW ANIMALS CONTROL WATER AND SALT LEVELS

The process of releasing useless, even poisonous, substances like waste nitrogen is called ***excretion***. It should not be confused with ***elimination*** (defecation). Whereas excretion is the release of wastes that have been inside the cells, tissue fluids, or blood of the organism, elimination is the release of wastes unabsorbed by the digestive tract.

In general, excretory mechanisms also serve a second important function: they help regulate the water and salt balance—processes that are in most instances inextricably intertwined.

■ PROBLEMS WITH LIVING IN DIRECT CONTACT WITH WATER

The first nitrogenous waste formed by deamination of amino acids is ammonia. Since ammonia is an exceedingly poisonous compound, no organism can survive if the ammonia concentration in the body fluids gets very high. However, since the small, highly soluble molecules of ammonia readily diffuse across cell membranes, there is no great difficulty in getting rid of them if an adequate supply of water is available. Thus many aquatic animals do not need to modify their waste ammonia since it can be promptly disposed of.

How marine invertebrates manage Many marine invertebrates lack special excretory systems; they simply release wastes across surface membranes. Such organisms seldom have any problem with water balance, because they are essentially isotonic with respect to the surrounding sea water, and hence neither take in or lose much water.

While maintaining a proper internal fluid environment is relatively simple for marine invertebrates as long as they remain in the sea, it is quite a different matter when they move into a hypotonic environment such as the brackish water of estuaries or the fresh waters of rivers and lakes. Many marine animals, including the spider crab (*Maia*; Fig. 31.7), are incapable of moving into such habitats, because their body fluids always lose salts until they have about the same salinity and osmotic concentration as their environment. These animals soon die because their cells cannot tolerate such changes.

Some marine animals, however, have evolved adaptations that enable them to move into hypotonic media. Some adaptations are merely mechanical: oysters and clams simply close their shells and exclude the external water during those parts of the tidal cycle when the water in the estuaries is very dilute. Others—those that played the principal role in the evolutionary movement of animals into fresh water—enable animals to regulate the osmotic concentrations of their body fluids and keep them constant despite fluctuations in the external medium. Such organisms are said to have the power of ***osmoregulation***.

The shore crab (*Carcinus*) has evolved sufficient osmoregulation to permit it to live in both sea water and brackish water (Fig. 31.7). In sea water the crab's body fluids are in osmotic equilibrium, but in brackish water they are hypertonic relative to the surrounding medium. To maintain the internal fluids near their normal concentration in brackish water, cells on the crab's gills remove salt from the surrounding water and actively secrete it into the blood, while the excretory organs eliminate the excess water that constantly pours in.

The strategies of freshwater animals Once the ancestors of the modern freshwater animals had made a permanent transition to freshwater environments, presumably by way of the estuaries, there was no longer any advantage in maintaining body fluids as concentrated as sea water. Such excessively hypertonic fluids would simply aggravate the problems of obtaining enough salt and bailing out excess water in a freshwater medium. Selection favored a reduction of the osmotic concentration of the body fluids; modern freshwater animals, both invertebrate and vertebrate, have osmotic concentrations decidedly lower than sea water (Table 31.2). Of course it would be impossible for body fluids, with their essential salts, nutrients, and wastes, to be as dilute as fresh water; the body fluids of freshwater animals are typically hypotonic relative to sea water, but hypertonic relative to fresh water (Table 31.2).

Since freshwater animals are hypertonic relative to the sur–

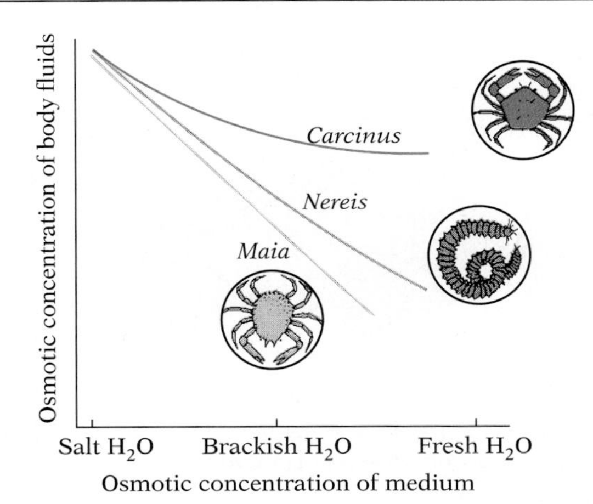

31.7 Variation of internal osmotic concentration with external osmotic concentration in three marine invertebrates

The spider crab (*Maia*) has no osmoregulatory capacity, and the concentration of its body fluids falls in direct proportion to the fall in the concentration of the external medium. The clam worm (*Nereis*) has very slight osmoregulatory capacity; the concentration of its body fluids does not bear a straight-line relationship to the concentration of the external medium. The shore crab (*Carcinus*) has considerable osmoregulatory capability and can maintain relatively concentrated body fluids even in a very dilute external medium.

Table 31.2 Concentrations of ions in the blood of freshwater animals compared with those of sea water and fresh water (millimoles/liter)

	Na^+	K^+	Ca^{++}	Mg^{++}	Cl^-
Sea water	470	9.9	10.2	53.6	548
Brown trout (*Salmo*)	144	6.0	5.3	?	151
Crayfish (*Cambarus*)	146	3.9	8.1	4.3	139
Mussel (*Anodonta*)	14	0.3	11.0	0.3	12
Fresh water [a]	0.65	0.01	2.0	0.2	0.5

SOURCE: Based on a larger table in C. L. Prosser and F. A. Brown, 1961. *Comparative Animal Physiology*, Saunders, Philadelphia.

[a] The values for fresh water are representative only; actual values vary greatly, as the water ranges from "soft" (low concentrations of dissolved minerals, such as 0.22 Ca^{++}) to "hard" (high concentrations of dissolved minerals, such as 5.0 Ca^{++}).

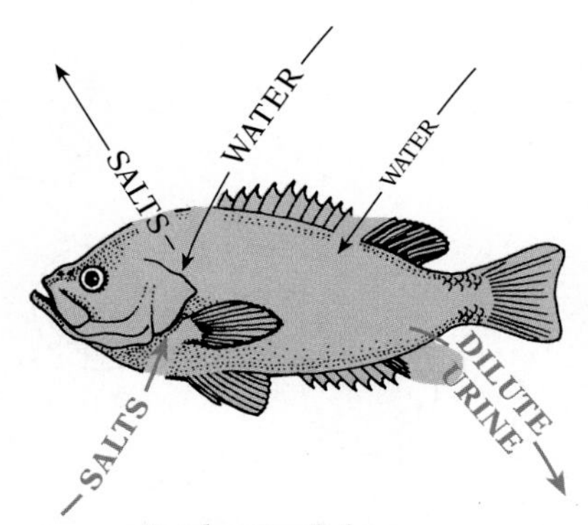

Freshwater fish
(hypertonic relative to medium)

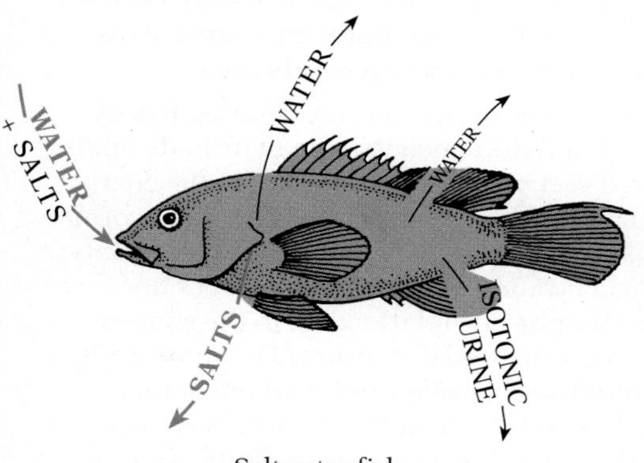

Saltwater fish
(hypotonic relative to medium)

Passive processes
Active processes

31.8 Osmoregulation in bony fish

Freshwater fish tend to take in excessive amounts of water and to lose too much salt. They compensate by seldom drinking, by actively absorbing salts through specialized cells in their gills, and by excreting copious dilute urine. Saltwater fish tend to lose too much water and to take in too much salt. They compensate by drinking constantly and by actively excreting salts across their gills. They cannot produce hypertonic urine; hence the kidneys are of little aid to them in osmoregulation.

rounding medium, there is a strong tendency for water to move into the organism and for salts to be lost to the surrounding water. A freshwater animal cannot solve this problem by resorting to impermeable membranes covering the entire body; some permeable membranes must be exposed to the water for gas exchange. (Because mammals that live in the water breathe air, never exposing respiratory membranes to the water, they *can* maintain an impermeable barrier between their body fluids and the water in which they live.) Thus freshwater animals must be able to carry out active osmoregulation to compensate for osmotic uptake and loss in the gills; this usually involves excretory organs that can remove the water as fast as it floods in—preferably through the production of urine more dilute than the body fluids—and/or special secretory cells somewhere on the body that can absorb salts from the environment and release them into the blood. Both tasks involve moving materials against concentration gradients, and therefore require energy.

The water and salt regulation of freshwater fish illustrates the problems and the usual solutions. The blood and tissue fluids of these fish are more concentrated than the environmental water (Table 31.2). Much of the body is covered by relatively impermeable skin and scales, and freshwater fish almost never drink. There is, however, a constant passive osmotic intake of water and loss of salt across the membranes of the gills and mouth. Correction of the resulting imbalance occurs in two ways: excess water is excreted in the form of very dilute and copious urine produced by the kidneys, and salts are actively absorbed by specialized cells in the gills (Fig. 31.8).

Why marine fish are at risk of drying out Curiously enough, most marine fish have the reverse problem: they steadily lose water to their environment and so are in constant danger of dehydration. The explanation is that the ancestors of the bony fish lived in fresh water, not in the sea; when some of their descendants moved back to the marine environment they retained their dilute body fluids. Marine bony fish are therefore hypotonic relative to the surrounding water, and have the problem of excessive water loss and salt intake. They use two corrective measures: they drink almost continuously to replace the water they are constantly losing, and, by means of specialized cells in the gills, they actively excrete the salts they unavoidably take in with the water (Fig. 31.8). Most of the nitrogenous wastes are excreted as ammonia through the gills; hence only a small quantity of urine is produced by the kidneys, so little water need be lost in this manner. Apparently the kidneys of fish have not evolved the capacity to produce urine more concentrated than their blood, and are consequently of no help in salt elimination.

The marine elasmobranch fish (sharks and their relatives) probably also evolved from freshwater ancestors, but they solved the osmotic problem in a very different way. Their blood contains about the same concentrations of salt as the blood of marine bony fishes, but their blood also contains high concentrations of urea, to which they are much more tolerant than most vertebrates. By conserving urea instead of excreting it, the marine elasmobranchs maintain a total osmotic concentration in their blood slightly greater than that of sea water. They therefore have no problem of water loss. Excess salt is excreted by special glandular cells in the rectum.

■ PROBLEMS WITH LIVING ON LAND

On land the greatest threat to life is desiccation. Water is lost by evaporation from the respiratory surfaces, by evaporation from the general body surface, by elimination in the feces, and by excretion in the urine. The lost water must be replaced if life is to continue. It is replaced by drinking, by eating foods containing water, and by the oxidizing of nutrients (water being one of the products of cellular respiration).

Though ammonia is a common nitrogenous excretory product for aquatic animals, where it can diffuse rapidly into the surrounding medium, this highly toxic substance would build up to fatal levels in terrestrial animals. Amphibians and mammals rapidly convert ammonia to urea, a compound that, though very soluble, is relatively nontoxic. Urea can remain in the body for some time before being excreted.

Though urea is a far more satisfactory excretory product than ammonia for land animals, it has the disadvantage of requiring critically needed water to keep it in solution. If, however, uric acid, a very insoluble compound, is excreted instead of urea, almost no water need be lost. Many terrestrial animals—most reptiles, birds, insects, and land snails—excrete uric acid or its salts. The excretion of this substance not only allows them to conserve water, but has another important advantage: each of these animals lays eggs enclosed within a relatively impermeable shell or membrane. Uric acid is so insoluble that it can be precipitated by the developing animal in almost solid form and stored in the egg without exerting harmful toxic or osmotic effects. In the nitrogen metabolism of fully terrestrial animals, uric acid excretion is correlated with egg laying, while urea excretion is correlated with viviparity (giving birth to living young).

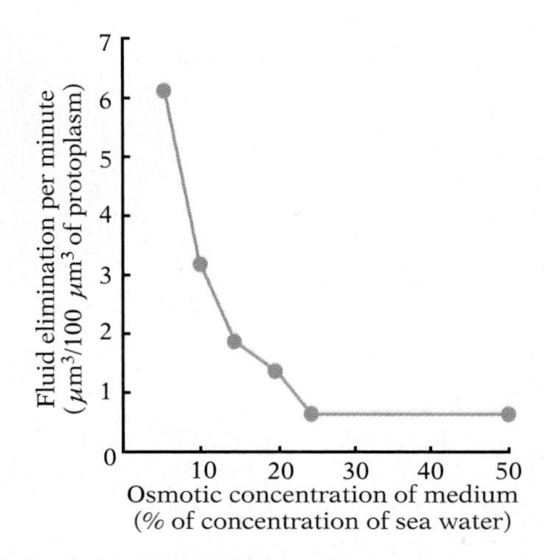

31.9 Rate of fluid elimination by contractile vacuole of *Amoeba lacerata* as a function of the osmotic concentration of the medium

Contractile-vacuole activity falls precipitously as the concentration of the medium goes up.

DISPOSING OF WASTES AND EXCESS WATER

■ EXPELLING WATER IN BURSTS: CONTRACTILE VACUOLES

Many unicellular and simple multicellular animals have no special excretory structures. Nitrogenous wastes are simply excreted across the general cell membranes into the surrounding water. Some protozoans do, however, have a special excretory organelle, the contractile vacuole. Each vacuole goes through a regular cycle: it fills with liquid and then contracts, ejecting its contents from the cell. Though some nitrogenous wastes are excreted, the primary function is elimination of excess water. They are much more common in freshwater protozoans than in marine forms, and their rate of fluid elimination becomes slower as the osmotic concentration of the environmental medium increases (Fig. 31.9).

How can a contractile vacuole take in dilute fluid when, by osmosis, water should move out of it, not in? In most protozoans the vacuole is surrounded by a layer of tiny vesicles, and these, in turn, are surrounded by a layer of mitochondria (Fig. 31.10). The vesi–

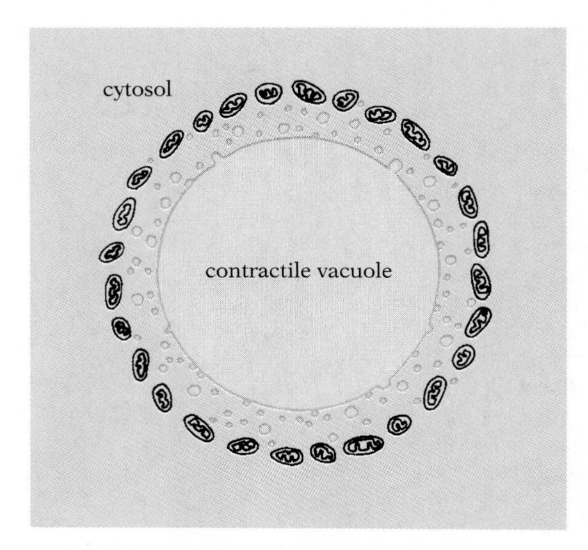

31.10 Contractile vacuole of *Amoeba proteus*

The numerous tiny vesicles around the vacuole fill with fluid; after most of the ions have been pumped out, the vesicles fuse with the vacuole and empty their contents into it. The layer of mitochondria just outside the vesicle layer presumably provides the ATP necessary to pump the ions out of the vesicles and then to expel the contents of the vacuole from the cell.

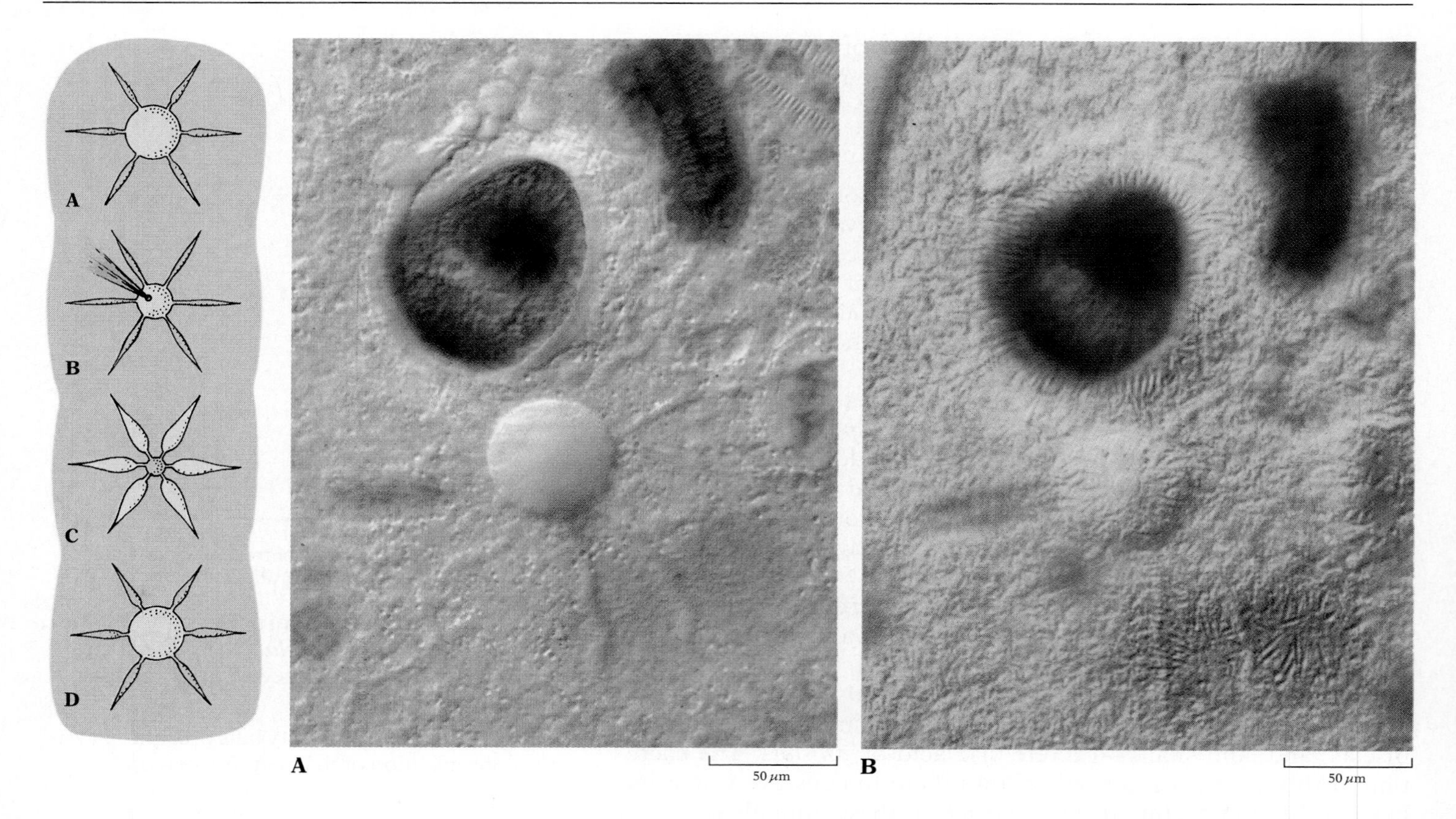

31.11 Contractile vacuole of *Paramecium caudatum*

The sequence shown diagrammatically above can also be observed in the three photographs of a live organism taken in the course of the expansion-contraction cycle of its contractile vacuoles. (Only one vacuole is seen here.) The large green object and the brown ones are remains of other microorganisms ingested by the Paramecium. (A) The vacuole is full. As shown in the diagram, a system of radiating canals brings fluid from the cytoplasm to the vacuole. (B) The vacuole is in the process of expelling its contents; in the photograph the opening to the outside can be seen as a small circle on the surface of the vacuole. (C) The vacuole is nearly empty, but the radiating canals are collecting more fluid from the cytoplasm to fill the reservoir again (D). Photographs taken with Nomarski optics.

cles initially contain a fluid isotonic with the cytosol, but actively pump out ions, using energy from ATP manufactured in the mitochondria. When the osmotic concentration of the vesicular fluid has fallen to about one-third that of the cytosol, the vesicles move to the contractile vacuole and fuse with it. The contractile vacuole grows larger as more and more vesicles merge with it and empty their fluid into it. The membrane of the vacuole itself is impermeable to water, which is held in the vacuole until its contraction expels the water from the cell. It does this through an exocytotic pore created when the vacuole and cell membranes fuse briefly.

In a few ciliate protozoans, fluid enters the contractile vacuole from radially oriented feeder canals rather than from vesicles (Fig. 31.11). The feeder canals, in turn, appear to collect fluid from a network of tiny tubules probably derived from the endoplasmic

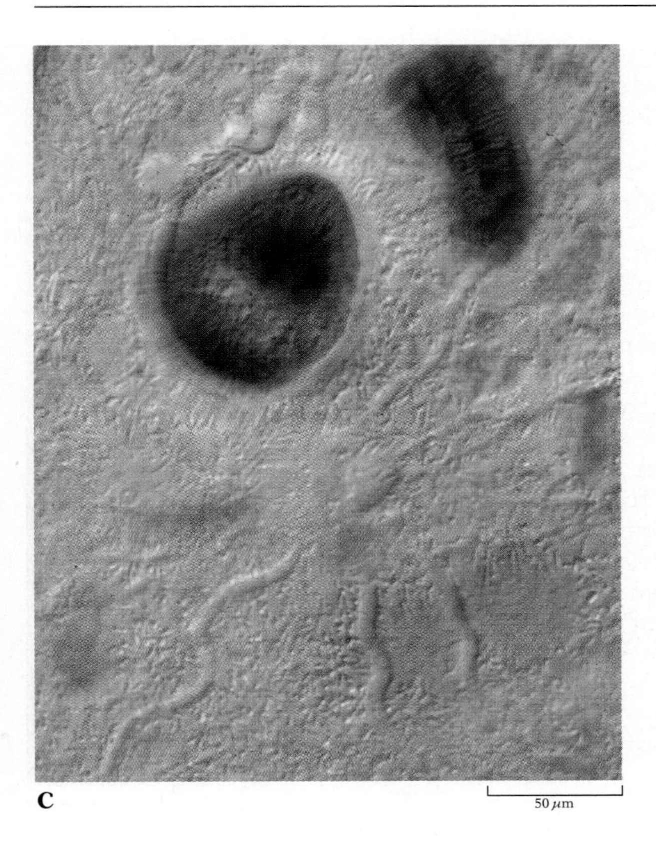

reticulum. Despite these anatomical differences, the mechanism of production of dilute fluid is probably similar to the one already described, in which ions are pumped out before the fluid reaches the contractile vacuole itself.

■ LEAKING WATER: FLAME-CELL SYSTEMS

The beginnings of a tubular excretory system are seen in the flatworms (planarians, flukes, tapeworms, and the like). Flatworm excretory systems usually consist of two or more longitudinal branching tubules running the length of the body (Fig. 31.12). In planarians and their relatives, the tubules open to the body surface through a number of tiny pores. In flukes the tubules unite to form an enlarged bladder that opens to the outside. The critical portions of the systems are many small bulblike structures located at the ends of side branches of the tubules. Each bulb has a hollow center into which a tuft of long cilia projects. The hollow centers of the bulbs are continuous with the cavities of the tubules. Water and some waste materials move from the tissue fluids into the bulbs. The constant undulating movement of the cilia creates a current that moves the collected liquid through the tubules to the excretory pores, where it leaves the body. The motion of the tuft of cilia resembles the flickering of a flame, so this type of excretory system is often called a flame-cell system.

Like the contractile vacuoles discussed earlier, flame-cell systems seem to function primarily in the regulation of water balance; most

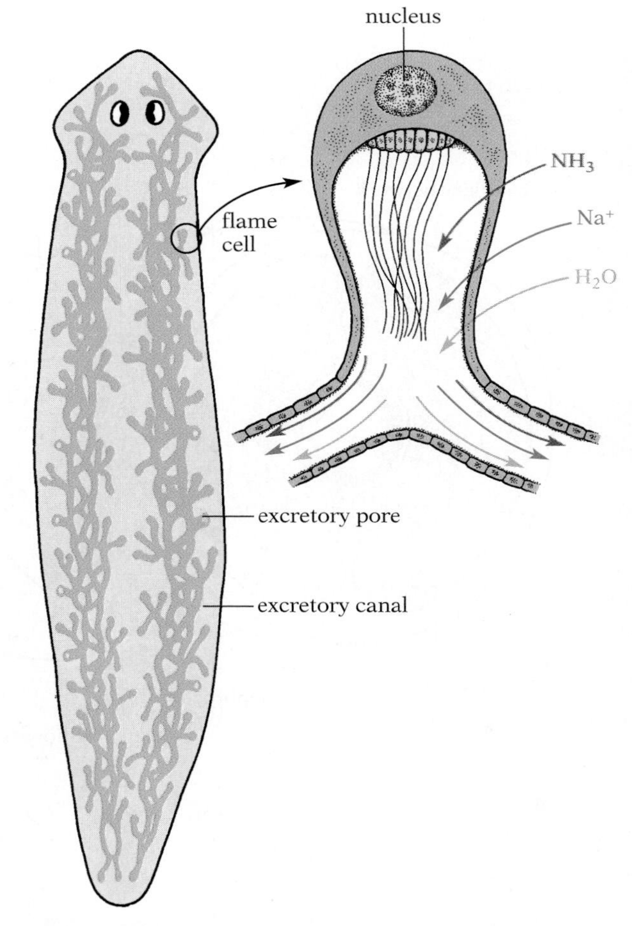

31.12 Flame-cell system of a planarian

Each of the two excretory canals consists of a longitudinal network of tubules, some ending in flame cells (one is shown enlarged at right) and others in excretory pores. The cilia in the flame cells create currents that move water and waste materials through the canals and out through the pores.

metabolic wastes of flatworms are excreted from the tissues into the gastrovascular cavity and eliminated from the body through the mouth.

■ CAPILLARY COOPERATION: NEPHRIDIA

Flame-cell systems, because they function in animals without a circulatory system, pick up substances only from the tissue fluids. In animals that have evolved a closed circulatory system, the blood vessels have become intimately associated with the excretory organs, making possible direct exchange of materials between the blood and the excretory system. This is especially clear in earthworms.

The earthworm's body is composed of a series of segments internally partitioned from each other by membranes. In general, each of the compartments thus formed has its own pair of excretory organs, called nephridia, which open independently to the outside. A typical nephridium (Fig. 31.13) consists of a nephrostome (an open ciliated funnel that corresponds functionally to the bulb in a flame-

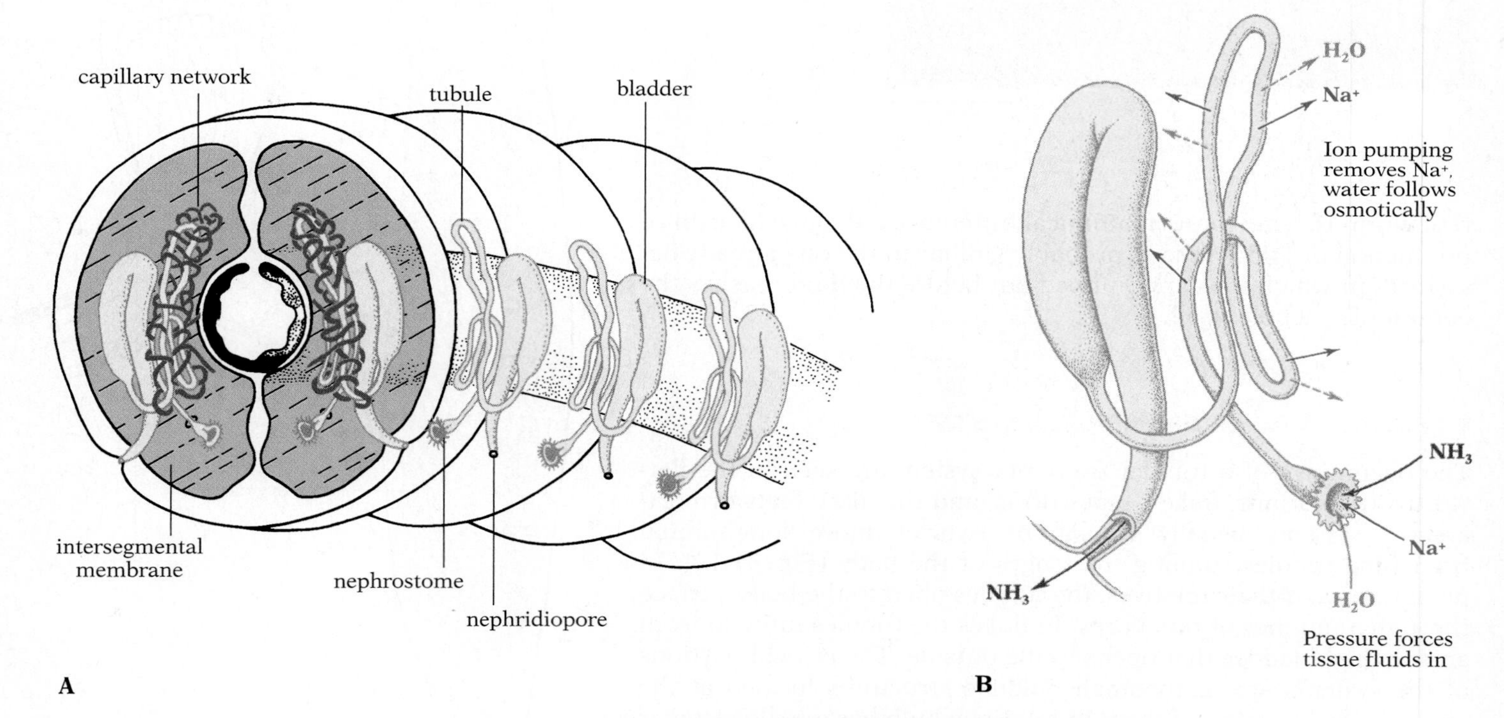

31.13 Nephridia of an earthworm

Each segment of the worm's body contains a pair of nephridia, one on each side. The open nephrostome of each nephridium is located in the segment ahead of the one containing the rest of the nephridium. The tubule from the nephrostome penetrates through the membranous partition between the segments and is then thrown into a series of coils, with which a network of blood capillaries is closely associated. (The capillaries are shown here only on the nephridia of one segment.) The coiled tubule empties into a storage bladder that opens to the outside through a nephridiopore. (The bladder and nephridiopore can be seen in the photograph of a cross section of an earthworm, Fig. 27.15, p. 776).

cell system), a coiled tubule running from the nephrostome, an enlarged bladder into which the tubule empties, and a nephridiopore, through which materials are expelled from the bladder to the outside. Blood capillaries form a network around the coiled tubule. Materials move from the body fluids into the nephridium through the open nephrostome, but some materials are also picked up by the coiled tubule directly from the blood in the capillaries. There is probably also some reabsorption of materials from the tubule into the blood capillaries. The principal advance of this type of excretory system over the flame cell, then, is the association of blood vessels with the coiled tubule.

■ THE INSECT ALTERNATIVE: MALPIGHIAN TUBULES

Insects do not have nephridia. The evolution of an open circulatory system in insects, with their consequent lack of blood capillaries, led to the evolution of entirely new excretory organs, called Malpighian tubules. They are diverticula (blind sacs) of the digestive tract located at the junction between the midgut and the hindgut (Fig. 31.14). They are bathed directly by the blood in the

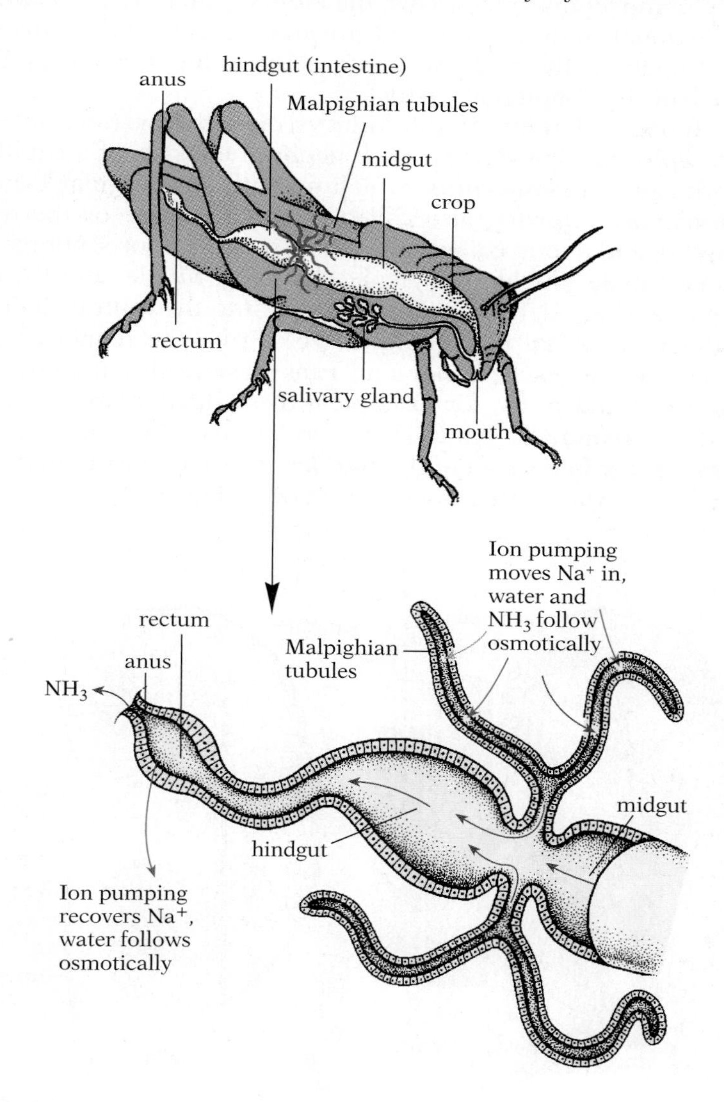

31.14 Malpighian tubules of an insect

These excretory organs arise as diverticula of the digestive system at the junction between the midgut and the hindgut.

open sinuses of the animal's body. Fluid is absorbed from the blood into the blind distal end of the Malpighian tubules. As the fluid moves through the proximal portion of the tubules, the nitrogenous material is precipitated as uric acid, and much of the water and various salts are reabsorbed. The concentrated, but still fluid, urine next passes into the hindgut and then into the rectum. The rectum has very powerful water-reabsorptive capacities, so the urine and feces leave the rectum as very dry material.

The highly effective role played in water conservation by the insect rectum is similar to that of the ***cloaca***, a common chamber in birds and some other vertebrates through which pass materials from the digestive, excretory, and reproductive systems. In particular, the urine passes through the cloaca directly to the posterior portion of the digestive tract. There it is subjected to the powerful reabsorption action of the rectum; the uric acid is therefore eliminated as a nearly dry powder or hard mass.

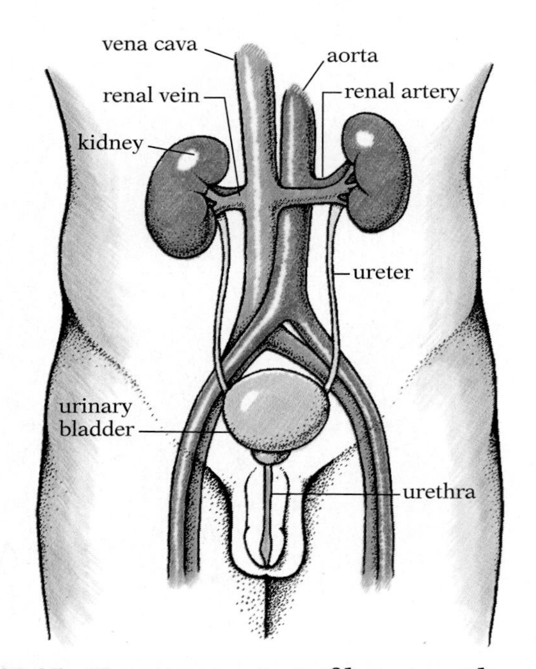

31.15 Excretory system of human males

The organs and vessels are shown larger relative to the body than they actually are.

■ THE VERTEBRATE ANSWER: THE KIDNEY

How the kidney is designed Higher vertebrates have typically evolved compact discrete organs, the kidneys (Fig. 31.15), in which the functional units of excretion are massed rather than distributed throughout the body. In humans the kidneys are located in the back of the abdominal cavity.

The functional units of the kidneys of higher vertebrates are called ***nephrons*** (Fig. 31.16). Each nephron consists of a capillary network called a ***glomerulus***, which fits into an invaginated bulb, the ***Bowman's capsule*** (also called Bowman's space or the renal corpuscle), and a long coiled tubule consisting of four sections: the proximal tubule, the loop of Henle, the distal tubule, and the collecting duct (Fig. 31.17). The ducts empty into the central cavity of the kidney, the renal pelvis (Fig. 31.16). From the renal pelvis, a large duct leaves each kidney and runs posteriorly. In some animals—frogs and birds, for example—these ducts empty into the cloaca. In mammals, which have no cloaca, the ducts (called ***ureters***) empty into the ***urinary bladder***. This storage organ drains to the outside via another duct, the ***urethra*** (Fig. 31.15).

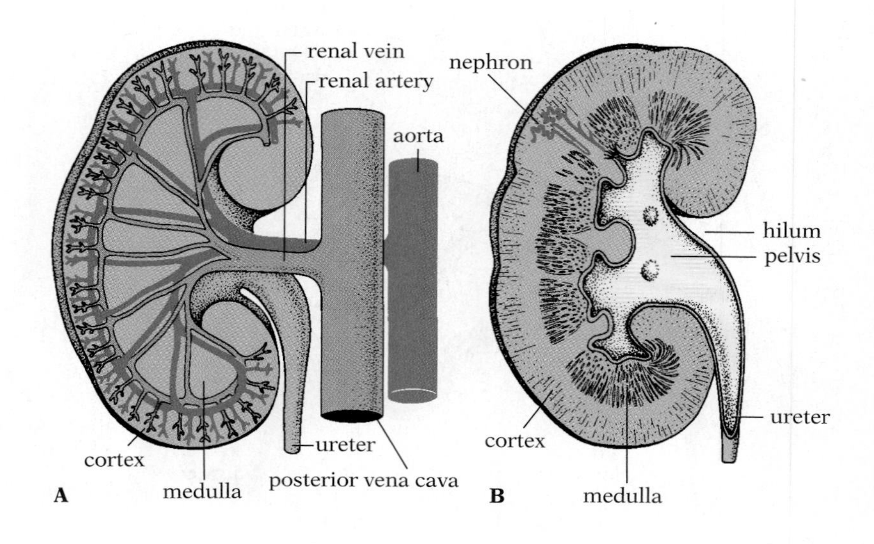

31.16 Sections of the human kidney

(A) The blood circulation of the kidney. (B) The cortex and the medulla, and the large renal pelvis into which the collecting tubules of the nephrons empty. One nephron is shown (pink).

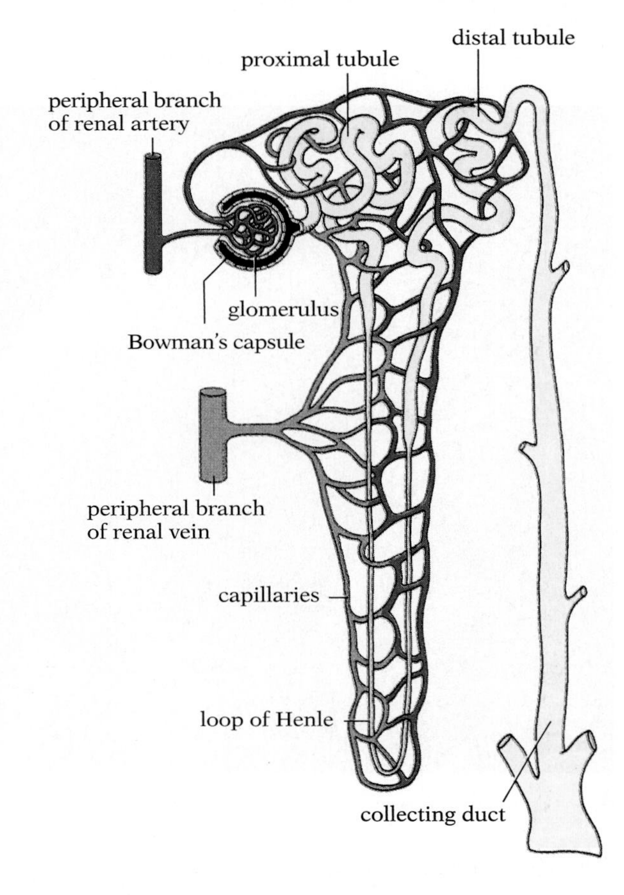

31.17 Human nephron

Each human kidney contains approximately 30–50 km of nephrons. The glomerulus and both the proximal and distal tubules are in the cortex, but the loop of Henle and the collecting duct run down into the medulla (compare with Fig. 31.16). The functioning of the nephron is explained in the text.

Blood capillaries and the capsules, tubules, and ducts of the nephrons are intimately associated in the modern vertebrate kidney. Blood reaches each kidney via a ***renal artery***, a short vessel leading directly from the aorta to the kidney (Fig. 31.16). There it breaks up into many smaller branches that run through the inner portion of the kidney (the medulla) into the outer kidney layer (the cortex), where each of the many tiny branch arterioles penetrates into a cuplike depression in the wall of a Bowman's capsule. Within each capsule, the arteriole breaks up into the glomerulus (Fig. 31.17). Blood leaves the glomerulus via an arteriole formed by the rejoining of the glomerular capillaries. After emerging from the capsule, the arteriole promptly divides again into many small capillaries that form a second dense network around the remaining elements of the nephron. Finally, these capillaries unite once more to form a small vein. The veins from the many nephrons then fuse to form the ***renal vein***, which leads to the posterior vena cava.

How urine is formed: an overview Urine formation involves three processes: ultrafiltration, reabsorption, and tubular secretion. The glomerulus acts as a simple mechanical filter: molecules small enough to pass through the basement membrane of the capillary walls and through the thin membranous walls of the capsule filter from the blood into the tubules as a result of the high hydrostatic pressure of the blood in the glomerulus. The filtration probably depends on pores in the endothelium lining the glomerular capillaries (Fig. 31.18) and between the epithelial cell processes in the wall

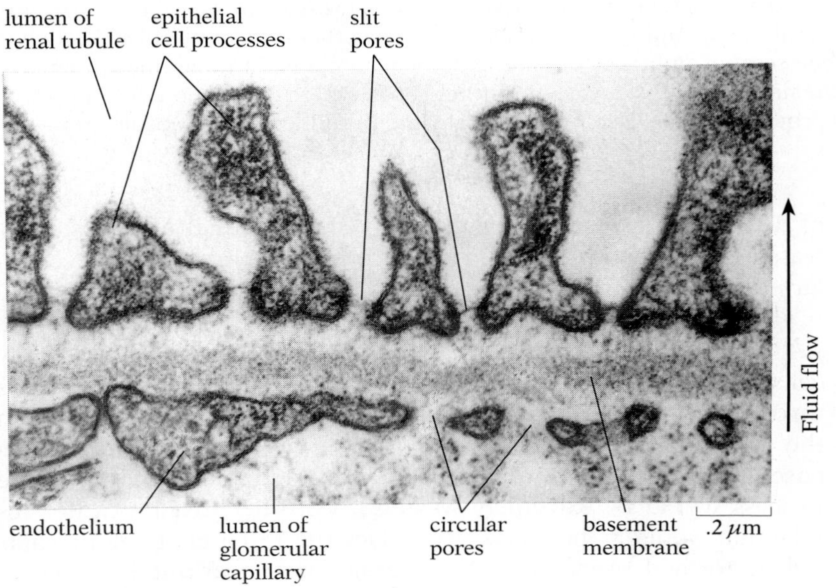

31.18 Section through the glomerular filtration barrier of a rat

The glomerular capillary wall has a complex structure consisting of three layers: a thin endothelium perforated with numerous circular pores, a continuous basement membrane made up of fine filamentous meshwork, and the epithelial cell processes, which are separated by elongated spaces called slit pores. The actual site of filtration is thought to be the basement membrane, which may contain smaller pores (not visible in this electron micrograph) just large enough to allow the passage of water, electrolytes, and other small molecules, and yet small enough to retain albumen and other plasma proteins. (Alternatively, the filtration barrier in the basement membrane may be a hydrated gel rather than a layer of discrete pores.)

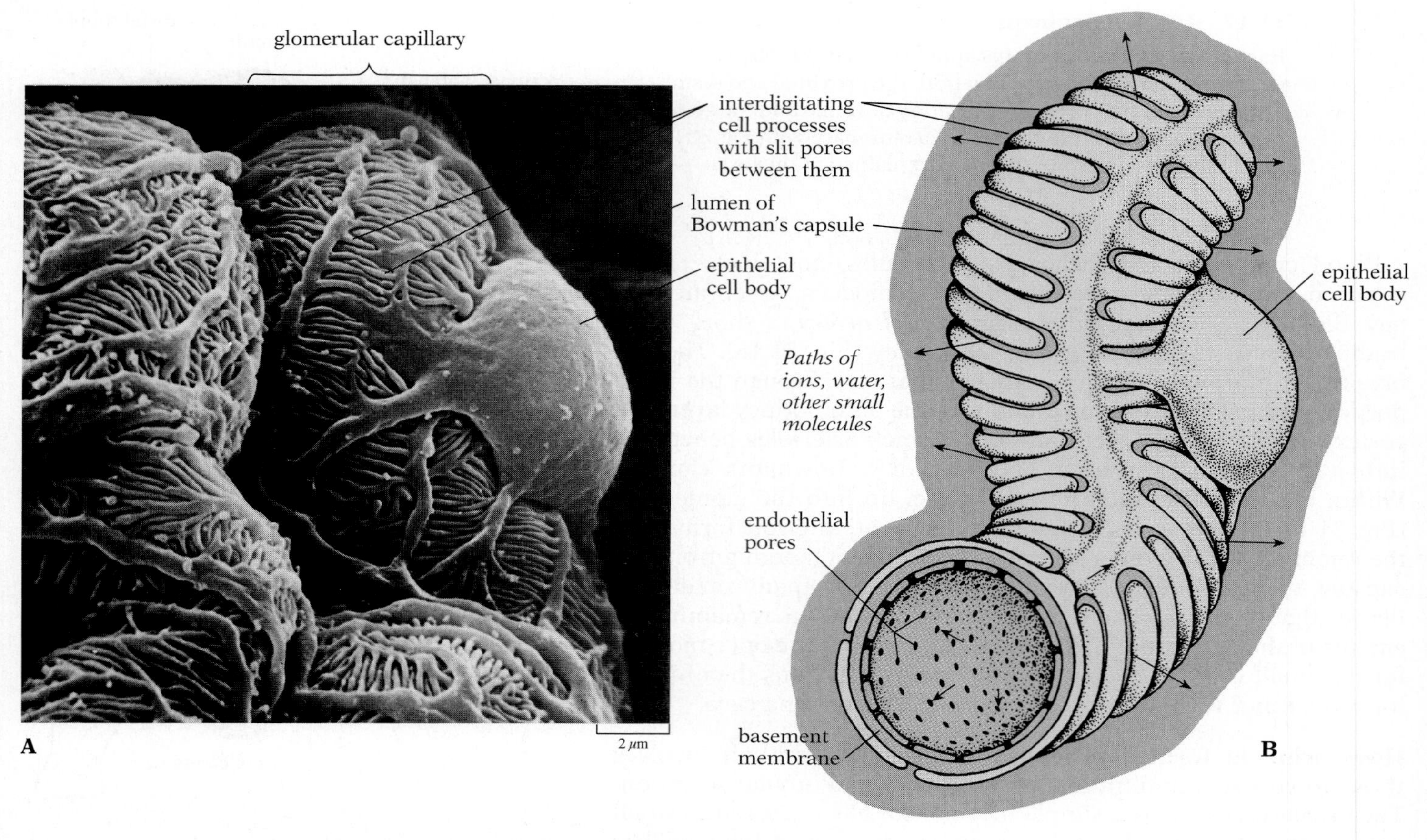

31.19 Part of a glomerulus from a rat kidney

(A) Scanning electron micrograph showing inner epithelial cells of a Bowman's capsule enveloping two glomerular capillaries. The individual cells that make up the inside layer of the cup of the capsule are so highly modified that they bear little resemblance to any other epithelial cells. Called podocytes, these cells consist of a cell body (of which only one is shown in this SEM) and numerous interdigitating processes that embrace the glomerular capillaries. The slit pores between the interdigitating processes provide an enormous area for the passage of the filtrated blood. Together, podocytes form the inner wall of the lumen of the capsule, through which the filtrate enters. From the lumen the filtrate moves into the tubule system of the nephron. (B) An interpretive drawing of one glomerular capillary with a podocyte cell body and interdigitating processes attached. Blood pressure in the capillary forces ions, water, and other small molecules through the endothelial pores and the basement membrane and out through the slit pores of the podocyte into the lumen of the capsule.

of Bowman's capsule (Fig. 31.19). Blood pressure forces ions, water, and other small molecules through the pores from the capillaries into the lumen of the capsule.

The liquid entering the lumen of the nephron has basically the same composition of dissolved substances as blood, lacking only cells, platelets, and plasma proteins, which are too large to pass through. However, if the filtrate in the nephron were expelled from the body without modification, many essential substances—glucose, amino acids, important ions—would be lost with it; the process would be extremely wasteful. Consider just the water intake that would be necessary to replace the 180 liters of glomerular filtrate formed every day in the average person's kidneys. Selective

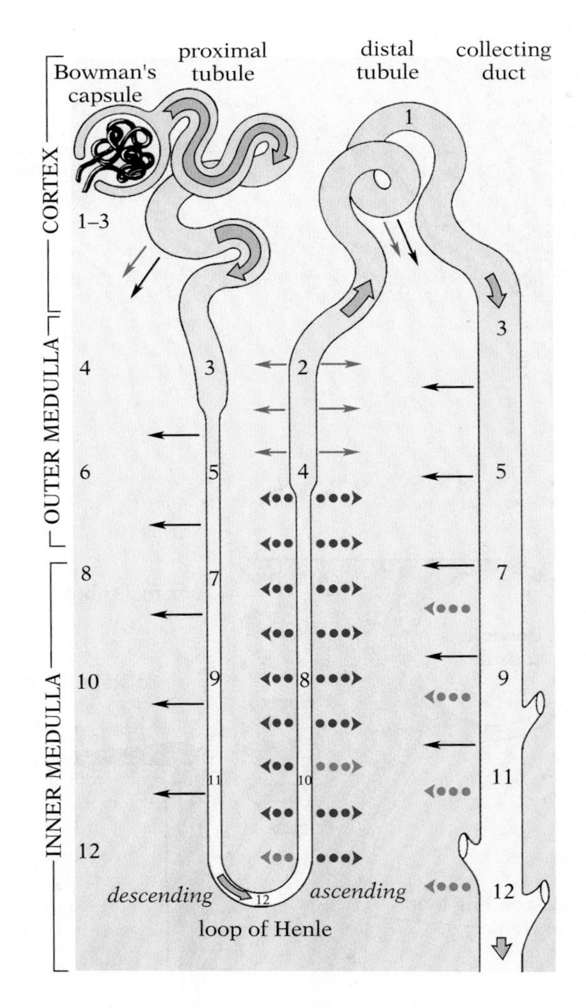

31.20 Operation of the nephron

The nephron serves to recover most of the water, salt, and useful substances in the filtrate entering from Bowman's capsule, while keeping urea and other wastes trapped in the fluids passing through. The most important component of the system is the concentration gradient of osmotically active substances (primarily Na^+, Cl^-, and urea) between the inner medulla (where it is highest) and the cortex. The creation of this gradient is described in the text. Filtrate passing through the nephron encounters a sequence of sodium pumps and selective permeabilities. In the proximal tubule, Na^+ is pumped out and water follows osmotically, greatly reducing the volume of filtrate without much changing its ionic concentration. The descending limb of the loop of Henle is impermeable to salt ions but can pass water; because of the increasing osmotic concentration in the medullar tissues through which it passes, the filtrate loses much more volume, and its ionic concentration rises. The ascending limb is permeable to salt but impermeable to water. Ions are lost passively in the inner medulla and actively pumped from the loop in the outer medulla; this serves to reduce the salt concentration without changing filtrate volume. In the distal tubules more salt is pumped out, and water follows osmotically, reducing filtrate volume further. The permeability of the collecting duct to water is hormonally controlled, with the result that filtrate volume can be maintained or greatly reduced depending on the body's need to recover water. The distal end of the duct (and, to a lesser extent, the ascending limb) is permeable to urea; the pumping of this substance into the tissues of the inner medulla (again, under hormonal control) helps create the intense osmotic concentration gradient essential to nephron function.

Figures indicate solute concentrations in hundreds of milliosmols per liter

reabsorption of most of the water, ions, and many of the other dissolved materials by the tubules of the nephrons is critical.

In humans the filtrate passes first through the proximal tubule, then through the long loop of Henle, then through the distal tubule, and finally into the collecting duct (Fig. 31.17). The Bowman's capsule and the proximal and distal tubules are in the kidney's cortex layer, while most of the loop of Henle and the collecting duct are in the medulla. As the filtrate moves through the tubules, as much as 99% of the water may be reabsorbed by the cells of the tubule walls and returned to the blood in the capillary network. In this way the human kidneys (and the kidneys of other mammals and of birds) produce concentrated urine—that is, urine that is hypertonic relative to the blood plasma—even though the initial filtrate was nearly isotonic.

The means by which the nephron succeeds in selectively concentrating the filtrate is both elegant and complex. The complexity does not stem from the mechanisms involved, which are familiar by now—ion pumping and osmosis. Instead, the complications arise from the organization of the system: the action of the nephron depends crucially on its remarkable anatomy. Moreover, events occurring later in the flow of filtrate control what is happening earlier. As a result, it makes more sense to describe the process from the end of the nephron back to the beginning. As we will see, the object of all the pumping and selective permeabilities in various regions of the nephron is to create a solute gradient, ranging from low solute concentration in the tissue of the cortex to a very high concentration in the tissue of the inner medulla; this gradient is indicated by the shading of the tissue in Figure 31.20. The solutes include Na^+, Cl^-, and urea; each is osmotically very active, and thus able to draw water into the tissue from the portions of the nephron that are permeable to water. Once we have seen how this all-important gradient is established, we will follow the filtrate through from the beginning (Bowman's capsule) to the end (the distal end of the collecting duct).

How the nephron works The filtrate leaving the collecting duct has a very high concentration of the waste product urea. This high concentration is created by a simple process: because the tubules are impermeable to urea, it remains in the filtrate as it passes through the nephron; at the same time, nearly all of the water is removed. The distal portion of the duct is the only part of the nephron with much permeability for urea. This permeability allows urea in the duct to diffuse into the inner medulla until it is at the same high level as the filtrate (Fig. 31.20), thus producing a high osmotic potential there. Since the urea in the inner medulla is not reabsorbed and removed by the capillaries, once this osmotic potential is established it tends to remain high—that is, only a small amount of urea is actually lost to the medulla.

As Figure 31.20 indicates, sodium ions are actively pumped out of the thick portion of the ascending loop of Henle, producing a lesser osmotic potential there. There is also some passive movement of salt out of the thin portion of the ascending loop of Henle; this movement, as we will see, results from the massive concentration of salt in the filtrate as it passes through the proximal tubule and the descending loop. The result of all this, then, is an osmotic gradient in the tissues surrounding the nephron: a very high concentration of solutes (ions and urea) in the inner medulla, a moderate concentration in the outer medulla, and a low concentration in the cortex—lower even than in the blood. With this gradient in place, we can now follow the filtrate through the nephron.

The filtrate entering the proximal tubule has a relatively low solute concentration. That value rises slightly as it passes through the proximal tubule, but the volume of filtrate drops dramatically (Fig. 31.21). The reason for this is that the proximal tubule pumps out about 75% of the Na^+ ions in the filtrate, and nearly three-quarters of the water in the filtrate follows osmotically. As we will see, both the water and salt removed at this point are recovered by the capillaries. What does not leave is urea and any other substances for which there are not specific pumps or permeabilities; these molecules remain trapped in the filtrate.

The 25% of the filtrate that remains now passes down the descending loop of Henle into the medulla. The descending loop is

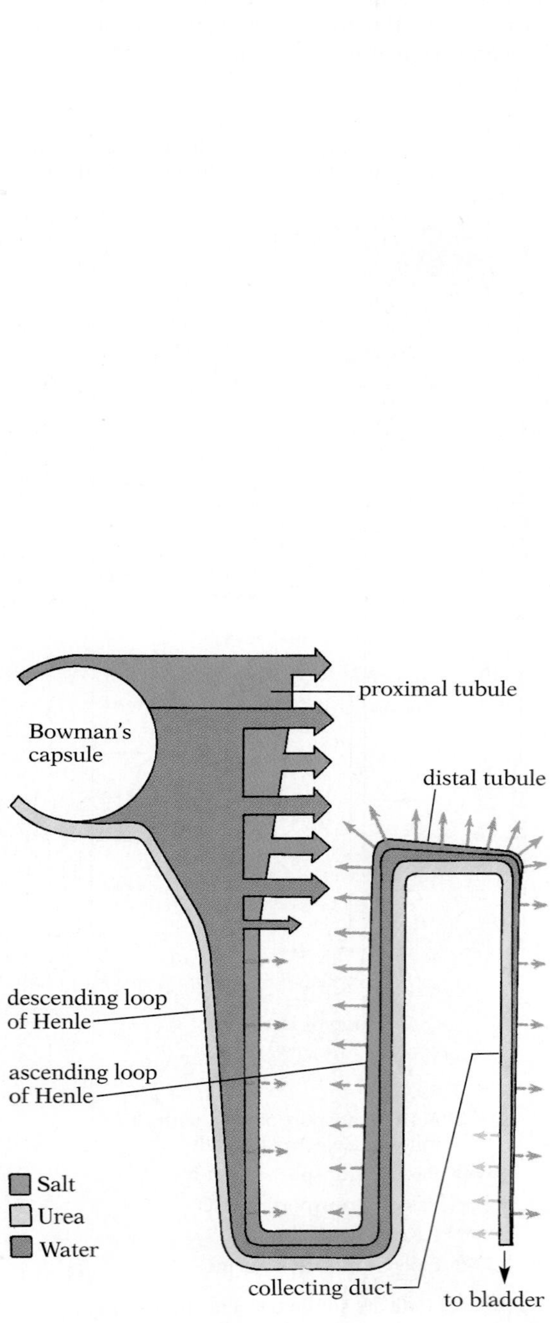

31.21 Fluid volume changes in the nephron

This schematic representation of the nephron distorts the width of the tubules to represent the relative amount of the initial filtrate still present at each point in the system, as well as each major component—water, salt ions, and urea. The selective permeability of each region and the location of sodium pumps are also indicated. About 75% of both salt and water is lost in the proximal tubule, about 75% of the remaining water is lost in the descending limb, about 80% of the remaining salt ions leave in the ascending limb, about 50% of the remaining water and salt is lost in the distal tubule, and up to 75% of the remaining water can be lost in the collecting duct.

permeable to water but not to salt. Because of the osmotic gradient we have already described, 75% of the remaining water leaves the descending tubule; about 6% of the original water now remains, and the salt concentration rises correspondingly (Figs. 31.20, 31.21).

The filtrate now enters the ascending loop. The entire ascending loop is impermeable to water, so there is no change in volume over this part of the nephron. The thin portion of the ascending loop is permeable to salt ions and (slightly) to urea, and some leaves osmotically in the inner medulla to join the urea already there, thus enhancing the osmotic potential of this tissue. In the thick portion of the ascending loop, Na^+ and Cl^- are actively pumped out (Fig. 31.20), helping to increase the osmotic potential of the outer medulla, as we mentioned earlier. The filtrate leaving the ascending portion of the loop, therefore, has had its salt concentration greatly reduced, and only about 4% of the salt entering from Bowman's capsule remains at this point. Most of the urea and all other waste products present initially are still trapped in the filtrate.

The distal tubule operates very much like the proximal tubule: sodium ions are pumped out and, because this part of the nephron is permeable to water, fluid follows. In consequence, the volume of filtrate is further reduced by another 50%; this leaves only about 3% of the original water that entered the nephron (Fig. 31.21). There is no loss of urea. The salt concentration at this point is as low as it ever gets in the filtrate.

The role of the collecting duct Now the remaining filtrate enters the collecting duct. Because the duct can be permeable to water, and because it plunges down into the medulla through tissues with an ever-increasing osmotic potential, as much as 75% of the remaining water can be lost at this stage, yielding highly concentrated urine. The degree of permeability, however, is controlled by various mechanisms designed to maintain blood volume and salt balance.

The need to recover nearly all of the water and salt originally present in the filtrate accounts for the close association of capillaries with the nephron. When these substances are pumped out or exit osmotically, they enter the tissues of the medulla and cortex. Salt is readily recovered in the medulla, where its concentration is far higher than in the blood passing through the capillaries. Water, on the other hand, is mainly picked up in the cortex, where the osmotic concentration of the blood, rich in plasma proteins unable to enter the filtrate, exceeds that of the tissues. Nearly all the glucose, amino acids, and ions are also reabsorbed and returned to the blood, though the exact recovery levels depend on the body's needs. Much of this reabsorption involves active transport, and thus energy expenditure, by the tubule cells.

In general, the kidney functions by forcing out of the blood in the glomerulus most molecules small enough to pass through the pores, and then reabsorbing into the capillaries surrounding the tubules and the loop of Henle only what is to be saved (Fig. 31.22). Hence the kidney effectively and automatically removes many chemicals by simply not transporting them back into the blood. This system is far safer than one requiring a special pump for ex-

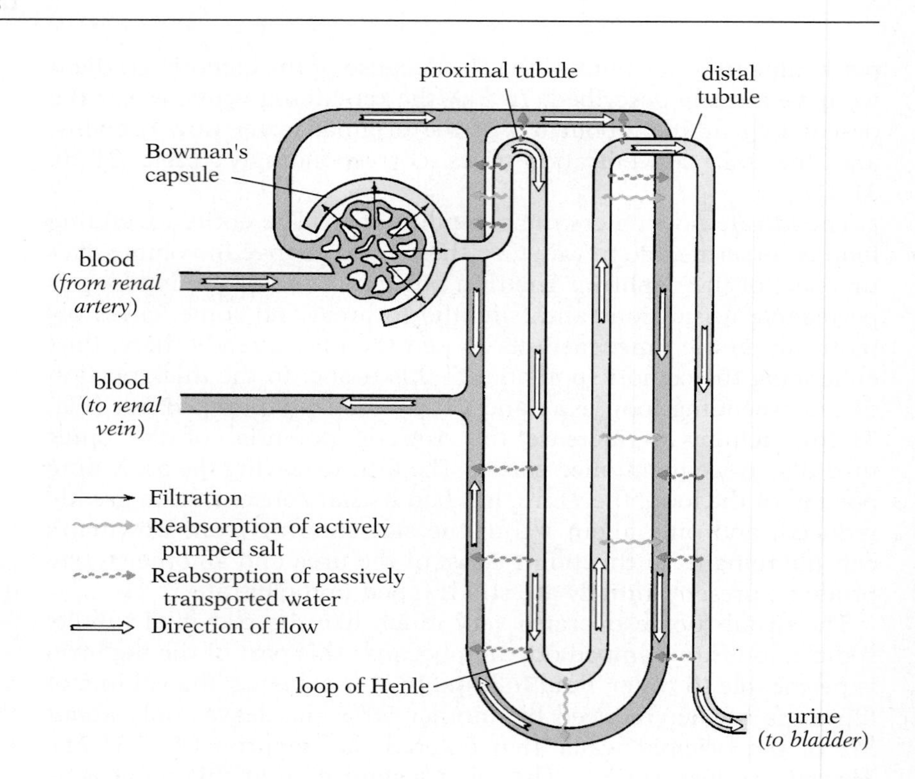

31.22 Summary of filtration and reabsorption in the nephron

The fluid in the blood arriving from the renal artery is forced through the walls of the capillaries in the Bowman's capsule. The water, ions, and small molecules that pass through this filter move through tubules leading to the pelvis of the kidney, and from there to the bladder. On the way, water leaving the tubules by osmosis, as well as ions and small molecules actively transported out of the tubules, can be reabsorbed by the capillaries closely associated with the tubules. Most of the reabsorption of water takes place near the proximal and distal tubules. Salt recovery occurs over the entire capillary circuit.

creting every undesirable element that might turn up in the blood.

Not all materials are subjected to the filtering and selective-reabsorption strategy we have described. Some larger molecules cannot pass through the pores of the glomerulus and are actively removed from the blood in the second bed of capillaries adjacent to the tubules. This tubular secretion supplements glomerular excretion and increases the efficiency of the overall excretory regulation of blood composition.

Special adaptations The kidneys of vertebrates vary considerably in their capacity to produce concentrated urine. Marine fish are incapable of producing urine more concentrated than their blood and must therefore excrete excess salt by another method—special cells in the gills. Without fresh water to drink, sea turtles and marine birds (albatrosses and penguins, for example) need to excrete the excess salt in the sea water they drink by some other mechanism; these animals have special glands near the nose that are capable of excreting salt in very concentrated solution.

Seals and some whales seldom drink; they get their water from the body fluids of the fish they eat, and thus benefit from the fishes' ability to excrete salt through the gills. A fish diet means much protein, however, and much urea to excrete. These animals have kidneys capable of excreting urine with a high urea concentration.

Kangaroo rats living in deserts almost never drink; nor do they get water by eating succulent food. Most of their water is metabolic water obtained during the respiratory breakdown of the dry grains they eat. They must, of course, conserve water extremely well: they are not active during the heat of the day; they do not sweat; they eliminate very dry feces; and they have extraordinarily efficient kidneys capable of producing extremely concentrated urine.

The human kidney is incapable of producing urine with a very high concentration of either salt or urea. Humans have no alternative excretory mechanisms like those of marine fish, turtles, and birds. Adrift at sea, humans are in serious danger indeed. Drinking sea water aggravates their condition because in the process of removing salt from their bodies they excrete more water than they drink. If they try to get their water by eating fish, as seals do, they excrete much water in the process of removing urea. Human kidneys are simply not adapted to life at sea or to life in very dry habitats.

■ HOW THE IONS ARE TRANSPORTED

Though animals tend to maintain a nonfluctuating fluid environment for their cells, one with approximately the same osmotic concentration as the cells, this does not mean that the extracellular and intracellular fluids have the same ionic composition. In fact, their compositions are very different. All cells, plant and animal, tend to accumulate certain ions in much higher concentrations than are found in the surrounding fluids, and to stabilize intracellular concentrations of other ions at levels far below those in the extracellular fluids. For example, the vast majority of cells maintain an internal concentration of sodium ions (Na^+) far below that in the fluids bathing them, while at the same time accumulating potassium ions (K^+) to a concentration many times that in the extracellular fluid.

Specific pumps in cell membranes use energy to move substances across the membranes against osmotic and/or electrical gradients. The most important of these in animals is the sodium-potassium pump (Fig. 31.23), a transmembrane protein that changes conformation when phosphorylated by ATP. In one conformation it is open to the outside of the cell, and has a high affinity for K^+ but a low affinity for Na^+. In the other conformation it is open to the inside, and has the opposite ion-binding properties—a low affinity for K^+ and a high affinity for Na^+. The result is that K^+ from outside the cell is bound and released inside, while Na^+ from inside the cell is bound and released outside. This ion pumping creates an electrochemical Na^+ gradient, which is then used to provide the energy needed to transport glucose into the cell (see Fig. 4.20, p. 104). A similar strategy is used to move amino acids into cells. The sodium-potassium pump is also responsible for maintaining the electrical activity of nerves and muscles, and for regulating the volume of cells. Roughly 30% of the ATP consumed by the body at rest is used to fuel the sodium-potassium pump system.[3]

The cells of the excretory and osmoregulatory organs use the same basic pump strategy, but the task they perform is more elaborate. The osmoregulatory cells in the gills of a marine fish, for example, must remove salt ions—Na^+ and Cl^-—from the blood and

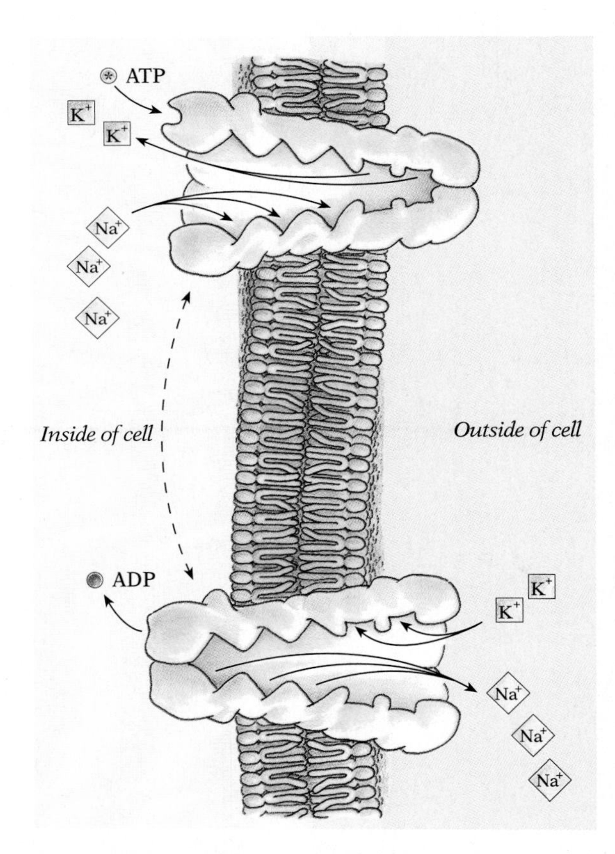

31.23 Sodium-potassium pump

The sodium pump uses energy from the cell and provides active transport of sodium against its gradient. Three sodium ions are exchanged for two potassium ions; both kinds of ions are already more concentrated on the side to which they are being moved. The net ionic effect is to pump positive charges out of the cell, so the inside of the cell becomes negatively charged with respect to the outside.

[3] The membranes of plant cells are distinctly different from those of animal cells in that the transmembrane ion pumps of plants mainly transport H^+ rather than Na^+. Apparently this allows plants to maintain a membrane potential about twice as large as that of animal cells. The main purpose of this ion gradient is to power the cotransport of sugars and amino acids into cells.

tissue fluid, and actively secrete the Na^+ into the surrounding sea water (the Cl^- follows electrostatically). The cells in the walls of the convoluted tubules in a mammalian kidney also remove Na^+ and Cl^- ions from the tubular fluid, actively reabsorbing Na^+ into the tissue fluid surrounding the nephron while Cl^- follows. (In the ascending limb of the loop of Henle the pump is a cotransporter, which moves both ions together.) In each instance the cells are doing more than simply expelling salt ions from their cytoplasm: they are picking up ions on one side of a cell and expelling them on the other side. The ions are being moved completely across the cell barrier that separates the tissue fluids of the fish from the sea water or that separates the contents of the mammalian nephron from the tissue fluid. This requires an asymmetry between the two sides of the cell.

The cross-cell movement could be accomplished by pumping salt into the cell on one side and pumping it out on the other. Thus the orientation of the cross-membrane pumps would differ between sides. Alternatively, maybe only one side of the cell has pumps: it is possible that salt ions diffuse passively into the cell on one side and are then actively expelled from the cell on the other side. This would work if the removal of ions from the cell lowered its salt-ion concentration below that in the environmental water, and if the membrane on the environmental side of the cell were permeable to Na^+ and Cl^-; in that case these ions would diffuse into the cell from the environmental water and then be actively expelled by the pump from the cell into the tissue fluid on the other side. In a marine fish the situation would be reversed: the pump would be in the membrane on the side of the cell exposed to the sea water, and the ions would diffuse passively into the cell on the tissue-fluid side. No one knows yet which of these cross-cell models is correct.

CHAPTER SUMMARY

HOW PLANTS CAN AVOID REGULATING THEIR FLUIDS

The stiff cell wall of plants keeps excess water from entering and bursting cells; the cuticle and stoma help limit water loss. (p. 879)

HOW THE LIVER REGULATES BLOOD CHEMISTRY

REGULATING BLOOD-SUGAR LEVEL The liver removes excess glucose from the blood stream and converts it into glycogen; the glycogen is converted back into glucose and released into the blood when the level of circulating glucose falls too low. (p. 882)

METABOLIZING LIPIDS AND AMINO ACIDS The liver processes lipids and amino acids. A limited quantity of excess amino acids is stored for later release into the circulation when the amino-acid level declines; beyond this, extra amino acids are deaminated and converted into glucose, glycogen, or fat; in most animals, the waste nitrogen group is converted into less toxic compounds like urea or uric acid. (p. 882)

A SUMMARY OF THE LIVER'S FUNCTIONS In addition to its regulation of glucose, amino acids, and fats, the liver converts excess carbohydrate to fat, detoxifies dangerous chemicals, synthesizes plasma proteins, stores important chemicals, destroys old corpuscles, excretes bile pigments, and synthesizes bile salts. (p. 883)

HOW ANIMALS CONTROL WATER AND SALT LEVELS

PROBLEMS WITH LIVING IN DIRECT CONTACT WITH WATER Most marine invertebrates have tissue fluids that are isotonic with sea water. Most marine vertebrates have hypotonic fluids, and thus risk losing water to the surrounding fluid and gaining salt; they minimize the problem by having only small regions (the gills) that are permeable, by drinking continuously, and by excreting salt. Freshwater organisms have hypertonic fluids, and thus risk losing salts and taking up too much water; they restrict exchange to the gills, excrete water, and pump in salts. (p. 884)

PROBLEMS WITH LIVING ON LAND Animals risk desiccation from breathing, evaporation from the skin, and loss of water in waste products. (p. 887)

DISPOSING OF WASTES AND EXCESS WATER

EXPELLING WATER IN BURSTS: CONTRACTILE VACUOLES Cytoplasmic fluid is collected into vesicles, which pump the salts and nutrients back into the cytoplasm, before adding their fluid contents to the vacuole. The vacuole later contracts, and the water is expelled from the cell. (p. 887)

LEAKING WATER: FLAME-CELL SYSTEMS In flatworms, water is collected in cilia-lined bulbs and gently pumped to the exterior of the body. (p. 889)

CAPILLARY COOPERATION: NEPHRIDIA In earthworms, water and wastes from tissue fluids and capillaries are collected in tubes that empty to the outside of the body. (p. 890)

THE INSECT ALTERNATIVE: MALPIGHIAN TUBULES Wastes are collected from the tissue fluids in tubules that empty into the digestive tract; excess water is recovered before the wastes are eliminated. (p. 891)

THE VERTEBRATE ANSWER: THE KIDNEY Fluid from the blood is forced into tubules in the kidney, and then the nutrients, as well as most of the water and salts, are recovered. The water is mostly removed by osmosis into hypertonic fluids in the kidney and recovered by capillaries; some salt is recovered this way, but most is actively pumped out to create the salt gradient that fuels the recovery of water. The key element in this system is the varying degree of permeability of different parts of the nephron. (p. 892)

HOW THE IONS ARE TRANSPORTED ATP-powered pumps in the cells lining parts of the nephron move Na^+ from the nephron fluid into the tissue fluid of the kidney. (p. 899)

STUDY QUESTIONS

1 There are artificial kidney machines, but nothing equivalent for replacing livers. Could one be designed for an individual eating a normal diet? How would the task be easier if an artificial diet were used? (pp. 881–884)

2 There are several species designed to migrate from freshwater to marine habitats (and back), such as salmon and certain eels. What kind of physiological changes (if any) must occur in the body of such a creature during the migration? (pp. 880, 884–887)

3 How could the kidneys be used to regulate blood volume (and thereby blood pressure) in the face of changing needs of the body? (pp. 892–899)

4 Is the elaborate filtration and concentration system of the kidney energetically less expensive than the more straightforward approach of just pumping urea and any excess salt out of the blood? If not, why not? If so, then what is the point of all this elaborate plumbing? (pp. 892–899)

SUGGESTED READING

SCHMIDT-NIELSEN, K., 1990. *Animal Physiology: Adaptation and Environment,* 4th ed. Cambridge University Press, New York.

VALTIN, H., 1983. *Renal function.* Boston, Little, Brown.

CHAPTER 32

DEVELOPMENT AND CHEMICAL CONTROL IN PLANTS

Like most organisms, plants respond to changes in their environment, but stimulus and response in plants is often not apparent. Plants do not see, hear, smell, or taste in the usual sense, although they do monitor many relevant chemical, tactile, and light stimuli. The responses of plants to these stimuli usually consist of changes in growth. In this chapter we will begin by outlining the patterns of plant development and growth, and then examine the signalling systems through which different parts of a plant coordinate growth and other activities.

FROM SEED TO PLANT

In higher plants, a diploid zygote develops from the fusion of haploid multicellular male and female gametes (Fig. 32.1). In angiosperms the zygote first develops as an embryo; its water is then largely removed, and it becomes a metabolically quiescent seed. Seeds are an evolutionary adaptation for dispersal and terrestrial survival, allowing the embryo to wait until conditions become favorable for growth.

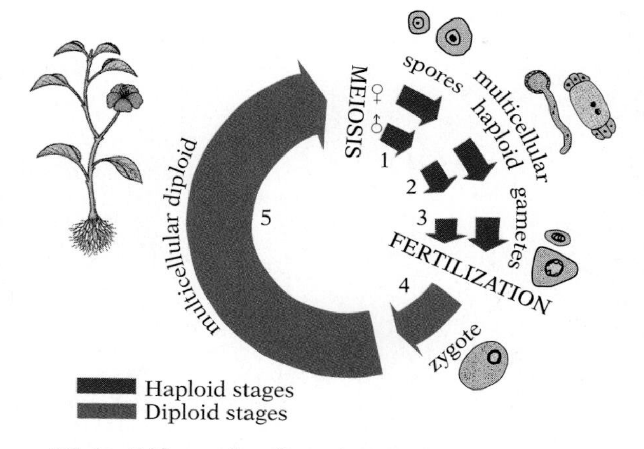

32.1 Life cycle of angiosperms

Once the seed germinates, the plant grows and then flowers, producing a new round of gametes (pollen and eggs). Temperate-zone plants thereafter follow one of four "pathways." Annuals invest everything in reproduction and die; biennials (which are rare) grow vegetatively the first year, overwinter as underground roots, regrow the second season, flower, and then die. Herbaceous perennials, like biennials, abandon their aboveground tissues during the winter, but they regrow them the next year using energy stored in roots, tubers, or bulbs. Woody perennials retain their stems and branches, and either keep or shed their leaves, depending on whether they are evergreen or deciduous. As we will see, one of the main roles of plant hormones is to coordinate the major changes in activity necessitated by germination, growth, flowering, and preparations for winter.

■ EVENTS INSIDE THE SEED

The egg cell of an angiosperm plant is retained within the ovary of the maternal plant and is fertilized there by a sperm nucleus from a pollen grain. After fertilization, the zygote undergoes a series of mitotic divisions and develops into a tiny ***embryo***. This embryo, together with a food-storage tissue called the ***endosperm***, becomes enclosed in a tough protective ***seed coat***. The resulting composite structure, made up of embryo, endosperm, and seed coat, is the seed. The embryo in some species, such as peas and beans, absorbs all the endosperm before the seed is released from the parent plant. In other species, such as corn, the embryo does not absorb significant quantities of the endosperm until the seed begins to germinate.[1] The embryonic stages of development usually do not last long, and by the time the ripe seed is released from the parent plant it is quite dry and its embryo has usually become quiescent.

How the embryo develops Soon after an egg cell has been fertilized, it begins to undergo a series of changes. Its wall, previously very thin, thickens; its endoplasmic reticulum and Golgi apparatus become more extensive; and new ribosomes are synthesized.

The first cell division may occur a day or so after fertilization. It invariably gives rise to two cells of unequal size: a smaller terminal cell and a larger basal cell containing vacuoles. The developmental fates of these two cells are very different. The terminal cell gives rise to the embryo proper. The basal cell, by means of about three transverse-division cycles, forms an elongate ***suspensor*** structure (Fig. 32.2A–B), which functions only while the embryo is in the seed—probably by pushing the embryo into the nutrients. We will trace the embryonic development of *Capsella*, a much-studied dicot commonly called shepherd's purse.

The terminal cell gives rise to a globular structure, in which three types of tissue begin to differentiate: a surface layer of ***protoderm***, which will form epidermal tissue; a middle layer of ***ground tissue***, which will form the cortex; and an inner core of ***provascular tissue***, which will form cambium and the vascular tissues (Fig. 32.2C). Shortly thereafter two mounds arise on the portion of the embryo opposite the suspensor (Fig. 32.2D). These mounds will become the ***cotyledons***, or embryo leaves (Fig. 32.2E). The part of the embryo proper below the point of attachment of the cotyledons is called the ***hypocotyl***; it will form the first part of the stem of the young plant.

As the cotyledons and hypocotyl of *Capsella* continue to elongate within the very limited space available to them in the seed, the embryo begins to curve back on itself (Fig. 32.2E). As most of the cells of the embryo take on more and more of the characteristics of the tissues to which they will give rise, small clumps of cells at each end of the embryonic axis remain relatively undifferentiated. One clump, located just beyond the point of attachment of the cotyledons, will become the ***apical meristem of the shoot***. The other clump, at the pole of the embryonic axis near the suspensor, will become the ***apical meristem of the root*** (Fig. 32.2E). In some species, cell divisions in these meristems during embryonic development give rise to an ***epicotyl***—a region of shoot above the point

[1] The form of the endosperm food reserve is of great significance to the human diet. For instance, beans store protein, grains store starch, and other seeds (commonly called nuts) rely on oils.

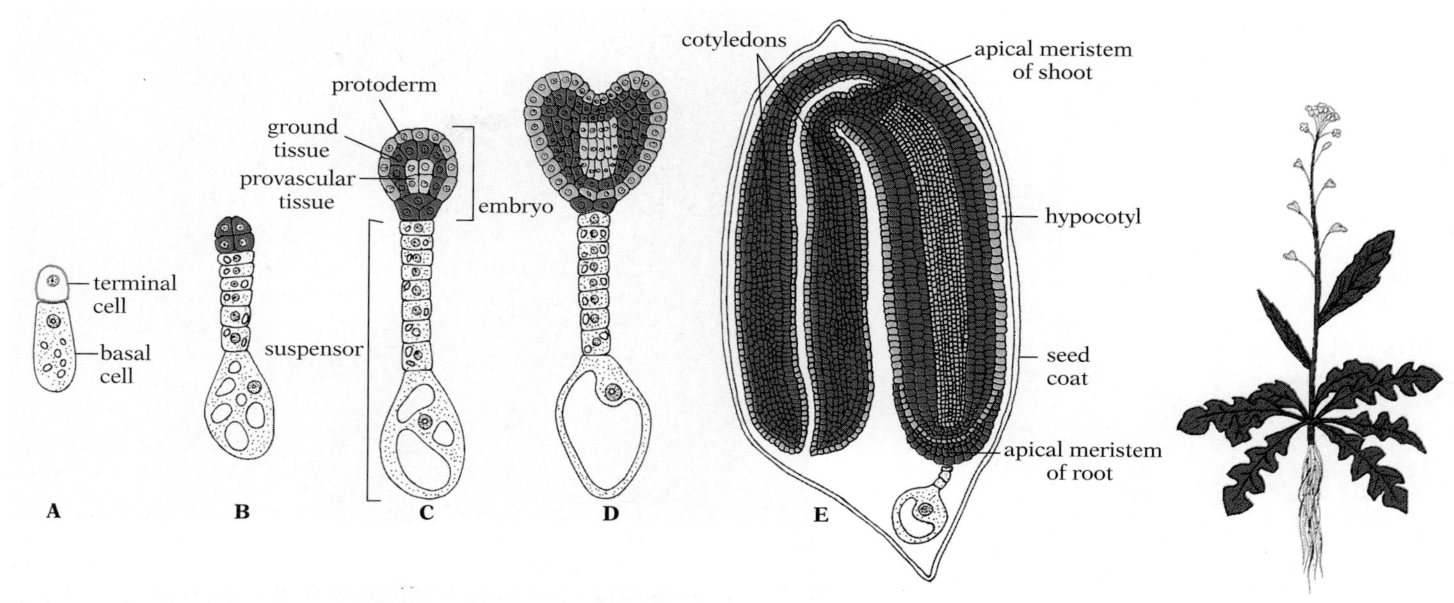

32.2 Embryonic development of shepherd's purse (*Capsella*)

(A) The first division of the zygote produces a smaller terminal cell (green) and a larger basal cell. (B–C) Divisions of the terminal cell give rise to a globular embryo, whereas divisions of the basal cell give rise to a stalklike suspensor. Cells of the globular embryo soon differentiate to form three tissue types: protoderm on the surface, provascular tissue in the center, and ground tissue in between. (D) The formerly globular embryo becomes heart-shaped as two mounds that will develop into cotyledons begin to form. (E) Elongation of the cotyledons and the hypocotyl within the confines of the seed causes the embryo to fold back on itself. A fully grown plant is also shown.

of attachment of the cotyledons, often bearing the first foliage leaves (the plumule)—and, at the other end of the embryonic axis, to a ***radicle***, which will develop into the primary root (Fig. 32.3).

There are several major differences between the embryonic development of an angiosperm plant and a typical animal:

1 Unlike animals, plants have no early developmental stage in which cleavage is unaccompanied by cell growth; the two processes—cell division and growth—typically occur together.

2 Plant cells do not migrate during morphogenesis. The cell walls constrain their shape, and the middle lamellae tend to connect each cell to its neighbors. Also, the plasmodesmata between adjacent cells would be ruptured if the cells were to slide along each other. The form and shape of the plant embryos are established not by movement of cells, as in animals, but by the patterns of cell division and growth.

3 In plants a few cells are early set aside to remain forever embryonic (meristematic).

4 The fully developed plant embryo does not possess, even in rudimentary form, all the organs of the adult plant; organogenesis (the formation of new organs) continues throughout the life of the

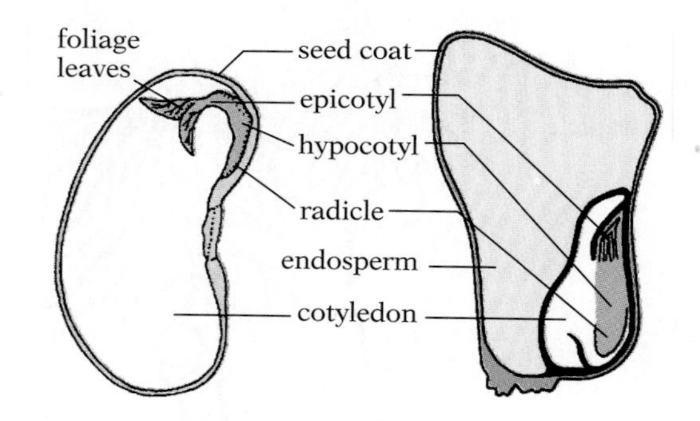

32.3 Dicot and monocot seeds

The embryos in these seeds consist of epicotyl, hypocotyl, radicle, and cotyledon. Left: A dicot seed (bean) in which one cotyledon has been removed to reveal the remaining parts of the embryo. Right: A monocot seed (corn) has only one cotyledon, but a large endosperm.

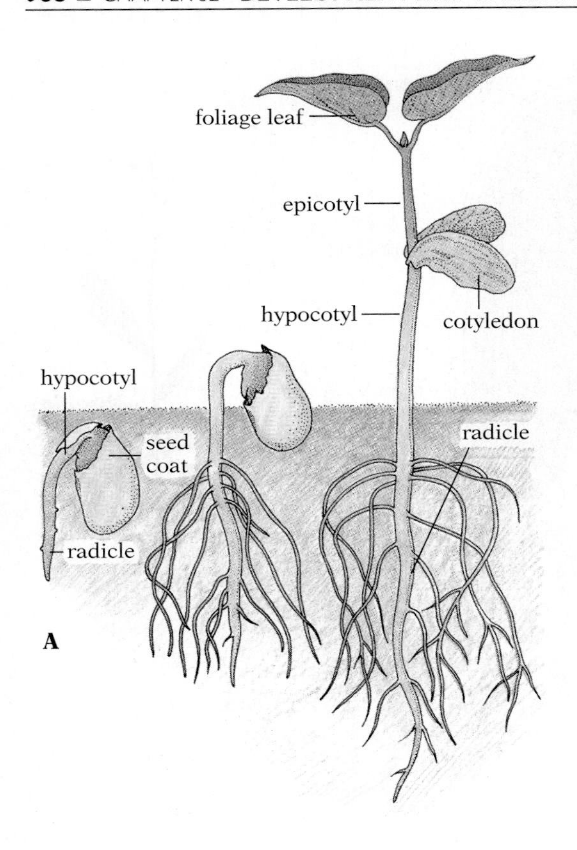

32.4 Germination and early development of a bean plant

(A) The hypocotyl and radicle emerge from the seed first (left). As the upper portion of the hypocotyl elongates, it forms an arch that pushes out of the soil into the air; the radicle gives rise to the first root system (middle). The hypocotyl straightens, pulling the cotyledons out of the ground as the epicotyl begins its development (right). (B) A germinating bean plant.

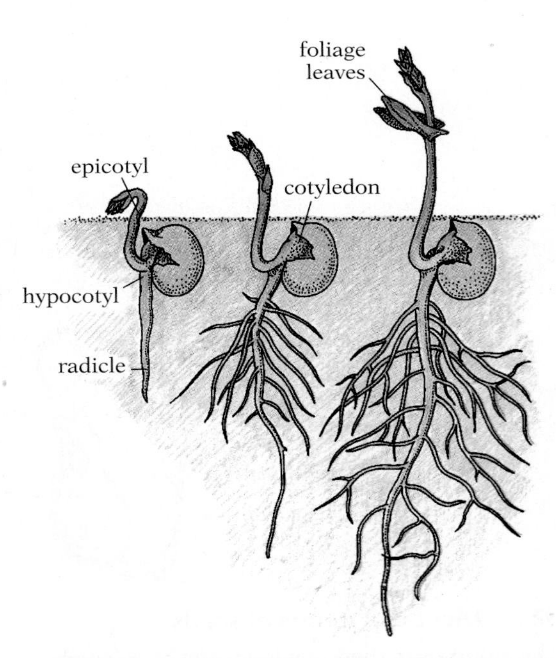

32.5 Germination and early development of a pea plant

The development of peas differs from that of beans in that no hypocotyl arch is formed and the cotyledons remain beneath the soil.

plant, as new roots, branches, leaves, and reproductive structures are formed during each growing season.

Breaking out of the seed Germination begins as a seed absorbs water, which increases its volume up to 200%. The resulting hydration of the protoplasm increases enzymatic activity, and the metabolic rate of the embryo rises markedly. Active cell division resumes, new protoplasm is synthesized, and cells increase in size by taking up water. The growing embryo bursts out of the seed coat, and forms a distinguishable shoot and root.

The hypocotyl (with attached radicle) is the first part of the embryo to emerge from the seed. The radicle promptly turns downward. By the time the epicotyl begins its rapid development, the radicle has already formed a young root system capable of anchoring the plant to the substrate and absorbing water and minerals.

In some dicots, the upper portion of the hypocotyl elongates and forms an arch, which pushes upward through the soil and emerges into the air (Fig. 32.4). Once the hypocotyl arch is exposed to light, it straightens, thus pulling the cotyledons and the epicotyl out of the soil. The epicotyl then begins to elongate. In these dicots, such as garden beans, the shoot of the mature plant is mostly of epicotyl origin, but a short region (usually little more than 1 cm) at the base of the stem is derived from the hypocotyl.

In other dicots, such as garden peas, the epicotyl begins to elongate soon after the young root system has begun to form; it grows upward and soon emerges from the soil (Fig. 32.5). In such plants

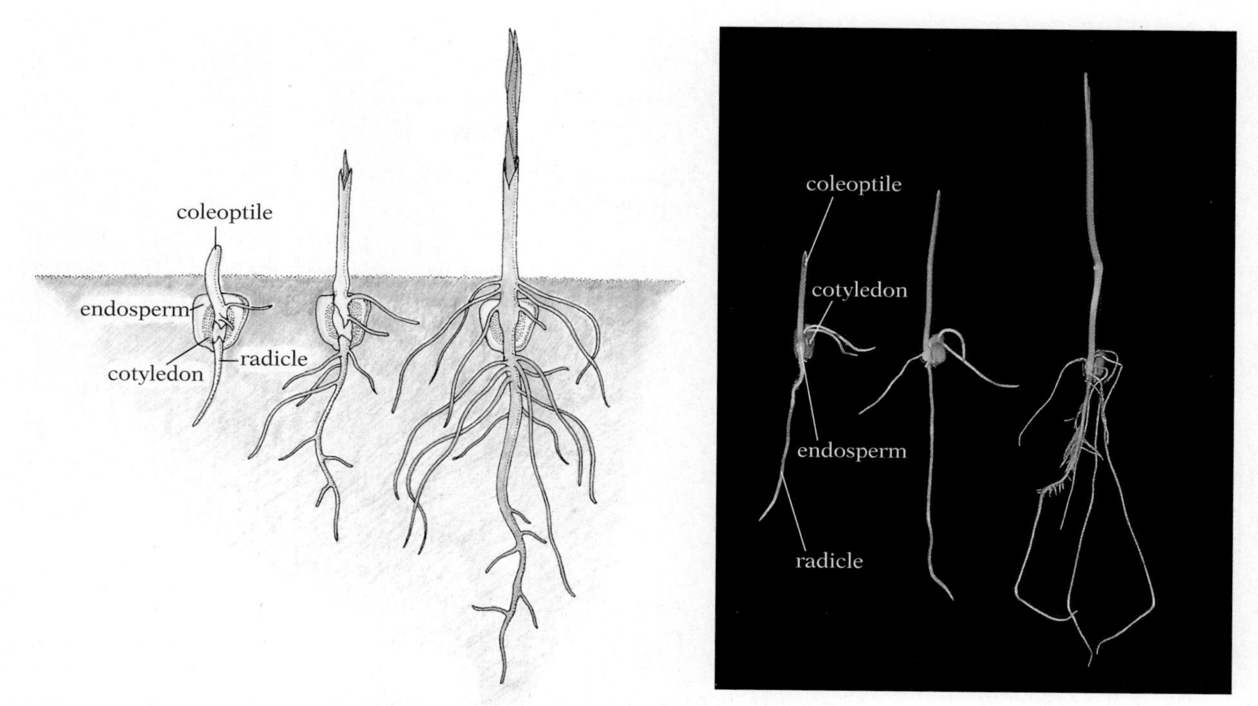

32.6 Germination and early development of a corn plant
As shown in the photograph and drawing, the young shoot is initially enclosed within a coleoptile, a tubular protective sheath.

the entire shoot is of epicotyl origin, and the food-storage tissue remains in the soil. A similar strategy is used by grass seedlings such as corn: the ***coleoptile*** (a cylindrical sheath enclosing the first leaves) emerges, but the cotyledon with its large endosperm remains underground (Fig. 32.6).

■ HOW PLANTS GROW

Growth of the shoot or root of a young plant involves both cell multiplication and cell elongation. In plants with only primary growth, both processes are ordinarily restricted to a limited region near the apex of the shoot or root. Let's look first at the root.

Root growth The extreme tip of a root is covered by a conical ***root cap*** consisting of a mass of nondividing cells (Fig. 32.7). These cells secrete a gelatinous substance that lubricates the surface of the root cap and facilitates the pushing of the root tip through the soil as the root elongates. As cells on the surface of the root cap break off, they are replaced by new cells added to the cap by the apical meristem. The apical meristem is usually 1 mm or less in length and is composed of relatively small, actively dividing cells. Most of the new cells are produced in the region of the meristem farther from the root cap. These cells are left behind as the

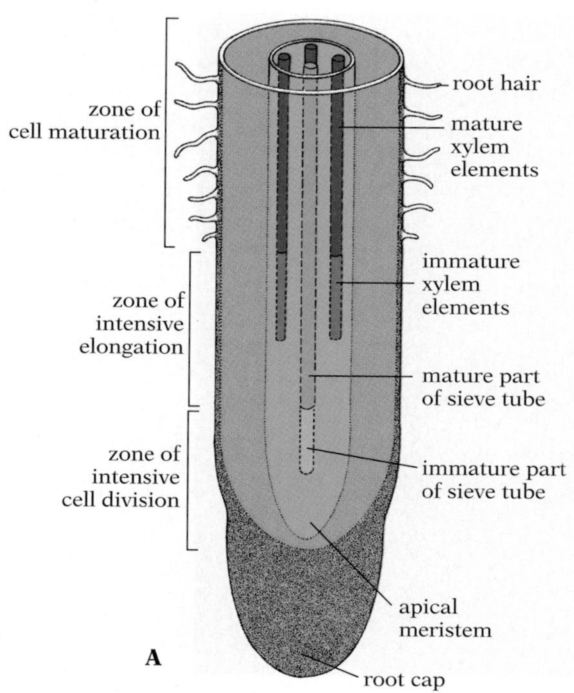

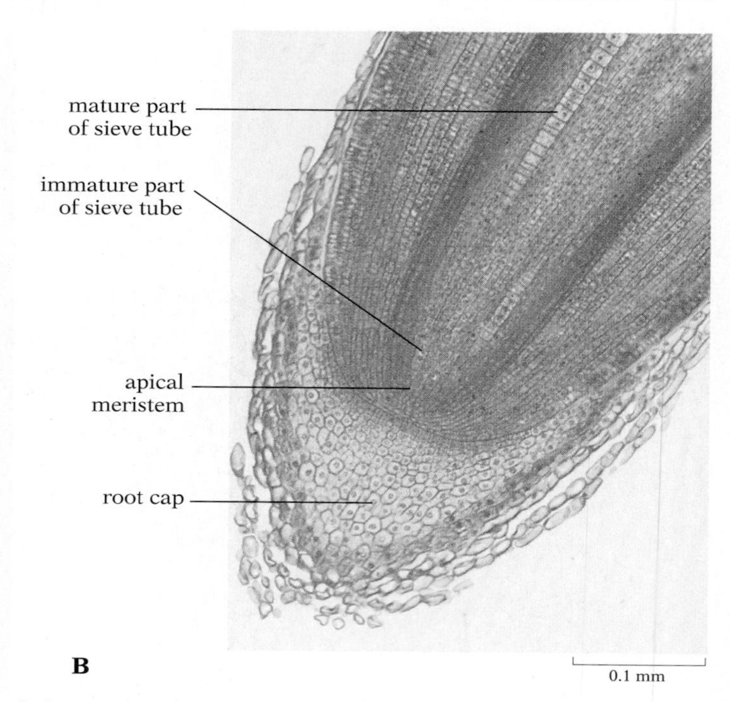

32.7 Longitudinal sections of root tips

(A) New cells are produced by mitotic divisions in the meristem region just behind the root cap, as shown in this drawing of a tobacco root. There is a zone of especially intensive cell elongation not far behind the meristem. Both phloem and xylem are well differentiated in the region of the root hairs, back of the zone of intensive elongation. (B) Many of the same structures are visible in the photograph of the distal part of the root tip of a corn plant.

meristem generates additional new cells in front of them and the tip continues to move through the soil. It is these new cells, derived from the apical meristem, that will form the primary tissues of the root.

Cell division and cell enlargement both occur in the meristem, but because the rate of division is high, the rate of enlargement is only sufficient to produce small cells. However, as the new cells become farther removed from the meristem by the addition of intervening cells, mitotic activity in most of them slows down; cell enlargement becomes the dominant process, and cell size increases. Most of the enlargement is elongation rather than increase in width. As we will see presently, cell elongation is under the control of hormones, particularly ***auxins***.

Since plant cells, unlike animal cells, are enclosed in a boxlike cell wall, cell growth is possible only if the walls can be extended. It is in this process—the enlarging of the walls—that auxin is so important. The walls are composed primarily of polysaccharides—mostly cellulose. In primary walls (those of growing cells) the cellulose is present in the form of long fibrils, each of which is thought to be a bundle of about 40 cellulose chains aligned parallel to one another and extensively cross-linked by hydrogen bonds

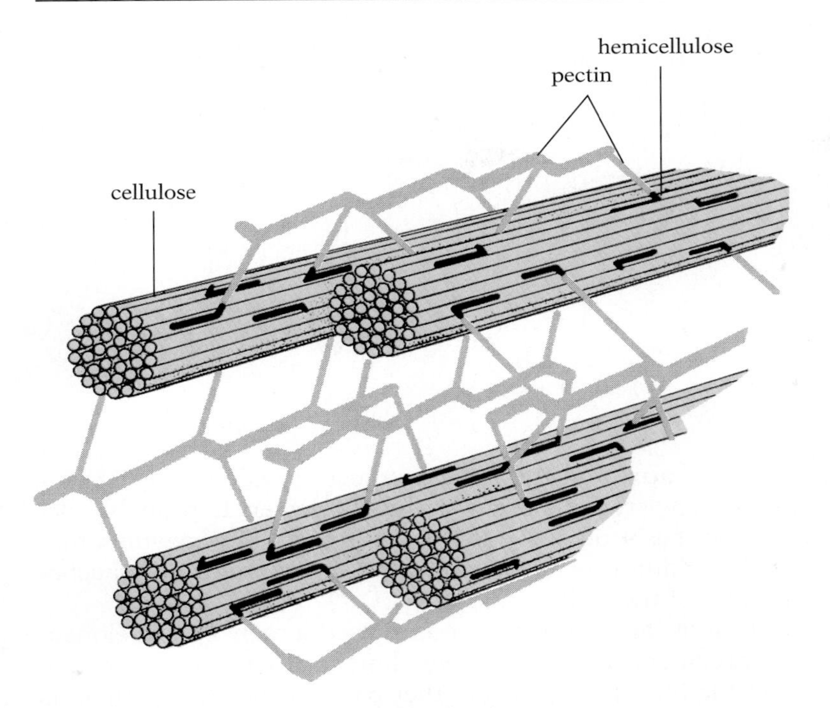

32.8 Molecular arrangement of polysaccharides in a primary cell wall

Each fibril is a bundle of parallel cellulose chains. The fibrils, which are cross-linked by pectin and hemicellulose, form a relatively rigid network.

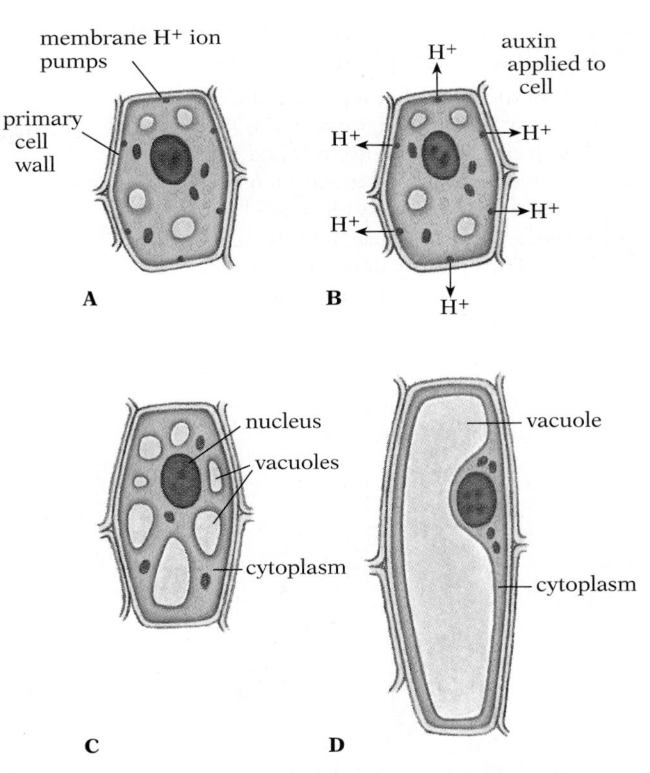

32.9 Elongation growth of a plant cell

Because the cytoplasm of plant cells is hypertonic relative to the surrounding fluid, osmotic pressure favors the entry of water. The stiff cell wall, however, prevents water from entering and swelling the cell. To grow, therefore, a plant cell must first soften its cell wall. The cell does this by pumping H^+ out of the cytoplasm, where these ions activate the enzymes that soften the wall (B). This ion pumping usually occurs as a response to the hormone auxin. Because the wall-softening enzymes are found only on the sides of plant cells, and not at their ends, the cell can only grow by elongating. As more and more water moves into the cell vacuoles, the wall is stretched, but in only one dimension (C). Almost no synthesis of new cytoplasm occurs during this kind of growth. The increased volume of the cell is taken up by the expanding vacuoles, which eventually fuse into a single large vacuole (D); in the mature cell the thin band of peripheral cytoplasm may constitute less than 10% of the cell's volume.

(Fig. 32.8). The cellulose fibrils, in turn, are linked to other polysaccharides, especially pectin and hemicellulose, which hold the fibrils in a more or less rigid pattern. For a cell to grow, two things must happen: some of the cross-linkages must be temporarily broken, so that the wall will become more plastic, and new wall material must be inserted.

Auxin seems to induce both greater wall plasticity and insertion of new wall material. Within minutes after auxin-activated proton (H^+) pumps in the membrane have transported hydrogen ions from the cytoplasm into the wall, some of the hydrogen bonds between the polysaccharides break, the fibrils loosen, and the walls become more pliable.

Because the cell contents are hypertonic relative to the extracellular fluids, and the sole constraint on further movement of water into the cell has been resistance by the wall, fluid begins to move into the cell as soon as the wall becomes less resistant to stretching. As more water floods into the cell, especially into the vacuole, the hydrostatic pressure inside the cell causes more stretching of the wall until the volume of the cell has increased as much as a hundredfold. This enormous increase in cell size is achieved with little or no synthesis of new cytoplasm: the cytoplasm is restricted to a thin layer next to the wall; the greater part of the cellular volume is filled by the expanded cell vacuole (Fig. 32.9). Cell growth in plants thus differs greatly from cell growth in animals, where most increase in size results from formation of new cytoplasm.

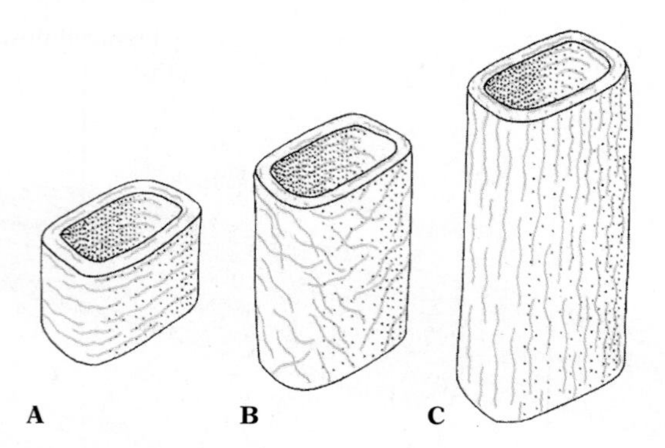

32.10 Change in the orientation of fibrils of the cell wall as a plant cell elongates

(A) Before the start of elongation, the cellulose fibrils are arranged horizontally. (B) As elongation begins and the wall is stretched, the fibrils are displaced and tipped toward a more vertical alignment. (C) In the fully elongated cell the fibrils in the outer, older wall layers are oriented vertically. The recently deposited fibrils in the inner, newer layers are horizontal.

The effect of auxin on the second aspect of wall enlargement—insertion of new wall material—is much slower. It probably depends on synthesis of new mRNA coding for the enzymes that catalyze the addition of further units to the complex polysaccharide structure of the wall.

Elongation of the root and stem is caused largely by cell elongation. (Enlargement of cells in other dimensions is, of course, important in the morphogenesis of other parts of the plant.) Auxin is targeted against the molecular attachments in the cell wall so that the stretching occurs in only one dimension—length, rather than width. The unidimensional stretching is reflected in the way the alignment of the cellulose fibrils changes during cell elongation (Fig. 32.10).

The zone where cell elongation predominates in the root is just behind the zone of intensive cell division in the meristem. The elongation zone usually extends only a few millimeters along the root (Fig. 32.7). In a corn seedling the fastest elongation occurs about 4 mm from the root tip; cells more than 10 mm behind the tip have completed their elongation. The situation is similar in bean seedlings (Fig. 32.11). Elongation in this zone has the effect of pushing the root tip through the soil faster than if it were driven only by the production of new cells in the zone of intensive cell division.

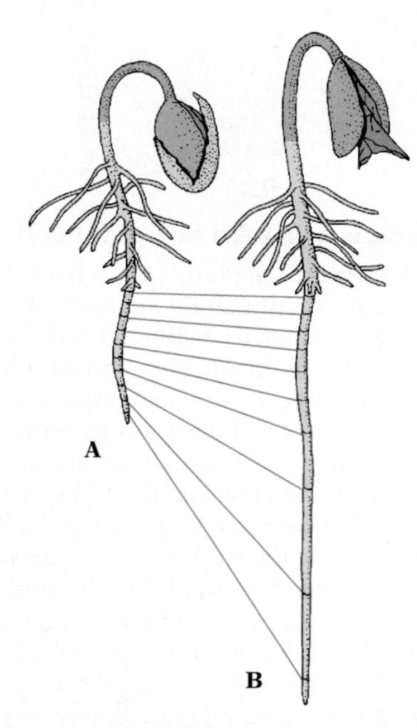

32.11 Growth of a bean root

Note that the parts of the root closer to the tip (but not at the tip) in the earlier stage (A) have undergone the most elongation by the later stage (B).

Though cell division has slowed down in the regions behind the apical meristem, it has not ceased. Indeed, persistent cell division is important to the development of the root. For example, it is persistent meristematic activity in the pericycle that gives rise to new lateral roots, which, as they grow, push out and through the overlying endodermal, cortical, and epidermal tissues to enter the soil (Figs. 32.12, 32.13).

The direction of root growth depends on several factors. The most important is gravity: roots normally grow down. (The mechanism of this downward growth will be described in a later section.) Another factor is touch: when a root contacts an obstacle (like a stone in the soil), its growth changes and the root takes a curving route to one side. A third factor in some (perhaps all) plants is a yet-to-be-identified chemical released by other roots of the plant. This substance causes roots to maintain an appropriate degree of spacing, allowing the network to operate efficiently and with a

32.12 Origin of a new lateral root from the pericycle of an older root

(A) A new lateral root starts to form when a group of cells in the pericycle (dark brown) begin to enlarge in a radial direction. (B) These cells then divide. (C) Continued divisions and elongations in this new direction produce a growing mass that pushes out and through the outer tissue layers of the old root, crushing the cells in its path. Note that a new root forms by a change in the orientation of cell elongation and cell division, not by the morphogenetic movements of cells expected in a comparable developmental event in an animal.

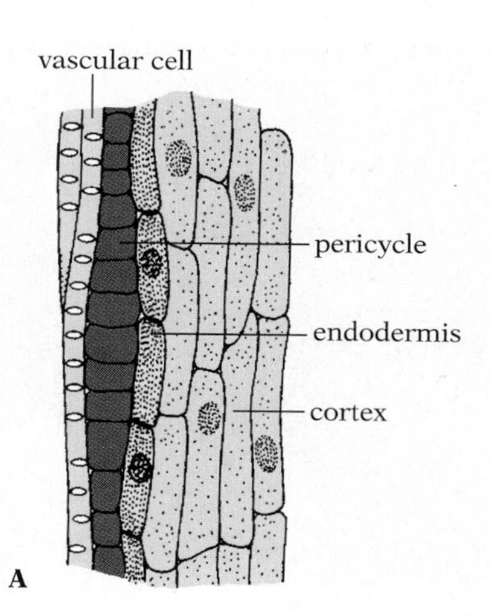

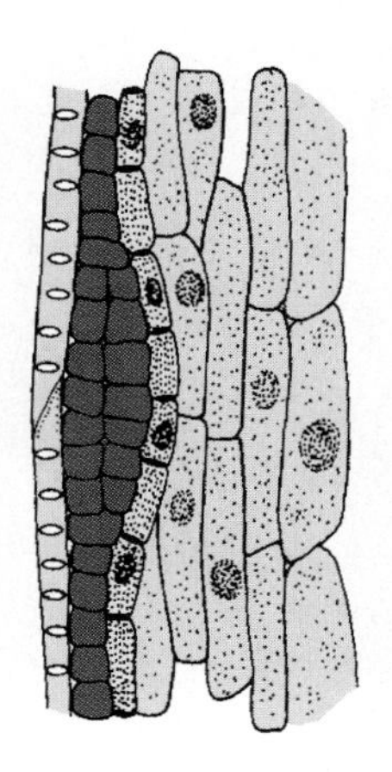

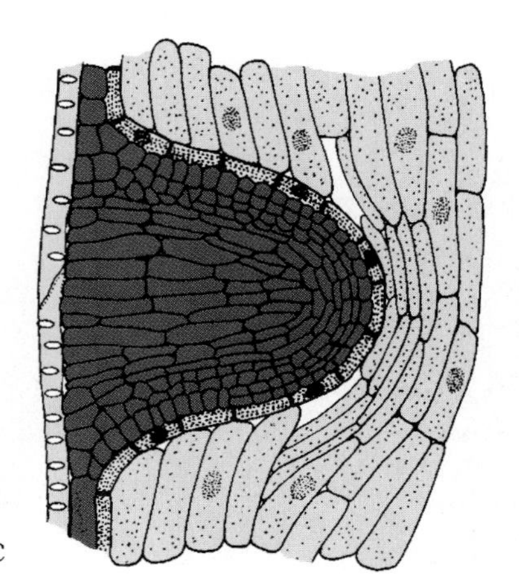

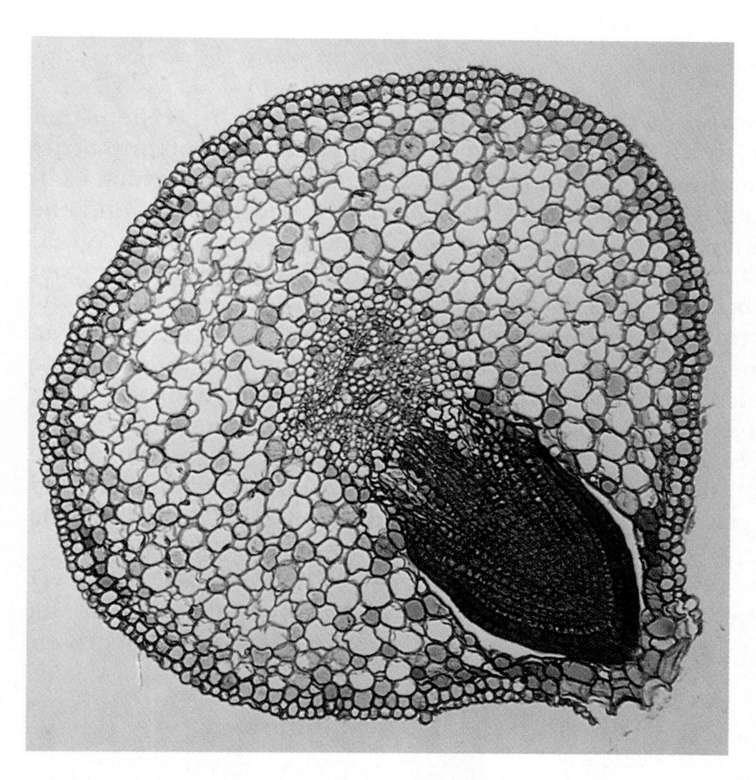

32.13 Origin of a lateral root in *Salix* (willow)

The new lateral root has pushed through the cortical and epidermal layers of the old root and is ready to enter the soil.

minimum of wasteful competition between roots. Temperature, water, and nutrient concentration also help orient and direct root growth.

Stem growth Growth of the main stem axis is basically similar to growth of the root. New cells are produced by an apical meristem, and these cells then elongate, pushing the apex upward. At regular intervals there is an increase in the rate of cell division under a lo-

A

apical meristem
leaf primordium

B

32.14 A leaf bud

(A) Leaf buds on a juniper branch. (B) Diagrammatic representation.

calized region of the sloping surface of the apical meristem that gives rise to a series of swellings that function as leaf primordia (the tissue that later develops into leaves). The point at which each leaf primordium arises from the stem is called a ***node***, and the length of stem between two successive nodes is an ***internode*** (see Fig. 22.54, p. 629). Most increase in length of the stem results from elongation of the cells in the young internodes.

At the tip of the stem is a series of internodes that have not yet undergone much elongation. The tiny leaf primordia that separate these internodes curve up and over the meristem, with the older, larger ones enveloping the younger, smaller ones (Figs. 32.14, 32.15). The resulting compound structure, consisting of the apical meristem and a series of unelongated internodes enclosed within the leaf primordia, is called a ***bud***. In shoots that have periodic

32.15 Sectioned *Elodea* bud

The stem tip forms a bud as upcurving leaf primordia enclose the apical meristem. Flower buds have a similar structure.

10 μm

(episodic) growth, the bud is protected on its outer surface by overlapping scales, modified leaves that grow from the base of the bud.

When a dormant bud "opens" in the spring, the scales curve away from the bud and then fall off, and the internodes that were contained within the bud begin to elongate rapidly. As the nodes become farther apart, mitotic activity in the leaf primordia gives rise to young leaves. The pattern of cell division, characteristic for each species, determines whether the leaves will be entire or lobed, simple or compound.

Before the leaf is fully formed, a small mound of meristematic tissue usually arises in the angle between the base of each leaf and the internode above it. Each of these new meristematic regions gives rise to a lateral or ***axillary bud*** with the same essential features as a ***terminal bud*** (Fig. 32.16). Elongation of the internodes of the lateral buds produces branch stems. In the root there are no surface meristems analogous to the lateral buds, and the lateral roots arise from cell division in deeper tissue layers.

The generation of nodes and the development of axillary buds is under hormonal control, (the mechanism of which we will discuss in a later section). The plant's response to these hormones depends in large part on the growth strategy of the species, a strategy that imparts a distinctive shape to most kinds of plants. Each shoot can be thought of as a developing structure that undergoes progressive differentiation. Young shoots usually have a distinctive internal anatomy and chemistry, and frequently produce leaves whose shape is readily discernible from those produced by older "adult" shoots.

The final step in the differentiation of a shoot is the generation of a reproductive structure. Some species simply produce shoots that grow, mature, and produce a flower or set of flowers at the end. In other species the shoots grow indefinitely, producing periodic side shoots that grow, mature, and produce flowering parts. Finally, there are plants in which the main shoot generates a terminal flower, at which point side shoots take over growth, producing more new side shoots when they themselves flower. A survey of the common plants in your area will quickly turn up examples of each strategy for branching and flowering.

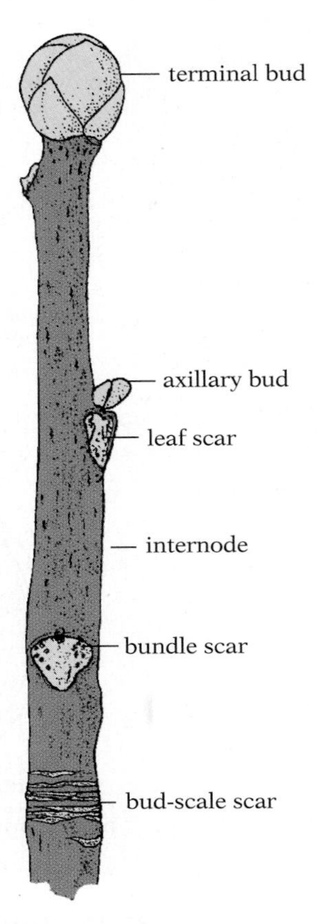

32.16 Portion of a stem

The bud-scale scar shows where the dormant terminal bud of the previous winter was; the length of stem between the bud-scale scar and the terminal bud is one year's growth. Leaf scars show where petioles were attached to the stem. The bundle scars (within each leaf scar) show where the vascular bundles passed into the petiole. The axillary bud will give rise to a branch stem during the next growing season.

How tissues differentiate All the new cells produced by the apical meristems are basically alike. Yet some of these cells will become xylem vessels, others sieve elements, and so forth. The process whereby a cell changes from its immature form to some one mature form is called ***differentiation***.

In the growing root or stem, cells begin to differentiate into the various tissues of the plant while they are still within the meristematic region. As the predominant activities of cell division and cell enlargement run their course, the cells are left to mature into their final form. Hence, though the pattern of histogenesis (formation of tissues) is laid out very close to the tip of the elongating axis, keep in mind the successive regions along the axis concerned with cell division, cell elongation, and cell maturation as shown in Fig. 32.7. It is useful to be able to recognize these regions in the different stages of growth.

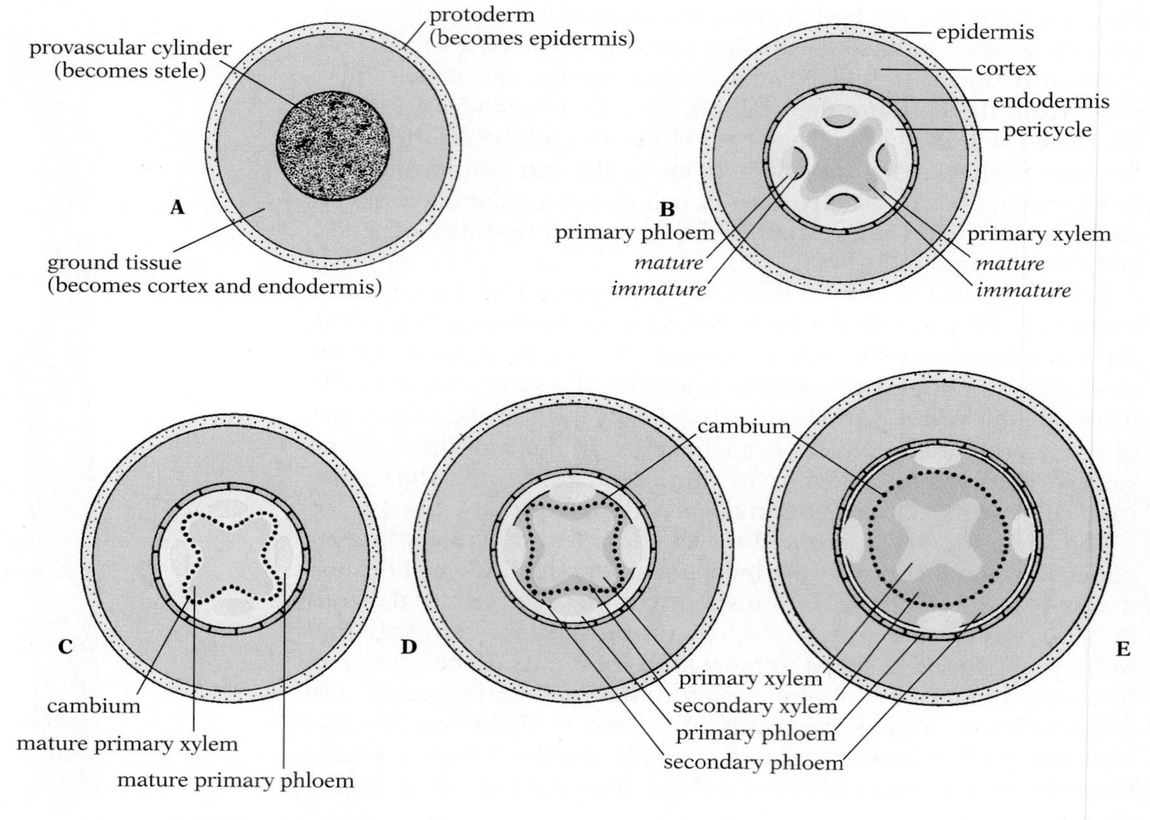

32.17 Differentiation in a young root

(A) Cross section just back of the meristem. Three distinct concentric areas can already be detected. (B) At a slightly later stage of development the protoderm has differentiated into epidermis; the ground tissue has differentiated into cortex and endodermis; and the provascular cylinder has begun to differentiate into primary xylem and primary phloem. (C) Differentiation in the provascular cylinder is complete, and the cambium is about to become active. (D) Divisions in the cambium have given rise to secondary xylem and secondary phloem, which are located between the primary xylem and primary phloem. (E) The areas of secondary tissue continue to thicken as more new cells are produced by the cambium.

Three concentric areas are visible even in the region just behind the meristem of a root (Fig. 32.17A): an outer protoderm, a wide area of ground tissue located beneath the protoderm, and an inner core of provascular tissue composed of particularly elongate cells. Just as in the embryo, the protoderm rapidly matures into the epidermis, the ground tissue into the cortex and endodermis, and the provascular core into the primary tissues of the stele: primary xylem, primary phloem, pericycle, and vascular cambium (Fig. 32.17B–C). Differentiation in the growing stem follows a similar pattern, except that there are usually two areas of ground tissue: one between the protoderm and the provascular cylinder, which gives rise to the cortex and endodermis, and a second inside the provascular cylinder, which becomes the pith.

As shown in Chapter 29, increase in circumference of the root or stem depends on formation of secondary tissues derived from lateral meristems, particularly the vascular cambium. As cells of the vascular cambium undergo mitosis, many new cells are produced on the inner face of the cambium that differentiate into secondary xylem, while other new cells are produced on the outer face of the cambium and differentiate into secondary phloem (Fig. 32.17D–E). As more and more secondary vascular tissue is formed and the circumference increases steadily, the old epidermis and cortex are broken and sloughed off. These are replaced by a secondary protective tissue, the cork, composed of cells derived from a new lateral meristem, the cork cambium. This cambium forms from a layer of the old cortex, from the pericycle, or even from the older phloem, depending on the species.

About 30–40 genes (many of which encode factors that regulate the transcription of other genes) are involved in establishing the radial and longitudinal patterns of roots and shoots, as well as orchestrating the development of embryos, leaves, and especially flowers. Complete flowers consist of four major concentrically arranged sets of parts mounted on a receptacle (Fig. 32.18A): the sepals (which enclose the flower bud), the petals, the pollen-bearing stamens, and the carpels (which provide the pathway by which growing pollen tubes reach the ovary). A variety of single-gene mutations alter the pattern of flower development. A major subset of these mutations are homeotic—that is, one organ or organ part is transformed into another. (Homeotic mutations in animal development are described in Chapter 14.) Three types of homeotic flower mutants are known (Fig. 32.18B–D). One converts sepals into carpels and petals into stamens; another transforms stamens into petals and carpels into sepals; the third turns petals into sepals and stamens into carpels. The anatomy of developing buds is basically concentric (Fig. 32.15), a series of layers surrounding the meristem.

Analysis of the three types of homeotic transformation suggests that, as in animals, developmental position is being "read" by cells on the basis of two overlapping chemical gradients. A change in the production rate of one of these chemical morphogens or a change in a cell's ability to measure the concentrations leads to these transformations. The similarity with the animal system runs still deeper: some of these genes contain a sequence sufficiently similar to the homeobox sequence of animals and fungi to suggest a common origin. (The role of homeobox sequences in organizing development is described in Chapter 14.)

The homeobox sequence turns up also near genes affecting leaf differentiation. Once a node becomes committed to producing a leaf (or a bud element is determined, and begins to form a sepal, petal, stamen, or carpel), another developmental sequence must take over to produce an appropriate structure. One well-studied homeobox mutation prevents the cells adjacent to leaf veins from maturing normally; instead they continue to divide, producing a knotty appearance where the leaf should be smooth. This suggests that there may be a morphogen disseminated from the developing veins. Given the enormous commercial potential of genetic redesign of plant growth and differentiation, this field promises to be one of the most active in modern biology.

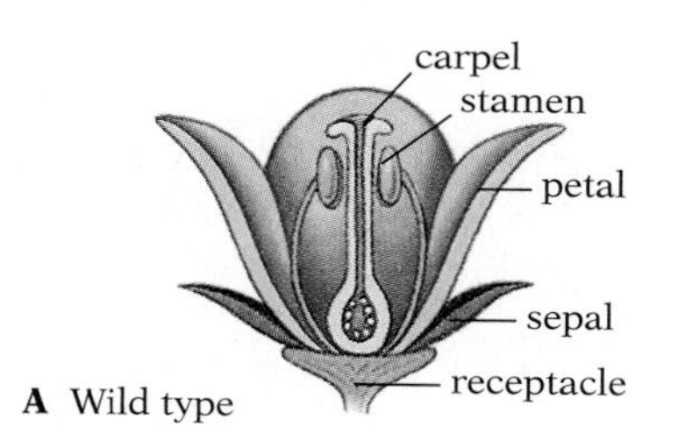

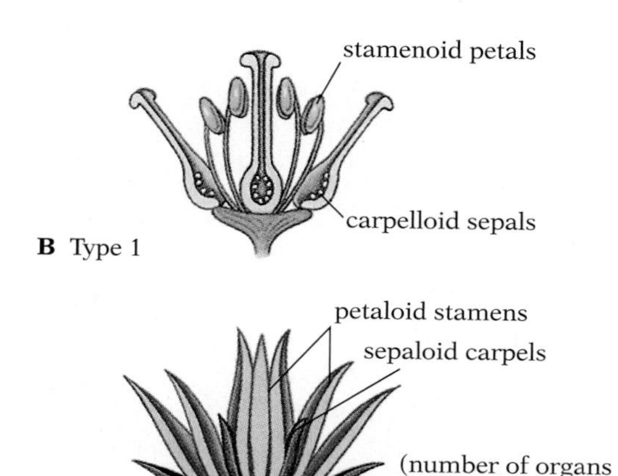

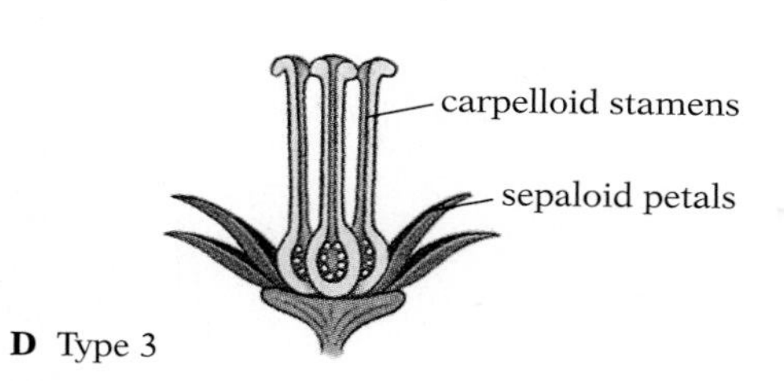

32.18 Homeotic mutation in flowers

Complete flowers consist of concentrically arranged sepals, petals, stamens, and carpels (A). Various homeotic mutations can create three basic patterns of transformation (B–D). The absence of other kinds of changes—of carpels into stamens, for example—makes it clear that the patterns are generated by at least two overlapping morphogenic gradients.

HOW HORMONES CONTROL PLANT GROWTH

We've seen that higher plants have specialized tissues—roots, stems, leaves, and reproductive organs—whose activity must be coordinated if the plant is to germinate, grow, and reproduce. Chemical control plays a role in orienting and regulating the growth of stems and roots, in timing both reproduction and shedding of leaves, in initiating the germination of seeds, and in regulating many other functions. When a chemical is secreted by tissues specialized for its production, moves to other parts of the plant (usually by means of the organism's internal transport system), and exerts an effect on a specific set of target cells, it is called a ***hormone***.

Plant hormones are produced most abundantly in the actively growing parts of the plant body, such as the apical meristems of the shoot and the root, young growing leaves, or developing seeds or fruits. There are no separate hormone-producing organs in plants analogous to the endocrine glands of higher animals.

■ THE MANY ROLES OF AUXIN

One group of plant hormones, the auxins, displays an amazing variety of effects on the growth of different plant tissues.

Growing toward light Most plants attempt to grow toward light. A potted plant indoors grows toward a window; you turn the plant and soon the shoot reorients toward the light of the window. Responding to light through differential growth is called phototropism, from the Greek words for "light" and "turning." (Other tropisms involve directed growth in response to other stimuli. Gravitropism, for example, is a response to gravity.) In plant shoots the phototropism is positive, a growth toward the stimulus; roots, on the other hand, exhibit negative phototropism, a growth away from light.

Around 1880, Charles Darwin and his son Francis became among the first to work on phototropism. They, like many who followed them, experimented on coleoptiles (see Fig. 32.6). The coleoptile grows principally by cell elongation. It exhibits a very strong positive phototropic response, bending toward light by means of differential growth. Cells on the side of the coleoptile away from the light elongate, forcing the tip toward the light. The Darwins showed that if the tip of the coleoptile is covered by a tiny black cap, the plant fails to grow toward light, while control coleoptiles with their tips exposed or covered with transparent caps bend normally (Fig. 32.19). Blocking light to the base of the coleoptile but not the tip fails to prevent bending. The tip of the coleoptile must play the key role in sensing the light, and then communicate this information to the growth zone (lying just below the tip), where the bending through differential growth occurs. Indeed, when the Darwins cut off the tip, the coleoptile failed to bend.

About 30 years later, P. Boysen-Jensen in Denmark showed that

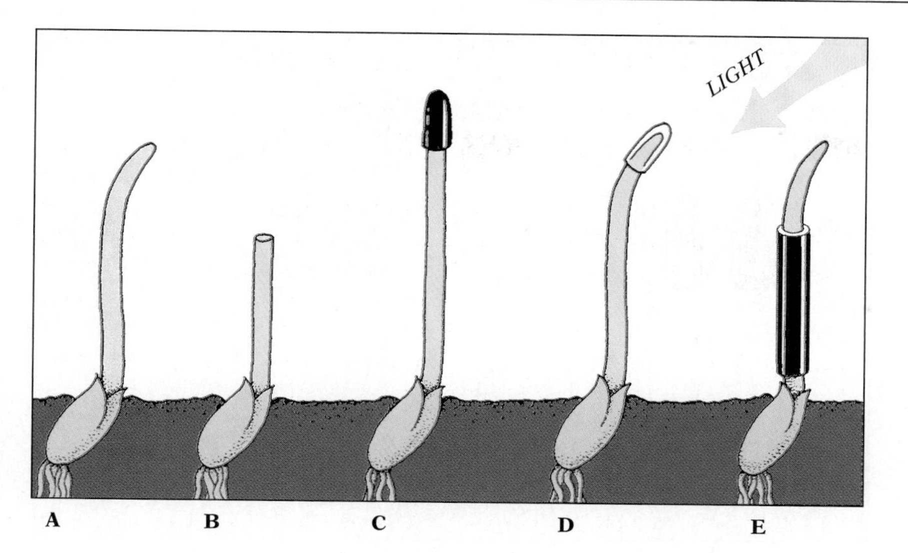

32.19 The Darwins' experiments on phototropism

(A) A coleoptile of canary grass grows toward the light. (B-C) The coleoptile does not bend if its tip is removed or is covered by an opaque cap. (D) The coleoptile does bend if its tip is covered by a transparent cap. (E) It also bends if its base is covered by an opaque substance (represented here as a tube; the Darwins used black sand).

the communication between tip and growth zone was probably chemical. He removed the tips of oat coleoptiles, spread a thin layer of gelatin on the cut end of each stump, and then replaced the tip (Fig. 32.20). When he illuminated the tip from the side, the coleoptile base grew toward the light; the message from the tip had moved across the gelatin barrier and induced bending in the base.

That the tip could cause the base of an oat coleoptile to bend even in the dark was demonstrated by A. Paál in Hungary in 1918. He cut off the tip and then replaced it off-center on the stump (Fig. 32.21). If he put the tip on the right side of the stump in the dark, the coleoptile bent to the left; if he put the tip on the left side, the coleoptile bent to the right. This asymmetric elongation of the coleoptile caused it to bend away from the side undergoing the greater elongation. This suggests that the tip continuously produces a substance that moves down the coleoptile stem and causes cells directly below it to grow. This would mean that light coming

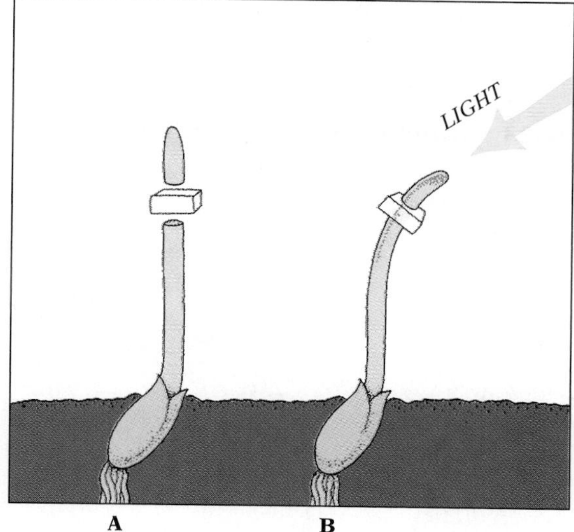

32.20 Boysen-Jensen's experiment

When the tip of an oat coleoptile is cut off, a layer of gelatin put on the end of the stump, and the tip replaced (A), the coleoptile will grow so that it turns toward the light (B). Presumably a chemical (red) moves from the tip, through the gelatin, and into the base, and stimulates the plant to turn.

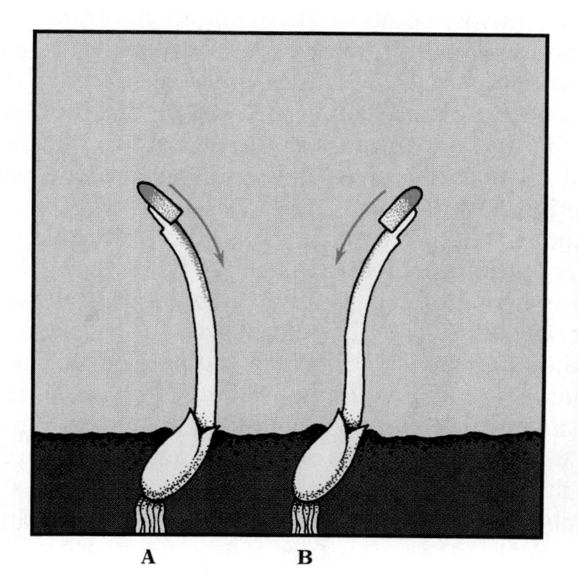

32.21 Paál's experiment

This experiment is performed in the dark. (A) If the tip of a coleoptile is cut off and then replaced right of center, the coleoptile will grow so that it bends to the left. (B) If the tip is placed left of center, it bends to the right. (Note that only the tip of the coleoptile is cut off; the rolled-up leaf inside is left intact.)

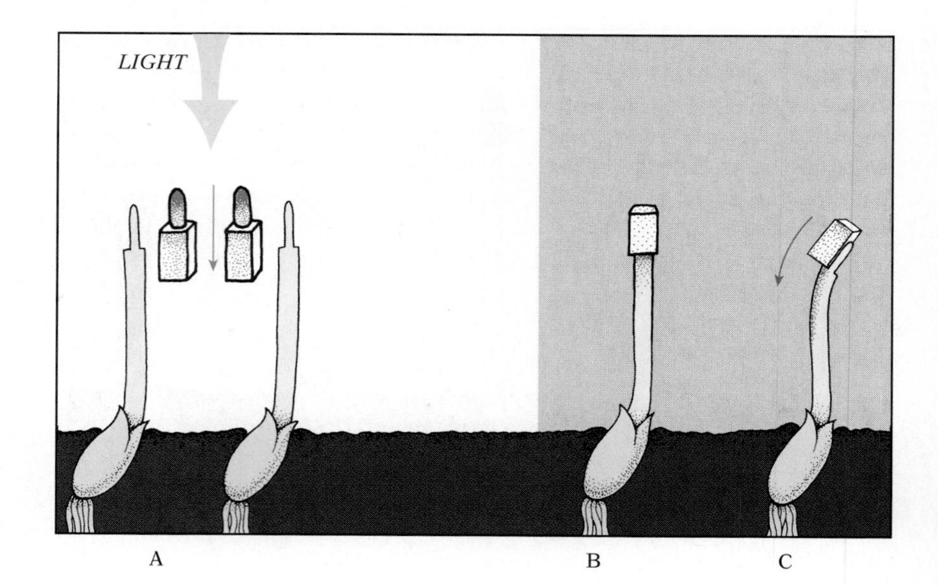

32.22 Went's experiment

When the tips of coleoptiles are cut off and placed on blocks of agar for about an hour (A), and one of the blocks (without the tip) is then put on a stump (B), the stump will resume growing even in the dark. If a block is placed off-center on a stump in the dark (C), the stump will grow so as to bend away from the side on which the block rests. Apparently a hormone has diffused from the tips into the blocks, and this hormone can then diffuse from the blocks into the stumps.

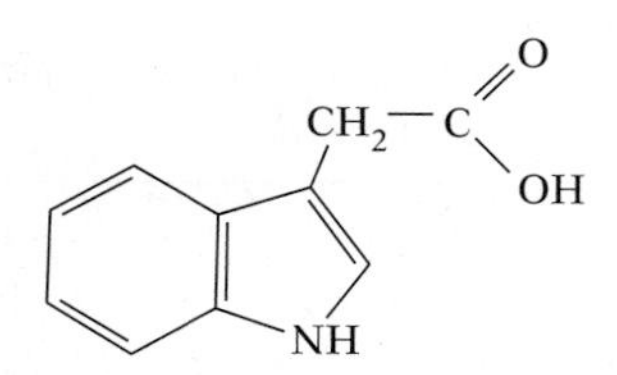

32.23 Structure of indoleacetic acid

from the side alters the relative amount of this substance moving down each side of the stem, thus causing more growth on one side than on the other.

The conclusive demonstration that a growth chemical moves downward from the tip was reported in 1926 by Frits Went, then in Holland. He removed the tips from oat coleoptiles and placed these isolated tips, base down, on blocks of agar (a gelatinous material) for about an hour (Fig. 32.22). He then put the blocks of agar, minus the tips, on the cut ends of the coleoptile stumps. The stumps resumed growth for a time. If the agar blocks were put on off-center, the stumps would bend, even in darkness. The growth-stimulating substance had been synthesized in the tips and had diffused into the blocks of agar while the tips were sitting on them. When these blocks were then placed on the stumps, the chemical moved down from them into the stumps and stimulated elongation. Went called this hypothetical diffusible hormone auxin (from a Greek word meaning "to grow").

Many chemicals, some of them found naturally in plants and some synthesized only in the laboratory, qualify as auxins. The one most thoroughly investigated is ***indoleacetic acid*** (Fig. 32.23), which has been isolated from numerous plants. Indoleacetic acid is the principal (perhaps the only) directly active natural auxin. The activity of other compounds may result from their conversion into indoleacetic acid by plants.

To summarize, these experiments indicate that there is little lateral movement of the auxin after it has been released from the tip; the hormone reaches and stimulates only those cells directly under the point of release. The differential growth that generates bending occurs when light strikes the tip of the plant from one side, somehow reducing the auxin supply on that side. Consequently the illuminated side of the plant grows more slowly than the shaded side, and this asymmetric growth produces bending toward the slower-growing illuminated side (Fig. 32.24).

Plant physiologists are still not certain how the tip detects the

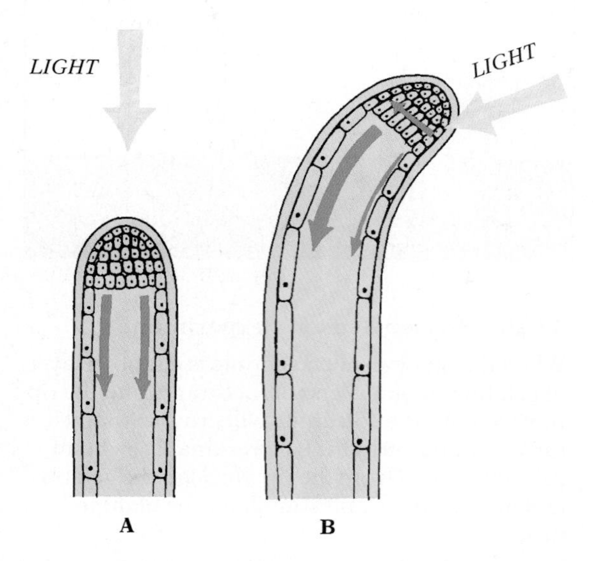

32.24 Auxin-mediated response to a light source in a young shoot

Light receptors in the growing tip of the plant are thought to respond to light by altering auxin distribution. (For clarity, only the cells in the growing tip and the periphery of the shoot are shown individually.) (A) With light from directly above, auxin is equally distributed to the growing tissues on all sides. (B) Light from the right causes a decrease in the supply of auxin to the right side of the plant, which results in a relative elongation of cells on the left side, and a bending toward the light.

light. Since the response is triggered only by blue light, the detection pigment ought to be orange (the apparent color of any substance that absorbs blue). But all candidates of the correct color have been conclusively ruled out. Nor do plant physiologists know how detection of light is coupled to asymmetric auxin distribution. The evidence indicates that there is active lateral transport of auxin in the tip from the illuminated to the shaded side (Fig. 32.25), but there is considerable debate on this point.

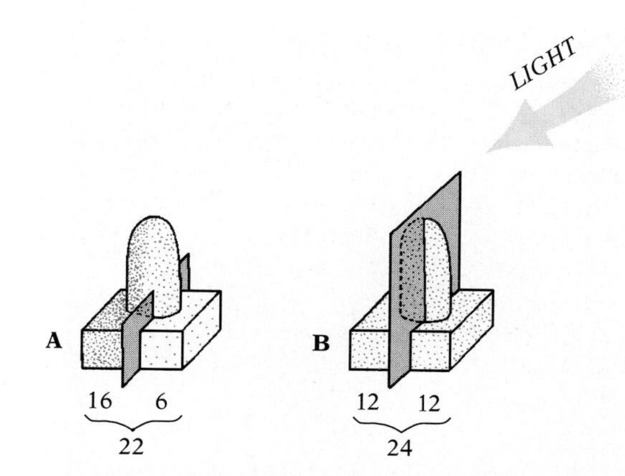

32.25 An experiment demonstrating lateral movement of auxin

The tips of two coleoptiles were placed on agar blocks partitioned by an auxin-impermeable barrier. (A) If the barrier extended only slightly into the base of the coleoptile tip, more auxin was later found in the side of the block away from the light (16 units) than in the side toward the light (6 units). (The units refer to degrees of bending that the block segments induce in decapitated coleoptiles in the dark, based on Went's quantification of his bending assay; the amount of bending, in turn, reflects the amount of auxin in the segment.) (B) Approximately the same total amount of auxin was found in the agar block under a coleoptile tip that was completely partitioned, but the amounts in the two sides of the block were the same. This experiment showed that light produces asymmetric auxin distribution by causing lateral movement of auxin, not by destroying auxin on the lighted side or by causing a differential production of this hormone.

How auxin acts Since auxin is produced continually in growing tips, it seems clear that whatever light receptors are involved respond to the light by triggering a redistribution of the hormone. The receptor molecules apparently direct the hormone to the side of the shoot away from the light. Horizontal movement and vertical movement of auxin probably involve identical processes. Neither depends on the vascular system, and movement from cell to cell takes place so quickly that active transport must be involved. This conclusion is reinforced by the observation that auxin can move against its concentration gradient, but cannot be transported in the presence of chemicals that inhibit metabolic processes.

According to one model, known as the ***acid-growth hypothesis***, once the auxin reaches the growing cells in the shoot, it activates pumps that transport H^+ ions from the cytoplasm to the cell wall, just outside the membrane. Here the lowered pH breaks the hydrogen bonds between the cellulose fibrils in the cell wall. As a result the walls become more extensible and the cells take in water and elongate. When there is more auxin on the side of the shoot away from the light, cells there elongate more than their counterparts on the lighted side, and the shoot bends. Cells outside the growing region are insensitive to auxin.

Growing up from and down into the ground Plants respond not only to light but to gravity as well. If you lay a potted plant on its side in the dark and leave it for a few hours, you will find that the shoot has begun to grow upward (Fig. 32.26). This is a negative gravitropic (also called a negative geotropic) response—the shoot turns away from the pull of gravity—while the roots show positive geotropism.

In a horizontally placed shoot, the concentration of auxin in the lower side increases while the concentration in the upper side decreases. This unequal distribution of auxin should stimulate the cells in the lower side to elongate faster than the cells in the upper side, and the shoot ought therefore to turn upward as it grows. Matters are not so simple, however: auxin may not be involved in dicot gravitropism, and it has at least some help from other hormones in monocots.

The situation in roots is also unclear. It seems likely that the dis-

32.26 Gravitropism of shoot and root

When a growing plant is left lying on its side, the shoot will grow upward and the roots will grow downward.

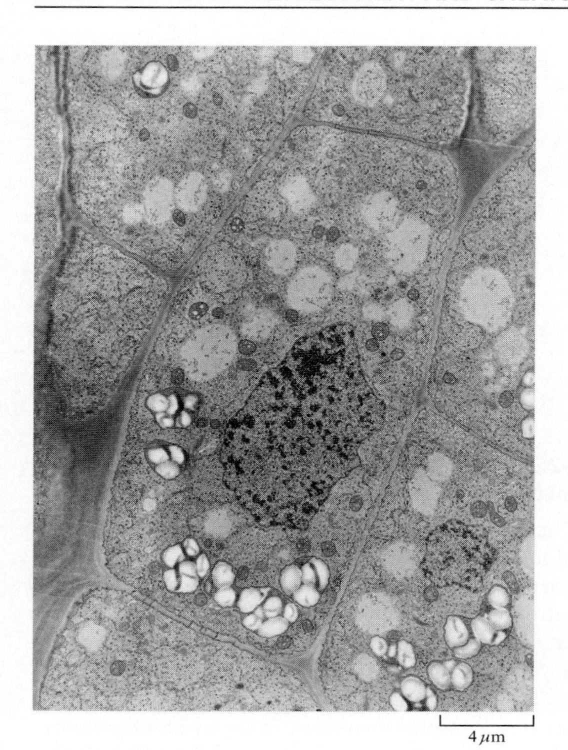

32.27 Amyloplasts in a root-cap cell

Since the starch they contain is denser than the surrounding cytoplasm, amyloplasts will move through the cytoplasm in response to gravity and accumulate within minutes at the bottom of a repositioned cell. Organelles of the same density as cytoplasm remain spread throughout the cell.

tribution of specialized starch plastids called ***amyloplasts*** (Fig. 32.27) in the root cap plays a major role, though even this is controversial. The amyloplasts (which are also found in shoot and meristem cells) settle rapidly to the bottom of root-cap cells, triggering the release of Ca^{++} ions. The calcium, in turn, seems to cause the redistribution of growth hormones between the top and bottom membranes of growing cells behind the root cap. Though auxin is clearly one of the substances transported in this reaction, there is no clear evidence that it plays a major role in the subsequent directed growth.

Inhibiting branching Auxin produced in the terminal bud moves downward in the shoot and inhibits development of lateral buds while it stimulates elongation of the plant's main stem. The terminal bud thus exerts ***apical dominance*** over the rest of the shoot, ensuring that the plant's energy for growth will be funneled into the main stem and produce a tall plant with relatively short lateral branches. Longer branches usually develop only from buds far enough below the terminal bud to be partly free of the apical dominance. If the terminal bud is removed, however, apical dominance is destroyed, and several of the upper lateral buds will begin to grow, producing branches whose terminal buds soon exert dominance over any buds below them (Fig. 32.28). Flower and shrub

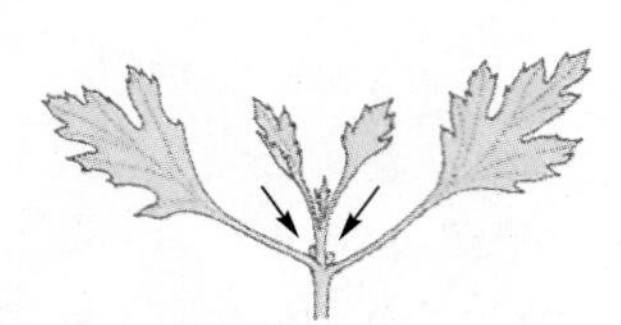

32.28 Inhibition of lateral buds by the terminal bud in chrysanthemum

As long as the terminal bud is intact, the lateral buds marked by arrows will grow very little, if at all. If the terminal bud is removed, those buds are released from inhibition and grow rapidly, forming the new leaders of the plant.

growers frequently pinch off the terminal buds of their plants one or more times each season in order to produce bushy, well-branched plants with many flowering points instead of tall spindly ones with sparse flowers. Pinching buds will not work, however, for plants in which the young leaves, not the terminal buds, exert control over the lateral buds. The different growth forms of various kinds of plants reflect the different degrees of apical dominance among species.

Once two or more branches have begun to develop, neither inhibits the other. Auxin secreted by the terminal bud of one branch does not reach the terminal bud of the other branch because it moves mostly downward in the stem (Fig. 32.29).

Forming fruit An organ whose normal development depends on the stimulatory effect of auxins is the fruit, which develops from the ovary or from the flower receptacle of the plant. In the absence of fertilization, fruit usually does not develop; instead, a weak layer of thin-walled cells, called an ***abscission layer***, forms at the base of the flower stalk (Fig. 32.30). This layer breaks, and the withered flower with its ovary falls off. If fertilization does occur, no abscission layer forms, and the ovary begins to grow rapidly. This period of rapid growth by the ovary (or by the receptacle in some plants), during which the ovary develops into the fruit, is initiated in most species by auxin released by the same pollen grains that bring about fertilization. The continued growth and development of the fruit in such plants depend on stimulation by auxins produced by the seeds contained within it.

Dropping leaves Unfertilized flowers drop off the plant because an abscission layer forms at the base of the flower stalk in the absence of high auxin production in the floral organs; ripe fruit also drops because of abscission-layer formation. Similarly, the shedding of leaves in autumn (or of diseased leaves at any time of year) by deciduous trees and shrubs usually (though not always) involves

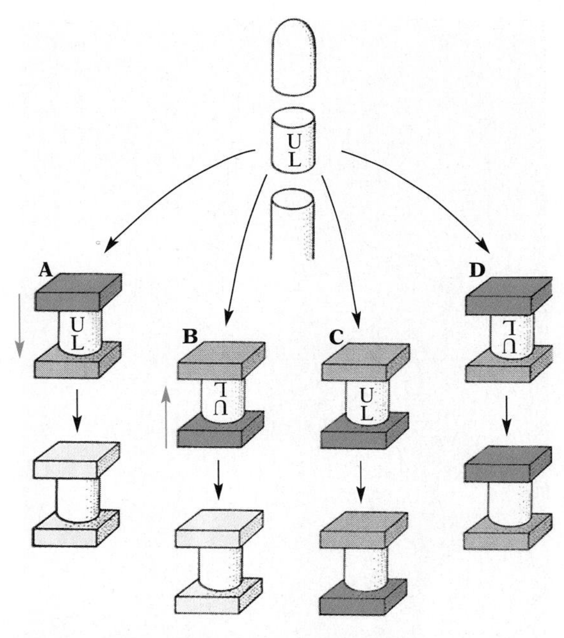

32.29 Experiment demonstrating the polarity of auxin movement

A segment is cut from a coleoptile or young stem (top). (A) An agar block containing auxin (red) is placed on the upper end of the segment and a block without auxin (gray) is placed on the lower end. Some auxin moves from the one block, through the coleoptile segment, into the other block. (B) The same thing happens even if the whole preparation is inverted, an indication that the movement is not a response to gravity. (C) When the agar block containing auxin is put on the lower end of the coleoptile segment and the block without auxin is put on the upper end, virtually no movement of auxin occurs, and inverting the group (D) makes no difference either. Hence auxin must move primarily in only one direction through a coleoptile or stem, and that direction is determined by properties of the cells themselves, not by the pull of gravity.

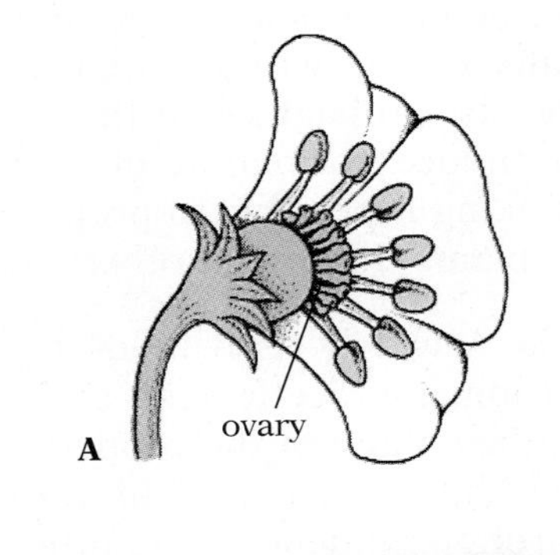

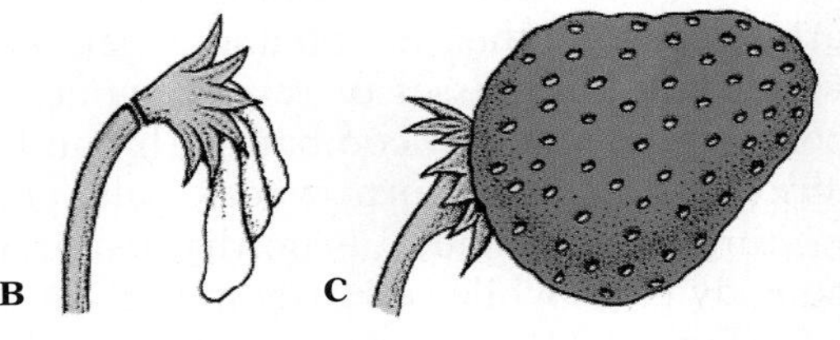

32.30 Formation of fruit

The formation of fruit from the ovary of a flower (A) depends on whether or not fertilization has occurred. If fertilization does not take place, a fragile abscission layer forms at the base of the flower stalk, and the withered flower soon drops off (B). During fertilization, pollen grains release auxins that initiate the growth and development of the ovary into the fruit (C).

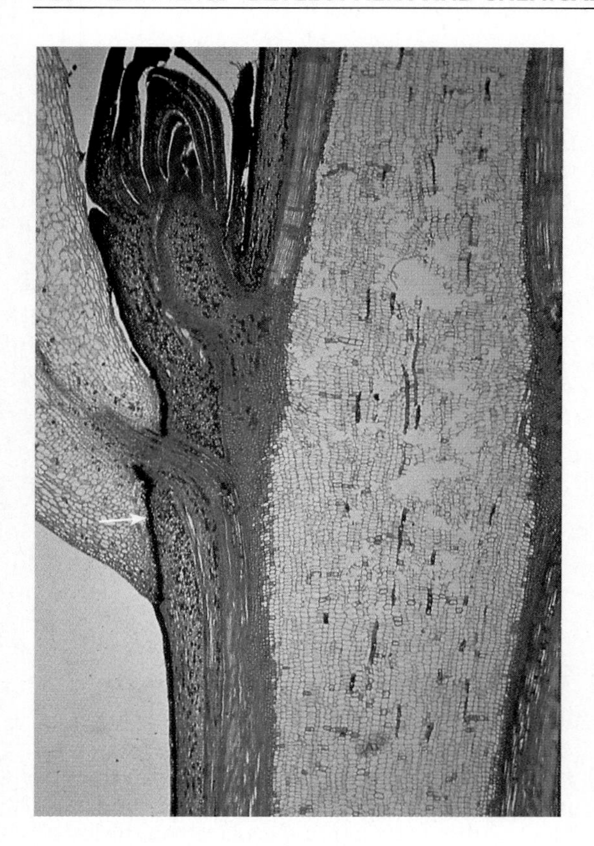

32.31 Abscission layer at base of petiole of a maple leaf
The arrow indicates the small cells of the abscission layer.

abscission-layer formation at the base of the petiole (Fig. 32.31), in part as a result of declining auxin production by the leaf blade. In short, auxin acts as an inhibitor of abscission; another hormone, ethylene (to be discussed shortly), is the principal promoter of abscission. The actual break in the abscission layer can be initiated by any slight strain like the pressure of a gentle wind, because the cell walls there have been weakened by an increasing concentration of cellulase, an enzyme that digests cellulose.

Controlling cell division Besides playing a part in cell elongation and abscission-layer formation, auxins are probably involved in cell division in certain tissues. Apparently it is auxin, moving downward from the buds in early spring, that stimulates renewed activity in the cambium, leading to production of new vascular tissue. As autumn approaches, auxin production by the buds and leaves declines, with the result that cambial activity also declines.

Auxins probably also initiate formation of lateral roots. As we saw earlier, such roots usually have their origin in the layer of relatively undifferentiated cells called the pericycle (see Fig. 32.12), which is just internal to the endodermis. Most of the time the cells of the pericycle show no meristematic activity. At intervals, however, a small group of cells in the pericycle changes into actively dividing meristematic tissue, giving rise to a new lateral root that bursts through the outer tissues of the main root and enters the soil. The stimulus initiating this meristematic activity in the pericycle probably comes from auxin. Auxin can, in fact, be applied to the roots of plants to induce lateral branching, but it is uncertain whether the auxin stimulates cell division directly; for instance, auxin could operate indirectly by eliciting increased production of another hormone (ethylene) that might then act as the direct stimulant.

Cuttings from some plants, such as geraniums and willows, will readily root in water or soil without application of hormone, but many plants cannot be propagated in this manner. Application of auxins will often induce formation of advantitious roots from leaves or stems, making it possible to propagate vegetatively many valuable strains of plants that might otherwise be lost.

Controlling weeds A widely used modern weed killer, or herbicide, is 2,4-dichlorophenoxyacetic acid, or 2,4-D (Fig. 32.32). This synthetic chemical has many of the properties of auxins. Artificial auxinlike chemicals have been used in vast quantities since the 1940s for the control of dandelions and other broad-leaved weeds. Because they are selective in their action and, when used in proper concentrations, will not kill grasses or related monocots, they are of enormous commercial value in combatting broad-leaved weeds in lawns, pastures, and fields of corn, wheat, oats, and rice. They kill plants by stimulating rapid, uncoordinated, and distorted growth of some body parts while seriously inhibiting the functioning of other parts (Fig. 32.33). The exact manner in which these re-

O
O — CH_2 — C
OH
Cl
Cl

32.32 Structure of 2,4-D

32.33 Effect of 2,4-D on a dandelion

sults are produced is not understood, nor do we understand why broad-leaved plants (dicots) are so much more susceptible than grasses (monocots).

32.34 Structure of gibberellic acid

■ STIMULATING GROWTH: THE GIBBERELLINS

The Japanese have long been familiar with a disease of rice that they call foolish-seedling disease. Afflicted plants grow unusually tall but cannot support their own weight. In 1926 a Japanese botanist, E. Kurosawa, found that all such plants are infected with the fungus *Gibberella fujikuroi.* He showed that when this fungus was moved to healthy seedlings, they developed the typical disease symptom of rapid stem elongation. He could also produce symptoms with an extract made from the fungus.

Several Japanese scientists, working on the problem of foolish-seedling disease during the 1930s, succeeded in isolating and crystallizing a substance from *Gibberella*, now known as ***gibberellin***, that produced typical disease symptoms when applied to rice plants. Apparently the gibberellin from the fungus instigates rapid growth of the host plant in the affected area; the fungus then uses the products of its host's elevated metabolism to support its own growth. More than 70 different gibberellins have been found in fungi and as natural hormones in higher plants. The gibberellin most often used in experimental work is ***gibberellic acid*** (Fig. 32.34).

The most dramatic effect of gibberellins is their stimulation of rapid stem elongation in dwarf plants and other plants that normally undergo little stem elongation (Fig. 32.35). Dwarf plants lack the last enzyme involved in the synthesis of the most active form of gibberellin. Administration of extra quantities of the hormone simply makes up for the deficiency and allows the plants to grow more normally.

Though both gibberellin and auxin stimulate stem elongation, they cannot substitute for each other in controlling plant growth. For one thing, the stages of development at which the plant is most sensitive to these two hormones often differ; in wheat coleoptiles, for example, responsiveness to gibberellin appears earlier than responsiveness to auxin. For another thing, gibberellin can move freely in the plant body through both the xylem and phloem, whereas auxin characteristically can move in only one direction and generally does not depend on the vascular system at all; hence gibberellin exerts systemic influences and cannot produce the bending movements that mark auxin-induced responses.

Gibberellins play a role in a host of developmental processes besides stem elongation. They can (1) often break seed and bud dor-

32.35 Effect of gibberellic acid on cabbage

The plants at left are normal. The one at right was treated with gibberellic acid once a week for 2 months. As discussed in Chapter 1, cabbage is one of several domesticated forms of a tall, spindly ancestor (see Fig. 1.14, p. 11). The loss of normal levels of gibberellic acid, with the consequent development of the desirable compact growth form of cabbage, was the result of intense selective breeding.

HOW WE KNOW:
HOW WITCHWEED TIMES ITS GERMINATION

Most grains use gibberellin as an internal signal to break dormancy. One species' "company memo," however, can become another species' meal ticket. Consider the widespread tropical plant parasite witchweed *(Striga,* see photo below*)*. This pest makes its living by attaching itself to the roots of grain plants and using the host's nutrients to fuel its own growth. Its parasitism is so successful that a single witchweed can produce up to 500,000 tiny seeds, and the species is said to destroy about 40% of the grain crop in Africa.

Even in dry areas the pest can flourish because its seeds are viable for at least two decades. The problem for the witchweed seed, though, is to synchronize its germination with that of a suitable host: a seed that begins growing too soon will quickly use up its minuscule supply of stored energy. Careful research reveals that witchweed counters with a two-step system for triggering germination. First, seeds wait for the soil to be moist for at least a week. Even when rain comes, however, the witchweed seed germinates only if it detects a grain seedling developing nearby. One of the signals it uses is a by-product of the gibberellin that initiates the host's own germination.

Because the basic chemistry of germination signals in grains is understood, agricultural scientists have developed a variety of strains that leak only small amounts of the germination compounds into the surrounding soil. Other strains lack sufficient levels of a second chemical signal that tells the developing witchweed it is close enough to its target to begin attempting to penetrate the root. Meticulous breeding, then, promises to help control the parasite. Certain herbicides also happen to stimulate weed germination. Applied after sufficient rain to a dormant field infected with the parasite seeds, it can induce a suicidal germination of virtually all of the witchweed seeds.

Witchweed

mancy; (2) induce the embryos in germinating seeds to produce an enzyme that hydrolyzes starch reserves in the seeds; (3) stimulate some biennials to flower during the first year of growth; (4) induce some summer-flowering plants to "bolt" and so produce a flowering stalk in the spring or fall when the day length is too short for them to flower normally; and (5) stimulate fruit set in some species.

Gibberellins are derived from the same biosynthetic pathway as vertebrate steroid hormones (like estrogen and testosterone) and are lipid-soluble; as a result they can cross cell membranes. In at least some cases, they function by turning particular genes on and off. In many seeds, for example, gibberellin produced in one part of a developing plant embryo in response to stimulation by water activates genes in another part that specify the synthesis of the enzyme α-amylase. This enzyme helps convert starch—the seed's supply of stored energy—into readily usable sugars.

■ MAINTAINING BALANCED GROWTH: THE CYTOKININS

The technique of ***tissue culture***—the growing of cells or bits of tissue on sterile nutrient media in the laboratory—has greatly facilitated research in both plant and animal developmental biology. It was thanks to this technique that a new class of hormones was discovered in the 1950s by Folke Skoog, Carlos O. Miller, and their associates.

These botanists developed methods of growing tissue from tobacco plants on tissue-culture media. The cells formed a tumorlike mass of tissue called a callus, in which the constituent cells often grew to huge size. In 1964 D. S. Letham and his associates isolated from corn seeds a compound called ***zeatin*** (Fig. 32.36), which is the most active known naturally occurring representative of the ***cytokinins***, a class of hormones promoting cell division.

The action of cytokinin on a callus in tissue culture depends on the presence of auxins. Indeed, the ratio of cytokinin to auxin appears to be of fundamental importance in determining the differentiation of new cells. When there is disproportionately more auxin than cytokinin, root growth is initiated; when the ratio favors cytokinin over auxin, stems and leaves are generated (Fig. 32.37). Since auxin is produced primarily by the meristems, while the major source of cytokinin is the roots, this interaction helps maintain a balance between a plant's investment in these two essential tissues. The changing ratio of these two hormones in a growing stem probably also determines the spatial extent of apical dominance. Both hormones influence cell growth, but auxin primarily stimulates elongation, whereas cytokinin promotes cell division. This sort of hormonal interaction is a recurrent theme in vertebrates as well.

Among their numerous other functions, cytokinins (1) stimulate conversion of proplastids into functional chloroplasts; (2) promote fruit development in some species; and (3) help retard the onset of senescence (aging), especially in leaves, by maintaining protein and nucleic acid synthesis and helping to preserve membrane integrity. As this list and the discussions of auxins and gibberellins suggest, all the major plant hormones seem to participate in some fashion in nearly all aspects of plant growth and development.

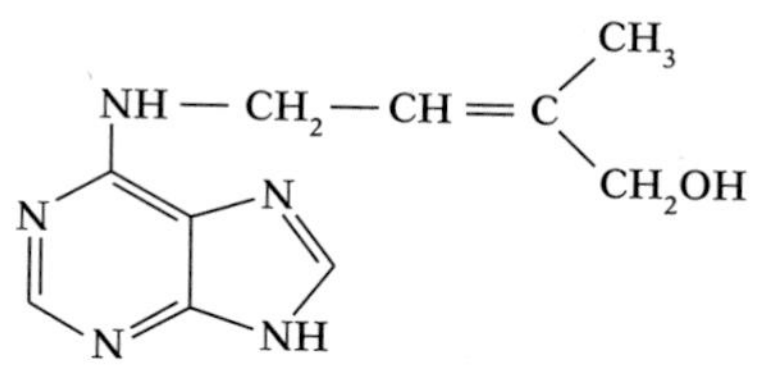

32.36 Cytokinin structure

The most common cytokinin is zeatin, shown here.

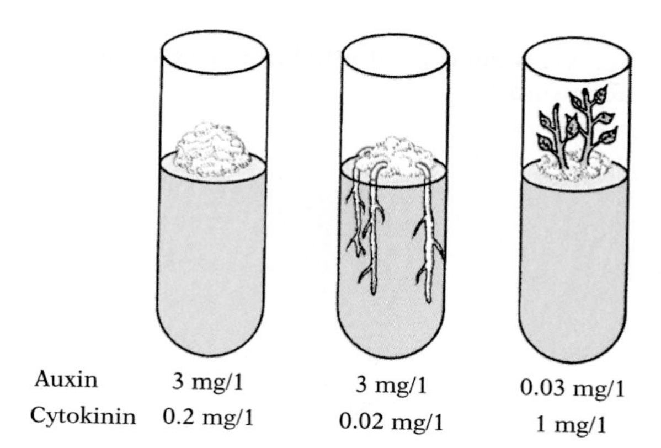

32.37 Effect of auxin and cytokinin on the development of cultured pith cells from a tobacco root

At certain concentrations of the hormones, only an amorphous callus is formed. At other concentrations, roots or shoots develop from the callus. Since all plant cells remain, at least in theory, developmentally totipotent (able to develop into any kind of tissue), growing a callus from a plant and then inducing root and shoot formation with appropriate hormone treatment has become a powerful method of propagating valuable species and variants, including those created through recombinant DNA techniques.

■ RESTRAINING GROWTH: THE INHIBITORS

Relatively little is known about growth inhibitors, which have effects opposite to those of auxins, gibberellins, and cytokinins. A few have been isolated and identified, but the existence of many others has simply been inferred.

The role of inhibitors in maintaining dormancy has attracted particular interest. Inhibitors probably block the activity of some buds and seeds in autumn, thus making sure that they will not begin to grow during a few warm days, only to be killed by the rigors of winter. Dormancy is broken—and the buds and seeds are set free to become active in the next growing season—when gradual breakdown over time, prolonged exposure to cold, or the leaching action of water has helped destroy the inhibitors. In addition, there may be a rise in another hormone (usually gibberellin) that opposes the inhibitors and helps break dormancy.

Inhibitors that must be leached out by water before the seeds can germinate constitute an important evolutionary adaptation in some desert plants. The seeds that fall to the ground will germinate only after long hard rains. Light showers, which might provide enough moisture for germination by seeds not adapted for life in the desert, do not leach out enough inhibitor to allow germination to begin in the desert-adapted seeds; hence no seedlings emerge to be killed by the dry conditions that may soon follow.

The most important known inhibitor is the hormone ***abscisic acid*** (Fig. 32.38). It helps induce dormancy in buds and seeds.

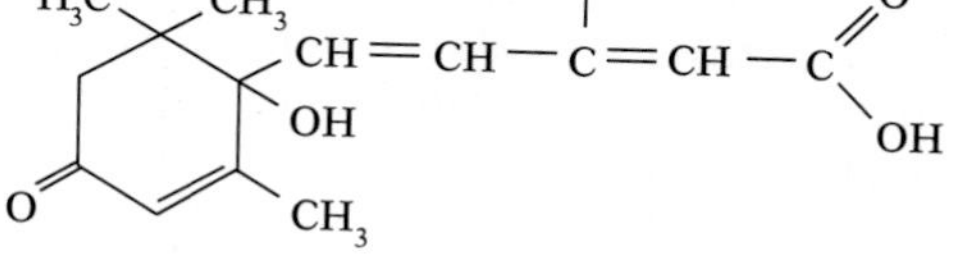

32.38 Structure of abscisic acid

When applied to an actively growing twig, it also induces many other changes (including reduced cell division, production of protective scales instead of foliage leaves, deposition of waterproofing substances, and so on) that prepare the plant for the winter. Despite its name, abscisic acid promotes leaf abscission in only a few species. Although it works mainly as an inducing signal for the expression of a variety of water-stress genes, abscisic acid also has some direct short-term effects; for example, it plays a role in controlling the stomata, the openings through which air enters and circulates in the leaves. The stomatal guard cells close when the plant begins to lose too much water, and it is abscisic acid that carries this message and thus initiates closing.

```
H       H
 \     /
  C = C
 /     \
H       H
```

32.39 Structure of ethylene, a gaseous hormone

■ CONTROLLING RIPENING: ETHYLENE GAS

"One rotten apple spoils the lot." That piece of folk wisdom rests on a familiar fact: when one apple in a barrel goes bad, the other apples in that barrel soon begin to rot. We now know that the bad apple affects other fruit by producing the highly volatile compound ethylene (Fig. 32.39), a plant hormone that plays a variety of roles in the life of plants.

One of the best-studied effects of ethylene is the stimulation of fruit ripening. Once fruit has attained its maximum size, it becomes sensitive to a host of chemical changes that cause it to ripen. The ripening process starts with a sudden sharp increase in carbon dioxide output, followed quickly by a sharp decline. This burst of metabolic activity, called the ***climacteric***, is triggered by a hundredfold increase in the concentration of ethylene. The change is autocatalytic: the more a fruit is stimulated to ripen by ethylene, the more ethylene it produces. Fruit is packaged in well-ventilated containers to prevent this reaction from getting out of hand. Inhibition of ethylene production, or removal of the ethylene as fast as the fruit produces it, prevents the climacteric, and no ripening occurs. Many commercial fruits are now picked and transported while they are still green and firm, and therefore resistant to damage. They can then be ripened with ethylene gas when ready for sale (Fig. 32.40).

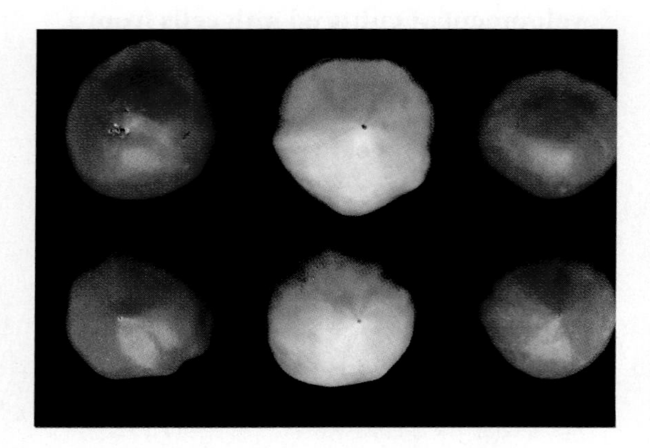

32.40 Effect of ethylene on tomato ripening

Tomatoes normally ripen in response to internally produced ethylene (left). Insertion of a genetically engineered gene coding for an antisense RNA that binds to and inactivates a mRNA involved in ethylene production produces tomatoes that fail to ripen (middle) unless treated artificially with ethylene (right). The ability to manipulate ripening enables growers to produce full-size green tomatoes that travel well and can be held in this state until needed.

Ethylene also contributes to leaf abscission (apparently by stimulating the production of cellulase) and to various other changes that characterize senescence in a plant or parts of a plant. In addition, it can sometimes stimulate radial growth of stems and roots; it can aid in breaking dormancy in the buds and seeds of some species; and it can help initiate flowering in some plants, such as the pineapple. Moreover, some effects usually attributed to auxin, such as lateral-bud inhibition, probably result in some cases from an auxin-induced increase in ethylene.

The use of a gaseous hormone has its costs: ethylene appears to be an important cue used by a variety of insects to help locate the ripening fruit into which they lay their eggs; the larvae that hatch then devour the fruit. A number of plants, however, use a gaseous hormone for defense: when herbivorous insects begin chewing on leaves, the damaged tissue releases volatile chemicals that cause other leaves on the plant (and, often, leaves of nearby plants of the same or even different species) to produce proteinase inhibitors, which are potent insecticides. Some parasitoid insects can home in on certain damage odors and lay their eggs in the herbivores.

HOW FLOWERING IS ORCHESTRATED

Flowering is not a random process. Some plants flower early in the spring, others flower in midsummer, and still others, like chrysanthemums, flower in the fall. Hormones are involved in these different flowering strategies.

■ THE ROLE OF DAY LENGTH

Around 1920, W. W. Garner and H. A. Allard discovered a mutant variety of tobacco, called Maryland mammoth, that grew unusually large (as much as 3 m tall), but would not flower. They propagated the new variety by cuttings and discovered that it would flower in the greenhouse in winter. Garner and Allard went on to show that day length is the critical factor (Fig. 32.41): the short days of late autumn and early winter induced flowering in Maryland mammoth tobacco.

Garner and Allard also experimented with Biloxi soybeans. They planted soybeans at 2-week intervals from early May through July, and found that all the plants flowered at the same time in September, even though their growing periods had differed by as much as 60 days. It was as though they were waiting for some signal from the environment. Garner and Allard were sure that the signal was short days.

Experiments with other species revealed that most plants can be placed in one of three groups: (1) short-day plants, which, like these unique strains of tobacco and soybeans, flower when the day length is below some critical value, usually in fall (examples include chrysanthemum, poinsettia, dahlia, aster, cocklebur, goldenrod, and ragweed); (2) long-day plants, which bloom when the day length exceeds some critical value, usually in summer (lettuce,

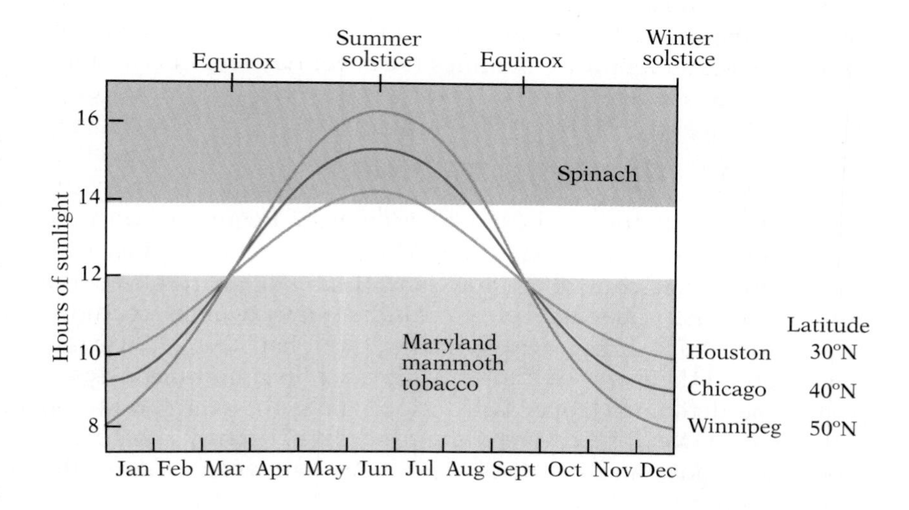

32.41 Day lengths during the year at different latitudes

Days are longest at the summer solstice (about June 21 in the northern hemisphere) and shortest at the winter solstice. The farther from the equator, the more extreme the variation in day length; above the polar circle, for instance, days are 24 hours long near the summer solstice. Some plants use day length as a cue for flowering. Maryland mammoth tobacco flowers when days are shorter than about 12 hours, while spinach requires days longer than 14 hours (which never occur in the tropics).

most grains, clover, larkspur, black-eyed Susan); and (3) day-neutral plants, which are independent of day length and can bloom whether the days are long or short (dandelion, sunflower, carnation, pansy, tomato, corn, string bean).

This response to the duration of days versus nights, which Garner and Allard called ***photoperiodism***, provides plants with a way of timing reproduction. Some species, for instance, set seed early; seedlings of these plants emerge the same year and overwinter as young plants. Others set seed late, and the seeds themselves overwinter. There are distinct costs and benefits to each strategy, including the availability of water and light for germinating seedlings at different times of year (trade-offs that vary depending on the climate and habitat involved). In addition, there is the cost of pollination: pollinators are in short supply in the spring, so plants must offer large quantities of nectar to lure them in; in the late summer it is a "seller's market," and pollination is cheap. Even if there is no best flowering time for a species, it may still be to the plants' advantage to bloom synchronously to maximize the chances of cross-pollination.

Despite the terms *long day* and *short day*, the critical parameter in flowering is actually night length (Fig. 32.42). If, for instance, a short-day plant is illuminated by a bright light for a few minutes, or even seconds, in the middle of the night during the normal flowering season, it will not bloom (Fig. 32.42F). The same sort of experiment will induce flowering at the wrong season by a long-day plant (Fig. 32.42C). Analogous experiments manipulating day length have no effect. Plants and many animals have a simple rule of thumb: whenever light is detected during the critical darkness "window" (which happens normally as the times of dawn and dusk shift with the seasons, but which can be made to occur with a pulse of light in experimental tests), genetic switches are altered, and a host of chemical changes follow.

The difference between long-day and short-day plants does not depend on the precise duration of darkness at the time of flowering. Rather, long-day (short-night) plants will flower only when the night is *shorter* than a critical value (Fig. 32.42A–B), whereas short-day (long-night) plants will flower only when the night is *longer* than a critical value (Fig. 32.42D–E). The specific critical night length for flowering in each species is thus a maximum value for long-day plants and a minimum value for short-day plants. In some species, temperature modifies the exact day-length criterion.

■ IS THERE A FLOWERING HORMONE?

How does the photoperiod exert its influence? Beginning with the 1936 experiments of M. H. Chailakhian, the evidence has supported hormonal control of flowering. Chailakhian removed the leaves from the upper half of chrysanthemums (which are short-day plants), but left the leaves on the lower half (Fig. 32.43). He then exposed the lower half to short days while simultaneously exposing the defoliated upper half to long days; the plants flowered (Fig. 32.43A). Next, he reversed the procedure, exposing the lower half to long days and the defoliated upper half to short days; the plants did not flower (Fig. 32.43B). He concluded that day length

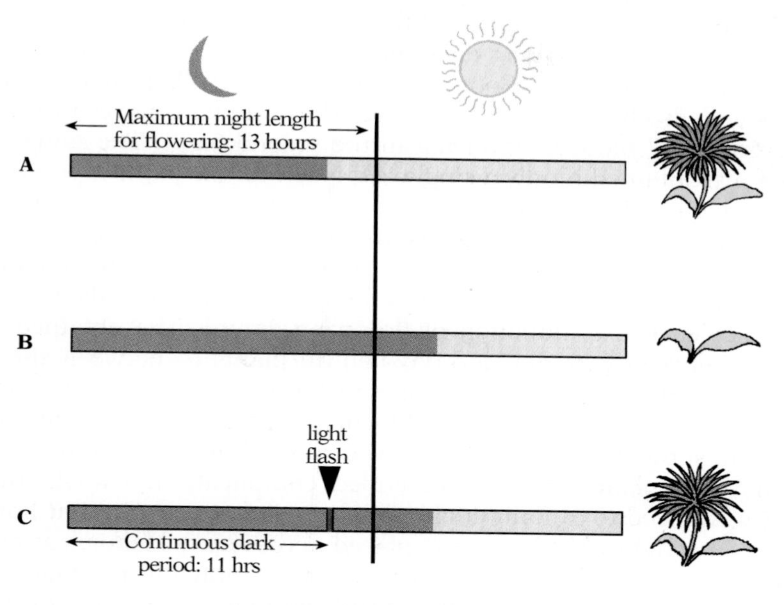

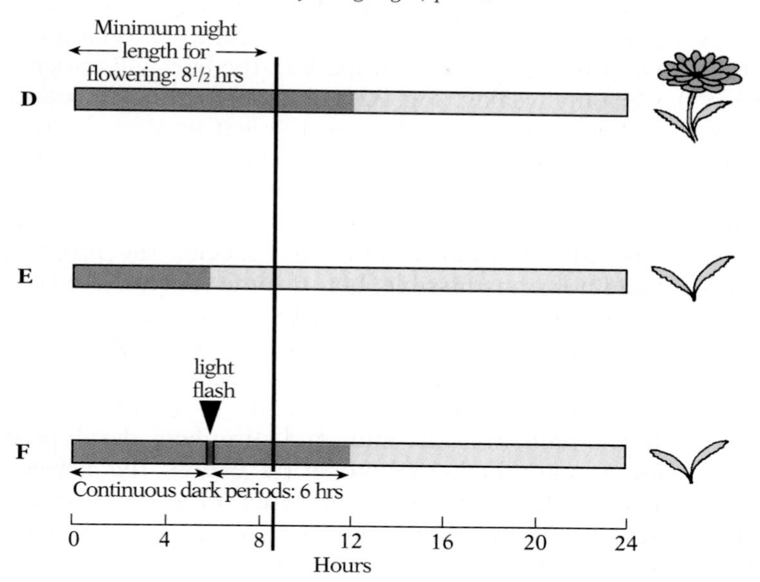

32.42 Comparison of long-day and short-day plants

Yellow bars indicate days and blue bars nights. The hypothetical long-day (short-night) plant shown here has a rather long critical night length of 13 hours, and the hypothetical short-day (long-night) plant has a rather short critical night length of $8^1/_2$ hours. In other words, in this example the critical night length for the short-night plant is actually longer than that for the long-night plant. The difference is that the critical night length is a *maximum* value for the short-night plant and a *minimum* value for the long-night plant. The short-night plant will flower when the night length is *below* the critical value (A) but will not flower when it is above the critical value (B); the plant will flower, however, if a long night is interrupted by a flash of light that reduces the period of *continuous* dark below the critical value (C). Conversely, the long-night plant will flower when the night length is *above* the critical value (D), but will not flower when it is below the critical value (E); the plant will not flower if a long night is interrupted by a flash of light that reduces the period of continuous dark below the critical value (F).

does not exert its effect directly on the flower buds, but causes the leaves to manufacture a hormone that moves from the leaves to the buds and induces flowering. This hypothetical hormone was named ***florigen***.

Further evidence for a flowering hormone comes from grafting experiments with cockleburs (which are also short-day plants; Fig.

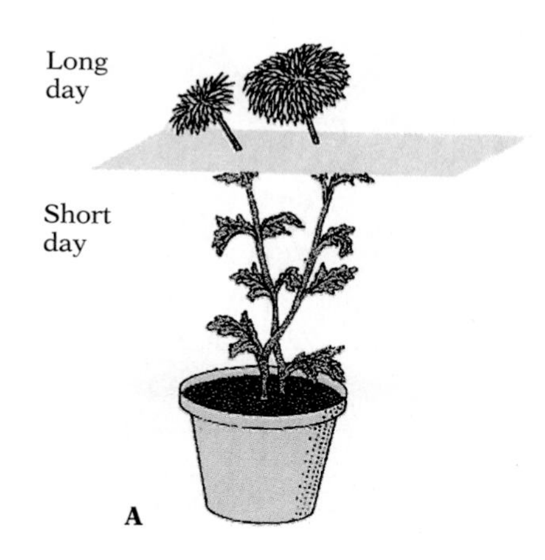

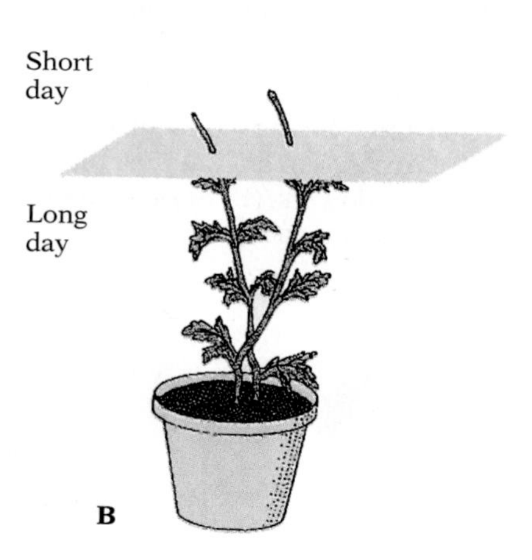

32.43 Chailakhian's experiment

(A) Chailakhian removed the leaves from the top half of a chrysanthemum (a short-day plant) and then exposed the top half of the plant to long days and the bottom half to short days. The plant flowered. (B) When he did the reverse experiment, the plant did not flower.

32.44 Grafting experiment with cockleburs

The two plants are separated by a light-tight barrier (black line), but are connected by a graft. (A) The plant exposed to an inducing photoperiod (short days) flowers. (B) Shortly thereafter the other plant begins to flower.

32.44). If one plant is grafted onto another through a light-tight partition, and if the first plant is exposed to an inducing photoperiod (short days/long nights) while the other is exposed to a noninducing photoperiod (long days/short nights), the plant exposed to short days will flower. Soon thereafter the plant exposed to long days will also flower. A stimulus from the first plant moves through the graft and induces flowering in the second plant. The same results are obtained if only a single leaf is left on the plant exposed to the inducing photoperiod.

The flower-inducing factor seems to be the same in both short-day and long-day plants: in many cases, if a long-day and a short-day plant are grafted together and then exposed to short days, both will flower. Cross induction of flowering can also be obtained in grafts between long-day and day-neutral plants or between short-day and day-neutral plants. Ringing experiments, in which a ring of outer tissue (including the phloem) is removed from a stem or branch, show that the stimulus is transported in the phloem.

In some plants there is an added complication, namely that leaves exposed to a noninducing photoperiod actively inhibit flowering. If a light-tight barrier is placed across a cocklebur leaf and the base of the leaf is exposed to a flower-inducing photoperiod while the tip of the leaf is exposed to a noninducing photoperiod, a nearby bud will flower (Fig. 32.45A). If the reverse experiment is run, with the tip of the leaf exposed to an inducing photoperiod and the base to a noninducing photoperiod, the bud will either not flower or flower only weakly (Fig. 32.45B). Any hormone produced under inducing conditions in the tip of the leaf may be destroyed as it passes through the noninduced base. The inhibiting effect can also be seen if one leaf is exposed to an inducing photoperiod and a noninduced leaf is located between the induced leaf and the bud (Fig. 32.45C–D). In cockleburs and other species the inhibition may be local and not transmissible, but in some species, like strawberries, there is evidence for a transmissible inhibitor.

The most obvious hypothesis to explain flowering that emerges from these various experiments is that an inducing photoperiod causes the leaves to increase the production of a hormone that then moves in the phloem to the buds and stimulates development of the flower. A noninducing photoperiod causes the leaves of many (but not all) plants to stop the production of this hormone. Under natural conditions, then, flower production is triggered when, as the seasons change, the photoperiod passes a critical value and production of the hormone by the leaves exceeds some inhibitory threshold.

Given the extraordinarily powerful biochemical techniques now available, the continued failure to isolate florigen is discouraging. Some biochemists working in this area are now inclined to doubt that a separate flowering hormone exists. It may be that flowering, like so many other plant functions, is actually controlled by the ratio of two or more other hormones, including (perhaps) one produced in the roots.

Whatever initiates the flowering process, it begins a complex developmental program. The blossom is an intricate structure. Its coloration and markings, time of opening, and the nature of its scent are adapted to particular classes of pollinators. Often no

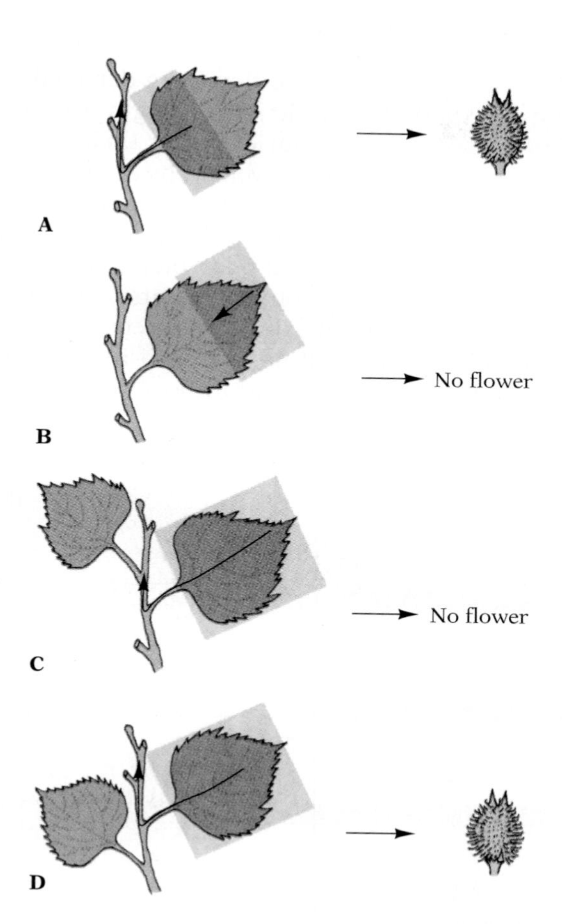

32.45 Experiments illustrating the inhibitory effect of long days on flowering in the cocklebur

(A) When the tip of the uppermost leaf is exposed to continuous illumination, but the base is covered with black paper to give it a short day, the plant flowers, presumably because some inducing substance synthesized in the shaded part of the leaf has moved (black arrow) to the flower bud. (B) The reverse procedure, shading the leaf tip and illuminating the leaf base, does not result in flowering, presumably because the inducing substance is destroyed as it passes through the illuminated part of the leaf. (C) When one leaf is shaded, but a leaf above it on the stem is illuminated, the plant does not flower; perhaps something at the base of the illuminated leaf destroys the inducing substance. (D) However, if the shaded leaf is located above the illuminated leaf, no such destruction can occur and the plant flowers.

pollen is produced during the period in which the ovules are capable of being fertilized, thereby reducing the chance of self-fertilization (inbreeding). One of the most unusual cases of bloom orchestration involves the voodoo lily (Fig. 32.46), whose complex flower generates fetid odors that attract hungry flies. These odors are volatilized by two bouts of massive heat production, each about 2 hours long, triggered by the hormone salicylic acid (a close relative of another plant product, acetylsalicylic acid, better known as aspirin). The lily's first release of odor and heat occurs in the late afternoon; flies enter and become trapped, depositing any pollen they may be carrying onto the female organs as they attempt to escape. The second rush of warmth (which, like the first, heats the flower 10–20°C above the ambient temperature) serves to awaken the blossom's torpid overnight guests, and to signal the male organs to shed pollen onto the flies, which are now allowed to escape and take the pollen to another lily.

32.46 Voodoo lily

The flower has been cut open to reveal the internal details. Pollen is dispensed from the white collar near the opening to the outside; the female organs are the white protuberances at the base of the chamber.

■ HOW IS DAY LENGTH MEASURED?

How is the photoperiod so critical to flowering detected and measured in the first place? What wavelengths of light are involved? In 1944, H. A. Borthwick, S. B. Hendricks, and their associates exposed Biloxi soybeans to different wavelengths of light. They found that red light (about 660 nm) is by far the most effective in inhibiting flowering in these short-day plants; the same red light is very effective in inducing flowering in long-day plants. It turns out that far-red light (infrared light of about 730 nm), invisible to the human eye, reverses the effect of exposure to red light (Fig. 32.47). A short-day (long-night) plant will not flower if its long night is interrupted by a bright flash of red light; if, however, the red flash is

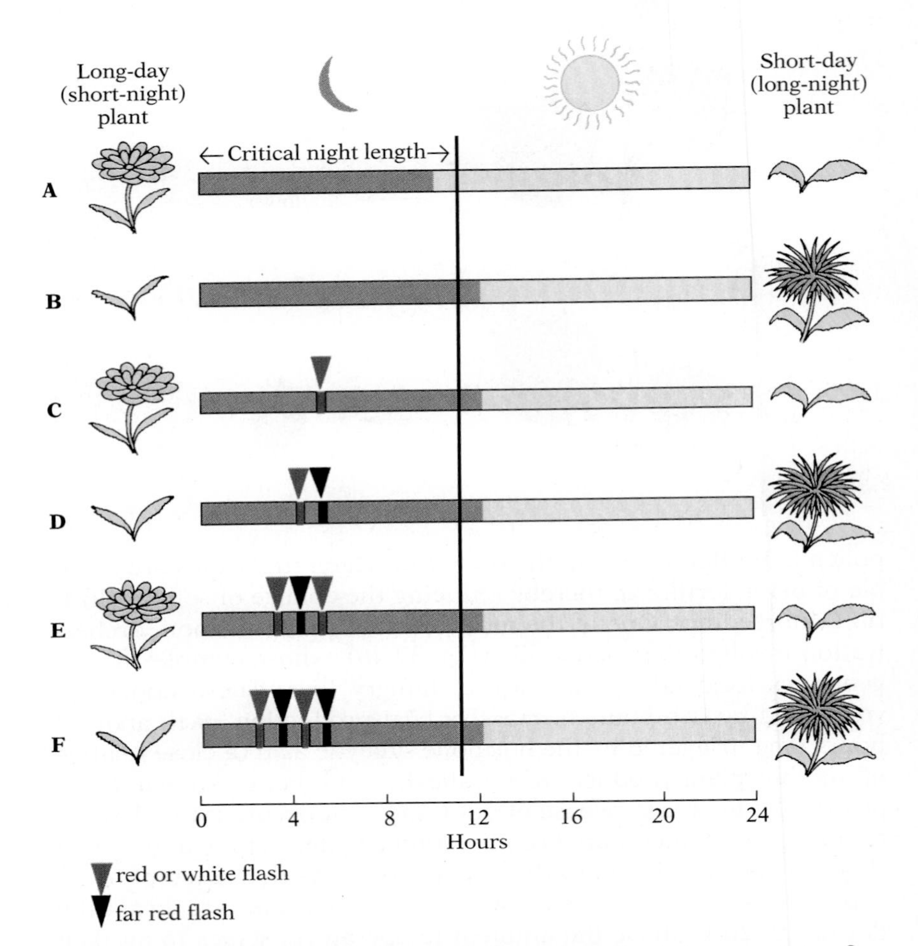

32.47 Reactions of long-day and short-day plants to a variety of light regimes

Yellow bars indicate days and blue bars nights. (A) Long-day (short-night) plants flower when the night is shorter than the critical value. (B) Short-day (long-night) plants flower when the night is longer than a critical value. (C) Either kind of plant can be tricked into treating a long night as a short one if the period of darkness is interrupted by an intense red or white flash. (D) If the red or white flash is followed by an infrared flash, the night-interrupting effect of the first flash is abolished. (E–F) The exact response depends on the color of the *last* flash of light.

followed immediately by a far-red flash, the plant flowers normally. Almost any number of successive flashes can be used, the final effect depending solely on whether the last flash was red or infrared.

The discovery that far-red light can reverse the effect of red or white light led Borthwick and Hendricks to conclude that a single receptor pigment, which they called ***phytochrome***, is involved; this pigment (since then identified as a protein with a prosthetic group; Fig. 32.48) exists in two forms: one that absorbs red light (P_r) and one that absorbs far-red light (P_{fr}). When P_r absorbs red light, it is rapidly converted into P_{fr}. Conversely, absorption of infrared light by P_{fr} rapidly converts it into P_r. The P_r form is apparently the more stable of the two; in darkness P_{fr} may revert to P_r over the course of several hours in some plants. In addition, P_{fr} is enzymatically destroyed more rapidly than P_r:

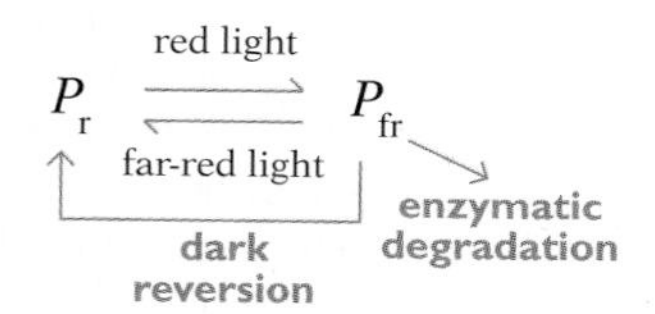

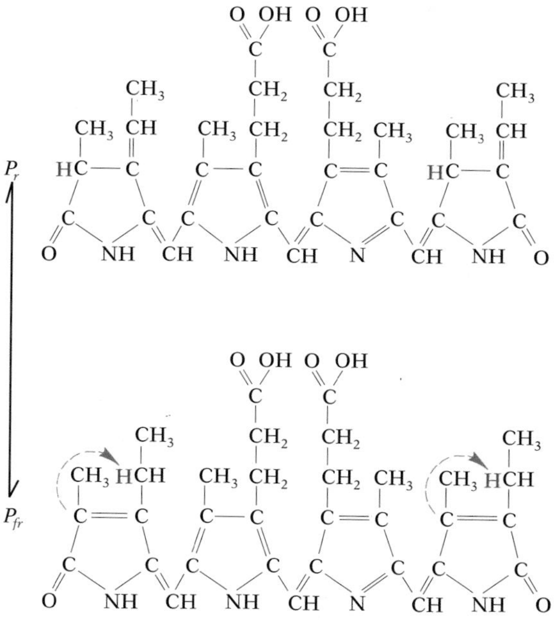

Though the P_{fr}-to-P_r pigment ratio ought to give plants a way of detecting the length of the photoperiod, it seems only to be involved in the rather mundane task of determining whether the plant is experiencing light or dark. The precise mechanism by which the plant measures the length of the dark period is apparently tied to a phenomenon—now believed to occur in all living cells—involving persistent and regular rhythms that depend on an "internal clock."

Once the phytochrome mechanism and the internal-clock mechanism together indicate to the plant when the photoperiod is appropriate for flowering, they trigger processes that no longer depend on light, phytochrome, or night length. This response begins in the leaves, which alter hormone production or sensitivity to hormone ratios as necessary to bring about actual flowering.

The phytochrome molecules can influence a variety of cell activities besides those associated with flowering. For example, germination of some types of seeds and spores requires exposure to red light, which is sensed by phytochrome. Gibberellin-controlled cell elongation, expansion of new leaves, breaking of dormancy in spring, and formation of plastids in cells also involve this pigment.

Recent evidence suggests that one of the major functions of phytochrome in plants is to detect the extent of competition and shading. Because foliage tends to absorb or reflect wavelengths below 700 nm, but to transmit wavelengths in the 700–800 nm (far-red) range, the ratio of red to far-red light in the sunlight reaching a leaf is an indication of the amount of shading by other leaves; the ratio reaching the stem, primarily by being reflected from adjacent plants, provides a measure of nearby competition. This is important information: it would be wasteful for a plant requiring direct sunlight to grow new leaves (or sometimes even to maintain existing ones) in the shade of other leaves, or to grow branches toward neighboring plants, or for a plant pollinated by insects to produce

32.48 Phytochrome prosthetic group

The color arrows indicate the shift in position of two hydrogen atoms that occurs when P_r is converted to P_{fr}.

flowers where insects would be unlikely to see them. Plants do indeed respond to shading and neighboring plants by modifying their normal patterns of stem elongation, amount of branching, leaf pigmentation, and flowering.

CHAPTER SUMMARY

FROM SEED TO PLANT

EVENTS INSIDE THE SEED After fertilization the embryo forms a three-layer structure consisting of future ectoderm, cortex, and vascular tissue. Next one or two cotyledons (the first leaves) develop along with two apical meristems, one for shoot and one for root growth. Active growth begins with massive water absorption at germination. (p. 903)

HOW PLANTS GROW Shoots and roots grow in length as apical meristems produce new cells, which then increase in length. Vascular tissue forms rapidly behind the growing meristems. Lateral meristem produces lateral growth: vascular cambium lying between the xylem and phloem generates new vascular tissue; pericycle generates lateral roots; cork cambium also produces lateral growth of the cortex. (p. 907)

HOW HORMONES CONTROL PLANT GROWTH

THE MANY ROLES OF AUXIN Auxin is produced in the apical meristem of shoots and causes both cell elongation and division. It is transported away from light and away from the tip, causing bending toward light. Auxin inhibits the formation of lateral branches, stimulates fruit growth after fertilization, inhibits leaf abscission, and stimulates lateral branching in roots. (p. 916)

STIMULATING GROWTH: THE GIBBERELLINS Gibberellins stimulate seed germination and seedling growth, stem growth, bolting and early flowering, and fruit set. (p. 923)

MAINTAINING BALANCED GROWTH: THE CYTOKININS Cytokinin is produced in root tips and, in conjunction with auxin, controls cell growth and development—especially, the division of resources between root and shoot growth. (p. 924)

RESTRAINING GROWTH: THE INHIBITORS Inhibitors block seed germination and induce the formation of protective structures in preparation for dormancy. (p. 925)

CONTROLLING RIPENING: ETHYLENE GAS Ethylene stimulates fruit ripening and leaf abcission. (p. 926)

HOW FLOWERING IS ORCHESTRATED

THE ROLE OF DAY LENGTH Many plants time their flowering by day length, which is a measure of the time of year. In fact, they actually measure night length. Long-day plants flower in the summer; short-day plants bloom in the fall. (p. 927)

IS THERE A FLOWERING HORMONE? Some systemic chemical or set of chemicals induces flowering; there may be a flowering hormone, or blooming may result from an interaction among other hormones. (p. 928)

HOW IS DAY LENGTH MEASURED? Phytochromes are converted by the red light in sunlight into an altered form, P_{fr}, which spontaneously returns to the original P_r form in the dark; plants monitor day length with phytochrome, and adjust their developmental program appropriately. (p. 932)

STUDY QUESTIONS

1 What might be the advantages favoring short-day, long-day, and day-neutral flowering in temperate-zone habitats? (With regard to dandelions, keep in mind that their relatively recent conversion to asexual reproduction makes pollinators irrelevant.) (pp. 927–928)

2 Why do plants grown in the dark become "leggy"? Explain the phenomenon both in terms of hormonal mechanisms and evolutionary logic. (pp. 923–924)

3 One indication that aphids are attacking a rose is that the sepals of very immature flower buds begin to enlarge and differentiate into leaves. (The bud itself never matures.) How does this observation fit in with what you know about a possible flowering hormone? (pp. 928–930)

4 In many species, seedlings exposed to mechanical contact (heavy rain, wind, or just being touched regularly by gardeners) grow about half as fast as undisturbed seedlings receiving equal amounts of light, water, and nutrients. What might be going on? How would you test your hypothesis?

SUGGESTED READING

EVANS, M. L., R. MOORE, AND K.-H. HASENSTEIN, 1986. How roots respond to gravity, *Scientific American* 255 (6).

GALSTON, A. W. 1994. *Life Processes of Plants*. W. H. Freeman, New York. *A beautifully illustrated introduction to plant growth and hormonal control.*

MEYEROWITZ, E. M., 1994. The genetics of flower development. *Scientific American* 271 (5). *On the developmental control of flower morphology.*

MOSES, P. B., AND N.-H. CHUA, 1988. Light switches for plant genes, *Scientific American* 258 (4).

QUAIL, P.H., et al., 1995. Phytochromes: photosensory perception and signal transduction. *Science* 268, pp. 675–680.

RAGHAVAN, V., 1992. Germination of fern spores. *American Scientist* 80 (2). *How the phytochromes trigger germination.*

CHAPTER 33

CHEMICAL CONTROL IN ANIMALS

Animal hormones are usually secreted by organs specialized for their production; they move through the circulatory system and bind to specific target tissue. There they help guide growth and development, and also aid in regulating metabolism and maintaining general homeostasis. In this chapter we will look at the role played by most of the hormones known to be important in humans, as well as chemical messengers that operate more locally; we will also explore how these various molecules exert their effects on target cells. We begin with a look at hormonal control in invertebrates.

HOW HORMONES CONTROL DEVELOPMENT IN INSECTS

Hormonal regulation has been found in a variety of invertebrates, including arthropods, annelid worms, molluscs, and echinoderms. Hormones probably play a role in all animals, but the question has not received much attention outside of mammals.

The action of invertebrate hormones is best understood in insects. Insects show a pattern of growth very different from that of vertebrates. Their body is encased in an exoskeleton, which limits growth until the exoskeleton is shed. V. B. Wigglesworth pioneered work on the mechanisms that control this periodic molting. Most of his experiments were performed on the bloodsucking bug *Rhodnius* (Fig. 33.1). This bug goes through five immature or nymphal stages, each separated by a molt, before it becomes an adult. During each nymphal stage it must obtain a blood meal, which engorges and stretches the abdomen. Repletion stimulates the release of hormones that cause molting at a fixed interval following the meal.

Ordinarily, the last molt (from the fifth nymphal stage to adult) occurs about 28 days after the blood meal. Wigglesworth showed that if *Rhodnius* is decapitated during the first few days after this meal, molting does not occur, even though the animal may continue to live for several months. Decapitation more than 8 days after the blood meal does not interfere with molting; a headless adult is produced. Furthermore, if the circulatory system of a bug decapitated 8 days after a blood meal is joined to that of a bug de-

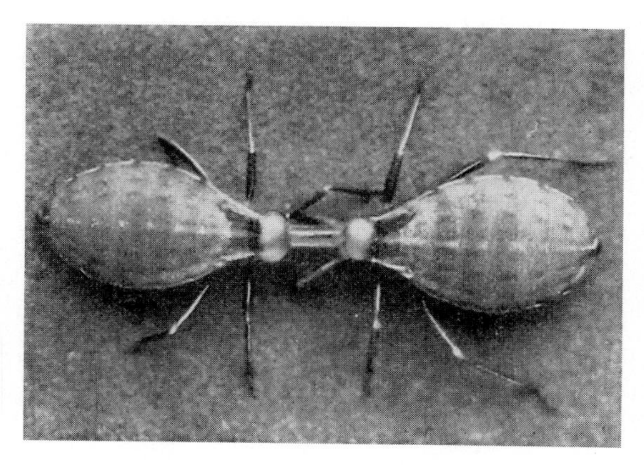

33.1 Decapitated bugs, joined experimentally

A fourth-stage larva of the bloodsucking bug *Rhodnius* was decapitated more than eight days after a blood meal and then connected by a glass tube to a larva decapitated one day after feeding. In repeated experiments of this type, both bugs molted into adults, indicating that a hormone secreted by the first insect passed via the blood to the second one.

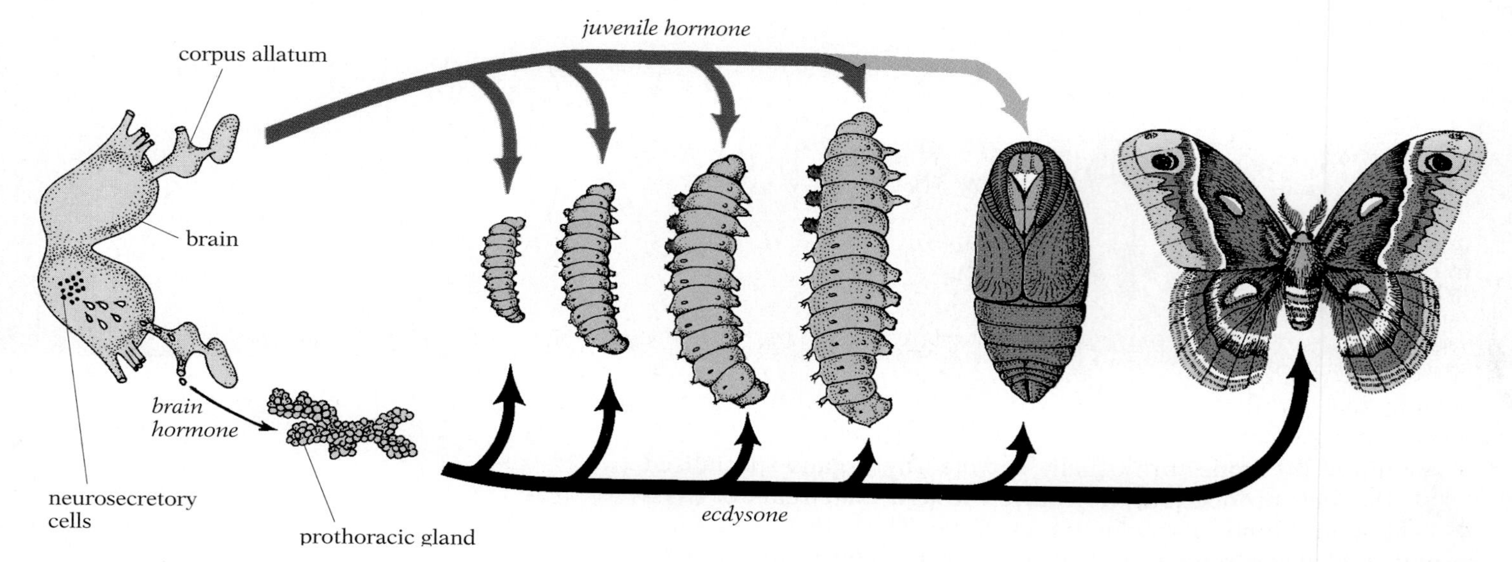

33.2 Interactions of juvenile hormone, brain hormone, and molting hormone (ecdysone) in the cecropia silkworm (*Hyalophora cecropia*)

If much juvenile hormone (JH) is present when the insect molts, it will molt into another larval stage. If a low concentration of JH is present, the larva will molt into a pupa. If no JH is present, the pupa will molt into an adult. A third substance, eclosion hormone (not shown), is required for shedding the pupal cuticle.

capitated soon after the meal, both bugs molt into adults. Clearly, some stimulus passes via the blood from one insect to the other and induces molting. That stimulus is a hormone secreted by the head 8 days after the blood meal.

This ***brain hormone***, a polypeptide, stimulates glands in the prothorax (the part of the body immediately behind the head, to which the first pair of legs is attached). The prothoracic glands, in turn, secrete a second hormone, ***ecdysone***, which induces molting (Fig. 33.2). Ecdysone, a lipid-soluble steroid, acts on the genes of several types of cells, stimulating the cells to grow and divide.

Wigglesworth became interested in the factors that determine whether a molt will result in an adult or in another immature stage. This is a particularly important question in insects like flies, beetles, and moths, which undergo a complete metamorphosis, which is a radical change from immature to adult form—from grub to fly or beetle, or from caterpillar to moth. Wigglesworth found that a third hormone, called ***juvenile hormone*** (JH), is involved. JH is produced by a pair of glands (corpora allata) located just behind the brain. When JH is present in high concentration at the time of molting, another immature stage follows the molt. The pupal stage—the changeover between the last larval stage and the adult in insects like flies and moths—results from a low concentration of JH. The hormone is absent in the pupa, and when it molts, an adult results. (A brief burst of ***eclosion hormone*** is essential at metamorphosis to trigger the behavior that leads to eclosion—the successful shedding of the pupal cuticle.)

Removal of the corpora allata from insects in the first or second immature stage results in pupation at the next molt, followed by a molt that results in a midget adult. Conversely, implantation of active corpora allata into insects about to undergo their final molt results in another immature stage instead of an adult; in this way several extra immature growth stages can be inserted into the insect's developmental sequence. These can be followed by pupation and a molt producing an unusually large adult when JH is finally eliminated. Certain trees, notably balsam fir and hemlock, defend themselves against insect pests by synthesizing artificial JH. As a result, most insects that attempt to feed on these trees never mature into egg-laying adults. Certain other plants produce ecdysone-like chemicals that initiate pupation prematurely; because of the hormonal imbalance, the insect develops abnormally and dies.

Many insect hormones have different functions in different periods of an organism's life. For example, though the molt from pupa to adult will proceed only if the quantity of juvenile hormone has been substantially reduced from previous stages, secretion of JH usually resumes after metamorphosis is complete. This is because a high concentration of JH is required in many species for females to deposit yolk in their eggs and for males to form mature sperm.

HOW WE KNOW:
THE DISCOVERY OF THE PAPER-TOWEL FACTOR

The idea of a JH mimic to fight insect pests grew from unlikely roots. When Czech scientist Karel Sláma moved his lab to Harvard University, the linden bugs he studied failed to mature in their new surroundings. In Prague they had moved into the adult stage after passing like clockwork through five larval instars. In Cambridge, however, all the bugs grew and molted into a sixth larval instar; some of these died, while the rest molted into a seventh instar, which then died.

After excluding a long list of possible differences such as the water, the linden seeds used as food, and so on, Sláma found the crucial variable: the paper towels he used at Harvard in place of the filter paper employed in Prague. Instead of dropping the issue now that he knew how to rear his bugs successfully in North America, however, he began a dogged search for the active factor.

Sláma discovered that almost any kind of paper around the Harvard lab—including napkins, magazine pages, and newspaper—seemed to block normal development. A visit to the library for international newspapers and periodicals was enlightening: European paper was benign, but Japanese paper was fatal. Since Japan imports its pulp from Canada, he began to investigate the main sources of North American pulp, including the balsam fir (see photo). It was in this species that he found juvabione—the chemical mimic of JH—that prevents insects from maturing. The juvabione tricks the JH receptors in the larvae into signaling that the corpora allata are still producing JH, and so adulthood is still at least one instar off.

Attempts to produce a novel insecticide using a JH mimic have yet to bear fruit. The most obvious difficulty with this approach is that JH, far from killing the ravenous larvae, actually extends their feeding phase by adding instars.

THE MANY ROLES OF HORMONES IN VERTEBRATES

The tissues that produce and release hormones are called ***endocrine*** tissues. The term *endocrine* (from the Greek *endon*, "within," and *krinein*, "to secrete") indicates that hormones are secreted internally into the blood supplying the endocrine tissues, with no special ducts involved. In fact, endocrine glands are often called the ductless glands. Most vertebrate hormones form weak bonds with plasma proteins and are transported in the blood in a bound form.

We will emphasize the mammalian hormonal system (especially the human system) in this chapter (Fig. 33.3; Table 33.1).

■ HOW HORMONES CONTROL DIGESTION

The first step toward understanding digestion came from experiments performed by the Russian physiologist Ivan Pavlov (1849–1936). He showed that both the secretion of saliva and the first phase of gastric secretion (secretion of gastric juice by the stomach) are under nervous control, and that gastric secretion actually begins before food reaches the stomach (when the animal tastes or even just sees or smells food). The partially digested food in the stomach in turn leads to the release of more gastric juice. Pavlov wondered whether the mere physical presence of food was sufficient to stimulate this second release.

When Pavlov inserted a piece of meat directly into a dog's stom-

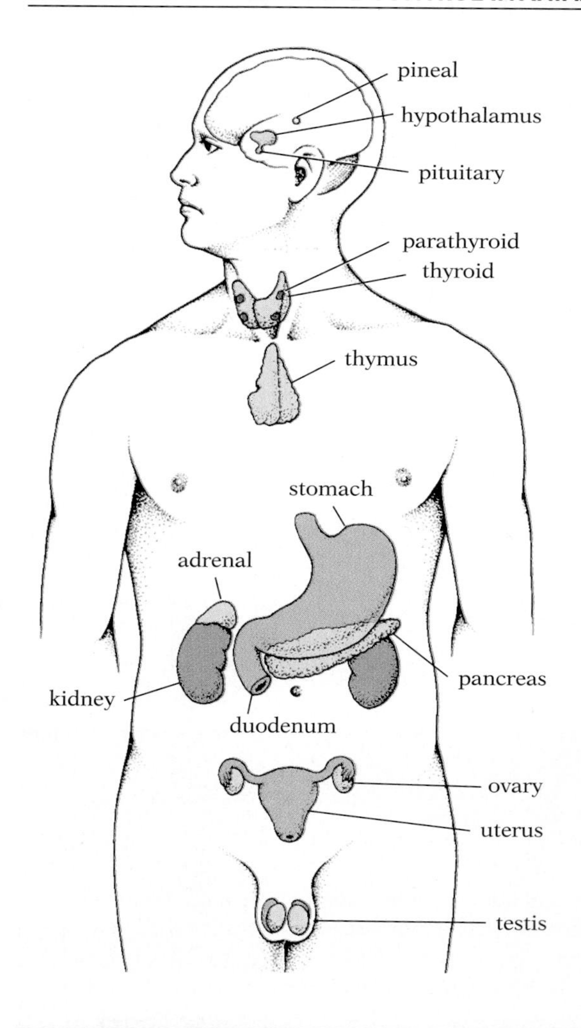

33.3 Major human endocrine organs

Table 33.1 Important mammalian hormones

SOURCE	HORMONE	PRINCIPAL EFFECTS
Pyloric mucosa of stomach	Gastrin	Stimulates secretion of gastric juice and pancreatic enzymes; increases intestinal motility
Mucosa of duodenum	Secretin	Stimulates flow of pancreatic enzymes
	Cholecystokinin	Stimulates release of bile from gallbladder; inhibits movement of food out of stomach
	Somatostatin	Inhibits digestion by counteracting effects of gastrin and secretin, and reducing secretion of cholecystokinin and gastrin
Pancreas	Insulin	Stimulates glycogen formation and storage, glucose oxidation, and synthesis of protein and fat; inhibits formation of new glucose
	Glucagon	Stimulates conversion of glycogen into glucose
Adrenal medulla	Adrenalin	Stimulates elevation of blood-glucose concentration and other fight-or-flight reactions
	Noradrenalin	Stimulates reactions similar to those produced by adrenalin

Table 33.1 Important mammalian hormones (continued)

SOURCE	HORMONE	PRINCIPAL EFFECTS
Adrenal cortex	Glucocorticoids (corticosterone, cortisol, cortisone, etc).	Stimulate formation of carbohydrate from protein and fat, thus helping maintain normal blood-sugar levels
	Mineralocorticoids (aldosterone, deoxycorticosterone, etc.)	Stimulate kidney tubules to reabsorb more sodium, chloride, and water and less potassium
	Cortical sex hormones	Stimulate development of secondary sexual characteristics, particularly those of the male
Heart	Atrial natriuretic factor (ANF)	Decreases blood volume and blood pressure
Thyroid	Thyroxin, triiodothyronine (together called TH)	Stimulate oxidative metabolism; help regulate growth and development
	Calcitonin	Prevents excessive rise in blood calcium
Parathyroids	Parathyroid hormone (PTH)	Regulates calcium-phosphate balance
Thymus	Thymosin	Stimulates immunologic competence in lymphoid tissues
Hypothalamus	Releasing hormones	Regulate hormone secretion by anterior pituitary
Posterior pituitary (storage of hypothalamic hormones)	Oxytocin	Stimulates contraction of uterine muscles; stimulates release of milk by mammary glands
	Vasopressin	Stimulates increased water reabsorption by kidneys; stimulates constriction of blood vessels (and other smooth muscle)
Anterior pituitary	Growth hormone (GH or STH)	Stimulates growth; stimulates protein synthesis, hydrolysis of fats, and increased blood-sugar concentrations
	Prolactin (PRL)	Stimulates milk secretion by mammary glands; participates in control of reproduction, osmoregulation, growth, and metabolism
	Melanophore–stimulating hormone (MSH)	Probably helps regulate salt and water balance; may influence certain types of behavior; controls cutaneous pigmentation in ectotherms
	Thyrotropic hormone (TSH)	Stimulates the thyroid
	Adrenocorticotropic hormone (ACTH)	Stimulates the adrenal cortex
	Follicle-stimulating hormone (FSH)	Stimulates growth of ovarian follicles and of seminiferous tubules of the testes
	Luteinizing hormone (LH)	Triggers ovulation; stimulates conversion of follicles into corpora lutea; stimulates secretion of sex hormones by ovaries and testes
Pineal	Melatonin	Helps regulate production of gonadotropins by anterior pituitary, perhaps by regulating hypothalamic releasing centers
Testes	Testosterone	Stimulates development and maintenance of male accessory reproductive structures, secondary sexual characteristics, and behavior; stimulates spermatogenesis
Ovaries	Estrogen	Stimulates development and maintenance of female accessory reproductive structures, secondary sexual characteristics, and behavior; stimulates growth of the uterine lining
	Progesterone	Prepares uterus for embryo implantation and helps maintain pregnancy

ach without allowing the animal to sense the food, no secretion of gastric juice occurred. However, when a piece of partly digested meat was inserted directly into the stomach, secretion began promptly. Compounds released from the partly digested meat trigger the secretion. Partial digestion during the first phase of gastric secretion, which is under nervous control, must release compounds that trigger the second phase of gastric secretion, thus continuing gastric digestion after nervous stimulation ceases.

How did the substances from partly digested meat stimulate release of gastric juice? They do not stimulate the gastric glands directly: Pavlov divided the stomach into two chambers and showed that partly digested meat in one chamber stimulates release of gastric juice by the glands in both chambers. Nor is the nervous system involved: isolating the stomach from nervous control does not stop the secretion.

Later researchers showed that the meat substances in the stomach trigger the release of a hormone, called ***gastrin***, into the blood; when it reaches the gastric glands, the hormone stimulates secretion. A similar pattern underlies the secretion of pancreatic enzymes: a hormone called ***secretin*** is released by the mucosal cells of the duodenum when they are stimulated by the acidity of food coming from the stomach. Another hormone, ***cholecystokinin***, produced by the duodenum under stimulation by acids and fats, triggers release of bile from the gallbladder. Cholecystokinin also acts on the nervous system to produce the drowsiness we often feel after a large meal. The hormone ***somatostatin*** helps regulate digestion by counteracting the effects of gastrin, secretin, and cholecystokinin. Thus the whole early process of digestion is hormonally regulated.

■ DIABETES AND THE PANCREAS

Diabetes mellitus, a disease in which much sugar is excreted in the urine, affects about 3.5% of the population. There are two major forms (Table 33.2); until quite recently the more common Type II diabetes was not recognized as a distinct disease. Both seem to be largely inherited.

Diabetes has been known for centuries, but its causes only began to be understood in the latter half of the 19th century. In 1889 two physicians, Johann von Mering and Oscar Minkowski, surgically removed the pancreas from a dog. A short time later they noticed that the dog's urine was attracting an unusual number of ants; it

Table 33.2 Types of diabetes

TYPE AND FREQUENCY	CAUSE
Type I (Juvenile-onset diabetes): *Frequency*—0.7% of population	Failure of islet cells to secrete insulin because of auto-immune destruction by T cells
Type II (Adult-onset diabetes): *Frequency*—3% of population	Although insulin production is normal, hormone either is removed by liver or fails to stimulate target cells

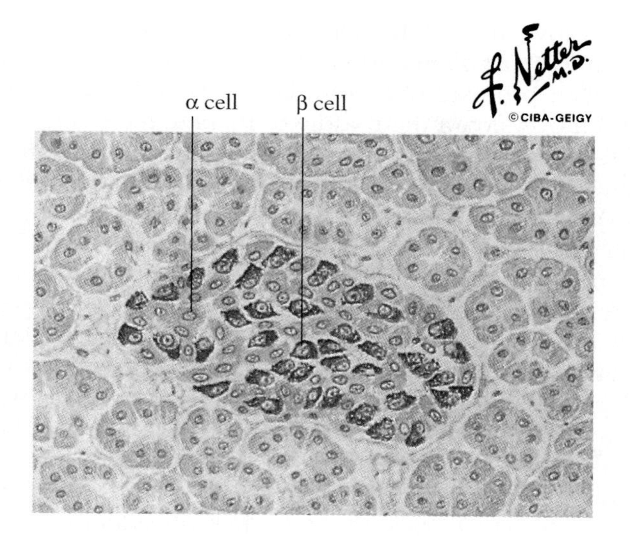

33.4 An islet of Langerhans

The endocrine cells of the pancreas form an islet, clearly distinct from the surrounding cells, whose function is secretion of digestive enzymes. The islet cells comprise about 1% of the pancreatic mass. The α islet cells secrete the hormone glucagon; the β islet cells, insulin. Two other minor classes of cells in the islets secrete hormones that, respectively, encourage and inhibit release of pancreatic enzymes. The pancreas contains about a million islets, each with approximately 3000 cells.

contained a high concentration of sugar. The dog soon developed other symptoms strikingly like those of human diabetes.

The pancreas, they felt certain, must secrete some substance that prevents diabetes, but all attempts at isolating the chemical failed. For example, grinding pancreatic tissue to produce extracts mixes the hormone with the pancreatic digestive enzymes, which destroy the hormone. This happens because the pancreas is a compound organ: it contains several types of cells that function independently. One kind produces and releases digestive enzymes. Other cells, called ***islet cells***, form into groups called islets of Langerhans, or simply islets (Fig. 33.4). It is in the β islet cells that the hormone that prevents diabetes is produced: if the pancreatic duct is tied off, most of the pancreas atrophies, but not the the islet cells; diabetes does not develop in animals treated this way.

The hormone whose absence leads to diabetes, ***insulin***, was finally isolated in 1922 by F. G. Banting and C. H. Best. They tied off the pancreatic ducts of a number of dogs. After the enzyme-producing tissue had atrophied, they removed the degenerated pancreas, froze and macerated it, and filtered the solution. When they injected the filtered material into dogs with experimentally induced diabetes, the dogs showed marked improvement. Banting and Best also obtained good results with extracts prepared from the pancreases of embryonic animals; since the islet cells develop in the embryo before the enzyme-producing cells, there are no enzymes to destroy the insulin during the extraction procedure. Banting shared a Nobel Prize in 1923 for this important work.

In most respects Banting and Best followed a procedure considered standard for demonstrating that a particular organ or tissue has an endocrine function:

1 Removal or destruction of the organ in question should result in predictable symptoms associated with absence of the hormone.

2 Administration of material prepared from the organ in question should relieve the symptoms.

3 The hormone should be present in both the organ and the blood, and the hormone should be extractable from each.

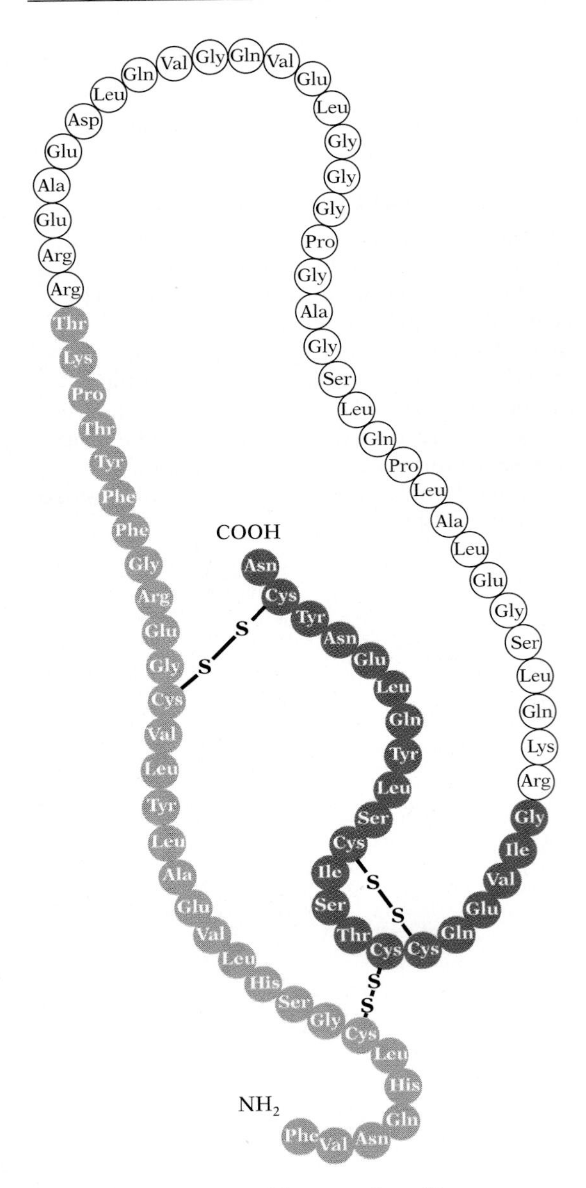

33.5 Activation of human insulin

The protein is synthesized initially as proinsulin, a single polypeptide zymogen (an inactive precursor) with three internal disulfide bonds. Removal of a long section (white circles) by an activating enzyme leaves the active hormone as two separate short polypeptide chains (color) held together by disulfide bonds.

We now know that the β islet cells of the pancreas first synthesize a much longer polypeptide zymogen—an inactive enzyme precursor—called proinsulin. Insulin becomes active when 35 amino acids are removed from the middle of its precursor (Fig. 33.5).

The presence of sugar in the urine of a diabetic means that the blood-sugar concentration is higher than normal and the kidneys are removing part of the excess. The liver plays a critical role in regulating blood-sugar levels: when blood coming to the liver from the intestines contains an abnormally high concentration of sugar, the liver removes much of the excess and stores it as glycogen. The muscles and adipose (fatty) tissue are also important elements in this regulatory system: when the blood-sugar concentration rises, part of the excess glucose is stored as glycogen in the muscles, and part is converted into fat and stored by adipose tissue.

This outline of the interplay between liver, muscles, and adipose tissues—all three target tissues of insulin—helps explain the actions of insulin, which have the net effect of reducing the concentration of glucose in the blood.

1 Insulin stimulates absorption of more glucose from the blood by muscle cells and adipose cells, and (indirectly) causes the liver to take up glucose and convert it into glycogen and fat.

2 It promotes both oxidation of glucose and conversion of glucose into glycogen in muscle as well as liver cells—actions whose effect is to reduce the supply of free glucose.

3 It inhibits metabolic breakdown of stored glycogen in liver and muscle cells.

4 It promotes synthesis of fats from glucose by adipose cells and also inhibits metabolic breakdown of fat.

5 It promotes uptake of amino acids by liver and muscle cells, and favors protein synthesis while inhibiting protein breakdown.

Though the first three actions concern carbohydrate metabolism, the last two deal with metabolizing fats and proteins. The promotion of fat and protein synthesis and the slowing of their catabolism force the cells to rely more on glucose as a source of metabolic energy; the result is a reduction in the supply of free glucose. The metabolic pathways for all nutrients form an interlocking system; alteration of one pathway influences the others.

Too much insulin in the system, whether from an overactive pancreas or from administration of too large a dose to a diabetic, can produce a severe reaction called insulin shock. The blood-sugar level falls so low that the brain, which stores almost no food, becomes overirritable; convulsions may result, followed by unconsciousness and often death. Far more common is a deficiency of insulin (as in all Type I diabetes, and hepatic-dysfunction Type II diabetes), or an insensitivity of the tissues to insulin (as in all other forms of Type II diabetes). The liver and muscles do not convert enough glucose into glycogen, the liver produces too much new glucose, and utilization of carbohydrate in cellular respiration is impaired. The blood-sugar level rises above normal, and part of the excess glucose begins to appear in the urine. More water must be excreted as a solvent for this glucose; the diabetic therefore tends to become dehydrated. The glycogen reserves become depleted as more and more glucose is poured into the blood and lost in the urine; yet the body still lacks sufficient energy, because of the impairment of carbohydrate metabolism. As the body begins to metabolize its reserves of proteins and fats—particularly the latter—the diabetic becomes emaciated, weak, and easily subject to infections. As if this were not enough, the excessive but incomplete metabolism of fats releases toxic substances that, seriously disturbing the delicately balanced pH of the body, often play a major part in the eventually fatal outcome of the untreated disease.

The islet cells of the pancreas secrete another polypeptide hormone called ***glucagon***. Glucagon has effects opposite to those of insulin; it causes an increase in blood-glucose concentrations. Fat cells have both insulin and glucagon receptors, and the balance between these contradictory hormones controls the metabolism of the cells. The normal functioning of an organism often depends on a delicate balance between opposing control systems; if one of these systems is disturbed and the proper balance destroyed, problems are inevitable.

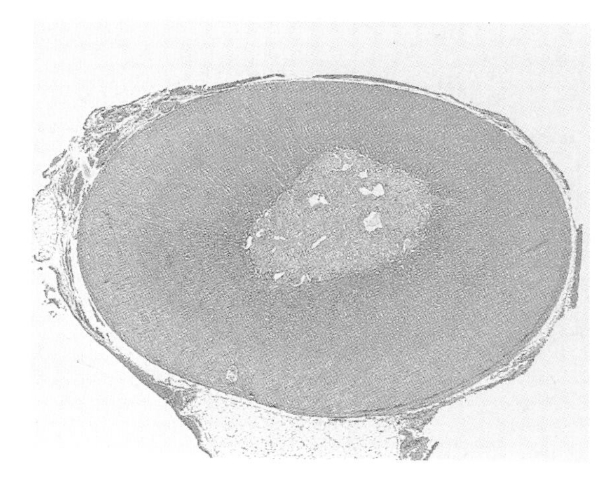

33.6 Cross section of the adrenal gland of a cat

The medulla (central region) is entirely surrounded by cortex.

■ TASKS OF THE ADRENAL GLANDS

The endocrine organs examined so far—stomach, duodenum, and pancreas—and the gonads (sex organs, which are discussed in Chapter 34), are all multipurpose organs: each is an important component in both the endocrine system and some other system. By contrast, the adrenals (and also the thyroid, the parathyroids, and the pituitary) do nothing but secrete hormones.

The two adrenal glands, as their name implies (*ad-* in Latin means "at" and *renal* refers to the kidney), lie very near the kidneys (Fig. 33.3). In mammals each adrenal is actually a double gland, composed of an inner corelike ***medulla*** and an outer barklike ***cortex*** (Fig. 33.6). The medulla and cortex arise in the embryo from different tissues, and their mature functions are unrelated. In fact, they remain as two separate pairs of glands in adult fish and amphibians.

Fight or flight: the adrenal medulla The adrenal medulla secretes two hormones, ***adrenalin*** (also known as epinephrine) and ***noradrenalin*** (norepinephrine), whose functions are similar but not identical. Both hormones have been isolated, identified, and synthesized in the laboratory (Fig. 33.7). They produce a great variety of effects on the body. For example, adrenalin causes

1 A rise in blood pressure

2 Acceleration of the heartbeat

3 Increased conversion of glycogen into glucose and release of glucose into the blood by the liver

4 Increased oxygen consumption

5 Release of reserve red blood corpuscles into the blood from the spleen

6 Vasodilation and increased blood flow in skeletal and heart muscle

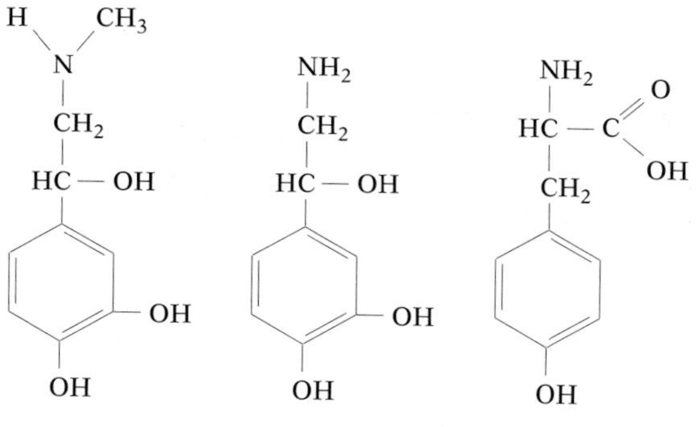

Adrenalin Noradrenalin Tyrosine

33.7 Structural formulas of adrenalin, noradrenalin, and tyrosine

The two hormones are derived from the amino acid, tyrosine.

7 Vasoconstriction and decreased blood flow in the skin and in the smooth muscle of the digestive tract

8 Inhibition of intestinal peristalsis

9 Erection of hairs

10 Production of "gooseflesh"

11 Dilation of the pupils.

This combination of reactions makes sense in response to intense physical exertion, pain, fear, anger, or other heightened emotional states. These so-called fight-or-flight reactions are in fact more often generated initially by the nervous system, with adrenalin (and to a lesser extent noradrenalin) helping to maintain them. These hormones aid in mobilizing the resources of the body in response to emergencies. They accomplish this by stimulating reactions that increase the supply of glucose and oxygen carried by the blood to the skeletal and heart muscles; they also help inhibit functions, such as digestion, that are not immediately important during the emergency and might otherwise compete with the skeletal muscles for oxygen.

It is probably as an antagonist to insulin that adrenalin fulfills its most important normal function. It elevates blood sugar by stimulating the liver to produce glucose from its glycogen reserves, and it acts on muscles to transform their glycogen stores into lactic acid, which is converted into glucose after being transported to the liver by the blood.

Remodeling cholesterol: the adrenal cortex The adrenal cortices are essential to life. If they are removed, death occurs after a host of symptoms: a severe disruption of ionic balance in the body fluids, lowered blood pressure, impairment of kidney function, impairment of carbohydrate metabolism with a marked decrease in both blood-glucose concentration and stored glycogen, loss of weight, general muscular weakness, and a peculiar browning of the skin. These symptoms are also seen to varying degrees in individuals whose adrenal cortices are insufficiently active, a condition known as Addison's disease.

The numerous symptoms of adrenal cortical insufficiency listed here are not related to a single hormone. The adrenal cortex produces many different hormones; the total number is unknown. All the cortical hormones are steroids manufactured through modifications of cholesterol; many differ from each other by only one or two atoms (Fig. 33.8). Despite these apparently minor differences,

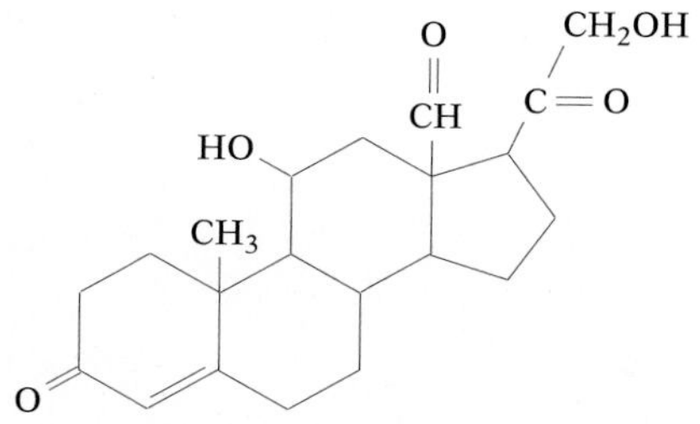

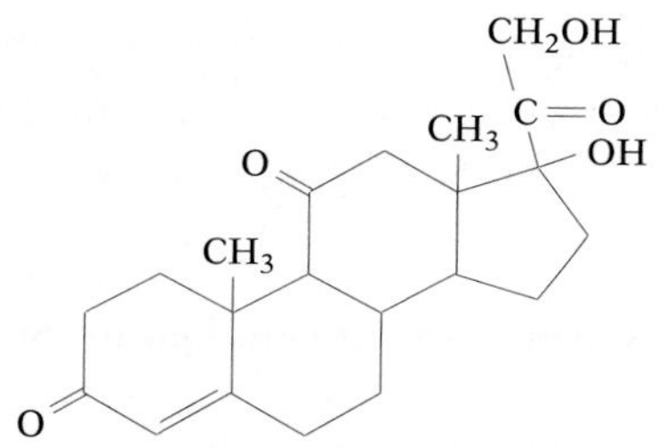

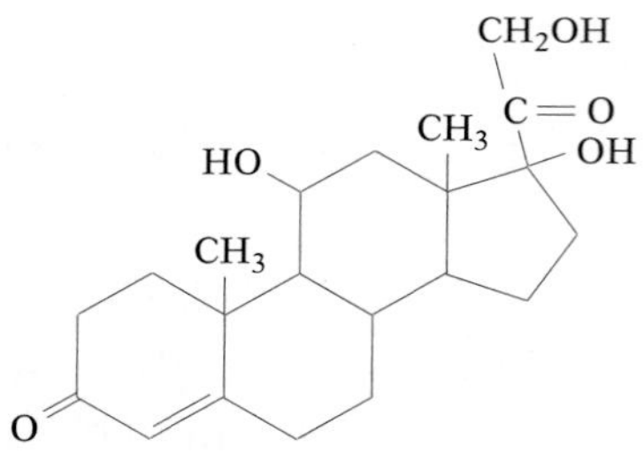

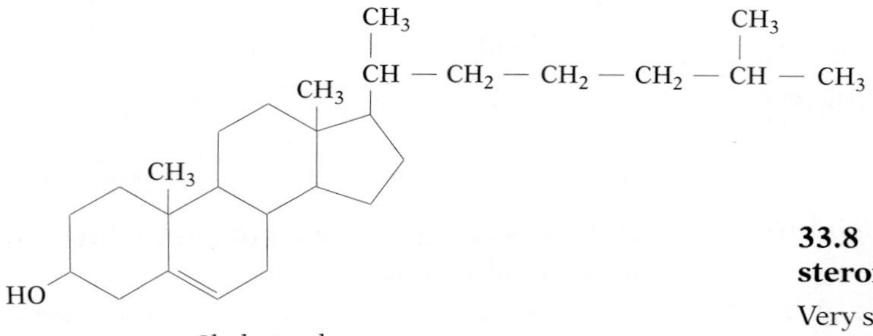

33.8 Some steroids secreted by the adrenal cortex, and the steroid—cholesterol—from which all are synthesized

Very slight differences in side chains can result in markedly different properties.

the various hormones have strikingly different functions: they bind to different receptors in target cells, and affect different sets of chemical reactions. In mammals, only the hormones of the adrenal cortex and those of the gonads and other reproductive structures are steroids. The hormones of the other endocrine organs are amino acids (or, like adrenalin and noradrenalin, compounds derived from amino acids), short polypeptide chains, or full-sized proteins (like insulin). Cortical steroids, which have enormous importance in vertebrates, can be grouped into three functional categories: (1) those that act primarily in regulating carbohydrate and protein metabolism, the glucocorticoids; (2) those that act primarily in regulating salt and water balance, called mineralocorticoids; and (3) those that function primarily as sex hormones.

Glucocorticoids (like cortisol, corticosterone, cortisone), cause a rise in blood sugar and an increase in liver glycogen; both effects probably result from an increased rate of conversion of protein into carbohydrate. The hormones also inhibit oxidation of glucose while promoting mobilization of fat reserves. In short, the glucocorticoids tend to elevate blood-sugar levels by stimulating the body to draw on its noncarbohydrate energy sources. When administered to a person with Addison's disease, they restore the blood-sugar level to normal. Thus they act as antagonists to insulin.

Mineralocorticoids (especially aldosterone) are part of a complex system for regulating blood volume, pressure, and salt balance. This remarkable homeostatic interaction will be described in the next section.

The sex hormones of the adrenals are very similar both chemically and functionally to the sex hormones produced by the gonads. Their normal role is not yet fully understood, but tumors or other disturbances of the adrenal cortex can cause excessive secretion of these hormones, especially of male hormones, resulting in masculinizing effects on females and early sexual development in males. Prenatal feminization is also possible if the mother's hormone balance is upset; the herbicide contaminant dioxin has this effect at extremely low levels, and can even be passed to newborns during lactation.

By and large, all cortical steroids pass through the cell membrane (some of them operating in conjunction with intercellular or intracellular receptors) and then act directly on the DNA, as we will see in more detail in a later section.

Soon after cortisone was isolated in 1935, it was found to have a remarkable ability to increase a test animal's resistance to exposure, cold, poisons, and other physiological stresses. Tests on patients suffering from severe rheumatoid arthritis yielded dramatic results: individuals nearly unable to move were essentially symptom-free after a week of hormone therapy; when injections were stopped, symptoms returned. Cortisone helps relieve the symptoms of a host of other diseases, including even pneumonia and tuberculosis, though it does not kill the bacteria responsible for the illness.

Cortisone and related hormones may produce these symptomatic effects by increasing the body's ability to endure stress. When administered over a long period of time, however, they can cause side effects as bad as or worse than the condition being treated. Among them are high blood pressure, excessive growth of hair, mental aberrations, lowered resistance to certain infections such as poliomyelitis and tuberculosis, peptic ulcers, cataracts, and brittle bones that are easily fractured. That the body's cortical hormones are not a universal panacea is not surprising. Being well and active is adaptive, so if the body could produce more of an existing hormone that would relieve illness without generating harmful side effects, the ability to secrete more of that hormone in response to illness would most likely have been reinforced through natural selection.

Cortisone and its chemical relatives are now most frequently used to facilitate healing where administration needs to be repeated only a few times, or to give partial relief from the symptoms of arthritis and other diseases of connective tissues (where they apparently cause changes in the collagen fibers). These hormones are also sometimes used to treat severe allergic diseases, particularly asthma, and some types of lymphatic diseases. Administered topically, cortisone can also reduce the severity of some rashes. And they are used for temporary relief of severe symptoms in emergency situations.

The dilemma presented by the cortical hormones serves as an example of a general problem faced by physicians every day. Most drugs—and other treatments, for that matter—have potentially harmful side effects. Physicians must therefore balance possible good against possible harm, and they must remember that even the safest drugs are dangerous when used in excessive quantity or at the wrong time. The body is, after all, a finely tuned machine, with interactions between its parts so intricate that they still largely defy analysis. There is a risk of damage to the machine when it is subjected to treatment with chemicals that almost always affect more different functions than can be predicted.

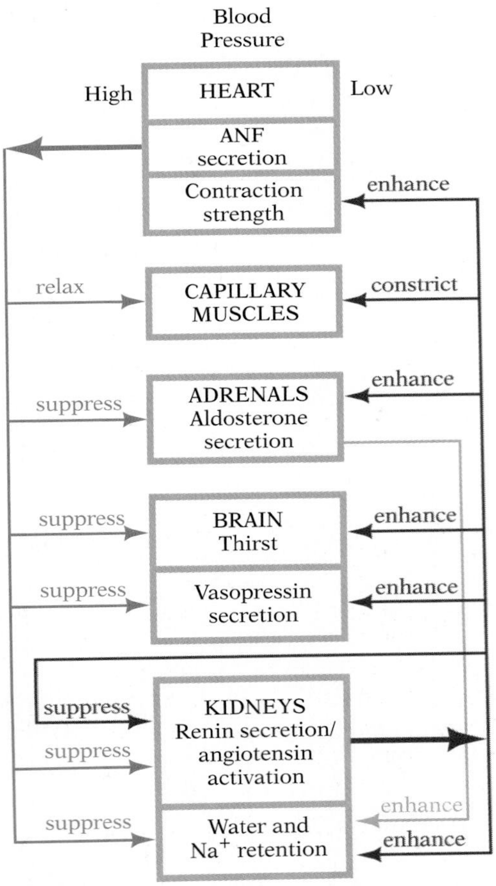

■ HOW HORMONES REGULATE THE BLOOD

As Figure 30.28 (p. 867) indicates, the circulatory system is doing many things at once, not all of which are always compatible with one another. Consider the difficulties in simply maintaining an optimal blood pressure: First, there is not enough blood to fill all the capillary beds simultaneously; when one set is opened, another must be closed or blood pressure will fall and shock may result. Second, blood pressure is in dynamic equilibrium with the volume of fluid in the tissues and lymph; if the osmotic activity of blood rises, as after a meal, water will be drawn into the capillaries from the tissues, and blood pressure will rise. Third, blood volume is constantly being affected by the kidneys, both directly by the removal of water, and indirectly through the extraction of salt (which alters the osmotic activity of the serum). Each meal, drink of fluid, bout of exercise, or major alteration in posture imposes changes that affect blood pressure, and each must be compensated for. The evolutionary solution to this problem involves not only the glucocorticoids, but hormones secreted by the heart, kidney, and brain.

Aldosterone, the main mineralocorticoid, has the simplest role in the drama. When the sodium ion concentration of the blood drops below a certain level, the adrenal cortex adds aldosterone to the blood. This hormone binds to cells in the kidney and promotes heightened reabsorption of Na^+ and water by the distal tubules

33.9 Hormonal regulation of blood pressure
The effects of vasopressin are not shown.

(Fig. 33.9). This negative-feedback system is itself modulated by another hormone, ***angiotensin***, which normally circulates in an inactive form in the blood. When the kidney senses a drop in blood pressure or sodium concentration, it releases the enzyme ***renin***, which activates angiotensin. Angiotensin not only promotes the secretion of more aldosterone, which helps solve the low-sodium problem, it also causes the smooth muscles of the vascular system to constrict. This reduces capillary volume, and so raises the blood pressure. Angiotensin also binds to receptors in the brain and stimulates thirst, which encourages the animal to drink more fluid (and so increase blood volume) and increases the strength of the heartbeat, which is another way to raise blood pressure. Finally, angiotensin stimulates the release of other hormones (ACTH, prolactin, and vasopressin, all described later in the chapter) and suppresses the further release of renin, which helps prevent an exaggerated response to the initial pressure drop (Fig. 33.9).

Given the pattern of counterbalancing hormones we have already seen in other contexts, it is not surprising that there is a chemical messenger designed to deal with high blood pressure. This hormone, ***atrial natriuretic factor (ANF)***, is released from granules in atrial muscle cells in response to atrial distension—that is, high blood pressure. It acts to suppress the secretion of renin, aldosterone, and vasopressin, relaxes smooth muscles, reduces thirst, and increases the kidneys' elimination of Na^+ and water by (as described in Chapter 31) closing channels in the collecting ducts (Fig. 33.8). The balance between ANF and the trio of hormones it suppresses keeps blood pressure, volume, and osmotic concentration at nearly optimal levels in the face of the inevitable challenges to the body.

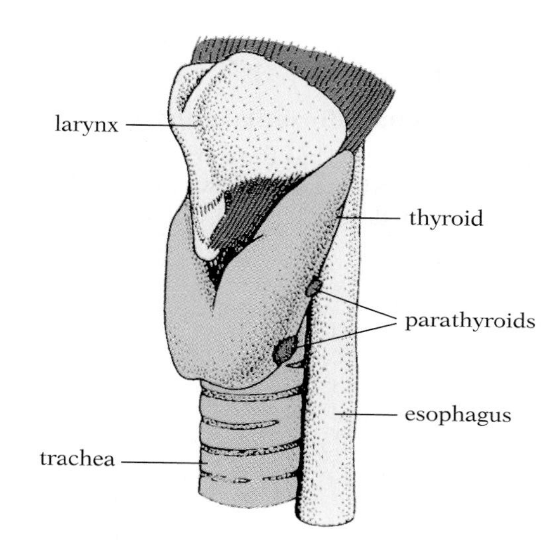

33.10 Thyroid and parathyroid glands

■ METABOLIC RATE AND THE THYROID

Most vertebrates have two thyroid glands, located in the neck; in humans the two have fused to form a single gland (Fig. 33.10). Years ago a condition known as goiter, in which the thyroid may become so enlarged that the whole neck looks swollen and deformed (Fig. 33.11), was very common in some areas of the world, such as the Swiss Alps and the Great Lakes region of the United States. Goiter is often associated with a group of other symptoms, including dry and puffy skin, loss of hair, obesity, a slower than normal heartbeat, physical lethargy, and mental dullness. No cause for this condition was known.

In 1883 a Swiss surgeon, who believed that the thyroid had no important function, removed the gland from a number of his patients. Most developed all the symptoms usually associated with goiter, except the swelling of the neck. This suggested that the normal thyroid secretes some chemical that prevents these symptoms. Patients with no thyroid and patients with the excessively large thyroid of a goiter show the same complex of symptoms because the malformed gland of the goiter, despite its large size, secretes too little hormone. By the 1890s patients with goiters or the other symptoms of hypothyroidism[1] were being successfully treated with

33.11 A Bangladeshi woman with goiter

[1] The prefix *hypo-* means "less than normal," while the prefix *hyper-* means "more than normal." Hence hypothyroidism means less than normal thyroid activity, and hyperthyroidism means more than normal thyroid activity.

injections of thyroid extract or with bits of sheep thyroid in their diets.

In 1905 David Marine noticed that in Cleveland, many people had goiters, as did a high percentage of the local dogs. Even many of the trout in the streams had goiters. Marine dicovered that goiters are caused by an insufficiency of iodine in the food and water. Though it took years to convince a skeptical public, the use of iodized salt finally became widespread, and hypothyroidism caused by insufficient iodine in the soil and water now seldom occurs in the United States or Europe.

33.12 Structural formula of thyroxin (T_4) Triiodothyronine (T_3) has the same formula, except that the iodine atom shown at upper left is replaced by a hydrogen.

The requirement for iodine became more understandable when a thyroid hormone, now called ***thyroxin*** or T_4, was isolated and synthesized. It proved to be an amino acid containing four atoms of iodine (Fig. 33.12). Later, another thyroid compound, identical to thyroxin except that it contains only three atoms of iodine, was found. This substance, called ***triiodothyronine*** or T_3, is 3–5 times more active than thyroxin, but is secreted in smaller amounts. Because T_4 and T_3 have virtually identical effects on target cells, they are usually considered together, under the designation "thyroid hormone" or TH.

The most characteristic effect of TH is stimulation of increased oxidative metabolism in most tissues. The hormone, which is lipid soluble, probably crosses the cell membrane directly and acts to alter gene expression. TH stimulates increased synthesis of certain enzymes, including mitochondrial respiratory enzymes, that facilitate an elevated basal metabolic rate (BMR). Hyperthyroidism—excessive secretion of TH—produces many symptoms that you might predict: a higher than normal body temperature, profuse perspiration, high blood pressure, loss of weight, irritability, and muscular weakness. It also produces one very characteristic but unexpected symptom: exophthalmia, a startling protrusion of the eyeballs (Fig. 33.13). Though hyperthyroidism can sometimes be controlled with antithyroid drugs, it is more often treated by surgical removal of part of the gland or by partial destruction of the gland with radioactive iodine.

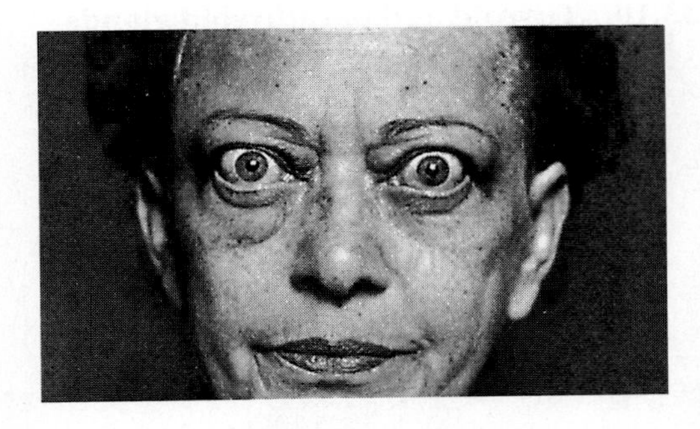

33.13 Exophthalmia

When hypothyroidism—the opposite of hyperthyroidism—is caused by malfunction of the thyroid gland itself rather than by dietary iodine insufficiency, it is treated by administration of thyroid hormone. The untreated condition is particularly serious when it occurs in newborns. These victims, grotesque in their development, are called cretins. They are dwarflike, and they never mature sexually. They have very low intelligence, seldom achieving a mental age of more than 4–5 years. Prevention of cretinism by early administration of hormone to babies showing deficiency symptoms is one of the triumphs of modern medicine.

Though many of the symptoms of hypothyroidism—such as slow heartbeat, obesity, physical lethargy, and mental dullness—may well be consequences of a low BMR, other symptoms, especially some seen in cretinism, cannot easily be explained in this way. An example is the abnormal distribution of protein in the bodies of cretins; they have an excessive amount of glycoprotein in the skin—the cause of their puffy appearance—and an unusually high concentration of protein in the blood plasma, but their kidneys and liver are severely deficient in protein and therefore markedly underdeveloped. Administration of TH relieves all these symptoms. It seems clear, then, that TH plays an important role in regulating the synthesis and distribution of protein.

The effect of TH on protein metabolism is only one manifestation of its more general role in the regulation of many aspects of development. Without the hormone, most vertebrates fail to develop normal adult form and function. TH is not only necessary for the protein synthesis required for proper growth; it is essential for functional maturation of the testes and ovaries, and it acts synergistically with growth hormone from the pituitary gland in promoting skeletal development. In many lower vertebrates, TH is necessary for metamorphosis and for molting.

Another thyroid hormone, ***calcitonin***, is functionally unrelated to TH; its chief effect is to prevent an excessive concentration of calcium in the blood. In lower vertebrates calcitonin is produced by separate glands (the ultimobranchial glands); the corresponding tissue in mammals becomes incorporated into the thyroid during embryonic development.

■ BONES AND THE PARATHYROID

The parathyroid glands in humans are small, pealike organs, usually four in number, located on the surface of the thyroid (Fig. 33.10). Both developmentally and functionally, they are totally independent of the thyroid.

The ***parathyroid hormone***, usually designated PTH, helps to regulate the calcium-phosphate balance between the blood and other tissues. Consequently it is an important element in maintaining the relative constancy of the internal fluids, a subject discussed at length in Chapter 31. We have already seen that insulin, glucagon, and calcitonin are important in this regard. PTH increases the concentration of calcium in the blood (thus functioning as a calcitonin antagonist), and decreases the concentration of phosphate, by acting on at least three organs: the kidneys, the intestines, and the bones. It inhibits excretion of calcium by the kidneys and intestines, and it stimulates release of calcium into the blood from the bones (which contain more than 98% of the body's calcium and 66% of its phosphate). Because calcium in bone is bonded with phosphate, the breakdown of bone releases phosphate as well as calcium. PTH compensates for this release of phosphate into the blood by stimulating the kidneys to excrete phosphate. Actually, the hormone overcompensates, causing more phosphate to be excreted than is added to the blood from bone; the result is that the concentration of phosphate in the blood drops as the secretion of PTH increases.

Naturally occurring hypoparathyroidism is very rare, but the parathyroids are sometimes accidentally removed during surgery on the thyroid. The result is a rise in the phosphate concentration in the blood and a drop in the calcium concentration (as more calcium is excreted by the kidneys and intestines and more is incorporated into bone). This change in the fluid environment of the cells produces serious disturbances, particularly of muscles and nerves. These tissues become very irritable, responding even to very minor stimuli with tremors, cramps, and convulsions. Complete absence of PTH is usually fatal unless large quantities of calcium are included in the diet. Injections of PTH are effective in preventing the symptoms.

Hyperparathyroidism sometimes occurs naturally when the glands become enlarged or develop tumors. PTH is then produced in such quantity that the opposing action of calcitonin is no longer able to maintain a proper balance. The most obvious symptom of this condition is bones that are weak and easily bent or fractured, because of the excessive withdrawal of calcium from them.

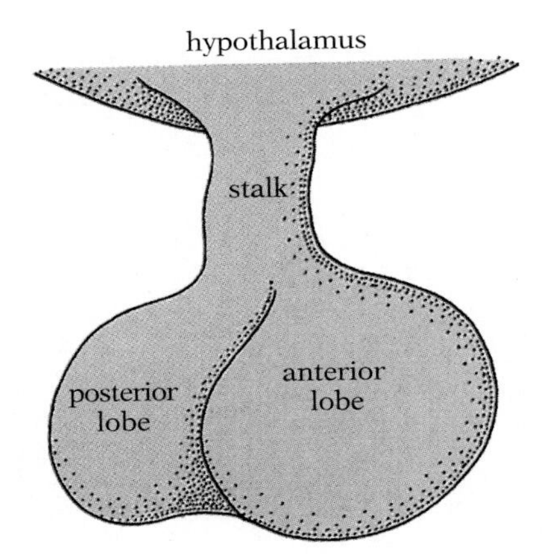

33.14 Pituitary gland

■ THE THYMUS AND THE IMMUNE SYSTEM

One of the functions of the thymus, a gland in the neck region particularly prominent in young animals, is production of the hormone ***thymosin***, which stimulates maturation of many immune-system cells. The role of the thymus in the immune system is described in Chapter 15.

■ THE PITUITARY AND THE HYPOTHALAMUS: REGULATING THE REGULATOR

The ***pituitary*** (also called the hypophysis) is a small gland lying just below the brain. Like the adrenals, the pituitary is a double gland (Fig. 33.14). It consists of an anterior lobe, which develops in the embryo as an outgrowth from the roof of the mouth, and a posterior lobe, which develops as an outgrowth from the lower part of the brain. The two lobes eventually contact each other as they grow, and the anterior lobe partly wraps itself around the posterior lobe. In time, the anterior lobe loses its connection with the mouth, but the posterior lobe retains a stalklike attachment to a part of the brain called the ***hypothalamus.*** The two lobes remain fully independent.

The posterior pituitary and water balance Two hormones, ***oxytocin*** and ***vasopressin,*** are released by the posterior pituitary. They are chemically very similar (Fig. 33.15). Each contains nine amino acids, seven of which occur in both compounds. Only two are different, yet they suffice to give the two hormones very different properties.

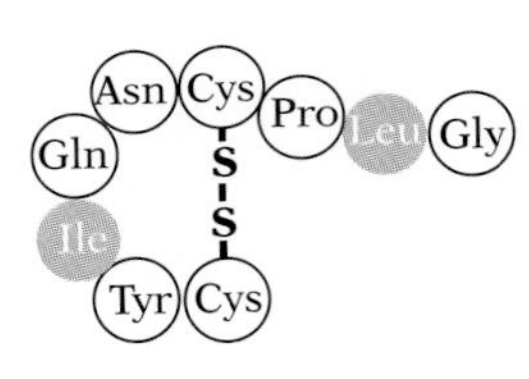

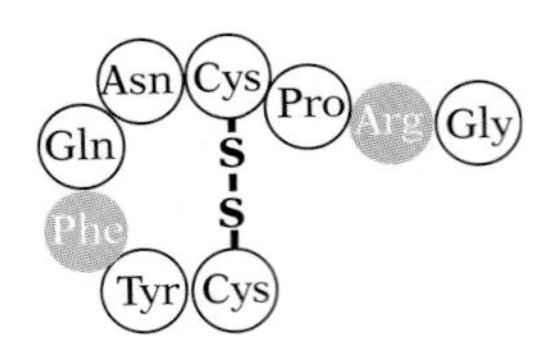

33.15 Oxytocin and vasopressin

The two hormones differ by only two amino acids, but this difference accounts for their distinct activities.

Oxytocin acts mainly on the muscles of the uterus, causing them to contract; hence a release of oxytocin generates labor in a pregnant woman. What causes this release is not yet understood. Injection of oxytocin can induce labor artificially.

Vasopressin causes constriction of the arterioles, and a consequent marked rise in blood pressure. It also stimulates the kidney tubules to reabsorb more water, in part by opening water channels in the collecting ducts. A human without vasopressin would excrete more than 20 liters of urine daily. The well-known diuretic effect of alcohol results from its tendency to suppress vasopressin release.

We said earlier that the posterior pituitary retains a stalklike connection with the hypothalamus (Fig. 33.14). Oxytocin and vasopressin do not originate in the posterior pituitary, but are produced by nerve cells in the hypothalamus and flow along their axons through the stalk to the posterior pituitary, where they are stored. The storage organ releases the hormones on stimulation by electrical signals from the hypothalamus.

Middle management: the anterior pituitary The anterior pituitary (also called the adenohypophysis) is an immensely important organ that produces hormones with widely varying and far-reaching effects. At least seven hormones are secreted by the anterior pituitary in humans.

Prolactin (***PRL***, also called lactogenic hormone) is the most versatile of all the pituitary hormones. It stimulates milk production by the female mammary glands shortly after the birth of a baby; its continued production depends on the mechanical stimulation provided by a suckling infant, and its absence causes milk production to cease. PRL also plays a variety of roles in reproduction, osmoregulation, growth, and the metabolism of carbohydrates and fats.

Another versatile pituitary hormone, ***growth hormone*** (***GH***, also called somatotropic hormone, or STH), plays a critical role in promoting normal growth, especially in combination with thyroid hormone. It is a powerful inducer of protein anabolism, favoring both cellular uptake of amino acids and their incorporation into protein. In this role it acts as an antagonist to the glucocorticoids, inhibiting their protein-catabolizing influence. In addition, GH is an insulin antagonist, acting to elevate blood-sugar concentration. Growth hormone also stimulates hydrolysis of fats in adipose tissues, thereby increasing fatty-acid concentrations in the blood.

If the supply of GH is seriously deficient in a child, growth will be stunted and the child will be a midget. Oversupply of the hormone in a child results in a giant. (The tallest pituitary giant on record reached a height of 2.72 m, or 8 feet 11 inches.) Both pituitary midgets and pituitary giants have relatively normal body proportions. However, if oversecretion of GH begins during adult life, only certain bones, such as those of the face, fingers, and toes, can resume growth. The result is a condition known as acromegaly, characterized by disproportionately large hands and feet and distorted features—a greatly enlarged and protruding jaw, enlarged cheekbones and eyebrow ridges, and a thickened nose.

The anterior pituitary also secretes a number of important hormones that control other endocrine organs. One of these hormones is ***thyrotropic hormone*** (also called thyrotropin or thyroid-stimulating hormone, TSH), which stimulates the thyroid (and which is over produced in infants whose breast milk is contaminated with even low concentrations of dioxin). Another is ***adrenocorticotropic hormone (ACTH)***, which stimulates the adrenal cortex. At least two ***gonadotropic hormones***, or gonadotropins—follicle-stimulating hormone (FSH), and luteinizing hormone (LH)—act on the gonads. Proper growth and development of these endocrine glands depend on adequate secretion of the appropriate tropic (stimulatory) hormone from the pituitary; if the pituitary is removed or becomes inactive, these organs atrophy. The pituitary is often called the master gland of the endocrine system.

Thyrotropic hormone illustrates the use of negative feedback. This two-chain glycoprotein is released by the pituitary when the

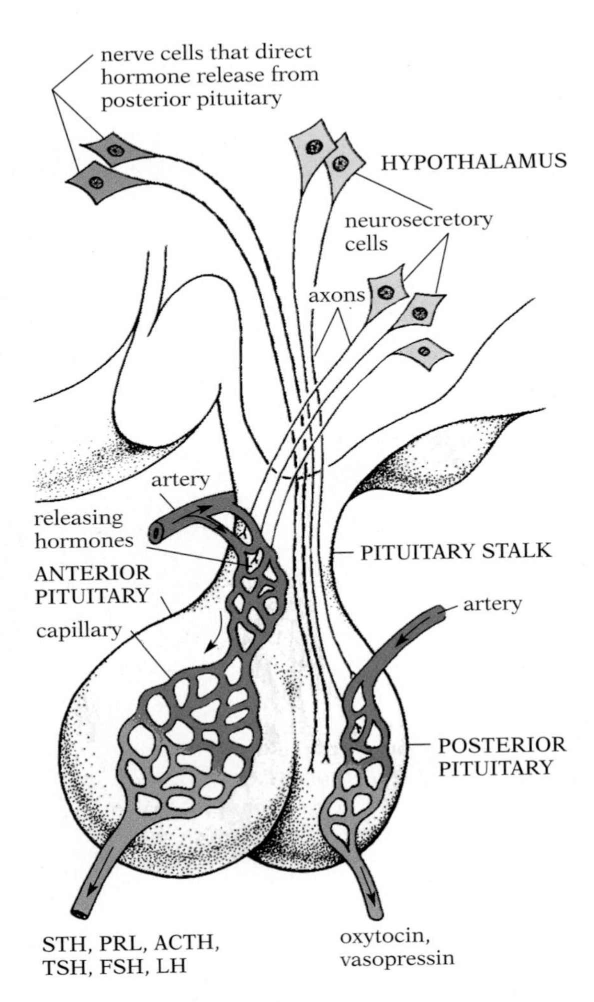

33.16 Connections between the pituitary and the hypothalamus

The pituitary and the hypothalamus are intimately associated. Certain nerve cells in the hypothalamus secrete hormones directly into a capillary bed, which then empties into another capillary bed in the anterior pituitary. The hormones transported in this manner regulate the release of the hormones synthesized in the anterior pituitary. Other nerve cells in the hypothalamus produce hormones that move down their axons to the posterior pituitary; the hormones are released directly into the capillaries there. The cells of the posterior pituitary capture these hormones from the blood and store them; electrical signals from another set of nerve cells in the hypothalamus later trigger their release into the bloodstream.

concentration of thyroxin in the blood is low; this release stimulates increased production of thyroxin by the thyroid. The resulting rise in concentration of thyroxin in the blood, however, inhibits secretion of more thyrotropic hormone by the pituitary. Thus there is negative feedback of information from the thyroid to the pituitary. Each sends chemical messengers to the other, but the message from the pituitary to the thyroid is stimulatory, tending to speed up the system while the return message from the thyroid to the pituitary is inhibitory, tending to slow the system down. This antagonistic interaction produces a delicately balanced control system.

The interaction of the pituitary with the adrenal cortex and with the gonads is similar to its interaction with the thyroid. The pituitary responds to low levels of cortical hormones by secreting more ACTH and to low levels of sex hormones by secreting additional quantities of the gonadotropic hormones. The resulting rise in concentration of cortical hormones or of sex hormones inhibits further secretion by the pituitary.

The anterior pituitary, important as it is for regulating other endocrine glands, does not seem to participate in controlling the pancreas, the adrenal medullae, or the parathyroids. Both the pancreas and the adrenal medullae are controlled in part by the nervous system (the pancreas is also influenced by adrenalin), and the parathyroids are thought to be regulated primarily by the concentration of calcium ions in the blood.

Releasing hormones: the role of the hypothalamus The delicately balanced feedback between the anterior pituitary and other endocrine glands is not the only factor that regulates the regulator. The desired norm must be changed from time to time. For example, many animals have annual reproductive cycles for which day length is the critical cue. Suppose an animal detects the lengthening of days in spring, and its gonads secrete more sex hormone. How is its perception of the stimulus—increasing day length—able to affect its endocrine glands? Perception of stimuli involves the nervous system, yet there are no nervous connections either to the anterior pituitary or to endocrine glands such as the gonads. However, nervous tissue is capable of producing hormones, as we saw in the interaction of the hypothalamus and the posterior pituitary. Moreover, it is the hypothalamus that acts as the main control center for gathering and integrating neural information bearing on the appropriate "setting" of the chemical thermostat. The suprachiasmatic nucleus of the hypothalamus in turn generates the rhythm that helps our bodies regulate daily activities.

As Fig. 33.16 indicates, hormones from the hypothalamus regulate hormone secretion by the anterior pituitary, though there is no direct physical connection between the hypothalamus and the anterior pituitary. Neurosecretory cells in the hypothalamus synthesize hormone, and transport it down their axons to capillary beds in the stalk. There is an unusual connection between the blood supplies of the stalk and the anterior pituitary. Arteries to the stalk break up into capillaries, and these capillaries eventually join to form several veins leading away from the hypothalamus. Unlike most veins, however, these do not run directly into a larger branch of the venous system; instead, they pass downward into the anterior pituitary and there break up into a second capillary bed. Thus hormones secreted into the stalk capillaries travel to the capillaries in the anterior pituitary (Fig. 33.16).

The hypothalamus is now known to produce a variety of special peptide hormones called ***releasing hormones***, which, carried directly to the anterior pituitary by the portal vessels, regulate the secretory activity of the anterior pituitary. Thus TSH-releasing hormone (also called thyrotropic releasing hormone or TRH)—a peptide only three amino acids long—stimulates the release of TSH; LH-releasing hormone (10 amino acids long) controls LH levels; corticotropic releasing hormone stimulates ACTH release; growth-hormone-releasing hormone stimulates release of growth hormone; and so on. A few of the hypothalamic hormones are inhibitory rather than stimulatory: prolactin-release-inhibiting hormone, for instance, inhibits release of prolactin by the anterior pituitary. As Chapter 35 details, nervous and hormonal control are each part of an integrated control system.

The sequence of responses triggered in our hypothetical animal by the perception of changing day length can now be reconstructed. The animal detects the light with its sense organs; nervous impulses are transmitted from the sense organs to the hypothalamus of the brain; the hypothalamus secretes gonadotropic-releasing hormone (GnRH) into the blood; the releasing hormone stimulates the anterior pituitary to increase its secretion of gonadotropic hormones; the gonadotropic hormones stimulate the gonads to secrete more sex hormone; and the sex hormone helps prepare the animal physiologically to begin the activities of the breeding season.

The hypothalamus is also one of the major points for the transmission of feedback information from the endocrine system. Consider again the negative-feedback effect of a rise in thyroxin on the anterior pituitary. To some extent, probably, the thyroxin exerts negative feedback on the pituitary directly, by inhibiting its secretion of TSH, but to a large extent it does so indirectly, by inhibiting the hypothalamus from secreting TSH-releasing hormone (Fig. 33.17). Similarly, much of the negative feedback action of other hormones is via the hypothalamus.

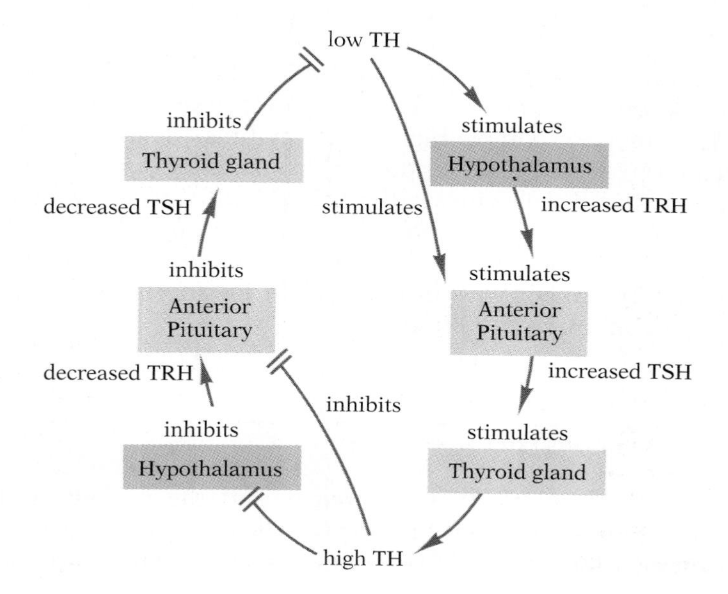

33.17 Feedback control of the thyroid gland

When the level of thyroxine (TH) in the blood is low, the hypothalamus secretes thyrotropic releasing hormone (TRH) into the blood; this stimulates the anterior pituitary to release thyrotropic hormone (TSH), which in turn stimulates the thyroid gland to secrete thyroxin (TH). The thyroxin stimulates target cells throughout the body, but it also inhibits both the hypothalamus and the anterior pituitary. (Arrows indicate stimulatory influences; barred lines, inhibitory influences.) Without the stimulatory effect of TSH on the thyroid, the secretion of thyroxin decreases. When the thyroxin level gets too low, the hypothalamus is again stimulated.

■ THE PINEAL AND RESPONSES TO LIGHT

The glandular appearance of the pineal, a lobe in the rear portion of the forebrain, has long intrigued investigators. Aristotle thought it must be the site of the soul, since, unlike the rest of the brain in which all structures exist as pairs (the right and left lobe of the hypothalamus, for example), the pineal is unpaired. In some lower vertebrates the pineal is eyelike and responds to light both by generating nervous impulses and by secreting a hormone called ***melatonin***, which lightens the skin by concentrating the pigment granules in melanophores (pigment-containing cells). In these animals the pineal is intimately involved (presumably through melatonin) in the control of circadian rhythms—cycles of activity repeated approximately every 24 hours.

The mammalian pineal, too, secretes melatonin, but it has no light-sensitive cells, and mammals have no melanophores. Though the pineal is a part of the brain, its principal (perhaps only) innervation originates outside the skull cavity in the neck. Hence the mammalian pineal and its hormone have long been a puzzle.

Current evidence suggests that the pineal functions as a neurosecretory transducer, converting neural information about light conditions into hormonal output. If this is correct, the conversion would take place as follows: Information about the light-dark cycle received by the eyes would go first (via the brain) to the nerve cells in the neck, and would then be conveyed to the pineal. The pineal would respond by secreting melatonin in inverse proportion to the amount of light. The melatonin, in turn, would influence the secretion of gonadotropic hormones by the anterior pituitary; it might do this directly, or by causing the secretion of a peptide in the hypothalamus that would then carry the message to the pituitary. The usual effect of melatonin is inhibitory; in autumn, for example, as the days get shorter, increased production of melatonin tends to turn off gonadotropic secretion, with the result that the gonads regress and the animal enters a nonreproductive phase.

The sequence of events is different in animals that breed in autumn. Moreover, melatonin may sometimes stimulate the testes instead of inhibiting them; whether the hormone is gonadotropic or antigonadotropic probably depends on when in the circadian cycle it is released.

Melatonin may affect parts of the brain other than the hypothalamus, particularly those concerned with locomotor rhythms, feeding rhythms, and other biological rhythms. The hormone is implicated in at least one serious syndrome: a substantial number of humans living well outside the tropics suffer from "winter depression" (also called seasonal affect disorder, or SAD), a consequence of overproduction of or heightened sensitivity to melatonin. Treatment with bright light in the morning appears to cure this debilitating condition.

HOW HORMONES ACT

The number of mammalian hormones and the variety of their functions is enormous. Despite this variety, only a limited number of mechanisms for intercellular communication are at work. Some of these strategies are very simple, some are well-understood components of the cell membrane, and some are unique to hormonal control.

Hormones gain entrance to target cells in various ways. Some, like the steroid hormones, cross the cell membrane either directly or bound to a receptor (Fig. 33.18); this receptor complex then binds to the DNA to exert the hormone's effect. A few hormones can pass into the cell through special channels, and some enter by

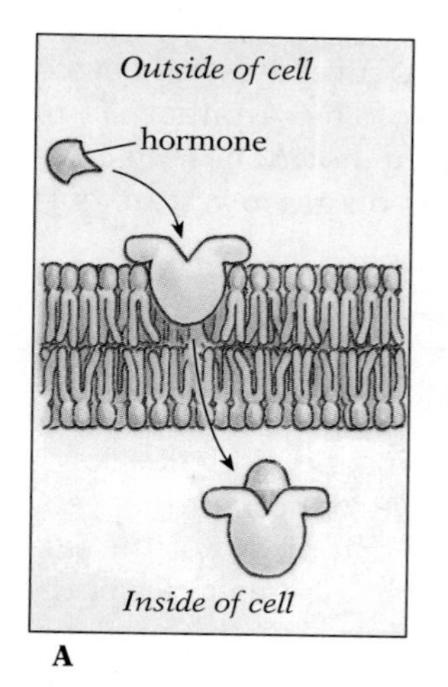

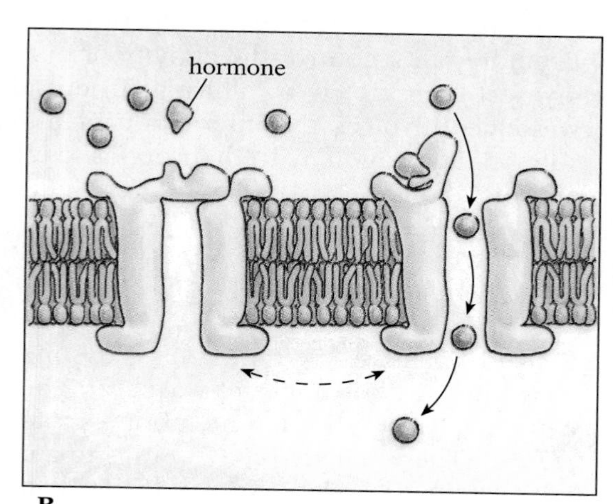

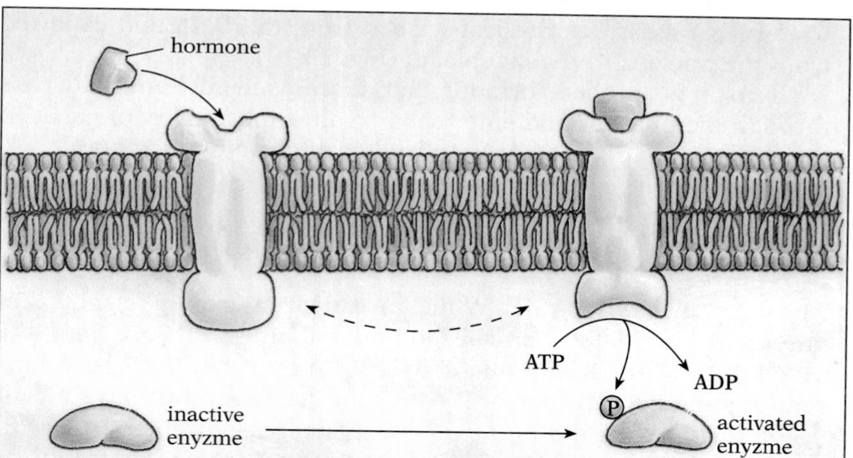

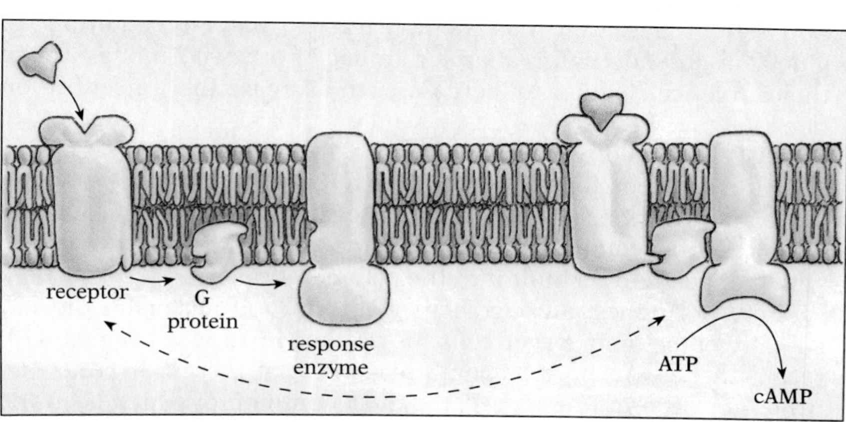

33.18 How hormones transmit messages
A hormone can act on its target cell in a variety of ways, several of which are illustrated here. (A) One type of hormone binds to a membrane receptor and enters the cytoplasm with it as a conjugated pair. Once inside, this type of hormone transmits its message by directly or indirectly altering gene expression or a specific chemical reaction. (B) Another type of hormone may transmit its message by facilitating the entrance into the target cell of a second substance, sometimes an ion, as in the case of the chemically gated ion channels by which most nerve cells communicate. Alternatively, a hormone may bind to and activate a membrane protein, which then causes another substance to be generated or activated on the inside of the membrane. An intracellular enzyme may be activated (usually by phosphorylation), as happens when insulin binds (C), or an intracellular messenger may be produced by activation of a response protein (D), usually by way of an intermediary (the G protein). In the version of this strategy shown here (and discussed further in the text) the binding of the hormone activates adenylate cyclase, which in turn catalyzes conversion of ATP into a so-called second messenger, cAMP.

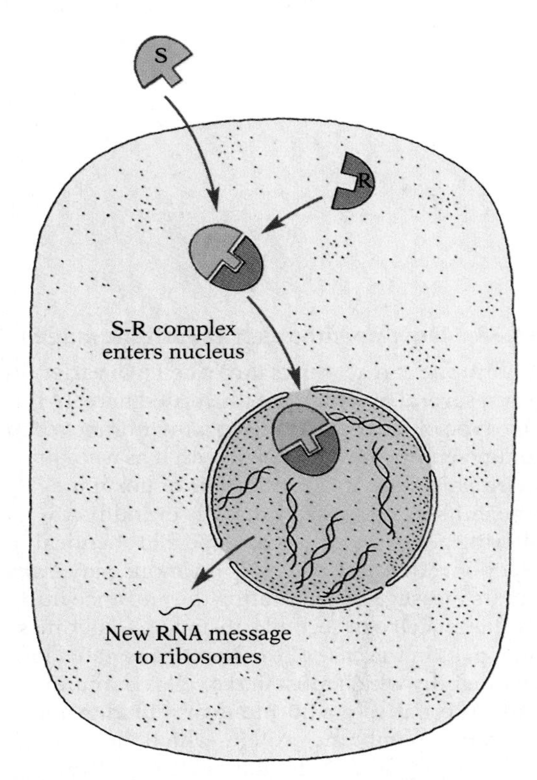

33.19 Model for the mode of action of steroid hormones

The hormone (S), being lipid-soluble, can penetrate the plasma membrane and enter the cytoplasm. There it binds with a receptor protein (R), and the complex then enters the nucleus, where it influences synthesis of RNA by the genes. A new message (in the form of new RNA) is thus sent to the ribosomes, which begin synthesizing the protein coded by the new RNA. The new protein, perhaps an enzyme, will then influence the chemical activity of the cell.

33.20 Cyclic AMP

being transported actively. Finally, most hormones do not enter the cell at all, but instead bind to extracellular receptors. Extracellular binding causes an intracellular effect, either by opening an ion channel in the membrane (Fig. 33.18B) or by activating an enzyme or second messenger inside the cell (Fig. 33.18C–D). In the next section we will consider in more detail how these various mechanisms work to control cell metabolism.

Within the cell, too, methods of hormonal control vary. Some, like testosterone, bind to the DNA and directly affect gene expression and, as a result, the production of specific enzymes. Others, like adrenalin, control the activity of enzymes already synthesized or exert their effects by altering structural proteins. Another set of strategies involves the interaction of different hormones, like insulin and glucagon, with their opposite effects on glucose transport and metabolism.

■ ALTERING GENE EXPRESSION

Steroid hormones, such as the ecdysone of insects and the corticoids and sex hormones of vertebrates, are hydrophobic; thus they can move easily through membranes into the cytoplasm of the target cell. There the steroid (S) binds to a specific cytoplasmic receptor protein (R). The complex (S-R) then moves into the nucleus, where it helps regulate the activity of specific genes, influencing which RNA messages are transcribed from the DNA and exported from the nucleus to the cytoplasm (Fig. 33.19).

Thyroxin is another hormone that, it seems likely, can easily pass through membranes and enter cells, to alter the activity of genes in the nucleus, influence mitochondrial enzyme activity, or both.

■ CHEMICAL MIDDLEMEN: SECOND MESSENGERS

The role of cyclic AMP While investigating how glucagon and adrenalin stimulate liver cells to release more glucose into the blood, E. W. Sutherland and T. W. Rall discovered that these hormones stimulate an increase in the intracellular concentration of ***cyclic adenosine monophosphate*** (usually shortened to ***cyclic AMP*** or ***cAMP***; Fig. 33.20). This compound leads, in turn, to activation of an enzyme necessary for breakdown of glycogen to glucose. Subsequent research has demonstrated that, in addition to glucagon and adrenalin, a large number of other hormones act on their target cells either to increase or to decrease the concentration of cAMP.

cAMP is related to ATP (adenosine triphosphate; see Fig. 6.3, p. 158, and has been found in almost all animal tissues studied (vertebrate and invertebrate) and in bacteria. It is synthesized from ATP by a reaction catalyzed by an enzyme called ***adenylate cyclase***, which appears to be built into the cell membrane.

Many hormones, among them glucagon and adrenalin, do not actually enter their target cells, but rather form weak bonds with receptor sites on the cell membrane. Most of these hormones depend on a ***second messenger***, the most common of which is cAMP.

The sequence is this: an extracellular first messenger, which is the hormone itself, goes from an endocrine gland to the target cell, where it binds and stimulates production of an intracellular second messenger, usually by activating adenylate cyclase (Fig. 33.18D). The cAMP then interacts with cytoplasmic enzyme systems and initiates the cell's characteristic responses to the hormonal stimulation. A signal-transduction protein, in the form of a three-unit guanine-binding protein (***G protein***; Fig. 33.18D), intervenes in this chain, and allows different hormone receptors to have different effects—to inhibit the cyclase, for instance, or to activate a different second messenger system. Mutations in G-protein genes can produce syndromes that were once thought to result from abnormal levels of hormone production.

The stimulation of glucose production in the liver by glucagon or adrenalin is an example of how cAMP influences the cell. In this case the cAMP activates a protein kinase enzyme. The activated kinase, in turn, activates a second enzyme, which activates a third enzyme, which catalyzes the first reaction in the breakdown of glycogen to glucose (Fig. 33.21).

This chain of enzyme-catalyzed reactions, an example of a cascade reaction, makes possible a very great ***amplification*** of the effects of the original hormone-binding event. Because an enzyme can be used over and over again, a single molecule of active adenylate cyclase can catalyze production of about 100 molecules of cAMP. Molecules of cAMP activate enzyme A. Each molecule of enzyme A activates roughly 100 molecules of enzyme B, and so on. The net result is that a single molecule of glucagon or adrenalin may lead to release of as many as 10^8 (100 million) molecules of glucose, within only 1 or 2 min. No wonder only very small quantities of hormone are needed.

We saw earlier that glucagon and adrenalin also raise blood-sugar levels by inhibiting glycogen synthesis. This action, like the one already discussed, depends on the protein kinase activated by cAMP. The protein kinase (enzyme A), which we previously saw activating an enzyme necessary for glucose production, is here inactivating an enzyme and hence blocking the conversion of glucose into glycogen (Fig. 33.22). Protein kinase is probably an essential link in all hormonal actions mediated by the adenylate cyclase system.

Glucagon and adrenalin can initiate the processes via adenylate cyclase even though they are chemically very different. This is possible, even though they bind to different receptor molecules on the membrane, because both types of receptor are capable of activating adenylate cyclase. Thus the two hormones have additive effects. However, this is true only for the liver; the two hormones do not have the same effect on muscle. Acting via adenylate cyclase, adrenalin can stimulate production of glucose from glycogen by muscle cells, but glucagon cannot do so, apparently because the membranes of muscle cells lack receptors for glucagon.

Clearly it is the presence or absence of receptors on the cell membrane that determines whether or not a particular cell type is influenced by a given hormone. Moreover, a cell may respond in the same way to two or more different hormones if it has cyclase-activating receptors for each of those hormones; the hormones

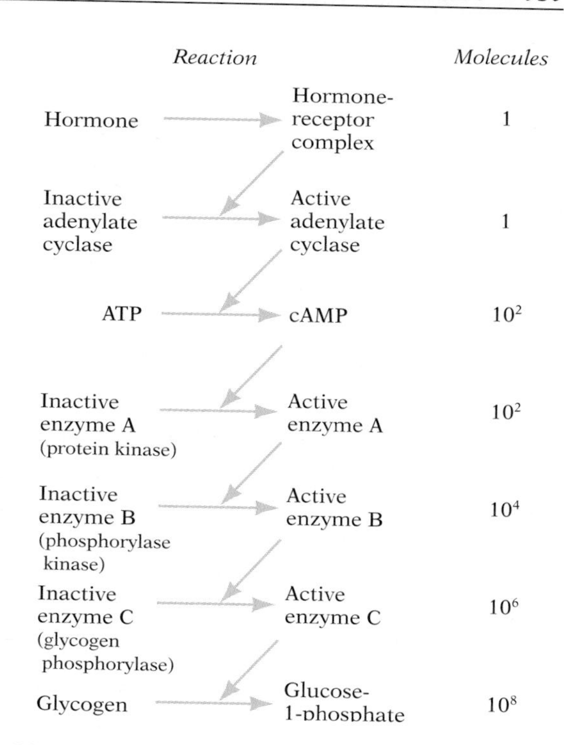

33.21 Hormonal stimulation of glucose production

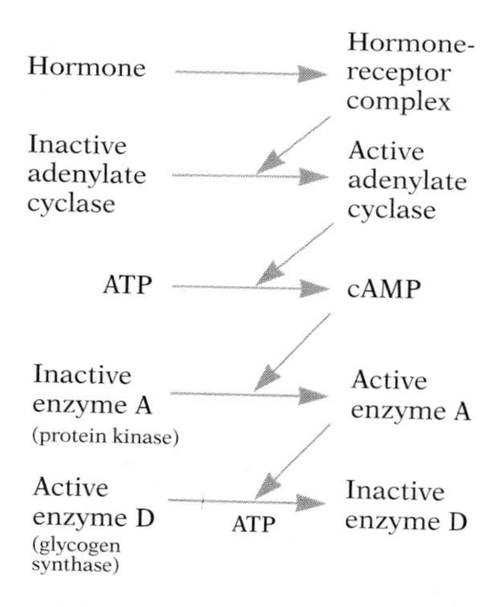

33.22 Hormonal inhibition of glycogen synthesis

themselves are specific only for the receptors, not for the reactions triggered through G-protein binding and second messenger activation.

The cell-specific distribution of hormone receptors helps explain how the many different hormones that depend on the adenylate cyclase system (including—in addition to glucagon and adrenalin—gastrin, secretin, parathyroid hormone, calcitonin, the tropic hormones of the anterior pituitary, and at least some of the hypothalamic releasing hormones) can have their own distinct sets of target cells, even though their immediate effect in every case is activation of adenylate cyclase. In addition, different cells respond differently to changes in their cAMP content depending on which molecular targets are synthesized in that cell.

Various chemicals other than hormones also influence cells by their effects on cAMP levels. Some of these, like histamine (an alerting chemical in the immune system), work through adenylate cyclase, but others exert their effect by acting on the enzyme ***phosphodiesterase***, which breaks down cAMP. In a normal cell a certain amount of this enzyme, which keeps the concentration of cAMP in check, is always present. Thus, if adrenalin induces a rise in the cAMP concentration, phosphodiesterase causes a return to more normal levels as soon as the hormone is no longer present. Any chemical capable of inhibiting this enzyme, like caffeine, or of activating it, like nicotine, can markedly influence the cAMP levels in the cell.

Ions as second messengers We said earlier that hormones that exert their effects on cell metabolism without entering the cell do so either by activating a second messenger or by opening an ion channel. In fact, the cAMP system and the ion-channel strategy are so intimately connected that it probably makes more sense to think of the ions as second messengers as well.

By far the most common second-messenger ion is Ca^{++}. Calcium ions can serve this function because the concentration of Ca^{++} in the cytosol is normally kept low: Ca^{++} ions are actively pumped both into the endoplasmic reticulum (ER) and out of the cell; they are also taken up and stored by the mitochondria, and are captured by various binding molecules free in the cytosol. As a result, when a hormone binds to its specialized receptor and thereby opens a Ca^{++} channel in the membrane, the strongly favorable electrochemical gradient causes Ca^{++} ions to rush in. Other hormones operate by activating (via the G-protein complex) a lipid second messenger—inositol trisphosphate—which diffuses to the ER and opens Ca^{++} channels there. In either case, these ions then bind to and activate particular intracellular enzymes, the most common of which is ***calmodulin***, a protein consisting of 148 amino acids. The Ca^{++}-calmodulin complex then activates enzymes of the particular system being controlled (Fig. 33.23). Which enzymes are regulated in a cell depends on the kind of cell it is—that is, on the particular genes active in the cell. Like cAMP, calmodulin controls different sets of enzymes in different sorts of cells.

The operation of the ion channel and the Ca^{++}-calmodulin complex is especially clear in muscle cells. In preparing skeletal muscles for high activity by stimulating the breakdown of glycogen to

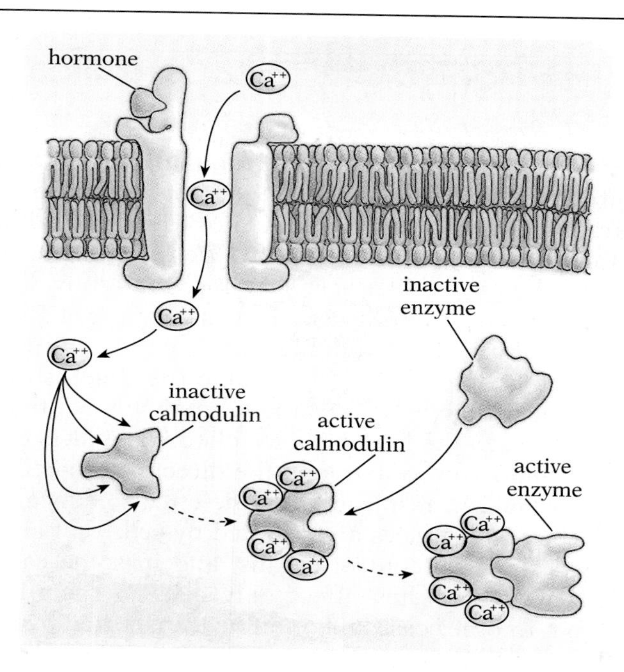

33.23 Calmodulin activation

The entrance of Ca^{++} ions in response to the binding of a hormone to its membrane receptor, or an intracellular second messenger to its receptor on the ER (not shown), activates calmodulin by filling its four Ca^{++} binding sites. The Ca^{++}-calmodulin complex in turn binds to and activates one or more specific enzymes in the cell.

glucose, adrenalin uses not only cAMP but also Ca^{++} as a second messenger. When adrenalin binds to its specialized receptor, a channel opens to Ca^{++}; the resulting calmodulin complex binds to phosphorylase kinase. This kinase in turn activates glycogen phosphorylase, the enzyme that converts glycogen into glucose–1–phosphate and sends it into the glycolytic pathway. Thus in muscles both cAMP and activated calmodulin are necessary to stimulate the production of glucose; the binding of the activated calmodulin to phosphorylase kinase is part of the kinase's activation. This interaction of cAMP, activated calmodulin, and the several enzymes gives us a glimpse of the elaborate interconnections underlying the chemistry of complex organisms.

■ WHY MOLECULAR ANTAGONISM IS ADAPTIVE

We have seen many cases in which one hormone triggers another, or in which two hormones act either synergistically or antagonistically to exert control. The antagonistic strategy, of which there are several examples in the next chapter, is relatively common: one aspect of metabolism (concentrations of particular substances in the blood, for instance) will be increased by one hormone and decreased by another. This type of control mechanism has at least two advantages over the use of a single hormone. The first is speed: secreting an antagonist to stop a process is faster than merely waiting for the stimulatory hormone to disappear. The second advantage is precision: a system that depends on the *ratio* of one hormone to another can be very precise in its response, even in regions of the body in which the absolute concentration of the hormones may be low because of factors such as poor circulation or depletion by receptors upstream.

NOT QUITE HORMONES: LOCAL CHEMICAL MEDIATORS

According to our working definition, animal hormones are substances produced by organs specialized for hormone synthesis; they are transported through the circulatory system, and they exert highly specific effects on target tissue. We have excluded neurotransmitters (chemicals, like acetylcholine, that convey messages from one nerve cell to another, or from a nerve cell to a muscle cell) because the cell that produces the communication chemical releases it directly onto the target cell, and the effect is to alter the electrical rather than the metabolic activity of that target. Somewhere between the long-distance chemical effects of the hormones of the endocrine system and the direct cell-to-cell electrical interactions of neurotransmitters are the effects of local chemical mediators. These substances are secreted by cells that are not part of specialized chemical-control organs, and in some cases are so rapidly destroyed that they affect only cells in their immediate neighborhood. In other cases they can be transported by the bloodstream to distant targets.

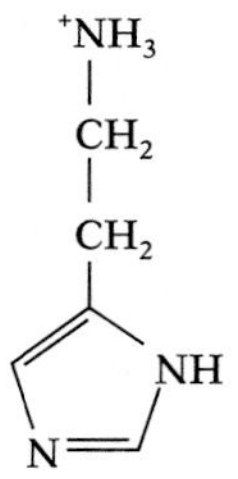

Histamine

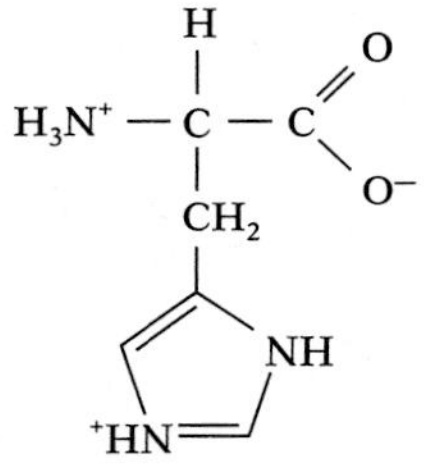

Histidine

33.24 Histamine and its precursor, histidine

■ HISTAMINE: THE MOLECULAR ALARM

Perhaps the best-known local chemical mediator is histamine, a derivative of the amino acid histidine (Fig. 33.24). As detailed in Chapter 15, histamine is produced primarily by the mast cells in connective tissue. These cells store histamine in large vesicles, and release it when they detect injury or a toxic substance in the vicinity. Histamine signals nearby capillaries to dilate and leak; this allows more blood to reach the site, and enables cells of the immune and cell-repair systems to leave the capillaries and reach the tissue in need of help. (Of course, the fluids that leak out also cause swelling.) Histamine attracts these repair cells, and they in turn secrete compounds that inactivate histamine.

One of the most obvious effects of a cold—the generalized swelling of nasal tissues—results from histamine release, and for this reason the antihistamines in many cold preparations provide some symptomatic relief. Most allergic reactions (among them the rare reaction to wasp or honey bee venom) are a consequence of uncontrolled histamine release after a malfunction of the immune system. Loss of blood fluids to the tissues causes blood pressure to fall, and swelling of the tissues may block the flow of air to the lungs. A severe allergic reaction can be fatal unless treated quickly by an injection of adrenalin, and even mild reactions such as hay fever can be debilitating.

■ JACKS OF ALL TRADES: THE PROSTAGLANDINS

The prostaglandins, though named for the prostate gland, where they were originally discovered, are now known to be secreted by most animal tissues. Prostaglandins exhibit a bewildering array of actions, including stimulation or relaxation of smooth muscle, dilation or constriction of blood vessels, stimulation of intestinal motility, modulation of synaptic transmission in the nervous system, stimulation of inflammation responses, and enhancement of the perception of pain. The effectiveness of aspirin in combating inflammation and pain results, at least in part, from inhibition of prostaglandin synthesis.

Prostaglandins (Fig. 33.25) are continually being synthesized from phospholipids in the cell membrane and released into the bloodstream or surrounding fluids. They exert control through changes in their rate of synthesis, which in turn is controlled by a great variety of stimuli, including hormones, nerve impulses, mechanical stimuli, oxygen deprivation, and inflammation. Interestingly, prostaglandins can affect the cells that manufacture them, as well as their neighbors. In some cases they circulate in the blood like hormones and exert their effects at distant locations. In others, they reach distant target sites without transport through the bloodstream; the prostaglandins in the semen secreted by male seminal vesicles, for instance, cause contractions of the uterine muscles in the female.

Prostaglandins bind to receptors on target cells. Studies of their effects on a variety of these cells have revealed that they often mimic the action of the corresponding stimulatory hormones, apparently with a variant of the cAMP mechanism. Like histamines, they are continually being destroyed by enzymes in the intercellular fluid and cell membranes.

33.25 Prostaglandin E

■ NITRIC OXIDE: A GAS AS CHEMICAL MEDIATOR

The most recently discovered local chemical mediator is nitric oxide (NO), a gas also used by macrophages in the immune system to help kill foreign cells. This versatile substance is also employed by the nervous system. Its discovery was difficult because it converts into other chemicals within a few seconds of release, a property well suited to both transmitters and short-range hormones. When its release is triggered, it moves from the cells lining the blood vessels to the surrounding smooth muscles and causes them to relax, thus lowering blood pressure; failures in the NO system lead to high blood pressure and hypertension. NO has its dark side as well: it is involved in the destruction of neurons following strokes.

■ ENDORPHINS: THE NATURAL OPIATES

Also close to the conceptual line separating hormones and neurotransmitters are the endorphins. Like hormones, they may be carried by the bloodstream to cells at some distance, but their targets are nerve cells, and they affect electrical rather than metabolic activity. The discovery of this group of compounds resulted from research on opiates, a group of highly addictive pain-killing drugs. Opiates such as morphine (Fig. 33.26) bind to specific receptors on nerve cells. Since receptors characteristically bind naturally occurring chemicals, the search was on for such chemicals, which would be internally synthesized pain-killing opiates.

The body is now known to synthesize more than half a dozen

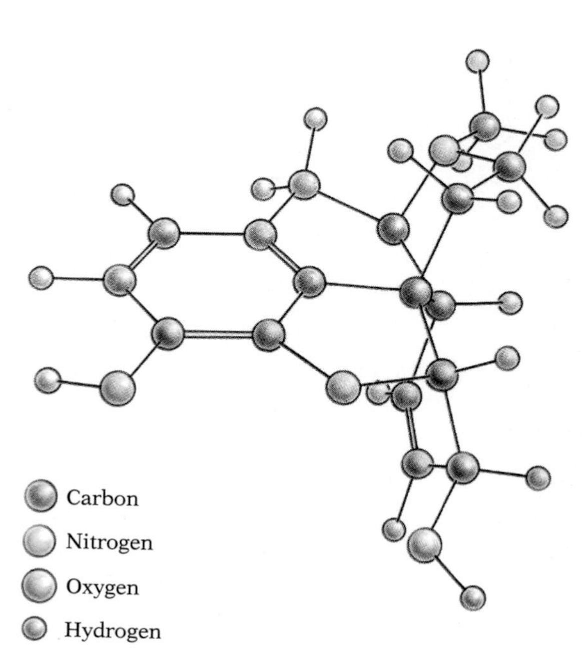

33.26 Morphine

The highly contorted molecule has four six-membered rings and one five-membered ring.

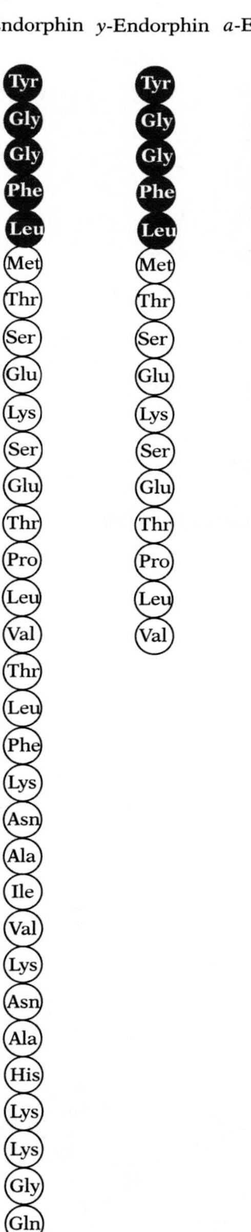

33.27 Endorphins
All four of these endorphins have the same sequence of amino acids at one end. Indeed, the sequences of all four are identical as far as they go. The differences among them result from their additional amino acids.

polypeptides that bind to the opiate receptors. Of these so-called endorphins, two short-chain polypeptides are produced by nerve cells throughout the nervous system, while the others (Fig. 33.27), which have the same terminal sequence as the first two, are produced in the anterior pituitary. The common terminal sequence probably explains why they all bind to the same receptors, while the differences in chain length give rise to the variations in their activity. Why there are so many different endorphins, and how they interact with their target cells, is not yet understood.

THE EVOLUTION OF HORMONES

One of the most pervasive trends in the organization of multicellular organisms is the specialization of particular cells and groups of cells. Two points about the evolution of such specialization are worth bearing in mind. First, many simple organisms face the same problems as more complex plants and animals with respect to transport, water balance, control of growth, and so on. Each cell in these unspecialized organisms, therefore, must perform many different duties. Second, evolution usually proceeds by modifying or building on systems and structures already present. Accordingly, it seems likely that some of the problems we share with less specialized organisms may have been solved by them in simpler ways; moreover, these solutions may in some cases have provided the raw materials for the elaborate mechanisms we now employ.

The evolution of hormonal control seems to underscore both of these points. Even animals with specialized endocrine organs also have unspecialized communication networks of the sort that would be expected in their simpler ancestors. We have already seen how various nonglandular cells can communicate through histamine, for instance. A more complex example involves insulin, which plays some roles in the body's information-processing system that we have not yet discussed. Insulin in the bloodstream is largely prevented from entering the brain by a thick membrane—the blood–brain barrier. The brain is nevertheless full of cells with insulin receptors, for insulin is also secreted by brain cells. This locally produced insulin serves a regulatory function in the brain quite independent of sugar metabolism in the body.

The discovery of the hormone in insects, protozoans, fungi, and even procaryotes like *E. coli* suggests that this unspecialized communication route is indeed a heritage from simpler forebears. The insulin of *E. coli* is so similar to human insulin that it promotes glucose oxidation in human fat cells—the standard test for insulin activity. In all likelihood, then, insulin originated as a nonglandular messenger before the vertebrate pancreas evolved, and its extended range of activities in complex organisms is an elaboration of its role in simpler organisms. Insulin and its receptor have, like many successful evolutionary systems, been duplicated in the genome and adapted to other jobs. In insects, for example, a hormone and receptor obviously derived from the insulin team regulate the secretion of ecdysone; in vertebrates it is clear that certain growth regulators and their receptors, such as epidermal growth factor, are also derived from this ancient combination.

Moreover, insulin is not the only hormone with a long pedigree: protozoans have biologically indistinguishable versions of β-endorphin and ACTH. As for receptors, adrenalin activates the adenylate cyclase system in protozoans just as it does in us, and is blocked by the same chemicals. Why microorganisms make and respond to what are for us specialized hormones remains to be understood. However, the possibility that the specialized hormones and receptor systems in higher plants and animals evolved from intracellular equivalents in unicellular organisms helps explain why there is no distinct functional line dividing the hormones from the many other chemicals made by cells for controlling metabolism and development.

CHAPTER SUMMARY

HOW HORMONES CONTROL DEVELOPMENT IN INSECTS

Hormones, especially ecdysone, control when insects molt. In species with complete metamorphosis, the hormone JH determines whether the next molt will produce another larva or an adult. Eclosion is triggered by yet another hormone. (p. 937)

THE MANY ROLES OF HORMONES IN VERTEBRATES

HOW HORMONES CONTROL DIGESTION Food in the stomach triggers release of gastrin, which causes gastric glands to secrete digestive enzymes. Secretin plays the same role for pancreatic enzymes. Cholecystokinin directs secretion from the gallbladder. Somatostatin regulates secretion by counteracting the effects of the other digestive hormones. (p. 940)

DIABETES AND THE PANCREAS The islet cells of the pancreas secrete insulin, which stimulates uptake and use of glucose; they also secrete glucagon, which has the opposite effect. Diabetes results from immunological destruction of the islet cells, reduced insulin secretion, or loss of sensitivity to insulin. (p. 942)

TASKS OF THE ADRENAL GLANDS The medulla secretes adrenalin and noradrenalin, which prepare the body for high activity. The cortex produces a variety of cholesterol-based hormones that regulate metabolism (glucocorticoids) and body fluids (mineralocorticoids), and act as sex hormones. (p. 945)

HOW HORMONES REGULATE THE BLOOD Aldosterone increases salt reabsorption by the kidney. Activated angiotensin amplifies this response and reduces capillary volume. Together they increase blood pressure. ANF counteracts these effects. (p. 948)

METABOLIC RATE AND THE THYROID Thyroid hormone (T_3 and T_4) controls basal metabolic rate, and thus activity and growth, as well as some aspects of protein synthesis. Calcitonin regulates blood-calcium level. (p. 949)

BONES AND THE PARATHYROID PTH regulates the balance between calcium and phosphate in body fluids, which affects calcium deposition in bones. (p. 951)

THE THYMUS AND THE IMMUNE SYSTEM Thymosin stimulates maturation of immune-system cells. (p. 952)

THE PITUITARY AND THE HYPOTHALAMUS: REGULATING THE REGULATORS Oxytocin acts on the uterus during labor; vasopressin helps increase blood pressure by acting on capillary muscles and the kidney, and also promotes water retention; both hormones are produced in the hypothalamus but stored and released from the posterior pituitary. The anterior pituitary produces prolactin (which regulates metabolism and plays a role in lactation), growth hormone (which stimulates growth), and a variety of stimulating hormones that act on other endocrine glands. Pituitary activity, in turn, is regulated by releasing hormones from the hypothalamus. (p. 952)

THE PINEAL AND RESPONSES TO LIGHT Melatonin controls skin shade in some animals, and appears to control annual reproductive cycles. (p. 956)

HOW HORMONES ACT

ALTERING GENE EXPRESSION Some hormones act by binding to DNA and altering gene expression. (p. 958)

CHEMICAL MIDDLEMEN: SECOND MESSENGERS Other hormones bind to membrane receptors, where they either open ion channels (usually Ca^{++}) or trigger production of other chemical messengers inside the cell. Typically the receptor activates an enzyme via a G-protein intermediate, which generates cAMP, which in turn activates the first step in an enzyme cascade. (p. 958)

WHY MOLECULAR ANTAGONISM IS ADAPTIVE Cellular reactions are faster and more accurate when activity is determined by ratios of two antagonistic hormones rather than by absolute levels of a single hormone. (p. 961)

NOT QUITE HORMONES: LOCAL CHEMICAL MEDIATORS

HISTAMINE: THE MOLECULAR ALARM Histamine is released by mast cells in response to injury or toxic chemicals; it causes capillaries to leak antibody-bearing fluid and immune-system cells into the tissues to speed healing. (p. 962)

JACKS OF ALL TRADES: THE PROSTAGLANDINS Prostaglandins have many functions, including regulating smooth-muscle activity and thus blood-vessel contraction, inflammatory responses, and pain sensitivity. (p. 962)

NITRIC OXIDE: A GAS AS CHEMICAL MEDIATOR This gas causes smooth muscle to relax. (p. 963)

ENDORPHINS: THE NATURAL OPIATES Endorphins reduce sensitivity to pain. (p. 963)

THE EVOLUTION OF HORMONES

Many hormones began as intracellular control chemicals, which were drafted into service for intercellular control. (p. 964)

STUDY QUESTIONS

1. What are some possible treatments for adult-onset diabetes? (pp. 942–945)
2. What are the costs and benefits of the hormone-antagonist system—that is, using hormones with opposite effects simultaneously? (p. 961)
3. Trace the hormonal consequences of a deep dive (which compresses all the tissues) and weightlessness. Which has more potential for problems: a rapid onset/offset of one of these circumstances, or a sustained period under such conditions? (pp. 866, 948–949)
4. How might the melatonin system have created the "winter-depression" syndrome as early humans moved out of the tropics about 250,000 years ago? Could a loss of skin pigment in nontropical populations be interpreted as a possible adaptation to extremes of day length? (p. 956)
5. How does adrenalin help counteract the effects of severe allergic reactions? (pp. 945–946)

SUGGESTED READING

ATKINSON, M.A., AND N. K. MACLAREN, 1990. What causes diabetes? *Scientific American* 263 (1).

BERRIDGE, M. J., 1985. The molecular basis of communication within the cell, *Scientific American* 253 (4). *Reviews the operation of second messengers.*

LIENHARD, G. E., J. W. SLOT, D. E. JAMES, AND M. M. MUECKLER, 1992. How cells absorb glucose, *Scientific American* 266 (1). *On the interactions between glucose channels and insulin.*

LINDER, M. E., AND A. G. GILMAN, 1992. G proteins. *Scientific American* 267 (1). *On the operation of these ubiquitous signal-transduction molecules.*

ORCI, L., J.-D. VASSALLI, AND A. PERRELET, 1988. The insulin factory, *Scientific American* 259 (3). *Traces the synthesis, sorting, packaging, activation, and secretion of insulin in great detail.*

RASMUSSEN, H., 1989. The cycling of calcium as an intracellular messenger, *Scientific American* 261 (4).

SNYDER, S. H., 1985. The molecular basis of communication between cells, *Scientific American* 253 (4). *An excellent review of hormones and local chemical mediators, and of the elaborate feedback system for controlling hormone levels.*

SNYDER, S. H., AND D. S. BREDT (1992). Biological roles of nitric oxide, *Scientific American* 266 (6). *On the many roles of this newly discovered transmitter, local chemical mediator, and toxic weapon.*

CHAPTER 34

HORMONES AND VERTEBRATE REPRODUCTION

THE EVOLUTIONARY FACTS OF LIFE

From an evolutionary standpoint, reproduction is the ultimate goal of life. All the other aspects of biology—nutrient procurement, gas exchange, internal transport, waste excretion, osmoregulation, growth, hormonal and nervous control, and behavior—are processes that enable organisms to survive to reproduce. Just as "the hen is the egg's way of producing another egg," humans too are elaborate devices for producing eggs and sperm, for bringing them together in the process of fertilization, and for giving birth to young. In this chapter we will examine the physiology of (primarily) mammalian reproduction.

THE TWO STRATEGIES FOR FERTILIZATION

Sexual reproduction in higher animals involves the union of two gametes—that is, an egg cell (ovum) and a sperm cell. Occasionally the same individual produces both gametes, which unite with one another. This process of self-fertilization is common among internal parasites such as tapeworms, whose chances of locating another individual for cross-fertilization are poor. However, most animals use cross-fertilization even when, as in earthworms, each individual is hermaphroditic (possessing both male and female sexual organs; Fig. 34.1). Among vertebrates, sexual reproduction always involves cross-fertilization; even in hermaphroditic species of vertebrates, such as some species of fish, individuals alternate between producing eggs and sperm, but never fertilize themselves.

34.1 Two earthworms mating

The worms are hermaphroditic, each possessing both male and female reproductive organs. Cross-fertilization nearly always involves coupling at two mating points: at the left point of joining, the upper worm acts as the male and the lower worm as the female, while at the right these roles are reversed.

There are two basic ways in which egg cells and sperm cells are brought together: (1) external fertilization, in which both types of gametes are shed into water, and the sperm swim or are carried by currents to the eggs; (2) and internal fertilization, in which the eggs are retained within the reproductive tract of the female until after they have been fertilized by sperm inserted into the female by the male.

External fertilization is limited essentially to animals living in aquatic environments; the flagellated sperm must have fluid in which to swim, and the eggs lack a protective coat or shell (so that sperm can penetrate and fertilize them), and would dry out in the air. Almost all aquatic invertebrates, most fish, and many amphibians use external fertilization. However, shedding eggs and sperm into the water is an uncertain method of fertilization: many of the sperm never locate an egg, and many eggs are never fertilized, even if both types of gametes are shed at the same time and in the same place, as is usually the case. Consequently, animals using external fertilization generally release vast numbers of eggs and sperm at one time (Fig. 34.2).

34.2 Toads spawning

The smaller male clasps the female in an embrace called amplexus, and sprays semen over the eggs she releases. Many hundreds of eggs are produced.

Most land animals, both invertebrate and vertebrate, use internal fertilization (Fig. 34.3). In effect, the sperm cells are provided with the sort of fluid environment that is no longer available to them outside the animals' bodies. The sperm can remain aquatic, swimming through the film of fluid present on the walls of the female reproductive tract. Once fertilized, the egg is either enclosed in a protective shell and released by the female, or held within the female's body until the embryonic stages of development have been completed. Internal fertilization requires close physiological and behavioral synchronization of the sexes, which involves extensive hormonal control.

The evolutionary trend in vertebrates has been toward internal fertilization. Fish almost always use external fertilization and hence lay eggs with no hard shell. Amphibians evolved from fish, and they too generally use external fertilization; they must return to the water or to a very moist place on land to lay their eggs. Some salamanders have evolved a behavioral sequence in which the male releases a membranous packet (spermatophore) containing sperm that the female picks up with her cloaca in a primitive type of internal fertilization.

The reptiles evolved from ancestral amphibians. They were the first vertebrates to be fully independent of the water for reproduction. They use internal fertilization, and lay eggs enclosed in tough membranes and shells. Since internal fertilization entails much less waste of egg cells than external fertilization does, only a few egg cells are released during each reproductive season. Birds evolved from one group of ancient reptiles, and they too employ internal fertilization and lay eggs with shells. Like reptiles and birds, mammals also use internal fertilization.

34.3 Internal fertilization by a terrestrial invertebrate

The male butterfly inserts sperm into the reproductive tract of the female. Fertilization occurs within the body of the female, and the eggs are coated with a protective shell before being laid.

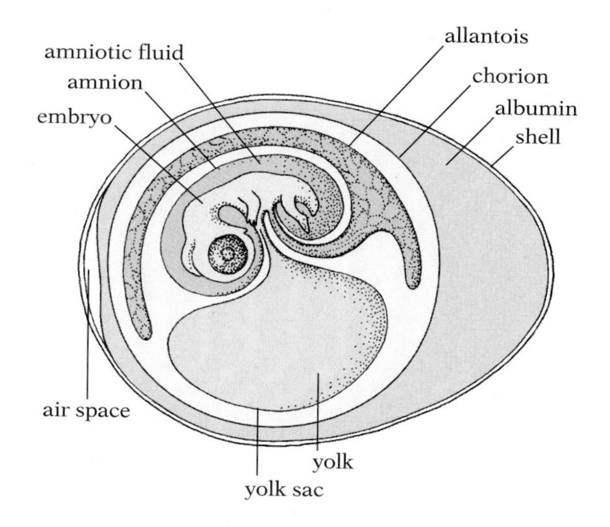

34.4 Embryonic membranes in a bird's egg

Though everything shown here is part of the "egg" in common parlance, the chorion is the outer boundary of structures derived from the true egg cell. The shell and thick layer of albumin (a protein) are outside the cell.

■ HOW THE EMBRYO IS PACKAGED

The eggs of land vertebrates such as reptiles and birds have four different membranes in addition to the shell. These are the amnion, the allantois, the yolk sac, and the chorion (Fig. 34.4). The ***amnion*** encloses a fluid-filled chamber housing the embryo, which can thus develop in an aquatic medium even though the egg as a whole may be laid on dry land. The ***allantois*** functions as a receptacle for the urinary wastes of the developing embryo, and its blood vessels, which lie near the shell, function in gas exchange. The ***yolk sac*** encloses the yolk, which is food for the developing embryo. The ***chorion*** is an outer membrane surrounding the embryo and the other membranes.

In mammals (with a few rare exceptions) no shell is formed and the egg is not laid. Instead, the early embryo with its membranes becomes implanted in a specialized chamber of the female genital tract, where embryonic development is completed. The young animal is then born alive.

HOW THE HUMAN REPRODUCTIVE SYSTEM WORKS

■ THE REPRODUCTIVE SYSTEM OF THE HUMAN MALE

The male gonads, or sex organs, are the ***testes*** (Figs. 34.5, 34.6)—oval glandular structures that form in the dorsal portion of the ab-

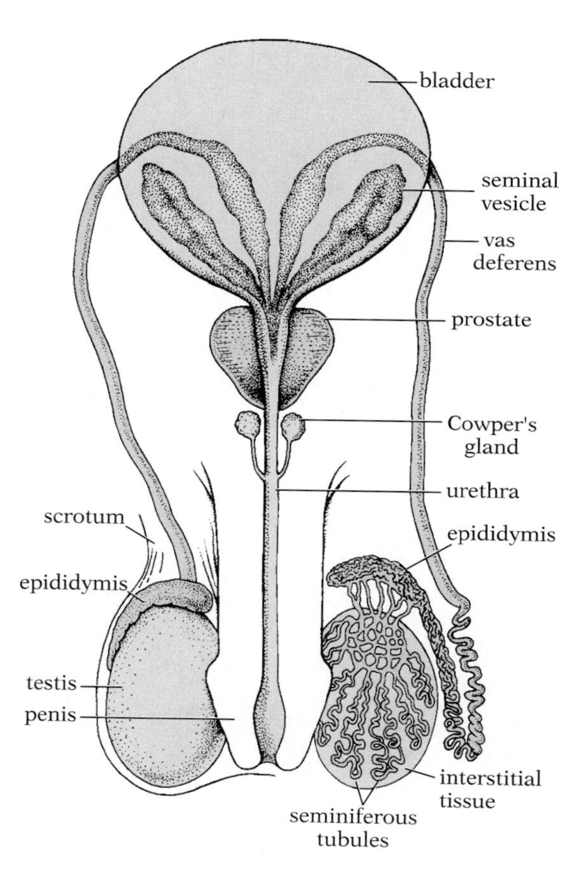

34.5 Reproductive tract of the human male: frontal view

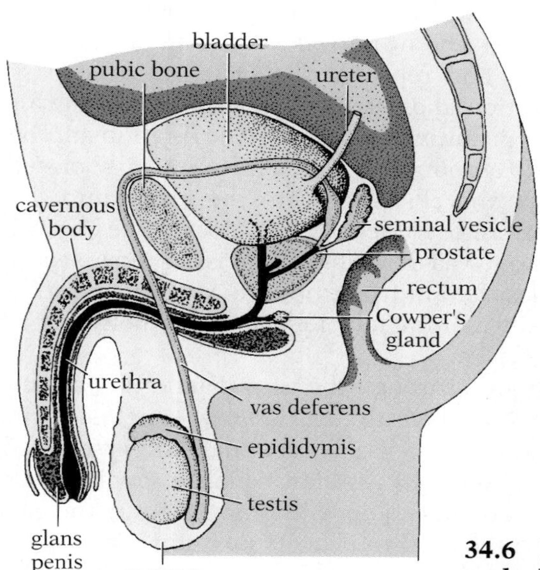

34.6 Reproductive tract of the human male: lateral view

dominal cavity from the same embryonic tissue that gives rise to the ovaries in females. In the human male the testes descend about the time of birth into the ***scrotal sac*** (scrotum), a pouch whose cavity is initially continuous with the abdominal cavity via a passageway called the ***inguinal canal***. After the testes have descended through the inguinal canal into the scrotum, the canal is slowly sealed off by growth of connective tissue.

Sometimes the inguinal canal fails to close properly; and even when it does, it remains a point of weakness that can open again when subjected to excessive strain, as when a man lifts a heavy object. The reopening of the canal is known as an inguinal hernia, or rupture. If the hernia is large, it must be repaired surgically to prevent a loop of the intestine from slipping through the opening into the scrotal sac, where the intestine may become caught so tightly that its blood supply is cut off and gangrene results. Inguinal hernia is largely a hazard of walking upright, which places much strain on the lower abdomen. In some mammals the inguinal canal remains partly open, and the testes move back into the abdominal cavity during the nonreproductive season.

Each testis has two functional components: the ***seminiferous tubules***, in which the sperm cells are produced, and the ***interstitial cells***, which secrete male sex hormone. The germinal epithelium of the seminiferous tubules becomes functional at the time of puberty. Mature sperm cells pass from the seminiferous tubules via many tiny ducts into a much-coiled tube, the ***epididymis*** (plural epididymides), which lies on the surface of each testis (Figs. 34.5, 34.6). Many sperm are stored in this organ until they are activated by secretions produced by it; they are then released during copulation. In addition, sperm are held in the vas deferens.

A long sperm duct, the ***vas deferens*** (plural vasa deferentia), runs from each epididymis through the inguinal canal and into the abdominal cavity, where it loops over the bladder and joins with the urethra just beyond the point where the urethra arises from the bladder. The urethra, in turn, passes through the ***penis*** and empties to the outside. The evolutionary trend in vertebrates is toward increasing independence of the reproductive system from the excretory system. The urethra in mammalian males is used by both the excretory and reproductive systems; urine passes through it during excretion and semen passes through it during sexual activity. In more primitive vertebrates the relationship between the excretory and reproductive systems is generally closer. In frogs, for example, sperm cells pass from the testes into the kidneys and down the excretory ducts to the cloaca; the reproductive system has no separate vas deferens. There is far more separation in mammalian males than in frogs, but the two systems still share the urethra. Only in mammalian females (and bony fish) has complete separation arisen.

As sperm pass through the vasa deferentia and urethra, seminal fluid is added to them to form ***semen***, which is a mixture of seminal fluid and sperm cells. The seminal fluid is secreted by three sets of glands: the ***seminal vesicles***, which empty into the vasa deferentia just before these join with the urethra; the ***prostate***, which empties into the urethra near its junction with the vasa deferentia; and the ***Cowper's glands***, which empty into the urethra at the base of the penis (Figs. 34.5, 34.6). Seminal fluid has a variety of functions:

1 It serves as a vehicle for transport of sperm.

2 It lubricates the passages through which the sperm must travel.

3 Its chemical buffers help protect the sperm from the acids in the female genital tract.

4 It contains much sugar which the active sperm need as a source of energy; sperm cells store very little food, and thus depend on external nutrients to supply power to their active flagella. The base of a sperm flagellum is packed with mitochondria, in which energy is extracted from the sugar absorbed from the seminal fluid (Fig. 34.7A).

During sexual excitement the arteries leading into the penis dilate, apparently in response to the release of nitric oxide from nerve cells; at the same time veins from the penis constrict. Much blood is pumped under considerable pressure through the arteries

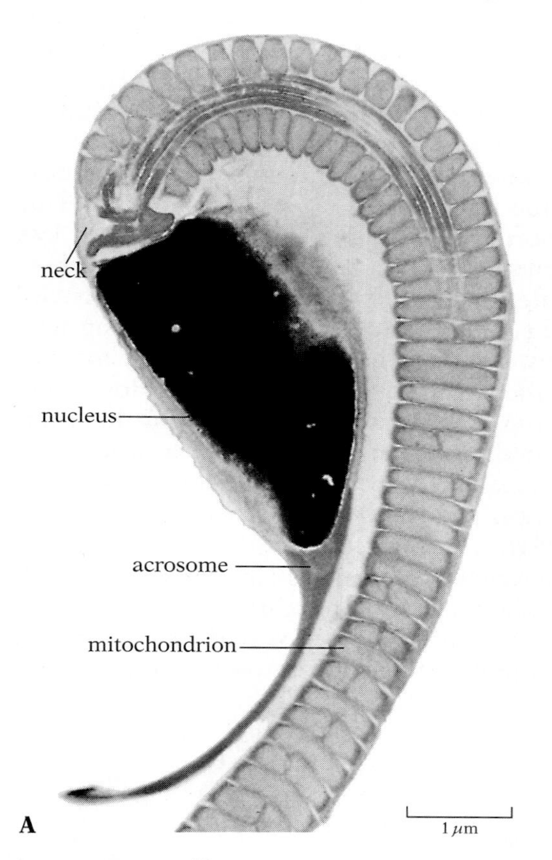

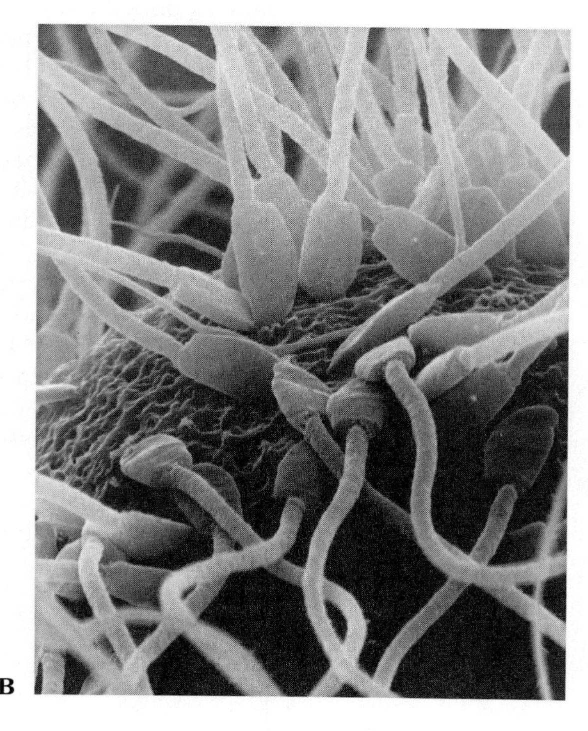

34.7 Mammalian spermatozoa

(A) Electron micrograph of longitudinal section of a sperm cell from a kangaroo rat. The sperm head, which contains the acrosome (an enzyme-filled organelle discussed in Chapter 13) and the nucleus, is connected by a short neck to a portion of the flagellum tightly packed with spirally arranged mitochondria. There are no mitochondria in the long distal portion of the flagellum, (not shown here). (B) Scanning electron micrograph of an Israeli sand rat egg surrounded by sperm. Only one sperm will fertilize the egg.

into the spaces in the ***cavernous body***, the spongy tissue of which the penis is largely composed (Fig. 34.6). The engorgement of the penis by blood under high arterial pressure causes it to increase greatly in size and to become hard and erect, thus preparing it for insertion into the female vagina during copulation (also called coitus). Erection of the penis does not involve activity of skeletal muscles; it is entirely a vascular phenomenon.

When the penis is sufficiently stimulated by friction during copulation, nervous reflexes cause waves of contraction in the smooth muscles of the walls of the epididymides, vasa deferentia, seminal glands, and urethra. These contractions move sperm from the epididymides and vasa deferentia, combine seminal fluid from the various glands with the sperm, and expel the semen from the urethra. An average of about 400 million sperm cells in about 3.5 ml of semen are released during one human ejaculation.

■ HOW HORMONES CONTROL MALE SEXUAL DEVELOPMENT

The testes begin secreting small amounts of the male sex hormone testosterone (Fig. 34.8) during embryonic development. The presence of this hormone in the embryo is crucial to the differentiation of male structures. The level of production remains low until the time of puberty, and no sperm cells are produced.

The factors governing onset of puberty are not well understood. One possibility is that the hypothalamus—which is inhibited by even low levels of the hormone during childhood—becomes less sensitive and begins sending more gonadotropic releasing hormone (GnRH) to the anterior pituitary, stimulating it to increase its secretion of the two gonadotropic hormones, luteinizing hormone (LH) and follicle-stimulating hormone (FSH). LH induces the interstitial cells of the testes to produce more testosterone. Together with FSH, testosterone induces maturation of the seminiferous tubules and stimulates them to begin sperm production (spermatogenesis). Maintenance of spermatogenesis over long periods requires the continued presence of testosterone (and the LH that induces it) and of FSH.

Once sufficient testosterone appears in the circulatory system, it stimulates maturation of the accessory reproductive structures and the complex of changes in the secondary sexual characteristics normally associated with puberty: growth of the beard and pubic

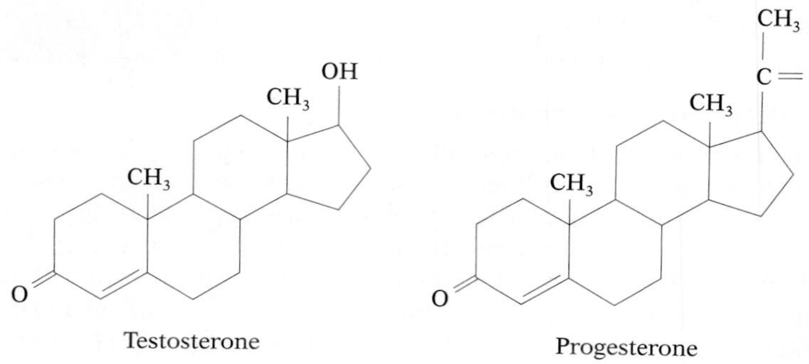

34.8 Structural formulas of male sex hormone (testosterone) and female sex hormone (progesterone)

Differing in only one side group, these two steroids have amazingly different effects on the body.

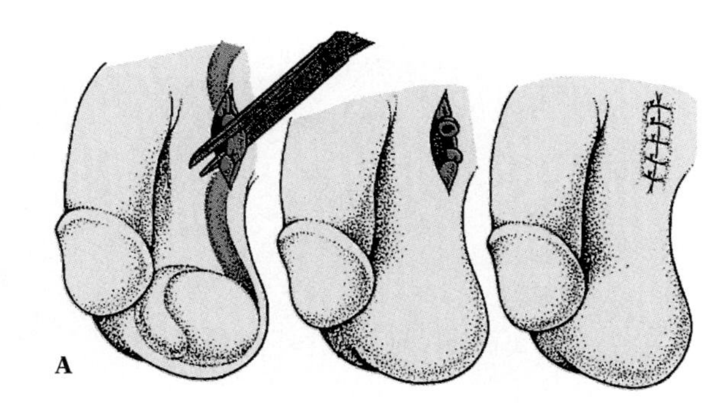

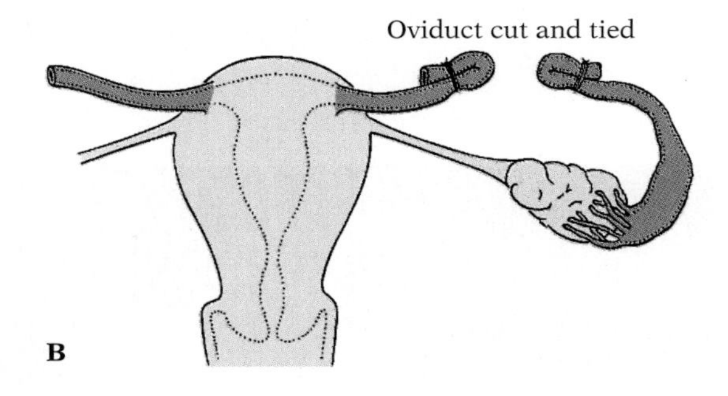

34.9 Surgical birth-control procedures

(A) Vasectomy. A small incision is made in the wall of the scrotum (on each side); the vas deferens (brown) is tied and cut, and the incision is sutured. (B) Tubal ligation. Each of the oviducts is cut and tied, so that eggs cannot descend and no fertilization can occur.

hair, deepening of the voice, development of larger and stronger muscles, and the like. If the testes are removed before puberty by castration, these changes do not occur; if castration occurs after puberty, there is some retrogression of the adult sexual characteristics. Castration after puberty abolishes the sex urge in many animals, but not in humans, where psychological factors are of great importance. Vasectomy, a surgical procedure often used as a birth-control method, involves cutting the vasa deferentia to prevent movement of sperm into the urethra (Fig. 34.9A). Vasectomy causes no regression of sexual characteristics because it does not alter hormone levels.

■ THE REPRODUCTIVE SYSTEM OF THE HUMAN FEMALE

The female gonads, the ***ovaries***, are located in the lower part of the abdominal cavity. Like the testes, the ovaries produce gametes (in this case egg cells) and secrete sex hormones. At the time of birth, a girl's ovaries already contain more than 100,000 primordial egg cells, or ***oocytes***. During the approximately 35 years of her reproductive life a woman ovulates about 13 times per year, producing one mature ovum each time, so fewer than 500 oocytes ever mature and leave the ovaries. The rest eventually degenerate, usually by age 50.

Each oocyte is enclosed within a cellular jacket called a ***follicle***. The oocyte fills most of the space in the small immature follicle (Fig. 34.10). During maturation, however, the follicle grows much

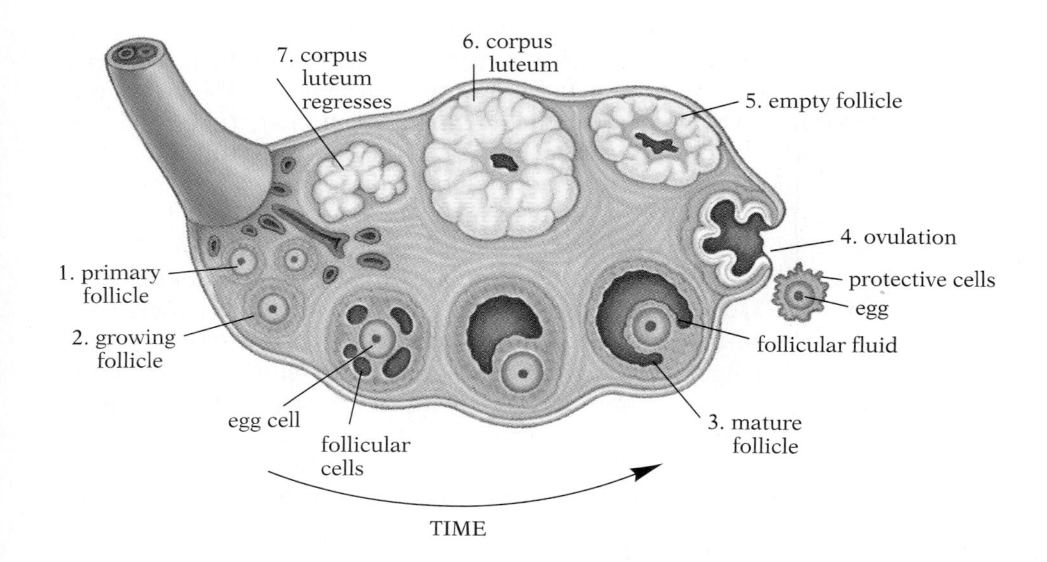

34.10 Schematic diagram of a human ovary showing various stages of egg development

Each oocyte is enclosed within a cluster of cells that forms a follicle (1). The oocyte fills most of the space in the small, immature follicle. As the follicle matures, it grows bigger relative to the oocyte and develops a large fluid-filled cavity (2,3). The oocyte, with its follicular cells, protrudes into the cavity. The outer wall ruptures and both the liquid and the detached oocyte with its surrounding cells are expelled during ovulation (4). The empty follicle is converted into a corpus luteum, which regresses at the end of the cycle (5–7).

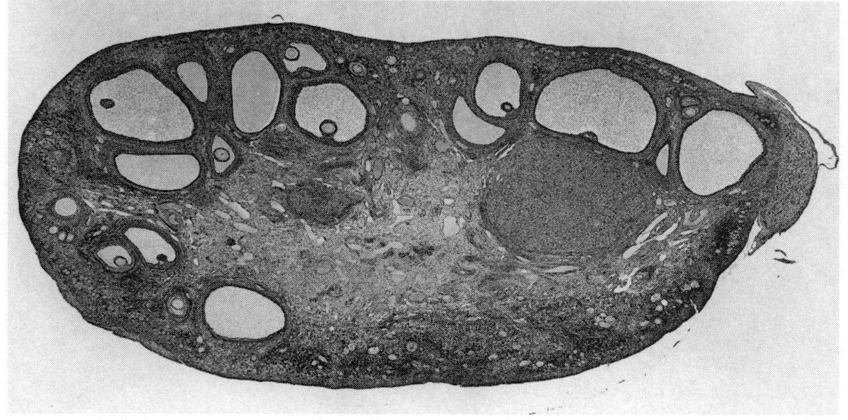

A

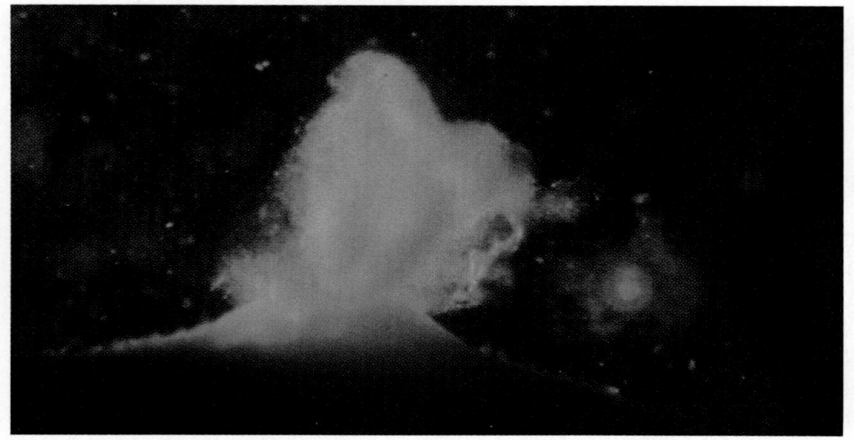

B

C

34.11 Sections of cat ovary

(A) Follicles of different sizes are shown in a section of the entire ovary. The more mature follicles each have a large cavity, with the egg cell embedded in a pedestal of epithelial cells that projects into the cavity. (B) An ovum (lower right) after being expelled from the ovary (center). (C) Enlarged view of a nearly mature follicle with its egg cell.

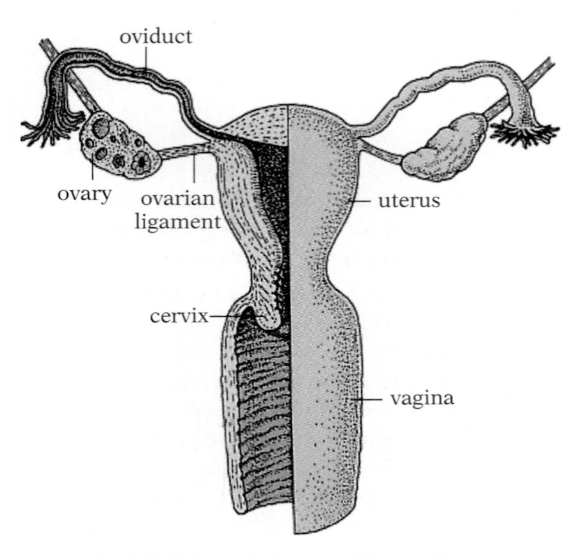

34.12 Reproductive tract of the human female

The wall of one side has been dissected away to reveal the interior structure.

bigger than the oocyte and develops a large fluid-filled cavity (Fig. 34.11); the oocyte, embedded in a mass of follicular epithelial cells, protrudes into the cavity. A ripe follicle bulges from the surface of the ovary; when ***ovulation*** occurs, the outer wall of the ovary ruptures and both the fluid and the detached oocyte are expelled. In the human, only one oocyte is normally released at each ovulation.

Ovulation releases the oocyte from the ovary into the abdominal cavity. From here, it is usually promptly drawn into the large funnel-shaped end of one of the ***oviducts*** (Fallopian tubes), which partly surround the ovaries (Fig. 34.12). Cilia lining the funnel of the oviduct produce currents that help move the oocyte into the oviduct. If sperm are present to meet the oocyte while it is still in the upper third of the oviduct, the penetration of a sperm through the membrane of the oocyte stimulates it to complete its maturation into a true egg cell (ovum), and almost immediately thereafter the nuclei of the sperm and ovum fuse in the process of fertilization. Tubal ligation, a method of permanent sterilization of females, is sometimes used as a birth-control measure. The oviducts are cut and tied so that sperm cannot reach the oocyte and no ovum can move down the oviduct into the uterus (Fig. 34.9B). Like vasectomy in the male, tubal ligation causes no change in hormone production.

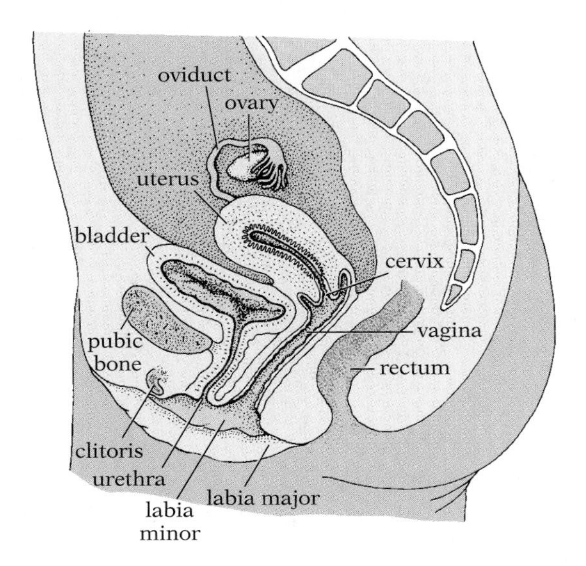

34.13 Reproductive tract of the human female: lateral view

The vestibule is the area (not labeled) into which the urethra and vagina open, and is bounded by the labia minor and labia major.

Each oviduct empties directly into the upper end of the ***uterus*** (womb). This organ, which is about the size of a fist, lies in the lower portion of the abdominal cavity just behind the bladder (Fig. 34.13). It has thick muscular walls and a mucous lining containing many blood vessels. If an egg is fertilized as it moves down the oviduct, it becomes implanted in the wall of the uterus, and there the embryo develops until the time of birth. Another method of birth control involves insertion of a plastic ring or spiral into the uterus (Fig. 34.14A). Such intrauterine devices (IUDs) seem to be very effective in preventing pregnancy, probably by preventing implantation in the uterus. However, these devices sometimes cause irritation and/or bleeding in the uterus; hence some women cannot tolerate them, and their safety for prolonged use is questionable.

At its lower end the uterus connects with a muscular tube, the ***vagina***, which leads to the outside. The vagina acts as the receptacle for the male penis during copulation. The great elasticity of its walls makes possible the passage of the baby during childbirth.

The uterus projects forward nearly at a right angle to the vagina, as shown in Figure 34.13. The ***cervix***, a muscular ring of tissue at the mouth of the uterus, protrudes into the vagina. Devices that block the mouth of the uterus by covering the cervix are widely used in birth control. One such device, the diaphragm, is a shallow rubber cup with a flexible spring around its rim. Inserted into the vagina and positioned so that it covers the entire cervical region (Fig. 34.14B), it is very effective in preventing sperm from entering the uterus, particularly if it is used in conjunction with a spermicidal jelly or cream. Another common contraceptive device, the condom, works by enclosing the penis, thus keeping sperm out of the vagina.

The opening of the vagina in young human females is partly closed by a thin membrane called the ***hymen***. Traditionally the hymen has been regarded as the symbol of virginity, to be destroyed the first time sexual intercourse takes place. Often, however, the membrane is ruptured during childhood by disease, a fall, or as a result of strenuous exercise.

The external female genitalia are collectively termed the ***vulva***. The vulvar region is bounded by two folds of skin, the labia minor and the labia major, which enclose the space known as the

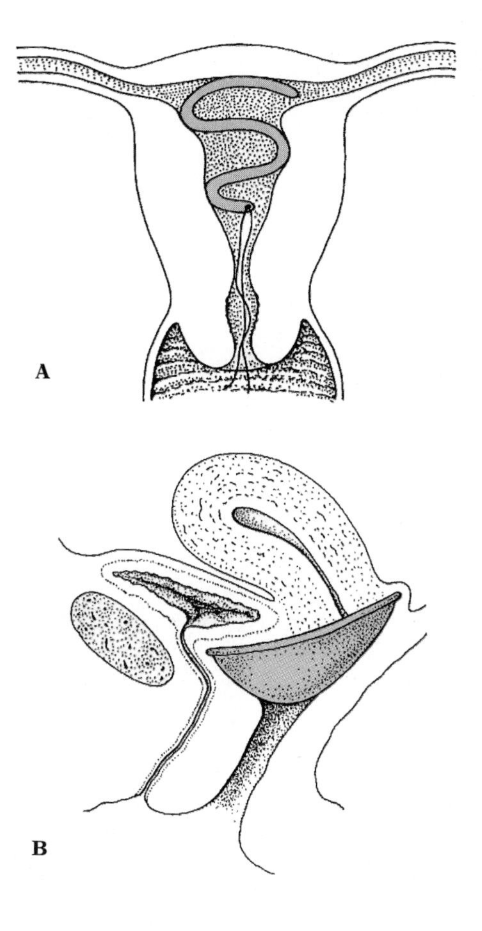

34.14 Two birth-control devices

(A) An IUD in place in the uterus. The strings that run through the cervix permit the woman to make sure the IUD has not been expelled.
(B) Diaphragm in position in the vagina. The device covers the mouth of the cervix. It is very effective in preventing sperm from entering the uterus when used with spermicidal jelly or cream.

vestibule (Fig. 34.13). The vagina opens into the rear portion of the vestibule, and the urethra opens into the midportion of the vestibule. In the adult mammalian female there is no interconnection between the excretory and reproductive systems. During embryonic development the vagina and urethra share a common opening, but as development proceeds, this opening becomes divided, so that the vagina and urethra have separate openings to the exterior.

In the anterior portion of the vestibule, in front of the opening of the urethra, is a small erectile organ, the ***clitoris***, which forms from the same embryonic tissue that gives rise to the penis in the male (see Fig. 14.9, p. 364). Like the penis, it becomes engorged with blood during sexual excitement and is the major site of stimulation during copulation.

■ HOW HORMONES DRIVE THE FEMALE REPRODUCTIVE CYCLE

As in the male, puberty in the female is thought to begin when the hypothalamus loses its sensitivity to the low levels of sex hormone in childhood and starts secreting more GnRH, which stimulates the anterior pituitary to release increased amounts of FSH and LH. These gonadotropic hormones cause maturation of the ovaries, which then begin secreting the female sex hormones, ***estrogen*** and ***progesterone***. The estrogen stimulates maturation of the accessory reproductive structures (increase in the size of the uterus and vagina, for example) and development of the female secondary sexual characteristics: broadening of the pelvis, development of breasts, change in the distribution of body fat, and some alteration of voice quality. The growth of pubic hair in both sexes depends on testosterone. The changing hormonal balance also triggers the onset of the ***menstrual cycle***.

For most species of mammals, rhythmic variation in the secretion of gonadotropic hormones in the females creates an ***estrous cycle***—periodic changes in the condition of the reproductive tract and in sexual readiness. The females of most species will copulate only near the time of ovulation when the uterine lining is thickest. During such periods the female is said to be "in heat," or in estrus. Many mammals have only one or a few estrous periods each year, but some, like rats, mice, and their relatives, may be in estrus as often as every 4 days. If fertilization does not occur, the thickened lining of the uterus is gradually reabsorbed by the female's body; ordinarily no discharge is associated with this process.

The reproductive cycle in humans and some other higher primates differs in several important ways. There is no distinct heat period, the female being to some degree receptive to the male throughout the cycle. And the thickened lining of the uterus is not completely reabsorbed if no fertilization occurs; instead, part of the lining is sloughed off during a period of discharge known as menstruation. The human menstrual cycle averages about 28 days, but there is extensive variation from person to person and from cycle to cycle in the same person.

Let's trace the sequence of events in a 28-day menstrual cycle,

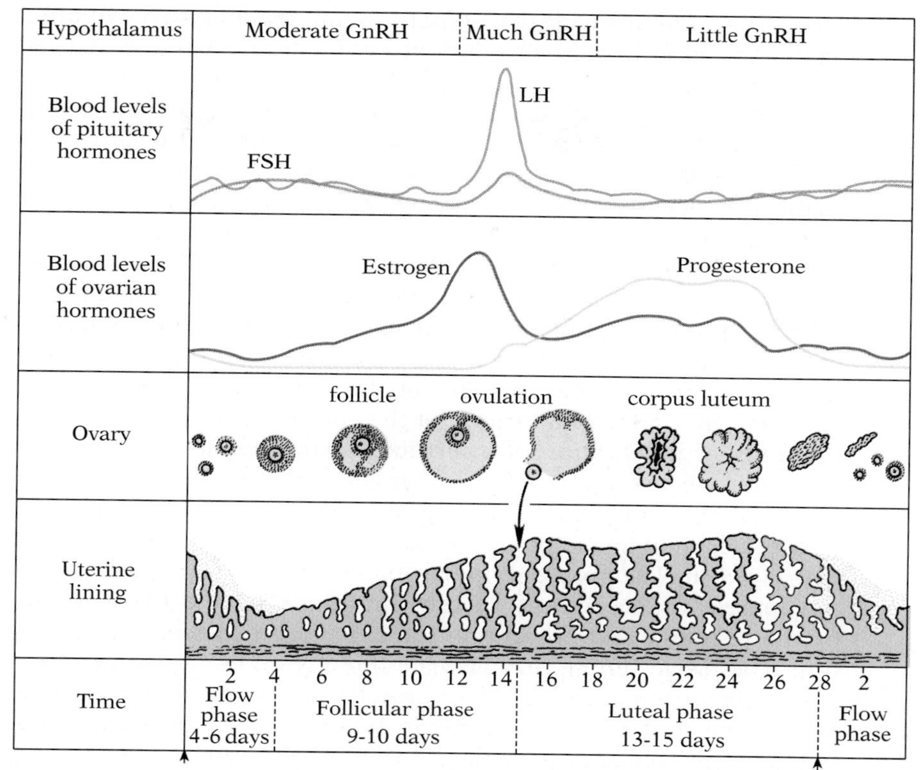

34.15 Sequence of events in the human menstrual cycle
The hormone levels shown are only approximate: they are based on radioimmunoassay determinations, the results of which vary considerably.

taking the first day of menstruation as the first day of the cycle (Fig. 34.15). When the period of discharge ends (day 4 in Fig. 34.15), the uterine lining is thin and there are no ripe follicles in the ovaries. Under the influence of FSH (follicle-stimulating hormone) from the anterior pituitary, several follicles in the ovaries begin growing. Influenced by the synergistic action of FSH and LH, the follicles begin secreting the first of the two female sex hormones, estrogen. One of the follicles soon becomes dominant: it continues to grow and secrete estrogen, while the others stop. The estrogen, in turn, stimulates the lining of the uterus to thicken. This follicular phase (growth phase) of the cycle lasts, on the average, about 9 to 10 days after the previous menstrual flow has stopped.

As the follicle grows, it produces more and more estrogen (Fig. 34.15). The eventually high level of estrogen (or, perhaps, the slowed rate of estrogen production as it reaches its peak) apparently stimulates (probably via GnRH from the hypothalamus) an abrupt surge of LH from the pituitary. This LH surge triggers ovulation by the dominant follicle, which by this time is mature and bulging from the surface of the ovary.[1] How ovulation is achieved is still unknown. Hormonally induced thinning of the follicular

[1] Some investigators think that a slight rise in progesterone just prior to ovulation acts synergistically with LH to trigger ovulation.

wall is one important factor. Whatever the mechanism, ovulation marks the end of the follicular (growth) phase of the menstrual cycle.

Following ovulation, LH induces changes in the follicular cells that convert the old follicle into a yellowish mass of cells rich in blood vessels. This new structure is called the ***corpus luteum*** (Latin for "yellow body"). From its location on the ovary, the corpus luteum continues secreting estrogen, though not as much as was secreted by the follicle just prior to ovulation.[2] The corpus luteum also secretes the second female sex hormone, progesterone.

Progesterone functions in preparing the uterus to receive the embryo. The uterine lining has already thickened substantially under the stimulation of estrogen during the follicular phase; now the progesterone causes maturation of the complex system of glands in the lining. Implantation of a fertilized ovum cannot occur unless progesterone has induced these changes in the uterine lining.

In addition, high levels of progesterone inhibit initiation of the next cycle, though the precise mechanism of this inhibition is not entirely clear. The progesterone probably acts on the hypothalamus to suppress GnRH release, thereby limiting FSH and LH secretion by the pituitary and preventing the LH surge necessary for ovulation. The progesterone may also act directly on the immature follicles in the ovary, inhibiting their growth and estrogen secretion. As long as progesterone is present in quantity in the system, then, there is little follicular growth.

This inhibiting action of progesterone is an important element in regulating the duration of the menstrual cycle. It is also the basis for the action of birth-control pills. These pills contain synthetic compounds similar to progesterone and estrogen. Taken regularly, they inhibit secretion of FSH and LH (probably by suppressing GnRH release by the hypothalamus) and thus prevent follicular growth and ovulation—and consequently conception. Like most contraceptive techniques, birth-control pills sometimes have undesirable side effects. These include migraine headaches, vascular disturbances, depression, reduced sexual drive, changes in skin pigmentation, and possibly increased risk of stroke and certain kinds of cancer (though there is evidence that they can also reduce the risk of some other forms of cancer).

If no fertilization occurs during a normal cycle, the corpus luteum begins to atrophy about 11 days after ovulation, and its secretion of progesterone falls. When this happens, the thickened lining of the uterus can no longer be maintained, and resorption of part of the lining begins. Human females reabsorb only part of the extra tissue laid down during the follicular and luteal phases; the rest is sloughed off during the 4–5 days of menstruation.

The fall of progesterone levels, resulting from atrophy of the corpus luteum, frees the hypothalamus from inhibition and allows it to stimulate the pituitary to increase secretion of FSH. The immature follicles are also freed from inhibition and, under the influ-

[2] In rats, maintenance of the corpus luteum requires prolactin in addition to LH, but current evidence strongly suggests that in most mammals, including humans, LH alone suffices for both the formation and the secretory activity of the corpus luteum.

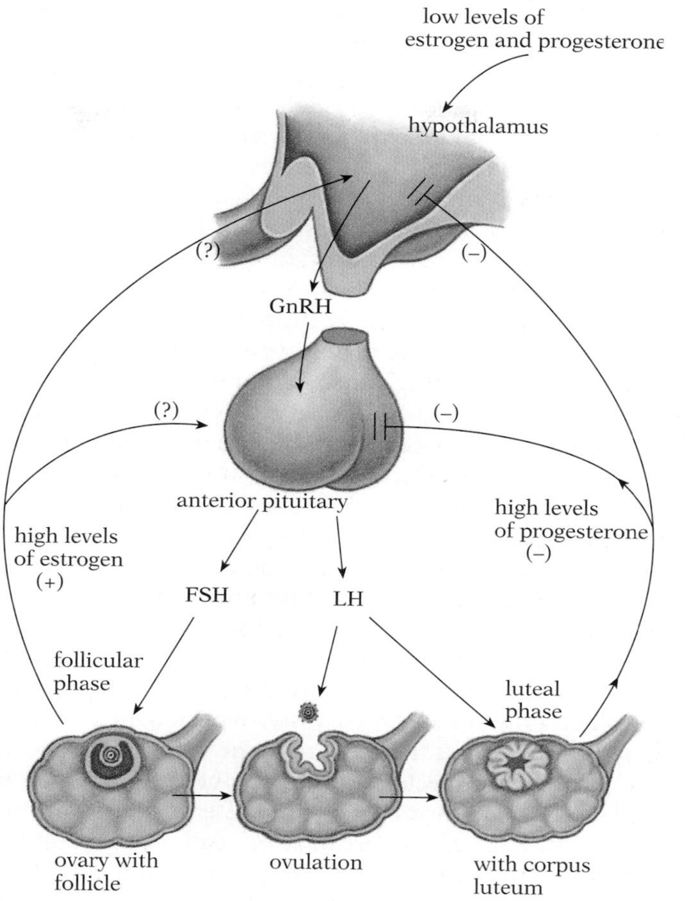

34.16 Hormonal control of the menstrual cycle

When the levels of sex hormones are low during the flow phase, the hypothalamus is stimulated to secrete GnRH, which stimulates the anterior pituitary to begin secreting FSH (and some LH). Under the influence of FSH the follicle begins to grow and produce estrogen. The rise in estrogen toward the end of the follicular phase causes a surge of LH secretion, thereby triggering ovulation. Under the influence of LH, the empty follicle is converted into a corpus luteum, which secretes high levels of progesterone and estrogen. These have a negative feedback effect on the hypothalamus and anterior pituitary, and FSH and LH secretion is inhibited, the corpus luteum atrophies, the levels of the sex hormones drop, and menstrual flow occurs. In humans it is not yet certain whether the feedback effects of estrogen and progesterone are directed at the hypothalamus or anterior pituitary. It is also unclear what causes the atrophy of the corpus luteum.

ence of the FSH, begin growing, and a new cycle begins (Fig. 34.15 and Fig. 34.16).

The critical event in resetting the system is probably the fall in progesterone levels as a result of atrophy of the corpus luteum. But the cause of this atrophy is unknown. In cows certain prostaglandins, produced by the nonpregnant uterus and carried to the ovary by a special system of portal blood vessels, act to cut off progesterone secretion. There is no such portal system in humans, nor is there solid evidence that the human uterus sends any inhibiting substance to the ovaries. Possibly the ovary itself produces prostaglandins that turn off progesterone secretion.

During the flow phase, particularly in the first few days, secretion of progesterone and estrogen by the old corpus luteum is at a low level, and the follicles of the new cycle have not yet begun producing significant amounts of estrogen. Their withdrawal at the end of the luteal phase of each menstrual cycle is often accompanied by physiological and psychological disturbance including irritability, depression, and sometimes nausea; abdominal cramps, caused by contractions of the uterus, are also common.

Emotional stress sometimes also accompanies the ***menopause***, a period lasting a year or two at the end of a woman's reproductive life. The menopause usually comes sometime between the ages of

40 and 55. It may result primarily from declining sensitivity of the ovaries to the stimulatory activity of gonadotropins. The ovaries atrophy, the remaining follicles disappear, the secretion of estrogen and progesterone falls to low levels, and menstruation stops. The changing hormonal balance during menopause may cause physiological and psychological disturbances until a new balance has been established. Sometimes physiological difficulties persist indefinitely; a fairly common condition is postmenopausal osteoporosis, a thinning and weakening of the bones thought to result from estrogen deficiency.

We have seen that ovulation in humans occurs roughly midway in the menstrual cycle. This ovulation is spontaneous; it does not depend on a copulatory stimulus. In some mammals, such as rabbits and cats, however, ovulation is reflex-controlled; the nervous stimulus of copulation is necessary for the hypothalamus to trigger release of LH by the pituitary, which in turn leads to ovulation. In such reflex ovulators it is possible to predict with great precision just when ovulation will occur; in the rabbit, for example, it occurs about 10½ hours after copulation. Such precision is not possible with spontaneous ovulators like humans. Yet predictions of the time of ovulation are important in the practice of the rhythm method of birth control.

The rhythm method is based on the premise that fertilization can take place only during a very short period in each menstrual cycle; copulation without risk of pregnancy should be possible during other parts of the cycle. Human egg cells begin to deteriorate about 12 hours (maximum 24 hours) after ovulation and can no longer be fertilized. Thus fertilization can occur only if fertile sperm are in the upper third of the oviduct during the 12–24 hours immediately following ovulation; ideally, conception would not be possible during the other 27 days of a 28-day cycle.

Immediately, the question of the fertile life of sperm in the female reproductive tract becomes pertinent. Conditions in the vagina are very inhospitable to sperm, and vast numbers are killed before they have a chance to pass the cervix. Many others die or become infertile in the uterus and oviducts, and many more go up the wrong oviduct or never find an oviduct at all. Despite the length and hazards of the journey, however, a significant number of sperm cells arrive in the upper part of the oviduct. If fact, they may be helped by contractions of the female genital tract, which seem to be accelerated by prostaglandins present in the seminal fluid and by oxytocin released by the posterior pituitary. Human sperm cells probably remain fertile only about 48 hours or less after their release, though the time is much longer for sperm in many other animals. For example, some bats mate in the fall, but do not ovulate until the spring, after hibernation; the sperm must therefore remain fertile for several months. In some invertebrates, such as bees and ants, fertile sperm may actually be stored in the body of the female for years. Thus one mating period provides a queen honey bee with enough sperm to last throughout her reproductive life, during which she produces tens or hundreds of thousands of offspring.

Since the fertile life of the human egg cell lasts at most 1 day and that of the human sperm at most 2 days, there is a period of about 3 days during which copulation can result in conception: the day when the egg is fertile and the two preceding days. But which 3 days? The inability to answer this question precisely makes the rhythm method of birth control unreliable. The 3 days come about midway through the menstrual cycle, but since the length of a given cycle cannot be predicted accurately, its midpoint can't be determined. The cycles of many women are very irregular, varying by as much as 8–15 days or more. Even women whose cycles are normally regular will have some that vary widely during the course of a year. Sickness or emotional upset frequently delays ovulation and prolongs the cycle by altering the hypothalamic control of gonadotropin secretion. There is also some evidence that copulation can advance ovulation, making fertilization more likely.

■ HOW HORMONES ORCHESTRATE PREGNANCY

So far we've assumed that conception does not occur. If the egg cell is fertilized sometime during the 12 hours after ovulation, a very different sequence of events unfolds. Only one of the millions of sperm cells released into the vagina actually penetrates the egg cell and fertilizes it. As soon as that one cell has fertilized the egg, the outer membrane of the egg becomes impenetrable. The fertilized egg, or zygote, moves down the oviduct, carried by fluid whose movement is caused by the beating of cilia and contractions of muscles in the oviduct. The rate of movement is partly controlled by estrogen. During the days of transit, cell division begins and an embryo is formed. (This process is outlined in Chapter 13.)

The human embryo becomes implanted in the wall of the uterus 8–10 days after fertilization. During the interval between fertilization and implantation, the embryo is nourished by its limited supply of yolk and by materials secreted by the glands of the female genital tract. The delay before implantation varies from species to species; it is about 20–22 days in sheep, 35 days in cows, and as long as 56 days in horses. In a few species there is a much longer delay, during which development of the embryo proceeds very slowly or even ceases; among mammals with delayed implantation are brown bears (about 5 months), pine martens (6 months), American badgers (2 months), and armadillos (14 weeks).

After implantation of the embryo in the uterine lining, the embryonic membranes form the ***umbilical cord.*** Blood vessels contributed by the allantois run through the cord to a large structure, the ***placenta,*** formed from the embryonic membranes (primarily the chorion) and from the adjacent uterine tissue (Fig. 34.17). Within the placenta the blood vessels of the embryo and those of the mother lie very close together, but they are not joined and there is no mixing of maternal and fetal blood. Exchange of materials takes place in the placenta by diffusion between the blood of the

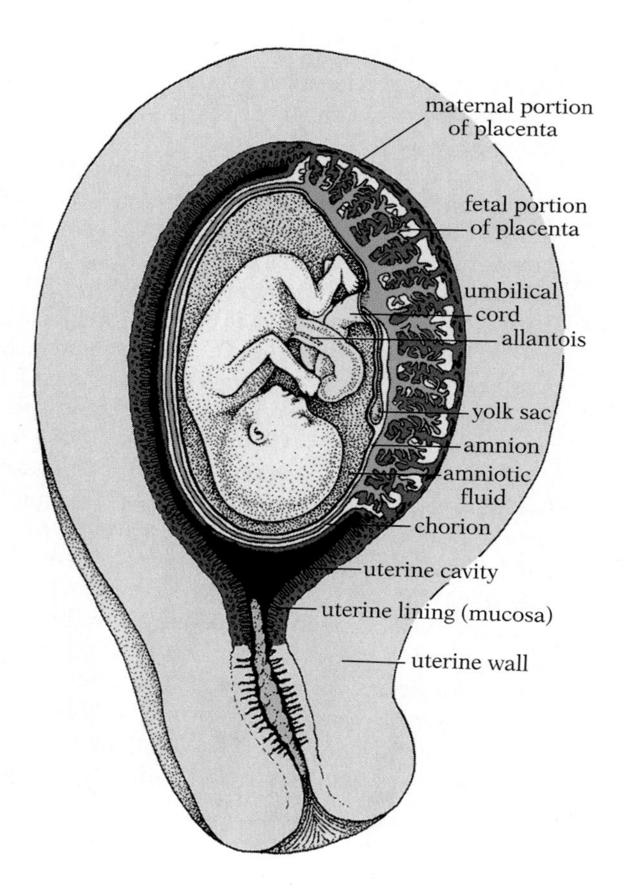

34.17 Uterus of a pregnant woman

The placenta consists of both maternal and fetal components, the latter derived from the chorion. Maternal and fetal capillaries lie side by side, allowing the diffusion of nutrients into the fetal circulation. Nutrients absorbed by the fetal capillaries are carried by the umbilical cord to the embryo, which lies in the amniotic sac, bathed by amniotic fluid.

mother and that of the embryo; nutritive substances and oxygen move from the mother to the embryo, and urinary wastes and carbon dioxide move from the embryo to the mother (see Fig. 30.7B, p. 846).

We saw earlier that progesterone is essential to maintain the uterine lining during implantation and pregnancy. When fertilization does not occur, the corpus luteum atrophies and cuts off the supply of progesterone, leading to menstruation. What prevents the corpus luteum from atrophying after fertilization? Apparently

EXPLORING FURTHER

DEVELOPMENT OF THE HUMAN EMBRYO

In humans the egg is fertilized soon after ovulation, while it is still in the upper portion of the oviduct. As it continues to move down the oviduct, the fertilized egg begins cleavage. By the time it becomes implanted in the uterine wall, 6–10 days after fertilization, it has reached the blastula stage. During the implantation process the amnion, which will enclose the embryo in a fluid-filled chamber throughout the rest of its development, begins to form.

By the 23rd day, two other embryonic membranes—the chorion and the allantois—have collaborated with maternal tissues of the uterus to give rise to a functional placenta; the chorion contributes the fingerlike villi of the fetal portion of the placenta, and the allantois contributes the blood vessels (see Fig. 34.17). By this time, the neural groove is complete; the mesoderm, too, is well developed, and individual segments (somites) are visible. The tubular embryonic heart has begun to pulsate weakly.

By the end of the first month of development the embryo, which now has arm and leg buds, is roughly 5 mm (1/5 inch) long. The hands and feet are still mittenlike, with no separations between the individual fingers and toes.

During the second month the embryo grows to a length of approximately 30 mm (about an inch) and weighs about 1 gram (0.03 ounce). After 8 weeks of development, the embryo, now called a ***fetus,*** is recognizable as human. It has a flat face with widely separated eyes, and the digits (fingers and toes) are well separated (Fig. 13.11B, p. 346). The brain shows structural organization, and sense organs are well developed. The muscles are sufficiently differentiated for some movement to be possible. Hardening of the skeleton has begun. The liver, which is the chief producer of blood cells during early pregnancy, is proportionately much larger than it will be later.

During the first 2 months of development the embryo is especially sensitive to a variety of factors that can cause serious malformations. For example, if the mother contracts rubella (German measles), this normally mild viral disease can produce a malformed heart, cataracts of the eyes, or deafness. The disastrous effects on the fetus of the tranquilizer thalidomide when taken by the mother early in pregnancy are well known. Fetal abnormalities induced by maternal consumption of alcohol (fetal-alcohol syndrome) and cocaine are all too common. These chemically induced problems can include severe retardation and personality disorders. Often the damage is done before the woman is even aware of her pregnancy.

By the end of 13 weeks (Fig. 13.11C, p. 346) the fetus is about 100 mm long, and the body proportions approach those expected at birth. The sex of the fetus is evident. The brain has begun to take its final shape, and the sense organs are almost complete. The heart too has taken on nearly its final form. Spontaneous movements are frequent.

At this point in development the basic pattern of all the physiological systems has been established. The remainder of fetal development is largely a combination of further refinement and elaboration and of growth in overall size (over 90% of fetal weight gain takes place in the last 4 months). In time, the bone marrow takes over from the liver the primary responsibility for producing blood cells; the eyes become light-sensitive, and later the ears respond to sound. Many new brain cells are produced; and numerous new neural circuits are established and become functional up to and past the time of birth.

By the end of 24 weeks the fetus has a fair chance of survival outside the uterus if it is given respiratory assistance and kept in an incubator. However, it is still so small—about 0.7 kg (1 1/2 pounds)—and so poorly developed that it is subject to medical problems for months (at least). If, on the other hand, birth occurs at full term (on the average, about 40 weeks or 280 days after the beginning of the mother's last regular menstrual period), the chances of survival are high.

Several important developmental changes in the circulatory system occur at the time of birth: the placental circulation is cut off; the ductus arteriosus (the shunt between the pulmonary artery and the aorta) is closed; the lungs are inflated for the first time; blood is forced into the pulmonary system; and production of fetal hemoglobin soon gives way to production of the adult type.

the chorionic portion of the placenta begins secreting a gonadotropic hormone (human chorionic gonadotropin, or HCG). This hormone preserves the corpus luteum, which continues to secrete progesterone and thus sustains the pregnancy. So much HCG is produced in a pregnant woman that much of it is excreted in the urine. Many commonly used tests for pregnancy, among them some available over the counter in drugstores, work by detecting HCG in the urine, and can yield reliable results quickly.

The necessity of early progesterone secretion by the corpus luteum is the basis of the abortion-inducing effects of a chemical known as RU 486, which works by blocking progesterone receptors on cells in both the uterus and the hypothalamus. The blocked receptors cannot bind to the DNA of those cells, with the result that certain genes critical for orchestrating pregnancy fail to be expressed. Taken in conjunction with prostaglandins within 9 weeks of fertilization, RU 486 almost always terminates pregnancy. Why this progesterone mimic fails to end pregnancy after the placenta takes over is not yet clear.

■ HORMONAL CONTROL DURING AND AFTER BIRTH

The complex interactions of hormones that control the birth process (parturition) are not well understood. Apparently a rise in prostaglandin production by the placenta plays a central role in initiating parturition. The major stimulus for this rise in late pregnancy appears to be an increase in estrogen secretion (also by the placenta), which is itself a result, at least in part, of increased secretion of estrogen precursors by the fetal adrenal cortex. In some mammals, but not in primates, the placenta in late pregnancy begins converting progesterone, which severely inhibits muscular contraction, into estrogen; the resulting fall in plasma progesterone levels may make the uterine muscles more susceptible to stimulation. Since oxytocin, released by the posterior pituitary (and perhaps also secreted by the placenta), is known to have a powerful stimulatory effect on uterine contractility, it seems likely that this hormone, too, contributes to the induction of labor.

Another hormone important in parturition is ***relaxin***, which is secreted during pregnancy by the ovaries and the placenta. This hormone has the effect of loosening the connections between the bones of the pelvis and thereby enlarging the birth canal and facilitating parturition. Relaxin also aids in softening and dilating the cervix. The activity of relaxin is enhanced by estrogens.

The hormonal control of milk secretion is also incompletely understood. Growth and development of the mammary glands in humans seem to be controlled by a complex interaction between estrogen, progesterone, thyroxin, insulin, growth hormone, prolactin, glucocorticoids, and human placental lactogen (a potent prolactin-like hormone produced by the placenta). Initiation and maintenance of lactation by mature mammary glands following parturition seem to be controlled primarily by prolactin and glucocorticoids. These hormones apparently become effective in inducing lactation when the high levels of sex hormones, which inhibit lactation, disappear at the time of parturition.

HOW WE KNOW:

HOW RU 486 STOPS PREGNANCY

The abortion-inducing drug RU 486 is chemically similar to progesterone, the hormone essential for maintaining pregnancy (Fig. A). It was created by researchers as one of many progesterone mimics whose action might throw some light on how the natural hormone works. Like progesterone, RU 486 is a lipid-soluble steroid: it is able to move through both the cell membrane and the nuclear envelope. More importantly, once in the nucleus this synthetic molecule binds to the progesterone receptor. From this point on, however, the action of RU 486 is different in illuminating and useful ways.

The first important difference is that RU 486 binds the hormone receptor better than progesterone itself. Thus when both the synthetic and natural hormone are present in the nucleus, most or all of the receptor molecules will be bound to RU 486. The unbound receptor is actually a complex, consisting of the receptor itself and a regulatory protein, *hsp,* which shields the site on the receptor that can recognize a specific DNA sequence. When progesterone binds to the receptor, the regulatory protein is released and the bound receptor attaches itself to the chromosome (Fig. B); this attachment promotes the transcription of genes essential for maintaining pregnancy. The critical difference in the action of RU 486 is that when it binds, either the *hsp* is not shed, or the conformation of the receptor is altered so that it cannot bind to the DNA. In either event, there is no transcription of the genes that encode proteins essential to pregnancy, and which are normally activated by the bound receptor.

Since even a short interruption in the supply of proges-

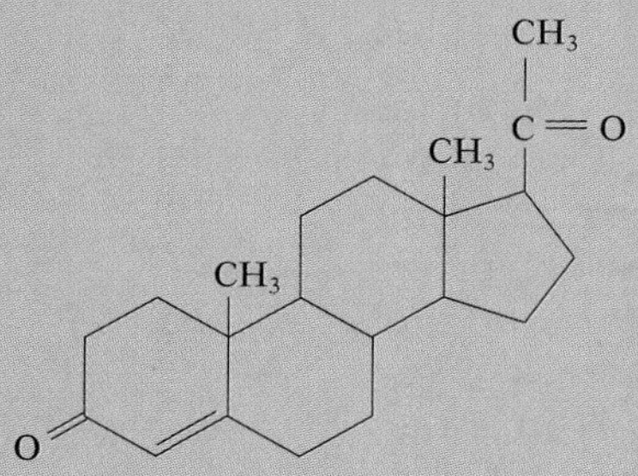

CH_3 C=O CH_3 CH_3 O

Progesterone

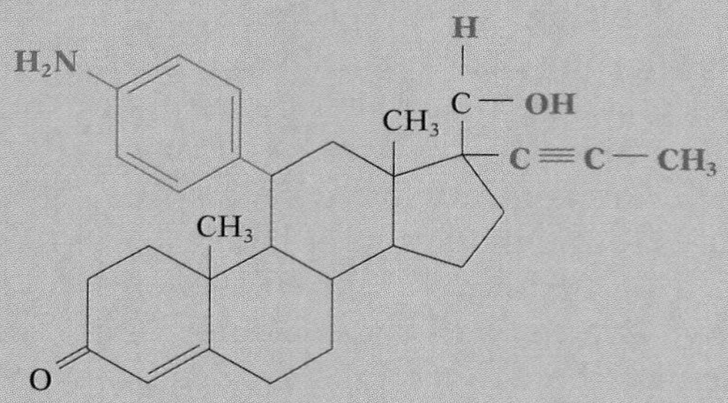

H C—OH H_2N CH_3 C≡C—CH_3 CH_3 O

RU486

A A comparison of progesterone and RU 486

The actual release of milk from the mammary glands involves both neural and hormonal mechanisms. Suckling causes nervous stimulation of that part of the hypothalamus that instigates release of oxytocin stored in the posterior pituitary. Even the sight of the young animal or, in cows, hearing rattling milk pails, can have the same effect. The oxytocin, in turn, induces constriction of the many tiny chambers in which the milk is stored in the mammary glands. The constriction forces the milk into ducts that lead to the nipple. Adrenalin inhibits this milk-ejection process.

terone damages the uterine lining, in about 96% of cases RU 486 causes a sloughing of the lining, and thus an end of pregnancy. This nonsurgical approach to inducing abortions has been used in more than 100,000 cases in Europe, and appears to be safe and effective. The potential utility of RU 486 in underdeveloped countries, where surgical procedures (even when available) are often fraught with a variety of risks, seems particularly great.

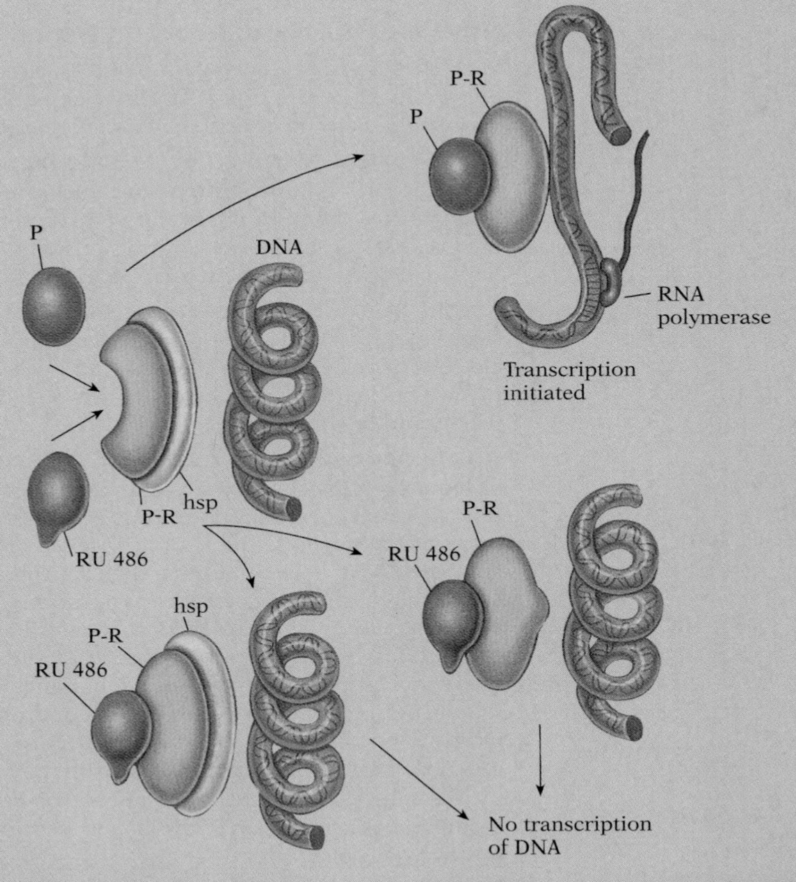

B Both progesterone (P) and RU 486 (RU) can bind to the progesterone-receptor proteins (P-R) in the nucleus of the cell. Attached to P-R is another protein (hsp). When progesterone binds to P-R (upper right), the hsp dissociates and the P-R complex binds to a specific site on the DNA, thereby facilitating transcription of DNA. When RU 486 binds, however, the RU 486-receptor complex either changes shape such that it cannot interact properly with the DNA to facilitate transcription (lower right) or, it may bind even more strongly to the hsp, effectively preventing an interaction with the DNA (lower left). In both cases the RU 486 occupies the binding site of the receptor without triggering a hormonal response.

CHAPTER SUMMARY

THE EVOLUTIONARY FACTS OF LIFE

THE TWO STRATEGIES FOR FERTILIZATION Most fertilization is either external (for example, gametes shed into water) or internal (with the gametes transferred directly into the female reproductive tract). Most fish and amphibians use external fertilization; reptiles, birds, and mammals use internal fertilization. (p. 969)

HOW THE EMBRYO IS PACKAGED The eggs of land vertebrates are surrounded by a series of protective membranes, which are involved in gas exchange, waste disposal, and water conservation. In reptiles and birds these membranes are enclosed by a shell. Nearly all mammalian embryos lack shells and are implanted, membranes and all, in the reproductive tract for development. (p. 971)

HOW THE HUMAN REPRODUCTIVE SYSTEM WORKS

THE REPRODUCTIVE SYSTEM OF THE HUMAN MALE Male gametes (sperm) are produced in the testes. During ejaculation, the sperm are mixed with various fluids to form semen, which is released through the penis. (p. 971)

HOW HORMONES CONTROL MALE SEXUAL DEVELOPMENT Hormones from the brain cause testosterone to be released into the blood by the testes; testosterone prompts the development of male structures and behavior, and causes the maturation of many sex-specific characters at puberty, when the level of testosterone rises. (p. 974)

THE REPRODUCTIVE SYSTEM OF THE HUMAN FEMALE Female gametes (eggs) are produced in the ovaries. A maturing oocyte, surrounded by follicle cells, is released from the ovary and passes through the oviduct. If fertilization occurs in the oviduct, the oocyte becomes implanted in the uterus and begins development. (p. 975)

HOW HORMONES DRIVE THE FEMALE REPRODUCTIVE CYCLE Maturation of female sex-specific traits occurs at puberty, when the levels of estrogen and progesterone rise. The estrous cycle of oocyte maturation, movement through the oviduct, and (in the absence of fertilization) menstruation is generated by a hormonal cycle that depends largely on hormones synthesized by the follicular cells. Early in oocyte development the follicle produces increasing amounts of estrogen until a burst of LH is triggered from the pituitary. This LH surge in turn causes the mature egg to leave the ovary (ovulation) and signals the remaining follicle (now called the corpus luteum) to reduce estrogen synthesis and produce increasing amounts of progesterone. If fertilization has not occurred, the corpus luteum degenerates, leading to menstruation. The drop in hormone production by the corpus luteum leads to increased release of FSH by the pituitary, and thus stimulates the next round of follicular development. (p. 978)

HOW HORMONES ORCHESTRATE PREGNANCY Implantation of the embryo after fertilization induces the formation of the placenta, which supplies nutrients to the developing embryo and carries away wastes. Hormones from the placenta keep the corpus luteum from atrophying. (p. 983)

HORMONAL CONTROL DURING AND AFTER BIRTH Hormones control nearly every aspect of labor, birth, and lactation; the details are poorly understood. (p. 985)

STUDY QUESTIONS

1 Since internal fertilization presumably evolved as a way of increasing the efficiency of sexual recombination, it is hard to understand why there are 400 million sperm in the average human ejaculate. Moreover, the egg to be fertilized is positioned at the end of an extremely (and, seemingly, unnecessarily) long, convoluted, and chemically hostile pathway, hardly the result of selection to increase fertilization efficiency. Given what you know about human reproductive physiology, under what circumstances would the male and female systems prove useful? (Be sure to consider a variety of social arrangements and the selfish perspective of each sex.)

2 Humans may be the only mammals without a period of estrus, or "heat." For what reasons might such an exceptional system have evolved in our species?

3 What might be the evolutionary logic of menopause? (Keep in mind there is no corresponding dramatic reduction in reproductive potential in males.)

4 What might be some of the reasons that the fetal blood supply is kept separate from that of the mother?

5 In most mammals (and all nonhuman primates) the female's breasts are not visible until she is well along in pregnancy. Even human females with little significant breast tissue before pregnancy nevertheless have mammary glands capable of providing as much milk as females with conspicuous prepregnancy breasts. What might be some of the plausible evolutionary scenarios for the evolution of these unusual structures?

SUGGESTED READING

RAHN, H., A. AR, AND C. V. PAGANELLI, 1979. How bird eggs breathe, *Scientific American* 240 (2). *On the development of bird embryos, with special attention to circulation, gas exchange, and waste disposal.*

ULMANN, A., G. TEUTSCH, AND D. PHILIBERT, 1990. RU 486, *Scientific American* 262 (6). *On the biochemistry of an abortion-inducing drug.*

CHAPTER 35

HOW NERVOUS SYSTEMS ARE ORGANIZED

THE UNUSUAL ANATOMY OF NEURONS

HOW SIMPLE NEURONAL PATHWAYS WORK

THE SIMPLEST NERVOUS SYSTEMS: NERVE NETS

WHY BRAINS EVOLVED IN THE HEAD

HOW THE INSECT NERVOUS SYSTEM IS ORGANIZED

WHAT'S SPECIAL ABOUT VERTEBRATES?

HOW NEURONS WORK

HOW AXONS CONDUCT

HOW NEURONS COMMUNICATE

COMMUNICATING WITH MUSCLES

HOW VERTEBRATE NEURAL PATHWAYS ARE ORGANIZED

RAPID AND AUTOMATIC RESPONSES: THE REFLEX ARC

UNCONSCIOUS BEHAVIOR: THE AUTONOMIC SYSTEM

COMPLEX BEHAVIOR: MOTOR PROGRAMS

NERVOUS CONTROL

Hormonal control works well for most tissues, but because it depends on the circulatory system, there are limits to the speed with which its messages can be transmitted. For virtually all animals, for many bacteria and protists, and even for some plants, hormonal communication is too slow for coordinating rapid responses. Electrical communication fills the need for speed. In this chapter we will look at how neurons work, and how they are wired together to create behavior.

HOW NERVOUS SYSTEMS ARE ORGANIZED

Almost every response to a stimulus, whether electrical or chemical, involves four stages: (1) detection of the stimulus, (2) conduction of a signal, (3) "processing" of the signal, and (4) response. These stages are evident in all organisms, from relatively simple procaryotes to complex multicellular organisms like ourselves.

■ THE UNUSUAL ANATOMY OF NEURONS

The specialized nerve cells of multicellular organisms evolved from early unicellular eucaryotes, which, like modern bacteria and motile protists, incorporated all four stages of information flow within a single cell. In contrast, in multicellular animals the anatomy of specialized nerve cells, or ***neurons***, often reflects their particular roles in detection, conduction, processing, and response control. Regardless of appearance, however, all neurons have the same functional organization, which enables them to collect information—from the environment directly (as sensory cells) or from other neurons—and to transmit it to target cells such as other neurons, muscle cells, and secretory cells. The typical nerve cell consists of a cell body, which contains the nucleus, and one or more long processes, or nerve fibers. In general, information is received along fibers called ***dendrites*** and transmitted to other cells along other fibers called ***axons*** (Fig. 35.1). Specialized terminals called ***synapses*** on the ends of axons convey the signals to the dendrites of target cells. In vertebrates the dendrites usually feed into the cell body (Fig. 35.1B–D), whereas in invertebrates the axon is most

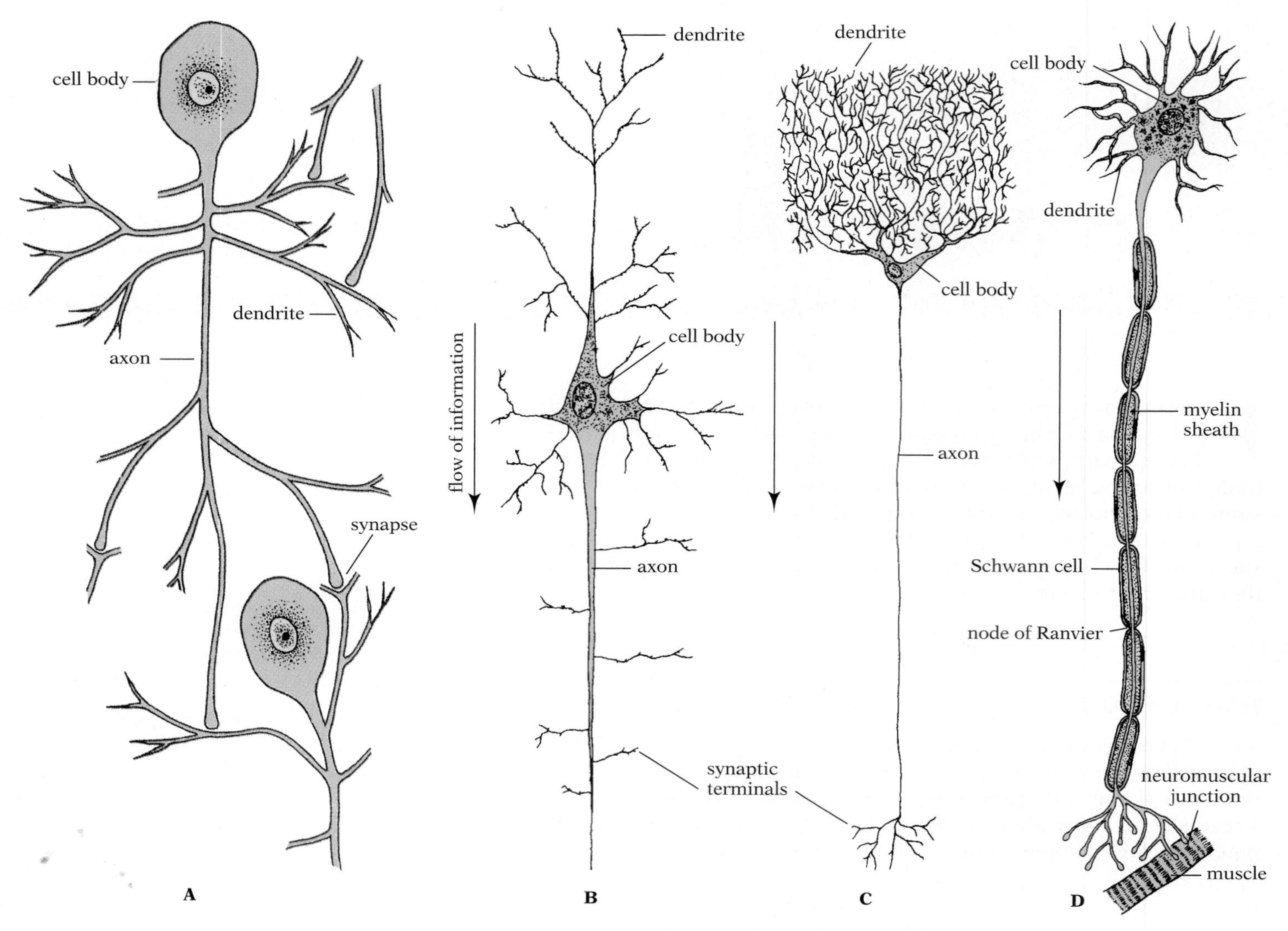

35.1 A variety of neuron morphologies

Information is normally collected by dendrites, conducted along an axon, and transmitted to other cells by terminals, at information links called synapses. In invertebrates, the cell body of the neuron usually lies outside the information pathway (A). In vertebrates (B–D), the cell body usually lies between the dendrites and the axon, and longer axons are frequently insulated by a series of myelin sheaths.

often connected directly to the dendrites, so that the cell body is out of the path of information flow (Fig. 35.1A). As we will see, it is on the fingerlike dendrites and (in vertebrates) on the cell body that much of neural processing takes place.

Dendrites are usually short, and neurons generally have many of them, receiving input from perhaps several thousand other cells. Most dendrites are profusely branched and have a spiny appearance. Their cytoplasm contains extensive arrays of flattened cisternae of the endoplasmic reticulum (ER), with a vast number of associated ribosomes, called Nissl substance (Fig. 35.2).

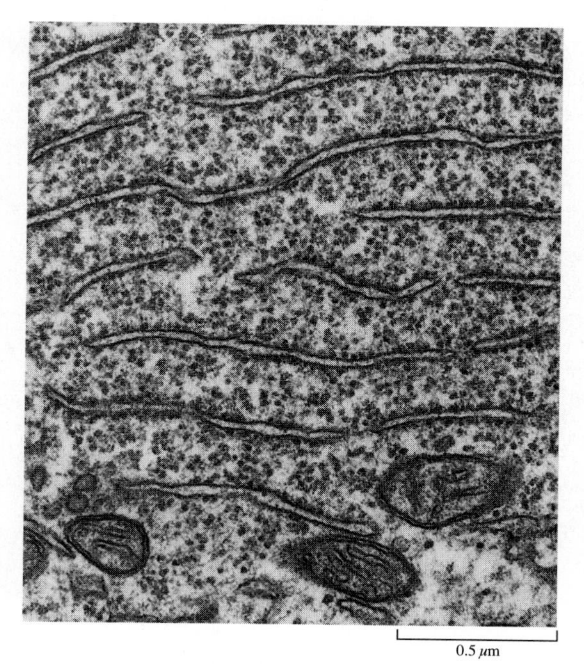

35.2 Nissl substance

Nissl substance, seen here in an electron micrograph, clearly consists of flattened cisternae of the ER plus numerous ribosomes. Many of the ribosomes are not directly attached to the cisternal membranes.

In contrast to the multiplicity of dendrites, there is usually only one axon per neuron, and it is usually longer and thicker than the dendrites; it may branch extensively, but it does not have a spiny appearance or contain ER. These histological differences reflect the basic functional distinction between dendrites and axons, mentioned earlier. As we will see, synapses are of great importance to the processing of information in the nervous system.

Within the ***central nervous system (CNS)*** of vertebrates, neurons are intimately associated with vast numbers of satellite cells called ***neuroglia*** (or simply glia). In the brain, glial cells outnumber neurons 10 to 1, and occupy about half the cranial volume. Some glia provide the neurons with nutrients, and may help maintain a homogeneous environment by absorbing substances secreted by the neurons. Much of this absorbed material is then cycled back to the neurons for reuse. In at least some areas of the CNS, glia provide a framework during development along which neurons migrate and along which axons grow to reach their targets. Certain glial cells, called astrocytes, appear to be able to communicate chemically with one another, slowly passing yet-to-be-understood messages over considerable distances.

One class of glia found in vertebrates is particularly well understood. The membranes of these specialized glial cells wrap around and around the axons of many neurons in the CNS to form a heavily lipid ***myelin sheath*** (Fig. 35.3). Many vertebrate axons peripheral to the CNS are enveloped by cells derived from the neural

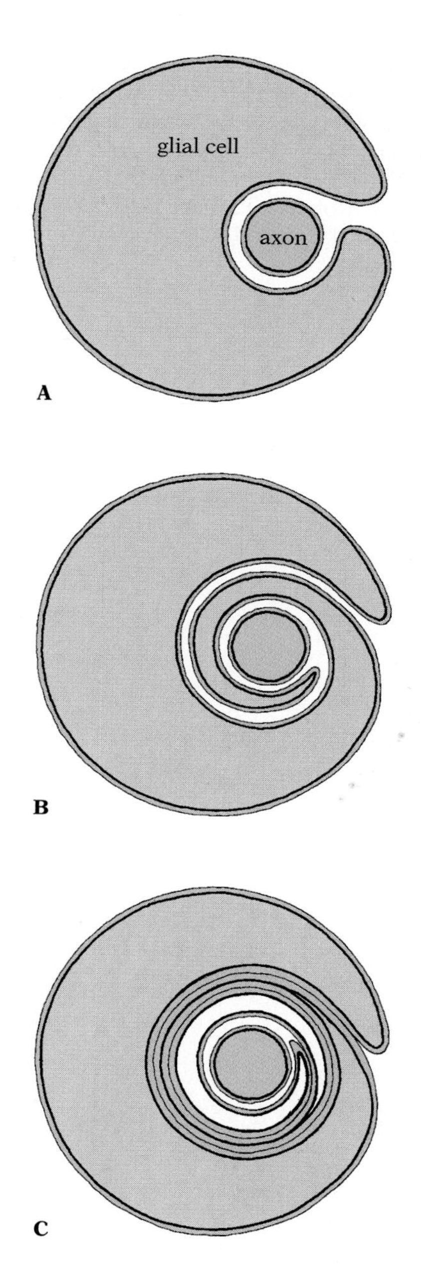

35.3 Development of the myelin sheath

(A) Initially the unmyelinated axon lies in a pocket of the glial cell. (B) The glial-cell membrane then begins to coil around the axon. (C) The membrane winds tightly around the axon, forming a myelin sheath.

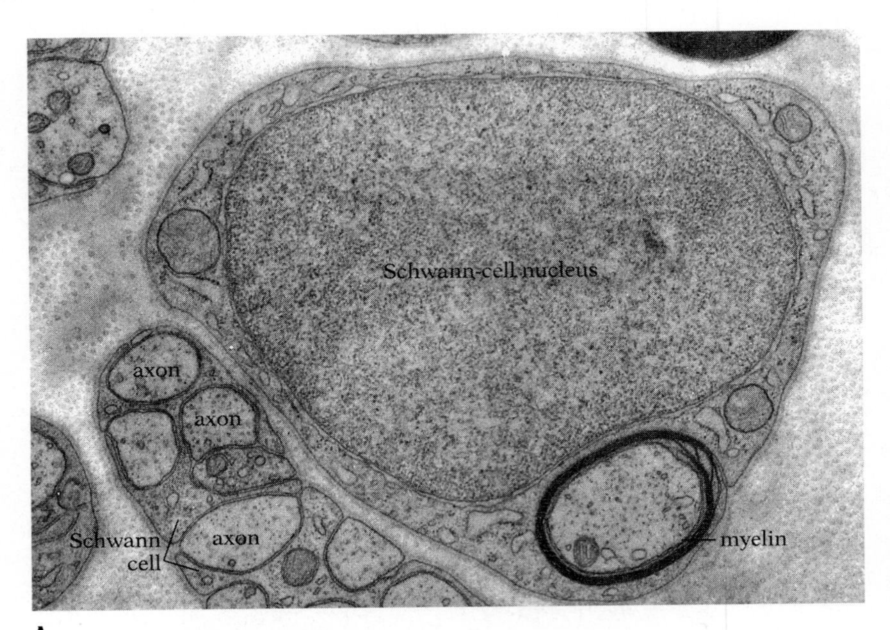

A

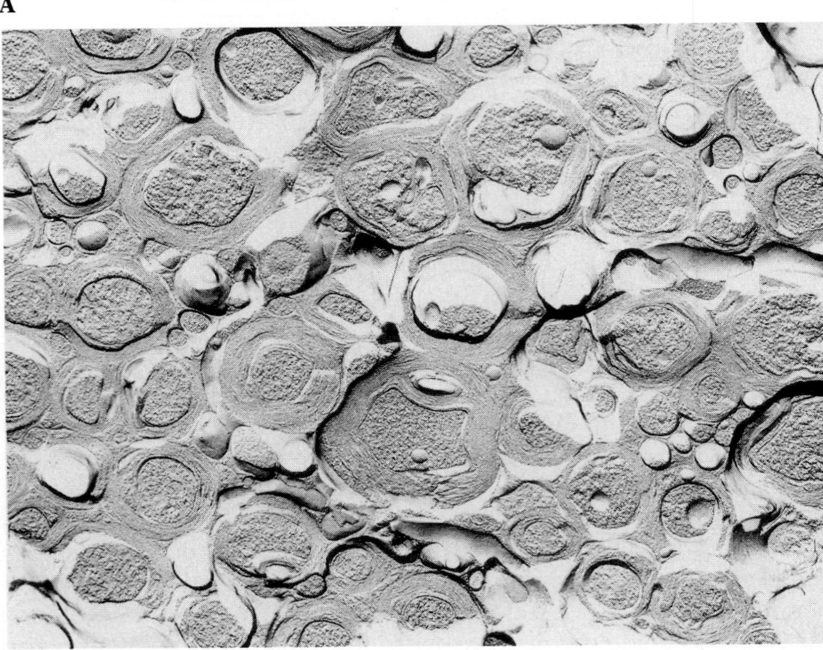

B

35.4 Nerves viewed in cross section

(A) In this transmission electron micrograph of part of a nerve from a guinea pig, the axons of the neurons are enveloped by Schwann cells; at lower left a single Schwann cell envelops several unmyelinated axons. The axon at lower right has a myelin sheath, formed from an invaginated coiled portion of the Schwann-cell membrane; note the large nucleus of the Schwann cell. (B) This freeze-fracture electron micrograph shows many transversely fractured myelinated axons in the CNS. Note the characteristic layering of the plasma membranes in the myelin sheaths. (From *Freeze Fracture Images of Cells and Tissues* by Richard L. Roberts, Richard G. Kessel, Hai-Nan Tung, Oxford University Press, New York, 1991)

crest, called ***Schwann cells***, which often give rise to myelin sheaths in much the same way as the glial cells within the CNS do (Fig. 35.4). Myelin sheaths reduce the area of neuron membrane that must be kept "charged" (thereby greatly reducing the associated metabolic cost), and speed up the conduction of impulses substantially in axons that have them; they are interrupted at regular intervals by ***nodes of Ranvier*** (Fig. 35.1D)—points where one glial or Schwann cell ends and another begins.

■ HOW SIMPLE NEURAL PATHWAYS WORK

All animal groups above the level of the sponges have some form of nervous system. In the tentacles of many coelenterates we see the simplest possible type of nervous pathway—one composed of only two specialized cells: a sensory receptor/conductor neuron, and an effector (response) cell (Fig. 35.5A). Such a pathway generates strictly automatic behavior—tentacle withdrawal, for instance—because the processing stage is essentially missing. Such automatic behavior is a pure ***reflex:*** there is only one input; there are no alternative pathways for the information to take; the behavioral response involves no coordination of effector cells; and there is no central neural control.

Simple as such a circuit looks, it does have one kind of flexibility: after repeated stimulation the sensory neuron can become less sensitive, a phenomenon known as ***sensory adaptation***. In coelenterates, for example, adaptation allows the tentacles to adjust to the constant background level of stimulation produced by water currents, so the defensive reflex is triggered only by an extraordinary stimulus. Sensory adaptation is a widespread phenomenon, observable even in bacteria.

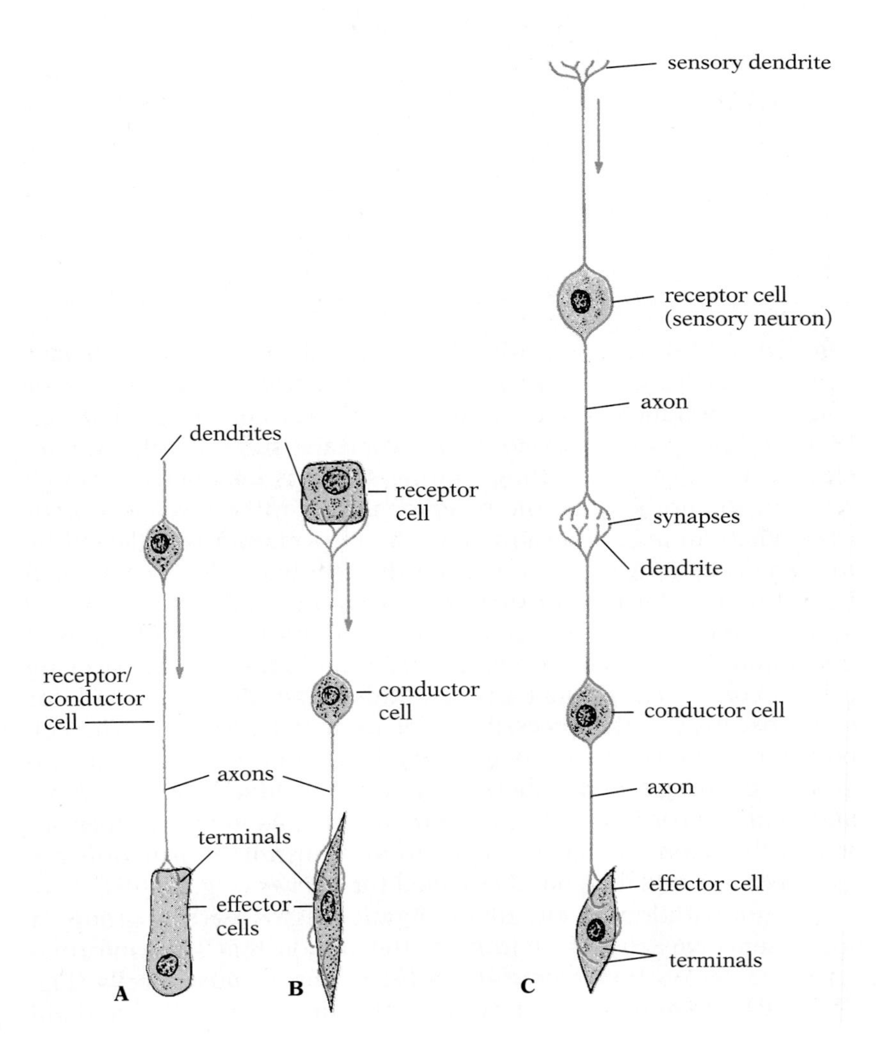

35.5 Simple nervous pathways

(A) In the simplest pathway—the one generating tentacle withdrawal in coelenterates, for example—only two cells are necessary. A single neuron acts as both sensory receptor and conductor of information; the other cell is the effector, and no processing stage is required. Such pathways are called reflex pathways. (B–C) Slightly more elaborate pathways may involve a specialized sensory receptor cell, itself sometimes a neuron, and a separate conductor cell. Again, however, there is no processing. The conductor cells shown here display typical neuron anatomy: a large cell body containing the nucleus, and two narrow projections—the base of the dendritic tree, which usually branches into many dendrites that collect information, and the axon for carrying it to other cells.

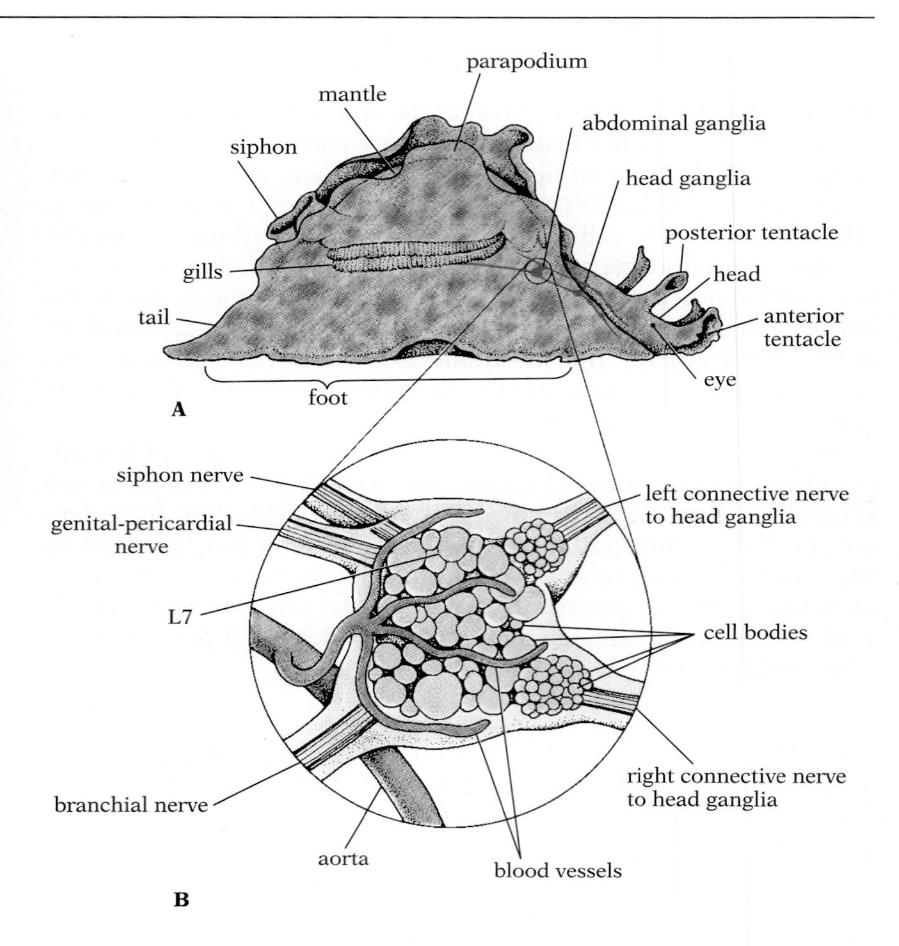

35.6 Sea slug, *Aplysia*

(A) *Aplysia* is basically a large, shell-less aquatic snail. The head has eyes and chemosensory tentacles. Under the mantle, which is the remnant of a shell, are the delicate gills, which extract oxygen from the water, and the siphon, which moves water across the gills. The mantle cavity also contains the so-called purple gland (not shown), from which the animal discharges a concealing ink when disturbed. The foot is the organ of movement. The bilaterally symmetrical nervous system (red) is organized as a series of paired ganglia. (B) The ganglia contain the cell bodies of the neurons. The long axons over which they transmit information are organized into cables called nerves. The abdominal ganglia, shown here, contain about 2000 neurons; they control circulation, respiration, and reproduction. The cell known as L7 is involved in defensive behavior, respiration, and circulation.

The gill-withdrawal circuit of *Aplysia* Most nervous pathways comprise at least three separate cells: a sensory receptor cell or neuron, a conductor neuron, and an effector cell (Fig. 35.5B–C). (Sensory receptors respond to appropriate stimulation with an electrical change; when the resulting signal is conducted through an axon to another cell, the receptor is said to be a ***sensory neuron***. When an axon is lacking and the information is collected by the dendrites of a separate neuron directly from the receptor-cell body, the receptor is a ***sensory cell***.) Neural organization involving separate receptors, conductors, and effectors is particularly well understood in the marine mollusc *Aplysia*, often called the sea slug (Fig. 35.6A). Eric Kandel selected this large shell-less snail for study because of the accessibility of its nervous system. The cell bodies of *Aplysia's* neurons are very large and are collected into groups called ***ganglia***, where they are individually recognizable and their electrical activity can be monitored. As in many other animals, the axons along which neurons transmit information are grouped into cablelike bundles called the ***nerves*** (Fig. 35.6B).

The gill-withdrawal circuit of *Aplysia* begins with a group of touch-sensitive sensory neurons in the siphon that send information to a set of ***motor neurons***, which control muscle cells (Fig. 35.7A–B). When touched, the animal withdraws its siphon and gill

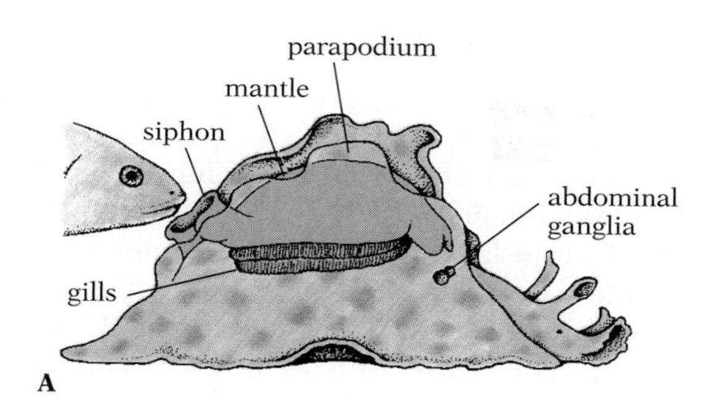

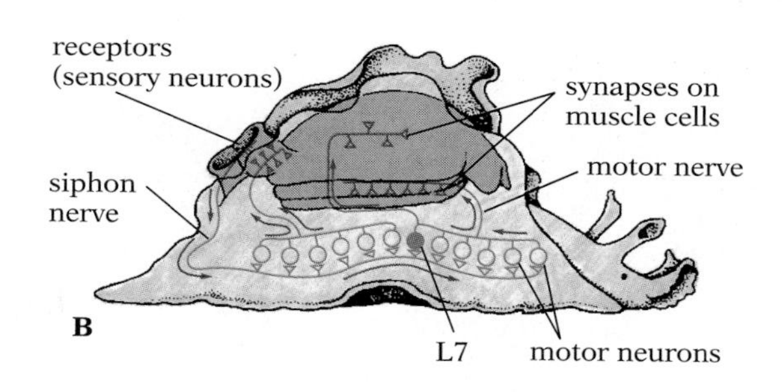

35.7 Gill-withdrawal behavior

(A) Normally, *Aplysia* pumps water across its exposed gills to obtain oxygen. (B) When the siphon is disturbed, however, the touch-sensitive receptors send signals through the siphon nerve to the abdominal ganglia; there the sensory information is communicated to a group of motor neurons that, when sufficiently stimulated, signal the muscles of the mantle, gill, and siphon. (The motor neurons, all of which are actually found in the abdominal ganglia, as well as the sensory and motor nerves, have been rearranged and enlarged here for clarity.) (C) As a result, a disturbed *Aplysia* withdraws these sensitive structures.

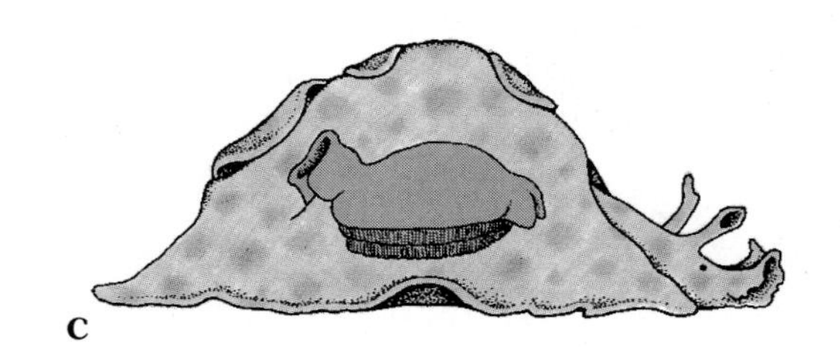

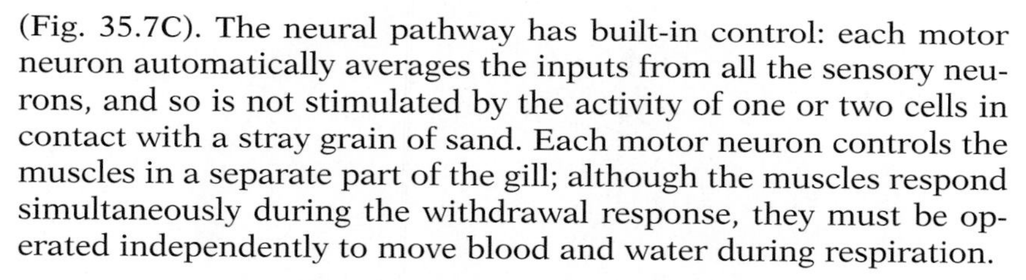

(Fig. 35.7C). The neural pathway has built-in control: each motor neuron automatically averages the inputs from all the sensory neurons, and so is not stimulated by the activity of one or two cells in contact with a stray grain of sand. Each motor neuron controls the muscles in a separate part of the gill; although the muscles respond simultaneously during the withdrawal response, they must be operated independently to move blood and water during respiration.

Why interneurons are important As more cells are added to a pathway, more information processing becomes possible. Cells specialized for processing are called ***interneurons***. The interneuron typically collects and processes input from many cells (often thousands), and passes on the resulting information to its target cells (Fig. 35.8). The signals an interneuron collects from the cells it monitors can be used to encourage or inhibit firing. Indeed, ***inhibition***, the ability to counteract excitatory input from other cells, is essential to information processing, since the nervous system often uses an antagonist strategy: contradictory signals are sent, and it is the ratio or difference between them that determines the nature and strength of the cell's response (Fig. 35.8). The importance of interneurons in this processing of contradictory signals would be hard to overemphasize: the brains of animals consist almost entirely of interneurons arranged in complex, highly specialized networks.

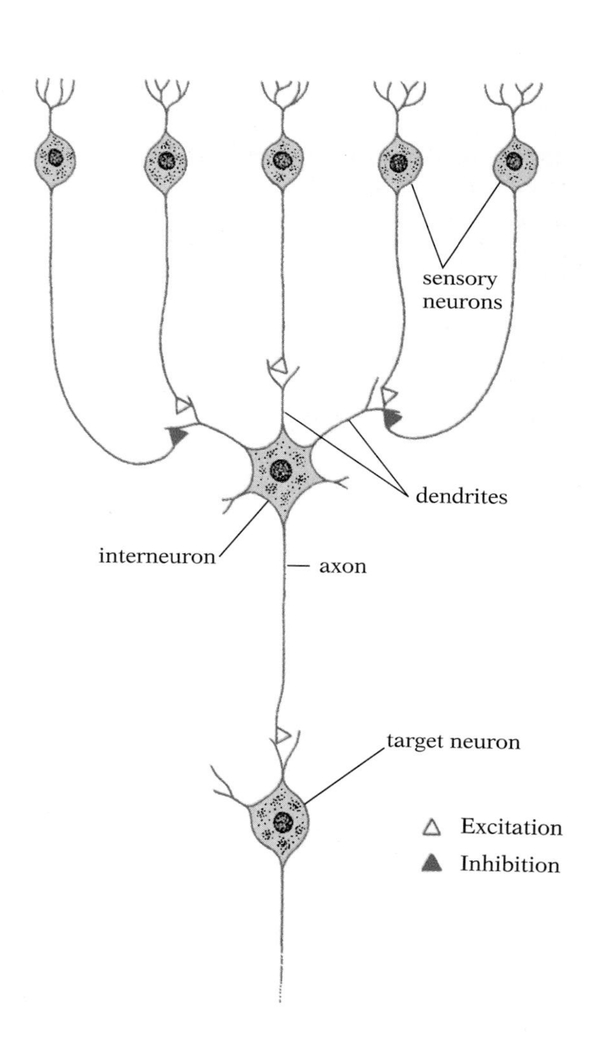

35.8 An interneuron

Interneurons combine information from many cells to produce a single output. In this example, five sensory neurons communicate with the interneuron, three that excite it and two that inhibit it. The relative strength of the various incoming signals determines whether excitation or inhibition predominates.

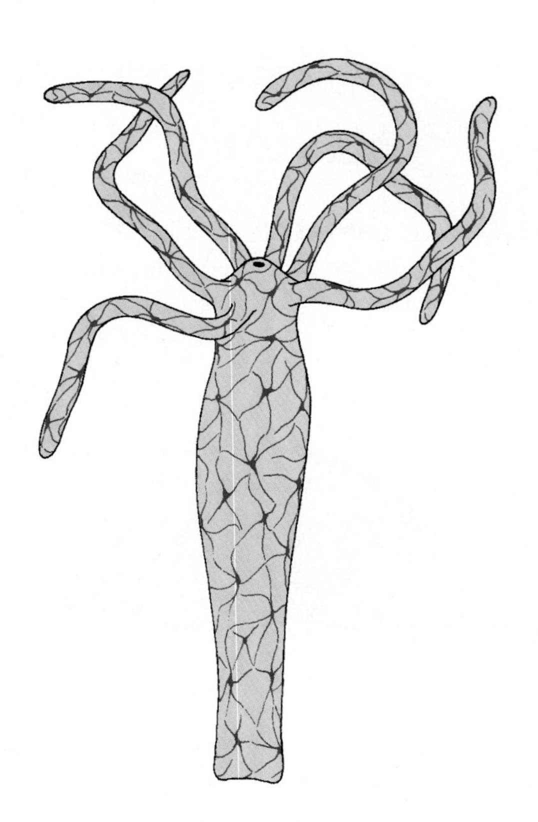

35.9 **Nerve-net system of hydra**

Conductor cells in organisms with nerve nets are not organized into specialized pathways. As a result, there can be no centralized control; only localized responses to stimuli are possible.

■ THE SIMPLEST NERVOUS SYSTEMS: NERVE NETS

Simple neural pathways are interconnected in most animals to form organized nervous systems. In coelenterates like hydra, the nervous system has distinct receptor, conductor, and effector cells (35.5B). The conductor cells, however, do not form definite pathways; instead they interlace to form a diffuse ***nerve net*** (Fig. 35.9). There is no central control: impulses simply spread from the region of initial stimulation to adjacent regions. This organization suits a sessile organism whose only means of escape involves moving the tentacle and the side of the body that has been touched. Without central coordination, an animal can produce only a narrow range of behavioral responses.

Other coelenterates display a degree of centralization. Jellyfish have a nerve ring in the "bell" portion of the body (Fig. 35.10). The other neurons tend to funnel into the ring, and conduction from one side of the animal to the other is thus fairly rapid and organized. This centralization is reflected in the jellyfish's rhythmic, coordinated swimming movements.

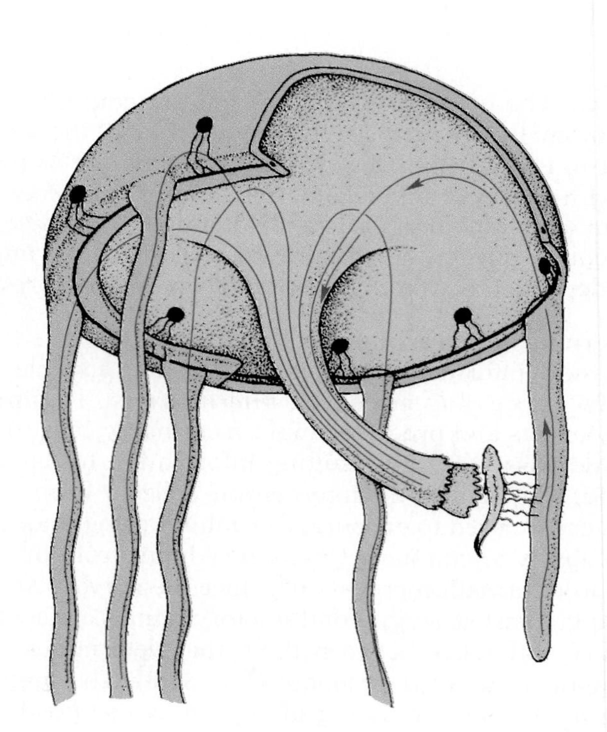

35.10 **Nervous system of jellyfish**

Jellyfish represent a degree of centralization among radially symmetrical animals. Here neurons (black) are organized into a primitive ring system that serves to synchronize the contractions of the swimming muscles of the bell. In addition, sensory neurons (red) on each of the peripheral tentacles send axons to muscles on the central stalk. When the firing of a nematocyst stimulates certain of these cells, the resulting signals direct the creature's mouth (at the base of the stalk) to the affected part of the tentacle for a possible meal.

■ WHY BRAINS EVOLVED IN THE HEAD

Nervous systems in bilaterally symmetrical animals show consistent evolutionary trends:

1 The nervous system becomes increasingly centralized by formation of major longitudinal nerve cords (the ***CNS***) through which most pathways between receptors and effectors pass.

2 Conduction along nervous pathways occurs in one direction only. (It is often bidirectional in nerve nets.) ***Afferent fibers***—sensory fibers leading toward the CNS—are distingiushed from motor fibers leading away from the CNS (***efferent fibers***).

3 Nervous pathways within the CNS become increasingly complex as large numbers of interneurons are included, a development that permits increased flexibility of response.

4 Cells performing different functions become increasingly segregated within the nervous system, so that distinct functional areas and structures become obvious.

5 An increasing ascendancy of the front end of the longitudinal cords leads to the formation of a ***brain***, which becomes more and more dominant—a process known as ***cephalization***.

6 The number and complexity of sense organs increases.

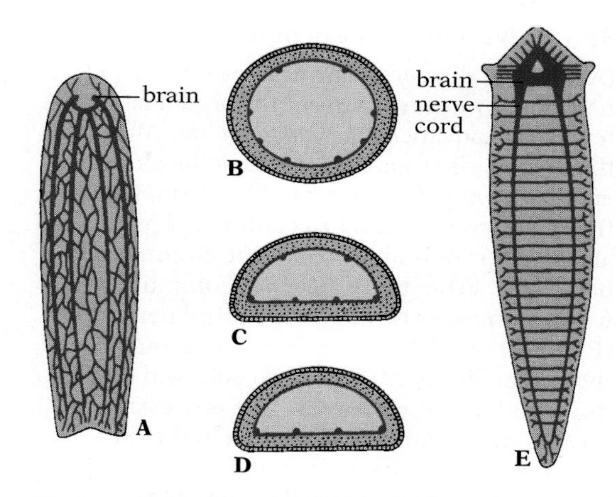

35.11 Nervous systems in flatworms

(A–B) Longitudinal and cross sections of some flatworms, showing some cords predominating in the neural net. (C–D) More advanced flatworms show fewer longitudinal cords. (E) The most advanced flatworms have only two cords.

These trends are not evident in the most primitive flatworms, which have a nerve net much like the hydra's. Slightly more advanced flatworms show the beginnings of major longitudinal cords formed by a grouping of neurons in the nerve nets (Fig. 35.11A). There can be as many as eight of these cords (Fig. 35.11B). Still more advanced flatworms show a reduction in the number of longitudinal cords (Fig. 35.11C–D), the most advanced having only two ventral cords (Fig. 35.11E).

Flatworms with the least advanced development of longitudinal cords have only slight swellings at the anterior end (Fig. 35.11A). More advanced flatworms have a much larger swelling (Fig. 35.11E), but this "brain" exerts only limited dominance over the rest of the CNS.

It's no accident that brains are usually located in the head. The anterior end of the body is usually the part of a bilateral animal that first encounters new stimuli as the animal moves. Natural selection therefore favored development of a particularly high concentration of sense organs in this region, which, in turn, led to an enlargement of the anterior end of the longitudinal nerve cord.[1]

The most primitive version of the brain was probably almost exclusively concerned with funneling impulses from the sense organs into the nerve cords that carried the information to the appropriate motor neurons. The selective advantage of comparing the various sensory inputs—processing the incoming information before passing it on to the muscles—must have led to an increase in the number of interneurons at the front, where the sensory receptors were

[1] Some dinosaurs, whose immense size made nerve transmission from brain to tail relatively slow, had auxiliary brainlike ganglia near the tail to facilitate rapid responses in the rear parts.

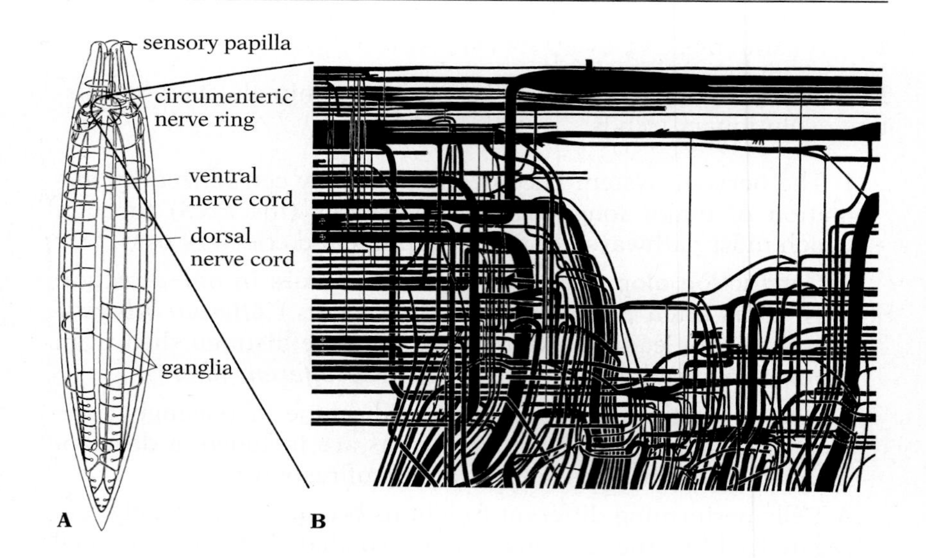

35.12 Nervous system of a nematode

(A) Though the nematode has fewer than 300 neurons, it displays many of the organizational features of advanced nervous systems. Most of the sensory neurons are located at the anterior end in a group of sensory papillae, from which they send information to a centralized processing area—a sort of brain—called the circumenteric nerve ring. From there, neurons communicate with the muscles through the ventral nerve cord. (B) A portion of the switchboard-like anatomy of the nematode "brain" has been enlarged to show its formidable organization. The processing of sensory information that goes on in these interneurons determines what actions will be performed, and where they will be directed.

already concentrated. In an organism like the nematode (Fig. 35.12), the brain became an area for analysis rather than a channel through which raw sensory data were funneled to the effectors farther back. With this increasing specialization for analysis, the brain became a coordination center as well, developing more and more dominance over the rest of the CNS.

The evolutionary trends whose beginnings can be seen so clearly in flatworms are most fully developed in the vertebrates—particularly the mammals—and in the higher invertebrates, including the annelids, molluscs, cephalopods, and arthropods. In all of these animals there is a high degree of centralization, and the older nerve-net strategy is represented by only a few vestiges in parts of the body where sluggish movements, such as the peristaltic contractions of the mammalian intestine, are controlled by slow, diffuse conduction.

■ HOW THE INSECT NERVOUS SYSTEM IS ORGANIZED

In the higher invertebrates, the CNS is a pair of ventrally located longitudinal cords in which the cell bodies of the neurons form ganglia, and the fibers are gathered into nerves that act as communication pathways between ganglia (Fig. 35.13). Even primitive annelids have prominent ganglionic masses, a pair in each body segment, which are connected by nerves running between the segments; almost all the cell bodies are located in ganglia. The brain is simply another ganglion that happens to be in the animal's head. Because the brain of invertebrates is normally formed by the fusion of the four most anterior ganglia, it is larger than the segmental ganglia, and has more sensory than motor neurons. Its dominance over the other ganglia is noticeable, but limited in comparison with that of the vertebrate brain.

More advanced arthropods, particularly some of the insects, show far more concentration of coordination in the front end. Moreover, many of the other segmental ganglia have fused, resulting in better integration of control between the segments. The

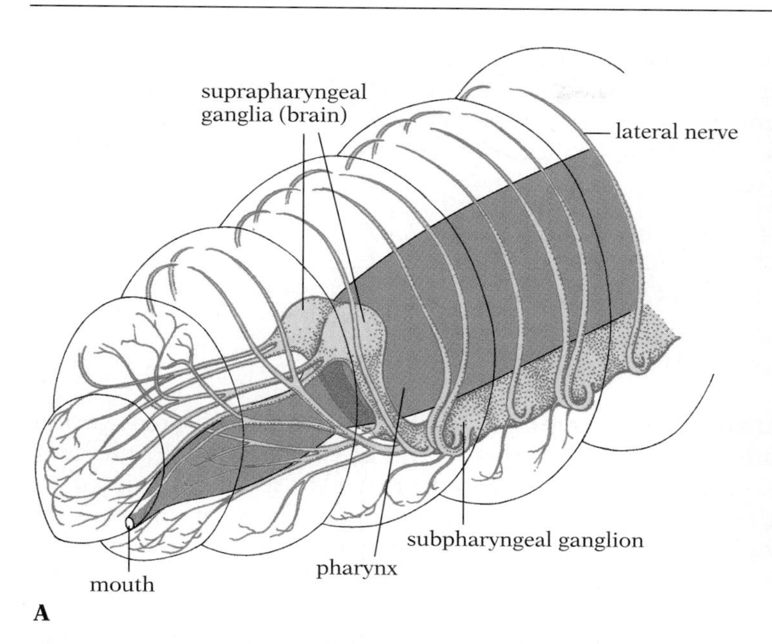

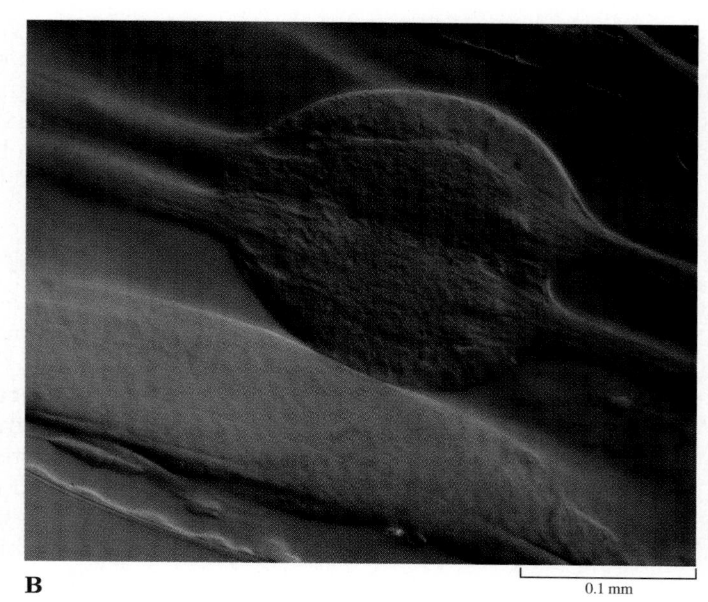

35.13 Earthworm and insect neural organization

(A) As in nematodes, many of the sensory receptors in the earthworm are located in the head, and carry information to the brain (the suprapharyngeal ganglia) for processing. From there information is conducted along the ventral cord and ultimately to the muscles. The brain of the earthworm serves as a device for deciding what action is to be taken and where to direct it. The brain of the earthworm is far more complex than that of nematodes. (B) Double ganglion and paired ventral nerve cord of a living fly larva, viewed with Nomarski optics.

brain, however, has remained relatively small, and the thoracic ganglia perform many vital coordinating functions.

The persistence of thoracic ganglia in insects is correlated with their anatomy. Since the legs and wings are attached to the thorax, a concentration of motor coordinating centers there is advantageous; and many of the sense organs are located on the legs or the thorax rather than on the head. (Flies have taste receptors on their feet, for instance, while many other insects have ears on the thorax or in their legs.) The neural "instructions" for such behavioral programs as walking, flying, courting, mating, and stinging are stored in the thoracic and abdominal ganglia, and the brain serves primarily to aim the behavior and to turn it on and off. Even headless flies and roaches can learn.

■ WHAT'S SPECIAL ABOUT VERTEBRATES?

The CNS of vertebrates (the spinal cord and brain) differs anatomically and functionally from that of annelids and arthropods:

1 The vertebrate spinal cord is single, is located dorsally, and forms in the embryo as a tube with a hollow central canal, a remnant of which survives in the adult (see Fig. 35.32). The cords of annelids and arthropods, on the other hand, are double (two cords lying side by side, though often partially fused); they are located ventrally and are always solid.

2 Though many simple coordinating functions in vertebrates are still performed in the spinal cord, the brain exerts far more dominance over the entire nervous system. Neural pathways that control complex behavior patterns are located exclusively in the brain.

HOW NEURONS WORK

■ HOW AXONS CONDUCT

Generating a nerve impulse Neurons respond to a great variety of stimuli such as electric shock, pressure, or abrupt pH changes. Various sensory neurons and sensory cells are specialized to respond to light, odors, movement, and so on. However, mild electrical stimuli are most often used in research because the intensity and duration of such stimuli can be precisely controlled. Consider the following experiment.

We are working with an isolated neuron. We place two electrodes several centimeters apart on the surface of the axon (Fig. 35.14A). The electrodes are connected to recording equipment, so we can detect any electrical changes. We apply an extremely mild electrical stimulus to the cell body; nothing happens (Fig. 35.14B). We increase the intensity of the stimulus and try again. This time

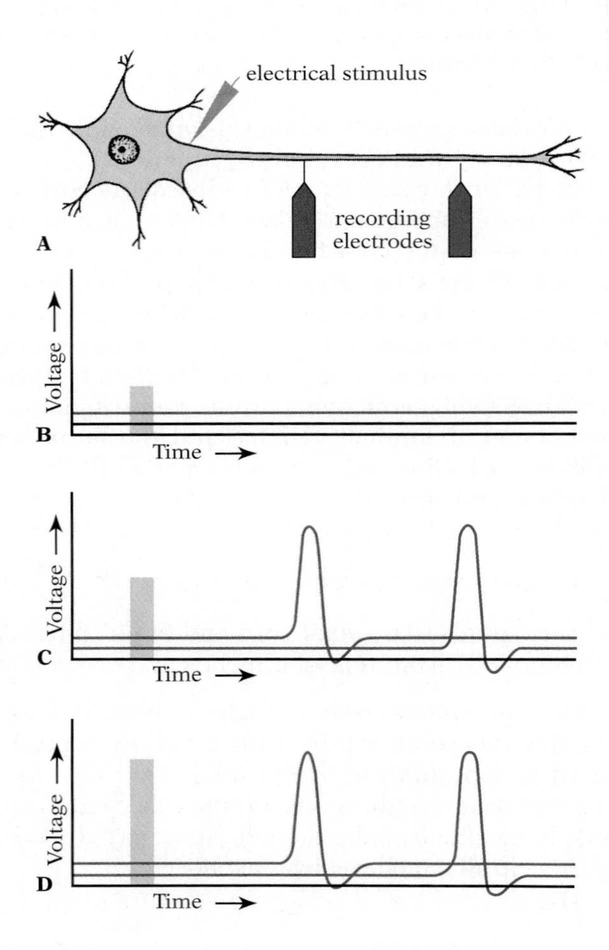

35.14 Initiation and propagation of an impulse

(A) By inserting an electrode into the cell body of a neuron to deliver precise amounts of current, and then monitoring the axon with recording electrodes, we can study the initiation and propagation of an impulse. In B–D the magnitude of the stimulus is shown in green, at left. (B) If the stimulus is below the cell's threshold for triggering an impulse, neither recording electrode registers any change. (C) If the stimulus is adequate, however, an impulse is registered as a spike first at electrode 1 (red) and then at electrode 2 (blue) as it moves along the axon. Note that the spike has the same magnitude at both places. (D) When a still larger stimulus is used, the speed of movement and the intensity of the impulse are not affected because the impulse is an all-or-none response.

our equipment tells us that an electrical change has occurred at the point of contact with the first electrode, and that a fraction of a second later a similar electrical change has taken place at the second electrode (Fig. 35.14C). We have stimulated the axon, and a wave of electrical activity has moved down the axon from the point of stimulation, passing first one electrode and then the other at a rate of 30–90 m/sec. Next we apply a still more intense stimulus, and again we record a wave of electrical change moving down the fiber (Fig. 35.14D), but the intensity and speed of this electrical activity are the same as those recorded from the previous milder stimulation.

We can draw several important conclusions from this experiment: (1) A nerve impulse can be detected as a wave of electrical activity moving along an axon. (2) A potential stimulus must be above a cell's ***threshold*** if it is actually to stimulate an axon. (3) Increasing the intensity of the stimulus above the threshold does not alter the intensity or speed of the impulse produced; the axon fires maximally or not at all; it is an ***all-or-none response***.

If the speed and intensity of an axon's response is all or none, how do animals detect the intensity of a stimulus? There are two ways to encode this: as the intensity of the stimulus increases, the number of impulses produced per unit time—the response frequency—usually goes up; and because neighboring cells usually have different thresholds, more and more neurons fire as stimulus intensity rises.

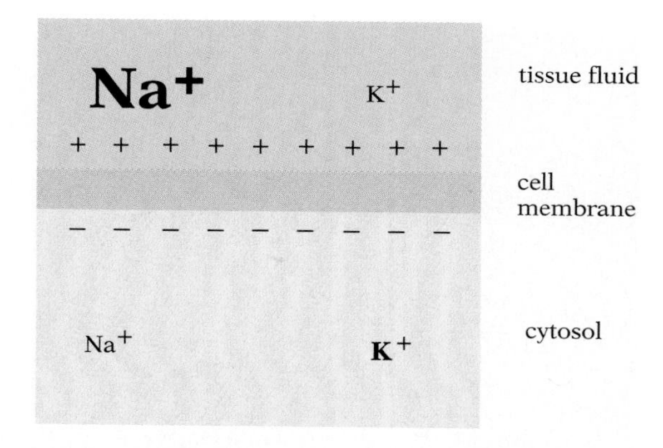

35.15 Polarization of the nerve-cell membrane

The concentration of sodium ions (Na^+) is much greater in the tissue fluid outside the cell, and the concentration of potassium ions (K^+) is greater inside the cell. Because the excess of Na^+ outside is larger than the excess of K^+ inside, the cell has a deficit of positively charged ions. As a result, the inside of the cell is negative (about –70 mv, or millivolts) relative to the outside. (The positive charge of the tissue fluid exerts an electrostatic force on the negative ions in the cell, drawing many of them to the membrane, as indicated here by the line of minus signs.)

Continuously regenerating the impulse The nerve impulse is not a simple electric current flowing through a cellular wire. The impulse travels too slowly, and the electrical resistance of cytoplasm is too high for a simple current even to reach the end of an axon, whereas the impulse is transmitted without any loss in strength. In fact, impulse conduction depends on an electrochemical change actively propagated along the neuron.

The basis of nerve conduction lies in the electrochemical gradient across the membrane and the characteristics of the various ion channels that participate in impulse propagation. The concentration of sodium (Na^+) is very low inside the cell and very high in the extracellular fluid; conversely, the concentration of potassium (K^+) is higher inside than outside, though the difference is not as great as with sodium. Negative ions (mostly charged proteins) are much more common inside cells. This unequal distribution of ions results in an electrostatic gradient across the membrane of about –70 mv (Fig. 35.15). This ***resting potential*** provides the power for impulse conduction.

In broad outline, impulses are propagated by changes in membrane permeability. During the passage of an impulse the selectivity of the membrane is momentarily altered; Na^+ and K^+ ions move across the membrane, and the electrostatic potential changes radically. This depolarization triggers the depolarization of adjacent patches. The details of this process were worked out using the giant nerve fibers of squid, which can be up to 1 mm in diameter. These fibers, which mediate the animal's escape response, innervate the muscles that propel the squid backwards by explosively expelling water from the mantle cavity (Fig. 35.16).

The great size of these squid nerve fibers allows them to conduct

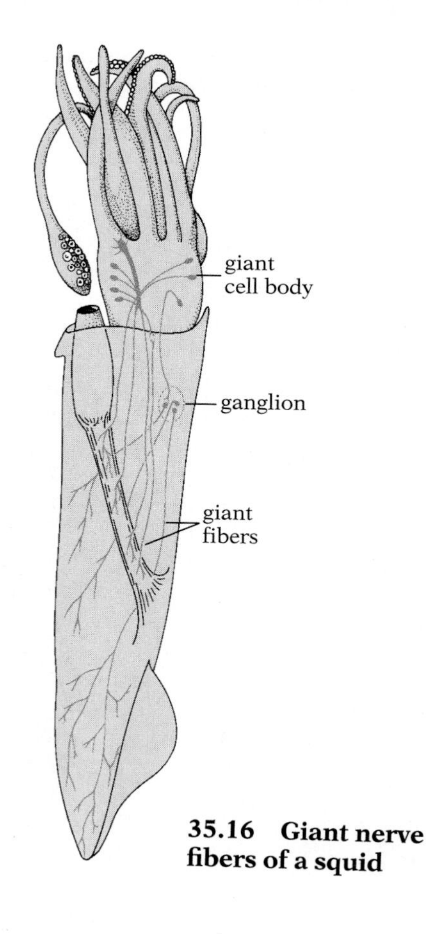

35.16 Giant nerve fibers of a squid

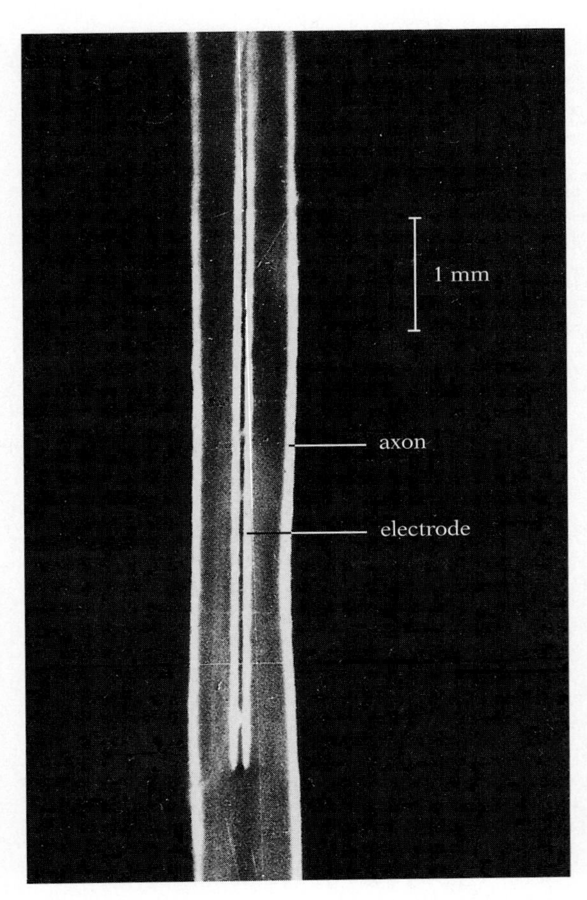

35.17 Giant axon of a squid with a glass-tube microelectrode inside it

nerve impulses very rapidly. (In general, the greater the diameter of a nerve fiber, the faster it conducts. As mentioned earlier, vertebrates evolved myelinated fibers as an alternative adaptation for increasing conduction speed.) In 1939 H. J. Curtis and K. S. Cole in the United States and A. L. Hodgkin and A. F. Huxley in England discovered how to insert a microelectrode into a squid giant nerve fiber. The microelectrode consisted of a glass tube filled with salt solution or metal (Fig. 35.17). With the microelectrode they could measure the electrostatic gradient across the membrane before and during the passage of a nerve impulse.

The patterns observed in the squid giant fiber (and essentially all neurons studied since) are accounted for by the now classic Hodgkin-Huxley model:

1 The membrane of the resting neuron is polarized, with the inside negative relative to the outside.

2 The concentration of Na^+ ions is much higher outside, and the concentration of K^+ ions is higher inside.

3 Stimulation causes the membrane to undergo a large but short-lived increase in permeability to Na^+ ions, which rush across the membrane into the cell, both because of their natural tendency to diffuse from regions of higher concentration to regions of lower concentration and because they are attracted by the net negative charge inside the cell. The inward flux of Na^+ is so great that for a moment the inside actually becomes positively charged relative to the outside.

4 A fraction of a second later, the permeability of the membrane to Na^+ returns to normal, while its permeability to K^+ increases greatly. The K^+ ions now rush out of the cell because their concentration is higher inside than out and because they are repelled by the momentarily high positive charge inside the cell. This exit of positively charged K^+ ions restores the charge inside the cell to its original negative state.

5 The impulse is propagated along the neuron because the cycle of changes at each point depolarizes the membrane at the adjacent point, and initiates a similar cycle of permeability changes there (Fig. 35.18); this in turn starts the cycle farther along the axon, and so on, like a chain of falling dominoes.

The nerve impulse does not decrease in strength as it moves along the fiber because it is constantly being regenerated; the electrical change, or ***action potential***, at each successive point is a new event, equal in magnitude to the electrical events at the preceding points. (Myelinated fibers conduct impulses faster than nonmyelinated fibers of the same diameter because the points at which the action potential is regenerated are not adjacent patches of membrane, but rather successive nodes with high densities of ion channels. By preventing leakage of ions, myelin allows the change in membrane potential at one node to be felt at the next.)

The role of gated channels The propagation of the action potential depends on an initial electrostatic gradient across the membrane, followed by a coordinated series of ion-specific changes in permeability. In turn, these permeability changes depend on gated

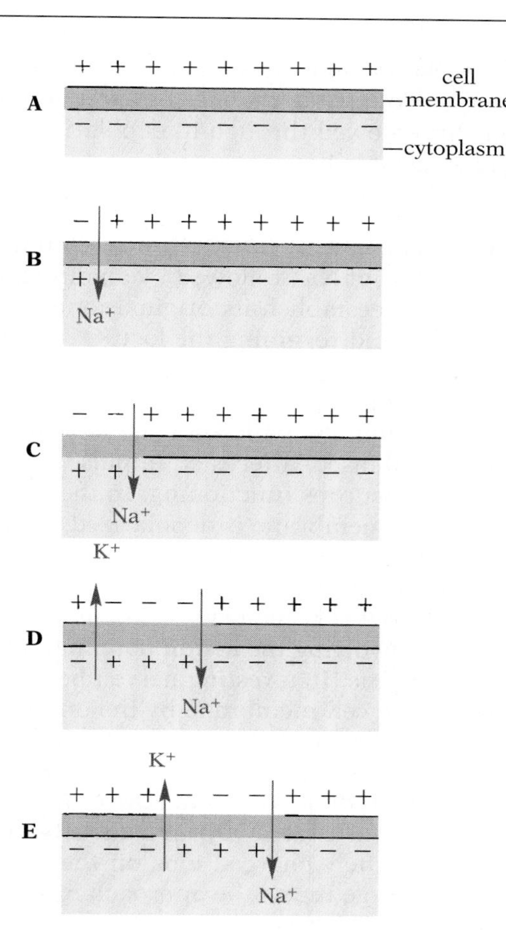

35.18 Model of propagation of a nerve impulse

(A) The interior of a resting nerve fiber is negative relative to the exterior because the concentration of positive ions is higher outside the cell. The interior has a high concentration of potassium ions (K^+), the exterior an even higher concentration of sodium ions (Na^+). (B) When a fiber is stimulated, the membrane, previously relatively impermeable to Na^+ ions, becomes highly permeable to them at the point of stimulation, and a large number rush into the cell (arrow). The result is a reversal of polarization at that point, the inside of the fiber becoming positive relative to the outside. (C) Meanwhile, because the change in membrane potential at the point initially stimulated has altered the potential at adjacent points, the same cycle of permeability changes is initiated, and Na^+ ions begin rushing inward at those points. (D) An instant later, the membrane at the initial point of stimulation becomes highly permeable to K^+ ions; a large number of these rush out of the cell, and the inside of the fiber once again becomes negative relative to the outside. (E) The cycle of changes at each point alters the potential (and hence, the permeability) of the membrane at adjacent points and initiates the same cycles of changes there; Na^+ ions rush into the cell, and K^+ ions rush out a moment later. The movement of this cycle of changes along the nerve fiber is called a nerve impulse.

channels; since both Na^+ and K^+ cross the membrane, there must be channels for each. The membrane proteins responsible for creating the action potential are ***voltage-gated channels***; that is, they open and close in response to changes in the electrostatic gradient across the membrane. Voltage-gated channels apparently have positively charged parts held in place by both the electrostatic repulsion of like-charged ions outside the membrane and the electrostatic attraction of oppositely charged ions inside. Once a stimulus has caused the membrane to depolarize by a precise amount and the electrostatic interactions are thereby weakened, the parts shift position, opening specific ion channels (Fig. 35.19).

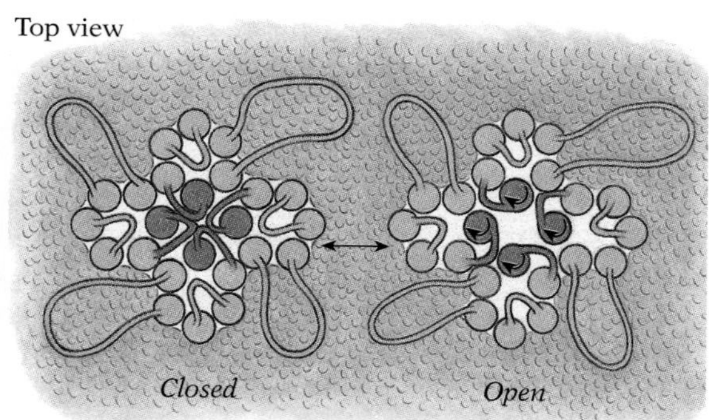

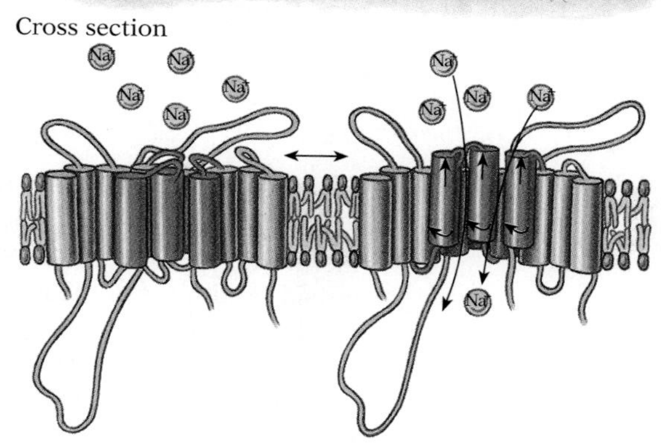

35.19 Voltage-gated sodium channel

The sodium channel is the most elaborate ion channel known. It includes 24 regions of α helix (represented here as cylinders) that penetrate the membrane and are organized into four clusters of six helices each. One helix in every cluster (shown in dark red) is positively charged, and is bound ionically to adjacent helices. When the membrane potential drops to threshold, these four helices slip outward about 0.5 nm, rotating about 60° in the process. This movement opens the channel to Na^+. When the normal potential is restored, the four charged helices are forced back down. The actual mechanism of channel blockage is unknown; it could be a simple physical hindrance, as suggested in the model depicted here.

The Na^+ channels open when the stimulus has reached the threshold for firing the neuron in question, while the K^+ channels do not open until enough Na^+ ions have flowed through to depolarize the membrane almost completely. As ionic flow causes the electrostatic gradient to change, the channels close.

The role of diffusion and electrostatic attraction If impulse conduction involves an inward flow of Na^+ followed by an outward flow of K^+, how does the neuron reestablish its original ionic balance, getting rid of the extra Na^+ and regaining the lost K^+? If the initial ionic distribution were not restored, the neuron would eventually lose its ability to conduct impulses, but neurons can continue to conduct impulses indefinitely with only a very brief refractory period (of about 0.5–2.0 msec) after each impulse.

Two mechanisms work to keep neurons functioning. In the short run, as an impulse passes and the membrane is depolarized, diffusion and electrostatic attraction seem to restore the electrochemical balance between Na^+ outside the cell and K^+ inside almost instantaneously. To understand how this happens, it is important to realize that virtually all events involving the action potential take place very close to the cell membrane. In a resting nerve fiber, free ions that have been attracted to the cell membrane by the opposite charges on the other side turn that part of the membrane into an area of concentrated charge; the electrostatic gradient across the thin membrane approaches an incredible 10^5 v/cm. Once the two surfaces of the membrane have been "coated" by ions attracted by the electrostatic force of the oppositely charged ions on the other side, this concentrated layer of charge repels the approach of additional ions from the same side (Fig. 35.20).

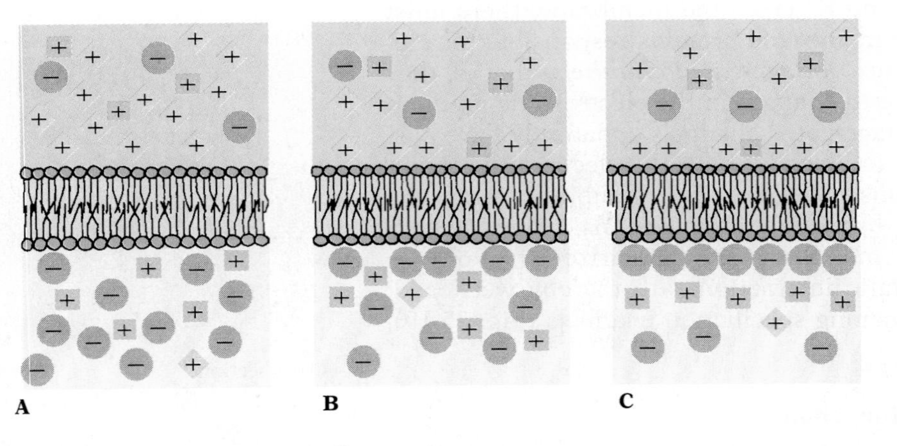

35.20 Charging of the membrane

(A–B) Electrostatic attraction across the cell membrane draws some of the positive ions on the outside and some of the negative ions on the inside to the respective surfaces of the membrane. (C) As a result, these surfaces become coated with oppositely charged ions. At some point, however, no more ions are attracted: the concentration of positive ions on the outer surface is so high that electrostatic forces repel the free-floating positive ions to the same degree as those ions are attracted to the negatively charged interior of the cell. On the other side of the membrane an equivalent situation develops with respect to negative ions.

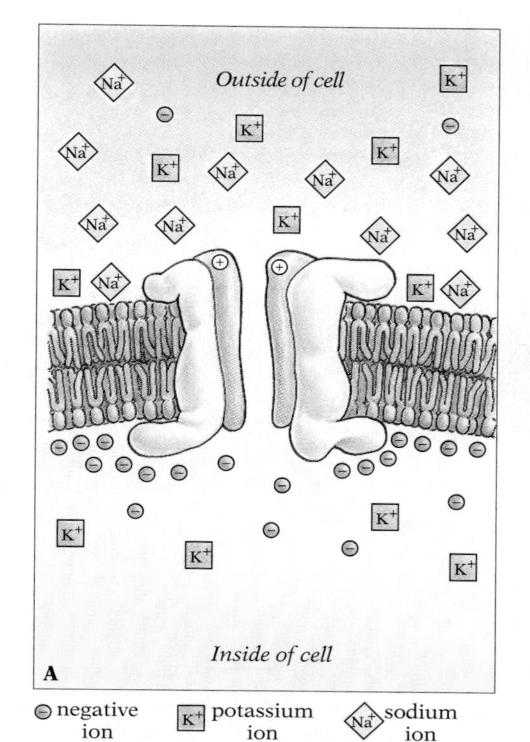

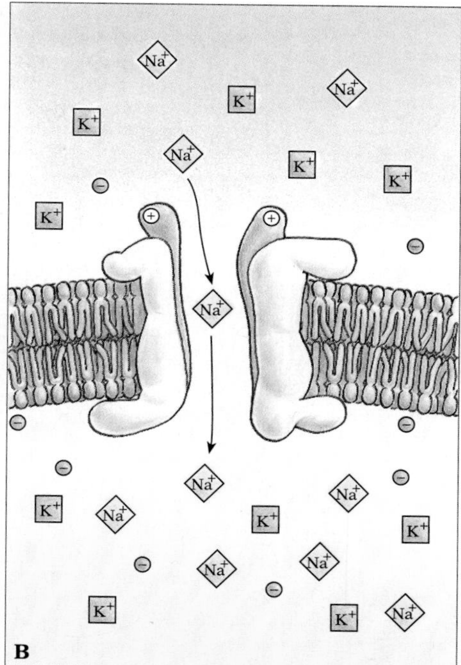

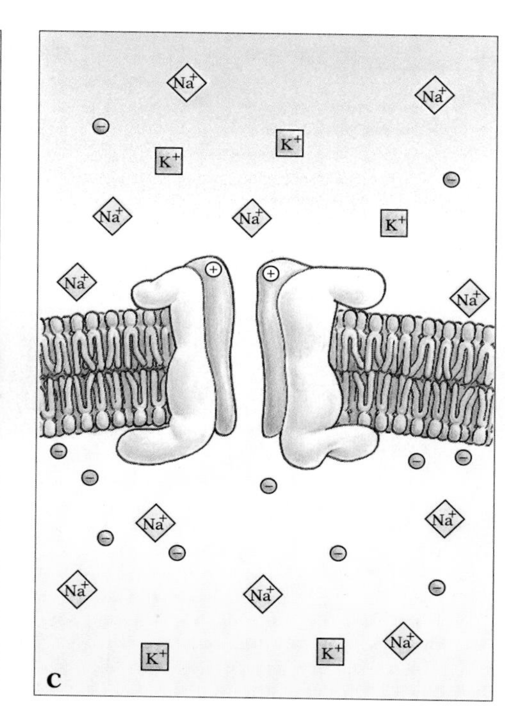

35.21 Diffusion after an action potential The ions that cross the neural membrane during an action potential rapidly diffuse away from the membrane into the abundant fluid inside and outside the cell. (A) Before the Na^+ channel shown here opens in response to partial membrane depolarization, the membrane is fully charged. (B) When the gate opens, Na^+ ions rush into the cell, reducing and then reversing the electrostatic gradient across the membrane. (C) After the channel closes, the excess Na^+ near the inner membrane of the neuron rapidly diffuses into the cytosol, and the membrane begins to recharge.

When an action potential allows ions to cross the membrane, only the ion concentrations near the highly charged inner and outer surfaces of the membrane are affected, simply because the permeability changes are so short-lived that only the closest ions have time to move to and through the membrane. The tiny resulting alteration in ion concentration is rapidly absorbed as the Na^+ ions that crossed into the neuron, and the K^+ ions that moved out, diffuse into the fluid on either side of the membrane. It is rather like adding a drop of ink to a pond: for a moment there is a dark patch, but it dissipates as the ink diffuses into the larger volume of water. The rapid diffusion of the small number of ions involved into the relatively enormous volume of the cell and the extracellular fluid allows the intense local electrostatic gradient at the cell membrane to reestablish itself almost immediately (Fig. 35.21).

Despite the dramatic nature of the events at the neural membrane, only a minute quantity—about 10^{-12} moles—of Na^+ enters the cell during an action potential. The net effect of a single action potential on the concentration of Na^+ in the cell as a whole is therefore negligible. In the long run, however, the potential will decline if the Na^+ that enters is not removed and the K^+ that leaves is not recovered.

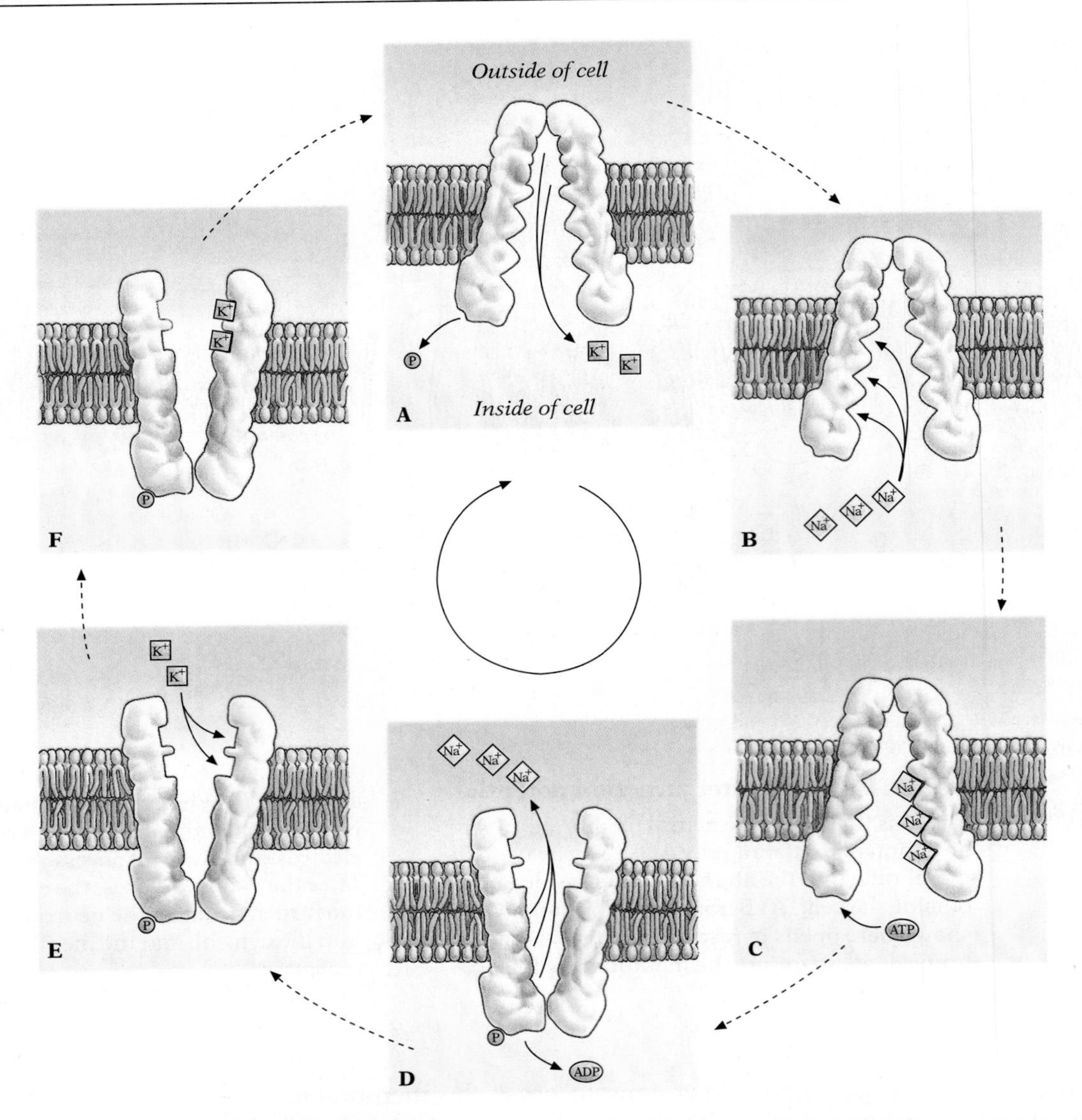

35.22 Model of the sodium-potassium pump

The pump consists of proteins embedded in the cell membrane. (A) When the pump complex is open to the inside of the cell, the K^+ binding sites no longer bind potassium, so those ions drift free into the cytosol, along with the phosphate group that powered the preceding cycle. (B) The Na^+ sites, on the other hand, become active. (C) Once the Na^+ ions have been bound, ATP binds to its site, thus phosphorylating the complex and (D) causing a conformational change that opens the pump to the outside; this change severely reduces the protein's affinity for Na^+ (which in consequence drifts free into the extracellular fluid), while (E) it activates the K^+ sites. The binding of K^+ ions causes the complex to return to its former conformation (A) and to release both the phosphate group obtained earlier from the ATP and the K^+ ions. (The extent of movement of the pump subunits has been greatly exaggerated for clarity.)

The sodium-potassium pump In the long run, the ionic "charge" is renewed by active pumping of Na^+ and K^+ ions across the membrane. Each neuron has a million or so ATP-powered sodium-potassium exchange pumps built into its membrane to enable it to keep conducting action potentials indefinitely. Each pump extrudes 200 Na^+ ions and takes up 135 K^+ ions every second (Fig. 35.22).

■ HOW NEURONS COMMUNICATE

Transmitters and synapses After an impulse reaches the end of an axon, the next step in neuronal communication begins. The axon usually synapses with the dendrites or cell body of other neurons. Since the terminal portion of an axon normally branches repeatedly, a single axon may synapse with many other neurons; moreover, it usually synapses at many points on each neuron (Fig. 35.23). Each tiny branch on an axon usually ends in a small buttonlike swelling called a ***synaptic terminal*** (Fig. 35.24).

In a few cases there is a gap junction between the membrane of the synaptic terminal and the membrane of the adjoining cell (see pp. 117–118). Such a junction permits direct electrical coupling between the two neurons, so that an impulse traveling down the axon of the first neuron can pass relatively unhindered to the next neuron. Since electrical synapses minimize the delay in transmission of impulses, they tend to occur in places in the nervous system where speed of conduction is of special importance. They also provide a high degree of certainty that an impulse in the first neuron will give rise to an impulse in the second neuron.

The vast majority of synapses, however, are not electrical, but chemical. A space about 20 nm wide—the ***synaptic cleft***—separates the synaptic terminal of the first (presynaptic) neuron and the membrane of the second (postsynaptic) neuron. Transmission across this cleft is effected by a diffusible ***transmitter*** chemical released from thousands of tiny ***synaptic vesicles*** in the terminal

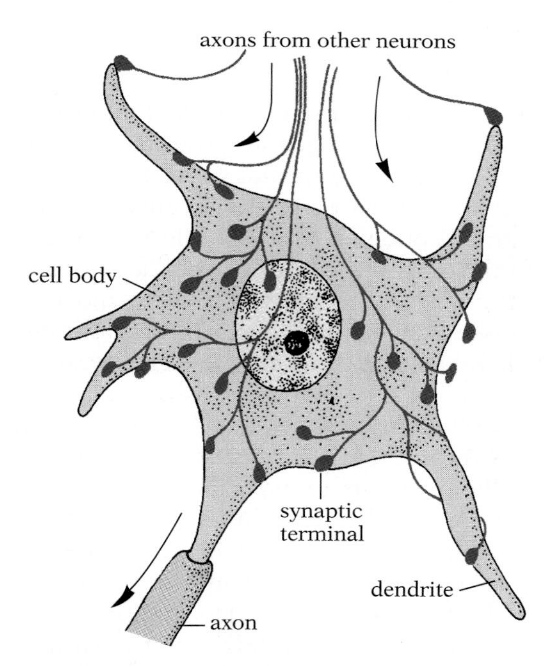

35.23 Synapses on a motor neuron

Many axons, each of which branches repeatedly, synapse on the dendrites and cell body of a single motor neuron. Each branch of an axon ends in a swelling called a synaptic terminal.

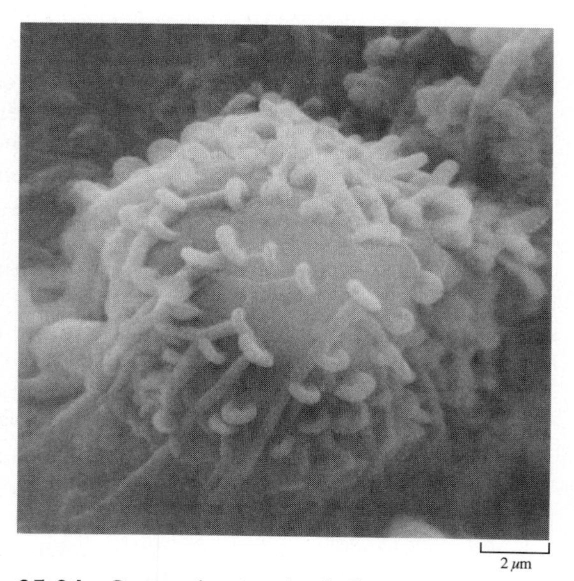

35.24 Synaptic terminals from *Aplysia*

The synaptic terminals of the numerous axons are in contact with the cell body of a postsynaptic neuron. Note in this scanning electron micrograph that it is the edge of the terminal, not its flattened end, that characteristically forms the synapse.

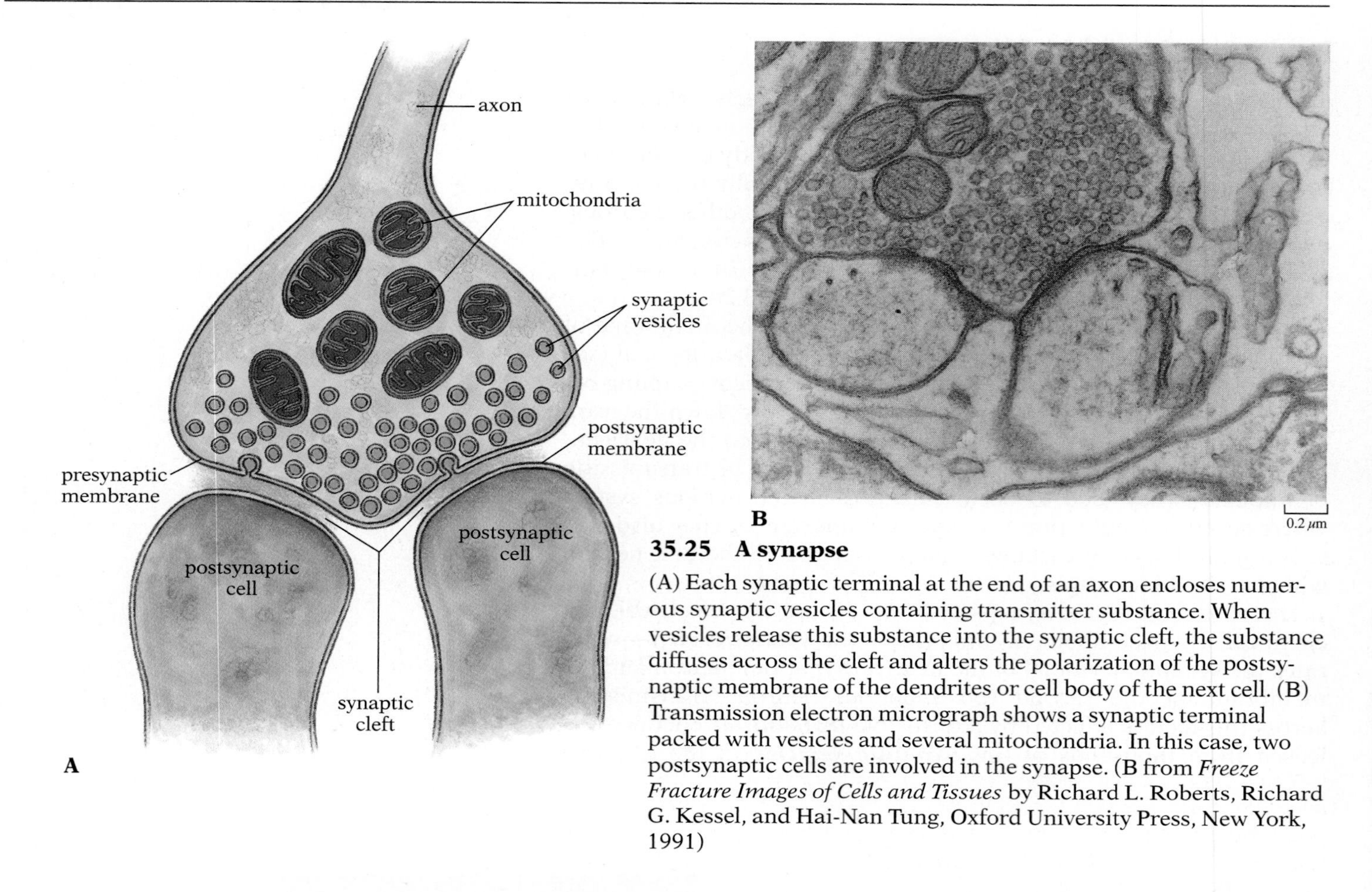

35.25 A synapse

(A) Each synaptic terminal at the end of an axon encloses numerous synaptic vesicles containing transmitter substance. When vesicles release this substance into the synaptic cleft, the substance diffuses across the cleft and alters the polarization of the postsynaptic membrane of the dendrites or cell body of the next cell. (B) Transmission electron micrograph shows a synaptic terminal packed with vesicles and several mitochondria. In this case, two postsynaptic cells are involved in the synapse. (B from *Freeze Fracture Images of Cells and Tissues* by Richard L. Roberts, Richard G. Kessel, and Hai-Nan Tung, Oxford University Press, New York, 1991)

(Fig. 35.25); each vesicle may contain as many as 10,000 molecules of transmitter. When an impulse traveling down the axon of the presynaptic neuron reaches the terminal, special voltage-gated calcium channels concentrated at the synapse open. Because they are 10,000 times more concentrated outside cells, Ca^{++} ions then diffuse into the terminal from the surrounding fluid. The Ca^{++} ions in some way stimulate synaptic vesicles in the terminal to move to the terminal membrane, fuse with it, and then rupture, thereby releasing the transmitter chemical into the synaptic cleft by exocytosis.

The transmitter molecules released into the cleft diffuse across it and bind by weak bonds to receptors on the ***postsynaptic membrane*** of the next neuron. The receptors are specific for a particular neurotransmitter. In the case of the transmitter ***acetylcholine*** (the chemical by which vertebrate motor neurons communicate with skeletal muscle cells), two molecules of transmitter must bind to a receptor to activate it; the contents of a single synaptic vesicle activate about 2000 of these receptors. The binding of transmitter to a receptor opens the gates of a channel, allowing a specific ion to pass through the postsynaptic membrane. This ion movement alters the target neuron's membrane potential. The channels with receptors for acetylcholine allow Na^+ to pass, thus partially depolarizing the cell near the channels. If the depolarization reaches threshold, the postsynaptic cell will fire.

The story of impulse transmission does not end with the diffusion of transmitter across the cleft. If the transmitter remained, the postsynaptic receptors would be stimulated indefinitely by a single action potential, so there must be a mechanism to destroy the transmitter. For instance, once acetylcholine has diffused across the synaptic cleft and exerted its effect on the postsynaptic membrane of the next cell, it is promptly destroyed by an enzyme called ***acetylcholinesterase***. Destruction of the transmitter makes it possible for the next impulse, with new information, to be transmitted. Many insecticides, such as the organophosphates (also known as nerve gases), are cholinesterase inhibitors. They block destruction of acetylcholine, with the predictable result that the synapses of an insect exposed to them become permanently active. Given in high enough doses, cholinesterase inhibitors will paralyze and kill an animal.

Acetylcholine is one of many transmitters found both outside and within the CNS. Other CNS transmitters in vertebrates include noradrenalin (which is also produced in the adrenal medulla as a hormone), serotonin, dopamine, nitric oxide, and gamma-aminobutyric acid (GABA). As we will see, certain disorders such as schizophrenia and severe depression are triggered by biochemical malfunctions of CNS transmitters, receptors, and neural hormones.

How transmitters work Most neurons—especially interneurons—collect and average information from many cells. This integration occurs on the membrane. In the case of transmitters like acetylcholine, the inward flow of Na^+ ions slightly decreases the polarization of the neuron so that the inside becomes less negative relative to the outside, creating an ***excitatory postsynaptic potential*** (***EPSP***, Fig. 35.26). If the EPSP is sufficiently large, it may spread to the base of the cell's axon (called the axon hillock), depolarize the membrane past its threshold, and trigger an impulse that will move down the axon to the next synapse.

The transmitter chemicals released by the terminals of some neurons have the opposite effect: they *increase* the polarization of the postsynaptic membrane and make the neuron harder to fire.

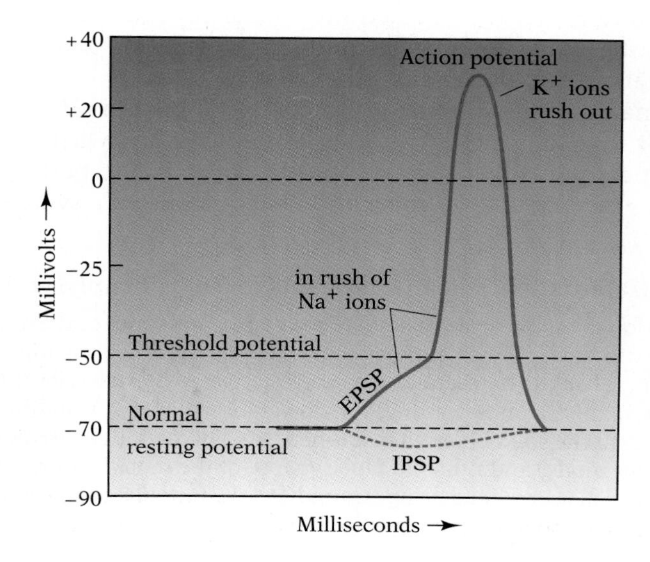

35.26 Effect of transmitter substance on the membrane potential of a neuron

The normal resting potential of a typical neuron is about −70 mv. An excitatory transmitter substance slightly reduces that polarization—that is, makes the inner surface of the membrane less strongly negative—thereby creating an excitatory postsynaptic potential (EPSP). If the EPSP reaches the threshold level (usually about −50 mv), an impulse (action potential) is triggered. If the transmitter substance had been inhibitory, the membrane could have become hyperpolarized (to perhaps −75 mv), a condition called an inhibitory postsynaptic potential (IPSP) (dashed curve), and no action potential would have resulted; the neuron would slowly have returned to its resting potential after release of the transmitter had ceased.

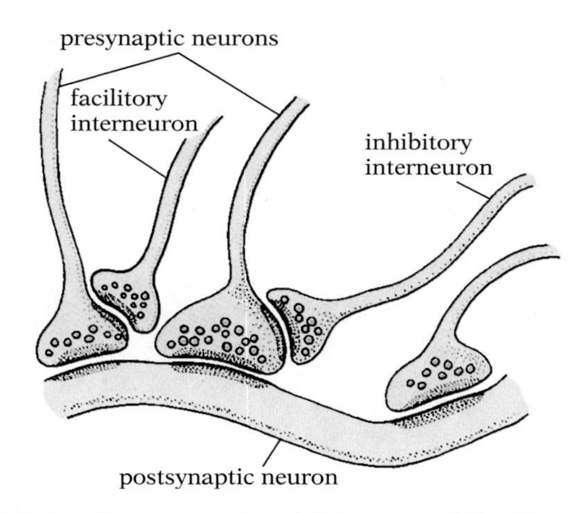

35.27 Presynaptic inhibition and facilitation

In presynaptic inhibition, an inhibitory interneuron synapses on the terminal of the presynaptic neuron. When the inhibitory cell releases its transmitter (often GABA), fewer vesicles in the terminal of the presynaptic neuron will release their own transmitter substance, with the result that the postsynaptic neuron receives less stimulation from this particular pathway. Similarly, in presynaptic facilitation, the facilitory interneuron releases its transmitter (often serotonin) onto the terminal of the presynaptic neuron, with the result that more vesicles in the terminal of the presynaptic neuron will release transmitter, and the postsynaptic neuron will receive increased stimulation from that pathway. Unlike postsynaptic inhibition, which reduces the responsiveness of the postsynaptic neuron to all excitatory inputs, presynaptic inhibition and facilitation affect specific inputs; the responsiveness of the postsynaptic neuron to inputs from other sources is left unchanged.

These transmitters usually produce their inhibitory effects by binding to and opening channels in the postsynaptic membrane that pass chloride ions (Cl^-) down their concentration gradient to the inside. Other inhibitory transmitters open potassium channels. When K^+ ions leave, or Cl^- ions enter, the cell membrane becomes hyperpolarized—the inside of the cell becomes more negative relative to the outside—and an ***inhibitory postsynaptic potential*** (***IPSP***) is produced (Fig. 35.26). Additional excitatory transmitters, and hence more than the usual number of excitatory impulses, are then needed to depolarize the membrane to threshold potential and fire the neuron. The balance between EPSPs and IPSPs underlies all neural processing.

The action of a transmitter depends on the ion specificity of the gated channels to which it binds in the membrane, not some character of the transmitter itself. For example, acetylcholine is excitatory at most neuromuscular junctions, but it is inhibitory on heart muscle. The channels associated with the acetylcholine receptors of heart muscle are quite different from those of other muscles.

Not all synaptic transmitters alter membrane potential directly. A few use a second-messenger strategy to alter the chemistry of the nerve cell. Indeed, a variety of transmitters—amino acid derivatives (called neuropeptides) such as serotonin and noradrenalin—activate the adenylate cyclase system of nerve cells, reminding us that the nervous and endocrine systems share a common evolutionary origin. In general, the second-messenger neurotransmitter system causes long-term changes in the excitability of the postsynaptic cell—changes of the sort that are thought to underlie learning and memory, for example. Transmitters that act directly on ion channels are responsible for short-term electrical events—the EPSPs and IPSPs.

Synaptic processing Both the excitatory and inhibitory synapses we have discussed so far determine whether or not a postsynaptic cell will fire an impulse. Synapses are also subject to ***presynaptic inhibition*** and ***presynaptic facilitation***. In such cases the terminals of inhibitory or excitatory interneurons synapse on the terminals of the presynaptic cell (Fig. 35.27) and act to alter the number of synaptic vesicles that release transmitter. This form of inhibition or facilitation has the decided advantage of modifying excitatory input to the postsynaptic cell from one information source without altering its responsiveness to other inputs.

Presynaptic interactions also underlie adaptive modifications of behavior such as the gill withdrawal of *Aplysia*; the wiring is shown schematically in Figure 35.28, omitting all but one pathway from sensory receptors to muscles. The solid-red synapse can

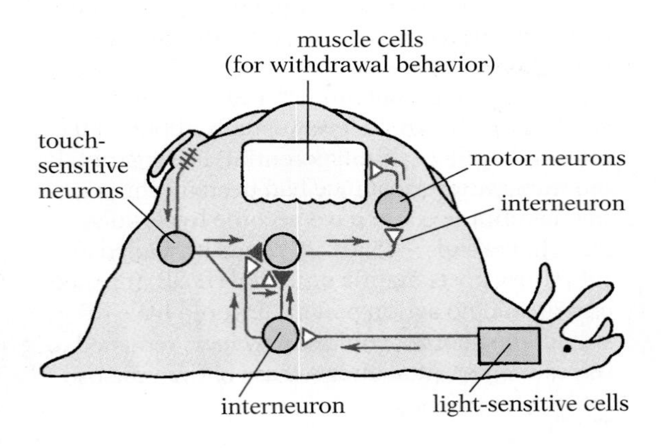

35.28 Habituation, sensitization, and learning in *Aplysia*

Habituation takes place at the solid red synapse, and reduces the animal's sensitivity to touch. Sensitization occurs when the presynaptic input (open blue triangle) on the habituated synapse is activated by an irrelevant stimulus (light, in this example; pathway of open blue synapses). Conditioning takes place at the solid blue synapse when firing of the touch circuit synapse (open red triangle) and the solid blue synapse of the sensitization circuit is correlated. This correlated firing strengthens the initially weak solid blue synapse enough that it can trigger the interneuron on its own.

habituate—that is, become less sensitive—with repeated stimulation. An *Aplysia* in rough water, for example, soon raises its withdrawal threshold as its synapses habituate, and so resumes its normal activities. Habituation is long lasting, but it can be instantly eliminated by ***sensitization:*** stimuli from other sensory modalities, such as a sudden change in illumination (Fig. 35.28), are relayed to critical synapses, where they release transmitters that undo any habituation.

Learning can take place at these special sensory junctions as well. The sensitization circuit (blue cells) sends an axon to the gill-withdrawal interneuron, but the connection (Fig. 35.28; blue synapse) is too weak to trigger firing. A second branch of the axon from the sensory cells (open red synapse) terminates on this ineffectual synapse, and strengthens it whenever the two fire simultaneously. Thus, if a reliable correlation develops between the firing of any of the several sensitization inputs and activity in the touch-receptor circuit, the synapse of the sensory system that can "predict" the imminent stimulation of the touch receptors soon becomes strong enough to trigger gill withdrawal on its own. This process—conditioning—is discussed in more detail in Chapter 38.

Synaptic integration Chemical synapses are points of resistance in neural circuits. An impulse may travel to the end of the axon of one neuron but, like the red synapse of the *Aplysia* circuit in Figure 35.28, die there because not enough excitatory transmitter is released to initiate the all-or-none response in the next neuron in the pathway. Indeed, it is rare that a single impulse from a single neuron can fire the next cell in the circuit. Ordinarily, excitatory transmitters from many different terminals must be released more or less simultaneously to fire a target cell. The individual EPSPs then combine to produce a large enough resultant EPSP to exceed the threshold of the target cell and so trigger an impulse. This additive phenomenon is called ***summation***. It can be spatial, as when several adjacent terminals from different neurons fire simultaneously, or it can be temporal, as when one cell fires at such high rate that individual EPSPs overlap in time. It can also be a combination of the two.

Summation on the postsynaptic membrane is algebraic: if both excitatory and inhibitory transmitters are released at the same time, the resulting EPSPs and IPSPs add according to sign. The result of this phenomenon, a kind of molecular "voting," is a postsynaptic potential that reflects the overall pattern of incoming information. This kind of neural processing, evident in most species of animals, is known as ***integration***. The cell integrates all the signals that converge on it (signals that, in the case of an interneuron or a motor neuron, may be coming from thousands of different interneurons or sensory neurons) and either fires an impulse or remains silent (Fig. 35.29). It is by combining many such simple yes-no, on-off decisions from a host of different cells that the nervous system processes the sensory information it receives.

Drugs and disease Synaptic malfunctions, particularly those involving transmitter irregularities, have been implicated in a variety of neurological diseases. Because synapses are responsible for information processing in the nervous system, and because their

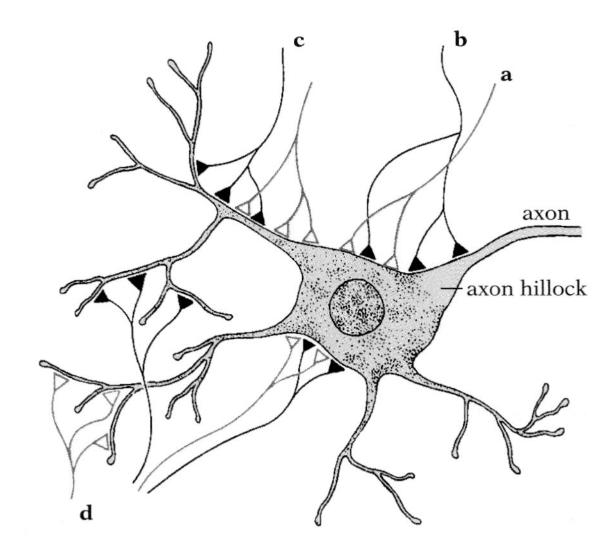

35.29 Integrative function of a neuron

The neuron receives both excitatory synapses (open triangles) and inhibitory synapses (solid triangles) from many different sources. The synapses vary in their distance from the base of the axon (the axon hillock), where impulses are generated. Whether or not the neuron will fire an impulse is determined at any given moment by the algebraic sum of all the individual EPSPs and IPSPs arriving at the axon hillock from the various synapses. However, unlike an axonic impulse, which shows no lessening with distance, depolarizations (EPSPs) or hyperpolarizations (IPSPs) decrease in magnitude as they spread along a dendrite or the cell body. Hence impinging interneurons such as those labeled **a** and **b**, which synapse near the axon hillock, can more easily influence the neuron's firing than interneurons such as **c** and **d**, which synapse on the cell at a distance from the hillock; only if these latter interneurons fire at a very high rate are they likely to have a major effect. In short, the geometry of the synapses on a neuron biases the integration process; inputs from some interneurons are given greater weight than inputs from other interneurons.

proper function depends on a very delicate balance among a variety of presynaptic enzymes, ion channels, transmitters, deactivating enzymes, and postsynaptic receptors and ion channels, it is not surprising that synaptic malfunctions can be devastating. Used with caution and informed medical supervision, some drugs that exert their effects at synapses can give relief from anxiety, severe neurological pain, or diseases involving biochemical disorders of the synapses. Used improperly, however, the same agents can induce symptoms strikingly similar to those seen in certain mental disorders, and in some cases the symptoms may be long-lasting or even permanent.

Neurological drugs alter synaptic function in a variety of ways. They can turn off certain synapses by interfering with synthesis of the appropriate transmitter, blocking uptake of the transmitter into synaptic vesicles, preventing release of transmitter from the vesicles into the cleft, or blocking the receptor sites on the postsynaptic membrane, so that the transmitter has no effect even when released. By contrast, other drugs can induce excessive and uncontrolled firing of postsynaptic cells by stimulating massive releases of transmitter, mimicking the effect of transmitter, or inhibiting the destruction of the transmitter once it has done its job, as in the case of the cholinesterase inhibitors mentioned earlier.

The physiological mode of action of a drug may help explain the behavioral symptoms the drug induces. Amphetamine, for example, acts as a stimulant because it increases the release of noradrenalin in the brain. Reserpine, on the other hand, acts as a tranquilizer because it blocks the uptake of noradrenalin into synaptic vesicles and hence prevents its release. Thus the contrast in the behavioral symptoms produced by these drugs is explained by their opposite effects on the same synapses.

Nicotine acts as a stimulant because it mimics the effect of acetylcholine. Chlorpromazine, a commonly used tranquilizer, inhibits transmission of impulses at both acetylcholine- and noradrenalin-mediated synapses by combining with receptor sites on the postsynaptic membranes and thereby blocking the transmitter chemicals. The antidepressant fluoxetine hydrochloride (Prozac) inhibits reabsorption of the presynaptic facilitator serotonin, thus enhancing its effect. Local anesthetics—cocaine, for instance—act by binding inside sodium channels and thereby blocking them. (Cocaine also inhibits the uptake of noradrenalin and enhances the action of the transmitter dopamine, thereby affecting the nervous system in other ways.)

LSD (lysergic acid diethylamide) produces its characteristic derangement of sensory experience and other mental functions by combining indiscriminately with receptor sites for serotonin. The benzodiazepines, of which the tranquilizer Valium is by far the most widely prescribed, interact synergistically with the inhibitory transmitter GABA to open chloride channels and thus inhibit synaptic transmission. The active ingredient of marijuana (tetrahydrocannabinol) binds to a G-protein-coupled receptor, which depresses CNS activity in complex ways.

Several neurological disorders have now been traced with more or less certainty to transmitter problems. Severe depression is firmly linked to a defect in the serotonin transport system: the brains of many suicide victims have about half of the normal serotonin level and only two-thirds of the usual number of binding sites. Abnormal dopamine levels are now tied to one form of schizophrenia, and abnormal dopamine receptors may be a major factor in one kind of alcoholism. The chemical basis of the manic-depressive syndrome (also called bipolar affective disorder) is sufficiently well understood to be treated chemically with lithium.

■ COMMUNICATING WITH MUSCLES

Just as there is a gap at the synapses between successive neurons in a neural pathway, there is also a gap between the terminals of an axon and the effector it innervates. When the effector is skeletal muscle, the gap is usually located within a specialized structure, the ***neuromuscular junction*** (or motor end plate), formed from the end of the axon and the adjacent portion of the muscle surface (Figs. 35.30, 35.31). Like most synapses between neurons, transmission across this gap is effected by transmitter chemicals, and the biochemical processes underlying neuromuscular transmission are very similar to those involved in ordinary chemical synapses.

Several kinds of neural networks in animals allow precise control of muscles. Typical of most invertebrates is a system that involves three classes of motor neurons: a fast fiber that produces large, rapid contractions; a slow fiber that generates slow, more finely controlled movement; and an inhibitory fiber that serves to block contraction. In higher invertebrates the muscles act like interneurons, integrating inhibitory and excitatory inputs. Many invertebrate muscle fibers display a graded response: they contract in proportion to the rate of stimulation.

In vertebrates the movement of skeletal muscles involves quite a different strategy. First, there are many more fibers per muscle, but each responds in an all-or-none fashion. Second, there are no inhibitory motor neurons; instead, muscles with opposite effects are paired—one that extends a finger, for instance, with one that retracts it. While in higher invertebrates muscle fibers integrate their various inputs themselves, in vertebrates this processing is accomplished by the CNS. There are far more motor-neuron fibers at the

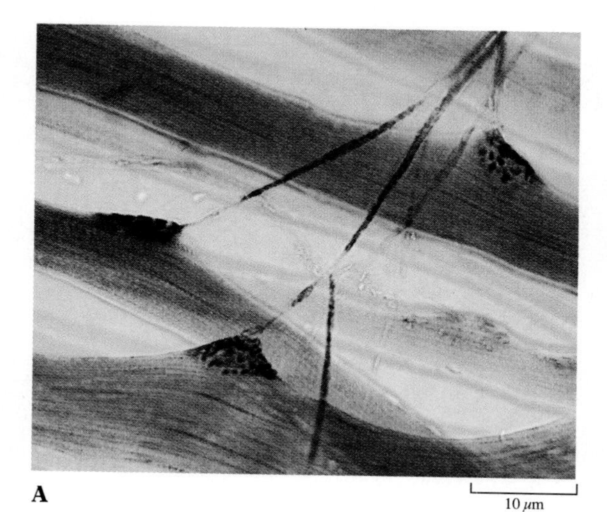

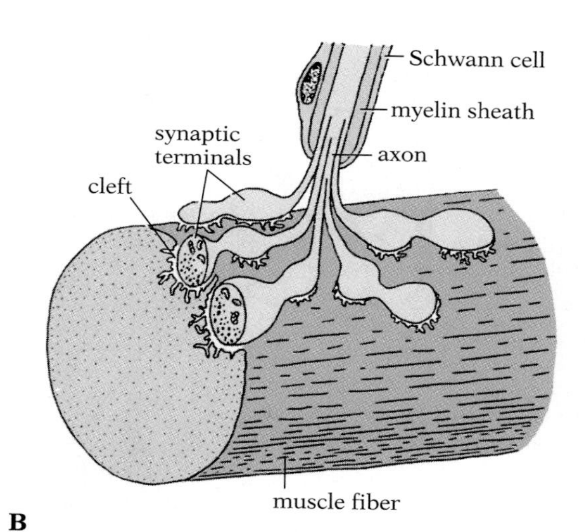

35.30 Neuromuscular junctions

(A) Toward its end, an axon supplying a muscle in a snake branches extensively and forms neuromuscular junctions on individual muscle fibers. (B) Close-up sketch of one neuromuscular junction. The junction is formed by branches of the axon, with their terminals, and the specialized adjacent portion of the muscle fiber. As in a synapse between two neurons, there is a cleft between muscle and nerve cells.

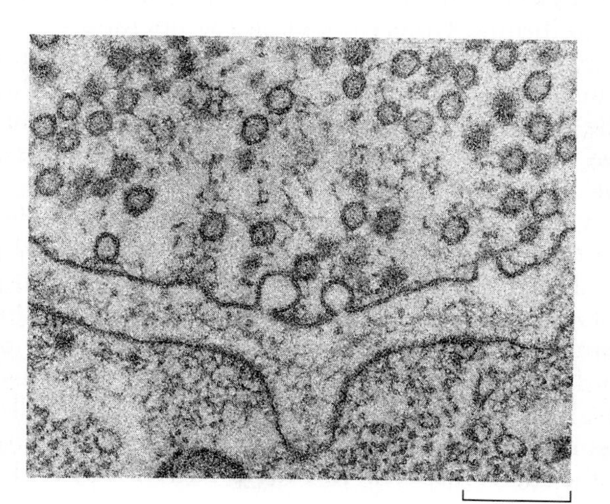

35.31 A neuromuscular junction of a frog

The upper half of the electron micrograph shows part of the terminal of an axon containing numerous synaptic vesicles, some of them releasing neurotransmitter. The lower half shows part of a muscle cell. Note the distinct cleft between the two cells.

muscles, each with correspondingly less potency, and the CNS exerts control by regulating the number that are active and their rate of firing, and by balancing the signals to paired (antagonist) muscles, so that they pull against one another.

The transmitter at the neuromuscular junctions of vertebrate skeletal muscle is acetylcholine. A variety of drugs that cause paralysis do so by blocking transmission between the motor neurons and the muscles. Examples are the poison produced by the bacterium responsible for botulism, and curare, the famous neuromuscular blocker used on the poison arrows of South American Indians. Curare is now known to act as an acetylcholine mimic: it binds to the receptors but fails to open the gates. To make matters worse, acetylcholinesterase cannot inactivate it. The receptor sites are permanently blocked, and neuromuscular transmission ceases. The action of botulin toxin, on the other hand, is not well understood. It first works presynaptically to cause terminals at neuromuscular junctions to be overactive, which results in muscular tremors and paralysis. Then it silences these same terminals, and death results. The neuromuscular junction is also the site of at least one deadly viral disease: rabies. The extreme virulence of rabies virus and the wide phylogenetic range of its hosts seem to be the consequence of the virus's specific affinity for acetylcholine receptors.

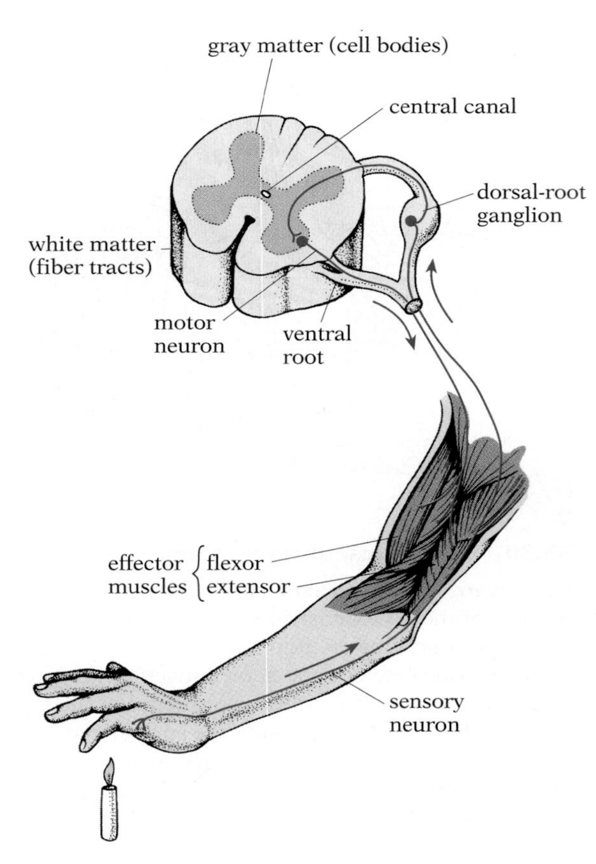

35.32 Hand-withdrawal circuit

When the hand touches something hot, signals from pain receptors reach motor neurons in the spinal cord, which in turn quickly retract the arm. These motor neurons both activate the flexor muscles and inhibit the extensors. (The central canal through the gray matter is a remnant of the hollow nerve cord evident early in development.)

HOW VERTEBRATE NEURAL PATHWAYS ARE ORGANIZED

Neural control depends on events at the level of the individual neuron and integrative processes between systems of neural cells. In vertebrates the simplest form of neural control is accomplished through the ***reflex arc***.

■ RAPID AND AUTOMATIC RESPONSES: THE REFLEX ARC

A sensory neuron connected to a motor neuron forms the simplest circuit in the nervous system. Circuits of this sort are seen in reflex arcs controlling behavioral responses that must occur quickly, such as emergency reactions and the automatic maintenance of some kind of equilibrium. Others fine-tune the operation of organs like the stomach. (A very local reflex arc, for instance, measures the pressure in the stomach and, using nitric oxide as its transmitter, relaxes the muscles of the stomach walls as appropriate.) In this section we will examine the organization of long-range reflex arcs in vertebrates.

A good example of a familiar emergency reaction is the withdrawal reflex. When we touch something hot, our hand jerks back automatically. The sensory neurons involved in this response run from the hand to the spinal cord; their cell bodies are located in a ***dorsal-root ganglion*** (or spinal ganglion) that lies just outside the spinal cord near its dorsal side (Fig. 35.32). The axons enter the cord dorsally and synapse with the dendrites or cell bodies of motor neurons within the gray matter of the spinal cord. The axons of the motor neurons exit the cord ventrally and run to the

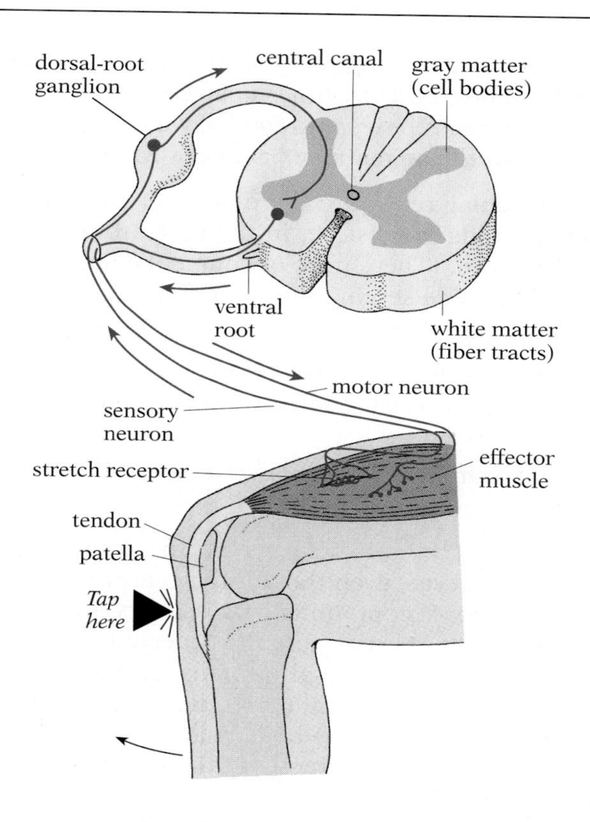

35.33 Knee-jerk reflex

This reflex, which is part of the automatic postural-control system, measures the load on the muscle (as reported by stretch receptors) and then adjusts the firing rate of the motor neurons to that muscle to compensate for any changes. The knee jerk itself results from the brief overloading of the muscle and its receptor when the tendon is tapped sharply.

muscles. In this reflex, a strong signal from the appropriate sensory cells both fires the flexor muscles and inhibits extensor muscles; this crucial motor response is well under way before the signals responsible for the conscious sensation of pain (which exit the reflex pathway in the spinal cord) ever reach the brain for analysis.

The kind of circuit that automatically maintains equilibrium is exemplified by the well-known knee-jerk reflex. Doctors test general nervous-system function by tapping a patient's knee with a rubber hammer and observing the reflex (Fig. 35.33). The sensory elements are stretch receptors, which measure the degree to which a particular muscle is stretched. As the force against which the muscle must act—the amount of weight on one leg in this case—increases, the muscle is stretched, and the receptors signal this fact through sensory neurons to the spinal cord. As in the previous example, the information is sent both to the brain for analysis and to motor neurons for immediate interim action. In this particular circuit, the motor neurons are those controlling the very muscles being monitored by the stretch receptors that have been activated. The arrival of signals from the receptor increases the firing rate of the motor neurons, and the muscles—extensors in this case—tighten to accommodate the added "load" it perceives from the hammer's tap.

This sort of automatic compensation illustrates negative feedback: a change in sensory input produces a self-correcting response that returns the system to its normal state. In this instance the tightening of the extensor in response to the stretch-receptor signal serves to shorten the muscle, thereby eliminating the stretch felt by the receptor, and so turning off the sensory signal. Thus the knee-jerk reflex automatically tunes posture.

Using these simple reflexes as a model, we can make several generalizations about the spinal reflex arcs of vertebrates:

1 For a particular reflex arc there is never more than one neuron, however long it must be, in the path from the sensory cell to the spinal cord. (There may, of course, be many such neurons running side by side serving the same function.)

2 The cell body of the sensory neuron is always outside the spinal cord in a dorsal-root ganglion.

3 The axons of sensory neurons always enter the spinal cord dorsally.

4 The axons of motor neurons always leave the spinal cord ventrally.

The sensory and motor neurons of the knee-jerk reflex run through the same nerve, even though they carry impulses in opposite directions. A nerve containing both sensory and motor fibers is called a mixed nerve. All the nerves connected to the spinal cord are mixed. There are 31 pairs in humans, all of which branch repeatedly after leaving the spinal cord, giving rise to smaller nerves that innervate most parts of the body below the head. Some nerves, on the other hand, connect directly to the brain rather than to the spinal cord. In humans there are 12 pairs of these cranial nerves, some purely motor, some purely sensory, and some mixed.

So far we have seen reflexes in their simplest isolated form. Very few reflex pathways, however, involve only two neurons in series. At least one interneuron is usually interposed between the sensory neuron and the motor neuron (Fig. 35.34), and many interneurons may be involved, their number depending on the balance between how quickly a reaction is needed and how much sensory integration and muscle coordination is desirable.

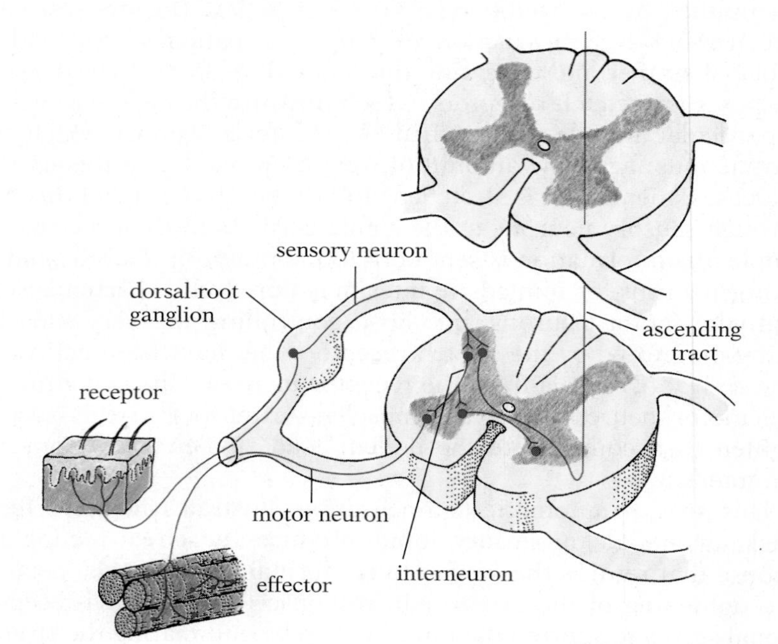

35.34 Reflex arc with interneurons

The sensory neuron synapses with several interneurons in the gray matter of the cord. Some of these interneurons may synapse directly with motor neurons on the same side, but some cross to the other side of the cord and synapse there with other motor neurons and with additional interneurons that run in ascending tracts through the cord to the brain.

It is important to keep in mind that a reflex arc, whether it includes few cells or many, makes two sorts of connections with other neural pathways. First, it almost always sends information to the brain, where instructions to counteract or augment the behavioral reaction can be issued. If you know that the doctor is going to strike your knee, for instance, you can inhibit or exaggerate the response.

The second sort of connection is with other reflex arcs. Consider for a moment the withdrawal reflex for a foot rather than a hand. Suppose you are walking barefoot and your right foot steps on a thorn. Impulses from pain receptors in the foot immediately ascend along sensory neurons to the spinal cord, activate the appropriate motor neurons, and descend to the muscles of the leg to excite the flexors and inhibit the extensors. At the same time, the interneurons send messages across the spinal cord to certain circuits for the other leg. These result in a reflex extension of the left leg in anticipation of the increased load it is about to bear. As the right foot is lifted from the ground, stretch receptors in the left leg take over control of their circuits to accommodate the extra weight more precisely. Now a new problem arises—balancing on one leg. Additional reflex arcs involving the muscles of the ankle must begin to operate, and so a cascade of independent circuits is recruited to handle the situation.

Meanwhile, of course, some of the interneurons have sent information to the brain. Part of the information has gone to portions of the brain concerned with our general level of awareness, and serve to make the victim of the thorn more alert: the process of sensitization. Still other messages have been sent to the part of the brain known as the cerebellum, where conscious motor control is coordinated. If the reflex arcs have proved inadequate for the situation or some other behavior seems called for, the cerebellum can issue commands to the muscles to take the necessary action.

Reflex circuitry is able to control and coordinate a variety of simple responses and automatically fine-tune behavior such as walking, whose details must be adjusted constantly as body weight shifts. Impressive as these interacting circuits are, however, the more complex behavior patterns of most animals stem from different pathways.

■ UNCONSCIOUS BEHAVIOR: THE AUTONOMIC SYSTEM

The central nervous system of vertebrates serves as a coordinating system for two kinds of pathways: somatic and autonomic. ***Somatic*** pathways, exemplified by reflex arcs and by more complex behavioral circuits to be discussed later, innervate skeletal muscle and include sensory and motor neurons lying largely outside the CNS. They involve, potentially, some conscious control of behavior, or at least an awareness that the behavior has occurred.

Autonomic pathways, in contrast, are basically unconscious internal reflexes; they are not ordinarily under conscious control and usually function without our being aware of them. They innervate the heart, some glands, and the smooth muscle in the walls of the digestive tract, respiratory system, excretory system, reproductive system, and blood vessels. Autonomic pathways are controlled

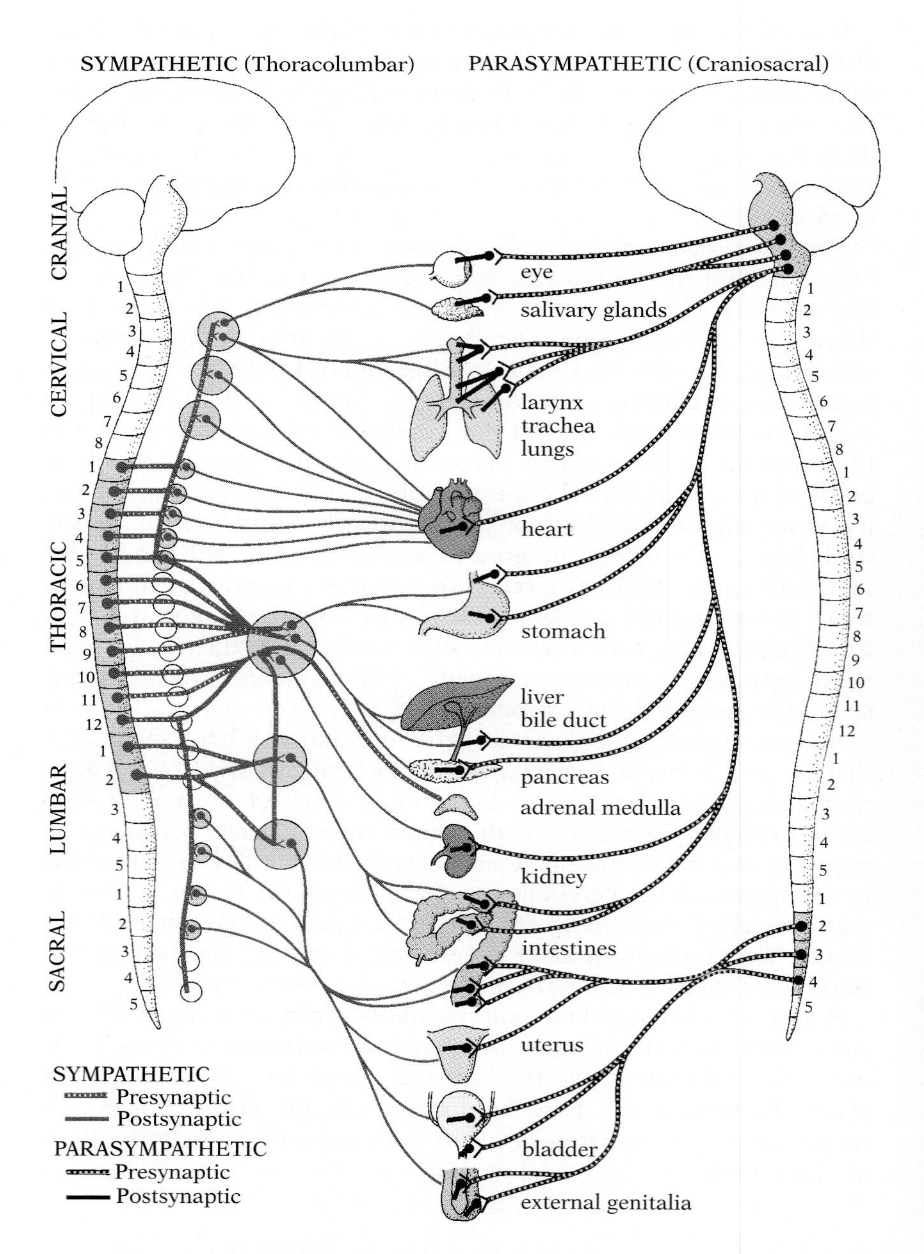

35.35 Autonomic nervous system

Of the 12 cranial and 31 spinal nerves, four cranial nerves (which emerge from the gray area of brain at upper right) and about half the spinal nerves (colored and gray segments emerging from the cord) contribute neurons to the autonomic nervous system, which innervates internal organs.

The ANS is customarily divided into two parts: the sympathetic and the parasympathetic systems. The pathways of both usually have two motor (efferent) neurons; a first (presynaptic) neuron exits from the CNS and synapses with a second (postsynaptic) neuron that innervates the target organ.

The presynaptic neurons of the sympathetic system exit from the thoracic and upper lumbar regions of the spinal cord, and synapse with the postsynaptic neurons in a series of small ganglia (circles) lying near the cord or in larger ganglia in the abdominal cavity; the postsynaptic neurons then run from the ganglia to the target organs. The presynaptic neurons of the parasympathetic system exit from the medulla of the brain and from the sacral region of the spinal cord. These are very long neurons that run all the way to the target organ, where they synapse with short postsynaptic neurons. Most internal organs are innervated by both the sympathetic and the parasympathetic system.

largely by the hypothalamus, the same part of the brain that orchestrates the many endocrine functions of the pituitary. The pathways of the autonomic nervous system differ structurally from somatic pathways in having two motor neurons instead of one. Somatic processing takes place largely in the CNS, while autonomic processing occurs in ganglia outside the CNS.

The sympathetic versus the parasympathetic system The autonomic nervous system (ANS) is composed of two parts, the sympathetic and the parasympathetic systems (Fig. 35.35), which differ both structurally and functionally. In the ***sympathetic system***, the cell bodies of the first motor neurons lie *inside* the thoracic and lumbar portions of the spinal cord. The axons of these neurons exit ventrally from the cord and run to ganglia lying near the cord;

there they synapse with the second motor neurons, whose cell bodies lie in the ganglia. Because the synapse between the first and second motor neurons occurs in a ganglion that is at a distance from the target organ, the axon of the second motor neuron is often quite long.

Two principal structural differences distinguish the ***parasympathetic system*** from the sympathetic system. First, the cell bodies of the first motor neurons in the parasympathetic system lie in the brain and in the sacral region of the spinal cord. Second, the synapses between the first and second motor neurons of the parasympathetic system occur quite near the target organs, or even inside the walls of those organs. As a result, the axons of the second motor neurons are relatively short.

Most internal organs are innervated by both sympathetic and parasympathetic fibers, with the two systems functioning largely in opposition to each other. Together they fine-tune the balance between what we might call active and passive behavior. At one extreme, the sympathetic system prepares an animal for emergency action during a crisis: it shuts down oxygen-consuming processes that are not immediately essential, such as digestion, and prepares the machinery that may be necessary for fighting or fleeing. At the other extreme, the parasympathetic system restores order or passivity after a crisis has ended, restarting the important but not immediately critical processes that have been shut down, and freeing the system of anxiety-producing chemicals. Normally, of course, animals function somewhere along a continuum between the active emergency behavioral state produced by the sympathetic system and the passive, almost vegetative behavior produced by the parasympathetic system.

In general, the sympathetic system gives rise to the same effects as the hormones of the adrenal medulla—the fight-or-flight reactions—but does so more rapidly. The nervous mechanism is far more important than the endocrine system in preparing an animal for emergency situations, as indeed it should be, considering how much faster neural transmission pathways are. The reason the sympathetic nervous system and the hormones of the adrenal medulla produce similar effects seems clear. While the transmitter at the neuromuscular junctions of both somatic and parasympathetic pathways is acetylcholine, the transmitter at the sympathetic terminals is noradrenalin (or, in a few cases, adrenalin). The effect on the target organs is identical because the transmitters of the sympathetic system and the hormones of the adrenal medulla are the same.

The reason the adrenal medulla releases the same substances as the sympathetic system apparently lies in a fascinating functional and evolutionary relationship between them. As we have seen, autonomic pathways normally include two motor neurons. There is a single exception: the sympathetic pathway to the adrenal medulla has only one motor neuron (Fig. 35.35). In the embryo, the adrenal medulla forms from presumptive nervous tissue (tissue that is destined to become neurons), and is itself actually a cluster of highly modified second motor neurons in this sympathetic pathway, specialized for neurosecretion rather than conduction. Evolution has converted what were probably once motor neurons of the sympa-

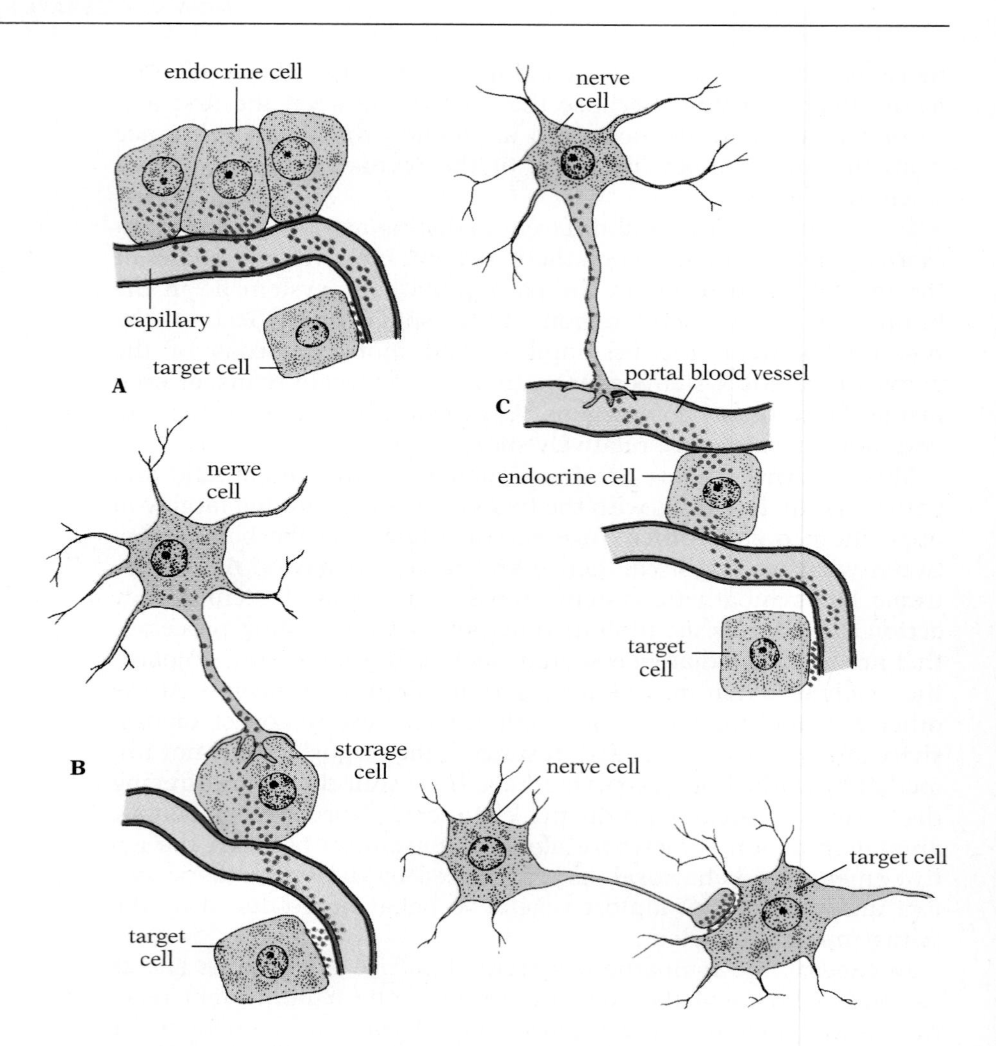

35.36 Four arrangements for chemical control

(A) Endocrine cells typically secrete their hormone directly into the blood, which carries the hormone to the target cells. (B) Some hormones secreted by nerve cells in the hypothalamus are stored in cells of the posterior pituitary, and later released into the blood. (C) Releasing hormones secreted by nerve cells in the hypothalamus are carried by the blood in a special portal system to endocrine cells of the anterior pituitary, which respond by secreting their own hormones into the general circulation. (D) A more typical nerve cell secretes its transmitter substance directly onto the target cell; no transport by the blood is involved.

thetic nervous system into an endocrine gland specialized for high-quantity secretion of the very substances that nearly all second motor neurons in this system produce.

Here, then, is another example of the parallels between the nervous system and the endocrine system, and of the similarity not only in their purpose but in their mechanisms of action (Fig. 35.36). Neurosecretion, after all, is fundamental to nerve action: impulse transmission across synapses depends on it. Natural selection seems to have favored a pair of parallel pathways: a high-speed nervous system and a low-speed endocrine system. This helps explain why many of the chemicals secreted by nervous tissue are basically hormones.

Autonomic control of heartbeat The nervous control of heart rate provides a good example of autonomic control, in which the effector—heart muscle in this case—is not under conscious control. (We may alter its rate of contraction indirectly, of course, by deliberately working into an emotional or excited state, or by consciously relaxing).

As detailed in Chapter 30, nervous impulses are not necessary to make the heart beat; beats are initiated directly by the S–A node in the wall of the heart itself. But the rhythm of the S–A node can be modified by impulses coming to it from both sympathetic nerves, which excite the S–A node, and parasympathetic nerves, which inhibit it. Though the sympathetic nerves to the heart emanate from the CNS in the thoracic region of the spinal cord, the impulses they carry originate in a cardiac-accelerating center in the medulla of the brain. Similarly, the impulses carried by the parasympathetic nerves to the heart originate in a cardiac-decelerating center in the medulla.

If blood pressure is high, pressure receptors in the carotid artery of the neck and in the arch of the aorta (and to a lesser extent in other arteries) are stimulated. Impulses travel from them to the medulla, where the sympathetic pathways from the cardiac-accelerating center are inhibited, while the parasympathetic pathways from the cardiac-decelerating center are activated. The result is that heart rate is slowed. As blood pressure falls, however, there is less stimulation of the pressure receptors, which consequently send fewer impulses to the medulla. The sympathetic pathways, freed of inhibition, begin carrying more impulses from the accelerating center to the S–A node, while the parasympathetic ones carry fewer. Impulses from chemoreceptors in the carotid artery and aorta, which respond to lowered O_2 and elevated CO_2 levels in the blood, can accelerate the firing rate, as can receptors in the wall of the right atrium, which respond to the filling of the atrium with blood. The result is an acceleration of heart rate. (As described in Chapter 33, a pair of hormones is also involved in modifying the strength of heart contractions in response to changes in blood pressure.)

The actual rate of heartbeat thus depends in part on the relative activity of the accelerating and decelerating centers in the medulla; the activity of these centers in turn reflects the amount of excitation they receive from the stretch receptors and the chemoreceptors in the arteries. These autonomic reflex circuits serve to fine-tune heart rate with a negative-feedback loop. The two centers in the medulla are also significantly influenced by signals from other parts of the brain. For example, when a person sees something frightening, impulses from processing centers in the brain that send signals to the medulla quicken the pulse. These two classes of control, one a feedback-regulated system that maintains bodily processes on an even keel, and the other an emergency system to take over in life-threatening situations, are analogous to the parasympathetic and sympathetic systems of the CNS. They also parallel the two kinds of reflexes discussed earlier: those that maintain equilibrium, such as the knee-jerk reflex, and those that handle emergencies, such as the withdrawal reflex.

■ COMPLEX BEHAVIOR: MOTOR PROGRAMS

Complex behavior—behavior that integrates various sensory inputs and is expressed through coordinated movements of several muscles in sequence—was once supposed to be organized through interacting reflex arcs. Each reflex arc presumably monitored a single sensor and controlled a single muscle (or a pair of antagonistic muscles, as in the hand-withdrawal reflex). Learned behavior, such as maze running in rats, was thought to come about through the linking of such reflex circuits into chains.

Insect flight The very plausible idea that complex behavior is built out of chains of reflexes was effectively disproven around 1960 by Donald Wilson, who was studying the circuitry underlying locust flight. Flight behavior seemed to be ideally suited for control by chains of reflexes: one set of flight muscles would raise the wings until the stretch receptors in the other set were sufficiently active to trigger the contractions that brought the wings down. As the wings came down, the stretch receptors in the "up" set of muscles would signal the increasing strain until what might be called the "up reflex" was triggered; the wings would then be pulled up until the receptors in the "down" muscles triggered the "down reflex"; and so on *ad infinitum*.

By tracing out the circuitry, Wilson discovered the requisite stretch receptors, and he cut the nerves running from the receptors to the thoracic ganglia. Wilson expected that once the triggering elements had been removed from the circuit, the behavior would collapse. To his surprise, the locusts continued to fly almost normally, though the wing beat rate dropped slightly and coordination suffered somewhat. Aside from the fine-tuning, the behavior remained intact.

Locust flight depends not on reflexes but rather on a special flying circuit—a ***motor program***—in the thoracic ganglia. This circuit generates all the commands necessary to produce flight (Fig. 35.37). Normally, of course, the circuit is fine-tuned by the stretch receptors.

Since Wilson's discovery, most of the complex rhythmic or sequential behaviors that have been investigated in detail in vertebrates and invertebrates have been found to rely on specialized self-contained circuits that orchestrate muscle movements in the proper order and with the correct timing; no complex behavior based on reflex chains is known. The list of behaviors controlled by

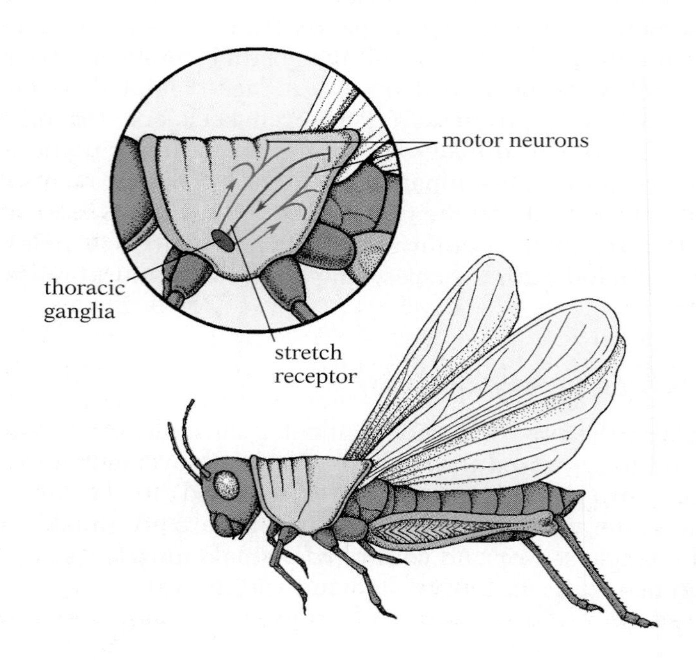

35.37 Flight circuitry in a locust

The motor neurons for the flight muscles are located in the thoracic ganglia, and the stretch receptors for the flight muscles send the information they collect about muscle position to the same ganglia. When flight is triggered by outside cues (loss of foot contact with the ground after jumping, for instance), a circuit (red) in the thoracic ganglia is activated that generates instructions for all the flight muscles. The information from the stretch receptors (blue) is used to fine-tune flight behavior.

motor programs includes feeding, walking, swimming, scratching, vocalizing, building, attack, courtship and mating, and so on.

Feeding in *Aplysia* As research techniques have improved, neurophysiologists have been able to map, cell by cell, some of the circuits responsible for complex behavior. A good example of an almost fully mapped behavioral pathway is the circuit controlling feeding in *Aplysia*. It illustrates the four steps in information flow—detection, conduction, processing, and response—in the context of relatively complex behavior.

Aplysia, which feeds on seaweed, has chemosensitive tentacles on its head that detect potential food. Signals from these receptors go to the brain for analysis. If the olfactory processing network determines that the animal is sensing suitable food (molluscs are able to learn food odors), it sends signals to the interneurons that control feeding behavior. However, these interneurons also receive input from other parts of the brain that may *prevent* feeding. Escape, for example, takes precedence over feeding, so when the animal senses danger, low-priority activities like feeding are instantly terminated.

When feeding is to proceed, a motor program is activated (Fig. 35.38); as in many invertebrate behaviors, the CNS simply stops inhibiting the appropriate circuit in a ganglion. In feeding, 22 muscles are controlled by six motor neurons that are themselves coordinated by two antagonistic groups of "command" interneurons, the protractors and the retractors. The protractor and retractor muscles are analogous to extensors and flexors, respectively. The protractor interneurons spontaneously produce rhythmic bursts of impulses every 3–4 sec. Signals from these cells activate in a particular order the muscles that open the mouth and bring its rasplike radular "teeth" down into position to bite its food. At the same time, these cells also inhibit the retractor motor neurons and their command interneurons.

When the rhythmic cycle of the protractor interneurons brings them into their quiet phase, they no longer inhibit the retractor interneurons. These cells, which fire continuously in the absence of inhibition, now activate the retractor motor neurons and inhibit those controlling the protractor muscles, thereby directing the sequence of muscle movements necessary to produce a biting movement and closing of the mouth.

The feeding circuit in *Aplysia* illustrates the three basic characteristics of a typical motor program: (1) The program is run by a self-contained circuit that coordinates the movement of several muscles. (2) Though self-contained, it takes advantage of sensory feedback to fine-tune the behavior automatically, in this instance by extending and strengthening the mouth-opening phase when appropriate. (3) It is under direct central control from the brain. In *Aplysia* the brain integrates information from chemosensory cells and "hunger" sensors, and then allows this motor program to be active when it is needed. Similarly, when other behavior becomes necessary—escape from a predatory starfish, for instance—the brain can switch off the feeding behavior. The brain can also order major changes in the rate of the behavior, lengthening or shortening the cycle time as appropriate.

The strategy of using two interacting sets of command cells, like

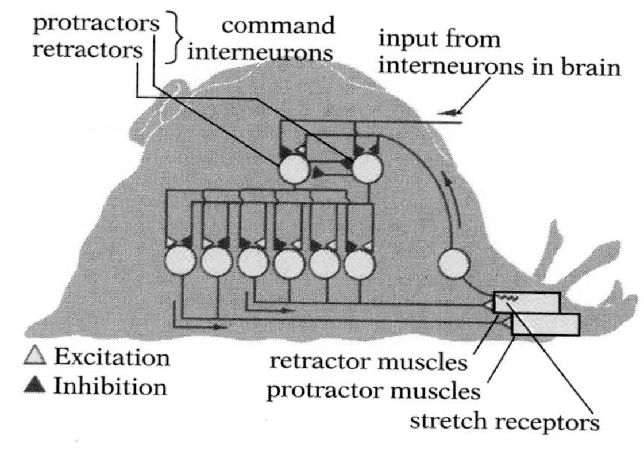

35.38 *Aplysia* feeding circuit

Aplysia feeds by opening its mouth with protractor muscles, and then closing it with retractor muscles. The protractor interneurons stimulate protractor motor neurons and simultaneously inhibit retractor motor neurons, while the retractor interneurons have the opposite effect. The two sets of interneurons coordinate their activity by taking turns (see text). The self-generated pattern of opening and closing can be modified by information arriving from stretch receptors in the retractor muscles. When the load on these muscles is unusually high, indicating that the piece of seaweed is too large or tough for the normal bite pattern, the stretch receptors cause the next opening phase to be extended. Input from interneurons in the brain keeps the circuit turned off except when it is needed.

Aplysia's protractors and retractors, is known as ***reciprocal inhibition***. It's probably the way virtually all rhythmic behavior is organized. Motor programs appear also to underlie nonrhythmic behavior such as swallowing and smiling. As described in Chapter 38, strategies have evolved for orchestrating even more complex behavior—the building of a bird nest, for instance, or the sequence by which a cat stalks, chases, catches, kills, disembowels, and eats a mouse: higher-level circuits coordinate the interactions of individual motor programs.

The realization that so much of behavior is accomplished by interacting groups of neural circuits similar to those of *Aplysia* is a major achievement for modern neurophysiology.

CHAPTER SUMMARY

HOW NERVOUS SYSTEMS ARE ORGANIZED

THE UNUSUAL ANATOMY OF NEURONS Neurons have dendrites to collect information and long axons to carry signals to distant targets. Communication from the axon of one cell to the dendrite of another occurs across synapses. Some glia serve to insulate axons, and thus speed conduction. (p. 991)

HOW SIMPLE NEURONAL PATHWAYS WORK Information is collected by sensory neurons and, in simple reflex circuits, passed to motor neurons. Most circuits, however, have interneurons that first integrate and process information from sensory neurons before passing it on to target cells. (p. 995)

THE SIMPLEST NERVOUS SYSTEMS: NERVE NETS Nerve nets carry general excitation throughout an organism. (p. 998)

WHY BRAINS EVOLVED IN THE HEAD Stimuli are often encountered by the front end of an organism; thus sensory receptors and their interneurons are usually concentrated there. (p. 999)

HOW THE NERVOUS SYSTEM IS ORGANIZED Most cell bodies and dendrites are localized ventrally in ganglia, with nerves containing many axons running between them. Insect ganglia are relatively self-sufficient: their wiring contains the information necessary to generate most behavior; the brain serves to trigger (or stop inhibiting) and orient behavior. (p. 1000)

WHAT'S SPECIAL ABOUT VERTEBRATES? The ganglia of vertebrates are located dorsally and are associated with the spinal cord; nearly all behavior is orchestrated in the brain. (p. 1001)

HOW NEURONS WORK

HOW AXONS CONDUCT Neurons have a resting membrane potential. When depolarized past a threshold, voltage-gated Na^+ channels open and close quickly, briefly admitting Na^+ ions that reverse the membrane potential. Voltage-gated K^+ channels open and close more slowly, allowing K^+ ions to escape and thus restoring the resting potential. This cycle of depolarization and repolarization occurs in successive patches of membrane, with each cycle triggering the cycle in the neighboring patch. The sodium-potassium pump maintains the resting potential. (p. 1002)

HOW NEURONS COMMUNICATE When an action potential arrives at a synapse, it triggers the release of transmitter chemicals, which bind to receptors on the target cell; there they cause changes in ion movement that either depolarize or hyperpolarize the target-cell membrane. Most target cells are muscle cells or dendrites of other neurons; axons can also synapse on other synapses. Target-cell response depends on the integration of incoming excitatory and inhibitory stimuli. Learning appears to involve both habituation and coordinated sensitization of synapses. Many neuro-active drugs act by altering the behavior of transmitter chemicals or their receptors. (p. 1009)

COMMUNICATING WITH MUSCLES Synapses between neurons and muscles are called neuromuscular junctions. Invertebrate muscles are controlled by the balance between excitatory and inhibitory synaptic activity; in vertebrates the relative stimulation of pairs of antagonistic muscles controls the rate and degree of contraction. (p. 1015)

HOW VERTEBRATE NEURAL PATHWAYS ARE ORGANIZED

RAPID AND AUTOMATIC RESPONSES: THE REFLEX ARC Vertebrate reflex arcs begin with sensory cells, which communicate through the spinal cord to muscles that effect the response; side branches usually carry excitation to the brain and to other reflex circuits as well. (p. 1016)

UNCONSCIOUS BEHAVIOR: THE AUTONOMIC SYSTEM Autonomic pathways control unconscious behavior via two sets of connections to each target, one from the sympathetic system, the other from the parasympathetic system. Sympathetic connections rouse animals in preparation for high activity; the parasympathetic system restores calm. Heartbeat, for instance, is under this antagonist control, with the sympathetic system accelerating the rate and the parasympathetic system slowing it. (p. 1019)

COMPLEX BEHAVIOR: MOTOR PROGRAMS Innate behavior is orchestrated by self-contained circuits rather than chains of reflexes; the details of the behavior may be controlled by feedback, and triggering of such a motor program is usually controlled by the brain. (p. 1023)

STUDY QUESTIONS

1 Most representations of nerve impulses show them as sharp spikes moving along axons. Given that the conduction rate is 30–90 m/sec, that the "spike" of the action potential lasts about 2 msec, and that axons range in length from 1 mm to 1 m, what does the spike really look like? (pp. 1002–1003)

2 Many circuits in organisms are designed to record the onset of a stimulus, and so are wired to suppress all but the first impulse. Design such a circuit.

3 When the kidneys or the hormonal systems that help regulate them begin to fail, the Na^+ concentration in the body's extracellular fluid can become too high or too low. What effect would high and low sodium concentrations have on neurons, both in the short run and the long term? (pp. 1002–1008)

4 Compare EPSP-IPSP integration on dendrites with the hormone-antagonist system as a way of maintaining fine control. (pp. 961, 1011–1012)

5 What sorts of behavioral and physiological experiments would be useful to prove that the song of mourning doves is an innate motor program? (pp. 1023–1026)

SUGGESTED READING

ALKON, D. L., 1989. Memory storage and neural systems, *Scientific American* 260 (1). *On the synaptic chemistry of learning.*

JACOBS, B. L., 1987. How hallucinogenic drugs work, *American Scientist* 75 (4). *On the interaction between drugs and the serotonin receptor.*

KIMELBERG, H. K., AND M. D. NORENBERG, 1989. Astrocytes, *Scientific American* 260 (4). *On an important kind of glia.*

LENT, C. M., AND M. H. DICKINSON, 1988. The neurobiology of feeding in leeches, *Scientific American* 258 (6).

NEHER, E., AND B. SAKMANN, 1992. The patch clamp technique, *Scientific American* 266 (3). *On the use of an isolated patch of membrane to study the operation of single ion channels.*

SNYDER, S. H., AND D. S. BREDT, 1992. Biological roles of nitric oxide, *Scientific American* 266 (6). *On the many roles of this newly discovered transmitter, local chemical mediator, and toxic weapon.*

STEVENS, C. F., 1979. The neuron, *Scientific American* 241 (3). *An excellent description of how nerve cells work.*

CHAPTER 36

HOW ENVIRONMENTAL CHANGES ARE DETECTED

HOW SENSORY STIMULI PRODUCE NEURAL IMPULSES

HOW RECEPTORS ENCODE STIMULUS STRENGTH

THE MANY SENSES OF TOUCH

MONITORING INTERNAL CONDITIONS

DETECTING TASTE AND SMELL

HOW ANIMALS SEE

HOW HEARING WORKS

MAINTAINING BALANCE AND SENSING MOVEMENT

SPECIALIZED AND UNUSUAL SENSES

MAKING SENSE OF SENSATIONS

HOW SENSORY DATA ARE PROCESSED

HOW THE VERTEBRATE BRAIN HAS EVOLVED

HOW THE FOREBRAIN IS ORGANIZED

SENSORY RECEPTION AND PROCESSING

Even simple procaryotes like *E. coli* collect information about their surroundings, "process" that data, and adjust their behavior appropriately. In this chapter we will examine the specialized receptor cells and analysis circuits that help higher animals make adaptive changes in their behavior.

HOW ENVIRONMENTAL CHANGES ARE DETECTED

HOW SENSORY STIMULI PRODUCE NEURAL IMPULSES

Each kind of sensory receptor is specialized for detecting one class of stimuli and converting the stimulus into a change in membrane polarization—the common currency of the nervous system. This process is called ***sensory transduction.*** The range of sensory specializations includes pressure, heat and cold, concentrations of particular chemicals, vibrations, light, electrical and magnetic fields, and so on.

In sensory cells the appropriate stimulus opens or closes gated channels, thus causing a change in membrane potential. For instance, when light is absorbed by visual pigments, the activated pigments close sodium channels in the membrane, thereby hyperpolarizing the sensory cells. Sodium channels in stretch-receptor cells are thought to open (and therefore to depolarize the membrane) simply as a result of the physical distortion of the membrane caused by stretching.

The receptor proteins in the membranes of sensory cells of the nose and tongue open Na^+ channels when the appropriate chemical substrates bind to them. When enough sugar molecules, for example, bind to receptor proteins on a sugar-sensitive cell, so many sodium channels open that the sensory cell depolarizes past threshold and fires. An adjacent cell that has receptor proteins for bitter chemicals like quinine will not respond, since its ion channels are not affected. In the presence of sufficient quinine, however, enough of its channels will open to depolarize *it* past threshold. The sensations of sweet and bitter are thus conveyed by identical means—the opening of Na^+ channels, followed by depolarization, followed by the generation of action potentials. They are distinguished because the different kinds of sensory cells are connected

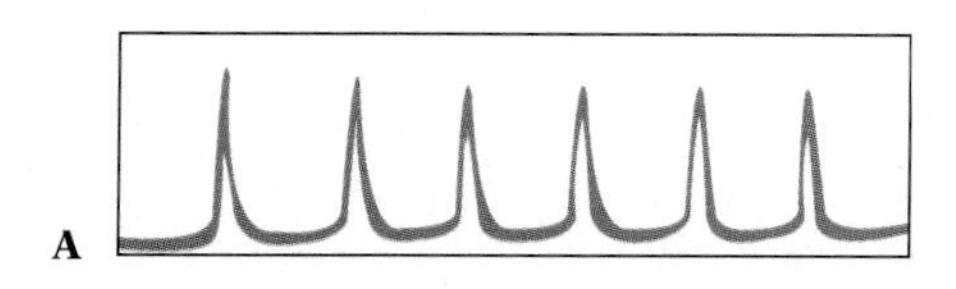

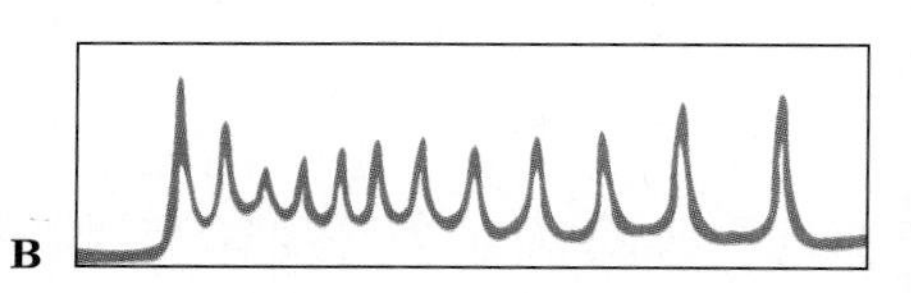

36.1 Relationship between impulse pattern and intensity of stimulation in a stretch-receptor cell

(A) A neuromuscular spindle is stretched slightly. The stimulus thus applied produces a generator potential (shown here as an upward shift in the baseline) that triggers a series of impulses (action potentials). (B) When the same neuromuscular spindle is stretched more, the generator potential (baseline) rises higher, and the frequency of the impulses increases. Thus the frequency of impulses is a function of the intensity of the generator potential, which in turn is a function of the strength of the stimulus.

to different targets in the brain. The specific sensations we experience are the brain's interpretation of incoming stimuli.

HOW RECEPTORS ENCODE STIMULUS STRENGTH

How do sensory cells encode the strength and (in many cases) the temporal pattern of the stimuli? The simplest example is the vertebrate stretch receptor.

Stretching a neuromuscular spindle produces a local depolarization of the receptor-cell membrane. When this depolarization, known as the ***generator potential***, reaches threshold level, it triggers an action potential in the nerve fiber (Fig. 36.1). An increase in the intensity of the stimulus (increased stretching in this case) causes a proportional rise in the generator potential. This rise, in turn, increases the frequency of the triggered impulses. Thus the frequency of output from the receptor is a direct measure of the strength of the stimulus.

Stretch-receptor cells differ in their behavior from most sensory cells: (1) They do not fire unless stimulated, whereas most sensory cells fire at a slow ***basal rate***, even when unstimulated. (2) Stretch-receptor cells continue to fire at a rate proportional to the stimulus intensity, whereas most sensory cells undergo ***adaptation***. For example, when a simple heat receptor in the skin is stimulated by warm water, it begins to fire at a higher rate; as time goes on, the sensory cell becomes less sensitive to temperature and the firing rate declines. Eventually the firing rate drops to the basal level, and we no longer sense that the water is warm. Adaptation differs from habituation (described in Chapter 35) in that adaptation occurs in the sensory cell rather than in or on an interneuron, and it is not abolished by sensitization.

This slow adaptation strategy makes good sense for most sorts of receptor cells because it allows them to operate over a far larger range of stimulus intensities than would otherwise be possible. For example, humans are sensitive to temperature variations over a range of at least 50°C. If a cell responded to these changes by varying its output frequency from, say, zero to 100 impulses per second, its ability to sense accurately the sorts of small temperature changes we normally experience and react to would be very poor: these small changes would produce only small changes in the cell's output frequency. Instead, however, the receptor cell responds accurately only over a range of 10°C, but through adaptation its sensitivity is centered around the current temperature. Hence it can report with high accuracy on minute changes in temperature.

Cells that adapt slowly are intermediate between the two extremes of receptor response: ***phasic*** cells, which adapt almost instantly, and ***tonic*** cells, which essentially never adapt (Fig. 36.2).

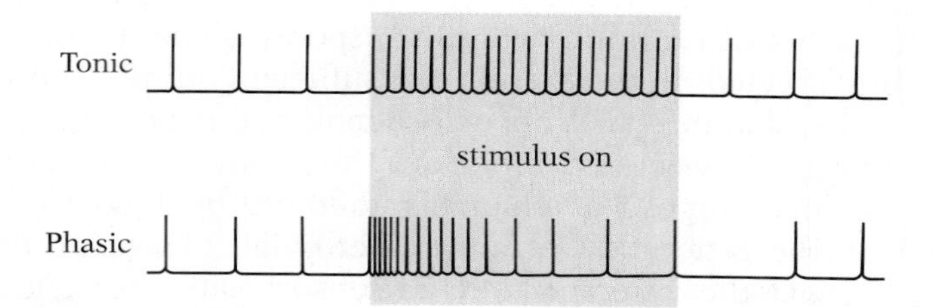

36.2 Varieties of receptor response

The response of most sensory receptor cells to a stimulus usually falls somewhere between the two extremes of tonic and phasic response. A highly tonic receptor fires at a continuous rate proportional to the strength of the stimulus (color) and immediately returns to its unstimulated state once the stimulus is removed. Since tonic cells respond efficiently only over a relatively narrow range of stimulus intensity, a series of receptors with differing ranges is required to monitor the strength of the stimulus effectively. By contrast, a highly phasic receptor returns to its basal firing rate soon after the onset of the stimulus, even while the stimulus continues to be applied. Removal of the stimulus on the highly phasic cell causes its firing rate to drop below its basal level until adaptation to the removal has taken place. As a result, phasic cells provide little information about the magnitude of a stimulus; instead, they provide precise information about the onset and end of stimulation, as well as indicating when subtle changes in stimulus strength occur.

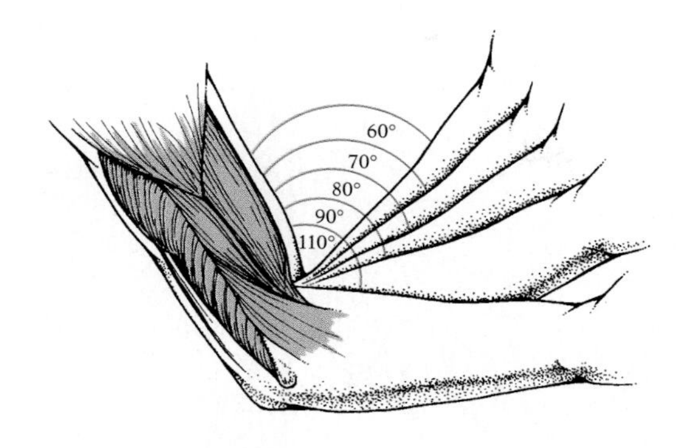

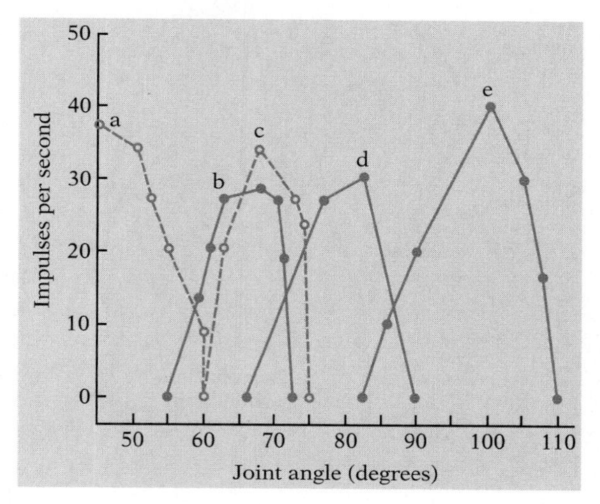

36.3 How joint receptors measure joint angles

Each receptor cell measures with great precision the degree to which a joint element is stretched (which in this case serves to define the angle between the upper and the lower arm); however, the limited maximum firing rate of each cell restricts the range over which its measurements can be made. As a result, many different receptor cells with overlapping sensitivity ranges come into play. Here five cells (out of a much larger population) with overlapping ranges encode the joint angle over a wide range. Notice how the pattern of activity shifts over a range of just 10°: at 60°, for example, spindle **b** responds with 21 impulses per second, while **c** is silent; at 65° **b** responds with 28 per second and **c** with 31 per second; at 70°, **b** responds with 20 per second and **c** with 28 per second. The relative firing rates are unique for each angle.

The vertebrate stretch receptor, described earlier, is tonic. A joint-receptor cell is another good example of a tonic cell: it continuously provides an accurate, unadapted measure of muscle loading (the amount of stress "loaded" onto the receptor). Accuracy in this task, of course, means that an individual sensory element can function only over a rather narrow range, but this problem is circumvented by a series of receptor cells with different but overlapping ranges (Fig. 36.3).

Phasic cells, on the other hand, are specialized to detect change. Invertebrate muscles, for instance, have two sorts of stretch-receptor cells: a nonadapting one virtually identical to the tonic cells seen in vertebrates, and a fast phasic cell that is sensitive to changes in muscle loading. The phasic cell produces a burst of impulses at the slightest alteration in muscle stretch, thus providing very precise information about when the change occurred, though very little about the actual *amount* of loading. Phasic cells are common in the vertebrate nervous system, but most sensory cells in both vertebrates and invertebrates fall somewhere between tonic and phasic extremes.

■ THE MANY SENSES OF TOUCH

There are several types of sensory receptors in vertebrate skin (see Fig. 25.12, p. 711), and each has an analogue in invertebrates. The vertebrate receptors are concerned with at least five different

36.4 Three receptors of the skin and the senses they mediate

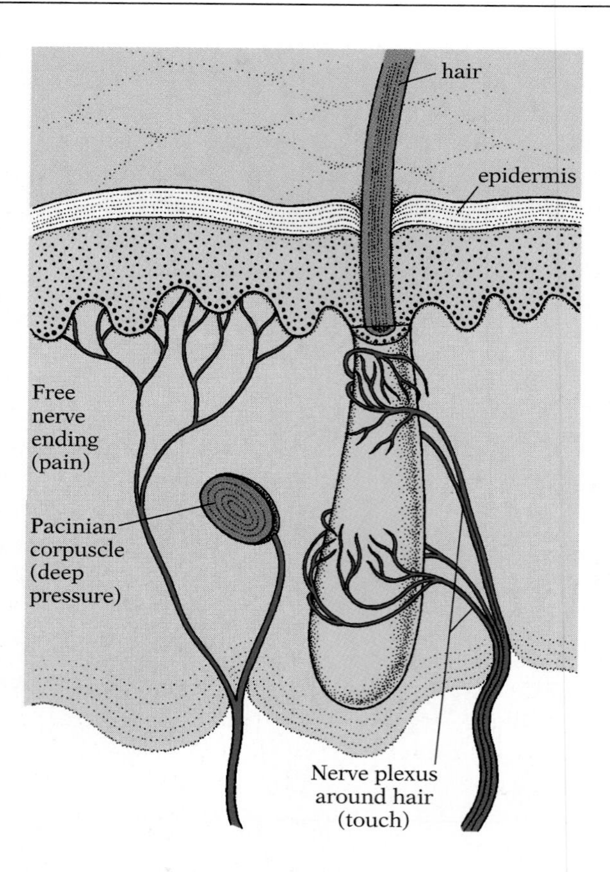

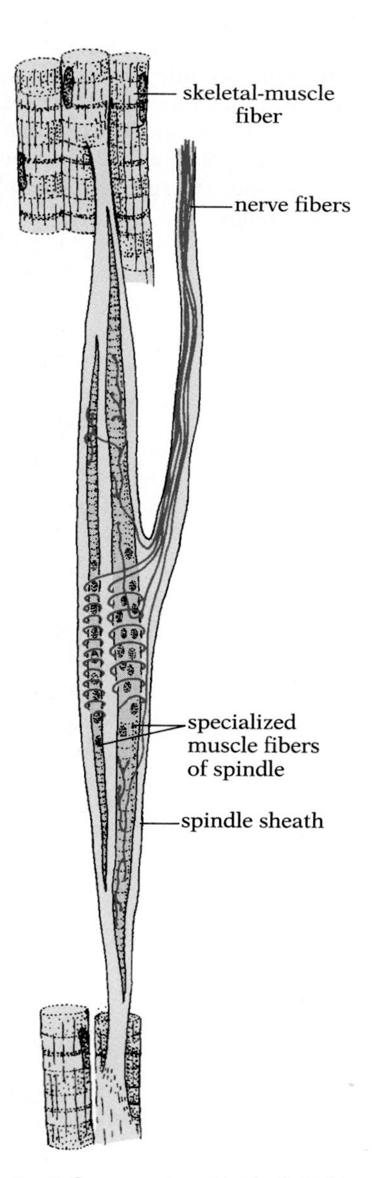

36.5 A stretch receptor in skeletal muscle

The terminal branches of sensory nerve fibers are intimately associated with several specialized muscle fibers that form the apparatus called a neuromuscular spindle.

senses: touch, pressure, heat, cold, and pain. Some of the skin receptors, particularly those concerned with pain, are simply the unmyelinated terminal branches of neurons (Fig. 36.4). Others are nets of nerve fibers surrounding, in mammals, the bases of hairs; these fibers, which are particularly important to the sense of touch, are stimulated by the slightest displacement of the tiny hairs present on most parts of the body. Still other skin receptors are more complex, consisting of nerve endings surrounded by a capsule of specialized connective-tissue cells (Fig. 36.4).

The relative abundance of the various types of receptors differs greatly in humans: pain receptors are about 30 times more abundant than cold receptors; cold receptors are about 10 times more abundant than heat receptors. Nor are the receptors distributed evenly over the body: touch receptors, for instance, are much more densely packed in the lips than on the back, while the fingertips have an intermediate concentration. The differences in density correlate well with the different functions of the body parts.

■ MONITORING INTERNAL CONDITIONS

Unlike the receptors in the skin, which receive information from the environment, certain other receptors function primarily in collecting data about the body itself. Among these immensely important monitors are the stretch receptors in the muscles and tendons (Fig. 36.5). They are sensitive to the changing tensions of muscles, and send impulses to the central nervous system (CNS) informing it of the position and movement of the various parts of the body. It is their well-tuned sensitivity to change that gives rise to the knee-jerk reflex (see Fig. 35.33, p. 1017).

Other internal receptors include those of the so-called visceral senses, located in the internal organs. Among those mentioned earlier are the receptors in the carotid artery sensitive either to elevated CO_2 concentration or high blood pressure. We seldom perceive the activity of such visceral receptors; responses to their messages are usually mediated by the autonomic system. Certain visceral receptors, though, help produce the conscious sensations of hunger, thirst, and nausea.

■ DETECTING TASTE AND SMELL

The receptors for taste and smell are chemoreceptors; they form weak bonds with particular chemicals. When we speak of a taste, we are usually referring to a compound sensation produced simultaneously by taste and olfactory (smell) receptors. Food often seems tasteless when we have a cold because the mucus coating our inflamed nasal passages blocks the olfactory receptors.

The four tastes The receptor cells for taste in our species are located in ***taste buds*** (Fig. 36.6) found mainly on the upper surface of the tongue. The receptor cells themselves are not neurons, but specialized cells with microvilli on their outer ends (Fig. 36.7). The ends of nerve fibers lie very close to these receptor cells, and when a receptor is stimulated, it generates impulses in the fibers.

There are four basic tastes: sweet, sour, salty, and bitter. The purest stimuli for these sensations are sucrose, H^+, NaCl, and quinine, respectively. In humans, receptors that respond maximally to the four basic taste stimuli are not evenly distributed: sweet and salt are most easily sensed at the front of the tongue, bitter at the

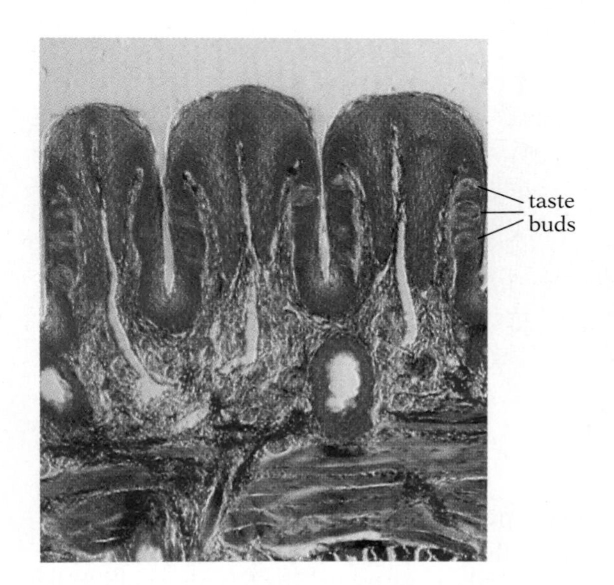

36.6 A section of mammalian tongue

The taste buds are oval structures located in the walls of the deep narrow pits.

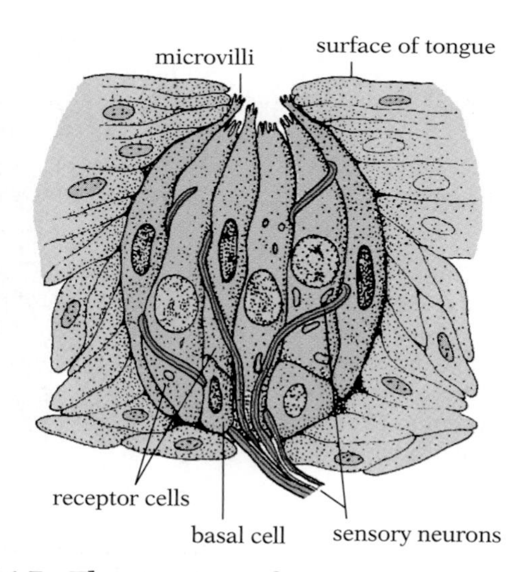

36.7 The structure of a taste bud

Each taste bud contains specialized receptor cells bearing sensory microvilli that are exposed in pits on the tongue surface. The ends of sensory neurons (blue) are closely associated with the receptor cells.

36.8 Taste receptors in a blowfly (*Calliphora vicina*)

(A) This scanning electron micrograph shows the proboscis of a blowfly. Most of the sensory hairs surrounding the labellum contain taste receptors. (B) The two small upward-curving hairs, one above each of the large claws on the lower part of the fly's foot, also contain taste receptors. Some of the forward-projecting hairs are probably touch receptors.

back, and sour at the sides. Unlike the four "pure" stimuli, though, most chemicals stimulate sensations of more than one category. The differential blending of these four sensations probably carries the information by which we distinguish tastes.

Some amount of blending is, in fact, inevitable. When neurophysiologists first succeeded in recording the activity of taste receptors in insects, where they are conveniently located near sensory hairs on the feet as well as on the mouthparts (Fig. 36.8), they usually found four or five taste-receptor cells per sensory hair. Each cell was neatly tuned to a particular taste class (though in many species one or another class was absent): one salt cell, one sugar cell, one cell for sour, one for bitter, and frequently one for water. In vertebrate taste buds, however, sharp distinctions are absent: a cell primarily responsive to salt, for instance, also responds somewhat to sweet, bitter, and sour. The advantage, if any, of this lack of absolute discrimination is not known.

How the nose knows The receptor cells for the sense of smell are located on the antennae of some insects and, for the most part, in the noses of terrestrial vertebrates. In humans the olfactory receptors are located in two clefts in the upper part of the nasal passages (Fig. 36.9). Olfactory receptors bear a cluster of modified cilia that contain the receptor sites (Fig. 36.9).

Humans are relatively insensitive to odors as compared with most other animals: a well-trained human can distinguish about

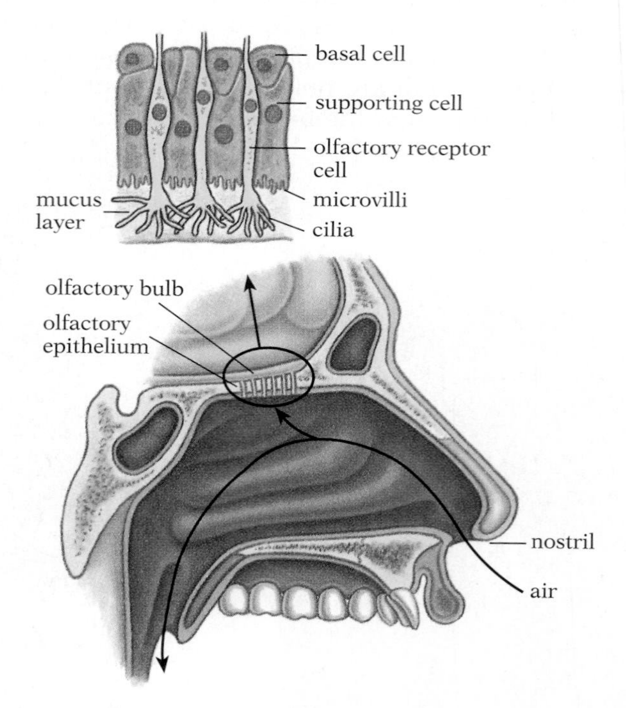

36.9 Human olfactory receptors

The receptor cells are located in the olfactory epithelium, which is located on the wall of the upper portion of the nasal cavity. The olfactory bulb of the brain receives information from the receptor cells. Inset: The cell bodies of most of the receptor cells are in the olfactory epithelium; their sensory cilia protrude into a layer of mucus on the surface of the epithelium.

36.10 Antennae bearing olfactory receptors of an adult male polyphemus moth (*Antheraea polyphemus*)

The thousands of sensory hairs extending perpendicularly from the combs of the male moth's antennae bear specialized cells that respond to minute amounts of the female's pheromone. While millions of molecules may be required to activate a human olfactory neuron, just a single molecule of pheromone at a receptor cell on the surface of a sensory hair triggers the moth's cell.

10,000 odors, albeit at high concentrations; for dogs, on the other hand, the number seems almost infinite, and the concentration necessary is very low.

The way in which odors are distinguished is not fully understood. Molecules stimulate olfactory receptors by binding to membrane proteins on the cilia. There appear to be many different kinds of receptors: for example, there are at least 30 specific kinds of odor "blindness" (anosmias), each of which must correspond to a receptor class; in addition, genes have been found for many different receptors, and estimates of the total number of genes involved range from 100 to 1000.

Even the existence of 1000 receptor classes, however, cannot alone explain the sensitivity of humans, much less of most mammals. The ability to distinguish odors arises because a single kind of molecule is usually able to bind with different degrees of fit to several kinds of receptors. The pattern of receptor stimulation thus produced—strong firing of certain receptor classes, less excitation of others, and no effect on most—is unique for each odor. It seems likely that dogs are more sensitive than humans because they have many more olfactory receptors of each class rather than a greater number of different classes.

Though most receptors respond to some degree to a variety of different odor molecules, some receptors are specialized for binding a single kind of molecule. Many animals are equipped to recognize, at very low concentrations, odors of special importance to their species. For example, some animals recognize predators chemically: the escape response of *Aplysia* is triggered by the species-specific odor of starfish, and some fish flee when they detect the odor produced by the damaged skin of conspecifics (members of the same species). Male moths ready to mate can detect the unique odor of the female of their species from enormous distances (Fig. 36.10). Organic chemists have been able to synthesize these ***pheromones*** to attract and trap male gypsy moths and Japanese beetles. The olfactory vocabularies of some social species include chemical messages for "attack," "flee," "come here," and the like. In general, there is a specific class of olfactory receptor, employing a highly specific binding protein, for each pheromone used by a species.

■ HOW ANIMALS SEE

Light-sensitive structures have evolved independently in a vast number of plants and animals. Though electromagnetic radiation takes many forms, ranging from low-energy radio waves to high-energy gamma rays, only the middle wavelengths have enough energy for vision, photosynthesis, and phytochrome conversion, but not so much that they can damage tissue. It is this relatively narrow band of wavelengths that so many unrelated kinds of organisms use to gather information about their environments, and it is this portion of the spectrum of electromagnetic radiation that we call light (see Fig. 7.3, p. 182).

Light receptors depend on plants Most animals respond to light. Even some protozoans react to changes in light intensity, moving toward or away from brightly lit areas. These organisms have a specialized region containing a pigment that reacts to light. The light-sensitive pigment common to many animals is a protein to which a portion of a carotenoid molecule is attached. Carotenoids, the yellow- or orange-light-sensitive pigments in plants, must be obtained by animals from their diets (in the form of vitamin A).

The light receptors of many invertebrates, like those of protozoans, do not function as eyes in the usual sense of the word. Some of these very simple receptors do nothing more than indicate the general light intensity; they do not form images (Fig. 36.11). In planaria, light receptors are arranged within a cup-shaped organ called an ***eye cup***; the shadow cast by the opaque edge of the cup defines the direction and elevation of the source of light (Fig. 36.12). Eye cups, however, are incapable of forming images.

Simple but grainy: compound eyes From this modest beginning, two considerably more complex strategies have evolved that allow animals to see images. These are the compound eye and the camera eye. The ***compound eye*** uses what is basically an array of tiny eye cups, each modified into a tube called an ***ommatidium***, which points out at the world in a slightly different direction from its

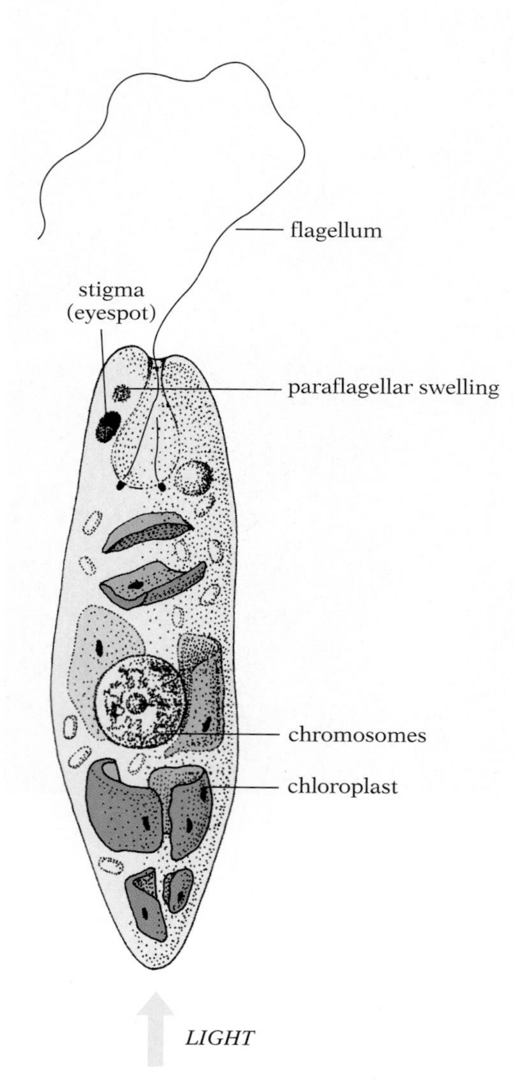

36.11 Light detection and response in *Euglena*

During part of the day the photosynthetic protist *Euglena* swims toward light; at other times it swims away. It points itself away from light, as shown here, by turning until its opaque stigma casts a shadow on the photosensitive paraflagellar swelling. To swim toward light, the organism turns until there is no shadow. The paraflagellar swelling, as its name implies, is involved in controlling flagellar movement.

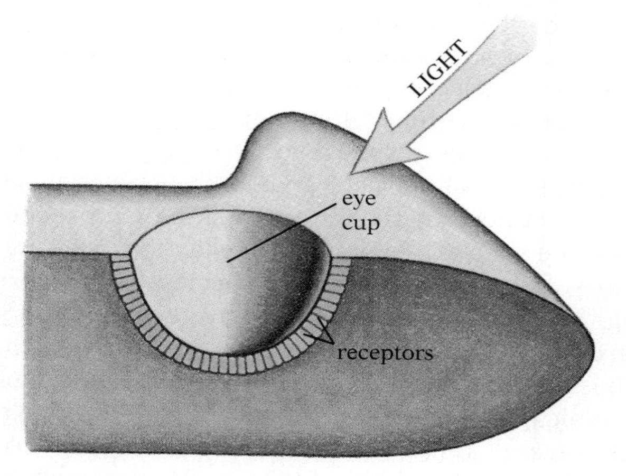

36.12 A planarian eye cup

The planarian eye cup provides directional information that the organism can use to maneuver in the environment; it does not form images. The direction of a light source is indicated by the location of the shadow cast by the cup's opaque edge onto receptors within the organ, as shown in this cross section.

36.13 Compound eyes of a horse fly

Each eye is composed of a huge number of ommatidia.

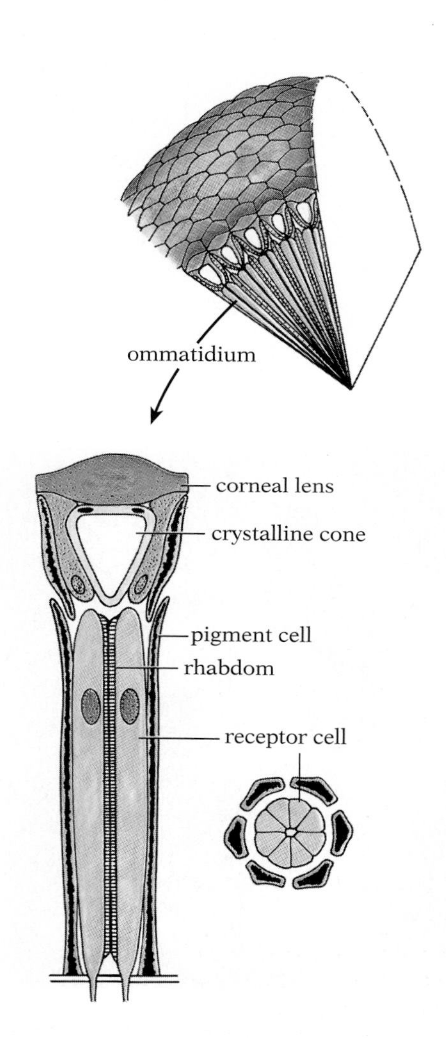

36.14 Ommatidia from the compound eye of a typical diurnal insect

Top: Section of a compound eye. Bottom: Longitudinal and cross sections of one ommatidium. The lens and crystalline cone focus incoming light rays into the rhabdom, a translucent cylinder formed by the highly specialized microvilli in the rhabdomeres of the eight receptor cells; photosensitive pigment is located in the microvilli. The pigment cells surrounding the ommatidium contain a dark pigment that prevents passage of light from one ommatidium to another.

neighbors (Figs. 36.13, 36.14). Light entering through the lens and crystalline cone of each ommatidium is focused onto an array of seven to nine elongated receptor cells. A specialized area called the ***rhabdomere*** runs the entire length of each cell and is located closest to the central axis of the ommatidium; together the rhadomeres form the rhabdom, which has layer upon layer of microvilli on which are spread light-sensitive pigments. With so many layers for light to pass through, compound eyes are more efficient at absorbing photons than camera eyes.

In many species with compound eyes the individual rhabdomeres have pigments that maximally absorb (and are therefore most sensitive to) light from particular parts of the spectrum. In honey bees, for instance, there are three pigments in the cells of each ommatidium: some cells have a pigment that absorbs green light most effectively, others absorb primarily blue light, and others respond best to ultraviolet (UV). The differing sensitivities of these three pigments form the basis of color vision in most insects.

Besides being very efficient at absorbing light, compound eyes are small and lightweight, which is important for a flying insect. In addition, most compound eyes enable arthropods to see details of movements that are far too rapid for our eyes. At the same time, the picture produced by an insect brain is probably a grainy mo-

A

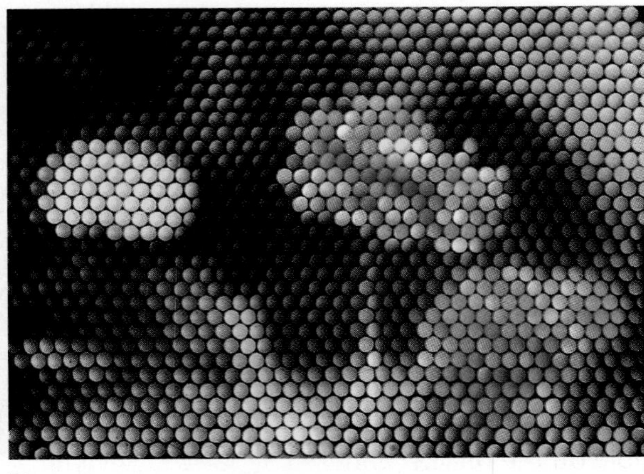
B

36.15 View through a compound eye

(A) Dahlias as seen by a human being from about 10 cm away. (B) The same scene as a honey bee might see it. The visual world of animals with compound eyes is thought to be broken up into a mosaic of spots. This picture, taken through an optical device, is not a perfect representation, however: it shows red but does not reproduce UV light, whereas bees see UV light and not red. Moreover, the circles should be vertically elongated ellipses, to reflect the peculiar anatomy of the bee's eye.

saic of the world, with far less precise delineation of objects in the visual field than we experience (Fig. 36.15).

Complex but crisp: camera eyes There are two versions of the ***camera eye***, the strategy employed by vertebrates and some molluscs. By far the rarer is the ***pinhole eye*** of organisms such as the chambered nautilus (Fig. 36.16). The pinhole eye is simply a covered eye cup with a tiny opening in its surface. Light from the world outside passes through the hole and is projected onto the array of receptors (the ***retina***) at the back (Fig. 36.17). Very little light can enter, and if the opening were widened to admit more light, the image, which is surprisingly sharp, would become blurred.

36.16 Cephalopods, with pinhole eyes

Cephalopods such as these chambered nautili, are the only organisms with pinhole eyes.

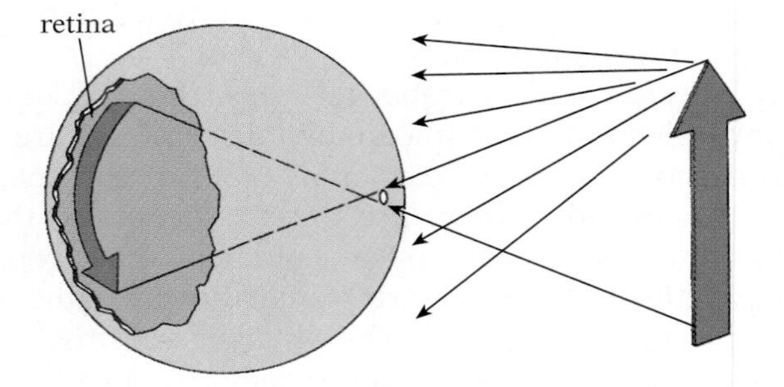

36.17 How a pinhole eye works

Light reflected from objects in the environment passes through the pinhole and projects a precise (though inverted) image on the retina. The main disadvantage of this eye type is that the image is relatively dim, since very little light is admitted.

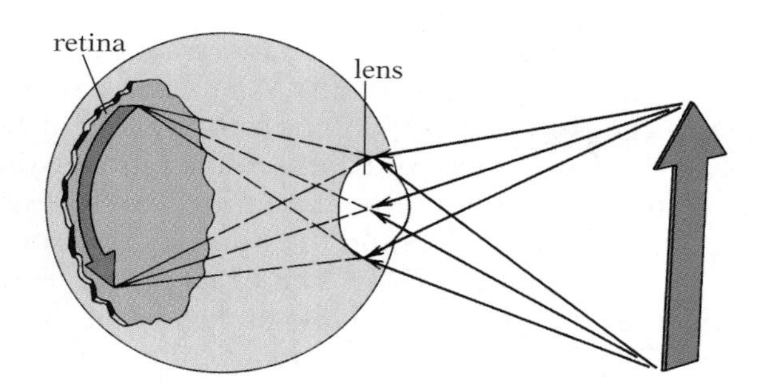

36.18 How a lens eye works

Light reflected from objects in the environment arrives at a lens and is focused on the retina. The lens eye admits much more light than the pinhole eye, but has great constraints on its focusing ability.

The ***lens eye***, on the other hand, can have a far larger opening for light, since the images are focused by a lens rather than a pinhole on the retina (Fig. 36.18). This more elaborate approach, however, is not without its problems. Unlike a pinhole eye, a lens eye can focus on objects at only one distance at any one time. Mammals and some birds and reptiles have muscles to change the shape (and thus the focus) of their lenses, while fish, amphibians, and other reptiles move the whole lens toward or away from the retina in the manner of a true camera. (Birds can also alter the shape of their cornea, the transparent structure through which light must pass to reach the lens.)

In addition, the retina and lens must have just the right spatial relationship: if the retina is too close to the lens, the animal will have difficulty seeing nearby objects (that is, it will be farsighted), whereas if the retina is too far away, distant objects cannot be brought into focus (the animal is nearsighted). In addition, irregularities in the lens itself may give rise to astigmatism, in which images of objects at the same distance from the lens but in different parts of the visual field come into focus at different distances. An insuperable disadvantage of the camera eye for many creatures is its bulk: a camera eye large enough to provide the same visual resolution as the compound eye of a honey bee would weigh more than the bee itself.

Design of the human eye The adult human eye is a globe-shaped lens eye with a diameter of about 2.5 cm (Fig. 36.19). It is encased in a tough but elastic coat of connective tissue, the ***sclera***. The anterior portion of the sclera, the ***cornea***, is transparent and more strongly curved. Just inside the sclera is a layer of darkly pigmented tissue, the ***choroid***, through which many blood vessels run. The choroid not only supplies blood to the eye; it also absorbs light, thus intercepting two kinds of light that would blur the image: light striking the eye but not entering through the lens, and internally reflected light. In nocturnal animals, by contrast, the choroid layer is usually highly reflective; while this still blocks light from entering except through the lens, it enhances internal reflection. The reflections reduce resolution, but increase sensitivity by sending unabsorbed light back through the receptor layer for another try. This mirrorlike layer accounts for the way a cat's eyes seem to glow in the dark.

Just behind the junction between the main part of the sclera and the cornea, the choroid becomes thicker and has smooth muscles embedded in it; this portion of the choroid is called the ***ciliary body***. Anterior to the ciliary body, the choroid leaves the surface of

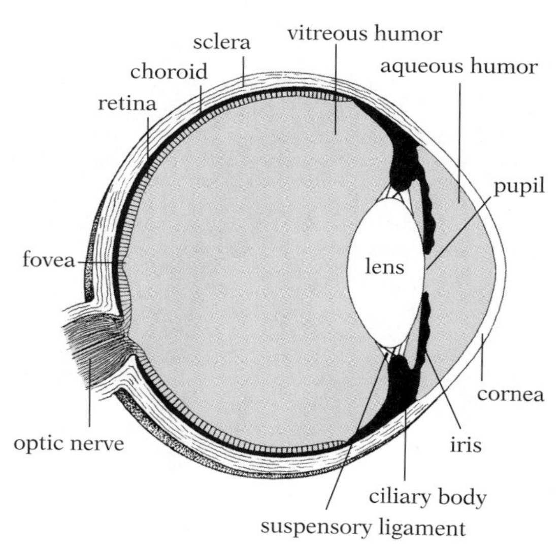

36.19 Human eye

The drawing shows a section through the left eye as seen from above.

the eyeball and extends into the cavity of the eye as a ring of pigmented tissue, the ***iris***. The iris contains smooth muscle fibers arranged both circularly and radially. When the circular muscle fibers contract, the ***pupil*** (the opening in the center of the iris) is reduced; when the radial muscles contract, the pupil is dilated. The iris thus regulates the amount of light admitted to the eye.

The ***lens*** is suspended just behind the pupil by a ***suspensory ligament*** attached to the ciliary body. The exact shape of the lens is controlled by the tension applied by an array of tiny muscles mounted here. The elasticity of the lens declines with age; by 50, most people can no longer focus on close objects, and they require reading glasses or bifocals. The lens and its suspensory ligament divide the cavity of the eyeball into two chambers. The chamber between the cornea and lens is filled with a watery fluid, the aqueous humor. The chamber behind the lens is filled with a gelatinous material, the vitreous humor.

The retina, which contains the receptor cells, is a thin tissue covering the inner surface of the choroid. It is composed of several layers of cells: the receptors, sensory neurons, and interneurons. The receptors are of two types, rods and cones (Fig. 36.20). The ***rod cells*** are more abundant toward the periphery of the retina, and are exceedingly sensitive to light; they enable us to see in dim

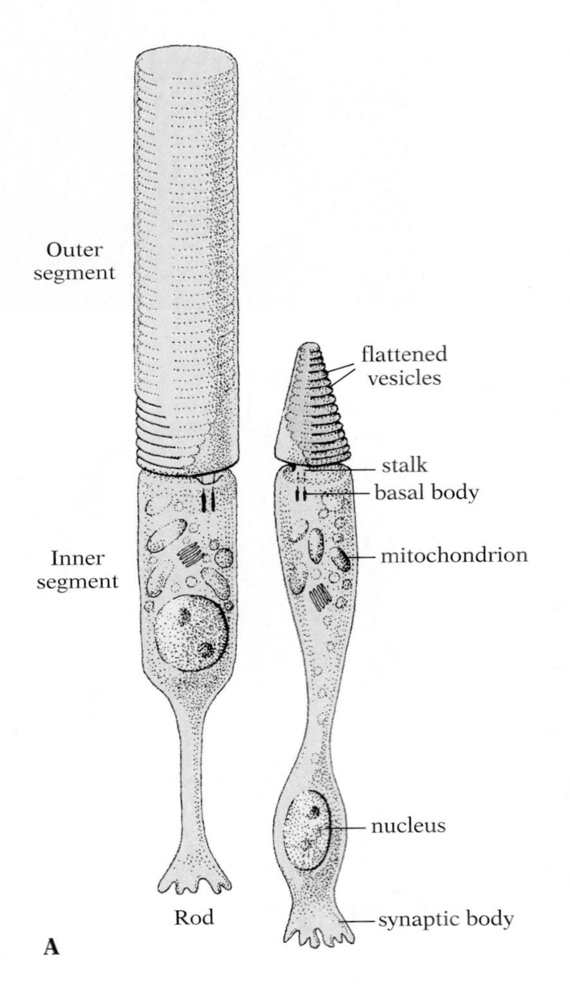

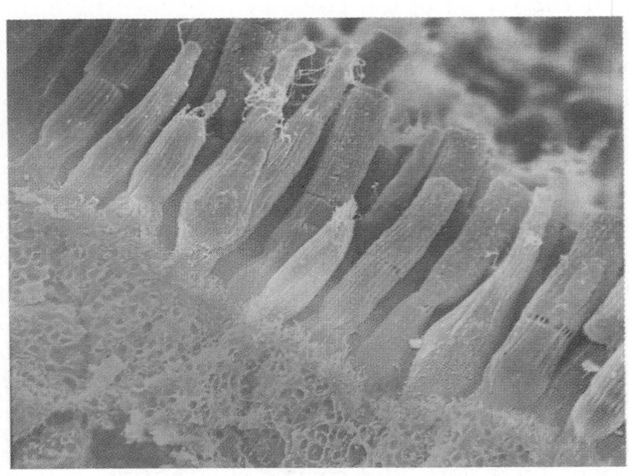

36.20 Rod and cone cells

(A) Cells from a human retina. The outer segment and the stalk connecting it to the inner segment develop as a highly specialized cilium. Electron microscopy reveals that a basal body is located at the inner end of the stalk, and that the nine peripheral microtubules characteristic of all cilia run from it through the stalk into the outer segment; as is true of most cilia that have lost their motile properties, the two central microtubules of motile cilia are absent. The visual pigment is located in the numerous membranous lamellae of the outer segment. The pigments of rods and cones are different: the pigments in rods detect low levels of light, while those of cones are responsible for color vision. (B) Scanning electromicrograph of rods (blue) and cones (greenish-blue). At their bases, the rods and cones synapse with various types of neurons (pink).

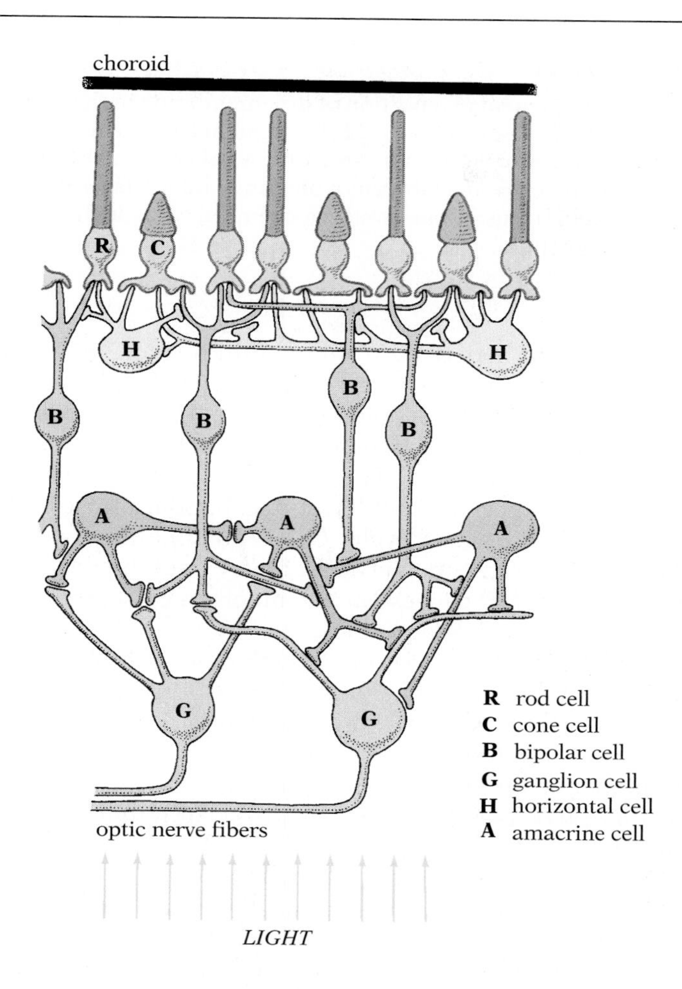

36.21 Cells of the human retina

The receptor cells (rods and cones) synapse at their bases with bipolar cells. The bipolar cells, in turn, synapse with ganglion cells; the axons of the ganglion cells form the optic nerve, which runs from the eye to the brain. Hence information that follows the most direct route to the brain moves from receptor cell to bipolar cell to ganglion cell to brain. Processing of information can occur within the retina because often several receptor cells synapse with a single bipolar cell and several bipolar cells synapse with a single ganglion cell. Besides convergence of information, lateral transfer of information from pathway to pathway also occurs: horizontal cells each receive synapses from many receptor cells and synapse onto many bipolar cells and other horizontal cells. Amacrine cells both receive synapses from and synapse onto bipolar cells; they also synapse onto many ganglion cells.

Note that the retina is arranged anatomically in reverse order from what might be expected; the receptor cells are in the back of the retina, and light must pass through the nerve cells to reach them.

light, but produce colorless, poorly defined images. The ***cone cells***, which are specialized for color vision in bright light, are especially abundant in the central portion of the retina, an area known as the ***fovea***. Because of the high density of receptors in the fovea, we are able to see the small area in the center of the visual field in fine detail.

The rods and cones synapse in the retina with short sensory neurons (bipolar cells); the bipolar cells in turn synapse with the retinal ganglion cells, whose axons, bundled together as the optic nerve, run to the visual centers of the brain (Fig. 36.21). The interconnection of neurons in the retina enables the eye to modify extensively the information transmitted from the 100 million or so receptor cells through the few million axons of the optic nerve to the brain. In most vertebrates these processing cells lie *between* the lens and the receptors, a location that requires a sizable hole in the retina (the blind spot) for the optic nerve fibers to pass through. In squid, octopus, and snakes, the processing cells are behind the receptors, an arrangement that creates no blind spot in their vision.

Sensitivity versus color: rods and cones The light-sensitive pigments in rods are embedded in the membranes of the flattened vesicles in the outer segment (Fig. 36.20). The rod pigment, ***rhodopsin***, consists of a protein (opsin) bonded to a prosthetic

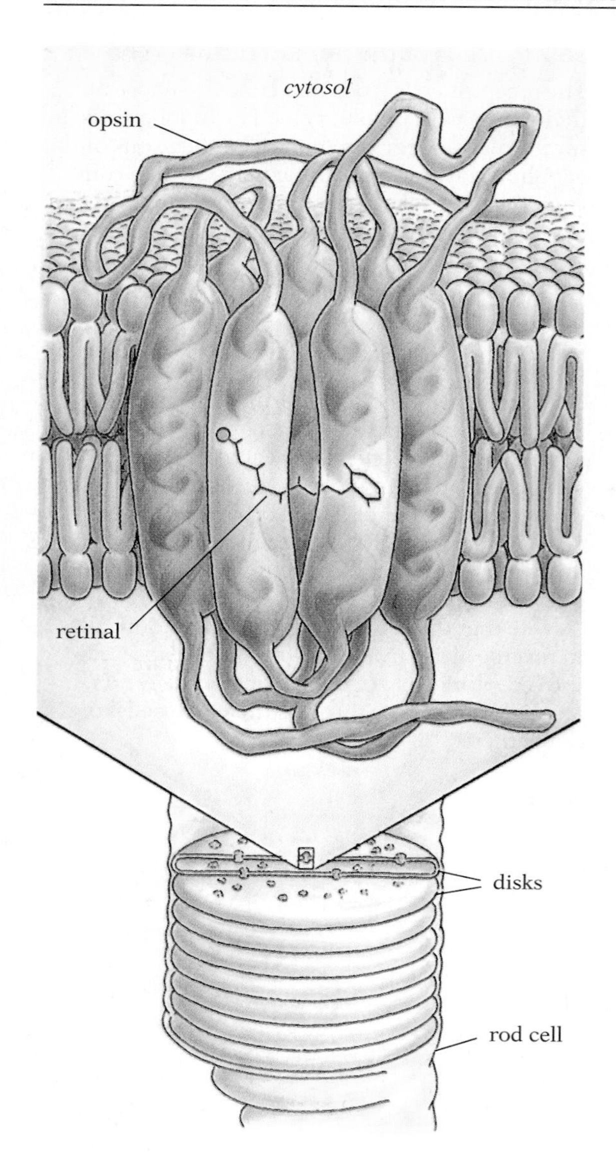

36.22 Rhodopsin

The top is an enlargement of a portion of disk membrane in the outer segment of a rod cell. Rhodopsin consists of a pigment (retinal) and a protein (opsin). Opsin has seven membrane-spanning α helices, which together hold the retinal. When retinal absorbs a photon, it interacts with another membrane protein (transductin) to start a chemical cascade that changes the membrane potential.

group (retinal) that is derived from vitamin A (Fig. 36.22). When a molecule of rhodopsin is struck by a photon of light, the retinal is converted briefly into an isomer (Fig. 36.23). The isomerization leads to conformational changes in the retinal, which activates an adjacent membrane protein. The second protein, transductin, releases a subunit that disinhibits a phosphodiesterase, which is thereby enabled to convert cyclic GMP (guanine monophosphate) to the noncyclic form. Since cGMP acts by opening sodium channels, the loss of cGMP causes Na^+ channels to close, thereby causing the membrane potential to increase.

The mechanism of cone vision is much more complex. There are three classes of cones, each containing a different pigment. All three human pigments have retinal as their prosthetic group, and all three absorb light over a wide range of wavelengths. Their protein components differ slightly; as a result, the range of wavelengths for each pigment centers on a different part of the spectrum (Fig. 36.24).

It may seem odd that we can distinguish colors so well when the absorption curves of the three pigments overlap so much: green-absorbing cones, for instance, may respond to orange, yellow, green, blue-green, or even blue light, and the "red" curve actually peaks in the yellow. The nervous system compares and contrasts the output of the cones in pairs to obtain the fine color distinctions we consciously perceive.

We are so accustomed to the human version of color vision that we tend to assume other animals must see the world the same way. Humans and other primates, however, are unusual among mammals: most mammals see the world in varying shades of gray. By contrast, most birds and many fish and reptiles appear to have color vision, and some even detect light in the UV range. Some of these color-sensitive vertebrates, however, have only two types of cones. Often in birds and reptiles, color sensitivity is further complicated by filtering: a colored oil droplet lies between the lens and the pigments so that only one part of the spectrum filters through to the receptor.

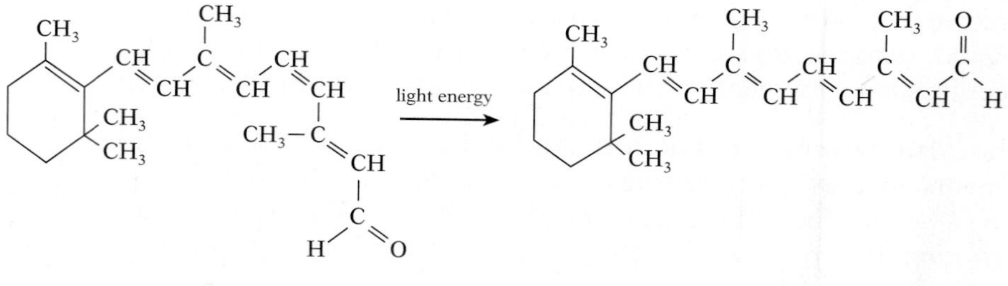

light energy

36.23 Light-driven change of retinal from one isomer into another

With the absorption of light energy, the shape of the side chain is altered.

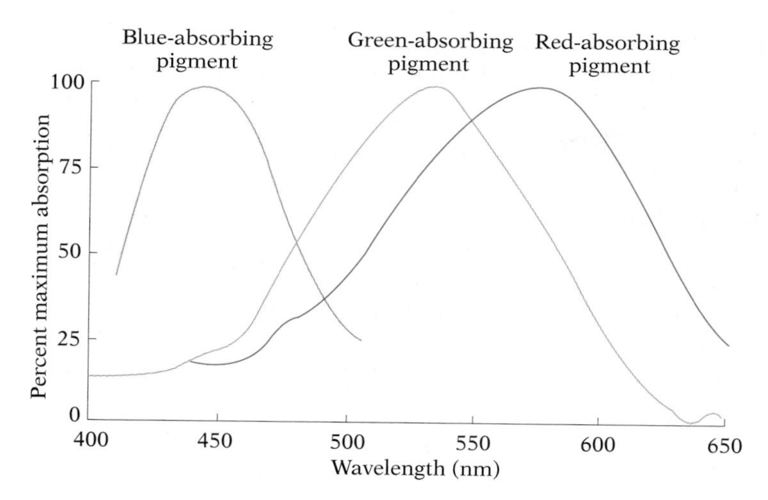

36.24 Absorption spectra of the three cone pigments in primates

The red-absorbing pigment actually has its maximum sensitivity in the yellow region of the spectrum. The relative narrowness of the absorption spectrum of the blue-absorbing cone is a result of faint yellow pigments in the lens and cornea, which absorb UV light.

■ HOW HEARING WORKS

Like vision, hearing is an important sensory ability among animals. For many species, sounds in the environment provide crucial information about predators, prey, and conspecifics. A variety of strategies for detecting them have evolved.

Sound is produced when an object vibrates, setting in motion the particles of the medium (usually air) and thus generating alternating bands of high and low pressure (Fig. 36.25). The number of vibrations per second is the sound's frequency, and is given in hertz (Hz). Many insects have long hairs that are moved back and forth by the air movement (Fig. 36.26A). Like tuning forks, these hairs

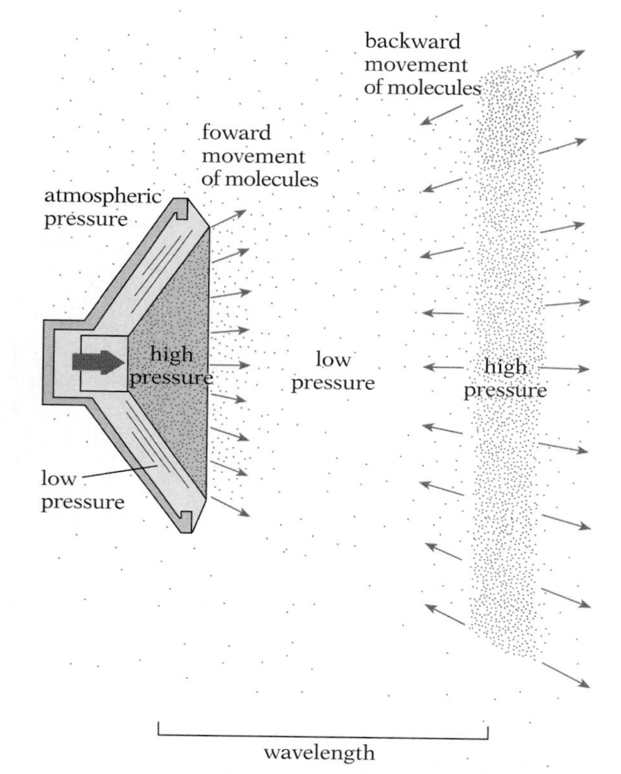

36.25 Sound generation by a loudspeaker

By vibrating and setting air molecules into motion, a loudspeaker creates waves of high and low pressure; these are the physical basis of sound. In this diagram, air pressure is indicated by the density of dots, while the directions of particle movement are shown by arrows. The distance between one band of high pressure and the next is one wavelength. Ears have evolved to respond either to particle movements (most invertebrate ears) or to waves of pressure (typical of higher vertebrates). What we hear is the result of the processing of sensory input in the nervous system.

A

B

36.26 Two sound detectors in insects

(A) The antennal hairs on the head of this male mosquito can resonate, and are therefore very sensitive to sounds of the proper frequencies. The hairs are particle-movement detectors. (B) By contrast, in katydids the pressure changes associated with sound are detected by membrane-covered organs on the forelegs.

are strongly resonant, so they respond best to vibrations within a narrow range—typically the frequencies emitted by other members of the species. In contrast, most vertebrates (including humans) and some invertebrates are sensitive to the pressure changes associated with sound waves (Fig. 36.26B).

Design of the human ear The human ear is divided into three parts: the ***outer ear***, the ***middle ear***, and the ***inner ear*** (Fig. 36.27A). The outer ear consists of the pinna (ear flap) and the auditory canal. At the inner end of the auditory canal is the ***tympanic membrane*** (eardrum).

On the other side of the tympanic membrane is the chamber of

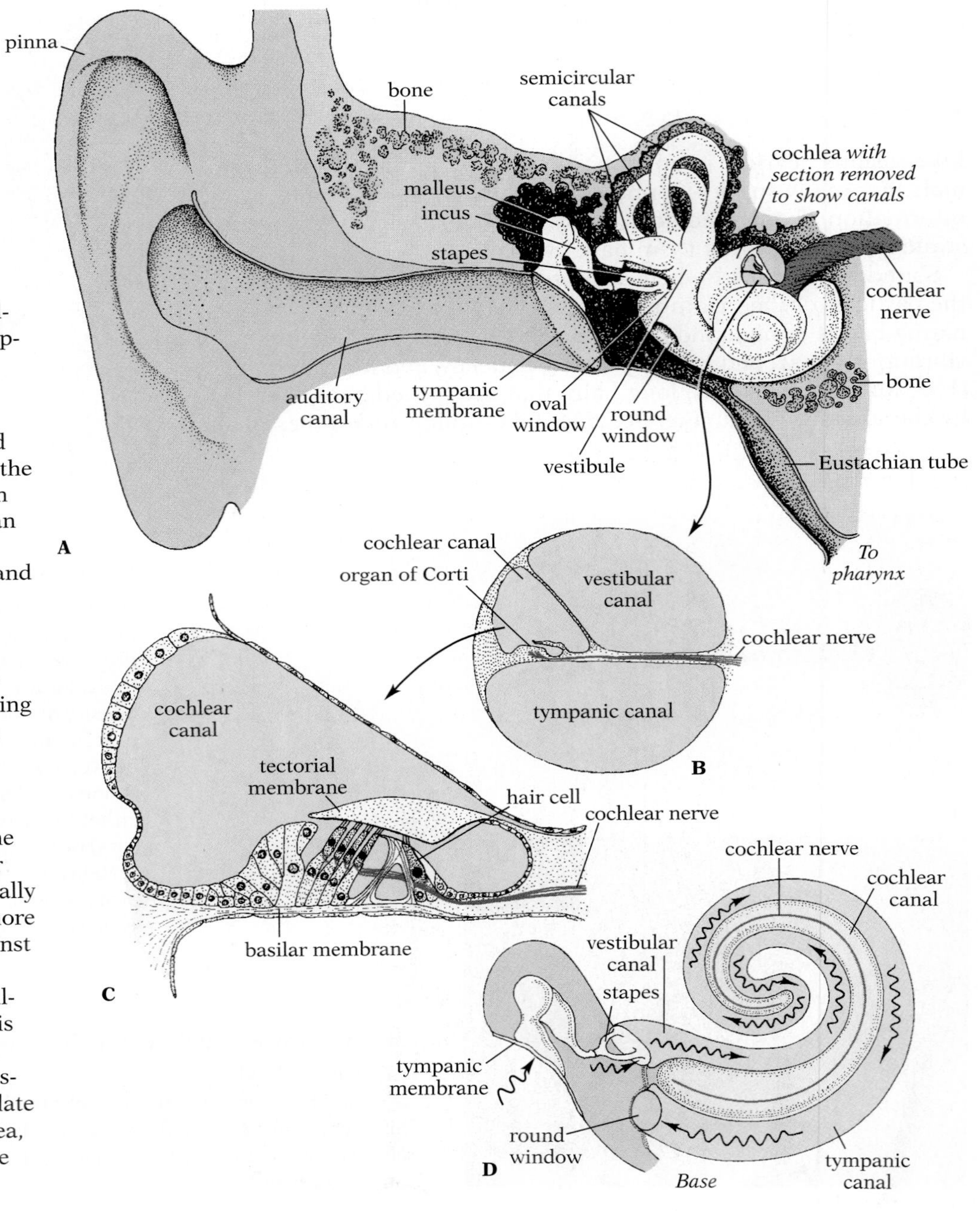

36.27 Human ear

(A) The major parts of the outer, middle, and inner ear (see text for description). (B) Enlarged cross section through one unit of the coil of the cochlea, showing the relationship between the vestibular, cochlear, and tympanic canals and the location of the organ of Corti. (C) Enlarged diagram of cochlear canals, showing the organ of Corti, which rests on the basilar membrane separating the cochlear and tympanic canals. When the basilar membrane vibrates and moves the sensory hair cells up and down, the hairs rub against the tectorial membrane overhanging them. The resulting deformation of the hairs produces a generator potential in the hair cells, which triggers impulses in sensory neurons running from the organ of Corti to the brain. (D) Diagram of the relationship between the middle ear and the cochlea, here pictured partially uncoiled to show its canal system more clearly. When the stapes moves against the oval window, the fluid in the vestibular and tympanic canals oscillates (wavy arrows); the oscillation is made possible by the flexible round window, which permits relief of pressure. High-frequency sounds stimulate hair cells near the base of the cochlea, and low-frequency sounds stimulate hair cells near the apex.

the middle ear. The air in this chamber is normally at the same pressure as that in the surrounding atmosphere. When the high-pressure part of a sound wave enters the ear, the pressure forces the eardrum to bulge slightly into the middle ear; when the low-pressure part of a wave enters, the higher-pressure air in the middle ear forces the eardrum to bulge into the outer ear. As a result, the eardrum vibrates, faithfully reproducing the pattern of pressure variation, and transforming it into mechanical movement.

The middle ear is connected to the pharynx via the ***Eustachian tube***, which opens when we swallow or yawn, allowing the pressure in the middle ear to equalize with the pressure outside. Periodic equalization is necessary because the external air pressure can change noticeably as the barometer rises or falls, and as we change altitude. During the course of the all-too-common cold, mucus may clog the Eustachian tube, destroying the ability to accommodate pressure changes.

Three small interconnected bones, the malleus, the incus, and the stapes (commonly known as the hammer, the anvil, and the stirrup), extend across the chamber of the middle ear from the tympanic membrane to a membrane called the ***oval window***. These bones amplify the sound waves received by the tympanic membrane. Another membrane, the ***round window***, lies just below the oval window.

On the inner side of the oval and round windows is the inner ear, a complicated labyrinth of interconnected fluid-filled chambers and canals. The upper group of chambers and canals is concerned with the sense of equilibrium. The lower portion of the inner ear, to which the upper chambers are connected, consists of a long coiled tube called the ***cochlea*** (Latin for "snail shell"). Inside the cochlea are three fluid-filled canals (Fig. 36.27B–D): the vestibular canal, which begins at the oval window; the tympanic canal, which begins at the round window and connects with the vestibular canal; and the cochlear canal, which lies between the other two.

The sensory portion of the cochlea, called the ***organ of Corti*** (Fig. 36.27C), projects into the cochlear canal from the ***basilar membrane***, which forms the lower boundary of the cochlear canal. The organ of Corti consists of a layer of epithelium on which lie rows of specialized receptor cells bearing bundles of sensory hairs at their apices (Fig. 36.28). Dendrites of sensory neurons terminate

36.28 A hair bundle from a cell in the ear of a bullfrog

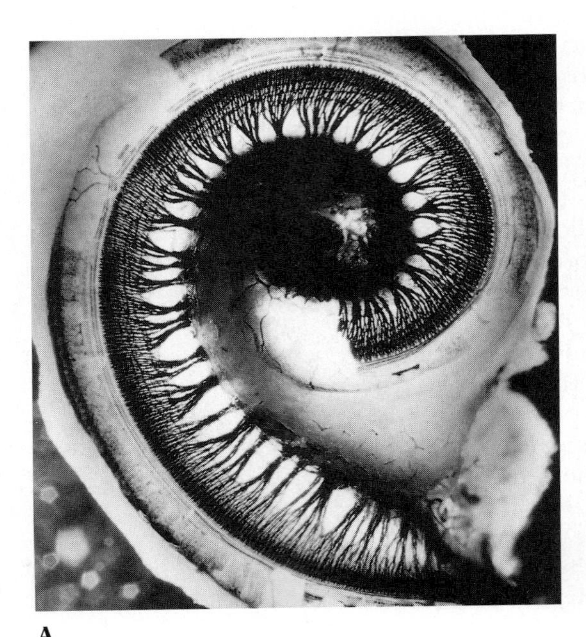

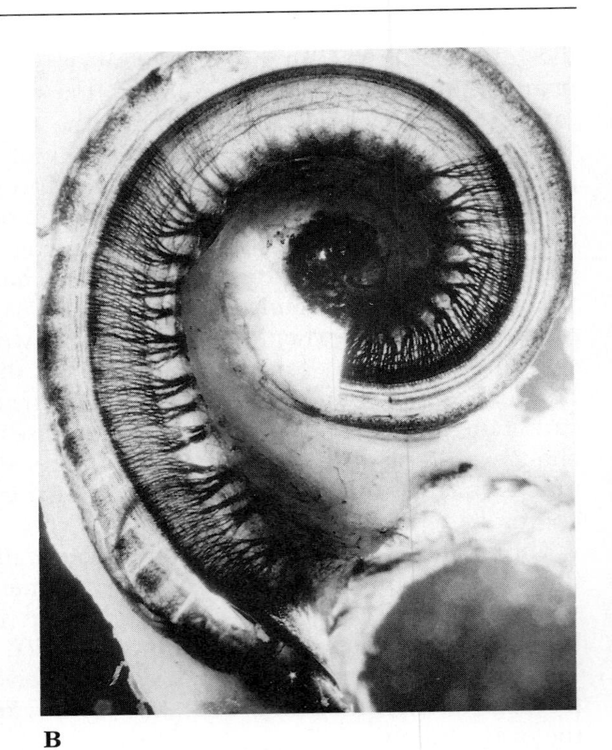

36.29 Normal and noise-damaged cochlea

(A) Longitudinal section of the cochlea showing normal distribution of the hair cells and associated sensory neurons. (B) Longitudinal section of noise-damaged cochlea. The hair cells and sensory neurons have been permanently damaged, resulting in a serious hearing loss. Individuals subject to excessive noise should wear ear protectors. Many musicians in rock bands have suffered loss of hearing from their occupation.

on the surfaces of the hair cells. Overhanging the hair cells is a gelatinous structure, the ***tectorial membrane***, into which the hairs project. Vibrations of the basilar membrane cause the sensory hairs to move up and down against the less mobile tectorial membrane, deforming the hairs; this movement physically opens K^+ channels in the hair cells (or in the hairs themselves), and so produces a degree of depolarization. Since the hairs on the hair cells are sheared off when the membrane moves violently, and since they cannot regrow, prolonged exposure to loud noises can result in some degree of permanent deafness (Fig. 36.29).

Operation of the ear Vibrations in the air pass down the auditory canal and cause the tympanic membrane to vibrate. These vibrations are transmitted across the middle ear to the oval window by the chain of middle-ear bones. Movement of the oval window moves the fluid in the canals of the cochlea: an inward push by the middle-ear bones sends some of the fluid in the vestibular canal into the tympanic canal, causing the round window to bulge outward. The process reverses itself each time the middle-ear bones draw back from the oval window. These waves in the fluid of the cochlea cause the basilar membrane to move up and down, rubbing the hair-cell bundles against the tectorial membrane. Thus stimulated, the hair cells activate the sensory neurons that carry impulses to the auditory centers of the brain.

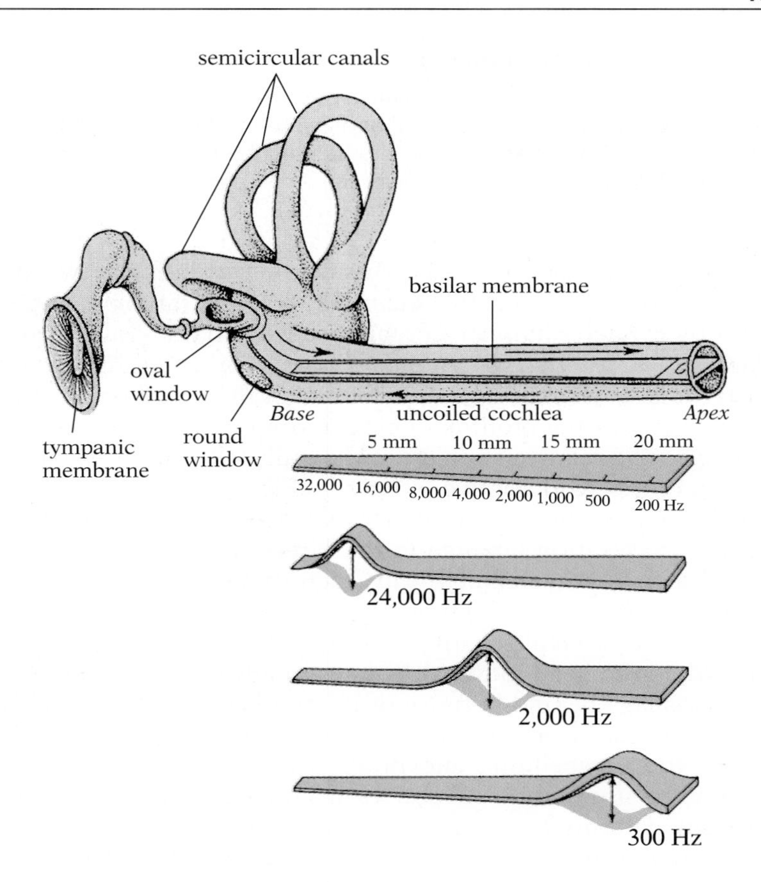

36.30 Mechanism of frequency detection in the human cochlea

The cochlea, here shown uncoiled, includes a basilar membrane that becomes progressively wider and thicker as it extends away from the oval window. The hairs also become progressively shorter and stiffer from base to apex. Sound waves of different frequencies give rise to pressure waves of corresponding frequencies in the cochlear fluid. These in turn cause the basilar membrane to vibrate at different points for different frequencies, thus stimulating surrounding sensory hairs to activate the sensory neurons that conduct impulses to the auditory centers of the brain. Processing of this auditory information in the brain results in discrimination between different frequencies, or pitches. With increasing age, the membrane begins to stiffen at the thin end, and so high-frequency sensitivity is progressively lost.

We are able to distinguish pitch (the frequency of sound waves) because the basilar membrane and its hair-cell bundles are tuned. For example, the basilar membrane becomes regularly wider and thicker as its distance from the oval window increases. Different parts of the membrane are stimulated to maximum movement by sound waves of different frequencies (Fig. 36.30). Just as thicker violin and guitar strings resonate to lower-frequency vibrations, so too the thicker parts of the basilar membrane vibrate when low-frequency vibrations stimulate them, giving rise to the sensation of low pitch; the thinner parts of the membrane resonate in the presence of high-frequency vibrations.

This broad resonance of the membrane is sharpened by the hair cells themselves. The hairs become progressively shorter and stiffer toward the thicker end, which endows them with a lower resonance. The opening and closing rates of the membrane channels in the hair cells declines toward the thick end, resulting in an appropriate electrical resonance. In addition, the hair cells are able to change their own shape (and thus their sensitivity and tuning) on the basis of feedback from the brain, thus selectively amplifying or attenuating particular sounds as the need arises. These several resonance gradients and active measures interact to maximize the tuning of the ear.

■ MAINTAINING BALANCE AND SENSING MOVEMENT

The upper portion of the labyrinth of the inner ear is composed of three ***semicircular canals*** and a large vestibule that connects

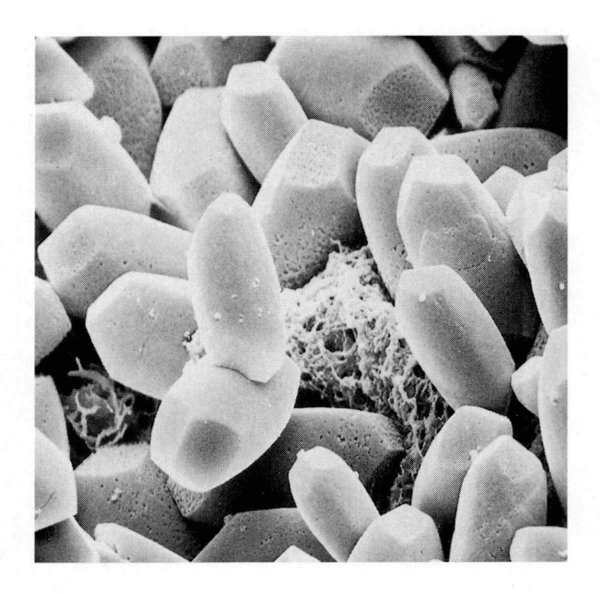

36.31 Otoliths in the inner ear

The crystals of calcium carbonate, called otoliths, shown in this scanning electron micrograph are surrounded by a gelatinous film in which hairs are embedded. When the head tilts, the altered pull of the otoliths on the hairs sends signals on head position to the brain.

them to the cochlea (Fig. 36.27). Inside the vestibule are two chambers, the ***utriculus*** and the ***sacculus***, oriented at right angles to each other. Each chamber contains sensory hairs embedded in a gelatinous layer that surrounds crystals of calcium carbonate called ***otoliths*** (Fig. 36.31). Any change in the speed or position of the head causes the otoliths to exert more pull on some hairs than on others. The relative strength of the pulls tells the cerebellum the position of the head relative to gravity (Fig. 36.32).

The design of the semicircular canals makes it possible to sense angular acceleration (change in the direction of motion). Like the axes of a three-dimensional grid, each of the three canals is oriented at right angles to the other two (Fig. 36.30). At the base of each canal is a small chamber containing a tuft of sensory hair cells. When the head moves or rotates in any direction, the fluid in the canals lags behind because of its inertia and viscosity. This results in increased pressure on the hair cells, which then send signals to the cerebellum. The brain, by integrating the different amounts of stimulation it receives from each of the three canals, can then determine the direction and rate of acceleration. The canals are not, however, perfect sensory organs. If the head rotates for an extended period and then stops abruptly, the fluid will continue to circulate. The result is the kind of dizziness we feel after being spun in a chair.

■ SPECIALIZED AND UNUSUAL SENSES

Just as bees and humans see different sets of colors, which suit each species' needs, so too does the auditory range of animals vary in adaptive ways. Bats, for instance, broadcast ultrasonic sound waves and use the echoes to locate obstacles and prey. They can hear sounds as high as 100,000 Hz or higher. Many insects that are hunted by bats, like moths and locusts, also hear ultrasonic sounds, and begin evasive maneuvers when they detect the bats' sonar pulses. Other animals, such as rats, use the ultrasonic range for communication. Elephants, on the other hand, are sensitive to sounds of extremely low frequency.

In addition to vision and hearing, many animals have senses that allow them access to information from other sources, including heat, electricity, and magnetism. Other sensory abilities may remain to be discovered.

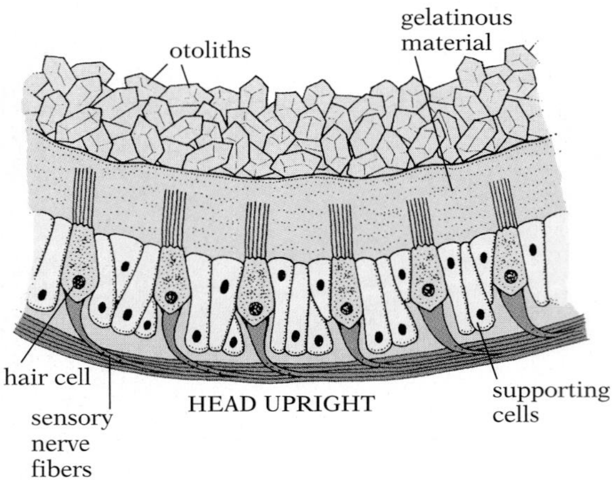

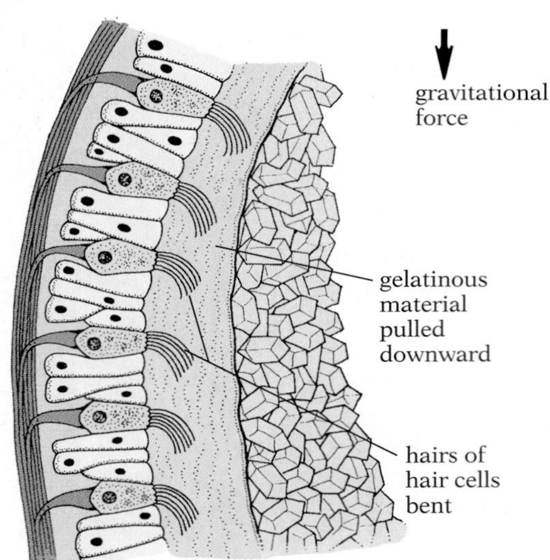

36.32 How otoliths function

When the head is in an upright position, the weight of the otoliths presses directly down on the sensitive hair cells, the tips of which are embedded in the membrane. When the head is tilted (bottom), the altered pull of the otoliths on the hairs generates signals to the brain.

Seeing heat We sense strong radiation in the infrared (IR) frequency range as heat. Some animals have specialized sensory structures that allow them to locate very faint IR sources. The most advanced of these structures are the infrared "eyes" of rattlesnakes and other pit vipers (Fig. 36.33). Intermediate between eye cups and pinhole eyes, these pit organs form crude images on a highly heat-sensitive "retina," enabling the snakes to hunt warm-blooded prey in the dark, while their conventional light-sensitive eyes enable them to see in daylight. The pit organs evolved from the heat sensors of the skin.

Sensing electricity Three classes of fish use electric fields. Members of the first class, the strongly electric fish, do not actually sense electric fields; instead they use highly modified muscle cells like a series of batteries to produce a strong electrical charge. Electric eels and some rays stun or kill both prey and potential predators with this charge.

Members of the second class, the weakly electric fish, are either nocturnal or live in murky water where vision is almost useless. They compensate for the lack of visual information by producing an electric field around themselves, which they monitor for signs of objects that disturb it (Fig. 36.34). They can also perceive the electrical signals broadcast by other members of their own species, an ability that is important in intraspecific communication and mating. The sensory organs of weakly electric fish are at the base of long, low-resistance, jelly-filled canals that radiate through the body from the head and monitor the electric field at points all over the body. The receptors are simply modified nerve cells that are ex-

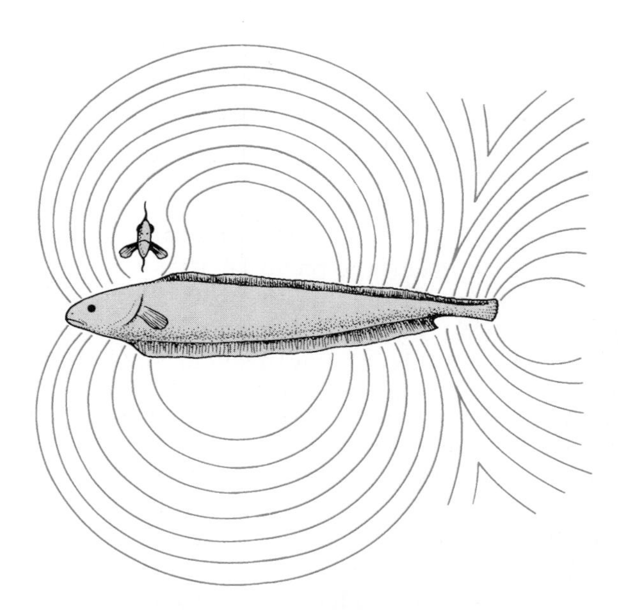

36.34 Electrolocation by a weakly electric fish

Nearby objects distort the field. Objects that are good conductors concentrate the lines of electric force on a small area of the fish's body, whereas poor conductors (here exemplified by a potential prey fish) spread the lines.

A

B

C

36.33 Infrared vision of pit vipers

(A) Pit vipers like the western rattlesnake, shown here, are known for their ability to see in daylight and to detect warm-blooded prey in the dark by means of a pit organ (arrow). (B) The organ is sensitive to the IR radiation of warm-blooded animals, forming crude images that probably resemble this thermal image of a gerbil. (C) The same animal in visible light.

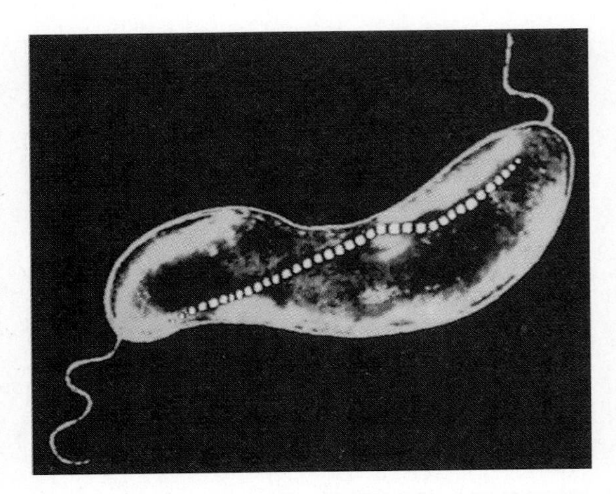

36.35 False-color micrograph of a magnetotactic bacterium

The chain of pale green magnetite crystals inside this mud-dwelling organism causes it to rotate into alignment with the lines of the earth's magnetic field. When the bacterium divides, each daughter cell gets part of the chain.

tremely sensitive to changes in the flow of current—in this case, the weak current generated by the fish itself, which then flows through the water and down the canals to the receptors.

Members of the third class, the passive electric fish, use their electrosensory apparatus to monitor their surroundings. These animals, which include the sharks, can sense the minute electric fields generated by the neuromuscular activity of their prey. Alerted by the odor of prey, a shark can use its electrical sense to detect a buried flounder, for example, from a distance of 1 m or more. Among higher vertebrates, only the platypus is known to have an electric sense, which it also uses to detect prey.

The electroreceptive organs of passively and weakly electric fish evolved from a line of tiny hairs (often just visible) running from head to tail along the sides of most other fish and many amphibians. These hairs are the sensory receptors of the ***lateral line organ***, a system of mechanoreceptors sensitive to water movements—especially those produced by nearby objects and individuals. Many species use these hairs to detect and localize prey.

Using the earth's magnetic field Various organisms are sensitive to the magnetic field that envelopes the earth. Magnetotactic bacteria (Fig. 36.35) build within themselves chains of magnetite (lodestone) that rotate them into alignment with the lines of the earth's magnetic field. These lines point north and down in the northern hemisphere, and so guide the many species of mud-dwelling, magnetite-bearing bacteria down to the bottom of the stagnant ponds and marshes they inhabit. In the southern hemisphere, where the lines of the magnetic field point south and down, the polarity of the chain is reversed. At least one species of alga also uses magnetite.

This kind of magnetic orientation probably evolved from more conventional weighting: most mud-dwelling bacteria have dense crystals at their front ends. The weight forces the front end down, and so aims a swimming bacterium toward the muddy bottom. Magnetite, the densest substance known to be synthesized by living organisms, makes an excellent weight, and the subsequent development of a chain of these crystals would have provided an even more effective way to orient the bacteria.

Behavioral studies indicate that many other organisms sense the earth's magnetic field. Sharks and most rays use their electrical sensitivity to detect the direction of this field. Among the other animals able to detect it are honey bees, homing pigeons, various migratory birds, tuna, and salmon. The basis of this ability in such

36.36 A miswired frog

In an experiment conducted by Marcus Jacobson and his colleagues, a patch of skin from the back of a tadpole was exchanged with a patch from its abdomen. The two patches continued to develop their original color patterns, and sent their sensory axons to their original targets in the CNS. As a result, when the transplanted patch on the back of the adult was tickled, the frog automatically scratched the place on its abdomen where the patch had originally begun to develop.

animals is not yet known, but localized deposits of magnetite have been found in association with the nervous system in each of them. In others, pigments in the retina seem to detect the field.

MAKING SENSE OF SENSATIONS

■ HOW SENSORY DATA ARE PROCESSED

When a sensory receptor is activated, the information it receives must be carried to the part of the CNS that is specialized to process it. The network of receptors, sensory neurons, interneurons, and brain cells is intricate and precisely ordered. The pattern arises during development and subsequent self-calibration.

The means by which the first axons destined to carry specific sorts of information "find" their proper targets during development are not well understood. During development, axons from sensory neurons seek out their targets: receptors transplanted from one location to another during development often send their axons to the part of the brain that would have been their destination if no transplantation had occurred. Moreover, the brain interprets input from these receptors as coming from the original location. For example, if a patch of skin with touch receptors from the back of a sufficiently developed tadpole is exchanged with a patch from its abdomen, the adult frog will scratch its abdomen when the skin on its back is irritated (Fig. 36.36).

The first breakthrough in understanding how animals process sensory information came from studies of a primitive invertebrate—the horseshoe crab, *Limulus*.

Enhancing contrast: lateral inhibition The activity of receptors in one ommatidium in the compound eye of *Limulus* has a dramatic effect on the behavior of receptors in other ommatidia. In the dark the receptors fire at a slow but regular basal rate. When a single ommatidium is illuminated, its receptors begin to fire at a more rapid rate; at the same time, however, the receptors in the adjacent ommatidia fall silent, inhibited from firing by their neighbors' excitation. This phenomenon is known as ***lateral inhibition***. The net result is that in the picture transmitted, the contrast between light and dark is enhanced at the edges: all the cells next to the stimulated receptors have ceased to fire even at their low basal rate.

Lateral inhibition is a kind of comparison-contrast strategy; it has been found in the visual wiring of every creature studied (though in vertebrates the receptor cells do not interact directly). It explains a familiar optical illusion: the contrast between two areas of differing shades of gray is enhanced at the border. The edge of the lighter area appears paler than the rest of that area, while the adjoining edge of the darker area appears even darker than the rest (Fig. 36.37). The effect of lateral inhibition is to pick out and enhance subtle features in the environment that might otherwise be missed.

A

B

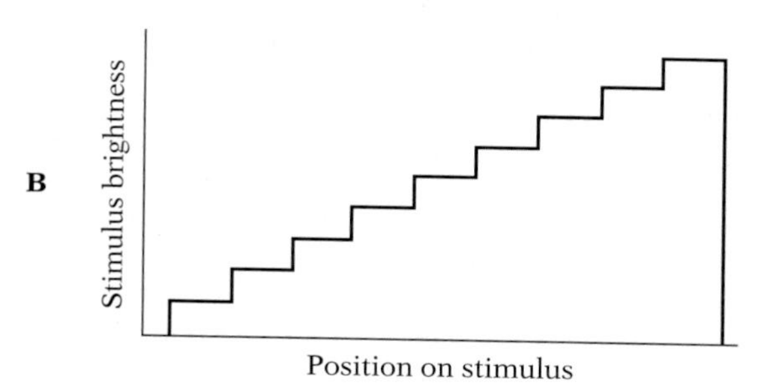

C

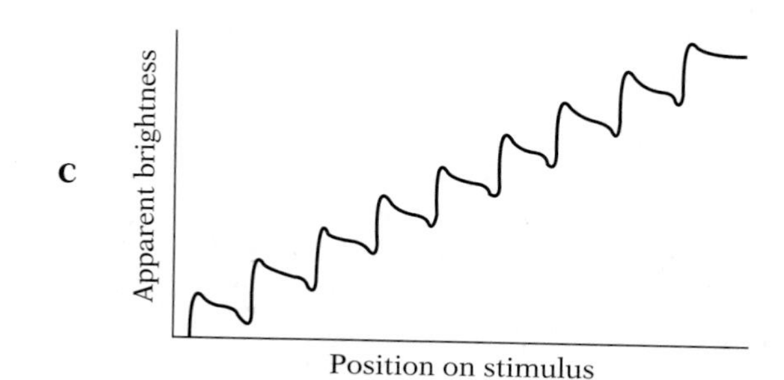

36.37 Mach bands

Lateral inhibition gives rise to a phenomenon known as Mach bands. For this series of grays (A), the actual brightness of each strip in relation to the others is shown (B), along with the perceived brightness (C). The phenomenon is evident as an exaggerated contrast between light and dark at the boundary between each two strips, and in the illusion that the right side of each strip is darkened.

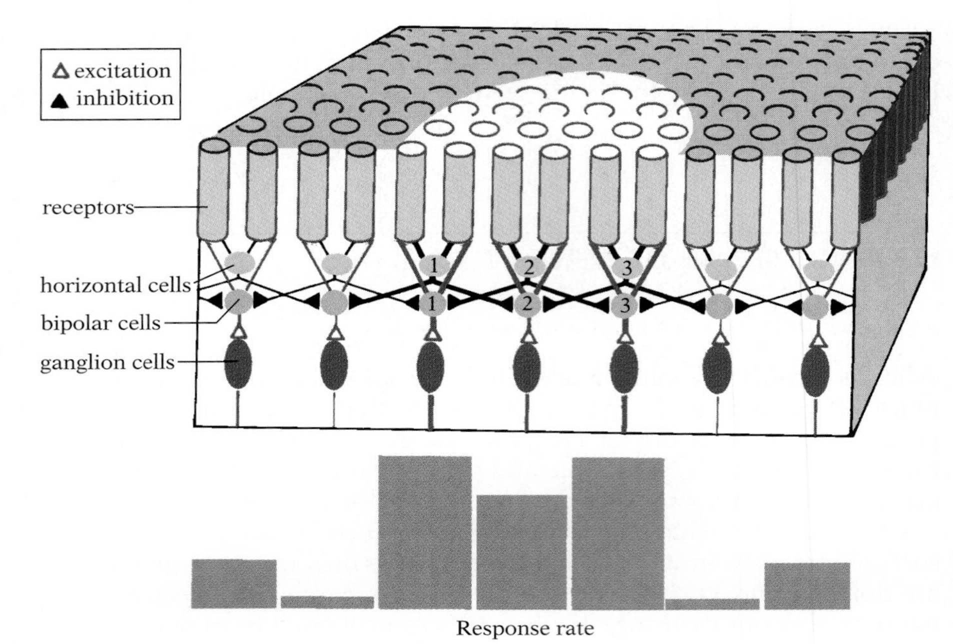

36.38 Lateral inhibition in the vertebrate retina

Top: A highly schematic representation of a section of vertebrate retina. Bipolar cells add the output of several receptors (only two of which are shown for each bipolar cell here) and excite the ganglion cells, which join together to form the optic nerve. In the phenomenon called lateral inhibition, each horizontal cell also adds the output of several receptors and sends out an axon that forms inhibitory synapses with neighboring bipolar cells and prevents them from firing maximally. (For simplicity, amacrine cells have been omitted.)

In this diagram a spot of light falls on several receptors; we will concentrate only on the six receptors in front. (The activity of the nerve fibers is indicated by their relative width.) Stimulated by input from receptors, bipolar cell 1 fires a ganglion cell. Input from the same receptors causes horizontal cell 1 to inhibit the bipolar cell to its left, causing its firing to drop below the basal rate; it also inhibits the cell to its right, bipolar cell 2, which is also inhibited by horizontal cell 3. Thus, even though bipolar cells 1 and 2 receive the same amount of excitation from receptors, bipolar cell 1 fires more strongly because it is inhibited by only one adjacent cell (horizontal cell 2), while bipolar cell 2 is inhibited by two such cells. Bipolar cell 3, like bipolar cell 1, is inhibited by only one adjacent horizontal cell, so it too fires more strongly than bipolar cell 2.

Bottom: Graph of the response rate of the ganglion cells. This pattern of lateral inhibition explains why the operation of the vertebrate retina accentuates the difference between light and dark along an edge. The cells just outside the illuminated area, receiving inhibition but no excitation, fire below their basal rate, thereby reporting their part of the visual field to be relatively darker than it is. Cells just inside the illuminated area, being inhibited by fewer adjacent cells than those toward the center, report their part of the visual field to be relatively lighter than it is. This strategy of lateral inhibition has been found to work in decoding a variety of features of the visual world in both invertebrates and vertebrates.

In the vertebrate retina, both horizontal cells and amacrine cells play an active role in lateral inhibition.[1] In many species, horizontal cells synapse on bipolar cells, which in turn synapse on ganglion cells. (In other species, horizontal cells synapse on the photoreceptors themselves.) Ganglion cells transmit visual information to the brain (Fig. 36.38). Since, in the simplest case, a ganglion cell is stimulated (via bipolar cells) by light falling on a small group of receptors, and is inhibited by light falling on the surrounding receptors, the ganglion cell responds best to a small spot of light on the activating receptors, surrounded by darkness. The ganglion cell, with its associated horizontal and bipolar and amacrine cells, is often called a ***spot detector*** (Fig. 36.39).

A second class of spot detector is wired to respond to the opposite situation, a small dark spot on a bright background. As a result, the picture falling on the retina is abstracted and transmitted to the brain as a pattern of light and dark spots.

Defining differences: opponent processing Other things are going on in the retina as well. Similar circuits are encoding color information by comparing the output of adjacent cone cells. This sort of comparison is necessary because the cones are very broadly tuned (see Fig. 36.24); a small hyperpolarization induced in a blue cone, for instance, may represent a dim blue light or a bright green one. The ambiguity is resolved by comparing the response of the blue cone with that of an adjacent green cone: if the green cone is unstimulated, the light cannot be green. Taken together, the relative responses of the three kinds of cone cells uniquely specify the wavelength of incoming light, thus providing the basis of our sensation of color. Comparing the output of two broadly tuned receptors in this way is called ***opponent processing***. Opponent

[1] For simplicity, this discussion focuses only on the activity of horizontal cells.

Class of spot detector

Stimulus	Response: A	Response: B
	++	--
	0	0
	0	0
	--	++

++ Strongly positive
0 Neutral
-- Strongly negative

36.39 Two kinds of spot detectors

The ganglion cells ultimately responsive to light falling on receptors in the retina belong to several classes. Two particular classes of spot detectors respond to spots of light in the receptive field. (In this diagram each class consists of 16 receptors, as well as bipolar, horizontal, and amacrine cells and a single ganglion cell, not shown.) Class A is most strongly excited when a spot of light falls in the center of its receptive field, and is inhibited when light falls on receptors to the outside. This can be seen in the table of responses ranging from strongly positive (++) for a focused spot of light, through relatively neutral (0) for an in-between stimulus, to strongly negative (--) when darkness is focused on the receptor cells at the center. Class (B) responds in exactly the opposite way.

processing is not confined to the retina: this comparison-contrast strategy is almost certainly used, for example, when the nervous system measures joint angles (Fig. 36.3) or pitch (Fig. 36.30).

■ HOW THE VERTEBRATE BRAIN HAS EVOLVED

Information from vertebrate sensory organs like the retina frequently undergoes some processing before it reaches the brain. Once in the brain, the information undergoes further processing that yields increasingly abstract versions of the original sensory input. How is this impressive neural computer organized, and how has it evolved?

Partly developed brains of vertebrate embryos, from primitive fish to humans, have striking structural similarities: all begin as three irregular swellings at the anterior end of the longitudinal nerve cord. In more advanced vertebrates these three regions undergo much modification in the course of development: walls thicken in some places, and distinctive outgrowths develop in others. The original three divisions of the brain—the ***forebrain***, the ***midbrain***, and the ***hindbrain*** (Fig. 36.40)—are still recognizable even in humans.

The vertebrate brain contains a series of hollow compartments, known as ventricles, that connect to the central canal of the spinal

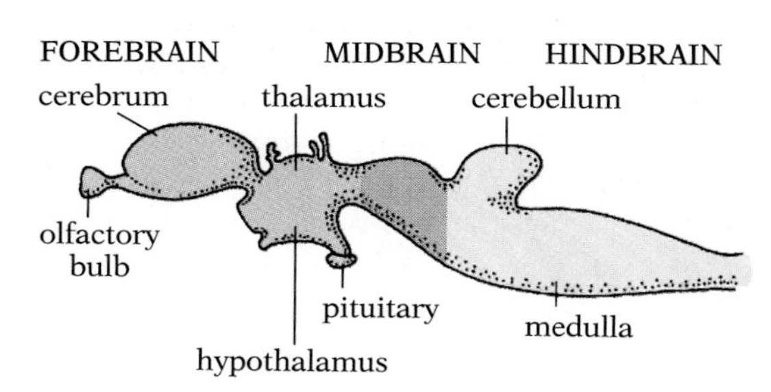

36.40 Principal divisions of the vertebrate brain

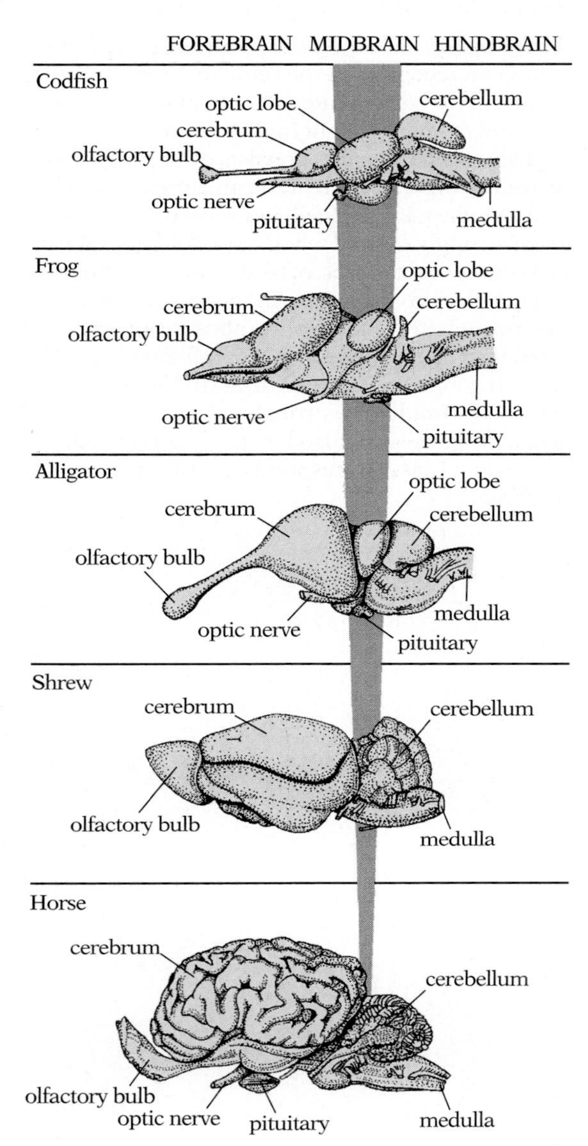

36.41 Evolutionary change in relative size of midbrain and forebrain in vertebrates

In this evolutionary sequence the relative size of the midbrain shows a marked decrease, and that of the forebrain a very considerable increase.

cord. The canal and the ventricles contain cerebrospinal fluid, which is kept circulating by the beating of cilia on the epithelial cells that line the ventricles. Both the brain and the cord are wrapped in three protective membranes, the ***meninges***. These are the pia, the innermost membrane, lying on the surface of the brain and spinal cord; the fragile arachnoid, just above the pia; and the tough dura, separating the other two from the inner surface of the skull and vertebrae. The spaces between the three meninges are filled with cerebrospinal fluid, which cushions the nervous tissue against damage.

Very early in its evolution, the vertebrate brain underwent modifications that set the stage for later evolutionary trends:

1 The ventral portion of the hindbrain, the ***medulla oblongata***, became specialized as a control center for some autonomic and somatic pathways concerned with visceral functions—breathing and heart rate, for instance—and as a connecting tract between the spinal cord and more anterior parts of the brain. In addition, the anterior dorsal portion of the hindbrain became much enlarged as the ***cerebellum***, a structure concerned with balance, equilibrium, and muscular coordination.

2 The dorsal part of the midbrain became specialized as the ***optic lobes***—the visual centers associated with the optic nerves.

3 The forebrain became divided into an anterior portion consisting of the ***cerebrum***, with its prominent olfactory bulbs, and a posterior portion consisting of the ***thalamus*** and ***hypothalamus***.

Continual evolution has made few changes in the hindbrain, though the cerebellum has become larger and more complex in many animals, particularly those with large bodies requiring fine muscle control. The most obvious evolutionary change has been the steady increase in the size and importance of the cerebrum, with a corresponding decrease in the relative size and importance of the midbrain (Fig. 36.41).

The ancestral cerebrum was probably only a pair of small smooth swellings concerned chiefly with olfaction. As in the spinal cord, the gray matter (cell bodies and synapses) was mostly internal. The synapses functioned predominantly as relays between the olfactory bulbs and more posterior parts of the brain; little processing of sensory information occurred in the cerebrum. The cerebrums of many modern fish are still little more than relay stations, though the areas of gray matter are more massive. In amphibians, which evolved from ancestral fish, there was an expansion of the gray matter and a multiplication of synapses between neurons that allowed it to process information coming from various sensory areas of the brain. Slowly, much of the gray matter moved outward from its initially internal position, until it came to lie on the surface of the cerebrum. This surface layer is known as the ***cerebral cortex.***

In certain advanced reptiles a new component of the cortex, the ***neocortex***, arose on the anterior surface of the cerebrum. Mammals, which evolved from reptiles of this type, show the greatest development of the neocortex. Even in primitive mammals the neocortex has expanded to form a surface layer covering most of the forebrain. This does not mean that the old cortex of the ancestral brain has been reduced; it has simply been covered over by the neocortex.

The neocortex became a major coordinating center for sensory and motor functions. As it continued to expand, both by increase in total size and by folding (which increases surface area), it came to dominate the other parts of the brain. The midbrain had been the chief control center in the earliest vertebrates. Then the thalamus portion of the forebrain became a major coordinating center, first sharing this function with the midbrain, then becoming dominant. Finally, the rise of the neocortex left the midbrain as a small connecting link between the hindbrain and the forebrain. The midbrain remains a control center for many unconscious mechanisms and some of the simpler visual functions; it also continues to play a major role in control of emotions.

Though brain size and complexity increases from fish through amphibians and reptiles to mammals, the fish brain did not stop evolving once the amphibian line had diverged, nor did the amphibian brain stop changing after reptiles appeared. The less advanced organization and complexity of fish and amphibian brains reflect the limited demands of their niches. Indeed, though the most primitive vertebrate brains are found in fish, the brains of some species of modern fish—the weakly electric fish in particular—are relatively large and complex, a result of continuing evolution to meet an unusual and neurally demanding way of life.

As specific areas of the vertebrate brain have increased in size, their internal organization has also become more complex. Originally brain areas were relatively unstructured: the neurons were scattered throughout. In areas of modest specialization, neurons are grouped into ***nuclei,*** with cell bodies and dendrites at the center and axons on the periphery. These axons bring information in and carry it out, and are channeled from one nucleus to another in nerve tracts. In even more highly evolved areas, these nuclei subdivide, and ultimately form regions called ***laminations***, in which loose layers of cell bodies and axons alternate. The most specialized brain nuclei have a tightly laminated structure with regular patterns of connections both within and between layers. This structural strategy makes possible complex synaptic organization and information processing.

In mammals, the increase in cortical area has resulted in a reduction in the relative amount of cortex devoted to strictly sensory and motor functions. The emphasis has shifted instead to the addition of association areas, in which information from different sensory systems converges and memory formation and storage occur. These make more flexible and complex behavior possible. The brain of our species represents the most extreme manifestation of this trend (Fig. 36.42).

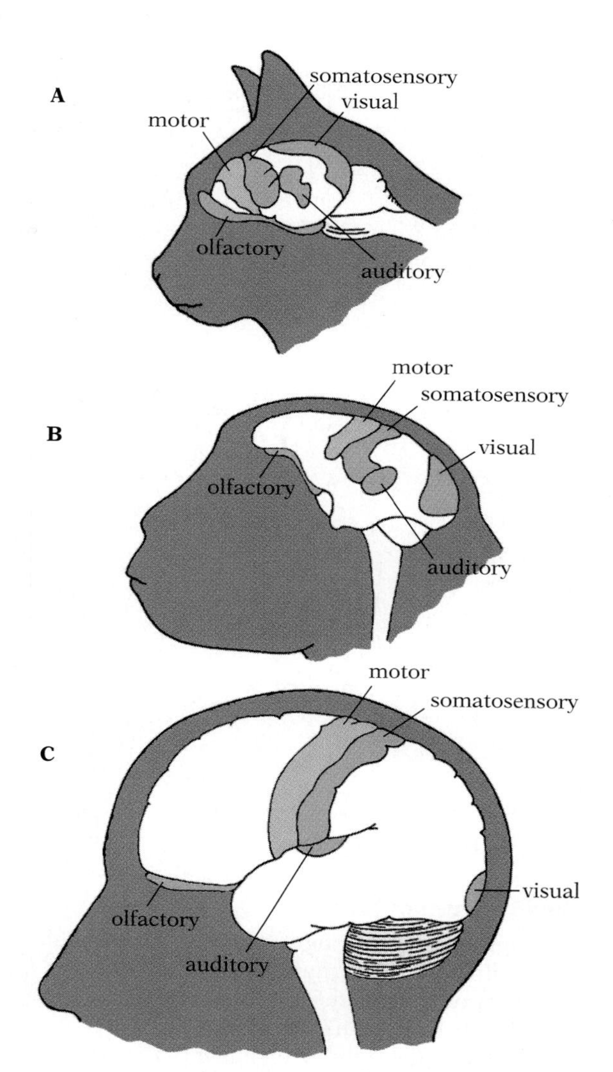

36.42 Proportion of cerebral cortex devoted to sensory and motor functions in three mammals

(A) In cats, sensory (tan) and motor (pink) areas constitute a major portion of the cortex. (B) In monkeys the proportion of cortex devoted to association areas (white) is much greater than in the cat. (C) In humans the sensory and motor areas occupy a relatively small percentage of the cortex, while most of the cortical area is devoted to association. The visual area shown here for each species is its primary visual cortex; as we will see in Figure 36.49, secondary areas, where more complex processing and associations occur, occupy large areas of cortex of humans and other primates.

■ HOW THE FOREBRAIN IS ORGANIZED

Much of our knowledge of vertebrate brain function is derived from research on rats, cats, monkeys, and chimpanzees. A considerable amount of information has accumulated about the human brain as well, most of it derived from electrical stimulation during brain surgery and from observations of the effects of tumors and

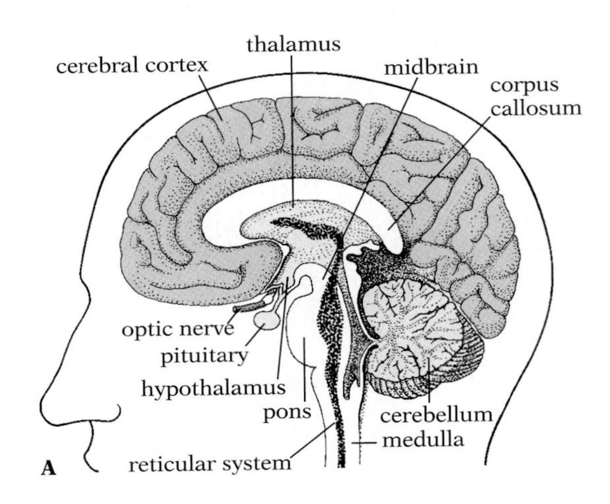

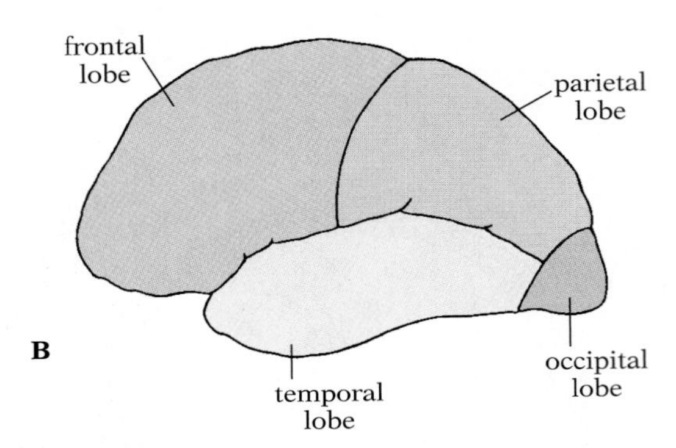

36.43 Human brain

(A) Sagittal section (longitudinal section through the midline) showing major parts of the brain. (B) Diagram of left side of the cerebral cortex showing its four main lobes. These lobes are duplicated in the right hemisphere of the brain.

accidental damage. The mammalian forebrain, as we have seen, is made up of the thalamus, the hypothalamus, and the cerebral cortex.

The thalamus The thalamus (Fig. 36.43A) is the major sensory-integration center in lower vertebrates; in higher vertebrates it has become in large part a sensory relay station on the way to the cerebrum, where complex integration takes place. Nevertheless, the thalamus continues to play an integrative role even in humans, and parts of it are intimately involved in memory.

The thalamus also contains part of the ***reticular system*** (Fig. 36.43A), a densely interconnected network of neurons that runs through the brainstem of the medulla and midbrain as well as the thalamus. Every sensory pathway running to higher centers of the brain sends side branches to the reticular system, as does every descending motor pathway. In this way, the reticular system is able to "listen in" on whatever is coming into or leaving the brain. It also sends a great many fibers of its own to areas of the cortex, brainstem, and spinal cord.

A major function of this curious area of the brain seems to be sensitization—the activation of other parts of the brain. It acts as the brain's arousal system, awakening and focusing an animal's attention on some change in the world around it. This adaptive response ensures that an animal will be alerted to anything that may be a predator, a potential mate, or food. One reason falling asleep is easiest in a dark, quiet bedroom is that fewer signals from the sensory receptors reach the reticular system, so it does less to arouse the brain. Barbiturates, which were once used in sleeping pills, block the reticular activating system and thus facilitate deep sleep. Destruction of the reticular system suspends the brain's ability to respond to sensory stimuli, and results in a permanent comatose state.

The reticular system does not, however, arouse animals indiscriminately. It selectively enhances or suppresses incoming sensory information on the way to the cortex to be processed. In a very real sense the reticular system "decides" what an animal will be most aware of—when it should pay more attention to sounds, say, than to its touch receptors, and vice versa. Such filtering is essential, since hundreds of millions of sensory receptors continually flood the brains of most mammals with irrelevant information. The system also modulates motor commands issued by the cortex, amplifying some and attenuating others.

The hypothalamus The hypothalamus is the part of the brainstem just ventral to the thalamus (Fig. 36.43A). As discussed in Chapter 33, its major functions include the synthesis of the hormones stored in the posterior pituitary and the secretion of the releasing hormones that help regulate the anterior pituitary. The hypothalamus is thus a crucial link between the neural and endocrine systems.

The hypothalamus is also the most important control center for the visceral and emotional responses of the body. Centers in the hypothalamus control hunger, thirst, body temperature, water balance, and blood pressure, as well as sexual desire, pleasure, pain, hostility, and so on. It is possible to induce behavior appropriate to

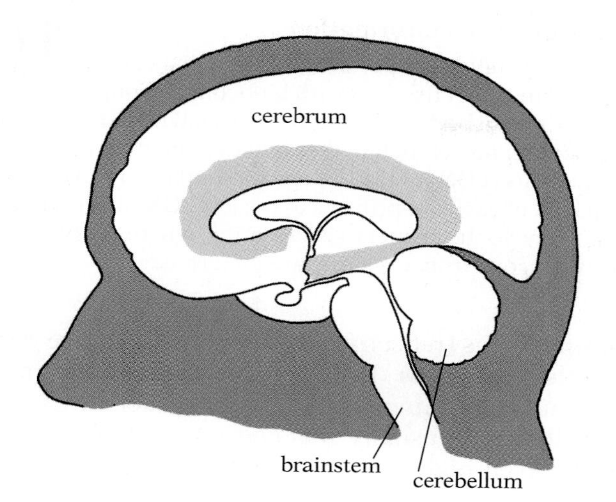

36.44 Limbic system of the human brain

The limbic system (green) is not an anatomically distinct structure, but rather a group of brain areas that are related functionally in giving rise to feelings and emotions.

each of these states by inserting microelectrodes surgically into the control centers of experimental animals and then stimulating the centers electrically. Rats with electrodes in their pleasure centers will spend virtually all their time pressing levers that turn on a tiny current in those regions, ignoring food and water almost to the point of starvation. Animals with electrodes in appropriate parts of the hypothalamus can be made to feel sated one moment and hungry the next, cold and then hot, angry and then calm.

The hypothalamus is the major integrating center for both visceral and emotional responses. It is not the only region controlling emotional responses, however, but rather one of a functionally related set of structures, known as the ***limbic system***, which surround the anterior end of the brainstem (Fig. 36.44).

The cerebral cortex In higher vertebrates, the cortex performs most of the complex processing of sensory information. Because the cortical cells are situated at the surface of the brain, electrical stimulation and recording are relatively convenient. As a result, we know a good deal about where and how sensory information is processed in the cortex.

The cortex contains discrete areas activated by different sensory systems (Fig. 36.42). These areas display a logical internal organization. In the somatosensory area, for instance, each part of the body has its own representation (Fig. 36.45A). To be sure, it is a distorted map. For example, the 4 cm^2 of lip surface are represented by a greater area of cortex than the 6000 cm^2 of neck, trunk,

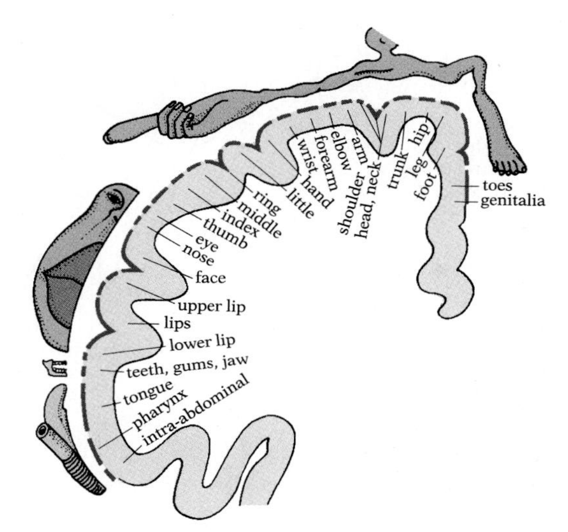

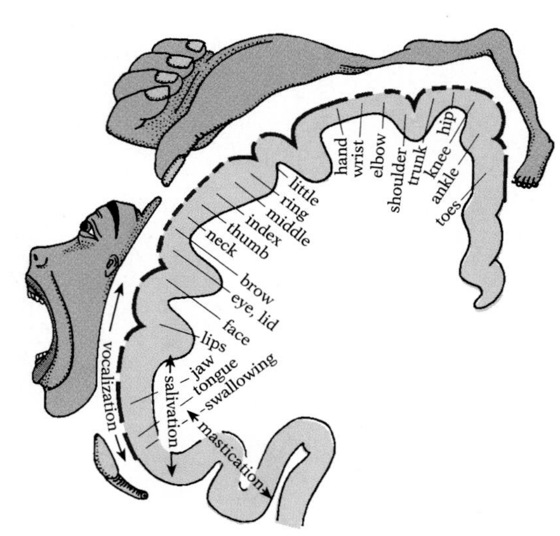

36.45 Functional map of the sensory and motor cortex

(A) The somatosensory strip of the human cerebral cortex (see Fig. 36.41C) receives sensory input from all parts of the body. On this strip adjacent parts of the body are usually represented next to each other. Parts of the body with large numbers of sensory receptors, such as the lips, tongue, and fingers, are represented by correspondingly large areas on the cortex. (B) A strip from the adjoining motor-control area of the cortex (Fig. 36.41C) shows similar organization: mapping proceeds from toe to head; adjacent parts of the body are usually represented next to each other; and those with many small muscles for fine motor control occupy especially large amounts of space.

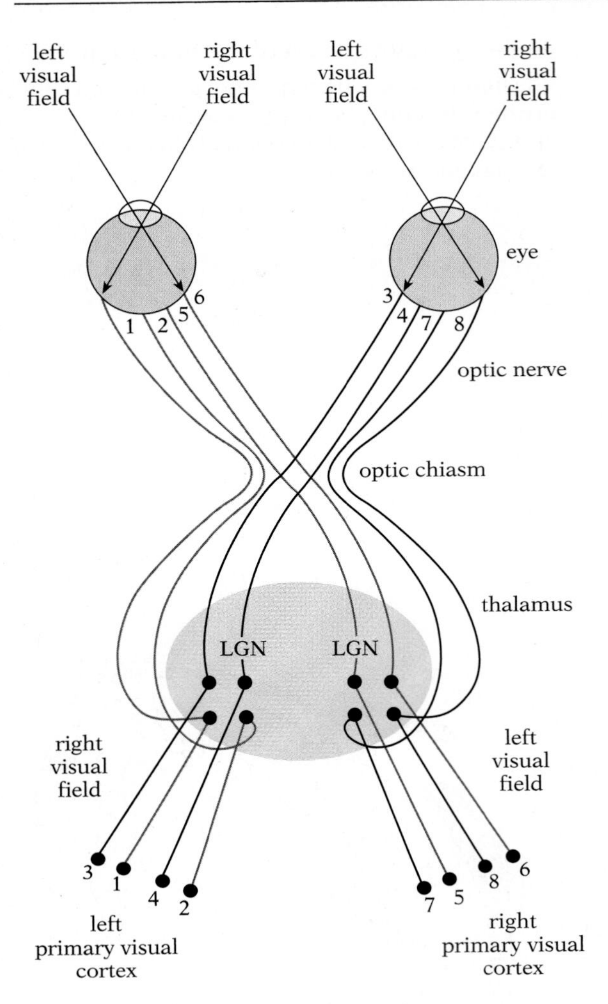

36.46 Flow of visual information

Axons from the ganglion cells in the retina run in the optic nerve to the optic chiasm. There, in creatures with binocular vision, the fibers from receptors looking out on the left half of the visual field in the left eye join the fibers representing the left half in the right eye and travel to the right lateral geniculate nucleus (LGN) of the thalamus. Similarly, information from the right visual field of each eye projects to the left LGN. From there each nucleus sends axons to the primary visual cortex, where the two images of its half of the world, one from each eye, are integrated.

hips, and upper legs (reflecting, of course, the relative density of receptors in these regions). However, the elements are arranged sequentially in an anatomically logical order, with contiguous regions on the body usually mapped on contiguous areas of the cortex. At an even higher level of detail within this strip of cortex, these positional relationships persist: an array of bristles on the face of a mouse is matched by an arrangement of cell groups on the cortex. The motor area, which lies alongside the somatosensory area, is mapped almost precisely in parallel (Fig. 36.45B).

Logical internal organization is seen also in the visual system. The optic nerve leaving one retina meets its counterpart from the other eye at the ***optic chiasm***. In animals with binocular vision (in which the fields of view of the two eyes overlap), the axons of each optic nerve separate into two parts there. Those from the cells in the left half of the retina of each eye (which look out at the same right half of the visual world) join to travel together to the left side of the thalamus. Those from cells in the right half of each retina go to the right side of the thalamus (Fig. 36.46). Hence the neurons from the left halves of both retinas synapse in the left lateral geniculate nucleus (LGN) of the thalamus, while those from the right halves of both retinas synapse in the right LGN; the fibers from the two eyes are thus brought together.

Fibers then leave the LGN and synapse in the left or right ***primary visual cortex*** consecutively: that is, fibers from the uppermost, farthest left patch of the left-eye retina are mapped onto the left visual cortex next to those from the uppermost, farthest left patch of the right-eye retina, and so on. Because of the high density of receptors in the fovea of the retina, the area devoted to the center of the visual field is greatly enlarged in the visual cortex. In birds and many lower vertebrates visual information is mapped onto an analogous area in the midbrain known as the optic tectum.

Visual processing in the cortex Lateral inhibition in the retina encodes the visual world into a series of spots of two classes: bright center/dark surround (Figs. 36.39A, 36.47A) and dark center/bright surround (Fig. 36.39B). The retina, thalamus, and tectum cortex add three additional classes of feature-detector circuits. (These classes are found in other vertebrates and in invertebrates as well.) One of these is spot-motion detectors. Each detector responds only to spots of a particular size moving in a specific direction at a certain rate (Fig. 36.47B). Different spot-motion detectors have different preferences, and virtually all contingencies appear to be represented. In frogs and toads such detectors are wired into circuits that aim and trigger the bug-catching behavioral responses.

Spot detector	Spot-motion detector	Line detector
Neural output	*Neural output*	*Neural output*

A B C

36.47 Feature detectors in the visual system

The response characteristics of three simple sorts of feature detectors are represented. In each case, the center of the receptive field is indicated by a red +. (A) This point is most responsive to a spot of light that is small and centered in the receptive field. (B) Spot-motion detectors respond to specific patterns of movement. The one shown here is stimulated by a small spot moving 135° to the right of vertical. Moreover, as the middle and bottom examples indicate, only a specific rate of movement stimulates a particular spot-motion detector maximally. (C) Line detectors respond to lines of specific width and direction. The one shown here requires a narrow bar oriented 45° to the right of vertical as its stimulus.

Spot-motion detector circuits work by comparing over time the output of a series of simple retinal spot-detector circuits (like those shown in Fig. 36.39): the motion-detecting feature detector is inhibited for all but one pattern of signals.

Feature detectors of another class respond only to lines of a particular orientation (Fig. 36.47C). There are detectors for all possible orientations of lines, but vertical and horizontal detectors seem to be especially numerous. In the cortex of higher vertebrates, the angle to which these line detectors respond best varies continuously across each small area of the cortex, and each unit of the grid thus established corresponds at least roughly to the receptive field of one spot detector in one eye (Fig. 36.48). A line detector works by comparing signals coming (via the LGN) from arrays of spot de-

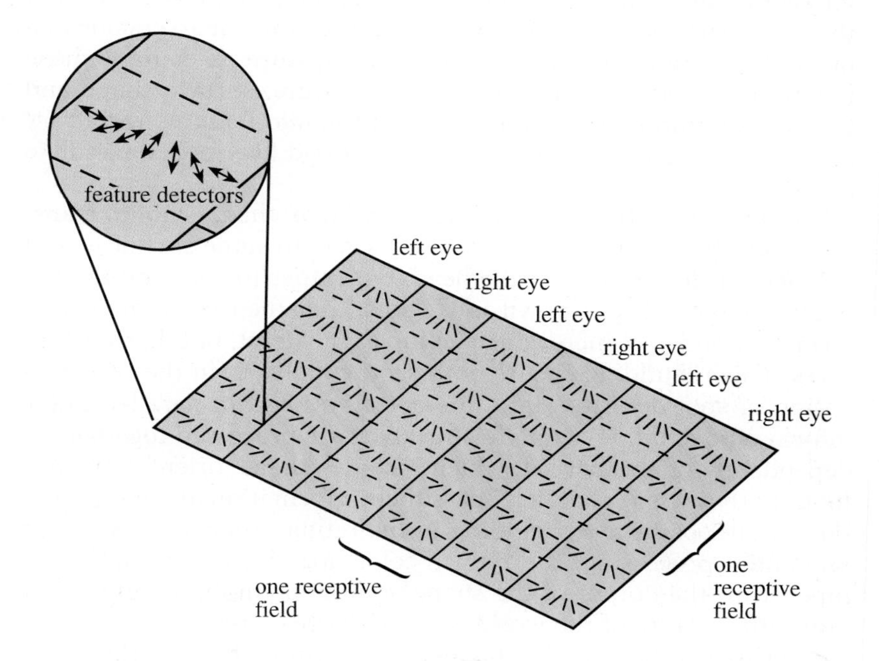

36.48 Diagram of the organization of a layer of visual cortex

Visual information arriving in the LGN is transmitted to specific cells that are arranged, several abreast, in narrow stripes. Each segment within a stripe contains cells responding to the receptive field of one group of receptors in one eye. Adjacent segments in the same stripe and in the next parallel stripe for the same eye code for contiguous receptors in the retina. Within these stripes are ordered sequences of line detectors, each maximally responsive to one angle. The pattern here is idealized.

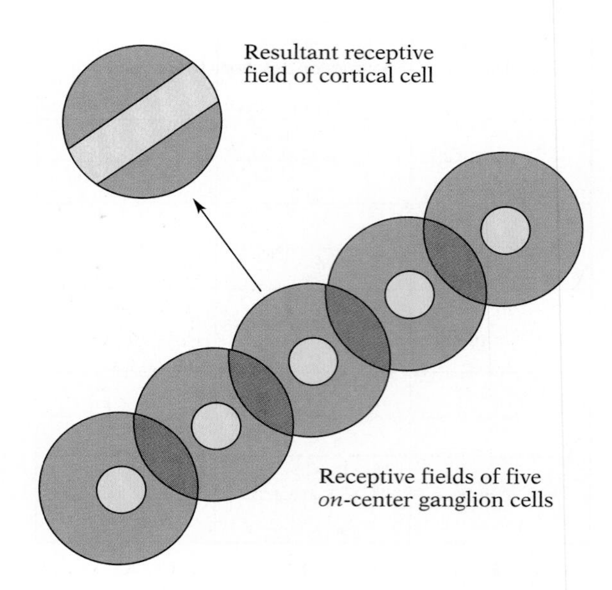

36.49 How line detectors utilize the input from spot detectors
Line detectors compare the signals coming from a series of spot detectors. Depending on which and how many spot detectors converge on a cell, a feature detector of this type can code for a line of any width, length, and orientation. Similarly, feature detectors have been found that respond to "hypercomplex" conformations such as points or curved edges or even certain shapes.

tectors in the retina. It is active only if the feature it recognizes is present (Fig. 36.49).

Feature detectors of still another class sort for rates and directions of motion in these lines. Each such detector has a well-defined specificity, and fires only when it detects motion at a particular speed *and* in a particular direction. Other feature detectors combine the input from the two eyes and, by comparing what the same parts of the two retinas are seeing, are able to encode distance. Vertebrates also have many other feature detectors, whose functions include detecting corners, curvature, surface shape and texture, specific rates of approach, and iconic features (arbitrary shapes, such as various sorts of T shapes and five-spoked handlike arrays).

The strategy of the vertebrate visual system, then, is not to transmit a faithful picture of the world passively to some sort of neural "TV screen" at the back of the head, but rather to sort through the data, first breaking it down into spots, and then rebuilding it in terms of a series of increasingly complex, abstract, but discrete features. This rebuilding occurs both sequentially (as in the recombination of spot detectors into line detectors) and in parallel. Color information and fine detail, for instance, are processed together independently of another set, which includes shape, orientation, and motion; these two "channels" pass their information along separate but parallel pathways beginning in the retina. After many steps of sequential processing, the refined-color and detail data are combined with the output of the shape/orientation/motion pathway to create the picture of the world we experience consciously.

In primates, at least, the much-edited picture of the world in the

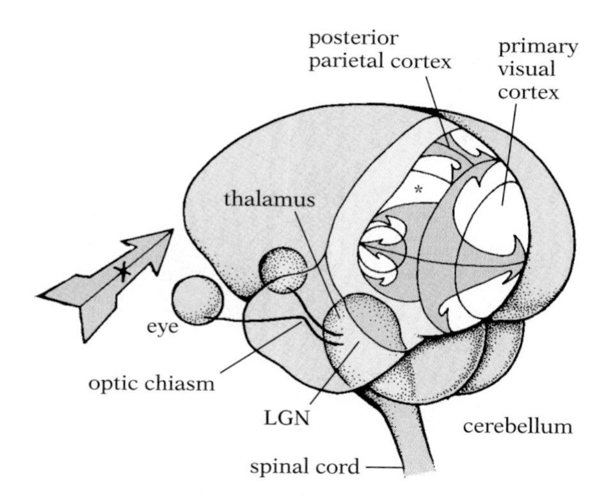

36.50 Projection of visual images in the cortex

The visual target in this example is a horizontal arrow; the X marks the center of the visual field. The right half of the visual field is shown here as it is transmitted from the retinas, via the LGN, to the cells in the left primary visual cortex of a squirrel monkey. (Part of this cortex is out of sight between the two hemispheres in this view.) The visual world we (and presumably the monkey) perceive is a spatially faithful picture except that the part seen by the fovea (the middle of the arrow) is magnified to allow high-resolution processing of the image in the center of the visual field. The picture is transmitted from the primary visual cortex to several other areas. In many of these the top and bottom halves of the visual field are separated. The organization of the visual area marked with an asterisk has not yet been traced.

left or right primary visual cortex is then passed to several nearby places in the cortex where further transformations are made (Fig. 36.50). Some of these areas may be involved in visual memory and visual association. One in particular—the posterior parietal cortex—seems to decide what in the visual field ought to be subject to the high-resolution scrutiny of the fovea, and then to issue instructions to aim the eyes appropriately. Another area concentrates on nearby stimuli—those within arm's reach. Others appear to bring the visual world at least roughly into register with other senses—touch and hearing most commonly—so that stimuli from a particular direction from any sensory modality map onto the same cells of a single array. The maps are particularly prominent in the midbrain. Such maps are both fine-tuned and brought into precise register with one another by early experience: the repeated correlated firing of two receptors in one map, or receptors in each of two maps, is taken as a sign that they represent similar locations, and synaptic strengths are modified accordingly. The parallels with the conditioning circuit of *Aplysia* (p. 1012) are striking.

Language processing Localized brain specialization is evident from studies of stroke victims who, because a clot has cut off the blood supply to a portion of the brain, have lost the use of specific neural regions. Some patients, for instance, can identify a dog as an animal, but cannot decide what kind of animal it is. Others cannot recognize familiar faces, but can recognize the same people by the sound of their voices. These syndromes involve consistent locations in the brain, suggesting that the processing or filing involved in at least certain categories of learned recognition is highly segregated and organized.

Complex but regular organization like that of the visual system appears to be the rule rather than the exception in the brain. From insects to primates, brains are specialized into nuclei consisting of relatively precise three-dimensional matrices of neurons. These nuclei receive input from particular sensory receptors and from other nuclei, and send information out to other nuclei, as well as to the appropriate motor areas.

Perhaps the most specialized pathway of this sort in humans is the one that underlies language acquisition and speech. The

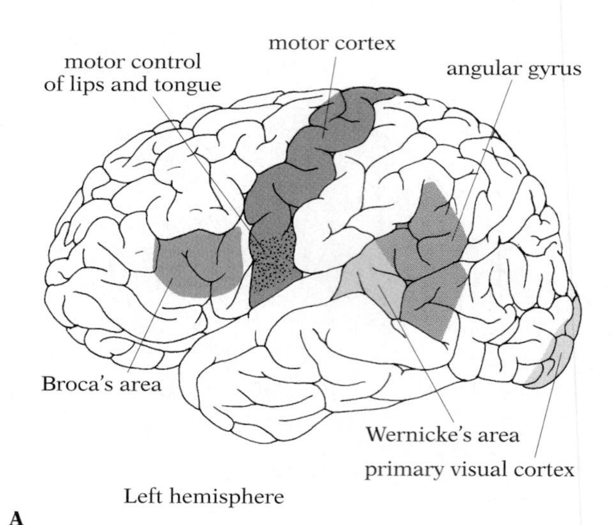

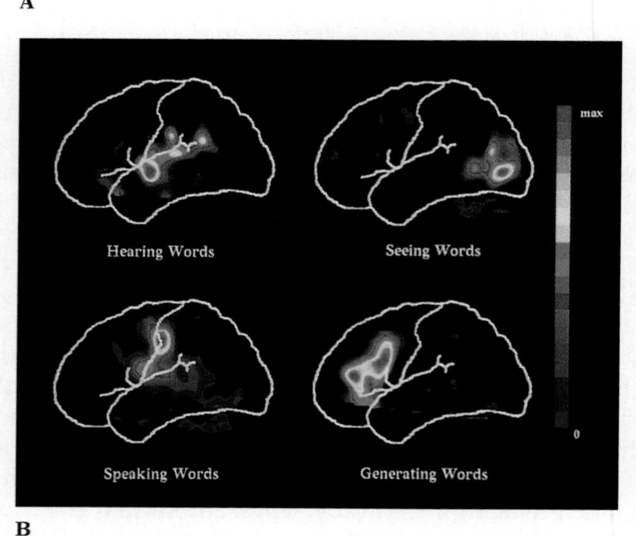

36.51 Anatomy of language

(A) The human brain contains many specialized centers concerned with language, particularly in the left cerebral hemisphere (shown here). Sounds are detected by the ears and processed in the midbrain (not shown). Spoken language is then sent to Wernicke's area. Written language is somehow abstracted from the other material processed in the primary visual cortex and sent to the angular gyrus, where it is translated into sound. It too is sent to Wernicke's area, which extracts meaning from incoming language, whatever the source. Language production also begins in Wernicke's area. Thoughts are encoded there into crude linguistic outlines, and these are sent to Broca's area, which refines them into grammatical sentences. Finally, Broca's area transmits directions to the adjacent motor cortex, which controls the organs of speech, to produce utterances. (B) Patterns of blood flow in the brain during language-related activity confirm the anatomical inferences drawn from studies of stroke and accident victims.

human brain is already wired at birth with the neural circuitry necessary for language. Linguistic meaning seems to be processed on one side (usually the left), in discrete, well-defined areas (Fig. 36.51). For written language the processing begins in the primary visual cortex, where what we see is analyzed before being sent to the ***angular gyrus***; there written words are translated into sounds, which are then passed to ***Wernicke's area***. Spoken language is sorted from other sounds in the midbrain, which sends the lowest frequencies of speech to Wernicke's area. Wernicke's area, then, is the destination for both written and spoken language. Higher frequencies go to the right side of the brain, where the emotional overtones of the speech are ascertained. So segregated are the intellectual and the emotional functions of the brain that many people with right-hemisphere damage can understand the *meaning* of a spoken sentence, but cannot say whether the speaker was happy or sad, angry or ironic. Conversely, people with left-hemisphere damage can often judge the mood and intention of a speaker, and yet have no idea of what has been said.

Wernicke's area is also the processing center most involved in the individual's own spoken and written expression. It is here that thoughts are formulated into crude linguistic structures before being sent to ***Broca's area*** for grammatical refinement. Damage in Broca's area often leaves a patient knowing what he wants to say, but unable to express it according to the accepted rules of tense, declension, number, gender, and so on; the damage may rob the speaker of such linguistic signposts as pronouns, conjunctions, and prepositions. Damage in Wernicke's area, on the other hand, leaves the patient talking in perfectly grammatical sentences that are nevertheless meaningless. Strikingly similar effects are seen in the sign-language communication of the congenitally deaf, indicating that these areas are specialized for both spoken and visually perceived language.

CHAPTER SUMMARY

HOW ENVIRONMENTAL CHANGES ARE DETECTED

HOW SENSORY STIMULI PRODUCE NEURAL IMPULSES Stimuli are detected from changes in membrane potential in sensory cells. (p. 1029)

HOW RECEPTORS ENCODE STIMULUS STRENGTH Most receptors have a basal firing rate and are relatively phasic, responding primarily to changes in stimulus intensity. Tonic receptors respond to absolute intensities; they usually cover only a limited part of the stimulus range, so the nervous system must rely on several different receptors to report the full range of intensities. (p. 1030)

THE MANY SENSES OF TOUCH Touch receptors include cells specialized to detect touch, pressure, heat, cold, and pain. (p. 1031)

MONITORING INTERNAL CONDITIONS Proprioceptors measure internal conditions and muscle tension. (p. 1032)

DETECTING TASTE AND SMELL Taste receptors in insects respond specifically to sweet, sour, salty, and bitter molecules; vertebrate taste receptors respond to the same but are less narrowly tuned. Odor receptor types are much more numerous, each kind responding when its membrane receptors bind to parts of odor molecules. (p. 1033)

HOW ANIMALS SEE Compound eyes view the world as a mosaic; in many species with compound eyes, each ommatidium has separate color receptors. Camera eyes focus an image on a retina. In many species with camera eyes, a fovea with a high density of receptors allows for finely detailed vision and (in many species) color sensitivity. Rods have high sensitivity to light but cannot distinguish colors; cones provide information about color only when the nervous system compares the activity of different cone types. (p. 1035)

HOW HEARING WORKS The human ear responds to the rapid variations in air pressure that accompany sound. The variations move the tympanic membrane, which transfers the vibrations to the cochlea, where they set the basilar membrane into motion. Different regions of the basilar membrane vibrate best at different frequencies, from which the nervous system infers pitch. (p. 1043)

MAINTAINING BALANCE AND SENSING MOVEMENT The semicircular canals register acceleration, while the otoliths indicate orientation relative to gravity. (p. 1047)

SPECIALIZED AND UNUSUAL SENSES Many species detect stimuli beyond our range of sensitivity—infrared and ultraviolet radiation, or infra- and ultrasonic sounds, for instance. Others detect stimuli to which we are insensitive, such as electric and magnetic fields. (p. 1048)

MAKING SENSE OF SENSATIONS

HOW SENSORY DATA ARE PROCESSED Lateral inhibition enhances sensory contrasts, and thus edges. Opponent processing uses the ratio of response of two differently tuned receptors to interpolate sensory value. (p. 1050)

HOW THE VERTEBRATE BRAIN HAS EVOLVED All vertebrate brains have a forebrain, midbrain, and hindbrain. The cerebellum of the hindbrain is specialized for muscular coordination. The forebrain has become increasingly specialized for sensory processing and decision making, particularly as its neocortex has mushroomed in size. Specialization in the neocortex has led to nuclei dedicated to particular functions, and laminations that process information sequentially. (p. 1053)

HOW THE FOREBRAIN IS ORGANIZED The thalamus is the main site of sensory processing in lower vertebrates; in mammals it relays information to the cerebral cortex for processing. The thalamus also contains the reticular formation, which controls general levels of alertness. The hypothalamus links the brain to the endocrine system, and plays a major role in controlling emotional states and drives. The cerebral cortex processes sensory information, analyzing it for increasingly specific features and integrating the results. The specificity of processing is illustrated by language use: decoding and encoding are accomplished in specific nuclei according to neural "rules" in place from birth. (p. 1055)

STUDY QUESTIONS

1 We saw in this chapter that most sensory cells adapt, and in Chapter 35 we saw that animals can habituate to sensation from a particular modality. Habituation takes place centrally in the nervous system, while adaptation occurs peripherally at the site of sensation. What simple behavioral test might allow you to determine whether an animal has stopped responding because of adaptation or habituation? Remember that a hand plunged into warm water soon gets used to the temperature, and the water begins to feel tepid. (pp. 995, 1012–1013, 1030–1031)

2 Color blindness occurs when one of the visual pigments is lost, leaving the brain unable to resolve the ambiguities inherent in a system that determines wavelength by comparing three broadly tuned receptors. Looking at Figure 36.24, identify the range of problems encountered by a red-blind person (the most common sort of color blindness). Would blue- or green-blind individuals be at more or less of a disadvantage? (pp. 1042–1043)

3 Lateral inhibition serves to emphasize contrast at the expense of reporting background or broad-area illumination. How does this phenomenon create the illusion of blackness in a TV picture? (Hint: Look at a TV screen when the set is switched off.) (pp. 1051–1052)

4 Individuals who have suffered damage to the visual area of the cortex are blind, and yet, when pushed to guess, they can often accurately describe objects in front of them—a phenomenon known as blindsight. How can this be? (pp. 1058–1060)

5 Most higher vertebrates determine the direction from which a sound is coming in two ways. For high frequencies, they compare the intensity heard by the two ears; the head acts as an acoustic shadow, so that the difference in intensity is greatest when the sound is coming directly from the side, but the loudness at the ears is the same when it comes from straight ahead. For low frequencies, animals usually compare the arrival time at the two ears, which again is most different when the sound is coming from the side, and is the same when the source is directly ahead. Design a circuit to make one of these distinctions, activating a specific cell in a linear array (encoding left to right) to indicate the direction of a sound source. (pp. 1051–1053)

SUGGESTED READING

GOULD, J. L., 1982. *Ethology: The Mechanisms and Evolution of Behavior*. W. W. Norton, New York. *Discusses animal senses and their role in behavior.*

HUBEL, D., 1988. *Eye, Brain, and Vision*. W. H. Freeman, New York. *Beautifully written and produced.*

KALIL, R. E., 1989. Synapse formation in the developing brain, *Scientific American* 261 (6). *On how correlated firing helps organize and tune brain circuits and maps.*

KIMURA, D., 1992. Sex differences and the brain, *Scientific American* 267 (3). *On sex-specific differences in information processing.*

KORETZ, J. F., AND G. H. HANDELMAN, 1988. How the human eye focuses, *Scientific American* 259 (1). *On the structure of the lens, how its shape is changed to allow focusing, and the process of aging.*

NEWMAN, E. A., AND P. H. HARTLINE, 1982. The infrared "vision" of snakes, *Scientific American* 246 (3). *A look at the neural basis of the ability of pit vipers to see in the dark.*

POSNER, M. I., AND M. E. RAICHLE, 1994. *Images of Mind.* W. H. Freeman, New York. *How the human brain processes information; well written and illustrated.*

SUGA, N., 1990. Biosonar and neural computation in bats, *Scientific American* 262 (6). *On the sensory processing underlying the ability of bats to locate prey in the dark.*

SZPIR, M., 1992. Accustomed to your face. *American Scientist* 80 (6). *On the highly localized and compartmentalized circuitry involved in face and name recognition*

TANAKA, K., 1993. Neuronal mechanisms of object recognition. *Science* 262, 685–688. *On the iconic model of visual processing.*

TREISMAN, A., 1986. Features and objects in visual processing, *Scientific American* 255 (5). *On parallel processing in the visual system.*

VAN ESSEN, D.C., C. H. ANDERSON, AND D. J. FELLEMAN, 1992. Information processing in the primate visual system. *Science* 255, 419–423. *On the serial- and parallel-processing pathways in the visual system of primates.*

CHAPTER 37

MUSCLES

Effectors are the parts of an organism that do things, that carry out its response to stimuli. Their actions range from phototropism to glandular secretion to muscle movement. In this chapter we will focus mainly on how vertebrate muscles operate and are controlled.

HOW PLANTS MOVE

As described in Chapter 32, plants respond to light and gravity through differential growth. Some plants are also capable of rapid movement: leaves, for example, may droop or fold at night and expand again in the morning. The flowers of many plants open and close in a regular fashion at different times of day. The leaves of the sensitive plant fold and droop within a few seconds after being touched (Fig. 37.1). The leaves of the Venus flytrap close rapidly around insects that have landed on them (see Fig. 26.17, p. 757). The seed or spore pods of some plants snap open at maturity, vigorously expelling their seeds (Fig. 22.33, p. 616).

These movements are far too rapid to depend on differential growth. They rely instead on changes in turgor pressure. Leaves droop when certain cells lose so much water that they are no longer turgid enough to keep the leaf rigid. Flowers fold when specially sensitive cells arranged in rows along the petals lose their turgidity; when these cells regain moisture, they open again. In the Venus flytrap, rapid changes in turgidity in special effector cells located along the hinge of the leaf close its "jaws" quickly. Similarly, in the sensitive plant, rapid turgidity changes in specialized effector cells at the bases of the leaflets and petioles make it seem to shy away from a touch.

Though active movement is not an exclusive characteristic of animals, the most elaborate mechanisms for producing locomotion are found in the animal kingdom. These effectors are usually controlled by the nervous system; unlike sensory receptors and conductor cells, however, effector cells are not part of the nervous system. Among the numerous effector systems not under nervous control are the nematocysts of coelenterates, the cilia and flagella of all animals, and some smooth muscles of vertebrates.

37.1 Rapid movement in plants

Within a few seconds of being touched, the sensitive plant (*Mimosa pudica*) can respond by folding its leaves. Rapid decreases in turgor pressure affect the rigidity of the leaves, causing them to droop or close.

MUSCLES AND SKELETONS

The most prominent effectors in all multicellular animals except the sponges are ***muscles***—tissues composed of specialized contractile cells.

■ WATER POWER: HYDROSTATIC SKELETONS

The first multicellular animals (disregarding the sponges) must have been no more than 1 mm long. They probably used cilia for swimming. Even today the smallest flatworms and the tiny larvae of many coelenterates depend primarily on cilia to get around. As animals evolved larger size, they evolved contractile tissues that first supplemented, and then supplanted the cilia as their chief means of locomotion. Coelenterates have very primitive contractile fibers, but their rhythmic contractions in the bell of a jellyfish enable it to swim weakly (see Fig. 35.10, p. 998), and contractions of other fibers allow jellyfish and hydra to move their tentacles. The hydra is even able to move by turning somersaults (Fig. 37.2).

37.2 Somersaulting locomotion of hydra

Some animals, such as flatworms and snails, glide forward slowly as waves of contraction pass along their longitudinal muscles. With each wave, points on the lower surface of their bodies advance a fraction of a millimeter; these points may then grip the substrate and, as the wave passes backward, act as anchors toward which more posterior parts of the lower surface are drawn. In this type of locomotion, little use is made of the circular muscles of the body or of the hydrostatic properties of the body contents (see next paragraph).

In many animals the muscle fibers of the body wall are arranged in prominent longitudinal and circular layers. The fibers in these two layers are antagonistic to each other—that is, they produce opposite actions: contraction of the longitudinal muscles shortens the animal; contraction of the circular muscles lengthens it. Because the semifluid body contents resist compression, and thus function as a ***hydrostatic skeleton***, the body volume remains constant and the shortening is accompanied by a compensating increase in diameter. The boneless appendages of many higher animals are moved in this way—the vertebrate tongue, for example, as well as the elephant's trunk and the squid's tentacles.

The most complete exploitation of the potentialities of hydrostatic skeletons is seen in certain annelid worms, such as earthworms. Here the body cavity is partitioned into a series of separate fluid-filled chambers (see Fig. 31.13, p. 890). Correlated with this ***segmentation*** of the body cavity is a similar segmentation of the musculature; each segment of the body has its own circular and longitudinal muscles. It is thus possible for the animal to elongate one part of the body while simultaneously shortening another part. The result is ***peristalsis***, a series of alternating waves of contraction, activating first longitudinal and then circular muscles. During the longitudinal contraction of a segment, the worm protrudes hard bristles known as setae that provide traction (Fig. 37.3).

For a worm with an unsegmented body cavity, it would not be so easy to perform a variety of localized movements, because changes

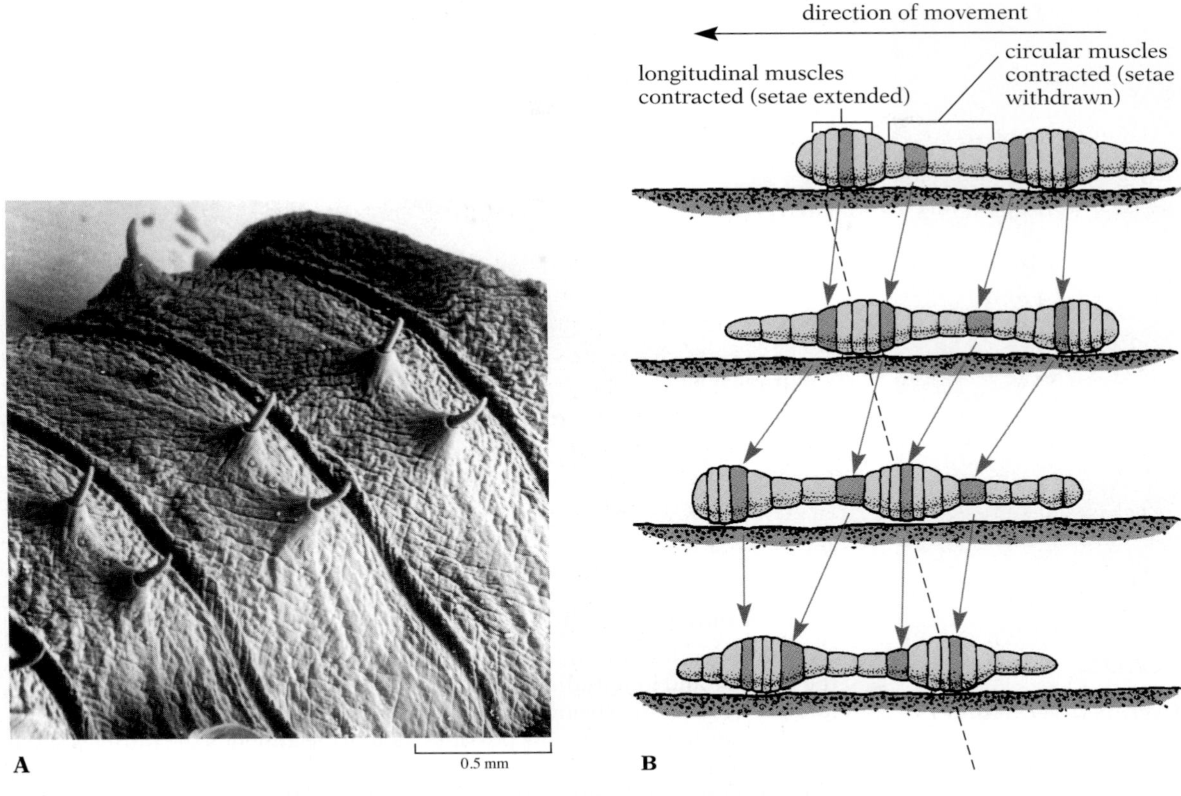

37.3 Use of setae, together with peristalsis, to create movement in earthworms

(A) This scanning EM of an earthworm shows the bristlelike setae of each segment. (B) The earthworm, represented here with 20 segments, uses its hydrostatic skeleton to generate movement. Some segments are shown in darker color and connected by arrows for easier identification. As the longitudinal muscles of a segment contract, the segment becomes short and thick, and its setae are extended to anchor the worm to the substrate. As the circular muscles of a segment contract, the opposite occurs; the segment becomes long and thin, extending forward as it loses contact with the substrate. Peristalsis—alternating waves of contraction of circular and longitudinal muscles—thus enables the earthworm to move. The progress of one wave of peristaltic contractions is indicated by the dashed line.

in the fluid pressure would be freely transmitted to all parts of the body. Segmentation usually carries with it a necessity for some degree of segmental organization of the nervous system and for serial repetition of other organs such as the nephridia.

Annelid segmentation probably evolved as an adaptation for burrowing. The compartmented hydrostatic skeleton aids movement by peristaltic waves, which can develop considerable thrust against the substrate and push aside the soil particles. Though there are many unsegmented worms that burrow by peristaltic wave motions, none can develop as much thrust or burrow so continuously and effectively as segmented worms.

Many marine annelids have lateral flaps on each segment called

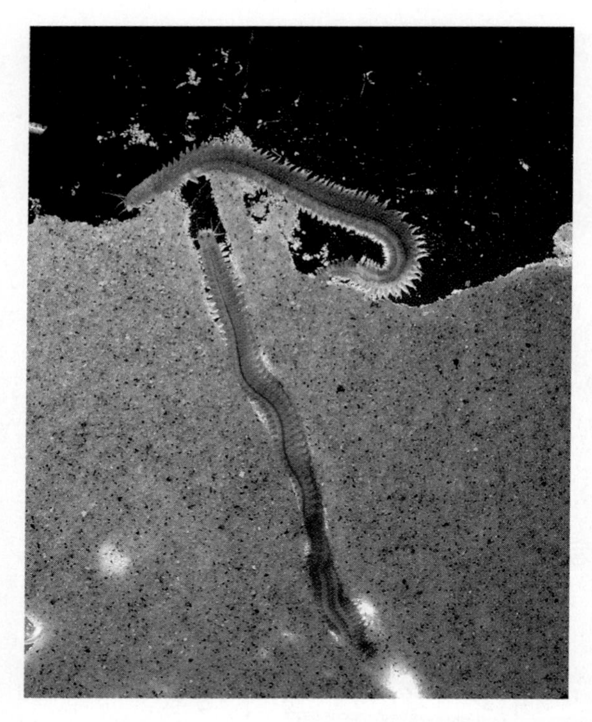

37.4 Marine worms (*Nereis diversicolor*) with parapodia

The worms get tiny particles of food and an adequate supply of oxygen as the waving parapodia produce water currents in the tubelike burrows the animals inhabit. Richly supplied with blood vessels, the parapodia also function as gills (see Fig. 28.10, p. 802).

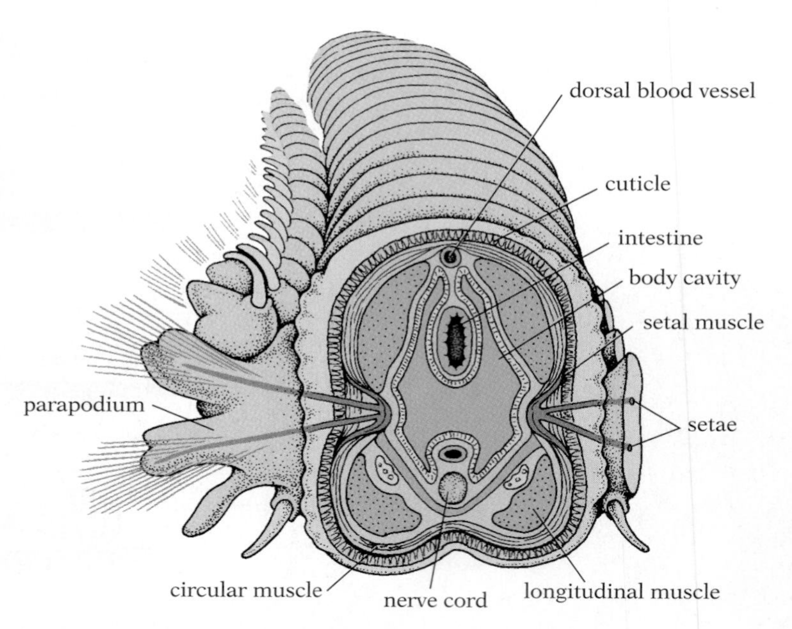

37.5 A marine annelid

This cross section through a segment of the marine annelid *Nereis* shows the circular and longitudinal muscles and the tough outer cuticle. The evolutionary loss of the strong cuticle made peristaltic movement in soft-bodied annelids like earthworms possible. In marine annelids the muscles serve instead to create an internal peristalsis that both pumps blood and turns the animal's body in rhythmic snakelike undulations that help it burrow into the ocean floor. The parapodia are moved from inside like oars by muscles that run from the cuticle to the inner ends of the hard setae.

parapodia (Fig. 37.4), which may have evolved originally to produce respiratory and feeding currents inside the tubes occupied by sedentary annelids; they often still function in this way and may also serve as gills and locomotory appendages. Annelids whose parapodia are particularly well adapted for locomotion have, in addition to muscles arranged in longitudinal and circular layers, many large muscles running at odd angles. Some of these muscles pull against the unusually tough body wall, and act to flex the body; the wall itself is so strong that these worms cannot create the alternating swelling and constriction that is essential for moving with peristaltic contractions (Fig. 37.5). Many of them also lack the high degree of internal segmentation of earthworms and their relatives. It was probably from this sort of annelid worms that the soft-bodied annelids (like earthworms) and the arthropods evolved.

■ WHY JOINTS ARE BETTER

Internal versus external skeletons The arthropods and the vertebrates are by far the most mobile of the multicellular animals. Both groups possess paired locomotory appendages—legs and sometimes wings. To largely replace a hydrostatic skeleton, each has evolved a hard, jointed skeleton, with most of the skeletal muscles so arranged that one end is attached to one section of the skeleton and the other end to a different section (Fig. 37.6). Hence when the muscle contracts, it causes the skeletal joint between its two points of attachment to bend. Though the skeletal and muscular systems of arthropods and vertebrates are functionally similar, there are many significant differences in structure and detail.

The most obvious difference is that arthropods have an ***exoskeleton***—a hard body covering with all the muscles and organs located inside it—whereas vertebrates have an ***endoskeleton***—a framework with attached muscles, all of which is embedded within the organism. Besides functioning as structures against which muscles can pull, both types of skeleton help provide shape and structural support, particularly for animals living on land, where the buoyancy of water is not available for support; in this respect they are analogous to the rigid xylem, which is a critical factor in enabling land plants to grow large. Exoskeletons, which are composed of noncellular materials secreted by the epidermis, function also as a protective armor for the softer body parts and, in terrestrial arthropods, as a waxy barrier preventing excessive water loss.

Exoskeletons create problems for growth; periodic molting of the exoskeleton and deposition of a new one are necessary to permit increase in size. The mechanics of exoskeletons also limit the maximum size of the animal. Since the weight of an animal is a function of its volume (length × width × height), doubling an animal's linear dimensions increases its weight by a factor of 8 (that is, 2^3). To support this added weight, the cylindrical exoskeletons of arthropods must become disproportionately thicker and heavier with increasing length. As a result, all large arthropods, such as lobsters, are aquatic: the buoyancy of the water they live in provides much of the support their weight requires.

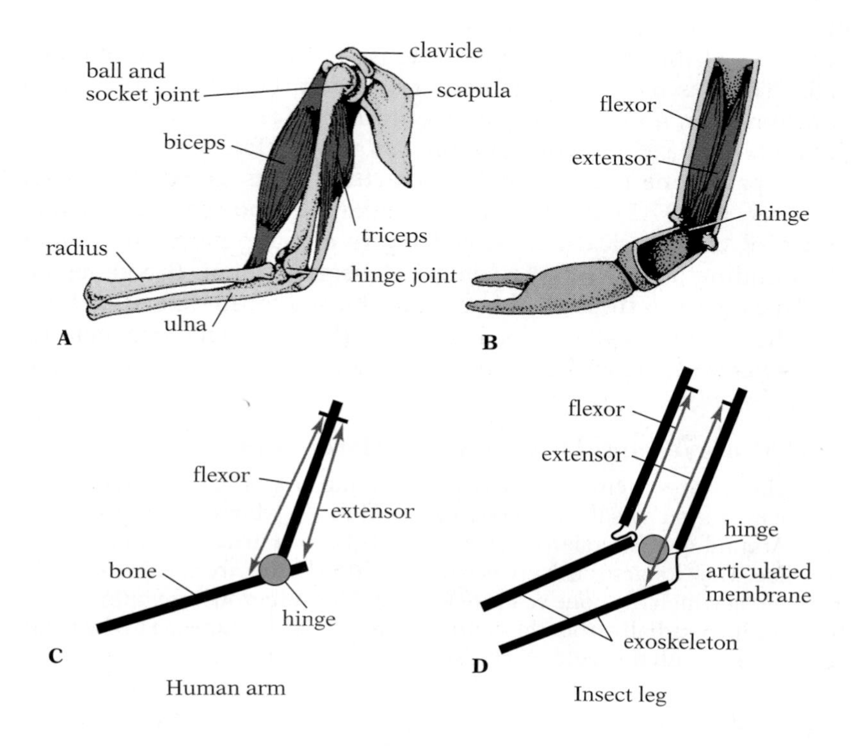

37.6 Mechanical arrangement of muscle and skeleton in a human arm and an insect leg

(A) When the biceps of the human arm contract, the arm is flexed (bent) at the elbow. The triceps has the opposite action; when it contracts, the lower arm is extended. (B) The comparable flexor and extensor muscles in the insect leg have the same action, even though the muscles are inside the skeleton. The joints in this example are so arranged that each segment, when bent, is perpendicular to its neighbors. Notice that only one of the two "fingers" of the claw can move. (C–D) The similarities between these two simple lever-joint strategies can be seen in these diagrams.

The same limitations apply in a somewhat less drastic way to vertebrates. The endoskeletons that provide support must be strong enough for the disproportionate increase of weight in larger animals. Elephants have much thicker bones than antelope, and the largest vertebrates, the whales, like the large arthropods, are aquatic. However, endoskeletons have two advantages: they can be made of stronger materials than exoskeletons, and they are inside the attached muscles instead of enclosing them. Unconfined by the walls of an exoskeleton (Fig. 37.6C–D), the muscles of vertebrates can be big enough to support a relatively large body. In small animals exoskeletons and endoskeletons are about equally effective, and in very small ones exoskeletons are probably stronger and more efficient.

How vertebrate skeletons work Vertebrate skeletons are composed primarily of bone and/or cartilage. Cartilage is firm, but not as hard or as brittle as bone. In all vertebrate embryos it is the primary component of the skeleton, and some adult vertebrates—notably the sharks, skates, and rays—have a permanent cartilaginous skeleton. In most vertebrates, bone progressively replaces cartilage as development proceeds; some cartilage is usually retained, however, where firmness combined with flexibility is needed: at the ends of ribs, on the articulating surfaces in skeletal joints, in the walls of the larynx and trachea, and in the external ear and the nose, for example.

Some bones are partly "spongy," consisting of a network of hardened bars with the spaces between them filled with marrow. Other bones are more compact, with only microscopic cavities in them. The shafts of typical long bones, like those of the upper arm and thigh, consist of compact bone surrounding a large central marrow cavity. In adults the marrow in the cavities of the shafts of long bones is primarily of the yellow fatty variety, while in the flat bones of the ribs and skull and in the ends of long bones, the marrow is primarily of the red variety and is active in the production of blood cells. There is no sharp distinction between the two types of marrow, however, and they may grade into each other. Even the most characteristic red marrow contains about 70% fat.

Compact bone is composed of structural units called ***Haversian systems*** (Fig. 37.7). Each unit is irregularly cylindrical, and is composed of concentrically arranged layers of hard inorganic matrix surrounding a microscopic central Haversian canal. Blood vessels and nerves pass through this canal. The scattered, irregular bone cells lie in small cavities located along the interfaces between adjoining concentric layers of the hard matrix. Exchange of materials

37.7 Cross section of bone, showing Haversian systems

Each Haversian system is seen as a nearly round area. The light circular core of each system is the Haversian canal, through which blood vessels pass. Around the Haversian canal is a series of concentrically arranged hard lamellae. The elongate dark areas between the lamellae are cavities, called lacunae, in which the bone cells are located. The numerous very thin dark lines running radially from the central canal across the lamellae to the lacunae are canaliculi through which tissue fluid can diffuse.

between the bone cells and the blood vessels in the Haversian canals occurs through radiating canaliculi ("little canals" in Latin), which penetrate and cross the layers of hard matrix.

Vertebrate skeletons have two components: (1) the axial skeleton, composed of the skull and the vertebral column with its associated rib cage; and (2) the appendicular skeleton, which includes the bones of the paired appendages (fins, legs, and wings) and their associated pectoral and pelvic girdles (Fig. 37.8). Some bones are joined together by immovable joints or sutures, as in the case of the numerous small bones that together constitute the skull. Many others are held together at movable joints by ***ligaments***. Skeletal muscles, attached to the bones by means of ***tendons***, produce their effects by bending the skeleton at these movable joints. The force causing the bending is always exerted as a pull by contracting muscles; muscles cannot push. Straightening or reversal of the direction in which a joint is bent must be accomplished by contraction of a different set of muscles.

If a given muscle is attached to two bones with one or more joints between them, contraction of the muscle generally causes movement of only one of the two bones, while the other is held relatively rigid by other muscles. The end of muscle attached to the essentially stationary bone (generally the proximal end in limb muscles) is called the ***origin***, and the end of the muscle attached to the bone that moves (generally the distal end in limb muscles) is called the ***insertion***. The movable bones behave like a lever system with the fulcrum at the joint. A single muscle sometimes has multiple origins and/or insertions, which may be on the same or on different bones. The action resulting from contraction of any specific muscle depends primarily on the exact positions of its origins and insertions and on the type of joint between them.

In general, vertebrate muscles operate in antagonistic groups—when one group of muscles is strongly contracted, an antagonistic group exerts a weaker opposing pull to fine-tune the movement. This strategy enables the antagonistic muscles to reverse the direction of movement instantly should that become necessary. In addition, other muscles (synergists) help guide and limit movement.

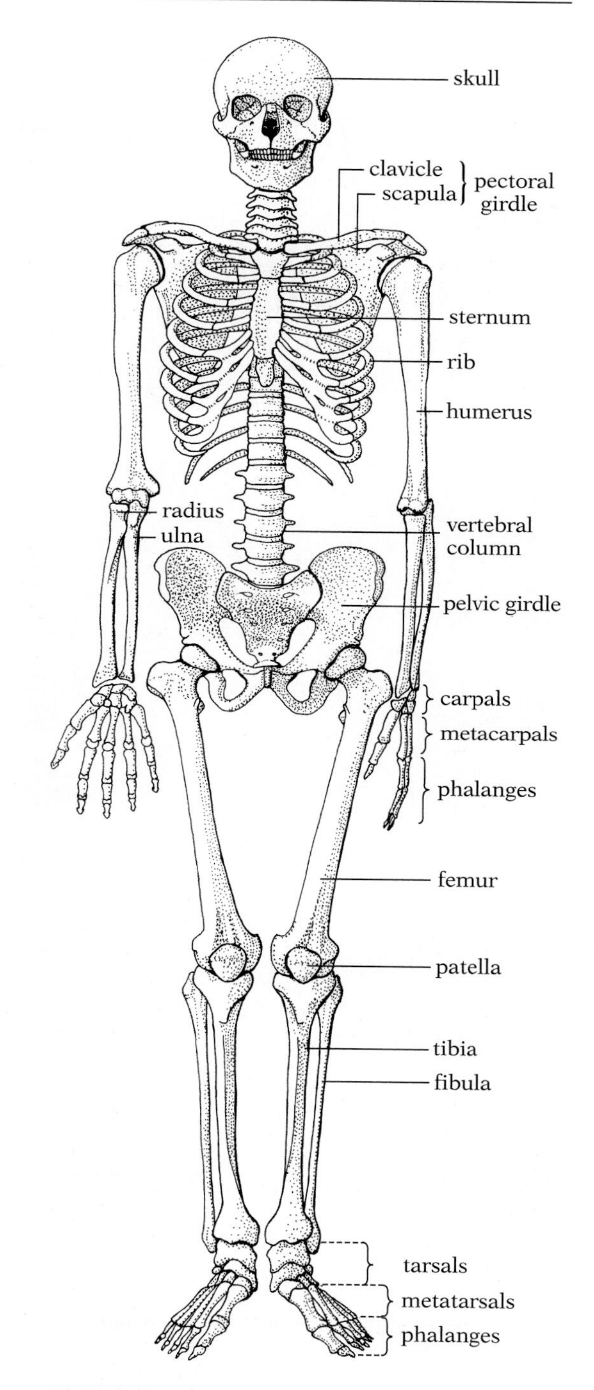

37.8 Human skeleton

HOW MUSCLES WORK

■ WHY ARE THERE SEVERAL KINDS OF MUSCLE?

Vertebrate muscles Three types of muscle tissue are recognized in vertebrates: skeletal muscle, smooth muscle, and cardiac (heart) muscle.

Skeletal muscle (also called voluntary or striated muscle) produces the movements of the limbs, trunk, face, jaws, and eyeballs. It is by far the most abundant tissue in the vertebrate body. Most of what we commonly call "meat" is skeletal muscle. Each skeletal-muscle cell—or fiber, as it is usually called—is roughly cylindrical, contains many nuclei, and is crossed by alternating light and dark

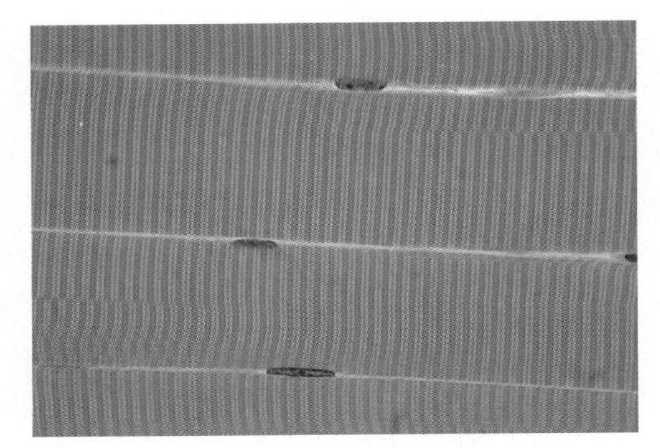

A

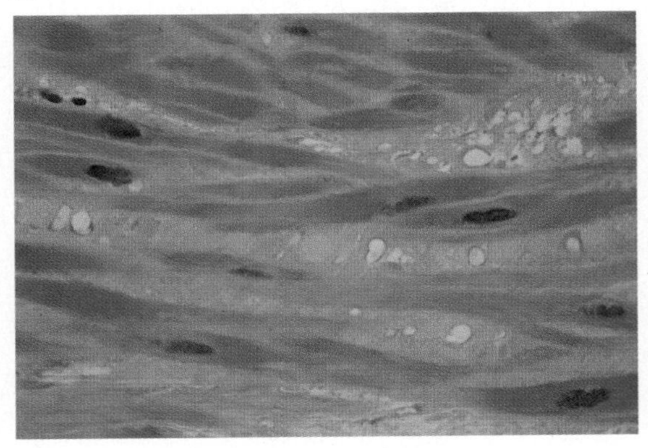

B

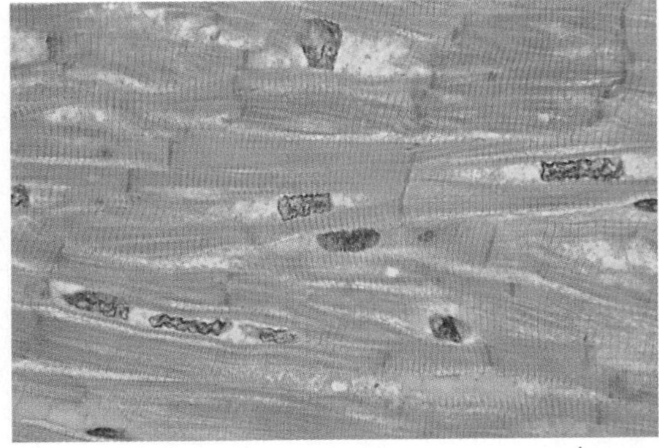

C

37.9 Three kinds of vertebrate muscle fiber

(A) Portions of four skeletal-muscle fibers from a monkey. Each fiber has several nuclei, located on its outer sheath, and is crossed by alternating light and dark bands, or striations. (B) Spindle-shaped smooth-muscle fibers from a human blood vessel. (C) Human cardiac muscle. The thick dark lines are intercalated disks, where one cell ends and another begins. Cardiac muscle cells often bifurcate, producing a complex three-dimensional network.

bands called ***striations*** (Fig. 37.9A). The fibers are usually bound together by connective tissue into bundles, which in turn are bound together by more connective tissue to form muscles. A muscle, then, is a composite structure made up of many bundles of muscle fibers, just as a nerve is composed of many nerve fibers bound together. Skeletal muscle is innervated by the somatic nervous system.

There are two types of skeletal muscle: red and white. ***Red muscle*** (or slow-twitch muscle) has a rich blood supply, numerous mitochondria, and much ***myoglobin***, a compound similar to hemoglobin that forms a loose combination with oxygen and stores it in the muscle. Red muscle oxidizes fatty acids as its primary source of energy. It contracts rather slowly, and is specialized for long-term activity without appreciable fatigue. By contrast, ***white muscle*** (also called fast-twitch muscle) has a more limited blood supply, fewer mitochondria, and a low myoglobin content. It depends almost entirely on anaerobic breakdown of glycogen for its energy supply. It is specialized for very fast contractions and can develop great tension, but only for a short time period, because it fatigues rapidly.

The light and dark meat of chicken provide a familiar example of these two types of muscle. The dark meat of the thigh is composed primarily of the slower, high-endurance red muscle needed for continuous walking, while the white meat of the breast is largely made up of the fast, low-endurance white muscle needed for an occasional escape from danger. By contrast, the breast and wing muscles of birds that fly a great deal consist of red fibers rather than white, while the leg muscles of relatively sedentary animals like rabbits, which use their legs primarily for rapid escape, consist predominantly of white fibers. In humans the fibers of leg muscles, which must be able to support the weight of the body for extended periods, are mostly red, while the fibers of arm muscles are white.

Smooth muscle (also called visceral muscle) forms the muscle layers in the walls of the digestive tract, bladder, various ducts, and other internal organs. It is also the muscle present in the walls of arteries and veins. The individual smooth-muscle cells are thin, elongate, and usually pointed at their ends (Fig. 37.9B). Each has a single nucleus. The fibers are not striated. They interlace to form sheets of muscle tissue rather than bundles. Smooth muscle is innervated by the autonomic nervous system.

The functional differences between skeletal muscle, which is primarily concerned with effecting adjustments to the vertebrate's external environment, and vertebrate smooth muscle, which brings about movements in response to internal changes, are reflected in differences in their physiological characteristics. For one thing, cells of skeletal muscle are innervated by only one nerve fiber; they contract when stimulated by nerve impulses and relax when no such impulses are reaching them. Smooth-muscle cells, by contrast, are usually innervated by two nerve fibers, one from the sympathetic system and one from the parasympathetic system; they contract in response to impulses from one of the fibers and are inhibited from contracting by impulses from the other. Secondly, skeletal muscles cannot function normally in the absence of nervous connections and actually degenerate when deprived of their innervation. In contrast, smooth muscle (and cardiac muscle) can often contract without any nervous stimulation, as is commonly the case in peristaltic contractions of the intestine; moreover, excitation spreads from one fiber to another, whereas skeletal fibers operate independently.

Smooth muscle has been retained even as the proportion of skeletal muscles in animals has steadily risen over the course of evolution, because it is efficient: smooth muscle is slow, but it uses only about 10 percent of the ATP required by skeletal muscle to produce the same strength of contraction.

Cardiac muscle, the tissue of which the heart is composed, shows characteristics of both skeletal and smooth muscle. Like skeletal muscle, its fibers are striated. Like smooth muscle, it is innervated by the autonomic nervous system, and its activity is more like that of smooth muscle. Where two separate cardiac-muscle fibers (cells) meet, their adjacent membranes are so tightly and complexly interdigitated, and have so many desmosomes and other fibrous reinforcements, that for many years these areas were not recognized as cellular junctions. The sites of these junctions are visible under light microscopes as dark-colored circles called intercalated disks (Fig. 37.9C).

Invertebrate muscles Muscles are structured somewhat differently among the invertebrates: all the muscles of insects are striated, even those in the walls of their internal organs; many other invertebrates have only smooth muscles. Some invertebrates have evolved arrangements enabling them to perform the same action in two different ways. Thus scallops, which swim by opening and closing their shells in a flapping motion, have two sets of shell-closing muscle fibers: a striated set, whose fast, short-term action is used in swimming, and a smooth set, whose slower but longer-lasting action is used for holding the shells tightly closed when the scallop is at rest or is attacked by a predator.

■ THE TWITCH: HOW MUSCLES CONTRACT

Individual muscle fibers, like individual nerve cells, fire only if stimulated above a threshold of intensity, duration, and rate. Relaxed vertebrate muscle fibers seem to exhibit the all-or-none property: when an excised vertebrate muscle fiber is stimulated at different values above the threshold, it shows the same degree of contraction as to a stimulus just at threshold.

While a relaxed, isolated fiber reacts in an all-or-none fashion, intact muscles—groups of fibers—do not. The same muscles can lift a pencil and a 10-kg weight; individual muscles give graded responses, depending on the strength of the stimulation. The same variability is evident in the contraction of frog leg muscle. If stimulated barely above threshold, the muscle gives a very weak twitch. A slightly stronger stimulus produces a slightly stronger twitch. Increasing the strength of the stimulus elicits an ever-stronger contraction from the muscle, until a point is reached beyond which further increases in the stimulus do not increase the strength of the response. The muscle has reached its maximal response (Fig. 37.10).

One reason for the graded response of a muscle composed of all-

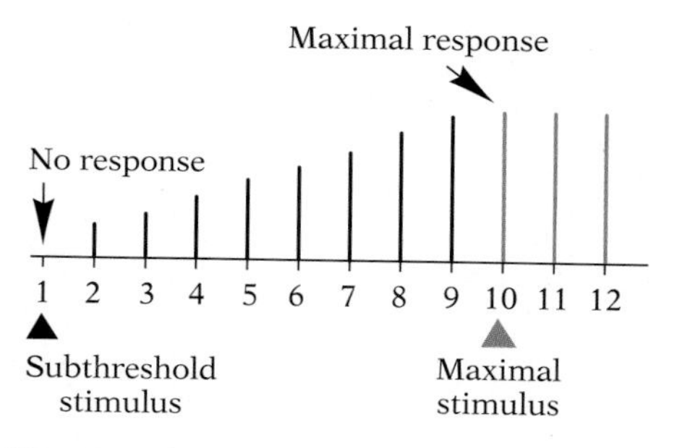

37.10 Response of a muscle to stimuli of various intensities

The numbers indicate the intensities at which stimuli are administered, and the height of the bars shows the strength of muscle response. Stimulus 1 is very weak and elicits no response; it is subthreshold. Stimulus 2 is somewhat stronger and proves to be above threshold, for the muscle contracts. Each stimulus from 3 to 10 is slightly stronger than the preceding one, and each elicits a correspondingly stronger muscle contraction. Stimuli 11 and 12 are stronger than 10, but the muscle gives no greater response, indicating that 10 elicits a maximal response.

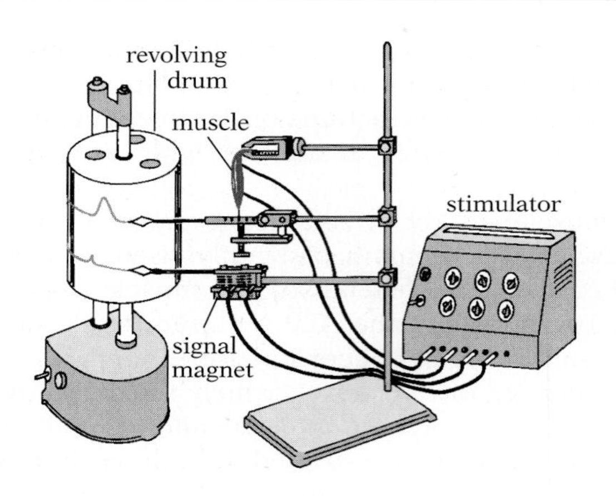

37.11 Kymograph apparatus for studying muscle contraction

The drum of the kymograph, which is covered with paper, revolves at a constant speed. The muscle is mounted in such a way that when it contracts, it raises a stylus that writes on the revolving drum. Wires lead from a stimulator to the muscle and also to a signal magnet that can deflect a second stylus. At the moment a stimulus is sent to the muscle, the signal stylus is deflected, producing a blip in the lower trace it is drawing on the revolving drum. This blip indicates exactly when the stimulus was administered. One such blip is shown here. The stimulus causes the muscle to contract, raising the stylus to which it is attached, and producing a corresponding rise in the upper trace on the revolving drum. As the muscle relaxes, the stylus is lowered and the trace falls. Thus the trace drawn on the kymograph drum gives a record of the contraction pattern of the muscle.

or-none fibers is that the threshold values of these fibers are not all the same; since different muscle fibers may be innervated by different nerve fibers, and since these do not all fire at the same time, an increase in the strength of the stimulus above the threshold level elicits a greater response from the whole muscle by stimulating more muscle fibers. Ultimately, however, as the intensity of stimulus is increased, all the fibers are stimulated to respond, at which point the muscle has reached its maximal response.

The response of a relaxed muscle is called a ***simple twitch***. If an isolated frog muscle is attached to a kymograph (Fig. 37.11), the characteristics of a simple twitch can be defined. Every time the muscle is stimulated by an electric shock, it moves a lever that writes on the revolving drum of the kymograph. The stronger the contraction of the muscle, the higher the lever will be pulled. When a single above-threshold stimulus is administered to this muscle, there is a brief ***latent period*** between stimulation of the muscle and the commencement of the shortening process—usually between 0.0025 and 0.004 sec (Fig. 37.12). The latent period is followed by the ***contraction period***; this phase is followed immediately by a ***relaxation period***.

The interaction of the various muscle fibers in a muscle, each exhibiting simple twitches at different thresholds, explains some of the graded response of muscles. It does not account for it completely, particularly if the fiber is not fully relaxed to begin with. If, for example, a pair of closely spaced stimuli is delivered to a relaxed excised muscle, the muscle may not have relaxed completely after contracting in response to the first stimulus by the time the second stimulus arrives. The result is a contraction that is greater

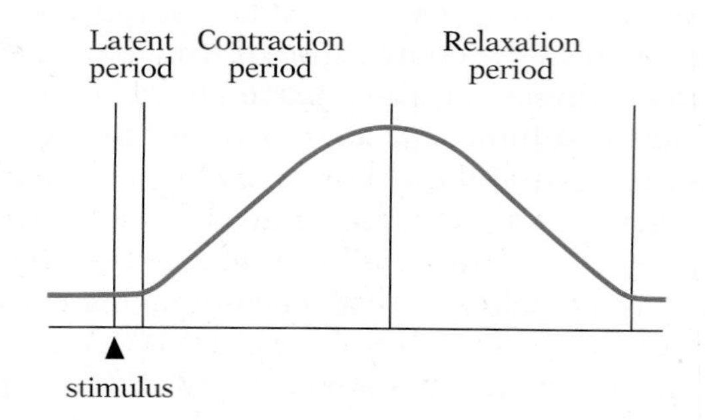

37.12 Kymograph record of a simple twitch

The duration of the latent period, which is too short to be measured even by the most sensitive kymograph, is exaggerated here for clarity.

than a single stimulus produces (Fig. 37.13). There has been a ***summation*** of contractions, the second adding to the first.

If the stimuli are submaximal, the summation results in part because the second stimulus recruits additional muscle fibers. Summation can also occur even if the individual stimuli are at maximal intensity, and thus all fibers in the muscle are activated by each stimulus. The physiological condition of the muscle fibers during activity changes, and thus increases the strength of subsequent contractions. The change results from the increase in temperature the contractions produce, and other factors related to the chemistry of contraction and its control, discussed in the next section.

When stimuli arrive very rapidly, a muscle can't relax at all between successive stimuli. The individual contractions become indistinguishable, and fuse into a single sustained contraction known as ***tetanus***. For the same reasons that apply to summation (of which tetanus is a form), a tetanic contraction is greater than a maximal simple twitch of the same muscle. Normally, a high percentage of our actions involve tetanic contractions rather than simple twitches, because a volley of nerve impulses is sent to the muscle. If, however, a tetanic contraction is maintained too long, the muscle will begin to fatigue, and the strength of its contraction will fall, even though the stimuli continue at the same intensity. Fatigue probably results from an accumulation of lactic acid, a depletion of stored energy reserves, and other chemical changes.

Some muscles are never completely relaxed, but are kept in a state of partial contraction called ***tonus*** (muscle tone). Tonus is maintained by alternate contraction of different groups of muscle fibers, so that no single fiber has a chance to fatigue.

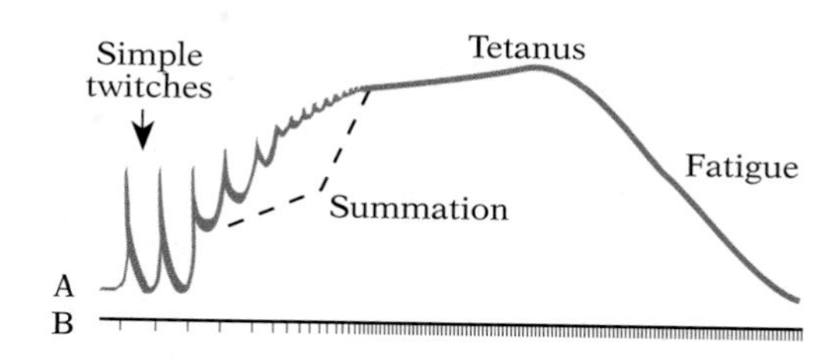

37.13 Summation and tetanus in muscle response

When the stimuli (line B) are widely spaced, the muscle has time to relax fully before the next stimulus arrives, and simple twitches result. (Because the drum is revolving much more slowly than in Fig. 37.12, each simple twitch is recorded as a sharp spike on the trace.) As the frequency of the stimuli increases, the muscle does not have time to relax fully from one contraction before the next stimulus arrives and causes it to contract again. The result is summation—contractions that are stronger (and hence produce taller spikes in the trace) than any single simple twitch. If the stimuli are very frequent, the muscle may not relax at all between successive stimulations; the resulting strong sustained contraction is called tetanus. If the very frequent stimulation continues, however, the muscle may fatigue and be unable to maintain the contraction.

■ THE MOLECULAR BASIS OF CONTRACTION

Supplying the energy The energy for muscle contraction comes from ATP. However, so little ATP is actually stored in the muscles that a few muscular twitches could quickly exhaust the supply, were it not for another high-energy phosphate that is stored in muscles. In vertebrates and some invertebrates, particularly echinoderms, this compound is creatine phosphate, formed by linkage of a high-energy phosphate group to the nitrogenous organic acid creatine. Many invertebrates use a similar compound, arginine phosphate. Creatine phosphate and arginine phosphate are called ***phosphagens***. Phosphagens cannot supply energy directly to the contraction mechanism of muscle, but they can pass their high-energy phosphate groups to ADP to form ATP. Enough high-energy phosphate is stored in the muscle to enable it to contract strongly during the several seconds' delay before the machinery of glycolysis and cellular respiration can be speeded up.

If the demands on the muscles are not great, much of the energy used to replenish the supply of phosphagens and ATP may come from the complete oxidation of nutrients to carbon dioxide and water. During any delay before adjustments of the respiratory and circulatory systems increase the oxygen supply to the active muscles, some of the O_2 for oxidative phosphorylation in red muscles may come from oxygenated myoglobin.

During violent muscular activity, such as strenuous exercise or the lifting of a very heavy object, the energy demands of the muscles (especially white muscles) may be greater than can be met by complete respiration, because oxygen cannot get to the tissues fast enough. Under these circumstances, lactic acid fermentation occurs; the muscles obtain the extra energy they need from the inefficient anaerobic processes of glycolysis alone, and they thus incur an ***oxygen debt***. Some of the lactic acid accumulates in the muscles, but much of it is transported by the blood to the liver. When the activity is over, hard breathing or panting helps supply the liver with the large quantities of O_2 it requires to reconvert the lactic acid into pyruvic acid. Some of the lactic acid is oxidized, and the resulting energy is used to resynthesize glycogen from the rest of the lactic acid. In this manner the oxygen debt is paid off.

Sliding filaments The major components of the contractile parts of muscles are two proteins, ***actin*** and ***myosin***. As described in Chapter 5, myosin uses ATP to fuel a conformational change; this change is harnessed to create movement. In muscle cells the myosin has a different form, and both proteins occur in arrays specialized for generating strong contractions.

Skeletal muscle has dramatic striations of light and dark bands; these bands reflect the specializations that underlie contraction. Each of the fibers that are joined together in a muscle is filled with numerous long thin myofibrils, each about 1–2 μm in diameter, with mitochondria in the cytoplasm between them. The myofibrils show the same pattern of cross striations as the fibers of which they are a part (Fig. 37.14). Wide light and dark bands—***I bands*** and ***A bands*** respectively—alternate. In the middle of each dark A band is a region called the ***H zone***, which is lighter than the rest of the A band, but darker than the I bands. In the middle of the light I band is a very dark, thin line called the ***Z line***. The entire region of a myofibril from one Z line to the next is called a ***sarcomere***. The sarcomeres are the functional units of muscular contraction.

Within each myofibril there are two types of filaments, thick

37.14 Skeletal muscle from a rabbit

The myofibrils run diagonally across this electron micrograph from lower left to upper right; each looks like a ribbon crossed by alternating light and dark bands. The wide light bands are I bands, with a dark narrow Z line in the middle of each. The wide dark bands are A bands, each with a lighter H zone across the middle.

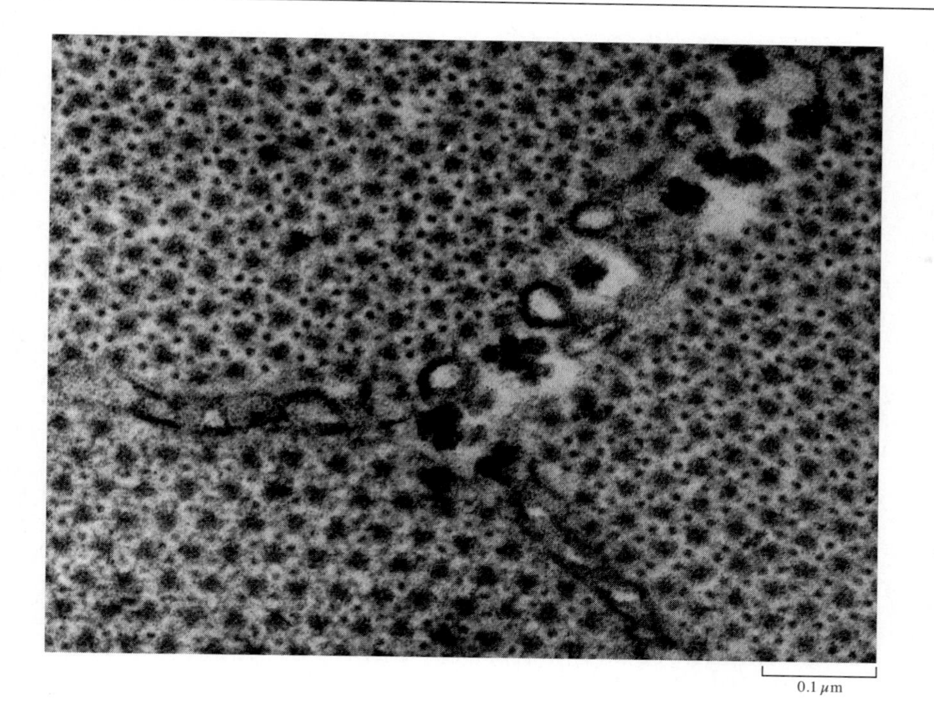

37.15 Cross section of frog skeletal muscle

Parts of three myofibrils are seen here in cross section in this electron micrograph, separated by a structure associated with the sarcomeres called the sarcoplasmic reticulum. Note the thin actin filaments that are arranged in a hexagonal pattern around the thick myosin filaments.

ones and thin ones, arranged in a precise pattern (Fig. 37.15). The two are interdigitated, with the thick ones located exclusively in the A bands and the thin ones primarily in the I bands, but extending some distance into the A bands. This distribution explains the different appearances of the A bands, I bands, and H zones. Each dark A band is precisely the length of one region of thick filaments; it is darkest near its borders, where the thick and thin filaments overlap, and lighter in its midregion, or H zone, where only the thick filaments are present (Fig. 37.16). Each light I band corresponds to a region where only the thin filaments are present. The Z line is a structure to which the thin filaments are anchored at their midpoints and against which they exert their pull during contraction; it also functions to hold the filaments in proper register. The protein that anchors the actin filaments to the Z line is called α-actinin, while another, known as desmin, keeps the actin in proper position.

Contraction occurs when the thick and thin filaments slide past each other. As they slide together, the zone of overlap between thick and thin filaments increases until the thin filaments actually meet and overlap slightly; this sliding together reduces the width of the H zone, and even obliterates it entirely if the thin filaments

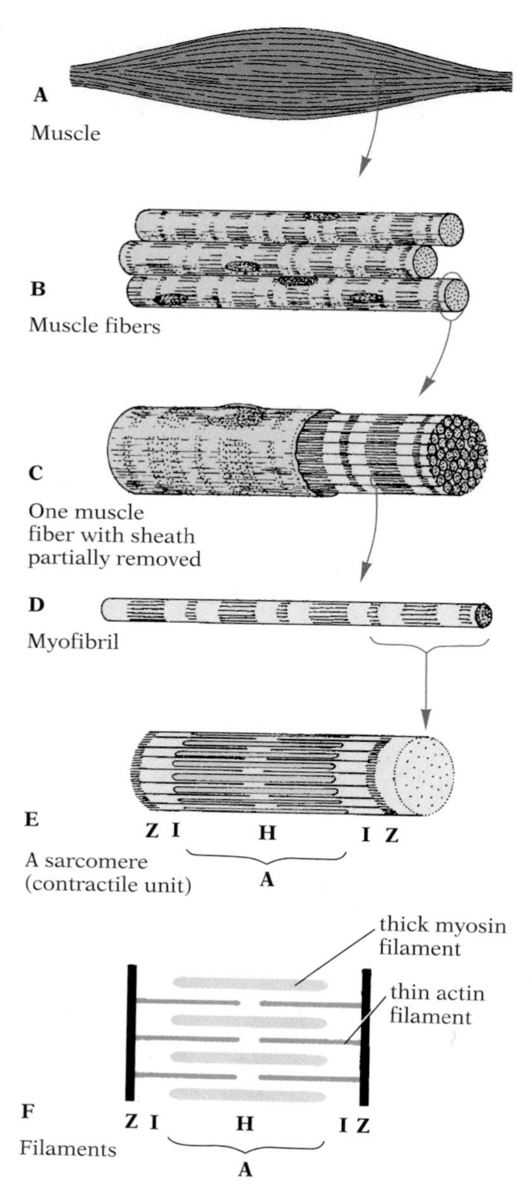

37.16 Component parts of skeletal muscle

The pattern of light and dark bands visible in myofibrils under high magnification (D; see also Fig. 37.14) results from the interdigitation of actin and myosin filaments in each myofibril. As shown in F, the A band corresponds to the length of the thick myosin filaments (pink); the lighter H zone is the region where only the thick filaments occur, while the darker ends of the A band are regions where thick and thin filaments overlap. The I band corresponds to regions where only thin actin filaments occur. The Z line is a structure to which the thin filaments are fastened at their midpoints.

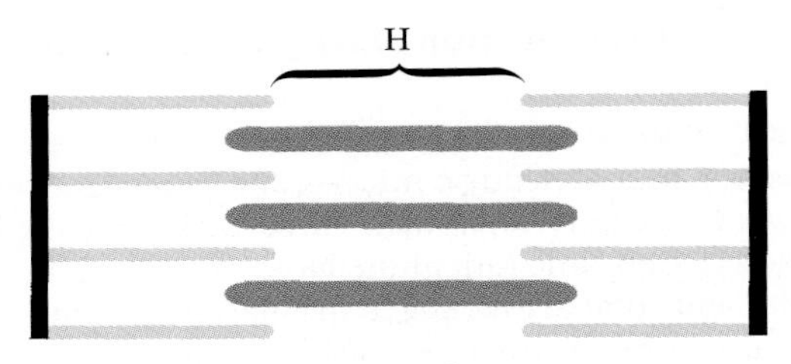

Relaxed

Moderately contracted

Strongly contracted

37.17 Arrangement of actin (lighter color) and myosin (darker color) filaments in a sarcomere in relaxed and contracted states

meet (Fig. 37.17). The sliding together also pulls the Z lines closer together and greatly reduces the width of the I bands. The width of the A bands, however, hardly changes, since these correspond to the full length of the thick filaments. These remain the same, except perhaps for a slight crumpling from contact with the Z lines under conditions of extreme contraction.

The thick filaments are composed of myosin and the thin filaments primarily of actin. They are connected by small cross bridges between them (Fig. 37.18) that arise from the thick filaments (Fig. 37.19A). Each thick filament is a bundle of myosin molecules, each of which is composed of an elongated tail portion and a pair of globular heads (Fig. 37.19B–C); the heads are the cross bridges.

The cross bridges act as hooks or levers that enable the myosin filaments to pull the actin filaments (Fig. 37.20). The cross bridges bend toward the actin, hook onto it at specialized receptor sites, and then bend in the other direction, pulling the actin with them; they then let go, bend back in the original direction, hook onto the actin at a new active site, and pull again. A ratchet mechanism powered by ATP slides the filaments together.

Though the remarkably precise arrangement of filaments found in striated-muscle cells is not found in smooth-muscle cells, the latter contracts by essentially the same mechanism.

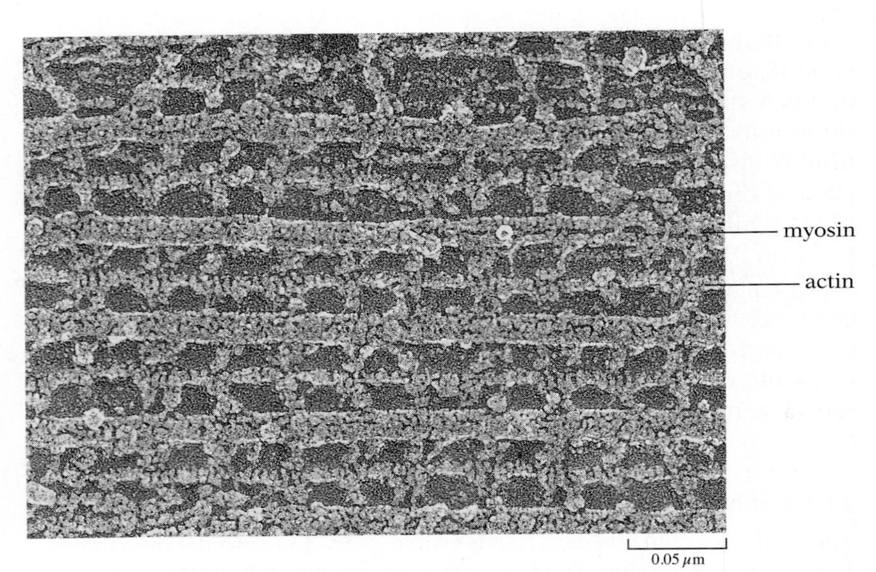

37.18 Insect flight muscle, showing cross bridges between filaments

The thick (myosin) and thin (actin) filaments are connected by cross bridges composed of myosin heads. Essentially the same structural pattern is seen in the myofibrils of skeletal muscle.

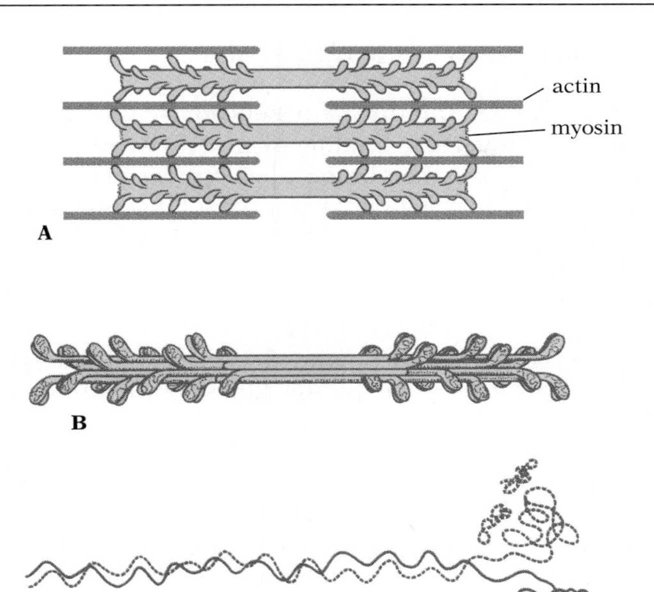

37.19 Molecular structure of myosin filaments

(A) Each myosin filament is linked to the adjacent thin filaments by numerous cross bridges. (B) The myosin filament is composed of a bundle of elongate molecules, each with a double club-shaped head, which acts as the cross bridge. Each thick filament has about 500 heads. (C) The tail of a myosin molecule is composed of two intertwined polypeptide chains. Each of its heads is formed by the coiled free end of one of the tail chains, plus two smaller polypeptides.

37.20 Action of cross bridges during contraction

Each of the 500 or so myosin heads of a thick filament acts independently. This drawing traces the action of two heads (dark color) through a cycle of movement on both sides of the H zone in one sarcomere. (A) On each side, an ATP-activated head is ready to bind to one of the spherical subunits in the actin filament. (Activation is indicated by an asterisk.) (B) Binding takes place. (C) The spent ATP is released in the form of ADP and phosphate, and this release triggers an allosteric change that enables the myosin cross bridges to bend and pull the actin filaments toward the H zone—to the left on the right side, and to the right on the left side. (D) As a result of the bending, the myosin heads lose their affinity for the actin and drift free. At the same time each becomes able to bind another molecule of ATP. (E) ATP has activated the heads by inducing an allosteric change that "cocks" them in preparation for another power stroke. The activated state is indicated by the asterisks. Research indicates that each myosin head goes through 5–10 cycles/sec during contraction.

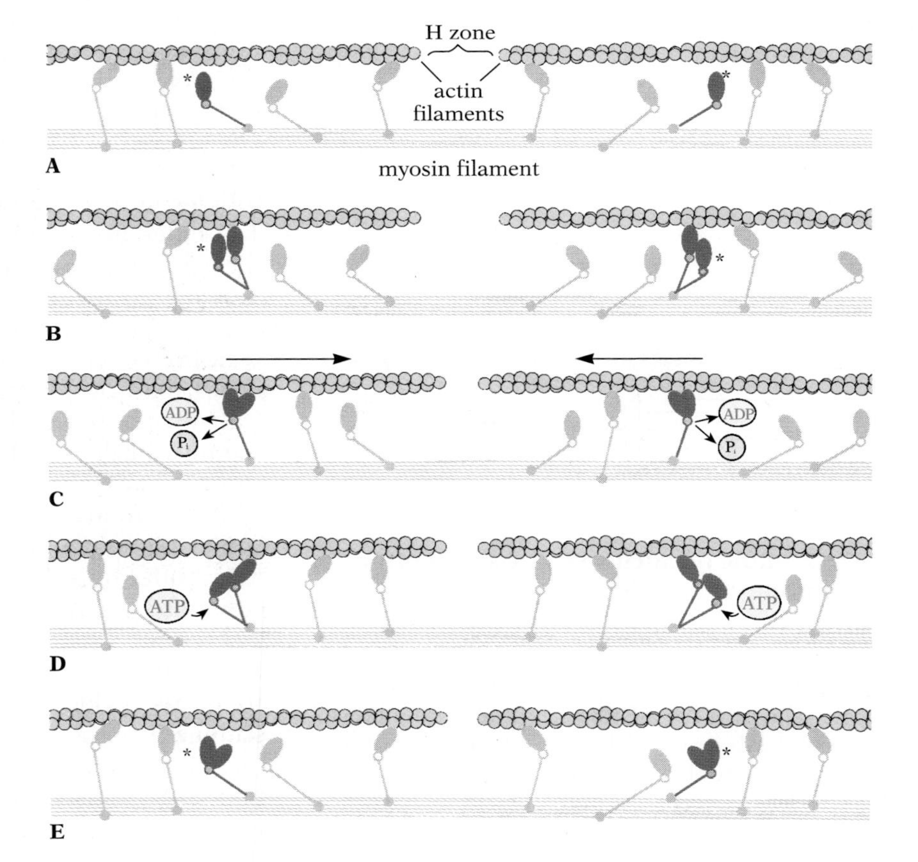

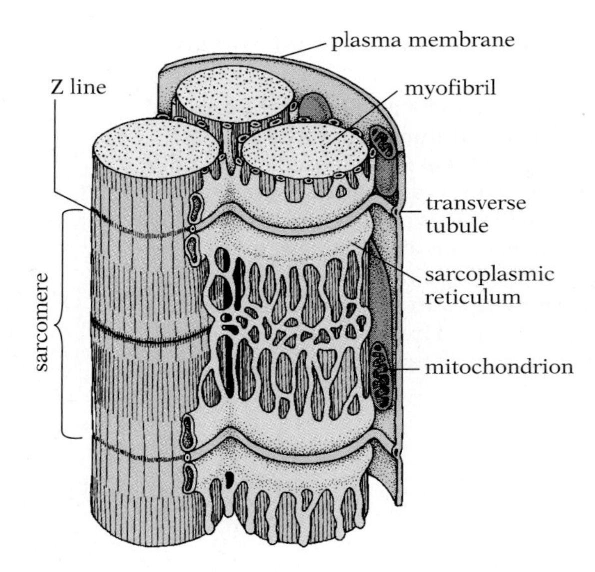

37.21 Sarcomeres with associated sarcoplasmic reticulum

Shown here are sarcomeres typical of amphibian skeletal muscle. The sarcomeres of mammalian skeletal muscle differ in having a separate transverse tubule (and associated sarcoplasmic reticulum) on each side of each Z line, rather than one tubule per line.

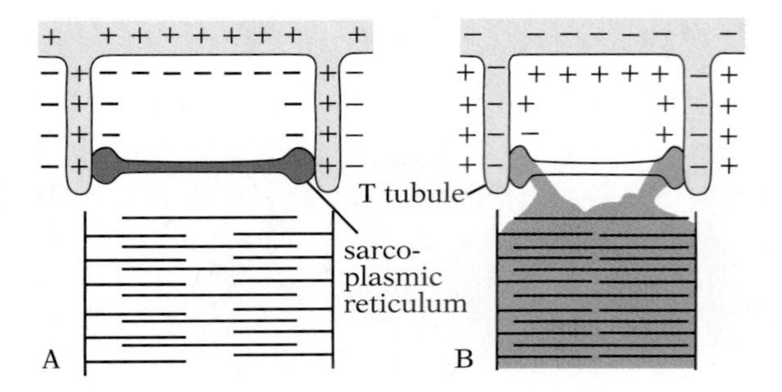

37.22 Role of calcium in the stimulation of muscle contraction

(A) A sarcomere in the resting (relaxed) condition. Ca^{++} ions (color) are stored in high concentration in the sarcoplasmic reticulum. (B) The polarization of the membranes of the T tubules is momentarily reversed during an action potential (impulse), and this reversal of polarization induces release of the Ca^{++} ions, which spread over the sarcomere and stimulate contraction.

■ HOW CONTRACTION IS TRIGGERED

How does a nervous impulse at a neuromuscular junction signal and control the molecular binding of myosin heads that results in contraction? Like the membrane of a resting neuron, the membrane of a resting muscle fiber is polarized, the outer surface being positively charged in relation to the inner one. Stimulatory transmitter substance released by a nerve axon at a neuromuscular junction (see Fig. 35.30, p. 1015) causes a momentary depolarization of the muscle membrane. If the depolarization reaches threshold, an impulse, or action potential, is triggered and propagates over the surface of the fiber by the same combination of voltage-gated Na^+ and K^+ channels found in the membrane of nerve cells. The action potential activates the contraction process indirectly, by means of Ca^{++} ions.

To contract, a vertebrate muscle fiber must shorten all of its many myofibrils simultaneously. The myofibrils in the center of a fiber are so far from the surface that Ca^{++} ions from outside could not possibly diffuse fast enough to reach them in the short interval between stimulation of the fiber and its contraction; thus some internal source of Ca^{++} is essential to rapid contraction. Muscle cells have two functionally related tubule systems that permit coordinated contraction: the ***sarcoplasmic reticulum***, which does not open to the exterior, and the ***T system***, or transverse tubule system, which is part of the plasma membrane surrounding the fiber.

The sarcoplasmic reticulum is the muscle cell's highly specialized version of the ubiquitous endoplasmic reticulum. Its membranous canals form a cufflike network around each sarcomere of the myofibrils (Fig. 37.21). The sarcoplasmic reticulum at the distal end of one sarcomere and that at the proximal end of the next sarcomere beyond it are very close together. Lying between them, at the level of the Z line, is usually a tubule of the T system (two tubules in mammals). Though the sarcoplasmic reticulum and the T tubules are in direct contact, there is no connection between their lumina (cavities), and hence no mixing of their contents.

When an action potential is propagated across the surface of a muscle cell, it also penetrates into the interior of the fiber via the membranes of the T tubules. The action potential moves much faster than diffusing ions, fast enough that the stimulus for contraction can reach all the myofibrils at nearly the same instant, and the myofibrils near the surface and those in the center of the fiber can contract together.

The intimate association between the T tubules and the sarcoplasmic reticulum allows action potentials moving along the membrane of a T tubule to trigger the reticulum. The reticulum contains large quantities of Ca^{++} ions. The action potential induces a sharp increase in the permeability of the reticular membranes to Ca^{++} ions, allowing these to escape in large numbers (Fig. 37.22). It is this suddenly released intracellular Ca^{++} that is the direct stimulant for contraction.

How does calcium actually trigger contraction of the muscle fibers? The answer lies in the structure of the thin filaments. The main protein in the thin filaments is actin; in addition, these filaments contain tropomyosin and the troponin complex, which are important regulatory proteins.

The subunits of the actin molecule are globular and form two helically intertwined rows, along which run the long thin molecules of the first regulatory protein, ***tropomyosin*** (Fig. 37.23). In the resting muscle tropomyosin prevents actin from binding to the cross bridges from the thick myosin filaments—probably by masking its binding sites for myosin. The molecules of the other regulatory proteins, the ***troponin complex***, are also globular and occur in triplets near every seventh pair of actin units; each complex has three binding sites: one for actin, one for tropomyosin, and one for Ca^{++} ions.

When Ca^{++} ions are released from the sarcoplasmic reticulum, they are picked up by the calcium-binding sites of the troponin complex. Troponin responds with a conformational change, which shifts its position. As a result tropomyosin no longer inhibits actin. The actin thus becomes free to bind with cross bridges from the myosin, and the contraction process is initiated.

In a resting muscle, then, the cross bridges—the globular myosin heads of the thick filaments—have been "cocked" (activated) by ATP, but they cannot bind to the thin actin filaments because tropomyosin is inhibiting the binding sites on the actin molecules (Fig. 37.24A). When stimulation from a motor nerve triggers an action potential and the action potential, transmitted along the T tubules, penetrates into the interior of the muscle fibers, the sarcoplasmic reticulum releases Ca^{++} ions. Some of these ions bind to the troponin complex, which then undergoes a conformational change, displacing the tropomyosin and exposing the myosin-binding sites of the actin to the cross bridges (Fig. 37.24B). The binding of the myosin causes the cross bridges to bend, thus initiating the power stroke that forces the filaments to slide along each other (Fig. 37.24C). When a new ATP molecule binds to a myosin cross bridge after the attachment to actin has been broken, the head is forced back to its original "cocked" conformation (Fig. 37.24D–E).

As long as free Ca^{++} ions (and ATP) remain available, as when the nerve continues to stimulate the muscle, the cycle of cross-bridge binding, power stroke, and recovery flip can occur over and over again, as the muscle continues to contract. When nervous

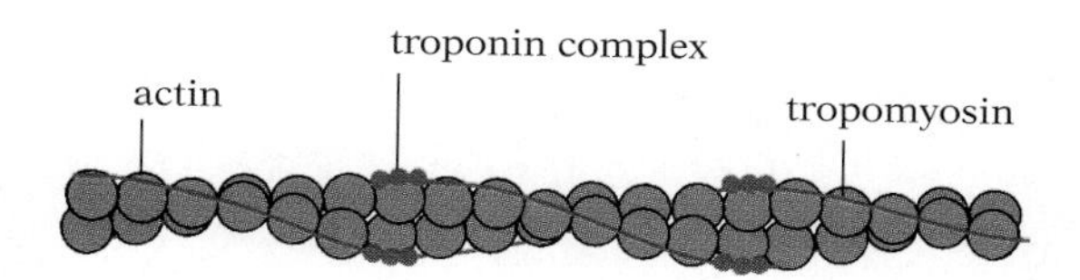

37.23 Molecular structure of a thin filament

Globular subunits of actin form two helically coiled rows. Molecules of the regulatory protein complex troponin (also globular) are evenly spaced along the rows of actin, and the long thin molecules of another regulatory protein, tropomyosin, run along the length of the rows.

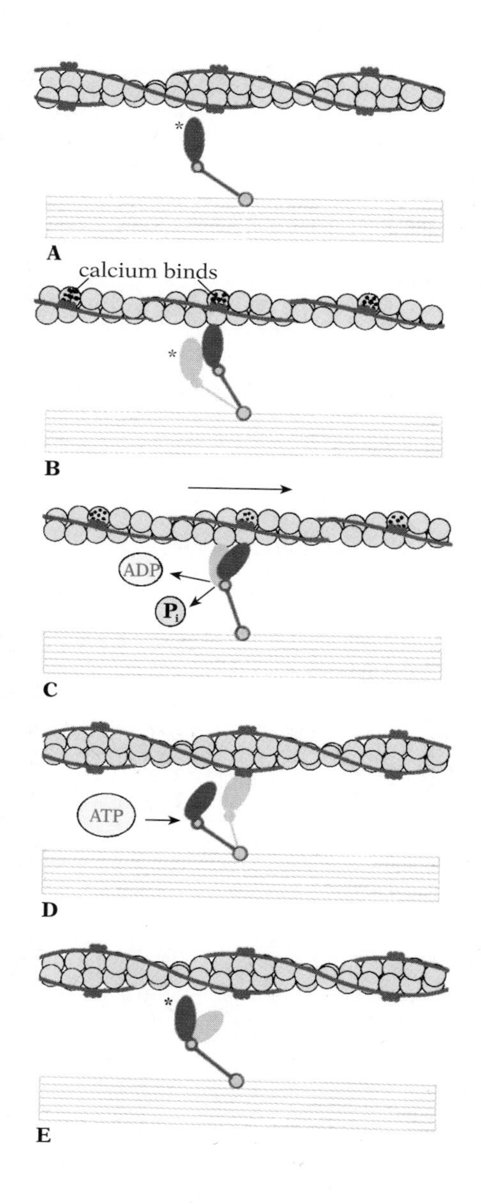

37.24 Model for the stimulation of muscle contraction

(A) In a resting muscle the myosin cross bridges that ATP has already activated (indicated by the asterisk) cannot bind to the actin in the thin filament, because the binding sites are masked by tropomyosin. (B) The binding of Ca^{++} ions to the troponin complex causes a conformational change that slightly displaces the tropomyosin. The active sites of the actin are thus exposed, and the cross bridges bind to the actin. (C) The binding of each myosin cross bridge to actin, with the concomitant release of ADP and phosphate, initiates a conformational change in the cross bridge, whose bending (the power stroke) forces the filaments to slide along each other. (D) The myosin head then dissociates from the actin, and its ATP-binding site becomes available once more. (E) ATP binds to the myosin head, which is thus "cocked" in preparation for a new stroke.

stimulation ceases, the muscle relaxes, because a calcium pump in the membrane of the sarcoplasmic reticulum quickly moves the Ca^{++} back into the reticulum. The troponin-tropomyosin system can then resume its inhibition of the myosin-binding sites on the actin. ATP, then, is not consumed by a muscle fiber unless release of Ca^{++} has set in motion the steps leading to a new contraction. As a result, there need be no cost to maintaining a contraction—a point well illustrated by the phenomenon of rigor mortis: after death, muscles initially relax as nervous stimulation ceases, but several hours later they contract as the sarcoplasmic reticulum breaks down and releases calcium, leaving the body rigid.

Smooth muscle, though similar in many ways, displays some interesting differences. In skeletal muscle ATP activates the myosin heads and Ca^{++} ions trigger movement by binding to the troponin complex of the actin filaments. In smooth muscle, Ca^{++} ions activate the myosin, through two intermediate enzymes, before ATP becomes involved. This helps explain why smooth muscle acts so slowly. It also accounts for the capacity of smooth muscle to be activated by hormones: as detailed in Chapter 33, hormones frequently work by opening membrane channels specific for Ca^{++} ions. These ions, in their role as second messengers, then bind to and activate an enzyme complex in the cytosol, such as calmodulin. This is precisely what happens in smooth muscle, where calmodulin takes the role played by troponin in skeletal muscle. In fact, one of the three subunits of the troponin complex in skeletal muscle is a modified calmodulin molecule.

There is yet another connection between the endocrine system and the smooth-muscle strategy of vertebrates: cyclic AMP acts as a second messenger to activate the myosin. The critical evolutionary steps from the slow-contraction system of smooth muscle to the highly organized fast-contraction system of skeletal muscle must have been (1) the evolution of the Z line dividing contraction units into discrete sarcomeres, (2) the minor modification of calmodulin into one of the three subunits of troponin, and (3) the evolution of the endoplasmic reticulum into a sarcoplasmic reticulum specialized for Ca^{++} transport to the myofibrils.

CHAPTER SUMMARY

HOW PLANTS MOVE

Plant movements depend on changes in turgor. (p. 1067)

MUSCLES AND SKELETONS

WATER POWER: HYDROSTATIC SKELETONS Some animals or their appendages are moved by muscles that narrow or shorten a sealed, fluid-filled compartment. (p. 1068)

WHY JOINTS ARE BETTER Arthropods and vertebrates generally attach their muscles to skeletal elements; muscle contraction pulls one element toward another. The movement is usually restricted by an intervening joint, which can make the movement precise and strong; additional control is generated when antagonistic muscles operating the same joint pull against one another. (p. 1070)

HOW MUSCLES WORK

WHY ARE THERE SEVERAL KINDS OF MUSCLE? Vertebrates have three kinds of muscle: skeletal, smooth, and cardiac. Skeletal muscles are of two types: white muscle is specialized for strong rapid contraction, but has little endurance; red muscle is slower but does not readily fatigue. Each skeletal muscle fiber is innervated separately, making precise control possible. Smooth-muscle activity spreads from one fiber to another, and thus is less precisely controlled; on the other hand, it is much more energy efficient. Cardiac muscle has properties intermediate between those of skeletal and smooth muscle, which suits it to its role of rhythmically pumping blood for decades. Invertebrate muscles do not fall readily into three types. (p. 1073)

THE TWITCH: HOW MUSCLES CONTRACT Individual vertebrate muscle fibers contract in an all-or-none fashion, but whole muscles shorten in graded proportion to the number of activated fibers. When a fiber is stimulated so often that it cannot relax between contractions, summation increases the degree of contraction. (p. 1075)

THE MOLECULAR BASIS OF CONTRACTION Phosphagens maintain a supply of ATP to fuel contraction; when energy demands are too great to be met by aerobic metabolism in muscles, they operate anaerobically, producing lactic acid, which creates an oxygen debt. ATP is used to move myosin heads relative to actin filaments; the sliding of myosin past actin generates contraction. (p. 1077)

HOW CONTRACTION IS TRIGGERED Impulses arriving at neuromuscular junctions depolarize the muscle membrane, leading to a massive release of Ca^{++} from storage in the sarcoplasmic reticulum within the fibers. The Ca^{++} binds to a regulatory protein, tropomyosin, on the actin, creating a conformational change that exposes the myosin-binding sites and thus permits contraction. (p. 1082)

STUDY QUESTIONS

1 Sketch a plausible design for the hydrostatic use of muscles in a tongue. Keep in mind that tongues can usually fold back on themselves (both on top and on bottom) and curl lengthwise. (pp. 1068–1070)

2 Compare the articulation of a human shoulder with that of a typical quadruped shoulder which can only move fore and aft. What might be the costs and benefits of having a limb constrained to move in only a single plane? (p. 1072)

3 Why do you suppose fish bones are so notoriously thin compared with, say, those of a small mammal the same size? (pp. 1071–1072)

4 In estimating time of death, pathologists must often use the onset or passing off of rigor, which varies from one part of the body to another, and can be influenced by external conditions. What is the most likely anatomical pattern of rigor, and what sorts of external conditions are likely to alter the timing, and in what way? (p. 1084)

SUGGESTED READING

COHEN, C., 1975. The protein switch of muscle contraction, *Scientific American* 233 (5). *A detailed account of the way calcium, troponin, and tropomyosin interact to control muscle contraction.*

RASMUSSEN, H., 1989. The cycling of calcium as an intracellular messenger, *Scientific American* 261 (4). *On the role of Ca^{++} in muscle contraction.*

SMITH, K. K., AND W. M. KIER, 1989. Trunks, tongues, and tentacles, *American Scientist* 77 (1). *On hydrostatic movement in vertebrates and cephalopods.*

WAYNE, R., 1993. Excitability in plant cells. *American Scientist* 81 (2). *On how plants create the rapid changes in turgidity that permit movement.*

CHAPTER 38

ANIMAL BEHAVIOR

The last few chapters have detailed how environmental stimuli are processed and transmitted to effectors to produce appropriate responses. In this chapter we will examine the *mechanisms* behind the behavior of animals—how behavioral repertoires are organized through coordination of sensory organs, neural pathways and processing networks, the endocrine system, and effector organs. The study of the evolution and mechanisms of animal behavior is known as ***ethology***.

The idea that behavior can be inherited has always been controversial. But by Darwin's time thousands of cases had been documented showing that behavior could be innate. The example most compelling to Darwin was the life history of the European cuckoo, which lays a single egg in the nests of other birds (Fig. 38.1). The cuckoo chick hatches in the nest of one of several possible host

38.1 Egg removal by a cuckoo

A cuckoo about to lay an egg in a reed warbler's nest first removes one of the warbler's eggs. The warbler apparently doesn't distinguish size differences among the eggs, but the proper count is crucial. If she found one egg too many, she would abandon the nest, and both her own eggs and the cuckoo's would perish.

A

B

38.2 Egg ejection by a fledgling cuckoo

(A) A newly hatched cuckoo rolls the eggs of its host out of the nest. As a result the young bird does not have to share any of the food its foster parents collect. (B) The unwitting foster parent continues to feed the cuckoo even when it has grown to several times the parents' size.

species, and even before its eyes are open, it ejects the hosts' own eggs and chicks (Fig. 38.2A). The foster parents, unable to recognize the cuckoo as an intruder, feed and care for the chick until it is fledged, by which time it is considerably larger than they are (Fig. 38.2B). Even though it has probably never seen or heard another cuckoo of either sex, the young bird is able to find and recognize a suitable mate the following spring, to court, and to copulate, all in a manner typical of its species.

How, Darwin wondered, is a cuckoo able to do precisely the right thing at the right time, without having had any opportunity to learn? When something must be learned—to recognize the appearance of the host species, for example, which a female cuckoo must do in order to locate the proper nest for her egg—how does the bird "know" to ignore a world full of distracting information and focus on what must be memorized? The baby cuckoo must inherit essential "instructions" in the genes that direct the wiring of its nervous system. The underlying instructions that direct learning and behavior like that of the cuckoo are known popularly as instinct.

As we will see, few behaviors are strictly innate or entirely learned: automatic behavior can usually be modified, and most learning appears to be guided by innate mechanisms. ***Instinct***, then, can be defined as the heritable, genetically specified neural circuitry that organizes and guides behavior. The behavior that is thereby produced can reasonably be said to be at least partially innate.

THE BUILDING BLOCKS OF BEHAVIOR

■ HOW DO ANIMALS SEE THE WORLD?

A major turning point in the study of behavior came around 1915, when Karl von Frisch discovered that honey bees have a range of sensory experience outside our own. Von Frisch had begun by wondering why flowers are colorful. The notion that insects might have color vision, and that colorful flowers might therefore appear attractive to their pollinators, was hardly taken seriously at the turn of the century. Nevertheless, von Frisch began a new experimental tradition by designing behavioral tests. First he trained honey bees to collect sugar solution from a dish placed on a blue card. Later he removed the card and food, and set out another blue card among cards of varying shades of gray, with an empty dish on each (Fig. 38.3). If bees have only black-and-white vision, they should confuse at least one shade of gray with blue. In fact, bees have no difficulty in distinguishing cards of colors they have been trained to recognize.

Subsequent work revealed that bees are blind to red, but their vision extends well into the ultraviolet (UV) range. Von Frisch therefore examined the world through UV filters, to see what bees are seeing, and discovered that bee-pollinated flowers show a distinctive bull's-eye pattern, with a dark center (Fig. 38.4). Later he and his students found that bees have several unusual capabilities: they see the direction of polarization of light; hear sounds inaudible to humans; smell carbon dioxide, humidity, and other odors too faint for our noses; and sense the earth's magnetic field.

38.3 A test for color vision

After being trained to go to a feeding dish on a blue card, honey bees were presented a varied array of cards, each under an empty food dish. The bees demonstrated their color vision by searching for food only on the blue card, where they had been trained to expect it.

Secret signals: sign stimuli The discoveries of von Frisch and others led Konrad Lorenz to realize that though the private sensory

A

B

38.4 Flowers in visible and UV light

(A) To us, this primrose appears to be a nearly uniform yellow. (B) To bees, which can see UV light, the center is marked by a dark bull's-eye.

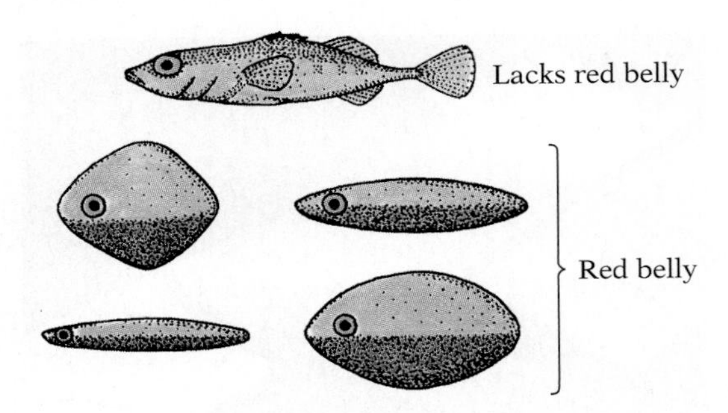

38.5 Models of male stickleback

The realistically shaped model lacking the red belly was attacked by territorial male sticklebacks much less often than were the oddly shaped models with red bellies. This indicates that color is more important than shape in male-male recognition. Other experiments show that males can recognize shapes in other contexts; for example, a male recognizes an egg-laden female by virtue of her swollen abdomen.

worlds of many animals overlap, their brains may interpret what they sense very differently. This discovery came in the early 1930s, when Lorenz noticed that he was attacked by his pet jackdaws whenever he carried something black hanging from his hand. The birds, which are themselves black, behaved as though any dangling black object was a fellow jackdaw in distress. In this situation the birds, though capable of recognizing each other as individuals in other situations, ignored most of what they could see and focused instead on a small (and in this case misleading) subset of cues.

Lorenz and the ethologist Niko Tinbergen examined this phenomenon further, and found that many animals are highly responsive to specific stimuli. For example, both males and females of the common minnowlike stickleback recognize breeding territorial males by the red stripe on their ventral surfaces. Sticklebacks are so thoroughly attuned to the red stripe that they are oblivious to additional cues that might otherwise be useful, such as the size and shape of the object displaying the color (Fig. 38.5). In behavioral terms a ***sign stimulus*** is any simple signal, such as the red stripe, that elicits a specific behavioral response. The specificity of a sign stimulus for a particular behavioral reaction is illustrated by Lorenz's observation that a mother hen can recognize a chick in trouble only by its special distress call, even when the chick is clearly visible (Fig. 38.6).

Sign stimuli seem to organize much of behavior. Ground-nesting birds such as gulls and geese instinctively rotate their eggs—a behavior that prevents the embryos from sticking to their shells—and then carefully roll any eggs that escape during this procedure back into their nests. These birds will also roll flashlight batteries, golf balls, beer bottles, and a variety of other rounded objects into their nests, apparently mistaking them for eggs. Similarly, male robins will attack anything red on their territories during the breeding season. In cases like these, certain stimuli trigger a particular behavior because the organism has an exaggerated neural sen-

38.6 Difference in response by a hen to her chick's visual and vocal distress signals

(A) The hen ignores the chick if she cannot hear its calls, even though its actions are clearly visible. (B) Distress calls elicit vigorous reaction from the hen even when she can't see the chick.

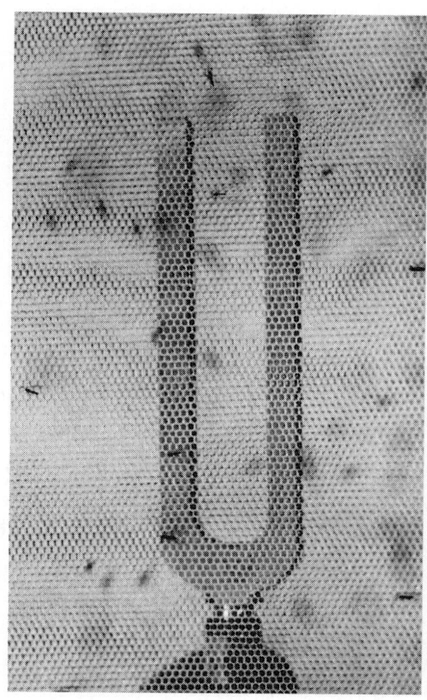
A

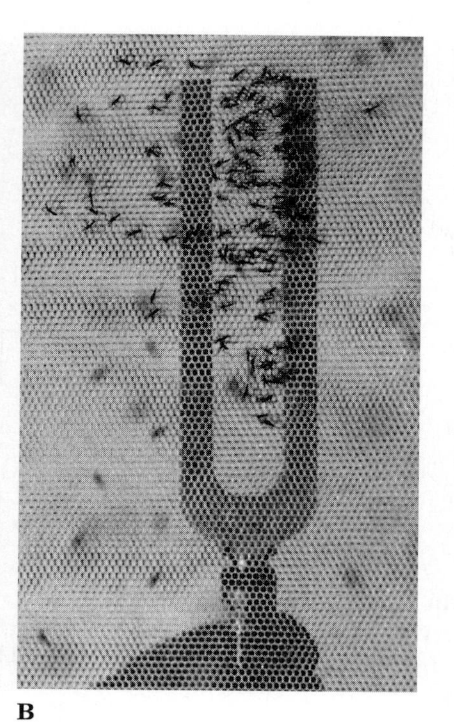
B

38.7 Response of male mosquitoes to a tuning fork

Left: The fork is silent. Right: The fork, which is vibrating at about the same frequency as the female mosquito's wingbeat, is attracting males.

sitivity to the stimulus. Because sign stimuli are said to "release" specific behaviors, they are frequently called ***releasers***.

Depending on the sensory world of the organism in question, behavior may be released by cues from many sensory modalities. ***Pheromones***, odors to which specialized receptor cells may be attuned, often serve as releasers. The odor of starfish that triggers an escape response in *Aplysia* is a pheromone. Sounds, too, may act as releasers. The high-frequency sounds produced by the wingbeat of a female mosquito attract males of the same species for mating (Fig. 38.7). The cries of bats release evasive maneuvering in certain moths. The bull's-eye pattern of many flowers in UV light, which attracts bees, is a visual releaser. The young of most species are born with the releasers necessary to direct and trigger parental care. The success of the cuckoo chick, for instance, is assured because it is born with the orange throat patch and specialized peeping that cause its surrogate parents to recognize and feed their own offspring.

Releasers have a great advantage: by initiating certain critical behavioral responses automatically, they can bypass the time-consuming and error-prone process of learning. Releasers have disadvantages as well—they may be triggered by crude and inappropriate stimuli, for instance. As we will see, a combination of learning and releasers frequently provides animals with the best features of both.

Recognizing sign stimuli Many—perhaps all—releasers depend on the feature-detector circuits described in Chapter 36. Working independently, Niko Tinbergen and Jack Hailman used models to determine which characteristics of the parent herring gull incite its chicks to peck at the adult's beak for food (Fig. 38.8). The baby

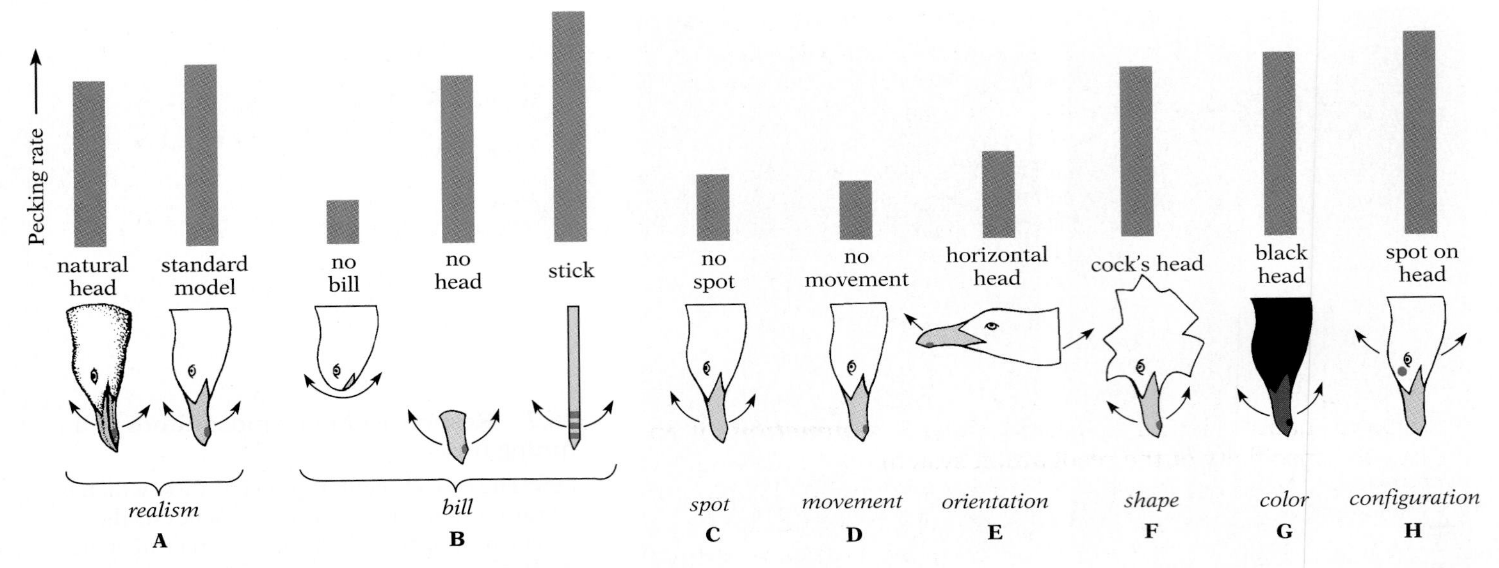

38.8 Releasers and pecking rate in herring gull chicks

To determine which characteristics of a herring gull's head release pecking in the chicks, Tinbergen and Hailman offered the chicks a series of models. Except as noted, all were held vertically and moved back and forth, as indicated by the arrows. (A) The standard flat cardboard model turned out to be slightly more effective than the head of a real bird. (B) A disembodied bill was almost as good. (C–E) Spot movement and bill orientation are crucial, as evidenced by their lack here. (F–G) However, head shape and color are not important. (G) The color of the bill and spot are likewise of little consequence, as long as they contrast well. (H) When the model with the misplaced spot was moved so that the spot traversed the same arc as the bill spot on the other models, chick response was high.

The two releasers for pecking, then, are a vertical bar moving horizontally, and a moving spot that contrasts with its background. A model emphasizing these features—a narrow stick with three spots that is waved back and forth (B)—is a supernormal stimulus.

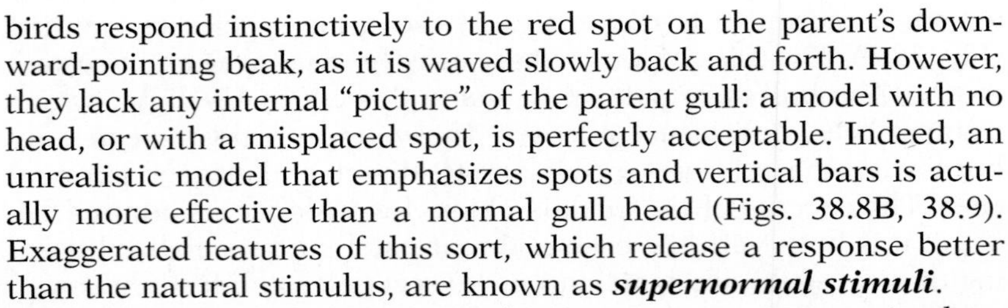

birds respond instinctively to the red spot on the parent's downward-pointing beak, as it is waved slowly back and forth. However, they lack any internal "picture" of the parent gull: a model with no head, or with a misplaced spot, is perfectly acceptable. Indeed, an unrealistic model that emphasizes spots and vertical bars is actually more effective than a normal gull head (Figs. 38.8B, 38.9). Exaggerated features of this sort, which release a response better than the natural stimulus, are known as ***supernormal stimuli***.

The two releasers operative in this instance—a spot moving horizontally (its color relatively unimportant as long as it contrasts with the background), and a vertical bar moving horizontally—combine in such a way that the releasing values of the two stimuli

38.9 Pecking by a herring gull chick

Young herring gull chicks peck at the red spots on their parents' bills from birth, and will peck at the cardboard model (on the right in this picture) as frequently as they will at a real bird. The chick's pecking behavior is so completely dictated by a set of releasers—one or more red spots on a vertical bar moved back and forth—that it will ignore a full representation of a herring gull in order to peck at the stick painted with three contrasting spots.

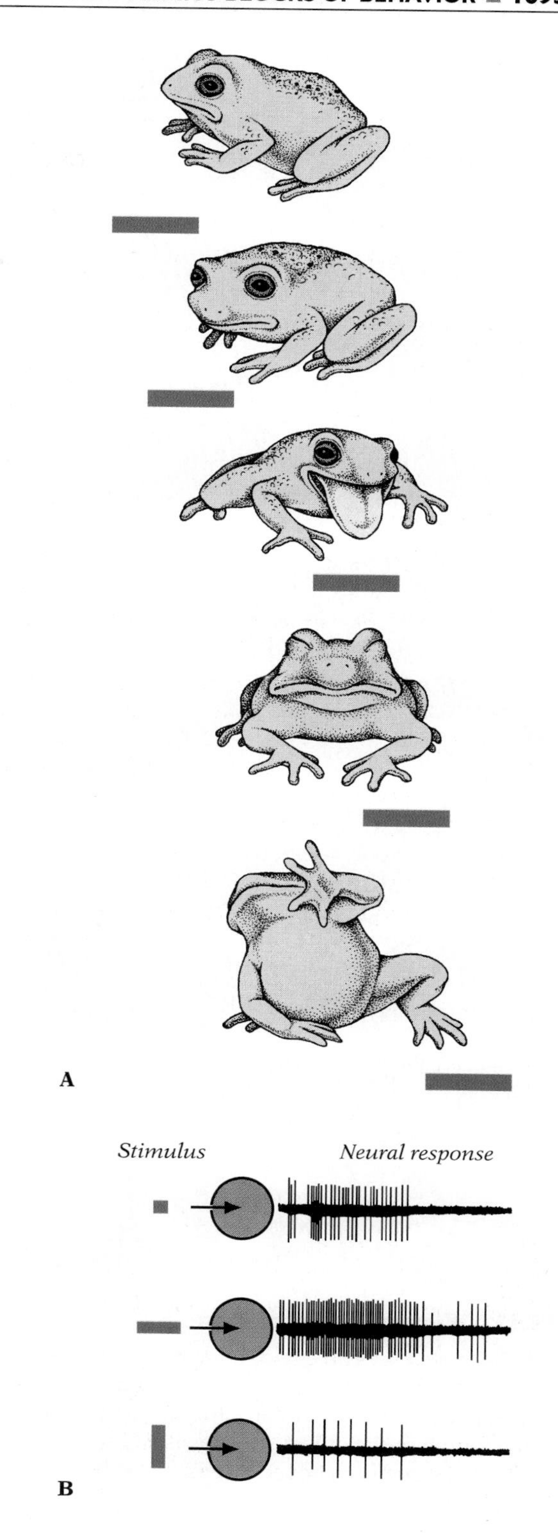

38.10 Prey-capture behavior by a toad

(A) A toad responds to an elongated stimulus moving lengthwise by turning toward it and striking at it with the tongue. The toad then "swallows" (with eyes shut) and wipes its mouth, even though it has not actually caught anything. (B) This behavior is controlled by feature detectors in the toad brain that respond preferentially to a wormlike stimulus (horizontal bar, middle row) moving into their receptive fields.

add together in the chick's central nervous system to produce an increased probability of response. This combination, common throughout nature, is called ***heterogeneous summation;*** it increases the specificity of the recognition system.

Hailman pointed out that the two releasers that enable chicks to recognize their parents correspond to two classes of feature detectors: a horizontal-motion spot detector, and a horizontal-motion vertical-line detector. A similar pattern is seen in the prey-recognition responses of the common European toad. Jörg-Peter Ewert has shown that the main stimulus that releases the prey-capture sequence (in which the toad turns toward an object and fires its long sticky tongue) is a bar moving along its long axis, regardless of direction. This is automatically provided by the worms and centipedes the toad favors (Fig. 38.10). This sign stimulus is even more effective in the presence of the odor of prey, which the toad recognizes innately.

Recording from the visual areas of the toad thalamus and tectum, Ewert found a sensory "map" of the toad's visual world, one layer of which contained feature-detector circuits tuned specifically to prey shape and movement in the lower half of the toad's visual field—the usual location of its terrestrial prey. Using microelectrodes, Ewert could sometimes stimulate an individual cell and thus cause the toad to turn to the approximate direction of the corresponding spot in the visual field and launch its tongue at an imaginary target. Ewert also found two other classes of feature detectors in toads—one for recognizing moving spots, presumably flying insects, and another for recognizing potential predators (essentially any large, moving, nearby object), to which the toad responds by crouching or fleeing.

Visual, auditory, and olfactory feature detectors also underlie much of the instinctive recognition of conspecifics that is necessary in most animal communication.

■ HOW RESPONSES ARE ORGANIZED

Fixed-action patterns Lorenz and Tinbergen noticed in the 1930s that some behaviors run to completion regardless of the situation. A dramatic example is the egg-rolling response of geese. If a goose notices an egg outside her nest, she rises, extends her neck until the underside of her bill is touching the egg, rolls it gently back into the nest, and then settles down to continue brooding (Fig. 38.11). To the casual observer this behavior appears to show

38.11 Egg rolling by a goose

When a goose sees an egg outside her nest, she rises, touches the egg with her beak, and then rolls it back in. She completes the same recovery behavior when the object she sees is a beer bottle, or when the egg is removed after she has begun to reach for it.

thought on the goose's part: she has recognized a problem and solved it. However, if the egg is removed while she is reaching for it, the goose will go on as if nothing had happened, rolling the nonexistent egg carefully into the nest. In short, egg rolling is an independent behavioral unit, a response to a releaser that, once triggered, proceeds to completion with little or no need for further feedback. Lorenz and Tinbergen called such units ***fixed-action patterns***.

Examples of fixed-action patterns are common. Among those we have discussed in this and previous chapters are the sequence by which the baby cuckoo disposes of its hosts' eggs and chicks, the prey-capture behavior of toads, the feeding, gill-withdrawal, and escape responses of *Aplysia*, locust flight, and swallowing in humans.

The fixed-action patterns that Lorenz and Tinbergen identified are what we now call motor programs (see pp. 1023–1026), the term we will use in the discussions that follow. Many researchers still employ the older term, however, to distinguish behavioral units (like egg rolling) that are almost completely independent of sensory feedback, from both feedback-dependent and learned motor programs such as shoe tying or piano playing.

Maturation versus motor learning It is often difficult to be sure whether or not a behavior has been learned if it is not actually exhibited at birth. Particularly among the invertebrates, though, opportunities for learning are so slight that many behaviors seen only in adults must be innate. A wasp that specializes in capturing honey bees, for instance, must be equipped from birth to spin a cocoon and emerge from it, dig out of its particular kind of burrow or chamber, groom itself, fly, court and mate, pounce on bees, sting them in an unarmored patch under the neck without being itself stung, squeeze the abdomen to obtain nectar the victim has collected, fly with its paralyzed prey to a place where it digs a burrow, lay an egg, seal the burrow, and so on. The wasp simply has no opportunity for trial-and-error learning between birth and its first task. Many other invertebrate behaviors, however, are varied and complex, and change with experience. Vertebrates, which usually lead longer lives than invertebrates, are in an even better position to take advantage of experience.

Nevertheless, even in vertebrates many behaviors that appear to be learned are actually innate. Eckhard Hess, for example, demonstrated the maturation of the pecking motor program in chicks. He fitted newborn chicks with tiny goggles that deflected their vision 7° to the right, and recorded the accuracy of their pecks by providing a target set in soft clay (the target was a nailhead, which, like a seed, acted as a sign stimulus for pecking). The pecks of both the normal chicks and those with the goggles were scattered, but the marks of the chicks with goggles were well to the side of the target (Fig. 38.12A, D). A few days later, chicks of both groups were able to produce a tight cluster of pecks, but those with the goggles were still missing the target as much as before (Fig. 38.12B, E). The chick's circuitry for aiming and pecking is already wired in at birth, and the improvement in accuracy is a consequence of increased nerve and muscle coordination rather than of learning.

Humans are no exception to this pattern. Smiling, for instance,

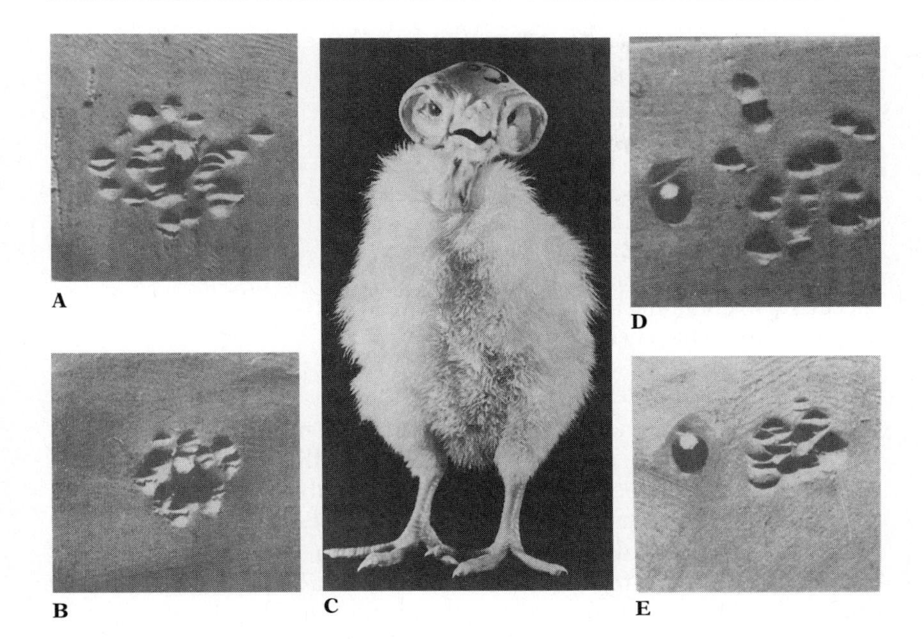

38.12 Maturation of pecking behavior in chicks

Newborn chicks peck at a target with fair accuracy (A), but their aim improves with age until at 4 days the pecks are tightly clustered (B). This improvement could be the result of some sort of maturation—better vision, perhaps, or strengthened neck muscles—or of learning, by which the chick recognizes and corrects its errors. Eckhard Hess pitted these alternatives against one another by raising chicks with goggles that deflected their vision to the right (C). As newborns, such birds produce the usual set of scattered pecks, but the pecking is well to the right of the target (D). By the fourth day, the pecks are tightly clustered but still misdirected (E), indicating that chicks are unable to learn to adjust their aim. The coordination of beak and eye involved in pecking must therefore be a wholly innate behavior, which matures without benefit of learning.

appears about a month after birth, but not as a result of learning: even blind infants start smiling on cue (Fig. 38.13). By contrast, many behaviors we know to be learned look like innate motor programs. Walking, for instance, which is innate in most species, must be learned initially in ours. Though the alternation of the legs and the interacting reflex arcs responsible for walking come prewired at birth—a properly supported human infant will perform walking motions on the delivery table—humans must learn to balance once they have matured enough to support their own weight. Yet after the difficult process of learning has been completed, simple walking becomes automatic. Swimming and bicycle riding develop the same way, and once painstakingly learned, neither is ever completely lost.

The same pattern may be seen in other animals: learned behavior can become stereotyped and largely automatic—that is, take on the characteristics of an innate motor program. This freeing of learned behavior from detailed conscious control allows conscious attention to be focused on new problems—an advantage for humans, as well as other animals. A bird that has learned to shell seeds automatically can devote its attention to watching for predators while it eats.

How behavior is orchestrated A remarkable degree of complexity and flexibility is possible without learning. For instance, all birds, as far as is known, follow innate instructions when building their nests, instinctively recognizing appropriate material and suitable locations. Bird nests vary in complexity from simple hollows scraped in the ground to elaborate multilayer structures with a different material for each layer—a robin will first use sticks, then twigs, then mud, and finally grass. The substances used to join these materials and to attach the nest to its support are as species-specific as the nests themselves, varying from dung to spider webs to special-adhesive saliva.

The means by which the many steps of nest construction are or-

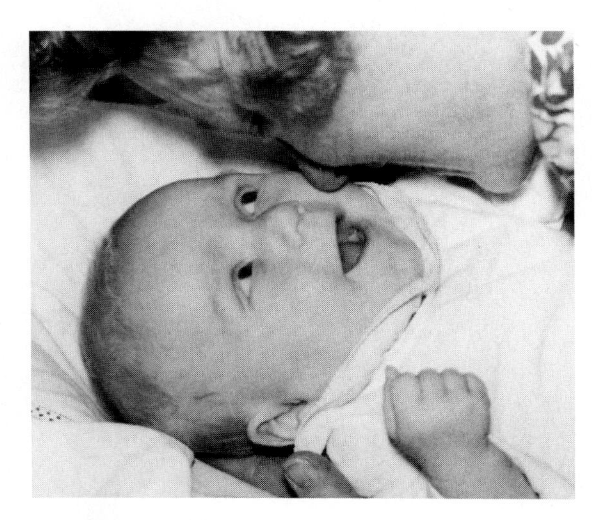

38.13 Smiling

Smiling appears spontaneously in human infants at about 4 weeks of age. The innate nature of this motor program is illustrated by the smile of this 11-week-old congenitally blind girl. Her eyes have fixated on the source of her mother's voice, a complex behavior that is also innate. Smiling helps cement a strong emotional attachment between parent and child.

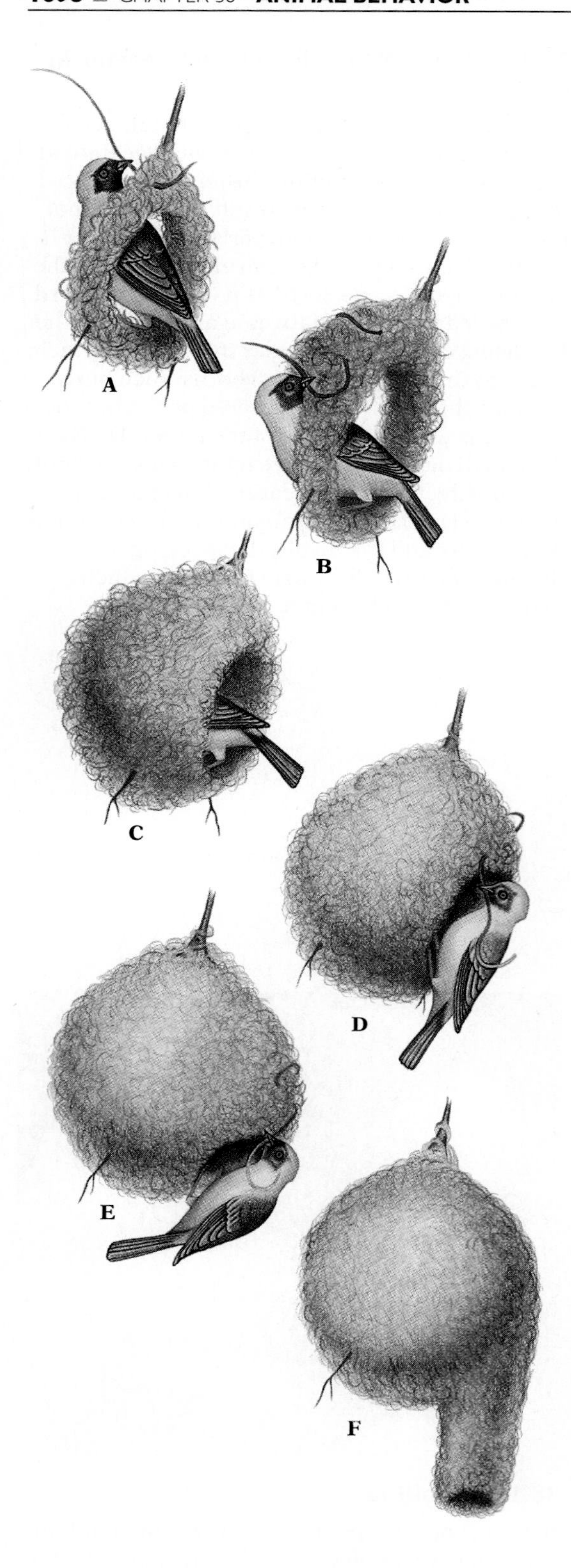

38.14 Construction of a village weaverbird nest

(A) A male village weaverbird begins with a suitable forked branch as a support and weaves around himself until he has a circular perch. (B–C) Then he stands facing out of the circle and weaves around himself until he has completed a roof and nest cup. (D–E) Next he weaves backward to create a doorway. (F) Finally, if he is successful in obtaining a mate, he weaves an entrance tube, which helps keep out nest predators.

chestrated are nowhere clearer than in the nest-building *tour de force* of weaverbirds, as observed by N. E. Collias and E. C. Collias. A male weaverbird begins by selecting a suitable branch, usually in the shape of an inverted Y, to support the nest (Fig. 38.14A). The male then begins collecting green vegetation, which he tears into long strips suitable for weaving. He next weaves the strips one by one over and around himself in an arc: he puts the tip of the strip through a narrow opening, vibrates his bill until the tip comes out the other side, and grasps the strip and pulls it through; he repeats this maneuver over and over until the strip is entirely woven in. This behavior resembles a kind of computer subroutine called a do loop—a series of steps repeated mechanically many times until a preset goal or criterion is reached. When all but the entrance to the nest is complete (Fig. 38.14E), the male breaks off building and begins an instinctive courtship display, in which he hangs upside down from the nest, vibrating his outspread wings and calling. If a female accepts the male, she lines the nest cup with a layer of soft grass and then a layer of feathers, while the male builds an entrance tube (Fig. 38.14F) that keeps out snakes and other nest predators.

The building of this complex structure is organized into relatively simple subroutines, each with its own motor program and clearly defined criteria for terminating the do loop. For example, the Colliases were able to end the thatching phase (during which wide strips are woven into the roof to make the nest waterproof) by covering the top of the nest with an opaque cloth; the bird is evidently programmed to stop thatching when an opaque layer is complete. Once the bird had begun his next "subroutine" (the stage that ordinarily comes after roof thatching) the Colliases could remove the cloth and the bird still would not resume thatching. Most complex innate behavior is probably structured along these lines: specific behavioral subroutines, cued by signals from the environment, start and stop in a preset ordered sequence, with little or no provision for creative problem solving in the face of setbacks.

■ WHAT MOTIVATES ANIMALS?

Different species behave differently: weaverbirds build nests quite unlike those of robins or geese. In addition, the same animal may behave in different ways at different times—birds are not always building nests, flying south, or courting potential mates. The force

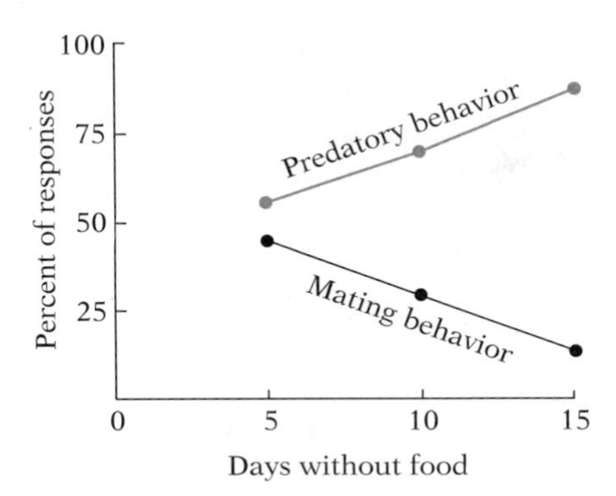

38.15 Change in response of a male jumping spider to a model as a function of the number of days he has gone without food

Among the prey of male spiders are some insects that closely resemble female spiders. It is therefore possible to construct models that can release either prey capture or mating, depending on the motivational state of the spider. The longer he has gone without food, the more likely he is to exhibit predatory behavior instead of mating behavior.

from within an animal that motivates it to do one thing now and another later is commonly known as ***drive***. A drive can have two basic effects: it can alter an animal's threshold to stimuli, thus making a particular behavior more or less likely, and it can substitute entirely new programs for old ones. Drives are shaped largely by the sorts of hormonal control, proprioceptive monitoring, and habituation we have examined in previous chapters.

Managing priorities In response to simultaneously active drives, an animal must choose among several behaviors, such as searching for food, attempting to attract a mate, grooming itself, repairing its nest or burrow, or even playing. At any given moment, some of these behaviors will be more important than others: escape behavior almost always takes precedence. Current priorities, established by the shifting thresholds of response to various drives, can help the animal select among the many behavioral possibilities that may be competing for its attention. These priorities determine, for instance, whether a spider will feed or mate (Fig. 38.15). Performing the behavior with the highest priority will lower the urgency of the drive that motivated it, and behavior that was formerly less important will then become the animal's highest priority. What behavior an animal chooses depends not only on the relative urgency of various drives, but on opportunity as well: an animal that is only mildly hungry but very thirsty will prefer water to food if both are present, but will probably eat if no water is available.

Changing behavioral programs In addition to altering thresholds to stimuli, drives can act by bringing in or retiring entire behavioral programs. A migratory bird, for example, must alter its dietary intake and metabolism dramatically to accumulate the fat reserves necessary for its annual journey, and then must set off in the correct direction at the appropriate time of year. How does the bird come to switch its behavior in anticipation of the need, even when it has never migrated before? And why do geese retrieve eggs outside their nests only in the period from about 1 week before laying until 1 week after hatching, while at other times they ignore the same cue? Animals often behave as though some sort of timer is switching their behavior patterns repeatedly from the set appropriate for one stage of their life cycle to the set appropriate for the next stage.

The brooding behavior of ring doves is a well-understood exam-

A

B

C

D

38.16 Behavior of a pair of ring doves

A pair of laboratory-reared doves court (A) and mate (B). They go on to establish their nest (in a ceramic dish in this case) (C) and produce crop milk for their offspring (D) in this artificial environment. Their remarkable willingness to continue this behavior outside of their natural environment has made the dove one of the standard laboratory animals.

ple of such automatic switching of behavior patterns (Fig. 38.16). In the normal course of events, ring doves court and pair, establish a nest in a shallow depression, lay two eggs in it (one day apart, both in the late afternoon), and begin to brood or incubate them. About 16 days later, the eggs hatch and the adults feed the young a liquid secretion known as crop milk, produced by special glands in their throats. If nesting ring doves are supplied with foreign eggs, they produce crop milk 16 days after eggs appeared in the nest, regardless of when the pair's own eggs were later laid, or when the chicks hatch. Evidently, the *sight of the eggs* initiates the ticking of a 16-day behavioral timer that controls the metabolic and physiological preparations necessary for feeding the young. The timer controls the level of the hormone prolactin in the parents. Prolactin binds to specific neurons in the brain and activates the neural circuits involved in parental behavior and in the production of crop milk.

How internal rhythms regulate behavior The most obvious of the internal timers that switch behavioral routines on and off and modulate drives is the nearly universal one that controls such daily behavioral cycles as the cycle of sleeping and waking. We know that this "clock" exists because an animal's cycles persist even in the absence of outside cues. A nocturnal animal like the flying squirrel, for instance, will continue to alternate about 10 hours of foraging with 14 hours of resting even in continuous darkness (Fig. 38.17).

Because the period of animal clocks is approximately 24 hours, the cycles they control are called ***circadian rhythms*** (from the Latin *circa*, "about," and *dies*, "day"). Under experimental conditions, activity rhythms drift systematically with respect to the 24-hour day (Fig. 38.17), but organisms can be reset by a species-specific hierarchy of cues, the most important of which is usually light. The clock's response to resetting (known as phase shifting or clock shifting) has a time schedule of its own. A flash of light seems to be taken by many animals as a sign stimulus for dawn; a flash in the six hours before dawn will advance the rhythm. If an animal is kept in the dark and a flash is administered in the first few hours *after* dawn ought to have occurred, the animal's schedule will be delayed. The process of resynchronizing the circadian clock to fit external cues may take several days, depending on the extent of the difference between internal and external time (Fig. 38.17). We experience this adjustment as "jet lag."

Clocks are found in virtually all living things, whether individual cells or whole multicellular plants or animals. Circadian clocks regulate enzyme activity, osmotic pressure, respiration rate, growth rate, membrane permeability, bioluminescence, sensitivity to light and temperature, and reactions to various drugs. Physicians are

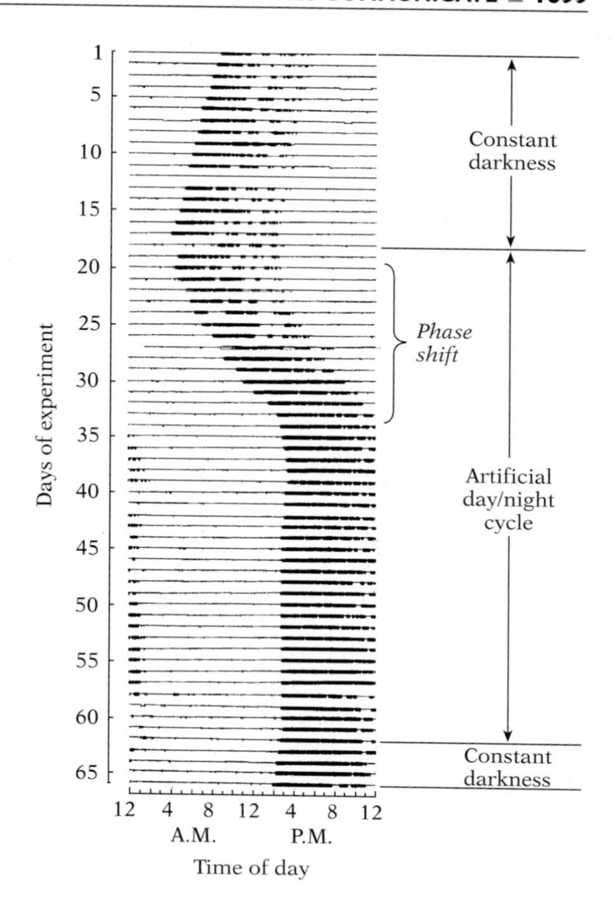

38.17 Record of a circadian-rhythm experiment

In constant darkness (days 1–17) a flying squirrel (*Glaucomys*) continued to spend about 10 hours in foraging activity (dark bars) and then 14 hours resting (light bars). The time of foraging drifted slowly with respect to a 24-hour day. When an artificial day/night cycle was imposed (days 18–61), the activity rhythm gradually shifted into phase with it. When constant darkness was resumed (day 62), the period of foraging again began to drift. Circadian rhythms in mammals are controlled by a region of the hypothalamus.

becoming increasingly aware that the proper dosage of a drug may be very different at different times of day; in some cases, what constitutes a beneficial dose at one time may actually be lethal at another.

There are other biological clocks as well. Many animals living near the seashore have a 13-hour tidal clock and a 27-day lunar rhythm that together enable them to anticipate the time and day of peak tides. The precision of this combination is obvious in a species of intertidal midge that emerges from the sand, mates, and lays eggs in the period between 2 and 5 a.m. during the month's lowest low tide. Such accuracy is essential since this species has only a 2-hour life span as adult. Many species—migratory birds, for instance—also display annual rhythms even under constant laboratory conditions.

HOW ANIMALS COMMUNICATE

Animals are frequently categorized as either solitary or social—that is, they either live by themselves, or live with conspecifics in pairs or groups. Those that are solitary come together with other members of their species only to mate. At this crucial point, each animal must find a reproductively ready member of the opposite sex of its own species. Communication probably first evolved to accomplish this goal, and diversified into the wide variety of signals that now communicate mood, intention, and all the information necessary to maintain order and stability in social groups.

■ GETTING THE MESSAGE THROUGH

Sensory channels The most primitive and widespread channel of communication is chemical. Many unicellular organisms depend on chemoreceptors to recognize when they have bumped into another individual of their species, while others, like the slime molds that aggregate periodically to reproduce, locate each other through pheromone (odor) trails. Only slightly more elaborate is the mating system of most moths: females release a species-specific pheromone, and males follow the odor upwind to its source. Pheromones also play an important role in groups as diverse as beetles, aquatic invertebrates, and mammals. In each case the role is basi-

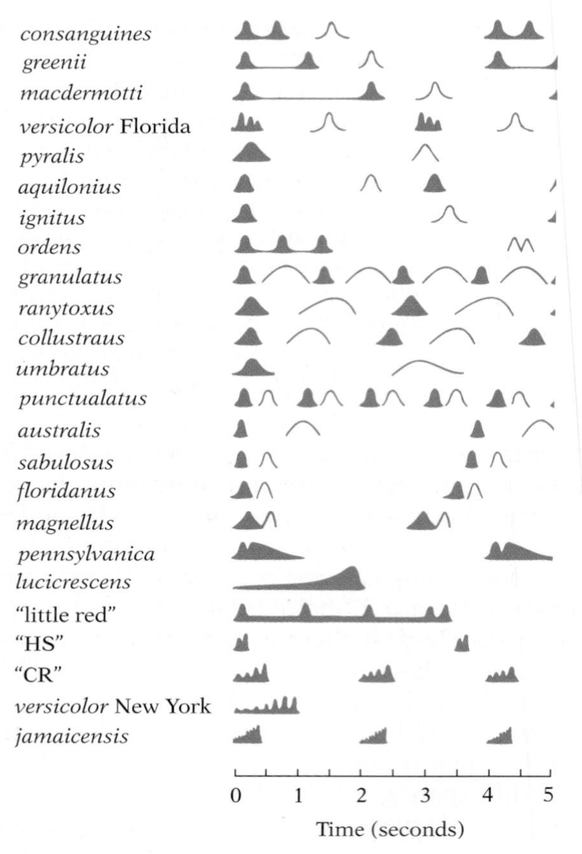

38.18 Firefly codes

The signals of various species of flying male fireflies (genus *Photuris*) are shown in solid color, while the female response is indicated in outline. Each female response occurs at a species-specific interval after the male flash. For some species (such as the last seven shown here), the female response is not yet known.

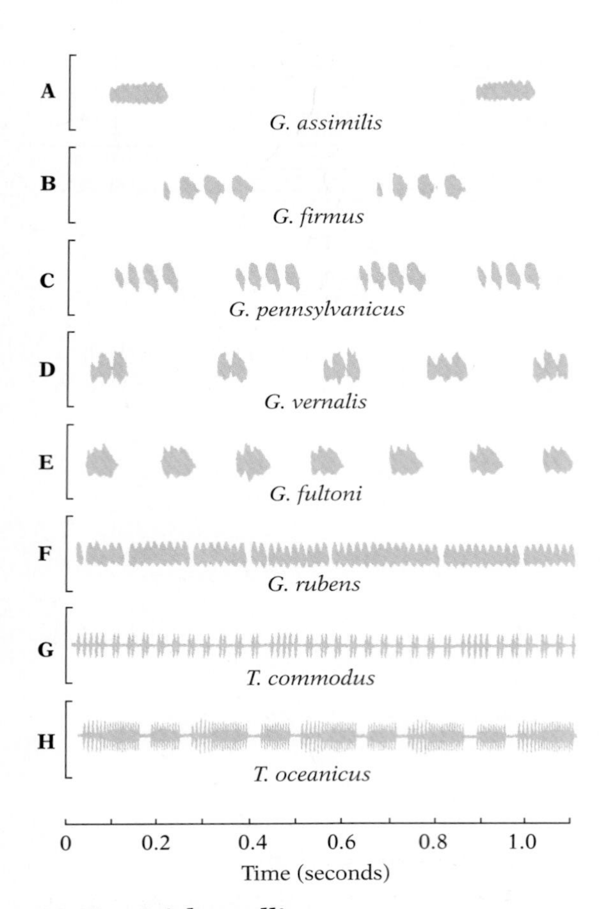

38.19 Cricket calling

Male crickets produce a pulsed calling song that attracts females of their species. The songs of the six species of *Gryllus* (A-F) are less complex than those of the two species of *Teleogryllus* (G–H). When a female approaches, the males switch to a courtship song.

cally the same: the odor informs potential mates of the species, sex, reproductive readiness, and location of an appropriate mate.

A number of species use other sensory modalities. Fireflies, for instance, produce pulsed signals that can be seen by other fireflies at great distances. Males fly about, flashing according to a species-specific code, while females wait on vegetation, flashing in answer (Fig. 38.18). The releaser in firefly communication is, in most cases, the interval between pulses. Both sexes are thoroughly tuned to a specific set of intervals; with a bit of experimentation you can lure males in with a penlight.

Other species employ auditory versions of the firefly system. Both crickets and frogs generate calling songs whose temporal characteristics minimize possible ambiguities of species or sex (Fig. 38.19). To our eyes and ears, the rhythmic pattern of pulses appears to be the most useful characteristic, but the feature detectors of the females seem to be tuned instead to the intervals between pulses. Hence a scrambled song that sounds totally different to our ears but faithfully preserves the intervals is as acceptable to females as the normal song.

Birds and mammals, which are capable of fine frequency (pitch) discrimination, can use different frequencies to convey separate messages. Among birds, species recognition is based not only on temporal intervals and sound frequency, but also on the rate of change of frequency with time. Even in species whose song is learned, feature detectors responsive to a specific combination of these three characteristics (and perhaps others) sensitize the animal to what it is to learn—what to sing, in males, and what to listen for, in females.

Specificity through multiple signals Much of the species specificity of animal communication depends on the *simultaneous* presence of several cues at once—rather in the manner of a fraternity handshake—to exclude all creatures that do not "belong." Obviously the multiplicity of cues reduces the potential for mistakes.

In a second and even more effective technique, the several cues must also appear in a particular order. The courtship sequence of queen butterflies (a species of milkweed butterfly closely related to the monarch) provides a good illustration of this more elaborate form of communication (Fig. 38.20). A male will chase any rapidly flapping object—including another butterfly in flight—until he

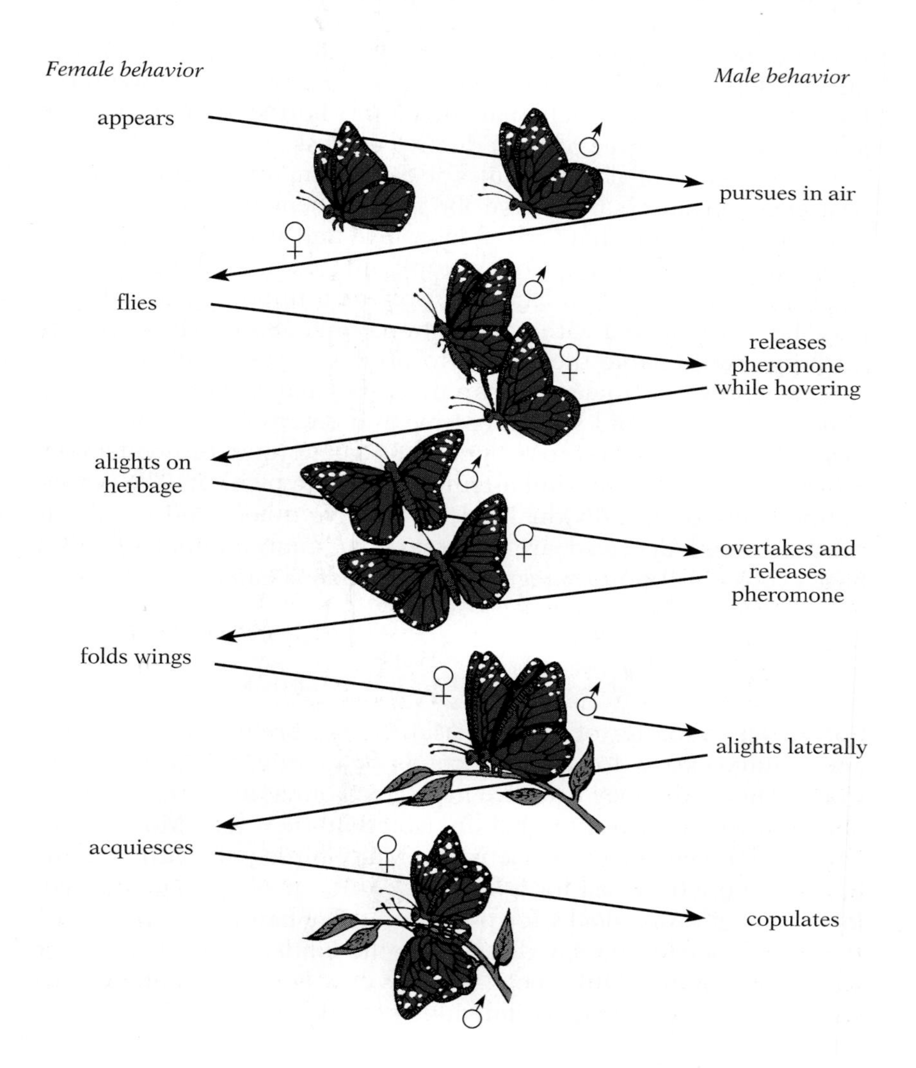

38.20 Butterfly courtship

Courtship of queen butterflies involves a series of signals from male to female and from female to male, presented in a particular order. If either individual fails to produce the right signal at the right time the courtship breaks off. This dependence on ritual ensures that only reproductively ready members of opposite sexes of the same species mate.

overtakes it. The flashing of the fluttering wings is the releaser that triggers the male's courtship behavior, causing him to extrude a pair of brushlike "hairpencils" from his abdomen, which emit a species-specific pheromone as he hovers over the object of his pursuit. If she is of the correct species and sex, and reproductively ready, she then alights on some nearby vegetation. (So, of course, might a falling leaf, which males often pursue.) The male hovers above the female, sweeping his hairpencils over her antennae. If all goes well, the female then closes her wings, thus signalling the male to alight and begin mating. Each step of the courtship sequence must be performed correctly for mating to take place. This strategy, involving a specific series of releasers, is almost universal.

38.21 Courtship among hangingflies

Males attract females with a pheromone, offer a gift of food (a dead fly in this case), and mate with the female while she is consuming it. Some males acquire the gift by hunting flies, while others steal it from a courting male by mimicking female behavior. Still other males suck out all the nutritive juices of their prey before offering it as a gift, although females that discover the deception will refuse to mate with the male.

Risks and deception Having to advertise for a mate carries the obvious risk of attracting predators as well. Bats, for example, target calling frogs and katydids, while parasitic flies home in on calling crickets. For males, then, mate attraction displays can be fatal, but celibacy is genetic suicide. In response to such pressures, many male frogs and crickets opt not to display; instead they hide near a male that does, and try to intercept the females he attracts. You might suppose that such cheaters would drive the conventional males extinct, but this is not the case: the disappearance of callers would doom the "parasitic" males. As described in Chapter 17, frequency-dependent selection usually leads to a balance between behavioral alternatives such that all males, honest or not, achieve about the same degree of reproductive success.

The existence of cheats makes it clear that sexual communication does not necessarily serve the mutual benefit of both parties. The human maxim that "all's fair in love and war" seems to have clear parallels in nature. For example, the male hangingfly hunts small insects, and offers captured prey as a nuptial gift to the female he has attracted with a pheromone (Fig. 38.21). The nutrients in the present enable the female to produce eggs. While a female eats the prey the male has captured, he mates with her. Some males, however, avoid spending time and energy looking for prey; instead they fly upwind to a signalling male, pretend to be female, take the proffered prey, and attempt to fly away with it. The potential benefits to an individual able to deceive others and the detriment to those deceived are so great that many animal societies have evolved elaborate safeguards to exclude cheaters.

■ THE CHALLENGES OF SOCIAL COMMUNICATION

Sociality is a matter of degree: crickets and fireflies, for instance, are solitary except for a few seconds of mating; at the other extreme, ants and honey bees live in large colonies, and are so dependent on community life that in isolation they die. Most social animals fall somewhere in between. Many birds pair with a member of the opposite sex for a few weeks to rear offspring, and may join winter feeding flocks for protection. The nature and degree of a species' sociality largely determines the kinds of social signals it will need. Among highly social animals new levels of complexity in communication emerge as individuals need signals to show mood or intention, to coordinate hunting or escape, to indicate social status, and so on.

Do honey bees have language? The ability of a forager bee to inform her hivemates of the distance and direction of a good source through a symbolic system of communication seems hardly less amazing now than it did when first discovered by Karl von Frisch in 1945. Foragers can be trained to use artificial food sources, marked with paint, and their dances (which normally occur in the darkness of the hive on vertical sheets of comb) can be observed in special glass-sided observation hives. Von Frisch decoded the dance by observing the differences between the dances of two groups trained to different locations.

In the significant central portion of the dance the forager vibrates, or "waggles," her body from side to side and simultaneously produces sound bursts by vibrating her folded wings. All the information that recruits need is conveyed in these acoustically emphasized waggle runs. "Up" on the comb is always the direction of the sun, and the angle the dance runs with respect to the vertical corresponds to the direction of the food with respect to the sun (Fig. 38.22). For example, if the food is 45° to the left of the sun, as seen from the hive, the dance will point 45° to the left of vertical. Even in the darkness of the hive the dance attenders can perceive the angle of the dance.

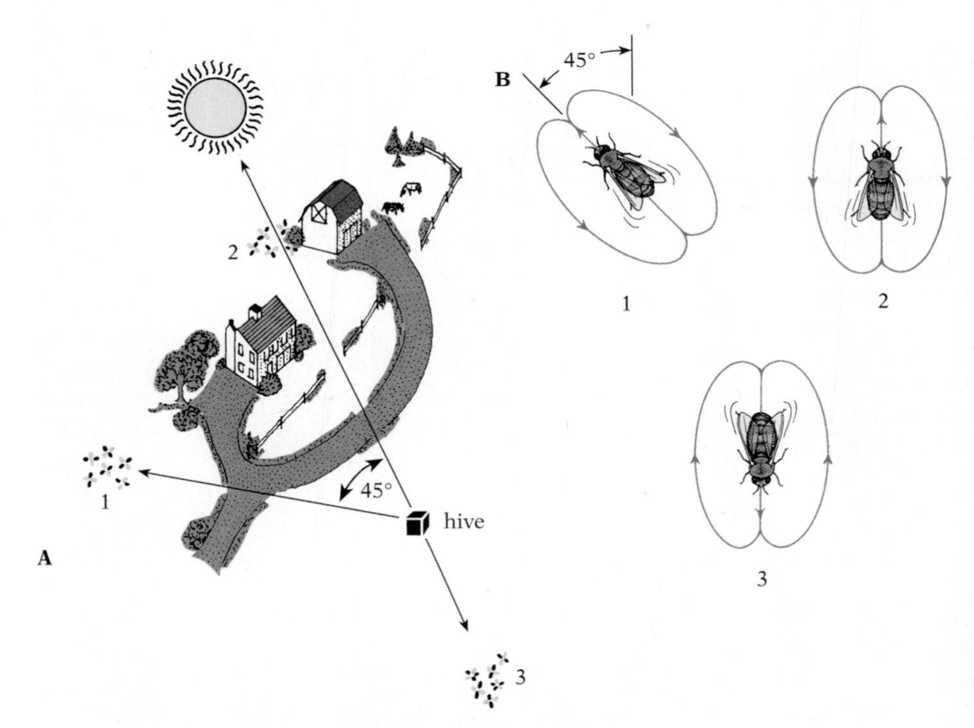

38.22 An example of bee language

(A) Three different sources of food were located: 1, 45° to the left of the sun as seen from the hive; 2, straight toward the sun; and 3, straight away from the sun. (B) The dances performed on the vertical comb in the darkened hive by forager bees from these food sources were oriented 45° to the left of vertical (1), straight up (2), and straight down (3). In short, the vertical direction on the comb symbolizes the sun. The dancing bee in the photograph, her legs bearing pollen, is closely surrounded by attenders.

HOW WE KNOW:
THE BEE LANGUAGE CONTROVERSY

The idea that a mere insect possesses the second most complex language in nature has always been difficult for humans to accept. In the late 1960s Adrian Wenner, Patrick Wells, and their colleagues in California looked closely at the phenomenon, and came to the striking conclusion that the bee's dance language, which was by then a cornerstone of behavioral wisdom, could be explained in other ways. They pointed out that any of Karl von Frisch's results could be accounted for if the recruited bees were merely searching for the floral odors they detected on dancing foragers—the very system used by other social bees. The dance correlations, they argued, were correlations only, and meant nothing to the dance attenders. They resurrected many long-forgotten studies showing that distance and direction correlations exist in the behavior of a variety of insects, such as beetles and ants, and that these behavior patterns clearly communicate nothing to other members of their species. Finally, Wenner and Wells attempted to repeat von Frisch's experiments using what they felt were improved controls for odor; they found that the recruited bees, when deprived of odor information, arrived in the field with little or no idea about the distance or direction of the food being advertised by the dancing. To make things more complex, the researchers showed that, under at least some circumstances, dancers could bring back locale odors on their bodies—the odors of vegetation near the food source, as opposed to floral odor alone. Dance attenders could then use these odors to gain location information without recourse to the so-called language. Wenner and Wells used unusual training and testing techniques, however, and many specialists remained skeptical.

The resulting dance-language controversy reawakened interest in honey bee communication, and led to dozens of new discoveries about bee navigation, colony organization, and decision making. It also produced unambiguous evidence that the language is real. The first clear-cut test, devised by one of us (James Gould), took advantage of two obscure facts about honey bees. The first is that if a bright light is directed onto the comb, the dancers will orient their dances to the light rather than to gravity; apparently they take the light to be the sun. By itself, this finding is of little use: although the dances can be reoriented, any directional interpretations being made by dance attenders would be similarly reoriented, and no misunderstanding would result.

The other curiosity, however, is that if the three simple eyes—ocelli—that lie between the large compound eyes are painted over, the bees' ability to detect light is greatly reduced. Apparently, the ocelli act in part as light-level meters, and adjust the sensitivity of the compound eyes. By painting the ocelli of foragers and using a bright light, Gould was able to reorient the directional interpretations of dance attenders without reorienting the dances of the ocelli-painted foragers. The dancers and attenders, if they share a language, would be using two different dialects. Gould was thus able to aim recruits at a site well away from the site where the dancers were collecting food. Moreover, recruits arrived preferentially at the arbitrary location. Since it is possible to make foragers "lie" to dance attenders, the dance language must be real.

Repetitions of the Wenner-Wells experiments by Gould demonstrated that their training conditions inhibited dancing and forced recruit bees to rely on odors. Thus, bees have two systems for mobilizing new foragers: one involves odor alone, while the other depends mainly on the dance language. Rather than being less complex than von Frisch had thought, Wenner and Wells showed that bee behavior is even more flexible and subtle than anyone had imagined.

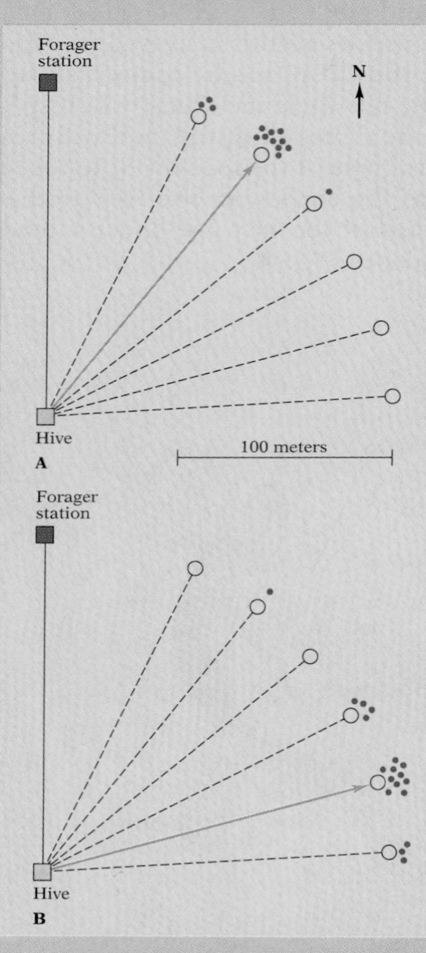

Ocelli-painted foragers were trained to a feeder north of the hive (forager station), and an array of recruit-capturing stations was set out in an arc to the right (open circles). (A) The bright light was used to aim the foragers at first one station in the array (heavy arrow), and later at another (B). The recruited bees (red dots) arrived mainly at the station indicated by the dancing.

Distance can be determined from the duration of the waggle run or the number of waggles in each run (Fig. 38.23). Each waggle specifies a particular increment of distance—about 40 m in the case of von Frisch's honey bees. The dance communication system is called a language because it refers to objects distant in both space and time (that is, the animal is not simply pointing and grunting) and because it is symbolic ("up," for instance, is an arbitrary symbol for the sun's direction; "down" or any other direction could have been used just as well). That the waggle as an indication of distance is a relatively arbitrary symbol is emphasized by the discovery that different subspecies of bees have different distance dialects (Fig. 38.23). These dialects are entirely instinctive: bees of one subspecies reared in a colony of a different subspecies misread the dances of their adopted hive.

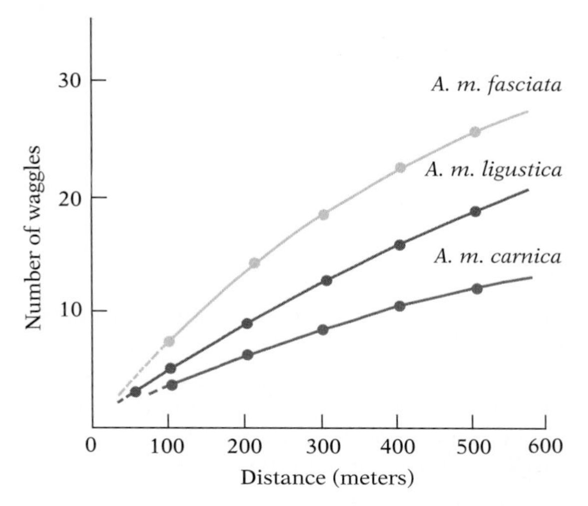

38.23 Distance codes in the bee dance

Different subspecies of honey bee use different dialects to indicate distance. For the German honey bee (*Apis mellifera carnica*), each waggle corresponds to about 40 m; for the Italian honey bee (*A. m. ligustica*), a waggle corresponds to 25 m; for the Egyptian honey bee (*A. m. fasciata*), a waggle represents about 15 m.

Learning and social communication As far as we know, the honey bee dance language is second only to human speech in its ability to convey complex information. Though used to relay information about water, nectar, pollen (which bees collect for its protein), tree sap (used by the bees to seal openings and entomb unwanted objects too large or awkward to remove from the hive), and new hive sites, it is basically a closed system under instinctive control: bees can perform or understand dances with no previous experience, but can use them only to specify the distance, the direction, and (in ways we have not discussed) the desirability of a location. In other highly social animals, however, communication seems to be more flexible.

Postural and facial cues play a role in many bird and mammalian communication systems. Dogs, for instance, solicit play with a half-crouch, tail up and wagging. This signal is species-specific; cats seem to misinterpret it, perhaps because their "vocabulary" includes no corresponding posture. Fear and aggressiveness are shown by facial cues easily read by other canids (Fig. 38.24). These may or may not be accompanied by acoustic signals such as short barks or growls. How do animals recognize and decode these visual messages? Unlike the acoustic and chemical signals we have discussed, complex visual information cannot readily be encoded as simple sign stimuli, and the neural mechanisms remain unknown.

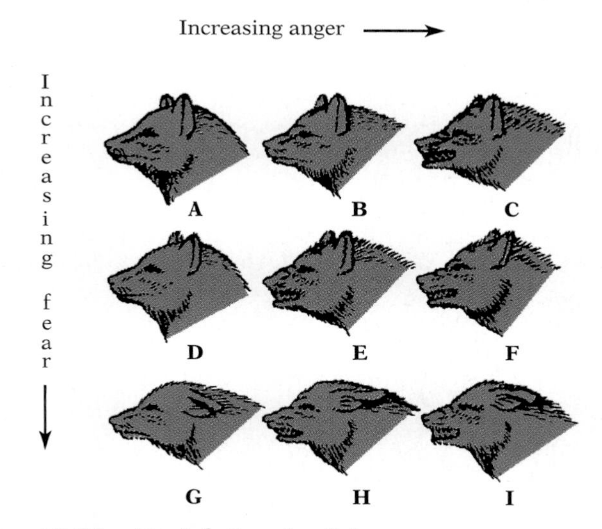

38.24 Facial signals of dogs

Dogs are among the many species of mammals and birds that use body postures along with other signals to communicate mood. Shown here is the simultaneous expression of varying degrees of anger and fear by dogs. Since these two states can vary independently, a dog can be purely angry (C), purely frightened (G), or both angry and frightened (I). Tail position also signals degrees of fear, as in the familiar tail-between-the-legs posture.

HOW ANIMALS LEARN

Innate behavior—nest building for example—can be amazingly complex. Purely instinctive behavior, however, is often rigid and unadaptable. The changing environmental and social conditions to which many species must adapt require the behavioral flexibility provided by learning.

■ THE PROCESS OF CONDITIONING

Many higher animals learn to recognize new objects and perform novel behavior. Such learning is often called conditioning.

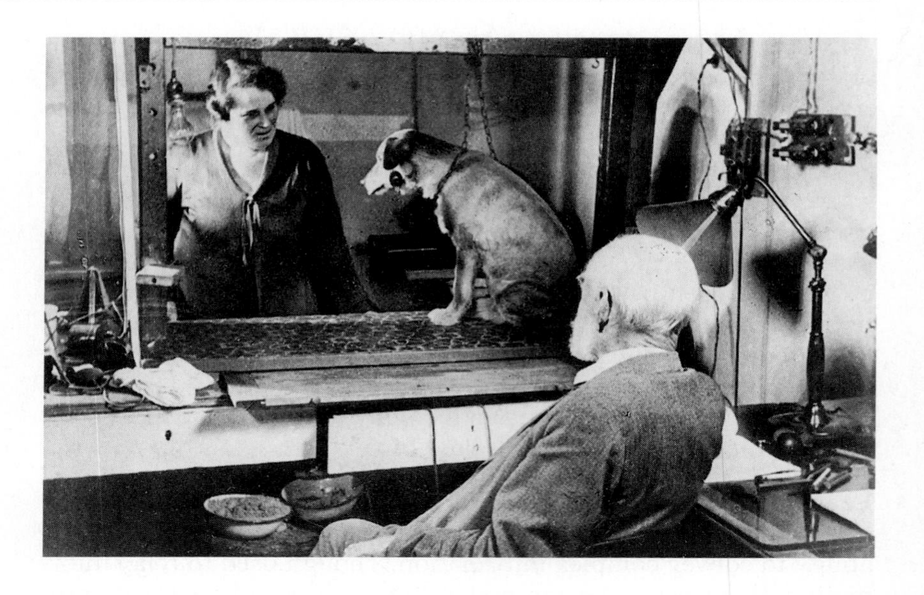

38.25 Pavlov's experiment showing classical conditioning

The device on the dog's cheek measures salivation, the unconditioned response; the dish at left contains meat powder, the unconditioned stimulus. The conditioning stimulus is the light.

Learning to recognize: classical conditioning Psychologists recognize two general forms of conditioning. The first, discovered by the Russian physiologist Ivan Pavlov (1849–1936), is called ***classical conditioning***. Pavlov encountered this form of learning during his pioneering studies of the physiology of digestion. While measuring the quantity of saliva produced by dogs when they see food, Pavlov noticed that if an irrelevant stimulus—a light or a ringing bell, for instance—appeared just before the dogs saw the food, the dogs came in time to associate the stimulus so closely with the expected food that the light or bell alone would trigger salivation (Fig. 38.25). They had been conditioned.

Classical conditioning begins with an unconditioned (innate) response (UR) that inevitably follows an unconditioned (instinctively recognized) stimulus (US). A novel conditioning stimulus (CS) is then paired with the US, and after this process has been repeated sufficiently, the CS alone will elicit the response (which at this point may be referred to as the CR, the conditioned response). In short, the animal comes wired with the sequence US $\longrightarrow$ UR; the environment (or the researcher) provides the relationship CS + US $\longrightarrow$ UR; and the animal generalizes from a series of individual experiences, so that CS $\longrightarrow$ UR. The gill-withdrawal behavior of *Aplysia*, described in Chapter 35, can be conditioned: the withdrawal, which ordinarily occurs after the siphon is touched, can be conditioned to a flash of light if the illumination consistently accompanies or slightly precedes the touch.

Experiments similar to Pavlov's show that classical conditioning is an instinctive process by which animals free themselves from exclusive dependence on sign stimuli for recognizing important objects and individuals in their environments. Species differ dramatically in the range of acceptable conditioned stimuli and in their ability to spot subtle CS-US connections.

Learning to do: trial-and-error learning Early psychologists believed that chains of linked conditioned stimuli and unconditioned responses could explain complex behavior. B. F. Skinner, however, recognized that classical conditioning cannot plausibly account for such behavior as the ability of a rat or an ant to learn a maze with 30 choice points. He pointed out that many animals seem instead to learn by doing—by experiencing the consequences of their behavior and altering it accordingly. This second learning strategy is called ***operant conditioning***, or ***trial-and-error learning***. In the laboratory, experimenters train animals to perform a behavior (Fig. 38.26); in nature, animals train themselves.

The capacity for trial-and-error learning confers a considerable advantage on animals, since it allows them to acquire motor behaviors that are not instinctive. A seed-eating bird, for instance, does not have innate motor programs for picking up the various kinds of seeds in its environment, cracking them open, and separating the kernels from the shells. Instead, the bird has an innate ability to recognize seedlike objects, along with a drive to experiment with anything that looks like a seed. Getting into the first sunflower seed may take a finch several minutes of manipulation with its beak and tongue, but the kernel the bird ultimately harvests provides the reward that motivates it to try another. By trial and error, the finch discovers which muscle movements help get at the seed and which are irrelevant; experience in opening a succession of seeds slowly shapes the bird's harvesting behavior into a quick and efficient series of movements that finally become automatic—a learned motor program.

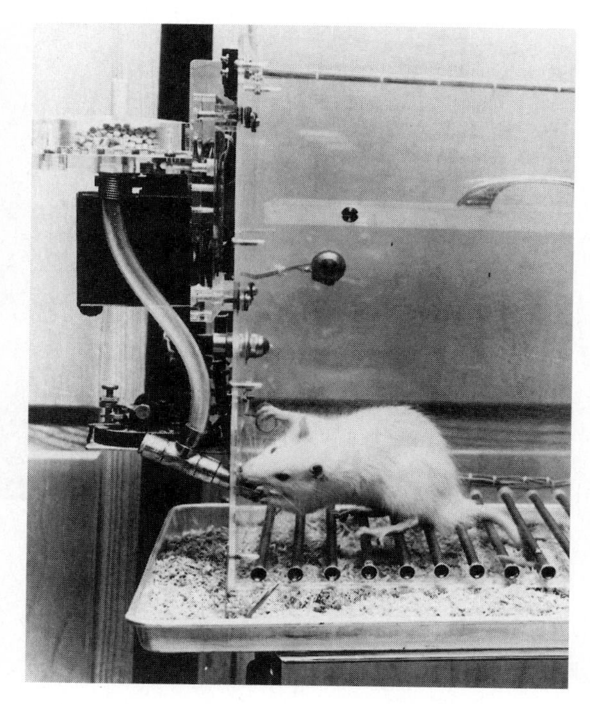

38.26 Trial-and-error learning in the laboratory

A pellet of food is dispensed from the apparatus when the rat presses the bar in response to the correct stimulus. This behavior is shaped by rewarding ever-closer approximations of the desired performance. The rat may be fed at first for being in the correct end of the box, and then only for accidentally touching the bar. Then the reward threshold may be raised to require actually pressing it. Finally the task of pressing in response to a particular stimulus is added.

Knowing what to learn: conditioning biases Most animals have strong species-specific "biases" that channel their learning along paths that are adaptive for them. Chicks, for instance, can learn to associate a particular sound with an impending shock and perform certain sorts of behavior (running or wing flapping, but not pecking) to avoid it. They also readily learn to associate color cues with a food reward and to peck (but not flap) in response. However, they are virtually unable to associate sound with food, or color with an impending shock. For birds, color is a better predictor of food than sound is, and pecking is a more appropriate behavioral "experiment" than flying or running away. Sound is a better predictor of the threat of predation, and flight is the appropriate response.

Animals, then, can be instinctively predisposed to recognize specific stimuli and to try specific sorts of behaviors in specific contexts. Clearly these biases reflect the cues and behaviors most likely to prove useful under natural conditions. Thus instinct can help focus and guide learning, so that animals can modify their behavior quickly and adaptively. Such examples demonstrate the oversimplification involved in dividing behavior into "instinctive" and "learned."

■ IS LEARNING PROGRAMMED?

The adaptive innate biases in classical conditioning and trial-and-error learning help explain how, through natural selection, specialized learning programs have evolved.

Parental imprinting Konrad Lorenz was fascinated by the phenomenon he subsequently labelled imprinting. Precocial young

A

B

38.27 Imprinted goslings

(A) Young goslings stay close to their own parent, whom they readily distinguish from the other geese. (B) Having been imprinted on Lorenz during their first day of life, these goslings follow him as if he were their parent.

birds—those able to walk from the moment of birth—must follow their parents if they are to survive, and must therefore be able to recognize them. Imagine a newly hatched gosling faced with the task of correctly identifying its parents. From birth, its visual field is full of an enormous variety of objects in the environment, and yet it identifies its parents and follows them; when offered a choice later, it is able to distinguish its parents from all other geese (Fig. 38.27A). Lorenz found that he could induce goslings to follow him instead if he removed the parents from view during the first day after hatching, and then walked away from the young birds while producing the appropriate species-specific call (Fig. 38.27B). The same trick, however, would not work if he waited until the third day.

The process by which the goslings follow their parents and memorize enough about them to ensure future recognition is now known as ***parental imprinting***. Lorenz discovered that parental imprinting is unlike conventional conditioning in several ways: (1) it involves a ***critical period***, or sensitive phase, during which the learning must take place (Fig. 38.28); (2) it involves neither reward nor punishment; and (3) it is irreversible.

Parental imprinting is seen in mammals as well as birds. The set of cues for parental recognition seems to be species-specific: auditory, visual, or (as is most often the case in mammals) olfactory. As with precocial birds, the critical period is normally early and brief, and the learning is generally not reversible.

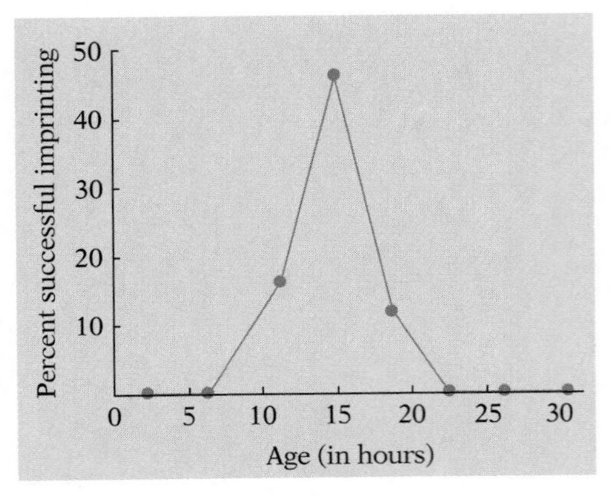

38.28 Sensitive phase for imprinting ducklings

Individual ducklings of various ages (in hours) were exposed to a moving decoy for 1 hour. They were then tested to see whether they had become imprinted on that object. The results showed that imprintability was at its peak when the birds were 15 hours old. Exposure to a suitable object too early or too late was ineffective.

Sexual imprinting The other sort of imprinting Lorenz explored is ***sexual imprinting***, the process by which many animals learn to recognize their species and, in many cases, their close relatives. Though most animals can identify reproductively ready members of the opposite sex of their own species by means of innately recognized cues, there are circumstances, particularly among birds and mammals, in which the animals require more detailed information than is provided by instinct alone. In the North Atlantic, for example, four closely related species of gull often nest together, but rarely interbreed. The only reliable morphological difference between them seems to be the color of the iris and of the fleshy ring that encircles the eye. By exchanging eggs of some birds between species and by painting the eye rings of others, N. G. Smith showed that the offspring use the eye color of their parents as a behavioral cue: when they have become adults, females choose mates whose eye ring and iris colors match those of the birds that raised them, whether natural or foster parents, and males will only copulate with females with the color combination they saw as young birds. The chicks, then, imprint on the parental eye color and later use it, in addition to other cues such as calls and postures common to gulls in general, in selecting a mate.

Other kinds of imprinting Many parasitic birds utilize imprinting. Hatched in strange nests by parents of another species, European cuckoos, for instance, memorize the songs of their host species and later use this knowledge to locate suitable hosts for their own young. Identifying the correct host is essential, for each cuckoo female lays eggs that pass for those of only one host species. A mismatch means her offspring may not survive, since many hosts eject eggs that do not resemble their own or, failing that, abandon the nest.

Many host birds imprint on their eggs. Species that live in close quarters, such as guillemots, memorize their own eggs so well that after a few hours even eggs of conspecifics are rejected. Similarly orioles, which are parasitized by cowbirds, will accept and imprint on a cowbird egg introduced *before* the host finishes laying her own. After her own eggs are laid, the oriole's imprinting program ceases, and she rejects any interlopers.

Another well-studied example of innately guided learning involves the way birds learn their species' songs. As Peter Marler has shown, most songbirds instinctively recognize only certain elements of their species' song. The elements they recognize function as acoustic releasers to trigger a detailed memorization of the father bird's song during a critical period. The sight of a singing bird near the nest is another releaser; in nature, the two occur together. As a result, a juvenile white-crowned sparrow memorizes only a white-crowned sparrow song. A bird that hears no song during the critical period, or only the songs of other species, learns nothing, and later produces a song that contains only the basic innate elements.

Various other types of imprinting occur. Female ducks, for instance, imprint on nest height on their second day of life, and subsequently build their own nests high or low as a consequence. Mice imprint on certain features of their birthplaces, and when they later disperse and choose their own home ranges, they pick those with similar features. This behavior is adaptive because it biases the animal toward areas that resemble the one that allowed it to survive, and are therefore likely to allow its progeny to survive. Salmon imprint on the odor of their home stream on the day they begin their journey to the sea, and use that memory years later on their way upstream from the ocean to track the tiny tributary in

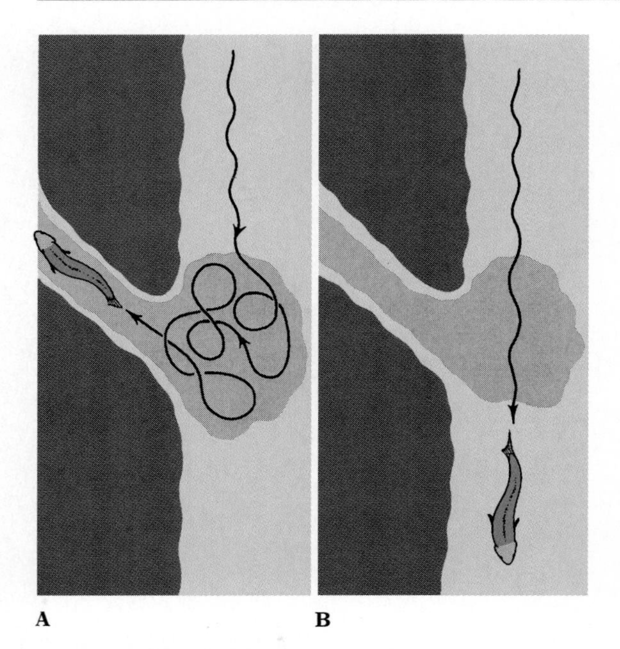

38.29 Tracks of salmon when they encounter morpholine

In one series of experiments (not discussed in the text), Arthur D. Hasler and his associates imprinted young salmon in the laboratory on the odor of a chemical called morpholine. Later, they used ultrasonic tracking to follow the adult salmon as they swam southward along the shore of Lake Michigan in search of a spawning stream. Morpholine was released in the area indicated in gray. (A) Fish previously imprinted on morpholine stopped their southward migration there, began to circle, and swam up the morpholine-scented stream. (B) Those not previously imprinted on the chemical typically swam through the morpholine-scented area without pausing.

which they were born (Fig. 38.29). Homing pigeons imprint on the location of their home loft as fledglings, and will return to it even after years of life in a cage hundreds of kilometers away. If the conditions are stable—that is, if the context, timing, and general sorts of cues that will be useful in the natural world are predictable—then an onboard set of instructions that channels learning into particular paths will be adaptive, and likely to evolve.

■ CULTURAL LEARNING

What makes our species unique is in large part our ability to pass on information from individual to individual and from generation to generation, thereby saving others the risky and time-consuming exercise of rediscovering by trial and error the lessons of the past. Our culture, then, is cumulative; each generation stands on the shoulders of the previous ones.

Among some other animals, the ability to pass novel and useful information from generation to generation has also evolved. Some of these new data are acquired initially through classical conditioning, some through trial-and-error learning, and some through a combination.

Learning about food Most specialist feeders (animals with diets confined to a limited number of substances) have innate mechanisms for recognizing their particular foods. Generalists, with more catholic tastes, are most often equipped with a mixture of innate guidance and a capacity to learn which foods are edible, and how best to handle them. The most thoroughly understood examples come from experimentally convenient ground-dwelling birds, such as jungle fowl, grouse, and domestic chickens. Work by Eckhard Hess has shown that chicks are born with strong innate preferences with regard to the size, shape, and color of food. Under natural conditions, however, the mother hen influences them by pecking at food, picking it up, dropping it, and uttering an innately recognized food call; these behaviors attract the chicks' attention, causing them to peck at whatever she has. The chicks soon learn from experience what is and is not edible. The strength of the bias introduced by the parent has been shown by experiments in which a mixture of grain dyed orange and green was offered to the chicks. At the same time they saw through a transparent barrier a hen trained to avoid one of the colors, or a crude model hen that pecked at only one color. Though the grain of each color was equally edible, the chicks developed a strong preference for the color selected by the hen.

Such experiments indicate that along with their innate preferences for certain foods, chicks have an innate predisposition to copy the food-selection behavior of the parent, with the result that knowledge is passed from generation to generation. Inborn predispositions and learning through experience act in parallel in the chicks for the transfer of cultural information, just as they do when a bird's innate song is modified and elaborated by learning. Virtually the same pattern is seen in a variety of birds and mammals, and is of special importance among primates (Fig. 38.30). It is a strategy that allows animals to discover novel food sources.

38.30 Termite hunting in chimpanzees

Tool using and social learning are crucial elements in the chimpanzee social system. Here an adult has selected and pruned a stick, which it uses to "fish" for ants or termites. Young chimpanzees observe their elders using this process, thus gaining cultural information on how to obtain food.

Recognizing enemies In recognizing their potential enemies, many animals display knowledge obtained through learning and cultural transmission. In at least some instances the learning is innately guided. Birds, for example, generally face two classes of threat: nest predators, such as owls, crows, and snakes, which attack eggs or chicks; and adult predators, such as hawks and cats, which capture and kill grown birds. The behavior in the face of these two types of enemies is quite different: most of the nesting birds in an area mob a nest predator (attack it en masse), but the same birds hide when a potential adult predator is seen. Birds appear to recognize a few animals of each class innately, but for the most part they must learn who their enemies are. This is accomplished without the young birds ever having to be subjected to life-threatening attacks directly.

By means of a clever series of experiments on European blackbirds, Eberhard Curio discovered the programming that underlies enemy recognition. Curio placed two cages of blackbirds on opposite sides of a hallway, in sight of each other. Between the two cages he installed a four-compartment box that allowed the occupants of each cage to see an object on their side, but not on the other side (Fig. 38.31). Curio then presented a stuffed owl to the birds in one cage and a harmless and unfamiliar bird, an Australian honey guide, to those in the other. The birds that saw the owl began at once to deliver the mobbing call and attempted to attack the stuffed figure through the wire of the cage. The birds on the other side, seeing only the honey guide (which unconditioned blackbirds ignore), and seeing and hearing the mobbing birds, began attempting to mob the honey guide (Fig. 38.31C).

The trained birds then passed on the practice of mobbing honey guides to other birds, and these passed it on to still others. Curio saw this mindless aversion for a creature that had never harmed a single blackbird transmitted through six generations in the laboratory. He was able to repeat this same blind but adaptive encultura-

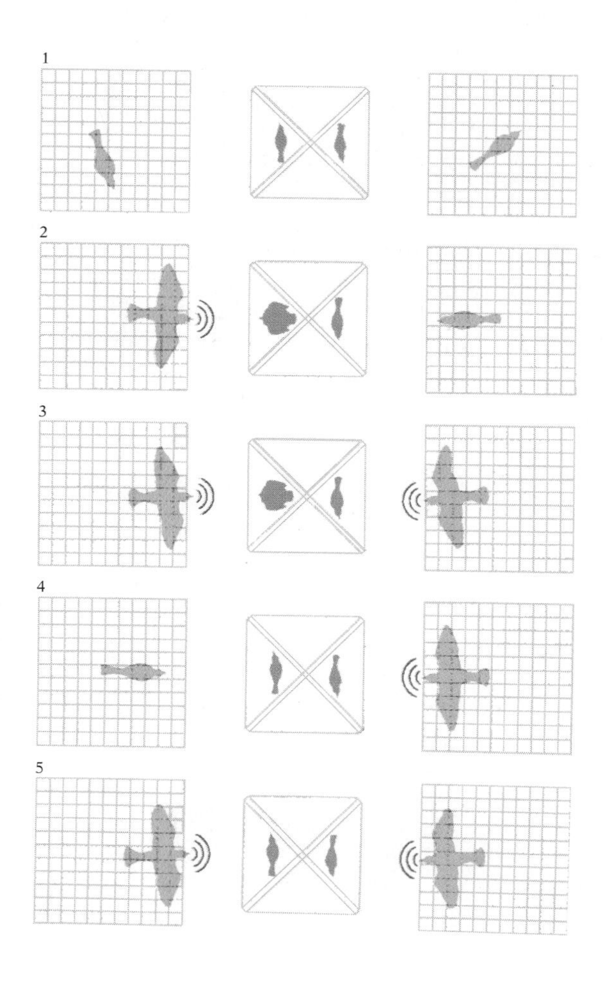

38.31 Enemy learning

Objects were presented to caged birds by means of a box (center column) that prevented the occupants of one cage (left column) from seeing the object viewed by the birds in the other cage (right column). A stuffed honey guide (1) elicited no reaction, but when one group of birds was shown a stuffed owl and attempted to mob it (2), the action caused the birds in the other cage to start attacking the honey guide (3). This latter group was then able to pass the aversion to the honey guide on to the occupants of the other cage (4, 5).

tion with a plastic bottle of laundry detergent as the object of official hatred. The innately recognized mobbing call is a sign stimulus for this piece of classical conditioning: when a bird hears the mobbing call, it automatically identifies the object of the call as an enemy. The warning signals of various other animals—the trumpeting of elephants, for instance—serve the same function, both alerting the experienced and teaching the young about danger before they come to harm.

HOW BIRDS NAVIGATE

Of all the astonishing feats resulting from the interplay of sign stimuli, motor programs, drives, and innately guided learning, perhaps none is more impressive than the ability of many creatures to find their way over great distances through unfamiliar territory. Animals as diverse as butterflies, sea turtles, and hummingbirds migrate thousands of miles to places they may never before have visited.

Among birds, two separate navigational strategies are evident. In one, the creatures are preprogrammed to fly a certain course. Wolfgang Wiltschko has shown that in the fall, garden warblers from northern Germany will use an internal compass to fly (or, in orientation cages, attempt to fly) southwest for several weeks, and then southeast for several more—following a course that, in the wild, carries them down through Spain, across Gibraltar, and into their winter ranges in Africa (Fig. 38.32). Timothy and Janet Williams discovered that many small birds in North America migrate to the East Coast, wait for a low-pressure front, and then fly southeast. In general, this course results in their catching winds that carry them to South America, though if the winds fail, they perish at sea by the millions.

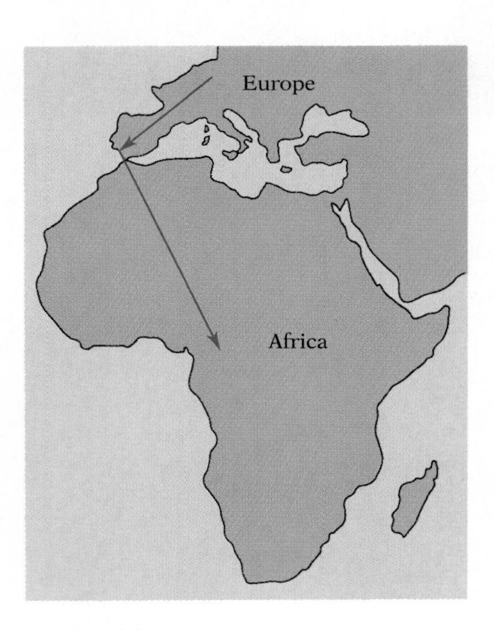

38.32 Warbler migration

Some European garden warblers reach their winter grounds after a two-leg journey. They know at birth the two flight bearings they need, and how long to fly in each direction.

This compass-and-timer strategy is good enough when the target is a continent, but will hardly serve when the goal is small. Many migrating animals need to know precisely where they are even when in an area for the first time. This need is filled by a mysterious but very real ability known as the ***map sense***, which represents something quite different from a mental map of a familiar area. An animal with a map sense behaves as though always aware of longitude and latitude. The nature of this map sense is one of the most intriguing mysteries in modern biology.

Though many migrating birds have a map sense, their journeys only twice a year make experimentation difficult. The animal many researchers prefer to work with, therefore, is the homing pigeon. A good homer can be taken from its loft and transported hundreds or even thousands of kilometers in total darkness (even anesthetized), and when released it will circle briefly and then fly off roughly in the direction of home (Fig. 38.33). To do this, homing pigeons use both a compass sense and a map sense.

The map sense can be surprisingly precise: Klaus Schmidt-Koenig and Charles Walcott released pigeons wearing translucent lenses that prevented them from perceiving shapes. These birds were able nevertheless to navigate to within a few kilometers of home (Fig. 38.34). Some researchers believe the map depends on

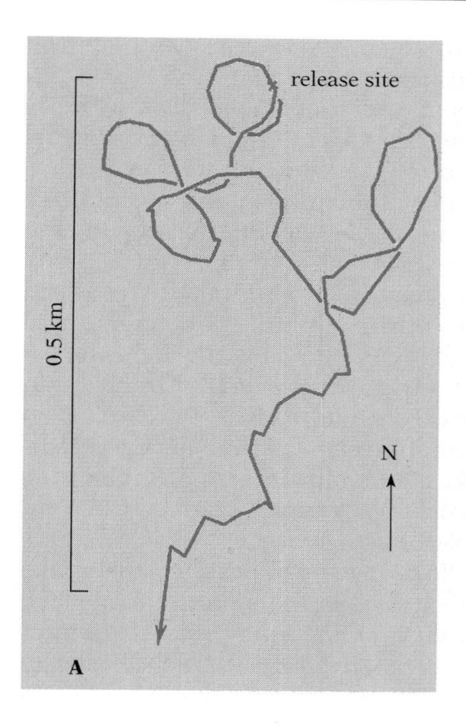

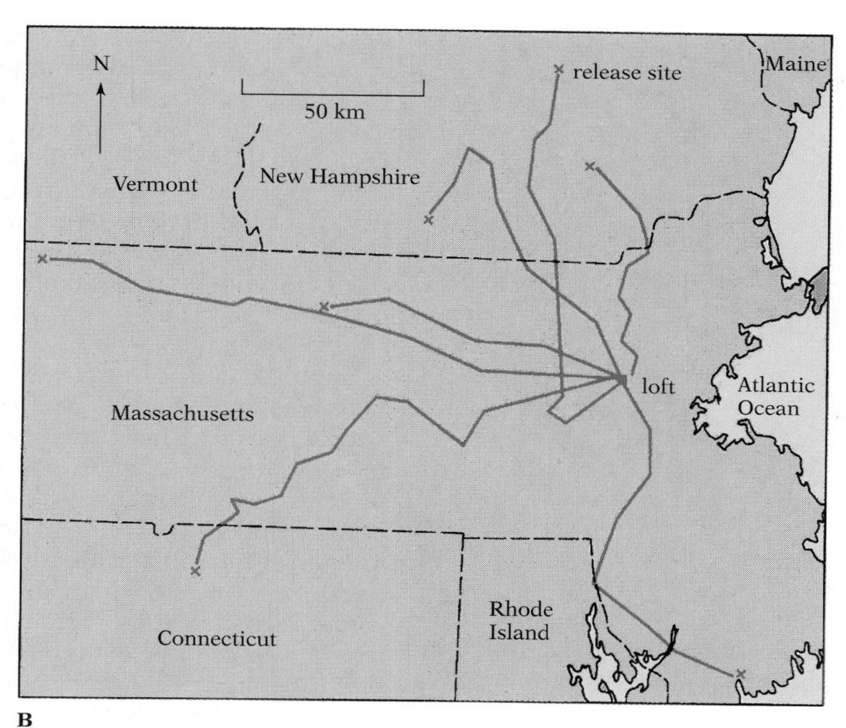

38.33 Pigeon homing

(A) Pigeons usually begin their journey home by circling the release site, but then they quickly set off along an irregular course for home. (B) The actual routes are rarely straight and direct, indicating that new map measurements must be taken from time to time to make midcourse corrections.

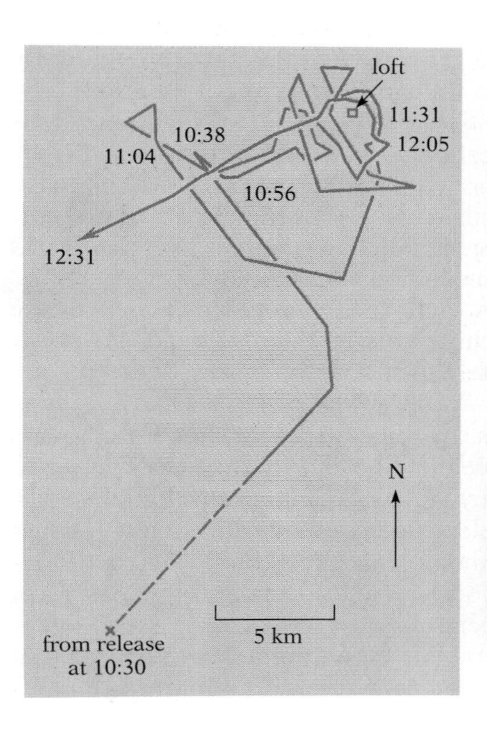

38.34 Flight of a pigeon that cannot see shapes

Pigeons that have arrived in the vicinity of home use visual landmarks to locate the loft. Those wearing translucent lenses (right) cannot see shapes, but nevertheless know when they are in the vicinity of home and fly wide circles nearby. The track of such a pigeon is shown (somewhat abridged) in a representative example here (left); it indicates a map sense accurate to within a few kilometers.

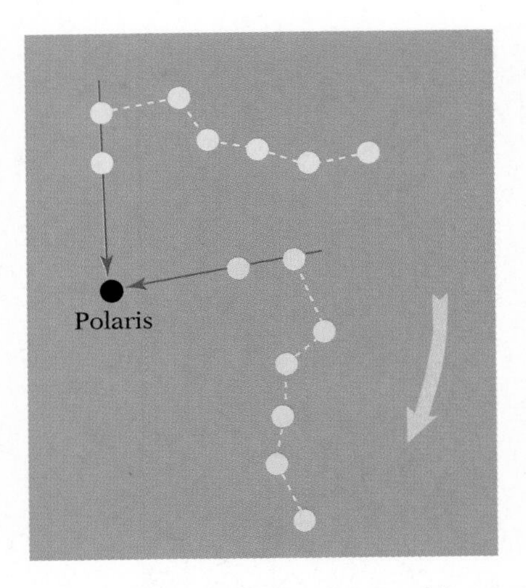

38.35 How north can be located by star patterns

Red arrows through the two end stars in the cup of the Big Dipper (Ursa Major) point toward Polaris, the North Star. Though the position of the constellation changes during the night, the same stars always determine an arrow pointing toward Polaris; hence directions can be determined without need of time compensation. Many different star patterns can be used for finding direction in this way.

olfactory cues, while others have concluded it relies on magnetic stimuli. Neither hypothesis explains all the data.

The compass sense in pigeons and migratory birds is now fairly well understood. Like many insects, pigeons and other diurnal birds use the position of the sun (or correlated patterns of polarized light) as their standard cue. Of course, the sun's position depends on the time of day, and birds have an internal time sense enabling them to allow for the westward movement of the sun from morning to night. Similarly, nocturnal migrants use a learned picture of the stars to set their course. The roles of the sun and stars in avian navigation are demonstrated by the behavior of caged birds under artificial skies: during migration season, the animals display an intense desire to escape in the direction their wild conspecifics are flying. A sudden shift of the artificial sun or pattern of stars results in an immediate compensatory change in the direction in which the caged birds are struggling to go.

Experiments by Stephen Emlen show that nocturnal migrants memorize the constellations while they are still nestlings, using the North Star, around which all other stars in the night sky appear to rotate, as their point of reference. As a result, they are able to infer north from even a small patch of sky (Fig. 38.35). Such birds can be raised under an artificial sky, with an arbitrary pattern of stars rotating about a pole at any chosen compass point. When the time to migrate arrives, they attempt to set off in the appropriate direction relative to the star patterns they observed during their infancy.

Homing pigeons demonstrate their use of the sun compass in an equally dramatic way. Correctly interpreting the sun's direction depends on an internal timer, which is sensitive to manipulations of the day/night cycle. For instance, a pigeon kept in a room whose lights go on, and later off, 6 hours early—on at midnight and off at noon—will misinterpret the sun's position accordingly. When such a bird is released at true noon, its internal clock reads 6 P.M. It sees the sun in the south, but because it has been clock-shifted, it interprets the sun's position as indicating west. Therefore, if its home is to the south, it will fly 90° to the left of the sun; attempting to fly south, it heads east (Fig. 38.36).

Pigeons, however, can also home under an overcast sky. If they use the sun as their compass, what guides them when it is invisible? William Keeton attacked this question by releasing both normal and clock-shifted birds on cloudy days (Fig. 38.37). The results are clear and dramatic: pigeons are able to home on overcast days, and they are then using cues that are not time-dependent, for the departure bearings of clock-shifted birds are not rotated. Obviously pigeons have a backup system, and Keeton guessed the second compass might be magnetic.

The clearest demonstration that pigeons use a magnetic compass was performed by Walcott. He fitted birds with tiny, head-mounted coils of wire. By passing a current from a battery through the coils, he was able to reverse the field sensed in the head. On cloudy days these birds flew away from home, while pigeons whose batteries were not connected to the coils homed normally (Fig. 38.38). On sunny days there was no effect.

The physiological and neural bases of animal behavior are fascinating topics of research, integrating several disciplines. A corre-

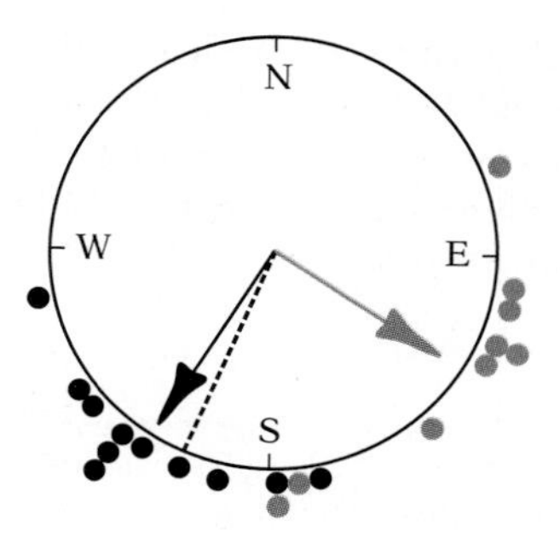

38.36 Effect of a 6-hour-fast clock shift on the initial bearings chosen by homing pigeons on a sunny day

Each dot indicates the bearing chosen by one bird; black dots represent control birds, color dots experimental birds. The dashed line marks the proper homeward direction. The arrows show the mean bearing (average direction) for each group. The length of each arrow is proportional to the degree of clustering of the dots: if all the birds flew off in the same direction, the resulting arrow would touch the circle; if the departure directions were widely scattered, the arrow would be very short. In this experiment the mean bearing of the clock-shifted birds (color arrow) is about 90° to the left of the mean bearing of the control birds (black arrow). The experimental birds have been clock-shifted a quarter of a day, and they have made an error of a quarter of a circle in reading the sun compass.

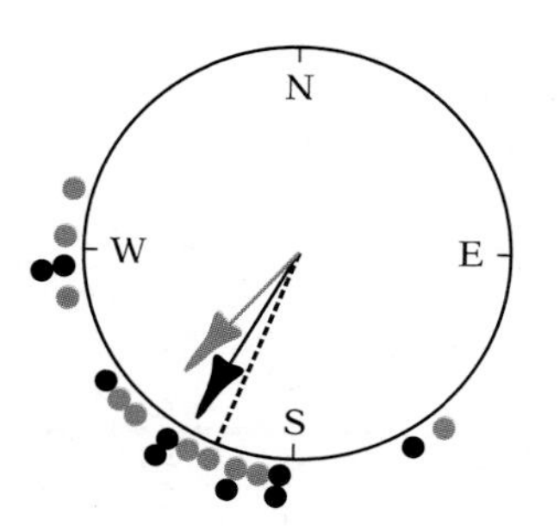

38.37 Effect of a 6-hr-fast clock shift on the initial bearings chosen by homing pigeons on a totally overcast day

When the sun is not visible, the experimental birds choose bearings (color dots) not significantly different from those of control birds (black dots); the 90° deflection of their bearings seen on sunny days (compare with Fig. 38.36) is not evident. In the absence of the sun compass, the birds appear to orient by some other system that does not require time compensation.

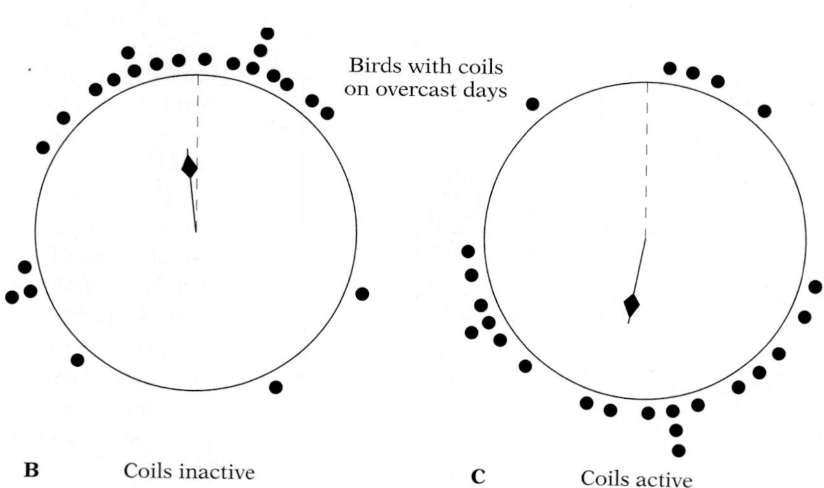

38.38 Effect of magnetic coils on pigeon homing

On overcast days, birds wearing active coils (A) that reversed the direction of the earth's field in their heads flew away from home (C), while those with inactive coils behaved normally (B). On sunny days there was no effect.

lated and equally rewarding endeavor is to try to understand the ecological factors that have shaped behavioral evolution, generating the patterns of social behavior and species interactions that create much of the remarkable diversity in nature. The ecological bases of behavior are discussed in Part VI.

CHAPTER SUMMARY

THE BUILDING BLOCKS OF BEHAVIOR

HOW DO ANIMALS SEE THE WORLD? Some species can sense stimuli of which we are unaware. The nervous system isolates species-specific sign stimuli that are used to guide or release behavior. Sign-stimulus recognition seems to depend on neural feature detectors. (p. 1089)

HOW RESPONSES ARE ORGANIZED Innate behavior usually consists of prewired motor programs, which may not mature until after birth. Many instances of complex behavior are organized as separate steps; each step must be completed before the next is initiated. (p. 1093)

WHAT MOTIVATES ANIMALS? Different behavioral alternatives have individual priorities, or drives, that depend on an organism's current needs. Some of these needs are anticipated by internal timers, which allow an animal to prepare in advance for an approaching contingency. (p. 1096)

HOW ANIMALS COMMUNICATE

GETTING THE MESSAGE THROUGH Most communication involves advertizing species identity and sex for mating. Species specificity is essential, and often depends on multiple cues transmitted either simultaneously or, more often, in a particular sequence. (p. 1099)

THE CHALLENGES OF SOCIAL COMMUNICATION Social species need an additional vocabulary to effect the coordination essential for group living. The most complex system known other than human speech is the dance language of honey bees, which communicates the location and quality of food. (p. 1102)

HOW ANIMALS LEARN

THE PROCESS OF CONDITIONING Classical conditioning involves substituting a reliable learned cue for a sign stimulus; it depends on recognizing a correlation between a conditioning stimulus and the innately recognized unconditioned stimulus. Operant conditioning involves learning a novel motor program (often assembling innate movements into a useful behavior pattern) based on trial-and-error experimentation. Both sorts of conditioning involve species-specific biases that usually speed learning by focussing an animal's attention on plausible cues or response elements. (p. 1105)

IS LEARNING PROGRAMMED? Some learning is largely innate. Imprinting, for example, proceeds at a specific time in response to predictable cues in the absence of reward. It can serve to guide an animal to memorize the distinctive features of its parents, its species, its birth area, or its mating signals. (p. 1107)

CULTURAL LEARNING Social learning can serve to accelerate the acquisition of information in groups by allowing older individuals to pass on knowledge to younger members of the group. Most cases appear to depend on innately recognized learning signals. (p. 1110)

HOW BIRDS NAVIGATE

In order to navigate in unfamiliar surroundings, birds need a compass and information about which direction to fly in. Inmost species the compass is celestial: the sun during the day or the stars at night; many have a magnetic compass that they use under overcast skies. In some species the information about which direction to fly comes from a map sense that tells them where they are relative to their goal. (p. 1112)

STUDY QUESTIONS

1 Compare and contrast the costs and benefits of heterogeneous summation versus a sequential series of releasers as alternative strategies for species and sex recognition in courtship. (pp. 1089–1093, 1099–1102)

2 Why might it be advantageous for migratory birds to have annual rhythms? (p. 1112)

3 Pavlov believed that the unconditioned stimulus in his experiments was the sight of food (meat powder in some cases). Is this likely? If not, what was probably going on, and what does it tell us about the potential power of classical conditioning? (pp. 1105–1106)

4 Mallard ducklings imprint on their mother. (The father, after helping to build the nest and incubate, leaves before hatching.) What does this mean for any possible sexual imprinting by sons versus daughters? In the laboratory the imprinting object is often a model of a male mallard; how might the imprinting experience affect the future mate choices of sons and daughters? (pp. 1107–1109)

5 Given that many animals learn about enemies using the blackbird system described in the text, what could be done to make species in wildlife parks that now flee when visitors approach less afraid of humans? (pp. 1111–1112)

SUGGESTED READING

GOULD, J. L., 1982. *Ethology: The Mechanisms and Evolution of Behavior*. W. W. Norton, New York. *An introductory textbook on animal behavior*.

GOULD, J. L., AND C. G. GOULD, 1988. *The Honey Bee*. W. H. Freeman, New York. *Well-illustrated descriptions of the honey bee's sensory world, communication, navigation, and learning*.

GOULD, J. L., AND C. G. GOULD, 1994. *The Animal Mind*. W. H. Freeman, New York. *On the interacting roles of instinct, learning, and insight in animal behavior.*

GOULD, J. L., AND P. MARLER, 1987. Learning by instinct, *Scientific American* 256 (1). *On innately guided learning*.

GWINNER, E., 1986. Internal rhythms in bird migration, *Scientific American* 254 (4). *On the role of annual rhythms and internal timers in guiding warbler migration*.

PART VI

ECOLOGY

◀ **Threatened reservoirs of diversity,** the earth's once-vast tropical rain forests are shrinking at an accelerating rate due to human activities, consigning countless species to extinction. This photograph, made in the Monteverde Cloud Forest Preserve in Costa Rica, shows the high canopy and vertical stratification characteristic of such forests. Epiphytic plants (plants that grow on other plants) are particularly abundant in this view.

CHAPTER 39

ECOLOGY OF POPULATIONS AND COMMUNITIES

Ecology is the study of interactions between organisms and their environments. Environment embraces everything outside the organism that impinges on it, including light, temperature, rainfall, humidity, and topography, as well as parasites, predators, mates, and competitors. An organism's response to this multifaceted environment depends on its anatomy, physiology, and behavioral repertoire; for instance, a species' enzymes and organ systems are at least as important as its predators in determining how individuals will be distributed over the range of potential habitats.

The interaction between an organism's biochemistry or general physiology and its environment is called ***physiological ecology***. There are three higher levels of ecological interactions: individuals interacting with other members of their species in a local ***population;*** the interactions of a population with other populations living in a local ***community;*** and interactions at a higher level still—the ***ecosystem***—which consists of a region's communities and their physical environments considered together. Each of these designations may be applied to a small local entity or to a large widespread one. Thus the sycamore trees in an isolated patch of forest may be regarded as a population, and so may all the sycamore trees in the eastern United States. Similarly, a small pond and its community of inhabitants, or the forest in which the pond is located, may be treated as an ecosystem.

The various ecosystems are linked to one another by biological, chemical, and physical processes. Inputs and outputs of energy, gases, inorganic chemicals, and organic compounds can cross ecosystem boundaries via wind and rain, in running water, and through the movement of animals and the dispersal of pollen and seeds. Thus the entire earth is itself an ecosystem, in that no part is fully isolated from the rest. The global ecosystem is ordinarily called the ***biosphere***.

The biosphere contains all living organisms and their environments. It forms a relatively thin shell around the earth, extending only a few kilometers above and below sea level. Except for energy, it is self-sufficient; all other requirements for life, such as water, oxygen, and nutrients, are supplied by utilization and recycling of materials already contained within the system.

In this chapter we will examine populations, including how they grow and regulate their size, and how they interact with other populations. In Chapter 40 we look at ecosystems and how they interact with communities.

HOW POPULATIONS ARE ORGANIZED AND GROW

We will look first at populations—groups of individuals belonging to the same species or to the same local subdivision of a species—and especially at the dynamics of populations and the environmental factors that help regulate them.

■ THE IMPORTANCE OF DENSITY AND DISTRIBUTION

Ecologists sometimes need to know the number of individuals in a population. In aiding an endangered species, for example, the size of the surviving population is a crucial factor in the design of proper management. Thus conservationists must have a good estimate of just how many whooping cranes or blue whales still exist.

More often ecologists are concerned not with the total numbers in a species, but with the density of the population in a given region—the number of individuals per unit area or volume (50 pine trees per hectare, for instance, or 5000 diatoms per liter of water). In some situations, especially when the size of the individuals in a population is extremely variable, ecologists find that biomass—the total weight of all the individuals per unit area or volume (or its energy equivalent in calories)—is a more useful index to the population's importance in the ecosystem.

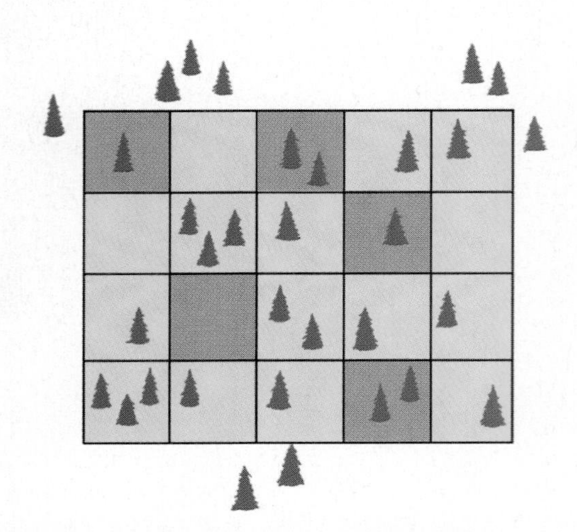

39.1 The quadrat sampling method

A grid is set up in the study area, and counts made in a sampling of squares. The results permit an estimate of the population of the entire study area. The five squares sampled here (brown) contain a total of six trees; hence it is estimated that the 20 squares of the entire grid will contain approximately 24 trees. In this example the selection of squares was random, but in some situations, as when obvious gradients run through the study area, other sampling designs may be more suitable.

How population size and density are estimated Measuring population parameters is not always simple. Counting all the individuals in an area may work if the organisms are conspicuous and not too abundant. An alternative procedure is to divide the area into a limited number of small sampling blocks (quadrats), and then estimate the density for the whole study area from these (Fig. 39.1). To avoid large errors in estimating population density, the sampling blocks must be as representative of the entire study area as possible.

Another method sometimes used in estimating the size of animal populations is the mark-and-recapture technique. A limited number of individuals (let's say 20) are captured at random, marked with a tag or dye, and then released back into the same population. At some later time a second group of animals is captured at random from the population, and the percentage of marked individuals determined. If 10% of the animals in this second group are marked, the investigator may conclude that the original 20 marked individuals represented 10% of the population in the study area, and hence that the total population is about 200.

Spacing: how organisms are distributed Within any given area the individuals of a population can be distributed in a variety of ways, which form a continuum ranging from a perfectly uniform distribution, through randomness, to a situation in which all individuals are found in groups or clumps (Fig. 39.2). Uniform distributions are not common; they occur only where environmental conditions are fairly uniform throughout the area and where, in addition, there is intense competition or antagonism between individuals. For example, creosote bushes are often spaced almost uniformly over a desert area because the roots of each bush give off toxic substances that prevent germination of seedlings in a circular zone around the base of the bush. Animals can create a nearly uniform distribution by defending private territories (Fig. 39.3).

Random distributions are also rare. They occur only where the environmental conditions are uniform, where there is no intense competition or antagonism between individuals, and where there is no tendency for individuals to aggregate. The chances that all three of these conditions will be met simultaneously are low. Thus ecologists usually work with nonrandom distributions, which complicates their sampling procedures and statistical tests. Hypothetical random distributions nevertheless represent a useful reference point, from which deviations toward uniform or clumped distributions can be measured.

Clumping, or aggregation, is by far the most common distribution pattern for both plants and animals in nature. There are several reasons for this:

1 Environmental conditions are seldom uniform throughout even a relatively small area. Variations in soil, in topography, in the distribution of other species, and in such microclimatic factors as moisture, temperature, and light may produce important habitat differences within the area. Organisms tend to occur in those spots where conditions are most favorable.

2 Reproductive patterns often favor clumping. This is particularly true in plants that reproduce vegetatively (asexually) and in animals whose young remain with the parent.

3 Animals often exhibit behavior patterns that lead to congregations ranging from loose groups to organized colonies, schools, flocks, or herds (Fig. 39.4).

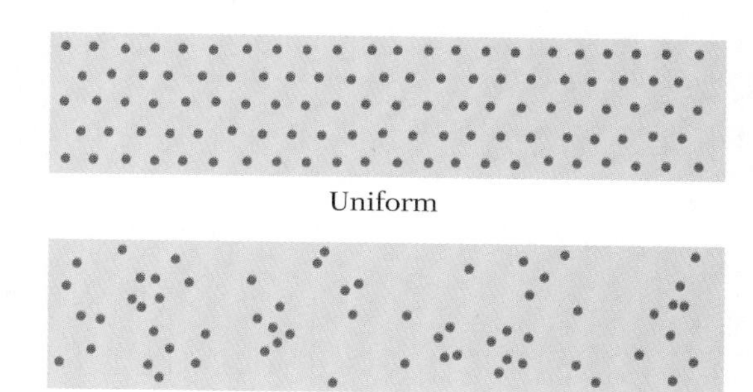

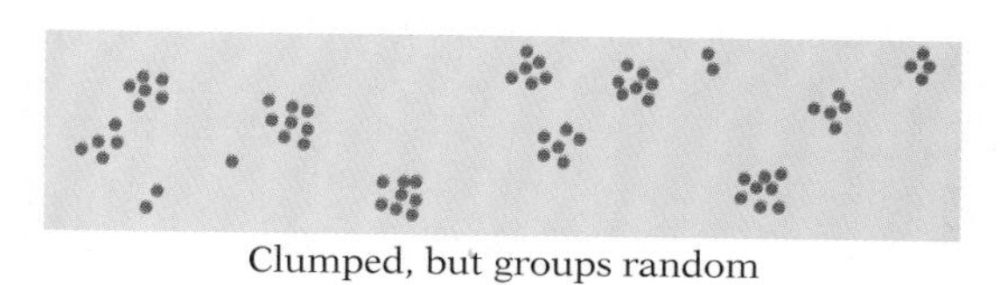

Clumped, but groups random

39.2 Uniform, random, and clumped distributions

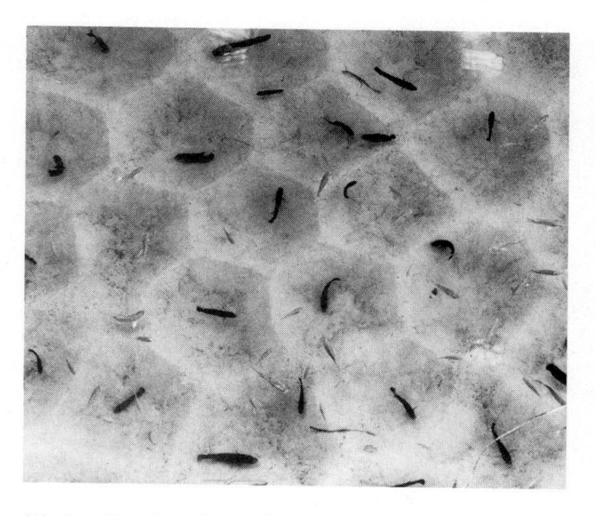

39.3 Territories of male cichlids

These male cichlids, known as black tilapias, establish clearly defined breeding territories in the bottom of their habitat. In this photograph, they have established breeding territories by digging in the sandy bottom of an artificial environment. In both cases, the result is an unusually uniform distribution of individuals.

39.4 A gannet colony

Like most gulls and other sea birds, gannets nest in dense colonies for mutual protection against predators. Since these birds hunt for food in the ocean, their territories are used only for nesting and rearing young.

A clumped distribution may increase competition for nutrients, food, space, or light, but this deleterious effect is often offset by some beneficial ones. For example, trees growing together in a hedgerow on the Great Plains may compete more intensely for nutrients and light than if they were widely separated, but because they shelter one another, they may be better able to withstand strong winds. Moreover, the clump, which has less surface area in proportion to mass than an isolated, exposed tree, may create its own more favorable microclimate. Within a clump of trees, for example, shading and protection from the wind help insulate against temperature extremes and reduce evaporation from the soil, fallen leaves are more likely to be trapped and help mulch and enrich the soil, and so on. Also, the existing trees may provide essential shelter that enables saplings on the periphery to grow, thus steadily enlarging the clump. A group of animals can have an advantage over isolated individuals in locating food and in withstanding attacks by predators. We will look in more detail at the costs and benefits of sociality in a later section.

■ HOW POPULATIONS GROW

Unrestrained Growth One way to understand the dynamics of real populations is to find out what to expect of a population under ideal conditions, and then to try to determine how actual conditions modify this expected pattern. Let's imagine a population with a stable age distribution; no predation, parasitism, or competition; no immigration or emigration; and living in an environment with unlimited resources. If a pair of house flies were to produce offspring, all of which survived long enough to produce a full complement of off-spring in their turn, and so on for a number of generations, one pair starting to breed in April could have almost 2×10^{20} (191,000,000,000,000,000,000) descendants by August—enough flies to cover the entire land surface of the earth to a depth of 10 cm. As discussed in Chapter 17, even elephants (the slowest-breeding animal known) would cover the planet in only a dozen centuries or so, if all survived to produce offspring.

All organisms, whether plant or animal, unicellular or multicellular, have the potential for explosive growth; in the absence of environmental limitations, their growth curve would be exponential (Fig. 39.5). If we define N as the number of individuals in the population, t the elapsed time, $\Delta N/\Delta t$ (where "Δ" means "change in") the rate of increase of the total number of individuals in the population, b the average birth rate per individual in the population, and d the average death rate per individual, then we can write the equation for population growth:

$$(1) \quad \frac{\Delta N}{\Delta T} = (b - d)N$$

An alternative way of expressing equation 1 is to define two separate times, t and $t+1$, and write the relationship as follows:[1]

[1] This form of the equation is favored by those familiar with the mathematics of iterated series.

$$(2) \quad N_{t+1} - N_t = (b - d)N_t$$

In either form of the equation, it should be clear that a population will grow only if the average birth rate exceeds the average death rate, so that the term $b - d$ is greater than zero (as in Fig. 39.5). Conversely, the population will decline if the birth rate is less than the death rate: $b - d < 0$. If births and deaths just balance out, the growth rate is zero: $b - d = 0$. It is the critical value of $b - d$, the difference between the birth and death rates, that determines whether a population will grow, be stable, or decline. This difference—the net rate of population change per individual at a given moment—is called r; in short, $b - d = r$. Thus we can rewrite equations (1) and (2) for exponential population change as follows:

$$(3) \quad \frac{\Delta N}{\Delta T} = rN \qquad \text{or} \qquad N_{t+1} - N_t = rN_t$$

In the hypothetical house-fly population living under unlimited environmental conditions, r is at its maximum for the species. With this minimum death rate and maximum birth rate, r is designated r_{max}; it represents the ***intrinsic rate of increase*** of the population. Clearly r_{max} varies among species; it is much larger for house flies than for elephants, for example.

In the exponential model (3), the rate of growth of the population as a whole is a function not just of r but also of N, the population size—in this context the number of individuals able to reproduce. Since N becomes larger with each successive generation, the rate of increase, $\Delta N/\Delta t$, also becomes larger with each generation. It is because of this accelerating rate of increase that the slope of the curve in Figure 39.5 becomes steeper and steeper.

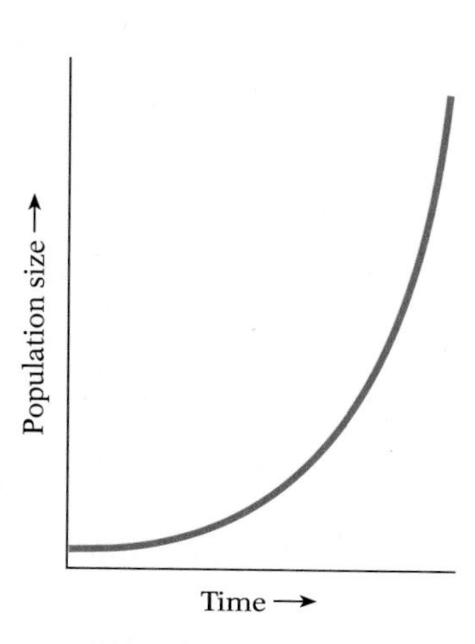

39.5 Exponential growth curve

The rate of increase accelerates steadily until, in theory, the population increases at an infinitely high rate. Clearly, no real population can continue increasing exponentially for long.

Growth to a limit No real population expands at infinite speed, or we would be buried in house flies and elephants. The exponential growth of real populations slows, stops, or declines. In many uncomplicated cases the actual growth curve shows initial rapid expansion of the population when it is at low densities, then decelerating growth at higher densities, and an eventual leveling off as the density approaches the ***carrying capacity*** (K) of the environment (Fig. 39.6). The carrying capacity is the maximum density of population that the environment can support over a sustained period.[2]

The population growth curve of Figure 39.6, called an S-shaped logistic curve, reflects a changing relationship between births and deaths. During the acceleration stage, births greatly exceed deaths. During the deceleration stage, the value of r is steadily falling (because the birth rate is declining, or the death rate is rising, or both), though it still is greater than zero. When the curve levels off, births and deaths are in balance and $r = 0$. However, for such changes in the relationship between births and deaths to occur—for the birth rate to decline (or the death rate to increase) as the population density rises—something must limit population growth, and this limitation becomes more severe as the density ap-

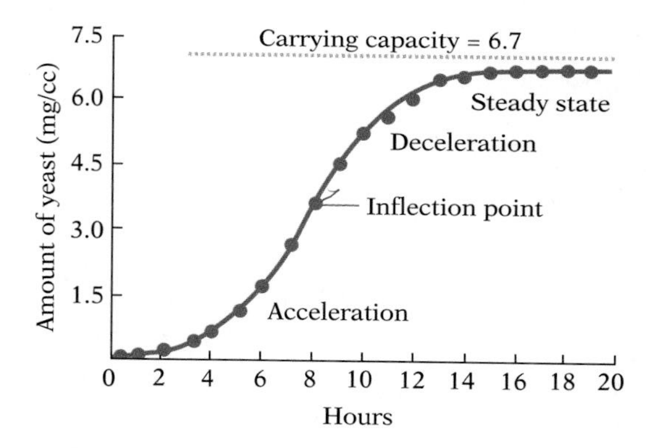

39.6 A representative logistic growth curve

Shown here is the growth curve of a laboratory population of yeast cells. Often called an S-shaped growth curve, the logistic curve exhibits an accelerating rate of growth at low densities, but eventually reaches an inflection point, where the rate of change shifts from acceleration to deceleration. The deceleration continues as the population density approaches the carrying capacity of the environment. When the carrying capacity is reached, there is no further increase in density, and the population continues in a steady state. This curve closely approximates the hypothetical logistic curve.

[2] More exactly, K is the maximum density of individuals that can coexist in a habitat; as we will see, antagonistic behavior often keeps the value of K below the density that a habitat could support.

Table 39.1 Growth of a population with r_{max} of 1.0, K of 100, and initial size N of 4[a]

$r_{max}\left(\frac{K-N_t}{K}\right)N_t$ POPULATION GROWTH	N_{t+1} NEW POPULATION	$t+1$ GENERATION
$1.0 \times \left(\frac{100-4.0}{100}\right) \times 4.0 = 3.8$	7.8	1
$1.0 \times \left(\frac{100-7.8}{100}\right) \times 7.8 = 7.3$	15.1	2
$1.0 \times \left(\frac{100-15.1}{100}\right) \times 15.1 = 12.8$	27.9	3
$1.0 \times \left(\frac{100-27.9}{100}\right) \times 27.9 = 20.1$	48.0	4
$1.0 \times \left(\frac{100-48.0}{100}\right) \times 48.0 = 24.9$	72.9	5
$1.0 \times \left(\frac{100-72.9}{100}\right) \times 72.9 = 19.8$	92.7	6
$1.0 \times \left(\frac{100-92.7}{100}\right) \times 92.7 = 6.8$	99.5	7
$1.0 \times \left(\frac{100-99.5}{100}\right) \times 99.5 = 0.5$	100	8

[a] Population totals assume that all individuals in the previous generation survive.

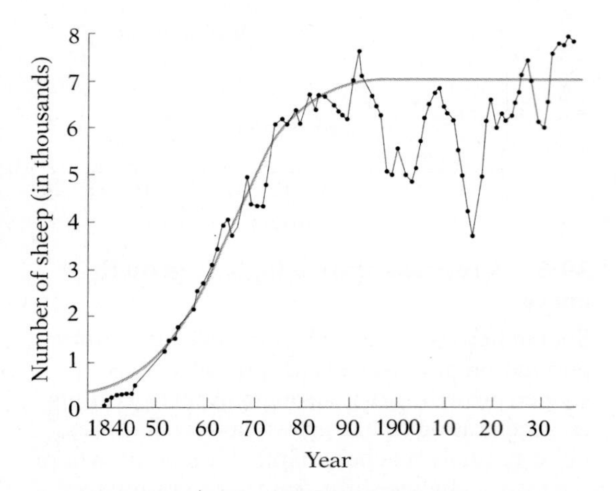

39.7 Growth curve of the sheep population introduced into South Australia

The smooth curve (color) is the hypothetical logistic curve about which the real curve fluctuates.

proaches the carrying capacity. We can express this ***density-dependent limitation*** as

$$(4) \quad \frac{K-N}{K}$$

This ratio represents the fraction of the total carrying capacity that remains to be filled. Inserting this limiting term into the two forms of the equation for maximum exponential growth (equation 3), we obtain

$$(5) \quad \frac{\Delta N}{\Delta t} = r_{max}\left(\frac{K-N}{K}\right)N \quad \text{or} \quad N_{t+1} - N_t = r_{max}\left(\frac{K-N_t}{K}\right)N_t$$

in which r_{max} and K are constants for a given species in a given environment; N, the population size, changes with the passage of time. Equation (5) is a formula for ***logistic growth:*** the growth is no longer a function solely of the intrinsic rate of increase, r_{max}, and the population size N, but also the ratio between $K - N$ and the carrying capacity K.

The S-shaped logistic growth curve derives its shape from the density-limiting term added in equation 5. At low population densities, where N is much smaller than the carrying capacity, the value of the limiting term is essentially K divided by K, or approximately 1; thus growth is primarily a function of $r_{max} \times N$ and increases almost exponentially. However, as growth continues, and N becomes larger and larger, the value of the limiting term declines steadily from 1 and acts as a brake on the rate of further growth (Table 39.1). Finally, when the carrying capacity is reached and N equals K, the value of the limiting term becomes zero and no further growth is possible:

$$(6) \quad \frac{\Delta N}{\Delta t} = r_{max}\left(\frac{K-N}{K}\right)N = r_{max} \times 0 \times N = 0$$

A population that has reached this steady-state level, in which births equal deaths, has ***zero population growth.***

The growth curves for some real populations approach the idealized logistic curve, but more often they fluctuate around the carrying capacity (Fig. 39.7). These variations around K usually represent some time lag between a change in density and its effect on the population. Note that whenever the population density fluctuates above the carrying capacity, so that N is greater than K, the limiting term $(K - N)/K$ becomes negative, and thus $\Delta N/\Delta t$ also becomes negative. This means that the population density will tend to decrease until it returns to the carrying capacity or below. In short, the limiting term provides feedback control, usually holding the population density near the steady-state level.

Harvesting populations Figure 39.6 shows us that the rate of increase of a population is greatest at the inflection point, rather than when the population density has reached its higher, steady-state level. The inflection point is the point of transition in the curve from an accelerating to a decelerating rate and in population studies is sometimes called the point of ***maximum sustainable yield.*** The density at which the inflection point occurs can be important in managing game animals and commercially valuable fish. If harvesting the organisms reduces their population density only to the point of maximum sustainable yield, there should be no lasting damage to the population. If, however, the population is overexploited, so that its density is reduced too far below this point, then the recovery of the population is slowed and perhaps endangered. In short, cropping resource populations down only to their point of maximum sustainable yield could theoretically be an optimum strategy in terms of both human benefit and perpetuation of the resource.

Unfortunately things are rarely quite this simple, though the same principles do apply. The problem is that the logistic growth curve (equation 5) says nothing about age and size distributions in the population. For example, in deciding on optimal cropping rates for resource populations, economic yield is critical. Though there may be a maximum yield in terms of numbers of animals caught if a fish population is cropped to the *K/2* level, economic yield may be greater if cropping is less severe, so that the fish will grow larger. In short, the point of optimal yield that managers of fish and game must strive to find is not necessarily the point of maximum sustainable yield.

Population crashes Not all growth curves of real populations are logistic. The population densities of many small short-lived organisms, or of organisms that live in disturbed or transient habitats, often crash abruptly (Fig. 39.8A). For example, a population of pea aphids in a field of alfalfa in spring may grow at an exponential rate if the weather is cool and moist, but if the weather then becomes hot and dry most of the aphids will die. This crash will occur when the weather changes even if the density of aphids is far below the carrying capacity of the alfalfa field. The weather exerts a ***density-independent limitation*** on the aphid population; that is, its effect does not depend significantly on the density of the aphids. In addition to the weather, density-independent limitations are created by sudden floods, fire, or physical disruption of the habitat such as clearing of a forest. Cycles and crashes can also be generated by the population-growth formula itself: a species with an r_{max} greater than 2.0 will overshoot K, then overcompensate and undershoot it, and so on; as K approaches 3.0, the curve becomes chaotic (Fig 39.8B)

Population profiles: mortality and survivorship Our discussion of growth curves so far has assumed that population growth depends only on birth and death rates. To understand a specific population, we must also consider the potential (maximum) life span, the average life expectancy, the average age of reproduction, and the age distribution of a population: most larger organisms, whether plants, animals, or fungi, do not reproduce at a steady rate. Instead, they live a significant part of their lives before beginning to reproduce, and after that their reproductive potential frequently varies with age—peaking and then declining in some species, such as ours, but increasing steadily in many trees and fish.

In attempting to understand the dynamics of a population, we often need to examine the mortality rates for the various age groups. Such data show what stages in the life cycle are most sus-

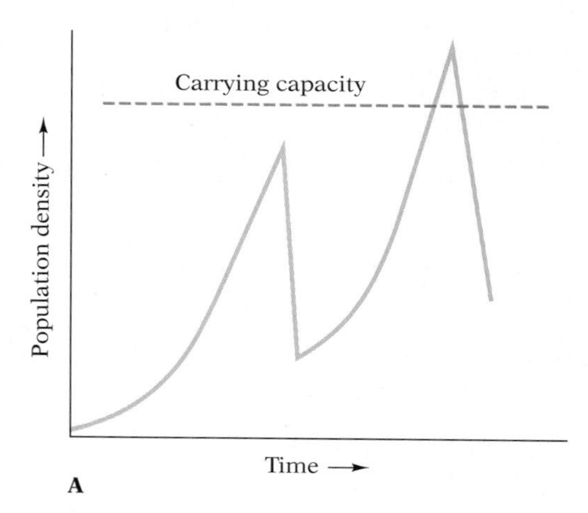

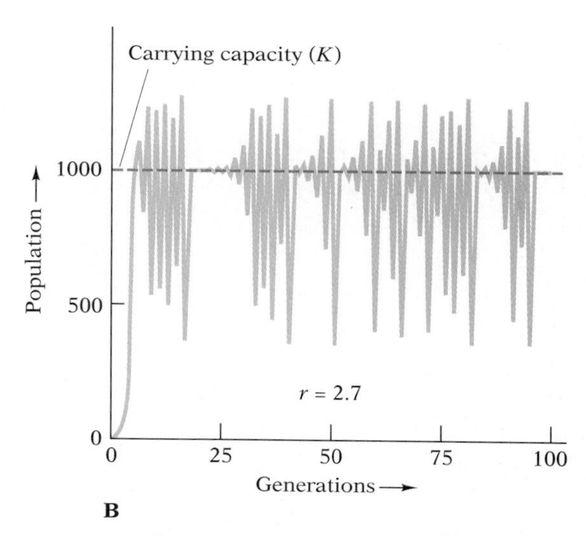

39.8 Unstable growth curves in which an exponentially growing population suddenly declines independent of its density

(A) This boom-and-bust growth curve of thrips is characteristic of populations of small organisms limited by density-independent factors such as weather. (B) When r_{max} is high, wildly fluctuating population sizes can occur independent of environmental conditions.

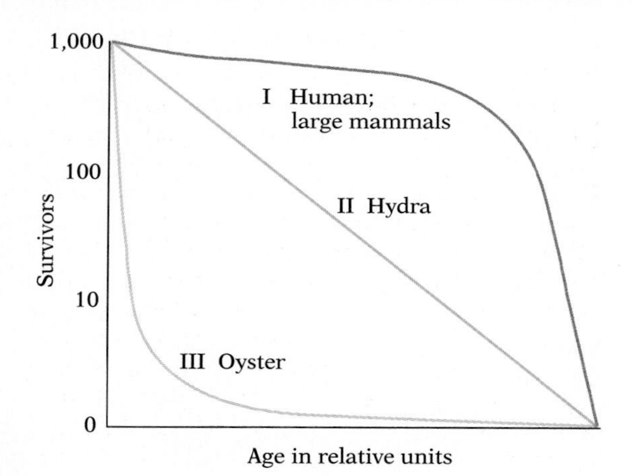

39.9 Three types of survivorship curve

For an initial population of 1000 individuals, the curves show the number of survivors at different ages from birth to the maximum possible age for that species. The curves represent three basic types of survivorship. The human curve is for societies with virtually no infant mortality.

ceptible to environmental control, and make it possible to compute the percentage of individuals likely to still be alive at the end of each age interval. The results may be graphed as a survivorship curve (Fig. 39.9).

The curves in Figure 39.9 illustrate several survivorship patterns. Curve I would be expected if all the individuals in the population lived almost as long as possible. There would be nearly full survival through all the early age intervals (as shown by the horizontal portions of the curve), with few premature deaths; mortality would result from old age. Curve III approaches the other extreme, where the mortality is exceedingly high among the very young, but any individual surviving the earliest life stages has a good chance of surviving for a long time thereafter. Between these two extremes is the condition represented by curve II, where the mortality rate of all ages is constant.

The survivorship curves for most wild-animal populations are probably intermediate between types II and III, and the curves for most plant populations are probably near the extreme of III. High mortality among the young is the general rule in nature.

Changes in environmental conditions may radically alter the shape of the survivorship curve for a population, and the altered mortality rates, in turn, may have profound effects on the dynamics of the population and on its future size. Consider the age distribution of three human populations. In Sweden (Fig. 39.10A) the distribution is nearly uniform, with the percentage of the population in each age class approximately equal except for the oldest classes. The survivorship curve of the Swedish population (human in Fig. 39.9) is close to type I. The birth rate equals the death rate, and the population size is stable; since each woman, on the average, has two children, individuals are only replacing themselves. The mortality rate is low up to about age 70, after which the number in each age class begins to drop steadily; note also the lower mortality rate for females beginning about age 60, which is typical of developed nations.

The population in the United States (Fig. 39.10B) is slightly different. Though the fertility rate is now close to replacement, this equilibrium has been achieved only recently. As a result, the population includes a disproportionate percentage of prereproductive and younger reproductive individuals, born when the average family size was greater than 2. If it were not for the high rate of immigration, the population of the United States would continue to grow only until the reproductive and prereproductive age classes neared equilibrium. This would happen about the year 2030, when the population would have reached about 300 million.

In India, by contrast (Fig. 39.10C), the age distribution is heavily weighted toward the bottom, with a relatively large percentage of the population in the younger age classes. The survivorship curve

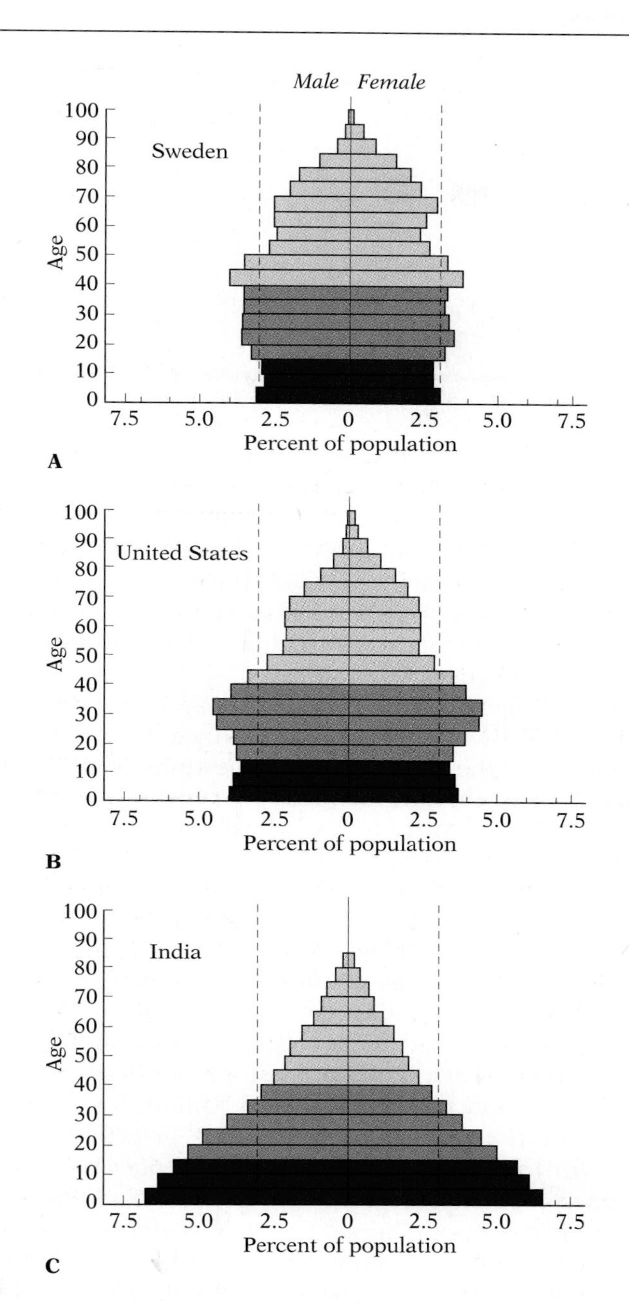

39.10 Age distribution in three human populations

(A) In Sweden, the size of each age class from prereproductive (dark blue) through early postreproductive ages (light blue) is roughly equal. Because the birth rate approximates the death rate, and has for many years, the population is not growing; reproductive individuals (medium blue) are simply replacing themselves. Were this not so, the percentage of the population in each of the prereproductive age classes (0–15 years) would be greater than the percentage in each of the reproductive age classes (15–40 years). For the life expectancy shown here, a stable age distribution would have about 3% of the population in each of the younger 5-year blocks, as indicated by the vertical dashed lines. Note that since older individuals of our species do not reproduce, any extension of the life span beyond about age 50 has little impact on the growth rate. (B) In the United States, the size of the youngest group (0–5 years) is roughly in equilibrium with that of the younger reproductive classes. However, since there is a bulge well beyond the 3% line in some of the age classes that have not finished reproducing, the population will continue to rise until at least 2000 in spite of the low birth rate. Immigration is also swelling the population. (C) In India, the population is heavily weighted toward the younger age classes as a result of a high birth rate. Even if the birth rate were to fall to two children per woman, the population would continue to grow for two generations.

of the Indian population is shifted toward type II, with higher infant mortality, a large average family size, and a rapidly growing population. The small numbers in the older age classes reflect the higher age-dependent mortality in India. In the absence of a dramatic reduction in the birth rate, advances in reducing infant and general mortality in developing countries (resulting from improvements in sanitation, nutrition, and medical care) inevitably result in greatly increased population growth.

Revolutions in medical and agricultural technologies have raised both infant survival rates and the carrying capacity of the earth. In industrialized societies, they have at the same time made efficient birth control possible. For many thousands of years the human population of the world increased very slowly, even though the

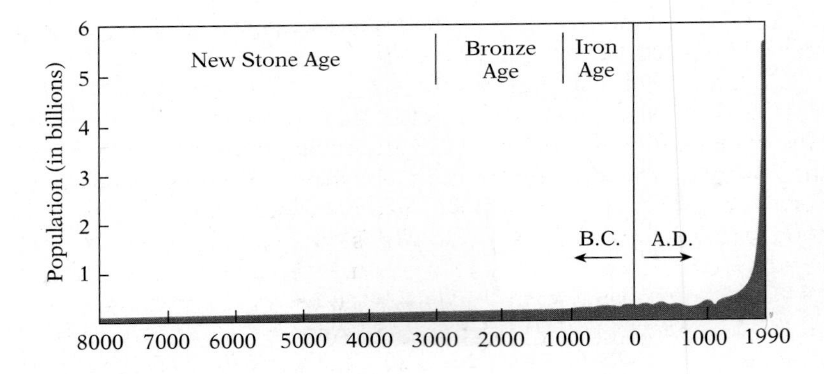

39.11 Growth of the human population of the world

Growth was slow for many thousands of years, but has become very rapid in the past century. The dip around 1350 corresponds to the outbreak of bubonic plague.

birth rate was probably high (Fig. 39.11). There were approximately 5 million people on earth 10,000 years ago; the number rose to only about 250 million by A.D. 1 and to 500 million by the year 1650. Until 300 years ago, then, the human population doubled approximately every 1600 years. At its present growth rate, the population of the world, estimated at 5.8 billion in 1996, will double in about 45 years.

How much longer the human species can continue with such a rate of increase—with such an imbalance between births and deaths—is one of the most pressing questions of our time. Many researchers believe that, in fact, it is *the* most important question. All other aspects of various ecological crises—hunger, poverty, crowding, pollution, accumulation of wastes, destruction of the environment on which all life depends—are the inexorable consequences of the increase in the number of humans.

There are at least two bright spots (if they can both be called that), in this otherwise bleak picture. The first is that in many cases a rise in standard of living has led, after a delay, to reduced growth rates: this is the so-called demographic transition. The second is that in areas that formerly had at least a modest standard of living (Eastern Europe and the former Soviet Union, for instance), economic deterioration has led to a negative growth rate (hardly a bright spot for those immediately concerned, but in a larger, demographic sense, a positive development nonetheless). Both instances suggest that human populations can, with sufficient experience of a degree of economic well-being, learn to judge the carrying capacity of their habitat (and even local changes in the value of K), and respond with sensible reproductive decisions.

■ HOW POPULATION SIZE IS REGULATED

The importance of growth rate One or more factors always work to limit the size of populations, human or otherwise. Most populations can be usefully described by the sort of growth curve that best characterizes them. Near one extreme of this conceptual continuum are populations that grow very quickly because they have a ***high*** $\boldsymbol{r_{max}}$. They have evolved this capacity to exploit transitional, disturbed, or otherwise unpredictable environments. The rapid multiplication of mosquitoes in springtime typifies the behavior of a high-r_{max} species; fortunately for us, the size of the mosquito population is generally held below saturation by environmental factors like drought and cold.

The high-r_{max} lifestyle is usually accompanied by a host of behavioral and physiological adaptations. In particular, individuals usually produce large numbers of small, quickly maturing offspring. In the absence of severe competition, the fitness of such individuals is heavily dependent on producing as many young as possible before some environmental disruption brings on a sudden crash. The survivorship curve of such a species usually resembles type III (Fig. 39.9). High-r_{max} species are relatively common, in part because directional selection often operates: as smaller size evolves, the organism's sensitivity to environmental fluctuations increases, which selects for a higher r_{max}, which in turn selects for yet smaller size, and so on.

Near the other end of this conceptual continuum are populations that are usually at or close to the carrying capacity K of their habitats, and so grow slowly or not at all. Selection has favored a ***low*** $\boldsymbol{r_{max}}$ in these species, of which most large mammals are examples. Low-r_{max} species are typically found in stable or predictable environments. Because their fitness depends less on rapid reproduction than on their ability to compete effectively for limiting resources, individuals of such species usually produce only a few large, longer-lived, slowly maturing young; they emphasize quality rather than quantity (Fig. 39.12). The survivorship curve for these species approximates type I. Again, directional selection appears to have operated to increase the relative number of low-r_{max} species: as larger size is favored, individuals become less sensitive to environmental fluctuations, which selects for lower r_{max}, which in turn selects for larger size, and so on.

39.12 A paradoxical example of low-r_{max} growth

Most insects have a relatively high-r_{max} lifestyle, characterized by small size and high reproductive rate. Honey bee colonies, on the other hand, are nearer the low-r_{max} end of the spectrum: they generally produce only one "offspring" a year, and invest heavily in it to help assure its competitive ability. That progeny is a swarm consisting of a queen and up to half of the parent colony's workforce (10,000–20,000 bees, weighing 1–3 kg in all); moreover, each individual in this swarm gorges itself with honey from the parent colony. This honey will be used to sustain the new colony and produce wax for making honeycomb once the swarm's scouts locate a suitable cavity for establishing a nest. Few high-r_{max} individuals invest up to half their "substance" in producing a single offspring.

The two extremes of the r_{max} spectrum illustrate dramatically the ways in which the various sorts of population controls operate. As we have seen, the density of the population in high-r_{max} species is generally too low for growth to be limited by the availability of essential resources like food, water, or nesting sites. Though they may be subject to severe predation, they reproduce so quickly and profusely that this factor is not greatly limiting. More important are the many density-*independent* factors such as weather, fire, or flooding, which can cause the population to crash periodically (see Fig. 39.8). In the rare instances that density-independent factors like environmental chance fail to crop the population of a high-r_{max} species, however, one of the density-dependent limitations (described shortly) inevitably comes into play.

The population of low-r_{max} species, on the other hand, is almost always kept from getting much larger than the carrying capacity by the limited availability of one or more critical resources. Thus, though chance can play a major role in altering population size, resource availability and other density-dependent factors are usually critical in limiting the population of most low-r_{max} species. Let's look at a few of these important factors, bearing in mind that they are as relevant to human populations as to those of microorganisms, plants, and nonhuman animals.

The roles of predation, parasitism, and disease Predation, parasitism, and disease usually influence the prey (or host) species in

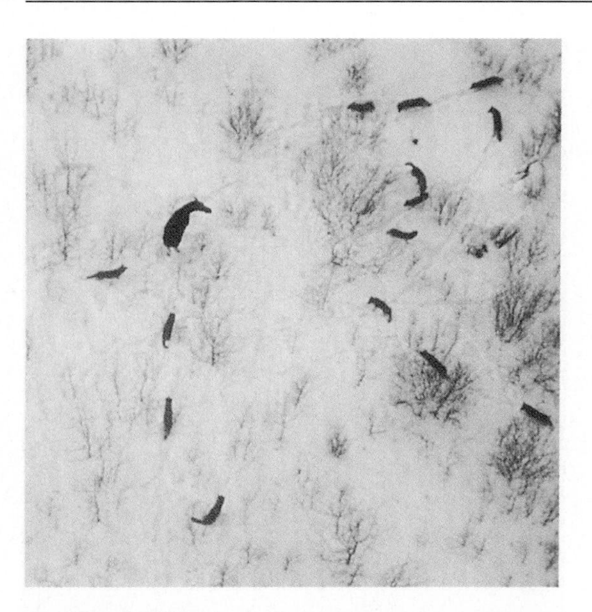

39.13 Population control by predation

A pack of wolves on Isle Royale corners an adult moose. Despite a long chase this individual was too strong and healthy to be worth the risk of attacking at close quarters. The wolves eventually left to find a more likely target.

a density-dependent manner. As the population of the host or prey increases, a higher percentage of the population is usually victimized, because the individuals—having perhaps been forced into less favorable situations or being weaker on account of the greater drain on available resources (or simply by virtue of being more common)—become easier to find and attack (Fig. 39.13). In the same way, the probability that a parasite will find a suitable host is also usually density-dependent. We've seen that in isolated species population size N depends ultimately on the species' growth rate r, as well as the carrying capacity of the habitat, K; when species interact, however, the population size of one (N_1) depends in large part on the growth rate of the other (r_2), and vice versa.

This intertwining of the population parameters of two species is especially obvious in predator-prey interactions: If the density of a prey species increases, the density of the predators feeding on it often increases also. This increase in predator number and attention can be one factor that causes the density of the prey to fall again. However, as the density of the prey falls, there is usually a corresponding but slightly delayed fall in the density of the predator species. The result can be a series of density fluctuations (Fig. 39.14). Such linked fluctuations of predator and prey suggest that the major limiting factor for some predators is the availability of its prey, and that predation is probably one significant limiting factor for the prey. (A note of caution is appropriate: prey species with high r_{max} values often display cycles in the absence of predators; moreover, cycles can sometimes be traced to the prey's own food supply, parasites, and weather.)

When the predator can switch to the most common of several potential prey species, or when environmental heterogeneity provides a refuge for rare individuals preyed on by specialist species, stable populations of both predator and prey can exist; for most interactions, however, population cycles are the rule.

Predator-prey balances can be destroyed inadvertently. For example, the application of insecticides to strawberries in an attempt to destroy cyclamen mites that were damaging the berries killed both the cyclamen mites and the carnivorous mites that preyed on them. However, because the cyclamen mites had a higher r_{max} value, once they were unconstrained by predation, they quickly reinvaded the strawberry fields, while the predatory mites returned

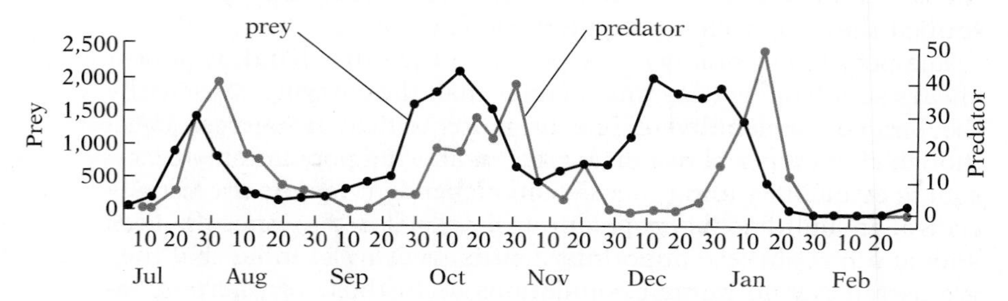

39.14 Linked fluctuations in predator and prey populations

The size of the population of a predatory mite tends to reflect (after a delay) fluctuations in the population of its prey (another mite). Predation is only one of many density-dependent limiting factors operating on the prey. The predator population, however, is limited mainly by the availability of prey.

much more slowly. The result was that the cyclamen mites, now free of their natural predators, rapidly increased in density and did more damage to the strawberries than if the insecticides had never been applied. Slower recovery of predatory (beneficial) arthropods than of prey (pest) arthropods is usual; moreover, the higher r_{max} values for pests means they are likely to evolve pesticide tolerance before their predators do. Together, these factors cast doubt on the long-term value of heavy pesticide use in agriculture.

The behavior of predators can greatly influence the reproductive strategy of prey. For instance, guppies living in some locales are subject to large predators specializing on adults, whereas in other habitats the threat is mainly from small fish that prey on the young. By way of compensating, guppies in the adult-predator habitat produce about twice as many fry as guppies in small-predator locales; the offspring in the former groups weigh about half as much as the fry in the latter habitat. These modifications in a species' r_{max} value can be created in the laboratory by introducing the same predators and waiting 30–60 generations.

Disease limits population size in a number of ways. Pathogens seem to take a constant toll of susceptible individuals, but there is a continual race between the host species (where selection favors immunity) and the parasite (for whom selection rewards better recognition of target-population cells). Because immunity also depends on the health of host individuals, disease can act to limit populations that are beyond the current carrying capacity of their habitat, culling the weaker members of the group. High population density also facilitates transmission of many diseases, adding to the efficacy of this means of control. Finally, disease can contribute a strong component of chance to population size when pathogens evolve new host specificities, as when one clone of the AIDS virus apparently accumulated a mutation enabling it to infect humans as well as its previous primate host (probably African green monkeys). Another example of this expansion of host range occurred in 1988 when a strain of canine distemper virus began to infect harbor seals off the coast of northwestern Europe, killing more than 12,000 in four months.

Disease can also exert a profound effect when preexisting geographic barriers are overcome, permitting the pathogen to reach a population that has not been subject to selection favoring immunity. It was this phenomenon that led to the massive mortality of the native inhabitants of the Americas when Europeans arrived in the 15th and 16th centuries, bringing with them a range of novel diseases like smallpox.

The impact of intraspecific competition Competition is one of the chief density-dependent limiting factors. The continued healthy existence of most low-r_{max} and some high-r_{max} organisms depends on utilization of some environmental resources that are in limited supply such as food, water, space, or light. Each member of a population shares the same basic requirements; each needs the same sort of food, shelter, and mates. Unless some other force, such as predation, holds a population below the carrying capacity of its environment, individuals in such populations must inevitably compete for resources.

Territorial birds illustrate the nature of this competition. Tawny

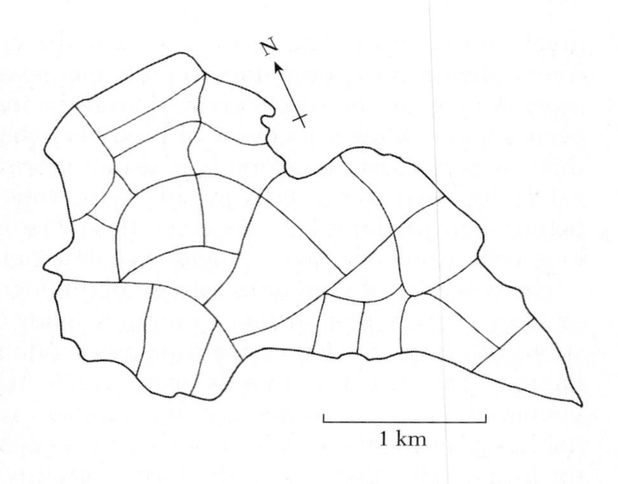

39.15 Owl territories

Pairs of tawny owls have divided this patch of habitat into a matrix of well-defined territories whose number and boundaries are relatively stable from year to year. The habitat supports an adult population of 50 individuals.

owls, for instance, live and hunt at night on well-defined territories that they occupy and defend year-round (Fig. 39.15). These territories are just large enough to supply a steady diet of rodents and, in some years, enough to allow reproduction. In a typical year in one habitat studied, eight of the 25 resident pairs did not even breed, and another nine cut their losses early by neglecting to incubate the eggs they had laid. Two more pairs allowed their chicks to starve, and the remaining six pairs laid an average of three eggs each rather than the species' maximum of four, so that in the end only 18 of the 100 potential offspring were fledged. Since in an average year only 11 adults died, creating 11 vacancies in the habitat (Fig. 39.16), even this number of surviving young was too high. Seven juveniles were unable to gain control of a territory, and so had to leave to "seek their fortunes" elsewhere. Clearly, intraspecific competition for limiting resources keeps tawny owl populations stable.

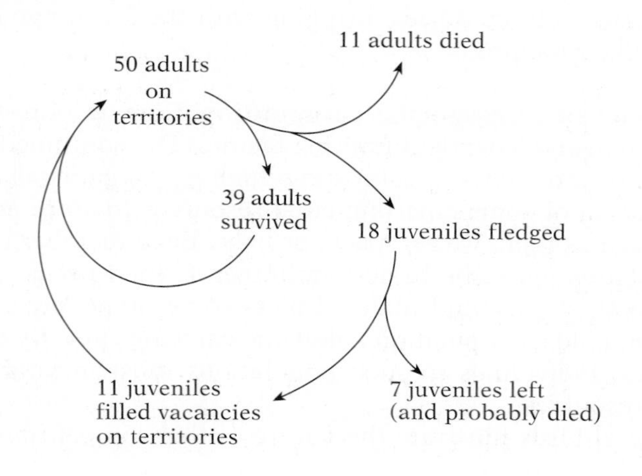

39.16 Typical yearly population flow in a group of tawny owls

A typical year for a tawny-owl population is shown here. On average, 18 juveniles were fledged (though in theory as many as 100 eggs could have been laid) to fill 11 vacancies left in the matrix of territories by the deaths of adults. Seven juveniles were therefore left to seek vacancies in other habitats; most did not survive.

Many plants also must compete for light, water, and nutrients, and a substantial surplus must be available if they are to produce seeds, which are metabolically costly. If flowers or vegetables are planted too close together in a garden, the densely packed plants will be weak and spindly, at least initially. If they are thinned, either artificially or by the natural death of the weakest individuals, the surviving plants will usually produce seeds or runners. Experiments varying the initial number of seeds or seedlings in a limited area show that natural thinning leads to a relatively constant level of production in many plants (Fig. 39.17). The same sort of competition for space, light, water, and nutrients operates in a forest and limits the density of the trees.

Why interspecific competition is important As detailed in Part III, competition between species can be a potent force in evolution. When two species compete for precisely the same limited resources, any systematic habitat-wide superiority of one species will inevitably lead to the extinction of the other. In any particular instance, the extent of this competition depends largely on the density of the competing species relative to the resources in question.

The ways in which an organism uses its environment to make a living define its ***niche***. A creature's niche is determined by its genetic endowment, and includes not merely what it eats, but how and where it finds and captures its food; what extremes of heat and cold, dry and wet, sun and shade, and other climatic factors it can withstand; and what values of these factors are optimal for it. It is also defined by such factors as what its parasites and predators are, and where, how, and when it reproduces. In addition, the time of year and the time of day when it is most active are important: the niche of honey bees, for instance, differs from that of small bumble bees even though they both harvest the same blossoms. Because honey bees overwinter as a group and regulate colony temperature, they are able to be active in early spring while the solitary bumble bee reproductives are still dormant. In late summer the situation is reversed: honey bees are at a disadvantage, having to forgo further reproduction to build up supplies of honey for later use in keeping the colony from freezing during the winter. The bumble bee colony, on the other hand, can channel all its earnings into explosive growth, sending out many reproductive individuals to found new colonies in the next season.

Since every aspect of an organism's existence helps define that organism's niche, we cannot completely describe or measure it; hence the concept of niche is an abstraction. One way to get a concrete grasp of a species' niche is by determining the acceptable limits and optimum values of variables that affect the species. Thus, for a particular species of caterpillar, we can begin with a variable such as the water content of its food plants, and determine the caterpillar's growth rate over a wide range of values. We can do the same for nitrogen content, and then plot the two sets of data against each other (Fig. 39.18). The result is the "adaptive space" the organism can occupy.

We could go on and determine the effects of a third variable, producing a three-dimensional adaptive volume. If we were to continue this procedure, adding every relevant variable, each determining a different dimension, we would obtain a hypervolume (a

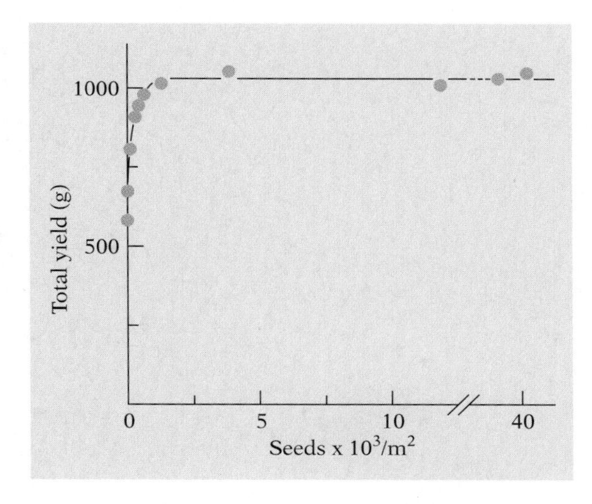

39.17 Effect of density-dependent competition in plants

For most plants, the amount of plant matter produced over the course of a growing season is virtually independent of the initial number of seeds or seedlings, as long as enough were present initially to exploit the patch. This Law of Constant Yield is illustrated here in an experiment with a species of grass (*Lolium loliaceum*) harvested 210 days after sowing.

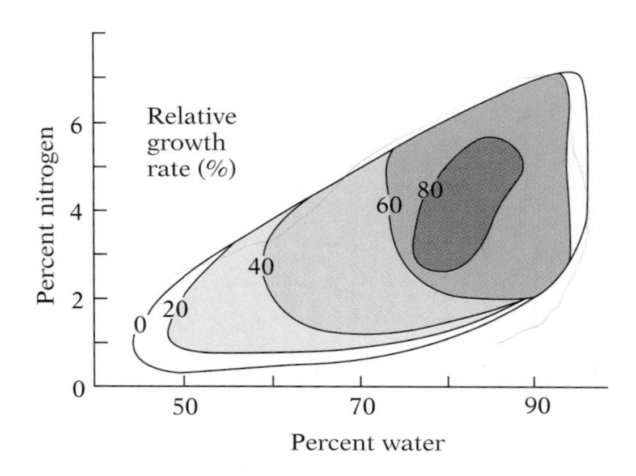

39.18 Graphical representation of a portion of the ecological niche of a species

The relationship between growth rate and two food variables is shown here for a species of caterpillar. If a third environmental variable were added to the analysis, a set of nested volumes would be defined. Though the procedure is difficult to carry further graphically, it can be continued mathematically to define a species' niche in a multidimensional hypervolume. The more the niche hypervolumes of two species overlap, the greater the potential for competition between the species.

multidimensional space) that would represent the niche of the species.

The more similar two niches are—the more the hypervolumes for two species overlap—the more likely it is that both species will compete for at least one limited resource (food, shelter, nesting sites, or the like). There is a limit on the amount of niche overlap compatible with coexistence. Competition for the one most limited resource therefore usually leads to one or two of four possible outcomes:

1 One species may simply become extinct. Usually this will be the result of the superiority of a rival species under ordinary circumstances, though luck and initial population size can be critical. Any systematic superiority need not lie in the ability to locate and harvest food; indeed, the species inferior in this regard may well be the one to survive if it is less susceptible to predation, disease, or extinction from the stresses of a fluctuating environment.

2 One species may be superior in some regions, and the second in other regions with different environmental conditions, with the result that one species is eliminated in some places and the second in other places. As a consequence, sympatry disappears, but both species survive in allopatric ranges. On a smaller scale, both may survive sympatrically in different microhabitats.

3 One species may be superior under normal conditions but at a strong disadvantage during periodic crises, which will reduce the population size of the superior species. Competition will then begin anew, as long as the generations overlap and crises are frequent enough to prevent the extinction of the inferior species.

4 Given time and slight differences in niche, selection may act to produce character displacement, which reduces competition. Interspecific competition for seeds in Darwin's finches on the larger islands led to character displacement in beak size, with the result that the several species tend to have small, medium, or large beaks, specialized for seeds of different sizes. Through character displacement two initially competing species may evolve greater differences between their niches.

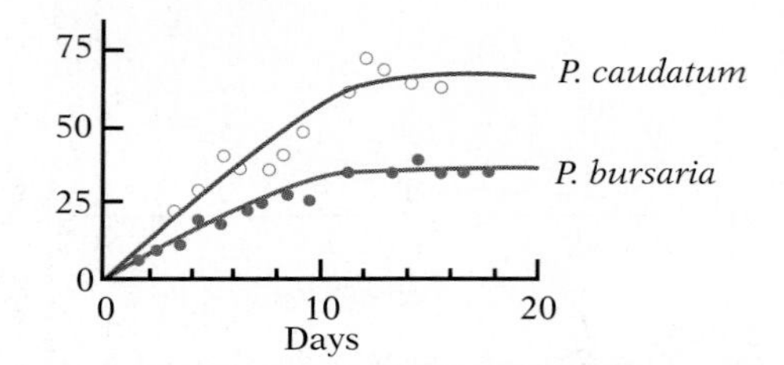

39.19 Coexistence based on microhabitat use

Although *Paramecium caudatum* competitively excludes some species of paramecia, and is itself excluded by others, it can coexist with *P. bursaria*, which is able to survive in low-oxygen parts of the habitat where *P. caudatum* is at a disadvantage. When by itself, *P. bursaria* takes over the entire container, reaching a density of about 150 on this scale. (The density numbers on the vertical scale give relative densities only, and hence have no units.)

Several classic studies illustrate the workings of competition in species interactions. As detailed in Chapter 18, two particular species of *Paramecium* cannot coexist in the same container, though either can survive on its own. In a related experiment, the loser of that competitive-exclusion encounter (*P. caudatum*) was pitted against another paramecium, *P. bursaria*, and both survived (Fig. 39.19). The reason was that *P. caudatum* could not flourish in the oxygen-poor water on the bottom, but *P. bursaria* was able to form a symbiotic association with bottom-living algae and thrive in the low-oxygen conditions.

When the flour beetles *Tribolium confusum* and *T. castaneum* are kept together in the same container of flour, one or the other species always goes extinct. The temperature and humidity greatly influence which species will win: *T. castaneum* usually wins under hot, wet conditions, while *T. confusum* favors cool, dry conditions. When intermediate conditions prevail, the outcome is unpre-

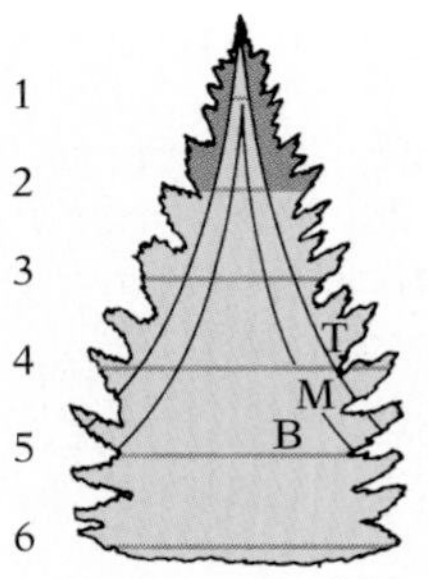

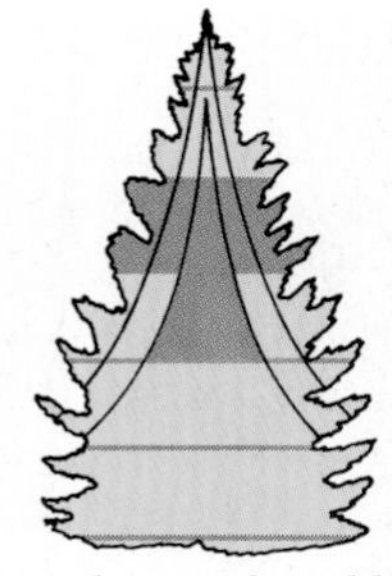

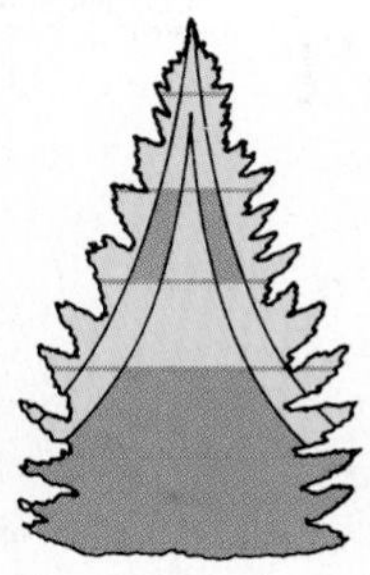

39.20 Differences in the feeding niches of three species of warblers living in the same community

The Cape May warbler feeds on the tips of the highest branches of the spruce tree; the bay-breasted warbler feeds in the middle part of the tree; and the yellow-rumped warbler feeds predominantly (though not exclusively) among the lower branches. Above: The dark green areas are the parts of the tree where the birds spend half their total feeding time. T, terminal parts of branches; M, middle; B, base. The height zones (1–6) are measured in 3-m units.

Cape May warbler

Bay-breasted warbler

Yellow-rumped warbler

dictable, but when twice as many beetles of one species are placed in the container at the outset, the variety with the initial numerical advantage usually wins. When a certain disease is present, a particular species wins. Chance, disease resistance, and priority are each important. Since nature is rarely uniform, both species survive in nature, though not in the same microhabitat.

It is not always easy to detect the differences between the niches of two or more closely related sympatric species, but close study usually reveals differences of fundamental importance. When Robert MacArthur studied a community where several closely related species of warblers (small insect-eating birds) occurred together, he found that their feeding habits were significantly different. Yellow-rumped warblers fed predominantly among the lower branches of spruce trees; bay-breasted warblers fed in the middle portions of the trees; and Cape May warblers fed toward the tops of the same trees and on the outer tips of the branches (Fig. 39.20).

Cormorants and shags, too, are closely related sympatric species of birds whose habits and ecological requirements appear very similar. This is deceptive, however: Though both nest on cliffs and feed on fish, cormorants nest on broad ledges and feed chiefly in shallow estuaries and harbors, whereas shags nest on narrow ledges and feed mainly at sea. Their niches, then, are very different, and there is little competition between them.

The role of emigration Emigration is common in many species, but in some, crowding induces physiological and behavioral changes that dramatically increase emigration. This is especially obvious in many species of aphids. When conditions are favorable, aphid populations consist mostly of wingless females reproducing parthenogenetically. When conditions deteriorate and competition becomes intense, winged females appear, and stand to gain a large

39.21 Locusts in the migratory phase, in Ethiopia

competitive advantage by moving out of the area in which they were born. However, given the high risk involved in searching for a new habitat—a suitable plant host—conditions must be intolerable before emigration makes sense. Many rodents (including the legendary lemming) are also programmed to emigrate in search of more favorable habitats when the population density—or, more often, some important secondary effect of crowding such as changes in hormone levels—exceeds a certain threshold.

The physiological and behavioral changes induced by crowding are particularly dramatic in several species of locusts. Individuals in the migratory phase have longer wings, a higher fat content, a lower water content, and a darker color than solitary-phase individuals; they are also much more gregarious and more readily stimulated to marching and flying by the presence of other individuals. The solitary phase is characteristic of low-density populations, and the migratory phase of high-density ones. As the density of a given population rises, the proportion of individuals developing into the migratory morph also rises; the sight and smell of other locusts seem to play an important role in triggering this line of development. When the proportion of migratory-phase individuals has risen sufficiently, enormous swarms emigrate from the crowded area, consuming most of the vegetation in their path and often completely devastating agricultural crops (Fig. 39.21).

How mutualism can play a part We have seen how the density of predators, prey, competing species, and parasites can affect each other. There are also density-dependent interspecific interactions in which the two species, instead of playing antagonistic roles, act in a mutually beneficial fashion. One of the more remarkable examples of such mutualism is the finely tuned cooperation between certain ants and the acacia trees in which they live. At first glance the ants of this species seem to be exploiting their hosts. Not only do they harvest the tree's nectar and leaves, but they also burrow into its thorns and stems to make their nests. A closer look, however, reveals that this acacia has evolved special adaptations that accommodate the ants. Compared to *Acacia* species not occupied by ants, its thorns are large and hollow (Fig. 39.22A), and the leaves bear nectaries at the base and extraordinarily nutritious structures called Beltian bodies at the tips (Fig. 39.22B).

Comparisons between normal "occupied" acacias and acacias from which the ants have been removed reveal that the ants are almost essential for the trees' survival. The ants keep acacias relatively free from plant-eating insects, and swarm out at the slightest disturbance to attack and repel browsing mammals as well. The ants prevent vines from damaging the tree, and defoliate nearby trees that shade their host.

Other examples of mutualism, such as the interdependence of many flowering plants and their insect pollinators, abound in nature. Among the most important and least recognized are the interactions between certain microorganisms and their plant and animal hosts. Chapter 26 in particular details two remarkable examples of micro- and macro-level mutualism. One involves nitrogen-fixing bacteria that are incorporated into nodules on the roots of legumes: The bacteria perform the extremely difficult and energetically expensive conversion of nitrogen gas into ammonia, thus supplying nitrogen, usually the most limiting of all mineral nutrients, to the host plant. In return, the legume supplies everything else the bacterium needs, including energy.

Equally remarkable is the meshwork of fungal hyphae, called mycorrhizae, that surround and radiate from the roots of most plants. The fungi themselves generally live as species-specific guests of the host plant, and in many cases absorb and pass on to the plant the vast majority of nutrients the host obtains. Most plants grown in the absence of mutualistic fungi fail to thrive, and many researchers believe that it was this kind of association between fungi and primitive plants that allowed plants to colonize the earth's surface. As detailed in Chapters 5 and 19, fungi and plants (as well as nearly all other eucaryotes) exist today only because of the ancient mutualistic relationship between the descendants of early procaryotes (now known as mitochondria and chloroplasts) and their primitive eucaryotic hosts.

■ HOW SOCIAL ORGANIZATION REGULATES POPULATIONS

Perhaps the single most important factor in regulating population size and distribution in many species of animals—birds and mammals especially—is social organization. A major challenge for ***behavioral ecology***—the study of how behavior has evolved in response to environmental pressures—is to understand why some species (honey bees, for instance) are social while others (like leafcutter bees) are not. Behavioral ecologists also seek to understand why sociality takes one form in, say, tricolor blackbirds, and quite another in a closely related species like redwing blackbirds.

Costs and benefits of sociality Most traits are the result of several countervailing selection pressures; they are adaptive on the whole because the *net* value of the trait is higher than that of any of the readily available alternatives. The same is true of sociality.

A

B

39.22 Mutualistic association of ants and acacia trees

(A) The acacia (*Acacia corigera*) provides overgrown hollow thorns that the ants (*Pseudomyrmex nigrocincta*) inhabit. (B) It also has specialized yellow protein-rich leaf tips called Beltian bodies, which the ants eat.

39.23 Cooperative hunting

By hunting together, these African wild dogs can capture much larger prey (here a wildebeest) than an individual could take on its own.

Its potential disadvantages are readily identified; the possible benefits that must amortize these costs are often less obvious.

One clear cost is competition: members of a species share the same basic niche, and so they inevitably compete to some degree for resources like food, water, mates, and breeding sites. To the extent that sociality brings members of a species into proximity, the frequency of such competition increases. It can take the form of time- and energy-consuming threats and fights, or simple interference as individuals get in one another's way, or decreased efficiency as several animals scour the same area for the same sort of food. Worse yet, concentrations of individuals also facilitate the transmission of diseases, and are often more obvious to predators.

Against these disadvantages are several possible benefits. Whether they are enough to turn the balance and lead to the evolution of some degree of sociality depends on ecological circumstances. For instance, group hunting may increase individual food intake. Lions, hyenas, and African wild dogs hunt cooperatively, bringing down prey too large for any single individual to capture alone (Fig. 39.13, 39.23). Groups can also cooperate to defend a feeding territory far larger than any individual could hold alone. This strategy, as Figure 39.24 illustrates, has been selected for

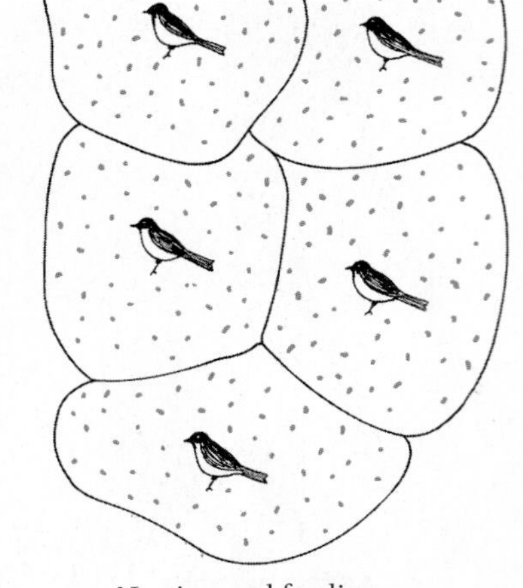

39.24 Effect of food distribution on blackbird social systems

(A) When food is distributed more or less evenly throughout the area, it may be energetically efficient for each individual to occupy and defend its own territory for both feeding and breeding. This pattern is typical of redwing blackbirds, which occupy territories in marshes and feed on emerging insects. (B) If food occurs in unpredictable patches, often far from good nesting sites, it may be more efficient for the individuals to occupy only small nesting territories in a colony and to forage as a group. This type of organization is seen in tricolor blackbirds, which nest together for mutual protection and feed as a flock. They forage on ripening grass and grain seeds.

39.25 Honey bees cooperate to regulate the internal temperature of the hive

These bees are fanning air out of the hive entrance, drawing cooler air in elsewhere. When the internal temperature is even higher, bees collect water and spread it on the comb and entrance, then fan it, cooling the hive evaporatively.

when the food supply—for instance, herds of grazing antelope, patches of ripening grasses, or trees unpredictably coming into fruit—is too rich for an individual to exploit completely, and too unpredictably dispersed to serve as a reliable ongoing supply of food for a single animal. Groups can sometimes have scouts that share information about the location of food, and thus monitor the habitat more efficiently. Finally, some groups cooperate in building nests and regulating their temperature (Fig. 39.25).

The most common pressure leading to sociality is defense. Though a group of animals is often more obvious to predators, it has more eyes available to watch its enemies, and early detection of a threat is usually the best defense. Antelope, though they accelerate more slowly than lions, can outrun them if they get a head start. Even the cheetah, the fastest of all land animals, is no threat if its target can start moving soon enough, since the cheetah has traded endurance for speed. Even when surprised, a multitude of running, crisscrossing animals in a group can confuse a predator. Two other factors can also play a role: with more eyes in the group, each individual can spend a greater proportion of its time feeding and less watching for danger (Fig. 39.26); in some cases (honey bees, colonies of nesting birds, and musk oxen, for example), cooperative defense is possible.

Controlling resources Group living, which will evolve when it benefits the individuals in the community, can have an enormous impact on population regulation. Territories, as we have already seen, clearly exclude some reproductively ready animals from prime habitats. The disenfranchised must either fight their way

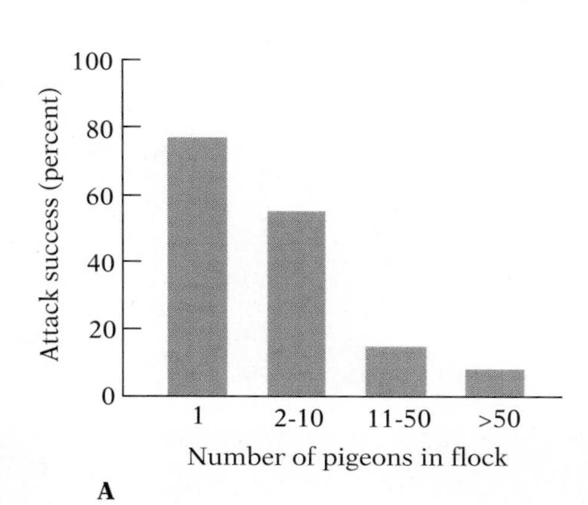

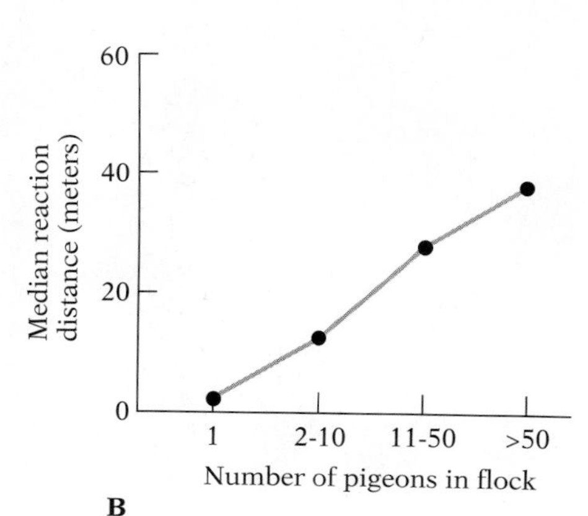

39.26 Defensive value of flocking in pigeons

(A) The more pigeons there are in a flock, the less chance a hunting hawk has of a successful attack. (B) The main reason for the lower success is that approaching hawks are spotted sooner when many eyes are watching for danger.

39.27 Peripheral territory in a wildebeest habitat

The risk this low-ranking male takes by defending his relatively poor territory is probably balanced by two possible benefits: there is always a chance that a female grazing on his territory may come into estrus while there and allow him to mate with her; and as a territory-holding male he is in a position to judge the fitness of males on adjacent territories, and can attempt to take over a better area if its owner begins to look vulnerable.

into the matrix, take a less desirable spot (Fig. 39.27), or forgo reproduction until they are older and stronger. Moreover, the ritualized trials of strength that have evolved to decide questions of dominance in many animal societies often benefit all: they establish the likely winner of an all-out fight without risk of injury, and save time that would be wasted if individuals had to contest every scrap of food or mating opportunity. Anarchy has been selected against because it reduces the fitness of all concerned; the innate rituals of social species bring efficiency and a measure of peace.

Group living in habitats where resources are thin or unpredictably distributed can still achieve the same effect. These societies usually have a dominance structure that limits the number of reproductive individuals of at least one sex. Surplus members are driven from the group, thus preventing depletion of resources. In many cases, group control falls to a single dominant individual; in lions and some primates, on the other hand, coalitions develop that are able to exert control even when no individual member of the ruling clique may be able to defeat a challenger single-handed (Fig. 39.28).

Conflicts to control resources critical to breeding—food, water, nesting sites—are the primary basis of male-contest sexual selection (discussed in Chapter 17), and have a profound effect on gene frequency, since certain phenotypes are more likely to succeed than others. Moreover, because in many species females exclude or harass other less vigorous females from the territory of even a polygynous male, there is frequently strong assortative mating between dominant males and dominant females. (As described in Chapter 17, assortative mating occurs when individuals do not mate randomly with respect to genotype.)

One remarkable consequence of this genetic sorting is that many species have evolved the ability to manipulate the sex ratios of their offspring. In species in which strong males can sire a disproportionate number of offspring, a dominant pair will often produce more males, whereas subordinate pairs will bias the sex ratio toward females. Similarly, dominant females may discriminate in favor of their male progeny by feeding them more often. In other species the sex ratio and postnatal biases depend on current food levels: when conditions are good, sons may be favored over daughters. The physiological basis of sex-ratio manipulation in birds and mammals is unknown.

What is biological altruism? It is essential to remember that each member in a group is acting in its own self-interest. Cooperation is selected for not because it is good for the species, but because it pays an individual to cooperate. ***Altruism*** as we commonly use the term—meaning self-sacrifice, or in biological terms, sacrifice of reproductive potential—probably does not exist; alleles

39.28 A primate alliance

The two baboons at the right are cooperating to threaten the baboon on the left, who could defeat either alone, but is unwilling to fight the two together. Such alliances permit small groups to control a sizable community.

for such a trait would quickly be selected against and disappear from the gene pool. Yet animals *do* help one another: birds care for their young, bees die defending the hive, primates share food and groom one another. In each case, however, the individuals involved are turning a genetic profit, or at least minimizing a potential loss. This is especially obvious in the context of parental care: the young are the parents' stake in the next generation, the sole index of evolutionary success. Investment in the young is an investment in immortality, the ultimate form of genetic selfishness. We will continue to refer to such behavior as altruistic, while keeping in mind that beneath the apparent selflessness lies genetic self-preservation.

The altruistic behavior of honey bees and other social insects is more subtle. W. D. Hamilton found the key to this apparent self-sacrifice when he realized that in order for the altruism that is evident in many social systems to evolve, alleles that code for neural circuits leading to altruism must have succeeded better even in the short run than genes that result in selfishness. He pointed out that the true measure of an allele's success, or fitness, is not whether a particular *individual* possessing it reproduces, but rather whether that allele is found in more individuals in the next generation. Alleles that cause one animal to be altruistic and forgo reproduction will still survive in the population if the resulting altruistic behavior sufficiently enhances the fitness of *other* individuals that carry the same alleles.

The easiest way to judge which other animals share alleles is by determining how closely individuals are related genetically: siblings, for instance, have on average half their alleles in common, so there is a 50% chance that any particular allele carried by one is possessed by the other. Similarly, each parent shares 1/2 its alleles with its offspring, but only 1/4 of its alleles with each grandchild. For genetically programmed altruism to evolve, the donor and recipient must be close kin, and the benefit must be much larger than the sacrifice; the *advantage* conferred on one individual by its sibling, for example, must be at least double the *loss* in fitness incurred by the altruistic sib.

Hamilton applied this idea of ***kin selection***—altruism that increases the fitness of kin—to the social insects. Members of the Hymenoptera (ants, bees, and wasps) have an unusual genetic characteristic: male hymenopterans develop from unfertilized eggs and have only one set of chromosomes. Under normal circumstances, when two individuals that are not hymenopterans mate, each contributes half of its chromosomes to each of its offspring. In contrast, male hymenopterans, having only 1/2 to begin with, pass *all* their chromosomes to each individual they sire. The result is that all of a male's daughters have exactly the same set of paternal alleles. Thus 50% of a daughter's alleles are identical to those of other daughters before we even consider the maternal contribution. Since the mother is diploid, there is a 50 : 50 chance of any given allele being shared between offspring; thus 1/4 (1/2 of the alleles shared times 1/2 of the total genome) of the offspring's alleles are identical through maternal inheritance.

In hymenopterans, therefore, the total relatedness is 3/4 — 1/2 from the father, and 1/4 from the mother (Fig. 39.29). (In conven-

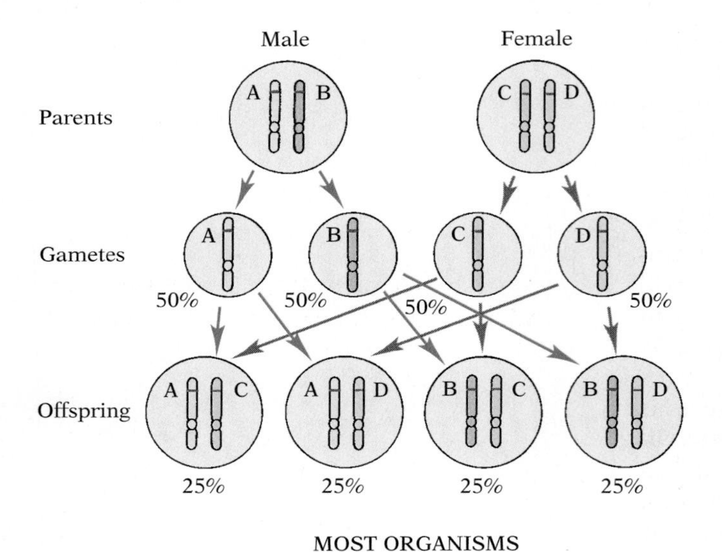

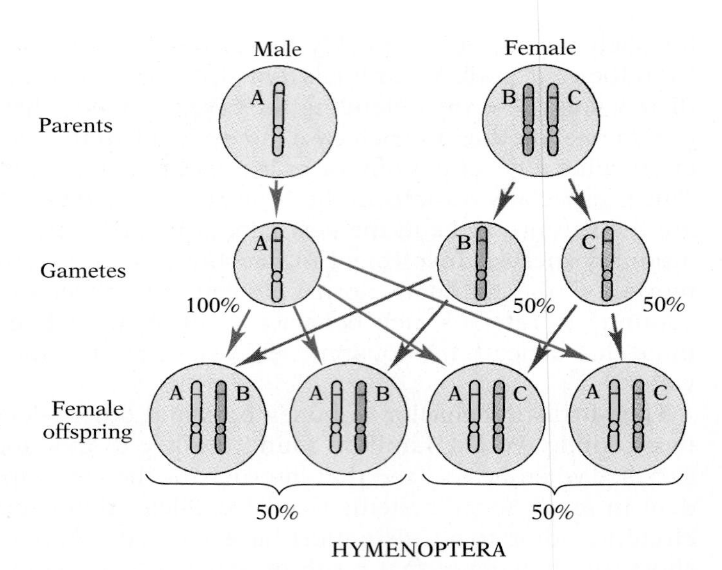

39.29 Inheritance in hymenopterans

In the Hymenoptera (ants, bees, and wasps) males have only half the normal number of chromosomes. All female offspring therefore receive exactly the same set of chromosomes from their father, while the chance of two daughters having a particular maternal chromosome in common is only 50%. As a result, sisters on average are more genetically similar to each other than to their mother.

39.30 A ground squirrel giving an alarm call

Alarm calls are sounded at the approach of airborne or terrestrial predators; there is a different call for each class of predator. The calls, which involve risk in that they draw attention to the individual giving the alarm, are produced primarily by females with close relatives living nearby.

tional diploid species, the equivalent values are 1/2 — 1/4 from the father and 1/4 from the mother.) Thus hymenopteran sisters share more alleles with each other and with any new offspring produced by their mother than they would share with their own daughters if they were to reproduce. Hence, alleles for altruism directed toward siblings instead of their own offspring could increase in frequency.

Even among more genetically conventional species, circumstances might cause selection for altruistic behavior toward kin. The best-understood case involves burrow-dwelling ground squirrels. These creatures hibernate for nearly two-thirds of the year, becoming fully active in their mountain habitats only during the short summer, when they eat grass and seed and reproduce. The females defend territories against intruding ground squirrels; when predators approach, certain females consistently give alarm calls (Fig. 39.30), thus increasing their chance of being attacked, while others never call. Juvenile females overcome their territoriality and clump together in small groups to hibernate near their birthplace, while males disperse and pass the winters alone.

In the light of kin selection, these behaviors make sense: the squirrels "sacrifice" themselves to aid the survival of their kin. The major source of mortality is the unpredictably long and cold winter of their mountain habitat. In addition, predation by coyotes, weasels, badgers, bears, and hawks takes about 5–10% of the population each year. Fighting kills another 5–10% of the adults, and adults kill roughly 10% of the young annually.

Surviving the first winter, controlling sufficient food resources, guarding against predators and aggressive conspecifics, and reproducing early are crucial to fitness in ground squirrels. The squirrels probably learn to recognize their relatives during their first day or two above ground. Any squirrel encountered by a youngster regu-

39.31 Mutual grooming in penguins
Each animal grooms an area the other cannot reach.

larly on its mother's territory is almost certainly a relation. A careful examination of the kin records derived from marking studies and blood tests shows that, in fact, the females do take risks to aid their kin. Their social organization serves to help kin face the challenges of winter survival, food scarcity, predation, and infanticide. Group hibernation is an efficient, low-risk way to survive winter by sharing warmth, and the programmed tolerance of kin means that only related females will cooperate in this way.

Only relatives—sisters, mothers, daughters, and their offspring—are allowed to share food during the warm season. All others are attacked. Thus the curious variability in the behavior of female ground squirrels—sharing territories and food with some squirrels while chasing out others—turns out to be based on kinship. Warning calls are produced much more frequently by older females with living kin than by males or childless females, and the advantage of such a warning to a female's nearby relations outweighs the risk she herself incurs. Finally, infants are never killed by relatives; rather, they are killed by homeless squirrels that would benefit from a reduction in competition. As a result, related females cooperate in driving off unrelated animals.

Though kin selection appears to be widespread and helps explain much of the behavior of social animals, there are many cases in which kinship is not a major factor. The motive force behind most of these instances is reciprocity: an animal confers a favor on another animal in the expectation of eventual repayment, an exchange called ***reciprocal altruism***. Two social animals may cooperate to groom places on each other that neither can reach, for example (Fig. 39.31). In most cases of reciprocal altruism, animals must be able to recognize and discriminate against cheaters, those chronically unwilling to repay favors. Failure to punish cheaters usually means that reciprocal altruism cannot persist in a population, since genes for the purely selfish strategy of nonreciprocation are then inevitably more fit. The necessity for keeping track of cheaters means that reciprocal altruism is most often found in small groups of animals capable of recognizing each other individually.

The realization that much of what appears to be altruistic behav-

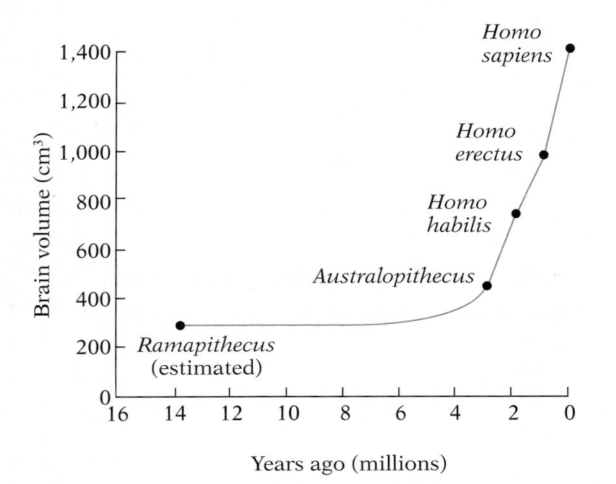

39.32 Evolution of brain size in humans

The brain capacity of the evolutionary line that led to humans has increased at an extraordinary rate over the last few million years. Brain volume is a rough measure of brain specialization when variables such as body size can be excluded. Therefore it seems likely that selection strongly favored the many specializations that form much of what we call intelligence during a brief, relatively recent period.

ior is ultimately selfish underscores the point that innate behavior, like all other specializations, enhances the fitness of the genes coding for it. The "good of the species" is, as far as we know, never selected for directly, though, like population regulation, it is frequently an incidental consequence of social organization.

Human ecology Our species evolved as hunter-gatherers, under strong selection pressure for high intelligence (Fig. 39.32). Studying the few groups of hunter-gatherers that are left may provide insights into the ecology of early humans. Until recently, these groups lived untouched by the technology and social adjustments necessary for crop cultivation and the domestication of animals that spread through most other human cultures 10,000–20,000 years ago. The best understood of the primitive tribes are the !Kung bush people of the Kalahari Desert in southern Africa. (The "!" represents a tongue click in their language.) Over the past few decades the number of !Kung groups depending exclusively on hunting and gathering has dwindled to zero. The following description of !Kung culture is based on information gathered over several decades and most recently synthesized by Richard B. Lee.

In the past, !Kung adults, like the members of other human societies, generally formed monogamous pair bonds and invested heavily in their offspring. Because breast feeding usually delays the resumption of the menstrual cycle, the interchild interval for couples was 3–4 years, which indicates an extremely low-r_{max} strategy. Unlike animals of predominantly monogamous nonhuman species, which tend to live as isolated pairs, the !Kung lived in groups of 20 to 30. These groups occupied traditional, undefended areas of about 500 km^2, centered on sources of water. Group interchange was frequent, and individuals often visited other bands for days or even weeks.

The !Kung divided the labor of obtaining food; women gathered fruits and vegetables, and men hunted game in small groups (Fig. 39.33). Since gathering is more efficient than hunting, women harvested about 60% of the protein and carbohydrates, yet worked fewer hours than the men. Though less efficient in terms of quantities obtained, hunting does provide essential amino acids, vita-

A

B

39.33 Hunter-gatherers

!Kung groups depended on both hunting (A) and gathering (B).

mins, and minerals. Food from both sources was seasonal, so the rhythm of the seasons controlled !Kung social organization. The limiting factor was water. Since about 95% of the rain fell during a 6-month period, there was an extended drought each year. During the rainy season a group would settle in a nut forest and eat their way out of it, collecting vegetables, fruit, and nuts over an ever-widening circle until the round trip for foraging reached 15–20 km. With the approach of the dry season, !Kung groups retreated to the most dependable water holes and made do with whatever food they could locate until the arrival of the next rainy season.

The controlling element in the behavioral ecology of the !Kung was the variability and density of the food supply. Since gathering was done from a base camp, at ever greater distances, it is clear that the larger the group, the larger the area that had to be harvested to support the population—and the farther individuals had to walk. Hence, if gathering had been the only activity, groups would have been very small. Hunting, however, tended to increase group size. The men normally hunted in pairs, and these teams frequently failed to find or to kill prey. The greater the number of pairs on the move, the more likely it was that at least one of them would be successful. Since a single wildebeest provides about 200 kg of meat, more than enough to feed a group of 30 for 2 weeks, an increase in the number of hunting pairs could dramatically improve the diet of the group as a whole.

!Kung social organization traditionally depended on reciprocal altruism within and between groups. Within a group, hunters and gatherers would share food daily, and the spottiness of rain meant that a group would have to borrow water from a neighboring group in some years and lend its own in others. The cement for this altruism appears in part to have been kinship. The groups were complex kin associations, and the choice of leaders was determined largely by kinship. Requests to use part of another group's territory or water were inevitably made through lines of kinship, and visits to neighboring groups always took the form of kin visiting kin. Strong cultural traditions were also of great importance in group cohesion. Both hunting and gathering depended on a sophisticated, culturally transmitted technology of digging sticks, woven nets, canteens, pouches, snares, spears, bows, poisoned arrows (Fig. 39.34), knives, fire, and so on.

39.34 A !Kung hunter preparing poisoned arrows.

It seems clear, then, that culture has been an enormously important component of human ecology for hundreds of thousands of years, and must have greatly influenced our species' evolution up until the discovery of domestication. How much of our present behavior is influenced by adaptations to this earlier lifestyle is a controversial but crucial question for behavioral ecologists.

HOW POPULATIONS FORM COMMUNITIES

In our discussion of populations, we considered species either alone or in pairs—as predator and prey, mutualists, or species with overlapping niches. Moreover, we viewed the environment as a more or less static backdrop against which various interspecific and intraspecific interactions take place. For many purposes, this narrow picture is reasonably accurate. In the broader view, however, each species usually interacts with several others, and is indirectly affected by many more. Furthermore, the environment is not static: habitats are constantly changing, and as a consequence different selection pressures are introduced. In this section we will consider how these complexities affect the biology of populations.

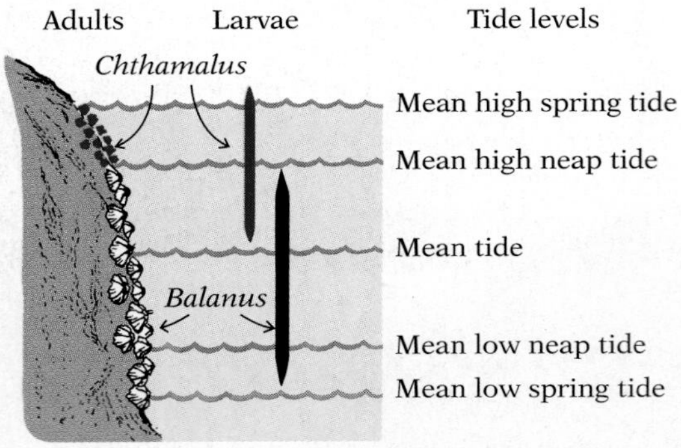

■ HOW DIVERSITY AFFECTS STABILITY

The species that make up a community influence one another in countless ways for both good and ill, through competition, predation, mutualism, and so on. Species interactions unnoticed by us in their natural state may still be of crucial importance to a community. Consider an experiment with an intertidal community of 15 species of marine invertebrates conducted at Mukkaw Bay, Washington, by R. T. Paine. The top predator in this community was a sea star, *Pisaster ochraceus*. When Paine excluded the sea star from one area but allowed it to remain in another, the result was a radical change in the species diversity of the first area. Of the original 15 species, only eight remained. One of two competing species of barnacle was eliminated, because the sites for its attachment were not cleared of other organisms by the sea star, and the other barnacle was a better competitor for what little space there was (Fig. 39.35). A sponge and its predator also disappeared; the sea star had some indirect influence on their existence, probably by clearing space for the sponge. It was impossible to predict beforehand the dramatic effects of removing *Pisaster*. In this community, the sea star is the ***keystone species***, controlling the community composition.

39.35 Effect of competition between two species of barnacles

Though there is a broad area in which the larvae of both species can settle successfully, competition eliminates most of the overlap by the time the adult stage is reached. *Chthamalus* is largely restricted to the zone above the level of the mean high neap tide.

Efforts to eliminate undesirable species from a community often reveal hidden linkages to other organisms, and demonstrate the complex interactions on which community stability rests. An unintended cautionary example was provided by the World Health Organization (WHO) in a campaign to eradicate malaria-carrying mosquitoes in the Borneo states of Malaysia, where as many as 90% of the population suffered from the disease. Mosquito control was achieved by spraying the inside of the village huts with DDT and dieldrin, two powerful insecticides, and malaria was indeed eradicated. However, soon the villagers began to notice that the thatch roofs of the sprayed huts were rotting and beginning to collapse. The deterioration was inflicted by the larvae of a moth that normally lives in small numbers in the thatch roofs. Whereas the thatch-eating moth larvae avoided food sprayed with DDT, the moth's natural enemy, a parasitic wasp, was very sensitive to it. The result was a substantial increase in the population of the larvae eating the thatch.

There was another more serious side effect. Cockroaches and a small house lizard, the gecko, are two normal inhabitants of the village huts. DDT-contaminated cockroaches were eaten by the geckos, which were in turn eaten by house cats. The cats, poisoned by the accumulation of the insecticide, died. The population of rats exploded, increasing the potential spread of leptospirosis, typhus, and plague. In an attempt to restore the cat population, WHO and the British Royal Air Force parachuted cats into the villages. With the cat population restored, the rat population plummeted.

Few pest-control efforts entail such complications, though unanticipated results are common. Insecticides that are rapidly broken down in the environment and are more selective in their toxicity have been developed, and these are less disruptive to biological communities. Much to be preferred, however, is biological control through the use of the natural predators, parasites, or pathogens of the pest (Fig. 39.36).

In each of the two examples just given, the removal of a single species caused major community upheavals. In the past, many ecologists held that in more complex communities, where there are more species and therefore more alternative interactions, there will be greater stability—a superior ability to withstand perturbations such as the removal of a species. In such a community, for example, the predators often have many prey species available and, because predators can switch from one to another, they were thought to be less sensitive to variations in the abundance of any one. Reciprocally, the prey may be subject to several different predators, its density regulation thus being less dependent on any single predator.

However, theoretical studies pioneered by Robert M. May show just the opposite trend: as species are added to a hypothetical community, stable configurations become progressively rarer. Moreover, as we have come to know more about the most complex and species-rich communities on earth—those in the tropics—it has become increasingly apparent that they are often no more resilient than temperate-zone communities; in some cases, in fact, they are less resilient, though for reasons that usually have more to do with nutrient retention and cycling than with species interactions.

Most ecologists now reject the notion that greater species diversity and complexity of interactions necessarily make for increased community stability, at least in the case of animals and large plants like trees. The response of a simple community to a disturbance may be more violent and immediate, but such a community may recover quickly and settle rapidly into a new functional mode. The complex community, by contrast, may sometimes respond less dramatically to the initial disturbance; however, because of the multiplicity of interactions within it, the effects may continue to ripple through the system for a long time, causing numerous smaller but still important dislocations and distortions.

A new line of evidence indicates that complex plant communities can sometimes enjoy an advantage when an external perturbation alters the availability of nutrients. As we know from studying natural selection, populations of a species often show considerable genetic variation. A community with a larger number of species is therefore more likely to have a useful variation that will help respond to any perturbation that requires an unusual genetic capacity (or preadaptation, as described in Chapter 17). Empirical data bear this idea out with particular force in studies of ecosystems responding to changes in CO_2 levels.

The complexity of the physical environment is also critical in predicting what will happen when species interact. In simple laboratory cultures of several competing species or of predator-prey systems, extinction of at least one of the species virtually always

A

B

39.36 Biological control of prickly pear cactus by a cactus moth

(A) Prickly pear cactus was introduced into Australia, where it spread and covered vast expanses of grazing land. (B) Within 3 years of introducing the cactus moth, the prickly pear population was virtually exterminated.

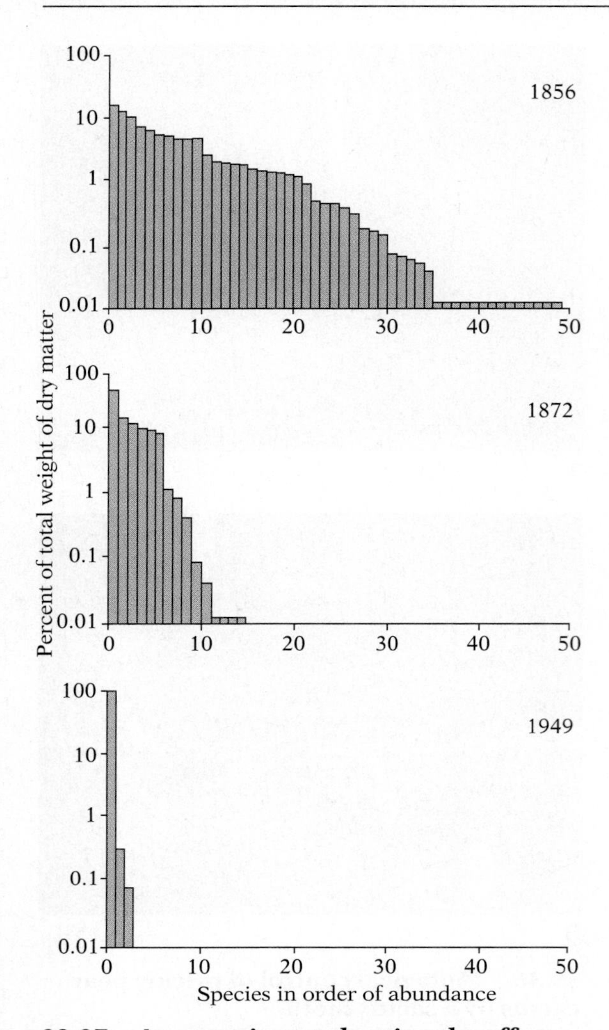

39.37 An experiment showing the effect of human intervention on species diversity in an ecosystem

Nitrogen fertilizer has been applied yearly since 1856 to an experimental grass plot at the Rothamsted Experimental Station in England. When the experiment began, 49 species were growing in the plot (top). By 1872 the number had fallen to 15 (middle). Only three species remained in 1949 (bottom). This sequence resembles an early succession run backward. The species diversity probably declined because one ecological factor—the ability to use exogenous nitrogen rapidly—was emphasized at the expense of all the others.

ensues (see Fig. 18.21, p. 505). If the physical environment in the culture is made more diverse, however, the wild fluctuations and oscillations typical of simple laboratory systems can often be reduced, and extinction can be postponed indefinitely (see Fig. 39.18). Thus structural diversity of the environment seems to favor community stability.

Human intervention in biological communities often has the effect of simplifying them, both by reducing species diversity (Fig. 39.37) and by diminishing the structural complexity. The result can make the communities far more prone to extreme fluctuations in response to changing conditions. The highly artificial communities created by modern agriculture, in particular, are notoriously unstable. Agricultural practice tends increasingly to emphasize monoculture—the planting of a single crop species in enormous fields from which all other plant species are systematically excluded. Such pure systems require constant vigilance to curb insect and mite infestations and outbreaks of strain-specific disease, to maintain soil fertility, and to clean out or kill invading weeds. As experiments have demonstrated, multiple-species crops have greater resistance to pests and disease, but the difficulty and expense of adapting conventional farm machinery to planting and harvesting such crops are prohibitive for large-scale farming.

■ SUCCESSION: WHY COMMUNITIES CHANGE

The dynamics of succession Succession is a more or less orderly process of community change. It involves the gradual replacement of the dominant species within a given area by other species. If a farmer's field is allowed to lie fallow, a crop of annual weeds will grow in it during the first year. Many perennial plants appear in the second year and become even more common in the third year. Soon, however, these are superseded as the dominant vegetation by woody shrubs, which may in turn be replaced eventually by trees.

What accounts for this change? Why does succession occur? The traditional view has been that the most important cause is the modification of the physical environment produced by the community itself: successional communities tend to alter the area in which they occur in such a way as to make it less favorable for themselves, and more favorable for other communities. In effect, each community in the succession "sows the seeds" of its own destruction.

Consider the alterations initiated by pioneer communities on newly formed land like sand dunes or lava. This process, known as ***primary succession***, begins with soil building: the first communities (often dominated by cyanobacteria and algae) will produce a layer of litter on the surface, creating the first traces of soil. This accumulation of litter affects the runoff of rainwater, the surface temperature, and the formation of humus (decomposed organic material). The humus, in turn, contributes to soil development and thus alters water retention, the pH and aeration of the soil, the availability of nutrients, and the sorts of soil organisms that will occur. However, the organisms characteristic of the pioneering communities that produced these changes may not prosper under the new conditions; they may be replaced by invading competitors that thrive in an area with more nutrients and better water retention.

Secondary succession involves reestablishment of a community after the original one is destroyed, such as when a forest is cleared, leading to the succession of weeds, herbs, shrubs, and then trees. No soil building is necessary. In fact, the best way to keep from confusing the two processes is to remember that the essential and dominant characteristic of primary succession is soil building, whereas, as we will see, secondary succession involves a series of competitive replacements of one species by another.

There has been growing recognition among researchers that modification of the habitat cannot be the entire explanation for ecological succession. To some degree, changes in vegetation merely demonstrate that some species are more easily dispersed and grow more rapidly than others. Annual herbs are far more important members of a pioneering community on recently abandoned farmland than are the seedlings of slow-growing trees, even though the trees may eventually become the dominant plants, because the herb seeds disperse farther and produce plants faster.

The course of succession One of the first examples of primary succession studied in detail was of sand-dune vegetation at the southern end of Lake Michigan. The lake once extended much farther south than it does today. As the lakeshore gradually receded northward after the last ice age (10,000 years ago), it left exposed a series of successively younger beaches and sand dunes. Hence someone who starts at the water's edge and walks south for several miles will pass through a series of communities (Fig. 39.38) that represent various successional stages beginning with bare beach and culminating in an old well-established forest. In some cases

39.38 Successional stages at the southern end of Lake Michigan

In a sequence of this sort, the distance from the water to the climax community—here a beech-maple forest—can be several miles.

these woods are dominated by beech and sugar maple trees; in others, there are different stable mixes such as oak and hickory.

The succession in this particular ecosystem can be described sequentially:

1 The lower beach near the water's edge has no land life because of the destructive action of waves.

2 The middle beach is ordinarily dry in summer, but is occasionally washed by the waves produced by strong winter storms. Conditions of life are very severe; a few succulent annuals grow there in summer.

3 Conditions on the upper beach are much less severe than on the lower and middle beaches, but vegetation is still very sparse.

4 Behind the upper beach is the foredune community, a pioneer community dominated by a perennial sand-binding beach grass. Tiger beetles, grasshoppers, and burrowing spiders are the characteristic animals of this community.

5 Higher up the beach, annuals have invaded, further enriching and stabilizing the soil.

6 Next appears a shrub-dominated region that includes dune willow, sand cherry, and bearberry. The plants in this community grow taller than the pioneering annuals, and so steal their light. The shrub community adds considerable humus to the soil.

7 The shrub region grades into a pine woods. However, because the pines create too much shade for their own seedlings, this stage is relatively brief.

8 The pine woods give way to hardwood forests growing in a deep, moist, humus-rich soil. These trees have taken over because their seedlings are able to survive the shade of both pines and their own parents, and their growth leads them eventually to overtop, shade, and kill the pines. The precise mix of beech, maple, oak, and hickory depends on a variety of local soil and climatic conditions, as well as the initial availability of seeds. Because the hardwood forest creates an environment in which only its own seedlings can survive, an apparent equilibrium (known as "climax") is reached.

There are many changes in the physical environment from the lakeshore to the forest—a progressive decrease in light intensity at ground level, a decrease in wind velocity and the rate of evaporation, an increase in soil moisture and relative humidity, and an increase in the amount of humus in the soil and of leaf mold on its surface. The series of communities and environmental changes seen from the lake to the forest duplicates the series of successional stages through which the area now covered by the forest must have passed since the time when it was a wave-washed beach.

Another well-understood succession is seen in ponds (Fig. 39.39). Sediments washed from the surrounding land begin to fill the pond, and the dead bodies of planktonic organisms add organic material. Soon pioneer submerged vascular plants appear in the shallower water near the margins of the pond. Their roots hold

A A newly formed pond near the beach has sandy borders bare of vegetation.

B After two years such a pond is ringed by low vegetation, including cottonwood saplings

C A 50-year-old pond is bordered by mature cottonwood trees. So much sediment is produced by organisms growing in the pond that only a small area of water, choked with weeds, remains.

D After 150–250 years an area that was once a pond has become a meadow.

39.39 Succession in ponds in Presque Isle, Pennsylvania, a peninsula in Lake Erie

the silt, and the pond bottom builds up faster where they are growing. As these plants die, their bodies accumulate faster than decomposers can break them down. Soon the water is shallow enough for broad-leaved floating pondweeds to become established, which displace any submerged species farther out into the pond. As the bottom continues to build up, the floating pondweeds are in their turn displaced by emergent species (plants that have their roots in the mud of the bottom, but their shoots extending into the air above the water), such as cattails, bulrushes, and reeds. These plants grow very close together and hold the sediment tightly, and their great bulk results in rapid accumulation of organic material. Soon conditions are dry enough for a few terrestrial plants to gain a foothold; an area that was formerly part of the pond is now dry land. The individual stages in this sequence can

39.40 An early successional stage on a bare rock surface

Lichens, shown here bearing fruiting bodies, are often the first multicellular organisms to gain a foothold on rock. Chemicals produced by the lichens corrode the rock surface and help prepare the way for later successional stages.

sometimes be seen as a series of zones girdling a pond or lake. With the passage of years, the pond becomes smaller and smaller as the zones move nearer and nearer its center. Eventually nothing of the pond remains.

Successions need not begin with land reclaimed from lakes or ponds. Consider the bare rock surfaces that were left by the volcanic explosion of Krakatoa or Mount Saint Helens, environments as inhospitable as the Galápagos must once have offered. The first pioneer organisms of the primary succession can be lichens (Fig. 39.40) and nitrogen-fixing cyanobacteria, which grow when the rock surface is wet and lie dormant when the surface is dry. The lichens release acids and other substances that corrode the rock. Dust particles and bits of dead bacteria may collect in these tiny crevices, and pioneer mosses can gain anchorage there. The mosses grow in tufts or clumps that trap more dust and debris and gradually form a thickening mat. A few fern spores or grass and herb seeds may land in the mat of soil and moss and germinate. These stages may be bypassed entirely if dead organic matter—seaweed, for example—is washed up and trapped in nooks and crannies, rapidly providing the soil needed by higher plants. Guano from seabirds can supply essential nitrogen. As more and more such plants survive and grow, they catch and hold still more mineral and organic material, and the new soil layer thus becomes thicker. Later, shrubs and even trees may start to grow in the soil that now covers what once was a bare rock surface.

Secondary succession on abandoned croplands, unused railway rights-of-way (Fig. 39.41), plowed grasslands, or cutover forests often proceeds relatively quickly in its initial stages; this is because the effects of the previous communities have not been wholly erased and the physical conditions are not as bleak as on a beach or a bare rock surface. During the first year a typical abandoned field will be covered with annual weeds, such as ragweed, horseweed, and crabgrass. In the second year ragweed, goldenrod, and asters will probably be common, and there will be much grass. The grass will usually be dominant for several years, and then more and more shrubs and tree seedlings will appear. As in many cases of pond and beach succession, the first tree seedlings to grow well in the unshaded field will be pines, and eventually a pine forest will replace the grass and shrubs. Because seedlings of oaks, hickories, and other deciduous trees are more shade-tolerant than pine

39.41 Secondary succession on an abandoned railway right-of-way

The ties are rotting, and vegetation is taking over in the formerly cleared area.

seedlings, these trees will gradually develop in the lower strata of the forest beneath the old pines, eventually replacing them. The deciduous forest thus formed is more stable, and will ordinarily maintain itself for a very long time.

Forests often become stratified into more or less distinct layers, each of which has its own species and interactions. The forest floor or herb layer, the shrub level, the short-tree level, and the canopy level are common strata of deciduous and tropical forests. The canopy species capture most of the sunlight, but much of the energy they assimilate must be used to build and maintain woody supporting tissues. The herb layer, on the other hand, receives as little as 1% of the available sunlight, but because the plants have no wood, they can use all the energy from photosynthesis for maintenance and reproduction.

As Figure 39.42 indicates, changes correlated with the succession of dominant plants also take place in the animal populations of the abandoned cropland community.

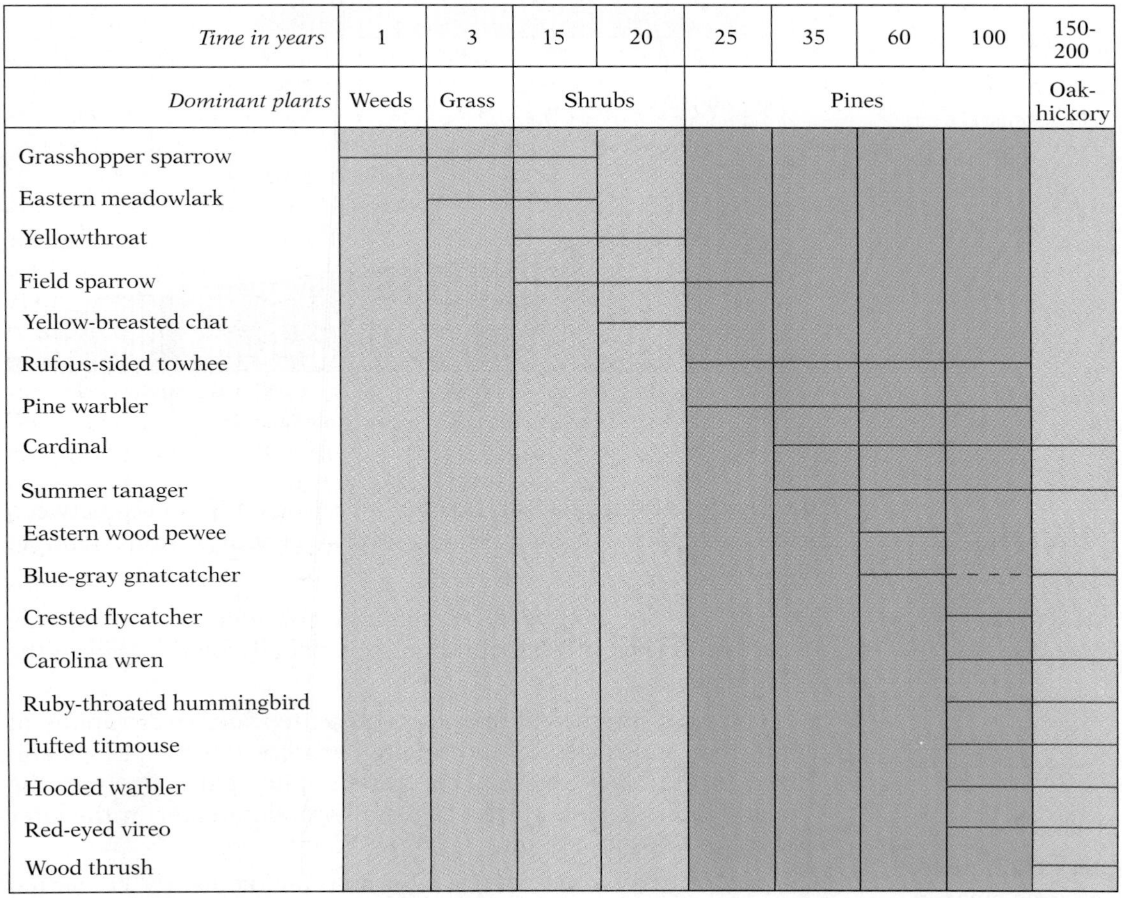

Time in years	1	3	15	20	25	35	60	100	150-200
Dominant plants	Weeds	Grass	Shrubs		Pines				Oak-hickory

Grasshopper sparrow
Eastern meadowlark
Yellowthroat
Field sparrow
Yellow-breasted chat
Rufous-sided towhee
Pine warbler
Cardinal
Summer tanager
Eastern wood pewee
Blue-gray gnatcatcher
Crested flycatcher
Carolina wren
Ruby-throated hummingbird
Tufted titmouse
Hooded warbler
Red-eyed vireo
Wood thrush

39.42 Bird succession on abandoned upland farmland in Georgia

The bars indicate when each bird species was present at a density of at least one pair per 10 acres. In the early stages (weed and grass), grasshopper sparrows and eastern meadowlarks were the dominant bird species. During the shrub stage yellowthroats and field sparrows became dominant. Pine warblers and rufous-sided towhees dominated the young pine forests, and red-eyed vireos, wood thrushes, and cardinals were the most common birds in the oak-hickory forests.

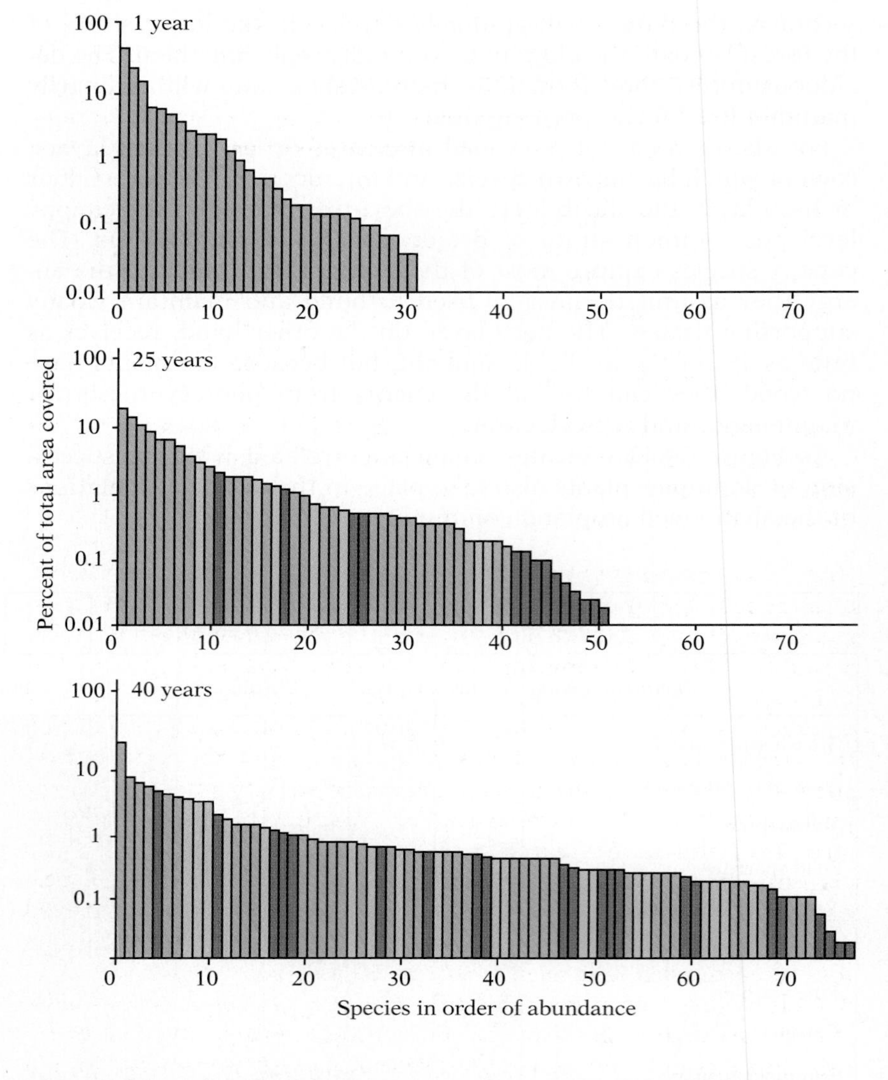

39.43 Changing pattern of species diversity and abundance in the early successional plant communities on an abandoned agricultural field in Illinois

After 1 year there were 31 species of plants, all herbs (green bars), growing in the field. After 25 years there were 51 species, including some shrubs (gold bars) and trees (purple bars). After 40 years there were 77 species, and shrubs and trees had greatly increased in relative abundance.

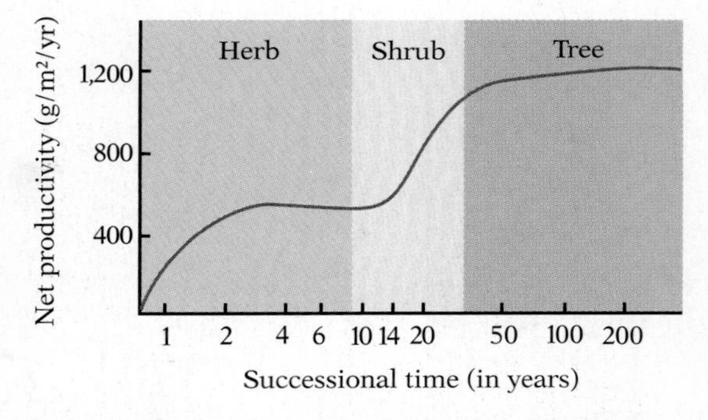

39.44 Change in net primary productivity during plant succession on an area cleared of an oak-pine forest in Brookhaven, New York

The first rise represents the invasion of the area by herbs. The later rise (after about 14 years) reflects the entrance of larger woody plants into the community.

The results of succession Despite numerous differences between various instances of succession, several generalizations tend to hold true:

1 The species composition changes continuously during the succession (Fig. 39.43), but change is usually more rapid in the earlier stages.

2 The total number of species represented increases rapidly at first, then more slowly, and finally becomes more or less stabilized in the older stages. This trend applies particularly to the heterotrophs, whose variety is usually much greater in the later stages of the succession.

3 Net primary productivity (the amount of energy converted into the products of photosynthesis by autotrophs, and available to heterotrophs) increases until it reaches a high level (Fig. 39.44).

4 The store of inorganic nutrients held in the organisms and soil of the ecosystem increases, and an increasing proportion of this store is held in the tissues of plants.

5 Both the total biomass in the ecosystem (Fig. 39.45) and the amount of nonliving organic matter increase during the succession until a more stable stage is reached.

6 The height and massiveness of the plants in the community increase and lead to greater differentiation of vertical strata into more and more microhabitats.

7 With more extensive aboveground plant cover, the microclimate within the community becomes increasingly determined by the community itself.

8 The food webs (discussed in Chapter 40) become more complex, and the relations between species in them become better defined or more specialized.

In summary, the trend of most successions is toward a more complex and longer-lasting ecosystem, in which less energy is wasted and hence a greater biomass can be supported without further increase in the supply of energy.

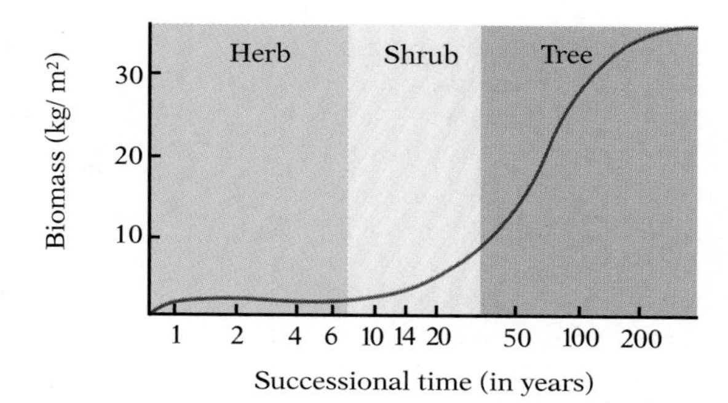

39.45 Change in biomass during succession in the Brookhaven study area

The total biomass remained low during the early years, when herbs were the dominant plants in the community, but increased later when shrubs and trees became more prominent.

Does succession ever stop? If some disruptive factor does not interfere, succession eventually reaches a stage far more stable than the stages that preceded. The important species populations attain a steady state, balancing births and deaths. Both energy flow and biomass reach equilibrium, with gross primary productivity equalled by total respiration. The community of this stage is called the ***climax community***. It has much less tendency than earlier successional communities to alter its environment in a manner injurious to itself. Its more complex organization, larger organic structure, and more balanced metabolism enable it to buffer its own physical environment to such an extent that it can be self-perpetuating. Consequently it may persist for centuries, unthreatened as long as climate, geography, and other major environmental factors remain essentially the same (Fig. 39.46).

A climax community is *not* static, however; it will change rapidly if there are major shifts in the environment, either physical or biotic. For example, 60 years ago chestnut trees were the most common species of tree in the climax forests of much of eastern North America. Since then they have been almost completely eliminated by a fungal blight, and today the climax forests of the region are dominated by other species. Apart from the time scale involved, there is no absolute distinction between climax and the other stages of succession.

Even in the absence of external change, climax communities slowly shift: the hardwood forest that appears after succession along Lake Michigan gradually becomes thinner and the proportion of oaks rises; this change results from the slow accumulation in the soil of H^+ from rainfall. Hydrogen ions compete with nutrient ions for ionic binding sites in the soil, thus lowering fertility. Acid rain accelerates this process.

It used to be thought that there is only one type of climax community for each region, and that any sites dominated by other communities have not yet reached climax, no matter how stable and long-lasting they may seem. Most modern ecologists, however, reject this view. They see the aggregation of species characterizing any given community as the product of local environmental condi-

39.46 Sequoia forest

The long-term stability of a climax forest is illustrated by this sequoia forest in south-central California. The exact age of the forest is not known, but some of the trees still standing are more than 2000 years old.

tions and of whatever plant and animal species happen to be available in the area. Since temperature, humidity, soil characteristics, topographic features, wind patterns, and other environmental conditions vary continuously in both space and time, vegetation likewise varies continuously. Boundaries between communities are therefore seldom distinct.

This view of communities as parts of a gradually changing continuum—whose characteristics at any specific place are uniquely determined by a combination of local physical conditions, local biotic factors, local species distributions, and a considerable element of chance—means that there is no absolute climax for any region. Climax, according to this interpretation, has meaning only in relation to the individual site and its environmental conditions. Ecologists holding this view study the gradients between the individual local climaxes in an attempt to learn how changes in the component parts of communities correlate with changing environmental conditions.

CHAPTER SUMMARY

HOW POPULATIONS ARE ORGANIZED AND GROW

THE IMPORTANCE OF DENSITY AND DISTRIBUTION Population sizes are often estimated through quadrat sampling or mark-and-recapture studies. Individuals in populations are usually clumped (as opposed to being uniformly or randomly distributed); often this is because the resources on which they depend are clumped, or because reproductive characteristics or behavioral needs bring or keep individuals in groups. (p. 1120)

HOW POPULATIONS GROW In the absence of restraint, populations grow exponentially at a rate r_{max}, the difference between the birth and death rates. This intrinsic rate of increase would lead to an infinite population size if one or more resources didn't limit the total sustainable population in the habitat, known as the carrying capacity *K*. In general, *r* declines from r_{max} to zero as population size increases to *K*. This density-dependent limitation creates a logistic growth pattern. Many populations grow exponentially and then fluctuate around *K*, because there is a delay between each change in population size and its impact on *r*—in particular, on the age structure of a population as well as on the delay between birth and reproduction. Other populations rarely reach *K* because chance or other events kill individuals before the habitat can be filled. (p. 1122)

HOW POPULATION SIZE IS REGULATED Intrinsic growth rates vary widely between species. Species with high r_{max} values usually specialize in exploiting unpredictable or highly variable habitats where density-independent factors control population size; individuals produce many small young when times are good. Low-r_{max} species are usually found in stable or predictable habitats where density-dependent factors are important, and produce relatively few but large young. Density-dependent factors controlling population size include predation, parasitism, disease, and intraspecific and interspecific competition for limiting resources. Interspecific competition occurs when the niches of two species overlap; when the overlap is very large, one species is usually driven extinct in the habitat. Interspecific mutualism can increase the carrying capacity of two cooperating species in a habitat. (p. 1128)

HOW SOCIAL ORGANIZATION REGULATES POPULATIONS Sociality can evolve if the net benefits of living in groups (especially hunting and foraging, resource control, and defense) exceed the costs (increased competition and fighting between members of a group, as well as greater susceptibility to disease). Group organization can lead to restricted reproduction when territorial behavior prevents some individuals from reproducing. Apparent altruism is observed most often in social species, but almost always involves kin selection or reciprocal altruism. Humans evolved as hunter-gatherers living in small territorial kin groups with low-r_{max} reproduction. (p. 1137)

HOW POPULATIONS FORM COMMUNITIES

HOW DIVERSITY AFFECTS STABILITY Populations interact in myriad ways—including competition, predation, and mutualism—to form diverse communities; these interactions sometimes create a degree of stability that can be greatly disturbed when species diversity is reduced. (p. 1146)

SUCCESSION: WHY COMMUNITIES CHANGE After a disturbance, the nature of the community in a habitat often changes in a regular way until a relatively stable community structure appears. This succession is often the result of an initial invasion of species that exploit disturbed habitats, followed by species that do well in the environment created by the earlier species, and so on until the climax community of populations changes only very slowly if at all. In general, the change in species composition is most rapid early in succession, while species diversity, primary productivity, and biomass increase as succession continues. (p. 1148)

STUDY QUESTIONS

1. In some species, an individual can opt for either a high-r_{max} or a low-r_{max} strategy, or even hedge its bets by dividing its investment between the two. Cite two unrelated examples, and analyze the circumstances under which the various alternatives would be favored.
2. Can males and females be thought of as occupying different points on the high-r_{max}–low-r_{max} continuum? How can the idea of differing life-history strategies be used to account for the phenomenon of sex-ratio manipulation? (pp. 1128–1129, 1140)
3. Analyze the effects of the various factors cited in the discussion of population regulation using human populations as your example. (pp. 1125–1137)
4. How do the various costs and benefits involved in sociality apply to plants? (pp. 1137–1144)
5. Coyotes are essentially solitary in some parts of North America, but relatively social in others. Formulate four different scenarios to account for this variability. How would you test your hypotheses against one another? (pp. 1137–1144)

SUGGESTED READING

CLUTTON-BROCK, T. H., 1985. Reproductive success in red deer, *Scientific American* 252 (2). *A classic example of a harem-based social system, with male contests and sex-ratio manipulation.*

COHEN, JOEL E., 1995. *How Many People Can the Earth Support?* W. W. Norton, New York. *An even-handed discussion of this all-important question.*

GOULD, J. L., AND C. G. GOULD, 1989. *Sexual Selection.* New York, W. H. Freeman. *A good introduction to behavioral ecology.*

MAY, R. M., 1978. The evolution of ecological systems, *Scientific American* 239 (3).

MAY, R. M., 1983. Parasitic infections as regulators of animal populations, *American Scientist* 71, 36–45.

POWER, J. F., AND R. F. FOLLETT, 1987. Monoculture, *Scientific American* 256 (3).

REICE, S. R., 1994. Nonequilibrium determinants of biological community structure, *American Scientist* 82 (5). *On the frequent circumstance in which succession is repeatedly disturbed before stability is reached, leading to greater biodiversity.*

WILKINSON, G. S., 1990. Food sharing in vampire bats, *Scientific American* 262 (2). *On a remarkable example of reciprocal altruism in bat roosts.*

CHAPTER 40

THE ECONOMY OF ECOSYSTEMS

HOW THE SUN CREATES CLIMATE

HOW THE SUN POWERS LIFE

HOW ENERGY MOVES THROUGH AN ECOSYSTEM

RECYCLING IN NATURE

THE IMPORTANCE OF SOIL

THE ECOLOGY OF THE PLANET

BIOMES: THE MAJOR FEATURES OF LIFE ON EARTH

HOW TODAY'S BIOMES CAME TO BE

HOW SPECIES PERSIST AND SPREAD

ECOSYSTEMS AND BIOGEOGRAPHY

Chapter 39 is concerned primarily with interactions between individuals within a population, between populations within a community, and between the community and its environment. The discussion of ecology in this chapter will explore some of the ways in which the movement of energy and materials binds the community and the physical environment together as a functioning system—in effect, the nutritional "economics" of ecosystems. We will also consider how the distribution and cycling of energy and materials affect the nature and distribution of communities on a global scale, a subject known as biogeography. We begin with the source of virtually all of the energy available to life, the sun, and consider how it both creates climate and fuels photosynthesis.

THE ECONOMY OF ECOSYSTEMS

HOW THE SUN CREATES CLIMATE

The sun is the ultimate energy source for life, and the distribution of its energy in large part determines the distribution of living things. The sun bathes the earth in warming radiation, but because the earth's surface curves away from the path of incident light, areas at different latitudes receive different amounts of sunlight, and consequently have different ranges of temperature. The tropics, for instance, receive almost five times as much energy per unit area as the polar latitudes. Moreover, since the earth's axis of rotation is tilted with respect to the orbit around the sun, mid-temperate latitudes receive more than twice as much solar energy at the beginning of summer than at the beginning of winter (Fig. 40.1).

In addition to a latitudinal gradient of temperature (Fig. 40.2A), the uneven distribution of sunlight causes the warm air of the tropics to rise, drawing along behind it the cooler surface air from the temperate zones. This affects the distribution of rain dramatically. The capacity of air to hold moisture decreases as the temperature falls, and the temperature of the atmosphere falls with increasing altitude. As a result, the moisture in the rising tropical air con-

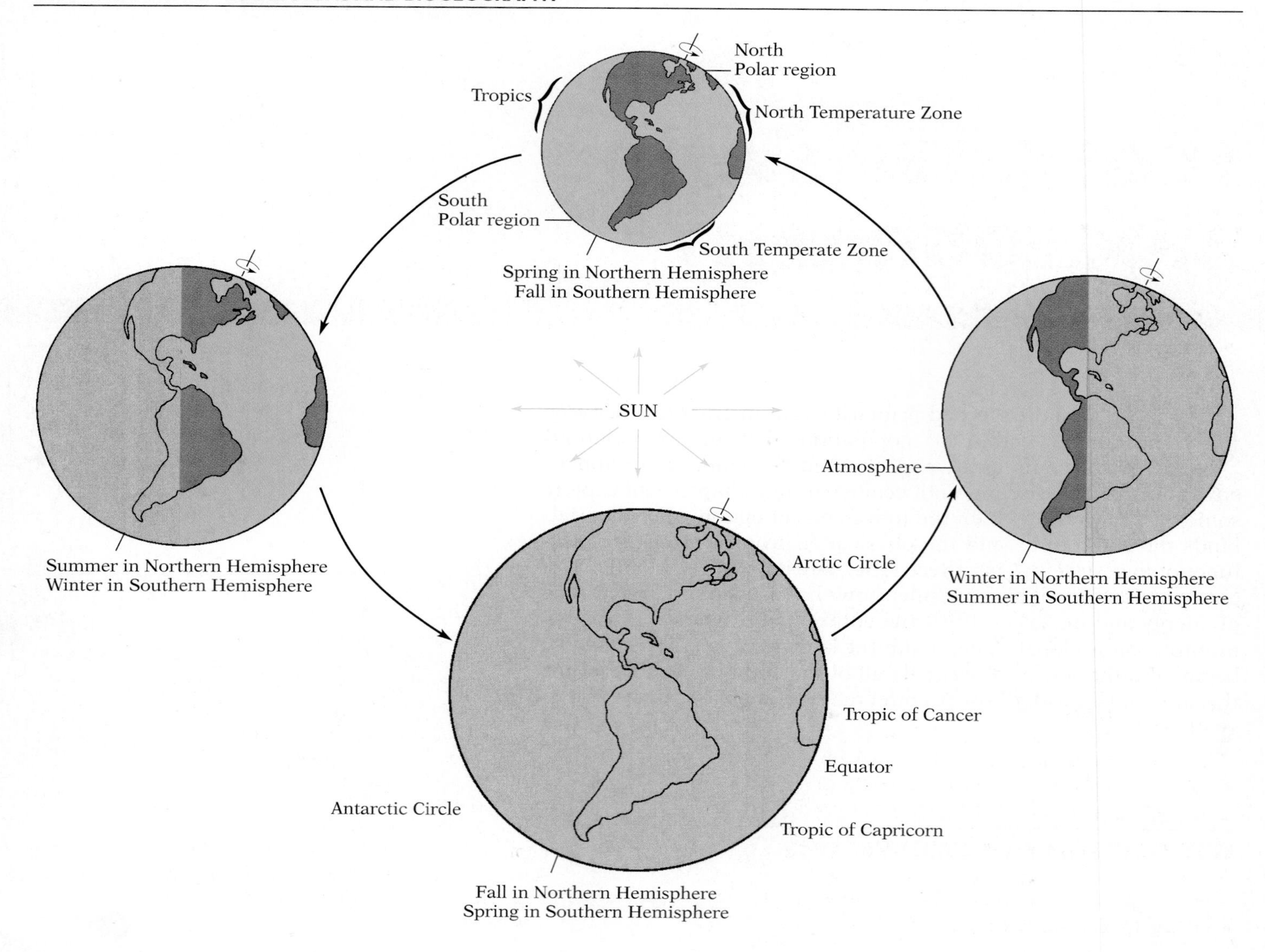

40.1 Source of the seasons
Because the earth's axis of rotation is inclined 23.5° with respect to its orbital plane, the Northern Hemisphere is tipped toward the sun during the summer (left) but away during the winter (right). The part of the earth tipped toward the sun is illuminated more vertically, so the energy received per unit area is higher in summer than in winter. The number of hours of sunlight is also greater in summer. The tropics, by contrast, receive strong, relatively vertical illumination all year, while the polar regions, on average, receive their sunlight more obliquely, spread out over more surface area. The absorption of sunlight by the atmosphere magnifies these seasonal and latitudinal differences in illumination: sunlight falling on the polar regions must travel through more air than sunlight in the tropics; hence more of its energy is dissipated before reaching the surface.

denses, bathing the tropics in rain. As this air cools and loses its moisture, it moves up and away from the equator, finally descending near the Tropic of Cancer or the Tropic of Capricorn (Fig. 40.2B). These areas of the world, which include Australia, Saudi Arabia, the veldt of South Africa, and the Sahara of northern Africa, tend as a result to be extremely dry.

Mountains can give rise to similarly radical variations in climate. Just as the moisture in tropical air condenses when it is carried up to cooler altitudes, so the moisture in winds blowing up and across a mountain tend to condense at higher altitudes on the windward side (the side facing the prevaling winds). As a result, that side is usually much more lush than the side swept by the dry descending air (Fig. 40.3). As we will see, direct solar energy and precipitation (which depends on solar energy to evaporate water into the air and then move the air to cooler regions, where the water condenses) are critical factors for understanding how terrestrial ecosystems work.

Temperature (°C)
30
20
10
0
−10
Annual Mean
Variation in monthly means
60° N
30°
0
30°
60° S
Latitude

A

Tropic of Cancer
Equator
Tropic of Capricorn
Cool, dry air descends
Warm, moist air rises — rain falls
Cool, dry air descends

B

40.2 Latitudinal patterns of temperature and rainfall

(A) More sunlight per unit area falls on tropical latitudes, generating the highest mean global temperatures; because of the seasons (see Fig. 40.1), the variation in mean monthly temperatures increases as you move from the equator into temperate and polar latitudes. (B) Because of these latitudinal temperature differences, warm, moist air rises near the equator, cools at higher altitudes, and—still over the tropics—releases its moisture as rain. As a result, when this air descends (roughly at the Tropic of Cancer and the Tropic of Capricorn), it is unusually dry, which contributes to the formation of deserts. This circulation of air is shown in cross section (at right) for the equinoxes; because the earth's angle to the sun varies with the seasons, the latitude of maximum solar irradiation moves north and south between the tropic lines, thus reducing the full impact of this wet-dry dichotomy. Its effect on the continent of Africa is evident in this NASA photograph (made with UV light). Vegetation is heaviest in the white areas, still abundant in the red areas, and somewhat less so in the green ones. The brown areas are very dry and often barren.

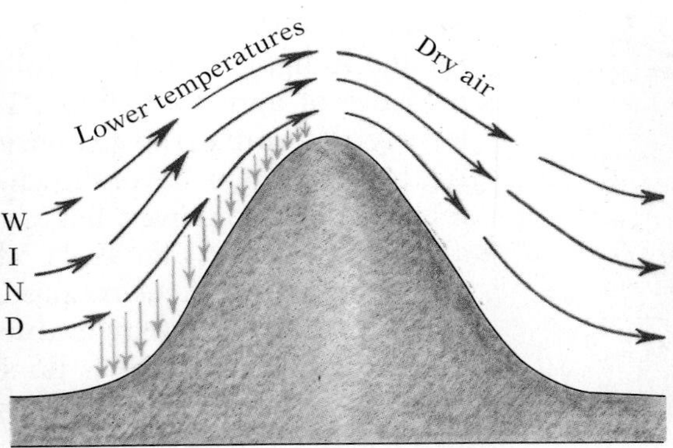

40.3 Effects of mountains on local climate

When moist air is forced up to cooler altitudes by mountains, the moisture frequently condenses as rain, and the result may be lush vegetation on the windward side, as found, for example, along the western border of the Sierra Nevada in California (left). The dry air descending on the other (leeward) side of the mountain creates a more arid environment there; the Mojave Desert, on the eastern border of the Sierra Nevada (right), is typical of the deserts that may occur downwind of a mountain range.

■ HOW THE SUN POWERS LIFE

All forms of life, with the exception of the chemosynthetic organisms, obtain their high-energy organic nutrients either directly or indirectly from photosynthesis. The total amount of energy converted into the products of photosynthesis is called ***gross primary productivity.*** Plants use 10–70% of their gross productivity in their own respiration; this respiratory "overhead cost" is usually greatest in large plants (especially trees), which have large quantities of vascular tissue to build annually. The rest is used to make new tissue, and is known as ***net primary productivity.*** The total net primary productivity of the biosphere, estimated at about 6×10^{20} calories of energy per year,[1] constitutes the energy base for heterotrophic life on earth. Heterotrophic organisms—most bacteria and protists, and virtually all fungi and animals[2]—obtain the energy they need by feeding on autotrophic organisms, on other heterotrophs that fed on autotrophs, or on the ***detritus*** (waste products or dead tissue) of other organisms.

As detailed in Chapter 7, photosynthesis can be summarized as follows:

$$6CO_2 + 6H_2O + \text{light} \longrightarrow C_6H_{12}O_6 + 6O_2$$

The energy of light is used to combine carbon dioxide gas and water to produce high-energy carbohydrates (glucose in this example) and oxygen gas. The carbohydrates in turn are used to build new tissue and to fuel metabolic processes. Metabolism requires respiration, a process that extracts the energy stored in carbohydrates and other energy-rich products of photosynthesis. Respiration reverses the photosynthetic reaction, producing ATP and heat instead of light.

At a glance, the reaction of photosynthesis tells us that light and water are critical to primary productivity. They are also critical for the process of transpiration: as detailed in Chapter 29, the heat of solar energy causes water in leaves to evaporate; the loss of leaf water draws nutrient-rich water up from the roots. (Chapter 26 describes the role of mineral nutrients in plant growth—particularly the nitrogen atoms required in every amino acid and nucleic acid.) Since, as shown in Figure 40.2A, temperature is usually strongly related to light intensity, it follows that temperature can be a good measure of light availability. Thus temperature and moisture may largely determine the primary productivity of a habitat. Helmut Lieth and others have shown that this intuitive guess has enormous predictive power: terrestrial productivity tracks precipitation and temperature (Fig. 40.4). They have also shown that transpiration rates are an excellent predictor of primary productivity.

Since, as we saw, temperature and precipitation are strongly correlated with latitude, it is no surprise that both transpiration and primary productivity are also correlated with latitude (Fig. 40.5). Tropical regions are generally the most productive on the globe,

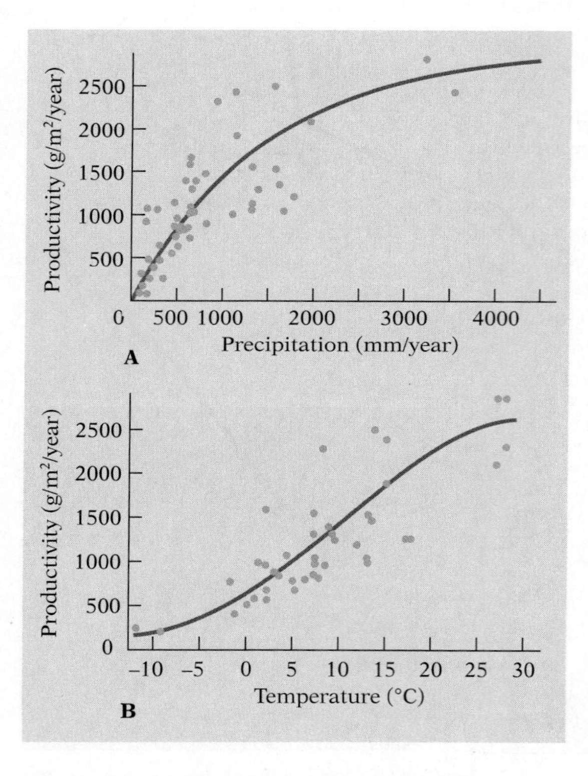

40.4 Patterns of precipitation and temperature on primary productivity

(A) The relationship between mean annual precipitation and primary productivity. (B) The relationship between mean annual temperature and primary productivity. Each curve is based on 52 sample areas around the globe.

[1] The energy content of organic materials is measured by the heat released when they are burned in pure oxygen. Organic matter has an energy content of about 4.25 kcal per dry gram of plant tissue and 5.0 kcal per dry gram of animal tissue. Some ecologists therefore give primary productivity values in dry-weight units. In these terms, the total net primary productivity for the biosphere is about 164 billion dry tons of organic matter per year.

[2] Certain fungi (in lichens) and animals (mostly certain corals) obtain at least some of their energy from photosynthetic symbionts.

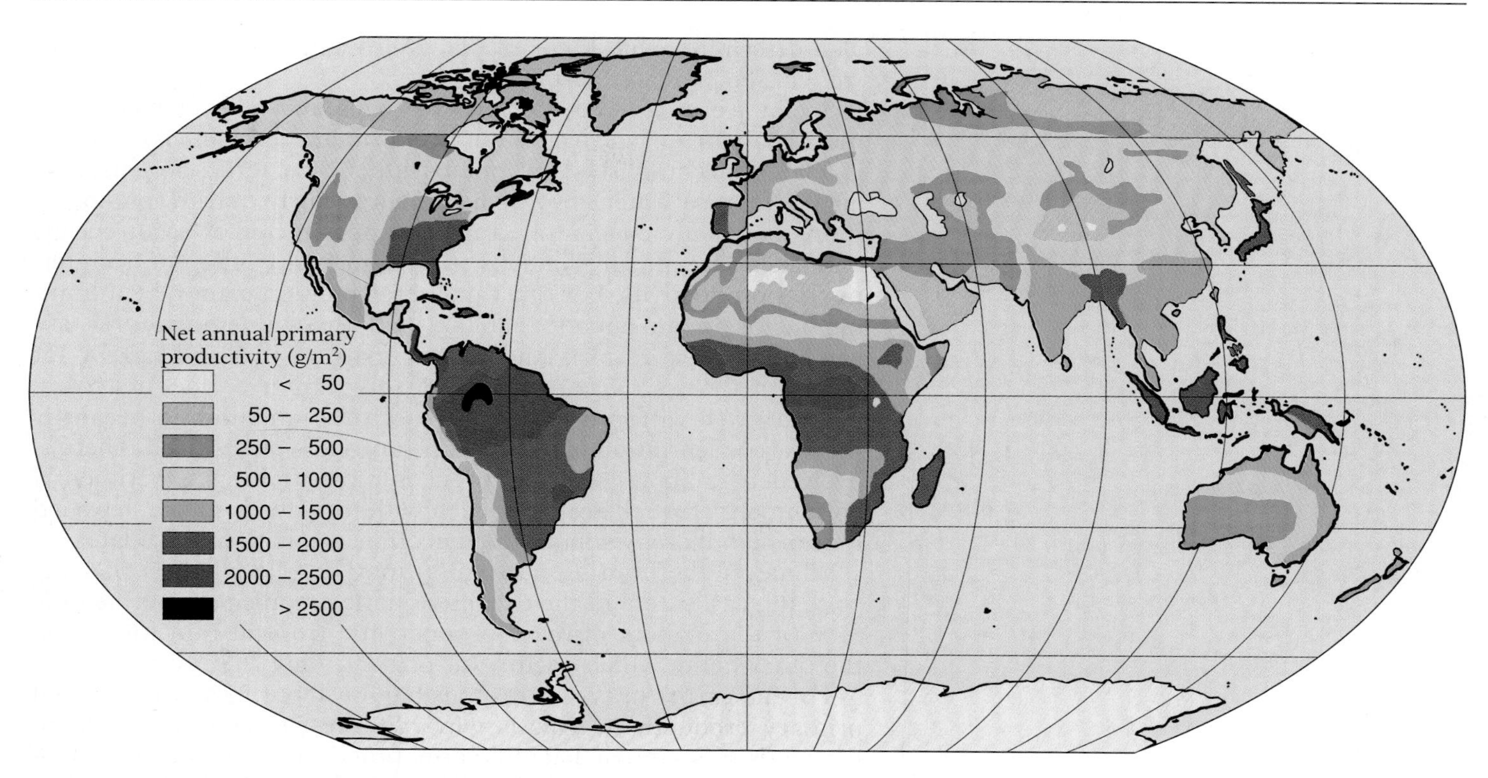

40.5 Global patterns of primary productivity

Transpiration data from around the world were used to construct this global productivity map for terrestrial habitats.

followed by the temperate regions that lie outside the arid belt (described earlier) created by atmospheric circulation and rainfall patterns. These general global patterns are further modified by ocean circulation, which brings unusually warm or cool water to the coasts (Fig. 40.6), as well as mountain ranges, which tend to remove moisture on the windward slopes. Temperature and rainfall

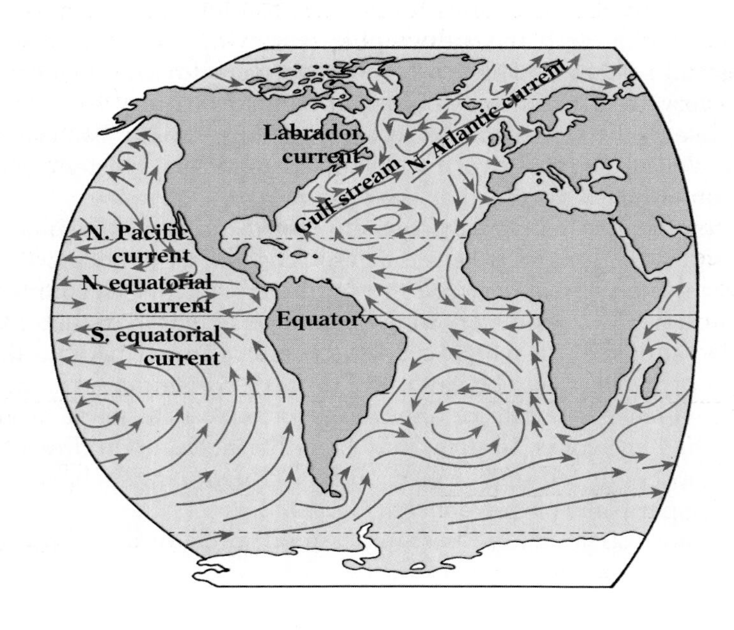

40.6 Surface ocean currents

Ocean currents can greatly modify coastal climate. Warm currents (red) increase coastal temperature; the Gulf Stream, for instance, makes the climate in Ireland and the United Kingdom much milder than we would expect in areas so far north. Cold currents (blue) have the opposite effect.

also determine what kinds of plants grow in different habitats, as we will see in a later section.

So far we've ignored the earth's oceans and lakes, which cover more than 70% of the earth's surface. Primary productivity in water is mainly accomplished by photosynthetic bacteria, protists, and algae. Because water absorbs light energy fairly quickly, photosynthesis is usually concentrated in the upper meter or so of oceans and lakes. Oceanic productivity is fundamentally different from terrestrial productivity. For one thing, precipitation cannot be a limiting factor in an aquatic habitat. And though temperatures are generally lower in lakes and oceans than on the land nearby (at least during the spring and summer), this parameter does not correlate very well with productivity. The critical limitation in oceans is the low concentration of many of the nutrients plants need. There is lively debate about which nutrient is most limiting: experimental enrichment of small areas of ocean with iron, for instance, have led to transient increases in productivity. Zinc is another candidate.

Whatever the limiting nutrient(s) may be, the result is that despite the vast extent of the oceans, aquatic organisms generate only 37% of the planet's primary productivity; tropical rain forests on the other hand, which occupy (at present) only slightly more than 3% of the earth's area, are responsible for about 22% of the global primary productivity. This is why rainforest ecology has become the focus of so much scientific attention. The relative lack of productivity in the oceans also underscores the importance of understanding in detail the ways nutrients and energy move through an ecosystem, and become available for organisms to use.

■ HOW ENERGY MOVES THROUGH AN ECOSYSTEM

The sequence of organisms through which energy and nutrients can move in a community is a ***food chain.*** In most communities, whether terrestrial or aquatic, there are many possible food chains intricately intertwined into a ***food web*** (Fig. 40.7). No matter how long a food chain or how complex a food web may be, however, certain basic characteristics are always present. Every food chain or web begins with the autotrophic organisms (usually plants in terrestrial systems), which are the ***producers*** for the community. And every food chain or web is punctuated at every level by ***decomposers,*** the organisms of decay, which are usually bacteria and fungi. Millipedes, earthworms, termites, some flies, lobsters, clams, and some catfish also feed at least partially on detritus.

Decomposers, like all heterotrophs, release simple substances—CO_2 and NH_3, for example—that are reusable as nutrients by the producers. The links between the producers and the decomposers are variable. The producers may die and be acted upon directly by the decomposers, in which event there are no intermediate links. Or the producers may be eaten by ***primary consumers,*** the herbivores, which is the fate of about 10% of terrestrial plant production. The herbivores, in turn, may be either acted upon directly by decomposers or fed upon by ***secondary consumers*** such as carnivores or parasites or scavengers (Fig. 40.8).

The most obvious (and best-understood) cycle of decomposition

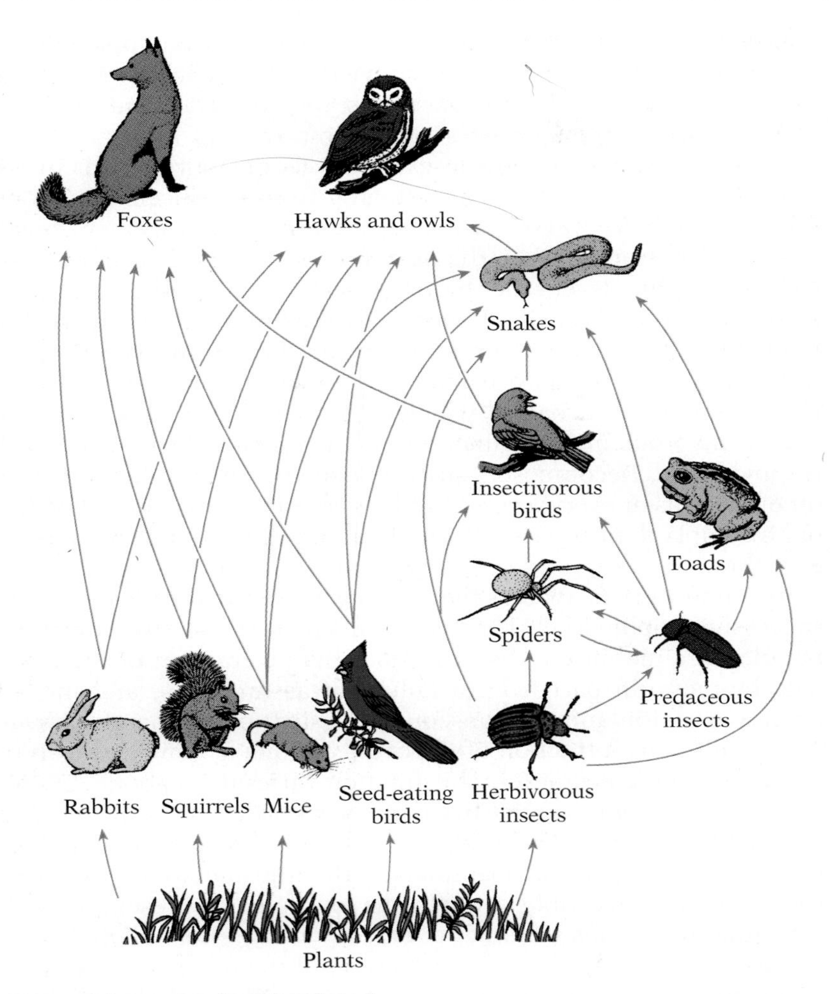

40.7 A hypothetical food web

This diagram illustrates the interdependence of living organisms, though no real food web would be as simple as this one. Parasites, disease-causing organisms, and decomposers are omitted.

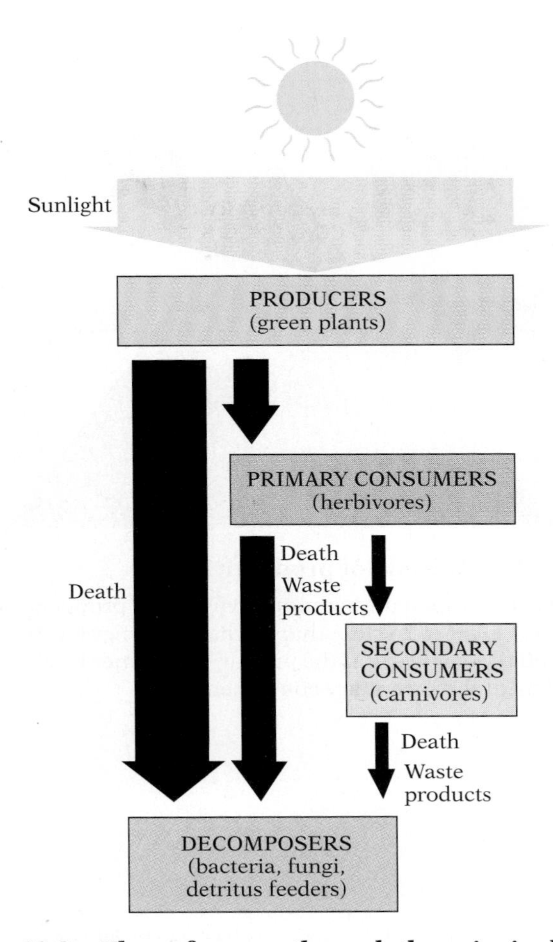

40.8 Flow of energy through the principal trophic levels in an ecosystem

The green plants are the producers, which are eaten by the herbivores, the primary consumers. The primary consumers may in turn be eaten by parasites or carnivores, the secondary consumers. When producers die, they become food for decomposer organisms; consumers generate waste products utilized by decomposers, and eventually "contribute" their own bodies when they die.

involves the leaf litter in forests. Rain leaches out about 20% of the dry mass of leaves in the form of minerals, salts, amino acids, simple sugars, and the like; the minerals and salts can be taken up by plant roots, and all can be fed on by bacteria and fungi. The remainder of the leaf consists of more complex molecules, mainly cellulose and lignin. Larger detritus feeders (particularly worms and some arthropods) are able to extract about a third of the remaining nutrients, but they cannot digest cellulose or lignin. By breaking up the leaf matter into smaller pieces, however, worms hasten the work of the main beneficiaries of leaf fall, the fungi, which can digest both cellulose and lignin. (A few species of bacteria can also digest cellulose, most often as symbionts in the digestive systems of termites and herbivores). As detailed in Chapter 23, fungi produce a meshwork of rootlike structures (hyphae) that release enzymes directly onto their food; the hyphae absorb the products of digestion. The role of fungi, which are busy consuming most of the vast quantities of dead plant material that accumulate

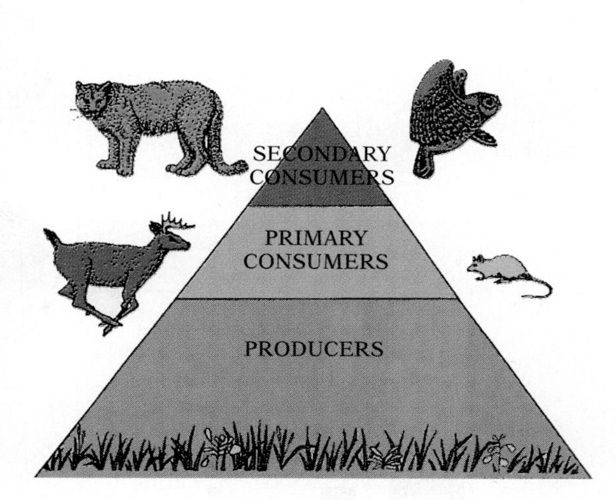

40.9 Pyramid of productivity

There is much more productivity at the producer level in an ecosystem than at the consumer levels, and there is more at the primary consumer level than at the secondary consumer level.

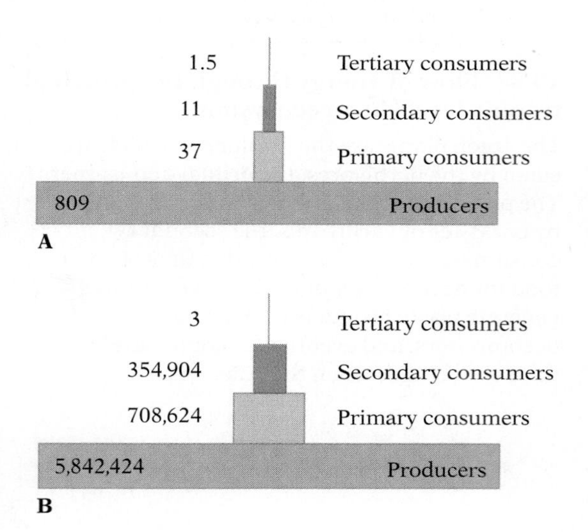

40.10 Examples of pyramids of biomass and numbers

(A) Pyramid of biomass in the aquatic ecosystem of Silver Springs, Florida. Figures represent grams of dry biomass per square meter. (B) Pyramid of numbers in a bluegrass field.

annually, is hard to appreciate. The network of microscopic hyphae in the soil and litter is generally invisible; the reproductive structures—mushrooms, for instance—are but the transient evidence of the massive ongoing process of decomposition.

Ecologists speak of the successive levels of nourishment in the food chains of a community, stretching from producers to decomposers, as ***trophic levels.*** Thus all the producers together constitute the first trophic level, the primary consumers (herbivores) the second trophic level, the herbivore-eating carnivores the third trophic level, and so on. The species that make up each trophic level differ from one community to another. Some species may function at two or more trophic levels within a single food web. For example, a chickadee, which eats seeds, herbivorous insects, and carnivorous insects, functions at the second, third, and fourth trophic levels. Decomposers stand outside this hierarchy; they feed on organisms at every level, as well as on each other, and thus defy our attempts to neatly categorize them in terms of nutrient and energy flow.

It's important to realize that energy is necessarily lost at each successive trophic level. The loss results in part from the consumer population's inability to harvest more than a fraction of the available biomass, in part from a failure of assimilation, and in part from respiration and the consequent dissipation of energy as heat. As a result, only a fraction of the energy at one trophic level can be passed on to the next level. This fraction varies from about 35% for the most efficient ectotherms consuming other animals, to less than 0.1% for some small endotherms feeding on plants. (The high values are realized when animals with little metabolic overhead feed on highly digestible animals; the low values represent animals with high overheads feeding on virtually indigestible plant matter. Values of 1–10% are the rule.) Nearly all of the rest of the energy—often more than 99%—is lost to decomposers or as waste heat. There is, therefore, far less productivity from the herbivores of a community than from the plants on which they feed, and there is still less productivity from the carnivores than from the herbivores.

The distribution of productivity within a community can be represented by a pyramid, with the first trophic level (producers) at the base and the last consumer trophic level at the apex (Fig. 40.9). Because of the rapid fall in productivity from one trophic level to the next, there are seldom more than four or five levels in a food chain; the fifth level rarely has more than about 0.0001% of the productivity of the first, and provides too low a density of potential food to support another level.

This ***pyramid of productivity*** is a characteristic of all ecosystems. Several other attributes of ecosystems may fit a pyramidal model because they are related to the flow of energy through the system. One example is the ***pyramid of biomass*** (Fig. 40.10A). In general, the decrease of energy at each successive trophic level means that less biomass can be supported at each level. Hence the total mass of carnivores in a given community is almost always less than the total mass of herbivores. However, the size, growth rate, and longevity of the species at the various trophic levels of a community are important in determining whether or not the pyramidal model applies to a community. For example, in some aquatic communities where the producers are small algae with high metabolic and reproductive rates, there may be a lower biomass of producers than of consumers at any given moment, but the total mass of all the algae that live during the course of a year is greater than the total mass of consumers that live during that year.

In general, carnivores are larger than their herbivorous prey, and secondary carnivores are often larger than the primary carnivores on which they feed. Since total biomass tends to decline at successive trophic levels, it follows that the number of individuals must decline at each level (except that decomposers outnumber all other groups combined). Consequently, some communities show a ***pyramid of numbers,*** with fewer individual herbivores than plants, and fewer individual carnivores than herbivores (Fig. 40.10B). ***Top predators*** (predators at the top of their food chains), such as lions, wolves, or killer whales, are not themselves preyed on because there are too few of them, they are too widely scattered, and they yield too little energy to make the effort worthwhile.

Some communities, however, have no pyramid of numbers. For example, there may be many more individual insect consumers than plants, even though their biomass may be less, because plant-eating insects are often far smaller than their food plants. A single large tree in spring can have hordes of leaf-eating caterpillars, as well as boring and sucking insects, feeding on it. Food chains involving parasites also tend to have inverted population-size relationships, because the parasites are smaller and usually more numerous than the hosts.

Given the inefficiency of energy transfer from one trophic level to the next, it might seem that the earth could support more humans if we lived on a wholly vegetable diet. This popular view has several flaws, however. One is that large areas of the world—much of Argentina, Australia, Africa, and western and southwestern United States, for example—can support only low-quality pasturage plants. They are unsuitable for human consumption but can sustain large herbivores adapted for just this kind of habitat. Another flaw involves human nutritional needs: most vegetarian diets require some supplemental protein, most commonly in the form of dairy products.

It is true, however, that most individuals in Western nations eat far more animal protein than they need, and cattle are fed on high-quality grain in the last few weeks before slaughter. In the United States, more than 60% of the total grain crop (mostly corn) is fed to livestock. Moreover, a substantial portion of the earth's cultivable land will always be needed to supply fodder for milk cows, which require high-quality grass to maintain lactation. Cattle raising has another controversial side: these and other ruminants are a major source of methane gas, which, as we will see in a later section, may be contributing to global warming.

■ RECYCLING IN NATURE

Energy is steadily drained from the ecosystem as it is passed along the links of a food chain. The system cannot continue functioning without a constant input of energy from the outside. Though there is no such thing as an energy cycle, most *materials* can and must

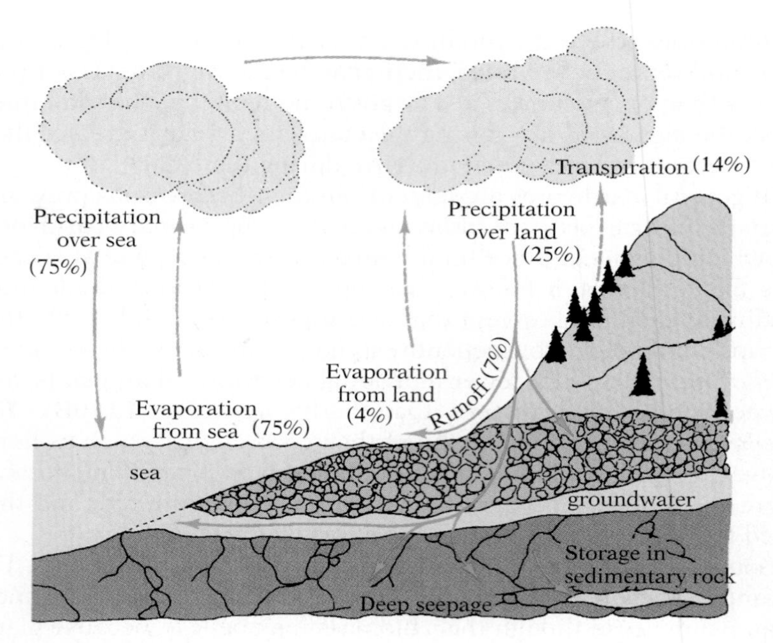

40.11 Water cycle

This diagram shows most of the major pathways of water movement through the ecosystem, although not the more recent ones created by human beings. Most of the evaporation from the land comes from plants; if the vegetation were removed, the majority of this water would join the runoff.

A

B

40.12 Hubbard Brook experiment

The pioneering Hubbard Brook experiment in New Hampshire involved careful measurements of water flow over a weir at the base of the watershed (A); this was followed by deforestation of 38 acres (B), after which water flow was monitored for several years. The volume of water runoff increased by about 50% after the watershed was clear-cut. Evaporation from the soil also increased dramatically. Previously this water—representing the majority of the precipitation (rain and snow) that entered the watershed—left this ecosystem through plant transpiration.

be used over and over again, and hence can be passed round through the ecosystem indefinitely.

As we've seen, the raw materials for photosynthesis are light, water, and carbon dioxide. In addition, growing plants need substantial quantities of nitrogen, phosphorus, potassium, sulfur, magnesium, and calcium (see Table 26.1, p. 746). Let's trace a few of these cycles here, and see how human activity is affecting the availability and distribution of these materials.

The water cycle When rain falls on the land, some of it evaporates back into the atmosphere. Of the rest, some is absorbed by plants or is consumed by animals, some runs off the surface of the land into streams and lakes, and some percolates down through the soil, to accumulate as groundwater (Fig. 40.11). Much of the water in the streams and lakes, as well as the subsurface groundwater, eventually finds its way to the ocean. There is constant evaporation from streams, lakes, and oceans, and also from the bodies of plants and animals. In the majority of ecosystems, most of the water exits through the leaves of plants as part of the process of transpiration (Fig. 40.12). The energy for transpiration and evaporation comes either directly or indirectly from solar radiation.

The endless cycling of water to earth as rain, back to the atmosphere through evaporation, and back again to earth as rain, maintains the various freshwater environments and supplies the vast quantities of water necessary for life on land. The water cycle is likewise a major factor in modifying temperatures; as we will see, it also transports many chemical nutrients through ecosystems.

Humans can have a major impact on the water cycle. Dams, for example, typically increase evaporation from entrapped water and thus reduce runoff, while forest cutting increases runoff at the expense of transpiration. When forest clearing occurs over a wide area, the ability of the forest itself to moderate local climate is lost. As the tropical rain forest of the Amazon is inexorably turned into farms, and then when it is played out into pasture land, for instance, climate models indicate that transpiration through the new constellation of plants will be 30% below present values. Because transpiration, like all evaporation, is a cooling process, the reduction in transpiration will cause the mean annual temperature to rise by 3°C. The increase in temperature will in turn reduce condensation in the atmosphere above the basin, and thus decrease the rainfall by about 25% (Fig. 40.13). The length of the dry season will increase, and the net change in climate will probably prevent regrowth of the forest. By altering the global water cycle, the loss of the Amazon rainforest is likely to change the climate in other parts of the world as well.

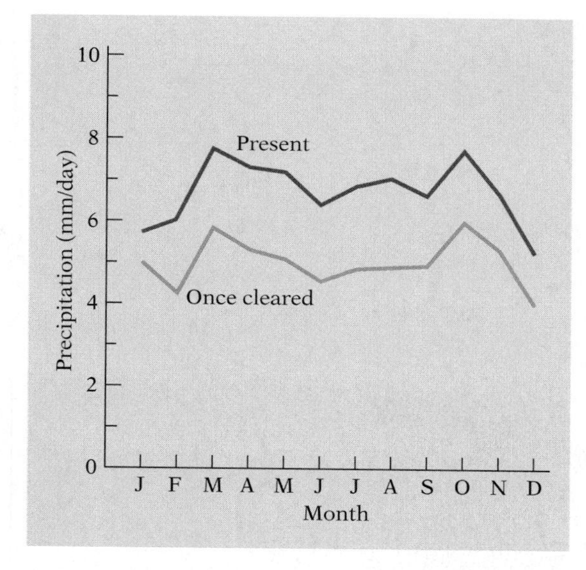

40.13 Effects of large-scale deforestation on rainfall

Conversion of the Amazon rain forest into pasture—a process already well under way—will probably reduce rainfall by about 1 m per year, about 25% below current levels. The reduced rainfall, in turn, will probably prevent regrowth of the forest.

The carbon cycle The carbon dioxide contained in the atmosphere or dissolved in water constitutes the reservoir of inorganic carbon from which almost all organic carbon is derived. Most of the rest is derived from sedimentary deposits of limestone ($CaCO_3$), which form from the shells of marine organisms; the erosive action of water on limestone leaches CO_2 into the water. Photosynthesis extracts the carbon from this inorganic reservoir and incorporates it into complex organic molecules (Fig. 40.14). Some of these organic molecules are soon broken down again, and their carbon is released as CO_2 by the plants in the course of their own respiration. However, much of it remains in the plant bodies until they

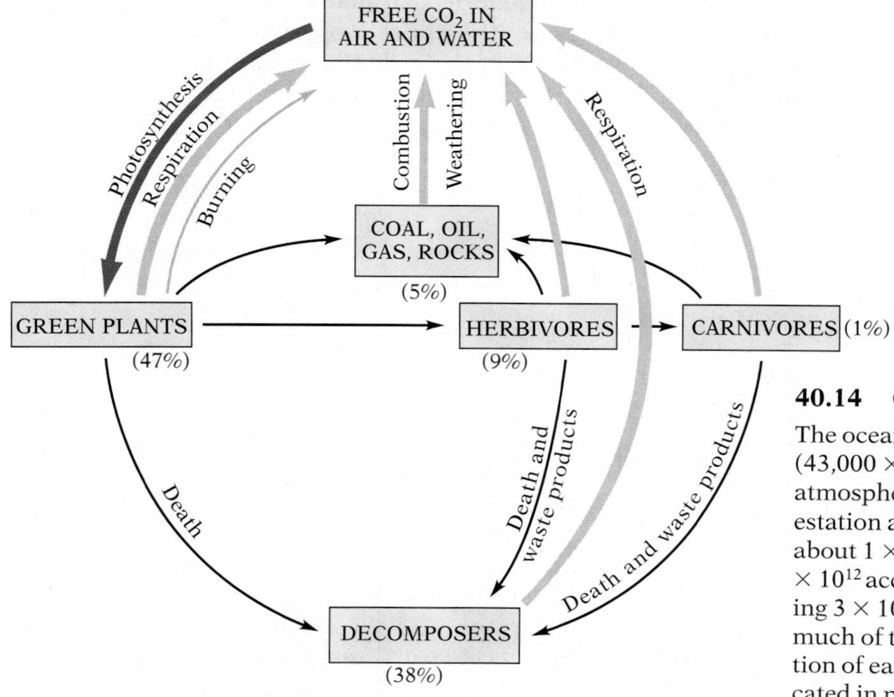

40.14 Carbon cycle

The oceans hold approximately 98% of the free carbon ($43,000 \times 10^{12}$ kg, mostly as bicarbonate ions), while the atmosphere contains only about 2% (740×1^{12} kg). Deforestation and burning of fossil fuels annually contribute about 1×10^{12} and 5×10^{12} kg respectively, of which about 3×10^{12} accumulates in the atmosphere. Some of the remaining 3×10^{12} kg is dissolved into the oceans, but the fate of much of this CO_2 has yet to be traced. The annual contribution of each reservoir of fixed carbon to the free CO_2 is indicated in parentheses. Note that the human input to the carbon cycle (from burning of plant matter and fossil fuels) is about 5% of the total.

die, are burned, or are eaten by animals. The carbon obtained from plants by animals may be released as CO_2 during respiration, be eliminated in more complex compounds in the body wastes, or remain in the animals until they die. Usually the wastes from animals and the dead bodies of both plants and animals are broken down (that is, their energy is extracted through respiration) by the decomposers, and the carbon is released as CO_2.

The pathways outlined here are all routes through which carbon moves rather rapidly, taking minutes or hours, or at most a few years. Alternative pathways can take much longer. Instead of decomposing promptly, the dead bodies of organisms are occasionally converted into coal, oil, gas, or rock (particularly limestone). Carbon in these forms may be removed from circulation for very long periods, but some of it may eventually return to the inorganic reservoir if the coal, oil, and gas (the fossil fuels) are burned or if the rocks weather. Humans have greatly accelerated the return of such carbon to the active cycle: about 5% of the current input to the carbon cycle comes from the burning of vegetation and fossil fuels. Where in the active cycle is this extra CO_2 going?

Of the CO_2 released by the burning of fossil fuels, about half remains in the atmosphere and about a third dissolves in the ocean, where it is available to phytoplankton for fixation. (The phytoplankton are then consumed by zooplankton with the inefficiency typical of grazers, so that most of the carbon sinks as feces or body parts, and vanishes from the cycle for many years.) The CO_2 reservoir in the atmosphere is also increasing as forests are burned and other areas are cleared for roads, buildings, or agriculture. (Agricultural crops usually fix less CO_2 than does the natural vegetation they displace, because they are highly productive for a relatively short time—see Figure 39.42, p. 1153; crops are represented by the portion of the curve from 0 to 1 year.) These human activities have increased atmospheric CO_2 and methane (CH_4) levels by 15% in the last 100 years; it is entirely possible that CO_2 and CH_4 concentrations will double in the next 100 years (Fig. 40.15).

Carbon dioxide and methane play an integral role in the regulation of temperatures on the surface of the earth. Heat radiated

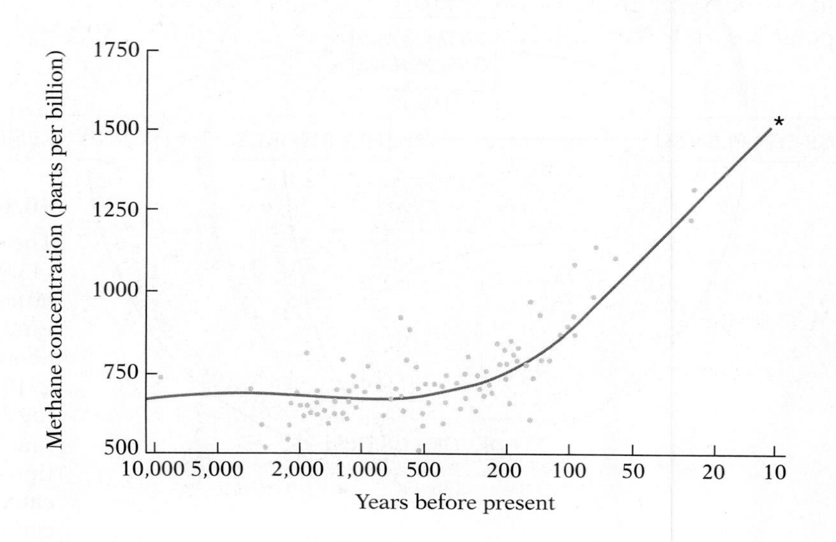

40.15 Increase in atmospheric methane over the past 10,000 years

Samples of air trapped in glaciers can be used to reconstruct the composition of the earth's atmosphere in the past. The methane level (which is strongly correlated with CO_2 concentration) has risen dramatically over the past 200 years. Most methane is released by anaerobic organisms in bogs, but ruminants, including domestic cattle, and the burning of fossil fuels are also major sources.

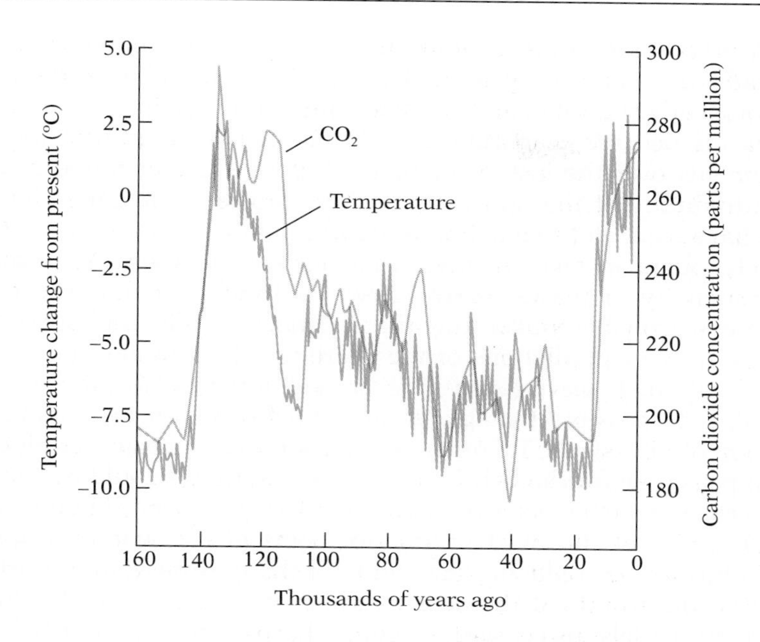

40.16 Correlation between atmospheric CO_2 concentration and average temperature

Global temperature (measured in terms of the amount of ice deposited in Antarctic glaciers) correlates well with atmospheric levels of CO_2 over the last 160,000 years. The interpretation of these data is controversial: in many cases the CO_2 levels seem to *follow* temperature changes; if there were a strict cause-and-effect relationship, CO_2 changes should precede temperature fluctuations.

from the earth is absorbed by CO_2 and CH_4 in the atmosphere; radiated back to the surface, it tends to warm the earth in what is called the ***greenhouse effect.*** Hence a rise in the CO_2 and CH_4 levels in the atmosphere should cause the temperature of the earth to increase. However, cloud distribution, total cloud cover, amount of water vapor, and amount of atmospheric particulate matter can alter the proportion of solar radiation reflected away from the earth. Because these factors are also changing as a result of human activities, it is difficult to predict what human activity is going to do to the climate. Historical evidence, however, indicates a strong link between CO_2 and CH_4 (which is about 20 times as active as CO_2) concentrations and temperature (Fig. 40.16). Most current estimates indicate that the average temperature of the earth will rise by at least 1–2°C over the next few decades. This minute increase in global temperature would be enough to melt a significant amount of polar and other ice, thereby raising the sea level by a few meters. The consequences for coastal areas, particularly major port cities like New York, could be disastrous. If warming continues, and the polar ice melts entirely, large areas would be flooded (Fig. 40.17).

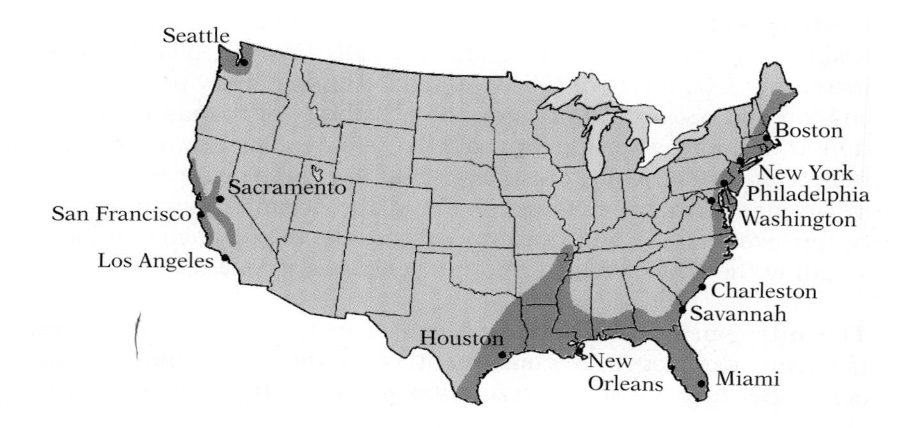

40.17 Effect on ocean level if the polar ice melts

Complete melting of the polar ice would raise sea level about 75 m, which would greatly alter the coastline of many countries.

At the same time, as CO_2 levels rise, global photosynthesis should be increasing—after all, CO_2 is a critical ingredient in photosynthesis. The rising global temperature should also be increasing oceanic evaporation, and thus increasing rainfall. Measurements over the last decade already reveal this expected rise in the humidity of the upper atmosphere (about 0.5% annually). As we have seen, elevated temperature and rainfall, like the rising CO_2 levels, ought to favor increased primary productivity. As primary productivity increases, more carbon is fixed, and thus the biosphere is probably moderating the rise in CO_2 levels. Global models differ in their predictions of the degree of this present and future moderation; however, since the human input is only 5% of the total, it is possible that plants and phytoplankton are taking up much of the slack. In other words, without primary producers compensating for the CO_2-induced changes, things could be getting worse a great deal faster than they are. Of course, the simultaneous destruction of the most productive areas of the world—tropical rain forests—is reducing the ability of the biosphere to fix carbon at just the historical moment in which it is most needed. Global climate models also disagree about where on the planet temperatures will rise the most.

It seems likely that major changes in primary productivity and plant type will occur in many habitats as temperatures and CO_2 levels rise. But what sorts of changes? In greenhouse tests plants simply grow better when there is more warmth and CO_2, but in nature the situation is more complex. In one technically impressive experiment, researchers encircled a test area with CO_2-releasing towers (so that CO_2 levels could be elevated regardless of wind direction); a central monitoring tower adjusted release rates to compensate for wind velocity (Fig. 40.18)

These Free Air Carbon Enrichment (FACE) studies show striking improvements in the growth of crop plants in response to elevated CO_2. But for unmanaged ecosystems, there is sometimes only a transient boost: apparently increased CO_2 availability doesn't help if water or a mineral nutrient is limiting. Often, even this view is too simple: in one fascinating study, researchers found that range grasses grow better with increased CO_2 only during dry years. The plants were able to conserve water for growth because the high CO_2 levels allowed them to keep their stomatal openings narrow; during wet years this advantage didn't matter.

Also a consideration is the question of how the community composition will change as climate and atmospheric CO_2 change. Rising temperatures, for instance, generally favor C_4 plants, while increasing CO_2 levels benefit C_3 plants. Another lively area of discussion and research centers on the possibility of human intervention: if we increase primary productivity in a major way, we should reduce the buildup of atmospheric CO_2. Are there better forest and range management techniques? Could the oceanic phytoplankton be fertilized? These questions underscore the importance of understanding the flow of energy and nutrients in ecosystems.

40.18 FACE experiment in a forest habitat

A central monitoring tower is surrounded by a circle of CO_2-releasing towers. Trees in this CO_2-enriched plot are growing faster than trees outside the experimental area.

The nitrogen cycle Another critical element in community metabolism is nitrogen, a constituent of amino acids and nucleic acids. Because most ionic binding sites in soil are negatively

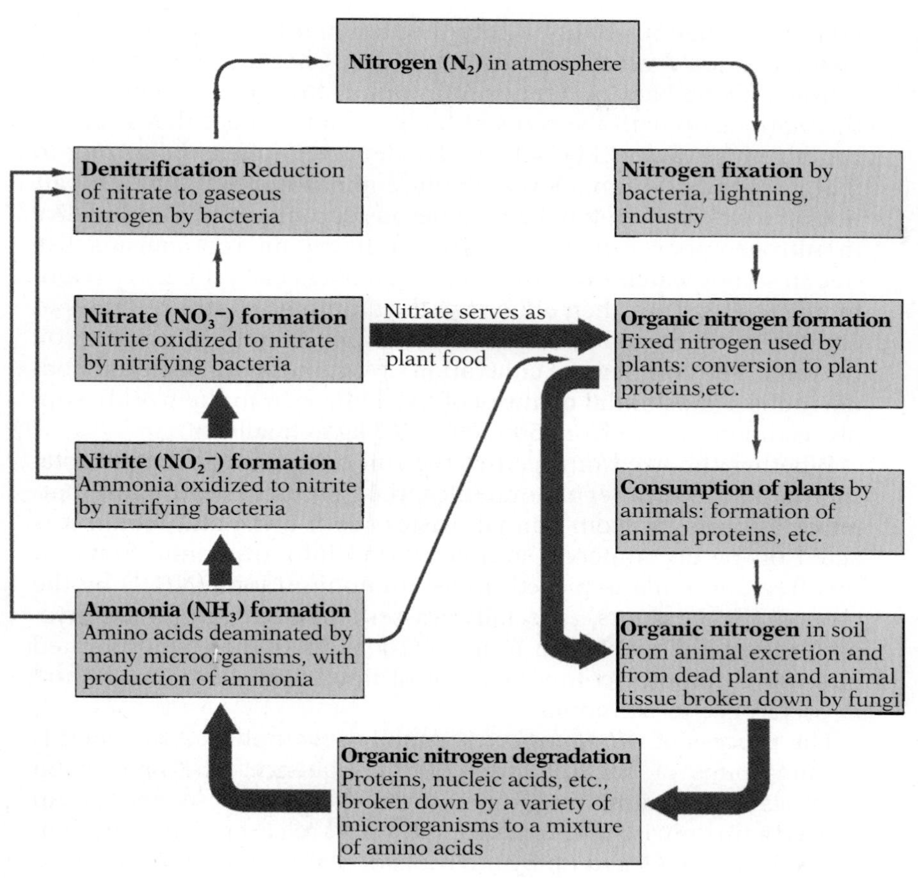

40.19 Nitrogen cycle prior to large-scale agriculture

This summary applies mainly to forests, grasslands, and other uncultivated areas of the world; where intensive agriculture is practiced, the extent of fixation by commercially grown legumes exceeds that of the other pathways for nitrogen fixation. Regardless of the source of fixed nitrogen, most of the nitrogen used by organisms was fixed long ago, and has cycled within the soil and ocean; it never enters the atmosphere. Not shown are minor long-term losses of organic nitrogen to sediments, and short-term losses of NH_3, NO_2^-, and NO_3^- to groundwater.

charged, nitrogen ions like NO_2^- and NO_3^- will wash through the soil quickly and be lost (Fig. 40.19). Thus plants must be very efficient at capturing nitrogen, or be able to obtain it from the few organisms able to fix nitrogen from air. (N_2 constitutes roughly 78% of the atmosphere.) Prior to the advent of large-scale human agriculture and industry, $100–250 \times 10^9$ kg of nitrogen were fixed annually on earth. About 50×10^9 kg were washed out of the atmosphere by rain; most of this fixation was a result of electrical discharges, such as lightning. The remaining $50\text{-}200 \times 10^9$ kg of nitrogen was fixed by bacteria, particularly the cyanobacteria, and provided most of the earth's usable nitrogen. Cyanobacteria are the most important nitrogen-fixing microorganisms that live free in the soil or water. These organisms, which may fix 10–20 kg of nitrogen per hectare annually, release ammonia into the surrounding medium, and when they die the fixed nitrogen in their cells is broken down to ammonia by decomposers. Through most of history, the rates of nitrogen fixation by cyanobacteria and the loss of fixed nitrogen as N_2 gas were roughly in balance.

The situation today is different, though estimates vary widely as to how different. Vast quantities of commercial fertilizer are being synthesized and applied to soils—probably more than 50×10^9 kg annually. Moreover, industrial processes are both fixing nitrogen from the atmosphere and releasing it back into the air. As a result, the washout of fixed nitrogen has increased by at least 100%,

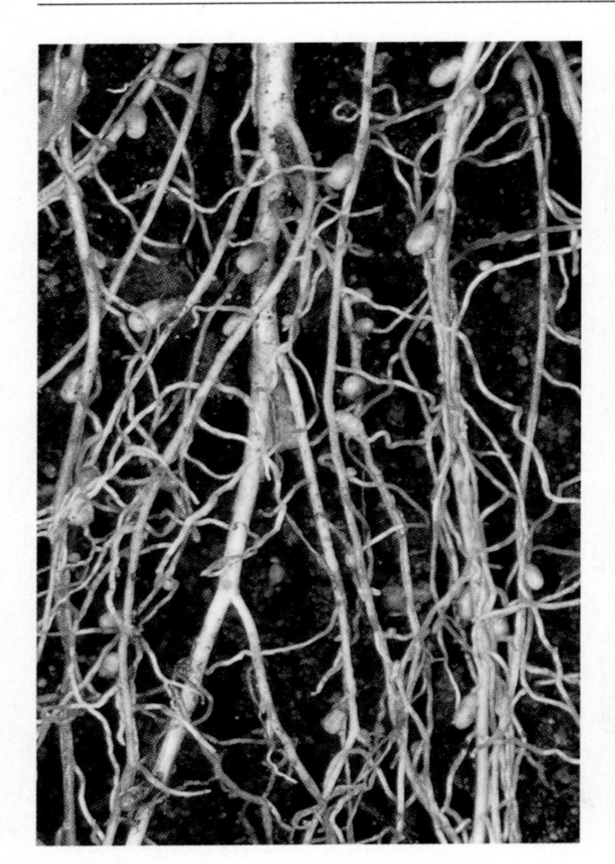

40.20 Roots of a clover plant, showing nodules containing nitrogen-fixing bacteria

adding another 50–100 × 10^9 kg. The amount fixed by bacteria has also increased dramatically. As detailed in Chapter 26, some of the nitrogen-fixing bacteria (genus *Rhizobium*) live in a close mutualistic relationship with the roots of higher plants, where they occur in prominent ***nodules*** (Fig. 40.20). The legumes (plants belonging to the pea family—bean, clover, alfalfa, lupine, and the like) are particularly well known for their numerous root nodules, and can live in nitrogen-poor soil. Other nitrogen-fixing microorganisms live free in soil or water. All of these nitrogen fixers can reduce N_2 to ammonia (NH_3). They then either use the ammonia in the synthesis of organic nitrogen-containing compounds, or excrete it into the soil or water. The commercial cultivation of legumes with bacterial nodules adds a substantial quantity of fixed nitrogen to the world's supply; estimates range from 50–200 × 10^9 kg annually.

Whether the fixed nitrogen in the soil originated in cyanobacteria, the bacteria in legume nodules, the tissues of dead plant matter or animals, or from animal wastes (both urine and feces), it is acted on by decomposers and converted into ammonia. Some of this free ammonia is picked up as ammonium ions (NH_4^+) by the roots of higher plants, especially grasses and trees, and by the mycorrhizae of the symbiotic fungi of roots. It is then incorporated into more complex compounds. Most flowering plants use nitrate in preference to ammonia.

The process of ***nitrification*** is usually accomplished by two different groups of bacteria, working in sequence. The first group converts ammonium ions into nitrite (NO_2^-); the second group converts this nitrite into nitrate (NO_3^-) and releases it into the soil, where it can be picked up by the roots of plants. Nitrogen can cycle repeatedly from plants to decomposers to nitrifying bacteria to plants without having to return to the gaseous N_2 state in the atmosphere. In this respect, the nitrogen cycle differs from the carbon cycle, where every turn of the cycle includes a return of CO_2 to the air or water.

Though nitrogen need not return to the atmosphere at every turn of the cycle, there is a steady drain of some of it away from the soil or water and back to the atmosphere: some bacteria carry out ***denitrification***, converting ammonia or nitrite or nitrate into N_2 and releasing it. The denitrifying bacteria remove nitrogen from the soil-and-organism part of the nitrogen cycle and return it to the atmosphere, while the nitrogen-fixing microorganisms do the reverse.

The phosphorus cycle Another mineral essential to life is phosphorus. Like nitrogen, it is one of the chief ingredients in commercial fertilizers. Phosphorus has its reservoir in rocks (Fig. 40.21).

Under natural conditions much less phosphorus is available to organisms than nitrogen; in natural waters, for example, the ratio of phosphorus to nitrogen is about 1:23. However, the mining of roughly 3 million tons each year has greatly accelerated the movement of phosphorus from rocks to the water-and-organism part of the cycle. This mineral, the normal limiting resource for algae in many freshwater lakes, is now being poured into the aquatic environment in enormous quantities in sewage and in runoff from in-

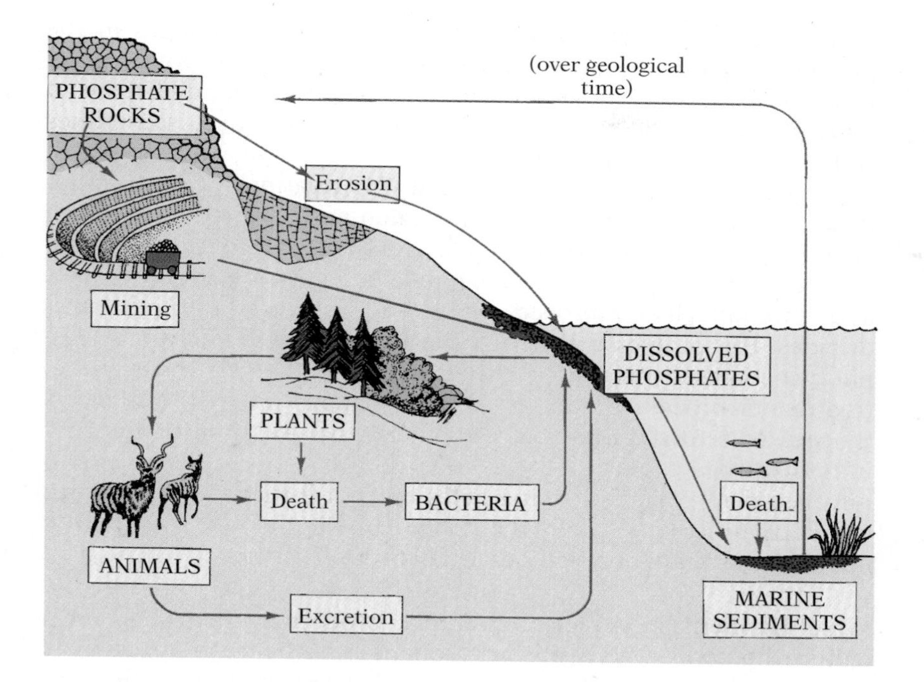

40.21 A simplified version of the phosphorus cycle

Phosphate from rock dissolves very slowly (unless the process is speeded up by human intervention). The dissolved phosphate can be used by plants, which may pass it to animals. Some of the phosphate is excreted by animals and goes immediately into the dissolved pool. When plants or animals die, phosphate is released by bacteria from organic compounds (like nucleic acids) that are present in the bodies. Each year huge quantities of dissolved phosphate are carried into the sea in runoff water. Though the formation of new rocks from marine sediments, where the phosphorus eventually comes to rest, is a very slow process, it is unlikely that we will soon run out of phosphate rock, because the known reserves are large. Nevertheless, supplies in soils are readily depleted, and so phosphorus is a limiting nutrient more often even than nitrogen.

organic fertilizers used in farming. One consequence is extensive algal or cyanobacterial "blooms" that cover the water with scum and foul shores with masses of rotting organic matter (Fig. 40.22).

The increased photosynthetic productivity associated with the algal blooms actually destroys many of the higher links in food webs. At the end of the growing season, many of the algae die and sink to the bottom, where they stimulate massive growth of heterotrophic bacteria the following year. The bacterial decomposers are so active that they consume most of the oxygen of the deeper, colder layers of lakes, with the result that cold-water fish such as trout, whitefish, pike, and sturgeon are asphyxiated, and are replaced by less valuable species such as carp and catfish.

40.22 A field experiment demonstrating the limiting nature of phosphorus in the eutrophication of a lake

The two basins of a lake were separated by a plastic curtain. The far basin was fertilized with phosphorus, carbon, and nitrogen. The near basin, used as a control, received only carbon and nitrogen. Within 2 months the far basin had a heavy bloom of algae, whereas the control basin showed no change in organic production.

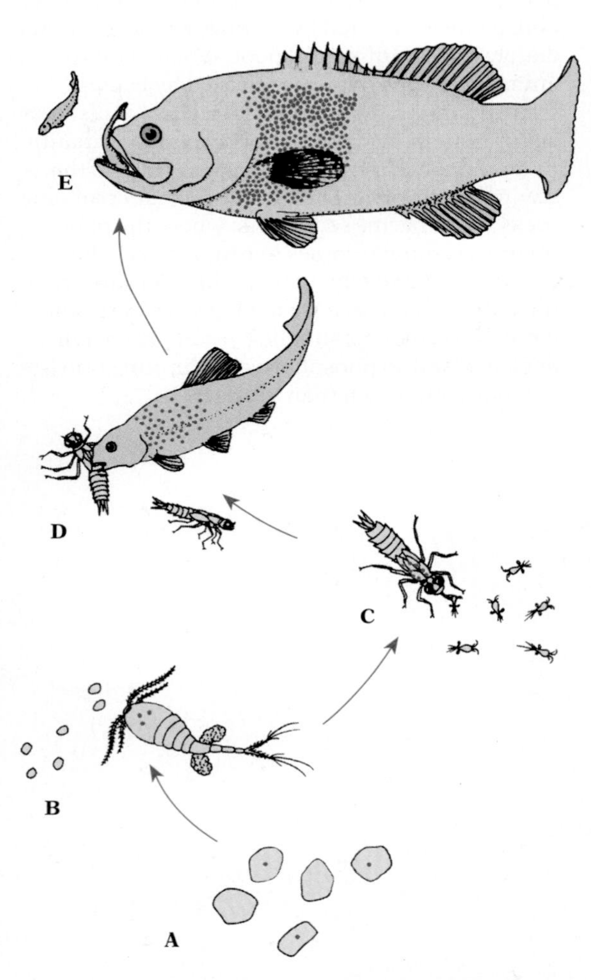

40.23 An example of biological magnification

(A) Some individuals of a single-celled plant species at the bottom of a food chain have picked up a small amount of a stable nonexcretable chemical (red). (B) *Cyclops*, a small crustacean, incorporates the chemical from the plants it eats into its own tissues. Like the other organisms in the chain, it lacks the biochemical pathways necessary to metabolize or excrete the novel substance. (C) A dragonfly nymph stores all the chemical acquired from the numerous *Cyclops* it eats. (D) Further magnification occurs when a minnow eats many of the dragonfly nymphs that have stored the chemical. (E) When a bass, the top predator in this food chain, eats many such minnows, the result is a very high concentration in its tissues of a chemical that was much less concentrated in the organisms lower in the chain.

Deoxygenation of the water also causes chemical changes in the bottom mud that produce increased quantities of odorous, sometimes toxic, gases. These changes lead to an accelerated form of ***eutrophication***, or nutrient enrichment of the lake, sometimes called "cultural eutrophication."[3]

Extensive use of advanced methods for treating sewage, the increasing reduction in the phosphate content of detergents, and a general reduction in industrial pollution have all slowed the undesirable change in the appearance and biotic composition of lakes, and dramatically improved water quality over the past two decades. But problems remain: at least 30% of the polluting phosphorus comes from agricultural sources, and advanced sewage-treatments processes are not able to remove all of the nitrates. Since fixed nitrogen is probably the major natural limiting nutrient in undisturbed estuaries, the change from phosphate detergents to nitrogen detergents is helping lakes but not estuaries. Only soaps, which clean less effectively than detergents in many cases, contain neither phosphate nor significant nitrogen.

What happens to novel chemicals? Modern industry and agriculture have been releasing vast quantities of new or previously rare chemicals into the environment. Their pathways through ecosystems are known for only a few. Most will probably be incorporated into the natural biogeochemical cycles and degraded to harmless simpler substances. However, many chemicals are so different from any naturally occurring substances that we have no idea as yet what their eventual fate or effects on the biosphere may be. The potent herbicide contaminant dioxin, for instance, turned out to mimic a naturally occurring steroid hormone that can feminize many animals and lower their reproductive fitness. Most of the by-products of industrial processes have not even been fully characterized chemically. Given the long list of substances whose dangers have been discovered over the last few decades, some of these will certainly prove harmful to life.

The matter is further complicated because a substance that is not harmful in the form in which it is released may be changed by microorganisms or natural physical processes into some other substance with vastly different properties. Mercury, for example, is released from plastics factories in an insoluble and nontoxic form that was once thought to be stable. When it settles in the bottom mud, however, microorganisms convert it into methyl mercury, a water-soluble compound that accumulates in higher organisms. The mercury poisoning that results is most severe in humans and other top predators.

The harmful effect on top predators of the mercury from plastics factories and from certain fungicides is an example of ***biological magnification***. If a persistent chemical (a biologically stable compound) is retained in the body when ingested, rather than excreted, that chemical will tend to become more and more concentrated as it is passed up the food chain (Fig. 40.23).

For instance, because of this biological magnification, DDT has

[3] Natural eutrophication refers to the accumulation of nutrients and organic matter as a normal part of the aging of lakes, and creates a typical successional series. Cultural eutrophication, by contrast, greatly decreases species diversity.

A

B

40.24 Effect of DDT on osprey

In 1950 there were over 200 mating pairs of osprey nesting at the mouth of the Connecticut River, where these pictures were taken. By 1970 only six mating pairs were observed. Decline in the local population is attributed to the detrimental effects of DDT and related hydrocarbons on the calcification of eggshells produced by these birds. The hydrocarbons had been introduced into the runoff of local streams and rivers in insecticides and consumed by the fish that were the osprey's prey. A high percentage of the weakened eggs broke during incubation; approximately 10 eggs, or two to three nestings, were needed to produce a single offspring. A nest with two eggs and a broken shell is seen in A; in B, a male and female osprey with some of the few young successfully hatched during this period. Since 1970 the local osprey population has grown considerably, largely because of a ban on the use of these chemicals.

had more severe effects on predatory birds such as the bald eagle, the peregrine falcon, and the osprey than on seed-eating birds. The reproductive rate of these birds has been calamitously reduced because DDT—and its metabolites, DDD and DDE—interfere with the deposition of calcium in eggshells; as a result, the thin-shelled eggs are easily broken and few birds hatch (Fig. 40.24). DDT is now largely outlawed, and some species are recovering.

In quite a different way, the man-made fluorocarbons (formerly used in the compressors of air conditioners and refrigerators and as an inert propellant in aerosol cans) has had the unexpected effect of depleting atmospheric ozone levels by about 1.0% annually (Fig. 40.25). Ozone absorbs most of the damaging ultraviolet (UV) light that would otherwise make the surface of the earth virtually uninhabitable. Strong international efforts to eliminate fluorocarbon use are slowing the rate of release dramatically, but fluorocarbons are very stable, and so will go on catalyzing the breakdown of ozone for decades. A decided rise in UV exposure (particularly outside the tropics) seems inevitable.

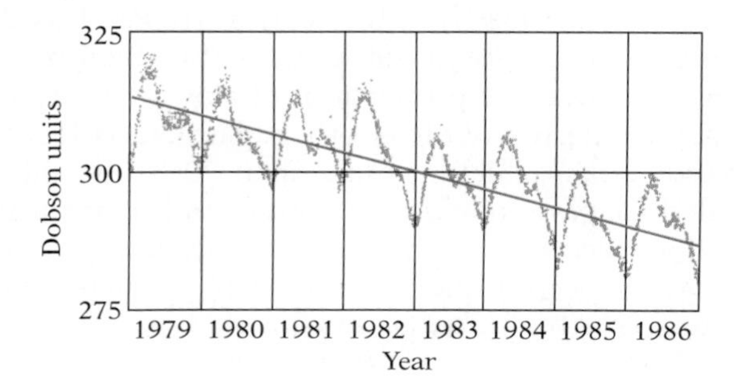

40.25 Decline in atmospheric ozone level

The global mean ozone concentration is shown for each day over the course of 8 years. There is both a seasonal rhythm and a long-term downward trend in the mid-latitudes of about 1% per year. The decline is largest in winter, and far more extreme at polar latitudes; tropical regions are incurring less ozone loss.

HOW WE KNOW: MAGNIFICATION FOR DEFENSE

While biological magnification generally damages animals higher on the food chain, there are cases in which it actually enhances the fitness of the organisms accumulating the toxin. This paradoxical phenomenon came to light when researchers sought to understand how puffer fish (*Fuga*) synthesize the potent nerve poison tetrodotoxin. Just 1 mg of this toxin is fatal to humans, yet the fish is considered a delicacy in the Orient. Careful removal of the liver, ovaries, skin, and intestinal tract generally renders the fish edible; despite these precautions, scores of people die annually from eating puffer fish.

Investigators had great difficulty isolating the sites of synthesis of tetrodotoxin in puffer fish, and for good reason: puffer fish do not make the poison themselves. The first major break came in the 1960s with the discovery by a Stanford chemist that certain newt eggs contain the toxin. Quickly researchers identified the substance in a variety of marine invertebrates as well as some algae. How could this apparently unique poison have evolved so many times? Japanese scientists guessed it might be a case of biological magnification: if some small organism produces it as chemical protection, specialized predators might then sequester the toxin for their own protection, and so on up the food chain to puffer fish, where the concentration would be enormously high.

Their guess turned out to be correct. They worked their way down the food chain, looking first for the animals puffer fish prey on. Tracing the chain was not as hard as it might sound: organisms with chemical protection generally advertise their defense with bright colors; they tend to be slow moving and conspicuous. Finally, at the end of the food chain they found a reef bacterium (*Alteromonas*) that is the original synthesizer of the poison. Subsequently, a few other microorganisms that produce the toxin have been identified. None of this work helps thrill-seeking gourmets, but it has identified alternative sources of tetrodotoxin, which is widely used in laboratories to study the activity of nerve cells.

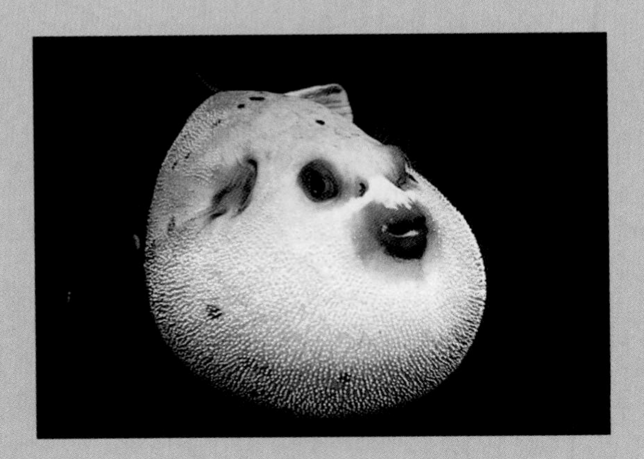

■ THE IMPORTANCE OF SOIL

Soil is essential to plants, not only as a substrate, but also as a reservoir for water and essential minerals. Each of these minerals comes to plants dissolved in soil water. The properties of soils, including particle size, amount of organic material, and pH, help determine the availability of water and minerals to the plants, and the rapidity with which these materials move through the soil.

Most soils are a combination of mineral particles, organic material, water, soluble chemical compounds, and air. In this complex the dominant components by far are the mineral particles, which (in good soils) are composed largely of silicon and aluminum compounds. They vary in size from tiny clay particles (less than 0.002 mm in diameter), through silt particles (0.002–0.05 mm), to sand grains (0.05–2.0 mm). The proportions of clay, silt, and sand in any given soil determine many of its other characteristics. For example, very sandy soils, which are less than 20% silt and clay, have many air-filled spaces. They are so porous and their particles have such a small surface area (compared with their volume) for binding water that moisture rapidly drains through them, and they are unsuitable for growth of most plants.

As the percentage of clay increases, the water retention of the soil also increases; in soils with much clay, the drainage is so poor and the water is held so tightly to the particles that the air spaces become filled with water; few plants are adapted to grow in such waterlogged soil. Though different species of plants are adapted to different soil types, most do best in ***loam***, which contains a fairly high proportion of particles of each size (for example, 24% clay, 29% silt, 30% fine sand, and 17% coarse sand). Drainage is good but not excessive, and there is good aeration; the soil particles are surrounded by (or contain) a shell of water, but there are numerous air-filled spaces between them.

Loams usually also contain considerable amounts of organic material (roughly 3–10%), mostly of plant origin. As this material decomposes, inorganic substances required for growth are released into the soil. The organic material thus contributes to soil fertility. It also helps to loosen soils with high clay contents and increases the proportion of pore spaces, thus promoting drainage and aeration. This is particularly true when the organic material is ***humus***, which consists mostly of decomposed cellulose and lignin (Fig. 40.26). Humus has the opposite effect on sandy soils, where it tends to reduce pore size by binding the sand grains together, thereby increasing the amount of water held in the soil. This is a clear case of how organisms—decomposers and the dead tissue they feed on—can modify their environment.

The proportion of clay particles also affects the amounts and availability to plants of certain minerals—in part because of the influence of the clay particles on water movement. If, for example, water percolates downward through the soil very rapidly and in large quantities, it tends to leach many important ions from the soil, carrying them too deep for roots to reach.

Leaching can be especially serious when it makes soil more acid, because few plants thrive in acid soils. The acidity of the soil is defined by the concentration of H^+ ions, which compete with nutrient

40.26 Plants growing in humus
Organic decomposition products containing high quantities of cellulose and lignin create humus. In this case, the underlying dense layer of humus, formed from decaying woody plants on a forest floor in Olympic Park Rain Forest in the state of Washington, improves the drainage and aeration of the soil and provides inorganic substances for the living plants the soil supports.

40.27 Forest damage attributed to acid rain

These trees are showing the effect of acid rain on forest soil in the Adirondack Mountains of northeastern New York State.

cations for negatively charged binding sites in the soil. Acidity reduces the availability of manganese, phosphate, and some other ions (see Fig. 26.11, p. 753), and leads to leaching of cations. High acidity also inhibits the activity of many soil organisms, which are important because they constantly work the soil, lightening and mixing it, and breaking down organic matter.

The availability of ions to plants depends not on the total amount of the various ions present in soils, but rather on the quantity of these ions that are free. A complex equilibrium generally exists between ions free in the soil water and ions adsorbed on the surface of colloidal clay and organic particles. Many factors, especially acidity, can shift this equilibrium, either increasing the proportion of ions bound to the particles, and thus reducing availability, or increasing the proportion of free ions available in the soil solution.

Even air conditions may influence the ionic makeup of soil. An extensive study of the experimental forest at Hubbard Brook (referred to earlier in Fig. 40.12), showed that rain often contains appreciable quantities of sulfuric acid, probably from sulfur dioxide in smoke. As we have seen, hydrogen ions from acids can displace nutrient cations, such as Ca^{++}, Mg^{++}, Mn^{++}, K^{+}, and Fe^{++}, from negatively charged sites on the soil particles; these nutrients are then leached more rapidly from the soil into streams and lakes. The loss of the cations caused by this acidity frees toxic aluminum ions (Al^{+++}), which compete with the remaining nutrient ions for uptake into plant roots. Soil acidity also favors the growth of certain mosses, which further acidify the water passing through the moss layer into the soil and kill the symbiotic fungi that help the roots of trees. The loss of soil fertility and the increase in toxicity as a consequence of this ***acid rain*** is increasingly widespread, and threatens forests in much of eastern North America and northern Europe (Fig. 40.27). The leaching of nutrients and toxic Al^{+++} into lakes can also accelerate eutrophication, and so lead to major changes in the character and abundance of aquatic species.

Control of sulfur emissions, however, is difficult: low-sulfur alternatives—nuclear-powered generating facilities, natural gas, and low-sulfur oil—are expensive and in short supply. To make matters worse, it seems likely that ground-level ozone (O_3), a by-product of even low-sulfur combustion, can drastically reduce the efficiency of photosynthesis, and may be a major factor in damaging forests.

Some of the effects of vegetation on soil were demonstrated dramatically in the Hubbard Brook study. As described earlier, the investigators first measured the nutrient input and output of a watershed for several years, and then cleared all the vegetation and again monitored input and output. They found an extraordinary loss of soil fertility. The runoff of nitrate rose steeply (as much as 45 times higher than in undisturbed watersheds; see Fig. 40.28), and there was a drastic increase in net losses of other nutrients as well. Removal of the vegetation so altered the chemistry of the soil that nutrients were bound less tightly to soil particles and thus were leached away.

Clearly, the wholesale human destruction of vegetation can result in increased erosion by wind and water as well as in severe loss of fertility in the soil that remains. The stability of the physical part of an ecosystem depends on the production and decomposi-

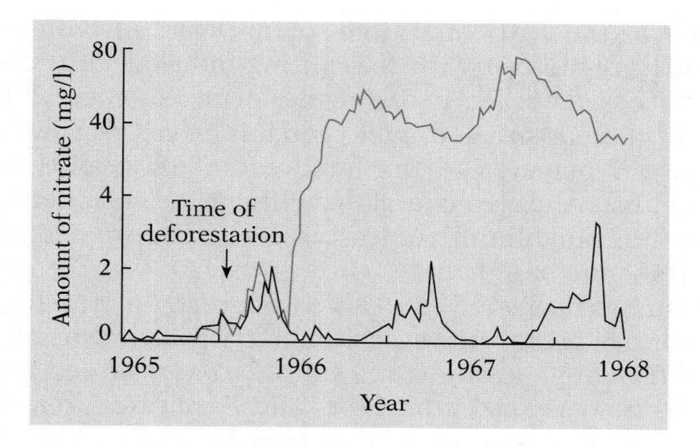

40.28 Change in the runoff of nitrate as a result of deforestation in the Hubbard Brook experimental forest

The colored curve indicates the output of nitrate in stream water in the deforested area, and the black curve the output in an undisturbed area. Shortly after deforestation the output in the experimental area rose to about 40 times its previous level, whereas the output in the undisturbed control area remained at about the same level. (Note that the vertical axis is not linear.) This shows how efficient plants and their associated root fungi are at capturing fixed nitrogen from the soil.

tion of organic matter, and on an orderly flow of nutrients between the living and the nonliving components of the system.

Cutting down forests is not the only way to ruin the soil; overgrazing and other poor farming practices have caused permanent damage in areas where there have never been forests. The valleys of the Tigris and the Euphrates rivers once supported the Sumerian civilization and the Babylonian Empire. But poor farming practices led to salt buildup in the soil (salinization) and extensive erosion; the amount of cultivable land today is less than 20% of what it once was. The ancient irrigation works are filled with silt, and so much soil has been washed into the Persian Gulf that the ancient seaport of Ur is now 240 km from the coast, its buildings buried under 10 m of silt.

In parts of the United States, overgrazing and the plowing under of native grasses, combined with a decade of drought, led to the Dust Bowl of the 1930s, when clouds of topsoil were blown hundreds of kilometers and vast areas were left barren (Fig. 40.29). Despite increasing knowledge of soil dynamics, wholesale destruction of the earth's soils and the vast numbers of species they support continues. The trees that make up the immense expanses of tropical rain forest are being cut down at a high rate; sometimes they are burned and plowed under in an effort to enrich the soil for agriculture. The land is planted for a few years at most, and finally

40.29 Dust Bowl in the United States

Drought, overgrazing, and poor farming practices led to the topsoil in larger areas of the American Midwest being blown away, leaving those regions infertile. In other areas, like the one pictured here, the blown topsoil accumulated, burying everything in dust.

40.30 Signs of salinization in Iran

A river on the island of Hormuz in the Persian Gulf shows obvious signs of salinization. Irrigation waters raised the water table of salty groundwater and brought in additional salt, much of which was eventually deposited in the topsoil. The salinization is made obvious here by evaporation, which has left an encrustation of salt on the banks.

abandoned as the nutrients in the soil are leached out. Because of the leaching, combined with the cutting-induced local changes in climate discussed earlier, many of the former forests, which occupied 7% of the earth's land, will probably never regrow. This vast feature of our planet, with its hundreds of thousands of unique species and its critical role in global climate, seems doomed to disappear within our lifetime unless vigorous and immediate efforts are undertaken to save it.

Irrigation has been viewed as a way of greatly increasing the productivity of dry areas. However, it is likely to be destructive in the long run. In many cases it leads simply to accelerated erosion. In other cases, as perhaps in the Tigris and Euphrates area, irrigation leads to rapid salinization until eventually there is so much salt that plants cannot grow. Salinization may occur because adding water to land overlying salty groundwater causes the water table to rise, pushing the salt into the topsoil. Salts originally present in low concentrations in the irrigation water also tend to accumulate in soil as water evaporates (Fig. 40.30).

The once fertile Indus Valley of Pakistan, the largest irrigated region in the world, now resembles an Arctic landscape because of the white crust of glistening salt. The irrigation system made possible by the Aswan High Dam in Egypt may produce similar salinization of the soil. Ecologists have speculated that while the Aswan High Dam was designed to bring another million acres of land under irrigation, it may well prove to be the ultimate disaster for Egypt.

One promising development is the use of drip irrigation, a method of reducing salinization by dripping water onto the soil from pipes at a rate calculated to ensure that virtually all of it penetrates into the soil instead of being lost by evaporation. Of particular value in drip irrigation is the use, where possible, of water from treated sewage; mineral nutrients are thus recycled to the land where they are needed, rather than released into streams and lakes where they are ecologically damaging.

THE ECOLOGY OF THE PLANET

The interactions of the various communities on earth are subtle and pervasive. The earth is itself an ecosystem, the biosphere; it is a complex array of communities—of populations and their environments—bound together by interrelated biological, chemical, and physical cycles and forces. In this section we will consider how the regional variations in climate discussed earlier profoundly influence the dispersion of species and the distribution of communities on earth.

■ BIOMES: THE MAJOR FEATURES OF LIFE ON EARTH

As we have seen, the major factors affecting primary productivity are mean annual temperature and precipitation. The same factors, along with associated parameters such as wind, humidity, latitude, altitude, and topography, result in large biotic regions called ***biomes*** (Fig. 40.31). Each biome is characterized by different climates, plants, and animals.

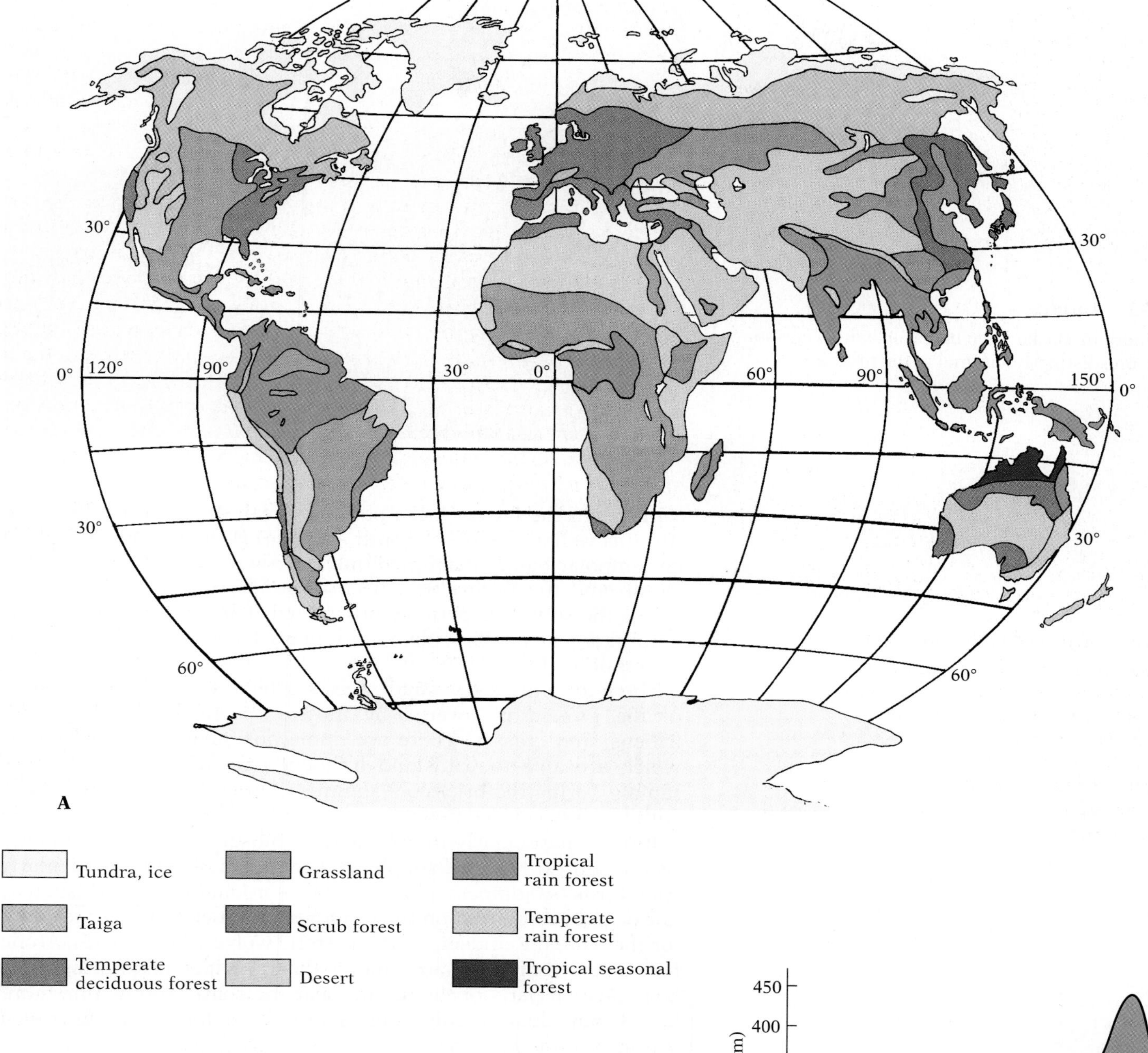

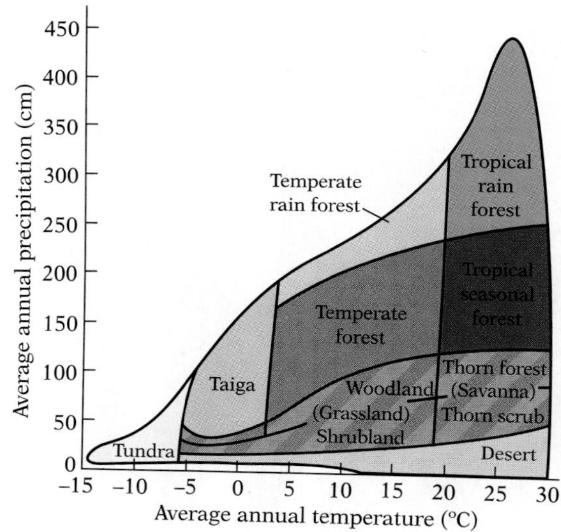

40.31 Major biomes of the world

The distribution of the earth's biomes (A), like primary productivity in general, depends on the combination of rainfall and temperature in each area (B). Grasslands and scrub forest can exist under the same combination of average conditions; which biome predominates depends on how the rain is distributed over the year as well as on other local conditions. The absence of any areas in the upper left reflects the inability of cold air to hold much water.

40.32 Tundra

A tundra in Alaska, seen in spring, with a caribou bull, one distinctive animal of the tundra.

Tundra In the far northern parts of North America, Europe, and Asia (as well as the tip of South America) is the tundra. It forms a circumpolar band interrupted only narrowly by the North Atlantic Ocean and the Bering Sea. It corresponds roughly to the region where the subsoil is permanently frozen. The land has the appearance of a gently rolling plain, with many lakes, ponds, and bogs in depressions (Fig. 40.32).

There are only a few small trees scattered on the tundra. Much of the ground is covered by mosses (particularly sphagnum), lichens, and grasses. There are numerous small perennial herbs, which are able to withstand frequent freezing and which grow rapidly during the brief cool summers, often carpeting the tundra with brightly colored flowers.

Insects, particularly flies (including mosquitoes), are abundant. As a result of this food supply, vast numbers of birds, particularly shorebirds (sandpipers, plovers, and so on) and waterfowl (such as ducks and geese), nest on the tundra in summer, but migrate south for the winter. Reindeer, caribou, Arctic wolves, Arctic foxes, Arctic hares, and lemmings are among the principal mammals; polar bears live on parts of the tundra near the coast. Though the number of individual organisms on the tundra is large, the number of species is quite limited.

Taiga South of the tundra is the taiga, a wide zone dominated by coniferous (evergreen) forests (Fig. 40.33), or boreal forest. Like the tundra, it is dotted by lakes, ponds, and bogs, and experiences very cold winters. The heavy snow cover, however, prevents the ground from freezing very deep. The taiga has longer summers, during which the subsoil thaws and vegetation grows abundantly.

The number of species living in the taiga is larger than on the tundra. Though conifers (including spruce, fir, and tamarack) are characteristic trees in the taiga, deciduous trees such as paper birch are also common. Moose, black bears, wolves, lynx, wolverines, martens, porcupines, and many smaller rodents are common

A

B

40.33 Taiga

(A) Coniferous forests, such as this one in Norway, cover extensive areas in the northern part of North America, Europe, and Asia. (B) Wolves, an important mammal of the taiga.

mammals in the taiga communities. Flying insects and birds are abundant in summer.

Deciduous forests The biomes south of the taiga do not form such definite circumglobal belts as the tundra and the taiga. There is more variation in the amount of rainfall at these latitudes, and consequently more longitudinal variation in the types of communities that predominate.

In those parts of the temperate zone where rainfall is abundant and the summers are relatively long and warm, as in most of the eastern United States, most of central Europe, and part of eastern Asia, the major communities are frequently dominated by broad-leaved trees. Such areas, in which the foliage changes color in autumn and drops, constitute the deciduous-forest biomes (Fig. 40.34). They characteristically include many more plant species than the taiga to the north, and show more vertical stratification of both plants and animals—that is, there are species suited to the ground, the low branches, and the treetops. Among the common mammals in this biome are squirrels, deer, foxes, and bears.

40.34 A temperate deciduous forest along the Housatonic River, Connecticut, in early autumn

A

B

40.35 Tropical rain forest

(A) A forest in Queensland, Australia. (B) Epiphytic bromeliads (distinctive plants of the tropics) growing on the branch of a tree in Costa Rica.

Tropical rain forests Tropical areas with abundant rainfall are (or, increasingly, were) usually covered by rain forests, which include some of the most productive communities on earth. As we pointed out earlier, although these forests now cover less than 4% of the land surface area, they account for more than 20% of the planet's net carbon fixation. By comparison, temperate forests are about half as productive per unit area, while boreal forests, grasslands, and woodlands are only one-quarter as productive.

Tropical rain forests are also the most complex communities on earth. The diversity of species is enormous; a temperate forest is composed of two or three (occasionally as many as 10) tree species, but a tropical rain forest may be composed of 400 or more. It may actually be difficult to find any two trees of the same species within an area of many hectares. A 13-km^2 rain-forest preserve in Costa Rica has 450 species of trees, more than 1000 other plant species, 400 species of birds, 58 species of bats, and 130 species of amphibians and reptiles. The accelerating loss of the earth's once-vast rain forests is condemning thousands of species to extinction.

The dominant trees in rain forests are usually very tall, and their interlacing tops form a dense canopy that intercepts much of the sunlight, leaving the forest floor only dimly lit even at midday (Fig. 40.35). The canopy likewise breaks the direct fall of rain, but water drips from it to the forest floor much of the time. It also shields the lower levels from wind, greatly reducing the rate of evaporation. The lower levels of the forest are consequently very humid. Temperatures near the forest floor are nearly constant. The pronounced differences in the microenvironmental conditions at different levels within such a forest result in a striking degree of vertical stratification; many species of animals and epiphytic plants (plants growing on the large trees) occur only in the canopy, others only in the middle strata, and still others only on the forest floor.

Grasslands Huge areas in both the temperate and the tropical regions of the world are covered by grassland biomes (Fig. 40.36).

40.36 A grassland in Tanzania, with black rhino and young.

These are typically areas where relatively low total annual rainfall (25–30 cm) or uneven seasonal occurrence of rainfall makes conditions inhospitable for forests but suitable for grasses. The grasslands of temperate regions characteristically undergo an annual warm-cold cycle, whereas the grasslands of the tropics undergo a wet-dry cycle instead. Different patterns of precipitation, wind, brush fires, and other local conditions can lead to scrub forests in such habitats (Fig. 40.31).

Temperate and tropical grasslands are remarkably similar in appearance, though the particular species inhabiting them may be very different. Both usually contain vast numbers of large herbivores, often including ungulates such as bison and pronghorn antelope in the United States, or wildebeest and gazelle in Africa. Burrowing rodents or rodentlike animals such as the prairie dogs of the western United States, are also common.

Deserts In places where rainfall is typically less than 25 cm (10 inches) per year, not even grasses can survive as the dominant vegetation. The resulting deserts are subject to the most extreme temperature fluctuations of any biome; during the day they are exposed to intense sunlight, and the temperature of the air may rise to 40°C or more, while surface temperature can exceed 70°C. In the absence of the moderating influence of vegetation, heat is rapidly lost at night, and a short while after sunset searing heat has usually given way to surprising cold.

Some deserts, such as parts of the Sahara, are nearly barren, but drought-resistant shrubs (sagebrush, creosote bush, and mesquite, for example) and succulent plants like cactus that can store much water (Fig. 40.37B) commonly grab a foodhold. Many small rapid-growing annual herbs bear seeds that will germinate only when there is a hard rain; once they germinate, the young plants shoot up, flower, set seed, and die, all within a few weeks.

Most desert animals are active primarily at night or during the

A

B

C

40.37 Deserts

(A) Monument Valley, Utah; vegetation is extremely sparse. (B) Cacti and other thorny, drought-resistant plants, abundant in many deserts, growing in Picacho Park, Arizona. (C) This desert toad of western Australia secretes fluid through its skin, which dries to form a moisture-proof film. It can then survive lengthy periods of drought buried in hardened mud.

brief periods in early morning and late afternoon when the heat is not so intense. During the day they remain in cool underground burrows or in cavities in plants or, in the case of some spiders and insects, in the shade of the plants. Among the animals often found in deserts are rodents like the kangaroo rat, snakes, lizards, a few birds, arachnids, and insects. Most show remarkable physiological and behavioral adaptations for life in this hostile environment (Fig. 40.37C).

The role of altitude We have seen a series of different biomes that change, largely as a result of temperature, as we move from north to south in the Northern Hemisphere. Changing altitude produces effects similar to changing latitudes: Higher altitudes, like higher latitudes, tend to be colder than lower ones. If rainfall and humidity are equivalent at the corresponding latitudes and altitudes, the changing pattern of vegetation seen on a mountainside as elevation increases resembles the pattern observed as one moves

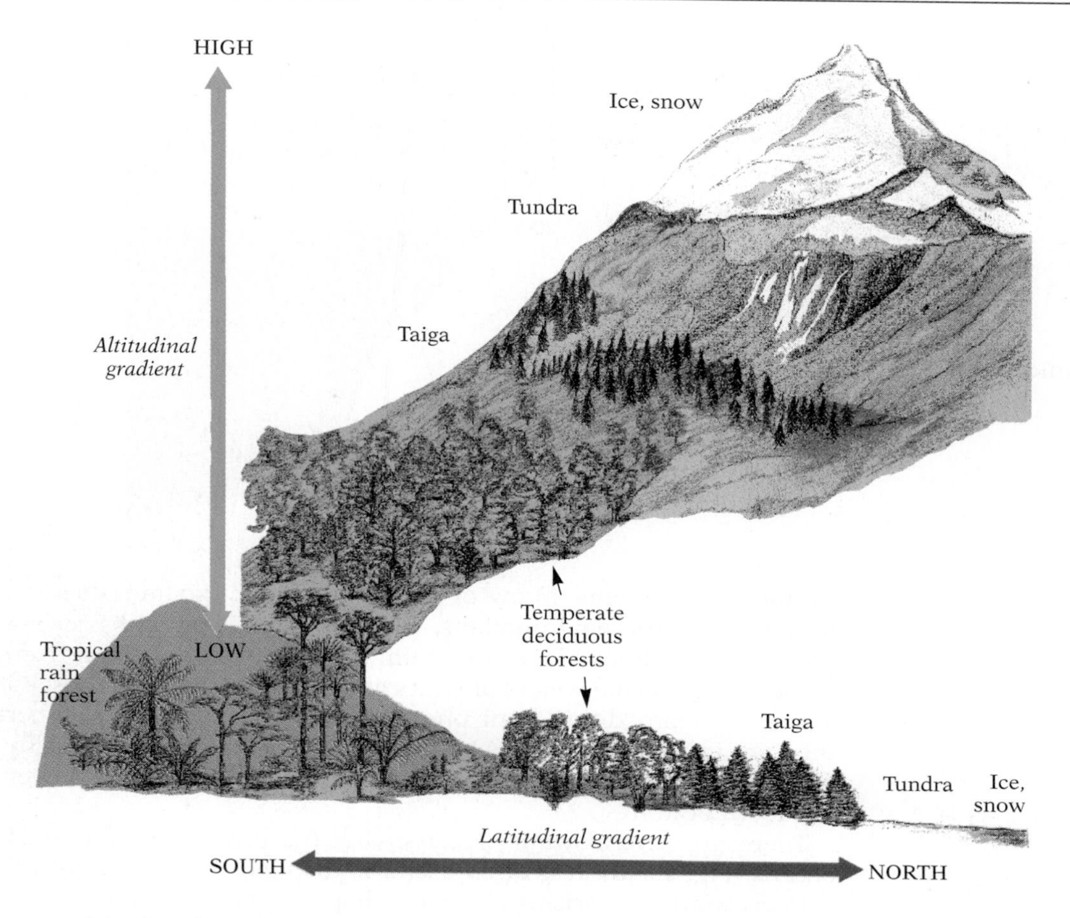

40.38 Similarity between latitudinal and altitudinal life zones in North America

toward higher latitudes (Fig. 40.38). Thus "arms" or isolated pockets of the taiga extend far south in the United States—on the slopes of the Appalachian Mountains in the east and of the Rockies and Coast Ranges in the west (Fig. 40.39). There are even tundra-like spots on the highest peaks. A 100-m increase in elevation is roughly equivalent to a 50-km increase in latitude.

40.39 An alpine meadow in the Sierra Nevada of California, with paintbrush in bloom

Trees can be seen in the lower, more protected areas. Most of the meadow is above the timberline. The upper limit of tree growth in mountainous regions (usually between 3000 and 3500 m) is determined largely by temperature and exposure to winter winds; it is also influenced by soil conditions and rainfall. More barren areas can be seen on the higher peaks in the distance.

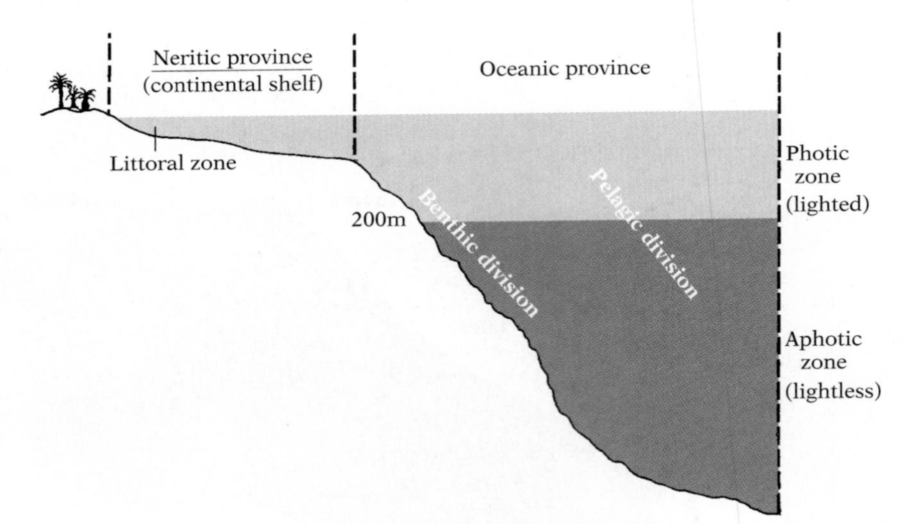

40.40 A classification of oceanic ecosystems

Aquatic ecosystems Many of the earth's biotic communities are found in aquatic environments, which vary in type with varying physical conditions. Thus the communities in lakes differ from those in the flowing waters of rivers and streams, and even those in a single stream differ from one another, depending on whether they are in rapids, where fast-flowing water is made turbulent by rocks and sudden falls, or in water flowing slowly and calmly over a smooth bottom.

There are several ways of classifying the ecosystems found in oceans (Fig. 40.40). It is often useful to distinguish between a ***benthic division***, comprising the ocean bottom together with all bottom-dwelling organisms, and a ***pelagic division***, consisting of the water above the bottom and all the organisms in it. Pelagic organisms include both the nekton (free-swimming animals) and the plankton (floating organisms carried passively with the water currents). When the distribution of light in the water is important, we distinguish between an upper, lighted ***photic zone*** and a deeper, lightless ***aphotic zone*** (usually below about 200 m); only in the ***euphotic zone*** (the upper 1–100 m depending on turbidity) is there enough light for net photosynthesis. Another distinction is made between the ***neritic province*** (the volume above the continental shelves) and the ***oceanic province*** (the volume of the main ocean basin).

The various subdivisions of the oceans are biologically more interdependent than the terrestrial biomes. For example, except for isolated colonies of chemosynthetic autotrophs found near underwater volcanic vents, benthic communities in the aphotic zone are made up of heterotrophs (including animals, fungi, and heterotrophic protists and bacteria) that receive their nutrients as a "rain" of dead organisms (detritus) from the water above. When the flow of energy and materials is considered, then, the communities in the dark aphotic zone cannot be understood apart from those in the euphotic zone, and the benthic division cannot be understood apart from the pelagic division.

The most complex oceanic communities occur in the shallow waters of the neritic province, especially the ***littoral zone***, which extends from the beach to a point where the water is deep enough

40.41 Distribution of present-day coral reefs

The intensity of solar radiation, which is greatest in the equatorial latitudes, is a major factor controlling the distribution of present-day coral communities in the earth's oceans. Reefs are rare where the water is so rich in nutrients that it can support other communities.

A

B

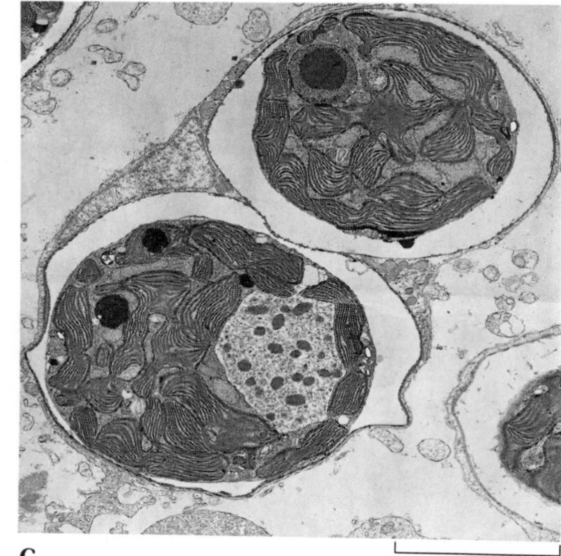

C

40.42 Corals

Hard corals growing in abundance in tropical waters provide diverse habitats for marine creatures. (A) Among the corals found in waters surrounding Fiji are leather and alapora corals. (B) All corals are colonial assemblies of individual polyps, as shown in this close-up view. (C) The formation of the coral skeleton is enhanced by the photosynthetic activities of single-celled endosymbiotic algae, or zooxanthellae, two of which can be seen in this electron micrograph of the tissues of a hard coral.

so that it is no longer completely stirred by the action of waves or tides. High primary productivity by both free-floating and bottom-anchored algae makes possible much niche diversification among herbivores in this zone. Because the littoral zone is subject to far more variation in temperature, water turbulence, salinity, and light than any other portion of the ocean, littoral communities vary greatly.

Coral reefs The intense sunlight falling on equatorial waters makes possible some of the most stable and diverse communities on earth (Figs. 40.41, 40.42). The cornerstones (almost literally) of these communities are diverse species of stony corals—colonial cnidarians that feed in a multitude of ways on zooplankton, detritus, organic solutions taken in through active transport, and the sugars and amino acids produced by symbiotic photosynthetic algae called zooxanthellae (Fig. 40.42C). Stony corals (as well as hard corals of other orders, and various species of algae) lay down calcium-based skeletons, which constitute the basic structure of a reef and provide a multitude of microhabitats for the organisms it supports. Like the rain forests to which they are often compared, coral reefs are inhabited by animals of virtually every phylum as well as the representatives of most of the plant phyla.

■ HOW TODAY'S BIOMES CAME TO BE

Though today's climate in the various biomes is a critical element in the distribution of living things, history is equally important. The earth and the organisms on it are constantly changing, and the present distribution is in large part the result of past, often quite different, conditions. For example, present conditions cannot explain why certain animals occur in South America, Africa, and southern Asia, but nowhere else. The fossil record combined with dramatic geological evidence of the past casts much light on the present geography of life.

Continental drift Early in the Mesozoic era, 225 million years ago, there was a single massive supercontinent called ***Pangaea*** (Fig. 40.43A). Pangaea broke up as the present-day continents drifted slowly apart. The first major break was an east-west one, separating a northern supercontinent called ***Laurasia*** (composed of the land that later became North America, Greenland, and Eurasia minus India) from a southern supercontinent called ***Gondwanaland*** (composed of the future South America, Africa, Madagascar, India, Antarctica, and Australia). Soon thereafter Gondwanaland began to break up as India drifted off to the north and an African–South American mass separated from an Antarctic-Australian mass. By the start of the Cretaceous period, roughly 135 million years ago, the continents were probably distributed as shown in Figure 40.43B.

By about 65 million years ago, South America had split from Africa and was drifting westward. India had moved farther northward but had not yet collided with the rest of Asia, and Australia had split from Antarctica and begun drifting northeastward (Fig. 40.43C). The Laurasian supercontinent was still intact; the split between North America and Eurasia was one of the last to take place, as the distribution of continents we know today slowly emerged (Fig. 40.43D).

As continents move in the course of millions of years, so that their distances from the earth's poles and from the equator change, their climates undergo major shifts. India, for example, has moved from a position next to Antarctica all the way across the equator to its present location in the tropics of the Northern Hemisphere. Australia, too, has moved steadily northward. Since shifts in climate would result in altered selection pressures, evolutionary

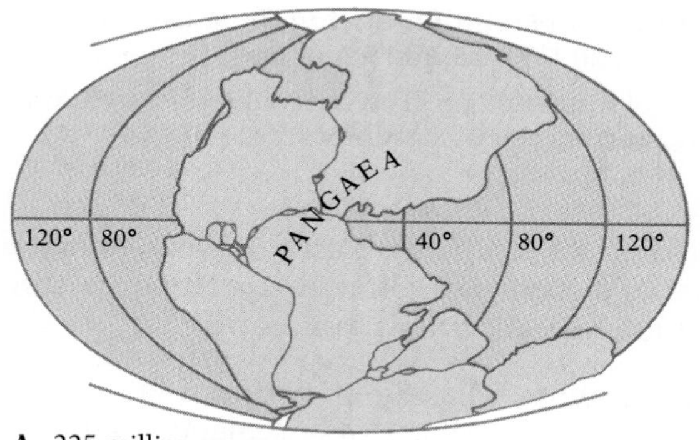

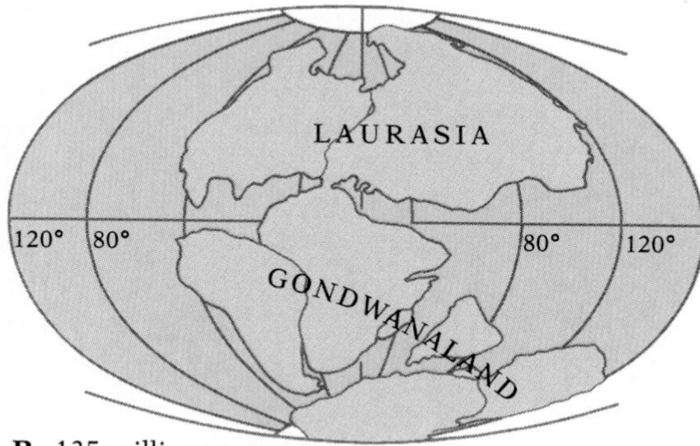

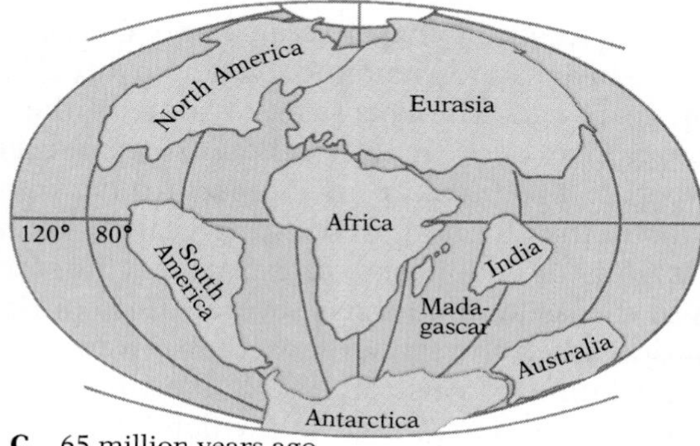

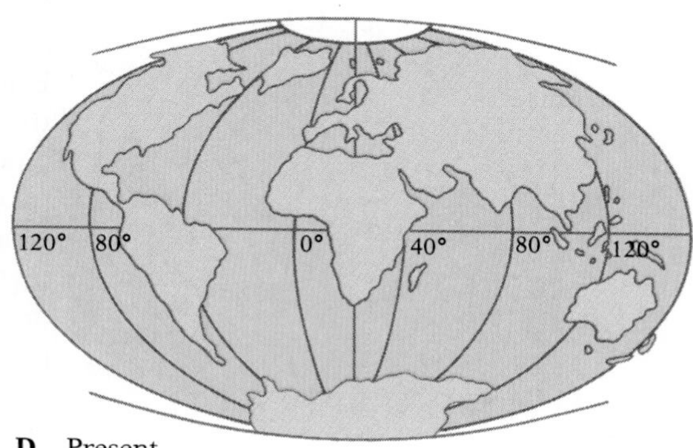

40.43 Origin of the modern distribution of continents through continental drift

(A) Early in the Mesozoic era, about 225 million years ago, all the earth's major landmasses were united in a single massive supercontinent called Pangaea. (B) About 135 million years ago, at the start of the Cretaceous period, Pangaea had broken into a northern supercontinent, Laurasia, and a southern supercontinent, Gondwanaland; Gondwanaland itself had also begun to break up. (C) By 65 million years ago, the breakup of Gondwanaland was complete, and the future South America, Africa, Madagascar, India, Antarctica, and Australia were drifting apart. (D) The present continental arrangement.

forces have changed the organisms in these regions gradually but dramatically.

Climatic changes have been brought about in other ways as well. Antarctica, for example, though probably always near the South Pole, has not always been covered by ice; fossils of amphibians and reptiles have been found there. During at least part of the Mesozoic, Antarctica must have been reasonably warm, and it was probably warm again about 50 million years ago, when tropical and subtropical climates were far more widespread on the earth than they are today (Fig. 40.44). By contrast, the earth was much colder only a few thousand years ago, during the periods of exten-

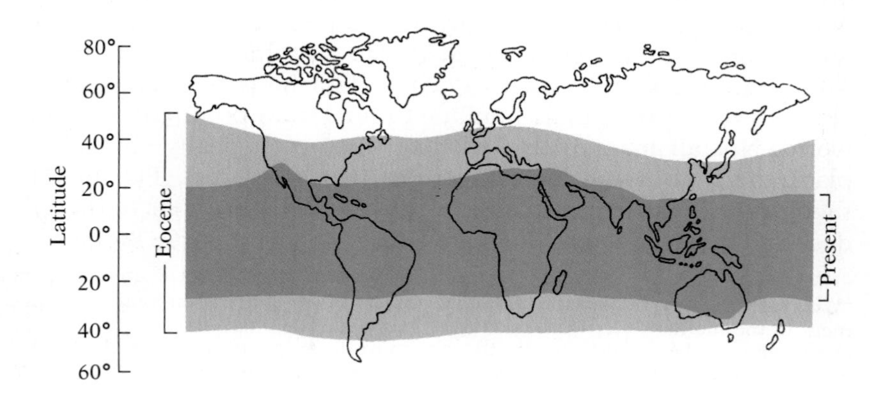

40.44 Distribution of tropical and subtropical forests today and during the Eocene, shown on a modern map

Warm conditions extended much farther north and south during the Eocene (about 50 million years ago). The possible causes of global warming and cooling cycles are hotly debated.

40.45 Distribution of glaciation during the most recent Pleistocene ice age, 20,000–15,000 years ago

The vast quantities of water tied up in glaciers (blue area) reduced the sea level dramatically, creating a new coastline (gray line).

sive glaciation (Fig. 40.45). Thus both the past configurations of the land masses and their past climates are important in explaining the distribution of organisms on the earth.

The island continents The biota (flora and fauna) of the ***Australasian region*** (Australia, New Zealand, and adjacent islands) is the most unusual found on any of the earth's major land masses. Many groups common in Australia occur nowhere else. Conversely, many taxa widespread in the rest of the world are absent from Australia. Australia has had no land connection with Eurasia or Africa since Gondwanaland began to break up more than 135 million years ago. The ancestors of some of the more ancient groups of organisms now living in the Australian region were probably there before the breakup of Pangaea; others must have reached Australia via the chain of intervening islands. In fact, most of the western islands of Indonesia were probably interconnected as an extension of the Asian land mass several times during recent periods of glaciation when the sea level was far lower.

The mammalian fauna of Australia is unlike that of any other continent. Except for wild dogs (dingoes), which humans imported in prehistoric times, the only placental mammals in Australia before European explorers landed there were a number of species of rodents belonging to a single family, and a variety of bats. Bats can, of course, fly across water barriers and would be expected to reach most oceanic islands. The rodents of Australia are apparently relatively recent arrivals, having come from Asia by island-hopping through Indonesia. Most of the ecological niches that on other continents would be filled by placental mammals are filled in Australia by marsupials.[4]

The marsupials probably reached Australia very early and, encountering no competition from placental mammals, radiated extensively. Since they were filling niches similar to those occupied elsewhere by placentals and were therefore subject to similar selection pressures, they evolved striking convergent similarities to the placentals. Certain of the marsupials resemble placental shrews, others resemble placental jumping mice, weasels, wolverines, wolves, anteaters, moles, rats, flying squirrels, groundhogs, bears, and so on (see Fig. 18.28, p. 518). The uninitiated visitor to an Australian zoo finds it hard to believe that he is not seeing close relatives of the mammals familiar to him from other parts of the world. Not all marsupials look like their ecologically equivalent placentals, however. Kangaroos are markedly different (Fig. 40.46), though some of them play an ecological role very similar to that of deer and other large placental grazers.

[4]Whereas the young of placental mammals undergo their entire embryonic development in the mother's uterus, the young of marsupial mammals develop for only a short while in the uterus; after birth, they move to a pouch on the mother's abdomen, attach to a nipple, and there complete their development.

40.46 Kangaroos

Though their appearance and behavior do not immediately suggest it, kangaroos are Australia's ecological equivalents of the ungulate grazers of other continents.

Another island continent, South America, which is known to biologists as the ***Neotropical region*** (meaning "new tropics"), has had a history similar to that of Australia. It, too, was isolated from other land masses through much of its history. And it, too, had an early mammalian fauna that included a variety of marsupials. The prevalence of marsupials in both Australia and South America probably means that these organisms were present throughout the region comprising Australia, Antarctica, and South America at a time when there were only small water breaks (if any) in this land mass (Fig. 40.43C), and when the climate was warmer than it is now.

Later a variety of placentals reached South America, probably during a short period of connection to North America via a Central American land bridge about 60 million years ago. After this land bridge disappeared, both the marsupials and the placentals of South America evolved, in isolation, many characteristics convergent to those evolved by placentals in Africa (Fig. 40.47). There

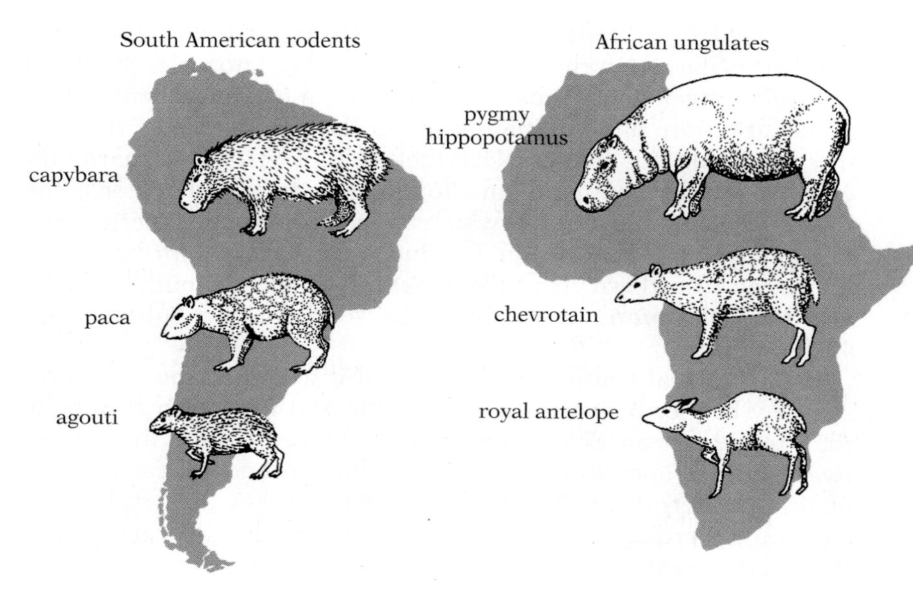

40.47 Convergence of South American rodents and African ungulates

There are remarkable similarities between the two groups of animals shown here, even though rodents and ungulates comprise separate mammalian orders that are not closely related.

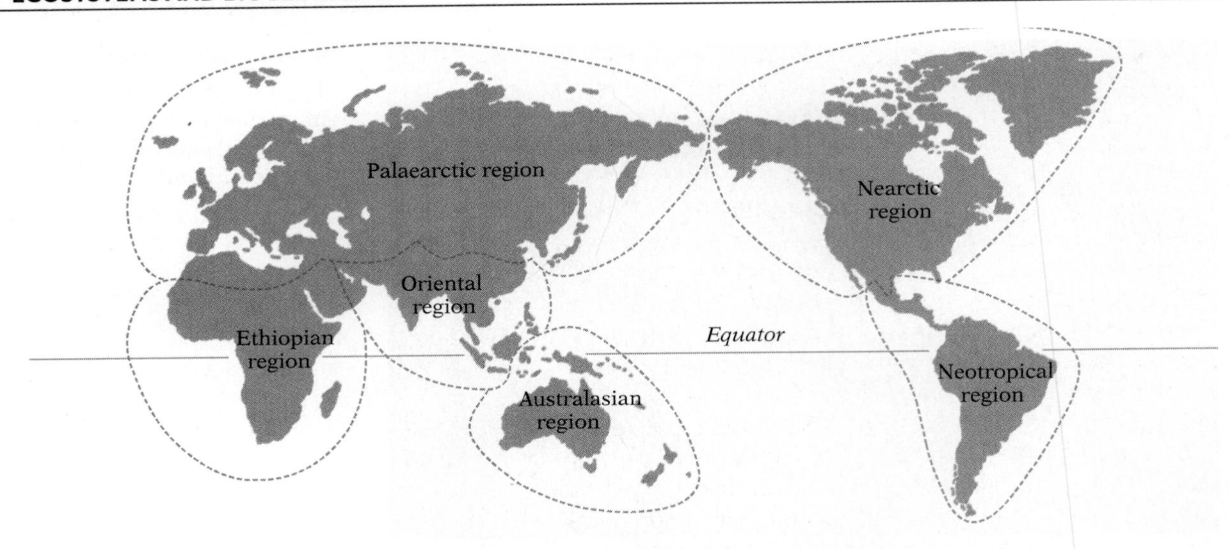

40.48 Biogeographic regions of the world

were times during this period of isolation when the water barrier between South America and northern Central America was not wide. A few additional placental mammals chanced to get across into South America then, including the ancestors of the modern New World monkeys and a number of rodents.

By five million years ago there were 23 families of mammals in South America; however, not one of these was represented in North America, where a different group of mammalian families lived. Then a land connection to North America was reestablished at the Isthmus of Panama (the two continents had not been connected since they were part of Pangaea). A few species moved from South America to North America, including the opossum, the porcupine, and the armadillo. However, most migrants moved in the opposite direction, from North America; much of the older South American fauna suffered extinction as a result of this invasion.

The World Continent Europe, Asia, Africa, and North America have formed a relatively continuous land mass, known as the World Continent, throughout most of geological time. Consequently their biotas are more alike in many aspects than they are like those of the two island continents. Nevertheless, biologists customarily divide the World Continent into four biogeographical regions: the ***Nearctic*** ("new northern"), which is most of North America; the ***Palaearctic*** ("old northern"), which is Europe, northernmost Africa, and northern Asia; the ***Oriental***, which is southern Asia; and the ***Ethiopian***, which is Africa south of the Sahara (Fig. 40.48).

The geological feature that looks as if it would be the main present-day barrier between the Palaearctic and Ethiopian regions is the Mediterranean Sea. Many species, however, have moved between Europe and North Africa by circling around the eastern end of the Mediterranean. The real barrier to species dispersal is the Sahara. Accordingly, Africa north of the Sahara is part of the Palaearctic region.

The Oriental region—tropical Asia—is separated from Palaearctic northern Asia in most places by east-west mountain ranges, of which the Himalayas between India and China are part. These mountains constitute important breaks between climatic regions; they act as both topographic and climatic barriers between cold-adapted and warm-adapted species. By contrast, north-south mountains such as those in North America tend to facilitate mixing of cold-adapted and warm-adapted species.

The Nearctic is considered part of the same land mass as the Palaearctic because North America and Eurasia remained connected through much of their geological history (Fig. 40.43A–C). Even after they broke apart as the northern part of the Atlantic Ocean formed, the Siberian land bridge between what are now Alaska and Siberia provided a link; part of this bridge is beneath water at the present time. Fossils of many temperate and even subtropical species of plants and animals are abundant in Alaska, suggesting that there was no climatic barrier at one time. Humans first crossed this bridge about 50,000 years ago. The Nearctic and Palaearctic regions remain biologically so similar that many biologists regard them as a single region, which they call the Holarctic.

The history of the major land masses and of their changing climates helps explain present distribution patterns of species. Consider the disjunct distribution referred to earlier, in which a species occurs in South America, Africa, and southern Asia. For species belonging to very ancient groups—the cockroaches, for example—this distribution indicates that the species occurred throughout the old Gondwanaland supercontinent and continued to survive in South America, Africa, and India after these land masses drifted apart. For species belonging to more recently evolved groups however—like the majority of modern mammalian and bird families, which arose after Gondwanaland had broken up—the species had to move between the New World and the Old World via either the North Atlantic or the Siberian land bridge, and between North and South America via Central America. The disjunct species later became extinct in the north, either because of climatic changes or competition. The fossil record shows that this pattern of dispersal between southern regions by way of the northern continents has occurred again and again. For example, members of the camel family occur today in South America (llamas, alpacas, vicuñas, and so on), northern Africa, and central Asia. However, fossils indicate that the family originated in North America, spread to South America via Central America and to the Old World via Siberia, and later became extinct in North America. The result is the disjunct distribution we see today.

■ HOW SPECIES PERSIST AND SPREAD

The many species of plants and animals inhabiting the earth do not often remain in the same location indefinitely; indeed, given the drastic changes that occur in climate, most terrestrial species would face inevitable extinction without the ability to adapt to new challenges or disperse to new areas. For a species to persist after a habitat change, it must have the ***physiological potential*** to survive and reproduce in the altered habitat. Before a species can successfully spread into a new area, it must meet the same physiological-potential criterion, plus at least two other major conditions: it must have the ***ecological opportunity*** to become established in the new area, and it must have ***physical access*** to it.

The importance of physiological potential Any new area into which colonizing members of a species move, as well as any habitat undergoing a ecological change, will differ to some degree from the area the animals are used to. Hence the population will immediately be subject to selection pressure for evolution of better adaptations for the new environment. But since such evolutionary improvement comes after ecological change or colonization, survival or colonization itself is possible only if individuals are already at least minimally preadapted for survival under the new environmental conditions. *Preadapted* does not imply any intentional preparation for an ecological change or the move into strange territory; it simply means that characteristics an organism has evolved in the previous habitat are at least minimally suited to the new habitat. For example, colonizers must be able to use some source of food in a new area, and they must be able to withstand the rigors of the new climate.

Let's see how climate limits the distribution of organisms. We have already touched on broad regional differences in temperature, humidity, rainfall, and other meteorological factors. The average pattern of such factors over an extended period constitutes the climate of a region.

Though the climate of a desert is dry, at any given time rain may be falling. Thus we distinguish between climate and weather, weather is the short-term pattern of meteorological events, and has obvious day-to-day effects on organisms. A late frost after a warm spell can destroy the potential productivity of a plant seedling, and the reproductive success of many animals will be very different in unusually dry and unusually wet years. Thus, though the climate of an area is important in determining distributions of species, the weather is what is important to the organism at any particular moment.

Indeed, the extremes of the weather—the highest or lowest temperatures of the year, for example, or the longest period without rainfall—are often most important in limiting the distribution of an organism. A plant that cannot tolerate temperatures below 0°C will be unable to survive in a region with a warm climate (an average annual temperature of, say, 25°C) if the temperature falls below 0°C even one day a year. All other conditions in the region may be optimal for the plant, but the one condition it cannot tolerate prevents it from growing there.

Ecologists must analyze the microenvironmental conditions—the ***microhabitat***—to understand the physical factors with which organisms must cope. Variations in the weather can be quite local. For example, within a single small area, temperature differences of several degrees may occur between sheltered and exposed places, or between the ground and various elevations above the ground. The same sorts of highly local variations may be found in humidity, wind velocity, amount of sunlight, soil type, and so on. Thus we should not expect to find exactly the same kinds of organisms living at all points within even a very small area. Plants and animals don't live under generalized regional conditions; they live within some range of microenvironmental conditions that may vary radically over a given region. Hence when ecologists say that plant species A ranges from South Carolina in the south to central New York in the north, and from the Atlantic coast to central Ohio in the west, they mean that, within the overall range, the species occurs in the microhabitats that support it and does not occur in any microhabitats outside that range.

The same sorts of considerations apply to other aspects of the environment. The climate of a region might be ideal for a particular species of plant, and the soil rich in nitrogen and potassium, but if there is less phosphorus than the plant requires, it cannot grow there. The importance of single environmental factors was recognized as long ago as 1840 by Justus von Liebig. Liebig's ***Law of the Minimum*** states that the growth of a plant will be limited by whichever requisite factor is most deficient in the local environment. V. E. Shelford expanded Liebig's Law, applying it also to animals and taking into account that too much may be as bad as too little. Shelford's ***Law of Tolerance*** states that the distribution of a species will be limited by its *range* of tolerance for local environmental factors (Fig. 40.49).

Though the principle behind both Liebig's and Shelford's laws is important in ecology, the assumption that a single factor is always limiting is potentially misleading. In nature, the various environmental factors interact in so many ways that it is often impossible to describe any one factor as the limiting one. As described in Chapter 39, when one condition is not optimal—though tolerable—for a species, the limits of tolerance for other factors may be reduced (the result being that an ellipsoidal hypervolume most accurately represents a species's ecological niche; see Figure 39.18, p. 1133). Moreover, unless the Law of Tolerance is extended to such biotic limiting factors as predation and competition, it has only restricted applicability. As an illustration, consider Klamath weed (*Hypericum perforatum*).

This plant, which is poisonous to cattle, was brought from Europe to the western United States, where it spread to millions of acres of valuable range land. To control Klamath weed, a species of flea beetle that feeds on it was introduced. The beetle soon eliminated the weed from the open range lands, though some small populations still persist in shaded places in forests, where the beetle does not feed on it as much. It is clearly the beetle that limits the range of the Klamath weed to forests; yet an observer who doesn't know the history of the weed-beetle interaction would hardly be able to determine what keeps the weed from spreading into the range lands, and might suppose that the plant requires a shaded habitat.

The need for ecological opportunity A colonizing species must encounter little competition or danger from natural enemies when it first reaches a new area. There must be underutilized resources it can exploit. The reason is simple: even if the colonizer has the physiological potential for surviving in the new habitat, chances are that it will be less well adapted to the new conditions than

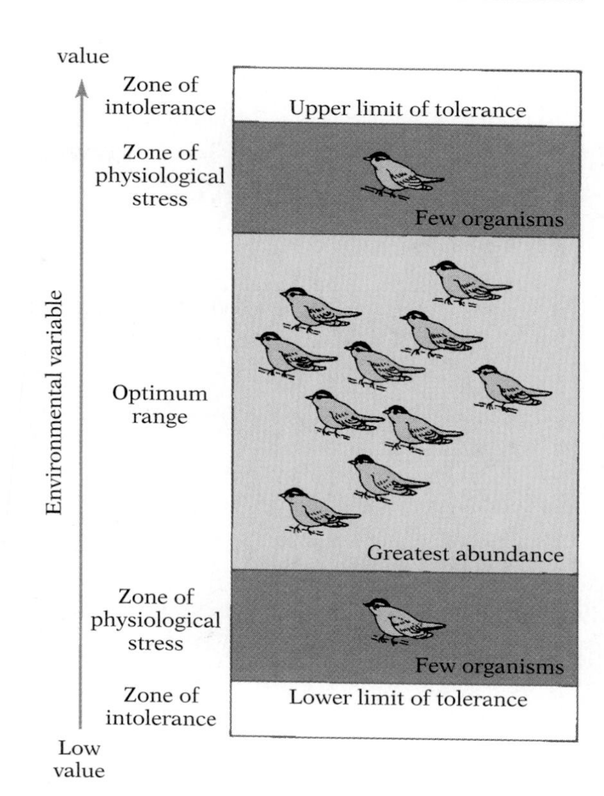

40.49 A diagrammatic illustration of the Law of Tolerance

The species in question is most abundant in areas where the environmental variable is within the optimum range for that species. The species is rare in areas where it experiences physiological stress because the environmental variable has either too high or too low a value. The species does not occur at all in areas beyond its upper and lower limits of tolerance.

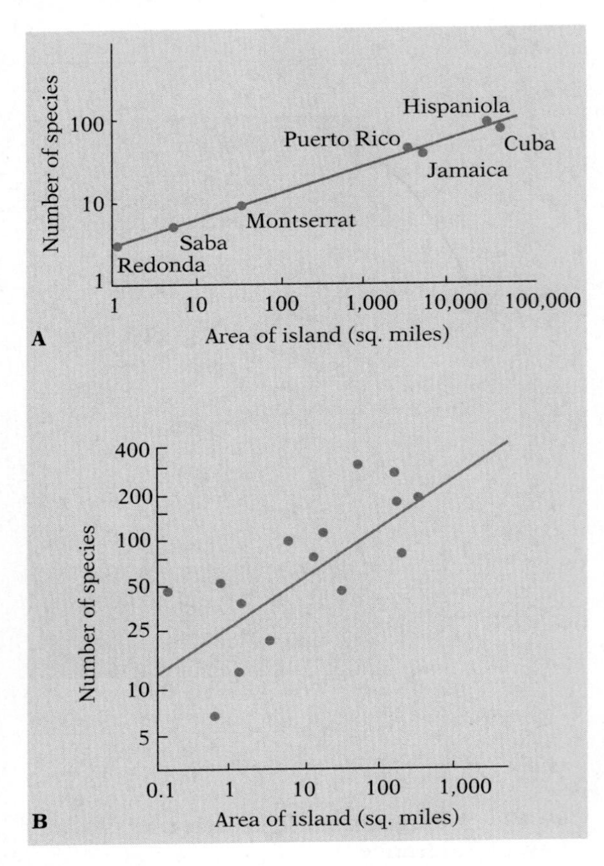

40.50 Two examples of area-species curves

(A) Number of species of amphibians and reptiles on islands in the West Indies. (B) Number of plant species on the Galápagos Islands. In each case the logarithms of the actual counts of species increase roughly as a linear function of the logarithms of the area of the islands.

species that have been in the area longer. If the niche of one of the established species is very similar to potential niches of the colonizer, the established species will probably have the competitive advantage and be able to prevent the colonizer from taking hold. But if conditions in the new range are very similar to those in the old, a colonizing species may sometimes be competitively superior to an established species and be able to supplant it.

Ecological opportunity is also affected by the size of the new area. The importance of habitat size was quantified by studies of island biogeography pioneered by Robert MacArthur and E. O. Wilson. We might reasonably expect a small island to support fewer species of any given group of organisms, and in fact, the number of species on an island is usually related to its area (Fig. 40.50). Though the slopes of area-species curves differ, depending on ecological conditions and on the taxonomic group of organisms studied, a rough generalization is that for each 10-fold increase in area the number of species approximately doubles. Thus an island with an area of 100 km^2 will, other things being equal, support roughly twice as many species as an island with an area of 10 km^2. In part this is because larger islands provide more diverse habitats and therefore a greater assortment of resources and opportunities for species interactions. However, species number on small islands is also kept down by the small size of the various populations. Chance, perhaps in the form of unusual weather, is more likely to wipe out every individual in a small population than in a larger one, and larger islands tend to support larger populations of each species.

Distance from the source of colonists profoundly affects the number of species found on an island as well. As we will see in the next section, species differ in their ability to cross climatic and geographical barriers. The higher species diversity of near-shore islands compared with mid-oceanic islands of similar size and topographic complexity is related to the relative accessibility of the near-shore islands. If that were all that mattered, however, we would expect that, given sufficient time, the diversity of organisms with good dispersal ability, such as birds, eventually would reach about the same level on similar islands, regardless of distance from the source of colonists. Yet even among birds and other mobile organisms, the "distance effect" is strong.

MacArthur and Wilson proposed that, just as a population with access to only limited resources must eventually reach an equilibrium between births and deaths, so the number of species on an island eventually reaches an equilibrium between the establishment of new species following immigration and the extinction of species already resident on the island. The size, climate, and topography of the island determine a probability of extinction that is independent of island isolation, but the rate of immigration of new species varies with the distance from the source of colonists. The inevitable result is that the equilibrium number of species on otherwise similar islands declines with increasing isolation, because the replacement of extinct species through immigration takes longer. Even when an island has reached equilibrium in terms of species diversity, however, there is still a slow but steady *turnover* of resident species.

Figure 40.51 shows how the rate of immigration of species not already established on an island falls as the number of established species on the island rises. Because more and more of the immigrants from source islands or continents represent species that have previously become established on the island, the percentage of immigrant individuals from species new to the island decreases. On the other hand, the rate of extinction rises as the number of species on the island increases. In part this is because the total number of species is greater. Another reason—which makes the extinction curve in Fig. 40.50 bend upward—is that, as the number of species on the island grows, interspecific competition also increases—a reflection of the increased overlap in resource use. The point at which the immigration/establishment and extinction curves cross represents a state of equilibrium for species diversity; both immigration and extinction continue, but the total number of resident species does not change significantly.

Figure 40.52 shows how the distance effect comes about. Two islands are compared: one is near the source of colonists, and so has a higher rate of immigration; the other is more distant, and consequently has a lower rate. The two immigration curves meet the horizontal axis at the same point: if there were no extinction on the islands, the maximum number of established species would eventually be the same for each, though that number would be reached more slowly on the distant island. Since the two islands are assumed to be similar in size, topography, and climate, a single extinction curve applies to both. However, the intersection of the extinction curve with the near-island immigration/establishment curve is to the right of its intersection with the distant-island curve. Thus, the equilibrium number of species on the near island is greater than the equilibrium number for the distant island.

Ecologists study island biogeography because it provides a clear model of the factors governing species dispersal in general. Freshwater lakes or springs can be considered biogeographic islands; so can forests separated from one another by grasslands, or deserts separated by wetter areas, or mountaintops, or even some parks. These habitat "islands," like real islands, obey the area-species rule: the greater the area, the greater the species diversity. In fact, habitat islands are more consistent with the theory than some real islands: Poorly dispersing organisms (lizards and freshwater fish, for example) have little chance of reaching oceanic islands, and little chance of establishing a thriving population on many of the less hospitable islands they do reach. The result is that chance effects dominate, and no regular pattern of distance or size will emerge from this "noisy" background. Habitat islands on the mainland, by contrast, are frequently easier to reach—the would-be immigrants have a land bridge and, often, freshwater routes in.

The ability to disperse It is useless for a species to have the physiological potential and ecological opportunity for surviving in a new range if it has no way of getting there. Doubtless many common North American mammals could survive and prosper in Australia, but unless they have some way of reaching that continent, this potential range extension will remain unrealized.

There are many ways organisms may disperse or be dispersed

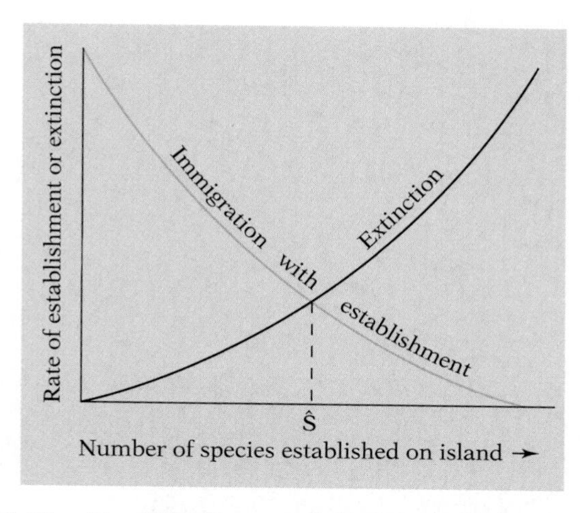

40.51 Equilibrium model of species diversity on an island

As the number of species established on the island increases, the rate of immigration and successful establishment of species not already established falls, while the rate of extinction of established species rises. When the point is reached at which establishment equals extinction (where the two curves cross), the island has the equilibrium number of species ($\hat{S}$); it cannot support more species unless the establishment rate increases or the extinction rate decreases.

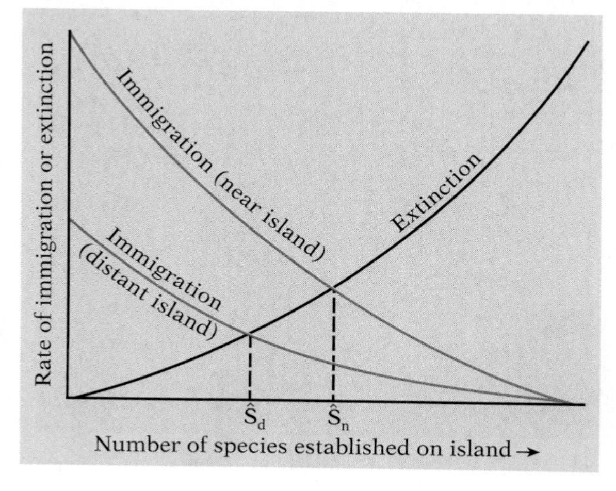

40.52 How the distance of an island from potential sources of colonizers affects the equilibrium number of species on the island

The rate of immigration of species not already established is lower for a distant island than for a near island. Hence species equilibrium is reached at a lower number of species ($\hat{S}_d$) for the distant island than for the near island ($\hat{S}_n$).

from one place to another. Most obvious for many animals is active locomotion: walking, crawling, swimming, or flying. Even many sedentary marine animals have a free-swimming larval stage. Active locomotion may carry the members of a single generation only a short distance from their point of origin, but over many generations the cumulative effect may spread the species over hundreds or thousands of miles. All the plant and animal species now present in the regions covered by glaciers 15,000 years ago (Fig. 40.45) recolonized these regions over the course of a few dozen centuries.

For plants and for many very small animals, passive transport is more important than active locomotion. For example, the seeds or spores of many plants may be blown great distances, and insects, spiders, and other invertebrates have been known to be blown hundreds of kilometers by storms. Pilots sometimes encounter large numbers of insects being swept along by fast-moving air currents at high altitudes. Aquatic organisms may similarly be swept along by water currents. Even some fairly large terrestrial plants and animals may be carried across many kilometers of water on floating logs or rafts of matted vegetation. Many such logs and rafts are swept out to sea by large rivers like the Amazon and the Congo, particularly during floods. A raft about 100 m^2, composed of soil and decaying organic matter laced together by roots, was sighted in the Atlantic Ocean off the coast of North America in 1892. Many shrubs and several trees 10 m tall were growing on it. This raft, which looked like a floating island, had drifted at least 1600 km.

A

B

40.53 Krakatoa

All life on the island of Krakatoa was destroyed by a volcanic eruption in 1883. (A) The effect of such an eruption can be seen in the photograph of an ash-covered beach on Krakatoa, largely barren of multicellular organisms, with the exception of an occasional sprouting coconut. (B) Now, however, large areas of the island have been recolonized.

Some plants and small animals are dispersed by birds and mammals. For example, the seeds of many plants pass through the digestive tracts of higher animals without being harmed, and may germinate and grow if the animals' feces are deposited in a favorable place; indeed, some seeds will not germinate without first passing through a vertebrate gut. Edible fruits tempt vertebrates to eat them, disperse their seeds and, when defecating, even supply fertilizer; the laxative effect of fruit may be adaptive in that it gets the seeds out of the gut before they are damaged by digestive enzymes. Birds sometimes transport seeds long distances in this way. Plant seeds and the eggs and larvae of some small aquatic animals may be transported on the feathers or feet of swimming or wading birds.

The recent history of Krakatoa, a small island between Java and Sumatra in the southwest Pacific, constitutes a natural experiment in dispersal and colonization. On August 27, 1883, a violent volcanic explosion destroyed much of the island and left the rest completely covered by a layer of hot ash and pumice 6–60 m deep. No life remained on the island. The island nearest to Krakatoa was 19 km away; most of the life on it was destroyed by toxic gases and a thick layer of ash produced by the explosion. The next-closest island was about 40 km away (Fig. 40.53).

Nine months after the eruption the sole living thing apparent on Krakatoa was a spider, which had probably been blown to the island. Less visible were dense colonies of photosynthetic cyanobacteria thriving in crevices near the shoreline. Only 3 years later, numerous plants were growing along the beaches, and several species of ferns and grasses were growing farther inland. The beach plants were of the kind found on the beaches of almost all tropical Pacific islands—plants whose seeds are highly resistant to sea water and are regularly carried long distances by ocean currents. The ferns reproduce by means of very light spores that can easily be carried by even gentle air currents, and can find a foothold in the thin soil.

Thirteen years after the explosion, the island was fairly well covered with vegetation, and about 9% of the species had probably reached the island on or in birds. By 1906 the island was densely covered with plants, and there were 263 species of animals living there; most were insects, but there were four species of land snails, two species of reptiles, and 16 species of birds. Many of the insects either flew or were blown to the island, but some of them (and perhaps the reptiles also) probably arrived on floating debris. The number of bird species on Krakatoa reached saturation by 1921, and has since undergone the slow turnover predicted by MacArthur and Wilson, with extinction of established species balanced by the establishment of immigrant species. The number of plant species continued to rise until at least the late 1930s.

Had Krakatoa been farther away from areas that could act as sources of colonizers, the species diversity would have increased more slowly. In addition, because of the distance effect, equilibrium would have been reached at a lower species number; as we have seen, the more distant an island is from a major source of new colonists, the lower its species diversity will be at equilibrium, because distance alone reduces the immigration rate.

Whatever the distance to be crossed, some species have readier physical access than others to a region that they might potentially colonize; thus it was easier for small animals that could be blown by wind or carried on rafts to get to Krakatoa than it was for large mammals. The geological or ecological zones that intervene between any two regions will be much more effective as barriers to dispersal for some species than for others. A wide expanse of ocean may almost completely prevent movement of horses or elephants, but coconut palms may cross it in fair numbers. A grassland separating two forested areas may be an almost insuperable barrier for some forest animals; other forest animals may have difficulty crossing the grassland, but manage to do so occasionally, and still others may cross freely. In short, what is a barrier to dispersal for one species may be a possible but difficult route for another and an easily negotiated path for a third.

Our increasing understanding of ecology underscores the problems the future holds. Once our species was a minor member of the biological cast on earth. Two technological breakthroughs—domestication, first of plants and then of animals—erased the major limiting factors that controlled our population size and distribution. The ability to manage crops and herds, and to store grain, allowed the development of cities and wealth and a self-perpetuating series of cultural and agricultural advances. As should be ominously clear from these last two chapters, the human population is continuing to increase exponentially; the effort to feed and otherwise support billions is leading inexorably to the rapid extinction of other species, and to the destruction of forests, croplands, water supplies, and even the air we breathe. Further scientific advances can ameliorate these problems and slow the destruction of the biosphere, but in the end only a dramatic end to human population growth offers any long-term hope for survival. Whether that end comes from a demoralizing and debilitating series of wars, diseases, and famines, or from conscious and informed decisions, will be the choice of the 21st century.

CHAPTER SUMMARY

THE ECONOMY OF ECOSYSTEMS

HOW THE SUN CREATES CLIMATE The differential warming of the earth by the sun creates the temperature differentials that create large-scale patterns of temperature and rainfall. Local features like oceans and mountains can dramatically influence local climate. (p. 1159)

HOW THE SUN POWERS LIFE The energy for life comes from the sun. The solar energy converted into photosynthetic products is gross primary productivity; the energy left after plant respiration is net primary productivity. To a first approximation, temperature and precipitation (or transpiration) predict primary terrestrial productivity. (p. 1162)

HOW ENERGY MOVES THROUGH AN ECOSYSTEM Energy passes through a food web from producers (the first trophic level) through primary consumers (herbivores, the second trophic level) various secondary consumers (mostly carnivores, third and higher trophic levels); energy from all levels sustains the decomposers. Because of the inefficiency of respiration, most of the energy in an organism is lost when it is consumed by another. Thus the energy flowing through each trophic level declines, leading to the pyramid of productivity. (p. 1164)

RECYCLING IN NATURE Although a constant input of energy is essential, most of the important nutrients for life are recycled. These nutrients include water, carbon (via CO_2), and nitrogen (mainly via NH_3 and other nitrogenous compounds). Phosphorus does not cycle in significant amounts. Burning of sequestered organic carbon (as the fossil fuels) is increasing the CO_2 concentration in the atmosphere, trapping more solar energy and increasing the global temperature. Some novel compounds, such as DDT, are not excreted or eliminated in organisms, and thus tend to build up through biological magnification at higher trophic levels. (p. 1167)

THE IMPORTANCE OF SOIL The physical and chemical composition of soils (especially the pH) largely determines their ability to retain water and make nutrients available to plants. The nature of the plant cover affects soil by controlling erosion, water retention, and humus input. (p. 1178)

THE ECOLOGY OF THE PLANET

BIOMES: THE MAJOR FEATURES OF LIFE ON EARTH Annual patterns of temperature and rainfall create a range of biomes, which include the tundra (in which the subsoil is frozen), the taiga (dominated by evergreen forests), deciduous forests, tropical rain forests, grasslands, and deserts. Increasing altitude can mimic higher latitudes. Several types of aquatic ecosystems also exist; temperature and light intensity are their most important parameters. (p. 1182)

HOW TODAY'S BIOMES CAME TO BE Many patterns of evolution and species distribution depend on where the drifting continents were in relation to each other millions of years ago. In particular, the flora and fauna on separated continents (Australia and, to a lesser extent, South America) evolved in isolation from those on the other continents. (p. 1192)

HOW SPECIES PERSIST AND SPREAD Species move into new habitats when they have the physiological potential to survive there, the ecological opportunity to make a living there in the face of competition by resident species, and physical access to the habitat. Habitat size and its distance from other habitats that can supply colonizers are important. Larger habitats can support larger populations and more species, with the result that the individual species go extinct in the habitat less often. Ecological vacancies created by local extinction in habitats distant from sources of colonizers usually remain unfilled longer than those in habitats close to sources of colonizing individuals. (p. 1197)

STUDY QUESTIONS

1 Compare and contrast the physiological challenges of deserts and marine shorelines. (pp. 1187–1188)

2 In the long run, what would happen to net productivity if there were no consumers? (pp. 1164–1167)

3 How would the earth's biomes and species be different if the earth's axis of rotation were not tilted relative to the sun? What changes would be evident if there were no oceans, or no continental drift? (pp. 1159–1164, 1182–1189, 1192–1196)

4 Given how fast carbonic anhydrase dissolves CO_2 in the blood, why couldn't the greenhouse warming be stopped by simply manufacturing enormous quantities of this enzyme and dumping it into the ocean? (pp. 872–873, 1169–1172)

SUGGESTED READING

BAZZAZ, F. A., AND E. D. FAJER, 1992. Plant life in a CO_2-rich world, *Scientific American* 266 (1). *On how plants will both gain and lose as CO_2 levels rise.*

CHARLSON, R. J., AND T. M. L. WIGLEY, 1994. Sulfate aerosol and climatic change, *Scientific American* 270 (2). *On how atmospheric sulfates moderate climate change, with the paradoxical prediction that reducing sulfur pollution would increase global warming.*

CULOTTA, E., 1995. Will plants profit from high CO_2? *Science* 268, 654-656. *A wide-ranging review of experimental approaches to this critical question.*

DALZIEL, I. W. D., 1995. Earth before Pangea, *Scientific American* 272 (1). *On the movements of the continents between 750 and 250 million years ago, and the biological consequences.*

MOHNER, V. A., 1988. The challenge of acid rain, *Scientific American* 259 (2).

POST, W. M., 1990. The global carbon cycle, *American Scientist* 78, 310-326. *An extremely detailed and wide-ranging treatment.*

WHITE, R. M., 1990. The great climate debate, *Scientific American* 263 (1). *On the controversy over the effects of increasing levels of carbon dioxide.*

WILSON, E. O., 1989. Threats to biodiversity, *Scientific American* 261 (3). *On human activities that are leading to large numbers of extinctions.*

WILSON, E. O., 1993. *The Diversity of Life*. W. W. Norton, New York. *An eloquent plea for the establishment of a new environmental ethic.**

*Available in paperback.

APPENDIX: A CLASSIFICATION OF LIVING THINGS

The classification given here is one of many in current use. Some other systems recognize more or fewer kingdoms and phyla (or divisions), and combine or divide classes in a variety of other ways. Chapters 19–25 discuss certain of the points at issue between advocates of different systems.

Botanists have traditionally used the term *division* for the major groups that zoologists have called phyla. A recent change in the botanical code allows the use of phyla for divisions, as we do here.

Most classes within a phylum are listed here, but where there is only one class, it is not named. For some classes—Insecta and Mammalia, for example—orders are given too. Except for a few extinct groups of particular evolutionary importance, like Placodermi, only groups with living representatives are included. A few of the better-known genera are mentioned as examples in most of the taxa.

Whenever possible, a (very) rough estimate of the number of living species is provided for higher taxa.

KINGDOM EUBACTERIA (3000)*

PHYLUM PROTEOBACTERIA Purple bacteria

- CLASS PURPLE SULFUR BACTERIA. *Pseudomonas, Escherichia*
- CLASS RHODOPSEUDOMONAS. *Rhodopseudomonas, Rhizobium*
- CLASS PURPLE NONSULFUR. *Sphaerotilus, Alcaligenes*
- CLASS DESULFOVIBRIO. *Desulfovibrio*
- CLASS RICKETTSIAE. *Rickettsia, Coxiella*
- CLASS MYXOBACTERIA. *Myxococcus, Chondromycos*

PHYLUM PROCHLOROPHYTA. *Chloroxybacteria, Prochloron*

PHYLUM CYANOBACTERIA. *Oscillatoria, Nostoc, Gloeocapsa, Microcystis*

PHYLUM GREEN SULFUR BACTERIA. *Chlorobium*

PHYLUM SPIROCHETES. *Treponema, Spirochaeta*

PHYLUM GRAM-POSITIVE BACTERIA

- CLASS CLOSTRIDIA. *Bacillus, Clostridium, Mycoplasma, Staphylococcus, Streptococcus*
- CLASS ACTINOMYCES. *Micrococcus, Actinomyces, Streptomyces*
- CLASS MYCOPLASMA. *Mycoplasma, Acholeplasma*

*There is no generally accepted classification for bacteria at the higher taxon level. One recent classification divides the bacteria into 17 distinct phyla, some without formal names. Another important classification assigns them to 19 "parts," most without formal names. This one, admittedly incomplete, is based for the most part on similarities in the sequences of ribosomal RNAs, as described in G. E. Fox et al., 1980, *Science* 209:457–463; and M. L. Sogin et al., 1989, *Science* 243:75–78.

KINGDOM ARCHAEBACTERIA (200)

PHYLUM METHANOGENS. *Methanobacterium*

PHYLUM HALOPHILES. *Halobacterium*

PHYLUM THERMOACIDOPHILES. *Thermoplasma*

PHYLUM SULFUR REDUCERS.

KINGDOM ARCHEZOA (400)†

PHYLUM METAMONADA. *Giardia, Spironucleus*

PHYLUM MICROSPORIDIA. *Vairimorpha, Nosema*

PHYLUM ARCHAEAMOEBA (or PELOBIONTA). *Pelomyxa, Mastigamoeba*

KINGDOM PROTISTA (27,000)†

KINETOPLASTID SUBKINGDOM

TRICHOMONADS

PHYLUM PARABASALA. Trichomonads and hypermastigids. *Trichomonas, Trichonympha, Calonympha*

EUGLENOIDS

PHYLUM EUGLENOZOA (600)

- **SUBPHYLUM KINETOPLASTIDA.** Trypanosomes and bodonids. *Trypanosoma, Bodo*
- **SUBPHYLUM EUGLENIDA.** Euglenids. *Euglena, Eutreptia, Phacus, Colacium*

†There are many alternative classification systems for protists and algae. The underlying uncertainty is reflected here by our use of colloquial groupings. The scheme used here, based primarily on rRNAs, is derived largely from Sogin et al. (1989); T Cavalier-Smith, 1989, *Nature* 339:100–101; and reviews for this edition by D. J. Patterson and T. Cavalier-Smith.

SARCODINE PROTISTS**

PHYLUM HETEROLOBOSEA (or PERCOLOZOA). Amoeboflagellates. *Percolomonas, Acrasis*

PHYLUM RIZOPODA. Amoebae with filose pseudopodia. *Euglypha*

PHYLUM AMOEBOZOA. Lobose amoebae. *Amoeba, Arcella, Entamoeba*

PHYLUM GRANULORETICULOSEA (or RETICULOSA). Foraminiferans. *Globigerina, Textularia*

SUBKINGDOM ACTINOPODA

PHYLUM HELIOZOA. Heliozoans. *Actinosphaerium, Actinophrys*

PHYLUM POLYCYSTINEA (or RADIOZOA.) Radiolarians. *Aulacantha, Acanthometron*

SUBKINGDOM AVEOLATA

PHYLUM DINOFLAGELLATA. Dinoflagellates (1000). *Gonyaulax, Gymnodinium, Ceratium, Oxyrrhis, Noctiluca*

PHYLUM APICOMPLEXA. Spore-forming parasitic protists. *Plasmodium, Eimeria*

PHYLUM CILIOPHORA. Ciliates (9000). *Paramecium, Stentor, Vorticella, Spirostomum*

AEROBIC ZOOFLAGELATE SUBKINGDOM

PHYLUM CHOANOFLAGELLATA (or CHOANOZOA). Choanoflagellates. *Daphanoeca, Monosiga*

PHYLUM OPALOZOA

- CLASS OPALINEA. Opalinids. *Opalina, Zelleriella*
- CLASS PLASMODIOPHOREA. Plasmodiophorids. *Plasmodiophora, Woronina*

SUBKINGDOM MYCETOZOA: Slime Molds

PHYLUM MYCETOZOA. Slime molds (500)

- CLASS MYXOGASTREA. True slime molds. *Physarum, Hemitrichia, Stemonitis*
- CLASS DICTYOSTELEA. Cellular slime molds. *Dictyostelium, Polysphondylium*
- CLASS PROTOSTELEA. Protostelids. *Protostelium*

** This classification, though useful, has no phylogenetic validity.

KINGDOM CHROMISTA†

SUBKINGDOM CRYPTISTA

PHYLUM CRYPTISTA

- CLASS CRYPTOMONADEA. Cryptomonads. *Cryptomonas, Chroomonas, Chilomonas, Hemiselmis*
- CLASS GONIOMONADEA. *Goniomones*

SUBKINGDOM HETEROKONTA: The Chrysophyta

PHYLUM SAGENISTA

- CLASS BICOSECIA. Biocosoecids. *Bicosoeca, Cafeteria*
- CLASS LABYRINTHULEA. Net slime molds. *Labyrinthula*

PHYLUM PSEUDOFUNGI

- CLASS OOMYCETES. Water molds, white rusts, downy mildews (400). *Saprolegnia, Phytophthora, Albugo*
- CLASS HYPHOCHYTRIOMYCETES. Hyphochytrids (25). *Rhizidiomyces*

PHYLUM OCHRISTA

- CLASS PHAEOPHYCEAE. Brown algae (1500). *Sargassum, Ectocarpus, Fucus, Laminaria*
- CLASS CHRYSOPHYCEAE. Golden-brown algae (650). *Chrysamoeba, Chromulina, Synura, Mallomonas*
- CLASS XANTHOPHYCEAE. Yellow-green algae (360). *Halosphaera, Tribonema*
- CLASS EUSTIGMATOPHYCEAE. Eustigmatophytes. *Pleurochloris, Vischeria*
- CLASS RAPHIDIOPHYCEAE. *Gonyostomum*
- CLASS BACILLARIOPHYCEAE (OR DIATOMEAE). Diatoms (10,000). *Pinnularia, Pleurosigma, Navicula, Melosira*

SUBKINGDOM HAPTOPHYTA

PHYLUM HAPTOPHYTA. Haptophytes and coccolithophores. *Isochrysis, Prymnesium, Phaeocystis, Coccolithus, Emiliania*

† There are many alternative classification systems for protists and algae. The underlying uncertainty is reflected here by our use of colloquial groupings. The scheme used here, based primarily on rRNAs, is derived largely from Sogin et al. (1989); T Cavalier-Smith, 1989, *Nature* 339:100–101; and reviews for this edition by D. J. Patterson and T. Cavalier-Smith.

KINGDOM PLANTAE

SUBKINGDOM BILIPHYTA†

PHYLUM RHODOPHYTA. Red algae (4000). *Nemalion, Polysiphonia, Dasya, Chondrus, Batrachospermum*

PHYLUM GLAUCOPHYTA. Glaucophytes (13)

SUBKINGDOM VIRIDIPLANTAE†

SECTION CHLOROPHYTA

PHYLUM CHLOROPHYTA. Green algae (7000). *Chlamydomonas, Volvox, Ulothrix, Oedogonium, Ulva*

SECTION STREPTOPHYTA

SUBSECTION CHAROPHYTA

PHYLUM CHAROPHYTA. Stoneworts (300). *Chara, Nitella, Tolypella, Spirogyra*

SUBSECTION EMBRYOPHYTA

PHYLUM BRYOPHYTA (23,600)

CLASS HEPATICAE. Liverworts. *Marchantia, Conocephalum, Riccia, Porella*

CLASS ANTHOCEROTAE. Hornworts. *Anthoceros*

CLASS MUSCI. Mosses. *Polytrichum, Sphagnum, Mnium*

PHYLUM TRACHEOPHYTA. Vascular plants

SUBPHYLUM PSILOPSIDA. *Psilotum, Tmesipteris*

SUBPHYLUM LYCOPSIDA. Club mosses (1500). *Lycopodium, Phylloglossum, Selaginella, Isoetes, Stylites*

SUBPHYLUM SPHENOPSIDA. Horsetails (25). *Equisetum*

SUBPHYLUM PTEROPSIDA. Ferns (10,000). *Polypodium, Osmunda, Dryopteris, Botrychium, Pteridium*

SUBPHYLUM SPERMOPSIDA. Seed plants

CLASS PTERIDOSPERMAE. Seed ferns. No living representatives

CLASS CYCADAE. Cycads (100). *Zamia*

CLASS GINKGOAE (1). *Gingko*

CLASS CONIFERAE. Conifers (500). *Pinus, Tsuga, Taxus, Sequoia*

CLASS GNETEAE (70). *Gnetum, Ephedra, Welwitschia*

CLASS ANGIOSPERMAE. Flowering plants

SUBCLASS DICOTYLEDONEAE. Dicots (225,000). *Magnolia, Quercus, Acer, Pisum, Taraxacum, Rosa, Chrysanthemum, Aster, Primula, Ligustrum, Ranunculus*

SUBCLASS MONOCOTYLEDONEAE. Monocots (50,000). *Lilium, Tulipa, Poa, Elymus, Triticum, Zea, Ophyrys, Yucca, Sabal*

† There are many alternative classification systems for protists and algae. The underlying uncertainty is reflected here by our use of colloquial groupings. The scheme used here, based primarily on rRNAs, is derived largely from Sogin et al. (1989); T Cavalier-Smith, 1989, *Nature* 339:100–101; and reviews for this edition by D. J. Patterson and T. Cavalier-Smith.

KINGDOM FUNGI

PHYLUM CHYTRIDIOMYCOTA. Chytrids (1000). *Olpidium, Rhizophydium, Diplophylctis, Cladochytrium*

PHYLUM ZYGOMYCOTA. Conjugation fungi (250)

CLASS ZYGOMYCETES. *Rhizopus, Mucor, Phycomyces, Choanephora, Entomophthora*

CLASS TRICHOMYCETES. *Stachylina*

PHYLUM ASCOMYCOTA. Sac fungi (12,000)

CLASS HEMIASCOMYCETES. Yeasts and their relatives. *Saccharomyces, Schizosaccharomyces, Endomyces, Eremascus, Taphrina*

CLASS PLECTOMYCETES. Powdery mildews, fruit molds, etc. *Erysiphe, Podosphaera, Aspergillus, Penicillium, Ceratocystis*

CLASS PYRENOMYCETES. *Sordaria, Neurospora, Chaetomium, Xylaria, Hypoxylon*

CLASS DISCOMYCETES. *Sclerotinia, Trichoscyphella, Rhytisma, Xanthoria, Pyronema*

CLASS LABOULBENIOMYCETES. *Herpomyces, Laboulbenia*

CLASS LOCULOASCOMYCETES. *Cochliobolus, Pyrenophora, Leptosphaeria, Pleospora*

PHYLUM BASIDIOMYCOTA. Club fungi (15,000)

CLASS HETEROBASIDIOMYCETES. Rusts and smuts. *Ustilago, Urocystis, Puccinia, Phragmidium, Melampsora*

CLASS HOMOBASIDIOMYCETES. Toadstools, bracket fungi, mushrooms, puffballs, stinkhorns, etc. *Coprinus, Marasmius, Amanita, Agaricus, Lycoperdon, Phallus*

KINGDOM ANIMALIA††

SUBKINGDOM PARAZOA

PHYLUM PORIFERA. Sponges (5000)

CLASS CALCAREA. Calcareous (Chalky) sponges. *Scypha, Leucosolenia, Sycon, Grantia*

CLASS HEXACTINELLIDA. Glass sponges. *Euplectella, Hyalonema, Monoraphis*

CLASS DEMOSPONGIAE. *Spongilla, Euspongia, Axinella*

CLASS SCLEROSPONGIAE. Coralline sponges. *Ceratoporella, Stromatospongia*

SUBKINGDOM EUMETAZOA

SECTION RADIATA

PHYLUM CNIDARIA (or COELENTERATA)

CLASS HYDROZOA. Hydrozoans (3700). *Hydra, Obelia, Gonionemus, Physalia*

CLASS CUBOZOA. Sea wasps (20). *Tripedalia*

CLASS SCYPHOZOA. Jellyfish (200). *Aurelia, Pelagia, Cyanea*

CLASS ANTHOZOA. Sea anemones and corals (6100). *Metridium, Pennatula, Gorgonia, Astrangia*

†† The classification of Animalia has been adapted from *Synopsis and Classification of Living Organisms*, 1982, ed. S. P. Parker, McGraw-Hill, New York.

PHYLUM PLACOZOA[@] (1). *Trichoplax*

PHYLUM CTENOPHORA. Comb jellies (90). *Pleurobrachia, Haeckelia*

SECTION MYXOZOA

PHYLUM MYXOZOA. Myxosporians (800). *Myxobolus, Myxidium, Ceratomyxa*

SECTION PROTOSTOMIA

PHYLUM PLATYHELMINTHES. Flatworms (10,000)

- CLASS TURBELLARIA. Free-living flatworms. *Planaria, Dugesia, Leptoplana*
- CLASS TREMATODA. Flukes. *Fasciola, Schistosoma, Prosthogonimus*
- CLASS CESTODA. Tapeworms. *Taenia, Dipylidium, Mesocestoides*

PHYLUM GNATHOSTOMULIDA (100). *Gnathostomula, Haplognathia*

PHYLUM NEMERTEA (or RHYNCHOCOELA). Proboscis or ribbon worms (650)

- CLASS ANOPLA. *Tubulanus, Cerebratulus*
- CLASS ENOPLA. *Amphiporus, Prostoma, Malacobdella*

PHYLUM ACANTHOCEPHALA. Spiny-headed worms (500). *Echinorhynchus, Gigantorhynchus*

PHYLUM MESOZOA (50)

- CLASS RHOMBOZOA. *Dicyema, Pseudicyema, Conocyema*
- CLASS ORTHONECTIDA. *Rhopalura*

PHYLUM ROTIFERA.[∞] Rotifers (1700). *Asplanchna, Hydatina, Rotaria*

PHYLUM GASTROTRICHA[∞] (2000). *Chaetonotus, Macrodasys*

PHYLUM KINORHYNCHA[∞] **(or ECHINODERA)** (100). *Echinoderes, Semnoderes*

PHYLUM NEMATA.[∞] Roundworms or nematodes (12,000). *Ascaris, Trichinella, Necator, Enterobius, Ancylostoma, Heterodera*

PHYLUM NEMATOMORPHA.[∞] Horsehair worms (230). *Gordius, Paragordius, Nectonema*

PHYLUM CHAETOGNATHA.[#] Arrow worms (70). *Sagitta, Spadella*

PHYLUM ENTOPROCTA (150). *Urnatella, Loxosoma, Pedicellina*

PHYLUM LORICIFERA (3, newly discovered and probably very common)

PHYLUM PRIAPULIDA (8). *Priapulus, Halicryptus*

@ Formerly considered a separate subkingdom.

∞ Formerly considered a class of Phylum Aschelminthes.

Formerly considered a phylum in the section Deuterostoma.

PHYLUM BRYOZOA[†] **(or ECTOPROCTA).** Bryozoans, moss animals (4000)

- CLASS GYMNOLAEMATA. *Paludicella, Bugula*
- CLASS PHYLACTOLAEMATA. *Plumatella, Pectinatella*
- CLASS STENOLAEMATA

PHYLUM PHORONIDA[†] (10). *Phoronis, Phoronopsis*

PHYLUM BRACHIOPODA.[†] Lamp shells (300)

- CLASS INARTICULATA. *Lingula, Glottidia, Discina*
- CLASS ARTICULATA. *Magellania, Neothyris, Terebratula*

PHYLUM MOLLUSCA. Molluscs

- CLASS CAUDOFOVEATA (70). *Chaetoderma*
- CLASS SOLENOGASTRES. Solenogasters (180). *Neomenia, Proneomenia*
- CLASS POLYPLACOPHORA. Chitons (600). *Chaetopleura, Ischnochiton, Lepidochiton, Amicula*
- CLASS MONOPLACOPHORA (8). *Neopilina*
- CLASS GASTROPODA. Snails and their allies (univalve molluscs) (25,000). *Helix, Busycon, Crepidula, Haliotis, Littorina, Doris, Limax*
- CLASS SCAPHOPODA. Tusk shells (350). *Dentalium, Cadulus*
- CLASS BIVALVIA. Bivalve molluscs (7500). *Mytilus, Ostrea, Pecten, Mercenaria, Teredo, Tagelus, Unio, Anodonta*
- CLASS CEPHALOPODA. Squids, octopuses, etc. (600). *Loligo, Octopus, Nautilus*

PHYLUM POGONOPHORA. Beard worms (100). *Siboglinum, Lamellisabella, Oligobrachia, Polybrachia*

PHYLUM SIPUNCULA (300). *Sipunculus, Phascolosoma, Dendrostomum*

PHYLUM ECHIURA (140)

- CLASS ECHIUROINEA. *Echiurus, Ikedella*
- CLASS XENOPNEUSTA. *Urechis*
- CLASS HETEROMYOTA. *Crisia, Tubulipora*

PHYLUM ANNELIDA. Segmented worms

- CLASS POLYCHAETA (INCLUDING ARCHIANNELIDA). Sandworms, tubeworms, etc. (8000). *Nereis, Chaetopterus, Aphrodite, Diopatra, Arenicola, Hydroides, Sabella*
- CLASS OLIGOCHAETA. Earthworms and many freshwater annelids (3100). *Tubifex, Enchytraeus, Lumbricus, Dendrobaena*
- CLASS HIRUDINOIDEA. Leeches (500). *Trachelobdella, Hirudo, Macrobdella, Haemadipsa*

PHYLUM ONYCHOPHORA (65). Peripatus. *Peripatopsis*

PHYLUM TARDIGRADA. Water bears (300). *Echiniscus, Macrobiotus*

PHYLUM ARTHROPODA (at least 2 million)

SUBPHYLUM TRILOBITA. No living representatives

SUBPHYLUM CHELICERATA

- CLASS EURYPTERIDA. No living representatives
- CLASS MEROSTOMATA. Horseshoe crabs (4). *Limulus*

† Bryozoa, Phoronida, and Brachiopoda are often referred to as the lophophorate phyla.

CLASS ARACHNIDA. **Spiders, ticks, mites, scorpions, whipscorpions, daddy longlegs, etc. (55,000; at least 500,000 undiscovered species of mites are thought to exist).** *Archaearanea, Latrodectus, Argiope, Centruroides, Chelifer, Mastigoproctus, Phalangium, Ixodes*

CLASS PYCNOGONIDA. **Sea spiders (1000).** *Nymphon, Ascorhynchus*

SUBPHYLUM CRUSTACEA (26,000). *Homarus, Cancer, Daphnia, Artemia, Cyclops, Balanus, Porcellio*

SUBPHYLUM UNIRAMIA

CLASS CHILOPODA. **Centipedes (2500).** *Scolopendra, Lithobius, Scutigera*

CLASS DIPLOPODA. **Millipedes (10,000; another 50,000 species thought to exist).** *Narceus, Apheloria, Polydesmus, Julus, Glomeris*

CLASS PAUROPODA **(500).** *Pauropus*

CLASS SYMPHYLA **(160).** *Scutigerella*

CLASS INSECTA. **Insects (900,000; another 1 million species are thought to exist)**

ORDER THYSANURA. **Bristletails, silverfish, firebrats.** *Machilis, Lepisma. Thermobia*

ORDER EPHEMERIDA. **Mayflies.** *Hexagenia, Callibaetis, Ephemerella*

ORDER ODONATA. **Dragonflies, damselflies.** *Archilestes, Lestes, Aeshna, Gomphus*

ORDER ORTHOPTERA. **Grasshoppers, crickets, etc.** *Schistocerca, Romalea, Nemobius, Megaphasma*

ORDER PHASMATOPTERA. **Walking sticks.** *Phyllium*

ORDER BLATTARIA. **Cockroaches.** *Blatta, Periplaneta*

ORDER MANTODEA. **Mantids.** *Mantis*

ORDER GRYLLOBLATTARIA. *Grylloblatta*

ORDER ISOPTERA. **Termites.** *Reticulitermes, Kalotermes, Zootermopsis, Nasutitermes*

ORDER DERMAPTERA. **Earwigs.** *Labia, Forficula, Prolabia*

ORDER EMBIIDINA (OR EMBIARIA OR EMBIOPTERA). *Oligotoma, Anisembia, Gynembia*

ORDER PLECOPTERA. **Stoneflies.** *Isoperla, Taeniopteryx, Capnia, Perla*

ORDER ZORAPTERA. *Zorotypus*

ORDER PSOCOPTERA. **Book lice.** *Ectopsocus, Liposcelis, Trogium*

ORDER MALLOPHAGA. **Chewing lice.** *Cuclotogaster, Menacanthus, Menopon, Trichodectes*

ORDER ANOPLURA. **Sucking lice.** *Pediculus, Phthirius, Haematopinus*

ORDER THYSANOPTERA. **Thrips.** *Heliothrips, Frankliniella, Hercothrips*

ORDER HEMIPTERA. **True bugs.** *Belostoma, Lygaeus, Notonecta, Cimex, Lygus, Oncopeltus*

ORDER HOMOPTERA. **Cicadas, aphids, leafhoppers, scale insects, etc.** *Magicicada, Circulifer, Psylla, Aphis, Saissetia*

ORDER NEUROPTERA. **Dobsonflies, alderflies, lacewings, mantispids, snakeflies, etc.** *Corydalus, Hemerobius, Chrysopa, Mantispa, Agulla*

ORDER COLEOPTERA. **Beetles, weevils.** *Copris, Phyllophaga, Harpalus, Scolytus, Melanotus, Cicindela, Dermestes, Photinus, Coccinella, Tenebrio, Anthonomus, Conotrachelus*

ORDER HYMENOPTERA. **Wasps, bees, ants, sawflies.** *Cimbex, Vespa, Glypta, Scolia, Bembix, Formica Bombus, Apis*

ORDER STREPSIPTERA. **Endoparasites**

ORDER MECOPTERA. **Scorpionflies.** *Panorpa, Boreus, Bittacus*

ORDER SIPHONAPTERA. **Fleas.** *Pulex, Nosopsyllus, Xenopsylla, Ctenocephalides*

ORDER DIPTERA. **True flies, mosquitoes.** *Aedes, Asilus, Sarcophaga, Anthomyia, Musca, Chironomus, Tabanus, Tipula, Drosophila*

ORDER TRICHOPTERA. **Caddisflies.** *Limnephilus, Rhyacophilia, Hydropsyche*

ORDER LEPIDOPTERA. **Moths, butterflies.** *Tinea, Pyrausta, Malacosoma, Sphinx, Samia, Bombyx, Heliothis, Papilio, Lycaena*

SECTION DEUTEROSTOMIA

PHYLUM ECHINODERMATA

SUBPHYLUM CRINOZOA

CLASS CRINOIDEA. **Crinoids, sea lilies (630).** *Antedon, Ptilocrinus, Comactinia*

SUBPHYLUM ASTEROZOA

CLASS STELLEROIDEA. **Sea stars, brittle stars (2600).** *Asterias, Ctenodiscus, Luidia, Oreaster, Asteronyx, Amphioplus, Ophiothrix, Ophioderma, Ophiura*

SUBPHYLUM ECHINOZOA

CLASS ECHINOIDEA. **Sea urchins, sand dollars, heart urchins (860).** *Cidaris, Arbacia, Strongylocentrotus, Echinanthus, Echinarachnius, Moira*

CLASS HOLOTHUROIDEA. **Sea cucumbers (900).** *Cucumaria, Thyone, Caudina, Synapa*

PHYLUM HEMICHORDATA (90)

CLASS ENTEROPNEUSTA. **Acorn worms.** *Saccoglossus, Balanoglossus, Glossobalanus*

CLASS PTEROBRANCHIA. *Rhabdopleura, Cephalodiscus*

CLASS PLANCTOSPHAEROIDEA

PHYLUM CHORDATA. Chordates

SUBPHYLUM TUNICATA (or UROCHORDATA). Tunicates (2000)

CLASS ASCIDIACEA. **Ascidians or sea squirts.** *Ciona, Clavelina, Molgula, Perophora*

CLASS THALIACEA. *Pyrosoma, Salpa, Doliolum*

CLASS APPENDICULARIA. *Appendicularia, Oikopleura, Fritillaria*

SUBPHYLUM CEPHALOCHORDATA. Lancelets, amphioxus (30). *Branchiostoma, Asymmetron*

SUBPHYLUM VERTEBRATA. Vertebrates

CLASS AGNATHA. **Jawless fish (50).** *Cephalaspis,* Pteraspis,* Petromyzon, Entosphenus, Myxine, Eptatretus*

CLASS ACANTHODII. **No living representatives**

CLASS PLACODERMI. **No living representatives**

CLASS CHONDRICHTHYES. **Cartilaginous fish, including sharks and rays (800).** *Squalus, Hyporion, Raja, Chimaera*

CLASS OSTEICHTHYES. **Bony fish (18,000)**

SUBCLASS SARCOPTERYGII

ORDER CERATODIFORMES. **Australian lungfish.** *Neoceratodus*

ORDER LEPIDOSIRENIFORMES. **Lungfish.** *Protopterus, Lepidosiren*

SUBCLASS ACTINOPTERYGII. Ray-finned fish. *Amia, Cyprinus, Gadus, Perca, Salmo*

CLASS AMPHIBIA **(3100)**

ORDER ANURA. **Frogs and toads.** *Rana, Hyla, Bufo*

ORDER CAUDATA (or URODELA). *Salamanders, Necturus, Triturus,*

* Extinct.

Plethodon, Ambystoma

ORDER GYMNOPHIONA(or APODA). *Ichthyophis, Typhlonectes*

CLASS REPTILIA (6500)

ORDER TESTUDINES. **Turtles.** *Chelydra, Kinosternon, Clemmys, Terrapene*

ORDER RHYNCHOCEPHALIA. *Tuatara, Sphenodon*

ORDER CROCODYLIA. **Crocodiles and alligators.** *Crocodylus, Alligator*

ORDER LEPIDOSAURIA. **Snakes and lizards.** *Iguana, Anolis, Sceloporus, Phrynosoma, Natrix, Elaphe, Coluber, Thamnophis, Crotalus*

CLASS AVES. **Birds (8600).** *Anas, Larus, Columba, Gallus, Turdus, Dendroica, Sturnus, Passer, Melospiza*

CLASS MAMMALIA. Mammals (4100)

SUBCLASS PROTOTHERIA

ORDER MONOTREMATA. **Egg-laying mammals.** *Ornithorhynchus, Tachyglossus*

SUBCLASS THERIA. Marsupial and placental mammals

ORDER METATHERIA (or MARSUPIALIA). **Marsupials.** *Didelphis, Sarcophilus, Notoryctes, Macropus*

ORDER INSECTIVORA. **Insectivores (moles, shrews, etc.).** *Scalopus, Sorex, Erinaceus*

ORDER DERMOPTERA. **Flying lemurs.** *Galeopithecus*

ORDER CHIROPTERA.** *Bats, Myotis, Eptesicus, Desmodus*

ORDER PRIMATA. **Lemurs, monkeys, apes, humans.** *Lemur, Tarsius, Cebus, Macacus, Cynocephalus, Pongo, Pan, Homo*

ORDER EDENTATA. **Sloths, anteaters, armadillos.** *Bradypus, Myrmecophagus, Dasypus*

ORDER PHOLIDOTA. **Pangolin.** *Manis*

ORDER LAGOMORPHA. **Rabbits, hares, pikas.** *Ochotona, Lepus, Sylvilagus, Oryctolagus*

ORDER RODENTIA. **Rodents.** *Sciurus, Marmota, Dipodomys, Microtus, Peromyscus, Rattus, Mus, Erethizon, Castor*

ORDER ODONTOCETA. **Toothed whales, dolphins, porpoises.** *Delphinus, Phocaena, Monodon*

ORDER MYSTICETA. **Baleen whales.** *Balaena*

ORDER CARNIVORA. **Carnivores.** *Canis, Procyon, Ursus, Mustela, Mephitis, Felis, Hyaena, Eumetopias*

ORDER TUBULIDENTATA. **Aardvark.** *Orycteropus*

ORDER PROBOSCIDEA. **Elephants.** *Elephas, Loxodonta*

ORDER HYRACOIEDEA. **Hyraxes, conies.** *Procavia*

ORDER SIRENIA. **Manatees.** *Trichechus, Halicore*

ORDER PERISSODACTYLA. **Odd-toed ungulates.** *Equus, Tapirella, Tapirus, Rhinoceros*

ORDER ARTIODACTYLA. **Even-toed ungulates.** *Pecari, Sus, Hippopotamus, Camelus, Cervus, Odocoileus, Giraffa, Bison, Ovis, Bos*

** Some authorities propose dividing the echo-locating bats and fruit-eating bats into separate orders.

CREDITS

About the cover Wendell Minor **Contents Part I** Bill Longcore/ Photoresearcher, Inc. **Part II** Illustration copyright of Irving Geis **Part III** Photo courtesy Bob Natalini **Part IV** Photo courtesy William E. Ferguson **Part V** M.P.L. Fogden/Bruce Coleman, Inc. **Part VI** © Art Wolfe/Tony Stone Images **About the author** Pryde Brown Photographs

1.1 Uniphoto Picture Agency, London. **1.3** Reunion des Musees Nationaux, Paris **1.5** National Portrait Gallery, London. **1.6** Rijksmuseum, Amsterdam. **1.7** Photo No. #326662. Courtesy Department Library Services, American Museum of Natural History. **1.8** (a) Sovfoto, New York. (b) F. Gohier, Photo Researchers, Inc. **1.9** Cop. Bibliotheque centrale M. N. H. N. Paris. **1.11** The New York Public Library, Astor, Tilden, and Lenox Foundations. **1.12** Tim Davis, Photo Researchers, Inc. **1.13** ©Lennart Nilsson, *Behold Man*, Little, Brown, and Company. **1.14** Modified from The Illustrated Origin of Species, by Charles Darwin, abridged and introduced by Richard E. Leakey, 1979; courtesy of Hill and Wang, a division of Farrar, Straus & Giroux, Inc. **1.15** Pages 36–37 of Darwin notebook "B," by permission of the Syndics of Cambridge University. **1.16** National Portrait Gallery, London. **1.17** (center) Kenneth W. Fink, Bruce Coleman, Inc. (all others) Louise B. Van der Meid. **1.18** Clockwise from left: (archaebacterium) H. W. Jannasch and C. O. Wirsen, *Bio Science*, vol. 29, 1979; copyright ©1979 by the American Institute of Biological Science; (archaezoan) Courtesy E. W. Daniels; (protist) © M. I. Walker, Science Source, Photo Researchers, Inc.; (fly agaric) © G. R. Roberts; (great egret) © M. P. Kahl, 1972, Photo Researchers, Inc.; (dahlias) Gene Ahrens, Bruce Coleman, Inc.; (kelp) © Jeff Rotman; (a true bacterium) CNRI, Photo Science Library, Photo Researchers, Inc. **1.19** Courtesy of Cold Spring Harbor Laboratory.

PART I: THE CHEMICAL AND CELLULAR BASIS OF LIFE

Powerhouse of the Cell: Bill Longcore/Photo Researchers, Inc.

2.1 NASA **2.3** Science Photo Library/Photo Researchers, Inc. **2.4** R. M. Feenstra and J. A. Stroscio, IBM Research **2.10** © David Newman/Visuals Unlimited **2.11a** © Dwight Kuhn **2.21** Photo by Ann F. Purcell **2.26** Herman Eisenbeiss/Photo Researchers, Inc. **2.27** © Dwight Kuhn, 1986. **2.29b** © Zefa Sauer, The Stock Market (#K8949) **2.30** Spenser Swanger/Tom Stack & Associates.

3.12 a Dwight Kuhn © 1980, b Courtesy W. Cheng, International Paper Company **3.14** Reproduced from R. G. Kessel and Randy H. Kardon, *Tissues and Organs: A Text-Atlas of Scanning Electron Microscopy*. W. H. Freeman, 1979. **3.24** Adapted by permission from *The Structure and Action of Proteins* by Richard E. Dickerson and Irving Geis, W. A. Benjamin, Inc., Menlo Park, Calif., Publisher; copyright © by Dickerson and Geis. **3.26** Adapted by permission from *The Structure and Action of Proteins* by Richard E. Dickerson and Irving Geis, W. A. Benjamin, Inc. Menlo Park, Calif., Publisher; copyright © by Dickerson and Geis. **3.27** Adapted by permission from *The Structure and Action of Proteins* by Richard E. Dickerson and Irving Geis, W. A. Benjamin, Inc. Menlo Park, Calif., Publisher; copyright © by Dickerson and Geis. **3.29** Courtesy John Kendrew, Medical Research Center of Molecular Biology **3.30** Adapted by permission from *The Structure and Action of Proteins* by Richard E. Dickerson and Irving Geis, W. A. Benjamin, Inc., Menlo Park, Calif., Publisher; copyright © by Dickerson and Geis. **3.38** Adapted from Bruce Alberts et al, *Molecular Biology of the Cell*, 3e, 1994, by permission of Garland Publishing, New York. **3.39** Jack Barrie/Bruce Coleman, Inc. **3.49** From Lehninger, *Biochemistry*, 2e, 1975. Worth Publishers, New York. **3.52** © K. Talaro/VU.

4.1a Science Photo Library, Photo Researchers, Inc. **4.3** Courtesy J. D. Pickett-Heaps, University of Melbourne. **4.4d** J. J. Cardamore and B. A. Phillips, University of Pittsburgh. **4.5** Reprinted with permission from M. Amrein, *Science*, 240:515. Copyright ©1988 American Association for the Advancement of Science. **4.6** Reprinted with permission from Dr. E. Henderson, from his article in *Science*, 257:1944. Copyright ©1992, American Association for the Advancement of Science. **4.11** © David M. Phillips/Visuals Unlimited. **4.13** The Liposome Company. **4.14** Courtesy J. David Robertson, Duke University. **4.16** Courtesy Daniel Branton. **4.25** Photo by Dorothy F. Bainton, M. D. **4.26** © R. L. Roberts **4.27a** J. Ross, J. Olmstead, and J. Rosenbaum, *Tissue and Cell*, vol. 7, 1975. **4.28** M. M. Perry and A. B. Gilbert, *J. Cell. Sci.*, vol. 39, 1979 by copyright permission of The Rockefeller University Press. **4.29a** N. Hirokawa and J. Heuser, *Cell*, vol. 30, 395–406. © 1982 by M. I. T., **b** courtesy of Dr. Richard G. W. Anderson, University of Texas Health Center, Dallas, Texas. **4.31b** V. Herzog, H. Sies, and F. Miller, J. *Cell. Biol.*, vol. 70, 1976, by copyright permission of the Rockefeller University Press. **4.32** © Biophoto Associates/Science Source, Photo Researchers, Inc. **4.33** Courtesy R. D. Preston and E. Frei, University of Leeds. **4.34b** H. Latta, W. Johnson, and T. Stanley, *J. Ultrastruct. Res.*, vol. 51, 1975.

5.1 Biophoto Associates/Science Source, Photo Researchers, Inc. **5.2** Dr. J. Nickerson, G. Krockmalinic, et al, *Science*, 259:1257, 1993. Reproduced by permission of the American Association for the Advancement of Science and the authors. **5.3** Photo courtesy of the Biology Department, Brookhaven National Laboratory. **5.4** Courtesy of Dr. Barbara Hamkalo and Dr. J. B. Rattner. **5.6** W. G. Whaley, H. H. Mollenhauer, and J. H. Leech, *Am. J. Bot.*, vol. 47, 1960. **5.7** Courtesy Daniel Branton, Harvard University. **5.8** Courtesy K. R. Porter, University of Colorado. **5.10** Micrograph courtesy D. S. Friend, University of California, San Francisco. **5.12** Courtesy D. S. Friend, University of California, San Francisco. **5.14** Courtesy D. S. Friend, University of California, San Francisco. **5.16** Micrograph by S. E. Frederick and E. H. Newcomb, J. Cell. Biol., vol. 43, 1969. **5.17** Courtesy D. S. Friend, University of California, San Francisco. **5.18top** Micrograph by W. P. Wergin, courtesy E. H. Newcomb, University of Wisconsin. **bottom** M. C. Ledbetter, Photo Researchers, Inc. **5.19** Courtesy Gordon Leedale, Biophoto Associates, Leeds, England. **5.22** Courtesy Elias Laxarides, California Institute of Technology. **5.23** Dr. Susumu Ito, Department of Neurobiology, Harvard Medical School. **5.25c** © Boehringer Ingelheim International GmbH, photo Lennart Nilsson. **d** Photograph by C. Lin, courtesy of Paul Forscher, Yale University. **5.26** Dr. John Hartwig et al., *Sci. Am.*, September 1994, used by permission of authors and W. H. Freeman & Co., Inc. **5.27a** H. Kim, L. I. Binder, J. L. Rosenbaum; The Journal of Cell Biology, vol. 80, 266–267, 1979 by copyright permission of the Rockefeller University Press. **b** Courtesy of Don Fawcett, Harvard Medical School. **5.30** Photo from A. Ashkin, K. Schutze, J. M. Dziedic, U. Euteneurt, and M. Schliwa, *Nature*, vol. 348; 1990, pp. 346–48. **5.32** U. Aebi, University of Basel, Switzerland. **5.33** Micrograph from J. Heuser and S. R. Salpeter, *J. Cell. Biol*. vol. 82, 1979 by copyright permission of the Rockefeller University Press. **5.34** M. McGill, D. P. Highfield, T. M. Monahan, and B. R. Brinkley, *J. Ultrastruct. Res.*, vol. 578, 1976. **5.35** Courtesy R. W. Linck, Harvard Medical School, and D. T. Woodrum. **5.36** Photo provided by Dr. E. R. Dirksen, University of California, Los Angeles. **5.37** Micrograph by K. Roberts, John Innes Institute, Norwich, England; from B. Alberts et al, Molecular Biology of the Cell, Garland Press, New York, 1983. **5.39** C. J. Brokaw, California Institute of Technology, from *Science*, vol. 178, 1972, copyright © by the American Association for the Advancement of Science. **5.41** Micrograph produced by J. W. Heuser, et al, *J. Cell Biol.*, vol. 95, p. 800, 1982. Reproduced by copyright permission of the Rockefeller University Press. **5.40** (mitochondrion) Courtesy K. R. Porter, University of Colorado; (lysosome) Courtesy A. B. Novikoff, Albert Einstein College of Medicine; (glycocalyx) Courtesy A. Ryter, Institut Pasteur, Paris; (Golgi apparatus) D. W. Fawcett/Visuals Unlimited; (centrioles) M. McGill, D. P. Highfield, T. M. Monohan, and B. R. Brinkley, *J. Ultrastruct. Res.*, vol. 57, 1976; (all other micrographs) Courtesy N. B. Gilula, Baylor College of Medicine. **5.41** (plasmodesma) Courtesy W. G. Whaley, et al., *J. Biophys. Biochem. Cytol.*, (*now J. Cell Biol.*) vol. 5, 1959; by copyright permission of the Rockefeller University Press; (nucleus, chloroplast, leucoplast, endoplasmic reticulum, mitochondrion, and Golgi apparatus) Courtesy A. Ryter, Institut Pasteur, Paris. **5.43** Courtesy J. Griffith, School of Medicine, University of North Carolina, Chapel Hill.

6.2 Courtesy Professor Dr. K. O. Stetter, Universitat Regensburg, Germany. **6.11a** © K. R. Porter, D. W. Fawcett/Visuals Unlimited. **c** photo by H. Fernandez-Moran, courtesy E. Valdivia, University of Wisconsin. From Fernandez-Moran, et al., J. *Cell Biol.*, 22:63-100, 1964. Reprinted by copyright permission of Rockefeller University Press. ***Exploring Further*** p. 183 Efraim Racker, Cornell University.

7.7 Courtesy of Professor C. N. Hunter, University of Sheffield, England. **7.12a** Micrograph by W. P. Wergin, courtesy E. H. Newcomb, University of Wisconsin. **7.16** (top left and right) Print Collection, Mirian and Ira D. Wallach Division of Art, Prints, and Photographs; (middle left and bottom left and right) General Research Division; (middle right) Rare Books and Manuscripts Division, The New York Public Library Astor, Lenox, and Tilden Foundations. **7.17** Courtesy John H. Troughton, New Zealand. **7.19** Courtesy Raymond Chollet, University of Nebraska. **7.22** Courtesy G. R. Roberts.

PART II: THE PERPETUATION OF LIFE

Molecule of inheritance: Illustration copyright by Irving Geis.

8.2 Carolina Biological Supply Company **8.3** M. McCarty, *Journal of Experimental Medicine*, vol. 79, 1944, 137–58. **8.5** Courtesy Lee D. Simon, Waksman Institute, Rutgers University. **8.12** From J. D. Watson, The Double Helix, Atheneum, New York, 1968. © J. D. Watson. Photo courtesy Cold Spring Harbor Lab Archives. **8.13** Courtesy Cold Spring Harbor Lab Archives. **8.16** From Alberts et al., *Molecular Biology of the Cell*, p. 247, fig. 3-36. Garland Press, New York, 1994. Based on V. Derbyshire and T. A. Stetix, Science, 260, 352–355, 1993. **8.18** Courtesy A. C. Arnberg, Biochemical Laboratory, State University, Groningen, The Netherlands. **8.19** Redrawn from M. S. Meselson and F. W. Stahl, Proc. Natl. Acad. Sci. U. S. A. vol. 44, 1958.

9.1 Courtesy David M. Prescott, from *Prog. Nucleic Acid*, vol. 3 (35), 1964. **9.3b** Jack R. Griffith, University of North Carolina, Chapel Hill. **9.6** Photo by B. Tagawa, courtesy F. Perrin and P. Chambin, *Sci. Am.*, 244 (1981) 60-71. Used by permission. **9.8** Jack R. Griffth, University of North Carolina, Chapel Hill. **9.9** Redrawn from Alberts et al., *Molecular Biology of the Cell* (1994), p. 232 fig. 6-19, Courtesy author and Garland Press, New York. Based upon figure from S. Stern, B. Weiser, and H. F. Noller, *J. Mol. Biol.*, 204, 447–481, 1988. **9.11** Adapted from R. Gupta, J. M. Lanter, and C. R. Woese, *Science*, vol. 221, 1993; copyright © 1983 A. A. A. S. **9.13** Micrograph from O. L Miller, Jr., B. A. Hamkalo, and C. A. Thomas, Jr., *Science*, vol. 169, 1970; copyright ©1970 by A. A. A. S. **9.14** Photograph courtesy Nigel Unwin, Stanford University School of Medicine. **9.16** Modified from B. Alberts et al., *Molecular Biology of the Cell*, Garland Press, New York, 1994. **9.17** Photo from M. A. Rould, J. J. Perona, D. Soll, and T. A. Steitz, *Science*, vol. 246, 1989. Copyright © 1989 A. A. A. S. **9.19** M. R. Hanson et al., *Mol. Gen. Genet.*, vol. 132, 1974. **9.21** Courtesy Bruce N. Ames, University of California, Berkeley.

10.1a,c Jack R. Griffith, University of North Carolina, Chapel Hill. **b** Courtesy S. N. Cohen, Stanford University. **10.2** Courtesy L. G. Caro, University of Geneva, and R. Curtiss, University of Alabama. **10.14** Courtesy Norman H. Olsen and Timothy S. Baxter, Purdue University. ***Exploring Futher*** p. 264 Photo courtesy Lark Sequencing. **10.16** Provided courtesy of Michael Simons, M. D., Cardiovascular Division, Beth Israel Hospital, Boston, MA. From Simons et al., *Nature*, vol. 359, 3 September 1992. **10.17** John D. Cunningham/ Visuals Unlimited. **10.18** Courtesy Dr. H. Anzai, Pharmaceutical Research Center, Meiji Seiki Kaisha Ltd. Yokohama, Japan. From *Biotechnology in Plant Disease Control*, Ilan Chet, ed. Copyright 1993 H. Anzai. Reprinted with permission of John Wiley & Sons, Inc. New York. **10.19** Courtesy Dr. John Sanford, Cornell University. **10.20** Courtesy Genentech, Inc. San Francisco, CA. ***How We Know*** p. 268 Photo compliments of Cellmark Diagnostics. **10.21** Photo courtesy Robert Devlin, Fisheries and Oceans, West Vancouver, B. C.

11.1a,b Courtesy Cold Spring Harbor Lab Archives. **11.2d** Courtesy J. Griffith, University of North Carolina, Chapel Hill. **11.5** Electron micrograph by G. F. Bahr and W. F. Engler, Armed Forces Institute of Pathology, Washington, D. C. **11.6** Courtesy Steven Henikoff, Fred Hutchinson Cancer Research Center, Seattle. **11.8** Photo courtesy J. G. Gall, Carnegie Institution, reproduced from M. B. Roth and J. G. Gall, *J. Cell Biol.*, 105, 1047–1054 (1987). Copyright 1987 by Rockefeller University Press. **11.9** Courtesy Michael Ashburner, Cambridge University. **11.11** Photograph courtesy O. L. Miller, Jr., and B. R. Beatty, *J. Cell Physiol.*, vol. 74, 1969. **11.13** Courtesy Gunter-Albrecht-Buehler, Cold Spring Harbor Laboratory, and Frank Solomon, Massachusetts Institute of Technology. **11.14** K. Porter, G. Fonte, and Weiss, *Cancer Res.*, vol. 34, 1974. **11.16** Curve for retinablastoma based on data from H. W. Hethcote and A. G. Knudson, *Proc. Natl. Acad. Sci. U. S. A.*, vol. 75, 1978; curves for prostate and skin cancer based on data from Japanese Cancer Association, *Cancer Mortality and Morbidity Statistics*, Japanese Scientific Press, Tokyo, 1981.

12.2 Micrograph from I. D. J. Burdett and R. G. E. Murray, Department of Microbiology and Immunology, The University of Western Ontario. **12.3** M. P. Marsden and U. K. Laemmli; *Cell*, 17:849-58, 1979, used by permission of MIT Press, Cambridge, MA. **12.4** Courtesy M. W. Shaw, University of Michigan, Ann Arbor. **12.6** Courtesy Andrew S. Bajer, University of Oregon. **12.10** Courtesy Andrew S. Bajer, University of Orgeon. **12.11** Adapted from Alberts et al., *Molecular Biology of the Cell*, 3e, Garland Press, New York, 1994. **12.12** Photo courtesy M. S. Fuller, from Fuller, *Mycologia*, vol. 60, 1968. **12.13** From H. W. Beams and R. G. Kessel, *Am. Sci.*, vol. 64, 1976; reprinted by permission of American Scientist, *Journal of Sigma Xi*, the Scientific Research Society. **12.15** All courtesy Andrew S. Bajer, University of Oregon. **12.16** From W. G. Whaley et al., *Am. J. Bot.*, vol. 47, 1960. **12.18** From D. von Wettstein, Proc. *Natl. Acad. Sci. U. S. A.* vol. 68, 1971. **12.20** Courtesy James Kezer, University of Orgeon. **12.23** Courtesy K. Izutsu, A. S. Bajer, and J. Mole-Bajer. **12.28** Courtesy Ed Reschke. **12.30** Photograph by L. Nilsson, from L. Nilsson, *Behold Man*, English translation © 1974, Albert Bonniers Forlag, Stockholm, and Little, Brown and Co., (Canada) Ltd.

13.1b Courtesy Dr. Sundstrom/Gamma Liaison **13.2f** D. M. Phillips, *J. Ultrastruct. Res.*, 72:1-12, 1980. **13.3** From *Science*, vol. 257, 10 July 1992; "Block of Ca + 2 Oscillation by Antibody..." fig. 3, p. 252. Courtesy Dr. Shun-ichi Miyazaki, Tokyo Women's Medical College. **13.8** Coutesy of Professors R. G. Kessel, University of Iowa, and C. Y. Shih. **13.11** Oxford Scientific Films **13.12** Photographs by L. Nilsson; from L. Nilsson, *Behold Man*, English translation © 1974, Albert Bonniers Forlag, Stockholm, and Little, Brown and Co. (Canada) Ltd. **13.14** Redrawn from G. J. Romanes, *Darwin and After Darwin*, Open Court Publishing Co., 1901. **13.15** Redrawn from D'A. W. Thompson, *On Growth and Form*, Cambridge, The University Press, 1942 (from G. Backman, after Stefanowska). **13.16** Redrawn from D'A. W. Thompson, *On Growth and Form*, Cambridge, The University Press, 1942 (after T. Ostwald). **13.17** Modified from V. B. Wigglesworth, *The Principles of Insect Physiology*, Methuen, 1947. **13.18** Redrawn from D'A. W. Thompson, *On Growth and Form*, Cambridge, The University Press, 1942 (from Quetelet's data). **13.20a-c,h** Photographs by D. Overcash, **d** photograph by L. West, **e–g** photos by E. R. Deginngger/Bruce Coleman, Inc. **13.22** From page 24 of *Sexual Selection* by James L. and Carol Grant Gould, W. H. Freeman & Co., Inc., based on A. Kornberg's *Replication*, which was published in 1980 by W. H. Freeman & Co., Inc.

14.5b Courtesy D. A. Melton, Harvard University **14.6** From L. Mann, *The Development of the Human Eye*, copyright © 1964 by Grune and Stratton, New York. **14.7** From W. Krommenhoek, J. Sebus, and G. J. van Esch, *Biological Structures*, copyright 1979 by L. C. G. Malmberg B. V., The Netherlands. **14.12b** Carolina Biological Supply Company. **14.15** Photographs by E. B. Lewis, California Institute of Technology. **14.16** Photograph by K. W. Tosney, as printed in N. K. Wessels, *Tissue Interactions and Development*, copyright © 1977 by Benjamin-Cummings, Menlo Park, California. **14.17** Photographs courtesy Dr. Dennis Summerbell, National Institute for Medical Research, London. **14.18d** Gregor Eichele, Baylor College of Medicine, Houston. **14.19** Adapted from M. Singer, *Sci. Am.*, October 1958, copyright © 1958 by Scientific American, Inc., all rights reserved. **14.21** Courtesy Corey S. Goodman and Michael J. Bastiani, from Goodman and Bastiani, *Sci. Am.*, December 1984, copyright © 1984 by Scientific American, Inc., all rights reserved. **14.22** Modified from J. E. Sulston and H. R. Horvitz, *Dev. Biol.*, vol. 56, 1977, copyright © 1977 by Academic Press.

15.3 Photo from R. G. Kessel and R. H. Kardon, *Tissues and Organs: A Text-Atlas of Scanning Electron Microscopy*, W. H. Freeman and Company, San Francisco, California, 1970. **15.4** Photo courtesy Peter Marks and Frederick Maxfield, from P. Marks and F. R. Maxfield, *J. Cell. Biol.*, 110:43-52 (1990). Reprinted by permission of the Rockefeller University Press. **15.5b** Computer graphic image by Arthur J. Olsen, © The Scripps Research Institute, 1986. **15.17** Reproduced from D. Lawson, C. Fewtrell, B. Gomperts, and M. Raff, *Journal of Experimental Medicine*, 142:391-402, 1975. Used by copyright permission of Rockefeller University Press. **15.19** E. Gueho/Photo Researchers, Inc. ***Exploring Further***, p. 402, photo courtesy Stephen T. Brentano.

16.5 Jane Burton/Bruce Coleman, Inc. **16.9** Courtesy M. C. Latham, Cornell University. **16.10** Modified from Francisco H. Alaya and John A. Kiger, Jr., *Modern Genetics*, Benjamin-Cummings, Menlo Park, California, 1980. **16.13** Courtesy S. B. Moore. **16.14** Modified from A. M. Winchester, *Genetics*, 5th ed., Houghton Mifflin, Boston, 1977. **16.16** Courtesy Marion I. Barnhart, Wayne State University Medical School, Detroit, Michigan. **16.19** K. R. Dronamraju, in E. J. Gardner, *Principles of Genetics*, copyright ©1975 by John Wiley & Sons, Inc.; reprinted with their permission. **16.20** Courtesy M. L. Barr, *Can. Cancer Conf.*, vol. 2, Academic Press, New York, 1957. **16.21** H. Chaumeton/Nature. **16.24** Modified from *Biological Science: An Ecological Apprach*, 4th ed., Houghton Mifflin, Boston, 1982. Used by permission. **16.26** E. J. Bingham, University of Wisconsin, Madison.

PART III: EVOLUTIONARY BIOLOGY

Diversity of Species: Photo courtesy Bob Natalini

17.2 F. Gohier/Photo Researchers, Inc. **17.3** Courtesy J. L. Gould **17.4a** George Holton/Photo Researchers, Inc.; **b** Masud Quraishy/Bruce Coleman, Inc. ***Exploring Further***, Fig. 2. © Kim Taylor/Bruce Coleman, Inc. **17.5** Adapted by permission from *The Structure and Action of Proteins* by Richard E. Dickerson and Irving Geis, W. A. Benjamin, Inc., Menlo Park, California, Publisher; copyright © 1969 by Dickerson and Geis. **17.7** A. J. Ribbink, *S. Afr. J. Zool.*, vol. 18, 1985. **17.8** Hans Reinhard/Bruce Coleman, Inc. **17.13** Data from C. M. Woodworth et al., *Agron, J.*, vol. 44, 1952. **17.15** Hans Reinhard/Bruce Coleman, Inc. **17.17** Photo courtesy Dr. Michio Hori, Department of Zoology, Kyoto University, Japan. **17.18** Animals Animals/ © Stouffer Prod. **17.19** John Wightman/Ardea London Ltd. **17.20** Photographs by John A. Endler, University of California at Santa Barbara. **17.21a** Courtesy Colin G. Beer, Rutgers University; **b** Courtesy John Sparks, BBC (Natural History). **17.20a** Courtesy D. J. Howell, Purdue University; **b** Courtesy M. Morcombe; **c** Bob & Clara Calhoun/Bruce Coleman, Inc.; **d** Courtesy E. S. Ross. **17.24** Courtesy Ken Paige, University of Illinois. **17.25** Derek Washington/Bruce Coleman, Inc. **17.26** Oxford Scientific Films/ Bruce Coleman, Inc. **17.27a** Courtesy Jeffrey L. Rotman; **b** Jane Burton/Bruce Coleman, Inc. **17.28a** David Overcash/Bruce Coleman, Inc.; **b** John Shaw/Bruce Coleman, Inc. **17.29** Peter Ward/ Bruce Coleman, Inc. **17.30** Jane Burton/Bruce Coleman, Inc. **17.31** Animals, Animals/© Breck Kent. **17.32** Art Wolfe/Tony Stone Images, New York. **17.33a** Courtesy James L. Castner, University of Florida; **b** Courtesy E. S. Ross; **c** Charles Angelo/Photo Researchers, Inc. **17.34a** D. Overcash; **b** J. Shaw/Bruce Coleman, Inc. **17.35** Courtesy Douglas Faulkner. **17.36a** © Steinhardt Aquarium/Tom McHugh/Photo Researchers, Inc.; **b** Joe McDonald/Bruce Coleman, Inc. **17.37** R. P. Carr/Bruce Coleman, Inc. **17.38** Photograph courtesy Department Library Services, American Museum of Natural History, from H. Kutter, *Neujahrsblatt herausgegeben von der Naturforschenden Gesellschaft in Zurich*, vol. 171, 1969. **17. 39** Courtesy E. S. Ross.

18.1 Photo by W. R. Eisenbohr, Courtesy Cies Sexton, U. S. Department of the Interior Geological Survey .**18.2** Irv Kornfield and Jeffrey N. Taylor, *Proc. Biol. Soc. Washington*, vol. 96, 1983. **18.3a** Modified from W. Auffenburg, *Tulane Stud. Zool. Bot.*, vol. 2, 1955; **b,d** Grant Heilman Photography; **c** Modified from R. E. Woodson, *Ann. Mo. Bot. Gard.*, vol. 34, 1947. **18.4** Adapted from Jens Clausen, David Keck, and William Hiesey, *Experimental Studies on the Nature of Species, III* (Carnegie Institution publication 581), 1948. **18.5a** Ed Degingger and **b** Josephy VanWormer/Bruce Coleman, Inc. **18.6** Redrawn from W. Auffenberg, *Tulane Stud. Zool. Bot.*, vol. 2, 1955. **18.7** Pat and Tom Leeson/Photo Researchers, Inc. **18.9** Modified from B. Wallace and A. M. Srb, *Adaptation*, copyright © 1964 by Prentice-Hall, Inc., Englewood Cliffs, New Jersey. **18.10** J. Shaw/ Bruce Coleman, Inc. **18.11** Courtesy E. S. Ross. **18.12** Phil Deginger/Bruce Coleman, Inc. **18.13** Courtesy Michael F. Whiting, from *Insect Homeotic Transformation*, p. 696, *Nature*, vol. 368, 21 April 1994 by Dr. Michael F. Whiting and Dr. Ward C. Wheeler, American Museum of Natural History. **18.14** Courtesy Thomas K. Wood, Department of Entomology, University of Delaware, Newark, Delaware. **18.15** Modified from D. Lack, *Darwin's Finches*, Cambridge, The University Press, 1947. **18.17** Drawing courtesy Sophie Webb **18.18** Modified from D. Lack, *Darwin's Finches*, Cambridge, The University Press, 1947. **18.19** Photograph by Irenaus Eibl-Eibesfeldt, Max Planck Institute for Behavioral Physiology. **18.20** Courtesy H. Douglas Pratt. **18.21** Modified from G. F. Gause, *Science*, vol. 79, 1934; copyright © 1934 by A. A. A. S. ***How We Know*** p. 506 Photos courtesy Peter Grant, Princeton University. **18.23** Illustrations by Marianne Collins are reproduced from *Wonderful Life: The Burgess Shale and the Nature of History*, by Stephen Jay Gould, with permission of W. W. Norton & Company, Inc.; copyright © 1989 Stephen Jay Gould. **18.24** Jane Burton/Bruce Coleman, Inc. **18.25** Redrawn from D. Lack, *Darwin's Finches*, Cambridge, The University Press, 1947. **18.26** Modified from G. G. Simpson, *Horses*, copyright © 1951 by Oxford University Press, Inc., renewed 1979 by G. G. Simpson. Used by permission. **18.28a** Chicago Zoological Park; photograph by Tom McHugh/Photo Researchers, Inc.; **b,c** Michael Morcombe; **d** Jack Fields/Photo Researchers, Inc.

PART IV: THE GENESIS AND DIVERSITY OF ORGANISMS

Major Milestones: Photo courtesy William E. Ferguson

19.1 Courtesy NASA **19.2** Courtesy David W. Deamer, University of California, Santa Cruz **19.3** Modified from R. E. Dickerson, *Sci. Am.*, September 1978; copyright © 1978 by Scientific American, Inc.; all rights reserved. **19.4** Courtesy Sigurgeir Jonasson. **19.6** Courtesy Sidney W. Fox, University of Miami, and Steven Brooke Studios, Coral Gables, Florida. **19.9** Courtesy J. W. Schopf, University of California, Los Angeles. **19.10** G. R. Roberts. **19.11a** Courtesy R. E. Lee, University of the Witwatersrand, South Africa; **b** MMJP Plant Research Lab. **19.12** D. A. Stetler and W. M. Laetsch, *Am. J. Bot.*, vol. 56, 1969, reprinted with permission from the Botanical Society of America. **19.13** W. J. Larsen, *J. Cell Biol.*, vol. 47, 1970, by copyright permission of the Rockefeller University Press. **19.14** Courtesy I. B. Dawid, National Intitutes of Health, Bethesda, Maryland, and D. R. Wolstenholme, University of Utah. **19.15** *Joenia annectens* adapted from drawing by Robert Golder, courtesy of Lynn Margulis and Karlene V. Schwartz, University of Massachusetts. **19.17a** Science Photo Library/Photo Researchers, Inc. **b** Biophoto Associates/Photo Researchers, Inc. **c** Courtesy of E. B. Daniels, from K. W. Jeon, ed. *The Biology of the Amoeba*, 1973, Academic Press, Orlando.

20.1a Courtesy M. Wurtz, University of Basel; **b** Courtesy M. Gomersall, McGill University; **c** Courtesy R. C. Williams, University of California, Berkeley. **20.3** Photo from K. Corbett, *Virology*, vol. 22, 1964; reprinted with permission of Academic Press, Inc., New York. **20.4** Photo courtesy T. O. Diener U.S.D.A. **20.5** From T. O. Diener, *Am. Sci.*, vol. 71, 1983. **20.6** Drawing adapted from one by Fred E. Cohen, which appeared on p. 33 of the January 1995 *Sci. Am.*, in the article by Stanley B. Prusiner, reprinted with permission of the author. **20.8b** Courtesy Esther R. Angert/Indiana University. Originally appeared on p. 1629 of *Science*, vol. 256, 19 June 1992. Reprinted with permission of A. A. A. S. **20.9a** Courtesy David Scharf/ Peter Arnold, Inc.; **b** Center for Disease Control, Atlanta, Georgia. **20.10** Turtox/Cambosco, Macmillan Science Co., Inc. **20.11** Courtesy Z. Skobe, Forsyth Dental Center/BPS. **20.12** Courtesy M.

Gomersall, McGill University. **20.13** S. Kimoto and J. C. Russ, *Am. Sci.*, vol. 57, 1969. **20.15** G. B. Chapman, *J. Bacteriol.*, vol. 71, 1956. **20.16** Photograph by Ginny Fonte, from H. C. Berg, *Sci. Am.*, August 1975; copyright © 1975 by Scientific American, Inc.; all rights reserved. **20.17** From R. C. Johnson, M. P. Walsh, B. Ely, and L. Shapiro, *J. Bacteriol.*, vol. 138, 1979. **20.18** Courtesy W. Burgdorfer, S. F. Hayes, and D. Corwin, Rocky Mountain Laboratories. **20.19** Courtesy David Chase, Veterans Hospital, Sepulveda, California. **20.20** Murti/Phototake, New York. **20.21** H. Reichenbach, Gesellschaft fur Biotechnologische Forschung, mbH. **20.22** Elliot Scientific Corp. **20.23** Elliot Scientific Corp. **20.24a** Science Source/Photo Researchers, Inc.; **b,c** © Thomas E. Adams/Visuals Unlimited. **20.25** Courtesy Malcolm Brown, Jr., University of Texas. **20.26** © S. Thompson/Visuals Unlimited. **20.27** This EM was made by Zell A. McGee in the course of research funded by the National Institute of Allergy and Infectious Diseases and was provided to the publisher without charge in the interest of science. **20.28** Photograph by Walther Stoeckenius, courtesy Carl Woese, University of Illinois. **20.29a** Courtesy of H. W. Jannasch, Woods Hole Oceanographic Institution; **b** From W. J. Jones, J. A. Leigh, F. Mayer, C. R. Woese, and R. S. Wolfe, *Arch. Microbiol.*, vol. 136, 1983, copyright © 1983 by Springer-Verlag, New York.

21.1 Photograph courtesy E. W. Daniels, art modified from K. W. Jeon, ed., *The Biology of the Ameoba*, 1973, Academic Press, Orlando. **21.2** From J. M. Jensen and S. R. Wellings, *J. Protozool.*, vol. 19, p. 297–305, 1971. **21.4** Courtesy E. V. Grave. **21.5** Photo courtesy Ed Reschke. **21.6** Photo courtesy E. V. Grave. **21.7** Courtesy E. V. Grave. **21.8** Edward Degginger/Bruce Coleman, Inc. **21.8** Courtesy E. V. Grave. **21.10** Courtesy E. V. Grave. **21.12a** Courtesy E. V. Grave; **b** Biophoto Associates/Photo Researchers, Inc.; **c** Courtesy Roman Vishniac; **d** Courtesy A. Fleury, Service d'Imagerie Cellulaire, Orsay, France. Originally appeared on the cover of *Science*, 9 December 1994. **21.13** Courtesy Thomas Eisner, Cornell University. **21.14** Manfred Kage/Peter Arnold, Inc. **21.16** Courtesy E. V. Grave. **21.19** Ray Simons/Photo Researchers, Inc. **21.22a,b** Courtesy K. B. Raper, *Proc. Am. Philos. Soc.*, vol. 104, 1960.; **c–f** Carolina Biological Supply Company.

22.2 B. S. C. Leadbeter, University of Birmingham. **22.4** Turtox/Cambosco, Macmillan Science Co., Inc. **22.5** Edward Degginger/Bruce Coleman, Inc. **22.6** H. Chaumeton/Nature. **22.7** Edward Degginger/Bruce Coleman, Inc. **22.8** Courtesy Sally Faulkner. **22.10b** Robert Carr/Bruce Coleman, Inc. **22.13b** Photo courtesy Susumu Honjo, Woods Hole Oceanographic Institution, Woods Hole, Massachusetts. **22.14** H. Chaumeton/Nature. **22.17** F. Sauer/Nature. **22.18** Courtesy Ed Reschke. **22.19** Courtesy Roman Vishniac. **22.20** Courtesy Roman Vishniac. **22.22** Courtesy Roman Vishniac. **22.27** Edward Degginger/Bruce Coleman, Inc. **22.30** Modified from H. J. Fuller and O. Tippo, *College Botany*, Holt, Reinhart & Winston, Inc., New York, 1954. **22.32** Jane Burton/Bruce Coleman, Inc. **22.33a–c** From G. Shih and R. Kessel, *Living Images: Biological Microstructures Revealed by Scanning Electron Microscopy*. © 1982 by Science Books International; **d** Stephen Dalton/Photo Researchers, Inc. **22.34** Adrian Davies/Bruce Coleman, Inc. **22.36** Adrian Davies/Bruce Coleman, Inc. **22.38** Courtesy Edward Degginger. **22.40** Carnegie Museum of Natural History, Pittsburgh, Pennsylvania. **22.41** Courtesy Field Museum of Natural History, Chicago. **22.42** Courtesy E. S. Ross. **22.43** John Shaw/Bruce Coleman, Inc. **22.44** Courtesy G. R. Roberts. **22.45** Stephen Parker/Photo Researchers, Inc. **22.49** Ed Reschke/Peter Arnold, Inc. **22.50** Courtesy Thomas Esiner, Cornell University. **22.51** Photo courtesy Nels R. Lersten, Iowa State University. **22.52a** Courtesy Victor B. Eichler; **b** Courtesy Thomas Eisner, Cornell University. **22.55** Redrawn from H. N. Andrews, *Science*, vol. 142, 1963; copyright © 1963 by A.A.A.S. **22.56** Courtesy E. S. Ross. **22.57** Courtesy Roman Vishniac. **22.58** From W. Krommenhoek, J. Sebus, and G. J. van Esch, *Biological Structures*, copyright © 1979 by L. C. G. Malmberg B. V., The Netherlands. **22.59a,b** Courtesy Thomas Eisner, Cornell University; **c** Courtesy E. S. Ross. **22.60** Modified from H. J. Fuller and O. Tippo, *College Botany*, Holt, Reinhart & Winston, Inc., New York, 1954. **22.64** M. I. Walker/Photo Researchers, Inc. **22.66** James H. Robinson/Photo Researchers, Inc. **22.68a** Jane Burton and **b** R. P. Carr/Bruce Coleman, Inc. **22.69** Modified from H. J. Fuller and O. Tippo, *College Botany*, Holt, Reinhart & Winston, Inc., New York, 1954.

23.1 Science Photo Library/Photo Researchers, Inc. **23.3** Photograph by W. H. Amos/Bruce Coleman, Inc. **23.5** M. P. L. Fogden/Bruce Coleman, Inc. **23.6** From L. W. Sharp, *Fundamentals of Cytology*, McGraw-Hill Book Co., New York, copyright © 1943; used by permission. **23.9a** Jane Burton and **b** L. West/Bruce Coleman; **c,d** V. Ahmadjian and J. B. Jacobs, *Nature*, vol. 289, 1981 ©1981, Macmillan Journals Ltd. **23.10a** Masana Izawa and **b** Satoshi Kuribyashi/Nature Production. **23.11** Photograph by B. A. Roy, reprinted by permission of *Nature*, vol. 362, page 57. **23.12** From L. W. Sharp, *Fundamentals of Cytology*, McGraw-Hill Book Co., New York, copyright © 1943; used by permission.

24.1 Courtesy Jeff Rotman. **24.4** Courtesy Oxford Scientific Films. **24.5** H. Chaumeton/Nature. **24.8** Carolina Biological Supply Company. **24.11** R. N. Mariscal/Bruce Coleman, Inc. **24.12** Courtesy Howard Hall. **24.15** After K. G. Grell, University of Tubingen (1974). **24.16** Courtesy Dr. A. Ruthmann, from J. Rassat and A. Ruthmann, *A. Zoomorphology* vol. 93, p. 59–72, Springer-Verlag, New York. **24.18** After K. G. Grell, University of Tubingen (1974). **24.20** H. Chaumeton/Nature. **24.24** CNRI/Science Photo Library/Photo Researchers, Inc. **24.26a** Courtesy Ed Reschke; **b** H. Chaumeton/Nature. **24.27** H. Chaumeton/Nature. **24.32** Adapted from *Life* by W. K. Purves and G. H. Orians, Sinauer Associates, Sunderland, Massachusetts; copyright © 1983. **24.33** Courtesy T. E. Adams. **24.35** H. Chaumeton/Nature. **24.36** H. Chaumeton/Nature. **24.38** Fred Bavendam/Peter Arnold, Inc. **24.39** Oxford Scientific Films. **24.41** H. Chaumeton/Nature. **24.42top** H. Chaumeton/Nature; **bottom** Rod Borland/Bruce Coleman, Inc. **24.43** Based in part on drawings by Louise G. Kingsbury. **24.44** F. Sauer/Nature. **24.45** H. Chaumeton/Nature. **24.46** Modified from W. Stempell, *Zoologie im Grundriss*, Borntraeger, 1926. **24.47** H. Chaumeton/Nature. **24.49** Photography by Joseph Pawlik, University of North Carolina at Wilmington. **24.50** Edward Degginger/Bruce Coleman, Inc. **24.51** Oxford Scientific Films. **24.52** Courtesy E. S. Ross. **24.53** Jane Burton/Bruce Coleman, Inc. **24.55** Courtesy Ed Reschke. **24.56** Scott Camazine/Photo Researchers, Inc. **24.57** Edward Degginger/Bruce Coleman, Inc. **24.58** H. Chaumeton/Nature. **24.59a** H. Chaumeton/Nature; **b** Hans Reinhard/Bruce Coleman, Inc. **24.60a** H. Chaumeton/Nature; **b** Robert Gossington/Bruce Coleman, Inc.; **c** Gilbert S. Grant/Photo Researchers, Inc.; **d** H. Chaumeton/Nature. **24.61** Oxford Scientific Films. **24.62** Courtesy G. R. Roberts. **24.63** A. Cosmos Blank, National Audobon Society/Photo Researchers, Inc. **24.65** Modified from T. I. Storer and R. L. Usinger, *General Zoology*, McGraw-Hill Book Co., New York, copyright © 1957; used by permission. **24.66** Courtesy Stephen Dalton/NHPA. **24.70** Modified from L. H. Hyman, *The Invertebrates*, McGraw-Hill Book Co., New York, copyright © 1955; used by permission. **24.71a** Robert Dunne and **b** Charlie Ott/Photo Researchers, Inc. **24.72** Andrew J. Martinez/Photo Researchers, Inc. **24.73** A. Kerstitch, Sea of Cortez Enterprises. **24.74b** Courtesy Smithsonian Institution, photo 62419. **24.75** H. Chaumeton/Nature.

25.1 Photograph by Bill Wood/Bruce Coleman, Ltd., London. **25.2** Turtox/Cambosco, Macmillan Science Co., Inc. **25.3** Photograph by H. Chaumeton/Nature. **25.4** Courtesy Ed Reschke. **25.6** Courtesy Ed Reschke. **25.7** Courtesy Ed Reschke. **25.8a** D. Claugher, courtesy of the Trustees, The British Museum (Natural History); **b** Courtesy Jerome Gross, Massachusetts General Hospital. **25.10a** Ed Reschke/Peter Arnold, Inc.; **b** Manfred P. Kage/Peter Arnold, Inc. **25.12** Photograph courtesy Ed Reschke. **25.14** Courtesy Heather Angel, Biofotos. **25.15** Chip Clark, National Museum of Natural History, Washington. **25.16** Modified from A. S. Romer, *The Vertebrate Body*, W. B. Saunders, New York, 1949. **25.17** Courtesy Howard Hall. **25.18** Courtesy the Royal Society, London. **25.19** Painting photograph by Chip Clark, courtesy National Museum of Natural History, Washington. **25.20a** S. C. Bisserot and **b** Hans Reinhard/Bruce Coleman, Inc. **25.21** Courtesy E. S. Ross. **25.22** Hans Pfletschinger/

Peter Arnold, Inc. **25.23a** G. R. Roberts; **b** Ferraro/Nature; **c,d** Hans Reinhard/Bruce Coleman, Inc.; **e** John Moss/Photo Researchers, Inc. **25.25** From a mural by Maidi Wiebe, courtesy Field Museum of Natural History, Chicago. **25.26a** Royal Tyrrell Museum, Alberta Culture and Muliculturalism; **b** Gregory S. Paul. **25.27** Gregory S. Paul. **25.28** Gregory S. Paul. **25.29** Photos courtesy Philippe Claeys, Department of Geology and Geophysics, University of California, Berkeley. **25.30** Courtesy D. G. Allen. **25.31** Warren Garst/ Tom Stack Associates. **25.32** Photo # ZWN0164, The Stock Market. **25.33** Tom McHugh/Photo Researchers, Inc. **25.34** Grospas/Nature. **25.36** M. P. L. Fogden/Bruce Coleman, Inc. **25.37a** Jan Lindblad/Photo Researchers, Inc.; **b** Peter Jackson/Bruce Coleman, Inc. **25.38** Jack Dermid/Bruce Coleman, Inc. **25.39** Monkey Jungle, Miami, Florida, R. P. Fontaine/Photo Researchers, Inc. **25.40** Brian Parker/Tom Stack and Associates. **25.41** Peter Davey/Bruce Coleman, Inc. **25.44** From R. E. Leakey and R. Lewin, *Origins*, Dutton, New York, 1977; reproduced by permission of Rainbird Publishing Group. **25.45** Margo Crabtree, courtesy A.A.A.S., used by permission of; **a** University of the Witwatersrand Medical School, and; **b-d** the National Museum of Kenya, Nairobi. **25.48** David Brill, 1985 National Geographic Society; photographed at the Institut du Quarternaire, Universite de Bordeaux, Talence, France.

PART V: THE BIOLOGY OF ORGANISMS

Internal Organs: M. P. L. Fogden/Bruce Coleman, Inc.

26.4 Courtesy G. R. Roberts. **26.5** Courtesy Ed Reschke. **26.8** Carolina Biological Supply Company. **26.12** Biophoto Associates, Science Source/Photo Researchers, Inc. **26.13** Dr. Jeremy Burgess/Photo Researchers, Inc. **26.13** Breck P. Kent/Animals Animlals, Earth Scenes. **26.14** U. N. Food and Agricultural Organization. **26.15** Both photos courtesy Hans W. Paerl, University of North Carolina; **b** is from Hans W. Paerl and Kathleen K. Gallucci, *Science*, vol. 227, 1985; copyright © 1985 by A.A.A.S. **26.16** Science Source/Photo Researchers, Inc. **26.17** Edward Degginger/Bruce Coleman, Inc. **26.18** Courtesy Edward Degginger.

27.1 Courtesy E. S. Ross. **27.2** Courtesy R. Farquharson, UNICEF. **27.4** From the *Vitamin Manual*, courtesy The Upjohn Company. **27.5** From the *Vitamin Manual*, courtesy The Upjohn Manual. ***How We Know***, p. 767; photo courtesy James L. Gould, from Ethology, 1e, 1982. Fig. 6-8, p. 94. W. W. Norton & Co., Inc. **27.7a–c** David Pramer, *Science*, vol. 144, 1964; copyright © 1964 by A.A.A.S.; **d** courtesy G. Barron and N. Allin, University of Guelph, Canada. **27.8** Courtesy K. G. Grell, University of Tubingen, Germany. **27.12** Modified from W. D. Russell-Hunter, *A Biology of Lower Invertebrates*, Macmillan Publishing Co., New York, 1968. **27.13** Photo courtesy E. V. Grave. **27.14** Oxford Scientific Films/Bruce Coleman, Inc. **27.15** Turtox, Cambosco, Macmillan Science Co., Inc. **27.17a** Courtesy Howard Hall; **b** courtesy Thomas Eisner; **c** courtesy William A. Watkins. **27.20** Adapted from an original painting by Frank H. Netter, M.D., from *The CIBA Collection of Medical Illustrations*, copyright © 1972 by CIBA Pharmaceutical Company, a division of CIBA-GEIGY Corp. **27.25** Adapted from Normal Kretckmer, *Sci. Am.*, October 1972; copyright © 1972 by Scientific American Inc.; all rights reserved. **27.26** From Warren Andrew, *Texbook of Comparative Histology*, Oxford, The University Press, 1959. **27.28a** Courtesy Susumu Ito, Department of Neurobiology, Harvard Medical School. **27.31** Courtesy Erwin and Peggy Bauer, Wildstock.

28.3 Courtesy Thomas Eisner, Cornell University. **28.4** Courtesy J. H. Troughton, Department of Scientific and Industrial Research, Wellington, New Zealand. **28.6** Courtesy J. H. Troughton, Department of Scientific and Industrial Research, Wellington, New Zealand. **28.8a** Courtesy Thomas Eisner, Cornell University; **b** courtesy G. R. Roberts. **28.10** Modified from Ralph Buschbaum, *Animals Without Backbones*, by permission of the University of Chicago Press, copyright © 1948 by the University of Chicago. **28.12** Courtesy James H. Carmichael, Jr. **28.16** © F. Sauer/Nature. **28.17** Bob Gossington/Bruce Coleman, Inc. **28.18a** Tom McHugh/Photo Researchers, Inc.; **b** Joe McDonald and; **c** Zig Leszczynski/Animals, Animals/ Earth Scenes; **d** S. J. Krasemann/Peter Arnold, Inc. **28.19** Photograph from R. G. Kessel and R. H. Kardon, *Tissues and Organs: A Text-Atlas of Scanning Electron Microscopy*, © W. H. Freeman and Company, 1979. **28.25** Courtesy H-R. Duncker, Justus Liebig University, Geissen, Germany. **28.27** Modified from R. Margaria et al., *J. Appl. Physiol.*, vol. 18, 1963. **28.28** Courtesy Roman Vishniac. **28.30a** Courtesy D. Claugher, by courtesy of the Trustees, The British Museum (Natural History); **b** from Warren Andrew, *Textbook of Comparative Histology*, Oxford, The University Press, 1959. **28.31** Kim Taylor/Bruce Coleman, Inc.

29.3 Photograph courtesy Thomas Eisner, Cornell University. **29.4** Courtesy Nels R. Lersten, Iowa State University. **29.5** Courtesy Ed Reschke. **29.7** Photograph courtesy Ed Reschke. **29.9** B. Bracegirdle; from W. Krommenhoek, J. Sebus, and G. J. van Esch, *Biological Structures*, copyright © 1979 by L. C. G. Malmberg B. V. The Netherlands. **29.10** Courtesy Thomas Eisner, Cornell University. **29.11** Modified from V. A. Greulach and J. E. Adams, *Plants: An Introduction to Modern Botany*, Wiley, New York, 1962. **29.14a** Courtesy J. H. Troughton, Department of Scientific and Industrial Research, Wellington, New Zealand; **b,c** courtesy B. G. Butterfield, Canterbury University, and B. A. Meylan, Department of Scientific and Industrial Research, Wellington, New Zealand. **29.15** Courtesy Thomas Eisner, Cornell University. **29.16** Courtesy Thomas Eisner, Cornell University. **29.19** Laura Riley/Bruce Coleman, Inc. **29.21** S. and O. Biddulph et al., *Plant Physiol.*, vol. 33, no. 4, 1958.

30.6 Micrograph from R. G. Kessel and R. H. Kardon, *Tissues and Organs: A Text-Atlas of Scanning Electron Microscopy*, © W. H. Freeman, San Francisco, 1979. **30.8** Modified from B. S. Guttman and J. W. Hopkins III, *Understanding Biology*; copyright © 1983 by Harcourt Brace Jovanovich, Inc.; used by permission of the publisher. **30.10** Albert Paglialunga/Phototake. ***Exploring Further***, p. 852; **a** courtesy Roman Vishniac; **b** courtesy Ed Reschke. **30.15a** Courtesy Thomas Eisner, Cornell University; **b** L. Nilsson, from L. Nilsson, *Behold Man*, English translation copyright © 1974 by Albert Bonniers Forlag, Stockholm, and Little, Brown and Co. (Canada) Ltd. **30.16a** Courtesy Don W. Fawcett, Harvard Medical School; **b** Micrograph from R. G. Kessel and R. H. Kardon, *Tissues and Organs: A Text-Atlas of Scanning Microscopy*, W. H. Freeman, San Francisco, copyright © 1979. **30.20** Manfred Kage/Peter Arnold, Inc. **30.22** Courtesy Turtox, Cambosco, Macmillan Science Co. **30.29** Courtesy Eila Kairinen, Gillette Research Institute. **30.30** Courtesy K. R. Porter and Ginny Fonte, University of Colorado.

31.1 Modified from E. G. Bollard, *J. Exp. Bot.*, vol. 4, 1953. **31.7** Modified from E. Baldwin, *An Introduction to Comparative Biochemistry*, Cambridge, The University Press, 1948. **31.9** After D. L. Hopkins, *Biol. Bull.*, vol. 90, 1946. **31.10** After E. H. Mercer, *Proc. Roy. Soc. Lond. B*, vol. 150, 1959. **31.11** Photographs courtesy Thomas Eisner, Cornell University. **31.12** Modified from Ralph Buschbaum, *Animals Without Backbones*, by permission of The University of Chicago Press, copyright © 1948 by the University of Chicago. **31.17** Modified from H. W. Smith, *The Kidney*, Oxford, The University Press, 1951. **31.18** M. G. Farquhar, University of California, San Diego. **31.19a** Courtesy F. Spinelli, CIBA-GEIGY Research Laboratory, Basel, Switzerland.

32.2e Redrawn from M. Schaffner, The Ohio Naturalist, 1906. **32.4b** Courtesy G. R. Roberts. **32.6** Photograph courtesy Donald Nevins, Iowa State University. **32.7a** Modified from K. Esau, *Plant Anatomy*, Wiley, New York, 1965; **b** Patrick J. Lynch/Photo Researchers, Inc. **32.8** Modified from Peter Albersheim, *Sci. Am.*, April, 1975; copyright © 1975 by Scientific American, Inc.; all rights reserved. **32.11** Modified from V. A. Greulach and J. E. Adams, *Plants: An Introduction to Modern Botany*, Wiley, New York, 1962. **32.12** Redrawn from *Biologie: Ein Lehrbuch*, edited by G. Czihak et al., 2nd ed., Springer-Verlag, 1978. **32.13** Courtesy Ed Reschke. **32.14a** Courtesy Edward Degginger; **b** Used by permission from J. D. Dodd, *Course Book in General Botany*, copyright © 1977 by the Iowa State University Press. **32.15** Carolina Biological Supply Company. **32.18** Redrawn after Z. Schwartz-Sommer et al., *Science*, vol. 250, 1990; copyright © 1990 by A.A.A.S. **32.27** Courtesy C. R. Hawes, Oxford Polytechnic. **32.31** ©Biophoto Associates, Science Source/Photo Researchers, Inc. **32.33** Courtesy Robert Newman, University of

Wisconsin. **32.35** Courtesy Sylvan Wittwer. **32.37** Redrawn from *Biologie: Ein Lehrbuch*, edited by G. Czihak et al., 2nd ed., Springer-Verlag, 1978. **32.40** Photograph by Stephen Gladfelter, Standford University, from research by P. W. Oeller, Lu M.-W., L. P. Taylor, D. A. Pike, and A. Theologis at the Plant Gene Expression Center, University of California, Berkeley. **32.45** After W. A. Jensen and F. B. Salisbury, *Botany: An Ecological Approach*; copyright © 1972 by Wadsworth Publishing Co., Inc., Belmont, California; used by permission of the publisher. **32.46** Photo by M. F. Nuttall, DuPont Experimental Station, courtesy Ilya Raskin, Rutgers State University.

33.1 Plate I from *Insect Hormones* by V. B. Wigglesworth, published in 1970 by Oliver & Boyd, Edinburgh, and W. H. Freeman, New York. **33.2** Modified from H. A. Schneiderman and L. I. Gilbert, *Science*, vol. 143, 1964; copyright © 1964 by A.A.A.S. ***How We Know***, p. 939. Photo from Michael Gadomski/Bruce Coleman, Inc. **33.4** Painting by Frank H. Netter, M.D.; reprinted with permission from *The CIBA Collection of Medical Illustrations*, copyright © 1965 by CIBA Pharmaceutical Company, division of CIBA-GEIGY Corporation; all rights reserved. **33.6** Carolina Biological Supply Company. **33.11** © John P. Kay/Peter Arnold, Inc. **33.13** From A. J. Carlson et al., *The Machinery of the Body*, University of Chicago Press, 1961. **33.16** Modified from H. Curtis, *Biology*, 4th ed.; copyright © 1983 by Worth Publishers, Inc. **33.26** Modified from an original model of a morphine molecule, courtesy Maitland Jones, Jr.

34.1 Hans Pfletschinger/Peter Arnold, Inc. **34.2** Michael Fogden/Bruce Coleman, Inc. **34.3** Ken M. Highfill/Photo Researchers, Inc. **34.7a** Courtesy D. M. Phillips, Population Council, New York; **b** David M. Phillips/Photo Researchers, Inc. **34.11a,c** Courtesy Thomas Eisner, Cornell University; **b** C. Edelman-La Villette/Photo Researchers, Inc.

35.2 Courtesy S. L Palay, Harvard Medical School. **35.4a** Courtesy H. deF. Webster, as printed in W. Bloom and D. W. Fawcett, *A Textbook of Histology*, 10th ed., W. B. Saunders Co., 1975; **b** R. L. Roberts, R. G. Kessel and H. N. Tung, *Freeze Fracture Images of Cells and Tissues*, Oxford University Press, 1991. **35.6a** H. Chaumeton/Nature; **b** adapted from E. R. Kandel, *Cellular Basis of Behavior*, W. H. Freeman, San Francisco, 1976. **35.7** Modified from E. R. Kandel, *Cellular Basis of Behavior*, W. H. Freeman, San Francisco, 1976. **35.10** Modified from T. H. Bullock and G. A. Horridge, *Structure and Function of the Nervous System of Invertebrates*, W. H. Freeman, San Francisco, 1965. **35.11** Modified from L. H. Hyman, *The Invertebrates*, vol. 2, McGraw-Hill Book Co., New York, copyright © 1951, and from Ralph Buschbaum, *Animals Without Backbones*, University of Chicago Press, copyright © 1948 by the University of Chicago, used by permission of the publisher. **35.12** Modified from R. Goldschmidt, *Z. Wiss. Zool., Abt. A.*, vol. 92, 1909. **35.13** Photograph courtesy Thomas Eisner, Cornell University. **35.17** Courtesy A. L. Hodgkin, *J. Physiol. (London)*, vol. 131, 1956. **35.19** Adapted from W. A. Catterall, *Science*, vol. 242, 1988; copyright © 1988 by A.A.A.S. **35.24** E. R. Lewis et al., *Science*, vol. 165, 1969; copyright © 1969 by A.A.A.S. **35.25** R. L. Roberts, R. G. Kessel and H. N. Tung, *Freeze Fracture Images of Cells and Tissues*, Oxford University Press, 1991. **35.30** Photograph courtesy Ed Reschke. **35.31** Courtesy J. E. Heuser, Washington University Medical Center.

36.1 Redrawn from B. Katz, *J. Physiol. (London)*, vol. 111, 1950. **36.3** Graph modified from S. Skoglund, *Acta Physiol. Scand., Suppl.* 124, vol. 36, 1956. **36.6** Courtesy Ed Reschke. **36.7** After Murray, 1973. **36.8** D. Claugher, by courtesy of the Trustees, The British National Museum of Natural History. **36.10** J. L. Lepore/Photo Researchers, Inc. **36.13** J. A. L. Cooke, Oxford Scientific Films. **36.14** Modified from R. E. Snodgrass, *Principles of Insect Morphology*, McGraw-Hill Book Co., New York, copyright © 1934; used with permission of McGraw-Hill Book Co. **36.15** Courtesy J. L. Gould and C. G. Gould. **36.16** Douglas Faulkner/Photo Researchers, Inc. **36.19** Photograph by L. Nilsson; from L. Nilsson, *Behold Man*, English translation © 1974, Albert Bonniers Forlag, Stockholm; Little Brown and Co. (Canada) Ltd. **36.20** Omikron/Photo Researchers, Inc. **36.24** After A. F. MacNichol. **36.26a** Courtesy Thomas Eisner, Cornell University; **b** courtesy E. S. Ross. **36.28** Courtesy A. J. Hudspeth and R. A. Jacobs from A. J. Hudspeth, *Nature*, 341:397-404 (1989). **36.29** Photos from Stuart Ira Fox, *Human Physiology*, 3rd ed., fig. 13.26, page 355, copyright © 1990 William C. Brown Communications, Inc. **36.30** Modified from G. von Bekesy, *Sump. Soc Exp. Biol.*, vol. 16, 1962; G. von Bekesy, *Experiments in Hearing*, McGraw-Hill Book Co., New York, copyright © 1960, used with the permission of McGraw-Hill Book Co.; M. S. Gordon, G. Bartholomew, A. D. Grinnell, C. B. Jorgensen, and F. N. White, *Animal Function: Principles and Adaptations*, Macmillan, New York, 1968, copyright © 1968 by M. S. Gordon. **36.31** L. Nilsson; from L. Nilsson, *Behold Man*, English translation 1974, Albert Bonniers Forlag, Stockholm; and Little, Brown and Co., (Canada) Ltd. **36.33a** William E. Ferguson; **b** S. L. Craig/Bruce Coleman, Inc.; **c** Edward Degginger/Bruce Coleman, Inc. **36.35** H. Vali, courtesy R. J. P. Williams, Oxford University. **36.37** Modified from W. M. Cornsweet, *Visual Perception*, Academic Press, New York, 1970; and S. Coren et al., *Sensation and Perception*, Academic Press, New York, 1978. **36.40** Modified from A. S. Romer and T. S. Parsons, *The Vertebrate Body*, 5th ed., copyright © 1977 by W. B. Saunders Company, reprinted by permission of CBS College Publishing. **36.41** Modified from A. S. Romer and T. S. Parsons, *The Vertebrate Body*, 5th ed., copyright © 1977 by W. B. Saunders Company, reprinted by permission of CBS College Publishing; and G. G. Simpson, C. S. Pittendrigh, and L. H. Tiffany, *Life: An Introduction to Biology*, copyright © 1957 by Harcourt Brace Jovanovich, Inc., used by permission of the publishers. **36.45** Modified from W. Penfield and T. Rasmussen, *The Cerebral Cortex of Man*, Macmillan, New York, 1950. **36.50** Based on data from John Allman, California Institute of Technology. **36.51** Institute of Medicine, National Academy of Sciences Press.

37.1 G. R. Roberts. **37.3** Photograph by D. Claugher, by courtesy of the Trustees, The British Museum of Natural History; drawing modified from J. Gray and H. W. Lissman from *J. Exp. Biol.*, vol. 15, 1938, by permission of Company of Biologists, Ltd. **37.4** H. Chaumeton/Nature. **37.7** © Biophoto Association, Photo Researchers, Inc. **37.9** Courtesy Ed Reschke. **37.14** Courtesy H. E. Huxley, Cambridge University. **37.15** Courtesy H. E. Huxley, Cambridge University. **37.18** Courtesy J. E. Heuser, Washington University Medical Center.

38.1 Ian Wyllie/Survival Anglia. London. **38.2a** Courtesy Paul Trotschel; **b** Stephen Dalton/Photo Researchers, Inc. **38.3** Courtesy John Sparks, BBC, Natural History. **38.4** Courtesy Thomas Eisner, Cornell University. **38.5** Modified from N. Tinbergen, *The Study of Instinct*, Oxford, The University Press, 1951. **38.6** Modified from N. Tinbergen, *The Study of Instinct*, Oxford, The University Press, 1951. **38.7** Courtesy E. R. Willis, Illinois State University. **38.8** Modified from N. Tinbergen and A. C. Perdeck, *Behaviour*, vol. 3, 1950; published by Oxford University Press. **38.9** Courtesy John Sparks, BBC, Natural History. **38.10** Modified from J.-P. Ewert, *Sci. Am.*, March 1974; copyright © 1974 by Scientific American; all rights reserved. **38.11** Adapted from K. Lorenz and N. Tinbergen, *Z. Tierpsychol.*, vol. 2, 1938. **38.12a,b,d,e** from E. H. Hess, *Sci. Am.*, July 1956; copyright © 1956 by Scientific American, Inc.,; all rights reserved; **c** Wallace Kirkland, copyright © 1954, Time, Inc. **38.13** Reprinted with permission of the author and publishers from D. G. Freedman, *Human Infancy: An Evolutionary Perspective*, Lawrence Erlbaum Associates, Hillsdale, New Jersey, 1975. **38.14** Adapted from N. E. Collias and E. C. Collias, *Auk*, vol. 79, 1962. **38.15** After Drees. **38.16** Courtesy R. Silver, Barnard College. **38.17** From K. Lorenz, *Symp. Soc. Exp. Biol.*, vol. 4, 1950. **38.18** Based on J. E. Lloyd, *Misc. Publ. Mus. Zool. Univ. Mich.*, vol. 130, 1966; and A. D. Carlson and J. Copeland, *Am. Sci.*, vol. 66, 1978, reprinted by permission of American Scientist, Journal of Sigma Xi, the Scientific Research Society. **38.19** D. R. Bentley, *Science*, vol. 174, 1971; copyright © 1971 by A.A.A.S. **38.20** From L. P. Brower, J. V. Z. Brower, and F. P. Cranston, *Zoologica*, vol. 50, 1965; used by permission of the New York Zoological Society. **38.21** Courtesy R. Thornhill, University of New Mexico. **38.22** Redrawn from K. von Frisch, *Bees: Their Vision, Chemical Senses, and Language*, copyright 1950 by Cornell University; used by permission of Cornell University Press; photograph courtesy Kenneth Lorenzen, University of California, Davis. **38.23** Based on data from

R. Boch, *Z. Vergl. Physiol.*, vol. 40, 1957. **38.24** From K. Lorenz, *Zool. Anz.*, vol. 17, 1953. **38.25** TASS from Sovfoto. **38.26** From W. R. Miles, *J. Comp. Psychol.*, vol. 10, 1930; copyright © 1930 by the Williams and Wilkins Co., Baltimore, Maryland. **38.27a** Eric Hosking/ Bruce Coleman, Inc.; **b** Nina Leen, *Life*, copyright © 1964 by Time, Inc. **38.28** From E. H. Hess, *Science*, vol. 130, 1959; copyright © 1959 by A.A.A.S. **38.30** Warren Gurst and Genny Gurst, Tom Stack and Associates. **38.31** Adapted by permission from *The Honey Bee* by James L. Gould and Carol Grant Gould, W. H. Freeman, New York, copyright © 1988. **38.32** Adapted by permission from E. Gwinner and W. Wiltschko, *J. Comp. Physiol.*, vol. 125, 1978; copyright © 1978 by Springer-Verlag, New York. **38.33a** From B. Elsner, *Animal Migration, Navigation, and Homing*, edited by K. Schmidt-Koenig and W. T. Keeton, Springer-Verlag, New York, 1978; **b** from M. Michener and C. Walcott, *J. Exp. Biol.*, vol. 47, 1967. **38.34**, left, Steve Johnson, courtesy Charles Walcott, Cornell University; right, from K. Schmidt-Koenig and C. Walcott, *Animal Behav.*, vol. 26, 1978. **38.36** Modified from N. Tinbergen and A. C. Perdeck, *Behaviour*, vol. 3, 1950. **38.38** Photograph by Russ Charig, courtesy Charles Walcott, Cornell University.

PART VI: ECOLOGY

Threatened reservoirs of diversity: © Art Wolfe/Tony Stone Images.

39.3 Photo by G. W. Barlow, from *Anim. Beh.*, vol. 22, 1974. **39.4** Francois Gohier/Photo Researchers, Inc. **39.6** Modified from T. Carlson, *Biochem. Z.*, vol. 57, 1913. **39.7** Modified from J. Davidson, *Trans. R. Soc. South Aust.*, vol. 62, 1938. **39.9** Modified from E. P. Odum, *Fundamentals of Ecology*, W. B. Saunders, 1959 after Deevey. **39.10** Office of Population Research, Washington, D. C. **39.12** Courtesy E. S. Ross. **39.13** Courtesy L. David Mech, USFWS Minnesota Research Group, St. Paul, MN. **39.15** Photograph by Hans Reinhard/Bruce Coleman, Inc.; drawing based on H. N. Southern, *J. Zool.*, vol. 162, 1970. **39.16** Based on H. N. Southern, *J. Zool.*, vol. 162, 1970. **39.17** From John L. Harper, *Population Biology of Plants*, Academic Press, 1977, p. 153, copyright © 1977 Academic Press, Orlando, Florida. **39.18** Redrawn after J. M. Scriber and G. Slankey, *Am. Rev. Entomology*, vol. 26, 1981. **39.19** Adapted from *Principles of Animal Ecology* by W. C. Allee, A. E. Emerson, O. Park, T. Park, and K. P. Schmidt, W. B. Saunders, Philadelphia; copyright © 1949. **39.20** Drawing adapted from R. H. MacArthur, *Ecology*, vol. 39, 1958, copyright © 1959 by the Ecological Society of America. Photos: **a** Philip Boyer and **b** R. Austing/Photo Researchers, Inc.; **c** Edgar T. Jones/Bruce Coleman, Inc. **39.21** G. Torlofi/F. A. O. Photo. **39.22** Carol Hughes/Bruce Coleman, Inc. **39.23** Norman Meyers/Bruce Coleman, Inc. **39.24** Redrawn by permission from E. O. Wilson, *Sociobiology*, Harvard University PRess, Cambridge, Massachusetts, 1975. **39.25** Hans Pfletschinger/Peter Arnold, Inc. **39.26** Adapted from R. E. Kenward, *J. Anim. Ecol.*, vol. 47, 1978. **39.27** Courtesy E. D. Estes, San Diego State University. **39.28** Coutesy L. T. Nash, Arizona State University. **39.30** George D. Lepp, Bio-Tec Images. **39.31** Roberto Bunge, Ardea London Ltd. **39.32** Redrawn with permission of Macmillan Publishing Co., Inc. from David R. Pilbeam, *The Ascent of Man: An Introduction to Human Evolution*; copyright 1972 by David R. Pilbeam. **39.33a** J. Tanaka and **b** M. Shostak/Anthro-Photo. **39.34** I. De Vore/Anthro-Photo. **39.35** Modified from J. H. Connell, *Ecology*, vol. 142, 1961; copyright © 1961 by the Ecological Society of America. **39.36** Photographs reproduced with permission of the Department of Lands, Queensland, Australia. **39.37** Redrawn after R. M. May, *Sci. Am.*, September, 1978; copyright 1978 by Scientific American, Inc., all rights reserved. **39.39** Courtesy E. J. Kormondy, *Smithsonian Magazine*, vol. 1, 1970. **39.40** Jeff Foote/Bruce Coleman, Inc. **39.41** S. Hurwitz/Bruce Coleman, Inc. **39.42** Based on data in E. P. Odum, *Fundamentals of Ecology*, W. B. Saunders, 1959. **39.43** Redrawn from R. M. May, *Sci. Am.*, September 1978; copyright 1978 by Scientific American, Inc.; all rights reserved. **39.44** Redrawn from R. H. Whittaker, *Communities and Ecosystems*, 2nd ed., Macmillan, New York, 1975; after B. Holt and G. M. Woodwell. **39.45** Redrawn from R. H. Whittaker, *Communities and Ecosystems*, 2nd ed., Macmillan, New York, 1975; after B. Holt and G. M. Woodwell. **39.46** Gene Ahrens/Bruce Coleman, Inc.

40.2a C. J. Tucker, J. R. G. Towshend, and T. E. Goff, *Science*, vol. 227, 1985. Redrawn with permission of the A.A.A.S.; **b** from Fig. 9-5 in R. E. Ricklefs, *Ecology*, 3rd ed., W. H. Freeman, 1990, p. 144. Based on data in H. H. Clayton and F. L. Clayton, 1947, World Weather Records 1931–1940, *Smithsonian Misc. Coll.* 105, 1–646. **40.3** Photographs courtesy E. S. Ross. **40.4** From fig. 12-3a and 12-4a, pp. 243–244 in Helmut Leith, Modeling the Primary Productivity of the World, in H. Leith and R. H. Whittaker, *Primary Productivity in the Biosphere*, Springer-Verlag, 1975. **40.5** From fig. 12.12c, in Helmut Leith, Modeling the Primary Productivity of the World, in Helmut Leith and R. H. Whittaker, *Primary Productivity in the Biosphere*, Springer-Verlag, 1975. **40.10** Modified from E. P. Odum, *Fundamentals of Ecology*, W. B. Saunders, 1959. **40.12** Photos courtesy James Hornbeck, U. S. Forest Service, Durham, New Hampshire. **40.13** From Shukla et al., *Science*, vol. 247, pp. 1322–25, fig. 3b, 1990. Copyright © 1990 A.A.A.S. **40.15** Redrawn after T. E. Graedel and P. J. Crutzen, *Sci. Am.*, September, 1989; copyright © 1989 by Scientific American, Inc.; all rights reserved. **40.16** Redrawn after S. H. Schneider, *Sci. Am.*, September, 1989; copyright © 1989 by Scientific American, Inc.; all rights reserved. **40.18** Photo by Will Owens, from Will Plants Profit From High CO_2? *Science*, vol. 268, 5 May 1995, p. 655. Copyright © 1995 by A.A.A.S. **40.19** Modified from M. H. Pelczar et al., *Microbiology*, 4th ed., McGraw-Hill Book Co., New York, copyright © 1977; used by permission. **40.20** Dr. Jeremy Burgess/Photo Researchers, Inc. **40.22** Courtesy D. W. Schindler, *Science*, col. 184, 1974; copyright © 1974 by A.A.A.S. **40.24** Courtesy Peter L. Ames. **40.25** Redrawn after K. P. Bowman, *Science*, vol. 239, 1988; copyright © 1988 by A.A.A.S. ***How We Know***, p. 1178 David Hall/Photo Researchers, Inc. **40.26** Larry Ditto/Bruce Coleman, Inc. **40.27** U.S.D.A. Forest Service, Washington, D.C. **40.28** Redrawn from G. E. Likens et al., *Ecol. Monogr.*, vol. 40, 1970; copyright © 1970 by the Ecological Society of America. **40.29** Courtesy National Archives. **40.30** Oxford Scientific Films/Animals, Animals, Earth Scenes. **40.31** After Whittaker (1970) Reprinted with permission from Macmillan Science Co., Inc., from *Communities and Ecosystems* by Robert Whittaker. Copyright © Robert Whittaker. **40.32** Photo by Tom Bean/DRK Photo, Sedona, Arizona. **40.33a** Paolo Koch/ Photo Researchers, Inc.; **b** Wolfgang Bayer/Bruce Coleman, Inc. **40.34** Zig Leszczynski, Animals, Animals, Earth Scenes. **40.35a** Ferrero/ Nature; **b** N. De Vore III/Bruce Coleman, Inc. **40.36** David Cayless/ Oxford Scientific Films. **40.37a** Ronald Toms/Oxford Scientific Films; **b,c** courtesy E. S. Ross. **40.39** John Elk III/Bruce Coleman, Inc. **40.41** After Schwarzbach, 1950. **40.42a** Courtesy Howard Hall; **b** courtesy James D. Jordan; **c** courtesy E. H. Newcomb and T. D. Pugh, University of Wisconsin/BPS. **40.44** Redrawn from John Napier, *The Roots of Mankind*, copyright © 1970, Smithsonian Institution Press, Washington, D. C.; used by permission. **40.46** Ferrero/Nature. **40.50a** Redrawn from R. H. MacArthur and E. O. Wilson, *The Theory of Island Biogeography*, copyright 1967 by Princeton University Press; used by permission; **b** redrawn by C. J. Krebs, *Ecology*, 2nd ed., Harper & Row, New York, 1978; after Preston. **40.53a** C. C. Reijnvaan from W. M. Doctors van Leeuwen, *Ann. Jard. Bot. Buitenzorg*, 1936; **b** courtesy Stephen Self, University of Texas, Arlington.

GLOSSARY

The Glossary gives brief definitions of the most important terms that recur in the text, excluding taxonomic designations. For fuller definitions, consult the index, where italicized page numbers refer you to explanations of key terms in context.

Of the basic units of measurement, some are tabulated on p. A16, others have their own alphabetical entries.

Interspersed alphabetically with the vocabulary are the main prefixes and combining forms used in biology. You will notice that, while they are generally of Greek or Latin origin, many of them have acquired a new meaning in biology (examples: ***blasto-, -cyte, caryo-, -plasm***). Familiarity with these forms will make it easier for you to learn and remember the numerous terms in which they are incorporated.

TABLE 1 *Standard prefixes of the metric system*

kilo- (k)	1.000	10^3
deci- (d)	0.1	10^{-1}
centi- (c)	0.01	10^{-2}
milli- (m)	0.001	10^{-3}
micro- (μ)	0.000001	10^{-6}
nano- (n)	0.000000001	10^{-9}

TABLE 2 *Common units of length, weight, and liquid capacity*

kilometer (km)	1,000 m	0.62137 mile
meter (m)		39.37 inches
centimeter (cm)	0.01 m	0.39 inch
millimeter (mm)	0.001 m	0.039 inch
micrometer* (μm)	10^{-6} m	
nanometer (nm)	10^{-9} m	
angstrom† (Å)	10^{-10} m	
kilogram (kg)	1,000 g	2.2 pounds
gram (g)		0.035 ounce
milligram (mg)	0.001 g	
microgram (μg)	10^{-6} g	
liter (l)	1,000 cm^3	1.057 quarts
milliliter (ml)	0.001 l	

*Formerly called micron.
†No longer used; nanometer used instead

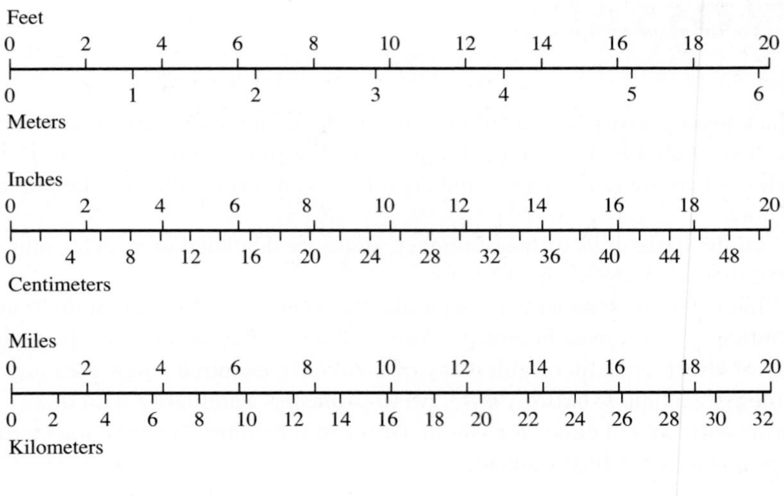

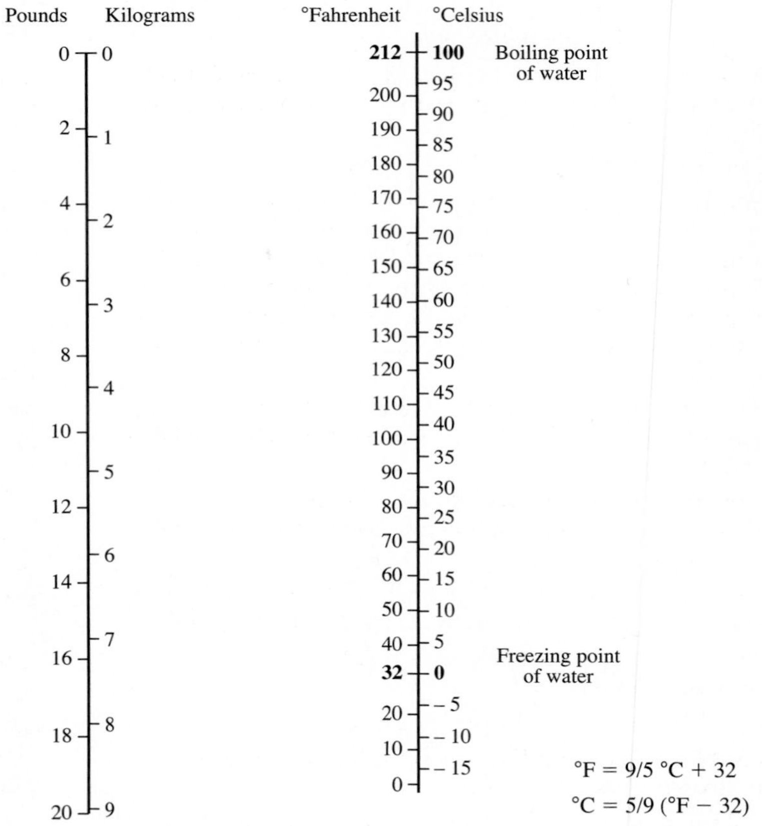

a- Without, lacking.

ab- Away from, off.

abdomen [L belly] In mammals, the portion of the trunk posterior to the thorax, containing most of the viscera except heart and lungs. In other animals, the posterior portion of the body.

absolute zero The temperature (–273°C) at which all thermal agitation ceases. The lowest possible temperature.

acellular Not constructed on a cellular basis.

acid [L *acidus* sour] A substance that increases the concentration of hydrogen ions when dissolved in water.

acoelomate A body plan in which there is no cavity between the digestive tract and the body wall.

ACTH *See* adrenocorticotropic hormone.

action potential *See* potential.

active site In an enzyme, the portion of the molecule that reacts with a substrate molecule.

active transport Movement of a substance across a membrane by a process requiring expenditure of energy by the cell.

ad- Next to, at, toward.

adaptation In evolution, any genetically controlled characteristic that increases an organism's fitness, usually by helping the organism to survive and reproduce in the environment it inhabits. In neurobiology, the process that results in a short-lasting decline in responsiveness of a sensory neuron after repeated firing; *cf.* habituation, sensitization.

adenosine diphosphate (ADP) A doubly phosphorylated organic compound that can be further phosphorylated to form ATP.

adenosine monophosphate (AMP) A singly phosphorylated organic compound that can be further phosphorylated to form ADP.

adenosine triphosphate (ATP) A triply phosphorylated organic compound that functions as "energy currency" for organisms.

adipose [L *adeps* fat] Fatty.

ADP *See* adenosine diphosphate.

adrenal [L *renes* kidneys] An endocrine gland of vertebrates located near the kidney.

adrenalin A hormone produced by the adrenal medulla that stimulates "fight-or-flight" reactions.

adrenocorticotropic hormone (ACTH) A hormone produced by the pituitary that stimulates the adrenal cortex.

adsorb [L *sorbēre* to suck up] Hold on a surface.

advanced New, unlike the ancestral condition.

aerobic [L *aer* air] With oxygen.

alcohol Any of a class of organic compounds in which one or more —OH groups are attached to a carbon backbone.

alkaline Having a pH of more than 7. *See* base.

all-, allo- [Gk *allos* other] Other, different.

allele Any of several alternative gene forms at a given chromosomal locus.

allopatric [L *patria* homeland] Having different ranges.

allosteric Of an enzyme: one that can exist in two or more conformations. *Allosteric control:* control of the activity of an allosteric enzyme by determination of the particular conformation it will assume.

altruism The willingness of an individual to sacrifice its fitness for the benefit of another. *Reciprocal altruism:* the performance of a favor by one individual in the expectation of a favor in return, as when two animals groom each other.

alveolus [L little hollow] A small cavity, especially one of the microscopic cavities that are the functional units of lungs.

amino acid An organic acid carrying an amino group (—NH_2); the building-block compound of proteins.

amnion [Gk caul] An extraembryonic membrane that forms a fluid-filled sac containing the embryo in reptiles, birds, and mammals.

amoeboid [Gk *amoibē* change] Amoebalike in the tendency to change shape by protoplasmic flow.

AMP *See* adenosine monophosphate.

amylase [L *amylum* starch] A starch-digesting enzyme.

an- Without.

anabolism [Gk *ana-* upward; *metabolē* change] The biosynthetic building-up aspects of metabolism.

anaerobic [L *aer* air] Without oxygen.

analogous Of characters in different organisms: similar in function and often in superficial structure but of different evolutionary origins.

anemia A condition in which the blood has lower than normal amounts of hemoglobin or red blood corpuscles.

angio-, -angium [Gk *angeion* vessel] Container, receptacle.

anisogamous Reproducing by the fusion of gametes that differ only in size, as opposed to gametes that are produced by oogamous species. Gametes of oogamous species, such as egg cells and sperm, are highly differentiated.

anterior Toward the front end.

antheridium [Gk *anthos* flower] Male reproductive organ of a plant; produces sperm cells.

antibody A protein, produced by the B lymphocytes of the immune system, that binds to a particular antigen.

antigen A substance, usually a protein or polysaccharide, that activates an organism's immune system.

anus [L ring] Opening at the posterior end of the digestive tract, through which indigestible wastes are expelled.

aorta The main artery of the systemic circulation.

apical At, toward, or near the apex, or tip, of a structure such as a plant shoot.

apo- Away from.

apoplast The network cell walls and intercellular spaces within a plant body; permits extensive extracellular movement of water within the plant.

aposematic [Gk *sēma* sign] Serving as a warning, with reference particularly to colors and structures that signal possession of defensive devices.

arch- [Gk *archein* to begin] Primitive, original.

archegonium [Gk *archegonos* the first of a race] Female reproductive organ of a higher plant; produces egg cells.

archenteron [Gk *enteron* intestine] The cavity in an early embryo that becomes the digestive cavity.

arteriole A small artery.

artery A blood vessel that carries blood away from the heart.

articulation A joint between bones. Articulating surfaces are those formed between bones and joints.

artifact A by-product of scientific manipulation rather than an inherent part of the thing observed.

ascus [Gk *askos* bag] The elongate spore sac of a fungus of the Ascomycota group.

asexual Without sex.

atmosphere (atm) (unit of pressure) The normal pressure of air at sea level: 101,325 newtons per square meter (approx. 14.7 pounds per square inch).

atom [Gk *atomos* indivisible] The smallest unit of an element, not divisible by ordinary chemical means.

atomic mass unit (amu) *See* dalton.

atomic weight The average weight of an atom of an element relative to ^{12}C, an isotope of carbon with six neutrons in the nucleus. The atomic weight of ^{12}C has arbitrarily been fixed as 12.

ATP *See* adenosine triphosphate.

auto- Self, same.

autonomic nervous system A portion of the vertebrate nervous system, comprising motor neurons that innervate internal organs and are not normally under direct voluntary control.

autosome [Gk *sōma* body] Any chromosome other than a sex chromosome.

autotrophic [Gk *trophē* food] Capable of manufacturing organic nutrients from inorganic raw materials.

auxin [Gk *auxein* to grow] Any of a class of plant hormones that promote cell elongation.

axon [Gk *axōn* axis] A fiber of a nerve cell that conducts impulses away from the cell body and can release transmitter substance.

bacteriophage [Gk *phagein* to eat] A virus that attacks bacteria, *abbrev.* phage.

basal At, near, or toward the base (the point of attachment) of a structure such as a limb.

basal body A structure, identical to the centriole, found at the base of cilia and eucaryotic flagella; consists of nine triplet microtubules arranged in a circle.

base (or alkali) A substance that increases the concentration of hydroxyl ions when dissolved in water. It has a pH higher than 7.

basidium The spore-bearing structure of Basidiomycota (club fungi).

bi- Two.

bilateral symmetry The property of having two similar sides, with definite upper and lower surfaces and anterior and posterior ends.

binary fision Reproduction by the division of a cell into two essentially equal parts by a nonmitotic process.

bio- [Gk *bios* life] Life, living.

biogenesis [Gk *genesis* source] Origin of living organisms from other living organisms.

biological magnification Increasing concentration of relatively stable chemicals as they are passed up a food chain from initial consumers to top predators.

biomass The total weight of all the organisms, or of a designated group of organisms, in a given area.

biome A large climatic region with characteristic sorts of plants and animals.

biotic Pertaining to life.

blasto- [Gk *blastos* bud] Embryo.

blastocoel [Gk *koilos* hollow] The cavity of a blastula.

blastopore [Gk *poros* passage] The opening from the cavity of the archenteron to the exterior in a gastrula.

blastula An early embryonic stage in animals, preceding the delimitation of the three principal tissue layers; frequently spherical and hollow.

B lymphocyte *See* lymphocyte.

buffer A substance that binds H^+ ions when their concentration rises and releases them when their concentration falls, thereby minimizing fluctuations in the pH of a solution.

C_3 plants Plants in which the Calvin cycle is the only pathway of CO_2 fixation. One of the products of photosynthesis is a three-carbon intermediary (C_3) of the Calvin cycle, from which the plants derive their name.

C_4 plants Also called **Kranz plants.** Plants in which one of the main early products of photosynthesis is a four-carbon compound (C_4). Kranz plants can carry out photosynthesis under conditions that are inhospitable to other plants.

caecum [L *caecus* blind] A blind diverticulum of the digestive tract.

calorie [L *calor* heat] The quantity of energy, in the form of heat, required to raise the temperature of one gram of pure water one degree from 14.5 to 15.5°C. The nutritionists' Calorie (capitalized) is 1,000 calories, or one kilocalorie.

cambium [L *cambiare* to exchange] The principal lateral meristem of vascular plants; gives rise to most secondary tissue.

cAMP *See* cyclic adenosine monophosphate.

capillarity [L *capillus* hair] The tendency of aqueous liquids to rise in narrow tubes with hydrophilic surfaces.

capillary [L *capillus*] A tiny blood vessel with walls one cell thick, across which exchange of materials between blood and the tissues takes place; receives blood from arteries and carries it to veins. Also, a similar vessel of the lymphatic system.

carbohydrate Any of a class of organic compounds composed of carbon, hydrogen, and oxygen in a ratio of about two hydrogens and one oxygen for each carbon; examples are sugar, starch, cellulose.

carbon fixation The process by which CO_2 is incorporated into organic compounds, primarily glucose; energy usually comes from the ATP and $NADP_{re}$ generated by photophosphorylation, and the metabolic pathway utilizing this energy is usually the Calvin cycle.

carboxyl group The —COOH group characteristic of organic acids.

cardiac [Gk *kardia* heart] Pertaining to the heart.

carnivore [L *carnis* of flesh; *vorare* to devour] An organism that feeds on animals.

carotenoid [L *carota* carrot] Any of a group of red, orange, and yellow accessory pigments of plants, found in plastids.

carrying capacity The maximum population that a given environment can support indefinitely.

cartilage A specialized type of dense fibrous connective tissue with a rubbery intercellular matrix.

caryo- [Gk *karyon* kernel] Nucleus.

Casparian strip A waterproof thickening in the radial and end walls of endodermal cells of plants.

cata- Down.

catabolism [Gk *katabolē* a throwing down] The degradational breaking-down aspects of metabolism, by which living things extract energy from food.

catalysis [Gk *katalyein* to dissolve] Acceleration of a chemical reaction by a substance that is not itself permanently changed by the reaction.

catalyst A substance that produces catalysis.

cation A positively charged ion.

caudal [L *cauda* tail] Pertaining to the tail.

cell cycle The cycle of cellular events from one mitosis through the next. Four stages are recognized, of which the last—distribution of genetic material to the two daughter nuclei—is mitosis proper.

cell sap *See* sap.

cellulose [L *cellula* cell] A complex polysaccharide that is a major constituent of most plant cell walls.

centi- [L *centum* hundred] One hundredth.

central nervous system A portion of the nervous system that contains interneurons and exerts some control over the rest of the nervous system. In vertebrates, the brain and the spinal cord.

centri- [L *centrum* center] Center.

centrifugation [L *fugere* to flee] The spinning of a mixture at very high speeds to separate substances of different densities.

centriole A cylindrical cytoplasmic organelle located just outside the nucleus of animal cells and the cells of some lower plants; associated with the spindle during mitosis and meiosis.

centromere [Gk *meros* part] A special region on a chromosome from which kinetochore microtubules radiate during mitosis or meiosis.

cephalization [Gk *kephalē* head] Localization of neural coordinating centers and sense organs at the anterior end of the body.

cerebellum [L small brain] A part of the hindbrain of vertebrates that controls muscular coordination.

cerebrum [L brain] Part of the forebrain of vertebrates, the chief coordination center of the nervous system.

channel *See* membrane channel.

character Any structure, functional attribute, behavioral trait, or other characteristic of an organism.

character displacement The rapid divergent evolution in sympatric species of characters that minimize competition and/or hybridization between them.

chemiosmotic gradient The combined electrostatic and osmotic concentration gradient generated by the electron-transport chains of mitochondria and chloroplasts; the energy in this gradient is used, for the most part, to synthesize ATP.

chemosynthesis Autotrophic synthesis of organic materials, energy for which is derived from inorganic molecules.

chitin [Gk *chitōn* tunic] Polysaccharide that forms part of the hard exoskeleton of insects, crustaceans, and other invertebrates; also occurs in the cell walls of fungi.

chlorophyll [Gk *chlōros* greenish yellow; *phyllon* leaf] The green pigment of plants necessary for photosynthesis.

chloroplast A plastid containing chlorophyll.

chrom-, -chrome [Gk *chrōma* color] Colored; pigment.

chromatid A single chromosomal strand.

chromatin The mixture of DNA and protein (mostly histones in the form of nucleosome cores) that comprises eucaryotic nuclear chromosomes.

chromatography Process of separating substances by adsorption on media for which they have different affinities.

chromosome [Gk *sōma* body] A filamentous structure in the cell nucleus (or nucleoid), mitochondria, and chloroplasts, along which the genes are located.

cilium [L eyelid] A short hairlike locomotory organelle on the surface of a cell (*pl* cilia).

cisterna [L cistern] A cavity, sac, or other enclosed space serving as a reservoir.

classical conditioning *See* conditioning.

cleavage Division of a zygote or of the cells of an early embryo.

climax (ecological) A relatively stable stage reached in some ecological successions.

cline [Gk *klinein* to lean] Gradual variation, correlated with geography, in a character of a species.

cloaca [L sewer] Common chamber that receives materials from the digestive, excretory, and reproductive systems.

clone [Gk *klōn* twig] A group of cells or organisms derived asexually from a single ancestor and hence genetically identical.

co- With, together.

codon The unit of genetic coding, three nucleotides long, specifying an amino acid or an instruction to terminate translation.

coel-, -coel [Gk *koilos* hollow] Hollow, cavity; chamber.

coelom A body cavity surrounded by mesoderm.

coenocytic [Gk *koinos* common] Having more than one nucleus in a single mass of cytoplasm.

coenzyme A nonproteinaceous organic molecule that plays an accessory role, but a necessary one, in the catalytic action of an enzyme.

coevolution Two or more organisms evolving, each in response to the other.

coleoptile [Gk *koleon* sheath, *ptilon* feather] A sheath around the young shoot of grasses.

collagen A fibrous protein; the most abundant protein in mammals.

collenchyma [Gk *kolla* glue] A supportive tissue in plants in which the cells usually have thickenings at the angles of the walls.

colloid [Gk *kolla*] A stable suspension of particles that, though larger than in a true solution, do not settle out.

colon The large intestine.

com- Together.

commensalism [L *mensa* table] A symbiosis in which one party is benefited and the other party receives neither benefit nor harm.

community In ecology, a unit composed of all the populations living in a given area.

competition In ecology, utilization by two or more individuals, or by two or more populations, of the same limited resource; an interaction in which both parties are harmed.

condensation reaction A reaction joining two compounds with resultant formation of water.

conditioning Associative learning. *Classical conditioning:* the association of a novel stimulus with an innately recognized stimulus. *Operant conditioning:* learning of a novel behavior as a result of reward or punishment; trial-and-error learning.

conformation (of a protein) [L *conformatio* symmetrical forming] The three-dimensional pattern according to which the polypeptide chains of a protein coil (secondary structure), fold (tertiary structure), and—if there is more than one chain—fit together (quarternary structure).

conjugation [L *jugare* to join, marry] Process of genetic recombination between two organisms (e.g., bacteria, algae) through a cytoplasmic bridge between them.

connective tissue A type of animal tissue whose cells are embedded in an extensive intercellular matrix; connects, supports, or surrounds other tissues and organs.

contractile vacuole An excretory and/or osmoregulatory vacuole in some cells, which, by contracting, ejects fluids from the cell.

cooperativity The phenomenon of enhanced reactivity of the remaining binding sites of a protein as a result of the binding of substrate at one site.

cork [L *cortex* bark] A waterproof tissue, derived from the cork cambium, that forms at the outer surfaces of the older stems and roots of woody plants; the outer bark or periderm.

corpus luteum [L yellow body] A yellowish structure in the ovary, formed from the follicle after ovulation, that secretes estrogen and progesterone (*pl.* corpora lutea).

cortex [L bark] In plants, tissue between the epidermis and the vascular cylinder of stems and roots. In animals, the outer barklike tissue of some organs, as *cerebral cortex, adrenal cortex,* etc.

cotyledon [Gk *kotyle* cup] A "seed leaf," a food-digesting and -storing part of a plant embryo.

countercurrent exchange A strategy in which two streams move past each other in opposite directions, facilitating the exchange of substances between them across a membrane. The gills in gas exchange and kidneys in the production of concentrated urine are two sites of countercurrent exchange.

covalent bond A chemical bond resulting from the sharing of a pair of electrons.

Crassulacean acid metabolism (CAM) A variation of photosynthesis found in some plants that grow in hot, dry environments. Plants such as succulents avoid water loss by closing their stomata during the day and opening them at night.

crossing over Exchange of parts between two homologous chromosomes.

cross section *See* section.

cryptic [Gk *kryptos* hidden] Concealing.

cuticle [L *cutis* skin] A waxy layer on the outer surface of leaves, insects, etc.

cyclic adenosine monophosphate (cyclic AMP or cAMP) Compound, synthesized in living cells from ATP, that functions as an intracellular mediator of hormonal action; also plays a part in neural transmission and some other kinds of cellular control systems.

cyst [Gk *kystis* bladder, bag] (1) A saclike abnormal growth. (2) Capsule that certain organisms secrete around themselves and that protects them during resting stages.

-cyte, cyto- [Gk *kytos* container] Cell.

cytochrome Any of a group of iron-containing enzymes important in electron transport during respiration or photophosphorylation.

cytokinesis [Gk *kinēsis* motion] Division of the cytoplasm of a cell.

cytoplasm All of a cell except the nucleus.

cytosol The relatively fluid, less structured part of the cytoplasm of a cell, excluding organelles and membranous structures.

dalton A unit of mass equal to one twelfth the atomic weight of ^{12}C, or 1.66024×10^{-24} gram. Formerly called atomic mass unit (amu).

deamination Removal of an amino group.

deciduous [L *decidere* to fall off] Shedding leaves each year.

dehydration reaction A condensation reaction.

deme [Gk *dēmos* population] A local unit of population of any one species.

dendr-, dendro- [Gk *dendron* tree] Tree; branching.

dendrite A short unsheathed fiber of a nerve cell—often spiny, usually branched and tapering—that receives many synapses and carries excitation and inhibition toward the cell body.

deoxyribonucleic acid (DNA) A nucleic acid found in most viruses, all bacteria, chloroplasts, mitochondria, and the nuclei of eucaryotic cells, characterized by the presence of a deoxyribose sugar in each nucleotide; the genetic material of all organisms except the RNA viruses.

-derm [Gk *derma* skin] Skin, covering; tissue layer.

di- Two.

dicot A member of a subclass of the angiosperms, or flowering plants, characterized by the presence of two cotyledons in the embryo, a netlike system of veins in the leaves, and flower petals in fours or fives; *cf.* monocot. *Herbaceous dicot:* a perennial whose aboveground parts die annually. *Woody dicot:* a perennial whose above ground parts—trunk and branches—remain alive and grow annually.

differentiation The process of developmental change from an immature to a mature form, especially in a cell.

diffusion The movement of dissolved or suspended particles from place to place as a result of their heat energy (thermal agitation).

digestion Hydrolysis of complex nutrient compounds into their building-block units.

diploid [Gk *diploos* double] Having two of each type of chromosome.

disaccharide A double sugar, one composed of two simple sugars.

distal [L *distare* to stand apart] Situated away from some reference point (usually the main part of the body).

diverticulum [L *devertere* to turn aside] A blind sac branching off a cavity or canal.

DNA *See* deoxyribonucleic acid.

dominant (1) Of an allele: exerting its full phenotypic effect despite the presence of another allele of the same gene, whose phenotypic expression it blocks or masks. *Dominant phenotype, dominant character:* one caused by a dominant allele. (2) Of an individual: occupying a high position in the social hierarchy.

dormancy [L *dormire* to sleep] The state of being inactive, quiescent. In plants, particularly seeds and buds, a period in which growth is arrested until environmental conditions become more favorable.

dorsal [L *dorsum* back] Pertaining to the back.

drive *See* motivation.

duodenum [From a Latin phrase meaning 12 *(duodecin)* finger's-breadths long] The first portion of the small intestine of vertebrates, into which ducts from the pancreas and gallbladder empty.

ecosystem [Gk *oikos* habitation] The sum of physical features and organisms occurring in a given area.

ecto- Outside, external.

ectoderm The outermost tissue layer of an animal embryo. Also, tissue derived from the embryonic ectoderm.

ectothermic *See* poikilothermic.

effector The part of an organism that produces a response, e.g., muscle, cilium, flagellum.

egg An egg cell or female gamete. Also a structure in which embryonic development takes place, especially in birds and reptiles; consists of an egg cell, various membranes, and often a shell.

electrochemical gradient Combined electrostatic and osmotic-concentration gradient, such as the chemiosmotic gradient of mitochondria and chloroplasts.

electron A negatively charged primary subatomic particle.

electronegativity The formal measure of an atom's attraction for free electrons. Atoms with few electron vacancies in their outer shell tend to be more electronegative than those with more. In covalent bonds, the shared electrons are, on average, nearer the more electronegative atom; this asymmetry, in part, gives rise to the polarity of certain molecules.

electronic charge unit The charge of one electron, or 1.6021×10^{-19} coulomb.

electron-transport chain A series of enzymes found in the inner membrane of mitochondria and (with somewhat different components) in the thylakoid membrane of chloroplasts. The chain accepts high-energy electrons and uses their energy to create a chemiosmotic gradient across the membrane in which it is located.

electrostatic force The attraction (also called *electrostatic attraction*) between particles with opposite charges, as between a proton and an electron, or between H^+ and OH^-; and the repulsion between particles with like charges, as between two H^+ ions.

electrostatic gradient The free-energy gradient created by a difference in charge between two points, generally the two sides of a membrane.

elimination (or defecation) The release of unabsorbed wastes from the digestive tract. *Cf.* excretion.

embryo A plant or animal in an early stage of development; generally still contained within the seed, egg, or uterus.

emulsion [L *emulsus* milked out] Suspension, usually as fine droplets, of one liquid in another.

-enchyma [Gk *parenchein* to pour in beside] Tissue.

end-, endo- Within, inside; requiring.

endergonic [Gk *ergon* work] Energy-absorbing; endothermic.

endocrine [Gk *krinein* to separate] Pertaining to ductless glands that produce hormones.

endocytosis The process by which the cell membrane forms an invagination which becomes a vesicle, trapping extracellular material that is then transported within the cell; in general, the invagination is triggered by the binding of membrane receptors to specific substances used by the cell.

endoderm The innermost tissue layer of an animal embryo.

endodermis A plant tissue, especially prominent in roots, that surrounds the vascular cylinder; all endodermal cells have Casparian strips.

endonuclease An enzyme that breaks bonds within nucleic acids, as opposed to an exonuclease, which can digest only a terminal group. *Restriction endonuclease:* an enzyme that breaks bonds only within a specific sequence of bases.

endoplasmic reticulum [L *reticulum* network] A system of membrane-bounded channels in the cytoplasm.

endoskeleton An internal skeleton.

endosperm [Gk *sperma* seed] A nutritive material in seeds.

endosymbiotic hypothesis Hypothesis that certain eucaryotic organelles—in particular mitochondria and chloroplasts—originated as free-living procaryotes that took up mutalistic residence in the ancestors of modern eucaryotes.

endothermic In thermodynamics, energy-absorbing (endergonic). In physiology, warm-blooded (homeothermic).

entropy Measure of the disorder of a system.

enzyme [Gk *zymē* leaven] A compound, usually a protein, that acts as a catalyst.

epi- Upon, outer.

epicotyl A portion of the axis of a plant embryo above the point of attachment of the cotyledons; forms most of the shoot.

epidermis [Gk *derma* skin] The outermost portion of the skin or body wall of an animal.

episome [Gk *sōma* body] Genetic element at times free in the cytoplasm, at other times integrated into a chromosome.

epithelium An animal tissue that forms the covering or lining of all free body surfaces, both external and internal.

equilibrium constant The ratio of products of a reaction to the reactants after the reaction has been allowed to proceed until there is no further change in these concentrations.

erythrocyte [Gk *erythros* red] A red blood corpuscle, i.e., a blood corpuscle containing hemoglobin.

esophagus [Gk *phagein* to eat] An interior part of the digestive tract; in mammals it leads from the pharynx to the stomach.

essential fatty acid A fatty acid an organism needs but cannot synthesize, and so must obtain preformed (or in a precursor form) from its diet.

estrogen [L *oestrus* frenzy] Any of a group of vertebrate female sex hormones.

estrous cycles [L *oestrus*] In female mammals, the higher primates excepted, a recurrent series of physiological and behavioral changes connected with reproduction.

estuary That portion of a river that is close enough to the sea to be influenced by marine tides.

eu- [Gk *eus* good] Most typical, true.

eucaryotic cell A cell containing a distinct membrane-bounded nucleus, characteristic of all organisms except bacteria.

evaginated [L *vagina* sheath] Folded or protruded outward.

eversible [L *evertere* to turn out] Capable of being turned inside out.

evolution [L *evolutio* unrolling] Change in the genetic makeup of a population with time.

ex-, exo- Out of, outside; producing.

excretion Release of metabolic wastes and excess water. *Cf.* elimination.

exergonic [Gk *ergon* work] Energy-releasing; exothermic.

exocytosis The process by which an intracellular vesicle fuses with the cell membrane, expelling its contents into its surroundings.

exon A part of a primary transcript (and the corresponding part of a gene) that is ultimately either translated (in the case of mRNA) or utilized in a final product, such as tRNA.

exoskeleton An external skeleton.

extrinsic External to, not a basic part of; as in *extrinsic isolating mechanism.*

fauna The animals of a given area or period.

feature detector A circuit in the nervous system that responds to a specific type of feature, such as a vertically moving spot or a particular auditory time delay.

feces [L *faeces* dregs] Indigestible wastes discharged from the digestive tract.

feedback The process by which a control mechanism is regulated through the very effects it brings about. *Positive feedback:* the process by which a small effect is amplified, as when a depolarization triggers an acton potential. *Negative feedback* (or feedback inhibition): the process by which a control mechanism is activated to restore conditions to their original state.

fermentation Anaerobic production of alcohol, lactic acid, or similar compounds from carbohydrates via the glycolytic pathway.

fertilization Fusion of nuclei of egg and sperm.

fetus [L *fetus* pregnant] An embryo in its later development, still in the egg or uterus.

fitness The probable genetic contribution of an individual (or allele or genotype) to succeeding generations. *Inclusive fitness:* the sum of an individual's personal fitness plus the fitness of that individual's relatives devalued in proportion to their genetic distance from the individual.

fixation (1) Conversion of a substance into a biologically more usable form, as the conversion of CO_2 into carbohydrate by photosynthetic plants or the incorporation of N_2 into more complex molecules by nitrogen-fixing bacteria. (2) Process of treating living tissue for microscopic examination.

flagellum [L whip] A long hairlike locomotory organelle on the surface of a cell.

flora The plants of a given area or period.

follicle [L *follis* bag] A jacket of cells around an egg cell in an ovary.

follicle-stimulating hormone (FSH) A gonadotropic hormone of the anterior pituitary that stimulates growth of follicles in the ovaries of females and function of the seminiferous tubules in males.

food chain Sequence of organisms, including producers, consumers, and decomposers, through which energy and materials may move in a community.

foot-candle Unit of illumination; the illumination of a surface produced by one standard candle at a distance of one foot; *cf.* lambert.

founder effect The difference between the gene pool of a population as a whole and that of a newly isolated population of the same species.

free energy Usable energy in a chemical system; energy available for producing change.

fruit A mature ovary or cluster of ovaries (sometimes with additional floral structures associated with the ovary).

fruiting body A spore-bearing structure (e.g., the above-ground portion of a mushroom).

FSH *See* follicle-stimulating hormone.

gamete [Gk *gamete(s)* wife, husband] A sexual reproductive cell that must usually fuse with another such cell before development begins; an egg or sperm.

gametophyte [Gk *phyton* plant] A multicellular haploid plant that can produce gametes.

ganglion [Gk tumor] A structure containing a group of cell bodies of neurons (*pl.* ganglia).

gastr-, gastro- [Gk *gastēr* belly] Stomach; ventral; resembling the stomach.

gastrovascular cavity An often branched digestive cavity, with only one opening to the outside, that conveys nutrients throughout the body; found only in animals without circulatory systems.

gastrula A two-layered, later three-layered, animal embryonic stage.

gastrulation The process by which a bastula develops into a gastrula, usually by an involution of cells.

gated channel A membrane channel that can open or close in response to a signal, generally a change in the electrostatic gradient or the binding of a hormone, transmitter, or other molecular signal.

gel Colloid in which the suspended particles form a relatively orderly arrangement; *cf.* sol.

-gen; -geny [Gk *genos* birth, race] Producing; production, generation.

gene [Gk *genos*] The unit of inheritance; usually a portion of a DNA molecule that codes for some product such as a protein, tRNA, or rRNA.

gene amplification Any of the strategies that give rise to multiple copies of certain genes, thus facilitating the rapid synthesis of a product (such as rRNA for ribosomes) for which the demand is great.

gene flow The movement of genes from one part of a population to another, or from one population to another, via gametes.

gene pool Thes sum total of all the genes of all the individuals in a population.

gene regulation Any of the strategies by which the rate of expression of a gene can be regulated, as by controlling the rate of transcription.

generator potential *See* potential.

genetic drift Change in the gene pool as a result of chance and not as a result of selection, mutation, or migration.

genome The cell's total complement of DNA: in eucaryotes, the nuclear and organelle chromosomes; in procaryotes, the major chromosome, episomes, and plasmids. In viruses and viroids, the total complement of DNA or RNA.

genotype The particular combination of genes present in the cells of an individual.

germ cell A sexual reproductive cell; an egg or sperm.

gibberellin A plant hormone—one of its effects is stem elongation in some dwarf plants.

gill An evaginated area of the body wall of an animal, specialized for gas exchange.

gizzard A chamber of an animal's digestive tract specialized for grinding food.

glucose [Gk *glykys* sweet] A six-carbon sugar; plays a central role in cellular metabolism.

glycocalyx The layer of protein and carbohydrates just outside the plasma membrane of an animal cell; in general, the proteins are anchored in the membrane, and the carbohydrates are bound to the proteins.

glycogen [Gk *glykys*] A polysaccharide that serves as the principal storage form of carbohydrate in animals.

glycolysis [Gk *glykys*] Anaerobic catabolism of carbohydrates to pyruvic acid.

Golgi apparatus Membranous subcellular structure that plays a role in storage and modification particularly of secretory products.

gonadotropic Stimulatory to the gonads.

gonadotropin A hormone stimulatory to the gonads, a gonadotropic hormone.

gonads [Gk *gonos* seed] The testes or ovaries.

gram molecule *See* mole.

granum [L grain] A stacklike grouping of photosynthetic membranes in a chloroplast (*pl.* grana).

guard cell A specialized epidermal cell that regulates the size of stoma of a leaf.

habit [L *habitus* disposition] In biology, the characteristic form or mode of growth of an organism.

habitat [L it lives] The kind of place where a given organism normally lives.

habituation The process that results in a long-lasting decline in the receptiveness of interneurons (primarily) to the input from sensory neurons or other interneurons; *cf.* sensitization, adaptation.

haploid [Gk *haploos* single] Having only one of each type of chromosome.

hem-, hemat-, hemo- [Gk *haima* blood] Blood.

hematopoiesis [Gk *poiēsis* making] The formation of blood.

hemoglobin A red iron-containing pigment in the blood that functions in oxygen transport.

hepatic [Gk *hēpar* liver] Pertaining to the liver.

herbaceous [L *herbaceus* grassy] Having a stem that remains soft and succulent; not woody.

herbaceous dicot *See* dicot.

herbivore [L *herba* grass; *vorare* to devour] An animal that eats plants.

Hertz A unit of frequency (as of sound waves) equal to one cycle per second.

hetero- [Gk *heteros* other] Other, different.

heterogamy [Gk *gamos* marriage] The condition of producing gametes of two or more different types.

heterotrophic [Gk *trophē* food] Incapable of manufacturing organic compounds from inorganic raw materials, therefore requiring organic nutrients from the environment.

heterozygous [Gk *zygōtos* yoked] Having two different alleles of a given gene.

Hg [L *hydrargyrum* mercury] The symbol for mercury. Pressure is often expressed in *mm Hg*—the pressure exerted by a column of mercury whose height is measured in millimeters (at 0° C, 1 mm Hg = 133.3 newtons per square meter).

hilum Region where blood vessels, nerves, ducts, enter an organ.

hist- [Gk *histos* web] Tissue.

histology The structure and arrangement of the tissues of organisms; the study of these.

histone One of a class of basic proteins serving as structural elements of eucaryotic chromosomes.

homeo-, homo- [Gk *homoios* like] Like, similar.

homeostasis The tendency in an organism toward maintenance of physiological and psychological stability.

homeothermic [Gk *thermē* heat] Capable of self-regulation of body temperature; warm-blooded, endothermic.

home range An area within which an animal tends to confine all or nearly all its activities for a long period of time.

homologous Of chromosomes: bearing genes for the same characters. Of characters in different organisms: inherited from a common ancestor.

homozygous [Gk *zygōtos* yoked] Having two copies of the same allele of a given gene.

hormone [Gk *horman* to set in motion] A control chemical secreted in one part of the body that affects other parts of the body.

hybrid In evolutionary biology, a cross between two species. In genetics, a cross between two genetic types.

hydr-, hydro- [Gk *hydor* water] Water; fluid; hydrogen.

hydration Formation of a sphere of water around an electrically charged particle.

hydrocarbon Any compound made of only carbon and hydrogen.

hydrogen bond A weak chemical bond formed when two polar molecules, at least one of which usually consists of a hydrogen bonded to a more electronegative atom (usually oxygen or nitrogen), are attracted electrostatically.

hydrolysis [Gk *lysis* loosing] Breaking apart of a molecule by addition of water.

hydrophilic Readily entering into solution by forming hydrogen bonds with water or other polar molecules.

hydrophobic Incapable of entering into solution by molecules that are neither ionic nor polar, and therefore cannot dissolve in water.

hydrostatic [Gk *statikos* causing to stand] Pertaining to the pressure and equilibium of fluids.

hydroxyl ion The $OH^{-2'}$ ion.

hyper- Over, overmuch; more.

hypertonic Of a solution (or colloidal suspension): tending to gain water from some reference solution (or colloidal suspension) separated from it by a selectively permeable membrane—usually because it has a higher osmotic concentration than the reference solution.

hypertrophy [Gk *trophē* food] Abnormal enlargement, excessive growth.

hypha [Gk *hyphē* web] A fungal filament.

hypo- Under, lower; less.

hypocotyl The portion of the axis of a plant embryo below the point of attachment of the cotyledons; forms the base of the shoot and the root.

hypothalamus [Gk *thalamos* inner chamber] Part of the posterior portion of the vertebrate forebrain, containing important centers of the autonomic nervous system and centers of emotion.

hypotonic Of a solution (or colloidal suspension): tending to lose water to some reference solution (or colloidal suspension) separated from it by a selectively permeable membrane—usually because it has a lower osmotic concentration than the reference solution.

imprinting A kind of associative learning in which an animal rapidly learns during a particular critical period to recognize an object, individual, or location in the absence of overt reward; distinguished from most other associative learning in that it is retained indefinitely, being difficult or impossible to reverse.

independent assortment The Principle of Independent Assortment is frequently referred to as Mendel's second law. Genes found on different chromosomes, so-called unlinked genes, assort independently in meiosis unless they are recombined by crossing over.

inducer In embryology, a substance that stimulates differentiation of cells or development of a particular structure. In genetics, a substance that activates particular genes.

inorganic compound A chemical compound not based on carbon.

in situ [L in place] In its natural or original position.

instinct Heritable, genetically specified neural circuitry that guides and directs behavior.

insulin [L *insula* island] A hormone produced by the β islet cells in the pancreas that helps regulate carbohydrate metabolism, especially conversion of glucose into glycogen.

integument [L *integere* to cover] A coat, skin, shell, rind, or other protective surface structure.

inter- Between (e.g., *interspecific,* between two or more different species).

interneuron A neuron that receives input from and synapses on other neurons, as distinguished from a sensory neuron (which receives sensory information) and a motor neuron (which synapses on a muscle).

intra- Within (e.g., *intraspecific,* within a single species).

intrinsic Inherent in, a basic part of; as in *intrinsic isolating mechanism.*

intron A part of a primary transcript (and the corresponding part of a gene) that lies between exons, and is removed before the RNA becomes functional.

invaginated [L *vagina* sheath] Folded or protruded inward.

invertebrate [L *vertebra* joint] Lacking a backbone, hence an animal without bones.

in vitro [L in glass] Not in the living organism, in the laboratory.

in vivo [L in the living] In the living organism.

ion An electrically charged atom.

ionic bond A chemical bond formed by the electrostatic attraction between two oppositely charged ions.

iso- equal, uniform

isogamy [Gk *gamos* marriage] The condition of producing gametes of only one type, with no distinction existing between male and female.

isolating mechanism An obstacle to interbreeding, either extrinsic, such as a geographic barrier, or intrinsic, such as structural or behavioral incompatibility.

isotonic Of a solution (or colloidal suspension): tending neither to gain nor to lose water when separated from some reference solution (or colloidal suspension) by a selectively permeable membrane—usually because it has the same osmotic concentration as the reference solution.

isotope [Gk *topos* place] An atom differing from another atom of the same element in the number of neutrons in its nucleus.

kilo- A thousand.

kin-, kino- [Gk *kinema* motion] Motion, action.

kinase An enzyme that catalyzes the phosphorylation of a substrate by ATP.

lactic acid A three-carbon organic acid produced in animals and some microorganisms by fermentation.

lambert In metric system, unit of brightness of a light source; approximately equivalent to 929 foot-candles.

lamella [L thin plate] A thin platelike structure; a fairly straight intracellular membrane.

larva [L ghost, mask] Immature form of some animals that undergo radical transformation to attain the adult form.

lateral Pertaining to the side.

lateral inhibition Process by which adjacent sensory cells or their targets interact to inhibit one another when excited, the result being an exaggerated contrast; neural basis of feature-detector circuits.

lenticel [L *lenticella* small lentil] A porous region in the periderm of a woody stem through which gases can move.

leukocyte [Gk *leukos* white] A white blood cell; *cf.* lymphocyte, macrophage.

LH *See* luteinizing hormone.

ligament [L *ligare* to bind] A type of connective tissue linking two bones in a joint.

ligase An enzyme that catalyzes the bonding between adjacent nucleotides in DNA and RNA.

lignin [L *lignum* wood] An organic compound in wood that makes cellulose harder and more brittle.

linkage The presence of two or more genes on the same chromosome, which, in the absence of crossing over, causes the characters they control to be inherited together.

lip- [Gk *lipos* fat] Fat or fatlike.

lipase A fat-digesting enzyme.

lipid Any of a variety of compounds insoluble in water but soluble in ethers and alcohols; includes fats, oils, waxes, phospholipids, and steroids.

locus [L place] In genetics, a particular location on a chromosome, hence often used synonymously with gene (*pl.* loci).

lumen [L light, opening] The space or cavity within a tube or sac (*pl.* lumina).

lung An internal chamber specialized for gas exchange in an animal.

luteinizing hormone (LH) A gonadotropic hormone of the pituitary that stimulates conversion of a follicle into corpus luteum and secretion of progesterone by the corpus luteum; also stimulates secretion of sex hormone by the testes.

lymph [L *lympha* water] A fluid derived from tissue fluid and transported in special lymph vessels to the blood.

lymphocyte A white blood cell that responds to the presence of a foreign antigen. *B lymphocyte:* a cell that upon stimulation by an antigen secretes antibodies. *T lymphocyte:* a cell that attacks infected cells and modulates the activity of B lympohocytes.

-lysis, lyso- [Gk *lysis* loosing] Loosening, decomposition.

lysogenic Of bacteria: carrying bacteriophage capable of lysing, i.e., destroying, other bacterial cells.

lysosome A subcellular organelle that stores digestive enzymes.

macro- Large.

macrophage A phagocytic white blood cell that ingests material—particularly viruses, bacteria, and clumped toxins—bound by circulating antibodies.

Malpighian tubule An excretory diverticulum of the digestive tract in insects and some other anthropods.

mast cell Cells that are specialized for the secretion of histamine and other local chemical mediators as part of the immune response.

matrix [L *mater* mother] A mass in which something is embedded, e.g., the intercellular substance of a tissue.

medulla [L marrow, innermost part] (1) The inner portion of an organ, e.g., *adrenal medulla,* (2) The *medulla oblongata,* a portion of the vertebrate hindbrain that connects with the spinal cord.

medusa [*after* Medusa, mythological monster with snaky locks] The free-swimming stage in the life cycle of a coelenterate.

mega- Large.

megaspore A spore that will germinate into a female plant.

meiosis [Gk *meiōsis* diminution] A process of nuclear division in which the number of chromosomes is reduced by half.

membrane A structure, formed mainly by a double layer of phospholipids, which surrounds cells and organelles.

membrane channel A pore in a membrane through which certain molecules may pass.

membrane pump A permease that uses energy, usually from ATP, to move substances across the membrane against their osmotic-concentration or electrostatic gradients.

meristematic tissue [Gk *meristos* divisible] A plant tissue that functions primarily in production of new cells by mitosis.

meso- Middle.

mesoderm The middle tissue layer of an animal embryo.

mesophyll [Gk *phyllon* leaf] The parenchymatous middle tissue layers of a leaf.

meta- Posterior, later; change in.

metabolism [Gk *metabolē* change] The sum of the chemical reactions within a cell (or a whole organism), including the energy-releasing breakdown of molecules (catabolism) and the synthesis of complex molecules and new protoplasm (anabolism).

metamorphosis [Gk *morphē* form] Transformation of an immature animal into an adult. More generally, change in the form of an organ or structure.

micro- Small. Male. In units of measurement, one millionth.

microfilament A long, thin structure, usually formed from the protein actin; when associated with myosin filaments, as in muscles, microfilaments are involved in movement.

microorganism A microscopic organism, especially a bacterium, virus, or protozoan.

microspore A spore that will germinate into a male plant.

microtubule A long, hollow structure formed from the protein tubulin; found in cilia, eucaryotic flagella, basal bodies/centrioles, and the cytoplasm.

middle lamella A layer of substance deposited between the walls of adjacent plant cells.

milli- One thousandth.

mineral In biology, any naturally occurring inorganic substance, excluding water.

mitochondrion [Gk *mitos* thread; *chondrion* small grain] Subcellular organelle in which aerobic respiration takes place.

mitosis [Gk *mitos*] Process of nuclear division in which complex movements of chromosomes along a spindle result in two new nuclei with the same number of chromosomes as the original nucleus.

modulator A control chemical that stabilizes an allosteric enzyme in one of its alternative conformations.

mold Any of many fungi that produce a cottony or furry growth.

mole The amount of a substance that has a weight in grams numerically equal to the molecular weight of the substance. One mole of a substance contains 6.023×10^{23} molecules of that substance; hence one mole of a substance will always contain the same number of molecules as a mole of any other substance.

molecular weight The weight of a molecule calculated as the sum of the atomic weights of its constituent atoms.

molecule A chemical unit consisting of two or more atoms bonded together.

mono- One.

monocot A member of a subclass of angiosperms, or flowering plants, characterized by the presence of a single cotyledon in the embryo, parallel veins in the leaves, and flower petals in threes; *cf.* dicot.

-morph, morpho- [Gk *morphē* form] Form, structure.

morphogenesis The establishment of shape and pattern in an organism.

morphology The form and structure of organisms or parts of organisms; the study of these.

motivation The internal state of an animal that is the immediate cause of its behavior; drive.

motor neuron A neuron, leading away from the central nervous system, that synapses on and controls an effector.

motor program A coordinated, relatively stereotyped series of muscle movements performed as a unit, either innate (as the movements of swallowing) or learned (as in speech); also, the neural circuitry underlying such behavior; *cf.* reflex.

mouthparts Structures or appendages near the mouth used in manipulating food.

mucosa Any membrane that secretes mucus (a slimy protective substance), e.g., the membrane lining the stomach and intestine.

muscle [L *musculus* small mouse, muscle] A contractile tissue of animals.

mutation [L *mutatio* change] Any relatively stable heritable change in the genetic material.

mutualism A symbiosis in which both parties benefit.

mycelium [Gk *mykēs* fungus] A mass of hyphae forming the body of a fungus.

myo- [Gk *mys* mouse, muscle] Muscle.

NAD *See* nicotinamide adenine dinucleotide.

NADP *See* nicotinamide adenine dinucleotide phosphate.

nano- [L *nanus* dwarf] One billionth.

natural selection Differential reproduction in nature, leading to an increase in the frequency of some genes or gene combinations and to a decrease in the frequency of others.

navigation The initiation and/or maintenance of movement toward a goal.

negative feedback *See* feedback.

nematocyst [Gk *nēma* thread; *kystis* bag] A specialized stinging cell in coelenterates; contains a hairlike structure that can be ejected.

neo- New.

neocortex Portion of the cerebral cortex in mammals, of relatively recent evolutionary origin; often greatly expanded in the higher primates and dominant over other parts of the brain.

nephr- [Gk *nephros* kidney] Kidney.

nephridium An excretory organ consisting of an open bulb and a tubule leading to the exterior; found in many invertebrates, such as segmented worms.

nephron The functional unit of a vertebrate kidney, consisting of Bowman's capsule, convoluted tubule, and loop of Henle.

nerve [L *nervus* sinew, nerve] A bundle of neuron fibers (axons).

nerve net A nervous system without any central control, as in coelenterates.

neuron [Gk nerve, sinew] A nerve cell.

neutron An electrically neutral subatomic particle with approximately the same mass as a proton.

niche The functional role and position of an organism in the ecosystem; the way an organism makes its living, including, for an animal, not only what it eats, but when, where, and how it obtains food, where it lives, etc.

nicotinamide adenine dinucleotide (NAD) An organic compound that functions as an electron acceptor, e.g., in respiration.

nicotinamide adenine dinucleotide phosphate (NADP) An organic compound that functions as an electron acceptor, e.g., in biosynthesis.

nitrogen fixation Incorporation of nitrogen from the atmosphere into substances more generally usable by organisms.

node (of plant) [L *nodus* knot] Point on a stem where a leaf or bud is (or was) attached.

nonhomologous Of chromosomes: two chromosomes that do not share the same genes and thus do not pair during meiosis.

notochord [Gk *nōtos* back; *chordē* string] In the lower chordates and in the embryos of the higher vertebrates, a flexible supportive rod running longitudinally through the back just ventral to the nerve cord.

nucleic acid Any of several organic acids that are polymers of nucleotides and function in transmission of hereditary traits, in protein synthesis, and in control of cellular activities.

nucleoid A region, not bounded by a membrane, where the chromosome is located in a procaryotic cell.

nucleolus A dense body within the nucleus, usually attached to one of the chromosomes; consists of multiple copies of the genes for certain kinds of rRNA.

nucleosome A complex consisting of several histone proteins, which together form a "spool," and chromosomal DNA, which is wrapped around the spool.

nucleotide A chemical entity consisting of a five-carbon sugar with a phosphate group and a purine or pyrimidine attached; building-block unit of nucleic acids.

nucleus (of cell) [L kernel] A large membrane-bounded organelle containing the chromosomes.

nutrient [L *nutrire* to nourish] A food substance usable in metabolism as a source of energy or of building material.

nymph [Gk *nymphē* bride, nymph] Immature stage of insect that undergoes gradual metamorphosis.

olfaction [L *olfacere* to smell] The sense of smell.

omnivorous [L *omnis* all; *vorare* to devour] Eating a variety of foods, including both plants and animals.

oncogene A gene that causes one of the biochemical changes that lead to cancer.

ontogeny [Gk *ōn* being] The course of development of an individual organism.

oo- [Gk *ōion* egg] Egg.

oogamy A type of heterogamy in which the female gametes are large nonmotile egg cells.

oogonium Unjacketed female reproductive organ of a thallophyte plant.

operant conditioning *See* conditioning.

operator A region on the DNA to which a control substance can bind, thereby altering the rate of transcription.

oral [L *oris* of the mouth] Relating to the mouth.

organ [Gk *organon* tool] A body part usually composed of several tissues grouped together into a structural and functional unit.

organelle A well-defined subcellular structure.

organic compound A chemical compound containing carbon.

organism An individual living thing.

orientation The act of turning or moving in relation to some external feature, such as a source of light.

osmol Measure of osmotic concentration; the total number of moles of osmotically active particles per liter of solvent.

osmoregulation Regulation of the osmotic concentration of body fluids in such a manner as to keep them relatively constant despite changes in the external medium.

osmosis [Gk *ōsmos* thrust] Movement of a solvent (usually water in biology) through a selectively permeable membrane.

osmotic potential The free energy of water molecules in a solution or colloid under conditions of constant temperature and pressure; since this free energy decreases as the proportion of osmotically active particles rises, a measure of the tendency of the solution or colloid to lose water.

osmotic pressure The pressure that must be exerted on a solution or colloid to keep it in equilibrium with pure water when it is separated from the water by a selectively permeable membrane; hence a measure of the tendency of the solution or colloid to take in water.

ov-, ovi- [L *ovum* egg] Egg.

ovary Female reproductive organ in which egg cells are produced.

ovulation Release of an egg from the ovary.

ovule A plant structure, composed of an integument, sporangium, and megagametophyte, that develops into a seed after fertilization.

ovum A mature egg cell (*pl.* ova).

oxidation Energy-releasing process involving removal of electrons from a substance; in biological systems, generally by the removal of hydrogen (or sometimes the addition of oxygen).

pancreas In vertebrates, a large glandular organ located near the stomach that secretes digestive enzymes into the duodenum and also produces hormones.

papilla [L nipple] A small nipplelike protuberance.

para- Alongside of.

parapodium [Gk *podion* little foot] One of the paired segmentally arranged lateral flaplike protuberances of polychaete worms.

parasitism [Gk *parasitos* eating with another] A symbiosis in which one party benefits at the expense of the other.

parasympathetic nervous system One of the two parts of the autonomic nervous system.

parathyroids Small endocrine glands of vertebrates located near the thyroid.

parenchyma A plant tissue composed of thin-walled, loosely packed, relatively unspecialized cells.

parthenogenesis [Gk *parthenos* virgin] Production of offspring without fertilization.

pathogen [Gk *pathos* suffering] A disease-causing organism.

pectin A complex polysaccharide that cross-links the cellulose fibrils in a plant cell wall and is a major constituent of the middle lamella.

pellicle [L *pellis* skin] A thin skin or membrane.

pepsin [Gk *pepsis* digestion] A protein-digesting enzyme of the stomach.

peptide bond A covalent bond between two amino acids resulting from a condensation reaction between the amino group of one acid and the acidic group of the other.

perennial A plant that lives for several years, as compared to annuals and biennials, which live for one and two years respectively.

peri- Surrounding.

pericycle A layer of cells inside the endodermis but outside the phloem of roots and stems.

periderm The corky outer bark of older stems and roots.

peristalsis [Gk *stalsis* contraction] Alternating waves of contraction and relaxation passing along a tubular structure such as the digestive tract.

permeable [L *permeare* to go through] Of a membrane: permitting other substances to pass through.

permease [L *permeare*] A protein that allows molecules to move across a membrane; *cf.* gated channel, membrane channel, membrane pump.

petiole [L *pediculus* small foot] The stalk of a leaf.

PGAL *See* phosphoglyceraldehyde.

pH Symbol for the logarithm of the reciprocal of the hydrogen ion concentration; hence a measure of acidity. A pH of 7 is neutral; lower values are acidic, higher values alkaline (basic).

phage *See* bacteriophage.

phagocytosis [Gk *phagein* to eat] The active engulfing of particles by a cell.

pharynx Part of the digestive tract between the oral cavity and the esophagus; in vertebrates, also part of the respiratory passage.

phenotype [Gk *phainein* to show] The physical manifestation of a genetic trait.

pheromone [Gk *pherein* to carry + hormone] A substance that, secreted by one organism, influences the behavior or physiology of other organisms of the same species when they sense its odor.

phloem [Gk *phloios* bark] A plant vascular tissue that transports organic materials; the inner bark.

-phore [Gk *pherin* to carry] Carrier.

phosphoglyceraldehyde (PGAL) A three-carbon phosphorylated carbohydrate, important in both photosynthesis and glycolysis.

phospholipid A compound composed of glycerol, fatty acids, a phosphate group, and often a nitrogenous group.

phosphorylation Addition of a phosphate group.

photo- [Gk *phōs* light] Light.

photon A discrete unit of radiant energy.

photoperiodism A response by an organism to the duration and timing of the light and dark conditions.

photophosphorylation The process by which energy from light is used to convert ADP into ATP.

photosynthesis Autotrophic synthesis of organic materials in which the source of energy is light; *cf.* photophosphorylation.

-phyll [Gk *phyllon* leaf] Leaf.

phylogeny [Gk *phylē* tribe] Evolutionary history of an organism.

physiology [Gk *physis* nature] The life processes and functions of organisms; the study of these.

-phyte, phyto- [Gk *phyton* plant] Plant.

phytochrome A protein pigment of plants sensitive to red and farred light.

pinocytosis [Gk *pinein* to drink] The active engulfing by cells of liquid or of very small particles.

pistil The female reproductive organ of a flower, composed of one or more megasporophylls.

pith A tissue (usually parenchyma) located in the center of a stem (rarely a root), internal to the xylem.

pituitary An endocrine gland located near the brain of vertebrates; known as the master gland because it secretes hormones that regulate the action of other endocrine glands.

placenta [Gk *plax* flat surface] An organ in mammals, made up of fetal and maternal components, that aids in exchange of materials between the fetus and the mother.

plasm-, plasmo-, -plasm [Gk *plasma* something formed or molded] Formed material; plasma; cytoplasm.

plasma Blood minus the cells and platelets.

plasma membrane The outer membrane of a cell.

plasmid A small circular piece of DNA free in the cytoplasm of a bacterial or yeast cell and replicated independently of the cell's chromosome.

plasmodesma [Gk *desma* bond] A connection between adjacent plant cells through tiny openings in the cell walls (*pl.* plasmodesmata).

plasmolysis Shrinkage of a plant cell away from its wall when in a hypertonic medium.

plastid Relatively large organelle in plant cells that functions in photosynthesis and/or nutrient storage.

pleiotropic [Gk *pleiōn* more] Of a gene: having more than one phenotypic effect.

poikilothermic [Gk *poikilos* various; *thermē* heat] Incapable of precise self-regulation of body temperature, dependent on environmental temperature; cold-blooded, ectothermic.

polar molecule A molecule with oppositely charged sections; the charges, which are far weaker than the charges on ions, a rise from differences in electronegativity between the constituent atoms.

pollen grain [L *pollen* flour dust] A microgametophyte of a seed plant.

poly- Many.

polycistronic Pertaining to the transcription of two or more adjacent cistrons (structural genes) into a single messenger RNA molecule.

polymer [Gk *meros* part] A large molecule consisting of a chain of small molecules bonded together by condensation reactions or similar reactions.

polymerase An enzyme complex that catalyzes the polymerization of nucleotides; examples are DNA polymerase, which is involved in replication, and RNA polymerase, which is involved in transcription.

polymorphism [Gk *morphē* form] The simultaneous occurrence of several discontinuous phenotypes in a population.

polyp [Gk *polypous* many-footed] The sedentary stage in the life cycle of a Cnidarian.

polypeptide chain A chain of amino acids linked together by peptide bonds.

polyploid Having more than two complete sets of chromosomes.

polysaccharide Any carbohydrate that is a polymer of simple sugars.

population In ecology, group of individuals belonging to the same species.

portal system [L *porta* gate] A blood circuit in which two beds of capillaries are connected by a vein (e.g., *hepatic portal system*).

positive feedback *See* feedback.

posterior Toward the hind end.

potential Short for *potential difference:* the difference in electrical charge between two points. *Resting p.:* a relatively steady potential difference across a cell membrane, particularly of a nonfiring nerve cell or a relaxed muscle cell. *Action p.:* a sharp change in the potential difference across the membrane of a nerve or muscle cell that is propagated along the cell; in nerves, identified with the nerve impulse. *Generator p.:* a change in the potential difference across the membrane of a sensory cell that, if it reaches a threshold level, may trigger an action potential along the associated neural pathway.

preadapted A structure that evolved for one function but can at least minimally perform another function when placed in a new environment.

predation [L *praedatio* plundering] The feeding of free-living organisms on other organisms.

presumptive Describing the developmental fate of a tissue that is not yet differentiated. Presumptive neural tissue, for example, is destined to become part of the nervous system once it has differentiated.

primary transcript Newly synthesized RNA—generally mRNA—before the introns are moved.

primitive [L *primus* first] Old, like the ancestral condition.

primordium [L *primus; ordiri* to begin] Rudiment, earliest stage of development.

pro- Before.

proboscis [Gk *boskein* to feed] A long snout; an elephant's trunk. In invertebrates, an elongate, sometimes eversible process originating in or near the mouth that often serves in feeding.

procaryotic cell A type of cell that lacks a membrane-bounded nucleus; found only in bacteria.

progesterone [L *gestare* to carry out] One of the principal female sex hormones of vertebrates.

promoter The region of DNA to which the transcription complex binds.

prot-, proto- First, primary.

protease A protein-digesting enzyme.

protein A long polypeptide chain.

proteolytic Protein-digesting.

proton A positively charged primary subatomic particle.

proto-oncogene A gene that can, after certain sorts of mutation or translocation, or after mutation or translocation in associated control regions, become an oncogene and cause one of the changes leading to cancer.

protoplasm Living substance, the material of cells.

provirus Viral nucleic acid integrated into the genetic material of a host cell.

proximal Near some reference point (often the main part of the body).

pseudo- False; temporary.

pseudocoelom A functional body cavity not entirely enclosed by mesoderm.

pseudogene An untranscribed region of the DNA that closely resembles a gene.

pseudopod, pseudopodium [L *podium* foot] A transitory cytoplasmic protrusion of an amoeba or an amoeboid cell.

pulmonary [L *pulmones* lungs] Relating to the lungs.

purine Any of several double-ringed nitrogenous bases important in nucleotides.

pyloric [Gk *pylōros* gatekeeper] Referring to the junction between the stomach and the intestine.

pyrimidine Any of several single-ringed nitrogenous bases important in nucleotides.

pyruvic acid A three-carbon compound produced by glycolysis.

race A subspecies.

radial symmetry A type of symmetry in which the body parts are arranged regularly around a central line (in animals, running through the oral-anal axis) rather than on the two sides of a plane.

radiation As an evolutionary phenomenon, divergence of members of a single lineage into different niches or adaptive zones.

receptor In cell biology, a region, often the exposed part of a membrane protein, that binds a substance but does not catalyze a reaction in the chemical it binds; the membrane protein frequently has another region that, as a result of the binding, undergoes an allosteric change and so becomes catalytically active.

recessive Of an allele: not expressing its phenotype in the presence of another allele of the same gene, therefore expressing it only in homozygous individuals. *Recessive character, recessive phenotype:* one caused by a recessive allele.

reciprocal altruism *See* altruistic behavior.

recombination In genetics, a novel arrangement of alleles resulting from sexual reproduction and from crossing over (or, in procaryotes and eucaryotic organelles, from conjugation). In gene evolution, a novel arrangement of exons resulting from a variety of processes that duplicate and transport segments of the chromosomes within the genome; these processes include transposition, unequal crossing over, and chromosomal breakage and fusion.

rectum [L *rectus* straight] The terminal portion of the intestine.

redox reaction [*from reduction-oxidation*] A reaction involving reduction and oxidation, which inevitably occur together; *cf.* reduction, oxidation.

reduction Energy-storing process involving addition of electrons to a substance; in biological systems, generally by the addition of hydrogen (or sometimes the removal of oxygen).

reflex [L *reflexus* bent back] An automatic act consisting, in its pure form, of a single simple response to a single stimulus, as when a tap on the knee elicits a knee jerk. Distinguished from a motor program, which involves a coordinated response of several muscles.

reflex arc A functional unit of the nervous system, involving the entire pathway from receptor cell to effector.

reinforcement (psychological) Reward for a particular behavior.

releaser *See* sign stimulus.

renal [L *renes* kidneys] Pertaining to the kidney.

respiration [L *respiratio* breathing out] (1) The release of energy by oxidation of fuel molecules. (2) The taking of O_2 and release of CO_2; breathing.

resting potential *See* potential.

restriction endonuclease *See* endonuclease.

reticulum [L little net] A network.

retina The tissue in the rear of the eye that contains the sensory cells of vision.

retrovirus An RNA virus that, by means of a special enzyme (reverse transcriptase), makes a DNA copy of its genome which is then incorporated into the host's genome.

rhizoid [Gk *rhiza* root] Rootlike structure.

ribonucleic acid (RNA) Nucleic acid characterized by the presence of a ribose sugar in each nucleotide. The primary classes of RNA are mRNA (messenger RNA, which carries the instructions specifying the order of amino acids in new proteins from the genes to the ribosomes where protein synthesis takes place), rRNA (ribosomal RNA, which is incorporated into ribosomes), and tRNA (transfer RNA, which carries amino acids to the ribosomes as part of protein synthesis).

ribosome A small cytoplasmic organelle that functions in protein synthesis.

RNA *See* ribonucleic acid.

salt Any of a class of generally ionic compounds that may be formed by reaction of an acid and a base, e.g., table salt, NaCl.

sap Water and dissolved materials moving in the xylem; less commonly, solutions moving in the phloem. *Cell sap:* the fluid content of a plant-cell vacuole.

saprophyte [Gk *sapros* rotten] A heterotrophoic plant or bacterium that lives on dead organic material.

sarcomere [Gk *sarx* flesh; *meros* part] The region of a skeletal-muscle myofibril extending from one Z line to the next; the functional unit of skeletal-muscle contraction.

sclerenchyma [Gk *sclēros* hard] A plant supportive tissue composed of cells with thick secondary walls.

section *Cross or transverse s.:* section at right angles to the longest axis. *Longitudinal s.:* section parallel to the longest axis. *Radial s.:* longitudinal section along a radius.

Sagittal s.: vertical longitudinal section along the midline of a bilaterally symmetrical animal.

seed A plant reproductive entity consisting of an embryo and stored food enclosed in a protective coat.

segmentation The subdivision of an organism into more or less equivalent serially arranged units.

selection pressure In a population, the force for genetic change resulting from natural selection.

semipermeable Permeable only to solvent (usually water); less strictly; selectively permeable, i.e., permeable to some substances but not to others.

sensitization The process by which an unexpected stimulus alerts an animal, reducing or eliminating any preexisting habituation; *cf.* adaptation, habituation.

sensory neuron A neuron, leading toward the central nervous system, that receives input from a receptor cell or is itself responsive to sensory stimulation.

septum [L barrier] A partition or wall (*pl.* septa).

sessile [L *sessilis* of sitting, low] Of animals, sedentary. Of plants, without a stalk.

sex-linked Of genes: located on the X chromosome.

sexual dimorphism Morphological differences between the two sexes of a species, as in the size of tails of peacocks as compared to peahens.

sexual selection Selection for morphology or behavior directly related to attracting or winning mates. *Male-contest sexual selection:* selection for morphology or behavior that enables a male to win fights or contests for access to females, gaining a high position in a dominance hierarchy, for example, or possession of a territory. *Female-choice sexual selection:* selection for morphology or behavior that enables a male to attract females directly.

shoot A stem with its leaves, flowers, etc.

sieve element A conductile cell of the phloem.

sign stimulus (or releaser) A simple cue that orients or triggers specific innate behavior.

sinus [L curve, hollow] (1) A channel for the passage of blood lacking the characteristics of a true blood vessel. (2) A hollow within bone or another tissue (e.g. the air-filled sinuses of some of the facial bones).

sol Colloid in which the suspended particles are dispersed at random; *cf.* gel.

solute Substance dissolved in another (the solvent).

solution [L *solutio* loosening] A homogeneous molecular mixture of two or more substances.

solvent Medium in which one or more substances (the solute) are dissolved.

-soma, somat-, -some [Gk *soma* body] Body, entity.

somatic Pertaining to the body; to all cells except the germ cells; to the body wall. *Somatic nervous system:* a portion of the nervous system that is at least potentially under control of the will; *cf.* autonomic nervous system.

specialized Adapted to a special, usually rather narrow, function or way of life.

speciation The process of formation of new species.

species [L kind] The largest unit of population within which effective gene flow occurs or could occur.

sperm [Gk *sperma* seed] A male gamete.

sphincter [Gk *sphinktēr* band] A ring-shaped muscle that can close a tubular structure by contracting.

spindle A microtubular structure with which the chromosomes are associated in mitosis and meiosis.

sporangium A plant structure that produces spores.

spore [Gk *spora* seed] An asexual reproductive cell, often a resting stage adapted to resist unfavorable environmental conditions.

sporophyll [Gk *phyllon* leaf] A modified leaf that bears spores.

sporophyte [Gk *phyton* plant] A diploid plant that produces spores.

stamen [L thread] A male sexual part of a flower; a microsporophyll of a flowering plant.

starch A glucose polymer, the principal polysaccharide storage product of vascular plants.

stele [Gk *stēlē* upright slab] The vascular cylinder in the center of a root or stem, bounded externally by the endodermis.

stereo- [Gk *stereos* solid] Solid; three-dimensional.

steroid Any of a number of complex, often biologically important compounds (e.g., some hormones and vitamins), composed of four interlocking rings of carbon atoms.

stimulus Any environmental factor that is detected by a receptor.

stoma [Gk mouth] An opening, regulated by guard cells, in the epidermis of a leaf or other plant part (*pl.* stomata).

stroma [Gk *stroma* bed, mattress] The ground substance within such organelles as chloroplasts and mitochondria.

subspecies A genetically distinct geographic subunit of a species.

substrate (1) The base on which an organism lives, e.g., soil. (2) In chemical reactions, a substance acted upon, as by an enzyme.

succession In ecology, progressive change in the plant and animal life of an area.

sucrose A double sugar composed of a unit of glucose and a unit of fructose; table sugar.

suspension A heterogeneous mixture in which the particles of one substance are kept dispersed by agitation.

sym-, syn- Together.

symbiosis [Gk *bios* life] The living together of two organisms in an intimate relationship.

sympathetic nervous system One of the two parts of the autonomic nervous system.

sympatric [L *patria* homeland] Having the same range.

symplast In a plant, the system constituted by the cytoplasm of cells interconnected by plasmodesmata.

synapse [Gk *haptein* to fasten] A juncture between two neurons.

synapsis The pairing of homologous chromosomes during meiosis.

synergistic [Gk *ergon* work] Acting together with another substance or organ to achieve or enhance a given effect.

systemic circulation The part of the circulatory system supplying body parts other than the gas-exchange surfaces.

-tactic Referring to a taxis.

taxis A simple continuously oriented movement in animals (e.g., phototaxis, geotaxis) (*pl.* taxes).

taxonomy [Gk *taxis* arrangement] The classification of organisms on the basis of their evolutionary relationships.

tendon [L *tendere* to stretch] A type of connective tissue attaching muscle to bone.

territory A particular area defended by an individual against intrusion by other individuals, particularly of the same species.

testis Primary male sex organ in which sperm are produced (*pl.* testes).

thalamus [Gk *thalamos* inner chamber] Part of the rear portion of the vertebrate forebrain, a center for integration of sensory impulses.

thallus [Gk *thallos* young shoot] A plant body exhibiting relatively little tissue differentiation and lacking true roots, stems, and leaves.

thorax [Gk *thōrax* breastplate] In mammals, the part of the trunk anterior to the diaphragm, which partitions it from the abdomen. In insects, the body region between the head and the abdomen, bearing the walking legs and wings.

thymus [Gk *thymos* warty excrescence] Glandular organ that plays an important role in the development of immunologic capabilities in vertebrates.

thyroid [Gk *thyreoeidēs* shield-shaped] An endocrine gland of vertebrates located in the neck region.

thyroxin A hormone, produced by the thyroid, that stimulates a speedup of metabolism.

tissue [L *texere* to weave] An aggregate of cells, usually similar in both structure and function, that are bound together by intercellular material.

T lymphocyte *See* lymphocyte.

toxin A proteinaceous substance produced by one organism that is poisonous to another.

trachea In vertebrates, the part of the respiratory system running from the pharynx into the thorax; the "windpipe." In land arthropods, an air duct running from an opening in the body wall to the tissues.

tracheid An elongate thick-walled tapering conductile cell of the xylem.

trans- Across; beyond.

transcription In genetics, the synthesis of RNA from a DNA template.

transduction [L *ducere* to lead] In genetics, the transfer of genetic material from one host cell to another by a virus. In neurobiology, the translation of a stimulus like light or sound into an electrical change in a receptor cell.

transformation The incorporation by bacteria of fragments of DNA released into the medium from dead cells.

translation In genetics, the synthesis of a polypeptide from an mRNA template.

translocation In botany, the movement of organic materials from one place to another within the plant body, primarily through the phloem. In genetics, the exchange of parts between nonhomologous chromosomes.

transpiration Release of water vapor from the aerial parts of a plant, primarily through the stomata.

transposition The movement of DNA from one position in the genome to another. *Transposon:* a mobile segment of DNA, usually encoding the enzymes necessary to effect its own movement.

-trophic [Gk *trophē* food] Nourishing; stimulatory.

tropic hormone A hormone produced by one endocrine gland that stimulates another endocrine gland.

tropism [Gk *tropos* turn] A turning response to a stimulus, primarily by differential growth patterns in plants.

turgid [L *turgidus* swollen] Swollen with fluid.

turgor pressure [L *turgēre* to be swollen] The pressure exerted by the contents of a cell against the cell membrane or cell wall.

tympanic membrane [Gk *tympanon* drum] A membrane of the ear that picks up vibrations from the air and transmits them to other parts of the ear; the eardrum.

urea The nitrogenous waste product of mammals and some other vertebrates, formed in the liver by combination of ammonia and carbon dioxide.

ureter The duct carrying urine from the kidney to the bladder in higher vertebrates.

urethra The duct leading from the bladder to the exterior in higher vertebrates.

uric acid An insoluble nitrogenous waste product of most land arthropods, reptiles, and birds.

uterus In mammals, the chamber of the female reproductive tract in which the embryo undergoes much of its development; the womb.

vaccine [L *vacca* cow] Drug containing an antigen, administered to induce active immunity in the patient.

vacuole [L *vacuus* empty] A membrane-bounded vesicle or chamber in a cell.

valence A measure of the bonding capacity of an atom, which is determined by the number of electrons in the outer shell.

vascular tissue [L *vasculum* small vessel] Tissue concerned with internal transport, such as xylem and phloem in plants and blood and lymph in animals.

vaso- [L *vas* vessel] Blood vessel.

vector [L *vectus* carried] Transmitter of pathogens.

vegetative Of plant cells and organs: not specialized for reproduction. Of reproduction: asexual. Of bodily functions: involuntary.

vein [L *vena* blood vessel] A blood vessel that transports blood toward the heart.

vena cava [L hollow vein] One of the two large veins that return blood to the heart from the systemic circulation of vertebrates.

ventral [L *venter* belly] Pertaining to the belly or underparts.

vessel element A highly specialized cell of the xylem, with thick secondary walls and extensively perforated end walls.

villus [L shaggy hair] A highly vascularized fingerlike process from the intestinal lining or from the surface of some other structure (e.g., a chorionic villus of the placenta) (*pl.* villi).

virus [L slime, poison] A submicroscopic noncellular, obligatorily parasitic entity, composed of a protein shell and a nucleic acid core, that exhibits some properties normally associated with living organisms, including the ability to mutate and to evolve.

viscera [L] The internal organs, especially those of the great central body cavity.

vitamin [L *vita* life] An organic compound, necessary in small quantities, that a given organism cannot synthesize for itself and must obtain prefabricated in the diet.

woody dicot *See* dicot.

X chromosome The female sex chromosome.

xylem [Gk *xylon* wood] A vascular tissue that transports water and dissolved minerals upward through the plant body.

Y chromosome The male sex chromosome.

yolk stored food material in an egg.

zoo- [Gk *zōion* animal] Animal, motile.

zoospore A ciliated or flagellated plant spore.

zygote [Gk *zygōtos* yoked] A fertilized egg cell.

zymogen [Gk *zymē* leaven] An inactive precursor of an enzyme.

Index

Page numbers in **boldface** refer to illustrations; those in *italics* identify definitions or main treatment of subjects mentioned in several parts of the book.